ARCTIC CLIMATE IMPACT ASSESSMENT

Rob
Pål Pr
Patricia
Snorri Balc
Elizabeth Bus
Terry V. Callagl

Paul Grabhorn
Susan Joy Hassol
Gordon McBean
Michael MacCracken
Lars-Otto Reiersen
Jan Idar Solbakken
Gunter Weller

Recommended Citation

CAMB
UNIVERS

CAMBRIDGE UNIVERSITY PRESS

Cambridge, New York, Melbourne, Madrid, Cape Town, Singapore, São Paulo

Cambridge University Press
40 West 20th Street, New York, NY 10011-4211, USA

Published in the United States of America by Cambridge University Press, New York

www.cambridge.org
Information on this title: www.cambridge.org/9780521865098

First published 2005

Printed in Canada by Friesens

A catalog record for this publication is available from the British Library.

ISBN-13 978-0-521- 86509 - 8 hardback
ISBN-10 0-521-86509 - 3 hardback

AMAP Secretariat
P.O. Box 8100 Dep.
N-0032 Oslo, Norway
Tel: +47 23 24 16 30
Fax: +47 22 67 67 06
http://www.amap.no

CAFF Internationa
Secretariat
Hafnarstraeti 97
600 Akureyri, Iceland
Tel: +354 461-3352
Fax: +354 462-3390
http://www.caff.is

IASC Secretariat
Middelthuns gate 29
P.O. Box 5156 Majorstu
N-0302 Oslo, Norway
Tel: +47 2295 9900
Fax: +47 2295 9901
http://www.iasc.no

Authors
Listed in each individual chapter

Project Production and Graphic Design
Paul Grabhorn, Joshua Weybright, Clifford Grabhorn (Cartography)

Editing
Carolyn Symon (lead editor), Lelani Arris, Bill Heal

Photography
Bryan and Cherry Alexander (Cover and Chapter 1)

Assessment Integration Team

ert Corell, Chair	American Meteorological Society, USA
estrud, Vice Chair	Centre for Climate Research in Oslo, Norway
A. Anderson	University of Alaska Fairbanks, USA
Jursson	Liaison for the Arctic Council, Iceland
h	Environment Canada, Canada
an	Abisko Scientific Research Station, Sweden
	Sheffield Centre for Arctic Ecology, UK
	Grabhorn Studio, Inc., USA
	Independent Scholar and Science Writer, USA
	University of Western Ontario, Canada
	Climate Institute, USA
	Arctic Monitoring and Assessment Programme, Norway
	Permanent Participants, Norway
	niversity of Alaska Fairbanks, USA

ACIA Secretariat

Gunter Weller, Executive Director
Patricia A. Anderson, Deputy Executive Director
Barb Hameister, Sherry Lynch
International Arctic Research Center
University of Alaska Fairbanks
Fairbanks, AK 99775-7740, USA
Tel: +907 474 5818
Fax +907 474 6722
http://www.acia.uaf.edu

ACIA, 2005. Arctic Climate Impact Assessment. Cambridge University Press, 1042p.

http://www.acia.uaf.edu

Preface

Earth's climate is changing, with the global temperature now rising at a rate unprecedented in the experience of modern human society. These climate changes, including increases in ultraviolet radiation, are being experienced particularly intensely in the Arctic. Because the Arctic plays a special role in global climate, these changes in the Arctic will also affect the rest of the world. It is thus essential that decision makers have the latest and best information available regarding ongoing changes in the Arctic and their global implications.

The Arctic Council called for this assessment and charged two of its working groups, the Arctic Monitoring and Assessment Programme (AMAP) and the Conservation of Arctic Flora and Fauna (CAFF), along with the International Arctic Science Committee (IASC), with its implementation. An Assessment Steering Committee (see page iv) was charged with the responsibility for scientific oversight and coordination of all work related to the preparation of the assessment reports.

This assessment was prepared over the past five years by an international team of over 300 scientists, other experts, and knowledgeable members of the indigenous communities. The lead authors were selected from open nominations provided by AMAP, CAFF, IASC, the Indigenous Peoples Secretariat, the Assessment Steering Committee, and several national and international scientific organizations. A similar nomination process was used by ACIA to select international experts who independently reviewed this report. The report has been thoroughly researched, is fully referenced, and provides the first comprehensive evaluation of arctic climate change, changes in ultraviolet radiation, and their impacts for the region and for the world. Written certification has been obtained from the ACIA leadership and all lead authors to the effect that the final scientific report fully reflects their expert views.

The scientific results reported herein provided the scientific foundations for the ACIA synthesis report, entitled "Impacts of a Warming Arctic", released in November 2004. This English language report is the only official document containing the comprehensive scientific assessment of the ACIA.

Recognizing the central importance of the Arctic and this information to society as it contemplates responses to the growing global challenge of climate change, the cooperating organizations are pleased to forward this report to the Arctic Council, the international science community, and others around the world.

Financial support for the ACIA Secretariat was provided by the U.S. National Science Foundation and National Oceanic and Atmospheric Administration. Support for ACIA-related workshops, participation of scientists and experts, and the production of this report was provided by the governments of the eight Arctic nations, several other governments, and the Secretariats of AMAP, CAFF, and IASC.

The Arctic Council

The Arctic Council is a high-level intergovernmental forum that provides a mechanism to address the common concerns and challenges faced by arctic people and governments. It is comprised of the eight arctic nations (Canada, Denmark/Greenland/Faroe Islands, Finland, Iceland, Norway, Russia, Sweden, and the United States of America), six Indigenous Peoples organizations (Permanent Participants: Aleut International Association, Arctic Athabaskan Council, Gwich'in Council International, Inuit Circumpolar Conference, Russian Association of Indigenous Peoples of the North, and Saami Council), and official observers (including France, Germany, the Netherlands, Poland, United Kingdom, non-governmental organizations, and scientific and other international bodies).

The International Arctic Science Committee

The International Arctic Science Committee is a non-governmental organization whose aim is to encourage and facilitate cooperation in all aspects of arctic research among scientists and institutions of countries with active arctic research programs. IASC's members are national scientific organizations, generally academies of science, which seek to identify priority research needs, and provide a venue for project development and implementation.

Assessment Steering Committee

Representatives of Organizations

Robert Corell, Chair	International Arctic Science Committee, USA
Pål Prestrud, Vice-Chair	Conservation of Arctic Flora and Fauna, Norway
Snorri Baldursson (to Aug. 2000)	Conservation of Arctic Flora and Fauna, Iceland
Gordon McBean (from Aug. 2000)	Conservation of Arctic Flora and Fauna, Canada
Lars-Otto Reiersen	Arctic Monitoring and Assessment Programme, Norway
Hanne Petersen (to Sept. 2001)	Arctic Monitoring and Assessment Programme, Denmark
Yuri Tsaturov (from Sept. 2001)	Arctic Monitoring and Assessment Programme, Russia
Bert Bolin (to July 2000)	International Arctic Science Committee, Sweden
Rögnvaldur Hannesson (from July 2000)	International Arctic Science Committee, Norway
Terry Fenge	Permanent Participants, Canada
Jan-Idar Solbakken	Permanent Participants, Norway
Cindy Dickson (from July 2002)	Permanent Participants, Canada

ACIA Secretariat

Gunter Weller, Executive Director	ACIA Secretariat, USA
Patricia A. Anderson	ACIA Secretariat, USA

Lead Authors[*]

Jim Berner	Alaska Native Tribal Health Consortium, USA
Terry V. Callaghan	Abisko Scientific Research Station, Sweden
	Sheffield Centre for Arctic Ecology, UK
Henry Huntington	Huntington Consulting, USA
Arne Instanes	Instanes Consulting Engineers, Norway
Glenn P. Juday	University of Alaska Fairbanks, USA
Erland Källén	Stockholm University, Sweden
Vladimir M. Kattsov	Voeikov Main Geophysical Observatory, Russia
David R. Klein	University of Alaska Fairbanks, USA
Harald Loeng	Institute of Marine Research, Norway
Gordon McBean	University of Western Ontario, Canada
James J. McCarthy	Harvard University, USA
Mark Nuttall	University of Aberdeen, Scotland, UK
	University of Alberta, Canada
James D. Reist (to June 2002)	Fisheries and Oceans Canada, Canada
Frederick J. Wrona (from June 2002)	National Water Research Institute, Canada
Petteri Taalas (to March 2003)	Finnish Meteorological Institute, Finland
Aapo Tanskanen (from March 2003)	Finnish Meteorological Institute, Finland
Hjálmar Vilhjálmsson	Marine Research Institute, Iceland
John E. Walsh	University of Alaska Fairbanks, USA
Betsy Weatherhead	University of Colorado at Boulder, USA

Liaisons

Snorri Baldursson (Aug. 2000 - Sept. 2002)	Conservation of Arctic Flora and Fauna, Iceland
Magdalena Muir (Sept. 2002 – May 2004)	Conservation of Arctic Flora and Fauna, Iceland
Maria Victoria Gunnarsdottir (from May 2004)	Conservation of Arctic Flora and Fauna, Iceland
Snorri Baldursson (from Sept. 2002)	Arctic Council, Iceland
Odd Rogne	International Arctic Science Committee, Norway
Bert Bolin (to July 2000)	Intergovernmental Panel on Climate Change, Sweden
James J. McCarthy (June 2001 – April 2003)	Intergovernmental Panel on Climate Change, USA
John Stone (from April 2003)	Intergovernmental Panel on Climate Change, Canada
John Calder	National Oceanic and Atmospheric Administration, USA
Karl Erb	National Science Foundation, USA
Hanne Petersen (from Sept. 2001)	Denmark

---

[*]Not all lead authors are members of the Assessment Steering Committee. For a full list of authors see Appendix A.

Contents

Chapter 1

An Introduction to the Arctic Climate Impact Assessment

Lead Authors
Henry Huntington, Gunter Weller

Contributing Authors
Elizabeth Bush, Terry V. Callaghan, Vladimir M. Kattsov, Mark Nuttall

Contents

I have heard it said by many Russians that their climate also is ameliorating! Will God, then, ... give them up even the sky and the breeze of the South? Shall we see Athens in Lapland, Rome at Moscow, the riches of the Thames in the Gulf of Finland, and the history of nations reduced to a question of latitude and longitude?
Astolphe de Custine, 14 July 1839 de Custine, 2002

1.1. Introduction

The *Arctic Climate Impact Assessment* (ACIA) is the first comprehensive, integrated assessment of climate change and ultraviolet (UV) radiation across the entire Arctic region. The assessment had three main objectives:

1. To provide a comprehensive and authoritative scientific synthesis of available information about observed and projected changes in climate and UV radiation and the impacts of those changes on ecosystems and human activities in the Arctic. The synthesis also reviews gaps in knowledge and the research required to fill those gaps. The intended audience is the international scientific community, including researchers and directors of research programs. The *ACIA Scientific Report* fulfills this goal.
2. To provide an accessible summary of the scientific findings, written in plain language but conveying the key points of the scientific synthesis. This summary, the *ACIA Overview Report* (ACIA, 2004a), is for policy makers and the general public.
3. To provide policy guidance to the Arctic Council to help guide the individual and collective responses of the Arctic countries to the challenges posed by climate change and UV radiation. The *ACIA Policy Document* (ACIA, 2004b) accomplishes this task.

An assessment of expected impacts is a difficult and long-term undertaking. The conclusions presented here, while as complete as present information allows, are only a step – although an essential first step – in a continuing process of integrated assessment (e.g., Janssen, 1998). There are many uncertainties, including the occurrence of climate regime shifts, such as possible cooling and extreme events, both of which are difficult if not impossible to predict. New data will continue to be gathered from a wide range of approaches, however, and models will be refined such that a better understanding of the complex processes, interactions, and feedbacks that comprise climate and its impacts will undoubtedly develop over time. As understanding improves it will be possible to predict with increasing confidence what the expected impacts are likely to be in the Arctic.

This assessment uses the definition of the Arctic established by the Arctic Monitoring and Assessment Programme, one of the Arctic Council working groups responsible for the ACIA. Each of the eight arctic coun-

Fig. 1.1. The four regions of the Arctic Climate Impact Assessment.

tries established the boundary in its own territory, and the international marine boundary was established by consensus. The definition of the arctic landmass used here is wider than that often used but has the advantage of being inclusive of landscapes and vegetation from northern forests to polar deserts, reflecting too the connections between the Arctic and more southerly regions. Physical, biological, and societal conditions vary greatly across the Arctic. Changes in climate and UV radiation are also likely to vary regionally, contributing to different impacts and responses at a variety of spatial scales. To strike a balance between overgeneralization and over-specialization, four major regions were identified based on differences in large-scale weather- and climate-shaping factors. Throughout the assessment, differences in climate trends, impacts, and responses were considered across these four regions, to explore the variations anticipated and to illustrate the need for responses targeted to regional and local conditions. The four ACIA regions are shown in Fig. 1.1. There are many definitions of the Arctic, such as the Arctic Circle, treeline, climatic boundaries, and the zone of continuous permafrost on land and sea-ice extent on the ocean. The numerous and complex connections between the Arctic and lower latitudes make any strict definition nearly meaningless, particularly in an assessment covering as many topics and issues as this one. Consequently, there was a deliberate decision not to define the Arctic for the assessment as a whole. Each chapter of this report describes the area that is relevant to its particular subject, implicitly or explicitly determining its own southern boundary.

1.2. Why assess the impacts of changes in climate and UV radiation in the Arctic?

1.2.1. Climate change

There are four compelling reasons to examine arctic climate change. First, the Arctic, together with the Antarctic Peninsula, experienced the greatest regional warming on earth in recent decades, due largely to various feedback processes. Average annual temperatures have risen by about 2 to 3 °C since the 1950s and in winter by up to 4 °C. The warming has been largest over the land areas (Chapman and Walsh, 2003; see also Figs. 1.2 and 1.3). There are also areas of cooling in southern Greenland, Davis Strait, and eastern Canada. The warming has resulted in extensive melting of glaciers (Sapiano et al., 1997), thawing of permafrost (Osterkamp, 1994), and reduction in extent of sea ice in the Arctic Ocean (Rothrock et al., 1999; Vinnikov et al., 1999). The warming has been accompanied by increases in precipitation, but a decrease in the duration of snow cover. These changes have been interpreted to be due at least in part to anthropogenic intensification of the global greenhouse effect, although the El Niño–Southern Oscillation and the inter-decadal Arctic Oscillation also affect the Arctic. The latter can result in warmer and wetter winters in its warm phases, and cooler, drier winters in its cool phases (see Chapter 2).

Second, climate projections suggest a continuation of the strong warming trend of recent decades, with the largest changes coming during winter months (IPCC, 1990, 1996, 2001a,b). For the B2 emissions scenario used by the Intergovernmental Panel on Climate Change (IPCC) and in the ACIA (see section 1.4.2), the five ACIA-designated general circulation models (GCMs; see section 1.4.2) project an additional warming in the annual mean air temperature of approximately 1 °C by 2020, 2 to 3 °C by 2050, and 4 to 5 °C by 2080; the three time intervals considered in this assessment (see Figs. 1.4 and 1.5). Within the Arctic, however, the models do show large seasonal and regional differences; in fact, the differences between individual models are greatest in the polar regions (McAvaney et al., 2001). The reduction in or loss of snow and ice has the effect of increasing the warming trend as reflective snow and ice surfaces are replaced by darker land and water surfaces that absorb more solar radiation. At one extreme, for example, the model of the Canadian Centre for Climate Modelling and Analysis projects near-total melting of arctic sea ice by 2100. Large winter warming in the Arctic is likely to accelerate already evident trends of a shorter snow season, retreat and thinning of sea ice, thawing of permafrost, and accelerated melting of glaciers.

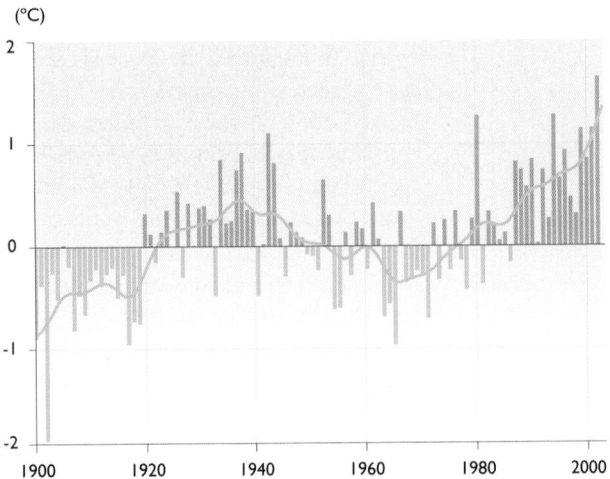

(°C)

Fig. 1.2. Annual average near surface air temperature from stations on land relative to the average for 1961–1990, for the region from 60° to 90° N (updated from Peterson and Vose, 1997).

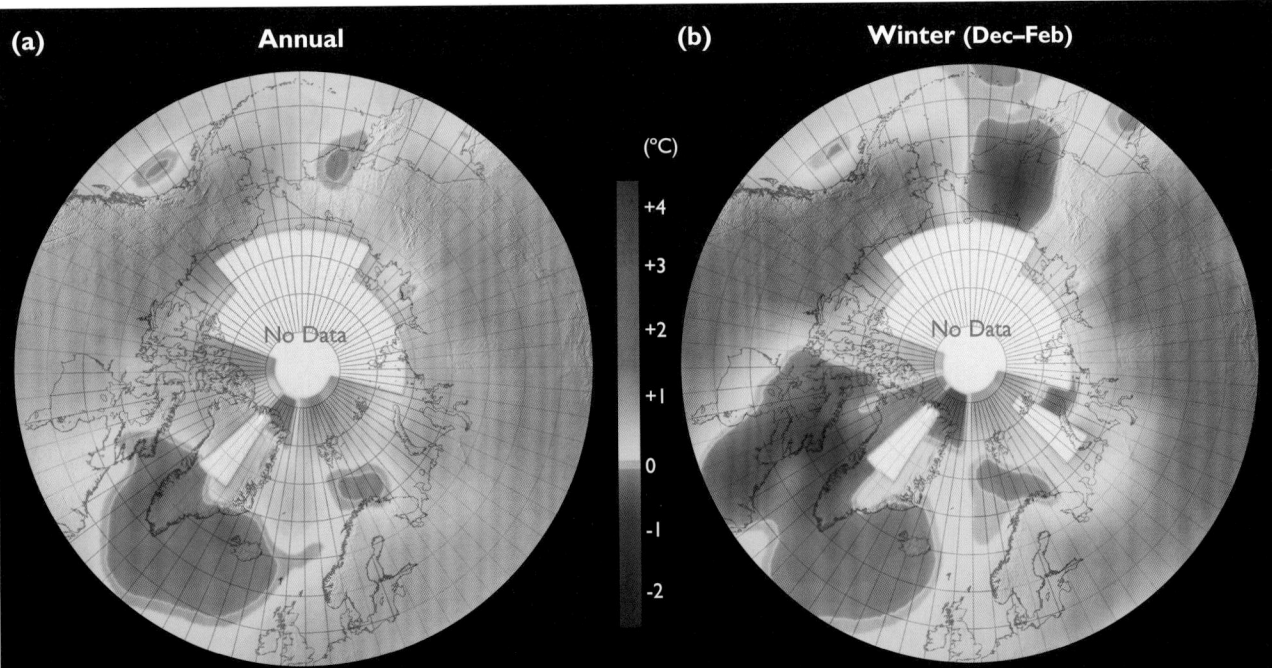

Fig. 1.3. Change in observed surface air temperature between 1954 and 2003: (a) annual mean; (b) winter (Chapman and Walsh, 2003, using data from the Climatic Research Unit, University of East Anglia, www.cru.uea.ac.uk/temperature).

Third, the changes seen in the Arctic have already led to major impacts on the environment and on economic activities (e.g., Weller, 1998). If the present climate warming continues as projected, these impacts are likely to increase, greatly affecting ecosystems, cultures, lifestyles, and economies across the Arctic (see Chapters 10 to 17). On land, the ecosystems range from the ecologically more productive boreal forest in the south to the tundra meadows and unproductive barrens in the High Arctic (Fig. 1.6). Reindeer herding and, to a lesser extent, agriculture are among the economic activities in terrestrial areas. Tourism is an increasing activity throughout the region. Some of the world's largest gas, oil, and mineral deposits are found in the Arctic. In the

marine environment, the Bering Sea, North Atlantic Ocean, and Barents Sea have some of the most productive fisheries in the world (Weller and Lange, 1999). As this assessment makes clear, all these systems and the activities they support are vulnerable to climate change.

In the Arctic there are few cities and many rural communities. Indigenous communities throughout the Arctic depend on the land, lakes and rivers, and the sea for food and income and especially for the vital social and cultural importance of traditional activities. The cultural diversity of the Arctic is already at risk (Freeman, 2000; Minority Rights Group, 1994), and this may be exacerbated by the additional challenge posed by climate change. The impacts of climate change will occur within the context of the societal changes and pressures that arctic indigenous residents are facing in their rapid transition to the modern world. The imposition of climate change from outside the region can also be seen as an ethical issue, in which people in one area suffer the consequences of actions beyond their control and in which beneficial opportunities may accrue to those outside the region rather than those within.

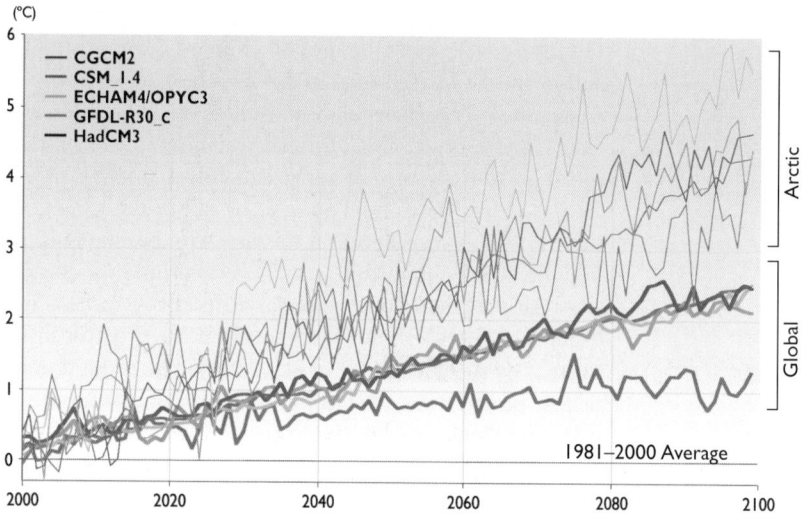

Fig. 1.4. Average surface air temperatures projected by the five ACIA-designated climate models for the B2 emissions scenario (see Chapter 4 for further details). The heavy lines are projected average *global* temperature increases and the thinner lines the projected average *arctic* temperature increases.

Fourth, climate change in the Arctic does not occur in isolation. The Arctic is an important part of the global climate system; it both affects and is affected by global climate change. Changes in climate in the Arctic, and in the environmental parameters such

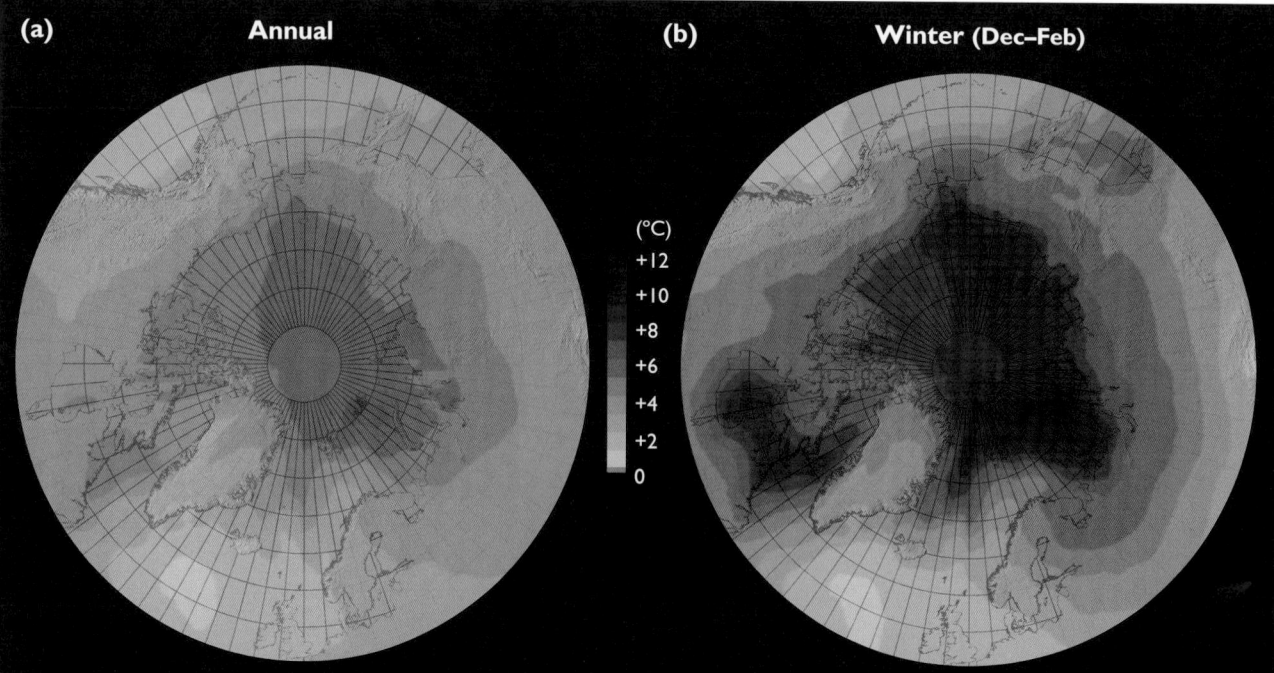

Fig. 1.5. (a) Projected annual surface air temperature change from the 1990s to the 2090s, based on the average change projected by the five ACIA-designated climate models using the B2 emissions scenario. (b) Projected surface air temperature change in winter from the 1990s to the 2090s, based on the average change projected by the five ACIA-designated climate models using the B2 emissions scenario.

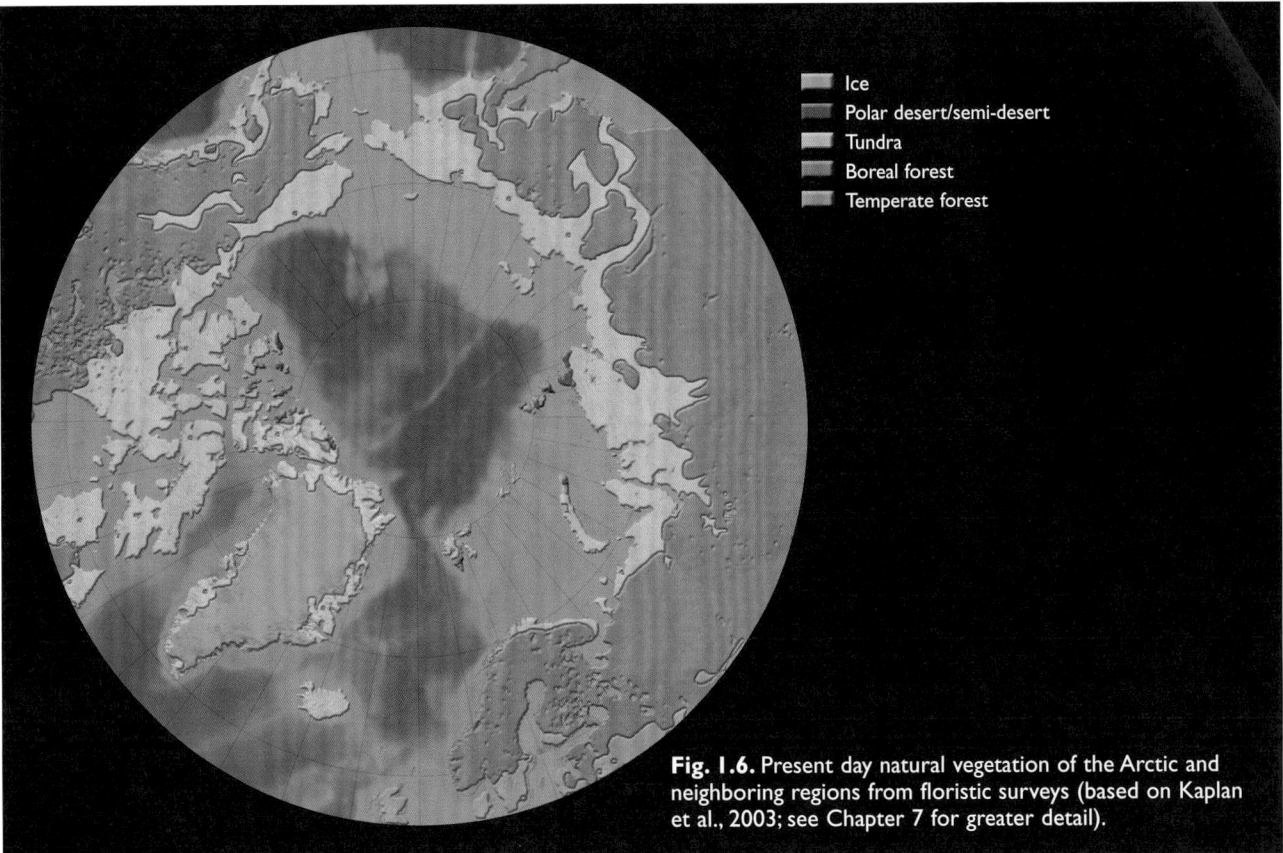

Fig. 1.6. Present day natural vegetation of the Arctic and neighboring regions from floristic surveys (based on Kaplan et al., 2003; see Chapter 7 for greater detail).

Legend:
- Ice
- Polar desert/semi-desert
- Tundra
- Boreal forest
- Temperate forest

as snow cover and sea ice that affect the earth's energy balance and the circulation of the oceans and the atmosphere, may have profound impacts on regional and global climates. Understanding the role of the Arctic and the implications of projected changes and their feedbacks, regionally and globally, is critical to assessing global climate change and its impacts. Furthermore, migratory species provide a direct biological link between the Arctic and lower latitudes, while arctic resources such as fish and oil play an economic role of global significance. Impacts on any of these may have global implications.

1.2.2. UV radiation

The case for assessing UV radiation is similarly compelling. Stratospheric ozone depletion events of up to 45% below normal have been recorded recently in the Arctic (Fioletov et al., 1997). Dramatic change in the thickness of the stratospheric ozone layer and corresponding changes in the intensity of solar UV radiation were first observed in Antarctica in the mid-1980s. The depletions of ozone were later found to be the result of anthropogenic chemicals such as chlorofluorocarbons reaching the stratosphere and destroying ozone. Ozone depletion has also been observed in the Arctic in most years since 1992. Owing to global circulation patterns, the arctic stratosphere is typically warmer and experiences more mixing than the antarctic stratosphere. The ozone decline is therefore more variable in the Arctic. For example, severe arctic ozone depletions were observed in most of the last ten springs, but not in 2002 owing to early warming of the stratosphere.

Although depletion of stratospheric ozone was expected to lead to increased UV radiation at the earth's surface, actual correlations have become possible only recently because the period of instrumental UV measurement is short. Goggles found in archaeological remains in the Arctic indicate that UV radiation has been a fact of human life in the Arctic for millennia. In recent years, however, UV radiation effects, including sunburn and increased snow blindness, have been reported in regions where they were not observed previously.

Future increases in UV-B radiation of 20 to 90% have been predicted for April for the period 2010 to 2020 (Taalas et al., 2000). Ultraviolet radiation can have a variety of harmful impacts on human beings, on plants and animals, and on materials such as paints, cloths, and plastics (Andrady et al., 2002). Ultraviolet radiation also affects many photochemical reactions, such as the formation of ozone in the lower atmosphere. In the Arctic, human beings and ecosystems have both adapted to the very low intensity of the solar UV radiation compared with that experienced at lower latitudes. The low intensity of UV radiation in the Arctic is a consequence of the sun never reaching high in the sky as well as the presence of the world's thickest ozone layer. The Arctic as a whole may therefore be particularly susceptible to increases in UV radiation.

Other factors that affect the intensity of UV radiation include cloudiness and the amount of light reflected by the surface. Climate change is likely to affect atmospheric circulation as well as cloudiness and the extent and duration of snow and ice cover, which in turn will

affect UV radiation. Thus, UV radiation is both a topic of concern in itself and also in relation to climate change (UNEP, 2003).

1.3. The Arctic Climate Impact Assessment

1.3.1. Origins of the assessment

The idea to conduct an assessment of climate and UV radiation in the Arctic grew from several initiatives in the 1990s. The International Arctic Science Committee (IASC) had been engaged in climate studies since it was founded in 1991, and conducted regional arctic impact studies throughout the 1990s. The Arctic Monitoring and Assessment Programme (AMAP) also conducted a preliminary assessment of climate and UV impacts in the Arctic, which was published in 1998. The need for a comprehensive and circum-Arctic climate impact study had been discussed by IASC for some time, and IASC invited AMAP and CAFF (Conservation of Arctic Flora and Fauna) to participate in a joint venture. A joint meeting between the three groups was held in April 1999 and the IASC proposal was used as the basis for discussion. A revised version of the proposal was then submitted to the Arctic Council and the IASC Council for approval. A joint project between the Arctic Council and IASC – the Arctic Climate Impact Assessment – was formally approved by the Arctic Council at its meeting in October 2000.

In addition to the work of the groups responsible for its production, the ACIA builds on several regional and global climate change assessments. The IPCC has made the most comprehensive and best-known assessment of climate change on a global basis (e.g., IPCC, 2001a,b), and has provided many valuable lessons for the ACIA. In addition, regional studies have examined, among other areas, Canada (Maxwell, 1997), the Mackenzie Basin (Cohen 1997a,b), the Barents Sea (Lange and the BASIS Consortium, 2003; Lange et al., 1999), and Alaska (Weller et al., 1999). (The results of these regional studies are summarized in Chapter 18.) Ozone depletion and UV radiation have also been assessed globally by the World Meteorological Organization (WMO,

2003) and the United Nations Environment Programme (UNEP, 2003). These assessments, and the research that they comprise, provide a baseline against which the findings of the ACIA can be considered.

1.3.2. Organization

The ACIA started in October 2000 and was completed by autumn 2004. Together, AMAP, CAFF, and IASC set up the organization for the ACIA, starting with an Assessment Steering Committee (ASC) to oversee the assessment. The members of the ASC included a chair, vice-chair, and executive director, all the lead authors for the ACIA chapters, several scientists appointed by the three sponsoring organizations, and three individuals appointed by the indigenous organizations in the Arctic Council. A subset of the ASC, the Assessment Integration Team, was created to coordinate the material in the various chapters and documents produced by the ACIA. The Arctic Council, including its Senior Arctic Officials, provided oversight through progress reports and documentation at all the Arctic Council meetings.

Funding was provided to the ACIA through direct and indirect support by each of the eight arctic nations. As the lead country for the ACIA, the United States provided financial support through the National Science Foundation and the National Oceanic and Atmospheric Administration, which allowed the establishment of an ACIA Secretariat at the University of Alaska Fairbanks. Contributions from the other arctic countries, as well as from the United Kingdom, supported the involvement of their citizens and provided in-kind support, such as hosting meetings and workshops.

Much of the credibility associated with an assessment comes from the reputation of the authors, who are well-recognized experts in their fields of study. Broad participation of experts from many different disciplines and countries in the writing of the ACIA documents was established through an extensive nomination process. From these nominations, the ASC selected lead and contributing authors for each chapter of the assessment. The chapters were drafted by around 180 lead and co-lead authors, contributing authors, and consulting authors from 12 countries, including all the arctic countries. The ultimate standard in any scientific publication is peer review. The scientific chapters of the ACIA were subject to a rigorous and comprehensive peer review process, which included around 200 reviewers from 15 countries.

1.3.3. Terminology of likelihood

Discussion of future events and conditions must take into account the likelihood that these events or conditions will occur. Often, assessments of likelihood are qualitative or cover a range of probabilities. To avoid confusion and to promote consistent usage, the ACIA has adapted a lexicon of terms from the US National Assessment Team (NAST, 2000) describing the likelihood of expected change. The stated likelihood of particular impacts occurring is based on expert evaluation of results from multiple lines of evidence including field and laboratory experiments, observed trends, theoretical analyses, and model simulations. Judgments of likelihood are indicated using a five-tier lexicon (see Fig. 1.7) consistent with everyday usage. These terms are similar to those used by the IPCC, though somewhat simplified, and are used throughout the ACIA.

Fig. 1.7. Five-tier lexicon describing the likelihood of expected change.

1.4. The assessment process

1.4.1. The nature of science assessment

The ACIA is a "science assessment" in the tradition of other major international assessments of current environmental issues. For example, the IPCC, the international body mandated to assess the relevant information for understanding the risk of human-induced climate change, recently released its Third Assessment Report (IPCC, 2001a,b). The WMO and UNEP jointly released their latest assessments of the issue of stratospheric ozone depletion (WMO, 2003; UNEP, 2003). Two Arctic Council working groups, AMAP and CAFF, have also recently completed science assessments of, respectively, pollution and biodiversity in the circumpolar Arctic (AMAP, 2002, 2003a,b, 2004a,b,c; CAFF, 2001). All of these, and indeed all other assessments, have in common the purpose of providing scientific advice to decision makers who need to develop strategies regarding their respective areas of responsibility. The ACIA responds directly to the request of the Arctic Council for an assessment that can provide the scientific basis for policies and actions.

The essence of a science assessment is to analyze critically and judge definitively the state of understanding on an issue that is inherently scientific in nature. It is a point-in-time evaluation of the existing knowledge base, highlighting both areas of confidence and consensus and areas of uncertainty and disagreement in the science. Another aim of an assessment is to stimulate research into filling emerging knowledge gaps and solving unresolved issues. A science assessment thus draws primarily on the available literature, rather than on new research. To be used within an assessment, a study must have been published according to standards of scientific excellence. (With regard to the incorporation of indigenous knowledge, see the discussion in section 1.4.3.) Publications in the open, peer-reviewed scientific literature meet this standard. Other resources, such as technical publications by government agencies, may be included if they have undergone review and are publicly available.

1.4.2. Concepts and tools in climate assessment

The arctic climate system is complex. The processes of climate and the ways in which various phenomena affect one another – the feedbacks in the system – are still not

fully understood. Specific feedbacks are introduced by the cryosphere and, in particular, by sea ice with its complex dynamics and thermodynamics. Other complex features include the internal dynamics of the polar atmosphere, stratification of both the lower troposphere and the ocean, and phenomena such as the dryness of the air and multiple cloud layers. All these add to the challenge of developing effective three-dimensional models and constructing climate scenarios based on the outcome of such models (Randall et al., 1998; Stocker et al., 2001).

"Climate scenario" means a plausible representation of the future climate that is consistent with assumptions about future emissions of greenhouse gases and other pollutants (emissions scenarios) and with the current understanding of the effects that increased atmospheric concentrations of these components have on climate (IPCC-TGCIA, 1999). Correspondingly, a "climate-change scenario" is the difference between conditions under a future climate scenario and those of today's climate. Being dependent on a number of assumptions about future human activities and their impact on the composition of the atmosphere, climate and climate-change scenarios are not predictions, but plausible descriptions of possible future climates.

Selection of climate scenarios for impact assessments is always controversial and vulnerable to criticism (Smith et al., 1998). The following criteria are suggested (Mearns et al., 2001) for climate scenarios to be most useful to impact assessors and policy makers: (1) consistency with global warming projections over the period 1990 to 2100 ranging from 1.4 to 5.8 °C (IPCC, 2001a); (2) physical plausibility; (3) applicability in impact assessments, providing a sufficient number of variables across relevant temporal and spatial scales; (4) representativeness, reflecting the potential range of future regional climate change; and (5) accessibility. It is preferable for impact researchers to use several climate scenarios, generated by different models where possible, in order to evaluate a greater range of possible futures. Practical limitations, however, typically mean researchers can only work with a small number of climate scenarios.

One starting point for developing a climate change scenario is to select an emissions scenario, which provides a plausible projection of future emissions of substances such as greenhouse gases and aerosols. The most recent IPCC emissions scenarios used in model simulations are those published in the Special Report on Emissions Scenarios (SRES, Nakićenović et al., 2000). The SRES emissions scenarios were built around four basic paths of development that the world may take in the 21st century. It should be noted that no probabilities were assigned to the various SRES emissions scenarios.

During the initial stage of the ACIA process, to stay coordinated with current IPCC efforts, it was agreed that the ACIA should work from IPCC SRES emissions scenarios (Källén et al., 2001). At that time, most of the available or soon-to-be-available simulations that allowed their own uncertainties to be assessed used the A2 and B2 emissions scenarios (Cubasch et al., 2001):

- The *A2 emissions scenario* assumes an emphasis on economic development rather than conservation. Population is projected to increase continuously.
- The *B2 emissions scenario* differs in having a greater emphasis on environmental concerns than economic concerns. It has intermediate levels of economic growth and a population that, although continuously increasing, grows at a slower rate than that in the A2 emissions scenario.

Both A2 and B2 can be considered intermediate scenarios. For reasons of schedule and limitations of data storage, ACIA had to choose one as the central emissions scenario. B2 was chosen because at the time it had been more widely used to generate scenarios, with A2 as a plausible alternative as its use increased.

Once an emissions scenario is selected, it must be used in a climate model (atmosphere–ocean general circulation model, or AOGCM; those used in this assessment are coupled atmosphere-land-ice-ocean models) to produce a climate scenario. Considering the large and increasing number of models available, selecting the models and model outputs for the assessment was not a trivial matter. The IPCC (McAvaney et al., 2001) concluded that no single model can be considered "best" and that it is important to utilize results from a range of coupled models.

Initially, a set of the most recent and comprehensive AOGCMs whose outputs were available from the IPCC Data Distribution Centre were chosen. Later, this set

was reduced to five AOGCMs (two European and three North-American) for practical reasons. The treatment of land surfaces and sea ice is included in all these models, but with varying degrees of complexity. The five ACIA-designated models and the institutes that run them are:

- CGCM2 (Canadian Centre for Climate Modelling and Analysis)
- CSM_1.4 (National Center for Atmospheric Research, USA)
- ECHAM4/OPYC3 (Max-Planck Institute for Meteorology, Germany)
- GFDL-R30_c (Geophysical Fluid Dynamics Laboratory, USA)
- HadCM3 (Hadley Centre for Climate Prediction and Research, UK).

In the initial phase of the ACIA, at least one simulation using the B2 emissions scenario and extending to 2100 was accomplished with each of the five ACIA-designated models. For climate change scenarios, the ACIA climate baseline is 1981–2000. Any differences from the more familiar IPCC baseline of 1961–1990 were small. Three 20-year time slices are the foci of the ACIA for the 21st century: 2011–2030, 2041–2060, and 2071–2090, corresponding to near-term, mid-term, and longer-term outlooks for climate change. A complete description and discussion of the modeling work under ACIA, as well as its limitations, are provided in Chapter 4.

Other types of scenario were also used by chapter authors or by the studies on which the chapters of the assessment are based. These include analogue scenarios of a future climate, based on past (instrumentally recorded) or paleo (geologically recorded) warm climates (i.e., temporal analogue scenarios) or current climates in warmer regions (i.e., spatial analogue scenarios). Although instrumental records provide relatively poor coverage for most of the Arctic, their use avoids uncertainties associated with interpreting other indicators, providing a significant advantage over other approaches. Overall, analogue scenarios were used widely in the ACIA, supplementing the scenarios produced by numerical models. No single impact model was used in the impacts chapters of the assessment; each chapter made use of its own approaches. Further work in this area might consider the need and ability to develop impact models that can be used to address the diversity of topics addressed in this assessment. Another need is for models and scenarios that are able to show more detailed regional and sub-regional variations and that can be used for local impact assessments.

1.4.3. Approaches for assessing impacts of climate and UV radiation

The study of climate and UV radiation involves detailed measurements of physical parameters and the subsequent analysis of results to detect patterns and trends and to create quantitative models of these trends and their interactions. As Chapters 2, 4, 5, and 6 show, this is not a trivial undertaking. The next step, using measurements and models to assess the likely impacts of changes in climate and UV radiation, is even more complex and uncertain. Ecosystems and societies are changing in ways great and small and are driven by many co-occurring factors regardless of variability in climate and UV radiation. Determining how changes in climate and UV radiation may affect dynamic systems relies on several sources of data and several approaches to analysis (see further discussion in Chapter 7).

Most experimental and empirical data can reveal how climate and UV radiation affect plants, animals, and human communities. Observational studies and monitoring can document changes in climate and UV radiation over time together with associated changes in the physical, biological, and social environment. The drawback to observational studies is that they are opportunistic and require that the correct parameters are tracked in a system in which change actually occurs. Establishing causal connections is harder, but can be done through studies of the physical and ecological processes that link environmental components. Experimental studies involve manipulations of small components of the environment, such as vegetation plots or streams. In these cases, the researcher determines the simulated climate or UV radiation change or changes, so there is great control over the conditions being studied. The drawback is that the range of climate and UV radiation conditions may not match that anticipated by various scenarios used for regional assessments, limiting the applicability of the experimental data to the assumptions of the particular assessment.

The use of analogues, as described at the end of the previous section, can help identify potential consequences of climate change. Looking at past climates and climate change events can help identify characteristic biota and how they change. Spatial analogues can be used to compare ecosystems that exist now with the ecosystems where similar climate conditions are anticipated in the future. A strength of analogues is that they enable an examination of actual changes over an ecosystem, rather than hypothetical changes or changes to small experimental sites. Their weakness is that perfect analogues cannot be found, making interpretation difficult because of the variety of factors that cannot be controlled.

For assessing impacts on societies, a variety of social and economic models and approaches can be used. Examining resilience, adaptation, and vulnerability (see further discussion in Chapter 17) offers a powerful means of understanding at least some of the dynamics and complexity associated with human responses to environmental and other changes. As with changes to the natural environment, examining societal dynamics can be achieved through models, observations, and the use of analogues.

These scientific approaches can be complemented by another source of information; indigenous and local knowledge[1]. This assessment makes use of such knowledge to an unprecedented degree in an exercise of this kind. Some extra attention to the topic is therefore warranted here. Indigenous residents of the Arctic have for millennia relied on their knowledge of the environment in order to provide food and other materials and to survive its harsh conditions. More recent arrivals, too, may have a wealth of local knowledge about their area and its environment. The high interannual variability in the Arctic has forced its residents to be adaptable to a range of conditions in climate and the abundance and distribution of animals. Although indigenous and local knowledge is not typically gathered for the specific purpose of documenting climate and UV radiation changes, it is nonetheless a valuable source of insight into environmental change over long periods and in great local detail, often covering areas and seasons in which little scientific research has been conducted. The review of documented information by the communities concerned is a crucial step in establishing whether the information contained in reports about indigenous and local knowledge reliably reflects community perspectives. This step of community review offers a similar degree of confidence to that provided by the peer-review process for scientific literature.

Determining how best to use indigenous knowledge in environmental assessments, including assessments of the impacts of climate and UV radiation, is a matter of debate (Howard and Widdowson, 1997; Stevenson, 1997), but the quality of information generated in careful studies has been established for many aspects of environmental research and management (e.g., Berkes, 1999; Huntington, 2000; Johannes, 1981). In making use of indigenous knowledge, several of its characteristics should be kept in mind. It is typically qualitative rather than quantitative, does not explicitly address uncertainty, and is more likely to be based on observations over a long period than on comparisons of observations taken at the same time in different locations. Identifying mechanisms of change can be particularly

---

[1]Many terms are used to refer to the type of knowledge referred to in this assessment as "indigenous knowledge". Among the terms in use in the literature are traditional knowledge, traditional ecological knowledge, local knowledge (often applied to the knowledge of non-indigenous persons), traditional knowledge and wisdom, and a variety of specific terms for different peoples, such as Saami knowledge or Inuit Qaujimajatuqangit. Within the context of this assessment, "indigenous knowledge" should be taken broadly, to include observations, interpretations, concerns, and responses of indigenous peoples. For further discussion see Chapter 3.

difficult. It is also important to note that indigenous knowledge refers to the variety of knowledge systems in the various cultures of the Arctic and is not merely another discipline or method for studying arctic climate.

Using more than one approach wherever possible can reduce the uncertainties inherent in each of these approaches. The ACIA has drawn on all available information, noting the limitations of each source, to compile a comprehensive picture of climate change and its impacts in the Arctic. Existing climate models project a wide range of conditions in future decades. Not all have been or can be studied empirically, nor can field studies examine enough sites to be fully representative of the range of changes across the Arctic. Instead, using data from existing studies to assess impacts from regional scenarios and models requires some extrapolation and judgment. In this assessment, the chapters addressing impacts may not be able to assess the precise conditions projected in the scenarios upon which the overall assessment is based. Instead, where necessary they will describe what is known and examine how that knowledge relates to the conditions anticipated by the scenarios.

1.5. The Arctic: geography, climate, ecology, and people

This section is intended for readers who are unfamiliar with the Arctic. Summaries and introductions to specific aspects of the Arctic can be found in reports published by AMAP (1997, 1998, 2002) and CAFF (2001), as well as the *Arctic Atlas* (State Committee of the USSR on Hydrometeorology and Controlled Natural Environments, 1985) published by the Arctic and Antarctic Research Institute in Russia. *The Arctic: Environment, People, Policy* (Nuttall and Callaghan, 2000) is an excellent summary of the present state of the Arctic, edited by two ACIA lead authors and with contributions from contributing ACIA authors.

1.5.1. Geography

The Arctic is a single, highly integrated system comprised of a deep, ice covered, and nearly isolated ocean surrounded by the land masses of Eurasia and North America, except for breaches at the Bering Strait and in the North Atlantic. It encompasses a range of land- and seascapes, from mountains and glaciers to flat plains, from coastal shallows to deep ocean basins, from polar deserts to sodden wetlands, from large rivers to isolated ponds. They, and the life they support, are all shaped to some degree by cold and by the processes of freezing and thawing. Sea ice, permafrost, glaciers, ice sheets, and river and lake ice are all characteristic parts of the Arctic's physical geography.

The Arctic Ocean covers about 14 million square kilometers. Continental shelves around the deep central basin occupy slightly more than half of the ocean's area — a significantly larger proportion than in any other ocean. The landforms surrounding the Arctic Ocean are of three major types: (1) rugged uplands, many of which were overrun by continental ice sheets that left scoured rock surfaces and spectacular fjords; (2) flat-bedded plains and plateaus, largely covered by deep glacial, alluvial, and marine deposits; and (3) folded mountains, ranging from the high peaks of the Canadian Rockies to the older, rounded slopes of the Ural Mountains. The climate of the Arctic, rather than its geological history, is the principal factor that gives the arctic terrain its distinctive nature (CIA, 1978).

1.5.2. Climate

The Arctic encompasses extreme climatic differences, which vary greatly by location and season. Mean annual surface temperatures range from 4 °C at Reykjavik, Iceland (64° N) and 0 °C at Murmansk, Russia (69° N) through -12.2 °C at Point Barrow, Alaska (71.3° N), -16.2 °C at Resolute, Canada (74.7° N), -18 °C over the central Arctic Ocean, to -28.1 °C at the crest of the Greenland Ice Sheet (about 71° N and over 3000 m elevation). Parts of the Arctic are comparable in precipitation to arid regions elsewhere, with average annual precipitation of 100 mm or less. The North Atlantic area, by contrast, has much greater average precipitation than elsewhere in the Arctic.

Arctic weather and climate can vary greatly from year to year and place to place. Some of these differences are due to the poleward intrusion of warm ocean currents such as the Gulf Stream and the southward extension of cold air masses. "Arctic" temperature conditions can occur at relatively low latitudes (52° N in eastern Canada), whereas forestry and agriculture can be practiced well north of the Arctic Circle at 69° N in Fennoscandia. Cyclic patterns also shape climate patterns, such as the North Atlantic Oscillation (Hurrell, 1995), which strongly influences winter weather patterns across a vast region from Greenland to Central Asia, and the Pacific Decadal Oscillation, which has a similar influence in the North Pacific and Bering Sea. Both may be related to the Arctic Oscillation (see Chapter 2).

1.5.3. Ecosystems and ecology

Although the Arctic is considered a single system, it is often convenient to identify specific ecosystems within that system. Such classifications are not meant to imply clear separations between these ecosystems. In fact, the transition zones between terrestrial, freshwater, and marine areas are often dynamic, sensitive, and biologically productive. Nonetheless, much scientific research, and indeed subsequent chapters in this assessment, use these three basic categories.

1.5.3.1. Terrestrial ecosystems

Species diversity appears to be low in the Arctic, and on land decreases markedly from the boreal forests to the polar deserts of the extreme north. Only about 3% (5900 species) of the world's plant species occur in the Arctic north of the treeline. However, primitive plant species of mosses and lichens are relatively abundant (Matveyeva and Chernov, 2000). Arctic plant diversity appears to be sensitive to climate. The temperature gradient that has such a strong influence on species diversity occurs over much shorter distances in the Arctic than in other biomes. North of the treeline in Siberia, for example, mean July temperature decreases from 12 to 2 °C over 900 km. In the boreal zone, a similar change in temperature occurs over 2000 km. From the southern boreal zone to the equator, the entire change is less than 10 °C (Chernov, 1995).

The diversity of arctic animals north of the treeline (about 6000 species) is similar to that of plants (Chernov, 1995). As with plants, the arctic fauna account for about 3% of the global total, and evolutionarily primitive species are better represented than advanced species. In general, the decline in animal species with increasing latitude is more pronounced than that of plants. An important consequence of this is an increase in dominance. "Super-dominant" species, such as lemmings, occupy a wide range of habitats and generally have large effects on ecosystem processes.

Many of the adaptations of arctic species to their current environments limit their responses to climate warming and other environmental changes. Many adaptations have evolved to cope with the harsh climate, and these make arctic species more susceptible to biological invasions at their southern ranges while species at their northern range limit are particularly sensitive to warming. During environmental changes in the past, arctic species have changed their distributions rather than evolving significantly. In the future, changes in the conditions in arctic ecosystems may affect the release of greenhouse gases to the atmosphere, providing a possibly significant feedback to climate warming although both the direction and magnitude of the feedback are currently very uncertain. Furthermore, vegetation type profoundly influences the water and energy exchange of arctic ecosystems, and so future changes in vegetation driven by climate change could profoundly alter regional climates.

1.5.3.2. Freshwater ecosystems

Arctic freshwater ecosystems are extremely numerous, occupying a substantial area of the arctic landmass. Even in areas of the Arctic that have low precipitation, freshwater ecosystems are common and the term "polar deserts" refers more to the impoverishment of vegetation cover than to a lack of groundwater. Arctic freshwater ecosystems include three main types: flowing water (rivers and streams), permanent standing water (lakes and ponds), and wetlands such as peatlands and bogs (Vincent and Hobbie, 2000). All provide a multitude of goods and services to humans and the biota that use them.

Flowing water systems range from the large, north-flowing rivers that connect the interiors of continents with the Arctic Ocean, through steep mountain rivers, to slow-flowing tundra streams that may contain water during spring snowmelt. The large rivers transport heat, water, nutrients, contaminants, sediment, and biota into the Arctic and together have a major effect on regional environments. The larger rivers flow throughout the year, but small rivers and streams freeze in winter. The biota of flowing waters are extremely variable: rivers fed mainly by glaciers are particularly low in nutrients and have low productivity. Spring-fed streams can provide stable, year-round habitats with a greater diversity of primary producers and insects.

Permanent standing waters vary from very large water bodies to small and shallow tundra ponds that freeze to the bottom in winter. By the time the ice melts in summer, the incoming solar radiation is already past its peak, so that the warming of lakes is limited. Primary production, by algae and aquatic mosses, decreases from the subarctic to the high Arctic. Zooplankton species are limited or even absent in arctic lakes because of low temperatures and low nutrient availability. Species abundance and diversity increase with the trophic status of the lake (Hobbie, 1984). Fish species are generally not diverse, ranging from 3 to 20 species, although species such as Arctic char (*Salvelinus alpinus*) and salmon (*Salmo salar*) are an important resource.

Wetlands are among the most abundant and productive aquatic ecosystems in the Arctic. They are ubiquitous and characteristic features throughout the Arctic and almost all are created by the retention of water above the permafrost. They are more extensive in the southern Arctic than the high Arctic, but overall, cover vast areas – up to 3.5 million km² or 11% of the land surface. Several types of wetlands are found in the Arctic, with specific characteristics related to productivity and climate. Bogs, for example, are nutrient poor and have low productivity but high carbon storage, whereas fens are nutrient rich and have high productivity. Arctic wetlands have greater biological diversity than other arctic freshwater ecosystems, primarily in the form of mosses and sedges. Together with lakes and ponds, arctic wetlands are summer home to hundreds of millions of migratory birds.

Arctic freshwater ecosystems are particularly sensitive to climate change because the very nature of their habitats results from interactions between temperature, precipitation, and permafrost. Also, species limited by temperature and nutrient availability are likely to respond to temperature changes and effects of UV radiation on dead organic material in the water column.

1.5.3.3. Marine ecosystems

Approximately two-thirds of the Arctic as defined by the ACIA comprises ocean, including the Arctic Ocean

and its shelf seas plus the Nordic, Labrador, and Bering Seas. These areas are important components of the global climate system, primarily because of their contributions to deepwater formation that influences global ocean circulation. Arctic marine ecosystems are unique in having a very high proportion of shallow water and coastal shelves. In common with terrestrial and freshwater ecosystems in the Arctic, they experience strong seasonality in sunlight and low temperatures. They are also influenced by freshwaters delivered mainly by the large rivers of the Arctic. Ice cover is a particularly important physical characteristic, affecting heat exchange between water and atmosphere, light penetration to organisms in the water below, and providing a biological habitat above (for example, for seals and polar bears (*Ursus maritimus*)), within, and beneath the ice. The marginal ice zone, at the edge of the pack ice, is particularly important for plankton production and plankton-feeding fish.

Some of these factors are highly variable from year to year and, together with the relatively young age of arctic marine ecosystems, have imposed constraints on the development of ecosystems that parallel those of arctic lands and freshwaters. Thus, in general, arctic marine ecosystems are relatively simple, productivity and biodiversity are low, and species are long-lived and slow-growing. Some arctic marine areas, however, have very high seasonal productivity (Sakshaug and Walsh, 2000) and the sub-polar seas have the highest marine productivity in the world. The Bering and Chukchi Seas, for example, include nutrient-rich upwelling areas that support large concentrations of migratory seabirds as well as diverse communities of marine mammals. The Bering and Barents Seas support some of the world's richest fisheries.

The marine ecosystems of the Arctic provide a range of ecosystem services that are of fundamental importance for the sustenance of inhabitants of arctic coastal areas. Over 150 species of fish occur in arctic and subarctic waters, and nine of these are common, almost all of which are important fishery species such as cod. Arctic marine mammals escaped the mass extinctions of the ice ages that dramatically reduced the numbers of arctic terrestrial mammal species, but many are harvested. They include predators such as the toothed whales, seals, walrus, sea otters, and the Arctic's top predator, the polar bear. Over 60 species of migratory and resident seabirds occur in the Arctic and form some of the largest seabird populations in the world. At least one species, the great auk (*Pinguinus impennis*), is now extinct because of overexploitation.

The simplicity of arctic marine ecosystems, together with the specialization of many of its species, make them potentially sensitive to environmental changes such as climatic change, exposure to higher levels of UV radiation, and increased levels of contaminants. Concomitant with these pressures is potential overexploitation of some marine resources.

1.5.4. Humans

Some two to four million people live in the Arctic today, although the precise number depends on where the boundary is drawn. These people include indigenous peoples (Fig. 1.8) and recent arrivals, herders and hunters living on the land, and city dwellers with desk jobs.

Humans have occupied large parts of the Arctic since at least the last ice age. Archeological remains have been found in northern Fennoscandia, Russia, and Alaska dating back more than 12 000 years (e.g., Anderson, 1988; Dixon, 2001; Thommessen, 1996). In the eastern European Arctic, Paleolithic settlements have been recorded from as early as 40 000 years ago (Pavlov et al., 2001). In Eurasia and across the North Atlantic, groups of humans have moved northward over the past several centuries, colonizing new lands such as the Faroe Islands and Iceland, and encountering those already present in northern Fennoscandia and Russia and in western Greenland (Bravo and Sorlin, 2002; Huntington et al., 1998).

In the 20th century, immigration to the Arctic has increased dramatically, to the point where non-indigenous persons outnumber indigenous ones in many regions. The new immigrants have been drawn by the prospect of developing natural resources, from fishing to gold to oil (CAFF, 2001), as well as by the search for new opportunities and escape from the perceived and real constraints of their home areas. Social, economic, and cultural conflicts have arisen as a consequence of competition for land and resources (Freeman, 2000; Minority Rights Group, 1994; Slezkine, 1994) and the incompatibility of some aspects of traditional and modern ways of life (e.g., Huntington, 1992; Nuttall, 2000). In North America, indigenous claims to land and resources have been addressed to some

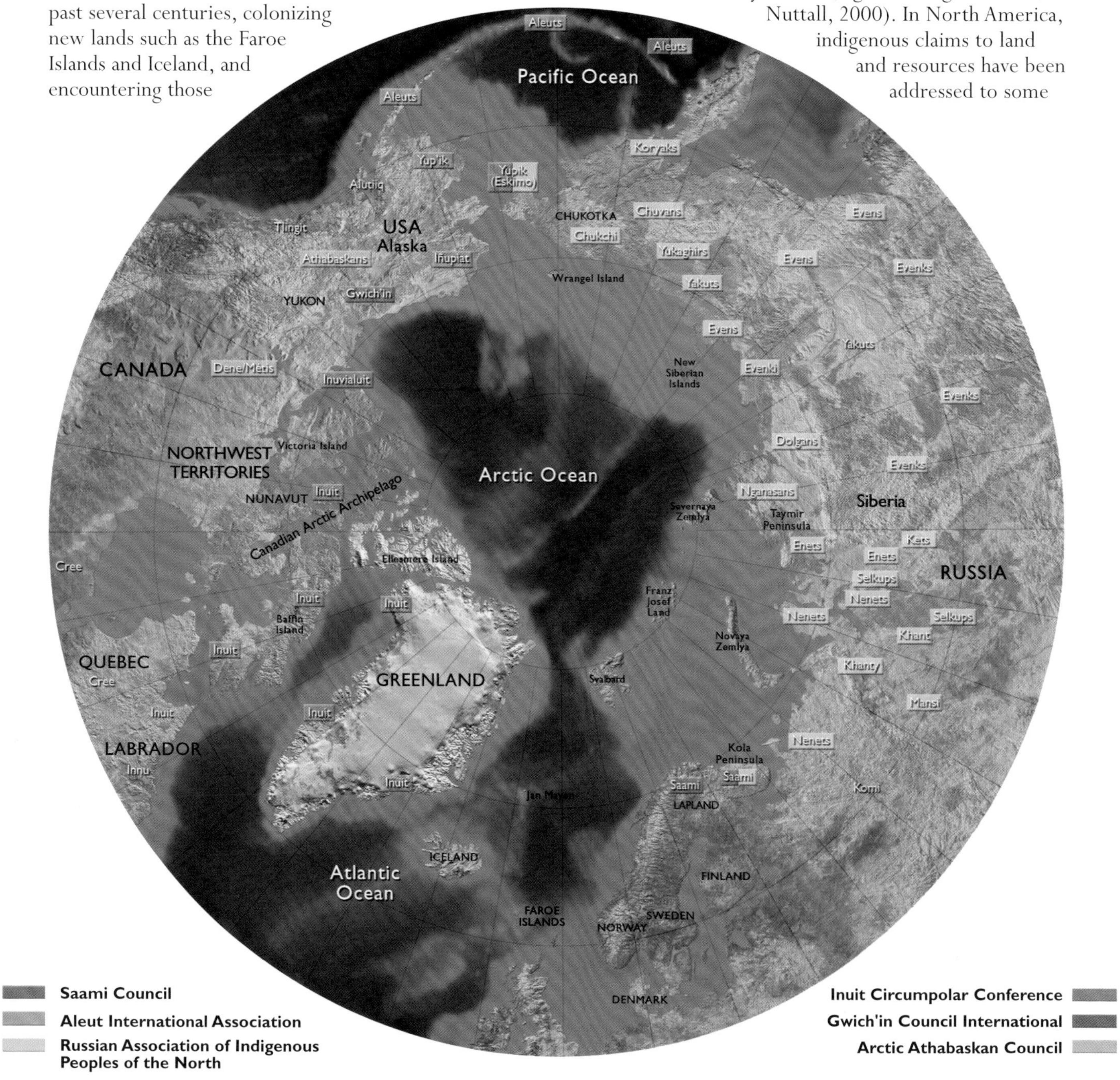

Fig. 1.8. Locations of indigenous peoples in the Arctic, showing affiliation to the Permanent Participants, the indigenous peoples' organizations that participate in the Arctic Council.

extent in land claim agreements, the creation of largely self-governed regions such as Nunavut and Greenland within nation states, and other political and economic actions. In Eurasia, by contrast, indigenous claims and rights have only recently begun to be addressed as matters of national policy (Freeman, 2000).

Many aspects of demography are also changing. Over the past decade, total population has increased rapidly in only three areas: Alaska, Iceland, and the Faroe Islands. Rapid declines in population have occurred across most of northern Russia, with lesser declines or modest increases in other parts of the North (see Table 1.1). Life expectancy has increased greatly across most of the Arctic in recent decades, but declined sharply in Russia in the 1990s. The prevalence of indigenous language use has decreased in most areas, with several languages in danger of disappearing from use. In some respects, the

disparities between northern and southern communities in terms of living standards, income, and education are shrinking, although the gaps remain large in most cases (Huntington et al., 1998). Traditional economies based on local production, sharing, and barter, are giving way to mixed economies in which money plays a greater role (e.g., Caulfield, 2000).

Despite this assimilation on many levels, or perhaps in response to it, many indigenous peoples are reasserting their cultural identity (e.g., Fienup-Riordan et al., 2000; Gaski, 1997). With this activism comes political calls for rights, recognition, and self-determination. The response of arctic indigenous groups to the presence of long-range pollutants in their traditional foods is a useful illustration of their growing engagement with the world community. In Canada particularly, indigenous groups led the effort to establish a national program to study

Table 1.1. Country population data (data sources as in table notes).

Country	Region	Total population	Indigenous population	Year of census/ estimate	Previous figure[a]	Previous indigenous figure[a]	Year of previous estimate
ALL	Arctic	3494107			3885798		
USA	Alaska (excluding Southeast)	553850	103000[b]	2000	481054	73235	1990
Canada	Total	105131	59685	2001	106705		1996
	Yukon Territory	28520	6540	2001	30766	6175	1996
	Northwest Territories	37100	18730	2001	39672	19000	1996
	Nunavut	26665	22720	2001	24730	20690	1996
	Nunavik, Quebec	9632	8750	2001	8715	7780	1996
	Northern Labrador[c]	3214	2945	2001	2822		1996
Denmark	Greenland	56542	49813[d]	2002	55419	48029[d]	1994
	Faroe Islands	47300	0	2002	43700	0	1995
Iceland		286275	0	2001	266783		1994
Norway	Finnmark, Troms, Nordland	462908		2002	468691		1990
	North of the Arctic Circle				379461	35000[e]	1990
Sweden	Norrbotten	254733	10000[ef]	2001	263735	6000[e]	1990
	North of the Arctic Circle	62000[g]			64000[g]		1990
Finland	Lapland	191768	4083[ei]	2000	200000[h]	4000[ei]	1995
Russia	Total	1535600		2002	1999711	67164[j]	1989
	Murmansk Oblast	893300		2002	1164586	1899[j]	1989
	Nenets Autonomous Okrug	41500		2002	53912	6468[j]	1989
	Yamalo-Nenets Autonomous Okrug	507400		2002	494844	30111[j]	1989
	Taimyr (Dolgano-Nenets) A.O.	39800		2002	55803	8728[j]	1989
	Sakha Republic (Arctic area)	k		2002	66632	3982[j]	1989
	Chukotka Autonomous Okrug	53600		2002	163934	15976[j]	1989

Data sources: AMAP, 1998; US Census Bureau, 2002 (www.census.gov); Statistics Canada, 2002 (www12.statcan.ca); Statistics Greenland, 2002 (www.statgreen.gl); Faroe Islands Statistics, 2002 (www.hagstova.fo); Statistics Iceland, 2002 (www.statice.is); Statistics Norway, 2002 (www.ssb.no); Statistics Sweden, 2002 (www.scb.se); Statistics Finland, 2002 (www.stat.fi); State Committee for Statistics, 2003 (www.eastview.com/all_russian_population_census.asp).

[a]Data from AMAP, 1998; [b]estimated by adding the number of Alaska Natives to a proportion of those listed as "mixed race" (calculated using the statewide figure for those of mixed race who are in part Alaska Native); [c]includes Davis Inlet, Hopedale, Makkovik, Nain, Postville, and Rigolet; [d]"indigenous" refers to people born in Greenland, regardless of ethnicity; [e]indigenous population is an estimate only; [f]estimate by the Saami Parliament for 1998 – the difference relative to the 1990 value probably reflects a difference in the method of estimate rather than an actual population increase; [g]estimate only, using the same percentage of the Norrbotten population in each case, rounded to the nearest thousand; [h]year of previous census/estimate unclear – population of Lapland reported as "slightly more than 200000"; [i]this value for the Saami population is for the four northernmost counties of Lapland (the "Saami Area"). There are an additional 3400 Saami elsewhere in Finland; [j]Indigenous figures refer only to the numerically-small peoples, i.e., not the Yakut, Komi, et al.; [k]for the districts of Anabarsk, Allaykhovsk, Bulun, Ust-Yansk, and Nizhnekolymsk.

contaminants, the results of which were used by those groups to advocate and negotiate international conventions to control persistent organic pollutants (Downie and Fenge, 2003). The arguments were often framed in terms of the rights of these distinct peoples to live without interference from afar. The use of international fora to make this case emphasizes the degree to which the indigenous groups think of themselves as participants in global, in addition to national, affairs.

At the same time that indigenous peoples are reaching outward, traditional hunting, fishing, herding, and gathering practices remain highly important. Traditional foods have high nutritional value, particularly for those adapted to diets high in fat and protein rather than carbohydrates (Hansen et al., 1998). Sharing and other forms of distributing foods within and between communities are highly valued, and indeed create a highly resilient adaptation to uncertain food supplies while strengthening social bonds (e.g., Magdanz et al., 2002). The ability to perpetuate traditional practices is a visible and effective way for many indigenous people to exert control over the pace and extent of modernization, and to retain the powerful spiritual tie between people and their environment (e.g., Fienup-Riordan et al., 2000; Ziker, 2002).

It is within this context of change and persistence in the Arctic today that climate change and increased UV radiation act as yet more external forces on the environment that arctic residents rely upon and know well. Depending on how these new forces interact with existing forces in each arctic society and each geographical region, the impacts and opportunities associated with climate change and UV radiation may be minimized or magnified (e.g., Hamilton et al., 2003). The degree to which people are resilient or vulnerable to climate change depends in part on the cumulative stresses to which they are subject through social, political, and economic changes in other aspects of their lives. It also depends in part on the sensitivity of social systems and their capacity for adaptation (see Chapter 17). The human impacts of climate change should be interpreted not in sweeping generalizations about the entire region, but as another influence on the already shifting mosaic that comprises each arctic community.

1.5.5. Natural resources and economics

In economic terms, the Arctic is best known as a source of natural resources. This has been true since the first explorers discovered whales, seals, birds, and fish that could be sold in more southerly markets (CAFF, 2001). In the 20th century, arctic minerals were also discovered and exploited, the size of some deposits of oil, gas, and metal ores more than compensating for the costs of operating in remote, cold regions (AMAP, 1998; Bernes, 1996). Military bases and other facilities were also constructed across much of the Arctic, providing employment but also affecting population distribution and local environments (e.g., Jenness, 1962). In recent decades, tourism has added another sector to the economies of

many communities and regions of the Arctic (Humphries et al., 1998). The public sector, including government services and transfer payments, is also a major part of the economy in nearly all areas of the Arctic, responsible in some cases for over half the available jobs (Huntington et al., 1998). In addition to the cash economy of the Arctic, the traditional subsistence and barter economies are major contributors to the overall well-being of the region, producing significant value that is not recorded in official statistics that reflect only cash transactions (e.g., Schroeder et al., 1987; Weihs et al., 1993).

The three most important economic resources of the Arctic are oil and gas, fish, and minerals.

1.5.5.1. Oil and gas

The Arctic has huge oil and gas reserves. Most are located in Russia: oil in the Pechora Basin, gas in the lower Ob Basin, and other potential oil and gas fields along the Siberian coast. Canadian oil and gas fields are concentrated in two main basins in the Mackenzie Delta/ Beaufort Sea region and in the Arctic Islands. In Alaska, Prudhoe Bay is the largest oil field in North America

and other fields have been discovered or remain to be discovered along the Beaufort Sea coast. Oil and gas fields also exist on Greenland's west coast and in Norway's arctic territories.

1.5.5.2. Fish

Arctic seas contain some of the world's oldest and richest commercial fishing grounds. In the Bering Sea and Aleutian Islands, Barents Sea, and Norwegian Sea annual fish harvests in the past have exceeded two million tonnes, although many of these fisheries have declined (in 2001 fish catches in the Bering Sea totaled 1.6 million tonnes). Important fisheries also exist around Iceland, Svalbard, Greenland, and Canada. Fisheries are important to many arctic countries, as well as to the world as a whole. For example, Norway is the world's biggest fish exporter with exports worth four billion US dollars in 2001.

1.5.5.3. Minerals

The Arctic has large mineral reserves, ranging from gemstones to fertilizers. Russia extracts the greatest quantities of these minerals, including nickel, copper, platinum, apatite, tin, diamonds, and gold, mostly on the Kola

Peninsula but also in Siberia. Canadian mining in the Yukon and Northwest Territories and Nunavut is for lead, zinc, copper, diamonds, and gold. In Alaska lead and zinc deposits in the Red Dog Mine, which contains two-thirds of US zinc resources, are mined, and gold mining continues. The mining activities in the Arctic are an important contributor of raw materials to the world economy.

1.6. An outline of the assessment

This assessment contains eighteen chapters. The seventeen chapters that follow this introduction are organized into four sections: climate change and UV radiation change in the Arctic, impacts on the physical and biological systems of the Arctic, impacts on humans in the Arctic, and future steps and a synthesis of the ACIA.

1.6.1. Climate change and UV radiation change in the Arctic

The arctic climate is an integral part of the global climate, and cannot be understood in isolation. Chapter 2 describes the arctic climate system, its history, and its connections to the global system. This description lays the foundation for the rest of the treatment of climate in this assessment. Chapter 3 lays another essential foundation for the assessment by describing how climate change appears from the perspective of arctic indigenous peoples, a topic also included in other chapters. Chapter 4 describes future climate projections, developed through use of emissions scenarios of greenhouse gases, and climate modeling. Several modeling simulations of future climates were developed specifically for this assessment, and these are described in detail. Chapter 5 provides the counterpart to Chapters 2 and 4 on observations and future projections of UV radiation and ozone, and their effects. The causes and characteristics of ozone depletion are discussed, together with models for the further depletion and eventual recovery of the ozone layer following international action.

1.6.2. Impacts on the physical and biological systems of the Arctic

The primary impacts of climate change and increased UV radiation in the Arctic will be to its physical and biological systems. Chapter 6 describes the changes that have already been observed, and the impacts that are expected to occur in the frozen regions of the Arctic, including sea ice, permafrost, glaciers, and snow cover. River discharge and river and lake ice break-up and freeze-up are also discussed. Chapter 7 discusses impacts on the terrestrial ecosystems of the Arctic, drawing on extensive research, experimental data, observations, and indigenous knowledge. Biodiversity, risks to species, including displacements due to climate change, UV radiation effects, and feedback processes as the vegetation and the hydrological regime change are discussed. Chapter 8 examines freshwater ecosystems in a similar fashion, including a discussion of freshwater fisheries in the Arctic. Chapter 9 covers the marine systems of the

Arctic, and includes topics from the physical ocean regime, including the thermohaline circulation, to sea ice, coastal issues, fisheries, and ecosystem changes.

1.6.3. Impacts on humans in the Arctic

The implications of climate change and changes in UV radiation for humans are many and complex, both direct and indirect. Chapter 10 addresses the challenges to bio-diversity conservation posed by climate change, especially given the relative paucity of data and the lack of circumpolar monitoring at present. Chapter 11 outlines the implications of climate change for wildlife conservation and management, a major concern in light of the substantial changes that are expected to impact upon ecosystems. Chapter 12 looks at traditional practices of hunting, herding, fishing, and gathering, which are also likely to be affected by ecosystem changes, as well as by changes in policies and society. Chapter 13 describes the commercial fisheries of the arctic seas, including seals and whales, with reference to climate as well as to fishing regulations and the socio-economic impacts of current harvests of fish stocks. Chapter 14 extends

the geographic scope of the assessment to the northern boreal forest, examining both that ecosystem and the implications of climate change for agriculture and forestry. Chapter 15 discusses the implications of climate and UV radiation on human health, both for individuals and for communities in terms of public health and cultural vitality. Chapter 16 explores the ways in which climate may affect man-made infrastructure in the Arctic, both in terms of threats to existing facilities such as houses, roads, pipelines, and other industrial facilities, and of future needs resulting from a changing climate.

1.6.4. Future steps and a synthesis of the ACIA

Chapter 17 presents an innovative way of examining societal vulnerability to climate change. It gives some initial results from current research but primarily illustrates prospects for applying this approach more broadly in the future. Chapter 18 contains a synthesis and summary of the main results of the ACIA, including implications for each of the four ACIA regions and directions for future research.

Acknowledgements

Many of the photographs used in this chapter were supplied by Bryan and Cherry Alexander.

References

ACIA, 2004a. Impacts of a Warming Arctic: Arctic Climate Impact Assessment. Cambridge University Press.

ACIA, 2004b. Arctic Climate Impact Assessment. Policy Document. Issued by the Fourth Arctic Council Ministerial Meeting, Reykjavík, 245 November 2004. [online at www.amap.no/acia].

AMAP, 1997. Arctic Pollution Issues: a State of the Arctic Environment Report. Arctic Monitoring and Assessment Programme, Oslo, 188pp.

AMAP, 1998. The AMAP Assessment Report: Arctic Pollution Issues. Arctic Monitoring and Assessment Programme, Oslo, xii + 859pp.

AMAP, 2002. Arctic Pollution 2002. Arctic Monitoring and Assessment Programme, Oslo, xi + 111pp.

AMAP, 2003a. AMAP Assessment 2002: Human Health in the Arctic. Arctic Monitoring and Assesment Programme, Oslo, xiii + 137pp.

AMAP, 2003b. AMAP Assessment 2002: The Influence of Global Change on Contaminant Pathways to, within, and from the Arctic. Arctic Monitoring and Assessment Programme, Oslo, xi + 65pp.

AMAP, 2004a. AMAP Assessment 2002: Persistent Organic Pollutants (POPs) in the Arctic. Arctic Monitoring and Assessment Programme, Oslo, xvi+310pp.

AMAP, 2004b. AMAP Assessment 2002: Radioactivity in the Arctic. Arctic Monitoring and Assessment Programme, Oslo, xi + 100pp.

AMAP, 2004c. AMAP Assessment 2002: Heavy Metals in the Arctic. Arctic Monitoring and Assessment Programme, Oslo.

Anderson, D.D., 1988. Onion portage: the archeology of a stratified site from the Kobuk River, northwest Alaska. Anthropological Papers of the University of Alaska Fairbanks, 22(1–2), 163pp.

Andrady, A.L., H.S. Hamid and A. Torikai, 2002. Effects of climate change and UV-B on materials. In: Environmental Effects of Ozone Depletion and its Interactions with Climate Change: 2002 Assessment, pp. 143–152. United Nations Environment Programme.

Berkes, F., 1999. Sacred Ecology: Traditional Ecological Knowledge and Resource Management. Taylor and Francis, xvi + 209pp.

Bernes, C., 1996. The Nordic Arctic Environment: Unspoilt, Exploited, Polluted? Nordic Council of Ministers, Copenhagen, 240pp.

Bravo, M. and S. Sorlin (eds.), 2002. Narrating the Arctic: a Cultural History of Nordic Scientific Practices. Science History Publications. 373pp.

CAFF, 2001. Arctic Flora and Fauna: Status and Conservation. Conservation of Arctic Flora and Fauna, Edita, Helsinki, 272pp.

Caulfield, R.A., 2000. Political economy of renewable resources in the Arctic. In: M. Nuttall and T.V. Callaghan (eds.). The Arctic: Environment, People, Policy, pp. 485–513. Harwood Academic Publishers.

Chapman, W.L. and J.E. Walsh, 2003. Observed climate change in the Arctic, updated from Chapman and Walsh, 1993: Recent variations of sea ice and air temperatures in high latitudes. Bulletin of the American Meteorological Society, 74(1):33–47.

Chernov, Y.I., 1995. Diversity of the Arctic terrestrial fauna. In: F.S. Chapin III and C. Korner (eds.). Arctic and Alpine Biodiversity: Patterns, Causes and Ecosystem Consequences, pp. 81–95. Springer-Verlag.

CIA, 1978. Polar Regions Atlas. Central Intelligence Agency, McLean, Virginia, 66pp. plus maps.

Cohen, S.J., 1997a. What if and so what in Northwest Canada: could climate change make a difference to the future of the Mackenzie Basin. Arctic, 50:293–307.

Cohen, S.J. (ed.), 1997b. Mackenzie Basin Impact Study (MBIS), Final Report. Environment Canada, Downsview, 372pp.

Cubasch, U., G.A. Meehl, G.J. Boer, R.J. Stouffer, M. Dix, A. Noda, C.A. Senior, S. Raper and K.S. Yap, 2001. Projections of future climate change. In: J.T. Houghton, Y. Ding, D.J. Griggs, M. Noguer, P.J. van der Linden, X. Dai, K. Maskell and C.A. Johnson (eds.). Climate Change 2001: The Scientific Basis. Contribution of Working Group I to the Third Assessment Report of the Intergovernmental Panel on Climate Change, pp. 525–582. Cambridge University Press.

de Custine, A., 2002. Letters from Russia. New York Review Books, xiii + 654pp.

Dixon, E.J, 2001. Human colonization of the Americas: timing, technology and process. Quaternary Science Reviews, 20:277–299.

Downie, D.L. and T. Fenge, 2003. Northern Lights against POPs: Combating Toxic Threats in the Arctic. McGill University Press, Montreal, xxv + 347pp.

Fienup-Riordan, A., W. Tyson, P. John, M. Meade and J. Active, 2000. Hunting Tradition in a Changing World. Rutgers University Press, New Jersey, xx + 310pp.

Fioletov, V.E., J.B. Kerr, D.I. Wardle, J. Davies, E.W. Hare, C.T. McElroy and D.W. Tarasick, 1997. Long-term ozone decline over the Canadian Arctic to early 1997 from ground-based and balloon observations. Geophysical Research Letters, 24(22):2705–2708.

Freeman, M.M.R. (ed.), 2000. Endangered Peoples of the Arctic. Greenwood Press, Connecticut, xix + 278pp.

Gaski, H. (ed.), 1997. Sami culture in a new era: the Norwegian Sami experience. Karasjok, Norway: Davvi Girji. 223pp.

Hamilton, L.C., B.C. Brown and R.O. Rasmussen, 2003. West Greenland's cod-to-shrimp transition: local dimensions of climate change. Arctic, 56(3):271–282.

Hansen, J.C., A. Gilman, V. Klopov and J.O. Ødland, 1998. Pollution and human health. In: AMAP. The Assessment Report: Arctic Pollution Issues, pp. 775–844. Arctic Monitoring and Assessment Programme, Oslo.

Hobbie, J.E., 1984. Polar limnology. In: F.B. Taub (ed.). Lakes and Rivers. Ecosystems of the World. Elsevier.

Howard, A. and F. Widdowson, 1997. Traditional knowledge threatens environmental assessment. Policy Options, 17(9):34–36.

Humphries, B.H., Å.Ø. Pedersen, P. Prokosch, S. Smith and B. Stonehouse (eds.), 1998. Linking Tourism and Conservation in the Arctic. Meddelelser No. 159. Norsk Polarinstittut, Tromsø, 140pp.

Huntington, H.P., 1992. Wildlife Management and Subsistence Hunting in Alaska. Belhaven Press, London, xvii + 177pp.

Huntington, H.P., 2000. Using traditional ecological knowledge in science: methods and applications. Ecological Applications, 10(5):1270–1274.

Huntington, H.P., J.H. Mosli and V.B. Shustov, 1998. Peoples of the Arctic. In: AMAP. The Assessment Report: Arctic Pollution Issues, pp. 141–182. Arctic Monitoring and Assessment Programme, Oslo.

Hurrell, J.W., 1995. Decadal trends in the North Atlantic Oscillation: regional temperatures and precipitation. Science, 269:676–679.

IPCC, 1990. Climate Change: the Scientific Assessment. J.T. Houghton, G.J. Jenkins and J.J. Ephraums (eds.). Intergovernmental Panel on Climate Change. Cambridge University Press, 364pp.

IPCC, 1996. Climate Change 1995: Impacts, Adaptations and Mitigation of Climate Change: Scientific-Technical Analysis Contributions of Working Group II to the Second Assessment Report of the Intergovernmental Panel on Climate Change. R.T. Watson, M.C. Zinyowera, R.H. Moss and D.J. Dokken (eds.). Cambridge University Press, 876pp.

IPCC, 2001a. Climate Change 2001: The Scientific Basis. Contribution of Working Group I to the Third Assessment Report of the Intergovernmental Panel on Climate Change. J.T. Houghton, Y. Ding, D.J. Griggs, M. Noguer, P.J. van der Linden, X. Dai, K. Maskell and C.A. Johnson (eds.). Cambridge University Press, 881pp.

IPCC, 2001b. Climate Change 2001: Impacts, Adaptation, and Vulnerability. Contribution of Working Group I to the Third Assessment Report of the Intergovernmental Panel on Climate Change. J.J. McCarthy, O.F. Canziani, N.A. Leary, D.J. Dokken and K.S. White (eds.). Cambridge University Press, 1032pp.

IPCC-TGCIA, 1999. Guidelines on the Use of Scenario Data for Climate Impact and Adaptation Assessment. Version 1. Prepared by T.R. Carter, M. Hulme and M. Lal, Intergovernmental Panel on Climate Change, Task Group on Scenarios for Climate Impact Assessment, 69pp.

Janssen, M., 1998. Modelling Global Changes. The Art of Integrated Assessment Modelling. Edward Elgar Publishing. 262pp.

Jenness, D., 1962. Eskimo administration: I. Alaska. Technical Paper No. 10, Arctic Institute of North America. 64pp.

Johannes, R.E., 1981. Words of the Lagoon: Fishing and Marine Lore in the Palau District of Micronesia. University of California Press, xiv + 245pp.

Källén, E., V. Kattsov, J. Walsh and E. Weatherhead, 2001. Report from the Arctic Climate Impact Assessment Modeling and Scenarios Workshop. Stockholm, 29–31 January 2001, 35pp.

Kaplan, J.O., N.H. Bigelow, I.C. Prentice, S.P. Harrison, P.J. Bartlein, T.R. Christensen, W. Cramer, N.V. Matveyeva, A.D. McGuire, D.F. Murray, V.Y. Razzhivin, B. Smith, D.A. Walker, P.M. Anderson, A.A. Andreev, L.B. Brubaker, M.E. Edwards and A.V. Lozhkin, 2003. Climate change and Arctic ecosystems: 2. Modeling, paleodata-model comparisons, and future projections. Journal of Geophysical Research, 108(D19):8171, doi:10.1029/2002JD002559.

Lange, M.A., B. Bartling and K. Grosfeld (eds.), 1999. Global changes in the Barents Sea region. Proceedings of the First International BASIS Research Conference, St. Petersburg, Feb. 22–25 1998, Institute for Geophysics, University of Münster, 470pp.

Lange, M.A. and the BASIS Consortium, 2003. The Barents Sea Impact Study (BASIS): Methodology and First Results. Continental Shelf Research, 23(17):1673–1694.

Magdanz, J.S., C.J. Utermohle and R.J. Wolfe, 2002. The Production and Distribution of Wild Food in Wales and Deering, Alaska. Technical Report 259. Alaska Department of Fish and Game, Division of Subsistence, Juneau, Alaska, xii + 136pp.

Matveyeva, N. and Y. Chernov, 2000. Biodiversity of terrestrial ecosystems. In: M. Nuttall and T.V. Callaghan (eds.). The Arctic: Environment, People, Policy, pp. 233–274. Harwood Academic Publishers.

Maxwell, B., 1997. Responding to Global Climate Change in Canada's Arctic. Vol. II of The Canada Country Study: Climate Impacts and Adaptation. Environment Canada, 82pp.

McAvaney, B.J., C. Covey, S. Joussaume, V. Kattsov, A. Kitoh, W. Ogana, A.J. Pitman, A.J. Weaver, R.A. Wood and Z.-C. Zhao, 2001. Model evaluation. In: J.T. Houghton, Y. Ding, D.J. Griggs, M. Noguer, P.J. van der Linden, X. Dai, K. Maskell and C.A. Johnson (eds.). Climate Change 2001: The Scientific Basis. Contribution of Working Group I to the Third Assessment Report of the Intergovernmental Panel on Climate Change, pp. 471–524. Cambridge University Press.

Mearns, L.O., M. Hulme, T.R. Carter, R. Leemans, M. Lal and P. Whetton, 2001. Climate Scenario Development. In: J.T. Houghton, Y. Ding, D.J. Griggs, M. Noguer, P.J. van der Linden, X. Dai, K. Maskell and C.A. Johnson (eds.). Climate Change 2001: The Scientific Basis. Contribution of Working Group I to the Third Assessment Report of the Intergovernmental Panel on Climate Change, pp. 739–768. Cambridge University Press.

Minority Rights Group, 1994. Polar Peoples. London: Minority Rights Group.

Nakićenović, N., J. Alcamo, G. Davis, B. de Vries, J. Fenhann, S. Gaffin, K. Gregory, A. Grübler, T.Y. Jung, T. Kram, E.L. La Rovere, L. Michaelis, S. Mori, T. Morita, W. Pepper, H. Pitcher, L. Price, K. Raihi, A. Roehrl, H.-H. Rogner, A. Sankovski, M. Schlesinger, P. Shukla, S. Smith, R. Swart, S. van Rooijen, N. Victor and Z. Dadi, 2000. IPCC Special Report on Emissions Scenarios. Cambridge University Press, 599pp.

NAST, 2000. Climate Change Impacts on the United States. National Assessment Synthesis Team, U.S. Global Change Research Program, Washington, DC. Cambridge University Press, 153pp.

Nuttall, M., 2000. Indigenous peoples, self-determination, and the Arctic environment. In: M. Nuttall and T.V. Callaghan (eds.). The Arctic: Environment, People, Policy, pp. 377–409. Harwood Academic Publishers.

Nuttall, M. and T.V. Callaghan (eds.), 2000. The Arctic: Environment, People, Policy. Harwood Academic Publishers, xxxviii + 647pp.

Osterkamp, T., 1994. Evidence for warming and thawing of discontinuous permafrost in Alaska. Eos, Transactions, American Geophysical Union, 75(44):85.

Pavlov, P., J.I. Svendsen and S. Indrelid, 2001. Human presence in the European Arctic nearly 40,000 years ago. Nature, 413:64–67.

Peterson, T.C. and R.S. Vose, 1997. An overview of the Global Historical Climatology Network temperature database. Bulletin of the American Meteorological Society, 78:2837–2849.

Randall, D., J. Curry, D. Battisti, G. Flato, R. Grumbine, S. Hakkinen, D. Martinson, R. Preller, J. Walsh and J. Weatherly, 1998. Status of and outlook for large-scale modeling of atmosphere-ice-ocean interactions in the Arctic. Bulletin of the American Meteorological Society, 79:197–219.

Rothrock, D., Y. Yu and G. Maykut, 1999. The thinning of the Arctic ice cover. Geophysical Research Letters, 26(23):3469–3472.

Sakshaug, E. and J. Walsh, 2000. Marine biology: biomass, productivity distributions and their variability in the Barents and Bering Seas. In: M. Nuttall and T.V. Callaghan (eds.). The Arctic: Environment, People, Policy, pp. 163–196. Harwood Academic Publishers.

Sapiano, J.J., W.D. Harrison and K.A. Echelmeyer, 1997. Elevation, volume and terminus changes of nine glaciers in North America. Journal of Glaciology, 44(146):119–135.

Schroeder, R.F., D.B. Andersen, R. Bosworth, J.M. Morris and J.M. Wright, 1987. Subsistence in Alaska: Arctic, Interior, Southcentral, Southwest, and Western Regional Summaries. Technical Paper 150. Alaska Department of Fish and Game, Division of Subsistence, Juneau, Alaska, 690pp.

Slezkine, Y., 1994. Arctic Mirrors: Russia and the Small Peoples of the North. Cornell University Press, xiv + 456pp.

Smith, J.B., M. Hulme, J. Jaagus, S. Keevallik, A. Mekonnen and K. Hailemariam, 1998. Climate Change Scenarios. In: J.F. Feenstra, I. Burton, J. Smith and R.S.J. Tol (eds.). Handbook on Methods for Climate Change Impact Assessment and Adaptation Strategies, Version 2.0, pp. 3-1 - 3-40. United Nations Environment Programme and Institute for Environmental Studies, Vrije Universiteit, Amsterdam.

State Committee of the USSR on Hydrometeorology and Controlled Natural Environments, 1985. Arctic atlas. Arctic and Antarctic Research Institute, Moscow. 204pp. (In Russian)

Stevenson, M.G., 1997. Ignorance and prejudice threaten environmental assessment. Policy Options, 18(2):25–28.

Stocker, T.F., G.K.C. Clarke, H. Le Treut, R.S. Lindzen, V.P. Meleshko, R.K. Mugara, T.N. Palmer, R.T. Pierrehumbert, P.J. Sellers, K.E. Trenberth and J. Willebrand, 2001. Physical climate processes and feedbacks. In: J.T. Houghton, Y. Ding, D.J. Griggs, M. Noguer, P.J. van der Linden, X. Dai, K. Maskell and C.A. Johnson (eds.). Climate Change 2001: The Scientific Basis. Contribution of Working Group I to the Third Assessment Report of the Intergovernmental Panel on Climate Change, pp. 417–470. Cambridge University Press.

Taalas, P., J. Kaurola, A. Kylling, D. Shindell, R. Sausen, M. Dameris, V. Grewe, J. Herman, J. Damski and B. Steil, 2000. The impact of greenhouse gases and halogenated species on future solar UV radiation doses. Geophysical Research Letters, 27(8):1127–1130.

Thommessen, T., 1996. The early settlement of northern Norway. In: L. Larsson (ed.). The Earliest Settlement of Scandinavia. Acta Archaeologica Lundensia, 8(24):235–240.

UNEP, 2003. Environmental Effects of Ozone Depletion and its Interactions with Climate Change: 2002 Assessment. United Nations Environment Programme. Journal of Photochemical and Photobiological Sciences, Special issue. DOI: 10.1039/b211913g.

Vincent, W.F. and J.E. Hobbie, 2000. Ecology of lakes and rivers. In: M. Nuttall and T.V. Callaghan (eds.). The Arctic: Environment, People, Policy, pp. 197–232. Harwood Academic Publishers.

Vinnikov, K.Y., A. Robock, R. Stouffer, J. Walsh, C. Parkinson, D. Cavalieri, J. Mitchell, D. Garrett and V. Zakharov, 1999. Global warming and northern hemisphere sea ice extent. Science, 286(5446):1934–1937.

Weihs, F.H., R. Higgins and D. Boult, 1993. A Review and Assessment of the Economic Utilizations and Potential of Country Foods in the Northern Economy. Report prepared for the Royal Commission on Aboriginal People, Canada.

Weller, G., 1998. Regional impacts of climate change in the Arctic and Antarctic. Annals of Glaciology, 27:543–552.

Weller, G. and M. Lange (eds.), 1999. Impacts of Global Climate Change in the Arctic Regions. Workshop on the Impacts of Global Change, 25–26 April 1999, Tromsø, Norway. Published for International Arctic Science Committee by Center for Global Change and Arctic System Research, University of Alaska Fairbanks, 59pp.

Weller, G., P. Anderson and B. Wang (eds.), 1999. Preparing for a Changing Climate: The Potential Consequences of Climate Change and Variability. A Report of the Alaska Regional Assessment Group for the U.S. Global Change Research Program. Center for Global Change and Arctic System Research, University of Alaska Fairbanks, 42pp.

WMO, 2003. Scientific Assessment of Ozone Depletion: 2002. Global Ozone Research and Monitoring Project – Report No. 47, World Meteorological Organization, Geneva.

Ziker, J.P., 2002. Peoples of the Tundra: Northern Siberians in the post-Communist Transition. Prospect Heights, IL, Waveland Press, x + 197pp.

Chapter 2

Arctic Climate: Past and Present

Lead Author
Gordon McBean

Contributing Authors
Genrikh Alekseev, Deliang Chen, Eirik Førland, John Fyfe, Pavel Y. Groisman, Roger King, Humfrey Melling, Russell Vose, Paul H. Whitfield

Contents

Summary

The arctic climate is defined by a low amount or absence of sunlight in winter and long days during summer, with significant spatial and temporal variation. The cryosphere is a prominent feature of the Arctic. The sensitivities of snow and ice regimes to small temperature increases and of cold oceans to small changes in salinity are processes that could contribute to unusually large and rapid climate change in the Arctic.

The arctic climate is a complex system with multiple interactions with the global climate system. The phase of the Arctic Oscillation was at its most negative in the 1960s, exhibited a general trend toward a more positive phase from about 1970 to the early 1990s, and has remained mostly positive since. Sea ice is a primary means by which the Arctic exerts leverage on global climate, and sea-ice extent has been decreasing. In terrestrial areas, temperature increases over the past 80 years have increased the frequency of mild winter days, causing changes in aquatic ecosystems; the timing of river-ice breakups; and the frequency and severity of extreme ice jams, floods, and low flows.

The observational database for the Arctic is quite limited, with few long-term stations and a paucity of observations in general, making it difficult to distinguish with confidence between the signals of climate variability and change. Based on the analysis of the climate of the 20th century, it is very probable that the Arctic has warmed over the past century, although the warming has not been uniform. Land stations north of 60° N indicate that the average surface temperature increased by approximately 0.09 °C/decade during the past century, which is greater than the 0.06 °C/decade increase averaged over the Northern Hemisphere. It is not possible to be certain of the variation in mean land-station temperature over the first half of the 20th century because of a scarcity of observations across the Arctic before about 1950. However, it is probable that the past decade was warmer than any other in the period of the instrumental record.

Evidence of polar amplification depends on the timescale of examination. Over the past 100 years, it is possible that there has been polar amplification, however, over the past 50 years it is probable that polar amplification has occurred.

It is very probable that atmospheric pressure over the Arctic Basin has been dropping, and it is probable that there has been an increase in total precipitation over the past century at the rate of about 1% per decade. Trends in precipitation are hard to assess because it is difficult to measure with precision in the cold arctic environment. It is very probable that snow-cover extent around the periphery of the Arctic has decreased. It is also very probable that there have been decreases in average arctic sea-ice extent over at least the past 40 years and a decrease in multi-year sea-ice extent in the central Arctic.

Reconstruction of arctic climate over the past thousands to millions of years demonstrates that arctic climate can vary substantially. There appears to be no natural impediment to anthropogenic climate change being very significant and greater in the Arctic than the change at the global scale. Especially during past cold periods, there have been times when temperature transitions have been quite rapid – from a few to several degrees change over a century.

2.1. Introduction

The Arctic is the northern polar component of the global climate system. The global climate system has been thoroughly examined in the recent reports of the Intergovernmental Panel on Climate Change (IPCC, 2001a,b,c), which include discussion of the impacts of climate change in the Arctic (IPCC, 2001a). Arctic climate is characterized by a low amount or absence of sunlight in winter and long days during summer. Although these solar inputs are a dominant influence, arctic climate exhibits significant spatial and temporal variability. As a result, the Arctic is a collection of regional climates with different ecological and physical climatic characteristics.

The cryosphere is a prominent feature of the Arctic, present as snow, ice sheets, glaciers, sea ice, and permafrost. The physical properties of snow and ice include high reflectivity, low thermal conductivity, and the high latent heat required to convert ice to liquid water; these contribute significantly to the regional character of arctic climate.

The arctic climate interacts with the climates of more southern latitudes through the atmosphere, oceans, and rivers. Because of these regionally diverse features, an exact geographic definition of the Arctic is not appropriate and this chapter focuses on the northernmost areas (usually north of 60° N), while acknowledging interactions with more southerly areas.

The observational database for the Arctic is quite limited, with few long-term stations and a paucity of observations in general. The combination of a sparse observational dataset and high variability makes it difficult to distinguish with confidence between the signals of climate variability and change.

With respect to the polar regions, the Intergovernmental Panel on Climate Change (IPCC, 2001a) stated:

> *Changes in climate that have already taken place are manifested in the decrease in extent and thickness of Arctic sea ice, permafrost thawing, coastal erosion, changes in ice sheets and ice shelves, and altered distribution and abundance of species in polar regions (high confidence).*

> *Climate change in polar regions is expected to be among the largest and most rapid of any region on the Earth,*

and will cause major physical, ecological, sociological, and economic impacts, especially in the Arctic, Antarctic Peninsula, and Southern Ocean (high confidence).

Polar regions contain important drivers of climate change. Once triggered, they may continue for centuries, long after greenhouse gas concentrations are stabilized, and cause irreversible impacts on ice sheets, global ocean circulation, and sea-level rise (medium confidence).

The arctic climate is a complex system and has multiple interactions with the global climate system. The sensitivities of snow and ice regimes to small temperature increases and of cold oceans to small changes in salinity, both of which can lead to subsequent amplification of the signal, are processes that could contribute to unusually large and rapid climate change in the Arctic. The Arctic Oscillation (AO) is an important feature of the arctic atmosphere and its connections with global climate (section 2.2). The phase of the AO was at its most negative in the 1960s, but from about 1970 to the early 1990s there was a general trend toward a more positive phase and it has remained mostly positive since. It is possible that this is the result of increased radiative forcing due to anthropogenic greenhouse gas (GHG) emissions, but it is also possible that it is a result of variations in sea surface temperatures. The Arctic Ocean (section 2.3) forms the core of the Arctic. Sea ice is the defining characteristic of the marine Arctic and is the primary means by which the Arctic exerts leverage on global climate. This leverage occurs through mediation of the exchange of radiation, sensible heat, and momentum between the atmosphere and the ocean. Terrestrial hydrology (section 2.4) and arctic climate are intricately linked. In terrestrial areas, temperature increases over the past 80 years have increased the frequency of mild winter days, causing changes in the timing of river-ice breakups; in the frequency and severity of extreme ice jams, floods, and low flows; and in aquatic ecosystems. The increased frequency of mild winter days has also affected transportation and hydroelectric generation.

There are both positive and negative feedback processes in the Arctic, occurring over a range of timescales. Positive feedbacks include snow and ice albedo feedback; reduction in the duration of time that sea ice insulates the atmosphere from the Arctic Ocean; and permafrost–methane hydrate feedbacks. Negative feedbacks can result from increased freshwater input from arctic watersheds, which makes the upper ocean more stably stratified and hence reduces temperature increases near the air–sea interface; reductions in the intensity of the thermohaline circulation that brings heat to the Arctic; and a possible vegetation–carbon dioxide (CO_2) feedback that has the potential to promote vegetation growth, resulting in a reduced albedo due to more vegetation covering the tundra. Polar amplification (greater temperature increases in the Arctic compared to the earth as a whole) is a result of the collective effect of these feedbacks and other process-

es. The Arctic is connected to the global climate, being influenced by it and vice versa (section 2.5).

Based on the analysis of the climate of the 20th century (section 2.6), it is very probable[2] that arctic temperatures have increased over the past century, although the increase has not been spatially or temporally uniform. The average surface temperature in the Arctic increased by approximately 0.09 °C/decade during the past century, which is 50% greater than the 0.06 °C/decade increase observed over the entire Northern Hemisphere (IPCC, 2001b). Probably as a result of natural variations, the Arctic may have been as warm in the 1930s as in the 1990s, although the spatial pattern of the warming was quite different and may have been primarily an artifact of the station distribution.

Evidence of polar amplification depends on the timescale of examination. Over the past 100 years, it is possible that there has been polar amplification, however, over the past 50 years it is probable that polar amplification has occurred.

It is very probable that atmospheric pressure over the Arctic Basin has been dropping, and it is probable that there has been an increase in total precipitation over the past century at the rate of about 1% per decade. Trends in precipitation are hard to assess because precipitation is difficult to measure with precision in the cold arctic environment. It is very probable that snow-cover extent around the periphery of the Arctic has decreased. It is also very probable that there have been decreases in average arctic sea-ice extent over at least the past 40 years and a decrease in multi-year sea-ice extent in the central Arctic.

Reconstruction of arctic climate over thousands to millions of years demonstrates that the arctic climate has varied substantially. There appears to be no natural impediment to anthropogenic climate change being very significant and greater in the Arctic than the change on the global scale. Section 2.7.2 examines the variability of arctic climate during the Quaternary Period (the past 1.6 million years) with a focus on the past 20 000 years. Arctic temperature variability during the Quaternary Period has been greater than the global average. Especially during past cold periods, there have been times when the variability and transitions in temperature have been quite rapid – from a few to several degrees change over a century. There have also been decadal-scale variations due to changes in the thermohaline circulation, with marked regional variations.

2.2. Arctic atmosphere

The arctic atmosphere is highly influenced by the overall hemispheric circulation, and should be regarded in this general context. This section examines Northern Hemisphere circulation using the National Centers for

---

[2]In this chapter, when describing changes in arctic climate, the words possible, probable, and very probable are used to indicate the level of confidence the authors have that the change really did occur, recognizing the limitations of the observing system and paleoclimatic reconstructions of arctic climate.

Environmental Prediction/National Center for Atmospheric Research reanalyses for the period from 1952 to 2003 (updated from Kalnay et al., 1996). Section 2.2.1 describes the main climatological features, while section 2.2.2 discusses the two major modes of variability: the AO (and its counterpart, the North Atlantic Oscillation) and the Pacific Decadal Oscillation. Because much of the observed change in the Arctic appears to be related to patterns of atmospheric circulation, it is important that these modes of atmospheric variability be described.

2.2.1. Climatology

Atmospheric circulation and weather are closely linked to surface pressure. Figure 2.1a shows the Northern Hemisphere seasonal mean patterns of sea-level pressure in winter and summer. The primary features of sea-level pressure in winter include the oceanic Aleutian and Icelandic Lows, and the continental Siberian High with its extension into the Arctic (the Beaufort High). The sea-level pressure distribution in summer is dominated by subtropical highs in the eastern Pacific and Atlantic Oceans, with relatively weak gradients in polar and subpolar regions. The seasonal cycle of sea-level pressure over the mid-latitude oceans exhibits a summer maximum and winter minimum. By contrast, the seasonal cycle of sea-level pressure over the Arctic and subarctic exhibits a maximum in late spring, a minimum in winter, and a weak secondary maximum in late autumn. The climatological patterns and seasonal cycle of sea-level pressure are largely determined by the regular passage of migratory cyclones and anticyclones, which are associated with storminess and settled periods, respectively. Areas of significant winter cyclonic activity (storm

tracks) are found in the North Pacific and North Atlantic. These disturbances carry heat, momentum, and moisture into the Arctic, and have a significant influence on high-latitude climate.

The Arctic is affected by extremes of solar radiation. The amount of solar radiation received in summer is relatively high due to long periods of daylight, but its absorption is kept low by the high albedo of snow and ice. The amount of solar radiation received in winter is low to non-existent. Figure 2.1b shows the seasonal mean patterns of surface air temperature. The Arctic is obviously a very cold region of the Northern Hemisphere, especially in winter when the seasonal mean temperature falls well below -20 °C. Temperature inversions, when warm air overlies a cold surface, are common in the Arctic. At night, especially on calm and clear nights, the ground cools more rapidly than the adjacent air because the ground is a much better emitter of infrared radiation than the air. The arctic winter is dominated by temperature inversions, due to the long nights and extensive infrared radiation losses. Arctic summers have fewer and weaker temperature inversions. On the hemispheric scale, there exist large north–south gradients of atmospheric temperature (and moisture). In winter, the continental landmasses are generally colder than the adjacent oceanic waters, owing to the influence of warm surface currents on the western boundaries of the Atlantic and Pacific Oceans.

2.2.2. Variability modes

2.2.2.1. Arctic/North Atlantic Oscillation

The North Atlantic Oscillation (NAO) has long been recognized as a major mode of atmospheric variability over the extratropical ocean between North America and Europe. The NAO describes co-variability in sea-level pressure between the Icelandic Low and the Azores High. When both are strong (higher than normal pressure in the Azores High and lower than normal pressure in the Icelandic Low), the NAO index is positive. When both are weak, the index is negative. The NAO is hence also a measure of the meridional gradient in sea-level pressure over the North Atlantic, and the strength of the westerlies in the intervening mid-latitudes. The NAO is most obvious during winter but can be identified at any time of the year. As the 20th century drew to a close, a series of papers were published (e.g., Thompson et al., 2000) arguing that the NAO should be considered as a regional manifestation of a more basic annular mode of sea-level pressure variability, which has come to be known as the Arctic Oscillation (AO). The AO is defined as the leading mode of variability from a linear principal component analysis of Northern Hemisphere sea-level pressure. It emerges as a robust pattern dominating both the intra-seasonal (e.g., month-to-month) and inter-annual variability in sea-level pressure.

Whether or not the AO is in fact a more fundamental mode than the NAO is a matter of debate. For example, Deser (2000) concluded that the correlation between the

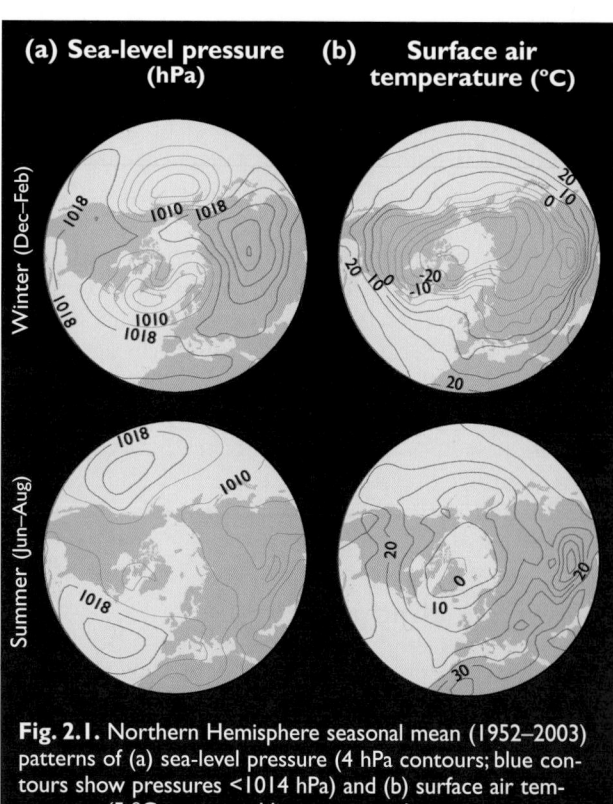

Fig. 2.1. Northern Hemisphere seasonal mean (1952–2003) patterns of (a) sea-level pressure (4 hPa contours; blue contours show pressures <1014 hPa) and (b) surface air temperature (5 °C contours; blue contours show temperatures <0 °C) (updated from Kalnay et al, 1996).

Pacific and Azores high-pressure areas was not significant, and that the AO cannot therefore be viewed as reflecting such a teleconnection. Ambaum et al. (2001) found that even the correlation between the Pacific and Icelandic–Arctic low-pressure centers was not significant. They argue that the AO is mainly a reflection of similar behavior in the Pacific and Atlantic basins. Regardless, the AO and NAO time series are very highly correlated, and for most applications (including this assessment), either paradigm can be used. Before proceeding with a description of the AO/NAO, two cautionary points must be mentioned. First, while the AO/NAO is obviously dominant, it explains only a fraction (i.e., 20 to 30%) of the total variability in sea-level pressure. Second, because the AO/NAO index is derived from a linear statistical tool, it cannot describe more general nonlinear variability. Monahan et al. (2003) have shown that hemispheric variability is significantly nonlinear, and the AO provides only the optimal linear approximation of this variability.

Figure 2.2a shows the AO/NAO time series obtained using monthly mean sea-level pressure for all months in the "extended winter" (November to April). There is considerable month-to-month and year-to-year variability, as well as variability on longer timescales. The AO/NAO index was at its most negative in the 1960s. From about 1970 to the early 1990s, there was a general increasing trend, and the AO index was more positive than negative throughout the 1990s. The physical origins of these long-term changes are the subject of considerable debate. Fyfe et al. (1999) and Shindell et al. (1999) have shown that positive AO trends can be obtained from global climate models using scenarios of increasing radiative forcing due to rising GHG concentrations. Rodwell et al. (1999) and Hoerling et al. (2001) have shown similar positive trends using global climate models run with fixed radiative forcing and observed annually varying sea surface temperatures. Rodwell et al. (1999) argued that slowly varying sea surface temperatures in the North Atlantic are locally communicated to the atmosphere through evaporation, precipitation, and atmospheric heating processes. On the other hand, Hoerling et al. (2001) suggested that changes in tropical sea surface temperatures, especially in the Indian and Pacific Oceans, may be more important than changes in sea surface temperatures in the North Atlantic. They postulated that changes in the tropical ocean alter the pattern and magnitude of tropical rainfall and atmospheric heating, which in turn produce positive AO/NAO trends. Regardless of the causes, it must be noted that AO/NAO trends do not necessarily reflect a change in the variability mode itself. As demonstrated by Fyfe (2003), the AO/NAO trends are a reflection of a more general change in the background, or "mean", state with respect to which the modes are defined.

Figure 2.2b shows the sea-level pressure anomaly pattern associated with the AO/NAO time series, as derived from a principal components analysis. The pattern shows negative anomalies over the polar and subpolar latitudes, and positive anomalies over the midlatitudes. The anomaly center in the North Atlantic,

while strongest in the vicinity of the Icelandic Low, extends with strength well into the Arctic Basin. Not surprisingly, these anomalies are directly related to fluctuations in cyclone frequency. Serreze et al. (1997) noted a strong poleward shift in cyclone activity during the positive phase of the AO/ NAO, and an equatorward shift during the negative phase. In the region corresponding to the climatological center of the Icelandic Low, cyclone events are more than twice as common during the positive AO/NAO extremes than during negative extremes. Systems found in this region during the positive phase are also significantly deeper than are their negative AO/NAO counterparts. McCabe et al. (2001) noted a general poleward shift in Northern Hemisphere cyclone activity starting around 1989, coincident with the positive trend in the AO/NAO time series. Figure 2.2c shows the pattern of surface air temperature anomalies associated with the AO/NAO time series. Negative surface air temperature anomalies centered in Davis Strait are consistent with southeasterly advection of cold arctic air by the AO/NAO-related winds. Easterly advection of warmer air, also linked to AO/NAO-related winds, accounts for the pattern of positive anomalies in surface air temperature over Eurasia.

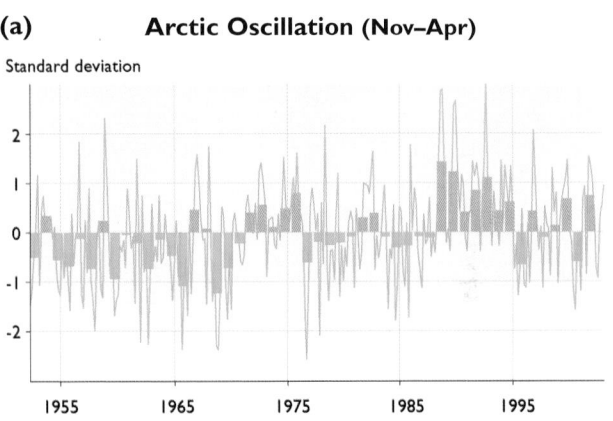

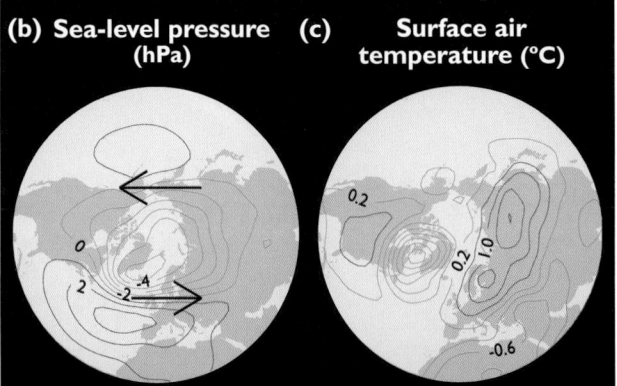

Fig. 2.2. Arctic Oscillation (a) time series based on anomalies of November to April monthly mean sea-level pressure, calculated relative to the 1952–2003 monthly mean (shading indicates November–April averages for each winter in the time series); and associated patterns of (b) sea-level pressure anomalies (1 hPa contours; arrows represent the anomalous wind direction) and (c) surface air temperature anomalies (0.4 °C contours) derived from a principal components analysis of the time series (updated from Kalnay et al, 1996).

2.2.2.2. Pacific Decadal Oscillation

The Pacific Decadal Oscillation (PDO) is a major mode of North Pacific climate variability. The PDO is obtained as the leading mode of North Pacific monthly surface temperature. Figure 2.3a shows the PDO time series obtained using monthly mean surface air temperature for all months in the extended winter (November to April). As with the AO/NAO time series, the PDO time series displays considerable month-to-month and year-to-year variability, as well as variability on longer timescales. The PDO was in a negative (cool) phase from 1947 to 1976, while a positive (warm) phase prevailed from 1977 to the mid-1990s (Mantua et al., 1997; Minobe, 1997). Major changes in northeast Pacific marine ecosystems have been correlated with these PDO phase changes. As with the AO/NAO, the physical origins of these long-term changes are currently unknown.

Figures 2.3b and 2.3c show the sea-level pressure and surface air temperature anomalies associated with the PDO time series, as derived from a principal components analysis. The sea-level pressure anomaly pattern is wave-like, with low sea-level pressure anomalies over the North Pacific and high sea-level pressure anomalies over western North America. At the same time, the surface air temperatures tend to be anomalously cool in the central North Pacific and anomalously warm along the west coast of North America. The PDO circulation anomalies extend well into the troposphere in a form similar to the Pacific North America pattern (another mode of atmospheric variability).

2.3. Marine Arctic

2.3.1. Geography

The Arctic Ocean forms the core of the marine Arctic. Its two principal basins, the Eurasian and Canada, are more than 4000 m deep and almost completely land-locked (Fig. 2.4). Traditionally, the open boundary of the Arctic Ocean has been drawn along the Barents Shelf edge from Norway to Svalbard, across Fram Strait, down the western margin of the Canadian Archipelago and across Bering Strait (Aagaard and Coachman, 1968a). Including the Canadian polar continental shelf (Canadian Archipelago), the total ocean area is 11.5 million km², of which 60% is continental shelf. The shelf ranges in width from about 100 km in the Beaufort Sea (Alaska) to more than 1000 km in the Barents Sea and the Canadian Archipelago. Representative shelf depths off the coasts of Alaska and Siberia are 50 to 100 m, whereas those in the Barents Sea, East Greenland, and northern Canada are 200 to 500 m. A break in the shelf at Fram Strait provides the only deep (2600 m) connection to the global ocean. Alternate routes to the Atlantic via the Canadian Archipelago and the Barents Sea block flow at depths below 220 m while the connection to the Pacific Ocean via Bering Strait is 45 m deep. About 70% of the Arctic Ocean is ice-covered throughout the year.

Like most oceans, the Arctic is stratified, with deep waters that are denser than surface waters. In a stratified ocean, energy must be provided in order to mix surface and deep waters or to force deep-water flow over obstacles. For this reason, seabed topography is an important influence on ocean processes. Sections 6.3 and 9.2.2 contain detailed discussions of the Arctic Ocean and sea ice.

The term "marine Arctic" is used here to denote an area that includes Baffin, Hudson, and James Bays; the Labrador, Greenland, Iceland, Norwegian, and Bering Seas; and the Arctic Ocean. This area encompasses 3.5 million km² of cold, low-salinity surface water and seasonal sea ice that are linked oceanographically to the Arctic Ocean and areas of the North Atlantic and North Pacific Oceans that interact with them. In this region, the increase in density with depth is dominated by an increase in salinity as opposed to a decrease in temperature. The isolated areas of the northern marine cryosphere, namely the Okhotsk and Baltic Seas and the Gulf of St. Lawrence, are not included in this chapter's definition of "marine Arctic".

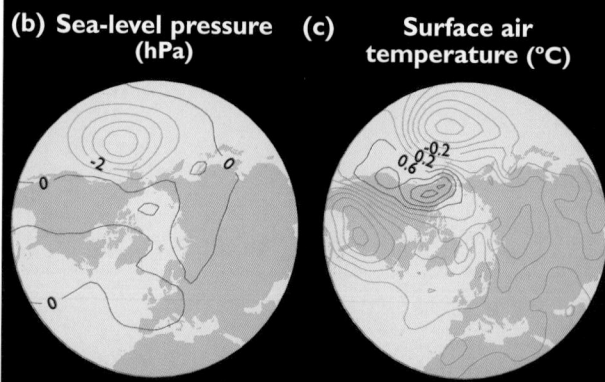

(a)   Pacific Decadal Oscillation (Nov–Apr)

Standard deviation

1955   1965   1975   1985   1995

(b) Sea-level pressure (c)   Surface air
(hPa)   temperature (°C)

Fig. 2.3. Pacific Decadal Oscillation (a) time series based on anomalies of November to April monthly mean surface air temperature in the North Pacific, calculated relative to the 1952–2003 monthly mean (shading indicates November–April averages for each winter in the time series); and associated patterns of (b) sea-level pressure anomalies (1.0 hPa contours) and (c) surface air temperature anomalies (0.2 °C contours) derived from a principal components analysis of the time series (updated from Kalnay et. al, 1996).

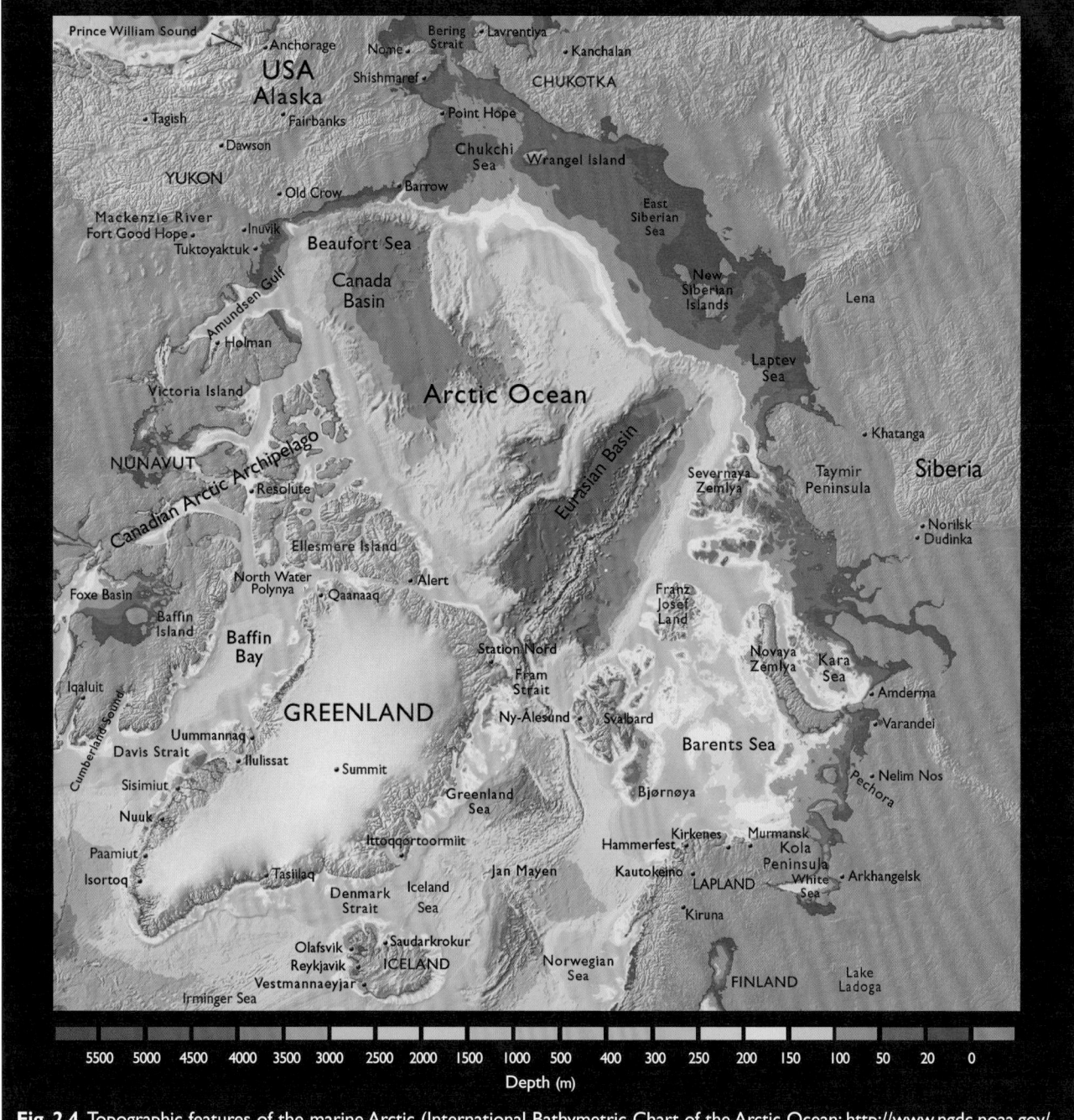

Fig. 2.4. Topographic features of the marine Arctic (International Bathymetric Chart of the Arctic Ocean; http://www.ngdc.noaa.gov/ mgg/bathymetry/arctic/arctic.html).

2.3.2. Influence of temperate latitudes

Climatic conditions in northern mid-latitudes influence the Arctic Ocean via marine and fluvial inflows as well as atmospheric exchange. The transport of water, heat, and salt by inflows are important elements of the global climate system. Warm inflows have the potential to melt sea ice provided that mixing processes can move heat to the surface. The dominant impediment to mixing is the vertical gradient in salinity at arctic temperatures. Therefore, the presence of sea ice in the marine Arctic is linked to the salt transport by inflows.

Approximately 11% of global river runoff is discharged to the Arctic Ocean, which represents only 5% of global ocean area and 1% of its volume (Shiklomanov et al.,

2000). In recognition of the dramatic effect of freshwater runoff on arctic surface water, the salt budget is commonly discussed in terms of freshwater, even for marine flows. Freshwater content in the marine context is the fictitious fraction of freshwater that dilutes seawater of standard salinity (e.g., 35) to create the salinity actually observed. For consistency with published literature, this chapter uses the convention of placing "freshwater" in quotes to distinguish the freshwater component of ocean water from the more conventional definition of freshwater.

The Arctic is clearly a shortcut for flow between the Pacific and Atlantic Oceans (Fig. 2.5). A flow of 800000 m³/s (0.8 Sv) follows this shortcut to the Atlantic via Bering Strait, the channels of the Canadian

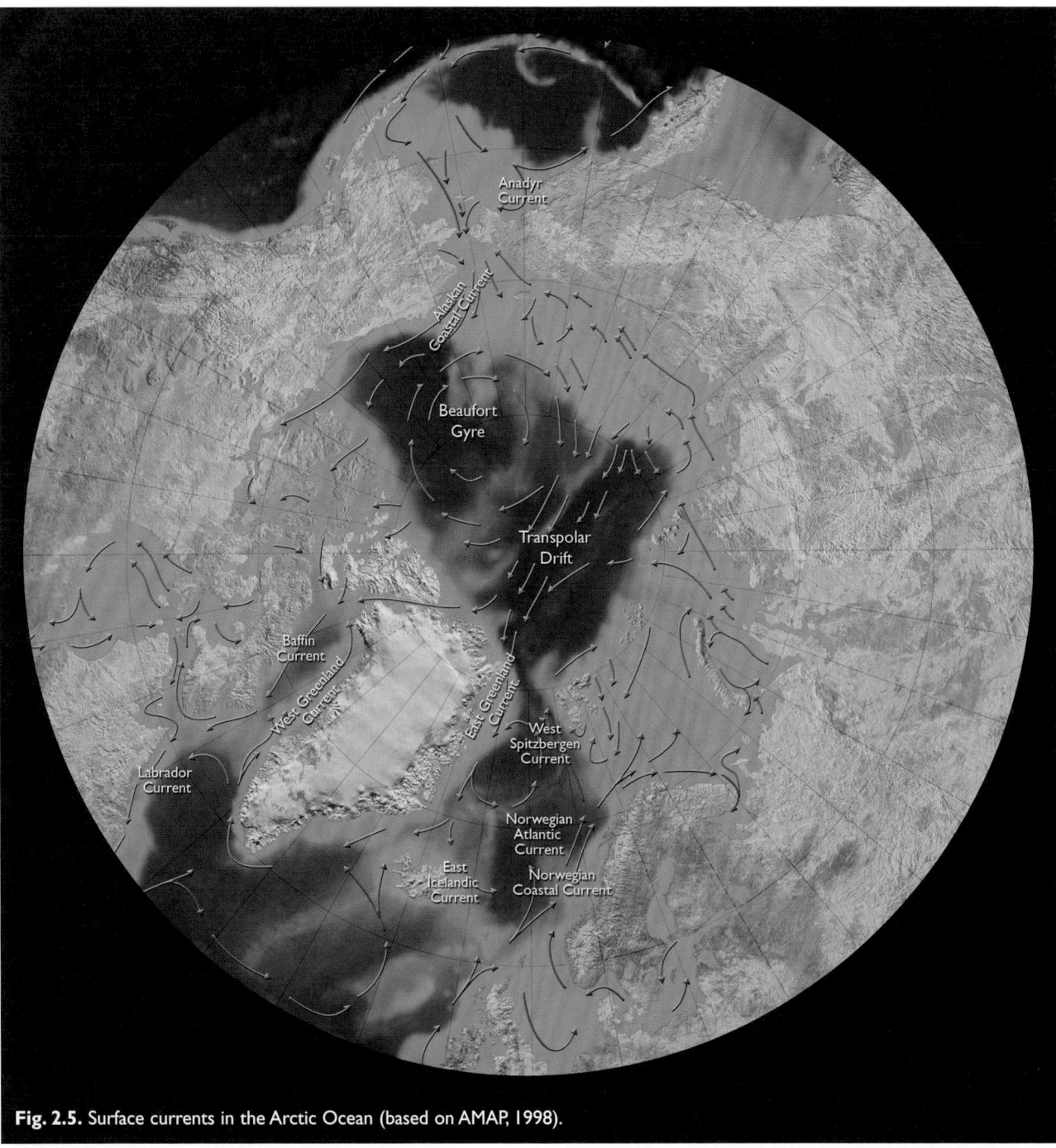

Fig. 2.5. Surface currents in the Arctic Ocean (based on AMAP, 1998).

Archipelago, and Fram Strait (Melling, 2000). The flow is driven by higher sea level (~0.5 m) in the North Pacific (Stigebrandt, 1984). The difference in elevation reflects the lower average salinity of the North Pacific, maintained by an excess of precipitation over evaporation relative to the North Atlantic (Wijffels et al., 1992). By returning excess precipitation to the Atlantic, the flow through the Arctic redresses a global-scale hydrologic imbalance created by present-day climate conditions. By transporting heat into the Arctic Ocean at depths less than 100 m, the flow influences the thickness of sea ice in the Canada Basin (Macdonald R. et al., 2002).

Much of the elevation change between the Pacific and the Atlantic occurs in Bering Strait. Operating like a weir in a stream, at its present depth and width the strait hydraulically limits flow to about 1 Sv (Overland and

Roach, 1987). Bering Strait is therefore a control point in the global hydrological cycle, which will allow more through-flow only with an increase in sea level. Similar hydraulic controls may operate with about 0.2 m of hydraulic head at flow constrictions within the Canadian Archipelago. The present "freshwater" flux through Bering Strait is about 0.07 Sv (Aagaard and Carmack, 1989; Fedorova and Yankina, 1964).

The Bering inflow of "freshwater" destined for the Atlantic is augmented from other sources, namely rivers draining into the Arctic Ocean, precipitation over ocean areas, and sea ice. The total influx to the marine Arctic from rivers is 0.18 Sv (Shiklomanov et al., 2000), about 2.5 times the "freshwater" flux of the Pacific inflow through Bering Strait. This estimate includes runoff from Greenland, the Canadian Archipelago, and the watersheds of the Yukon River (carried through Bering Strait by the Alaskan Coastal Current), Hudson Bay, and James Bay. The average annual precipitation minus evaporation north of 60° N is 0.16 m/yr (Barry and Serreze, 2000), corresponding to a freshwater flux of 0.049 Sv over marine areas. The combined rate of freshwater supply to the marine Arctic is 0.3 Sv.

Sea ice has a high "freshwater" content, since it loses 80% of its salt upon freezing and all but about 3% through subsequent thermal weathering. Although about 10% of sea-ice area is exported annually from the Arctic Ocean through Fram Strait, this is not a "freshwater" export from the marine Arctic, since the boundary is defined as the edge of sea ice at its maximum extent.

Freezing segregates the upper ocean into brackish surface (ice) and salty deeper components that circulate differently within the marine Arctic. The melting of sea ice delivers freshwater to the surface of the ocean near the boundary of the marine Arctic. The flux of sea ice southward through Fram Strait is known to be about 0.09 Sv (Vinje, 2001), but the southward flux of seasonal sea ice formed outside the Arctic Ocean in the Barents, Bering, and Labrador Seas; the Canadian Archipelago; Hudson and Baffin Bays; and East Greenland is not known.

The inflows to the marine Arctic maintain a large reservoir of "freshwater" (i.e., diluted seawater and brackish sea ice). Aagaard and Carmack (1989) estimated the volume of "freshwater" stored within the Arctic Ocean to be 80 000 km^3. A rough estimate suggests that there is an additional reservoir of approximately 50 000 km^3 in the marginal seas described in the previous paragraph. The total reservoir of "freshwater" equals the accumulation of inflow over about 15 years.

The "freshwater" reservoir feeds two boundary currents that flow into the western North Atlantic – the East Greenland Current and the Labrador Current (Aagaard and Coachman, 1968a,b). The former enters the Greenland Sea via Fram Strait and the latter enters the Labrador Sea via Davis Strait, gathering a contribution from Hudson Bay via Hudson Strait.

Northbound streams of warm saline water, the Norwegian Atlantic Current and the West Greenland Current, counter the flow of low-salinity water toward the Atlantic. The Norwegian Atlantic Current branches into the West Spitzbergen Current and the Barents Sea through-flow. The former passes through Fram Strait with a temperature near 3 °C and follows the continental slope eastward at depths of 200 to 800 m as the Fram Strait Branch (Gorshkov, 1980). The latter, cooled to less than 0 °C and freshened by arctic surface waters, enters the Arctic Ocean at depths of 800 to 1500 m in the eastern Barents Sea (Schauer et al., 2002). The West Greenland Current carries 3 °C seawater to northern Baffin Bay, where it mixes with arctic outflow and joins the south-flowing Baffin Current (Melling et al., 2001).

The inflows via the West Spitzbergen Current and Barents Sea through-flow are each about 1 to 2 Sv. The West Greenland Current transports less than 0.5 Sv. The associated fluxes of "freshwater" are small because salinity is close to 35. All fluxes vary appreciably from year to year.

The Fram Strait and Barents Sea branches are important marine sources of heat and the most significant sources of salt for arctic waters subjected to continuous dilution. The heat loss to the atmosphere in the ice-free northeastern Greenland Sea averages 200 W/m^2 (Khrol, 1992). The average heat loss from the Arctic Ocean is 6 W/m^2 of which 2 W/m^2 comes from the Atlantic-derived water. The impact of the incoming oceanic heat on sea ice is spatially non-uniform because the upper-ocean stability varies with the distribution of freshwater storage and ice cover.

2.3.3. Arctic Ocean

The two branches of Atlantic inflow interleave at depths of 200 to 2000 m in the Arctic Ocean because of their high salinity, which makes them denser than surface waters despite their higher temperature. They circulate counter-clockwise around the basin in narrow (50 km) streams confined to the continental slope by the Coriolis Effect. The streams split where the slope meets mid-ocean ridges, creating branches that circulate counter-clockwise around the sub-basins (Rudels et al., 1994). The delivery of new Atlantic water to the interior of basins is slow (i.e., decades).

The boundary currents eventually return cooler, fresher, denser water to the North Atlantic via Fram Strait (Greenland side) and the Nordic Seas. The circuit time varies with routing. The role of arctic outflow in deep convection within the Greenland Sea and in the global thermohaline circulation is discussed in section 9.2.3. In the present climate, Atlantic-derived waters in the Arctic Ocean occur at depths too great to pass through the Canadian Archipelago.

Inflow from the North Pacific is less saline and circulates at a shallower depth than Atlantic inflow. It spreads north from Bering Strait to dominate the upper ocean of the western Arctic – the Chukchi and Beaufort Seas, Canada Basin, and the Canadian Archipelago. An oceanic front presently located over the Alpha-Mendeleyev Ridge in Canada Basin separates the region of Pacific dominance from an "Atlantic domain" in the eastern hemisphere. A dramatic shift of this front from the Lomonosov Ridge in the early 1990s flooded a wide area of former Pacific dominance with warmer and less stratified Atlantic water (Carmack et al., 1995).

The interplay of Atlantic and Pacific influence in the Arctic Ocean, the inflows of freshwater, and the seasonal cycle of freezing and melting create a layered structure in the Arctic Ocean (Treshnikov, 1959). These layers, from top to bottom, include snow; sea ice; surface seawater strongly diluted by precipitation, river discharge, and ice melt; warm summer intrusions from ice-free seas (principally the Bering Sea); cold winter intrusions from freezing seas; cool winter intrusions from ice-free seas (principally the Barents Sea); warm intrusions of the Fram Strait Branch; cool intrusions of the Barents Sea Branch; recently-formed deep waters; and relict deep waters. The presence and properties of each layer vary with location across the Arctic Ocean.

The cold and cool winter intrusions form the arctic cold halocline, an approximately isothermal zone wherein salinity increases with depth. The halocline isolates sea ice from warm deeper water because its density gradient inhibits mixing, and its weak temperature gradient minimizes the upward flux of heat. The cold halocline is a determining factor in the existence of year-round sea ice in the present climate. Areas of seasonal sea ice either lack a cold halocline (e.g., Baffin Bay, Labrador Shelf, Hudson Bay) or experience an intrusion of warm water in summer that overrides it (e.g., Chukchi Sea, coastal Beaufort Sea, eastern Canadian Archipelago). The stability of the cold halocline is determined by freshwater dynamics in the Arctic and its low temperature is maintained by cooling and ice formation in recurrent coastal polynyas (Cavalieri and Martin, 1994; Melling, 1993; Melling and Lewis, 1982; Rudels et al., 1996). Polynyas are regions within heavy winter sea ice where the ice is thinner because the oceanic heat flux is locally intense or because existing ice is carried away by wind or currents. The locations and effectiveness of these "ice factories" are functions of present-day wind patterns (Winsor and Björk, 2000).

2.3.4. Sea ice

Sea ice is the defining characteristic of the marine Arctic. It is the primary method through which the Arctic exerts leverage on global climate, by mediating the exchange of radiation, sensible heat, and momentum between the atmosphere and the ocean (see section 2.5). Changes to sea ice as a unique biological habitat are in the forefront of climate change impacts in the marine Arctic.

The two primary forms of sea ice are seasonal (or first-year) ice and perennial (or multi-year) ice. Seasonal or first-year ice is in its first winter of growth or first summer of melt. Its thickness in level floes ranges from a few tenths of a meter near the southern margin of the marine cryosphere to 2.5 m in the high Arctic at the end of winter. Some first-year ice survives the summer and becomes multi-year ice. This ice develops its distinctive hummocky appearance through thermal weathering, becoming harder and almost salt-free over several years. In the present climate, old multi-year ice floes without ridges are about 3 m thick at the end of winter.

The area of sea ice decreases from roughly 15 million km² in March to 7 million km² in September, as much of the first-year ice melts during the summer (Cavalieri et al., 1997). The area of multi-year sea ice, mostly over the

Arctic Ocean basins, the East Siberian Sea, and the Canadian polar shelf, is about 5 million km² (Johannessen et al., 1999). A transpolar drift carries sea ice from the Siberian shelves to the Barents Sea and Fram Strait. It merges on its eastern side with clockwise circulation of sea ice within Canada Basin. On average, 10% of arctic sea ice exits through Fram Strait each year. Section 6.3 provides a full discussion of sea ice in the Arctic Ocean.

Sea ice also leaves the Arctic via the Canadian Archipelago. Joined by seasonal sea ice in Baffin Bay, it drifts south along the Labrador coast to reach New-foundland in March. An ice edge is established in this location where the supply of sea ice from the north balances the loss by melt in warm ocean waters. Sea-ice production in the source region in winter is enhanced within a polynya (the North Water) formed by the persistent southward drift of ice. Similar "conveyor belt" sea-ice regimes also exist in the Barents and Bering Seas, where northern regions of growth export ice to temperate waters.

First-year floes fracture easily under the forces generated by storm winds. Leads form where ice floes separate under tension, exposing new ocean surface to rapid freezing. Where the pack is compressed, the floes buckle and break into blocks that pile into ridges up to 30 m thick. Near open water, notably in the Labrador, Greenland, and Barents Seas, waves are an additional cause of ridging. Because of ridging and rafting, the average thickness of first-year sea ice is typically twice that achievable by freezing processes alone (Melling and Riedel, 1996). Heavily deformed multi-year floes near the Canadian Archipelago can average more than 10 m thick.

Information on the thickness of northern sea ice is scarce. Weekly records of land-fast ice thickness obtained from drilling are available for coastal locations around the Arctic (Canada and Russia) for the 1940s through the present (Melling, 2002; Polyakov et al., 2003a). Within the Arctic Ocean, there have been occasional surveys of sea ice since 1958, measured with sonar on nuclear submarines (Rothrock et al., 1999; Wadhams, 1997; Winsor, 2001). In Fram Strait and the Beaufort Sea, data have been acquired continuously since 1990 from sonar operated from moorings (Melling, 1993; Melling and Moore, 1995; Melling and Riedel, 1996; Vinje et al., 1998). The average thickness of sea ice in the Arctic Ocean is about 3 m, and the thickest ice (about 6 m) is found along the shores of northern Canada and Greenland (Bourke and Garrett, 1987). There is little information about the thickness of the seasonal sea ice that covers more than half the marine Arctic.

Land-fast ice (or fast ice) is immobilized for up to 10 months each year by coastal geometry or by grounded ice ridges (stamukhi). There are a few hundred meters of land-fast ice along all arctic coastlines in winter. In the present climate, ice ridges ground to form stamukhi in depths of up to 30 m, as the pack ice is repeatedly crushed against the fast ice by storm winds. In many areas, stamukhi stabilize sea ice for tens of kilometers from shore. Within the Canadian Archipelago in late winter, land-fast ice bridges channels up to 200 km wide and covers an area of 1 million km^2. Some of this ice is trapped for decades as multi-year land-fast ice (Reimnitz et al., 1995). The remobilization of land-fast ice in summer is poorly understood. Deterioration through melting, flooding by runoff at the coast, winds, and tides are contributing factors.

Many potential impacts of climate change will be mediated through land-fast ice. It protects unstable coastlines and coastal communities from wave damage, flooding by surges, and ice ride-up. It offers safe, fast routes for travel and hunting. It creates unique and necessary habitat for northern species (e.g., ringed seal (*Phoca hispida*) birth lairs) and brackish under-ice migration corridors for fish. It blocks channels, facilitating the formation of polynyas important to northern ecosystems in some areas, and impeding navigation in others (e.g., the Northwest Passage).

2.4. Terrestrial water balance

The terrestrial water balance and hydrologic processes in the Arctic have received increasing attention, as it has been realized that changes in these processes will have implications for global climate. There are large uncertainties concerning the water balance of tundra owing to a combination of:

- the sparse network of *in situ* measurements of precipitation and the virtual absence of measurements of evapotranspiration in the Arctic;
- the difficulty of obtaining accurate measurements of solid precipitation in cold windy environments, even at manned weather stations;
- the compounding effects of elevation on precipitation and evapotranspiration in topographically complex regions of the Arctic, where the distribution of observing stations is biased toward low elevations and coastal regions; and
- slow progress in exploiting remote sensing techniques for measuring high-latitude precipitation and evapotranspiration.

Uncertainties concerning the present-day distributions of precipitation and evapotranspiration are sufficiently large that evaluations of recent variations and trends are problematic. The water budgets of arctic watersheds reflect the extreme environment. Summer precipitation plays a minor role in the water balance compared to winter snow, since in summer heavy rains cannot be absorbed by soils that are near saturation. In arctic watersheds, precipitation exceeds evapotranspiration, and snowmelt is the dominant hydrologic event despite its short duration. In the boreal forest, water balance dynamics are dominated by spring snowmelt; water is stored in wetlands, and evapotranspiration is also a major component in the water balance (Metcalfe and Buttle, 1999). Xu and Halldin (1997) suggested that the effects of climate variability and change on streamflow will depend on the ratio of annual runoff to annual precipitation, with the greatest sensitivity in watersheds with the lowest ratios.

2.4.1. Permanent storage of water on land

The great ice caps and ice sheets of the world hold 75% of the global supply of freshwater; of these, the Greenland Ice Sheet contains 2.85 million km^3 of freshwater (IPCC, 2001c). The northern portions of mid-latitude cyclones carry most of the water that reaches arctic ice caps, with the result that precipitation generally decreases from south to north. Runoff often exceeds precipitation when ice caps retreat. The behavior of glaciers depends upon climate (see section 6.5).

Temperature and precipitation variations influence the arctic ice caps; for example, temperature increases coupled with decreased precipitation move the equilibrium line (boundary between accumulation and ablation) higher, but with increased precipitation, the line moves lower (Woo and Ohmura, 1997). Small shifts in precipitation could offset or enhance the effect of increasing temperatures (Rouse et al., 1997). Water is also stored in permanent snowfields and firn (compact, granular snow that is over one year old) fields, perched lakes (lakes that are raised above the local water table by permafrost), and as permafrost itself. Whitfield and Cannon (2000) implicated shifts between these types of storage as the source of increases in arctic streamflow during recent warmer periods. The IPCC (2001b) stated: "Satellite data show that there are very likely to have been decreases of about 10% in the extent of snow cover since the late 1960s, and ground-based observations show that there is very likely to have been a reduction of about two weeks in the annual duration of lake and river ice cover in the mid- and high latitudes of the Northern Hemisphere, over the 20th century".

2.4.2. Hydrology of freshwater in the Arctic

The Arctic has four hydrologic periods: snowmelt; outflow breakup period (several days in length but accounting for 75% of total annual flow); a summer period with no ice cover and high evaporation; and a winter period where ice cover thicker than 2 m exists on lakes. Four types of arctic rivers show different sensitivity to climatic variations:

- Arctic–nival: continuous permafrost where deep infiltration is impeded by perennially frozen strata, base flow and winter flow are low, and snowmelt is the major hydrologic event.
- Subarctic–nival: dominated by spring snowmelt events, with peak water levels often the product of backwater from ice jams. Groundwater contributions are larger than those in arctic–nival systems. In some areas, complete winter freezing occurs.

- Proglacial: snowmelt produces a spring peak, but flows continue throughout the summer as areas at progressively higher elevations melt. Ice-dammed lakes are possible.
- Muskeg: large areas of low relief characterized by poor drainage. Runoff attenuation is high because of large water-holding capacity and flow resistance.

Fens (peatlands) are wetlands that depend upon annual snowmelt to restore their water table, and summer precipitation is the most important single factor in the water balance (Rouse, 1998). Actual evapotranspiration is a linear function of rainfall. If summer rainfall decreases, there would be an increase in the severity and length of the water deficit. Water balance has a significant effect on the carbon budget and peat accumulation; under drier conditions, peatlands would lose biomass, and streamflows would decrease. Krasovskaia and Saelthun (1997) found that monthly flow regimes in Scandinavia have stable average patterns that are similar from year to year. They demonstrated that most rivers are very sensitive to temperature rises on the order of 1 to 3 °C, and that nival (snow-dominated) rivers become less stable while pluvial (rain-dominated) rivers become more stable. Land storage of snow is important in the formation of the hydrograph in that the distributed nature of the snow across the land "converts" the daily melt into a single peak. Kuchment et al. (2000) modeled snowmelt and rainfall runoff generation for permafrost areas, taking into account the influence of the depth of thawed ground on water input, water storage, and redistribution.

Where they exist, perennial snow banks are the major source of runoff, and as little as 5% of watershed area occupied by such snow banks will enhance runoff compared to watersheds without them. The resulting stream discharge is termed "proglacial", and stored water contributes about 50% of the annual runoff. During winter, when biological processes are dormant, the active layer freezes and thaws. Spring snowmelt guarantees water availability about the same time each year, at a time when rainfall is minimal but solar radiation is near its maximum. Summer hydrology varies from year to year and depends upon summer precipitation patterns and magnitudes. Surface organic soils, which remain saturated throughout the year (although the phase changes), are more important hydrologically than deeper mineral soils. During dry periods, runoff is minimal or ceases. During five years of observations at Imnavait Creek, Alaska, an average of 50 to 66% of the snowpack moisture became runoff, 20 to 34% evaporated, and 10 to 19% added to soil moisture storage (Kane et al., 1989). All biological activity takes place in the active layer above the permafrost. Hydraulic conductivity of the organic soils is 10 to 1000 times greater than silt. Unlike the organic layer, the mineral layer remains saturated and does not respond to precipitation events. Soil properties vary dramatically over short vertical distances. The snowmelt period is brief, lasting on the order of 10 days, and peak flow happens within 36 hours of the onset of flow.

Evapotranspiration is similar in magnitude to runoff as a principal mechanism of water loss from a watershed underlain by permafrost. Water balance studies indicate that cumulative potential evaporation is greater than cumulative summer precipitation.

Snowmelt on south-facing slopes occurred one month earlier than on north-facing slopes in subarctic watersheds (Carey and Woo, 1999). On south-facing slopes, the meltwater infiltrated and recharged the soil moisture but there was neither subsurface flow nor actual runoff. The north-facing slopes had infiltration barriers, thus meltwater was impounded in the organic layer and produced surface and subsurface flows. Permafrost slopes and organic horizons are the principal controls on streamflow generation in subarctic catchments. Seppälä (1997) showed that permafrost is confining but not impermeable. Quinton et al. (2000) found that in tundra, subsurface flow occurs predominantly through the saturated zone within the layer of peat that mantles hill slopes, and that water flow through peat is laminar.

Beltaos (2002) showed that temperature increases over the past 80 years have increased the frequency of mild winter days, which has augmented flows to the extent that they can affect breakup processes. There are several implications of this change, including increases in the frequency of mid-winter breakup events; increased flooding and ice-jam damages; delayed freeze-up dates; and advanced breakup dates. Prowse and Beltaos (2002) suggested that climate change may alter the frequency and severity of extreme ice jams, floods, and low flows. These climate-driven changes are projected to have secondary effects on fluvial geomorphology; river modifying processes; aquatic ecology; ice-induced flooding that supplies water and nutrients to wetlands; biological templates; dissolved oxygen depletion patterns; transportation and hydroelectric generation; and ice-jam damage.

The hydrology and the climate of the Arctic are intricately linked. Changes in temperature and precipitation directly and indirectly affect all forms of water on and in the landscape. If the storage and flux of surface water changes, a variety of feedback mechanisms will be affected, but the end result is difficult to project. Snow, ice, and rivers are considered further in Chapters 6 and 8.

2.5. Influence of the Arctic on global climate

2.5.1. Marine connections

Although the marine Arctic covers a small fraction of the globe, positive feedback between the Arctic Ocean and the climate system has the potential to cause global effects. The thermohaline circulation is the global-scale overturning in the ocean that transports significant heat via a poleward flow of warm surface water and an equatorward return of cold, less saline water at depth. The overturning crucial to this transport in the Northern Hemisphere occurs in the Greenland, Irminger, and

Labrador Seas (Broecker et al., 1990). The occurrence and intensity of overturning is sensitive to the density of water at the surface in these convective gyres, which is in turn sensitive to the outflow of low-salinity water from the Arctic. An increase in arctic outflow is very likely to reduce the overturning and therefore the oceanic flux of heat to northern high latitudes. The overturning also moderates anthropogenic impacts on climate because it removes atmospheric CO_2 to the deep ocean. A comprehensive description of the dynamics and consequences of the marine connections is given in section 9.2.3.

2.5.1.1. Ice-albedo feedback to warming and cooling

Sea ice is an influential feature of the marine Arctic. It reflects a large fraction of incoming solar radiation and insulates the ocean waters against loss of heat and moisture during winter. Sea ice also inhibits the movement and mixing of the upper ocean in response to wind. By stabilizing the upper ocean through melting, it may control the global heat sink at high northern latitudes (Manabe et al., 1991; Rind et al., 1995).

The global impact of ice-albedo feedback is predicated on the existence of a strong relationship between atmospheric temperature increases and sea-ice extent. The seasonal analogue of climate change effects on the marine cryosphere is the dramatic expansion of sea-ice extent in winter and its retreat in summer, in tune with (at a lag of several months) the seasonal variation in air temperature. Another relevant analogue is the seasonal progression from frequently clear skies over the marine cryosphere in winter to dominance by fog and stratiform cloud in summer. The increased moisture supply at the melting surface of the ice pack promotes the formation of low clouds that reflect most of the incoming solar radiation in summer, replacing the weakened reflecting capability of melting sea ice. Thus, cloud cover is an important partner to sea ice in the albedo feedback mechanism.

2.5.1.2. Freshwater feedback to poleward transport of heat and freshwater

Deep convection in the northwest Atlantic Ocean is a crucial part of the global thermohaline circulation. Water freshened by arctic outflow is cooled, causing it to sink deep into the ocean, from where it flows either south to the North Atlantic or north into the Arctic Basin (Aagaard and Greisman, 1975; Nikiforov and Shpaikher, 1980). Deep convection has considerable interannual variability controlled by atmospheric circulation. It operates to link the stochastic effects of atmospheric variability to slow oscillations in the ocean–atmosphere system via the oceanic transports of heat and "freshwater" in the global thermohaline circulation (Broecker, 1997, 2000).

The Greenland Sea is one region where new deep water forms (Swift and Aagaard, 1981). Here, warm and saline water of Atlantic origin meets cold arctic water of lower salinity. Extremely low temperatures cause rapid cooling of the sea surface, which may trigger either deep convective mixing or intensified ice formation, depending on the density of waters at the sea surface. Convection can reach depths of about 2000 m (Visbeck et al., 1995) and the temperature change in the water at that depth is an indicator of the intensity of deep-water formation, with warmer temperatures indicating less deep-water formation. Observations show periods of deep-water temperature increases in the Greenland Sea in the late 1950s and between 1980 and 1990, and temperature decreases in the early 1950s and in the late 1960s. A large increase (0.25 °C) in deep-water temperature occurred in the 1990s (Alekseev et al., 2001). The decrease in deep-water formation implied by increasing deep-water temperatures has weakened the thermohaline circulation, leading to a decreased overflow of deep water through the Faroe-Shetland channel (Hansen et al., 2001).

A reduction in the vertical flux of salt and reduced deep-water formation is likely to trigger a prolonged weakening of the global thermohaline circulation. With less bottom-water formation, there is likely to be a reduction in upwelling at temperate and subtropical latitudes. Paleoclimatic shifts in the thermohaline circulation have caused large and sometimes abrupt changes in regional climate (section 2.7). Dickson et al. (2002) demonstrated that the flows of dense cold water over sills in the Faroe–Shetland Channel and in Denmark Strait are the principal means of ventilating the deep waters of the North Atlantic. Both the flux and density structure of "freshwater" outflow to the North Atlantic are critical to the arctic influence on global climate via the thermohaline circulation (Aagaard and Carmack, 1989).

2.5.2. Sea level

Global average sea level rose between 0.1 and 0.2 m during the 20th century (IPCC, 2001b), primarily because of thermal expansion of warming ocean waters. Although the thermal expansion coefficient for seawater is small, integrated over the 6000 m depth of the ocean the resulting change in sea level can generate changes of significance to ecosystems and communities near coastlines. The warming of arctic seawater will have a negligible impact on local sea level because cold (<0 °C) seawater expands very little with an increase in temperature. However, arctic sea level will respond to changes in the levels of the Atlantic and Pacific Oceans via dynamic links through Bering Strait, Fram Strait, and the Canadian Archipelago. In many areas of the Arctic, sea level is also changing very rapidly as a result of postglacial rebound of the earth's crust. For example, the land at Churchill, Canada (on the western shore of Hudson Bay), rose one meter during the 20th century. In many parts of the Arctic, changes in the elevation of the shoreline due to crustal rebound are likely to exceed the rise in sea level resulting from oceanic warming.

The Arctic Ocean stores a large volume of "freshwater". Arctic sea level is sensitive to "freshwater" storage and will rise if this inventory increases, or fall if "freshwater" storage declines. Changes in northern hydrology are therefore likely to have an important effect on arctic sea level by changing "freshwater" storage in the Arctic.

On a timescale of centuries, and with a sufficient increase in temperature, accumulation or ablation of terrestrial ice caps in Greenland and Antarctica are very likely to be the dominant causes of global changes in sea level. There is an interesting aspect to sea-level change in the vicinity of these ice caps: a sea-level increase caused by the ice cap melting, distributed globally, may be offset by changes in the local gravitational anomaly of the ice, which pulls the sea level up towards it. As a result, it is possible that sea level could actually drop at locations within a few hundred kilometers of Greenland, despite an average increase in sea level worldwide. Sections 6.5 and 6.9 provide further details related to ice caps, glaciers, and sea-level rise.

2.5.3. Greenhouse gases

Arctic ecosystems are characterized by low levels of primary productivity, low element inputs, and slow element cycling due to inhibition of these processes by very cold climatic conditions. However, arctic ecosystems still tend to accumulate organic matter, carbon (C), and other elements because decomposition and mineralization processes are equally inhibited by the cold, wet soil environment (Jonasson et al., 2001). Owing to this slow decomposition, the total C and element stocks of wet and moist arctic tundra frequently equal or exceed stocks of the same elements in much more productive ecosystems in temperate and tropical latitudes. Methane (CH_4) production, for example, is related to the position of the water table in the active layer, which will be affected by changes in active-layer depth and/or permafrost degradation. Natural gas hydrates are also found in the terrestrial Arctic, although only at depths of several hundred meters. Currently, arctic and alpine tundra is estimated to contain 96 x 10^{12} kg of C in its soil and permafrost. This is roughly 5% of the world's soil C pool (IPCC, 2001c). An additional 5.7 x 10^{12} kg of C is stored in arctic wetland, boreal, and tundra vegetation, for a total of 102 x 10^{12} kg of terrestrial C (Jonasson et al., 2001). This is fully discussed in section 7.4.2.1. Thawing of permafrost has the potential to release large stores of CO_2 and CH_4 that are presently contained in frozen arctic soils, both as a direct consequence of thawing and as an indirect consequence of changes in soil wetness (Anisimov et al., 1997; Fukuda, 1994). Although it is not clear whether the Arctic will be a net source or sink of C in the future, the large amounts of C that could be taken up or released make improved understanding of arctic processes important.

The Arctic Ocean was not initially believed to be a significant sink of C because its sizeable ice cover prevents atmosphere–ocean exchange and biological production in the central ocean was believed to be small. Under warmer climate conditions, however, the amount of C that the Arctic Ocean can sequester is likely to increase significantly. In the northern seas, hydrated CH_4 is trapped in solid form at shallow depths in cold sediments. Gas hydrates are likely to decompose and release CH_4 to the atmosphere if the temperature of water at the seabed rises by a few degrees (Kennett et al., 2000) over centuries to a millennium. This is discussed further in section 9.5.5.

2.6. Arctic climate variability in the twentieth century

2.6.1. Observing systems and data sources

All arctic countries maintain programs of synoptic observations to support their economic activity and the sustainability of communities in the Arctic. Due to the harsh environment and the sparseness of the observation network, the need for meteorological observations is often a major (or even the only) reason for the existence of many arctic settlements. Systematic *in situ* arctic meteorological observations started in the late 18th century in the Atlantic sector (Tabony, 1981). In Fennoscandia, the oldest systematic climatic observations north of 65° N were made in Tornio, Finland, between 1737 and 1749, and regular weather stations were established around 1850. At Svalbard, the first permanent weather station was established in 1911. The first station in the Russian north was established at Arkhangelsk in 1813. Most of the meteorological network in central and northern Alaska was established in the 1920s, with the first station, Kotzebue, opening in 1897. The first meteorological observations in southern Alaska (Sitka at 57° N) were made in 1828. In northern Canada, systematic meteorological observations started in the 1940s.

Meteorological observations in the Arctic Ocean began with the first research voyage of Fridtjof Nansen onboard *Fram* (1894–1896). Additional observations were made during the 1920s and 1930s by ships trapped in the pack ice. A new phase of Arctic Ocean observations began in the mid-1930s with the establishment of North Pole ice stations (Arctic Climatology Project, 2000).

Economic issues led to a significant reduction in the existing meteorological network in northern Russia and Canada in the 1990s. Thus, during the past decade, the number of arctic meteorological stations has noticeably decreased, and the number of the stations conducting atmospheric measurements using balloons has decreased sharply.

The national meteorological services of the Nordic countries, Canada, Russia, and the United States maintain extensive archives of *in situ* observations from their national networks. The station density in these networks varies substantially, from 2 per 1000 km² in Fennoscandia to 1 per 100 000 km² in Canada north of

60° N, northern Alaska, and (since the 1990s) in northern regions of Siberia.

In seeking to assemble a high-quality record, different levels of quality assurance, data infilling, and homogenization adjustments are required. The Global Historical Climatology Network (GHCN) dataset includes selected quality controlled long-term stations suitable for climate change studies, while the Global Daily Climatology Network dataset goes through a more limited screening. The Integrated Surface Hourly Dataset incorporates all synoptic observations distributed through the Global Telecommunication System during the past 30 years. All rawindsonde observations in the Arctic (300 stations north of 50° N and 135 stations north of 60° N) are currently collected in the Comprehensive Aerological Reference Data Set.

The sea-ice boundaries in the Atlantic sector of the Arctic Ocean have been documented since the beginning of the 20th century. Since the late 1950s, sea-ice observations have been conducted throughout the year. The development of shipping along the coast of Siberia in the mid-1920s led to sea-ice monitoring in Siberian Arctic waters. By the late 1930s, aviation had become the main observation tool; since the 1970s, satellite remote sensing has been used. The notes of seamen and the logs of fishing, whaling, and sealer vessels operating in arctic waters serve as an important source of information about changes in the state of arctic sea ice throughout the 19th and 20th centuries (Vinje, 2001). A significant amount of historical data on sea ice near the shores of Iceland was preserved and generalized in many studies (Ogilvie and Jónsdóttir, 2000). Information on sea-ice thickness is scarce; observations of ice draft using upward-looking sonar from submarines and stationary systems are the primary source of this information (see section 2.3.4).

Sea-ice data are concentrated at two World Data Centers. Datasets of satellite observations of sea-ice and snow-cover extent (Ramsay, 1998), snow water equivalent from the Special Sensor Microwave Imager (Armstrong and Brodzik, 2001; Grody and Basist, 1996), cloudiness (Rossow and Schiffer, 1999), and the radiation budget (Wielicki et al., 1995) are available from the National Aeronautics and Space Administration (Goddard Institute for Space Studies, Langley Atmospheric Sciences Data Center) and the National Oceanic and Atmospheric Administration (National Climatic Data Center, National Snow and Ice Data Center – NSIDC). A suite of arctic-related datasets is available from the NSIDC (http://nsidc.org/index.html).

Oceanographic measurements in the central Arctic were initiated in 1894 by Nansen (1906). They were restarted in the 1930s, interrupted during the Second World War, and resumed in the late 1940s with the help of aviation and drifting stations. Since 1987, icebreakers with conductivity, temperature, and depth sondes have been used to make observations.

The hydrologic network in the Arctic is probably the weakest of the arctic observation networks, which makes information about the arctic water budget quite uncertain (Vörösmarty et al., 2001). A circumpolar river discharge dataset is available online (R-ArcticNET, 2003).

2.6.2. Atmospheric changes

2.6.2.1. Land-surface air temperature

Although several analyses (e.g., Comiso, 2003; Polyakov et al., 2003b) have examined large-scale temperature variations in the Arctic, no study has evaluated the spatial and temporal variations in temperature over all land areas in the zone from 60° to 90° N for the entire 20th century. Consequently, temperature trends are illustrated for this latitude band using the Climatic Research Unit (CRU) database (Jones and Moberg, 2003) and the GHCN database (updated from Peterson and Vose, 1997). Both databases were used in the IPCC Third Assessment Report (IPCC, 2001c) to summarize the patterns of temperature change over global land areas since the late 19th century. While the impact of urbanization on large-scale temperature trends in the Arctic has not been assessed, the results of Jones et al. (1990), Easterling et al. (1997), Peterson et al. (1999), and Peterson (2003) indicate that urbanization effects at the global, hemispheric, and even regional scale are small (<0.05 °C over the period 1900 to 1990).

Figure 2.6 depicts annual land-surface air temperature variations in the Arctic (north of 60° N) from 1900 to 2003 using the GHCN dataset. The CRU time series is virtually identical to the GHCN series, and both document a statistically significant warming trend of 0.09 °C/decade during the period shown (Table 2.1). The arctic trend is greater than the overall Northern Hemisphere trend of 0.06 °C/decade over the same period (IPCC, 2001b). In general, temperature increased from 1900 to the mid-1940s, then decreased

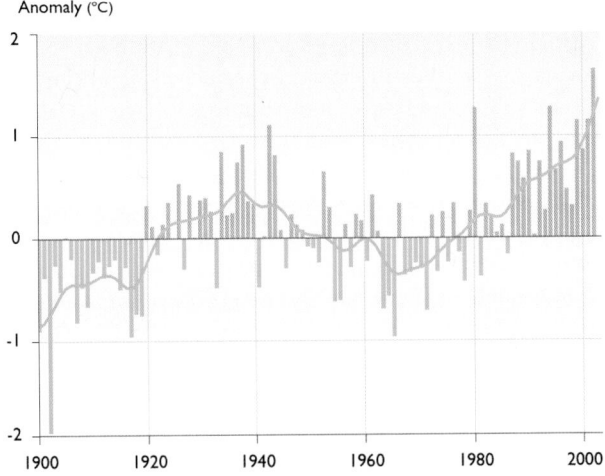

Fig. 2.6. Annual anomalies of land-surface air temperature in the Arctic (60° to 90° N) for the period 1900 to 2003 using the GHCN dataset (updated from Peterson and Vose, 1997). Anomalies are calculated relative to the 1961–1990 average. The smoothed curve was created using a 21-point binomial filter, which approximates a 10-year running mean.

Table 2.1. Least-squares linear trends in annual anomalies of arctic (60° to 90° N) land-surface air temperature (°C/decade) from the GHCN (updated from Peterson and Vose, 1997) and CRU (Jones and Moberg, 2003) datasets. Anomalies are calculated relative to the 1961–1990 average.

	1900–2003	1900–1945	1946–1965	1966–2003
GHCN dataset	0.09[a]	0.29[a]	-0.14	0.40[a]
CRU dataset	0.09[a]	0.31[a]	-0.20	0.40[a]

[a]value significant at the 1% level of confidence or better

until about the mid-1960s, and increased again thereafter. The general features of the arctic time series are similar to those of the global time series, but decadal trends and interannual variability are greater in the Arctic. Comparisons of the arctic and global time series are discussed later in this section.

Figure 2.7 illustrates the patterns of land-surface air temperature change in the Arctic between 1900 and 2003 and for three periods therein. Trends were calculated from annually averaged gridded anomalies using the method of Peterson et al. (1999) with the requirement that annual anomalies include a minimum of ten months of data. For the period 1900 to 2003, trends were calculated only for those 5° x 5° grid boxes with annual anomalies in at least 70 of the 104 years. The minimum number of years required for the shorter time periods (1900–1945, 1946–1965, and 1966–2003) was 31, 14, and 26, respectively. The three periods selected for this figure correspond approximately to the warming–cooling–warming trends seen in Fig. 2.6. The spatial coverage of the region north of 60° N is quite varied. During the first period (1900–1945), there are only three stations meeting the requirement of 31 years of data in Region 4 and only four in Region 3 (see section 18.3 for map and description of the ACIA regions). The highest density of stations is found in Region 1. The coverage for the second and third periods is more uniform.

Although it is difficult to assess trends in some areas, air temperature appears to have increased throughout the Arctic during most of the 20th century. The only period characterized by widespread cooling was 1946 to 1965, and even then large areas (e.g., southern Canada and southern Eurasia) experienced significant increases in temperature. Temperatures in virtually all parts of the Arctic increased between 1966 and 2003, with trends exceeding 1 to 2 °C/decade in northern Eurasia and northwestern North America. The average trend between 1966 and 2003 over the Arctic was 0.4 °C/decade, approximately four times greater than the average for the century. While most pronounced in winter and spring, all seasons experienced an increase in temperature over the past several decades (Fig. 2.8). The trends shown in Fig. 2.8 were calculated from seasonally-averaged gridded anomalies using the method of Peterson et al. (1999) with the requirement that the calculation of seasonal anomalies should include all three months. Trends were calculated only for those 5° x 5° grid boxes containing seasonal anomalies in at least 26 of the 38 years. The updated analysis by Chapman and Walsh (2003) showed generally similar

patterns of change over the period 1954 to 2003, but with areas of cooling around southern Greenland.

The instrumental record of land-surface air temperature is qualitatively consistent with other climate records in the Arctic (Serreze et al., 2000). For instance, temperatures in the marine Arctic (as measured by coastal land stations, drifting ice stations, and Russian North Pole stations) increased at the rate of 0.05 °C/decade during the 20th century (Polyakov et al., 2003b). As with the land-only record, the increases were greatest in winter and spring, and there were two relative maxima during the century (the late 1930s and the 1990s). For periods since 1950, Polyakov et al. (2003b) found the rate of temperature increase in the marine Arctic to be similar to that noted for the GHCN dataset. Because of the scarcity of data prior to 1945, it is very difficult to say whether the Arctic as a whole was as warm in the 1930s and 1940s as it was during the 1990s. In the Polyakov et al. (2003b) analysis, only coastal stations were chosen and most of the stations contributing to the high average temperatures in the 1930s were in Scandinavia. Interior stations, especially those between 60° N (the southern limit for the analysis in this section) and 62° N (the southern limit for the Polyakov et al. (2003b) study), have warmed more than coastal stations over the past few decades. As discussed in section 2.6.3, arctic sea-ice extent contracted from 1918 to 1938 and then expanded between 1938 and 1968 (Zakharov, 2003). The expansion after 1938 implies that the Arctic was cooling during that period.

Polyakov et al. (2002) presented the temperature trends (°C/yr) for their dataset (as described above) and for the Jones et al. (1999) Northern Hemisphere dataset. For comparison purposes, the temperature trends (°C/yr) for the land-surface temperatures (GHCN dataset) were computed for the latitude bands 60° to 90° N and 0° to 60° N (Fig. 2.9). The trend shown for any given year before present is the average trend from that year through 2003; for example, the value corresponding to 60 years before present is the average trend for the period 1944 to 2003. In both latitudinal bands, the trend over any period from 120 years ago to the present is positive (i.e., the Arctic is warming). Although the trends for both bands have been increasing over the past 60 years, the trend in the 60° to 90° N band is larger. The rate of temperature increase in the Arctic (as defined here) exceeds that of lower latitudes. Due to natural variability and sparse data in the Arctic, the arctic trend shows more variability and the confidence limits are wider. Over the past 40 years, the arctic warming trend was about 0.04 °C/yr (0.4 °C/decade) compared to a trend of 0.025 °C/yr for the lower latitudes.

Likewise, surface temperatures derived from satellite thermal infrared measurements, which provide circumpolar coverage from 1981 to 2001, exhibited statistically significant warming trends in all areas between 60° to 90° N except Greenland (Comiso, 2003). The warming trends were 0.33 °C/decade over the sea ice, 0.50 °C/decade over Eurasia, and 1.06 °C/decade over North America. In addition, the recent reduction in sea-ice thickness (Rothrock et al., 1999), the retreat of sea-ice cover (Parkinson et al., 1999), and the decline in peren-

nial sea-ice cover (Comiso, 2002) are consistent with large-scale warming in the Arctic. In view of this evidence, it is very probable that the Arctic has warmed over the past few decades.

Global climate model simulations generally indicate that increasing atmospheric GHG concentrations will result in greater temperature increases in the Arctic than in other parts of the world. As stated by the IPCC (2001c), model experiments show "a maximum warming in the high latitudes of the Northern Hemisphere". In reference to warming at the global scale, the IPCC (2001b) also concluded, "There is new and stronger evidence that

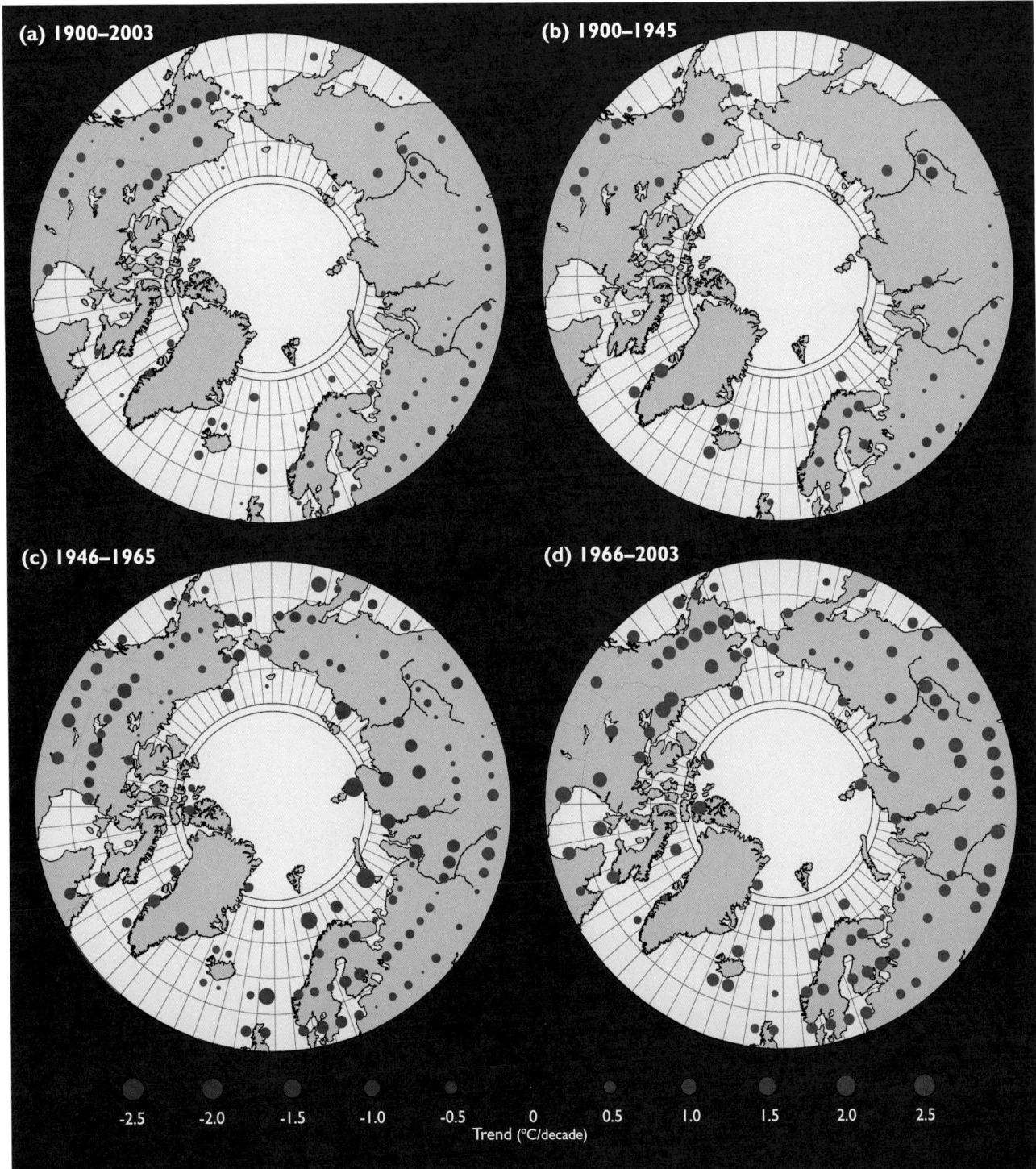

Fig. 2.7. Annual land-surface air temperature trends for (a) 1900 to 2003; (b) 1900 to 1945; (c) 1946 to 1965; and (d) 1966 to 2003, calculated using the GHCN dataset (updated from Peterson and Vose, 1997).

most of the warming observed over the past 50 years is attributable to human activities".

The question is whether there is definitive evidence of an anthropogenic signal in the Arctic. This would require a direct attribution study of the Arctic, which has not yet been done. There are studies showing that an anthropogenic warming signal has been detected at the regional scale. For example, Karoly et al. (2003) concluded that temperature variations in North America during the second half of the 20th century were probably not due to natural variability alone. Zwiers and Zhang (2003) were able to detect the combined effect of changes in

GHGs and sulfate aerosols over both Eurasia and North America for this period, as did Stott et al. (2003) for northern Asia (50°–70° N) and northern North America (50°–85° N). In any regional attribution study, the importance of variability must be recognized. In climate model simulations, the arctic signal resulting from GHG-induced warming is large but the variability (noise) is also large. Hence, the signal-to-noise ratio may be lower in the Arctic than at lower latitudes. In the Arctic, data scarcity is another important issue.

A related question is whether the warming in the Arctic is enhanced relative to that of the globe (i.e., polar

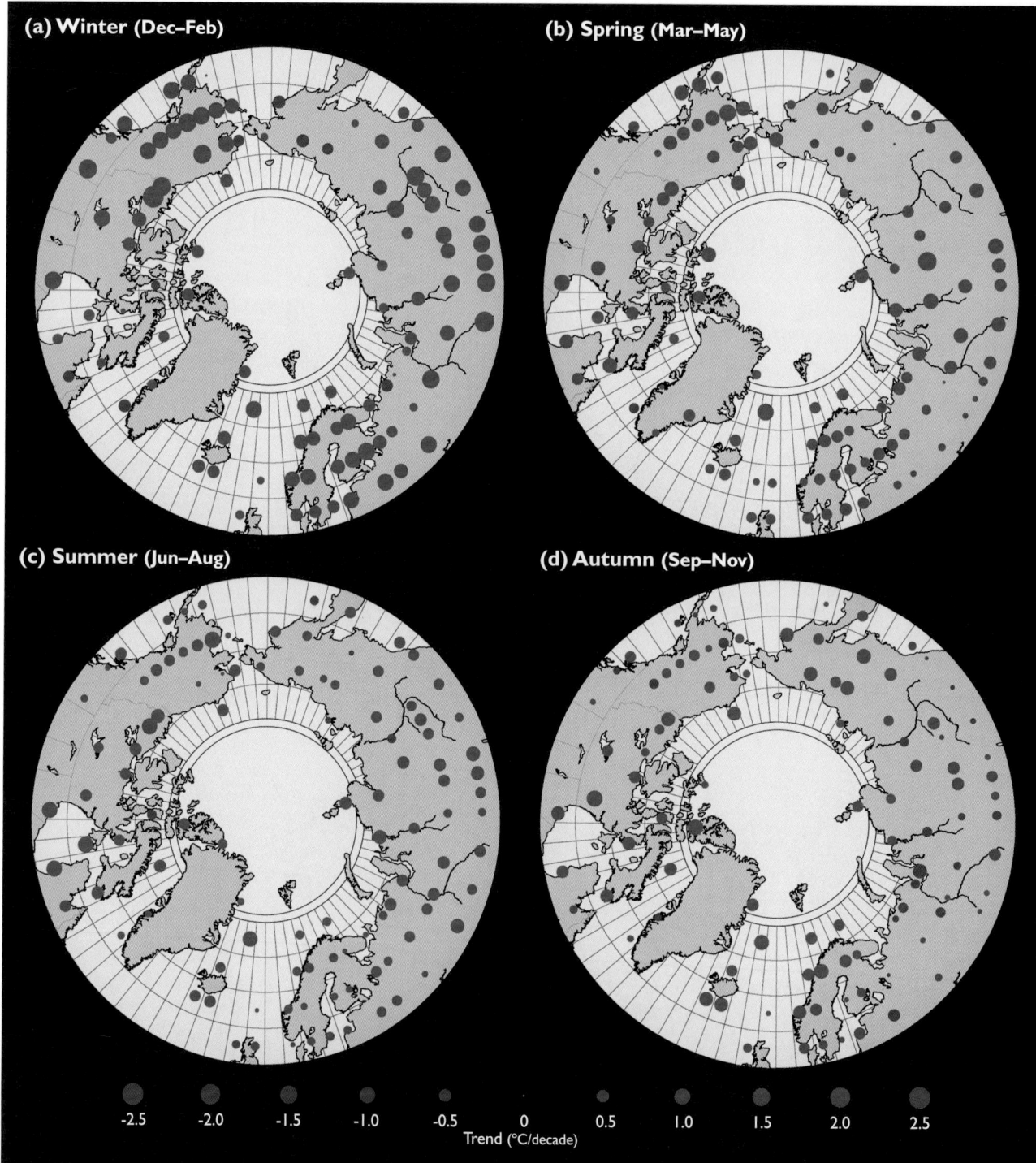

Fig. 2.8. Seasonal land-surface air temperature trends for the period 1966 to 2003 calculated using the GHCN dataset (updated from Peterson and Vose, 1997).

amplification). For example, Polyakov et al. (2002) concluded that observed trends in the Arctic over the entire 20th century did not show polar amplification. A number of studies (e.g., Comiso, 2003; Thompson and Wallace, 1998) suggested that much of the recent warming resulted from changes in atmospheric circulation; specifically, during the past two decades when the AO was in a phase that brought relatively warm air to the same areas that experienced the greatest increases in temperature (i.e., northern Eurasia and northwestern North America). Serreze et al. (2000), however, noted that the changes in circulation were "not inconsistent" with anthropogenic forcing.

It is clear that trends in temperature records, as evidence of polar amplification, depend on the timescale chosen. Over the past 100 years, it is possible that there has been polar amplification but the evidence does not allow a firm conclusion. As noted previously, the analysis presented here of land stations north of 60° N does show more warming over the past 100 years (0.09 °C/decade for the Arctic compared to 0.06 °C/decade for the

globe). However, an analysis of coastal stations north of 62° N (Polyakov et al., 2002) found arctic warming of 0.05 °C/decade over the past 100 years. Johannessen et al. (2004) found, with a more extensive dataset, that the "early warming trend in the Arctic was nearly as large as the warming trend for the last 20 years" but "spatial comparison of these periods reveals key differences in their patterns". The pattern of temperature increases over the past few decades is different and more extensive than the pattern of temperature increases during the 1930s and 1940s, when there was weak (compared to the present) lower-latitude warming.

In conclusion, for the past 20 to 40 years there have been marked temperature increases in the Arctic. The rates of increase have been large, and greater than the global average. Two modeling studies have shown the importance of anthropogenic forcing over the past half century for modeling the arctic climate. Johannessen et al. (2004) used a coupled atmosphere–ocean general circulation model (AOGCM) to study the past 100 years and noted, "It is suggested strongly that whereas the earlier warming was natural internal climate-system variability, the recent SAT (surface air temperature) changes are a response to anthropogenic forcing". Goosse and Renssen (2003) simulated the past 1000 years of arctic climate with a coarser resolution AOGCM and were able to replicate the cooling and warming until the mid-20th century. Without anthropogenic forcing, the model simulates cooling after a temperature maximum in the 1950s. There is still need for further study before it can be firmly concluded that the increase in arctic temperatures over the past century and/or past few decades is due to anthropogenic forcing.

2.6.2.2. Precipitation

Precipitation assessments in the Arctic are limited by serious problems with the measurement of rainfall and snowfall in cold environments (Goodison et al., 1998). Methods for correcting measured precipitation depend upon observing practices, which differ between countries; in addition, metadata (information about the data) needed to perform corrections are often inadequate. The IPCC (2001b) concluded, "It is very likely that precipitation has increased by 0.5 to 1% per decade in the 20th century over most mid- and high latitudes of the Northern Hemisphere continents".

An updated assessment of precipitation changes north of 60° N is presented in Figs. 2.10 and 2.11. These figures provide estimates of linear trends in annual and seasonal precipitation based on available data in the GHCN database (updated from Peterson and Vose, 1997). Trends were calculated from annually-averaged gridded anomalies (baseline 1961–1990) using the method of Peterson et al. (1999) with the requirement that annual anomalies include a minimum of ten months of data (Fig. 2.10) and that seasonal anomalies must contain all three months (Fig. 2.11). In Fig. 2.10, for the period 1900 to 2003 trends were calculated only for those 5° x 5° grid boxes containing annual

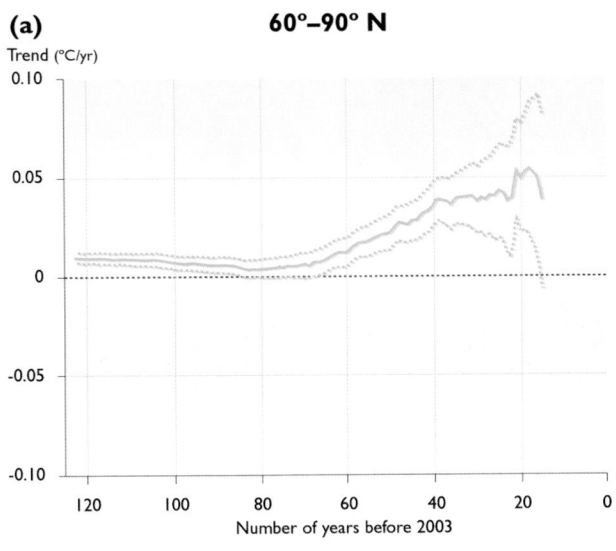

(a) 60°–90° N

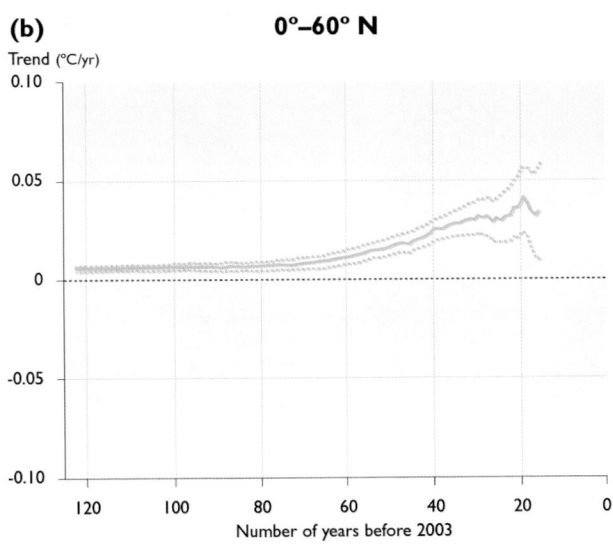

(b) 0°–60° N

Fig. 2.9. Trends in anual mean land-surface air temperatures (solid lines) and their 95% significance levels (dashed lines) over the past 120 years for (a) 60° to 90° N and (b) 0° to 60° N (data from the GHCN dataset, updated from Peterson and Vose, 1997).

anomalies in at least 70 of the 104 years; the minimum number of years required for the shorter time periods (1900–1945, 1946–1965, and 1966–2003) was 31, 14, and 26 respectively. In Fig. 2.11, trends were calculated only for those 5° x 5° grid boxes containing seasonal anomalies in at least 26 of the 38 years.

For the entire period from 1900 to 2003, Table 2.2 indicates a significant positive trend of 1.4% per decade. New et al. (2001) used uncorrected records and found that terrestrial precipitation averaged over the 60° to 80° N band exhibited an increase of 0.8% per decade over the period from 1900 to 1998. In gen-

eral, the greatest increases were observed in autumn and winter (Serreze et al., 2000). Fig. 2.10 shows that most high-latitude regions have experienced an increase in annual precipitation. During the arctic warming in the first half of the 20th century (1900–1945), precipitation increased by about 2% per decade, with significant positive trends in Alaska and the Nordic region. During the two decades of arctic cooling (1946–1965), the high-latitude precipitation increase was roughly 1% per decade. However, there were large regional contrasts with strongly decreasing values in western Alaska, the North Atlantic region, and parts of Russia. Since 1966, annual precipitation

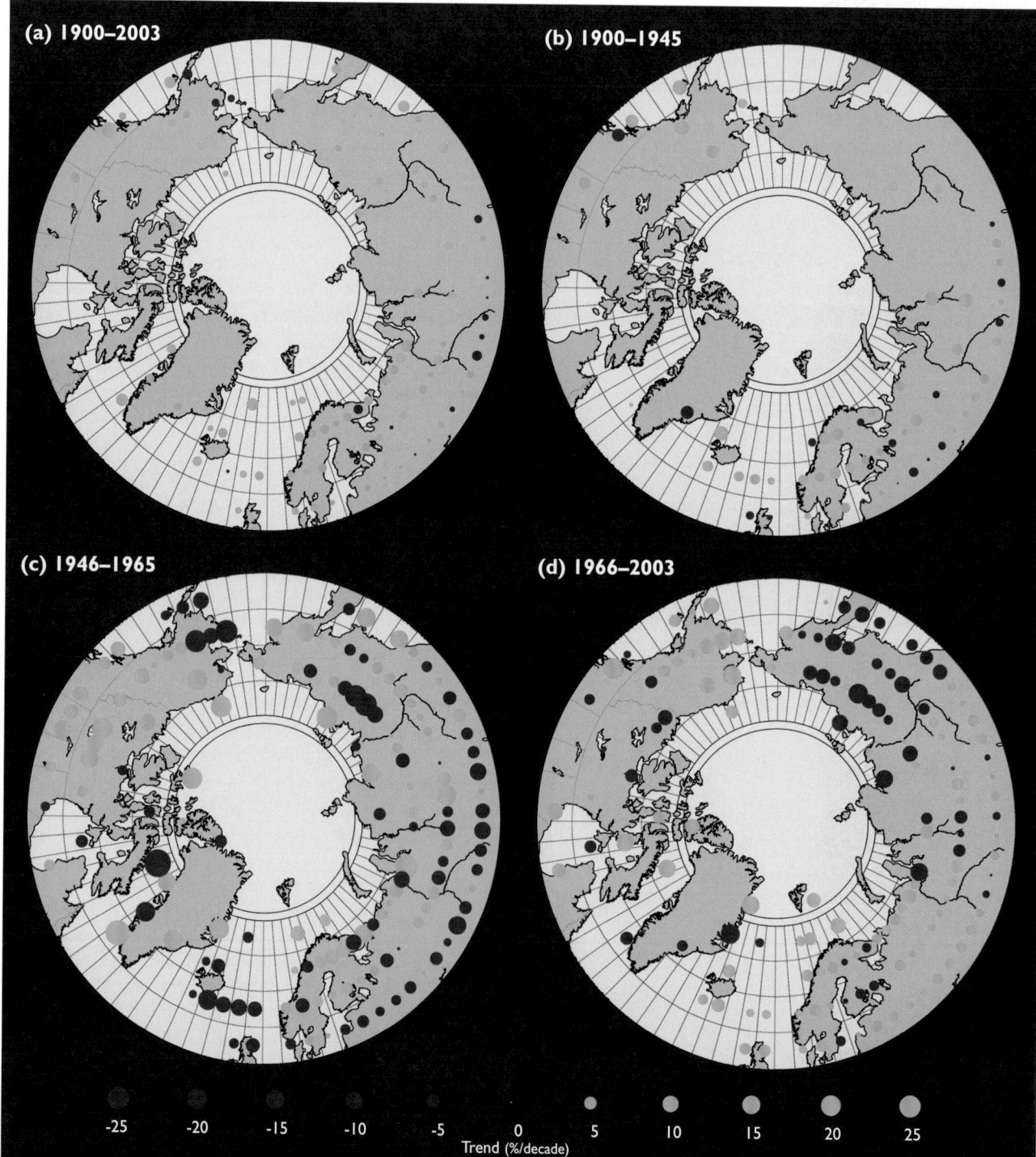

Fig. 2.10. Annual land-surface precipitation trends for (a) 1900 to 2003; (b) 1900 to 1945; (c) 1946 to 1965; and (d) 1966 to 2003, calculated using the GHCN database (updated from Peterson and Vose, 1997).

has increased at about the same rate as during the first half of the 20th century. In eastern Russia, annual precipitation decreased, mostly because of a substantial decrease during winter (Fig. 2.11). Average Alaskan precipitation has increased in all seasons.

These trends are in general agreement with the results of a number of regional studies. For example, there have been positive trends in annual precipitation (up to a 20% increase) during the past 40 years over Alaska (Karl et al., 1993), Canada north of 55° N (Mekis and Hogg, 1999), and the Russian permafrost-free zone (Groisman and Rankova, 2001). Siberian precipitation exhibited little change during the second half of the 20th century (Groisman et al., 1991, updated). Hanssen-Bauer et al. (1997) documented that annual precipitation in northern Norway increased by 15% during the 20th century. Førland et al. (1997) and Hanssen-Bauer and Førland (1998) found precipitation increases of about 25% over the past 80 years in Svalbard. Lee et al. (2000) described details of specific periods and changes in annual precipitation, which generally increased between 1880 and 1993 in the Arctic.

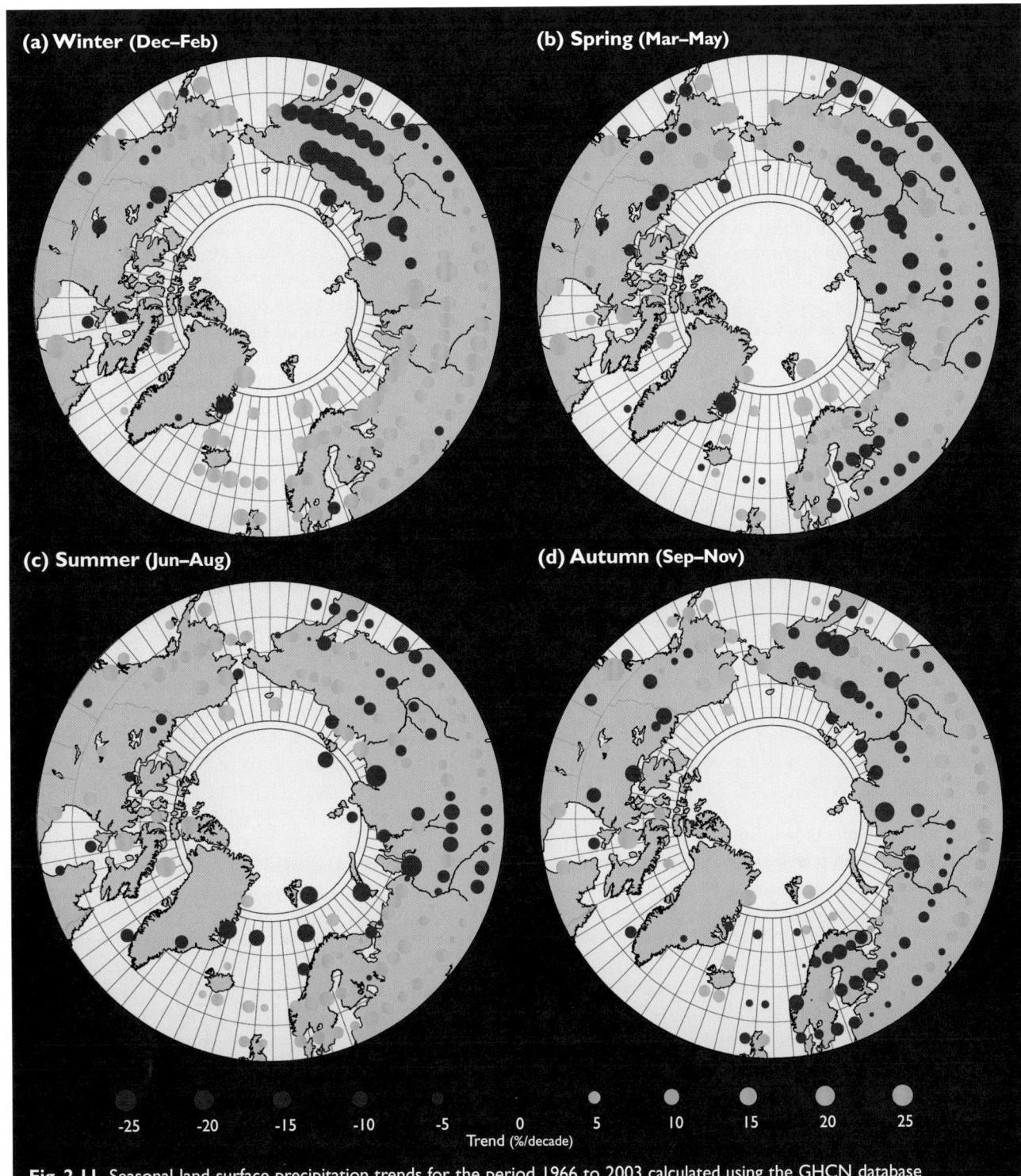

Fig. 2.11. Seasonal land-surface precipitation trends for the period 1966 to 2003 calculated using the GHCN database (updated from Peterson and Vose, 1997).

Table 2.2. Least-squares linear trends in annual anomalies of arctic (60° to 90° N) precipitation (% per decade) from the GHCN database (updated from Peterson and Vose, 1997). Anomalies are calculated relative to the 1961–1990 average.

	1900–2003	1900–1945	1946–1965	1966–2003
GHCN	1.4[a]	2.3[a]	1.3	2.2

[a]value significant at the 1% level of confidence or better

Overall, it is probable that there was an increase in arctic precipitation over the past century.

A few studies suggest that the fraction of annual precipitation falling as snow has diminished, which is consistent with widespread temperature increases in the Arctic. For the period 1950 to 2001, Groisman et al. (2003) found an increase of 1.2% per decade in annual rainfall north of 50° N, an increase that was partially due to an additional fraction of liquid precipitation during spring and autumn. In addition, Førland and Hanssen-Bauer (2003) found that the fraction of solid precipitation diminished at all stations in the Norwegian Arctic between 1975 and 2001. If the fraction of precipitation falling in solid form has decreased, it is possible that the arctic trend in total precipitation may be overestimated because gauge under-catch is less problematic during liquid precipitation events (Førland and Hanssen-Bauer, 2000).

Reporting changes in precipitation frequency in cold environments is difficult because many events result in small precipitation totals and many gauges have accuracy problems during light precipitation events. To avoid uncertainties in the measurement of small amounts of precipitation, Groisman et al. (2003) assessed the climatology and trends in the annual number of days with precipitation >0.5 mm for the period 1950 to 2001. They found no significant change in the number of wet days over North America and northern Europe (including western Russia) but a significant decrease in wet days over eastern Russia. The latter was first reported by Sun et al. (2001) for eastern Russia south of 60° N and is remarkable because it is accompanied by an increase in heavy precipitation in the same region (Groisman et al., 1999a,b; Sun et al., 2001).

The IPCC (2001c) concluded that it was likely that there had been a widespread increase in heavy and extreme precipitation events in regions where total precipitation had increased (e.g., in the high latitudes of the Northern Hemisphere). Subsequent to the study of Groisman et al. (1998, 1999a), which reported that heavy precipitation events appear to have increased at high latitudes, a series of analyses demonstrated that changes in heavy precipitation over Canada are spatially diverse (Akinremi et al., 1999; Mekis and Hogg, 1999; Stone et al., 2000). Mekis and Hogg (1999) noted that the magnitude of heavy precipitation events making up the top 10% of such events increased in northern Canada but decreased in the south during the 20th-century period of record. Zhang X. et al. (2001) reported a marked decadal increase in heavy snowfall events in northern Canada. Stone et al. (2000) showed that during the second half of the 20th century, heavy and intermediate daily precipitation events became significantly more frequent in over one-third of all districts of Canada, indicating a shift to more intense precipitation. Groisman et al. (2002) reported an increase in the frequency of "very heavy" (upper 0.3%) rain events over the European part of Russia during the past 65 years. Comparing two multidecadal periods in the second half of the 20th century, Frich et al. (2002) reported an increase in most indicators of heavy precipitation events over western Russia, eastern Canada, and northern Europe, while Siberia showed a decrease in the frequency and/or duration of heavy precipitation events.

2.6.2.3. Sea-level pressure

Sea-level pressure data from the Arctic Buoy Program (1979–1994) show decreasing pressures in the central arctic that are greater than decreases in sea-level pressure found anywhere else in the Northern Hemisphere (Walsh et al., 1996). These decreases are greatest and statistically significant in autumn and winter, and are compensated by pressure increases over the subpolar oceans. It is very probable that sea-level pressure has been dropping over the Arctic during the past few decades. A useful review of other evidence for these and related changes, and their linkages to the AO, was undertaken by Moritz et al. (2002). A similar pattern of sea-level pressure change is simulated by some climate models when forced with observed variations in GHGs and tropospheric sulfate aerosols (Gillett et al., 2003). However, for unknown reasons, the magnitude of the anthropogenically forced pressure change in climate models is generally much less than the observed pressure change. Longer-term (>100 years) observations, based in part on recently released Russian meteorological observations from coastal stations, show that the decrease in central arctic sea-level pressure during recent decades is part of a longer-term (50 to 80 year) oscillation that peaked in the late 1800s and again in the 1950s and 1960s (Polyakov et al., 2003b). This suggests that the decrease in central arctic sea-level pressure during recent decades may be a result of the combined effects of natural and anthropogenic forcing.

2.6.2.4. Other variables

Groisman et al. (2003) assessed a set of atmospheric variables derived from daily temperatures and precipitation that have applications in agriculture, architecture, power generation, and human health management. This section provides examples of these derived variables, including a temperature derivative (frequency of thaws) and a temperature and precipitation derivative (rain-on-snow-events). A day with thaw (i.e., snowmelt) can be defined as a day with snow on the ground when the daily mean temperature is above -2 °C (Brown R., 2000). During these days, snow deteriorates, changes its physical properties, and (eventually) disappears. In win-

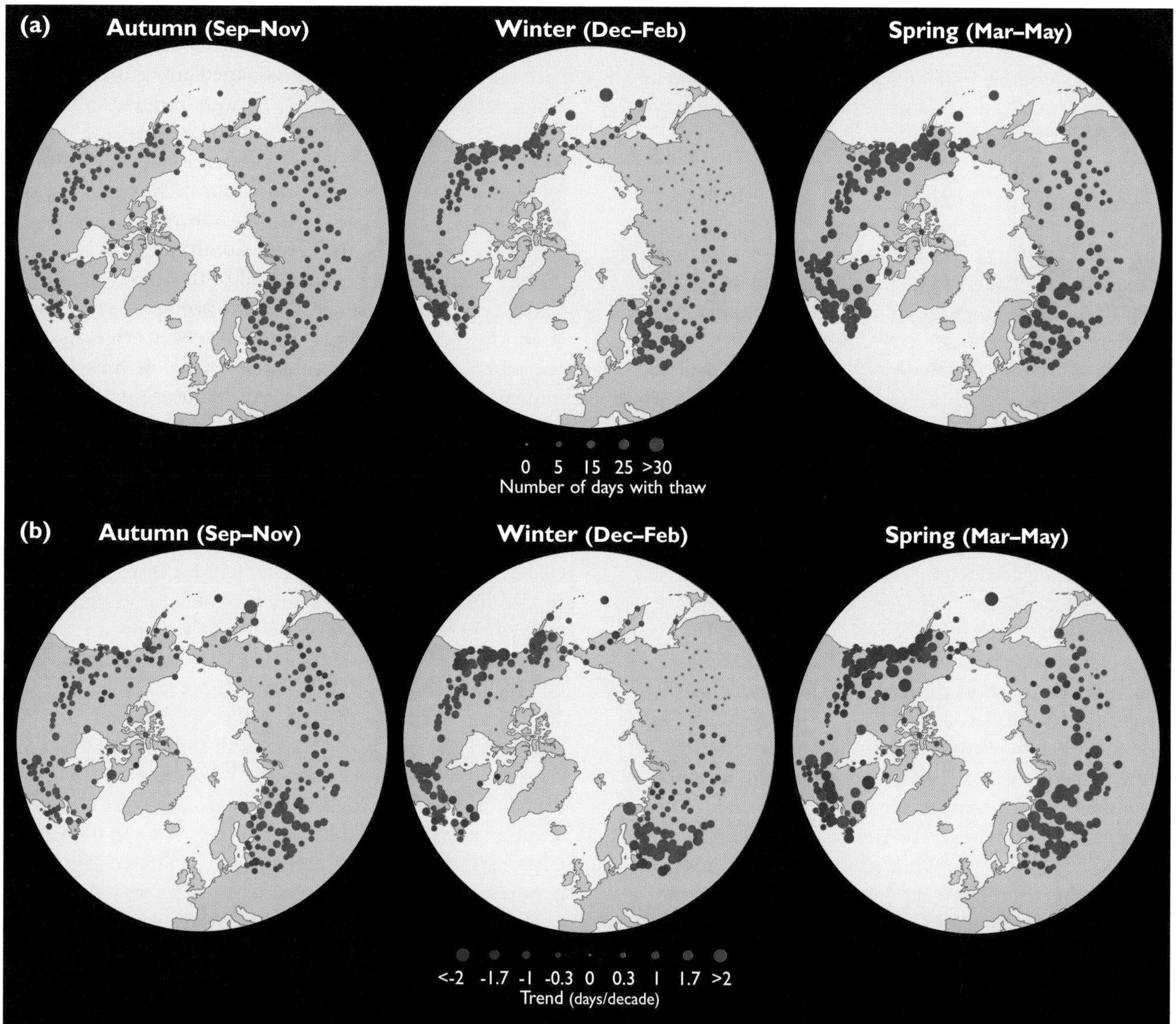

(a)

| Autumn (Sep–Nov) | Winter (Dec–Feb) | Spring (Mar–May) |

0   5   15   25   >30
Number of days with thaw

(b)

| Autumn (Sep–Nov) | Winter (Dec–Feb) | Spring (Mar–May) |

<-2   -1.7   -1   -0.3   0   0.3   1   1.7   >2
Trend (days/decade)

Fig. 2.12. Seasonal frequency of days with thaw: (a) climatology and (b) trends for the period 1950 to 2000 (Groisman et al., 2003).

ter and early spring at high latitudes, thaws negatively affect transportation, winter crops, and the natural environment including vegetation and wild animals. In late spring, intensification of thaw conditions leads to earlier snow retreat and spring onset (Brown R., 2000; Cayan et al., 2001; Groisman et al., 1994). Gradual snowmelt during the cold season affects seasonal runoff in northern watersheds, reducing the peak flow of snowmelt origin and increasing the mid-winter low flow (Vörösmarty et al., 2001). Figures 2.12 and 2.13 show the climatology of thaws in North America and Russia and quantitative estimates of change during the past 50 years. The time series shown in Fig. 2.13 demonstrate statistically significant increasing trends for winter and autumn of 1.5 to 2 days per 50 years. This change constitutes a 20% (winter) to 40% (autumn) increase in the thaw frequency during the second half of the 20th century.

Rain falling on snow causes a faster snowmelt and, when the rainfall is intense, may lead to flash floods. Along the western coast of North America, rain-on-snow events are the major cause of severe flash floods (Wade et al., 2001). Groisman et al. (2003) found a significant

increase in the frequency of winter rain-on-snow events in western Russia (50% over the 50 years from 1950 to 2000) and a significant reduction (of similar magnitude) in western Canada. During spring, there has been a

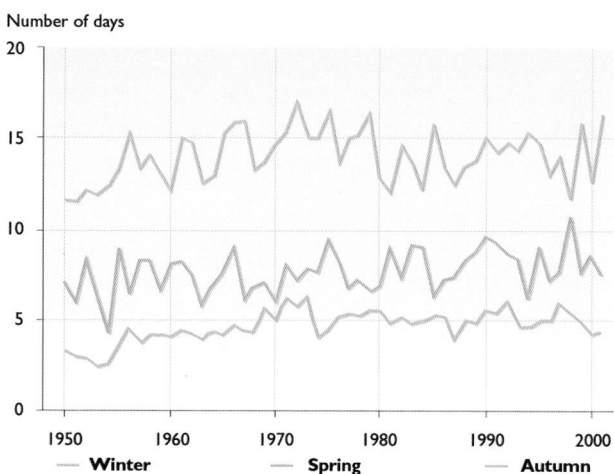

Fig. 2.13. Time series of variations in the frequency of days with thaw, averaged over the arctic land area for winter, spring, and autumn (Groisman et al., 2003).

significant increase in rain-on-snow events in Russia (mostly in western Russia), but there has also been a significant decrease in the frequency of these events in western Canada. The decrease in western Canada is mostly due to reduced snow-cover extent.

2.6.3. Marine Arctic

The 20th-century record of sea-ice extent reveals phases of expansion between 1900 and 1918 and between 1938 and 1968, and of contraction between 1918 and 1938 and from 1968 to the present (Zakharov, 2003). The IPCC (2001b) concluded, "Northern Hemisphere spring and summer sea-ice extent has decreased by about 10 to 15% since the 1950s. It is likely that there has been about a 40% decline in Arctic sea-ice thickness during late summer to early autumn in recent decades and a considerably slower decline in winter sea-ice thickness". Since 1978, the average rate of decrease has been 2.9±0.4% per decade (Fig. 2.14; Cavalieri et al., 1997). The extent of perennial sea ice has decreased at more than twice this rate (Johannessen et al., 1999). The contraction has been greatest in August and September, while in January and February there has been a small expansion (Serreze et al., 2000). Recent trends vary considerably by region. In eastern Canadian waters, ice cover has actually increased (Parkinson, 2000). Interdecadal variations are large and dominated by anomalies of opposite sign in the Eurasian and American sectors (Deser et al., 2000). These variations are associated with anomalies in atmospheric pressure and temperature that closely resemble those associated with the AO (section 2.2).

The thickness of sea ice has decreased by 42% since the mid-1970s within a band across the central Arctic Ocean between the Chukchi Sea and Fram Strait (Rothrock et al., 1999; Wadhams, 2001). In the 1990s, the average thickness in this zone at the end of summer was only 1.8 m. However, the extent to which this change is a result of greater melt and reduced growth

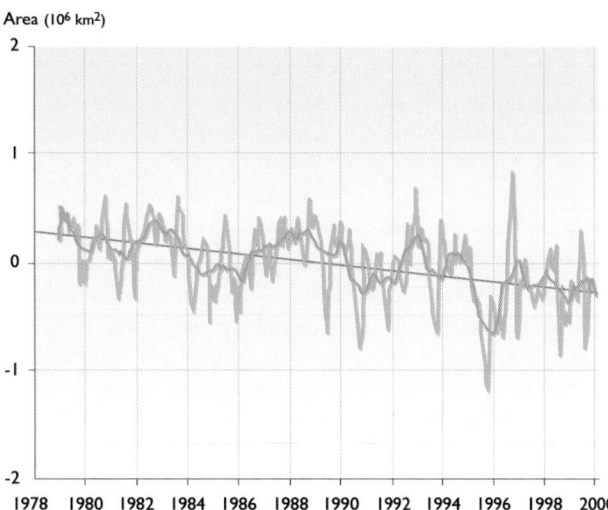

Area (10^6 km²)

Fig. 2.14. Time series of monthly sea-ice extent anomalies in the Arctic Basin (blue line) with respect to the 1979–2001 average, with 12-month running mean (lavender line) and least-squares linear fit (grey line), based on data from 1979 to 2001 (National Snow and Ice Data Center, Boulder, Colorado).

(namely a warming climate) remains to be resolved. Thick sea ice observed before the 1990s may simply have been driven out of the band of observation by wind changes in the 1990s associated with the AO (Holloway and Sou, 2001; Polyakov and Johnson, 2000; Zhang T. et al., 2000). The impacts of the concurrent inflow of warmer Atlantic water and disappearing cold halocline (see below) are also unclear (Winsor, 2001). There are no systematic trends over the past 50 years in the thickness of coastal land-fast ice in either northern Canada or Siberia (Brown R. and Coté, 1992; Polyakov et al., 2003a). Interannual variability in the thickness of coastal land-fast ice is dominated by changes in winter snow accumulation, not air temperature.

Sea-ice drift in the central Arctic Ocean has been monitored since 1979 under the direction of the International Arctic Buoy Program. The pattern of sea-ice drift changed in the late 1980s coincident with the change in the pattern of surface winds caused by the AO. In the first decade of observations, the transpolar drift of sea ice moved directly from the Laptev Sea to Fram Strait. The Beaufort Gyre extended from the Canadian Archipelago to Wrangel Island. During the 1990s, the source of the transpolar drift moved eastward into the East Siberian Sea and the Beaufort Gyre shrank eastward into the Beaufort Sea (Kwok, 2000; Mysak, 2001; Polyakov and Johnson, 2000). The change in circulation moved the thickest sea ice toward the Canadian Archipelago and may have facilitated increased export of perennial sea ice across the north of Greenland and out through Fram Strait (Maslowski et al., 2000). Section 6.3 provides additional information regarding sea ice.

There are few systematic long-term oceanographic observations in the marine Arctic, except perhaps in the European sector. Russian airborne surveys are the primary source of information for the central Arctic prior to 1990 (Gorshkov, 1980). Because most records are of short duration, confidence in discriminating between trends and variability in the marine Arctic is low.

Most evidence of change in ice-covered waters is recent, emerging from observations made in the 1990s. Temperature and salinity at intermediate depths in the arctic basins were stable between 1950 and 1990. In the early 1990s, the temperature of the Fram Strait Branch north of the Barents Sea increased by 2 °C. The temperature increase has since spread around the perimeter of the basin, reaching the Beaufort Sea north of Alaska in 2001 (Carmack et al., 1997; Ivanov, 2000; Morison et al., 1998; Rudels et al., 1997; Schauer et al., 2002). The spatial pattern has remained consistent with that of past decades, with the highest temperatures in the eastern sector (Alekseev et al., 1998, 1999). However, the ocean front that separates waters of Atlantic and Pacific origin in the Arctic moved 600 km eastward along the Siberian slope in the early 1990s, shifting its alignment from the Lomonosov Ridge to the Alpha-Mendeleyev Ridge (Carmack et al., 1995; McLaughlin et al., 1996;

Morison et al., 1998). This shift reflects a 20% increase in the volume of Atlantic water within the Arctic Ocean. The cold halocline separating sea ice from warm Atlantic water in the Eurasian Basin virtually disappeared in the 1990s (Steele and Boyd, 1998). In general, the pattern of spreading and the greater thickness of the Atlantic layer are consistent with increases in the rate of inflow and temperature of Atlantic water entering the Arctic via Fram Strait (Dickson et al., 2000; Grotefendt et al., 1998).

The magnitude and temperature of inflow to the Arctic via the Barents Sea is highly variable, in response to wind changes associated with the AO (Dickson et al., 2000). Both the inflow and its temperature have increased since 1970, mirroring trends observed during the arctic warming of 1910 to 1930 (Loeng et al., 1997).

The properties of the surface layer of the Arctic Ocean vary greatly by season and from year to year. Russian data reveal an average 1.3 m decrease in surface (upper 50 m) "freshwater" storage in the Arctic Ocean in winter between 1974 and 1977 (Alekseev et al., 2000). Surface-layer variability is greatest in the area of seasonal sea ice because varying ice conditions strongly modulate the temperature and salinity of the upper ocean (e.g., Fissel and Melling, 1990). Possible temporal trends in most peripheral seas of the marine Arctic remain obscured by high variability.

In the decade following 1963 there was an obvious freshening of the upper waters of the North Atlantic (the Great Salinity Anomaly: Dickson et al., 1988) which is thought to have originated in the Arctic Ocean (e.g., Hakkinen, 1993; Mysak et al., 1990). Unusually warm summers from 1957 to 1962 in Baffin Bay and the Canadian Archipelago may have promoted melting of land ice and an enhanced freshwater outflow (Alekseev et al., 2000). The Great Salinity Anomaly, first evident north of Iceland (Steffansson, 1969), peaked in the North Atlantic in 1969 (Malmberg and Blindheim, 1994), at the same time that unusually fresh and cold water was observed on the other side of Greenland (Drinkwater, 1994). The simultaneous development of freshening in both areas indicates paths for "freshwater" transport via both Fram Strait and the Canadian Archipelago. A new freshwater pulse may appear in the North Atlantic in response to lengthening of the arctic summer melt period in the 1990s (Belkin et al., 1998). For further discussion of change in the Atlantic sector, refer to section 9.2.4.2.

In the central Greenland Sea, change in the temperature of water at the 2000 m depth is an indicator of varying deep-water formation (Alekseev et al., 2001; Boenisch et al., 1997). The observation of a large temperature increase (0.25 °C) in the central Greenland Sea in the 1990s is consistent with the weakening of the thermohaline circulation implied by decreasing deep-water flow toward the Arctic through the Faroe-Shetland Strait (Hansen et al., 2001).

2.6.4. Terrestrial system

About 24.5% of the exposed land surface in the Northern Hemisphere is underlain by permafrost (Zhang T. et al., 2000), and any increase in temperature is very likely to cause this area to shrink, in turn affecting arctic groundwater supply. Section 6.6.1.2 provides a more detailed description of recent changes in permafrost. Permafrost is very sensitive to climate change, and is dependent upon ambient temperatures for its existence and distinctive properties (Lunardini, 1996). Brown J. et al. (2000) indicated that only the thin active (seasonally thawed) layer of permafrost responds immediately to temperature changes, making it both an indicator and a product of climate change, as its depth is dependent on temperature. Vourlitis and Oechel (1996) reported an annual increase of 10 to 22 cm in the depth of the active layer between June and August in some permafrost areas. Climate change is very likely to accelerate this depth increase, due to both higher temperatures and increased evapotranspiration, which makes the soil drier throughout the active layer and causes a descent of the permafrost table (Vourlitis and Oechel, 1996).

Along the land-sea margin of the Arctic, over the past few decades there has been an increase in the magnitude of oscillations of the permafrost bottom (deepest extent of permafrost), and a greater variability in thickness during the warmer part of longer-term cycles, indicating an overall degradation of coastal permafrost (Romanovskii and Hubberten, 2001). Along a north–south transect in Alaska, permafrost temperatures at 15 to 20 m depths have increased between 0.6 and 1.5 °C over the past 20 years, mean annual ground surface temperatures have increased by 2.5 °C since the 1960s, and discontinuous permafrost has begun thawing downward at a rate of about 0.1 m/yr at some locations (Osterkamp, 2003; Osterkamp and Romanovsky, 1999; Romanovsky and Osterkamp, 2000; Romanovsky et al., 2003). The effects of increased permafrost temperatures on groundwater could possibly include the development of large, unfrozen near-surface aquifers with groundwater flow throughout the year; more active recharge and discharge of aquifers and increased groundwater flow would affect the base flow volumes and chemistry of many rivers.

Romanovsky and Osterkamp (2000) took a series of precise permafrost temperature measurements from four observation sites in Alaska, and found that unfrozen water has the greatest effect on the ground thermal regime immediately after freeze-up and during the cooling of the active layer, but is less important during the warming and thawing of the active layer. Jorgenson et al. (2001) showed widespread permafrost degradation in the Tanana Valley due to increases in air and permafrost temperatures. Permafrost temperatures have increased in Alaska, and decreased in eastern Canada (Serreze et al., 2000). Increased permafrost temperatures have also been observed in Siberia (Isaksen et al., 2001). Climate change is likely to drastically affect wildlife habitat, as permafrost is essential to the existence and functioning of arctic wetlands.

2.7. Arctic climate variability prior to 100 years BP

This section examines the record of past climate change in the Arctic with the objective of providing a context for evaluating evidence of more recent climate change and the possible impacts of future climate change. This review focuses on the past two million years (approximately), and particularly the past 20 000 years. Over geological time periods, the earth's natural climate system has been forced or driven by a relatively small number of external factors. Tectonic processes acting very slowly over millions of years have affected the location and topography of the continents through plate tectonics and ocean spreading. Changes in the orbit of the earth occur over tens to hundreds of thousands of years and alter the amount of solar radiation received at the surface by season and by latitude. These orbital changes drive climate responses, including the growth and decay of ice sheets at high latitudes. Finally, changes in the emissivity of the sun have taken place over billions of years, but there have been shorter-term variations that occurred over decades, centuries, and millennia. As a consequence of changes in these external forces, global climate has experienced variability and change over a variety of timescales, ranging from decades to millions of years.

The sparseness of instrumental climate records prior to the 20th century (especially prior to the mid-19th century) means that estimates of climate variability in the Arctic during past centuries must rely upon indirect "proxy" paleoclimate indicators, which have the potential to provide evidence for prior large-scale climatic changes. Typically, the interpretation of proxy climate records is complicated by the presence of "noise" in which climate information is immersed and by a variety of possible distortions of the underlying climate information (see Bradley, 1999 for a comprehensive review; IPCC, 2001). Careful calibration and cross-validation procedures are necessary to establish a reliable relationship between a proxy indicator and the climatic variable or variables of interest, providing a "transfer" function that allows past climatic conditions to be estimated. Of crucial importance is the ability to date different proxy records accurately in order to determine whether events occurred simultaneously, or whether some events lagged behind others.

Sources of paleoclimatic information in the Arctic are limited to a few, often equivocal types of records, most of which are interpreted as proxies for summer temperature. Little can be said about winter paleoclimate (Bradley, 1990). Only ice cores, tree rings, and lake sediments provide continuous high-resolution records. Coarsely resolved climate trends over several centuries are evident in many parts of the Arctic, including:

- the presence or absence of ice shelves deduced from driftwood frequency and peat growth episodes and pollen content;
- the timing of deglaciation and maximum uplift rates deduced from glacio-isostatic evidence as well as glacial deposits and organic materials over-ridden by glacial advances or exposed by ice recession;
- changes in the range of plant and animal species to locations beyond those of today (Bradley, 1990); and
- past temperature changes deduced from the geothermal information provided by boreholes.

In contrast, large-scale continuous records of decadal, annual, or seasonal climate variations in past centuries must rely upon sources that resolve annual or seasonal climatic variations. Such proxy information includes tree-ring width and density measurements; pollen, diatom, and sediment changes from laminated sediment cores; isotopes, chemistry, melt-layer stratigraphy, acidity, pollen content, and ice accumulation from annually resolved ice cores; and the sparse historical documentary evidence available for the past few centuries. Information from individual paleoclimate proxies is often difficult to interpret and multi-proxy analysis is being used increasingly in climate reconstructions. Taken as a whole, proxy climate data can provide global-scale sampling of climate variations several centuries into the past, and have the potential to resolve both large-scale and regional patterns of climate change prior to the instrumental period.

2.7.1. Pre-Quaternary Period

Over the course of millions of years, the Arctic has experienced climatic conditions that have ranged from one extreme to the other. Based on the fossil record, 120 to 90 million years before present (My BP) during the mid-Cretaceous Period, the Arctic was significantly warmer than at present, such that arctic geography, atmospheric composition, ocean currents, and other factors were quite different from at present. In contrast, as recently as 20 ky BP, the Arctic was in the grip of intense cold and continental-scale glaciation was at its height. This was the latest of a series of major glacial events, which, together with intervening warm periods, have characterized the past two million years of arctic environmental history.

The most recent large-scale development and build-up of ice sheets in the Arctic probably commenced during the Late Tertiary Period (38 to 1.6 My BP). Ice accumulation at that time may have been facilitated initially by plate tectonics (Ruddiman et al., 1989). According to Maslin et al. (1998), the onset of Northern Hemisphere glaciation began with a significant build-up of ice in southern Greenland. However, progressive intensification of glaciation does not seem to have begun until 3.5 to 3 My BP, when the Greenland Ice Sheet expanded to include northern Greenland. Maslin et al. (1998) suggested that the Eurasian Arctic, northeast Asia, and Alaska were glaciated by about 2.7 My BP and northeast

America by 2.5 My BP. Maslin et al. (1998) suggested that tectonic changes, including the deepening of the Bering Strait, were too gradual to be responsible for the speed of Northern Hemisphere glaciation, but suggested that tectonic changes brought the global climate to a critical threshold while relatively rapid variations in the orbital parameters of the earth triggered the glaciation.

2.7.2. Quaternary Period

The Quaternary Period (the last 1.6 My) has been characterized by periodic climatic variations during which the global climate system has switched between interglacial and glacial stages, with further subdivision into stadials (shorter cold periods) and interstadials (shorter mild episodes). Glacial stages are normally defined as cold phases with major glacier and ice sheet expansion. Interglacials are defined as warm periods when temperatures were at least as high as during the present Holocene interglacial.

This interglacial–glacial–interglacial climate oscillation has been recurring with a similar periodicity for most of the Quaternary Period, although each individual cycle appears to have had its own idiosyncrasies in terms of the timing and magnitude of specific events. It has been estimated that there have been between 30 and 50 glacial/interglacial cycles during the Quaternary Period (Ruddiman et al., 1989; Ruddiman and Kutzbach, 1990), primarily driven by changes in the orbit of the earth (Imbrie and Imbrie, 1979).

One of the earliest (and certainly the most well known) hypotheses concerning the effects of orbital configuration on glacial cycles is described by Milankovitch (1941), who presents the argument that a decrease in summer insolation is critical for glacial initiation. Low summer insolation occurs when the tilt of the axis of rotation of the earth is small; the poles are pointing less directly at the sun; the Northern Hemisphere summer solstice is farthest from the sun; and the earth's orbit is highly eccentric. The key orbital parameters involved include changes in the eccentricity of the orbit of the earth with a period of 100 ky; the tilt of the axis of rotation of the earth, which oscillates between 22.2° and 24.5° with a period of 41 ky; and the position of the earth within its elliptical orbit during the Northern Hemisphere summer, which changes over a period of 23 ky. Bradley (1990) pointed out that such orbitally induced radiation anomalies are especially significant at high latitudes in summer when daylight persists for 24 hours. The latitude most sensitive to low insolation values is 65° N, the latitude at which most ice sheets first formed and lasted longest during the last glaciation. The amount of summer insolation reaching the top of the atmosphere at 65° N and nearby latitudes can vary by as much as ±12% around the long-term mean value. Changes in winter insolation also occur with exactly the opposite timing, but they are not considered important to ice-sheet survival. When summer insolation is weak, less radiation is delivered to the surface at high latitudes,

resulting in lower regional temperatures. Lower temperatures reduce the summer ablation of the ice sheets and allow snow to accumulate and ice sheets to grow. Once ice sheets are created, they contribute to their own positive mass balance by growing in elevation, reaching altitudes of several kilometers where prevailing temperatures favor the accumulation of snow and ice.

Other hypotheses concerning the causes of glacial initiation include that of Young and Bradley (1984), who argued that the meridional insolation gradient is a critical factor for the growth and decay of ice sheets through its control over poleward moisture fluxes during summer and resultant snowfall. Snowfall could increase in a cooler high-latitude climate through enhanced storm activity forced by a greater latitudinal temperature gradient. Ruddiman and McIntyre (1981) and Miller and de Vernal (1992) suggested that the North Atlantic Ocean circulation is important for modulating ice volume. There is evidence that the North Atlantic remained warm during periods of ice growth and they proposed that enhanced meridional temperature gradients and evaporation rates during the winter enhanced snow delivery to the nascent ice sheets of northeastern Canada. An active thermohaline circulation in the North Atlantic could increase the moisture supply to sites of glacial initiation through its ability to maintain warmer sea surface temperatures, limit sea-ice growth, and allow for greater evaporation from the ocean surface.

These external forcing mechanisms in turn cause responses and chain reactions in the internal elements of the earth's system (e.g., Bradley, 1985). Changes in one internal element of the system can cause responses in other elements. These lead to feedback effects that can amplify or attenuate the original signal. Thus, ice sheets, ice caps, and glaciers play an important role in the global climate system. Glacial advance and retreat may therefore be both a consequence and a cause of climate change (Imbrie et al., 1993a,b).

Given the inherent errors in dating techniques, gaps in the stratigraphic record, and the varying rates of response of different biological proxy indicators, there is considerable uncertainty about the timing of specific events and whether climate changes were truly synchronous in different regions. The errors and uncertainties tend to be amplified farther back in the paleoclimatic record, particularly in the Arctic, where much of the paleoclimatic evidence from earlier parts of the Quaternary Period has been removed or obfuscated as a result of later glaciations. Consequently, this review of climate variability in the Arctic during the Quaternary Period focuses first on the more complete and reliable evidence of climate conditions immediately prior to the onset of the most recent glacial–interglacial oscillation (~130 ky BP). This is followed by a brief review of conditions during the Last Glacial Maximum (~20 ky BP) and the subsequent period of deglaciation, and culminates in a review of the evidence of climatic variability during the present interglacial, the Holocene.

2.7.3. Last interglacial and glaciation

2.7.3.1. Last interglacial: The Eemian

Climatic conditions during interglacial periods are generally considered to be broadly comparable to present-day conditions. The most recent interglacial, the Eemian, extended from the end of the penultimate glaciation about 130 ky BP until about 107 ky BP when the last glacial period began (see IPCC, 2001c for further discussion of the Eemian). The Eemian is often regarded as a typical interglacial event with characteristics including relatively high sea level, a retreat to minimum size of global ice sheets, and the establishment of biotic assemblages that closely parallel those at present. According to most proxy data, the last interglacial was slightly warmer everywhere than at present (IPCC, 2001c). Brigham-Grette and Hopkins (1995) reported that during the Eemian the winter sea-ice limit in Bering Strait was at least 800 km farther north than today, and that during some summers the Arctic Ocean may have been ice-free. The northern treeline was more than 600 km farther north, displacing tundra across all of Chukotka (Lozhkin and Anderson, 1995). Western European lake pollen records show deciduous forests (characteristic of warmer conditions) across much of Western Europe that were abruptly replaced by steppic taxa characteristic of colder conditions; this shift is associated with a cold event at 107 ky BP. This relatively prolonged warm period is also detected in northeast Atlantic marine sediments, but is not evident throughout the North Atlantic. Faunal and lithic records from the polar North Atlantic (Fronval and Jansen, 1997) indicate that the first abrupt cooling occurred around 118 to 117 ky BP (Adkins et al., 1997; Keigwin et al., 1994).

Evidence of warmer conditions in the Arctic than exist at present is provided by a re-evaluation of the oxygen isotope ratio (δ^{18}O) record obtained from Greenland ice core samples (Cuffey and Marshall, 2000). These authors suggest that the Greenland Ice Sheet was considerably smaller and steeper during the Eemian than at present and probably contributed 4 to 5.5 m to global sea level during that period. This implies that the climate of Greenland during the Eemian was more stable and warmer than previously thought, and the consequent melting of the Greenland Ice Sheet more significant.

Some researchers suggest that a paleoclimatic reconstruction of the Eemian provides a means of establishing the mode and tempo of natural climate variability with no anthropogenic influence. However, a general lack of precise absolute timescales and regionally-to-globally synchronous stratigraphic markers makes long-distance correlation between sites problematic, and inferred terrestrial changes are difficult to place within the temporal framework of changes in ice volume and sea level (Tzedakis, 2003).

2.7.3.2. Last glaciation: Wisconsinan/ Weichselian

Although the timing of the end of the Eemian interglacial is subject to some uncertainty, high-resolution North Atlantic marine sediment records indicate that the Eemian ended with abrupt changes in deep-water flow occurring over a period of less than 400 years (Adkins et al., 1997; IPCC, 2001c). Evidence in marine sediments of an invasion of cold, low salinity water in the Norwegian Sea at this time has been linked by Cortijo et al. (1994) to a reduction in warm-water transport by the North Atlantic Drift and the thermohaline circulation. It seems likely that this contributed to the onset of widespread arctic glaciation in sensitive areas.

Following the initial cooling event (~107 ky BP), climatic conditions often changed suddenly, followed by several thousand years of relatively stable climate or even a temporary reversal to warmth. Overall, however, there was a decline in global temperatures. The boundaries of the boreal forests retreated southward and fragmented as conditions grew colder. Large ice sheets began to develop on all the continents surrounding the Arctic Ocean. The point at which the global ice extent was at its greatest (~24 to 21 ky BP) is known as the Last Glacial Maximum (LGM – Clark and Mix, 2000).

At its maximum extent, the Laurentide Ice Sheet extended from the Arctic Ocean in the Canadian Archipelago to the midwestern United States in the south, and from the Canadian Cordillera to the eastern edge of the continent. Local ice sheets also covered the Alaska and Brooks Ranges in Alaska. The Eurasian and Laurentide Ice Sheets were responsible for most of the glacio-eustatic decrease in sea level (about 120 m) during the LGM. The pattern of postglacial isostatic rebound suggests that the ice was thickest over Hudson Bay. The different parts of the Laurentide Ice Sheet reached their maximum extent between 24 and 21 ky BP (Dyke et al., 2002). The Innuitian ice buildup appears to have culminated in the east after 20.5 ky BP. Dyke et al. (2002) suggested that the entire ice sheet system east of the Canadian Cordillera responded uniformly to changes in climate. In contrast, the Cordilleran Ice Sheet did not reach its maximum extent until 15.2 to 14.7 ky BP, well after the LGM and the insolation minimum at approximately 21 ky BP (Dyke et al., 2002). This out-of-phase response may be attributable to the effects of growth of the Cordilleran Ice Sheet, which would have intercepted moisture transport to the interior plains at the expense of the Laurentide Ice Sheet. During its maximum extent, the Laurentide Ice Sheet was more than twice the size of the Eurasian Ice Sheet. Changes in climate during the LGM are discussed by the IPCC (2001c).

The Eurasian Ice Sheet initiation began 28 ky BP as a result of temperature changes that lowered equilibrium line altitudes across the Scandinavian mountains, Svalbard, Franz Josef Land, and Novaya Zemlya. Ice flow north from Scandinavia and south from Svalbard, in conjunction with eustatic sea-level decreases, caused the margin of the ice sheet to migrate into the Barents Sea. Complete glaciation of the Barents Sea by a grounded ice sheet was achieved by 20 ky BP. The ice sheet at its maximum extent covered Scandinavia and the Barents Shelf and included a marine-based margin along the northern Barents Shelf, the western Barents Sea, western Scandinavia, and northern Great Britain and Ireland. The eastern margin of the ice sheet is generally thought to have been located west of the Taymir Peninsula (Mangerud et al., 2002). It appears that at the LGM, cold arid conditions persisted across much of eastern Siberia and that an East Siberian Sea Ice Sheet did not exist (Brigham-Grette et al., 2003), as some have claimed (Grosswald and Hughes, 2002). Glaciers in the Urals at the LGM appear to have been confined to the mountain valleys, rather than coalescing with the Barents-Kara Ice Sheet, as happened during previous glaciations (Mangerud et al., 2002).

2.7.4. Last glacial/interglacial transition through to mid-Holocene

2.7.4.1. Last glacial/interglacial transition

The most extreme manifestation of climate change in the geological record is the transition from full glacial to full interglacial conditions. Following the LGM (~24 to 21 ky BP), temperatures close to those of today were restored by approximately 10 ky BP (IPCC, 2001c).

The inception of warming appears to have been very rapid (NRC, 2002). The rate of temperature change during the recovery phase from the LGM provides a benchmark against which to assess rates of temperature change in the late 20th century. Available data indicate an average warming rate of about 2 °C per millennium between about 20 and 10 ky BP in Greenland, with lower rates for other regions. On the other hand, very rapid temperature increases at the start of the Bølling-Allerød period (14.5 ky BP; Severinghaus and Brook, 1999) or at the end of the Younger Dryas (~11 ky BP) may have occurred at rates as large as 10 °C per 50 years over substantial areas of the Northern Hemi-

sphere. Almost synchronously, major vegetation changes occurred in Europe and North America and elsewhere (Gasse and van Campo, 1994). There was also a pronounced warming of the North Atlantic and North Pacific (Webb et al., 1998).

Oxygen isotope measurements from Greenland ice cores demonstrate that a series of rapid warm and cold oscillations, called Dansgaard–Oeschger (D-O) events, punctuated the last glaciation, often taking Greenland and northwestern Europe from a full-glacial climate to conditions about as warm as at present (Fig. 2.15). For the period between 115 and 14 ky BP, 24 of these short-lived warm events have been identified in the Greenland ice core data, although many lesser warming events also occurred (Dansgaard et al., 1993). According to Lang et al. (1999), associated temperature changes may have been as great as 16 °C. The D-O oscillations are correlated with sea surface temperature variations derived from several North Atlantic deep-sea cores (Bond et al., 1993). From the speed of the climate changes recorded in the Greenland Ice Sheet (Dansgaard et al., 1989), it is widely thought that the complete change in climate occurred, at least regionally, over only a few decades. These interstadials lasted for varying periods of time, usually a few centuries to about 2000 years, before equally rapid cooling returned conditions to their previous state. The evidence from high-resolution deep sea cores from the North Atlantic (Bond et al., 1997) suggests that during at least the past 30 000 years, interstadials tended to occur at the warmer points of a background North Atlantic temperature cycle that had a periodicity of approximately 1500 years.

Heinrich events appear to be the most extreme of a series of sudden, brief cold events that seem to have occurred very frequently over the past 115 000 years, apparently tending to start at the low point of the same 1500-year temperature cycle. Heinrich events occurred during times of decreasing sea surface temperatures in the form of brief, exceptionally large discharges of icebergs in the North Atlantic from the Laurentide and European Ice Sheets that left conspicuous layers of detrital material in deep-sea sediments. Accompanying the Heinrich events were large decreases in the oxygen isotope ratio of planktonic foraminifera, providing evidence of lowered surface salinity probably caused by melting of drifting ice (Bond et al., 1993). Heinrich events appear at the end of a series of saw-tooth-shaped temperature cycles known as Bond cycles. During the Pleistocene Epoch, each cycle was characterized by the relatively warm interstadials becoming progressively cooler.

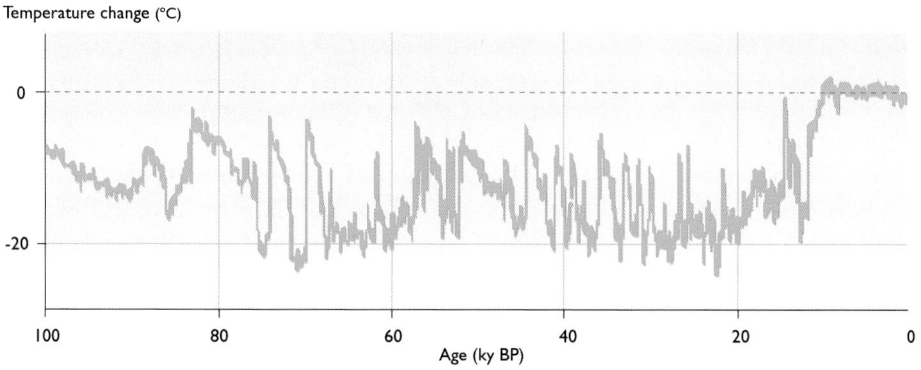

Temperature change (°C)

Fig. 2.15. Temperature change over the past 100 ky (departure from present conditions) reconstructed from a Greenland ice core (Ganopolski and Rahmstorf, 2001).

Deep-sea cores also show the presence of ice-rafting cycles in the intervals between Heinrich events (Bond and Lotti, 1995). The duration of these ice-rafting cycles varies between 2000 and 3000 years and they closely coincide with the D-O events. A study of the ice-rafted material suggests that, coincident with the D-O cooling, ice within the Icelandic ice caps and within or near the Gulf of Saint Lawrence underwent nearly synchronous increases in rates of calving. The Heinrich events reflect the slower rhythm of iceberg discharges into the North Atlantic, probably from Hudson Strait.

Air temperature, sea surface temperature, and salinity variations in the North Atlantic are associated with major changes in the thermohaline circulation. A core from the margin of the Faroe-Shetland Channel covering the last glacial period reveals numerous oscillations in benthic and planktonic foraminifera, oxygen isotopes, and ice-rafted detritus (Rasmussen et al., 1996). These oscillations correlate with the D-O cycles, showing a close relationship between the deep-ocean circulation and the abrupt climatic changes during the last glaciation. It is increasingly apparent that large, globally linked environmental feedbacks were involved in the generation of the large, rapid temperature increases that occurred during glacial termination and the onset of D-O events.

During the last glacial–interglacial transition, the movements of the North Atlantic Polar Front have been described as pivoting around locations in the western North Atlantic. Iceland, situated in the middle of the North Atlantic, has glaciers sensitive to changes in oceanic and atmospheric frontal systems (Ingolfsson et al., 1997). The late-glacial (subsequent to the LGM) records from Iceland indicate that relatively warm Atlantic water reached Iceland during the Bølling–Allerød Interstadial, with a short cooling period corresponding with the Older Dryas. Karpuz et al. (1993) suggested that the marine polar front was located close to Iceland during the Bølling–Allerød, and Sarnthein et al. (1995) concluded that sea-surface circulation was mainly in Holocene interglacial mode after 12.8 ky BP.

Warm episodes were also associated with higher sea-surface temperatures and the presence of oceanic convection in the Norwegian Greenland Sea. Cold episodes were associated with low sea surface temperatures, low salinity, and no convection (Rasmussen et al., 1996). Subsequent to the initial phase of deglaciation on Spitsbergen (~13 to 12.5 ky BP), most of the eastern and northern Barents Sea was deglaciated during the Bølling–Allerød Interstadial (Vorren and Laberg, 1996). Vorren and Laberg (1996) speculate on the existence of a close correlation between summer air temperatures and the waxing and waning phases of the northern Fennoscandian and southern Barents Sea Ice Sheets. Most of the ice-sheet decay is attributed by these authors to increases in air temperature, which caused thinning of the ice sheets, making them susceptible to decoupling from the seabed and increased calving.

After a few thousand years of recovery, the Arctic was suddenly plunged back into a new and very short-lived cold event known as the Younger Dryas (see NRC, 2002, for a detailed discussion) leading to a brief advance of the ice sheets. The central Greenland ice core record (from the Greenland Ice Core Project and the Greenland Ice Sheet Project 2) indicates that the return to the cold conditions of the Younger Dryas from the incipient interglacial warming 13 ky BP took place in less than 100 years. The warming phase at the end of the Younger Dryas, which took place about 11 ky BP and lasted about 1300 years, was very abrupt and central Greenland temperatures increased by 7 °C or more in a few decades.

2.7.4.2. Early to mid-Holocene

Following the sudden end of the Younger Dryas, the Arctic entered several thousand years of conditions that were warmer and probably moister than today. Peak high-latitude summer insolation (Milankovitch orbital forcing) occurred during the earliest Holocene, with a maximum radiation anomaly (approximately 8% greater than at present) attained between 10 and 9 ky BP (Berger and Loutre, 1991). Although most of the Arctic experienced summers that were warmer (1–2 °C) than at present during the early to middle Holocene, there were significant spatial differences in the timing of this warm period. This was probably due to the effects of local residual land-ice cover and local sea surface temperatures (Overpeck et al., 1997). In most of the ice cores from high latitudes, the warm period is seen at the beginning of the Holocene (about 11 to 10 ky BP). In contrast, central Greenland (Dahl-Jensen et al., 1998) and regions downwind of the Laurentide Ice Sheet, including Europe (COHMAP Members, 1988) and eastern North America (Webb et al., 1998), did not warm up until after 8 ky BP. The rapid climatic events of the last glacial period and early Holocene are best documented in Greenland and the North Atlantic and may not have occurred throughout the Arctic. This early Holocene warm period appears to have been punctuated by a severe cold and dry phase about 8200 years ago, which lasted for less than a century, as recorded in the central Greenland ice cores (Alley et al., 1997).

Glacio-isostatic evidence indicates deglaciation was underway by the beginning of the Holocene and maximum uplift occurred between 8 and 7 ky BP and even earlier in many areas. By 9 ky BP, Spitsbergen glaciers had retreated to or beyond their present day positions (Sexton et al., 1992) and the marine faunal evidence suggest that this period was as warm if not warmer than at present along the west and north coasts of Svalbard (Salvigsen et al., 1992). The retreat of the largest of the glaciers along the Gulf of Alaska began as early as 16 to 14 ky BP (Mann D. and Hamilton, 1995). Although much of the high and mid-Canadian Arctic remained glaciated, warm summers are clearly registered by enhanced summer melting of the Agassiz Ice Cap (Koerner and Fisher, 1990). Following the large, abrupt change in stable-isotope ratios marking the end of the last glaciation, $\delta^{18}O$ profiles from Agassiz Ice Cap cores show a rapid warming trend that reached a maximum between approximately 9 and 8 ky BP.

Deglacial marine sediments in Clements Markham Inlet on the north coast of Ellesmere Island resemble those characteristic of temperate (as opposed to polar) tide-water glaciers, suggesting that climatic conditions in the early Holocene were significantly warmer there than today (Stewart, 1988). Glaciers had retreated past present-day termini in some areas by 7.5 ky BP. Increasing sea surface temperatures in Baffin Bay enhanced precipitation on Baffin Island (Miller and de Vernal, 1992), leading to a widespread early Holocene glacial advance along the east coast. Marine mammals and boreal mollusks were present far north of their present-day range by 7.5 to 6.5 ky BP, as were many species of plants between 9.2 and 6.7 ky BP (Dyke et al., 1996, 1999). Caribou were able to survive in the northernmost valleys of Ellesmere Island and Peary Land by 8.5 ky BP or earlier. Such evidence indicates very warm conditions early in the Holocene (before 8 ky BP).

Early Holocene summer temperatures similar to those at present have been reconstructed in Arctic Fennoscandia by numerous studies using a range of proxies and multi-proxy analyses (Rosén et al., 2001). However, abrupt climatic variations were characteristic of the early Holocene, with distinct cool episodes around 9.2, 8.6, and 8.2 ky BP (Korhola et al., 2002). The most recent of these events might be connected to the widely known "8.2 ky event", which affected terrestrial and aquatic systems in northern Fennoscandia (Korhola et al., 2002; Nesje et al., 2000; Snowball et. al., 1999). Hantemirov and Shiyatov (2002) report that open larch forests were already growing in the Yamal Peninsula of northwestern Siberia 10.5 to 9 ky BP and that the most favorable period for tree growth lasted from 9.2 to 8 ky BP. During the early Holocene, reconstructed mean temperature anomalies for the warmest month, based on pollen data across Northern Europe, show temperatures comparable to those at present (Davis et al., 2003). Temperatures then increased around 6 ky BP, with the onset of this increase delayed to around 9 ky BP in the east.

Boreal forest development across northern Russia (including Siberia) commenced by 10 ky BP (MacDonald G. et al., 2000). Over most of Russia, forests advanced to or near the current arctic coastline between 9 and 7 ky BP, and retreated to their present position by between 4 and 3 ky BP. Forest establishment and retreat were roughly synchronous across most of northern Russia, with the exception of the Kola Peninsula, where both appear to have occurred later. During the period of maximum forest extension, the mean July temperature along the northern coastline of Russia may have been 2.5 to 7.0 °C warmer than present.

The Arctic appears to have been relatively warm during the mid-Holocene, although records differ spatially, temporally, and by how much summer warmth they suggest (relative to the early Holocene insolation maximum; Bradley, 1990; Hardy and Bradley, 1996).

A review of the Holocene glaciation record in coastal Alaska (Calkin, 1988) suggests that glacier fluctuations in arctic, central interior, and southern maritime Alaska were mostly synchronous. Ager (1983) and Heusser (1995) report that pollen records indicate a dramatic cooling about 3.5 ky BP and suggest an increase in precipitation and storminess in the Gulf of Alaska accompanied by a rejuvenation of glacial activity. In northern Iceland, the Holocene record of glacier fluctuations indicates two glacial advances between 6 and 4.8 ky BP (Stötter, 1991).

In the Canadian Arctic, interior regions of Ellesmere Island appear to have retained extensive Innuitian and/or plateau ice cover until the mid-Holocene (Bell, 1996; Smith, 1999), after which ice margins retreated to positions at or behind those at present. Restricted marine-mammal distributions imply more extensive summer sea ice between 8 and 5 ky BP, and hence cooler conditions (Dyke et al., 1996, 1999). However, marine conditions at 6 ky BP warmer than those at present are suggested by analyses of marine microfossils (Gajewski et al., 2000) performed over a broad area from the high Arctic to the Labrador Sea via Baffin Bay. A multi-proxy summary of marine and terrestrial evidence from the Baffin sector (Williams et al., 1995) suggests that warming began around 8 ky BP, intensified at 6 ky BP, and that conditions had cooled markedly by 3 ky BP.

Dugmore (1989) demonstrated that, in Iceland, the Sólheimajökull glacier extended up to 5 km beyond its present limits between 7 and 4.5 ky BP. Major ice sheet advances also occurred before 3.1 ky BP and between 1.4 and 1.2 ky BP. In the 10th century (1 ky BP), this glacier was also larger than during the period from AD 1600 to 1900, when some other glaciers reached their maximum Holocene extent. Stötter et al. (1999) suggested that major glacier advances in northern Iceland occurred at around 4.7, 4.2, 3.2–3.0, 2.0, 1.5, and 1.0 ky BP.

Evidence for a mid-Holocene thermal maximum in Scandinavia is considerable, and based on a wide range of proxies (Davis et al., 2003). Treelines reached their maximum altitude (up to 300 m higher than at present; Barnekow and Sandgren, 2001), and glaciers were greatly reduced or absent (Seierstad et al., 2002). Pollen and macrofossil records from the Torneträsk area in northern Swedish Lapland indicate optimal conditions for Scots pine (*Pinus sylvestris*) from 6.3 to 4.5 ky BP (Barnekow and Sandgren, 2001) and records of treeline change in northern Sweden show high-elevation treelines around 6 ky BP (Karlén and Kuylenstierna, 1996). These data indicate an extended period in the early to mid-Holocene when Scandinavian summer temperatures were 1.5 to 2 °C higher than at present.

Tree-ring data from the Torneträsk area indicate particularly severe climatic conditions between 2.6 and 2 ky BP

(600–1 BC). This period includes the greatest range in ring-width variability of the past 7400 years in this area, indicating a highly variable but generally cold climate (Grudd et al., 2002). This period is contemporary with a major glacial expansion in Scandinavia when many glaciers advanced to their Holocene maximum position (Karlén, 1988) with major effects on human societies (Burenhult, 1999; van Geel et al., 1996).

Especially severe conditions in northern Swedish Lapland occurred 2.3 ky BP (330 BC), with tree-ring data indicating a short-term decrease in mean summer temperature of about 3 to 4 °C. A catastrophic drop in pine growth at that time is also reported by Eronen et al. (2002), who state that this was the most unfavorable year for the growth of treeline pines in Finnish Lapland in the past 7500 years. Reconstructed Holocene summer temperature changes in Finnish Lapland, based on proxy climate indicators in sediments from Lake Tsuolbmajavri, show an unstable early Holocene between 10 and 8 ky BP in which inferred July air temperatures were about the same as at present most of the time, but with three successive cold periods at approximately 9.2, 8.6, and 8.2 ky BP, and a "thermal maximum" between approximately 8 and 5.8 ky BP, followed by an abrupt cooling (Korhola et al., 2002). Dated subfossils (partially fossilized organisms) show that the pine treeline in northwestern Finnish Lapland retreated a distance of at most 70 km during this cooling, but that the shift was less pronounced in more easterly parts of Lapland (Eronen et al., 2002).

Hantemirov and Shiyatov (2002) reported that the most favorable period for tree growth in the Yamal Peninsula of northwestern Siberia lasted from 9.2 to 8 ky BP. At that time, the treeline was located at 70° N. Then, until 7.6 ky BP, temperatures decreased but this did not result in any significant shift in the treeline. The treeline then moved south until, by 7.4 ky BP, it was located at approximately 69° N. It remained here until 3.7 ky BP when it rapidly retreated (~20 km) to within 2 to 3 km north of its present position south of the Yamal Peninsula. This retreat in the space of only 50 years coincides with an abrupt and large cooling as indicated in the tree-ring data. This cooling event may have been associated with the eruption of the Thera (Santorini) volcano in the southern Aegean around 3.6 ky BP.

Tree-ring data from the Kheta-Khatanga plain region and the Moyero-Kotui plateau in the eastern part of the Taymir Peninsula indicate climatic conditions more favorable for tree growth around 6 ky BP, as confirmed by increased concentrations of the stable carbon isotope ^{13}C in the annual tree rings (Naurzbaev et al., 2002). The growth of larch trees at that time was 1.5 to 1.6 times greater than the average radial growth of trees during the last 2000 years, and the northern treeline is thought to have been situated at least 150 km farther north than at present, as indicated by the presence of subfossil wood of that age in alluvial deposits of the Balakhnya River. During the past 6000 years, the eastern Taymir tree-ring chronologies show a significant and progressive decrease in tree growth and thus temperature.

2.7.5. Last millennium

Over the last millennium, variations in climate across the Arctic and globally have continued. The term "Medieval Warm Period", corresponding roughly to the 9th to the mid-15th centuries, is frequently used but evidence suggests that the timing and magnitude of this warm period varies considerably worldwide (Bradley and Jones, 1993; Crowley and Lowery, 2000; IPCC 2001c). Current evidence does not support a globally synchronous period of anomalous warmth during that time frame, and the conventional term of "Medieval Warm Period" appears to have limited utility in describing trends in hemispheric or global mean temperature changes.

The Northern Hemisphere mean temperature estimates of Mann M. et al. (1999), and Crowley and Lowery (2000), show that temperatures during the 11th to the 14th centuries were about 0.2 °C higher than those during the 15th to the 19th centuries, but somewhat below the temperatures of the mid-20th century. The long-term hemispheric trend is best described as a modest and irregular cooling from AD 1000 to around 1850 to 1900, followed by an abrupt 20th-century warming.

Regional evidence is, however, quite variable. Crowley and Lowery (2000) show that western Greenland exhibited local anomalous warmth only around AD 1000 (and to a lesser extent, around AD 1400), and experienced quite cold conditions during the latter part of the 11th century. In general, the few proxy temperature records spanning the last millennium suggest that the Arctic was not anomalously warm throughout the 9th to 14th centuries (Hughes and Diaz, 1994).

In northern Swedish Lapland, Scots pine tree-ring data indicate a warm period around AD 1000 that ended about AD 1100 when a shift to a colder climate occurred (Grudd et al., 2002). In Finnish Lapland, based on a 7500-year Scots pine tree-ring record, Helama et al. (2002) reported that the warmest non-overlapping 100-year period in the record is AD 1501 to 1600, but AD 1601 was unusually cold. Other locations in Fennoscandia and Siberia were also cold in AD 1601, and Briffa et al. (1992, 1995) linked the cold conditions to the AD 1600 eruption of the Huaynaputina volcano in Peru. In northern Siberia, and particularly east of Taymir where the most northerly larch forests occur, long-term temperature trends derived from tree rings indicate the occurrence of cool periods during the 13th, 16th to 17th, and early 19th centuries. The warmest periods over the last millennium in this region were between AD 950 and 1049, AD 1058 and 1157, and AD 1870 and 1979. A long period of cooling began in the 15th century and conditions remained cool until the middle of the 18th century (Naurzbaev et al., 2002).

For the most part, "medieval warmth" appears to have been restricted to areas in and around the North Atlantic, suggesting that variability in ocean circulation may have played a role. Keigwin and Pickart (1999) suggested that the temperature contrasts between the North Atlantic and other areas were associated with changes in ocean currents in the North Atlantic and may to a large extent reflect century-scale changes in the NAO.

By the middle of the 19th century, the climate of the globe and the Arctic was cooling. Overall, the period from 1550 to 1900 may have been the coldest period in the entire Holocene (Bradley, 1990). This period is usually called the "Little Ice Age" (LIA), during which glaciers advanced on all continents. The LIA appears to have been most clearly expressed in the North Atlantic region as altered patterns of atmospheric circulation (O'Brien et al., 1995). Unusually cold, dry winters in central Europe (e.g., 1 to 2 °C below normal during the late 17th century) were very probably associated with more frequent flows of continental air from the northeast (Pfister, 1999; Wanner et al., 1995). Such conditions are consistent with the negative or enhanced easterly wind phase of the NAO, which implies both warm and cold anomalies over different regions of the North Atlantic sector. Although the term LIA is used for this period, there was considerable temporal and spatial variability across the Arctic during this period.

Ice shelves in northwestern Ellesmere Island probably reached their greatest extent in the Holocene during this interval. On the Devon Island Ice Cap, 1550 to 1620 is considered to have been a period of net summer accumulation, with very extensive summer sea ice in the region. There is widespread evidence of glaciers reaching their maximum post-Wisconsinan positions during the LIA, and the lowest $\delta^{18}O$ values and melt percentages for at least 1000 years are recorded in ice cores for this interval. Mann M. et al. (1999) and Jones et al. (1998) supported the theory that the 15th to 19th centuries were the coldest of the millennium for the Northern Hemisphere overall. However, averaged over the Northern Hemisphere, the temperature decrease during the LIA was less than 1 °C relative to late 20th-century levels (Crowley and Lowery, 2000; Jones et al., 1998; Mann M. et al., 1999). Cold conditions appear, however, to have been considerably more pronounced in particular regions during the LIA. Such regional variability may in part reflect accompanying changes in atmospheric circulation. Overpeck et al. (1997) summarized arctic climate change over the past 400 years.

There is an abundance of evidence from the Arctic that summer temperatures have decreased over approximately the past 3500 years. In the Canadian Arctic, the melt record from the Agassiz ice core indicates a decline in summer temperatures since approximately 5.5 ky BP, especially after 2 ky BP. In Alaska, widespread glacier advances were initiated at approximately 700 ky BP and continued through the 19th century (Calkin et al., 2001). During this interval, the majority of Alaskan glaciers reached their Holocene maximum extensions. The pattern of LIA glacier advances along the Gulf of Alaska is similar on decadal timescales to that of the well-dated glacier fluctuations throughout the rest of Alaska.

There is a general consensus that throughout the Canadian Archipelago, the late Holocene has been an interval of progressive cooling (the "Neoglacial", culminating in the LIA), followed by pronounced warming starting about 1840 (Overpeck et al., 1997). According to Bourgeois et al. (2000), the coldest temperatures of the entire Holocene were reached approximately 100 to 300 years ago in this region. Others, working with different indicators, have suggested that Neoglacial cooling was even greater in areas to the south of the Canadian Archipelago (Johnsen et al., 2001). Therefore, even if the broad pattern of Holocene climatic evolution is assumed to be coherent across the Canadian Archipelago, the available data suggest regional variation in the amplitude of temperature shifts.

The most extensive data on the behavior of Greenland glaciers apart from the Greenland Ice Sheet come from Maniitsoq (Sukkertoppen) and Disko Island. Similar to the inland ice-sheet lobes, the majority of the local glaciers reached their maximum Neoglacial extent during the 18th century, possibly as early as 1750. Glaciers started to retreat around 1850, but between 1880 and 1890 there were glacier advances. In the early 20th century, glacier recession continued, with interruptions by some periods of advance. The most rapid glacial retreat took place between the 1920s and 1940s.

In Iceland, historical records indicate that Fjallsjökull and Breidamerkurjökull reached their maximum Holocene extent during the latter half of the 19th century (Kugelmann, 1991). Between 1690 and 1710, the Vatnajökull outlet glaciers advanced rapidly and then were stationary or fluctuated slightly. Around 1750 to 1760 a significant re-advance occurred, and most of the glaciers are considered to have reached their maximum LIA extent at that time (e.g., Grove, 1988). During the 20th century, glaciers retreated rapidly. During the LIA, Myrdalsjökul and Eyjafjallsjökull formed one ice cap, which separated in the middle of the 20th century into two ice caps (Grove, 1988). Drangajökull, a small ice cap in northwest Iceland, advanced across farmland by the end of the 17th century, and during the mid-18th century the outlet glaciers were the most extensive known since settlement of the surrounding valleys. After the mid-19th century advance, glaciers retreated significantly. On the island of Jan Mayen, some glaciers reached their maximum extent around 1850. The glaciers subsequently experienced an oscillating retreat, but with a significant expansion around 1960 (Anda et al., 1985).

In northeastern Eurasia, long-term temperature trends derived from tree rings close to the northern treeline in east Taymir and northeast Yakutia indicate decreasing temperatures during the LIA (Vaganov et al., 2000).

Variations in arctic climate over the past 1000 years may have been the result of several forcing mechanisms. Bond et al. (2001) suggested variations in solar insolation. Changes in the thermohaline circulation or modes of atmospheric variability, such as the AO, may also have been primary forcing mechanisms of century- or millennial-scale changes in the Holocene climate of the North Atlantic. It is possible that solar forcing may excite modes of atmospheric variability that, in turn, may amplify climate changes. The Arctic, through its linkage with the Nordic Seas, may be a key region where solar-induced atmospheric changes are amplified and transmitted globally through their effect on the thermohaline circulation. The resulting reduction in northward heat transport may have further altered latitudinal temperature and moisture gradients.

2.7.6. Concluding remarks

Natural climate variability in the Arctic over the past two million years has been large. In particular, the past 20 000-year period is now known to have been highly unstable and prone to rapid changes, especially temperature increases that occurred rapidly (within a few decades or less). These temperature increases occurred during glacial terminations and at the onset of D-O interstadials. This instability implies rapid, closely linked changes within the earth's environmental system, including the hydrosphere, atmosphere, cryosphere, and biosphere. Not only has the climate of the Arctic changed significantly over the past two million years, there have also been pronounced regional variations associated with each change.

The Arctic is not homogeneous and neither is its climate, and past climate changes have not been uniform in their characteristics or their effects. Many of these changes have not been synchronous nor have they had equal magnitudes and rates of change. Climate changes in one part of the Arctic may trigger a delayed response elsewhere, adding to the complexity. The paleoenvironmental evidence for the Arctic suggests that at certain times, critical thresholds have been passed and unpredictable responses have followed. The role that anthropogenic changes to the climate system might play in exceeding such thresholds and the subsequent response remains unclear.

It is clear that between 400 and 100 years BP, the climate in the Arctic was exceptionally cold. There is widespread evidence of glaciers reaching their maximum post-Wisconsinan positions during this period, and the lowest $\delta^{18}O$ values and melt percentages for at least 1000 years are recorded in ice cores for this interval. The observed warming in the Arctic in the latter half of the 20th century appears to be without precedent since the early Holocene (Mann M. and Jones, 2003).

2.8. Summary and key findings

This chapter has described the arctic climate system; the region's impact on the global climate system; recent climatic change depicted by the instrumental record; and the historical/paleoclimatic perspective on arctic climatic variability. Features of the arctic climate system that are unique to the region include the cryosphere, the extremes of solar radiation, and the role of salinity in ocean dynamics.

The climate of the Arctic is changing. Trends in instrumental records over the past 50 years indicate a reasonably coherent picture of recent environmental change in northern high latitudes. The average surface temperature in the Arctic increased by approximately 0.09 °C per decade over the past century, and the pattern of change is similar to the global trend (i.e., an increase up to the mid-1940s, a decrease from then until the mid-1960s, and a steep increase thereafter with a warming rate of 0.4 °C per decade). It is very probable that the Arctic has warmed over the past century, at a rate greater than the average over the Northern Hemisphere. It is probable that polar amplification has occurred over the past 50 years. Because of the scarcity of observations across the Arctic before about 1950, it is not possible to be certain of the variation in mean land-station temperature over the first half of the 20th century. However, it is probable that the past decade was warmer than any other in the period of the instrumental record. The observed warming in the Arctic appears to be without precedent since the early Holocene.

It is very probable that atmospheric pressure over the Arctic Basin has been dropping and it is probable that there has been an increase in total precipitation at the rate of about 1% per decade over the past century. It is very probable that snow-cover extent around the periphery of the Arctic, sea-ice extent averaged over the Arctic (during at least the past 40 years), and multi-year sea-ice extent in the central Arctic have all decreased.

These climate changes are consistent with projections of climate change by global climate models forced with increasing atmospheric GHG concentrations, but definitive attribution is not yet possible.

Natural climate variability in the Arctic over the past two million years has been substantial. In particular, the past 20 000-year period is now known to have been highly unstable and prone to large rapid changes, especially temperature increases that occurred quickly (within a few decades or less). It is clear that between 400 and 100 years BP the climate in the Arctic was exceptionally cold, and there is widespread evidence of glaciers reaching their maximum post-Wisconsinan positions during this period.

Changes in the Arctic are very likely to have significant impacts on the global climate system. For example, a reduction in snow-cover extent and a shrinking of the marine cryosphere would increase heating of the surface, which is very likely to accelerate warming of the Arctic and reduce the equator-to-pole temperature gradient. Freshening of the Arctic Ocean by increased precipitation and runoff is likely to reduce the formation of cold deep water, thereby slowing the global thermohaline circulation. It is likely that a slowdown of the thermohaline circulation would lead to a more rapid rate of rise of global sea level, reduce upwelling of nutrients, and exert a chilling influence on the North Atlantic region as Gulf Stream heat transport is reduced. It would also decrease the rate at which CO_2 is transported to the deep ocean. Finally, temperature increases over permafrost areas could possibly lead to the release of additional CH_4 into the atmosphere; if seabed temperatures rise by a few degrees, hydrated CH_4 trapped in solid form could also escape into the atmosphere.

Although it is possible to draw many conclusions about past arctic climate change, it is evident that further research is still needed. The complex processes of the atmosphere, sea-ice, ocean, and terrestrial systems should be further explored in order to improve projections of future climate and to assist in interpreting past climate. Reconstructions of the past have been limited by available information, both proxy and instrumental records. The Arctic is a region of large natural variability and regional differences and it is important that more uniform coverage be obtained to clarify past changes. In order for the quantitative detection of change to be more specific in the future, it is essential that steps be taken now to fill in observational gaps across the Arctic, including the oceans, land, ice, and atmosphere.

Acknowledgements

I would like to offer a special acknowledgement for David Wuertz, NOAA National Climate Data Center. I would also like to thank Mark Serreze, Inger Hanssen-Bauer, and Thorstein Thorsteinsson who helped by participating in meetings and providing comments on the text. Supporting me in my work have been Jaime Dawson and Anna Ziolecki.

References

Aagaard, K. and E.C. Carmack, 1989. The role of sea ice and other fresh water in the Arctic circulation. Journal of Geophysical Research, 94(C10):14485–14498.

Aagaard, K. and L.K. Coachman, 1968a. The East Greenland Current north of Denmark Strait, I. Arctic, 21:181–200.

Aagaard, K. and L.K. Coachman, 1968b. The East Greenland Current north of Denmark Strait, II. Arctic, 21:267–290.

Aagaard, K. and P. Greisman, 1975. Toward new mass and heat budgets for the Arctic Ocean. Journal of Geophysical Research, 80:3821–3827.

Adkins, J.F., E.A. Boyle, L. Keigwin and E. Cortijo, 1997. Variability of the North Atlantic thermohaline circulation during the last interglacial period. Nature, 390:154–156.

Ager, T.A., 1983. Holocene vegetation history of Alaska. In: H.E. Wright Jr. (ed). Late Quaternary Environments of the United States. Vol. 2, The Holocene, pp. 128–140. University of Minnesota Press.

Akinremi, O.O., S.M. McGinn and H.W. Cutforth, 1999. Precipitation trends on the Canadian Prairies. Journal of Climate, 12:2996–3003.

Alekseev, G.V., L.V. Bulatov, V.F. Zakharov and V.V. Ivanov, 1998. Heat expansion of Atlantic waters in the Arctic Basin. Meteorologiya i Gidrologiya, 7:69–78.

Alekseev, G.V., Ye.I. Aleksandrov, R.V. Bekryaev, P.N. Svyaschennikov and N.Ya. Harlanienkova, 1999. Surface air temperature from meteorological data. In: Detection and Modelling of Greenhouse Warming in the Arctic and sub-Arctic, INTAS Project 97-1277, Technical Report on Task 1. Arctic and Antarctic Research Institute, St. Petersburg, Russia.

Alekseev, G.V., L.V. Bulatov and V.F. Zakharov, 2000. Freshwater melting/freezing cycle in the Arctic Ocean. In: E.L. Lewis, E.P. Jones, P. Lemke, T.D. Prowse and P. Wadhams (eds.). The Freshwater Budget of the Arctic Ocean, pp. 589–608. NATO Science Series, Kluwer Academic Publishers.

Alekseev, G.V., O.M. Johannessen, A.A. Korablev, V.V. Ivanov and D.V. Kovalevsky, 2001. Interannual variability in water masses in the Greenland Sea and adjacent areas. Polar Research, 20(2):201–208.

Alley, R.B., P.A. Mayewski, T. Sowers, M. Stuiver, K.C. Taylor and P.U. Clark, 1997. Holocene climatic instability: a prominent, widespread event 8200 years ago. Geology, 25:483–486.

AMAP, 1998. AMAP Assessment Report: Arctic Pollution Issues. Arctic Monitoring and Assessment Programme, Oslo, Norway, xii+859 pp.

Ambaum, M.H., B.J. Hoskins and D.B. Stephenson, 2001. Arctic Oscillation or North Atlantic Oscillation? Journal of Climate, 14:3495–3507.

Anda, E., O. Orheim and J. Mangerud, 1985. Late Holocene glacier variations and climate at Jan Mayen. Polar Research, 3:129–140.

Anisimov, O.A., N.I. Shiklomanov and F.E. Nelson, 1997. Effects of global warming on permafrost and active layer thickness: results from transient general circulation models. Global and Planetary Change, 61:61–77.

Arctic Climatology Project, 2000. Environmental Working Group Arctic Meteorology and Climate Atlas. F. Fetterer and V. Radionov (eds.). National Snow and Ice Data Center, Boulder, Colorado. CD-ROM.

Armstrong, R.L. and M.J. Brodzik, 2001. Recent northern hemisphere snow extent: A comparison of data derived from visible and microwave satellite sensors. Geophysical Research Letters, 28:3673–3676.

Barnekow, L. and P. Sandgren, 2001. Palaeoclimate and tree-line changes during the Holocene based on pollen and plant macrofossil records from six lakes at different altitudes in northern Sweden. Review of Palaeobotany and Palynology, 117:109–118.

Barry, R.G. and M.C. Serreze, 2000. Atmospheric components of the Arctic Ocean freshwater balance and their interannual variability. In: E.L. Lewis, E.P. Jones, P. Lemke, T.D. Prowse and P. Wadhams (eds.). The Freshwater Budget of the Arctic Ocean, pp. 345–351. NATO Science Series, Kluwer Academic Publishers.

Belkin, I.M., S. Levitus, J. Antonov and S.-A. Malmberg, 1998. Great salinity anomalies in the North Atlantic. Progress in Oceanography, 41:1–68.

Bell, T., 1996. The last glaciation and sea level history of Fosheim Peninsula, Ellesmere Island, Canadian High Arctic. Canadian Journal of Earth Sciences, 33:1075–1086.

Beltaos, S., 2002. Effects of climate on mid-winter ice jams. Hydrological Processes, 16:789–804.

Berger, A. and M.F. Loutre, 1991. Insolation values for the climate of the last 10 million years. Quaternary Science Reviews, 10:297–317.

Boenisch, G., J. Blindheim, J.L. Bullister, P. Schlosser and D.W.R. Wallace, 1997. Long term trends of temperature, salinity, density and transient tracers in the central Greenland Sea. Journal of Geophysical Research, 102(C8):18553–18571.

Bond, G.C. and R. Lotti, 1995. Iceberg discharges into the North Atlantic on millennial timescales during the last glaciation. Science, 267:1005–1010.

Bond, G.C., W. Broecker, S. Johnsen, J. McManus, L. Labeyrie, J. Jouzel and G. Bonani, 1993. Correlations between climate records from North Atlantic sediments and Greenland ice. Nature, 365:143–147.

Bond, G.C., W. Showers, M. Cheseby, R. Lotti, P. Almasi, P. deMenocal, P. Priore, H. Cullen, I. Hajdas and G. Bonani, 1997. A pervasive millennial-scale cycle in North Atlantic Holocene and glacial climates. Science, 278:1257–1266.

Bond, G., B. Kromer, J. Beer, R. Muscheler, M.N. Evans, W. Showers, S. Hoffmann, R. Lotti-Bond, I. Hajdas and G. Bonani, 2001. Persistent solar influence on North Atlantic climate during the Holocene. Science, 294:2130–2136.

Bourgeois, J.C., R.M. Koerner, K. Gajewski and D.A. Fisher, 2000. A Holocene ice-core pollen record from Ellesmere Island, Nunavut, Canada. Quaternary Research, 54:275–283.

Bourke, R.H. and R.P. Garrett, 1987. Sea ice thickness distribution in the Arctic Ocean. Cold Regions Science and Technology, 13:259–280.

Bradley, R.S., 1985. Quaternary Paleoclimatology: Methods of Paleoclimatic Reconstruction. Allen & Unwin, 490pp.

Bradley, R.S., 1990. Holocene paleoclimatology of the Queen Elizabeth Islands, Canadian high Arctic. Quaternary Science Reviews, 9:365–384.

Bradley, R.S., 1999. Paleoclimatology: Reconstructing Climates of the Quaternary. Academic Press, 610pp.

Bradley, R.S. and P.D. Jones, 1993. 'Little Ice Age' summer temperature variations: their nature and relevance to recent global warming trends. The Holocene, 3:367–376.

Briffa, K.R., P.D. Jones, T.S. Bartholin, D. Eckstein, F.H. Schweingruber, W. Karlén, P. Zetterberg and M. Eronen, 1992. Fennoscandian summers from AD 500: temperature changes on short and long timescales. Climate Dynamics, 7:111–119.

Briffa, K.R., P.D. Jones, F.H. Schweingruber, S.G. Shiyatov and E.R. Cook, 1995. Unusual twentieth-century summer warmth in a 1,000-year temperature record from Siberia. Nature, 376:156–159.

Brigham-Grette, J. and D.M. Hopkins, 1995. Emergent marine record and paleoclimate of the last interglaciation along the Northwest Alaskan Coast. Quaternary Research, 43:159–173.

Brigham-Grette, J., L.M. Gualtieri, O.Y. Glushkova, T.D. Hamilton, D. Mostoller and A. Kotov, 2003. Chlorine-36 and ^{14}C chronology support a limited last glacial maximum across central Chukotka, northeastern Siberia, and no Beringian ice sheet. Quaternary Research, 59:386–398.

Broecker, W.S., 1997. Thermohaline circulation, the Achilles heel of our climate system: Will man-made CO_2 upset the current balance? Science, 278:1582–1588.

Broecker, W.S., 2000. Was a change in thermohaline circulation responsible for the Little Ice Age? Proceedings of the National Academy of Sciences, 97:1339–1342.

Broecker, W.S., G. Bond and M.A. Klas, 1990. A salt oscillator in the glacial Atlantic? 1: the concept. Paleoceanography, 5:469–477.

Brown, J., K.M. Hinkel and F.E. Nelson, 2000. The Circumpolar Active Layer Monitoring (CALM) program: Research designs and initial results. Polar Geography, 24(3):165–258.

Brown, R.D., 2000. Northern Hemisphere snow cover variability and change, 1915–97. Journal of Climate, 13:2339–2355.

Brown, R.D. and P. Coté, 1992. Interannual variability of landfast ice thickness in the Canadian High Arctic, 1950–89. Arctic, 45(3):273–284.

Burenhult, G. (ed.), 1999. Arkeologi i Norden, Volume 1. Natur och Kultur, Stockholm, 540pp.

Calkin, P.E., 1988. Holocene glaciation of Alaska (and adjoining Yukon Territory, Canada). Quaternary Science Reviews, 7:159–184.

Calkin, P.E., G.C. Wiles and D.J. Barclay, 2001. Holocene coastal glaciation of Alaska. Quaternary Science Reviews, 20:449–461.

Carey, S.K. and M.-K. Woo, 1999. Hydrology of two slopes in subarctic Yukon, Canada. Hydrological Processes, 13:2549–2562.

Carmack, E.C., R.W. Macdonald, R.G. Perkin, F.A. McLaughlin and R.J. Pearson, 1995. Evidence for warming of Atlantic water in the southern Canadian Basin of the Arctic Ocean: results from the Larsen-93 expedition. Geophysical Research Letters, 22(9):1061–1064.

Carmack, E.C., K. Aagaard, J.H. Swift, R.W. Macdonald, F.A. McLaughlin, E.P. Jones, R.G. Perkin, J.N. Smith, K.M. Ellis and L.R. Killius, 1997. Changes in temperature and tracer distributions within the Arctic Ocean: Results from the 1994 Arctic Ocean section. Deep-Sea Research II, 44:1487–1502.

Cavalieri, D.J. and S. Martin, 1994. The contribution of Alaskan, Siberian and Canadian coastal polynyas to the cold halocline layer of the Arctic Ocean. Journal of Geophysical Research, 99(C9):18343–18362.

Cavalieri, D.J., P. Gloersen, C.L. Parkinson, J.C. Comiso and H.J. Zwally, 1997. Observed hemispheric asymmetry in global sea ice changes. Science, 278:1104–1106.

Cayan, D.R., S. Kammerdiener, M.D. Dettinger, J.M. Caprio and D.H. Peterson, 2001. Changes in the onset of spring in the western United States. Bulletin of the American Meteorological Society, 82:399–416.

Chapman, W.L. and J.E. Walsh, 2003. Observed climate change in the Arctic, updated from Chapman and Walsh, 1993. Recent variations of sea ice and air temperatures in high latitudes. Bulletin of the American Meteorological Society, 74(1):33–47. http://arctic.atmos.uiuc.edu/CLIMATESUMMARY/2003/

Clark, P.U. and A.C. Mix, 2000. Global change: Ice sheets by volume. Nature, 406:689–690.

COHMAP Members, 1988. Climatic changes of the last 18,000 years: observations and model simulations. Science, 241:1043–1052.

Comiso, J.C., 2002. A rapidly declining perennial sea ice cover in the Arctic. Geophysical Research Letters, 29(20):1956, doi:10.1029/2002GL015650.

Comiso, J., 2003. Warming trends in the Arctic from clear sky satellite observations. Journal of Climate, 16:3498–3510.

Cortijo, E., J. Duplessy, L. Labeyrie, H. Leclaire, J. Duprat and T. van Weering, 1994. Eemian cooling in the Norwegian Sea and North Atlantic ocean preceding ice-sheet growth. Nature, 372:446–449.

Crowley, T.J. and T. Lowery, 2000. How warm was the Medieval warm period? Ambio, 29:51–54.

Cuffey, K.M. and S.J. Marshall, 2000. Substantial contribution to sea-level rise during the last interglacial from the Greenland ice sheet. Nature, 404:591–594.

Dahl-Jensen, D., K. Mosegaard, N. Gunderstrup, G.D. Clow, S.J. Johnsen, A.W. Hansen and N. Balling, 1998. Past temperatures directly from the Greenland ice sheet. Science, 282:268–271.

Dansgaard, W., J.W. White and S.J. Johnsen, 1989. The abrupt termination of the Younger Dryas climate event. Nature, 339:532–534.

Dansgaard, W., S.J. Johnsen, H.B. Clausen, D. Dahl-Jensen, N.S. Gundestrup, C.U. Hammer, C.S. Hvidberg, J.P. Steffensen, A.E. Sveinbjörnsdottir, J. Jouzel and G. Bond, 1993. Evidence for general instability of past climate from a 250-kyr ice-core record. Nature, 364:218–220.

Davis, B.A.S., S. Brewer, A.C. Stevenson, J. Guiot and Data Contributors, 2003. The temperature of Europe during the Holocene reconstructed from pollen data, Quaternary Science Reviews, 22:1701–1716.

Deser, C., 2000. On the teleconnectivity of the 'Arctic Oscillation'. Geophysical Research Letters, 27:779–782.

Deser, C., J.E. Walsh and M.S. Timlin, 2000. Arctic sea ice variability in the context of recent atmospheric circulation trends. Journal of Climate, 13:617–633.

Dickson, R.R., J. Meincke, S.-A. Malmberg and A.J. Lee, 1988. The 'Great Salinity Anomaly' in the Northern North Atlantic 1968–1982. Progress in Oceanography, 20:103–151.

Dickson, R.R., T.J. Osborn, J.W. Hurrell, J. Meincke, J. Blindheim, B. Adlandsvik, T. Vinje, G. Alekseev and W. Maslowski, 2000. The Arctic Ocean response to the North Atlantic Oscillation. Journal of Climate, 13:2671–2696.

Dickson, R.R., I. Yashayaev, J. Meincke, B. Turrell, S. Dye and J. Holfort, 2002. Rapid freshening of the deep North Atlantic Ocean over the past four decades. Nature, 416:832–837.

Drinkwater, K.F., 1994. Climate and oceanographic variability in the Northwest Atlantic during 1980s and early 1990s, NAFO Scientific Council Research Document 94/71. Northwest Atlantic Fisheries Organization, Dartmouth, Nova Scotia, Canada, 39pp.

Dugmore, A.J., 1989. Tephrochronological studies of Holocene glacier fluctuations in south Iceland. In: J. Oerlemans (ed.). Glacier Fluctuations and Climatic Change, pp. 37–55. Kluwer Academic Publishers.

Dyke, A.S., J. Hooper and J.M. Savelle, 1996. A history of sea ice in the Canadian Arctic Archipelago based on postglacial remains of the bowhead whale (*Balaena mysticetus*). Arctic, 49(3):235–255.

Dyke, A.S., J. Hooper, C.R. Harington and J.M. Savelle, 1999. The Late Wisconsinan and Holocene record of walrus (*Odobenus rosmarus*) from North America: a review with new data from Arctic and Atlantic Canada. Arctic, 52(2):160–181.

Dyke, A.S., J.T. Andrews, P.U. Clark, J.H. England, G.H. Miller, J. Shaw and J.J. Veillette, 2002. The Laurentide and Innuitian ice sheets during the Last Glacial Maximum. Quaternary Science Reviews, 21:9–31.

Easterling, D.R., B. Horton, P.D. Jones, T.C. Peterson, T.R. Karl, D.E. Parker, M.J. Salinger, V. Razuvayev, N. Plummer, P. Jamason and C.K. Folland, 1997. Maximum and minimum temperature trends for the globe. Science, 277:364–367.

Eronen, M., P. Zetterberg, K.R. Briffa, M. Lindholm, J. Meriläinen and M. Timonen, 2002. The supra-long Scots pine tree-ring record for Finnish Lapland: Part 1, chronology construction and initial inferences. The Holocene, 12(6):673–680.

Fedorova, Z.P. and Z.S. Yankina, 1964. The passage of Pacific Ocean water through the Bering Strait into the Chukchi Sea. Deep-Sea Research and Oceanographic Abstracts, 11:427–434.

Fissel, D.B. and H. Melling, 1990. Interannual variability of oceanographic conditions in the southeastern Beaufort Sea. Canadian Contractor Report of Hydrography and Ocean Sciences No 35. Institute of Ocean Sciences, Sidney, British Columbia, Canada, 102pp.

Førland, E.J. and I. Hanssen-Bauer, 2000. Increased precipitation in the Norwegian Arctic: True or False? Climatic Change, 46:485–509.

Førland, E.J and I. Hanssen-Bauer, 2003. Past and future climate variations in the Norwegian Arctic: overview and novel analyses. Polar Research, 22(2):113–124.

Førland, E.J., I. Hanssen-Bauer and P. Nordli, 1997. Climate Statistics and Long-term Time Series of Temperature and Precipitation at Svalbard and Jan Mayen. Norwegian Meteorological Institute Report 21/97, 72pp.

Frich, P., L.V. Alexander, P. Della-Marta, B. Gleason, M. Haylock, A.M.G. Klein Tank and T. Peterson, 2002. Observed coherent changes in climatic extremes during the second half of the twentieth century. Climate Research, 19:193–212.

Fronval, T. and E. Jansen, 1997. Eemian and early Weichselian (140–60 ka) paleoceanography and paleoclimate in the Nordic seas with comparisons to Holocene conditions. Paleoceanography, 12(3):443–462.

Fukuda, M., 1994. Methane flux from thawing Siberian permafrost (ice complexes) results from field observations. EOS, Transactions of the American Geophysical Union, 75:86.

Fyfe, J.C., 2003. Separating extratropical zonal wind variability and mean change. Journal of Climate, 16:863–874.

Fyfe, J.C., G.J. Boer and G.M. Flato, 1999. The Arctic and Antarctic oscillations and their projected changes under global warming. Geophysical Research Letters, 26:1601–1604.

Gajewski, K., R. Vance, M. Sawada, I. Fung, L.D. Gignac, L. Halsey, J. Johm, P. Maisongrande, P. Mandell, P.J. Mudie, P.J.H. Richard, A.G. Sherin, J. Soroko and D.H. Vitt, 2000. The climate of North America and adjacent ocean waters ca. 6 ka. Canadian Journal of Earth Sciences, 37:661–681.

Ganopolski, A. and S. Rahmstorf, 2001. Rapid changes of glacial climate simulated in a coupled climate model. Nature, 409:153–158.

Gasse, F. and E. van Campo, 1994. Abrupt post-glacial events in west Asia and North Africa monsoon domains. Earth and Planetary Science Letters, 126:453–456.

Gillett, N.P., F.W. Zwiers, A.J. Weaver and P.A. Stott, 2003. Detection of human influence on sea-level pressure. Nature, 422:292–294.

Goodison, B.E., P.Y.T. Louie and D. Yang, 1998. WMO Solid Precipitation Intercomparison, Final Report. Instruments and Observing Methods Report 67, WMO/TD 872. World Meteorological Organization, Geneva, 87pp. + Annexes.

Goosse, H. and H. Renssen, 2003. Simulating the evolution of the Arctic climate during the last millennium. In: Proceedings of the Seventh Conference on Polar Meteorology and Oceanography and Joint Symposium on High-Latitude Climate Variations, May 12–16, 2003, Hyannis, Massachusetts, Abstract 1.5. American Meteorological Society.

Gorshkov, S.G. (ed.), 1980. Atlas of the Oceans. Arctic Ocean. USSR Ministry of Defense, Voenno-Morskoy Flot., Moscow, 199pp.

Grody, N.C. and A.N. Basist, 1996. Global identification of snowcover using SSM/I measurements. IEEE Transactions on Geoscience and Remote Sensing, 34:237–249.

Groisman, P.Ya. and E.Ya. Rankova, 2001. Precipitation trends over the Russian permafrost-free zone: removing the artifacts of pre-processing. International Journal of Climatology, 21:657–678.

Groisman, P.Ya., V.V Koknaeva, T.A. Belokrylova and T.R. Karl, 1991. Overcoming biases of precipitation: a history of the USSR experience. Bulletin of the American Meteorological Society, 72:1725–1733.

Groisman, P.Ya., T.R. Karl and T.W. Knight, 1994. Observed impact of snow cover on the heat balance and the rise of continental spring temperatures. Science, 263:198–200.

Groisman, P.Ya., T.R. Karl, D.R. Easterling, R.W. Knight, P.B. Jamason, K.J. Hennessy, R. Suppiah, C.M. Page, J. Wibig, K. Fortuniak, V.N. Razuvaev, A. Douglas, E. Førland and P.-M. Zhai, 1998. Heavy rainfall in a changing climate. In: Proceedings of the Ninth Symposium on Global Change Studies, 11–16 January, 1998, Phoenix, Arizona, pp. j5–j10. American Meteorological Society.

Groisman, P.Ya., T.R. Karl, D.R. Easterling, R.W. Knight, P.B. Jamason, K.J. Hennessy, R. Suppiah, Ch.M. Page, J. Wibig, K. Fortuniak, V.N. Razuvaev, A. Douglas, E. Førland and P.-M. Zhai, 1999a. Changes in the probability of heavy precipitation: Important indicators of climatic change. Climatic Change, 42(1):243–283.

Groisman, P.Ya., E.L. Genikhovich, R.S. Bradley and B.-M. Sun, 1999b. Trends in turbulent heat fluxes over Northern Eurasia. In: M. Tranter, R. Armstrong, E. Brun, G. Jones, M. Sharp and M. Williams (eds.). Interactions Between the Cryosphere, Climate and Greenhouse Gases, pp. 19–25. Proceedings of the International Union of Geodesy and Geophysics 1999 Symposium HS2, Birmingham, UK, July 1999. International Association of Hydrological Sciences Publication No. 256. IAHS Press.

Groisman, P.Ya., R.W. Knight and T.R. Karl, 2002. Very heavy precipitation over land: estimates based on a new global daily precipitation data set. In: Proceedings of the 13th Symposium on Global Change Studies, Orlando, Florida, 13–17 January, 2002, pp. 88–90. American Meteorological Society.

Groisman, P.Ya., B. Sun, R.S. Vose, J.H. Lawrimore, P.H. Whitfield, E. Førland, I. Hanssen-Bauer, M.C. Serreze, V.N. Razuvaev and G.V. Alekseev, 2003. Contemporary climate changes in high latitudes of the northern hemisphere: daily time resolution. In: Proceedings of the 14th AMS Symposium on Global Change and Climate Variations. CD-ROM with Proceedings of the Annual Meeting of the American Meteorological Society, Long Beach, California, 9–13 February, 2003. American Meteorological Society, 10pp.

Grosswald, M.G. and T.J. Hughes, 2002. The Russian component of an Arctic ice sheet during the Last Glacial Maximum. Quaternary Science Reviews, 21:121–146.

Grotefendt, K., K. Logemann, D. Quadfasel and S. Ronski, 1998. Is the Arctic Ocean warming? Journal of Geophysical Research, 103(C12):27679–27687.

Grove, J.M., 1988. The Little Ice Age. Methuen, London, 498pp.

Grudd, H., K.R. Briffa, W. Karlén, T.S. Bartholin, P.D. Jones and B. Kromer, 2002. A 7400-year tree-ring chronology in northern Swedish Lapland: natural climatic variability expressed on annual to millennial timescales. The Holocene, 12(6):657–666.

Hakkinen, S., 1993. An Arctic source for the Great Salinity Anomaly: a simulation of the Arctic ice-ocean system for 1955–1975. Journal of Geophysical Research, 98(C9):16397–16410.

Hansen, B., W.R. Turrell and S. Osterhus, 2001. Decreasing overflow from the Nordic Seas into the Atlantic Ocean through the Faroe Bank Channel since 1950. Nature, 411:927–930.

Hanssen-Bauer, I. and E.J. Forland, 1998. Long-term trends in precipitation and temperature in the Norwegian Arctic: can they be explained by changes in atmospheric circulation patterns? Climate Research, 19:143–153.

Hanssen-Bauer, I., E.J. Førland, O.E. Tveito and P.O. Nordli, 1997. Estimating regional trends – comparisons of two methods. Nordic Hydrology, 28:21–36.

Hantemirov, R.M. and S.G. Shiyatov, 2002. A continuous multimillennial ring-width chronology in Yamal, northwestern Siberia. The Holocene, 12(6):717–726.

Hardy, D.R. and R.S. Bradley, 1996. Climatic change in Nunavut. Geoscience Canada, 23:217–224.

Helama, S., M. Lindholm, M. Timonen, J. Meriläinen and M. Eronen, 2002. The supra-long Scots pine tree-ring record for Finnish Lapland: Part 2, interannual to centennial variability in summer temperatures for 7500 years. The Holocene, 12(6):681–687.

Heusser, C.J., 1995. Late Quaternary vegetation response to climatic-glacial forcing in North Pacific America. Physical Geography, 16:118–149.

Hoerling, M.P., J.W. Hurrell and T. Xu, 2001. Tropical origins for recent North Atlantic climate change. Science, 292:90–92.

Holloway, G. and T. Sou, 2001. Has Arctic sea ice rapidly thinned? Journal of Climate, 15:1692–1701.

Hughes, M.K. and H.F. Diaz, 1994. Was there a 'Medieval Warm Period' and if so, where and when? Climatic Change, 265:109–142.

Imbrie, J. and K.P. Imbrie, 1979. Ice Ages: Solving the Mystery. Harvard University Press, 224pp.

Imbrie, J., A. Berger, E.A. Boyle, S.C. Clemens, A. Duffy, W.R. Howard, G. Kukla, J. Kutzbach, D.G. Martinson, A. McIntyre, A.C. Mix, B. Molfino, J.J. Morley, L.C. Peterson, N.G. Pisias, W.L. Prell, M.E. Raymo, N.J. Shackleton and J.R. Toggweiler, 1993a. On the structure and origin of major glaciation cycles. 2: The 100,000 year cycle. Paleoceanography, 8:699–735.

Imbrie, J., A. Berger and N.J. Shackleton, 1993b. Role of orbital forcing: a two-million-year perspective. In: J.A. Eddy and H. Oeschger (eds.). Global Changes in the Perspective of the Past, pp. 263–277. John Wiley.

Ingolfsson, O., S. Björck, H. Haflidason and M. Rundgren, 1997. Glacial and climatic events in Iceland reflecting regional North Atlantic climatic shifts during the Pleistocene-Holocene transition. Quaternary Science Reviews, 16:1135–1144.

IPCC, 2001a. Climate Change 2001: Impacts, Adaptation, and Vulnerability. Contribution of Working Group II to the Third Assessment Report of the Intergovernmental Panel on Climate Change. McCarthy, J.J., O.F. Canziani, N.A. Leary, D.J. Dokken and K.S. White (eds.). Cambridge University Press, 1032 pp.

IPCC, 2001b. Climate Change 2001: Synthesis Report. A Contribution of Working Groups I, II, and III to the Third Assessment Report of the Intergovernmental Panel on Climate Change. Watson, R.T., and the Core Writing Team (eds.). Cambridge University Press, 398 pp.

IPCC, 2001c. Climate Change 2001: The Scientific Basis. Contribution of Working Group I to the Third Assessment Report of the Intergovernmental Panel on Climate Change. Houghton, J.T., Y. Ding, D.J. Griggs, M. Noguer, P.J. van der Linden, X. Dai, K. Maskell and C.A. Johnson (eds.) Cambridge University Press, 881 pp.

Isaksen, K., P. Holmlund, J.L. Sollid and C. Harris, 2001. Three deep alpine-permafrost boreholes in Svalbard and Scandinavia. Permafrost and Periglacial Processes, 12:13–25.

Ivanov, V.V., 2000. Akademik Fedorov Cruise. Arctic Climate System Study, ACSYS Arctic Forecast, 2:2–3. Tromso.

Johannessen, O.M., E. Shalina and M. Miles, 1999. Satellite evidence for an Arctic sea ice cover in transformation. Science, 286:1937–1939.

Johannessen, O.M., L. Bengtsson, M.W. Miles, S.I. Kuzmina, V.A. Semenov, G.V. Alekseev, A.P. Nagurnyi, V.F. Zakharov, L.P. Bobylev, L.H. Pettersson, K. Hasselmann and H.P. Cattle, 2004. Arctic climate change: observed and modelled temperature and sea-ice variability. Tellus A, 56:328–341.

Johnsen, S.J., D. Dahl-Jensen, N. Gunderstrup, J.P. Steffensen, H.B. Clausen, H. Miller, V. Masson-Delmotte, A.E. Sveinbjornsdottir and J. White, 2001. Oxygen isotope and palaeotemperature records from six Greenland ice-core stations: Camp Century, Dye-3, GRIP, GISP, Renland and NorthGRIP. Journal of Quaternary Science, 16:299–307.

Jonasson, S.E., G.R. Shaver and F.S. Chapin III, 2001. Biogeochemistry in the Arctic: patterns, processes and controls. In: E.D. Schulze, M. Heimann, S.P. Harrison, E.A. Holland, J.J. Lloyd, I.C. Prentice and D. Schimel (eds.). Global Biogeochemical Cycles in the Climate System, pp. 139–150. Academic Press.

Jones, P.D. and A. Moberg, 2003. Hemispheric and large-scale surface air temperature variations: an extensive revision and an update to 2001. Journal of Climate, 16:206–223.

Jones, P.D., P.Ya. Groisman, M. Coughlan, N. Plummer, W.C. Wang and T.R. Karl, 1990. Assessment of urbanization effects in time series of surface air temperature over land. Nature, 347:169–172.

Jones, P.D., K.R. Briffa, T.P. Barnett and S.B. Tett, 1998. High-resolution palaeoclimatic records for the last millennium: interpretation, integration and comparison with General Circulation Model control run temperatures. The Holocene, 8:455–471.

Jones P.D., M. New, D.E. Parker, S. Martin and I.G. Rigor, 1999. Surface air temperature and its changes over the past 150 years. Reviews of Geophysics, 37(2):173–199.

Jorgenson, M.T., C.H. Racine, J.C. Walters and T.E. Osterkamp, 2001. Permafrost degradation and ecological changes associated with a warming climate in central Alaska. Climatic Change, 48:551–579.

Kalnay, E.M., M. Kanamitsu, R. Kistler, W. Collins, D. Deaven, L. Gandin, M. Iredell, S. Saha, G. White, J. Woollen, Y. Zhu, A. Leetmaa, B. Reynolds, M. Chelliah, W. Ebisuzaki, W. Higgins, J. Janowiak, K.C. Mo, C. Ropelewski, J. Wang, R. Jenne and D. Joseph, 1996. The NCEP/NCAR Reanalysis Project. Bulletin of the American Meteorological Society, 77:437–471.

Kane, D.L., L.D. Hinzman, C.S. Benson and K. R. Everett, 1989. Hydrology of Imnavait Creek, an arctic watershed. Holarctic Ecology, 12:262–269.

Karl, T.R., R.G. Quayle and P.Ya. Groisman, 1993. Detecting climate variations and change: new challenges for observing and data management systems. Journal of Climate, 6:1481–1494.

Karlén, W., 1988. Scandinavian glacial and climate fluctuations during the Holocene. Quaternary Science Reviews 7:199–209.

Karlén, W. and J. Kuylenstierna, 1996. On solar forcing of Holocene climate: evidence from Scandinavia. The Holocene, 6:359–365.

Karoly, D.J., K. Braganza, P.A. Stott, J.M. Arblaster, G.A. Meehl, A.J. Broccoli and K.W. Dixon, 2003. Detection of a human influence on North American climate. Science, 302:1200–1203.

Karpuz, N., E. Jansen and H. Haflidason, 1993. Paleoceanographic reconstruction of surface ocean conditions in the Greenland, Iceland and Norwegian Seas through the last 14 ka based on diatoms. Quaternary Science Reviews, 12:115–140.

Keigwin, L.D. and R.S. Pickart, 1999. Slope water current over the Laurentian Fan on interannual to millennial time scales. Science, 286:520–523.

Keigwin, L.D., W.B. Curry, S.J. Lehman and S. Johnson, 1994. The role of North Atlantic climate change between 70 and 130 kyr ago. Nature, 371:323–326.

Kennett, J.P., K.G. Cannariato, I.L. Hendy and R.J. Behl, 2000. Carbon isotopic evidence for methane hydrate instability during Quaternary interstadials. Science, 288:128–133.

Khrol, V.P. (ed.), 1992. Atlas of the Energy Budget of the Northern Polar Region. Gidrometeoizdat, St. Petersburg, 52pp.

Koerner, R.M. and D.A. Fisher, 1990. A record of Holocene summer climate from a Canadian high-Arctic ice core. Nature, 343:630–631.

Korhola, A., K. Vasko, H.T. Toivonen and H. Olander, 2002. Holocene temperature changes in northern Fennoscandia reconstructed from chironomids using Bayesian modeling. Quaternary Science Reviews, 21:1841–1860.

Krasovskaia, I. and N.R. Saelthun, 1997. Sensitivity of the stability of Scandinavian river flow regimes to a predicted temperature rise. Hydrological Sciences, 42:693–711.

Kuchment, L.S., A.N. Gelfan and V.N. Demidov, 2000. A distributed model of runoff generation in the permafrost regions. Journal of Hydrology, 240:1–22.

Kugelmann, O., 1991. Dating recent glacier advances in the Svarfadardalur-Skidadalur area of northern Iceland by means of a new lichen curve. In: J.K. Maizels and C. Caseldine (eds.). Environmental Change in Iceland: Past and Present, pp. 203–217. Kluwer Academic Publishers.

Kwok, R., 2000. Recent changes in Arctic Ocean sea ice motion associated with the North Atlantic Oscillation. Geophysical Research Letters, 27:775–778.

Lang, C., M. Leuenberger, J. Schwander and J. Johnsen, 1999. 16 °C rapid temperature variation in central Greenland 70 000 years ago. Science, 286:934–937.

Lee, S.E., M.C. Press and J.A. Lee, 2000. Observed climate variations during the last 100 years in Lapland, Northern Finland. International Journal of Climatology, 20:329–346.

Loeng, H., V. Ozhigin and B. Ådlandsvik, 1997. Water fluxes through the Barents Sea. ICES Journal of Marine Science, 54:310–317.

Lozhkin, A.V. and P.M. Anderson, 1995. The last interglaciation in Northeast Siberia. Quaternary Research, 43:47–158.

Lunardini, V.J., 1996. Climatic warming and the degradation of warm permafrost. Permafrost and Periglacial Processes, 7:311–320.

MacDonald, G.M., A.A. Velichko, C.V. Kremenetski, O.K. Borisova, A.A. Goleva, A.A. Andreev, L.C. Cwynar, R.T. Riding, S.L. Forman, T.W.D. Edwards, R. Aravena, D. Hammarlund, J.M. Szeicz and V.N. Gattaulin, 2000. Holocene treeline history and climate change across northern Eurasia. Quaternary Research, 53:302–311.

Macdonald, R.W., F.A. McLaughlin and E.C. Carmack, 2002. Fresh water and its sources during the SHEBA drift in the Canada Basin of the Arctic Ocean. Deep-Sea Research I, 49:1769–1785.

Malmberg, S.A. and J. Blindheim, 1994. Climate, cod and capelin in northern waters. ICES Marine Science Symposia, 198:297–310.

Manabe, S., R.J. Stouffer, M.J. Spelman and K. Bryan, 1991. Transient response of a coupled ocean-atmosphere model to gradual changes of atmospheric CO_2. Part I. Annual mean response. Journal of Climate, 4:785–818.

Mangerud, J., V. Astakhov and J.-I. Svensen, 2002. The extent of the Barents-Kara ice sheet during the Last Glacial Maximum. Quaternary Science Reviews, 21:111–119.

Mann, D.H. and T.L. Hamilton, 1995. Late Pleistocene and Holocene paleoenvironments of the North Pacific coast. Quaternary Science Reviews, 14:449–471.

Mann, M.E. and P.D. Jones, 2003. Global surface temperatures over the past two millennia. Geophysical Research Letters, 30:1820–1824, doi: 10.1029/2003GL017814.

Mann, M.E., R.S. Bradley and M.K. Hughes, 1999. Northern Hemisphere temperatures during the past millennium: inferences, uncertainties, and limitations. Geophysical Research Letters, 26:759–762.

Mantua, N.J., S.R. Hare, Y. Zhang, J.M. Wallace and R.C. Francis, 1997. A Pacific decadal climate oscillation with impacts on salmon. Bulletin of the American Meteorological Society, 78:1069–1079.

Maslin, M.A., X.S. Li, M.-F. Loutre and A. Berger, 1998. The contribution of orbital forcing to the progressive intensification of northern hemisphere glaciation. Quaternary Science Reviews, 17:411–426.

Maslowski, W., B. Newton, P. Schlosser, A. Semtner and D. Martinson, 2000. Modelling recent climate variability in the Arctic Ocean. Geophysical Research Letters, 27(22):3743–3746.

McCabe, G.J., M.P. Clark and M.C. Serreze, 2001. Trends in Northern Hemisphere surface cyclone frequency and intensity. Journal of Climate, 14:2763–2768.

McLaughlin, F.A., E.C. Carmack, R.W. Macdonald and J.K. Bishop, 1996. Physical and geochemical properties across the Atlantic/Pacific water mass front in the southern Canadian Basin. Journal of Geophysical Research, 101(C1):1183–1197.

Mekis, E. and W.D. Hogg, 1999. Rehabilitation and analysis of Canadian daily precipitation time series. Atmosphere-Ocean, 37:53–85.

Melling, H., 1993. The formation of a haline shelf front in an ice-covered arctic sea. Continental Shelf Research, 13:1123–1147.

Melling, H., 2000. Exchanges of freshwater through the shallow straits of the North American Arctic. In: E.L. Lewis, E.P. Jones, P. Lemke, T.D. Prowse and P. Wadhams (eds.). The Freshwater Budget of the Arctic Ocean. Proceedings of a NATO Advanced Research Workshop, Tallinn, Estonia, April 1998, pp. 479–502. Kluwer Academic Publishers.

Melling, H., 2002. Sea ice of the northern Canadian Arctic Archipelago. Journal of Geophysical Research, 107(C11):3181, doi:10.1029/2001JC001102.

Melling, H. and E.L. Lewis, 1982. Shelf drainage flows in the Beaufort Sea and their effect on the Arctic Ocean pycnocline. Deep-Sea Research A, 29:967–985.

Melling, H. and R.M. Moore, 1995. Modification of halocline source waters during freezing on the Beaufort Sea shelf: Evidence from oxygen isotopes and dissolved nutrients. Continental Shelf Research, 15(1):81–113.

Melling, H. and D.A. Riedel, 1996. Development of seasonal pack ice in the Beaufort Sea during the winter of 1991–92: A view from below. Journal of Geophysical Research, 101(C5):11975–11992.

Melling, H., Y. Gratton and G. Ingram, 2001. Ocean circulation within the North Water Polynya of Baffin Bay. Atmosphere-Ocean, 39(3):301–325.

Metcalfe, R.A. and J.M. Buttle, 1999. Semi-distributed water balance dynamics in a small boreal forest basin. Journal of Hydrology, 226:66–87.

Milankovitch, M., 1941. Canon of insolation and the ice age problem. Special Publication 132, Koniglich Serbische Akademie, Belgrade. (English translation by the Israel Program for Scientific Translations, Jerusalem, 1969).

Miller, G.H. and A. de Vernal, 1992. Will greenhouse warming lead to Northern Hemisphere ice-sheet growth? Nature, 355:244–246.

Minobe, S., 1997. A 50–70 year climatic oscillation over the North Pacific and North America. Geophysical Research Letters, 24:683–686.

Monahan, A.H., J.C. Fyfe, and L. Pandolfo, 2003. The vertical structure of wintertime climate regimes of the northern hemisphere extratropical atmosphere. Journal of Climate, 16:2005–2021.

Morison, J., M. Steele and R. Andersen, 1998. Hydrography of the upper Arctic Ocean measured from the nuclear submarine U.S.S. Pargo. Deep-Sea Research I, 45:15–38.

Moritz, R.E., C.M. Bitz and E.J. Steig, 2002. Dynamics of recent climate change in the Arctic. Science, 297:1497–1501.

Mysak, L.A., 2001. Patterns of Arctic circulation. Science, 293:1269–1270.

Mysak, L.A., D.K. Manak and R.F. Marsden, 1990. Sea ice anomalies observed in the Greenland and Labrador Seas during 1901–1984 and their relation to an interceded Arctic Climate cycle. Climate Dynamics, 5:111–132.

Nansen, F., 1906. Northern waters: Captain Roald Amundsen's oceanographic observations in the Arctic Seas 1901. Videnskabs-Selskabets Skrifter, I, Matematisk-Naturvidenskabelig Klasse, No. 3, 145pp.

Naurzbaev, M.M., E.A. Vaganov, O.V. Sidorova and F.H. Schweingruber, 2002. Summer temperatures in eastern Taimyr inferred from a 2427-year late-Holocene tree-ring chronology and earlier floating series. The Holocene, 12:727–736.

Nesje, A., Ø. Lie and S.O. Dahl, 2000. Is the North Atlantic Oscillation reflected in Scandinavian glacier mass balance records? Journal of Quaternary Science, 15:587–601.

New, M., M. Todd, M. Hulme and P. Jones, 2001. Precipitation measurements and trends in the twentieth century. International Journal of Climatology, 21:1899–1922.

Nikiforov, Ye.G. and A.O. Shpaikher, 1980. On the regular features of formation of large-scale oscillations of the hydrological regime of the Arctic Ocean. Gidrometeoizdat, St. Petersburg, 269pp. (In Russian)

NRC, 2002. Abrupt Climate Change: Inevitable Surprises. National Research Council, National Academy Press, Washington, D.C., 230 pp.

O'Brien, S., P.A. Mayewski, L.D. Meeker, D.A. Meese, M.S. Twickler and S.I. Whitlow, 1995. Complexity of Holocene climate as reconstructed from a Greenland ice core. Science, 270:1962–1964.

Ogilvie, A.E.J. and I. Jónsdóttir, 2000. Sea ice, climate, and Icelandic fisheries in the eighteenth and nineteenth centuries. Arctic, 53(4):383–394.

Osterkamp, T.E., 2003. A thermal history of permafrost in Alaska. In: M. Phillips, S. Springman and L.U. Arenson (eds.). Permafrost, pp. 863–868. Swets & Zeitlinger, Lisse.

Osterkamp, T.E. and V.E. Romanovsky, 1999. Evidence for warming and thawing of discontinuous permafrost in Alaska. Permafrost and Periglacial Processes, 10:17–37.

Overland, J.E. and A.T. Roach, 1987. Northward flow in the Bering and Chukchi Seas. Journal of Geophysical Research, 92:7097–7105.

Overpeck, J., K. Hughen, D. Hardy, R. Bradley, R. Case, M. Douglas, B. Finney, K. Gajewski, C. Jacoby, A. Jennings, S. Lamoureux, A. Lasca, G. MacDonald, J. Moore, M. Retelle, S. Smith, A. Wolfe and G. Zielinski, 1997. Arctic environmental change of the last four centuries. Science, 278:1251–1256.

Parkinson, C.L., 2000. Recent trend reversals in Arctic sea ice extents: Possible connections to the North Atlantic Oscillation. Polar Geography, 24:1–12.

Parkinson, C.L., D.J. Cavalieri, P. Gloersen, H.J. Zwally and J.C. Comiso, 1999. Arctic sea ice extents, areas, and trends, 1978–1996. Journal of Geophysical Research, 104(C9):20837–20856.

Peterson, T.C., 2003. Assessment of urban versus rural *in situ* surface temperatures in the contiguous United States: no difference found. Journal of Climate, 16:2941–2959.

Peterson, T.C. and R.S. Vose, 1997. An overview of the Global Historical Climatology Network temperature database. Bulletin of the American Meteorological Society, 78:2837–2849.

Peterson, T.C., K.P. Gallo, J. Livermore, T.W. Owen, A. Huang and D.A. McKittrick, 1999. Global rural temperature trends. Geophysical Research Letters, 26:329–332.

Pfister, C., 1999. Wetternachsage: 500 Jahre Klimavariationen und Naturkatastrophen 1496–1995. Paul Haupt, Bern, 304pp.

Polyakov, I.V. and M.A. Johnson, 2000. Arctic decadal and interdecadal variability. Geophysical Research Letters, 27:4097–4100.

Polyakov, I.V., G.V. Alekseev, R.V. Bekryaev, U. Bhatt, R.L. Colony, M.A. Johnson, V.P. Karklin, A.P. Makshtas, D. Walsh and A.V. Yulin, 2002. Observationally based assessment of polar amplification of global warming. Geophysical Research Letters, 29(18):1878, doi:10.1029/2001GL011111.

Polyakov, I.V., G.V. Alekseev, R.V. Bekryaev, U.S. Bhatt, R. Colony, M.A. Johnson, V.P. Karklin, D. Walsh and A.V. Yulin, 2003a. Long-term ice variability in Arctic marginal seas. Journal of Climate, 16(12):2078–2085.

Polyakov, I.V., R.V. Bekryaev, G.V. Alekseev, U.S. Bhatt, R.L. Colony, M.A. Johnson, A.P. Makshtas and D. Walsh, 2003b. Variability and trends of air temperature and pressure in the maritime Arctic, 1875–2000. Journal of Climate, 16:2067–2077.

Prowse, T.D. and S. Beltaos, 2002. Climatic control of river-ice hydrology: a review. Hydrological Processes, 16:805–822.

Quinton, W.L., D.M. Gray and P. Marsh, 2000. Subsurface drainage from hummock-cover hillslopes in the Arctic tundra. Journal of Hydrology, 237:113–125.

R-ArcticNET, 2003. A Regional, Electronic, Hydrographic Data Network for the Arctic Region. www.r-arcticnet.sr.unh.edu.

Ramsay, B.H., 1998. The interactive multi-sensor snow and ice mapping system. Hydrological Processes, 12:1537–1546.

Rasmussen, T.L., E. Thomsen, L.D. Labeyrie and T.C.E. van Weering, 1996. Circulation changes in the Faeroe-Shetland Channel correlating with cold events during the last glacial period (58–10 ka). Geology, 24:937–940.

Reimnitz, E., H. Eicken and T. Martin, 1995. Multiyear fast ice along the Taymyr Peninsula, Siberia. Arctic, 48(4):359–367.

Rind, D., R. Healy, C. Parkinson and D. Martinson, 1995. The role of sea ice in 2xCO$_2$ climate model sensitivity. Part I: The total influence of sea ice thickness and extent. Journal of Climate, 8:449–463.

Rodwell, M.J., D.P. Rowell and C.K. Folland, 1999. Oceanic forcing of the wintertime North Atlantic Oscillation and European climate. Nature, 398:320–323.

Romanovskii, N.N. and H.W. Hubberten, 2001. Results of permafrost modelling of the lowlands and shelf of the Laptev Sea Region, Russia. Permafrost and Periglacial Processes, 12:191–202.

Romanovsky, V.E. and T.E. Osterkamp, 2000. Effects of unfrozen water on heat and mass transport processes in the active layer and permafrost. Permafrost and Periglacial Processes, 11:219–239.

Romanovsky, V.E., D.O. Sergueev and T.E. Osterkamp, 2003. Temporal variations in the active layer and near-surface permafrost temperatures at the long-term observatories in Northern Alaska. In: M. Phillips, S. Springman and L.U. Arenson (eds.). Permafrost, pp. 989–994. Swets & Zeitlinger, Lisse.

Rosén, P., U. Segerstrom, L. Eriksson, I. Renberg and H.J.B. Birks, 2001. Holocene climate change reconstructed from diatoms, chironomids, pollen and near-infrared spectroscopy at an alpine lake (Sjuodjijaure) in northern Sweden. The Holocene, 11:551–562.

Rossow, W.B. and R.A. Schiffer, 1999. Advances in understanding clouds from ISCCP. Bulletin of the American Meteorological Society, 80:2261–2287.

Rothrock, D.A., Y. Yu and G.A. Maykut, 1999. Thinning of the Arctic sea-ice cover. Geophysical Research Letters, 26:3469–3472.

Rouse, W.R., 1998. A water balance model for a subarctic sedge fen and its application to climatic change. Climatic Change, 38:207–234.

Rouse, W.R., M.S.V. Douglas, R.E. Hecky, A.E. Hershey, G.W. Kling, L. Lesack, P. Marsh, M. McDonald, B.J. Nicholson, N.T. Roulet and J.P. Smol, 1997. Effects of climate change on the freshwaters of Arctic and subArctic North America. Hydrological Processes, 11:873–902.

Ruddiman, W.F. and J.E. Kutzbach, 1990. Late Cenozoic plateau uplift and climate change. Transactions of the Royal Society of Edinburgh: Earth Sciences, 81:301–314.

Ruddiman, W.F. and A. McIntyre, 1981. The North Atlantic during the last deglaciation. Palaeogeography, Palaeoclimatology, Palaeoecology, 35:145–214.

Ruddiman, W.F., M.E. Raymo, D.G. Martinson, B.M. Clement and J. Backman, 1989. Pleistocene evolution: Northern Hemisphere ice sheets and North Atlantic Ocean. Paleoceanography, 4:353–412.

Rudels, B., E.P. Jones, L.G. Anderson and G. Kattner, 1994. On the intermediate depth waters of the Arctic Ocean. In: O.M. Johannessen, R.D. Muench and J.E. Overland (eds.). The Polar Oceans and their Role in Shaping the Global Environment. Geophysical Monograph Series, 85:33–46. American Geophysical Union.

Rudels, B., L.G. Anderson and E.P. Jones, 1996. Formation and evolution of the surface mixed layer and halocline of the Arctic Ocean. Journal of Geophysical Research, 101(C4):8807–8821.

Rudels, B., C. Darnel, J. Gunn and E. Zakharchuck, 1997. CTD observations. In: E. Rachor (ed.). Scientific Cruise report of the Arctic Expedition ARK - XI/1 of RV 'Polarstern' in 1995. Berichte zur Polarforschung, 226:22–25.

Salvigsen, O., S.L. Forman and G.H. Miller, 1992. The occurrence of thermophilous mollusks on Svalbard during the Holocene and their paleoclimatic implications. Polar Research, 11:1–10.

Sarnthein, M., E. Jansen, M. Weinelt, M. Arnold, J.C. Duplessy, T. Johannessen, S. Jung, N. Koc, L. Labeyrie, M. Maslin, U. Pflaumann and H. Schulz, 1995. Variations in Atlantic surface ocean paleoceanography, 50°–80° N: a time-slice record of the last 30,000 years. Paleoceanography, 10:1063–1094.

Schauer, U., B. Rudels, E.P. Jones, L.G. Anderson, R.D. Muench, G. Bjork, J.H. Swift, V. Ivanov and A.-M. Larsson, 2002. Confluence and redistribution of Atlantic Water in the Nansen, Amundsen and Makarov basins. Annales Geophysicae, 20:257–273.

Seierstad, J., A. Nesje, S.O. Dahl and J.R. Simonsen, 2002. Holocene glacier fluctuations of Grovabreen and Holocene snow-avalanche activity
reconstructed from lake sediments in Groningstolsvatnet, western
Norway. The Holocene, 12:211–222.

Seppälä, M., 1997. Piping causing thermokarst in permafrost, Ungava
Peninsula, Quebec, Canada. Geomorphology, 20:313–319.

Serreze, M.C., F. Carse, R.G. Barry and J.C. Rogers, 1997. Icelandic Low
cyclone activity: climatological features, linkages with the NAO, and
relationships with recent changes in the northern hemisphere circulation. Journal of Climate, 10:453–464.

Serreze, M.C., J.E. Walsh, F.S. Chapin III, T. Osterkamp, M. Dyurgerov,
V. Romanovsky, W.C. Oechel, J. Morison, T. Zhang and R.G. Barry,
2000. Observational evidence of recent change in the northern high
latitude environment. Climatic Change, 46:159–207.

Severinghaus, J.P. and E. Brook, 1999. Abrupt climate change at the end
of the last glacial period inferred from trapped air in polar ice.
Science, 286:930–934.

Sexton, D.J., J.A. Dowdeswell, A. Solheim and A. Elverhøi, 1992. Seismic
architecture and sedimentation in north-west Spitsbergen fjords.
Marine Geology, 103:53–68.

Shiklomanov, I.A., A.I. Shiklomanov, R.B. Lammers, B.J. Peterson and
C.J. Vörösmarty, 2000. The dynamics of river water inflow to the
Arctic Ocean. In: E.L. Lewis, E.P. Jones, P. Lemke, T.D. Prowse and
P. Wadhams (eds.). The Freshwater Budget of the Arctic Ocean, pp.
281–296. Kluwer Academic Publishers.

Shindell, D.T., R.L. Miller, G.A. Schmidt and L. Pandolfo, 1999.
Simulation of recent northern winter climate trends by greenhouse-gas
forcing. Nature, 399:452–455.

Smith, I.R., 1999. Late Quaternary glacial history of Lake Hazen Basin
and eastern Hazen Plateau, northern Ellesmere Island, Nunavut,
Canada. Canadian Journal of Earth Sciences, 36:1547–1565.

Snowball, I., P. Sandgren and G. Petterson, 1999. The mineral magnetic
properties of an annually laminated Holocene lake-sediment sequence
in northern Sweden. The Holocene, 9:353–362.

Steele, M. and T. Boyd, 1998. Retreat of the cold halocline layer in the Arctic
Ocean. Journal of Geophysical Research, 103(C5):10419–10435.

Steffansson, U., 1969. Temperature variation in the North Icelandic coastal
area during recent decades. JoKull, 19:18–28.

Stewart, T.G., 1988. Deglacial-Marine Sediments from Clements
Markham Inlet, Ellesmere Island, N.W.T., Canada. Ph.D. Thesis.
University of Alberta, 229pp.

Stigebrandt, A., 1984. The North Pacific: a global-scale estuary. Journal of
Physical Oceanography, 14:464–470.

Stone, D.A., A.J. Weaver and F.W. Zwiers, 2000. Trends in Canadian
precipitation intensity. Atmosphere-Ocean, 38:321–347.

Stott, P.A., G.S. Jones and J.F.B. Mitchell, 2003. Do models underestimate the solar contribution to recent climate change? Journal of
Climate, 16:4079–4093.

Stötter, J., 1991. New observations on the postglacial history of
Tröllaskagi, northern Iceland. In: J.K. Maizels and C. Caseldine (eds.).
Environmental Change in Iceland: Past and Present, pp. 181–192,
Kluwer Academic Publishers.

Stötter, J., M. Wastl, C. Caseldine and T. Häberle, 1999. Holocene palaeoclimatic reconstruction in Northern Iceland: approaches and results.
Quaternary Science Reviews, 18:457–474.

Sun, B., P.Ya. Groisman and I.I. Mokhov, 2001. Recent changes in cloud-
type frequency and inferred increases in convection over the United
States and the former USSR. Journal of Climate, 14:1864–1880.

Swift, J.H. and K. Aagaard, 1981. Seasonal transitions and water mass
formation in the Iceland and Greenland seas. Deep-Sea Research A,
28:1107–1129.

Tabony, R.C., 1981. A principal component and spectral analysis of
European rainfall. Journal of Climatology, 1:283–294.

Thompson, D.W.J. and J.M. Wallace, 1998. The Arctic Oscillation signature in the wintertime geopotential height and temperature fields.
Geophysical Research Letters, 25:1297–1300.

Thompson, D.W.J., J.M. Wallace and G.C. Hegerl, 2000. Annular modes
in the extratropical circulation. Part II: Trends. Journal of Climate,
13:1018–1036.

Treshnikov, A.F., 1959. Surface waters in the Arctic Basin. Problemy
Arktiki, 7:5–14.

Tzedakis, C., 2003. Timing and duration of Last Interglacial conditions in
Europe: A chronicle of a changing chronology. Quaternary Science
Reviews, 22(8–9):763–768.

Vaganov, E.A., K.R. Briffa, M.M. Naurzbaev, F.H. Schweingruber,
S.G. Shiyatov and V.V. Shishov, 2000. Long-term climatic changes in
the arctic region of the Northern Hemisphere. Doklady Earth
Sciences, 375:1314–1317.

van Geel, B., J. Buurman and H.T. Waterbolk, 1996. Archaeological and
palaeoecological indications of an abrupt climate change in the
Netherlands and evidence for climatological teleconnections around
2650 BP. Journal of Quaternary Science, 11:451–460.

Vinje, T., 2001. Fram Strait ice fluxes and atmospheric circulation:
1950–2000. Journal of Climate, 14:3508–3517.

Vinje, T., N. Nordlund and Å. Kvambekk, 1998. Monitoring ice thickness
in Fram Strait. Journal of Geophysical Research, 103(C5):10437–
10450.

Visbeck, M.J., J. Fischer and F. Schott, 1995. Preconditioning the
Greenland Sea for deep convection. Ice formation and ice drift. Journal
of Geophysical Research, 100(C9):18489–18502.

Vörösmarty, C.J., L.D. Hinzman, B.J. Peterson, D.H. Bromwich, L.C.
Hamilton, J. Morison, V.E. Romanovsky, M. Sturm and R.S. Webb,
2001. The Hydrologic Cycle and Its Role in Arctic and Global
Environmental Change: A Rationale and Strategy for Synthesis Study.
Arctic Research Consortium of the U.S., Fairbanks, Alaska, 84pp.

Vorren, T.O. and J.S. Laberg, 1996. Late glacial air temperature, oceano-
graphic and ice sheet interactions in the southern Barents Sea region.
In: J.T. Andrews, W.E.N. Austin, H. Bergsten and A.E. Jennings (eds.).
Late Quaternary Palaeoceanography of the North Atlantic Margins.
Geological Society Special Publication, 111:303–321.

Vourlitis, G.L. and W.C. Oechel, 1996. The heat and water budgets in the
active layer of the arctic tundra at Barrow, Alaska. Journal of
Agricultural Meteorology, 52:293–300.

Wade, N.L., J. Martin and P.H. Whitfield, 2001. Hydrologic and climatic
zonation of Georgia Basin, British Columbia. Canadian Water
Resources Journal, 26:43–70.

Wadhams, P., 1997. Ice thickness in the Arctic Ocean: the statistical reliability of experimental data. Journal of Geophysical Research,
102(C13):27951–27959.

Wadhams, P., 2001. Sea ice thinning in the Nansen Basin and Greenland
Sea. Second Wadati Conference on Global Change and the Polar
Climate, March 7–9, 2001, Tsukuba, Japan. Extended Abstracts, pp.
17–21.

Walsh, J.E., W.L. Chapman and T.L. Shy, 1996. Recent decrease of sea
level pressure in the Central Arctic. Journal of Climate, 9:480–488.

Wanner, H., C. Pfister, R. Bràzdil, P. Frich, K. Fruydendahl, T. Jonsson,
J. Kington, H.H. Lamb, S. Rosenorn and E. Wishman, 1995.
Wintertime European circulation patterns during the Late Maunder
Minimum Cooling Period (1675–1704). Theoretical and Applied
Climatology, 51:167–175.

Webb, I., D.W.J. Thompson and J.E. Kutzbach, 1998. An introduction to
Late Quaternary climates: data syntheses and model experiments.
Quaternary Science Reviews, 17:465–471.

Whitfield, P.H. and A.J. Cannon, 2000. Recent climate moderated shifts in
Yukon territory. In: D.L. Kane (ed.). Proceedings of the American
Water Resources Association Spring Specialty Conference, Water
Resources in Extreme Environments, Anchorage, Alaska, May 1–3,
2000, pp. 257–262.

Wielicki, B.A., R.D. Cess, M.D. King, D.A. Randall and E.F. Harrison,
1995. Mission to planet Earth: role of clouds and radiation in climate.
Bulletin of the American Meteorological Society, 76:2125–2153.

Wijffels, S.E., R.W. Schmitt, H.L. Bryden and A. Stigebrandt, 1992.
Transport of freshwater by the oceans. Journal of Physical
Oceanography, 22:155–162.

Williams, K.M., S.K. Short, J.T. Andrews, A.E. Jennings, W.N. Mode and
J.P.M. Syvitski, 1995. The Eastern Canadian Arctic at ca. 6 ka BP: a
time of transition. Géographie physique et Quaternaire, 49:13–27.

Winsor, P., 2001. Arctic sea ice thickness remained constant during the
1990s. Geophysical Research Letters, 28(6):1039–1041.

Winsor, P. and G. Björk, 2000. Polynya activity in the Arctic Ocean from
1958–1997. Journal of Geophysical Research, 105(C4):8789–8803.

Woo, M.-K. and A. Ohmura, 1997. The Arctic Islands. In: W.G. Bailey,
T.R. Oke and W.R. Rouse (eds.). The Surface Climates of Canada.
Canadian Association of Geographers Series in Canadian Geography
4:172–197.

Xu, C.-Y. and S. Halldin, 1997. The effect of climate change on river flow
and snow cover in the NOPEX area simulated by a simple water balance model. Nordic Hydrology, 28:273–282.

Young, M.A. and R.S. Bradley, 1984. Insolation gradients and the paleocli-
matic record. In: A.L. Berger, J. Imbrie, J. Hays, G. Kukla and B.
Saltzman (eds.). Milankovitch and Climate: Understanding the
Response to Astronomical Forcing, Part 2, pp. 707–713. D. Reidel.

Zakharov, V.F., 2003. Sea ice extent changes during XX century.
Meteorology and Hydrology, 5:75–86 (in Russian).

Zhang, T., J.A. Heginbottom, R.G. Barry and J. Brown, 2000. Further
statistics on the distribution of permafrost and ground ice in the
Northern Hemisphere. Polar Geography, 24(2):126–131.

Zhang, X., K.D. Harvey, W.D. Hogg and T.R. Yuzyk, 2001. Trends in
Canadian streamflow. Water Resources Research, 37:987–998.

Zwiers, F.W. and X. Zhang, 2003. Towards regional climate change
detection. Journal of Climate, 16:793–797.

Chapter 3

The Changing Arctic: Indigenous Perspectives

Lead Authors
Henry Huntington, Shari Fox

Contributing Authors
Fikret Berkes, Igor Krupnik

Case Study Authors
Kotzebue: Alex Whiting
The Aleutian and Pribilof Islands Region, Alaska: Michael Zacharof, Greg McGlashan, Michael Brubaker, Victoria Gofman
The Yukon Territory: Cindy Dickson
Denendeh: Chris Paci, Shirley Tsetta, Chief Sam Gargan, Chief Roy Fabian, Chief Jerry Paulette, Vice-Chief Michael Cazon, Sub-Chief Diane Giroux,
    Pete King, Maurice Boucher, Louie Able, Jean Norin, Agatha Laboucan, Philip Cheezie, Joseph Poitras, Flora Abraham, Bella T'selie, Jim Pierrot,
    Paul Cotchilly, George Lafferty, James Rabesca, Eddie Camille, John Edwards, John Carmichael, Woody Elias, Alison de Palham, Laura Pitkanen,
    Leo Norwegian
Nunavut: Shari Fox
Qaanaaq, Greenland: Uusaqqak Qujaukitsoq, Nuka Møller
Sapmi: Tero Mustonen, Mika Nieminen, Hanna Eklund
Climate Change and the Saami: Elina Helander
Kola: Tero Mustonen, Sergey Zavalko, Jyrki Terva, Alexey Cherenkov

Consulting Authors
Anne Henshaw, Terry Fenge, Scot Nickels, Simon Wilson

Contents

We cannot change nature, our past, and other people for that matter, but we can control our own thoughts and actions and participate in global efforts to cope with these global climate changes. That I think is the most empowering thing we can do as individuals. George Noongwook, St. Lawrence Island Yupik, Savoonga, Alaska, as quoted in Noongwook, 2000

Summary

Indigenous peoples in the Arctic have for millennia depended on and adapted to their environment. Their knowledge of their surroundings is a vital resource for their well-being. Their knowledge is also a rich source of information for others wishing to understand the arctic system. Many of the aspects of climate change and its impacts considered in this chapter are also considered in other chapters. To avoid excessive disruption to the flow of the text, cross-references are included only where other chapters contain extended discussion or additional material. Within the context of climate change, indigenous observations and perspectives offer great insights not only in terms of the nature and extent of environmental change, but also in terms of the significance of such change for those peoples whose cultures are built on an intimate connection with the arctic landscape.

This chapter reviews the concept of indigenous knowledge, summarizes those indigenous observations of environmental and climatic change that have been documented to date, and presents a series of case studies, largely from hunting and herding societies, examining the perspectives of specific communities or peoples. Although idiosyncratic, the case studies each attempt to convey the sense of how climate change is seen, not in the form of aggregate statistics or general trends, but in specific terms for particular individuals and communities. The case studies provide the basis for a discussion of resilience, or protecting options to increase the capacity of arctic societies to deal with future change, and a review of further research needs.

The observations and case studies contain some common themes. One such observation is that the weather has become more variable and thus less predictable by traditional means. Social changes, such as less time spent on the land, may influence this observation, but there are climatological implications that merit further study. In terms of perceptions of the significance of climate change, there are few, if any, areas where climate change is regarded as the most pressing problem being faced. Nonetheless, most arctic residents are aware of climate change, have experience of the types of changes being seen and anticipated, and are concerned about the implications for themselves, their communities, and the future.

Several of the general conclusions drawn in this chapter are likely to be applicable to all communities affected by climate change, whether the impacts are on balance beneficial or harmful. Climate change is not an isolated phenomenon, but one that is connected to the web of activities and life surrounding indigenous peoples. Thus, it must be understood and assessed in terms of its interactions with other phenomena and with current and future societal and environmental changes. Responses to climate change will not be effective unless they reflect the particular circumstances of each place. Increasing resilience is a useful way to consider the merits of various response options, which are best developed and evaluated iteratively to promote adjustment and improvement as experience and knowledge increase. Indigenous perspectives on climate change offer an important starting point for collaborative development of effective responses.

3.1. Introduction

The indigenous peoples of the Arctic have adapted to great environmental variability, cold, extended winter darkness, and fluctuations in animal populations, among many other challenges posed by geography and climate. Although the arctic climate has always undergone change, current and projected changes make it timely and important to reflect on the ways that such changes affect arctic residents, particularly the indigenous residents whose way of life is so closely linked to their surroundings. It is also important to consider how these indigenous residents observe and feel about the changes that are occurring. Together, such perspectives can help the global community understand what is at stake in a changing Arctic.

Much of the Arctic has been inhabited since at least the end of the last ice age, and some areas for far longer (Pitulko et al., 2004). During this time, human groups have come and gone, and evolved and adapted, their patterns of settlement changing, often abruptly, in response not only to climate but also to regional patterns such as resource availability, relations with neighbors, landscape change, hunting and fishing technology, and the rise of reindeer husbandry (Krupnik, 1993). In recent centuries and in particular the twentieth century, forces from outside the region have shaped human patterns in the Arctic, as the modern world has extended its reach and influence. Today, the Arctic is home to a large number of indigenous peoples with distinct cultures, languages, traditions, and ways of interacting with their environment (Freeman, 2000; Nuttall, 1998). They have in common a close connection to their surroundings, an intimate understanding of their environment (e.g., Fienup-Riordan et al., 2000), complex relationships with national and sub-national governments and non-indigenous migrants to the Arctic (Minority Rights Group, 1994; Nuttall, 1992; Pika, 1999), a way of life that mixes modern and traditional activities, and a major stake in the future of the region (CAFF, 2001; Huntington et al., 1998; Nuttall and Callaghan 2000; Slezkine, 1994). An overview of humans in the Arctic is given in Chapter 1.

This chapter attempts to show some of the observations of change that indigenous inhabitants of the Arctic consider to be related to climate change. In doing so, the comments and perspective also show what climate change means to them and their communities within the context of the other forces affecting their lives and cultures. Although little material is available concerning indigenous perspectives on ultraviolet (UV) radiation and ozone depletion, the chapter includes a short summary of some related observations (Box 3.1). Other chapters describe impacts on specific components of the environment and areas of human activity and so draw extensively on indigenous knowledge and perspectives, a level of inclusion that is unprecedented in an assessment of this type and scope.

This chapter addresses the impacts of climate change and variability on those affected most directly: the people whose ways of life are based on their use of the land and waters of the Arctic. This has been achieved using a series of case studies drawn from existing research projects that have been selected to give, through specific examples rather than general summaries, a sense of the variety of indigenous perspectives on climate change in the Arctic. The case studies are idiosyncratic, reflecting differences in the communities they describe as well as differences in the aims and methods of the studies from which they derive. Because they are examples, the case studies cannot reflect all the views held within arctic communities. Some communities, such as those in Greenland that fish

for cod, may see benefits from climate change if fish stocks increase, a perspective that may be missing from case studies focusing more on the negative impacts of climate change. Nonetheless, the case studies are intended to give a human face to some of the impacts of weather and climate change observed by arctic residents.

Although people plan around expectations that reflect the climate of their area, their daily activities are affected more by the day's weather. Many of the statements and perspectives contained in this chapter reflect perceptions of weather and changes in weather patterns and variability, which are also of interest to climatologists examining the ways that climate change is manifested in the Arctic (Overland et al., 2002; J. Walsh, International Arctic Research Center, University of Alaska, Fairbanks, pers. comm., 2003). The distinction between weather change and climate change is not simple, and observations about weather may indicate something significant about the arctic climate. It is also likely that the publicity surrounding climate change has led many people in the Arctic as elsewhere to interpret observations in the light of climate, whether or not this is appropriate. This chapter presents the connections indigenous peoples draw between their own observations and the general phenomenon of climate change.

Box 3.1. Indigenous observations concerning the sun and ultraviolet radiation

Many people in the Arctic have observed changes in the characteristics of the sun and its effects since the early 1990s (Fox, 1998; McDonald et al., 1997). While not discussed in terms of UV radiation, many indigenous observations do include the same concerns as UV scientists (see Chapters 5 and 15). Most commonly expressed is the perception that the sun is stronger or more "stinging" and "sharp" feeling (e.g., Fox, 1998). The sun's heat seems to have become more intense and northern residents report unusual sunburns, eye irritations, and skin rashes (Kassi, 1993; Fox, 1998).

The direct heat from the sun is warmer, it is not the same anymore and you can't help but notice that. It is probably not warmer overall, but the heat of the sun is stronger. G. Kappianaq, Igloolik, 1997 as quoted in Fox, 1998

The reason why I mention the fact that the sun seems warmer is because another [piece of] evidence to that is that we get some skin diseases or some skin problems. Because I think in the past when Peter [a Clyde River elder] was a young boy they never seemed to have these skin problems and I see them more and more these days. J. Qillaq, Clyde River, 2001 as quoted in Fox, 2004

Humans are not the only ones affected by a more intense sun. Inuit in Nunavut link other environmental changes to the sun. In some areas, for example, although the overall temperature may not be warmer, elders claim that the heat of the sun is causing small ponds to be warmer than usual or to dry up altogether. In some places, meat hung out to dry seems to get burned by the sun, and caribou skins seem to rip more easily around the neck area, a new condition elders link to skins possibly being burnt or becoming too hot from the sun (Fox, 2004).

Archaeological sites in the Arctic have contained sun goggles, indicating that indigenous peoples have, for a long time, made an effort to shield their eyes from the blinding light of sunshine on snow. These days, indigenous peoples are doing more to protect themselves from sun damage. High quality sunglasses and goggles are popular and many people who spend time on the land are now using sun lotion and lip balm. In Igloolik, Nunavut, for example, the nursing station has had more requests for sun lotion in recent years, but it is unclear whether this is due to more sunburn, or a greater awareness of the damage caused by sun exposure (Fox, 1998). Still, elders and older hunters who have grown up on the land and spent decades on the sea ice and snow say they are only now beginning to experience sunburn. While rates of skin cancer remain low in the Arctic (see section 15.3.3.2), community members note it will be important to monitor how serious the new sun-related skin ailments become. Residents also want to monitor how a more intense sun may affect arctic animals and plants over time.

In describing the significance of climate change for indigenous peoples, it is important to remember that there are many forms of environmental change in the Arctic, as well as extensive social changes related to modernization and globalization (e.g., AMAP, 1998, 2002; CAFF, 2001; Freeman, 2000; Gaski, 1997; Nuttall, 2000). The challenges these pose often require great attention and effort by indigenous peoples and organizations. From negotiating the creation of Nunavut in Canada to responding to threats from oil and gas development in northern Russia, arctic indigenous peoples have had to organize themselves to articulate and fight for their values and ways of life. In some cases, they have been successful in promoting global action. The Stockholm Convention on Persistent Organic Pollutants was adopted in 2001, in no small part resulting from concerns about contaminants in the Arctic and their impacts on indigenous peoples and cultures (Downie and Fenge, 2003). More recently, Inuit leaders have framed climate change as a human rights issue (Sheila Watt-Cloutier as quoted in Brown P., 2003). Climate change is a topic about which indigenous peoples have a great deal to share with the world.

3.2. Indigenous knowledge

Indigenous peoples have long depended on their knowledge and skills for survival, including their ability to function in small, independent groups by dividing labor and maintaining strong social support and mutual ties both within and between their immediate communities (e.g., Burch, 1998; Krupnik, 1993; Freeman, 2000; Usher et al., 2003). Knowledge about the environment is equally important. Understanding the patterns of animal behavior and aggregation is necessary for acquiring food. Successful traveling and living in a cold-dominated landscape requires the ability to read subtle signs in the ice, snow, and weather. Gradual shifts in social patterns and environmental conditions make this a continuous process of learning and adapting. In the past, sudden shifts in physical conditions, such as abrupt warming or cooling, led to radical changes including the abandonment of large areas for extended periods that is apparent from the archeological record (Fitzhugh, 1984; McGhee, 1996). Knowing one's surroundings was an often-tested requirement, one that remains true today for those who travel on and live off the land and sea (Berkes, 1999; Berkes et al., 2000; Fox, 1998; Huntington et al., 1999; Inglis, 1993; Krupnik and Jolly, 2002).

3.2.1. Academic engagement with indigenous knowledge

Those outside indigenous communities have not always recognized or respected the value of this knowledge. Occasionally used and less frequently credited prior to and during most of the twentieth century, indigenous knowledge from the Arctic has received increasing attention over the past couple of decades (e.g., Freeman, 1976; Inglis, 1993; Nadasdy, 1999; Stevenson, 1996). This interest, arising from research in the ethnosciences,

has taken the form of studies to document indigenous knowledge about various aspects of the environment (Ferguson and Messier, 1997; Fox, 2002; Huntington et al., 1999; Kilabuck, 1998; McDonald et al., 1997; Mymrin, et al., 1999; Riedlinger and Berkes, 2001), the increasing use of cooperative approaches to wildlife and environmental management (Berkes, 1998, 1999; Freeman and Carbyn, 1988; Huntington 1992a,b; Pinkerton, 1989; Usher, 2000), and a greater emphasis on collaborative research between scientists and indigenous people (Huntington, 2000a; Krupnik and Jolly, 2002). This section describes some of the characteristics of indigenous knowledge and its relevance for studies of climate change and its implications.

The topic of indigenous knowledge is not without disputes and controversy. In fact, agreement has not even been found on the appropriate term – "traditional knowledge", "traditional ecological knowledge", "traditional knowledge and wisdom", "local and traditional knowledge", "indigenous knowledge", and various combinations of these words and their acronyms are among those that have been used (e.g., Huntington, 1998; Kawagley, 1995; Turner et al., 2000). Terms specific to particular peoples are also common, such as "Saami knowledge" or "Inuit Qaujimajatuqangit". Although their definitions largely overlap, each raises difficulties. The term "indigenous" in this context excludes long-term arctic residents not of indigenous descent, implies that all indigenous persons hold this knowledge, and emphasizes ancestry over experience. "Traditional" has a connotation of being static and from past times, whereas this knowledge is current and dynamic. "Local" fails to capture the sense of continuity and the practice of building on what was learned by previous generations. "Knowledge" by itself omits the insights learned from experience and application, which are better captured by "wisdom". All of these terms neglect the spiritual dimensions of knowledge and connection with the environment that are often of greatest importance to those who hold this knowledge. Some groups, such as the World Intellectual Property Organization, identify "indigenous knowledge" as a subset of "traditional knowledge", with the latter incorporating folklore (WIPO, 2001). The issue of terminology will not be resolved here, but the term "indigenous knowledge" is used in a broad sense, encompassing the various systems of knowledge, practice, and belief gained through experience and culturally transmitted among members and generations of a community (Berkes, 1999; Huntington, 1998).

By any term, indigenous knowledge plays a vital role in arctic communities, and its perpetuation is important to the future of such communities. It has also become a popular research topic. Scholars within and outside the indigenous community discuss its nature, the appropriate ways in which it should be studied and used, how it can be understood, and how it relates to other ways of knowing such as the scientific. Many agree that indigenous knowledge offers great insight from people who live close to and depend greatly on the local environment and its ecology (Berkes, 1998, 1999; Freeman and Carbyn, 1988; Huntington, 2000a; Inglis, 1993; Mailhot, 1993). Most of these scholars also recognize, however, that gaining access to and using this knowledge must be done with respect for community rights and interests, and with awareness of the cultural contexts within which the knowledge is gathered, held, and communicated (e.g., Krupnik and Jolly, 2002; Wenzel, 1999). Successful efforts are typically built on trust and mutual understanding. It takes time for knowledge holders to feel comfortable sharing what they know, for researchers to be able to understand and interpret what they see and hear, and for both groups to understand how indigenous knowledge is represented and for what purpose.

The legal and political contexts of indigenous knowledge must also be taken into account. The intellectual property aspects of indigenous knowledge are being explored (WIPO, 2001). Some jurisdictions in the Arctic require that it be considered in processes such as resource management and environmental impact assessment (e.g., Smith D., 2001). Throughout the Arctic, there is increasing political pressure to use indigenous knowledge, but often without clear guidance on exactly how this should be achieved. Most existing ethical guidelines or checklists for community involvement in research identify the areas to be addressed in research agreements, but do not resolve how the controversial questions are best answered (e.g., Grenier, 1998; IARPC, 1992). Such uncertainty may lead to reluctance on the part of some researchers to engage in studies of indigenous knowledge, but at present there are many good examples of collaborative projects that have benefited both the communities involved and those conducting the research (e.g., Huntington et al., 1999, 2002; Kilabuck 1998; McDonald et al., 1997).

3.2.2. The development and nature of indigenous knowledge

Careful observation of the world combined with interpretation in various forms is the foundation for indigenous knowledge (Cruikshank, 2001; Huntington, 2000a; Johnson, 1992; Kilabuck, 1998; Krupnik and Jolly, 2002). The ability to thrive in the Arctic depends in large part on the ability to anticipate and respond to dangers, risks, opportunities, and change. Knowing where caribou are likely to be is as important as knowing how to stalk them. Sensing when sea ice is safe enough for travel is an essential part of bringing home a seal. The accuracy and reliability of this knowledge has been repeatedly subjected to the harshest test as people's lives have depended on decisions made on the basis of their understanding of the environment. Mistakes can lead to death, even for those with great experience. Thus, information of particular relevance to survival has been valued and refined through countless generations, as individuals combine the lessons of their elders with personal experience (e.g., Ingold and Kurtilla, 2000).

Indigenous knowledge is far more than a collection of facts. It is an understanding of the world and of the human place in the world (Agrawal, 1995; Berkes, 1999; Berkes et al., 2000; Fehr and Hurst, 1996; Kawagley, 1995). From observations, people everywhere find patterns and similarities and associations, from which they develop a view of how the world works, a view that explains the mysteries surrounding them, that gives them a sense of place (Berkes, 1999; Brody, 2000; Nelson, 1983). In the Arctic, parallels may be drawn, for example, in the migrations of caribou, cranes, and whales (Huntington et al., 1999). Systems of resource use are developed to make efficient use of available resources (Berkes, 1998, 1999; Berkes et al., 2000). Hunters develop rituals and practices that reflect their view of the world (Cruikshank, 1998; Fienup-Riordan, 1994). Stories, dances, songs, and artwork express this worldview (Cruikshank, 1998). In turn, culture shapes perception, and the world is interpreted according to the way it is understood. When personal memories and stories are retold to family members, relatives, neighbors, and others, as is common practice across the Arctic, an extensive local record is built. Non-verbal transmission of knowledge and skills, for example through observation and imitation, is also common. It often extends over several generations and represents the accumulated knowledge of many highly experienced and respected persons. Learning the knowledge of one's people involves absorbing the stories and lessons, then watching closely to figure out exactly what is meant and how to use it, and adapting it to one's own needs and experiences. In these ways, indigenous knowledge is continually evolving (Ingold and Kurtilla, 2000).

3.2.3. The use and application of indigenous knowledge

Studies of indigenous knowledge often make comparisons with scientific knowledge in an effort to determine the "accuracy" of indigenous knowledge as measured on a scale that is intended to be objective. Other studies use indigenous knowledge in the generation of new hypotheses or for the identification of geographic locations for research (Albert, 1988; Huntington, 2000a; Johannes, 1993; Nadasdy, 1999; Riedlinger and Berkes, 2001). While this can be worthwhile, the value of indigenous knowledge lies primarily within the group and culture in which it developed. Holders of this knowledge use it when making decisions or in setting priorities, and an understanding of the nature of this knowledge can help explain the rationale behind these processes (Cruikshank, 1981, 1998; Feldman and Norton, 1995; Kublu et al., 1999). "Accuracy" in this context depends on the uses to which the knowledge is put, not on an external evaluation.

The emphasis on the cultural aspects of indigenous knowledge in this assessment is not intended to detract from the great utility it has in ecological and environmental research and management (Berkes, 1998, 1999; Fox, 2004; Freeman and Carbyn, 1988; Krupnik and Jolly, 2002; Riedlinger and Berkes, 2001). In this setting, accuracy as evaluated externally may be a key concern

Box 3.2. Place names as indicators of environmental change

Indigenous peoples use a variety of cultural mechanisms to pass on climatic and environmental knowledge and its "attendant adaptive behavior" from one generation to the next (Gunn, 1994; Henshaw, 2003). These mechanisms include place names, which reflect perceptions of the environment and can serve as a repository for accumulated knowledge. When conditions change, place names can serve as indicators of environmental change. Place names show how perceptions of physical geography, ecology, and climate transform observations and experiences to memory shared among members of a particular group (Cruikshank, 1990; Müller-Wille, 1983, 1985; Peplinski, 2000; Rankama, 1993).

As environmental conditions change, place names (or toponyms) may change or persist, providing insight into the nature of those changes and the adaptations that accompany them. For example, near Iqaluit, Nunavut, there is a site called Pissiulaaqsit, which translates as "a place where there is an absence of guillemots" (Peplinski, 2000). Local residents explain that the name is significant because guillemots nested there in the past. In the Sikusilarmiut land-use area, covering most of the Foxe Peninsula in Nunavut, Henshaw (2003) has documented more than 300 toponyms around the community of Kinngait (Cape Dorset). The extensive naming of places, often using descriptive terms, creates an important frame of reference for navigation, with crucial implications for safety, travel, and hunting. Many of the place names refer to features or phenomena that may be highly sensitive to environmental change.

For example, Ullivinirkallak is a place that used to be used for storing walrus (_Odobenus rosmarus_) meat. That it is no longer used for that purpose may indicate change to permafrost. Qimirjuaq is a large plateau with ice and snow even in summer. The area watered by melting snow produces abundant berries, and so the size and condition of the snowfield is monitored closely. The berry pickers quickly note changes in the persistence and characteristics of the snowfield. Seasonal features such as polynyas or migratory routes are also named, as are patterns of currents and sea-ice movements. Documenting these names and the conditions that occur at these locations can provide a means of monitoring and identifying future environmental change.

---

because the information is being applied for a purpose that may be very different from that for which it was originally generated. There are many instances where indigenous knowledge of the habits of an animal such as the bowhead whale (_Balaena mysticetus_) (Albert, 1988) or the interactions within an ecosystem such as sea-ice phenomena (Norton, 2002) were – and are – far in advance of scientific understanding, and in fact were used by scientists to make significant progress in ecology and biology (Freeman, 1992; Krupnik and Jolly, 2002). This is especially true in the Arctic, where scientific inquiry is a relatively recent phenomenon, and where researchers often depend on the knowledge and skills of their indigenous guides.

To apply indigenous knowledge to environmental research and management, consideration must be given to the ways in which it is acquired, held, and communicated. Indigenous knowledge is the synthesis of innumerable observations made over time (Agrawal, 1995; Huntington, 1998). Added weight is often given to anomalous occurrences, in order to be better prepared for surprises and extremes. It is typically qualitative; when quantities are noted, they are more often relative than absolute. Indigenous knowledge evolves with changing social, technological, and environmental conditions (Krupnik and Vakhtin, 1997), and thus observations of change over time can be influenced by these as well as by the vagaries of memory. Indeed, one of the main challenges in evaluating observations of environmental change is that of addressing the many factors that influence the ways in which people remember and describe events. In addition, some communities today are experiencing erosion of indigenous knowledge and the esteem in which it is held, which has emotional and practical impacts on individuals and communities (Fox, 2002).

Indigenous knowledge has been documented on various topics in various places in the Arctic, largely in North America. These efforts have rarely focused on climate change or even included climate change as an explicit topic of discussion. Nonetheless, substantial information is available, including evidence from place names (Box 3.2) and the archaeological record (Box 3.3). Further documentation is highly desirable, both for increasing the understanding of climate dynamics and as a means of engaging arctic residents in the search for appropriate responses to the impacts of climate change.

3.3. Indigenous observations of climate change

Indigenous peoples have only recently been engaged in climate change research and only through a relatively small number of projects. However, these projects have amassed a large collection of indigenous knowledge and observations about climate and environmental change, reflecting the depth of knowledge held by these peoples. Figures 3.1 and 3.2 present examples of observations documented in these projects, and highlight five major topic areas: changes in weather, seasons, wind, and sea ice (Fig. 3.1), and changes in animals and insects

(Fig. 3.2). This information is organized by community and region across the Arctic, but is derived from projects conducted in different ways, with different objectives, and at different times. This compilation provides a useful introduction to changes experienced by indigenous peoples, but should not be used for detailed comparisons across regions without referring to the original reports. Also, some of the changes were not necessarily considered by the observers to be climate-driven, and this is particularly true for information in Fig. 3.2, while some do have connections to climate. The original reports should be used for clarification.

Many of the topics addressed by indigenous observations in Figs. 3.1 and 3.2 are discussed in other chapters of this assessment. There are many links between indigenous and scientific observations of arctic climate and environmental change and many opportunities for complementary perspectives on the nature of various phenomena and their impacts. For example, Chapter 7 reports that biologists connect a changing climate to changing animal migration patterns, such as caribou (see section 7.3.5). Indigenous knowledge is cited as

helping to explain how caribou migrations may be triggered by seasonal cues such as day length, air temperature, or ice thickness (Thorpe et al., 2001). Also, scientific descriptions of changes in the arctic climate (such as those reported in Chapter 2) are often consistent with indigenous observations. For example, observational data from the scientific record indicate that the Arctic is warming in western Canada, Alaska, and across Eurasia, but experiencing no change or cooling in eastern Canada, Greenland, and the northwestern Atlantic (see section 2.6.2.1). This is supported by indigenous observations by comparing those from communities in Alaska with those from Igloolik and Iqaluit in Nunavut, Canada.

Indigenous and scientific observations do not always agree, however. For example, in the Kitikmeot region of Nunavut, Inuit have observed more abundant and new types of shrubs and lichens (Thorpe et al., 2001). While the increased abundance of shrubs corresponds with aerial photography of vegetation change, experimental evidence suggests that lichens should decrease under the changing environmental conditions seen in the Kitikmeot (see section 7.3.3.1). There are probably

Box 3.3. Archaeology and past changes in the arctic climate

The documentation of indigenous observations of climate change has focused primarily on recent decades. But the arctic environment has long been recognized for its extreme variability and rapid fluctuations. Several past examples of both extreme warming and cooling events are documented in Chapter 2. In addition to the work by climatologists and physical scientists, social scientists, archaeologists, and ethnohistorians have accumulated a large body of evidence concerning past changes in the environment. They often use proxy data, such as rapid shifts in human subsistence practice, change in settlement areas, substantial population moves, or certain migration patterns, as indicators of rapid transitions in arctic ecosystems.

Since the 1970s, archaeologists have developed detailed scenarios of how past climate changes have affected human life, local economies, and population distribution within the Arctic. One of the clearest examples of such links was the expansion of indigenous bowhead whaling and the rapid spread of the whaling-based coastal Eskimo cultures from northern Alaska across the central Canadian Arctic to Labrador, Baffin Island, and eventually to Greenland around 1000 years ago. Based upon recent radiocarbon dating and paleoenvironmental data, this enormous shift in population and economy took place within less than 200 years, caused at least in part by the rapidly changing sea-ice and weather conditions in the western and central Arctic (Bockstoce, 1976; Maxwell, 1985; McCartney and Savelle, 1985; McGhee, 1969/70, 1984; Stoker and Krupnik, 1993; Whitbridge, 1999). When around 300 to 400 years later the arctic climate shifted to the next cooling phase, Inuit were forced to abandon whaling over most of the central Canadian Arctic. This extreme cooling trend around 400 to 500 years ago left many Inuit communities isolated and under heavy environmental stresses that triggered population declines and loss of certain subsistence skills and related knowledge.

One well-known example of these impacts illustrates that not all responses are effective, and that people may not be able to adapt to all types of change. The Polar Eskimo (Inughuit) of northwest Greenland lost the use of their skin hunting boats (kayaks), bows and arrows, and fish spears when they became isolated from other communities by expanded glaciers and heavier sea ice during the Little Ice Age. In consequence, open water hunting for seals and walruses declined, and hunting for caribou and ptarmigan (*Lagopus* spp.) was completely abandoned, to the extent that their meat was considered unfit for human consumption (Gilberg 1974-75, 1984; Mary-Rousselière, 1991). As game animals were labeled "unclean" or "unreachable", the whole body of related expertise about animal habits, observation practices, the pursuit and capture of animals, and the butchering and storing of meat was reduced dramatically or completely lost. Some shifts of this kind, as well as stories about hardship caused by environmental change, have been preserved in indigenous oral traditions, folklore, and myths (Cruikshank, 2001; Gubser, 1965; Krupnik, 1993; Minc, 1986). However, few systematic attempts have been made so far to use indigenous knowledge to track historical or pre-historical cases of arctic climate change.

Elim, Alaska, USA
Weather: Heavy storms washing up timber onto shorelines; not seen before. More warm days, sometimes several in a row (Charles Saccheus as quoted in Krupnik, 2000).
Sea Ice: Ice no longer stable in spring. Fast ice melting faster (Charles Saccheus as quoted in Krupnik, 2000).

Yukon Territory, Canada
Weather: Year of "no real summer" sometime in the middle of the 19th century recorded in oral tradition (Cruikshank, 1981). Summers getting hotter, winters getting warmer (Kassi, 1993).

Barrow, Alaska, USA
Sea Ice: Differences in quality of sea ice; less salty, easier to chop, breaks up sooner. Fast ice retreats early; breaks up and retreats 20 to 30 miles and does not come back. No longer ice coming in during autumn, now water freezes in place and ice floes no longer seen drifting to shore. Multi-year ice does not arrive until later (Charles Brower as quoted in Krupnik, 2000).

Arviat, Nunavut, Canada
Weather: Winters warmer; in 1940s and 1950s frostbite only took seconds; it is not that cold anymore (GN, 2001).
Sea Ice: Sea ice forms later and overall thickness is reduced. People are less confident in winter ice travel and sea ice breaks up earlier and more quickly, for example, in June the last few years (GN, 2001).

Northwestern Hudson Bay, Canada
Weather: Weather highly variably since the 1940s; by 1990s weather changes are quick, unexpected, and difficult to predict. Used to be more clear calm days, winters were colder, and low temperatures persisted longer. Cooler summers in the early 1990s (McDonald et al., 1997).

Western Hudson Bay, Canada
Weather: Longer winters and colder springs (McDonald et al., 1997).

Western James Bay, Canada
Weather: Winters shorter and warmer (McDonald et al., 1997).
Wind: Winds shift several times per day (McDonald et al., 1997).

Eastern James Bay, Canada
Weather: In early 1990s autumn weather changed quickly. Cold weather arrives earlier but lakes freeze later (McDonald et al., 1997). Shorter spring and autumn seasons and colder winters in reservoir areas (McDonald et al., 1997).
Sea Ice: Salinity changing along the northeast coast with more freshwater ice forming in the bay. Ice less solid in La Grande River area. Sea ice freezes later and breaks up earlier (McDonald et al., 1997).

Eastern Hudson Bay, Canada
Weather: Persistence of cold weather into spring with spring and summer cooling trend (McDonald et al., 1997).
Wind: Since 1984, April and May winds in the Belcher Islands have blown mostly from the north, reducing the size of Canada geese flocks, slowing the spring melts, and contributing to the spring and summer cooling trend in eastern Hudson Bay (McDonald et al., 1997).
Sea Ice: Between the 1920s and 1970s the "ice bridge" between the Belcher Islands and eastern Hudson Bay mainland occasionally froze by late February or March. In the 1970s it began to freeze earlier and by late 1980s started freezing as soon as the early freezing season began. During the 1950s, 35 polynyas were open all winter in the Belcher Islands archipelago; in the 1960s to 1970s, 13, and in the early 1990s, only three (McDonald et al., 1997). Sea ice freezes faster and solid ice cover is larger and thicker with fewer polynyas. The flow edge melts before breaking up (McDonald et al., 1997).

Hudson Strait, Canada
Weather: Cooling trend observed – spring and early summer used to be warm, cold weather now returns following a March or April warm spell and persists into May, June, and July (McDonald et al., 1997).
Sea Ice: Timing of autumn freeze-up unchanged but sea ice freezes faster than in the past and quality deteriorated in some areas since more slush ice develops in early freezing (McDonald et al., 1997). Since the late 1980s, fast ice in the Lake Harbour, Ivujivik, and Salluit areas has been extending farther into Hudson Strait. In early 1990s, it froze over completely (McDonald et al., 1997). Large recurring polynya used by Ivujivik and Salluit sea ice hunters started to freeze over in the 1980s and no longer opened during early 1990s spring tides (McDonald et al., 1997). Floe edge melts before breaking (McDonald et al., 1997).

Iqaluit, Nunavut, Canada
Weather: Weather more unpredictable since the 1990s. More unusually hot days in summer but temperatures cooler overall (Fox, 1998).
Wind: Winds change suddenly, weather and wind changes were more subtle in the past (Fox, 1998).
Sea Ice: Ice conditions becoming more unpredictable through the 1990s with several accidents occurring in late 1990s. In mid/late 1990s, sea ice breaks up more quickly near Iqaluit and earlier than usual. Less sea ice near shorelines (Fox, 1998).

Igloolik, Nunavut, Canada
Weather: Weather increasingly unpredictable in recent years. Fewer periods of extended clear weather and more sudden storms (Fox, 1998).
Wind: Winds are stronger and occur more often. The winds are now responsible for ice break-up, as opposed to the ice thinning first. Recent changes in the winds have had an impact on sea-ice formation and decay processes (Fox, 1998).

Aleuts

Yup'ik

Alutiiq

Tlingit

USA
Alaska

Athabaskans   Iñupiat

YUKON   Gwich'in

CANADA   Dene/Métis

Inuvialuit

NORTHWEST
TERRITORIES   Victoria Island

NUNAVUT   Inuit

Canadian Arctic Archipelago

Cree

Ellesmere Island

Inuit   Inuit
Baffin
Island

QUEBEC   Inuit
Cree   GREENLAND

Inuit   Inuit

LABRADOR
Innu   Inuit

ICELAND

Atlantic
Ocean

Saami Council

Russian Association of Indigenous
Peoples of the North

Aleut International Association

Inuit Circumpolar Conference

Gwich'in Council International

Arctic Athabaskan Council

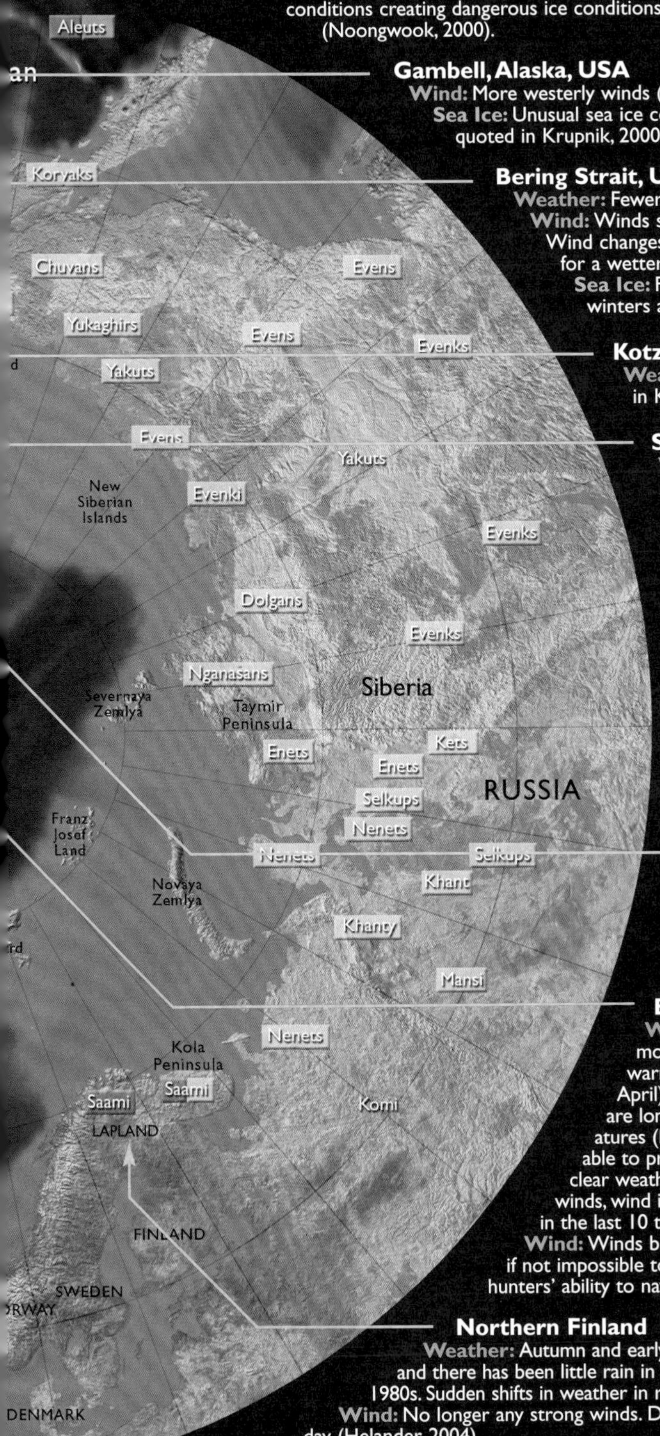

St. Lawrence Island, Alaska, USA

Weather: More extreme weather conditions in the last 10 to 20 years (e.g., winds and storms that cause dangerous ice conditions that impede hunting). More warm weather (Noongwook, 2000).

Wind: More intense storms which last longer than in the past. Winds constantly change from one direction to another and with more intensity (Noongwook, 2000).

Sea Ice: Delays in ice packing and freeze-up due to changing winds (Noongwook, 2000). Increased frequency of windy and warm conditions creating dangerous ice conditions such as thin ice, snow-covered small open leads, and very rough ice (Noongwook, 2000).

Gambell, Alaska, USA

Wind: More westerly winds (Conrad Oozeva as quoted in Krupnik, 2000).

Sea Ice: Unusual sea ice conditions in winter 1999/2000; no icebergs that autumn (Edmond Apassingok as quoted in Krupnik, 2000).

Bering Strait, USA

Weather: Fewer calm days (Pungowiyi, 2000).

Wind: Winds stronger. Winds may shift direction but stay strong for long periods in spring. Wind changes distribution of sea ice. Mid-July to September more wind from the south makes for a wetter season, also more erosion due to wave action (Pungowiyi, 2000).

Sea Ice: Formation of sea ice is later in the autumn. Ice thinner than usual due to warmer winters and winds. Different formation processes and earlier break-up (Pungowiyi, 2000).

Kotzebue, Alaska, USA

Weather: Weather patterns change too fast to predict (Caleb Pungowiyi as quoted in Krupnik, 2000).

Sachs Harbour, Nunavut, Canada

Weather: Increased weather variability; more sudden and intense changes in weather; changes most noticeable in transition months; more extreme weather; weather more unpredictable. Longer, warmer summers. Spring melt is faster and spring comes earlier. Shorter, warmer winters. August is a warm month now, used to be the "cooling off month". Autumn comes later. Longer duration of "hot" days, now a whole week rather than one to two days (Jolly et al., 2002). More "bad" weather with blowing snow and whiteouts (Berkes and Jolly, 2001).

Wind: Changes in velocity and direction with more intense wind storms and more wind in summer (Jolly et al., 2002).

Sea Ice: Less/no multi-year ice in July and August; more open water and rougher water; more ice movement than before; not able to see the permanent ice pack to the west; ice breaks up earlier and freezes later; rate of ice break-up has increased; seasonal ice in harbor is thinner (not safe); less and thinner fast ice; changes in distribution and extent of local pressure ridges; leads farther away from shore; ice pans do not push up on shore anymore; open water in winter is closer than before; changes in ice color and texture (Jolly et al., 2002).

Kitikmeot Region, Canada

Weather: Warmer temperatures; unpredictable weather; late autumn; early spring; more extreme hot days; sporadic extreme heat days; spring melt came earlier than in the past in the 1990s; earlier snow melt (Thorpe, 2000).

Sea Ice: Unusually high number of cracks in sea ice in early spring around Hope Bay (Thorpe, 2000).

Baker Lake, Nunavut, Canada

Weather: Winters were colder and longer in the past, one used to get frostbite more quickly in the 1940s. Early and quick spring melt periods in recent years with warmer springs (e.g., snow "gets shiny" (i.e., a melt layer) earlier (March instead of April) and does not refreeze at night which is what is expected) (Fox, 2004). Summers are longer, caching meat has to be put off until later in the year to wait for lower temperatures (Fox, 2004). Increased weather variability; experienced hunters and elders no longer able to predict weather; sudden storms; more cloudy periods; and less long stretches of clear weather (Fox, 2002). Unstable weather recognized in temperature fluctuations, shifting winds, wind intensity, and storm behavior (GN, 2001). Summers warmer and winters less severe in the last 10 to 15 years (GN, 2001). More extreme hot days in summer (GN, 2001).

Wind: Winds blow stronger resulting in the snow being packed unusually hard. Igloos are difficult if not impossible to build because of this. The direction of the dominant wind has changed affecting hunters' ability to navigate using wind-formed ridges in the snow (Fox, 2002, 2004).

Northern Finland

Weather: Autumn and early winter are warmer. In recent years, the ground has not frozen properly in autumn and there has been little rain in September. Winters no longer have long cold periods as in the 1960s and even 1980s. Sudden shifts in weather in recent years and the weather is becoming more difficult to predict (Helander, 2004).

Wind: No longer any strong winds. Direction of the wind can shift fast, blowing from different directions on the same day (Helander, 2004).

Clyde River, Nunavut, Canada

Weather: Weather highly variable and more difficult to predict in recent years. Warmer springs with quicker spring melt (Fox, 2002, 2004).

Wind: Winds have changed in direction, frequency, and intensity. Winds also change suddenly (Fox, 2002, 2004).

Sea Ice: Usual leads have not formed in last few years and new ones have opened in unusual and unexpected areas. Ice is thinner and dangerous for travel in some areas, especially in spring (e.g., near Home Bay) and some residents have observed a trend in recent years of more icebergs (Fox, 2002). The texture of sea ice is different in some areas. For example, it can be mushy or too soft. Sea ice and icebergs are expected to chip off in a certain way when struck with an object. The ice is not chipping off as expected in some areas (Fox, 2004).

Fig. 3.1. Indigenous observations of changes in weather, seasons, wind, and sea ice.

Arviat, Nunavut, Canada

Caribou: Antlers not as thick (GN, 2001).
Bears: Polar bears and grizzly bears common in areas and during times not normal compared to the past, e.g., polar bears still around in July and August, and in December (GN, 2001).
Birds: Harder snow too difficult for ptarmigan, they are seeking out new habitat in willows where there is softer snow and better access to food. Sandhill cranes have changed with many observed eating carrion (GN, 2001).

Western Hudson Bay, Canada

Whales: Fewer whales now visit mouths of Nelson and Churchill rivers. Large numbers reported near Severn and Winisk rivers where abundant whitefish (McDonald et al., 1997).
Walrus: Decline in walrus populations in James Bay and southwestern Hudson Bay associated with changing shorelines and habitat alteration (overgrown with willow). Walrus also used to inhabit an island in the Winisk area until it began merging with the coastal shoreline in the early 1980s. Now walrus only return to visit in groups of two or three (McDonald et al., 1997). Large numbers reported near Severn and Winisk rivers where abundant whitefish (McDonald et al., 1997).
Moose: Change in taste of meat. Greater number drowning. No moose at Marsh Point (McDonald et al., 1997).
Caribou: Increase in numbers. Pin Island herd is mixing with Woodland herd (McDonald et al., 1997).
Bears: Thin-looking bears in York Factory area. Drink motor oil. Change in behavior (McDonald et al., 1997).
Birds: More snow geese migrating to and from the west. Habitat changes at March Point staging area. Earlier and shorter autumn migration.
Fish: Mercury contamination. Loss of habitat including spawning grounds. Change in taste of fish, some are inedible (McDonald et al., 1997).

Western James Bay, Canada

Walrus: Decrease in numbers in Attawapiskat area (McDonald et al., 1997).
Bears: Recent increase in reproduction rates. Fearless of humans (McDonald et al., 1997).
Birds: Habitat changes in Moose Factory area. More snow geese flying in from the west. Canada geese arrive from the north first part of June. Change in autumn migration pattern (McDonald et al., 1997).
Fish: Morphological changes in sturgeon. Dried river channels (McDonald et al., 1997).

Eastern James Bay, Canada

Whales: Decrease in beluga numbers (McDonald et al., 1997).
Walrus: No longer present in Wemindji area (McDonald et al., 1997).
Moose: Loss of habitat. Decrease in numbers. Change in body condition. Change in taste of meat (McDonald et al., 1997).
Caribou: Change in body condition and behavior. Increase in number of diseased livers and intestines. Change in diet. Change in taste of meat. More caribou along coast (McDonald et al., 1997).
Birds: Coastal and inland habitat changes for snow geese and Canada geese. Coastal flyways shifted eastward. Fewer geese being harvested in spring and autumn. Large flocks of non-nesting/molting geese along coastal flyway (McDonald et al., 1997).
Fish: Mercury contamination. Loss of adequate habitat for several species, e.g. white-fish, sturgeon, and pike. Morphological changes in sturgeon (McDonald et al., 1997).

Eastern Hudson Bay, Canada

Whales: Fewer beluga seen along eastern coast of Hudson Bay and James Bay; more offshore of east coast Hudson Bay (McDonald et al., 1997).
Walrus: Shift away from Belcher Islands (McDonald et al., 1997).
Moose: In-migration from southeastern James Bay (McDonald et al., 1997).
Caribou: Caribou from different areas mingle together. Very large herds. Traveling closer to coast. Change in diet. Change in taste of the meat (McDonald et al., 1997).
Bears: Polar bear numbers increased since the 1930s, and more rapidly since 1960s. Inuit think they are relocating to eastern Hudson Bay in response to effects of an abundance of ringed seals, the extended floe edge, and hunting quotas in effect since the 1970s (McDonald et al., 1997).
Birds: Canada and snow goose migrations have shifted east, from the Quebec coast towards the mountains of mid-northern Quebec. Along the Manitoba and northwestern Ontario coasts, more geese are entering and leaving Hudson Bay from the west and fewer taking north-south coastal route. Cree attribute this to combination of factors in last 20 years; weather system changes, coastal and land habitat changes, disturbance from inexperienced hunters, and wildlife management practices, e.g., aircraft and banding (McDonald et al., 1997). Smaller flocks of Canada geese arrive in Belcher Islands since 1984. Increase in non-nesting/molting geese in Belcher and Long islands (McDonald et al., 1997).
Fish: Fish habitat reduced due to lowering water levels. Decrease in Arctic char and Arctic cod in Inukjuak area (McDonald et al., 1997).

Hudson Strait, Canada

Whales: Decrease in beluga numbers in Salluit area (McDonald et al., 1997).
Walrus: Increase in numbers around Nottingham Island (McDonald et al., 1997).
Caribou: Increase in numbers and shifts in distribution, perhaps as part of long-term cycles (Ferguson et al., 1998). Increase in abnormal livers (e.g., spots and lumps). Change in diet (McDonald et al., 1997).
Bears: Decrease in numbers in Ivujivik area (McDonald et al., 1997).
Birds: New snow goose migration routes. Increase in number of molting snow geese. Canada geese no longer nest in Soper River area (McDonald et al., 1997).

Iqaluit, Nunavut, Canada

Birds: Robins sighted in Iqaluit and Kangirsuk (Northern Quebec) in 1999 (George, 1999).

Clyde River, Nunavut, Canada

Seals: Seals not as healthy in last few years; missing patches of fur, molting at unusual times of the season, suffering from skin rashes, white pustules on the meat (Fox, 2004).
Bears: Increased numbers of polar bears in the area and polar bears arrive at unexpected times of the season (Fox, 2004).

Aleuts

Yup'ik

Alutiiq

Tlingit

USA
Alaska

Athabaskans   Iñupiat

YUKON   Gwich'in

CANADA   Dene/Métis

Inuvialuit

NORTHWEST   Victoria Island
TERRITORIES

Inuit

NUNAVUT   Inuit

Canadian Arctic Archipelago

Ar

Ellesmere Island

Cree

Inuit   Inuit

Baffin
Island

Inuit

QUEBEC   GREENLAND
Cree

Inuit

Inuit

LABRADOR
Innu   Inuit

ICELAND

Atlantic
Ocean

Saami Council

Russian Association of Indigenous
Peoples of the North

Aleut International Association

Inuit Circumpolar Conference

Gwich'in Council International

Arctic Athabaskan Council

Bering Strait, USA

Seals: Spotted seals declined from late 1960s/early 1970s to present. 1996 and 1997 spring break-up came early resulting in more strandings of baby ringed seals on the beach. Fewer seals in Nome area these days, perhaps as result of less ice for ringed seal dens (Pungowiyi, 2000).

Walrus: Physical condition of walrus generally poor 1996-1998 – animals skinny and productivity low. One cause was reduced sea ice, which forces walrus to swim farther between feeding areas. Walrus in good condition in spring 1999 after cold winter and good ice in Bering Sea (Pungowiyi, 2000).

Birds: Spring migrations early. Geese and songbirds arrive in late April, earlier than in past. August 1996 and 1997 large die-offs of kittiwakes and murres. Other birds doing well, little snow (and lower hare population, food competitor) has been good for ptarmigan (Pungowiyi, 2000).

Fish: Chum salmon crashed in Norton Sound in early 1990s and have been down since (Pungowiyi, 2000).

Insects: New insects appearing not seen before. Mosquitoes still the same (Pungowiyi, 2000).

Sachs Harbour, Nunavut, Canada

Seals: Increasing occurrence of skinny seal pups at spring break-up (Jolly et al., 2002).

Caribou: Increased forage availability for caribou. Changes in the timing of intra-island caribou migration (Jolly et al., 2002).

Bears: Less polar bears seen in autumn due to lack of ice (Berkes and Jolly, 2001).

Muskox: Increased forage availability for muskox (Jolly et al., 2002).

Birds: Difficult to hunt geese in spring because of quick melt (Berkes and Jolly, 2001). Robins have been observed; previously unknown small birds (Jolly et al., 2002).

Fish: Different species observed. More least cisco (locally called "herring") caught now (Berkes and Jolly, 2001). Two species of Pacific salmon caught near the community for the first time (Jolly et al., 2002).

Kitikmeot Region, Canada

Caribou: Caribou changing migration routes due to early cracks in sea ice. Changes in vegetation types and abundance affecting caribou foraging strategies. Massive caribou drownings increasing due to thinner ice, e.g., massive drowning observed in 1996. Lower water levels may mean caribou can save energy by not having to swim as far, however, changing shorelines due to dropping water levels are affecting caribou forage (though unclear how). Caribou deaths due to exhaustion from extreme heat and attempts to escape more mosquitoes (Thorpe, 2000).

Seals: Seals come up through the unusually high number of cracks in sea ice in early spring around Hope Bay, which attracts polar bears (Thorpe, 2000).

Bears: Grizzly bears seen for the first time crossing from the mainland northward to Victoria Island in 1999. Spring 2000, unusually high numbers of grizzly bears and grizzly tracks (Thorpe, 2000).

Birds: New birds seen for the first time such as the robin and unidentified yellow songbird (Thorpe, 2000).

Insects: Number of mosquitoes increasing with temperature, but this occurs only to a threshold then the mosquitoes cannot survive (Thorpe, 2000).

Baker Lake, Nunavut, Canada

Caribou: Caribou not as healthy in recent years. Meat is tough and skin around the neck area tears too easily. More liquid in joints and more white pustules on meat (Fox, 2004). Caribou less fat and undernourished due to heat and dryness in summer. Caribou skins are weak and tear easily during field dressing. More diseased caribou, e.g., sores in mouth and on tongue (GN, 2001). Links between changing caribou condition and climate not always clear (GN, 2001; Fox, 2004).

Bears: More grizzly bear sightings and encounters around Baker Lake area (GN, 2001; Fox, 2004).

Muskox: More muskox sightings around Baker Lake area (Fox, 2004).

Birds: Birds seem smaller and not as happy. Redpolls and white-throated sparrows more common, lapland longspurs hardly seen any more (Fox, 2002). More ravens (GN, 2001).

Fish: Changes in fish (mainly char and trout); trout darker in color, little fat observed between meat layers when boiled, less fish in usual fishing spots, fish eating things they are not supposed to eat, fish too skinny, smell different – "like earth", mushy meat (Fox, 2004).

Insects: Mosquitoes decreasing in numbers in some areas with increasing summer temperatures since there is less standing water. At least ten new kinds of insects in the area, all winged insects, some recognized from treeline area (Fox, 2004). Strange kinds of flying insects being observed. Warmer temperatures may be responsible for arrival of flying insects from the south and for insects being active longer in the year (GN, 2001).

Northwestern Hudson Bay, Canada

Whales: Decrease in numbers in Arviat and Repulse Bay area (McDonald et al., 1997).

Walrus: Decrease in numbers near Arviat and Whale Cove. Increase in numbers near Coral Harbour and Chesterfield Inlet (McDonald et al., 1997).

Caribou: Increase in numbers. Not intimidated by exploration activity. Feed close to exploration camps. Change in diet (McDonald et al., 1997).

Bears: Inuit in northwestern and eastern Hudson Bay report increasing numbers of polar bears. Appear leaner and more aggressive (McDonald et al., 1997).

Birds: More Canada geese in Repulse Bay area during summers of 1992 and 1993 (McDonald et al., 1997).

Fish: Decrease in Arctic cod in near-shore areas. Arctic cod no longer found in near-shore areas off Cape Smith and Repulse Bay (McDonald et al., 1997).

Northern Finland

Birds: Many types of bird have declined in numbers including crows, buzzards, and some falcons. Arctic terns, long-tailed duck, and osprey have disappeared in some areas (Helander, 2004).

Fish: Fish populations have gone into decline in many lakes, partly to due overharvesting, but also due to factors unknown to local people. For example, in Rievssatjavri, in the reindeer village of Kaldoaivi, perch have disappeared but pike survive (Helander, 2004).

Insects: The number of insects has decreased, mosquito populations among others (Helander, 2004).

Map labels: Aleuts, Koryaks, Chuvans, Evens, Yukaghirs, Evens, Evenks, Yakuts, Evens, Yakuts, New Siberian Islands, Evenki, Evenks, Dolgans, Evenks, Nganasans, Siberia, Severnaya Zemlya, Taymir Peninsula, Enets, Kets, Enets, RUSSIA, Selkups, Nenets, Franz Josef Land, Nenets, Selkups, Novaya Zemlya, Khant, Khanty, Mansi, Kola Peninsula, Nenets, Saami, Saami, Komi, LAPLAND, FINLAND, SWEDEN, WAY, DENMARK

Fig. 3.2. Indigenous observations of changes in animals and insects.

other disagreements between indigenous and scientific knowledge. Examining the reasons for these differences, however, may drive interesting questions for further research on environmental change. Trying to link different perspectives may result in meaningful insights into the nature and impacts of arctic environmental change (Huntington et al., 2004).

The spatial scale of the observations in Figs. 3.1 and 3.2 is significant. Models of arctic climate provide information on regional scales. Indigenous observations, by contrast, are more localized. A major challenge is to refine model outputs to finer scales, which requires the connection of large- and small-scale processes and information. A corresponding challenge is to combine indigenous observations from various areas to create a regional picture of environmental change. Using these different sources of information across different scales may help to identify the local components of regional processes as well as the regional processes that account for locally observed change.

The information in Figs. 3.1 and 3.2 provides a starting point for studies of the link between indigenous knowledge and other research, for example by cross-referencing different perspectives on climate and environmental change. In this context, several important points about the figures should be noted. First, the information is not comprehensive. The projects cited and even the observations taken from particular projects are only examples. There is not the space to record all documented observations here. Second, each observation is from a particular person, from a particular place, and with a particular history and point of view. Such details are lost when the information is reduced to fit this type of format, and so the informa-

tion presented here is out of context. The condensed format is valuable for certain purposes, such as a broad comparison across regions or with scientific findings, but the original sources should always be consulted when using the information presented here.

3.4. Case studies

Indigenous perspectives on the changing Arctic vary widely over time and space, as may be expected given the differences between the histories, cultures, ways of life, social and economic situations, geographical locations, and other characteristics of the many peoples of the region. These perspectives cannot be illustrated by generalizations nor, in the space allotted and with the materials currently available, comprehensively for the entire Arctic. The case studies used in this section were chosen as illustrations of indigenous perspectives on climate change, and were drawn from the limited number of studies that have been done on this and related topics. Such a sample of opportunity inevitably results in omissions, such as the lack of indigenous fishers' voices and the absence of case studies across most of the Russian Arctic. It is also important to note that climate change cannot be neatly separated from the many factors that affect the relationship of people with their environment. Many of the observations and interpretations given in the case studies reflect an interaction between climate change and other factors, rather than being the result of climate change in isolation.

Each of the case studies comes from an existing project, whose researchers were willing to contribute material to this chapter. The formats for the case studies vary greatly and were chosen by the authors to reflect the type of

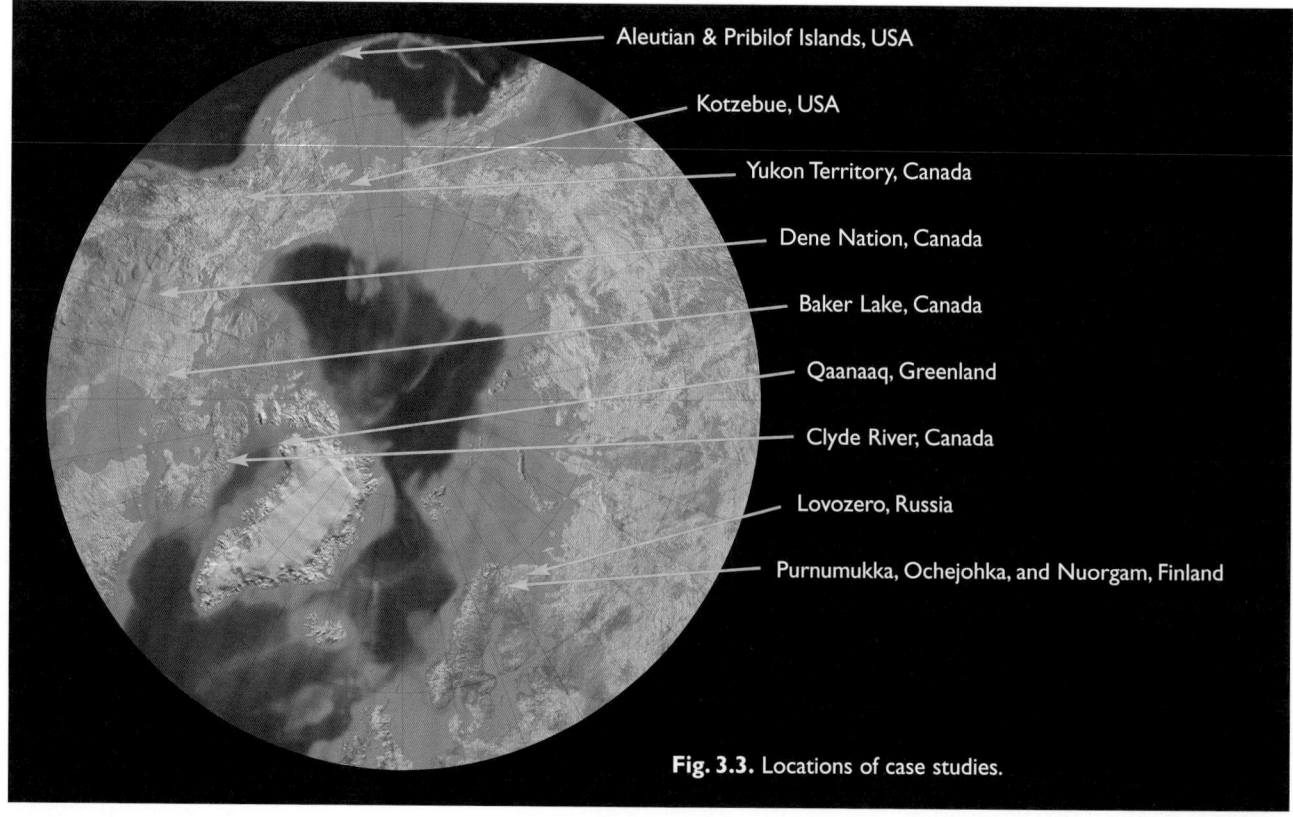

Fig. 3.3. Locations of case studies.

material they gathered and the way in which the study was conducted. It was felt that the resulting inconsistencies in style were preferable to imposing a uniform approach to very different materials generated in very different ways. Of course, each study is selective in that it cannot cover all that a given people or community has to say about climate change. The case studies describe those aspects of climate change and related topics that the authors and the communities represented find most significant. Figure 3.3 shows the locations of the case studies discussed in this chapter.

The projects from which the case studies are drawn have, in most cases, produced a separate report or reports elsewhere, which contain more thorough discussions of methods, approaches, and results. They also acknowledge the support that was required from funding agencies, collaborators, and, most importantly, indigenous communities and people in conducting and reporting each study.

In addition to the longer case studies presented, three short case studies are included to give perspectives from other parts of the Arctic or to emphasize a particular point of view. The Greenland case study (section 3.4.6) and the second case study from Finland (3.4.8) are drawn from interviews with individuals. The Aleut case study (section 3.4.2) describes the background and plans for a project to be carried out in the village of Nelson Lagoon, Alaska, with additional observations from other communities in the region. As is the case for the longer case studies, the three short case studies are illustrative rather than representative, and are given as examples.

3.4.1. Northwest Alaska: the Qikiktagrugmiut

The Native Village of Kotzebue, the tribal government of the community, conducted a study to document traditional knowledge of environmental change in their region from the 1950s to the present. The study was conceived, developed, and carried out by members and employees of the tribe. Interviewers used a semi-directive interview to engage elders in conversations about environmental change. The results were compiled in a report (Whiting, 2002), which included a discussion of the implications of the various observations recorded. The documentation of this knowledge will be valuable to future tribal members for historic preservation purposes and comparative analysis. The case study reported here draws largely on the discussion contained in the project's final report.

Kotzebue, a community of 3000 inhabitants, the majority of whom are Iñupiaq, is located on the Baldwin Peninsula 30 miles above the Arctic Circle in northwest Alaska. Its Iñupiaq name is *Qikiktagruk*, which means "almost an island". The people indigenous to this locality are the *Qikiktagrugmiut*, "people of Qikiktagruk", and the tribal government representing them is the Native Village of Kotzebue. This case study documents the perspectives of the Qikiktagrugmiut, who have lived in the area as a distinct group for centuries (Burch, 1998).

The subject of climate change, and environmental change in general, is of personal interest to a people that continue to interact directly with the natural world for their spiritual, cultural, and nutritional sustenance. The Qikiktagrugmiut, while giving some thought to the broader definition and future implications of environmental change mostly focus on present and near-future effects. They are specifically concerned with change as it relates to climate and day-to-day weather and how these compare to what is considered "normal" based on personal experiences over the last half of the twentieth century. Some significant changes in weather types and patterns have occurred in recent decades, in some cases creating conditions that have not been observed previously by the current generations of Qikiktagrugmiut. The implications of these changes are hard to predict.

Local hunter discussion focuses on the elements of the climate and "country" (the environment outside the settlements) on a daily basis. Their interest is not primarily comfort, although that is usually referred to in conversation, but on whether conditions are conducive to travel and how they will affect the movement of animals. Hunter mobility is critical to the ability to access a migratory and widespread animal resource. During summer, high winds and fog, along with the break-up of the ice at the end of May and beginning of June, heavily influence the ability to travel. In winter, high winds are more an inconvenience than an impediment to travel, unless the winds are associated with blizzards and whiteout conditions. While these will stop travel at times, it is still possible to travel under the most adverse winter conditions, unlike in summer. More commonly, extremely low temperatures restrict winter travel. In recent years, however, the period of time temperatures stay extremely low has been shortening. The timing of freeze-up, the thickness of the newly forming ice, and the timing and amount of snow are also key factors affecting travel. In other words, the weather is immediately important for daily activity.

There is little doubt that the weather and the entire environment that Qikiktagrugmiut depend on to support their way of life are changing. Environmental change is not necessarily bad. For instance, at the beginning of the twentieth century, caribou (*Rangifer tarandus*) were rare in the locality of Kotzebue. If not for the region's reindeer (*Rangifer tarandus*) herds imported at this time from the Chukotsk Peninsula, Russia, many people would have had a hard time finding enough meat and skins for clothing to survive the winter. Now the Western Arctic Caribou Herd that travels through the area numbers about half a million (attributed to natural variability in caribou/lichen cycles and a decrease in reindeer competition due to a loss of interest in herding and from caribou herds absorbing many reindeer and even reindeer herds), and local reindeer herds are absent. In addition, moose (*Alces alces*) also began to appear in the region from around the 1950s onward, again attributed to natural variability. These types of change contribute to the quality of life of the region's residents.

Weather has also changed over the last half century. Seasons are now less consistent. For example, higher temperatures have become more common during autumn and winter, sometimes creating mid-winter fog, a new phenomenon. At the same time, lower temperatures in the summer have also become more common. Daily changes are now more extreme. It is relatively common for the temperature to change from -35 °C one day to 0 °C the next or vice versa. Unusual swings in weather occur not just during winter. An exceptional example is the snowstorm during July 2000, which left snow covering the tundra for a day and reduced the berry crop dramatically that year. Another example includes having the least amount of precipitation recorded for the month of November during 1999 and again in 2001, 0.25 mm, or 100 times less than the average amount of precipitation recorded for that month over the last fifty years. According to Qikiktagrugmiut, the increased variability and unpredictability in weather appears to have started during the 1970s and has continued through the 1980s and 1990s and into the new century.

The relationship between weather and the Qikiktagrugmiut is more intimate than for most people in the United States. Their daily traditional activities are almost entirely dictated by the weather and other environmental conditions, such as snow depth and animal distribution. For most urban communities the concern with weather has more to do with comfort level and recreation. For the Qikiktagrugmiut, the weather determines if daily activities can be carried out safely and productively (for instance water and ice travel and being able to dry meat and fish successfully). The weather is also tied to the ability of the land to produce natural crops of fur, meat, and berries.

This disparity in how weather is perceived by the rural Alaskan communities versus the urban mainstream is apparent from watching weather forecasters across the country, including urban centers in Alaska, such as Anchorage. The premise of these forecasters is based on the urban view that "good" weather should be sunny and warm. For the Qikiktagrugmiut, however, weather is "good" if it is favorable to the country's productivity and the ability of people to access the land's resources. Thus, "good" weather may include rain in July to produce a bountiful berry crop and extremely low temperatures during early autumn so that Kotzebue Sound and the surrounding rivers and lakes freeze quickly and reliably for safe travel. These two conditions, rain and extreme low temperatures, are almost universally portrayed as "bad" weather in urban settings. In addition, the Qikiktagrugmiut's ability to cope with extreme weather events differs from that of most urban communities across the nation. Blizzards that would shut down entire cities and be portrayed as mini-disasters by the urban media are looked upon favorably by rural Alaskan communities as a means by which travel is improved (through additional snow filling in willow stands, tundra tussocks, creeks/gullies, and compacting snow cover by the associated winds), allowing greater access to the country for travel and harvesting animals. Even in the town of Kotzebue, extreme weather events have relatively little direct impact. Schools and businesses, for example, are rarely closed due to weather.

These characteristics of the people appear to show an ability to successfully adapt and live in an inherently variable local environment. The real challenge with assessing the impacts of climate change, however, is in trying to understand the interconnectedness and the wide-ranging impacts that collectively work to change the shape of the web of activities and life in this part of the world.

Some Qikiktagrugmiut live out in the country outside the communities. Their ability to travel and obtain the necessary requirements of life is dependent on the length and quality of the freeze-up and the length of the break-up, which are determined by the weather conditions during autumn and spring. In addition, many people who are at the fringes of production, the young and the elderly, depend on favorable weather to be able to participate in the limited harvesting activities available to them. Ice fishing in front of Kotzebue, for instance, supplies people with traditional autumn food (saffron cod (*Eleginus gracilis*) and smelt (*Osmerus mordax*)), and is an important social activity that binds the community and gives the elderly and young people one of their few chances during the year to harvest traditional foods. During autumns with a late freeze-up, ice fishing is limited or less productive. Thus, a single climate variable in one season disproportionately affects this segment of the population by substantially reducing their annual harvesting opportunity.

3.4.1.1. The impacts of late freeze-up

A closer look at the Qikiktagrugmiut understanding of one event and its impacts, such as late freeze-up, can show how they see consequences that are widespread and varied yet still intertwined, so that it is impossible to look at any one thing in isolation. Late freeze-up is one likely consequence of regional climate warming, and hence a relevant example for considering the impacts of climate change. To illustrate the complexity of determining whether overall changes are positive or negative and how this depends on context and perception, this section uses the example of late freeze-up and its impacts on people, spotted seals (*Phoca largha*), caribou, and red foxes (*Vulpes vulpes*). The impacts are those that the Qikiktagrugmiut would immediately associate with late freeze-up, showing both the scope of their environmental knowledge and the patterns of interconnection that they see in their surroundings. This exercise shows how the timing, quality of ice, speed of complete freezing, associated weather, and ecological effects all combine to produce the many and varied impacts of a late freeze-up.

Impacts on humans and their way of life

The impacts of late freeze-up on humans vary widely and include better whitefish (*Coregonus* spp.) harvests, better clamming (*Macoma* spp.), better spotted seal hunting, better access to caribou, better arctic fox (*Alopex lagopus*) harvests, better access to driftwood, a shorter ice-fishing season, poor access to Kotzebue for people living out in the country, rough ice conditions, more danger from thin ice, and more erosion and flood problems.

- People living outside Kotzebue at remote campsites have an extended period for whitefish harvesting. Late season storm surges can reach the beach, piling porous sand across the mouth of a major harvesting river, trapping the fish behind the sand dam from where they are easily caught.
- Late season storm surges wash clams onto the beach at Sisaulik (a peninsula across the sound from Kotzebue where some of the Qikiktagrugmiut live during the summer and autumn), which can then be collected and stored in cool saltwater for many days of clambakes.
- Hunters have a longer period for using boats to hunt spotted seals, which are present in prolific numbers feeding on large schools of fish. Also, a long period of thin ice enables the seals to feed far into the sound. When the ice thickens overnight, many may try to return to open water by crawling on top of the ice, where they are easily reached by hunters now able to travel on the ice.
- Caribou hunters have a longer period in which to use boats to reach caribou (conversely snow-machine access will be delayed to later in the winter). There is, however, an increased risk during extended freeze-ups that boats will get caught in young ice and have to be abandoned for the winter. This happened during the late freeze-up of 2000.
- Arctic foxes are concentrated along the coast during the long season of open water, unable to get out onto the sea ice.
- More logs are washed up on the mud flats by late-season high water, for use by people living out in the country for their early autumn fuel supply.
- Ice for autumn fishing is missing, so the ice-fishing season is shorter in front of Kotzebue. In many cases, the ice fishers will then miss the largest runs of smelt and saffron cod, which tend to come past Kotzebue in large concentrations earlier, rather than later, in the autumn.
- People living out in the country must wait for a longer period before they can reach Kotzebue for expendable supplies such as gas, propane, medical needs and other necessities or must risk traveling under very dangerous conditions, which has caused the loss of life in some cases.
- Repeated incomplete freezing and thawing of the northern sound means that the ice that does appear can be piled up by the wind, creating very rough conditions and many obstacles to travel by snowmachine and dogs which begin once the ice freezes permanently.
- Snow can pile up on thin ice which makes such areas less likely to freeze completely and thus more dangerous once travel begins. There is often much snow on the ground during autumns with late freeze-ups because the low pressure conditions that contribute to slow ice growth are also associated with snow and storm fronts.
- Late season storm surges, unimpeded by ice, can create erosion and flood problems along the beach and road in front of Kotzebue.

Impacts on spotted seal

The impacts of late freeze-up on spotted seals include better access to inshore waters and the fishes that congregate there, better haul outs for resting, and greater risk of being trapped.

- Owing to the absence or patchiness of ice, spotted seals have increased access to the extreme inshore waters where smelt and saffron cod, and other food fishes, congregate in large numbers in early autumn. The seals force the fish into concentrated groups next to shore during the open water period, which is probably the most efficient way for them to catch the fish easily and in large numbers. Also, late freeze-up would allow seals increased access to the Noatak River, which holds large char (*Salvelinus malma*) and chum salmon (*Oncorhynchus keta*) at this time.
- Thin or patchy ice is better for hauling out on, allowing the seals to rest close to their major food source at this time of year, thus increasing the net amount of energy gained from this seasonal activity.
- Because the seals are able to haul out and breathe through the thin ice, they have a greater risk of becoming trapped too far from open water when the ice begins to thicken. Once temperatures drop well below freezing and stay there, which can happen rapidly at this time of year, the ice can become too solid and extensive for the seals to reach open water, which will force them to travel out over the ice (and in some cases over land) in order to reach the open water of the Chukchi Sea, leaving them vulnerable to starvation and predation.

Impacts on caribou

One of the impacts of late freeze-up on caribou is slower movements.

- The warm weather associated with late freeze-up makes caribou less likely to travel long distances thus slowing the autumn migration. In addition to being slowed by the warm weather and their own lack of initiative to move, extended thin ice conditions hamper movement, because the ice does not support the animals when they try to cross water bodies in their path. Although the consequences of this are unclear, they are probably many and varied, such as being forced to stay for extended periods of time on less productive ranges and increased vulnerability to predators such as wolves (*Canis lupus*) that are lighter and able to take advantage of the thin ice that is an obstacle for the caribou.

Impacts on red foxes

The impacts of late freeze-up on red foxes include better feeding and increased competition with Arctic foxes.

- A longer period of late season open water allows more storm surges to reach the shore, closing off coastal rivers with porous sand that allows large amounts of whitefish to become trapped and frozen into the ice at coastal river outlets. These provide a substantial food resource for many of the foxes along the coast. In addition, late season storms result in more sources of fox food in the form of enormous schools of baitfish and marine mammal carcasses that are deposited on the beach by the waves. Also, a longer hunting season for spotted seals and caribou by boat hunters means that more caribou gut piles and lost seals become available prior to the long period of beach foraging. Almost all foxes within the vicinity of the coast rely heavily on beach scavenging during the time around freeze-up, which also coincides with low human traffic along the coast. A particularly good year for late season beach foraging allows the foxes to accumulate critical amounts of fat to survive the long winter months ahead.
- An extended period of open water along the coast can impede the movement of arctic foxes onto the pack ice, which results in increased competition with the red foxes that rely on coastal food sources. If this occurs during a high in the four-year arctic fox population cycle, the effect is multiplied by the large numbers of arctic foxes migrating south and being stopped by the open water along the coast.

While this list of impacts arising from late freeze-up is not exhaustive, the examples indicate the interconnectedness that complicates an effort to understand the changes that occur from year to year as well as the long- and short-term effects of changes to the various combinations of environmental elements.

The challenge posed by climate change to indigenous peoples is their ability to respond and adapt to changes in the local environment, while continuing to prosper. Since the history of indigenous peoples is replete with change, it is important to ask whether they and their cultures are threatened by continued change, or whether change is just a threat to current understanding of the environment, which in any case is continually changing, slowly and on a daily basis. For example, seal hunting in leads during winter has decreased in importance and participation each year, due in part to the cultural economy's changing dependency on the seal for food and domestic utilitarian purposes, and in part to the unpredictable, and thus more dangerous, ice conditions of late. It is an activity that relies on the most extreme form of specialized knowledge of the environment that needs to be taught and learned over many years. More rapid environmental change is generally

harder to adapt to. Recently, two experienced seal hunters were lost on the ice while hunting. Local interpretation of the event concluded that climate change has resulted in unusual and unpredictable ice conditions and that this must have been the cause of the tragedy, as the two men would not have had trouble traveling over ice under normal circumstances.

Even if processes are in motion that will change the entire ecosystem, whether this will result in circumstances that are not conducive to human existence, or in a new ecosystem with resources available for human consumption following some degree of adaptation, is unknown. Archaeologists have found this to have occurred in the past, with arctic societies having changed from terrestrial-based cultures to marine-based cultures and back again. The best that can be done at this point is to continue to observe, document, and discuss the changing environment and to hope that indigenous peoples will be able to adapt to whatever future environments may evolve in their traditional homelands.

3.4.2. The Aleutian and Pribilof Islands region, Alaska

The Aleut International Association (AIA) and the Aleutian and Pribilof Islands Association (APIA) prepared this summary of current observations, concerns, and plans related to climate change in their region. Michael Zacharof is President of AIA and lives on St. Paul Island in the Bering Sea. Greg McGlashan is the Tribal Environmental Programs Director on St. George Island. Michael Brubaker is the Community Services Director for APIA. Victoria Gofman is Executive Director of AIA.

There are several examples of how climate change is affecting people and communities in the Aleutian and Pribilof Islands region. The Nelson Lagoon Tribal Council has for several years been concerned about the effect of changing weather patterns on the narrow spit of sand they occupy between Nelson Lagoon (a prime nesting habitat for Steller's eider (*Polysticta stelleri*)) and the Bering Sea. The changing climate is having dramatic effects on the security of the village and the local infrastructure.

Like many Alaskan coastal communities, Nelson Lagoon has been battling the effects of winter storms for years, most notably by building increasingly strong breakwalls along the shore. The increasing violence of the storms and changing winter sea-ice patterns have exacerbated the problem, reducing sections of a structure they hoped would provide decades of protection to kindling within just a few seasons. This is because their breakwall was designed to brace the shore ice, which would in turn provide the real buffer from winter storm wave action. As the winters have been warmer over the past six years, the buffer provided by the shore ice has been lost, allowing the full force of the waves to surge against the wall and the village.

In addition to the breakwall, other vital infrastructure has been disrupted by the changing weather patterns. In 2000, the pipeline that provides the village's drinking water was threatened when storm waves eroded cover soil and caused a breach in the line. US Public Health Service engineers made an emergency trip to Nelson Lagoon to help repair the damage.

Based on these events, the Nelson Lagoon Tribal Council applied for funding from the US Environmental Protection Agency to establish a tribal environmental program with a focus on community planning and increasing understanding about the long-term impacts of climate change. This program is currently under development.

Other climate-related observations from the Aleutian Islands include the presence of non-indigenous warm-water fish species. One example is salmon sharks (*Lamna ditropis*), which have been observed as far north as False Pass, an important migratory passage for whales, salmon, and other marine life between the North Pacific and Bering Sea. It is likely that salmon sharks will soon be seen in the Bering Sea, competing for resources with marine mammals such as the Steller sea lion (*Eumetopias jubatus*).

On the Pribilof Islands in the Bering Sea, many changes have been seen. On St. George Island, the gravel of Staraya Artil Beach is being washed away, leading to shoreline erosion. As the gravel recedes and sea level rises, various artifacts and animal bones are appearing. No one on the island has seen anything like this before. On neighboring St. Paul Island, the 2002/2003 winter was warm with little snow. As a result, there were few eiders, which reduced subsistence hunting opportunities. On the other hand, northern fur seals (*Callorhinus ursinus*) and Steller sea lions remained on the island throughout the winter.

For such reasons, the Aleut are very interested in finding better ways to monitor and study climate-related changes in the region, as well as to affect global policies. The Aleut International Association recognizes the need for comprehensive monitoring of the Bering Sea region from the Alaska Peninsula in the east to the Commander Islands in Russia in the west. Climate change should be an integral part of such monitoring. Although climate change observations in the region are numerous, almost no systematic records have been analyzed to document climate change, either in Alaska or Russia.

The Aleut, who live in the region all year round, have intimate knowledge of the land, sea, and climate. They are an invaluable resource and important partners in research. Currently, Aleut tribes in Alaska are developing their marine research capacity, which includes working with villages all along the Aleutian chain and engaging the region's small-boat fishing fleet. This is a large untapped resource, which could provide an effective link between traditional and western knowledge, involve the region's people in efforts to understand sci-

entific research, and empower those most affected by climate change and its impacts. The current tribal structure and environmental program capacity provide excellent opportunities for research focused on climate change and other emerging issues. The Aleut International Association enables international research and monitoring in the Bering Sea by connecting peoples and governments on both sides of the Bering Sea and by helping the Aleut people address the most vital problems that they face today, one of which is climate change.

3.4.3. Arctic Athabaskan Council: Yukon First Nations

The Council of Yukon First Nations held a series of workshops in February 2003 to address climate change. On February 12 and 13, twenty elders representing the Elders Council of the Council of Yukon First Nations took part in the Elders Climate Change Workshop. This workshop happened as a result of suggestions given by community elders. They said that the changing northern climate was an important issue because it had implications for human health and cultural survival. Following a round-table discussion, the elders divided themselves into three groups – representing the north, central, and south Yukon regions. Each elder shared his or her knowledge and concerns about changes in weather and how these changes have affected the way of life in their nation's traditional territory. A second Yukon First Nations Climate Change Forum was held February 26 and 27. Elders, community representatives, scientists, and government representatives participated in the forum. Elders listened to explanations of government programs and research and provided suggestions as to how federal and territorial governments and researchers could work in partnership with Yukon First Nations communities to develop a long-term strategy for climate change.

Yukon First Nations elders share a growing concern about the changes that are taking place on their lands and in their way of life. Climate change is responsible for some but not all of these changes. Contaminants from local sources and long-range transport have been a source of worry for cultural activities such as trapping, hunting, and eating traditional foods, resulting in changes to community social life. As the environment has changed, people have had to learn to cope with the change. Understanding why things have changed and why this is happening helps people to adjust. Yukon elders want to share what they know, and to learn as much as they can from others about climate change and its impacts. Only through knowledge and the ability to take action for themselves will Yukon First Nations be able to respond effectively to climate change.

At the Elders Climate Change Workshop and the Yukon First Nations Climate Change Forum, elders and community representatives described the changes that they have seen. In the northern Yukon, freezing rains in November have meant that animals cannot eat. Birds that usually migrate south in August and September are now being seen in October and November. In some areas, thawing permafrost has caused the ground to drop and in some cases has made the area smell foul. In more southerly communities, rings around the moon are no longer seen, although they are still visible in the northernmost community. There are increased sightings of new types of insect and an increase in cougar (*Puma concolor*) and mule deer (*Odocoileus hemionus*). People used to be able to predict when it would get colder by looking at tree leaves. It is difficult to do that now. Lakes and streams are drying up, or are becoming choked with weeds, making the water undrinkable. Many animals are changing their distribution and behavior. Bears used to go into their dens in October and November, but are now out until December. One bear was spotted in winter sleeping under a tree but above ground, rather than in a den, which was regarded as an exceptional and unprecedented sighting.

Many of these changes have been observed since the 1940s, although some began earlier. These changes have produced much concern and anxiety. People are concerned about the future. Habits have changed, so that people now depend more and more on market foods and eat fewer traditional foods. In the face of such changes, people often mobilize to take action. One result in the Yukon is that elders are willing to assist in developing a strategy and are prepared to find ways to cope with the changes, as difficult as that may be. In many statements, the elders expressed their fears about these changes. One said that he had never expected to see the day when people would worry about water, but now he hears about that all the time. Another remembered, "We were able to read the weather, what it was going to be like".

At the workshops, Yukon First Nations elders made it clear that partnerships must be developed between the federal government, the territorial government, and Yukon First Nations, as well as with scientists and non-governmental organizations. Each community must have the ability to prepare for change and to be a part of designing the strategies that are being developed in relation to climate change. At the international level, Yukon First Nations elders have shared their traditional knowledge and cultural values to help shape policies through the Arctic Athabaskan Council. The elders and community representatives concluded that their workshops were just a start. More is needed. To that end, an elders panel of five members has been created. This is to be consulted and to have a representative who will work with the Council of Yukon First Nations and the Arctic Athabaskan Council.

3.4.4. Denendeh: the Dene Nation's Denendeh Environmental Working Group

This case study is the first official publication of Dene observations and knowledge on climate change. There have been documents by newcomers, missionaries, fur traders, and others, in which it was observed and noted that the Dene knew much about their relationship with the land. There have been articles and books written on the problems of relying on these observations (Brown J. and Vibert, 1996; Krech, 1999). The Dene are continuing to write their histories and curriculum for their schools, teaching us about their world, but these are beyond this discussion (Erasmus et al.,2003; Watkins 1977).

Raven is known as a trickster. He said let's see who can kill a moose. So everyone ran into the bush. The Raven said he killed a moose. The people asked what are we going to do with the intestines? They said to string it along the creek. They were melting the grease into the intestine. (This was the first pipeline). They were wondering why this grease was never filling up. So a couple of kids followed him. They found the Raven at the end drinking the grease. So they threw a bone in the grease. The Raven swallowed the bone. That's why the Raven today says, KAA! Bella T'selie, Liidlii Kue, Denendeh, March 12, 2003

My mother used to get the best water and spruce bough for the floor. It was natural. Even if we spilt something it would seep through the spruce bough. So when we lived in a house we didn't know how to make the transition. People got sick. There's not many people who can live in both worlds. We have sacred lands that we don't go into. We knew of this one area to be a bad place (Deline). After development, we found out how bad it really was. Scientists like to talk about things apart. We think in holistic terms and cannot think about things separately. Dene spirituality is in traditional knowledge. Dene ways are very formal. We cannot separate spirituality in Dene, but scientists think this is ridiculous. Bella T'selie, Liidlii Kue, Denendeh, March 12, 2003

We survive by caribou. When you hunt caribou it can take up to three weeks for the trip. That's why we need to protect our caribou. That's why I brought some caribou for you to taste. Eddie Camille, Liidlii Kue, Denendeh, March 11, 2003

Like all environmental issues, climate change is understood and talked about by the Dene (Chief Roy Fabian in Dene Nation, 2002, Chief Sam Gargan in Dene Nation, 2003) as it relates to the people and the land (Dene Nation, 1984). Denendeh will face significant change based on current climate change scenarios and models. The issues of changing climate are different for Dene than they are for other indigenous peoples and each has much to contribute to this discussion. Dene knowledge speaks to the past, and explains the now, as well as what may occur in the future. This knowledge is different from what scientists know about climate change (see for example the discussion on climate and the high Arctic in Dick, 2001). Each form of knowledge can be gathered together, not necessarily to create a single synthesis, but to allow each to appreciate and increase what is known about climate change from both perspectives. The Denendeh Environmental Working Group workshops are a first step in a larger effort to bring Dene views and voices into climate change discussions in the north, in Canada, and into international discussions such as the SnowChange Conference (see section 3.4.7) and this assessment.

3.4.4.1. Dene Nation

Dene Nation is a non-profit Aboriginal governmental organization mandated to retain sovereignty by strengthening Dene spiritual beliefs and cultural values in Denendeh, which encompasses five culturally and geographically distinct areas, six language groups, and is home to over 25 000 Dene in twenty-nine communities. The settlement of communities has varied as has the population and composition of each community. As indigenous peoples their cultures, languages, and title come from time immemorial. In the international arena the Dene use the Arctic Athabaskan Council, a Permanent Participant to the Arctic Council, and the Assembly of First Nations. These links enable the Dene to tackle difficult science and policy issues of climate change by maintaining activities at the local to international level.

3.4.4.2. Climate change policies and programs in Denendeh

The Dene have always observed the climate and have stories that speak about the way things were before time and as they are meant to be in the future. Climate change, as discussed here, is concerned with the phenomenon that has been speeding up following industrialization. Since the 1970s, a great deal of what has entered into Dene thinking has come from international discussions of greenhouse gas emissions and global warming (e.g., Brown et al., 2000). Changing climate is indeed being experienced as local changes on the land; however, policies and programs dealing with these changes often have little to do with the needs of the people on the land.

In order to discuss Dene knowledge, impacts, and adaptations of changes in climate, it is first important to place into context the reason why Dene observations and knowledge are being documented. The development of national government policy and programs has had a significant influence on Dene understanding of climate change and a brief summary of government policy and programs provides the context from which to understand the development of the Denendeh Environmental Working Group.

The Canadian government's initiatives fund a variety of activities. The general direction for the bulk of these programs, toward national objectives, means that there is a limited engagement for Dene in these activities. Work at the local level must serve some national priority or it is not funded. For example, taken to its full extent the development of the Hub Pilot Advisory Team and the Public Education Outreach gives the appearance of democratic and public institutions responding to national information needs surrounding climate change.

Likewise, the Canadian Climate Impacts and Adaptation Research Network facilitates a transparent network of researchers and government scientists working collectively on climate change. These and other activities appear to be developing a critical mass of information, but are not focused on the specific or local realities of climate change in any single location; how it is being experienced. The Joint Ministers of Energy and Environment are not concerned specifically with the fact that indigenous peoples across the north are experiencing climate change differently. In order to counter the homogenizing influence of federal programs, limited attempts, in particular the Northern Ecosystem Initiative under Environment Canada, have evolved, responding to particular realities of northern Canada and the indigenous peoples living there.

Northern indigenous peoples' organizations in Canada have approached climate change independently and the result has been a piece-meal approach to bringing forward indigenous views and knowledge. Hearing from the people themselves is important (Dene Nation, 2002, 2003; Nunavut Tunngavik Inc. et al., 2001). Equally important is finding ways to interpret what is being said by these people to those who lack the historical and anthropological knowledge necessary to understand fully what is being said (Krupnik and Jolly, 2002). This is particularly pronounced in climate change research because of its heavy reliance on physical scientists who may wrongly conclude that indigenous knowledge is anecdotal and therefore without scientific value (Dene Nation, 1999). Dene Nation decided that the most efficient way of contributing to discussions on climate change was by sharing some of what was gathered during workshops where Dene knowledge could be shared and documented.

3.4.4.3. Denendeh Environmental Working Group

The Denendeh Environmental Working Group (DEWG) developed out of two complementary forces. First, a number of climate change issues were of interest to the Dene and had broad policy/program implications for Denendeh. Second, the Environment and Lands division of Dene Nation had previous experience with a Denendeh environment committee. Dene Nation's objective in forming the DEWG was to work on specific areas of collective interest, to educate about the issues and best apply the policies and programs that exist, and to lobby in all arenas regarding Dene sovereignty, spiritual beliefs, and cultural values.

The DEWG is a non-political forum where Dene and invited guests from government, academia, and non-governmental organizations can gather to share climate change knowledge and observations. The first workshop on climate change was held in Thebachaghe (Fort Smith) in 2002. The second, held in Liidlii Kue (Fort Simpson) in 2003, examined climate change and forests. Themes of future workshops include water and fish.

The DEWG membership changes with each meeting, but includes one technical staff member working on environment and lands issues from each region and a regional elder. There are a number of concerns about changes in climate that are specific and unique, as well as others that are common to each Dene community and region. These differences and commonalities are recognized and the issues and knowledge brought forward during each meeting are documented. The mixture of tradition and modernity find common hearing and attempts are made to improve what is known from both traditional knowledge and practices and science and government policy and programs. In the future, transcripts and tapes could serve a number of educational and other research purposes. The work of documenting Dene concerns and responses to climate change is still in its infancy. Conclusions should not be drawn on impacts and adaptations until a critical mass of information is gathered, the definition of which was an important consideration during the DEWG meetings during 2003 and 2004.

The DEWG facilitates more than the documentation of Dene views and knowledge on climate change issues. A significant goal is to facilitate the sharing among regions of climate change knowledge in a systematic way. The workshops are proving to be an opportunity for people from each region to hear from one another about changes they have observed.

The broad nature of climate change research and programs is being complemented and improved by focusing at the regional and local levels. The DEWG is important for education and outreach. In particular, each meeting challenges the participants to find ways to communicate the sometimes-complex science of climate change in a way that is accessible and free from technical jargon. It challenges elders and technical staff to speak with scientists and build, when necessary, research alliances and networks of researchers.

The DEWG has been shaped by four basic discussion questions:

1. Is there a difference today in Denendeh and is climate change having a role in these changes, what else may be causing it?
2. What climate change programs are there and how can our communities be more involved in research and communication about these changes?
3. If it is important to document Dene climate change views/knowledge, how should we communicate this knowledge with each other and to policymakers, governments, and others outside the north?
4. Is the DEWG a good mechanism to discuss climate change, what should we be talking about, and what else do we need to do?

Answers to these questions continue to evolve. Elders, in particular, find that there is a difference in how the climate is changing in Denendeh, and that many changes

on the land are attributed to climate change, or that climate change is at least having a role in the land and animals being different. The behavior of Dene themselves, including increased use of transportation like vehicles and skidoos, is blamed in part for increased changes in the climate and overall health of Denendeh.

Change is manifest in how animals behave, such as wolves acting unpredictably. Invasive species, such as moose moving further north and buffalo (*Bison bison*) moving into Monfwi region, are being observed. Birds never before seen and increasing variations in insects are also being noted. A problem identified for trees was increased pine and spruce parasites and diseases.

The overall health of trees and their ability to fight disease and withstand the increased frequency of insect infestations is of concern for forests in Denendeh. For the past five to six years, trees have been dying in greater numbers. Elders noted that trees were soggy and not frozen through so that in cases of emergency one could not easily save oneself by making a fire. This was serious as ice was unpredictable in places and there were increased instances of people falling through.

In the Mackenzie Delta, many channels had changed with some widening, making winter land travel impossible. Groundwater is down in some areas because of increased levels of vegetation, especially willows. Changes in vegetation were not the cause but rather the effect of climate change. Changes in water also lead to increased disease in wildlife. Elders are observing how many freshwater fish populations and runs are less healthy, there is increased occurrence of "unhealthy fish".

Interconnectedness among all parts of the environment is a feature repeated by elders during the workshops. The links between activities on the land, development impacts, lack of capacity to deal with change, and the overall adaptive ability of Dene cultures are important considerations. For the Dene, it is wrong to separate climate change from human and governance issues. Dene talk about climate change in part as "how things grow" (Dene Nation, 2003).

Climate change is affecting how traditions are maintained. It is difficult to demonstrate to non-Dene the impacts of climate change on traditions and to identify those impacts that have other sources, for example television. The changes and associated impacts are known by elders and others who continue Dene traditions, and stem from the overall unpredictability of climate and increased warming which alters the availability of species and so on. An indication of the erosion of traditions is that, "in the old days, everything was dried; meat and fish". Less food is dried now. For example, the organizers of the DEWG workshops could not secure an adequate supply of dry fish and dry meat for participants, even though these foods were once abundant in Dene communities. There are very good lunches but typically few traditional foods are served. In the future the supply of food and food preparation methods will be questioned in more detail. In this point lies the strength of having a series of workshops by and for the Dene on climate change and associated issues; to continue to learn and to share/protect Dene observations and knowledge.

Traditional knowledge teaches Dene about relationships, to know how things are related to each other. So for example, when asking how trees are affected by changes in climate, it may be appropriate to consider what is happening with drinking water. The relationship may be that there are different trees now, the willows having replaced spruce and other trees dying off, while the water tastes bad because of warming. Seen in this way, the entire world relates to all other parts, including the Dene.

Dene explain both the causes of climate change and the future in terms of their daily lives and cultural understandings. The physical connections Dene have with the land have been much discussed; always with an underlying concern about more than just the physical and immediate concerns of everyday life. What a person does affects everyone else, whether it is throwing garbage into the river and affecting those downstream, or the way that cities create huge ecological footprints from car exhaust, factories, and industries. Also, what was done in the past affects now and into the future.

The integrity of culture and land is essential to Dene. They have made many observations and see climate change as more than the weather becoming warmer. Some discussion was placed on the weather and how it was becoming unpredictable. In the past, Dene elders could predict the weather, but this is no longer the case. Warmer temperatures and changing precipitation patterns, although seasonal and spatially variable, cause concern for animal migrations, in particular the lack of snow causing caribou to wander all over the place whereas in the past they would break trail for each other and stay together. There is a Dene legend about people in the past who were able to control the weather. These people would predict the weather. They could tell what was going to come before it happened by watching the color of the sky and connecting this to cloud patterns. Not many people can predict weather anymore.

Holism of ecosystems means all Dene, elders and youth, working together equally. When Dene talk about trees, they talk as well about porcupine (*Erethizon dorsatum*) and fish. It is important to understand Dene concern for the overall health and faith all Dene have for the land. They are not as confident today with the natural environment. It is being changed by anthropogenic development far removed from Denendeh. Adapting to these impacts will depend, in part, on strengthening Dene teachings and traditional knowledge in order to provide a solid cultural foundation for actions that are taken. In this way, adaptations and other responses can be developed that are consistent with Dene cultural values but that may draw upon other resources such as outside technology.

Proceedings of the workshops have been produced which summarize the discussions and outcomes of the meetings. To protect the interests of the Dene, the detailed content of these reports is protected and cannot be reviewed or used without the full, involved, and meaningful consent of the Dene. In addition, Dene Nation has developed a web page summarizing the key findings of the DEWG (www.denenation.com). Digital photos and audio recordings, in each of the languages spoken during the meeting, were also collected and are housed in the resource center/archives of Dene Nation. The Dene want both to share and protect their knowledge. At the first workshop the working group asked that a book be written. Each region and each person is writing a chapter in this book on Dene observations and knowledge of climate change. Each meeting brings out something different from what is going on and what collectively is the traditional knowledge and practical projects the Dene are engaged in. Much is being learned from Dene elders, but it is often not written, as this is not how it has traditionally been conveyed, learned, and taught. An important issue raised was how intellectual property rights work and that knowledge holders had rights that had to be respected. Guests and observers to the workshops were asked not to present information verbatim, as expert testimony, and for profit. The ownership of traditional knowledge and stories, the benefits accrued by appropriating the voices or representations made without permission is a serious consideration. Once written, the form of knowledge that is conveyed is the lowest (crudest) level of understanding. Dene oral tradition and practice is the authority and so even what is written here is only one telling of the climate change story.

Regional technical staff spoke about changes in climate, and causes and responses to these changes, including traditional knowledge and stories passed down from their relatives. Dene have always been able to adapt to change. For example, Gwich'in learned to cope with and understand change. In the Sahtú, the people lived off the land for generations, and many recalled parents and grandparents teaching how the climate was changing and that the sun had changed the most. The strength of the moon was observed to be "not as bright" as it was in the past.

Dene demonstrate extensive knowledge about climate change in their daily lives and during the workshop a small amount was documented. This was done to share Dene views and observations among regions and at the international level. Dene knowledge is shared to educate non-Dene. There is significant concern that traditional knowledge, observations, and activities on the land be made apparent to Canadians and to the international community so that what is known can be improved, so that better decisions can be made.

> *All the problems are man-made. We have to make a lot of noise to be heard. There's some places down south you can't even go fishing but we can still go fishing up here. All the stuff going into the water from down south is coming up here. We need to put the fire out at the source.*

We can't forget our traditional knowledge and to use our Elders. I want our children a hundred years from now to say "My god, they did a good job!" Leo Norwegian, Liidlii Kue, Denendeh, March 13, 2003

3.4.5. Nunavut

In 1995, Shari Fox began a research project with the communities of Iqaluit and Igloolik to help document and communicate Inuit observations and perspectives of climate and environmental change. In 2000, the communities of Qamani'tuaq (Baker Lake) and Kangiqtugaapik (Clyde River) joined the project. A long-term, multi-phase approach has driven the research and the integration of multiple techniques such as interviews, focus groups, videography, and mapping were used to collect, analyze, and communicate information (Fox, 2002). Close collaboration with individuals and communities has been central to the project. The case study presented here draws on two examples from the project to show how Inuit in Nunavut are observing and experiencing climate and environmental changes and their associated impacts and hazards. Comprehensive findings and discussion are presented by Fox (2004).

Inuit have a traditional juggling game. The weather is sort of like that now. The weather is being juggled; it is changing so quickly and drastically. N. Attungala, Baker Lake, 2001

The people of Nunavut have reported many observations of environmental change (e.g., Fox, 2002; Thorpe et al., 2002). Two examples that appear to be related to climate change are increased weather variability across the region and changing water levels in the Baker Lake area.

3.4.5.1. Increased weather variability

Participants in all four communities in this project have noted an increase in weather variability and unpredictability since the early 1990s, an observation shared by many other communities in several parts of the Arctic (Riedlinger et al., 2001; Whiting, 2002; sections 3.4.1 and 3.4.8). The weather has become unpredictable, with more extremes, and elders can no longer predict it using their traditional skills.

The weather has changed. For instance, elders will predict that it might be windy, but then it doesn't become windy. And then it often seems like its going to be very calm and then it suddenly becomes windy. So their predictions are never correct anymore, the predictions according to what they see haven't been true. P. Kunuliusie, Clyde River, 2000

The weather when I was young and vulnerable to the weather, according to my parents, was more predictable, in that we were able to tell where the wind was going to be that day by looking at the cloud formations... Now, in the 1990s and prior to that, the weather patterns seem to have changed a great deal. Contrary to our beliefs and

ability to predict [by] looking at the sky, especially the cloud formations, looking at the stars, everything seems to be contrary to our training from the hunting days with our fathers. The winds could pick up pretty fast now. Very unpredictable. [The winds] could change directions, from south [to] southeast in no time. Whereas before, before 1960s when I was growing up to be a hunter, we were able to predict. L. Nutaraluk, Iqaluit, 2000

Right now the weather is unpredictable. In the older days, the elders used to predict the weather and they were always right, but right now, when they try to predict the weather, it's always something different. Z. Aqqiaruq, Igloolik, 2000

When I lived out on the land I would always know when the weather would be bad from the clouds but nowadays when you look at the clouds and they say there is going to be bad weather that evening it doesn't always happen. You could wake up and it would be nice and through the whole evening it would still be nice. The indicators that we would use before don't always happen. J. Nukik, Baker Lake, 2001

Sudden storms and unpredictable winds make for hazardous travel conditions. No longer confident in their weather predictions, experienced hunters in Clyde River pack for several extra days when heading out on the land, expecting to be caught in unpredictable weather. In Baker Lake, erratic weather has proved especially dangerous in winter. According to residents, changing wind patterns have changed the snow structure around Baker Lake and, in combination with unpredictable weather, dangerous situations have occurred in recent years. For example, changing wind patterns in the area have packed the snow extremely hard. As a result, hunters and travel parties are unable to build igloos, still commonly relied upon as temporary and emergency shelters. A number of accidents and deaths on the land in the last few years have been blamed on sudden storms and those involved not being able to find good snow with which to build shelters (N. Attungala, Baker Lake, 2001).

There used to be different layers of snow back then. The wind would not blow as hard, not make the snow as hard as it is now... But nowadays, the snow gets really hard and it's really hard to tell [the layers] and it's really hard to make shelters with that kind of snow because it's usually way too hard right to the ground. T. Qaqimat, Baker Lake, 2001

For many elders, increased weather variability and unpredictability also have an emotional and personal impact. For much of their lives they have been able to advise the people around them confidently about when and where to travel, providing weather predictions. As their skills no longer work, some elders are now less confident and feel sadness that their advisory roles have changed.

Some of the ways people are coping with more variable weather have already been mentioned – for example, packing extra supplies on trips. Longer-term strategies are more difficult to design. Inuit are careful to note that the weather is always changing, it has always changed – the weather is always different. However, according to many people, the recent climate and environmental changes are outside the expected variability. In turn, while there have always been accidents on the land, some people are concerned that recent environmental changes are to blame for recent accidents and this requires further investigation so that precautions can be taken.

3.4.5.2. Changing water levels in Baker Lake

Baker Lake is the only inland Inuit community in Nunavut. Many of the groups that settled here brought a heritage closely tied to survival from lakes and rivers (Webster, 1999). For example, Harvaqtuurmiut depended on caribou and specialized in hunting at autumn river-crossings. Ukkuhiksalingmiut, from the Back River area, relied mainly on fish. Groups such as the Qairngnirmiut have always hunted and fished in the Baker Lake region. After settling in the community in the 1950s, these groups and others often returned to their traditional hunting and fishing grounds, but also used their skills in local waters.

Hunters explain that in the 1940s and 1950s they traveled without difficulty by boat and outboard motor through the lakes and rivers around Baker Lake. In the 1960s, residents began to observe that water levels were gradually dropping. Since the 1990s, water levels have dropped dramatically, with extremely low levels observed since 1998. Between 1998 and 2002, travel routes on rivers were blocked by shallow water and hunters were unable to get to the caribou hunting grounds and their usual camping areas. Large amounts of equipment made portaging difficult or impossible through these newly shallow areas and many hunting parties had to turn back. Accessible only by boat, these summer caribou hunting grounds can no longer be reached.

The shallow water has also affected fish. There are fewer fish in areas where they are traditionally expected, and when present are often small or skinny. Lake trout (*Salvelinus namaycush*) are darker than usual. Fish also smell different, "like earth", and no longer have white fat layers between the meat. While Inuit note that there are other possibilities for changes in fish (e.g., pollution), there are connections between the observed changes and recent water levels.

> There is a lot less water, around all these islands [in Baker Lake]. The shore is getting closer to Sadluq Island, they are almost joined together. There used to be a lot of water. We could go through with our outboard motors and boats, but now there is getting to be less and less water all over… At the mouth of Prince River there used to be a lot of fish and you used to be able to get char [Salvelinus alpinus]. There's been a lot less fish because there's not as much water anymore. And we used

> to be able to get a lot of fish all the time at Qikiqtaujaq and all the other places where you can get fish. The fish were more plentiful and they used to be bigger. Now you hardly get char anymore at Prince River or any of these fishing places because the water level has gone down.
> L. Arngna'naaq, Baker Lake, 2001

> The lakes and the rivers are getting less water and a lot of them are getting more shallow and some places don't have any water left. They're not as healthy anymore. Things are not as healthy because there is not as much water and there were a lot places that probably in the '50s there was a lot of good water around [and] you could travel all over the place. [There used to be] a lot of water in the lakes and rivers and anything that had water was a lot cleaner, but now some of the waters have things that cause illnesses and that has really affected the food in the water and also things that eat things that are in the water. That is quite dangerous, the level of the water going down, because of the effect on the things in the water and the things that use them. Like we eat fish or anything that gets things out of the water because the water is not as healthy as it used to be and there is less water. N. Attungala, Baker Lake, 2001

In the community, these observations are shared between hunting parties, families, and at meetings of hunters and trappers. Different individuals and groups make assessments about lake and river conditions based on their own knowledge and the knowledge and information shared by others. People understand lake and river changes by linking experiences on the water to observations of other factors such as precipitation, temperature, history of the water body, condition of animals and fish populations, and vegetation. In the case of recent low water levels around Baker Lake, residents identify the situation as unprecedented, both in terms of water levels and the condition of the fish populations. While unsure as to the cause, residents suspect lower water levels are related to changes in the climate over the last decade, particularly warmer, drier summers that last longer than normal. Others explain that the land is "growing", which may be another way of looking at permafrost thaw. Scientists have studied how thawing permafrost can cause water levels to drop since the soil is able to hold more water (Rouse et al., 1997).

Like many of the environmental changes being experienced in the Arctic, low water levels and subsequent impacts are a relatively recent phenomenon (or have not been experienced in a long time). Coping strategies are focused primarily on the day-to-day. Hunters adjust their travel routes and hunting grounds, looking for caribou in other areas. Residents change fishing areas looking for more abundant and healthier fish populations. While these strategies allow residents to continue day-to-day life, it is unclear what will happen if the water condition persists or worsens. In the community, long-term plans are yet to be developed, but are of interest.

3.4.5.3. Discussion

Unpredictable weather and lower water levels are two examples from a number of environmental changes observed by Inuit in Nunavut in recent years. In some cases, these changes are clearly linked to climate and in others they are not. Sometimes the observations with the least obvious linkages are the ones that spark the most interest for scientists and offer the most opportunity for advancing knowledge of climate change. For example, although unpredictable weather could be a matter of knowledge erosion (i.e., elders no longer live on the land and thus have lost prediction skills), something recognized by Inuit themselves, the fact that indigenous observers seem to consistently cite recent variable weather patterns across Nunavut and other parts of the Arctic has raised some scientific eyebrows that the phenomenon could be linked in some way to climate change (Fox et al., in prep). However, it is probably neither knowledge erosion nor climate change exclusively. As with many environmental changes in Nunavut, multiple factors including climate and cultural factors, interact to create the impacts felt by Inuit. For Nunavut, these factors and their interactions are just beginning to be uncovered. Also, many of the climate and environmental changes observed by Inuit refer to the last decade and so ties to long-term climate change are still to be made – although many useful indicators have been identified. What is clear is that Inuit will play a future role in such investigations and that their observations will help guide, inform, and challenge scientific efforts to understand arctic climate and environmental change.

3.4.6. Qaanaaq, Greenland

Uusaqqak Qujaukitsoq is a hunter from Qaanaaq, North Greenland. He has served in the Greenland Parliament and on the Executive Committee of the Inuit Circumpolar Conference. Uusaqqak has been involved in many natural resource use and conservation issues, including the Greenland Home Rule Government's seal skin campaign. His comments were made in response to an invitation to describe climate changes that have occurred in his region.

Change has been so dramatic that during the coldest month of the year, the month of December 2001, torrential rains have fallen in the Thule region so much that there appeared a thick layer of solid ice on top of the sea ice and the surface of the land. The impact on the sea ice can be described in this manner: the snow that normally covers the sea ice became *nilak* (freshwater ice), and the lower layer became *pukak* (crystallized ice), which was very bad for the paws of our sled dogs.

In January 2002, our outermost hunting grounds were not covered by sea ice because of shifting wind conditions and sea currents. We used to go hunting to these areas in October only four or five years ago. It is hard to tell what impact such conditions will have to the land animals. Since I haven't been out to see the feeding grounds of the arctic hares (*Lepus arcticus*), musk ox (*Ovibos moschatus*), and reindeer this year, I can't tell how it is, but I can guess that it will be difficult for the animals to find anything to feed on because of the layer of ice that covers everything. It is hard to say what can be done about these conditions.

Sea-ice conditions have changed over the last five to six years. The ice is generally thinner and is slower to form off the smaller forelands. The appearance of *aakkarneq* (ice thinned by sea currents) happens earlier in the year than normal. Also, sea ice, which previously broke up gradually from the floe-edge towards land, now breaks off all at once. Glaciers are very notably receding and the place names are no longer consistent with the appearance of the land. For example, *Sermiarsussuaq* ("the smaller large glacier"), which previously stretched out to the sea, no longer exists.

3.4.7. Sapmi: the communities of Purnumukka, Ochejohka, and Nuorgam

This case study comes from a project carried out as part of the SnowChange program organized by the Environmental Engineering Department at Tampere Polytechnic in Finland (Mustonen and Helander, 2004). SnowChange sponsored conferences in 2002 and 2003 at which arctic indigenous peoples, interested researchers, and other parties discussed their observations and concerns regarding their cultures and the effects of climate change and other aspects of the modern world. In addition, SnowChange researchers and students have conducted several projects around the Arctic, documenting indigenous observations and perspectives. The study took place in two locations: the small reindeer herding community of Purnumukka, in central Lapland, and the Saami communities of Nuorgam and Ochejohka (Utsjoki) in the northeast corner of Finland, the only part of the country where Saami represent the majority of the population. In Purnumukka, initial community contacts, networking, and interviews took place in September 2001. In March and April 2002, Mika Nieminen spent a month in the region, living and practicing reindeer herding with Pentti Nikodemus and Riitta Lehvonen and interviewing active herders, hunters, elders, and fishermen. SnowChange researchers have been in active communication with Purnumukka residents on a monthly basis since then. In Nuorgam and Ochejohka, SnowChange researchers spoke with elders, reindeer herders, fishermen, and cultural activists about the changes taking place in the local area and communities. The interviews were conducted in March 2002 in the Skallovaara reindeer corral area, in Ochejohka village, in Nuorgam, in the remote area of Lake Pulmanki, and in the village of Sirma, which is on the Norwegian side of the Deatnu (Teno or Tana) River.

Sapmi is the Saami homeland that extends across northern Norway, Sweden, Finland, and Russia. Residents of the Finnish part of Sapmi have many concerns about changes that are taking place in their region. Not all changes are due to climate. The biggest environmental impacts in the region to date result from the construction for hydropower purposes of the Lokka reservoir in 1967 and the Porttipahta reservoir in 1970, and the massive clear cuts preceding the construction of the reservoirs. The best lichen areas were flooded and the herders had to move 5000 reindeer from the grazing grounds because of the construction. These particular grazing grounds were excellent autumn grazing areas. Many people continue to refer to the creation of the reservoirs as a marker of great change, including changes in weather patterns, snowfall, and ice formation.

This case study presents comments by elders, as they carry the most extensive knowledge, with additions from the younger generation of Saami living in the region. Many more people were interviewed than are quoted here. The quotes used are considered to best illustrate the themes and ideas that typify what was learned from the Saami interviewed and provides insight into what the changes mean for people in the area. The interviews covered more material than is presented in this brief case study. Some of the additional material was reported in section 3.3.

The following people are quoted in this case study.

- Veikko Magga – a reindeer herder for over 50 years and a member of the reindeer herders association of Lapin Paliskunta
- Niila Nikodemus – an 86-year-old elder and the oldest reindeer herder in Purnumukka
- Heikki Hirvasvuopio – 65 years old and still active in reindeer herding
- Aslak Antti Länsman – 55 years old and a local reindeer herder belonging to the Polmak Lake Siida
- Niillas Vuolab – an elder and the oldest reindeer herder in Ochejohka. Niillas Vuolab was born in the Saami Community of Angel in 1916 and came to Ochejohka in the autumn of 1916
- Ilmari Vuolab – the son of Niillas Vuolab. Ilmari Vuolab was a 51-year-old reindeer herder from the reindeer herding area of Kaldoaivi. He passed away in April 2003
- Taisto Länsman – a reindeer herder from Lake Pulmanki with extensive written records of the ice break-ups in the lake

3.4.7.1. Weather, rain, and extreme events

Heikki Hirvasvuopio reflected on the seasonal changes and the autumn weather.

> *Temperatures used to be well below freezing in autumn and winter came when it was supposed to. It was not mild autumn like now. It used to be longer, snow would fall. Now sleet and rain will fall. Summers used to be to the "standard form", this means that fair weather would stay longer. We would have the reindeers in the big fell areas because especially the beginning of the summer would be very hot. Insects would be there as well with the rein-*

> *deers. Now this has changed. The summers are very unstable. Reindeer are staying in the forests now; they do not go up to the fells any more.*

Veikko Magga reported that the amount and consistency of snowfall has fluctuated in Purnumukka and the Vuotso region.

> *It is springtime nowadays when the snow actually comes. Some comes in the autumn, but there is no proper freezing, only so that the lowermost snow freezes. The lichen freezes solid into this layer and the reindeer cannot get proper food because of this. For a couple of years in a row there has been less snow in the autumn than previously but I think sometimes it falls earlier, sometimes later. It has always been this way.*

Niila Nikodemus discussed snow and ice cover based on his extensive experience and a lifetime of observations.

> *There is normal fluctuation in the amounts of snow. However, snow falls later. In the 1950s and 1960s, there used to be a permanent snow cover always in October. Starting in the 1970s, it can be November, middle of November. Ice has thinned, especially in the small rivers and ditches here. I wonder if the reservoirs affected this? It used to be that we could just drive away on the ice. There used to be a proper ice cover on the small rivers.*

Heikki Hirvasvuopio outlined the impacts of climate change on the reindeer herding.

> *During autumn times, the weather fluctuates so much, there is rain and mild weather. This ruins the lichen access for the reindeer. In some years this has caused massive loss of reindeers. It is very simple — when the bottom layer freezes, reindeer cannot access the lichen. This is extremely different from the previous years. This is one of the reasons why there is less lichen. The reindeer has to claw to force the lichen out and the whole plant comes complete with roots. It takes, as you know, extremely long for a lichen to regenerate when you remove the roots of the lichen. This current debate of loss of grazing grounds is not therefore connected with a larger number of reindeers, this is not the case. In previous years, the numbers of the herds were much bigger. The main reason for the loss of grazing lichen areas is the bad weather conditions that contribute to the bigger impact on the area. I do not think many people have observed this reason in their thinking.*

Aslak Antti Länsman said that for the past ten years the autumns have been mild with sleet, which has frozen the surface of the ground.

> *This has affected the reindeer economy for sure. As the ice cover was there this had a negative impact on the reindeer herding. It is a big question mark why these changes are taking place. Is it so called greenhouse effect, or air pollution? Or is it connected with the same events that melted the last ice age? Who knows?*

Ilmari Vuolab stated:

> Yes, the weather conditions have changed overall so it is not possible to be certain that this amount of snow will be at this time. This is a problem.

3.4.7.2. Birds

Heikki Hirvasvuopio talked about the disappearance of birds in Kakslauttanen.

> … especially with ground birds, we could be talking about a near extermination when compared to the previous amounts. I used to hunt them quite a lot while reindeer herding, so I have a good idea of the stocks. We cannot even talk now about the same amounts during the same day. This affects especially ptarmigans, capercaillie, and ground birds. With small singing birds, the same trend is visible. Nowadays it is silent in the forest – they do not sing in the same way anymore. It used to be that your ears would get blocked as the singing was so powerful before. They have disappeared completely as well.

Taisto Länsman was concerned that there were far fewer small birds compared to his childhood. Niillas Vuolab shared the concern and stated that:

> Nature has grown much poorer. For example during summers migratory birds are fewer today. We see the usual species, but the numbers are down. Especially marine birds, such as long-tailed ducks [Clangula hyemalis], white-winged scoters [Melanitta fusca], black scoters [Melanitta nigra]; we do not see these anymore. After the war and even later great big flocks would come here. I feel ptarmigans have diminished as well.

3.4.7.3. Insects

Heikki Hirvasvuopio said:

> Both mosquitoes and gadflies have disappeared. Especially we can see gadflies disappearing. And the reason is that in the olden times reindeers had to move up to the big fells as the vermin were plentiful then.

Niila Nikodemus pointed out that:

> The [number of insects] really depends on the particular summer. If it is a rainy springtime they will be plenty, but if it will be a dry spring, hardly any will appear.

3.4.7.4. Traditional calendar and knowledge

Veikko Magga stated:

> Traditional knowledge has changed like reindeer herding in a negative direction. We have to feed the reindeers with hay and fodder quite much now. But I would not advocate that the traditional Saami calendar is mixed up yet. But traditional weather reading skills cannot be trusted anymore. In the olden times one could see beforehand what kind of weather it will be. These signs and skills holds true no more. Old markers do not hold true, the world has changed too much now. We can say nature is mixed up now. An additional factor is that reindeer herding is being pressured from different political, social, and economic fronts at all times now. Difficulties are real. A way of living that used to support everything is now changing and people do not find employment enough.

Heikki Hirvasvuopio discussed traditional forecasts.

> The periods of weather are no longer the norm. We had certain stable decisive periods of the year that formed the traditional norms. These are no longer at their places. Certain calendar days, like Kustaa Vilkuna [a Finnish folk historian of weather and culture from the mid-twentieth century] wrote, held really true. But these are no longer so. Today we can have almost 30 degrees of variation in temperature in a very small time period. In the olden days the Saami would have considered this almost like an apocalypse if similar drastic changes had taken place so rapidly. Before I spent all of my winters in the forest and was at home for one week at the most. Nowadays the traditional weather forecasting cannot be done anymore as I could before. Too many significant and big changes have taken place. Certainly some predictions can be read from the way a reindeer behaves and this is still a way to look ahead, weather-wise. But for the markers in the sky we look now in vain. Long term predictions cannot be done anymore.

Ilmari Vuolab stated that:

> The traditional markers of the nature do not hold true anymore. Ecosystem seems to have changed. It is a very good question, as what has contributed to the change. It cannot be all because of cyclic weather patterns of different years. I believe the changes we have seen are a long-term phenomenon. The wise people say that changes will be for the next 100 years even if we act now to reduce emissions. It is obvious that the reindeer herding is the basis of the Saami culture here and other subsistence activities that are related to the nature, such as picking cloudberries or ptarmigan hunting. I feel the Saami have been always quite adaptive people and we adapt to the changes as well. After all, climate changes in small steps, not in one year.

3.4.8. Climate change and the Saami

Elina Helander is a researcher at the Arctic Centre of the University of Lapland as well as a Saami from Ochejohka. She took part in the preparation and community activities for the documentation work that produced the Sapmi case study in section 3.4.7.

The Saami have an ecological knowledge of their own, rooted in the traditional way of life. They have their own knowledge derived from experience, long-term observation, and the utilization of natural resources. This knowledge is best expressed and transmitted through the Saami language. Saami ecological knowledge goes beyond observation and documentation because it is a precondition for their survival. Particularly interesting is the fact that indigenous people like the Saami have long-term experience in adaptation. People in the villages are worried as they face global changes. The Saami are used to combining different economic activities, such as berry picking, reindeer herding, fishing, hunting, trapping, and handicraft. If the changes are sudden, accumulate rapidly, and have impacts on all or most of local resources, and if the resource base is scarce, then the problems start to show themselves immediately. Many claim that the weather has become warmer, and especially the fall and early winter are warm. During the recent years, the ground has not frozen properly in the fall, and there has been little rain in September. There are many salmon rivers and lakes in the Utsjoki area where I come from. When the ground does not freeze in the fall, and there is little snow during the winter, there is very little water in late May and early summer in the rivers and lakes. Then, of course, with little rain during June, the rivers are almost dry and the fish cannot go upriver. But during the recent years, it has happened that in July there are heavy rains. Consequently, the amount of water increases enormously and it becomes impossible to fish in small salmon (*Salmo salar*) rivers. Many herders and subsistence hunters claim that there are no winds anymore. Wind has some positive effects. For instance, wind gathers the snow to certain spots. In other spots there is little snow and it is then easy for the reindeer to dig through where the amount of snow is small. The wind can also make the snow soft, but on the other hand, the extremely strong wind, *guoldu* in Saami, makes the snow hard. During the recent years, the weather has started to change rapidly, so that sudden shifts take place. There are no longer stable periods of a cold weather type. It has also become more difficult to predict the coming weather. People are more careful when moving across lakes and rivers. In our area the moose migrate in early November from north to south, but they can be hindered from doing this if there is no ice in the rivers and lakes. There must come about a radical change regarding the ecological awareness in humanity if we want to do something positive regarding the changes that occur and are predicted to come. When talking about the snow change, we should not only monitor and accept the changes. We have to resist the global changes when resistance is imperative, i.e., when the changes made by man cause serious damage to nature, societies, and people.

3.4.9. Kola: the Saami community of Lovozero

This case study is also drawn from research carried out as part of the SnowChange initiative. SnowChange cooperates fully with the Russian Association of Indigenous Peoples of the North, the national indigenous organization that is also a Permanent Participant of the Arctic Council. For this case study, researchers spoke with elders, reindeer herders, cultural activists, and other local people. Researchers selected from the material gathered in interviews the comments most relevant to the changes local people see in the local ecological and climatic situation. The interviews were recorded and edited by Jyrki Terva, Tero Mustonen, Sergey Zavalko, and several indigenous and non-indigenous students of ecology at the Murmansk Humanities Institute, Murmansk State Technical University, and Tampere Polytechnic. The community visits functioned as story-telling and ecological teaching experiences, especially for the indigenous students who participated.

The tundra is like my dear mother to me! We herded reindeer with the whole family. How else should we do it? We took care of the shelter. We knitted, we washed, we smoothed down clothes. What did we do? We baked bread. When it is warm, it is warm. When it is not warm, it is cold. I spent my whole life on the tundra. Even after I retired, I spent a year in the tundra. Life was easy; the only thing we missed was the television. Before that all we did was to stay in the earth hut. Summer or winter, always living in the shelter in the tundra. Maria Zakharova, Lovozero Elder

The Murmansk Oblast (province) is located in northwestern Russia on the Kola Peninsula. It borders the Republic of Karelia (Russia) in the south and Lapland (Finland) and Finnmark (Norway) in the west. The oblast covers 144 900 square kilometers and has a population of around 900 000. The capital is the city of Murmansk, the largest city north of the Arctic Circle, with a population of around 350 000. The town of Lovozero (in Saami, *Luujavre*) is the main Saami community on the Kola Peninsula, with approximately 800 Saami among its 3000 inhabitants.

Climate change research among indigenous peoples is only just beginning in Russia. Russia's indigenous communities, as well as mainstream society, have many other, often more urgent, social, economic, and political issues to deal with. However, the massive differences in ecosystems and regions across the vast expanse of northern Russia make local assessments of climate change imperative in order to understand their real and specific impacts. Documenting change in the Russian indigenous communities is a vital and much-needed process. Even as the reclaiming of economic, ecological, social, and cultural rights continues, local voices in the remote regions of northern Russia are often not heard. The people consulted during this case study delivered a clear and coherent message – local people should be heard.

3.4.9.1. Observations of change in Lovozero

Documentation of change cannot be separated from broader questions of the development of Russian territories and their indigenous peoples. The people interviewed stated that there are many other concerns in addition to climate change, such as the state of Russian society, economic hardship, and lack of resources. But climate change has had a definite impact on the traditional lifestyle. Larisa Avdeyeva, director of the Saami Culture Center in Lovozero, stated that:

> Reindeer herders especially have observed change. They talk about the changes in the behavior of reindeer. People have to travel with the reindeer and navigate differently. Bogs and marshes do not freeze immediately, rhythms change, and we have to change our routes of movement and this means the whole system of living is under change. Everything has become more difficult. I have conversed with reindeer herders and they have told me of these kinds of observations. They have seen as well that in areas where it was possible to collect a lot of cloudberries [Rubus chamaemorus] before, now the berries are not ripe because of climatic warming and melting of glaciers. Changes are very visible.

Climate change, particularly changes in local weather, has become an increasingly important issue for the reindeer herders and others in Lovozero. Avdeyeva continued,

> Nowadays snows melt earlier in the spring time. Lakes, rivers, and bogs freeze much later in the autumn. Reindeer herding becomes more difficult as the ice is weak and may give away. The rhythm of the yearly cycle of herding and slaughtering of reindeer is disrupted and the migration patterns of the reindeer change as well.

Broader social and environmental awareness penetrated into Russian society from the mid-1980s onward. Olga Anofrieva discussed how this has been seen in terms of local industrial pollution.

> The biggest benefit of Perestroika was that the industries no longer polluted as much as before. Here, locally, nature rested. Now the arms race is being taken down, as are the amounts of military ships and military personnel. One could say that we have now a positive era, quite difficult, but nevertheless so. We find ourselves now in a situation where soon decisive moves have to be made. We must develop ourselves differently. I think changes in the weather are more to do with God's influences.

3.4.9.2. Weather, rain, and extreme events

Avdeyeva reported that climate variation had been witnessed locally and had caused alarm.

> I would say the climate is warming globally, we have already observed this here. For example the reindeer herders coming out off the tundra have said that last year the bogs and rivers stayed open for a long time and it was hard to gather the reindeer. If this event was previously due in November, now we have it in December or January. Bogs stay unfrozen for a long time and it is very difficult to try to catch reindeer in such conditions. The herders say the climate has warmed and everything is a result of that.

Avdeyeva pointed out that change is more than general warming.

> Extreme events have been seen mostly in the spring time. This year we had thunder in May, and usually this occurs in July. Monthly mean temperatures have increased and spring has warmed up. During the winter of 2001 to 2002, there was little snow and that is why there was little water in rivers and lakes. The low water levels have affected negatively boating. Of course we understand all of these events are related. There is way too little rain and storms. There is certainly thunder though. Lighting was rare here before the 1990s, as well as heavy rains. Now there are more of those.

Arkady Khodzinsky, a reindeer herder from Lovozero, provided more details.

> The weather has changed to worse and to us it is a bad thing. It affects mobility at work. In the olden days [the 1960s and 1970s], the permanent ice cover came in October and even people as old as myself remember how on 7 of November we would go home to celebrate the anniversary of the Great Socialist Revolution. These days you can venture to the ice only beginning in December. This is how things have changed. This year the ice came and froze a little early but for sure the weather has changed very much. All began about six years ago. Everything went haywire. Yes, six years ago! Now it can rain in January. Once three years ago I came back from the tundra, there was a full winter there. I was here in the community for some time, resting and lo and behold! On the tundra spring had arrived because it had rained!

Vladimir Lifov, another reindeer herder in Lovozero, concurred.

> Oh, it is warmer. Before when going to the tundra we had to take a lot of warm clothes, otherwise we would freeze. But nowadays you can sleep with just one malitsa [reindeer-skin coat] on during the whole night. It is all right with that one malitsa. Previously we were using as well boots made out of reindeer skin. You never froze your feet in those. But now you do not need them any more.

Lifov described recent winters.

> Yes, it is very interesting. First it snows, then it melts, like it would be summertime. And this all over again. First there is a big snowfall, then it warms up, and then it freezes. During winter now it can rain, as happened last New Year. Before it never rained during wintertime. Rain in the middle of winter? To the extent that snow disappears? Yes, it is true. Rain, and the snow melts!

3.4.9.3. Rivers, lakes, and ice

The herders traditionally use waterways, such as many rivers and lakes of the Kola Peninsula, in their transportation routes. Recent changes have caused uncertainty and the fear that the routes cannot be traveled on safely. Arkady Khodzinsky described the situation.

> *Rivers do not freeze at all; they are only covered in snow. Ice arrives, but the surface of the water drops so that ice is like on top of empty space and then it is covered with snow. Of course the stream flows like it should. But the changes have taken place in the last six to seven years. Before we saw none of that. Well, for the past few years the weather has been different. No decent ice comes anymore. When the freeze-up occurs, the ice sometimes melts right away.*

Vasily Lukov, a reindeer breeder from Lovozero, saw both impacts and a possible explanation.

> *The River Virma grows shallower every year. Now there is hardly any water left and it can freeze all the way to the bottom. There used to be a lot of fish, but now they are almost all gone. I think it is due to the drying of the bogs and marshes, improvements of the ground. Now the melt is slow. First the water gets on top of the ice and the river melts first from the middle. Steep riverbanks are still frozen but gradually they melt as well. Nowadays there is no actual ice melting event like before.*

3.4.9.4. Plants, birds, and insects

In September 2001, Larisa Avdeyeva spoke about the changes in plant life.

> *New species of plants have arrived. We never saw them before. This is what we have observed. New plants have arrived here and on tundra. Even there are arrival species in the river, previously known in middle parts of Russia. This past summer and the previous were very hot here. Rivers and lakes are filled with small-flowered a kind of duckweed [Lemnaceae] and the lake started to bloom. Life for the fish is more difficult and likewise people's fishing opportunities as lakes are closed up by the new plants. We have observed that the trees in our village grow much faster. New unknown plant species have arrived here in great numbers. New bird species have arrived here. As well, the birds stay in our village longer than before. Some new beautiful never-before-seen birds have arrived.*

Reindeer herders have witnessed the changes as well, as Arkady Khodzinsky reported.

> *The birds are about the same as they have always been but their numbers are decreasing all the time. Yes, there are very few birds nowadays. It used to be that there were ptarmigan on top of every bush. Nowadays it is not like that anymore and it feels bad. To give you an example, in earlier times I was sitting and watching the herd. I tapped my foot to the ground and a ptarmigan would fly to me. When I would say "Kop, kop" to it, it would come so close I could even hold it. Then I said again "Kop, kop" and it took off. Now there are very few goose. It used to be that they were all*

> *over. Before, when we were at the camp and we would see geese, we would know the spring is coming. All people enjoy the arrival of spring. Nowadays we see no geese. Occasionally one or two flocks fly over, but this is a rare event. There are no birds of prey any more. Very small numbers of those remain. Every one has disappeared somewhere. We used to see northern goshawks, they would fly high and scream. It was nice to follow them in the sky. All of them have disappeared and I do not know where.*

Khodzinsky described how the presence of insects had also changed dramatically.

> *I cannot comprehend that there are no mosquitoes. I think for two years now there have been no mosquitoes. In recent times they have not troubled us at all. Here in Lovozero it will be soon like down south. Before there were insects and they would sting you, but we no longer need mosquito hats even. The biting midges come in August usually. This year there have not been biting midges or mosquitoes at all. Of course this is bad. I think they have disappeared from the northland altogether.*

3.4.9.5. Traditional calendar and knowledge

The legends, stories, and traditional knowledge of living off the land have taught the Saami to notice changes and to adapt locally. At the core of their knowledge is the Saami calendar, a system of local traditional knowledge of marker days, seasons, and certain activities tied to seasonal cycles. Now there is concern. Avdeyeva noted that the calendar is off balance, adding to the burdens of observed changes.

> *When we ask the elders and reindeer herders for example what kind of summer it will be, how many berries to expect, or what kind of fish and how much to expect, they answer us that they cannot predict anything because our Saami calendar of the yearly cycle has collapsed completely with the changes that have taken place in nature. They cannot foresee accurately and with precision. Before we would ask the reindeer herders and the answers would be right to the mark, but now the predicted times keep on moving and changing. Two days ago I had a conversation with my cousin who is a reindeer herder and had just returned from the tundra. I told him: "Look how nice the sky is, see those clouds, what a nice weather!" He would tell me: "Well you say the weather is good now when you are here in the village, but out on the tundra they do not know what is to come. You should not say this sort of thing". He told me a Saami saying: "Do not predict today something that an old lady can tell for sure tomorrow". The Saami weather calendar is not accurate with the changes that we are witnessing. Yes, the reindeer herders see it and keep on discussing this at all times. We talk and discuss the changes. It is difficult to make use of the elders' knowledge because the climate has changed. People have it hard. Yes, it is so.*

3.4.9.6. Reindeer

The most vocal and urgent messages of change relate to the reindeer, which remains a key species for the community in cultural, social, economic, and ecological terms. There is concern. Reindeer are acting differently, and herders spend less time with the herds on the tundra. Mixing herds with the wild or feral reindeer is another concern. People such as Maria Zakharova have spoken at length about the often-emotional relationships and concerns regarding the reindeer.

> On the tundra the reindeers used to run towards people, but now they run away. The reindeers are our children. In the olden times when we used to have just the reindeers the air was clean. How should I explain? Now they drive around in skidoos and you can smell the gasoline, yuck! What did they herd with? The reindeer! Now they have started to herd with skidoos. Why on earth? They should rather train the reindeers like our fathers and forefathers did. Now everything is in ruins. There used to be many young reindeers. Yes, at the time the herds were bigger as well.

As Vladimir Lifov put it:

> Our income diminishes because of climate change, of course, and in a very drastic way. Even my wife has said that it would be time to forget the reindeer. But I tell her always: "Tamara, we depend on these reindeer. If there are no reindeer, we have nothing to do here either".

3.4.9.7. Overall concerns

Larisa Avdeyeva stressed the main points felt by many.

> The cycle of the yearly calendar has been disturbed greatly and this affects the reindeer herding negatively for sure. We should start working differently in a new way. We still have not thought this and we still have not pondered this – we try to start from the needs of the people and be flexible. We Saami have an anecdote, rather it is a legend, which has the law of the Saami life in it. People tell this onwards always. The Saami say: "We are not reapers, we are not field-plowers, we are reindeer herders. The reindeer are our bread. Everybody should cherish the land. The green land with its flowers and lichens was given to us so that we should pass it on to our children". We try to follow this Saami law because there are laws that the Saami follow. And the Saami guide other people to follow those laws in our land. It is true. This is the truth.

In 2002 as the documentation teams returned to the community, she continued:

> We feel some unexpected changes are taking place in the tundra. But in the recent years it has been wonderful to follow what our youth have been doing. They have understood that they are needed here. We need well-trained and strong-spirited youth here. Nowadays the young people are very different from 20 to 30 years ago. They are more self-confident, stronger in character and very proud of their nation. When I was 18, we could not even imagine that.

She looks to the future with a positive sense. She concluded the interview session in June by saying that "Yes, respect is coming back. And the consideration for reindeer herding is increasing as well".

3.5. Indigenous perspectives and resilience

The Arctic is by nature highly variable. The availability of many resources is cyclical or unpredictable. For example, there are always uncertainties in caribou migrations. The lesser snow geese may arrive early or late; the nesting success varies considerably from year to year. Weather is changeable and inconsistent. There are large natural variations in the extent of sea-ice cover from year to year, and in freeze-up and break-up dates. Erosion, uplift, thermokarst, and plant succession change landscapes and coastlines. The peoples of the Arctic are familiar with these characteristics of their homeland, and recognize that surprises are inherent in their ecosystems and ways of life. That is why indigenous peoples tend to be flexible in their ways and to have cultural adaptations, such as mobile hunting groups and strong sharing ethics, that help deal with environmental variability and uncertainty (e.g., Berkes and Jolly, 2001; Krupnik, 1993; Smith E., 1991). Many of the points raised in this section are discussed further in Chapter 17.

Against this backdrop, especially in a time of rapid social change, climate change may be regarded as simply another aspect of the variable and challenging Arctic (see Box 3.4). Just as easily, it can be dismissed from further thought as a vast, slow, and unstoppable force to be accommodated over time, in contrast to rapid, worrisome, and potentially tractable political, economic, and social problems. Some of these larger socio-economic changes have had significant impacts on lifestyles and culture, including the erosion of indigenous knowledge and the social standing of its holders (e.g., Dorais, 1997). While some observers hold that impacts of climate change will be devastating for indigenous peoples over the course of the next several decades, others argue that climate change in the shorter term will be less important than existing and ongoing economic, cultural, and social changes. It is important, however, to be cautious in making comparisons between impacts from the very different phenomena of climate change and social change.

The case studies in this chapter show that while both views have some legitimacy, by themselves they are simplistic and inaccurate portrayals of how climate change is perceived by arctic residents. The chapter title itself is plural, reflecting the diversity of ways in which climate change and the arctic environment are seen by indigenous peoples of the Arctic. Even within one group, there is a range of views, and differences in the perceived

Box 3.4. Political relations, self-determination, and adaptability

Social pathologies resulting from social change clearly weigh more heavily on the minds of northerners now than the effects of climate change. It is important, however, to be careful when comparing social change apples with climate change oranges. The ACIA climate change scenarios point to very significant environmental changes across most of the Arctic within a couple of generations. Notwithstanding high levels of suicide and other social pathologies in many northern communities, many indigenous peoples in the Arctic are interacting with, and adapting to, changing economic and social circumstances and the adoption of new technologies, in short, to "globalization". In the midst of globalization, arctic indigenous peoples still identify themselves as arctic indigenous peoples. But the sheer magnitude of projected impacts resulting from climate change raises questions of whether many of the links between arctic indigenous peoples and the land and all it provides will be eroded or even severed.

Certainly arctic indigenous peoples are highly skilled in and accustomed to adapting, as the archaeological and historical records and current practices illustrate. Adapting to climate change in the modern age, however, may be a very different prospect than adaptations in the past. It is clear that support from regional and national governments will be important for the effectiveness of the adaptations required. Herein lies a crucial point: the policies and programs of regional and national governments can encourage, enable, and equip northerners to adapt to climate change, although it is important to note that the projected magnitude of change in the Arctic may, eventually, overwhelm adaptive capacity no matter what policies and programs are in place. On the other hand, policies and programs of national governments could, conceivably, make adaptation more strained and difficult by imposing further constraints at levels from the individual to the regional.

Empowering northern residents, particularly indigenous peoples, through self-government and self-determination arrangements, including ownership and management of land and natural resources, is a key ingredient that would enable them to adapt to climate change. Indigenous peoples want to see policies that will help them protect their self-reliance, rather than become ever more dependent on the state. There are compelling reasons for the national governments of the arctic states to provide northerners, specifically indigenous peoples, with the powers, resources, information, and responsibilities that they need to adapt to climate change, and to do so on their own terms.

Berkes and Jolly (2001) provide a practical and positive example of what needs to be done. The ability of Inuvialuit of the Canadian Beaufort Sea region to adapt to climate change is grounded in the Inuvialuit Final Agreement, which recognizes rights of land ownership, co-operative management, protected areas, cash, and economic development opportunities. For indigenous peoples themselves, their own institutions and representative organizations must learn quickly from the well-documented adaptive efforts of hunters as well as from positive examples such as the Inuvialuit Final Agreement. National governments, often slow moving and ill equipped to think and act in the long term, must also understand the connections between empowerment and adaptability in the north if their policies and programs are to succeed in helping people respond to the long-term challenge posed by climate change.

importance of various threats, of which climate change is only one (Fox, 2002). A comprehensive survey of the many and varied perspectives of all arctic residents has not been possible, and it is not clear in any case how valuable such a survey would be. The case studies herein illustrate that while generalizations are possible, the particular circumstances, location, economic base, and culture of a particular group, as well as each individual's personal history and experiences, are crucial factors in determining how and what people think about climate change, how climate change may or may not affect them, and what can or cannot be done in response.

The archeological record reveals that, with or without modern anthropogenic influences, the arctic climate has experienced sudden shifts that have had severe consequences for the people who live there (McGhee, 1996). In some cases, people have simply died as resources dwindled or became inaccessible. In other cases, communities moved location, or shifted their hunting and gathering patterns to adapt to environmental change. Indigenous peoples today have more options than in the

past, but not all of these allow for the retention of all aspects of their cultures or for maintaining their ways of life. For example, many of these options have become available at the cost of dependency on the outside world. A wider range of foods is available, but communities are less self-sufficient than before. Settled village life provides educational opportunities, but indigenous knowledge is eroded because its transmission requires living on the land (Ohmagari and Berkes, 1997). Considerable infrastructure has been built over the past century, bringing improvements in the material standard of living in the Arctic. At the same time, the settled way of life has reduced both the flexibility of indigenous peoples to move with the seasons to obtain their livelihoods and the extent of their day-to-day contact with their environment, and thus the depth of their knowledge of precise environmental conditions. Instead, they have become dependent on mechanized transportation and fossil fuels to carry out their seasonal rounds while based in one central location. Together, these dependencies have increased the vulnerability of arctic communities to the impacts of climate change.

Resilience is the counterpart to vulnerability. It is a systems property; in this case, a property of the linked system of humans and nature, or socio-ecological systems, in the Arctic. Resilience is related to the magnitude of shock that a system can absorb, its self-organization capability, and its capacity for learning and adaptation (Folke et al., 2002; Resilience Alliance, 2003). Resilience is especially important to assess in cases of uncertainty, such as anticipating the impacts of climate change. Managing for resilience enhances the likelihood of sustaining linked systems of humans and nature in a changing environment in which the future is unpredictable. More resilient socio-ecological systems are able to absorb larger shocks without collapse. Building resilience means nurturing options and diversity, and increasing the capability of the system to cope with uncertainty and surprise (Berkes et al., 2003; Folke et al., 2002). Examining climate change and indigenous peoples in this way can illuminate some of the reasons that indigenous perspectives are a critical element in responding to climate change.

Life in the Arctic requires great flexibility and resilience in this technical sense. Many of the well known cultural adaptations in the Arctic, such as small group and individual flexibility and the accumulation of specialist and generalist knowledge for hunting and fishing, may be interpreted as mechanisms providing resilience (Berkes and Jolly, 2001; Fox, 2002). Such adaptations enhance options and were (and still are) important for survival. If the caribou or snow geese do not show up at a particular time and place, the hunter has back-up options and knows where to go for fish or ringed seals instead. However, cultural change and loss of some knowledge and sensitivity to environmental cues, and developments such as fixed village locations with elaborate infrastructure that restrict options, may reduce the adaptive capacity of indigenous peoples.

One approach to improve the situation is to develop policy measures that can help build resilience and add options. For example, Folke et al. (2002) have suggested that one such policy direction may be the creation of flexible multi-level governance systems that can learn from experience and generate knowledge to cope with change. In this context, the significance of indigenous observations includes their relevance for understanding the processes by which people and communities adapt to climate change.

While the scale of the impacts of climate change over the long-term is projected to be very significant, it is not yet clear how quickly these changes will take place or their spatial variation within the circumpolar Arctic. Be that as it may, human societies will attempt to adapt, constrained by the cultural, geographic, climatic, ecological, economic, political, social, national, regional, and local circumstances that shape them. As with all adaptations, those that are developed in the Arctic in response to climate change will protect some aspects of society at the expense of others. The overall success of the adapta-

tions, however, will be determined by arctic residents, probably based in large part on the degree to which they are conceived, designed, developed, and carried out by those who are doing the adapting.

Nevertheless, support from regional and national governments may be important for the effectiveness of the adaptations required. For example, Berkes and Jolly (2001) have argued that co-management institutions in the Canadian Western Arctic under the Inuvialuit Final Agreement have been instrumental in relaying local concerns across multiple levels of political organization. They have also been important in speeding up two-way information exchange between indigenous knowledge holders and scientists, thus enhancing local adaptation capabilities by tightening the feedback loop between change and response (e.g., Smith D., 2001).

In this regard, response by the community itself, through its own institutions, is crucial to effective adaptation. Directives from administrative centers or solutions devised by outsiders are unlikely to lead to the specific adaptations necessary for each community. Indigenous perspectives are needed to provide the details that arctic-wide models cannot provide. Indigenous knowledge perspectives can help identify local needs, concerns, and actions. This is an iterative rather than a one-step solution because there is much uncertainty about what is to come. Thus, policies and actions must be based on incomplete information, to be modified iteratively as the understanding of climate change and its impacts evolve.

Indigenous perspectives are also important in that indigenous peoples are experts in learning-by-doing. Science can learn from arctic indigenous knowledge in dealing with climate change impacts, and build on the adaptive management approach – which, after all, is a scientific version of learning-by-doing (Berkes et al., 2003). Multi-scale learning is key – learning at the level of community institutions such as hunter-trapper committees, regional organizations, national organizations, and international organizations such as the Arctic Council. The use of adaptive management is a shift from the conventional scientific approach, and the creation of multi-level governance, or co-management systems, is a shift from the usual top-down approach to management.

One significant aspect of the indigenous perspectives introduced in this chapter is that they help illustrate that the vulnerability and resilience of each group or community differ greatly from place to place and from time to time. In considering the impacts of climate change in the Arctic and the options for responding to those changes, it is essential to understand the nature of the question. It is also essential to consider what is at stake. The indigenous peoples of the Arctic are struggling to maintain their identity and distinctive cultures in the face of national assimilation and homogenization, as well as globalization (Freeman, 2000; Nuttall, 1998). The response to climate change can exacerbate or mitigate the impacts of that climate change itself. For policy-makers, taking the nature and diversity of indigenous perspectives into account is essential in the effort to help those groups adapt to a changing climate.

For indigenous peoples themselves, an understanding of the ways in which they are resilient and the ways in which they are vulnerable is an essential starting point in determining how they will respond to the challenges posed by climate change. As noted, physical, ecological, and social forces interact to shape these characteristics for each group of people. In times of rapid change, the dynamics of this interplay are particularly difficult for a society to track. An assessment of individual and collective perspectives of arctic indigenous peoples on the challenges ahead can help determine strengths, weaknesses, and priorities. This chapter is a first step in the direction of such an assessment, and shows the need for further work to enable indigenous communities in the Arctic to reflect on the implications of climate change for themselves and for their future.

3.6. Further research needs

This chapter reviews observations of the environmental changes occurring in the Arctic as well as the ways in which people view those changes. In both cases, there is a growing but still insufficient body of research to draw on. For some areas, such as the central and eastern Russian Arctic, few or no current records of indigenous observations are available. To detect and interpret climate change, and to determine appropriate response strategies, more research is clearly needed.

In terms of indigenous observations, documentation of existing knowledge about changes that have occurred and prospective monitoring for future change are both important (Riedlinger and Berkes, 2001; Huntington, 2000b). More research on knowledge documentation has taken place in Canada, particularly among Inuit, regarding indigenous knowledge of climate change than elsewhere (Fox, 1998, 2002, 2004; Furgal et al., 2002; Jolly et al., 2002, 2003; Nickels et al., 2002; Thorpe et al., 2001, 2002), but even there a great deal more can be done. In Eurasia and Greenland, little systematic work of this kind has been done, and research in these regions is clearly needed (Mustonen, 2002). Indigenous observation networks have been set up in Chukotka, Russia (N. Mymrin, Eskimo Society of Chukotka, Provideniya, Russia, pers. comm., 2002), and some projects have taken place in Alaska (Huntington et al., 1999, Krupnik, 2002; Whiting, 2002), but again, little systematic work has been done to set up, maintain, and make use of the results from such efforts.

In terms of indigenous perspectives and interpretations of climate change, most research has taken place in Canada (e.g., Krupnik and Jolly, 2002; McDonald et al., 1997), building largely on the documentation of observations noted above. To date, however, little has been done to connect these perspectives to potential response strategies (see Table 3.1). Some research on responses

Table 3.1. Indigenous responses to climate change in the Inuvialuit Settlement Region of Canada's Northwest Territories (adapted from Nickels et al., 2002).

Observation	Effect	Response/adaptation
Erosion of the shoreline	Relocation of homes and possibly community considered	Stone breakwalls and gravel have been placed on the shoreline to alleviate erosion from wave action
Warmer temperatures in summer	Not able to store country food properly and thus not able to store it for use in winter	Community members travel back to communities more often in summer to store country food. This is expensive as it requires more fuel and time
Warmer temperatures in summer	Can no longer prepare dried/smoked fish in the same way, it gets cooked in the heat	People are building thicker roofs on the smoke houses to keep some heat out, tarpaulins and other materials are used to shelter country foods from heat
Lower water levels and some brooks drying up	Not as many good natural sources of drinking water available	Bottled water now taken on trips
Changing water levels and the formation of shifting sand bars	More difficult to plan travel in certain areas	Community members are finding new (usually longer and therefore more costly) routes to their usual camps and hunting grounds or are flying, incurring still greater expense
Warmer weather in winter	Animal fur is shorter and not as thick, changing the quality of the fur/skin used in making clothing, decreasing the money received when sold	Some people do not bother to hunt/trap, while others buy skins from the store that are not locally trapped, are usually not as good quality, and are expensive
Water warmer at surface	Kills fish in nets	Nets are checked and emptied more frequently so that fish caught in nets do not perish in the warm surface water and spoil
More mosquitoes and other biting insects	Getting bitten more	Use insect repellent lotion or spray as well as netting and screens for windows and entrances to houses
Changing animal travel/migration routes	Makes hunting more expensive, requires more fuel, gear, and time – high costs mean some residents (particularly elders) cannot afford to go hunting	Initiation of a community program for elders, through which younger hunters can provide meat to elders who are unable to travel or hunt for themselves

has been undertaken recently or is underway in Alaska (e.g., Brunner et al., 2004; George et al., 2004), but more is needed to determine the needs of those designing response strategies, the ways in which information is used in the process of designing them, and the ways in which researchers and indigenous peoples can contribute. An essential component of this line of research is to understand how various actors see the issue of climate change, why they see it the way they do, and what can be done to arrive at a common understanding of the threat posed by climate change, the need for responses, and the needs and capabilities of local residents. Consideration of regional similarities and differences across the Arctic may help communities to learn from each other's experiences, too, as well as to incorporate greater understanding of the cumulative impacts of various factors influencing communities (see Chapter 17).

In working toward this goal, an often neglected topic is the linking of indigenous and scientific observations of climate change and the interpretation of these observations. Proponents of the documentation and use of indigenous knowledge often stress both similarities and differences with scientific knowledge (e.g., Stevenson, 1996), but little is done to bridge the gap (e.g., Agrawal, 1995). While the two approaches differ in substantive ways, there are examples of how interactions between them can benefit both and produce a better overall understanding of a given topic (e.g., Albert, 1988; Fox, 2004; Huntington, 2000a; Norton, 2002). Part of the problem is in determining how indigenous knowledge can best be incorporated into scientific systems of knowledge acquisition and interpretation. Part of the problem is in finding ways to involve indigenous communities in scientific research as well as in communicating scientific findings to indigenous communities. And a large part of the problem is in establishing the trust necessary to find appropriate solutions to both goals. Collaborative research is the most promising model for addressing these challenges, as demonstrated by the projects through which the case studies in this chapter were produced, as well as by the results reported from other projects associated with the ACIA, particularly the vulnerability approach described in Chapter 17. Further development of the collaborative model, from small projects to large research programs and extending from identifying research needs to designing response strategies, is an urgent need.

3.7. Conclusions

The case studies and the summary of indigenous observations presented in this chapter were drawn from a variety of studies, conducted in many arctic communities and cultures, and translated from a number of languages. From this diversity of sources, some common themes emerge. While the specifics of these themes and how they are dealt with will depend on circumstances particular to each community, indigenous peoples around the Arctic nonetheless have some shared experiences with climate change.

One topic that stands out across all regions is increased weather variability and unpredictability. Experienced hunters and elders from around the Arctic express concern that they cannot predict the weather like they used to: the weather changes more quickly and in unexpected ways. Arctic residents recognize that the climate is inherently variable. However, many indigenous observers identify the unpredictable and unseasonable weather of the last decade or so as unprecedented. It is true that many factors, such as less frequent time on the land and a tendency to remember the past in rosier terms than are justified, could contribute to changing perceptions of weather predictability even in the absence of actual changes in weather patterns. Nonetheless, similar observations have been made independently by many people in all areas around the Arctic. Such widespread observations of and concern over increased weather variability point, at a minimum, to an important and interesting area for further research, particularly in collaboration with meteorologists and climatologists. Fox et al. (in prep) have done the initial work to link indigenous and scientific observations of weather variability for one community on Baffin Island. Further investigation covering a larger region would be useful and desirable.

While increased weather variability clearly stands out as the most common observation of change across arctic communities, changes in wind and changes in sea ice are also important and widespread. The details of both, however, depend on the location of the observation. In some communities, residents are concerned about changes in wind direction, in others wind strength and the frequency of high winds have changed, and in some places both trends have been seen. Changes in sea ice are similarly variable in time and space. Sea ice may be of the usual thickness but lesser extent in one area in a given year, and the usual extent but reduced thickness in a different area or in another year. The common theme is the prevalence of unusual characteristics and patterns in winds and sea ice. This leads to another insight from analyzing indigenous observations, which is the stress on interconnections between impacts from climate and environmental changes.

Many of the examples of indigenous perspectives of climate change presented here illustrate how the impacts of climate change are connected, interacting to produce further changes. For example, the Nunavut case study (section 3.4.5) shows how wind changes in the Baker Lake area have packed the snow unusually hard, making igloo building difficult or impossible. When these wind-driven snow changes interact with the recent unpredictable weather conditions, dangerous situations arise as travelers are unable to build emergency snow shelters. The Kotzebue case study (section 3.4.1) offers several examples of the different and interacting consequences of change in a single variable in that region.

The climate and environmental changes observed by arctic indigenous peoples produce impacts through their interactions with one another, and through the ways in which they play out in social, political, and cultural contexts. Indigenous perceptions of climate change do not arise in isolation, but are shaped by these contexts as well as the context of the overall climate change debate. This is best demonstrated in the case study from the Kola Peninsula (section 3.4.9). While the Saami of that region have observed the impacts of climate change and are concerned about the long-term implications, they are far more concerned about their immediate economic and political circumstances. When people are concerned about making a living and providing food for their families, it is not surprising that less immediate concerns do not rate as highly. In Nunavut, the situation is not as dire, but people are nonetheless very concerned about issues such as poverty, housing, and cultural preservation. Here, too, climate change may not be regarded as a top priority issue.

The contexts within which indigenous peoples observe, assess, interact, and respond to the impacts of climate change are extremely important, especially as individuals and communities begin to develop ways to adapt to these changes. Political or economic situations will play a role in constraining or enabling people to adapt. For example, Chapter 17 discusses how reindeer herders in Finnmark are hindered in their ability to deal with icy grazing areas in the autumn. In the past, herders could move the reindeer to other pastures that had not iced over. Owing to changes in land use and new boundaries, however, herders are now prevented from moving their herds and are thus vulnerable to localized freezing events.

Two areas in particular need further development to enhance the abilities of indigenous peoples to cope with the impacts of climate change. First, increasing flexibility and the response options available will allow a broader array of potential responses. This entails devolving authority and capacity to more local levels so that people and communities can choose for themselves the responses that make the most sense in their particular situation, given the costs and benefits of those responses. Such responses range from changing regulations concerning resource use to moving settlements to more favorable locations. Second, more information about the potential types of changes that may be seen will help identify particular areas of vulnerability. The common themes and concerns in this chapter – increased variability in weather, changes in wind patterns, changes in sea ice and snow, more freeze-thaw cycles, more and stronger storms – are topics that are not well addressed in typical climate models (see Chapter 4). Greater attention to the climate parameters that affect local people and ecosystems directly will help to identify critical areas for local and regional action.

These steps can and should flow from the documentation and presentation of indigenous perspectives on climate change. Indigenous knowledge and perspectives are a foundation upon which individuals, communities, and regions can design responses and take action. Other

information and expertise are also essential to this process, and collaborative approaches are thus the most likely to be effective in identifying and addressing the challenges and opportunities posed by climate change. Randall Tetlichi, a Vuntut Gwitchin leader from Old Crow, Yukon Territory, referred to the need to draw on scientific and traditional knowledge as the need to "double understand" (quoted in Kofinas et al., 2002). For the peoples of the Arctic, whose future is at stake, having the ability to make choices and changes is a matter of survival, to which all available resources must be applied.

Acknowledgements

We are grateful to all those who contributed to the studies on which the case studies are based, who are too numerous to be listed here.

References

Agrawal, A., 1995. Dismantling the divide between indigenous and scientific knowledge. Development and Change, 26(3):413–439.

Albert, T.F., 1988. The role of the North Slope Borough in Arctic environmental research. Arctic Research of the United States, 2:17–23.

AMAP, 1998. AMAP Assessment Report: Arctic Pollution Issues. Arctic Monitoring and Assessment Programme, Oslo, xii + 859p.

AMAP, 2002. Arctic Pollution 2002. Arctic Monitoring and Assessment Programme, Oslo, xi + 111p.

Berkes, F., 1998. Indigenous knowledge and resource management systems in the Canadian subarctic. In: F. Berkes and C. Folke (eds.). Linking Social and Ecological Systems, pp. 98–128. Cambridge University Press.

Berkes, F., 1999. Sacred Ecology: Traditional Ecological Knowledge and Resource Management. Taylor & Francis, xvi + 209p.

Berkes, F. and D. Jolly, 2001. Adapting to climate change: social-ecological resilience in a Canadian Western Arctic community. Conservation Ecology, 5(2):18 [online: www.consecol.org/vol5/iss2/art18].

Berkes, F., J. Colding and C. Folke, 2000. Rediscovery of traditional ecological knowledge as adaptive management. Ecological Applications, 10(5):1251–1262.

Berkes, F., J. Colding and C. Folke (eds.), 2003. Navigating Social-Ecological Systems: Building Resilience for Complexity and Change. Cambridge University Press.

Bockstoce, J.R., 1976. On the development of whaling in the Western Thule Culture. Folk, 18:41–46.

Brody, H., 2000. The other side of Eden: hunters, farmers and the shaping of the world. Douglas & McIntyre, 368p.

Brown, J. and E. Vibert (eds.), 1996. Reading beyond Words: Contexts for Native History. Broadview Press, xxvii + 519p.

Brown, L.R., C. Flavin, H. French, J. Abramovitz, S. Dunn, G. Gardner, A. Mattoon, A. Platt McGinn, M. O'Meara, M. Renner, C. Bright, S. Postel, B. Halweil and L. Starke (eds.), 2000. State of the World 2000: A Worldwatch Institute Report on Progress Toward a Sustainable Society. W.W. Norton, ix + 276p.

Brown, P., 2003. Global warming is killing us too, say Inuit. The Guardian (UK), December 11, 2003.

Brunner, R.D., A.H. Lynch, J. Pardikes, E.N. Cassano, L. Lestak and J. Vogel, 2004. An Arctic disaster and its policy implications. Arctic, 57(4):336–346.

Burch, E.S. Jr., 1998. The Inupiaq Eskimo nations of northwest Alaska. University of Alaska, xviii + 473p.

CAFF, 2001. Arctic Flora and Fauna: Status and Conservation. Conservation of Arctic Flora and Fauna, Helsinki, 272p.

Cruikshank, J., 1981. Legend and landscape: convergence of oral and scientific traditions in the Yukon Territory. Arctic Anthropology, 18(2):67–94.

Cruikshank, J., 1990. Getting the words right: perspectives on naming and places in Athapascan oral history. Arctic Anthropology, 27(1):52–65.

Cruikshank, J., 1998. The social life of stories: narrative and knowledge in the Yukon Territory. University of Nebraska Press, xvii + 211p.

Cruikshank, J., 2001. Glaciers and climate change: perspectives from oral tradition. Arctic, 54(4):377–393.

Dene Nation, 1984. Denendeh: a Dene celebration. Dene National Office, Yellowknife, Northwest Territories, 144p.

Dene Nation, 1999. TK for Dummies: The Dene Nation Guide to Traditional Knowledge. Dene National Office, Yellowknife, Northwest Territories, 13p.

Dene Nation, 2002. The Denendeh Environmental Working Group: Climate Change Workshop. [online: www.denenation.com]

Dene Nation, 2003. Report of the Second Denendeh Environmental Working Group: Climate Change and Forests Workshop. [online: www.denenation.com]

Dick, L., 2001. Muskox Land: Ellesmere Island in the Age of Contact. University of Calgary Press, xxv + 615p.

Dorais, L.-J., 1997. Quaqtaq: Modernity and Identity in an Inuit Community. University of Toronto Press, ix + 132p.

Downie, D.L. and T. Fenge (eds.), 2003. Northern Lights against POPs: Combating Toxic Threats in the Arctic. McGill University Press, xxv + 347p.

Erasmus, B., C.J. Paci and S. Irlbacher Fox, 2003. History and Development of the Dene Nation. Indigenous Nations Studies Journal, 4(2).

Fehr, A. and W. Hurst (eds.), 1996. A seminar on two ways of knowing: indigenous and scientific knowledge. Aurora Research Institute, Inuvik, Northwest Territories, 93p.

Feldman, K.D. and E. Norton, 1995. Niqsaq and napaaqtuq: issues in Inupiaq Eskimo life-form classification and ethnoscience. Etudes/Inuit/Studies, 19(2):77–100.

Ferguson, M.A.D., R.G. Williamson and F. Messier, 1998. Inuit knowledge of long-term changes in a population of Arctic tundra caribou. Arctic, 51(3):201–219.

Ferguson, M.A.D. and F. Messier, 1997. Collection and analysis of traditional ecological knowledge about a population of arctic tundra caribou. Arctic, 50(1):17–28.

Fienup-Riordan, A., 1994. Boundaries and Passages: Rule and Ritual in Yup'ik Eskimo Oral Tradition. University of Oklahoma Press, xxiv + 389p.

Fienup-Riordan, A., W. Tyson, P. John, M. Meade and J. Active, 2000. Hunting Tradition in a Changing World. Rutgers University Press, xx + 310p.

Fitzhugh, W.W., 1984. Paleo-Eskimo cultures of Greenland. In: D. Damas (ed.). Arctic. Handbook of North American Indians, vol. 5, pp. 528–539. Smithsonian Institution, Washington D.C.

Folke, C., S. Carpenter, T. Elmqvist, L. Gunderson, C.S. Holling, B. Walker, J. Bengtsson, F. Berkes, J. Colding, K. Danell, M. Falkenmark, L. Gordon, R. Kasperson, N. Kautsky, A. Kinzig, S. Levin, K.-G. Maler, F. Moberg, L. Ohlsson, P. Olsson, E. Ostrom, W. Reid, J. Rockstrom, H. Savenije and U. Svedin, 2002. Resilience for Sustainable Development: Building Adaptive Capacity in a World of Transformations. Environmental Advisory Council, Ministry of the Environment, Stockholm, 74p.

Fox, S., 1998. Inuit Knowledge of Climate and Climate Change. M.A. Thesis, University of Waterloo, Canada.

Fox, S., 2002. These are things that are really happening: Inuit perspectives on the evidence and impacts of climate change in Nunavut. In: I. Krupnik and D. Jolly (eds.). The Earth is Faster Now: Indigenous Observations of Arctic Environmental Change, pp. 12–53. Arctic Research Consortium of the U.S., Fairbanks, Alaska.

Fox, S., 2004. When the Weather is Uggianaqtuq: Linking Inuit and Scientific Observations of Recent Environmental Change in Nunavut, Canada. Ph.D. Dissertation, University of Colorado.

Fox, S., M. Pocernich and J.A. Miller, in prep. Climate and weather variability in the Eastern Canadian Arctic: linking Inuit observations and meteorological data.

Freeman, M.M.R., (ed.), 1976. Report of the Inuit land use and occupancy project. 3 vols. Ottawa: Indian and Northern Affairs Canada.

Freeman, M.M.R., 1992. The nature and utility of traditional ecological knowledge. Northern Perspectives, 20(1):9–12.

Freeman, M.M.R. (ed.), 2000. Endangered Peoples of the Arctic: Struggles to Survive and Thrive. The Greenwood Press, xix + 278p.

Freeman, M.M.R. and L.N. Carbyn (eds.), 1988. Traditional Knowledge and Renewable Resource Management in Northern Regions. Boreal Institute for Northern Studies, Alberta, 124p.

Furgal, C., D. Martin and P. Gosselin, 2002. Climate change and health in Nunavik and Labrador: lessons from Inuit knowledge. In: I. Krupnik and D. Jolly (eds.). The Earth is Faster Now: Indigenous Observations of Arctic Environmental Change, pp. 266–299. Arctic Research Consortium of the U.S., Fairbanks, Alaska.

Gaski, H. (ed.), 1997. Sami culture in a new era: the Norwegian Sami experience. Davvi Girji, Karasjok, Norway, 223p.

George, J., 1999. Global warming brings red-breasted robins to Iqaluit. Nunatsiaq News. September 30.

George, J.C., H.P. Huntington, K. Brewster, H. Eicken, D.W. Norton and R. Glenn, 2004. Observations on shorefast ice failures in Arctic Alaska and the responses of the Inupiat hunting community. Arctic, 57(4):363–374.

Gilberg, R., 1974–75. Changes in the life of the polar Eskimos resulting from a Canadian immigration into the Thule District, North Greenland in the 1860s. Folk, 16–17:159–170.

Gilberg, R., 1984. Polar Eskimo. In: D. Damas (ed.). Arctic. Handbook of North American Indians vol. 5, pp. 577–594. Smithsonian Institution, Washington D.C.

GN, 2001. Inuit Qaujimajangit Hilap Alanguminganut/Inuit Knowledge of Climate Change: A Sample of Inuit Experiences of Climate Change in Nunavut. Baker Lake and Arviat, Nunavut. January–March 2001. Government of Nunavut, Department of Sustainable Development, Environmental Protection Services.

Grenier, L., 1998. Working with indigenous knowledge: a guide for researchers. International Development Research Center, Ottawa, 115p.

Gubser, N.J., 1965. The Nunamiut Eskimo: Hunters of Caribou. Yale University Press.

Gunn, J., 1994. Global climate and regional biocultural diversity. In: C. Crumley (ed.). Historical Ecology: Cultural Knowledge and Changing Landscapes, pp. 67–98. School of American Research Press.

Helander, E., 2004. Global change – climate observations among the Sami. In: T. Mustonen and E. Helander (eds.). Snowscapes, Dreamscapes: SnowChange book on community voices of change, pp. 302–309. Tampere Polytechnic Institute, Finland.

Henshaw, A., 2003. Climate and culture in the North: The interface of archaeology, paleoenvironmental science and oral history. In: S. Strauss and B. Orlove (eds.). Weather, Climate, Culture. Berg Press.

Huntington, H.P., 1992a. The Alaska Eskimo Whaling Commission and other cooperative marine mammal management organizations in Alaska. Polar Record, 28(165):119–126.

Huntington, H.P., 1992b. Wildlife management and subsistence hunting in Alaska. Belhaven Press, xvii + 177p.

Huntington, H.P., 1998. Observations on the utility of the semi-directive interview for documenting traditional ecological knowledge. Arctic, 51(3):237–242.

Huntington, H.P., 2000a. Using traditional ecological knowledge in science: methods and applications. Ecological Applications, 10(5):1270–1274.

Huntington, H.P. (ed.), 2000b. Impacts of Changes in Sea Ice and Other Environmental Parameters in the Arctic. Report of the Marine Mammal Commission workshop, Girdwood, Alaska, 15–17 February 2000. Marine Mammal Commission, Bethesda, Maryland, iv + 98p.

Huntington, H.P., J.H. Mosli and V.B. Shustov, 1998. Peoples of the Arctic: characteristics of human populations relevant to pollution issues. In: AMAP Assessment Report: Arctic Pollution Issues, pp. 141–182. Arctic Monitoring and Assessment Programme, Oslo.

Huntington, H.P. and the communities of Buckland, Elim, Koyuk, Point Lay and Shaktoolik, 1999. Traditional knowledge of the ecology of beluga whales (*Delphinapterus leucas*) in the eastern Chukchi and northern Bering seas, Alaska. Arctic, 52(1):49–61.

Huntington, H.P., P.K. Brown-Schwalenberg, M.E. Fernandez–Gimenez, K.J. Frost, D.W. Norton and D.H. Rosenberg, 2002. Observations on the workshop as a means of improving communication between holders of traditional and scientific knowledge. Environmental Management, 30(6):778–792.

Huntington, H.P., T. Callaghan, S. Fox and I. Krupnik, 2004. Matching traditional and scientific observations to detect environmental change: a discussion on Arctic terrestrial ecosystems. Ambio, 33(7):18–23.

Inglis, J.T. (ed.), 1993. Traditional Ecological Knowledge: Concepts and Cases. International Program on Traditional Ecological Knowledge and International Development Research Centre, Ottawa, 142p.

Ingold, T. and T. Kurtilla, 2000. Perceiving the environment in Finnish Lapland. Body and Society, 6(3–4):183–196.

IARPC, 1992. Principles for the conduct of research in the Arctic. Interagency Arctic Research Policy Committee, Arctic Research of the United States, 6:78–79.

Johannes, R.E., 1993. Integrating traditional ecological knowledge and management with environmental impact assessment. In: J.T. Inglis (ed.). Traditional Ecological Knowledge: Concepts and Cases, pp. 33–39. International Program on Traditional Ecological Knowledge and International Development Research Centre, Ottawa.

Johnson, M. (ed.), 1992. Lore: capturing traditional environmental knowledge. Dene Cultural Institute and International Development Research Centre, Ottawa, 190p.

Jolly, D., F. Berkes, J. Castleden, T. Nichols and the Community of Sachs Harbour, 2002. We can't predict the weather like we used to: Inuvialuit observations of climate change, Sachs Harbour, Western Canadian Arctic. In: I. Krupnik and D. Jolly (eds.). The Earth is Faster Now: Indigenous Observations of Arctic Environmental Change, pp. 92–125. Arctic Research Consortium of the U.S., Fairbanks, Alaska.

Jolly, D., S. Fox and N. Thorpe, 2003. Inuit and Inuvialuit knowledge of climate change. In: J. Oakes, R. Riewe, K. Wilde, A. Edmunds and A. Dubois, (eds.), pp. 280–290. Native Voices in Research. Aboriginal Issues Press.

Kassi, N., 1993. Native perspective on climate change. In: G. Wall (ed.). Impacts of Climate Change on Resource Management in the North, pp. 43–49. Department of Geography Publication Series, Occasional Paper No. 16. University of Waterloo, Ontario.

Kawagley, A.O., 1995. A Yupiaq Worldview: A Pathway to Ecology and Spirit. Waveland Press, 166p.

Kilabuck, P., 1998. A Study of Inuit Knowledge of Southeast Baffin Beluga. Nunavut Wildlife Management Board, Iqaluit, Northwest Territories, iv + 74p.

Kofinas, G.P. and the communities of Aklavik, Arctic Village, Old Crow, and Fort McPherson, 2002. Community contributions to ecological monitoring: knowledge co-production in the U.S.-Canada Arctic borderlands. In: I. Krupnik and D. Jolly (eds.). The Earth is Faster Now: Indigenous Observations of Arctic Environmental Change, pp. 55–91. Arctic Research Consortium of the U.S., Fairbanks, Alaska.

Krech, S. III, 1999. The Ecological Indian: Myth and History. W.W. Norton, 318p.

Krupnik, I., 1993. Arctic Adaptations: Native Whalers and Reindeer Herders of Northern Eurasia. University Press of New England, xvii + 355p.

Krupnik, I., 2000. Native perspectives on climate and sea-ice changes. In: H.P. Huntington (ed.). Impacts of Changes in Sea Ice and other Environmental Parameters in the Arctic, pp. 25–39. Report of the Marine Mammal Commission workshop, Girdwood, Alaska, 15–17 February 2000. Bethesda, Maryland.

Krupnik, I., 2002. Watching ice and weather our way: some lessons from Yupik observations of sea ice and weather on St. Lawrence Island, Alaska. In: I. Krupnik, and D. Jolly (eds.). The Earth is Faster Now: Indigenous Observations of Arctic Environmental Change, pp. 156–197. Arctic Research Consortium of the U.S., Fairbanks, Alaska.

Krupnik, I. and D. Jolly (eds.), 2002. The Earth is Faster Now: Indigenous Observations of Arctic Environmental Change. Arctic Research Consortium of the U.S., Fairbanks, Alaska, xxvii + 356p.

Krupnik, I. and N. Vakhtin, 1997. Indigenous knowledge in modern culture: Siberian Yupik ecological legacy in transition. Arctic Anthropology, 34(1):236–252.

Kublu, A., F. Laugrand and J. Oosten, 1999. Introduction. In: J. Oosten and F. Laugrand (eds.), pp. 1–12. Interviewing Inuit elders. Vol. 1. Iqaluit: Nunavut Arctic College.

Mailhot, J., 1993. Traditional Ecological Knowledge: The Diversity of Knowledge Systems and Their Study. Great Whale Public Review Support Office, Montreal, 48p.

Mary-Rousselière, G., 1991. Qitdlarssuaq: The Story of a Polar Migration. Wuerz Publishing.

Maxwell, M.S., 1985. Prehistory of the Eastern Arctic. Academic Press.

McCartney, A.P. and J.M. Savelle, 1985. Thule Eskimo whaling in the central Canadian Arctic. Arctic Anthropology, 22(2):37–58.

McDonald, M., L. Arragutainaq and Z. Novalinga, 1997. Voices from the Bay: Traditional Ecological Knowledge of Inuit and Cree in the James Bay Bioregion. Canadian Arctic Resources Committee and Environmental Committee of the Municipality of Sanikiluaq, Ottawa, 90p.

McGhee, R., 1969/70. Speculations on climatic change and Thule Culture development. Folk, 11–12:173–184.

McGhee, R., 1984. Thule prehistory of Canada. In: D. Damas (ed.). Arctic. Handbook of North American Indians, vol. 5, pp. 369–376. Smithsonian Institution, Washington, D.C.

McGhee, R., 1996. Ancient People of the Arctic. University of British Columbia Press, xii + 244p.

Minc, L.D., 1986. Scarcity and survival: the role of oral tradition in mediating subsistence crises. Journal of Anthropological Archaeology, 5:39–113.

Minority Rights Group (ed.), 1994. Polar Peoples: Self Determination and Development. London: Minority Rights Group.

Müller-Wille, L., 1983. Inuit toponymy and cultural sovereignty. In: L. Müller-Wille (ed.). Conflict in the Development in Nouveau-Quebec, pp. 131–150. Centre of Northern Studies and Research at McGill University, Montreal.

Müller-Wille, L., 1985. Une methodologie pour les enuêtes toponymiques autochtones: le répertoire Inuit de la region de Kativik et de sa zone côtière. Etudes/Inuit/Studies, 9(1):51–66.

Mustonen, T., 2002. Snowchange 2002: indigenous views on climate change: a circumpolar perspective. In: I. Krupnik and D. Jolly (eds.). The Earth is Faster Now: Indigenous Observations of Arctic Environmental Change, pp. 350–356. Arctic Research Consortium of the U.S., Fairbanks, Alaska.

Mustonen, T. and E. Helander (eds.), 2004. Snowscapes, Dreamscapes: SnowChange book on community voices of change. Tampere Polytechnic Institute, Finland, 562p.

Mymrin, N.I., the communities of Novoe Chaplino, Sireniki, Uelen, and Yanrakinnot and H.P. Huntington, 1999. Traditional knowledge of the ecology of beluga whales (Delphinapterus leucas) in the northern Bering Sea, Chukotka, Russia. Arctic, 52(1):62–70.

Nadasdy, P., 1999. The politics of TEK: power and the 'integration' of knowledge. Arctic Anthropology, 36(1):1–18.

Nelson, R.K., 1983. Make prayers to the raven. University of Chicago Press, xvi + 292p.

Nickels, S., C. Furgal, J. Castleden, P. Moss-Davies, M. Buell, B. Armstrong, D. Dillion and R. Fonger, 2002. Putting a human face on climate change through community workshops: Inuit knowledge, partnerships, and research. In: I. Krupnik and D. Jolly (eds.). The Earth is Faster Now: Indigenous Observations of Arctic Environmental Change, pp. 300–333. Arctic Research Consortium of the U.S., Fairbanks, Alaska.

Noongwook, G., 2000. Native observations of local climate changes around St. Lawrence Island. In: H.P. Huntington (ed.). Impacts of Changes in Sea Ice and other Environmental Parameters in the Arctic, pp. 21–24. Marine Mammal Commission, Maryland.

Norton, D.W., 2002. Coastal sea ice watch: private confessions of a convert to indigenous knowledge. In: I. Krupnik and D. Jolly (eds.). The Earth is Faster Now: Indigenous Observations of Arctic Environmental Change, pp. 126–155. Arctic Research Consortium of the U.S., Fairbanks, Alaska.

Nunavut Tunngavik Inc., Association, Kitikmeot Inuit, and Canada, Indian and Northern Affairs, 2001. Elders' Conference on Climate Change, Cambridge Bay 2001, March 29–31. Workshop report, 92p.

Nuttall, M., 1992. Arctic Homeland: Kinship, Community and Development in Northwest Greenland. Belhaven Press, 194p.

Nuttall, M., 1998. Protecting the Arctic: Indigenous Peoples and Cultural Survival. Harwood Academic Publishing, 204p.

Nuttall, M., 2000. Indigenous peoples, self-determination, and the Arctic environment. In: M. Nuttall and T.V. Callaghan (eds.). The Arctic: Environment, People, Policy, pp. 377–409. Harwood Academic Publishers.

Nuttall, M. and T.V. Callaghan (eds.), 2000. The Arctic: Environment, People, Policy. Harwood Academic Publishers, xxxviii + 647p.

Ohmagari, K. and F. Berkes, 1997. Transmission of indigenous knowledge and bush skills among the Western James Bay Cree women of subarctic Canada. Human Ecology, 25:197–222.

Overland, J.E., W. Muyin and N.A. Bond, 2002. Recent temperature changes in the western Arctic during spring. Journal of Climate, 15:1702–1716.

Peplinski, L., 2000. Public resource management and Inuit toponymy: implementing policies to maintain human-environmental knowledge in Nunavut. M.A. Thesis, Royal Roads University, British Columbia.

Pika, A. (ed.), 1999. Neotraditionalism in the Russian North. Indigenous Peoples and the Legacy of Perestroika. Edited in English by B. Grant. Circumpolar Research Series 6. Canadian Circumpolar Institute and University of Washington Press.

Pinkerton, E. (ed.), 1989. Cooperative Management of Local Fisheries. University of British Columbia Press, xiii + 299p.

Pitulko, V.V., P.A. Nikolsky, E.Yu. Girya, A.E. Basilyan, V.E. Tumskoy, S.A. Koulakov, S.N. Astakhov, E.Yu. Pavlova and M.A. Anisimov, 2004. The Yana RHS Site: humans in the Arctic before the last glacial maximum. Science, 303:52–56.

Pungowiyi, C., 2000. Native observations of change in the marine environment of the Bering Strait region. In: H.P. Huntington (ed.). Impacts of Changes in Sea Ice and other Environmental Parameters in the Arctic, pp. 18–20. Report of the Marine Mammal Commission workshop, Girdwood, Alaska, 15–17 February 2000. Bethesda, Maryland.

Rankama, T., 1993. Managing the landscape: a study of Sami place names in Utsjoki, Finnish Lapland. Etudes/Inuit/Studies, 17(1):47–67.

Riedlinger, D. and F. Berkes, 2001. Contributions of traditional knowledge to understanding climate change in the Canadian Arctic. Polar Record, 37(203):315–328.

Riedlinger, D., S. Fox and N. Thorpe, 2001. Inuit and Inuvialuit knowledge of climate change in the Northwest Territories and Nunavut. In: J. Oakes and R. Riewe (eds.). Native voices in research: Northern and native studies, pp. 21–48. Winnipeg: University of Manitoba.

Resilience Alliance, 2003. The Resilience Alliance. [online: www.resalliance.org]

Rouse, W.R., M.S.V. Douglas, R.E. Hecky, A.E. Hershey, G.K. Kling, L. Lesack, P. Marsh, M. McDonald, B.J. Nicholson, N.T. Roulet and J.P. Smol, 1997. Effects of climate changes on the freshwaters of Arctic and Subarctic North America. Hydrological Processes, 11:873–902.

Slezkine, Y., 1994. Arctic Mirrors: Russia and the Small Peoples of the North. Cornell, xiv + 456p.

Smith, D., 2001. Co-management in the Inuvialuit Settlement Region. In: Arctic Flora and Fauna: Status and Trends, pp. 64–65. Conservation of Arctic Flora and Fauna, Helsinki.

Smith, E.A., 1991. Inujjuamiut Foraging Strategies: Evolutionary Ecology of an Arctic Hunting Economy. Aldine de Gruyter, xx + 455p.

Stevenson, M.G., 1996. Indigenous knowledge and environmental assessment. Arctic, 49(3):278–291.

Stoker, S. and I.I. Krupnik, 1993. Subsistence whaling. In: J.J. Burns, J.J. Montague and C.J. Cowles (eds.). The Bowhead Whale, pp. 579–629. The Society for Marine Mammalogy, Kansas.

Thorpe, N., 2000. Contributions of Inuit ecological knowledge to understanding the impacts of climate change on the Bathurst Caribou Herd in the Kitikmeot Region, Nunavut. Unpublished Masters Thesis, School of Resource and Environmental Management. Project No. 268. Simon Fraser University, 111p.

Thorpe, N., N. Hakongak, S. Eyegetok and the Kitikmeot Elders, 2001. Thunder on the tundra: Inuit qaujimajatuqangit of the Bathurst caribou. Generation Printing, xv + 208p.

Thorpe, N., S. Eyegetok, N. Hakongak and the Kitikmeot Elders, 2002. Nowadays it is not the same: Inuit qaujimajatuqangit, climate, and caribou in the Kitikmeot region of Nunavut, Canada. In: I. Krupnik and D. Jolly (eds.). The Earth is Faster Now: Indigenous Observations of Arctic Environmental Change, pp. 198–239. Arctic Research Consortium of the U.S., Fairbanks, Alaska.

Turner, N.J., M. Boelscher Ignace and R. Ignace, 2000. Traditional ecological knowledge and wisdom of aboriginal peoples in British Columbia. Ecological Applications, 10(5):1275–1287.

Usher, P.J., 2000. Traditional ecological knowledge in environmental assessment and management. Arctic, 53(2):183–193.

Usher, P.J., G. Duhaime and E. Searles, 2003. The household as an economic unit in Arctic aboriginal communities, and its measurement by means of a comprehensive survey. Social Indicators Research, 61(2):175–202.

Watkins, M. (ed.), 1977. Dene Nation: The Colony Within. University of Toronto Press, xii + 189p.

Webster, D., 1999. Harvaqtuurmiut heritage: the heritage of the Inuit of the Lower Kazan River. Artisan Press.

Wenzel, G., 1999. Traditional ecological knowledge and Inuit: reflections on TEK research and ethics. Arctic, 52(2):113–124.

Whitbridge, P., 1999. The prehistory of Inuit and Yupik whale use. Revista de Arqueologia Americana, 16:99–154.

Whiting, A., 2002. Documenting Qikiktagrugmiut knowledge of environmental change. Native Village of Kotzebue, Alaska.

WIPO, 2001. Intellectual property needs and expectations of traditional knowledge holders. World Intellectual Property Organization, Geneva.

Chapter 4

Future Climate Change: Modeling and Scenarios for the Arctic

Lead Authors
Vladimir M. Kattsov, Erland Källén

Contributing Authors
Howard Cattle, Jens Christensen, Helge Drange, Inger Hanssen-Bauer, Tómas Jóhannesen, Igor Karol, Jouni Räisänen, Gunilla Svensson, Stanislav Vavulin

Consulting Authors
Deliang Chen, Igor Polyakov, Annette Rinke

Contents

Summary

Increased atmospheric concentrations of greenhouse gases (GHGs) are very likely to have a larger effect on climate in the Arctic than anywhere else on the globe. Physically based, global coupled atmosphere-land-ocean climate models are used to project possible future climate change. Given a change in GHG concentrations, the resulting changes in temperature, precipitation, seasonality, etc. can be projected. Future emissions of GHGs and aerosols can be estimated by making assumptions about future demographic, socioeconomic, and technological changes. The Intergovernmental Panel on Climate Change (IPCC) prepared a set of emissions scenarios for use in projecting future climate change. This assessment uses the A2 and B2 emissions scenarios, which are in the middle of the range of scenarios provided by the IPCC. Projections from the IPCC climate models indicate a global mean temperature increase of 1.4 °C by the mid-21st century compared to the present climate for both the A2 and B2 scenarios (IPCC, 2001). Toward the end of the century, the global mean temperature increase is projected to be 3.5 °C and 2.5 °C for the two scenarios, respectively.

Over the Arctic, the ACIA-designated models project a larger mean temperature increase: for the region north of 60° N, both emissions scenarios result in a 2.5 °C increase by the mid-21st century. By the end of the 21st century, arctic temperature increases are projected to be 7 °C and 5 °C for the A2 and B2 scenarios, respectively, compared to the present climate. By then, in the B2 scenario, the models project temperature increases of around 3 °C for Scandinavia and East Greenland, about 2 °C for Iceland, and up to 5 °C for the Canadian Archipelago and Russian Arctic. The five-model mean warming over the central Arctic Ocean is greatest in autumn and winter (up to 9 °C by the late 21st century in the B2 scenario), as the air temperature reacts strongly to reduced ice cover and thickness. Average autumn and winter temperatures are projected to rise by 3 to 5 °C over most arctic land areas by the end of the 21st century. By contrast, summer temperature increases over the Arctic Ocean are projected to remain below 1 °C throughout the 21st century. The contrast between greater projected warming in autumn and winter and lesser warming in summer also extends to the surrounding land areas but is less pronounced there. In summer, the projected warming over northern Eurasia and northern North America is greater than that over the Arctic Ocean, while in winter the reverse is projected. All of the models suggest substantially smaller temperature increases over the northern North Atlantic sector than in the other parts of the Arctic.

By the late 21st century, projected precipitation increases in the Arctic range from about 5 to 10% in the Atlantic sector to as much as 35% in certain high Arctic locations (for the B2 scenario). As for temperature, the projected increase in precipitation is generally greatest in autumn and winter and smallest in summer.

A slight decrease in pressure in the polar region is projected for throughout the year. While impact studies would benefit from projections of wind characteristics and storm tracks in the Arctic, available analyses in the literature are insufficient to justify firm conclusions about possible changes in the 21st century.

The models also project a substantial decrease in snow and sea-ice cover over most of the Arctic by the end of the 21st century.

The projected increase in arctic temperatures is accompanied by large between-model differences and considerable interdecadal variability. Dividing the average projected temperature change by the magnitude of projected variability suggests that, despite the large warming projected for the Arctic, the signal-to-noise ratio is actually lower in the Arctic than in many other areas.

The Arctic is a region characterized by complex and insufficiently understood climate processes and feedbacks, contributing to the challenge that the Arctic poses from the view of climate modeling. Several weaknesses of the models related to descriptions of high-latitude surface processes have been identified, and these are among the most serious shortcomings of present-day arctic climate modeling.

Local and regional climate features, such as enhanced precipitation close to steep mountains, are not well represented in global climate models due to the limited horizontal resolution of the models. To describe local climate, physical modeling or statistically based empirical links between the large-scale flow and local climate can be used. Despite rapid developments in arctic regional climate modeling, the current status of developments in this field did not allow regional models to be used as principal tools for the ACIA. Therefore, the ACIA used projections from coupled global models, either directly or in combination with statistical downscaling techniques.

A model simulation provides one possible climate scenario. This is not a prediction of future climate change, but a projection based on a prescribed change in the concentration of atmospheric GHGs. A climate shift can be caused by natural variability as well as by changes in GHG concentrations. Natural variability in the Arctic is large and could mask or amplify a change resulting from increased atmospheric GHG concentrations. To assess the relative importance of natural variability versus a prescribed climate forcing, an ensemble of differently formulated climate models should be used. For this assessment, five different models are used to give an indication of simulation uncertainty versus forced changes, although greater numbers of simulations would provide a better estimate of climate change probability distributions, and perhaps allow the estimation of changes in the frequency of winter storms, and temperature and precipitation extremes, etc.

While the level of uncertainty in climate simulations can probably be reduced with improved model formulations, it will never be certain that all physical processes relevant to climate change have been included in a model simulation. There can still be surprises in the understanding of climate change. The projections presented here are based on the best knowledge available today about climate change; as climate-change science progresses there will always be new results that may change the understanding of how the arctic climate system works.

4.1. Introduction

To assess climate change impacts on societies, ecosystems, and infrastructure, possible changes in physical climate parameters must first be projected. The physical climate change projections must in turn be calculated from changes in external factors that can affect the physical climate. Examples of such factors include atmospheric composition, particularly atmospheric concentrations of GHGs and aerosols, and land-surface changes (e.g., deforestation). This chapter describes the options available to make such projections and their application to the Arctic. The main emphasis is on physically based models of the climate system and the relationship between global climate change and regional effects in the Arctic.

Physically based climate models are used to obtain climate scenarios – plausible representations of future climate that are consistent with assumptions about future emissions of GHGs and other pollutants (i.e., emissions scenarios) and with present understanding of the effects of increased atmospheric concentrations of these components on the climate (IPCC-TGCIA, 1999). Correspondingly, by using a climate change scenario, the difference between the projected future climate and the current climate is described. Being dependent on sets of prior assumptions about future human activities, demographic and technological change, and their impact on atmospheric composition, climate change scenarios are not predictions, but rather plausible, internally consistent descriptions of possible future climates.

In addition to physical climate modeling, there are alternative methods for providing climate scenarios for use in impact assessments. These include synthetic scenarios (also referred to as arbitrary or incremental scenarios) and analogue scenarios. None of the alternatives provide a physically consistent climate change scenario including both atmospheric composition changes and physically coupled changes in temperature, precipitation, and other climate variables. Nevertheless, due to their relative simplicity they can be useful and adequate for some types of impact studies. There are also climate scenarios that do not fall into any of these categories, which primarily employ extrapolation of either ongoing trends in climate, or future regional climate, on the basis of projected global or hemispheric mean climate

change. A separate group of scenarios is based on expert judgments. All of the methods have their limitations, but each has some particular advantages (see Carter et al., 2001; Mearns et al., 2001).

Synthetic scenarios are based on incremental changes in climatic variables, particularly air temperature (e.g., +1, +2, +3 °C) and precipitation (e.g., +5, +10, +15%). Such scenarios often assume a uniform annual change in the variables over the region under consideration; however, some temporal and spatial variability may be introduced as well. Synthetic scenarios provide a framework for conducting sensitivity studies of potential impacts of climate change using impact models. Careful selection of the range and combinations of changes (e.g., using knowledge based on climate model projections), can facilitate "guided" sensitivity analysis, enabling an examination of both the modeled behavior of a system under a plausible range of climatic conditions and the robustness of impact models applied under changed and often unprecedented environmental conditions. Synthetic scenarios can provide a useful context for understanding and evaluating responses to more complex scenarios based on climate model outputs. Transparency to users and limited computational resource requirements, which allow examination of a wide range of potential climate changes (the range is further increased by the possibility of changing individual variables independent of one another), are among the advantages of synthetic scenarios. Their main disadvantage is the lack of internal consistency in applying uniform changes over large and highly variable areas such as the Arctic. Arbitrary changes in different variables may also lead to inconsistencies in synthetic scenarios that can limit their applicability and appropriateness. In addition, synthetic scenarios are not directly related to GHG forcing.

Analogue scenarios of a future climate are of two types: temporal analogue scenarios, which are based on previous warm climate conditions (determined either by instrumental or proxy data), and spatial analogue scenarios, which are based on current climate conditions in warmer regions. The use of historic instrumental records is an apparent advantage of the past climate analogues over other approaches. However, the availability of historic observational data for the Arctic is extremely limited. Proxy climate data, while representing in some cases a physically plausible climate different from the current climate to a degree similar to that of the climate projected for the 21st century, are also not available for many locations. The quality of geological records is often uncertain, and the resolution coarse. Furthermore, the paleoclimate changes are unlikely to have been driven by an increase in GHG concentrations. Spatial analogues are also unrelated to GHG forcing and are often physically implausible. The lack of availability of proper analogues is the major problem for the analogue scenario approach. The IPCC recommends that analogue scenarios are not used, at least not independently of other types of scenario (Carter et al., 1994).

Physical climate models are based on the laws of physics and discrete numerical representations of these laws that allow computer simulations. Trenberth (1992) describes how climate models can be constructed and their underlying physical principles. Of the hierarchy of climate models (Box 4.1), global coupled atmosphere-ocean general circulation models (AOGCMs) are widely acknowledged as the principal, and most promising rapidly developing tools for simulating the response of the global climate system to increasing GHG concentrations. In its Third Assessment Report, the IPCC (2001) concluded that state-of-the-art AOGCMs in existence at the turn of the century provided "credible simulations of climate, at least down to subcontinental scales and over temporal scales from seasonal to decadal", and as a class were "suitable tools to provide useful projections of the future climate" (McAvaney et al., 2001). The IPCC (2001) identified the following primary sources of uncertainty in climate scenarios based on AOGCM projections: uncertainties in future emissions of GHGs and aerosols (emissions scenarios), and in conversion of the emissions to atmospheric concentrations and to radiative forcing of the climate; uncertainties in the global and regional climate responses to emissions simulated by different AOGCMs; and uncertainties due to inaccurate representation of regional and local climate. A disadvantage of the AOGCMs as a tool for constructing scenarios is their high demand for computational resources, which makes it expensive and time-consuming to carry out calculations for multiple emissions scenarios.

The selection of climate scenarios for impact assessments is always controversial and vulnerable to criticism (Smith et al., 1998). Mearns et al. (2001) suggested that, to be useful for impact assessments and policy makers, climate scenarios should be consistent with global projections at the regional level (i.e., projected changes in regional climate may lie outside the range of global mean changes but should be consistent with theory and model-based results); be physically plausible and realistic; provide a sufficient number of variables and appropriate temporal and spatial scales for impact assessments; be representative, reflecting the potential range of future regional climate change; and be accessible.

Compared to the other methods of constructing climate change scenarios, only AOGCMs (possibly in combination with dynamic or statistical downscaling methods)

Box 4.1. Climate model hierarchy

Climate models have very different levels of complexity with respect to resolution and comprehensiveness. Available computing resources may limit model complexity for practical reasons, but the scientific question to be addressed is the main factor determining the required model complexity. Different levels of reduction (or simplification) create a hierarchy of climate models (McAvaney et al., 2001).

Simple climate models of the energy-balance type, with zero (globally averaged) to two (latitude and height) spatial dimensions, belong to the lowest level of the hierarchy. Based upon parameters derived from more complex climate models, they are useful in studies of climate sensitivity to a particular process over a wide range of parameters (e.g., in a preliminary analysis of climate sensitivity to various emissions scenarios, see section 4.4.1). Simple climate models can also be used as components of integrated assessment models, for example, in analyses of the potential costs of emission reductions or impacts of climate change (see Mearns et al., 2001).

Earth system models of intermediate complexity (EMICs) bridge the gap between the simple models and the comprehensive three-dimensional climate models (see Claussen et al., 2002). These models explicitly simulate interactions between different components of the climate system; however, at least some of the components have a reduced complexity, potentially limiting their applicability. These models are computationally efficient, allow for long-term climate simulations measured in thousands and tens of thousands of years, and are primarily used for studies of particular climate processes and feedbacks that are not believed to be affected by the dynamical simplifications introduced.

Comprehensive three-dimensional coupled atmosphere-ocean general circulation models (AOGCMs) occupy the top level of the hierarchy. The term "general circulation" refers to large-scale flow systems in the atmosphere and oceans, and the associated redistribution of mass and energy in the climate system. General circulation models (GCMs) simulate the behavior of these systems and the interactions between them and with other components of the climate system, such as sea ice, the land surface, and the biosphere. Atmosphere-ocean general circulation models are widely acknowledged as the most sophisticated tool available for global climate simulations, and particularly for projecting future climate states.

Atmosphere-ocean general circulation models were preceded by far less computationally demanding atmospheric GCMs coupled to simple parameterizations of the upper mixed layer of the ocean (AGCM/OUML), which still play an important role in studies of processes and feedbacks in the climate system (see also section 4.2.1) and in paleoclimate simulations.

have the potential to provide spatially and physically consistent estimates of regional climate change due to increased atmospheric GHG concentrations (IPCC-TGCIA, 1999). The AOGCM projections are available for a large number of climate variables, at a variety of temporal scales, and for regular grid points all over the world, which should be sufficient for many impact assessments. Employing an ensemble of different models increases the representativeness of AOGCM-based scenarios. When AOGCMs are used to provide the central scenarios, they can be combined with other types of scenarios (e.g., with synthetic scenarios applied at the regional level, for which the AOGCMs provide a physically plausible range of climate changes).

For this assessment, five AOGCMs (referred to as the ACIA-designated models, see section 4.2.7) were selected for constructing future climate change scenarios for the Arctic (see section 1.4.2). The ACIA-designated models are drawn from the generation of climate models evaluated by the IPCC (2001). This chapter begins with a brief description of the state-of-the-art in AOGCM development at the time of the IPCC assessment (section 4.2), followed by an evalua-

tion of the ACIA-designated models' performance in simulating the current climate of the Arctic (section 4.3). Projections of future climate change in the Arctic using the ACIA-designated models are the central focus of this chapter (section 4.4). An assessment of possible climate change at scales smaller than subcontinental, such as the scale considered by the ACIA, requires the application of a downscaling technique to the AOGCM output (see Box 4.1). In this assessment, two methods of downscaling AOGCM projections have been considered: regional climate modeling (section 4.5), and statistical downscaling (section 4.6). Finally, section 4.7 presents the outlook for improving AOGCM-based climate change projections for the Arctic.

4.2. Global coupled atmosphere–ocean general circulation models

The atmosphere, oceans, land surface, cryosphere, and associated biology and chemistry form interactively coupled components of the total climate system. Climate models are primary tools for the study of climate, its sensitivity to external and internal forcing factors, and the mechanisms of climate variability and

---

While the resolution of AOGCMs used for projections of future climate is rapidly improving, it is still insufficient to capture the fine-scale structure of climatic variables in many regions of the world that is necessary for impact assessment studies (Giorgi et al., 2001; Mearns et al., 2001). Hence, a number of techniques exist to enhance the resolution of AOGCM outputs. These techniques fall into three categories:

- High- or variable-resolution stand-alone AGCM simulations initialized using atmospheric and land-surface conditions interpolated from the corresponding AOGCM fields and driven with the sea surface temperature and sea-ice distributions projected by the AOGCM.

- High-resolution regional (or limited-area) climate models (RCMs) restricted to a domain with simple lateral boundaries, at which they are driven by outputs from GCMs or larger-scale RCMs.

- Statistical downscaling methods that are based on empirically derived relations between observed large-scale climate variables and local variables, and which apply these relations to the large-scale variables simulated by GCMs (or RCMs).

Each of the regionalization techniques is characterized by its own set of advantages and disadvantages. Giorgi et al. (2001) provided details on the high- and variable-resolution AGCMs, while RCMs and statistical downscaling are discussed in sections 4.5 and 4.6.

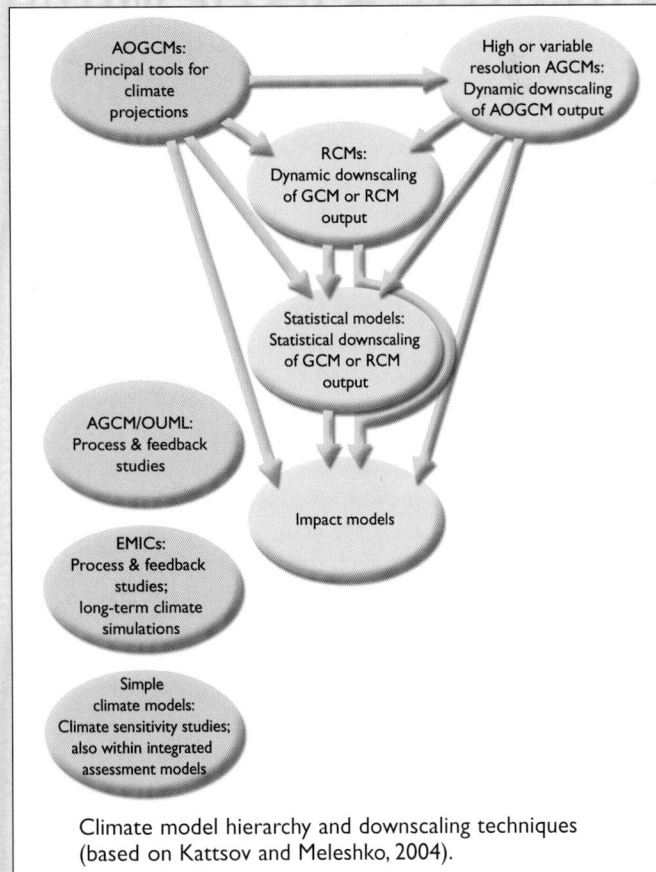

Climate model hierarchy and downscaling techniques (based on Kattsov and Meleshko, 2004).

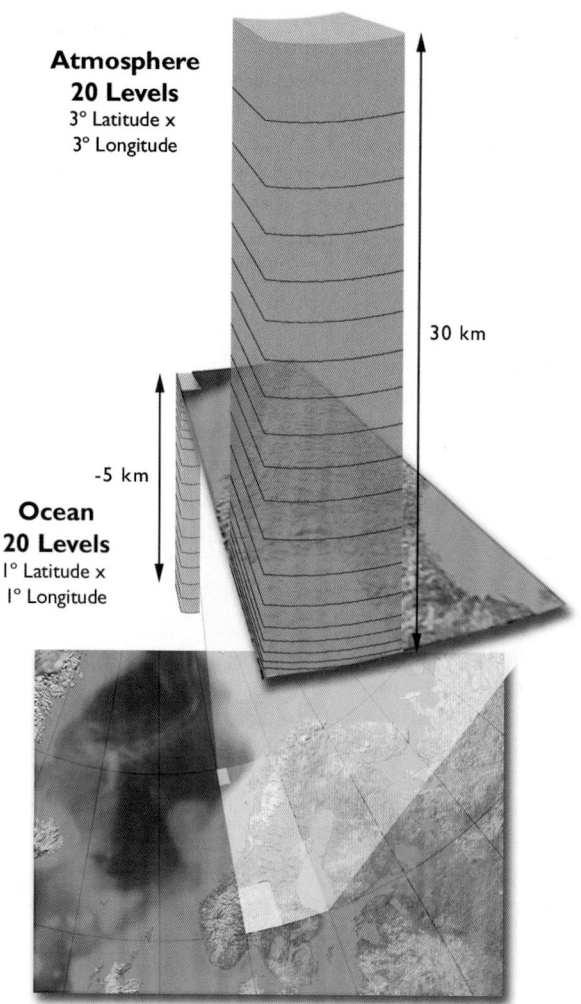

**Atmosphere
20 Levels**
3° Latitude x
3° Longitude

30 km

-5 km

**Ocean
20 Levels**
1° Latitude x
1° Longitude

Fig. 4.1. Schematic illustrating the representation of the earth system by a coupled atmosphere–ocean general circulation model. Actual grid size and number of levels may vary.

change. These models attempt to take into account the various processes important for climate in the atmosphere, the oceans, the land surface, and the cryosphere, as well as the interactions between them (Fig. 4.1). In addition, models are increasingly incorporating components that describe the role of the biosphere and chemistry in order to provide a comprehensive description of the total earth system. Because physical processes and feedbacks play a key role in the arctic climate system, this section focuses primarily on the physical components of climate models. However, land-surface biology is an important factor in determining the key thermal and radiative properties of the land surface, surface hydrology, turbulent heat and gas exchanges, and other processes. Likewise, the interaction of ocean biology with physical processes is important for air-sea gas exchange, including key processes related to cloud formation such as dimethyl sulfide exchange.

Coupled AOGCMs are made up of component models of the atmosphere, ocean, cryosphere, and land surface that are interactively coupled via exchange of data across the interfaces between them. For example, the ocean component is driven by the atmospheric fluxes of heat, momentum, and freshwater simulated by the atmospheric com-

ponent. These heat and freshwater fluxes are themselves functions of the sea surface temperatures simulated by the ocean model. Other driving fluxes for the ocean are produced by the brine rejection that occurs during sea-ice formation, freshwater from sea-ice melt, and freshwater river discharge at the continental boundaries.

Atmosphere-ocean general circulation models are continually evolving. The state-of-the-art climate modeling described in this section refers to the generation of models from the late 1990s and very early 2000s, and is close to that evaluated by the IPCC (2001).

4.2.1. Equilibrium and transient response experiments

Early climate simulations were conducted using atmospheric models coupled to highly simplified representations of the ocean. In these models, only the upper ocean was normally represented and then only as a simple fixed-depth slab of water some tens of meters deep in which the temperature responded directly to changes in atmospheric heat fluxes. Such models are still useful for short sensitivity experiments, such as exploring the impact of new representations of physical processes on climate change in experiments in which the concentration of atmospheric GHGs is instantaneously doubled in the model atmosphere. These models enable a quick assessment of the "equilibrium response" of climate to a given perturbation. The equilibrium response is the change in climate resulting from a perturbation (e.g., a specified increase in effective carbon dioxide (CO_2) concentration) after a period long enough for the climate to reach an equilibrium state. However, such models assume that vertical and horizontal heat transports in the ocean do not change when the climate changes.

Many centers have developed models with full dynamic deep-ocean components over the past decade. The dynamic oceans introduce the long timescales (multicentury to millennia) associated with the equilibration of the abyssal ocean. Such long timescales are absent in the models that represent the ocean as a shallow slab of water. In particular, this development has enabled the exploration of the "transient response" of climate to changing concentrations of GHGs, as well as the examination of many aspects of natural climate variability. The transient response is the change over time as the perturbation (e.g., a continuous change in GHG concentrations) is applied. In the case of GHG-induced temperature change, the transient response is smaller than the equilibrium response because the large thermal inertia of the oceans slows the rate of warming.

4.2.2. Initialization and coupling issues

Owing to embodied feedbacks between ocean and atmosphere, an AOGCM-simulated climate is less constrained than climates simulated by stand-alone atmospheric or oceanic general circulation models (GCMs). Upon coupling, an AOGCM-simulated climate typically undergoes a so-called coupling shock (fast drift due to imbalances in the initial conditions between the component models at the time of coupling) and then, after a close-to-balance state between the interacting components of the AOGCM has been achieved, a gradual drift toward the model's equilibrium climatic state. The presence of climate drift, if it is significant, can complicate the study of a possible climate change signal. For example, large drifts can potentially distort the behavior of various feedback processes present in the climate system and, dependent on the mean state of the model, distort the calculated climatic response to a given change in forcing.

The climate drift problem can introduce many technical considerations into the application of AOGCMs. To limit the influence of climate drift (especially the fast-drift component), careful initialization of AOGCMs is very important. This has led to a relatively wide array of initialization methods (e.g., Stouffer and Dixon, 1998). Initialization techniques often include a sequence of runs of component models separately, and in the coupled mode the components are constrained by observations at their interfaces. This makes it possible to reduce the climate drift and, particularly, the coupling shock.

Until recently, it has been necessary to use so-called "flux adjustments" (or "flux corrections", Sausen et al., 1987) to prevent drift in the climate of the coupled system that arises from inadequacies in the component models and in the simulated fluxes at their interfaces. These adjustments are normally derived as fields of spatially varying "corrections" to the heat and freshwater fluxes between the atmosphere and ocean components of the model. They are often derived during a calibration run of the AOGCM in which the sea surface temperatures and surface salinities are constrained to observed climatological values of these quantities. The flux adjustments are then applied to succeeding runs of the model to provide improved simulation of the coupled system. In some cases, flux adjustments have also been applied to momentum fluxes. While flux adjustments have not been applied over land, it has in the past been necessary to flux-adjust the fields of sea-ice concentration and thickness. A driver in the development of coupled models has been to improve models to the stage where they can run without flux adjustments, as is now the case for some AOGCMs. In the AOGCMs that continue to use this technique, flux adjustments have become smaller as models have improved. Interestingly, the IPCC did not find systematic differences in the simulation of internal climate variability between flux-adjusted and non-flux adjusted AOGCMs (McAvaney et al., 2001), thus supporting the use of both types of model in the detection and attribution of climate change.

4.2.3. Atmospheric components of AOGCMs

The atmospheric component of AOGCMs enables simulation of the evolution with time of the spatial distributions of the vector wind, temperature, humidity, and surface pressure. This is done by discretization of the basic equations governing the behavior of the atmosphere, and implementation of these discretized equations on an appropriate computer. The equations are time-stepped forward at intervals that typically vary from a few minutes to tens of minutes, depending on the model formulation and resolution, to produce an evolving simulation of the behavior of the atmospheric flow and associated temperature, humidity, and surface-pressure fields.

The model dynamics are usually represented either as periodic functions defined as the sum of several waves (spectral models) or on a grid of points (finite-difference models) covering the globe for various levels of the atmosphere. Typically, the atmospheric components of the generation of climate models evaluated by the IPCC (2001) operated on grids with a horizontal spacing of 200 to 300 km and 10 to 20 vertical levels. Various schemes are available for the specification of vertical coordinates (e.g., Kalnay, 2003).

Simulation of some climatic variables in high latitudes (e.g., atmospheric moisture) using global models presents certain problems. Finite-difference GCMs require undesirable filtering operations in order to avoid computational instability when a reasonable time step is used in regions of converging meridians such as the Arctic. While polar filtering is not needed in spectral models, these models produce fictitious negative moisture amounts in the dry high-latitude atmosphere, thus calling for correction procedures. Both problems are apparently overcome by the application of semi-Lagrangian schemes for moisture advection, which are used in a number of atmospheric general circulation models (AGCMs). However, in semi-Lagrangian schemes, the advantages of large time steps and the absence of spurious negative moisture values are partially offset by the lack of exact moisture conservation. New schemes have recently started to appear that combine the semi-Lagrangian approach with mass conservation (e.g., Zubov et al., 1999), but have other disadvantages. Hopes for improved climate simulations of the polar regions are also associated with spherical geodesic grids, which allow for approximately uniform discretization of the sphere. Such grids are already used in some global numerical weather-prediction models (Majewski et al., 2002).

A key issue is the simulation of the basic physical processes that take place in the atmosphere and determine many of the feedbacks for climate variability and change. Examples include the representation of clouds and radiation; dry and moist convective processes; the formation of precipitation and its deposition on the surface as rain or snow; the interactions between the atmosphere and the land-surface orography (including the drag on the atmosphere caused by breaking gravity waves); and atmospheric boundary-layer processes and their interaction with the surface. Because these processes take place on scales much smaller than the model grid, they must be represented in terms of the large-scale variables in the model (vector wind, temperature, humidity, and surface pressure). Key atmospheric processes from an arctic surface climate perspective include the representations of the planetary boundary layer, clouds, and radiation.

Energy, momentum, and moisture from the free troposphere are transferred via the atmospheric boundary layer (ABL) to the surface and vice versa. Atmospheric general circulation models have difficulty with the proper representation of turbulent mixing processes in general, which has implications for the representation of boundary-layer clouds (IPCC, 2001). The ABL in the Arctic differs significantly from its mid-latitude counterpart, so parameterizations based on mid-latitude observations tend to perform poorly in the Arctic. Parameterizations of the surface fluxes are usually based on the Monin-Obukhov similarity theory. These parameterizations work reasonably well for cases where the vertical stratification of the atmosphere is weakly stable, but simulate surface fluxes of momentum, heat, and water vapor that are too small in the very stable stratified conditions (Poulus and Burns, 2003) common in the high Arctic. In the very stable cases, turbulence may not be stationary, local, and continuous (Mahrt, 1998) – assumptions used in ABL parameterizations of surface fluxes. In addition, vertical resolution is a critical issue because the very thin stable surface layer is usually shallower than the first vertical model layer. Deviations from observations in the ABL during winter, found in simulations with a regional climate model for the Arctic (section 4.5.1), indicate the necessity of improvements in the atmospheric parameterization that better describe the vertical stratification and atmosphere–surface energy exchange (Dethloff et al., 2001). The mean monthly turbulent heat-flux distribution at the surface strongly depends on different ABL parameterizations and leads to different spatial distributions of temperature, wind, moisture, and other variables throughout the arctic atmosphere. The greatest changes are found in the ABL above the sea-ice edge in January.

Model resolution, both horizontal and vertical, is a problem in simulating the arctic ABL. The vertical discretization of current AGCMs cannot resolve the large temperature gradients and inversions that exist in the arctic ABL. Insufficient resolution gives rise to sensible heat fluxes in the models that tend to be too large. However, simply increasing the resolution will not solve the problem. Even if the very stable ABL can be simulated in finer detail, the fundamental problem of current theories predicting turbulent fluxes that are too small will still remain.

Specific cloud types observed in the arctic ABL present a serious challenge for atmospheric models. Parameterizing low-level arctic clouds is particularly difficult because of complex radiative and turbulent interactions with the surface (e.g., Randall et al., 1998).

The atmospheric components of AOGCMs usually focus on representation of tropospheric processes and the effects of stratospheric processes on the troposphere, while their descriptions of stratospheric processes are less satisfactory. For example, the insufficient vertical resolution in the stratosphere (as compared to that in the troposphere) prevents the atmospheric components of AOGCMs from properly representing important stratospheric phenomena, such as the quasi-biennial oscillation and sudden stratospheric temperature increases (Takahashi, 1999).

Current AOGCMs generally do not include interactive atmospheric chemistry models (Austin et al., 2003). Most of the atmospheric photochemical processes are therefore simulated with chemical transport models (CTMs) that use atmospheric wind velocities and temperature prescribed either from observational data or from GCM simulations. In the latter case, CTMs can be used to project the evolution of the atmospheric content of ozone, other radiatively active gases (e.g., methane and nitrous oxide), and aerosols (Austin et al., 2003; WMO, 2003). Projections of the distributions of tropospheric ozone and aerosols (sulfates, soot, sea salt, and mineral dust collectively known as "arctic haze") are particularly important to climate change projections (IPCC, 2001).

4.2.4. Ocean components of AOGCMs

The oceanic component in AOGCMs has improved substantially over the past decade. These models now include representation of the full dynamics and thermodynamics of the global ocean basins and allow simulation of the full three-dimensional current, temperature, and salinity structure of the ocean and its evolution. Important physical processes are associated with the upper-ocean mixed layer and diffusive processes in the ocean. The freezing, melting, and dynamics of sea ice and ice–ocean interactions are also taken into account. Until recently, because of limitations in available computing power, AOGCMs typically had similar horizontal resolution in the ocean and atmospheric components. Such ocean models poorly represent the large-scale ocean current structure, not only because of the lack of resolution of narrow boundary currents such as the Gulf Stream and the Kuroshio, but also because of the high viscosity coefficients necessary for computational stability (e.g., Bryan et al., 1975). However, as available computing power has increased, the resolution of the ocean component of AOGCMs has increased to roughly one degree of latitude and longitude. Although this resolution does not allow explicit representation of ocean eddies (a resolution of one-third of a degree is considered "eddy permitting", and one-ninth of a degree or better, "eddy-resolving"), it does result in a much-improved representation of ocean current structure.

The Arctic Ocean has always been and still remains one of the weak spots in AOGCMs. This is partly due to specific numerical problems such as the singularity of the longitude-latitude spherical coordinates (converging meridians) at the North Pole (see Randall et al., 1998).

Until recently, filtering, or even inserting an artificial island at the North Pole, have been among the usual, but undesirable, ways to overcome the pole problem. Rotating grids or introducing alternative grids, for example, geodesic grids providing approximately uniform discretization of the sphere (e.g., Sadourny et al., 1968) or using curvilinear generalized coordinates (Murray, 1996), are now being pursued in order to eliminate the converging meridian problem. Such features are now starting to appear in oceanic components of AOGCMs (e.g., Furevik et al., 2003). However, a greater challenge is insufficient understanding of some phenomena related to the general circulation of the Arctic Ocean and subarctic seas. In particular, improvement is needed in representing ocean/atmosphere/sea-ice interaction processes in order to better evaluate their importance within the context of natural variability and anthropogenically forced change in the climate system. A particular problem for the oceanic component of AOGCMs is the treatment of air–ice–ocean interactions and water-mass formation (creation of water bodies with a homogenous distribution of temperature and salinity) over the shallow continental shelves, which requires adequate resolution of shallow water layers, water-mass formation and mixing processes, continental runoff, and ice processes.

4.2.5. Land-surface components of AOGCMs

The land-surface components of climate models include representation of the thermal and soil-moisture storage properties of the land surface through modeling of its upper layers. Key properties include surface roughness and albedo, which are normally specified from global datasets, although models with interactive land-surface properties are now being developed.

Possible changes in vegetation and the effects that these changes may have on future climate are not often taken into account in climate change projections. These effects may be substantial and would be manifested in the local fluxes of water, heat, and momentum controlled by surface roughness, albedo, and surface moisture. The arctic land types have special features that are not well represented in the present generation of climate models (Harding et al., 2001). This is particularly true for winter conditions where snow distribution and its interaction with vegetation are poorly understood and modeled.

The discharge of river water to the ocean, especially to the Arctic Ocean whose freshwater budget is much more influenced by terrestrial water influx than are the budgets of other oceans, is of potential importance to climate change. The land-surface components of AOGCMs usually include river-routing schemes, in which the land surface is represented as a set of watersheds draining the runoff (integrated over their territories at each time step) into the grid boxes of the ocean model closest to the grid points specified as river mouths in the land-surface model. Such schemes are able to provide reasonable annual means of the dis-

charge, but shift and sharpen its seasonal cycle, especially for the Arctic Ocean terrestrial watersheds with their high seasonality of discharge. More comprehensive river-routing schemes (e.g., Hagemann and Dümenil, 1998), allowing for simulations of horizontal transport of the runoff within model watersheds, are usually not used interactively in AOGCMs.

4.2.6. Cryospheric components of AOGCMs

Snow cover and sea ice are the two primary elements of the cryosphere represented interactively in AOGCMs, although some models now incorporate explicit parameterizations of permafrost processes. The large ice sheets are represented, although non-interactively, by land-surface topography and surface albedo (typically fixed at a value of around 0.8). Likewise, there is usually no explicit representation of glaciers.

The insulating effects and change in surface albedo due to snow cover are of particular importance for climate change projections. AOGCMs demonstrate varying degrees of sophistication in their snow parameterization schemes. For example, some can represent snow density, liquid water storage, and wind-blown snow (see Stocker et al., 2001). Advanced albedo schemes incorporate dependencies on snow age or temperature. However, a major uncertainty exists regarding the ability of AOGCMs to simulate terrestrial snow cover (McAvaney et al., 2001; see also section 6.4), particularly its albedo effects and the masking effects of vegetation that are potentially important in determining the surface energy budget (see section 7.5).

Sea-ice components of AOGCMs usually include parameterizations of the accumulation and melting of snow on the ice, and thermodynamic energy transfers between the ocean and atmosphere through the ice and snow. Most of the AOGCMs evaluated by the IPCC (2001) employed simplistic parameterizations of sea ice. Recent advances in stand-alone sea-ice modeling, including those in modeling sea-ice thermodynamics (e.g., introducing the effects of subgrid-scale parameterizations with multiple thickness categories – the so-called "ice-thickness distribution"), are now being incorporated into AOGCMs. However, understanding is still insufficient for treating some atmosphere–ice–ocean interaction issues (e.g., heat distribution between concurrent lateral and vertical ice melt or accumulation). The primary differences among the various representations relate to treatment of internal stresses in calculating sea-ice model dynamics. An evaluation of the different treatment of sea-ice rheologies (relationships between internal stresses and deformation) was the core task for the Sea-Ice Model Intercomparison Project (SIMIP) initiated in the late 1990s. Having considered a hierarchy of stand-alone sea-ice models with different dynamic parameterizations, SIMIP found the viscous-plastic rheology to provide the best simulation results and adopted it as a starting point for further optimizations (Lemke et al., 1997). Other developments, including the elastic-viscous-plastic rheology (Hunke and Dukowicz, 1997), are helpful in achieving high computational efficiency. However mature the status of stand-alone sea-ice dynamics modeling, some AOGCMs still employ a simple, so-called "free drift" scheme that only allows ice to be advected with ocean currents. There is a large range in the ability of AOGCMs to simulate the position of the ice edge and its seasonal cycle (McAvaney et al., 2001). However, there is no obvious connection between the fidelity of simulated ice extent and the inclusion of an ice-dynamics scheme. This is apparently due to the additional impact of simulated wind-field errors (e.g., Bitz et al., 2002; Walsh et al., 2002), which may offset improvements from the inclusion of more realistic ice dynamics. Conversely, the importance of improved sea-ice dynamics and thermodynamics has become apparent, and the AOGCM community is responding by including more sophisticated treatments of sea-ice physics.

4.2.7. AOGCMs selected for the ACIA

Selecting AOGCM simulation results to be used in an impact assessment is not a trivial task, given the variety of models. The IPCC (McAvaney et al., 2001) concluded that the varying sets of strengths and weaknesses that AOGCMs display means that, at this time, no single model can be considered "best" and it is important to utilize results from a range of coupled models in assessment studies. The choice of AOGCMs for this assessment used the criteria suggested by Smith et al. (1998): vintage, resolution, validity, representativeness of results, and accessibility of the model outputs.

While models do not necessarily improve with time, later versions (often with higher resolution) are usually preferred to earlier ones. An important criterion for selecting an AOGCM to be used in constructing regional climate scenarios is its validity as evaluated by analyses of its performance in simulating present-day and past climates (the evolution of 20th century climate in particular). The validity is evaluated by comparing the model output with observed climate, and with output from other models for the region of interest and larger scales, to determine the ability of the model to simulate large-scale circulation patterns. Well-established systematic comparisons of this type are provided by international model intercomparison projects (MIPs, see Box 4.2). Finally, when several AOGCMs are to be selected for use in an impact assessment, the model results should span a representative range of changes in key variables in the region under consideration.

Section 1.4.2 provides details of the procedure for selecting AOGCMs for the ACIA. Initially, a set of the most recent and comprehensive AOGCMs whose outputs were available from the IPCC Data Distribution Center was chosen. This set was later reduced to five AOGCMs (two European and three North American),

Box 4.2. Model intercomparison projects

Model intercomparison projects (MIPs) allow comparison of the ability of different models to simulate current and perturbed climates, in order to identify common deficiencies in the models and thus to stimulate further investigation into possible causes of the deficiencies (Boer, 2000a,b). This is currently the only way to increase the credibility of future climate projections. Participation in MIPs is an important prerequisite for an AOGCM to be employed in constructing climate scenarios (e.g., for the ACIA).

In MIPs, models of the same class (AOGCMs, stand-alone AGCMs or oceanic GCMs, RCMs) are run for the same period using the same forcings. Typically, diagnostic subprojects are established that concentrate upon analyses of specific variables, phenomena, or regions. Occasionally, experimental subprojects are initiated, aimed mainly at answering questions related to model sensitivity.

Of the many international MIPs conducted in the past decade, two are of primary importance for the ACIA: the Atmospheric Model Intercomparison Project (AMIP: Gates, 1992; Gates et al., 1998), and the Coupled Model Intercomparison Project (CMIP: Meehl et al., 2000). Both included subprojects devoted specifically to model performance at high latitudes among their numerous diagnostic subprojects.

Thirty AGCMs were included in the second phase of the AMIP (AMIP-II, concluded in 2002). All of these were forced with the same sea surface temperatures (SSTs) and sea-ice extents prescribed from observations, and a set of constants, including GHG concentrations. The AMIP-II simulations span the period from 1979 to 1996. AMIP findings related to AGCM performance in the Arctic have been reported since the early 1990s (e.g., Bitz et al., 2002; Frei et al., 2003a; Kattsov et al., 1998, 2000; Tao et al., 1996; Walsh et al., 1998, 2002). Coupled Model Intercomparison Project experiments belong to the class of idealized (e.g., 1% per year increase in CO_2) transient experiments with AOGCMs. Räisänen (2001) discussed some results of the second phase of the CMIP (CMIP2) related to the Arctic (see also section 4.4.5).

The Climate of the 20th Century project was initiated in order to determine to what extent stand-alone AGCMs are able to simulate observed climate variations of the 20th century against a background of natural variability primarily due to the accessibility of model output, as well as storage and network limitations. By the initial phase of the ACIA, at least one Special Report on Emissions Scenarios (SRES: Nakićenović and Swart, 2000) B2 simulation (see section 4.4.1) extending to 2100 had been generated by each of the ACIA-designated models. All of the models are well documented, participate in major international MIPs, and have had their pre-SRES simulations (see Box 4.2) analyzed for the Arctic and the results published (e.g., Walsh et al., 2002). The five ACIA-designated models listed in Table 4.1, together with information on their formulations, provided the core data for constructing the ACIA climate change scenarios.

4.2.8. Summary

Atmosphere-ocean general circulation models are widely acknowledged to be the primary tool for projecting future climate. As understanding of the earth's climate system increases and computers become more sophisticated, the scope of processes and feedbacks simulated by AOGCMs is steadily increasing. In addition to representing the general circulation of the atmosphere and the ocean, the AOGCMs include interactive components representing the land surface and cryosphere. The biosphere and the carbon and sulfur cycle components of AOGCMs are evolving, while the atmospheric chemistry component is currently being developed off-

line. The ability to increase confidence in model projections of arctic climate is limited by the need for further advances in the representation of the arctic climate system in the AOGCMs (see section 4.7).

4.3. Simulation of observed arctic climate with the ACIA-designated models

Model-based scenarios of future climate are only credible if the models simulate the observed climate (present-day and past) realistically – both globally and in the region of interest. While an accurate simulation of the present-day climate does not guarantee a realistic sensitivity to an external forcing (e.g., higher GHG concentrations), a grossly biased present-day simulation may lead to weakening or elimination of key feedbacks in a simulation of change, or conversely may cause key feedbacks to be exaggerated. The ability of the models to reproduce climate states in the past – under external forcings differing from those at present – can therefore help to add to the credibility of their future climate projections.

Boer (2000a) distinguishes three major categories of model evaluation: the morphology of climate, including spatial distributions and structures of means, variances, and other statistics of climate variables; budgets, balances, and cycles in the climate system; and process studies of climate. A comprehensive assessment of recent AOGCM simulations of observed global climate is pro-

(Folland et al., 2002). In this MIP, the AGCMs are forced with observed SSTs and sea-ice extents and prescribed changes in radiative forcing (GHGs, trace gases, stratospheric and tropospheric ozone, direct and indirect effects of sulfate aerosols, solar variations, and volcanic aerosols).

The outputs of models archived at the IPCC Data Distribution Center provide an additional opportunity for AOGCM intercomparison (IPCC-TGCIA, 1999). The archived outputs have a limited set of variables, but include at least two scenarios (A2 and B2) from the IPCC Special Report on Emissions Scenarios (SRES: Nakićenović and Swart, 2000) and at least two pre-SRES (IS92a) emissions scenarios (GHGs only and GHGs plus sulfate aerosols). The simulation results that are available usually span the 20th and 21st centuries. The selection of these AOGCMs by the IPCC for use in its Third Assessment Report (IPCC, 2001) was an indication that these models provide the most viable basis for climate change assessment.

The foci of the Arctic Regional Climate Model Intercomparison Project (ARCMIP) include the surface energy balance over ocean and land, clouds and precipitation processes, stable planetary boundary layer turbulence, ice-albedo feedback, and sea-ice processes (Curry J. and Lynch, 2002; see also section 4.5.1). Another international effort – the Arctic Ocean Model Intercomparison Project (AOMIP) – aims to identify strengths and weaknesses in Arctic Ocean models using realistic forcing (Proshutinsky et al., 2001; see also section 4.5.2). The major goals of the project are to examine the ability of Arctic Ocean models to simulate variability at seasonal to decadal scales, and to qualitatively and quantitatively understand the behavior of the Arctic Ocean under changing climate forcing.

Other GCM MIPs of relevance to the ACIA include the Ocean Model Intercomparison Project (WCRP, 2002), which is designed to stimulate the development of ocean models for climate research, and the Paleoclimate Modeling Intercomparison Project (Braconnot, 2000), which compares AGCM/OUML models (see Box 4.1) and AOGCMs in simulations of paleoclimate conditions during periods that were significantly different from the present-day climate. There are also a number of MIPs devoted to intercomparison of specific parameterizations employed in GCMs, including the Sea-Ice Model Intercomparison Project (Lemke et al., 1997), the Snow Models Intercomparison Project (Etchevers et al., 2002), and polar clouds (IGPO, 2000).

Table 4.1. Key features of the ACIA-designated AOGCMs.

	Atmospheric resolution[a]	Ocean resolution[b]	Land-surface scheme[c]	Sea-ice model[d]	Flux adjustment[e]	Primary reference
CGCM2						
Canadian Centre for Climate Modelling and Analysis, Canada	T32 (3.8° × 3.8°) L10	1.8° × 1.8° L29	M, BB, F, R	T, R	H, W	Flato and Boer, 2001
CSM_1.4						
National Center for Atmospheric Research, United States	T42 (2.8° × 2.8°) L18	2.0° × 2.4° L45	C, F	T, R	-	Boville et al., 2001
ECHAM4/OPYC3						
Max-Planck Institute for Meteorology, Germany	T42 (2.8° × 2.8°) L19	2.8° × 2.8° L11	M, BB, R	T, R	H*, W*	Roeckner et al., 1996
GFDL-R30_c						
Geophysical Fluid Dynamics Laboratory, United States	R30 (2.25° × 3.75°) L14	2.25° × 1.875° L18	B, R	T, F	H, W	Delworth et al., 2002
HadCM3						
Hadley Centre for Climate Prediction and Research, United Kingdom	2.5° × 3.75° L19	1.25° × 1.25° L20	C, F, R	T, F	-	Gordon et al., 2000

[a]Horizontal resolution is expressed either as degrees latitude by longitude or as a spectral truncation (either triangular (T) or rhomboidal (R)) with a rough translation to degrees latitude and longitude. Vertical resolution (L) is the number of vertical levels; [b]Horizontal resolution is expressed as degrees latitude by longitude, while vertical resolution (L) is the number of vertical levels; [c]B=standard bucket hydrology scheme (single-layer reservoir of soil moisture which changes with the combined action of precipitation (snowmelt) and evaporation, and produces runoff when the water content reaches the prescribed maximum value); BB=modified bucket scheme with spatially varying soil moisture capacity and/or surface resistance; M=multilayer temperature scheme; C=complex land-surface scheme usually including multiple soil layers for temperature and moisture, and an explicit representation of canopy processes; F=soil freezing processes included; R=river routing of the discharge to the ocean (land surface is represented as a set of river drainage basins); [d]T=thermodynamic ice model; F="free drift" dynamics; R=ice rheology included; [e]H=heat flux; W=freshwater flux; asterisks indicate annual mean flux adjustment only.

vided by McAvaney et al. (2001), who, in particular, regarded as well-established the ability of the AOGCMs "to provide credible simulations of both the annual mean climate and the climatological seasonal cycle over broad continental scales for most variables of interest for climate change". In this context, clouds and humidity were mentioned as major sources of uncertainty, in spite of incremental improvements in their modeling.

In this section, the first two categories of model evaluation (Boer, 2000a) are addressed for the five ACIA-designated AOGCM simulations of the observed arctic climate. The primary focus is on the evaluation of representations of surface air temperature and precipitation as reproduced by the AOGCMs for the ACIA climatological baseline period (1981–2000). The evaluation of individual ACIA-designated model simulations compared to historical data is also considered.

In most cases, the area between 60° and 90° N is used as a reference region for model evaluation. In some cases, however, smaller areas are used for consistency with observational data (e.g., precipitation, see section 4.3.1). In cases where a variable was missing from one of the five model outputs, a subset of four models was evaluated for that variable.

4.3.1. Observational data and reanalyses for model evaluation

A considerable number of datasets are available for the Arctic, including remotely sensed and *in situ* data, observations from the arctic buoy program, historical data, and field experiments (see section 2.6). However, for

evaluation of three-dimensional AOGCMs, observational data readily available at regularly spaced grid points are the most useful. *In situ* observations are not representative of conditions covering an area the size of an average model grid box, thus a comprehensive analysis is required to match model simulations and observations.

A good opportunity for model evaluation is provided by reanalyses employing numerical weather prediction models to convert irregularly spaced observational data into complete global, gridded, and temporally homogeneous data (presently available for periods of several decades). Reanalyses include both observed (assimilated) variables (e.g., temperature, geopotential height) and derived fields (e.g., precipitation, cloudiness). For some of the derived fields, direct observations are nonexistent (e.g., evaporation). The quality of a reanalysis is not the same for different variables; it may also vary regionally for the same variable, depending on the availability of observations. In areas where observations are sparse, each reanalysis primarily represents the quality of the model's simulation. For variables that are not observed, the reanalysis may not be realistic. Errors in a model's physical parameterizations can also adversely affect the reanalysis. However, despite these problems, reanalyses provide the best gridded, self-consistent datasets available for model evaluation.

It is worthwhile noting that direct point-to-point and time-step-to-time-step comparison of a climate GCM output against observations, reanalyses, or another climate model simulation is not methodologically correct. Only spatial and temporal statistics can be used for the evaluation. For state-of-the-art AOGCMs, spatial aver-

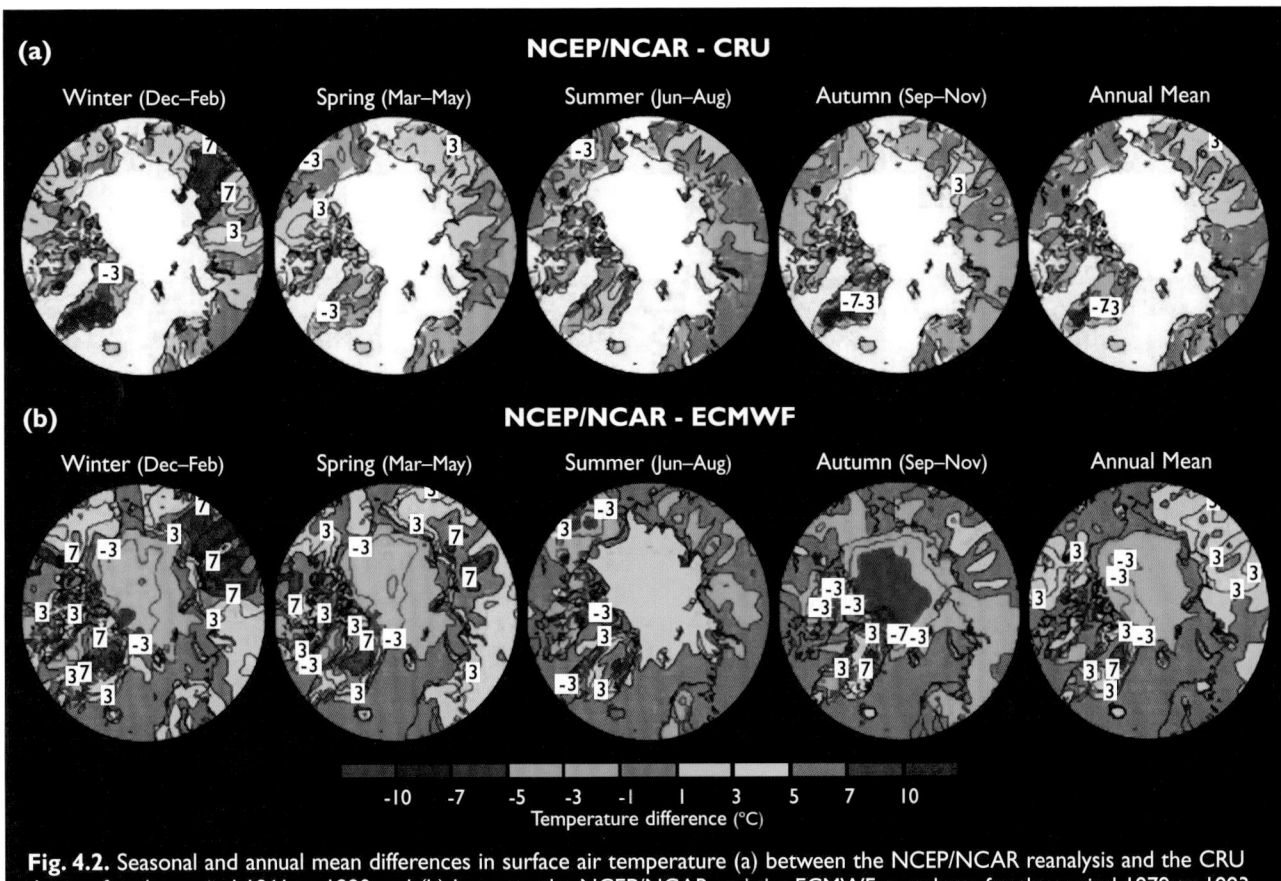

Fig. 4.2. Seasonal and annual mean differences in surface air temperature (a) between the NCEP/NCAR reanalysis and the CRU dataset for the period 1961 to 1990 and (b) between the NCEP/NCAR and the ECMWF reanalyses for the period 1979 to 1993.

ages should be at subcontinental or greater scales, such as the Arctic Ocean; the four ACIA regions (see section 1.1) including their marine parts; or the watersheds of major rivers.

Observational data for validating AOGCM performance in the Arctic (particularly the central Arctic) are characterized by a comparatively high level of uncertainty. Because of the sparsity of direct observations, even the temperature climatology in the Arctic is imperfectly known. Model-simulated surface air temperature and atmospheric pressure have primarily been compared with the National Centers for Environmental Prediction/National Center for Atmospheric Research (NCEP/NCAR) reanalysis (Kistler et al., 2001). To estimate the accuracy of the NCEP/NCAR reanalysis, its pattern of surface air temperatures was compared against two other datasets (Fig. 4.2). The first, compiled at the Climatic Research Unit (CRU), University of East Anglia (New et al., 1999, 2000), is based on the interpolation of weather station observations. It is therefore expected to be accurate where the station density is sufficient, but it covers only land areas. The second dataset used for comparison is the European Centre for Medium-Range Weather Forecasts (ECMWF) reanalysis (ERA-15; Gibson et al., 1997). Neither of the two reanalyses should be considered as "truth" but their differences provide some information about the probable magnitude of errors in them. The ECMWF reanalysis is only available for the period since 1979; the difference between the ECMWF and NCEP/

NCAR reanalyses shown in Fig. 4.2 was calculated for the overlapping interval (1979–1993).

The differences between the NCEP/NCAR reanalysis and the CRU dataset for the period 1961 to 1990 are smallest in summer (generally within ±1 °C, and almost everywhere within ±3 °C) and largest in winter. In winter, temperatures in the NCEP/NCAR reanalysis are higher than in the CRU dataset over most of northern Siberia and North America, but lower over the northeastern Canadian Archipelago and Greenland. Locally, the differences are as great as 15 °C in northern Siberia (NCEP/NCAR warmer than CRU) and Greenland (NCEP/NCAR colder than CRU). Despite these very large regional differences, the NCEP/NCAR and CRU mean temperatures over the entire arctic land area are in all seasons within 2 °C of each other.

The differences between the NCEP/NCAR and ECMWF reanalyses over land follow the NCEP/NCAR minus CRU differences in most, but not all, respects. Substantial differences also occur between the ECMWF reanalysis and the CRU dataset, most notably in spring when the ECMWF temperatures show a widespread cold bias compared to the CRU dataset. Over the central Arctic Ocean, temperatures in the NCEP/NCAR reanalysis are lower than temperatures in the ECMWF reanalysis throughout most of the year, with the greatest differences (up to 5–7 °C) in autumn. In summer, however, NCEP/NCAR temperatures are slightly higher than ECMWF temperatures.

The quality of precipitation climatologies for high latitudes derived from reanalyses is lower than that for temperature (or, e.g., atmospheric pressure). On the other hand, assessments of simulated precipitation at high latitudes are confounded by the uncertainties in observational estimates, which suffer from errors in gauge measurements of solid precipitation, especially when the solid precipitation occurs under windy conditions. Depending on the partitioning between falling snow (precipitation) and wind-blown snow from the surface, the error can range from a significant "undercatch" to a significant "overcatch". Because different observational climatologies incorporate varying types and degrees of adjustment, there is some variance among the observational estimates, more so in the monthly means than in the annual means (for details see sections 2.6.2.2 and 6.2.1). The primary observational dataset used here is an outgrowth of an arctic climatology originally compiled by Bryazgin (1976), whose monthly mean fields were extended for inclusion in Khrol (1996), and subsequently updated and enhanced by additional corrections. This compilation includes data from the Russian drifting ice stations and high-latitude land-surface stations, and it is gridded over the 65° to 90° N domain. Additional precipitation climatologies used in this assessment are Legates and Willmott (1990) and Xie and Arkin (1998). Both the Bryazgin and Legates-Willmott climatologies are based on gauge-corrected *in situ* data only. The two climatologies are multi-year averages over periods that do not coincide with each other or with the ACIA climatological baseline (see section 4.3.2). The Xie and Arkin climatology is a blend of *in situ* and satellite data, and reanalysis where *in situ* and satellite data are not available. It differs significantly from the other two not only in spatial distributions, but also in areal averages. The Xie and Arkin dataset provides monthly means for individual years over a period that includes the ACIA baseline.

4.3.2. Specifying the ACIA climatological baseline

A climatological baseline is a period of years representing the current climate in terms of the mean and variability over the period. To satisfy widely adopted IPCC (1994) criteria, a baseline period should:

- be representative of the present-day or recent average climate in the region considered;
- be of sufficient duration to encompass a range of climatic variations;
- cover a period for which data on all major climatological variables are abundant, adequately distributed in space, and readily available;
- include data of sufficiently high quality for use in evaluating impacts; and should
- be consistent or readily comparable with baseline climatologies used in other impact assessments.

Until recently, the most widely used baseline period has been the "classical" 30-year period defined by the World Meteorological Organization (WMO). Usually the period 1961 to 1990 is used (as was the case for the first three IPCC Assessment Reports). In some cases, an earlier period (1951–1980) was used. It is expected that the IPCC Fourth Assessment Report will use the climatological baseline 1971–2000 (IPCC-TGCIA, 1999).

The 20-year period 1981–2000 was selected as the ACIA climatological baseline. While shorter than the 30-year WMO standard, the ACIA baseline is linked to the period of high-quality (satellite) observations of sea-ice extent and concentration (important climatological variables for the Arctic), which have been available only since the late 1970s (IPCC, 2001; see also section 6.3). The precise coincidence of the baseline duration with the ACIA future time slices (also 20 years, see section 1.4.2) is also methodologically consistent. Another technical reason for selecting the 1981–2000 baseline period, rather than 1971–2000, was the availability of the former (but not the latter) in the outputs of all five B2 simulations stored in the ACIA archive.

A serious concern is that the ACIA baseline duration is insufficient to reflect natural climatic variability on a multi-decadal timescale. Indeed, the ACIA climatological baseline includes at least ten of the warmest years globally since the mid-19th century when the instrumental record began (IPCC-TGCIA, 1999). However, considering the large interdecadal variability of arctic climate during the entire period of the instrumental record (e.g., Bengtsson et al., 2003; Polyakov and Johnson, 2000; Polyakov et al., 2002a,b), any particular 30-year (or even longer) period of the 20th century could exhibit a similar limitation.

Table 4.2. Multi-year means of surface air temperature and precipitation derived from the NCEP/NCAR reanalysis and averaged over the WMO standard (1961–1990) and ACIA (1981–2000) climatological baselines (Kattsov et al., 2003).

	Global		60°–90° N	
	Surface air temperature (°C)	Precipitation (mm/d)	Surface air temperature (°C)	Precipitation (mm/d)
Winter (Dec–Feb)				
WMO	12.3	2.69	-22.1	1.10
ACIA	12.4	2.70	-21.5	1.12
Spring (Mar–May)				
WMO	13.7	2.71	-11.6	1.03
ACIA	13.8	2.72	-10.9	1.05
Summer (Jun–Aug)				
WMO	15.3	2.91	5.5	1.65
ACIA	15.5	2.89	5.7	1.67
Autumn (Sep–Nov)				
WMO	13.8	2.69	-9.1	1.31
ACIA	13.9	2.69	-8.6	1.31
Annual				
WMO	13.8	2.75	-9.3	1.28
ACIA	13.9	2.75	-8.8	1.28

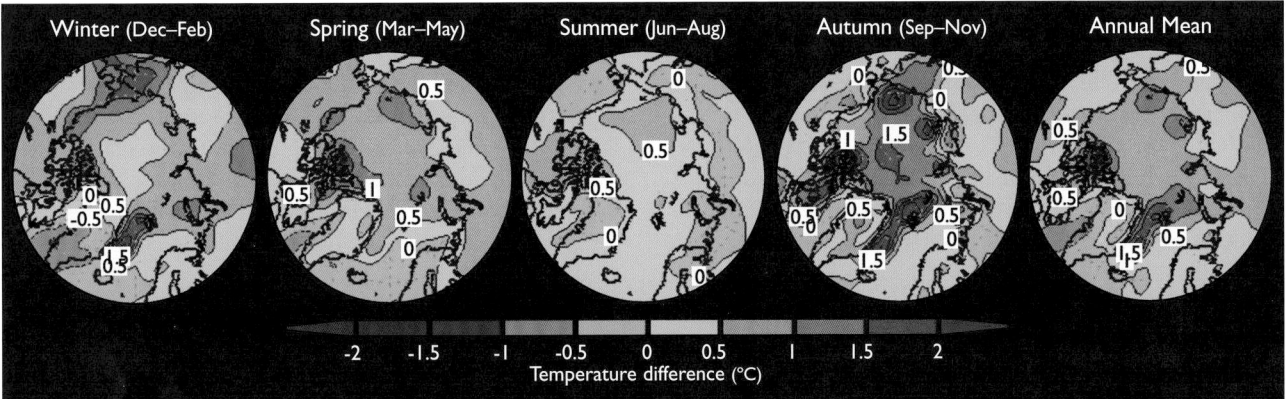

Fig. 4.3. Differences in seasonal and annual multi-year mean surface air temperature in the northern polar region (60°–90° N) between the ACIA (1981–2000) and the standard WMO (1961–1990) climatological baselines, obtained from the NCEP/NCAR reanalysis.

Following the recommendations of the IPCC Task Group on Scenarios for Climate and Impact Assessment, the ACIA climatological baseline (1981–2000) was compared with the standard 1961–1990 baseline, using surface air temperature and precipitation data from the NCEP/NCAR reanalysis (Kistler et al., 2001). Table 4.2 provides seasonal and annual multi-year means of the atmospheric variables for the two baseline periods.

The differences in the global means between the two baseline periods are systematic, but small. Globally, the ACIA baseline period is warmer by 0.1 to 0.2 °C in all seasons. The differences in global precipitation are negligible. For the polar region (60°–90° N), the differences between the two baselines are larger. The difference in the surface air temperature is at a maximum in winter (0.6 °C) and is smallest in summer (0.2 °C). The ACIA baseline annual mean precipitation is the same as the 1961–1990 mean.

Geographically, the differences between the two baseline periods in the NCEP/NCAR reanalysis are more pronounced (Fig. 4.3). The Arctic is generally warmer during the ACIA baseline period, especially in autumn. The strongest warming is evidently associated with the marginal sea-ice zone, particularly along the east coast of Greenland. Mean sea-level pressure (SLP) is generally lower (by up to about 1.5 hPa) over the central Arctic

and northern North Atlantic in the ACIA baseline period compared to the 1961–1990 baseline period.

Because 1961–1990 is expected to be superseded in the near future by 1971–2000 as the new standard 30-year averaging period, it is worth comparing the ACIA climatological baseline against the latter. Figure 4.4 shows the spatial distributions of seasonal and annual differences in surface air temperature between the 1981–2000 and 1971–2000 periods.

In summary, for surface air temperature, precipitation, and atmospheric pressure, the ACIA baseline period has systematic but generally small differences in comparison with the WMO standard baseline (1961–1990). The differences can easily be taken into account when a comparison between climate change scenarios employing the different baselines is required. An advantage of the ACIA climatological baseline period is that it is more "current" than the 1961–1990 period. The duration of the ACIA baseline period is exactly the same as that adopted for the ACIA future time slices. There are only minor geographical differences in seasonal temperature means between the ACIA baseline and the new standard baseline (1971–2000). While the ACIA climatological baseline period (1981–2000) satisfies the IPCC (1994) selection criteria, its relative shortness (compared to the standard 30-year period) may not provide an adequate representation of the probability of extreme events.

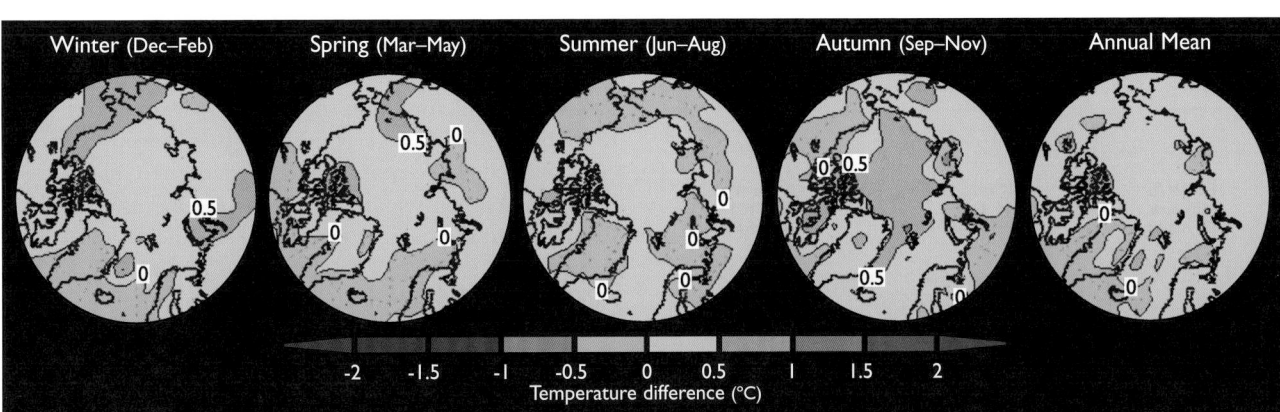

Fig. 4.4. Differences in seasonal and annual multi-year mean surface air temperature in the northern polar region (60°–90° N) between the ACIA (1981–2000) and the new standard (1971–2000) climatological baselines, obtained from the NCEP/NCAR reanalysis.

4.3.3. Surface air temperature

The seasonal cycle of the simulated and analyzed air temperatures at the 2 m height (1.5 m height for HadCM3) for the period 1981 to 2000 is illustrated in Fig. 4.5a, which shows means for the entire area north of 65° N. While there are differences between the five-model mean and the NCEP/NCAR reanalysis, these are relatively small compared with the range of model results. In particular, the area mean temperatures during the greater part of the year differ by about 5 °C between the models simulating the highest and lowest temperatures, and local differences are even larger. In late winter (February–March), all the models simulate slightly lower temperatures than those in the NCEP/NCAR reanalysis. It should be noted that the 1.5 and

2 m temperatures in the models are not prognostic variables – they are derived from the prognostic temperatures at the lowest model level (typically a few tens of meters) using the models' ABL parameterizations; as a result, the biases may be partly due to the diagnostic schemes. In addition, surface elevation differences between the models and the reanalysis, as a result of differences in spatial resolution, could be contributing to the apparent biases.

The large-scale spatial distribution of annual mean temperature in the Arctic is, on average, reasonably well reproduced by the ACIA-designated models. The simulations tend to be slightly colder than the NCEP/NCAR reanalysis in northern Eurasia and somewhat warmer over the western Arctic Ocean and northern North America (Fig. 4.6). A sharp local maximum in the five-model bias (Fig. 4.6c) in southern Greenland probably reflects the relatively smooth model topographies. However, the biases vary substantially between the individual models. Figure 4.6d shows the number of the five models that simulate lower mean annual temperatures than those in the NCEP/NCAR reanalysis. There are a few areas in the western Arctic where all five models simulate higher mean annual temperatures than those in the NCEP reanalysis and a few areas in the eastern Arctic where all five models simulate lower mean annual temperatures. The seasonal distribution of the biases in the five-model mean temperature is shown in Fig. 4.6e-h. The cold bias in northern Eurasia is most pronounced in winter and spring, whereas the warm bias in northern North America persists for most of the year and is largest in autumn. The simulated temperatures in the central Arctic Ocean tend to exceed the NCEP/NCAR reanalysis estimate in spring and especially in autumn. The five-model mean simulated summer temperatures in the central Arctic are lower than in the NCEP/NCAR reanalysis and the model-to-model variation is relatively small.

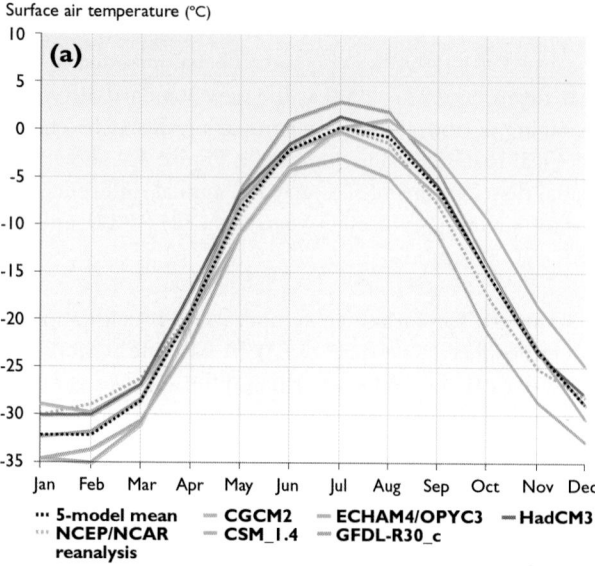

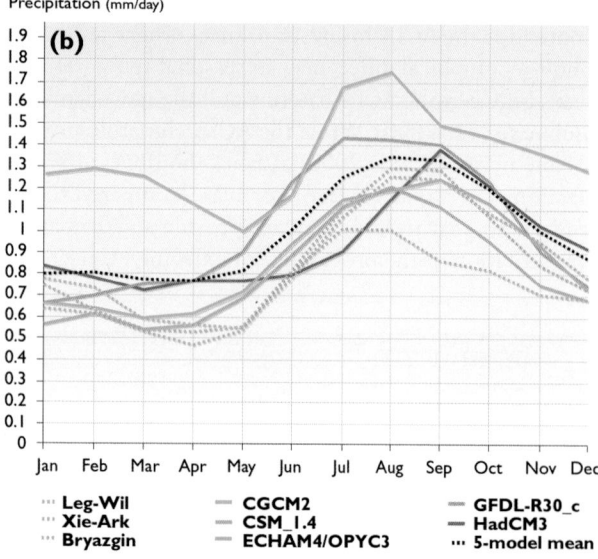

Fig. 4.5. Seasonal cycles of (a) surface air temperature and (b) precipitation simulated by the five ACIA-designated models for the period 1981–2000 and averaged over the area 65°–90° N. For comparison, data for the same area are included in (a) from the NCEP/NCAR reanalysis for the same time period and in (b) from three climatologies: Bryazgin (1936–1990), Legates-Willmott (1920–1980), and Xie-Arkin (1981–2000) (Khrol, 1996; Legates and Willmott, 1990; Xie and Arkin, 1998).

The ability of AOGCMs to provide credible projections of future climates is strongly supported by their ability to simulate the evolution of the climate during past centuries. One of the five-member ensemble 20th century simulations with the GFDL-R30_c model driven by historical changes in GHG and sulfate aerosol concentrations demonstrated an impressive resemblance to the observed warming that occurred in two distinct periods in the first and second halves of the 20th century with a pronounced maximum in the Arctic (Delworth and Knutson, 2000). The early 20th-century warming was not obtained in the other simulations of the GFDL-R30_c ensemble, which highlights the role of internal variability in the climate evolution, and therefore proves the necessity of ensemble simulations, rather than single runs, in order to better delineate the associated uncertainties. In the HadCM3 model, a good fit to 20th century observations was only obtained when natural (varying) forcing from solar and volcanic activity was also included (Stott et al., 2000; Stott, 2003; Tett et al., 2000; see also IPCC, 2001a).

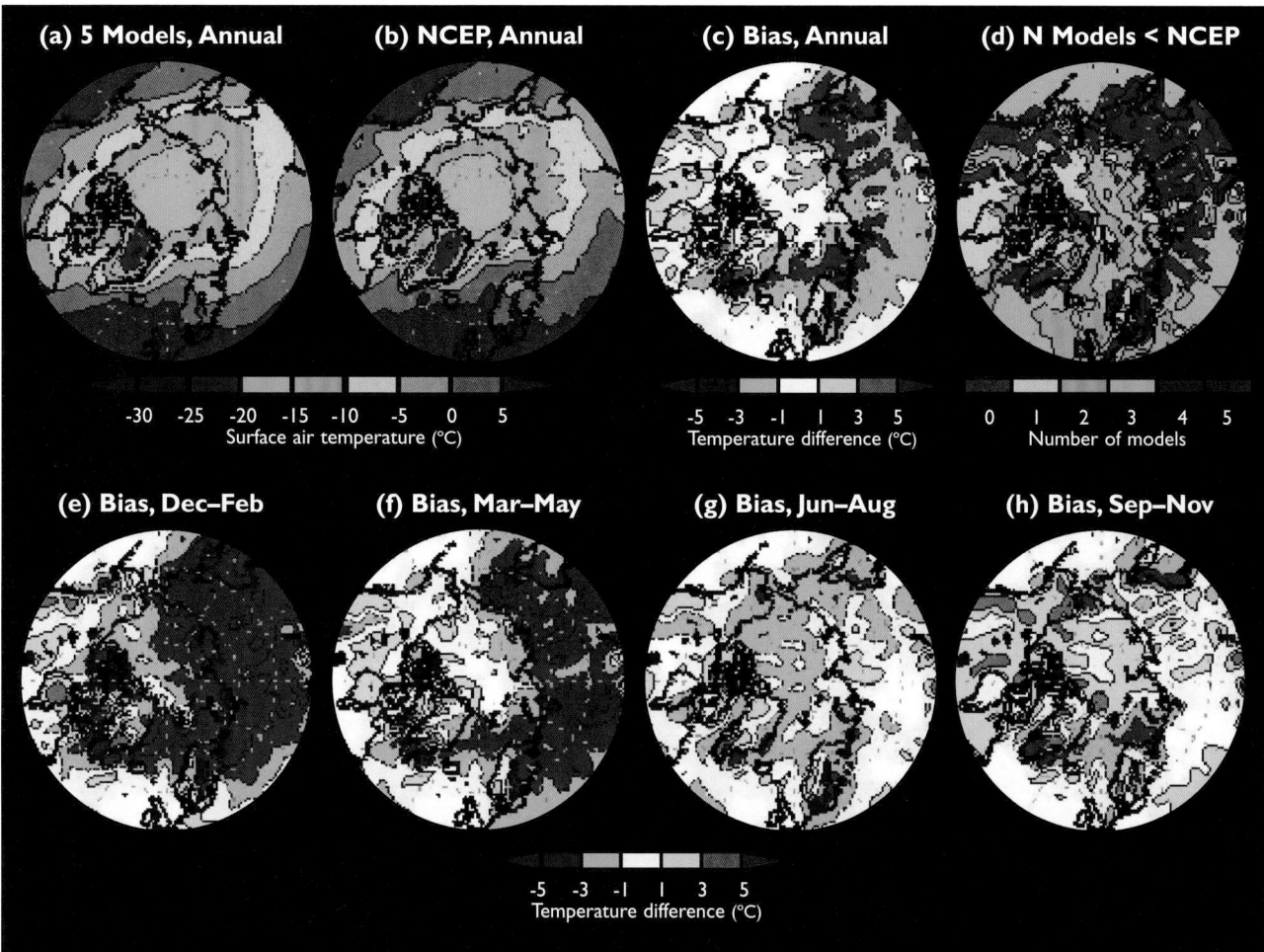

Fig. 4.6. Comparison of surface air temperature simulated by the ACIA-designated models for 1981–2000 and the NCEP/NCAR reanalysis temperature for the same period. The top row shows (a) the simulated annual mean temperature averaged over the five ACIA-designated models; (b) the NCEP/NCAR reanalysis for 1981–2000; (c) the difference between (a) and (b); and (d) the number of models (out of five) in which the simulated annual mean temperature is lower than that of the NCEP/NCAR reanalysis. The bottom row shows seasonal differences between the five-model mean and the NCEP/NCAR reanalysis for (e) winter; (f) spring; (g) summer; and (h) autumn.

4.3.4. Precipitation

More so than with temperature, there are major systematic differences between precipitation in the ACIA 1981–2000 simulations and in observational datasets. The five-model mean seasonal cycle of precipitation in the area 65° to 90° N is in qualitative agreement with the climatologies (Fig. 4.5b). While the range between the individual simulations is substantial throughout the year, it is noteworthy that the observational climatologies demonstrate a comparable scatter, at least in summer and autumn.

As shown in Fig. 4.7a-d, the average simulated annual precipitation generally exceeds the Bryazgin estimate (Bryazgin, 1976; Khrol, 1996) and other observational estimates in most of the Arctic; in some areas by a factor of two. The reverse is true, however, in the northeastern North Atlantic and parts of northwestern Eurasia, probably because simulated cyclone activity in this area tends to be too weak (see section 4.3.5). The same geographical pattern of biases persists throughout the year, although the magnitude of these biases varies with season. The positive biases relative to the Bryazgin climatology are generally greatest in spring and smallest in summer (Fig. 4.7e-h).

The differences between the simulated precipitation and the observational estimates may be partly due to measurement errors that lead to underestimation of the actual precipitation, particularly when it falls in solid form. However, this clearly cannot explain all the differences in the spatial and seasonal distributions of precipitation. For example, the difference between the five-model area mean and the observational estimates is substantially larger in spring than in winter, in contrast to what might be expected from measurement errors alone.

The ability of AOGCMs (including three of the ACIA-designated models: HadCM3, ECHAM4/OPYC3, and CGCM2) to reproduce the 20th century increase in arctic precipitation has been demonstrated (e.g., Kattsov and Walsh, 2002). In all of the ACIA-designated model simulations, positive linear trends in 20th century arctic precipitation agree with available observational estimates (see section 2.6.2.2).

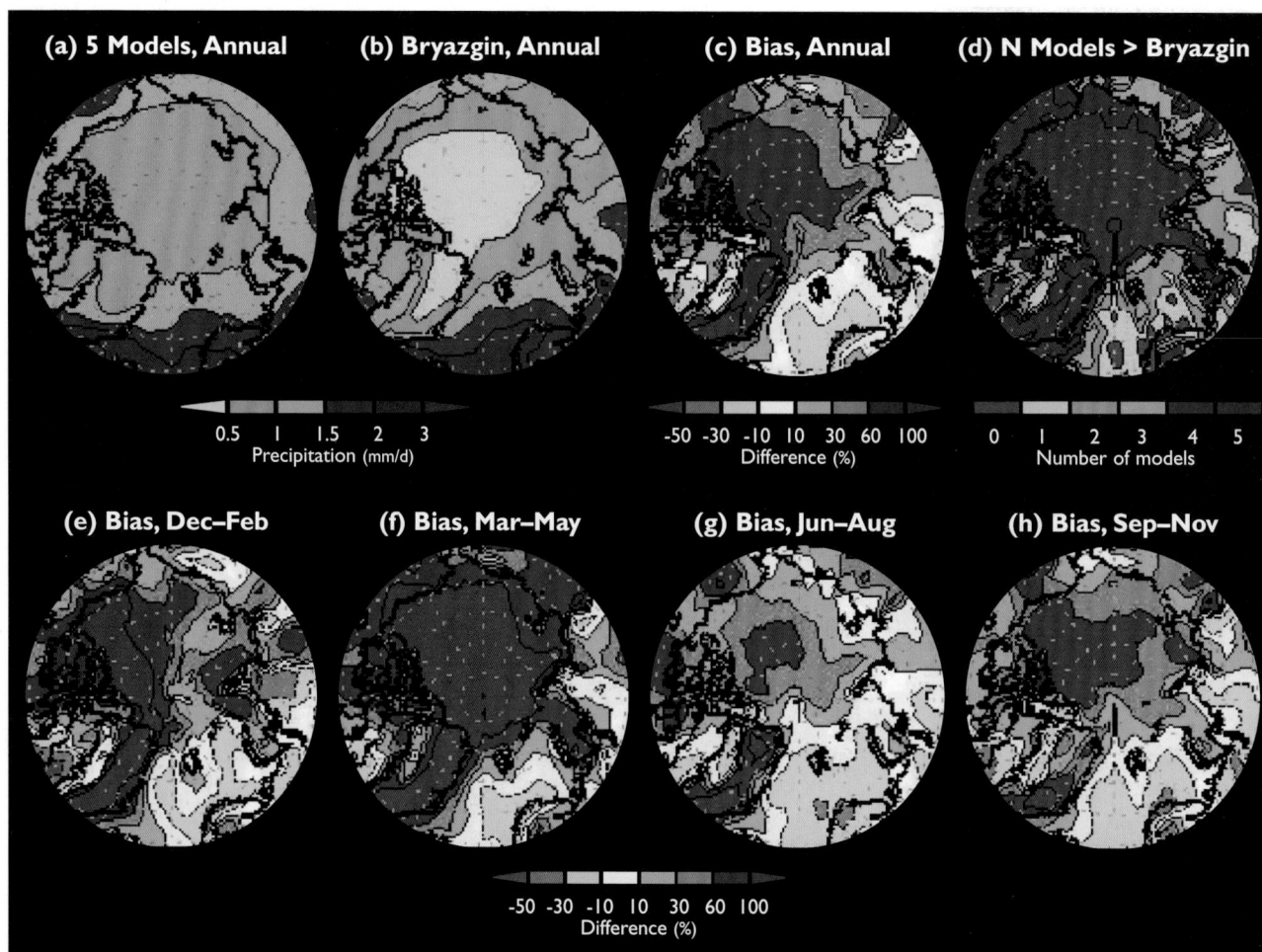

Fig. 4.7. Comparison of precipitation simulated by the ACIA-designated models for 1981–2000 and precipitation from the Bryazgin climatology (Khrol, 1996) for 1936–1990. The top row shows (a) the simulated annual mean precipitation averaged over the five ACIA-designated models; (b) the Bryazgin climatology for 1936–1990; (c) the percentage difference between (a) and (b); and (d) the number of models (out of five) in which the simulated annual mean precipitation exceeds that of the Bryazgin climatology. The bottom row shows seasonal percentage differences between the five-model mean and the Bryazgin climatology for (e) winter; (f) spring; (g) summer; and (h) autumn.

Further details on the evaluation of the five ACIA-designated models with respect to simulation of precipitation and other components of arctic hydrology and climatology can be found in Chapter 6.

4.3.5. Other climatic variables

The distribution of the 1981–2000 mean SLP is qualitatively similar between the mean of four of the ACIA-designated models (for GFDL-R30_c, only surface pressure fields were available) and the NCEP/NCAR reanalysis, but there are important differences in details (Fig. 4.8). Surface pressure is the prognostic variable in GCMs, while SLP is diagnosed using different reduction schemes. Some of the variations in SLP between the different models, and between the models and the reanalysis, may be due to the use of different SLP reduction schemes.

The models simulate the main lobes of the annual mean Icelandic and Aleutian lows quite well, but the simulated extension of the Icelandic low towards the Barents Sea is too weak. The positive pressure bias over the eastern Arctic Ocean and the negative bias in western

Eurasia (Fig. 4.8c) suggest that the path of cyclone activity is too far south in the simulations. There also tends to be a slight negative pressure bias in the western Canadian Arctic, which suggests that the simulated cyclone activity is too strong in that region. Winter pressure biases make the greatest contribution to the four-model annual mean pressure biases. However, the positive bias over the eastern Arctic Ocean persists throughout the year. As with temperature and precipitation, the pressure biases also vary between the individual models. The shift of the arctic air mass relative to observations is a well-known feature of both AOGCMs and stand-alone AGCMs (e.g., AMIP; see Box 4.2). The pressure biases contribute to significant differences in wind forcing of sea ice between the AOGCM simulations and the real world, and lead to distortions in simulated spatial distributions of sea ice (see Bitz et al., 2002; Walsh, in press; Walsh et al., 2002).

Cloudiness and the radiative properties of clouds, particularly in the Arctic, remain a major challenge to simulate. Figure 4.9 shows the dramatic scatter between the total cloud amounts simulated by four of the ACIA-designated models (the results from CGCM2 were not

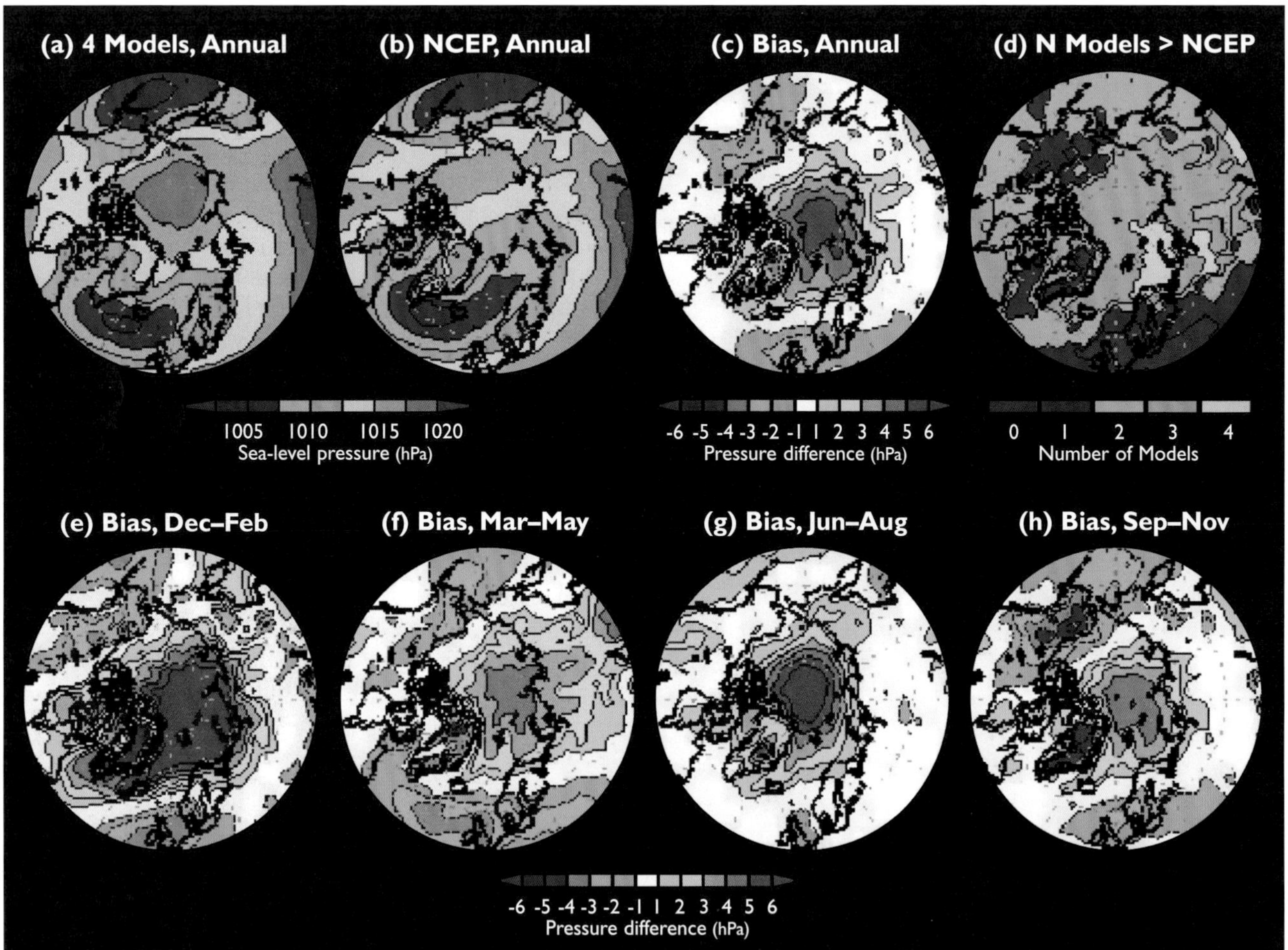

Fig. 4.8. Comparison of mean sea-level pressure (SLP) simulated by four of the ACIA-designated models for 1981–2000 and the NCEP/NCAR reanalysis SLP for the same period. The top row shows (a) the simulated annual mean SLP averaged over the four ACIA-designated models; (b) the NCEP reanalysis for 1981–2000; (c) the difference between (a) and (b); and (d) the number of models (out of four) in which the simulated annual mean SLP exceeds the NCEP/NCAR reanalysis. The bottom row shows seasonal differences between the four-model mean and the NCEP/NCAR reanalysis for (e) winter; (f) spring; (g) summer; and (h) autumn.

available) over the Arctic Ocean between 70° and 90° N. Two observational estimates, one based primarily on Television and Infrared Observation Satellite Operational Vertical Sounder (TOVS) data (Schweiger et al., 1999) and the other obtained primarily from surface-based observations (Hahn et al., 1995), diverge substantially in late summer and early autumn, but give an idea of the seasonality of arctic cloud cover. While the inter-model scatter is quite large (e.g., the difference between the highest (CSM_1.4) and lowest (GFDL-R30_c) simulations approaches 60% in winter), the four-model mean underestimates cloudiness in the warm season and overestimates it in winter. Of the four

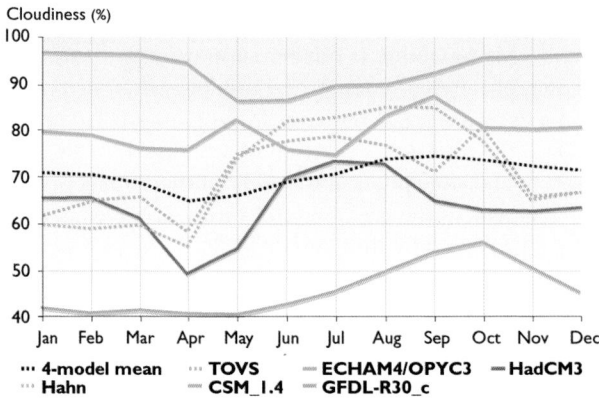

Fig. 4.9. Annual cycle of monthly mean cloudiness over the Arctic Ocean (70°–90° N) for 1981–2000 simulated by four of the ACIA-designated models, and observational estimates from TOVS satellite data and surface observations (Hahn et al., 1995).

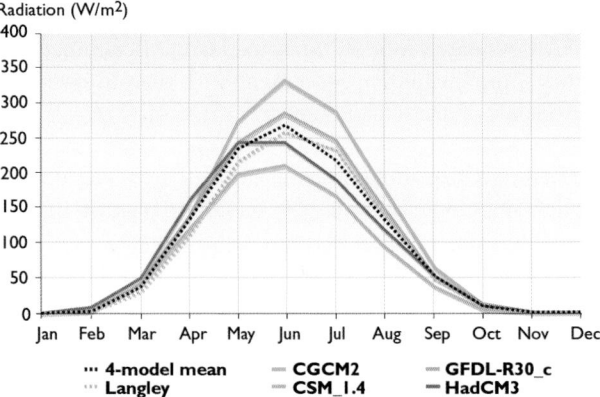

Fig. 4.10. Annual cycle of incident solar radiation at the surface of the Arctic Ocean (70°–90° N) for 1981–2000 simulated by four of the ACIA-designated models, and the observationally based estimate obtained using data from the Langley Atmospheric Sciences Data Center (1983–1991).

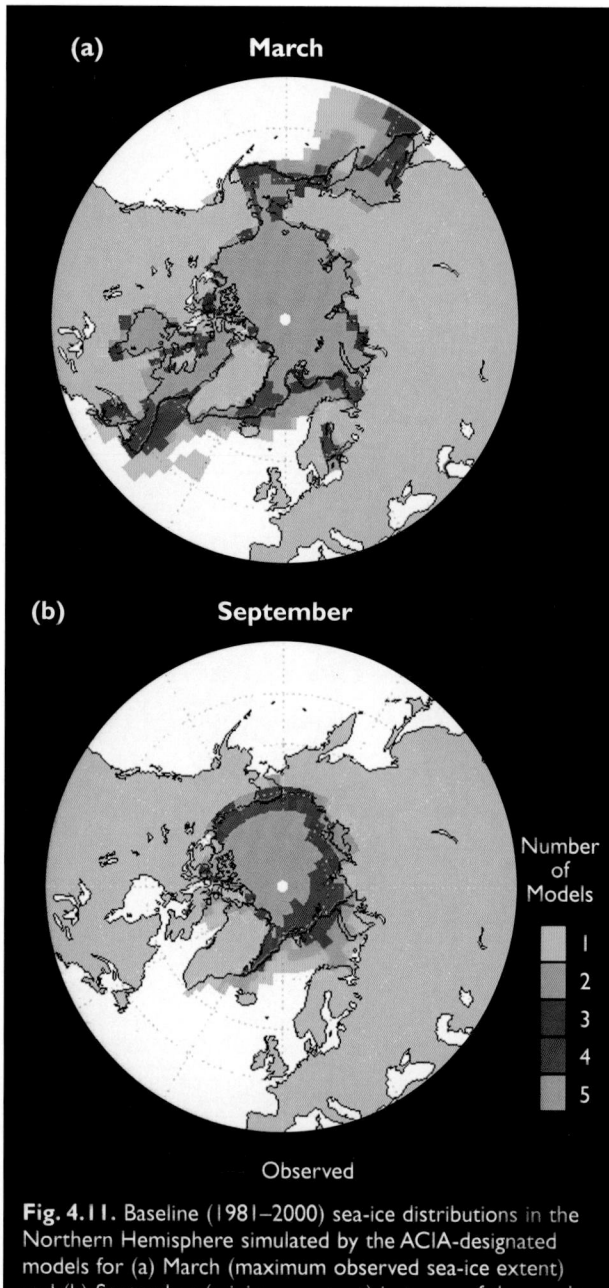

Fig. 4.11. Baseline (1981–2000) sea-ice distributions in the Northern Hemisphere simulated by the ACIA-designated models for (a) March (maximum observed sea-ice extent) and (b) September (minimum extent) in terms of the number of models simulating sea ice in each grid cell (based on Meleshko et al., 2004).

ACIA-designated models, only the HadCM3 simulation shows some qualitative agreement with the observed seasonality of arctic cloud cover.

Surface radiative fluxes also vary widely between the models. Figure 4.10 shows the seasonal variation in incident solar radiation as simulated by four of the ACIA-designated models (ECHAM4/OPYC3 data were not available) for the Arctic Ocean between 70° and 90° N and the lone observational estimate obtained from the Langley Atmospheric Sciences Data Center. The four-model mean seasonal cycle is close to the observed one. In summer, however, the difference between the highest and lowest simulated values (CGCM2 and CSM_1.4, respectively) reaches a maximum of up to 125 W/m².

The large inter-model scatter in the ACIA-designated model simulations of baseline (1981–2000) sea-ice and terrestrial snow-cover distributions reflects problems in modeling the cryosphere as discussed in sections 4.2.5 and 4.2.6. Figure 4.11 presents an integrated picture of sea-ice distributions in the Northern Hemisphere as simulated by the five ACIA-designated models for March (maximum observed sea-ice extent) and September (minimum extent). The distribution is represented by the number of models simulating sea ice in each of the 2.5° x 2.5° grid cells. For each model, sea ice is defined as present in a grid cell if its quantity exceeds one of the *ad hoc* critical values (depending on what variable is available from the model output): 5 cm thickness, 45 kg/m² mass, or 10% areal coverage. For the greater part of the Arctic Ocean, all five models simulate sea ice in both March and September; however, major differences between the models occur along the margins of the ice cover.

A detailed evaluation of the five ACIA-designated models with respect to simulation of the baseline sea-ice and terrestrial snow-cover distributions in the Northern Hemisphere is provided in sections 6.3.3 and 6.4.3.

4.3.6. Summary

A key characteristic of the simulations of the arctic climate from the ACIA-designated models is their large inter-model scatter. Biases in surface air temperature and SLP spatial distributions and simulation of excessive arctic precipitation are among the most important systematic errors. Significant uncertainty is introduced by the simulated cloudiness – one of the key variables in climate system feedbacks (this is also a problem in the current generation of AOGCMs outside of the Arctic). The large inter-model scatter in reproducing sea-ice and terrestrial snow-cover extent limits the credibility of future climate projections obtained with the models. Conversely, compared to the five individual simulations, the five-model ensemble means show reasonable agreement with available observations, at least for the area averages. The evaluation of the ability of the ACIA-designated AOGCM ensemble to simulate observed climate conditions supports use of the ensemble for constructing 21st-century climate change scenarios for the Arctic. This suitability is further supported by the ability of some of the ACIA-designated models, when driven by estimates of historical radiative forcing, to satisfactorily emulate the observed evolution of arctic surface air temperature and precipitation throughout the 20th century, which enhances the credibility of future arctic climate change projected by these models.

As a consequence of the biases and inter-model scatter in AOGCM simulations of the present-day arctic climate, the ACIA has chosen to append (add or subtract) the simulated changes to observed baseline climates as was done, for example, by the National Assessment Synthesis Team (NAST, 2001) rather than to use the model-simulated climates directly in impact studies. The climate changes should be expressed either as absolute differences (e.g., temperature), or as ratios (e.g., precipitation).

4.4. Arctic climate change scenarios for the 21st century projected by the ACIA-designated models

The IPCC Third Assessment Report (IPCC, 2001), based on a set of AOGCM projections, has provided the following global context for the ACIA. For the last three decades of the 21st century (2071–2100), the IPCC (2001) projects a mean increase in globally averaged surface air temperature, relative to the period 1961–1990, of 3.0 °C (with a range of 1.3–4.5 °C derived from the nine models used by the IPCC) for the A2 emissions scenario and 2.2 °C (with a range of 0.9–3.4 °C) for the B2 emissions scenario. Most of the spatial patterns of the AOGCM-projected responses to the SRES emissions scenarios are similar to other emissions scenarios, including the idealized 1% per year CO_2 increase. The following list summarizes the key projections of climate change over the 21st century by the AOGCMs used in the IPCC (2001) assessment.

- It is very likely that nearly all land areas will warm more rapidly than the global average, particularly during the cold season in northern high latitudes.
- Models project a decrease in the diurnal temperature range in many areas, with nighttime lows increasing more than daytime highs.
- Models project a decrease in Northern Hemisphere snow cover and sea-ice extent, and continued retreat of glaciers and ice caps.
- Projected increases in mean precipitation are likely to lead to increases in interannual precipitation variability.
- Increases in the lowest daily minimum temperatures are projected to occur over nearly all land areas and are generally greatest in areas where snow and ice retreat.
- Frost days and cold waves are very likely to become fewer.
- High extremes of precipitation are projected to increase more than the mean, and the intensity of precipitation events is projected to increase.
- The frequency of extreme precipitation events is projected to increase almost everywhere.
- For some other extreme phenomena, many of which may have important impacts on the environment and society, the confidence in model projections and understanding is currently inadequate to make firm projections.
- No clear consensus has been reached about how extratropical cyclones are likely to change, as the results differ between the relatively few studies that have been conducted.
- Most AOGCMs project a weakening of the Northern Hemisphere thermohaline circulation, which would contribute to a reduction of surface warming in the subarctic North Atlantic.

- There is no clear agreement on likely changes in the probability distribution or structure of natural modes of variability, like the North Atlantic Oscillation (NAO) or the Arctic Oscillation (AO), whose magnitude and character changes vary across the models.

Box 4.3 reviews some of the major uncertainties associated with AOGCM projections.

In this section, the Arctic is considered in the context of the 21st-century global climate change projections listed previously, focusing on surface air temperature and precipitation in the Arctic and the inter-model differences for the A2 and B2 emissions scenarios. All changes are compared to the ACIA climatological baseline (1981–2000). A wider context of climate change simulations is provided by comparing the behavior of the five ACIA-designated AOGCMs in Phase 2 of the Coupled Model Intercomparison Project (CMIP2: Meehl et al., 2000) with the other 14 models included in that project.

4.4.1. Emissions scenarios

Emissions scenarios are plausible representations of the future development of emissions of radiatively active substances (GHGs, aerosols), based on a coherent and internally consistent set of assumptions about demographic, socioeconomic, and technological changes and their key relationships in the future. Emissions scenarios are converted into concentration scenarios that are used as input for climate model projections.

In idealized transient experiments (e.g., CMIP2, see Box 4.2), the atmospheric CO_2 concentration increases gradually, usually at a rate of 1% (compound) per year, which results in a doubling of its concentration in 70 years. The 1% idealized scenario lies at the high end of the SRES scenarios.

In transient experiments with a detailed forcing scenario, the concentrations of CO_2 and other GHGs such as methane and nitrous oxide are prescribed as a function of time, based on an emissions scenario for these gases. Frequently, sulfate aerosols are also included. Examples of the scenarios used in model simulations include the IPCC IS92 scenarios (Leggett et al., 1992) and the more recent SRES scenarios (Nakićenović and Swart, 2000).

The SRES emissions scenarios were built around four narrative storylines that describe the evolution of the world in the 21st century. Altogether, 40 different emissions scenarios were constructed. Six of these (A1B, A1T, A1FI, A2, B1, and B2) were chosen by the IPCC as illustrative "marker" scenarios. The SRES scenarios include no additional mitigation initiatives, which means that no scenarios are included that explicitly assume the implementation of the United Nations Framework Convention on Climate Change or the emission targets of the Kyoto Protocol.

Box 4.3. Uncertainties in climate change scenarios based on AOGCM simulations

Uncertainties in future GHG and aerosol emissions, their conversion to atmospheric concentrations, and their contribution to radiative forcing of the climate. Different assumptions about future social and economic development, and hence future GHG and aerosol emissions, comprise one of the major uncertainties in the climate change scenarios. For example, the IPCC Special Report on Emissions Scenarios (Nakićenović and Swart, 2000; see also section 4.4.1) presents 40 different emissions scenarios. Uncertainty is also associated with the conversion of emissions into atmospheric GHG and aerosol concentrations. Additional uncertainty arises from the calculation of radiative forcing associated with given concentrations, which occurs implicitly within AOGCMs, but is problematic in particular for aerosols.

Uncertainties in the global and regional climate responses to a radiative forcing from different AOGCM simulations. Due to different representations of processes and feedbacks in the climate system (e.g., Stocker et al., 2001), or by excluding some of them, AOGCMs differ in their sensitivity to the same radiative forcing. Sometimes, the difference in AOGCM sensitivity manifests itself only in projected regional climate change patterns, while the magnitudes of projected global mean changes remain similar between models. At long timescales, the forcing uncertainty (differences between emissions scenarios) and model uncertainty are of approximately equal importance.

Uncertainties due to insufficient AOGCM resolution and different methods of regionalizing (downscaling) AOGCM results. Insufficient AOGCM resolution limits direct use of their outputs in impact assessments. In most cases, a climate scenario for a certain region requires a combination of the simulated variables and observed data, which may be accomplished with different methods. In addition, observational data often fail to capture the full range of decadal-scale natural variability. Finally, gridding the observational data in order to create baseline climatologies can introduce errors. Employing regional climate models to enhance the spatial and temporal resolution of AOGCM outputs introduces further uncertainties arising from individual features of the RCMs. In principle, by employing a range of downscaling methods, quantification of this class of uncertainties is possible, but this is seldom done (Mearns et al., 2001).

Uncertainties due to forced and unforced natural variability. In addition to anthropogenic forcing, climate change in the real world is affected by largely unpredictable natural variability. Part of the natural variability is thought to be due to variations in solar and volcanic activity, but a substantial part is unforced, resulting from the internal dynamics of the climate system. Climate models also simulate unforced natural variability, such that, when the same model is run with the same forcing scenario but different initial conditions, there are non-negligible differences in the results, particularly in regional details that are affected by internal variability much more than global means. When different models are run using the same forcing scenario, some of the differences in their results arise from different realizations of natural variability.

The greatest difference between the SRES scenarios and the earlier IS92 scenarios relates to sulfur emissions and hence sulfate aerosol concentrations. The commonly adopted intermediate IS92 scenario (IS92a) assumed a doubling of anthropogenic sulfur emissions between 1990 and 2050, and little change in emissions thereafter. In contrast, all six illustrative SRES scenarios project lower sulfur emissions in 2100 than at present, although some include an increase over the next few decades. The lower sulfur emissions together with higher GHG emissions in some of the SRES scenarios are the main reason for the upward shift in the IPCC projections of the increase in global mean temperature between 1990 and 2100 from 1.0 to 3.5 °C in the Second Assessment Report (IPCC, 1996) to 1.4 to 5.8 °C in the Third Assessment Report (IPCC, 2001). The CO_2 emissions and the derived atmospheric CO_2 concentrations, the sulfur dioxide (SO_2) emissions, and projections of global mean temperature increases for the SRES marker scenarios and for the IS92a scenario are shown in Fig. 4.12.

No probabilities are assigned to the various SRES scenarios. During the initial stage of the ACIA process, to stay coordinated with current IPCC efforts, it was agreed that the ACIA projections would be based on the IPCC SRES scenarios (Källén et al., 2001). By that time, most of the available (and expected to be shortly available) AOGCM simulations to be relied upon had been forced by two emissions scenarios: A2 and B2 (Cubasch et al., 2001). Globally, the model mean transient climate responses to the A2 and B2 emissions scenarios are close to each other for each of the different models through the first half of the 21st century and only diverge significantly after that. Given the schedule for producing this assessment and the limits of resources for data storage, the B2 emissions scenario was chosen as the primary scenario for use in ACIA impact analyses.

In a number of studies (e.g., Carter et al., 2000; Ruosteenoja et al., 2003), a "pattern-scaling" technique is applied to represent a wider range of possible future

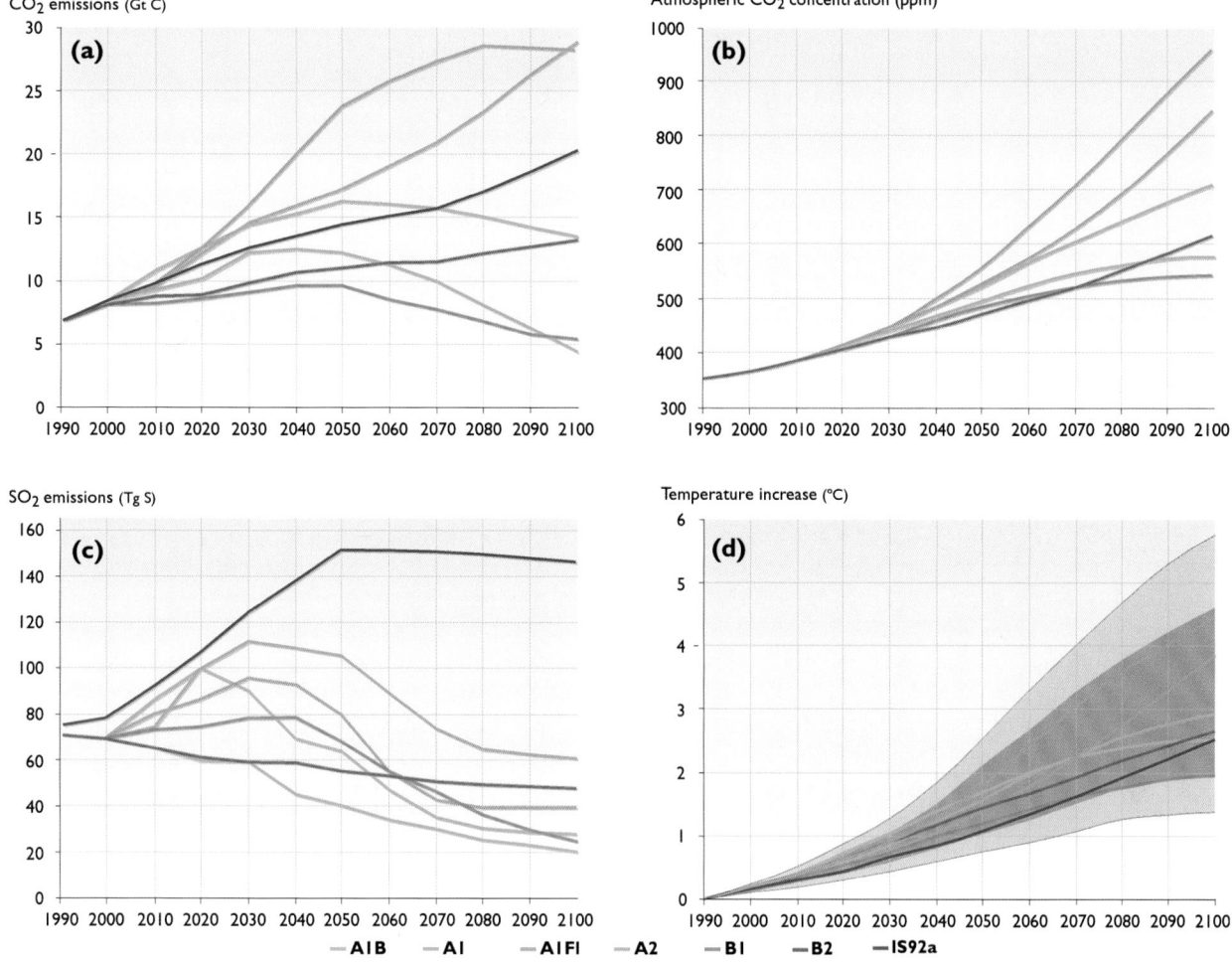

Fig. 4.12. Comparison of the six SRES marker scenarios and the IS92a scenario showing (a) CO_2 emissions; (b) the resulting atmospheric CO_2 concentrations; (c) sulfur dioxide (SO_2) emissions; and (d) projections of global mean temperature increases relative to 1990. In (d), the projected temperature increase for each emissions scenario is an average of the results of the seven climate models used by IPCC (2001). The dark shading represents the range of average warming across all 40 SRES scenarios, and the light shading the range across all scenarios and models (based on IPCC, 2001).

forcings than are available from AOGCM simulations alone. In particular, Ruosteenoja et al. (2003) extrapolated available A2 and B2 AOGCM projections of temperature and precipitation for the SRES B1 and A1FI scenarios over 32 world regions, including the Arctic. The application of this technique is based on the assumption that the spatial pattern of the response is independent of the forcing, while the amplitude of the response at each location is linearly proportional to the global mean change in surface air temperature. The global mean temperature changes for the entire range of SRES scenarios (as shown by shading in Fig. 4.12d) were calculated using a simple climate model system (Box 4.1) calibrated to be consistent with each AOGCM being emulated. Additional assumptions of this approach, first suggested by Santer et al. (1990), are that the patterns of the climate response to anthropogenic forcing can be adequately defined from AOGCM simulations and that they are stable through time and across a representative range of possible anthropogenic forcings. Uncertainties due to scaling climate response patterns increase for scenarios that include substantial, regionally differentiated aerosol forcings and in regions where there is an enhanced

response, for example, near sea-ice and snow margins (Carter et al., 2000).

Box 4.4 discusses specific issues related to the use of the ACIA-designated model projections as a basis for local scenarios.

4.4.2. Changes in surface air temperature

Figure 4.13 displays the evolution of arctic annual mean temperature during the 21st century projected by the five ACIA-designated model simulations using the B2 emissions scenario. The projections for the 60° to 90° N polar region are shown along with the range of global mean temperature changes projected by the same simulations. In the five ACIA-designated model projections, by the late 21st century (2071–2090), the global mean temperature increase (from the 1981–2000 baseline) varies from 1.4 °C (CSM_1.4) to 2.1 °C (ECHAM4/ OPYC3 and CGCM2), with a five-model average of 1.9 °C. In the Arctic, the increase in mean annual temperature projected by the five models is significantly larger, reaching 3.7 °C (twice the increase in the global mean) for the area north of 60° N. The projected

Table 4.3. Increases in mean annual surface air temperature in the Arctic (60°–90° N) compared to the 1981–2000 baseline, as projected by the five ACIA-designated models forced with the B2 emissions scenario (Kattsov et al., 2003).

	Temperature change (°C)					
	CGCM2	CSM_1.4	ECHAM4/OPYC3	GFDL-R30_c	HadCM3	Five-model mean
2011–2030	1.2	1.5	1.3	1.0	1.1	1.2
2041–2060	2.5	2.2	3.2	2.5	2.2	2.5
2071–2090	3.7	2.8	4.6	3.8	4.0	3.7

Box 4.4. Specific issues related to the use of ACIA-designated AOGCM projections as a basis for local scenarios

There are two basic approaches for applying climate change projections in impact studies: deterministic and probabilistic. In the deterministic approach, a single best-guess projection of climate change is used for impact modeling. The simplest way of making a best-guess projection, which is based on the assumption that all models give equally likely results, is to use the arithmetic mean of all available model results (such as the mean of the five ACIA simulations). In the probabilistic approach, in contrast, the projections from different climate models are treated as giving a probability distribution of future climate changes (Giorgi and Mearns, 2003; Palmer and Räisänen, 2002; Räisänen and Palmer, 2001). With this approach, each climate model projection is used separately in an impact analysis. If averaging of different scenarios is performed, it is done in the last phase: the calculated impacts, rather than the climate change projections, are averaged. Because the impacts may depend nonlinearly on changes in climate, the average impact scenario derived by the probabilistic method may differ from the impact scenario obtained by using the average climate change projection. For this reason and because information on uncertainty of the impacts is also important, the probabilistic approach is, in principle, preferable to the deterministic approach. However, it is also more demanding in terms of the computations required.

The simplest variants of both deterministic and probabilistic methods assume that all available climate model results are equally likely. In some situations, this simple assumption may be questionable. For example, when models have serious problems in their control climates, it may be best to exclude them from the calculations. Furthermore, for some situations different models may give such widely divergent results that a deterministic averaging may be misleading. Examples of situations in which local scenario construction requires special care include the following.

- Climate changes in the North Atlantic depend to a high degree on the state of the ocean circulation. Some models project a significantly reduced thermohaline circulation as a consequence of climate change, while others do not. It is currently not possible to determine which scenario is most likely. In this case, arithmetic averaging of the ACIA-designated model simulations may not be meaningful. It makes more sense to consider all or at least two scenarios separately: one with and one without a significantly reduced thermohaline circulation (see also Box 4.5).

- The Barents Sea is currently ice-free throughout the year, mainly as a result of the northward flow of warm Atlantic water into the region. Models that simulate an ice-covered Barents Sea for the present-day climate and near ice-free conditions in a climate change scenario will therefore highly overestimate the amplitude of local warming in the region. In this situation, only model realizations with a realistic sea-ice distribution in the baseline simulation should be used for a scenario of future climate change.

- Sea ice is generally not well handled by AOGCMs, and this is also the case for the ACIA-designated models. Specifically, problems relate to their simulation of major polynyas and differences in larger-scale ice cover. For these areas, future climate change cannot be projected using any specific model or a combination of models. Given the present state of modeling, expert judgment has to be used in combination with the model projections.

- Snow cover during winter is reasonably well represented in the Arctic in all of the ACIA-designated models. However, in both of the transition seasons, as well as during summer, some models exhibit an unrealistically extensive snow pack for the present-day climate, and then complete absence of snow for some regions in a climate change scenario. As a consequence, these simulations project temperature increases that are too large in these regions. In this situation, only model realizations with a realistic seasonal distribution of snow cover in the baseline simulation should be used for scenarios of future climate change.

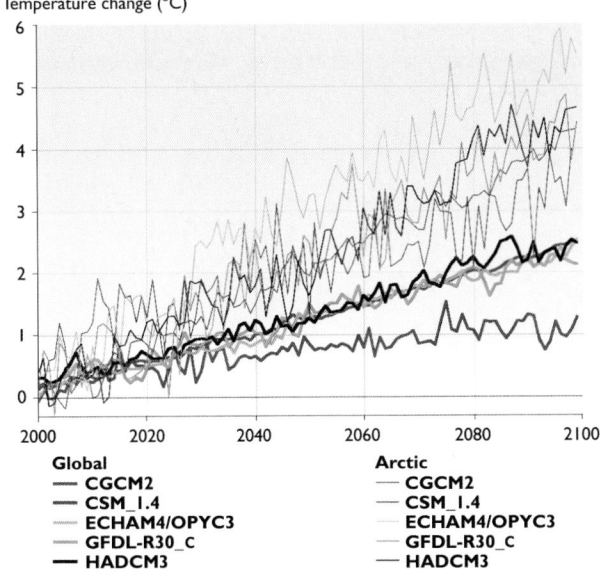

Temperature change (°C)

Global
— CGCM2
— CSM_1.4
— ECHAM4/OPYC3
— GFDL-R30_C
— HADCM3

Arctic
— CGCM2
— CSM_1.4
— ECHAM4/OPYC3
— GFDL-R30_C
— HADCM3

Fig. 4.13. Global and arctic (60°–90° N) changes in annual mean surface air temperature relative to the baseline period 1981–2000 as projected by the five ACIA-designated models forced with the B2 emissions scenario.

temperature change varies by about ±25% about the mean of the five models. For the Arctic north of 60° N, the projected area mean temperature increase by 2071–2090 ranges from 2.8 °C (CSM_1.4) to 4.6 °C (ECHAM4/OPYC3), with the other models within the 3.7 to 4.0 °C range (Table 4.3). The projected arctic mean temperature increase exceeds the projected global mean temperature increase in all of the models.

In Fig. 4.14, spatial patterns of projected increases in the annual mean temperature in the Arctic are put in a global perspective for the three 21st-century time slices. On average, the models project a greater temperature increase at high northern latitudes than anywhere else in the world (Fig. 4.14, left column). By 2071– 2090, the five-model average projected increase in the mean annual temperature in the central Arctic is more than 5 °C (about three times the global mean). By that time, the mean annual temperature is projected to increase by around 3 °C in Scandinavia and East Greenland, about 2 °C in Iceland, and up to 5 °C in the Canadian Archipelago and Russian Arctic. However, the variation in projected temperature change between the individual models is also generally much larger over the Arctic than for most other regions of the globe (Fig. 4.14, middle column). The standard deviation of the mean temperature change averaged over the five models also varies substantially across the Arctic, but these variations are difficult to interpret because there are only five models in the sample. The relative agreement between the different projections is measured by the ratio between the five-model mean change and the inter-projection standard deviation (Fig. 4.14, right column). Because the large standard deviations compensate for the large average warming, this signal-to-noise ratio in the Arctic is not exceptional. Very high relative agreement (mean exceeding the standard deviation by a factor of six) occurs by 2071–2090 in some (but not all) arctic regions, but this is also the case in lower latitudes where both the mean and the standard deviation are smaller.

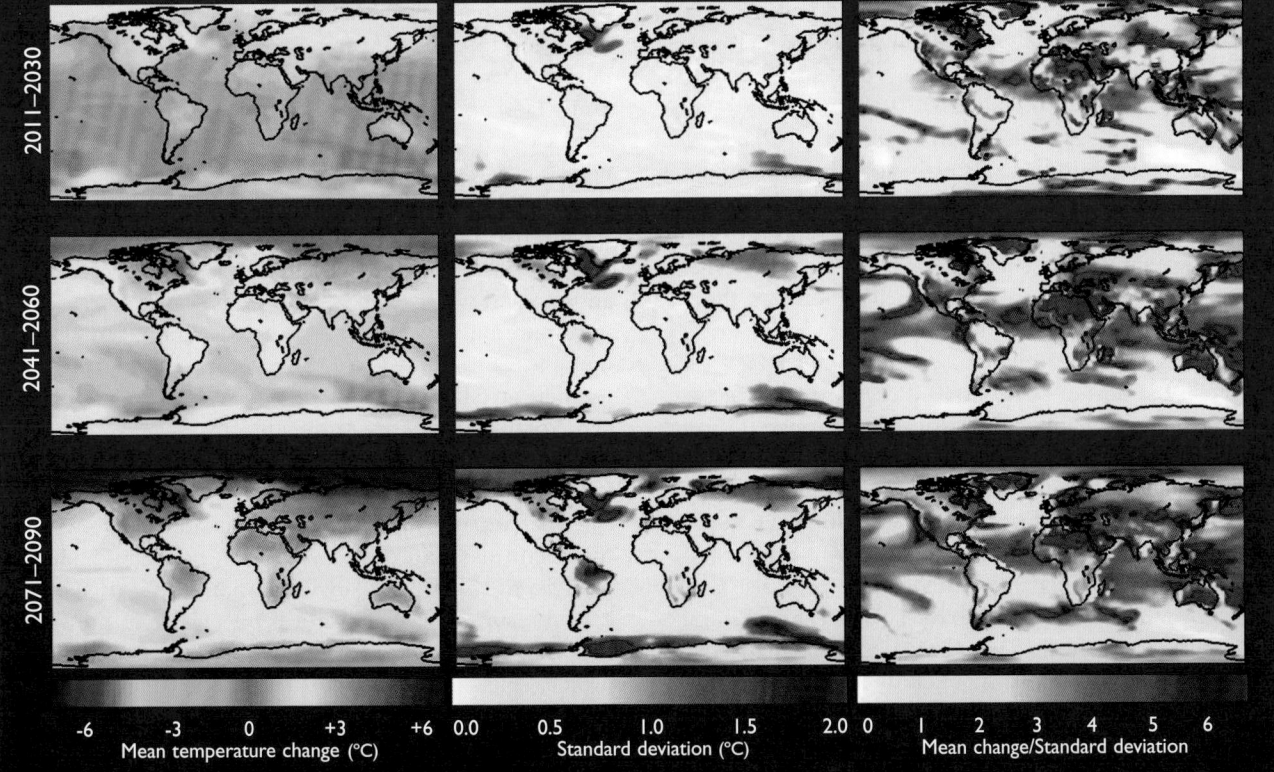

-6   -3   0   +3   +6   0.0   0.5   1.0   1.5   2.0   0   1   2   3   4   5   6
Mean temperature change (°C)   Standard deviation (°C)   Mean change/Standard deviation

Fig. 4.14. Changes in annual mean temperature projected by the ACIA-designated models for the early (top row), middle (center row), and late (bottom row) 21st century, as compared to the ACIA baseline (1981–2000). From left to right: five-model mean change; the inter-projection standard deviation; and the ratio between the mean and the standard deviation.

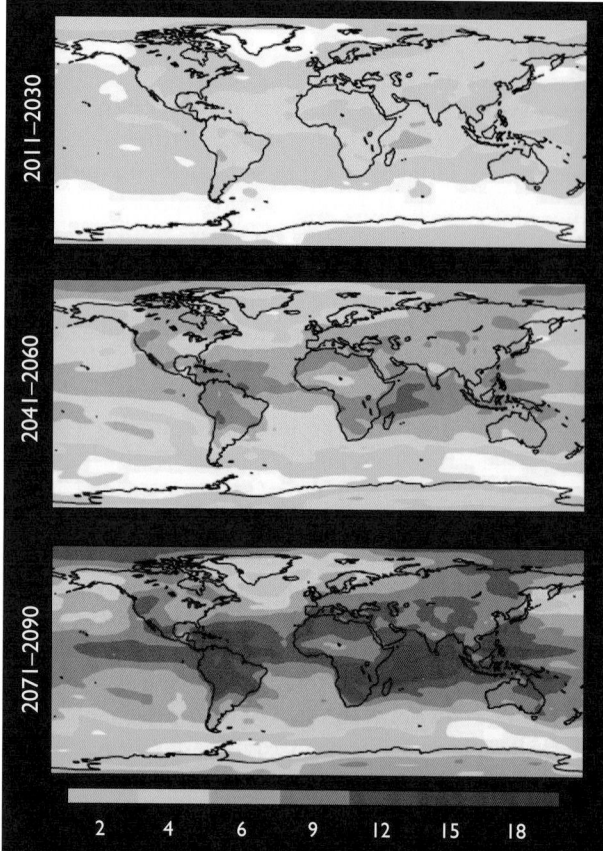

Fig. 4.15. The ratio between the ACIA five-model average projected change in mean annual temperature and √2 times the standard deviation of 20-year mean temperatures in the CMIP2 control simulations.

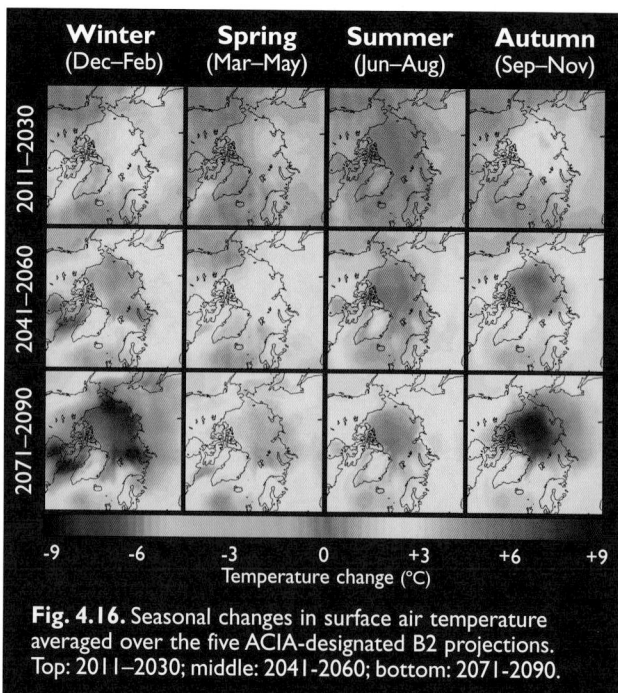

Fig. 4.16. Seasonal changes in surface air temperature averaged over the five ACIA-designated B2 projections. Top: 2011–2030; middle: 2041-2060; bottom: 2071-2090.

The differences in temperature change between the five ACIA-designated simulations are caused primarily by model differences and by noise associated with internal variability. To investigate the role of the latter factor alone, the five-model mean temperature changes were compared with the variability of 20-year mean temperatures generated in CMIP2 control simulations. The signal-to-noise ratio that was estimated by dividing the average warming by √2 times the standard deviation of 20-year means is shown in Fig. 4.15. The ratios are generally higher than those in Fig. 4.14, because internal variability does not cause all differences between the ACIA-designated simulations. Importantly, however, the ratio is still relatively low (<2) in some parts of the Arctic in the period 2011–2030. Because a ratio of two defines the lower limit of statistical significance, this suggests that, at least in some parts of the Arctic, the GHG-induced temperature increase may remain difficult to differentiate from natural variability over the next few decades. Indeed, the real signal-to-noise ratio might be even lower than calculated here, because the CMIP2 simulations exclude external sources of natural variability. Later in the 21st century, the signal-to-noise ratio increases, but remains generally lower in the Arctic than at low latitudes, where the internal variability is much smaller. Thus, despite the large average warming suggested by these simulations, the Arctic might not be the area where anthropogenic climate changes are easiest to detect.

Together, the results in Figs. 4.14 and 4.15 emphasize the importance of improving understanding of how best to interpret the relatively large arctic warming in the light of simulation and projection uncertainties. This issue is discussed further in section 4.7.

Figure 4.16 displays spatial distributions of seasonal temperature changes in the Arctic for the three 21st-century time slices. The five-model mean projected temperature increase over the central Arctic Ocean is greatest in autumn (up to 9 °C by 2071–2090), when the air temperature reacts strongly to reduced sea-ice cover and thickness. Average autumn and winter temperatures are projected to rise by 3 to 5 °C over most arctic land areas. By contrast, projected temperature increases over the Arctic Ocean in summer remain below 1 °C, because the temperature is held close to the freezing point by the presence of melting ice in both the control and the climate change simulations. The contrast between larger temperature increases in autumn and winter and smaller temperature increases in summer also extends to the surrounding land areas, but is less pronounced there. In summer, projected temperature increases over northern Eurasia and northern North America are larger than over the Arctic Ocean, while in winter the reverse is projected.

The spatial patterns of projected climate change within the Arctic also differ markedly between the individual models, so that, at any single location, the scatter of the model results is larger than it is for change in the arctic area mean temperature. However, all of the models project substantially smaller temperature increases over the northern North Atlantic sector than for other parts of the Arctic. In this area of the sinking branch of the Atlantic thermohaline circulation, the ocean is well mixed to a great depth. Therefore, much of the GHG-induced heating is devoted to warming the deeper ocean, rather than to warming the surface water. The warming is further reduced because the thermohaline circulation weakens in most of the model projections, transporting less warm water from the subtropical regions to the northern North Atlantic. The different degree of projected weakening of the thermohaline circulation in the models presents a special problem for scenario development for the North Atlantic area (see Boxes 4.4 and 4.5).

A comparison between changes in mean annual surface air temperature in the area north of 60° N projected by four of the ACIA-designated models (CGCM2, ECHAM4/OPYC3, GFDL-R30_c, and HadCM3) for the A2 and B2 emissions scenarios is shown in Fig. 4.17. The difference between the two scenarios is not dramatic during the first half of the 21st century, but becomes more systematic and significant during the second half of the century. These differences between the projected A2 and B2 arctic area-averaged temperature increases do not exceed the differences between the highest and lowest projections from the different ACIA-designated AOGCMs.

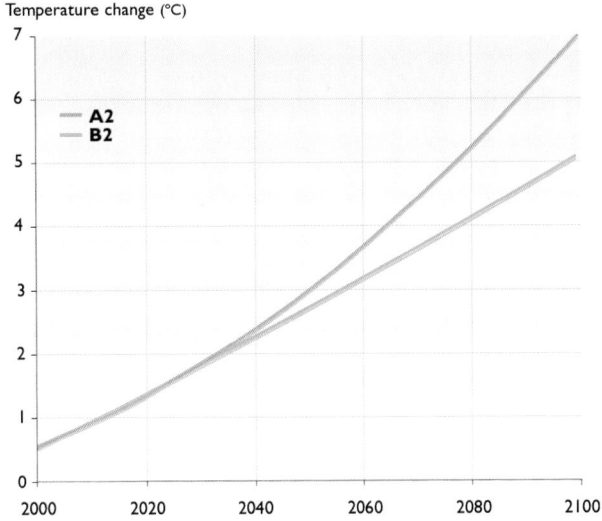

Fig. 4.17. Projected changes in mean annual surface air temperature in the Arctic (60°–90° N) for the A2 and B2 emissions scenarios, relative to 1981–2000. A binomial approximation is used to smooth the original mean of the four ACIA-designated model projections (CGCM2, ECHAM/OPYC3, GFDL-R30_c, and HadCM3) for each emissions scenario.

Box 4.5. The Atlantic meridional overturning circulation in the 21st century

Evidence from paleoclimate records indicates that the earth underwent large and rapid climate changes during the last glacial and early postglacial periods. The origin of the abrupt changes is being closely studied. A number of studies suggest that the Atlantic meridional overturning circulation (MOC), as part of the global thermohaline circulation, played an active and important role in these rapid climate transitions (e.g., Broecker, 1997; Ganopolski and Rahmstorf, 2001). Furthermore, modeling studies agree that the Atlantic surface freshwater balance is a key control parameter for the strength and variability of the Atlantic MOC. A common finding from idealized models and AOGCMs is that the strength of the Atlantic MOC decreases if the net flux of freshwater to the high northern latitudes increases (Manabe and Stouffer, 1997; Otterå et al., 2003; Rind et al., 2001; Schiller et al., 1997; Vellinga et al., 2002).

Increased precipitation at high northern latitudes is commonly projected by AOGCM simulations forced with increasing GHG concentrations (e.g., Räisänen, 2001). There is also recent observational evidence of intensification of the hydrological cycle at high northern latitudes (Curry R. et al., 2003; Dickson et al., 2002; Peterson et al., 2002). This raises questions about whether the Atlantic MOC will weaken in the 21st century, and what the likelihood is of a full shutdown of the Atlantic MOC.

Most AOGCMs forced with prescribed scenarios of the major GHG concentrations and aerosol particle distributions project a weakening of the MOC. The projected changes in the maximum strength of the Atlantic MOC by the end of the 21st century range from about zero to a reduction of 30 to 50% (Cubasch et al., 2001). The projected changes, if any, typically start around 2000 and show a quasi-linear trend thereafter. Irrespective of model differences in the sensitivity of the simulated Atlantic MOC to global climate change, no AOGCM has yet projected a shutdown of the Atlantic MOC by 2100 (Cubasch et al., 2001).

The present generation of AOGCMs used for these simulations does not include freshwater runoff from melting ice sheets and glaciers; therefore, it is possible that the model MOC sensitivity to global climate change is too weak. However, sensitivity experiments with freshwater artificially added to high northern latitudes indicate that fluxes corresponding to several times the present-day freshwater input are required to significantly alter the Atlantic MOC (Manabe and Stouffer, 1997; Otterå et al., 2003; Schiller et al., 1997; Vellinga et al., 2002), so this shortcoming may not be significant. As a result, the IPCC Third Assessment Report (Cubasch et al., 2001) concluded that it is unlikely that the Atlantic MOC will experience a shutdown in the 21st century. It is also likely that the major part of the North Atlantic–Nordic Seas region will experience warming throughout the 21st century, even with a weakened Atlantic MOC.

4.4.3. Changes in precipitation

Increases in arctic and global mean annual precipitation over the 21st century projected by the five ACIA-designated models forced with the B2 emissions scenario are displayed in Fig. 4.18. By the end of the 21st century (2071–2090), the projected change in global mean precipitation varies from 1.4% (ECHAM4/OPYC3) to 4.7% (GFDL-R30_c), with a mean of

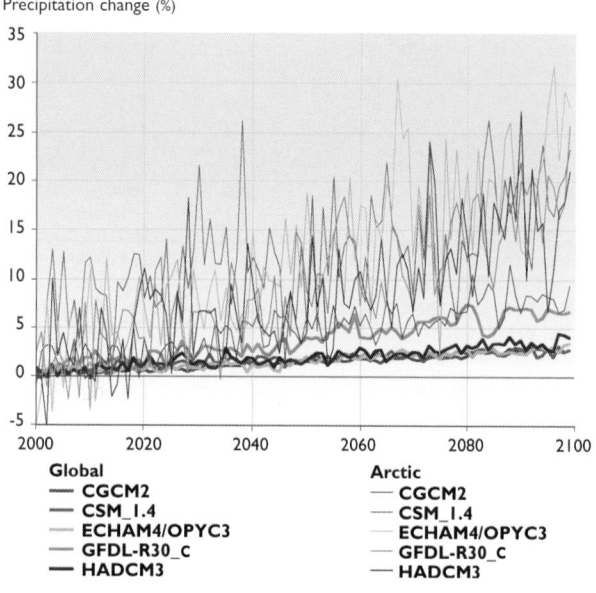

Precipitation change (%)

Fig. 4.18. Global and arctic (60°–90° N) percentage change in mean annual precipitation relative to the baseline period (1981–2000) projected by the five ACIA-designated models forced with the B2 emissions scenario.

2.5%. The Arctic Ocean and terrestrial arctic regions of North America and Eurasia are among the areas with the greatest projected percentage increase in precipitation. The general increase in high-latitude precipitation is a robust and qualitatively well-understood result from climate change experiments. With increasing temperature, the ability of the atmospheric circulation to transport moisture from lower to higher latitudes increases, leading to an increase in precipitation in the polar areas where the local evaporation is relatively small (e.g., Manabe and Wetherald, 1975).

For the area north of 60° N, all five models simulate an increase in annual precipitation by 2071–2090, which varies from 7.5% (CGCM2) to 18.1% (ECHAM4/OPYC3), and from 12 to 14% in the other models (Table 4.4). The differences between the models increase rapidly as the spatial domain becomes smaller.

The spatial distribution of the projected mean annual precipitation changes (five-model average) is shown in the left column of Fig. 4.19. By 2071–2090, projected precipitation increases in the Arctic vary from about 5 to 10% in the Atlantic sector to up to 35% locally in the high Arctic. Most of the projected increase in precipitation is due to increasing atmospheric water vapor convergence, which apparently results from the ability of a warmer atmosphere to transport more water vapor from lower to higher latitudes. The increased water vapor convergence is particularly important for precipitation in the high Arctic, where the local evaporation is small (see section 6.2). Although evaporation is projected to increase

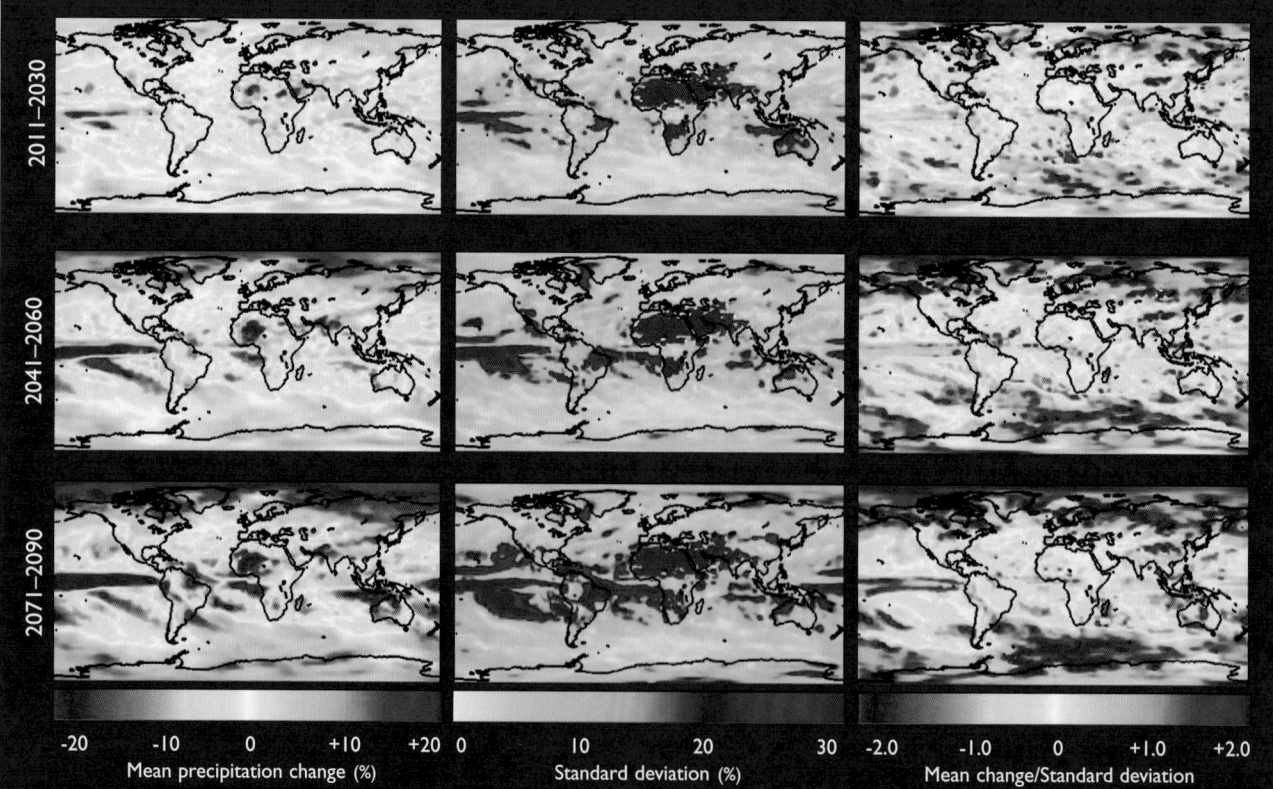

Fig. 4.19. Changes in mean annual precipitation projected by the ACIA-designated models for the early (top), middle (center), and late (bottom) 21st century, as a percentage of baseline (1981–2000) values. From left to right: five-model mean change; the inter-projection standard deviation; and the ratio between the mean and the standard deviation.

Table 4.4. Percentage increases in mean annual precipitation in the Arctic (60°–90° N) compared to the 1981–2000 baseline, as projected by the five ACIA-designated models forced with the B2 emissions scenario (from Kattsov et al., 2003).

	Precipitation increase (%)					
	CGCM2	CSM_1.4	ECHAM4/OPYC3	GFDL-R30_c	HadCM3	Five-model mean
2011–2030	2.3	8.3	5.1	3.0	4.0	4.3
2041–2060	4.6	8.3	12.3	7.4	7.3	7.9
2071–2090	7.5	14.0	18.1	11.9	12.9	12.3

slightly in most of the Arctic, the change is near zero (or even negative) in the Atlantic sector, where the change in sea surface temperature is small. Unlike surface air temperature, the ratio between the five-model mean precipitation change and the inter-model standard deviation (middle column) is larger in the Arctic than almost anywhere else (right column), with comparable agreement only at high southern latitudes. Nevertheless, the relative agreement between models is worse for precipitation than temperature changes.

Similar to projected temperature increases, the projected increase in precipitation is generally greatest in autumn and winter and smallest in summer (Fig. 4.20), but the summer minimum is less pronounced than that of temperature change.

The four-model projected changes in mean annual precipitation for the area north of 60° N (Fig. 4.21) behave similarly to temperature. The difference between the A2 and B2 emissions scenarios is small in the first half of the 21st century. Later in the century, the difference increases and is systematic (i.e., shown in all models), but does not exceed the inter-model scatter.

4.4.4. Changes in other variables

The average projected changes in mean seasonal and annual sea-level pressure are shown in Fig. 4.22.

Throughout the year, there is a slight projected decrease in pressure in the polar region, suggesting a shift toward the positive phase of the AO. However, the changes are small. Even in winter, when the projected decrease in sea-level pressure over the central Arctic is greatest, this amounts to only 4 hPa. The changes projected by individual models are larger, but vary widely, especially in winter. While many impact studies would benefit from projections of wind characteristics and storm tracks in the Arctic, the available analyses in the literature are insufficient to justify any firm conclusions about their possible changes in the 21st century.

Changes in cloud cover over the Arctic are small but systematic. By 2071–2090, the five-model average projects an increase in mean annual cloud cover. This is accompanied by a projected decrease (four-model average) in the mean incident short wave radiation. In summer, this flux is projected to decrease across the Arctic by more than 10 W/m² by 2071–2090 compared to the baseline (1981–2000).

The ACIA-designated models agree in their projections of decreases in sea-ice and terrestrial snow extents during the 21st century, as well as general increases both in precipitation minus evaporation over the marine Arctic and in river discharge to the Arctic Ocean from the surrounding terrestrial watersheds. The projected cryospheric and hydrological changes are quantified and discussed in detail in Chapter 6.

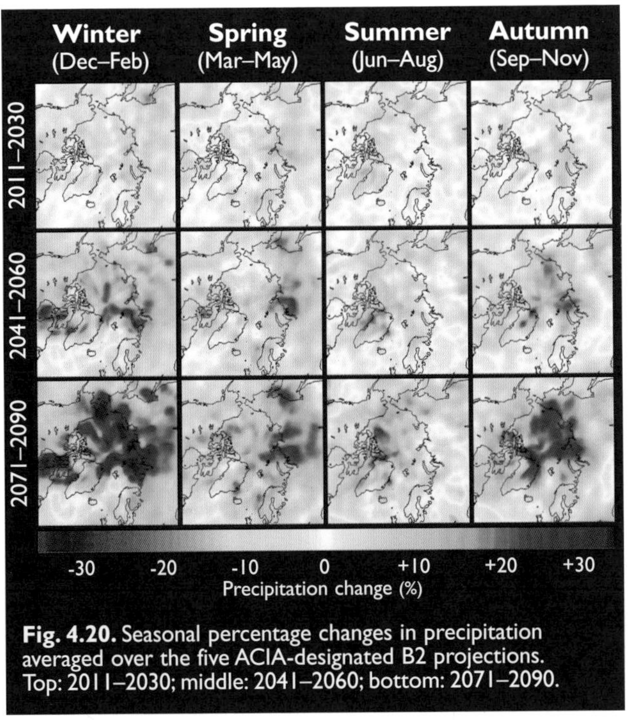

Fig. 4.20. Seasonal percentage changes in precipitation averaged over the five ACIA-designated B2 projections. Top: 2011–2030; middle: 2041–2060; bottom: 2071–2090.

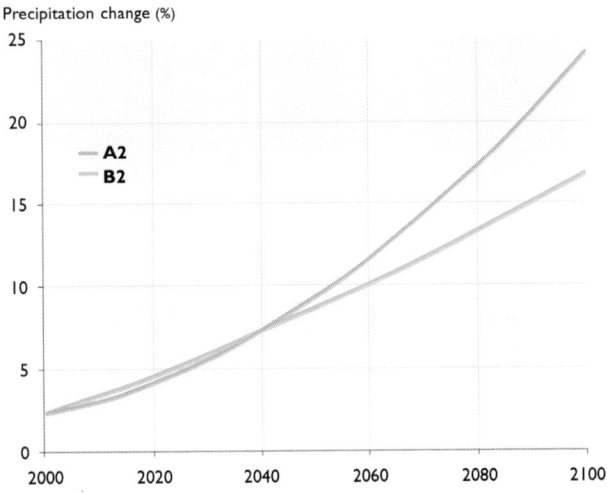

Fig. 4.21. Projected percentage changes in mean annual precipitation in the Arctic (60°–90° N) for the A2 and B2 emissions scenarios, relative to 1981–2000. A binomial approximation has been applied to the original mean of the four ACIA-designated model projections (CGCM2, ECHAM/OPYC3, GFDL-R30_c, and HadCM3) for each emissions scenario.

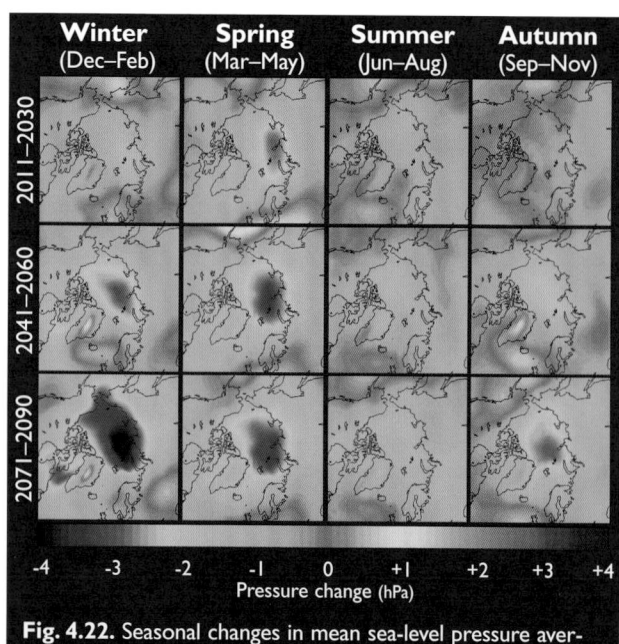

Fig. 4.22. Seasonal changes in mean sea-level pressure averaged over the five ACIA-designated B2 projections. For GFDL-R30_c, the variable used is surface pressure change, rather than sea-level pressure.

4.4.5. ACIA-designated models in the CMIP2 exercise

The five ACIA-designated models also participated in the CMIP2 intercomparison (Meehl et al., 2000), together with 14 other models. The model versions used in the CMIP2 simulations are in two cases (CGCM and CSM) slightly different from the ACIA-designated model ver-

sions, but this is unlikely to have any substantial influence on the comparison. The CMIP2 intercomparison helps to place the ACIA-designated model results in the broader context of model behavior. The climate changes in the CMIP2 simulations (Table 4.5) are examined for the 20-year period centered on the year when atmospheric CO_2 doubles in the simulations, which takes 70 years in these idealized experiments.

Figure 4.23a shows the global and arctic (60°–90° N) mean temperature changes in the CMIP2 simulations. Models located above the diagonal dashed line project greater temperature increases in the Arctic than globally, while models below the dashed line project the reverse. The global temperature increase varies from 1.1 °C (MRI2) to 3.1 °C (CCSR2), with a mean of 1.75 °C. The mean value for the five ACIA-designated models is very similar but the spread is smaller, 1.4 to 2.0 °C. The 19-model average projected increase in mean annual temperature in the Arctic (60°–90° N) is 3.4 °C (twice the global mean), and for the five ACIA-designated models the projected arctic temperature increase is 3.6 °C. Again, the range of the five ACIA-designated model projections (3.1–4.1 °C) is much smaller than the total range (1.5–7.6 °C) of all 19 CMIP2 simulations. However, individual outliers contribute substantially to the large range in the CMIP2 results.

Similar conclusions are valid for the projected change in arctic mean precipitation (Fig. 4.23b). The full range of projected precipitation change in the 19 models (4–24%) is wider than the range in the five ACIA-

Table 4.5. The 19 models participating in the CMIP2 intercomparison, with the five ACIA models shown in bold. The first column provides the symbol used in Figure 4.23.

	Model acronym	Institution	Reference
A	BMRC	Bureau of Meteorology Research Center	Power et al., 1993
B	**CCCma[a]**	**Canadian Centre for Climate Modelling and Analysis**	**Flato et al., 2000**
C	CCSR1	Center for Climate System Research	Emori et al., 1999
D	CCSR2	Center for Climate System Research	Nozawa et al., 2000
E	CERFACS	Centre Européen de Reserche et de Formation Avancée en Calcul Scientifique	Barthelet et al., 1998
F	CSIRO	Commonwealth Scientific and Industrial Research Organization	Hirst et al., 2000
G	ECHAM3	Max-Planck Institute for Meteorology	Voss et al., 1998
H	**ECHAM4/OPYC3**	**Max-Planck Institute for Meteorology**	**Roeckner et al., 1999**
I	GFDL-R15	Geophysical Fluid Dynamics Laboratory	Manabe et al., 1991
J	**GFDL-R30_c**	**Geophysical Fluid Dynamics Laboratory**	**Knutson et al., 1999**
K	GISS	Goddard Institute for Space Studies	Russell and Rind, 1999
L	HadCM2	Hadley Centre for Climate Prediction and Research	Johns et al., 1997
M	**HadCM3**	**Hadley Centre for Climate Prediction and Research**	**Gordon et al., 2000**
N	IAP/LASG	Institute of Atmospheric Physics, Chinese Academy of Sciences	Zhang et al., 2000
O	LMD/IPSL	Laboratoire de Météorologie Dynamique, Institut Pierre Simon Laplace	Braconnot et al., 1997
P	MRI1	Meteorological Research Institute	Tokioka et al., 1995
Q	MRI2	Meteorological Research Institute	Yukimoto et al., 2000
R	**NCAR-CSM[a]**	**National Center for Atmospheric Research**	**Boville and Gent, 1998**
S	NCAR/DOE-PCM	National Center for Atmospheric Research/ Department of Energy	Washington et al., 2000

[a]An earlier model version than that used by the ACIA participated in the CMIP2 intercomparison.

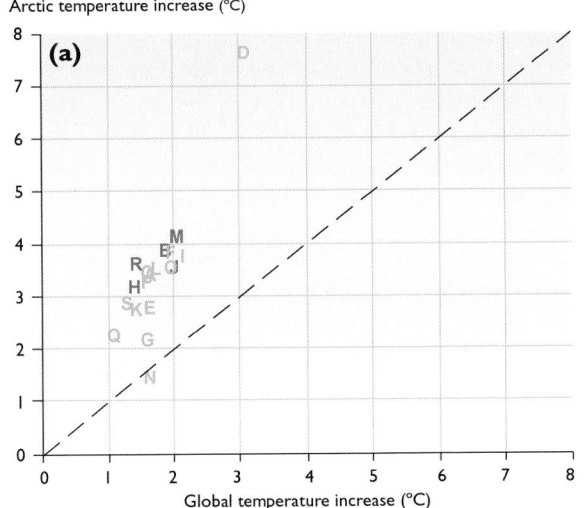

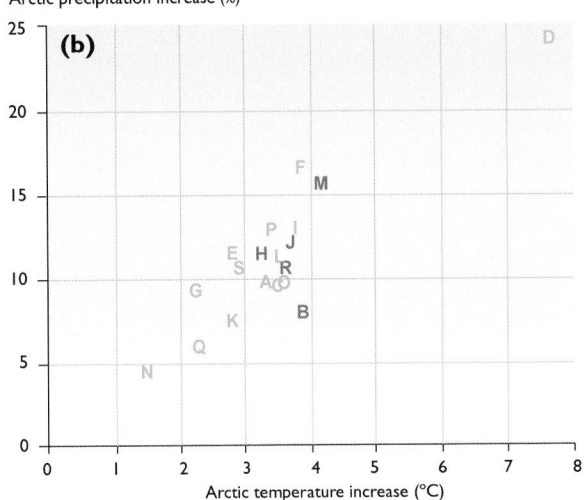

Fig. 4.23. Projections from the 19 CMIP2 models of (a) changes in global mean annual temperature (horizontal axis) and arctic (60°–90° N) mean annual temperature (vertical axis) and (b) changes in arctic mean annual temperature (horizontal axis) and precipitation (vertical axis). Table 4.5 lists the model associated with each letter. The five ACIA-designated models are shown in red and the others in blue.

designated models (8–15%), but the 5-model and the 19-model means are both 11%. The larger set of models shows an expected tendency for the projected precipitation change to be greatest for the models with the greatest projected temperature increase, which is not visible in the results from the five ACIA-designated models alone. The same tendency is seen in a comparison of projected global mean temperature and precipitation change (see Cubasch et al., 2001).

Figure 4.24 shows the spatial distribution of the projected change in mean annual temperature from the CMIP2 experiments, as averaged over all 19 models and over the five ACIA-designated models. The basic patterns are very similar, with the greatest temperature increases over the central Arctic in both cases. A similar comparison of projected precipitation changes (Fig. 4.25) also indicates broad agreement between the 19-model mean and the 5-model mean. However, some of the spatial details in the 5-model mean, such as the maxima over northeastern Greenland and eastern Siberia, are smoothed out when the change is averaged over all 19 models.

4.4.6. Summary

Projections of arctic climate change in the 21st century from all of the ACIA-designated AOGCMs are qualitatively consistent and in line with the IPCC conclusions listed at the beginning of section 4.4. The across-model scatter of the arctic climate change scenarios is significant, but smaller than the scatter between the climates simulated by the different models for the baseline period. Even the difference between the two single-model projections driven with both the A2 and B2 emissions scenarios is comparable to the range of corresponding changes projected by the five ACIA-designated models forced with the B2 emissions scenario.

In summary, the five ACIA-designated AOGCMs appear to be a representative sample of climate models, at least in terms of the average response of arctic temperature and precipitation to the B2 emissions scenario. However, this set of projections does not capture the full range of uncertainty associated with model and emissions scenario differences, at least in the second half of the 21st century.

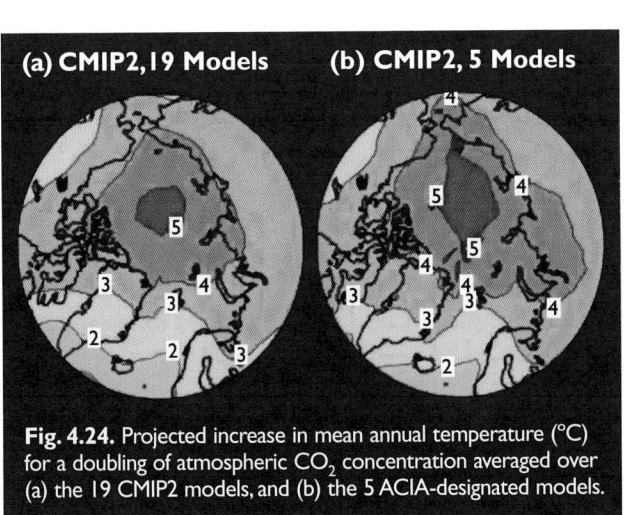

Fig. 4.24. Projected increase in mean annual temperature (°C) for a doubling of atmospheric CO_2 concentration averaged over (a) the 19 CMIP2 models, and (b) the 5 ACIA-designated models.

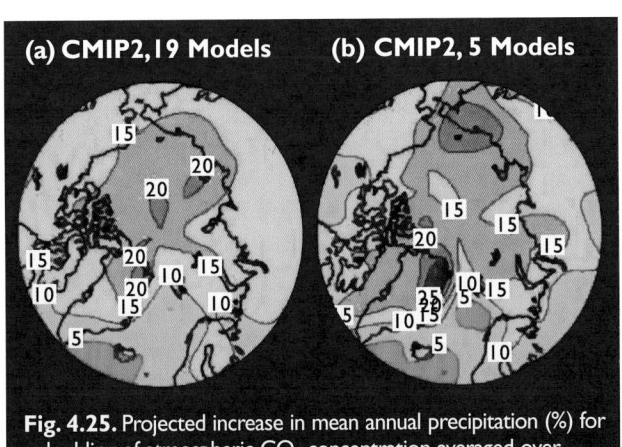

Fig. 4.25. Projected increase in mean annual precipitation (%) for a doubling of atmospheric CO_2 concentration averaged over (a) the 19 CMIP2 models, and (b) the 5 ACIA-designated models.

4.5. Regional modeling of the Arctic

An improved understanding of the arctic climate system is necessary to provide better quantitative assessments of the magnitude of potential global change and to clarify the role of the Arctic in the global climate system. The deficiencies of AOGCMs in describing arctic climate are partly due to inadequate parameterizations of physical processes. Equally important, AOGCMs are characterized by a rather coarse horizontal resolution, which fails to capture atmospheric mesoscale features caused by coastlines, ice sheets, sea-ice margins, and mountains. To some extent, this failure is overcome when the resolution is increased (Giorgi et al., 2001). The most obvious step – to simply increase resolution in AOGCMs – has until now been impractical due to the computational capacities required. Experience with high- or variable-resolution AGCMs has been limited and no results are available that focus on the performance of these models for the Arctic.

Another approach to obtaining enhanced regional details is the use of a nested limited-area model. This technique has garnered considerable interest in recent years because it requires less computational capacity than a global model with comparable resolution. These models can be used for process studies and for model validation studies using lateral boundaries from observation-based analysis (e.g., Giorgi and Mearns, 1999; Giorgi et al., 2001). However, the technique can also be used as the basis for dynamic downscaling of AOGCM simulations. Models used in this way are often referred to as regional climate models (RCMs; see Box 4.1).

4.5.1. Regional climate models of the arctic atmosphere

4.5.1.1. General

Regional models show considerable skill in short-term (few hour to few day) weather forecasting and are used worldwide for this purpose. This application is basically an initial-value problem, and most of the success of this approach can be ascribed to the high resolution of the models and the use of very recent observations.

Likewise, at timescales of a few years to decades and beyond, regional models have shown their strength in comparison with coarse-resolution global models, as they are capable of capturing the fine-scale details of climatic processes – such as the presence of complex topography and small-scale weather features such as tropical cyclones and even polar lows – much more realistically than global models (Giorgi et al., 2001). Moreover, it has been shown that the statistics of extreme precipitation are realistically simulated by high-resolution regional models (Frei et al., 2003b) and that they show skill even at their grid scale (Huntingford et al., 2003). This is a boundary-value problem and the assessment of the skill of the model is based on statistics of model performance over long time periods (several years or more).

Regional model projections are limited by the quality of the global model projections used for the lateral boundary conditions. In this respect, a potential problem is the mismatch between scales in the coarse-resolution global model and the high-resolution regional model. Recently, this has been demonstrated not to be a fundamental problem if proper boundary condition procedures are used (Denis et al., 2002). In fact, it has been demonstrated that, when driven by realistic (observed) boundary conditions, regional climate models are capable of capturing the overall observed regional climatic evolution and can add realistic spatial and temporal information to the information provided by the driving model (Dethloff et al., 2002; Frei et al., 2003b; Giorgi et al., 2001). When nested within a GCM, it is important, however, to stress that the large scale circulation is imposed by the lateral boundaries. The regional model is not able to, nor is it intended to, correct the large-scale errors made by the global model. The role of the regional model is instead to add regional detail and fine spatial and temporal scales to the simulation, not to improve the large-scale simulation.

Atmospheric regional modeling systems for the Arctic have been developed by Lynch et al. (1995) and Dethloff et al. (1996). Dethloff et al. (1996) applied the regional atmospheric climate model HIRHAM (Christensen J. et al., 1996) to the entire Arctic north of 65° N. This model has so far been the only RCM used with a circumpolar focus for climate simulations, although the recent Arctic Regional Model Intercomparison Project (ARCMIP; see section 4.5.1.2) initiative has increased the interest in arctic RCMs. Many other groups have developed RCMs, but mostly with a more southerly region of interest, although in some cases parts of the Arctic have been included in simulations performed with these models. For example, several models have been applied over most of the European continent including the Scandinavian Peninsula and parts of the North Atlantic Ocean extending all the way to the ice margins; even parts of Greenland have been included in such simulations (Christensen J. et al., 1997; Rummukainen et al., 2001). Likewise, experiments with a Canadian RCM have been applied to the whole of Canada (Laprise et al., 1998) including a fair proportion of the Arctic. The development of these models continues, but limited results with an arctic focus have been published.

Giorgi et al. (2001) documented how much insight has been provided into fundamental issues concerning the nested regional modeling technique. For example, multi-year to multi-decadal simulations must be used for climate change studies to provide meaningful climate statistics, to identify significant systematic model errors and climate changes relative to internal model and observed climate variability, and to allow the atmospheric model to equilibrate with the land surface conditions (e.g., Christensen O., 1999; Jones et al., 1997; Machenhauer et al., 1998; McGregor et al., 1999). In addition, the choice of domain size is not a trivial question. The influence of the boundary forcing is reduced as region size increases (Jacob and Podzun, 1997; Jones et al., 1995) and may be dominated by the internal model physics for certain variables and seasons (Noguer et al., 1998). This can lead to the RCM solution significantly departing from the driving data, which can make the interpretation of downscaled regional climate changes more difficult (Jones et al., 1997). For most experiments with very high resolution, the domain size is limited by practical considerations and the large-scale flow is, therefore, constrained substantially by the driving model (Christensen J. and Kuhry, 2000). Denis et al. (2002) demonstrated that when the discrepancy in resolution between the model providing the lateral boundary conditions and the nested RCM does not exceed a factor of 10, the RCM is able to generate added value high-resolution information. With this limitation, they also showed that for a typically sized domain, the RCM does not introduce any spurious developments due to the nesting technique, and is fully capable of a consistent development within the model domain.

Configurations of RCM model physics are derived either from a pre-existing (and well-tested) limited-area model system with modifications suitable for climate applications or are implemented directly from a GCM (see Giorgi et al., 2001). In the first approach, each set of parameterizations is developed and optimized for the respective model resolutions. However, this makes interpreting differences between the nested model and the driving GCM more difficult, as these will result from more than just changes in resolution. Also, the different model physics schemes may result in inconsistencies near the boundaries (Machenhauer et al., 1998; Rummukainen et al., 2001). The second approach maximizes compatibility between the models. However, physics schemes developed for coarse-resolution GCMs may not always be adequate for the high resolutions used in nested regional models and may at a minimum require recalibration. Overall, both strategies have shown performance of similar quality and either is acceptable (Giorgi and Mearns, 1999).

4.5.1.2. Simulations of present-day climate with regional climate models

Regional climate models have been used for a wide variety of research worldwide, and have generated a sizeable research community. However, very few groups have focused on the Arctic to date, although several new initiatives have recently been undertaken. Intercomparison projects entailing participation of different international research groups have been conducted or are currently underway (Christensen J. et al., 1997, 2002; Machenhauer et al., 1998; Takle et al., 1999; Rinke et al., 2000; www:awi-potsdam.de/www-pot/atmo/glimpse/index.html). This section describes the existing applications of RCMs for simulating present-day arctic climate conditions.

Advances in regional climate modeling must be based on an analysis of physical processes and comparison with observations. In data-poor regions such as the Arctic, this procedure may be complemented by a community-based approach (i.e., through collaborative analysis by several research groups). To illustrate this approach, simulations of the Arctic Basin north of 65° N have been performed and planned with different RCMs, driven at their lateral boundaries by observationally based analyses (Rinke et al., 2000). Motivated by this, an international intercomparison of regional model simulations for the Arctic, ARCMIP, has been organized under the auspices of the World Climate Research Programme. The foci of the evaluation include the surface energy balance over ocean and land, clouds and precipitation processes, stable planetary boundary layer turbulence, ice-albedo feedback, and sea-ice processes. The preliminary results from this project indicate that the participating regional models are able to reproduce reasonably well the main features of the large-scale flow and the surface parameters in the Arctic. However, in order to reach definitive conclusions in an RCM intercomparison, ensemble simulations with adequate spin-up (time for regional model processes to equilibrate with prescribed external forcings) and equivalent initialization of surface fields are required (Rinke et al., 2000). Several aspects of the intercomparison are difficult due to the lack of adequate data for model validation. However, some aspects of the models, such as the hydrological cycle, compare well with each other (see Rinke et al., 2000).

Determining the primary causes of model biases, deficiencies, and uncertainties in atmosphere-only and coupled atmosphere–ice–ocean climate models for the Arctic is of vital importance in order to model the arctic climate adequately. Participants in the ARCMIP project are seeking to improve model representations through an intercomparison of simulations by different models and comparison with observations made during the Surface Heat Budget of the Arctic Ocean field-- experiment year (October 1997 to October 1998).

For the Arctic, only very limited multi-year RCM experiments have been conducted. For example, Kiilsholm et al. (2003) assessed the uncertainty in regional accumulation rates for the Greenland Ice Sheet due to model resolution. They used the HIRHAM RCM (50 km resolution) with boundary conditions from a 30-year control simulation with the ECHAM4/OPYC3 model (~300 km horizontal resolution; Roeckner et al., 1996). Figure 4.26 compares the resulting accumulation and ablation zones as simulated by the RCM and the GCM. Table 4.6 illustrates the ability of the models to simulate present regional (northern Greenland) changes in accumulation rates. It appears that the RCM simulation is in better agreement with the observational evidence than the GCM simulation.

Non-arctic applications have shown that regional models may have significant potential for use as dynamic interpolators, yielding useful data for a wide range of times and locations where *in situ* observations are not available. As regional models become increasingly accu-

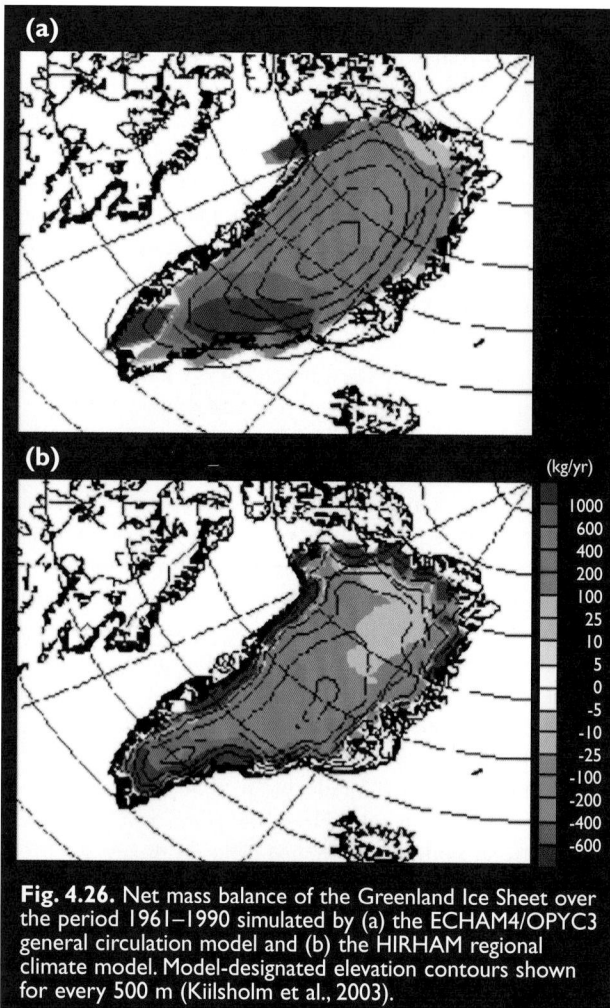

Fig. 4.26. Net mass balance of the Greenland Ice Sheet over the period 1961–1990 simulated by (a) the ECHAM4/OPYC3 general circulation model and (b) the HIRHAM regional climate model. Model-designated elevation contours shown for every 500 m (Kiilsholm et al., 2003).

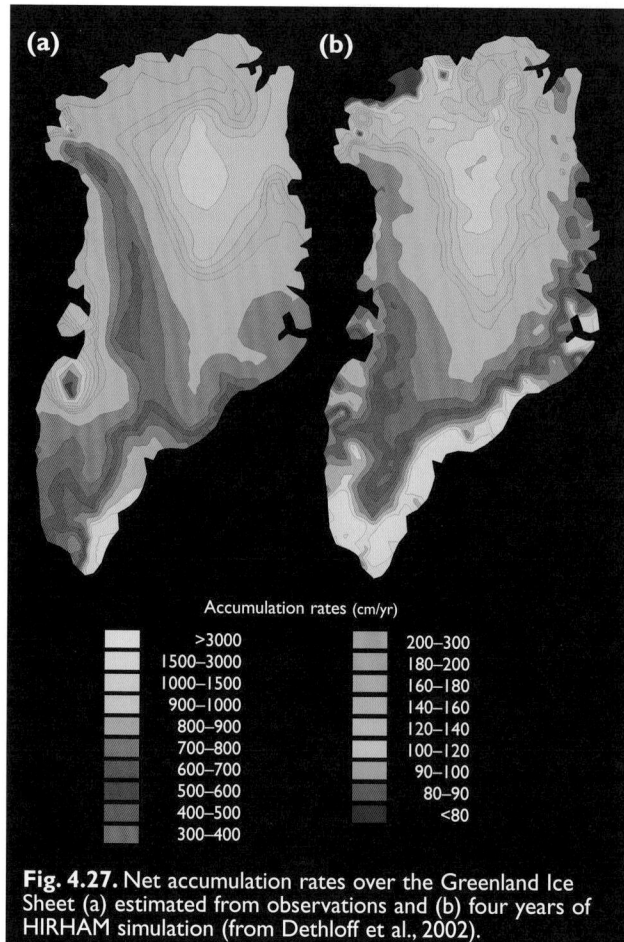

Fig. 4.27. Net accumulation rates over the Greenland Ice Sheet (a) estimated from observations and (b) four years of HIRHAM simulation (from Dethloff et al., 2002).

rate, they could become valuable tools for glaciological research. A primary application in glaciology is the investigation of mass balance changes in continental ice sheets. Determining the climatic conditions affecting ice sheets is important because major changes in ice-sheet dimensions affect climate and sea level throughout the world. Much recent work has gone into the validation of RCM simulations (Bromwich et al., 2001; Cassano et al., 2001) using observational data analyses (Hanna and Valdes, 2001) for the Greenland Ice Sheet. Dethloff et al. (2002) carried out a detailed RCM validation study on the basis of multi-year ensemble simulations, selected annual simulations, and derived results for the mass balance of the Greenland Ice Sheet. Figure 4.27 illus-

trates net accumulation rates from the HIRHAM simulations. Compared with available results from earlier work, these results indicate a high degree of skill in spatial representation.

Another promising application was identified by Christensen J. and Kuhry (2000), who analyzed the ability of an RCM to simulate permafrost zonation at very high spatial resolution. Based on a simple permafrost index (Nelson and Outcalt, 1987), but applied to the subsurface model layers rather than near-surface air temperatures, they documented that at sufficiently high resolution the permafrost zonation in complex regions can be quite accurately modeled.

4.5.1.3. Time-slice projections from atmospheric RCMs

In its Third Assessment Report (IPCC, 2001) the IPCC considered the concept of using RCMs for climate change projections at some length, and concluded that it is essential that information from a transient AOGCM simulation be available. In such studies, the RCM is used to provide a reinterpretation of the overall AOGCM behavior, including its response to external forcing (from GHGs and aerosols). Sometimes ocean modeling is also done as part of the regional model simulation (Räisänen et al., 2003). In a typical experiment (Christensen J. et al., 2002; Jones et al., 1995; Kiilsholm et al., 2003; Machenhauer et al., 1996;

Table 4.6. Comparison of observed changes in net accumulation rates for the north Greenland Ice Sheet (76°–79° N) with simulations by a general circulation model (ECHAM4/OPYC3) and a regional climate model (HIRHAM). Model uncertainties are estimated as one standard deviation of the 30-year interannual variability in the simulation.

	Change in net accumulation rate (mm/yr)	
	27°–30° W	60°–65° W
ECHAM4/OPYC3[a]	170±40	-1390±560
HIRHAM[a]	75±24	-223±189
Observed[b]	97±84	-310±107

[a]1961–1990 (Kiilsholm et al., 2003); [b]1954–1995 (Paterson and Reeh, 2001)

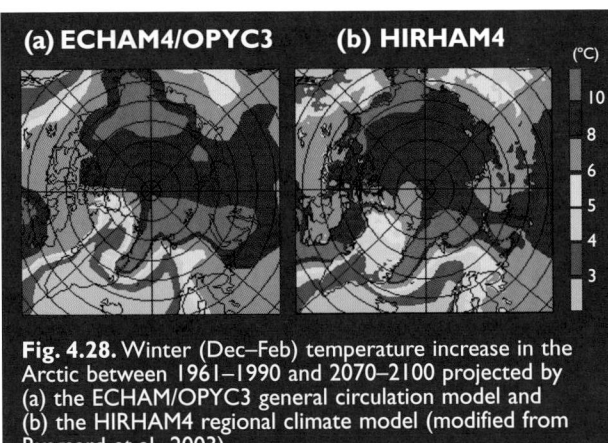

Fig. 4.28. Winter (Dec–Feb) temperature increase in the Arctic between 1961–1990 and 2070–2100 projected by (a) the ECHAM/OPYC3 general circulation model and (b) the HIRHAM4 regional climate model (modified from Rysgaard et al., 2003).

Räisänen et al., 1999; Rummukainen et al., 2001), two time slices (e.g., 1961–1990 and 2071–2100) are selected from a transient AOGCM simulation. The RCM simulations include prescribed time-dependent GHG and aerosol concentrations for the corresponding periods of the AOGCM run. The time-dependent sea surface temperatures and sea-ice distributions simulated by the AOGCM are also prescribed as lower boundary conditions, although some models also incorporate an interactive ocean/sea-ice model (Räisänen et al., 2003). The RCM simulations are typically initialized using atmospheric and land-surface conditions interpolated from the corresponding AOGCM fields, and there may be a considerable spin-up period before the actual simulation is started (e.g., Christensen O., 1999).

Only a few studies of this type have been conducted for time slices of durations long enough to encompass the large interannual variability in Arctic. Only two sets of experiments exist to date that cover the entire Arctic (Dorn et al., 2003; Kiilsholm et al., 2003), while Haugen et al. (2000) reported simulations that only cover the Atlantic sector, with the main focus of the experiment being Norwegian land territories (mainland Norway and Svalbard). Kiilsholm et al. (2003) and Dorn et al. (2003) have conducted a set of such experiments using information from transient simulations with the ECHAM4/OPYC3 model (Stendel et al., 2000).

This section highlights the experiments conducted by Kiilsholm et al. (2003) using the HIRHAM4 model forced with the B2 emissions scenario, as they are the only ones with complete multi-year integrations that are generally consistent with the characteristics of the ACIA scenarios. In this study, the AOGCM had a resolution of approximately 300 km, while the resolution of the RCM was approximately 50 km. This discussion focuses on differences in the climate projections of the RCM and the driving AOGCM.

Figure 4.28 depicts the winter temperature change as simulated by the AOGCM and the RCM. In general, the patterns of warming as well as their amplitude are quite similar. However, the RCM depicts some regional patterns that can be ascribed to the higher resolution.

Along the ice margin in the Greenland Sea and the Barents Sea, where sea ice retreats in the simulations (sea ice in the RCM is interpolated from the AOGCM), the RCM shows greater temperature increases. This is due to a stronger response to sea-ice changes resulting from a better description of the nonlinear energy cascade connected with mesoscale weather developments (e.g., stronger cyclonic developments) in the RCM than in the AOGCM. Conversely, temperature increases projected by the RCM tend to be lower over most of the central Arctic and all of Siberia, particularly during summer. This is due to a more realistic simulation of the present-day snow pack by the RCM than by the AOGCM (see Box 4.4).

Figure 4.29 shows the winter change in simulated precipitation minus evaporation. As with temperature, the large-scale agreement is striking. However, regional details are evident along the North Atlantic storm track and close to complex topography. In general, the RCM shows a stronger increase in precipitation minus evaporation upwind of major topographical obstacles and a corresponding decrease in precipitation minus evapotranspiration downwind (e.g., Scandinavia, the Rocky Mountains, and Siberia). This is partly explained by the increased topographical gradients due to the higher resolution of the model.

These apparently small differences may have substantial effects on the assessment of future changes in various geophysical systems, such as the mass balance of glaciers and the Greenland Ice Sheet in particular (Kiilsholm et al., 2003) and the depth of the permafrost active layer (Walsh, in press; see also section 6.6). The RCM is better suited for simulating climate change at the regional level, particularly for areas with complex topography and coastlines. This has been confirmed in multiple applications of the model outside of the Arctic (e.g., Giorgi et al., 2001; Christensen J. and Christensen O., 2003; Huntingford et al., 2003).

Projected changes in arctic climate due to anthropogenic GHG emissions will occur together with natural dynamic processes in the climate system. In order to improve projections of the evolution of arctic climate, the effects

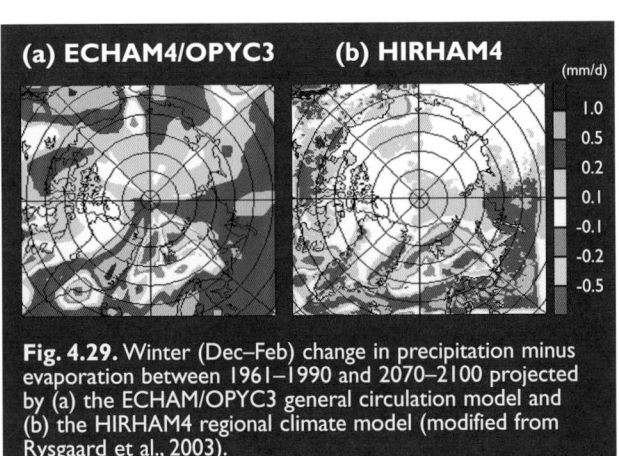

Fig. 4.29. Winter (Dec–Feb) change in precipitation minus evaporation between 1961–1990 and 2070–2100 projected by (a) the ECHAM/OPYC3 general circulation model and (b) the HIRHAM4 regional climate model (modified from Rysgaard et al., 2003).

of natural climate variability and in particular their regional dimensions must be taken into account.
One major phenomenon contributing to the natural variability of the climate of the Northern Hemisphere is the NAO, which is also associated with the AO, and is described in more detail in section 2.2.2. In general, the influence of the NAO on arctic temperatures is directly opposed in the western and eastern Arctic, and is stronger over land areas than over ocean areas or sea ice. Dorn et al. (2003) investigated the combined effects of varying phases of the NAO and increasing GHG and aerosol concentrations on arctic winter temperatures (Fig. 4.30). In this study, different phases of the NAO in a transient coupled AOGCM simulation were considered in time slice simulations using an RCM. Two future periods, 2013–2020 and 2039–2046, representing a positive and a negative phase of the NAO, respectively, were analyzed. The level of GHGs and aerosols in the simulation is higher during the negative NAO phase (2039–2046) than during the positive phase (2013–2020). Although mean arctic winter temperatures are projected to be approximately 1.3 °C higher during the negative phase compared to the positive phase, regions of warming and cooling between the two periods can be observed in Figs. 4.30a and 4.30b. Subsequent to the positive phase, a strong warming of more than 6 °C is simulated over some areas of Alaska, the Labrador Sea, and Baffin Island, whereas a similar strong cooling is simulated only over the northern Barents Sea. The temperature effect of the NAO is altered by the general temperature increase resulting from enhanced GHG and aerosol concentrations, but the influence of the NAO is still clearly evident at the regional scale. Although the statistical significance of these differences has not been assessed, the results clearly show that regional changes in the Arctic at decadal timescales may, at least for several more decades, be dominated by changes in the overall atmospheric flow rather than by the temperature increases due to rising GHG concentrations.

Circulation in an RCM is determined by the boundary conditions provided by the driving AOGCM. As noted in section 4.3, simulations from different global models can be very different in terms of circulation patterns, and the small-scale response in a regional model can amplify such

differences. An example is given in Fig. 4.31. Here, the Rossby Centre regional model (RCAO: Rummukainen et al., 2001) is driven by 30-year global climate simulations from the HadAM3H and ECHAM4/OPYC3 global climate models forced with the A2 and B2 emissions scenarios. Both regional simulations driven by the HadAM3H scenarios (Fig. 4.31a,b) show only moderate increases in precipitation while both ECHAM4/OPYC3 simulations (Fig. 4.31c,d) show dramatic precipitation increases, particularly on the western side of the Scandinavian mountains. The difference between the A2 and B2 emissions scenarios is quite small relative to the inter-model differences, which are due to a clear difference in the circulation regime change simulated by the two AOGCMs. The HadAM3H model projects a relatively small change in the north-south pressure gradient across the Nordic region, in sharp contrast to the ECHAM4/OPYC3 model, which projects a substantial strengthening of the north–south pressure gradient. This difference is in turn connected with a difference in the projected shifts in the main storm track regions over the North Atlantic. The physical reason behind the different responses in

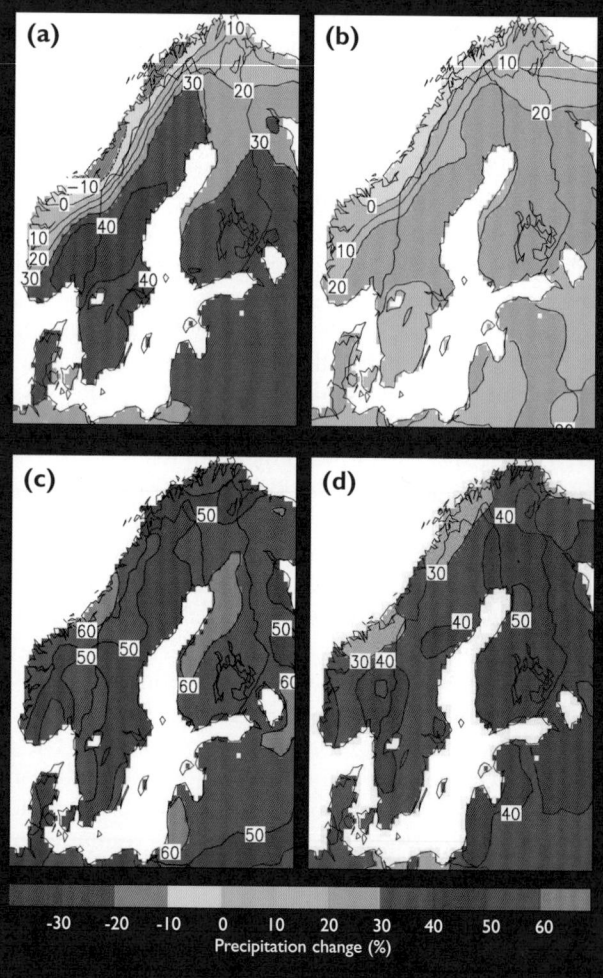

Fig. 4.31. Percentage changes in winter (Dec–Feb) precipitation over the Scandinavian region between 1961–1990 and 2071–2100 as simulated by a regional climate model (RCAO) driven by (a) the HadAM3H model forced with the A2 emissions scenario; (b) the HadAM3H model forced with the B2 emissions scenario; (c) the ECHAM4/OPYC3 model forced with the A2 emissions scenario; and (d) the ECHAM4/OPYC3 model forced with the B2 emissions scenario.

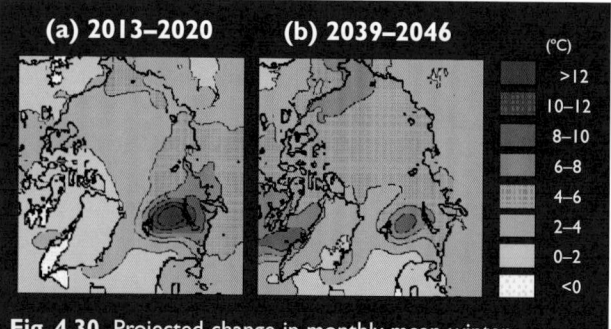

Fig. 4.30. Projected change in monthly mean winter (Dec–Mar) temperature at 2 m height compared to the control climate for (a) 2013–2020, when the simulated North Atlantic Oscillation (NAO) was in a positive phase; and (b) 2039–2046, when the simulated NAO was in a negative phase (Dorn et al., 2003).

storm track regions for the two global models is at present unclear. Keeping in mind the results of Dorn et al. (2003), a plausible explanation is that part of the difference may be due to different simulations of decadal variability in the NAO by the two AOGCMs.

The lack of experiments with different RCMs using boundary conditions from more than one GCM so far prevents the practical use of RCM information for general impacts work. This is also the situation even for better-studied regions, such as continental Europe, due to a lack of appropriate coordination of these efforts; however, see Christensen J. et al. (2002) for an update on recent progress.

4.5.2. Regional Arctic Ocean models

Proper treatment of the Arctic Ocean requires a spatial resolution high enough to account for reduced Rossby length scales (the smallest scale at which the rotation of the Earth has a dominating influence on flow dynamics); to permit the important flows through Bering Strait and the Canadian Archipelago; and to accurately represent the complex bottom topography steering the currents, as well as the continental slopes and the large continental shelves where the thermohaline, wind-driven, and tidal dynamics interact.

Currently, about a dozen Arctic Ocean models exist, most of which are participating in the Arctic Ocean Model Intercomparison Project (AOMIP: Proshutinsky et al., 2001; see also Box 4.2). Most of the Arctic Ocean models are derived from global oceanic GCMs – in many cases from the Geophysical Fluid Dynamics Laboratory (GFDL) Modular Ocean Model (MOM: Pacanowski and Griffies, 1999). The Arctic Ocean models represent a wide spectrum of numerical approaches, employing either finite-difference (in most cases) or finite-element approximations, and three types of vertical coordinates: z (constant geopotential surfaces), sigma (bathymetry-following), and isopycnal (constant potential density referenced to a given pressure). The models differ in their spatial resolution (from 40–50 km down to 16–20 km in the horizontal and typically 25–30 levels in the vertical); specifications of surface and lateral fluxes; and formulations of the surface mixed and bottom boundary layers. Regional models of the Arctic Ocean are usually coupled to comprehensive sea-ice models, although these often differ in their treatment of dynamics and thermodynamics.

To ensure a degree of fidelity to the simulations, an artificial constraint known as "climate restoring" (i.e., relaxation to the observed climate) based on the surface salinity is often introduced. This constraint prevents a model from drifting significantly from observations. Restoring time constants vary across models from several months (strong restoring) to several years (weak restoring). Some models do not employ restoring at all. However, the biases in simulations monotonically increase with the value of the restoring constant, reaching their highest levels in the models without restoring (Proshutinsky et al., 2001). This emphasizes that there are some processes and feedbacks crucial for representing the Arctic Ocean general circulation that are still neither sufficiently understood nor properly represented in the regional models, just as is the case for the global models.

From the viewpoint of their employment in downscaling AOGCM outputs or in constructing scenarios of future climate change, regional ocean models lag behind the atmospheric RCMs.

4.5.3. Coupled arctic regional climate models

The construction of coupled RCMs is a recent development. These models couple atmospheric RCMs to other models of climate system components, such as lake, ocean/sea-ice, chemistry/aerosol, and land biosphere/hydrology models (Bailey and Lynch, 2000a,b; Bailey et al., 1997; Hostetler et al., 1994; Lynch et al., 1995, 1997, 1998; Mabuchi et al., 2000; Maslanik et al., 2000; Qian and Giorgi, 1999; Rinke et al., 2003; Roed et al., 2000; Rummukainen et al., 2001; Small et al., 1999a,b; Tsvetsinskaya et al., 2000; Weisse et al., 2000). These initial efforts provide a path toward the development of coupled "regional climate system models". For some parts of the Arctic, coupled mesoscale atmosphere–ice–ocean models already exist, although they are restricted to small domains and short integration times (e.g., Lynch et al., 1997, 2001; Roed et al., 2000; Schrum et al., 2001).

For the circumpolar Arctic, Maslanik et al. (2000) presented the first results of a coupled atmosphere–ice–ocean RCM called ARCSyM. The oceanic component in this RCM was a simple mixed layer model. Rinke et al. (2003) presented a more complex coupled RCM (i.e., a fully coupled atmosphere–ice–ocean circulation model system) called HIRHAM-MOM. The ability to simulate conditions over the Arctic Ocean during April to September 1990, a period of anomalous atmospheric circulation and sea-ice conditions, was investigated with both models. A common result was found: neither model was able to correctly reproduce the large retreat of sea ice in the eastern Eurasian Basin and the adjacent shelf sea observed during the summer of 1990 (Maslanik et al., 1996). The sea ice in the Chukchi and East Siberian Seas does not retreat as completely in the model simulations as it does in observations inferred from satellite data.

The HIRHAM-MOM coupled model (Rinke et al., 2003) reproduced the general sea-level pressure patterns for the summer of 1990, based on a comparison with ECMWF analyses. However, discrepancies appeared in late summer that significantly affected variables such as wind flow and sea-ice transport. Similar to the ARCSyM results (Maslanik et al., 2000), HIRHAM-MOM simulated too much sea ice in the Bering, Chukchi, and East Siberian Seas during the summer of

1990. This is also the case for the ocean/sea-ice models driven by ECMWF atmospheric data.

The results from both coupled models highlight the importance of regional atmospheric circulation in driving interannual variations in arctic sea-ice extent, and illustrate the level of model performance required to simulate such variations. Such studies are valuable because they indicate improvements needed in the models by evaluating the results against observations, and the roles of key processes and feedbacks by comparing the results to those of the uncoupled atmospheric model. While results for the Baltic Sea imply improved model performance when an ocean model is coupled to the RCM (Räisänen et al., 2003), high-resolution coupled model systems for the Arctic have not provided improved performance to date.

4.5.4. Summary

The current status of arctic regional climate modeling did not allow RCMs to be employed as principal tools for the ACIA. Present scenarios of future arctic climate change are therefore based on results from global AOGCMs. However, presently available global coupled models have a coarse spatial resolution that limits their ability to capture many important aspects of climate change. In particular, intense storms, the effects of topography, and fundamental aspects of regional ocean circulation cannot be represented adequately. To improve the modeling of such phenomena, development of regional coupled ocean-ice-atmosphere climate models should receive a high priority in the efforts of the climate modeling community. Such developments should go hand in hand with developments in global modeling, in particular development of high-resolution AOGCMs.

4.6. Statistical downscaling approach and downscaling of AOGCM climate change projections

Statistical downscaling (also called empirical downscaling) is a tool for downscaling climate information from coarse spatial scales to finer scales. It may be applied as an alternative, or as a supplement, to dynamic downscaling (i.e., regional modeling). The underlying concept is that local climate is conditioned by large-scale climate and by local physiographical features such as topography, distance to a coast, and vegetation (von Storch, 1999). At a specific location, therefore, links should exist between large-scale and local climatic conditions. Statistical downscaling consists of identifying empirical links between large-scale patterns of climate elements (predictors) and local climate (the predictand), and applying them to output from global or regional models. Successful statistical downscaling is thus dependent on long reliable series of predictors and predictands. Giorgi et al. (2001) provide a survey of statistical downscaling studies with emphasis on studies published between 1995 and 2000.

4.6.1. Approach

4.6.1.1. Predictands

Although mean temperature and precipitation (seasonal, monthly, or daily) are the most commonly used local predictands, statistical downscaling has also been applied to generate local scenarios of cloud cover, daily temperature range, extreme temperatures, relative humidity, sunshine duration, snow-cover duration, and sea-level anomalies (Enke and Spekat, 1997; Heyden et al., 1996; Kaas and Frich, 1995; Martin et al., 1997; Schubert, 1998; Solman and Nuñez, 1999). Even sea ice (Omstedt and Chen, 2001), ocean salinity and oxygen concentrations (Zorita and Laine, 2000), zooplankton (Heyden et al., 1998), and phytoplankton spring blooms in a Swedish lake (Blenckner and Chen, 2003) have been used as predictands.

4.6.1.2. Predictors

The large-scale predictors should satisfy certain conditions: they should be reproduced realistically by the particular global model; they should (alone or combined) be able to account for most of the observed variations in the predictand; the statistical relationships should be physically interpretable and temporally stationary; and, when applied to a changing climate, predictors that "carry the climate change signal" should be included (Giorgi et al., 2001).

The optimal choice of predictors depends upon the predictand. For downscaling local temperature, large-scale fields of geopotential height or air temperature might be used as the "signal-carrying" predictors (e.g., Huth, 1999). For precipitation, absolute or specific humidity may be used (Crane and Hewitson, 1998; Hellström et al., 2001; Wilby and Wigley, 2000). In maritime regions, air temperature can sometimes serve as a proxy for humidity (Wilby and Wigley, 2000). In addition, some indicator of atmospheric circulation (e.g., sea-level pressure or a geopotential height field) is usually included (e.g., Chen and Chen, 2003).

4.6.1.3. Methods

Surveys of methods for establishing links between large-scale predictors and local predictands are provided by Hewitson and Crane (1996), Zorita and von Storch (1997), Wilby and Wigley (1997), Xu (1999), Giorgi et al. (2001), and Mearns et al. (2001). The choice of method should depend on predictand, time resolution, and also on the application of the scenario. Linear methods such as canonical correlation analysis (CCA), singular value decomposition (SVD), and multiple linear regression analysis (MLR) can, in most cases, be used to generate scenarios of monthly or seasonal values (e.g., Busuioc et al., 1999; Corte-Real et al., 1995; Huth and Kysely, 2000; Sailor and Li, 1999). To generate scenarios for variables such as daily precipitation, however, nonlinear techniques such as weather classification (e.g.,

Conway and Jones, 1998; Enke and Spekat, 1997; Goodess and Palutikof, 1998, Palutikof et al., 2002; Schnur and Lettenmaier, 1997), neural nets (Cavazoz, 1999; Clair and Ehrman, 1998; Crane and Hewitson, 1998; Schoof and Pryor, 2001), or analogues (Zorita and von Storch, 1999) are most useful. Weather generators (e.g., Semenov and Barrow, 1997; Semenov et al., 1998; Wilby et al., 1998; Wilks, 1999) can also be applied for generating scenarios with daily resolution, starting from monthly climate-change scenarios generated by one of the above methods.

4.6.1.4. Comparison of statistical downscaling and regional modeling

Several studies have compared results from statistical and regional modeling (Cubasch et al., 1996; Hellström et al., 2001; Kidson and Thompson, 1998; Murphy, 1999, 2000). The main impression from these studies is that results from the two downscaling techniques are usually quite similar for present-day climate, while differences in future climate projections are found more frequently. These differences can, to a large degree, be explained by the unwise choice of predictors in the statistical downscaling, for example, predictors that carry the climate signal (Murphy, 2000). It has also been suggested that results from statistical downscaling may be misleading because the projected climate change exceeds the range of data used to develop the model (Mearns et al., 2001). However, differences between results from statistical downscaling and regional modeling may also result from the ability of statistical downscaling to reproduce local features that are not resolved in the regional models (Hanssen-Bauer et al., 2003).

Some disadvantages of statistical downscaling versus regional modeling are as follows.

- The major weakness of statistical downscaling is the assumption that observed links between large-scale predictors and local predictands will persist in a changed climate.
- A problem when applying statistical downscaling techniques to daily values is that the observed autocorrelation between the weather at consecutive time steps is not necessarily reproduced. If it is essential to reproduce this, a suitable method (e.g., weather generators; Katz and Parlange, 1996; Wilks, 1999) should be used.
- Statistical downscaling does not necessarily reproduce a physically sound relationship between different climate elements. Using a downscaling method based on weather classification for several predictands (e.g., Enke and Spekat, 1997) can minimize this problem.
- Successful statistical downscaling depends on long, reliable observational series of predictors and predictands.

Some advantages of statistical downscaling versus regional modeling are as follows.

- Statistical downscaling is less technically demanding than regional modeling. It is thus possible to downscale from several GCMs and several different emissions scenarios relatively quickly and inexpensively (Benestad, 2002).
- It is possible to tailor scenarios for specific localities, scales, and problems. The spatial resolution applied in regional climate modeling is still too coarse for many impact studies, and some variables are either not available or not realistically reproduced by regional models. For example, Omstedt and Chen (2001) applied statistical downscaling to infer sea-ice extent in the Baltic Sea.
- In most cases, the development of statistical downscaling models includes an evaluation of AOGCM performance in simulating the climate of a specific region (Busuioc et al., 1999, 2001a). Methods applied in statistical downscaling have been used to evaluate large-scale fields of single variables as well as the links between different fields (Busuioc et al., 2001b; Hanssen-Bauer and Førland, 2001; Wilby and Wigley, 2000). The stability of these links under global change has also been investigated (e.g., Chen and Hellström, 1999). Such analyses can indicate which variables serve as the best predictors.

4.6.2. Statistical downscaling of AOGCM climate change projections in the Arctic

A survey of statistical downscaling studies up to 2000 is provided by the IPCC (Giorgi et al., 2001). However, very few of these studies considered arctic sites. During the 1990s, a few statistical downscaling studies in the Atlantic/European sector of the Arctic were performed at the Danish Meteorological Institute. Kaas and Frich (1995) downscaled monthly means of diurnal temperature range (DTR) and cloud cover at ten synoptic stations from Greenland in the west to Finland in the east. The 500 hPa height and the 500/1000 hPa thickness anomaly fields were used as predictors in an MLR-based model. The model predictors were taken from the final 30 years of a control simulation of the 20th century and the final 30 years of a scenario "A" (business as usual) simulation of the 21st century generated by the ECHAM1 model (Cubasch et al., 1992). Kaas and Frich (1995) found that statistically significant negative trends in DTR were projected for Fennoscandia, especially in central and eastern areas, and especially during winter. For Greenland and Iceland, only minor trends in DTR were projected. Positive trends in cloud cover were projected over most of the area; these were most significant in northeastern areas of Fennoscandia.

In Canada, artificial neural networks (ANNs) have been applied to model hydrological variables (Clair and Ehrman, 1998; Clair et al., 1998; Ehrman et al., 2000). Clair et al. (1998) present a scenario for changed runoff from different Canadian ecozones (including arctic areas) under conditions of doubled atmospheric CO_2

concentrations. A doubled-CO_2 equilibrium climate change scenario produced by the Canadian climate model CCC (Boer et al., 1992) was used to generate scenarios for individual basins. Temperature and precipitation scenarios were fed into the ANN to produce scenarios of changes in runoff. In the arctic ecozones that were investigated in the study, the projected changes in annual runoff were between 0 and +10%, the spring melt advanced by between a couple of weeks and one month, and there was a tendency for reduced runoff during summer. Qualitatively similar findings are reported in section 6.8, where annual discharge from various North American rivers to the Arctic Ocean is projected to increase by 10 to 25% during the 21st century.

Recently, most of the statistical downscaling studies for the Arctic have been performed for the European sector as part of the Norwegian Regional Climate Development Under Global Warming (RegClim) project or the Swedish Regional Climate Modelling Programme (SWECLIM), both of which use regional modeling and statistical downscaling to generate climate scenarios. Statistical downscaling using results from several global models and for various emissions scenarios has been completed (Benestad, 2002, 2004; Chen et al., 2001). The primary global model used in the RegClim project was ECHAM4/OPYC3 (Roeckner et al., 1996, 1999), forced with the IS92a emissions scenario. The primary

case studied was GSDIO, which included changes in GHGs, tropospheric ozone, and direct as well as indirect sulfur aerosol forcing (Roeckner et al., 1999). In SWECLIM, the main global models were HadCM2 (Johns, 1996; Johns et al., 1997) and ECHAM4/OPYC3. Hanssen-Bauer and Førland (2001) evaluated the ECHAM4/OPYC3 simulation of present-day climate over Norway and Svalbard, while Räisänen and Döscher (1999) evaluated the HadCM2 simulation of the present-day climate of northern Europe.

Sea-level pressure has proven to be a good indicator for the Scandinavian climate (Busuioc et al., 2001b; Chen, 2000; Chen and Hellström, 1999; Hanssen-Bauer and Førland, 2001), and was therefore used as a large-scale predictor in the statistical downscaling models. Depending on predictands, additional predictors included air temperature at 2 m height, humidity, and precipitation. The models were developed by linear techniques: CCA, SVD, or MLR. Monthly precipitation and temperatures at selected stations were the main predictands (Benestad, 2002, 2004; Busuioc et al., 2001a; Chen and Chen, 1999; Hanssen-Bauer et al., 2003; Hellström et al., 2001). Although not used for scenario estimation, some non-standard climate variables such as annual maximum sea ice extent over the Baltic Sea (Chen and Li, 2004; Omstedt and Chen, 2001), sea level near Stockholm (Chen and Omstedt, 2002), and spring phytoplankton blooms in a Swedish lake (Blenckner and Chen, 2003) were also linked to atmospheric circulation and may thus be projected by statistical downscaling.

The statistically downscaled temperature scenario based upon the GSDIO integration projects increases in mean annual temperature of 0.2 to 0.5 °C per decade in Norway up to 2050 (Hanssen-Bauer et al., 2003), and 0.6 °C per decade in Svalbard (Hanssen-Bauer, 2002). Cumulative frequency (relative number of years that

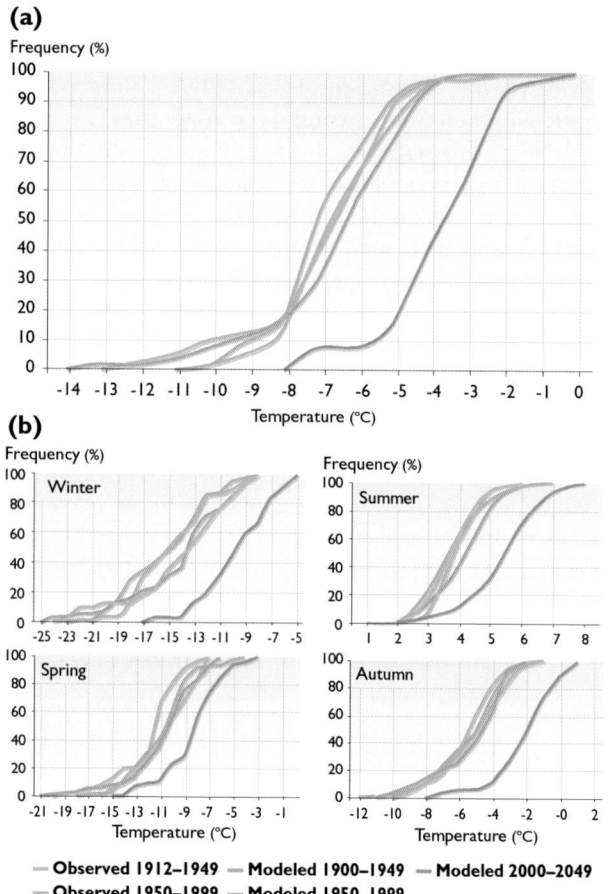

Fig. 4.32. Cumulative frequency plots of observed and modeled (a) mean annual temperature and (b) mean seasonal temperature at Svalbard Airport for various time slices (Hanssen-Bauer, 2002).

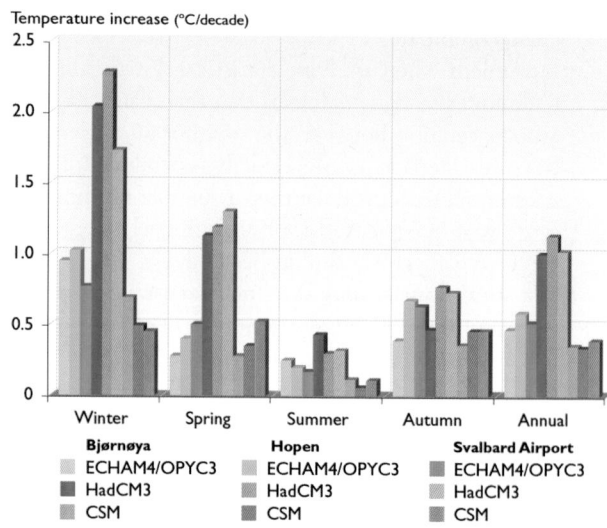

Fig. 4.33. Projections of seasonal and annual temperature increases between 1981–2000 and 2030–2050 for the arctic stations Bjørnøya, Hopen, and Svalbard Airport, based on statistical downscaling of three different general circulation models: ECHAM4/OPYC3, HadCM3, and CSM (Benestad et al., 2002).

temperatures go below a certain threshold given as a function of the threshold value) plots of annual and seasonal mean temperatures based on this study (Fig. 4.32) show good correspondence between observations and model output for time slices within the 20th century. The smallest warming rates were simulated in southern Norway along the coast; the rates increase when moving inland and northward. Along the coast of southern Norway, the modeled warming rates are similar in all seasons. Farther north and in the interior, considerably larger warming rates are projected for winter than for summer. Comparing these results to the results from dynamic downscaling of the same integration (Bjørge et al., 2000) shows only minor differences in summer and autumn. In winter and spring, on the other hand, the statistical downscaling projects greater warming rates for locations that are exposed to temperature inversions. Hanssen-Bauer et al. (2003) argue that it is reasonable to expect weaker winter inversions in the future, and that greater winter warming rates should be expected in valleys compared to mountains.

Benestad (2002, 2004) shows that statistical downscaling from several different climate models gives different local warming rates over Fennoscandia. Still, the temperature signal is robust in some respects: all models simulate warming; the warming is larger inland than along the coast; and the seasonal patterns are similar. A comparison of statistically downscaled temperature scenarios for Svalbard based on the GSDIO integration and HadCM3, both forced with the IS92a emissions scenario, and NCAR's CSM forced with a 1% per year increase in atmospheric CO_2 concentration, revealed differences (Fig. 4.33) that to a large degree can be explained in terms of different descriptions of sea-ice extent (Benestad et al., 2002). The HadCM3 model, which projects significantly stronger warming in this area than the other simulations, projects a substantial retreat of sea ice in the Barents Sea. The CSM model, which projects the most moderate warming rates, shows no melting of sea ice this area. Local temperature sce-

narios in the high Arctic are thus closely related to the projected changes in regional sea-ice cover. If the AOGCM fails to reproduce either the present sea-ice border or future melting in the region, the local temperature projections will be suspect.

Hellström et al. (2001) compared two dynamically and statistically downscaled precipitation scenarios for Sweden. The precipitation climates of the GCMs, dynamic models (i.e., RCMs), and statistical models from the control runs were also compared with respect to their ability to reproduce the observed seasonal cycle (Fig. 4.34). Improvements in the representation of the seasonal cycle by the downscaling models compared to the GCMs significantly increase the credibility of the downscaling models.

Chen et al. (2001) applied statistically downscaled scenarios from 17 CMIP2 AOGCMs to quantify AOGCM-related uncertainty in the estimation of precipitation scenarios. The result shows that there is an overall projected increase in annual precipitation over the 21st century throughout Sweden. The projected increase is greater in northern than southern Sweden. The precipitation in autumn, winter, and spring is projected to increase throughout the country, whereas decreasing summer precipitation is projected for the southern part of the country. The estimates for winter have a higher level of confidence than the estimates for summer. A statistically downscaled precipitation scenario based upon the GSDIO integration (Hanssen-Bauer et al., 2003) also projects increased annual precipitation in Norway. The projected rates of increase are smallest in southeastern Norway, where they are not statistically significant, and greatest along the northwestern and western coast, where they are highly significant (Fig. 4.35). In winter and autumn, statistically significant positive trends are projected for most of Norway, while most of the modeled changes in spring and summer precipitation are not statistically significant.

To date, statistical downscaling has primarily concentrated on monthly and annual scales. However, Linderson et al. (2004) developed downscaling models for daily statistics based upon monthly precipitation values for southern Sweden. Future scenarios of selected daily precipitation statistics were downscaled from a GCM developed at the Canadian Centre for Climate Modelling and Analysis (CGCM1; e.g., Flato et al., 2000). The downscaling models use large-scale precipitation, relative humidity, and circulation indices as predictors. The models are skillful in reproducing the variability of mean precipitation and the frequency of days with no precipitation, but less skillful concerning extremes and statistics of days with precipitation. By the time that atmospheric CO_2 doubles in the model, the CGCM1 projects an increase of 10% in annual mean precipitation (statistically significant at the 95% level), and an insignificant reduction in the annual frequency of days with precipitation of 1.4%. An increase in precipitation intensity is projected throughout most of the year, especially during

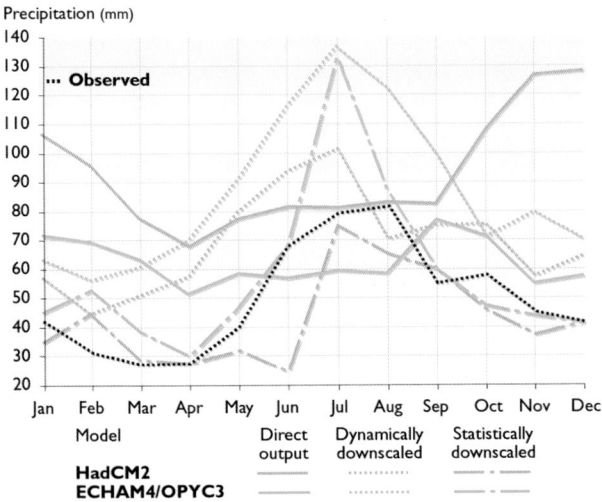

Fig. 4.34. Seasonal cycle of observed and modeled control period (1921–1950) precipitation for Kvikkjokk in northern Sweden (Hellström et al., 2001).

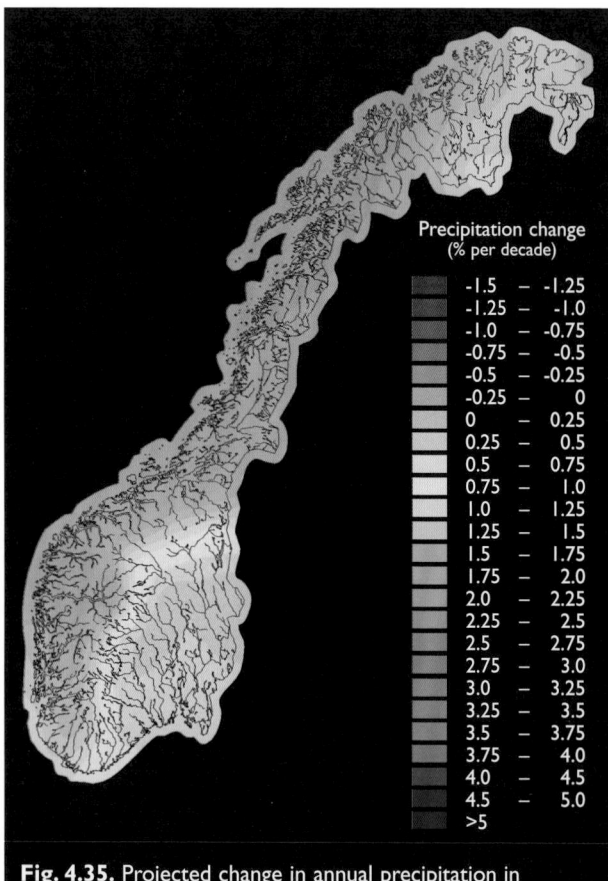

Fig. 4.35. Projected change in annual precipitation in Norway between 1961–1990 and 2031–2050 based on statistically downscaled output from the ECHAM4/OPYC3 GSDIO scenario (Hanssen-Bauer et al., 2003).

winter. The major changes include a substantial increase in winter precipitation, a delay in the timing of the summer maximum, and prolonged duration of the winter maximum relative to present-day climate. This indicates a more maritime precipitation climate in the scenario climate compared to the control climate.

Statistically downscaled climate scenarios for the North American and Russian parts of the Arctic have not yet been published. However, the Canadian Climate Impacts Scenarios (CCIS) Project has validated a downscaling tool, the Statistical DownScaling Model (SDSM; Wilby et al., 2002). The CCIS Project is providing predictor variable data to support the SDSM.

4.6.3. Summary

Although only a few statistical downscaling studies have been performed for arctic localities so far, results from the available studies indicate that these methods are able to resolve projected changes in temperature gradients between the coast and inland, and changes in valley temperature inversions. They also generate a more realistic representation of the annual precipitation cycle than the AOGCMs. If careful attention is given to the choice of predictors, scenarios from statistical downscaling and regional modeling seem to be consistent. However, statistical downscaling does not necessarily reproduce a physically based relationship between dif-

ferent climate elements. Conversely, the ability to downscale from several models and integrations is useful for assessing uncertainty. Results from downscaling temperature scenarios from several models underscore that the local temperature projections in the high Arctic depend on how well changes in sea-ice extent are represented in the AOGCMs.

4.7. Outlook for improving climate change projections for the Arctic

To provide more reliable climate change scenarios for the Arctic, several aspects of numerical climate models need further development. The most challenging aspects of model development are the physical parameterization schemes: much of the uncertainty in arctic climate change projections can be attributed to an insufficient knowledge of many of the physical processes active in the arctic domain. There is also substantial natural variability in the arctic climate system and this part of the uncertainty cannot be eliminated simply by model development and a refinement of the descriptions of physical processes. The large-scale flow is dominated by variability patterns such as the AO and the NAO (see section 2.2.2). In climate change simulations, both the frequency and nature of these flow patterns may be altered. To assess changes in the flow patterns, there must be a greater focus on the climate predictability problem to probe the inevitable natural uncertainty through a systematic search in probability space. To do this, ensemble projections are required where both initial states and uncertain model parameters are varied within a realistic range associated with a probability distribution. The development of more sophisticated physical parameterization schemes and the introduction of ensemble climate-change scenarios will both require considerable computing resources.

Historically, physical parameterization schemes have primarily been based on process descriptions and measurements from mid- and low latitudes (e.g., Randall et al., 1998). Assuming that the same physical processes are relevant to the Arctic, the developments have "propagated" from lower to higher latitudes in global models, and from AOGCMs to RCMs. In recent years, the Arctic has received particular attention from the climate modeling community, motivated by the strong arctic response to an increased GHG forcing in climate models. This has been demonstrated in the northern high latitudes along with a tremendous inter-model scatter, both in sensitivity to the forcing and in simulating the observed climate in this region. In particular, the amplification of global-model systematic errors in regional arctic models presents a serious challenge to future regional model developments.

In this section, research and model development priorities are summarized, aimed at an improvement of AOGCM performance in the Arctic and, particularly, at an increase in the credibility of AOGCM-based projections of future climate.

4.7.1. The Arctic part of the climate system – a key focus in developing AOGCMs

Surface air temperature and precipitation are variables of central interest from the viewpoint of AOGCM-based climate change scenarios. The level of confidence that can be placed on the projected changes depends on the accuracy and adequacy of the representations of many physical processes, particularly boundary-layer fluxes of heat, moisture, and momentum; clouds; and radiative fluxes.

Sea ice plays a dominant role in determining the intensity of these fluxes in the Arctic and to a large extent determines the climate sensitivity of the Arctic, in particular to GHG forcing. Description of sea ice is thus of central importance in the arctic climate system, and there is a considerable scope for improvement of the sea-ice components of current AOGCMs. More sophisticated treatments of sea-ice dynamics and thermodynamics can be included – up to the level of stand-alone regional Arctic Ocean/sea-ice models. However, even in the most comprehensive present-day sea-ice models, some important processes are not properly represented, including heat distribution between concurrent lateral and vertical melt or growth of the ice and convective processes inherent in sea ice (melt-pond and brine convection).

Improvements in the performance of AOGCM sea-ice components are hampered by errors in the forcing fields that determine sea-ice distribution. For example, the systematic bias in the arctic surface atmospheric pressure and the associated bias in the wind forcing of sea-ice as simulated by atmospheric components of AOGCMs (and stand-alone AGCMs) prevents even sea-ice models with advanced dynamics from properly simulating spatial distributions of sea ice (Bitz et al., 2002). The causes of the atmospheric pressure biases are not clear. Possible linkages include topographic (resolution) effects on atmospheric dynamics, lower boundary fluxes, as well as atmospheric chemistry and dynamics of the upper atmosphere (Walsh, in press).

The atmospheric boundary layer in the Arctic is poorly represented in current AOGCMs. It is unlikely that the representation can be improved just by increasing model vertical resolution. Insufficient understanding of the physics of the atmospheric boundary layer in the Arctic and the inappropriate parameterizations used in the current generation of AOGCMs call for further research in this field. To a certain extent, the same can be said about radiative transfer parameterizations, which should account for specific features of the arctic atmosphere and the underlying surface, including both the vertical and the horizontal heterogeneity of this complex system (Randall et al., 1998).

From a global perspective, clouds have been identified as the most serious source of uncertainty in present-day climate models (McAvaney et al., 2001). This is also true for the Arctic. In particular, the multilayer arctic clouds with their specific complexities associated with mixed phases and low temperatures need to be represented better. Other uncertain aspects of clouds involve the radiative properties of ice crystals and the concentration of various types of crystals, which have very different properties with respect to their interaction with electromagnetic radiation.

Present climate-change modeling efforts largely focus upon effects in the atmosphere, including effects on air temperatures and precipitation. Modeling potential climate change in the marine Arctic has received less attention, although changes in the thermohaline circulation have been extensively studied, primarily with low resolution, uncoupled models. Due to the lack of coordination among modeling studies, few definitive projections can be made about changes to such variables as Arctic Ocean temperatures and salinities, stratification, and circulation (including the thermohaline circulation). In light of this, future modeling efforts should attempt to more fully address changes in the ocean. This will require better resolution in the ocean models and improved coupling between the dynamic atmosphere and dynamic ocean components, particularly in the presence of sea ice.

The freshwater budget of the Arctic Ocean (and its possible link to the intermittency of North Atlantic deep-water formation) is affected by the hydrological cycle not only in the region, but also far beyond it, including the vast terrestrial watersheds of the Arctic Ocean. For satisfactory simulations, river discharge into the Arctic Ocean needs to be properly represented in order to maintain the observed stratification and sea-ice distribution and transport. It is not clear whether the simple river discharge schemes used in current AOGCMs are sufficient, although it appears that incorporating more comprehensive river routing schemes will help ensure proper seasonality of the discharge and result in an improvement in the representation of Arctic Ocean general circulation. Accounting for the freshwater influx into the ocean from glaciers and the Greenland Ice Sheet will require more advanced parameterizations than those employed today and, ideally, require introducing dynamics into the ice-sheet components.

Processes and feedbacks associated with vegetation may also play an important role in the terrestrial Arctic, affecting heat, water, and momentum fluxes. The effects of vegetation on terrestrial snow cover and surface albedo, evapotranspiration processes, and the possible expansion of boreal forests into regions currently occupied by tundra are among many processes that may potentially be crucial in the context of climate change. Developing comprehensive interactive dynamic vegetation components of AOGCMs should eventually increase confidence in AOGCM-based projections of future climate.

Climatic changes of special concern for indigenous communities include weather variability and predictability;

the extent, thickness, and quality of sea ice; the extent, duration, and hardness of snow cover; freeze-thaw cycles (particularly in autumn, when a layer of ice on the ground may be produced and last all winter, blocking access to forage for grazing animals); sudden changes in wind direction; and changes in the strength and frequency of winds and storms (Chapter 3). While most of these quantities are either directly available, or easily derivable from standard model outputs, representation of a few of these variables will require additional efforts from the modeling community in the future.

4.7.2. Improved resolution of arctic processes

To model climatically important processes in the Arctic, models with a high spatial resolution are required. To achieve this with present-day computing resources, regional models are required. In the future, global models may also have adequate resolution, but for the foreseeable future regional models will be required to complement the global simulations, because their results are closer to actual local climatic conditions and can more easily be translated into impacts than global model results.

When nested within a GCM, the large-scale circulation is imposed by the lateral boundaries of the RCM. Regional models are not able to, nor intended to, correct large-scale errors made by the global model from which conditions are drawn. The role of the regional model is to add regional detail and fine spatial and temporal scales to the simulation, not to improve the large-scale simulation. An alternative to regional models is presented by the evolving global variable-resolution stretched-grid approach that provides additional spatial detail over a region of interest (e.g., Giorgi et al., 2001). This technique allows for a feedback to the global scale from the region with high resolution. While this may seem appealing at first, it raises the question of whether this feedback is preferable when similar feedbacks from other regions are represented at a lower resolution.

The need for high resolution is not restricted to the atmospheric model component. To simulate the coupled atmosphere–ice–ocean system in the Arctic, a high-resolution ocean component is also required. In particular, the coupling processes occur on small horizontal and vertical scales, thus a high-resolution regional coupled atmosphere–ocean model is needed. Some early versions of such coupled models already exist, but much additional development work is required.

A further increase in atmospheric resolution (to <10 km horizontally) will require the use of non-hydrostatic model equations. New parameterizations of physical processes such as cloud formation and turbulence are also necessary at these scales. With a very high resolution (<1 km), non-hydrostatic models start to resolve individual clouds, thus necessitating further changes in cloud

parameterizations. A special emphasis needs to be put on cloud microphysics, including the ice crystals and aerosols that provide nuclei for the condensation process.

The arctic climate depends on the unique high-latitude characteristics of processes such as ice dynamics and persistent low-level clouds. Simulation deficiencies are partly due to coarse model resolution and partly due to inadequate model process descriptions. As mentioned previously, most model formulations are based on low-latitude observations that do not cover the extreme conditions occurring in the Arctic. To validate coupled high-resolution models in the Arctic, improved and extended observational datasets are required. *In situ* observations exist for a few locations and restricted time periods, but more such datasets are needed. To obtain better coverage in space and time, remote sensing instruments are necessary. Several satellite missions are planned that hopefully will provide observational datasets with a much better coverage.

4.7.3. Better representation of the stratosphere in AGCMs

Most current AGCMs are aimed at simulating tropospheric processes, and the stratosphere is only included with a limited resolution. On the other hand, many of the middle atmosphere three-dimensional circulation models describe only the stratosphere and the mesosphere, having a lower boundary at the tropopause (IPCC, 2001). Such models are primarily intended to simulate processes that are internal to the stratosphere, and it is assumed that the interaction with the troposphere can be neglected.

To model current arctic climate and stratospheric and tropospheric ozone concentrations, as well as to project their future changes, AGCMs must describe the troposphere and the stratosphere in comparable detail. Most models assume that all ozone-related processes are located in the stratosphere: ozone and ozone-related species, as well as their photochemical sources and sinks and air transport in the ozone layer (20–30 km average height). However, some of the stratospheric transport features have a tropospheric origin. Two important processes in this regard are planetary wave propagation in the northern mid-latitudes and gravitational wave destruction in the middle and upper stratosphere. While the planetary waves are well-resolved by climate models, gravity-wave drag occurs at small scales and is therefore difficult to simulate with the coarse grids of most current AGCMs. In many AGCMs, these dynamic factors are roughly parameterized by Rayleigh friction at upper model levels. This parameterization also serves the purpose of preventing spurious reflection of vertically propagating gravity waves at the upper boundary. This feature is necessary in a climate model, but in order to resolve vertically propagating waves realistically, more resolution is needed in the middle and upper stratosphere. Austin et al. (1997) demonstrated that the shift from 19 to 49 levels in an AGCM with coupled chemistry considerably improved the ozone and temperature simulation in the winter stratosphere.

The AO is another important feature affecting the stratosphere (Wallace, 2000). The AO is a naturally occurring phenomenon but difficult to project with current GCMs. A gradual increase in the AO positive phase persistence has been observed in recent years (Hoerling et al., 2001; Shindell et al., 2001). While the change may be a natural fluctuation in the AO, it may also be a result of increased atmospheric GHG concentrations. A better resolution of the stratosphere in models is required to determine whether this is the case, and, if it is, whether further increases in GHG concentrations are likely to exert a greater influence on the AO. A change in the AO would also influence the ozone distribution in the arctic stratosphere, giving rise to additional climate-relevant feedbacks in the Arctic. One example could be a change in the latitudinal heating gradient in the stratosphere caused by a change in the ozone distribution. An altered heating gradient would result in a changed temperature gradient, which in turn would change the zonal wind distribution. The zonal wind distribution determines the vertical planetary wave propagation characteristics that in turn affect ozone distribution.

4.7.4. Coupling chemical components to GCMs

Ozone is an important GHG and moderates fluxes of ultraviolet radiation at ground level. In addition to an adequate description of dynamic processes, GCMs must incorporate detailed photochemical components for better simulation of ozone formation and destruction in the atmosphere. Due to the complicated character of ozone photochemistry in the arctic stratosphere, which has significant input from heterogeneous reactions on polar stratospheric cloud (PSC) particle surfaces, the inclusion of the microphysics of particle formation and destruction must be considered. This is omitted in the photochemical components of most present-day GCMs. Instead, the observed spectra of PSC particles is assumed to appear immediately when the air temperature drops below a certain threshold temperature and to disappear at once when the temperature rises above the threshold. The actual delay in the observed PSC effects compared to those modeled indicates the importance of considering PSC microphysics in models, especially in the simulation of arctic ozone "mini-holes" and their rapid evolution in space and time (Austin et al., 2003).

The denitrification of cold polar air in winter is another process in the microphysics of PSC formation and the chemistry that activates ozone-depleting chlorine radicals and repartitioning of bromine species. Polar stratospheric cloud particles that contain liquid nitric acid are supercooled ternary solutions of nitric acid, sulfuric acid, and water. They grow in the stratosphere to nitric acid dihydrate (NAD) and nitric acid trihydrate (NAT) large particles, which remove nitric acid from the stratosphere by gravitational sedimentation and con-

tribute to the denitrification (removal of nitric acid) of the arctic stratosphere. Both NAD and NAT particles are formed intensively at temperatures of 190 to 192 K – the "nucleation window" (Tabazadeh et al., 2001). These "window temperature" belts are persistent at the periphery of winter polar vortices in the Antarctic for several months, and in the Arctic for about a month, and produce significant denitrified stratospheric layers (Tabazadeh et al., 2001).

This phenomenon as well as the PSC microphysics and chemistry have spatial and temporal scales finer than current GCM and CTM grids can resolve. A suitable parameterization of these effects is needed in addition to an elaboration of the whole photochemical computation scheme. Together with the necessary refinement of the simulation of dynamic processes, these requirements make the problem of arctic ozone modeling computationally demanding and scientifically challenging.

Another atmospheric chemistry aspect of the Arctic is the production of cloud condensation nuclei near the surface and the possible involvement of naturally occurring dimethyl sulfide (DMS) in this process. Dimethyl sulfide particles originate from arctic seawater, and the flux to the atmosphere is thus strongly coupled to the existence of sea ice. It may be that the local arctic production of DMS is a determining factor for droplet size distributions in low clouds and thus may have a significant effect on low-cloud radiative properties. If this is the case, cloud properties would be sensitive to the occurrence of sea ice and a dramatic change in sea-ice distribution would affect the arctic radiative balance. This type of effect, as well as arctic haze effects, whose radiative forcing has been estimated from observations (e.g., Herber et al., 2002; Quinn et al., 2002), need to be included in AGCMs.

4.7.5. Ensemble simulations

A more ambitious strategy for ensemble climate simulations is needed in order to better understand natural climate variability in the Arctic and how it may be affected by global climate change. In discussing the impacts of climate change, changes in the distribution of climatic events are as interesting as changes in the mean. The ACIA used the results of five climate models to study future changes in arctic climate. In order to increase the accuracy of the different error estimates, a larger scenario sample is needed. In numerical weather prediction, experience has shown that a sample involving 50 to 100 simulations with identical models but different initial states gives a reasonable estimate of forecast uncertainties. For arctic climate change, error estimates based on a sample of this size could be adequate. In addition, it would be advantageous to increase model resolution to better capture physical processes and to better describe sharp spatial gradients (fronts), which are often the regions where extreme events occur. Both types of improvement require large additional computing resources. Further research is needed to find a reasonable balance between ensemble size, model resolution, and the complexity of physical process descriptions. For climate simulation ensembles, it is also necessary to perturb model parameters and external forcings. The uncertainty aspects to be addressed thus include natural variability, uncertainties in model sensitivity to prescribed forcings, and uncertainties in the forcings.

Estimates of extreme events and their frequency of occurrence also require ensemble simulations. For precipitation in particular, extreme events are often more interesting than changes in the mean (Palmer and Räisänen, 2002). To obtain reliable estimates of changes in the frequency of extreme events, ensemble simulations are necessary. This is thus an added benefit of climate-change projection ensembles and is required to make projections of changes in storm frequencies or other extreme events. It has recently been shown that an increased GHG forcing could contribute to an increase in intense storm events in particular areas of the North Atlantic Ocean and Western Europe (Van den Brink et al., 2003). At the same time, a decrease in storm frequencies is projected for other regions of the North Atlantic. To arrive at this result, a very large ensemble was used, and in order to achieve that, a simplified atmospheric model (quasi-geostrophic, three vertical levels, and a coarse horizontal resolution) had to be utilized. The drawback of using such a simplified model is that storm dynamics are not described in full detail and storm characteristics have to be derived from empirically based, statistical methods similar to the downscaling technique discussed in section 4.6. Other studies using more advanced model tools but smaller ensembles have not been able to simulate significant increases in storm frequencies (Carnell and Senior, 1998; Knippetz et al., 2000; Lunkeit et al., 1996; Ulbrich and Christoph, 1999).

4.7.6. Conclusions

The general increase in computing resources that have become available for climate system modeling in recent years favors progress in developing new generations of AOGCMs – mostly by adding new components, increasing resolution, and extending ensembles of simulations. Conversely, the Arctic is one of the regions of the world with limited availability of observational data necessary for model validation and evaluation (e.g., Walsh, in press). Nevertheless, it has been shown that model performance can be improved with systematic model improvements and better resolution. For the Arctic, it is necessary to perform climate change simulations involving the entire globe; however, spatial resolution in the Arctic could be improved with the use of regional models driven by global simulations. The ultimate goal is to use as high a resolution as possible over the entire globe. In simulating arctic climate change, sea-ice processes are of primary importance. Boundary-layer fluxes and clouds are closely linked with sea-ice processes. All components require improvements to increase confidence in climate change projections.

Expectations for model improvement are increasing because of increasing international activity in the field of model intercomparison exercises (e.g., Puri, 2002; Box 4.2), allowing the identification of model errors, their causes, and how they may be reduced.

Some scientists doubt that AOGCMs can provide realistic scenarios of future climate change. However, even if present day models have major shortcomings and need to be improved, they still provide useful information about possible changes in the future climate. The models are based on a physical understanding of the climate system and, as such, provide a physically coherent picture of likely climate change. There are very few other methods, if any, which can be used to provide such credible climate change estimates. Statistical methods, other than simple extrapolation of present trends, require a physical model in the background to provide a basis to generate statistically representative estimates of variables that cannot be deduced directly from the physical model. The authors of this chapter are thus confident that future model improvements will provide better estimates of the arctic climate change that may occur as a result of increasing atmospheric GHG concentrations.

There will always be uncertainties in the estimates and some of these uncertainties cannot be reduced below a certain level. These include, for example, uncertainties associated with the lack of observations to provide an accurate initial state for a model simulation, model parameter uncertainties, and the inherent limited predictability of any atmospheric/oceanic simulation (Lorenz, 1963). While the level of uncertainty can be lowered, it will never be certain that all physical processes relevant to climate change have been included in a model simulation. There could still be surprises to come in the understanding of climate change. Solar variability, the effects of cosmic rays, and volcanic eruptions may all contribute more to arctic climate change than is presently thought, but this remains to be seen. As climate change science progresses there will always be new results that could significantly change understanding of how the arctic climate system works; however, the present estimates are based on the best knowledge available today about climate change.

References

Austin, J., N. Butchart and R. Swinbank, 1997. Sensitivity of ozone and temperature to vertical resolution in a GCM with coupled stratospheric chemistry. Quarterly Journal of the Royal Meteorological Society, 123:1405–1431.

Austin, J., D. Shindell, S.R. Beagley, C. Bruhl, M. Damers, E. Manzini, T. Nagashima, P. Newman, S. Pawson, G. Pitari, E. Rozanov, C. Schnadt and T.A. Shepherd, 2003. Uncertainties and assessments of chemistry-climate models of the stratosphere. Atmospheric Chemistry and Physics, 3:1–27.

Bailey, D.A. and A.H. Lynch, 2000a. Development of an Antarctic Regional Climate System Model: Part 1. Sea ice and large-scale circulation. Journal of Climate, 13:1337–1350.

Bailey, D.A. and A.H. Lynch, 2000b. Development of an Antarctic Regional Climate System Model: Part 2. Station validation and surface energy balance. Journal of Climate, 13:1351–1361.

Bailey, D.A., A.H. Lynch and K.S. Hedström, 1997. The impact of ocean circulation on regional polar climate simulations using the Arctic Region Climate System Model. Annals of Glaciology, 25:203–207.

Barthelet, P., L. Terray and S. Valcke, 1998. Transient CO$_2$ experiment using the ARPEGE/OPAICE non flux corrected coupled model. Geophysical Research Letters, 25:2277–2280.

Benestad, R., 2002. Empirically downscaled multi-model ensemble temperature and precipitation scenarios for Norway. Journal of Climate, 15:3008–3027.

Benestad, R., 2004. Tentative probabilistic temperature scenarios for northern Europe. Tellus A: Dynamic Meteorology and Oceanography, 56:89–101.

Benestad, R., E.J. Forland and I. Hanssen-Bauer, 2002. Empirically downscaled temperature scenarios for Svalbard. Atmospheric Science Letters, 3(2–4):71–93.

Bengtsson, L., V.A. Semenov and O. Johannessen, 2003. The Early Century Warming in the Arctic – A Possible Mechanism. Rep. 345. Max Planck Institute for Meteorology, Hamburg, Germany, 31pp.

Bitz, C., G. Flato and J. Fyfe, 2002. Sea ice response to wind forcing from AMIP models. Journal of Climate, 15:523–535.

Bjørge, D., J.E. Haugen and T.E. Nordeng, 2000. Future Climate in Norway. Dynamical downscaling experiments within the RegClim project. Research Report 103, Norwegian Meteorological Institute, Oslo.

Blenckner, T. and D. Chen, 2003. Comparison of the impact of regional and north-Atlantic atmospheric circulation on an aquatic ecosystem. Climatic Research, 23:131–136.

Boer, G.J., 2000a. Analysis and verification of model climate. In: P. Mote and A. O'Neill (eds.). Numerical Modeling of the Global Atmosphere in the Climate System. NATO Science Series C-550. Kluwer Academic Publishers.

Boer, G.J., 2000b. Climate model intercomparison. In: P. Mote and A. O'Neill (eds.). Numerical Modeling of the Global Atmosphere in the Climate System. NATO Science Series C-550. Kluwer Academic Publishers.

Boer, G.J., N.A. McFarlane and M. Lazare, 1992. Greenhouse gas-induced climate change simulated with the CCC second-generation general circulation model. Journal of Climate, 5:1045–1077.

Boville, B.A. and P.R. Gent, 1998. The NCAR Climate System Model, Version One. Journal of Climate, 11:1115–1130.

Boville, B.A., J.T. Kiehl, P.J. Rasch and F.O. Bryan, 2001. Improvements to the NCAR CSM-1 for transient climate simulations. Journal of Climate, 14:164–179.

Braconnot, P. (ed.), 2000. Paleoclimate Modeling Intercomparison Project (PMIP): Proceedings of the Third PMIP Workshop, La Huardière, Canada, 4–8 October 1999. ICPO Publications Series No. 34, PAGES 2000–1, WMO/TD No. 1007, WCRP Series Report No. 111, 271pp.

Braconnot, P., O. Marti and S. Joussaume, 1997. Adjustment and feedbacks in a global coupled ocean-atmosphere model. Climate Dynamics, 13:507–519.

Broecker, W.S., 1997. Thermohaline circulation, the achilles heel of our climate system: will man-made CO$_2$ upset the current balance? Science, 278:1582–1588.

Bromwich, D., J. Cassano, T. Klein, G. Heinemann, K. Hines, K. Steffen and J. Box, 2001. Mesoscale modeling of katabatic winds over Greenland with the Polar MM5. Monthly Weather Review, 129(9):2290–2309.

Bryan, K., S. Manabe and R.C. Pacanowski, 1975. A global ocean-atmosphere climate model. II: the oceanic circulation. Journal of Physical Oceanography, 5:30–46.

Bryazgin, N.N., 1976. Yearly mean precipitation in the Arctic region accounting for measurement errors. Proceedings of the Arctic and Antarctic Research Institute, 323:40–74. (In Russian)

Busuioc, A., H. von Storch and R. Schnur, 1999. Verification of GCM-generated regional seasonal precipitation for current climate and of statistical downscaling estimates under changing climate conditions. Journal of Climate, 12:258–272.

Busuioc, A., D. Chen and C. Hellström, 2001a. Performance of statistical downscaling models in GCM validation and regional climate estimates: application for Swedish precipitation. International Journal of Climatology, 21:557–578.

Busuioc, A., D. Chen and C. Hellström, 2001b. Temporal and spatial variability of precipitation in Sweden and its link with the large scale atmospheric circulation. Tellus, 53A(3):348–367.

Carnell, R.E. and C.A. Senior, 1998. Changes in mid-latitude variability due to increasing greenhouse gases and sulphate aerosols. Climate Dynamics, 14:369–383.

Carter, T.R., M.L. Parry, H. Harasawa and S. Nishioka, 1994. IPCC Technical Guidelines for Assessing Climate Change Impacts and Adaptations. Department of Geography, University College, London.

Carter, T.R., M. Hulme, J.F. Crossley, S. Malyshev, M.G. New, M.E. Schlesinger and H. Tuomenvirta, 2000. Climate Change in the 21st Century – Interim Characterizations based on the New IPCC Emissions Scenarios. The Finnish Environment 433, Finnish Environment Institute, Helsinki, 148pp.

Carter, T.R., E.L. La Rovere, R.N. Jones, R. Leemans, L.O. Mearns, N. Nakićenović, A.B. Pittock, S.M. Semenov and J. Skea, 2001. Developing and applying scenarios. In: J.J. McCarthy, O.F. Canziani, N.A. Leary, D.J. Dokken and K.S. White (eds.). Climate Change 2001: Impacts, Adaptation, and Vulnerability. Pp. 145–190. Contribution of Working Group II to the Third Assessment Report of the Intergovernmental Panel on Climate Change. Cambridge University Press.

Cassano, J.J., T.R. Parish and J. King, 2001. Evaluation of turbulent surface flux parameterizations for the stable boundary layer over Halley, Antarctica. Monthly Weather Review, 129:26–46.

Cavazos, T., 1999: Large-scale circulation anomalies conductive to extreme precipitation events and derivation of daily rainfall in northeastern Mexico and southeastern Texas. Journal of Climate, 12:1506–1523.

Chen, D., 2000. A monthly circulation climatology for Sweden and its application to a winter temperature case study. International Journal of Climatology, 20:1067–1076.

Chen, D. and Y. Chen, 1999. Development and verification of a multiple regression downscaling model for monthly temperature in Sweden. In: D. Chen, C. Hellström, Y. Chen (eds.). Preliminary Analysis and Statistical Downscaling of Monthly Temperature in Sweden, pp. 41–55. Report C16, Earth Sciences Centre, University of Gothenburg, Sweden.

Chen, D., and Y. Chen, 2003. Association between winter temperature in China and upper air circulation over East Asia revealed by Canonical Correlation Analysis. Global and Planetary Change, 37:315–325.

Chen, D. and C. Hellström, 1999. The influence of the North Atlantic Oscillation on the regional temperature variability in Sweden: spatial and temporal variations. Tellus, 51A(4):505–516.

Chen, D. and X. Li, 2004. Scale dependent relationship between maximum ice extent in the Baltic Sea and atmospheric circulation. Global and Planetary Change, 41:275–283.

Chen, D. and A. Omstedt, 2002. Using statistical downscaling to quantify the GCM-related uncertainty in regional climate change scenarios: A case study of Swedish precipitation. SWECLIM Newsletter, Swedish Meteorological and Hydrological Institute, Norrköping, Sweden.

Chen, D., C. Achberger, J. Räisänen and C. Hellström, 2001. Using statistical downscaling to quantify the GCM-related uncertainty in regional climate change scenarios: A case study of Swedish precipitation. SWECLIM Newsletter, Swedish Meteorological and Hydrological Institute, Norrköping, Sweden.

Christensen, J.H. and O.B. Christensen, 2003. Severe summer flooding in Europe. Nature, 421:805–806.

Christensen, J.H. and P. Kuhry, 2000. High resolution regional climate model validation and permafrost simulation for the East-European Russian Arctic. Journal of Geophysical Research, 105:29647–29658.

Christensen, J.H., O.B. Christensen, P. Lopez, E. van Meijgaard and M. Botzet, 1996. The HIRHAM4 Regional Atmospheric Climate Model. Danish Meteorological Institute, Scientific Report 96-4.

Christensen, J.H., B. Machenhauer, R.G. Jones, C. Schär, P.M. Ruti, M. Castro and G. Visconti, 1997. Validation of present-day regional climate simulations over Europe: LAM simulations with observed boundary conditions. Climate Dynamics, 13:489–506.

Christensen, J.H., T.R. Carter and F. Giorgi, 2002. PRUDENCE employs new methods to assess European climate change. Eos, Transactions, American Geophysical Union, 83(13):147.

Christensen, O.B., 1999. Relaxation of soil variables in a Regional Climate Model. Tellus, 51A:674–685.

Clair, T.A. and J. Ehrman, 1998. Using neural networks to assess the influence of changing seasonal climates in modifying discharge, dissolved organic carbon, and nitrogen export in eastern Canadian rivers. Water Resources Research, 34(3):447–455.

Clair, T.A., J. Ehrman and K. Higuchi, 1998. Changes to the runoff of Canadian ecozones under a doubled CO$_2$ atmosphere. Canadian Journal of Fisheries and Aquatic Sciences, 55(11):2464–2477.

Claussen, M., L.A. Mysak, A.J. Weaver, M. Crucifix, T. Fichefet, M.-F. Loutre, S.L. Weber, J. Alcamo, V.A. Alexeev, A. Berger, R. Calov, A. Ganopolski, H. Goosse, G. Lohmann, F. Lunkeit, I.I. Mokhov, V. Petoukhov, P. Stone and Z. Wang, 2002. Earth system models of intermediate complexity: closing the gap in the spectrum of climate system models. Climate Dynamics, 18:579–586.

Conway, D. and P.D. Jones, 1998. The use of weather types and air flow indices for GCM downscaling. Journal of Hydrology, 212:348–361.

Corte-Real, J., X. Zhang and X. Wang, 1995. Large-scale circulation regimes and surface climate anomalies over the Mediterranean. International Journal of Climatology, 15:1135–1150.

Crane, R.G. and B.C. Hewitson, 1998. Doubled CO$_2$ precipitation changes for the Susquehanna Basin: downscaling from the genesis General Circulation Model. International Journal of Climatology, 18:65–76.

Cubasch, U., K. Hasselmann, H. Höck, E. Maier-Raimer, U. Mikolajewicz, B.D. Santer and R. Sausen, 1992. Time-dependent greenhouse warming computations with a coupled ocean-atmosphere model. Climate Dynamics, 9:55–69.

Cubasch, U., H. von Storch, J. Waszkewitz and E. Zorita, 1996. Estimates of climate change on southern Europe derived from dynamical climate model output. Climate Research, 7:129–149.

Cubasch, U., G.A. Meehl, G.J. Boer, R.J. Stouffer, M. Dix, A. Noda, C.A. Senior, S. Raper and K.S. Yap, 2001. Projections of future climate change. In: J.T. Houghton, Y. Ding, D.J. Griggs, M. Noguer, P.J. van der Linden, X. Dai, K. Maskell and C.A. Johnson (eds.). Climate Change 2001: The Scientific Basis, pp. 526–582. Contribution of Working Group I to the Third Assessment Report of the Intergovernmental Panel on Climate Change. Cambridge University Press.

Curry, J.A. and A.H. Lynch, 2002. Comparing Arctic Regional Climate Models. Eos, Transactions, American Geophysical Union, 83:87.

Curry R., B. Dickson and I. Yashayaev, 2003. A change in the freshwater balance of the Atlantic Ocean over the past four decades. Nature, 426:826–829.

Delworth, T.L. and T.R. Knutson, 2000. Simulation of early 20th century global warming. Science, 287:2246–2250.

Delworth, T.L., R.J. Stouffer, K.W. Dixon, M.J. Spelman, T.R. Knutson, A.J. Broccoli, P.J. Kushner and R.T. Wetherald, 2002. Simulation of climate variability and change by the GFDL R30 coupled climate model. Climate Dynamics, 19(7):555–574.

Denis, B., R. Laprise and D. Cava, 2002. Downscaling ability of one-way nested regional climate models: the big-brother experiment. Climate Dynamics, 18:627–646.

Dethloff, K., A. Rinke, R. Lehmann, J.H. Christensen, M. Botzet and B. Machenhauer, 1996. A regional climate model of the Arctic atmosphere. Journal of Geophysical Research, 101:23401–23422.

Dethloff, K., C. Abegg, A. Rinke, I. Hebestadt and V.F. Romanov, 2001. Sensitivity of Arctic climate simulations to different boundary-layer parameterizations in a regional climate model. Tellus, 53A:1–26.

Dethloff, K., M. Schwager, J.H. Christensen, S. Kiilsholm, A. Rinke, W. Dorn, F. Jung-Rothenhäusler, H. Fischer, S. Kipfstuhl and H. Miller, 2002. Greenland precipitation from ice core estimates and regional climate model simulations. Journal of Climate, 15:2821–2832.

Dickson, B., I. Yashayaev, J. Meincke, B. Turrell, S. Dye and J. Holfort, 2002. Rapid freshening of the deep North Atlantic Ocean over the past four decades. Nature, 416:832–837.

Dorn, W., K. Dethloff, A. Rinke and E. Roeckner, 2003. Competition of NAO regime changes and increasing greenhouse gases and aerosols with respect to Arctic climate projections. Climate Dynamics, 21(5–6): 447–458.

Ehrman, J., K. Higuchi and T.A. Clair, 2000. Backcasting to test the use of neural networks for predicting runoff in Canadian rivers. Canadian Water Resources Journal, 25(3):279–291.

Emori, S., T. Nozawa, A. Abe-Ouchi, A. Numaguti, M. Kimoto and T. Nakajima, 1999. Coupled ocean-atmosphere model experiments of future climate change with an explicit representation of sulfate aerosol scattering. Journal of the Meteorological Society of Japan, 77:1299–1307.

Enke, W. and A. Spekat, 1997. Downscaling climate model outputs into local and regional weather elements by classification and regression. Climate Research, 8:195–207.

Etchevers, P., E. Martin, R. Brown, C. Fierz, Y. Lejeune, E. Bazile, A. Boon, Y.-J. Dai, R. Essery, A. Fernandez, Y. Gusev, R. Jordan, V. Koren, E. Kowalczyck, R. Nasonova, D. Pyles, A. Schlosser, A. Shmakin, T.G. Smirnova, U. Strasser, D. Verseghy, T. Yamazaki and Z.-L. Yang, 2002. SnowMIP, an intercomparison of snow models: first results. In: Proceedings of the International Snow Science Workshop, Penticton, Canada, 29 September – 4 October 2002, 8pp.

Flato, G.M. and G.J. Boer, 2001. Warming asymmetry in climate change experiments. Geophysical Research Letters, 28:195–198.

Flato, G.M., G.J. Boer, W.G. Lee, N.A. McFarlane, D. Ramsden, M.C. Reader and A.J. Weaver, 2000. The Canadian Centre for Climate Modelling and Analysis global coupled model and its climate. Climate Dynamics, 16:451–467.

Folland, C., J. Shukla, J. Kinter and M. Rodwell, 2002. The climate of the twentieth century project. Exchanges (Newsletter of CLIVAR), 7(2):37–39.

Frei, A., J.A. Miller and D.A. Robinson, 2003a. Improved simulations of snow extent in the second phase of the Atmospheric Model Intercomparison Project (AMIP-2). Journal of Geophysical Research, 108:(D12), 4369, doi:10.1029/2002JD003030.

Frei, C., J.H. Christensen, M. Deque, D. Jacob, R.G. Jones and P.L. Vidale, 2003b. Daily precipitation statistics in regional climate models: evaluation and intercomparison for the European Alps. Journal of Geophysical Research, 108: 10.1029/2002JD002287.

Furevik, T., M. Bentsen, H. Drange, I.K.T. Kindem, N.G. Kvamstø and A. Sorteberg, 2003. Description and validation of the Bergen Climate Model: ARPEGE coupled with MICOM. Climate Dynamics, 21:27–51.

Ganopolski, A. and S. Rahmstorf, 2001. Rapid changes of glacial climate simulated in a coupled climate model. Nature, 409:153–158.

Gates, W.L., 1992. AMIP: The Atmospheric Model Intercomparison Project. Bulletin of the American Meteorological Society, 73:1962–1970.

Gates, W.L., J.S. Boyle, C.C. Covey, C.G. Dease, C.M. Doutriaux, R.S. Drach, M. Fiorino, P.J. Gleckler, J.J. Hnilo, S.M. Marlais, T.J. Phillips, G.L. Potter, B.D. Santer, K.R. Sperber, K.E. Taylor and D.N. Williams, 1998. An overview of the results of the Atmospheric Model Intercomparison Project (AMIP). PCMDI Rep. No. 45, UCRL-JC-129928. Program for Climate Model Diagnosis and Intercomparison, Lawrence Livermore National Laboratory, Livermore, CA, 29 pp. + figs.

Gibson, J.K., P. Kållberg, S. Uppala, A. Hernandez, A. Nomura and E. Serrano, 1997. ERA description. ECMWF Reanalysis Project Rep. Series 1. European Centre for Medium Range Weather Forecasts, Reading, UK, 66pp.

Giorgi, F. and L.O. Mearns, 1999. Introduction to special section: Regional climate modeling revisited. Journal of Geophysical Research, 104:6335–6352.

Giorgi, F. and L.O. Mearns, 2003. Probability of regional climate change calculated using the Reliability Ensemble Averaging (REA) method. Geophysical Research Letters, 30(12):311–314.

Giorgi, F., B. Hewitson, J. Christensen, M. Hulme, H. von Storch, P. Whetton, R. Jones, L. Mearns and C. Fu, 2001. Regional climate information – evaluation and projections. In: J.T. Houghton, Y. Ding, D.J. Griggs, M. Noguer, P.J. van der Linden, X. Dai, K. Maskell and C.A. Johnson (eds.). pp. 583–638. Climate Change 2001: The Scientific Basis. Contribution of Working Group I to the Third Assessment Report of the Intergovernmental Panel on Climate Change. Cambridge University Press.

Goodess, C.M. and J.P. Palutikof, 1998. Development of daily rainfall scenarios for southeast Spain using a circulation-type approach to downscaling. International Journal of Climatology, 10:1051–1083.

Gordon, C., C. Cooper, C.A. Senior, H. Banks, J.M. Gregory, T.C. Johns, J.F.B. Mitchell and R.A. Wood, 2000. The simulation of SST, sea ice extents and ocean heat transports in a version of the Hadley Centre coupled model without flux adjustments. Climate Dynamics, 16:147–168.

Hagemann, S. and L. Dümenil, 1998. A parameterization of the lateral water flow for the global scale. Climate Dynamics, 14:17–31.

Hahn, C.J., S.G. Warren and J. London, 1995. The effect of moonlight on observations of cloud cover at night, and applications to cloud climatology. Journal of Climate, 8:1429–1466.

Hanna, E. and P. Valdes, 2001. Validation of ECMWF (re)analysis surface climate data, 1979–1998 for Greenland and implications for mass balance modelling of the ice sheet. International Journal of Climatology, 21(2):171–195.

Hanssen-Bauer, I., 2002. Temperature and precipitation at Svalbard 1900–2050: measurements and scenarios. Polar Record, 38(206):225–232.

Hanssen-Bauer, I. and E.J. Førland, 2001. Verification and analysis of a climate simulation of temperature and pressure fields over Norway and Svalbard. Climate Research, 16:225–235.

Hanssen-Bauer, I., E.J. Førland, J.E. Haugen and O.E. Tveito, 2003. Temperature and precipitation scenarios for Norway: Comparison of results from dynamical and empirical downscaling. Climate Research, 25:15–27.

Harding, R.J., S.-E. Gryning, S. Halldin and C.R. Lloyd, 2001. Progress in understanding of land surface/atmosphere exchanges at high latitudes. Theoretical and Applied Climatology, 70:5–18.

Haugen, J.E., D. Bjørge and T.E. Nordeng, 2000. Dynamic downscaling: further results. RegClim General Technical Report, No.4. Available from the Norwegian Institute for Air Research.

Hellström, C., D. Chen, C. Achberger and J. Räisänen, 2001. Comparison of climate change scenarios for Sweden based on statistical and dynamical downscaling of monthly precipitation. Climate Research, 19:45–55.

Herber, A., L. Thomason, H. Gernandt, U. Leiterer, D. Nagel, K.-H. Schulz, J. Kaptur, T. Albrecht and J. Notholt, 2002. Continuous day and night aerosol optical depth observations in the Arctic between 1991 and 1999. Journal of Geophysical Research, 107(D10):4097, doi: 10.1029/2001JD000536.

Hewitson, B.C. and R.G. Crane, 1996. Climate downscaling: techniques and application. Climate Research, 7:85–95.

Heyden, H., E. Zorita and H. Von Storch, 1996. Statistical downscaling of monthly mean North Atlantic air-pressure to sea level anomalies in the Baltic Sea. Tellus, 48A:312–323.

Heyden, H., H. Fock and W. Greve, 1998. Detecting relationships between interannual variability in ecological time series and climate using a multivariate statistical approach – a case study on Helgoland Road zooplankton. Climate Research, 10:179–191.

Hirst, A., S.P. O'Farrell and H.B. Gordon, 2000. Comparison of a coupled ocean-atmosphere model with and without oceanic eddy-induced advection. Part I: Ocean spinup and control integrations. Journal of Climate, 13:139–163.

Hoerling, M.P., J.W. Hurrell and T. Xu, 2001. Tropical origin for recent North Atlantic climate change. Science, 292:90–92.

Hostetler, S.W., F. Giorgi, G.T. Bates and P.J. Bartlein, 1994. Lake-atmosphere feedbacks associated with paleolakes Bonneville and Lahontan. Science, 263:665–668.

Hunke, E.C. and J.K. Dukowicz, 1997. An elastic-viscous-plastic model for sea ice dynamics. Journal of Physical Oceanography, 27:1849–1867.

Huntingford, C., R.G. Jones, C. Prudhomme, R. Lamb and J.H.C. Gash, 2003. Regional climate model predictions of extreme rainfall for a changing climate. Quarterly Journal of the Royal Meteorological Society, 129:1607–1621.

Huth, R., 1999. Statistical downscaling in central Europe: Evaluation of methods and potential predictors. Climate Research, 13:91–101.

Huth, R. and J. Kysely, 2000. Constructing site-specific climate change scenarios on a monthly scale using statistical downscaling. Theoretical and Applied Climatology, 66:13–27.

IGPO, 2000. GEWEX Cloud System Study (GCSS) Second Science and Implementation Plan. Global Energy and Water Cycle Experiment (GEWEX). International GEWEX Project Office, Document No. 34, 52pp.

IPCC, 1994. IPCC Technical Guidelines for Assessing Climate Change Impacts and Adaptations. Prepared by Working Group II. T.R. Carter, M.L. Parry, H. Harasawa and S. Nishioka (eds.). WMO/UNEP, CGER-IO15-94. Intergovernmental Panel on Climate Change, 59pp.

IPCC, 1996. Climate Change 1995: The Science of Climate Change. Contribution of Working Group I to the Second Assessment Report of the Intergovernmental Panel on Climate Change. J.T. Houghton, L.G. Meira Filho, B.A. Callander, N. Harris, A. Kattenberg and K. Maskell (eds.). Intergovernmental Panel on Climate Change. Cambridge University Press, 572pp.

IPCC, 2001. Climate Change 2001: The Scientific Basis. Contribution of Working Group I to the Third Assessment Report of the Intergovernmental Panel on Climate Change. J.T. Houghton, Y. Ding, D.J. Griggs, M. Noguer, P.J. van der Linden, X. Dai, K. Maskell and C.A. Johnson (eds.). Intergovernmental Panel on Climate Change. Cambridge University Press, 881pp.

IPCC-TGCIA, 1999. Guidelines on the Use of Scenario Data for Climate Impact and Adaptation Assessment. Version 1. Prepared by T.R. Carter, M. Hulme and M. Lal. Intergovernmental Panel on Climate Change, Task Group on Scenarios for Climate and Impact Assessment, 69pp.

Jacob, D. and R. Podzun, 1997. Sensitivity studies with the Regional Climate Model REMO. Meteorology and Atmospheric Physics, 63:119–129.

Johns, T.C., 1996. A description of the second Hadley Centre coupled model (HADCM2). Report No. 71. Hadley Centre for Climate Prediction and Research, UK.

Johns, T.C., R.E. Carnell, J.F. Crossley, J.M. Gregory, J.F.B. Mitchell, C.A. Senior, S.F.B. Tett and R.A. Wood, 1997. The second Hadley Centre coupled atmosphere-ocean GCM: model description, spin-up and validation. Climate Dynamics, 13:103–134.

Jones, R.G., J.M. Murphy and M. Noguer, 1995. Simulation of climate change over Europe using a nested regional climate model. I: Assessment of control climate, including sensitivity to location of lateral boundaries. Quarterly Journal of the Royal Meteorological Society, 121:1413–1449.

Jones, R.G., J.M. Murphy, M. Noguer and A.B. Keen, 1997. Simulation of climate change over Europe using a nested regional-climate model. II: comparison of driving and regional model responses to a doubling of carbon dioxide. Quarterly Journal of the Royal Meteorological Society, 123:265–292.

Kaas, E. and P. Frich, 1995. Diurnal temperature range and cloud cover in the Nordic countries: observed trends and estimates for the future. Atmospheric Research, 37:211–228.

Kalnay, E., 2003. Atmospheric Modeling, Data Assimilation and Predictability. Cambridge University Press, 341pp.

Kattsov, V.M. and V.P. Meleshko, 2004. Evaluation of atmosphere-ocean general circulation models used for projecting future climate change. Izvestia, Russian Academy of Sciences – Atmospheric and Oceanic Physics, 40(6):647–658.

Kattsov, V.M. and J.E. Walsh, 2002. Reply to comments on 'Twentieth-century trends of Arctic precipitation from observational data and a climate model simulation' by H. Paeth, A. Hense, and R. Hagenbrock. Journal of Climate, 15:804–805.

Kattsov, V.M., V.P. Meleshko, V.M. Gavrilina, V.A. Govorkova and T.V. Pavlova, 1998. Freshwater budget of the polar regions as simulated with current atmospheric general circulation models. Izvestia Russian Academy of Sciences Phys. Atmos. Ocean, 34:479–489.

Kattsov, V.M., J.E. Walsh, A. Rinke and K. Dethloff, 2000. Atmospheric climate models: simulations of the Arctic Ocean fresh water budget components. In: E.L. Lewis, E.P. Jones, P. Lemke, T.D. Prowse and P. Wadhams (eds.). The Freshwater Budget of the Arctic Ocean, pp. 209–247. Kluwer Academic Publishers.

Kattsov, V.M., S.V. Vavulin, V.A. Govorkova and T.V. Pavlova, 2003. Scenarios of the Arctic climate change in the 21st century. Russian Meteorology and Hydrology, 10:5–19.

Katz, R.W. and M.B. Parlange, 1996. Mixtures of stochastic processes: applications to statistical downscaling. Climate Research, 7:185–193.

Källén, E., V. Kattsov, J. Walsh and E. Weatherhead, 2001. Report from the Arctic Climate Impact Assessment Modeling and Scenarios Workshop. Stockholm, January 29–31 2001. 35pp.

Khrol, V.P. (ed.), 1996. Atlas of Water Balance of the Northern Polar Area. Gidrometeoizdat, St. Petersburg, 81pp.

Kidson, J.W. and C.S. Thompson, 1998. Comparison of statistical and model-based downscaling techniques for estimating local climate variations. Journal of Climate, 11:735–753.

Kiilsholm, S., J.H. Christensen, K. Dethloff and A. Rinke, 2003. Net accumulation of the Greenland Ice Sheet: Modelling Arctic regional climate change. Geophysical Research Letters, 30, 10.1029/2002GL015742.

Kistler, R., E. Kalnay, W. Collins, S. Saha, G. White, J. Woollen, M. Chelliah, W. Ebisuzaki, M. Kanamitsu, V. Kousky, H. van den Dool, R. Jenne and M. Fiorino, 2001. The NCEP-NCAR 50-year reanalysis: Monthly means CD-ROM and documentation. Bulletin of the American Meteorological Society, 82:247–268.

Knippetz, P., U. Ulbrich and P. Speth, 2000. Changing cyclones and surface wind speeds over the North Atlantic and Europe in a transient GHG experiment. Climate Research, 15:109–122.

Knutson, T.R., T.L. Delworth, K.W. Dixon and R.J. Stouffer, 1999. Model assessment of regional surface temperature trends (1949–1997). Journal of Geophysical Research, 104:30981–30996.

Langley Atmospheric Sciences Data Center, 1983–1991. Langley Eight-Year Shortwave and Longwave Surface Radiation Budget Dataset, July 1983–June 1991. NASA. (CD-ROM)

Laprise, R., D. Caya, M. Giguère, G. Bergeron, H. Côte, J.-P. Blanchet, G.J. Boer and N.A. McFarlane, 1998. Climate and climate change in western Canada as simulated by the Canadian regional climate model. Atmosphere-Ocean, 36:119–167.

Legates, D.R. and C.L. Willmott, 1990. Mean seasonal and spatial variability in gauge-corrected global precipitation. International Journal of Climatology, 10:111–133.

Leggett, J.W., J. Pepper and R.J. Swart, 1992. Emission scenarios for the IPCC: an update. In: J.T. Houghton, B.A. Callander and S.K. Varney (eds.). Climate Change 1992. The Supplementary Report to the IPCC Scientific Assessment, pp. 69–95. Cambridge University Press.

Lemke, P., W.D. Hibler III, G.M. Flato, M. Harder and M. Kreyscher, 1997. On the improvement of sea-ice models for climate simulations: The sea-ice model intercomparison project. Annals of Glaciology, 25:183–187.

Linderson, M-L., C. Achberger and D. Chen, 2004. Statistical downscaling and scenario construction of precipitation in Scania, southern Sweden. Nordic Hydrology, 35(3):261–278.

Lorenz, E.N., 1963. The predictability of hydrodynamic flow. Transactions of the New York Academy of Sciences, Series II, 25:409–432.

Lunkeit, F., M. Ponater, R. Sausen, M. Sogalla, U. Ulbrich and M. Windelband, 1996. Cyclonic activity in a warmer climate. Beitrage zur Physik der Atmosphare, 69:393–407.

Lynch, A.H., W.L. Chapman, J.E. Walsh and G. Weller, 1995. Development of a regional climate model of the western Arctic. Journal of Climate, 8:1555–1570.

Lynch, A.H., M.F. Glück, W.L. Chapman, D.A. Bailey and J.E. Walsh, 1997. Remote sensing and climate modeling of the St. Lawrence Is. Polynya. Tellus, 49A:277–297.

Lynch, A.H., D.L. McGinnes and D.A. Bailey, 1998. Snow-albedo and the spring transition in a regional climate system model: influence of land surface model. Journal of Geophysical Research, 103:29037–29049.

Lynch, A.H., J.A. Maslanik and W. Wu, 2001. Mechanisms in the development of anomalous sea ice extent in the western Arctic: A case study. Journal of Geophysical Research, 106:28097–28105.

Mabuchi, K., Y. Sato and H. Kida, 2000. Numerical study of the relationships between climate and the carbon dioxide cycle on a regional scale. Journal of the Meteorological Society of Japan, 78:25–46.

Machenhauer, B., M. Windelband, M. Botzet, R. Jones and M. Deque, 1996. Validation of present-day regional climate simulations over Europe: Nested LAM and variable resolution global model simulations with observed or mixed layer ocean boundary conditions. MPI Report 191. Max Planck Institute for meteorology, Hamburg.

Machenhauer, B., J. Windelband, M. Botzet, J.H. Christensen, M. Deque, R. Jones, P.M. Ruti and G. Visconti, 1998. Validation and analysis of regional present-day climate and climate change simulations over Europe. MPI Report 275. Max Planck Institute for meteorology, Hamburg.

Majewski, D., D. Liermann, P. Prohl, B. Ritter, M. Buchhold, T. Hanisch, G. Paul, W. Wergen and J. Baumgardner, 2002. The operational global Icosahedral-Hexagonal Gridpoint Model GME: description and high-resolution tests. Monthly Weather Review, 130:319–338.

Mahrt, L., 1998: Stratified atmospheric boundary layers and breakdown of models. Theoretical and Computational Fluid Dynamics, 11:263–279.

Manabe, S. and R.J. Stouffer, 1997. Coupled ocean-atmosphere model response to freshwater input: comparision to Younger Dryas event. Paleoceanography, 12:321–336.

Manabe, S. and R.T. Wetherald, 1975. The effects of doubling the CO_2 concentration on the climate of a general circulation model. Journal of Atmospheric Sciences, 32:3–15.

Manabe, S., R.J. Stouffer, M.J. Spelman and K. Bryan, 1991. Transient responses of a coupled ocean-atmosphere model to gradual changes of atmospheric CO_2. Part I: Annual mean response. Journal of Climate, 4:785–818.

Martin, E., B. Timbal and E. Brun, 1997. Downscaling of general circulation model output: simulation of snow climatology of the French Alps and sensitivity to climate change. Climate Dynamics, 13:45–56.

Maslanik, J.A., M.C. Serreze and R.G. Barry, 1996: Recent decreases in Arctic summer ice cover and linkage to atmospheric circulation anomalies. Geophysical research Letters, 23:1677–1680.

Maslanik, J.A., A.H. Lynch, M.C. Serreze and W. Wu, 2000. A case study of regional climate anomalies in the Arctic: performance requirements for a coupled model. Journal of Climate, 13:383–401.

McAvaney, B.J., C. Covey, S. Joussaume, V. Kattsov, A. Kitoh, W. Ogana, A.J. Pitman, A.J. Weaver, R.A. Wood and Z.-C. Zhao, 2001. Model evaluation. In: J.T. Houghton, Y. Ding, D.J. Griggs, M. Noguer, P.J. van der Linden, X. Dai, K. Maskell and C.A. Johnson (eds.). pp. 471–524. Climate Change 2001: The Scientific Basis. Contribution of Working Group I to the Third Assessment Report of the Intergovernmental Panel on Climate Change. Cambridge University Press.

McGregor, J.L., J.J. Katzfey and K.C. Nguyen, 1999. Recent regional climate modelling experiments at CSIRO. In: H. Ritchie (ed.). Research Activities in Atmospheric and Oceanic Modelling, pp. 7.37–7.38. CAS/JSC Working Group on Numerical Experimentation Report 28. WMO/TD - No. 942. World Meteorological Organization, Geneva.

Mearns, L.O., M. Hulme, T.R. Carter, R. Leemans, M. Lal and P. Whetton, 2001. Climate scenario development. In: J.T. Houghton, Y. Ding, D.J. Griggs, M. Noguer, P.J. van der Linden, X. Dai, K. Maskell and C.A. Johnson (eds.). pp. 739–768. Climate Change 2001: The Scientific Basis. Contribution of Working Group I to the Third Assessment Report of the Intergovernmental Panel on Climate Change. Cambridge University Press.

Meehl, G.A., G.J. Boer, C. Covey, M. Latif and R.J. Stouffer, 2000. The Coupled Model Intercomparison Project (CMIP). Bulletin of the American Meteorological Society, 81:313–318.

Meleshko, V.P., V.M. Kattsov, V.A. Govorkova, S.P. Malevsky-Malevich, E.D. Nadyozhina and P.V. Sporyshev, 2004. Anthropogenic climate changes in the 21st century in Northern Eurasia. Russian Meteorology and Hydrology, 7:5–26.

Murphy, J., 1999. An evaluation of statistical and dynamical techniques for downscaling local climate. Journal of Climate, 12:2256–2284.

Murphy, J., 2000. Predictions of climate change over Europe using statistical and dynamical downscaling techniques. International Journal of Climatology, 20:489–501.

Murray, R.J., 1996. Explicit generation of orthogonal grids for ocean models. Journal of Computational Physics, 126:251–273.

Nakićenović, N. and R. Swart (eds.), 2000. Intergovernmental Panel on Climate Change, Special Report on Emissions Scenarios. Cambridge University Press, 599pp.

NAST, 2001. Climate Change Impacts on the United States: The Potential Consequences of Climate Variability and Change. National Assessment Synthesis Team, Report for the US Global Change Research Program. Cambridge University Press, 620pp.

Nelson, F.E. and S.I. Outcalt, 1987. A computational method for prediction and regionalization of permafrost. Arctic and Alpine Research, 19:279–288.

New, M., M. Hulme and P. Jones, 1999. Representing twentieth-century space-time climate variability. Part I: Development of a 1961–90 mean monthly terrestrial climatology. Journal of Climate, 12:829–856.

New, M., M. Hulme and P. Jones, 2000. Representing twentieth-century space-time climate variability. Part II: Development of 1901–96 monthly grids of terrestrial surface climate. Journal of Climate, 13:2217–2238.

Noguer, M., R. Jones and J. Murphy, 1998. Sources of systematic errors in the climatology of a regional climate model over Europe. Climate Dynamics, 14:691–712.

Nozawa, T., S. Emori, T. Takemura, T. Nakajima, A. Numaguti, A. Abe-Ouchi and M. Kimoto, 2000. Coupled ocean-atmosphere model experiments of future climate change based on IPCC SRES scenarios. Preprints, Eleventh Symposium on Global Change Studies, 9–14 January 2000, Long Beach, California, pp. 352–355.

Omstedt, A. and D. Chen, 2001. Influence of atmospheric circulation on the maximum ice extent in the Baltic Sea. Journal of Geophysical Research, 106(C3):4493–4500.

Otterå, O.H., H. Drange, M. Bentsen, N.G. Kvamsto and D. Jiang, 2003: The sensitivity of the present day Atlantic meridional overturning circulation to freshwater forcing. Geophysical Research Letters, 30: 1898, doi:101029/2003GL017578.

Pacanowski, R.C. and S.M. Griffies, 1999. The MOM 3 Manual. Geophysical Fluid Dynamics Laboratory/NOAA, Princeton, New Jersey, 680pp.

Palmer, T.N. and J. Räisänen, 2002. Quantifying the risk of extreme seasonal precipitation events in a changing climate. Nature, 415:512–514.

Palutikof, J.P., C.M. Goodess, S.J. Watkins and T. Holt, 2002. Generating daily rainfall and temperature scenarios at multiple sites: Examples from the Mediterranean. Journal of Climate, 15:3529–3548.

Paterson, A.B. and N. Reeh, 2001. Thinning of the ice sheet in northwest Greenland over the past forty years. Nature, 414:60–62.

Peterson, B.J., R.M. Holmes, J.W. McClelland, C.J. Vörösmarty, R.B. Lammers, A.I. Shiklomanov, I.A. Shiklomanov and S. Rahmstorf, 2002. Increasing river discharge to the Arctic Ocean. Science, 298:2171–2173.

Polyakov, I. and M. Johnson, 2000. Arctic decadal and interdecadal variability. Geophysical Research Letters, 24:4097–4100.

Polyakov, I., S.-I. Akasofu, U. Bhatt, R. Colony, M. Ikeda, A. Makshtas, C. Swingley, D. Walsh and J. Walsh, 2002a. Trends and variations in Arctic climate system. Eos, Transactions, American Geophysical Union, 83(47):547–548.

Polyakov, I., G. Alekseev, R. Bekryaev, U. Bhatt, R. Colony, M. Johnson, V. Karklin, A. Makshtas, D. Walsh and A. Yulin, 2002b. Observationally based assessment of polar amplification of global warming. Geophysical Research Letters, 29(18):1878–1881.

Poulus, G.S., and S.P. Burns, 2003. An evaluation of bulk Ri-based surface layer flux formulas for stable and very stable conditions with intermittent turbulence. Journal of Atmospheric Science, 60:2523–2537.

Power, S.B., R.A. Colman, B.J. McAvaney, R.R. Dahni, A.M. Moore and N.R. Smith, 1993. The BMRC coupled atmosphere/ocean/sea-ice model. Bureau of Meteorology Research Centre, Research Rep. 37. Melbourne, 58pp.

Proshutinsky, A., M. Steele, J. Zhang, G. Holloway, N. Steiner, S. Hakkinen, D. Holland, R. Gerdes, C. Koeberle, M. Karcher, M. Johnson, W. Maslowski, W. Walczowski, W. Hibler and J. Wang, 2001. Multinational effort studies differences among Arctic Ocean models. Eos, Transactions, American Geophysical Union, 82(637):643–644.

Puri, K., 2002. Activities of the CAS/JSC Working Group on Numerical Experimentation (WGNE). In: H. Ritchie (ed.). Research Activities in Atmospheric and Oceanic Modelling. Report No. 32, WMO/TD-No. 1105.

Qian, Y. and F. Giorgi, 1999. Interactive coupling of regional climate and sulfate aerosol models over East Asia. Journal of Geophysical Research, 104:6501–6514.

Quinn, P.K., T.L. Miller, T.S. Bates, J.A. Ogren, E. Andrews and G.E. Shaw, 2002. A 3-year record of simultaneously measured aerosol chemical and optical properties at Barrow, Alaska. Journal of Geophysical Research, 107(D11):4130, doi:10.1029/2001JD001248.

Randall, D., J. Curry, D. Battisti, G. Flato, R. Grumbine, S. Hakkinen, D. Martinson, R. Preller, J. Walsh and J. Weatherly, 1998. Status of and outlook for large-scale modeling of atmosphere-ice-ocean interactions in the Arctic. Bulletin of the American Meteorological Society, 79:197–219.

Räisänen, J., 2001. CO_2-induced climate change in the Arctic area in the CMIP2 experiments. SWECLIM Newsletter, 11:23–28.

Räisänen, J. and R. Döscher, 1999. Simulation of present-day climate in Northern Europe in the HadCM2 GCM. Reports on Meteorology and Climatology, No. 48. Swedish Meteorological and Hydrological Institute.

Räisänen, J. and T.N. Palmer, 2001. A probability and decision-model analysis of a multi-model ensemble of climate change simulations. Journal of Climate, 14:3212–3226.

Räisänen, J., M. Rummukainen, A. Ullerstig, B. Bringfelt, U. Hansson and U. Willén, 1999. The first Rossby Centre regional climate scenario – dynamical downscaling of CO_2-induced climate change in the HadCM2 GCM. Reports on Meteorology and Climatology, No. 85, Swedish Meteorological and Hydrological Institute, 56pp.

Räisänen, J., U. Hansson, A. Ullerstig, R. Döscher, L.P. Graham, C. Jones, M. Meier, P. Samuelsson and U. Willén, 2003. GCM driven simulations of recent and future climate with the Rossby Centre coupled atmosphere Baltic Sea regional climate model RCAO. Reports on Meteorology and Climatology, No. 101, Swedish Meteorological and Hydrological Institute, 61pp.

Rind, D., P. deMenocal, G. Russell, S. Sheth, D. Collins, G. Schmidt and J. Teller, 2001. Effects of glacial meltwater in the GISS coupled atmosphere-ocean model 1. North Atlantic Deep Water response. Journal of Geophysical Research, 16:27335–27353.

Rinke, A., A.H. Lynch and K. Dethloff, 2000. Intercomparison of Arctic regional climate simulations: case studies of January and June 1990. Journal of Geophysical Research, 105:29669–29683.

Rinke, A., R. Gerdes, K. Dethloff, T. Kandlbinder, M. Karcher, F. Kauker, S. Frickenhaus, C. Köberle and W. Hiller, 2003. A case study of the anomalous Arctic sea ice conditions during 1990: insight from coupled and uncoupled regional climate model simulations. Journal of Geophysical Research, 108 (D9),4275, doi:10.1029/2002JD003146.

Roeckner, E., J.M. Oberhuber, A. Bacher, M. Christoph and I. Kirchner, 1996. ENSO variability and atmospheric response in a global coupled atmosphere-ocean GCM. Climate Dynamics, 12:737–754.

Roeckner, E., L. Bengtsson, J. Feichter, J. Lelieveld and H. Rodhe, 1999. Transient climate change simulations with a coupled atmosphere-ocean GCM including the tropospheric sulfur cycle. Journal of Climate, 12:3004–3032.

Roed, L.P., X.B. Shi and B. Hackett, 2000. The Importance of Allowing Turbulent and Diffusive Diapycnal Mixing in Isopycnic Coordinate Ocean Models. RegClim General Tech. Rep. 4, 139–148. Available from the Norwegian Institute for Air Research.

Rummukainen, M., J. Räisänen, B. Bringfelt, A. Ullerstig, A. Omstedt, U. Willén, U. Hansson and C. Jones, 2001. A regional climate model for northern Europe: model description and results from the downscaling of two GCM control simulations. Climate Dynamics, 17:339–359.

Ruosteenoja, K., T.R. Carter, K. Jylhä and H. Tuomenvirta, 2003. Future climate in world regions: an intercomparison of model-based projections for the new IPCC emissions scenarios. The Finnish Environment 644, Finnish Environment Institute.

Russell, G.L. and D. Rind, 1999. Response to CO_2 transient increase in the GISS coupled model: regional coolings in a warmer climate. Journal of Climate, 12:531–539.

Rysgaard, S., T. Vang, M. Stjernholm, B. Rasmussen, A. Windelin and S. Kiilsholm, 2003. Physical conditions, carbon transport and climate change impacts in a NE Greenland fjord. Arctic, Antarctic and Alpine Research, 35(3):301–312.

Sadourny, R., A. Arakawa and Y. Mintz, 1968. Integration of the nondivergent barotropic equation with an icosahedral hexagonal grid on the sphere. Monthly Weather Review, 96:351–356.

Sailor, D.J. and X. Li, 1999. A semiempirical downscaling approach for predicting regional temperature impacts associated with climatic change. Journal of Climate, 12:103–114.

Santer, B.D., T.M.L. Wigley, M.E. Schlesinger and J.F.B. Mitchell, 1990. Developing climate scenarios from equilibrium GCM results. MPI Rep. 47, Max Planck Institute for Meteorology, Hamburg, 29pp.

Sausen, R., K. Barthel and K. Hasselmann, 1987. A flux correction method for removing the climate drift of climate models. Report No. 1. Max Planck Institute for Meteorology, 39pp.

Schiller, A., U. Mikolajewicz and R. Voss, 1997. The stability of the North Atlantic thermohaline circulation in a coupled ocean-atmosphere general circulation model. Climate Dynamics, 13:325–347.

Schnur, R. and D. Lettenmaier, 1997. A case study of statistical downscaling in Australia using weather classification by recursive partitioning. Journal of Hydrology, 211:362–379.

Schoof, J.T. and S.C. Pryor, 2001. Downscaling temperature and precipitation: A comparison of regression-based methods and artificial neural networks. International Journal of Climatology, 21:773–790.

Schrum, C., U. Huebner, D. Jacob and R. Podzun, 2001. A coupled atmosphere/ice/ocean model for the North Sea and the Baltic Sea. Berichte aus dem Zentrum für Meeres- und Klimaforschung Nr. 41, Reihe B: Ozeanographie. Inst. Meereskunde, Hamburg.

Schubert, S., 1998. Downscaling local extreme temperature changes in south-eastern Australia from the CSIRO MARK2 GCM. International Journal of Climatology, 18:1419–1439.

Schweiger, A.J., R.W. Lindsay, J.R. Key and J.A. Francis, 1999. Arctic clouds in multiyear satellite data sets. Geophysical Research Letters, 26:1845–1848.

Semenov, M.A. and E. Barrow, 1997. Use of stochastic weather generator in the development of climate change scenarios. Climatic Change, 35:397–414.

Semenov, M.A., R.J. Brooks, E.M. Barrow and C.W. Richardson, 1998. Comparison of the WGEN and LARS-WG stochastic weather generators for diverse climates. Climate Research, 10:95–107.

Shindell, D.T., G.A. Schmidt, R.L. Miller and D. Rind, 2001. Northern Hemisphere winter climate response to greenhouse gas, ozone, solar, and volcanic forcing. Journal of Geophysical Research, 106:7193–7210.

Small, E.E., F. Giorgi and L.C. Sloan, 1999a. Regional climate model simulation of precipitation in central Asia: Mean and interannual variability. Journal of Geophysical Research, 104:6563–6582.

Small, E.E., L.C. Sloan, S. Hostetler and F. Giorgi, 1999b. Simulating the water balance of the Aral Sea with a coupled regional climate-lake model. Journal of Geophysical Research, 104:6583–6602.

Smith, J.B., M. Hulme, J. Jaagus, S. Keevallik, A. Mekonnen and K. Hailemariam, 1998. Climate change scenarios. In: J.F. Feenstra, I. Burton, J. Smith and R.S.J. Tol (eds.). Handbook on Methods for Climate Change Impact Assessment and Adaptation Strategies, pp. 3-1–3-40. Version 2.0. United Nations Environment Programme and Institute for Environmental Studies, Vrije Universiteit, Amsterdam.

Solman, S.A. and M.N. Nuñez, 1999. Local estimates of global climate change: A statistical downscaling approach. International Journal of Climatology, 19:835–861.

Stendel, M., T. Schmith, E. Roeckner and U. Cubasch, 2000. The climate of the 21st century: transient simulations with a coupled atmosphere-ocean general circulation model. Danish Climate Centre Report 00-6, 51pp.

Stocker, T.F., G.K.C. Clarke, H. Le Treut, R.S. Lindzen, V.P. Meleshko, R.K. Mugara, T.N. Palmer, R.T. Pierrehumbert, P.J. Sellers, K.E. Trenberth and J. Willebrand, 2001. Physical climate processes and feedbacks. In: J.T. Houghton, Y. Ding, D.J. Griggs, M. Noguer, P.J. van der Linden, X. Dai, K. Maskell and C.A. Johnson (eds.), pp. 418–470. Climate Change 2001: The Scientific Basis. Contribution of Working Group I to the Third Assessment Report of the Intergovernmental Panel on Climate Change. Cambridge University Press.

Stott, P., 2003: Attribution of regional-scale temperature changes to anthropogenic and natural causes. Geophysical Research Letters, 30(14): 1728, doi:10.1029/2003GL017324.

Stott, P.A., S.F.B. Tett, G.S. Jones, M.R. Allen, J.F.B. Mitchell and G.J. Jenkins, 2000. External control of twentieth century temperature variations by natural and anthropogenic forcings. Science, 15:2133–2137.

Stouffer, R.J. and K.W. Dixon, 1998. Initialization of coupled models for use in climate studies. In: A. Staniforth (ed.). Research Activities in Atmospheric and Oceanic Modelling, pp. I.1–I.15. Rep. 27, WMO/TD-No. 865.

Tabazadeh, A., E.J. Jensen, O.B. Toon, K. Drdla and M.R. Schoeberl, 2001. Role of the stratospheric polar freezing belt in denitrification. Science, 291:2591–2594.

Takahashi, M., 1999. Simulation of the quasibiannial oscillation in a general circulation model. Geophysical Research Letters, 26, 1307–1310.

Takle, E.S., W.J. Gutowski, R.W. Arritt, Z. Pan, C.J. Anderson, R.S. da Silva, D. Caya, S.-C. Chen, F. Giorgi, J.H. Christensen, S.-Y. Hong, H.-M.H. Juang, J. Katzfey, W.M. Lapenta, R. Laprise, P. Lopez, G.E. Liston, J. McGregor, A. Pielke and J.O. Roads, 1999. Project to intercompare regional climate simulation (PIRCS): Description and initial results. Journal of Geophysical Research, 104:19443–19461.

Tao, X., J.E. Walsh and W.L. Chapman, 1996. An assessment of global climate model simulations of Arctic air temperatures. Journal of Climate, 9:1060–1076.

Tett, S.F.B., G.S. Jones, P.A. Stott, D.C. Hill, J.F.B. Mitchell, M.A. Allen, W.J. Ingram, T.C. Johns, C.E. Johnson, A. Jones, D.L. Roberts, D.M.H. Sexton and M.J. Woodage, 2000. Estimation of Natural and Anthropogenic Contributions to 20th Century. Hadley Centre Tech Note 19. Hadley Centre for Climate Prediction and Response, UK, 52pp.

Tokioka, T., A. Noda, A. Kitoh, Y. Nikaidou, S. Nakagawa, T. Motoi, S. Yukimoto and K. Takata, 1995. A transient CO_2 experiment with the MRI CGCM. Quick Report. Journal of the Meteorological Society of Japan, 73(4):817–826.

Trenberth, K. (ed.), 1992. Climate System Modelling. Cambridge University Press, 788pp.

Tsvetsinskaya, E., L.O. Mearns and W.E. Easterling, 2000. Effects of plant growth and development on interannual variability in mesoscale atmospheric simulations. In: Proceedings of the Tenth International Offshore and Polar Engineering Conference. Vol. I, pp. 729–736. International Society of Offshore and Polar Engineers, Cupertino, California.

Ulbrich, U. and M. Christoph, 1999. A shift of the NAO and increasing storm track activity over Europe due to anthropogenic greenhouse gas forcing. Climate Dynamics, 15:551–559.

Van den Brink, H.W., G.P. Konnen and J.D. Opsteegh, 2003. The reliability of extreme surge levels, estimated from observational records of order of hundred years. Journal of Coastal Research, 19:376–388.

Vellinga, M., R.A. Wood and J.M. Gregory, 2002. Processes governing the recovery of a perturbed thermohaline circulation in HadCM3. Journal of Climate, 15:764–780.

von Storch, H., 1999. The global and regional climate system. In: H. von Storch and G. Flöser (eds.). Anthropogenic Climate Change, pp. 3–36. Springer Verlag.

Voss, R., R. Sausen and U. Cubasch, 1998. Periodically synchronously coupled integrations with the atmosphere-ocean general circulation model ECHAM3/LSG. Climate Dynamics, 14:249–266.

Wallace, J.M., 2000. North Atlantic Oscillation/annular mode: Two paradigms – one phenomenon. Quarterly Journal of the Royal Meteorological Society, 126:729–805.

Walsh, J.E., in press. Summary of a workshop on modeling the Arctic atmosphere, Madison, Wisconsin, 20–22 May 2002. Bulletin of the American Meteorological Society.

Walsh, J.E., V. Kattsov, D. Portis and V. Meleshko, 1998. Arctic precipitation and evaporation: model results and observational estimates. Journal of Climate, 11:72–87.

Walsh, J.E., V. Kattsov, W. Chapman, V. Govorkova and T. Pavlova, 2002. Comparison of Arctic climate simulations by uncoupled and coupled global models. Journal of Climate, 15:1429–1446.

Washington, W.M., J.W. Weatherly, G.A. Meehl, A.J. Semtner Jr., T.W. Bettge, A.P. Graig, W.G. Strand Jr., J. Arblaster, V.B. Wayland, R. James and Y. Zhang, 2000. Parallel climate model (PCM) control and transient simulations. Climate Dynamics, 16:755–774.

WCRP, 2002. WOCE/CLIVAR, 2002: WOCE/CLIVAR Working Group on Ocean Model Development. Report of the Third Session. World Climate Research Programme, Informal Rep. No. 14/2002.

Weisse, R., H. Heyen and H. von Storch, 2000. Sensitivity of a regional atmospheric model to a sea state-dependent roughness and the need for ensemble calculations. Monthly Weather Review, 128:3631–3642.

Wilby, R.L. and T.M.L. Wigley, 1997. Downscaling general circulation model output: a review of methods and limitations. Progress in Physical Geography, 21:530–548.

Wilby, R.L. and T.M.L. Wigley, 2000. Precipitation predictors for downscaling: observed and general circulation model relationships. International Journal of Climatology, 20:641–661.

Wilby, R.L., T.M.L. Wigley, D. Conway, P.D. Jones, B.C. Hewitson, J. Main and D.S. Wilks, 1998. Statistical downscaling of general circulation model output: A comparison of methods. Water Resources Research, 34:2995–3008.

Wilby, R.L., C.W. Dawson and E.M. Barrow, 2002. SDSM – a decision support tool for the assessment of regional climate change impacts. Environmental Modelling and Software, 17:145–157.

Wilks, D., 1999. Multisite downscaling of daily precipitation with a stochastic weather generator. Climate Research, 11:125–136.

Yukimoto, S., M. Endoh, Y. Kitamura, A. Kitoh, T. Motoi and A. Noda, 2000. ENSO-like interdecadal variability in the Pacific Ocean as simulated in a coupled GCM. Journal of Geophysical Research, 105(D10):13945–13963.

Chapter 5

Ozone and Ultraviolet Radiation

Lead Authors

Betsy Weatherhead, Aapo Tanskanen, Amy Stevermer

Contributing Authors

Signe Bech Andersen, Antti Arola, John Austin, Germar Bernhard, Howard Browman, Vitali Fioletov, Volker Grewe, Jay Herman, Weine Josefsson, Arve Kylling, Esko Kyrö, Anders Lindfors, Drew Shindell, Petteri Taalas, David Tarasick

Consulting Authors

Valery Dorokhov, Bjorn Johnsen, Jussi Kaurola, Rigel Kivi, Nikolay Krotkov, Kaisa Lakkala, Jacqueline Lenoble, David Sliney

Contents

Summary

Depletion of stratospheric ozone over the Arctic can reduce normally high winter and spring ozone levels and allow more ultraviolet (UV) radiation to reach the surface of the earth. Arctic ozone levels exhibit high natural seasonal and interannual variability, driven primarily by atmospheric dynamics that govern the large-scale meridional transport of ozone from the tropics to high latitudes. The spatial distribution of total column ozone over the Arctic is less symmetric around the pole than is the case for ozone over the Antarctic. The large natural variability in arctic ozone complicates the interpretation of past changes and the projection of future ozone levels. Observations have shown substantial late winter and early spring reductions in arctic total column ozone over the last two decades. These reductions have been directly linked to chemical reactions occurring at low temperatures in the presence of anthropogenic chlorine and bromine compounds. Between 1979 and 2000, the trend in mean annual total column ozone over the Arctic was about -3% per decade (7% accumulated loss), while the trend in mean spring total column ozone was about -5% per decade (11% accumulated loss). Arctic ozone depletion is also strongly affected by stratospheric temperatures and polar stratospheric cloud formation. Climate change leading to lower temperatures in the stratosphere is likely to increase the frequency and severity of ozone-depletion episodes.

Ozone levels directly influence the amount of UV radiation reaching the surface of the earth. Surface UV radiation levels are also strongly affected by clouds, aerosols, altitude, solar zenith angle, and surface albedo. These different factors contribute to high variability in UV radiation levels and make it difficult to identify changes that result from ozone depletion. Because of the low solar elevation in the Arctic, the region is subject to an increased proportion of diffuse UV radiation, from scattering in the atmosphere as well as from reflectance off snow and ice. Reflectance off snow can increase the biologically effective irradiance by more than 50%. Changes in global climate are likely to result in changes in arctic snow cover and sea ice. Snow and ice cover strongly attenuate UV radiation, protecting organisms underneath. A reduction in snow and ice cover on the surface of rivers, lakes, or oceans is likely to increase the exposure of many organisms to damaging UV radiation. Loss of snow or ice cover earlier in the spring, when UV radiation is very likely to be at increased levels, would be stressful for both aquatic and terrestrial life.

Ground-based measurements of UV radiation levels are conducted in all arctic countries. However, the current monitoring network does not provide sufficient coverage over vast regions. Available individual measurements suggest localized increases in UV radiation levels reaching the surface, but the measurement time series are not yet long enough to allow trends to be detected.

Reconstructed time series of surface UV radiation levels based on total column ozone, sunshine duration, and cloud cover suggest distinct increases, but reconstruction methods are less certain than direct measurements because they involve assumptions about the spectral characteristics of cloud and aerosol attenuation and surface reflectivity. The increases in UV radiation levels occur primarily in the spring, when ozone depletion reaches a maximum, and can result in spring UV radiation levels that are higher than those measured during the summer.

Atmospheric sampling indicates that the Montreal Protocol and its amendments have already resulted in a leveling off of some atmospheric halogen concentrations. However, climate change and other factors are likely to complicate the recovery of the ozone layer. Changes in both the overall meteorology of the region and in atmospheric composition may delay or accelerate the recovery of the arctic ozone layer. Ozone levels are projected to remain depleted for several decades and thus surface UV radiation levels in the Arctic are likely to remain elevated in the coming years. The elevated levels are likely to be most pronounced in the spring, when ecosystems are most sensitive to harmful UV radiation. Exposure to UV radiation has been linked to skin cancers, corneal damage, cataracts, immune suppression, and aging of the skin in humans, and can also have deleterious effects on ecosystems and on materials.

5.1. Introduction

Ultraviolet radiation levels reaching the surface of the earth are directly influenced by total column ozone amounts and other geophysical parameters. In the Arctic, UV radiation is of particular concern, particularly during the spring and summer when the region experiences more hours of sunshine compared to lower latitudes. Goggles found in archaeological remains suggest that indigenous peoples had developed protection from sunlight long before the onset of anthropogenic ozone depletion (e.g., Hedblom, 1961; Sliney, 2001). Although systematic measurements of UV radiation levels have been performed for little more than decade, analysis of fossil pigments in leaf sediments suggests that past UV radiation levels in the Arctic may have been similar to modern-day (pre-depletion) levels (Leavitt et al., 1997). In recent years, however, Arctic ozone depletion (which has sometimes been severe) has allowed more UV radiation to reach the surface. In the years since ozone depletion was first observed over the Arctic, UV radiation effects such as sunburn and increased snow blindness have been reported in regions where they were not previously observed (Fox, 2000).

Less than 10% of the solar energy reaching the top of the atmosphere is in the UV spectral region, with wavelengths between 100 and 400 nm. The shortest wavelengths (100–280 nm) are referred to as UV-C radiation. Radiation at these wavelengths is almost entirely absorbed by atmospheric oxygen and ozone, preventing it from reaching the surface. Wavelengths between 280 and 315 nm comprise the UV-B portion of the spectrum (while some communities use 320 nm to mark the division between UV-B and UV-A radiation, it is the convention in this report to use 315 nm). Ultraviolet-B radiation is absorbed efficiently but not completely by atmospheric ozone. Wavelengths between 315 and 400 nm are referred to as UV-A radiation. Absorption of UV-A radiation by atmospheric ozone is comparatively small.

Although the intensity of solar UV-B radiation is low, the energy per photon is high. Due to this high energy, UV-B radiation can have several harmful impacts on human beings (i.e., DNA damage, skin cancers, corneal damage, cataracts, immune suppression, aging of the skin, and erythema), on ecosystems, and on materials (e.g., UNEP, 1998, 2003). These effects are discussed in detail in sections 7.3, 7.4, 8.6, 9.4, 14.12, 15.3.3, 16.3.1, and 17.2.2.3. Ultraviolet-B radiation also affects many photochemical processes, including the formation of tropospheric ozone from gases released into the environment by motor vehicles or other anthropogenic sources.

The amount of UV radiation reaching the surface of the earth is expressed in terms of irradiance, denoting the radiant power per unit area reaching a surface. Figure 5.1 shows typical spectral irradiance in the UV-A and UV-B wavelengths for the Arctic. The values were obtained using a radiative transfer model with a solar zenith angle of 56.5°, total column ozone of 300 Dobson units (DU), and surface albedo of 0.6.

The exposure necessary to produce some biological effect, such as erythema (skin reddening), at each

wavelength in the UV spectral region is given by an action spectrum. In general, shorter UV-B wavelengths have greater biological effects than longer UV-A wavelengths, and action spectra account for this wavelength dependence. The action spectra are used to provide biological weighting factors to determine sensitivities to UV radiation exposure. The action spectrum often used to estimate human health effects is the McKinlay-Diffey erythemal response spectrum (McKinlay and Diffey, 1987). This curve is shown in Fig. 5.1 and represents the standard erythemal action spectrum adopted by the Commission Internationale de l'Eclairage (CIE) to represent the average skin response over the UV-B and UV-A regions of the spectrum (CIE, 1998).

The biological response is determined by multiplying the spectral irradiance at each wavelength by the biological weighting factor provided by the action spectrum. As ozone levels decrease, the biological response increases (see Fig. 5.1). Integrating the product of the spectral irradiance and the biological weighting factor over all wavelengths provides a measure of the biologically effective UV irradiance, or dose rate, with units W/m². This dose rate is scaled to produce a UV index value (WHO, 2002), which is made available to the public to provide an estimate of the level of UV radiation reaching the surface in a particular area at a particular time. Summing the dose rate over the exposure period (e.g., one day) results in a measure of the biologically effective radiation exposure, or dose, with units J/m². In the Arctic, the extended duration of sunlight during the summer can result in moderately large UV radiation doses. When considering biological impacts, it is important to distinguish that the definition of dose presented here differs slightly from that used by biologists, who refer to dose as the amount actually absorbed by the receptor. In addition, for some biological effects the cumulative dose model outlined above is too simple, because dose history also plays a role. In many cases, repair mechanisms cause the dose received over a longer time period to have less effect than a single, intense exposure.

Although some exposure to UV radiation can be beneficial, increases in surface UV radiation doses can have detrimental effects on humans and organisms in the Arctic. The levels of UV radiation reaching the surface are affected not only by total column ozone and solar zenith angle, but also by cloudiness, surface reflectance (albedo), and atmospheric aerosol concentrations. Climate change is likely to affect both future cloudiness and the extent of snow and ice cover in the Arctic, in turn leading to local changes in the intensity of solar UV radiation. It is very likely that climate change is already influencing stratospheric dynamics, which are very likely to in turn affect ozone depletion and surface UV radiation levels in the future. This chapter addresses some of the factors influencing total column ozone and surface UV irradiance, and describes both observed and projected changes in arctic ozone and UV radiation levels.

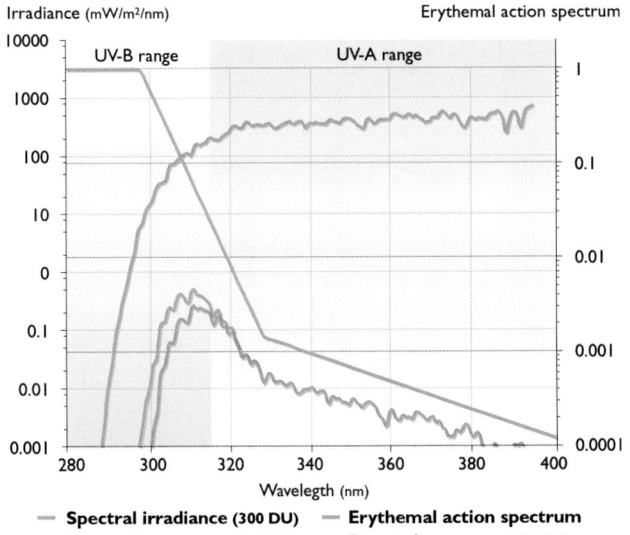

Fig. 5.1. Spectral UV irradiance in the UV-A and UV-B wavelengths (for total column ozone of 300 DU), the CIE erythemal action spectrum, and biological response curves for total column ozone of 300 and 400 DU.

5.2. Factors affecting arctic ozone variability

Ozone in the atmosphere prevents most harmful UV radiation from reaching the biosphere. About 90 to 95% of atmospheric ozone is found in the stratosphere; the remaining 5 to 10% is in the troposphere. Most of the stratospheric ozone is produced by photochemical reactions in equatorial regions; at high latitudes, there is less photochemical ozone production and much of the stratospheric ozone is imported from low latitudes by the Brewer-Dobson circulation. This diabatic circulation also distributes ozone to lower altitudes in the high latitude regions, where, owing to a longer photochemical lifetime, it accumulates. For these reasons, total column ozone tends to exhibit global maxima near the poles. The atmospheric circulation varies seasonally, and oscillations in the circulation patterns explain some of the natural spatial, seasonal, and annual variations in the global total ozone distribution. In the Northern Hemisphere, the maximum total column ozone usually occurs in spring and the minimum in autumn. Solar activity also causes small fluctuations in total column ozone in phase with the solar cycle.

In addition to natural ozone production and destruction processes (WMO, 1995, 1999, 2003), stratospheric ozone is destroyed by heterogeneous chemical reactions involving halogens, particularly chlorine and bromine, which are derived from chlorofluorocarbons (CFCs) and other ozone-depleting substances. In the presence of solar radiation, extremely low stratospheric temperatures facilitate ozone depletion chemistry. Thus, ozone depletion can occur in relatively undisturbed polar vortices (see section 5.2.2, Box 5.1) with the return of sunlight in early spring. The fundamental processes governing ozone levels over the Arctic and Antarctic are the same, however, relative rates of production and destruction can differ. Low temperatures within the stable Antarctic vortex and the presence of ozone-depleting gases have led to an area of large-scale ozone depletion, the "ozone hole", which has been observed every spring since the 1980s. In contrast, the arctic polar vortex is less stable, result-

ing in arctic air masses that are on average warmer than air masses over the Antarctic. However, chemical ozone depletion has been observed over the Arctic during springs when temperatures in the arctic stratosphere were lower than normal. The decreases over the Arctic and Antarctic have both been sizeable (Fig. 5.2), although climatological spring ozone levels over the Arctic tend to be higher than those over the Antarctic, so that total column ozone after depletion events is higher in the Arctic than in the Antarctic. The depletion observed over the Antarctic in spring 2002 was not as severe as in previous years, but this was due to exceptional meteorological conditions and does not indicate an early recovery of the ozone layer.

Since the late 1980s, much attention has been directed to studying ozone depletion processes over the Arctic. Arctic ozone levels have been significantly depleted in the past decade, particularly during the late winter and early spring (seasons when pre-depletion ozone levels were historically higher than at other times of the year). Several studies (Austin, 1992, 1994; Austin and Butchart, 1992; Austin et al., 1995; Guirlet et al., 2000) have focused on both the chemical and dynamic factors contributing to this depletion. These factors have combined to change the overall concentrations and distribution of ozone in the arctic stratosphere (e.g., Weatherhead, 1998; WMO, 1995, 1999, 2003), and the observed changes have not been symmetric around the North Pole. The greatest changes in ozone levels have been observed over eastern Siberia and west toward Scandinavia.

Ozone depletion can increase the level of UV radiation reaching the surface. These increased UV doses, particularly when combined with other environmental stressors, are very likely to cause significant changes to the region's ecosystems. Ozone depletion has been greatest in the spring, when most biological systems are particularly sensitive to UV radiation. The depletion has not been constant over time: very strong ozone depletion has been observed in some years while very little depletion has been observed in other years.

Transport of low-ozone air masses from lower latitudes can result in a few days of very low ozone and high UV radiation levels (Weatherhead, 1998). This transport of low-ozone air masses is often observed in late winter or early spring and is likely to have occurred naturally for decades. Climate change is likely to change transport patterns and is therefore likely to alter the frequency and severity of these events (Schnadt and Dameris, 2003).

5.2.1. Halogens and trace gases

Chlorine and bromine compounds cause chemically induced ozone depletion in the arctic stratosphere (E.C., 2003; Solomon, 1999; WMO, 1999, 2003). The source gases for these halogens are predominantly anthropogenic (E.C., 2003; WMO, 1999, 2003) and

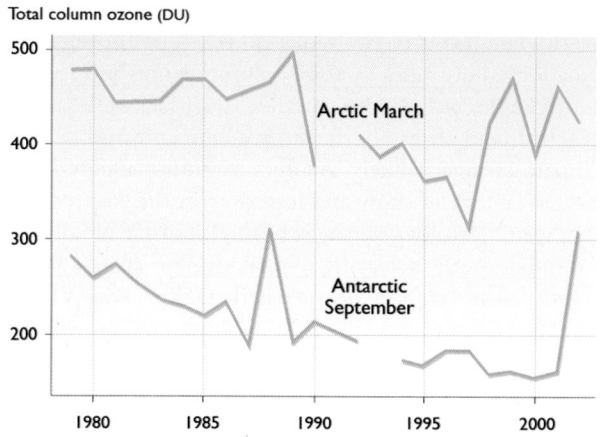

Fig. 5.2. Spring ozone depletion over the Antarctic and the Arctic between 1979 and 2002.

are transported to the polar stratosphere over a period of 3 to 6.5 years (Harnisch et al., 1996; Schmidt and Khedim, 1991; Volk et al., 1997). In the stratosphere, the source gases are converted through photolysis and reaction with the hydroxyl radical to inorganic species of bromine, chlorine, and fluorine. The halogens are normally present as reservoir species (primarily hydrogen chloride – HCl, chlorine nitrate – $ClONO_2$, and bromine nitrate), which are efficiently converted into photochemically active species in the presence of sulfate aerosols or polar stratospheric clouds (WMO, 1999, 2003). Subsequently, in the presence of sunlight, reactive compounds (e.g., chlorine monoxide, bromine monoxide) are formed that react with and destroy stratospheric ozone in catalytic cycles.

The concentrations of chlorine measured in the stratosphere correspond well with the concentrations of CFCs and related gases that have been measured in the troposphere (Chang et al., 1996; Russell et al., 1996; Zander et al., 1996). From the mid-1980s to the early 1990s, the atmospheric chlorine concentration increased approximately 3 to 4% per year (WMO, 1990, 1992), while between 1995 and 1997, the rate of stratospheric chlorine increase was estimated at $1.8\pm0.3\%$ per year (WMO, 1999). An analysis of long-term trends in total column inorganic chlorine through 2001, based on 24 years of HCl and $ClONO_2$ data, showed a broad plateau in inorganic chlorine levels after 1996 (Rinsland et al., 2003). Some uncertainty remains concerning the time lag between reductions in emissions of chlorine-containing compounds at the surface and chlorine concentrations in the stratosphere (e.g., Waugh et al., 2001), although this lag is thought to be between 3 and 5 years on average. Other studies report an estimated total organic bromine growth rate of 2.2% per year (Butler et al., 1998; Wamsley et al., 1998), although errors in the experimental method make the stratospheric bromine mixing ratios more difficult to determine. More recently, Montzka et al. (2003) reported that total organic bromine amounts in the troposphere have decreased since 1998.

Changing concentrations of the trace gases nitrous oxide (N_2O), methane (CH_4), water vapor, and carbon dioxide (CO_2) directly affect ozone chemistry and also alter local atmospheric temperatures by radiative cooling or heating, influencing the reaction rates of ozone depletion chemistry and the formation of ice particles. All of these gases emit radiation efficiently to space from the stratosphere (although CO_2 and water vapor are the most important), so increases in the abundances of these gases are very likely to lead to stratospheric cooling. In the polar regions, this cooling is very likely to lead to ozone depletion through heterogeneous chemistry. Lower temperatures facilitate the formation of polar stratospheric cloud particles, which play a role in transforming halogens to reactive compounds that can destroy ozone very rapidly. Small changes in temperature have been shown to have a significant effect on ozone levels (e.g., Danilin et al., 1998; Tabazadeh et al., 2000).

The trace gases N_2O, CH_4, and water vapor are also important chemically. In the stratosphere, CH_4 acts as an important source of water vapor and is also a sink for reactive chlorine. In addition, stratospheric water vapor is an important source of hydrogen oxide radicals, which play an important role in ozone destruction. Evans et al. (1998), Dvortsov and Solomon (2001), Shindell (2001), and Forster and Shine (2002) have studied the effects of water vapor on homogeneous chemistry. Their model results suggest that increases in water vapor reduce ozone levels in the upper stratosphere, increase ozone levels in the middle stratosphere, and reduce ozone levels in the lower stratosphere. Ozone levels in the lower stratosphere dominate total column ozone, and the model results differ most in this region. In the simulations of Evans et al. (1998), reductions in lower-stratospheric ozone levels occur only in the tropics when water vapor increases, while in the other simulations, the reductions extend to the mid-latitudes or the poles. The models of Dvortsov and Solomon (2001) and Shindell (2001) projected a slower recovery of the ozone layer as a result of increased stratospheric water vapor, and a 1 to 2% reduction in ozone levels over the next 50 years compared to what would be expected if water vapor did not increase.

Water vapor affects heterogeneous chemistry by enhancing the formation of polar stratospheric clouds (PSCs). This effect may be much more important than the relatively small impacts of water vapor on homogeneous chemistry. Kirk-Davidoff et al. (1999) projected a significant enhancement of arctic ozone depletion in a more humid atmosphere. Much of this projected effect is based on the radiative cooling of the stratosphere assumed to be induced by water vapor, a value that is currently uncertain. A smaller value would imply a reduced role for water vapor in enhancing PSC formation. Even using a smaller cooling rate, however, the impact on ozone is likely to be large, as the $\sim 3\ ^{\circ}C$ cooling of the stratosphere projected to occur if CO_2 concentrations double is of comparable magnitude to the cooling that would be caused by a water vapor increase of only ~ 2 ppmv. Although precise quantification of radiative forcing due to water vapor is difficult, an estimate by Tabazadeh et al. (2000) suggests that an increase of 1 ppmv in stratospheric water vapor (with constant temperature) would be equivalent to a $\sim 1\ ^{\circ}C$ decrease in stratospheric temperature and would cause a corresponding increase in PSC formation. This comparison suggests that the radiative impact of water vapor is larger than its effects on chemistry or microphysics. Given the potential for atmospheric changes in the Arctic, and the large ozone losses that could result from a slight cooling (Tabazadeh et al., 2000), it is important both to understand trends in stratospheric water vapor, and to resolve differences in model projections of the radiative impact of those trends.

Changing concentrations of the trace gases N_2O and CH_4 may also affect ozone levels. Nitrous oxide breaks down to release nitrogen oxide radicals, which are extremely reactive and play an important role in ozone chemistry. Increases in CH_4 concentrations lead to an increase in hydrogen oxide radicals but at the same time increase the sequestration of chlorine radicals into HCl. The effects of increases in these gases on ozone depletion are thought to be relatively small (e.g., Shindell et al., 1998b; WMO, 1999), although a recent study by Randeniya et al. (2002) suggests that increasing concentrations of N_2O may have a larger impact than previously thought.

5.2.2. Arctic ozone depletion and meteorological variability

Partitioning the transport and chlorine chemistry contributions to arctic ozone variability is a subject of much discussion (Shepherd, 2000). The degree of ozone depletion in the Arctic depends strongly on air temperatures and PSC formation. Several methods have been used to estimate the total column ozone depletion in the arctic polar vortex based on meteorological measurements (e.g., Goutail et al., 1999; Manney et al., 1996; Müller et al., 1997; Rex et al., 1998), and comparisons between the different studies show good agreement (Harris et al., 2002). Since 1988–1989, three winters (1994–1995, 1995–1996, and 1999–2000) have had particularly low stratospheric temperatures and were characterized by PSC formation in both the early and late parts of the season (Braathen et al., 2000; Pawson and Naujokat, 1999). Some of the most severe arctic ozone losses (up to 70% at 18 km altitude) were observed during those winters (Knudsen et al., 1998; Rex et al., 1999; 2002).

Chipperfield and Pyle (1998) used models to investigate the sensitivity of ozone depletion to meteorological variability, chlorine and bromine concentrations, denitrification, and increases or decreases in stratospheric water vapor. Although the models tended to underestimate observed rates of arctic ozone depletion, their results agreed at least qualitatively with empirical estimates of ozone depletion, which suggest that substantial arctic ozone depletion is possible when both early and late winter temperatures in the stratosphere are extremely low. Cold early winters or cold late winters alone are not enough to produce extensive ozone depletion, but can still cause depletion to occur. During the winters of 1993–1994 and 1996–1997, temperatures in the arctic stratosphere were very low in late winter compared to earlier in the season. Ozone losses at specific altitudes during these years were of the order of 40 to 50% (Braathen et al., 2000; Schulz et al., 2000).

Dynamic processes dominate the short-term (day-to-day) variability in winter and spring total column ozone at mid- and high latitudes. Local changes in total column ozone of the order of 100 DU have been frequently reported (e.g., Peters et al., 1995) and are linked to three main transport processes:

1. A shift in the location of the polar vortex leads to changes in total column ozone, because the polar vortex air masses are characterized by low ozone levels compared to air masses outside the vortex.
2. Tropical upper-tropospheric high-pressure systems moving to higher latitudes cause an increase in the height of the tropopause at those latitudes, and thus a reduction in the overall depth of the stratospheric air column, as a result of divergence, resulting in ozone redistribution and a decline in total column ozone (e.g., James, 1998).
3. Tropical lower-stratospheric or upper-tropospheric air masses may be mixed into the stratosphere at higher latitudes. Referred to as "streamers" (e.g., Kouker et al., 1999), these phenomena introduce lower ozone content to the high-latitude air masses.

These three transport processes are not independent and can occur simultaneously, potentially increasing total column ozone variability.

Box 5.1. The polar vortex and polar stratospheric clouds

Winter and early spring ozone levels in the Arctic are influenced by the *polar vortex*, a large-scale cyclonic circulation in the middle and upper troposphere. This circulation keeps ozone-rich mid-latitude air from reaching the vortex region and can also lead to very cold air temperatures within the vortex.

Cold temperatures allow the formation of *polar stratospheric clouds* (PSCs), which play two important roles in polar ozone chemistry. First, the particles support chemical reactions leading to active chlorine formation, which can catalytically destroy ozone. Second, nitric acid removal from the gas phase can increase ozone loss by perturbing the reactive chlorine and nitrogen chemical cycles in late winter and early spring (WMO, 2003).

As the stratosphere cools, two types of PSCs can form. *Type-1* PSCs are composed of frozen nitric acid and water and form at temperatures below 195 K. At temperatures below 190 K, *Type-2* PSCs may form. Type-2 PSCs are composed of pure frozen water and contain particles that are much larger than the Type-1 PSC particles. Both types of PSCs occur at altitudes of 15 to 25 km and can play a role in ozone depletion chemistry, although Type-2 PSCs are quite rare in the Northern Hemisphere.

The occurrence of ozone minima or ozone "mini-holes" at northern mid- and high latitudes caused by tropopause lifting (process 2) exhibits high interannual variability. James (1998) found no detectable trend in mini-hole occurrences using Total Ozone Mapping Spectrometer (TOMS) satellite data for the period from 1979 to 1993. However, an analysis of satellite data by Orsolini and Limpasuvan (2001) found an increase in the frequency of ozone mini-holes in the late 1980s and early 1990s. The increase may be linked to the positive phase of the North Atlantic Oscillation (NAO; see section 2.2.2.1), which displaces the westerly jet to higher latitudes, allowing pronounced northward intrusions of high-pressure systems (processes 2 and 3). A similar link between the NAO and the frequency of ozone mini-holes has been found in ground-based measurements (Appenzeller et al., 2000).

Coupled chemistry-climate models are currently able to simulate these meteorological phenomena (Eyring et al., 2003; Stenke and Grewe, 2003). Stenke and Grewe (2003) compared simulations from a coupled chemistry-climate model with TOMS data and showed that ozone minima were fairly well represented in the simulations. Such simulations suggest that the processes affecting PSC formation can significantly increase chemical ozone depletion, leading to mini-hole occurrences or other substantial ozone minima.

5.2.3. Large-scale dynamics and temperature

The Arctic is highly affected by atmospheric processes, and mid- and high-latitude dynamics can play an important role in arctic ozone depletion. The Northern Hemisphere is characterized by large landmasses and several high mountain ranges at middle and high latitudes. These geographic features generate planetary-scale atmospheric waves that disturb the northern polar vortex. As a result, the polar vortex tends to be less stable and less persistent over the Arctic than over the Antarctic. Ozone depletion over the Arctic has therefore been less severe than that over the Antarctic, but is still greater than the depletion observed at tropical or mid-latitudes. Ozone depletion in the Arctic is characterized by large interannual variability, depending largely on the strength of the polar vortex and on air temperatures within it. During years when the polar vortex was especially strong, substantial (up to 40%) total column ozone depletion was observed (Weatherhead, 1998; WMO, 2003).

Changes in the dynamics of the stratosphere play a role in long-term trends as well as in inter- and intra-annual variability in arctic ozone levels. The stratospheric circulation determines how much ozone is transported from the lower-latitude production regions, as well as the extent, strength, and temperature of the winter polar vortex. The variability of polar vortex conditions is strongly influenced by fluctuations in the strength of the planetary-wave forcing

of the stratosphere. There is evidence from both observations and modeling studies that long-term trends in arctic ozone levels are not solely driven by trends in halogen concentrations, but are also a function of changes in wave-driven dynamics in the stratosphere (Fusco and Salby, 1999; Hartmann et al., 2000; Hood et al., 1999; Kodera and Koide, 1997; Kuroda and Kodera, 1999; Pitari et al., 2002; Randel et al., 2002; Shindell et al., 1998a; Waugh et al., 1999). During years in which planetary waves penetrate effectively to the stratosphere, the waves enhance the meridional Brewer-Dobson circulation, which brings more ozone from the low-latitude middle and upper stratosphere to the polar region and then down to the arctic lower stratosphere. At the same time, the planetary waves are likely to disrupt the polar vortex, reducing the occurrence of temperatures low enough for PSC formation. Increased planetary-wave activity is thus highly correlated with greater ozone levels, but projections of future wave forcing remain uncertain (WMO, 2003).

Extremely low stratospheric temperatures (below 190 K) in the polar regions can lead to the formation of PSCs (Box 5.1). Polar stratospheric clouds contribute significantly to ozone chemistry, leading to accelerated ozone destruction. Over the Antarctic, stratospheric temperatures are routinely lower than these thresholds every spring. Over the Arctic, stratospheric temperatures are often near these critical temperature thresholds, such that during periods when the temperatures are slightly lower than average, accelerated ozone depletion is observed, while during periods when the temperatures are slightly higher than average, ozone levels can appear climatologically normal. Current climate models suggest that stratospheric temperatures are likely to decrease in the coming decades as a result of increasing atmospheric concentrations of greenhouse gases, thus, it is likely that there will be more periods when accelerated ozone destruction could occur. The combination of dynamics, interannual variability, and the coupling between chemistry and radiative forcing makes projecting future arctic stratospheric temperatures and ozone depletion extremely challenging.

5.3. Long-term change and variability in ozone levels

In the early 1970s, scientists began projecting that anthropogenic emissions of CFCs and other halocarbons would lead to stratospheric ozone depletion (Molina and Rowland, 1974). These projections were confirmed when the Antarctic ozone hole was discovered in 1985 (Farman et al., 1985), and subsequent work (e.g., Anderson et al., 1989) identified and refined the chemical mechanisms that are responsible for ozone depletion. Since that time, decreases in stratospheric and total column ozone have been reported over both poles and in the mid-latitudes in both hemispheres.

5.3.1. Monitoring stratospheric ozone over the Arctic

Ground-based and satellite-borne instruments are used to monitor the concentrations and vertical distributions of stratospheric ozone. Ground-based Dobson spectroradiometers have been used since the 1920s, and Brewer spectroradiometers have been introduced more recently to provide both ozone and UV radiation monitoring. Currently, more than 30 Dobson and Brewer instruments are operated in or near the Arctic. In Russia, total ozone is monitored using filter radiometers. In addition to these measurements of total column ozone, the vertical ozone distribution or ozone profile can be measured using ozonesondes (balloon-borne measuring devices). Figure 5.3 shows the current network of regularly reporting total ozone and ozonesonde stations in or near the Arctic. The ground-based monitoring network provides the longest and most accurate record of stratospheric ozone levels. In addition to ground-based monitoring, various satellite-borne instruments have been in orbit since the 1970s and are able to provide global spatial coverage not available from ground-based networks. Because the ground-based monitoring network does not cover all parts of the Arctic, monitoring arctic ozone levels relies on a combination of ground-based and satellite-borne instruments.

5.3.2. Total column ozone on a global scale

Total column ozone is a measure of the total number of ozone molecules in a column of atmosphere above a particular location. Total column ozone is important because of its direct, measurable effect on the amount of UV radiation reaching the surface. The variability in

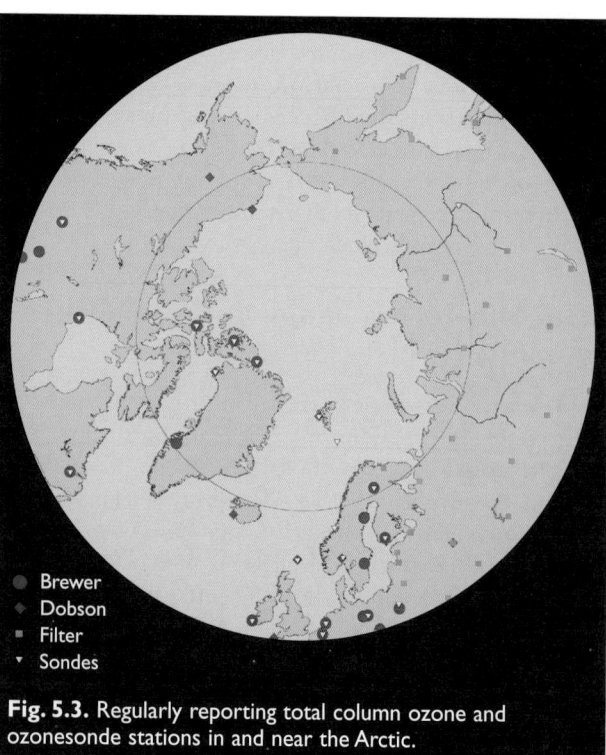

Fig. 5.3. Regularly reporting total column ozone and ozonesonde stations in and near the Arctic.

total column ozone at a single location is strongly influenced by the movement of air from one region to another. Thus, total column ozone averages over the entire globe, or over large regions, often show less variability than total column ozone at a specific location (Bodeker et al., 2001). Although ozone measurements have been made by satellite-borne instruments since the late 1970s, orbits and instrument capabilities have not always ensured year-round monitoring of conditions in the arctic stratosphere.

Instrument drift, problems with calibration, and other issues influencing data continuity can all affect estimates of ozone levels derived from satellite data. Careful comparison with well-calibrated ground-based instruments has helped resolve many of these difficulties, and the satellite data have been used in many analyses of ozone depletion (e.g., Herman and Larko, 1994; McPeters et al., 1996; Newman et al., 1997; Reinsel et al., 1994; Staehelin et al., 2002; Weatherhead et al., 2000). The results indicate strong downward trends in stratospheric ozone amounts, particularly during the late winter and spring. The data show strong latitudinal variability as well as observable longitudinal variations.

Several datasets of zonal total column ozone values were compared and used to estimate long-term changes in total column ozone. The datasets were prepared by different groups and are based on TOMS, Solar Backscatter Ultraviolet (SBUV, SBUV/2), Global Ozone Monitoring Experiment, and ground-based measurements (Fioletov et al., 2002). To avoid problems of missing data at high latitudes, and to estimate global total ozone, it was assumed that deviations from the long-term mean over regions with no data (such as over the poles) were the same as the deviations in the surrounding latitude belts. The results suggest that global average total column ozone in the late 1990s was 3% lower than in the late 1970s.

5.3.3. Total column ozone trends

The decline in total column ozone is a function of the solar cycle, atmospheric dynamics, chemistry, and temperatures. In general, the agreement between the long-term trends in total column ozone obtained from satellite and ground-based data is very good: both indicate a latitudinal variation in the trends, with values close to zero over the equator and substantial declines outside the 35° S to 35° N zone.

Satellite data indicate that variations in the total column ozone trends are predominantly latitudinal, with some smaller longitudinal differences. The greatest decrease in total column ozone over the Northern Hemisphere high latitudes (7% per decade) occurred in the spring (March–May) over the subpolar regions of Siberia, northern Europe, and the Canadian Arctic. These longitudinal differences correspond at least partially to large relative decreases during the winter and spring, which occur when air masses with relatively low ozone concentrations typical of the polar vortex are transported over regions with high climatological ozone values. In these situations, the decrease in total column ozone is not limited to the polar vortex area alone (WMO, 1999, 2003). Unlike the winter and spring depletion, the summer and autumn decrease in total column ozone over the Northern Hemisphere has been smaller and more uniform with longitude.

5.3.4. Variations in arctic total column ozone

Variations and trends in total column ozone over the Arctic are similar to those over mid-latitudes. However, a strong polar vortex in late winter and early spring leads to an additional decrease in total column ozone. Extremely large decreases in total column ozone over the Arctic were observed in certain years,

for example, in 1993 and 1997 (Fig. 5.4), which have been partly attributed to a strong polar vortex during those years. Because of the large interannual variability in the strength of the vortex, ozone decreases in the late 1990s and early 2000s were not as large. Decreases in total column ozone associated with the polar vortex can be as large as 45% over vast areas and can last longer than two weeks (Weatherhead, 1998). These traits make vortex-related decreases different from local anomalies or mini-holes, which are caused by advections of tropical and polar air into the mid-latitudes. Mini-holes can be as deep as 35 to 40%, but last only a few days (Weatherhead, 1998).

The trend in mean annual total column ozone over the Arctic was approximately -3% per decade for the period from 1979 to 2000 (a total decrease of about 7%). Trends depend on season; the trend in mean spring total column ozone was approximately -5% per decade for the period from 1979 to 2000 (a total decrease of 11%). Large mean monthly decreases in total column ozone (30–35% below pre-depletion levels) were reported in March 1996 and 1997. Some of the daily total column ozone values during these months were below 270 DU, or 40 to 45% below pre-depletion levels.

5.3.5. Ozone profiles

The vertical distribution of ozone within the column plays a lesser role than the total column ozone in determining surface UV radiation levels. At the present time, approximately 20 stations measure vertical ozone profiles during the winter and spring. Measurements of the vertical profile of ozone concentration using ozonesondes have been made weekly since 1980 at several sites in Canada (Edmonton, Goose Bay, Churchill, and Resolute), since 1987 at Alert, and since 1992 at Eureka. Ozone soundings are also performed regularly at Sodankylä, Finland; Ny Ålesund, Norway; Scoresbysund and Thule, Greenland; and Yakutsk, Russia; and occasionally at Bear Island, Norway. In 1988, Europe, Canada, and Russia coordinated an ozonesonde network to measure ozone amounts within the polar vortex. The network consists of 19 stations and has provided assessments of chemical ozone loss for almost every winter since 1988–1989 (Rex et al., 2002). Preliminary analysis of the profiles suggests that trends in ozone concentrations as a function of altitude are most significant in the lower and middle stratosphere, at pressure altitudes of approximately 100 to 25 hPa.

5.4. Factors affecting surface ultraviolet radiation levels in the Arctic

The factors that affect UV radiation levels in the Arctic are generally well established (WMO, 2003), and are illustrated in Fig. 5.5. Atmospheric ozone levels, solar zenith angle, clouds, aerosols, and altitude are all major factors affecting UV radiation levels reaching the surface of the earth. In the Arctic, snow and ice cover add

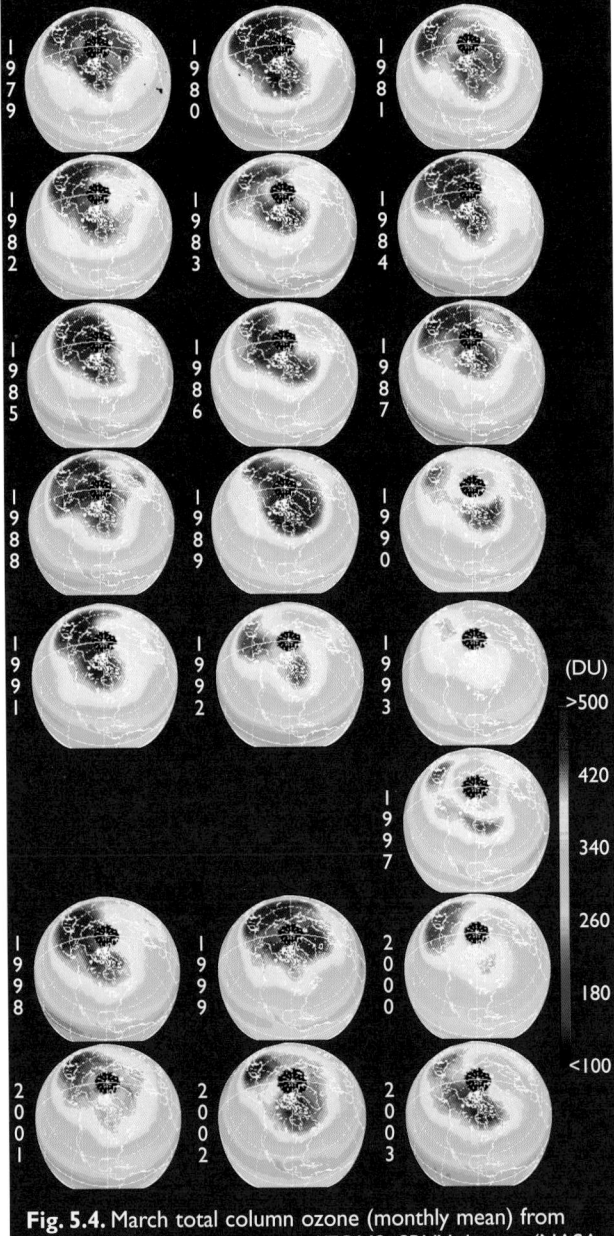

Fig. 5.4. March total column ozone (monthly mean) from 1979 to 2003, from the merged TOMS+SBUV dataset (NASA Goddard Space Flight Center, 2004).

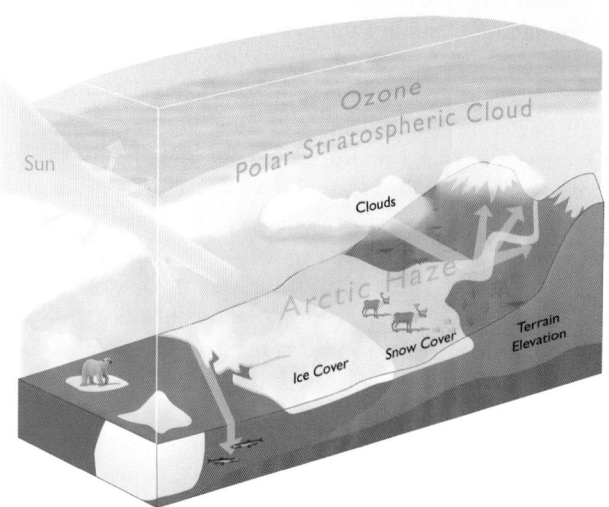

Fig. 5.5. Factors affecting UV radiation in the Arctic.

further complexity to the estimation of UV radiation exposure. When UV radiation passes through the atmosphere it is partially absorbed by ozone, and scattered by air molecules, aerosol particles, and clouds. Attenuation of UV-B radiation as it passes through the ozone layer is primarily a consequence of the sharp increase in the ozone absorption cross section at shorter wavelengths. The ratio of diffuse to global (direct and diffuse) radiation is greater in the UV than in the visible spectrum, primarily due to the wavelength dependence of Rayleigh scattering. Moreover, the ratio is usually

higher in the Arctic than at lower latitudes due to large solar zenith angles and frequent snow cover.

Many of the factors affecting UV radiation have large natural variations, which makes it difficult to discern changes in UV radiation levels that result from ozone depletion. Furthermore, the factors are not independent but interact in complex ways. For example, enhancement of surface UV irradiance by multiple scattering depends on both surface albedo and cloud conditions. These features make polar regions, including the Arctic, unique and complex in terms of their UV radiation environments.

Table 5.1 summarizes the factors that affect surface UV radiation levels in the Arctic.

5.4.1. Extraterrestrial solar spectrum

The radiation output of the sun varies over a range of timescales. Over the last century, the largest variation has been the 11-year solar cycle, which can be estimated by the average number of sunspots. The variation in solar irradiance is dependent on wavelength, with greater variability at shorter wavelengths (Solanki and Unruh, 1998). It has been estimated, using models and data from the Upper Atmosphere Research Satellite Solar Stellar Irradiance Comparison Experiment instrument, that although the total solar irradiance varies by only about 0.1% over the 11-year solar cycle, the amplitude of variation is as high as 8.3% for wave-

Table 5.1. Factors affecting surface UV irradiance in the Arctic.

Factor	Correlation with UV doses	Summary remarks
Solar activity	Negative	In the past century, changes in solar activity have caused fluctuations in surface UV irradiance on the order of a few percent.
Solar zenith angle	Negative	Diurnal and seasonal changes in solar zenith angle depend on latitude. In the Arctic, seasonal variations are extreme while diurnal variations are smaller than those at lower latitudes.
Atmospheric ozone	Negative	The amount of ozone in the stratosphere directly affects the amount of UV radiation reaching the troposphere and the surface of the earth.
Cloudiness	Negative/Positive	Thick clouds can attenuate UV radiation reaching the surface of the earth by tens of percent. Multiple reflections between clouds and snow-covered surfaces can lead to increases in surface UV irradiance, also of the order of tens of percent.
Atmospheric aerosols	Negative	Aerosols can attenuate UV radiation reaching the surface of the earth.
Altitude	Positive	Estimated changes in erythemal UV irradiance with altitude vary from 7 to 25% per 1000 m altitude gain.
Surface albedo	Positive	Reflection off snow can increase surface UV doses by more than 50%.
Snow and ice cover	Negative/Positive	Changes in the extent and duration of snow or ice cover can expose organisms currently shielded from UV radiation. Organisms living above the snow or ice cover will receive lower UV doses as melting snow or ice reduces the surface albedo.
Water quality	Not applicable	The amount of UV radiation penetrating through water is affected by UV-absorbing dissolved organic carbon. Organisms in the near-surface layer experience the greatest exposure to UV radiation.
Receptor orientation	Not applicable	The UV radiation doses received by a vertical surface (such as eyes or face) in the Arctic can be substantially higher than those that are received by a horizontal surface.

lengths in the 200 nm range and 0.85% for wavelengths in the 300 nm range (Lean, 2000).

Fligge and Solanki (2000) reconstructed solar spectral irradiance from 1700 to the present using a model of the magnetic features of the surface of the sun. Their results suggest that since the Maunder solar activity minimum in 1700, solar irradiance has increased by approximately 3% at wavelengths shorter than 300 nm. According to Rozema et al. (2002), the increased solar activity since 1700 has led to enhanced atmospheric ozone production and reduced surface UV-B irradiance. Thus, while the 11-year solar cycle has only a small effect on surface UV-B irradiance, longer-term variations in solar activity have the potential to affect future UV radiation levels.

The amount of UV radiation reaching the earth also depends on the distance between the earth and the sun. Due to the eccentricity of the orbit of the earth, this distance varies throughout the year. The earth is closest to the sun on 3 January (perihelion) and farthest away on 4 July (aphelion). The difference between the perihelion and aphelion distances is about 3%, and therefore extraterrestrial irradiance is about 7% higher during the austral (Southern Hemisphere) summer than it is during the boreal (Northern Hemisphere) summer.

5.4.2. Solar zenith angle

The solar zenith angle (SZA) is the angle between zenith and the position of the sun. Its cosine is approximately inversely proportional to the path length that the direct solar beam has to travel through the atmosphere to reach the surface of the earth. At large SZAs, when the sun appears low in the sky, atmospheric gases and aerosols absorb more UV radiation owing to the longer path length that photons must travel. Variations in the SZA cause clear diurnal and annual variations in surface UV radiation levels. The SZA is also responsible

for most of the latitudinal variation in surface UV radiation levels. The percentage change between summer and winter UV radiation levels is higher in the Arctic than at lower latitudes, while diurnal variations in the SZA are smaller at higher latitudes. In general, SZAs are large in the Arctic and therefore, arctic UV irradiances are typically lower than those at lower latitudes. However, when daily integrated doses are compared, the length of arctic summer days somewhat compensates for the effect of large SZAs. The annual variation of the clear-sky daily erythemal dose at latitudes of 50°, 70°, and 80° N is shown in Fig. 5.6. The values are based on radiative transfer calculations assuming moderate polar ozone levels (300 DU), snow-free conditions with a surface albedo of 0.03, and clear skies. The seasonal variation in erythemal dose is caused solely by the seasonal variation in the SZA.

5.4.3. Ozone levels

Absorption by ozone causes attenuation of UV-B irradiance. It has been repeatedly demonstrated that a decrease in total column ozone leads to an increase in UV radiation levels (WMO, 2003). The relationship depends somewhat on the vertical distribution of ozone in the atmosphere. At small SZAs, a redistribution of ozone from the stratosphere to the troposphere leads to a decrease in UV-B radiation levels at the surface (Brühl and Crutzen, 1989). At very large SZAs, this redistribution leads to an increase in UV-B radiation levels (Krotkov et al., 1998). Lapeta et al. (2000) and Krzyscin (2000) further quantified this effect, and concluded that the erythemally weighted UV dose rate varies by a maximum of 5% owing to changes in the ozone profile.

The change in surface UV irradiance as a result of a change in total column ozone depends highly on the wavelength of the radiation. Traditionally, radiation amplification factors (RAFs) have been used to quantify the change in biologically effective irradiances as a result of a change in total column ozone (e.g., Booth and Madronich, 1994; van der Leun et al., 1989; WMO, 1989). These factors can also be used to indicate the sensitivity of a particular UV radiation effect to a change in total column ozone. Values of RAFs depend largely on the biological effect and vary between 0.1 and approximately 2.5 (Madronich et al., 1998). The RAF for the standard erythemal action spectrum (CIE, 1998) is 1.1 at small SZAs (Madronich et al., 1998), indicating that a 1% decrease in total column ozone leads to a 1.1% increase in erythemal UV radiation. For large changes in total ozone, the relationship is nonlinear, and a more complex relationship is required to estimate the corresponding changes in biologically effective UV radiation (Booth and Madronich, 1994). In the Arctic, where SZAs are often large, RAFs should be used with caution due to their pronounced dependence on the SZA and on total column ozone at large SZAs (Micheletti et al., 2003). For example, at an 80° SZA and total column ozone of

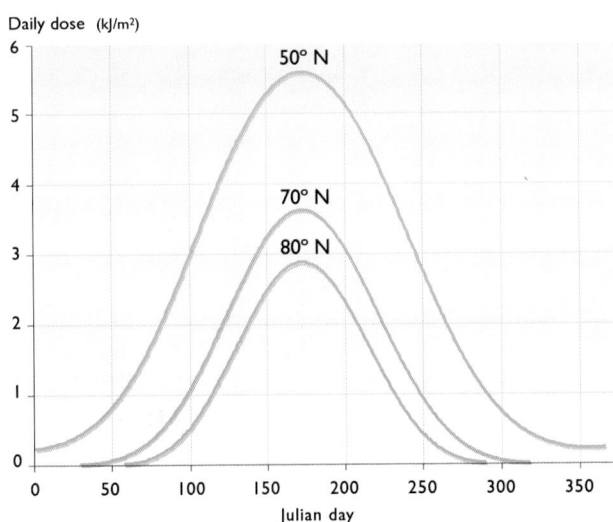

Fig. 5.6. Modeled clear-sky daily erythemal UV radiation dose at latitudes of 50°, 70°, and 80° N.

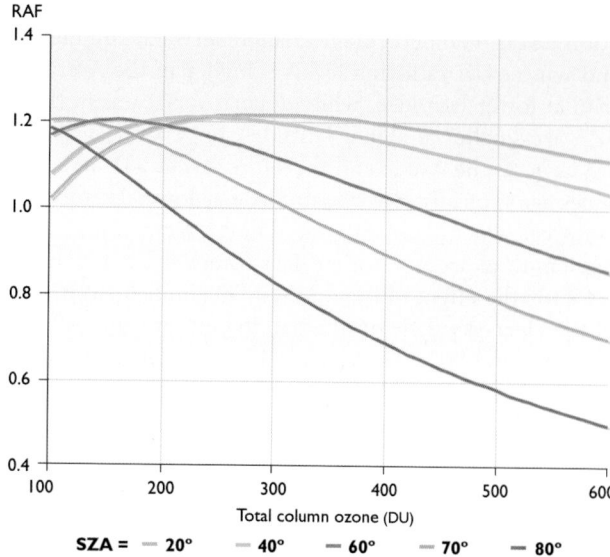

Fig. 5.7. Erythemal radiation amplification factor (RAF) as a function of total column ozone and solar zenith angle (SZA).

300 DU, the erythemal RAF is reduced to approximately 0.8, which is about 27% lower than that for smaller SZAs typical of the mid-latitudes (Fig. 5.7).

5.4.4. Clouds

The effect of clouds on UV radiation is difficult to quantify because of their complex three-dimensional character and rapid temporal variation. A uniform cloud layer generally leads to a decrease in irradiance at the surface of the earth, because part of the radiation that is reflected upward by the cloud layer escapes into space. However, local surface UV irradiance can be increased if clouds are not obstructing the disk of the sun and additional radiation is reflected from the side of a broken cloud field toward the ground (Mims and Frederick, 1994; Nack and Green, 1974). In meteorology, cloud cover is traditionally measured in "octas". The sky is divided into eight sectors and the octa number, between zero and eight, is based on the number of observed sectors containing clouds. Bais et al. (1993) and Blumthaler et al. (1994a) showed that when the solar disk is clear of clouds, cloud amounts up to six octas have little effect on irradiance compared to clear-sky situations. Thiel et al. (1997) and Josefsson and Landelius (2000) have further parameterized the attenuation of UV irradiance as a function of cloud cover and type.

Cloud transmittance of UV radiation depends on wavelength (Frederick and Erlick, 1997; Kylling et al., 1997; Seckmeyer et al., 1996). The maximum transmittance occurs at approximately 315 nm, although the actual location of this maximum depends on the cloud optical depth, the amount of tropospheric ozone, and the SZA (Mayer et al., 1997). In general, clouds in the Arctic tend to be optically thinner than clouds at lower latitudes owing to reduced atmospheric water vapor content. When the ground is covered by snow, attenua-

tion of UV radiation by clouds is further diminished owing to multiple scattering between the ground surface and the cloud base (Nichol et al., 2003).

5.4.5. Aerosols

Aerosols are solid or liquid particles suspended in the atmosphere, found primarily in the lower part of the troposphere. The attenuation of surface UV irradiance by aerosols depends on the aerosol optical depth (AOD), single scattering albedo, asymmetry factor, and aerosol profile. Measurements of AOD are routinely carried out at visible and UV-A wavelengths (e.g., Holben et al., 1998). The AOD is generally assumed to follow Ångström's law, which states that AOD is proportional to $\lambda^{-\alpha}$, where λ is wavelength and α is the Ångström coefficient. Converting the AOD measured at longer wavelengths to an AOD value for the UV-B spectrum is not straightforward, however, because α is not easy to measure and is likely to have some wavelength dependence. The single scattering albedo is the ratio of the scattering cross section of the aerosol to its extinction cross section, and is typically greater than 0.95 in relatively unpolluted areas of the Arctic (d'Almeida et al., 1991).

Episodes of long-range transport of pollutants have been observed in the Arctic. These episodes, combined with the lower rates of particle and gas removal in the cold and stable arctic atmosphere, can lead to a phenomenon called "arctic haze" (Shaw, 1985, 1995). Arctic haze events result in increased aerosol concentrations and mostly occur in winter and spring.

Relatively few studies have addressed the role of aerosols in attenuating solar UV radiation in the Arctic. Wetzel et al. (2003) conducted field investigations at Poker Flat, Alaska, and sampled different air mass types originating from sources outside the region. The measured AOD at 368 nm ranged from 0.05 to 0.25, and

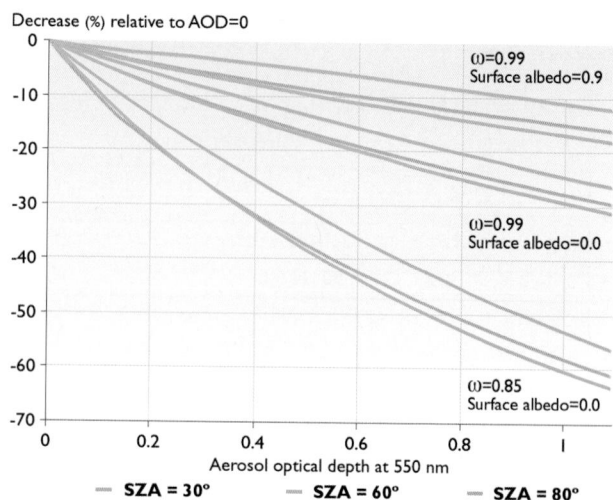

Fig. 5.8. Percentage decrease in erythemal UV irradiance as a function of aerosol optical depth (AOD), based on theoretical calculations with a radiative transfer model (ω=aerosol single scattering albedo).

estimates for the single scattering albedo varied from 0.63 to 0.95, the former being for spring air masses originating from Asia and the latter for cleaner air masses of marine origin. Herber et al. (2002) summarized eight years of measurements of AOD at the Koldeway station in Ny Ålesund, Norway, and reported strong arctic haze events, mainly in late winter and spring. The mean AOD at 371 nm during arctic haze conditions was about 0.18, while in the autumn the average AOD was only 0.05. Quinn et al. (2002) presented results from three years of simultaneous measurements of aerosol chemical composition and light scattering and absorption at Barrow, Alaska. They found that sulfate concentrations were highest at Barrow and decreased with latitude from Poker Flat to Denali to Homer, suggesting a north–south gradient. Ricard et al. (2002) studied the chemical properties of aerosols in northern Finland, and found that, compared to other arctic sites, the aerosols reflect smaller contributions from arctic haze and marine events in winter and larger contributions from biogenic sources in summer. For the range of aerosols sampled at Poker Flat, Alaska, Wetzel et al. (2003) found that the attenuation of UV radiation at 305 nm and 368 nm ranged from a few percent up to about 11%.

Figure 5.8 illustrates the decrease in erythemal UV irradiance as a function of AOD, based on theoretical calculations with a radiative transfer model (Mayer et al., 1997). The figure indicates that in the Arctic, where SZAs tend to be high, the reduction of erythemal UV irradiance by aerosols depends strongly on aerosol properties, including the single scattering albedo, and on surface properties, including surface albedo. In practice, AOD and single scattering albedo cannot be directly translated into UV attenuation, as the asymmetry factor, vertical distribution, and other factors must also be taken into account.

5.4.6. Altitude

Ultraviolet radiation levels increase with altitude for several reasons. At higher elevations, the atmosphere is optically thinner, and therefore fewer particles exist to absorb or scatter radiation. Higher elevations also experience a reduced influence from tropospheric ozone or aerosols in the boundary layer. In the Arctic and in mountainous regions, the ground is more likely to be covered by snow at higher altitudes, which leads to higher albedo and increased UV reflectance. Clouds below a mountain summit have a reflective effect similar to snow-covered ground, and will therefore increase UV radiation levels at the summit. In contrast, the same cloud may reduce UV radiation levels in a valley below the mountain. The variation of UV radiation levels with altitude depends on several factors, all of which have different wavelength dependencies; therefore, this variation cannot be expressed by a simple relationship. Changes in erythemal UV irradiance with altitude reported in the literature vary between 7 and 25% per 1000 m of altitude gain

(Blumthaler et al., 1994b, 1997; Gröbner et al., 2000; McKenzie et al., 2001a).

5.4.7. Surface albedo

The extent and duration of snow cover in the Arctic has a significant effect on surface UV radiation doses. An increase in surface albedo leads to an increase in downwelling UV radiation, as part of the radiation that is reflected upward is backscattered by air molecules or clouds. Snow is particularly efficient at reflecting UV radiation; multiple reflections between snow-covered ground and clouds, therefore, can lead to a significant increase in surface UV radiation levels compared to a snow-free situation (Kylling et al., 2000a).

Surface albedo at UV wavelengths is generally low, except in the presence of snow cover. Blumthaler and Ambach (1988) measured erythemally weighted surface albedos for various snow-free surfaces and reported values ranging between 0.01 and 0.11. Spectral measurements by Feister and Grewe (1995) and McKenzie and Kotkamp (1996) confirm these low values. For snow-covered surfaces, the measurements suggest values ranging from 0.50 to 0.98. In general, dry new snow has the highest albedo, which ranges from 0.90 to 0.98 (Grenfell et al., 1994). The albedo of a snow-covered surface depends not only on snow depth and condition, but also on topography, vegetation, and man-made structures (Fioletov et al., 2003). Albedo is an important factor affecting UV radiation levels in the Arctic, where the ground is covered by snow for extensive periods of the year. Figure 5.9 shows the spectral amplification of surface UV irradiance by surface albedo for clear-sky conditions. The figure indicates that snow cover, with an albedo that can be greater than 0.8, can increase erythemal irradiance by up to 60% compared to a snow-free case (albedo 0.2 or less). The amplification is greatest at short UV wavelengths, and thus increases the ratio of UV-B to UV-A radiation.

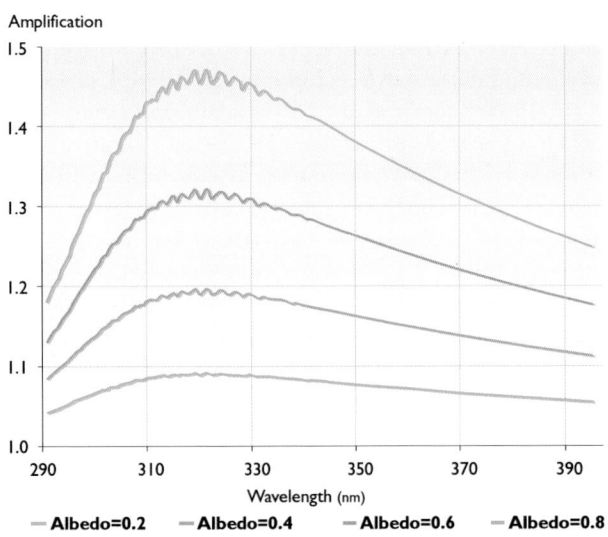

Fig. 5.9. Spectral amplification of surface UV radiation by surface albedo under clear-sky conditions (adapted from Lenoble, 1998).

Scattering in the atmosphere may occur far away from the location of interest; therefore, the ground properties of a large area around the measurement site must be considered. The regionally averaged albedo is often referred to as the "effective albedo" (Gröbner et al., 2000; Kylling et al., 2000b), and can be considered the albedo estimate that gives the best agreement between measured and modeled irradiances when used in a radiative transfer model. Three-dimensional radiative transfer models have shown that the area of significance, defined by an increase in UV irradiance of more than 5% when effective albedo is taken into account, can extend more than 40 km around the point of interest (Degünther et al., 1998; Lenoble, 2000; Ricchiazzi and Gautier, 1998).

5.4.8. Snow and ice cover

Many arctic ecosystems are shielded from UV radiation for much of the year by snow or ice cover. The transmission of UV radiation through snow or ice depends on wavelength, the thickness of the cover, and the optical properties of the snow or ice. In general, radiation is attenuated by a factor that changes exponentially with the thickness of the snow or ice cover. Shorter wavelengths are more strongly attenuated than longer wavelengths. Field measurements conducted at Alert, Canada, suggest that a 10 cm snow-cover depth reduces the amount of transmitted 321 nm UV radiation by two orders of magnitude (King and Simpson, 2001). According to Perovich (1993), approximately 1.3 m of white ice would be required to achieve a similar attenuation of transmitted 300 nm UV radiation. It is difficult to project how changes in snow and ice cover will affect the amount of UV radiation to which terrestrial and aquatic life forms in the Arctic are exposed. Observed and projected changes in sea-ice and snow cover are discussed in sections 6.3 and 6.4, respectively.

5.4.9. Water quality

The water quality parameters that are known to affect underwater UV radiation levels are dissolved organic carbon and chlorophyll a (Kuhn et al., 1999; Laurion et al., 1997; Morris et al., 1995; Scully and Lean, 1994). A general optical characterization of water columns is obtained from diffuse attenuation coefficients ($K_d(\lambda)$), which are calculated from measurements of spectral irradiance at various depths. For comparative purposes, wavelength-specific 10% depths (the depth to which 10% of the below-surface irradiance penetrates) are often derived from the $K_d(\lambda)$ values. It is important to note that the choice of 10% depth is arbitrary and is not based upon any correlation with biological effects. Figure 5.10 shows 10% penetration depths in Lake Cromwell, the St. Lawrence River estuary, and the Gulf of St. Lawrence, Canada. In clear ocean water, 10% of the radiation at longer UV wavelengths can penetrate to a depth of nearly 100 m. In shallower water, this depth may be of the

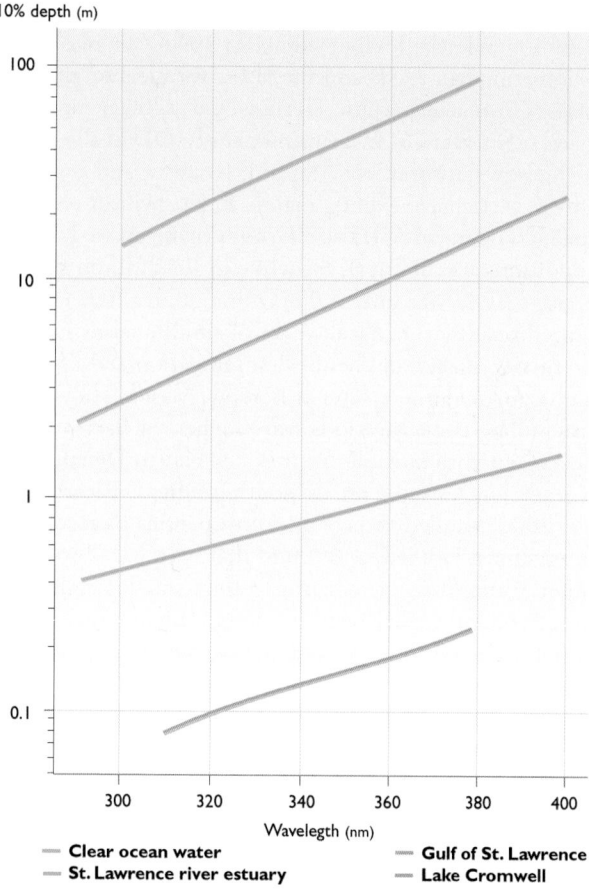

Fig. 5.10. Ten percent depth penetrations at selected locations in eastern Canada (data from Booth and Morrow, 1997; Kuhn et al., 1999; Scully and Lean, 1994; and Smith and Baker, 1979).

order of only a meter, but organisms living within this 1 m layer would still be at risk. At 310 nm, 10% penetration depths are 20 m for clear ocean water (Smith and Baker, 1979); 1 to 4 m for the Gulf of St. Lawrence and for coastal zones (Booth and Morrow, 1997; Kuhn et al., 1999); 0.5 m for estuarine waters; and 0.1 m for Lake Cromwell, Québec (Scully and Lean, 1994). Measurements made in arctic waters suggest that 10% penetration depths are typically less than 5 m (Aas et al., 2001). Ultraviolet-A radiation generally reaches greater depths. Organisms residing in the near-surface layer experience the greatest exposure to UV radiation.

5.4.10. Receptor orientation

Ultraviolet irradiance has traditionally been measured on a flat, horizontal surface. While this approach has sound physical merit, it does not accurately represent the UV irradiance that reaches many biological receptors. The amount of UV radiation incident on a vertical (as opposed to horizontal) surface has important biological implications, particularly in terms of effects on the eye (Meyer-Rochow, 2000; Sliney, 1986, 1987). Recent studies have explored both the effect of high snow reflectivity (e.g., McKenzie et al., 1998; Schmucki et al., 2001) and the orientation of the receptor on UV radiation doses. Some investigators have measured the amount of UV radiation incident on a surface oriented perpendicular to the rays of the sun (e.g., Philipona et al., 2001) while others have measured the amount incident on a vertical surface (Jokela et al., 1993; Weatherhead, 1998; Webb et al., 1999). As reported by Webb et al. (1999), the relationship between irradiance measured on a horizontal surface and that measured on a vertical surface depends on the orientation of the vertical surface, the SZA, and the wavelength. For wavelengths shorter than 400 nm, Webb et al. (1999) found distinct maxima in the vertical to horizontal irradiance ratios during the morning and afternoon. Under cloudless, snow-free conditions, the maximum ratios ranged from 1.4 at 300 nm to 7 at 500 nm. Snow cover, which increases surface albedo, may substantially increase these ratios. For example, in the presence of fresh snow cover and at SZAs greater than 60°, Philipona et al. (2001) reported a 65% increase in erythemal UV irradiance on a surface oriented perpendicular to the sun compared to the irradiance observed on a horizontal surface under the same conditions.

Similar results were obtained by Jokela et al. (1993), who pointed UV radiometers azimuthally South, North, West, and East to assess UV dose rates on vertical surfaces in Saariselkä, Finland. The results indicated a snow albedo of 0.83, in good agreement with data presented by Blumthaler and Ambach (1988). The ratios of vertical to horizontal dose rates varied from about 0.25 to 1.4, depending on direction and on whether the ground was barren or covered with fresh snow. In general, the observations indicated that spring ozone depletion could greatly increase ocular UV radiation doses because of the significant effect of snow reflection. The measurements by Jokela et al. (1993) show that ocular doses of UV radiation in Saariselkä can be higher at the end of April than at any other time of the year. These high doses suggest that the amount of UV radiation incident on the eye when looking toward the horizon can be equivalent or greater than the amount of UV radiation incident on the eye when looking directly upward.

For many biological systems, the actinic flux (the radiation incident at a point) is a more relevant quantity than the horizontal irradiance. Only recently have measurements of the spectral actinic flux become more common (Hofzumahaus et al., 1999; Webb et al., 2002). Webb et al. (2002) found that the ratio of actinic flux to horizontal irradiance varied between 1.4 and 2.6 for UV wavelengths, and depended on wavelength, SZA, and the optical properties of the atmosphere.

5.5. Long-term change and variability in surface UV irradiance

Several instruments and methods have been used to determine surface UV irradiance in the Arctic. Spectral and broadband radiometers, as well as narrowband multi-filter instruments, are used to measure surface UV irradiance. Quality-controlled measurements of UV irradiance have been available for little more than a decade in the Arctic, with limited spatial coverage. In addition to the direct ground-based UV irradiance measurements, surface UV irradiance can be reconstructed using satellite data, or by using observed total column ozone combined with commonly available meteorological data. In addition, historic UV-B radiation levels can be reconstructed using biological proxies.

5.5.1. Ground-based measurements

Surface UV irradiance measurements have been made in the Arctic for many years (Hisdal, 1986; Stamnes et al., 1988; Wester, 1997), and a discernible improvement in the quality of these measurements occurred during the 1990s. Lantz et al. (1999) and Bais et al. (2001) reported that similarly calibrated spectroradiometers typically differ by less than 5% in the UV-A spectrum, and by 5 to 10% in the UV-B spectrum. However, reliable data have only been available since the 1990s, which is inadequate for long-term trend analyses (Weatherhead et al., 1998). Nevertheless, the measured surface UV irradiance time series illustrate the variability of UV radiation and the role of different processes that affect UV irradiance at each of the monitoring sites. The ground-based UV irradiance records are also crucial for validating the indirect methods of estimating UV irradiance.

The current network of ground-based UV irradiance measurement stations in the Arctic is shown in Fig. 5.11. Only those installations operated on a regular basis are shown. The measuring instruments fall

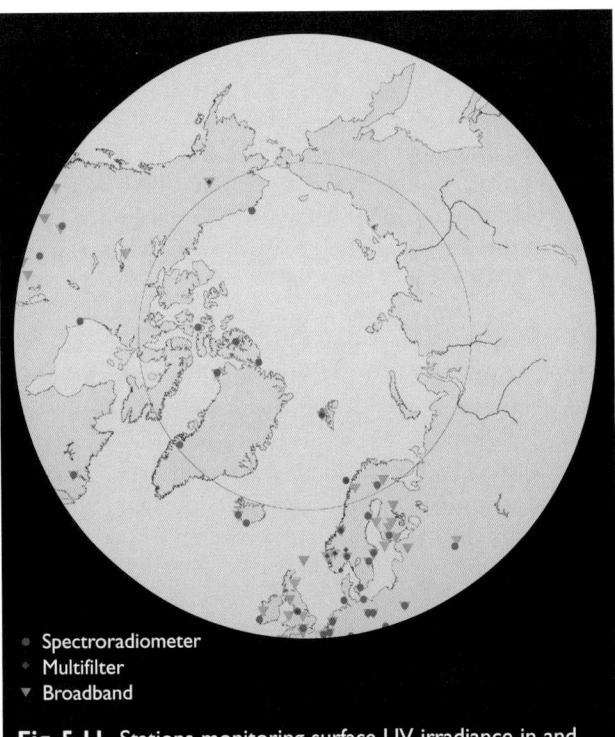

Fig. 5.11. Stations monitoring surface UV irradiance in and near the Arctic.

- • Spectroradiometer
- ✚ Multifilter
- ▼ Broadband

into three categories: spectroradiometers, multi-filter instruments, and broadband instruments. Spectroradiometers scan radiation in narrow wavelength bands with a typical resolution of 1 nm or less (Seckmeyer et al., 2001). They provide the greatest measurement accuracy and the spectral information enables versatile data use. However, spectroradiometers are costly and require trained personnel for maintenance and operation. Multi-channel or multi-filter instruments typically consist of several filtered photodetectors that measure radiation in selected wavelength bands (Bigelow et al., 1998; Dahlback, 1996; Harrison et al., 1994). They provide much faster sampling than conventional scanning spectroradiometers, allowing the evaluation of both short- and long-term changes in UV irradiance. Methods have also been developed to reconstruct the high-resolution spectrum from these multi-filter instrument measurements (Dahlback, 1996; Fuenzalida, 1998; Min and Harrison, 1998). Broadband instruments collect radiation over a portion of the spectrum and apply a weighting function to the measurements. Many of these instruments are designed to measure the erythemal irradiance. Broadband instruments are comparatively inexpensive and easy to maintain. However, their spectral response often deviates from the one that they are attempting to estimate. Thus, the calibration of broadband instruments depends on solar zenith angle and total column ozone: determining the calibration (Bodhaine et al., 1998; Mayer and Seckmeyer, 1996) and maintaining the long-term stability of the broadband instruments can be very demanding (Borkowski, 2000; Weatherhead et al., 1997). In addition to the instruments described previously, the UV radiation dose (the biological dose rate integrated over time), can be measured using biological UV dosimeters, which directly quantify the biologically effective solar irradiance affecting certain processes by allowing biological systems to act as UV radiation sensors. A number of different sensors have been developed, based on triggers that include direct DNA damage, the inactivation of bacterial spores and bacteriophages, photochemical reactions involving vitamin D photosynthesis, and the accumulation of polycrystalline uracil. Agreement with weighted spectroradiometer measurements has been shown for some of these systems (Bérces et al., 1999; Furusawa et al., 1998; Munakata et al., 2000).

The Canadian UV-monitoring network includes four arctic measurement sites: Churchill, Resolute, Eureka, and Alert. Monitoring at these sites began in 1992, 1991, 1997, and 1995, respectively. There are reports of increased UV-B radiation levels in the Arctic during six winters in the 1990s (Fioletov and Evans, 1997; Kerr and McElroy, 1993), but according to Tarasick et al. (2003), the maximum UV indices (section 5.1) measured at the Canadian Arctic monitoring sites were 8 in Churchill, 4 in Resolute and Eureka, and 3 in Alert. These values are relatively small compared to the UV indices measured at Canadian monitoring sites outside of the Arctic. However, the maximum observed UV index may not be the best indicator of UV radia-

tion doses received in the Arctic because it does not take into account day length or receptor orientation. Moreover, the Canadian UV measurements imply that snow is a major factor affecting UV irradiance at high latitudes, and may enhance the dose received by a horizontal surface by 40%.

Measurements of spectral UV irradiance at Sodankylä, Finland began in 1990. Lakkala et al. (2003) analyzed the Sodankylä data for the period from 1990 to 2001, incorporating corrections for temperature, cosine error, noise spikes, and wavelength shifts. No statistically significant changes in any month were found over the 12-year period, which may be a result of the relatively short period of analysis coupled with the high natural interannual variability of UV irradiance. The lack of a distinct trend over this particular period may also be linked to ozone levels, which decreased in the early 1990s, reached a minimum in the middle of the analysis period, and then increased in the very late 1990s and early 2000s. Arola et al. (2003) applied methods to separate the effects of the different factors that affect short- and long-term changes in UV irradiance to the Sodankylä data and found that in some cases, ozone levels accounted for nearly 100% of the short-term variability in monthly mean irradiance, although on average they accounted for about 35% of the variability. The effects of clouds were smaller, accounting for a maximum of 40% and an average of 12% of the short-term irradiance variability. Albedo-related effects were strongest at Sodankylä during the month of May, accounting for a maximum of 21% and an average of 7% of the short-term irradiance variability.

Measurements are also available from Tromsø, Norway and were recently used by Kylling et al. (2000a) to analyze the effects of albedo and clouds on UV irradiance. The results, shown in Fig. 5.12, indicate that snow

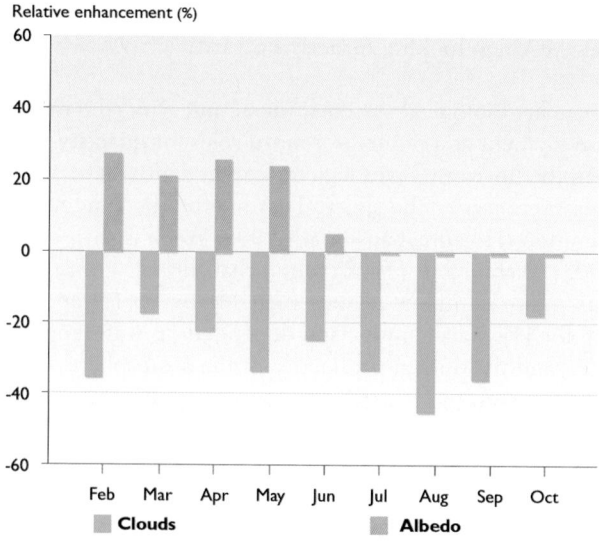

Fig. 5.12. Relative change in monthly erythemal UV radiation doses owing to the presence of clouds and albedo from snow cover (adapted from Kylling et al., 2000a).

increases monthly erythemal UV radiation doses by more than 20% in spring and early summer. The albedo effect at Tromsø is relatively small because the area is influenced by the Gulf Stream, which prevents the adjacent ocean from freezing over. In addition, the terrain around Tromsø is inhomogeneous, consisting of open fjords surrounded by high mountains. Results from a three-dimensional radiative transfer model (Kylling and Mayer, 2001) suggest that the inhomogeneous surface albedo reduces the effective albedo to only 0.55 to 0.60. On specific days, clouds can reduce daily erythemal UV radiation doses at Tromsø by up to 85%. Clouds reduce average monthly erythemal UV radiation doses by 20 to 40% (Fig. 5.12) and this attenuation is fairly constant throughout the year. Multiple scattering between the ground and cloud base links the effects of surface albedo and cloud cover. Attempts to separate the two factors (e.g., Arola et al., 2003 and Kylling et al., 2000a) are only approximations.

Spectroradiometer measurements at Barrow, Alaska commenced in January 1991 (Booth et al., 2001). Increased levels of UV radiation were observed in 1993, which were probably due to low ozone levels related to the injections of aerosols into the stratosphere from the eruption of Mt. Pinatubo (Gurney, 1998). Photochemically induced ozone depletion events at Barrow have primarily occurred during March and April. The most notable case occurred on 18 April 1997, when the daily erythemal UV radiation dose was 36% higher than the climatological average dose. This increase is still small compared to spikes seen in UV radiation measurements at sites affected by the Antarctic ozone hole. By comparing measured and modeled spectra, Bernhard et al. (2003) estimated that Barrow's effective UV albedo (surface albedo for UV wavelengths) during the winter and early spring is 0.85±0.10. The albedo is greater than that of Tromsø because the ocean adjacent to Barrow is frozen until late in the spring, and the treeless tundra surrounding the measurement site is covered by snow until June. The higher albedo leads to a 40 to 55% enhancement of erythemal UV radiation when skies are clear. No statistically significant changes in UV irradiance were found at Barrow for the period from 1991 to 2001 (Booth et al., 2001), for reasons similar to those for Sodankylä. It should be noted that the observed reductions in ozone levels have been much greater in the European and Russian sectors of the Arctic than over Alaska. UV radiation increases at Barrow are therefore less pronounced compared to those measured at the other arctic sites.

5.5.2. Reconstructed time series

In order to assess variations in UV irradiance over longer time periods, various empirical methods for reconstructing UV irradiance data have been developed. These methods are usually based on total column ozone or other commonly available weather data, including global (direct and diffuse) radiation levels.

Bodeker and McKenzie (1996), for example, presented a model based on total column ozone, broadband radiation measurements, and radiative transfer calculations. Similar methods were developed and used by McArthur et al. (1999), Kaurola et al. (2000), and Feister et al. (2002) for estimating UV irradiance at various locations in Canada and Europe. Gantner et al. (2000) estimated clear-sky UV irradiance using an empirically determined statistical relationship between clear-sky UV irradiance and total column ozone. The results obtained with reconstruction methods generally compare well with observations, but there are some sources of uncertainty. For example, accounting for the influence of aerosols on the amount of UV radiation reaching the surface of the earth is difficult, as is accounting for sulfur dioxide from volcanic and anthropogenic sources, although Fioletov et al. (1998) reported that sulfur dioxide had a negligible influence on erythemal UV irradiance at 12 out of 13 Canadian and Japanese sites. The importance of examining the homogeneity of the data cannot be overestimated when using long time series of measurements as input data for estimating past UV radiation levels. Likewise, it is of great importance to properly validate the empirical methods against independent measurement data.

Fioletov et al. (2001) developed a statistical model using global radiation levels, total column ozone, dew point temperature, and snow cover data to reconstruct past UV irradiance at several Canadian sites, including Churchill. The model is based on previous efforts to estimate UV-A irradiance from pyranometer measurements (McArthur et al., 1999). Fioletov et al. (2001) produced UV irradiance estimates for the period from 1965 to 1997, and found a statistically significant increase in the erythemally weighted UV irradiance (~11% per decade) over the period from 1979 to 1997. This increase follows a reported decline in UV radiation levels from 1965 to 1980. The increase in UV irradiance between 1979 and 1997 is more than twice the increase that would be expected to result from the observed decline in total column ozone, because concurrent changes in surface albedo and cloud conditions have enhanced the increase in surface UV irradiance over this period. Fioletov et al. (2001) also examined the frequency of extreme values of UV irradiance at Churchill, and reported that the number of hours when the UV index exceeds five increased from an average of 26 per year during the period 1970 to 1979 to an average of 58 per year during the period 1990 to 1997.

Díaz et al. (2003) estimated narrowband (303.030–307.692 nm) irradiances for Barrow from pyranometer data and satellite-measured total column ozone, and found that spring UV irradiance increased between 1979 and 2000. Lindfors et al. (2003) presented a method for estimating daily erythemal UV radiation doses using total column ozone, sunshine duration, and snow depth as input data. They estimated UV radiation doses at Sodankylä for the period from 1950 to 1999, and found a statistically significant March increase

(3.9% per decade). April also showed an increasing trend. For both March and April, the trend was more pronounced during the latter part of the period (1979–1999), suggesting a connection to stratospheric ozone depletion. For July, a statistically significant decreasing trend of 3.3% per decade was found, attributed to changes in total column ozone and sunshine duration (defined by the World Meteorological Organization as the time during which direct solar radiation exceeds 120 W/m²). June and August exhibited decreasing trends, although not statistically significant, and trends for May and September were negligible.

In addition to atmospheric observations, biological proxies can be used to estimate past UV irradiance. For example, Leavitt et al. (1997) proposed that the past UV radiation environment of freshwater ecosystems could be reconstructed by studying fossil pigments in lake sediments, because some algae and other aquatic organisms produce photoprotective pigments when exposed to UV radiation. However, the long-term variation of underwater UV irradiance is primarily controlled by the amount of dissolved organic matter, which limits the use of fossil pigment sediments for reconstructing surface UV irradiance without ancillary information about dissolved organic matter (Leavitt et al., 2003).

Archaeological findings offer further evidence that UV radiation has long been a concern in the Arctic. Goggles found in the ancestral remains of indigenous peoples have thin slits, which allow the wearer to see but limit the amount of sunlight reaching the eyes (Hedblom, 1961; Sliney, 2001). The indigenous peoples of the Arctic have historically relied heavily on the environment for their livelihood and well-being, and have therefore been acutely aware of changes in climate-related variables. Fox (2000) documented some of these observations, which include reports of sunburn and increased incidence of snow blindness in populations not normally experiencing these effects. The existence of goggles confirms that arctic peoples have long sought protection from the sun and its glare, but the reports of sunburn and other effects of UV radiation suggest that the recent changes in UV irradiance are unusual when considered in this longer context. This information serves as a useful proxy of UV irradiance over time, and provides firsthand evidence of UV radiation impacts resulting from ozone depletion in the Arctic. The information also helps to corroborate available UV radiation and ozone measurements.

5.5.3. Surface estimates from satellite data

Estimates of surface UV irradiance derived from satellite data use a radiative transfer model, with input parameters determined from the satellite measurements. In addition to total column ozone, satellite-derived information about clouds, aerosols, and surface albedo is needed to accurately model these parameters, and various sources of data have been used for that pur-

pose (Herman et al., 1999; Krotkov et al., 2001; Li et al., 2000; Lubin et al., 1998; Meerkoetter et al., 1997; Verdebout, 2000). The advantages of satellite measurements include global or near-global spatial coverage and long-term data continuity. However, there are limitations: the best spatial resolution of the currently available data from ozone-monitoring satellite instruments is 40 × 40 km and satellites usually provide only a single overpass per day. Satellite instruments also have difficulty probing the lower atmosphere, where UV-absorbing aerosols or tropospheric ozone can significantly affect surface UV irradiance. Fortunately, neither absorbing aerosols nor tropospheric ozone are major factors affecting UV radiation in the Arctic, with the exception of air masses occasionally transported from lower latitudes (section 5.4.5).

Several studies have assessed the accuracy and limitations of satellite-based UV irradiance estimates by comparing them to ground-based data (Arola et al., 2002; Chubarova et al., 2002; Fioletov et al., 2002; Kalliskota et al., 2000; McKenzie et al., 2001b; Slusser et al., 2002; Wang et al., 2000). For noon irradiance, the root mean square difference between the TOMS-derived estimates and ground-based measurements is of the order of ±25%. Estimating surface UV irradiance in the Arctic is more uncertain than that at lower latitudes: when the ground is snow-covered, the satellite-derived UV irradiance estimates have been systematically lower than the ground-based measurements. The bias originates from underestimates of surface albedo, which is a critical parameter for accurately estimating surface UV irradiance in the Arctic. Some studies have addressed surface albedo variations related to snow and ice cover (Arola et al., 2003; Lubin and Morrow, 2001; Tanskanen et al., 2003), but none of the proposed methods have completely resolved the issue. Snow or ice cover also complicates the determination of cloud optical depth properties (Krotkov et al., 2001), and the radiative transfer models are less accurate at high solar zenith angles. Because of these limitations, the accuracy of current satellite retrieval algorithms is not adequate for monitoring arctic surface UV irradiance, although some information on variations in UV irradiance over time can be inferred. The satellite-derived daily erythemal UV radiation dose in and near the Arctic on 21 March 2000 is shown in Fig. 5.13.

Long-term trends in surface UV irradiance were determined using time series derived from TOMS data. The analyses were performed for clear-sky estimates (i.e., the modulation of surface UV radiation by clouds or aerosols was not considered). The trend analysis methods applied were similar to those used to study long-term changes in total column ozone. The trend models take into account changes in solar activity and oscillations in the atmospheric circulation that affect stratospheric ozone levels. Using TOMS Version 7 ozone and reflectivity data from 1979 to 1992 and the corresponding TOMS surface UV retrieval algorithm, Herman et al. (1996) estimated that zonally averaged

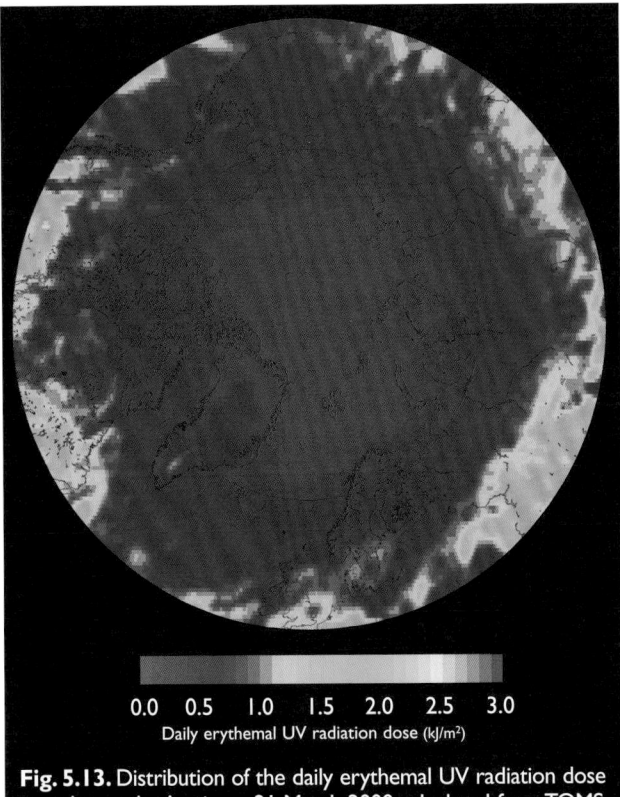

Fig. 5.13. Distribution of the daily erythemal UV radiation dose in and near the Arctic on 21 March 2000 calculated from TOMS total ozone data (Finnish Meteorological Institute, 2004, using data from the National Aeronautics and Space Administration).

mean annual surface UV irradiances at 300, 310, and 320 nm increased by 15%, 6%, and 2% per decade, respectively, for the latitude band 60° to 70° N. The greatest increases were found in winter and spring. Ziemke et al. (2000) studied the spatial distribution of the trends in erythemally weighted surface UV irradiance at northern latitudes using TOMS/Nimbus-7 data and found positive trends exceeding 10% per decade at high latitudes in Eastern Siberia. Unfortunately, the studies based on long-term satellite data did not include the region north of 65° N, and the retrieval algorithm used for the calculations is known to underestimate surface UV irradiance over snow-covered terrain.

5.6. Future changes in ozone

Stratospheric ozone levels over the polar regions are very different from levels over the mid-latitudes. Total column ozone over the Arctic in winter and spring is usually higher than that over the equator and northern mid-latitudes. Ozone levels over the Arctic are marked by a strong annual cycle, with a peak in the spring, and a decrease in late summer and throughout the autumn. Low stratospheric temperatures provide the potential for substantial ozone depletion in the winter and early spring, reducing ozone levels at a time when they would normally be high and when reproduction and new growth leave ecosystems particularly vulnerable. The same physical and chemical processes govern ozone levels and ozone depletion over both the Arctic and Antarctic. However, a stronger, less dis-

turbed polar vortex over Antarctica and thus uniformly lower stratospheric temperatures, have resulted in greater percentage ozone losses over the past two decades in the Antarctic compared to the Arctic. In the years where dynamic conditions allow for similarly cold stratospheric temperatures in the Arctic, significant ozone losses have been observed at northern high latitudes as well.

The Montreal Protocol and its amendments have already resulted in a decrease in the atmospheric concentrations of some ozone-depleting substances (Anderson et al., 2000; Montzka et al., 1999). Although scientific understanding of the dynamics and other factors influencing ozone depletion remains incomplete, most projections suggest that mid-latitude ozone levels will gradually recover over the next 50 years (WMO, 2003). Confirming either a change in ozone trends or an actual increase in ozone levels is likely to require some time because of natural variability and intrinsic measurement errors (Reinsel et al., 2002; Weatherhead et al., 1998, 2000). In polar regions, the projections of recovery are complicated by the effects of dynamic processes and climate change.

Recovery of the ozone layer is likely to occur in stages. The first signs of recovery should be a reduction in the downward trend followed by an increase in ozone levels. Final recovery may be defined as an overall return to pre-depletion ozone levels or as the determination that ozone levels are no longer being affected by anthropogenic ozone-depleting substances. Newchurch et al. (2003) reported evidence of a reduction in the downward trend in ozone levels based on satellite estimates averaged over 60° S to 60° N. Their analysis indicates that since 1997 there has been a slowdown in mid- and low-latitude stratospheric ozone losses at altitudes of 35 to 45 km. These changes in loss rates are consistent with the slowdown in total stratospheric chlorine increases, and, if they continue, will represent the first stage of a mid-latitude ozone recovery. No evidence of this change in loss rates has been reported for polar latitudes, and it is also important to note that ozone at the altitudes where these reduced loss rates are being observed plays a lesser role than total column ozone in absorbing UV radiation.

Several models have been used to project future ozone levels (WMO, 1999, 2003). Intercomparison of these models (WMO, 2003) shows qualitative agreement between the projections, although specific projections of recovery rates can disagree significantly. Two-dimensional models have been used for global ozone level projections, while three-dimensional chemistry–climate models are useful for simulating polar processes (Austin et al., 2003; WMO, 2003). Three-dimensional models can provide multi-year time slice simulations, which have the advantage that several realizations are available for a single year, allowing a better assessment of the projections. Three-dimensional models are also able to address dynamic changes in well-mixed greenhouse gases and provide a more detailed evolution of ozone levels based on the mechanisms that are likely to occur in the atmosphere. These models can also provide information on the expected range of interannual variability.

Austin et al. (2003) compared several chemistry–climate models used in recent ozone assessments (WMO, 2003). These include the Unified Model with Eulerian Transport and Chemistry (UMETRAC; Austin, 2002), the Canadian Middle Atmosphere Model (de Grandpre et al., 2000), the Middle Atmosphere European Centre Hamburg Model (ECHAM) with chemistry (Manzini et al., 2002; Steil et al., 1998, 2003), the ECHAM model with chemistry run at Deutschen Zentrum für Luft- und Raumfahrt (E39/C; Hein et al., 2001; Schnadt et al., 2002), the University of Illinois at Urbana-Champaign (UIUC) model (Rozanov et al., 2001; Yang et al., 2000), the Center for Climate System Research/National Institute for Environmental Studies (CCSR/NIES) model (Nagashima et al., 2002; Takigawa et al., 1999), the Goddard Institute for Space Studies (GISS) model (Shindell et al., 1998b), and the Università degli Studi dell'Aquila (ULAQ) model (Pitari et al., 2002). The models were compared based on their ability to simulate ozone climatologies for the current atmosphere. For the Northern Hemisphere, all the models tended to overestimate the area-weighted hemispheric total column ozone, by an average of 7.2%. The models were unable to simulate the observed loss rates within the arctic polar vortex (Becker et al., 1998, 2000; Bregman et al., 1997; Hansen et al., 1997; Woyke et al., 1999), so the modeled ozone depletion is less than that observed. Uncertainties in the model projections include temperature biases, leading to the "cold pole problem" (Pawson et al., 2000); these biases are worse in some models than in others. In the Northern Hemisphere, the biases are sometimes positive at certain altitudes, resulting in projections of insufficient ozone depletion in early winter, but excess depletion in spring. The cold pole problem is due largely to the absence of gravity-wave forcing, which many models now include. Other uncertainties include the inability of the models to accurately simulate PSCs and to account for all aspects of constituent (chemical) transport, including processes occurring at the upper boundary of the model. Changes in planetary waves and heat flux also pose uncertainties, and are discussed in greater detail in section 5.6.2.

5.6.1. Considerations for projecting future polar ozone levels

The chemical contributions to ozone depletion are generally understood well enough to describe the annual ozone losses observed over Antarctica as a result of efforts to understand the ozone hole observed there. Over the Arctic, however, ozone depletion processes are often much more complicated and depend greatly on climate conditions and climate change. For example, when potential increases in stratospheric water vapor

and corresponding stratospheric cooling resulting from climate change are included in models, the resulting projected mid-latitude ozone decrease in the 2030s surpasses that resulting from the projected amounts of CFC-derived halogens (Shindell and Grewe, 2002). At high latitudes, the effects of stratospheric water vapor and stratospheric cooling on the ozone column are anticipated to be even larger due to the effects of PSCs. Separating the chemical and dynamic/climate-related contributions to ozone depletion is not a simple task, and many questions concerning the future of ozone over the Arctic remain unanswered.

Most model projections suggest small but continuing ozone losses over the Arctic for at least the next two decades (Austin et al., 2000; WMO, 2003). Ozone depletion in the Arctic is strongly influenced by the dynamics of the polar atmosphere: changes in circulation, and particularly changes that affect air temperatures in the polar region, can have a substantial effect. For example, a strong polar vortex results in decreasing stratospheric temperatures, which further strengthen the polar vortex. This positive feedback effect contributes to increased ozone depletion, and is likely to be exacerbated by the stratospheric cooling projected to occur as a result of future climate change.

While dynamics determine the onset of ozone depletion and also influence the rate and severity of the depletion processes, the main driver for upper stratospheric (~40 km) ozone loss and for the spring losses in the polar stratosphere is the chemistry associated with chlorine and bromine (Solomon, 1999; WMO, 1999, 2003). The Montreal Protocol and its amendments have led to a reduction in atmospheric chlorine concentrations, and concentrations of ozone-depleting halogens are expected to continue to decrease between 2000 and 2050. The decreases were first reported in the troposphere (Montzka et al., 1996, 1999), but have also been observed in the upper stratosphere (Anderson et al., 2000). Bromine is another halogen particularly effective at destroying ozone, and its overall levels may increase or remain high because of shorter-lived substances, such as bromoform (Dvortsov et al., 1999). However, the magnitude of ozone loss will depend greatly on dynamic and climate conditions (section 5.2), with low temperatures contributing to the formation and persistence of PSCs. Over the polar regions, heterogeneous chemistry in or on these clouds converts stable chlorine and bromine reservoirs to more active forms that can deplete ozone. Future volcanic eruptions could also change stratospheric ozone levels worldwide for at least several years, and could have a large effect on arctic ozone levels as long as halogen loading remains large (Tabazadeh et al., 2002).

5.6.2. The role of climate change in arctic ozone recovery

Projections of the recovery of ozone levels in the Arctic depend on projections of global climate change.

Understanding long-term changes (natural and anthropogenic), will be essential to improving assessments and projections of the dynamic structure of the stratosphere (E.C., 2003). Current chemical and dynamics models project that climate change resulting from increased atmospheric concentrations of carbon dioxide and other greenhouse gases will warm the troposphere, but will cool the stratosphere. In the Arctic, this cooling is likely to lead to increased ozone destruction, as lower temperatures are likely to result in the formation and persistence of PSCs, which aid in the activation of ozone-depleting compounds and can therefore accelerate ozone depletion. Stratospheric cooling resulting from climate change is therefore likely to lead to an increased probability of larger and longer-lasting ozone holes in the Antarctic and extensive, more severe ozone losses over the Arctic (Dameris et al., 1998). On the other hand, climate change could possibly trigger an increase in planetary waves, enhancing the transport of warm, ozone-rich air to the Arctic (Schnadt et al., 2002). This increased transport would counter the effects of heterogeneous chemistry and possibly hasten recovery of the ozone layer. Understanding this "dynamic effect on chemistry" requires improved information about the effects of increasing greenhouse gas concentrations so that the balance between dynamics and radiation can be deduced. If radiative effects dominate, planetary wave activity would be more likely to decrease, resulting in more ozone depletion at arctic latitudes.

Another climate feedback affecting ozone is a potential increase in stratospheric water vapor due to changes in tropopause temperatures (Evans et al., 1998). Few long-term datasets of water vapor concentrations are available, but previous studies of existing observations have suggested that stratospheric water vapor has been increasing (Oltmans and Hofmann, 1995; Oltmans et al., 2000; Randel et al., 1999). Analyses of 45 years of data (1954–2000) by Rosenlof et al. (2001) found a 1% per year increase in stratospheric water vapor concentrations. Analyses of satellite data, however, have shown less evidence of a water vapor increase (Randel et al., 2004). Increased water vapor is likely to contribute to increased ozone destruction by affecting the radiation balance of the stratosphere (Forster and Shine, 2002; Shindell, 2001). Greater water vapor concentrations in the stratosphere can raise the threshold temperatures for activating heterogeneous chemical reactions on PSCs, and can cause a decrease in the temperature of the polar vortex (Kirk-Davidoff et al., 1999).

Ozone itself is central to climate change science: it is an important greenhouse gas in the infrared part of spectrum and is the primary absorber of solar UV radiation. Ozone is critical to the radiation balance of the atmosphere, and to the dynamics of the stratosphere. Indeed, recent observational findings confirm that "the stratosphere is a major player in determining the memory of the climate system" (Baldwin et al., 2003). Stratospheric ozone levels play a role in determining

many properties of the polar atmosphere, including the strength of the polar vortex. Observations show that the strengths of the polar vortices affect surface temperatures in the polar regions and at mid-latitudes in both hemispheres. Connections between ozone levels and other properties of the stratosphere can alter weather processes in the troposphere, with an effect whose magnitude is comparable to that of the El Niño–Southern Oscillation (Gillett and Thompson, 2003; Hartmann et al., 2000).

Projected future changes in ozone levels over the polar regions differ from projected changes over the rest of the globe, where stratospheric temperatures do not reach the low thresholds necessary for the formation of PSCs. In recent years, the arctic polar vortex has increased in strength and has become more persistent (Waugh et al., 1999; Zhou et al., 2000). A strong polar vortex can enhance the amount of depletion experienced over the Arctic. If these strong polar vortex conditions continue in future years, arctic ozone recovery is likely to be substantially delayed (Shindell et al., 1998b). For example, in an analysis of approximately 2000 ozonesonde measurements, Rex et al. (2004) found that each 1 °C cooling of the arctic stratosphere resulted in an additional 15 DU of chemical ozone loss. Their findings indicate that over the past four decades, the potential for the formation of PSCs increased by a factor of three, resulting in stratospheric conditions that have become significantly more favorable for large arctic ozone losses. This relationship between potential amounts of PSCs and ozone loss is not well-represented in current chemistry–climate models. If the arctic stratosphere continues to cool as a result of climate change, the region is likely to continue to experience severe ozone depletion until chlorine and bromine loadings have returned to background levels. Any delay in the recovery of the ozone layer over polar regions means a longer-lasting, and perhaps more severe, threat of ecosystem damage due to increased UV irradiance.

5.6.3. Projected changes in ozone amounts

A number of two-dimensional models using specified scenarios of atmospheric halocarbon concentrations were used to estimate future ozone levels for the most recent Scientific Assessment of Ozone Depletion (WMO, 2003). These included the Atmospheric and Environmental Research (AER) model (Weisenstein et al., 1998), the Max-Planck Institute (MPI) model (Grooß et al., 1998), the Goddard Space Flight Center (GSFC) model (Fleming et al., 1999), GSFC-INT (interactive version of the GSFC model; Rosenfield et al., 1997), the National Oceanic and Atmospheric Administration/National Center for Atmospheric Research (NOCAR) model (Portmann et al., 1999), the University of Oslo (OSLO) model (Stordal et al., 1985), the National Institute for Public Health and the Environment (RIVM) model (Velders, 1995), the State University of New York – St. Petersburg (SUNY-SPB) model (Smyshlyaev et al., 1998), UIUC (Wuebbles et

al., 2001), and ULAQ (Pitari and Rizi, 1993).
Figure 5.14 shows the spring (March–May) changes in ozone for the latitude band 60° to 90° N (relative to 1980 levels) projected by these two-dimensional models. The spring is interesting because ozone depletion reaches its most severe levels and UV irradiance can also be relatively high during what is the beginning of the growth period for many biological systems. The model results shown are for the greenhouse gas scenario MA2 and baseline halocarbon scenario AB (WMO, 2003). All the models except RIVM include arctic chemistry, while only the MPI and UIUC models include the 11-year solar cycle. The projected spring changes in arctic ozone levels are about twice as large as those projected for the Northern Hemisphere mid-latitudes and about three times as large as projected changes in the 60° N to 60° S annual average. Generally, the models simulate local minimums in arctic ozone levels in the late 1990s, followed by a gradual increase. The majority of the models project significantly lower ozone levels in 2020 compared to 1980.

Because the two-dimensional models are unable to incorporate dynamic effects, their results are considered very rough projections for the polar regions, where ozone levels are strongly influenced by atmospheric dynamics. The model simulations used in the 2002 assessment (WMO, 2003) differed from those used in prior assessments in that they incorporate a lower level of stratospheric aerosols and thus project a more rapid recovery of the ozone layer. About half of the models project recovery to 1980 levels by 2050. A two-dimensional model simulation with the GSFC-INT model by Rosenfield et al. (2002) projects that arctic ozone recovery will be slowest in the spring, with total column ozone returning to 1980 levels after 2050, and earliest in the autumn, with total column

ozone returning to 1980 levels before 2035. The results from the two-dimensional models project a range of arctic ozone recovery rates, from about 0.5% per decade to about 2% per decade.

Three-dimensional model simulations for the Arctic are also presented in the assessment (WMO, 2003). These models offer greater insight into the dynamic factors affecting current and future arctic ozone levels. Figure 5.15 shows the spring (March–May) average change in ozone relative to 1980 for the latitude band 60° to 90° N projected by the UMETRAC and GISS models, which are transient simulations, and E39/C, which is a time-slice simulation. In general, the three-dimensional models simulate larger ozone depletion over the Arctic between 1980 and 2000 than do the two-dimensional models. The different three-dimensional models project quite different future ozone levels. The UMETRAC model provides projections through 2020; these projections indicate slow recovery (a few percent) between 2000 and 2020. The E39/C model provides simulations for 1960, 1980, and 1990, and a projection for 2015. The ozone levels simulated by the E39/C model for 1980 are about 6% lower than the 1960 levels, but this large decrease between 1960 and 1980 is not corroborated by observations. The E39/C model simulates the same rate of decrease between 1980 and 1990, while the projections for 2015 show ozone levels above those of 1980 but still lower than 1960. The GISS model is the only three-dimensional model that provides projections beyond 2020; these projections indicate further ozone depletion between 1995 and 2015, with only modest recovery in 2045.

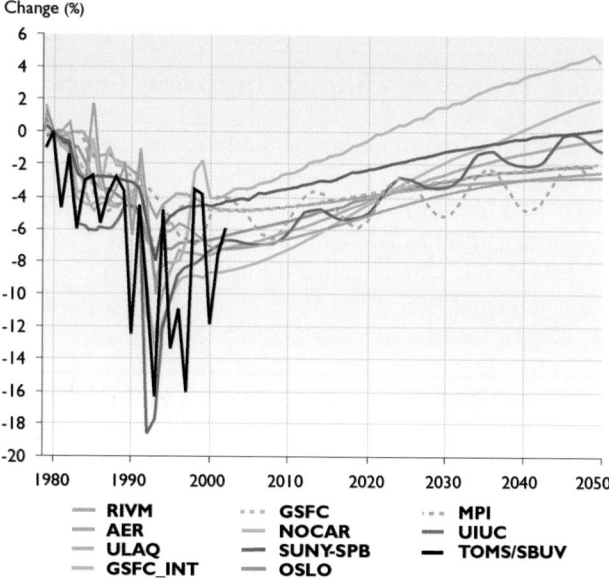

Fig. 5.14. Two-dimensional model projections of the change in spring (March–May) total column ozone relative to 1980; averaged over the band from 60° to 90° N. TOMS/SBUV observations are shown for comparison.

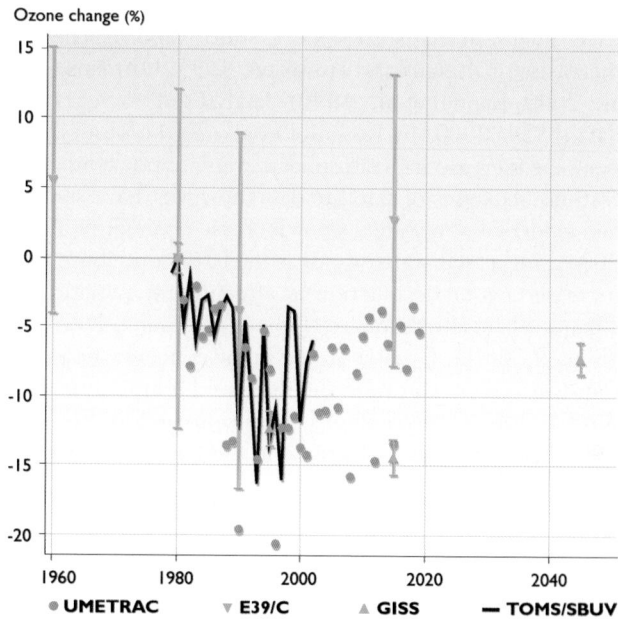

Fig. 5.15. Three-dimensional model projections of the change in spring (March–May) total column ozone relative to 1980; averaged over the band from 60° to 90° N. Error bars represent variability in model projections averaged over 5 years for the GISS model, and twice the standard deviation of the 20 individual years within each model sample for the E39/C model. TOMS/SBUV observations are shown for comparison.

Total column ozone (DU)

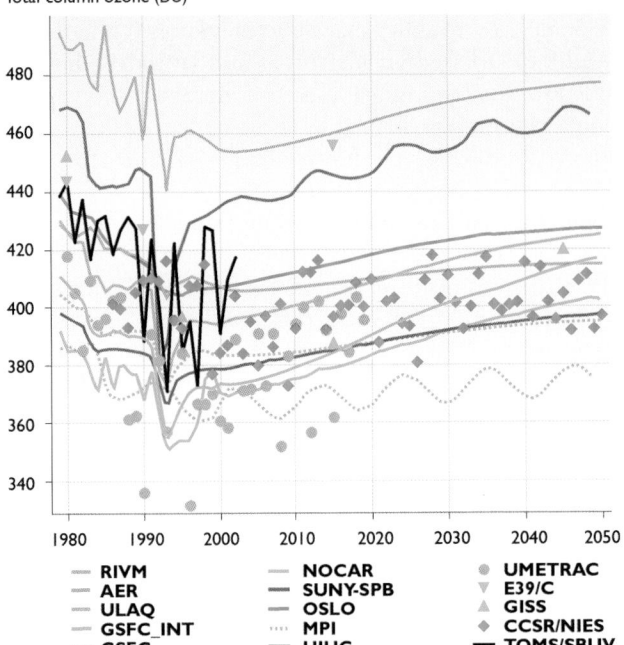

Fig. 5.16. Two- and three-dimensional model projections of spring (March–May) total column ozone averaged over the band from 60° to 90° N. TOMS/SBUV observations through 2002 are shown for comparison.

The two- and three-dimensional model projections of spring total column ozone for the 60° to 90° N band are shown in Fig. 5.16. The results indicate large differences in the projected total ozone column amounts, with most models projecting lower ozone levels than have been observed. All of the models project that ozone levels will remain substantially below pre-depletion levels for at least the next two decades.

The three-dimensional models can offer insight into the spatial distribution of ozone depletion and recovery. As Fig. 5.17 shows, both the E39/C and UMETRAC models project greater changes near Greenland than elsewhere in the Arctic, although the UMETRAC model projects continued depletion and the E39/C model projects earlier recovery for this region. These

differences indicate the uncertainty in the spatial distribution of future ozone levels. Zonally symmetric dynamics in the GISS model result in near-zonally symmetric ozone loss and recovery.

Austin et al. (2003) summarized the uncertainties in many of the chemistry–climate models that are used to project future ozone levels. Some of the most important uncertainties related to arctic projections are due to the cold temperature biases in the arctic winter that most models have, and that the models are forced with less than half the observed trend in stratospheric water vapor. Differences in gravity-wave and planetary-wave simulations as well as model resolution can lead to very different projections of polar temperatures and transport of ozone to the poles. Arctic ozone depletion is also subject to large natural variability, complicating definitive projections of how ozone levels will evolve (Austin et al., 2003; WMO, 2003).

As this section suggests, modeling past and future ozone levels, particularly in the Arctic, is challenging. One of the primary challenges is the difficulty of simulating observed polar temperatures, which are essential for determining the severity of chemical ozone depletion by anthropogenic chlorine and bromine. Many of the current chemistry–climate models do not reproduce the observed occurrence of PSCs or the large observed increase in PSC occurrence since the 1960s (Austin et al., 2003, Pawson and Naujokat, 1999; WMO, 2003). Some models are also unable to accurately reproduce the observed ozone loss rates within the arctic polar vortex. As reported by Rex et al. (2004), the limitations of accurately simulating the relationship between potential amounts of PSCs and ozone depletion may be leading to more optimistic projections of arctic ozone levels than are likely to occur given the influences of climate change. The difficulties in simulating arctic stratospheric temperatures stem partly from the strong influence of polar dynamics, and until these processes are better understood, future changes in arctic dynamics and ultimately in arctic ozone levels will be difficult to project.

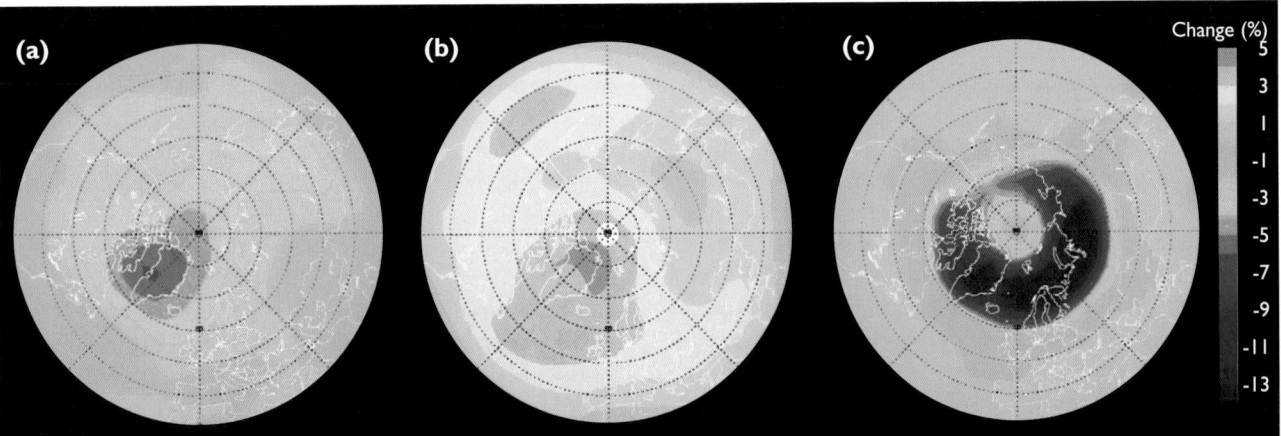

Fig. 5.17. Percentage change in average annual total column ozone projected by (a) the UMETRAC model for the period 2010–2019 (average over the time slice) relative to 1980–1984; (b) the E39/C model for 2015 relative to 1980; and (c) the GISS model for the period 2010–2019 (average over the time slice) relative to 1980.

Because of the anticipated decline in stratospheric chlorine and bromine concentrations resulting from the Montreal Protocol and its amendments, an increase in arctic ozone levels is expected to occur eventually. However, any quantitative statements concerning the timing and magnitude of arctic ozone layer recovery are highly uncertain.

5.7. Future changes in ultraviolet radiation

While there are early signs that the Montreal Protocol and its amendments are working, a return to normal ozone levels is not likely to occur for several decades. Scientists primarily concerned with chemical contributions may be interested in the earlier signs of ozone recovery (for example, a reduction in the downward trend in ozone levels), while those studying UV radiation and its effects are likely to focus on an overall recovery from depleted values. Delayed recovery of total column ozone in the polar regions implies the possibility of increased UV irradiance at northern latitudes for another one to five decades. Although there are many uncertainties in the models, current projections indicate that ozone depletion over the Arctic is likely to continue for 50 years or more (WMO, 1999, 2003).

Because ozone depletion at polar latitudes is expected to persist, and possibly worsen, over the next few decades (WMO, 2003), UV radiation reaching the surface is likely to remain at levels greater than those observed in the past. Using ozone projections from the GISS chemistry–climate model, which projects the most severe, longest-lasting ozone depletion of any of the models, Taalas et al. (2000) estimated that the worst-case spring erythemal UV radiation doses averaged over the period from 2010 to 2020 will increase by up to 90% relative to average 1970–1992 conditions. By comparison, the annual UV radiation dose increases for the entire Northern Hemisphere are estimated to be 14% for 2010 to 2020, and 2% for 2040 to 2050. Ultraviolet radiation projections depend on future ozone levels, which are highly uncertain. Future UV radiation levels will also be complicated by changes in snow and ice cover and albedo and are likely to vary locally as well as regionally. Reuder et al. (2001) simulated ozone levels using various CFC emissions scenarios, as well as changes in future temperatures and dynamics and used these results to project UV radiation conditions over central Europe. Their results indicate a slight increase in spring UV radiation levels between the present and 2015, and the potential for continued above-normal late winter and spring UV radiation levels through approximately 2050, although future climate-induced forcings of arctic ozone recovery still need to be better explored.

Continued increased UV levels are likely to have profound effects on human health. Slaper et al. (1996) estimated that reaching minimum ozone levels around the year 2000 (based on the Copenhagen Amendments

to the Montreal Protocol) would be likely to result in a 10% increase in skin cancer 60 years later. While these results were based on analyses at mid-latitudes, they illustrate the long-range effects of increased UV radiation levels on human health. These human health effects are discussed in more detail in section 15.3.3.

While humans can choose clothing, sunglasses or goggles, and other protection to reduce their exposure to UV radiation, plants and animals in the Arctic must adapt to their environment through slower, biological means or by migrating or seeking shelter to reduce their exposure. Section 7.3.2 describes plant and animal adaptations to UV radiation exposure in greater detail. These adaptations include thick leaves and protective pigments in plants and reflective white feathers and fur in arctic animals. Sections 8.6 and 9.4 also address potential adaptations and protections for organisms experiencing increased UV radiation exposure. If UV radiation levels in the Arctic remain high into the 21st century, the relative incidence and time frame of effects on human health and on plants and animals would also be extended.

Ozone levels and future changes in these levels are not the only factors affecting anticipated UV irradiance in the Arctic; aerosol concentrations and cloud cover also play a role (see section 5.4). These factors are likely to change, at least on a regional basis, in the future. A more active hydrological cycle in the Arctic, projected to occur as a result of climate change (IPCC, 2001), is likely to result in changes in cloud cover. In general, increases in cloud cover will reduce UV irradiance at the surface, except in certain conditions when multiple reflections between clouds and a snow-covered surface may enhance the UV radiation dose. Current model projections suggest that cloud cover over the Arctic is likely to increase in some areas and decrease in others. Uncertainties in projections of future aerosol or cloud changes and in the understanding of aerosol–cloud–UV radiation interactions complicate projections of future surface UV irradiance.

Sea ice and snow cover are also likely to be affected by climate change, and can have major effects on incident UV radiation, both by reflecting radiation and by protecting organisms buried beneath it. Almost all climate models project increases in precipitation in the Arctic (IPCC, 2001). Model projections indicate that overall temperatures in the Arctic are likely to be warmer, suggesting that for late spring through early autumn, much of the precipitation increase is likely to be in the form of rain, or of rain falling on existing snow cover, possibly enhancing the rate of snowmelt. The extent and duration of snow cover in the Arctic is important in part because of its relation to UV radiation doses. In the polar regions, UV radiation doses affecting biological organisms depend greatly on the local surface albedo. Reflection off snow, for example, can increase the amount of UV radiation reaching an organism's face, eyes, or other surface. These amplified doses can be particularly pronounced at low solar elevations, or in the presence of increased multiple scattering by non-absorbing aerosols. Any shift in the extent or duration of snow cover, particularly during the critical spring months, is likely to amplify the biologically effective UV radiation doses received by ecosystems potentially already stressed by climate change. In areas normally covered by snow, early spring snowmelt, such as has been observed by Stone et al. (2002), is very likely to leave organisms at ground level vulnerable to increased UV irradiance during periods of spring ozone depletion.

UV radiation exposure can have a range of effects on humans and on the overall arctic environment. Human health concerns include skin cancers, corneal damage, cataracts, immune suppression, and aging of the skin. Ultraviolet radiation can also have deleterious effects on both terrestrial and aquatic ecosystems, and is known to affect infrastructure through damage to plastics, wood, and other materials. Many of these effects require increased study. The combination of future climate change and the likelihood of prolonged increases in arctic UV radiation levels present a potentially challenging situation for the people and environment of the Arctic. These effects and some of their expected consequences are discussed in greater detail in sections 7.3, 7.4, 8.6, 9.4, 14.12, 15.3.3, 16.3.1, and 17.2.2.3.

5.8. Gaps in knowledge, future research, and observational needs

Four key areas of research activity will improve the ability of the scientific community to assess the changes in and effects of ozone depletion and UV radiation in the Arctic: addressing unanswered scientific questions concerning variability and long-term changes in both ozone levels and UV irradiance; ensuring accurate and comprehensive monitoring of ozone and UV radiation levels; improving analysis of emerging data and incorporating this new understanding into modeling efforts; and undertaking cross-disciplinary studies to determine the effects of changes in UV irradiance. All four areas are important both to improve scientific understanding and to provide relevant information for policy decisions.

There are a number of unanswered scientific questions concerning the sources of variability in ozone and UV radiation levels in the Arctic. Improved knowledge is needed to quantify the effects of trace gases, dynamics, and temperature on arctic ozone levels. The influence of climate change on both ozone and UV radiation levels needs to be better understood. Understanding the controls on and interactions between various processes will greatly improve projections of future ozone levels. In addition to ozone levels, several other factors, including cloud conditions, aerosol concentrations, and surface albedo, affect surface UV irradiance. The interactions between and overall influence of these factors are still the subject of much uncertainty, but

future changes in any one parameter – for example, in cloudiness or snow melt timing – could substantially affect UV radiation levels in the Arctic. Quantifying these factors across the Arctic will provide opportunities to more realistically assess changes in UV irradiance and their effects.

Although many of the questions regarding the cause of ozone depletion have been addressed and confirmed in a number of studies both within and outside the Arctic, questions still remain concerning the future of ozone and UV radiation levels in the Arctic. Current model projections vary widely in terms of future ozone and UV radiation levels, with the large differences due mainly to uncertainties regarding the roles of dynamics, temperature, and trace gases. Because of these uncertainties, continued monitoring of both ozone and UV radiation levels in the Arctic is important. Ozone levels and the stratosphere are also known to play a key role in influencing and modulating climate. Monitoring efforts are necessary both to document the evolution of ozone and UV radiation levels over time and to validate model projections. Past monitoring of ozone and UV radiation levels has shown that changes occur on a regional basis even within the Arctic, indicating that regional observations are critical for assessing the overall status of the Arctic. Satellite monitoring of the Arctic for most times of the year has been ongoing for the past few decades, but continued observations will be necessary to understand the evolution of arctic ozone. The continuation of ground-based monitoring of ozone levels depends on available funding and is highly uncertain at this time. Adding UV radiation monitoring in the Russian Arctic and coordinating the existing surface UV radiation and ozone level monitoring throughout the Arctic would allow for a more accurate assessment of the changes that are occurring.

Analyses of emerging ozone and UV radiation data continue to reveal new information concerning the relative importance of trace gases, dynamics, and temperatures in the Arctic. Continued studies are likely to add to the understanding of UV radiation levels in the Arctic. Campaigns with intensive measurements of a variety of parameters as well as detailed monitoring of trace gases and vertically resolved ozone concentrations are important for advancing this understanding. Analyzing these measurements and using the available information in conjunction with model results will help achieve the best possible insight into future ozone and UV radiation levels, as well as the impacts of specific changes in these levels. Recent studies support the idea that advanced three-dimensional models will be fundamental for obtaining improved projections of future ozone levels over the Arctic.

One of the most important issues to address in the Arctic is determining the impacts of increased UV irradiance, particularly as levels are likely to remain higher than normal in the coming decades. This work requires coordinated cooperation between the UV radiation monitoring community and the biological and impacts communities. Few cross-disciplinary efforts have been implemented so far, but the collaboration that has taken place has resulted in many of the impact studies cited in other chapters of this assessment. Many questions remain concerning the impacts of UV radiation on individual species, ecosystems, and human health. Because the Arctic is likely to have elevated spring UV radiation levels for some time, understanding the magnitude of potential impacts will be critical for future policy decisions. Because of projected future changes in emissions and atmospheric concentrations of various trace gases, future arctic ozone levels are highly uncertain, not only for the next few decades but throughout the rest of the 21st century. Policy decisions regarding trace gas emissions are likely to directly influence ozone levels in the Arctic, and should be based on not only improved understanding of how ozone and UV radiation levels are likely to evolve, but also on increased knowledge of how the ecosystems, infrastructure, industries, and people of the Arctic will be affected.

In addition to the scientific requirements outlined above, there is ongoing concern about maintaining the existing international legislation to protect the ozone layer. Most developed countries have ratified the Montreal Protocol and its amendments, but economic and political pressures have led some countries to suggest that they cannot meet their obligations under the current agreements. The atmospheric concentrations of most ozone-depleting substances are decreasing due to compliance with the Protocol and its amendments, but future compliance is uncertain and requires vigilance and cooperation. In addition, because arctic ozone and UV radiation levels are strongly influenced by climate change, including the effects of changes in temperature, trace gas concentrations, and dynamics, international legislation regarding climate change is likely to directly affect arctic ozone levels in the coming decades. Any climate change policies need to be considered in light of the impacts both on arctic climate and on arctic ozone and UV radiation levels.

References

Aas, E., J. Høkedal, N.K. Hørjerslev, R. Sandvik and E. Saksnaug, 2001. Spectral properties and UV-attenuation in Arctic marine waters. In: D.O. Hessen (ed.). UV-Radiation and Arctic Ecosystems, pp. 23–56. Springer-Verlag.

Anderson, J.G., W.H. Brune and M.H. Proffitt, 1989. Ozone destruction by chlorine radicals within the Antarctic vortex – the spatial and temporal evolution of ClO-O$_3$ anticorrelation based on in situ ER-2 data. Journal of Geophysical Research, 94(D9):11465–11479.

Anderson, J., J.M. Russell III, S. Solomon and L.E. Deaver, 2000. HALOE confirmation of stratospheric chlorine decreases in accordance with the Montreal Protocol. Journal of Geophysical Research, 105:4483–4490.

Appenzeller, C., A.K. Weiss and J. Staehelin, 2000. North Atlantic Oscillation modulates total ozone winter trends. Geophysical Research Letters, 27(8):1131.

Arola, A., S. Kalliskota, P.N. den Outer, K. Edvardsen, G. Hansen, T. Koskela, T.J. Martin, J. Matthijsen, R. Meerkoetter, P. Peeters, G. Seckmeyer, P. Simon, H. Slaper, P. Taalas and J. Verdebout, 2002. Four UV mapping procedures using satellite data and their validation against ground-based UV measurements. Journal of Geophysical Research, 107, (D16):4310 doi:10.1029/2001JD000462.

Arola, A., J. Kaurola, L. Koskinen, A. Tanskanen, T. Tikkanen, P. Taalas, J.R. Herman, N. Krotkov and V. Fioletov, 2003. A new approach to estimating the albedo for snow-covered surfaces in the satellite UV method. Journal of Geophysical Research, 108(D17):4531 doi:10.1029/2003JD003492.

Austin, J., 1992. Toward the four dimensional assimilation of stratospheric chemical constituents. Journal of Geophysical Research, 97(D2):2569–2588.

Austin, J., 1994. The influence of climate change and the timing of stratospheric warming on ozone depletion. Journal of Geophysical Research, 99(D1):1127–1145.

Austin, J., 2002. A three-dimensional coupled chemistry-climate model simulation of past stratospheric trends. Journal of Atmospheric Sciences, 59:218–232.

Austin J. and N. Butchart, 1992. A three-dimensional modelling study of the influence of planetary wave dynamics on polar ozone photochemistry. Journal of Geophysical Research, 97(2):2569–2588.

Austin, J., D.J. Hofmann, N. Butchart and S.J. Oltmans, 1995. Mid stratospheric ozone minima in polar regions. Geophysical Research Letters, 22(18):2489–2492.

Austin, J., J. Knight and N. Butchart, 2000. Three-dimensional chemical model simulations of the ozone layer:1979–2015. Quarterly Journal of the Royal Meteorological Society, 126(565 Part B):1533–1556.

Austin, J., D. Shindell, S.R. Beagley, C. Bruhl, M. Dameris, E. Manzini, T. Nagashima, P. Newman, S. Pawson, G. Pitari, E. Rozanov, C. Schnadt and T.G. Shepherd, 2003. Uncertainties and assessments of chemistry-climate models of the stratosphere. Atmospheric Chemistry and Physics, 3:1–27.

Bais, A.F., C.S. Zerefos and C. Meleti, 1993. Spectral measurements of solar UV-B radiation and its relations to total ozone, SO$_2$, and clouds. Journal of Geophysical Research, 98(D3):5199–5204.

Bais, A.F., B.G. Gardiner, H. Slaper, M. Blumthaler, G. Bernhard, R. McKenzie, A.R. Webb, G. Seckmeyer, B. Kjeldstad, T. Koskela, P.J. Kirsch, J. Gröbner, J.B. Kerr, S. Kazadzis, K. Leszczynski, D. Wardle, C. Brogniez, D. Gillotay, H. Reinen, P. Weihs, T. Svenoe, P. Eriksen, F. Kuik, A. Redondas, 2001. SUSPEN intercomparison of ultraviolet spectroradiometers. Journal of Geophysical Research, 106(D12):12509–12525.

Baldwin, M.P., D.W. J. Thompson, E.F. Shuckburgh, W.A. Norton and N.P. Gillett, 2003. Weather from the stratosphere? Science, 301:317.

Becker, G., R. Müller, D.S. McKenna, M. Rex and K.S. Carslaw, 1998. Ozone loss rates in the Arctic stratosphere in the winter 1991/1992: Model calculations compared with Match results. Geophysical Research Letters, 25:4325–4328.

Becker, G., R. Müller, D.S. McKenna, M. Rex, K.S. Carslaw and H. Oelhauf, 2000. Ozone loss rates in the Arctic stratosphere in the winter 1994/1995: Model simulations underestimate results of the Match analysis. Journal of Geophysical Research, 105:14184–15175.

Bérces, A., S. Fekete, P. Gáspá, P. Gróf, P. Rettberg, G. Horneck and G. Rontó, 1999. UV dosimeters in the assessment of the biological hazard from environmental radiation. Journal of Photochemistry and Photobiology B-Biology, 53:36–43.

Bernhard, G., C.R. Booth and R. McPeters, 2003. Calculation of total column ozone from global UV spectra at high latitudes. Journal of Geophysical Research, 108(D17):4532 doi: 10.1029/2003JD003450.

Bigelow, D.S., J.R. Slusser, A.F. Beaubien and J.H. Gibson, 1998. The USDA Ultraviolet Radiation Monitoring Program. Bulletin of the American Meteorological Society, 79:601–615.

Blumthaler, M. and W. Ambach, 1988. Solar UVB-albedo of various surfaces. Photochemistry and Photobiology, 48(1):85–88.

Blumthaler, M., W. Ambach and M. Salzgeber, 1994a. Effects of cloudiness on global and diffuse UV irradiance in a high-mountain area. Theoretical and Applied Climatology, 50:23–30.

Blumthaler, M., A.R. Webb, G. Seckmeyer, A.F. Bais, M. Huber and B. Mayer, 1994b. Simultaneous spectroradiometry: a study of solar UV irradiance at two altitudes. Geophysical Research Letters, 21(25):2805–2808.

Blumthaler, M., W. Ambach and R. Ellinger, 1997. Increase in solar UV radiation with altitude. Journal of Photochemistry and Photobiology B Biology, 39:130–134.

Bodeker, G.E. and R.L. McKenzie, 1996. An algorithm for inferring surface UV irradiance including cloud effects. Journal of Applied Meteorology, 35:1860–1877.

Bodeker, G.E., J.C. Scott, K. Kreher and R.L. McKenzie, 2001. Global ozone trends in potential vorticity coordinates using TOMS and GOME intercompared against the Dobson network. Journal of Geophysical Research, 106(D19):23029–23042.

Bodhaine, B.A., E.G., Dutton, R.L. McKenzie and P.V. Johnston, 1998. Calibrating broadband UV instruments: ozone and solar zenith angle dependence. Journal of Atmospheric and Oceanic Technology, 15:916–926.

Booth, C.R. and S. Madronich, 1994. Radiation amplification factors: improved formulation accounts for large increases in ultraviolet radiation associated with Antarctic ozone depletion. In: C.S. Weiler and P.A. Penhale (eds.). Antarctic Research Series, 62, pp. 39–42.

Booth, C.R. and J.H. Morrow, 1997. The penetration of UV into natural waters. Photochemistry and Photobiology, 65(2):254–257.

Booth, C.R., G. Bernhard, J.C. Ehramjian, V.V. Quang, and S.A. Lynch, 2001. NSF Polar Programs UV Spectroradiometer Network 1999–2000 Operations Report. Biospherical Instruments, San Diego, 219pp.

Borkowski, J.L., 2000. Homogenization of the Belsk UV-B series (1976–1997) and trend analysis. Journal of Geophysical Research, 105(D4):4873–4878.

Braathen, G., M. Müller, B.-M. Sinnhuber, P. von der Gathan, E. Kyrö, I.S. Mikkelsen, B. Bojkov, V. Dorokhov, H.Fast, C. Parrondo and H. Kanzawa, 2000. Temporal evolution of ozone in the Arctic vortex from 1988–89 to 1999–2000. Results from the Third European Stratospheric Experiment on Ozone (THESEO), Palermo, Italy, 2000.

Bregman, A., M. van den Brock, K.S. Carslaw, R. Müller, T. Peter, M.P. Scheele and J. Lelieveld, 1997. Ozone depletion in the late winter lower Arctic stratosphere: observations and model results. Journal of Geophysical Research, 102:10815–10828.

Brühl, C. and P.J. Crutzen, 1989. On the disproportionate role of tropospheric ozone as a filter against solar UV-B radiation. Geophysical Research Letters, 16(7):703–706.

Butler, J.H., S.A. Montzka, A.D. Clarke, J.M. Lobert and J.W. Elkins, 1998. Growth and distribution of halons in the atmosphere. Journal of Geophysical Research, 103(D11):1503–1511.

Chang, A.Y., R.J. Salawitch and H.A. Michelson, 1996. A comparison of measurements from ATMOS and instruments aboard the ER-2 aircraft: halogenated gases. Geophysical Research Letters, 23(17):2393–2396.

Chipperfield, M.P. and J.A. Pyle, 1998. Model sensitivity studies of Arctic ozone depletion. Journal of Geophysical Research, 103(D21):28389–28403.

Chubarova, N.Y., A.Y. Yurova, N.A. Krotkov, J.R. Herman and P.K. Bhartia, 2002. Comparisons between ground based measurements of broadband UV irradiance (300–380 nm) and TOMS UV estimates at Moscow for 1979–2000. Optical Engineering, 41(12):3070–3081.

CIE, 1998. Erythema Reference Action Spectrum and Standard Erythema Dose. Joint ISO/CIE Standard, ISO 17166:1999/CIE S007–1998, Commission Internationale de l'Eclairage.

Dahlback, A., 1996. Measurements of biologically effective UV doses, total ozone abundances, and cloud effects with multichannel, moderate bandwidth filter instruments. Applied Optics, 35(33):6514–6521.

d'Almeida, G.A., P. Koepke and E.P. Shettle, 1991. Atmospheric Aerosols: Global Climatology and Radiative Characteristics. A. Deepak Publishing, 561pp.

Dameris, M., V. Grewe, R. Hein and C. Schnadt, 1998. Assessment of the future development of the ozone layer. Geophysical Research Letters, 25:3579–3582.

Danilin, M.Y., N.-D. Sze, M.K.W. Ko, J.M. Rodrigues and A. Tabazadeh, 1998. Stratospheric cooling and Arctic ozone recovery. Geophysical Research Letters, 25:2141–2144.

de Grandpre, J., S.R. Beagley, V. Fomichev, E. Griffion, J.C. McConnell, A.S. Medvedev and T.G. Shepherd, 2000. Ozone climatology using interactive chemistry: results from the Canadian Middle Atmosphere Model. Journal of Geophysical Research, 105:26475–26491.

Degünther, M., R. Meerkoetter, A. Albold and G. Seckmeyer, 1998. Case study on the influence of inhomogeneous surface albedo on UV irradiance. Geophysical Research Letters, 25:3587–3590.

Díaz. S., D. Nelson, G. Deferrari and C. Camilión, 2003. A model to extend spectral and multiwavelength UV irradiances time series: model development and validation. Journal of Geophysical Research, 108(D4): 10.1029/2002JD002134.

Dvortsov, V.L. and S. Solomon, 2001. Response of the stratospheric temperatures and ozone to past and future increases in stratospheric humidity. Journal of Geophysical Research, 106:7505–7514.

Dvortsov, V.L., M.A. Geller, S. Solomon, S.M. Schauffler, E.L. Atlas and D.R. Blake, 1999. Rethinking reactive halogen budgets in the midlatitude lower stratosphere. Geophysical Research Letters, 26(12):1699–1702.

E.C., 2003. Ozone-climate interactions: air pollution research report No 81. Directorate-General for Research, Environment, and sustainable development programme, European Commission.

Evans, S.J., R. Toumi, J.E. Harries, M.P. Chipperfield and J.M. Russell, 1998. Trends in stratospheric humidity and the sensitivity of ozone to these trends. Journal of Geophysical Research, 103:8715–8725.

Eyring, V., M. Dameris, V. Grewe, I. Langbein and W. Kouker, 2003. Climatologies of subtropical mixing derived from 3D models. Atmospheric Chemistry and Physics, 3:1007–1021.

Farman, J.C., B.G. Gardiner and J.D. Shanklin, 1985. Large losses of total ozone in Antarctica reveal seasonal ClO_x/NO_x Interaction. Nature, 315:207–210.

Feister, U. and R. Grewe, 1995. Spectral albedo measurements in the UV and visible region over different types of surfaces. Photochemistry and Photobiology, 62(4):736–744.

Feister, U., E. Jakel and K. Geridee, 2002. Parameterization of daily solar global ultraviolet radiation Photochemistry and Photobiology, 76(3):281–293.

Fioletov, V.E. and W.F.J. Evans, 1997. The influence of ozone and other factors on surface radiation. In: D.I. Wardle, J.B. Kerr, C.T. McElroy and D.R. Francis (eds.). Ozone Science: a Canadian Perspective on the Changing Ozone Layer, pp. 73–90. Environment Canada Report CARD 97-3.

Fioletov, V.E., E. Griffioen and J.D. Kerr, 1998. Influence of volcanic sulfur dioxide on spectral UV irradiance as measured by Brewer spectroradiometer. Geophysical Research Letters, 25(10): 1665–1668.

Fioletov, V.E., L.J.B. McArthur, J.B. Kerr and D.I. Wardle, 2001. Long-term variations of UV-B irradiance over Canada estimated from Brewer observations and derived from ozone and pyranometer measurements. Journal of Geophysical Research, 106(D19):23,009–32,027.

Fioletov, V.E., J.B. Kerr, D.I. Wardle, N. Krotkov and J.R. Herman, 2002. Comparison of Brewer UV irradiance measurements with TOMS satellite retrievals. Optical Engineering, 41(12): 3051–3061.

Fioletov, V.E., J.B. Kerr, L.J.B. McArthur, D.I. Wardle and T.W. Mathews, 2003. Estimating UV Index climatology over Canada. Journal of Applied Meteorology, 42:417–433.

Fleming, E.L., C.H. Jackman, R.S. Stolarski and D.B. Considine, 1999. Simulation of stratospheric tracers using an improved empirically based two-dimensional model transport formulation. Journal of Geophysical Research, 104:23911–23934.

Fligge, M. and S. Solanki, 2000. The solar spectral irradiance since 1700. Journal of Geophysical Research, 27(14):2157–2160.

Forster, P.M. and K.P. Shine, 2002. Assessing the climate impact of trends in stratospheric water vapor. Geophysical Research Letters, 29(6):1029/2001GL013909.

Fox, S.L., 2000. Arctic climate change: observations of Inuit in the Eastern Canadian Arctic. In: F. Fetterer and V. Radionov (eds.). Arctic Climatology Project, Environmental Working Group Arctic Meteorology and Climate Atlas. U.S. National Snow and Ice Data Center, Boulder. CD-ROM.

Frederick, J.E. and C. Erlick, 1997. The attenuation of sunlight by high latitude clouds: spectral dependence and its physical mechanisms. Journal of Atmospheric Sciences, 54:2813–2819.

Fuenzalida, H.A., 1998. Global ultraviolet spectra derived directly from observations with multichannel radiometers. Applied Optics, 37(33):7912–7919.

Furusawa, Y., L.E. Quintern, H. Holtschmidt, P. Koepke and M. Saito, 1998. Determination of erythema-effective solar radiation in Japan and Germany with a spore monilayer film optimized for the detection of UVB and UVA – results of a field campaign. Applied Microbiology and Biotechnology, 50:597–603.

Fusco, A.C. and M.L. Salby, 1999. Interannual variations of total ozone and their relationship to variations of planetary wave activity. Journal of Climate, 12:1619–1629.

Gantner, L., P. Winkler and U. Kohler, 2000. A method to derive long-term time series and trends of UV-B radiation (1968–1997) from observations at Hohenpeissenberg (Bavaria). Journal of Geophysical Research, 105(D4):4879–4880.

Gillett, N.P. and D.W.J. Thompson, 2003. Simulation of recent Southern Hemisphere climate change. Science, 302:273.

Goutail, F., J.P. Pommereau and C. Philips, 1999. Depletion of column ozone in the Arctic during the winters of 1993/94 and 1994/95. Journal of Atmospheric Chemistry, 32:1–34.

Grenfell, T.C., S.G. Warren and P.C. Mullen, 1994. Reflection of solar radiation by the Antarctic snow surface at ultraviolet, visible, and near-infrared wavelengths. Journal of Geophysical Research, 99:18,669–18,684.

Gröbner, J., A. Albold, M. Blumthaler, T. Cabot. A. De la Casiniere, J. Lenoble, T. Martin, D. Masserot, M. Müller, R. Philipona, T. Pichler, E. Pougatch, G. Rengarajan, D. Schmucki, G. Seckmeyer, C. Sergent, M.L. Toure and P. Weihs, 2000. Variability of spectral solar ultraviolet irradiance in an Alpine environment. Journal of Geophysical Research, 105(D22):26,991–27,003.

Grooß, J.-U., C. Brühl and T. Peter, 1998. Impact of aircraft emissions on tropospheric and stratospheric ozone, I, Chemistry and 2-D model results. Atmospheric Environment, 32:3173–3184.

Guirlet, M., M.P. Chipperfield, J.A. Pyle, F. Goutail, J.P. Pommereau and E. Kyro, 2000. Modeled Arctic ozone depletion in winter 1997/1998 and comparison with previous winters. Journal of Geophysical Research, 105(D17):22185–22200.

Gurney, K.R., 1998. Evidence for increasing ultraviolet irradiance at Point Barrow, Alaska. Geophysical Research Letters, 25:903–906.

Hansen, G., T. Svenoe, M.P. Chipperfield, A. Dahlbade and V.P. Hoppe, 1997. Evidence of substantial ozone depletion in winter 1995/96 over northern Norway Geophysical Research Letters, 24:799–804.

Harnisch, J., R. Borchers and P. Fabian, 1996. Tropospheric trends for CF4 and C2F6 since 1982 derived from SF6 dated stratospheric air Geophysical Research Letters, 23(10):1099–1102.

Harris, N.R.P., M. Rex, F. Goutail, B.M. Knudsen, G.L. Manney, R. Muller and P. von der Gathen, 2002. Comparison of empirically derived ozone loss rates in the Arctic. Journal of Geophysical Research, 107(D20), doi:10.1029/2001JD000482.

Harrison, L., J. Michalsky and J. Berndt, 1994. Automated multifilter rotating shadow-band radiometer and instrument for optical depth and radiation measurements. Applied Optice, 33:5118–5125.

Hartmann, D.L., J.M. Wallace, V. Limpasuvan, D.W.J. Thompson and J.R. Holton, 2000. Can ozone depletion and global warming interact to produce rapid climate change. Proceedings of the National Academy of Sciences, 97:1412–1417.

Hedblom, E.E., 1961. Snowscape eye protection. Archives of Environmental Health, 2:685–704.

Hein, R., M. Dameris and C. Schnadt, 2001. Results of an interactively coupled atmospheric chemistry-general circulation model: comparison with observations. Annales Geophysicae, 19(4): 435–437.

Herber, A., L.W. Thomason, H. Gernandt, U. Leiterer, D. Nagel, K.-H. Schulz, J. Kaptur, T. Albrecht and J. Not, 2002. Continuous day and night aerosol optical depth observations in the Arctic between 1991 and 1999. Journal of Geophysical Research, 107(D10): 10.1029/2001JD000536.

Herman, J.R. and D. Larko, 1994. Low ozone amounts during 1992 and 1993 from Nimbus 7 and Meteor 3 total ozone mapping spectrometers. Journal of Geophysical Research, 99:3483–3496.

Herman, J.R., P.K. Bhartia and J. Ziemke, 1996. UV-B increases (1979–1992) from decreases in total ozone. Geophysical Research Letters, 23(16):2117–2120.

Herman, J.R., N. Krotkov, E. Celarier, D. Larko and G. Labow, 1999. The distribution of UV radiation at the Earth's surface from TOMS measured UV-backscattered radiances. Journal of Geophysical Research, 104:12059–12076.

Hisdal, V., 1986. Spectral distribution of global and diffuse solar radiation at Ny Alesund, Spitzbergen. Polar Research, 5:1–27.

Hofzumahaus, A., A. Kraus and M. Muller, 1999. Solar actinic flux spectroradiometry: a technique for measuring photolysis frequency in the atmosphere. Applied Optics, 38(21):4443–4460.

Holben, B.N., T.F. Eck and I. Slutsker, 1998. AERONET-A federated instrument network and data archive for aerosol characterization. Remote Sensing of the Environment, 66(1):1–16.

Hood, L., S. Rossi and M. Beulen, 1999. Trends in lower stratospheric zonal winds, Rossby wave breaking behavior, and column ozone at northern midlatitudes. Journal of Geophysical Research, 104:24,321–24,339.

IPCC, 2001. Climate Change 2001: Impacts, Adaptation, and Vulnerability. Contribution of Working Group I to the Third Assessment Report of the Intergovernmental Panel on Climate Change. J.J McCarthy, O.F. Canziani, N.A. Leary, D.J. Dokken and K.S. White (eds.). Cambridge University Press, 1032pp.

James, P.M., 1998. A climatology of ozone mini-holes over the Northern Hemisphere. International Journal of Climatology, 18(12):1287–1303.

Jokela, K., K. Leszczynski and R. Visuri, 1993. Effects of Arctic ozone depletion and snow on UV exposure in Finland. Photochemistry and Photobiology, 58(4):559–566.

Josefsson, W. and T. Landelius, 2000. Effect of clouds on UV irradiance: as estimated from cloud amount, cloud type, precipitation, global radiation, and sunshine duration. Journal of Geophysical Research, 105(D4):4927–4935.

Kalliskota, S., J. Kaurola, P. Taalas, J.R. Herman, E.A. Celarier and N. Krotkov, 2000. Comparison of daily UV doses estimated from Nimbus 7/TOMS measurements and ground-based spectroradiometric data. Journal of Geophysical Research, 105:5059–5067.

Kaurola, J., P. Taalas and T. Koskela, 2000. Long-term variations of UV-B doses at three stations in northern Europe. Journal of Geophysical Research, 105(D16):20813–20820.

Kerr, J.B. and C.T. McElroy, 1993. Evidence for large upward trends of ultraviolet-B radiation linked to ozone depletion. Science, 262(5136):1032–1034.

King, M.D. and W.R. Simpson, 2001. Extinction of UV radiation in Arctic snow at Alert, Canada (82° N). Journal of Geophysical Research, 106(D12):12499–12507.

Kirk-Davidoff, D.B., E.J. Hintsa, J.G. Anderson and D.W. Keith, 1999. The effect of climate change on ozone depletion through changes in stratospheric water vapor. Nature, 402:399–401.

Knudsen, B.M., N. Larsen, I.S. Mikkelsen, J-J. Morcrette, G.O. Braathen, E. Kyro, H. Fast, H. Gernandt, H. Kanzawa, H. Nakane, V. Dorokhov, V. Yushkov, G. Hansen, M. Gil and R.J. Shearman, 1998. Ozone depletion in and below the arctic vortex for 1997. Geophysical Research Letters, 25:627–630.

Kodera, K. and H. Koide, 1997. Spatial and seasonal characteristics of recent decadal trends in the Northern Hemisphere troposphere and stratosphere. Journal of Geophysical Research, 102: 19,433–19,447.

Kouker, W., D. Offermann, V. Küll, T. Reddmann, R. Ruhnke and A. Franzen, 1999. Streamers observed by the CRISTA experiment and simulated in the KASIMA model. Journal of Geophysical Research, 104:16405–16418.

Krotkov, N.A., P.K. Bhartia, J.R. Herman, V. Fioletov and J. Kerr, 1998. Satellite estimation of spectral surface UV irradiance in the presence of tropospheric aerosols - 1. Cloud-free case. Journal of Geophysical Research, 103(D8):8779–8793.

Krotkov, N.A., J.R. Herman, P.K. Bhartia, V. Fioletov and Z. Ahmad, 2001. Satellite estimation of spectral surface UV irradiance 2. Effects of horizontally homogeneous clouds and snow. Journal of Geophysical Research, 106:11743–11759.

Krzyscin, J.W., 2000. Total ozone influence on the surface UV-B radiation in late spring-summer 1963–1997: An analysis of multiple timescales Journal of Geophysical Research, 105(D4):4993–5000.

Kuhn, P., H. Browman and B. McArthur, 1999. Penetration of ultraviolet radiation in the waters of the estuary and Gulf of St. Lawrence. Limnology and Oceanography, 44(3):710–716.

Kuroda, Y. and K. Kodera, 1999. Role of planetary waves in the stratosphere-troposphere coupled variability in the Northern Hemisphere winter. Geophysical Research Letters, 26:2375–2378.

Kylling, A. and B. Mayer, 2001. Ultraviolet radiation in partly snow covered terrain: observations and three-dimensional simulations. Geophysical Research Letters, 27(9):1411–1414.

Kylling, A., A. Albold and G. Seckmeyer, 1997. Transmittance of a cloud is wavelength-dependent in the UV-range: physical interpretation. Geophysical Research Letters, 24(4):397–400.

Kylling, A., A Dahlback and B. Mayer, 2000a. The effect of clouds and surface albedo on UV irradiances at a high latitude site. Geophysical Research Letters, 27(9):1411–1414.

Kylling, A., T. Persen, B. Mayer and T. Svenøe, 2000b. Determination of an effective spectral surface albedo from ground based global and direct UV irradiance measurements. Journal of Geophysical Research, 105:4949–4959.

Lakkala, K., E. Kyro and T. Turunen, 2003. Spectral UV measurements at Sodankylä during 1990–2001. Journal of Geophysical Research, 108(D19): doi:10.1029/2002JD003300.

Lantz, K.O., P. Disterhoft, J.J. DeLuisi, E. Early, A. Thompson, D. Bigelow and J. Slusser, 1999. Methodology for deriving clearsky erythemal calibration factors for UV broadband radiometers of the U.S. Central Calibration Facility. Journal of Atmospheric and Oceanic Technology, 16:1736–1752.

Lapeta, B., O. Engelsen and Z. Litynska, 2000. Sensitivity of surface UV radiation and ozone column retrieval to ozone and temperature profiles. Journal of Geophysical Research, 105(D4): 5001–5007.

Laurion, I., W.F. Vincent and S Lean, 1997. Underwater ultraviolet radiation: Development of spectral models for northern high latitude lakes. Photochemistry and Photobiology, 65(1):107–114.

Lean, J., 2000. Evolution of the Sun's spectral irradiance since the Maunder minimum. Journal of Geophysical Research, 27(16):2425–2428.

Leavitt, P.R., R.D. Vinebrooke, D.B. Donald, J.P. Smol and D.W. Schindler, 1997. Past ultraviolet radiation environments in lakes derived from fossil pigments. Nature, 388:457–459.

Leavitt, P.R., B.F. Cumming and J.P. Smol, 2003. Climatic control of ultraviolet radiation effects on lakes. Limnology and Oceanography, 48(5):2062–2069.

Lenoble, J., 1998. Modeling of the influence of snow reflectance on ultraviolet irradiance for cloudless sky. Applied Optics, 37(12):2441–2447.

Lenoble, J., 2000. Influence of the environment reflectance on the ultraviolet zenith radiance for cloudless sky. Applied Optics, 39:4247–4254.

Li, Z., P. Wang and J. Cihlar, 2000. A simple and efficient method for retrieving surface UV radiation dose rate from satellite. Journal of Geophysical Research, 105:5027–5036.

Lindfors, A.V., A. Arola, J. Kaurola, P. Taalas and T. Svenøe, 2003. Long-term erythemal UV doses at Sodankyla estimated using total ozone, sunshine duration, and snow depth. Journal of Geophysical Research, 108(D16):10.1029/2002JD003325.

Lubin, D. and E. Morrow, 2001. Ultraviolet radiation environment of Antarctica. 1. Effect of sea ice on top-of-atmosphere albedo and on satellite retrievals. Journal of Geophysical Research, 106: 33453–33461.

Lubin, D., E.H. Jensen and H.P. Gies, 1998. Global surface ultraviolet radiation climatology from TOMS and ERBE data. Journal of Geophysical Research, 103:26061–26091.

Madronich, S., R.L. McKenzie, L.O. Björn and M.M. Caldwell, 1998. Changes in biologically active radiation reaching the Earth's surface. Journal of Photochemistry and Photobiology B, 46:5–19.

Manney, G., L. Froidevaux and J.W. Waters, 1996. Arctic ozone depletion observed by URAS MLS during the 1994/95 winter. Geophysical Research Letters, 23:85–88.

Manzini, E., B. Steil, C. Bruhl, M. Giorgetta and K. Kruger, 2002. A new interactive chemistry climate model. 2: sensitivity of the middle atmosphere to ozone depletion and increase in greenhouse gases: implications for recent stratospheric cooling. Journal of Geophysical Research, 108:10.1029/2002JD002977.

Mayer, B. and G. Seckmeyer, 1996. All-weather comparison between spectral and broadband (Robertson-Berger) UV measurements. Photochemistry and Photobiology, 64(7):792–799.

Mayer, B., G. Seckmeyer and A. Kylling, 1997. Systematic long-term comparison of spectral UV measurements and UVSPEC modeling results. Journal of Geophysical Research, 102(D7):8755–8768.

McArthur, L.J.B., V.E. Fioletov, J.B. Kerr, C.T. McElroy and D.I. Wardle, 1999. Derivation of UV-A irradiance from pyranometer measurements. Journal of Geophysical Research, 104(D23): 30139–30151.

McKenzie, R.L. and M. Kotkamp, 1996. Upwelling UV spectral irradiance and surface albedo measurements at Lauder, New Zealand. Geophysical Research Letters, 23(14):1757–1760.

McKenzie, R.L., K.J. Paulin and S. Madronich, 1998. Effects of snow cover on UV irradiance and surface albedo: a case study. Journal of Geophysical Research, 103(D22):28785–28792.

McKenzie, R.L., P.V. Johnston, D. Smale, B.A. Bodhaine and S. Madronich, 2001a. Altitude effects on UV spectral irradiance deduced from measurements at Lauder, New Zealand, and at Mauna Loa Observatory, Hawaii. Journal of Geophysical Research, 106(D19):22845–22860.

McKenzie, R.L., G. Seckmeyer, A.F. Bais, J.B. Kerr and S. Madronich, 2001b. Satellite retrievals of erythemal UV dose compared with ground-based measurements at northern and southern midlatitudes. Journal of Geophysical Research, 106(D20): 24051–24062.

McKinlay, A.F. and B.L. Diffey, 1987. A reference action spectrum for ultraviolet induced erythema in human skin. CIE (Commission International de l'Éclairage) Research Note, 6(1):17–22.

McPeters, R.D., P.K. Bhartia, A.J. Krueger, J.R. Herman, B.M. Schlesinger, C.G. Wellemeyer, C.J. Seftor, G. Jaross, S.L. Taylor, T. Swissler, O. Torres, G. Labow, W. Byerly and R.P. Cebula, 1996. Total Ozone Mapping Spectrometer (TOMS) Data Products User's Guide. NASA Ref. Publ. No 1384.

Meerkoetter, R., B. Wissinger and G. Seckmeyer, 1997. Surface UV from ERS-2/GOME and NOAA/AVHRR data: a case study. Geophysical Research Letters, 24:1939–1942.

Meyer-Rochow, V.B., 2000. Risks, especially for the eye, emanating from the rise of solar UV-radiation in the Arctic and Antarctic regions. International Journal of Circumpolar Health, 59:38–51.

Micheletti, M.I., R.D. Piacentini and S. Madronich, 2003. Sensitivity of biologically active UV radiation to stratospheric ozone changes: effects of action spectrum shape and wavelength range. Photochemistry and Photobiology, 78(5):465–461.

Mims, F.M. III and J.E. Frederick, 1994. Cumulus clouds and UV-B. Nature, 371:291.

Min, Q. and L. Harrison, 1998. Synthetic spectra for terrestrial ultraviolet from discrete measurements. Journal of Geophysical Research, 105:17,033–17,039.

Molina, M.J. and F.S. Rowland, 1974. Stratospheric sink for chlorofluoromethanes: chlorine atom-catalysed destruction of ozone. Nature, 249(5460):810–812.

Montzka, S.A., J.H. Butler, R.C. Myers, T.M. Thompson, T.H. Swanson, A.D. Clarke, L.T. Lock and J.W. Elkins, 1996. Decline in the tropospheric abundance of halogen from halocarbons: implications for stratospheric ozone depletion. Science, 272(5266): 1318–1322.

Montzka, S.A., J.H. Butler, J.W. Elkins, T.M. Thompson, A.D. Clarke and L.T. Lock, 1999. Present and future trends in the atmospheric burden of ozone-depleting halogens. Nature, 398:690–694.

Montzka, S.A., J.H. Butler, B.D. Hall, D.J. Mondeel and J.W. Elkins, 2003. A decline in tropospheric organic bromine. Geophysical Research Letters, 30(15):doi:10.1029/2003GL017745.

Morris, D.P., H. Zagarese and C.E. Williamson, 1995. The attenuation of solar UV radiation in lakes and the role of dissolved organic carbon. Limnology and Oceanography, 40(8):1381–1391.

Müller, R., P.J. Crutzen, J.U. Gross, C. Bruhl, J.M. Russell, H. Gernant, D.S. McKenna and A.F. Tuck, 1997. Severe chemical loss in the Arctic during the winter of 1995–96. Nature, 389: 709–712.

Munakata, N., S. Kazadzis, A. Bais, K. Hieda, G. Rontó, P. Rettberg and G. Horneck, 2000. Comparisons of spore dosimetry and spectral photometry of solar-UV radiation at four sites in Japan and Europe. Photochemistry and Photobiology, 72:739–745.

Nack, M.L. and A.E.S. Green, 1974. Influence of clouds, haze, and smog on the middle ultraviolet reaching the ground. Applied Optics, 13:2405–2415.

Nagashima, T., M. Takahashi, M. Takigawa and H. Akiyoshi, 2002. Future development of the ozone layer calculated by a general circulation model with fully interactive chemistry. Geophysical Research Letters, 29(8) doi:10.1029/2001GL014026.

Newchurch, M.J., E-S.Yang, D.M. Cunnold, G.C. Reinsel, J.M. Zawodny and J.M. Russell III, 2003. Evidence for slowdown in stratospheric ozone loss: first stage of ozone recovery. Journal of Geophysical Research, 108(D16):4507 doi:10.1029/2003JD003471.

Newman, P., J. Gleason, R. McPeters and R. Stolarski, 1997. Anomalously low ozone over the Arctic, Geophysical Research Letters, 24:2689–2692.

Nichol, S.E., G. Pfister, G.E. Bodeker, R.L. McKenzie, S.W. Wood and G. Bernhard, 2003. Moderation of cloud reduction of UV in the Antarctic due to high surface albedo. Journal of Applied Meteorology, 42(8):1174–1183.

Oltmans, S.J. and D.J. Hofmann, 1995. Increase in lower-stratospheric water vapor at a mid-latitude Northern Hemisphere site from 1981 to 1994. Nature, 374:146–149.

Oltmans, S.J., H. Vomel, D.J. Hofmann, K.H. Rosenlof and D. Kley, 2000. The increase in stratospheric water vapor from balloon borne, frost point hygrometer measurements at Washington, DC, and Boulder, Colorado. Geophysical Research Letters, 27(21):3453–3456.

Orsolini, Y.J. and V. Limpasuvan, 2001. The North Atlantic Oscillation and the occurrences of ozone miniholes. Geophysical Research Letters, 28(21):4099–4102.

Pawson, S. and B. Naujokat, 1999. The cold winters in the middle 1990s in the northern lower stratosphere. Journal of Geophysical Research, 104(D12):14209–14222.

Pawson, S., K. Kodera, K. Hamilton, (plus 37 others), 2000. The GCM-Reality Intercomparison Project for SPARC (GRIPS): scientific issue and initial results. Bulletin of the American Meteorological Society, 81(4):781–796.

Perovich, D.K., 1993. A theoretical model of ultraviolet light transmission through Antarctic sea ice. Journal of Geophysical Research, 98:22579–22587.

Peters, D., J. Egger and G. Entzian, 1995. Dynamical aspects of ozone mini-hole formation. Meteorology and Atmospheric Physics, 55:205–214.

Philipona, R., A. Schilling and D. Schmucki, 2001. Albedo-enhanced maximum UV irradiance measured on a surface oriented normal to the sun. Photochemistry and Photobiology, 73(4):366–369.

Pitari, G. and V. Rizi, 1993. An estimate of the chemical and radiative perturbation of stratospheric ozone following the eruption of Mt. Pinatubo. Journal of Atmospheric Sciences, 50:3260–3276.

Pitari, G., E. Mancini, V. Rizi and D.T. Shindell, 2002. Impact of future climate and emission changes on stratospheric aerosols and ozone. Journal of Atmospheric Sciences, 59:414–440.

Portmann, R.W., S.S. Brown, T. Gierczak, R.K. Talukdar, J.B.
    Burkholder and A.R. Ravishankara, 1999. Role of nitrogen oxides
    in the stratosphere: A reevaluation based on laboratory data.
    Geophysical Research Letters, 26:2387–2390.
Quinn, P.K., T.L. Miller, T.S. Bates, J.A. Ogren, E. Andrews and G.E.
    Shaw, 2002. A 3-year record of simultaneously measured aerosol
    chemical and optical properties at Barrow, Alaska. Journal of
    Geophysical Research, 107(D11):4130,
    doi:10.1029/2001JD001248.
Randel, W.J., F. Wu, J.M. Russell III and J. Waters, 1999. Space-time
    patterns of trends in stratospheric constituents derived from UARS
    measurements. Journal of Geophysical Research, 104:3711–3727.
Randel, W.J., F. Wu and R. Stolarski, 2002. Changes in column ozone
    correlated with the stratospheric EP flux. Journal of the
    Meteorological Society of Japan, 80(4B):849–862.
Randel, W.J., F. Wu, S.J. Oltmans, K. Rosenlof and G. Nedoluha,
    2004. Interannual changes of stratospheric water vapor and corre-
    lations with tropical tropopause temperatures. Journal of the
    Atmospheric Sciences, 61:2133–2148.
Randeniya, L.K., P.F. Vohralik and I. Plumb, 2002. Stratospheric
    ozone depletion at northern mid latitudes in the 21st century: the
    importance of future concentrations of greenhouse gases nitrous
    oxide and methane. Geophysical Research Letters, 29(4):
    1051, 2002.
Reinsel, G.C., G.C. Tiao, D.J. Wuebbles, J.B. Kerr, A.J. Miller, R.M.
    Nagatani, L. Bishop and L.H. Ying, 1994. Seasonal trend analysis of
    published ground-based and TOMS total ozone data through 1991.
    Journal of Geophysical Research, 99:5449–5464.
Reinsel, G.C., E. Weatherhead, G.C. Tiao, A.J. Miller, R.M. Nagatani,
    D.J. Wuebbles and L.E. Flynn, 2002. On detection of turnaround
    and recovery in trend for ozone. Journal of Geophysical Research,
    107: 10.1029/2001JD000500.
Reuder, J., M. Dameris and P. Koepke, 2001. Future UV radiation in
    Central Europe modeled from ozone scenarios. Journal of
    Photochemistry and Photobiology, 61:94–105.
Rex, M., P. von der Gathen, N.R.P. Harris, B.M. Knudsen, G.O.
    Brathen, H. de Backer, R. Fabian, H. Fast, M. Gil, E. Kyro, I.S.
    Mikkelsen, M. Rummukainen, J. Stahelin and C. Varotsos, 1998.
    In situ measurements of stratospheric ozone depletion rates in the
    Arctic winter 1991/1992: a Lagrangian approach. Journal of
    Geophysical Research, 103:5843–5853.
Rex, M., P. von der Gathen, G.O. Braathen, N.P.L. Harris, E. Reimer,
    A. Beck, R. Alfier, R. Kruger-Carstensen, M.P. Chipperfield, H.
    DeBacker, F. O'Connor, H. Dier, V. Dorokhov, H. Fast, A. Gamma,
    M. Gil, E. Kyro, Z. Litynska, I.S. Mikkelsen, M. Molyneux, G.
    Murphy, S.J. Reid, M. Rummukainen, C. Zerefos, 1999. Chemical
    ozone loss in the Arctic winter 1994/95 as determined by the Match
    technique. Journal of Atmospheric Chemistry, 32:35–99.
Rex, M., R.J. Salawitch, N.R.P. Harris (plus 61 others), 2002.
    Chemical loss of Arctic ozone. Journal of Geophysical Research,
    107(D20): doi:10.1029/2001JD000533.
Rex, M., R.J. Salawitch, P. von der Gathen, N.R.P. Harris, M.P.
    Chipperfield and B. Naujokat, 2004. Arctic ozone loss and climate
    change. Geophysical Research Letters, 31:L04116,
    doi:10.1029/2003GL018844.
Ricard, V., J.-L. Jaffrezo, V.-M. Kerminen, R.E. Hillamo, M. Sillanpää,
    S. Ruellan, C. Liousse and H. Cashier, 2002. Two years of continu-
    ous aerosol measurements in northern Finland. Journal of
    Geophysical Research, 107(D11):4129–4139.
Ricchiazzi, P.J. and C. Gautier, 1998. Investigation of the effect of sur-
    face heterogeneity and topography on the radiation environment of
    Palmer Station, Antarctica with a hybrid 3-D radiative transfer
    model. Journal of Geophysical Research, 103, 6161–6176.
Rinsland, C.P., E. Mathieu, R. Zander, N.B. Jones, M.P. Chipperfield,
    A. Goldman, J. Anderson, J.M. Russell, P. Demoulin, J. Notholt,
    G.C. Toon, J.F. Blavier and B. Sen, 2003. Long-term trends of inor-
    ganic chlorine from ground-based infrared solar spectra: past
    increases and evidence for stabilization. Journal of Geophysical
    Research, 108(D8): 4252, doi:10.1029/2002JD003001.
Rosenfield, J.E., J.B. Considine, P.E. Meade, J.T. Bacmeister, C.H.
    Jackman and M.R. Schoeberl, 1997. Stratospheric effects of Mount
    Pinatubo aerosol studied with a coupled two-dimensional model.
    Journal of Geophysical Research, 102:3649–3670.
Rosenfield, J.E., A.R. Douglas and D.B. Considine, 2002. The impact
    of increasing carbon dioxide on ozone recovery. Journal of
    Geophysical Research, 107(D6): doi:10.1029/2001JD000824.
Rosenlof, K.H., S.J. Oltmans, D. Kley, J.M. Russell, E.W. Chiou, W.P.
    Chu, D.G. Johnson, K.K. Kelly, H.A. Michelsen, G.E. Nedoluha,
    E.E. Remsberg, G.C. Toon and M.P. McCormick, 2001.
    Stratospheric water vapor increases over the past half-century.
    Geophysical Research Letters, 28(7):1195–1198.

Rozanov, E.V., M.E. Schlesinger and V.A. Zubov, 2001. The University
    of Illinois at Urbana-Champaign three-dimensional stratosphere-
    troposphere general circulation model with interactive ozone pho-
    tochemistry: fifteen-year control run climatology. Journal of
    Geophysical Research, 106:27233–27254.
Rozema, J., B. van Geel, L. Björn, J. Lean and S. Madronich, 2002.
    Toward solving the UV Puzzle. Science, 296:1621–1622.
Russell, J.M., M.Z. Luo and R.J. Cicerone, 1996. Satellite confirma-
    tion of the dominance of chlorofluorocarbons in the global strato-
    spheric chlorine budget. Nature, 379(6565):526–529.
Schmidt, U. and A. Khedim, 1991. In situ measurements of carbon
    dioxide in the winter Arctic vortex and mid-latitudes – an indicator
    of the age of stratospheric air. Geophysical Research Letters,
    18(4):763–766.
Schmucki, D., S. Voigt, R. Philipona, C. Frohlich, J. Lenoble,
    A. Ohmura and C. Wehrli, 2001. Effective albedo derived from UV
    measurements in the Swiss Alps. Journal of Geophysical Research,
    106(D6):5369–5383.
Schnadt, C. and M. Dameris, 2003. Relationship between North
    Atlantic Oscillation changes and stratospheric ozone recovery in
    the Northern Hemisphere in a chemistry-climate model.
    Geophysical Research Letters, 30(9): 1487,
    doi:10.1029/2003GL017006.
Schnadt, C., M. Dameris, M. Ponater, R. Hein, V. Grewe and B. Steil,
    2002. Interaction of atmospheric chemistry and climate and its
    impact on stratospheric ozone. Climate Dynamics, 18:501–517.
Schulz, A., M. Rex, J. Steger (plus 27 others), 2000. Match observa-
    tions in the Arctic winter 1996/97: high stratospheric ozone loss
    rates correlate with low temperatures deep inside the polar vortex.
    Geophysical Research Letters, 27:205–208.
Scully, N.M. and D.S. Lean, 1994. The attenuation of ultraviolet radia-
    tion in temperate lakes. Archiv für Hydrobiologie, 43:135–144.
Seckmeyer, G., A. Albold and R. Erb, 1996. Transmittance of a cloud is
    wavelength-dependent in the UV-range. Geophysical Research
    Letters, 23(20):2753–2755.
Seckmeyer, G., A. Bais, G. Bernhard, M. Blumthaler, C.R. Booth,
    P. Disterhoft, P. Eriksen, R.L. McKenzie, M. Miyauchi and C. Roy,
    2001. Instruments to measure solar ultraviolet radiation. Part 1:
    Spectral instruments. Global Atmospheric Watch Publication No.
    125, WMO TD No. 1066. World Meteorological Organization.
Shaw, G.E., 1985. On the climatic relevancy of Arctic haze: static
    energy balance considerations. Tellus B, 37(1):50–52.
Shaw, G.E., 1995. The Arctic haze phenomenon. Bulletin of the
    American Meteorological Society, 76:2403–2413.
Shepherd, T.G., 2000. The middle atmosphere. Journal of Atmospheric
    and Terrestrial Physics, 62:1587–1601.
Shindell, D.T., 2001. Climate and ozone response to increased strato-
    spheric water vapor. Geophysical Research Letters, 28:1551–1554.
Shindell, D.T. and V. Grewe, 2002. Separating the influence of halogen
    and climate changes on ozone recovery in the upper stratosphere.
    Journal of Geophysical Research, 107(D12):
    doi:10.1029/2001JD000420.
Shindell, D.T., D. Rind and P. Lonergan, 1998a. Climate change and
    the middle atmosphere: Part IV: ozone response to doubled CO_2.
    Journal of Climate, 11(5):895–918.
Shindell, D.T., D. Rind and P. Lonergan, 1998b. Increased polar strato-
    spheric ozone losses and delayed eventual recovery owing to
    increasing greenhouse gas concentrations. Nature, 392:589–592.
Slaper, H., G.J.M. Velders, J.S. Daniel, F.R. deGruijl and J.C.
    vanderLeun, 1996. Estimates of ozone depletion and skin cancer
    incidence to examine the Vienna Convention achievements. Nature,
    384(6606):256–258.
Sliney, D.H., 1986. Physical factors in cataractogenesis: ambient ultra-
    violet radiation and temperature. Investigative Ophthalmology and
    Visual Science, 27:781–790.
Sliney, D.H., 1987. Estimating the solar ultraviolet radiation exposure
    to an intraocular lens patient. Journal of Cataract and Refractive
    Surgery, 13:269–301.
Sliney, D.H., 2001. Photoprotection of the eye – UV radiation and
    sunglasses. Photochemistry and Photobiology, 64:166–175.
Slusser, J., W. Gao and R. McKenzie, 2002. Advances in UV ground-
    and space-based measurements and modeling. Optical Engineering,
    41(12):3006–3007.
Smith, R.C. and K.S. Baker, 1979. Penetration of UV-B and biological-
    ly effective dose rates in natural water. Photochemistry and
    Photobiology, 29(2):311–323.
Smyshlyaev, S.P., V.L. Dvortsov, M.A. Geller and V. Yudin, 1998.
    A two-dimensional model with input parameters from a general
    circulation model: ozone sensitivity to different formulations for
    the longitudinal temperature variation. Journal of Geophysical
    Research, 103:28373–28387.

Solomon, S., 1999. Stratospheric ozone depletion: a review of concepts and history. Reviews of Geophysics, 37:275–316.

Solanki, S. and Y. Unruh, 1998. A model of the wavelength dependence of solar irradiance variations. Astronomy and Astrophysics. 329:747–753.

Staehelin, J., J. Mader, A.K. Weiss and C. Appenzeller, 2002. Long-term trends in Northern mid-latitudes with special emphasis on the contribution of changes in dynamics. Physics and Chemistry of the Earth, 27(6–8):461–469.

Stamnes, K., K Henriksen and P. Ostensen, 1988. Simultaneous measurement of UV radiation received by the biosphere and total ozone amount. Geophysical Research Letters, 15:4418–4425.

Steil, B., M. Dameris, C. Bruhl, P.J. Crutzen, V. Grewe, M. Ponater and R. Sausen, 1998. Development of a chemistry module for GCMs: first results of a multiannual integration. Annales Geophysicae, 16(2):205–208.

Steil, B., C. Bruhl, E. Manzini, P.J. Crutzen, J. Lelieveld, P.J. Rasch, E. Roeckner and K. Kruger, 2003. A new interactive chemistry climate model. 1: Present day climatology and interannual variability of the middle atmosphere using the model and 9 years of HALOE/UARS data. Journal of Geophysical Research, 108:10.1029/2002JD002971.

Stenke, A. and V. Grewe, 2003. Impact of ozone mini-holes on the heterogeneous destruction of stratospheric ozone. Chemosphere, 50(2):177–190.

Stone, R.S., F.G., Dutton, J.M. Harris and D. Longenecker, 2002. Earlier spring snowmelt in northern Alaska as an indicator of climate change. Journal of Geophysical Research, 107(D10): 10.1029/2000JD000286.

Stordal, F., I.S.A. Isaksen and K. Horntveth, 1985. A diabatic circulation two-dimensional model with photochemistry: simulations of ozone and long-lived tracers with surface sources. Journal of Geophysical Research, 90:5757–5776.

Taalas, P., J. Kaurola, A. Kylling, D. Shindell, R. Sausen, M. Dameris, V. Grewe, J. Herman, J. Damski and B. Steil, 2000. The impact of greenhouse gases and halogenated species on future solar UV radiation doses. Geophysical Research Letters, 27:1127–1130.

Tabazadeh, A., M.L. Santee, M.Y. Danilin, H.C. Pumphrey, P.A. Newman, P.J. Hamill and J.L. Mergenthaler, 2000. Quantifying denitrification and its effect on ozone recovery. Science, 288: 1407–1411.

Tabazadeh, A, K. Drdla, M.R. Schoeberl, P. Hamill and O.B. Toon, 2002. Arctic 'ozone hole' in a cold volcanic stratosphere. Proceedings of the national Academy of Sciences of the United States of America, 99(5):2609–2612.

Takigawa, M., M. Takahashi and H. Akiyoshi, 1999. Simulation of ozone and other chemical species using a Center for Climate Systems Research/National Institute for Environmental Studies atmospheric GCM with coupled stratospheric chemistry. Journal of Geophysical Research, 104:14003–14018.

Tanskanen, A., A. Arola and J. Kujanpää, 2003. Use of the moving time-window technique to determine surface albedo from TOMS reflectivity data. In: Proceedings of the International Society for Optical Engineering Vol. 4896.

Tarasick, D.W., V.E. Fioletov, D.I. Wardle, J.B. Kerr, J.B. McArthur and C.A. McLinden, 2003. Climatology and trends in surface UV radiation survey article. Atmosphere-Ocean, 41(2) 121–138.

Thiel, S., K. Steiner and H.K. Seidlitz, 1997. Modification of global erythemal effective irradiance by clouds. Photochemistry and Photobiology, 65(6):969–973.

UNEP, 1998. Environmental Effects of Ozone Depletion. United Nations Environment Programme.

UNEP, 2003. Environmental Effects of Ozone Depletion: 2002 Assessment. Photochemistry and Photobiology, 2:1–72. United Nations Environment Programme

Van der Leun, J.C., Y. Takizawa and J.D. Longstreth, 1989. Human Health. In: Environmental Effects Panel Report. United Nations Environment Programme, 19pp.

Velders, G.J.M., 1995. Scenario Study of the Effects of CFC, HCFC, and HFC Emissions on Stratospheric Ozone. RIVM Report 722201006, National Institute of Public Health and the Environment, Netherlands.

Verdebout, J., 2000. A method to generate surface UV radiation maps over Europe using GOME, Meteosat, and ancillary geophysical data, Journal of Geophysical Research, 105:5049–5058.

Volk, C.M., J.W. Elkins, D.W. Fahey, G.S. Dutton, J.M. Gilligan, M. Loewenstein, J.R. Podolske, K.R. Chan and M.R. Gunson, 1997. Evaluation of source gas lifetimes from stratospheric observations. Journal of Geophysical Research, 102(D21):25543–25564.

Wamsley, P.R., J.W. Elkins, D.W. Fahey, G.S. Dutton, C.M. Volk, R.C. Myers, S.A. Montzka, J.H. Butler, A.D. Clarke, P.J. Fraser, L.P. Steele, M.P. Lucarelli, E.L. Atlas, S.M. Schauffler, D.R. Blake, F.S. Rowland, R.M. Stimpfle, K.R. Chan, D.K. Weisenstein and M.K.W. Ko, 1998. Distribution of halon-1211 in the upper troposphere and lower stratosphere and the 1994 total bromine budget. Journal of Geophysical Research, 103(D1):1513–1526.

Wang, P., Z. Li and J. Cihlar, 2000. Validation of an UV inversion algorithm using satellite and surface measurements. Journal of Geophysical Research, 105:5037–5048.

Waugh, D.W., W.J. Randel, S. Pawson, P.A. Newman and E.R. Nash, 1999. Persistence of the lower stratospheric polar vortices. Journal of Geophysical Research, 104:27,191–27,201.

Waugh, D.W., D.B. Considine and E.L. Fleming, 2001. Is upper stratospheric chlorine decreasing as expected? Geophysical Research Letters, 28(7):1187–1190.

Weatherhead, E.C. (Ed.), 1998. Climate change, ozone, and ultraviolet radiation. In: AMAP Assessment Report: Arctic Pollution Issues, pp. 717–774. Arctic Monitoring and Assessment Programme, Oslo.

Weatherhead, E.C., G.C. Tiao, G.C. Reinsel, J.E. Frederick, J.J. DeLuisi, D. Choi and W.K. Tam, 1997. Analysis of long-term behavior of ultraviolet-B radiation measured by Robertson-Berger meters at 14 sites in the United States. Journal of Geophysical Research, 102(D7):8737–8754.

Weatherhead, E.C., G.C. Reinsel, G.C. Tiao, J.E. Frederick, X.L. Meng, D. Choi, W.K. Cheang, T. Keller, J. DeLuisi, D. Wuebbles, J. Kerr and A.J. Miller, 1998. Factors affecting the detection of trends: statistical considerations and applications to environmental data. Journal of Geophysical Research, 103(D14):17,149–17,161.

Weatherhead, E.C., G.C. Reinsel, G.C. Tiao, C.H. Jackman, L. Bishop, S.M. Hollandsworth Frith, J. DeLuisi, T. Keller, S.J. Oltmans, E.L. Fleming, D.J. Wuebbles, J.B. Kerr, A.J. Miller, J. Herman, R. McPeters, R.M. Nagatani and J.E. Frederick, 2000. Detecting the recovery of total column ozone. Journal of Geophysical Research, 105:22201–22210.

Webb, A.R., P. Weihs and M. Blumthaler, 1999. Spectral UV irradiance on vertical surfaces: a case study. Photochemistry and Photobiology, 69:464–470.

Webb, A.R., A.F. Bais and M. Blumthaler, 2002. Measuring spectral actinic flux and irradiance: experimental results from the Actinic Flux Determination from Measurements of Irradiance (ADMIRA) project. Journal of Atmospheric and Oceanic Technology, 19(7):1049–1062.

Weisenstein, D.K., M.K.W. Ko, I.G. Dyominov, G. Pitari, L. Ricciardulli, G. Visconti and S. Bekki, 1998. The effects of sulphur emissions from HSCT aircraft: A 2-D model intercomparison. Journal of Geophysical Research, 103:1527–1547.

Wester, U., 1997. A portable lamp system circulated between Nordic solar UV laboratories. In: B. Kjeldstad, B. Johnsen and T. Koskela (eds.). The Nordic Intercomparison of Ultraviolet and Total Ozone Instruments at Izana, Oct. 1996: Final Report, pp. 25–42. Yliopistopaino, Helsinki.

Wetzel, M.A., G.E. Shaw, J.R. Slusser, R.D. Borys and C.F. Cahill, 2003. Physical, chemical, and ultraviolet radiative characteristics of aerosol in central Alaska. Journal of Geophysical Research, 108:10.1029/2002JD003208.

WHO, 2002. Global Solar UV Index: A Practical Guide. A joint recommendation of the World Health Organization, World Meteorological Organization, United Nations Environment Programme, and the International Commission on Non-Ionizing Radiation Protection.

WMO, 1989. Scientific Assessment of Stratospheric Ozone: 1989. Global Ozone Research and Monitoring Project, Report No. 20, vol. 1, World Meteorological Organization, 387pp.

WMO, 1990. Report of the International Ozone Trends Panel: 1988. Global Ozone Research and Monitoring Project, Report No. 18, World Meteorological Organization.

WMO, 1992. Scientific Assessment of Ozone Depletion, 1991. WMO Ozone Report 25, World Meteorological Organization.

WMO, 1995. Scientific Assessment of Ozone Depletion: 1994. WMO Ozone Report 37, World Meteorological Organization.

WMO, 1999. Scientific Assessment of Ozone Depletion: 1998. WMO Ozone Report 44, World Meteorological Organization.

WMO, 2003. Scientific Assessment of Ozone Depletion: 2002. Global Ozone Research and Monitoring Project – Report No. 47. World Meteorological Organization Geneva, 498pp.

Woyke, T., R. Müller, F. Stroh, D.S. McKenna, A. Engel, J.J. Margitan, M. Rex and K.S. Carslaw, 1999. A test of our understanding of the ozone chemistry in the Arctic polar vortex based on in situ measurements of ClO, BrO and O_3 in the 1994/1995 winter. Journal of Geophysical Research, 104:18755–18768.

Wuebbles, D.J., K.O. Patten, M.T. Johnson and R. Kotamarthi, 2001.
New methodology for Ozone Depletion Potentials of short-lived
compounds: n-propyl bromide as an example. Journal of
Geophysical Research, 106(D13):14551–14571.

Yang, P.C., X.J. Zhou and J.C. Bien, 2000. A nonlinear regional
prediction experiment on a short-range climate process of the
atmospheric ozone. Journal of Geophysical Research, 105(D10):
12253–12258.

Zander, R., E. Mahieu and M.R. Gunson, 1996. The 1994 northern
mid latitude budget of stratospheric chlorine derived from
ATMOS/ATLAS-3 observations. Geophysical Research Letters,
23(17):2357–2360.

Zhou, S.T., M.E. Gelman, A.J. Miller and J.P. McCormack, 2000.
An inter-hemisphere comparison of the persistent stratospheric
polar vortex. Geophysical Research Letters, 27(8):1123–1126.

Ziemke, J.R., S. Chandra and J. Herman, 2000. Erythemally weighted UV trends over northern latitudes derived from Nimbus 7
TOMS measurements. Journal of Geophysical Research,
105(D6):7373–7382.

Chapter 6

Cryosphere and Hydrology

Lead Author
John E. Walsh

Contributing Authors
Oleg Anisimov, Jon Ove M. Hagen, Thor Jakobsson, Johannes Oerlemans, Terry D. Prowse, Vladimir Romanovsky, Nina Savelieva, Mark Serreze, Alex Shiklomanov, Igor Shiklomanov, Steven Solomon

Consulting Authors
Anthony Arendt, David Atkinson, Michael N. Demuth, Julian Dowdeswell, Mark Dyurgerov, Andrey Glazovsky, Roy M. Koerner, Mark Meier, Niels Reeh, Oddur Sigurðsson, Konrad Steffen, Martin Truffer

Contents

Summary

Recent observational data present a generally consistent picture of cryospheric change shaped by patterns of recent warming and variations in the atmospheric circulation. Sea-ice coverage has decreased by 5 to 10% during the past few decades. The decrease is greater in the summer; new period-of-record minima for this season were observed several times in the 1990s and early 2000s. The coverage of multi-year ice has also decreased, as has the thickness of sea ice in the central Arctic. Snow-covered area has diminished by several percent since the early 1970s over both North America and Eurasia. River discharge over much of the Arctic has increased during the past several decades, and on many rivers the spring discharge pulse is occurring earlier. The increase in discharge is consistent with an irregular increase in precipitation over northern land areas. Permafrost temperatures over most of the subarctic land areas have increased by several tenths of a degree to as much as 2 to 3 °C during the past few decades. Glaciers throughout much of the Northern Hemisphere have lost mass over the past several decades, as have coastal regions of the Greenland Ice Sheet. The glacier retreat has been especially large in Alaska since the mid-1990s. During the past decade, glacier melting resulted in an estimated sea-level increase of 0.15 to 0.30 mm/yr. Earlier breakup and later freeze-up have combined to lengthen the ice-free season of rivers and lakes by up to three weeks since the early 1900s throughout much of the Arctic. The lengthening of the ice-free season has been greatest in the western and central portions of the northern continents. While the various cryospheric and atmospheric changes are consistent in an aggregate sense and are quite large in some cases, it is likely that low-frequency variations in the atmosphere and ocean have played at least some role in forcing the cryospheric and hydrological trends of the past few decades.

Model projections of climate change indicate a continuation of recent trends throughout the 21st century, although the rates of the projected changes vary widely among the models. For example, arctic river discharge is likely to increase by an additional 5 to 25% by the late 21st century. Trends toward earlier breakup and later freeze-up of arctic rivers and lakes are likely if the projected warming occurs. Models project that the wastage of arctic glaciers and the Greenland Ice Sheet will contribute several centimeters to global sea-level rise by 2100. The effects of thermal expansion and isostatic rebound are superimposed on the glacial contributions to sea-level change, all of which combine to produce a spatially variable pattern of projected sea-level rise of several tens of centimeters in some areas (the Beaufort Sea and much of the Siberian coast) and sea-level decrease in other areas (e.g., Hudson Bay and Novaya Zemlya). Increased inflow of cold, fresh water to the Arctic Ocean has the potential for significant impacts on the thermohaline circulation and global climate.

Models project that summer sea ice will decrease by more than 50% over the 21st century, which would extend the navigation season in the Northern Sea Route by between two and four months. Snow cover is projected to continue to decrease, with the greatest decreases projected for spring and autumn. Over the 21st century, permafrost degradation is likely to occur over 10 to 20% of the present permafrost area, and the southern limit of permafrost is likely to move northward by several hundred kilometers. Arctic coastal erosion and coastal permafrost degradation are likely to accelerate this century in response to a combination of arctic warming, sea-level rise, and sea-ice retreat.

6.1. Introduction

The term "cryosphere" is defined (NRCC, 1988) as: "That part of the earth's crust and atmosphere subject to temperatures below 0 °C for at least part of each year". For purposes of monitoring, diagnosis, projection, and impact assessment, it is convenient to distinguish the following components of the cryosphere: sea ice, seasonal snow cover, glaciers and ice sheets, permafrost, and river and lake ice. Sections 6.3–6.7 address each of these variables separately. In addition, section 6.2 addresses precipitation and evapotranspiration, which together represent the net input of moisture from the atmosphere to the cryosphere. Section 6.8 addresses the surface flows that are the primary hydrological linkages between the terrestrial cryosphere and other parts of the arctic system. These surface flows will play a critical role in determining the impact of cryospheric change on the terrestrial and marine ecosystems of the Arctic, as well as on arctic and perhaps global climate. Finally, section 6.9 addresses sea-level variations that are likely to result from changes in the cryosphere and arctic hydrology.

The different components of the cryosphere respond to change over widely varying timescales, and some of these are not in equilibrium with today's climate. The following sections examine recent and ongoing changes in each cryospheric component, as well as changes projected for the 21st century. Summaries of the present distributions of each variable precede the discussions of change. Each section also includes brief summaries of the impacts of the projected changes, although these summaries rely heavily on references to later chapters that cover many of the impacts in more detail. Each section concludes with a brief description of the key research needs that must be met to reduce uncertainties in the diagnoses and projections discussed. Relevant information from indigenous peoples on cryospheric and hydrological variability is given in Chapter 3.

6.2. Precipitation and evapotranspiration

6.2.1. Background

The cryosphere and hydrological system will respond not only to changes in the thermal state of the Arctic, but also to available moisture. For example, higher temperatures will alter the phase of precipitation, the length of the melt season, the distribution of permafrost, and the depth of the active layer, with consequent impacts on river discharge, subsurface storage, and glacier mass balance. However, these systems also depend on the balance between precipitation (P) and evapotranspiration/sublimation (collectively denoted as E).

The distribution of P and E in the Arctic has been a subject of accelerating interest in recent years. Two factors account for this surge of interest. The first is the realization that variations in hydrological processes in the Arctic have major implications not only for arctic terrestrial and marine ecosystems, but also for the cryosphere and the global ocean. The second arises from the large uncertainties in the distribution of P and E throughout the Arctic. Uncertainties concerning even the present-day distributions of P and E are sufficiently large that evaluations of recent variations and trends are problematic. The uncertainties reflect:

- the sparse network of *in situ* measurements of P (several hundred stations, with very poor coverage over northern Canada and the Arctic Ocean), and the virtual absence of such measurements of E (those that do exist are mostly from field programs of short duration);
- the difficulty of obtaining accurate measurements of solid P in cold windy environments, even at manned weather stations;
- the compounding effects of elevation on P and E in topographically complex regions of the Arctic, where the distribution of observing stations is biased toward low elevations and coastal regions; and
- slow progress in exploiting remote sensing techniques for measuring high-latitude P and E owing to the heterogeneous emissivity of snow- and ice-covered surfaces, difficulties with cloud/snow discrimination, and the near-absence of coverage by ground-based radar.

Progress in mapping the spatial and seasonal distributions of arctic P has resulted from the use of information on gauge bias adjustment procedures, for example, from the World Meteorological Organization (WMO) Solid Precipitation Measurement Intercomparison (Goodison et al., 1998). Colony et al. (1998), Yang (1999), and Bogdanova et al. (2002) recently completed summaries of P over the Arctic Ocean, where only measurements from coastal and drifting ice stations are available. The Bogdanova et al. (2002) study, which accounts for all the major systematic errors in P measurement, found the

mean annual bias-corrected P for the central Arctic Ocean to be 16.9 cm – 32% higher than the uncorrected value. The spatial pattern shows an increase from minimum values of <10 cm/yr over Greenland and 15 to 20 cm/yr over much of the Arctic Ocean, to >50 cm/yr over parts of the North Atlantic subpolar seas.

Estimates of evaporation over the Arctic Ocean are scarce. The one-year Surface Heat Budget of the Arctic Ocean (SHEBA) project collected some of the best measurements during 1997 and 1998 in the Beaufort Sea. These observations showed that evaporation was nearly zero between October and April, and peaked in July at about 7 mm/month (Persson et al., 2002).

Serreze et al. (2003) compiled estimates of P and E for the major terrestrial watersheds of the Arctic using data from 1960 to 1989. Table 6.1 presents basin-averaged values of mean annual P, precipitation minus evapotranspiration (P-E), runoff (R), and E (computed in two ways). In this study, P was derived from objectively analyzed fields of gauge-adjusted station measurements; P-E from the atmospheric moisture flux convergences in the National Centers for Environmental Prediction/National Center for Atmospheric Research (NCEP/NCAR) reanalysis; and R from gauges near the mouths of the major rivers. E was computed in two ways: E1 is the difference between the independently derived P and P-E, and E2 is the difference between basin-averaged P and R.

The two estimates of E differ by as much as 20%, providing a measure of the uncertainty in the basin-scale means of the hydrological quantities. At least some, and probably most, of the uncertainty arises from biases in measurements of P. All basins show summer maxima in P and E, and summer minima in P-E (Fig. 6.1).

Precipitation minus evapotranspiration is essentially zero during July and August in the Mackenzie Basin, and negative during June and July in the Ob Basin, illustrating the importance of E in the hydrological budget of arctic terrestrial regions. In addition, about 25% of July P in the large Eurasian basins is associated with the recycling of moisture from E (Serreze et al., 2003). The relatively low ratios of R to P (R/P, Table 6.1) in the Ob Basin are indicative of the general absence of permafrost (19% coverage in this basin, see section 6.8.2), while the relatively high ratios (and smaller E values) in the Lena and Mackenzie Basins are consistent with larger proportions of permafrost, which reduces infiltration and enhances R.

Table 6.1. Mean annual water budget components in four major drainage basins based on data from 1960 to 1989 (Serreze et al., 2003).

	P (mm)	P-E (mm)	E1 (mm)	E2 (mm)	R (mm)	R/P
Ob	534	151	383	396	138	0.26
Yenisey	495	189	306	256	239	0.48
Lena	403	179	224	182	221	0.55
Mackenzie	411	142	269	241	171	0.41

P: mean annual precipitation; P-E: precipitation minus evapotranspiration; E1: difference between the independently-derived P and P-E; E2: difference between basin-averaged P and R; R: runoff.

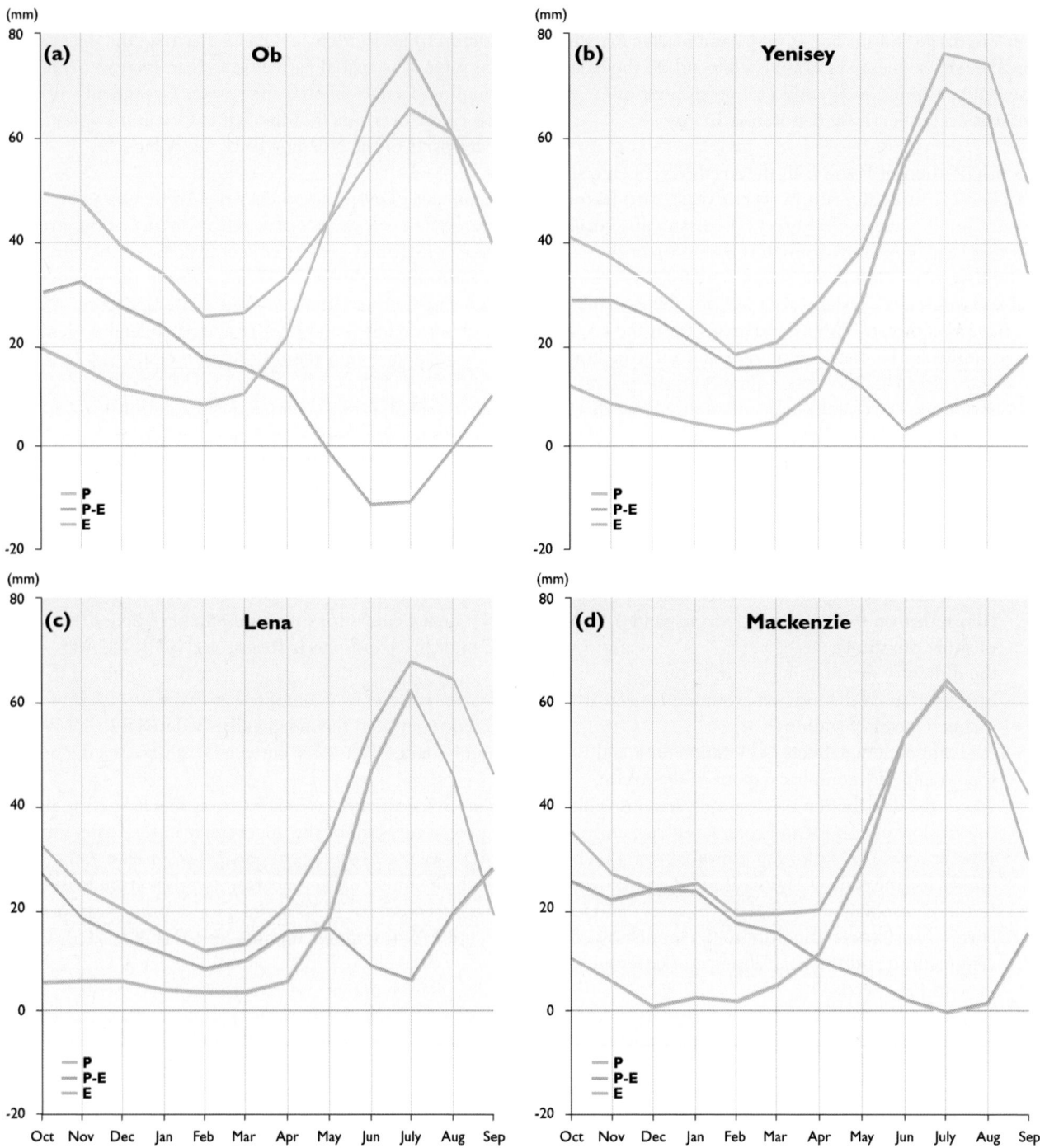

Fig. 6.1. Mean monthly precipitation (P), precipitation minus evapotranspiration (P-E), and evapotranspiration (E) for the four major arctic watersheds, using data from 1960 to 1989. E is the calculated difference between P and P-E (E1 in text). Seasonal cycle corresponds to water year (Serreze et al., 2003).

Additional estimates of the freshwater budget components of arctic and worldwide rivers, using data for earlier years, are provided by Oki et al. (1995). Rouse et al. (2003) provide a more detailed analysis of the Mackenzie Basin water cycle. The present-day hydrological regimes of the various arctic subregions are discussed further in sections 6.4 and 6.8.

6.2.2. Recent and ongoing changes

Given the uncertainties in the climatologies of arctic P and E, it is not surprising that information on recent variations and trends in these variables is limited. Time series obtained from reanalyses are subject to inhomogeneities resulting from changes in the input data over multi-decadal timescales, while trends computed using station data are complicated by measurement errors. Changes in the rain/snow ratio during periods of warming or cooling at high-latitude sites further complicate the use of *in situ* measurements for trend determination (Forland and Hanssen-Bauer, 2000).

The Intergovernmental Panel on Climate Change (IPCC, 1996, 2001) has consistently reported 20th-century P increases in northern high latitudes (55°–85° N; see Fig. 3.11 of IPCC, 1996). The increase is similar to that in Karl's (1998) "Arctic region", which includes the area poleward of 65° N but excludes the waters surrounding southern Greenland. In both cases, the greatest increase appears to have occurred during the first half of the 20th century. However, the time series are based on data from the synoptic station network, which is unevenly distributed and has undergone much change. Nevertheless, the increase in the early 20th century is reproduced by some model simulations of 20th-century climate (Kattsov and Walsh, 2000; Paeth et al., 2002).

Groisman and Easterling (1994) present data showing an increase in P over northern Canada (poleward of 55° N) since 1950. For the period since 1960, the gauge-adjusted and basin-averaged data of Serreze et al. (2003) show no discernible trends in mean annual P over the Ob, Yenisey, Lena, and Mackenzie Basins. However, summer P over the Yenisey Basin decreased by 5 to 10% over the four decades since 1960. The variations in P in these basins are associated with variations in the atmospheric circulation.

Although they are subject to the caveats that accompany trends of derived quantities in a reanalysis, trends of annual E (determined primarily by summer E) in the NCEP/NCAR reanalysis are negative in the Ob Basin and positive in the Yenisey and Mackenzie Basins. Serreze et al. (2003) suggest that recent increases in winter discharge from the Yenisey Basin may have been associated with permafrost thawing within the basin in recent decades.

Further discussion of recent trends in variables associated with P may be found in sections 6.4.2 and 6.8.2.

6.2.3. Projected changes

The five ACIA-designated climate models (section 1.4.2), forced with the B2 emissions scenario (section 4.4.1), were used to project 21st-century change in P, E, and P-E. The models are the CGCM2 (Canadian Centre for Climate Modelling and Analysis), CSM_1.4 (National Center for Atmospheric Research), ECHAM4 OPYC3 (Max-Planck Institute for Meteorology), GFDL-R30_c (Geophysical Fluid Dynamics Laboratory), and HadCM3 (Hadley Centre for Climate Prediction and Research). Model projections are presented as averages for the Arctic Ocean and for the five largest arctic river basins: the Ob,

Yenisey, Pechora, Lena, and Mackenzie. The models differ widely in their simulations of baseline (1981–2000) values of P, E, and P-E (Table 6.2). For each of the three variables (P, E, and P-E), the projected changes by 2071–2090 are generally smaller than the range in baseline values simulated by the different models.

In general, the models project modest increases in P by the end of the 21st century. Figure 6.2 illustrates the changes projected for the 2071–2090 time slice as percentages of the baseline (1981–2000) values simulated by the models. The values of P, E, and P-E projected for the earlier time slices are generally between the models' baseline values and those for the 2071–2090 time slice, although sampling variations result in some instances of non-monotonicity, especially when the changes are small. As indicated in Fig. 6.2, there is a wider across-model range in projected changes in E than in projected changes in P. There is even considerable disagreement among the models concerning the sign of the changes in E: in every region, at least one model projects a decrease, although most of the projected changes are positive. However, the baseline values for E from which the changes occur are much smaller than the corresponding baseline values for P (Table 6.2), so the projected unit changes in E are generally smaller than the projected unit changes in P.

Table 6.2. Ranges in baseline (1981–2000) values of mean annual precipitation and evapotranspiration simulated by the five ACIA-designated models for the Arctic Ocean and major arctic river basins.

	P (mm)	E (mm)
Arctic Ocean	220[a] – 504[b]	39[b] – 92[c]
Ob	708[c] – 1058[d]	302[a] – 426[d]
Yenisey	604[c] – 898[b]	224[a] – 276[b]
Lena	552[c] – 881[b]	200[c] – 312[d]
Pechora	493[c] – 1080[b]	144[c] – 246[d]
Mackenzie	670[c] – 958[d]	330[c] – 557[a]

P: Precipitation; E: Evapotranspiration.
[a]ECHAM4/OPYC3; [b]CGCM2; [c]CSM_1.4; [d]HadCM3.

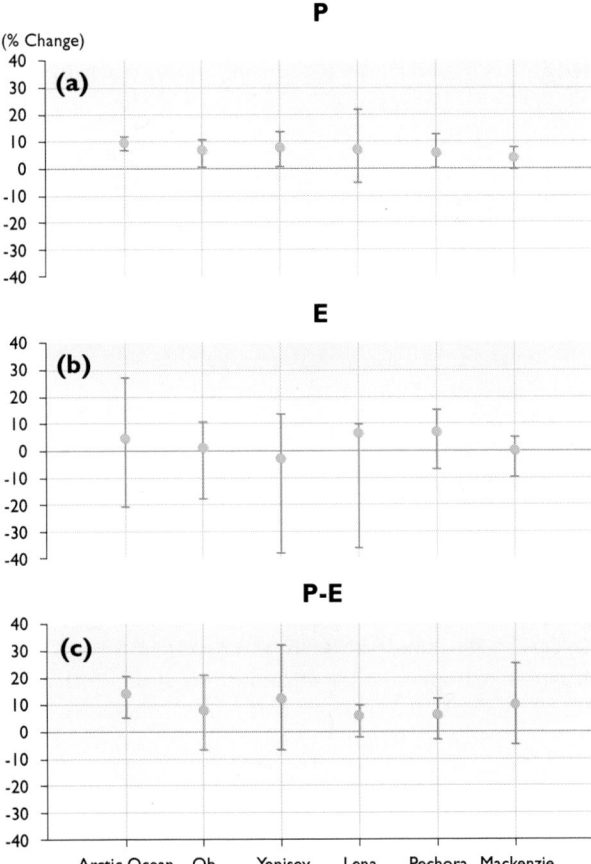

Fig. 6.2. Percentage change in (a) precipitation, (b) evapotranspiration, and (c) precipitation minus evapotranspiration between 1981–2000 and 2071–2090 projected by the five ACIA-designated models for the Arctic Ocean and five major arctic river basins. Solid circles are five-model means; vertical line segments denote the ranges of the five model projections.

Of the variables considered here, the one with the greatest relevance to other parts of the arctic system is P-E, which represents the net moisture input to the surface from the atmosphere. With one exception (the CSM_1.4, which projects the least warming of the five models), the projected changes in P-E are positive (Fig. 6.2). The greatest increase, 14% (averaged across all models), is projected to occur over the Arctic Ocean, where even the CSM_1.4 projects an increase in P-E. Over the terrestrial watersheds, the projected increases range from 6 to 12% (averaged across all models). These changes are considerably smaller than the departures from the means occurring during individual years and even during multi-year periods in the model simulations. Since the projected changes in P and P-E are generally positive, it is likely that the most consequential changes in these variables will be increases in the frequency and/or duration of wet periods. However, the annual averaging of the variables shown in Fig. 6.2 obscures a potentially important seasonality. The projected changes in P-E are generally smaller, and occasionally negative, over the major river basins during the warm season. This relative decrease in projected P-E during summer is the result of two factors: an increase in E due to projected temperature increases; and a longer season with a snow-free surface and above-freezing temperatures in the upper soil layers, resulting in greater projected E. Consequently, the model projections point to the distinct possibility that increased river flow rates during winter and spring will be accompanied by decreased flow rates during the warm season. The latter is consistent with the results of the Mackenzie Basin Impact Study (Cohen, 1997).

6.2.4. Impacts of projected changes

On other parts of the physical system

The projected increases in P, and more importantly in P-E, imply an increase in water availability for soil infiltration and runoff. The increases in P-E projected to occur by 2071–2090 over the major terrestrial watersheds imply that the mean annual discharge to the Arctic Ocean will increase by 6 to 12%. Since the mean annual P-E over the Arctic Ocean is projected to increase by 14% over this period, a substantial increase in the freshwater supplied to the Arctic Ocean is projected to occur by the later decades of the present century. If there is an increase in the supply of fresh water to the Arctic Ocean, it will increase the stratification of the Arctic Ocean, facilitate the formation of sea ice, and enhance freshwater export from the Arctic Ocean to the North Atlantic (sections 6.5.4 and 6.8.4). In addition, increased aquatic transport and associated heat fluxes across the coastal zone are likely to accelerate the degradation of coastal permafrost in some areas.

The projected increases in P and P-E imply generally wetter soils when soils are not frozen, increased surface flows above frozen soils, wetter active layers in the summer, and greater ice content in the upper soil layer during winter. To the extent that the projected increase in P occurs as an increase in snowfall during the cold season (section 6.4), the Arctic Ocean and its terrestrial watersheds will experience increases in snow depth and snow water equivalent, although the seasonal duration may be shorter if warming accompanies the increase in P. Moreover, the projected increase in mean annual P-E obscures important seasonality. Recent trends of increasing E in the Yenisey and Mackenzie Basins (section 6.2.2) raise the possibility that P-E will actually decrease during the summer when E exceeds P, resulting in a drying of soils during the warm season.

On ecosystems

The projected increase in P-E over the terrestrial watersheds will increase moisture availability in the upper soil layers, favoring plant growth in regions that are presently moisture-limited. However, as previously noted, projected increases in E during the summer are likely to lead to warm-season soil drying and reduced summer river levels. Thawing of permafrost, which could increase the subsurface contribution to streamflow and possibly mitigate the effect of increased E during summer, is another complicating factor.

The projected increase in river discharge is likely to increase nutrient and sediment fluxes to the Arctic Ocean, with corresponding impacts on coastal marine ecosystems (section 9.3.2). If P increases during winter and ice breakup accelerates, an increase in flood events is likely. Higher flow rates in rivers and streams caused by such events are likely to have large impacts on riparian regions and flood plains in the Arctic. Wetland ecosystems are likely to expand in a climate regime of increased P-E, with corresponding changes in the fluxes of trace gases (e.g., carbon dioxide and methane) across the surface–atmosphere interface.

Projected increases in P and P-E will result in generally greater availability of surface moisture for arctic residents. In permafrost-free areas, water tables are likely to be closer to the surface, and moisture availability for agriculture will increase. During the spring period when enhanced P and P-E are likely to increase river levels, the risk of flooding will increase. Lower water levels during the summer would affect river navigation, increase the threat from forest fires, and affect hydropower generation.

6.2.5. Critical research needs

It is apparent from Table 6.2 and Fig. 6.2 that models differ widely in their simulations of P and P-E in baseline climate simulations and in projections of future climate. The result is a very large range in uncertainty for future rates of moisture supply to the arctic surface. There is an urgent need to narrow the range in uncertainty by determining the reasons for the large across-model variances in P and E, and by bringing the models' baseline simulations of P and E into closer agreement with observational data. That the observational data are also uncertain indicates a need for collaboration between the observational and modeling communities, including the remote sensing community, in reconciling models and data.

The most problematic variable of those considered is E. Despite its direct relevance to the surface moisture budget and to terrestrial ecosystems, very few observational data are available for assessing model simulations of E. The 21st-century simulations summarized here show that the models do not agree even on the sign of the changes in E in the Arctic. Improved model parameterizations of E will need to address factors such as the effects of vegetation change and simulation of transpiration rates using more realistic vegetation parameters, such as leaf area index instead of a single crop factor. Datasets for validating and calibrating model-simulated E (including better use of satellite data) are one of the most urgent needs for developing scenarios of arctic hydrology.

6.3. Sea ice

6.3.1. Background

Sea ice has long been regarded as a key potential indicator and agent of climate change. In recent years, sea ice has received much attention in the news media and the scientific literature owing to the apparent reduction in coverage and thickness of sea ice in the Arctic. Since the potential impacts of these changes on climate, ecosystems, and infrastructure are large, sea ice is a highly

important variable in an assessment of arctic change. Section 6.9.1 discusses sea ice within the context of coastal stability and sea-level rise.

Owing to the routine availability of satellite passive microwave imagery from the Scanning Multichannel Microwave Radiometer and the Special Sensor Microwave/Imager (SSM/I) sensors, sea-ice coverage has been well monitored since the 1970s. Figure 6.3 shows mean sea-ice concentrations for the months of the climatological maximum (March) and minimum (September) for the period 1990 to 1999 derived from SSM/I data. The accuracy of passive microwave-derived sea-ice concentrations varies from approximately 6% during winter to more than 10% during summer. The sea-ice variable most compatible with pre-satellite information (based largely on ship reports) is sea-ice extent, defined as the area of ocean with an ice concentration of at least 15%. Arctic sea-ice extent, including all subpolar seas except the Baltic, ranges from about 7 million km² at its September minimum to about 15 million km² at its March maximum. The areal coverage of sea ice (excluding open water poleward of the ice edge) ranges from 5 to 6 million km² in late summer to about 14 million km² in the late winter (Parkinson et al., 1999). Interannual variability in the position of the sea-ice edge is typically one to five degrees of latitude for a particular geographic region and month. The departures from normal at a particular time vary regionally in magnitude and in sign.

While ice extent and areal coverage have historically been used to monitor sea ice, ice thickness is an equally important consideration within the context of the sea-

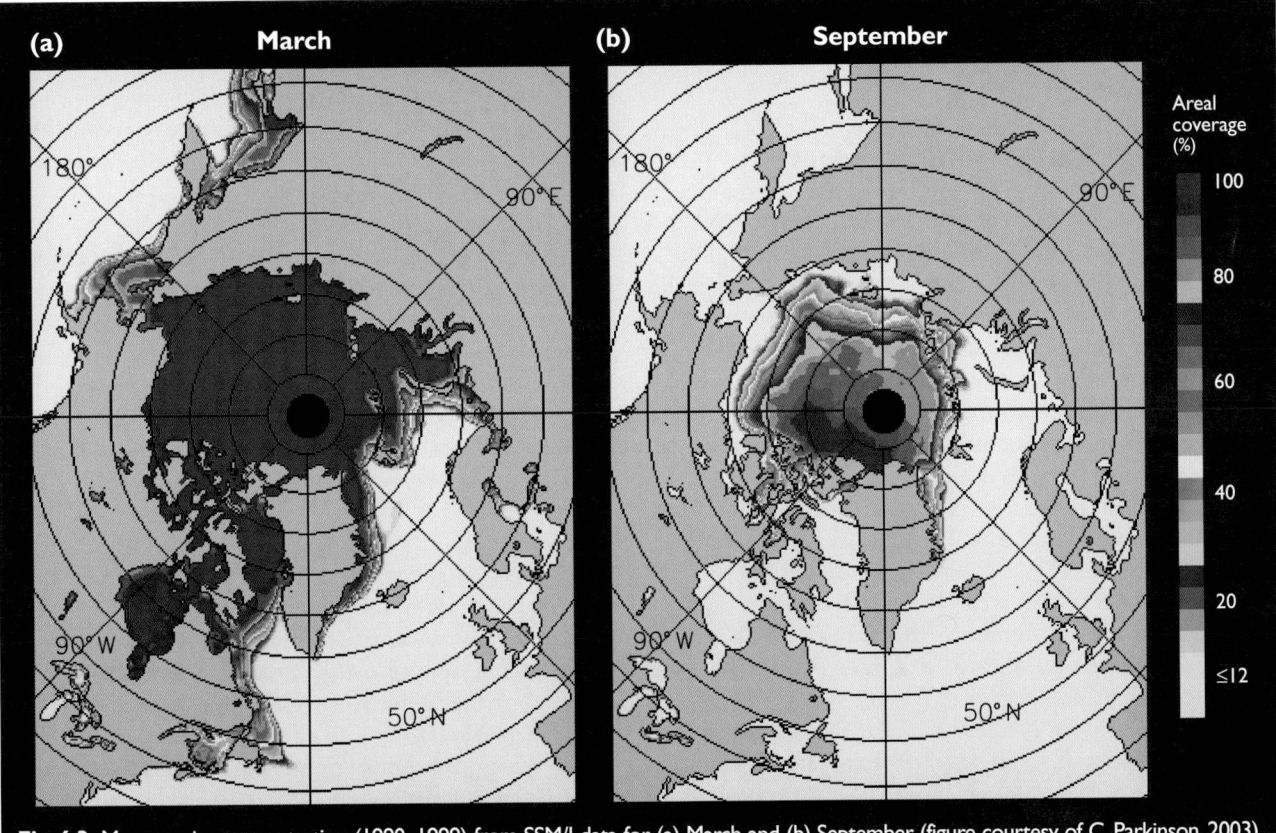

Fig. 6.3. Mean sea-ice concentration (1990–1999) from SSM/I data for (a) March and (b) September (figure courtesy of C. Parkinson, 2003).

ice mass budget. Unfortunately, sea-ice thickness measurements are less routine, consisting largely of upward-looking sonar measurements from occasional and irregular submarine cruises and, in recent years, from moored sonar on or near the continental shelves. In addition, direct measurements of fast-ice (sea ice attached to the shore) thickness have been made for several decades in some coastal regions, and occasional direct measurements have been made in the central Arctic at manned ice camps. The general pattern of sea-ice thickness has been determined, but it is subject to variations and uncertainties that have not been well quantified. Sea-ice thickness generally increases from the Siberian side of the Arctic to the Canadian Archipelago, largely in response to the mean pattern of sea-ice drift and convergence (although air temperatures are also generally lower on the Canadian side of the Arctic Ocean). In areas of perennial sea ice, the seasonal cycle of melt and ablation has an amplitude of about 0.5 to 1.0 m.

The albedo of sea ice is of critical importance to the surface energy budget and to the ice-albedo feedback, both of which can accelerate sea-ice variations over timescales ranging from the seasonal to the decade-to-century scale of interest in the context of climate change. The albedo of sea ice and snow-covered sea ice has been measured throughout the annual cycle at a local scale (e.g., at ice stations such as SHEBA). However, the albedo of sea ice over scales of 10 to 100 km^2 is strongly dependent on the surface state (snow-covered versus bare ice, melt-pond distribution, and the proportion of open water, i.e., leads and polynyas). Robinson et al. (1992) summarized several years of interannual variations in surface albedo in the central Arctic Ocean. Similar compilations depicting decadal or longer-scale variations, or variations outside the Arctic Ocean, do not exist despite the potential value of such datasets for assessing the ice-albedo–temperature feedback.

6.3.2. Recent and ongoing changes

There has been an apparent reduction in sea ice over the past several decades, although this varies by region, by season, and by the sea-ice variable measured. Figure 6.4 shows the time series of Northern Hemisphere sea-ice extent, in terms of the seasonal cycle and the interannual variations (departures from climatological mean daily ice extent), for the period 1972 to 2002. Passive microwave imagery was available almost continuously during this period. Arctic sea-ice extent decreased by 0.30±0.03 x 10^6 km^2/10 yr between 1972 and 2002, but by 0.36±0.05 x 10^6 km^2/10 yr between 1979 and 2002, indicating a 20% acceleration in the rate of decrease (Cavalieri et al., 2003). Over the full 31-year period, the trend in summer (September) is -0.38± 0.08 x 10^6 km^2/10 yr, whereas in winter (March) the trend is -0.27±0.05 x 10^6 km^2/10 yr. For the 24-year period (1979–2002), the corresponding summer and winter trends are -0.48±0.13 x 10^6 km^2/10 yr and -0.29±0.06 x 10^6 km^2/10 yr, respectively (Cavalieri et al., 2003). These trends contrast with those of

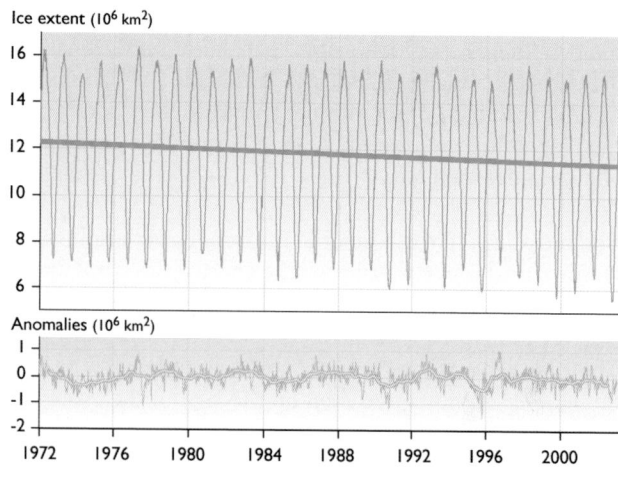

Fig. 6.4. Daily arctic sea-ice extent (upper) and anomalies (lower) between 1972 and 2002. A linear trend line is superimposed on the daily extents and a 365-day running mean has been applied to the daily anomalies (Cavalieri et al., 2003).

Southern Hemisphere sea ice, where the trends are either close to zero or slightly positive, depending on the period of analysis.

The recent trend of decreasing sea ice has also been identified in the coverage of multi-year sea ice in the central Arctic Ocean. An analysis of passive microwave-derived coverage of multi-year sea ice in the Arctic showed a 14% decrease in winter multi-year sea ice between 1978 and 1998 (Johannessen et al., 1999). Comiso (2002) analyzed trends in end-of-summer minimum ice cover for 1979 to 2000. Figure 6.5 contrasts the sea-ice concentrations at the time of ice minima during the first and second halves of the study period. The decrease is especially large north of the Russian and Alaskan coasts. The rate of decrease in perennial sea ice (9% per decade) computed by Comiso (2002) is consistent with the trend in multi-year sea-ice coverage found by Johannessen et al. (1999), and is slightly greater than the rate of decrease in total ice-covered area in recent decades (Cavalieri et al., 2003).

The decrease in sea-ice extent over the past few decades is consistent with reports from indigenous peoples in various coastal communities of the Arctic. In particular, the themes of a shortened ice season and a deteriorating sea-ice cover have emerged from studies that drew upon the experiences of residents of Sachs Harbor, Canada and Barrow, Alaska, as well as communities on St. Lawrence Island in the Bering Sea (Krupnik and Jolly, 2002).

Vinnikov et al. (1999) extended the record back to the 1950s using data from ships, coastal reports, and aircraft surveys, and found that the trends are comparable to those of the satellite period and are statistically significant. This study also compared the observed trends of the past several decades with estimates of natural (low-frequency) variability generated by a Geophysical Fluid Dynamics Laboratory (GFDL) climate model and showed that the decrease in arctic sea-ice extent is highly unlikely to have occurred as a result of natural vari-

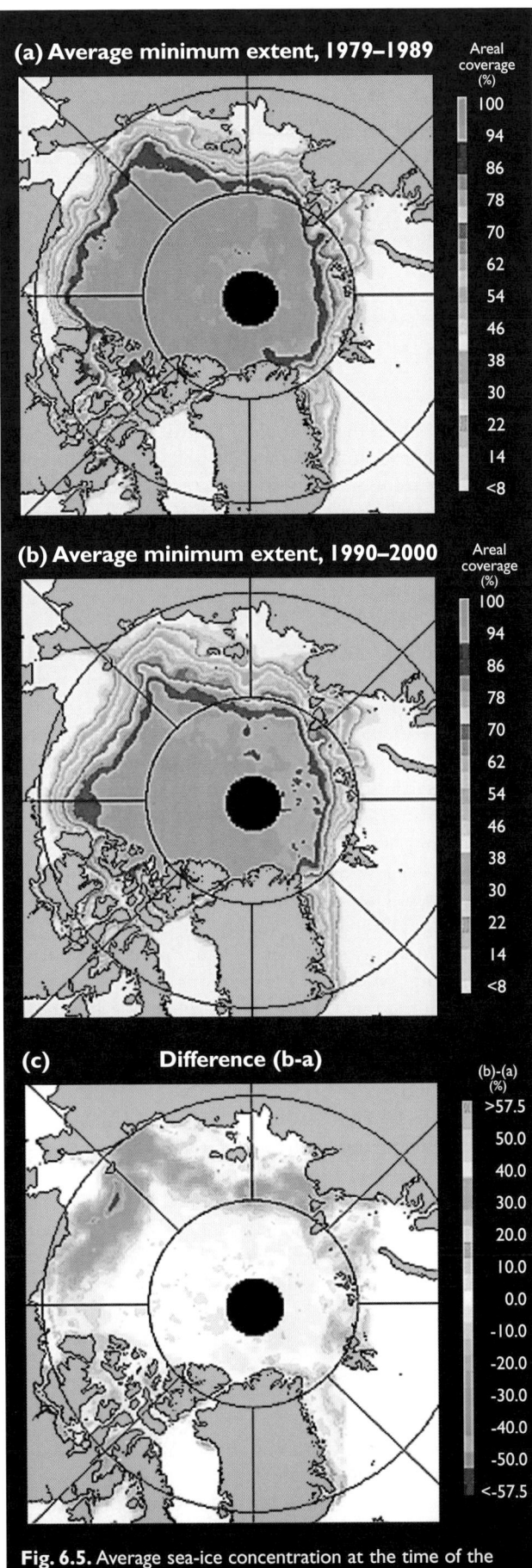

Fig. 6.5. Average sea-ice concentration at the time of the summer ice minimum for two 11-year periods: (a) 1979–1989; and (b) 1990–2000. (c) depicts the loss of ice between the two periods (Comiso, 2002).

ability alone. However, this conclusion is based on the assumption that the natural variability of sea ice can be reliably inferred from climate model simulations.

For longer timescales, the lack of sea-ice data limits estimates of hemispheric-scale trends. However, sufficient data are available for portions of the North Atlantic subarctic, based largely on historical ship reports and coastal observations, to permit regional trend assessments over periods exceeding 100 years. Perhaps the best-known record is the Icelandic sea-ice index, compiled by Thoroddsen (1917) and Koch (1945), with subsequent extensions (e.g., Ogilvie and Jonsson, 2001). The index combines information on the annual duration of sea ice along the Icelandic coast and the length of coastline affected by sea ice. Figure 6.6 shows several periods of severe sea-ice conditions, especially during the late 1800s and early 1900s, followed by a long interval (from about 1920 to the early 1960s) in which sea ice was virtually absent from Icelandic waters. However, an abrupt change to severe ice conditions in the late 1960s serves as a reminder that decadal variability is a characteristic of sea ice. Since the early 1970s, sea-ice conditions in the vicinity of Iceland have been relatively mild.

In an analysis that drew upon ship reports from the ocean waters east of Iceland, Vinje (2001) found that the extent of ice in the Nordic Seas during April had decreased by about 33% since the 1860s (Fig. 6.7). However, this dataset and longer versions spanning the past several centuries indicate large variations in trends over multi-decadal periods. Some earlier multi-decadal periods show trends comparable to those of the past several decades.

A widely cited study by Rothrock et al. (1999), based on a comparison of upward-looking sonar data from submarine cruises during 1958–1976 and 1993–1997, found a decrease of about 40% (1.3 m) in the sea-ice draft (proportional to thickness) in the central Arctic Ocean from the earlier to the later period. Wadhams and Davis (2000) provide further submarine-measured evidence of sea-ice thinning in the Arctic Ocean.

While the findings concerning ice draft and multi-year sea-ice coverage are compatible, the trends in ice draft have been evaluated using data from a relatively small subset of the past 45 years. Anisimov et al. (2003)

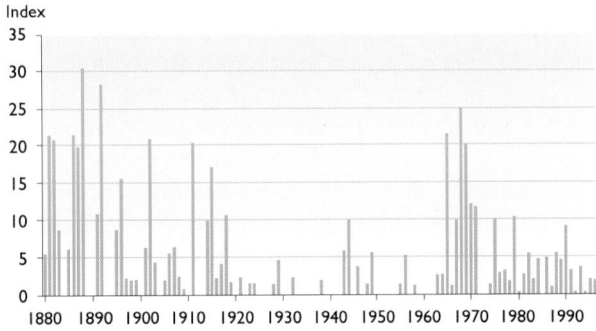

Fig. 6.6. Annual values of the Icelandic sea-ice index (T. Jakobsson, 2003).

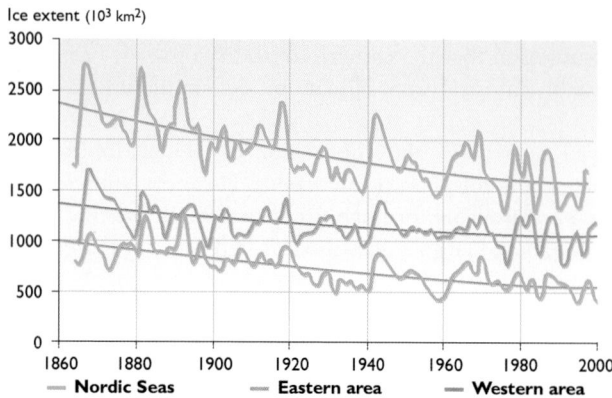

Fig. 6.7. Historical record of April sea-ice extent (two-year running means) in the Nordic Seas and in their eastern and western subregions (Vinje, 2001).

showed that a one-year shift in the sample of years examined by Rothrock et al. (1999) results in a much weaker trend in sea-ice draft. There are also indications that at least some of the decrease in ice thickness is a consequence of variations in the wind-driven advection of sea ice and that increases in ice thickness in unsampled regions (e.g., offshore of the Canadian Archipelago) may partially offset the decreases in the central Arctic Ocean detected in the 1990s (Holloway and Sou, 2002). Specifically, the sea-ice drafts in the western Arctic Ocean (Beaufort sector) appear to have decreased by about 1.5 m between the mid-1980s and early 1990s when the Beaufort Gyre weakened considerably in association with a change in the Arctic Oscillation (AO), altering the ice drift and dynamics in the region near the North Pole (Tucker et al., 2001). Proshutinsky and Johnson (1997) show that the pattern of arctic sea-ice drift has historically varied between two regimes, characterized by relatively strong and weak phases of the Beaufort anticyclone.

The association between the AO (or the North Atlantic Oscillation – NAO) and arctic sea ice is increasingly used to explain variations in arctic sea ice over the past several decades (e.g., Kwok, 2000; Parkinson, 2000; Rigor et al., 2002). Research has related the wind forcing associated with this atmospheric mode to sea-ice export from the Arctic Ocean through Fram Strait to the North Atlantic Ocean (Kwok and Rothrock, 1999), and to ice conditions along the northwestern coastline of the Canadian Archipelago (Agnew et al., 2003). However, studies of longer periods suggest that such associations with Fram Strait sea-ice export may not be temporally robust because of relatively subtle shifts in the centers of action of the NAO (Hilmer and Jung, 2000). Cavalieri (2002) reveals a consistent relationship over decadal timescales between Fram Strait sea-ice export and the phase of atmospheric sea-level pressure wave 1 at high latitudes. The phase of this wave appears to be a more sensitive indicator of Barents Sea low-pressure systems that drive sea ice through Fram Strait than the NAO index. In general, the role of sea-ice motion in diagnoses of historical change and projections of future change is largely unexplored.

6.3.3. Projected changes

This section summarizes the changes in sea ice projected for the 21st century by the five ACIA-designated models. In the case of the CGCM2 model, an ensemble of three different 21st-century simulations was available. The models all project decreases in sea-ice extent during the 21st century, although the time series contain sufficient variability that increases are found over occasional intervals of one to ten years, especially when coverage in specific regions of the Arctic is examined.

Two factors hamper quantitative comparisons of the projected changes in sea ice. First, the sea-ice variables archived by the various modeling centers vary from model to model, ranging from the presence of ice (binary 1/0) to concentration, thickness, and grid-cell mass. Since all of these variables permit evaluations of sea-ice extent (defined as the area poleward of the ice edge), ice extent is used for comparisons between the various models. Second, the sea ice simulated by these models for the baseline climate (1981–2000) is generally not in agreement with observed coverage (e.g., Fig. 6.3), especially when coverage in specific regions is considered. These biases in the baseline climate will confound interpretations of the model-derived coverage for a future time (e.g., the ACIA time slices centered on 2020, 2050, and 2080), since changes from a biased initial state are unlikely to result in a projected state that is free of biases. In an attempt to optimize the informational content of the projections of sea ice, the future sea-ice states projected by each model have been crudely adjusted by adding to each projection the baseline climate bias of sea ice for the particular model, month, and longitude. The need for this type of ad hoc adjustment will be eliminated as coupled atmosphere–ocean–ice model simulations become more realistic. The following synthesis of projections includes examples of both the raw (unadjusted) projections and the adjusted projections.

Figure 6.8 shows the 21st-century time series of total Northern Hemisphere sea-ice extent for March and September projected by the five models. The upper panels show the raw (unadjusted) time series and the lower panels show the adjusted time series. While the trends and variations are the same in both panels for a particular model, the starting points in 2000 are generally not, owing to the biases in the baseline climate simulations. Many of the differences between the models' unadjusted projections are due to the differences in the simulated baseline (1981–2000) sea-ice extent. For example, the unadjusted March sea-ice extents simulated for 1981–2000 range from approximately 13 to 20 million km², while the corresponding observational value, averaged over the entire month of March for the period 1990–1999, is about 14.5 million km². The models' raw projections show an even greater range in September, varying from about 2 to 11 million km², compared to the observational value of approximately 8 million km². The CSM_1.4 model consistently projects the greatest

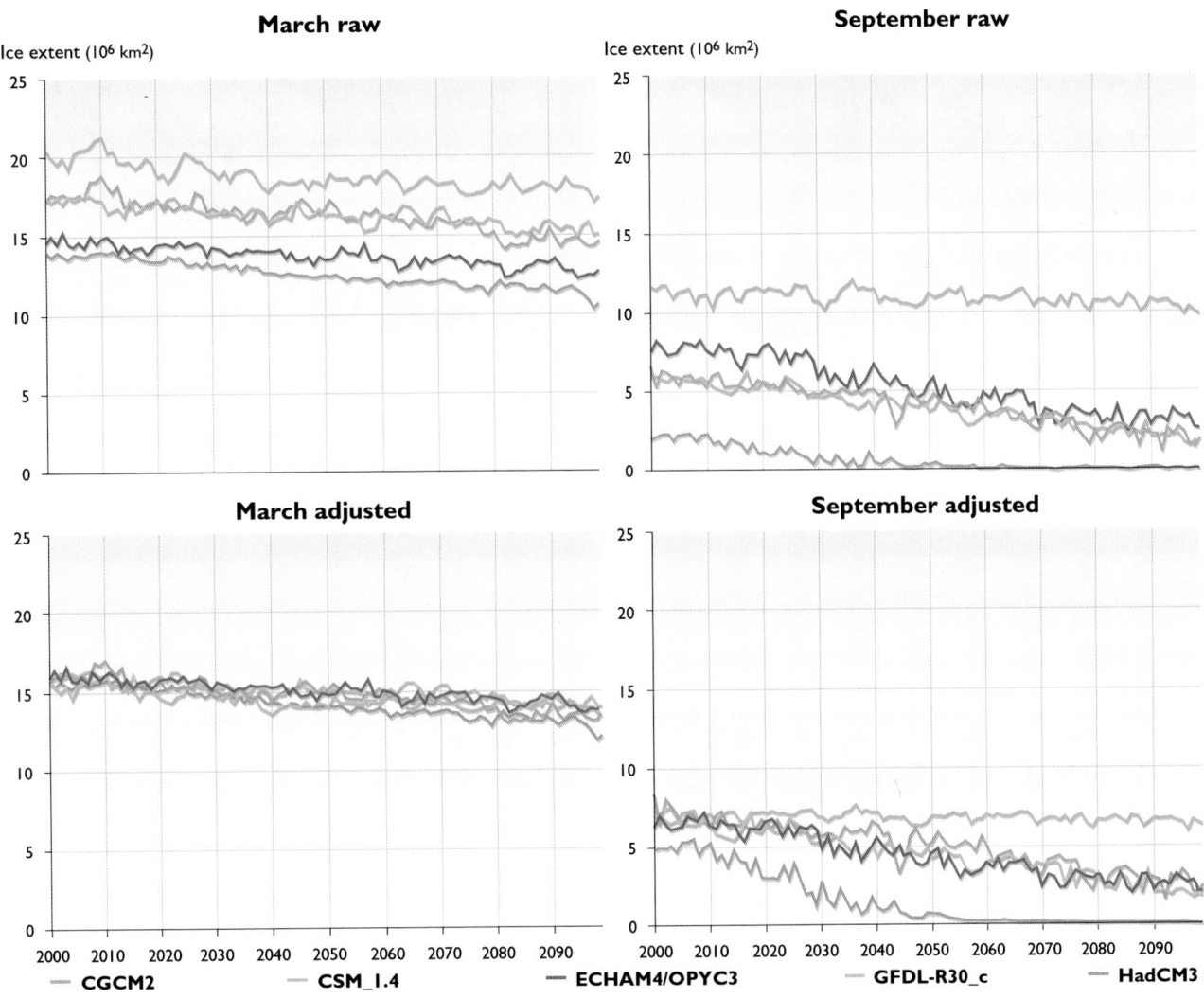

Fig. 6.8. 21st-century total Northern Hemisphere sea-ice extent projected by the five ACIA-designated models for March (left panels) and September (right panels). Upper panels show raw (unadjusted) model output; lower panels show projections adjusted for biases in simulated baseline (1981–2000) sea ice.

sea-ice extent, while the CGCM2 model consistently projects the least ice extent.

The raw projections from the CGCM2 model indicate an ice-free Arctic during September by the mid-21st century, but this model simulated less than half of the observed September sea-ice extent at the start of the 21st century. There is very little difference between the three ensemble simulations from the CGCM2 model, indicating that the initial conditions are less important than the choice of the model. None of the other models projects ice-free summers in the Arctic by 2100, although the sea-ice extent projected by the HadCM3 and ECHAM4/OPYC3 models decreases to about one-third of initial (2000) and observed September values by 2100.

For March, the projected decreases in sea-ice extent by 2100 vary from about 2 to 4 million km². Unlike September, none of the model projections for 2100 is close to ice-free in March, although the sea-ice extent projected by the CGCM2 model is only about 10 million km², which is about two-thirds of the initial (2000) March extent. A large proportion of the differences between the projected March sea-ice extents in 2100 is

attributable to the differences in the initial (2000) ice extent simulated by the models.

Table 6.3 summarizes the 21st-century changes in mean annual sea-ice extent projected by the models. The greatest reductions in sea-ice cover, both as actual areas and as percentage reductions, are projected by the model with the least initial (2000) sea ice, while the smallest losses are projected by the model with the most initial (2000) sea ice. Insofar as sea-ice extent and mean ice thickness are positively correlated, this relationship is not surprising, i.e., the models projecting the greatest ice extent also project the thickest ice, which is more difficult to lose in a climate change scenario. However, the association found here between the initial sea-ice extent and the rate of ice retreat does not seem to be present in the Coupled Model Intercomparison Project suite of coupled global models (Bitz, pers. comm., 2003). Flato (2004) illustrated this lack of association.

When projections are examined on the basis of the the four ACIA regions (section 18.3), some spatial variations in the model-projected sea-ice retreat are apparent. However, the regional differences are generally small, and

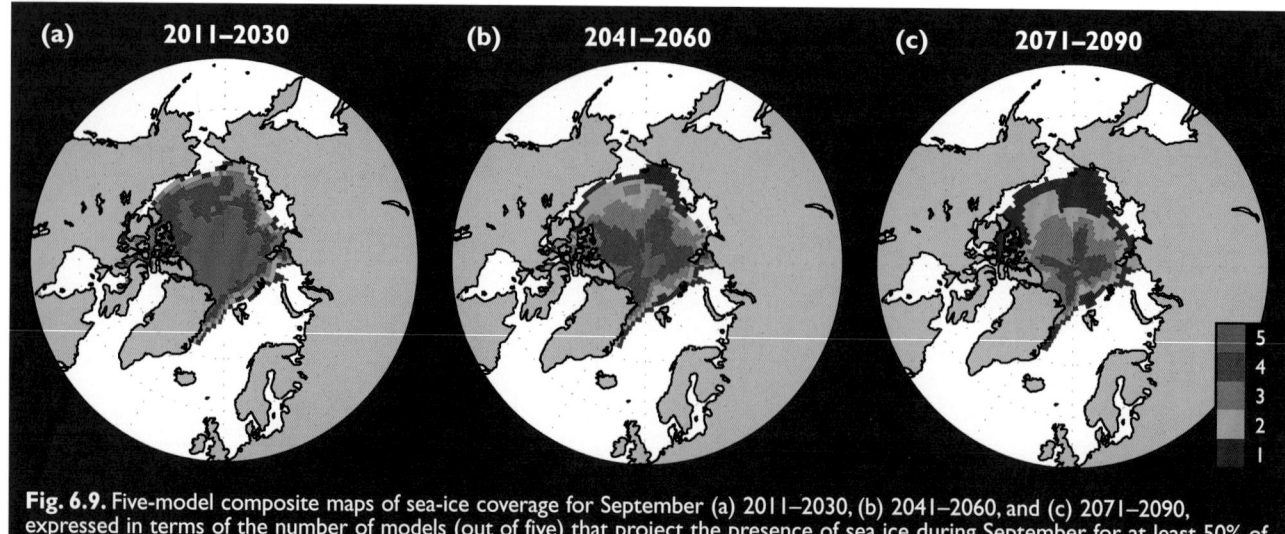

Fig. 6.9. Five-model composite maps of sea-ice coverage for September (a) 2011–2030, (b) 2041–2060, and (c) 2071–2090, expressed in terms of the number of models (out of five) that project the presence of sea ice during September for at least 50% of the years in the time slice.

are considerably less than the differences between the models. Winter sea-ice retreat, as measured by the changes in projected March ice extent, is greatest in Region 3 (150° E–120° W) for three of the model simulations (GFDL-R30_c, HadCM3, CSM_1.4). In the CGCM2 simulation, the March retreat is greatest in Region 1 (30° W–60° E). For the summer, the models show more regional variation in their projections of the greatest retreat. The GFDL-R30_c model projects the greatest summer sea-ice loss in Region 3, which is projected to become ice-free in September by the end of the 21st century. The HadCM3 and CGCM2 models project the most rapid retreat in Region 1, which is projected to become ice-free by 2100 using the unadjusted results from both simulations. The CSM_1.4 model projects little sea-ice loss in any region during the summer.

"Best estimates" of the sea-ice distributions in the ACIA time slices (2011–2030, 2041–2060, and 2071–2090) can be obtained by compositing the adjusted fields of sea ice from the five models. Figures 6.9 and 6.10 show these fields for September and March, respectively, expressed in terms of the number of models (out of five) that project the presence of sea ice during the specified month for at least 50% of the years in the time slice. Comparisons with Fig. 6.3 provide measures of the changes from 1990–1999 observed values. The distributions in Figs. 6.9 and 6.10 illustrate the

tendency for the projected reductions in sea ice to be greater, especially as a percentage of the initial (2000) values, in September than in March. The September values for all of the time slices are less than the maximum of five (models projecting the presence of sea ice) over much of the Arctic Ocean (Fig. 6.9), which at present is largely ice-covered in September.

The projected reduction in sea-ice extent in winter (March, Fig. 6.10) is less than in summer, especially when expressed as a percentage of the present coverage. Most of the Arctic Ocean is projected to remain ice-covered in March, although the March sea-ice edge is projected to retreat substantially in the subpolar seas. However, the models that simulate sea-ice thickness or mass per grid cell project that the ice becomes thinner in the central Arctic Ocean throughout the 21st century.

6.3.4. Impacts of projected changes

On other parts of the physical system

The projected changes in sea ice extent and thickness are sufficiently large that their impact on the surface energy and moisture budgets will be substantial, affecting climate at least locally and regionally. For the five ACIA-designated models, the amount by which sea-ice extent is projected to decrease is correlated with the amount by

Table 6.3. Changes in mean annual Northern Hemisphere sea-ice extent between 2000 and 2100 projected by the five ACIA-designated models.

	Unadjusted projections			Adjusted projections		
	Ice extent (10⁶ km²)		Change (%)	Ice extent (10⁶ km²)		Change (%)
	2000	2100		2000	2100	
CGCM2	9.7	5.6	-42	12.3	6.6	-46
CSM_1.4	16.5	14.2	-14	12.3	10.8	-12
ECHAM4/OPYC3	11.9	8.9	-25	12.3	9.3	-24
GFDL-R30_c	11.9	8.5	-29	12.3	8.6	-30
HadCM3	12.8	9.4	-27	12.3	9.1	-26

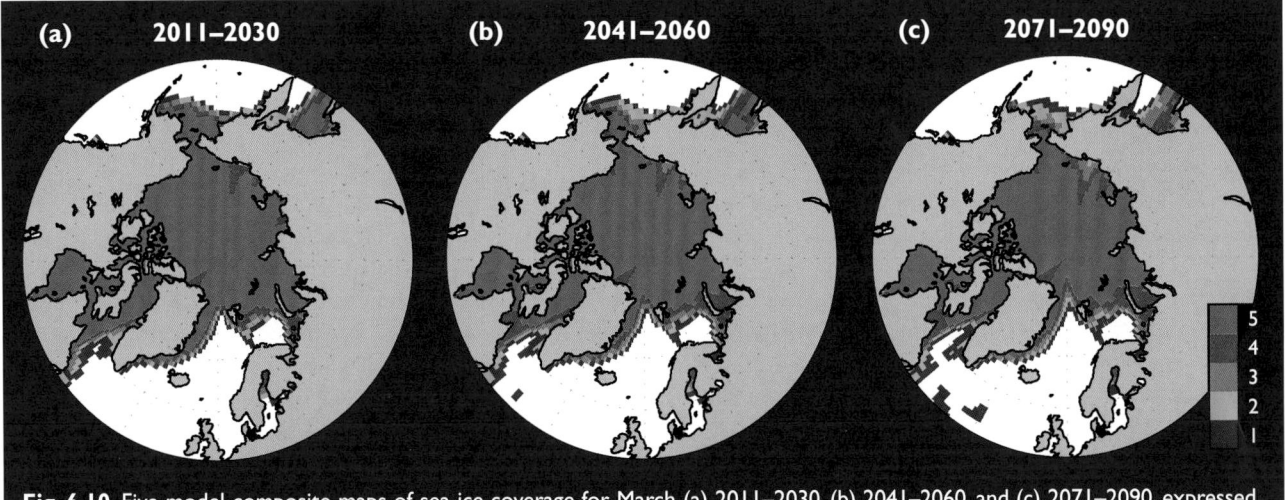

Fig. 6.10. Five-model composite maps of sea-ice coverage for March (a) 2011–2030, (b) 2041–2060, and (c) 2071–2090, expressed in terms of the number of models (out of five) that project the presence of sea ice during March for at least 50% of the years in the time slice.

which the Arctic is projected to warm (section 4.4.2) throughout the 21st century. This ranges from the relatively weak warming and small sea-ice retreat projected by the CSM_1.4 model to the strongest warming and greatest retreat projected by the CGCM2 model. For many months, especially in autumn and early winter, the projected loss of sea ice is unambiguously associated with the degree of warming projected by a particular model. Interestingly, the projected loss of sea ice is also consistently related to the models' projected global surface air warming.

A loss of sea ice is likely to enhance atmospheric humidity and cloudiness, and the general increase in precipitation noted in section 6.2 is at least partially attributable to the projected reduction in sea ice in the 21st-century scenarios, especially over and near the areas of sea-ice retreat. In areas of sea-ice retreat, ocean temperature and salinity near the surface will change, as will the upper-ocean stratification. Biogenic aerosol fluxes are also likely to increase.

There is potential for feedback between meteorological conditions and oceanographic conditions in that greater expanses of open water (at above-freezing temperatures) could strengthen low-pressure systems as they move across the arctic seas. More intense low-pressure systems will increase sea level and storm-surge height owing to the hydrostatic effect. Changes in sea-ice concentrations will also affect wave generation through the magnitude of the wind stress acting directly on the ocean.

On ecosystems

Light penetration in the upper ocean will increase in areas of sea-ice retreat, affecting phytoplankton blooms and the marine food web. Changes in ocean temperature accompanying a retreat in sea ice are likely to affect the distribution of fish stocks (Chapter 13). Marine mammals (e.g., walrus and polar bears) that rely on sea ice as a platform will be forced to find new habitats, and whale migration routes are likely to change as sea ice retreats.

On people

If the projected changes in sea ice occur, commercial navigation opportunities (section 16.3.7), and opportunities for offshore mineral extraction (section 16.3.10) will increase. Fish and mammal harvests are likely to be affected, and tourism activities are likely to increase. The absence of sea ice in previously ice-covered areas will have impacts on some types of military operations. Vulnerability to storms is likely to increase in low-lying coastal areas as the ice-free season lengthens, with corresponding impacts on residents and infrastructure (section 16.2.4.2). The stability of coastal sea ice for travel and other purposes will be reduced, with negative impacts on traditional subsistence activities.

6.3.5. Critical research needs

The discussion in section 6.3.3 focused on the large-scale sea-ice properties that can be simulated by models. The importance of small/subgrid-scale processes on large-scale behavior should also be emphasized. Among the main challenges involved in modeling ocean mixing in ice-covered seas is a representation of the effects of small-scale inhomogeneities in sea-ice cover (primarily lead fraction and distribution). This affects the surface exchange fluxes of momentum, heat, freshwater, and greenhouse gases (GHGs), and mixing processes under the ice. Processes specific to the surface boundary layer include the radically different surface fluxes in ice-covered versus ice-free fractions of a climate model grid cell; the strongly asymmetrical behavior of ice basal melting versus freezing; the interaction of tides and currents with ice-bottom morphology; and the modification of momentum transfer mechanisms as surface wave effects are replaced by stress transfer through the sea-ice cover. These are all subgrid-scale effects. Their successful representation in a climate model requires a combination of detailed observations, mathematical and physical process modeling, stochastic analysis, and numerical modeling at a range of resolutions and physical complexity. The understanding and modeling of these processes are critical to more consistent and accurate simulations of sea-ice cover and climate. The inadequate treatment of small-scale processes may have contributed to the systematic errors in the model simulations discussed in section 6.3.3. These errors are limited to some regions and seasons for a few of the models, but are more pervasive in others. The errors increase the uncertainty in the projected rates of change in sea-ice variables. Thus, reducing or eliminating these errors is a high priority for assessments of future change in the arctic marine environment.

Model resolution is presently inadequate to capture changes in sea ice in coastal areas and in geographically complex areas such as the Canadian Archipelago. For example, finer resolution is required to address the types of sea-ice change that will affect navigability in the Northwest Passage. Section 16.3.7 addresses changes affecting the Russian Northern Sea Route (the Northeast Passage).

Data on surface albedo, particularly its seasonal, interannual, and interdecadal variations, are needed for a more rigorous assessment of the albedo–temperature feedback, including its magnitude in the present climate and the validity of its treatment in climate models. Field programs have made local measurements of surface albedo, radiative fluxes, and associated cloud parameters, but such data have not been fully exploited for model simulations of climate change. Also, the albedo–temperature feedback almost certainly involves changes in cloudiness, yet the nature and magnitude of these cloud-related effects are unknown.

Systematically compiled data on sea-ice thickness are needed to provide a spatial and temporal context for the recent decrease in sea ice observed in the central Arctic Ocean. The possibility that compensating increases in sea-ice thickness have occurred in other (unmeasured) areas of the Arctic Ocean raises fundamental questions about the nature and significance of the decreases detected in the vicinity of the submarine measurements. Satellite techniques for measuring sea-ice thickness throughout the Arctic would be particularly valuable. Moreover, the apparent redistribution of sea ice in recent decades indicates the importance of including ice motion in model-derived scenarios of change.

Finally, the role of sea-ice variations in the thermohaline circulation of the North Atlantic and the global ocean (section 2.5.1) must be clarified. While the potential exists for sea-ice variations to have significant global impacts (Mauritzen and Hakkinen, 1997), variations in the temperature and salinity of ocean water advected poleward from lower latitudes may explain much of the variability in deep convection in the subpolar seas. A better understanding of the relationship between sea ice and ocean circulation is perhaps the highest priority for assessments of arctic–global interactions, given the potential for sea ice to have a substantial effect on the thermohaline circulation, which in turn has the poten-

tial to change the climate of northern Europe and much of the Arctic Ocean.

6.4. Snow cover

6.4.1. Background

Terrestrial snow cover is the most rapidly varying cryospheric variable on the surface of the earth. An individual frontal cyclone can change the area of snow-covered land (or sea ice) by 0.1 to 1.0 million km² in a matter of days. Snow cover also displays large spatial variability in response to wind, and to topographic and vegetative variations. Yet it is the spatially integrated accumulation of snow over one to two seasons that has important hydrological implications for arctic terrestrial regions and the polar oceans, and hence for terrestrial and marine ecosystems. Snow also represents the fundamentally important accumulation component of ice sheets and glaciers (section 6.5). Finally, snow cover influences the ground thermal regime and therefore the permafrost changes (section 6.6) that have additional hydrological implications (section 6.8).

Before the availability of satellite imagery in the 1960s, snow cover was determined from occasional aerial photographs and from point measurements, often made at weather stations spaced irregularly over the land surface. In cold and windy environments such as the Arctic, point measurements are inaccurate because snow gauges are inefficient and drifting snow contaminates the measurements (Goodison and Yang, 1996). In addition, even accurate point measurements may not be representative of large-area or regional snow-cover conditions. The inaccuracy of the point measurements makes them inadequate for mapping the detailed spatial structure of snow coverage and depth, especially in regions of significant topography. Because snow cover is easily identified in visible and near-infrared wavelength bands, owing to its high reflectance, satellites have proven valuable in monitoring variations in snow cover at various scales over the past three to four decades. Unfortunately, most sensors cannot measure snow depth or water equivalent (Dankers and De Jong, 2004).

The present distribution of snow cover in the Northern Hemisphere, excluding permanently glaciated areas such as Greenland, varies from <1 million km² in late August to 40 to 50 million km² in February (Ramsay, 1998; Robinson, 1993). The large range in the February values indicates the interannual variability. Figure 6.11 shows the frequency of snow cover on the land areas of the Northern Hemisphere from 1966 to 2000 during winter (December), early and late spring transition months (February and May), and an autumn transition month (October). It is apparent that snow is a quasi-permanent feature of the arctic terrestrial landscape during winter. The variability inherent in subarctic land areas during the spring (Fig. 6.11b), when insolation is relatively strong, implies that the timing of the snowmelt, which reduces the surface albedo by 20 to

60%, can strongly affect surface absorption of solar radiation. Models have demonstrated the importance of snow cover for the surface energy budget, soil temperature, and the permafrost active layer (Ling and Zhang, 2003; Sokratov and Barry, 2002). Snow cover is also highly variable in October (Fig. 6.11c), when insolation and hence the potential for snow to affect the surface absorption of solar radiation is weaker. Snow is rarely present over the subarctic land areas in July and August.

The hydrologically important characteristic of snow cover is its water equivalent, since this moisture is eventually released to the atmosphere by sublimation or evaporation, or to the polar oceans by runoff. Some of the snow water is siphoned off for human use prior to its eventual release to the atmosphere or ocean. Estimated fields of snow water equivalent (or snow depth) can be derived from satellite passive microwave measurements (Armstrong and Brodzik, 2001; Chang et al., 1987;

Goodison and Walker, 1995). Although the spatial coverage of these measurements is complete and their broad spatial patterns are correct, there are large uncertainties and errors for areas in which vegetative masking (vegetation obscuring the underlying snow, making the ground appear darker) is significant (e.g., the boreal forests of the subarctic). Even allowing for the uncertainties, the derived snow water equivalents represent large water supplies that are released to other parts of the climate system during spring melt.

Station-derived climatologies of snow depth represent alternatives to the satellite-derived estimates and their associated uncertainties. Such climatologies have been compiled for Canada (e.g., Brown R. and Braaten, 1998) and for Russia (e.g., Ye H. et al., 1998). However, these compilations are subject to elevation- and location-related biases in the station networks. Section 6.4.2 summarizes broad-scale variations in these trends.

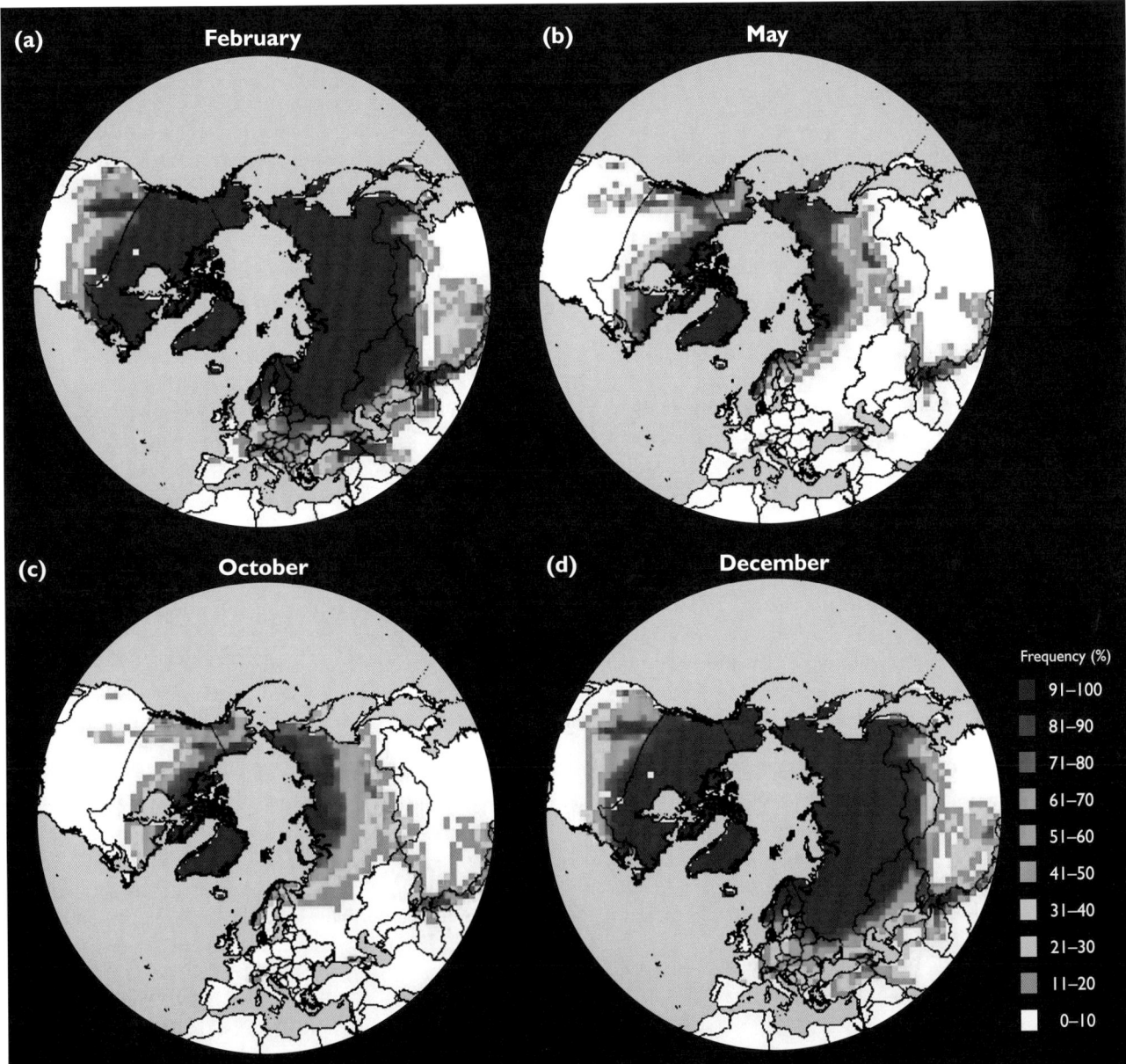

Fig. 6.11. Frequency of snow cover on the land areas of the Northern Hemisphere during early and late spring transition months (February and May), an autumn transition month (October), and winter (December). Frequency is determined by the percentage of weeks in the specified month over the 35-year period 1966 to 2002 that a location had snow cover. In all panels, the 50% contour (in the green zone) represents the approximate climatological mean position of the snow boundary (figure courtesy of D. Robinson, 2003).

Radionov et al. (1997) compiled statistics of snow depth and duration over the central Arctic Ocean using measurements obtained at the Russian drifting ice stations, primarily between the 1950s and 1990.

Over glaciated regions, where snow stakes and altimetry are major sources of information, the problem of spatial integration of snow measurements is quite different from that over regions of seasonal snow cover. Section 6.5 discusses mass-balance measurements for glaciers and ice caps on a regional basis.

Albedo is highly relevant to the role of snow in the surface thermal regime. The albedo of snow-covered land areas is highly variable, depending on snow depth, snow age, and the masking characteristics of vegetation. There are few systematic compilations of surface albedo over snow cover, so the climatology and variability of surface albedo are not well documented at the circumpolar scale. Winther (1993) and others have studied albedo variations on a regional basis.

Snow is a key variable in the rates of soil warming and permafrost thawing. Because snow effectively insulates the upper soil layers during winter, increases in snow depth generally result in higher soil temperatures during the cold season, while an absence of snow results in more rapid and greater cooling of the soil. If snowfall changes substantially as climate changes, warming and thawing or cooling and freezing may significantly affect the upper soil layers. Permafrost models (section 6.6) require information on snow cover in addition to air temperature if they are to provide valid simulations of variations in the temperature and water phase in the upper soil layers. In general, climate models treat the subsurface effects of snow rather crudely, particularly with regard to the freeze-thaw cycle of soils over seasonal to centennial timescales.

6.4.2. Recent and ongoing changes

Over a few months, snow cover in the Northern Hemisphere (excluding sea ice, Greenland, and glaciers) varies between the 0 to 5 million km² typical of summer and the more than 40 million km² typical of winter. The rapidity of the expansion and retreat (melt) is comparable in the autumn and spring, and indicates the short timescales for variations in snow cover relative to other cryospheric variables. Interannual variations are also rapid. Figure 6.12, for example, shows the 12-month running means of snow extent between 1972 and 2003, the period of homogeneous visible satellite data. While these fluctuations complicate the detection and interpretation of trends, least-squares fits to the time series in Fig. 6.12 indicate that the areal coverage of snow has decreased over the past few decades. The decrease for the Northern Hemisphere is nearly 10% over the period 1972 to 2003. Both North America and Eurasia show decreases, although the decrease is greatest for Eurasia. However, the decrease is highly seasonal, varying from no significant change in autumn and winter to decreases

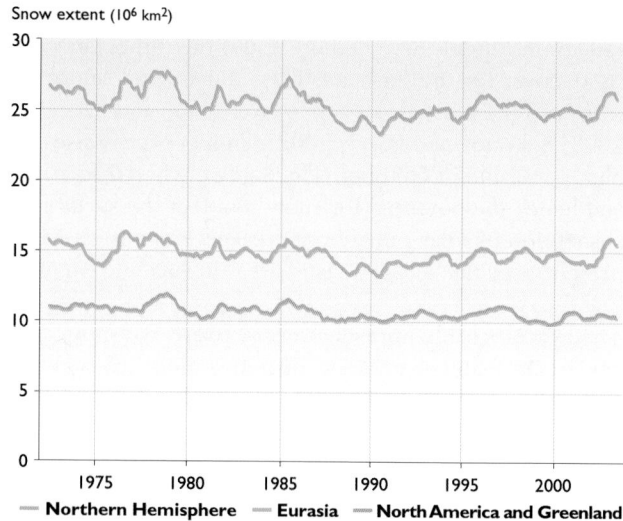

Fig. 6.12. Twelve-month running mean of snow extent in the Northern Hemisphere from 1972 to 2003, showing the entire hemisphere, North America and Greenland, and Eurasia (data from D. Robinson, 2003).

greater than 10% in spring and summer (Fig. 6.13). The large areal decrease in spring is correlated with the large spring warming over the northern land areas (section 2.6.2). The summer trend, to which the arctic land areas are probably making key contributions, has received little attention in the literature.

For the period before satellite data were available, variations and trends in snow cover have been assessed largely on a regional basis. North American snow cover shows a general decrease in spring (March–April) extent since the 1950s (Fig. 6.13a), although there are indications that spring snow extent increased during the earlier part of the 20th century. The spring decrease is also apparent in the Eurasian data (Fig. 6.13b). The total extent of Northern Hemisphere snow during spring and summer was lower in the 1990s than at any time in the past 100 years (IPCC, 2001). However, the longer records shown in Fig. 6.13 do not indicate a systematic decrease in snow cover during autumn or early winter for either landmass.

Recent variations in snow depth are more difficult to assess because of measurement and remote-sensing difficulties in vegetated areas. For the pre-satellite era, the sparseness of the synoptic station network precluded systematic mapping of snow depth in many high-latitude areas. Nevertheless, there have been compilations and analyses of snow-depth data for particular regions. Snow depth appears to have decreased over much of Canada since 1946, especially during spring (Brown R. and Braaten, 1998). Winter snow depths have decreased over European Russia since 1900 (Meshcherskaya et al., 1995), but have generally increased elsewhere over Russia during the past few decades (Fallot et al., 1997), in agreement with the increase in precipitation noted in section 6.2.2. Ye H. (2001) reports a small (several day) increase in the snow-season length, due primarily to later snowmelt, over north-central and northwest Asia

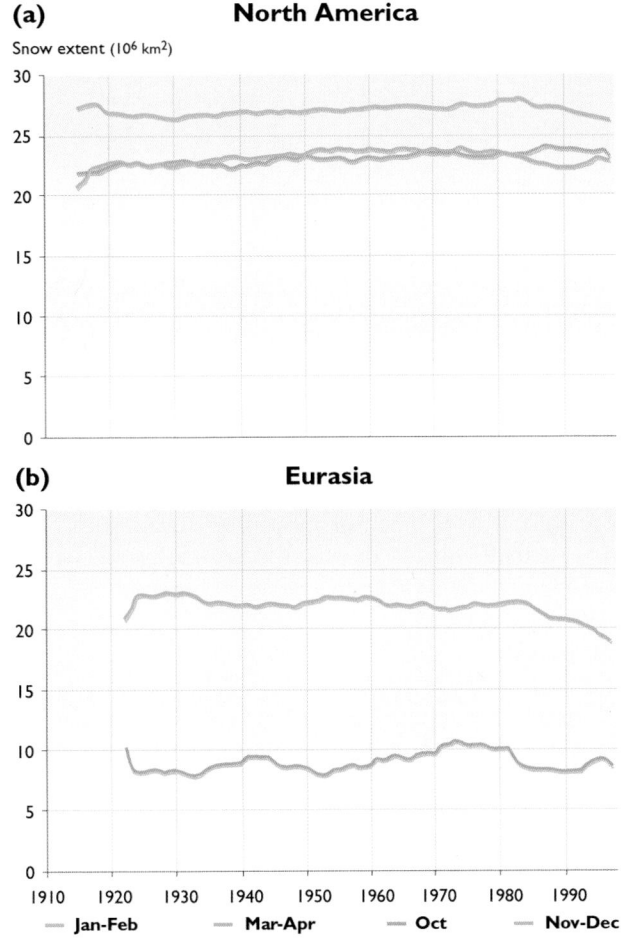

(a)   North America

Snow extent (10⁶ km²)

(b)   Eurasia

1910 1920 1930 1940 1950 1960 1970 1980 1990
— Jan-Feb   — Mar-Apr   — Oct   — Nov-Dec

Fig. 6.13. Eleven-year running mean of snow cover extent from 1915 to 1997 for (a) North America and (b) Eurasia (data from R. Brown, 2000).

for the period 1937 to 1994. However, this study excludes the years since 1994, during which northern Asia has been relatively warm in winter and spring.

Over the central Arctic Ocean, measurements at Russian North Pole ice stations from the mid-1950s to 1990 suggest a decrease in snow depth, although considerable variability is superimposed on this decrease (Radionov et al., 1997). Measurements of total precipitation at the same sites show little indication of any systematic trend (Radionov et al., 1997).

Thus, there are consistent indications of spring warming of arctic terrestrial regions (section 2.6.2.1) and earlier

disappearance of snow cover. Associated changes are the earlier breakup of ice-covered lakes and rivers (section 6.7) and a seasonal advance (from early spring to late winter) in the timing of the primary pulse of river discharge in the Arctic (section 6.8).

6.4.3. Projected changes

Simulations of 21st-century climate from the five ACIA-designated climate models (section 4.4) were examined for projected changes in snow cover. These changes are complicated by the conflicting effects of higher temperatures, which will result in a poleward retreat of the snow margin, but which are also likely to contribute to an acceleration of the hydrological cycle and – in those regions that remain consistently below freezing – an increase in snowfall and possibly snow depth (water equivalent).

Table 6.4 summarizes the changes in snow cover projected by the five ACIA-designated models, including snow cover simulated for the baseline climate (1981–2000) and the changes averaged over each of the three ACIA time slices: 2011–2030, 2041–2060, and 2071–2090. It is apparent that the projected changes in mean annual snow cover are not large, even by 2071–2090 when the changes range from -9 to -18%.

However, there is a notable seasonality to the changes in snow cover. Table 6.5 summarizes the changes projected to occur between the baseline (1981–2000) and 2071–2090 by season. While the percentage changes are greatest in summer (when areal coverage is very small in the present climate), the actual decrease in snow-covered area is greatest in spring (April and May).

A more detailed evaluation (Table 6.6) shows that the months with the largest projected reductions in snow extent are April and November, followed by May, March, and December. The changes most relevant to arctic hydrology are those that occur in spring, when a reduction in snow cover implies an earlier pulse of river discharge to the Arctic Ocean and coastal seas. That the greatest projected changes occur during the spring, late autumn, and early winter indicates a shortened snow season in the model simulations for the late 21st century. Figure 6.14 illustrates the distribution of snow cover during the 2071–2090 time slice for March, May,

Table 6.4. Northern Hemisphere mean annual snow cover simulated by the five ACIA-designated models for baseline climate (1981–2000) and percentage change from the baseline projected for each of the ACIA time slices.

	Snow cover (10⁶ km²)	Percentage change from baseline (1981–2000)		
	1981–2000	2011–2030	2041–2060	2071–2090
CGCM2	27.8	-7	-13	-17
CSM_1.4	23.8	-3	-5	-9
ECHAM4/OPYC3	18.6	-6	-13	-18
GFDL-R30_c	31.9	-4	-7	-10
HadCM3	23.0	-4	-8	-10
Observed	23.2			

Table 6.5. Seasonal change in snow extent (10^6 km²) between 1981–2000 and 2071–2090 projected by the five ACIA-designated models.

	Winter (Dec–Feb)	Spring (Mar–May)	Summer (Jun–Aug)	Autumn (Sep–Nov)
CGCM2	-5.8	-6.8	-3.2	-3.2
CSM_1.4	-2.8	-3.2	-0.4	-2.2
ECHAM4/OPYC3	-5.5	-6.2	-0.6	-3.4
GFDL-R30_c	-2.3	-5.1	-0.6	-4.6
HadCM3	-2.8	-3.1	-0.5	-3.0
Five-model mean	-3.8	-4.9	-1.1	-3.3

October, and December. A comparison with the present-day frequency distributions for May, October, and December (Fig. 6.11) shows that the snow retreat is modest but visually noticeable, especially in May.

6.4.4. Impacts of projected changes

On other parts of physical system

The primary effects of a reduction in snow cover will be on the surface energy budget (hence soil temperature and permafrost) and on the surface moisture budget (runoff, evaporation). However, the effects of changes in snow cover will vary seasonally. During winter, snow insulates the ground, so a reduction in snow cover or depth will lead to cooling of the underlying ground. During spring, a decrease in snow cover will lower the surface albedo, leading to enhanced absorption of solar radiation and warming of the ground. For sea ice, decreased snow cover will accelerate ice melt in the spring while increased snow cover will retard ice melt.

Changes in snow cover also have the potential to influence significantly the distribution of vegetation (Bruland and Cooper, 2001), which can then influence the atmosphere through changes in vegetative masking, surface albedo, and the surface energy budget. Snow cover also affects the exchange of GHGs between the land surface and the atmosphere, as documented in the Land Arctic Physical Processes experiment (Anon, 1999).

On ecosystems

The growth season of high-latitude vegetation, and hence primary production and carbon dioxide (CO_2) uptake, depends strongly on the timing of snow disappearance, which in turn depends on antecedent snow accumulation. Snow insulates underlying vegetation and other biota (e.g., mammals, insects). The runoff pulse produces large biogeochemical fluxes from terrestrial to marine ecosystems, and changes in the spring snowmelt will affect both the timing and intensity of this pulse. An important characteristic of snow cover is its structure, especially the presence of ice layers that can result from thaw–freeze

cycles or from freezing rain events. The presence of ice layers can severely hamper winter grazing by wildlife (section 12.2.4). Unfortunately, neither the observational database nor the model output can provide useful information on the presence of ice layers in snow, although it is reasonable to assume that a warming climate will increase the frequency of winter freeze–thaw cycles and freezing rain events in arctic terrestrial regions.

On people

Changes in the amount of snow will have impacts on transportation (e.g., feasibility, safety, costs); recreation activities and the businesses dependent on them (e.g., ski resorts, snow machines); snow loading on structures and removal costs; avalanche hazards in areas with steep topography; and water supplies for various population sectors. Because the costs of clearing snow from roads are significant in many mid- and high-latitude communities, economic consequences of changes in the amount of snow are very likely. In addition, changes in the length of the snow-free season would affect agricultural, industrial, and commercial activities as well as transportation in many high-latitude communities. Changes in the amount of snow and the length of the snow season will also directly affect the subsistence activities of indigenous communities.

6.4.5. Critical research needs

Global climate model simulations of snow extent have shown some improvement over the past decade (Frei et al., 2003). However, a climatology of the spatial distribution of snow water equivalent in each month is a critical need for model validation and hydrological simulations. This is especially urgent for high latitudes, where there are few *in situ* measurements of snow water equivalent to complement the estimates derived from passive microwave measurements. Information on snow albedo over northern terrestrial regions, especially for vegetated areas and for the late winter and spring seasons when the timing of snowmelt is hydrologically critical, is an additional requirement. Global daily snow-albedo products derived from the Moderate-

Table 6.6. Five-model monthly means of the projected change in snow extent (10^6 km²) between 1981–2000 and 2071–2090.

	Jan	Feb	Mar	Apr	May	Jun	Jul	Aug	Sep	Oct	Nov	Dec
Change in extent	-3.7	-3.8	-4.2	-5.4	-5.0	-2.1	-0.9	-0.3	-1.2	-3.6	-5.1	-4.0

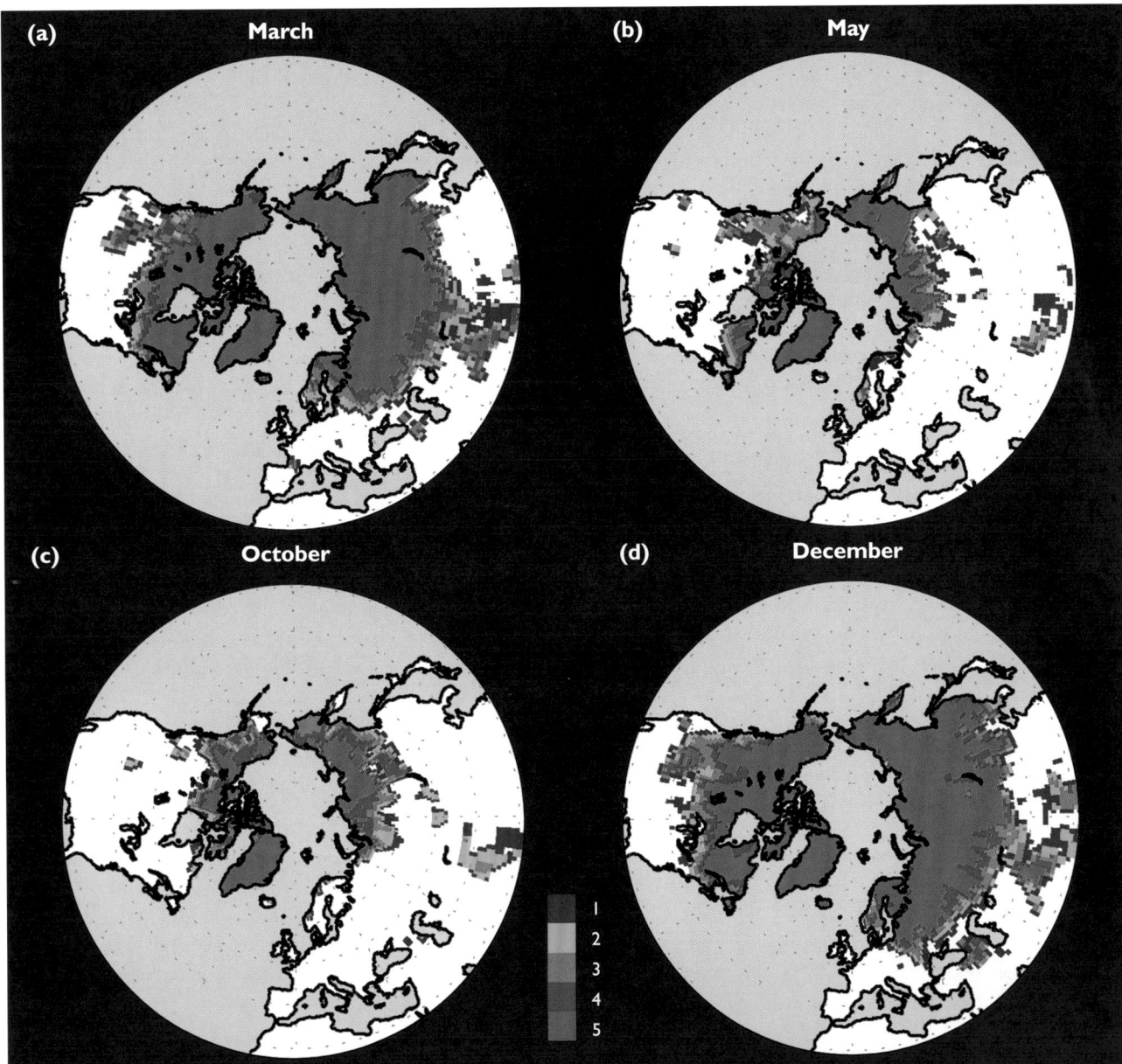

Fig. 6.14. Distribution of snow cover during the 2071–2090 time slice for March, May, October, and December, using as the measure of coverage the number of models (out of five) that project the presence of snow during the specified month for at least half of the years in the time slice.

Resolution Imaging Spectroradiometer are now available (Hall et al., 2002); such products should be used for quantitative assessments of large-scale albedo variations and for climate model validation.

The ability of models to simulate the snowmelt process also needs further investigation within the context of arctic hydrology. This should result in evaluations of feedbacks between the timing of snowmelt and broader changes in terrestrial ecosystems.

A potentially important but often overlooked process is the sublimation of snow, especially when enhanced by blowing snow. Sublimation can be a key part of the hydrological cycle locally and regionally (Pomeroy and Li, 2000), yet climate models do not include the enhancement of sublimation by blowing snow, and some models do not include even the direct sublimation of snow from the surface.

6.5. Glaciers and ice sheets

6.5.1. Background

Dowdeswell and Hagen (2004) estimated that the total volume of land ice in the Arctic is about 3.1 million km^3, which corresponds to a sea-level equivalent of about 8 m. In terms of volume and area, the largest feature is the Greenland Ice Sheet, which covers about four times the combined area of the glaciers and ice caps of Alaska, the Canadian Arctic, Iceland, Svalbard, Franz Josef Land, Novaya Zemlya, Severnaya Zemlya, and northern Scandinavia (Table 6.7). However, unlike most small glaciers and ice caps, more than half the surface of the Greenland Ice Sheet is at altitudes that remain well below freezing throughout the year. Hence, relative to the Greenland Ice Sheet, the smaller ice caps and glaciers are susceptible to greater percentage changes of mass and area in response to changes in temperature and precipitation.

The arctic glaciers and ice caps are irregularly distributed in space (Fig. 6.15), and are located in very different climatic regimes. The glaciers in southern Alaska and Iceland are subject to a maritime climate with a relatively small annual temperature range and high precipitation rates (a few meters per year). Conversely, the glaciers in the Canadian High Arctic are in a very continental climate. The summer is short, the annual temperature range is very large, and precipitation is about 0.25 m/yr. The conditions on Svalbard and the Russian Arctic islands fall between these two climatic regimes.

The Greenland Ice Sheet covers a wide latitude belt. The climate is dry and cold in the north, although summer temperatures can be high, with mean July temperatures of up to 5 to 6 °C (Ohmura, 1997). The North Atlantic storm track directly influences the southeastern part of the ice sheet. Maritime air masses are pushed onto the ice sheet and release large amounts of moisture. The accumulation rates are greatest in this part of the Greenland Ice Sheet.

The morphology of arctic glaciers shows great variety (e.g., Williams R. and Ferrigno, 2002). Some ice caps are dome-shaped, with lobes and outlet glaciers in which the ice drains away from the accumulation area to the melting regions or calving bays. Examples occur in the Canadian Arctic, Iceland, and the Russian Arctic islands. In other regions, large glaciers originate from ice fields

that cover the area between mountain ranges (e.g., in southern Alaska). Many regions (e.g., Svalbard) also have a large number of individual valley glaciers.

There are many surging glaciers in the Arctic. In a surging event, glacier fronts can move forward many kilometers (sometimes more than 10 km) in a matter of years. After a surge, a build-up phase starts and the glacier accumulates mass for the next surge. Depending on the size of the glacier, the duration of the build-up phase ranges from a few decades to a few centuries. Surging glaciers occur in Alaska, Canada, Svalbard, and Iceland, and have also been observed in other areas of the Arctic. A surge event may change the flow and geometry of the glacier. While an individual surge is not directly related to climate change, increased melting may have an effect on the periodicity of surging.

Glaciers, ice caps, and ice sheets respond to climate changes over very different timescales depending on their size, shape, and temperature condition. The smaller glaciers are likely to respond quickly, with shape, flow, and front position changing over a few years or a few decades, while the Greenland Ice Sheet responds to climate changes over timescales of up to millennia. Parts of the Greenland Ice Sheet may still be responding to climate variations that occurred thousands of years ago.

Many glaciers in dry regions have low accumulation rates. Consequently, it takes a long time before the climate signal penetrates into these glaciers, and over a 100-year timescale, the effects are unlikely to be very large. However, in areas where meltwater penetration increases, the effect of latent heat release is likely to cause a faster response in the thermal regime.

Because arctic glaciers have such a wide variety of morphological and climatic regimes, the most difficult task in this assessment is to extrapolate results for a few glaciers and ice caps to all ice masses in the Arctic. Mass-balance measurements have been conducted on some glaciers for shorter or longer periods (Fig. 6.15), but only a small fraction of the glaciated area is moni-

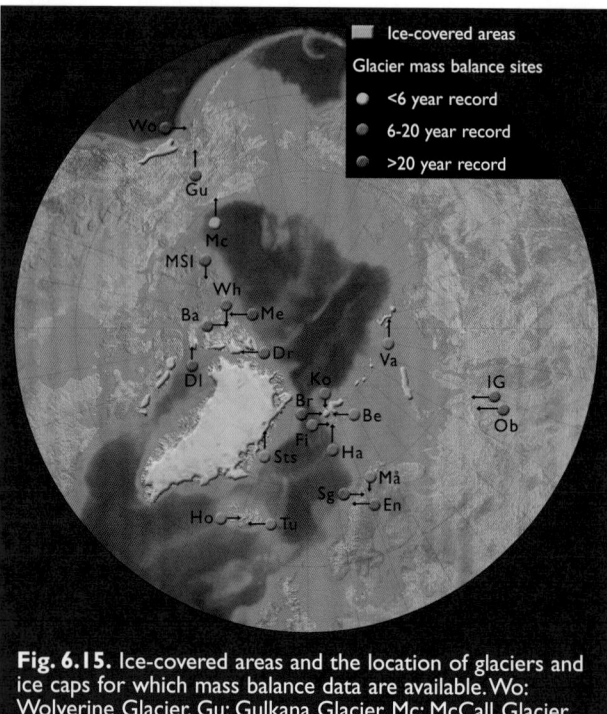

Fig. 6.15. Ice-covered areas and the location of glaciers and ice caps for which mass balance data are available. Wo: Wolverine Glacier, Gu: Gulkana Glacier, Mc: McCall Glacier, MSI: Melville South Ice Cap, Ba: Baby Glacier, Me: Meighen Ice Cap, Dl: Devon Ice Cap, Dr: Drambui Glacier, Ho: Hofsjökull, Tu: Tungnarjökull, Br: Austre Brøggerbreen, Ko: Kongsvegen, En: Engabreen, Sg: Storglaciären, IG: Igan, Ob: Obruchev, Va: Vavilov, Ha: Hansbreen, Wh: White, Be: Bear Bay, Fi: Finsterwalderbreen, Ma: Storglaciären, Sts: Storstrommen (modified from Dowdeswell et al., 1997).

Table 6.7. Ice coverage in arctic regions with extensive glaciation (Dowdeswell and Hagen, 2004).

	Glacier area (10^3 km²)
Greenland Ice Sheet	1640.0
Canadian Arctic (>74° N)	108.0
Canadian Arctic (<74° N)	43.4
Alaska	75.0
Iceland	10.9
Svalbard	36.6
Franz Josef Land	13.7
Novaya Zemlya	23.6
Severnaya Zemlya	18.3
Norway/Sweden	3.1

tored. Attempts have been made to extrapolate measurements, parameters, and models from a small number of glaciers to obtain regional estimates (e.g., Dyurgerov and Meier, 1997), although extrapolation introduces considerable uncertainty into conclusions about ongoing and future changes in the area and volume of land ice and associated changes in sea level (section 6.5.3).

Glaciers gain mass from snowfall and lose mass mainly through iceberg calving, surface melting and runoff, and melting under floating ice shelves. The specific net balance is the net annual change in mass per square meter, often expressed in kg/m^2 or meter water equivalent (mwe). The mass balance is positive in the accumulation zone and negative in the ablation zone. The equilibrium line separates the accumulation and ablation zones.

Meltwater formed at the surface may percolate into the snowpack and refreeze to form ice lenses and glands. Eventually the meltwater freezes onto the ice surface below the snowpack to form superimposed ice. Part of this ice subsequently melts in the summer, but the remainder survives. Refreezing and the formation of superimposed ice can have a significant influence on the energy budget of the melt process (Ambach, 1979) and can decrease the altitude of the equilibrium line. The quantitative effects of refreezing on the mean specific balance of a glacier are not well understood, and are rarely treated well in mass-balance models. With respect to relatively rapid climate change, an important question is whether the increasing amounts of meltwater will add to runoff or be retained in cold firn (compact, granular snow that is over one year old) fields.

Iceberg calving is another process for which a universal model does not exist. For many glaciers in the Arctic, the amount of ice lost by meltwater runoff is larger than the amount lost by calving, but calving is signifi-

cant for many glaciers (typically 15 to 40% of the total mass loss for glaciers on the islands in the Eurasian Arctic sector) (Dowdeswell and Hagen, 2004). For the Greenland Ice Sheet, the IPCC (2001) estimated that the losses from meltwater runoff and calving are of the same order of magnitude. Researchers have attempted to determine a linear relationship between the calving rate and water depth or ice thickness at the glacier front. This seems to work for many individual glaciers, but the coefficients show large spatial variation.

The mass-balance sensitivities vary widely among glaciers. Differences in the sensitivity to annual anomalies of temperature and precipitation reach one order of magnitude. In the high Arctic, where winter temperatures are consistently below freezing, only the summer temperature affects the mass balance. In a maritime climate such as for Iceland, sensitivity to temperature change is much greater, and temperature anomalies in other seasons are also important. For all glaciers in maritime climates, where most of the summer precipitation falls as rain, the sensitivity to precipitation anomalies shows a marked seasonality.

The sensitivity of a glacier to atmospheric forcing can be estimated using a mass-balance model and the field measurements that are required for calibrating the model. The potential contribution to sea-level rise from an entire region is obtained by multiplying the annual climate sensitivities (expressed in millimeters) by the total glacier area in the region. However, the mass-balance sensitivity for a region must be extrapolated from the calculations for specific glaciers. Figure 6.16 summarizes mass-balance sensitivities to temperature for the Arctic regions in Table 6.7 and shows the corresponding consequences for global sea level.

Recent events have demonstrated the potential for calving glaciers to undergo very rapid change. Calving glaciers that have retreated over large distances during the last hundred years, or even the last few decades, exist throughout the Arctic and subarctic. Well-documented examples include Jakobshavn Isbrae, West Greenland (Weidick et al., 1992; Williams R.S., 1986), Breidamerkurjökull, Iceland (Björnsson et al., 2001), Columbia Glacier, Alaska (Pfeffer et al., 2000), and Kronebreen and Hansbreen in Svalbard (Jania and Kaczmarska, 1997, Lefauconnier et al., 1994).

Although understanding of the processes that control calving is limited (e.g., Van der Veen, 1996, 1997), a clear relation to climatic forcing is not evident. Many internal mechanisms play a role, including sedimentary and erosive processes below and in front of a glacier tongue. During retreat, some calving glaciers tend to "jump" from pinning point to pinning point; other glaciers retreat steadily over a rather simple bed. While climatic factors will affect the extent of calving glaciers over the long term, the response of many glaciers is of an irregular and episodic nature, and is therefore unpredictable.

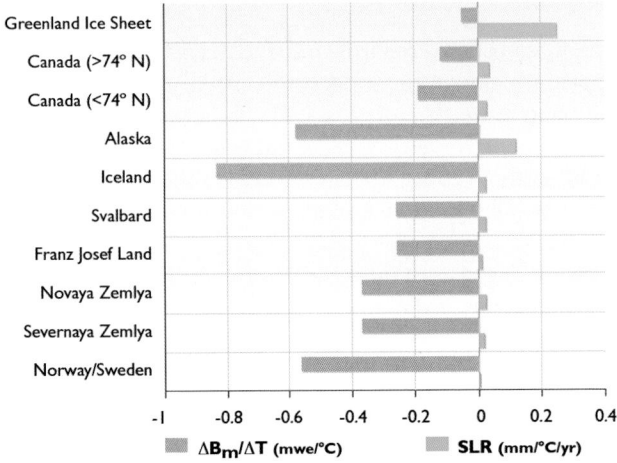

Fig. 6.16. Mass-balance sensitivity to temperature change (lavender) and potential sea-level rise (blue) for the arctic regions with extensive glaciation listed in Table 6.7. $\Delta B_m/\Delta T$ is the change in mass balance (meter water equivalent) per °C of temperature change; SLR is sea-level rise (J. Oerlemans, 2003).

6.5.2. Recent and ongoing changes

The general pattern of glacier and ice-cap variations in the Arctic (apart from the Greenland Ice Sheet) is a retreat of glacier fronts, indicating a volume decrease since about 1920 that follows a period of general temperature increase throughout the Arctic. However, there are large regional variations in the magnitude of this retreat, and it is not known whether thickening in the accumulation areas may be compensating for some or all of the frontal retreat. Long-term mass-balance investigations have been conducted for only a few glaciers (Fig. 6.15), which occupy less than 0.1% of the total glaciated area in the Arctic. For the measured glaciers, no clear trends are discernible in the mass-balance parameters, winter accumulation, or summer melting prior to 1990 (Dowdeswell et al., 1997; Jania and Hagen, 1996). Several of the glaciers had a negative mass balance, but with no acceleration in the melt rate. However, changes in these trends since 1990 have been observed. Arendt et al. (2002) observed increased and accelerating melting of Alaskan glaciers, and the same trend has been reported for the Devon Ice Cap in northern Canada (Koerner, pers. comm., 2003). In other parts of the Arctic (e.g., Svalbard), no accelerated melting has been observed (Hagen and Liestøl, 1990; Lefauconnier et al., 1999). In subarctic areas (i.e., Scandinavia), increased precipitation and positive mass balance were observed from 1988 to 1998, although the mass balances have generally been negative since 1998 (Dowdeswell and Hagen, 2004; Dyurgerov and Meier, 1997; Jania and Hagen, 1996).

Figure 6.17 presents a spatially integrated picture of arctic ice caps and mountain glaciers, obtained by grouping glaciers into geographic regions and assuming that glaciers in the same region have similar mass balances (Church J. et al., 2001; Dyurgerov and Meier, 1997). The figure shows the accumulated ice volume change

since 1960 in three areas: the North American Arctic (Alaska and Canada); the Russian Arctic (arctic islands, northeast Siberia, and polar Urals); and the European Arctic (Scandinavia, Svalbard, Iceland, and Jan Mayen). The arctic-wide volume change, which is negative and dominated by changes in the North American Arctic, is also shown. The net accumulation in the European Arctic (primarily Scandinavia) is due to the increased precipitation accompanying the northward shift of the Atlantic storm track in recent decades, when the NAO has been in a predominantly positive phase.

6.5.2.1. Alaska

The total area of Alaskan glaciers is approximately 75 000 km². The largest glaciers in Alaska occur along the southern and western shores of the Pacific Mountain System. Despite the vast number and size of Alaskan glaciers, mass-balance data are available for only a very few.

Arendt et al. (2002) estimated the volume changes in 67 Alaskan glaciers between the mid-1950s and the mid-1990s using airborne laser altimetry measurements, and found that the glaciers had thinned at an average rate of 0.52 m/yr. Extrapolating this thinning rate to all glaciers in Alaska results in an estimated volume change of -52±15 km³/yr, which is equivalent to a sea-level rise of 0.14±0.04 mm/yr. Additional measurements from 28 of these glaciers between the mid-1990s and 2000–2001 indicate that the rate of thinning has increased to -1.8 m/yr. When this rate of thinning is extrapolated to all Alaskan glaciers, the equivalent sea-level rise is 0.27±10 mm/yr, which is nearly double the estimated contribution from the Greenland Ice Sheet during the same period (Rignot and Thomas, 2002). This rapid wastage of Alaskan glaciers represents about half the estimated loss of mass by glaciers worldwide (Meier and Dyurgerov, 2002), and the largest glacial contribution to sea-level rise yet deduced from measurements (Arendt et al., 2002).

6.5.2.2. Canadian Arctic

The mass-balance records in the eastern Canadian Arctic are among the longest in existence, with many covering more than 40 years. The larger ice masses had slightly negative mass balances between the early 1960s and the mid-1980s (Koerner, 1996), although the balances became increasingly negative both with diminishing size of the ice caps and/or with a more westerly location. No persistent trends were observed in any of the data prior to the mid-1980s. However, the mass balances have become increasingly negative since the mid-1980s (Koerner and Lundgaard, 1995). Summer mass-balance trends determine the annual balance trends; the winter balances have shown no significant trend over the entire 40-year period. This indicates that, as in other parts of the Arctic, summer temperature drives variations in the annual mass balance. At present, there are no systematic observations of mass balance in the western Canadian Arctic.

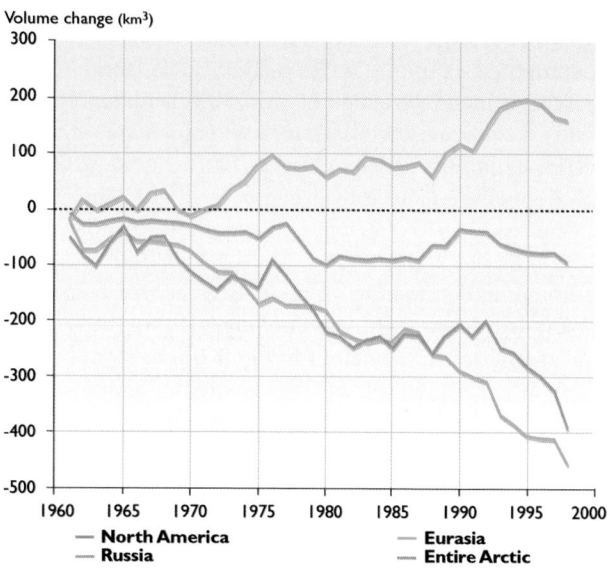

Fig. 6.17. Accumulated annual volume change in ice caps and glaciers in the North American Arctic, the Russian Arctic, the Eurasian Arctic, and the entire Arctic (M. Dyurgerov, 2003).

Ice core records show that, while the present mass balance is negative, it was more negative in the early part of the present interglacial, when substantial glacier retreat occurred (Koerner, 2002). Balances that were more positive occurred over the past 3000 years, terminating with the onset of the modern warming period about 150 years ago.

In the subarctic areas of the Canadian Cordillera, Demuth et al. (2002) found a period of declining glacier-derived discharge during the last half of the 20th century, despite a general warming trend. This decline appears to be due to the substantial contraction of outlet glaciers since the Neoglacial maximum stage (ca. 1850).

6.5.2.3. Greenland Ice Sheet

The Greenland Ice Sheet ($1\,640\,000$ km^2) is the largest ice mass in the Arctic. Two factors contribute to the difficulty of measuring the total mass balance of the Greenland Ice Sheet: short-term (interannual to decadal) fluctuations in accumulation and melt rate cause variations in surface elevation that mask the long-term trend; and climate changes that occurred hundreds or even thousands of years ago still influence ice flow, as do changes that are more recent.

The geological and historical records show that the marginal zone of the Greenland Ice Sheet has thinned and retreated over the past hundred years (Weidick, 1968). Whether this mass loss was compensated, partly or fully, by thickening in the interior is unknown. Although several expeditions have crossed the ice sheet since the late 19th century, the earliest measurements of sufficient precision to permit calculation of surface-elevation change are those made by the British North Greenland Expedition (BNGE), which crossed the ice sheet during 1953 and 1954. Comparing these data to modern surface elevations measured by satellite radar altimetry and airborne laser altimetry shows that between 1954 and 1995, ice thickness along the BNGE traverse changed little on the northeast slope, whereas ice on the northwest slope thinned at a rate of up to 30 cm/yr (Paterson and Reeh, 2001). Height measurements repeated in 1959, 1968, and 1992 along a profile across the ice sheet in central Greenland showed thickening on the western slope between 1959 and 1968, but subsequent thinning between 1968 and 1992 (Möller, 1996), probably reflecting decadal-scale fluctuations in accumulation rates.

The IPCC (2001) provides estimated of the individual terms of the mass budget of the Greenland Ice Sheet, consisting of: accumulation (520 ± 26 km^3/yr); runoff (329 ± 32 km^3/yr); and iceberg calving (235 ± 33 km^3/yr). There are large uncertainties in these estimates, but they show that calving and surface melting are of the same order of magnitude.

Rignot and Thomas (2002) mapped estimated thickening rates in Greenland by synthesizing airborne laser altimeter and satellite-borne radar altimeter surveys, mass-budget calculations, and direct measurements of changes in surface elevation. The higher-elevation areas appear to be in balance to within 1 cm/yr, although temporal variations in snow accumulation rates create local thickening or thinning rates of up to 30 cm/yr. In contrast, the coastal regions appear to have thinned rapidly between the 1993–1994 and 1998–1999 laser altimeter surveys (Krabill et al., 1999). A conservative estimate of the rate of net ice loss (~50 km^3/yr) corresponds to a sea-level rise of 0.13 mm/yr. Since variations in summer temperatures do not explain the rapid thinning of many outlet glaciers, the coastal thinning is apparently a result of glacier dynamics rather than a response to atmospheric warming (Rignot and Thomas, 2002). Alternatively,

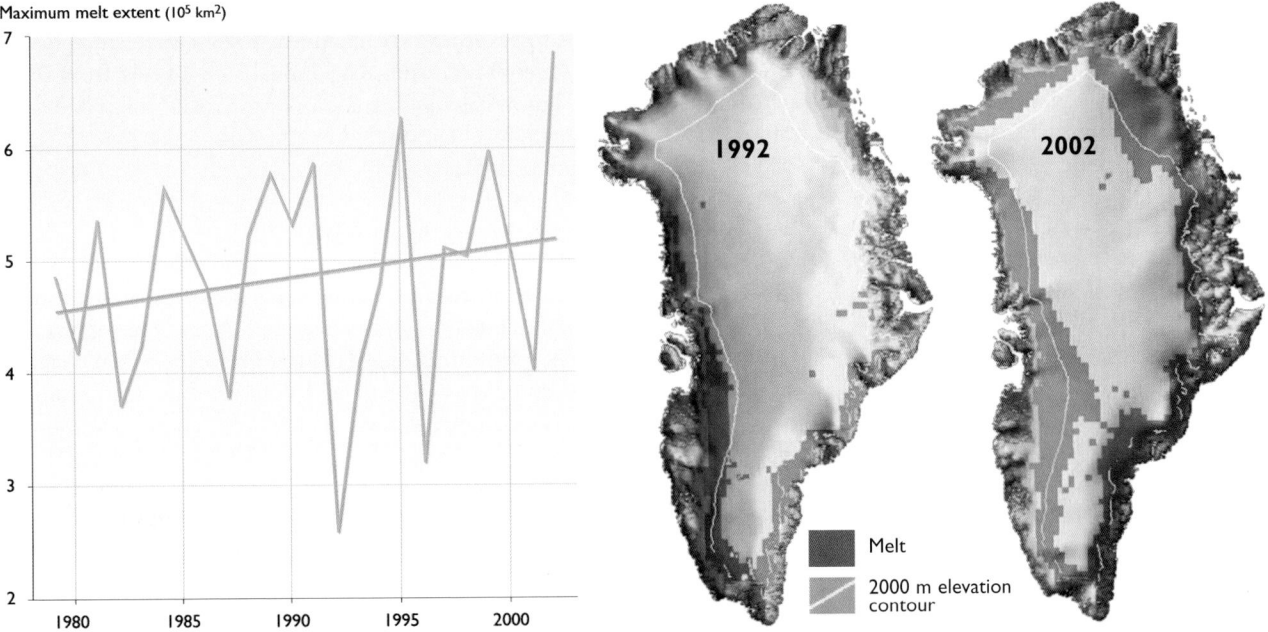

Fig. 6.18. Time series of maximum summer melt extent over Greenland from 1979 to 2002 (left) and the melt extent in 1992 and 2002 (right) (figure provided by K. Steffen, 2003).

Zwally et al. (2002) suggested that increased basal lubrication, due to additional surface meltwater reaching the glacier beds via crevasses and moulins, may play a role in outlet glacier thinning.

The extent of surface melt over Greenland increased between 1979 and 2002, although large interannual variations are superimposed on this increase. Figure 6.18 shows the time series of the maximum summer melt extent, together with maps of the melt areas in 1992 (year of minimum melt) and 2002 (year of maximum melt).

6.5.2.4. Iceland

Most Icelandic glaciers are subject to a maritime climate, with annual precipitation of up to 7 m at the highest elevations. Annual glacier-front variations are monitored at about 50 sites, and complete records from 1930 to the present exist for some of the glaciers (Sigurðsson, 1998), which show a clear response to variations in climate during this period (Jóhannesson and Sigurðsson, 1998). Mass-balance measurements have been made annually at several ice caps for 10 to 15 years (Björnsson et al., 2002; Sigurðsson, 2002). The glaciers in Iceland, particularly those on the south and southeast coasts, have a very great annual turnover of freshwater. During a year of strongly negative mass balance the glaciers can add more than 10 km^3 (equivalent to 0.03 mm global sea-level rise) to the normal precipitation runoff from the glaciated areas. Between 1991 and 2001, Vatnajökull Ice Cap lost 0.6% of its mass (Björnsson et al., 2002), equivalent to 24 km^3 of runoff or a global sea-level rise of 0.06 mm.

Icelandic glaciers advanced almost continuously during the Little Ice Age (c.1400–1900), and reached a maximum extent about 1890 (Sigurðsson, 2005). During the first quarter of the 20th century, glacier fronts retreated slightly. The rate of retreat increased significantly in the 1930s, but decreased after 1940. About 1970, most non-surging glaciers in Iceland started advancing. This period of advance was more or less continuous until the late 1990s. By 2000, all Icelandic glaciers were retreating, owing to consistently negative mass balances after 1995 (Sigurðsson, 2002).

6.5.2.5. Svalbard

The ice masses of Svalbard cover an area of approximately 36 600 km^2. Annual mass-balance measurements have been made on several Svalbard glaciers for up to 30 years (Hagen and Liestøl, 1990; Lefauconnier et al., 1999). No significant changes in mass balance have been observed during the past 30 years. The measured mean net mass balance has been negative, with no discernible change in trend. The winter accumulation is stable, and annual variations are small. The mean summer ablation is also stable with no significant trend. However, there are large interannual variations, and summer ablation drives the variations in the annual net mass balance. The low-altitude glaciers are shrinking steadily but with a slightly smaller negative net balance than those observed three decades ago. Glaciers with high-altitude accumulation areas have mass balances close to zero.

6.5.2.6. Scandinavia

Glaciers in the western maritime region of southern Scandinavia grew slowly between 1960 and 1988. Mass balances became even more positive (or less negative) between 1988 and 1998. The increase in net mass balance was due to greater winter snowfall, in contrast to Svalbard where summer ablation drives the variability in net balance. Positive mass balances due to greater winter snowfall were also observed in northern Scandinavia, at least at latitudes below 68° N.

However, low accumulation rates and high ablation rates resulted in negative mass balances for all Scandinavian glaciers between 1999 and 2003. Reichert et al. (2001) used model-based calculations to show that natural variability (e.g., the NAO) can explain many of the shorter-term variations in the mass balance of Scandinavian glaciers.

6.5.2.7. Novaya Zemlya

The glacier area of Novaya Zemlya is about 23 600 km^2 (Glazovsky, 1996; Koryakin, 1997). No direct mass-balance measurements have been made on Novaya Zemlya, but several studies of glacier extent indicate a general retreat and thus negative mass balance. Koryakin (1997) reported reductions in glacier area on Novaya Zemlya during each of four periods spanning 1913 to 1988. Zeeberg and Forman (2001) reported that tidewater calving glaciers on north Novaya Zemlya receded rapidly (>300 m/yr) during the first half of the 20th century. However, 75 to 100% of the net 20th-century retreat occurred by 1952; between 1964 and 1993, half the studied glaciers were stable, while the remainder retreated <2.5 km.

Mass losses from calving appear to be greater than mass losses from recession. The annual iceberg flux from the 200 km of calving fronts on Novaya Zemlya has been estimated to be about 2 km^3/yr (Glazovsky, pers. comm., 2003).

6.5.2.8. Franz Josef Land

The primary measurements available for the Franz Josef Land archipelago (glacier area 13 700 km^2) are of glacier extent rather than mass balance. Franz Josef Land glaciers receded between 1953 and 1993, resulting in an estimated change in glacier area of -210 km^2, and a corresponding volume change of -42 km^3 (Glazovsky, 1996). The largest changes appear to have occurred in southern parts of the ice caps on the different islands of the archipelago.

6.5.2.9. Severnaya Zemlya

Very few measurements of mass balance have been made in the Russian Arctic islands (Glazovsky, 1996). The only comprehensive data are those from the Vavilov Ice Cap (1820 km^2) on October Revolution Island (79° N) in Severnaya Zemlya, where the mass balance did not differ significantly from zero over a ten-year period beginning in 1975. Observations of changing ice-front positions suggest that the glaciers of the Vavilov Ice Cap have generally been retreating during the 20th century, providing a qualitative indication that the mean mass balance has been negative over this period, but less negative than in the other Russian Arctic islands.

6.5.3. Projected changes

In view of the limited knowledge of arctic glaciers and the uncertainties discussed in section 6.5.2, a mass-balance sensitivity approach (Dyurgerov and Meier, 1997; Oerlemans and Reichert, 2000) was used to project future change in glaciers and ice sheets. The monthly anomalies in surface temperature and precipitation generated by the five ACIA-designated models (section 4.4) were used to calculate projected changes in glacier mass balance. Projected regional changes were extrapolated from the sensitivities of glaciers for which mass-balance data exist. These projections assume that the glaciers are in balance with the baseline climates (temperature and precipitation) simulated by the models, although this assumption is unlikely to be correct (see section 6.5.5).

This approach to projecting changes in mass balance does not include glacier or ice-sheet dynamics, calving, or an explicit treatment of internal accumulation (refreezing of meltwater that percolates into the glacier); other types of mass-balance models would provide different results. Nevertheless, the use of a single mass-balance model implies that the range in projected mass-balance changes described in this section can be attributed solely to differences in the projections of temperature and precipitation generated by the five ACIA-designated models.

The output from each mass-balance model run (using input from the different ACIA-designated models) was first averaged over the regions listed in Table 6.7 (the Greenland Ice Sheet was split into four parts). The results are summarized in Fig. 6.19, which shows the projected contribution of arctic land ice to sea-level rise between 2000 and 2100. The results from the different models diverge significantly over time, ranging from close to zero to almost six centimeters by 2100. The result using output from the CSM_1.4 model is an outlier. This model projects a large increase in precipitation for the Arctic, which compensates for the enhanced ablation associated with the modest temperature rise projected by the model. The effects of temperature on glacier and ice cap mass balances projected by the different models are generally similar. The differences in the projected changes in sea level are therefore primarily due to differences in the modeled precipitation rates.

If the CSM_1.4 outlier is not included, the mean of the projected changes in sea level is an increase of approximately 4 cm between 2000 and 2100. This change is

somewhat smaller than the 70-year (2000–2070) increase of 5.7 cm estimated by Van de Wal and Wild (2001). However, the model used by Van de Wal and Wild was forced with doubled CO_2 concentrations throughout the 70-year simulation, rather than the gradual increase of GHGs in the B2 emissions scenario used to force the ACIA-designated models (section 4.4.1). In addition, the 70-year simulation used prescribed sea surface temperatures, while the coupled ACIA-designated models generate sea surface temperatures as part of the simulation process.

At a regional scale, the differences between the model projections are even greater. The projections do not even agree on the sign of the contribution of arctic land ice to sea-level change resulting from precipitation changes in some regions (e.g., Svalbard), indicating that model projections of changes in glacier mass balance at a regional scale are highly uncertain.

Of the five ACIA-designated models, the ECHAM4/ OPYC3 model projects the greatest temperature effects on the mass balance of glaciers and ice caps. Earlier versions of this model were used by Ohmura et al. (1996) to assess possible changes in the Greenland Ice Sheet driven by climate change, and by Van de Wal and Wild (2001) as noted previously. Figure 6.20 compares the projected contributions to sea-level change (due to temperature effects alone) from various regions using output from the ECHAM4/OPYC3 model. The Greenland Ice Sheet is projected to make the largest contribution, which is a direct consequence of its size. Although the glaciers in Alaska cover a much smaller area, they are also projected to make a large contribution, in agreement with recent analyses (Arendt et al., 2002; Meier and Dyurgerov, 2002). For Alaskan glaciers, the relative-

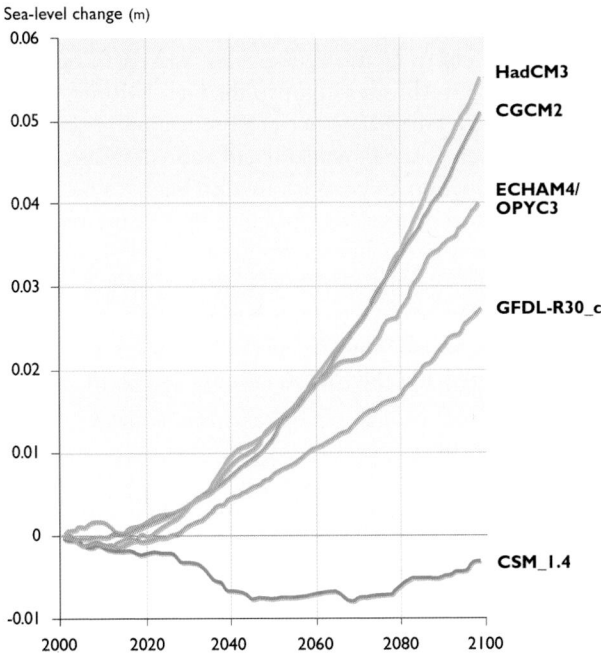

Fig. 6.19. Projected contribution of arctic land ice to sea-level change between 2000 and 2100, calculated using output from the five ACIA-designated models (J. Oerlemans, 2003).

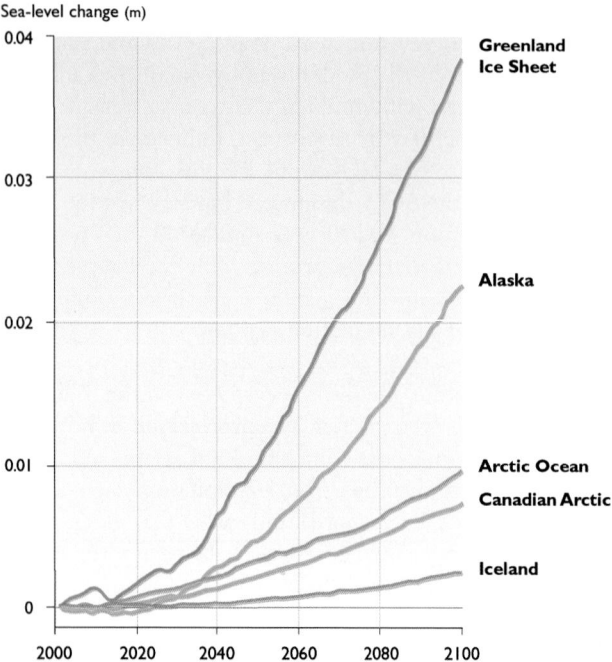

Fig. 6.20. Projected contribution of arctic land ice to sea-level change between 2000 and 2100 for various regions, calculated from ECHAM4/OPYC3 model output using temperature effects only (J. Oerlemans, 2003).

ly large sensitivity to temperature change drives the regional changes. Glaciers and ice caps in the Eurasian Arctic Ocean (Svalbard, Franz Josef Land, Severnaya Zemlya, and Novaya Zemlya) are projected to contribute about the same amount as those in the Canadian Arctic. Section 6.9 reviews all of the variables that may contribute to future sea-level change.

6.5.4. Impacts of projected changes

On other parts of the physical system

The greatest impact of changes in the mass of arctic land ice over decade-to-century timescales is likely to be a change in the freshwater input to the high-latitude oceans, which will change ocean stratification in sensitive areas such as the Greenland and Labrador Seas. Sea-ice production and export are also likely to be affected if more freshwater goes into the oceans and stabilizes the water column (section 9.2.3.1). In some areas, increased freshwater flux is likely to increase the formation of sea ice. Over longer timescales, changes in glacial ice (especially the Greenland Ice Sheet) may affect the geoid and the rotation rate of the earth.

On ecosystems

To the extent that changes in freshwater influx affect upper-ocean stratification (and possibly sea ice), impacts on marine ecosystems are likely. Riparian ecosystems are also likely to be affected by changes in river flow and aufeis (ice formed when water from a stream emerges and freezes on top of existing ice) production. Any significant change in sea level will have impacts on ecosystems in low-lying coastal areas.

On people

The greatest direct impacts on humans from changes in arctic land ice are likely to result from changes in global sea level, which will affect coastal communities in many parts of the Arctic. Other possible impacts include changes in hydropower production and water supply from glacier-fed lakes and reservoirs. Changes in iceberg production will increase or decrease hazards to shipping and navigation.

6.5.5. Critical research needs

The compilation of an up-to-date global glacier inventory is a critical research need. For some regions, existing inventories are sparse; inventories also need to be updated where glacier areas have changed. A global satellite-derived dataset of exposed ice areas is a minimum requirement. Ideally, a complete glacier database describing individual glacier locations, areas, and geometries should be compiled, so that mass-balance measurements on individual benchmark glaciers can be extrapolated to unmeasured glaciers with greater certainty.

For future projections, it would be useful to develop additional mass-balance models so that spatial variations can be better depicted and so that the records can be extended back in time at locations for which atmospheric data are available. For this purpose, additional mass-balance observations should be obtained in regions where existing data are particularly sparse, in order to provide credibility and a sense of the uncertainties in model projections of future trends. It is also important to continue the ongoing monitoring of glacier mass balance with *in situ* measurements on selected glaciers in order to improve understanding of the response of glaciers to climate change, improve model projections of future change, and calibrate remote-sensing data.

Improved climate model projections of temperature and precipitation over ice-covered regions are a top priority for improving projections of changes in glaciers, ice sheets, and sea level. As the results in section 6.5.3 imply, the ranges in the projected changes in temperature and precipitation among the ACIA-designated models are far greater than the mean changes projected for glaciated areas. The application of the mass-balance sensitivity approach described in section 6.5.3 also requires analyses of the likely discrepancies between mass balances calculated from the models' baseline simulations of temperature and precipitation and the present-day glacier mass balances.

With regard to sea level, a critical need is to determine whether the recent negative mass balance and increasing summer melt area of the Greenland Ice Sheet (equivalent to at least 0.13 mm/yr sea-level rise) are part of a long-term trend (Rignot and Thomas, 2002). Continued monitoring of the ice sheet, using, for example, radar and laser altimeters such as those planned for the Cryosphere Satellite (CryoSat) and the Ice, Cloud and

Land Elevation Satellite (ICESat), is likely to improve understanding of the current mass balance of the Greenland Ice Sheet.

In order to improve projections of *future* mass-balance changes, the following studies should be given high priority:

- improving understanding of albedo changes and feedback mechanisms;
- studies of outlet glacier dynamics with emphasis on their potential for triggering persistent, rapid changes in ice-sheet volume;
- improving ice-dynamic models for determining the long-term response of the ice sheet to past climate change;
- improving parameterization and verification of internal-accumulation models; and
- improving understanding of the relationships between climate change, meltwater penetration to the bed, and changes in iceberg production.

Future changes in mass balance are strongly dependent on future changes in climate. Consequently, the ability to project changes in the mass balance of the Greenland Ice Sheet is linked closely to the ability of atmosphere–ocean general circulation models (AOGCMs) to project changes in regional climate over Greenland. For example, recent AOGCM model runs project a greater increase in the accumulation rate over Greenland associated with a temperature increase than did previous studies (Van de Wal et al., 2001). If the latest projections prove to be accurate, increases in accumulation would largely compensate for the increased runoff resulting from projected temperature increases.

6.6. Permafrost

6.6.1. Terrestrial permafrost

6.6.1.1. Background

Permafrost is soil, rock, sediment, or other earth material with a temperature that has remained below 0 °C for two or more consecutive years. Permafrost underlies most of the surfaces in the terrestrial Arctic. Permafrost extends as far south as Mongolia (Sharkhuu, 2003), and is present in alpine areas at even lower latitudes. Figure 6.21 shows the distribution of permafrost in the Northern Hemisphere, classified into continuous, discontinuous, and sporadic zones. In the continuous zones, permafrost occupies the entire area (except below large rivers and lakes). In the discontinuous and sporadic zones, the percentage of the surface underlain by permafrost ranges from 10 to 90%. Discontinuous permafrost underlies a larger percentage of the landscape than does sporadic permafrost, although there is not a standard definition of the boundary between the two zones (Fig. 6.21 uses 30% coverage as the boundary). In the Northern Hemisphere, permafrost zones occupy approximately 26 million km² or about 23% of

the exposed land area, but permafrost actually underlies 13 to 18% of the exposed land area (Anisimov and Nelson, 1997; Zhang T. et al., 2000). Distinctions are made between permafrost that is very cold (temperatures of -10 °C and lower) and thick (500–1400 m), and permafrost that is warm (within 1 or 2 °C of the melting point) and thin (several meters or less). Ground ice (0–20 m depth) in permafrost exhibits large spatial variability, with generally much more ice in lowland permafrost than in mountain permafrost (Brown J. et al., 1998).

The role of permafrost in the climate system is threefold (Anisimov and Fitzharris, 2001). First, because it provides a temperature archive, permafrost is a "geoindicator" of environmental change. At depths below 15 to 20 m, there is generally little or no annual cycle of temperature, so seasonality does not influence warming or cooling. Second, permafrost serves as a vehicle for transferring atmospheric temperature changes to the hydrological and biological components of the earth system. For example, the presence of permafrost significantly alters surface and subsurface water fluxes, as well as vegetative functions. Third, changes in permafrost can feed back to climate change through the release of trace gases such as CO_2 and methane (CH_4), linking climate change in the Arctic to global climate change (Anisimov et al., 1997; Fukuda, 1994).

The active layer is the seasonally thawed layer overlying permafrost. Most biogeochemical and hydrological

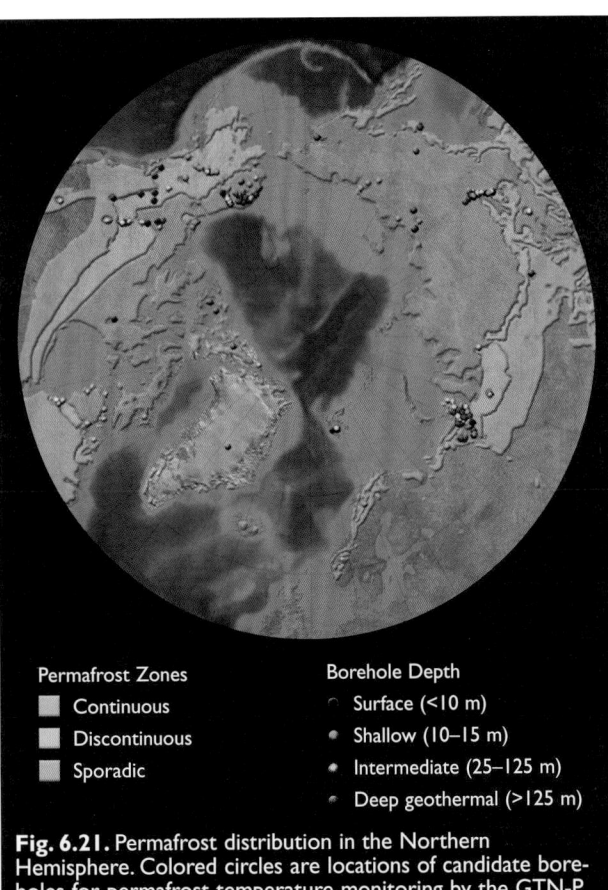

Permafrost Zones
- Continuous
- Discontinuous
- Sporadic

Borehole Depth
- Surface (<10 m)
- Shallow (10–15 m)
- Intermediate (25–125 m)
- Deep geothermal (>125 m)

Fig. 6.21. Permafrost distribution in the Northern Hemisphere. Colored circles are locations of candidate boreholes for permafrost temperature monitoring by the GTN-P (based on Romanovsky et al., 2002).

processes in permafrost are confined to the active layer, which varies from several tens of centimeters to one to two meters in depth. The rate and depth of active-layer thaw are dependent on heat transfer through layers of snow, vegetation, and organic soil. Snow and vegetation (with the underlying organic layer) have low thermal conductivity and attenuate annual variations in air temperature. During summer, the thermal conductivity of the organic layer and vegetation is typically much smaller than in winter. This leads to lower heat fluxes in summer and ultimately keeps permafrost temperatures lower than they would be in the absence of vegetation and the organic layer. The latent heat associated with evapotranspiration and with melting and freezing of water further complicates the thermodynamics of the active layer. Over longer timescales, the thawing of deep permafrost layers can lag considerably (decades or centuries) behind a warming of the surface because of the large latent heat of fusion of ice (Riseborough, 1990). Moreover, thermal conductivity is typically 20 to 35% lower in thawed mineral soils than in frozen mineral soils. Consequently, the mean annual temperature below the level of seasonal thawing can be 0.5 to 1.5 °C lower than on the ground surface.

Thawing of permafrost can lead to subsidence of the ground surface as masses of ground ice melt, and to the formation of uneven topography known as thermokarst. The development of thermokarst in some areas of warm and discontinuous permafrost in Alaska has transformed some upland forests into wetlands (Osterkamp et al., 2000). Recent thaw subsidence has also been reported in areas of Siberia (Fedorov, 1996) and Canada (Smith et al., 2001). Climate-induced thermokarst and thaw subsidence may have detrimental impacts on infrastructure built upon permafrost (Anisimov and Belolutskaia, 2002; Nelson, 2003; Nelson et al., 2001), as section 16.3 discusses in more detail. Permafrost degradation can also pose a serious threat to arctic biota through either oversaturation or drying (Callaghan and Jonasson, 1995). The abundance of ground ice is a key factor in subsidence, such that areas with little ice (e.g., the Canadian Shield or Greenland bedrock masses) will suffer fewer subsidence effects when permafrost degrades.

Seasonal soil freezing and thawing are the driving forces for many surficial processes that occur in areas with permafrost or seasonally frozen soils. Cryoturbation, a collective term for local vertical and lateral movements of the soil due to frost action, is one of these potentially important cryogenic processes (see Washburn (1956) for a review). Cryoturbation typically occurs in the permafrost zone, but also occurs in soils that freeze only seasonally. Cryoturbation can cause the downward displacement of organic material from the near-surface organic horizons to the top of the permafrost table, resulting in sequestration of organic carbon in the upper permafrost layer (Williams P. and Smith, 1989). During the past several thousand years, a significant amount of organic carbon has accumulated in permafrost due to this process.

6.6.1.2. Recent and ongoing changes

Because surface temperatures are increasing over most permafrost areas (section 2.6.2), permafrost is receiving increased attention within the context of past and present climate variability. Measurements of ground temperature in Canada, Alaska, and Russia have produced a generally consistent picture of permafrost warming over the past several decades. Lachenbruch and Marshall (1986) were among the first to document systematic warming by using measurements from permafrost boreholes in northern Alaska to show that the surface temperature increased by 2 to 4 °C between the beginning of the 20th century and the mid-1980s. Measurements conducted by Clow and Urban (see Nelson, 2003) in the same Alaska borehole network indicated further warming of about 3 °C since the late 1980s. Figure 6.22 confirms this warming with results from a site-specific permafrost model driven by observed air temperatures and snow depths for the period 1930 to 2003, calibrated using measurements of permafrost temperatures between 1995 and 2000. While warming has predominated since 1950, considerable interannual variability is also apparent.

Data from northwestern Canada, indicating that temperatures in the upper 30 m of permafrost have increased by up to 2 °C over the past 20 years (Burn, see Couture et al., 2003; Nelson, 2003), provide further evidence of warming. Although cooling of permafrost in the Ungava Peninsula of eastern Canada in recent decades has been widely cited as an exception to the dominant warming trend, Brown J. et al. (2000) and Allard et al. (see Nelson, 2003) indicated that shallow permafrost temperatures in the region have increased by up to nearly 2 °C since the mid-1990s. Smith et al. (2003) reported warming in the upper 30 m of permafrost in the Canadian High Arctic since the mid-1990s. Smaller temperature increases, averaging 1 °C or less, have been reported in northwestern Siberia (Chudinova et al., 2003; Pavlov and Moskalenko, 2002). Measurements from a network of recently drilled boreholes in mountainous areas of Europe indicate warming of a degree or less (Harris and Haeberli, 2003), while Isaksen et al.

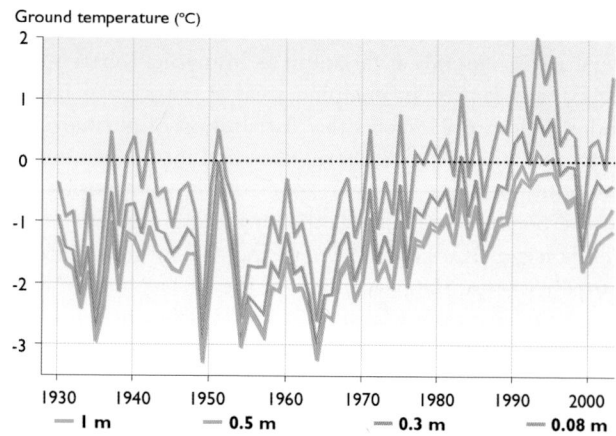

Ground temperature (°C)

Fig. 6.22. Simulated mean annual ground temperature at Fairbanks (Bonanza Creek), Alaska, from 1930 to 2003 (V. Romanovsky, 2004).

(2001) have reported warming of Scandinavian permafrost. Table 6.8 summarizes recent trends in permafrost temperatures in terms of region, time period, and the approximate temperature change over the period of record. In general, the changes in permafrost temperature are consistent with other environmental changes in the circumpolar Arctic (Anisimov et al., 2003; Serreze et al., 2000).

Most of the boreholes mentioned in this section are included in an emerging system for comprehensive monitoring of permafrost temperatures (Fig. 6.21), the Global Terrestrial Network for Permafrost (GTN-P), established with the assistance of the International Permafrost Association. Burgess et al. (2000) provide an overview of the GTN-P.

6.6.1.3. Projected changes

At present, land-surface parameterizations used in global climate models such as those designated by the ACIA (section 4.2.7) do not adequately resolve the soil, and the models do not archive the soil output needed to assess changes in permafrost distribution and active-layer characteristics. The more viable approach to date has been the use of AOGCM output as input to soil modules run in an off-line mode, often in combination with baseline climatic data obtained from meteorological observations

(Anisimov and Poliakov, 2003). This section summarizes results from several studies that used this approach with output from the ACIA-designated models to show spatially distributed fields of projected changes in permafrost for three different times in the 21st century.

Circumpolar projections

At the circumpolar scale, climate change can be expected to reduce the area occupied by frozen ground and to cause shifts between the zones of continuous, discontinuous, and sporadic permafrost, comparable to the changes that occurred during warm epochs in the past (Anisimov et al., 2002). Such changes can be projected using a relatively simple frost-index-based model of permafrost driven by scenarios of climate change. Anisimov and Nelson (1997) used this method to calculate areas occupied by near-surface permafrost in the Northern Hemisphere under present-day climatic conditions and climatic conditions projected for the 2041–2060 time slice. Scenarios of climate change used in these calculations were based on the results from several transient and equilibrium experiments with general circulation models. These results have been updated (Anisimov, unpubl. data, 2003) using output from the five ACIA-designated models (section 4.4). Table 6.9 presents projections of the area occupied by different permafrost zones in 2030, 2050, and 2080. Results for

Table 6.8. Recent trends in permafrost temperature (Romanovsky et al., 2002, updated).

Region	Depth (m)	Period of record	Permafrost temperature change[a] (°C)	Reference
United States				
Trans-Alaska pipeline route	20	1983–2000	+0.6 to +1.5	Osterkamp, 2003; Osterkamp and Romanovsky, 1999
Barrow Permafrost Observatory	15	1950–2001	+1	Romanovsky et al., 2002
Russia				
East Siberia	1.6–3.2	1960–1992	+0.03/yr	Romanovsky, pers. comm., 2003
Northwest Siberia	10	1980–1990	+0.3 to +0.7	Pavlov, 1994
European north of Russia, continuous permafrost zone	6	1973–1992	+1.6 to +2.8	Pavlov, 1994
European north of Russia, discontinuous permafrost zone	6	1970–1995	up to +1.2	Oberman and Mazhitova, 2001
Canada				
Alert, Nunavut	15–30	1995–2000	+0.15/yr	Smith et al., 2003
Northern Mackenzie Basin, Northwest Territories	28	1990–2000	+0.1/yr	Couture et al., 2003
Central Mackenzie Basin, Northwest Territories	15	1985–2000	+0.03/yr	Couture et al., 2003
Northern Québec	10	late 1980s– mid-1990s	-0.1/yr	Allard et al., 1995
Norway				
Juvvasshøe, southern Norway	~5	past 60–80 years	+0.5 to +1.0	Isaksen et al., 2001
Svalbard				
Janssonhaugen	~5	past 60–80 years	+1 to +2	Isaksen et al., 2001

[a]Temperature change over period of record, unless otherwise noted.

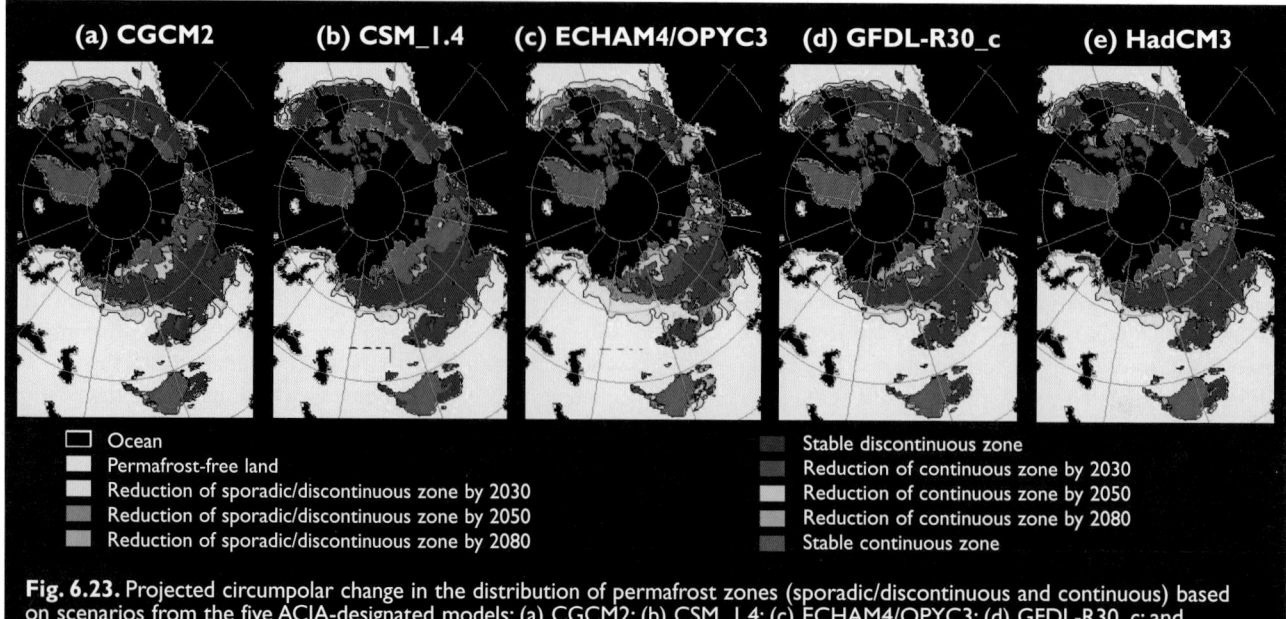

Fig. 6.23. Projected circumpolar change in the distribution of permafrost zones (sporadic/discontinuous and continuous) based on scenarios from the five ACIA-designated models: (a) CGCM2; (b) CSM_1.4; (c) ECHAM4/OPYC3; (d) GFDL-R30_c; and (e) HadCM3 (adapted from Anisimov and Belolutskaia, in press a).

2030 rather than 2020 are shown because the latter show little change from present-day distributions.

The projected reductions in the area occupied by near-surface permafrost (the uppermost few meters of frozen ground; conditions in the deeper layers are not addressed in these projections) vary substantially depending on the scenario used, indicating that the uncertainties in the forcing data are large. Among the five model scenarios, the two outliers are the ECHAM4/OPYC3-based scenario, which projects the greatest contraction of the area occupied by near-surface permafrost, and the CSM_1.4 scenario that projects only modest changes. Projections from the three other scenarios (CGCM2, GFDL-R30_c,

and HadCM3) are relatively close to each other. According to the "median" GFDL-R30_c-based scenario, the total area occupied by near-surface permafrost is projected to decrease by 11, 18, and 23% by 2030, 2050, and 2080, respectively. The projected contractions of the continuous near-surface permafrost zone for the same years are 18, 29, and 41%, respectively. Figure 6.23 shows the projected changes in the distribution of permafrost zones (continuous and sporadic/discontinuous) calculated using output for 2030, 2050, and 2080 from the five ACIA-designated models.

A progressive increase in the depth of seasonal thawing could be a relatively short-term reaction to climate

Table 6.9. Projected area occupied by permafrost zones in 2030, 2050, and 2080 calculated using output from the five ACIA-designated models.

		Total permafrost		Continuous permafrost	
		Area (10⁶ km²)	% of present-day value	Area (10⁶ km²)	% of present-day value
CGCM2	2030	23.72	87	9.83	79
	2050	21.94	81	8.19	66
	2080	20.66	76	6.93	56
ECHAM4/OPYC3	2030	22.30	82	9.37	75
	2050	19.31	71	7.25	58
	2080	17.64	65	5.88	47
GFDL-R30_c	2030	24.11	89	10.19	82
	2050	22.38	82	8.85	71
	2080	20.85	77	7.28	59
HadCM3	2030	24.45	90	10.47	84
	2050	23.07	85	9.44	76
	2080	21.36	78	7.71	62
CSM_1.4	2030	24.24	89	10.69	86
	2050	23.64	87	10.06	81
	2080	21.99	81	9.14	74

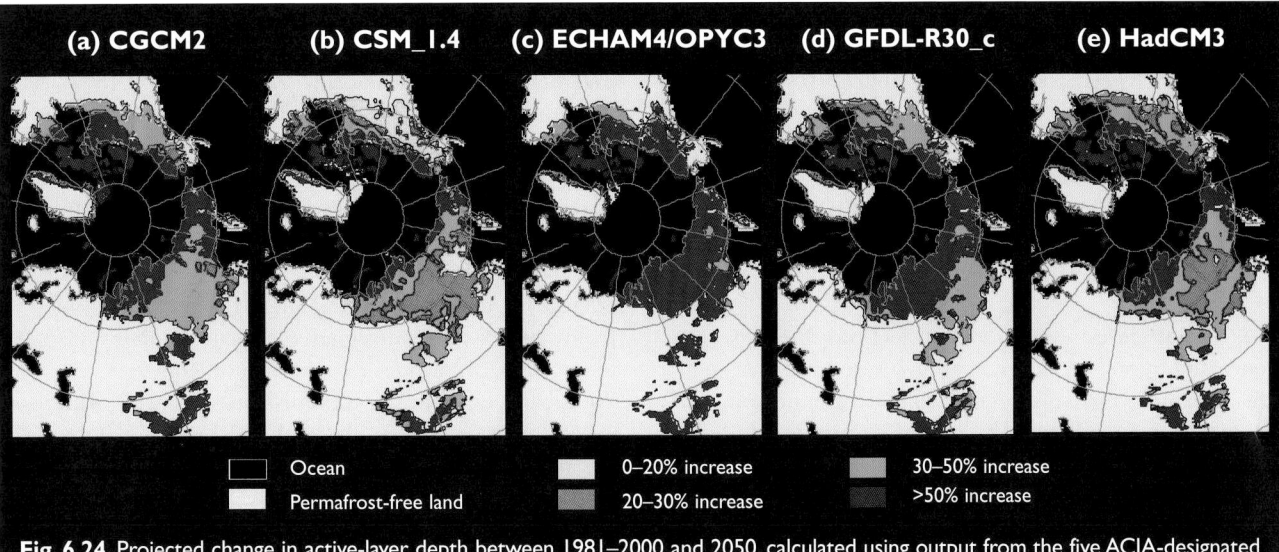

Fig. 6.24. Projected change in active-layer depth between 1981–2000 and 2050, calculated using output from the five ACIA-designated models: (a) CGCM2; (b) CSM_1.4; (c) ECHAM4/OPYC3; (d) GFDL-R30_c; and (e) HadCM3 (Anisimov et al., 1997, updated).

change in permafrost regions, since it does not involve any lags associated with the thermal inertia of the climate/permafrost system. One of the most successful and frequently used approaches to active-layer mapping is based on semi-empirical methods developed primarily for the practical needs of cold-regions engineering, but adjusted for use at the hemispheric scale. The fundamentals of these methods were formulated by Russian geocryologists (Garagulya, 1990; Kudryavtsev, 1974) and have been used by other investigators (e.g., Anisimov et al., 1997; Romanovsky and Osterkamp, 1995; Sazonova and Romanovsky, 2003). Monthly temperature and precipitation simulated by the ACIA-designated models for baseline (1981–2000) and year 2050 climate conditions were used as input to the Kudryavtsev (1974) model to calculate projected changes in seasonal thaw depth during the first half of the 21st century (Fig. 6.24). The projected increases in active-layer depth range from 0 to 20% to more than 50%.

The calculations require several assumptions about soil, organic layer, vegetation, and snow-cover properties (see Anisimov et al., 1997). Digital representation of soil properties for each cell was obtained from the Global Ecosystems Database (Staub and Rosenzweig, 1987). Over much of the permafrost area, the calculations were made for silt covered with a 10 cm organic layer. The cal-

culations assume that vegetation does not change as temperatures increase, although a study by Anisimov and Belolutskaia (in press b) indicates that climate-induced vegetation changes are likely to both largely offset the effects of warming in the northernmost permafrost regions and enhance the degradation of sporadic and discontinuous permafrost. The projected changes represent the behavior of permafrost with highly generalized properties averaged over 0.5° by 0.5° grid cells. Owing to the effects of local environmental factors, including topography and vegetation variations, seasonal thaw depth is characterized by pronounced spatial and temporal variability that cannot be resolved at the scale of the model calculations. More details on the spatial variability of environmental features, including low-level vegetation, would improve the spatial accuracy of the projected changes in permafrost (e.g., Smith and Burgess, 1999).

The maps in Fig. 6.24 provide a broad picture of the projected hemispheric-scale changes in seasonal thaw depth under changing climate conditions. These projections are generally consistent with the results obtained from the more detailed regional studies for Alaska and Siberia. Although the results of the calculations are model-specific, there is a general consensus among the models that seasonal thaw depths are likely to increase by more than 50% in the northernmost permafrost loca-

Table 6.10. Projected regional increases in mean annual air temperature (ΔT_a), permafrost temperature (ΔT_s), and depth of seasonal thaw (ΔZ) between 1981–2000 and 2050, calculated using output from the five ACIA-designated models.

	ΔT_a (°C)	ΔT_s (°C)	ΔZ (%)
Arctic coast of Alaska and Canada	2.0–3.0	2.0–2.5	≥50
Central Canada	1.5–2.5	1.0–2.0	≤30
West coast of Canada	1.0–2.0	0.5–1.5	≤10
Northern Scandinavia	1.5–2.0	1.0–2.0	≤10
Siberia	2.0–3.0	2.0–2.5	≥50
Yakutia	1.5–2.5	1.5–2.0	≤30
Russian Arctic coast	2.0–3.0	2.0–2.5	≥50

tions (including much of Siberia, the Far East, the North Slope of Alaska, and northern Canada) and by 30 to 50% in most other permafrost regions. Table 6.10 shows, by region, the range of changes in mean annual air temperature, permafrost temperature, and depth of seasonal thaw projected to occur between 1981–2000 and 2050 by the ACIA-designated models.

Regional projections

This section describes regional projections that were generated by soil models run at high resolution for an area in which relatively detailed information on soil properties was available. The soil information included soil temperatures used for model calibration. While this section presents results for northern Alaska, model-based evaluations of the sensitivities of permafrost to warming in other areas, including Canada, are also available (Smith and Burgess, 1999; Wright et al., 2000).

Detailed projections of future changes in permafrost in northern Alaska were obtained from a soil model (Zhuang et al., 2001) calibrated using observational data from three sites on the North Slope of Alaska. The two major types of vegetation in northern Alaska are tundra and taiga (boreal forest). Permafrost is continuous north of the Brooks Range and discontinuous in much of Interior Alaska to the south. Permafrost is >600 m thick in northern areas but is only one to sev-

eral meters thick near its southern limits. In the lowlands of the southern discontinuous zone, where the mean annual air temperature ranges from -7 to 0 °C, the temperature of the permafrost below the layer of seasonal temperature variation ranges from -5 to -1 °C. In the continuous permafrost zone north of the Brooks Range, permafrost temperatures typically range from -11 to -4 °C.

Surface air temperature and snow-cover projections from the five ACIA-designated climate models and the older HadCM2 model (Sazonova and Romanovsky, 2003), forced with the B2 emissions scenario (section 4.4.1), were used as input to the soil model to project the active-layer and mean annual ground-temperature dynamics in northern Alaska between 2000 and 2100. The across-model average projected increase in mean annual air temperature between 2000 and 2100 ranges from 8 to 10 °C in the north to 4 to 6 °C in the southern part of the region.

In the central and northern areas of Alaska, projected increases in mean annual ground temperatures between 2000–2010 and 2100 range from 1 to 2 °C using the CGCM2 climate scenario to 5 °C using the HadCM3 and ECHAM4/OPYC3 scenarios (Fig. 6.25). The HadCM3, HadCM2, and ECHAM4/OPYC3 scenarios generate significant projected increases in mean annual ground temperatures over the entire area.

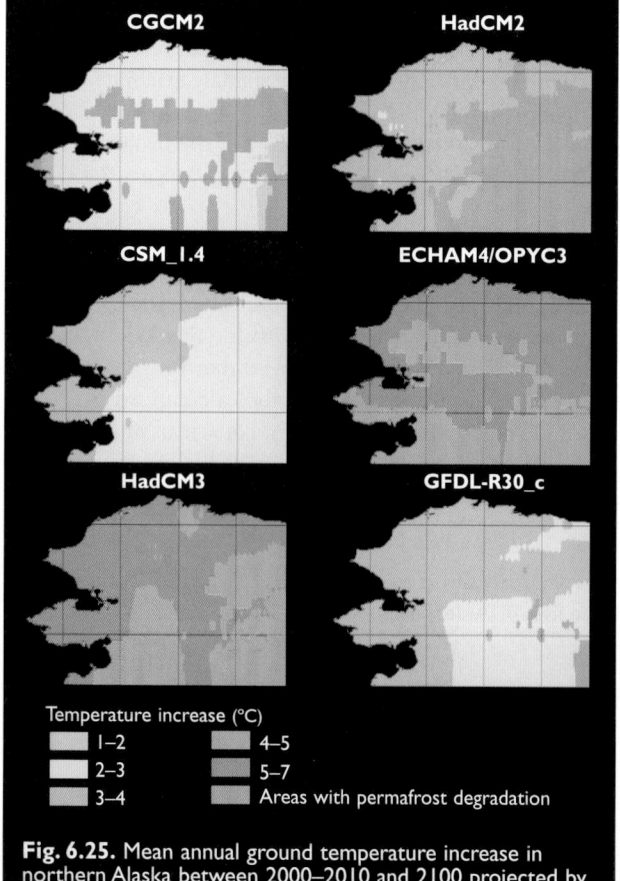

Fig. 6.25. Mean annual ground temperature increase in northern Alaska between 2000–2010 and 2100 projected by a soil model forced with output from the ACIA-designated models. A reference case using forcing from the HadCM2 model is shown at the upper right (Sazonova et al., 2004).

CGCM2   HadCM2
CSM_1.4   ECHAM4/OPYC3
HadCM3   GFDL-R30_c

Temperature increase (°C)
1–2   4–5
2–3   5–7
3–4   Areas with permafrost degradation

CGCM2   HadCM2
CSM_1.4   ECHAM4/OPYC3
HadCM3   GFDL-R30_c

Increase in active layer thickness (m)
0.0–0.3   1.2–1.8
0.3–0.6   1.8–2.4
0.6–0.8
0.8–1.2   Areas with permafrost degradation

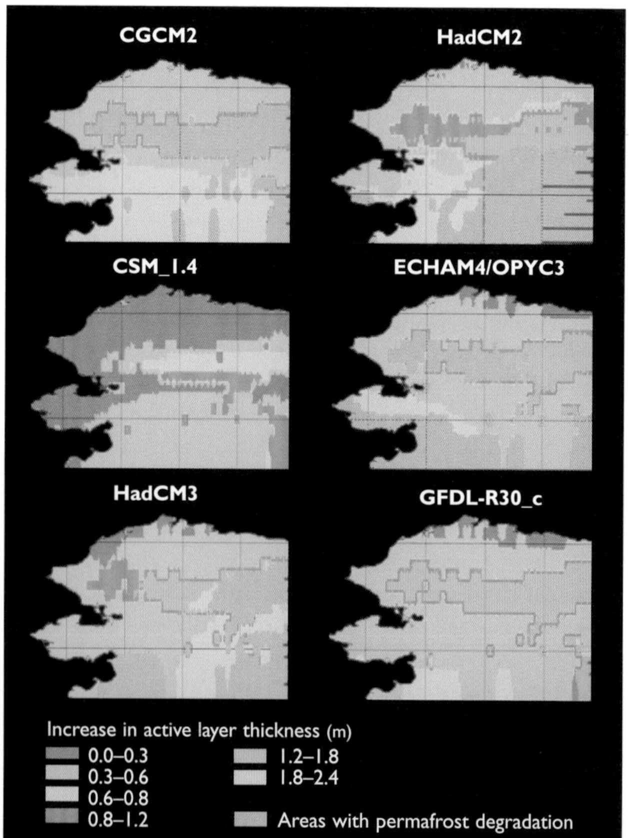

Fig. 6.26. Increase in active-layer thickness in northern Alaska between 2000 and 2100 projected by a soil model forced with output from the ACIA-designated models. A reference case using forcing from the HadCM2 model is shown at the upper right (Sazonova et al., 2004).

An analysis of the maximum active-layer thickness was also performed. All the scenarios project that, by 2100, the active-layer thickness is likely to increase by up to 1 m in areas occupied by coarse-grained material and rocks with high thermal conductivity, and by up to 0.5 m throughout the rest of the region (Fig. 6.26). The HadCM2, HadCM3, and ECHAM4/OPYC3 scenarios generate the greatest projected increases in active-layer thickness.

By 2100, all the scenarios except for the CSM_1.4 project that a zone with permafrost degradation (failure of some portion of the former active layer to refreeze during winter) will exist in northern Alaska (Fig. 6.26). The HadCM2 and GFDL-R30_c scenarios project that this zone will occupy the southeastern part of the modeled area. The HadCM2 scenario projects a relatively constant increase in this zone throughout the years, with almost one-third of the modeled area degraded by 2100. The ECHAM4/OPYC3 scenario projects the second largest zone of degradation by 2100, but the development of this zone throughout the century is not uniform. The CGCM2 scenario projects that this zone will be located in the southeastern and central parts of the modeled area and in the Brooks Range. The HadCM3 and ECHAM4/OPYC3 scenarios project that the zone will occupy the southeastern and southwestern parts of the modeled area and some parts of the Brooks Range (the eastern part in HadCM3, the western part in ECHAM4/OPYC3). The GFDL-R30_c scenario projects that the zone of permafrost degradation will occupy less than 5 to 7% of the modeled area. The CSM_1.4 scenario is an outlier in that it projects no permafrost degradation between 2000 and 2100.

Similar dependencies on climate model forcing scenarios have been found by Malevsky-Malevich et al. (2003), who used output from the same climate models to drive a different type of soil model. The results of these simulations showed that the projected active-layer response in Siberia would be greater in southern and western regions than in eastern and northern regions, indicating the potential importance of snow cover to projections of permafrost change (Stieglitz et al., 2003). The decrease in snow-cover duration is projected to be greater in southern and western Siberia than in northern and eastern Siberia, and greater in the spring season (section 6.4) when insolation is relatively high.

The scenarios of permafrost change clearly vary with the choice of climate model, and they contain many examples of decadal-scale variations that can complicate the detection of change. Nevertheless, the projected changes are substantial in nearly all cases, and terrestrial permafrost is likely to remain one of the more useful indicators of global change because large regions of the arctic terrestrial system now have mean annual temperatures close to 0 °C.

6.6.1.4. Impacts of projected changes

On other parts of the physical system

Projected climate change is very likely to increase the active-layer thickness and the thawing of permafrost at greater soil depths. The impacts of permafrost degradation include changes in drainage patterns and surface wetness resulting from subsidence and thermokarst formation, especially where soils are ice-rich. Thawing of ice-rich permafrost can trigger mass movements on slopes, and possibly increase sediment delivery to watercourses. Thawing of permafrost in peatlands and frozen organic matter sequestered by cryoturbation is likely to accelerate biochemical decomposition and increase the GHGs released into the atmosphere.

On ecosystems

Changes in surface drainage and wetness are likely to result in vegetative changes (e.g., shallow-rooted versus deeper-rooted vegetation, changes in plant density); the development of thermokarst has transformed some upland forests into extensive wetlands. Microbial, insect, and wildlife populations are likely to evolve over time as soil drainage and wetness change (section 7.4.1). Changes in drainage resulting from changes in the distribution of permafrost are also likely to affect terrestrial ecosystems, and will determine the response of peatlands and whether they become carbon sources or sinks (section 7.5.3).

On people

Permafrost degradation is likely to cause instabilities in the landscape, leading to surface settlement and slope collapse, which may pose severe risks to infrastructure (e.g., buildings, roads, pipelines). The possibilities for land use change with soil wetness. Offshore engineering (e.g., for resource extraction) is highly affected by coastal permafrost and its degradation (section 16.3.10). Containment structures (e.g., tailing ponds, sewage lagoons) often rely on the impermeable nature of frozen ground; thawing permafrost would reduce the integrity of these structures. Over the very long term, the disappearance of permafrost coupled with infrastructure replacement will eliminate many of the above concerns.

6.6.1.5. Critical research needs

In order to improve the credibility of model projections of future permafrost change throughout the Arctic, the soil/vegetation models must be validated in a more spatially comprehensive manner. In particular, there is a need for intercomparison of permafrost models using the same input parameters and standardized measures for quantifying changes in permafrost boundaries. Global models do not yet use such regionally calibrated permafrost models, nor do they treat the upper soil layers in sufficient detail to resolve the active layer. The need for additional detail is particularly great for areas with thin permafrost (e.g., Scandinavia). Enhanced model resolution, and validation and calibration at the circumpolar scale, will be necessary before fully coupled simulations by global models will provide the information required for assessment activities such as the ACIA.

There is likely to be a significant linkage between changes in terrestrial permafrost and the hydrology of arctic drainage basins. Long-term field data are required to increase understanding of permafrost–climate interactions and the interaction between permafrost and hydrological processes, and for model improvement and validation. The active-layer measurements from the Circumpolar Active Layer Monitoring Program and the borehole measurements from the GTN-P will be especially valuable in this regard, if the numbers of sites are increased.

6.6.2. Coastal and subsea permafrost

6.6.2.1. Background

The terms subsea (or offshore) and coastal permafrost refer to geological materials that have remained below 0 °C for two or more years and that occur at or below sea level. At present, the thermal regime of subsea permafrost is controlled partially or completely by seawater temperature. Subsea permafrost has formed either in response to negative mean annual sea-bottom temperatures or as the result of inundation of terrestrial permafrost. Coastal permafrost includes the areas of permafrost that are near a coastline (offshore or onshore) and that are affected, directly or indirectly, by marine processes. Direct marine influences include seawater temperature, sea-ice action, storm surges, wave action, and tides. Indirect marine influences include the erosion of cliffs and bluffs. This review focuses on those parts of the permafrost environment found below the storm tide line, since the thermal and chemical environments that affect them are substantially different from those affecting terrestrial permafrost.

The development and properties of subsea permafrost are largely dependent on the detailed history of postglacial relative sea level. Coastal permafrost conditions are influenced by a range of oceanographic and meteorological processes, ranging from sea-ice thickness to storm-surge frequency. During the transition from terrestrial to submarine, permafrost is subjected to a set of intermediary environments that affect its distribution and state.

Based on the strict definition above, not all permafrost is frozen, since the freezing point of sediments may be depressed below 0 °C by the presence of salt or by capillary effects in fine-grained material. In the marine environment, non-frozen materials do not present serious problems for engineering activities, so modifiers are used to further define frozen permafrost as either ice-bonded, ice-bearing, or both (see Sellmann and Hopkins, 1984). Ice-bearing material refers to permafrost or seasonally frozen sediments that contain some ice. Ice-bonded sediments are mechanically cemented by ice. While the ice

component of permafrost usually consists of pore or interstitial ice that fills the small spaces between individual grains of sand, silt, or gravel, it sometimes occurs in much larger forms referred to as "massive ice". Unfrozen fluids may be present in the pore spaces in both ice-bearing and ice-bonded materials. As with terrestrial permafrost, some subsea permafrost has an active (seasonally thawed) layer. Hubberten and Romanovskii (2003) discussed the characteristics of permafrost in one particular offshore environment, the Laptev Sea.

While the stability of terrestrial permafrost depends directly on atmospheric forcing (temperature and precipitation), the effect of atmospheric forcing on the stability of subsea permafrost is a second- or third-order impact mediated through oceanographic and sea-ice regimes. Most subsea permafrost formed during past glacial cycles, when continental shelves were exposed to low mean annual temperatures during sea-level lowstands, thus it is restricted to those parts of the Arctic that were not subjected to extensive glaciation during the Late Quaternary Period (Mackay, 1972). Permafrost that developed on exposed continental shelves during glacial epochs subsequently eroded when sea-level rise submerged the shelves during interglacial warm intervals and regraded the land surface to a quasi-equilibrium seabed profile. Positive mean annual sea-bottom temperatures degrade upland permafrost as it passes through the coastal zone, but with continued sea-level rise, the sea-bottom water temperature falls to negative values and permafrost degradation slows. Thus, in any locality, the distribution of relict subsea permafrost is a function of its original distribution on land, and the depth to ice-bonded or ice-bearing permafrost is a function of the time spent in the zone of positive sea-bottom temperatures along with other variables (e.g., volumetric ice content, salt content, etc.) (Mackay, 1972; Osterkamp and Harrison, 1977; Vigdorchik, 1980).

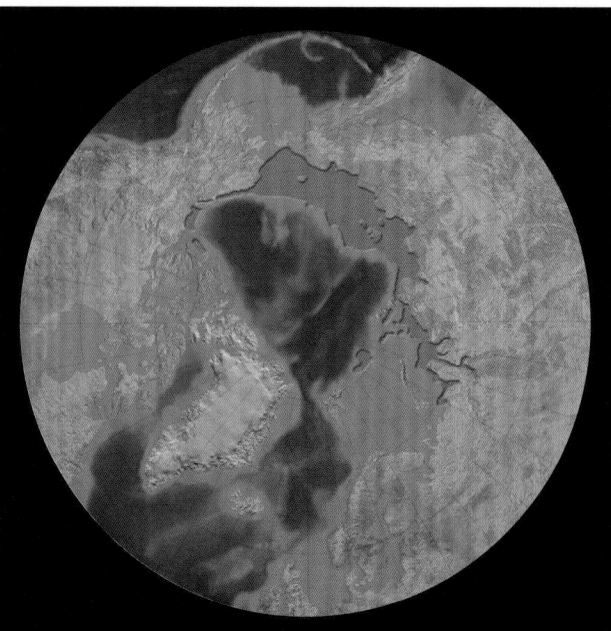

Fig. 6.27. Subsea permafrost distribution in the Arctic (based on Brown J. et al., 1998).

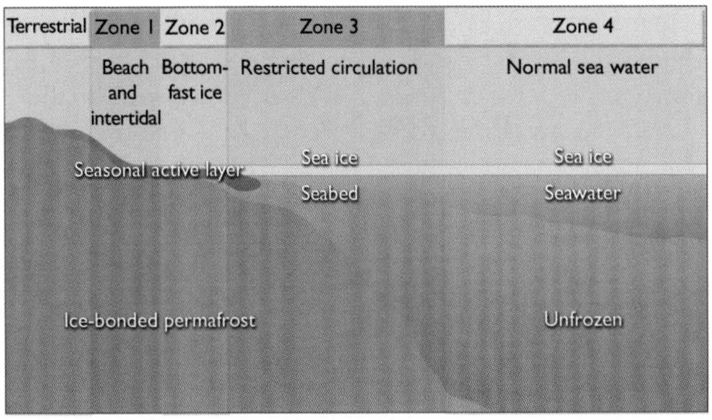

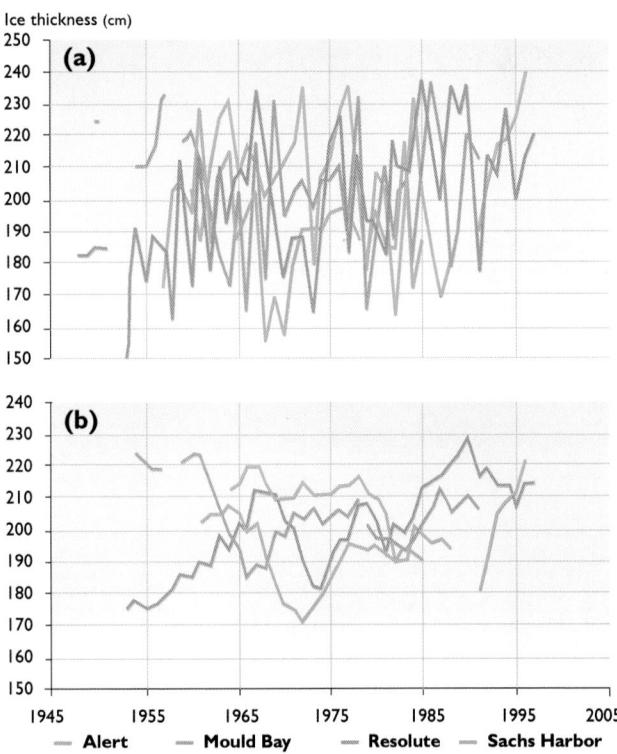

Fig. 6.28. The coastal and offshore permafrost zones (based on Osterkamp, 2001).

Subsea relict permafrost is thought to contain or overlie large volumes of CH_4 in the form of gas hydrates at depths of up to several hundred meters. Degradation of gas hydrates resulting from climate change (section 6.6.2.3) could increase the flux of CH_4 to the atmosphere (Judge and Majorowicz, 1992; Kvenvolden, 1988).

A combination of observations and models has been used to estimate the distribution of subsea permafrost (Fig. 6.27). The distribution is largely inferred from glacial extent during the last glacial maximum, water temperature, and the location of the 100 m depth contour (approximate minimum sea level during the past 100000 years). Narrow zones of coastal permafrost are probably present along most arctic coasts.

Coastal and subsea permafrost can be subdivided into four zones (Fig. 6.28), based primarily on water depth and on the dominant processes that operate in those depth zones (after Osterkamp, 2001). Zone 1 covers the inter- and supra-tidal environments of the beaches and flats. Seaward of the intertidal zone, in Zone 2, the seasonal ice cover freezes to the seabed each year, allowing cold winter temperatures to penetrate the water column and reach the sediments. This occurs in water depths of 1.5 to 2 m. Zone 3 covers areas where water depths are too great for the sea ice to freeze to the seabed; however, under-ice circulation may be restricted, with attendant higher salinities and lower seabed temperatures. In Zone 4, "normal" seawater salinity and temperatures prevail, providing a more or less constant regime. Sea-bottom temperatures on the arctic shelves range from -1.5 to -1.8 °C; salinities range from 30 to 34 (Arctic Climatology Project, 1997, 1998).

6.6.2.2. Recent and ongoing changes

There are no ongoing programs to monitor the state of coastal and subsea permafrost, although some effort is being devoted to monitoring the forcing variables and coastal erosion (Brown J. and Solomon, 1999; Rachold et al., 2002; see also Table 16.8). Therefore, most publications addressing changes in the state of subsea permafrost are model-based and speculative. Zones 1 and 2 are the most dynamic, especially in locations where erosion is rapid (e.g., the Laptev and Beaufort

Sea coasts). In these areas, erosion rates of several meters per year cause a rapid transition from terrestrial to nearshore marine conditions. High rates of erosion caused by exposure to waves and storm surges during the open water season lead to deep thermal notch development in cliffs, block failure in the backshore (area reached only by the highest tides), melting of sea-level-straddling massive ice, and possible offshore thermokarst development (e.g., Dallimore et al., 1996; Mackay, 1986; Wolfe et al., 1998). The rate at which destabilization of permafrost in these zones occurs is dependent on the erosion rate, which in turn varies according to storm frequency and severity (Solomon et al., 1994) and the presence or absence of sea ice. The degree to which permafrost destabilization affects erosion remains conjectural. Thaw subsidence in Zones 2 and 3 that accompanies melting of excess ice (ice that is not in thermal equilibrium with the existing soil–ice–air configuration) in the nearshore provides accommodation space for sediments produced by erosion. Thus, there is a potential feedback between high erosion rates and thaw subsidence, but there are few observations to support this hypothesis. An analysis of time series of erosion measurements and environmental forcing (e.g., weather, storms, freeze-thaw cycles) in the Beaufort Sea area did not reveal any trends, but showed pronounced interannual and decadal-scale variability (e.g., Solomon et al., 1994).

Sea-ice thickness plays a major role in the development of subsea permafrost within Zone 2. However, none of the recent analyses of historic data on sea-ice thickness

Ice thickness (cm)

Fig. 6.29. Maximum thickness of landfast ice measured at four coastal locations in the Canadian Arctic, plotted as (a) annual values and (b) five-year moving averages (data from the Canadian Ice Service, 2003).

in the Arctic (e.g., Rothrock et al., 1999; Winsor, 2001, see also section 6.3.2) addresses the state of the sea ice that forms very close to the coast (Manson et al., 2002), since the coastal waters are too shallow for submarines. Time series of ice thickness measurements from several coastal locations extending back to the late 1940s are available from the Canadian Ice Service (Wilson, pers. comm., 2003). Polyakov et al. (2003) describe a similar dataset from Russia. Neither dataset shows any trend over the period of record, which is dominated by large interannual fluctuations. Figure 6.29 illustrates the variability in the annual maximum thickness of landfast ice measured at four coastal locations in the Canadian Arctic. Smoothing the data with a five-year moving average reveals some similarities between the stations.

Air-temperature changes in the Arctic are well documented, and many studies have examined the impact of these changes on the active-layer thickness and temperature of terrestrial permafrost, however, there are no equivalent multi-year studies for coastal permafrost.

Seabed temperature is a critical upper boundary condition for subsea permafrost. Decadally averaged temperatures (1950s–1980s) for various water depths in the Arctic Ocean are available from the National Snow and Ice Data Center in Boulder, Colorado (Arctic Clima-

tology Project, 1997, 1998). These data indicate that there have been decadal-scale changes of a degree or more in the temperatures of shallow water, and smaller changes in deeper water (Fig. 6.30). Interdecadal variability is apparent along the Beaufort Shelf (warmer in the 1960s and 1980s) and the Laptev Sea Shelf (cooler in the 1980s).

6.6.2.3. Projected changes

The stability of coastal and subsea permafrost in a changing climate depends directly on the magnitude of changes in water temperature and salinity, air temperature, sea-ice thickness, and coastal and seabed stability. It is difficult to extract the relevant projections of environmental forcing (subsurface and seabed water temperatures in particular) from any of the scenarios generated by climate models. In general terms:

• The projected increase in air temperature (section 4.4.2) is likely to increase backshore thermokarst development, resulting in more rapid input of material to, and sediment deposition in, the coastal environment. Increased air temperatures will also tend to increase permafrost instability in Zone 1, especially in the supra-tidal environment, and will probably result in increased coastal water temperatures. These factors, cou-

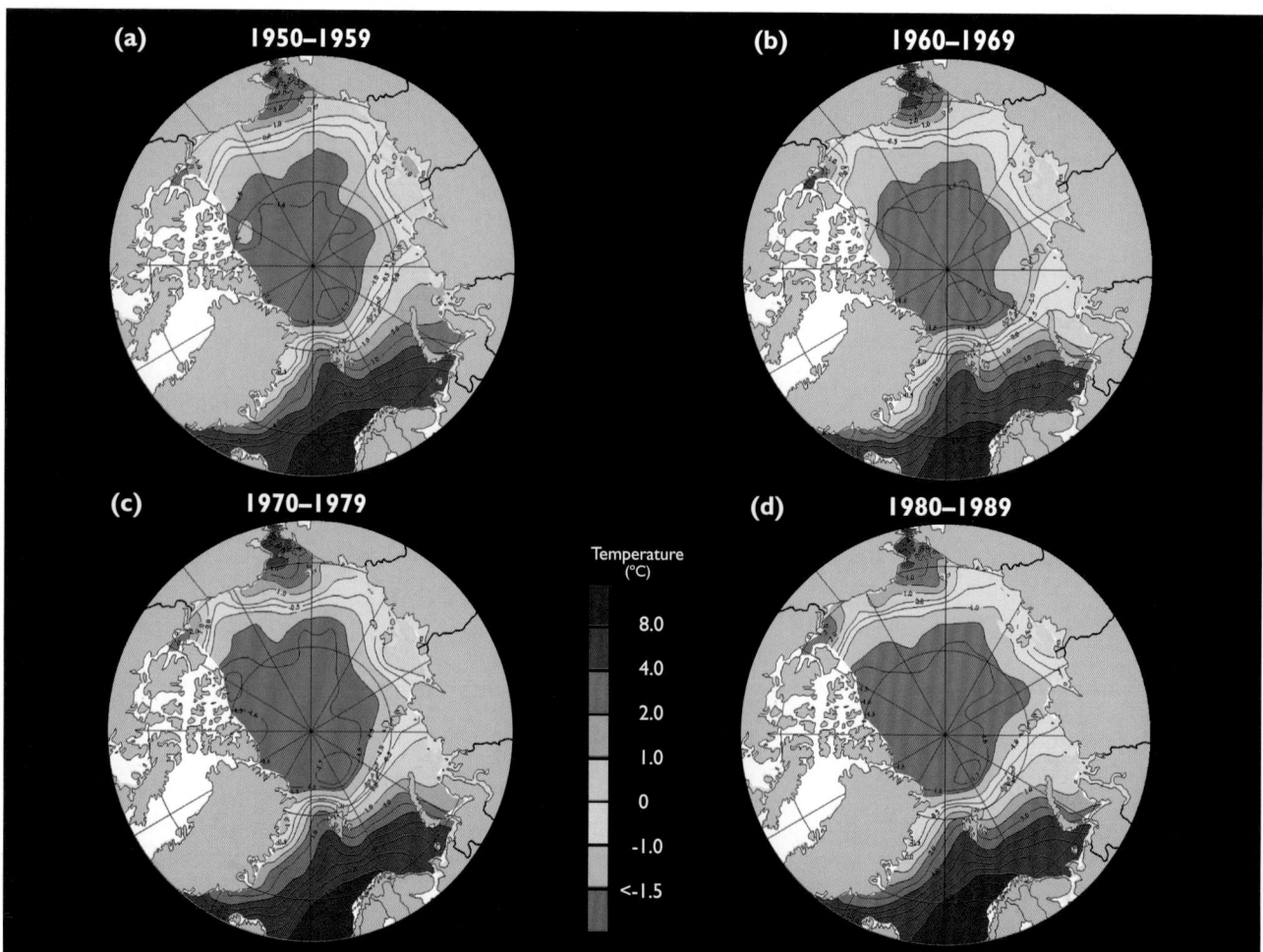

Fig. 6.30. Average decadal summer ocean temperature at 10 m depth from (a) the 1950s to (d) the 1980s (data from the Arctic Climatology Project, 1998).

pled with decreased sea-ice thickness, are likely to cause more rapid warming of permafrost in Zones 1 and 2.

- Projected increases in sea level (section 6.9.3) are likely to shift the location of Zones 1 and 2 to higher elevations, resulting in increased rates of erosion and attendant increases in instability of coastal permafrost.
- Longer open-water seasons, resulting from projected decreases in sea ice cover (section 6.3.3), are likely to expose coastal environments to more storms, resulting in increased rates of backshore erosion. This will lead to higher rates of nearshore deposition in localized areas, with attendant possible permafrost aggradation (e.g., Dyke and Wolfe, 1993). However, rapid backshore erosion is also likely to increase the rate at which terrestrial permafrost is exposed to coastal conditions, leading to warming of terrestrial permafrost in the backshore environment.
- Projected changes in fluvial inflow patterns (section 6.8.3) are likely to change the nearshore and coastal salinity and temperature regimes. Lower flow rates and changes in the timing and duration of floods will affect rates of erosion and deposition, and also the salinity and temperature of the coastal ocean. Higher water temperatures will increase permafrost destabilization; lower flow rates could result in higher coastal salinity, which also would increase rates of permafrost thaw.
- Changes in the thickness and extent of sea ice are likely to affect Zone 3 in that there may be less restriction of circulation as sea-ice conditions become less severe. This would inhibit brine formation in bays, reducing rates of permafrost thawing by an unknown amount. Changing sea-ice regimes will also affect pressure-ridge development, which will change under-ice circulation, but the direction of change is uncertain. It is likely that the effects will be local.
- Changes in sea-bottom water temperature and salinity could possibly occur in Zones 3 and 4, although most AOGCMs do not explicitly project values for these variables. Thawed sediments above the permafrost surface will buffer the effect: a change of 1 °C would take several decades to propagate from the seabed surface to the upper permafrost boundary 10 to 100 m below the seabed surface (Taylor, pers. comm., 2002). Gas hydrates within and beneath the subsea permafrost will also be buffered from the immediate effects of changing seabed conditions and in the near term may be relatively unaffected. Over longer time periods (100 years or more), there is potential for increased instability of subsea gas hydrates in shallow waters of the Arctic. In water deeper than 200 m, other gas hydrates (unrelated to permafrost) that are also susceptible to destabilization by increased bottom-water temperatures may exist close to the seabed surface.

6.6.2.4. Impacts of projected changes

On other parts of the physical system

Decreases in the stability of coastal permafrost are likely to result in greater nearshore thaw subsidence and increased rates of coastal erosion. This will introduce greater sediment loads to the coastal system; higher levels of suspended sediment and changes in depositional patterns may ensue. Increased erosion rates will also result in greater emissions of CO_2 from coastal and nearshore sources, and increased emission rates of CH_4 from terrestrial permafrost. Over the long-term, destabilization of intra-permafrost gas hydrates is likely to enhance climate change.

On ecosystems

Changing deposition patterns and suspended sediment loads along the coast are likely to have impacts on marine ecosystems, including anadromous fish migration, phytoplankton blooms, and benthic communities. Negative or positive impacts are possible. Increased suspended material may increase nutrients, resulting in higher productivity in nutrient-limited systems. Conversely, increased suspended material lowers light levels, resulting in lower productivity. The potential impacts of changes in subsea and coastal permafrost on marine ecosystems are discussed further in Chapter 9.

On people

Decreases in the stability of coastal permafrost will have an impact on coastal infrastructure. Increased erosion rates, caused in part by nearshore thaw subsidence, are likely to affect communities and industrial facilities situated close to the coast. Permafrost thawing and subsidence could affect pipelines in nearshore and coastal environments in excess of their design specifications. Warming and/or thawing permafrost is likely to reduce the foundation strength of wharves and associated pilings. In deeper water (Zones 3 and 4), permafrost warming could affect design considerations for hydrocarbon production facilities, including casing strings and platforms anchored to the seabed. Chapter 16 addresses specific infrastructure issues associated with changes in coastal permafrost.

6.6.2.5. Critical research needs

A circumpolar program to monitor changes in the coastal and offshore cryosphere is required, as is a better understanding of the processes that drive those changes. The Arctic Coastal Dynamics (ACD) project, sponsored by the International Arctic Science Committee and the International Permafrost Association, is promoting the need for such studies. At present, there is no monitoring of coastal and subsea permafrost, and this lack represents a critical gap in the understanding of coastal stability in the Arctic. The absence of monitoring is a result of the difficulty in working in arctic coastal environments, particularly in Zones 1 and 2. Equipment for measuring temperatures throughout the year must be placed in such a way that cables are not jeopardized by storms and sea ice. The technology exists, but it is more expensive than that used for similar measurements on land.

A comprehensive understanding of coastal permafrost processes, including the interaction between storms and permafrost, is needed. Heat convection is thought to play a major role in coastal permafrost thawing during storms after the thawed overlying material is removed (Kobayashi et al., 1999). However, given the difficulty of making measurements at the shoreline under storm conditions, there are no observations supporting this hypothesis. Laboratory studies could play a role in this regard. Thaw subsidence rates can exceed the rate of eustatic sea-level rise (rise due to changes in the mass of ocean water, see section 6.9.1), and are therefore thought to contribute to coastal erosion, at least at a local scale. However, there are few documented observations of the magnitude of thaw subsidence and/or its role in coastal erosion.

The role of brine exclusion and convection in enhancing coastal and subsea permafrost degradation requires further investigation.

Finally, the gas hydrates in coastal and subsea permafrost require further study in order to evaluate their stability over the range of future climate change scenarios produced by climate models.

6.7. River and lake ice

6.7.1. Background

Ice cover plays a fundamental role in the biological, chemical, and physical processes of arctic freshwater systems (e.g., Adams, 1981; Magnuson et al., 1997; Prowse, 2001; Schindler et al., 1996; Willemse and Tornqvist, 1999; see also section 8.2). In particular, freshwater ice is integral to the hydrological cycle of northern systems. The duration and composition of lake ice, for example, controls the seasonal heat budget of lake systems. This in turn determines the magnitude and timing of evaporation from these systems, ranging from small ponds to large lakes; storage levels in the latter also control the flow of some of the major arctic rivers, such as the Mackenzie. Similarly, river ice has a significant influence on the timing and magnitude of extreme hydrological events, such as low flows (Prowse and Carter, 2002) and floods (Prowse and Beltaos, 2002). Many of these hydrological events are due more to in-channel ice effects than to landscape runoff processes (Gerard, 1990; Prowse, 1994).

Lake ice and river ice also serve as climate indicators, and long-term records of these variables provide useful proxy climate data. For some sites in northern Finland and Siberia, the dates of ice formation and breakup have been recorded since the 16th century (Magnuson et al.,

2000). Given the proxy potential of freshwater ice, considerable recent research has focused on the use of remote sensing for documenting ice phenology (e.g., Wynn and Lillesand, 1993) and on the hemispheric process-based modeling of lake-ice patterns for assessing the effects of climate variability and change (e.g., Walsh et al., 1998).

Freeze-over processes differ significantly between lakes and rivers, although the timing of each depends on the magnitude of the open-water heat storage and the autumn rate of cooling. The stratigraphy of lake ice is relatively simple, usually consisting of clear columnar ice, an intermediate layer of translucent granular ice, and a surface layer of snow (e.g., Adams, 1981). In contrast, river ice forms by the dynamic accumulation of various ice forms and is characterized by a more complex vertical and horizontal ice structure (e.g., Prowse, 1995). The thickness and physical characteristics (e.g., optical and mechanical) of the ice cover exert significant control over the thermal and mass balance of the underlying water bodies, and produce important hydrological responses. These in turn affect a number of chemical and biological processes as discussed in section 8.2.

Freeze-up is controlled by a combination of atmospheric heat fluxes. Of all meteorological variables, freeze-up timing correlates best with air temperature in the preceding weeks to months (e.g., Palecki and Barry, 1986; Reycraft and Skinner, 1993). In the case of lakes, area and depth are important determinants of the heat budget and therefore the timing of freeze-up (Stewart and Haugen, 1990). Depth is not as important a factor in determining the timing of lake-ice breakup as it is for freeze-up (Vavrus et al., 1996; Wynn et al., 1996); the timing of lake breakup is determined more by the energy balance characterizing the melt period leading up to the event (e.g., Anderson et al., 1996; Assel and Robertson, 1995). Ice breakup on rivers is a more complex process than that on lakes and, because it is not as strongly related to a single meteorological variable such as air temperature, it is less valuable as a climatic indicator (Barry, 1984). For example, the primary determinant of mechanical strength is insolation-induced decay, which is dependent on solar radiation and ice-cover composition (Prowse and Demuth, 1993). The latter, which can be influenced by snow loading and the generation of surface "snow-ice", also controls the surface albedo and the effectiveness of insolation in reducing ice strength. Ice breakup on northern rivers usually coincides with spring melt of the catchment snow cover and produces the major hydrological event of the year (Church M., 1974; Prowse, 1994).

The most common approach to projecting breakup timing on lakes and rivers employs an air temperature index, such as accumulated degree-days, which reflects the amount of ice deterioration and, in the case of rivers, the magnitude of the snowmelt flood wave (Prowse, 1995). While recognizing that numerous physical and climatological factors influence the rates and timing of ice formation and decay processes, Magnuson et al. (2000) estimated the change in freeze-up and breakup dates relative to a change in air temperature in the preceding weeks or months to be approximately 5 days/°C based on results from a number of lake- and river-ice case studies. Estimating the severity of breakup requires the use of a wider range of meteorological factors (e.g., Gray and Prowse, 1993).

Ice jams frequently form on rivers, and, because of their high hydraulic resistance, produce flood levels that often far exceed those for equivalent discharge under open-water conditions, usually with a high recurrence interval (Beltaos, 1995; Prowse and Beltaos, 2002). Dredging, blasting, and aerial bombing have been used to try to dislodge ice jams, with varying degrees of success, but these approaches often produce local environmental damage (Burrell, 1995).

One of the most persistent effects of ice on river hydrology is its influence on water levels. Although the additional hydraulic resistance of a stable ice cover tends to elevate channel water levels, the greatest effect occurs when the ice cover is hydraulically rough, as it often is following a dynamically active freeze-up period. In combination with rapid freezing, hydraulic staging of the ice cover can extract significant amounts of water such that a period of low flow prevails. The release of water stored during this period can significantly augment the spring freshet (Prowse and Carter, 2002).

6.7.2. Recent and ongoing changes

Although changes in the thickness and composition of freshwater ice or in the severity of freeze-up and breakup events can have significant implications for numerous physical and ecological processes, available documentation of past changes is largely limited to simple observations of the timing of freeze-up and breakup. Such data are, however, useful indicators of climate change, although long-term records are relatively scarce. Magnuson et al. (2000) assessed freeze-up and breakup trends for Northern Hemisphere lakes and rivers that had records spanning at least 100 years within the period from 1846 to 1995 (only three sites had records beginning prior to 1800). Of the 26 rivers and lakes included in the study, most are located south of 60° N. Over the 150-year period, average freeze-up dates were delayed by 5.8 d/100 yr and average breakup dates advanced by 6.3 d/100 yr, corresponding to an increase in air temperature of about 1.2 °C/100 yr. Magnuson et al. (2000) further observed that the few available longer time series indicate that a trend of reduced ice cover began as early as the 16th century, although rates of change increased after approximately 1850. Figure 6.31 shows the time series of freeze-up and breakup dates for a sample of the rivers and lakes studied. The high-latitude water bodies in Fig. 6.31 (the Mackenzie River in Canada and Kallavesi Lake in Finland) both show general trends toward later freeze-up dates, although decadal-scale variations make trends sensitive to the

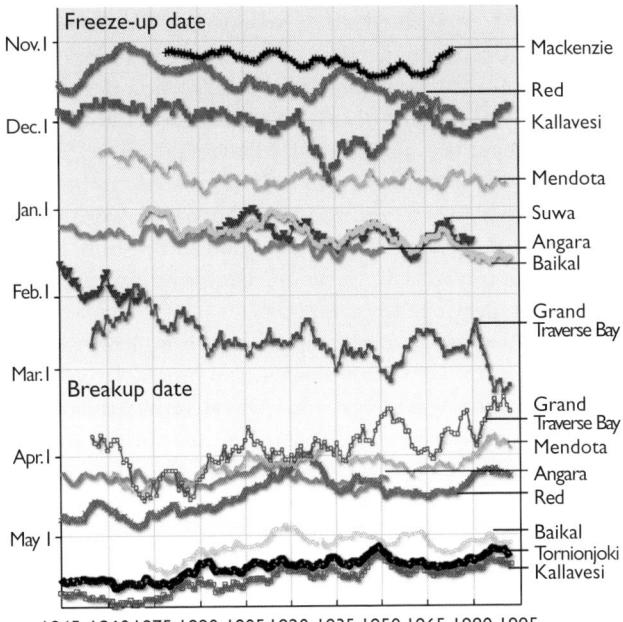

Fig. 6.31. Ten-year running means of freeze-up (top) and breakup (bottom) dates of selected lakes and rivers in the Northern Hemisphere: Mackenzie River (Canada), Red River (Canada), Kallavesi Lake (Finland), Lake Mendota (U.S.), Lake Suwa (Japan), Angara River (eastern Russia), Lake Baikal (eastern Russia), Grand Traverse Bay (Lake Michigan, United States), and Tornionjoki River (Finland) (Magnuson et al., 2000).

beginning and end dates of the calculations. Similar sensitivity is found in the trend toward earlier breakup of Kallavesi Lake, and in the breakup date of the Tanana River at Nenana, Alaska, for which the time series in Fig. 6.32 shows interannual and decadal-scale variations superimposed on the trend toward earlier breakup.

Most of the very long-term records analyzed by Magnuson et al. (2000) are geographically diverse and give little insight into potential regional trends. Moreover, few sites are even located above 60° N. To gain a better understanding of recent and ongoing changes at high latitudes, records shorter than 150 years must be examined. The most comprehensive regional evaluation of freeze-up and breakup dates that

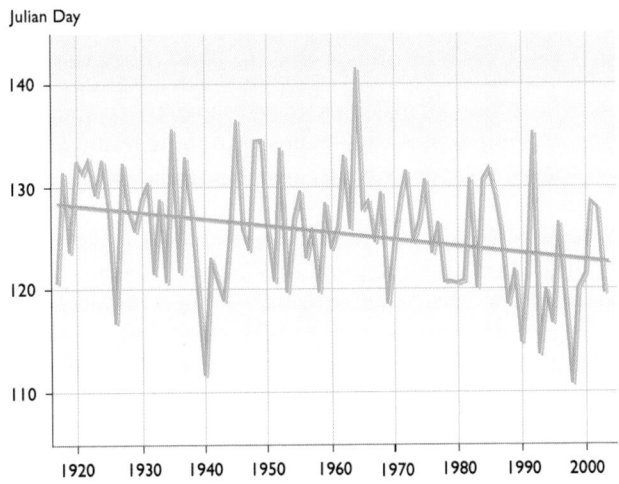

Fig. 6.32. Time series of the breakup date of the Tanana River at Nenana, Alaska (modifed from W. Chapman using data from the National Snow and Ice Data Center, 2003).

includes areas of the Arctic and subarctic was conducted by Ginzburg et al. (1992) and Soldatova (1993), using data from about 1893 to 1985 for homogenous hydrological regions of the Former Soviet Union. Although appreciable interdecadal variability was found, significant long-term spatial patterns and temporal trends in freeze-up and breakup dates were identified for the period. The most significant regional trend was toward later river-ice freeze-up dates in the European part of the Former Soviet Union and western Siberia. A weaker but still significant trend toward earlier freeze-up dates was found for portions of rivers (e.g., the Yenisey and Lena) in central and eastern Siberia (Ginzburg et al., 1992). A similar broad-scale spatial pattern is evident for breakup dates (Soldatova, 1993). Breakup on major rivers in the European part of the Former Soviet Union and western Siberia advanced by an average of 7 to 10 d/100 yr, resulting in a reduction in ice-season duration of up to a month. However, some rivers in central and eastern Siberia exhibited an opposing trend: later breakup dates and hence an increase in ice-season duration. Freeze-up and breakup dates were well correlated ($r^2=0.6$–0.7) with the mean air temperature in the preceding autumn and spring months, respectively. A gradual advancement in the date of breakup has been documented for lakes in southern Finland (Kuusisto and Elo, 2000) and rivers in northern Sweden/Finland and Latvia (Kuusisto and Elo, 2000; Zachrisson, 1989).

In northwestern North America, studies of the Tanana River (1917–2000; Sagarin and Micheli, 2001) and the Yukon River (1896–1998; Jasek, 1998) indicate that the average date of breakup has advanced by approximately 5 d/100 yr. This trend is characterized by a number of interdecadal cycles. Zhang X. et al. (2001) conducted a Canada-wide assessment of river freeze-up, breakup, and ice duration using records spanning 50 years or less. The major spatial distinction in breakup timing was between eastern and western sites, with the western sites (e.g., the Yukon and other western rivers) showing trends towards earlier breakup dates. Notably, there was also a nation-wide trend to earlier freeze-up dates.

Smith (2000) conducted a study of shorter-term records from nine major arctic and subarctic rivers in Russia. Some trends were opposite to those found for the longer-term and broader regional studies of Ginzburg et al. (1992) and Soldatova (1993), possibly because of the shorter record lengths (54 to 71 years) or differences resulting from site-specific factors. In particular, earlier rather than later freeze-up dates were found for rivers west of and including the Yenisey, whereas later freeze-up dates were observed for rivers in far eastern Siberia. Although Smith (2000) found no statistically significant shifts in breakup timing, there were significant shifts toward an earlier melt onset, producing a trend toward a longer period of pre-breakup melt. According to breakup theory, a longer melt period favors "thermal" breakups – low-energy events that

are less likely to produce floods and related disturbances (e.g., Gray and Prowse, 1993). Similar analyses, or studies that focus on breakup characteristics beyond simple timing, have not been conducted elsewhere.

6.7.3. Projected changes

Although many case studies of existing data show relationships between the timing of freeze-up and breakup and the preceding autumn and spring air temperatures, such relationships are not necessarily temporally stable. For example, Livingstone (1999) found that the influence of April air temperatures on Lake Baikal breakup dates has varied considerably over the past 100 years, and accounts for only 12 to 39% of the variance in breakup dates. Furthermore, there is no guarantee that such empirical relationships will hold for future climatic conditions, particularly if they are characterized by significant changes in the composition of the major heat fluxes (e.g., Bonsal and Prowse, 2003).

Considering only the projected changes in air temperature, the general pattern of change in freshwater ice will be a general reduction in ice cover on arctic rivers and lakes. This reduction will be greatest in the regions of greatest warming. The warming varies somewhat from model to model, but is generally larger in the northernmost land areas than in the subpolar land areas. None of the five ACIA-designated models projects a cooling over northern terrestrial regions. However, the reduction in river and lake ice may be modified by changes in precipitation (including snowfall) over the 21st century, and the projected changes in precipitation vary substantially from model to model.

Freeze-up and breakup dates are projected to respond more strongly to warming than to cooling because of albedo–radiation feedbacks (e.g., Vavrus et al., 1996). However, changes in winter precipitation will modify this pattern. Increased snowfall should lead to a delay in breakup owing to additions of white ice and longer-lasting higher albedo. Conversely, decreased snowfall should advance breakup owing to lower spring albedo, although reduced insulation in winter could also lead to enhanced ice growth. Projecting specific regional responses requires detailed physical modeling (employing multi-variable meteorological input) of changes not only in winter snow and ice conditions, but also in the open-water heat budgets that strongly influence freeze-up timing and subsequent ice growth. Projections for river ice are even more complex because the heat budgets of contributing catchment flow must be considered, together with changes in the timing and magnitude of flow that control many of the important river-ice hydrological extremes. Prowse and Beltaos (2002) reviewed a range of the complex interacting hydraulic, mechanical, and thermal changes that could result from shifts in temperature and precipitation. In general, as for lake ice, the duration and composition of river-ice cover would change, as would the potential for extreme conditions during freeze-up and breakup.

6.7.4. Impacts of projected changes

On other parts of the physical system

A number of physical, biological, and chemical changes will result from changes in the timing, composition, and duration of lake- and river-ice cover. Some of the most direct changes will be shifts in the thermal and radiation regimes, which can have indirect effects on freshwater habitat and quality (e.g., water temperature and dissolved oxygen). Hydrological processes (e.g. discharge timing, evaporation) are also likely to be affected. For northern peatlands, ice-induced changes in open-water evaporation and resultant water levels are likely to determine whether they become sources or sinks of CO_2 and CH_4 (section 7.5.3). For regions with extensive lake cover and substantial winter water storage, changes in the timing and magnitude of winter snowfall will produce corresponding changes in the winter pulsing of river discharge. Changes in the timing and severity of freeze-up and breakup will alter the hydrological extremes (e.g., low flows and floods) that dominate the flow regime of northern systems. A change in breakup intensity will also alter channel-forming processes, as well as levels of suspended sediment ultimately carried to the Arctic Ocean.

On ecosystems

Biological and chemical changes are likely to result from changes in the timing and duration of lake- and river-ice cover (section 8.4). A change in breakup intensity will affect the supply of floodwater, organic carbon, and nutrients to riparian zones; the ecosystem health of river deltas is particularly dependent on such fluxes.

On people

Winter roads that use the ice cover of interconnected lakes and rivers service extensive areas of the north, particularly those areas being explored or developed by resource extraction industries. Any changes in the thickness, composition, and/or mechanical strength of such ice will have major transportation and financial implications. The greatest economic impact is likely to stem from a decrease in ice thickness and bearing capacity, which could severely restrict the size and load limit of vehicular traffic. Changing ice regimes will also affect shipping operations, particularly on large Russian rivers where icebreakers are employed to extend the shipping season to northern towns and industries. Considering the high operational costs of ice breaking, any reduction in the duration of the ice season or breakup severity should translate into significant cost savings.

Changes in the ice regime will also require changes in operating strategies for hydroelectric installations, both at the generating facility to reduce impacts from ice (e.g., accumulations of frazil, the slushy ice-water mixture that develops when turbulent water starts to freeze), and for management of downstream flows to minimize negative impacts on river ecology and infrastructure. Major economic savings are likely to accrue to hydroelectric facilities if climate change reduces the length of the ice season. However, if the length of the ice season is reduced, the increased time required for freeze-up to a hydraulically stable ice cover will have at least some negative economic consequences (e.g., the necessary reductions in peak generating capacity during the freeze-up period).

6.7.5. Critical research needs

Critical research needs with regard to river and lake ice include improved understanding of the interacting hydrological and meteorological controls on freeze-up and breakup, reliable projections of changes in these controls over the 21st century, and further refinement of models of lake-ice growth and ablation (e.g., Duguay et al., 2003) and river-ice dynamics (e.g., Petryk, 1995) for use in forecasting future conditions. There is a particular need for more credible model projections of precipitation and surface solar radiative fluxes. Snowfall can influence river and lake ice by changing the composition of the ice cover and, through its effects on insulation and insolation, ice growth and ablation rates. Accumulated winter precipitation also determines the magnitude of the spring runoff, which controls the severity of breakup and associated ice-jam flooding. Surface radiative fluxes are key controls of river and lake ice, affecting both rates of ablation and changes in mechanical strength of the ice cover. However, model projections of future radiative fluxes, which will depend strongly on changes in cloudiness, are highly uncertain.

6.8. Freshwater discharge

6.8.1. Background

Many of the linkages between the arctic system and global climate involve the hydrological cycle. Theoretical arguments and models both suggest that net high-latitude precipitation increases in proportion to increases in mean hemispheric temperature (Manabe and Stouffer, 1993; Rahmstorf and Ganopolsky, 1999). Section 6.2.3 showed that precipitation and precipitation minus evapotranspiration (P-E) are projected to increase in the Arctic as GHG concentrations increase. This is supported by the nearly linear relationship between temperature and ice accumulation found in Greenland ice cores over the past 20000 years (Van der Veen, 2002). At the same time, increased freshwater export from the Arctic Ocean may reduce North Atlantic Deep Water formation and Atlantic thermohaline circulation (Broecker, 1997). These changes in Atlantic thermohaline circulation may trigger major climatic shifts. Terrestrial discharge, or river runoff, to the Arctic Ocean may therefore have global implications.

To analyze the variability of the Arctic Ocean's freshwater budget, both the Arctic Ocean watershed and the adjacent territories from which the runoff originates (Fig. 6.33) must be considered. The total area of the

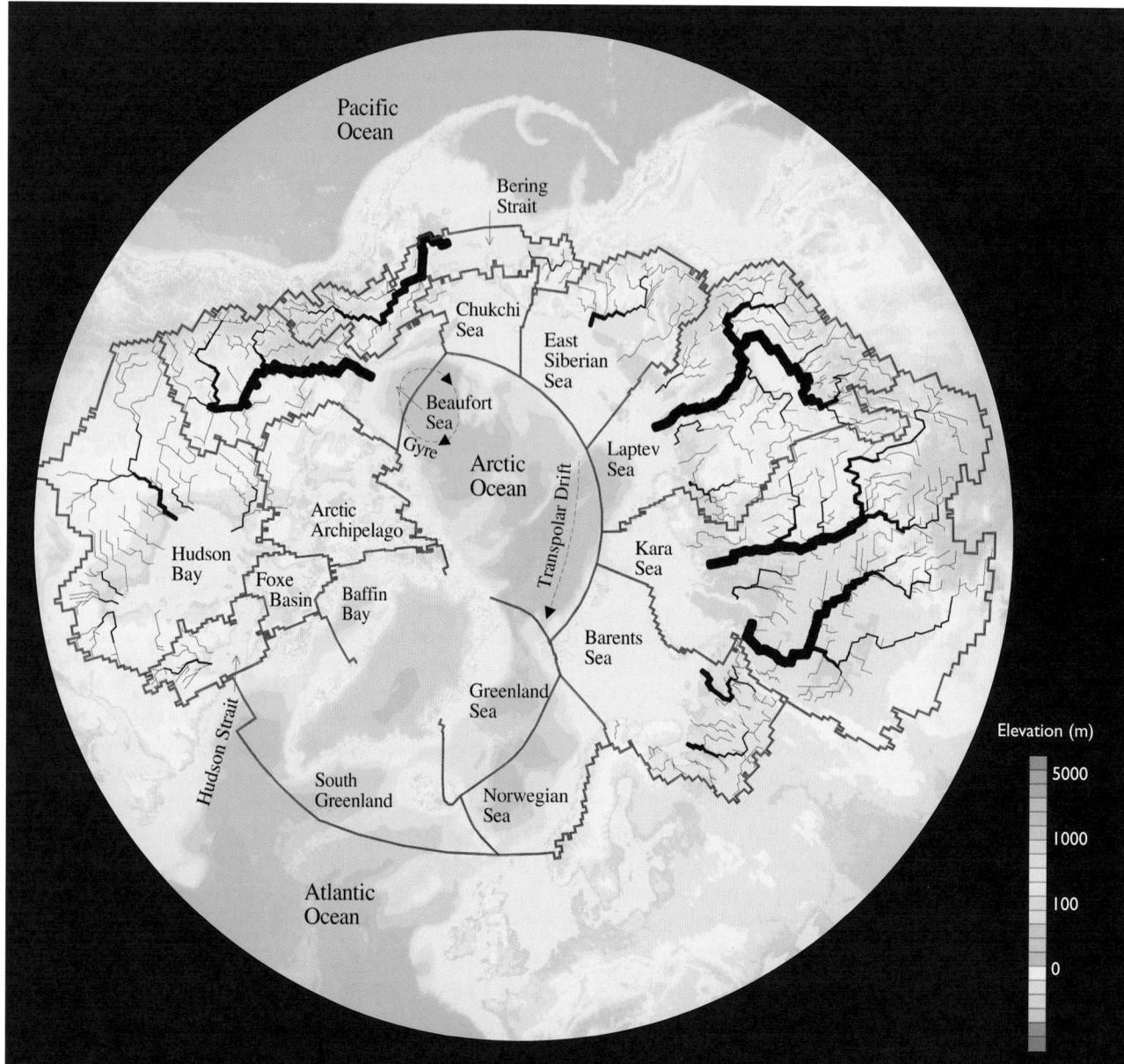

Fig. 6.33. The Arctic "half-hemisphere" showing oceans and shelf seas with catchment areas. Blue lines show river networks with line thickness representing relative discharge. Table 6.11 provides freshwater discharge statistics for different regions of the Arctic Ocean (figure prepared by A. Shiklomanov, 2003).

Arctic Ocean drainage basin, together with the adjacent Hudson Bay and Bering Sea drainage basins, is about 24 million km² (Forman et al., 2000; Shiklomanov I. et al., 2000). This huge area includes a wide variety of surface types and climate zones, from semiarid regions in the south to polar deserts in the north.

Table 6.11. Mean annual discharge of freshwater to the Arctic Ocean for the period 1921 to 2000 (Shiklomanov I. et al., 2000, updated by A. Shiklomanov, 2003).

Basin	Discharge (km³/yr)	Coefficient of variation	Maximum discharge km³	year	Minimum discharge km³	year
Bering Strait	301	0.09	362	1990	259	1999
Hudson Bay and Strait	946	0.09	1140	1966	733	1989
North America (Arctic Ocean drainage basin only)	1187	0.09	1510	1996	990	1953
North America including Hudson Bay basin	2133	0.07	2475	1996	1800	1998
Europe	697	0.08	884	1938	504	1960
Asia (Arctic Ocean drainage basin only)	2430	0.06	2890	1974	2100	1953
Arctic Ocean drainage basin	4314	0.05	4870	1974	3820	1953
Arctic Ocean drainage basin and Hudson Bay basin	5250	0.04	5950	1974	4700	1953

Number of gauges

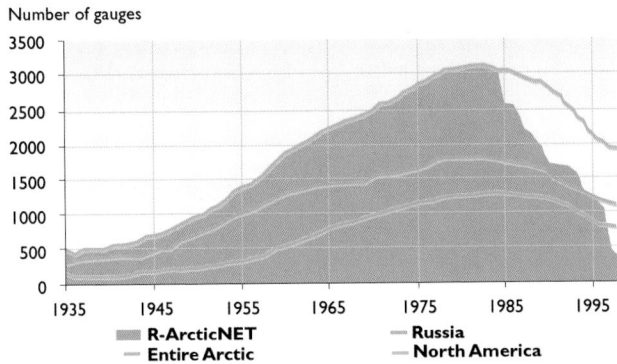

Fig. 6.34. Time series of the number of river discharge gauges in the Arctic Ocean drainage basin, the Russian Arctic, and the North American Arctic. Shaded area represents the number of stations with data that have been included in the R-ArcticNET database of the Arctic-RIMS project (Shiklomanov A. et al., 2002).

The spatial distribution of the hydrological monitoring network in the Arctic is extremely uneven. The greatest numbers of stations are located in Europe, the southern part of western Siberia, and the Hudson Bay drainage basin, while the northern part of the Arctic, including Greenland, the arctic islands, and coastal regions, is essentially unmonitored. Monitoring capacity in both North America and Eurasia peaked in 1985, when the percentage of Arctic Ocean drainage area monitored by gauges was 50.2% in North America, 85.1% in Asia, and 70.7% in Europe (but 0% in Greenland). Since 1985, the number of hydrometric stations has decreased significantly, owing to budget constraints and (in Siberia) population losses. The total number of gauges throughout the Arctic is now 38% lower than in 1985 (Fig. 6.34); in 1999, the discharge monitoring network had the same number of gauges as in 1960 (Shiklomanov A. et al., 2002).

The total freshwater discharge from the land area into the Arctic Ocean is the sum of river discharge into the ocean, glacier and ice sheet discharge, subsurface water flows (mainly from the freeze–thaw cycle in the active layer of permafrost soils), and groundwater flows. Most of the glacier streamflow enters the subpolar seas (e.g., Greenland Sea, Baffin Bay/Davis Strait, Gulf of Alaska) rather than the Arctic Ocean. Subsurface and groundwater flow to the Arctic Ocean is considered to be orders of magnitude lower than river discharge (Grabs et al., 2000), but is very important during winter, when other sources of discharge are substantially reduced. Seasonally frozen ground also affects the interactions between surface runoff, subsurface runoff, and subsurface storage (Bowling et al., 2000).

Estimates of river discharge to the Arctic Ocean are highly dependent on the specification of the contributing area (Prowse and Flegg, 2000). According to a contemporary assessment based on available hydrometric and meteorological information (Shiklomanov I. et al., 2000), the total long-term river discharge from the entire Arctic Ocean drainage basin (including Hudson Bay but not Bering Strait) is 5250 km³/yr (46% from

Asia, 41% from North America and Greenland, and 13% from Europe). Excluding Hudson Bay, direct river discharge to the Arctic Ocean is 4320 km³/yr (1190, 2430, and 700 km³/yr from North America, Asia, and Europe respectively). Table 6.11 presents statistics of the annual discharge to the Arctic Ocean from the different drainage areas. The World Climate Research Programme (WCRP, 1996) provided a similar estimate of freshwater flux from land areas to the Arctic Ocean (4269 km³/yr with 41% attributed to unmeasured discharge). Earlier estimates of total river discharge to the Arctic Ocean were lower: 3300 km³/yr (Aagaard and Carmack, 1989); 3500 km³/yr (Ivanov, 1976); and 3740 km³/yr (Gordeev et al., 1996). These earlier studies underestimated the freshwater discharge to the Arctic Ocean because they did not fully consider the unmonitored discharges from the mainland and/or arctic islands. Shiklomanov I. et al. (2000) discussed methods for estimating runoff from unmonitored areas. The total river discharge to the Arctic Ocean (~4300 km³/yr) is several times the net input of freshwater from P-E over the Arctic Ocean (assuming an average P-E of 10 to 15 cm/yr (section 6.2) and an area of about 10 million km²).

The spatial distribution of runoff across the Arctic is highly heterogeneous. When standardized to units of runoff volume per area of a drainage basin, the smallest values (<10 mm/yr) are found in the prairies of Canada and the steppes of West Siberia. The highest values (>1000 mm/yr) are found in Norway, Iceland, and the mountain regions of Siberia (Lammers et al., 2001).

Rivers flowing into the Arctic Ocean are characterized by very low winter runoff, high spring flow rates driven by snowmelt, and rain-induced floods in the summer and autumn. The degree of seasonality depends on climate conditions, land cover, permafrost extent, and level of natural and artificial runoff regulation. Snowmelt contributes up to 80% of the annual runoff in regions with a continental climate and continuous permafrost, such as the northern parts of central and eastern Siberia, and contributes about 50% of the annual runoff in northern Europe and northeastern Canada (AARI, 1985). Most eastern Siberian and northern Canadian rivers with drainage areas smaller than 10⁵ km² that flow through the continuous permafrost zone have practically no runoff during winter because the supply of groundwater is so low.

6.8.2. Recent and ongoing changes

Shiklomanov I. et al. (2000) calculated time series of river discharge to the Arctic Ocean from the individual drainage areas between 1921 and 1999 based on hydrometeorological observations. Figure 6.35 summarizes the temporal variations by region. Cyclical discharge variations with relatively small positive trends are evident for the Asian and Northern American regions, while river discharge to Hudson Bay decreased by 6% over the period. According to this assessment, annual

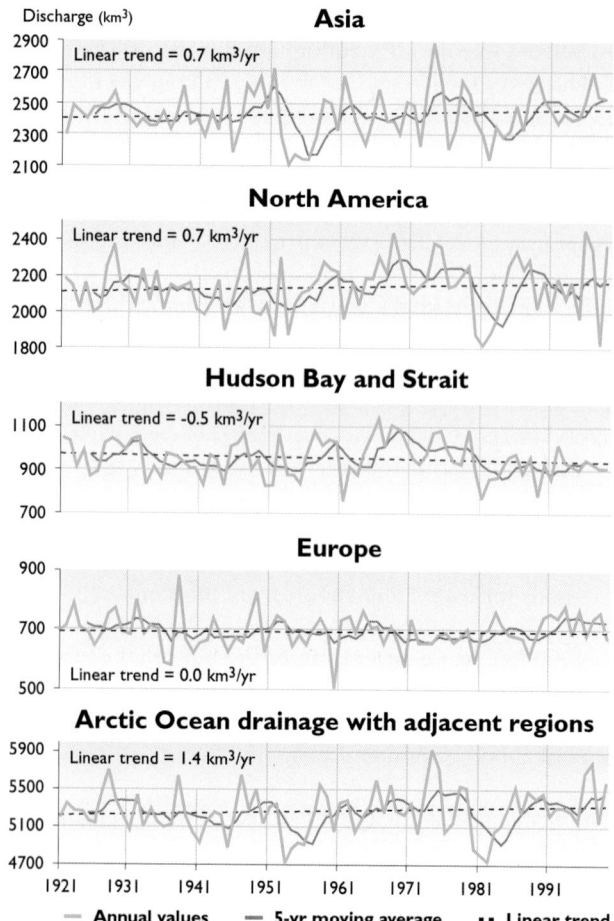

Fig. 6.35. Time series of river discharge to the Arctic Ocean from different parts of the drainage basin between 1921 and 1999 (Shiklomanov I. et al., 2000).

freshwater discharge to the Arctic Ocean increased by 112 km³ between 1921 and 1999.

In a study that used long-term hydrometric observations for the Eurasian Arctic, Peterson et al. (2002) found that the annual discharge to the Arctic Ocean from the six largest Eurasian rivers increased by 7% between 1936 and 1999. Although the increase is not monotonic (Fig. 6.36), it is statistically significant. The 7% increase in the discharge of these rivers implies that the annual freshwater inflow to the Arctic Ocean is now 128 km³ greater than it was in the mid-1930s.

Variations in river discharge occur primarily in response to variations in atmospheric forcing, particularly air temperature and precipitation. The ten largest arctic river basins all show an increase in air temperature during the past 30 years (New et al., 2000), as shown in Table 6.12. The greatest runoff increase is observed in the large European rivers (e.g., Severnaya Dvina, Pechora), where significant increases in precipitation have occurred. The largest Siberian river basins (e.g., the Yenisey and Lena), in which permafrost is widespread, show an increase in runoff despite a decreasing trend in precipitation. Factors that may have contributed to this increase include a shorter winter period, faster spring snowmelt (reducing evaporation and infiltration losses),

thawing permafrost, and saturated soils resulting from an increase in groundwater storage. However, there are large uncertainties in the precipitation data and the calculated changes in precipitation. The relationship between changes in precipitation and river discharge clearly requires additional investigation.

In addition to its correlation with regional temperature, discharge from the large Siberian rivers is correlated with global mean surface air temperature and with the NAO index (Peterson et al., 2002), as shown in Fig. 6.36. The linkage between the NAO (or its broader manifestation, the AO) and Eurasian temperature and precipitation has been documented by Thompson and Wallace (1998) and Dickson et al. (2000). The strengthened westerlies characteristic of the positive phase of the NAO enhance the transport of moisture and relative warmth across northern Europe and northern Asia (Semiletov et al., 2002). It is apparent from Fig. 6.36 that the NAO (or the AO) should be considered in diagnoses of variations in Eurasian river discharge over interannual to decadal timescales.

Savelieva et al. (2000) related changes in the seasonality of Siberian river discharge in the second half of the 20th century to a climate shift that occurred in the 1970s. A shift in climatic conditions over the Pacific Ocean and Siberia around 1977 has been well-documented (Hare and Mantua, 2000; Minobe, 2000; Overland et al., 1999).

When analyzing seasonal trends in river discharge, the downstream gauges of most of the large rivers cannot be used because of the impoundments within their basins. Even small reservoirs can have a significant impact during times of low flow (winter). For example, Ye B. et al. (2003) showed that a relatively small reservoir in the Lena Basin significantly changed the winter

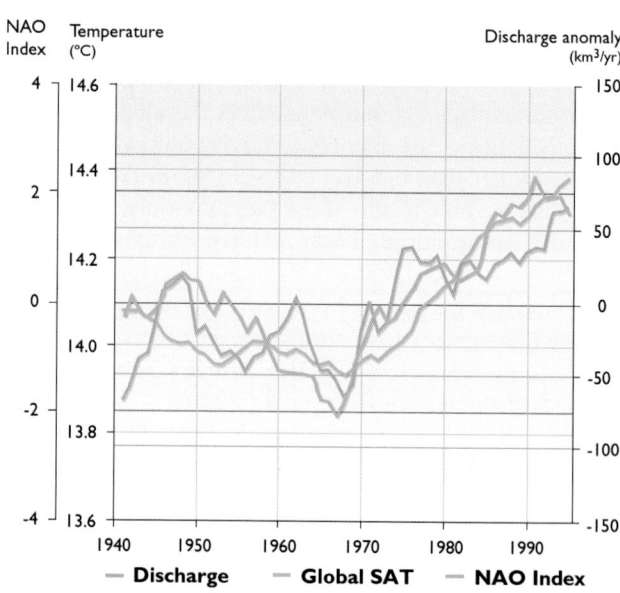

Fig. 6.36. Ten-year running averages of the Eurasian Arctic river discharge anomaly (departure from the 1936–1999 mean), global mean surface air temperature (SAT), and the winter (Dec–Mar) NAO index (Peterson et al., 2002).

discharge regime at downstream locations. Therefore, in a recent attempt to identify seasonal variations and changes in the hydrological regime of the Eurasian Arctic, Georgievsky et al. (2002) analyzed data from 97 rivers with monthly discharge records exceeding 50 years and no significant human influence. The results show that between 1978 and 2000, winter river runoff increased relative to its longer-term mean across most of the region (Fig. 6.37c). Significant increases (10 to 30%) in winter and summer–autumn runoff occurred in rivers located in the part of European Russia that drains into the Arctic Ocean (Fig. 6.37b, c). Even greater changes occurred in Siberia. Winter runoff increased by up to 40 to 60% in the Irtysh Basin and in southeastern Siberia, and by up to 15 to 35% in northern Siberia.

Figure 6.37d shows that the annual runoff in the Eurasian part of the arctic drainage basin has been significantly higher during the last 20 to 25 years, excluding south-central Siberia where annual runoff has been lower than the longer-term average. The most significant increase (>30%) has occurred in European Russia and the western part of the Irtysh Basin. The runoff in the Lena Basin has increased by up to 15 to 25% in the south and by 5 to 15% in the north.

6.8.3. Projected changes

General conclusions about the influence of projected climate change on river discharge to the Arctic Ocean are drawn from a synthesis of studies by various investigators from different countries (e.g., Shiklomanov A., 1994, 1997; Van der Linden et al., 2003). Climate change scenarios used in all of these estimates were generated by various general circulation models (GCMs) forced with doubled atmospheric CO_2 concentrations, with transient increases of CO_2, and/or with other forcing from paleoclimatic reconstructions. The climate scenarios generated by the GCMs have been used to force hydrological models of varying complexity (Shiklomanov I. and Lins, 1991). The projections obtained with this approach go beyond the projections of changes in P-E (section 6.2.3), since they include the effects of changing temperature and changes in the atmospheric circulation (winds) projected by the GCMs. The projections are summarized in Table 6.13.

Miller and Russell (1992) performed one of the first such assessments, using scenarios for doubled atmospheric CO_2 concentrations from the Goddard Institute for Space Studies (GISS) and Canadian GCMs as input to a simple water-budget model. This assessment projected increases of 10 to 45% in the discharges of the large Eurasian and North American rivers (Table 6.13).

Shiklomanov A. (1994) projected the impact of climate change on the annual and seasonal discharges of the rivers in the Yenisey drainage basin using a number of GCM scenarios and paleoclimate reconstructions as input to the detailed hydrological model developed by

Table 6.12. Changes in air temperature, precipitation, and runoff in the largest arctic river basins between 1936 and 1996, computed based on a linear trend (compiled by A. Shiklomanov using data from New et al., 2000).

	Period	Change for the period			Permafrost extent (% of total area)
		Air temperature (°C)	Precipitation (mm/yr)	River runoff[a] (mm/yr)	
Severnaya Dvina	1936–1996	0.3	24	37	0
	1966–1996	1.3	62	44	
Pechora	1936–1996	0.5	60	53	31
	1966–1996	1.7	27	30	
Ob	1936–1996	1.2	3.8	6	19
	1966–1996	2.2	-4	1	
Yenisey	1936–1996	1.2	-11	13	71
	1966–1996	2.5	0	27	
Lena	1936–1996	1.1	-5	22	94
	1966–1996	2.1	-24	10	
Indigirka	1936–1996	0.0	-34	17	100
	1966–1996	1.0	-42	1	
Kolyma	1936–1996	0.0	-29	-5	100
	1966–1996	0.6	-36	15	
Yukon	1957–1996	1.6	19	6	90
	1966–1996	2.2	43	13	
Mackenzie	1966–1996	1.4	-6	-5	55
Back	1966–1996	1.7	6	6	100

[a]Change in river runoff is presented as the net change (mm/yr) in precipitation minus evapotranspiration, which is equivalent to total basin runoff (km³/yr) divided by the area (km²) of the drainage basin.

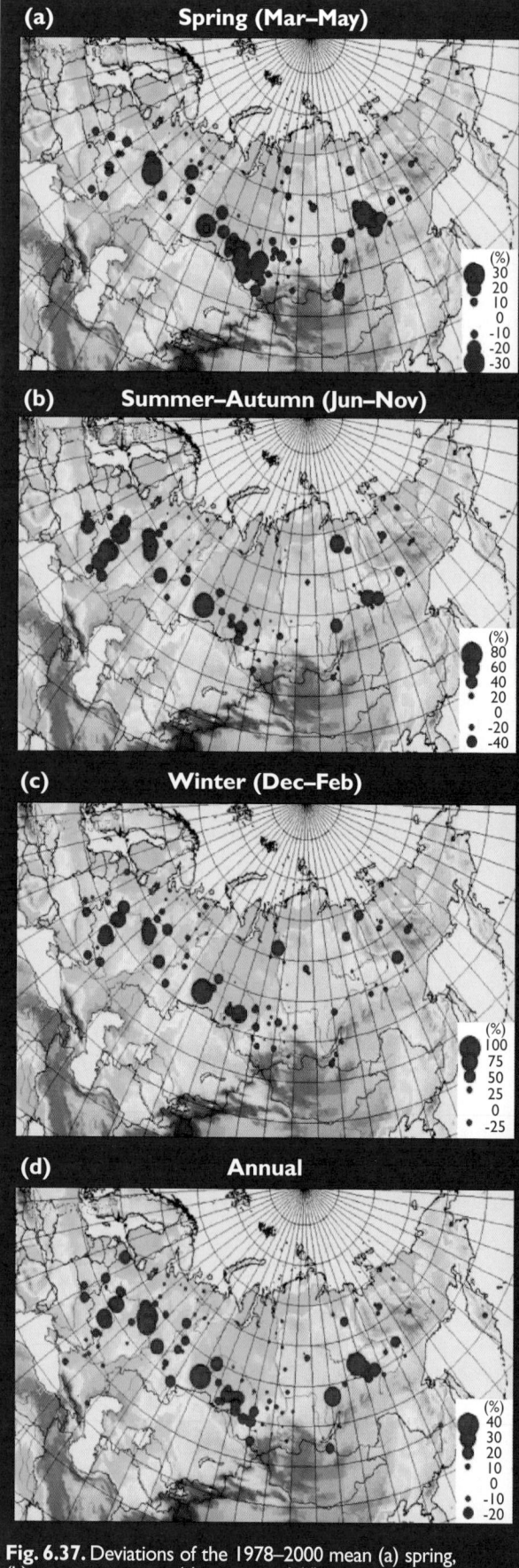

Fig. 6.37. Deviations of the 1978–2000 mean (a) spring, (b) summer–autumn, (c) winter, and (d) annual runoff expressed as a percentage of the long-term mean for each location and season. Red and blue circles denote positive and negative deviations, respectively. The long-term mean is calculated using the entire gauge station history (at least 50 years for all plotted locations). Stations with no deviation in a specific period are not shown. Tan, orange, and brown indicate progressively higher elevations; green indicates low elevations (Georgievsky et al., 2002).

the State Hydrological Institute (Russia) (Table 6.13). According to the more plausible climate scenarios, the mean annual discharge of the Yenisey River is likely to increase by 15 to 20%, and the winter discharge is likely to increase by 50 to 60%. Similar evaluations of river discharge to the Barents Sea drainage basin (Georgievsky et al., 1996) project increases of 14 to 35% and 25 to 46% in the mean annual and winter discharge, respectively (Table 6.13).

Arnell (1999) used six GCM scenarios developed by the Hadley Centre to project changes in runoff from the world's largest rivers between 1961–1990 and 2050. This study projected discharge increases ranging from 3–10% to 30–40% for the largest rivers in the Arctic Ocean drainage basin (Table 6.13). Other assessments (Arora and Boer, 2001; Miller and Russell, 2000; Mokhov et al., 2003) also project significant increases in both the discharge of the largest rivers and in the total river discharge to the Arctic Ocean. Georgievsky et al. (2003) provided projections for the Lena Basin based on a water balance model with a three-layer environment (the active soil layer and two layers of groundwater reservoirs) using input from the HadCM3 model forced with the A2 emissions scenario (section 4.4.1). Annual runoff is projected to increase by 27 mm (12.5%), with higher percentage increases in winter and spring runoff, which would increase the probability of extreme flooding.

The results presented above indicate that, if atmospheric CO_2 concentrations double and the model projections of runoff changes are correct, the total annual discharge to the Arctic Ocean from arctic land areas can be expected to increase by 10 to 20%. The increase in winter discharge is likely to be as high as 50 to 80%. At the same time, 55 to 60% of annual discharge is likely to enter the ocean during the peak runoff season (April–July). It must be emphasized, however, that the B2 emissions scenario (section 4.4.1) used to force the ACIA-designated models does not lead to a doubling of CO_2 concentrations during the 21st century. Relative to the atmospheric CO_2 concentration in 2000 (~370 ppm), the CO_2 concentrations in the B2 emissions scenario increase by about 30% by 2050 and 65% by 2100. (In the A2 emissions scenario, the corresponding increases are about 40% by 2050 and 120% by 2100.)

For this reason, it is not surprising that the 10 to 20% increase in discharge cited above is larger than the ACIA-designated model projections of increases in precipitation and P-E, which are about 10% or slightly less (Fig. 6.2). The projected changes in temperature, precipitation, and river discharge obtained from other forcing scenarios (e.g., Table 6.13) must be tempered accordingly.

6.8.4. Impacts of projected changes

On other parts of physical system

Changes in freshwater runoff are likely to affect upper-ocean salinity, sea-ice production, export of

Table 6.13. Projected change in the discharge of the largest arctic rivers using different climate models and forcing scenarios.

	GCM and forcing scenario	Discharge change (%)		Reference
		Annual discharge	Winter discharge	
Yenisey, Lena, Ob, Kolyma	Canadian[a], GISS[b] $2\times CO_2$	10–45		Miller and Russell, 1992
Yenisey	SHI[c] 1 °C	9	34	Shiklomanov A., 1994, 1997
	SHI[c] 2 °C	9	61	
	SHI[c] 4 °C	15	325	
	GFDL[d] $2\times CO_2$	19	70	
	UKMO[e] $2\times CO_2$	45	80	
Inflow into the Barents Sea	GFDL[d] $2\times CO_2$; UKMO[e] $2\times CO_2$	14–35	25–46	Georgievsky et al., 1996
Yenisey	HadCM2[f]; HadCM3[f]; 6 scenarios by 2050	6–14		Arnell, 1999
Lena		12–25		
Ob		3–10		
Kolyma		30–40		
Mackenzie		12–20		
Yukon		20–30		
Arctic total	GISS[b]	12		Miller and Russell, 2000
Eurasian rivers	CO_2: +0.5%/yr to 2100	9		
North American rivers		23		
Lena	HadCM3[f] $2\times CO_2$	12		Georgievsky et al., 2003
Yenisey	HadCM3[f] $2\times CO_2$	8		Mokhov and Khon, 2002; Mokhov et al., 2003
Lena		24		
Ob		4		
Yenisey	ECHAM4/OPYC3[g] $2\times CO_2$	8		Mokhov and Khon, 2002; Mokhov et al., 2003
Lena		22		
Ob		3		
Yenisey	CGC[a] $2\times CO_2$	18		Arora and Boer, 2001
Lena		19		
Ob		-12		
Mackenzie		20		
Yukon		10		
Usa (Pechora basin)	HadCM2[f] (2080)	-16		Van Der Linden et al., 2003
	HadCM2[f] (2230)	10		

[a]Canadian Centre for Climate Modelling and Analysis; [b]Goddard Institute for Space Studies (United States); [c]State Hydrological Institute (Russia); [d]Geophysical Fluid Dynamics Laboratory (United States); [e]United Kingdom Met Office; [f]Hadley Centre (United Kingdom); [g]Max-Planck Institute for Meteorology (Germany).

freshwater to the North Atlantic subpolar seas, and possibly the thermohaline circulation. In particular, Steele and Boyd (1998) have argued that recent changes in the upper layers of the Arctic Ocean are attributable to altered pathways of Siberian river runoff in the Arctic Ocean.

On ecosystems

Changes in extreme runoff events (floods) are likely to alter biological production and biodiversity in riparian systems (section 8.4). The area of ponds and wetlands, for which water levels are critical, can determine whether vast northern peatlands will become sources or sinks of CO_2 and CH_4 (section 7.5.3). Changes in the fluxes of water and hence nutrients to and from ponds and wetlands will affect aquatic ecology (section 8.4).

On people

Traditional lifestyles are likely to be affected by changes in recharging of ponds and wetlands. Navigability of arctic rivers will be affected if runoff levels change substantially, especially during the warm season. Increases in the frequency and magnitude of extreme discharge events will result in catastrophic floods and are likely to require revisions of current construction requirements (section 16.4). Changes in the thickness and/or duration of freshwater ice cover will affect transportation in northern regions (i.e., ice roads) and will influence expanding development (e.g., oil and gas, diamond, and gold exploration).

6.8.5. Critical research needs

Perhaps the most critical need pertaining to surface flows in the Arctic concerns the network of gauge stations for monitoring discharge rates. This network has degraded seriously in the past decade, such that present measurements of surface flows in the Arctic are much less complete than in the recent past. Many of the monitoring station closures have occurred in Russia and Canada, and to a lesser extent in Alaska. It is important to reopen some of these stations or otherwise enhance the current monitoring network.

Better estimation of subsurface flows is also required. If permafrost thaws in regions of significant discharge to the Arctic Ocean, a quantitative understanding of subsurface flows will become increasingly essential for closing the arctic hydrological budget. A related need is an improved understanding of the relationship between net atmospheric moisture input (P-E), river discharge, and changes in permafrost. Recent findings by Serreze et al. (2003) and Savelieva et al. (2002) suggested that changes in permafrost may be affecting the linkage between precipitation and river discharge, in terms of the relationships between the water-year means of the two quantities and between the seasonality of the two quantities.

6.9. Sea-level rise and coastal stability

6.9.1. Background

Sea-level rise is one of the most important consequences of climate change and has the potential to cause significant impacts on ecosystems and societies. Changes in sea level will directly affect coastal stability. While section 6.5 discussed glacier wastage as a contributor to sea-level change, the present section addresses sea-level change in a broader context. The consequences for arctic coastal stability serve as the motivation and target of the present discussion of sea level and its variations. Sea level is discussed further in section 2.5.2 and 16.2.4.

Vulnerability to sea-level change varies substantially among arctic coastal regions. Figure 6.38 shows the areas of the Arctic that are presently less than ten meters above mean sea level. Substantial portions of the coasts of Siberia, Alaska, and Canada are low-lying and hence vulnerable to sea-level rise, although the rate of isostatic rebound in eastern Canada is substantial, as discussed below.

"Mean sea level" at the coast is defined as the height of the sea with respect to a local land benchmark, averaged over a long enough period (e.g., a month or a year) that fluctuations caused by waves and tides are largely removed. Changes in mean sea level measured by coastal tide gauges are called "relative sea-level changes", because they can be caused by movement of the land on which the tide gauge is situated or by climate-driven oceanic changes affecting the height of the adjacent sea surface. These two causes of relative sea-level change can have similar rates (several millimeters per year) over decadal or longer timescales. In addition, sea level is affected by changes in the spatial distribution of the water in the ocean resulting from atmosphere–ocean

processes, for example, atmospheric (hydrostatic) pressure and storm winds, tides, and changes in the ocean circulation (currents). To infer sea-level changes arising from changes in the atmosphere or the ocean, the movement of the land must be subtracted from the records of tide gauges and geological indicators of past sea level. Because the processes contributing to sea-level change all have significant spatial variability, there will be considerable geographic variability in changes in the rate of relative sea-level rise.

6.9.1.1. Vertical motions of the land surface (isostatic changes)

Widespread land movements are caused by glacio-isostatic adjustment (a slow response to the melting of large ice sheets), and by tectonic land movements that include rapid displacements (earthquakes) and slow movements (associated with mantle convection and sediment transport). Glacio-isostatic adjustment and tectonic movements both vary widely in space. Therefore, sea-level change is not expected to be geographically uniform, and information about its distribution is needed to inform assessments of the impacts on coastal regions.

Glacio-isostatic adjustment is a response of the earth to loading and unloading by glaciers during the last major glaciation and the subsequent deglaciation. Regions that hosted thick accumulations of ice experienced subsidence followed by rebound when the ice retreated. Other parts of the earth experienced both subsidence and uplift or subsidence alone. Because the response of the earth's crust and mantle is slow, recovery from this loading and unloading is still occurring (Peltier, 2001, 2004). Glacio-isostatic adjustment varies considerably around the Arctic, from uplift in the Canadian Archipelago, Greenland, and Norway to subsidence along the Beaufort Sea and Siberian coasts. Tectonic motion is caused by movements of the crustal plates that result in rapid changes (earthquakes) or slow, gentle uplift or subsidence. Local loading of the crust by sediment (e.g., in deltas) can also cause subsidence.

6.9.1.2. Climate-driven oceanic changes affecting the height of the sea surface

In the absence of vertical motions of the land surface, there are two main components of sea-level rise:

- "steric rise", which refers to processes that cause an increase in ocean volume without a change in mass, primarily through changes in temperature (thermal expansion) and salinity (freshening); and
- "eustatic rise", which refers (at least in the oceanographic community) to changes resulting from an increase in the mass of water. Increased runoff from terrestrial regions, including glaciers and ice sheets, contributes directly to eustatic rise.

Changes in the amount of water stored on land will alter the mass of the ocean. (Sea level will be unaffected by

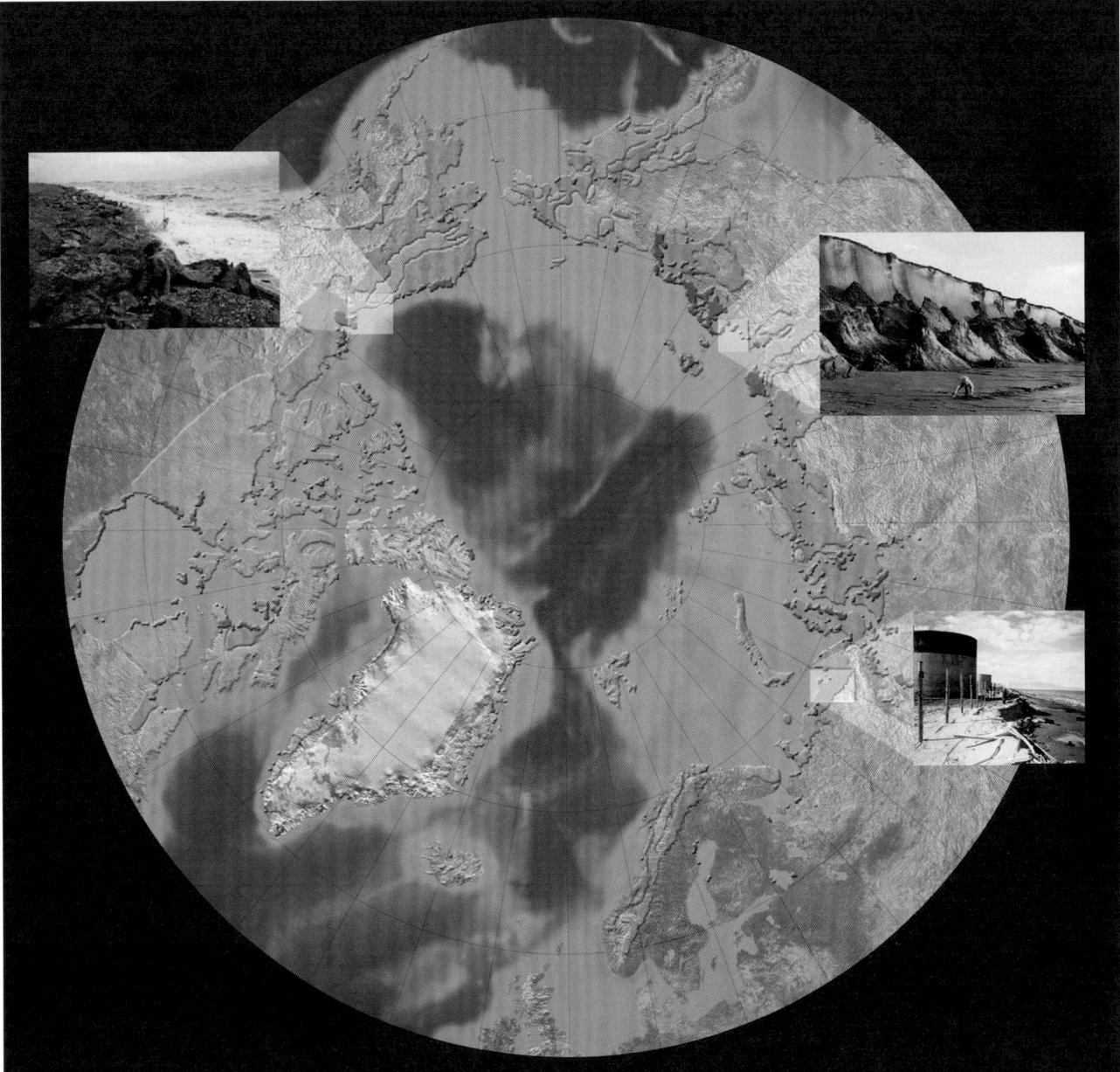

Fig. 6.38. Vulnerability of arctic coasts to sea-level rise and erosion, showing land areas of the Arctic with elevations less than 10 m above mean sea level (red), regions with unlithified coasts (green), and regions with lithified coasts (brown) (data from Brown J. et al., 1997 and Peltier, 2004). Examples of unstable coastal environments are shown in the insets from the Pechora (photo: S. Ogorodov, courtesy of the Arctic Coastal Dynamics website), Laptev (photo: M. Grigoriev, courtesy of the Arctic Coastal Dynamics website), and Beaufort (photo: S. Solomon, Geological Survey of Canada) Sea coasts.

the melting of sea ice, which displaces the volume of ocean water equivalent to its mass.) Climate change is projected to reduce the amount of water frozen in glaciers and ice caps owing to increased melting and evaporation (section 6.5.3). Greater melting of, and evaporation from, the Greenland and Antarctic ice sheets are also projected, but might be offset by increased precipitation, especially over Antarctica. Recent laser altimetry work by the National Aeronautics and Space Administration (NASA) shows that increased precipitation over Greenland does not presently balance the increased melting at lower elevations (Zwally et al., 2002). Increased discharge of ice from the ice sheets into the ocean is also possible. The ice sheets respond to climate change over timescales of up to millennia, so they could still be gaining or losing mass in response to climate variations that occurred as far back as the last glacial period, and they will continue to change for thousands of years after climate stabilizes. In addition to changes in glaciers and ice sheets, groundwater extraction and impounding of water in reservoirs can affect sea level.

6.9.1.3. Variations in sea level arising from atmosphere–ocean processes (including sea ice)

A variety of atmosphere–ocean processes contribute to spatial and temporal variations in sea level. The frequency, direction, magnitude, and duration of winds are the important variables which, when combined with water depth and coastal morphology, determine wave and storm-surge forcing at the coast. Storm surges form in response to lowered atmospheric pressure in cyclonic weather systems and accumulation of water in shallow coastal areas due to wind stress. Storm surges can cause increases in water level many times greater than the normal tidal range. Associated waves can directly impact coastal bluffs, resulting in rapid coastal retreat.

Changes in meteorological forcing over time can be identified using data from the terrestrial and ice-island weather observation networks that have been active in the circumpolar coastal and ocean region since the 1940s. Of the parameters observed at a weather station, winds represent the single most important forcing agent for the coastal regime, driving waves and surges and affecting sea-ice formation and presence. It is important to consider the distributions of wind speeds rather than the averages, because the greatest impacts are typically associated with high-magnitude winds. A strong wind event can also break up existing sea ice, further exposing the coast to wave activity.

Wind speeds in excess of 10 m/s are considered strong enough to have an impact on the coastal regime (Solomon et al., 1994) when occurring during the open water period (broadly defined as July through October, although there is regional variability). Serreze et al. (2000) summarized a number of studies that have identified a decreasing trend in sea-level pressure over the Arctic Ocean. This has been linked to an increasing trend in both the frequency (trend significant in all seasons) and intensity (trend significant in summer) of cyclonic activity (McCabe et al., 2001; Serreze et al., 2000), although these trends are not apparent at all sites (e.g., Barrow, Alaska; Lynch et al., 2004). At many sites in the circumpolar coastal station network, there was also a decrease between 1950 and 2000 in the observed time between cyclonic events during the open-water season (Atkinson, pers. comm., 2003).

Wind forcing of oceanographic processes in the Arctic is moderated by the presence of sea ice throughout much of the year. Severe winter storms, which have such a devastating effect on temperate coasts, have little impact on arctic shores because sea ice protects the Arctic Ocean coastline from waves for eight months or more of the year. The impacts of changes in the duration of the open-water season depend on not only the magnitude, duration, and direction of winds, but also on the extent and concentration of sea ice. Sea ice affects wave generation and, to a lesser extent, the magnitude of storm surges. The duration of the open-water season varies considerably both spatially and temporally. Some parts of the Canadian Archipelago are never ice-free, while sea ice is rare in the vicinity of northern Norway (under the influence of the Gulf Stream). In some locations, interannual variations in the duration and extent of the open water range from a few weeks and several kilometers of fetch to eight to ten weeks and hundreds of kilometers of fetch, as in the Beaufort Sea in late summer and early autumn. Changes in the duration of the open-water season will be critical to the future impacts of coastal storms in the Arctic. There are various examples of Inuit peoples having difficulty coping with thinning ice, retreating

ice-floe edges, and increasing storm frequency during the last decade (Kerr, 2002). These experiences indicate that, while sea-level rise is one of the most well known possible consequences of climate change, the fate of sea ice may be equally or more important to natural and human coastal systems.

Sea-ice pressure-ridge keels and icebergs scour the seabed, resulting in a characteristic roughened or ploughed seabed morphology that may affect resuspension rates and could change the degree of consolidation of the seabed surface. This process may be responsible for enhanced coastal retreat along parts of the Canadian Beaufort Sea coast.

Compounding the problem of sea-level change are concerns about arctic coastal stability, which directly affects human settlements and development. The stability of any coast is a function of the interaction between meteorological and oceanographic forcing and the physical properties of coastal materials. Short-term (hours to days) meteorological and oceanographic events are superimposed on the medium-term elevation of the sea surface, which changes seasonally and interannually due to natural climate variability, and over the longer term due to the combination of vertical motion of the land and the volume and distribution of the global ocean.

High-latitude coastal environments differ from their temperate counterparts in several important ways. The interaction between the atmosphere and the ocean that produces waves and storm surges is mediated by the presence and concentration of sea ice, while coastal materials are either strengthened or destabilized by the presence of permafrost, the abundance of ground ice, and associated temperature regimes (Are, 1988; Kobayashi et al., 1999).

Arctic coasts are as variable as coasts in temperate regions. The most unstable coasts are those that are composed of unlithified sediments (Fig. 6.38). These sediments were deposited during the past glacial and interglacial periods, are affected by permafrost erosion, and contain variable, sometimes significant, amounts of ground ice. Unlithified sediments are concentrated along much of the Russian coast and along the Beaufort Sea coast. Bedrock coasts are located in the Canadian Archipelago, and along Greenland, Norway, and extreme western Russia. The bedrock coasts generally coincide with those areas that experienced extensive glaciation during the Holocene. Even along bedrock and fjord-dominated coasts, beaches and deltas do occur, and those are usually where human settlements and infrastructure are concentrated (section 16.3).

6.9.2. Recent and ongoing changes

According to the IPCC (2001), sea level has risen more than 120 m over the past 20000 years as a result of mass loss from melting ice sheets. The IPCC also reports that, based on geological data, global mean sea level may have increased at an average rate of about 0.5 mm/yr over the last 6000 years and at an average rate of 0.1 to 0.2 mm/yr over the last 3000 years. In addition, vertical land movements are still occurring today.

The IPCC (2001) consensus value for the current rate of sea-level rise is 1 to 2 mm/yr. Recent results from global tide gauge analyses (length <70 years) show that the present global mean rate of sea-level rise is closer to 2 mm/yr (Douglas, 2001; Peltier, 2001 as reported in Cabanes et al., 2001). Complex short-term variations are superimposed on this rise, complicating the evaluation of trends (Fig. 6.39).

While both steric rise and eustatic rise are thought to be responsible for the current increase in sea level (Douglas et al., 2001), Cabanes et al. (2001) recently used a combination of ocean temperature measurements and satellite observations of sea level to argue that nearly all the present sea-level rise can be accounted for by steric processes alone. Others (e.g., Meier and Wahr, 2002; Munk, 2002) have countered that, due to melting glaciers, the eustatic contribution is significant. Munk (2003) and others report observed recent freshening of ocean water, which also suggests an increase in freshwater discharge from the land. These arguments, combined with new data from satellite-borne altimeters and gravity-measuring instruments and new calculations of glacio-isostatic adjustments, make this a complex subject for which there is no firm consensus, but rather an evolving spectrum of views.

The mass of the ocean, and thus sea level, changes as water is exchanged with glaciers and ice caps. Observational and modeling studies of glaciers and ice caps (section 6.5) indicate a contribution to sea-level rise of 0.2 to 0.4 mm/yr averaged over the 20th century. Modeling studies suggest that climate changes during the 20th century have led to contributions of -0.2 to 0.0 mm/yr from Antarctica (the result of increasing precipitation) and 0.0 to 0.1 mm/yr from Greenland (from changes in both precipitation and runoff).

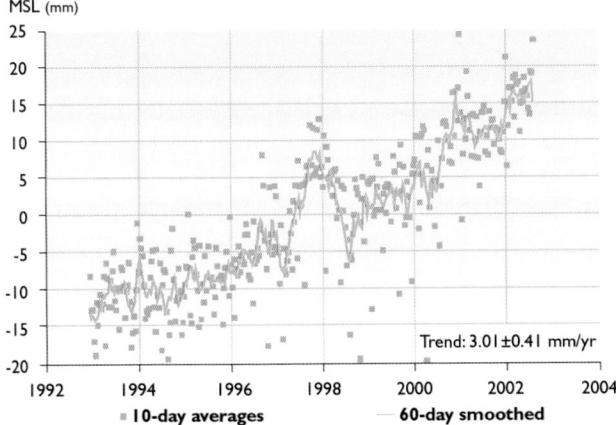

Fig. 6.39. Temporal variations in global mean sea level (MSL) computed from TOPEX/POSEIDON measurements between December 1992 and July 2002 (Leuliette et al., 2004; http://sealevel.colorado.edu).

Rates of change in the Arctic similar to the global values have been documented. Sea-level curves for the Beaufort Sea shelf suggest a rapid rise (~7 to 8 mm/yr) between 10 000–8000 and 3000 years BP, and a slower average rise (~1.1 to 2.5 mm/yr) from 3000 years BP to the present (Campeau et al., 2000; Hill et al., 1993). These rates refer to relative sea-level rise and therefore include the isostatic contribution. Multi-decadal tide gauge data from Russian Arctic stations show relative sea-level rise of 0.3 to 3.2 mm/yr (Fig. 6.40; Proshutinsky et al., 2001). Data from additional stations in Alaska (Zervas, 2001) show mostly falling sea levels in the Pacific region (probably related to tectonic activity) and a single station in the Canadian Arctic showing rising sea level (Douglas et al., 2001). Analyses reported by Proshutinsky (pers. comm., 2003) indicate that in the Russian sector of the Arctic Ocean, sea level increased by an average of 1.85 mm/yr over the past 50 years (after correction for glacio-isostatic adjustments). Most of the increase is attributed to a combination of steric effects (0.64 mm/yr), decreasing sea-level pressure (0.56 mm/yr), and increasing cyclonic curvature of the mean wind field (0.19 mm/yr). For comparison, the IPCC (2001) reports observational estimates of the steric effect of about 1 mm/yr over recent decades, similar to values of 0.7 to 1.1 mm/yr simulated by AOGCMs over a comparable period. Averaged over the 20th century, AOGCM simulations suggest thermal expansion rates of 0.3 to 0.7 mm/yr.

Sea-ice conditions, especially greater amounts of open water, also contribute to increasing coastal instability. Figure 6.40 contrasts the present-day September sea-ice extent with that projected by the ACIA models for the 2071–2090 time slice. The significant projected increase in open-water extent is likely to create more energetic wave and swell conditions at the coast resulting in greater coastal instability.

The ACD project (section 6.6.2.5) was initiated in 1999 to compile circumpolar coastal change data, to develop a comprehensive set of monitoring sites, and to synthesize information about historical coastal environmental forcing. Based in part on information from the ACD project, it is clear that large regions of the arctic coast are undergoing rapid change. For example, Rachold et al. (2002) reported that the average rate of retreat of the Laptev Sea coast is 2.5 m/yr, a rate that contributes more sediment and organic carbon to the sea than does the Lena River. Retreat also predominates along the coast of the Beaufort Sea and large portions of the Russian coast. However, most arctic coastal stability monitoring records are too short and infrequent to identify trends in rates of retreat. The records show considerable annual and decadal variation in recent rates of retreat, which are attributed to variability in the frequency and severity of coastal storms and variations in open-water extent.

6.9.3. Projected changes

IPCC (2001) projections of the components contributing to sea-level change between 1990 and 2100 are as

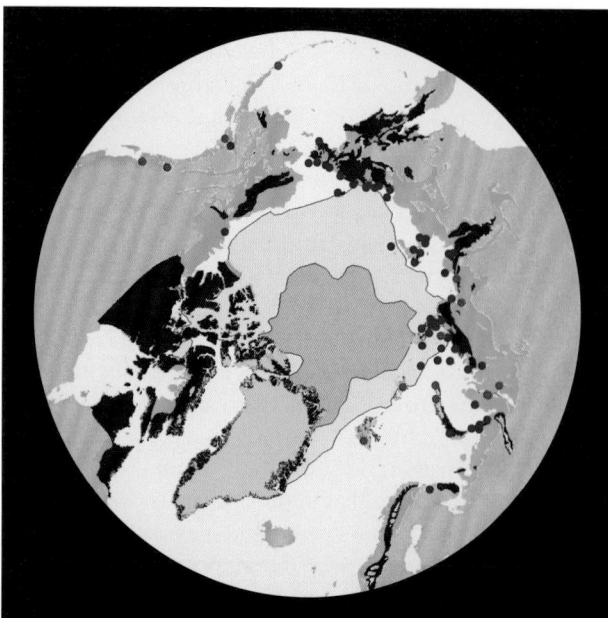

Fig. 6.40. Factors affecting coastal stability in the Arctic. Blue dots show stations where relative sea level (RSL) is rising; red dots show stations where RSL is falling. The light green area in the Arctic Ocean represents present-day September sea-ice extent (concentrations exceeding 15% in at least half the Septembers from 1980 to 1999). The pink area is the corresponding ACIA model-projected sea-ice extent (ice present in September in more than 50% of the years from 2071–2090; five-model average). Regions with unlithified coasts are shown in dark green, while regions with lithified coasts are shown in dark brown (data from Proshutinsky et al., 2001; Proshutinsky, pers. comm., 2003; Zervas, 2001; and the Hadley Centre's HadISST dataset).

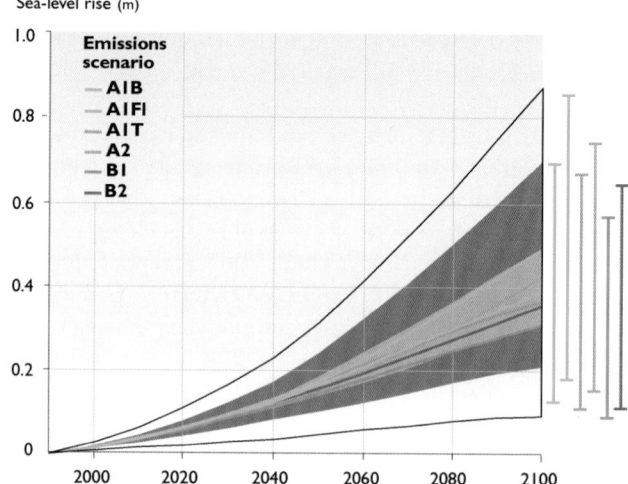

Fig. 6.41. Projected global sea-level rise between 1990 and 2100 using 35 different emissions scenarios. Thermal expansion and land ice changes were calculated using a simple climate model calibrated separately for each of seven AOGCMs. Contributions from changes in permafrost, the effect of sediment deposition, and the long-term adjustment of the ice sheets to past climate change were added. Each of the six colored lines is the average of all the model projections for one of six illustrative emissions scenarios; vertical bars at the right show the range of these projections. The region of light shading shows the range of the seven-model average projection for all 35 scenarios. The dark shading shows the entire range of projections by the seven models for the 35 scenarios. The outermost lines delineate the range of projections from all models and scenarios with the additional uncertainties due to changes in land ice, permafrost, and sediment deposition included (IPCC, 2001).

follows: thermal expansion of 0.11 to 0.43 m, accelerating throughout the 21st century; a glacier contribution of 0.01 to 0.23 m; a Greenland contribution of -0.02 to 0.09 m; and an Antarctic contribution of -0.17 to 0.02 m. The glacier and Greenland contributions are consistent with the estimate of 0.04 m derived from the ACIA-designated models forced with the B2 emissions scenario (section 6.5). Including thawing of permafrost, sediment deposition, and the ongoing contributions from ice sheets as a result of climate change since the Last Glacial Maximum, projected global average sea-level rise between 1990 and 2100 ranges from 0.09 to 0.88 m over the 35 emissions scenarios used by the IPCC (Fig. 6.41).

The wide range in the projections shown in Fig. 6.41 reflects systematic uncertainties in modeling. The central value of 0.48 m represents more than a doubling of the mean rate of global sea-level rise over the 20th century. Based on the summation of the components contributing to sea-level rise, the IPCC projects that global sea level will rise by 0.11 to 0.77 m between 1990 and 2100 (Church J. et al., 1991; IPCC, 2001). However, the large variation among the models is an outstanding feature of the IPCC projections of sea-level rise. Moreover, sea-level rise will depend strongly on the actual GHG emissions (for which any scenario beyond the 21st century is highly uncertain) and the behavior of glaciers and ice sheets, particularly the Greenland and the West Antarctic Ice Sheets. The range in the projections given above

makes no allowance for instability of the West Antarctic Ice Sheet, although it is now widely agreed that major loss of grounded ice and accelerated sea-level rise due to West Antarctic Ice Sheet instability are very unlikely during the 21st century.

One of the important results of these model studies is that projected sea-level rise is not globally uniform. In particular, the greatest sea-level increases are projected for the Arctic, based on output from seven of nine models (Gregory, pers. comm., 2003). Figure 6.42 shows the IPCC (2001) model-projected sea-level changes after transformation to the same polar projection, providing a better depiction of spatial variability in the Arctic.

The IPCC (2001) stated that confidence in the regional distribution of sea-level change projected by AOGCMs is low because there is little similarity between model projections for various regions. One of the reasons for these regional differences may be the across-model variations in projected freshening of the Arctic Ocean owing to increased runoff or precipitation over the ocean (Bryan, 1996; Miller and Russell, 2000). Using the NASA–GISS atmosphere–ocean model, Miller and Russell (2000) project that arctic sea level will rise by 0.73 m between 2000 and 2100, of which 0.42 m is due to thermal expansion and about 0.31 m is from increased freshwater input, which also reduces the salinity. The mean global sea-level rise projected by the NASA–GISS model is 0.45 m during the same period.

Land movements, both isostatic and tectonic, will continue throughout the 21st century at rates that are unaffected by climate change. It is possible that by 2100, many regions currently experiencing decreases in relative sea level will instead have a rising relative sea level. Extreme high-water levels will occur with increasing frequency as a result of projected mean sea-level rise. Their frequency is likely to be further increased if storms become more frequent or severe as a result of climate change.

Projected decreases in sea-ice extent (section 6.3.3; see also section 16.2.5) will result in longer open-water seasons and an increased probability that severe storm events will occur without the protection of winter ice cover.

6.9.4. Impacts of projected changes

On other parts of the physical system

Increases in sea level are likely to increase coastal erosion rates in low-elevation areas and affect sediment transport in coastal regions. The salinity of bays, estuaries, and low-lying coastal areas is likely to increase as sea level rises. Spatial variations in sea-level changes may alter ocean currents and sediment transport. An increase in the duration of the open-water season will increase the frequency of wave-induced mixing, sediment transport, and storm surges during shoulder seasons, especially the autumn.

On ecosystems

Extensive coastal lowlands and immense deltas host ecosystems that will be affected by increases in sea level. Wetlands are likely to move farther inland, and coastal flood events will increase. Salinity will increase in coastal marine ecosystems that are now freshened by terrestrial discharge. If storm surges and coastal flood events increase in frequency and/or intensity, bird and fishery habitats in the affected areas are likely to be adversely affected. Decreased sea-ice coverage, which increases the probability that arctic storms will occur over open water, may change the timing and extent of mixing regimes and water–sediment interaction in coastal waters, with potential effects on benthic and coastal ecosystems.

On people

In the Arctic, many communities and much of the infrastructure are located on the coast and are therefore vulnerable to projected changes. Coastal erosion will have impacts on infrastructure and other human activities in the vicinity of coastlines (section 16.4.3). The frequency

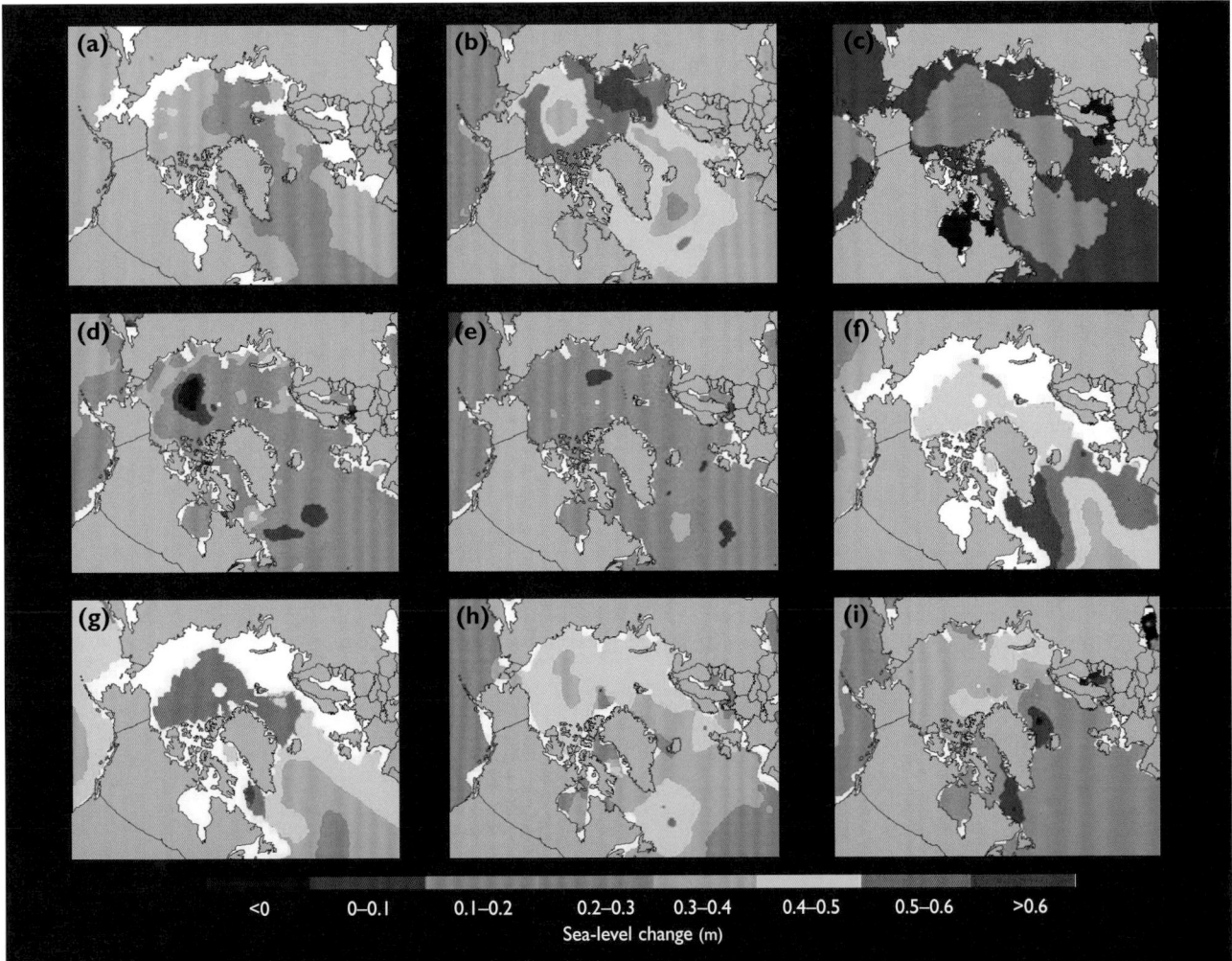

<0	0–0.1	0.1–0.2	0.2–0.3	0.3–0.4	0.4–0.5	0.5–0.6	>0.6

Sea-level change (m)

Fig. 6.42. Projected sea-level rise between 2000 and 2100 based on output from nine global climate models: (a) CSIRO Mk2 GS; (b) GFDL-R3_c; (c) MRI2 GS; (d) HadCM3 GSIO; (e) HadCM2 GS; (f) CMC2 GS; (g) SMCI GS; (h) GFDL r15 b; and (i) ECHAM4/OPYC3 G (figure courtesy of J. Gregory, 2003; see IPCC (2001) for description of models and scenarios used).

and severity of coastal flood events are likely to increase, and the severity may be compounded by the loss of sea ice along the coast during the autumn and spring seasons during which, under present climate conditions, coastal waters are usually ice-covered.

6.9.5. Critical research needs

The IPCC (2001) estimated that eustatic sea-level rise over the past 100 years was 0.10 to 0.20 m, which is higher than the estimated total sea-level rise over the same period. This discrepancy in estimates of historical change points to the imperfect state of current knowledge of sea-level variations. Similarly, the relative importance of eustatic and steric contributions to recent sea-level rise is a subject of current disagreement, as noted in section 6.9.2. Attribution is a key issue in understanding and projecting sea-level change, and must be regarded as a key research need.

Uncertainties in the projected sea-level changes are very large, although improved observations and modeling have reduced some uncertainties since publication of the Second Assessment Report (IPCC, 1996). The present wide range of projections is partially a consequence of uncertainties in the scenarios of future GHG forcing, as well as uncertainties in the response of different global climate models to this forcing. It is also a consequence of how different models simulate (or do not simulate) the response of glaciers, ice sheets, and the oceans to climate change. In order to more realistically capture glacier and ice sheet responses, models will need to resolve the topographic features that surround most glaciers (and the topography of the larger ice sheets). Topographic resolution is essential for simulation of the precipitation and melt regimes, including the location of the equilibrium line, of significant glaciers. Other research needs for projecting the response of glaciers and ice sheets are discussed in section 6.5.5. The discharge of glacial meltwater to the ocean creates spatial gradients in sea level and upper-ocean salinity, both of which will trigger oceanic responses. These responses are not well understood and are poorly simulated, yet are crucial to projecting regional variations in sea-level rise.

The contribution of Antarctica to future changes in sea level remains a major uncertainty. Whether or not an increase in precipitation will offset the effects of enhanced melt in a warmer climate is a first-order question that needs to be addressed. Changes in the West Antarctic Ice Sheet have the potential to contribute significantly to future changes in global sea level.

For model simulations to be useful in assessments of coastal stability, enhanced resolution of coastlines is required in order to simulate realistically the sharp near-coastal gradients that are likely to characterize GHG-induced changes in temperature and precipitation. Credible simulations of changes in storminess and surface winds will also require improved model resolution and parameterizations of surface fluxes in high latitudes.

Finally, assessments of coastal vulnerability in the Arctic suffer from the limitations of the observational network for monitoring coastal surface winds and coastal stability. Data from the existing station network are of limited utility for driving numerical wave and surge models, which require offshore wind fields. The availability of surface wind data from the Arctic Ocean is poorer still. Evaluations of current trends in storm events and the associated coastal vulnerabilities will require additional sources of reliable surface wind data from coastal and offshore areas in the Arctic. Observations of coastal responses to wind and sea-level forcing are also essential in order to quantify the relationships between environmental processes and coastal impacts.

Acknowledgements

The authors thank William Chapman of the University of Illinois for providing information on the model-derived scenarios.

Personal communications and unpublished data

Anisimov, O., 2003. State Hydrological Institute, St. Petersburg, Russia.
Arctic Coastal Dynamics program, www.awi-potsdam.de/www-pot/geo/acd.html.
Atkinson, D., 2003. National Research Council of Canada.
Bitz, C., 2003. University of Washington.
Canadian Ice Service, http://ice-glaces.ec.gc.ca/App/WsvPageDsp.cfm?ID=210&LnId=6&Lang=eng.
Chapman, W., 2003. University of Illinois.
Dyurgerov, M., 2003. Cooperative Institute for Research in Environmental Sciences, University of Colorado.
Glazovsky, A.F., 2003. Russian Academy of Sciences.
Gregory, J., 2003. University College of Reading, United Kingdom.
Hadley Centre for Climate Prediction and Research, United Kingdom, http://badc.nerc.ac.uk/data/hadisst/.
Jakobsson, T., 2003. Icelandic Meteorological Office.
Koerner, R.M., 2003. Geological Survey of Canada.
National Snow and Ice Data Center, Boulder, Colorado.
Oerlemans, J., 2003. Utrecht University, the Netherlands.
Parkinson, C., 2003. NASA Goddard Space Flight Center.
Proshutinsky, A., 2003. Woods Hole Oceanographic Institution, Massachusetts.
Robinson, D., 2003. Snow Data Resource Center, Rutgers University Climate Laboratory, Piscataway, New Jersey.
Romanovsky, V., 2003, 2004. Geophysical Institute, University of Alaska, Fairbanks.
Shiklomanov, A., 2003. University of New Hampshire.
Steffen, K., 2003. Cooperative Institute for Research in Environmental Sciences, University of Colorado.
Taylor, A.E., 2002. Geological Survey of Canada.
Wilson, K.J., 2003. Canadian Ice Service.

References

Aagaard, K. and E.C. Carmack, 1989. The role of sea ice and other freshwater in the Arctic circulation. Journal of Geophysical Research, 94:14485–14498.
AARI, 1985. Arctic Atlas. Arctic and Antarctic Research Institute, Moscow, 204pp. (In Russian)
Adams, W.P., 1981. Snow and ice on lakes. In: D.M. Gray and D.H. Male (eds.). Handbook of Snow, pp. 437–474. Pergamon Press.
Agnew, T.A., B. Alt, R. De Abreu and S. Jeffers, 2003. The loss of decades old sea ice plugs in the Canadian Arctic islands. Sixth Conference on Polar Meteorology and Oceanography, American Meteorological Society, paper 1.5.
Allard, M., B. Wang and J.A. Pilon, 1995. Recent cooling along the southern shore of Hudson Strait, Quebec, Canada, documented from permafrost temperature measurements. Arctic and Alpine Research, 27:157–166.

Ambach, W., 1979. Zur Wärmehaushalt des Grönlandischen Inlandeises: Vergleichende Studie im Akkumulationsgebiet- und Ablationsgebiet. Polarforschung, 49:44–54.

Anderson, W.L., D.M. Robertson and J.J. Magnuson, 1996. Evidence of recent warming and El Niño-related variations in ice breakup of Wisconsin lakes. Limnology and Oceanography, 41:815–821.

Anisimov, O.A. and M.A. Belolutskaia, 2002. Assessment of the impacts of climate change and degradation of permafrost on infrastructure in the northern Russia. Meteorology and Hydrology, 9:15–22. (In Russian)

Anisimov, O.A. and M.A. Belolutskaia, in press a. Impacts of changing climate on permafrost: predictive modeling approach and evaluation of uncertainties. In: Yu. Izrael (ed.). Problems of Ecological Monitoring and Ecosystems Modeling. Moscow. (In Russian)

Anisimov, O.A. and M.A. Belolutskaia, in press b. Modeling climate-permafrost interaction: effects of vegetation. Meteorology and Hydrology. (In Russian)

Anisimov, O.A. and B. Fitzharris, 2001. Polar Regions (Arctic and Antarctic). In: J. McCarthy, O. Canziani, N.A. Leary, D.J. Dokken and K.S. White (eds.). Climate Change 2001: Impacts, Adaptation, and Vulnerability, pp. 801–841. Contribution of Working Group II to the Third Assessment Report of the Intergovernmental Panel on Climate Change. Cambridge University Press.

Anisimov, O.A. and F.E. Nelson, 1997. Permafrost zonation and climate change: Results from transient general circulation models. Climatic Change, 35:241–258.

Anisimov, O.A. and V.Yu. Poliakov, 2003. GIS assessment of climate-change impacts in permafrost regions. Proceedings of the Eighth International Permafrost Conference, vol. 1, pp. 9–14.

Anisimov, O.A., N.J. Shiklomanov and F.E. Nelson, 1997. Effects of global warming on permafrost and active layer thickness: results from transient general circulation models. Global and Planetary Change, 15:61–77.

Anisimov, O.A., A.A. Velichko, P.F. Demchenko, E.V. Eliseev, I.I. Mokhov and V.P. Nechaev, 2002. Impacts of climate change on permafrost in the past, present, and future. Proceedings of Russian Academy of Science, Physics of Atmosphere and Ocean, 38(7): 23–51.

Anisimov, O.A., M.A. Beloloutskaia and V.A. Lobanov, 2003. Observed climatic and environmental changes in the high latitudes of the Northern Hemisphere. Meteorology and Hydrology, 2:18–30. (In Russian)

Anon, 1999. Final Report. Land Arctic Physical Process (LAPP). Contract No. ENV4-CT95-0093. 1 April 1996 to 31 March 1999. 134pp.

Arctic Climatology Project, 1997. Environmental Working Group Joint U.S.-Russian Atlas of the Arctic Ocean - Winter Period. L. Timokhov and F. Tanis (eds.). Environmental Research Institute of Michigan in association with the National Snow and Ice Data Center, Ann Arbor, Michigan. CD-ROM.

Arctic Climatology Project, 1998. Environmental Working Group Joint U.S.-Russian Atlas of the Arctic Ocean - Summer Period. L. Timokhov and F. Tanis (eds.). Environmental Research Institute of Michigan in association with the National Snow and Ice Data Center, Ann Arbor, Michigan. CD-ROM.

Are, F.A., 1988. Thermal abrasion of sea coasts (Parts I and II). Polar Geography and Geology, 12:1–157.

Arendt, A.A., K.E. Echelmeyer, W.D. Harrison, C.A. Lingle and V.B. Valentine, 2002. Rapid wastage of Alaska glaciers and their contribution to rising sea level. Science, 297:382–386.

Armstrong, R. and M. Brodzik, 2001. Validation of passive microwave snow algorithms. In: M. Owe, K. Brubaker, J. Ritchie and A. Rango (eds.). Remote Sensing and Hydrology 2000. International Association of Hydrological Sciences, Publ. No. 267, pp. 87–92.

Arnell, N.W., 1999. Climate change and global water resources. Global Environmental Change, 9:31–49.

Arora, V.K. and G.J. Boer, 2001. Effects of simulated climate change on the hydrology of major river basins. Journal of Geophysical Research, 106(D4):3335–3348.

Assel, R.A. and D.M. Robertson, 1995. Changes in winter air temperatures near Lake Michigan, 1851–1993, as determined from regional lake-ice records. Limnology and Oceanography, 40:165–176.

Barry, R.G., 1984. Possible CO_2-induced warming effects on the cryosphere. In: N.-A. Morner and W. Karlin (eds.). Climatic Changes on a Yearly to Millennial Basis, pp. 571–604. D. Reidel.

Beltaos, S. (ed.), 1995. River Ice Jams. Water Resources Publications, Colorado, 372pp.

Björnsson, H., F. Pálsson and S. Gudmundsson, 2001. Jökulsárlón at Breidamerkursandur, Vatnajökull, Iceland: 20th century changes and future outlook. Jökull, 50:1–18.

Björnsson, H., F. Pálsson and H.H. Haraldsson, 2002. Mass balance of Vatnajökull (1991–2001) and Langjökull (1996–2001), Iceland. Jökull, 51:75–78.

Bogdanova, E.G., B.M. Ilyin and I.V. Dragomilova, 2002. Application of an improved bias correction model to precipitation measured at Russian North Pole drifting stations. Journal of Hydrometeorology, 3:700–713.

Bonsal, B.R. and T.D. Prowse, 2003. Trends and variability in spring and autumn 0 °C-isotherm dates over Canada. Climatic Change, 57: 342–358.

Bowling, L.C., D.P. Lettenmaer and B.N. Matheussen, 2000. Hydroclimatology of the Arctic Drainage Basin. In: E.L. Lewis, E.P. Jones, P. Lemke, T.D. Prowse and P. Wadhams (eds.). The Freshwater Budget of the Arctic Ocean, pp. 47–90. NATO Meeting/NATO ASI Series, Kluwer Academic Publishers.

Broecker, W.S., 1997. Thermohaline circulation, the Achilles heel of our climate system: will man-made CO_2 upset the current balance? Science, 278:1582–1588.

Brown, J. and S.M. Solomon, 1999. Arctic Coastal Dynamics. Report of an International Workshop. Geological Survey of Canada, Open File 3929.

Brown, J., O.J. Ferrians, J.A. Heginbottom and E.S. Melnikov, 1997. Circum-Arctic Map of Permafrost and Ground-ice Conditions. U.S.G.S. Circum-Pacific Map series, Map CP-45. U.S. Geological Survey.

Brown, J., O.J. Ferrians Jr., J.A. Heginbottom and E.S. Melnikov, 1998. Circum-Arctic Map of Permafrost and Ground-Ice Conditions. National Snow and Ice Data Center/World Data Center for Glaciology, Boulder, Colorado. Digital Media.

Brown, J., K.M. Hinkel and F.E. Nelson, 2000. The Circumpolar Active Layer Monitoring (CALM) program: Research designs and initial results. Polar Geography, 24:163–258.

Brown, R.D., 2000. Northern Hemisphere snow cover variability and change, 1915–1997. Journal of Climate, 13:2339–2355.

Brown, R.D. and R.O. Braaten, 1998. Spatial and temporal variability of Canadian monthly snow depths. Atmosphere-Ocean, 36:37–45.

Bruland, O. and E. Cooper, 2001. Snow distribution and vegetation. In: P. Kuhry (ed.). Proceedings of Arctic Feedbacks to Global Change. Rovaniemen Paintuskeskus Oy, Rovaniemi, Finland, pp. 110.

Bryan, K., 1996. The steric component of sea level rise associated with enhanced greenhouse warming: a model study. Climate Dynamics, 12:545–555.

Burgess, M.M., S.L. Smith, J. Brown, V. Romanovsky and K. Hinkel, 2000. Global Terrestrial Network for Permafrost (GTNet-P): Permafrost monitoring contributing to global climate observations. Geological Survey of Canada, Current Research 2000-E14, 8pp.

Burrell, B.C., 1995. Mitigation. In: S. Beltaos (ed.). River Ice Jams, pp. 205–239. Water Resources Publications, Colorado.

Cabanes, C., A. Cazenave and C. Le Provost, 2001. Sea level rise during past 40 years determined from satellite and *in situ* observations. Science, 294:840–842.

Callaghan, T.V. and S. Jonasson, 1995. Implications for changes in Arctic plant biodiversity from environmental manipulation experiments. In: F.S. Chapin III and C.H. Korner (eds.). Arctic and Alpine Biodiversity: Patterns, Causes and Ecosystem Consequences, pp. 151–164. Springer-Verlag.

Campeau, S., A. Hequette and R. Pienitz, 2000. Late Holocene diatom biostratigraphy and sea-level changes in the southeastern Beaufort Sea. Canadian Journal of Earth Sciences, 37:63–80.

Cavalieri, D.J., 2002. A link between Fram Strait sea ice export and atmospheric planetary wave phase. Geophysical Research Letters, 20(12):10.1029/2002GL014684.

Cavalieri, D.J., C.L. Parkinson and K.Y. Vinnikov, 2003. 30-year satellite record reveals contrasting Arctic and Antarctic decadal sea ice variability. Geophysical Research Letters, 30(18), 1970, doi:10.1029/2003GL018031.

Chang, A.T.C., J.L. Foster and D.K. Hall, 1987. Nimbus-7 derived global snow cover parameters. Annals of Glaciology, 9:39–44.

Chudinova, S.M., S.S. Bykhovets, V.A. Sorokovikov, D.A. Gilichinsky, T-J. Zhang and R.G. Barry, 2003. Could the current warming endanger the status of frozen ground regions of Eurasia? In: W. Haeberli and D. Brandova (eds.). Permafrost. Extended Abstracts reporting current research and new investigation, pp. 21–22. Eighth International Conference on Permafrost, University of Zurich.

Church, J.A., J.S. Godfrey, D.R. Jackell and T.J. McDougall, 1991. A model of sea level rise caused by ocean thermal expansion. Journal of Climate, 4:438–456.

Church, J.A., J.M. Gregory, P. Huybrechts, M. Kuhn, L. Lambeck, T. Nhuan, M.Y. Qin and P.L. Woodworth, 2001. Changes in sea level. In: J.T. Houghton, Y. Ding, D.J. Griggs, M. Nogeur, P.J. Van der Linden, X.Dai, K. Maskell, C.A. Johnson (eds). Climate Change 2001. The Scientific Basis, pp. 641–693. Intergovernmental Panel on Climate Change, Cambridge University Press.

Church, M., 1974. Hydrology and permafrost with reference to northern North America. In: Permafrost Hydrology, Proceedings of Workshop Seminar 1974, pp. 7–20. Canadian National Committee for the International Hydrological Decade, Ottawa.

Cohen, S.J. (ed.), 1997. Mackenzie Basin Impact Study (MBIS): Final Report. Environment Canada.

Colony, R., V.F. Radionov and F.J. Tanis, 1998. Measurements of precipitation and snow pack at Russian North Pole drifting stations. Polar Record, 34:3–14.

Comiso, J.C., 2002. A rapidly declining perennial sea ice cover in the Arctic. Geophysical Research Letters, 29(20):1956, doi:10.1029/2002GL015650.

Couture, R., S. Smith, S.D. Robinson, M.M. Burgess and S. Solomon, 2003. On the hazards to infrastructure in the Canadian North associated with thawing of permafrost. Proceedings of Geohazards 2003, Third Canadian Conference on Geotechnique and Natural Hazards, pp. 97–104. Canadian Geotechnical Society.

Dallimore, S.R., S.A. Wolfe and S.M. Solomon, 1996. Influence of ground ice and permafrost on coastal evolution, Richards Island, Beaufort Sea, N.W.T. Canadian Journal of Earth Sciences, 33: 664–675.

Dankers, R. and S.M. De Jong, 2004. Monitoring snow-cover dynamics in northern Fennoscandia with SPOT Vegetation Images. International Journal of Remote Sensing, 25(15):2933–2949.

Demuth, M.N., A. Pietroniro, T. Ouarda and J.Yetter, 2002. The Impact of Climate Change on the Glaciers of the Canadian Rocky Mountain Eastern Slopes and Implications for Water Resource Adaptation in the Canadian Prairies. Geological Survey of Canada Open File 4322.

Dickson, R.R., T.J. Osborn, J.W. Hurrell, J. Meincke, J. Blindheim, B. Adlandsvik, T. Vinje, G. Alekseev and W. Maslowski, 2000. The Arctic Ocean response to the North Atlantic Oscillation. Journal of Climate, 13:2671–2696.

Douglas, B.C., 2001. Sea level changes in the era of the recording. In: B. Douglas, M. Kearney and S. Leatherman (eds.). Sea Level Rise: History and Consequences, pp. 37–64. Academic Press.

Douglas, B.C., M. Kearney and S. Leatherman (eds.), 2001. Sea Level Rise: History and Consequences. Academic Press, 232pp.

Dowdeswell, J.A. and J.O. Hagen, 2004. Arctic ice masses. Chapter 15. In: J.L. Bamber and A.J. Payne (eds.). Mass Balance of the Cryosphere. Cambridge University Press, 712pp.

Dowdeswell, J.A., J.O. Hagen, H. Bjornsson, A.F. Glazovsky, W.D. Harrison, P. Holmlund, J. Jania, R.M. Koerner, B. Lefauconnier, C.S.L. Ommanney and R.H. Thomas, 1997. The mass balance of circum-Arctic glaciers and recent climate change. Quaternary Research, 48:1–14.

Duguay, C.R., G.M. Flato, M.O. Jeffries, P. Menard, K. Morris and W.R. Rouse, 2003. Ice-cover variability on shallow lakes at high latitudes: model simulations and observations. Hydrological Processes, 17:3465–3483.

Dyke, L.D. and S. Wolfe, 1993. Ground temperatures and recent coastal change at the north end of Richards Island, Mackenzie Delta, Northwest Territories. In: Current Research, Part E, Geological Survey of Canada Paper 93-1E, pp. 83–91.

Dyurgerov, M.B. and M.F. Meier, 1997. Year-to-year fluctuations of global mass balance of small glaciers and their contribution to sea-level change. Arctic and Alpine Research, 29:392–401.

Fallot, J.-M., R.G. Barry and D. Hoogstrate, 1997. Variations of mean cold season temperature, precipitation and snow depths during the last 100 years in the Former Soviet Union (FSU). Hydrological Sciences Journal, 42:301–327.

Fedorov, A.N., 1996. Effects of recent climate change on permafrost landscapes in central Sakha. Polar Geography, 20:99–108.

Flato, G.M., 2004. Sea-ice and its response to CO_2 forcing as simulated by global climate models. Climate Dynamics, 23:229–241.

Forland, E.J. and I. Hanssen-Bauer, 2000. Increased precipitation in the Norwegian Arctic: True or false? Climatic Change, 46:485–509.

Forman, S.L., W. Maslowski, J.T. Andrews, D. Lubinski, M. Steele, J. Zhang, R.B. Lammers and B.J. Peterson, 2000. Researchers explore arctic freshwater's role in ocean circulation. Eos, Transactions, American Geophysical Union, 81(16):169–174.

Frei, A., J.A. Miller and D.A. Robinson, 2003. Improved simulations of snow extent in the second phase of the Atmospheric Model Intercomparison Project (AMIP-2). Journal of Geophysical Research, 108(D12):4369, doi:10.1029/2002JD003030.

Fukuda, M., 1994. Methane flux from thawing Siberian permafrost (ice complexes) results from field observations. Eos, Transactions, American Geophysical Union, 75:86.

Garagulya, L.S., 1990. Application of Mathematical Methods and Computers in Investigations of Geocryological Processes. Moscow State University Press, Moscow, 124pp.

Georgievsky, V.Yu., A.V. Ezhov, A.L. Shalygin, I.A. Shiklomanov and A.I. Shiklomanov, 1996. Evaluation of possible climate change impact on hydrological regime and water resources of the former USSR rivers. Meteorology and Hydrology, 11:89–99. (In Russian)

Georgievsky, V.Yu., I.A. Shiklomanov and A.L. Shalygin, 2002. Long-term Variations in the Runoff over the Russian Territory. Report of the State Hydrological Institute, St. Petersburg, Russia, 85pp.

Georgievsky, V.Yu., I.A. Shiklomanov and A.L. Shalygin, 2003. Possible Changes of Water Resources and Water Regimes in the Lena Basin Due to Global Climate Warming. Report of the State Hydrological Institute, St. Petersburg, Russia.

Gerard, R., 1990. Hydrology of floating ice. In: T.D. Prowse and C.S.O. Ommanney (eds.). Northern Hydrology: Canadian Perspectives, pp. 103–134. NHRI Science Report No. 1. National Hydrology Research Institute, Environment Canada.

Ginzburg, B.M., K.N. Polyakova and I.I. Soldatova, 1992. Secular changes in dates of ice formation on rivers and their relationship with climate change. Soviet Meteorology and Hydrology, 12:57–64.

Glazovsky, A.F., 1996. Russian Arctic. In: J. Jania and J.O. Hagen (eds.). Mass Balance of Arctic Glaciers, pp. 44–53. International Arctic Science Committee Report No. 5.

Goodison, B.E. and A.E. Walker, 1995. Canadian development and use of snow cover information from passive microwave satellite data. In: B.T. Choudhury, Y.H. Kerr, E.G. Njoku and P. Pampaloni (eds.). Passive Microwave Remote Sensing of Land-Atmosphere Interactions, pp. 245–261. VSP BV Zeitt, Netherlands.

Goodison, B.E. and D.Q. Yang, 1996. In-situ measurements of solid precipitation in high latitudes: the need for correction. Proceedings of the workshop on the ACSYS Solid Precipitation Climatology Project, WCRP-93, WMO/TD No.739 3–17pp.

Goodison, B.E., P.Y.T. Louie and D. Yang, 1998. WMO Solid Precipitation Measurement Intercomparison, Final Report. WMO TD-No. 872. World Meteorological Organization, 212pp.

Gordeev, V.V., J.M. Martin, I.S. Sidorov and M.V. Sidorova, 1996. A reassessment of the Eurasian river input of water, sediment, major elements, and nutrients to the Arctic Ocean. American Journal of Science, 296:664–691.

Grabs, W.E., F. Portmann and T. de Couet, 2000. Discharge observation networks in Arctic regions: Computations of the river runoff into the Arctic Ocean, its seasonality and variability. In: E.L. Lewis, E.P. Jones, P. Lemke, T.D. Prowse and P. Wadhams (eds.). The Freshwater Budget of the Arctic Ocean, NATO Science Series, pp. 249–268. Kluwer Academic Publishers.

Gray, D.M. and T.D. Prowse, 1993. Snow and floating ice. In: D. Maidment (ed.). Handbook of Hydrology, pp. 7.1–7.58. McGraw-Hill.

Groisman, P.Y. and D.R. Easterling, 1994. Variability and trends of total precipitation and snowfall over the United States and Canada. Journal of Climate, 7:184–205.

Hagen, J.O. and O. Liestøl, 1990. Long term glacier mass balance investigations in Svalbard 1950–1988. Annals of Glaciology, 14:102–106.

Hall, D.K., G.A. Riggs, V.V. Salomonson, N.E. DiGirolamo and K.J. Bayr, 2002. MODIS snow-cover products. Remote Sensing of Environment, 83:181–194.

Hare, S.R. and N.J. Mantua, 2000. Empirical evidence for North Pacific regime shifts in 1977 and 1989. Progress in Oceanography, 47(2–4):103–147.

Harris, C. and W. Haeberli, 2003. Warming Permafrost in the Mountains of Europe. World Meteorological Organization Bulletin, 52(3), 6pp.

Hill, P.R., A. Héquette and M.H. Ruz, 1993. Holocene sea level history of the Canadian Beaufort Shelf. Canadian Journal of Earth Sciences, 30:103–108.

Hilmer, M. and T. Jung, 2000. Evidence for a recent change in the link between the North Atlantic Oscillation and Arctic sea ice export. Geophysical Research Letters, 27:989–992.

Holloway, G. and T. Sou, 2002. Has Arctic sea ice rapidly thinned? Journal of Climate, 15:1691–1701.

Hubberten, H.-W. and N.N. Romanovskii, 2003. The main features of permafrost in the Laptev Sea, Russia – a review. In: M. Phillips (ed.). ICOP 2003: Permafrost. Proceedings of the Eighth International Permafrost Conference, Zurich. pp. 431–436. A.A. Balkema.

IPCC, 1996. Climate Change 1995: The Science of Climate Change. J.T. Houghton, L.G. Meira Filho, B.A. Callender, N. Harris, A. Kattenberg and K. Maskell (eds.). Intergovernmental Panel on Climate Change, Cambridge University Press, 572pp.

IPCC, 2001. Climate Change 2001: The Scientific Basis. Contribution of Working Group I to the Third Assessment Report of the Intergovernmental Panel on Climate Change. J.T. Houghton, Y. Ding, D.J. Griggs, M. Noguer, P.J. van der Linden, X. Dai, K. Maskell and C.A. Johnson (eds.). Cambridge University Press, 881pp.

Isaksen, K., P. Holmlund, J.L. Sollid and C. Harris, 2001. Three deep alpine-permafrost boreholes in Svalbard and Scandinavia. Permafrost and Periglacial Processes, 12:13–25.

Ivanov, V.V., 1976. Fresh water balance of the Arctic Ocean. Proceedings of the Arctic and Antarctic Research Institute, 323:138–147.

Jania, J., and J.O. Hagen (eds.), 1996. Mass Balance of Arctic Glaciers. International Arctic Science Committee Report No. 5, 62pp.

Jania, J. and M. Kaczmarska, 1997. Hans Glacier – a tidewater glacier in southern Spitsbergen: summary of some results. In: C.J. Van der Veen (ed.). Calving Glaciers: Report of a workshop, February 28–March 2, 1997, pp. 95–104. Byrd Polar Research Center, Report No. 15. Ohio State University.

Jasek, M.J., 1998. 1998 break-up and flood on the Yukon River at Dawson – Did El Niño and climate play a role? In: Proceedings of the Fourteenth International Ice Symposium, Potsdam, New York Vol. 2, pp. 761–768.

Johannessen, O.M., E.V. Shalina and M.W. Miles, 1999. Satellite evidence for an Arctic sea ice cover in transformation. Science, 286:1937–1939.

Jóhannesson, T. and O. Sigurðsson, 1998. Interpretation of glacier variations in Iceland 1930–1995. Jökull, 45:27–33.

Judge, A.S. and J.A. Majorowicz, 1992. Geothermal conditions for gas hydrate stability in the Beaufort-Mackenzie area: the global change aspect. Palaeogeography, Palaeoclimatology, Palaeoecology (Global and Planetary Change Section), 98:251–263.

Karl, T., 1998. Regional trends and variations of temperature and precipitation. In: R.T. Watson, M.C. Zinyowera, R.H. Moss and D.J. Dokken (eds.). The Regional Impacts of Climate Change: An Assessment of Vulnerability, pp. 412–425. Intergovernmental Panel on Climate Change, Cambridge University Press.

Kattsov, V.M. and J.E. Walsh, 2000. Twentieth-century trends of Arctic precipitation from observational data and a climate model simulation. Journal of Climate, 13:1362–1370.

Kerr, R., 2002. A warmer Arctic means change for all. Science, 297:1490–1492.

Kobayashi, N., J.C. Vidrine, R.B. Nairn and S.M. Solomon, 1999. Erosion of frozen cliffs due to storm surge on the Beaufort Sea Coast. Journal of Coastal Research, 15:332–344.

Koch, L., 1945. The east Greenland ice. Meddelelser om Grønland, 130(3):1–374.

Koerner, R.M., 1996. Canadian Arctic. In: J. Jania and J.O. Hagen (eds.). Mass Balance of Arctic Glaciers. International Arctic Science Committee, Rep. No. 5.

Koerner, R.M., 2002. Glaciers of the High Arctic Islands. In: R.S. Williams and J.G. Ferrigno (eds.). Satellite Image Atlas of Glaciers of the World: Glaciers of North America, pp. J111–J146. U.S. Geological Survey Professional Paper 1386–J.

Koerner, R.M. and L. Lundgaard, 1995. Glaciers and global warming. Geographie physique et Quaternaire, 49:429–434.

Koryakin, V.S., 1997. Glaciers of Novaya Zemlya. Zemlya i Vselennaya (Earth and Universe), 1:17–24. (In Russian)

Krabill, W., W. Abdalati, E. Frederick, S. Manizade, C. Martin, J. Sonntag, R. Swift, R. Thomas, W. Wright and J. Yungel, 1999. Greenland Ice Sheet: High-elevation balance and peripheral thinning. Science, 289:428–430.

Krupnik, I. and D. Jolly (eds.), 2002. The Earth is Faster Now: Indigenous Observations of Arctic Environmental Change. Arctic Research Consortium of the United States, Fairbanks, Alaska, 384pp.

Kudryavtsev, V.A., 1974. Fundamentals of Frost Forecasting to Geological Engineering Investigations. Nauka, Moscow, 222pp.

Kuusisto, E. and A.-R. Elo, 2000. Lake and river ice variables as climate indicators in Northern Europe. Internationale Vereinigung fur Theoretische und Angewandte Limnologie: Verhandlungen, 27:2761–2764.

Kvenvolden, K.A., 1988. Methane hydrates and global climate. Global Biogeochemical Cycles, 2:221–229.

Kwok, R., 2000. Recent changes in Arctic Ocean sea ice motion associated with the North Atlantic Oscillation. Geophysical Research Letters, 27:775–778.

Kwok, R. and D.A. Rothrock, 1999. Variability of Fram Strait ice flux and North Atlantic Oscillation. Journal of Geophysical Research, 104:5177–5189.

Lachenbruch, A.H. and B.V. Marshall, 1986. Changing climate: geothermal evidence from permafrost in the Alaskan Arctic. Science, 234:689–696.

Lammers, R.B., A.I. Shiklomanov, C.J. Vorosmarty, B.M. Fekete and B.J. Peterson, 2001. Assessment of contemporary arctic river runoff based on observation discharge records. Journal of Geophysical Research, 106(D4):3321–3334.

Lefauconnier, B., J.O. Hagen and J.P. Rudant, 1994. Flow speed and calving rate of Kongsbreen Glacier, 79° N, Spitsbergen, Svalbard, using SPOT images. Polar Research, 13(1):59–65.

Lefauconnier, B., J.O. Hagen, J.B. Ørbeck, K. Melvold and E. Isaksson, 1999. Glacier balance trends in the Kongsfjorden area, western Spitsbergen, Svalbard, in relation to the climate. Polar Research, 18(2):307–313.

Leuliette, E.W., R.S. Nerem and G.T. Mitchum, 2004. Calibration of TOPEX/Poseidon and Jason altimeter data to construct a continuous record of mean sea level change. Marine Geodesy, 27(1–2):79–94.

Ling, F. and T. Zhang, 2003. Impact of the timing and duration of seasonal snow cover on the active layer and permafrost in the Alaskan Arctic. Permafrost and Periglacial Processes, 14:141–150.

Livingstone, D.M., 1999. Ice break-up on southern Lake Baikal and its relationship to local and regional air temperatures in Siberia and to the North Atlantic Oscillation. Limnology and Oceanography, 44:1486–1497.

Lynch, A.H., J.A. Curry, R.D. Brunner and J.A. Maslanik, 2004. Towards an integrated assessment of the impacts of extreme wind events on Barrow, Alaska. Bulletin of the American Meteorological Society, 85:209–221.

Mackay, J.R., 1972. Offshore permafrost and ground ice, Southern Beaufort Sea, Canada. Canadian Journal of Earth Sciences, 9:1550–1561.

Mackay, J.R., 1986. Fifty years (1935 to 1985) of coastal retreat west of Tuktoyaktuk, District of Mackenzie. Geological Survey of Canada, Paper 86-1A:727–735.

Magnuson, J.J., K.E. Webster, R.A. Assel, C.J. Bowser, P.J. Dillon, J.G. Eaton, H.E. Evans, E.J. Fee, R.I. Hall, L.R. Mortsch, D.W. Schindler and F.H. Quinn, 1997. Potential effects of climate changes on aquatic systems: Laurentian Great Lakes and Precambrian Shield Region. In: C.E. Cushing (ed.). Freshwater Ecosystems and Climate Change in North America, pp. 7–53. John Wiley and Sons.

Magnuson, J.J., D.M. Robertson, R.H. Wynne, B.J. Benson, D.M. Livingstone, T. Arai, R.A. Assel, R.D. Barry, V. Card, E. Kuusisto, N.G. Granin, T.D. Prowse, K.M. Stewart and V.S. Vuglinski, 2000. Ice cover phenologies of lakes and rivers in the Northern Hemisphere and climate warming. Science, 289(5485):1743–1746.

Malevsky-Malevich, S.P., E.K. Molkentin, E.D. Nadyozhina, T.V. Pavlova and O.B. Shklyarevich, 2003. Possible changes of active layer depth in the permafrost areas of Russia in the 21st century. Russian Meteorology and Hydrology, (12):80–88.

Manabe, S. and R.J. Stouffer, 1993. Century-scale effects of increased atmospheric CO_2 on the ocean-atmosphere system. Nature, 364(6434):215–218.

Manson, G.K., S.M. Solomon, J.J. van der Sanden, D.L. Forbes, I.K. Peterson, S.J. Prinsenberg, D. Frobel, T.L. Lynds and T.L. Webster, 2002. Discrimination of nearshore, shoreface and estuarine ice on the north shore of Prince Edward Island, Canada, using Radarsat-1 and airborne polarimetric c-band SAR. In: Proceedings of the Seventh International Conference on Remote Sensing for Marine and Coastal Environments. Miami, Florida. 20–22 May 2002. Document 0043, 7pp.

Mauritzen, C. and S. Hakkinen, 1997. Sensitivity of thermohaline circulation to sea ice forcing in an arctic-North Atlantic model. Journal of Geophysical Research, 102:3257–3260.

McCabe, G.J., M.P. Clark and M.C. Serreze, 2001. Trends in northern hemisphere surface cyclone frequency and intensity. Journal of Climate, 14:2763–2768.

Meier, M.F. and M.B. Dyurgerov, 2002. How Alaska affects the world. Science, 297:350–351.

Meier, M.F. and J.M. Wahr, 2002. Sea level is rising: Do we know why? Proceedings of the National Academy of Sciences, 99(10):6524–6526.

Meshcherskaya, A.V., I.G. Belyankina and M.P. Golod, 1995. Monitoring tolshching cnozhnogo pokprova v osnovioi zerno proizvodyaschei zone Byvshego SSSR za period instrumental'nykh nablyugenii, Izvestiya Akad. Nauk SSR, Sser. Geograf., 101–110.

Miller, J.R. and G.L. Russell, 1992. The impact of global warming on river runoff. Journal of Geophysical Research, 97(D3):2757–2764.

Miller, J.R. and G.L. Russell, 2000. Projected impact of climate change on the freshwater and salt budgets of the Arctic Ocean by a global climate model. Geophysical Research Letters, 27:1183–1186.

Minobe, S., 2000. Spatio-temporal structure of the pentadecadal variability over the North Pacific. Progress in Oceanography, 47(2–4):381–407.

Mokhov, I.I. and V.Ch. Khon, 2002. Hydrological regime in basins of Siberian rivers: Model estimates of changes in the 21st century. Meteorology and Hydrology, 8:77–92. (In Russian)

Mokhov, I.I., V.A. Semenov and V.Ch. Khon, 2003. Estimates of possible regional hydrologic regime changes in the 21st century based on Global Climate Models. Izvestiya, Atmospheric and Oceanic Physics, 39(2):130–144.

Möller, D., 1996. Die Höhen und Höhenänderungen des Inlandeises. Die Weiterfürung der geodätischen Arbeiten der Internationalen Glaziologischen Grönland-Expedition (EGIG) durch das Institut für Vermessungskunde der TU Braunschweig 1987–1993. Deutsche Geodätische Kommission bei der Bayrischen Akademie der Wissenschaften, Reihe B, Angewandte Geodäsie, Heft Nr. 303. Verlag der Bayrischen Akademie der Wissenschaften: 49–58.

Munk, W., 2002. Twentieth century sea level: An enigma. Proceedings of the National Academy of Sciences, 99:6550–6555.

Munk, W., 2003. Ocean freshening, sea level rising. Science, 300: 2041–2043.

Nelson, F.E., 2003. Geocryology: (Un)frozen in time. Science, 299:1673–1675.

Nelson, F.E., O.A. Anisimov and N.I. Shiklomanov, 2001. Subsidence risk from thawing permafrost. Nature, 410:889–890.

New, M., M. Hulme and P. Jones, 2000. Representing twentieth-century space-time climate variability. Part II: Development of 1901–96 monthly grids of terrestrial surface climate. Journal of Climate, 13:2217–2238.

NRCC, 1988. Glossary of Permafrost and Related Ground-ice Terms. Permafrost Subcommittee, National Research Council of Canada, Technical Memorandum 142, 156pp.

Oberman, N.G. and G.G. Mazhitova, 2001. Permafrost dynamics in the north-east of European Russia at the end of the 20th century. Norwegian Journal of Geography, 55:241–244.

Oerlemans, J. and B.K. Reichert, 2000. Relating glacier mass balance to meteorological data using a Seasonal Sensitivity Characteristic (SSC). Journal of Glaciology, 46:1–6.

Ogilvie, A.E.J. and T. Jonsson, 2001. 'Little Ice Age' research: A perspective from Iceland. Climatic Change, 48:9–52.

Ohmura, A., 1997. New temperature distribution maps for Greenland. Zeitschrift fur Gletscherkunde und Glazialgeologie, 23:1–45.

Ohmura, A., M. Wild and L. Bengtsson, 1996. A possible change in mass balance of Greenland and Antarctic ice sheets. Journal of Climate, 9:2124–2137.

Oki, T., K. Musiake, H. Matsuyama and K. Masuda, 1995. Global atmospheric water balance and runoff from large river basins. Hydrological Processes, 9:655–678.

Osterkamp, T.E., 2001. Sub-sea permafrost. In: Encyclopedia of Ocean Sciences, pp. 2902–2912. Academic Press.

Osterkamp, T.E., 2003. A thermal history of permafrost in Alaska. Proceedings of Eighth International Conference on Permafrost, Zurich, pp. 863–868.

Osterkamp, T.E. and W.D. Harrison, 1977. Sub-sea permafrost regime at Prudhoe Bay, Alaska, U.S.A. Journal of Glaciology, 19:627–637.

Osterkamp, T.E. and V.E. Romanovsky, 1999. Evidence for warming and thawing of discontinuous permafrost in Alaska. Permafrost and Periglacial Processes, 10(1):17–37.

Osterkamp, T.E., L. Viereck, Y. Shur, M.T. Jorgenson, C. Racine, A. Doyle and R.D. Boone, 2000. Observations of thermokarst and its impact on boreal forests in Alaska, U.S.A. Arctic, Antarctic and Alpine Research, 32:303–315.

Overland, J.E., J.M. Adams and M.A. Bond, 1999. Decadal variability of the Aleutian Low and its relation to high latitude circulation. Journal of Climate, 12:1542–1548.

Paeth, H., A. Hense and R. Hagenbrock, 2002. Comments on 'Twentieth-century trends of Arctic precipitation from observational data and a climate model simulation.' Journal of Climate, 15: 800–803.

Palecki, M.A. and R.G. Barry, 1986. Freeze-up and break-up of lakes as an index of temperature changes during the transition seasons: a case study for Finland. Journal of Climate and Applied Meteorology, 25:893–902.

Parkinson, C.L., 2000. Recent trend reversals in Arctic sea ice extents: Possible connections to the North Atlantic Oscillation. Polar Geography, 24:1–12.

Parkinson, C.L., D.J. Cavalieri, P. Gloersen, H.J. Zwally and J.C. Comiso, 1999. Arctic sea ice extents, areas and trends. Journal of Geophysical Research, 104:20837–20856.

Paterson, W.S.B. and N. Reeh, 2001. Thinning of the ice sheet in north-west Greenland over the past forty years. Nature, 414:60–62.

Pavlov, A.V., 1994. Current changes of climate and permafrost in the Arctic and Sub-Arctic of Russia. Permafrost and Periglacial Processes, 5:101–110.

Pavlov, A.V. and N.G. Moskalenko, 2002. The thermal regime of soils in the north of western Siberia. Permafrost and Periglacial Processes, 13:43–51.

Peltier, W.R., 2001. Global glacial isostatic adjustment and modern instrumental records of relative sea level history. In: B. Douglas, M. Kearney and S. Leatherman (eds.). Sea Level Rise: History and Consequences. International Geophysics Series, 75:65–95.

Peltier, W.R., 2004. Global glacial isostasy and the surface of the ice-age Earth: the ICE-5G(VM2) model and GRACE. Annual Review of Earth and Planetary Sciences, 32:111–149.

Persson, P.O.G., C.W. Fairall, E.L. Andreas, P.S. Guest and D.K. Perovich, 2002. Measurements near the atmospheric surface flux group tower at SHEBA: Near surface conditions and surface energy budget. Journal of Geophysical Research, 107(C10), doi: 10.1029/2002JC000705.

Peterson, B.J., R.M. Holmes, J.W. McClelland, C.J. Vorosmarty, R.B. Lammers, A.I. Shiklomanov, I.A. Shiklomanov and S. Rahmstorf, 2002. Increasing river discharge to the Arctic Ocean. Science, 298:2171–2173.

Petryk, S., 1995. Numerical modeling. In: S. Beltaos (ed.). River Ice Jams, pp. 147–172. Water Resources Publications, Colorado.

Péwé, T.L., 1983. Alpine permafrost in the contiguous United States: A review. Arctic and Alpine Research, 15:145–156.

Pfeffer, W.T., J. Cohn, M.F. Meier and R.M. Krimmel, 2000. Alaskan glacier beats a rapid retreat. Eos, Transactions, American Geophysical Union, 81:48.

Polyakov, I., G. Alexeev, R. Bekryaev, U.S. Bhatt, R. Colony, M. Johnson, V. Karklin, D. Walsh and A. Yulin, 2003. Long-term variability of ice in the arctic marginal seas. Journal of Climate, 16:2078–2085.

Pomeroy, J.W. and L. Li, 2000. Prairie and Arctic areal snow cover mass balance using a blowing snow model. Journal of Geophysical Research, 105(D21):26610–26634.

Proshutinsky, A. and M. Johnson, 1997. Two circulation regimes of the wind-driven Arctic Ocean. Journal of Geophysical Research, 102:12493–12514.

Proshutinsky, A., V. Pavlov and R.H. Bourke, 2001. Sea level rise in the Arctic Ocean. Geophysical Research Letters, 28:2237–2240.

Prowse, T.D., 1994. The environmental significance of ice to cold-regions streamflow. Freshwater Biology, 32(2):241–260.

Prowse, T.D., 1995. River ice processes. In: S. Beltaos (ed.). River Ice Jams, pp. 29–70. Water Resources Publications, USA.

Prowse, T.D., 2001. River-ice ecology: Part A) Hydrologic, geomorphic and water-quality aspects. Journal of Cold Regions Engineering, 15(1):1–16.

Prowse, T.D. and S. Beltaos, 2002. Climatic control of river-ice hydrology: a review. Hydrological Processes, 16(4):805–822.

Prowse, T.D. and T. Carter, 2002. Significance of ice-induced hydraulic storage to spring runoff: a case study of the Mackenzie River. Hydrological Processes 16(4):779–788.

Prowse, T.D. and M.N. Demuth, 1993. Strength variability of major river-ice types. Nordic Hydrology, 24(3):169–182.

Prowse, T.D. and P.O. Flegg, 2000. The magnitude of river flow to the Arctic Ocean: dependence on contributing area. Hydrological Processes, 14(16–17):3185–3188.

Rachold, V., J. Brown and S.M. Solomon, 2002. Arctic Coastal Dynamics. Report of an International Workshop, Potsdam (Germany) 26–30 November 2001. Report on Polar Research 413, 27 extended abstracts, 103 pp.

Radionov, V.F., N.N. Bryazgin and E.I. Alexandrov, 1997. The Snow Cover of the Arctic Basin. Tech. Rep. APL-UW TR 9701. Applied Physics Laboratory, University of Washington, Seattle, 95pp.

Rahmstorf, S. and A. Ganopolski, 1999. Long-term global warming scenarios computed with an efficient coupled climate model. Climatic Change, 43:353–367.

Ramsay, B.H., 1998. The interactive multisensor snow and ice mapping system. Hydrological Processes, 12:1537–1546.

Reichert, B.K., L. Bengtsson and J. Oerlemans, 2001. Midlatitude forcing mechanisms for glacier mass balance investigated using general circulation models. Journal of Climate, 14:3767–3784.

Reycraft, J. and W. Skinner, 1993. Canadian lake ice conditions: An indicator of climate variability. Climatic Perspectives, 15:9–15.

Rignot, E. and R.H. Thomas, 2002. Mass balance of the polar ice sheets. Science, 297:1502–1506.

Rigor, I.G., J.M. Wallace and R.L. Colony, 2002. Response of sea ice to the Arctic Oscillation. Journal of Climate, 15:2648–2663.

Riseborough, D.W., 1990. Soil latent heat as a filter of the climate signal in permafrost. In: Proceedings of the Fifth Canadian Permafrost Conference, Université Laval, Quebec, Collection Nordicana No. 54, pp. 199–205.

Robinson, D.A., 1993. Hemispheric snow cover from satellites. Annals of Glaciology, 17:367–371.

Robinson, D.A., M.C. Serreze, R.G. Barry, G. Scharfen and G. Kukla, 1992. Large-scale patterns of snow melt and parameterized surface albedo in the Arctic Basin. Journal of Climate, 5:1109–1119.

Romanovsky, V.E. and T.E. Osterkamp, 1995. Interannual variations of the thermal regime of the active layer and near-surface permafrost in northern Alaska. Permafrost and Periglacial Processes, 6:313–335.

Romanovsky, V.E., M. Burgess, S. Smith, K. Yoshikawa and J. Brown, 2002. Permafrost temperature records: Indicators of climate change. Eos, Transactions, American Geophysical Union, 83:589–594.

Rothrock, D.A., Y. Yu and G.A. Maykut, 1999. Thinning of the Arctic sea-ice cover. Geophysical Research Letters, 26:3469–3472.

Rouse, W.R., E.M. Blyth, R.W. Crawford, J.R. Gyakum, J.R. Janowicz, B. Kochtubajda, H.G. Leighton, P. Marsh, L. Martz, A. Pietroniro, H. Ritchie, W.M. Schertzer, E.D. Soulis, R.E. Stewart, G.S. Strong and M.K. Woo, 2003. Energy and water cycles in a high-latitude north-flowing river system. Bulletin of the American Meteorological Society, 84:73–87.

Sagarin, R. and F. Micheli, 2001. Climate change in nontraditional data sets. Science, 294:811.

Savelieva, N.I., I.P. Semiletov, L.N. Vasilevskaya and S.P. Pugach, 2000. A climate shift in seasonal values of meteorological and hydrological parameters for Northeastern Asia. Progress in Oceanography, 47:279–297.

Savelieva, N.I., I.P. Semiletov, L.N. Vasilevskaya and S.P. Pugach, 2002. Climatic variability of the Siberian rivers seasonal discharge. In: I.P. Semiletov (ed.). Hydrometeorological and Biogeochemistry Research in the Arctic. Trudy Arctic Regional Center, 2:9–22. (In Russian)

Sazonova, T.S. and V.E. Romanovsky, 2003. A model for regional-scale estimation of temporal and spatial variability of active-layer thickness and mean annual ground temperatures. Permafrost and Periglacial Processes, 14(2):125–139.

Sazonova, T.S., V.E. Romanovsky, J.E. Walsh and D.O. Sergueev, 2004. Permafrost dynamics in the 20th and 21st centuries along the East Siberian transect. Journal of Geophysical Research, 109, doi:10.1029/2003JD003680.

Schindler, D.W., S.E. Bayley, B.R. Parker, K.G. Beaty, D.R. Cruikshank, E.J. Fee, E.U. Schindler and M.P. Stainton, 1996. The effects of climate warming on the properties of boreal lakes and streams at the Experimental Lakes Area, northwestern Ontario. Limnology and Oceanography, 41:1004–1017.

Sellman, P.V. and D.M. Hopkins, 1984. Subsea permafrost distribution on the Alaskan Shelf. In: Final Proceedings of the Fourth International Permafrost Conference, 17–22 July 1983, pp. 75–82. National Academy Press, Washington, D.C.

Semiletov, I.P., N.I. Savelieva and G.E. Weller, 2002. Cause and effect linkages between atmosphere, the Siberian rivers and conditions in the Russian shelf seas. In: I.P. Semiletov (ed.). Changes in the Atmosphere-Land-Sea System in the American Arctic. Proceedings of the Arctic Regional Center, Vladivostok, 3:63–97.

Serreze, M.C., J.E. Walsh, F.S. Chapin III, T. Osterkamp, M. Dyurgerov, V. Romanovsky, W.C. Oechel, J. Morison, T. Zhang and R.G. Barry, 2000. Observational evidence of recent changes in the northern high-latitude environment. Climatic Change, 46:159–207.

Serreze, M.C., D.H. Bromwich, M.P. Clark, A.J. Etringer, T. Zhang and R. Lammers, 2003. Large-scale hydro-climatology of the terrestrial Arctic drainage system. Journal of Geophysical Research, 108(D2). doi:10.1029/2001JD000919.

Sharkhuu, N., 2003. Recent changes in permafrost of Mongolia. Proceedings of the Eighth International Conference on Permafrost, pp. 1029–1034.

Shiklomanov, A.I., 1994. Influence of anthropogenic changes in global climate on the Yenisey River Runoff. Meteorology and Hydrology, 2:84–93. (In Russian)

Shiklomanov, A.I., 1997. On the effect of anthropogenic change in the global climate on river runoff in the Yenisei basin. In: Runoff Computations for Water Projects. Proceedings of the St. Petersburg Symposium 30 Oct.–03 Nov. 1995, pp. 113–119. IHP-V UNESCO Technical Document in Hydrology N9.

Shiklomanov, A.I., R.B. Lammers and C.J. Vorosmarty, 2002. Widespread decline in hydrological monitoring threatens pan-Arctic research. Eos, Transactions, American Geophysical Union, 83(2):13,16,17.

Shiklomanov, I.A. and H. Lins, 1991. Effect of climate change on hydrology and water management. Meteorology and Hydrology, 4:51–65. (In Russian)

Shiklomanov, I.A., A.I. Shiklomanov, R.B. Lammers, B.J. Peterson and C.J. Vorosmarty, 2000. The dynamics of river water inflow to the Arctic Ocean. In: E.L. Lewis, E.P. Jones, P. Lemke, T.D. Prowse and P. Wadhams (eds.). The Freshwater Budget of the Arctic Ocean, pp. 281–296. Kluwer Academic Publishers.

Sigurðsson, O., 1998. Glacier variations in Iceland 1930–1995 – From the database of the Iceland Glaciological Society. Jökull, 45:3–25.

Sigurðsson, O., 2002. Jöklabreytingar 1930–1960,1960–1990 og 1999–2000. (Glacier variations 1930–1960, 1960–1990 and 1999–2000.) Jökull, 51:79–86.

Sigurðsson, O., 2005. Variations of termini of glaciers in Iceland in recent centuries and their connection with climate. In: C. Caseldine, A. Russell, J. Hardardóttir and O. Knudsen (eds.). Iceland - Modern Processes and Past Environments, pp. 180–192. Elsevier.

Smith, L.C., 2000. Trends in Russian Arctic river-ice formation and breakup, 1917–1994. Physical Geography, 21:46–56.

Smith, S.L. and M.M. Burgess, 1999. Mapping the sensitivity of Canadian permafrost to climate warming. In: M. Tranter, R. Armstrong, E. Brun, G. Jones, M. Sharp and M. Williams (eds.). Interactions Between the Cryosphere, Climate and Greenhouse Gases, pp. 71–80. IAHS Publication No. 256.

Smith, S.L., M.M. Burgess and F.M. Nixon, 2001. Response of active-layer and permafrost temperatures to warming during 1998 in the Mackenzie Delta, Northwest Territories and at Canadian Forces Station Alert and Baker Lake, Nunavut. Geological Survey of Canada Current Research 2001-E5, 8pp.

Smith, S.L., M.M. Burgess and A.E. Taylor, 2003. High Arctic permafrost observatory at Alert, Nunavut – analysis of a 23-year data set. Proceedings of the Eighth International Conference on Permafrost, 1073–1078.

Sokratov, S.A. and R.G. Barry, 2002. Intraseasonal variations in the thermoinsulation effect of snow cover on soil temperatures and energy balance. Journal of Geophysical Research, 107(D9–10), ACL 13 1–7.

Soldatova, I.I., 1993. Secular variations in river break-up dates and their relationship with climate variation. Russian Meteorology and Hydrology, 9:70–76.

Solomon, S.M., D.L. Forbes and B. Kierstead, 1994. Coastal Impacts of Climate Change: Beaufort Sea Erosion Study. Geological Survey of Canada, Open File 2890, 85pp.

Staub, B. and C. Rosenzweig, 1987. Global Gridded Data Sets of Soil Type, Soil Texture, Surface Slope and Other Properties. National Center for Atmospheric Research, Boulder, Colorado. (In digital format available on Global Ecosystem's Database CD, version 1, 1991)

Steele, M. and T. Boyd, 1998. Retreat of the cold halocline layer in the Arctic Ocean. Journal of Geophysical Research, 103:10419–10435.

Stewart, K.M. and R.K. Haugen, 1990. Influence of lake morphometry on ice dates. Internationale Vereinigung fur Theoretische und Angewandte Limnologie: Verhandlungen, 24:122–127.

Stieglitz, M., S.J. Dery, V.E. Romanovsky and T.E. Osterkamp, 2003. The role of snow cover in the warming of arctic permafrost. Geophysical Research Letters, 30(13), doi:10.1029/2003GL017337.

Thompson, D.W. and J.M. Wallace, 1998. The Arctic Oscillation signature in the wintertime geopotential height and temperature fields. Geophysical Research Letters, 25:1297–1230.

Thoroddsen, T., 1917. Arferdi a Islandi i thusund ar, Hid islenzka fraedafelag, Copenhagen.

Tucker, W.B. III, J.W. Weatherly, D.T. Eppler, L.D. Farmer and D.L. Bentley, 2001. Evidence for rapid thinning of sea ice in the western Arctic Ocean at the end of the 1980s. Geophysical Research Letters, 28:2851–2854.

Van de Wal, R.S.W. and M. Wild, 2001. Modelling the response of glaciers to climate change by applying volume-area scaling in combination with a high-resolution GCM. Climate Dynamics, 18:359–366.

Van de Wal, R.S.W., M. Wild and J. de Wolde, 2001. Short-term volume changes of the Greenland ice sheet in response to doubled CO_2 conditions. Tellus, 53B:94–102.

Van der Linden, S., T. Virtanen, N. Oberman and P. Kuhry, 2003. Sensitivity analysis of discharge in the arctic Usa basin, East-European Russia. Climatic Change, 57:139–161.

Van der Veen, C.J., 1996. Tidewater calving. Journal of Glaciology, 41:375–385.

Van der Veen, C.J. (ed.), 1997. Calving Glaciers: Report of a workshop, February 28 – March 2, 1996. BPRC Report No. 15. Byrd Polar Research Center, Ohio State University, 194pp.

Van der Veen, C.J., 2002. Polar ice sheets and global sea level: how well can we predict the future? Global and Planetary Change, 32:165–194.

Vavrus, S.J., R.H. Wynne and J.A. Foley, 1996. Measuring the sensitivity of southern Wisconsin lake ice to climate variations and lake depth using a numerical model. Limnology and Oceanography, 41:822–831.

Vigdorchik, M.E., 1980. Arctic Pleistocene History and the Development of Submarine Permafrost. Westview Press, Colorado, 286pp.

Vinje, T., 2001. Anomalies and trends of sea ice extent and atmospheric circulation in the Nordic Seas during the period 1864–1998. Journal of Climate, 14:255–267.

Vinnikov, K.Y., A. Robock, R.J. Stouffer, J.E. Walsh, C.L. Parkinson, D.J. Cavalieri, J.F.B. Mitchell, D. Garrett and V.F. Zakharov, 1999. Global warming and Northern Hemisphere sea ice extent. Science, 286:1934–1937.

Wadhams, P. and N.R. Davis, 2000. Further evidence of ice thinning in the Arctic Ocean. Geophysical Research Letters, 27:3973–3975.

Walsh, S.E., S.J. Vavrus, J.A. Foley, A. Fisher, R.H. Wynne and J.D. Lenters, 1998. Global patterns of lake ice phenology and climate: model simulations and observations. Journal of Geophysical Research, 103:28825–28837.

Washburn, A.L., 1956. Classification of patterned ground and review of suggested origins. Geological Society of America Bulletin, 67:823–865.

WCRP, 1996. Proceedings of the workshop on the Implementation of the Arctic Precipitation Data Archive (APDA) at the Global Precipitation Climate Centre (GPCC), Offenbach, Germany. World Climate Research Programme. Report No. WCRP-98, 44pp.

Weidick, A., 1968. Observation on some Holocene glacier fluctuations in West Greenland. Meddelelser om Grønland, 165(6) 202pp.

Weidick, A., C.E. Bøggild and O.B. Olesen, 1992. Glacier Inventory and Atlas of West Greenland. Report 158. Grønlands Geologiske Undersøgelse, 194pp.

Willemse, N.W. and T.E. Tornqvist, 1999. Holocene century-scale temperature variability from West Greenland lake records. Geology, 27(7):580–584.

Williams, P.J. and M.W. Smith, 1989. The Frozen Earth: Fundamentals of Geocryology. Cambridge University Press, 306pp.

Williams, R.S. Jr., 1986. Glaciers and glacial landforms. Chapter 9. In: N.M. Short and R.W. Blair Jr. (eds.). Geomorphology from Space. A Global Overview of Regional Landforms. NASA Special Publication SP-486, pp. 521–596.

Williams, R.S. Jr. and J.G. Ferrigno (eds.), 2002. Satellite Image Atlas of Glaciers of the World: Glaciers of North America. U.S. Geological Survey, Washington, D.C., 405pp.

Winsor, P., 2001. Arctic sea ice thickness remained constant during the 1990s. Geophysical Research Letters, 28(6):1039–1041.

Winther, J.-G., 1993. Short- and long-term variability of snow albedo. Nordic Hydrology, 24:199–212.

Wolfe, S.A, S.R. Dallimore and S.M. Solomon, 1998. Coastal permafrost investigations along a rapidly eroding shoreline, Tuktoyaktuk, N.W.T. In: A.G. Lewkowicz and M. Allard (eds.). Proceedings of the Seventh International Conference on Permafrost, pp. 1125–1131.

Wright, J.F., M.W. Smith and A.E. Taylor, 2000. Potential changes in permafrost distribution in the Fort Simpson and Norman Wells Area. In: L.D. Dyke and G.R. Brooks (eds.). The Physical Environment of the Mackenzie Valley, Northwest Territories: A Base Line for the Assessment of Environmental Change, pp. 197–207. Geological Survey of Canada Bulletin 547.

Wynn, R.H. and T.M. Lillesand, 1993. Satellite observation of lake ice as a climatic indicator: initial results from state-wide monitoring in Wisconsin. Photogrammetric Engineering and Remote Sensing, 59:1023–1031.

Wynn, R.H., J.J. Magnuson, M.K. Clayton, T.M. Lillesand and D.C. Rodman, 1996. Determinants of temporal coherence in the satellite-derived 1987–1994 ice breakup dates of lakes on the Laurentian Shield. Limnology and Oceanography, 41:831–838.

Yang, D., 1999. An improved precipitation climatology for the Arctic Ocean. Geophysical Research Letters, 26:1625–1628.

Ye, B., D. Yang and D. Kane, 2003. Changes in Lena River streamflow hydrology: Human impacts versus natural variations. Water Resources Research, 39(7), doi:10.1029/2003WR001991.

Ye, H., 2001. Increases in snow season length due to earlier first snow and later last snow dates over north central and northwest Asia during 1937–94. Geophysical Research Letters, 28:551–554.

Ye, H., H. Cho and P.E. Gustafson, 1998. The changes in Russian winter snow accumulation during 1936–1983 and its spatial patterns. Journal of Climate, 11:856–863.

Zachrisson, G., 1989. Climate variation and ice conditions in the River Torneälven. In: Proceedings of the Conference on Climate and Water, Helsinki. Publications of the Academy of Finland, Vol. 1:353–364.

Zeeberg, J.J. and S.L. Forman, 2001. Changes in glacier extent on north Novaya Zemlya in the twentieth century. The Holocene, 11(2):161–175.

Zervas, C. 2001. Sea Level Variations of the United States 1854–1999. NOAA Technical Report NOS CO-OPS 36. National Oceanic and Atmospheric Administration, Silver Spring, Maryland, 65pp+appendices.

Zhang, T., J.A. Heginbottom, R.G. Barry and J. Brown, 2000. Further statistics on the distribution of permafrost and ground ice in the Northern Hemisphere. Polar Geography, 24(2):125–131.

Zhang, X., K.D. Harvey, W.D. Hogg and T.R. Yuzyk, 2001. Trends in Canadian streamflow. Water Resources Research, 37:987–998.

Zhuang, Q., V.E. Romanovsky and A.D. McGuire, 2001. Incorporation of a permafrost model into a large-scale ecosystem model: Evaluation of temporal and spatial scaling issues in simulating soil thermal dynamics. Journal of Geophysical Research, 106(24):33649–33670.

Zwally, H.J., W. Abdalati, T. Herring, K. Larson, J. Saba and K. Steffen, 2002. Surface melt-induced acceleration of Greenland ice-sheet flow. Science, 297:218–222

Chapter 7

Arctic Tundra and Polar Desert Ecosystems

Lead Author
Terry V. Callaghan

Contributing Authors
Lars Olof Björn, F. Stuart Chapin III, Yuri Chernov, Torben R. Christensen, Brian Huntley, Rolf Ims, Margareta Johansson, Dyanna Jolly Riedlinger, Sven Jonasson, Nadya Matveyeva, Walter Oechel, Nicolai Panikov, Gus Shaver

Consulting Authors
Josef Elster, Heikki Henttonen, Ingibjörg S. Jónsdóttir, Kari Laine, Sibyll Schaphoff, Stephen Sitch, Erja Taulavuori, Kari Taulavuori, Christoph Zöckler

Contents

Summary

The dominant response of current arctic species to climate change, as in the past, is very likely to be relocation rather than adaptation. Relocation possibilities vary according to region and geographic barriers. Some changes are occurring now.

Some groups such as mosses, lichens, and some herbivores and their predators are at risk in some areas, but productivity and number of species is very likely to increase. Biodiversity is more at risk in some ACIA regions than in others: Beringia (Region 3) has a higher number of threatened plant and animal species than any other ACIA region.

Changes in populations are triggered by trends and extreme events, particularly winter processes.

Forest is very likely to replace a significant proportion of the tundra and this will have a great effect on the composition of species. However, there are environmental and sociological processes that are very likely to prevent forest from advancing in some locations.

Displacement of tundra by forest will lead to a decrease in albedo, which will increase the positive feedback to the climate system. This positive feedback is likely to dominate over the negative feedback of increased carbon sequestration. Forest development is very likely to also ameliorate local climate, for example, by increasing temperature.

Warming and drying of tundra soils in parts of Alaska have already changed the carbon status of this area from sink to source. Although other areas still maintain their sink status, the number of source areas currently exceeds the number of sink areas. However, geographic representation of research sites is currently small. Future warming of tundra soils is likely to lead to a pulse of trace gases into the atmosphere, particularly from disturbed areas and areas that are drying. It is not known if the circumpolar tundra will be a carbon source or sink in the long term, but current models suggest that the tundra is likely to become a weak sink for carbon because of the northward movement of vegetation zones that are more productive than those they displace. Uncertainties are high.

Rapid climate change that exceeds the ability of species to relocate is very likely to lead to increased incidence of fires, disease, and pest outbreaks.

Enhanced carbon dioxide concentrations and ultraviolet-B radiation levels affect plant tissue chemistry and thereby have subtle but long-term impacts on ecosystem processes that reduce nutrient cycling and have the potential to decrease productivity and increase or decrease herbivory.

7.1. Introduction

The Arctic is generally recognized as a treeless wilderness with cold winters and cool summers. However, definitions of the southern boundary vary according to environmental, geographic, or political biases. This chapter focuses on biota (plants, animals, and microorganisms) and processes in the region north of the northern limit of the closed forest (the taiga), but also includes processes occurring south of this boundary that affect arctic ecosystems. Examples include animals that migrate south for the winter and the regulation of the latitudinal treeline. The geographic area defined in this chapter as the present-day Arctic is the area used for developing scenarios of future impacts: the geographic area of interest will not decrease under a scenario of replacement of current arctic tundra by boreal forests.

7.1.1. Characteristics of arctic tundra and polar desert ecosystems

The southern boundary of the circumpolar Arctic as defined in this chapter is the northern extent of the closed boreal forests (section 14.2.3). This is not a clear boundary but a transition from south to north consisting of the sequence: closed forest, forest with patches of tundra, tundra with patches of forest, and tundra. The transition zone is relatively narrow (30 to 150 km) when compared to the width of the forest and tundra zones in many, but not all areas. Superimposed on the latitudinal zonation of forest and tundra is an altitudinal zonation from forest to treeless areas to barren ground in some mountainous regions of the northern taiga. The transition zone from taiga to tundra stretches for more than 13 400 km around the lands of the Northern Hemisphere and is one of the most important environmental transition zones on Earth (Callaghan et al., 2002a,b) as it represents a strong temperature threshold close to an area of low temperatures. The transition zone has been called forest tundra, subarctic, and the tundra–taiga boundary or ecotone. The vegetation of the transition zone is characterized by an open landscape with patches of trees that have a low stature and dense thickets of shrubs that, together with the trees, totally cover the ground surface.

The environmental definition of the Arctic does not correspond with the geographic zone delimited by the Arctic Circle (66.5° N), nor with political definitions. Cold waters in ocean currents flowing southward from the Arctic depress the temperatures in Greenland and the eastern Canadian Arctic whereas the northward-flowing Gulf Stream warms the northern landmasses of Europe (section 2.3). Thus, at the extremes, polar bears and tundra are found at 51° N in eastern Canada whereas agriculture is practiced north of 69° N in Norway. Arctic lands span some 20° of latitude, reaching 84° N in Greenland and locally, in eastern Canada, an extreme southern limit of 51° N.

The climate of the Arctic is largely determined by the relatively low solar angles with respect to the earth. Differences in photoperiod between summer and winter become more extreme toward the north. Beyond the Arctic Circle, the sun remains above the horizon at midnight on midsummer's day and remains below the horizon at midday on midwinter's day.

Climatically, the Arctic is often defined as the area where the average temperature for the warmest month is lower than 10 °C (Köppen, 1931), but mean annual air temperatures vary greatly according to location, even at the same latitude (see Chapter 2). They vary from -12.2 °C at Point Barrow, Alaska (71.3° N) to -28.1 °C at the summit of the Greenland Ice Sheet (about 71° N) (Weller, 2000) and from 1.5 °C at 52° N in subarctic Canada to 8.9 °C at 52° N in temperate Europe. The summer period, or period of most biological activity, progressively decreases from about 3.5 to 1.5 months from the southern boundary of the Arctic to the north, and mean July temperature decreases from 10–12 °C to 1.5 °C. In general, annual precipitation in the Arctic is low, decreasing from about 250 mm in southern areas to as low as 45 mm in the northern polar deserts (Jonasson et al., 2000), with extreme precipitation amounts in subarctic maritime areas (e.g., 1100 mm at 68° N in Norway). However, owing to low rates of evaporation the Arctic cannot be considered arid: even in the polar deserts, air humidity is high and the soils are moist during the short growth period (Bovis and Barry, 1974). In the Arctic context, "desert" refers to extreme poverty of life.

The Arctic is characterized by the presence of continuous permafrost (section 6.6.1), although there are exceptions such as the Kola Peninsula. Continuous and deep (>200 m) permafrost also exists south of the treeline in large areas of Siberia extending south to Mongolia. The depth of the active (seasonally frozen) layer of the soil during the growing season depends on summer temperatures and varies from about 80 cm near the treeline to about 40 cm in polar deserts. However, active-layer depth varies according to local conditions within landscapes according to topography: it can reach 120 cm on south-facing slopes and be as little as 30 cm in bogs even

in the southern part of the tundra zone. In many areas of the Arctic, continuous permafrost occurs at greater depths beneath the soil surface and degrades into discontinuous permafrost in the southern part of the zone. Active-layer depth, the extent of discontinuous permafrost, and coastal permafrost are very likely to be particularly sensitive to climatic warming (section 6.6). Permafrost and active-layer dynamics lead to topographic patterns such as polygons in the landscape. Topography plays an important role in defining habitats in terms of moisture and temperature as well as active-layer dynamics (Brown et al., 1980; Webber et al., 1980), such that arctic landscapes are a mosaic of microenvironments. Topographic differences of even a few tens of centimeters (e.g., polygon rims and centers) are important for determining habitats, whereas larger-scale topographic differences (meters to tens of meters) determine wind exposure and snow accumulation that in turn affect plant communities and animal distribution. Topographic differences become more important as latitude increases.

Disturbances of ecosystems are characteristic of the Arctic. Mechanical disturbances include thermokarst induced by permafrost thaw (section 6.6.1); freeze–thaw processes; wind, sand, and ice blasts; seasonal ice oscillations; slope processes; snow load; flooding during thaw; changes in river volume; and coastal erosion and flooding. Biological disturbances include insect pest outbreaks, peaks of grazing animals that have cyclic populations, and fire. These disturbances operate at various spatial and temporal scales (Fig. 7.1) and affect the colonization and survival of organisms and thus ecosystem development.

Arctic lands are extensive beyond the northern limit of the tundra–taiga ecotone, encompassing an area of approximately 7 567 000 km², including about

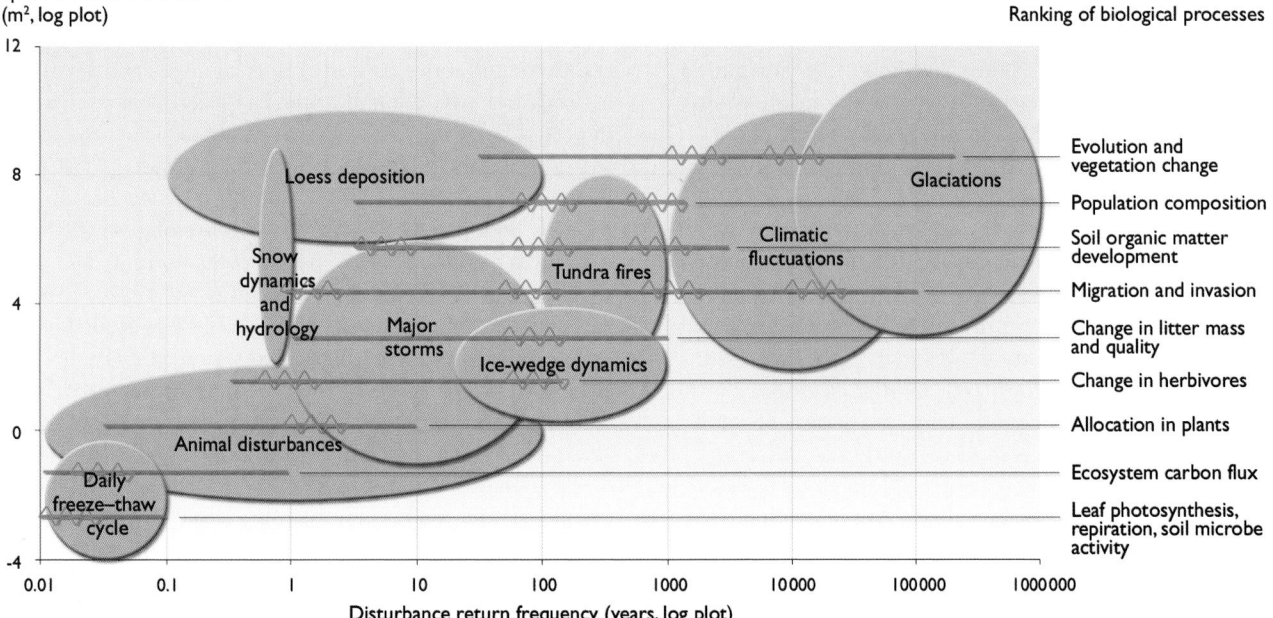

Fig. 7.1. Timescale of ecological processes in relation to disturbances (shown as breaks in horizontal lines) in the Arctic. The schematic does not show responses projected as a result of anthropogenic climate change (based on Oechel and Billings, 1992; Shaver et al., 2000; Walker D. and Walker, 1991).

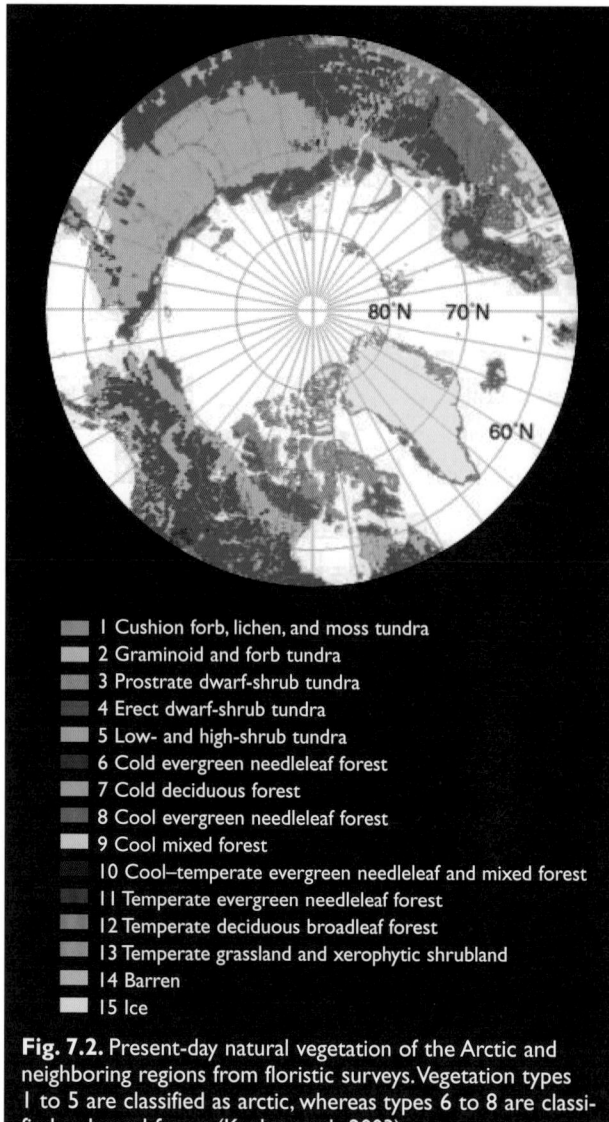

1 Cushion forb, lichen, and moss tundra
2 Graminoid and forb tundra
3 Prostrate dwarf-shrub tundra
4 Erect dwarf-shrub tundra
5 Low- and high-shrub tundra
6 Cold evergreen needleleaf forest
7 Cold deciduous forest
8 Cool evergreen needleleaf forest
9 Cool mixed forest
10 Cool–temperate evergreen needleleaf and mixed forest
11 Temperate evergreen needleleaf forest
12 Temperate deciduous broadleaf forest
13 Temperate grassland and xerophytic shrubland
14 Barren
15 Ice

Fig. 7.2. Present-day natural vegetation of the Arctic and neighboring regions from floristic surveys. Vegetation types 1 to 5 are classified as arctic, whereas types 6 to 8 are classified as boreal forest (Kaplan et al., 2003).

2 560 000 km^2 in the former Soviet Union and Scandinavia, 2 480 000 km^2 in Canada, 2 167 000 km^2 in Greenland and Iceland, and 360 000 km^2 in Alaska (Bliss and Matveyeva, 1992). Figure 7.2 shows the distribution of arctic and other vegetation types based on a classification by Walker D. (2000) and mapped by Kaplan et al. (2003). The distribution of arctic landmasses is often fragmented: seas separate large arctic islands (e.g., Svalbard, Novaya Zemlya, Severnaya Zemlya, New Siberian Islands, and Wrangel Island) and the landmasses of the Canadian Archipelago and Greenland. Similarly, the Bering Strait separates the arctic lands of Eurasia and North America. Large mountains such as the east–west running Brooks Range in Alaska and the Putorana Plateau in Siberia separate tundra and taiga. Such areas of relief contain outposts of boreal species on their southern major slopes that are likely to expand northward and higher-elevation areas that are likely to act as refuges for arctic-alpine species. The Taymir Peninsula is the only continuous landmass that stretches 900 km from the northern tundra limit to taiga without geographic barriers to the dispersal of animals and plants (Matveyeva and Chernov, 2000). The width of the tundra zone varies greatly in different parts of its circumpolar distribution.

On average, it does not exceed 300 km, and in some regions (e.g., the lower reaches of the Kolyma River), the tundra zone extends only 60 km from the treeline to the coast. In such areas, the tundra zone is very likely to be highly vulnerable to climate warming.

The vegetation of the Arctic varies from forest tundra in the south, where plant communities have all the plant life forms known in the Arctic and have continuous canopies in several layers extending to more than 3 m high, to polar deserts in the north, where vegetation colonizes 5% or less of the ground surface, is less than 10 cm high, and is dominated by herbs, lichens (symbionts of algae and fungi), and mosses (Fig. 7.3). Species richness in the Arctic is low and decreases toward the north: there are about 1800 species of vascular plants, 4000 species of cryptogams, 75 species of terrestrial mammals, 240 species of terrestrial birds, 3000 species of fungi, 3300 species of insects (Chernov, 2002; Matveyeva and Chernov, 2000), and thousands of prokaryotic species (bacteria and Archaea) whose diversity in the tundra has only recently started to be estimated. However, the Arctic is an important global pool of some groups such as mosses, lichens, springtails (and insect parasitoids: Hawkins, 1990; Kouki et al., 1994, Price et al., 1995) because their abundance in the Arctic is higher than in other biomes. Net primary production (NPP), net ecosystem production (NEP), and decomposition rates are low. Food chains are often short and typically there are few representatives at each level of the chain. Arctic soils are generally shallow and underdeveloped with low productivity and immature moor-type humus (Brown et al., 1980). Substantial heterogeneity of the soil cover, owing to numerous spatial gradients, has an important influence on the microtopographical distribution of the soil biota (invertebrates, fungi, and bacteria) that will possibly amplify any negative effects of climate change.

The Arctic has a long history of human settlement and exploitation, based initially on its rich aquatic biological resources and more recently on its minerals and fossil hydrocarbons. At the end of the last glacial stage, humans migrated from Eurasia to North America across the ice-free Bering land bridge and along the southern coast of Beringia (ca. 14 000–13 500 years BP; Dixon, 2001). As early as about 12 200 years BP, areas north of the Fennoscandian Ice Sheet in northernmost Finnmark (Norway) had been settled (Thommessen, 1996). Even earlier Paleolithic settlements (ca. 40 000 years BP) have been recorded in the eastern European Arctic (Pavlov et al., 2001). The impacts of these peoples on terrestrial ecosystems are difficult to assess but were probably small given their small populations and "hunter-gatherer" way of life. The prey species hunted by these peoples included the megafauna, such as the woolly mammoth, which became extinct. The extent to which hunting may have been principally responsible for these extinctions is a matter of continuing debate (Stuart et al., 2002) but this possibility cannot be excluded (Alroy, 2001). It is also uncertain to what extent the extinction of the megafauna may have contributed to, or been at least partly a result

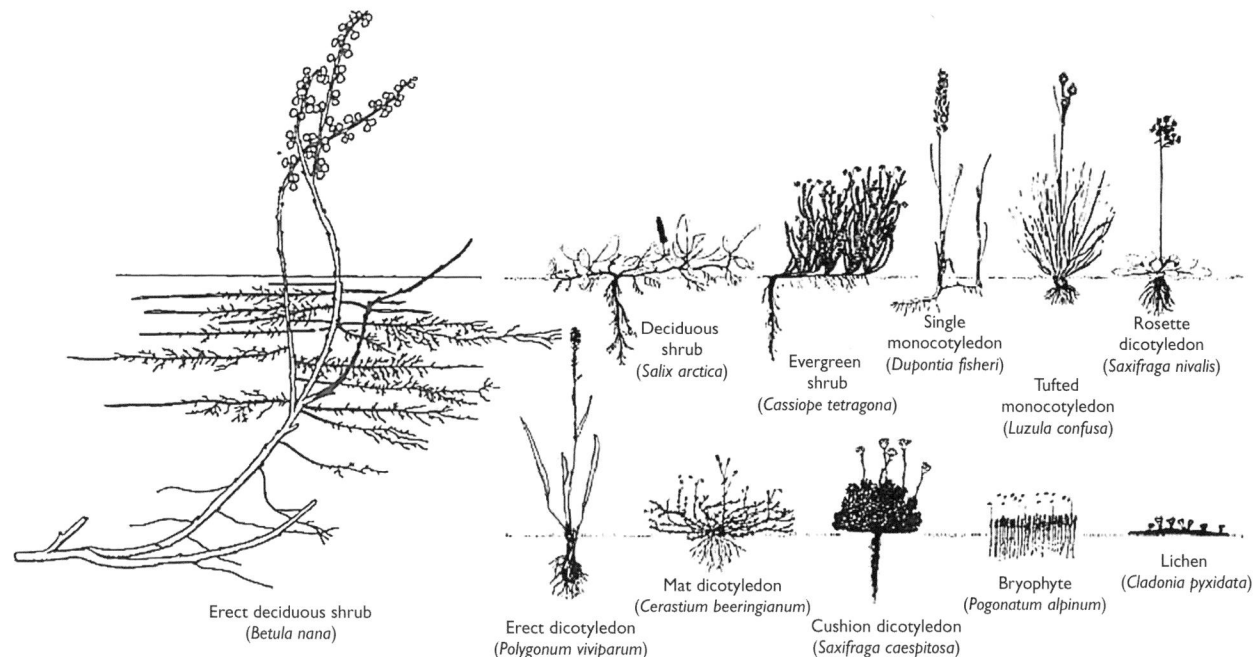

Fig. 7.3. Growth forms of arctic plants (modified from Webber et al., 1980 and T. Polozova, pers. comm., 2005).

of, the accelerated northward movement of trees and shrubs and consequent changes in vegetation structure (section 7.2). Although estimates of the population density of megafaunal species have large uncertainties, it seems unlikely that megafaunal populations were sufficient to constrain the spread of woody taxa in response to favorable environmental change.

During the last 1000 years, resources from terrestrial ecosystems have been central to the mixed economies of the Arctic: many inland indigenous communities still derive most of their protein from subsistence activities such as caribou/reindeer hunting (Berkes and Fast, 1996). During this period, increasing trade between peoples of temperate latitudes and arctic indigenous peoples is likely to have affected a few target animal species, such as the reindeer that was domesticated in Fennoscandia and Russia, ermine hunted for fur, and birds of prey used for hunting as far away as the eastern Mediterranean. However, the most dramatic impacts occurred after World War II as a result of the exploitation of minerals and oil and fragmentation of the arctic landscape by infrastructure (Nellemann et al., 2001). Vlassova (2002) suggested that industrial activities and forestry have displaced the Russian forest tundra southward by deforesting 470 000 to 500 000 km² of land that now superficially resembles tundra. Although this estimate has been challenged as greatly exaggerated (because northern taiga areas have been included in that estimate's definition of forest tundra), such effects have occurred locally in the Yamal Peninsula and the estimate highlights a need for reappraisal. Knowledge of possible past interactions between humans and the environment that may have shaped present-day arctic ecosystems is limited, but shows that any future increases in population density and human activity are likely to modify the projected responses of arctic ecosystems to changes in climate and ultraviolet (UV) radiation levels.

7.1.2. Raison d'être for the chapter

The Arctic is experiencing dramatic environmental changes that are likely to have profound impacts on arctic ecosystems. The Arctic is outstanding among global biomes in that climate change dominates the major factors affecting biodiversity (Sala and Chapin, 2000). Present-day arctic biota are also relatively restricted in range and population size compared with their Quaternary situation. For example, when the treeline advanced northward during the early Holocene warming, a lowered sea level allowed a belt of tundra to persist around the Arctic Basin, whereas any future northward migration of the treeline is very likely to further restrict tundra areas because sea level is projected to rise. Arctic ecosystems are known to be vulnerable to disturbances (Crawford, 1997b; Forbes et al., 2001; Walker D. and Walker, 1991) and to have long recovery times: subarctic birch forest defoliated by insects can take 70 years to recover (Tenow and Bylund, 2000). Current and projected environmental changes are likely to create additional stresses and decrease the potential for ecosystem recovery from natural disturbances, while providing thresholds for shifts to new states (e.g., disturbance opening gaps for invasion of species new to the Arctic).

Changes in arctic ecosystems and their biota are important to arctic residents in terms of food, fuel, and culture (Chapter 12) and are likely to have global impacts because of the many linkages between the Arctic and more southerly regions. Several hundreds of millions of birds migrate to the Arctic each year and their success in the Arctic determines their success and impacts at lower latitudes (section 7.3.1.2). Physical and biogeochemical processes in the Arctic affect atmospheric circulation and the climate of regions outside of the Arctic (section 7.5). It is known that ecosystems have responded to past environmental changes (section 7.2) and that environmental changes are presently occurring in the Arctic (Chapman and Walsh, 1993 as quoted in Weller, 2000; Dye, 2002; Fioletov et al., 1997; Chapters 2, 5, and 6). This understanding indicates that there are very likely to be responses of arctic ecosystems to projected future and ongoing climate change. It is also known that current levels of ultraviolet-B (UV-B) radiation, as well as higher levels, can affect subarctic plants (Gwynn-Jones et al., 1997; Johanson et al., 1995; Phoenix et al., 2000). Arctic plants may be particularly sensitive to increases in UV-B irradiance because UV-B radiation damage is not dependent on temperature whereas enzyme-mediated repair of DNA damage could be constrained by low temperatures (Björn, 2002; Li et al., 2002a,b; Paulsson, 2003).

For all of these reasons, understanding the relationships between ecosystems and the arctic environment is important. Although many aspects of its environment are changing concurrently (e.g., climate, pollution, atmospheric nitrogen deposition, atmospheric concentrations of carbon dioxide (CO_2), UV-B radiation levels, and land use), the specific mission of this chapter is to focus on the impacts of changes in climate and UV-B radiation levels on arctic terrestrial ecosystems and their species and processes.

7.1.3. Rationale for the structure of the chapter

The effects of climate are specific to species, the age and developmental stages of individuals, and processes from metabolism to evolution (Fig. 7.1). Although there are many ways in which to organize an assessment of climate and UV-B radiation impacts, this chapter follows a logical hierarchy of increasing organizational biological complexity to assess impacts on species, the structure of ecosystems, the function of ecosystems, and landscape and regional processes. A basic understanding of biological processes related to climate and UV-B radiation is required before the impacts of changes in these factors on terrestrial ecosystems can be assessed (Smaglik, 2002). Consequently, this chapter progresses from a review of climate and UV radiation controls on biological processes to an assessment of the potential impacts of changes in climate and UV-B radiation levels on processes at the species and regional levels. Some effects of climate change on ecosystems may be beneficial to humans, while others may be harmful.

The changes in climate and UV-B radiation levels that are used in this chapter to assess biological impacts are of two types: those already documented (section 2.6) and those projected by scenarios of future change in UV-B radiation levels (section 5.7) and climate (section 4.4) derived from models. Mean annual and seasonal temperatures have varied considerably in the Arctic since 1965 (Chapman and Walsh, 1993 as quoted in Weller, 2000; section 2.6.2.1). Mean annual temperatures in western parts of North America and central Siberia have increased by about 1 °C (up to 2 °C in winter) per decade between 1966 and 1995 while temperatures in

West Greenland and the eastern Canadian Arctic have decreased by 0.25 to 1 °C per decade (Chapman and Walsh, 1993, quoted in Weller, 2000). Over a longer period, from 1954 to 2003, the annual increase and decrease in temperatures have been slightly less: about 2 to 3 °C for the whole period (Chapter 1, Fig. 1.3). Temperature increases in Fennoscandia over the past century have been small, ranging from about 1 °C in the west to near 0 °C in the east (Lee et al., 2000).

Precipitation has also changed. The duration of the snow-free period at high northern latitudes increased by 5 to 6 days per decade and the week of the last observed snow cover in spring advanced by 3 to 5 days per decade between 1972 and 2000 (Dye, 2002). Stratospheric ozone has been depleted over recent decades (e.g., by a maximum of 45% below normal over the high Arctic in spring; Fioletov et al., 1997). This has probably led to an increase in surface UV-B radiation levels in the Arctic, although the measurement period is short (section 5.5). Scenarios of future change project that mean annual temperatures in the Arctic will increase by nearly 4 °C by 2080 (section 4.4.2) and that spring (April) UV-B radiation levels will increase by 20 to 90% in much of the Arctic by 2010–2020 (Taalas et al., 2000).

The assessment of impacts on terrestrial ecosystems presented in this chapter is based on existing literature rather than new research or ACIA modeling activities. Existing long-term experimental manipulations of temperature and/or UV-B radiation relied on earlier scenarios of climate and UV-B radiation change (IPCC, 1990). However, the most recent scenarios (Chapters 4, 5, and 6) are used to provide a context for the assessment in this chapter, and to modify projections of ecosystem responses based on earlier scenarios where appropriate. The ACIA climate scenarios (section 4.4) are also used directly to illustrate the responses of some species to projected climate changes.

7.1.4. Approaches used for the assessment: strengths, limitations, and uncertainties

This chapter assesses information on interactions between climate, UV-B radiation levels, and ecosystems from a wide range of sources including experimental manipulations of ecosystems and environments in the field; laboratory experiments; monitoring and observation of biological processes in the field; conceptual modeling using past relationships between climate and biota (paleo-analogues) and current relationships between climate and biota in different geographic areas (geographic analogues) to infer future relationships; and process-based mathematical modeling. Where possible, indigenous knowledge (limited to published sources) is included as an additional source of observational evidence. Relevant information from indigenous peoples on arctic tundra and polar desert ecosystems is given in Chapter 3.

Each method has uncertainties and strengths and these are discussed in section 7.7. By considering and comparing different types of information, it is hoped that a more robust assessment has been achieved. However, the only certainties in this assessment are that there are various levels of uncertainty in the projections and that even if an attempt is made to estimate the magnitude of these uncertainties, surprise responses of ecosystems and their species to changes in climate and UV-B radiation levels are certain to occur.

7.2. Late-Quaternary changes in arctic terrestrial ecosystems, climate, and ultraviolet radiation levels

In order to understand the present biota and ecosystems of the Arctic, and to project the nature of their responses to potentially rapid future climate change, it is necessary to examine at least the last 21 000 years of their history. This period, which is part of the late Quaternary Period, extends from the present back to the last glacial maximum (LGM), encompassing the Holocene, or postglacial period, that spans approximately the last 11 400 years. A review of this period of the history of the biota and ecosystems found in the Arctic today also must examine a spatial domain that is not restricted to the present arctic regions. At the LGM, many of these regions were submerged beneath vast ice sheets, whereas many of the biota comprising present arctic ecosystems were found at lower latitudes.

7.2.1. Environmental history

At the LGM, vast ice sheets accumulated not only on many high-latitude continental areas but also across some relatively shallow marine basins. The beds of relatively shallow seas such as the North Sea and Bering Sea were exposed as a result of a global sea-level fall of approximately 120 m, resulting in a broad land connection between eastern Siberia and Alaska and closure of the connection between the Pacific and Arctic Oceans. The reduction in sea level also exposed a broad strip of land extending northward from the present coast of Siberia. Most, if not all, of the Arctic Ocean basin may have been covered by permanent sea ice.

Although details of the extent of some of the ice sheets continue to be a controversial matter (see e.g., Astakhov, 1998; Grosswald, 1988, 1998; Lambeck, 1995; Siegert et al., 1999), it is certain that the majority of land areas north of 60° N were ice-covered. The principal exceptions were in eastern Siberia, Beringia, and Alaska, although there is some geological evidence to suggest that smaller ice-free areas also persisted in the high Arctic, for example in the northernmost parts of the Canadian Archipelago (Andrews, 1987) and perhaps even in northern and northeastern Greenland (Funder et al., 1998). This evidence is supported by recent molecular genetic studies of arctic species; for example, a study of the dwarf shrub *Dryas integrifolia* indicates glacial occurrences in the high Arctic (Tremblay and Schoen, 1999) as well as in Beringia, and a study of the collared lemming *Dicrostonyx groenlandicus* indicates separate glacial populations east and west of the Mackenzie River (Ehrich et al., 2000; Fedorov and Goropashnaya, 1999), the latter most probably in the Canadian Archipelago. The latter conclusion is supported by the phylogeography (relationship between genetic identity and geographic distribution) of the *Paranoplocephala arctica* species complex, a cestode parasite of *Dicrostonyx* spp., indicating that two subclades probably survived the LGM with their host in the Canadian High Arctic (Wickström et al., 2003). More controversial are suggestions that elements of the arctic flora and fauna may have survived the LGM on nunataks (hills or mountains extending above the surface of a glacier) in glaciated areas of high relief such as parts of Greenland, Svalbard, and Iceland (Rundgren and Ingolfsson, 1999). Although a recent molecular genetic study of the alpine cushion plant *Eritrichium nanum* (Stehlik et al., 2001) provides strong evidence for survival of that species on nunataks within the heart of the European Alps, similar studies of arctic species have so far not supported the hypothesis of survival on nunataks in areas such as Svalbard (Abbott et al., 2000) that experienced extreme climatic severity as ice sheets extended to margins beyond the current coast during the LGM.

Direct evidence of the severity of the full glacial climate in the Arctic comes from studies of ice cores from the Greenland Ice Sheet and other arctic ice sheets (section 2.7) that indicate full glacial conditions with mean annual temperatures 10 to 13 °C colder than during the Holocene (Grootes et al., 1993). Paleotemperature reconstructions based upon dinoflagellate cyst assemblages indicate strong seasonal temperature fluctuations, with markedly cold winter temperatures (de Vernal and Hillaire-Marcel, 2000; de Vernal et al., 2000).

The LGM was, however, relatively short-lived; within a few millennia of reaching their maximum extent many of the ice sheets were decaying rapidly and seasonal temperatures had increased in many parts of the Arctic. Deglaciation was not, however, a simple unidirectional change; instead a series of climatic fluctuations occurred during the period between about 18 000 and 11 400 years BP that varied in intensity, duration, and perhaps also in geographic extent. The most marked and persistent of these fluctuations, the Younger Dryas event (Alley, 2000; Peteet, 1993, 1995), was at least hemispheric in its extent, and was marked by the reglaciation of some regions and readvances of ice-sheet margins in others. Mean annual temperatures during this event fell substantially; although not as low as during the glacial maximum, they were nonetheless 4 to 6 °C cooler than at present over most of Europe (Walker M.J., 1995), and as much as 10 to 12 °C colder than at present in the northern North Atlantic and the Norwegian Sea (Koç et al., 1996), as well as in much of northern Eurasia (Velichko, 1995). The end of the Younger Dryas was marked by a very rapid rise in temperatures. At some locations, mean annual temperature rose by more than 5 °C in less than 100 years (Dansgaard et al., 1989). The most rapid changes probably were spatially and temporally transgressive, with the global mean change thus occurring much less rapidly. Nonetheless, in many areas summer temperatures during the early Holocene rose to values higher than those at present. Winter conditions remained more severe than today in many higher-latitude areas, however, because the influence of the decaying ice sheets persisted into the early millennia of the Holocene.

Despite higher summer temperatures in the early to mid-Holocene in most of the Arctic, Holocene climate has not differed qualitatively from that at present. Following the general thermal maximum there has been a modest overall cooling trend throughout the second half of the Holocene. However, a series of millennial and centennial fluctuations in climate have been super-imposed upon these general longer-term patterns (Huntley et al., 2002). The most marked of these occurred about 8200 years BP and appears to have been triggered by the catastrophic discharge of freshwater into the northern North Atlantic from proglacial lakes in North America (Barber et al., 1999; Renssen et al., 2001). A reduction in strength, if not a partial shut-down, of the thermohaline circulation in the northern North Atlantic and Norwegian Sea was also associated with this event, as well as with the series of less severe climatic fluctuations that continued throughout the Holocene (Bianchi and McCave, 1999).

The most recent of these climatic fluctuations was that of the "Little Ice Age" (LIA), a generally cool interval spanning approximately the late 13th to early 19th centuries (section 2.7.5). At its most extreme, mean annual temperatures in some arctic areas fell by several degrees. Sea ice extended around Greenland and in some years filled the Denmark Strait between Greenland and Iceland (Lamb H.H., 1982; Ogilvie, 1984; Ogilvie and Jonsdottir, 2000; Ogilvie and Jonsson, 2001), the Norse settlement of Greenland died out (Barlow et al., 1997; Buckland et al., 1996), and the population of Iceland was greatly reduced (Ogilvie, 1991; Sveinbjarnardóttir, 1992). Although there was great temporal climate variability (on decadal to centennial timescales) within the LIA, and spatial variability in the magnitude of the impacts, it was apparently a period of generally more severe conditions in arctic and boreal latitudes; the marked impacts upon farming and fisheries (Lamb H.H., 1982) imply similar impacts on other components of the arctic ecosystem. Since the early 19th century, however, there has been an overall warming trend (Overpeck et al., 1997), although with clear evidence of both spatial variability and shorter-term temporal variability (Maxwell, 1997). The magnitude of this recent warming is comparable to that of the warmest part of the Holocene, at least in those parts of the Arctic that have experienced the most rapid warming during the last 30 years or so.

The solar variability thought to be responsible for the LIA, and for other similar centennial to millennial climatic fluctuations, probably also affected the ozone layer and UV-B radiation levels. Ultraviolet-B irradiance at ground level absorbed by DNA could have been

between 9 and 27% higher during periods of low solar output (cool periods) than during periods of high solar output (Rozema et al., 2002; see also section 5.4.1).

7.2.2. History of arctic biota

During the LGM, when most land areas in the Arctic were ice-covered, biomes able to support the elements of the arctic biota, including some species that are now extinct, were extensive south of the Fennoscandian Ice Sheet in Europe (Huntley et al., 2003). Similar biomes apparently were extensive south of the Eurasian ice sheets of northern Russia, eastward across Siberia and the exposed seabed to the north, and via Beringia into Alaska and the northern Yukon (Ritchie, 1987), although they were much more restricted south of the Laurentide Ice Sheet in central and eastern North America (Lister and Bahn, 1995). The most extensive and important of these glacial biomes, the steppe–tundra, has been interpreted and referred to by various authors as "tundra–steppe" or "Mammoth steppe" (Guthrie, 2001; Walker D. et al., 2001; Yurtsev, 2001). The vegetation of this biome comprised a no-analogue combination of light-demanding herbaceous and dwarf-shrub taxa that are found today either in arctic tundra regions or in the steppe regions that characterize central parts of both North America and Eurasia (Yurtsev, 2001). Evidence of an abundance of grazing herbivores of large body mass, some extant (e.g., reindeer/caribou – *Rangifer tarandus*; muskox – *Ovibos moschatus*) and others extinct (e.g., giant deer or "Irish elk" – *Megaloceros giganteus*; woolly mammoth – *Mammuthus primigenius*; woolly rhinoceros – *Coelodonta antiquitatis*), associated with this biome suggests that it was much more productive than is the contemporary tundra biome. This productive biome, dominated by non-tree taxa, corresponded to a no-analogue environment that was relatively cold throughout the year, with a growing season short enough to exclude even cold-tolerant boreal trees from at least the majority of the landscape. The "light climate", however, was that of the relatively lower latitudes (as low as 45° N in Europe) at which this biome occurred, rather than that of the present arctic latitudes; the greater solar angle and consequent higher insolation intensities during the summer months probably made an important contribution to the productivity of the biome.

The productive steppe–tundra and related biomes were much more spatially extensive during the last glacial stage than is the tundra biome today (Fig. 7.4). The last glacial stage was thus a time when many elements of the present arctic biota thrived, almost certainly in greater numbers than today. Fossil remains of both arctic plants (see e.g., West, 2000) and mammals (see e.g., FAUN-MAP Working Group, 1996; Lundelius et al., 1983; Stuart, 1982) found at numerous locations attest to their widespread distribution and abundance. Similar conclusions have been reached on the basis of phylogeographic studies of arctic-breeding waders (Kraaijeveld and Nieboer, 2000). Species such as red knot (*Calidris canutus*) and ruddy turnstone (*Arenaria interpres*) are

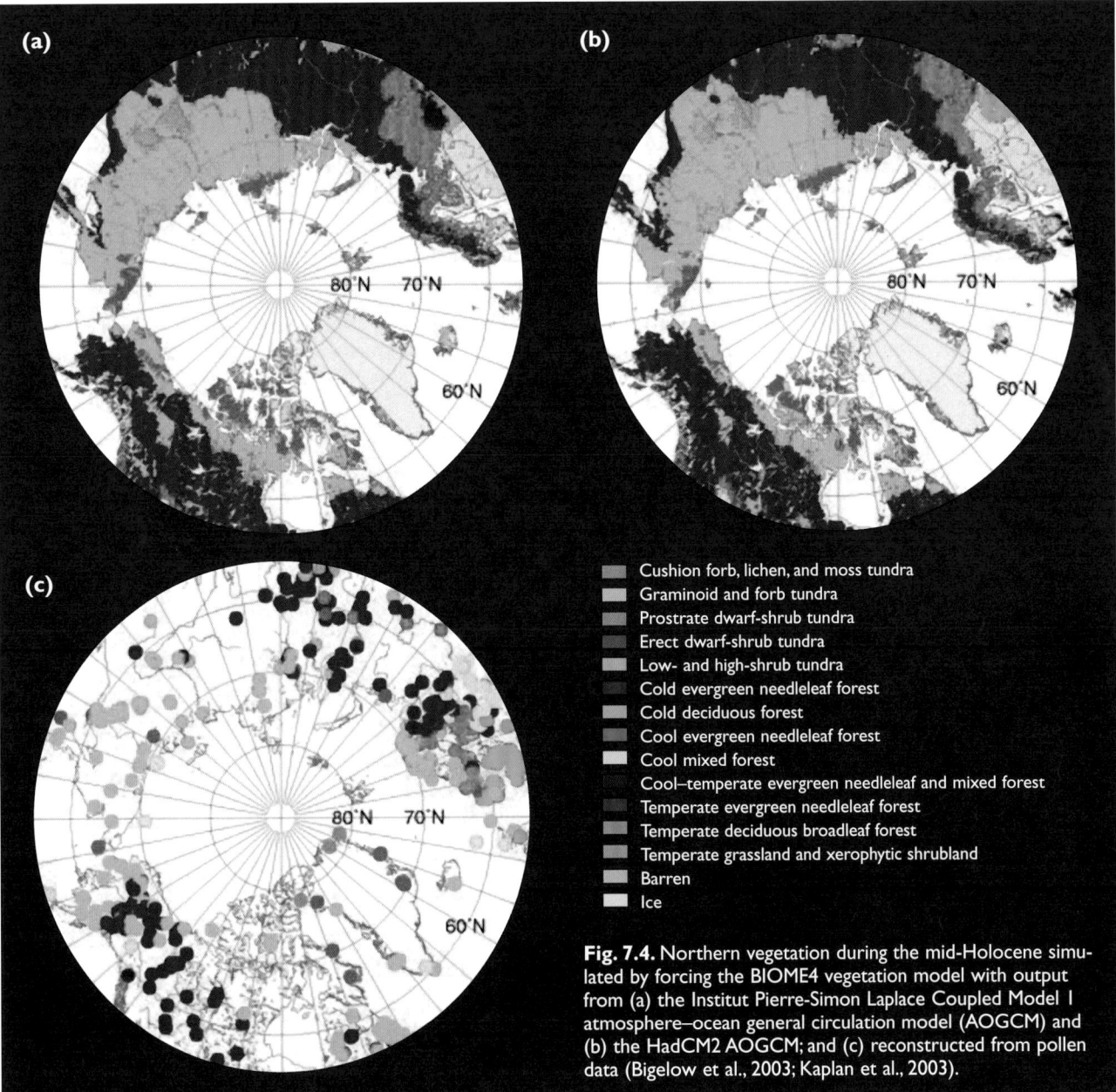

Fig. 7.4. Northern vegetation during the mid-Holocene simulated by forcing the BIOME4 vegetation model with output from (a) the Institut Pierre-Simon Laplace Coupled Model I atmosphere–ocean general circulation model (AOGCM) and (b) the HadCM2 AOGCM; and (c) reconstructed from pollen data (Bigelow et al., 2003; Kaplan et al., 2003).

Legend:
- Cushion forb, lichen, and moss tundra
- Graminoid and forb tundra
- Prostrate dwarf-shrub tundra
- Erect dwarf-shrub tundra
- Low- and high-shrub tundra
- Cold evergreen needleleaf forest
- Cold deciduous forest
- Cool evergreen needleleaf forest
- Cool mixed forest
- Cool–temperate evergreen needleleaf and mixed forest
- Temperate evergreen needleleaf forest
- Temperate deciduous broadleaf forest
- Temperate grassland and xerophytic shrubland
- Barren
- Ice

inferred to have had much larger populations and more extensive breeding areas during glacial stages, although others, such as dunlin (*C. alpina*), exhibit evidence of range fragmentation during glacial stages leading to the evolution of distinct geographically restricted infraspecific taxa. Phylogeographic studies of other arctic taxa show individualistic responses (see Weider and Hobaek, 2000 for a recent review). Some species, such as Arctic char (*Salvelinus alpinus*; Brunner et al., 2001), and genera, such as whitefish (*Coregonus* spp.; Bernatchez et al., 1999), exhibit evidence of sub-taxa whose origins are apparently related to recurrent isolation of populations throughout the alternating glacial and interglacial stages of the Pleistocene. Collared lemmings (*Dicrostonyx* spp.), however, apparently parallel *C. alpina* in exhibiting genetic differentiation principally as a consequence of the relatively recent geographic isolation of populations during the last glacial stage (Fedorov and Goropashnaya, 1999; Fedorov et al., 1999b). Other species, such as the polar bear (*Ursus maritimus*; Paetkau et al., 1999), exhibit

little or no evidence of genetic differentiation that might indicate past population fragmentation, and Fedorov et al. (1999a) inferred that Eurasian true lemmings (*Lemmus* spp.) experienced no effective reduction in population size during recent glacial–interglacial cycles.

In the context of their late-Quaternary history, the arctic biota at present are relatively restricted in range and population size. Although tundra areas were of even smaller extent during the early part of the Holocene than at present, as a result of greater northward extension of the treeline (Huntley, 1997; Huntley and Bradshaw, 1999; MacDonald et al., 2000), that reduction in extent was small in magnitude compared to that experienced at the end of the last glacial stage, during which they were much more extensive than at any time since. Similarly, while extant arctic taxa at the lower taxonomic levels often exhibit considerable diversity that can be related to their late-Quaternary history, the biota as a whole has suffered a recent reduction in overall diversity owing to the extinctions of many species, and some genera, that did not survive into the Holocene. Of at least 12 large herbivores and six large carnivores present in steppe–tundra areas at the LGM (Lister and Bahn, 1995; Stuart, 1982), only four and three, respectively, survive today. Of the surviving species, only two herbivores (reindeer/caribou and muskox) and two carnivores (brown bear – *Ursus arctos* and wolf – *Canis lupus*) occur today in the arctic tundra biome. Present arctic geography also imposes extreme migratory distances upon many tundra-breeding birds owing to the wide separation between their breeding and wintering areas (Davidson N. et al., 1986; Wennerberg, 2001), rendering many of them, in common with much of the arctic biota, extremely vulnerable to any further climatic warming (Evans, 1997).

7.2.3. Ecological history

Although relatively few in overall number, paleo-ecological studies of the late Quaternary Period have been conducted in many parts of the Arctic (see e.g., Anderson and Brubaker, 1993, 1994; Lamb H.F. and Edwards, 1988; MacDonald et al., 2000; Ritchie, 1987). In areas that were by then ice free, the transition to the Holocene was marked by evidence of rapid ecological response. Elsewhere, in proximity to the decaying ice sheets, there was a lag between the global changes and the ecological changes because of the regional influence of the ice sheets. Although the precise nature of the ecological changes depended upon location, the overall picture was one of widespread rapid replacement of the open, discontinuously vegetated tundra and polar desert that characterized most ice-free areas during the late-glacial period by closed tundra. This was in turn replaced by shrub tundra and subsequently by arctic woodlands or northern boreal forest in southern areas of the Arctic. In areas that were unglaciated at the LGM (e.g., Alaska), the ecological transition began earlier, coinciding with the first rapid climatic warming recorded in Greenland about 14700 years BP (Björck et al., 1998; Stuiver et al., 1995). In Alaska, tundra was replaced by shrub tundra during the late-glacial stage, and the first forest stands (of balsam poplar – *Populus balsamifera*) were already present before the transition to the Holocene (Anderson and Brubaker, 1994). South of the Arctic, the extensive areas of steppe–tundra that were present at the LGM were rapidly replaced by expanding forests. Only in parts of northernmost Siberia may fragments of the steppe–tundra biome have persisted into the Holocene, supporting the last population of woolly mammoths that persisted as recently as 4000 years BP (Vartanyan et al., 1993).

The early Holocene was characterized by higher summer insolation intensities at northern latitudes than at present. The warmer summer months enabled trees to extend their ranges further northward than at present; positive feedback resulting from the contrasting albedo of forest compared to tundra (sections 7.4.2.4 and 7.5.4.2) probably enhanced this extension of the forest (Foley et al., 1994). Boreal forest trees expanded their ranges at rates of between 0.2 and 2 km/yr (Huntley and Birks, 1983; Ritchie and McDonald, 1986). They exhibited individualistic responses with respect to their distributions and abundance patterns in response to climatic patterns that differed from those of today. Milder winters and more winter precipitation in western Siberia during the early Holocene, for example, allowed Norway spruce (*Picea abies*) to dominate in areas where Siberian fir (*Abies sibirica*) and Siberian stone pine (*Pinus sibirica*) have become important forest components during the later Holocene (Huntley, 1988, 1997; Huntley and Birks, 1983). Throughout northern Russia, the arctic treeline had advanced more or less to the position of the present arctic coastline by about 10200 years BP, although the lower sea level at that time meant that a narrow strip of tundra, up to 150 km wide at most, persisted north of the treeline (MacDonald et al., 2000). Subsequently, as sea level continued to rise during the early Holocene, tundra extent reached a minimum that persisted for several millennia. For tundra species, including tundra-breeding birds, the early Holocene thus seems likely to have been a time of particular stress. This stress may, however, have been in part relieved by enhanced productivity in these areas, compared to modern tundra ecosystems, as a consequence of the warmer summers and higher insolation intensity.

In glaciated areas of the Arctic, such as northern Fennoscandia and much of arctic Canada, peatlands became extensive only after the mid-Holocene (see e.g., Lamb H.F., 1980; Vardy et al., 1997) in response to the general pattern of climatic change toward cooler and regionally moister summer conditions. The same cooling trend led to the southward retreat of the arctic treeline, which reached more or less its present location in most regions by about 4500 years BP (MacDonald et al., 2000). The consequent increase in tundra extent probably relieved the stress experienced by tundra organisms during the early Holocene, although the cooler, less productive conditions, and the increasing extent of seasonally waterlogged tundra peatlands, may have offset this at least in part. While the early Holocene was a time of permafrost decay and thermokarst development, at least in some regions (Burn, 1997), the extent of permafrost has increased in many areas during the later Holocene (see e.g., Kienel et al., 1999; Vardy et al., 1997).

7.2.4. Human history related to ecosystems

Recently discovered evidence (Pavlov et al., 2001) shows that Paleolithic "hunter-gatherers" were present about 40000 years BP (long before the LGM) as far north as 66°34' N in Russia, east of the Fennoscandian Ice Sheet. Although it seems likely that humans did not range as far north during the glacial maximum, it is clear that they expanded rapidly into the Arctic during the deglaciation.

Humans entered North America via the Bering "land bridge" and along the southern coast of Beringia about 14000 to 13500 years BP (Dixon, 2001). These so-called Clovis hunters were hunter-gatherers who had developed sophisticated ways of working stone to produce very fine spear- and arrowheads. Over the next few millennia, they expanded their range and population rapidly, occupying most of the North American continent. Their prey apparently included many of the large vertebrate species that soon became extinct. The extent to which human hunting may have been principally responsible for these extinctions is a matter of continuing debate, but recent simulations for North America indicate that this possibility cannot be excluded (Alroy, 2001). However, these extinctions also coincide with an environmental change that caused the area of the biome with which the large arctic vertebrates were associated to be reduced to an extent that was apparently unprecedented during previous glacial–interglacial cycles (Sher, 1997). It thus is more probable that the hunting pressure exerted by humans was at most an additional contributory factor leading to the extinctions, rather than their primary cause.

In Eurasia, Paleolithic hunter-gatherers shifted their range northward into the Arctic at the end of the last glacial stage, as did their large vertebrate prey. To the south, they were replaced by Mesolithic peoples who occupied the expanding forests. By the early Holocene these Mesolithic peoples had expanded well into the Arctic (Thommessen, 1996), where they probably gave rise to the indigenous peoples that in many cases continued to practice a nomadic hunter-gatherer way of life until the recent past or even up to the present day in some regions. The arrival of later immigrants has had major impacts upon indigenous peoples and their way of life (Chapters 3, 11, and 12). In turn, land use and natural resource exploitation by the immigrants, as well as the changes that they have brought about in the way of life of indigenous peoples, have had negative impacts on many arctic ecosystems. These impacts in some cases have possibly increased the vulnerability of these ecosystems to the pressures that they now face from climate change and increased exposure to UV-B radiation.

7.2.5. Future change in the context of late-Quaternary changes

The potential changes for the next century can be put into context by comparing their rates and magnitudes to those estimated for the changes documented by paleoecological and other evidence from the late Quaternary Period (Table 7.1).

It is apparent from Table 7.1 that projected future changes have several characteristics that pose a particular threat to the biota and ecosystems of the Arctic. First, climatic changes over the next century are likely to be comparable in magnitude to the changes that occurred between full glacial conditions and present conditions, and greater than the maximum changes that occurred during the Holocene. Second, the global increase in mean annual temperature is projected to occur at rates that are

Table 7.1. Comparison of key aspects of projected future environmental changes with late-Quaternary changes.

	Late Quaternary	Projected Future
Sea level	ca. 120 m lower at LGM; increased at a maximum rate of ca. 24 mm/yr (Fairbanks, 1989)	0.09–0.88 m higher by 2100; 3–10 m higher in 1000 years increasing at a rate of 1–9 mm/yr (IPCC, 2001)
Climate		
Mean annual temperature	full glacial: global mean ca. 5 °C lower; regionally in the Arctic 10–13 °C lower	2100: global mean 1.5–5.8 °C higher; regionally in the Arctic 2.1–8.1 °C higher (IPCC, 2001)
	Holocene: global mean <1 °C higher at maximum; regionally in the Arctic similar to present	
Winter temperature	full glacial: >15 °C cooler regionally in the Arctic	2100: 4–10 °C higher regionally in the Arctic (IPCC, 2001)
	Holocene: ca. 2–4 °C warmer regionally in the Arctic at maximum	
Rate of increase in mean annual temperature	global: ≤1°C per millennium; regionally in the Arctic: >5 °C in a century	global: 1.5–5.8 °C per century; regionally in the Arctic: 2.1–8.1 °C in a century (IPCC, 2001)
Ecosystem responses		
Treeline displacement	full glacial: >1000 km southward; Holocene: 50–200 km northward at maximum (Kaplan, 2001)	2100: >500 km northward. It is possible that anthropogenic disturbance might result in an opposite response (see section 7.5.3.2)
Range margin displacement rates	early Holocene: rates of 0.2–2 km/yr estimated for trees from pollen data (Huntley, 1988)	21st century: potential rates of 5–10 km/yr estimated from species–climate response models (Huntley et al., 1995)
Area of tundra	full glacial: 197% (ranging from 168 to 237%) of present; Holocene: 81% (ranging from 76 to 84%) of present at minimum	2100: 51% of present (J. Kaplan, pers. comm., 2002; see Kaplan et al., 2003)
UV-B radiation levels	No long-term trend known. Due to solar variability, levels of DNA-active UV-B wavelengths may have varied by up to 27% within a period of ca. 150 years (Rozema et al., 2002)	In addition to natural solar cycles, it is very likely that anthropogenic cooling of the stratosphere will delay recovery of the ozone layer

higher than the rate of global temperature increase during the last deglaciation; in parts of the Arctic the rate of warming is likely to match the most rapid regional warming of the late Quaternary Period. Third, as a consequence of this temperature increase, and the accompanying rise in sea level, tundra extent is likely to be less than at any time during the late Quaternary Period. Fourth, global mean temperatures and mean annual temperatures in the Arctic are very likely to reach levels unprecedented in the late Quaternary Period; this is very likely to result in a rapid reduction in the extent of permafrost, with associated thermokarst development in areas of permafrost decay leading to potentially severe erosion and degradation of many arctic peatlands (section 7.5.3.1). The combination of projected future climate change with other anthropogenic effects (including enhanced levels of UV-B radiation, deposition of nitrogen compounds from the atmosphere, heavy metal and acidic pollution, radioactive contamination, and increased habitat fragmentation) suggests that the future is very likely to be without a past analogue and will pose unprecedented challenges to arctic ecosystems and biota that evolved in response to global cooling throughout the last five million years or so (the late Tertiary and Quaternary Periods), during which our own species also evolved.

7.2.6. Summary

At the LGM, vast ice sheets covered many continental areas. The beds of some shallow seas were exposed, connecting previously separated landmasses. Although some areas were ice-free and supported a flora and fauna, mean annual temperatures were 10 to 13 °C colder than during the Holocene. Within a few millennia of the glacial maximum, deglaciation started but was not a simple unidirectional change: a series of climatic fluctuations occurred between about 18 000 and 11 400 years BP. During the Younger Dryas event, mean annual temperatures fell substantially in some areas and reglaciation occurred. At the end of the event, mean annual temperatures rose by more than 5 °C in less than 100 years in at least some parts of the Arctic. Following the general thermal maximum in the Holocene, there has been a modest overall cooling trend. However, superimposed upon the general longer-term patterns have been a series of millennial and centennial fluctuations in climate, the most marked of which occurred about 8200 years BP. The most recent of these climatic fluctuations was that of the LIA, a generally cool interval spanning approximately the late 13th to early 19th centuries. At its most extreme, mean annual temperatures in some arctic areas fell by several degrees, with impacts on human settlements in the north.

In the context of at least the last 150 000 years, arctic ecosystems and biota have been close to their minimum extent within the last 10 000 years. They suffered loss of diversity as a result of extinctions during the rapid, large-magnitude global warming at the end of the last glacial stage. Consequently, arctic ecosystems and biota are already stressed; some are extremely vulnerable to current and projected future climate change. For example, migratory arctic-breeding birds today face maximal migration distances between their wintering and breeding areas.

Evidence from the past indicates that arctic species, especially larger vertebrates, are very likely to be vulnerable to extinction if climate warms. The treeline is very likely to advance, perhaps rapidly, into tundra areas of northern Eurasia, Canada, and Alaska, as it did during the early Holocene, reducing the extent of tundra and contributing to the pressure upon species that makes their extinction possible. Species that today have more southerly distributions are very likely to extend their ranges north, displacing arctic species. Permafrost is very likely to decay and thermokarst develop, leading to erosion and degradation of arctic peatlands. Unlike the early Holocene, when lower relative sea level allowed a belt of tundra to persist around at least some parts of the Arctic Basin when treelines advanced to the present coast, sea level is very likely to rise in the future, further restricting the area of tundra and other treeless arctic ecosystems.

The negative response of arctic ecosystems in the face of a shift to global climatic conditions that are apparently without precedent during the Pleistocene is likely to be considerable, particularly as their exposure to co-occurring environmental changes (i.e., enhanced levels of UV-B radiation, deposition of nitrogen compounds from the atmosphere, heavy metal and acidic pollution, radioactive contamination, increased habitat fragmentation) is also without precedent.

7.3. Species responses to changes in climate and ultraviolet-B radiation in the Arctic

The individual of a species is the basic unit of ecosystems that responds to changes in climate and UV-B radiation levels. Individuals respond to environmental changes over a wide range of timescales: from biochemical, physiological, and behavioral processes occurring in less than a minute to the integrative responses of reproduction and death (Fig. 7.1). Reproduction and death drive the dynamics of populations while mutation and environmental selection of particular traits in individuals within the population lead to changes in the genetic composition of the population and adaptation.

Current arctic species have characteristics that have enabled them to pass various environmental filters associated with the arctic environment (Körner, 1995; Walker M.D., 1995), whereas species of more southerly latitudes either cannot pass these filters or have not yet arrived in the Arctic. Changes in arctic landscape processes and ecosystems in a future climatic and UV-B radiation regime will depend upon the ability of arctic species to withstand or adapt to new environments and upon their interactions with immigrant species that can pass through less severe environmental filters. This section focuses on the attributes of current arctic species that constrain or facilitate their responses to a changing climate and UV-B radiation regime.

Soil characteristics will determine to some extent the responses of vegetation to climate change. Arctic soils (and particularly moisture content) vary from the forest tundra to the polar deserts and within each of these vegetation zones.

In the high-arctic polar deserts, skeletal soils and stony ground predominate (Aleksandrova, 1988). Materials range from boulders to gravel and the dominant erosion process is physical weathering (e.g., freeze–thaw cracking) rather than chemical and biochemical weathering, which are strongly suppressed by lack of heat. Freeze–thaw cycles lead to a sorting of stones by size and formation of patterned ground consisting, for example, of stone nets. An organic layer is missing from the soil profile and organic material is restricted to small pockets under sparse plant cover or in cracks. The soils are neutral or only weakly acidic and the soil complexes are almost completely saturated with moisture although this differs between the polar deserts of Canada, Greenland, and the Russian Arctic. Gleys are almost absent and the active-layer depth is about 30 cm.

In the tundra biome, soil profiles are characterized by an organic layer that is often less than 10 cm deep on dry ridges, is deeper in moist and mesic habitats, and extends to deep deposits of peat in wet areas (Nadelhoffer et al., 1992). Below the organic layer is a mineral layer. The active layer is deepest in the dry areas (~1 m) owing to the lack of summer insulation, and is shallowest in wet areas (~20 cm) due to efficient insulation by continuous vegetation cover and organic soil. The pH of tundra soils

is generally acidic. Chemical and biochemical processes are important but sorting of materials and patterned ground are still evident, leading to landscapes with larger polygons than those found in the polar deserts. In both polar-desert and tundra soils, the permafrost is generally continuous. In contrast, in soils of the forest tundra, the permafrost generally becomes discontinuous and the depth of the organic layer decreases except for waterlogged depressions where peat bogs are found.

Soil formation processes in the Arctic are slow and the type of soil is very likely to constrain potential rates of colonization by southern species.

7.3.1. Implications of current species distributions for future biotic change

7.3.1.1. Plants

Species diversity

About 3% (~5900 species) of the global flora occurs in the Arctic as defined in this chapter (0.7% of the angiosperms (flowering plants), 1.6% of the gymnosperms (cone-bearing plants), 6.6% of the bryophytes, and 11% of the lichens) (Table 7.2). There are more species of primitive taxa (cryptogams), that is, mosses, liverworts, lichens, and algae, in the Arctic than of vascular plants (Matveyeva and Chernov, 2000). Less than half (about 1800) of arctic plant species are vascular plants. There are about 1500 species of vascular plants common to both Eurasia (Matveyeva and Chernov, 2000; Sekretareva, 1999) and North America (Murray, 1995). Similar numbers of non-vascular plants probably occur in the Arctic on both continents, although their diversity has been less

Table 7.2. Biodiversity estimates in terms of species richness (number of species) for the Arctic north of the latitudinal treeline and percentage of world biota (Chernov, 2002; Matveyeva and Chernov, 2000).

	Animals			Plants			Fungi	
Group	Number of species	% of world biota	Group	Number of species	% of world biota	Group	Number of species	% of world biota
Mammals	75	1.7	Angiosperms	1735	0.7	Fungi	2500	2.3
Birds	240	2.9	monocotyledons	399	0.6			
Insects	3300	0.4	dicotyledons	1336	0.7			
Diptera	1600	0.9	Gymnosperms	12	1.6			
Beetles	450	0.1	Pteridophytes	62	0.6			
Butterflies	400	0.3	Mosses	600	4.1			
Hymenoptera	450	0.2	Liverworts	250	2.5			
Others	400		Lichens	2000	11.0			
Springtails	400	6.0	Algae	1200	3.3			
Spiders	300	1.7						
Mites	700	1.9						
Other groups[a]	600	--						
Total estimate	6000	--		5859	3.0			

[a]Amphibians and reptiles (7 species), centipedes (10 species), terrestrial mollusks (3 species), oligochaetes (earthworms and enchytraeids) (70 species), and nematodes (~500 species).

thoroughly documented. In the Russian Arctic, for example, 735 bryophyte species (530 mosses and 205 liverworts) and 1078 lichen species have been recorded (Afonina and Czernyadjeva, 1995; Andreev et al., 1996; Konstantinova and Potemkin, 1996). In general, the North American and Eurasian Arctic are similar to one another in their numbers of vascular and non-vascular plant species, of which a large proportion (about 80%) of vascular plant species occurs on at least two continents. An even larger proportion (90%) of bryophyte species occurs in both the North American and Eurasian Arctic.

About 40% of arctic vascular plants (and a much higher percentage of mosses and lichens) are basically boreal species that now barely penetrate the Arctic (Table 7.3). They currently occur close to the treeline or along large rivers that connect the subarctic with the Arctic. These boreal species within the Arctic will probably be the primary boreal colonizers of the Arctic in the event of continued warming. Polyzonal (distributed in several zones), arctoboreal (in taiga and tundra zones), and hypoarctic (in the northern taiga and southern part of the tundra zone) species have even greater potential to widen their distribution and increase their abundance in a changing climate. The majority of cryptogams have wide distributions throughout the Arctic. Such species are likely to survive a changing climate, although their abundance is likely to be reduced (sections 7.3.3.1 and 7.4.1.2).

In contrast to the low diversity of the arctic flora at the continental and regional scales, individual communities (100 m² plots) within the Arctic have a diversity similar to or higher than those of boreal and temperate zones. These diversities are highest in continental parts of the Arctic such as the Taymir Peninsula of Russia, where there are about 150 species of plants (vascular plants, lichens, and mosses) per 100 m² plot, 40 to 50 species per square meter, and up to 25 species per square decimeter (Matveyeva, 1998).

Latitudinal gradients of species diversity

Latitudinal gradients suggest that arctic plant diversity is sensitive to climate. The number of vascular plant species declines five-fold from south to north on the Taymir Peninsula (Matveyeva, 1998). Summer temperature is the environmental variable that best predicts plant diversity in the Arctic (Young, 1971). Other factors are also important, however: regions at different latitudes that have a similar maximum monthly temperature often differ in diversity. Taymir biodiversity values are intermediate between the higher values in Chukotka and Alaska, which have a more complicated relief, geology, and floristic history, and the lower values in the eastern Canadian Arctic with its impoverished flora resulting from relatively recent glaciation. All diversity values on the Yamal Peninsula are even lower than in Canada because of a wide distribution of sandy soils and perhaps its young age. Similar patterns are observed with butterflies (Fig. 7.5c) and spiders (Chernov, 1989, 1995). Therefore, latitudinal gradients of species diversity are

best described as several parallel gradients, each of which depends on summer heat, but which may differ from one geographic region to another. This must be taken into consideration when projecting future changes in biodiversity. Figure 7.5b illustrates how current bioclimatic distributions are related to climate change scenarios by plotting the likely changes in the number of ground beetles for three time slices of mean July tem-

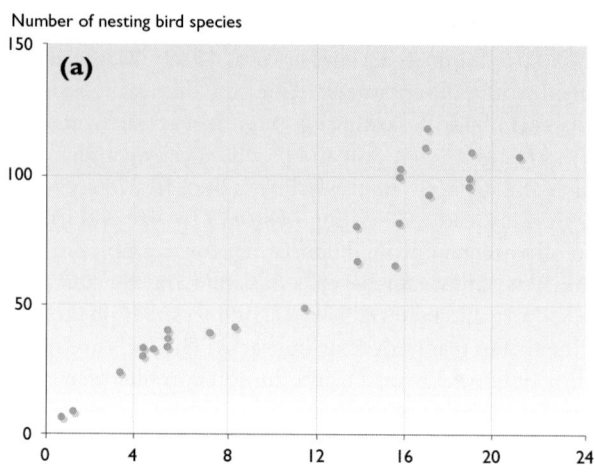

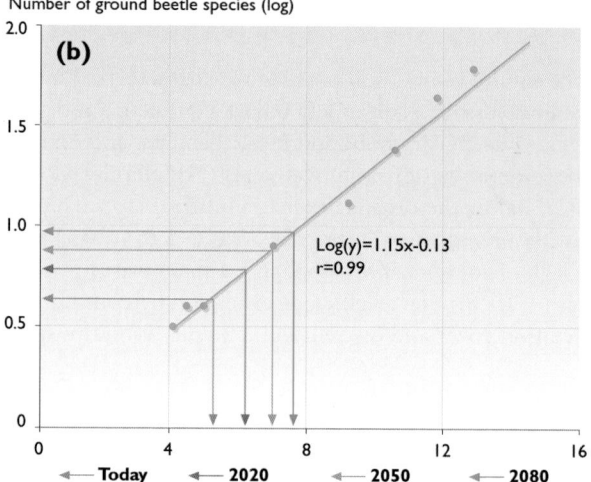

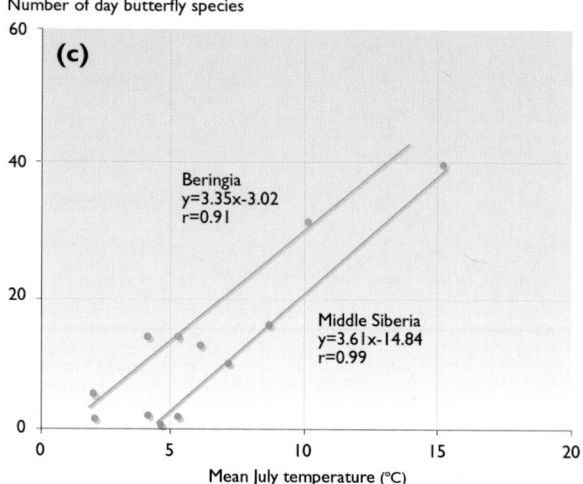

Fig. 7.5. Relationship between mean July temperature and (a) the number of nesting bird species in western and middle Siberia; (b) the number of ground beetle species in local faunas of the Taymir Peninsula; and (c) the number of day butterfly species in the middle Siberian and Beringian areas of the Arctic (Chernov, 1989, 1995; Matveyeva and Chernov, 2000).

perature derived from the mean of the scenarios generated by the five ACIA-designated models.

At the level of the local flora (the number of species present in a landscape of about 100 km²), there is either a linear or an "S"-shaped relationship between summer temperature and the number of species (Fig. 7.6). The number of species is least sensitive to temperature near the southern margin of the tundra and most sensitive to temperatures between 3 and 8 °C. This suggests that the primary changes in species composition are very likely to occur in the northern part of the tundra zone and in the polar desert, where species are now most restricted in their distribution by summer warmth and growing-season length. July temperature, for example, accounts for 95% of the variance in number of vascular plant species in the Canadian Arctic (Rannie, 1986), although extreme winter temperatures are also important (section 7.3.3.1). Summer warmth, growing-season length, and winter temperatures all affect the growth, reproduction, and survival of arctic plants. The relative importance of each varies from species to species, site to site, and year to year.

The steep temperature gradient that has such a strong influence on species diversity occurs over much shorter distances in the Arctic than in other biomes. North of the treeline in Siberia, mean July temperature decreases from 12 to 2 °C over 900 km, whereas mean July temperature decreases by 10 °C over 2000 km in the boreal zone, and decreases by less than 10 °C from the southern boreal zone to the equator (Chernov, 1995).

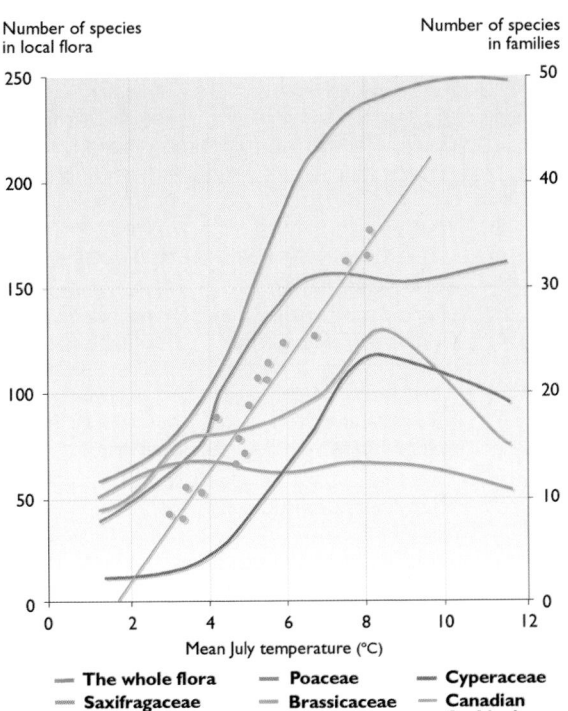

Fig. 7.6. Relationship between mean July temperature and the number of vascular plant species in local floras of the Taymir Peninsula (Matveyeva and Chernov, 2000) and the number of vascular species in the local flora of the Canadian Archipelago (Rannie, 1986).

Comparing the temperature decrease of 10 °C with the projected 2.5 °C increase in mean July temperature in the Arctic by 2080 (mean of the two extremes – 1.1 and 4.2 °C – projected by the five ACIA-designated models) suggests that much of the Arctic is very likely to remain within the arctic summer climate envelope (although the increase in winter temperature is projected to be higher).

Because of the steep latitudinal temperature gradients in the Arctic, the distance that plants must migrate in response to a change in temperature is much less in the Arctic than in other biomes, particularly where topographic variations in microclimate enable plants to grow far north of their climatic optima. The low solar angle and presence of permafrost make topographic variations in microclimate and associated plant community composition particularly pronounced in the Arctic. Thus, both the sensitivity of arctic species diversity to temperature and the short distance over which this temperature gradient occurs suggest that arctic diversity is very likely to respond strongly and rapidly to high-latitude temperature change.

Latitudinal patterns of diversity differ strikingly among different groups of plants (Table 7.3). Many polyzonal, boreal, and hypoarctic species have ranges that extend into the Arctic. Some of these (e.g., the moss *Hylocomium splendens* and the sedges *Eriophorum angustifolium* and *E. vaginatum*) are important dominants within the Arctic. Tussocks of *E. vaginatum* structure the microtopography of broad areas of tussock tundra (Bliss and Matveyeva, 1992), and *H. splendens* exerts a control over nutrient cycling (Hobbie, 1996). Tall willow (*Salix* spp.) and alder (*Alnus fruticosa*) shrubs as well as dwarf birch (*Betula exilis* and *B. nana*) form dense thickets in the southern part of the tundra zone and often have outlier populations that extend far to the north in favorable habitats (Matveyeva and Chernov, 2000).

Species that are important community dominants are likely to have a particularly rapid and strong effect on ecosystem processes where regional warming occurs. Hemiarctic species are those that occur throughout the Arctic. Many of these species are common community dominants, including *Carex bigelowii/arctisibirica*, *C. stans*, *Dryas octopetala/punctata*, *Cassiope tetragona*, and the moss *Tomentypnum nitens*. Due to their current widespread distribution, their initial responses to climatic warming are likely to be increased productivity and abundance, probably followed by northward extension of their ranges. The species most vulnerable to climate change are likely to be euarctic (polar willow – *Salix polaris*) and hyperarctic species that now have the greatest abundance and widest ecological amplitude in the northernmost part of the tundra zone and in polar deserts, respectively. These groups of species are best adapted to the climate conditions of the high Arctic where they are distributed in a wide range of habitats where more competitive southern species are absent. In the more southerly regions of the tundra zone, they

Table 7.3. Current diversity changes with latitude in the Arctic, excluding limnic and marine animals (compiled and modified from Matveyeva and Chernov, 2000). General information about how species within the various categories are likely to respond to changes in climate and UV radiation levels is presented in the text, but insufficient information is available for most of the species in the table.

Category	Optimum of distribution	Examples		
		Plants	Birds	Mammals and invertebrates
Polyzonal	Different zones in the Holarctic; within the tundra zone usually in local habitats and wet depressions	Soil algae; the mosses *Hylocomium splendens* sensu lato, *Aulacomnium turgidum*, and *Racomitrium lanuginosum*; the liverwort *Ptilidium ciliare*; the lichens *Cetraria islandica*, *Psora decipiens*, and *Cladina rangiferina*; the vascular species *Cardamine pratensis*, *Chrysosplenium alternifolium*, and *Eriophorum angustifolium*; the sedge *Carex duriuscula*[a]; the herb *Helictotrichon krylovii*[a]; and the moss *Tortula ruralis*[a]	Common raven (*Corvus corax*), peregrine falcon (*Falco peregrinus*), white wagtail (*Motacilla alba*), northern wheatear (*Oenanthe oenanthe*)	Wolf (*Canis lupus*), ermine (*Mustela erminea*), weasel (*M. nivalis*), voles (*Microtus gregalis* and *M. oeconomus*), and the mite *Chiloxanthus pilosus*[b]
Zonal boreal	Not abundant and constrained to the southern Arctic in benign habitats such as river valleys, south-facing slopes, and wet areas	Tree species of *Larix*; the orchid *Corallorrhiza*; the shrub *Salix myrtilloides*; the sedge *Carex chordorrhiza*; the herbs *Allium schoenoprasum*, *Cortusa matthioli*, *Galium densiflorum*, and *Sanguisorba officinalis*; and the forest mosses *Climacium dendroides*, *Pleurozium shreberi*, and *Rhytidiadelphus triquetrus*	The forest birds *Turdus iliacus* and *T. pilaris* (thrushes); the leaf warblers Arctic warbler (*Phylloscopus borealis*) and yellow-browed warbler (*P. inornatus*); and the "river" ducks *Anas acuta*, *A. penelope*, and *A. crecca*	Reindeer/caribou (*Rangifer tarandus*), wolverine (*Gulo gulo*), and brown bear (*Ursus arctos*)

Zonal Arctic

Category	Optimum of distribution	Plants	Birds	Mammals and invertebrates
Hypoarctic	Optima in the southern tundra subzone	The shrubs *Betula nana* and *B. exilis* and the sedge *Eriophorum vaginatum*[c]	Ptarmigan (*Lagopus lagopus*), spotted redshank (*Tringa erythropus*), little bunting (*Emberiza pusilla*), bar-tailed godwit (*Limosa lapponica*)	The vole *Microtus middendorffi*, the ground beetle *Carabus truncaticollis*, the bumblebee *Bombus cingulatus*, and the spider *Alopecosa hirtipes*
Hemiarctic	Throughout the tundra zone but most frequent in the middle	Most of the dominant species, including the grasses *Arctophila fulva* and *Dupontia fisheri*; the sedges *Carex bigelowii/arctisibirica* and *C. stans*; the shrub willow *Salix reptans*; the dwarf shrubs *Dryas punctata/octopetala* and *Cassiope tetragona*; the mosses *Tomentypnum nitens*, *Drepanocladus intermedius*, and *Cinclidium arcticum*; the herbs *Lagotis minor* and *Pedicularis hirsuta*; and the moss *Polytrichum juniperinum*	Lapland longspur (*Calcarius lapponicus*), lesser golden plover (*Pluvialis dominica*), Pacific golden plover (*P. fulva*), and the dunlins *Calidris alpina* and *C. minuta*	The lemming *Lemmus sibiricus*, the bumblebee *Bombus balteatus*, the ground beetles *Curtonotus alpinus* and *Pterostichus costatus*, and the flower-fly *parasyrphus tarsatus*
Euarctic	Northern part of the tundra zone, rare in the southern part	The dwarf shrubs *Salix polaris* and *S. arctica*[d]	Black-bellied plover (*Pluvialis squatarola*), curlew sandpiper (*Calidris ferruginea*), snowy owl (*Nyctea scandiaca*), snow bunting (*Plectrophenax nivalis*) and several more	The lemming *Dicrostonyx torquatus*, the bumblebees *Bombus hyperboreus* and *B. polaris*, and the crane fly *Tipula carinifrons*
Hyperarctic	Polar desert and the northernmost part of the tundra zone	Almost no plants are restricted to these zones; the following have their highest frequencies there: the grasses *Phippsia algida* and *Poa abbreviata*; the herbs *Cerastium regelii*, *Draba oblongata*, *D. subcapitata*, *Saxifraga hyperborea*, and *S. oppositifolia*; the mosses *Dicranoweisia crispula*, *Bryum cyclophyllum*, *Orthothecium chryseon*, and *Seligeria polaris*; and the lichens *Cetrariella delisei*, *Arctocetraria nigricascens*, *Dactylina ramulosa*, *D. madreporiformis*, and *Thamnolia subuliformis*	The wader species *Calidris alba* and *C. canutus*	No terrestrial mammal species are restricted to this zone. The collembolan *Vertagopus brevicaudus*

[a]"steppe" species; [b]bizonal steppe and tundra species; [c]this group characterizes the southern tundra subzone; [d]this group is relatively small, but is valuable in subdividing the tundra zone into subzones.

are able to grow only (or mainly) in snow beds (depressions in the landscape where snow accumulates and plant growth and diversity is limited by particularly short snow-free periods). It is likely that their ecological amplitude will narrow and their abundance decrease during climate warming.

Thus, responses to climate change will be different in various groups of plants. Some currently rare boreal species are likely to move further north and the relative abundance and range of habitats occupied by more common species are likely to increase. When southern species with currently narrow niches penetrate into the harsher ecosystems at high latitudes, therefore, there is likely to be a broadening of their ecological niches there. In contrast, some true arctic species (endemics) that are widespread at high latitudes are likely to become more restricted in their local distribution within and among ecosystems. They could possibly disappear in the lower latitudes where the tundra biome is particularly narrow. Only few high-arctic plants in Greenland are expected to become extinct, for example, *Ranunculus sabinei*, which is limited to the narrow outer coastal zone of North Greenland (Heide-Jørgensen and Johnsen, 1998). However, temperature is not the only factor that currently prevents some species from being distributed in the Arctic. Latitude is important, as life cycles depend not only on temperature but on the light regime as well. It is very likely that arctic species will tolerate warmer summers whereas long day lengths will initially restrict the distribution of some boreal species (section 7.3.3.1). New communities with a peculiar species composition and structure are therefore very likely to arise, and will be different from those that exist at present.

7.3.1.2. Animals

Species diversity

The diversity of arctic terrestrial animals beyond the latitudinal treeline (6000 species) is nearly twice as great as that of vascular plants and bryophytes (Chernov, 1995, 2002; Table 7.2). As with plants, the arctic fauna accounts for about 2% of the global total, and, in general, primitive groups (e.g., springtails, 6% of the global total) are better represented in the Arctic than are advanced groups such as beetles (0.1% of the global total; Chernov, 1995; Matveyeva and Chernov, 2000). There are about 322 species of vertebrates in the Arctic, including approximately 75 species of mammals, 240 species of birds, 2 species of reptiles, and 5 species of amphibians. Insects are the most diverse group of arctic animals; of approximately 3300 arctic species, about 50% are Diptera, 10% are beetles (Coleoptera), 10% are butterflies (Lepidoptera), and 10% are Hymenoptera. The Arctic has about 300 species of spiders (Arachnida), 700 species of mites (Acarina), 400 species of springtails (Collembola), 500 species of nematodes, 70 species of oligochaetes (of which most are Enchytraeidae), only a few mollusks, and an unknown number of protozoan species.

In the Arctic as defined by CAFF (Conservation of Arctic Flora and Fauna) (which includes forested areas), some 450 species of birds have been recorded breeding. Some species extend their more southern breeding areas only marginally into the Arctic. Others are not migratory and stay in the Arctic throughout the year. About 280 species have their main breeding distribution in the Arctic and migrate regularly (Scott, 1998). An estimate of the total number of individuals involved is not possible, as too little is known about the population size of most species or their arctic proportion. However, a rough first approximation suggests that there are at least several hundred million birds. Population sizes of water birds are better known and the Arctic is of particular importance for most water birds, such as divers, geese, and waders. Twelve goose species breed in the Arctic, eleven almost entirely and eight exclusively. These comprise about 8.3 million birds. The total number of arctic-breeding sandpipers (24 species) exceeds 17.5 million birds (Zöckler, 1998). The total number of water birds, including other wader species, divers, swans, ducks, and gulls is estimated to be between 85 and 100 million birds.

Latitudinal gradients of species diversity

Latitudinal patterns of diversity in arctic animals are similar to those described for arctic plants. Species diversity declines in parallel with decreasing temperature in most animal groups (Fig. 7.5), including birds, ground beetles, and butterflies (Chernov, 1995). However, in some groups (e.g., peatland birds and sawflies at local sites in the European Arctic), concentrations of both species diversity and density per unit area can increase compared with more southern territories, perhaps because the habitat types appropriate to these groups are more diverse in the tundra than in the boreal forest. In general, the latitudinal decline in the number of animal species is more pronounced (frequently greater than 2.5-fold) than in vascular plants. As with plants, at a given temperature there are more animal species in Beringia, with its complicated relief, geology, and biogeographic history than in the Taymir Peninsula. Many animal species are restricted to the boreal zone because they depend on the crown, wood, roots, or litter of trees, which are absent in the tundra zone. These groups include wood-boring insects and wood-decaying fungi and their predators (Chernov and Matveyeva, 1997), as well as mammals and birds that specialize on tree seeds and leaves. Other important animals, including the raven (*Corvus corax*), wolf, red fox (*Vulpes vulpes*), and ermine (*Mustela erminea*), are primarily boreal in distribution but remain an important component of many arctic ecosystems. There are a few terrestrial animals restricted to the high Arctic such as the sanderling (*Calidris alba*) and a common Collembola, *Vertagopus brevicaudus*. Other arctic species have their centers of distribution in the northern, mid- or southern Arctic (Table 7.3). The more diverse patterns of animal distribution compared to plants make it more difficult to project how animals will respond to climatic warming. Some herbivores have distributions that are more limited than those of their host plants (Strathdee and Bale, 1998), so it is possible that warming will allow these species to extend northward relatively rapidly.

As with plants, latitudinal patterns of diversity differ strikingly among different groups of animals (Table 7.3). The common species tend to be more broadly distributed in the far north. In northern Taymir, there are only 12 species of springtails but 80% of these occur in all microsites and topographic locations investigated (Chernov and Matveyeva, 1997). Some boreal birds, such as the American robin (*Turdus migratorius*), penetrate only into the southern part of the tundra while others can occur far from the area of their continuous distribution (climatic optimum): in the vicinity of Dixon (Taymir), the forest thrushes *T. pilaris* and *T. iliacus* form populations in the northernmost part of the tundra zone, 400 km from the last outposts of the forests. At the southern limits of the tundra, there is greater specialization among microhabitats. Many more species occur in intrazonal habitats, occupying relatively small and isolated sites, than in zonal habitats that contain only a small proportion of the regional fauna. Warming is therefore likely to lead to more pronounced habitat and niche specialization.

An important consequence of the decline in numbers of species with increasing latitude is an increase in dominance. For example, one species of collembolan, *Folsomia regularis*, may constitute 60% of the total collembolan density in the polar desert (Babenko and Bulavintsev, 1997). These "super-dominant" species are generally highly plastic, occupy a wide range of habitats, and generally have large effects on ecosystem processes. Lemmings (*Lemmus* spp. and *Dicrostonyx* spp.) are super-dominant species during peak years of their population cycles (Stenseth and Ims, 1993) and have large effects on ecosystem processes (Batzli et al., 1980; Laine and Henttonen, 1983; Stenseth and Ims, 1993).

7.3.1.3. Microorganisms

Species diversity

Microbial organisms are critically important for the functioning of ecosystems, but are difficult to study and are poorly understood compared with other species. However, the International Biological Programme (IBP; 1964–1974) significantly advanced understanding of arctic microorganisms, compared with those of other biomes, when an inventory of microbial communities was undertaken in the tundra (Heal et al., 1981). At the beginning of the 21st century, the knowledge of microbial diversity in the tundra remains little better than 30 to 40 years ago, and recent outstanding progress in molecular microbial ecology has rarely been applied to arctic terrestrial studies.

Presently there are 5000 to 6000 named bacterial species globally and about the same number of fungi (Holt et al., 1994), compared with more than one mil-

lion named plant and animal species (Mayr, 1998; Wilson E., 1992). Some scientists have interpreted this difference to mean that the bacteria are not particularly diverse (Mayr, 1998). However, there are several reasons (section 7.7.1.1) to believe that the apparent limited diversity of microbes is an artifact.

Recent progress in molecular biology and genetics has revolutionized bacterial classification and the understanding of microbial phylogeny ("family trees") and biodiversity in general. The DNA sequencing technique has reorganized bacterial classification and brought order to microbial taxonomy (Wayne et al., 1987). Moreover, the microbial inventory can now be done without isolation and cultivation of the dominant microorganisms, because it is enough to extract the total community DNA from the soil and amplify, clone, and sequence the individual genes. This culture-independent approach has been applied occasionally for analysis of microbial communities in subarctic and arctic soils, most often to study relatively simple communities in hot springs, subsoils, and contaminated aquifers (section 7.3.5.2). Analysis of Siberian subsurface permafrost samples (Gilichinsky, 2002; Tsapin et al., 1999) resulted in the formation of a clone library of 150 clones which has been separated into three main groups of Eubacteria. From 150 clones so far analyzed, the authors have identified several known species (*Arthrobacter*, *Clostridium*, and *Pseudomonas*), while the most abundant phylotypes were represented by completely unknown species closely affiliated with iron-oxidizing bacteria.

Another area of intensive application of molecular tools was northern wetlands (cold, oligotrophic (nutrient poor), and usually acidic habitats) related to the methane cycle (sections 7.4 and 7.5). The most challenging tasks were to determine what particular microbial organisms are responsible for the generation and uptake of methane (so-called methanogens and methanotrophs) in these northern ecosystems and what their reaction might be to warming of arctic soils. It was found that most of the boreal and subarctic wetlands contain a wide diversity of methanogens (Hales et al., 1996; Galand et al., 2002) and methanotrophs (McDonald et al., 1999; Radajewsky et al., 2000), most of them distantly related to known species. Only recently, some of these obscure microbes were obtained in pure culture or stable consortia (Dedysh et al., 1998; Sizova et al., 2003). The novel microbes of methane cycles are extreme oligotrophic species that evolved to function in media with very low concentrations of mineral nutrients. Taxonomically, these methanogens form new species, genera, and even families within the Archaea domain (Sizova et al., 2003). The acidophilic methanotrophs form two new genera: *Methylocapsa* and *Methylocella* (Dedysh et al., 1998, 2000), the latter affiliating with heterotrophic *Beijerinckia indica*.

A DNA-based technique allows determination of the upper limit for variation of microbial diversity in the Arctic as compared with other natural ecosystems, that is, how many species (both cultured and unculturable) soils contain. This technique is called DNA reassociation (how quickly the hybrid double helix is formed from denatured single-stranded DNA).

Arctic polar desert and tundra contain considerable microbial diversity, comparable with boreal forest soils and much higher than arable soils. Although extreme environmental conditions restrain the metabolic activity of arctic microbes, they preserve huge potential that is likely to display the same activity as boreal analogues during climate warming.

There is a much higher degree of genomic diversity in prokaryotic communities (prokaryotes such as cyanobacteria have a simple arrangement of genetic material whereas eukaryotes such as microalgae have genetic material arranged in a more advanced way, in that the DNA is linear and forms a number of distinct chromosomes) of heterogeneous habitats (virgin soils, pristine sediments) as compared with more homogeneous samples: the DNA diversity found in 30 to 100 cc of heterogeneous samples corresponds to about 10^4 different genomes, while in pond water and arable soils the number of genomes decreases to 10^0 to 10^2. Based on extrapolation and taking into account that listings of species can significantly overlap for microbial communities of different soils, a rough estimate is that there could be from 10^4 to 10^9 prokaryotic species globally (Staley and Gosink, 1999; Torsvik et al., 2002).

The conventional inventory approach based on cultivation suggests that at present in the Arctic, it is possible to identify in any particular soil no more than 100 prokaryotic species from the potential of 1000 to 3000 "genome equivalents" (Table 7.4) and no more than 2000 species of eukaryotes. In the broadly defined Russian Arctic, 1750 named fungi species (not including yeast and soil fungi) have been identified (Karatygin et al., 1999). About 350 of these are macromycetes. However, the number of fungi species in the Arctic proper is 20 to 30% less, but these data are far from complete. The Arctic has fewer species of bacteria, fungi, and algae than other major biomes; actinomycetes are rare or absent in most tundra sites (Bunnell et al., 1980). While most major phyla of microflora are represented in tundra

ecosystems, many species and genera that are common elsewhere, even in subarctic ecosystems, are rare or absent in tundra. Gram-positive bacteria, including gram-positive spore forms, are absent or rare in most tundra sites. *Arthrobacter* and *Bacillus* can rarely be isolated and then only from drier areas. *Azotobacter*, the free-living nitrogen-fixing bacterium, is extremely rare in tundra, and the moderate rate of nitrogen fixation observed *in situ* is mainly due to the activity of cyanobacteria. Sulfur-oxidizing bacteria are also reported to be rare or absent. Even using enrichment techniques, Bunnell et al. (1980) rarely found chemoautotrophic sulfur-oxidizing bacteria. Photosynthetic sulfur bacteria were not found in any IBP tundra biome sites and were reported in only one subarctic site (Bunnell et al., 1980, Dunican and Rosswall, 1974), although they are common in coastal areas of the west and south coasts of Hudson Bay. Sulfur-reducing bacteria, while not abundant in tundra sites, have been reported in sites in the Arctic and Antarctic. Iron-oxidizing bacteria are very rare in tundra sites. Despite ample iron substrate in tundra ponds and soils, chemoautotrophic ferrous iron oxidizers were not found in IBP tundra sites (Dunican and Rosswall, 1974). In contrast, methanotrophic and methanogenic bacteria appear to be widespread in tundra areas.

As with bacteria, many generally common fungi are conspicuous by their rare occurrence or absence in tundra areas. *Aspergillus*, *Alternaria*, *Botrytis*, *Fusarium*, and *Rhizopus* simply do not occur and even *Penicillium* are rare (Flanagan and Scarborough, 1974). Yeasts can be isolated readily but there is very low species diversity in culture media. Only three different species were reported for Point Barrow tundra (Bunnell et al., 1980). Aquatic fungi show high diversity, especially Chytridiales and Saprolegniales. However they may not be endemic, and reflect the annual migration into the Arctic of many avian species, especially waterfowl. The so-called higher fungi, Basidiomycetes and Ascomycetes, also have low diversity. They are reduced in the Arctic to 17 families, 30 genera, and about 100 species. In comparison, subarctic and temperate regions contain at least 50 families, not less than 300 genera, and anywhere up to 1200 species (Miller and Farr, 1975). Mycorrhizal symbionts on tundra plants are common. Arbuscular, ecto-, ericoid, arbutoid, and orchid mycorrhizal fungi are associated with plants in arctic ecosystems

Table 7.4. Microbial genome size in the Arctic as compared with other habitats (after Torsvik et al., 2002).

DNA source	Number of cells per cm³	Community genome complexity (base pairs)[a]	Genome equivalents[b]
Arctic desert (Svalbard)	7.5×10^9	$5–10 \times 10^9$	1200–2500
Tundra soil (Norway)	37×10^9	5×10^9	1200
Boreal forest soil	4.8×10^9	25×10^9	6000
Forest soil, cultivated prokaryotes	0.014×10^9	0.14×10^9	35
Pasture soil	18×10^9	$15–35 \times 10^9$	3500–8800
Arable soil	21×10^9	$0.57–1.4 \times 10^9$	140–350
Salt-crystallizing pond, 22% salinity	0.06×10^9	0.029×10^9	7

[a]the number of nucleotides in each strand in the DNA molecule; [b]measure of diversity specified at a molecular level.

(Michelsen et al., 1998). The ectomycorrhizal symbionts are important as they form mycorrhizal associations with *Betula*, *Larix*, *Pinus*, *Salix*, *Dryas*, *Cassiope*, *Polygonum*, and *Kobresia*. Based on fungal fruitbodies, Borgen et al. (in press) estimate that there are 238 ectomycorrhizal fungal species in Greenland, which may increase to around 250 out of a total of 855 basidiomyceteous fungi in Greenland when some large fungal genera such as *Cortinarius* and *Inocybe* have been revised. With the exception of *Eriophorum* spp., Flanagan (unpub. data, 2004) found endotrophic *Arbuscula*-like mycorrhizae on all ten graminoid plants examined. The number of fungal species involved in other mycorrhizal symbioses is not clear.

Tundra algae exhibit the same degree of reduction in species diversity seen among the fungi and bacteria (Bunnell et al., 1980; Elster, 2002; Fogg, 1998), which documents a diversity much reduced from that of the microflora of temperate regions. Cyanobacteria and microalgae are among the oldest (in evolutionary terms) and simplest forms of life on the planet that can photosynthesize. Unicellular and filamentous photosynthetic cyanobacteria and microalgae are among the main primary colonizers adapted to conditions of the arctic terrestrial environment. They are widespread in all terrestrial and shallow wetland habitats and frequently produce visible biomass. Terrestrial photosynthetic microorganisms colonize mainly the surface and subsurface of the soil and create the crust (Elster et al., 1999). Algal

communities in shallow flowing or static wetlands produce mats or mucilaginous clusters that float in the water but are attached to rocks underneath (Elster et al., 1997). Terrestrial and wetland habitats represent a unique mosaic of cyanobacteria and algae communities that occur up to the highest and lowest possible latitudes and altitudes as long as water (liquid or vapor) is available for some time during the year (Elster, 2002). The arctic soil and wetland microflora is composed mainly of species from Cyanobacteria, Chrysophyceae, Xanthophyceae, Bacillariophyceae, Chlorophyceae, Charophyceae, Ulvophyceae, and Zygnemaphyceae. A wide range of species diversity (between 53 and 150 to 160 species) has been reported from various arctic sites (Elster, 2002).

Latitudinal gradients of species diversity

Arctic soils contain large reserves (standing crops) of microbial (mainly fungal) biomass, although the rate of microbial growth is generally lower than in the boreal zone. Surprisingly, under severe arctic conditions, soil microbes fail to produce spores and other dormant structures that are adaptations to harsh environments (Fig. 7.7). The species diversity of all groups of soil microorganisms is lower in the Arctic than further south, decreasing from about 90 species in Irish grassland, through about 50 species in Alaskan birch forest, to about 30 species in Alaskan tundra (Flanagan and Scarborough, 1974). As with plants and animals, there are large reductions in numbers of microbial species with increasing latitude, although these patterns are less well documented. A correlate of the decreasing number of species with increasing latitude is increasing dominance of the species that occur, as with plants and animals. One yeast (*Cryptococcus laurentii*), for example, constitutes a large proportion of yeast biomass across a range of community types in the northern Taymir Peninsula (Chernov, 1985).

The hyphal length of fungi in the Arctic shows a latitudinal trend in which the abundance of fungi, as measured by hyphal length, decreases toward the north. Although it is not known if this trend also applies to the species diversity of fungal mycelia (the belowground network of fungal filaments or hyphae), it is clear that the amount of fungal hyphae is low in the Arctic (Robinson et al., 1996). In the high Arctic, fungal hyphal length was 23±1 m/g in a polar semi desert on Svalbard (78°56' N), 39 m/g on a beach ridge, and 2228 m/g in a mesic meadow on Devon Island (75°33' N). At Barrow, Alaska, hyphal length was 200 m/g. In a subarctic mire in Swedish Lapland, hyphal length was 3033 m/g. These values can be compared with 6050 to 9000 m/g in temperate uplands in the United Kingdom and 1900 to 4432 m/g in temperate woodland soils.

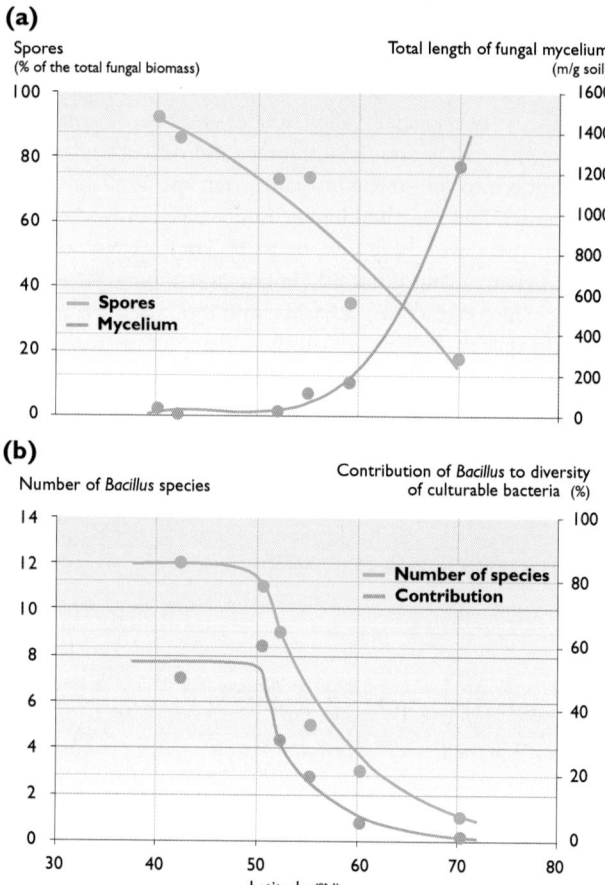

Fig. 7.7. Latitudinal distribution of (a) soil fungi and (b) bacilli; curves are hand-fitted (recalculated from data in Mirchink, 1988).

7.3.1.4. Summary

Species diversity appears to be low in the Arctic, and decreases from the boreal forests to the polar deserts of the extreme north. Only about 3% (~5900 species) of the global flora (excluding algae) occur in the Arctic north of the treeline. However, primitive plant species (mosses and lichens) are particularly abundant. Although the number of plant species in the Arctic is low in general, individual communities of small arctic plants have a diversity similar to or higher than those of boreal and temperate zones: there can be up to 25 species per square decimeter. Latitudinal gradients suggest that arctic plant diversity is sensitive to climate, and species number is least sensitive to temperature near the southern margin of the tundra. The temperature gradient that has such a strong influence on species diversity occurs over much shorter distances in the Arctic than in other biomes.

The diversity of arctic animals beyond the latitudinal treeline (~6000 species) is nearly twice as great as that of vascular plants and bryophytes. As with plants, the arctic fauna accounts for about 3% of the global total, and in general, primitive groups (e.g., springtails) are better represented in the Arctic than are advanced groups such as beetles. In general, the decline in animal species with increasing latitude is more pronounced than that of plants (frequently greater than 2.5-fold). An important consequence of the decline in numbers of species with increasing latitude is an increase in dominance. "Super-dominant" plant and animal species occupy a wide range of habitats, and generally have large effects on ecosystem processes.

Microbial organisms are more difficult to enumerate. Arctic soils contain large reserves of microbial biomass, although diversity of all groups of soil microorganisms is lower in the Arctic than further south. Many common bacteria and fungi are rare or absent in tundra areas. As with plants and animals, there are large reductions in numbers of microbial species with increasing latitude, and increasing dominance of the species that occur.

The latitudinal temperature gradient within tundra is steeper than for any other biome, and outlier populations of more southerly species frequently exist in favorable microenvironments far to the north of their centers of distribution. Consequently, migration of southerly taxa is very likely to occur more rapidly in the Arctic than in other biomes. Temperature-induced biotic change is likely to be most pronounced at the northern extreme of tundra, where species distributions are most temperature-sensitive.

The initial response of diversity to warming is likely to be an increase in the diversity of plants, animals, and microbes, and reduced dominance of species that are currently widespread. Taxa most likely to expand into tundra are boreal taxa that currently exist in river valleys and could spread into the uplands, or animal groups such as wood-boring beetles that are presently excluded due to a lack of food resources. Although current extreme environmental conditions restrain the metabolic activity of arctic microbes, they preserve huge potential that is ready to display the same activity as boreal analogues during climate warming. Warming could possibly cause the extinction of a few arctic plants that currently occur in narrow latitudinal strips of tundra adjacent to the sea. Some animals are arctic specialists and could possibly face extinction. Those plant and animal species that have their centers of distribution in the high or middle Arctic are most likely to show reduced abundance in their current locations should projected warming occur.

7.3.2. General characteristics of arctic species and their adaptations in the context of changes in climate and ultraviolet-B radiation levels

7.3.2.1. Plants

For the past 60 years, arctic plant ecologists have been concerned with the adaptations and traits of arctic plants that enable them to survive in harsh climates (e.g., Billings and Mooney, 1968; Bliss, 1971; Porsild, 1951; Russell, 1940; Savile, 1971; Sørensen, 1941). It is now important to consider how plants that are adapted to harsh environments will respond to climatic warming, and particularly how former adaptations may constrain their survival in competition with more aggressive species that are projected to immigrate from the south. Only in the past 20 years have ecologists considered arctic plant adaptations to UV-B radiation (e.g., Björn, 2002; Robberecht et al., 1980).

Plant adaptations to the arctic climate are relatively few compared with adaptations of plants to more southerly environments (Porsild, 1951; Savile, 1971) for the following reasons (Jonasson et al., 2000):

- arctic plants have inhabited the Arctic (except for ice-free refugia) for a relatively short period of time, particularly in Canada and the Yamal Peninsula;
- life spans and generation times are long, with clonal reproduction predominating;
- flowering and seed set are relatively low and insecure from year to year; and
- the complexity of the plant canopy is relatively small and the canopy is low, so that climbing plants with tendrils, thorns, etc. are not present.

Annuals and ephemeral species are very few (e.g., cold eyebright – *Euphrasia frigida* and Iceland purslane – *Koenigia islandica*). Many arctic plants are pre-adapted to arctic conditions (Crawford et al., 1994) and have migrated to the Arctic along mountain chains (Billings, 1992) or along upland mires and bogs. Although specific adaptations to arctic climate and UV-B radiation levels are absent or rare, the climate and UV-B radiation regimes of the Arctic have selected for a range of plant characteristics (Table 7.5).

The first filter for plants that can grow in the Arctic is freezing tolerance, which excludes approximately 75% of the world's vascular plants (Körner, 1995). However, many temperature effects on plants, particularly those

Table 7.5. Summary of major characteristics of current arctic plants related to climate and UV-B radiation.

Climatic factor	General effects on plants	Adaptations/characteristics of arctic plants	References
Aboveground environment			
Freezing temperatures	Plant death	Evergreen conifers tolerate temperatures between -40 and -90 °C; arctic herbaceous plants tolerate temperatures between -30 and -196 °C	Larcher, 1995
Ice encapsulation	Death through lack of oxygen	Increased anoxia tolerance	Crawford et al., 1994
Low summer temperatures	Reduced growth	Increased root growth, nutrient uptake, and respiration	Shaver and Billings, 1975; Chapin F., 1974; Mooney and Billings, 1961
		Minimized coupling between the vegetation surface and the atmosphere: cushion plants can have temperature differentials of 25 °C	Mølgaard, 1982
		Occupation of sheltered microhabitats and south-facing slopes	Bliss, 1971; Walker M.D. et al., 1991
Short, late growing seasons	Constraint on available photosynthetically active radiation and time for developmental processes	Long life cycles	Callaghan and Emanuelsson, 1985
		Slow growth and productivity	Wielgolaski et al., 1981
		Dependence on stored resources	Jonasson and Chapin, 1985
		Long flowering cycles with early flowering in some species	Sørensen, 1941
		Increased importance of vegetative reproduction	Bell and Bliss, 1980
		Clonal growth; clones surviving for thousands of years	Jónsdóttir et al., 2000
		Long-lived leaves maximizing investment of carbon	Bell and Bliss, 1977
Interannual variability	Sporadic seed set and seedling recruitment	Dependence on stored resources	Jonasson and Chapin, 1985
		Long development processes buffer effects of any one year	Sørensen, 1941
		Clonal growth	Callaghan and Emanuelsson, 1985; see also Molau and Shaver, 1997; Brooker et al., 2001; Molau and Larsson, 2000
Snow depth and duration	*Negative*: constrains length and timing of growing season	Where snow accumulates, snow beds form in which specialized plant communities occur	Gjærevoll, 1956
		Where snow is blown off exposed ridges (fellfields), plants are exposed to summer drought, winter herbivory, and extreme temperatures	Billings and Bliss, 1959; Savile, 1971
	Exerts mechanical pressure on plants	Responses and adaptations not measured	-
	Positive: Insulation in winter (it is seldom colder than -5 °C under a 0.5 m layer)	Low plant stature	Crawford, 1989
	Reduction of plant temperature extremes and freeze–thaw cycles	Low stature to remain below winter snow cover reduces the risk of premature dehardening	Ögren, 1996, 1997
	Protection from wind damage, abrasion by ice crystals, and some herbivory	Low stature to remain below winter snow cover; growth in sheltered locations	Sveinbjörnsson et al., 2002
	Protection from winter desiccation when water loss exceeds water supply from frozen ground	Low stature to remain below winter snow cover; deciduous growth	Barnes et al., 1996; Havas, 1985; Ögren, 1996; Taulavuori E. et al., 1997; Taulavuori K. et al., 1997b
	Protection from chlorophyll bleaching due to light damage in sunny habitats	Low stature to remain below winter snow cover; deciduous growth	Curl et al., 1972
	Source of water and nutrients late into the growing season	Zonation of plant species related to snow depth and duration	Fahnestock et al., 2000; Gjærevoll, 1956
Increased UV-B radiation levels	Damage to DNA that can be lethal or mutagenic	Reflective/absorptive barriers such as thick cell walls and cuticles, waxes, and hairs on leaves; physiological responses such as the induction or presence of UV-B radiation absorbing pigments (e.g., flavonoids) and an ability to repair some UV-B radiation damage to DNA	Robberecht et al., 1980; Semerdjieva et al., 2003
		Repair is mediated through the enzyme photolyase that is induced by UV-A radiation. There is so far no indication of adaptations to UV-B radiation that are specific to plants of the Arctic	Li et al., 2002a,b

Climatic factor	General effects on plants	Adaptations/characteristics of arctic plants	References
Aboveground environment			
Variable atmospheric CO_2 concentrations	Increased atmospheric CO_2 concentrations usually stimulate photosynthesis and growth if other factors are non-limiting; an increased ratio of carbon to nitrogen in plant tissues	Photosynthesis in Alaskan graminoids acclimated to high CO_2 concentrations within six weeks, with no long-term gain	Tissue and Oechel, 1987
		The dwarf willow (*Salix herbacea*) has been able to alter its carbon metabolism and morphology in relation to changing CO_2 concentrations throughout the last 9000 years	Beerling and Rundgren, 2000
		Species such as the moss *Hylocomium splendens* are already adapted to high CO_2 concentrations; frequently experiencing 400–450 ppm, and sometimes over 1100 ppm, to compensate for low light intensities under mountain birch woodland. These high CO_2 concentrations are caused by both soil and plant respiration close to the forest understory surface in still conditions and when light intensities are low	Sonesson et al., 1992
Soil environment			
Low rates of nutrient availability, particularly nitrogen	Reduced growth and reproduction	Conservation of nutrients in nutrient-poor tissues	Wielgolaski et al., 1975
		Long nitrogen retention time resulting from considerable longevity of plant organs and resorption of nutrients from senescing tissues and retention of dead leaves within plant tufts and cushions	Berendse and Jonasson, 1992
		Substantial rates of nutrient uptake at low temperatures	Chapin F. and Bloom, 1976
		Increased surface area for nutrient uptake by increased biomass of roots relative to shoots (up to 95% of plant biomass can be below ground)	Chapin F., 1974; Shaver and Cutler, 1979
		Associations with mycorrhizal fungi	Michelsen et al., 1998
		Nitrogen uptake by rhizomes	Brooker et al., 1999
		Some arctic plants can take up nutrients in organic forms, bypassing some of the slow decomposition and mineralization processes	Chapin F. et al., 1993
		Dependence on atmospheric nutrient deposition in mosses and lichens	Jónsdóttir et al. 1995
Soil movement at various spatial scales resulting from freeze–thaw cycles, permafrost dynamics, and slope processes	Freeze–thaw cycles heave ill-adapted plants from the soil and cause seedling death	Areas of active movement select for species with elastic and shallow roots or cryptogams without roots	Perfect et al., 1988; Wager, 1938; Jonasson and Callaghan, 1992; Chapter 5
Shallow active layer	Limits zone of soil biological activity and rooting depth; shallow rooting plates of trees can lead to falling	Shallow-rooting species, rhizome networks	Shaver and Cutler, 1979
Biotic environment			
Herbivory	Removal of plant tissue sometimes leading to widespread defoliation and death	Arctic plants do not have some morphological defenses (e.g., thorns) found elsewhere	Porsild, 1951
		Many plants have secondary metabolites that deter herbivores; some substances are induced by vertebrate and invertebrate herbivores	Haukioja and Neuvonen, 1987; Seldal et al. 1994
		Protected growing points; continuous leaf growth in summer; rapid modular growth in some graminoids; regeneration from torn fragments of grass leaves, mosses, and lichens	Savile, 1971; Sørensen, 1941
Competition	Suppression of some species and increased dominance of others leading to changes in community structure	Secondary metabolites in some arctic species inhibit the germination and growth of neighboring species	Michelsen et al., 1995; Zackrisson and Nilsson, 1992
Facilitation	Mutual benefits to plant species that grow together	Positive plant interactions are more important than plant competition in severe physical environments	Brooker and Callaghan, 1998; Callaway et al., 2002
		Nitrogen-fixing species in expanding glacial forefields facilitate the colonization and growth of immigrant plant species	Chapin F. et al., 1994
		Plant aggregation can confer advantages of shelter from wind	Carlsson and Callaghan, 1991
		Hemiparasites can stimulate nutrient cycling of potential benefit to the whole plant community	Quested et al., 2002

with roots and in the long term, are indirect (Chapin F., 1983). Plant nutrients in arctic soils, particularly nitrogen, are available to higher plants (with roots) at low rates (Russell R., 1940) because of slow microbial decomposition and mineralization rates of organic matter constrained by low temperatures (Heal et al., 1981). Arctic plants use different strategies for nutrient uptake (Callaghan et al., 1991), and different sources of nitrogen, which reduces competition among plants and facilitates greater plant diversity (McKane et al., 2002).

Many of the adaptations of arctic species to their current environments, such as slow and low growth, are likely to limit their responses to climate warming and other environmental changes. If changes in climate and UV radiation levels adversely affect species such as mosses that play an important role in facilitation, normal community development and recovery after disturbance are likely to be constrained. Many arctic plant characteristics are likely to enable plants to cope with abiotic selective pressures (e.g., climate) more than biotic pressures (e.g., interspecific competition). This is likely to render arctic organisms more susceptible to biological invasions at their southern distributional limits, while populations at their northern range limit (e.g., boreal species in the tundra) are likely to respond more than species at their southern limit to warming *per se*. Thus, as during past environmental changes, arctic species are very likely to change their distributions rather than evolve significantly.

Summary

Plant adaptations to the arctic climate are absent or rare: many species are pre-adapted. The first filter for arctic plants is freezing tolerance, which excludes approximately 75% of the world's vascular plants. Short growing seasons and low solar angles select for long life cycles in which slow growth often uses stored resources while development cycles are often extended over multiple growing seasons. Some plant species occupy microhabitats, or exhibit behavior or growth forms that maximize plant temperatures compared with ambient temperatures. Low soil temperatures reduce microbial activity and the rates and magnitude of nutrient availability to the roots of higher plants. Mechanisms to compensate for low nutrient availability include the conservation of nutrients in nutrient-poor tissues, resorption of nutrients from senescing tissues, enhanced rates of nutrient uptake at low temperatures, increased biomass of roots relative to shoots, associations with mycorrhizal fungi, uptake of nutrients in organic forms, and uptake of nitrogen by rhizomes. Temperature fluctuations around 0 °C cause frost-heave phenomena that can uproot ill-adapted plants.

Snow distribution determines the period over which plants can intercept solar radiation and grow. Snow cover insulates plants against low air temperatures in winter and extremes of temperature in spring, protects plants from physical damage from abrasion by ice crystals, and provides a source of water often late into the growing season. Where snow cover is thin (e.g., on exposed ridges), growing seasons are usually long but water can become limiting; where snow accumulates in sheltered depressions, snow beds form in which specialized plant communities occur.

Many arctic plants are pre-adapted to relatively high levels of UV-B radiation. They exhibit various mechanisms to protect DNA and sensitive tissues from UV-B radiation and an ability to repair some UV-B radiation damage to DNA. Thick cell walls and cuticles, waxes, and hairs on leaves, and the presence or induction of UV-B radiation absorbing chemical compounds in leaves, protect sensitive tissues. There appear to be no specific adaptations of arctic plant species to high atmospheric CO_2 concentrations.

Arctic plant species do not show the often-complex interactions with other organisms prevalent in southern latitudes. Arctic plants are adapted to grazing and browsing mainly through chemical defenses rather than the possession of spines and thorns. Facilitation increases in importance relative to competition at high latitudes and altitudes.

Thus, many of the adaptations of arctic species to their current environments are likely to limit their responses to climate warming and other environmental changes. Many characteristics are likely to enable plants to cope with abiotic selective pressures (e.g., climate) more than biotic pressures (e.g., inter-specific competition). This is likely to render arctic organisms more susceptible to biological invasions and they are very likely to change their distributions rather than evolve significantly in response to warming.

7.3.2.2. Animals

Classical arctic zoology typically focused on morphological and physiological adaptations to life under conditions of extremely low winter temperatures (Schmidt-Nielsen, 1979; Scholander et al., 1950). Physiological studies contribute to a mechanistic understanding of how arctic animals cope with extreme environmental conditions (especially low temperatures), and what makes them different from their temperate counterparts. Ecological and evolutionary studies focus on how life-history strategies of arctic animals have evolved to tolerate environmental variation in the Arctic, how flexible life histories (in terms of both phenotypic plasticity and genetic variation) are adapted to environmental variation, and how adjustments in life-history parameters such as survival and reproduction translate into population dynamics patterns.

Adaptations to low temperatures

Arctic animals have evolved a set of adaptations that enable them to conserve energy in low winter temperatures. Warm-blooded animals that persist throughout the arctic winter have thick coats of fur or feathers that often turn white (Scholander et al., 1950). The body shapes of high-arctic mammals such as reindeer/caribou, collared lemmings, Arctic hares (*Lepus arcticus*), and Arctic foxes (*Alopex lagopus*) are rounder and their extremities shorter than their temperate counterparts (Allen's rule). Body size within some vertebrate taxa increases toward the north (Bergman's rule), but there are several notable arctic exceptions to this (e.g., reindeer/caribou; Klein, 1999; muskox; Smith P. et al., 2002). There are few physiological adaptations in homeotherms (warm-blooded animals) that are unique to arctic animals. However, several adaptations may be considered to be typical of the Arctic, including fat storage (e.g., reindeer/caribou and Arctic fox; Prestrud and Nilssen, 1992) and reduced body-core temperature and basal metabolism in the winter (e.g., Arctic fox; Fuglei and Øritsland, 1999). While hibernation during the winter is found in a few arctic mammals such as the Arctic ground squirrel (*Spermophilus parryii*), most homeothermic animals are active throughout the year. Small mammals such as shrews (*Sorex* spp.), voles (*Microtus, Clethrionomys* spp.), and lemmings with relatively large heat losses due to a high surface-to-volume ratio stay in the subnivean space (a cavity below the snow) where they are protected from low temperatures during the winter. Even medium-sized birds and mammals such as ptarmigan (*Lagopus mutus*) and hares seek thermal refuges in snow caves when resting. In the high Arctic, the normal diurnal activity patterns observed at more southerly latitudes are replaced by activity patterns that are independent of the time of the day (e.g., Svalbard ptarmigan – *Lagopus mutus hyperboreus*; Reierth and Stokkan, 1998).

In heterothermic (cold-blooded) invertebrates, hairiness and melanism (dark pigmentation) enable them to warm up in the summer season. Invertebrates survive low winter temperatures in dormancy mainly due to two cold-hardiness strategies: freeze tolerance and freeze avoidance (Strathdee and Bale, 1998). Typically, supercooling points are lower in arctic than in temperate invertebrates. Freeze tolerance, which appears to be an energetically less costly strategy than extended supercooling, is a common strategy in very cold regions. Wingless morphs occur frequently among arctic insects, probably because limited energy during the short growth season is allocated to development and reproduction, rather than in an energetically costly flight apparatus. A short growth season also constrains insect body size and number of generations per year. Life cycles are often extended in time and/or simplified because invertebrates may need several seasons to complete their life cycles. Small body size in arctic insects seems to be a strategy to shorten generation time (Strathdee and Bale, 1998). Moreover, individuals from arctic populations are able to grow faster at a given temperature than southern conspecifics (e.g., Birkemoe and Leinaas, 2000). Thus, arctic invertebrates may be particularly efficient in utilizing relatively short warm periods to complete life-cycle stages.

A short breeding season also underlies several life-history adaptations in birds and mammals, such as synchronized breeding, shortened breeding season, specif-

ic molting patterns, and mating systems (Mehlum, 1999). Although adjustments to low temperatures and short growth seasons are widespread in arctic animals, successful species cannot be generalized with respect to particular life-history traits (Convey, 2000). Both flexible and programmed life cycles are common in polar arthropods (Danks, 1999).

While there are many examples that show that winter temperatures lower than species-specific tolerance limits set the northern borders of the geographic distribution of animals, there are hardly any examples that demonstrate that high temperatures alone determine how far south terrestrial arctic animals are found. Southern range borders are typically set by a combination of abiotic factors (e.g., temperature and moisture in soil invertebrates) or, probably most often, by biotic factors such as food resources, competitors, and natural enemies.

Migration and habitat selection

Many vertebrates escape unfavorable conditions through movement (either long-distance migration or more short-range seasonal movement) between different habitats in the same landscape. Seasonal migration to southern overwintering areas is almost the rule in arctic birds. Climate may interfere in several ways with migrating birds, such as mismatched migration timing, habitat loss at stopover sites, and weather en route (Lindström and Agrell, 1999), and a mismatch in the timing of migration and the development of invertebrate food in arctic ponds (section 8.5.6). Many boreal forest insects invade the low-arctic tundra in quite large quantities every summer (Chernov and Matveyeva, 1997), but few of these are likely to return in the autumn. Birds residing in the tundra throughout the year are very few and include species such as Arctic redpoll (*Carduelis hornemanni*), willow grouse (*Lagopus lagopus*), ptarmigan, raven, gyrfalcon (*Falco rusticolus*), and snowy owl (*Nyctea scandiaca*). Like several other arctic predators that specialize in feeding on lemmings and Arctic voles, the snowy owl emigrates when cyclic lemming populations crash to seek high-density prey populations elsewhere in the Arctic and subarctic. A similar nomadic lifestyle is found in small passerine seed-eating birds such as redpolls and crossbills (*Loxia* spp.) in the forest tundra. These birds move between areas with asynchronous mast years (years with exceptionally abundant seed production) in birch and conifers. A substantial fraction of the Arctic fox population migrates after lemming peaks and sometimes these migrations may extend far into the taiga zone (Hersteinsson and Macdonald, 1992). Most reindeer and caribou populations perform seasonal migrations from coastal tundra in summer to continental areas of forest tundra and taiga in the winter. Inuit ecological knowledge explains caribou migrations as triggered by seasonal "cues", such as day length, temperature, or ice thickness (Thorpe et al., 2001). Reindeer on isolated arctic islands are more sedentary without pronounced seasonal migrations (Tyler and Øritsland, 1989). Lemmings and ptarmigans shift habitat seasonally within the same landscapes (Kalela, 1961). In peak population years, the seasonal habitat shifts of the Norway lemming (*Lemmus lemmus*) may become more long-distance mass movements (Henttonen and Kaikusalo, 1993). For small mobile animals (e.g., wingless soil invertebrates such as Collembola and mites), habitat selection on a very small spatial scale (microhabitat selection) enables individuals to find spatial refuges with temperature and moisture regimes adequate for survival (Hodkinson et al., 1994; Ims et al., 2004). The variability in microclimatic conditions may be extremely large in the high Arctic (Coulson et al., 1995).

Adaptations to the biotic environment

Generalists in terms of food and habitat selection seem to be more common among arctic animals than in communities further south (e.g., Strathdee and Bale, 1998). This may be either due to fewer competitors and a less tightly packed niche space in arctic animal communities and/or because food resource availability is less predictable and the appropriate strategy is to opt for more flexible diets. Notable exceptions to food resource generalism are lemming predators (e.g., least weasels – *Mustela nivalis*, and several owls – *Asio*, *Nyctea* spp., raptors – *Buteo* spp., and skuas – *Stercorarius* spp.) and a number of host-specific phytophagous insects (e.g., aphids – *Acyrthosiphon* spp. and sawflies – Symphyta). Many water birds, such as geese and sandpipers (*Calidris* spp.) with 75 and 90%, respectively, of species breeding in the Arctic, are habitat specialists. Some species exhibit a large flexibility in their reproductive strategy based on food resources. Coastal populations of Arctic foxes with a relatively predictable food supply from the marine ecosystem (e.g., seabird colonies) have smaller litter sizes than inland "lemming foxes" that rely on a highly variable food supply (Tannerfeldt and Angerbjörn, 1998). Specialists on highly fluctuating food resources such as seeds from birch and conifers, as well as lemmings and voles, respond to temporary superabundant food supplies by having extraordinarily high clutch or litter sizes.

High Arctic environments contain fewer natural enemy species (e.g., predators and parasites) and some animals seem to be less agile (e.g., Svalbard reindeer; Tyler and Øritsland, 1989) and are possibly less resistant to disease (Piersma, 1997).

Ultraviolet-B radiation

Little is known about animal adaptations to UV-B radiation. Non-migrant species, such as reindeer/caribou, Arctic foxes, hares, and many birds, have white fur or feathers that presumably reflect some UV-B radiation. There is some evidence, however, that feathers can be affected by high UV-B radiation levels (Bergman, 1982), although this early research needs to be repeated. There is also a possibility that fur absorbs UV-B radiation. Eyes of arctic vertebrates experience extremes of UV-B radiation, from dark winter conditions to high UV-B radiation environments in spring. However, mechanisms of toler-

ance are unknown. Invertebrates in general have DNA that is robust to UV-B radiation damage (Koval, 1988) and various adaptations to reduce the absorption of UV-B radiation. Some subarctic caterpillars possess pigmented cuticles that absorb in the UV-B wavelengths, while pre-exposure to UV-B radiation can induce pigmentation (Buck and Callaghan, 1999). Collembolans and possibly other invertebrates have dark pigmentation that plays a role in both thermoregulation and UV-B radiation protection (Leinaas, 2002).

Patterns of population dynamics

In tundra habitats, population cycles in small- to medium-sized birds and mammals are the rule, with few exceptions. The period of the population cycle in lemmings and voles varies geographically, and is between three and five years. Cyclicity (e.g., spatial synchronicity and period between peak population years) seems to be associated with spatial climate gradients (coast to inland and south to north) in Fennoscandia (Hansson and Henttonen, 1988; Strann et al., 2002), although the biotic mechanisms involved are still much debated (Hanski et al., 2001). Some lemming populations show geographic variation in the cycle period within arctic Siberia; and also (for example) exhibit a long period (5 years on Wrangel Island) and a relatively short period (3 years in Taymir) between peak population years (Chernov and Matveyeva, 1997). Within regions (e.g., northern Fennoscandia), small rodent cycles may show distinct interspecific synchrony over large spatial scales (Myrberget, 1973). However, recent spatially extensive surveys in northern Canada (Predavec et al., 2001) and Siberia (Erlinge et al., 1999) have indicated that the spatial synchrony of lemming populations is not as large-scale as the snowshoe hare (*Lepus americanus*) cycles in boreal North America (Elton and Nicholson, 1942). This is at least partly due to the geographically variable cycle period.

Populations of small- and medium-sized bird and mammal predators follow the dynamics of their lemming and vole prey species (Wiklund et al., 1999). The signature of lemming and vole population dynamics can also be found in the reproductive success and demography of mammals and birds, for example, waders and geese (e.g., Bety et al., 2002), that serve as alternative prey for lemming predators. Among northern insects, population cycles are best known in geometrid moths, particularly the autumnal moth (*Epirrita autumnata*), which exhibits massive population outbreaks with approximately 10-year intervals that extend into the forest tundra (Tenow, 1972, 1996). In the tundra biome, no herbivorous insects are known to have population cycles (Chernov and Matveyeva, 1997). However, the population dynamics of tundra invertebrates is poorly known due to the lack of long-term data. Soil invertebrates such as Collembola (Birkemoe and Sømme, 1998; Hertzberg et al., 2000) sometimes exhibit large interannual fluctuations in population density. Large fluctuations in numbers have also been observed in arctic ungulate populations (reindeer/caribou and muskox), and seem to be the result of several biotic factors in combination with climatic variation (Klein, 1996, 1999; Morneau and Payette, 2000).

Summary: Implications for animal responses to climate change

Terrestrial arctic animals possess many adaptations that enable them to persist in the arctic climate. Physiological and morphological traits in warm-blooded vertebrates (mammals and birds) include thick fur and feather plumages, short extremities, extensive fat storage before winter, and metabolic seasonal adjustments, while cold-blooded invertebrates have developed strategies of cold hardiness, high body growth rates, and pigmented and hairy bodies. Arctic animals can survive under an amazingly wide range of temperatures, including high temperatures. The short growing season represents a challenge for most arctic animals and life-history strategies have evolved to enable individuals to fulfill their life cycles under time constraints and high environmental unpredictability. The biotic environment (e.g., the ecosystem context) of arctic species is relatively simple with few enemies, competitors, and available food resources. For those reasons, arctic animals have evolved fewer traits related to competition for resources, predator avoidance, and resistance to diseases and parasites than have their southern counterparts. Life cycles that are specifically adjusted to seasonal and multi-annual fluctuations in resources are particularly important because such fluctuations are very pronounced in terrestrial arctic environments. Many arctic animals possess adaptations for escaping unfavorable weather, resource shortages, or other unfavorable conditions through either winter dormancy or by selection of refuges at a wide range of spatial scales, including microhabitat selection at any given site, seasonal habitat shifts within landscapes, and long-distance seasonal migrations within or across geographic regions.

Based on the above general characteristics, if climate changes, terrestrial arctic animals are likely to be most vulnerable to the following conditions: higher summer temperatures that induce desiccation in invertebrates; climatic changes that interfere with migration routes and staging sites for long-distance migrants; climatic events that alter winter snow conditions and freeze–thaw cycles resulting in unfavorable temperature, oxygen, and CO_2 conditions for animals below the snow and limited resource availability (e.g., vegetation or animal prey) for animals above the snow; climatic changes that disrupt behavior and life-history adjustments to the timing of reproduction and development that are currently linked to seasonal and multi-annual peaks in food resource availability; and the influx of new competitors, predators, parasites, and diseases.

7.3.2.3. Microorganisms

As a group, microorganisms are highly mobile, can tolerate most environmental conditions, and have short generation times that can facilitate rapid adaptation to new environments associated with changes in climate and UV-B radiation levels.

Adaptations to cold

The ability to resist freezing (and to restore activity after thawing) and the ability to metabolize below the freezing point are fundamental microbial adaptations to cold climates prevailing at high latitudes.

Cell viability depends dramatically on the freezing rate, which defines the formation of intracellular water crystals (Kushner, 1981; Mazur, 1980). Cold-adapted microbial species are characterized by remarkably high resistance to freezing due to the presence of specific intracellular compounds (metabolic antifreeze), stable and flexible membranes, and other adaptations. Lichens are extreme examples (Kappen, 1993): the moist thalli of such species as *Xanthoria candelaria* and *Rhizoplaca melanophthalma* fully tolerated gradual or rapid freezing to -196 °C, and even after being stored for up to several years, almost immediately resumed normal photosynthetic rates when warmed and wetted. For five to seven months of cold and continuous darkness, they remain green with intact photosynthetic pigments. However, freeze resistance is not a unique feature of arctic organisms.

The ability of microorganisms to grow and metabolize in frozen soils, subsoils, or water is generally thought to be insignificant. However, microbial growth and activity below the freezing point has been recorded in refrigerated food (Larkin and Stokes, 1968) as well as in arctic and antarctic habitats such as sea ice, frozen soil, and permafrost (Kappen et al., 1996; Schroeter et al., 1994). Such activity has important implications for ecosystem function. Field measurements of gas fluxes in Alaska and northern Eurasia revealed that winter CO_2 emissions can account for up to half of annual CO_2 emissions (Oechel et al., 1997; Panikov and Dedysh, 2000; Sommerfeld et al., 1993; Zimov et al., 1993), implying significant cold-season activity in psychrophilic ("cold-loving") soil microbes. Soil fungi (including mycobionts, the symbiotic fungal component of lichens) have been considered as the most probable candidates for the majority of the tundra soil respiration occurring at temperatures below 0 °C (Flanagan and Bunnell, 1980) because their live biomass was estimated to be ten times larger than that of cohabiting bacteria.

Winter CO_2 emissions have been also explained by other mechanisms (e.g., the physical release of summer-accumulated gases or abiotic CO_2 formation due to cryoturbation; Coyne and Kelley, 1971). Most recent studies (Finegold, 1996; Geiges, 1996; Mazur, 1980; Rivkina et al., 2000; Russell N., 1990) agree that microbial growth is limited at about -12 °C and that occasional reports of microbial activity below -12 °C (e.g., continuous photosynthesis in arctic and antarctic lichens down to -17 °C; Kappen et al., 1996; Schroeter et al., 1994, and photosynthetic CO_2 fixation at -24 °C; Lange and Metzner, 1965) were not carefully recorded and confirmed. Under laboratory conditions, Rivkina et al. (2000) quantified microbial growth in permafrost samples at temperatures down to -20 °C. However, the data points below -12 °C turned out to be close to the detection limits of the highly sensitive technique that they employed. The authors concluded that nutrient uptake at -20 °C could be measured, but only transiently, "whereas in nature (i.e., under stable permafrost conditions)... the level of activity, if any, is not measurable ..." (Rivkina et al., 2000).

Recently, a new precise technique was applied to frozen soil samples collected from Barrow, Alaska, and incubated at a wide range of subzero temperatures under laboratory conditions (Panikov et al., 2001). The rate of CO_2 production declined exponentially with temperature and unfrozen water content when the soil was cooled below 0 °C, but it remained surprisingly positive and measurable (e.g., 8 ng C/d/kg) at -39 °C. A range of experimental results and treatments confirmed that this CO_2 production at very low temperatures was due to microbial respiration, rather than to abiotic processes. The demonstration that microorganisms can survive low temperatures suggests the possibility that ancient bacteria of distinctive genotypes trapped in permafrost will be released and become active during permafrost thawing (Gilichinsky, 1994). However, the period over which ancient permafrost is likely to thaw will be significantly longer than the next 100 years (section 6.6.1.3).

Dark pigmentation causes higher heat absorption in lichens, which is especially favorable in the cold polar environment (Kershaw, 1983; Lange, 1954).

Adaptations to drought

Freezing is always associated with a deficiency of available water. Thus, true psychrophilic organisms must also be "xerotolerant" (adapted to extremely dry environments). A number of plants and microorganisms in polar deserts, such as lichens, are termed "poikilohydrous", meaning that they tend to be in moisture equilibrium with their surroundings (Blum, 1974). They have high desiccation tolerance and are able to survive water loss of more than 95% and long periods of drought. Rapid water loss inactivates the thallus, and in the inactive state the lichen is safe from heat-induced respiratory loss and heat stress (Kappen, 1974; Lange, 1953). In unicellular microorganisms, drought resistance can also be significant, although mycelial forms of microbial life (fungi and actinomycetes) seem to have a much higher resistance to drought due to their more efficient cytoplasm compartmentalization and spore formation.

Adaptations to mechanical disturbance

Wind, sand, and ice blasts, and seasonal ice oscillations, are characteristic features of arctic environments that affect the colonization and survival of organisms. Most lichens are adapted to such effects by forming a mechanically solid thallus firmly attached to the substrate.

Windswept habitats such as hillsides can be favorable if they provide a suitably rough substrate and receive sufficient moisture from the air. In contrast, shallow depressions or small valleys, although more sheltered, are bare of lichens because snow recedes from them only for very brief periods each season or persists over several years. This phenomenon is one reason for the so-called trimline effect (a sharp delineation on rocks between zones with and without lichens: Corner and Smith, 1973; Koerner, 1980; Smith R., 1972). The abrasive forces of the ice at the bottoms of glaciers may destroy all epilithic (rock-attached) lichen vegetation, but lichens once established are able to survive long periods of snow cover, and even glacial periods (Kappen, 1993).

Adaptations to irradiance

Strong pigmentation is typical for numerous microorganisms inhabiting tundra and polar deserts, especially for those that are frequently or permanently exposed to sunlight at the soil surface (lichens and epiphytic bacteria). Pigments (melanin, melanoids, carotenoids, etc.) are usually interpreted as a protection against strong irradiation. Pigmentation may be constitutive for particular species or appear as a plastic response to irradiance, for example, originally colorless *Cladonia* and *Cladina* lichens quickly develop dark-pigmented thalli after exposure to higher levels of solar radiation (Ahmadjian, 1970). Buffoni Hall et al. (2002) demonstrated that in *Cladonia arbuscula* ssp. *mitis* an increase in phenolic substances is specifically induced by UV-B radiation, and that this increase leads to attenuation of the UV-B radiation penetrating into the thallus. The accumulation of the protective pigment parietin in *Xanthoria parietina* is induced specifically by UV-B radiation (Gauslaa and Ustvedt, 2003), while in *Cladonia uncialis* and *Cladina rangiferina* only UV-A radiation had a stimulating effect on the accumulation of usnic acid and atranorin, respectively. Photorepair of radiation-damaged DNA in *Cladonia* requires not only light, but also high temperature and a hydrated thallus (Buffoni Hall et al., 2003). As in higher plants, carotenoids protect algae, fungi, and lichens from excessive photosynthetically active radiation (MacKenzie et al., 2002) and perhaps also have a role in protecting them from ultraviolet radiation. In contrast to higher plants, flavonoids do not act as screening compounds in algae, fungi, and lichens.

Braga et al. (2001a,b) surveyed the UV radiation sensitivity of conidia (spore-forming bodies) of 30 strains belonging to four species of the fungus *Metarhizium*, an important biological insecticide. Exposing the fungus to UV-B radiation levels within an ecologically relevant range revealed great differences between the strains, with strains from low latitudes generally more tolerant of UV-B radiation than strains from high latitudes.

Algae

Seven inter-related stress factors (temperature, water, nutrient status, light availability and/or UV radiation, freeze–thaw events, and growing-season length and unpredictability) are important for life in arctic terrestrial and shallow wetlands (Convey, 2000). Cyanobacteria and algae have developed a wide range of adaptive strategies that allow them to avoid or at least minimize injury. Three main strategies for coping with life in arctic terrestrial and wetland habitats are avoidance, protection, and the formation of partnerships with other organisms (Elster and Benson, 2004). Poikilohydricity (tolerance of desiccation) and shelter strategies are frequently interconnected, and when combined with cell mobility and the development of complex life cycles, afford considerable potential for avoidance. The production of intracellular protective compounds, which control the cell solute composition and viscosity (changes in the carbohydrate and polyols composition of the cell), together with changes in cell wall structures (production of multi-layered cell walls and mucilage sheets) are very common phenomena. The association of cyanobacteria and algae with fungi in lichens provides the benefit of physical protection.

Summary

Arctic microorganisms are not only resistant to freezing, but some can metabolize at temperatures down to -39 ºC. During winter, this process could be responsible for up to 50% of annual CO_2 emissions from tundra soils. Cold-tolerant microorganisms are usually also drought-tolerant. Microorganisms are tolerant of mechanical disturbance and high irradiance. Pigmentation protects organisms such as lichens from high irradiance, including UV radiation, and pigments can be present in considerable concentrations. Cyanobacteria and algae have developed a wide range of adaptive strategies that allow them to avoid, or at least minimize, UV injury. However, in contrast to higher plants, flavonoids do not act as screening compounds in algae, fungi, and lichens.

As a group, microorganisms are highly adaptive, can tolerate most environmental conditions, and have short generation times that can facilitate rapid adaptation to new environments associated with changes in climate and UV-B radiation levels.

7.3.3. Phenotypic responses of arctic species to changes in climate and ultraviolet-B radiation

Species responses to climate change are complex. They respond individualistically to environmental variables such as temperature (Chapin F. and Shaver, 1985a), and various processes within a given species (e.g., reproductive development, photosynthesis, respiration, leaf phenology in plants) respond individualistically to a given environmental change. Knowledge of species responses to changes in temperature comes from many sources including indigenous knowledge, current species distributions related to climate, and experimental manipulations of temperature in the laboratory and field.

7.3.3.1. Plants

The information presented in this section relates to individual plant species and how they have responded to changes in various aspects of climate and UV radiation. The information is taken mainly from experiments in which climate variables or UV radiation levels were modified and the responses of the individual species determined while they were growing in natural communities. Indigenous knowledge is also included.

Responses to current changes in climate

Indigenous knowledge studies in Canada describe poor vegetation growth in eastern regions associated with warmer summers and less rain (Fox, 2002), but describe increased plant biomass and growth in western regions, particularly in riparian areas and of moisture-tolerant species such as shrubs (Riedlinger, 2001; Thorpe et al., 2001), due to lengthening of the growing season, marked spring warming, and increased rainfall.

Inuit participating in the Tuktu and Nogak Project in the Kitikmeot region of Nunavut (Thorpe, 2000; Thorpe et al., 2001) observed that vegetation was more lush, plentiful, and diverse in the 1990s compared to earlier decades. Willows and alders were described as taller, with thicker stem diameters and producing more branches, particularly along shorelines. Other indigenous communities have also reported increases in vegetation, particularly grasses and shrubs – stating that there is grass growing in places where there used to be only gravel. On Banks Island, in the western Canadian Arctic, Inuvialuit point to observations that the *umingmak* (muskox) are staying in one place for longer periods of time as additional evidence that vegetation is richer (Riedlinger, 2001). In addition, Riedlinger (2001) has documented Inuvialuit observations of an increase in forbs such as *qungalik* (Arctic sorrel – *Oxyria digyna*), which is described as coming out earlier in the spring, and noticeably "bigger, fresher, and greener".

The Arctic Borderlands Ecological Knowledge Co-op monitors the annual quality and quantity of salmonberries locally called *akpiks* (cloudberry – *Rubus chamaemorus*), and has documented recent observations of high temperatures early in the year that "burn" berry plant flowers, early spring melt that results in inadequate moisture for the plants later in the year, and intense summer sun that "cooks" the berries before they can be picked (Kofinas et al., 2002). On Banks Island, local residents report years where the grass remained green into the autumn, leaving it vulnerable to freezing (Riedlinger, 2001). This corresponds to experiments that show a similar effect on Svalbard (Robinson et al., 1998).

In northern Finland, marshy areas are said to be drying up. Sami reindeer herders from Kaldoaivi in Utsjoki have observed that berries such as bog whortleberry (*Vaccinium uliginosum*) have almost disappeared in some areas. Other berries such as cloudberry and lingonberry

(*V. vitis-idaea*) are said to have declined in the last 30 years (Helander, 2002). Indigenous peoples' observations of declining cloudberry production are supported by experiments that postulate declines in growth in warm winters (Marks and Taylor, 1978) and provide detailed mechanisms of fruit production (Jean and Lapointe, 2001; Korpelainen, 1994; Wallenius, 1999).

Indigenous knowledge also records changes in species distribution: some existing species have become more widespread and new species have been seen. In addition to increased shrub abundance, Thorpe et al. (2001, 2002) documented reports of new types of lichens and flowering plants on Victoria Island in Nunavut and more individual plants of the same species (Thorpe, 2000). The increases in shrubs in this area correspond to aerial photographic evidence of increases in shrub abundance in Alaska (Sturm et al., 2001b). However, the reports of new types of plants, and lichens in particular, contrast with experimental evidence that shows a decrease in lichens and some mosses when flowering plant biomass increases (Cornelissen et al., 2001; Potter et al., 1995). A possible reason for this is that results from warming experiments cannot be extrapolated throughout the Arctic because of variations in recent and projected climate owing to both cooling and warming (sections 2.6.2.1 and 4.4.2): warming experiments in continuous vegetation show declines of lichens, whereas lichens may expand their distribution during warming in the high Arctic where vascular plant competitors are sparse (Heide-Jørgensen and Johnsen, 1998).

In contrast to observed responses of plants to recent warming, remote sensing by satellites has shown that the start dates of birch pollen seasons have been delayed at high altitudes and in the northern boreal regions of Fennoscandia (Høgda et al., 2002). In the Faroe Islands, there has been a lowering of the alpine zone in response to a 0.25 °C cooling over the past 50 years (Fossa, 2003).

Projected responses to future temperature changes

Warming *per se* is very likely to be favorable to the growth, development, and reproduction of most arctic plant species, particularly those with high phenotypic plasticity (flexible/responsive growth and development). However, other limiting factors such as nutrients and moisture or competition from immigrant species are likely to modify plant responses to warming. In some cases, the direct and indirect effects of warming are projected to generate negative responses:

- Increased respiration relative to photosynthesis can result in negative carbon balances, particularly in clonal plants that accumulate old tissues, for example, the cushion form ecotype (Fig. 7.3) of purple saxifrage (*Saxifraga oppositifolia*; Crawford, 1997a) and some species of the herb *Ranunculus* (Cooper, 1996).

- It is possible that cushion forms of arctic plants, including mosses, that have low atmospheric coupling and experience high temperatures will experience thermal death during warming, particularly when combined with reduced cooling by evapotranspiration under drought conditions.

- Exposure to high levels of solar radiation and increases in temperature could possibly cause damage and death to some species, particularly those occupying shady and wet habitats, that have low thermal tolerances (as low as 42 °C in Arctic sorrel; Gauslaa, 1984).

- During warming, arctic species with conservative nutrient-use strategies, slow growth, and particularly inflexible morphologies such as those of cushion and mat plants, are likely to be at a competitive disadvantage to more responsive, faster growing, taller species immigrating from southern latitudes. After six years of shading (simulating competition), increasing temperatures, and fertilizing a heath and a fellfield community in Swedish Lapland, shading was found to have the greatest effect on aboveground growth (Graglia et al., 1997). In another experiment, flowering of the dwarf heather-like shrub *Cassiope tetragona* stopped when it was shaded (Havström et al., 1995a). In contrast, a meta-analysis by Dormann and Woodin (2002) found no significant effect of shading on biomass.

Populations at the most environmentally extreme boundary of their distributions (in terms of latitude, altitude, and habitat mosaics within landscapes) tend to be responsive to amelioration of physical environmental factors such as temperature that limit their distributions; these populations have the potential to expand their distribution. In contrast, populations at the most environmentally benign boundary of their distribution tend to be constrained by competition with more responsive species of more benign environments (Wijk, 1986) and tend to be displaced by environmental amelioration.

An International Tundra Experiment (ITEX; Henry and Molau, 1997) meta-analysis of arctic vascular plant species responses to simulated summer warming (1.2 to 1.8 °C mean daily near-surface and soil temperature increase) using standard open-top chambers compared key species from 13 sites over a period of one to four years (Arft et al., 1999). The simulated temperature increase is comparable to the projected increase in mean arctic summer air temperature of 1.8 °C by 2050 (mean of the five ACIA-designated model scenarios). In ITEX and earlier experiments, phenology (bud burst and flowering) was advanced in warming treatments at some sites (Wookey et al., 1993, 1995). In Swedish Lapland, growth accelerated and the period between thawing and anther appearance advanced by two weeks (Welker et al., 1997). In contrast, there was little change in growth cessation at the end of the season in response to higher temperatures. However, nutrient addition prolonged the growth period of polar semi-desert species on Svalbard in autumn but reduced frost hardening, leading to dramatic loss of aboveground biomass during November 1993, which was extremely warm and wet (Robinson et al., 1998). This corresponds with the indigenous observations noted above.

Experiments conducted by ITEX showed that initial increases in vegetative growth were generally, but not always, reduced in later years, probably because temperature increases stimulated the use of stored resources more than the uptake of new resources. Similarly, growth responses of subarctic dwarf shrubs to soil warming increased initially but soon returned to former levels. This response followed an initial increase in nitrogen mineralization as a result of soil warming, but increases in mineralization did not persist (Hartley et al., 1999). In contrast, reproductive success improved in later years in the ITEX experiments (Arft et al., 1999) due to the extended period between flower bud initiation and seed set in arctic flowering plants. Similarly, over an 18-year period, flowering of the widespread sedge *Carex bigelowii* was strongly correlated with July temperature of the previous year (Brooker et al., 2001). *Eriophorum* species exhibited even more dramatic interannual variation in flowering than *Carex* species, but there was no simple correlation with weather in the flowering year or the previous year (Shaver et al., 1986).

The ITEX experiments showed that responses of growth and reproduction to temperature increases varied among vascular plant life forms. Herbaceous species responded more strongly and consistently to warming than did woody forms over a four-year period (Arft et al., 1999). Over longer time periods, the growth form, number, and position of meristems in some woody plants such as *Betula nana* (Fig. 7.3) allowed a much greater response that completely changed the height and structure of the whole canopy (Bret-Harte et al., 2001; but see Graglia et al., 2001a for a different response). In the subarctic, Graglia et al. (2001a) showed that initial plant responses (abundance) to temperature increases and other treatments persisted throughout a ten-year period. Graminoids were particularly responsive to fertilizer additions in the subarctic and their increased growth and litter production suppressed the growth of mosses and lichens (Cornelissen et al., 2001; Graglia et al., 2001a; Molau and Alatalo, 1998; Potter et al., 1995). Evergreens were more responsive to nutrient addition and temperature increases than deciduous species (Arft et al., 1999).

Mosses and lichens appear to be particularly vulnerable to climate warming, at least in areas of continuous vegetation cover. A meta-analysis of lichen responses to warming experiments across the Arctic showed that lichen biomass decreased as vascular plant biomass increased following warming (Cornelissen et al., 2001). This group of plants is particularly important because a large proportion of global lichen diversity is found in the Arctic, some species are important winter forage for reindeer/caribou, and some are important nitrogen fixers in strongly nitrogen-limited systems. A 22-year study

of the lichen flora of the Netherlands showed changes that researchers suggest are related to an increase in temperature, although the subtropical species might be more sensitive to nitrogen (Palmqvist et al., 2002). Fifty percent of the arctic–alpine/boreal–montane lichen species were declining while subtropical species were invading (van Herk et al., 2002). The widespread moss *Hylocomium splendens* shows a complex response to warming (Callaghan et al., 1999): in warming experiments growth is reduced (Graglia et al., 2001a; Potter et al., 1995), whereas growth increases in relation to increases in mean annual temperature throughout its arctic distribution range (section 7.3.5.2, Fig. 7.15; Callaghan et al., 1997). This suggests some limitation in the experimental simulation of natural/anthropogenic warming. If, however, moss growth and abundance are reduced by higher temperatures, soil thermal regimes, biogeochemical cycling, and energy and heat exchange between the biosphere and atmosphere will be significantly affected (Hobbie, 1996).

Plant species respond differently to warming according to previous temperature history related to latitude, altitude, interannual temperature variations, and interactions among species. Phenological responses to warming are greatest at cold sites in the high Arctic (Arft et al., 1999; Wookey et al., 1993), whereas growth responses to warming are greatest at sites in the low Arctic. Growth responses of *Cassiope tetragona* to warming were greatest at a site in the high Arctic and a high-altitude site in the low Arctic when compared with the warmest low-altitude site in the low Arctic (Havström et al., 1993). Over a period of five years, shoot elongation responses to warming were greatest in cold summers (Molau, 2001; Richardson, 2000). Laine (1988) showed that the reproduction of bilberry (*Vaccinium myrtillus*) depended to some extent on the climate in the previous years (see section 14.7.3 for examples of this in trees), whereas Shevtsova et al. (1995, 1997) showed no such response for co-occurring lingonberry and crowberry (*Empetrum nigrum*).

Most information on plant responses to climate warming is limited to the short term and small plot – even if the short term is two decades. Because of the great longevity and clonal growth of arctic plants, it is difficult to extrapolate plant responses from an individual plant to the population. However, the impacts of climate change (temperature, nutrients, CO_2) on demographic parameters and population growth statistics were determined for the sedge *Carex bigelowii* by Carlsson and Callaghan (1994) and Callaghan and Carlsson (1997), who showed that climate change increased tiller size, vegetative production of young tiller generations, survival of young tillers, and flowering, and reduced the age of a tiller at flowering and tiller life span. Two mathematical models showed that the changes in demographic parameters led to an increase in the population growth rate, with young tillers dominating this increase. The rate of vegetative spread more than doubled, while cyclical trends in flowering and population growth decreased substantially.

Responses to precipitation changes

Precipitation in the Arctic is extremely variable between seasons and from place to place, but the amount of snow is difficult to measure (section 2.4). Precipitation varies from over 1000 mm in coastal areas (e.g., Norway and Iceland) to less than 45 mm in the polar deserts, where most of the annual precipitation occurs as snow. The interaction between precipitation and temperature is extremely important for plant growth and ecosystem processes and it is difficult to separate their effects.

Observations show that precipitation has increased by up to 20% in northern latitudes within the last 40 years (section 2.6.2.2), although there has been a 10% decrease in snow-cover extent in the Northern Hemisphere in the last 20 years (section 2.4.1). The most recent climate scenarios for the North Atlantic region suggest increased mean annual temperatures and precipitation for the entire region over the next 100 years (IPCC, 2001; section 4.4).

Effects of changes in snow depth, duration, and timing of the snow-free period

The interaction between snow amount and temperature determines the start and duration of the snow-free period. The duration of the snow-free period at high northern latitudes has increased by five to six days per decade and the week of the last observed snow cover in spring has advanced by three to five days per decade between 1972 and 2000 (Dye, 2002). Even if precipitation increases, therefore, temperature increases may still result in shorter duration of snow and less snow cover (sections 2.6 and 4.4.3). In contrast, the start of the growing season has been delayed by up to one week over the last 20 years in the high-altitude and

northern boreal regions of Fennoscandia (Høgda et al., 2001). Hydrological models applied to the Tana River Basin of northernmost Finland project increases in growing-season length, from 30 days in the mountains to 70 days near the coast of the Barents Sea, by 2100 (Dankers, 2002). This change is associated with an earlier start to the growing season of about three weeks and a delayed end of two to three weeks.

The timing of the start of the snow-free period is of critical importance, and more important than the timing of autumn snowfall, because solar angles are already high when plants start growth and each extra snow-free day at the beginning of the growing season will enable plants to access high levels of photosynthetically active radiation (Fig. 7.8; see also section 7.5). In an alpine area, productivity decreased by about 3% per day that the snow release date was delayed (Ostler et al., 1982). The timing of snowmelt has also been found to have considerable effects on plant phenology (more so than temperature in some cases: Hollister and Webber, 2000), with a contracting of development time that is associated with a decrease in productivity and reproductive output (Callaghan, 1974). Some plant species, such as the deciduous shrub species *Salix pulchra* and *Betula nana*, can respond to changes in growing-season timing (Pop et al., 2000), but others, particularly evergreen and early-flowering species appear to be particularly vulnerable (Kudo, 1991, 1993).

An experiment that manipulated snow conditions by using snow fences at Toolik Lake, Alaska, showed that drifts increased winter temperatures and CO_2 flux (Jones et al., 1998, Walker M.D. et al., 1999; Welker et al., 2000). Under the drifts, temperatures were more constant than in control plots. Plant growth increased despite a shorter growing season, although this was thought to be a transitory response and contrasts with the reduced growth of plants associated with late snow beds (Fig. 7.9).

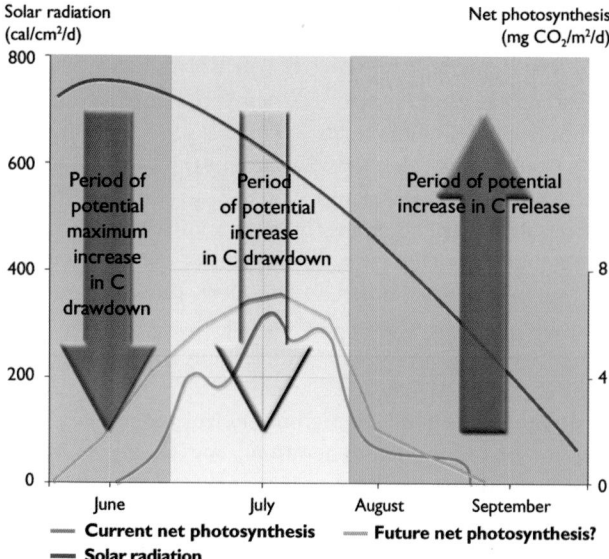

Fig. 7.8. Relationship between timing of the growing season and the seasonal pattern of irradiance together with an indication of where transient switches from carbon (C) sink to carbon source could occur (modified from Chapin F. and Shaver, 1985b).

Fig. 7.9. Snow bank vegetation showing increasing vegetation development with increasing growing-season length represented as distance from the snow patch, Disko Island, West Greenland (photo: T.V. Callaghan).

Frost resistance and avoidance

Changes in snow depth and duration are likely to negatively affect the frost resistance and avoidance of many plants at high latitudes. Damage to foliage and apical meristems occurs when they are "triggered" to premature bud burst and development by an earlier onset of the growing season (resulting from early snowmelt) when the annual hardening/dehardening process is at its most sensitive phase, and when there is a risk of short periods of cold weather. Bilberry is a species whose requirement for cool temperatures to enable it to break dormancy (i.e., chilling requirement) is fulfilled early (Havas, 1971). Accelerated dehardening of bilberry was found as consequence of a minor (2 to 3 °C) increase in temperature (Taulavuori K. et al., 1997b), suggesting that climatic warming is very likely to entail a real risk of early dehardening and subsequent frost damage of shoots. The explanation for this may be the higher but fluctuating temperatures, which increase the cryoprotectant-consuming freeze–thaw cycles (Ögren, 1996, 1997; Ögren et al., 1997). In addition to frost resistance, frost avoidance is likely to be disturbed by a thin or lacking snow cover. The risk is likely to be highest at high latitudes, where plants that are genetically adapted to the presence of snow may have lost some potential for frost resistance during their evolution. Subarctic provenances of bilberry, for example, have shown reduced frost resistance compared to provenances from southern Finland (Havas, 1971).

Other global change factors might affect frost resistance, but few, sometimes conflicting, reports have been published of studies performed at high latitudes. Nitrogen pollutant (or fertilizer) can impair the frost resistance of plants. Such an effect was demonstrated for mountain avens (*Dryas octopetala*) on Svalbard during a warm period in early winter (Robinson et al., 1998; see previous subsection on projected temperature responses). However, recent studies with the heather *Calluna vulgaris* (Caporn et al., 1994), bilberry (Taulavuori E. et al., 1997), and lingonberry (Taulavuori K. et al., 2001) have demonstrated improved frost resistance caused by extra nitrogen, probably because these ericaceous species are plants adapted to low-nutrient habitats, such as those at high latitudes.

Snow depth and duration vary greatly with topography at the landscape level. High summer temperatures are very likely to decrease the abundance and size of snow beds. Changes in snow patches observed by indigenous peoples are already causing concern in Baker Lake, Clyde River, and Iqaluit; Fox (2002) describes *aniuvak* (permanent snow patches) that are melting in the hills around those communities. *Aniuvak* are good areas for caching meat and provide a sanctuary for caribou to evade flying insects. Indigenous peoples' explanations for the melting snow patches relate more to changes in precipitation and mean relative humidity than to temperature increases. The specialized plants characteristic of late snow beds (Gjærevoll, 1956) are very likely to be at particular risk as temperatures increase.

Summer precipitation

Altered timing and rate of snowmelt are very likely to differentially alter the availability of water in different facies of the tundra landscape mosaic, which are very likely to in turn significantly affect the predominant vegetation type and its growth dynamic through the active season (Molau, 1996). Artificial increases in summer precipitation produced few responses in arctic plants compared with manipulations of other environmental variables (Dormann and Woodin, 2002). However, mosses benefited from moderate summer watering (Phoenix et al., 2001; Potter et al., 1995; Sonesson et al., 2002) and nitrogen fixation rates by blue-green algae associated with the moss *Hylocomium splendens* increased (Solheim et al., 2002). Addition of water to a polar semi-desert community in summer produced surprisingly few responses (Press et al., 1998a). Also in the high Arctic, comparisons were made between sites with high and low plant densities. Although there was little difference in soil moisture and plant water relations, and water availability did not constrain the adult vascular plants, surface water flow in snow-flush areas (accumulations of snow that provide water during summer) allowed greater development of cyanobacterial soil crusts, prolonged their nitrogen fixing activity, and resulted in greater soil nitrogen concentrations (Gold and Bliss, 1995). Because of their importance in facilitating vascular plant community development, Gold and Bliss (1995) projected that the effects of climate change on non-vascular species are very likely to be of great consequence for high-arctic ecosystems.

Responses to increased atmospheric carbon dioxide concentrations

There are very few arctic experiments that have manipulated atmospheric CO_2 concentrations in the field (Gwynn-Jones et al., 1996, 1999; Tissue and Oechel, 1987), but more laboratory experiments have been conducted on arctic vascular plants (Oberbauer et al., 1986), mosses, and lichens (Schipperges and Gehrke, 1996; Sonesson et al., 1992, 1995, 1996).

The first experiment that manipulated CO_2 in the Arctic concluded that elevated CO_2 concentrations had no long-term effects because photosynthetic acclimation (i.e., down-regulation, the physiological adjustment of photosynthetic rate so that no differences are found between plants grown at ambient and elevated CO_2 levels) in cottongrass (*Eriophorum vaginatum*) was apparent within three weeks and biomass did not increase. However, there was prolonged photosynthetic activity in autumn and more biomass was allocated to roots (Tissue and Oechel, 1987). The lack of response and enhanced root biomass were attributed to nutrient limitation (Oechel et al., 1997). Although increases in tiller production of cottongrass were not considered to be an important response, this can lead to long-term increases in population growth (Carlsson and Callaghan, 1994).

Longer-term CO_2 enrichment experiments in the subarctic also show that growth responses are dominated by early, transient responses (Gwynn-Jones et al., 1997). Four dwarf shrubs were studied over the first three years of the experiment; one, the deciduous bilberry, showed increased annual stem growth (length) in the first year whereas two other evergreen dwarf shrubs (mountain crowberry – *Empetrum hermaphroditum* and lingonberry) showed reduced growth. In the seventh year, increased CO_2 concentrations significantly increased the leaf ice nucleation temperature (i.e., reduced frost resistance, which can be harmful during the growing season) in three of the four species tested (Beerling et al., 2001). Bog whortleberry (or bog bilberry – *Vaccinium uliginosum*), lingonberry, and mountain crowberry showed increases in leaf ice nucleation temperature exceeding 2.5 °C whereas bilberry showed no significant effect, as in another study (Taulavuori E. et al., 1997). Increased CO_2 concentrations interacted with high UV-B radiation levels to increase leaf ice nucleation temperature by 5 °C in bog whortleberry. This effect coincides with indigenous knowledge and other experiments that show increased frost sensitivity of some arctic plants to changes in climate and UV-B radiation levels (see responses to cloudiness and photoperiod in this section).

An expected (and subsequently observed) response to increased atmospheric CO_2 concentrations was a change in leaf chemistry (e.g., an increase in the carbon to nitrogen ratio) that was expected to affect herbivory (Fajer et al., 1989) and decomposition (Robinson et al., 1997). Surprisingly, herbivory was not affected. However, increased CO_2 concentrations were found to play a role in nutrient cycling by altering the composition of microbial communities after five years (Johnson et al., 2002; section 7.3.3.3), suggesting that chemical changes are occurring in plants exposed to high CO_2 concentrations but the changes have not yet been identified.

In laboratory studies, the moss *Hylocomium splendens* that naturally experiences high CO_2 levels in the birch woodlands of the Swedish subarctic was shown to have photosynthetic rates that were limited by light, temperature, and water for most of the growing season (Sonesson et al., 1992). Enhanced CO_2 concentrations for five months decreased photosynthetic efficiency, light compensation point, maximum net photosynthesis, and surprisingly, growth (Sonesson et al., 1996). Similar experiments with three lichen species, *Cladonia arbuscula*, *Cetraria islandica*, and *Stereocaulon paschale*, failed to show any response of fluorescence yield to enhanced CO_2 concentration (1000 ppm), although there was an interaction between CO_2 and UV-B radiation levels (Sonesson et al., 1995). Perhaps the lack of response in moss and lichens reflects their adaptation to the currently high levels of CO_2 that they experience close to the ground surface (Sonesson et al., 1992) via the process of down-regulation.

In contrast to some views that responses of plants (mainly growth) to increased CO_2 concentrations are relatively small and by inference insignificant (Dormann and Woodin, 2002), recent results show that in the long term, increased CO_2 concentrations can have the wide-ranging and important effects discussed previously (Beerling et al., 2001; Johnson et al., 2002).

Responses to increased ultraviolet-B radiation levels

One common method for simulating the effects of ozone depletion has been to irradiate organisms and ecosystems with artificial UV-B radiation. Results are often reported in relation to the equivalent percentage of ozone depletion. It should be noted, however, that the radiation spectrum from the lamps used in experiments differs from the spectrum of the radiation increase that would ensue from real ozone depletion. Therefore, the degree of simulated ozone depletion depends on the "weighting func-

Table 7.6. Summary of UV-B radiation effects on subarctic dwarf shrubs (based on Phoenix, 2000 and other sources referred to in the text).

	Bilberry (*Vaccinium myrtillus*)	Bog whortleberry (*Vaccinium uliginosum*)		Lingonberry (*Vaccinium vitis-idaea*)		Mountain crowberry (*Empetrum hermaphroditum*)	
	Ambient to enhanced UV-B	Zero to ambient UV	Ambient to enhanced UV-B	Zero to ambient UV	Ambient to enhanced UV-B	Zero to ambient UV	Ambient to enhanced UV-B
Stem length	-	0	0	0	0	0	0
Branching	0	0	0	0	0	0	0
Leaf thickness	+	0	0	0	0		
Flowering	+	+		0	0		
Berry production	+	0		0	0	0	0
Phenology	0	0	0	0	0	0	0
Total UV-B radiation-absorbing compounds	-	-	-	-	0	0	0
Leaf ice nucleation temperature	0		+		+	0	+

+ indicates an increase; - indicates a decrease; 0 indicates no effect compared to control; blank indicates no information

tion" applied in the calculations, and the knowledge of the appropriate weighting function is very incomplete. Weighting functions are also species-specific: a certain amount of applied artificial radiation does not correspond to the same degree of ozone depletion for a plant and a tadpole, for example. The information in the following sections should be read with this in mind.

Relatively little is known about plant responses to changes in UV-B radiation levels. Field experiments on subarctic (Table 7.6) and high-arctic ecosystems show species-specific responses to ambient UV-B radiation levels and to enhanced UV-B radiation levels equivalent to a 15% decrease in stratospheric ozone from 1990 levels. (The 15% decrease is equivalent to losses of ozone projected to occur throughout much of the Arctic by 2015. However, the values do not apply to Beringia for April and October 2015; Taalas et al., 2000.) On the whole, the effects of increased UV-B radiation levels are relatively few compared with effects of increased temperature and nitrogen (Dormann and Woodin, 2002).

A global meta-analysis of plant responses to increased UV-B radiation levels showed that there was a small but significant reduction in biomass and plant height (Searles et al., 2001). In the subarctic, measurements of stem length, branching, leaf thickness, flowering, berry production, phenology, and total UV-B radiation absorbing compounds were affected significantly by ambient UV-B radiation levels in only two of three dwarf shrubs (i.e., bog whortleberry and lingonberry; Phoenix, 2000). Mountain crowberry and lingonberry showed no responses to enhanced UV-B radiation levels after seven years of exposure whereas bog whortleberry and bilberry showed few responses (Table 7.6). Enhanced UV-B radiation levels have been shown to reduce the height growth, but not biomass, of the moss-

es *Sphagnum fuscum* and *Hylocomium splendens* in the subarctic (Gehrke et al., 1995).

The UV-B radiation studies (Table 7.6) showed that arctic species were more tolerant of enhanced UV-B radiation levels than previously thought, and that the production of UV-B radiation absorbing compounds did not show the simple relationship with UV-B radiation dose expected from laboratory studies. Another surprise effect was the responsiveness of frost hardiness in some arctic dwarf shrubs to increased UV-B radiation levels. Dunning et al. (1994) pioneered investigation of the relationship between UV-B radiation levels and frost resistance in a *Rhododendron* species and concluded that increased exposure to UV-B radiation increases (although only marginally) cold resistance. In contrast, K. Taulavuori et al., (unpubl. data, 2004) found decreased frost resistance in bilberry in response to elevated UV-B radiation levels and Beerling et al. (2001) showed decreased frost resistance in bog whortleberry, lingonberry, and mountain crowberry. A combination of elevated CO_2 and UV-B radiation levels increased late-season frost sensitivity of leaves of bog whortleberry from -11.5 to -6 °C. Increased frost sensitivity at the beginning and/or end of the short arctic growing season is likely to curtail the season even further. As some models of vegetation redistribution related to temperature change use the critical freezing temperatures for leaf damage in temperate trees and shrubs (Prentice et al., 1992), modeled past and future northward migration of temperate vegetation should be reconsidered in relation to changing CO_2 and UV-B radiation levels.

The resilience of the subarctic dwarf shrubs to enhanced UV-B radiation levels probably reflects pre-adaptation to higher levels than are currently experienced in the Arctic (Phoenix, 2000). These species currently extend southward to about 40° N and they probably existed even further south in a higher UV-B radiation regime during the early Holocene. The increased UV-B radiation levels currently applied in experiments are equivalent to the difference in ambient UV-B radiation levels between the site of the experiment (68° N) and Helsinki (59° N) (Fig. 7.10). In addition, many arctic plants have thick leaves that might attenuate UV-B radiation entering leaf tissues. However, one particular climate–UV radiation interaction that could possibly increase the damage experienced by plants is the combination of possible earlier snow-free periods (Dankers, 2002) with higher spring UV-B radiation levels at the surface of the earth (Taalas et al., 2000). Such a combination of effects would expose young, potentially sensitive plant shoots and flower buds to particularly high UV irradiation (Zepp et al., 2003).

Responses to changes in cloudiness and photoperiod

An important characteristic of the arctic environment is the daily and seasonal patterns of the light period or photoperiod. At midwinter, intermediate latitudes (40°

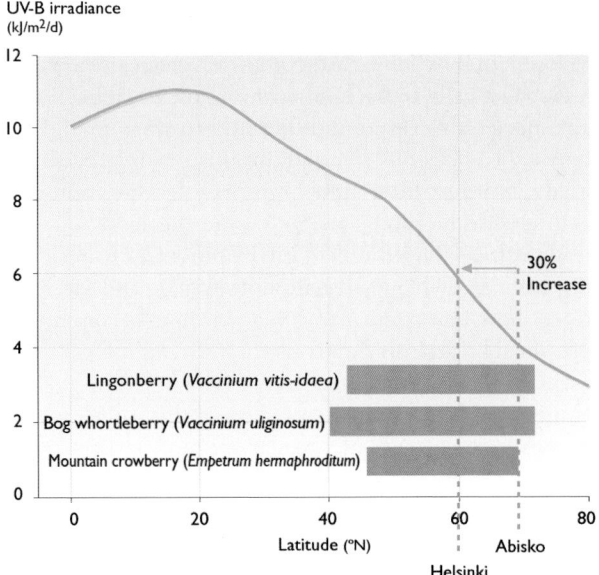

Fig. 7.10. Dwarf shrub distributions in relation to latitude and solar UV-B radiation incident at the surface of the earth (Gwynn-Jones et al., 1999; Hultén, 1964; Phoenix, 2000).

to 50° N) have about an eight-hour day length, whereas a polar night without sunrise prevails north of the Arctic Circle (66.5° N). Consequently, day-length change during spring and autumn occurs much faster at high latitudes.

Frost-resistance patterns change seasonally and are environmentally controlled, mainly by temperature and day length; which of these factors predominate depends on the seasonal growth cycle (Fuchigami et al., 1982). The development of frost resistance by almost all woody plants at high latitudes is characterized by strong dependency on the photoperiod for growth cessation and cold hardening. Scots pine (*Pinus sylvestris*) seedlings from the northern boreal forest develop a high degree of frost resistance during the late summer as a consequence of the shortening days (Taulavuori K. et al., 1997a). The frost-hardening process is initiated even at high temperatures (20 °C) in experimental conditions that mimic the ambient photoperiod (Taulavuori K. et al., 2000). Given the marked photoperiodic control of the frost-hardening process in woody species at high latitudes, it is understandable that they harden more extensively compared to populations at lower latitudes under similar temperatures. For example, the lowest survival temperature of bilberry in the central Alps (~50° N) at midwinter is around -35 °C (Sakai and Larcher, 1987 and references therein), while the same level of frost resistance is already achieved at the end of September in northern Finland (65° N) (Taulavuori E. et al., 1997).

In a changing climate, photoperiod will not change, but species that are migrating will experience changes in photoperiod. It is unlikely however, that this will constrain species initially. Many northern boreal species, for example, experienced arctic photoperiods earlier in the Holocene before they were displaced southward by climate cooling (section 7.2). If and when species with a more southerly distribution migrate into the Arctic, photoperiod constraints could possibly affect growth and flowering but this is largely unknown. However, experiments with transplanting herbs between the Austrian Alps, Abisko, and Svalbard showed that allocation of biomass in some species such as glacier buttercup (*Ranunculus glacialis*) was affected by photoperiod and this constrained any potential increases of vigor that might have occurred due to climate warming (Prock and Körner, 1996). In contrast, herbs such as *Geum* (Prock and Körner, 1996) and some grasses (Heide et al., 1995) that are not sensitive to photoperiod could possibly benefit from climate warming.

It has been suggested that increased UV-B radiation effects might be small in the future because of increased cloudiness (section 4.4.4) that is likely to counteract to some extent the effect of decreasing ozone levels (Dormann and Woodin, 2002). However, projections of increased cloudiness, and particularly future cloud types, are uncertain. It is more likely that UV-B radiation effects will be reduced by decreases in albedo as snow

and ice distribution and seasonal duration decline, and as the boreal forest displaces part of the current tundra.

Arctic plants differ in the degree to which they gain or lose carbon in photosynthesis at "night" (22:00 to 04:00 hours when light intensity is low but it is not necessarily dark). Under conditions of cloudy nights, those species that gain carbon at night, for example, Arctic dryad (*Dryas integrifolia*), alpine foxtail (*Alopecurus alpinus*), glaucous willow (*Salix glauca*), and arctic willow (*Salix arctica*) (25–30% of diurnal carbon gain; Semikhatova et al., 1992), are likely to have reduced competitive ability compared with species that do not. In contrast, increased cloudiness during the day probably favors those species that gain carbon at night. Those species that lose carbon at night (e.g., *Eriophorum angustifolium*; Semikhatova et al., 1992) are likely to be disadvantaged by warming.

Responses to potential changes in pollinator abundance and activity

The rapid phenological changes that have been observed in response to simulated climate change have the potential to disrupt the relationships that plants have with animal, fungal, and bacterial species that act as pollinators, seed dispersers, herbivores, seed predators, and pathogens (Dunne et al., 2003). These disruptions are likely to have the strongest impact if the interacting species are influenced by different abiotic factors or if their relative responses to the same factors (e.g., elevated temperatures) are different. However, wind and self-pollination are more widespread among arctic flowering plants, so any mismatch between pollinator activity and flowering phenology is likely to be of greater significance to any plants immigrating to the Arctic as temperatures increase. Little appears to be known about these processes.

Summary

Species responses to changes in temperature and other environmental variables are complex. Species respond individualistically to each environmental variable. Plant species also respond differently to warming according to previous temperature history related to latitude, altitude, interannual temperature variations, and interactions among species. Some species are already responding to recent environmental changes. Indigenous knowledge, aerial photographs, and satellite images show that some arctic vegetation is becoming more shrubby and productive.

Summer warming experiments showed that initial increases in the growth of vascular species were generally reduced with time, whereas reproductive success improved in later years. Over short periods (four years), herbaceous plants responded more than woody plants, but over longer periods, woody plant responses were dominant and could change the canopy height and structure. Mosses and lichens were generally disadvantaged by higher-plant responses to warming.

Responses to warming are critically controlled by moisture availability and snow cover. Already, indigenous observations from North America and Lapland show a drying trend with reduced growth of economically important berries. However, experimental increases in summer precipitation produced few responses in arctic plants, except for mosses, which showed increased growth. An experiment that manipulated snow conditions showed that drifts increased winter temperatures and CO_2 flux and, surprisingly, that plant growth increased despite a shorter growing season. In general, however, any earlier onset of the snow-free period is likely to stimulate increased plant growth because of high solar angles, whereas an increase in the snow-free period in autumn, when solar angles are low, will probably have little impact.

Carbon dioxide enrichment experiments show that plant growth responses are dominated by early, transient responses. Surprisingly, enhanced CO_2 did not affect levels of herbivory, but significantly increased the leaf ice-nucleation temperature (i.e., increased frost sensitivity) of three of four dwarf shrub species, and altered the composition of microbial communities after five years. A general lack of responses of mosses and lichens reflects their adaptation to the currently high levels of CO_2 that they experience close to the ground surface.

Ambient and supplemental UV-B radiation levels produced complex, individualistic, and somewhat small responses in species. Overall, arctic species were far more tolerant of enhanced UV-B radiation levels than previously thought, and the production of UV-B absorbing compounds did not show the simple relationship with UV-B radiation dose expected from laboratory studies. Some arctic dwarf shrubs exhibited increased frost sensitivity under increased UV-B radiation levels. The arctic photoperiod is unlikely to be a general constraint to species migrations from the south, as trees and southern species previously occurred further north than at present.

7.3.3.2. Animals

In contrast to plants, there are relatively few experiments that have addressed how animal populations respond to simulated climate change and UV-B radiation levels in the Arctic. The few experiments have focused on invertebrates (e.g., insects and soil animals) for which the microclimate can be manipulated on small experimental plots. Experiments on free-ranging vertebrate populations may not be feasible for logistical reasons. On the other hand, more time series of population data are available for conspicuous vertebrates such as reindeer/caribou and lemmings than, for example, soil invertebrates. Time series can be analyzed with respect to the influence of current climate variability (including recent changes).

Responses to current changes in climate and ultraviolet radiation levels

Ice-crust formation on the tundra as a result of freeze–thaw events during the winter affects most terrestrial arctic animals. Dense snow and ice severely limit forage availability for large ungulates such as reindeer/caribou and muskox (Klein, 1999). Dramatic reindeer population crashes resulting from periodic ice crusting have been reported from the western coastal part of the Russian Arctic, Svalbard, and Fennoscandia (Aanes et al., 2000; Putkonen and Roe, 2003; Reimers, 1982; Syroechovski and Kuprionov, 1995). Similar events have been reported for muskox in the southern parts of their range in Greenland (Forchhammer and Boertmann, 1993). Inuit in Nunavut report that caribou numbers decrease in years when there are many freeze–thaw cycles (Thorpe et al., 2001) and the probability of such freeze–thaw events is said to have increased as a result of more short-term fluctuations in temperature. In central Siberia, where winter climate is colder and more stable, reindeer population dynamics are less climate-driven (Syroechovski and Kuprionov, 1995). Swedish Saami note that over the last decade, autumn snow lies on unfrozen ground rather than on frozen ground in the summer grazing areas and this results in poor-quality spring vegetation that has rotted (Nutti, pers. comm., 2004); certain microfungi seem to be responsible for this (Kumpula et al., 2000).

Long and accurate time series data on Svalbard reindeer populations (Aanes et al., 2000; Solberg et al., 2001) show that the amount of precipitation during the winter, which is highly variable and is well described by the Arctic Oscillation index (Aanes et al., 2002), provides the most important check on the reindeer population growth rate in concert with population density. Winters with freezing rain were associated with severe population crashes both in one reindeer population (although the natural dynamics of an introduced herd may have contributed to this) and in an introduced population of sibling voles (*Microtus rossiaemeridionalis*; Fig. 7.11).

Episodes of mild weather and wet snow lead to a collapse of the subnivean space and subsequent frost encapsulates food plants in ice, making them unavailable to small mammal herbivores, and even killing plants in some cases (Callaghan et al., 1999; Robinson et al., 1998). Accordingly, the survival rate of tundra voles (*Microtus oeconomus*) decreases dramatically in winters with many freeze–thaw cycles (Aars and Ims, 2002; Fig. 7.12). For example, the lemming increases observed at Kilpisjärvi (northwest Finnish Lapland) in 1997 and 2001 were probably curtailed by warm spells and rain in January that resulted in freezing of the ground layer (Henttonen, unpubl. data, 2004). Inuit residents of the western Canadian Arctic are also concerned with the impacts of thaw slumping on lemming populations and their predators (owls). Thaw slumps at lake edges have been occurring more extensively and at a faster rate in recent years, linked to warmer temperatures and an increase in wind activity and rain; thawing of ice-bound soil destroys lemming burrows (IISD and the community of Sachs Harbour, 2000).

There has been speculation about whether the recent dampened amplitude of population cycles and more spatially asynchronous dynamics of voles and lemmings in northern Fennoscandia may be the result of occasionally unfavorable winters disrupting the normal population dynamics (Yoccoz and Ims, 1999). Figure 7.13 illustrates changes since the beginning of the 1990s in populations of the formerly cyclic and numerically dominant grey-sided vole (*Clethrionomys rufocanus*) and other vole species. In long qualitative time series (up to 100 years), periods with loss of cyclicity and synchrony are evident (Angerbjörn et al., 2001; Steen et al., 1990; Stenseth and Ims, 1993), but it is unclear whether this is related to fluctuations in climate. There is a correlation between sunspot activity and snowshoe hare cycles in North America (Sinclair et al., 1993), but no such relationship has been found for the mountain hare (*Lepus timidus*) in northern Finland (Ranta et al., 1997). There are no relationships between sunspot activity and outbreak years in the autumnal moth in Fennoscandia (Ruohomäki et al., 2000), although there are few studies of the role of climatic variability in arctic insect and soil arthropod populations because of the lack of long quantitative time series.

Arctic indigenous peoples are rich sources of information about recent changes in animal health and behavior, in particular reindeer/caribou. Increases in vegetation (longer grass, riparian areas with denser vegetation) are linked to increased forage availability and more mosquitoes and flies, resulting in increased insect harassment of reindeer/caribou (Riedlinger, 2001). Changes in "the warmth of the sun", day length, and the timing of the growing season may trigger reindeer/caribou to cross a frozen lake or river when the ice is no longer thick enough to support their weight (Thorpe et al., 2001). However, some of the environmental changes may be

beneficial. Stronger and more frequent winds are said to provide reindeer/caribou with relief from insect harassment, meaning they can spend more time inland and not in coastal areas (Riedlinger, 2001). Qitirmiut in Nunavut know that caribou adapt to the heat by staying near coastal areas and shorelines, lying on patches of snow, drinking water, standing in the water, eating moist plants, and sucking mushrooms (Thorpe et al., 2001). However, increases in the number of extremely hot days, combined with changing water levels and vegetation patterns, are likely to affect the ability of reindeer/caribou to respond in these ways.

Climatic cooling has to some extent caused habitat degradation in some coastal areas as a result of grubbing by snow geese on their staging ground. The lesser snow goose (*Anser caerulescens caerulescens*) breeds in coastal areas of the Hudson Bay region, which has experienced climatic cooling since the mid-1970s. This has delayed migration of the breeding populations (Hansell et al., 1998). Huge aggregations of staging and local geese in the coastal marshes have led to intense grubbing and degradation of salt-marsh sward (Srivastava and Jefferies, 1996). Long-term observations and modeling have shown that goose reproductive variables are directly and indirectly dependent on selected climatic variables, particularly those relating to spring (Skinner W. et al., 1998). Nest initiation date, hatching date, and clutch size were associated with the date of the last snow on the ground and mean daily temperature between 6 and 20 May. Early snowmelt allows geese to forage and females to build up nutrient stores before nest initiation. Goslings that hatch earlier in the spring have a higher probability of survival than those hatching later. Inclement weather, such as cumulative snowfall, freezing rain, and northerly and easterly winds can result in nest abandonment by females and even adult starvation while incubating eggs.

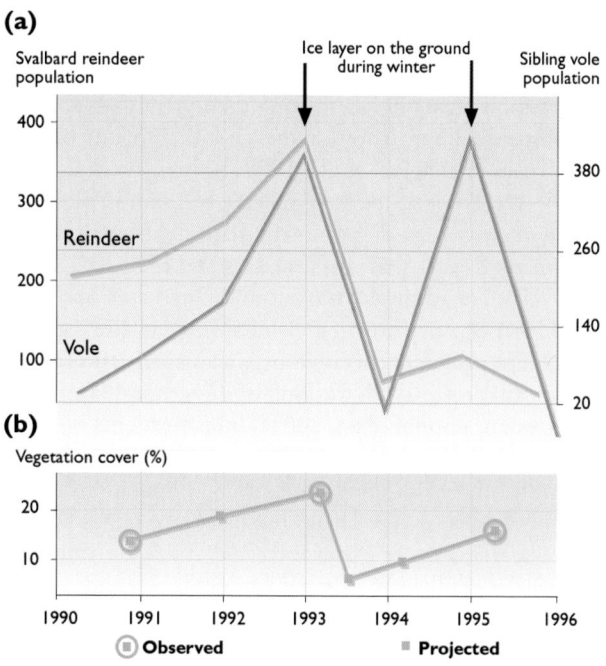

Fig. 7.11. (a) Population dynamics of Svalbard reindeer at Brøggerhalvøya and sibling voles at Fuglefjella on Svalbard (Aanes et al., 2000; Yoccoz and Ims, 1999); (b) observed and projected changes in vegetation (Robinson et al., 1998; Callaghan et al., 1999).

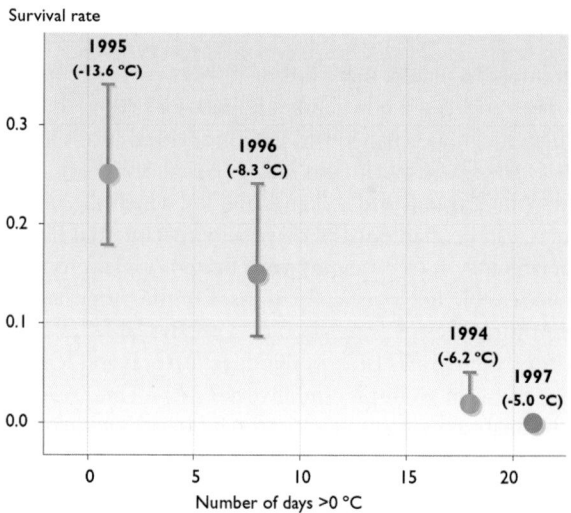

Fig. 7.12. Yearly winter survival rate (with 95% confidence intervals) of experimental tundra vole populations plotted against the number of days with temperatures above 0 °C during December through February. Mean winter temperature and the year are denoted above the survival rate estimates (Aars and Ims, 2002).

Responses to projected changes in climate

Despite adaptations to low temperatures, warming experiments have shown that temperatures higher than normal do not present any physiological problem for arctic arthropods provided that water is available (Hodkinson et al., 1998). Arctic aphids were more successful in terms of the number of completed generations through the summer when temperature was experimentally elevated (Strathdee et al., 1993). The effects of experimental warming were more pronounced in the high Arctic at Svalbard than in subarctic Abisko (Strathdee et al., 1995). The combination of high temperatures and drought seem to be very problematic for terrestrial invertebrates (Strathdee and Bale, 1998), but the hydrological aspect of climate change in tundra habitats is an important issue that has rarely been addressed in studies of arctic animals (Hodkinson et al., 1999).

Some of the most important effects of higher summer temperatures on arctic terrestrial animals are likely to be mediated through intensified interspecific interactions (parasitism, predation, and competition). Higher temperatures in the Arctic are very likely to lead to invasions of species with more southerly distributions. Such range expansions are projected to be particularly rapid in those species for which food resources (e.g., host plants) are already present (Hodkinson and Bird, 1998). For example, the mountain birch *Betula pubescens* ssp. *czerepanovii*, the main food plant of the autumnal moth, occurs in the continental parts of the Fennoscandian forest tundra where winter temperatures are occasionally lower than the tolerance limit for over-wintering eggs (Tenow, 1972); however, warmer winters are likely to lead to the exploitation of this existing food source. Many insects belonging to the boreal forest already invade the low-arctic tundra in quite large quantities every summer (Chernov and Matveyeva, 1997) and the Arctic is subject to a "steady rain" of wind-dispersed small invertebrates (Elton, 1925) that are likely to rapidly become established when environmental conditions are adequate. Due to the lack of long-term monitoring programs, there are presently no arctic equivalents of the detailed and quantitative documentation of the northward spread of insects in Europe (e.g., Parmesan et al., 1999). Several generalist predators not yet present in the Arctic are likely to spread northward with increased ecosystem productivity due to warming. The red fox has already expanded into the Arctic, probably at the expense of the Arctic fox (Hersteinsson and MacDonald, 1992).

Winter warming will alter snow cover, texture, and thickness. A deeper snow cover is likely to restrict reindeer/caribou access to winter pastures, their ability to flee from predators, and energy expenditure traveling across snow. Changes in snow depth and texture are very likely to also determine whether warm-blooded small vertebrates will find thermal refuges for resting in snow dens (ptarmigan and hares) or for being active in the subnivean space (Pruitt, 1957). Ice-crust formation reduces the insulating properties of the snowpack (Aitchinson, 2001) and makes vegetation inaccessible to herbivores. There is ample observational evidence that the current incidence and degree of winter ice crusting clearly affects the population dynamics patterns of both large and small mammal herbivore species (see previous subsection). Moreover, there is experimental evidence that population densities of numerically dominant tundra Collembola (springtail) species such as *Folsomia quadrioculata* and *Hypogastrura tullbergi* can be halved following an episode of freezing rain on Spitzbergen (Coulson et al., 2000). The projected winter temperature increase of 6.3 °C by 2080 (mean of the five ACIA-designated model scenarios) is very likely to result in an increase of alternating periods of melting and freezing (section 6.4.4). Putkonen and Roe (2003) found that episodes with rain-on-snow in the winter presently occur over an area of 8.4 x 10^6 km^2 in the Arctic and they projected that this area would increase 40% by 2080–2089. The projected increase in the frequency of winter warming is very likely to severely suppress population densities; distort the cyclic dynamics and degree of geographic synchrony in lemmings, voles, and geometrid moths; and in some cases even lead to population extinctions.

Responses to projected increases in ultraviolet-B radiation levels

The extent to which animals are adapted to incident UV-B radiation levels must be inferred in most cases. Hairs and feathers necessary for insulation against low temperatures also presumably protect the skins of mammals and birds from UV-B radiation, while white winter hair and feathers reflect UV-B radiation to some extent. The eyes of non-migratory animals must be extremely well adapted to UV-B radiation in order to be effective in the dark arctic winter yet also cope with high UV-B radiation levels in the bright, snowy spring. Invertebrates have coloring that may serve many functions. Melanic forms of invertebrates might have advantages in thermoregulation and UV-B radiation protection (Leinaas, 2002). If white coloration, insulation, and melanistic thermal regulation decrease due to reduced snow cover and higher temperatures, sensitivity to increased UV-B radiation levels is likely to increase.

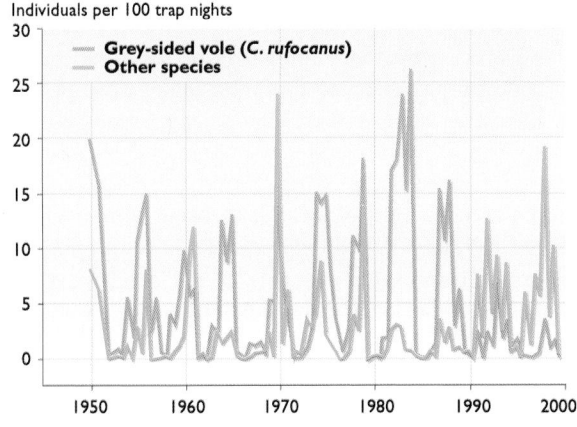

Fig. 7.13. Population dynamics of the grey-sided vole and other vole species (combined) at Kilpisjärvi, northern Finland between 1950 and 2000 (Henttonen and Wallgren, 2001).

Four species of Collembola on Svalbard were investigated by Leinaas (2002) with respect to UV-B radiation tolerance: *Hypogastrura viatica*, *Folsomia sexoculata*, *Onychiurus groenlandicus*, and *O. arcticus*. The first three species coexist in wet shore habitats, with the very heavily pigmented *H. viatica* on the surface and *F. sexoculata*, which as an adult is also very heavily pigmented, lower down. *O. groenlandicus* is a soil-dwelling, unpigmented species. Although *O. arcticus* is most commonly found under small stones and in rock crevices, and is thus rather unexposed, it has some pigmentation. In an experiment with enhanced UV-B radiation levels (0.5 W/m² in the 300 to 320 nm band for 12 to 14 hours per day, approximately equivalent to clear sky summer conditions in southern Norway) the unpigmented *O. groenlandicus* experienced 100% mortality within one week, while the heavily pigmented *H. viatica* was not affected.

Caterpillars of subarctic moths have skins that absorb UV-B radiation to varying extents and the degree of absorption may depend on previous exposure to high UV-B radiation levels (Buck and Callaghan, 1999). However, UV-B radiation levels affect animals indirectly via the quality and quantity of food that is available to them as a result of UV-B radiation impacts on plant growth and secondary metabolite production (section 7.4.1.4).

It is possible to infer some responses of animals to future increases in UV-B radiation levels by comparing them to the effects of natural UV-B radiation levels on animals along latitudinal gradients. Along a south-to-north gradient starting at 55.7° N, ambient UV-B radiation levels reduced hatchling size in frogs at sites up to 66° N, with no latitudinal gradient in UV-B radiation tolerance (Pahkala et al., 2002). Surprisingly, for a given time of the year, although UV-B radiation levels decrease with increasing latitudes, the frogs were exposed to higher UV-B radiation levels during the sensitive stages of their life cycles (egg and tadpole) at high latitudes than at low latitudes (Merilä et al., 2000). These studies suggest that an increase in UV-B radiation levels due to anthropogenic ozone depletion is likely to reduce the populations of those amphibians that have distribution ranges extending into the Arctic.

Enhanced UV-B radiation levels are thought to improve the immune system of the autumnal moth in the subarctic and to destroy the polyhedrosis virus. As this virus and the parasitoid wasp *Cotesia jucunda* are both important controllers of the survival of moth caterpillars, increased UV-B radiation levels could possibly lead to increased moth populations and birch forest defoliation. However, no direct effects of enhanced UV-B radiation levels on moth fecundity or survival have been detected (Buck and Callaghan, 1999).

Summary

Evidence for animal responses to climate change is scarcer than for plants because field experiments are less feasible for mobile animals, especially vertebrates. In many cases inferences are made based on time-series analyses of population abundance data for a few conspicuous species such as ungulates and lemmings.

Winter climate impacts, especially those events that affect properties of snow and ice, are particularly important. Freeze–thaw cycles leading to ice-crust formation have been shown to severely reduce the winter survival rate of a variety of species, ranging from soil-dwelling springtails (Collembola) to small mammals (lemmings and voles) to ungulates (in particular reindeer/caribou). Such icing induces conditions of anoxia that affect invertebrates, creates unfavorable thermal conditions for animals under the snow, and renders vegetation unavailable for herbivores. A deeper snow cover is likely to restrict reindeer/caribou access to winter pastures and their ability to flee from predators. The projected increase in the frequency of freeze–thaw cycles is very likely to disrupt the population dynamics of many terrestrial animals, and indications that this is already happening to some extent are apparent in the recent loss of the typical three-to-four year population cycles of voles and lemmings in subarctic Europe.

Experimental elevation of summer temperature has shown that many invertebrates respond positively to higher temperatures in terms of population growth, as long as desiccation is not induced. Many invertebrates, such as insects, are very likely to rapidly expand their ranges northward into the Arctic if climate warming occurs, because they have vast capacities to become passively or actively dispersed and host species (both plants and animals) are already present north of their present range borders.

Little is known about the responses in arctic animals to expected increases in UV-B radiation levels. However, there are some indications that arctic animals are likely to be more exposed and susceptible to such changes than their southern counterparts. The effects of increased UV-B radiation levels on animals are likely to be subtle and indirect, such as reduced food quality for herbivores and increased disease resistance in insect pest species.

7.3.3.3. Microorganisms

Recent experiments that manipulate the environment (e.g., soil heating, changing the water table, atmospheric CO_2 enrichment, and UV-B radiation supplementation and attenuation) have added new information about the effects of environmental change on the soil microbial community at the species level. In general, climate change is likely to alter microbial community composition and substrate utilization (Lipson et al., 1999). Tundra soil heating, atmospheric CO_2 enrichment, and amendment with mineral nutrients generally accelerate microbial activity (leading to a higher growth rate). Higher CO_2 concentrations tend to intensify root exudation, which is the main source of available carbon for soil and rhizosphere bacteria. Much less is known about the transient changes in the species composition of soil microorganisms induced by manipulation of UV radiation levels, although supplementation of UV-B radiation in the field resulted in changes in the composition of microbial communities (Johnson et al., 2002). Laboratory incubation of tundra soils from Barrow, Alaska, at different temperatures had strong effects on community composition assessed from a molecular biology approach, but only after a temperature shift of more than 10 °C (Panikov, 1999).

A mathematical simulation of the changes in tundra microbial community structure (Panikov, 1994, 1997) showed, surprisingly, that the effects of temperature on the soil microbial community were less significant compared with effects on the plant community. Probable reasons for this include strong metabolic interactions between individual populations within the microbial community (in which the product of one organism is used as a nutrient substrate by other organisms) that stabilize community structure in a wide range of environmental conditions; the wide temperature tolerance of microbial species; and the lower resolution power of microbial taxonomy as compared with plant taxonomy.

The model (Panikov, 1994, 1997) generated realistic patterns of mass and energy flow (primary productivity, decomposition rates, and soil respiration) under present-day conditions and in response to warming, pollution, fertilization, drying/rewetting of soil, etc. (Fig. 7.14). Figure 7.14a shows that L-selected species (*Bacillus*) display only sporadic occurrence under normal cold tundra conditions, in agreement with observations, and attain high population density after soil warming. Simulated soil warming accelerated both primary productivity and

organic matter decomposition, but the latter was more affected. Soil warming also led to a negative carbon balance in the soil, as respiration exceeded photosynthesis leading to a decline in accumulated organic carbon (Fig. 7.14b; sections 7.4.2.1, 7.4.2.2, 7.5.1.1, and 7.5.4).

Conidia (spores) of the fungus *Metarhizium* are sensitive to UV-B radiation. There are great differences between strains, but strains from high latitudes are less tolerant than those from lower latitudes (Braga et al., 2001a,b). In one species (*M. anisopliae*) it was shown that UV-A also had a negative effect and, when comparing strains, the sensitivity to UV-A radiation did not correlate with the sensitivity to UV-B radiation (Braga et al., 2001c). Several groups have studied the effects of UV-B radiation on phylloplane (leaf surface-dwelling) fungi and litter-decomposing fungi. Moody et al. (1999) found that five of the investigated species were sensitive and seven relatively insensitive to UV-B radiation. The spore production in the litter decomposers was generally inhibited by UV-B radiation (except for one species), while that in phylloplane species was unaffected. However, the sensitivity of spores is not equivalent to the sensitivity of the metabolic machinery of the vegetative body of a fungus (i.e., the thallus or mycelium) that produces the spores.

In the subarctic (Abisko), a study of the decomposition rates of a standard litter type showed that there was a change in the composition of fungal species resulting from elevated UV-B radiation levels (Moody et al., 2001). These results to some extent resemble those from an earlier experiment studying the decomposition of dwarf-shrub litter at the same site (Gehrke et al., 1995).

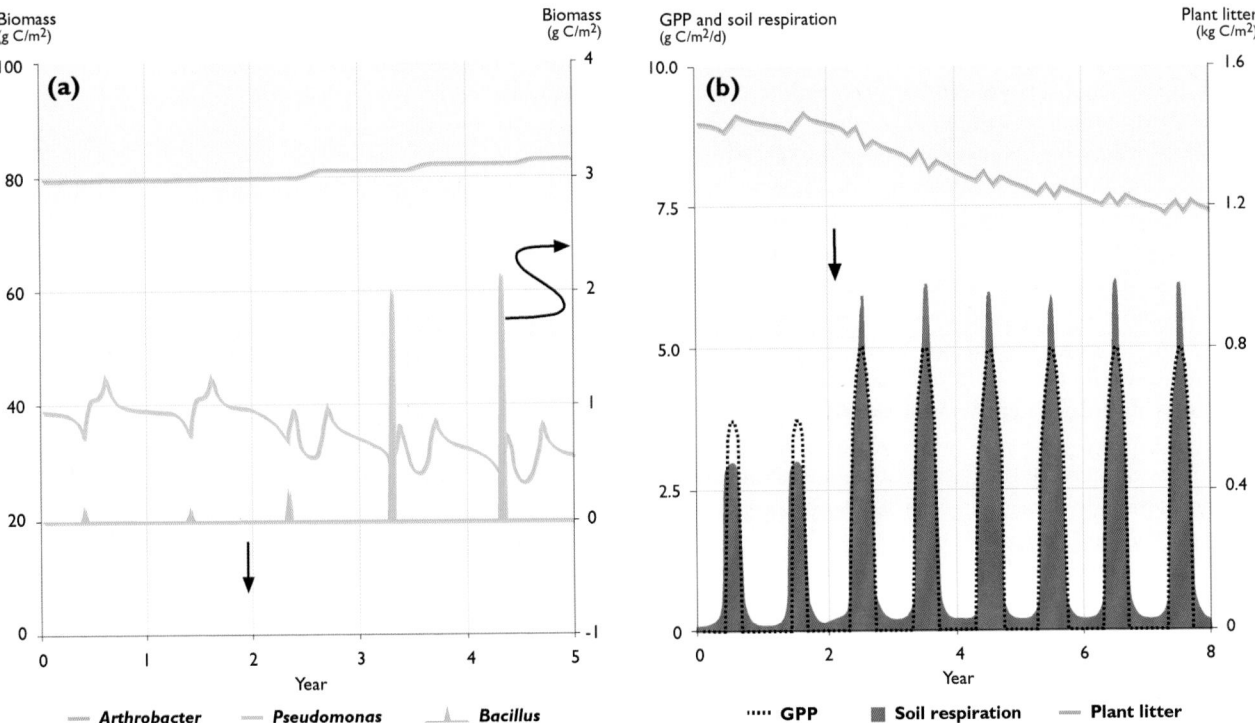

Fig. 7.14. Simulation of changes in a tundra microbial community (Barrow, Alaska) induced by climate warming: (a) population dynamics of dominant soil bacteria (right axis applies to *Bacillus*); (b) carbon budget including gross primary production (GPP), soil respiration, and litter dynamics. The simulation assumed that average air temperature instantly increased by 10 °C in year two of the simulation, indicated by the vertical arrows in each panel (Panikov, 1994).

The arctic periglacial environment represents a unique mosaic of unstable habitats (gradation between terrestrial and shallow wetland environments) where large variations in cyanobacterial and algal diversity, productivity, and life strategy exist (Elster and Svoboda, 1995, 1996; Elster et al., 1997, 2002; Kubeckova et al., 2001). Prokaryotic cyanobacteria and eukaryotic algae have different life strategies with respect to their susceptibility to severe and unstable conditions (Elster, 2002). Cyanobacteria are well adapted to changeable conditions involving low and high radiation levels (including UV-B radiation), and cycles of desiccation and rehydration, increasing and decreasing salinity, and freezing and thawing. This gives them a great ecological advantage and allows them to be perennial. In contrast, eukaryotic algae have higher rates of photosynthesis and lower resistance to changes in irradiation, desiccation, rehydration, and freeze–thaw cycles. These features predetermine their annual character. If the arctic terrestrial environment becomes colder, the cyanobacteria are very likely to become the dominant community. In contrast, if temperatures become warmer, the eukaryotic algae are very likely to start to predominate. In addition, the ongoing temperature increase in the Arctic is very likely to influence cyanobacteria and algal production, as well as the balance between cyanobacteria and algae and invertebrate herbivore activity. Invertebrate grazing pressure is likely to increase and much of the visible cyanobacteria and algae biomass could possibly disappear from arctic locales (Elster et al., 2001).

Summary

Tundra soil heating, CO_2 enrichment, and amendment with mineral nutrients generally accelerate microbial activity. Higher CO_2 concentrations tend to intensify root exudation, which is the main source of available carbon for soil and rhizosphere bacteria. Supplementation of UV-B radiation in the field resulted in changes in the composition of microbial communities. Laboratory incubation of tundra soils had strong effects on community composition after a temperature shift of more than 10 °C. Surprisingly, the effects of many factors on the soil microbial community were less significant compared with effects on the plant community. However, a mathematical simulation of the changes in microbial community structure in the tundra showed that soil warming resulted in stimulation of bacterial growth.

The effects of increased UV-B radiation levels on microorganisms include damage to high-latitude strains of fungal spores, and damage to some species of leaf-dwelling fungi and soil-dwelling decomposer fungi that resulted in a change in the composition of the fungal communities.

Cyanobacteria are better adapted to changeable and harsh conditions than algae, and in milder climates are likely to be dominated by algae. However, herbivory of both cyanobacteria and algal biomass is likely to increase in a warmer climate.

7.3.4. Genetic responses of species to changes in climate and ultraviolet-B radiation levels

Many widely distributed arctic species show large ecological amplitude (broad niches) and are taxonomically complex, often with many subspecies, while species with a narrower distribution range often show a more restricted ecological amplitude. It is necessary to know the extent of genetic variation in arctic species and the underlying causes of differentiation or homogenization (biogeography, historical bottlenecks, reproductive biology, and demography) in order to assess responses of species to climate change.

7.3.4.1. Plants

In spite of rapid development in recent years of different molecular techniques suited for population genetic studies, there are still few studies of arctic plants, most of which have focused on biogeographical and phylogeographical questions related to vascular plant species. Such studies may reveal the migratory potential of the species in response to climate change. During the Pleistocene glaciations, arctic plants were restricted to refugia within or south of the present-day Arctic, from which they could re-colonize areas as conditions improved during interglacial periods (Abbott et al., 2000; Tremblay and Schoen, 1999). The rate of colonization by different species during the Holocene probably depended on the location of their closest refugia, their dispersal biology, and their genetic makeup. Genetic phylogeographical studies provide evidence for relatively fast migration rates in most vascular species (Abbott and Brochmann, 2003; Bennike, 1999; Brochmann et al., 2003) and possibly bryophytes as well (Derda and Wyatt, 1999). However, in the modern context of rapid climate change, migration rates need to be considered on somewhat shorter timescales than thousands of years.

The level of genetic variation within and between populations indicates the potential for local adaptation to environmental change and hence population resilience to environmental change. Based on the relatively young age of populations and low recruitment of sexually reproduced offspring, it was long believed that genetic variation in arctic plants would be low. However, the number of genetic studies is limited and no such general pattern of genetic variation has been identified. Arctic plants show the same range of genetic variation as temperate plants, ranging from comparatively high levels (Bauert, 1996; Gabrielsen and Brochmann, 1998; Jefferies and Gottlieb, 1983; Jonsson et al., 1996; Philipp, 1997; Stenström et al., 2001) to very low levels (Bayer, 1991; Max et al., 1999; Odasz and Savolainen, 1996; Philipp, 1998; Stenström et al., 2001) of variation. However, genetic variation among arctic plants may be of greater value in terms of biodiversity than in other biomes due to much lower species diversity. Furthermore, high levels of polyploidy in many arctic vascular plant species may promote the proportion of the genetic variation partitioned within individuals, which may be important when passing through evolutionary bottlenecks (Brochmann and Steen, 1999).

By comparing 19 different populations of three rhizomatous *Carex* taxa, distributed among 16 sites within arctic Eurasia, ranging from northern Scandinavia in the west to Wrangel Island in the east, Stenström et al. (2001) showed that the levels of genetic variation were not related to climate, but were to a large extent explained by differences in glaciation history at the sampling sites. Populations in areas deglaciated about 10 000 years BP had significantly lower genetic variation than populations in areas deglaciated 60 000 years BP or those in areas not glaciated at all during the Weichselian. Relatively young population age may also be responsible for a low genetic variation in some other populations (e.g., Bayer, 1991; Max et al., 1999), while in yet others, breeding systems apparently play a large role (e.g., Odasz and Savolainen, 1996; Philipp, 1998). In general, populations of insect- or self-pollinated plant species have lower genetic variation than populations of wind-pollinated species (Hamrick and Godt, 1990), and this seems to apply to arctic plants as well.

Those plant species representing populations with relatively high levels of genetic variation usually have a large geographic distribution, for example purple saxifrage (Abbott et al., 1995), nodding saxifrage (*Saxifraga cernua*; Gabrielsen and Brochmann, 1998), moss campion (*Silene acaulis*; Abbott et al., 1995; Philipp, 1997), *Carex bigelowii* sensu lato (Jonsson et al., 1996; Stenström et al., 2001), and *C. stans* (Stenström et al., 2001). In these species, the genetic variation among populations (G_{ST}) is a relatively small proportion of the total genetic variation (i.e., they show low degrees of population differentiation). Large variation within populations, however, increases possibilities for ecotypic differentiation. In the Arctic, extremely steep environmental gradients are frequent on a microtopographical scale and ecotypic differentiation has been demonstrated over such short distances for alpine timothy (*Phleum alpinum*; Callaghan, 1974), *Carex aquatilis* (Shaver et al., 1979), mountain avens (McGraw and Antonovics, 1983), and purple saxifrage (Crawford and Smith, 1997), all widely distributed plant species in the Arctic. Ecotypic differentiation in response to this small-scale heterogeneity may preserve genetic variation and in that way contribute to resilience to change at the species rather than the population level. Thus, an initial response to climate change in such species is likely to be a change in the distribution and abundance of ecotypes within a species distribution (Crawford and Smith, 1997). In addition, many arctic plants show large phenotypic plasticity, which is likely to further increase their resilience (Stenström et al., 2002; Table 7.5).

If the degree of genetic variation can be used as an indication of resilience of populations to change, it is likely that this resilience will be greatest among plants in old populations of widely distributed, wind-pollinated vascular species (e.g., rhizomatous *Carex* populations in eastern Siberia). However, generation time and seedling recruitment may affect the adaptation rate. Many of the dominant arctic plants such as the rhizomatous *Carex* species are clonal, that is, they do not rely on seed production through sexual reproduction for short-term population maintenance. Genetic individuals of these plant species may live to be thousands of years old (Jónsdóttir et al., 2000), which may decrease the adaptation rate. However, experiments with plants from outside of the Arctic have shown that UV-B radiation may increase the rate of genetic change. Exposure to high UV-B radiation levels can activate mutator transposons that amplify the mutation effect beyond the immediate UV-B radiation damage (Walbot, 1999), and increased levels of UV-B radiation may lead to an increased tendency to mutations in future generations (Ries et al., 2000).

For plants with long-lived seed, further genetic variation is preserved in the seed banks. Dormant seed populations may be genetically different from the aboveground populations (McGraw, 1995) and potentially able to better exploit a new climate.

Genetic variation has been studied in fewer moss and lichen species than in vascular plants. Boreal and antarctic bryophytes usually show high levels of variation (Cronberg et al., 1997; Derda and Wyatt, 1999; Skotnicki et al., 1998, 1999), but the partitioning of genetic variation among and within populations depends on species. Scandinavian populations of the widely distributed moss *Hylocomium splendens*, including two subarctic alpine populations, showed high genetic variation within populations and low G_{ST}, a pattern similar to widely distributed, wind-pollinated vascular species (Cronberg et al., 1997). In contrast, North American and European populations of *Polytrichum commune* have low within-population variation and high G_{ST} (Derda and Wyatt, 1999).

7.3.4.2. Animals

The genetics of arctic terrestrial animals have been thoroughly studied mainly for a few well-known mammal species such as reindeer/caribou (Flagstad and Røed, 2003), lemmings (Ehrich et al., 2000; Fedorov et al., 1999a,b), and Arctic fox (Dalén et al., 2005). These studies have focused on phylogeographical patterns and the relative roles of present gene flow and historic processes (especially glacial–interglacial cycles; see section 7.2) based on neutral genetic markers (especially mitochondrial DNA). The present genetic differentiation reflects to a large extent historic processes and the presence of current migration barriers. For mammals with relatively restricted mobility such as lemmings, even small-scale barriers (e.g., large rivers) can form the borders between subspecies (Fedorov et al., 1999a,b), while a very mobile animal such as the Arctic fox, which readily moves between continents and islands on sea ice, appears to be relatively panmictic (i.e., shows little genetic structuring) at the circumpolar scale (Dalén et al., 2005).

Current gene flow (an indication of mobility) and population history (origin and differentiation) indicate the ability of a species to track the location of its habitats through time (i.e., a species is able to relocate its distributional range according to any changes in the distribution of its habitat through the process of dispersal). A mobile species will have better prospects for survival than a relatively sedentary species. Moreover, a species with high genetic/racial diversity has proved an ability to adapt to different environmental conditions in the past and is likely to do the same in the future. It should be noted, however, that markers of genetic variation/differentiation currently used (e.g., mitochondrial DNA) may have little bearing on the genetic variation in morphology and life-history traits (see Flagstad and Røed, 2003). It is these latter traits that decide whether a species or a morph will be able to adapt to future changes. Currently, there are few studies of arctic animals using a quantitative genetics approach (Roff, 1997) that address the potential for rapid adaptations to climatic change. Elsewhere, using a quantitative genetic research protocol, Réale et al. (2003) showed that northern boreal red squirrels (*Tamiasciurus hudsonicus*) were able to respond genetically within a decade to increased spring temperatures.

7.3.4.3. Microorganisms

Assessment of genetic responses of microorganisms to climate change is based on laboratory models, as observations made within arctic terrestrial ecosystems are absent. Short generation times and the impressive genetic plasticity of bacteria make them a favorite topic in theoretical studies of general population genetics. Because most mutations are deleterious, mutation rates are generally thought to be low and, consequently, mutator alleles should be selected against. However, up to 1% of natural bacterial isolates have been found to be mutators. A mutator can be viewed as behaving altruistically because, although it reduces individual fitness, it increases the probability of an adaptive mutation appearing. These results may help to explain observations that associate high mutation rates with emerging pathogens that cause spontaneous epidemic outbreaks (Sniegowski et al., 1997; Wilke et al., 2001).

In the arctic environment, intensive mutagenic effects are likely to result from increased UV radiation levels and also from aerosols and volatile chemical mutagens transported to the cool polar atmosphere from the mid- and low latitudes. The direct mutagenic effects are very likely to be weak, especially if the protective shielding effects of soil particles and impressive genetic plasticity of bacteria are taken into account. However, it is possible that mutants could lead to epidemic outbreaks that could have profound and unexpected consequences for the whole ecosystem.

7.3.4.4. Summary

Arctic plants show the same range of genetic variation as temperate plants, ranging from comparatively high levels to very low levels. In widespread *Carex* taxa, levels of genetic variation were not related to climate, but were to a large extent explained by differences in glaciation history at the sampling sites: populations in areas deglaciated approximately 10000 years BP had significantly lower genetic variation than populations in areas deglaciated 60000 years BP.

Plant species representing populations with relatively high levels of genetic variation usually have a large geographic distribution. On a microtopographical scale, extremely steep environmental gradients are frequent and ecotypic differentiation has been demonstrated over short distances for several widespread species. This heterogeneity, together with large phenotypic plasticity, is likely to contribute to resilience to change at the population and species levels. For plants with long-lived seed, further genetic variation related to former environments is preserved in the seed banks. Thus, there are several mechanisms for widespread arctic plant species to respond to environmental change.

Experiments with plants from outside the Arctic have shown that increased levels of UV-B radiation can speed up genetic change and may lead to an increased tendency for mutations in future generations.

The present genetic differentiation of arctic terrestrial animals that have been studied thoroughly (e.g., reindeer/caribou, lemmings, Arctic fox) to a large extent reflects historic processes and the presence of current migration barriers. For mammals with relatively restricted mobility such as lemmings, even small-scale barriers (e.g., large rivers) can form the borders between subspecies, while a very mobile animal such as the Arctic fox shows little genetic structuring at the circumpolar scale. A species with high genetic/racial diversity has proved an ability to adapt to different environmental conditions in the past and is likely to do the same in the future.

There is a paucity of studies of arctic animals that have addressed the potential for rapid adaptations to climatic change. Elsewhere, it was shown that northern boreal red squirrels were able to respond genetically within a decade to increased spring temperatures.

Up to 1% of natural bacterial isolates have been found to be mutators, and high mutation rates are associated with emerging pathogens causing spontaneous epidemic outbreaks. In the Arctic, intensive mutagenic effects are likely to result from increased UV radiation levels and also from aerosols and volatile chemical mutagens. Although the effects are very likely to be weak, it is possible that mutants could lead to epidemic outbreaks that could have profound and unexpected consequences for the whole ecosystem.

7.3.5. Recent and projected changes in species distributions and potential ranges

Paleoecological research (section 7.2) and observations over many decades demonstrate that the geographic ranges of terrestrial species in general are well correlated with bioclimatic variables. Furthermore, the strength of these relationships is independent of trophic level (Huntley et al., 2003). Major climate-related species distributions at the large scale include the limit of trees, which is associated with the isoline for mean July air temperatures of about 10 °C (Brockmann-Jerosch, 1919 as discussed in Körner, 1999) and soil temperature of 7 °C (Körner, 1998), and the limit of woody plants such as dwarf shrubs that are one indicator of the boundary of the polar desert biome (Edlund and Alt, 1989). Such relationships suggest that species distributions at the macrogeographical and landscape scale are very likely to change as temperature changes. This section assesses the effects of climate change on recent changes in species distributions and those projected to occur in the future.

7.3.5.1. Recent changes

Indigenous knowledge projects have documented recent changes in the ranges of caribou in relation to changes in weather, based on hunters' understanding of how environmental conditions affect seasonal caribou distribution patterns (Kofinas et al., 2002). Hunters' explanations of caribou distributions may provide indications of potential range changes given projected climate change. For example, in the El Niño year of 1997–1998, several thousand Porcupine Caribou overwintered on the Yukon Coast in arctic Canada. Hunters in Aklavik, Northwest Territories, explained this phenomenon in terms of the Beaufort Sea ice pack, which was farther from the Yukon North Slope than in most years, resulting in warmer coastal temperatures and thus more abundant forage for caribou. In July 1997, as the caribou moved into Canada from their Alaskan calving grounds, several large groups remained on the coast, taking advantage of the rich forage opportunities. A mild autumn and the lack of icing events that push the caribou south for the winter kept the caribou in the area into October, as the animals could continue to access summer forage. The herd that remained on the coast for the winter was reported to be in better condition than the herd wintering in the usual locations.

Indigenous knowledge has also documented recent changes in the ranges of other animals in relation to changes in the weather. In the Canadian Arctic, Inuit in communities such as Baker Lake have reported insects previously associated with areas south of the treeline (Fox, 2002). In more western regions, there have been more frequent sightings of "mainland ducks" such as pintail ducks (*Anas acuta*) and mallard (*A. platyrhynchos*; Riedlinger, 2001).

Working in the Canadian Arctic using a conventional scientific approach, Morrison et al. (2001) summarized the trends in population data for breeding waders, and found that almost all arctic-breeding species were declining. The reasons for the trends were not always clear and were probably of multiple origins. Long-term monitoring in Finland has shown a substantial decline in the populations of many arctic and subarctic bird species over the past 20 years (Väisänen et al., 1998), but the trend in bird populations is not always negative. Zöckler et al. (2003) found that almost half of the arctic-breeding, long-distance migrants studied are presently in decline. For many species there are still insufficient data available, and only a few species (8%) show an increasing trend. In most cases, it is not easy to correlate trends with climate change. As the trends in some species are different outside and inside the Arctic, there is an indication that factors of a more global nature are involved. An example is the drastic decline of the ruff (*Philomachus pugnax*) in almost all breeding sites outside the Arctic in contrast to their stable or even increasing populations in some (but not all) northern arctic areas (Zöckler, 2002). This coincides with the recent northern expansion of other wet-grassland waders, such as the common snipe (*Gallinago gallinago*) in the Bolshemelzkaya tundra (Morozov, 1998), the black-tailed godwit (*Limosa limosa*), and the northern lapwing (*Vanellus vanellus*) in northern Russia concomitant with a northward expansion of agriculture including sown meadows (Lebedeva, 1998). Several other bird species have recently been recorded in more northerly locations in the Arctic (Zöckler et al., 1997), suggesting that some species are shifting their distribution in response to alteration of habitats by climate change. The emerging picture is that the ruff is being forced to retreat to its core arctic habitats owing to the effects of global climate change in combination with increasing nutrient enrichment on the quality of wet grassland habitats (Zöckler, 2002).

A recent global meta-analysis of plants claims that a climate change signal has been identified across natural ecosystems (Parmesan and Yohe, 2003). Range shifts of plants averaging 6.1 km per decade toward the poles and 6.1 m per decade in altitude have been identified in response to a mean advancement of spring (initiation of greening) by 2 to 3 days per decade. Although some northern treeline data were included in the analysis, little information was available for arctic ecosystems.

7.3.5.2. Projected future changes in species distributions

Models of species–climate response surfaces based upon correlations between species ranges and bioclimatic variables are able to simulate the recently observed range changes of at least some species of birds (Zöckler and Lysenko, 2000) and butterflies (Hill et al., 1999; 2003; Virtanen and Neuvonen, 1999). Related studies have shown that, at least in the case of butterflies, the extent to which species have realized their projected range changes over the last 30 to 50 years is strongly related to their degree of habitat restriction: generalist species are much more able to achieve the projected range expansions than are specialist species (Warren et al., 2001).

Such models (Hill et al., 2003; Huntley et al., 1995) project future ranges of arctic species that are often markedly reduced in spatial extent compared to the species' present ranges. The range limits of boreal and temperate species shift poleward in response to the same future climate scenarios. However, the large magnitude of the shifts in projected range margins results in potential reductions in the ranges of many boreal species because they are limited to the north by the Arctic Ocean.

The extent to which arctic plant species experience the rapid range reductions simulated by such models will depend principally upon two factors. First, such reductions are likely to happen most rapidly in species that experience some physiological constraint at their southern range margin, for example, the winter thermal constraint postulated for cloudberry (Marks, 1978; Marks and Taylor, 1978) or the summer thermal constraints postulated for the great skua (*Catharacta skua*; Furness, 1990). Species whose southern range margin is determined by biotic interactions are likely to experience less

rapid range reductions. Second, such reductions are very likely to happen more rapidly where the northward migration of boreal or temperate species is not limited either by habitat availability or propagule (dispersal stage of a plant or animal, such as fertilized eggs, larvae, or seeds) dispersal. "Fugitive" species of the early successional communities that characteristically follow disturbance of the boreal forests have the required dispersal ability to achieve rapid poleward range expansions. Unless other factors (e.g., herbivore pressure or a lack of microsites for successful seedling establishment) exclude them, these species are likely to extend into the Arctic rapidly, forming transient ecosystems that will persist until the arrival of the more slowly expanding late-successional boreal species.

Loss of habitat is a particularly important possibility that would constrain species ranges. The most dramatic change in habitat for many water birds is the projected loss of tundra habitat, which varies between 39 and 57% by the end of the 21st century (Harding et al., 2002; Haxeltine and Prentice, 1996). Vegetation models (Neilson and Drapek, 1998) combined with maps of water-bird distributions show a large variation in the impact of projected vegetation changes on 25 selected

Table 7.7. Loss of breeding area habitat projected by two different general circulation models for arctic water-bird species, and their globally threatened status (based on Zöckler and Lysenko, 2000).

		Loss of habitat (%)[a]		Red List[d]
		HadCM2GSa1[b]	UKMO[c]	
Tundra bean goose	*Anser fabalis rossicus/serrirostris*	76	93	
Red-breasted goose	*Branta ruficollis*	67	85	VU
Spoon-billed sandpiper	*Eurynorhynchus pygmaeus*	57	57	VU/EN
Emperor goose	*Anser canagicus*	54	54	!
Ross's gull	*Rhodostethia rosea*	51	73	
Red-necked stint	*Calidris ruficollis*	48	68	
Sharp-tailed sandpiper	*Calidris acuminata*	46	74	
Little stint	*Calidris minuta*	45	65	
Curlew sandpiper	*Calidris ferruginea*	41	70	
Pectoral sandpiper	*Calidris melanotos*	38	60	
Dunlin	*Calidris alpina*	36	58	
White-fronted goose	*Anser albifrons*	36	57	
Long-billed dowitcher	*Limnodromus scolopaceus*	31	54	
Great knot	*Calidris tenuirostris*	31	42	
Lesser white-fronted goose	*Anser erythropus*	28	29	VU
Barnacle goose	*Branta leucopsis*	21	27	
Western sandpiper	*Calidris mauri*	19	21	
Brent goose	*Branta bernicla*	16	44	
Red knot	*Calidris canutus*	16	33	
Greater snow goose	*Anser caerulescens*	14	46	
Canada goose	*Branta canadensis*	13	22	
Pink-footed goose	*Anser brachyrhynchus*	10	10	
Sanderling	*Calidris alba*	5	25	

[a]Value could be substantially higher as unclassified areas in these analyses may contain different tundra types; [b]moderate warming; [c]extreme warming; [d]VU=vulnerable as a globally threatened species (BirdLife International, 2001), EN=suggested for upgrading to endangered as a globally threatened species, !=suggested for inclusion in the Red List.

species (Zöckler and Lysenko, 2000). Vegetation scenarios derived from the HadCM2GSa1 model project that 76% of tundra bean goose (*Anser fabalis rossicus/serrirostris*) habitat will be affected by the alteration of tundra vegetation, while only 5% of sanderling habitat will be affected (Neilson and Drapek, 1998). However, the sanderling, similar to many other high-arctic breeders, might be affected even more strongly, as southern tundra habitat types are projected to replace their specific high-arctic habitats. Whereas the more southerly breeding species can shift northwards, it is likely to be increasingly difficult for high-arctic breeders to compete. For two of the three water-bird species that are considered globally threatened, namely the red-breasted goose (*Branta ruficollis*) and the spoon-billed sandpiper (*Eurynorhynchus pygmaeus*), 67 and 57% of their current breeding range is projected to change from tundra to forest, respectively (Table 7.7). This additional loss of habitat is likely to place these two species at a higher risk of extinction. The emperor goose (*Anser canagicus*), already in decline and with 54% of its small range projected to be affected, is highlighted as needing further conservation attention.

Geographic ranges of plants

Strong relationships between growth and temperature in the circumpolar ericaceous dwarf shrub *Cassiope tetragona* and the feather moss *Hylocomium splendens* can be used to model range changes. The growth of *C. tetragona* is strongly related to mean July temperature (Havström et al., 1995b) and that of *H. splendens* is related to mean annual temperature (Callaghan et al., 1997) throughout their northern ranges (Fig. 7.15a). Mean July and mean annual temperatures are to some extent representative of latitude, as they decrease toward the north. The natural climatic warming from the beginning of the Little Ice Age to the present is the equivalent of only a minor shift in latitude for *C. tetragona*. On the other hand, projected future warming is likely to produce a greater latitudinal displacement that, at the northern limit of the current ranges of the two species, is very likely to result in a northern range extension (Fig. 7.15a). In contrast, at the southern edge of the ranges, future warming is very unlikely to increase growth beyond the genetic capabilities of the species, and the dynamics of the species in this part of their ranges are very likely to be determined by the responses of competi-

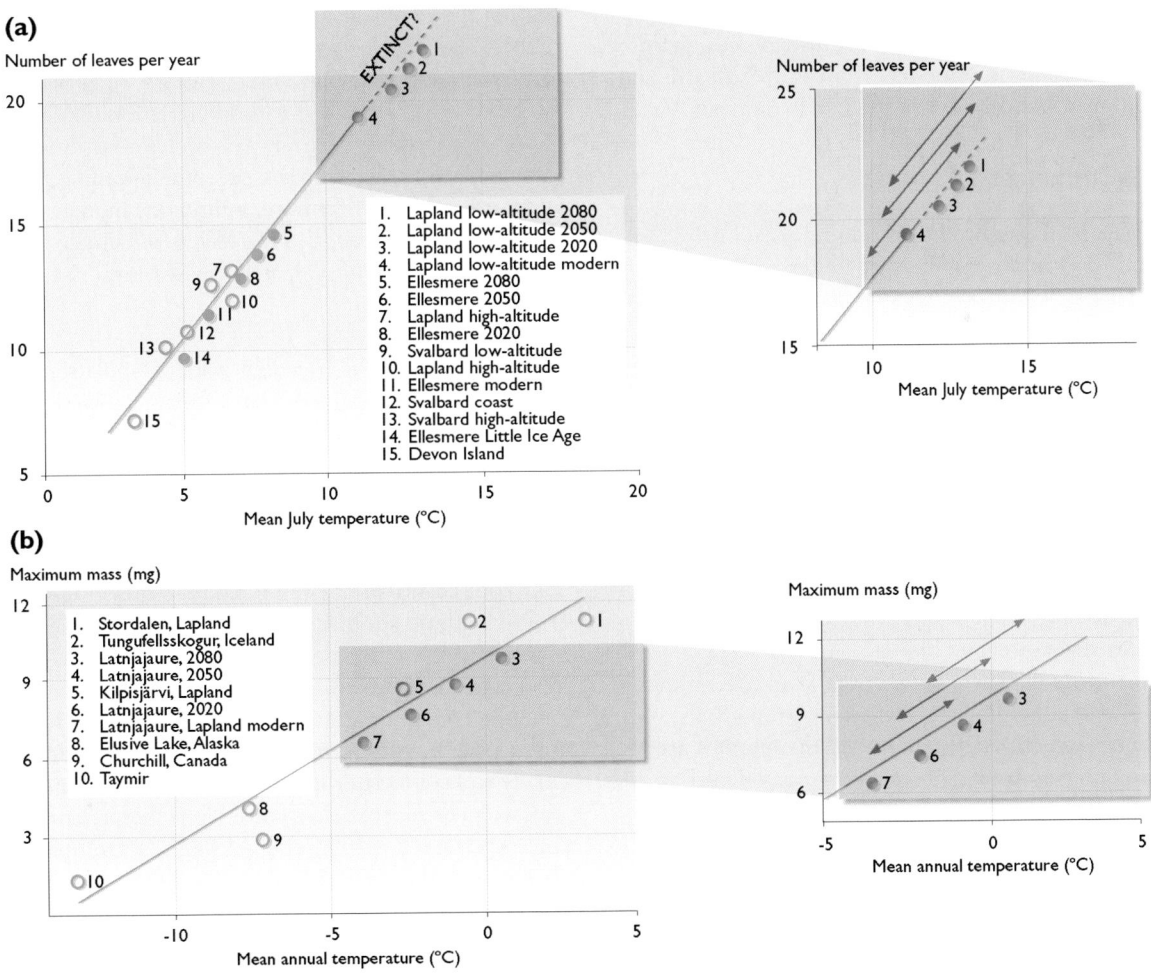

Fig. 7.15. (a) Number of leaves produced per year in shoots of *Cassiope tetragona* as a function of mean July temperature and (b) correlation between a growth parameter in *Hylocomium splendens* and mean annual temperature at seven arctic and subarctic sites. The boxes on the regression lines contain scenarios of growth in ACIA time slices resulting from temperature increases projected by the five ACIA-designated models. The hatched line in (a) depicts projected growth that is probably outside the capability of the species. The expanded boxes to the right of each panel depict uncertainty ranges associated with each of the projections (doubled-ended arrows). Open circles represent sequences of present-day relationships between climate and growth, while filled circles represent present-day and projected growth at three sites (based on Callaghan et al., 1997 and Havström et al., 1995b).

tors to warming. A similar analysis for *H. splendens* shows how a current alpine population is very likely to resemble a population from a lowland forested area under projected future climate conditions (Fig. 7.15b).

At the landscape scale, plants are distributed in mosaics associated with microhabitats, and the larger-scale latitudinal range changes are very likely to be associated with initial changes in landscape mosaics. Cushion plants and other species characteristic of wind-exposed patches are likely to become restricted in distribution by increased snow cover. In contrast, plants of snow beds might become more restricted if snow duration decreases. Wetland species will become restricted by drying and species in drier areas will become more restricted by increased soil moisture. Plants currently restricted to south-facing slopes and warm springs (to some extent analogues of future warmer habitats and hot spots of biodiversity) north of their main distribution areas are likely to provide an "inoculum" for rapid colonization of surrounding habitats when climate becomes warmer, although they themselves are likely to be displaced from their current niches by less diverse shrub–thicket communities. Examples include orchids, ferns, and herbs in warm springs on West Greenland (although orchids and ferns are unlikely to become widely distributed), ericaceous dwarf shrubs in some inner fjords of Svalbard, and the large shrubs/small trees of the North Slope of Alaska.

Geographic ranges of animals

Often, causes of observed trends in migrant bird population numbers cannot easily be attributed to local, site-related factors within and outside the Arctic, such as drainage, land use change, hunting and persecution by humans, and predation. Even among global factors, climate change is one of an array of drivers, such as eutrophication, often working in synergy with climate change and reinforcing the effect. In addition, migratory birds are also strongly affected by climate change outside of their arctic breeding grounds. Desertification, droughts, and wetland loss; eutrophication of staging and wintering wetlands; changes in land use; and application of chemicals and nutrients on wintering grounds lead to changes in vegetation and biomass on coastal staging and wintering grounds. The effects of sea-level rise on the extent of coastal staging and wintering grounds are very likely to be particularly harmful, and the hunting pressure on wintering waders in certain areas will also reduce bird populations.

There are few studies of the impacts of climate change on migratory species, although recent trends in some species (e.g., arctic geese) are well known (Madsen et al., 1999). Very little can be concluded from the observed impacts of current climate variability on migratory birds, as existing monitoring programs are few (e.g., Soloviev et al., 1998) and often began only recently.

An analysis of Hadley Centre spring and summer temperature and precipitation data over the last 50 years, inter-polated over the currently known arctic distribution areas of the white-fronted goose (*Anser albifrons*) and the Taymir population of the red knot (*Calidris canutus canutus*), demonstrates a significant correlation between the mean June temperature and the percentage of juveniles in the population as a measure of breeding success. The Nearctic population of the red knot (*C. c. islandica*) and the curlew sandpiper (*Calidris ferruginea*) breeding on the Taymir Peninsula do not show such a correlation (Zöckler and Lysenko, 2000). The HadCM2GSa1 model, forced with a 1% per year increase in atmospheric CO_2 concentrations, projects a moderate increase in mean June temperature (by the time CO_2 concentrations double) in the Taymir breeding area of the white-fronted goose, which is likely to favor the goose population. The conditions for the Taymir population are projected to be particularly favorable in the period around 2020. However, a considerable initial cooling and lack of warming over present-day values by 2080 in the breeding grounds of the West Greenland goose population is likely to lead to a drop in the size of the fragile Greenland population. Although the ACIA-designated model projections differ from those used in Zöckler and Lysenko (2000), possible decreases in temperature in ACIA Region 1 (section 18.3.1) are within the range of the projections (section 7.6, Table 7.14). The Zöckler and Lysenko (2000) study must be interpreted in relation to other factors, such as other weather parameters and natural predation that often fluctuates in three- to four-year cycles according to the abundance of the main prey (i.e., the lemming; see section 7.4.1.4). Furthermore, hunting by humans, mainly outside of the Arctic, and the effects of climate change outside of the Arctic, in particular sea-level rise, need to be taken into account.

Investigations of the breeding wader population in northeast Greenland for over 30 years showed that spring snow cover is the main factor governing initiation of egg laying in high-arctic waders, such as the red knot and other sandpipers, while June temperature does not appear to be important (Meltofte, 1985, 2000; pers. comm., 2004). In fact, waders breed earlier in the arid but cool far north of Greenland than they do in the "mild", humid southern areas of the high Arctic, where snow cover is much deeper and more extensive. Projections of future climate for northeast Greenland include cooler summers, later snowmelt, and less snow-free space for the arriving waders to feed, which are likely to lead to later breeding and smaller populations. Snow cover must still be considered the prime regulating factor for initiation of egg laying, but temperature – so important for determining invertebrate food availability (section 8.5.6) – is important as well when sufficient snow-free habitat is already present.

Although global climate change in synergy with global eutrophication is likely to lead to an increase in biomass in the Arctic, a change in vegetation height and density, and a general change in vegetation structure with shifts in species distribution that are very likely to have an enormous impact on water birds (which are highly dependent on open landscapes and lightly vegetated breeding sites), global climate change and eutrophication are very likely to provide opportunities for other birds with more southerly distributions, such as owls and woodpeckers. Some birds, including most goose species and a few waders, have demonstrated a certain ability to adjust to new and changing habitats (Lugert and Zöckler, 2001), but the majority of birds breeding in the high Arctic are likely to be to be pushed to the edge of survival with little habitat left.

Geographic ranges of microorganisms

Studies of geographic ranges of microbes related to extremely cold environments such as the Arctic, and also to climate change, are in their infancy. Unlike plant and animal ecology, soil microbiology still does not have a solution to the central biogeographical problem of whether soil microorganisms are cosmopolitan (widely distributed) or endemic (restricted to one location) species. Until the ranges of species are known, the bacteria that might be threatened by climate change cannot be identified (Staley, 1997).

The prevailing hypothesis for bacterial biogeography is based on the axiom of the Dutch microbiologists Baas-Becking and Beijerinck, who stated, "Everything is everywhere, but the environment selects" (Beijerinck, 1913). This hypothesis assumes that free-living bacteria are cosmopolitan in their geographic distribution; they are readily disseminated from one location on earth to another by water and air currents or by animal vectors such as birds that migrate between regions. Only recently has it been possible to rigorously test the cosmopolitan distribution of bacteria with unbiased molecular biology approaches. Studies outside of the Arctic demonstrate that the cyanobacterium *Microcoleus chthonoplastes* is a cosmopolitan species (Garcia-Pichel et al., 1996). Using different molecular biology techniques, Stetter et al. (1993) discovered that hyperthermophilic (heat-loving) Archaea isolated from Alaskan oil reservoirs showed a high degree of DNA–DNA reassociation with selected *Archaeoglobus*, *Thermococcus*, and *Pyrococcus* species, and concluded that the species were the same as those from European thermal marine sources. In a separate study, DNA–DNA reassociation of a strain of *Archaeoglobus fulgidus* isolated from North Sea crude oil fields showed 100% relatedness to an *Archaeoglobus fulgidus* strain from Italian hydrothermal systems (Beeder et al., 1994). These two studies comprise some of the best evidence to date supporting the cosmopolitan hypothesis of Baas-Becking.

However, 3-chlorobenzoate-degrading bacteria isolated from soils in six regions on five continents (Fulthorpe et al., 1998) were found to have restricted or unique ranges. Also, plant species have been reported to harbor their own unique symbiotic species of fungi associated with leaves, bark, roots, etc (Hawksworth, 1991), so, by definition, the existence of endemic plants should imply the existence of respective microbial symbionts. Therefore, arctic microbial communities may consist of a mixture of species: some that are endemic and some that are cosmopolitan.

7.3.5.3. Summary

Monitoring of distribution ranges with a spatial representation as good as for temperate latitudes is not available for the terrestrial Arctic. Indigenous knowledge projects have documented recent changes in caribou ranges in relation to changes in weather. Hunters' explanations of caribou distributions may provide indications of potential range changes under scenarios of warming temperatures, such as overwintering of caribou in coastal areas during warm winters. Other arctic indigenous observations include insects previously associated with areas south of the treeline and more frequent sightings of "mainland ducks". In contrast, almost all arctic-breeding species are declining. The reasons for the trends are not always clear and probably of multiple origins, although there are suggestions that some species are shifting their distribution in response to alteration of habitats by climate change.

Quantitative monitoring of conspicuous and popular species such as birds and butterflies has demonstrated that many formerly southern species are rapidly approaching arctic regions and some have already entered. Arctic birds, especially arctic-breeding water birds and waders that can be counted on staging and wintering grounds, show mostly declining population trends; some species have declined dramatically. It is likely that these changes result from the combined action of eutrophication and habitat loss on wintering and staging sites as well as concurrent climate change, although separating the relative contributions of these factors is difficult. Based on climate models, dramatic reductions in the populations of tundra birds are projected as a generally warmer climate is likely to increase vegetation height and decrease the Arctic's landmass.

Species–climate response-surface models are able to simulate the recently observed range changes of at least some species of both birds and butterflies. At least in the case of butterflies, the extent to which species have realized their projected range changes over the last 30 to 50 years is strongly related to their degree of habitat restriction: generalist species are much more able to achieve the projected range expansions than are specialist species. Simulated potential future ranges are often markedly reduced in spatial extent compared to present ranges. The range limits of boreal and temperate species shift poleward but the large magnitude of the shifts in projected range margins results in potential reductions in the ranges of many boreal species because they are limited to the north by the Arctic Ocean. Species that experience some physiological constraint at their southern range margin are likely to be affected sooner than those that are affected by biotic relationships such as competition from immigrant species. Loss of habitat, such as tundra ponds for many arctic birds, is a particularly important possibility that would constrain species ranges. In contrast, plant populations that are outliers of more southerly regions and restricted to particularly favorable habitats in the Arctic, are likely to spread rapidly during

Fig. 7.16. Forest tundra vegetation represented by the Fennoscandian mountain birch forest, Abisko, northern Sweden (photo: T.V. Callaghan).

Fig. 7.17. Zonal tussock tundra near Toolik Lake, Alaska, with large shrubs/small trees of *Salix* in moist sheltered depressions (photo: T.V. Callaghan).

warming. Models of moss and dwarf-shrub growth along latitudinal gradients show considerable potential for range expansion in the north, but considerable uncertainty, in relation to ACIA scenarios of warming.

Most microorganisms detected in northern ecosystems, such as free-living bacteria, are probably cosmopolitan in their geographic distribution and readily disseminated from one location to another, and the environment selects those that can proliferate. However, some species, particularly symbionts with endemic plants, can themselves be candidates for endemic status.

7.4. Effects of changes in climate and UV radiation levels on structure and function of arctic ecosystems in the short and long term

Section 7.3 assessed the responses of individual species to changes in climate and UV-B radiation levels. The present section assesses the responses of species aggregated into communities and ecosystems. The two main attributes of ecosystems that respond to environmental change are structure and function: each is assessed separately although the two attributes strongly interact.

In this section, ecosystem structure is defined in terms of spatial structure (e.g., canopy structure and habitat), trophic interactions, and community composition in terms of biodiversity; while ecosystem function is defined in terms of carbon and nutrient cycling including dissolved organic carbon export, soil processes, controls on trace gas exchange processes, primary and secondary productivity, and water and energy balance.

Although ecosystem structure and function are closely interconnected, this section focuses on the two aspects separately for clarity, and limits the discussion here to plot (single square meter) scales: processes at the landscape and regional scales are covered in sections 7.5 and 7.6. Community responses to climate and UV radiation change presented in this section include effects on the

diversity of plant growth forms in terms of biomass contribution, but the details of impacts on biodiversity in terms of organism survival and population dynamics are included in section 7.3.

7.4.1. Ecosystem structure

7.4.1.1. Local and latitudinal variation

The Arctic is characterized by ecosystems that lack trees. There is a broad diversity in ecosystem structure among these northern treeless ecosystems that follows a latitudinal gradient from the treeline to the polar deserts. Typical communities for a particular latitude are called "zonal", but local variation at the landscape level occurs and these "intrazonal communities" are frequently associated with variations in soil moisture and snow accumulation (Chernov and Matveyeva, 1997; Walker M.D. et al., 1989).

According to Bliss and Matveyeva (1992), zonal communities south of the arctic boundary near the mean July isotherms of 10 to 12 °C consist of taiga (i.e., the northern edge of the boreal forest). This is characterized by a closed-canopy forest of northern coniferous trees with mires in poorly drained areas. To the north of this transition zone is the forest tundra. It is characterized by white spruce (*Picea glauca*) in Alaska, mountain birch (*Betula pubescens* ssp. *czerepanovii*) in Fennoscandia (Fig. 7.16), by birch and Norway spruce in the European Russian Arctic (Kola Peninsula and the Pechora lowlands), by Dahurian larch (*Larix dahurica*) in central and eastern Siberia, and by evergreen coniferous trees in Canada (Hustich, 1983). The vegetation of the forest tundra is characterized by sparse, low-growing trees with thickets of shrubs. North of this zone is the low Arctic, which is characterized by tundra vegetation in the strict sense (Fig. 7.17), consisting of communities of low, thicket-forming shrubs with sedges, tussock-forming sedges with dwarf shrubs, and mires in poorly drained areas. To the north of this zone is the high Arctic, which consists of polar semi-desert communities (Fig. 7.18) in the south, characterized by

Fig. 7.18. Polar semi-desert dominated by mountain avens (*Dryas octopetala*), Ny Ålesund, Svalbard (photo: T.V. Callaghan).

Fig. 7.19. Polar desert, Cornwallis Island, Northwest Territories, Canada (photo: J. Svoboda).

cryptogam–herb, cushion plant–cryptogam, and, to a limited extent, mire communities. To the extreme north is the polar desert where only about 5% of the ground surface is covered by herb–cryptogam communities (Fig. 7.19). In this zone, the mean July temperature is below 2 °C and precipitation, which falls mainly as snow, is about 50 mm per year.

The tundra zone can be further subdivided into three subzones: the southern tundra with shrub–sedge, tussock–dwarf shrub, and mire communities; the typical tundra with sedge–dwarf shrub and polygonal mire communities (Fig. 7.20); and the northern arctic tundra that consists of dwarf shrub–herb communities. The northern end of the latitudinal gradient, occurring primarily on islands and on the mainland only at Cape Chelyuskin (Taymir Peninsula), is occupied by polar deserts where woody plants are absent, and forbs and grasses with mosses and lichens are the main components of plant communities (Matveyeva and Chernov, 2000).

This vegetation classification has geographic connotations and cannot be applied easily to reconstructions of past vegetation throughout the circumpolar Arctic (Kaplan et al., 2003). A recent classification of tundra vegetation at the biome level (Walker D., 2000) has been proposed by Kaplan et al. (2003; Table 7.8, Fig. 7.2).

Within the biomes or zonal vegetation types, there are intrazonal habitats that are frequently associated with variations in soil moisture and snow accumulation, and that have a microclimate that deviates from the general macroclimate associated with flat surfaces. The intrazonal habitats form a mosaic of communities. Each of these tend to have fewer species than the "plakor," or zonal, communities. For example, poorly drained areas are often dominated by sedges with an understory of mosses and liverworts, but lack fruticose lichens (Matveyeva and Chernov, 2000). Although each intrazonal community has relatively few species, together they are more differentiated and diverse than zonal ones, and are responsible for about 80% of total species diversity in the regional flora and fauna. Disturbances, particularly freeze–thaw cycles and thermokarst (Fig. 7.21) that form patterned ground, also create landscape mosaics (Fig. 7.20). Diversity "focal points/hot spots" (Walker M.D., 1995) and "oases" (Edlund and Alt, 1989; Svoboda and Freedman, 1994) enrich landscapes by possessing a larger number of species, including those of more southerly distribution. Examples include dense willow thickets

Table 7.8. Circumpolar tundra biome classification (Walker D., 2000; Kaplan et al., 2003).

Biome	Definition	Typical taxa
Low- and high-shrub tundra	Continuous shrubland, 50 cm to 2 m tall, deciduous or evergreen, sometimes with tussock-forming graminoids and true mosses[a], bog mosses, and lichens	*Alnus, Betula, Salix, Pinus pumila* (in eastern Siberia), *Eriophorum, Sphagnum*
Erect dwarf-shrub tundra	Continuous shrubland 2 to 50 cm tall, deciduous or evergreen, with graminoids, true mosses[a], and lichens	*Betula, Cassiope, Empetrum, Salix, Vaccinium,* Poaceae, Cyperaceae
Prostrate dwarf-shrub tundra	Discontinuous "shrubland" of prostrate deciduous dwarf-shrubs 0 to 2 cm tall, true mosses[a], and lichens	*Salix, Dryas, Pedicularis,* Asteraceae, Caryophyllaceae, Poaceae, true mosses[a]
Cushion forb, lichen, and moss tundra	Discontinuous cover of rosette plants or cushion forbs with lichens and true mosses[a]	Saxifragaceae, Caryophyllaceae, *Papaver, Draba,* lichens, true mosses[a]
Graminoid and forb tundra	Predominantly herbaceous vegetation dominated by forbs, graminoids, true mosses[a], and lichens	*Artemisia, Kobresia,* Brassicaceae, Asteraceae, Caryophyllaceae, Poaceae, true mosses[a]

[a]"true" mosses exclude the genus *Sphagnum*

Fig. 7.20. Polygonal wet tundra near Prudhoe Bay, Alaska (photo: T.V. Callaghan).

Fig. 7.21. Thermokarst in the Russian tundra, New Siberian Islands (photo: T.V. Callaghan).

two meters in height in sheltered valleys at 75° N in Taymir and stands of balsam poplar north of the treeline in the northern foothills of the Brooks Range, Alaska, that are likely to respond rapidly to warming. There are numerous other types of plant communities, such as the moss-dominated tundra of Iceland (Fig. 7.22).

The vertical structure of arctic ecosystems is as important as horizontal structure in explaining their current and future functioning. This structure is most pronounced in low-arctic shrub communities, where there is a well-developed shrub canopy and an understory of mosses, similar to the vertical structure of boreal forests. Vertical structure is also pronounced below ground, with mosses and lichens lacking roots, some species rooted in the moss layer, others rooted just beneath the mosses, and a few species rooted more deeply.

The most striking latitudinal trend in plant functional types is the decrease in height of woody plants (from trees to tall shrubs, to low and prostrate shrubs, to dwarf shrubs, and eventually the loss of woody plants with increasing latitude). These functional types often occur in low abundance in zones north of their main areas of dominance, suggesting that they are likely to expand rapidly in response to warming through vegetative reproduction (Bret-Harte et al., 2001; Chapin F. et al., 1995) and sexual reproduction (Molau and Larsson, 2000), although range expansion will depend on geographic barriers such as mountains and seas (section 7.6, Table 7.14). Recent warming in Alaska has caused a substantial increase in shrub density and size in the moist tundra of northern Alaska (Sturm et al., 2001b). In areas where shrubs are absent, shrubs are likely to exhibit time lags in migrating to new habitats (Chapin F. and Starfield, 1997). Shrubs colonize most effectively in association with disturbances such as flooding in riparian zones, thermokarst, and frost boils (patterned ground formation caused by soil heave) throughout their latitudinal range, so migration may be

strongly influenced by climate- or human-induced changes in the disturbance regime. Woody species affect ecosystem structure and function because of their potential to dominate the canopy and reduce light availability to understory species (Bret-Harte et al., 2001; Chapin F. et al., 1996) and to reduce overall litter quality (Hobbie, 1996) and rates of nutrient cycling.

A similar latitudinal decline in abundance occurs with sedges, which are absent from polar deserts, suggesting that this group is also likely to expand northward with warming (Matveyeva and Chernov, 2000). *Carex stans* and *C. bigelowii* now mark the northernmost boundary of the tundra zone and might be a sensitive indicator of species responses to warming. Sedges have important effects on many ecosystem processes, including methane flux, because of their transport of oxygen to soils, transport of methane to the atmosphere, and inputs of labile carbon to the rhizosphere (Joabsson and Christensen, 2001; Torn and Chapin, 1993). Prostrate and dwarf shrubs such as *Dryas* spp., arctic willow, and polar willow are likely to decline in abundance with warming in the

Fig. 7.22. *Racomitrium/Empetrum* heath in Iceland showing erosion (photo: T.V. Callaghan).

low Arctic, due to competition with taller plants, but are likely to increase in abundance in the current polar deserts. These changes in distribution are very likely to substantially reduce the extent of polar desert ecosystems (section 7.5.3.2), which are characterized by the absence of woody plants.

7.4.1.2. Response to experimental manipulations

Experimental manipulation of environmental factors projected to change at high latitudes (temperature, snow, nutrients, solar radiation, atmospheric CO_2 concentrations, and UV-B radiation levels) has substantial effects on the structure of arctic ecosystems, but the effects are regionally variable. The effects of these variables on individual species are discussed in section 7.3; while this section focuses on overall community structure and species interactions.

Plant communities

Nutrient addition is the environmental manipulation that has the greatest effect on the productivity, canopy height, and community composition of arctic plant communities (Jonasson et al., 2001; Press et al., 1998a; van Wijk et al., 2004; Fig. 7.23). Fertilization also increases biomass turnover rate, so eventual biomass may or may not change in response to nutrient addition. In northern

Sweden, for example, nutrient addition to a mountain birch (*Betula pubescens* ssp. *czerepanovii*) site (cf. a Swedish treeline heath and fellfield) caused an initial biomass increase. This biomass increase was not maintained over the long term, however, because expansion of the grass *Calamagrostis lapponica* negatively affected the growth of mosses and evergreen shrubs, leading to a negligible change in community biomass (Parsons et al., 1994; Press et al., 1998b). Similarly, addition of nitrogen and phosphorus at a site in northern Alaska increased productivity and turnover within three years (Chapin F. et al., 1995). There was, however, little change in biomass because the rapidly growing sedges, forbs, and deciduous shrubs responded most strongly, whereas evergreen shrubs and mosses declined in abundance (Fig. 7.24). After 9 and 15 years, competitive interactions altered the relative abundance of plant functional types, with the tallest species (the deciduous shrub *Betula nana*; Fig. 7.3) responding most strongly (Bret-Harte et al., 2001; Chapin F. et al., 1995; Shaver et al., 2001). Litter and/or shade from this species reduced the growth of lichens, mosses, and evergreen shrubs. In vegetation types without any pronounced change in relative proportions of dominant species or life forms following fertilizer addition, as in Swedish treeline and high-altitude heaths and in Alaskan wet-sedge tundra, the biomass of most dominant life forms increased. This resulted in up to a doubling of biomass after five to nine years of treatment

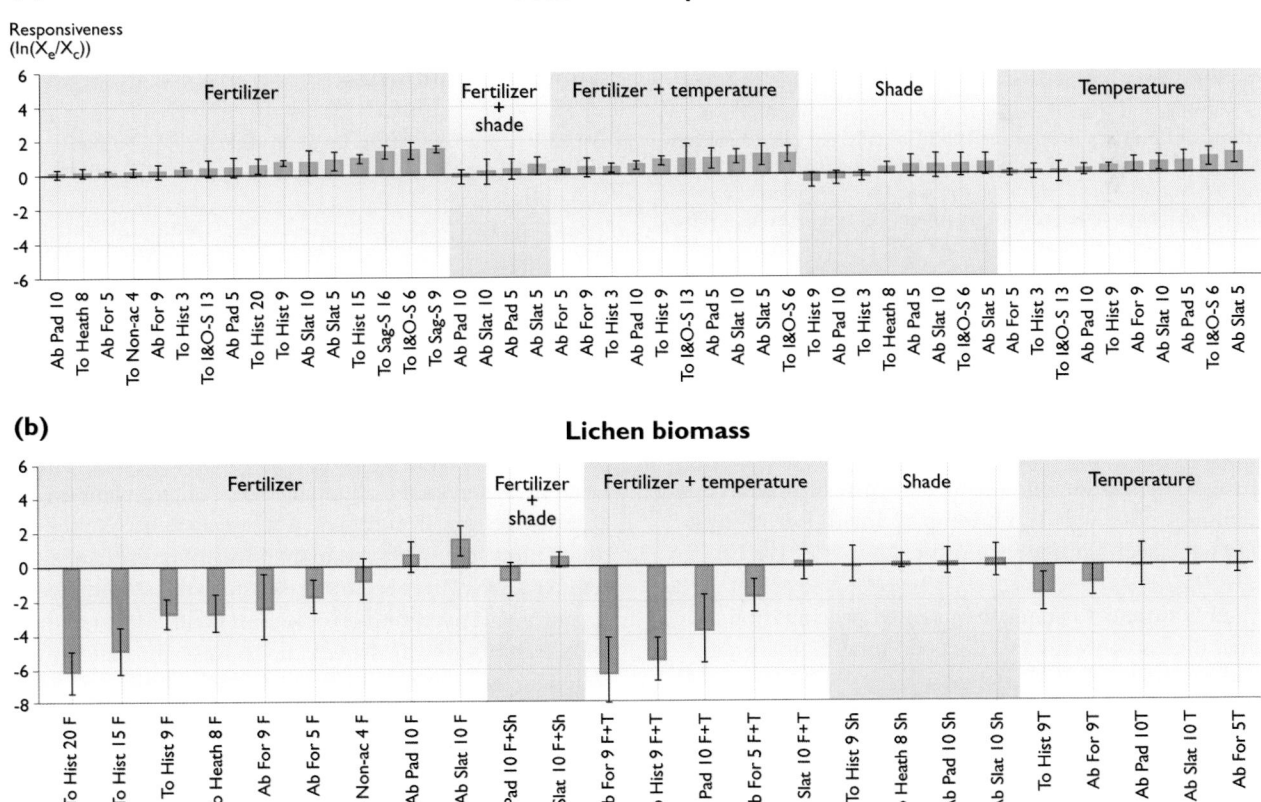

Fig. 7.23. Results of long-term (generally 10 years or more) experiments in a range of habitats at Toolik Lake, Alaska, and Abisko, Sweden, showing the responsiveness of aboveground biomass ordered by treatment and degree of responsiveness (X is the mean value of the analyzed characteristic for the experimental (X_e) and control (X_c) groups). Data are given for (a) total vascular plant biomass and (b) lichen biomass. Codes relate to the geographical region (To=Toolik, Ab=Abisko), the site name, and the duration of the experiment (van Wijk et al., 2004).

(a) **1983**

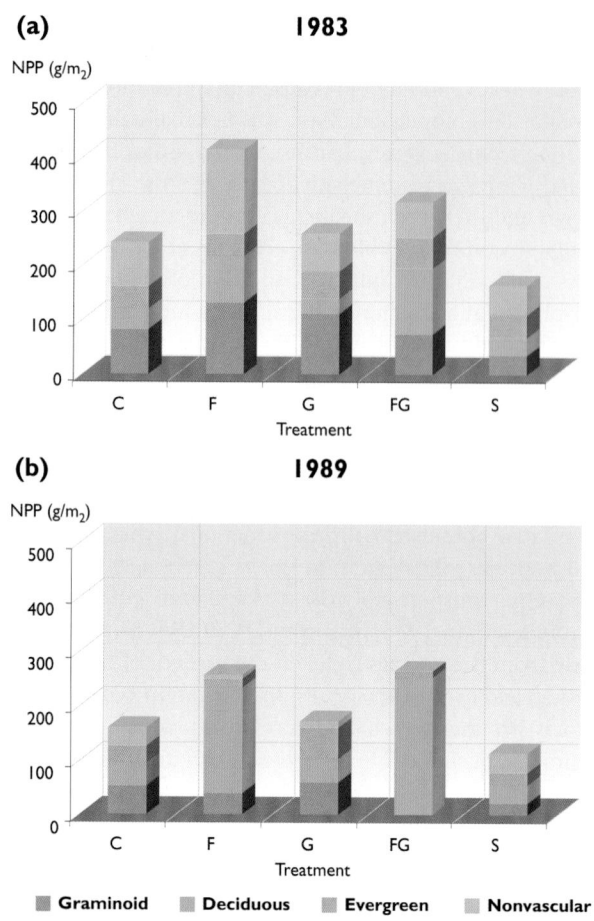

(b) **1989**

Graminoid   Deciduous   Evergreen   Nonvascular

C=control (unmanipulated) plots; F=annual N+P fertilizer addition;
G=warming in a plastic greenhouse during the growing season;
FG=fertilizer plus greenhouse treatment; S=50% light reduction
(by shading) during the growing season.

Fig. 7.24. Effects of long-term fertilizer addition and experimental warming and shading during the growing season on aboveground net primary production (NPP) of different plant functional types at Toolik Lake, Alaska, showing NPP by functional type and treatment (a) in 1983, after three years of treatment, and (b) in 1989, after nine years of treatment (Chapin F. et al., 1995).

(Jonasson et al., 1999b; Shaver et al., 1998). In polar semi-deserts, nutrient addition generally had a negative effect on vascular plants, due to enhanced winterkill, but stimulated the growth of mosses (Robinson et al., 1998), an effect opposite to that in low-arctic tundra. This difference is probably due to the immigration of nitrogen-demanding mosses from nearby bird-cliff communities in the high Arctic compared with loss of existing moss species in the low Arctic.

Water additions to simulate increased precipitation have generally had only minor effects on total biomass and production (Press et al., 1998a).

Experimental summer warming of tundra vegetation within the range of projected temperature increases (2 to 4 °C over the next 100 years) has generally led to smaller changes than fertilizer addition (Arft et al., 1999; Jonasson et al., 2001; Shaver and Jonasson, 2000; van Wijk et al., 2004; Fig. 7.23). For example, temperature enhancement in the high-arctic semi-desert increased plant cover within growing seasons but the

effect did not persist from year to year (Arft et al., 1999; Robinson et al., 1998). In the low Arctic, community biomass and nutrient mass changed little in response to warming of two Alaskan tussock sites (Chapin F. et al., 1995; Hobbie and Chapin, 1998) and two wet-sedge tundra sites (Shaver et al., 1998), coincident with relatively low changes in soil nutrient pools and net mineralization. Tussock tundra showed little response to warming, as some species increased in abundance and others decreased (Chapin F. and Shaver, 1985a,b; Chapin F. et al., 1995), similar to a pattern observed in subarctic Swedish forest floor vegetation (Press et al., 1998b). The responses to warming were much greater in Swedish treeline heath and in fellfield (Jonasson et al., 1999b). Biomass in the low-altitude heath increased by about 60% after air temperatures were increased by about 2.5 °C, but there was little additional effect when temperatures were further increased by about 2 °C. In contrast, biomass approximately doubled after the first temperature increase (2.5 °C) and tripled after the higher temperature increase (an additional 2 °C) in the colder fellfield. Hence, the growth response increased from the climatically relatively mild forest understory through the treeline heath to the cold, high-altitude fellfield where the response to warming was of the same magnitude as the response to fertilizer addition (Jonasson et al., 2001). A general long-term (10 years or more) response to environmental manipulations at sites in subarctic Sweden and in Alaska was a decrease in total nonvascular plant biomass and particularly the biomass of lichens (van Wijk et al., 2004: Fig. 7.23).

Animal communities

Air-warming experiments at Svalbard (79° N) had greater effects on the fauna above ground than below ground, probably because the soil is more buffered against fluctuations in temperature and moisture than the surface (Hodkinson et al., 1998). Species with rapid life cycles (aphids and Collembola) responded demographically more quickly than species (e.g., mites) with slow life cycles (Coulson et al., 1996). Responses to warming differed among sites. The abundance of Collembola declined at barren sites where higher temperatures also caused drought and mortality due to desiccation, whereas the abundance of Collembola increased at moister sites. In summer, water availability is probably much more important to many invertebrates than is temperature. Mites are more resistant than Collembola to summer desiccation (Hodkinson et al., 1998) and to anoxic conditions in winter due to ice-crust formation following episodes of mild weather (Coulson et al., 2000). Ice-crust formation during the winter may increase winter mortality by 50% in Collembola (Coulson et al., 2000). Freeze–thaw events in spring may also cause differential mortality among species, thus altering community composition (Coulson et al., 1995). In experiments conducted simultaneously at several sites and over several years, the natural spatial and temporal variability in community structure and population density of soil invertebrates was larger than the effects of the experimental manipulation within years and sites. This demonstrates that there is a large variability in the structure and function of high-arctic invertebrate communities due to current variation in abiotic conditions. It also indicates that arctic invertebrate communities can respond rapidly to change.

Compared to the high Arctic, subarctic invertebrate communities at Abisko responded less to experimental temperature increases (Hodkinson et al., 1998). However, nematode population density increased substantially, and the dominance changed in favor of plant- and fungal-feeding species with elevated summer temperatures and nitrogen (N), phosphorus (P), and potassium (K) fertilization, indicating a shift in the decomposition pathway (Ruess et al., 1999a,b).

Microbial communities

Although the biomass of microorganisms is a poor predictor of the productivity and turnover of microorganisms and their carbon, alternative methods focusing on population dynamics of microbial species within communities are extremely difficult to employ in the field. Microbial biomass has therefore been used to quantify microbial processes within ecosystems. The sensitivity of microbial biomass, generally measured as biomass carbon (C), and nutrient content to changed environmental conditions in the Arctic has not been well studied. Long-term addition of easily processed C generally increases the microbial biomass; and addition of inorganic nutrients generally, but not always, increases microbial nutrient content without appreciable effect on the biomass (Jonasson et al., 1996, 1999b; Michelsen et al., 1999; Schmidt et al., 2000). In some cases, however, a combination of C and nutrient addition has led to a pronounced increase in both microbial biomass and nutrient content (Schmidt et al., 2000). This suggests a general C limitation of microbial biomass production, and increased sink strength for soil nutrients (i.e., increased sequestration of nutrients) if the amounts of both labile C and nutrients increase, but relatively weak effects from increased nutrient availability alone. In the widespread drier ecosystem types in the Arctic, the soil microbial biomass is likely to be further limited by low water supply. Water addition to a high-arctic semi-desert led to a substantial increase in microbial biomass C and microbial activity (Illeris et al., 2003).

Data on the effects on ecosystems of growing-season temperature increases of 2 to 4 °C over five (Jonasson et al., 1999b; Ruess et al., 1999a,b) and ten (Jonasson and Michelsen, unpub. data, 2005) years have not shown appreciable long-term changes in microbial biomass and nutrient stocks. This suggests that an increase in growing-season temperature alone is unlikely to have any strong impact on microbial C and nutrient sequestration, and that changes in soil nutrient availability are likely to lead to greater changes than the direct effects of increased temperature. Temperature effects on ecosystem processes are likely, however, to be different from the observed relatively small effects on microbial biomass and nutrient stocks, because temperature changes are likely to affect rates of decomposition and nutrient mineralization, rather than pool sizes, resulting in altered C balance and rates of nutrient supply to the plants (section 7.4.2.1).

Appreciable seasonality in microbial biomass and nutrient mass have been reported, however, that seemingly are independent of ambient temperature. In general, the masses change little, or fluctuate, during summer (Giblin et al., 1991; Jonasson et al., 1999b; Schmidt et al., 1999, 2002). In contrast, pronounced increases in both biomass and nutrient mass have been reported in autumn (Bardgett et al., 2002; Jaeger et al., 1999; Lipson et al., 1999), probably as a function of increased input of labile C and nutrients from plants as they senesce, although these data are from mountain and alpine, rather than from arctic soils. The increase seems to continue through winter, although at a slower rate (Grogan and Jonasson, 2003; Schmidt et al., 1999), despite soil temperatures below 0 °C (Clein and Schimel, 1995; see section 7.3.2.3). It is followed by a sharp biomass decline in the transition between winter and spring (Bardgett et al., 2002; Brooks et al., 1998; Grogan and Jonasson, 2003), which may (Brooks et al., 1998) or may not (Grogan and Jonasson, 2003; Larsen et al., 2002) coincide with a decrease in microbial N and an increase in mineralized N, indicating a pronounced transformation of microbial N to soil inorganic N (Giblin et al., 1991; Schmidt et al., 1999). Indeed, this seasonal pattern suggests a temporal partitioning of resource uptake with low competition between plants and microbes for nutrients, as microbes absorb most nutrients in autumn and plants in spring, coincident with the nutrient release from declining microbial populations. However, it may also be an indication that plants compete well for nutrients during the growing season (Schimel and Chapin, 1996), and microbes access nutrients efficiently only when the sink strength for nutrients in plants is low (Jonasson et al., 1999b).

Laboratory experiments have shown that the spring decline of microbial mass is likely to be an effect of repeated freeze–thaw cycles (Schimel and Clein, 1996). Indeed, Larsen et al. (2002) reported a microbial decline in soils subjected to repeated freezing and thawing but not in the same soils kept constantly frozen before thawing. The seasonal dynamics of microbial biomass and microbial and inorganic soil nutrients therefore suggests that "off growing-season" changes in climate during the transition between winter and spring (e.g., changed frequency of freeze–thaw events and warmer winters) are likely to have greater impacts on nutrient transformations between microbes, soils, and plants than changes during the growing season.

Manipulations simulating enhanced UV-B radiation levels (equivalent to a 15% reduction in stratospheric ozone levels) and a doubling of atmospheric CO_2 concentrations for seven years altered the use of labile C sub-

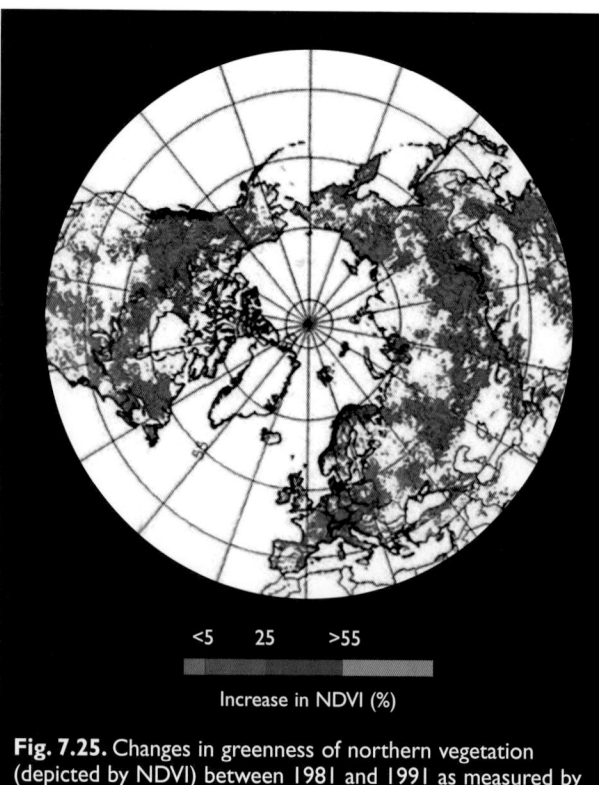

Fig. 7.25. Changes in greenness of northern vegetation (depicted by NDVI) between 1981 and 1991 as measured by satellite instruments (Myneni et al., 1997).

strates by gram-negative bacteria (Johnson et al., 2002). Although these rhizosphere bacteria are a relatively small component of the belowground microbial biomass, they are likely to be particularly responsive to environmentally induced changes in belowground plant C flow.

Ultraviolet-B radiation also affects the structure of fungal communities. Microcosms of subarctic birch forest floor litter exposed to enhanced UV-B radiation levels showed a reduction in fungal colonization of leaf veins and lamina (Gehrke et al., 1995). Fungal composition was also altered in the UV-B radiation treatments, with a reduction in *Mucor hiemalis* and a loss of *Truncatella truncata*. Similar findings of fungal community change were obtained in subarctic Abisko, in an ecosystem that was the source of the litter used by Gehrke et al. (1995). In this field study of the decomposition rates of a standard litter type, there was also a change in the composition of the fungal community associated with litter resulting from elevated UV-B radiation levels (Moody et al., 2001). So far, no change in plant community structure has been found in the Arctic in response to artificially enhanced or reduced UV-B radiation levels and CO_2 concentrations.

7.4.1.3. Recent decadal changes within permanent plots

Satellite measurements suggest a widespread increase in indices of vegetation greenness (e.g., the normalized difference vegetation index – NDVI) and biomass at high latitudes (Myneni et al., 1997, 2001; Fig. 7.25), although changes in satellites and sensor degradation may have contributed to this trend (Fung, 1997). Aerial photo-

graphs show a general increase in shrubbiness in arctic Alaska (Sturm et al., 2001b) and indigenous knowledge also reports an increase in shrubbiness in some areas. These observations are consistent with the satellite observations. However, it has been difficult to corroborate these with studies of permanent plots, because of the paucity of long-term vegetation studies in the Arctic. In arctic Alaska, for example, a trend toward reduced abundance of graminoids and deciduous shrubs during the 1980s was reversed in the 1990s (Shaver et al., 2001). In Scandinavia, decadal changes in vegetation were affected more strongly by the cyclic abundance of lemmings than by climatic trends (Laine and Henttonen, 1983).

7.4.1.4. Trophic interactions

Trophic-level structure is simpler in the Arctic than further south. In all taxonomic groups, the Arctic has an unusually high proportion of carnivorous species and a low proportion of herbivores (Chernov, 1995). As herbivores are strongly dependent on responses of vegetation to climate variability, warming is very likely to substantially alter the trophic structure and dynamics of arctic ecosystems. The herbivore-based trophic system in most tundra habitats is dominated by one or two lemming species (Batzli et al., 1980; Oksanen et al., 1997; Wiklund et al., 1999) while the abundance of phytophagous (plant-eating) insects relative to plant biomass is low in arctic tundra (Strathdee and Bale, 1998). Large predators such as wolves, wolverines, and bears are less numerous in the tundra than in the boreal forest (Chernov and Matveyeva, 1997) and predation impacts on tundra ungulates are usually low. Thus, the dynamics and assemblages of vertebrate predators in arctic tundra are almost entirely based on lemmings and other small rodent species (*Microtus* spp. and *Clethrionomys* spp.; Batzli, 1975; Wiklund et al., 1999), while lemmings and small rodents consume more plant biomass than other herbivores. Climate has direct and indirect impacts on the interactions among trophic levels, but there is greater uncertainty about the responses to climate change of animals at the higher trophic levels.

Plant-herbivore interactions

Plant tissue chemistry and herbivory

Arctic and boreal plant species often contain significant concentrations of secondary metabolites that are important to the regulation of herbivory and herbivore abundance (Haukioja, 1980; Jefferies et al., 1994). Secondary compounds also retard decomposition of leaves after litter fall (Cornelissen et al., 1999). These secondary metabolites are highly variable in their chemical composition and in their antiherbivore effects, both within and among species. One hypothesis about the regulation of these compounds that has received widespread discussion is the carbon–nutrient balance hypothesis of Bryant et al. (1983; Coley et al., 1985), which attempts to explain this variation in part on the basis of C versus nutrient limitation of plant growth. Although many other factors in addition to carbon–nutrient balance are probably important to the regulation of plant–herbivore interactions in the Arctic (e.g., Iason and Hester, 1993; Jefferies et al., 1994; Jonasson et al., 1986), the abundance of secondary chemicals is often strongly responsive to changes in the environment including temperature, light, and nutrient availability (e.g., Graglia et al., 2001b; Haukioja et al., 1998; Laine and Henttonen, 1987). In a widespread arctic shrub species, *Betula nana* (Fig. 7.3), Graglia et al. (2001b) found that fertilization and shading generally led to decreased condensed and hydrolyzable tannin concentrations in leaves, whereas warming in small field greenhouses increased condensed tannins and decreased hydrolyzable tannins. There was also a large difference in both the average concentrations and the responsiveness of the concentrations of phenolics in plants from northern Alaska versus northern Sweden, with the plants from Sweden having generally higher concentrations but being less responsive to environmental changes. Such data suggest that the effects of climate change on plant–herbivore interactions are likely to be highly variable, species-specific, and also dependent on the nature of the change and on ecotypic or subspecific differences, perhaps related to local evolution in the presence or absence of herbivores.

Plant exposure to UV-B radiation has the ability to change the chemistry of leaf tissues, which has the potential to affect the odor that herbivores such as reindeer/caribou use to detect food, and the quality of food in terms of palatability and digestibility (Gwynn-Jones, 1999). In general, enhanced UV-B radiation levels can reduce soluble carbohydrates and increase phenolic compounds and flavonoids. Such changes are expected to reduce forage quality.

Plant exposure to increased CO_2 concentrations can also affect plant tissue quality and consequently herbivory (Agrell et al., 1999). Enriched CO_2 concentrations may lead to the accumulation of carbohydrates and phenolic compounds while reducing N concentrations in leaves. However, these phytochemical responses can be significantly modified by the availability of other resources such as nutrients, water, and light. Unfortunately, little information about the impacts of increased CO_2 concentrations on herbivory is available for the Arctic.

Herbivore abundance and vegetation production

Invertebrates

Insect population outbreaks seldom extend into the tundra. However, in the forest near the treeline, insect defoliators can have devastating impacts on the ecosystem. Climate change is very likely to modify the population dynamics of such insects in several ways (Bylund, 1999; Neuvonen et al., 1999). In the autumnal moth, eggs laid on birch twigs in autumn cannot tolerate winter temperatures lower than -36 °C. For this reason, the moth is destroyed in parts of the terrain (e.g., depressions) where winter temperatures drop below this criti-

cal minimum (Tenow and Nilssen, 1990; Virtanen and Neuvonen, 1999). Warmer winters are very likely to reduce winter mortality and possibly increase outbreak intensity. Moreover, lower minimum temperatures are likely to allow the autumnal moth and the related, less cold-tolerant winter moth (*Operophtera brumata*) to extend their geographic distributions into continental areas with cold winters (Tenow, 1996). However, predicting the effect of a changing climate is not straightforward because moth responses are season-specific. For instance, increasing spring temperatures are likely to cause a mismatch between the phenology of birch leaves and hatching of larvae that are currently synchronized (Bale et al., 2002). Moreover, natural enemies such as parasitoid wasps and ants are likely to increase their abundances and activity rates if summer temperature rises. Currently, there is cyclicity in the populations of the autumnal moth and outbreak proportions occur approximately every 10 to 11 years (Tenow, 1972, 1996). The defoliated forests require about 70 years to attain their former leaf area, although insect outbreaks in subarctic Finland followed by heavy reindeer browsing of regenerating birch shoots have led to more or less permanent tundra (Kallio and Lehtonen, 1973; Lehtonen and Heikkinen, 1995). There are no population outbreaks in the autumnal moth further south in Fennoscandia, most likely due to the high abundance of generalist parasitoids that keep moth populations below outbreak levels (Tanhuanpää et al., 2001). However, the border between outbreaking and non-outbreaking populations of geometrid moths is likely to move northward if climate changes.

Enhanced UV-B radiation levels applied to birch leaves alters the chemistry or structure of the leaves such that caterpillars eat three times as much leaf biomass to maintain body development (Buck and Callaghan, 1999; Lavola et al., 1997, 1998). There is also a tendency for enhanced UV-B radiation levels to increase the immunocompetence of the caterpillars, which could possibly make them more tolerant to the wasp parasitoid (Buck, 1999). Although the effects of winter warming on eggs, increased UV-B radiation levels on leaves, and immunocompetence on caterpillars are likely to increase future damage to subarctic birch forests, it is not known to what extent other processes susceptible to spring and summer climate variability may alleviate these effects.

Vertebrates

The herbivore-based trophic system in most tundra habitats is dominated by one or two lemming species (Batzli et al., 1980; Oksanen et al., 1997; Wiklund et al., 1999). Lemming abundance is the highest in coastal tundra, especially in moist sedge meadows that are the optimum habitat for *Lemmus*. Collared lemmings (*Dicrostonyx*) usually do not reach as high densities in their preferred habitats on drier ridges where herbs and dwarf shrubs dominate. Voles (*Microtus* and *Clethrionomys* spp.) are likely to become more abundant than lemmings in some low-arctic tundra habitats and forest tundra (Chernov and Matveyeva, 1997). At the landscape scale, lemmings and voles are very patchily distributed according to the abundance of their preferred food plants, as well as the distribution of snow (Batzli, 1975; section 7.3.2.2). Lemming peak densities exceed 200 individuals per hectare in the most productive *Lemmus* habitats in both Siberia and North America (Batzli, 1981) and the standing crop of lemmings may approach 2.6 kg dry weight per hectare. The population builds up during the winter (due to winter breeding) and peak densities may be reached in late winter/early spring when the standing crop of food plants is minimal. The diet of *Lemmus* consists mainly of mosses and graminoids, while *Dicrostonyx* prefers herbs and dwarf shrubs (Batzli, 1993). Lemmings have a high metabolic rate, and *Lemmus* in particular has a low digestive efficiency (about 30%, compared to 50% in other small rodents). Consequently, their consumption rate and impact on the vegetation exceeds that of all other herbivores combined (with the exception of local effects of geese near breeding colonies). Moreover, lemmings destroy much more vegetation than they ingest and after population peaks typically 50% of the aboveground biomass has been removed by the time the snow melts (Turchin and Batzli, 2001). In unproductive snow beds, which are favored winter habitats of the Norway lemming (Kalela, 1961), up to 90–100% of the mosses and graminoids present during the winter may be removed (Koskina, 1961). If winters become so unfavorable for lemmings that they are unable to build up cyclic peak densities, the species-rich predator community relying on lemmings is likely to collapse (see next subsection). Moreover, their important, pulsed impact on vegetation as a result of grazing and nutrient recycling is likely to cease. Changes in snow conditions, relative abundances of preferred food plants, and climate impacts on primary production are all very likely to affect lemming populations, and are likely to result in a northward displacement of the climatically determined geographic borders between cyclic and non-cyclic populations of small herbivores (small rodents and moths), as well as the species distributions *per se*.

Wild populations of other herbivorous mammal species in the tundra, such as hares, squirrels, muskox, and reindeer/caribou, never reach population densities or biomass levels that can compare with peak lemming populations (Chernov and Matveyeva, 1997). Moving herds of reindeer/caribou represent only patchy and temporary excursions in numbers, biomass, and impacts on vegetation; averaged over space and time some of the largest herds approach only 0.01 individuals and 0.5 kg of dry weight per hectare (Batzli, 1981) on their summer pastures and usually take less than 10% of the vegetation (Jefferies et al., 1994). The only cases where reindeer/caribou have been shown to have large impacts on vegetation seem to be in unusual circumstances (stranding on islands; Klein, 1968) or under human intervention (e.g., removing top predators or introductions to islands) where overshooting reindeer/caribou populations have led to vegetation destruction, habitat degradation, and subsequent population crashes.

Although the cooling since the mid-1970s in the Hudson Bay region has affected the reproduction of snow geese (*Anser caerulescens*), the mid-continental population is currently growing by 5% per year (Skinner W. et al., 1998). This, in combination with the staging of snow geese in La Pérouse Bay, Manitoba, because of bad weather further north, leads to increasing foraging for roots and rhizomes of the graminoids *Puccinellia phryganodes* and *Carex subspathacea* (Jefferies et al., 1995). The rate of removal of belowground organs in the salt marshes combined with intense grazing of sward during summer exceeds the rate of recovery of the vegetation. It is estimated that geese have destroyed 50% of the salt marsh graminoid sward of La Pérouse Bay since 1985. This loss of vegetation cover exposes the sediments of the salt marshes, which have become hypersaline (salinities exceeding 3.2) as a result of increased evapotranspiration. This further reduces plant growth and forage availability to the geese. In turn, this is reducing goose size, survivorship, and fecundity. Other factors that are affected by the trophic cascades initiated by the geese include reduced N mineralization rates and declines in the populations of soil invertebrates, waders, and some species of duck such as the widgeon (*Anas americana*).

Cyclic populations

Herbivore–plant interactions have been proposed to produce population cycles in arctic herbivores through several mechanisms including nutrient recycling (Schultz, 1969), production cycles inherent in food plants (Tast and Kalela, 1971), induced chemical defense in plants (Haukioja, 1991), and recurrent overgrazing (Oksanen et al., 1981). The empirical evidence is mixed. There is at least partly supporting evidence for induced chemical defense in the *Epirrita*–birch system (Ruohomäki et al., 2000) and for overgrazing in the *Lemmus*–plant system in unproductive tundra habitats (Turchin et al., 2000). There is little evidence, however, for mechanisms involving nutrient cycling and chemical defense in the case of lemmings and voles (Andersson and Jonasson, 1986; Jonasson et al., 1986). Climate is somehow involved in all the hypotheses of population cycles related to plant–herbivore interactions. For example, allocation strategies in plants and the amount of secondary compounds (induced chemical defense hypothesis) depend on temperature and growing-season length (see plant tissue chemistry subsection). Plant production and biomass are also controlled by temperature (overgrazing hypothesis). Climate change is thus likely to modify the population dynamics patterns and roles of key herbivores such as lemmings and moths because the dynamics of herbivore-plant interactions are likely to change. As early as 1924, Charles Elton pointed out the potentially decisive role of climate in determining the generation of cycles in northern animal populations (Elton, 1924).

Mathematical modeling shows that specialist resident predators such as small mustelids and the Arctic fox can also impose prey population cycles due to sufficiently strong numerical and adequate functional responses (Gilg et al., 2003; Turchin and Hanski, 1997). Moreover, nomadic specialists such as birds of prey can dampen lemming cycles and decrease the degree of regional asynchrony if their predation rates are sufficiently high (Ims and Andreassen, 2000; Ims and Steen, 1990). The impacts of bird predators have a strong seasonal component since most migrate south for the winter (Ims and Steen, 1990). Reliable estimates of predation rates on cyclic lemming populations are rare. Indirect estimates based on the energy requirements of predators at Point Barrow, Alaska, indicated that avian predators could account for 88% of the early summer mortality, but it was concluded that neither this nor winter predation by weasels could stop lemming population growth under otherwise favorable winter conditions (Batzli, 1981). In the Karup Valley, Greenland, the combined impact of different predators both limited population growth and caused population crashes in collared lemmings (Gilg et al., 2003). In a declining lemming population in an alpine area in Norway, almost 50% predation was demonstrated by following the fates of radio-tagged individuals (Heske et al., 1993). Using the same methodology, Reid et al. (1995), Wilson D. et al. (1999), and Gilg (2002) showed that predation was the predominant mortality factor in populations of collared lemmings at various locations in northern Canada and eastern Greenland.

Predator–prey interactions

The dynamics and assemblages of vertebrate predators in arctic tundra are almost entirely based on lemmings and other small rodent species (*Microtus* spp. and *Clethrionomys* spp.; Batzli, 1975; Wiklund et al., 1999). Birds of prey such as snowy owls, short-eared owls (*Asio flammeus*), jaegers (skuas – *Stercorarius* spp.), and rough-legged buzzards (*Buteo lagopus*) are lemming and vole specialists that are only able to breed at peak lemming densities and which aggregate in areas with high lemming densities. Since lemming cycles are not synchronized over large distances (Erlinge et al., 1999; Predavec et al., 2001), the highly mobile avian predators can track lemming population peaks in space. Mammal lemming and vole specialists in the Arctic, such as the least weasel (*Mustela nivalis*) and the ermine, are less mobile than birds but both have high pregnancy rates and produce large litters in lemming peak years (MacLean et al., 1974). In lemming low years, weasel and ermine reproduction frequently fails and mortality rates increase (Gilg et al., 2003; Hanski et al., 2001). In coastal and inland tundra habitats where bird colonies are lacking, the Arctic fox also exhibits the population dynamics typical of a lemming specialist (Angerbjörn et al., 1999). The lemming cycles also impose cyclic dynamics in other animals such as geese and waders because they serve as alternative prey for predators in lemming crash years (Bety et al., 2002; Sutherland, 1988). Recently observed increased predation pressure on water birds in various arctic regions might reflect a change of the lemming cycle in response to climate change, with secondary effects on predators and water birds as an alternative prey (Soloviev et al., 1998; Summers and Underhill, 1987). Thus, a large part of the tundra vertebrate community cycle is in a rhythm dictated by the lemming populations (Chernov and Matveyeva, 1997; Stenseth and Ims, 1993).

This rhythm is likely to be disrupted by projected future variations in snow properties (e.g., snow-season length, snow density, and snow-cover thickness; Yoccoz and Ims, 1999). For small mammals living in the subnivean space, snow provides insulation from low temperatures as well as protection from most predators such as foxes and raptors (Hansson and Henttonen, 1988) and increases in snow are likely to be beneficial. The effect on large mammal prey species (ungulates) is likely to be the opposite, as deeper snow makes reindeer/caribou and moose (*Alces alces*) more vulnerable to predators such as wolves (Post et al., 1999), but more extensive snow patches provide relief from insect pests (section 7.3.3.2). If climate change results in more frequent freeze-thaw events leading to a more shallow and icy snowpack (section 6.4.4), this is likely to expose small mammals to predators, disrupt population increases, and thereby prevent cyclic peak abundances of lemmings and voles. For nomadic predators whose life-history tactic is based on asynchronous lemming populations at the continental scale, an increased frequency of large-scale climatic anomalies that induces continent-wide synchrony (the "Moran effect"; Moran, 1953) is very likely to have devastating effects.

Long-term monitoring (>50 years) of small rodents near the treeline at Kilpisjärvi in subarctic Finland has shown a pronounced shift in small rodent community structure and dynamics since the early 1990s (Henttonen and Wallgren, 2001; see Fig. 7.13). In particular, the previously numerically dominant and cyclically fluctuating grey-sided vole has become both less abundant and less variable in abundance. The Norway lemming and *Microtus* voles also have lower peak abundances, and the small rodent community is currently dominated by the relatively more stable red-backed vole (*Clethrionomys rutilus*). Similar changes took place in the mid-1980s in the northern taiga (Hanski and Henttonen, 1996; Henttonen, 2000; Henttonen et al., 1987) and are still prevailing. For predators that specialize in feeding on small rodents, the lack of cyclic peak abundances of small rodents, especially in the spring (Oksanen et al., 1997), is likely to have detrimental consequences, as they need to breed successfully at least every three to four years to sustain viable populations. At Kilpisjärvi, the least weasel has become rare. Moreover, the severe decline of the Arctic fox and the snowy owl in Fennoscandia, both of which prey on *Microtus* voles and lemmings in mountain and tundra habitats, may be due to lower peak abundances of small mammal prey species in their habitats (Angerbjörn et al., 2001). In Alaska, a similar decrease in lemming cyclicity occurred in the 1970s (Batzli et al., 1980).

Large predators such as wolves, wolverines, and bears are less numerous in the tundra than in the boreal forest (Chernov and Matveyeva, 1997). Consequently, predation impacts on tundra ungulates are usually low. While 79% of the production in small herbivores (voles, lemmings, ptarmigan, and Arctic hares) was consumed by predators averaged across a number of sites in arctic Canada, the corresponding number was only 9% for large herbivores (caribou and muskox; Krebs et al., 2003).

Insect pests, parasites, and pathogens

Plants

Disease in plants is likely to increase in those parts of species distribution ranges where a mismatch between the rate of relocation of the species and the northward/upward shift of climatic zones results in populations remaining in supra-optimal temperature conditions. Under these conditions, species can experience thermal injury (particularly plants of wet and shady habitats; Gauslaa, 1984), drought, and other stresses that make plants more susceptible to disease.

Very little is known about the incidence and impacts of plant diseases in arctic ecosystems. However, recent work has shown that a fungal pathogen (*Exobasidium*) of *Cassiope tetragona* and *Andromeda polifolia* reduces host plant growth, reproductive investment, and survival (Skinner L., 2002). As the incidence of disease increases with an increase in temperature downward along an altitudinal gradient, climate warming is likely to increase the incidence of at least this naturally occurring disease in the Arctic. The incidence of new diseases from increasing mobility of pathogens with a southern distribution is a possibility.

Animals

Ultraviolet-B radiation can reduce the impact of viral and fungal pathogens on insects. The nuclear polyhedrosis virus is a major cause of death of the defoliating autumnal moth. However, this virus is killed by UV-B radiation (Killick and Warden, 1991). Species and strains of the fungus *Metarhizium* are important agents of insect disease, but some, particularly high-latitude strains, are sensitive to UV-B radiation (Braga et al., 2001a,b).

Parasitism is perhaps the most successful form of life, but until recently has been underestimated, especially in the Arctic (Henttonen and Burek, 2001; Hoberg et al., 2003). Parasitism in the Arctic has been poorly studied with respect to both taxonomy and biodiversity as well as the ecological impact parasites may have on animal species and communities.

Recent research on the evolution and phylogeography of typical arctic animals like lemmings has revealed how greatly the alternating glacial and interglacial periods have influenced their distribution and genetic diversity (Fedorov et al., 1999a,b). The impact seems to be at least as profound on the helminth parasites of arctic rodents (Haukisalmi et al., 2004; Hoberg et al., 2003). Such impacts of past climatic fluctuations can be used to project some possible consequences of the present warming. If the arctic host populations become fragmented due to the northward expansion of southern biogeographic elements, extinction of parasites in small host populations and/or cryptic speciation (isolation events seen in parasites, often only by using molecular methods, that are not evident in host populations) in refugia are likely to follow. Phylogeographic structure (often cryptic speciation) can be seen in rodent cestodes in the Arctic even if there is no such structure in the host. This is true also for ruminant parasites.

Phylogenetic studies have shown that host switches have occurred in many clades of rodent cestodes (Wickström et al., in press). It seems plausible that host switches have been promoted by climatic events that force host assemblages, earlier separated by geography or habitat, to overlap in their distribution.

Macroparasites, such as intestinal worms, often have complicated life cycles. In the main host, in which the parasite reproduces, parasites are controlled by host immunity. On the other hand, the free-living intermediate stages (eggs and larvae), and those in intermediate hosts, are subject to extrinsic environmental conditions like temperature and humidity. Temperature strongly affects the development rate of parasite larvae. For example, a small increase in temperature has a clear effect on the development of the muskox lungworm *Umingmakstrongylus pallikuukensis* in its gastropod intermediate hosts (Kutz et al., 2002). Therefore, a slight increase in temperature and in growing-season length is very likely to profoundly affect the abundance and geographic distribution of potentially harmful parasites such as lungworms. Lungworm infections have become conspicuous in recent years as summer temperatures in the Arctic have increased.

The free-living stages of parasites are prone to desiccation. In addition to temperature effects on their development, the survival and abundance of free-living intermediate stages depend greatly on humidity. In addition, the same factors affect drastically the abundance, survival, and distribution of the intermediate hosts of parasites, like insects, gastropods, and soil mites. Haukisalmi and Henttonen (1990) found that precipitation in early summer was the most important factor affecting the prevalence of common nematodes and cestodes in *Clethrionomys* voles in Finnish Lapland. Temperature and humidity also affect the primary production and development of the free-living stages of abomasal nematodes of reindeer/caribou (Irvine et al., 2000). Recently, Albon et al. (2002) showed that abomasal nematodes affect the dynamics of Svalbard reindeer through fecundity. Consequently, even slight climatic changes are likely to have surprising effects on the large ungulates, and possibly on humans exploiting them, through enhanced parasite development (Chapter 15).

The complicated life cycles of parasites cause intrinsic lags in their capacity to track the changes in the population density of their hosts, and these lags are further retarded by unfavorable arctic conditions. Any climatic factor promoting the development of a parasite, so that it can respond in a density-dependent way to host dynamics, is likely to alter the interaction between parasite and host, and the dynamics of both.

There is considerable uncertainty about the possibilities for invasion of pathogens and parasites into the Arctic as a result of climate warming (but see section 15.4.1.2). However, increased tourism combined with a warmer climate could possibly increase the risk of such invasions.

Climate change is likely to affect the important interaction between parasitic insects and reindeer/caribou. Insect harassment is already a significant factor affecting the condition of reindeer/caribou in the summer (section 7.3.3.2). These insects are likely to become more widespread, abundant, and active during warmer summers while many refuges for reindeer/caribou on glaciers and late snow patches are likely to disappear.

Microbe–plant and microbe–microbivore interactions

Although data on the dynamics and processes in arctic microbial communities and on processes in the soil–microbial–plant interface are accumulating rapidly, it is not yet possible to reach firm conclusions about how the dynamics and processes will change in a changing climate. However, the following can be stated. First, short-term (seasonal) changes in microbial processes may not necessarily have major influences on longer-term (annual to multi-annual) processes. Second, microbes and plants share common nutrient resources, although they may not be limited by the same resource. For example, while nutrient supply rates generally control plant productivity, microbial productivity may be constantly or periodically controlled by the abundance of labile C. Third, the nutrient supply rate to the pool available to plants may not be controlled principally by continuous nutrient mineralization, but rather by pulses of supply and sequestration of nutrients linked to microbial population dynamics and abiotic change, such as freeze–thaw cycles.

Jonasson et al. (1999b) showed that despite no appreciable effect on the microbial biomass and nutrient mass, warming increased plant productivity. Because plant productivity was limited principally by a low rate of N supply, it appears that the mineralization of litter or soil organic matter, or microbial solubilization of organic N, increased, and that the plants rather than the microbes sequestered the "extra" N in inorganic or organic form. However, microbes increased their nutrient content in cases when the sink strength for nutrients in the plants decreased (e.g., after shading) at the same time as soil inorganic N also increased. This suggests either that plants compete successfully with microbes for nutrients, or that the microbial requirement for nutrients was satisfied, and the microbes absorbed a "surplus" of nutrients, which is likely if they

were limited by C rather than nutrients. This does not fully preclude nutrient competition, however, because it is possible that the plants accessed the nutrients from pulse releases from microbes during periods of population dieback. If so, seasonal changes in the frequency of such pulses are of importance for projecting changes in ecosystem function and need further investigation. This is particularly obvious, considering that the microbial N and P content typically exceeds the amounts annually sequestered by plants several-fold, and should constitute an important plant nutrient source (Jonasson et al., 1999a, 2001).

The plant–microbe interaction may also be mutalistic through the mycorrhiza by which the fungal partner supplies nutrients to the plant in exchange for C supplied by the plant. A large proportion of the plant species in shrubby vegetation, common in the Arctic, associate with ecto- or ericaceous mycorrhizal fungi. These mycorrhiza types have enzyme systems able to break down complex organic molecules and thereby supply the plant partner with N (Read et al., 1989), the most common production-limiting element for plants. Changes in plant species composition as a consequence of climatic changes are very likely to substantially affect microbial community composition, including that of mycorrhizal fungi. Unfortunately, studies of the effects of projected climate change on mycorrhizal associations in the Arctic are virtually nonexistent. However, a decade of warming of a fellfield led to a substantial increase in willow (*Salix*) biomass, but few changes in the community of associated ecto-mycorrhizal fungi (Clemmensen and Michelsen, unpub. data, 2004).

The effects of microbivores on the microbial community are yet poorly explored and can only be listed as potentially important for projecting effects of global change. It appears, however, that populations of nematodes increase strongly with warming. Because nematodes are the main predators of fungi and bacteria, it may be that increased biomass production of microbes is masked in a warmer environment because of predation by strongly responding microbivores (Ruess et al., 1999a,b). If so, the release rate of plant-available nutrients is likely to increase (e.g., Ingham et al., 1985), which may explain the enhanced nutrient sequestration by plants in warmer soils (rather than pulse sequestration after microbial dieback).

7.4.1.5. Summary

Changes in climate and UV radiation levels are very likely to affect three important attributes of ecosystem structure: spatial structure (e.g., canopy structure and habitat), trophic interactions, and community composition in terms of biodiversity. Ecosystem structure varies along a latitudinal gradient from the treeline to the polar deserts of the high Arctic. Along this gradient there is a decreasing complexity of vertical canopy structure and ground cover ranging from the continuous and high canopies (>2 m) of the forest tundra in the south to the low canopies (~5 cm) that occupy less than 5% of the ground surface in the polar deserts. Within each arctic vegetation zone, there are often outliers of more southerly zones. Changes in vegetation distribution in relation to climate warming are likely to occur by local expansion of these intrazonal communities and northward movement of zones. Satellite measurements, aerial photographs, and indigenous knowledge show a recent increase in shrubbiness in parts of the Arctic.

Experimental manipulations of environmental factors projected to change at high latitudes show that some of these factors have strong effects on the structure of arctic ecosystems, but the effects are regionally variable. Nutrient addition has the greatest effect on the productivity, canopy height, and community composition of arctic plant communities. Nutrients also increase biomass turnover, so biomass may or may not respond to nutrient addition. Summer warming of tundra vegetation within the range of projected temperature increases (2 to 4 °C over the next 100 years) has generally led to smaller changes compared with fertilization and always to greater responses compared with irrigation. Plant growth response increased from a climatically relatively mild forest understory through a treeline heath to a cold, high-altitude fellfield. Total nonvascular plant biomass and particularly the biomass of lichens decreased in response to 10 years or more of environmental manipulations at sites in subarctic Sweden and in Alaska. Warming experiments in the high Arctic had a greater effect on the fauna above ground compared with fauna below ground and in the subarctic. Spring freeze–thaw events are important, and will probably cause differential mortality among species, thus altering community composition. In general, arctic invertebrate communities are very likely to respond rapidly to change. In contrast, long-term data on the effects of summer warming (2–4 °C) of ecosystems have not shown appreciable changes in microbial biomass and nutrient stocks. This suggests that a temperature increase alone is unlikely to have any strong impact on microbial carbon and nutrient sequestration. Manipulations simulating enhanced UV-B radiation levels and a doubling of atmospheric CO_2 concentration for seven years altered the use of labile carbon substrates by gram-negative bacteria, suggesting a change in community composition. UV-B radiation also affects the structure of fungal communities. So far, no change in plant community structure has been found in the Arctic in response to manipulations of UV-B radiation levels and CO_2 concentration.

Trophic interactions of tundra and subarctic forest plant-based food webs are centered on a few dominant animal species, which often have cyclic population fluctuations that lead to extremely high peak abundances in some years. Small herbivorous rodents of the tundra (mainly lemmings) are the main trophic link between plants and carnivores. Small-rodent population cycles with peak densities every three to five years induce strong pulses of disturbance, energy, and nutrient flows,

and a host of indirect interactions throughout the food web. Lemming population cycles are crucial for nutrient cycling, structure and diversity of vegetation, and for the viability of a number of predators and parasites that are specialists on rodent prey/hosts. Trophic interactions are likely to be affected by climate change. Ice crusting in winter is likely to render vegetation inaccessible for lemmings, deep snow is likely to render rodent prey less accessible to predators, and increased plant productivity due to warmer summers is likely to dominate food-web dynamics. Long-term monitoring of small rodents at the border of arctic Fennoscandia provides evidence of pronounced shifts in small rodent community structure and dynamics that have resulted in a decline in predators (including Arctic fox, snowy owls, buzzards, and skuas) that specialize in feeding on small rodents.

In subarctic forests, a few insect defoliators such as the autumnal moth that exhibit cyclic peak densities at approximately 10-year intervals are dominant actors in the forest food web. At outbreak densities, insects can devastate large tracts of birch forest and play a crucial role in forest structure and dynamics. Trophic interactions with either the mountain birch host or its insect parasitoids are the most plausible mechanisms generating cyclic outbreaks in *Epirrita*. Climate is likely to alter the role of *Epirrita* and other insect pests in the birch forest system in several ways. Warmer winters are likely to increase egg survival and expand the range of the insects into areas outside their present outbreak ranges. However, the distribution range and activity of natural enemies are likely to keep the insect herbivore populations below outbreak densities in some areas.

Climate change is likely to also affect the important interaction between parasitic insects and reindeer/caribou. Insect harassment is already a significant factor affecting the condition of reindeer in the summer. These insects are likely to become more widespread, abundant, and active during warmer summers while refuges for reindeer/caribou on glaciers and late snow patches are likely to disappear. There are large uncertainties about the outcome of the potential spread of new trophic interactants, especially pests and pathogens, into the Arctic.

Disease in plants is likely to increase in those parts of species distribution ranges where a mismatch between the rate of relocation of the species and the northward/upward shift of climatic zones results in populations remaining in supra-optimal temperature conditions. The incidence of new diseases from increasing mobility of pathogens with a southern distribution is a possibility, but increases in UV-B radiation levels could possibly reduce the impact of viral and fungal pathogens.

Microbe–plant interactions can be competitive for nutrients and also mutualistic through mycorrhizal associations. Warming will probably affect both types of relationship, but information is scarce.

7.4.2. Ecosystem function

7.4.2.1. Biogeochemical cycling: dynamics of carbon and nutrients

Arctic ecosystems are characterized by low primary productivity, low element inputs, and slow element cycling, yet they tend to accumulate organic matter, C, and other elements because decomposition and mineralization processes are even more strongly limited than productivity by the arctic environment, particularly the cold, wet soil environment (Jonasson et al., 2001). Because of this slow decomposition, the total C and element stocks of wet and moist arctic tundra frequently equal and may exceed the stocks of the same elements in the much more productive systems of temperate and even tropical latitudes (Table 7.9).

Low-arctic sites with warmer and dryer soils, and extremely unproductive high-arctic polar deserts and semi-deserts, have smaller accumulations of organic matter (Table 7.10). Most of the organic matter and element accumulation occurs in soils, while large accumulations of biomass are limited by a lack of tall woody plant forms such as trees; by selection for slow-growing, low, compact plant forms; and by low productivity and low availability of soil-available elements such as N or P. Typically, the majority of the biomass consists of roots and belowground stems, with aboveground plant mass accounting for less than one-third, and sometimes only 5 to 10%, of the total.

In addition to the large C stocks within the seasonally thawed active layer of the soil (Table 7.10), an equally large pool of organic C may be held in the upper permafrost, within 1 to 2 m of the surface (Michaelson et al., 1996). While these frozen C stocks are not actively involved in C cycling on a seasonal or yearly basis, in the long term, they represent an important C sink, and they are likely to be of particular importance if climate change leads to greater soil thawing or to loss of permafrost (section 6.6.1.3).

The largest body of information on organic matter, C, and nutrient budgets of a wide range of arctic ecosystems comes from the IBP Tundra Biome program, which took place during the late 1960s and early 1970s (Bliss et al., 1981). Since then, research on arctic element cycling has tended to focus on controls over individual biogeochemical processes rather than on comparisons of overall budgets and element stocks. The recent surge of interest in climate change and feedbacks from the Arctic to the globe has highlighted the relevance and utility of those earlier studies, particularly as currently only a few sites are being studied at the whole-system level.

Microbes in arctic soils contain only one or a few percent of the ecosystem C pool. However, the proportions of ecosystem N and P are appreciably higher due to high concentrations of N and P in the microbial tissue compared to the concentrations in plants and soil organic matter (Jonasson et al., 1999a). As a proportion of the total soil organic matter, microbial biomass and nutrient content are similar to ecosystems outside the Arctic, but as a proportion of the total organic matter in the ecosystem (soil plus vegetation) microbial biomass and nutrient content are high in comparison with other systems due to the relatively small vegetation component in the Arctic. Data from various arctic and subarctic sites have shown that microbes commonly contain appreciably less C, slightly less or comparable amounts of N, and much higher amounts of P than the entire plant biomass (Cheng and Virginia, 1993; Hobbie and Chapin, 1996; Jonasson et al., 1996, 1999a; Schmidt et al., 2002).

Nutrient mineralization rates are low, however: typically ten-fold lower than in the boreal region. The low rate is mainly due to low soil temperatures, and it leads to low supply rates of nutrients to the plant-available pool and nutrient-constrained plant productivity in most arctic ecosystems (Nadelhoffer et al., 1992). The combination of low mineralization rates and high proportions of nutrients in microbes compared to plants leads to possible competition for nutrients between microbes and plants during periods of rapid microbial growth (Kaye and Hart, 1997). However, microbes are likely to also release a pulse of nutrients during periods of population decline when the cells are lysed and nutrients are leached (Giblin et al., 1991; Jonasson et al., 2001). To project microbial effects on nutrient-constrained plant productivity as a result of environmental changes, it is essential to understand not only how the microbial processing rate of organic matter will change, but also the controls on microbial population sizes and how

Table 7.9. Average carbon (C) pools and total C in arctic and alpine tundra, the neighboring boreal zone, and all terrestrial ecosystems. The soil pools do not include organic C in permafrost beneath the seasonally thawed active layer (from Jonasson et al., 2001, after data in McGuire et al., 1997).

	Area (10^6 km^2)	Soil (g C/m^2)	Vegetation (g C/m^2)	Soil: Vegetation ratio	Total C (10^{12} kg)		
					Soil	Vegetation	Soil and vegetation
Arctic and alpine tundra	10.5	9200	550	17	96	5.7	102
Boreal woodlands[a]	6.5	11750	4150	2.8	76	27	103
Boreal forest	12.5	11000	9450	1.2	138	118	256
Global terrestrial	130.3	5900	7150	0.8	772	930	1702

[a]comparable to forest tundra

Table 7.10. Soil organic matter, plant biomass, and net primary production (NPP) in the primary arctic ecosystem types (after Jonasson et al., 2001; based on data from Bliss and Matveyeva, 1992 and Oechel and Billings, 1992).

	Soil organic matter (g/m²)	Vegetation biomass (g/m²)	NPP (g/m²/yr)	Soil: Vegetation Ratio	Soil: NPP Ratio (yr)	Vegetation: NPP Ratio (yr)	Percentage of total area
High Arctic							
Polar desert	20	2	1	10	20	2.0	15
Semi-desert	1030	250	35	4.1	29	7.1	8
Wet sedge/mire	21000	750	140	28	150	5.4	2
Low Arctic							
Semi-desert	9200	290	45	32	204	6.4	6
Low shrub	3800	770	375	4.9	10	2.1	23
Wet sedge/mire	38750	959	220	40	176	4.3	16
Tall shrub	400	2600	1000	0.2[a]	0.4	2.6	3
Tussock/sedge dwarf shrub	29000	3330	225	8.7	129	16	17

[a]source data for this value, which is surprisingly low, have not been found

changes in the populations affect nutrient cycling and interact with plant processes (see also section 7.4.1.4).

Spatial variability

Although the productivity of the most productive tundra may rival that of highly productive shrub and marsh systems at lower latitudes, most arctic systems lie at the low end of the global productivity range. What is striking is the wide range of variation (about three orders of magnitude) in NPP and standing stocks of organic matter in soils and vegetation within the Arctic (Table 7.10). In general, productivity and organic matter stocks decrease with temperature and precipitation from south to north, but local variation in productivity in relation to topography is dramatic (often 10- to 100-fold). Among the most important correlates of topographic variation in productivity are the duration and depth of winter snow cover and degree of protection from winter wind damage, as well as variation in soil moisture, soil thaw, and soil temperature. Local variation in these factors can be nearly as great as that across a wide range of latitudes (Billings, 1973; Jonasson et al., 2001; Shaver et al., 1996). Local variation in productivity is also associated with dramatic shifts in the relative abundance of plant functional types including both vascular and nonvascular plants (sections 7.3 and 7.4.1). Because of this dramatic local variability, primary production and organic matter accumulation are distributed in a mosaic fashion across the Arctic, with a higher frequency of more productive sites (usually wet or moist lowlands) at lower latitudes.

The spatial distribution of productivity and organic matter in the Arctic is broadly predictable in relation to temperature, soil moisture, and other soil factors such as pH, topography, and snow cover (Walker D. et al., 1998, 2002; Walker M.D. et al., 1989, 1994). The proximate controls on C cycling in these ecosystems, how-

ever, are much more closely tied to the inputs and turnover of other elements, especially N and P (Shaver et al., 1992). Because N and P inputs (by deposition, fixation, or weathering) are low, even where rapid photosynthesis is possible C cannot be stored in organic matter any faster than the rate of N or P accumulation. Thus, for example, in the Canadian High Arctic where a large portion of the surface is bare ground, C fixation and accumulation is closely tied to low, wet areas where anaerobic soil conditions favor N fixation (Gold and Bliss, 1995). In other arctic systems, such as Alaskan wet and moist tundras, the total amounts of N and P in soil organic matter may be very high but, due to slow rates of decomposition, the availability of these elements to plants is low, leading to low productivity despite high element and soil organic matter stocks (Giblin et al., 1991).

Temporal variability

Interannual variation in biogeochemical cycles has received little attention in the Arctic, although a few multi-year records of ecosystem C exchange, N deposition, and C and N losses at the watershed or catchment level do exist (Hershey et al., 1997; Kling et al., 2000; Oechel et al., 2000a; Steiglitz et al., 2000). Clearly, the trend over the Holocene, at least, has been one of overall accumulation of elements in organic matter since the loss of the glacial ice cover, but the variation in rates of accumulation (or loss) at the scale of years to decades is particularly poorly understood.

The net C balance of arctic ecosystems, as in any terrestrial ecosystem, may be positive or negative depending on the timescale over which it is measured and the environmental conditions during the measurement period. This dynamic balance is called net ecosystem production (NEP[3]) and is defined as the difference between two large, opposing fluxes: gross ecosystem production

---

[3]The biological definition of NEP results in positive values when carbon accumulates in the biosphere, whereas measurements of carbon fluxes and the concept of carbon sinks and sources use a convention of negative fluxes when carbon accumulates in the biosphere.

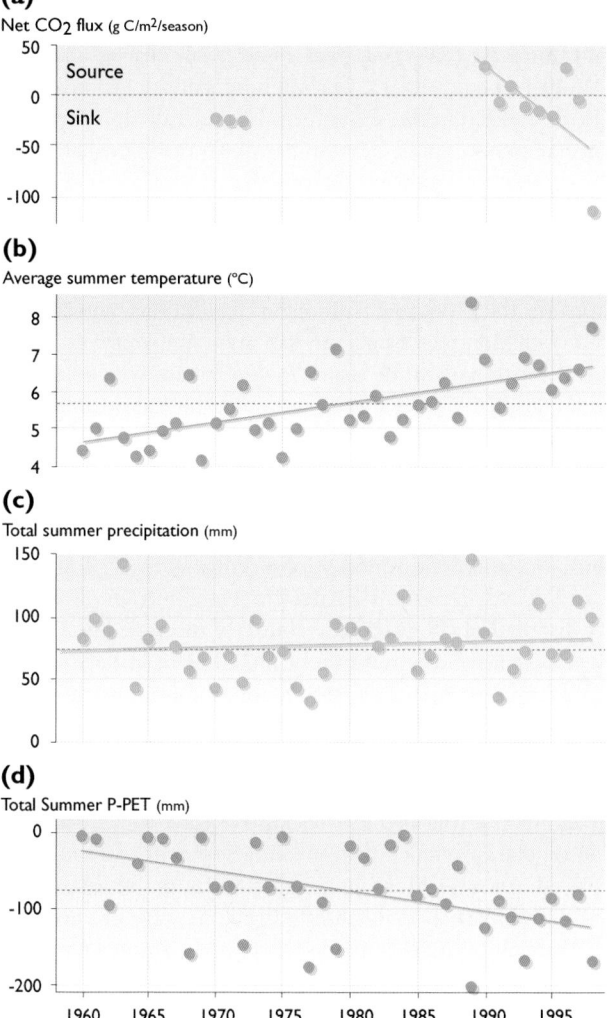

(a)
Net CO₂ flux (g C/m²/season)

(b)
Average summer temperature (°C)

(c)
Total summer precipitation (mm)

(d)
Total Summer P-PET (mm)

Fig. 7.26. Long-term trends in (a) summer net CO₂ flux, (b) average summer temperature, (c) total summer precipitation, and (d) summer precipitation (P) less potential evapotranspiration (PET) for Alaskan coastal wet-sedge tundra (Oechel et al., 1993, 2000b).

(GEP, or gross ecosystem photosynthesis) and ecosystem respiration (R_E), both measured in units of C (mass or moles) per unit area and time. Ecosystem respiration has two major components, R_A (autotrophic or plant respiration) and R_H (heterotrophic or animal plus microbial respiration). Each of these components of NEP has different relationships to current temperature, moisture, and light conditions. Because they are measured at the whole-system level, all three components are also a function of the current functional mass or surface area of the organisms as well as their current nutritional status. Thus, for example, even though long-term NEP must be positive for the large C accumulations in tundra ecosystems to occur, on a daily or seasonal basis NEP swings from strongly negative (at "night", even under the midnight sun, or in winter) to strongly positive (at midday in midsummer). These daily and seasonal fluctuations have been measured at an increasing number of arctic sites in recent years (Nordström et al., 2001; Oechel et al., 2000a; Søgaard et al., 2000; Vourlitis and Oechel, 1997, 1999).

Carbon balance may also vary sharply among years, and may be either positive or negative on an annual basis. Recent work in Alaska (Oechel et al., 2000a) indicates that although C was accumulating in wet and moist tundras in the 1960s and 1970s (i.e., "negative" C balance), during much of the 1980s and 1990s there was a net loss of C from these ecosystems in both winter *and* summer (i.e., "positive" C balance). In the late 1990s, the summer C balances changed again so that the ecosystems were net sinks, but it is not yet clear whether the C balances for the full year have returned to net C accumulation (Fig. 7.26; see also section 7.5.1.1). It is also unclear whether the shifts in NEP (from C sink to C source) that have occurred over the past 40 years are related in any direct way to weather, because the entire period has been one of general warming in northern Alaska (section 7.5). Modeling studies (e.g., Clein et al., 2000; McGuire et al., 2000; McKane et al., 1997) suggest that in the short term (within one or a few years) the response of R_E (both R_A and R_H) to temperature is more rapid than the response of GEP, leading to a short-term loss of C with warming. In the long term, however, the interaction between temperature and soil nutrient availability might increase GEP sufficiently to cause an eventual return to net C accumulation. There is also evidence from manipulation experiments (Shaver and Chapin, 1991; Shaver et al., 1998) and latitudinal gradients (Callaghan and Jonasson, 1995) that increases in air temperature can result in soil cooling after long periods, as higher air temperatures lead to increased leaf area indices that intercept a greater proportion of incoming radiation before it reaches the soil surface, thereby leading to soil cooling.

Species and functional type composition of the vegetation are keys to long-term change in productivity, because of differences in nutrient use and allocation, canopy structure, phenology, and relative growth rates among plants (Chapin F. et al., 1993; Hobbie, 1995). Large differences exist, for example, in the *rate* at which tundra plants can respond to changes in weather and climate, due to differences in allocation to stems versus leaves or to secondary chemistry versus new growth (Shaver et al., 2001), in the ability to add new meristems (Bret-Harte et al., 2001), and in the constraints on the amount of growth that can be achieved by a single meristem within a single year (i.e., determinate versus indeterminate growth). Species and functional types also differ in their growth phenology and thus in their ability to take advantage of a change in the timing and duration of the growing season. For example, moss-dominated ecosystems in Iceland have limited ability to respond to climate change without a complete change to a vascular plant-dominated community (Jónsdóttir et al., 1995), whereas shrubs and small trees already present in sheltered, moist depressions on the North Slope of Alaska seem to be already expanding their distribution (Sturm et al., 2001b) and therefore productivity.

The chemical composition of primary productivity (leaves versus wood, secondary chemistry, species composition) is important as a long-term feedback to productivity and its responsiveness to climate change. It is also important in terms of both animal community composition and secondary (herbivore) productivity. Carbon to nitrogen ratios; lignin and protein content; and tannin, resin, and phenolic content are all important in determining forage quality (section 7.4.1.4) and the susceptibility of plant litter to decomposition, and thus the remineralization of essential limiting nutrients like N and P (Hobbie, 1996).

Despite the critical importance of NPP together with NEP and the considerable research already conducted on these parameters, additional field measurements and focused process studies are needed to resolve issues relating to the different methodologies used for measuring NPP and NEP (Hobbie et al., 2000; Williams et al., 2000). Results from different methodologies also need to be reconciled (section 7.7.1.1).

Budgets of N and P were developed for several arctic sites during the IBP studies 30 years ago (Chapin F. et al., 1980) but complete documentation of inputs and outputs of any element other than C has not been attempted since then for any arctic site. Part of the problem is that individual N and P inputs and outputs in arctic ecosystems, such as N fixation, N deposition, denitrification, rock weathering, or losses in streamflow (see next subsection), are even smaller than the amounts annually recycled by mineralization of organic matter (Fig. 7.27; Nadelhoffer et al., 1991; Peterson et al., 1992; Shaver et al., 1992), except perhaps in the high Arctic (Gold and Bliss, 1995). Thus, very long-term records are needed to evaluate the significance of interannual variation in N and P budgets, while most studies of the component processes last only one to three years.

Inputs/outputs, primary production, and net ecosystem production

The dominant C input to arctic ecosystems is from photosynthesis in vascular and nonvascular plants, which in total comprises GEP. The relative (apparent) impor-

tance of various controls on primary production differs depending on the level (leaf, canopy, or whole vegetation) and timescale (daily, seasonal, or decadal) of examination (Williams et al., 2001). Carbon inputs at the leaf level are clearly limited in the short term by generally low irradiance and consequent low temperatures during usually short, and late, growing seasons (Fig. 7.8), despite a wide range of specific photosynthesis-related adaptations to the arctic environment (section 7.3). Photosynthesis in arctic plants is also often sensitive to changes in CO_2 concentrations (in the short term), moisture conditions, and snow (UV radiation effects are variable and comparatively small). Although arctic plants in general are well adapted to the arctic climate, there is considerable variation in the responses of photosynthesis to microclimate among plant functional types. In the longer term and at the level of whole vegetation canopies, however, C inputs are limited by generally low canopy leaf areas, leaf phenology and duration, and light interception (Williams et al., 2001). Canopy leaf area is low because low soil nutrient availability, particularly N, limits the ability of the vegetation to develop a large, photosynthetically efficient leaf area (Williams and Rastetter, 1999), and it also limits the ability of the vegetation to use newly fixed C for new growth, because growth requires adequate supplies of multiple elements in addition to C (Jonasson et al., 2001; Shaver et al., 1992). It is also low because of the low stature of the vegetation, which prevents development of a multilayered canopy. Other environmental factors such as wind and soil disturbance also limit C gain. Storage of photosynthate and nutrients acquired in previous years plays a key role in determining the current year's productivity (Chapin F. and Shaver, 1985b).

Carbon outputs from arctic ecosystems occur via a wider array of processes and are regulated very differently from C inputs (section 7.4.2.2). The dominant form of C loss is as CO_2, produced by both plants and soil biota. Autotrophic or plant respiration (R_A) typically accounts for about half of GEP on an annual basis (Williams et al., 2000, 2001) but follows a very different seasonal and daily pattern (discussed previously).

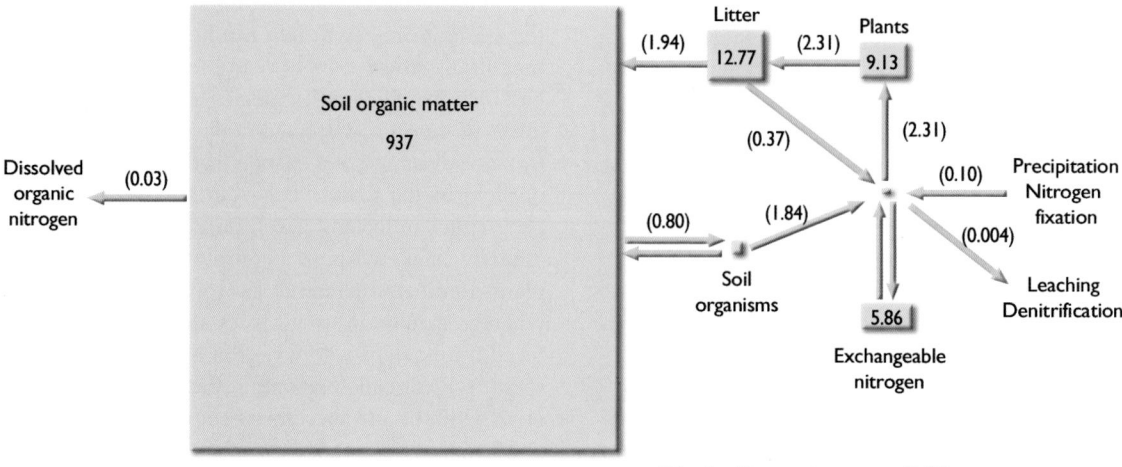

Fig. 7.27. Nitrogen budget for wet-sedge tundra at Barrow, Alaska. Numbers in boxes are N stocks in g/m²; numbers in parentheses are N fluxes in g/m²/yr (Shaver et al., 1992; adapted from Chapin F. et al., 1980).

Heterotrophic respiration (R_H), mostly by soil organisms, accounts for most of the other half of GEP, although in the long term the sum of R_A and R_H must be slightly less than GEP if C is to accumulate in soil organic matter. Heterotrophic respiration produces both CO_2 and methane (CH_4), the latter produced anaerobically in wet soils (section 7.4.2.2). Much of the CH_4 produced in arctic soils is oxidized to CO_2 before it reaches the atmosphere; net CH_4 emissions thus are normally only a fraction of CO_2 emissions from arctic soils (less than 5%), but CH_4 is a much more powerful greenhouse gas than CO_2. Other aspects of C balance are important yet difficult to quantify. Examples are plant root respiration, the sloughing of dead material from roots, root exudation, and the growth and respiration of microorganisms intimately associated with plant roots.

Most of the respiratory CO_2 and CH_4 losses from arctic systems move directly to the atmosphere. Significant fractions of these gases, however, travel in dissolved forms in soil water, eventually reaching streams and lakes where they are released to the atmosphere (Kling et al., 1991, 1992). In addition, soil and surface waters contain significant amounts of dissolved organic forms of C, much of which is eventually consumed by aquatic microbes, producing more CO_2 (Chapter 8). Together, these losses to aquatic systems may add up to a significant component of the net C balance of arctic systems. Synoptic, simultaneous analysis of aquatic C losses at the same time, place, and scale as direct atmospheric exchanges has not been completed, but estimates of aquatic C losses suggest that these may equal as much as 20 to 30% of GEP.

Winter CO_2 losses are a second major gap in the understanding of C losses from arctic ecosystems. Although winter CO_2 losses have long been recognized (Coyne and Kelley, 1971), more recent research indicates that these losses are greater than was previously thought and may be the product of significant respiratory activity during the winter (Fahnestock et al., 1999; Hobbie et al., 2000; Oechel et al., 1997; Welker et al., 2000; section 7.3.2.3), when recently fixed C is respired (Grogan et al., 2001).

There are few studies of inputs and outputs of N in the Arctic, largely because early work suggested that they were small relative to standing N stocks and internal recycling, and thus were less important, at least on a short-term basis (Shaver et al., 1992). In the long term (several decades or more), however, understanding of N inputs and outputs is essential to understanding how the total pool sizes of N change over time. Changes in standing stocks of N are closely tied to the accumulation or loss of organic matter and C in the Arctic (Gold and Bliss, 1995).

Nitrogen enters arctic ecosystems through atmospheric deposition and microbially mediated N fixation (Fig. 7.27). Nitrogen deposition rates are low in the Arctic relative to other parts of the world, mostly because the atmosphere is cold enough that it cannot hold the high concentrations of N species such as nitrate (NO_3) that are deposited at lower latitudes. Thus, N deposition can account for only about 5% or less of the annual plant N uptake requirement in Alaskan wet-sedge tundra (Chapin F. et al., 1980), although this might increase with increased industrial activity at lower latitudes. In regions such as northern Scandinavia that are subject to N deposition from lower-latitude anthropogenic sources, however, N deposition may be greater than 0.1 g/m²/yr, which if continued for many years is sufficient to affect plant growth and productivity (Back et al., 1994). Nitrogen fixation rates are usually assumed to be of similar magnitude, although the only relatively recent studies (Chapin D. and Bledsoe, 1992; Lennihan et al., 1994) indicate that, at least in the high Arctic, N fixation might account for more than 10% of plant requirements with the remaining 90% supplied by recycling from the soils.

Nitrogen losses are also poorly understood. There have been no recent, published studies of denitrification in the Arctic; although anaerobic soils might be expected to have high potential for denitrification, the generally low rates of NO_3 production in tundra soils suggest that this is also a small component of the annual N budget. Possible spring losses of N in the form of nitrous oxide (N_2O) have been suggested (Christensen T. et al., 1999a) but not yet verified in the Arctic. Nitrogen losses in streams have been monitored at several locations, and are of roughly the same magnitude as N deposition (Hershey et al., 1997).

Responses to climate change

Responses of element cycles in arctic ecosystems to climate change factors have been studied in multi-year manipulation experiments in several contrasting ecosystem types (Dormann and Woodin, 2002; Shaver and Jonasson, 2000; Fig. 7.28). These experiments include manipulations of air temperature, CO_2 concentration, light, water (both excess and deficit), nutrients, and UV-B radiation levels. One common observation from these experiments is that although short-term responses to single factors like changing CO_2 concentrations or temperature increases are measurable and often significant, these responses are often not sustained due to other limitations. A general conclusion is that nutrient limitation dominates the multi-year responses and is linked to changes in other factors (e.g., temperature and water) through their indirect effects on nutrient mineralization and availability to plants. In wet systems, water-table depth and soil drainage are critical variables limiting nutrient turnover in the soil; increases in C turnover in these systems are not linked to increases in C accumulation because increased C accumulation requires increased N and/or P supply (Oechel et al., 1998, 2000a; Shaver et al., 1998). A large pool of nutrients exists in organic matter, and may drive large changes in organic matter stocks if the nutrients can be mineralized and not leached from the system. Similarly, short-term increases in photosynthesis and growth in response to high CO_2 concentrations are often not sustained due to nutrient limitation (Grulke et al., 1990; Hartley et al., 1999; Oechel et al., 1994; section 7.3).

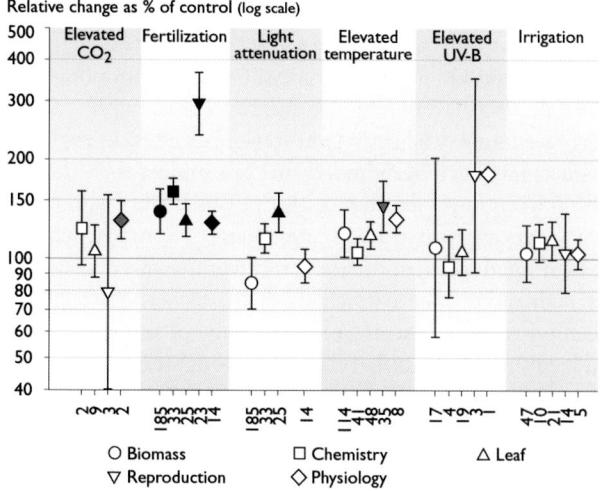

Relative change as % of control (log scale)

O Biomass   □ Chemistry   △ Leaf
▽ Reproduction   ◇ Physiology

Fig. 7.28. Mean response of plants to environmental perturbations, shown as a percentage of untreated controls (vertical lines depict ±1 standard deviation). Black and grey symbols indicate that means differ significantly from 100% (P<0.05 and 0.1>P>0.05, respectively; two-tailed t-test). Numbers along the horizontal axis refer to corrected sample size (df+1) (Dormann and Woodin, 2002).

The results of several manipulation experiments indicate that nutrient mineralization stimulated by increased soil temperatures is unlikely to be sustained in the long term. Soil temperature increases cause an immediate increase in soil respiration in laboratory studies, but few arctic field studies have shown increased mineralization rates in response to increased air temperatures (but see Schmidt et al., 1999) and longer-term field studies show an acclimation to increased soil temperature (Hartley et al., 1999; Luo et al., 2001). Some studies have shown soil *cooling* in response to increasing air temperatures in experiments (Shaver and Chapin, 1991) and along latitudinal gradients (Callaghan and Jonasson, 1995). Higher air temperatures stimulate leaf area development (Myneni et al., 1997) and a greater leaf area index would be expected to intercept thermal radiation before it reaches the soil, leading to soil cooling. In addition, organic matter in deeper soil profiles is less responsive to temperature increases than that in surface layers (Christensen T. et al., 1999b; Grogan et al., 2001), again suggesting that any temperature-induced mineralization is likely to be transitory.

Overall, however, multi-year experiments suggest that the arctic ecosystems most responsive to climate change are very likely to be those in which the environmental change is linked to a large change in nutrient inputs or soil nutrient turnover, and/or large changes in leaching or erosional losses of soil-available nutrients (Fig. 7.27). The effects of UV radiation on overall organic matter cycling are generally unknown, but not unimportant. Recent work by Niemi et al. (2002) showed that increased UV-B radiation decreased CH_4 emissions from a peatland in northern Finland, while three studies show UV-B radiation effects on *Sphagnum* growth (Gehrke et al., 1995; Searles et al., 1999; Sonesson et al., 2002) with potential implications for C sequestration. Long-term responses of biogeochemical cycling to increased

CO_2 concentrations and UV-B radiation levels are small in magnitude but are likely to lead to longer-term changes in biogeochemical cycling and ecosystem structure (sections 7.3.3.3 and 7.4.1.2). However, most of the current understanding of UV radiation responses is based on species- and tissue-level research.

Biodiversity and species effects on biogeochemistry

It is important to determine if species or growth form composition of the vegetation have any impact on biogeochemistry of arctic ecosystems, or if biogeochemistry is largely regulated by climate and resource availability irrespective of species composition. Although only partial answers to this question are currently available, there are at least five main mechanisms by which species composition are likely to have important consequences for biogeochemistry. These are:

1. Species composition is likely to affect the rate of change in ecosystems in response to environmental change, through differences in species growth, reproduction, and dispersal rate potential (e.g., Bret-Harte et al., 2001).
2. Species are likely to affect nutrient availability and C cycling through differences in the turnover of elements in their living tissues and in the decomposability of their dead parts (Hobbie and Chapin, 1998; Quested et al., 2003).
3. Species are likely to affect element accumulations in living plants through differences in their biomass allocation patterns and in their biomass element concentrations and element ratios (Bret-Harte et al., 2002; Quested et al., 2003; Shaver et al., 2000).
4. Species are likely to differ in their effects on snow accumulation and snowmelt, surface energy balance, and soil temperature regimes, with important feedbacks to element cycles (McFadden et al., 2001; Sturm et al., 2001a).
5. Physiological mechanisms, for example, wetland species that act as conduits for CH_4 transport from the soil to the atmosphere (Fig. 7.29; Joabsson and Christensen, 2001; Niemi et al., 2002; Öquist and Svensson, 2002).

All five of these species effects have been documented in arctic systems, although it is often uncertain how to scale up from small experimental communities to larger units of the landscape.

Species richness or diversity itself is likely to also affect the biogeochemistry of arctic ecosystems, although the magnitude of the effect is hard to judge. There is a weak positive correlation between productivity and vascular species richness in arctic vegetation, but, like most vegetation, richness declines when productivity is increased artificially by fertilizer addition or other disturbance (Gough et al., 2000). Recent evidence suggests that arctic plants obtain their N from diverse sources in the soil (Michelsen et al., 1996, 1998; Nadelhoffer et al., 1996) and that the

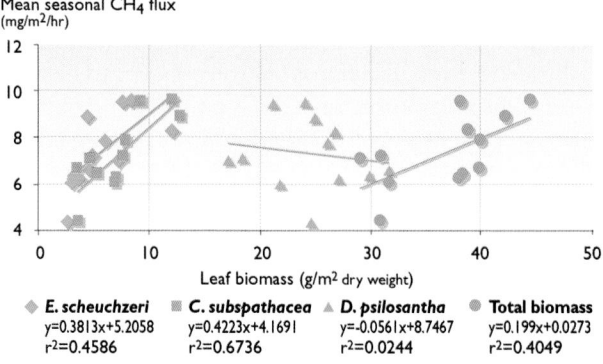

Mean seasonal CH₄ flux
(mg/m²/hr)

Leaf biomass (g/m² dry weight)

◆ *E. scheuchzeri*	■ *C. subspathacea*	▲ *D. psilosantha*	● Total biomass
y=0.3813x+5.2058	y=0.4223x+4.1691	y=-0.0561x+8.7467	y=0.199x+0.0273
r²=0.4586	r²=0.6736	r²=0.0244	r²=0.4049

Fig. 7.29. Mean seasonal methane emissions from a high-arctic fen in Zackenberg, northeast Greenland, as a function of leaf biomass of the grasses *Eriophorum scheuchzeri, Dupontia psilosantha*, and *Carex subspathacea*, and total leaf biomass of the three species (Joabsson and Christensen, 2001).

relative abundance of different species reflects different abilities to acquire the different forms of N (McKane et al., 2002). These latter studies suggest that diversity will probably increase productivity in arctic vegetation by increasing total uptake of different forms of a strongly limiting element, N. Partial support for this conclusion comes from experiments involving removal of individual species from arctic vegetation, in which the remaining species failed to increase in abundance (Fetcher, 1985; Jonasson, 1992; Shevtsova et al., 1995, 1997).

Vascular plants directly affect the substrate availability for methanogens and have the capability to transport gases between the anaerobic parts of soils and the atmosphere (Bubier and Moore, 1994; Joabsson and Christensen, 2001; Joabsson et al., 1999; Niemi et al., 2002). Different vascular plant species have different effects, however, and the vascular plant species composition in wet tundra ecosystems may be a key determinant of the scale of CH_4 emissions; for example, Fig. 7.29 shows that the minor constituents of the total vascular plant biomass (*Carex* and *Eriophorum*) seem to be "driving" net CH_4 emissions from the site, suggesting that shifts in vascular plant species composition alone could lead to significant effects on trace-gas exchange (Joabsson and Christensen, 2001; Ström et al., 2003). Changes in species composition *per se* caused by climate warming and increased UV-B radiation levels are likely to cause a change in CH_4 emissions, adding to the direct effect of a changing soil climate on these emissions (Niemi et al., 2002; see section 7.4.2.2).

Role of disturbance

Disturbances are expected to increase with climatic warming, mainly through thermokarst (section 7.5.2) and possibly also through increased fire in some northern ecosystems and insect pest outbreaks in subarctic forests (section 7.4.1.4). In general, physical disturbance of arctic ecosystems results in greater soil warming and permafrost thawing, which tend to increase soil organic matter and nutrient turnover. Typically, vegetation productivity increases dramatically although soil respiration also

increases. It is not yet clear whether the increased plant growth is sufficient to compensate for losses of soil organic matter. In the long term, however, arctic landscapes should gain organic matter in both soils and vegetation on disturbed sites. The timing and trajectory of these changes are key unknowns requiring further research.

7.4.2.2. Soil processes and controls over trace-gas exchanges

During the last decade, trace-gas production, emissions, and assimilation have attracted considerable attention from the scientific community, for the following reasons:

- most of these gases are "radiatively active" (i.e., they affect the heat balance and contribute to the "greenhouse effect" responsible for climate instability and change);
- the atmospheric concentration (mixing ratios) of these gases underwent remarkably rapid changes after the industrial revolution (e.g., the concentration of atmospheric CH_4 has been growing at an annual rate of 0.8 to 1%, which can have a significant impact on the biosphere apart from the greenhouse effect); and
- most trace gases are intermediate or end products/substrates of key biogeochemical processes, such that monitoring these gas species can be used for early detection of any anomaly in ecosystem functioning.

Table 7.11 lists the major trace gases and their potential impacts on ecosystems, although not all of the listed gases are of primary importance for arctic terrestrial ecosystems. This review is restricted to CO_2 (section 7.4.2.1), CH_4, and N_2O.

Soil and ecosystem processes responsible for trace-gas emissions

Trace-gas exchange with the atmosphere occurs through a set of coupled soil/ecosystem processes, including:

- production of substrate(s) for processing by trace gas-producing organisms;
- conversion of substrate(s) to respective gaseous species in parallel with gas consumption (Table 7.12); and
- mass transfer of produced gas to the free atmosphere, which includes three main mechanisms: molecular diffusion, vascular gas transfer (i.e., through plant "conduits"), and ebullition (i.e., bubble formation).

Substrates are formed by one of three processes: decomposition (hydrolytic breakdown of plant litter, oxidation, fermentation); N mineralization, and photosynthesis and photorespiration. However, the starting point for almost all substrates is the primary production of organic matter by plant photosynthesis or (occasionally) bacterial chemosynthesis. There are two main flows of C substrates from plants: plant litter formation with lignocellulose as a main component resistant to microbial breakdown; and the continuous supply of readily available C monomers (root and foliage exudation). The chain of events leading to the formation of immediate precursors of trace gases can be long and intricate (Panikov, 1999). It is worthwhile to note that the most successful model simulations of trace gas emissions include vegetation or primary productivity modules.

Trace-gas transport

There are three main transport mechanisms: molecular diffusion, vascular transport of gas through plant roots, and ebullition. Vascular transport can be described as a diffusion process through plant root aerochyma (parenchyma containing large air spaces typical of emergent and marginal wetland species), which is a continuous network of gas-filled channels. Vascular transport is two or three orders of magnitude more rapid than diffusion in water. Ebullition is probably the most difficult process to simulate and describe mathematically due to its stochastic nature. In northern soils, ebullition and vascular transport were shown to be the major transport mechanisms, accounting for up to 98% of total CH_4 emissions (Christensen T. et al., 2003a).

Environmental controls on methane fluxes

Methane is produced from anaerobic decomposition of organic material in waterlogged, anaerobic parts of the soil. Wet and moist tundra environments are known to be significant contributors to atmospheric CH_4 (Bartlett and Harriss, 1993; Fung et al., 1991). Methane is formed through the microbial process of methanogenesis, which is controlled by a range of factors: most notably temperature, the persistence of anaerobic conditions, gas transport by vascular plants, and the supply of labile organic substrates (Joabsson and Christensen, 2001; Schimel, 1995; Ström et al., 2003; Whalen and Reeburgh, 1992). Figure 7.30 shows the variety of controls on CH_4 formation rates at different spatial and

Table 7.11. Trace gases produced in tundra soils and their potential impacts on terrestrial ecosystems.

Gas species	Main soil sources	Main soil sinks	Environmental control on "source–sink" balance	Impact[a]	Reference related to the Arctic
Carbon dioxide (CO_2)	Organic matter decomposition	Photosynthetic uptake; formation of carbonates	Temperature; moisture; available nutrients	GHG; productivity of plant community (indirectly)	Nordström et al., 2001; Oechel et al, 2000b; Søgaard et al., 2000;
Methane (CH_4)	Methanogenesis (anaerobic decomposition of organic matter)	Uptake by methanotrophic bacteria	Temperature; moisture; nutrients; plant community structure	GHG; highly combustible	Christensen T. et al., 2000; Friborg et al., 2000; Whalen and Reeburgh 1990
Nitrous oxide (N_2O)	Denitrification; nitrification	Uptake by aerobic and anaerobic soil bacteria	Temperature; moisture; N fertilizers	GHG	Christensen T. et al., 1999a
Carbon monoxide (CO)	Anaerobic decomposition of organic matter; airborne pollution	Uptake by carboxydobacteria	Temperature; moisture	Atmospheric photochemical reactions; toxicity	Conrad, 1996; Whalen and Reeburgh, 2001
Molecular hydrogen (H_2)	Fermentation, Pollution	Uptake by methanogens, acetogens, sulfite reducers, and H_2-oxidizing bacteria	Temperature; moisture	Atmospheric photochemical reactions; toxicity	Conrad, 1996; no specific arctic works
non-methyl hydrocarbons (terpene and isoprene derivatives)	Plants and microorganisms (bacteria, fungi)	Photooxidation; uptake by aerobic soil bacteria	Temperature; moisture; plant phenology and physiological state	GHG; ozone and aerosol formation within plant canopy	Guenther et al., 1993; Isidorov et al., 1983; Isidorov and Jdanova, 2002
Methylated halogens (methyl bromide, methyl iodide, etc.)	Soil affected by oceanic water	Bacterial decomposition	Soil hydrology; location within landscape relative to oceanic shoreline	Ozone depletion; phytotoxicity	Dimmer et al., 2001 (data for Irish peatland ecosystems)
Dimethyl sulfide (DMS)		Hydrolysis; uptake as a sulfur source by plants	Soil hydrology and location within landscape relative to oceanic shoreline	Phytotoxicity	Legrand, 1995
Sulfur oxides (SO_2; SO)	Fuel combustion: airborne contamination of soil	Sulfate and sulfite formation; ion exchange; plant and microbial uptake	Temperature; moisture	Acid rain; phytotoxicity	Conrad, 1996
Ammonia (NH_3)	Airborne soil contamination	Conversion to ammonium (NH_4^+); ion exchange; plant and microbial uptake	Temperature; moisture	Plant productivity	Conrad, 1996

[a]GHG=greenhouse gas

temporal scales. Methane is not only produced but also consumed in the aerobic parts of the soil through the microbial process of methanotrophy, which can even take place in dry soils with the bacteria utilizing atmospheric CH_4 (Christensen T. et al., 1999a; Panikov, 1999; Sjögersten and Wookey, 2002; Whalen and Reeburgh, 1992). Methanotrophy is responsible for the oxidation of approximately 50% of the CH_4 produced at depth in the soil and therefore is as important to net CH_4 emissions as methanogenesis. The anaerobic process of methanogenesis is much more responsive to temperature than CH_4 uptake, so soil warming in the absence of any other changes is very likely to accelerate emissions (the difference between production and consumption), in spite of the simultaneous stimulation of the two opposing processes.

Apart from temperature, water regime, and plant cover, methanogenic bacteria are strongly affected by biological interactions within the soil community. Competition with acetogenic and sulfidogenic bacteria for molecular hydrogen (the outcome of which depends on the affinity to hydrogen, the temperature, and density of the various populations) determines the pattern of gas formation not only quantitatively, but also in qualitative terms. For example, ecosystems can be a source of CH_4 (if methanogenic bacteria prevail), or hydrogen sulfide and other sulfides (if sulfate-reducing bacteria dominate), or acetic acid (if a large population of acetogens is present).

Early empirical models of northern wetland/tundra CH_4 exchanges suggested sensitivity to climate change (Harriss et al., 1993; Roulet et al., 1992). A simple mechanistic model of tundra CH_4 emissions that includ-

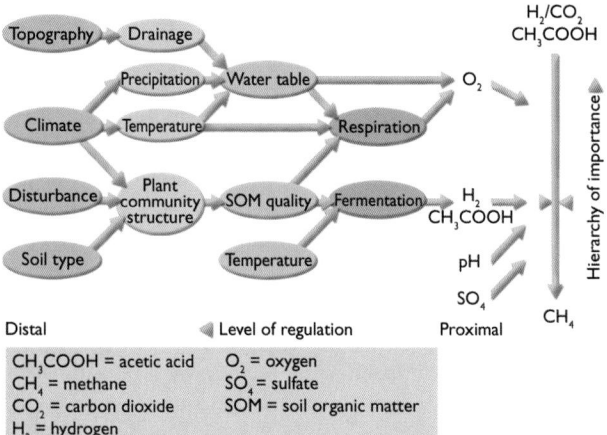

Fig. 7.30. Controls on methanogenesis (redrawn from Davidson E. and Schimel, 1995).

ed the combined effects of temperature, moisture, and active-layer depth also suggested significant changes in CH_4 emissions as a result of climate change (Christensen T. and Cox, 1995). More complex wetland CH_4 emission models suggest that winter processes have a strong influence on net annual CH_4 emissions (Panikov and Dedysh, 2000). Variations in CH_4 emissions at the regional and global scale are driven largely by temperature (Crill et al., 1992; Harriss et al., 1993) with the important modulating effects of vascular plant species composition superimposed (Christensen T. et al., 2003b; Ström et al., 2003). An initial warming is, hence, projected to lead to increased CH_4 emissions, the scale of which will depend on associated changes in soil moisture conditions and the secondary effects of changes in vegetation composition (section 7.5.1.2).

Table 7.12. Examples of soil processes where trace gases are formed or consumed.

Gas species	Source reaction	Sink reaction
CO_2	Oxidative decomposition of dead organic matter (OM): $OM + O_2 \rightarrow CO_2 + H_2O + $"ash" Oxidation of root exudates Fermentation: $(CH_2O)_n$ [a] $\rightarrow CO_2 + VOC^b + H_2$	sulfur[c] (S) oxidizing bacteria Photosynthesis (plant and green bacteria): $nCO_2 + nH_2O \rightarrow nCH_2O^a + nO_2$ Soil chemosynthesis (nitrifying, hydrogen-oxidizing, iron (Fe^{2+})-oxidizing, and sulfur[c] (S) bacteria) Carbonate formation and leaching
CO	Fermentation O_2-limited oxidation reactions	Activity of carboxydobacteria: $CO + 1/2\ O_2 \rightarrow CO_2$ Spontaneous chemical oxidation
CH_4	Methanogenesis: $CO_2 + H_2 \rightarrow CH_4$ or CH_3COO^-(acetate)$\rightarrow CH_4$	Methanotrophy: $CH_4 \rightarrow CO_2 + H_2O$ Photochemical reactions with hydroxyl and chlorine radicals
N_2O	Nitrification: $NH_4^+ \rightarrow \approx 99\%\ NO_3^- + \approx 1\%\ N_2O$ Denitrification $NO_3^- \rightarrow N_2O \rightarrow N_2$ (molecular nitrogen)	The N_2O is finally converted to stable end products such as NO_3^- (nitrification, aerobic conditions) or N_2 and NH_3 (denitrification, anaerobic conditions) upon completion of the respective processes
Nitric oxide (NO)	Nitrate reduction $NO_3^- \rightarrow NO$	NO oxidation
Isoprene (C_5H_8)	Secondary metabolic reactions in plants: pyruvate $\rightarrow ... \rightarrow$ isoprene Fermentation	Microbial oxidation Formation of phytogenic aerosols and sedimentation

[a]carbohydrate; [b]volatile organic carbon (acetic, propionic, or butyric acid; aldehydes; alcohols; ketones; ethers; etc.); [c]these bacteria can use several S-containing reduced compounds (S, S^{2-}, SO_3^{2-}, etc.) as an energy source.

Controls on nitrous oxide fluxes

The simulation of N_2O emissions requires consideration
of the combined N and C cycles, because the substrates
of denitrification include electron acceptors (NO_3) as
well as oxidizable C substrates. The ecosystem–
atmosphere fluxes of N_2O are associated with fundamental transformations of N in the soil, namely the processes
of nitrification and denitrification. Very few field studies
of N_2O fluxes are available from the Arctic (Christensen
T. et al., 1999a), but very small releases of N_2O are
expected from arctic soils due to their general nutrient
limitation. However, one issue of potentially great
importance in the Arctic is early spring fluxes of N_2O.
Denitrification has been found to take place even below
the freezing point (Dorland and Beauchamp, 1991;
Malhi et al., 1990). During freezing and thawing, C is
liberated and this may increase denitrification activity in
the soil (Christensen S. and Christensen, 1991). During
the spring thaw of the soil, a significant percentage of
annual N_2O emissions can take place (Papen and
Butterbach-Bahl, 1999). The early spring fluxes may also
explain the significant N_2O production measured in fertilized plots on a subarctic heath (Christensen T. et al.,
1999a) on soils that showed no emissions from control
plots. Hence, as with CH_4, the winter and "shoulder"
season processes are generally very important but also
the least well understood.

7.4.2.3. Water and energy balance

Arctic ecosystems exhibit the greatest seasonal changes
in energy exchange of any terrestrial ecosystem because
of the large changes in albedo from late winter, when
snow reflects most incoming radiation, to summer when
the ecosystem absorbs most incoming radiation. About
90% of the energy absorbed during summer is transferred to the atmosphere, with the rest transferred to
the soil in summer and released to the atmosphere in
winter (Eugster et al., 2000). Consequently, arctic
ecosystems have a strong warming effect on the atmosphere during the snow-free season.

Vegetation profoundly influences the water and energy
exchange of arctic ecosystems. In general, ecosystems
with high soil moisture have greater evapotranspiration
than dry ecosystems, as in any climatic zone. Arctic
ecosystems differ from those at lower latitudes, however,
in that there is no consistent relationship between CO_2
flux and water vapor flux, because vascular plants
account for most CO_2 flux, whereas mosses account for
most water vapor flux (McFadden et al., 2003). This
contrasts with other major biomes on Earth, where
these two fluxes are strongly correlated (Kelliher et al.,
1995; Schulze et al., 1994).

Within tundra, vegetation strongly influences the winter
energy budget through its effects on snow depth and
density. Shrubs increase snow depth by reducing the
velocity of blowing snow and reducing sublimation rates;
models suggest that in northern Alaska this shrub-

induced reduction in sublimation can increase ecosys-
tem-scale winter snow accumulation by 20% (Sturm et
al., 2001a). By acting as a "snow fence", shrubs also
cause snow to accumulate within shrub patches and to
be depleted from shrub-free zones, increasing the spatial
heterogeneity of snow depth. Snow within shrub
canopies is deeper and less dense, which reduces heat
transfer through the snowpack and increases winter soil
temperatures by 2 °C relative to adjacent shrub-free tundra. Warmer soil temperatures beneath shrubs may
increase winter decomposition and enhance nutrient
availability, creating a positive feedback that promotes
shrub growth (Sturm et al., 2001a).

Midsummer vegetation feedbacks to regional climate are
determined largely by midsummer patterns of water and
energy exchange (Chapin F. et al., 2000a). Midsummer
albedo is greatest in sedge communities, whose standing
dead leaves reflect much of the incoming radiation
(Chapin F. et al., 2000a; McFadden et al., 1998).
Evergreen forests and forest tundra, in contrast, have a
particularly low albedo because of the dark absorptive
nature of evergreen leaves and the effectiveness of complex forest canopies in capturing light (Chapin F. et al.,
2000b; Eugster et al., 2000). Tundra and forest tundra
canopies dominated by deciduous plants are intermediate in albedo and therefore in the quantity of energy that
they absorb and transfer to the atmosphere (Baldocchi et
al., 2000; Chapin F. et al., 2000b). A larger proportion
of the energy transfer to the atmosphere occurs as sensible heat flux in forests, forest tundra, and shrub tundra
than in wet tundra (Boudreau and Rouse, 1995; Chapin
F. et al., 2000a; Eugster et al., 2000; Lafleur et al.,
1992; McFadden et al., 1998).

All arctic ecosystems exhibit greater ground heat flux
during summer (5 to 15% of net radiation) than do temperate ecosystems (generally close to zero), due to the
strong thermal gradient between the ground surface and
permafrost and the long hours of solar radiation (Chapin
F. et al., 2000a). Ground heat fluxes are reduced in tundra ecosystems with a large leaf area, which shades the
ground surface (McFadden et al., 1998), or where the
ground cover is highly insulative, as with *Sphagnum*
mosses (Beringer et al., 2001a). Grazing and other
processes causing surface disturbance increase ground
heat flux and thaw depth (Walker D. et al., 1998).
Future changes in vegetation driven by climate change
are very likely to profoundly alter regional climate.

7.4.2.4. Summary

Arctic ecosystems tend to accumulate organic matter
and elements despite low inputs because organic matter
decomposition is very slow. As a result, soil-available
elements like N and P place key limits on increases in C
fixation and further biomass and organic matter accumulation. Key issues for projecting whole-system responses
to climate change include the importance of carbon–
nutrient interactions; the interactions of C and nutrient
cycles with temperature, water, and snow cover; the

magnitude of dissolved organic and inorganic C losses in soil water; and the magnitude and role of wintertime processes. Most disturbances are expected to increase C and element turnover, particularly in soils. This is likely to lead to initial losses of elements but eventual, slow recovery. Individual species and species diversity have clear impacts on element inputs and retention in arctic ecosystems, but their magnitude relative to climate and resource supply is still uncertain. Similarly, the current information about the long-term effects of increasing CO_2 concentrations and UV-B radiation levels on whole ecosystems indicates that direct effects of these variables will probably be small relative to changes in soil resources and element turnover. Indirect effects of increasing CO_2 concentrations and UV-B radiation levels are likely to be more important at the ecosystem level (e.g., through changes in species composition).

The most important trace gases in arctic ecosystems are CO_2 and CH_4. Trace-gas exchange with the atmosphere occurs through a set of coupled soil ecosystem processes. Wet and moist tundra environments are known to be significant contributors to atmospheric CH_4. However, CH_4 is also consumed in aerobic parts of the soil. Methane emissions from the ecosystems are a balance between production and consumption, with production more responsive to warming than consumption. Soil warming in the absence of any other changes is very likely to accelerate emissions. Winter processes and vegetation type also affect CH_4 emissions. Nitrous oxide emissions are also sensitive to winter conditions and potential winter warming.

Arctic ecosystems exhibit the largest seasonal changes in energy exchange of any terrestrial ecosystem because of the large changes in albedo from late winter, when snow reflects most incoming radiation, to summer when the ecosystem absorbs most incoming radiation. Vegetation profoundly influences the water and energy exchange of arctic ecosystems. Vascular plants account for most CO_2 flux, whereas mosses account for most water vapor flux. Albedo during the period of snow cover declines from tundra to forest tundra to deciduous forest to evergreen forest. Shrubs and trees increase snow depth, which in turn increases ground heat fluxes; ecosystems with a large leaf area and insulating moss carpets reduce ground heat fluxes and conserve permafrost. Future changes in vegetation driven by climate change are very likely to profoundly alter regional climate.

7.5. Effects of climate change on landscape and regional processes and feedbacks to the climate system

Biological and physical processes and phenomena in the arctic system operate at various temporal and spatial scales to affect large-scale feedbacks and interactions with the earth system. Understanding these processes at multiple scales is critical because the effects of the complex interactions between physical, biological, and

human dimensions on system performance cannot be projected by simply applying a different scale to existing results. Therefore, a multidisciplinary and quantitative approach is necessary to understand and project the response of the arctic system to variability in temperature and moisture. The large scale, inter-related processes described in this section include:

- ecosystem processes extrapolated to the landscape or regional scale (e.g., trace-gas exchange, water and energy exchange, and disturbance);
- changes in ecosystem distribution and abundance in the landscape;
- changes in vegetation zonation (e.g., treeline movement);
- interactions between terrestrial and freshwater ecosystems; and
- regional feedbacks.

Paleoclimate studies and studies of the contemporary Arctic together have identified four potential feedback mechanisms between the impacts of climate change on the Arctic and the global climate system; these are:

- albedo (reflectivity);
- greenhouse gas emissions and/or uptake through biological responses to warming;
- greenhouse gas emissions from methane hydrates released from thawing permafrost; and
- freshwater fluxes that affect thermohaline circulation.

In the past, three of the potential feedbacks have been generally positive and only one negative.

Some of the feedbacks such as energy and water exchange operate at local to regional scales whereas others, particularly trace-gas fluxes, have the potential to operate at regional to global scales. This section assesses the impacts of changes in climate (but not UV radiation levels, for which data are lacking) on ecosystem processes at the larger scale. The section explores the implications of these changes for feedbacks from terrestrial ecosystems to the climate system, but does not calculate changes in forcing (section 4.7.1). Nor does it consider freshwater discharge (sections 6.8.3 and 8.4.2) or methane hydrate feedbacks (section 6.6.2).

7.5.1. Impacts of recent and current climate on carbon flux

There are two complementary approaches to solve the carbon-flux inventory problem: "bottom-up" and "top-down". The first is based on the long-term monitoring of gas emissions within networks of field stations or sites that cover the main types of habitats. At its simplest, total circumpolar emissions are estimated from the number and area of the types of northern ecosystems differentiated in terms of easily mapped features (e.g., vegetation, soil properties, relief, and geomorphology) and the characteristic annual exchange of CO_2 and CH_4 from each ecosystem. The data on CO_2 and CH_4 fluxes come from three main groups of available techniques that operate at different spatial scales: closed and open-top chambers (0.1 to 1 m^2), micrometeorological towers based on eddy covariance and gradient methods (10 to 10 000 m^2), and aircraft sensing (up to tens and hundreds of square kilometers). All three groups of techniques have their advantages and disadvantages. However, continuous measurements with towers seem to be the most appropriate for providing reliable information on the temporal variation of gas emissions at the ecosystem and landscape spatial levels.

7.5.1.1. Recent changes in carbon dioxide fluxes

Recent variations in arctic climate have had profound effects on some ecosystem- and regional-level C fluxes and, in general, these fluxes reflect the recent spatial variability in climate change. The assessment in this section is restricted to C in the active layer of soils and in plants, and does not consider C in permafrost and methane hydrates (section 6.6).

The North Slope of Alaska has experienced a rise in temperature (Fig. 7.26; Weller, 2000), an increase in growing-season length, and a decrease in available soil moisture (Oechel et al., 1993, 1995, 2000a; Serreze et al., 2000) over the last three to four decades. This has resulted in North Slope ecosystems changing from a sink for C throughout the Holocene (Marion and Oechel, 1993) to a source of C to the atmosphere beginning in the mid-1970s to early 1990s (Oechel et al., 1993, 1995, 2000a; Fig. 7.26). However, as there has been a change in climate, with progressive warming, drying, and lengthening of the growing season, there has been physiological, community, and ecosystem level adjustment that has reduced the rate of C loss from North Slope ecosystems (Fig. 7.26). In addition, wetter areas of the North Slope are not showing the same increase in CO_2 emissions (Harazono et al., 2003). The interannual variations in the C balance measured on the coastal plain of Alaska are very large, ranging from a net summer CO_2 uptake of about 25 g C/m^2/yr to a summer loss of over 225 g C/m^2/yr. If the latter fluxes held worldwide for wet coastal and moist tussock tundra, this would result in a net loss of up to 0.3 Pg C/yr from these two ecosystem types alone.

In northeast Greenland, the recent climatic history is different than that of Alaska: there has been no significant trend toward higher temperatures (Weller, 2000) and integrating over all vegetation types shows that the Zackenberg valley is a small net sink (2.3 g C/m^2/yr) with a large uncertainty range (±16.2 g C/m^2/yr). This integrated study of the valley shows that Landsat-derived C flux estimates are in good agreement with ground-based eddy correlation flux measurements covering all the dominant vegetation types in the valley. The Landsat method estimated a midday uptake rate in August 1997 of 0.77 g C/m^2/d for the valley as a whole whereas the ground-based measurements showed the

uptake rate to be 0.88 g C/m^2/d (Søgaard et al., 2000). The measured annual balance in the valley varies from significant uptake in the intensively studied fen areas (on the order of 18.8±6.7 g C/m^2/yr) to net C losses in the dry heath (Christensen T. et al., 2000; Nordström et al., 2001; Søgaard et al., 2000).

Like Alaska, northern Scandinavian areas have experienced warming in recent years. The ecosystem C balance for a subarctic Swedish peatland was found to be a sink of between 15 and 25 g C/m^2/yr (Friborg, pers. comm., 2005). Similarly, in Finland, a net annual uptake of about 20 g C/m^2/yr was reported for a subarctic fen at Kaamanen (Aurela et al., 2002). Six years of continuous measurements at this fen show marked interannual variation in the CO_2 balances (sinks ranging from 4 to 52 g C/m^2/yr), which mainly reflect the variations in spring temperatures and the timing of the snow melt (Aurela et al., 2004). Studies of fluxes in high-arctic barren tundra on Svalbard show a very limited source of around 1 g C/m^2/yr (Lloyd C., 2001). Overall, the synthesis of regional C flux information from measurements at several sites in northern Europe and Greenland (the Land Arctic Physical Processes project; Laurila et al., 2001) indicates that arctic landscapes are remarkably similar in their C fluxes during midsummer, but the length of the growing season and the shoulder season fluxes are the key determinants for the net annual fluxes. This causes substantial interannual variability at the individual sites, and general uncertainty about whether the circumpolar Arctic is presently a source or a sink for C.

Recent work in East European tundra indicates a substantial current source of C in the northeastern European tundra areas (Heikkinen et al., 2004). When combined with the areas of the northern Alaska tundra mentioned above that are also a source of C, source areas (East European tundra, Svalbard, and Alaska) may exceed sink areas (northeast Greenland, northern Scandinavia). However, data are available for only a small part of the Arctic.

There may be a correlation between C balance and recent climatic history in areas that have seen a significant warming and drying: these areas experienced at least a temporary release of CO_2, while areas that have not experienced the same extent of warming and drying, or have possibly experienced a warming and wetting, remain atmospheric CO_2 sinks and could possibly even become large sinks. A complete synthesis of the available information from the circumpolar Arctic is underway but not yet available.

New models and approaches make estimation of current and future global C balances possible. The modeling approach has been used to explore potential changes in arctic terrestrial ecosystems from C sink to source status (Clein et al., 2000; McGuire et al., 2000). The Terrestrial Ecosystem Model (Marine Biological Laboratory, Woods Hole) has been used to estimate current and future C fluxes, while the model Hybrid v4.1

(Institute of Terrestrial Ecology, United Kingdom: Friend et al., 1997) has been used to simulate vegetation and carbon-pool changes at high latitudes for the period from 1860 to 2100 (White et al., 2000). For the present day, the models simulate a mix of C sinks and sources that reflect variations in current and past climate. McGuire et al. (2000) estimate that average circumpolar C fluxes presently constitute a small sink of 17 g C/m²/yr with a standard deviation of 40 g C/m²/yr that crosses the boundary between sink and source status. This uncertainty range is comparable to the Lund-Potsdam-Jena Dynamic Global Vegetation Model (LPJ) outputs (section 7.5.4.1; Sitch et al., 2003) and the calculation of current sink status corresponds to the projections by White et al. (2000).

Although it can be concluded that source areas currently exceed sink areas, there is great uncertainty about the current CO_2 balance of the Arctic, owing to geographically sparse measurements and inadequate representation of ecosystem dynamics in current models.

7.5.1.2. Current circumpolar methane fluxes

Probably the most intensive studies and the longest observations of CH_4 fluxes were performed in North America, mainly within the central Alaskan and North Slope sites at Barrow, Atqasuk, Toolik Lake, and Prudhoe Bay (Christensen T., 1993; Morrissey and Livingston, 1992; Vourlitis et al., 1993; Whalen and Reeburgh, 1990, 1992). In northern Eurasia including Russia, extensive measurements of gas emissions were initiated in the late 1980s, either as short-term measurements across geographic transects or as long time series of flux measurements at individual sites. The first approach is illustrated by chamber measurements of CH_4 (and CO_2) fluxes across the Russian Arctic (Christensen T. et al., 1995, 1998). The second approach is found at a number of field stations where gas fluxes are measured mainly during the summer season (Panikov et al., 1995, 1997; Wagner et al., 2003; Zimov et al., 1993).

The general characteristics of spatial and temporal flux variations can be formulated as follows. First, there are evident temperature-related variations: even within northern wetlands, the highest net fluxes occur in warmer soils, with maximum values attained in the boreal zone. This trend is especially evident with respect to CH_4 emissions, which increase along the latitudinal sequence Barrow–Toolik Lake–Fairbanks or Taymir–Surgut–Tomsk. Seasonal variations also follow a temperature curve, although winter, autumn, and spring emissions are often measurable (sections 7.3.2.3 and 7.4.2.1). Seasonal measurements of CH_4 emissions from five different wetland sites along a transect from northeast Greenland across Iceland and Scandinavia to Siberia also showed a clear positive relationship to the mean seasonal temperatures of the sites (Christensen T. et al., 2003b). Second, there are always enhanced emissions from wetland patches covered by vascular plants (e.g., *Eriophorum*, *Carex*, and *Menyanthes*) as compared with pure

Sphagnum lawn (section 7.4.2.2). Third, variations in the water table affect CH_4 (and CO_2) emissions in opposite ways, with CH_4 fluxes stimulated and CO_2 suppressed by an increase in the water table. However, the range of fluxes varies so widely that uncertainty in regional and global estimates remains too large and is very dependent on the site-specific features of a particular study. For example, extensive measurements using various techniques over the Hudson Bay Lowland (Roulet et al., 1994) led to the conclusion that northern wetlands are modest sources of atmospheric CH_4 (average July emissions as low as 10 to 20 mg CH_4/m²/d). On the other hand, Alaskan wet meadow and shrub/tussock tundra have average summer emissions as high as 100 to 700 mg CH_4/m²/d (Christensen T., 1993; Whalen and Reeburgh, 1992). The uncertainty in regional and global estimates that follows from these differences in measured fluxes is very frustrating and calls for alternative ways to solve the problem of scaling up fluxes.

One such alternative is the inverse modeling approach. In this top-down approach, information about the temporal and spatial variation of CH_4 and CO_2 emissions from soils is deduced from observations of gas mixing ratios in the atmosphere (obtained from a network of National Oceanic and Atmospheric Administration/Climate Monitoring and Diagnostics Laboratory field stations scattered over the globe, mainly in oceanic regions far from industrial effects). These data are fitted to a three-dimensional atmospheric transport model, which is combined with a tropospheric background chemistry module and accounts for all essential sources and sinks of gases. The model is validated against an "internal standard" such as methyl chloroform. Presently available results from inverse modeling (Hein et al., 1997) do not deviate significantly from data obtained by the bottom-up approach. The contribution of high-latitude regions (>60° N) to the global CH_4 source was less than 13% (70 Tg CH_4/yr), and northern wetlands were responsible for emissions of less than 30 Tg CH_4/yr. This estimate appears to contradict the latitudinal gradient of atmospheric CH_4 concentration that has a well-expressed maximum in the north, however, the higher concentration of CH_4 in the high-latitude atmosphere can be explained by low concentrations of the hydroxyl radical and, hence, lower rates of photochemical reactions that break down atmospheric CH_4.

7.5.1.3. Relative contributions of methane and carbon dioxide to the carbon budget

The formation of CO_2 and CH_4 is a result of aerobic and anaerobic decomposition, respectively. The ratio of respired CO_2 to CH_4 is hence an indication of how reduced the soil environment is. An increasingly reduced soil environment (i.e., higher CH_4/CO_2 ratio) also leads to slower overall decomposition rates, as anaerobic decomposition is less efficient in absolute C terms (i.e., less C is released compared to aerobic decomposition). This generally leads to a buildup of stored organic C in wet tundra soils, as net primary production is not normally limited by wet soil conditions to the same extent as respiration.

The net CH_4/CO_2 ratio of the total respiration is also a function of the amount of CH_4 that is oxidized in the aerobic soil layers above a given anaerobic zone of production and even the atmospheric CH_4 uptake that takes place in some dry tundra soil environments. The CH_4/CO_2 ratio or the percentage contribution of CH_4 to total respired C varies from <1% in dry ecosystems to >20% in extreme cases in wet tundra ecosystems. Typical annual average contributions of CH_4 to the total C flux range from 2 to 10% for wet tundra and northern wetlands (e.g., Christensen T. et al., 1996; Clymo and Reddaway, 1971; Klinger et al., 1994; Svensson, 1980; Svensson et al., 1999).

In the context of climate change, it is very important to note that the relative contribution of CH_4 to total radiative forcing is much greater on a per molecule basis than CO_2 (IPCC, 2001). The "global warming potential" (GWP) indicates how many times stronger a given greenhouse gas is compared to CO_2 on a per molecule basis, and this is dependent on the time horizon of interest. For example, over a 100-year time horizon, the GWP of CH_4 is 23 and over a 20-year horizon it is 63 (IPCC, 2001).

From the perspective of climate change, it is, hence, not very informative to look only at the C balance of an ecosystem if there are fluxes of CH_4 or other greenhouse gases such as N_2O (Christensen T. and Keller, 2003). Calculations have shown that ecosystems such as the huge western Siberian lowlands, despite being strong sinks for C, are sources of radiative forcing due to considerable CH_4 emissions (Friborg et al., 2003). Data are, however, scarce when it comes to full annual budgets of both CO_2 and CH_4 fluxes from tundra regions. Figure 7.31 shows calculations based on accumulated continu-

ous eddy correlation measurements of CO_2 and CH_4 fluxes in the Zackenberg Valley, northeast Greenland during the summer of 1997 (Friborg et al., 2000; Søgaard et al., 2000). The figure illustrates that the effect of a net C accumulation ("minus" in the budget) during the season is completely offset if the CO_2 equivalent of CH_4 is calculated (using a 20-year time horizon) and added to the budget. Using the 100-year time horizon, the ecosystem is still a small sink of CO_2 equivalent at the end of the growing season. However, if the autumn and winter fluxes (which are entirely sources) were included, the annual total would probably add up to a source as well.

In general, there are significant CH_4 emissions from the most productive tundra areas due to the predominantly wet soil conditions. It is very likely that, at the landscape, regional, and global scales, the tundra represents a source of radiative forcing due to emissions of CH_4, which is the most important greenhouse gas driving the ecosystem influence on atmospheric radiative forcing.

7.5.2. Current circumpolar water and energy balances

Arctic ecosystems exhibit the greatest seasonal changes in energy exchange of any terrestrial ecosystem because of the large changes in albedo from late winter, when snow reflects most incoming radiation (albedo about 0.7), to summer when the ecosystem absorbs most incoming radiation (albedo about 0.15). This change in albedo combined with greater incoming solar radiation in summer than in winter causes much greater energy absorption in summer than in winter. About 90% of the energy absorbed during summer is transferred to the atmosphere, with the rest transferred to the soil in summer and released to the atmosphere in winter (Eugster et al., 2000). Also, snow within shrub canopies is deeper and less dense, which reduces heat transfer through the snowpack and increases winter soil temperatures by 2 °C relative to adjacent shrub-free tundra. Consequently, arctic ecosystems have a strong warming effect on the atmosphere during the snow-free season, and any increase in the duration of snow-free conditions results in a strong positive feedback to regional climate warming (Bonan et al., 1995; Foley et al., 1994).

Climate influences the partitioning of energy between sensible and latent heat fluxes. Cold moist air from coastal oceans, for example, minimizes latent heat flux (evapotranspiration), as does extremely warm dry air, which can induce stomatal closure (Eugster et al., 2000; Rouse, 2000); evapotranspiration is therefore greatest at intermediate temperatures. Conversely, sensible heat flux is a larger proportion of the energy transfer to the atmosphere when air is cold and moist or when drought limits stomatal conductance under dry conditions. Heat that is conducted into the ground during summer is released to the atmosphere in winter, with any seasonal imbalance causing changes in permafrost temperature leading to a probability of thermokarst (Osterkamp and Romanovsky, 1999).

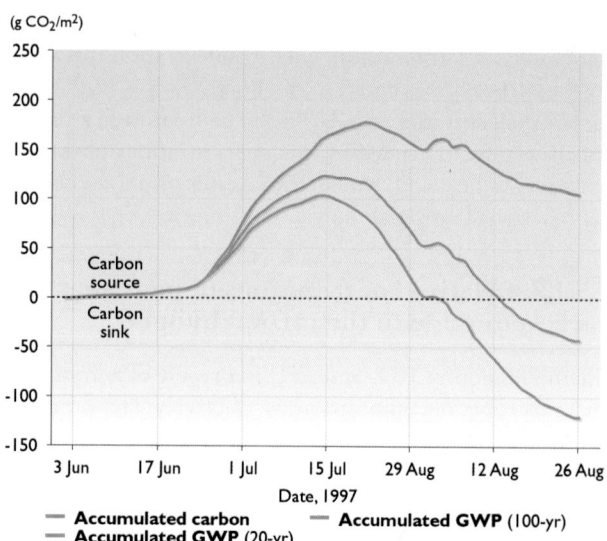

Fig. 7.31. Accumulated carbon and global warming potential (GWP, calculated as CO_2 equivalent) from CO_2 and CH_4 fluxes measured throughout the summer of 1997 at Zackenberg, northeast Greenland (data from Friborg et al., 2000; Søgaard et al., 2000).

There are large regional differences among arctic ecosystems in energy exchange and partitioning. Albedo during the period of snow cover is extremely high in tundra and declines with increasing development of a plant canopy above the snow, from tundra to shrub tundra to forest tundra to deciduous forest to evergreen forest (Betts A. and Ball, 1997). These differences in albedo are an important feedback to climate during spring, when the ground is snow-covered, and incoming radiation is high. As a result of differences in albedo and sensible heat flux, forests at the arctic treeline transfer about 5 W/m^2 more energy to the atmosphere than does adjacent tundra (Beringer et al., 2001b). This difference in energy transfer to the atmosphere depending on vegetation type is an order of magnitude less than the heating contrast that had been hypothesized to be required for treeline to regulate the position of the Arctic Front (Pielke and Vidale, 1995). Thus, the location of the Arctic Front is more likely to govern the position of the treeline than the other way around (Bryson, 1966).

7.5.3. Large-scale processes affecting future balances of carbon, water, and energy

This section assesses the effects of climate change on permafrost degradation and vegetation redistribution as a prerequisite for assessing changes in feedbacks from future terrestrial ecosystems to the climate system.

7.5.3.1. Permafrost degradation

Soil C storage is greatest where the drainage is slight and the limited precipitation is held near the surface by permafrost and modest topography. This results in ponds, wetlands, and moist tundra with a saturated seasonal active layer that limits microbial activity. Increases in active-layer depth can cause subsidence at the surface, a lowering of the soil water table (Hinzman et al., 2003), and, potentially, thermokarst erosion (Hinzman et al., 1997). This can drain surrounding areas, often increasing the decomposition rate of soil organic matter, which accelerates the loss of belowground C stores (Oberbauer et al., 1991, 1996) and results in a change in plant communities and their abilities to sequester atmospheric CO_2. Initially, increased soil decomposition rates can increase mineralization rates (Rastetter et al., 1997) and result in increased NPP (section 7.4.2.1). However, continued thawing of permafrost and increased drainage of surface water in areas with low precipitation are likely to lead to drier soils, a decrease in NPP, and possibly even desertification (section 7.5.3.2).

Full permafrost disintegration in subarctic discontinuous permafrost regions may in some cases show a different response. Monitoring of changes in permafrost distribution in subarctic Sweden as part of the Circumpolar Active Layer Monitoring Program (Brown et al., 2000) shows that permafrost loss causes mires to shift from ombrotrophic moss- and shrub-dominated systems to wetland systems dominated by minerotrophic vascular plants (Christensen T. et al., 2004; Svensson et al.,

1999). This, in turn, leads to a significant lowering of soil redox potentials, an increase in anaerobic decomposition, and increased CH_4 emissions. Wet minerotrophic soils and vegetation are in general associated with the highest CH_4 emissions in subarctic and arctic tundra environments. Discontinuous permafrost regions are considered some of the most vulnerable to climate warming; given projected temperature increases over the next 100 years, effects such as the changes in vegetation composition discussed here are likely.

Permafrost degradation and disintegration are therefore very likely to have major effects on ecosystem C balances and CH_4 emissions. The rate of permafrost thawing, the amount of ground surface subsidence, and the response of the hydrological regime to permafrost degradation will depend on numerous site characteristics. Changes in the hydrological regime will also alter the soil thermal regime. In areas with significant topographic variations, flowing water can carry heat into drainage channels, causing increased soil temperatures and increased active-layer thickness (Hastings et al., 1989; Kane et al., 2001). In regions with minor topographic variations, subtle differences in elevation can create cooler, saturated wetlands (as mentioned in the previous paragraph) or markedly drier, warmer uplands (Rovansek et al., 1996).

7.5.3.2. Changes in circumpolar vegetation zones

While climate-driven changes in the structure and distribution of plant communities affect trace gas fluxes and water and energy balances at the landscape scale (section 7.4.2), changes in the location and extent of broad vegetation zones is a longer-term integrative process that is likely to lead to regional and even global impacts on feedbacks to the climate system (Betts R., 2000; Chapin F. et al., 2000a; Harding et al., 2002; McGuire et al., 2002). Such vegetation zone changes are also likely to affect permafrost dynamics (section 6.6), biodiversity (section 7.3.1), and ecosystem services (Vlassova, 2002). Past climate-driven changes in vegetation zones such as forest and tundra (section 7.2; Payette et al., 2002) lead to the expectation that future climate warming is very likely to result in vegetation and ecosystem change, but projecting future changes is complex and relies on modeling.

Dynamics of the treeline and changes in the areas of tundra and taiga vegetation

The latitudinal treeline or tundra–taiga boundary is an exceptionally important transition zone in terms of global vegetation, climate feedbacks, biodiversity, and human settlement.

The treeline stretches for more than 13 000 km around the Northern Hemisphere and through areas that are experiencing different types of environmental change, for example, cooling, warming, or marginal temperature change and increasing or decreasing land use. Climate is only one of a suite of environmental factors that are now changing, and a critically important challenge is to determine how human impacts in the ecotone will modify its projected response to climate change (Vlassova, 2002).

The lack of standardized terminology and the wide variation in methodology applied to locate, characterize, and observe changes in the boundary have resulted in a poor understanding of even the current location and characteristics of the boundary. Particular areas of uncertainty include the Lena Delta of Siberia (Callaghan et al., 2002a) and forests in Iceland that have been subjected to major environmental and land-use changes since colonization by people about 1100 years BP (section 14.3.4.4). One of the major problems with current studies of the latitudinal "treeline" is the concept of "line" inappropriately applied to the transition from forest, through an area dominated by forest in which patches of tundra occur, to tundra in which patches of forest occur, and then eventually to tundra without trees. Often there are east–west gradients related to the presence of a river valley, bogs, mires, uplands, etc. that also confound the concept of a linear boundary.

Dynamics of the boundary

Current and projected changes in the location of the tundra–taiga boundary should be seen in the context of the longer-term past cooling trend during which the treeline has been at its lowest latitudinal and altitudinal locations for several thousands of years (section 7.2.4). Examples of recent treeline advance include a 40 m increase in the elevation of the subarctic treeline in northern Sweden during the 20th century (Emanuelsson, 1987; Kullman, 1979; Rapp, 1996; Sonesson and Hoogesteger, 1983), an increase in shrub growth in Alaska (Sturm et al., 2001b), and an increase in shrubbiness and larch advance in the northeast Russian European Arctic (Katenin, unpub. data, 2004). In contrast, other studies show a surprising southward displacement of the tundra–taiga boundary (Crawford et al., 2003; Kozhevnikov, 2000; Vlassova, 2002). Part of this is a counter-intuitive response to warming in which an increasingly oceanic climate (e.g., in the western Russian Arctic: see section 8.4.1, Fig. 8.9) together with permafrost thawing have led to paludification (waterlogging and peat formation) and the death of treeline trees (Skre et al., 2002). Part is associated with human activities, including mining, farming, and forestry, that have led to ecosystem degradation in the forest tundra zone and the movement of its northern boundary southward in some locations (Vlassova, 2002). In the Archangelsk region and the Komi Republic, the southern border of the forest tundra zone now lies 40 to 100 km further south than when previously surveyed. One report claims that anthropogenic tundra now covers about 470 000 to 500 000 km² of the forest tundra stretching from Archangelsk to Chukotka (Vlassova, 2002), although it is likely that this estimate includes deforestation in some of the northern boreal forest zone.

Although records of recent changes in the location of the latitudinal treeline are surprisingly rare, there is good evidence of increased growth in current northern forests. Comparisons of the greenness index (NDVI) from satellite images show that May to September values for the Northern Hemisphere between 55° and 75° N increased by around 5 to 50% from 1982 to 1990 (Myneni et al., 1997; Fig. 7.25), with greater increases in North America than in Eurasia. The increased greenness was associated with an increase in growing-season length of between 3.8 and 4.3 days for the circumpolar area, mainly due to an earlier start of the growing season.

Projecting future changes in the tundra–taiga boundary

In order to model changes in the location of the tundra–taiga ecotone and estimate future areas of tundra to the north and taiga to the south, it is necessary to understand the causes of the treeline. Opinions on the mechanisms controlling the location of the treeline vary greatly. Some researchers see the limit of tree growth as a universal mechanism related to a specific process such as sink (C and nutrient) limitation (Körner, 1998, 1999) or C limitation (Nojd and Hari, 2001). Others see a range of possible mechanisms that operate in different places and at different times (Sveinbjörnsson et al., 2002). These mechanisms are in turn affected by environmental factors such as incident radiation, temperature, wind, moisture, and soil nutrients, which affect tree reproduction, seedling establishment, and the growth and physiology of mature trees. Extreme conditions such as ice-crystal abrasion and soil movement also directly damage tree tissues (e.g., conifer needles) and displace individuals. Diseases, pests, fires, and human activities all exert some control on the treeline at certain places and at certain times (Chapter 14).

Models of vegetation redistribution resulting from global change utilize more general driving variables, related to, for example, biogeography and biogeochemistry. Most current global vegetation models and regional models project that a major part of the tundra (between 11 and 50% according to location) will be displaced by a northward shift of the boreal forest over the period in which atmospheric CO_2 concentrations double (Harding et al., 2002; Kaplan et al., 2003; Skre et al., 2002; Table 7.13; section 7.6, Table 7.14; Fig. 7.32). The treeline is projected to move north in all sectors of the Arctic, even in Greenland and Chukotka where only fragments of forest exist today (Kaplan et al., 2003). However, to date there have been fewer observations of this type of forest response than projections suggest, even though temperature has already risen dramatically in some areas.

The observations of the latitudinal treeline that show a recent *southern* displacement suggest that there is very unlikely to be a general northward displacement of the latitudinal treeline throughout the circumpolar region as the models project. In addition to potential paludification (Crawford et al., 2003) and local human activities displacing treelines southward, permafrost thawing, surface-water drainage, and soil drying in areas of low precipitation are likely to lead to the formation of vegetation similar to the tundra–steppe (Yurtsev, 2001; section 7.2.2). Increased disturbances such as pest outbreaks, thermokarst, and fire are also likely to locally affect the direction of treeline response. In addition, some tree species show reduced responsiveness to increases in temperature with increasing continentality of their location and decreased precipitation (Linderholm et al., 2003; section 14.7). This suggests that increased temperatures with no comparable increase in precipitation are likely to lead to reduced tree growth and/or changes in species and lack of treeline advance. Even in areas projected to undergo warming with none of the moderating factors listed above, it is uncertain if the rate of tree migration can keep up with the rate of projected temperature increases. Past tree migration rates were generally of the order of 0.2 to 0.4 km/yr but sometimes reached 4.0 km/yr (section 7.2; Payette et al., 2002). These rates suggest that in those areas of the Arctic that have warmed substantially in the last 30 years, treeline should already have advanced by about 6 to 120 km. Such shifts have not been recorded in the Arctic, although Parmesan and Yohe (2003) claim to have identified a poleward displacement of species ranges of 6.1 km per decade globally.

Overall, it is likely that treeline will show many different responses throughout the circumpolar north according to different degrees of temperature change associated with various changes in precipitation, permafrost dynamics, land use, and tree species migration potential.

Projecting future changes in the areas of tundra and polar desert

The LPJ model was used to project vegetation changes in the northern areas of the Arctic (Sitch et al., 2003; Table 7.13; section 7.6, Table 7.14), using climate scenarios generated by four of the five ACIA-designated models (CGCM2, GFDL-R30_c, HadCM3, and ECHAM4/OPYC3; section 4.2.7) forced with the B2 emissions scenario (section 4.4.1). Although the results and interpretations are preliminary, LPJ simulations are consistent in projecting a decrease in the area of polar desert that will be replaced by northward-moving tundra (Table 7.13). The area of the Arctic covered by polar desert is projected to decrease by 17.6% (range 14 to 23%) between 1960 and 2080. In this model, the two vegetation zones were defined by plant functional types: woody species for the tundra, and absence of woody species for the polar desert. In the BIOME4 (coupled biogeography and biogeochemistry model) simulations by Kaplan et al. (2003), driven by output from the HadCM2GSa1 model forced with the IS92a emissions scenario, five tundra biomes were constructed (Table 7.8). The most significant changes appear to be a significant northward advance of the cold evergreen needleleaf forest that is particularly dramatic in the region of arctic Russia between Chukotka and the Taymir Peninsula. This greatly reduces the area of tundra. However, low- and high-shrub tundra in the Canadian Archipelago remains as a wide zone and displaces prostrate dwarf-shrub tundra (Figs. 7.2 and 7.32). Earlier modeling by White et al. (2000) projected that the area of tundra would be halved by forest expansion by 2100.

Table 7.13. Averages and ranges of the drivers and responses of a leading dynamic global vegetation model (LPJ) driven with the outputs for the terrestrial Arctic (>60° N) from four different climate models (CGCM2, GFDL-R20_c, HadCM3, ECHAM4/OPYC3) forced with the B2 emissions scenario (Sitch et al., 2003; Sitch et al., submitted).

	Average	Range
Temperature change (°C)		
2000–2100	5.0	4.7 to 5.7
Precipitation change (mm/yr)		
2000–2100	42.9	9.0 to 78.0
NPP (Pg C/yr)		
1960s	2.83	2.77 to 2.88
2080s	4.87	4.57 to 5.19
Change (%)	72.4	60.9 to 87.4
Change in C storage 1960–2080 (Pg C)		
Vegetation	5.73	3.59 to 7.65
Soil	6.98	1.6 to 15.6
Litter	5.6	3.4 to 9.6
Total	18.3	12.2 to 31.3
Percent areal vegetation change[a]		
Increase in taiga area[b]		
1960–2020	4.4	3.1 to 5.3
1960–2050	7.4	6.4 to 8.4
1960–2080	11.3	9.8 to 14.4
Change in polar desert area[c]		
1960–2020	-7.5	-13.3 to -4.2
1960–2050	-13.2	-18.5 to -10.6
1960–2080	-17.6	-23.0 to -14.2

[a]Only a proxy as the change is derived from functional characteristics of the vegetation produced by the model rather than projections of specific vegetation composition *per se*. For a proper vegetation distribution estimate, it would be more appropriate to use a coupled biogeography and biogeochemistry model such as BIOME4; [b]based on the percentage increase in woody plants simulated by LPJ; [c]based on the percentage reduction in bare ground simulated by LPJ

7.5.4. Projections of future balances of carbon, water, and energy exchange

Because the Arctic contains huge stores of C in the soil and permafrost (section 7.4.2), and because the Arctic has the capacity for unlimited additional storage or significant loss (Billings et al., 1982, 1984), it can be a major positive or negative feedback to increasing trace-gas concentrations in the atmosphere and to global climate change. Loss of CO_2 from arctic ecosystems could potentially lead to enormous positive feedbacks to global climate change by release to the atmosphere of the estimated 250 Pg C stored in the large arctic soil pool, although this is unlikely to happen (Billings, 1987;

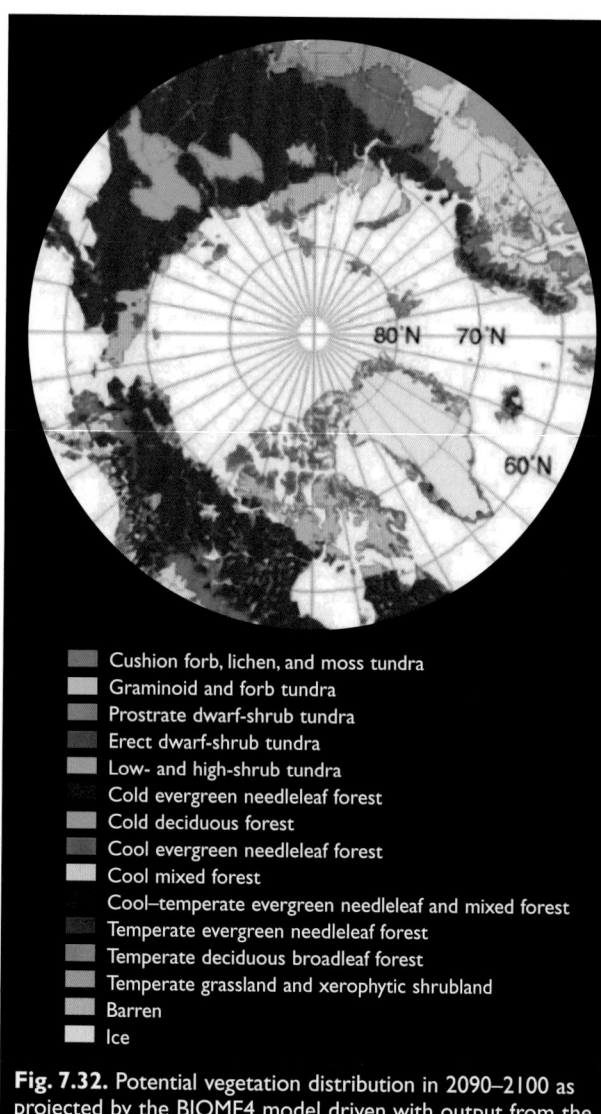

Fig. 7.32. Potential vegetation distribution in 2090–2100 as projected by the BIOME4 model driven with output from the HadCM2GSa1 model forced with the IS92a emissions scenario (Kaplan et al., 2003).

Billings et al., 1982; Lal et al., 2000; Oechel and Billings, 1992; Oechel et al., 1993, 1995, 2000a). In addition, an increasing snow-free period (Engstrom et al., 2002; Keyser et al., 2000), increasing shrub cover (Silapaswan et al., 2001; Sturm et al., 2001a,b), and the northward migration of the treeline (Lloyd A. et al., 2003) are very likely to decrease arctic albedo and further increase regional warming (Bonan et al., 1992; Chapin F. et al., 2000a,b; Foley et al., 1994; McFadden et al., 1998; Thomas and Rowntree, 1992). This section assesses likely changes in the C balance and water and energy exchange in relation to vegetation change.

7.5.4.1. Carbon balance

The vegetation distribution model BIOME3, forced with present-day and doubled atmospheric CO_2 concentrations, projected decreases in the extent of the Scandinavian, central northern Siberian, and Eurasian tundra areas ranging from 10 to 35% as a result of displacement by taiga (Harding et al., 2002). This vegetation change was projected to significantly increase CO_2 uptake and reduce CH_4 emissions, with net C seques-

tration in the biosphere of a magnitude (4.6 Pg C) that would alter the radiative forcing of the earth. Using another model, McGuire et al. (2000) estimated that circumpolar mean C uptake would increase from a current 12 g $C/m^2/yr$ to 22 g $C/m^2/yr$ by 2100, because NPP increased more than respiration throughout the period (McGuire et al., 2000). It should be noted, however, that throughout the 200-year model run, the standard deviation of C uptake always crosses the zero line (McGuire et al., 2000). White et al. (2000) obtained comparable results from their Hybrid v4.1 model, projecting that high-latitude terrestrial ecosystems would remain a sink for C.

The LPJ model (Sitch et al., 2003; Box 7.1) was used to produce estimates of future changes in arctic C storage and fluxes for this assessment, based on four different general circulation model (GCM) outputs. The results and analyses are preliminary but indicate a consistent net additional sink in the Arctic in 2080 compared to 2000, with increases in accumulated arctic C storage varying between 12 and 31 Pg C depending on the climate scenario used. Figure 7.33 shows the C storage anomalies projected by the LPJ model, and Table 7.14 (section 7.6) shows further details of the regional subdivision of these projections.

There are great uncertainties associated with these estimates due to the complex differential response of NPP and respiration to the climate drivers (temperature, precipitation), which themselves are highly spatially variable and interact. However, the general response of the model seems to be as follows. In areas with little or no vegetation (e.g., polar desert), increasing CO_2 concentrations and temperature (e.g., increasing growing-season length) lead to increased vegetation growth and northward plant migration, resulting in an increase in C stocks. This seems to be a general pattern acting through increased productivity throughout the Arctic, all else being equal (see NPP projections in Table 7.13). However, increased temperature leads also to increased heterotrophic (soil microorganism) respiration. Therefore, areas that at present contain large soil C stocks are likely to release larger amounts of C from the soil as respiration responds to a warmer climate. Whether these areas become net sources or sinks depends on the balance between temperature-enhanced respiration and increased productivity (hence increased biomass and litterfall) due to increased CO_2 concentrations and longer growing seasons. When the LPJ model was forced with the climate projections of the ECHAM4/OPYC3 model, which projects very large temperature increases, projected respiration was enhanced more than productivity. Over the entire Arctic, C storage is projected by the LPJ model to balance, due to northward migration of plants, etc., with C loss in areas that experience large temperature changes and have large stocks of soil C. On the whole, all runs of the LPJ model project that C storage will increase. The scenario with the highest projected temperature change (ECHAM4/OPYC3) results in the low-

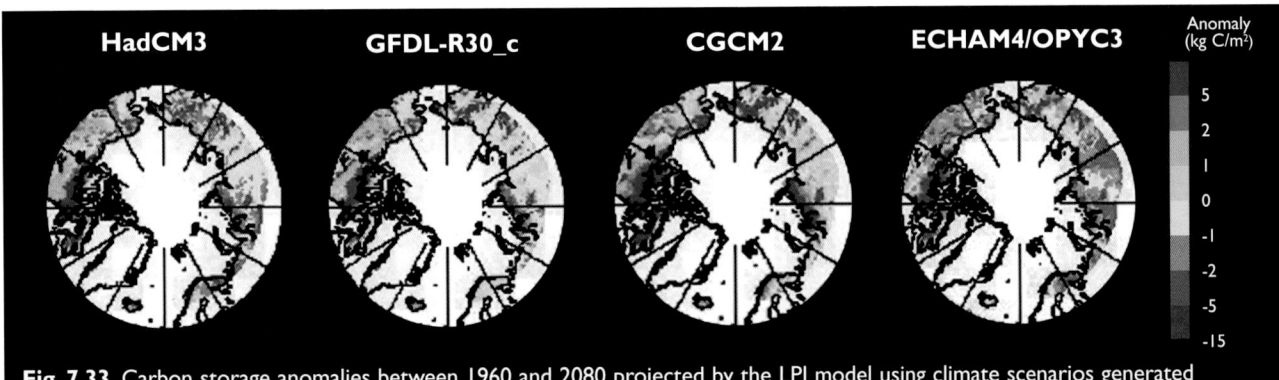

Fig. 7.33. Carbon storage anomalies between 1960 and 2080 projected by the LPJ model using climate scenarios generated by four of the five ACIA-designated models (Sitch et al., 2003).

est overall C gain, while the scenario with the lowest projected temperature change (CGCM2) results in the highest C gain.

The current estimated circumpolar emissions of CH_4 are in the range of 20 to 60 Tg CH_4/yr. These have a significant potential for feedback to a changing climate. Large-scale CH_4 flux models are currently not as advanced as general carbon-cycle models and few allow for projections of future change based on climate change scenarios. Early attempts to assess and model tundra CH_4 emissions driven by climate change all indicated a potential increase in emissions (Christensen T. and Cox, 1995; Harriss et al., 1993; Roulet et al., 1992) but more recent improved mechanistic models (Granberg et al., 2001; Walter and Heimann, 2000) have not yet been followed up by full coupling to GCM projections to assess future circumpolar CH_4 emissions. Another critical factor is the geographic extent of wetlands and how these may change in the future. There is, however, little doubt that with climate scenarios of warming and wetting of arctic soils, CH_4 emissions are very likely to increase, while with warming and drying, emissions are likely to change little or decline relative to current emissions.

Lakes and streams cover large portions of many arctic landscapes and, due to low evapotranspiration, runoff is a major component of arctic water budgets. These surface freshwaters contain large amounts of dissolved organic and inorganic C that is carried into them by soil and groundwater flow from the terrestrial portions of their watersheds (Kling et al., 1992, 2000). The inorganic C is largely CO_2 produced by soil and root respiration. Organic C concentrations in soil water, groundwater, and surface waters are typically several times greater than inorganic C concentrations and are a major source of respiratory CO_2 produced in lakes and streams, thus adding to their already high dissolved inorganic C content.

Because the dissolved CO_2 in surface waters is typically supersaturated with respect to the atmosphere, and the surface area and flow of freshwater is large, surface waters of arctic landscapes emit large amounts of CO_2 to the atmosphere (Kling et al., 1991; section 8.4.4.4). Estimates of CO_2 emissions from surface waters are as large as 20 to 25% of gross landscape CO_2 fixation and thus may be a major component of landscape C balance that is not accounted for in studies that include terrestri-

Box 7.1. The Lund-Potsdam-Jena model

The Lund-Potsdam-Jena (LPJ) dynamic global vegetation model combines process-based, large-scale representations of terrestrial vegetation dynamics and land–atmosphere carbon and water exchanges in a modular framework. Features include feedback through canopy conductance between photosynthesis and transpiration, and interactive coupling between these rapid processes and other ecosystem processes including resource competition, tissue turnover, population dynamics, soil organic matter and litter dynamics, and fire disturbance. Ten plant functional types (PFTs) are differentiated by physiological, morphological, phenological, bioclimatic, and fire-response attributes. Resource competition and differential responses to fire between PFTs influence their relative fractional cover from year to year. Photosynthesis, evapotranspiration, and soil water dynamics are modeled on a daily time step, while vegetation structure and PFT population densities are updated annually.

Within the biosphere model (Sitch et al., 2003), the raw GCM (CGCM2, GFDL-R30_c, HadCM3, and ECHAM4/OPYC3 forced with the B2 emissions scenario) climatologies for 1900 to 2100 were not used directly. The climate simulated by present-day GCMs is not spatially detailed enough to directly drive a biosphere model, therefore the anomaly approach was used. The data were downscaled from the GCM-specific grid onto a grid with 0.5° resolution. Climate anomalies projected by the GCMs were normalized to the 1961–1990 observed average monthly Climatic Research Unit climatology (CRU CL 1.0: New et al., 1999).

al CO_2 fluxes only. Similar large CO_2 losses also occur in freshwaters of boreal, temperate, and tropical landscapes (Cole et al., 1994), but they are generally not considered in landscape-level C budgets. At present, little is known about controls over these CO_2 losses or how they might change with changes in climate or water balance. Attempts to measure the losses directly have yielded inconsistent results (Eugster et al., 2003).

7.5.4.2. Energy and water exchange

Many of the likely changes in water and energy exchange that occur in response to projected future warming are likely to act as a positive feedback to warming. Earlier disappearance of snow from the tundra is very likely to lead to a decline in albedo and an increase in regional warming (Bonan et al., 1992; Thomas and Rowntree, 1992). Similarly, an expansion of forest will lead to a reduction in albedo, because trees mask a snow-covered surface. In areas where forest expansion occurs, this is very likely to lead to significant heating of the lower atmosphere (sections 7.4.2.3 and 7.5.2). Paleoclimate modeling experiments have shown that the northward movement of the treeline 6000 years BP accounted for half of the climatic warming that occurred at that time (Foley et al., 1994). Although the current arctic treeline appears relatively stable or retreating in some areas of human impact (Callaghan et al., 2002a; Vlassova, 2002; section 7.5.3.2), any future northward advance of the treeline is likely to contribute to regional warming, while treeline retreat is likely to contribute to regional cooling, particularly in late spring due to the large differences in albedo between snow-covered tundra and adjacent forest.

A positive feedback (leading to increased warming) of displacement of tundra by trees and shrubs could possibly offset the negative feedback (leading to cooling) due to increased C sequestration at the local level (Harding et al., 2002), but the climate forcing by energy and water exchange operates primarily at the regional scale, where the energy exchange occurs, whereas the negative feedback due to sequestration of atmospheric C is likely to vary between regions and contribute to warming through changes in the globally mixed pool of atmospheric CO_2. Models suggest that forests in the eastern Canadian Arctic would have a net negative feedback through sequestration of C whereas forests in arctic Russia would have a net positive feedback to climate through decreased albedo (Betts R., 2000; Betts A. and Ball, 1997). This complex balance between opposing feedbacks indicates that encouraging forest to displace tundra as an appropriate strategy to mitigate *global* climate change should take into account the local feedback.

An important contributing factor to the effect of vegetation change on albedo is the characteristics of the plant canopy in terms of canopy height relative to snow height, leaf duration, and leaf optical properties. The greatest changes in albedo are likely to occur after increases in area relative to tundra vegetation of the fol-

lowing vegetation types (in decreasing order of effect): dark, evergreen boreal trees such as pine and spruce, deciduous conifer trees such as larch, deciduous angiosperm trees such as birch, and low shrubs such as willows and dwarf birch.

The vegetation changes projected to occur in northern Alaska in response to climatic warming are calculated to increase summer heating of the atmosphere by 3.7 W/m² (Chapin F. et al., 2000a). This warming is equivalent to the unit-area effect of a doubling of atmospheric CO_2 concentrations or a 2% increase in the solar constant (i.e., the difference that caused a switch from a glacial to an interglacial climate), two forcings that are known to have large climatic effects (Kattenberg et al., 1996). Regional climate simulations suggested that a conversion from moist tussock tundra to shrub tundra would cause a 1.5 to 3.5 °C increase in mean July temperature on the Alaskan North Slope, reflecting greater sensible heat fluxes to the atmosphere from the shrub-dominated ecosystem. Thus vegetation changes of the sort that have recently been observed (Sturm et al., 2001b) are very likely to have large positive feedbacks to regional warming if the increased shrub cover is extensive. This vegetation–climate feedback requires only modest increases in shrub density to enhance sensible heat flux (McFadden et al., 1998).

The transition from tundra to forest also affects evapotranspiration and the water storage capacity of the biosphere, such that freshwater runoff via rivers to the Arctic Ocean could possibly decrease (Harding et al., 2002).

Other human activities also have impacts on the local climate of the forest tundra. Deforestation, as a result of industrial activities or forestry, increases wind speeds; pollution leads to earlier snowmelt and increased temperatures, and the northward extension of farming and settlements in general induces permafrost thawing (Vlassova, 2002).

7.5.5. Summary

Biological and physical processes in the Arctic operate at various temporal and spatial scales to affect large-scale feedbacks and interactions with the earth system. There are four main potential feedback mechanisms between the impacts of climate change on the Arctic and the global climate system: albedo, greenhouse gas emissions or uptake by ecosystems, greenhouse gas emissions from methane hydrates, and freshwater fluxes that affect the thermohaline circulation. All of these feedbacks are controlled to some extent by changes in ecosystem distribution and character and particularly by large-scale movement of vegetation zones. However, it is difficult to assess the consequences of the interacting feedbacks or even of individual feedbacks.

There are currently too few full annual measurements available to give a solid answer to the question as to whether the circumpolar Arctic is an atmospheric source or a sink of CO_2 at the landscape scale (Box 7.2). Indications are, however, that source areas currently exceed sink areas. Measurements of CH_4 sources are also inadequate, but the available information indicates emissions at the landscape level that are of great importance to the total greenhouse gas balance of the circumpolar north. In addition to the effect of greenhouse gases, the energy and water balances of arctic landscapes encompass important feedback mechanisms in a changing climate. Increasing density and spatial expansion of the vegetation cover will reduce albedo and cause more energy to be absorbed on the ground; this effect is likely to exceed the negative feedback of increased C sequestration via greater primary productivity. The degradation of permafrost has complex consequences. In areas of discontinuous permafrost, warming is very likely to lead to a complete loss of the permafrost. Depending on local

hydrological conditions, this may in turn lead to a wetting or drying of the environment with subsequent implications for greenhouse gas fluxes. Models projecting vegetation change in response to scenarios of future climate change indicate a 7 to 18% decrease in the area occupied by polar desert and a 4 to 11% increase in the area occupied by taiga over the next 80 years. This is projected to lead to increased carbon storage over this same period due to productivity being stimulated more than respiration. However, the balance depends on the degree of projected warming. With greater temperature increases, heterotrophic respiration is stimulated more and the projected carbon gain is less. Very few models are available for projections of future CH_4 emissions, but these emissions will be extremely important to the total greenhouse gas balance and functioning of the circumpolar Arctic.

Box 7.2. Will the Arctic become a source of carbon, or remain a sink?

There is not yet a definitive answer to this question, although past opinions favored the hypothesis that the tundra will switch from being a sink for carbon during recent millennia to becoming a source under future warming scenarios, based mainly on the response of increased soil respiration to warming relative to increases in photosynthesis. At a recent meeting of experts on carbon dynamics in the Arctic, the following authoritative consensus statements were made.

Consensus statements on arctic carbon source/sink functioning from the Synthesis Workshop on "Current and future status of carbon storage and ecosystem–atmosphere exchange in the circumpolar north: Processes, budgets and projections", Skogar, Iceland, 21 June 2003.

- The available carbon flux data show large interannual variability. Arctic terrestrial ecosystems are a patchwork, with some regions being sources of carbon to the atmosphere (mostly dry and mesic ecosystems) and some regions being sinks (mostly wet tundra). Current indications are that source areas exceed sink areas.
- The available data indicate that when considering both CO_2 and CH_4, the Arctic is a source of radiative forcing.
- Contrary to the data from ground-based measurements, current global carbon models indicate that the Arctic is a small carbon sink. This apparent discrepancy is, however, within the uncertainty range of the data and model outputs.
- The recent remote-sensing record indicates a greening of the Arctic, suggesting increased photosynthetic activity and net primary productivity, but does not address the belowground processes (i.e., respiration).
- Projections from global models of vegetation and soil responses to climate change suggest that enhanced vegetation production will exceed increases in decomposition, thereby resulting in net carbon sequestration. However, there are large uncertainties, including the response of heterotrophic respiration, nutrient cycling, permafrost dynamics, land-cover change, and scaling issues.
- Experimental data indicate that warming enhances carbon loss under dry soil conditions and enhances carbon sequestration under wet soil conditions. Thus, future responses to regional warming will depend substantially on changes in soil moisture.
- Recent observations indicate that cold-season greenhouse gas emissions contribute substantially to the annual budget.

The new, but uncertain, insights from these statements are that the Arctic is already a source of carbon and radiative forcing, and that it is likely to become a weak sink of carbon during future warming. Other uncertainties, not addressed in the consensus statements, include:

- potential complex flux patterns over long time periods during which acclimation of carbon fluxes to warming leads to transient trends within millennia-long smoothing of shorter-term dynamics;
- probable increased physical and biotic disturbance during warming that might increase carbon emissions, particularly if carbon storage shifts from tundra soils to invading forests; and
- various feedbacks between changing vegetation, soil temperatures, quality and quantity of litter, and biodiversity of decomposer organisms.

Many uncertainties remain, and it should be noted that the carbon considered in this discussion is that in the active layer of the soil and in vegetation, and not that trapped in permafrost and methane hydrates.

Table 7.14. Summary baseline information for the four ACIA regions.

	Region 1 Arctic Europe, East Greenland, European Russian North, and North Atlantic	Region 2 Central Siberia	Region 3 Chukotka, Bering Sea, Alaska, and western Arctic Canada	Region 4 Northeast Canada, Labrador Sea, Davis Strait, and West Greenland
Projected environmental changes				
Mean Annual Temperature[a] (°C)				
Baseline 1981–2000	-17 to 16	-8 to 4	-8 to 12	-20 to 12
2020 (change from baseline)	-1 to 3	0 to 2.5	-0.5 to 3.0	-1 to 4
2050 (change from baseline)	-0.5 to 4.0	0.5 to 4.0	0 to 4	0 to 7
2080 (change from baseline)	1 to 7	1 to 6	0.5 to 6	0 to 8
Precipitation[a] (mm/month)				
Baseline 1981–2000	10 to 150	10 to 70	10 to 150	5 to 130
2020 (change from baseline)	-10 to 12	-8 to 5	-10 to 8	-10 to 20
2050 (change from baseline)	-20 to 20	-2 to 5	-4 to 12	-8 to 35
2080 (change from baseline)	-20 to 25	0 to 10	-3 to 12	-15 to 35
Increase in UV-B radiation levels from 1979–1992 baseline (%)[b]				
Mean for 2010–2020	0 to 10	0 to 8	0 to 2	0 to 14
Mean for 2040–2050	0 to 2	0 to 2	0 to 2	0 to 2
Change in albedo[c] (due to vegetation change)				
2050	-0.10 to 0.05	-0.050 to 0.025	-0.10 to 0.025	-0.050 to 0.025
2080	-0.10 to 0.01	-0.050 to 0.025	-0.10 to 0.025	-0.10 to 0.025
Ecosystem processes projected by the LPJ model[d]				
NPP (Pg C/yr)		See Table 7.13 for the arctic total		
1960s	1.2	4.0	4.6	1.5
2080s	1.8	6.5	7.5	3.8
Change (%)	46.3	62.3	63.8	144.4
NEP: Change in C storage 1960–2080 (Pg C)		See Table 7.13 for the arctic total		
Vegetation	0.2	1.7	2.5	1.3
Soil	-0.1	0.5	1.9	4.7
Litter	-0.02	0.5	1.8	3.4
Total	0.04 (-0.7 to 0.8)	2.8 (-0.9 to 7.1)	6.2 (4.1 to 9.5)	9.3 (6.5 to 14.0)
Landscape processes projected by the LPJ model[d]				
Increase in taiga area (%)[e,f]				
1960–2020	1.1 (-1.1 to 2.3)	6.1 (4.2 to 8.6)	4.2 (1.9 to 5.7)	3.7 (2.9 to 4.4)
1960–2050	3.2 (1.6 to 4.3)	9.4 (7.5 to 10.2)	8.2 (6.5 to 9.7)	5.0 (3.8 to 6.2)
1960–2080	5.0 (3.7 to 5.9)	13.7 (11.1 to 17.3)	11.9 (9.8 to 15.1)	9.5 (6.8 to 12.5)
Change in polar desert area (%)[e,g]				
1960–2020	-2.3 (-3.1 to -1.1)	-6.9 (-11.0 to -3.9)	-5.3 (-10.8 to -2.4)	-12.7 (-23.0 to -7.2)
1960–2050	-3.5 (-4.8 to -2.7)	-9.9 (-13.2 to -7.7)	-11.0 (-15.7 to -8.9)	-23.6 (-33.5 to -16.3)
1960–2080	-4.2 (-5.9 to -3.2)	-11.4 (-14.6 to -10)	-13.2 (-16.6 to -11.4)	-35.6 (-47.8 to -25.9)
Biodiversity				
Rare endemic vascular plant species[h]	2	18	69	8
Threatened vascular plant species (occurring at a single unprotected location in each region)[h]	1	4	11	0
Threatened animal species[i]	2	4	6	1

[a]from the ACIA-designated model simulations; [b]from Taalas et al. (2000); [c]Betts R. (2001) using IS92a emissions scenario; [d]averages and ranges (in parentheses) of the responses of the LPJ model (Sitch et al., 2003) driven with outputs from four different climate models (CGCM2, GFDL-R30_c, HadCM3, ECHAM4/OPYC3) forced with the B2 emissions scenario (section 4.4); [e]only a proxy as the change is derived from functional characteristics of the vegetation simulated by the model rather than projections of specific vegetation composition per se. For a proper vegetation distribution estimate, it would be more appropriate to use a coupled biogeography and biogeochemistry model such as BIOME4; [f]based on the percentage increase in woody plants simulated by LPJ; [g]based on the percentage reduction in bare ground simulated by LPJ; [h]from Talbot et al. (1999); [i]IUCN (2003).

7.6. Synthesis: Scenarios of projected changes in the four ACIA regions for 2020, 2050, and 2080

This synthesis draws on information in this chapter that can be assessed within the four ACIA regions. Most of the information is therefore based on model output. Details of the regions and the logic determining them are presented in section 18.3, while details of the models that generate the UV-B radiation and climate scenarios are presented in sections 5.6 and 4.2.7, respectively. Many of the details relating to vegetation and carbon dynamics are derived specifically for this section from the LPJ model (Sitch et al., 2003), details of which are presented in Box 7.1. Other aspects of the assessment that cannot currently be divided into the ACIA regions are summarized within and at the end of the various sections of this chapter.

7.6.1. Environmental characteristics

The four ACIA regions (section 18.3) differ greatly in their geography and climatology, which leads to variation in future possibilities for the relocation of species and ecosystems, and differences in scenarios of future changes in climate and UV-B radiation levels (Table 7.14).

Geographically, Region 4 has a far greater extent of land at high latitudes compared with the other regions. This is likely to support northward migration of arctic biota even if the Canadian Archipelago and the glacial landscape of Greenland together with lack of suitable soils will, to some extent, pose barriers to migration. Relatively narrow tundra zones in some parts of Regions 1 and 3 could possibly, with sea-level rise and northward boreal forest expansion, disappear and forest reach the shore of the Arctic Ocean (Fig. 7.32). Region 1 contains the relatively isolated high-arctic islands of Svalbard, and the islands of Iceland and the Faroe Islands, that are likely to experience delayed immigration of southern species during warming. Both Iceland and the Faroe Islands have equivocal positions within classifications of the Arctic: the northern part of Iceland and the alpine zones of the Faroe Islands (Fossa, 2003) have the strongest arctic characteristics and climate warming is likely to lead to altitudinal displacement of tundra-like vegetation in both areas, and displacement from the northern coastal area in Iceland. The imbalance of species loss and replacement by species migrating more slowly to islands is projected to lead to an initial loss in diversity (Heide-Jørgensen and Johnsen, 1998).

The scenarios of temperature change generated by the five ACIA-designated models show complex temporal patterns, in some areas shifting from initial cooling to substantial warming. The data used for the modeling of vegetation zone displacement and carbon storage used a different baseline period (1961–1990) than that used for the ACIA-designated model projections (1981–2000) and excluded output from the CSM_1.4 model. In addition, the output from the LPJ model is provided for 2100,

rather than the 2071–2090 time slice used in the ACIA scenarios (section 7.5). It is therefore difficult to compare the results, even though both approaches had four GCMs in common and used the same emissions scenario.

Changes in UV-B radiation levels are projected to vary among regions, but only over the next 20 years. By 2050, recovery of the stratospheric ozone layer is projected to reduce UV-B radiation to relatively low levels above present-day levels, with no differences among the ACIA regions. Of course, this recovery depends entirely on the success of management and regulation of ozone-depleting substances. In the near future, however, increases in UV-B radiation levels are projected to be greatest in Region 4, followed by Regions 1 and 2 (Taalas et al., 2000).

Active-layer thickness is projected to increase by 20 to 60% by 2071–2100 (compared to the IPCC baseline, 1961–1990). The greatest percentage increases are projected to occur in northern Siberia and the interior of the Alaska–Yukon area. In general, the greatest relative changes in the active layer are projected to occur in those regions where the active layer is presently shallow (section 6.6.1.3). Degradation of continuous permafrost to discontinuous permafrost and the disappearance of discontinuous permafrost are projected to occur at the southern boundaries of all of the ACIA regions.

7.6.2. Vegetation zones and carbon balance

Region 1 (Arctic Europe, East Greenland, European Russian North, and North Atlantic) includes many high-arctic areas but these are separated from terrestrial ecosystems at lower latitudes by barriers of open sea. The possibilities for future species relocation are limited, even though moderate warming is projected here (Table 7.14). In contrast, Region 2 (Central Siberia) has continuous landmasses from the tropics to the high Arctic. This region is currently warming, and scenarios project that future warming will be greater here than elsewhere. The possibilities for responses in ecosystem distribution, structure, and C balance are therefore considerable. This is illustrated by large projected increases in taiga that displaces tundra in particular, and also in projected decreases in polar desert that is displaced to some extent by northward movement of the tundra (Table 7.14). There is also a projected northward displacement and reduction in prostrate dwarf-shrub tundra, particularly in Yakutia and the Taymir Peninsula, together with a displacement of erect dwarf-shrub tundra from much of the Russian Arctic by low- and high-shrub tundra that is projected to expand markedly there (Figs. 7.2 and 7.32). Region 3 (Chukotka, Bering Sea, Alaska, and the western Canadian Arctic) has little high-arctic area and a large maritime influence. Increases in temperature and precipitation are projected to be moderate, as are changes in vegetation (Table 7.14). Region 4 (Northeast Canada, Labrador Sea, Davis Strait, and West Greenland) is a region of fragmented landmasses that are often extensively glaciated or have recently become deglaciated. This area has experienced recent cooling, but a warming trend is projected to occur between now and 2100. Increases in temperature and precipitation are projected to lead to relatively small increases in taiga (compared with other regions) but a particularly large (~36%) loss of polar desert by 2080.

In terms of C storage, all regions are projected to accumulate C, largely because of the replacement of bare ground by tundra. Consequently, the greatest carbon gain is projected to occur in Region 4 (Table 7.14; Fig. 7.33). In contrast, the smallest gains – but still gains – are projected to occur in Region 1, which has the smallest projected increase in temperature.

7.6.3. Biodiversity

Biodiversity is affected by habitat fragmentation. Scenarios of the impacts of projected human infrastructure development on arctic flora and fauna suggest that in the Arctic, these impacts are likely to extend for 4 to 10 km away from the infrastructure (Nellemann et al., 2001). This is a much wider zone of influence than in other regions of the earth. Nellemann et al. (2001) calculated that 50 to 80% of the Arctic is likely to be affected by infrastructure development by 2050. Of course, infrastructure development varies among the ACIA regions and this remains to be characterized. However, threats to flora and fauna are likely to be increased by the additive or possibly interactive effects of infrastructure development and climate change.

The number of rare endemic vascular plant species in the Arctic varies greatly between the regions (Table 7.14). Region 1 has relatively little land mass and supports only two of the rare endemic vascular plant species. Region 4, which encompasses a significant proportion of the high Arctic, contains 8% of the species and Region 2 contains 18%. In contrast, over 70% of the rare endemic vascular plant species are found in Beringia (Region 3). Twenty-four species are found on Wrangel Island (Talbot et al., 1999). A recent modification (Talbot and Murray, 2001) of the list of threatened arctic plant species (Talbot et al., 1999) adds a further 63 plant species, but data have not yet been compiled on their distributions within the ACIA regions. Although Table 7.14 shows clear regional differences in the distribution of rare and endemic plant species, and also a surprisingly high number of these species, it should be noted that the taxonomic treatment of species is likely to vary from region to region and there is uncertainty about the taxonomic status of some of the species.

It is not clear to what extent the rarity of the species listed in Tables 7.7 and 7.14 will be affected by climate change, as many other factors determine rarity. However, species concentrated in small areas such as Wrangel Island are particularly at risk from any future climate warming and species invasion.

The likely impacts of climate change on biodiversity in terms of threatened species require new concepts of "threatened species" and "protection" of currently per-

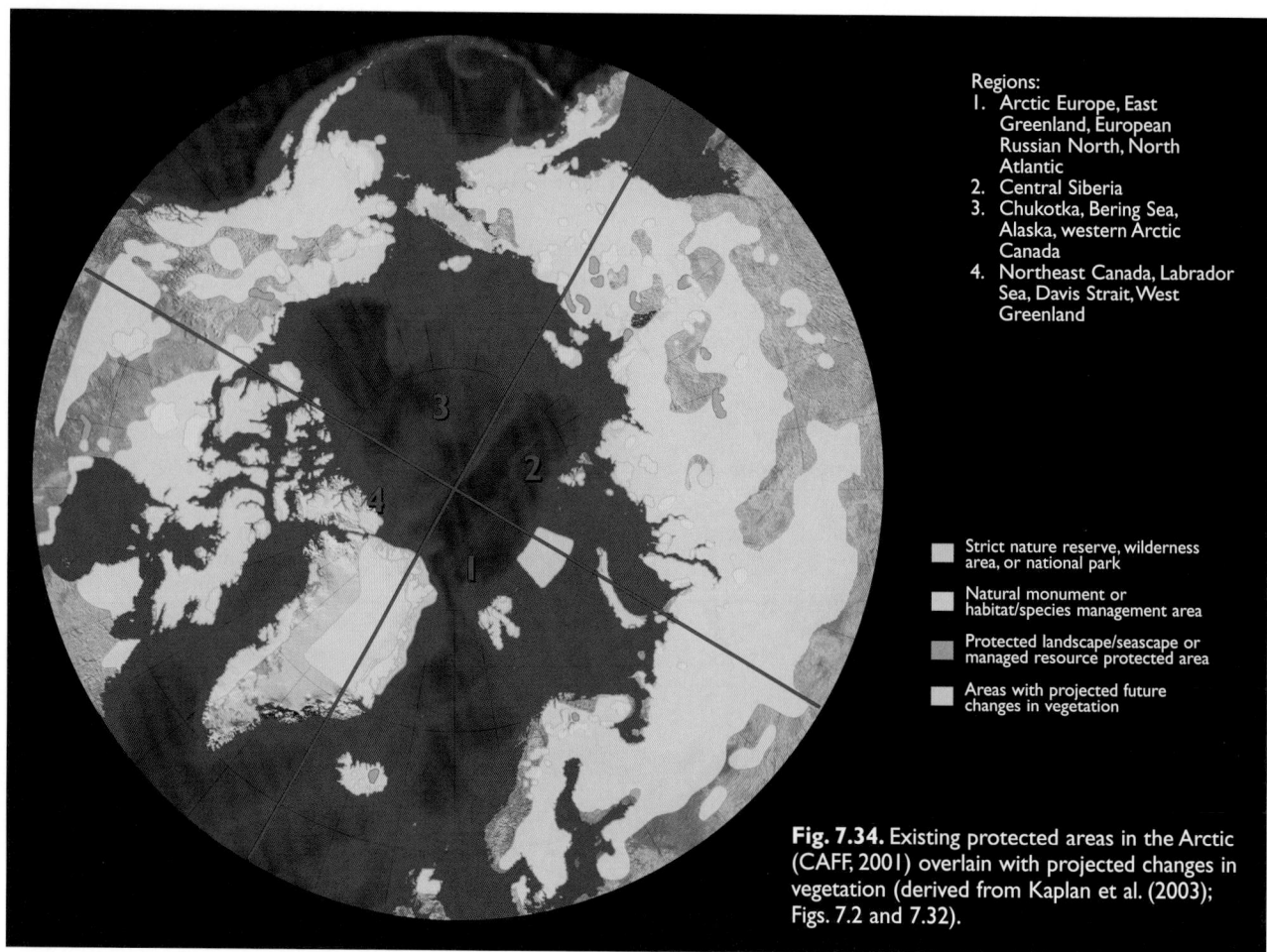

Regions:
1. Arctic Europe, East Greenland, European Russian North, North Atlantic
2. Central Siberia
3. Chukotka, Bering Sea, Alaska, western Arctic Canada
4. Northeast Canada, Labrador Sea, Davis Strait, West Greenland

■ Strict nature reserve, wilderness area, or national park

■ Natural monument or habitat/species management area

■ Protected landscape/seascape or managed resource protected area

■ Areas with projected future changes in vegetation

Fig. 7.34. Existing protected areas in the Arctic (CAFF, 2001) overlain with projected changes in vegetation (derived from Kaplan et al. (2003); Figs. 7.2 and 7.32).

ceived threatened species (Fig. 7.34; Chapter 11). The numbers of species currently perceived as threatened vary between regions. Region 3 contains significantly more rare plant species and threatened animal and plant species than other regions. Although temperature and precipitation changes are likely to be less in this region than in others, the vulnerability of the biodiversity of this area is likely to be considerable. Northward expansion of dwarf-shrub and tree-dominated vegetation into an area such as Wrangel Island that is rich in rare endemic species could possibly result in the loss of many plant species. Although some of these might not be considered vulnerable because they are currently in "protected" areas, this protection is against local human activities (e.g., hunting, infrastructure development, etc.) and protection cannot extend to changes in climate and UV-B radiation levels. It is possible that some plant species, particularly outliers of more southerly distributions, might experience population expansion or reproduction and recruitment to populations, leading to initial expansion in response to warming. However, displacement of herbaceous species by woody immigrants is possible in the long term in mesic areas. In contrast to the possibility that some threatened species will proliferate in a warmer climate, some currently widespread species are very likely to become less abundant and even "threatened".

The greatest long-term threat to arctic diversity is the loss of arctic habitat (section 7.2.6). In locations where the tundra zone is narrow, boreal forest moves northward, and the ocean moves southward due to sea-level rise, there is very likely to be, over a period of centuries, a loss of arctic ecosystems and the species that characterize them.

7.7. Uncertainties and recommendations

7.7.1. Uncertainties

Current understanding of ecological processes and changes driven by climate and UV-B radiation is strong in some geographic areas and in some disciplines, but is weak in others. Although the ability to make projections has recently increased dramatically with increased research effort in the Arctic and the introduction of new technologies, current understanding is constrained by various uncertainties. This section focuses on these uncertainties and recommends ways in which they can be reduced.

7.7.1.1. Uncertainties due to methodologies and conceptual frameworks

Methods of projecting impacts on species and ecosystems

Each method has advantages and strengths and has led to the important and extensive current knowledge base. However, each method also has uncertainties that need to be identified so that methods can be refined and uncertainties quantified.

The use of *paleo analogues* to infer future changes underrepresents the differences between past changes and those likely to occur in the future due to differences in the starting state of the environment and biota, and the different nature of past and likely future changes. Major differences include anthropogenic effects (e.g., extent of land-use change, current and future stratospheric ozone depletion, and trans-boundary pollution) that are probably unprecedented.

Using *geographic analogues* can indicate where communities and species should be in a warmer world, but they do not indicate at what rate species can relocate or if new barriers to dispersal such as fragmented habitats will prevent potential distributions from being achieved.

Observations and monitoring provide essential data about changes as they occur, and can be used to test hypotheses and model projections, but they have little predictive power in a time of changing climate during which many biotic responses are nonlinear.

Experiments that simulate future environmental conditions (e.g., CO_2 concentrations, UV-B radiation levels, temperature, precipitation, snow depth and duration, etc.) all have artifacts, despite attempts to minimize them. It is difficult in field experiments to include simulations of all likely eventualities: in warming experiments, it is very difficult to identify separate effects of seasonal warming and extreme events because most experiments are small in spatial extent, and are short-term in the context of the life cycles of arctic plants and animals. It is also difficult to identify the complex interactions among all the co-occurring environmental change variables, and ecological processes determined from experiments in one geographic area may not relate sufficiently to other areas because of different ecological conditions and histories.

Indigenous knowledge, although a valuable contributor to current understanding of ecological changes (section 7.7.2.2), is more qualitative than quantitative, and often characterized by relatively coarse measures (i.e., monthly and seasonal change rather than daily or weekly). The information available is sometimes limited to phenomena that fall within the cycle of subsistence resource use, and is more likely to be diachronic (long time series of local information) and not synchronic (simultaneously observed). It is often difficult to assign particular environmental changes to individual changes in biota, to determine mechanisms of change, and distinguish climate-related change from other changes occurring in the environment. Indigenous knowledge is variable between and within communities, and interpretation and verification processes are as important as collection and documentation. It is a knowledge system.

Uncertainties can be reduced when information from several methods converges. This chapter accepts all methodologies, knowing their limitations, and qualifies the information presented by the methodologies used to obtain it.

Measuring primary production and controlling factors

Key unknowns about primary productivity in the Arctic include root production and turnover and belowground allocation processes in general, including allocation to mycorrhizae and exudation. Also poorly understood are long-term (multi-year to multi-decade) interactions between the carbon cycle and nutrient cycles, in which relatively slow changes in soil processes and nutrient availability interact with relatively rapid changes in photosynthesis in response to climate change. One major unknown is the control on dispersal, establishment, and rate of change in abundance of species and functional types that are more productive than current arctic species but are not now present or common in arctic vegetation (e.g., trees and tall shrubs).

There are two major approaches to assessing NEP: classic biomass weighing and year-round CO_2 flux recording, but these approaches are not always compatible. One particular gap in current estimates is the lateral transport of organic C from one ecosystem to another. The two methodologies give opposite results when accounting for the input of allochthonous (produced outside) organic matter to a particular ecosystem: CO_2 flux measurement gives negative NEP due to increased CO_2 emission from soil to atmosphere, while weighing gives a higher accumulation of organic C in the soil. In addition, current estimates of buried C released to ecosystems due to thermokarst and soil erosion (Figs. 7.21 and 7.22) are poor.

Difficulties in studying microorganisms

Understanding of microbes that are critically important in many ecosystem processes is limited. Knowledge of microbial diversity and function has been strongly constrained by lack of development in methodology and conceptual frameworks.

Bacteria and even the more advanced microscopic yeast and fungi cannot be characterized by visual observation alone due to their very simple shapes (rods, spheres, filaments). Typically, microbial strains must be cultivated in pure culture to reveal their various functional features, and an appreciable amount of laboratory work is required to differentiate a microbe from close relatives. Only a small fraction of soil microorganisms are able to grow on artificial laboratory media, and less than 1% of the cells observed with a microscope form colonies on the plate. The main reasons for this "Great Plate Count Anomaly" (Staley and Konopka, 1985) include the metabolic stress of the "famine-to-feast" transition occurring when cells are brought from soil to artificial, nutrient-rich media; inadequacy of cultivation conditions compared with the natural environment; and metabiotic interactions/cooperation in natural communities that are broken after cells have been separated by plating (Panikov, 1995). This technical problem has resulted in an underestimation of diversity in natural habitats.

Fortunately, new cultivation approaches are being developed that are helping to overcome this problem (Staley and Gosink, 1999). However, it is not presently possible to make a fair comparison between the numbers of species of animals and plants versus bacteria given that these groups are defined differently (Staley, 1997).

Lack of feasible technologies to measure microbial population dynamics in the field has lead to the use of microbial biomass, which is convenient but a poor predictor of productivity and C and nutrient turnover. More sensitive measurements of microbial activity need to be developed for field application.

Incomplete databases

Length of time series of data: Although many long time series of relevant data (e.g., species performance and phenology) exist, most information relates to short time series. This is a particular problem in the Arctic where complex population dynamics (e.g., cycles) need to be understood over periods long enough to allow trends to be separated from underlying natural dynamics. Observations of trace gas emissions also require annual observations over time periods long enough to encompass significant climate variability. Experiments are usually too brief to capture stable responses to environmental manipulations and to avoid artifacts that are disturbance responses. Long time series of data are also necessary in order to identify extreme events and nonlinear system changes.

Geographic coverage and spatial scaling: The ecosystems and environments in the Arctic are surprisingly variable yet generalizations to the circumpolar Arctic are often made from a few plot-level studies. Sometimes, particular experiments (e.g., CO_2 concentration and UV-B radiation manipulations) or observations are restricted to a few square meters of tundra at just one or two sites. Uncertainties due to generalizing and scaling up are thus significant. The IBP and the ITEX are exceptional examples of how standardized experiments and observations can be implemented throughout the Arctic.

Coverage of species and taxa: Chapin F. and Shaver (1985a) and others have demonstrated the individualistic responses of species to experimental environmental manipulations, including climate, while Dormann and Woodin (2002) have shown the inadequacy of the concept of "plant functional types" in generalizing plant responses to such experiments. An approach has to be developed to measure responses of a relevant range of species to changes in climate, and particularly UV-B radiation levels. Plants studied in the UV-B radiation manipulation experiments were generally at their northernmost distributional limits and well adapted to high UV-B radiation levels characteristic of southern parts of their ranges. Greater responses would be expected from species at their southern distributional limits where increased UV-B radiation would exceed levels in the plants' recent "memory".

Some species and taxonomic groups are particularly difficult to study, or have little socioeconomic value, and so are underrepresented in databases. Examples include mosses, lichens, soil fauna and flora, and microorganisms (section 7.7.1.1).

Nomenclature and concepts

The restricted use of appropriate language often generates uncertainties. The nomenclature of vegetation and plant community types allows changes in the *distribution* of these assemblages of species in a changed climate to be modeled, but constrains understanding of changes in the *structure* of the assemblages that are likely to occur because assemblages of species do not move *en bloc*. This problem limits understanding of novel future communities (Chapin F. and Starfield, 1997) and no-analogue communities of the past, and emphasizes the uncertainties due to the inability of quantitative models to project qualitative changes in systems. Similarly, the concept of "line" to denote the limit of species distributions (e.g., treeline) is inadequate to express the gradient of changes from one zone to another that can occur over tens of kilometers.

The concept of "species" is particularly difficult in the context of microorganisms as discussed previously, and even as applied to flowering plants. The traditional view that there are few rare and endemic arctic plant species is challenged by recent studies of the flora of Wrangel Island and Beringia (Table 7.14) but it is not known to what degree plant taxonomy is problematic (although the Pan Arctic Flora project is addressing this problem). Such problems need to be resolved before the impacts of climate change on biodiversity can be assessed.

7.7.1.2. Uncertainties due to surprises

Perhaps the only certainty in the assessment of impacts of changes in climate and UV radiation levels on terrestrial ecosystems in this chapter is that there will be surprises. By definition, it is difficult to predict surprises. However, the possibility that climate *cooling* will occur because of a change in thermohaline circulation is potentially the most dramatic surprise that could occur.

Regional cooling

The potential for a negative feedback arising from an increased freshwater flux to the Greenland, Icelandic, and Norwegian seas and the Arctic Basin, leading to a partial or complete shut down of the thermohaline circulation of the global oceans, remains an area of considerable uncertainty (sections 6.8.4 and 9.2.3.4). Such an event would be likely to lead to marked and rapid regional cooling in at least northwest Europe. This region at present enjoys an anomalously warm climate given its latitude (50°–72° N), enabling agriculture to be practiced and substantial settlements maintained at far higher latitudes than in any other arctic or subarctic region. Such cooling would be likely to qualitatively alter terrestrial

ecosystems (Fossa, 2003), agriculture, and forestry over very large areas of Fennoscandia and Europe.

Mutations

Mutations are projected to occur as a result of increased levels of UV radiation and also as a result of aerosols and volatile chemical mutagens transported to the cool polar atmosphere from the mid- and low latitudes. The direct mutagenic effect is probably not strong, especially if the protecting shielding effects of soil particles and adaptive mechanisms are taken into account. However, possible microbial mutants could lead to epidemic outbreaks that could have profound and unexpected consequences for the Arctic and elsewhere.

Desertification

Several approaches suggest that climate warming will lead to an increase in the productivity of arctic vegetation and long-term net sequestration of CO_2. However, the complex interactions among warming, permafrost dynamics, hydrology, precipitation, and soil type are poorly understood. Desertification is a plausible outcome in some areas where scenarios suggest that permafrost will thaw, drainage will increase, temperatures will increase, and precipitation will *not* increase substantially. In areas of sandy soil and loess deposition, such as parts of eastern Siberia, there is a particular risk of desertification. In the polar deserts, herb barrens, and heaths of northern Greenland, plant productivity is strongly correlated with precipitation and increased evapotranspiration is likely to lead to a similar process (Heide-Jørgensen and Johnsen, 1998). Locally, the impacts of overgrazing and anthropogenic disturbance can accelerate the process. A clear example of the effect of warming and drying on the C balance of Alaskan tundra is provided in section 7.5, and an example of boreal coniferous forest loss in North America is given in section 14.7.3.1. However, the possible wider geographic scale of this process is unknown.

Changes in current distributions of widespread and rare species

Climate change could possibly have counter-intuitive impacts on species distributions. Currently rare arctic plant species, particularly those that are northern outliers of species with more southerly distributions, could possibly expand during initial phases of climate warming. In contrast, currently widespread species, particularly lichens and mosses, could possibly become more restricted in their abundance during warming. It is necessary to reassess the concept of "threatened species" in the context of climate and UV radiation change (see section 7.6 and Chapter 11).

7.7.1.3. Model-related uncertainties

During the IBP period (late 1960s and early 1970s), tundra research was characterized by extensive field observations but a general lack of modeling capability. Currently, a technological revolution has stimulated model generation and remote sensing of ecosystem change. However, in some cases, validation is insufficient. Models that project NPP at a global or circumpolar level are insufficiently validated, as recent measurements of NPP are rare and restricted to a few localities. The lack of inter-comparison between models and existing observations leads to potential projection errors: modeled displacement of the tundra by the boreal forest contradicts current observations of the southward retreat of the treeline in some areas and the expansion of "pseudotundra" in parts of Russia due to permafrost degradation, paludification, and human activities.

The climate models used to provide future climate scenarios for the ACIA were forced with the B2 emissions scenario (section 4.4.1), but projections based on the A2 emissions scenario have been used to a limited extent as a plausible alternative (IPCC, 2001; section 1.4.2). The A2 scenario assumes an emphasis on economic development rather than conservation, while the B2 scenario assumes a greater emphasis on environmental concerns: each has considerable uncertainties (section 4.4.1). In this chapter, projections based mainly on the B2 emissions scenario were used to model changes in vegetation and carbon storage. Projections based on the A2 emissions scenario result in higher temperatures for a particular time period than those based on the B2 emissions scenario. The changes projected by models forced with the A2 emissions scenario for the 2041–2060 and 2071–2090 time slices occur 5 to 10 and 10 to 20 years earlier, respectively, compared with projections based on the B2 scenario. Potential impacts on ecosystems would thus occur faster should emissions follow the A2 scenario.

The major implication for ecosystems of a faster rate of temperature change is an increased mismatch between the rate of habitat change and the rate at which species can relocate to occupy new habitats in appropriate climate envelopes. The overall, generalized, difference between projections based on the B2 and A2 emissions scenarios is likely to be an increased risk of disturbance and disease in species that, under projected conditions based on the A2 scenario, cannot relocate quickly enough. Projections based on the A2 scenario also imply an increased mismatch between initial stimulation of soil respiration and longer-term vegetation feedbacks that would reduce C fluxes to the atmosphere.

7.7.2. Recommendations to reduce uncertainties

7.7.2.1. Thematic recommendations and justification

This section reviews important thematic topics that require particular research. For each topic, the state of knowledge and important gaps are summarized, and recommendations to fill these gaps are suggested (in italic font).

Mechanisms of species responses to changes in climate and UV-B radiation levels. Changes in microbe, animal, and plant populations are triggered by trends in climate and UV radiation levels exceeding thresholds, and by extreme events, particularly during winter. However, information is uneven and dominated by trends in summer climate. *Appropriate scenarios of extreme events are required, as is deployment of long-term experiments simulating extreme events and future winter processes in particular. A better understanding of thresholds relevant to biological processes is also required.*

Biodiversity changes. Some groups of species are very likely to be at risk from climate change impacts, and the biodiversity of particular geographic areas such as Beringia are at particular risk. It is not known if currently threatened species might proliferate under future warming, nor which currently widespread species might decrease in abundance. *The nature of threats to species, including microbes, must be reassessed using long-term climate and UV-B radiation change simulation experiments. There is also a need to identify and monitor currently widespread species that are likely to decline under climate change, and to redefine conservation and protection in the context of climate and UV radiation change.*

Relocation of species. The dominant response of current arctic species to climate change, as in the past, is very likely to be relocation rather than adaptation. Relocation possibilities are very likely to vary according to region and geographic barriers. Some changes are already occurring. However, knowledge of rates of relocation, impact of geographic barriers, and current changes is poor. *There is a need to measure and project rates of species migration by combining paleo-ecological information with observations from indigenous knowledge, environmental and biodiversity monitoring, and experimental manipulations of environment and species.*

Vegetation zone redistribution. Forest is very likely to replace a significant proportion of the tundra and is very likely to have a great effect on species composition. However, several processes (including land use and permafrost dynamics) are expected to modify the modeled response of vegetation redistribution related to warming. *Models of climate, hydrology (permafrost), ecosystems, and land use need to be developed and linked. These models need to be based on improved information about the current boundaries of major vegetation zones, defined and recorded using standardized protocols.*

Carbon sinks and sources in the Arctic. Current models suggest that arctic vegetation and active-layer soils will be a sink for C in the long term because of the northward movement of vegetation zones that are more productive than those they displace. Model output needs to be reconciled with observations that tundra areas that are C sources currently exceed those that are C sinks, although the measurements of circumpolar C balance are very incomplete. To what extent disturbance will reduce the C sink strength of the Arctic is also unknown.

There is a need to establish long-term, annual C monitoring throughout the Arctic; to develop models capable of scaling ecosystem processes from plot experiments to landscapes; to develop observatories, experiments, and models to relate disturbance such as desertification to C dynamics; and to improve the geographic balance of observations by increasing high-arctic measurements. There is also a need to combine estimates of ecosystem carbon flux with estimates of C flux from thawing permafrost and methane hydrates.

Ultraviolet-B radiation and CO₂ impacts.

Enhanced CO_2 concentrations and UV-B radiation levels have subtle but long term impacts on ecosystem processes that reduce nutrient cycling and have the potential to decrease productivity. However, these are generalizations from very few plot-scale experiments, and it is difficult to understand impacts that include large herbivores and shrubs. *There is a need for long-term experiments on CO_2 and UV-B radiation effects interacting with climate in a range of arctic ecosystems; short-term experiments stimulating repeated episodes of high UV radiation exposure; long-term experiments that determine the consequences of high CO_2 concentrations and UV-B radiation levels for herbivores; and short-term screening trials to identify the sensitivity of a wide range of species, including soil microbes, to current and projected UV-B radiation levels.*

Local and regional feedbacks.

Displacement of tundra by forest is very likely to lead to a decrease in albedo with a potential for local warming, whereas C sequestration is likely to increase with potential impacts on global concentrations of greenhouse gases. However, the timing of the processes and the balance between the processes are very uncertain. How local factors such as land use, disturbance, tree type, and possible desertification will affect the balance is also uncertain. *There is a need for long-term and annual empirical measurements, analysis of past remotely sensed images, and collection of new images together with the development and application of new models that include land use, disturbance, and permafrost dynamics.*

7.7.2.2. Recommendations for future approaches to research and monitoring

No single research or monitoring approach is adequate, and confidence is increased when results from different approaches converge (section 7.7.1.1). Current approaches should be maintained, and new approaches and even paradigms (e.g., when defining "threatened species" and "protected areas") developed. Some important approaches are highlighted below.

Reducing uncertainty by increasing and extending the use of indigenous knowledge

Arctic indigenous peoples retain strong ties to the land through subsistence economies and they are "active participants" in ecosystems (*sensu* Bielawski, 1997). Unlike a scientist, a hunter is not bound in his observations by a project timeline, budget, seasonality, or logistical constraint (Krupnik, 2000; Riedlinger and Berkes, 2001).

Subsistence activities occur on a daily basis, year after year, and throughout the winter period when many scientists are south in home institutions. Indigenous peoples of the Arctic therefore possess a substantial body of knowledge and expertise related to both biological and environmental phenomena. Such local expertise can highlight qualitative changes in the environment and provide pictures of regional variability across the Arctic that are difficult to capture using coarser-scale models.

This chapter presents some of the first efforts at linking western science and indigenous knowledge to expand the range of approaches that inform this assessment. However, the potential is far greater, including, for example, local-scale expertise, information on climate history, generation of research hypotheses, community monitoring, and community adaptation (Riedlinger and Berkes, 2001).

Monitoring

Long-term environmental and biological monitoring have been undervalued but are becoming increasingly necessary to detect change, to validate model projections and results from experiments, and to substantiate measurements made from remote sensing. Present monitoring programs and initiatives are too scarce and are scattered randomly. Data from the Arctic on many topics are often not based on organized monitoring schemes, are geographically biased, and are not long-term enough to detect changes in species ranges, natural habitats, animal population cycles, vegetation distribution, and C balance. More networks of standardized, long-term monitoring sites are required to better represent environmental and ecosystem variability in the Arctic and particularly sensitive habitats. Because there are interactions among many co-varying environmental variables, monitoring programs should be integrated. Observatories should have the ability to facilitate campaigns to validate output from models or ground-truth observations from remote sensing. There should be collaboration with indigenous and other local peoples' monitoring networks where relevant. It would be advantageous to create a decentralized and distributed, ideally web-based, meta-database from the monitoring and campaign results, including relevant indigenous knowledge.

Monitoring also requires institutions, not necessarily sited in the Arctic, to process remotely sensed data. Much information from satellite and aerial photographs exists already on vegetation change, such as treeline displacement, and on disturbances such as reindeer/caribou overgrazing and insect outbreaks. However, relatively little of this information has been extracted and analyzed.

Monitoring C fluxes has gained increased significance since the signing of the Kyoto Protocol. Past temporal and spatial scales of measurement used to directly measure C flux have been a poor match for the larger scale of arctic ecosystem modeling and extrapolation. It remains a challenge to determine if flux measurements and model output are complementary. The technological difficulties in extrapolating many non-linear, complex, interacting factors that comprise fluxes at hundreds to thousands of square kilometers over time, space, and levels of biological and environmental organization in the Arctic have been significant (Oechel et al., 2000b, 1998). Research is needed to better understand how the complex system behaves at the meter scale related to larger spatial scales that can be efficiently modeled and evaluated at the regional and circumpolar scale. To do this, extensive long-term and year-round eddy covariance sites and other long-term flux sites, including repeated aircraft flux measurements and remote sensing (Oechel et al., 2000a), provide the basis for estimating circumpolar net ecosystem CO_2 exchanges. Currently, the circumpolar Arctic is disproportionately covered by current and recent measurements, with Canadian and high-arctic regions particularly poorly represented.

Long-term and year-round approach to observations and experiments

Many observations and experiments are short term (<5 years) and they are biased toward the summer period often because researchers have commitments to institutions outside the Arctic during winter. However, throughout this assessment it has become clear that long-term and year-round measurements and experiments are essential to understand the slow and complex responses of arctic organisms and ecosystems to climate change.

Long-term (>10 years) observations and experiments are required in order to enable transient responses to be separated from possible equilibrium responses; increase the chances that disturbances, extreme events, and significant interannual variation in weather are included in the observations; and allow possible thresholds for responses to be experienced.

Year-round observations are necessary to understand the importance of winter processes in determining the survival of arctic species and the function of arctic ecosystems. Such observations are necessary to recognize the projected amplification of climate warming in winter and to redress the current experimental bias toward summer-only warming. For microbes, it is particularly important to understand changes in winter respiration and nutrient mobilization during freeze–thaw events in spring and late autumn.

It is important to improve the appropriateness of the timing of observations and experiments. For example, current information about the impacts of increased UV-B radiation levels is mainly derived from general summer enhancements or filtration of UV-B radiation, although future increases in UV-B radiation levels are likely to be highest in spring and during specific stratospheric ozone depletion events. The frequency of observations should be fitted to the rate of change of the species or processes of interest, for example, decadal measurements may suffice for some variables such as treeline movement.

Increasing the complexity and scale of environmental and ecosystem manipulation experiments

Single-factor manipulation experiments now have limited applicability because it is clear that there are many interactive effects among co-occurring environmental change variables. There is need for well-designed, large, mutifactorial environmental (e.g., climate, UV-B radiation levels, and CO_2 concentrations) *and* ecosystem (e.g., species removal and addition) manipulation experiments that are long-term and seek to understand annual, seasonal, and event-based impacts of changing environments. The complexity of appropriate treatments and timescales is vast and the spatial scale is also a significant challenge, as it is important to have manipulations that can be related to larger plants (e.g., trees, shrubs) and animals (e.g., reindeer/caribou).

Assessing the impacts of cooling on ecosystems

Scenarios of increased temperature dominate the approaches to projecting responses of ecosystems to future climate. However, cooling in some areas remains a possibility. As the impacts of cooling on terrestrial ecosystems and their services to people are likely to be far more dramatic than the impacts of warming, it is timely to reassess the probabilities of cooling projected by GCMs and the appropriateness of assessing cooling impacts on ecosystems.

Modeling responses of arctic ecosystems to climate and UV-B radiation change and communicating results at appropriate geographic scales

High-resolution models are needed at the landscape scale for a range of landscape types that are projected to experience different future envelopes of climate and UV-B radiation levels. Modeling at the landscape scale will simulate local changes that relate to plot-scale experiments and can be validated by results of experiments and field observations. Visualization of model results presented at the landscape scale will also enhance the understanding of the changes and their implications by local residents and decision-makers. A particular challenge is to provide scenarios for changes in climate and UV-B radiation levels at the scale of tens of meters.

7.7.2.3. Funding requirements

It is inappropriate here to comment on levels of funding required to fulfill the recommendations discussed above. However, it is appropriate to highlight two essential aspects of funding.

First, current short-term funding is inappropriate to support research into long-term processes such as ecosystem responses to climate change and UV-B radiation impacts. *A stable commitment to long-term funding is necessary.*

Second, funding possibilities that are restricted to single nations, or at best a few nations, make it extremely difficult to implement coordinated research that covers the variability in ecosystems and projected climate change throughout the circumpolar north, even though the instruments for coordination exist (e.g., within the International Arctic Sciences Committee, the International Council of Scientific Unions, and the International Geosphere–Biosphere Programme). Limitation of international funding possibilities leads to geographic biases and gaps in important information. *Circumpolar funding is required so that coordinated projects can operate at geographically appropriate sites over the same time periods.*

Acknowledgements

Many individuals and organizations kindly contributed information to this assessment and we are grateful to all. In particular, we thank the Steering Committees of the IASC projects on the Tundra-Taiga Interface and Feedback from Arctic Terrestrial Ecosystems. ITEX participants also gave support. We thank Dr. Anders Michelsen, University of Copenhagen, for information on arctic mycorrhizal associations and Jed Kaplan (MPI Jena) for providing unpublished estimates of the past and potential future extent of tundra derived from simulations made using the BIOME 4 model. We are grateful to Håkan Samuelsson and Chris Callaghan for help with some of the figures. Terry Callaghan and Margareta Johansson are very grateful to the Swedish Environmental Protection Agency (Naturvårdsverket) for funding to allow them to contribute to the ACIA process. Participation of K. Laine and E. Taulavuori have been facilitated by financial support from the Academy of Finland and Thule Institute, University of Oulu. Phycology research in the Arctic by J. Elster and colleagues has been sponsored by two grants; the Natural Environment Research Council (LSF-82/2002), and the Grant Agency of the Ministry of Education of the Czech Republic (KONTAKT - ME 576). We are grateful to A. Katenin for providing unpublished data and T. Polozova for providing part of Figure 7.3. Finally, we are grateful to the ACIA leadership, integration team, and chapter liason Pål Prestrud for guidance and encouragement.

Personal communications and unpublished data

Callaghan, T.V. Abisko Scientific Research Station, Sweden.
Clemmensen, K.E. and A. Michelsen, 2004. Institute of Biology, University of Copenhagen.
Flanagan, P.W., 2004. Global Environmental Enterprises, Inc., Chester, Virginia.
Friborg, T., 2005. Institute of Geography, University of Copenhagen, Denmark.
Henttonen, H., 2004. Thule Insitute, University of Oulu, Finland.
Jonasson, S. and A. Michelsen, 2005. Institute of Biology, University of Copenhagen.
Kaplan, J., 2002. Canadian Centre for Climate Modelling and Analysis.
Katenin, A., 2004. Komorov Botanical Institute, St. Petersburg, Russia.
Meltofte, H., 2004. National Environment Research Institute, Denmark.
Nutti, E., 2004. Karesuando, Sweden.
Polozova, T., 2005. Komorov Botanical Institute, St. Petersburg, Russia.
Svoboda, J. Department of Botany, University of Toronto, Canada.
Taulavuori, K., E. Taulavuori and K. Laine, 2004. Thule Institute, University of Oulu, Finland.

References

Aanes, R., B.-E. Sæther and N.A. Øritsland, 2000. Fluctuations of an introduced population of Svalbard reindeer: the effects of density dependence and climatic variation. Ecography, 23:437–443.
Aanes, R., B.-E. Sæther, F.M. Smith, E.J. Cooper, P.A. Wookey and N.A. Øritsland, 2002. The Arctic Oscillation predicts effects of climate change in two trophic levels in a high-arctic ecosystem. Ecology Letters, 5:445–453.

Aars, J. and R.A. Ims, 2002. Intrinsic and climatic determinants of population demography: the winter dynamics of tundra voles. Ecology, 83:3449–3456.
Abbott, R.J. and C. Brochmann, 2003. History and evolution of the arctic flora: in the footsteps of Eric Hultén. Molecular Ecology, 12:299–313.
Abbott, R.J., H.M. Chapman, R.M.M. Crawford and D.G. Forbes, 1995. Molecular diversity and derivations of populations of *Silene acaulis* and *Saxifraga oppositifolia* from the high Arctic and more southerly latitudes. Molecular Ecology, 4:199–207.
Abbott, R.J., L.C. Smith, R.I. Milne, R.M.M. Crawford, K. Wolff and J. Balfour, 2000. Molecular analysis of plant migration and refugia in the Arctic. Science, 289:1343–1346.
Afonina, O.M. and I.V. Czernyadjeva, 1995. Mosses of the Russian Arctic: check-list and bibliography. Arctoa, 5:99–142.
Agrell, J., E.P. McDonald and R.L. Lindroth, 1999. Responses to defoliation in deciduous trees: effects of CO_2 and light. In: A. Hofgaard, J.P. Ball, K. Danell and T.V. Callaghan (eds.). Animal Responses to Global Change in the North. Ecological Bulletin 47:84–95.
Ahmadjian, V., 1970. Adaptations of Antarctic terrestrial plants. In: H.W. Holdgate (ed.). Antarctic Ecology, pp. 801–811. Academic Press.
Aitchinson, C.W., 2001. The effect of snow cover on small animals. In: H.G. Jones, J.W. Pomeroy, D.A. Walker and R.W. Hoham (eds.). Snow Ecology: An Interdisciplinary Examination of Snow-Covered Ecosystems, pp. 229–265. Cambridge University Press.
Albon, S.D., A. Stien, R.J. Irvine, R. Langvatn, E. Ropstad and O. Halvorsen, 2002. The role of parasites in the dynamics of a reindeer population. Proceedings of the Royal Society of London B, 269:1625–1632.
Aleksandrova, V.D., 1988. Vegetation of the Soviet Polar Deserts. Cambridge University Press, 228pp.
Alley, R.B., 2000. The Younger Dryas cold interval as viewed from central Greenland. Quaternary Science Reviews, 19:213–226.
Alroy, J., 2001. A multispecies overkill simulation of the end-Pleistocene megafaunal mass extinction. Science, 292:1893–1896.
Anderson, P.M. and L.B. Brubaker, 1993. Holocene vegetation and climate histories of Alaska. In: H.E. Wright Jr., J.E. Kutzbach, T. Webb III, W.F. Ruddiman, F.A. Street-Perrott and P.J. Bartlein (eds.). Global Climates since the Last Glacial Maximum, pp. 386–400. University of Minnesota Press.
Anderson, P.M. and L.B. Brubaker, 1994. Vegetation history of north central Alaska: A mapped summary of late-Quaternary pollen data. Quaternary Science Reviews, 13:71–92.
Andersson, M. and S. Jonasson, 1986. Rodent cycles in relation to food resources on an alpine heath. Oikos, 46:93–106.
Andreev, M.P., Yu.V. Kotlov and I.I. Makarova, 1996. Checklist of lichens and lichenicolous fungi of the Russian Arctic. The Bryologist, 99(2):137–169.
Andrews, J.T., 1987. The late Wisconsin glaciation and deglaciation of the Laurentide ice sheet. In: W.F. Ruddiman and H.E. Wright Jr. (eds.). North America and Adjacent Oceans during the Last Deglaciation, pp. 13–37. The Geological Society of America, Boulder, Colorado.
Angerbjörn, A., M. Tannerfeldt and S. Erlinge, 1999. Predator-prey relationships: arctic foxes and lemmings. Journal of Animal Ecology, 68:34–49.
Angerbjörn, A., M. Tannerfeldt and H. Lundberg, 2001. Geographical and temporal patterns of lemming population dynamics in Fennoscandia. Ecography, 24:298–308.
Arft, A.M., M.D. Walker, J. Gurevitch, J.M. Alatalo, M.S. Bret-Harte, M. Dale, M. Diemer, F. Gugerli, G.H.R. Henry, M.H. Jones, R.D. Hollister, I.S. Jónsdóttir, K. Laine, E. Lévesque, G.M. Marion, U. Molau, P. Mølgaard, U. Nordenhäll, V. Raszhivin, C.H. Robinson, G. Starr, A. Stenström, M. Stenström, Ø. Totland, P. L. Turner, L.J. Walker, P.J. Webber, J.M. Welker and P.A. Wookey, 1999. Responses of tundra plants to experimental warming: meta-analysis of the International Tundra Experiment. Ecological Monographs, 69:491–511.
Astakhov, V., 1998. The last ice sheet of the Kara Sea: Terrestrial constraints on its age. Quaternary International, 45/46:19–28.
Aurela, M., T. Laurila and J.-P. Tuovinen, 2002. Annual CO_2 balance of a subarctic fen in northern Europe: Importance of the wintertime efflux. Journal of Geophysical Research, 107(D21):4607, doi: 10.1029/2002JD002055.
Aurela M., T. Laurila and J.-P. Tuovinen, 2004. The timing of snow melt controls the annual CO_2 balance in a subarctic fen. Geophysical Research Letters, 31:L16119, doi:10.1029/2004GL020315.
Babenko, A.B. and V.I. Bulavintsev, 1997. Springtails (Collembola) of Eurasian polar deserts. Russian Journal of Zoology, 1(2):177–184.
Back, J., S. Neuvonen and S. Huttunen, 1994. Pine needle growth and fine structure after prolonged acid rain treatment in the subarctic. Plant, Cell and Environment, 17:1009–1021.

Baldocchi, D., F.M. Kelliher, T.A. Black and P.G. Jarvis, 2000. Climate and vegetation controls on boreal zone energy exchange. Global Change Biology, 6(S1):69–83.

Bale, J.S., G.J. Masters, I.D. Hodkinson, C. Awmack, T.M. Bezemer, V.K. Brown, J. Butterfield, A. Buse, J.C. Coulson, J. Farrar, J.E.G. Good, R. Harrington, S. Hartley, T.H. Jones, R.L. Lindroth, M.C. Press, I. Symrnioudis, A.D. Watt and J.B. Whittaker, 2002. Herbivory in global climate change research: direct effects of rising temperature on insect herbivores. Global Change Biology, 8:1–16.

Barber, D.C., A. Dyke, C. Hillaire-Marcel, A.E. Jennings, J.T. Andrews, M.W. Kerwin, G. Bilodeau, R. McNeely, J. Southon, M.D. Morehead and J.-M. Gagnon, 1999. Forcing of the cold event of 8,200 years ago by catastrophic drainage of Laurentide lakes. Nature, 400:344–348.

Bardgett, R.D., T.C. Streeter, L. Cole and I.R. Hartley, 2002. Linkages between soil biota, nitrogen availability, and plant nitrogen uptake in a mountain ecosystem in the Scottish Highlands. Applied Soil Ecology, 19:121–134.

Barlow, L.K., J.P. Sadler, A.E.J. Ogilvie, P.C. Buckland, T. Amorosi, J.H. Ingimundarson, P. Skidmore, A.J. Dugmore and T.H. McGovern, 1997. Interdisciplinary investigations of the end of the Norse Western Settlement in Greenland. The Holocene, 7(4):489–499.

Barnes, J.D., M.R. Hull and A.W. Davison, 1996. Impacts of air pollutants and elevated carbon dioxide on plants in wintertime. In: M. Iqbal and M. Yunus (eds.). Plant Response to Air Pollution, pp. 135–166. John Wiley & Sons.

Bartlett, K.B. and R.C. Harriss, 1993. Review and assessment of methane emissions from wetlands. Chemosphere, 26:261–320.

Batzli, G.O., 1975. The role of small mammals in arctic ecosystems. In: F.B. Golley, K. Petrusewicz and L. Ryszkowski (eds.). Small Mammals: Their Productivity and Population Dynamics, pp. 243–267. Cambridge University Press.

Batzli, G.O., 1981. Populations and energetics of small mammals in the tundra ecosystem. In: L.C. Bliss, O.W. Heal and J.J. Moore (eds.). Tundra Ecosystems: a Comparative Analysis, pp. 377–396. Cambridge University Press.

Batzli, G.O., 1993. Food selection by lemmings. In: N.C. Stenseth and R.A. Ims (eds.). The Biology of Lemmings, pp. 281–301. Academic Press.

Batzli, G.O., R.G. White, S.F. MacLean Jr., F.A. Pitelka and B.D. Collier, 1980. The herbivore-based trophic system. In: J. Brown, P.C. Miller, L.L. Tieszen and F.L. Bunnell (eds.). An Arctic Ecosystem: The Coastal Tundra at Barrow, Alaska, pp. 335–410. Dowden, Hutchinson and Ross, Pennsylvania.

Bauert, M.R., 1996. Genetic diversity and ecotypic differentiation in arctic and alpine populations of *Polygonum viviparum*. Arctic and Alpine Research, 28:190–195.

Bayer, R.J., 1991. Allozymic and morphological variation of *Antennaria* (Asteraceae: Inuleae) from the low arctic of northwestern North America. Systematic Botany, 16:492–509.

Beeder, J., R.K. Nilsen, J.T. Rosnes, T. Torsvik and T. Lien, 1994. *Archaeoglobus fulgidus* isolated from hot North Sea oil field waters. Applied and Environmental Microbiology, 60:1227–1231.

Beerling, D.J. and M. Rundgren, 2000. Leaf metabolic and morphological responses of dwarf willow (*Salix herbacea*) in the sub-Arctic to the past 9000 years of global environmental change. New Phytologist, 145:257–269.

Beerling, D.J., A.C. Terry, P.L. Mitchell, T.V. Callaghan, D. Gwynn-Jones and J.A. Lee, 2001. Time to chill: effects of simulated global change on leaf ice nucleation temperatures of subarctic vegetation. American Journal of Botany, 88(4):628–633.

Beijerinck, M.W., 1913. De infusies en de ontdekking der backteriën. In: Jaarboek van de Koninklijke Akademie v. Wetenschappen. Müller, Amsterdam.

Bell, K.L. and L.C. Bliss, 1977. Overwinter phenology of plants in a polar semi-desert. Arctic, 30:118–121.

Bell, K.L. and L.C. Bliss, 1980. Plant reproduction in a high arctic environment. Arctic and Alpine Research, 12:1–10.

Bennike, O., 1999. Colonisation of Greenland by plants and animals after the last ice age: a review. Polar Record, 35:323–336.

Berendse, F. and S. Jonasson, 1992. Nutrient use and nutrient cycling in northern ecosystems. In: F.S. Chapin III, R.L. Jefferies, J.F. Reynolds, G.R. Shaver and J. Svoboda (eds.). Arctic Ecosystems in a Changing Climate: An Ecophysiological Perspective, pp. 337–356. Academic Press.

Bergman, G., 1982. Why are wings of *Larus fuscus* so dark? Ornis Fennica, 59:77–83.

Beringer, J., A.H. Lynch, F.S. Chapin III, M. Mack and G.B. Bonan, 2001a. The representation of Arctic soils in the Land Surface Model: the importance of mosses. Journal of Climate, 14:3324–3335.

Beringer, J., N.J. Tapper, I. McHugh, F.S. Chapin III, A.H. Lynch, M.C. Serreze and A. Slater, 2001b. Impact of Arctic treeline on synoptic climate. Geophysical Research Letters, 28:4247–4250.

Berkes, F. and H. Fast, 1996. Aboriginal peoples: the basis for policy-making towards sustainable development. In: A. Dale and J.B. Robinson (eds.). Achieving Sustainable Development, pp. 204–264. University of British Columbia Press.

Bernatchez, L., A. Chouinard and G. Lu, 1999. Integrating molecular genetics and ecology in studies of adaptive radiation: whitefish, *Coregonus* sp., as a case study. Biological Journal of the Linnean Society, 68(1–2):173–194.

Betts, A.K. and J.H. Ball, 1997. Albedo over the boreal forest. Journal of Geophysical Research, 102(D24):28901–28909.

Betts, R.A., 2000. Offset of the potential carbon sink from boreal forestation by decreases in surface albedo. Nature, 408:187–190.

Betts, R.A., 2001. The boreal forest and surface albedo change. In: Proceedings of the International Conference on Arctic Feedbacks to Global Change, Rovaniemi, Finland, 25–26 October 2001, pp. 15–19. Arctic Centre, University of Lapland, Rovaniemi.

Bety, J., G. Gauthier, E. Korpimäki and J.F. Giroux, 2002. Shared predators and indirect trophic interactions: lemming cycles and arctic-nesting geese. Journal of Animal Ecology, 71:88–98.

Bianchi, G.G. and I.N. McCave, 1999. Holocene periodicity in North Atlantic climate and deep-ocean flow south of Iceland. Nature, 397:515–517.

Bielawski, E., 1997. Aboriginal participation in global change research in the Northwest Territories of Canada. In: W.C. Oechel, T. Callaghan, T. Gilmanov, J.I. Holten, B. Maxwell, U. Molau and B. Sveinbjornsson (eds.). Global Change and Arctic Terrestrial Ecosystems, pp. 475–483. Springer-Verlag.

Bigelow, N.H., L.B. Brubaker, M.E. Edwards, S.P. Harrison, I.C. Prentice, P.M. Anderson, A.A. Andreev, P.J. Bartlein, T.R. Christensen, W. Cramer, J.O. Kaplan, A.V. Lozhkin, N.V. Matveyeva, D.F. Murray, A.D. McGuire, V.Y. Razzhivin, J.C. Ritchie, B. Smith, D.A. Walker, K. Gajewski, V. Wolf, B.H. Holmqvist, Y. Igarashi, K. Kremenetskii, A. Paus, M.F.J. Pisaric and V.S. Volkova, 2003. Climate change and Arctic ecosystems: 1. Vegetation changes north of 55°N between the last glacial maximum, mid-Holocene, and present. Journal of Geophysical Research, 108(D19):8170, doi:10.1029/2002JD002558.

Billings, W.D., 1973. Arctic and Alpine vegetations: similarities, differences, and susceptibility to disturbance. Bioscience, 23:697–704.

Billings, W.D., 1987. Carbon balance of Alaskan tundra and taiga ecosystems: Past, present, and future. Quaternary Science Reviews, 6:165–177.

Billings, W.D., 1992. Phytogeographic and evolutionary potential for the Arctic flora and vegetation in a changing climate. In: F.S. Chapin III, R.L. Jefferies, J.F. Reynolds, G.R. Shaver and J. Svoboda (eds.). Arctic Ecosystems in a Changing Climate: An Ecophysiological Perspective, pp. 91–109. Academic Press.

Billings, W.D. and L.C. Bliss, 1959. An alpine snowbank environment and its effects on vegetation, plant development and productivity. Ecology, 40(3):388–397.

Billings, W.D. and H.A. Mooney, 1968. The ecology of Arctic and Alpine plants. Biological Reviews, 43:481–529.

Billings, W.D., J.O. Luken, D.A. Mortensen and K.M. Peterson, 1982. Arctic tundra: a source or sink for atmospheric carbon dioxide in a changing environment? Oecologia, 53:7–11.

Billings, W.D., K.M. Peterson, J.O. Luken and D.A. Mortensen, 1984. Interaction of increasing atmospheric carbon dioxide and soil nitrogen on the carbon balance of tundra microcosms. Oecologia, 65:26–29.

BirdLife International, 2001. Threatened Birds of Asia: The BirdLife International Red Data Book. N.J. Collar, A.V. Andreev, S. Chan, M.J. Crosby, S. Subramanya and J.A. Tobias (eds.). BirdLife International, Cambridge, UK, 3038pp.

Birkemoe, T. and H.P. Leinaas, 2000. Effects of temperature on the development of an arctic Collembola (*Hypogastrura tullbergi*). Functional Ecology, 14:693–700.

Birkemoe, T. and L. Sømme, 1998. Population dynamics of two collembolan species in an Arctic tundra. Pedobiologia, 42:131–145.

Björck, S., M.J.C. Walker, L.C. Cwynar, S. Johnsen, K.-L. Knudsen, J.J. Lowe and B. Wohlfarth, 1998. An event stratigraphy for the Last Termination in the North Atlantic region based on the Greenland ice-core record: a proposal by the INTIMATE group. Journal of Quaternary Science, 13(4):283–292.

Björn, L.O., 2002. Effects of UV-B radiation on terrestrial organisms and ecosystems with special reference to the Arctic. In: D.O. Hessen (ed.). UV Radiation and Arctic Ecosystems. Ecological Studies, 153:93–121.

Bliss, L.C., 1971. Arctic and alpine plant life cycles. Annual Review of Ecology and Systematics, 2:405–438.

Bliss, L.C. and N.V. Matveyeva, 1992. Circumpolar Arctic vegetation. In: F.S. Chapin III, R.L. Jefferies, J.F. Reynolds, G.R. Shaver, and J. Svoboda (eds.). Arctic Ecosystems in a Changing Climate: An Ecophysiological Perspective, pp. 59–89. Academic Press.

Bliss, L.C., O.W. Heal and J.J. Moore (eds.), 1981. Tundra Ecosystems: A Comparative Analysis. Cambridge University Press, 813pp.

Blum, O.B., 1974. Water relations. In: V. Ahmadjian and M.E. Hale (eds.). The Lichens, pp. 381–400. Academic Press.

Bonan, G.B., D. Pollard and S.L. Thompson, 1992. Effects of boreal forest vegetation on global climate. Nature, 359:716–718.

Bonan, G.B., F.S. Chapin III and S.L. Thompson, 1995. Boreal forest and tundra ecosystems as components of the climate system. Climatic Change, 29:145–167.

Borgen, T., S.A. Elborne and H. Knudsen. A Checklist of the Greenland Basidiomycota. Meddelelser om Grønland, Bioscience (in press.).

Boudreau, L.D. and W.R. Rouse, 1995. The role of individual terrain units in the water balance of wetland tundra. Climate Research, 5:31–47.

Bovis, M.J. and R.G. Barry, 1974. A climatological analysis of north polar desert areas. In: T.L. Smiley and J.H. Zumberge (eds.). Polar Deserts and Modern Man, pp. 23–31. University of Arizona Press.

Braga, G.U.L., S.D. Flint, C.D. Miller, A.J. Anderson and D.W. Roberts, 2001a. Variability in response to UV-B among species and strains of *Metarhizium* isolated from sites at latitudes from 61° N to 54° S. Journal of Invertebrate Pathology, 78:98–108.

Braga, G.U.L., S.D. Flint, C.L. Messias, A.J. Anderson and D.W. Roberts, 2001b. Effects of UVB irradiance on conidia and germinants of the entomopathogenic hyphomycete *Metarhizium anisopliae*: A study of reciprocity and recovery. Photochemistry and Photobiology, 73:140–146.

Braga, G.U.L., S.D. Flint, C.D. Miller, A.J. Anderson and D.W. Roberts, 2001c. Both solar UVA and UVB radiation impair conidial culturability and delay germination in the entomopathogenic fungus *Metarhizium anisopliae*. Photochemistry and Photobiology, 74:734–739.

Bret-Harte, M.S., G.R. Shaver, J.P. Zoerner, J.F. Johnstone, J.L. Wagner, A.S. Chavez, R.F. Gunkelman, S.C. Lippert and J.A. Laundre, 2001. Developmental plasticity allows *Betula nana* to dominate tundra subjected to an altered environment. Ecology, 82:18–32.

Bret-Harte, M.S., G.R. Shaver and F.S. Chapin III, 2002. Primary and secondary stem growth in arctic shrubs: implications for community response to environmental change. Journal of Ecology, 90:251–267.

Brochmann, C. and S.W. Steen, 1999. Sex and genes in the flora of Svalbard – implications for conservation biology and climate change. In: I. Nordal and V.Y. Razzhivin (eds.). The Species Concept in the High North - a Panarctic Flora Initiative. Det Norske Videnskaps – Akademi. I. Matematisk Naturvitenskapelig Klasse, Skrifter, Ny serie 38:33–72. Novus Forlag, Oslo.

Brochmann, C., T.M. Gabrielsen, I. Nordal, J.Y. Landvik and R. Elven, 2003. Glacial survivial or *tabula rasa*? The history of North Atlantic biota revisited. Taxon, 52:417–450.

Brockmann-Jerosch, H., 1919. Baumgrenze und Klimacharakter. Pflanzengeographische Kommission der Schweiz Naturforschenden Gesellschaft, Beiträge zur geobotanischen Landesaufnahmne, Vol 6. Rascher, Zürich.

Brooker, R. and T.V. Callaghan, 1998. The balance between positive and negative plant interactions and its relationship to environmental gradients: a model. Oikos, 81:196–207.

Brooker, R., T.V. Callaghan and S. Jonasson, 1999. Nitrogen uptake by rhizomes of the clonal sedge *Carex bigelowii*: a previously overlooked nutritional benefit of rhizomatous growth. New Phytologist, 142:35–48.

Brooker, R.W., B.Å. Carlsson and T.V. Callaghan, 2001. *Carex bigelowii* Torrey ex Schweinitz (*C. rigida* Good., non Schrank; *C. hyperborea* Drejer). Journal of Ecology, 89:1072–1095.

Brooks, P.D., M.W. Williams and S.K. Schmidt, 1998. Inorganic nitrogen and microbial biomass dynamics before and during spring snowmelt. Biogeochemistry, 43:1–15.

Brown, J., K.R. Everett, P.J. Webber, S.F. MacLean Jr. and D.F. Murray, 1980. The coastal tundra at Barrow. In: J. Brown, P.C. Miller, L.L. Tieszen and F.L. Bunnell (eds.). An Arctic Ecosystem: The Coastal Tundra at Barrow, Alaska, pp. 1–29. Dowden, Hutchinson and Ross, Pennsylvania.

Brown, J., K.M. Hinkel and F.E. Nelson, 2000. The Circumpolar Active Layer Monitoring (CALM) Program: research designs and initial results. Polar Geography, 24(3):165–258.

Brunner, P.C., M.R. Douglas, A. Osinov, C.C. Wilson and L. Bernatchez, 2001. Holarctic phylogeography of Arctic char (*Salvelinus alpinus* L.) inferred from mitochondrial DNA sequences. Evolution, 55(3):573–586.

Bryant, J.P., F.S. Chapin III and D.R. Klein, 1983. Carbon/nutrient balance of boreal plants in relation to vertebrate herbivory. Oikos, 40:357–368.

Bryson, R.A., 1966. Air masses, streamlines, and the boreal forest. Geographical Bulletin, 8:228–269.

Bubier, J.L. and T.R. Moore, 1994. An ecological perspective on methane emissions from northern wetlands. Trends in Ecology and Evolution, 9:460–464.

Buck, N.D., 1999. The direct and indirect effects of enhanced UV-B on larvae of the moth *Epirrita autumnata*. Ph.D Thesis, University of Sheffield, 206pp.

Buck, N.D. and T.V. Callaghan, 1999. The direct and indirect effects of enhanced UV-B on the moth caterpillar *Epirrita autumnata*. In: A. Hofgaard, J.P. Ball, K. Danell and T.V. Callaghan (eds.). Animal Responses to Global Change in the North. Ecological Bulletin 47:68–76.

Buckland, P.C., T. Amorosi, L.K. Barlow, A.J. Dugmore, P.A. Mayewski, T.H. McGovern, A.E.J. Ogilvie, J.P. Sadler and P. Skidmore, 1996. Bioarchaeological and climatological evidence for the fate of Norse farmers in medieval Greenland. Antiquity, 70(267):88–96.

Buffoni Hall, R.S., J.F. Bornman and L.O. Björn, 2002. UV-induced changes in pigment content and light penetration in the fruticose lichen *Cladonia arbuscula* ssp. mitis. Journal of Photochemistry and Photobiology B: Biology, 66:13–20.

Buffoni Hall, R.S., M. Paulsson, K. Duncan, A.K. Tobin, S. Widell and J.F. Bornman, 2003. Water- and temperature-dependence of DNA damage and repair in the fruticose lichen *Cladonia arbuscula* ssp. *mitis* exposed to UV-B radiation. Physiologia Plantarum, 118:371–379.

Bunnell, F., O.K. Miller, P.W. Flanagan and R.E. Benoit, 1980. The microflora: composition, biomass and environmental relations. In: J. Brown, P.C. Miller, L.L. Tieszen and F.L. Bunnell (eds.). An Arctic Ecosystem: The Coastal Tundra at Barrow, Alaska, pp. 255–290. Dowden, Hutchinson and Ross, Pennsylvania..

Burn, C.R., 1997. Cryostratigraphy, paleogeography, and climate change during the early Holocene warm interval, western Arctic coast, Canada. Canadian Journal of Earth Sciences, 34:912–925.

Bylund, H., 1999. Climate and the population dynamics of two insect outbreak species in the north. In: A. Hofgaard, J.P. Ball, K. Danell and T.V. Callaghan (eds.). Animal Responses to Global Change in the North. Ecological Bulletin 47:54–62.

CAFF, 2001. Arctic Flora and Fauna: Status and Conservation. Conservation of Arctic Flora and Fauna, Edita, Helsinki, 272pp.

Callaghan, T.V., 1974. Intraspecific variation in *Phleum alpinum* L. with special reference to polar regions. Arctic and Alpine Research, 6:361–401.

Callaghan, T.V. and B.A. Carlsson, 1997. Impacts of climate change on demographic processes and population dynamics in Arctic plants. In: W.C. Oechel, T.V. Callaghan, T. Gilmanov, J.I. Holten, B. Maxwell, U. Molau and B. Sveinbjornsson (eds.). Global Change and Arctic Terrestrial Ecosystems, pp. 129–152. Springer Verlag.

Callaghan, T.V. and U. Emanuelsson, 1985. Population structure and processes of tundra plants and vegetation. In: J. White (ed.). The Population Structure of Vegetation, pp. 399–439. Junk, Dordrecht.

Callaghan, T.V. and S. Jonasson, 1995. Arctic terrestrial ecosystems and environmental change. Philosophical Transactions of the Royal Society of London A, 352:259–276.

Callaghan, T.V., A.D. Headley and J.A. Lee, 1991. Root function related to the morphology, life history and ecology of tundra plants. In: D. Atkinson (ed.). Plant Root Growth: An Ecological Perspective, pp. 311–340. Blackwell.

Callaghan, T.V., B.A. Carlsson, M. Sonesson and A. Temesvary, 1997. Between-year variation in climate-related growth of circumarctic populations of the moss *Hylocomium splendens*. Functional Ecology, 11:157–165.

Callaghan, T.V., M.C. Press, J.A. Lee, D.L. Robinson and C.W. Anderson, 1999. Spatial and temporal variability in the responses of Arctic terrestrial ecosystems to environmental change. Polar Research, 18(2):191–197.

Callaghan, T.V., R.M.M. Crawford, M. Eronen, A. Hofgaard, S. Payette, W.G. Rees, O. Skre, B. Sveinbjörnsson, T.K. Vlassova and B.R. Werkman, 2002a. The dynamics of the tundra-taiga boundary: An overview and suggested coordinated and integrated approach to research. Ambio Special Report, 12:3–5.

Callaghan, T.V., B.R. Werkman and R.M.M. Crawford, 2002b. The tundra-taiga interface and its dynamics: concepts and applications. Ambio Special Report, 12:6–14.

Callaway, R.M., R.W. Brooker, P. Choler, Z. Kikvidze, C.J. Lortie, R. Michalet, L. Paolini, F.I. Pugnaire, B. Newingham, E.T. Aschehoug, C. Armas, D. Kikodze and B.J. Cook, 2002. Positive interactions among alpine plants increase with stress. Nature, 417:844–848.

Caporn, S.J.M., M. Risager and J.A. Lee, 1994. Effect of nitrogen supply on frost hardiness in *Calluna vulgaris* (L.) Hull. New Phytologist, 128:461–468.

Carlsson, B.A. and T.V. Callaghan, 1991. Positive plant interactions in tundra vegetation and the importance of shelter. Journal of Ecology, 79:973–983.

Carlsson, B.A. and T.V. Callaghan, 1994. Impact of climate change factors on the clonal sedge *Carex bigelowii*: implications for population growth and vegetative spread. Ecography, 17:321–330.

Chapin, D.M. and C.S. Bledsoe, 1992. Nitrogen fixation in arctic plant communities. In: F.S. Chapin III, R.L. Jefferies, J.F. Reynolds, G.R. Shaver and J. Svoboda (eds.). Arctic Ecosystems in a Changing Climate: An Ecophysiological Perspective, pp. 301–319. Academic Press.

Chapin, F.S. III, 1974. Morphological and physiological mechanisms of temperature compensation in phosphate absorption along a latitudinal gradient. Ecology, 55:1180–1198.

Chapin, F.S. III, 1983. Direct and indirect effects of temperature on Arctic plants. Polar Biology, 2:47–52.

Chapin, F.S. III and A.J. Bloom, 1976. Phosphate absorption: adaptation of tundra graminoids to a low temperature, low phosphorus environment. Oikos, 26:111–121.

Chapin, F.S. III and G.R. Shaver, 1985a. Individualistic growth response of tundra plant species to environmental manipulations in the field. Ecology, 66:564–576.

Chapin, F.S. III and G.R. Shaver, 1985b. Arctic. In: B.F. Chabot and H.A. Mooney (eds.). Physiological Ecology of North American Plant Communities, pp. 16–40. Chapman and Hall.

Chapin, F.S. III and A.M. Starfield, 1997. Time lags and novel ecosystems in response to transient climatic change in Arctic Alaska. Climatic Change, 35:449–461.

Chapin, F.S. III, P.C. Miller, W.D. Billings and P.I. Coyne, 1980. Carbon and nutrient budgets and their control in coastal tundra. In: J. Brown, P.C. Miller, L.L. Tieszen, and F.L. Bunnell (eds.). An Arctic Ecosystem: The Coastal Tundra at Barrow, Alaska, pp. 458–482. Dowden, Hutchinson and Ross, Pennsylvania.

Chapin, F.S. III, L. Moilanen and K. Kielland, 1993. Preferential use of organic nitrogen for growth by a non-mycorrhizal Arctic sedge. Nature, 361:150–153.

Chapin, F.S. III, L.R. Walker, C.L. Fastie and L.C. Sharman, 1994. Mechanisms of primary succession following deglaciation at Glacier Bay, Alaska. Ecological Monographs, 64:149–175.

Chapin, F.S. III, G.R. Shaver, A.E. Giblin, K.J. Nadelhoffer and J.A. Laundre, 1995. Responses of Arctic tundra to experimental and observed changes in climate. Ecology, 76:694–711.

Chapin, F.S. III, M.S. Bret-Harte, S.E. Hobbie and H. Zhong, 1996. Plant functional types as predictors of transient responses of Arctic vegetation to global change. Journal of Vegetation Science, 7:347–358.

Chapin, F.S. III, W. Eugster, J.P. McFadden, A.H. Lynch and D.A. Walker, 2000a. Summer differences among Arctic ecosystems in regional climate forcing. Journal of Climate, 13:2002–2010.

Chapin, F.S. III, A.D. McGuire, J. Randerson, R. Pielke Sr., D. Baldocchi, S.E. Hobbie, N. Roulet, W. Eugster, E. Kasischke, E.B. Rastetter, S.A. Zimov and S.W. Running, 2000b. Arctic and boreal ecosystems of western North America as components of the climate system. Global Change Biology, 6(S1):211–223.

Chapman, W.L. and J.E. Walsh, 1993. Recent variations of sea ice and air temperatures in high latitudes. Bulletin of the American Meteorological Society, 74(1):33–47.

Cheng, W. and R.A. Virginia, 1993. Measurement of microbial biomass in arctic tundra soils using fumigation-extraction and substrate-induced respiration procedures. Soil Biology and Biochemistry, 25:135–141.

Chernov, Y.I., 1985. Synecological analysis of yeasts in the Taimyr tundra. Ekologiya, 1:54–60.

Chernov, Y.I., 1989. Heat conditions and Arctic biota. Ekologiya, 2:49–57. (In Russian)

Chernov, Y.I., 1995. Diversity of the Arctic terrestrial fauna. In: F.S. Chapin III and C. Körner (eds.). Arctic and Alpine Biodiversity: Patterns, Causes and Ecosystem Consequences. Ecological Studies, 113:81–95.

Chernov, Y.I., 2002. Arctic biota: taxonomic diversity. Zoologicheski Zhurnal, 81(12):1411–1431. (In Russian with English summary)

Chernov, Y.I. and N.V. Matveyeva, 1997. Arctic ecosystems in Russia. In: F.E. Wielgolaski (ed.). Polar and Alpine Tundra. Ecosystems of the World, 3:361–507. Elsevier.

Christensen, S. and B.T. Christensen, 1991. Organic matter available for denitrification in different soil fractions: Effect of freeze/thaw cycles and straw disposal. Journal of Soil Science, 42:637–647.

Christensen, T.R., 1993. Methane emission from Arctic tundra. Biogeochemistry, 21(2):117–139.

Christensen, T.R. and P. Cox, 1995. Response of methane emission from Arctic tundra to climatic change: results from a model simulation. Tellus B:47:301–309.

Christensen, T.R. and M. Keller, 2003. Element interactions and trace gas exchange. In: J.M. Melillo, C.B. Field and B. Moldan (eds.). Interactions of the Major Biogeochemical Cycles: Global Changes and Human Impacts. SCOPE 61, pp. 247–258. Island Press.

Christensen, T.R., S. Jonasson, T.V. Callaghan and M. Havström, 1995. Spatial variation in high-latitude methane flux along a transect across Siberian and European tundra environments. Journal of Geophysical Research, 100(D10):21035–21046.

Christensen, T.R., I.C. Prentice, J. Kaplan, A. Haxeltine and S. Sitch, 1996. Methane flux from northern wetlands and tundra: an ecosystem source modelling approach. Tellus B:48:652–661.

Christensen, T.R., S. Jonasson, A. Michelsen, T.V. Callaghan and M. Havström, 1998. Environmental controls on soil respiration in the Eurasian and Greenlandic Arctic. Journal of Geophysical Research, 103(D22):29015–29021.

Christensen, T.R., A. Michelsen and S. Jonasson, 1999a. Exchange of CH_4 and N_2O in a subarctic heath soil: effects of inorganic N and P and amino acid addition. Soil Biology and Biochemistry, 31:637–641.

Christensen, T.R., S. Jonasson, T.V. Callaghan and M. Havström, 1999b. On the potential CO_2 releases from tundra soils in a changing climate. Applied Soil Ecology, 11:127–134.

Christensen, T.R., T. Friborg, M. Sommerkorn, J. Kaplan, L. Illeris, H. Søgaard, C. Nordstrøm and S. Jonasson, 2000. Trace gas exchange in a high-arctic valley. 1: Variations in CO_2 and CH_4 flux between tundra vegetation types. Global Biogeochemical Cycles, 14:701–714.

Christensen, T.R., N. Panikov, M. Mastepanov, A. Joabsson, A. Stewart, M. Oquist, M. Sommerkorn, S. Reynaud and B. Svensson, 2003a. Biotic controls on CO_2 and CH_4 exchange in wetlands - a closed environment study. Biogeochemistry, 64:337–354.

Christensen, T.R., A. Ekberg, L. Ström, M. Mastepanov, N. Panikov, M. Öquist, B.H. Svensson, H. Nykänen, P.J. Martikainen and H. Oskarsson, 2003b. Factors controlling large scale variations in methane emissions from wetlands. Geophysical Research Letters, 30(7):1414, doi:10.1029/2002GL016848.

Christensen, T.R, T. Johansson, H.J. Åkerman, M. Mastepanov, N. Malmer, T. Friborg, P. Crill and B. Svensson, 2004. Thawing sub-arctic permafrost: effects on vegetation and methane emissions. Geophysical Research Letters, 31(4):L04501, doi:10.1029/2003GL018680.

Clein, J.S. and J.P. Schimel, 1995. Microbial activity of tundra and taiga soils at sub-zero temperatures. Soil Biology and Biochemistry, 27:1231–1234.

Clein, J.S., B.L. Kwiatkowski, A.D. McGuire, J.E. Hobbie, E.B. Rastetter, J.M. Melillo and D.W. Kicklighter, 2000. Modelling carbon responses of tundra ecosystems to historical and projected climate: A comparison of a plot- and a global-scale ecosystem model to identify process-based uncertainties. Global Change Biology, 6(S1):127–140.

Clymo, R.S. and E.J.F. Reddaway, 1971. Productivity of *Sphagnum* (bog-moss) and peat accumulation. Hydrobiologia, 12:181–192.

Cole, J.J., N.F. Caraco, G.W. Kling and T.K. Kratz, 1994. Carbon dioxide supersaturation in the surface waters of lakes. Science, 265:1568–1570.

Coley, P.D., J.P. Bryant and F.S. Chapin III, 1985. Resource availability and plant antiherbivore defense. Science, 230:895–899.

Conrad, R., 1996. Soil microorganisms as controllers of atmospheric trace gases (H_2, CO, CH_4, OCS, N_2O, and NO). Microbiological Reviews, 60(4):609–640.

Convey, P., 2000. How does cold constrain life cycles of terrestrial plants and animals? CryoLetters, 21:73–82.

Cooper, E.J., 1996. An ecophysiological investigation of some species of Arctic temperature *Ranunculus L.* with respect to climate warming. Responses of above- and below-ground growth and carbon dioxide exchange to season and temperature. Ph.D Thesis, University of Bradford.

Cornelissen, J.H.C., N. Perez-Harguindeguy, S. Diaz, J. Grime, B. Marazano, M. Cabido, F. Vendramini and B. Cerabolini, 1999. Leaf structure and defence control litter decomposition rate across species and life forms in regional floras on two continents. New Phytologist, 143:191–200.

Cornelissen, J.H.C., T.V. Callaghan, J.M. Alatalo, A. Michelsen, E. Graglia, A.E. Hartley, D.S. Hik, S.E. Hobbie, M.C. Press, C.H. Robinson, G.H.R. Henry, G.R. Shaver, G.K. Phoenix, D. Gwynn-Jones, S. Jonasson, F.S. Chapin III, U. Molau, J.A. Lee, J.M. Melillo, B. Sveinbjornsson and R. Aerts, 2001. Global change and Arctic ecosystems: is lichen decline a function of increases in vascular plant biomass? Journal of Ecology, 89:984–994.

Corner, R.W.M. and R.I.L. Smith, 1973. Botanical evidence of ice recession in the Argentine Islands. British Antarctic Survey Bulletin, 35:83–86.

Coulson, S.J., I.D. Hodkinson, A.T. Strathdee, W. Block, N.R. Webb, J.S. Bale and M.R. Worland, 1995. Thermal environments of Arctic soil organisms during the winter. Arctic and Alpine Research, 27:364–370.

Coulson, S.J., I.D. Hodkinson, N.R. Webb, W. Block, J.S. Bale, A.T. Strathdee, M.R. Worland and C. Wooley, 1996. Effects of experimental temperature elevation on high-arctic soil microarthropod populations. Polar Biology, 16:147–153.

Coulson, S.J., H.P. Leinaas, R.A. Ims and G. Søvik, 2000. Experimental manipulation of winter surface ice layer: The effects on a High Arctic soil microarthropod community. Ecography, 23:299–306.

Coyne, P.I. and J.J. Kelley, 1971. Release of carbon dioxide from frozen soil to the Arctic atmosphere. Nature, 234:407–408.

Crawford, R.M.M., 1989. Studies in Plant Survival. Blackwell.

Crawford, R.M.M., 1997a. Habitat fragility as an aid to long-term survival in Arctic vegetation. In: S.J. Woodin and M. Marquiss (eds.). Ecology of Arctic Environments, pp. 113–136. Blackwell Scientific Ltd.

Crawford, R.M.M. (ed.), 1997b. Disturbance and Recovery in Arctic Lands: An Ecological Perspective. Kluwer Academic Publishers, 621pp.

Crawford, R.M.M. and L.C. Smith, 1997. Responses of some high Arctic shore plants to variable lengths of growing season. Opera Botanica, 132:201–214.

Crawford, R.M.M., H.M. Chapman and H. Hodge, 1994. Anoxia tolerance in high Arctic vegetation. Arctic and Alpine Research, 26:308–312.

Crawford, R.M.M., C.E. Jeffree and W.G. Rees, 2003. Paludification and forest retreat in northern oceanic environments. Annals of Botany, 91:213–226.

Crill, P., K.B. Bartlett and N.T. Roulet, 1992. Methane flux from boreal peatlands. Suo, 43:173–182.

Cronberg, N., U. Molau and M. Sonesson, 1997. Genetic variation in the clonal bryophyte Hylocomium splendens at hierarchical geographical scales in Scandinavia. Heredity, 78:293–301.

Curl, H. Jr., J.T. Hardy and R. Ellermeier, 1972. Spectral absorption of solar radiation in alpine snowfields. Ecology, 53:1189–1194.

Dalén, L., E. Fuglei, P. Hersteinsson, C. Kapel, J. Roth, G. Samelius, M. Tannerfeldt and A. Angerbjörn, 2005. Population history and genetic structure of a circumpolar species: the Arctic fox. Biological Journal of the Linnean Society, 84:79–89.

Dankers, R., 2002. Sub-arctic Hydrology and Climate Change. A Case Study of the Tana River Basin in Northern Fennoscandia. Netherlands Geographical Studies, 304. Royal Dutch Geographical Society/ Faculteit Ruimtelijke Wetenschappen Universiteit Utrecht, 240pp.

Danks, H.V., 1999. Life cycles in polar arthropods – flexible or programmed? European Journal of Entomology, 96:83–102.

Dansgaard, W., J.W.C. White and S.J. Johnsen, 1989. The abrupt termination of the Younger Dryas climate event. Nature, 339:532–534.

Davidson, E.A. and J.P. Schimel, 1995. Microbial processes of production and consumption of nitric oxide, nitrous oxide and methane. In: P.A. Matson and R.C. Harriss (eds.). Biogenic Trace Gases: Measuring Emissions from Soil and Water, pp. 327–357. Blackwell Science.

Davidson, N.C., K.B. Strann, N.J. Crockford, P.R. Evans, J. Richardson, L.J. Standen, D.J. Townshend, J.D. Uttley, J.R. Wilson and A.G. Wood, 1986. The origins of Knots Calidris canutus in arctic Norway in Spring. Ornis Scandinavica, 17(2):175–179.

de Vernal, A. and C. Hillaire-Marcel, 2000. Sea-ice cover, sea-surface salinity and halo/thermocline structure of the northwest North Atlantic: modern versus full glacial conditions. Quaternary Science Reviews, 19(1–5):65–85.

de Vernal, A., C. Hillaire-Marcel, J.-L. Turon and J. Matthiessen, 2000. Reconstruction of sea-surface temperature, salinity, and sea-ice cover in the northern North Atlantic during the last glacial maximum based on dinocyst assemblages. Canadian Journal of Earth Sciences, 37(5):725–750.

Dedysh, S.N., N.S. Panikov, W. Liesack, R. Grosskopf, J. Zhou and J.M. Tiedje, 1998. Isolation of acidophilic methane-oxidizing bacteria from northern peat wetlands. Science, 282:281–284.

Dedysh, S.N., W. Liesack, V.N. Khmelenina, N.E. Suzina, Y.A. Trotsenko, J.D. Semrau, A.M. Bares, N.S. Panikov and J.M. Tiedje, 2000. Methylocella palustris gen. nov., sp. nov., a new methane-oxidizing acidophilic bacterium from peat bogs, representing a novel subtype of serine-pathway methanotrophs. International Journal of Systematic and Evolutionary Microbiology, 50:955–969.

Derda, G.S. and R. Wyatt, 1999. Levels of genetic variation and its partitioning in the wide-ranging moss Polytrichum commune. Systematic Botany, 24:512–528.

Dimmer, C.H., P.G. Simmonds, G. Nickless and M.R. Bassford, 2001. Biogenic fluxes of halomethanes from Irish peatland ecosystems. Atmospheric Environment, 35:321–330.

Dixon, E.J., 2001. Human colonization of the Americas: timing, technology and process. Quaternary Science Reviews, 20:277–299.

Dorland, S. and E.G. Beauchamp, 1991. Denitrification and ammonification at low soil temperatures. Canadian Journal of Soil Science, 71:293–303.

Dormann, C.F. and S.J. Woodin, 2002. Climate change in the Arctic: using plant functional types in a meta-analysis of field experiments. Functional Ecology, 16:4–17.

Dunican, L.K. and T. Rosswall, 1974 Taxonomy and physiology of tundra bacteria in relationship to site characteristics. In: A.J. Holding, O.W. Heal, S.F. MacLean Jr. and P.W. Flanagan (eds.). Soil Organisms and Decomposition in Tundra, pp. 79–92. International Biological Programme Tundra Biome Steering Committee, Stockholm.

Dunne, J.A., J. Harte and K.J. Taylor, 2003. Subalpine meadow flowering phenology responses to climate change: integrating experimental and gradient methods. Ecological Monographs, 73:69–86.

Dunning, C.A., L. Chalker-Scott and J.D. Scott, 1994. Exposure to ultraviolet-B radiation increases cold hardiness in Rhododendron. Physiologia Plantarum, 92:516–520.

Dye, D.G., 2002. Variability and trends in the annual snow-cover cycle in Northern Hemisphere land areas, 1972–2000. Hydrological Processes, 16:3065–3077.

Edlund, S.A. and B.T. Alt, 1989. Regional congruence of vegetation and summer climate patterns in the Queen Elizabeth Islands, Northwest Territories, Canada. Arctic, 42:3–23.

Ehrich, D., V.B. Fedorov, N.C. Stenseth, C.J. Krebs and A. Kenney, 2000. Phylogeography and mitochondrial DNA (mtDNA) diversity in North American collared lemmings (Dicrostonyx groenlandicus). Molecular Ecology, 9(3):329–337.

Elster, J., 2002. Ecological classification of terrestrial algal communities in polar environments. In: L. Beyer and M. Bolter (eds.). Geoecology of Antarctic Ice-free Coastal Landscapes. Ecological Studies 154:303–326.

Elster, J. and E.E. Benson, 2004. Life in the polar terrestrial environment: a focus on algae. In: B. Fuller, N. Lane and E.E. Benson (eds.). Life in the Frozen State, pp. 111–150. CRC Press.

Elster, J. and J. Svoboda, 1995. In situ simulation and manipulation of a glacial stream ecosystem in the Canadian High Arctic. In: A. Jenkins, R.C. Ferrier and C. Kirby (eds.). Ecosystem Manipulation Experiments: Scientific Approaches, Experimental Design and Relevant Results. Ecosystem Research Report 20, pp. 254–263. Commission of the European Communities, Brussels.

Elster, J. and J. Svoboda, 1996. Algal diversity, seasonality and abundance in, and along glacial stream in Sverdrup Pass 79° N, Central Ellesmere Island, Canada. Memoirs of National Institute of Polar Research, Special Issue 51:99–118.

Elster, J., J. Svoboda, J. Komárek and P. Marvan, 1997. Algal and cyanoprocaryote communities in a glacial stream, Sverdrup Pass, 79° N, Central Ellesmere Island, Canada. Archiv fuer Hydrobiologie. Supplementband. Algological Studies, 85:57–93.

Elster, J., A. Lukesova, J. Svoboda, J. Kopecky and H. Kanda, 1999. Diversity and abundance of soil algae in the polar desert, Sverdrup Pass, central Ellesmere Island. Polar Record, 35(194):231–254.

Elster, J., J. Svoboda and H. Kanda, 2001. Controlled environment platform used in temperature manipulation study of a stream periphyton in the Ny-Ålesund, Svalbard. In: J. Elster, J. Seckbach, W.F. Vincent and O. Lhotsky (eds.). Algae and Extreme Environments – Ecology and Physiology. Nova Hedvigia Beihefte, 123:63–75.

Elster, J., J. Svoboda, S. Ohtani and H. Kanda, 2002. Feasibility studies on future phycological research in polar regions. Polar Bioscience, 15:114–122.

Elton, C.S., 1924. Periodic fluctuations in the number of animals: their causes and effects. British Journal of Experimental Biology, 2:119–163.

Elton, C.S., 1925. The dispersal of insects to Spitsbergen. Transactions of the Entomological Society of London 1925, 73:289–299.

Elton, C. and M. Nicholson, 1942. The ten-year cycle in numbers of the lynx in Canada. Journal of Animal Ecology, 11:215–244.

Emanuelsson, U., 1987. Human influence on vegetation in the Torneträsk area during the last three centuries. In: M. Sonesson (ed.). Research in Arctic Life and Earth Sciences: Present Knowledge and Future Perspectives. Ecological Bulletin, 38:95–111.

Engstrom, R.N., A.S. Hope, D.A. Stow and H. Kwon, 2002. Characteristics of the spatial distribution of surface moisture in an eddy flux tower footprint in Arctic Coastal Plain ecosystems. EOS, Transactions of the American Geophysical Union, 83(2002 Fall Meeting Supplement): Abstract #H61B-0764.

Erlinge, S., K. Danell, P. Frodin, D. Hasselquist, P. Nilsson, E.-B. Olofsson and M. Svensson, 1999. Asynchronous population dynamics of Siberian lemmings across the Palaearctic tundra. Oecologia, 119:493–500.

Eugster, W., W.R. Rouse, R.A. Pielke, J.P. McFadden, D.D. Baldocchi, T.G.F. Kittel, F.S. Chapin III, G.E. Liston, P.L. Vidale, E. Vaganov and S. Chambers, 2000. Land-atmosphere energy exchange in Arctic tundra and boreal forest: available data and feedbacks to climate. Global Change Biology, 6(S1):84–115.

Eugster, W., G. Kling, T. Jonas, J.P. McFadden, A. Wust, S. MacIntyre and F.S. Chapin III. 2003. CO_2 exchange between air and water in an Arctic Alaskan and midlatitude Swiss lake: Importance of convective mixing. Journal of Geophysical Research, 108(D12):4362, doi:10.1029/2002JD002653.

Evans, P.R., 1997. Migratory birds and climate change. In: B. Huntley, W. Cramer, A.V. Morgan, H.C. Prentice and J.R.M. Allen (eds.). Past and Future Rapid Environmental Changes: The Spatial and Evolutionary Responses of Terrestrial Biota, pp. 227–238. Springer Verlag.

Fahnestock, J.T., M.H. Jones and J.M. Welker, 1999. Wintertime CO_2 efflux from Arctic soils: Implications for annual carbon budgets. Global Biogeochemical Cycles, 13:775–780.

Fahnestock, J.T., K.L. Povirk and J.M. Welker, 2000. Ecological significance of litter redistribution by wind and snow in arctic landscapes. Ecography, 23:623–631.

Fairbanks, R.G., 1989. A 17,000-year glacio-eustatic sea level record: Influence of glacial melting rates on the Younger Dryas event and deep-ocean circulation. Nature, 342:637–642.

Fajer, E.D., M.D. Bowers and F.A. Bazzaz, 1989. The effects of enriched carbon dioxide atmospheres on plant-insect herbivore interactions. Science, 243:1198–1200.

FAUNMAP Working Group, 1996. Spatial response of mammals to late Quaternary environmental fluctuations. Science, 272:1601–1606.

Fedorov, V.B. and A.V. Goropashnaya, 1999. The importance of ice ages in diversification of Arctic collared lemmings (Dicrostonyx): evidence from the mitochondrial cytochrome b region. Hereditas, 130(3):301–307.

Fedorov, V.B., K. Fredga and G.H. Jarrell, 1999a. Mitochondrial DNA variation and the evolutionary history of chromosome races of collared lemmings (Dicrostonyx) in the Eurasian Arctic. Journal of Evolutionary Biology, 12(1):134–145.

Fedorov, V., A. Goropashnaya, G.H. Jarrell and K. Fredga, 1999b. Phylogeographic structure and mitochondrial DNA variation in true lemmings (Lemmus) from the Eurasian Arctic. Biological Journal of the Linnean Society, 66(3):357–371.

Fetcher, N., 1985. Effects of removal of neighboring species on growth, nutrients, and microclimate of Eriophorum vaginatum. Arctic and Alpine Research, 17:7–17.

Finegold, L., 1996. Molecular and biophysical aspects of adaptation of life to temperatures below the freezing point. Advances in Space Research, 18(12):87–95.

Fioletov, V.E., J.B. Kerr, D.I. Wardle, J. Davies, E.W. Hare, C.T. McElroy and D.W. Tarasick, 1997. Long-term ozone decline over the Canadian Arctic to early 1997 from ground-based and balloon observations. Geophysical Research Letters, 24(22):2705–2708.

Flagstad, Ø. and K.H. Røed, 2003. Refugial origins of reindeer (Rangifer tarandus L.) inferred from mitochondrial DNA sequences. Evolution, 57:658–670.

Flanagan, P.W. and F.L. Bunnell, 1980. Microfloral activities and decomposition. In: J. Brown, P.C. Miller, L.L. Tieszen and F.L. Bunnell (eds.). An Arctic Ecosystem: The Coastal Tundra at Barrow, Alaska, pp. 291–335. Dowden, Hutchinson, and Ross, Pennsylvania.

Flanagan, P.W. and W. Scarborough, 1974. Physiological groups of decomposer fungi in tundra plant remains. In: A.J. Holding, O.W. Heal, S.F. MacLean Jr. and P.W. Flanagan (eds.). Soil Organisms and Decomposition in Tundra, pp. 159–181. International Biological Programme Tundra Biome Steering Committee, Stockholm.

Fogg, G.E., 1998. The Biology of Polar Habitats. Oxford University Press, 274pp.

Foley, J.A., J.E. Kutzbach, M.T. Coe and S. Levis, 1994. Feedbacks between climate and boreal forests during the Holocene epoch. Nature, 371:52–54.

Forbes, B.C., J.J. Ebersole and B. Strandberg, 2001. Anthropogenic disturbance and patch dynamics in circumpolar arctic ecosystems. Conservation Biology, 15(4):954–969.

Forchhammer, M.C. and D.M. Boertmann, 1993. The muskoxen Ovibos moschatus in north and northeast Greenland: population trends and the influence of abiotic parameters on population dynamics. Ecography, 16:299–308.

Fossa, A.M., 2003. Mountain vegetation in the Faroe Islands in a climate change perspective. Ph.D Thesis, University of Lund, 119pp.

Fox, S., 2002. These are things that are really happening: Inuit perspectives on the evidence and impacts of climate change in Nunavut. In: I. Krupnik and D. Jolly (eds.). The Earth is Faster Now: Indigenous Observations of Arctic Environmental Change, pp. 12–53. Arctic Research Consortium of the United States, Fairbanks, Alaska.

Friborg, T., T.R. Christensen, B.U. Hansen, C. Nordstrøm and H. Søgaard, 2000. Trace gas exchange in a high arctic valley 2: Landscape CH_4 fluxes measured and modeled using eddy correlation data. Global Biogeochemical Cycles, 14:715–724.

Friborg, T., H. Soegaard, T.R. Christensen, C.R. Lloyd and N.S. Panikov, 2003. Siberian wetlands: Where a sink is a source. Geophysical Research Letters, 30(21):2129, doi:10.1029/2003GL017797.

Friend, A.D., A.K. Stevens, R.G. Knox and M.G.R. Cannell, 1997. A process-based, terrestrial biosphere model of ecosystem dynamics. (Hybrid v3.0). Ecological Modelling, 95:249–287.

Fuchigami, L.H., C.J. Weiser, K. Kobayshi, R. Timmis and L.V. Gusta, 1982. A degree growth stage (°GS) model and cold acclimation in temperate woody plants. In: P.H. Li and A. Sakai (eds.). Plant Cold Hardiness and Freezing Stress: Mechanisms and Crop Implications, Vol. 2, pp. 93–116. Academic Press.

Fuglei, E. and N.A. Øritsland, 1999. Seasonal trends in body mass, food intake and resting metabolic rate, and induction of metabolic depression in arctic foxes (Alopex lagopus) at Svalbard. Journal of Comparative Physiology B, 169:361–369.

Fulthorpe, R.R., A.N. Rhodes and J.M. Tiedje, 1998. High levels of endemicity of 3-chlorobenzoate-degrading soil bacteria. Applied and Environmental Microbiology, 64:1620–1627.

Funder, S., C. Hjort, J.Y. Landvik, S.-I. Nam, N. Reeh and R. Stein, 1998. History of a stable ice margin - East Greenland during the Middle and Upper Pleistocene. Quaternary Science Reviews, 17(1–3):77–123.

Fung, I., 1997. A greener north. Nature, 386:659–660.

Fung, I., J. John, J. Lerner, E. Matthews, M. Prather, L.P. Steele and P.J. Fraser, 1991. Three-dimensional model synthesis of the global methane cycle. Journal of Geophysical Research, 96:13033–13065.

Furness, R.W., 1990. Evolutionary and ecological constraints on the breeding distributions and behaviour of skuas. In: R. van den Elzen, K.-L. Schuchmann and K. Schmidt-Koenig (eds.). Current Topics in Avian Biology. Proceedings of the International Centennial Meeting of the Deutsche Ornithologen-Gesellschaft, pp. 153–158. Bonn.

Gabrielsen, T.M. and C. Brochmann, 1998. Sex after all: high levels of diversity detected in the arctic clonal plant Saxifraga cernua using RAPD markers. Molecular Ecology, 7:1701–1708.

Garcia-Pichel, F., L. Prufert-Bebout and G. Muyzer, 1996. Phenotypic and phylogenetic analyses show Microcoleus chthonoplastes to be a cosmopolitan cyanobacterium. Applied and Environmental Microbiology, 62:3284–3291.

Galand, P.E., S. Saarnio, H. Fritze and K. Yrjala, 2002. Depth related diversity of methanogen Archaea in Finnish oligotrophic fen. FEMS Microbiology Ecology, 42:441–449.

Gauslaa, Y., 1984. Heat resistance and energy budget in different Scandinavian plants. Holarctic Ecology, 7:1–78.

Gauslaa, Y. and E.M. Ustvedt, 2003. Is parietin a UV-B or a blue-light screening pigment in the lichen Xanthoria parietina? Photochemical and Photobiological Sciences, 2:424–432.

Gehrke, C., U. Johanson, T.V. Callaghan, D. Chadwick and C.H. Robinson, 1995. The impact of enhanced ultraviolet-B radiation on litter quality and decomposition processes in Vaccinium leaves from the Subarctic. Oikos, 72:213–222.

Geiges, O., 1996. Microbial processes in frozen food. Advances in Space Research, 18(12):109–118.

Giblin, A.E., K.J. Nadelhoffer, G.R. Shaver, J.A. Laundre and A.J. McKerrow, 1991. Biogeochemical diversity along a riverside toposequence in arctic Alaska. Ecological Monographs, 61:415–435.

Gilg, O., 2002. The summer decline of the collared lemming, Dicrostonyx groenlandicus, in high Arctic Greenland. Oikos, 99:499–510.

Gilg, O., I. Hanski and B. Sittler, 2003. Cyclic dynamics in a simple vertebrate predator-prey community. Science, 302:866–868.

Gilichinsky, D. (ed.), 1994. Viable Microorganisms in Permafrost. Russian Academy of Science, Pushchino Scientific Centre, 115pp.

Gilichinsky, D., 2002. Permafrost as a microbial habitat. In: G. Bitton (ed.). Encyclopedia of Environmental Microbiology, Vol 6, pp. 932–956. Wiley.

Gjærevoll, O., 1956. The plant communities of the Scandinavian alpine snowbeds. Det Kongelige Norske Videnskabernes Selskabs Skrifter, 1956(1):1–405.

Gold, W.G. and L.C. Bliss, 1995. The nature of water limitations for plants in a high arctic polar desert. In: T.V. Callaghan, W.C. Oechel, T. Gilmanov, J.I. Holten, B. Maxwell, U. Molau, B. Sveinbjornsson and M.J. Tyson (eds.). Global Change and Arctic Terrestrial Ecosystems, pp. 149–155. Ecosystems Research Report 10. European Commission, Brussels.

Gough, L., G.R. Shaver, J. Carroll, D.L. Royer and J.A. Laundre, 2000. Vascular plant species richness in Alaskan arctic tundra: the importance of soil pH. Journal of Ecology, 88:54–66.

Graglia, E., S. Jonasson, A. Michelsen and I.K. Schmidt, 1997. Effects of shading, nutrient application and warming on leaf growth and shoot densities of dwarf shrubs in two arctic-alpine plant communities. Ecoscience, 4:191–198.

Graglia, E., S. Jonasson, A. Michelsen, I.K. Schmidt, M. Havström and L. Gustavsson, 2001a. Effects of environmental perturbations on abundance of subarctic plants after three, seven and ten years of treatments. Ecography, 24:5–12.

Graglia, E., R. Julkunen-Tiitto, G. Shaver, I.K. Schmidt, S. Jonasson and A. Michelsen, 2001b. Environmental control and intersite variations of phenolics in *Betula nana* in tundra ecosystems. New Phytologist, 151:227–236.

Granberg, G., M. Ottosson-Lofvenius, H. Grip, I. Sundh and M. Nilsson, 2001. Effect of climatic variability from 1980 to 1997 on simulated methane emission from a boreal mixed mire in northern Sweden. Global Biogeochemical Cycles, 15(4):977–991.

Grogan, P. and S. Jonasson, 2003. Controls on annual nitrogen cycling in the understory of a subarctic birch forest. Ecology, 84(1):202–218.

Grogan, P., L. Illeris, A. Michelsen and S. Jonasson, 2001. Respiration of recently-fixed plant carbon dominates mid-winter ecosystem CO_2 production in sub-arctic heath tundra. Climatic Change, 50:129–142.

Grootes, P.M., M. Stuiver, J.W.C. White, S. Johnsen and J. Jouzel, 1993. Comparison of oxygen isotope records from the GISP2 and GRIP Greenland ice cores. Nature, 366:552–554.

Grosswald, M.G., 1988. An Antarctic-style ice sheet in the Northern Hemisphere: Toward a new global glacial theory. Polar Geography and Geology, 12:239–267.

Grosswald, M.G., 1998. Late-Weichselian ice sheets in Arctic and Pacific Siberia. Quaternary International, 45/46:3–18.

Grulke, N., G.H. Riechers, W.C. Oechel, U. Hjelm and C. Jaeger, 1990. Carbon balance in tussock tundra under ambient and elevated atmospheric CO_2. Oecologia, 83:485–494.

Guenther, A., P. Zimmerman and M. Wildermuth, 1993. Natural volatile organic compound emission rate estimates for U.S. woodland landscapes. Atmospheric Environment, 28:1197–1210.

Guthrie, R.D., 2001. Origin and causes of the mammoth steppe: a story of cloud cover, woolly mammal tooth pits, buckles, and inside-out Beringia. Quaternary Science Reviews, 20:549–574.

Gwynn-Jones, D., 1999. Enhanced UV-B radiation and herbivory. In: A. Hofgaard, J.P. Ball, K. Danell and T.V. Callaghan (eds.). Animal Responses to Global Change in the North. Ecological Bulletin, 47:77–83.

Gwynn-Jones, D., L.O. Björn, T.V. Callaghan, C. Gehrke, U. Johanson, J.A. Lee and M. Sonesson, 1996. Effects of enhanced UV-B radiation and elevated concentrations of CO_2 on a subarctic heathland. In: C. Korner and F.A. Bazzaz (eds.). Carbon Dioxide, Populations and Communities, pp. 197–207. Academic Press.

Gwynn-Jones, D., J.A. Lee and T.V. Callaghan, 1997. Effects of enhanced UV-B radiation and elevated carbon dioxide concentrations on a sub-Arctic forest heath ecosystem. Plant Ecology, 128:243–249.

Gwynn-Jones, D., U. Johanson, G.K. Phoenix, C. Gehrke, T.V. Callaghan, L.O. Björn, M. Sonesson and J.A. Lee, 1999. UV-B impacts and interactions with other co-occurring variables of environmental change: an Arctic perspective. In: J. Rozema (ed.). Stratospheric Ozone Depletion: The Effects of Enhanced UV-B Radiation on Terrestrial Ecosystems, pp. 187–201. Backhuys Publishers.

Hales, B.A., C. Edwards, D.A. Ritchie, G. Hall, R.W. Pickup and J.R. Saunders, 1996. Isolation and identification of methanogen-specific DNA from blanket bog peat by PCR amplification and sequence analysis. Applied and Environmental Microbiology, 62:668–675.

Hamrick, J.L. and M.J.W. Godt, 1990. Allozyme diversity in plant species. In: A.H.D. Brown, M.T. Clegg, A.L. Kahler and B.S. Weir (eds.). Plant Population Genetics, Breeding, and Genetic Resources, pp. 43–63. Sinauer Associates Inc.

Hansell, R.I.C., J.R. Malcom, H. Welch, R.L. Jefferies and P.A. Scott, 1998. Atmospheric change and biodiversity in the Arctic. Environmental Monitoring and Assessment, 49:303–325.

Hanski, I. and H. Henttonen, 1996. Predation on competing rodent species: a simple explanation of complex patterns. Journal of Animal Ecology, 65:220–232.

Hanski, I., H. Henttonen, E. Korpimäki, L. Oksanen and P. Turchin, 2001. Small-rodent dynamics and predation. Ecology, 82:1505–1520.

Hansson, L. and H. Henttonen, 1988. Rodent dynamics as community processes. Trends in Evolution and Ecology, 3:195–200.

Harazono, Y., M. Mano, A. Miyata, R.C. Zulueta and W.C. Oechel, 2003. Inter-annual carbon dioxide uptake of a wet sedge tundra ecosystem in the Arctic. Tellus B, 55:215–231.

Harding, R., P. Kuhry, T.R. Christensen, M.T. Sykes, R. Dankers and S. van der Linden, 2002. Climate feedbacks at the tundra-taiga interface. Ambio Special Report, 12:47–55.

Harriss, R., K. Bartlett, S. Frolking and P. Crill, 1993. Methane emissions from northern high-latitude wetlands. In: R.S. Oremland (ed.). Biogeochemistry of Global Change: Radiatively Active Trace Gases, pp. 449–486. Chapman & Hall.

Hartley, A.E., C. Neill, J.M. Melillo, R. Crabtree and F.P. Bowles, 1999. Plant performance and soil nitrogen mineralization in response to simulated climate change in subarctic dwarf shrub heath. Oikos, 86:331–343.

Hastings, S.J., S.A. Luchessa, W.C. Oechel and J.D. Tenhunen, 1989. Standing biomass and production in water drainages of the foothills of the Philip Smith Mountains, Alaska. Holarctic Ecology, 12:304–311.

Haukioja, E., 1980. On the role of plant defences in the fluctuation of herbivore populations. Oikos, 35:202–213.

Haukioja, E., 1991. Induction of defenses in trees. Annual Review of Entomology, 36:25–42.

Haukioja, E. and S. Neuvonen, 1987. Insect population dynamics and induction of plant resistance: the testing of hypotheses. In: P. Barbosa and J.C. Schultz (eds.). Insect Outbreaks: Ecological and Evolutionary Perspectives, pp. 411–432. Academic Press.

Haukioja, E., V. Ossipov, J. Koricheva, T. Honkanen, S. Larsson and K. Lempa, 1998. Biosynthetic origin of carbon-based secondary compounds: cause of variable responses of woody plants to fertilization? Chemoecology, 8:133–139.

Haukisalmi, V. and H. Henttonen, 1990. The impact of climatic factors and host density on the long-term population dynamics of vole helminths. Oecologia, 83:309–315.

Haukisalmi, V., L.M. Wickström, H. Henttonen, J. Hantula and A. Gubányi, 2004. Molecular and morphological evidence for multiple species within *Paranoplocephala omphalodes* (Hermann, 1783) (Cestoda: Anoplocephalidae) in Microtus-voles (Arvicolinae). Zoologica Scripta, 33(3):277–290.

Havas, P., 1971. The water economy of the bilberry (*Vaccinium myrtillus*) under winter conditions. Reports from the Kevo Subarctic Research Station, 8:41–52.

Havas, P., 1985. Winter and the boreal forests. Aquilo Series Botanica, 23:9–16.

Havström, M., T.V. Callaghan and S. Jonasson, 1993. Differential growth responses of *Cassiope tetragona*, an arctic dwarf shrub, to environmental perturbations among three contrasting high- and subarctic sites. Oikos, 66:389–402.

Havström, M., T.V. Callaghan and S. Jonasson, 1995a. Effects of simulated climate change on the sexual reproductive effort of *Cassiope tetragona*. In: T.V. Callaghan, W.C. Oechel, T. Gilmanov, U. Molau, B. Maxwell, B. Tyson, B. Sveinbjörnsson and J.I. Holten (eds.). Global change and Arctic terrestrial ecosystems, pp 109–114. Proceedings of papers contributed to the international conference, 21–26 August 1993, Oppdal, Norway. European Commission Ecosystems research report 10.

Havström, M., T.V. Callaghan, S. Jonasson and J. Svoboda, 1995b. Little ice age temperature estimated by growth and flowering differences between sub fossil and extant shoots of *Cassiope tetragona*, an arctic heather. Functional Ecology, 9:650–654.

Hawkins, B.A., 1990. Global patterns of parasitoid assemblage size. Journal of Animal Ecology, 59:57–72.

Hawksworth, D.L., 1991. The fungal dimension of biodiversity: magnitude, significance, and conservation. Mycological Research, 95:641–655.

Haxeltine, A. and I.C. Prentice, 1996. BIOME3: An equilibrium terrestrial biosphere model based on ecophysiological constraints, resource availability, and competition among plant functional types. Global Biogeochemical Cycles, 10:693–710.

Heal, O.W., P.W. Flanagan, D.D. French and S.F. MacLean Jr., 1981. Decomposition and accumulation of organic matter in tundra. In: L.C. Bliss, O.W. Heal and J.J. Moore (eds.). Tundra Ecosystems: A Comparative Analysis, pp. 587–634. Cambridge University Press.

Heide, O.M., R.K.M. Hay and H. Baugerod, 1985. Specific daylength effects on leaf growth and dry-matter production in high-latitude grasses. Annals of Botany, 55:579–586.

Heide-Jørgensen, H.S. and I. Johnsen, 1998. Ecosystem vulnerability to climate change in Greenland and the Faroe Islands. Miljønyt, 33:1–266.

Heikkinen, J.E.P., T. Virtanen, J.T. Huttunen, V. Elsakov and P.J. Martikainen, 2004. Carbon balance in East European tundra. Global Biogeochemical Cycles, 18(1), GB1023, doi:10.1029/2003GB002054.

Hein, R., P.J. Crutzen and M. Heimann, 1997. An inverse modeling approach to investigate the global atmospheric methane cycle. Global Biogeochemical Cycles, 11:43–76.

Helander, E., 2002. Global change – climate change observations among the Sami. Proceedings of the Snowchange Conference, February 21–24, 2002, Tampere, Finland.

Henry, G.H.R. and U. Molau, 1997. Tundra plants and climate change: the International Tundra Experiment (ITEX). Global Change Biology, 3(S1):1–9.

Henttonen, H., 2000. Long-term dynamics of the bank vole *Clethrionomys glareolus* at Pallasjarvi, northern Finnish taiga. In: G. Bujalska and L. Hansson (eds.). Bank Vole Biology: Recent Advances in the Population Biology of a Model Species. Polish Journal of Ecology, 48(Suppl.):87–96.

Henttonen, H. and K.A. Burck, 2001. Parasitism: an underestimated stressor of arctic fauna? In: Arctic Flora and Fauna: Status and Conservation, p. 145. Conservation of Arctic Flora and Fauna, Helsinki.

Henttonen, H. and A. Kaikusalo, 1993. Lemming movements. In: N.C. Stenseth and R.A. Ims (eds.). The Biology of Lemmings, pp. 157–186. Academic Press.

Henttonen, H. and H. Wallgren, 2001. Rodent dynamics and communities in the birch forest zone of northern Fennoscandia. In: F.E. Wielgolaski (ed.). Nordic Mountain Forest Ecosystems. Man and the Biosphere Series 27:261–278. UNESCO, Paris and Parthenon Publishing Group.

Henttonen, H., T. Oksanen, A. Jortikka and V. Haukisalmi, 1987. How much do weasels shape microtine cycles in the northern Fennoscandian taiga? Oikos, 50:353–365.

Hershey, A.E., W.B. Bowden, L.A. Deegan, J.E. Hobbie, B.J. Peterson, G.W. Kipphut, G.W. Kling, M.A. Lock, R.W. Merritt, M.C. Miller, J.R. Vestal and J.A. Schuldt, 1997. The Kuparuk River: A long-term study of biological and chemical processes in an arctic river. In: A.M. Milner and M.W. Oswood (eds.). Freshwaters of Alaska: Ecological Syntheses. Ecological Studies 119:107–130. Springer-Verlag.

Hersteinsson, P. and D.W. MacDonald, 1992. Interspecific competition and the geographical distribution of red and arctic foxes *Vulpes vulpes* and *Alopex lagopus*. Oikos, 64:505–515.

Hertzberg, K., N.G. Yoccoz, R.A. Ims and H.P. Leinaas, 2000. The effects of spatial habitat configuration on recruitment, growth and population structure in arctic Collembola. Oecologia, 124:381–390.

Heske, E.J., R.A. Ims and H. Steen, 1993. Four experiments on a Norwegian subalpine microtine rodent assemblage during a summer decline. In: N.C. Stenseth and R.A. Ims (eds.). The Biology of Lemmings, pp. 411–424. Academic Press.

Hill, J.K., C.D. Thomas and B. Huntley, 1999. Climate and habitat availability determine 20th century changes in a butterfly's range margin. Proceedings of the Royal Society of London: Biological Sciences, 266:1197–1206.

Hill, J.K., C.D. Thomas and B. Huntley, 2003. Modelling present and potential future ranges of European butterflies using climate response surfaces. In: C.L. Boggs, W.B. Watt and P.R. Ehrlich (eds.). Butterflies: Ecology and Evolution Taking Flight, pp. 149–167. University of Chicago Press.

Hinzman, L.D., D.J. Goering, S. Li and T.C. Kinney, 1997. Numeric simulation of thermokarst formation during disturbance. In: R.M.M. Crawford (ed.). Disturbance and Recovery in Arctic Lands: An Ecological Perspective, pp. 199–212. Kluwer Academic Publishers.

Hinzman, L.D., M. Fukuda, D.V. Sandberg, F.S. Chapin III and D. Dash, 2003. FROSTFIRE: An experimental approach to predicting the climate feedbacks from the changing boreal fire regime. Journal of Geophysical Research, 108(D1):8153, doi:10.1029/2001JD000415.

Hobbie, S.E., 1995. Direct and indirect species effects on biogeochemical processes in arctic ecosystems. In: F.S. Chapin III and C. Körner (eds.). Arctic and Alpine Biodiversity: Patterns, Causes, and Ecosystem Consequences. Ecological Studies, 113:213–224. Springer-Verlag.

Hobbie, S.E., 1996. Temperature and plant species control over litter decomposition in Alaskan tundra. Ecological Monographs, 66:503–522.

Hobbie, S.E. and F.S. Chapin III, 1996. Winter regulation of tundra litter carbon and nitrogen dynamics. Biogeochemistry, 35:327–338.

Hobbie, S.E. and F.S. Chapin III, 1998. The response of tundra plant biomass, aboveground production, nitrogen, and CO_2 flux to experimental warming. Ecology, 79:1526–1544.

Hobbie, S.E., J.P. Schimel, S.E. Trumbore and J.R. Randerson, 2000. Controls over carbon storage and turnover in high-latitude soils. Global Change Biology, 6(S1):196–210.

Hoberg, E.P., S.J. Kutz, K.E. Galbreath and J. Cook, 2003. Arctic biodiversity: from discovery to faunal baselines – revealing the history of a dynamic ecosystem. Journal of Parasitology, 89: S84–S95.

Hodkinson, I.D. and J. Bird, 1998. Host-specific insect herbivores as sensors of climate change in arctic and alpine environments. Arctic and Alpine Research, 30(1):78–83.

Hodkinson, I.D., S.J. Coulson, N.R. Webb, W. Block, A.T. Strathdee and J.S. Bale, 1994. Feeding studies on Onychiurus arcticus (Tullberg) Collembola: Onychiuridae on West Spitsbergen. Polar Biology, 14:17–19.

Hodkinson, I.D., N.R. Webb, J.S. Bale, W. Block, S.J. Coulson and A.T. Strathdee, 1998. Global change and high arctic ecosystems: conclusions and predictions from experiments with terrestrial invertebrates on Spitsbergen. Arctic and Alpine Research, 30(3):306–313.

Hodkinson, I.D., N.R. Webb, J.S. Bale and W. Block, 1999. Hydrology, water availability and tundra ecosystem function in a changing climate: the need for a closer integration of ideas? Global Change Biology, 5:359–369.

Høgda, K.A., S.R. Karlsen and I. Solheim, 2001. Climate change impact on growing season in Fennoscandia studied by a time series of NOAA AVHRR NDVI data. Geoscience and Remote Sensing Symposium, 2001. IGARSS '01, 2001 IEEE International, 3:1338–1340.

Høgda, K.A., S.R. Karlsen, I. Solheim, H. Tommervik and H. Ramfjord, 2002. The start dates of birch pollen seasons in Fennoscandia studied by NOAA AVHRR NDVI data. Geoscience and Remote Sensing Symposium, 2002. IGARSS '02, 2002 IEEE International, 6:3299–3301.

Hollister, R.D. and P.J. Webber, 2000. Biotic validation of small open-top chambers in a tundra ecosystem. Global Change Biology, 6:835–842.

Holt, J.G., N.R. Krieg, P.H.A. Sneath, J.T. Staley and S.T. Williams (eds.), 1994. Bergey's Manual of Determinative Bacteriology, 9th Edition. Williams and Wilkins, 787pp.

Hultén, E., 1964. The Circumpolar Plants. 1, Vascular Cryptogams, Conifers, Monocotyledons. Almqvist & Wiksell, 275pp.

Huntley, B., 1988. Glacial and Holocene vegetation history: Europe. In: B. Huntley and T. Webb III (ed.). Vegetation History. Handbook of Vegetation Science, 7:341–383. Kluwer Academic Publishers.

Huntley, B., 1997. The responses of vegetation to past and future climate changes. In: W.C. Oechel, T. Callaghan, T. Gilmanov, J.I. Holten, B. Maxwell, U. Molau and B. Sveinbjornsson (eds.). Global Change and Arctic Terrestrial Ecosystems, pp. 290–311. Springer-Verlag.

Huntley, B. and H.J.B. Birks, 1983. An Atlas of Past and Present Pollen Maps for Europe: 0–13000 Years Ago. Cambridge University Press, 667pp.

Huntley, B. and R. Bradshaw, 1999. Palaeoecological evidence and opportunities in determining environmental change and ecological responses. In: M. Turunen, J. Hukkinen, O.W. Heal, J.I. Holten and N.R. Saelthun (eds.). A Terrestrial Transect for Scandinavia/Northern Europe: Proceedings of the International SCANTRAN Conference. Ecosystems Research Report 31:153–157. European Commission, Brussels.

Huntley, B., P.M. Berry, W. Cramer and A.P. McDonald, 1995. Modelling present and potential future ranges of some European higher plants using climate response surfaces. Journal of Biogeography, 22:967–1001.

Huntley, B., M. Baillie, J.M. Grove, C.U. Hammer, S.P. Harrison, S. Jacomet, E. Jansen, W. Karlen, N. Koc, J. Luterbacher, J. Negendank and J. Schibler, 2002. Holocene paleoenvironmental changes in north-west Europe: Climatic implications and the human dimension. In: G. Wefer, W.H. Berger, K.-E. Behre, and E. Jansen (eds.). Climate Development and History of the North Atlantic Realm, pp. 259–298. Springer-Verlag.

Huntley, B., M.J. Alfano, J.R.M. Allen, D. Pollard, P.C. Tzedakis, J.-L. de Beaulieu, E. Gruger and B. Watts, 2003. European vegetation during Marine Oxygen Isotope Stage-3. Quaternary Research, 59(2):195–212.

Hustich, I., 1983. Tree-line and tree growth studies during 50 years: some subjective observations. Collection Nordicana, 47:181–188.

Iason, G.R. and A.J. Hester, 1993. The response of heather (*Calluna vulgaris*) to shade and nutrients - predictions of the carbon-nutrient balance hypothesis. Journal of Ecology, 81:75–80.

IISD and the community of Sachs Harbour, 2000. Sila Alangotok: Inuit Observations on Climate Change. Video production. International Institute for Sustainable Development, Winnipeg.

Illeris, L., A. Michelsen and S. Jonasson. 2003. Soil plus root respiration and microbial biomass following water, nitrogen, and phosphorus application at a high arctic semi desert. Biogeochemistry, 65:15–29.

Ims, R.A. and H.P. Andreassen, 2000. Spatial synchronization of vole population dynamics by predatory birds. Nature, 408:194–196.

Ims, R.A. and H. Steen, 1990. Geographical synchrony in cyclic microtine populations: a theoretical evaluation of the role of nomadic avian predators. Oikos, 57:381–387.

Ims, R.A., H.P. Leinaas and S. Coulson, 2004. Spatial and temporal variation in patch occupancy and population density in a model system of an arctic Collembola species assemblage. Oikos, 105:89–100.

Ingham, R.E., J.A. Trofymow, E.R. Ingham and D.C. Coleman, 1985. Interactions of bacteria, fungi, and their nematode grazers: effects on nutrient cycling and plant growth. Ecological Monographs, 55:119–140.

IPCC, 1990. Climate Change: The IPCC Scientific Assessment. J.T. Houghton, G.J. Jenkins and J.J. Ephraums (eds.). Intergovernmental Panel on Climate Change, World Meteorological Organization/United Nations Environment Programme. Cambridge University Press, 364pp.

IPCC, 2001. Climate Change 2001: The Scientific Basis. Contribution of Working Group I to the Third Assessment Report of the Intergovernmental Panel on Climate Change. J.T. Houghton, Y. Ding, D.J. Griggs, M. Noguer, P.J. van der Linden, X. Dai, K. Maskell and C.A. Johnson (eds.). Cambridge University Press, 881pp.

Irvine, R.J., A. Stien, O. Halvorsen, R. Langvatn and S.D. Albon, 2000. Life-history strategies and population dynamics of abomasal nematodes in Svalbard reindeer (*Rangifer tarandus platyrhynchus*). Parasitology, 120:297–311.

Isidorov, V.A. and M. Jdanova, 2002. Volatile organic compounds from leaves litter. Chemosphere, 48(9):975–979.

Isidorov, V.A., I.G. Zenkevich and B.V. Ioffe, 1983. Methods and results of gas chromatographic-mass spectrometric determination of volatile organic substances in an urban atmosphere. Atmospheric Environment, 17(7):1347–1353.

IUCN, 2003. The 2003 IUCN Red List of Threatened Species. World Conservation Union, Species Survival Commission, Gland, Switzerland. http://www.iucnredlist.org/

Jaeger, C.H., R.K. Monson, M.C. Fisk and S.K. Schmidt, 1999. Seasonal partitioning of nitrogen by plants and soil microorganisms in an alpine ecosystem. Ecology, 80:1883–1891.

Jean, D. and L. Lapointe, 2001. Limited carbohydrate availability as a potential cause of fruit abortion in *Rubus chamaemorus*. Physiologia Plantarum, 112(3):379–387.

Jefferies, R.L. and L.D. Gottlieb, 1983. Genetic variation within and between populations of the asexual plant Puccinellia x phryganodes. Canadian Journal of Botany, 61:774–779.

Jefferies, R.L., D.R. Klein and G.R. Shaver, 1994. Herbivores and northern plant communities: Reciprocal influences and responses. Oikos, 71:193–206.

Jefferies, R.L., F.L. Gadallah, D.S. Srivastava and D.J. Wilson, 1995. Desertification and trophic cascades in arctic coastal ecosystems: a potential climatic change scenario? In: T.V. Callaghan, W.C. Oechel, T. Gilmanov, J.I. Holten, B. Maxwell, U. Molau, B. Sveinbjornsson and M.J. Tyson, (eds.). Global Change and Arctic Terrestrial Ecosystems, pp. 201–206. Ecosystems Research Report 10. European Commission, Brussels.

Joabsson, A. and T.R. Christensen, 2001. Methane emissions from wetlands and their relationship with vascular plants: an Arctic example. Global Change Biology, 7:919–932.

Joabsson, A., T.R. Christensen and B. Wallén, 1999. Vascular plant controls on methane emissions from northern peatforming wetlands. Trends in Ecology and Evolution, 14:385–388.

Johanson, U., C. Gehrke, L.O. Bjorn and T.V. Callaghan, 1995. The effects of enhanced UV-B radiation on the growth of dwarf shrubs in a subarctic heathland. Functional Ecology, 9:713–719.

Johnson, D., C.D. Campbell, J.A. Lee, T.V. Callaghan and D. Gwynn-Jones, 2002. Arctic soil microorganisms respond more to elevated UV-B radiation than CO$_2$. Nature, 416:82–83.

Jonasson, S., 1992. Plant responses to fertilization and species removal in tundra related to community structure and clonality. Oikos, 63:420–429.

Jonasson, S. and T.V. Callaghan, 1992. Root mechanical properties related to disturbed and stressed habitats in the Arctic. New Phytologist, 122:179–186.

Jonasson, S. and F.S. Chapin III, 1985. Significance of sequential leaf develop¬ment for nutrient balance of the cotton sedge, *Eriophorum vaginatum* L. Oecologia, 67:511–518.

Jonasson, S., J.P. Bryant, F.S. Chapin III and M. Andersson, 1986. Plant phenols and nutrients in relation to variations in climate and rodent grazing. American Naturalist, 128:394–408.

Jonasson, S., A. Michelsen, I.K. Schmidt, E.V. Nielsen and T.V. Callaghan, 1996. Microbial biomass C, N, and P in two arctic soils and responses to addition of NPK fertilizer and sugar: Implications for plant nutrient uptake. Oecologia, 106:507–515.

Jonasson, S., A. Michelsen and I.K. Schmidt, 1999a. Coupling of nutrient cycling and carbon dynamics in the Arctic, integration of soil microbial and plant processes. Applied Soil Ecology, 11:135–146.

Jonasson, S., A. Michelsen, I.K. Schmidt and E.V. Nielsen, 1999b. Responses in microbes and plants to changed temperature, nutrient and light regimes in the Arctic. Ecology, 80:1828–1843.

Jonasson, S., T.V. Callaghan, G.R. Shaver and L.A. Nielsen, 2000. Arctic Terrestrial ecosystems and ecosystem function. In: M. Nuttall and T.V. Callaghan (eds.). The Arctic: Environment, People, Policy, pp. 275–313. Harwood Academic Publishers.

Jonasson, S., F.S. Chapin III and G.R. Shaver, 2001. Biogeochemistry in the Arctic: patterns, processes and controls. In: E.-D. Schulze, M. Heimann, S.P. Harrison, E.A. Holland, J.J. Lloyd, I.C. Prentice and D. Schimel (eds.). Global Biogeochemical Cycles in the Climate System, pp. 139–150. Academic Press.

Jones, M.H., J.T. Fahnestock, D.A Walker, M.D. Walker and J.M. Welker, 1998. Carbon dioxide fluxes in moist and dry arctic tundra during the snow-free season: Responses to increases in summer temperature and winter snow accumulation. Arctic and Alpine Research, 30:373–380.

Jónsdóttir, I.S., T.V. Callaghan and J.A. Lee, 1995. Fate of added nitrogen in a moss-sedge Arctic community and effects of increased nitrogen deposition. Science of the Total Environment, 161:677–685.

Jónsdóttir, I.S., M. Augner, T. Fagerström, H. Persson and A. Stenström, 2000. Genet age in marginal populations of two clonal Carex species in the Siberian Arctic. Ecography, 23:402–412.

Jonsson, B.O., I.S. Jonsdottir and N. Cronberg, 1996. Clonal diversity and allozyme variation in populations of the arctic sedge Carex bigelowii (Cyperaceae). Journal of Ecology, 84:449–459.

Kalela, O., 1961. Seasonal change of the habitat in the Norwegian lemming, *Lemmus lemmus*. Annales Academiae Scientiarium Fennicae (A, IV), 55:1–72.

Källen, E., V. Kattsov, J. Walsh and E. Weatherhead, 2001. Report from the Arctic Climate Impact Assessment Modeling and Scenarios Workshop. January 29–31, 2001, Stockholm. Available from: ACIA Secretariat, P.O. Box 757740, Fairbanks, Alaska 99775 USA or www.acia.uaf.edu, 28pp.

Kallio, P. and J. Lehtonen, 1973. Birch forest damage caused by *Oporinia autumnata* (Bkh.) in 1965–66, in Utsjoki, N Finland. Reports from the Kevo Subarctic Research Station, 10:55–69.

Kane, D.L., K.M. Hinkel, D.J. Goering, L.D. Hinzman and S.I. Outcalt, 2001. Non-conductive heat transfer associated with frozen soils. Global and Planetary Change, 29:275–292.

Kaplan, J.O., 2001. Geophysical applications of vegetation modelling. Ph.D. Thesis, Lund University.

Kaplan, J.O., N.H. Bigelow, I.C. Prentice, S.P. Harrison, P.J. Bartlein, T.R. Christensen, W. Cramer, N.V. Matveyeva, A.D. McGuire, D.F. Murray, V.Y. Razzhivin, B. Smith, D.A. Walker, P.M. Anderson, A.A. Andreev, L.B. Brubaker, M.E. Edwards and A.V. Lozhkin, 2003. Climate change and Arctic ecosystems: 2. Modeling, paleodata-model comparisons, and future projections. Journal of Geophysical Research, 108(D19):8171, doi:10.1029/2002JD002559.

Kappen, L., 1974. Response to extreme environments. In: V. Ahmadjian and M.E. Hale (eds.). The Lichens, pp. 311–380. Academic Press.

Kappen, L., 1993. Lichens in the Antarctic region. In: E.I. Friedmann (ed.). Antarctic Microbiology, pp. 433–490. Wiley-Liss, Inc.

Kappen, L., B. Schroeter, C. Scheidegger, M. Sommerkorn and G. Hestmark, 1996. Cold resistance and metabolic activity of lichens below 0 °C. Advances in Space Research, 18(12):119–128.

Karatygin, I.V., E.L. Nezdoiminogo, Y.K. Novozhilov and M.P. Zhurbenko, 1999. Russian Arctic Fungi: A Checklist. St. Petersburg. 212pp.

Kattenberg, A., F. Giorgi, H. Grassl, G.A. Meehl, J.F.B. Mitchell, R.J. Stouffer, T. Tokioka, A.J. Weaver and T.M.L. Wigley, 1996. Climate models - projections of future climate. In: J.T. Houghton, L.G. Meira Filho, B.A. Callander, N. Harris, A. Kattenberg and K. Maskell (eds.). Climate Change 1995. The Science of Climate Change. Contribution of Working Group I to the Second Assessment Report of the Intergovernmental Panel on Climate Change, pp. 285–357. Cambridge University Press.

Kaye, J.P. and S.C. Hart, 1997. Competition for nitrogen between plants and soil microorganisms. Trends in Ecology and Evolution, 12:139–143.

Kelliher, F.M., R. Leuning, M.R. Raupach and E.-D. Schulze, 1995. Maximum conductances for evaporation from global vegetation types. Agricultural and Forest Meteorology, 73:1–16.

Kershaw, K.A., 1983. The thermal operating environment of a lichen. The Lichenologist, 15:191–207.

Keyser, A.R., J.S. Kimball, R.R. Nemani and S.W. Running, 2000. Simulating the effects of climate change on the carbon balance of North American high-latitude forests. Global Change Biology, 6(S1):185–195.

Kienel, U., C. Siegert and J. Hahne, 1999. Late Quaternary palaeoenvironmental reconstructions from a permafrost sequence (North Siberian Lowland, SE Taymyr Peninsula) - a multidisciplinary case study. Boreas, 28(1):181–193.

Killick, H.J. and S.J. Warden, 1991. Ultraviolet penetration of pine trees and insect virus survival. Entomophaga, 36:87–94.

Klein, D.R., 1968. The introduction, increase, and crash of reindeer on St. Matthew Island. Journal of Wildlife Management, 32:350–367.

Klein, D.R., 1996. Arctic ungulates at the northern edge of terrestrial life. Rangifer, 16:51–56.

Klein, D.R., 1999. The roles of climate and insularity in establishment and persistence of *Rangifer tarandus* populations in the high Arctic. In: A. Hofgaard, J.P. Ball, K. Danell and T.V. Callaghan (eds.). Animal Responses to Global Change in the North. Ecological Bulletin 47:96–104.

Kling, G.W., G.W. Kipphut and M.C. Miller. 1991. Arctic lakes and streams as gas conduits to the atmosphere: Implications for tundra carbon budgets. Science, 251:298–301.

Kling, G.W., G.W. Kipphut and M.C. Miller, 1992. The flux of CO_2 and CH_4 from lakes and rivers in arctic Alaska. Hydrobiologia, 240:23–36.

Kling, G.W., G.W. Kipphut, M.C. Miller and W.J. O'Brien, 2000. Integration of lakes and streams in a landscape perspective: the importance of material processing on spatial patterns and temporal coherence. Freshwater Biology, 43:477–497.

Klinger, L.F., P.R. Zimmerman, J.P. Greenberg, L.E. Heidt and A.B. Guenther, 1994. Carbon trace gas fluxes along a successional gradient in the Hudson Bay lowland. Journal of Geophysical Research, 99(D1):1469–1494.

Koç, N., E. Jansen, M. Hald and L. Labeyrie, 1996. Late glacial–Holocene sea surface temperatures and gradients between the North Atlantic and the Norwegian Sea: implications for the nordic heat pump. In: J.T. Andrews, W.E.N. Austin, H.E. Bergsten and A.E. Jennings (eds.). Late Quaternary Palaeoceanography of the North Atlantic Margins. Geological Society Special Publication 111:177–185.

Koerner, R.M., 1980. The problem of lichen-free zones in Arctic Canada. Arctic and Alpine Research, 12:87–94.

Kofinas, G. with the communities of Aklavik, Arctic Village, Old Crow and Fort McPherson, 2002. Community contributions to ecological monitoring: knowledge co-production in the U.S.-Canada arctic borderlands. In: I. Krupnik and D. Jolly (eds.). The Earth is Faster Now: Indigenous Observations of Arctic Environmental Change, pp. 54–91. Arctic Research Consortium of the United States, Fairbanks, Alaska.

Konstantinova, N.A. and A.D. Potemkin, 1996. Liverworts of the Russian Arctic: an annotated check-list and bibliography. Arctoa, 6:125–150.

Köppen, W., 1931. Grundriss der Klimakunde. Walter de Gruyter & Co., 388pp. (In German)

Körner, C.H., 1995. Alpine plant diversity: A global survey and functional interpretations. In: F.S. Chapin III and C. Körner (eds.). Arctic and Alpine Biodiversity: Patterns, Causes and Ecosystem Consequences. Ecological Studies, 113:45–62.

Körner, C., 1998. A re-assessment of high elevation treeline positions and their explanation. Oecologia, 115:445–459.

Körner, C., 1999. Alpine Plant Life. Functional Plant Ecology of High Mountain Ecosystems. Springer, 338pp.

Korpelainen, H., 1994. Sex ratios and resource allocation among sexually reproducing plants of *Rubus chamaemorus*. Annals of Botany, 74(6):627–632.

Koskina, T.V., 1961. New data on the nutrition of Norwegian lemming (*Lemmus lemmus*). Bulletin of the Moscow Society of Naturalists, 66:15–32.

Kouki, J., P. Niemel and M. Viitasaari, 1994. Reversed latitudinal gradient in species richness of sawflies (Hymenoptera, Symphyta). Annales Zoologici Fennici, 31:83–88.

Koval, T.M., 1988. Enhanced recovery from ionizing radiation damage in a lepidopteran insect cell line. Radiation Research, 115:413–420.

Kozhevnikov, Y.P., 2000. Is the Arctic getting warmer or cooler? In: B.S. Ebbinge, Y.L. Mazourov and P.S. Tomkovich (eds.). Heritage of the Russian Arctic: Research, Conservation and International Co-operation, pp.145–157. Ecopros.

Kraaijeveld, K. and E.N. Nieboer, 2000. Late Quaternary paleogeography and evolution of Arctic breeding waders. Ardea, 88(2):193–205.

Krebs, C.J., K. Danell, A. Angerbjorn, J. Agrell, D. Berteaux, K.A. Bråthen, O. Danell, S. Erlinge, V. Fedorov, K. Fredga, J. Hjältén, G. Högstedt, I.S. Jónsdóttir, A.J. Kenney, N. Kjellén, T. Nordin, H. Roininen, M. Svensson, M. Tannerfeldt and C. Wiklund, 2003. Terrestrial trophic dynamics in the Canadian Arctic. Canadian Journal of Zoology, 81:827–843.

Krupnik, I., 2000. Native perspectives on climate and sea-ice changes. In: H.P. Huntington (ed.). Impacts of Changes in Sea Ice and other Environmental Parameters in the Arctic. Final Report of the Marine Mammal Commission Workshop, Girdwood, Alaska, 15–17 February 2000, pp. 35–52. Marine Mammal Commission, Maryland.

Kubeckova, K., J. Elster and H. Kanda, 2001. Periphyton ecology of glacial and snowmelt streams, Ny-Ålesund, Svalbard: presence of mineral particles in water and their erosive activity. In: J. Elster, J. Seckbach, W. Vincent and O. Lhotsky (eds.). Algae and Extreme Environments. Nova Hedwigia Beihefte, Beiheft 123:141–172.

Kudo, G., 1991. Effects of snow-free period on the phenology of alpine plants inhabiting snow patches. Arctic and Alpine Research, 23:436–443.

Kudo, G., 1993. Relationship between flowering time and fruit set of the entomophilous alpine shrub, *Rhododendron aureum* (Ericaceae), inhabiting snow patches. American Journal of Botany, 80:1300–1304.

Kullman, L., 1979. Change and stability in the altitude of the birch tree-limit in the southern Swedish Scandes 1915–1975. Acta Phytogeographica Suecica, 65:1–121.

Kumpula, J., P. Parikka and M. Nieminen, 2000. Occurrence of certain microfungi on reindeer pastures in northern Finland during winter 1996–97. Rangifer, 20:3–8.

Kushner, D., 1981. Extreme environments: Are there any limits to life? In: C. Ponnamperuma (ed.). Comets and the Origin of Life, pp. 241–248. D. Reidel.

Kutz, S.J., E.P. Hoberg, J. Nishi and L. Polley, 2002. Development of the muskox lungworm *Umingmakstrongylus pallikuukensis* (Protostrongylidae), in gastropods in the Arctic. Canadian Journal of Zoology, 80:1977–1985.

Lafleur, P.M., W.R. Rouse and D.W. Carlson, 1992. Energy balance differences and hydrologic impacts across the northern treeline. International Journal of Climatology, 12:193–203.

Laine, K., 1988. Long-term variations in plant quality and quantity in relation to cyclic microtine rodents at Kilpisjärvi, Finnish Lapland. Acta Universitatis Ouluenis Series A, 198. 33pp.

Laine, K. and H. Henttonen, 1983. The role of plant production in microtine cycles in northern Fennoscandia. Oikos, 40:407–418.

Laine, K. and H. Henttonen, 1987. Phenolics/nitrogen ratios in the blueberry *Vaccinium myrtillus* in relation to temperature and microtine density in Finnish Lapland. Oikos, 50:389–395.

Lal, R., J.M. Kimble and B.A. Stewart (eds.), 2000. Global Climate Change and Cold Regions Ecosystems. Lewis Publishers, 265pp.

Lamb, H.F., 1980. Late Quaternary vegetational history of southeastern Labrador. Arctic and Alpine Research, 12:117–135.

Lamb, H.F. and M.E. Edwards, 1988. Glacial and Holocene vegetation history: the Arctic. In: B. Huntley and T. Webb III (eds.). Vegetation History. Handbook of Vegetation Science, 7:519–555.

Lamb, H.H., 1982. Climate, History and the Modern World. Methuen, 387pp.

Lambeck, K., 1995. Constraints on the Late Weichselian ice sheet over the Barents Sea from observations of raised shorelines. Quaternary Science Reviews, 14:1–16.

Lange, O.L., 1953. Hitze- und Trockenresistenz der Flechten in Beziehung zu ihrer Verbreitung. Flora, 140:39–97.

Lange, O.L., 1954. Einige Messungen zum Warmehaushalt poikilohydrer Flechten und Moose. Archiv fur Meteorologie, Geophysik und Bioklimatologie, Serie B: Allgemeine und Biologische Klimatologie, 5:182–190.

Lange, O.L. and H. Metzner, 1965. Lichtabhängiger Kohlenstoff-Einbau in Flechten bei tiefen Temperaturen. Naturwissenschaften, 52:191–192.

Larcher, W., 1995. Physiological Plant Ecology. 3rd Edition. Springer Verlag, 506pp.

Larkin, J.M. and J.L. Stokes, 1968. Growth of psychrophilic micro-organisms at subzero temperatures. Canadian Journal of Microbiology, 14:97–101.

Larsen, K.S., S. Jonasson and A. Michelsen, 2002. Repeated freeze–thaw cycles and their effects on biological processes in two arctic ecosystem types. Applied Soil Ecology, 21:187–195.

Laurila, T., H. Soegaard, C.R. Lloyd, M. Aurela, J.-P. Tuovinen and C. Nordström, 2001. Seasonal variations of net CO_2 exchange in European Arctic ecosystems. Theoretical and Applied Climatology, 70:183–201.

Lavola, A., R. Julkunen-Tiitto, P. Aphalo, T. de la Rosa and T. Lehto, 1997. The effect of UV-B radiation on UV-absorbing secondary metabolites in birch seedlings grown under simulated forest soil conditions. New Phytologist, 137:617–621.

Lavola, A., R. Julkunen-Tiitto, H. Roininen and P. Aphalo, 1998. Host-plant preference of an insect herbivore mediated by UV–B and CO_2 in relation to plant secondary metabolites. Biochemical Systematics and Ecology, 26:1–12.

Lebedeva, E., 1998. Waders in agricultural habitats of European Russia. In: H. Hotker, E. Lebedeva, P.S. Tomkovich, J. Gromadzka, N.C. Davidson, J. Evans, D.A. Stroud and R.B. West (eds.). Migration and International Conservation of Waders: Research and Conservation on North Asian, African and European Flyways. International Wader Studies, 10:315–324.

Lee, S.E., M.C. Press and J.A. Lee, 2000. Observed climate variations during the last 100 years in Lapland, Northern Finland. International Journal of Climatology, 20:329–346.

Legrand, M., 1995. Atmospheric chemistry changes versus past climate inferred from polar ice cores. In: R.J. Charlson and J. Heintzenberg (eds.). Aerosol Forcing of Climate, pp. 123–151, Wiley.

Lehtonen, J. and R.K. Heikkinen, 1995. On the recovery of mountain birch after *Epirrita* damage in Finnish Lapland, with a particular emphasis on reindeer grazing. Ecoscience, 2:349–356.

Leinaas, H.P., 2002. UV tolerance, pigmentation and life forms in high Arctic collembola. In: D. Hessen (ed.). UV Radiation and Arctic Ecosystems. Ecological Studies, 153:123–134.

Lennihan, R., D.M. Chapin and L.G. Dickson, 1994. Nitrogen fixation and photosynthesis in high arctic forms of Nostoc commune. Canadian Journal of Botany, 72:940–945.

Li, S., M. Paulsson and L.O. Björn, 2002a. Temperature-dependent formation and photorepair of DNA damage induced by UV-B radiation in suspension-cultured tobacco cells. Journal of Photochemistry and Photobiology B:66(1):67–72.

Li, S., Y. Wang and L.O. Björn, 2002b. Temperature effects on the formation of DNA damage in *Nicotiana tabacum* leaf discs induced by UV-B irradiation. Ecologic Science, 21:115–117.

Linderholm, H.W., B.Ö. Solberg and M. Lindholm, 2003. Tree-ring records from central Fennoscandia: the relationship between tree growth and climate along a west-east transect. The Holocene, 13:887–895.

Lindström, Å. and J. Agrell, 1999. Global change and possible effects on the migration and reproduction of arctic-breeding waders. In: A. Hofgaard, J.P. Ball, K. Danell and T.V. Callaghan (eds.). Animal Responses to Global Change in the North. Ecological Bulletin 47:145–159.

Lipson, D.A., S.K. Schmidt and R.K. Monson, 1999. Links between microbial population dynamics and nitrogen availability in an alpine ecosystem. Ecology, 80:1623–1631.

Lister, A. and P. Bahn, 1995. Mammoths. Boxtree, London, 168pp.

Lloyd, A.H., T.S. Rupp, C.L. Fastie and A.M. Starfield, 2003. Patterns and dynamics of treeline advance on the Seward Peninsula, Alaska. Journal of Geophysical Research, 108(D2):8161, doi:10.1029/2001JD000852.

Lloyd, C.R., 2001. The measurement and modelling of the carbon dioxide exchange at a high Arctic site in Svalbard. Global Change Biology, 7:405–426.

Lugert, J. and C. Zockler, 2001. The bird fauna of the Yakut horse pastures in northeast Siberia. In: B. Gerken and M. Görner (eds.). Neue Modelle zu Maßnahmen der Landschaftsentwicklung mit großen Pflanzenfressern -Praktische Erfahrungen bei der Umsetzung. Natur- und Kulturlandschaft, 4:458–461.

Lundelius, E.L. Jr., R.W. Graham, E. Anderson, J. Guilday, J.A. Holman, D.W. Steadman, and D.S. Webb, 1983. Terrestrial vertebrate faunas. In: S.C. Porter (ed.). Late Quaternary Environments of the United States. Vol 1: The Late Pleistocene, pp. 311–353. University of Minnesota Press.

Luo, Y., S. Wan, D. Hui and L.L. Wallace, 2001. Acclimatization of soil respiration to warming in a tall grass prairie. Nature, 413:622–625.

MacDonald, G.M., A.A. Velichko, C.V. Kremenetski, O.K. Borisova, A.A. Goleva, A.A. Andreev, L.C. Cwynar, R.T. Riding, S.L. Forman, T.W.D. Edwards, R. Aravena, D. Hammarlund, J.M. Szeicz and V.N. Gattaulin, 2000. Holocene treeline history and climate change across northern Eurasia. Quaternary Research, 53(3):302–311.

MacKenzie, T.D.B, M. Krol, N.P.A. Huner and D.A. Campbell, 2002. Seasonal changes in chlorophyll fluorescence quenching and the induction and capacity of the photoprotective xanthophyll cycle in Lobaria pulmonaria. Canadian Journal of Botany, 80:255–261.

MacLean, S.F., B.M. Fitzgerald and F.A. Pitelka, 1974. Population cycles in arctic lemmings: winter reproduction and predation by weasels. Arctic and Alpine Research, 6:1–12.

Madsen, J., G. Cracknell and A.D. Fox (eds.), 1999. Goose Populations of the Western Paleoartic: A Review of Status and Distribution. Wetlands International Publications 48. National Environment Research Institute, Ronde, Denmark, 344pp.

Malhi, S.S., W.B. McGill and M. Nyborg, 1990. Nitrate losses in soils: effect of temperature, moisture and substrate concentration. Soil Biology and Biochemistry, 22:733–737.

Marion, G.M. and W.C. Oechel, 1993. Mid- to late-Holocene carbon balance in Arctic Alaska and its implications for future global warming. The Holocene, 3:193–200.

Marks, T.C., 1978. The carbon economy of *Rubus chamaemorus* L. II. Respiration. Annals of Botany, 42:181–190.

Marks, T.C. and K. Taylor, 1978. The carbon economy of *Rubus chamaemorus* L. I. Photosynthesis. Annals of Botany, 42:165–179.

Matveyeva, N.V., 1998. Zonation in Plant Cover of the Arctic. Proceedings of the Komarov Botanical Institute, 21. Russian Academy of Sciences, Moscow, 219pp. (In Russian)

Matveyeva, N. and Y. Chernov, 2000. Biodiversity of terrestrial ecosystems. In: M. Nuttall and T.V. Callaghan (eds.). The Arctic: Environment, People, Policy, pp. 233–274. Harwood Academic Publishers.

Max, K.N., S.K. Mouchaty and K.E. Schwaegerle, 1999. Allozyme and morphological variation in two subspecies of *Dryas octopetala* (Rosaceae) in Alaska. American Journal of Botany, 86:1637–1644.

Maxwell, B., 1997. Recent climate patterns in the Arctic. In: W.C. Oechel, T. Callaghan, T. Gilmanov, J.I. Holten, B. Maxwell, U. Molau and B. Sveinbjornsson (eds.). Global Change and Arctic Terrestrial Ecosystems, pp. 21–46. Springer-Verlag.

Mayr, E., 1998. Two empires or three? Proceedings of the National Academy of Sciences, 95:9720–9723.

Mazur, P., 1980. Limits to life at low temperatures and at reduced water contents and water activities. Origins of Life, 10:137–159.

McDonald, I.R., M. Upton, G. Hall, R.W. Pickup, C. Edwards, J.R. Saunders, D.A. Ritchie and J.C. Murrel, 1999. Molecular ecological analysis of methanogens and methanotrophs in blanket bog peat. Microbial Ecology, 38:225–233.

McFadden, J.P., F.S. Chapin III and D.Y. Hollinger, 1998. Subgrid-scale variability in the surface energy balance of Arctic tundra. Journal of Geophysical Research, 103(D22):28947–28961.

McFadden, J.P., G.E. Liston, M. Sturm, R.A. Pielke Sr. and F.S. Chapin III, 2001. Interactions of shrubs and snow in arctic tundra: Measurements and models. In: A.J. Dolman, A.J. Hall, M.L. Kavvas, T. Oki and J.W. Pomeroy (eds.). Soil-Vegetation-Atmosphere Transfer Schemes and Large-Scale Hydrological Models. IAHS Publication 270:317–325. International Association of Hydrological Sciences, UK.

McFadden, J.P., W. Eugster and F.S. Chapin III, 2003. A regional study of the controls on water vapor and CO_2 exchange in Arctic tundra. Ecology, 84:2762–2776.

McGraw, J.B., 1995. Patterns and causes of genetic diversity in arctic plants. In: F.S. Chapin III and C. Körner (eds.). Arctic and Alpine Biodiversity: Patterns, Causes and Ecosystem Consequences. Ecological Studies, 113:33–43.

McGraw, J.B. and J. Antonovics, 1983. Experimental ecology of *Dryas octopetala* ecotypes. I. Ecotypic differentiation and life-cycle stages of selection. Journal of Ecology, 71:879–897.

McGuire, A.D., J.M. Melillo, D.W. Kicklighter, Y. Pan, X. Xiao, J. Helfrich, B.M. Moore III, C.J. Vorosmarty and A.L. Schloss, 1997. Equilibrium responses of global net primary production and carbon storage to doubled atmospheric carbon dioxide: sensitivity to changes in vegetation nitrogen concentration. Global Biogeochemical Cycles, 11:173–189.

McGuire, A.D., J.S. Clein, J.M. Melillo, D.W. Kicklighter, R.A. Meier, C.J. Vorosmarty and M.C. Serreze, 2000. Modeling carbon responses of tundra ecosystems to historical and projected climate: sensitivity of pan-arctic carbon storage to temporal and spatial variation in climate. Global Change Biology, 6(S1):141–159.

McGuire, A.D., C. Wirth, M. Apps, J. Beringer, J. Clein, H. Epstein, D.W. Kicklighter, J. Bhatti, F.S. Chapin III, B. de Groot, D. Efremov, W. Eugster, M. Fukuda, T. Gower, L.D. Hinzman, B. Huntley, G.J. Jia, E.S. Kasischke, J. Melillo, V.E. Romanovsky, A. Shvidenko, E. Vaganov and D.A. Walker, 2002. Environmental variation, vegetation distribution, carbon dynamics, and water/energy exchange in high latitudes. Journal of Vegetation Science, 13:301–314.

McKane, R.B., E.B. Rastetter, G.R. Shaver, K.J. Nadelhoffer, A.E. Giblin, J.A. Laundre and F.S. Chapin III, 1997. Climatic effects on tundra carbon storage inferred from experimental data and a model. Ecology, 78:1170–1187.

McKane, R.B., L.C. Johnson, G.R. Shaver, K.J. Nadelhoffer, E.B. Rastetter, B. Fry, A.E. Giblin, K. Kielland, B.L. Kwiatkowski, J.A. Laundre and G. Murray, 2002. Resource-based niches provide a basis for plant species diversity and dominance in arctic tundra. Nature, 415:68–71.

Mehlum, F., 1999. Adaptation in arctic organisms to a short summer season. In: S.-A. Bengtsson, F. Mehlum and T. Severinsen (eds.). The Ecology of the Tundra in Svalbard. Norsk Polarinstitutt, Meddeleser, 150:161–169.

Meltofte, H., 1985. Populations and breeding schedules of waders, *Charadrii*, in high Arctic Greenland. Meddelelser om Gronland Bioscience, 16:1–43.

Meltofte, H., 2000. Birds. In: K. Caning and M. Rasch (eds.). Zackenberg Ecological Research Operations, ZERO, 5th Annual Report, 1999, pp. 32–39. Danish Polar Center, Ministry of Research and Information Technology.

Merilä, J., M. Pahkala and U. Johanson, 2000. Increased ultraviolet-B radiation, climate change and latitudinal adaptation - a frog perspective. Annales Zoologici Fennici, 37:129–134.

Michaelson, G.J., C.-L. Ping and J.M. Kimble, 1996. Carbon storage and distribution in tundra soils in Arctic Alaska, U.S.A. Arctic and Alpine Research, 28:414–424.

Michelsen, A., I.K. Schmidt, S. Jonasson, J. Dighton, H.E. Jones and T.V. Callaghan, 1995. Inhibition of growth, and effects on nutrient uptake on Arctic graminoids by leaf extracts – allelopathy or resource competition between plants and microbes? Oecologia, 103:407–418.

Michelsen, A., I.K. Schmidt, S. Jonasson, C. Quarmby and D. Sleep, 1996. Leaf ^{15}N abundance of subarctic plants provides field evidence that ericoid, ectomycorrhizal and non- and arbuscular mycorrhizal species access different sources of soil nitrogen. Oecologia, 105:53–63.

Michelsen, A., C. Quarmby, D. Sleep and S. Jonasson, 1998. Vascular plant 15N abundance in heath and forest tundra ecosystems is closely correlated with presence and type of mycorrhizal fungi in roots. Oecologia, 115:406–418.

Michelsen, A., E. Graglia, I.K. Schmidt, S. Jonasson, D. Sleep and C. Quarmby, 1999. Differential responses of grass and a dwarf shrub to long-term changes in soil microbial biomass C, N and P following factorial addition of NPK fertilizer, fungicide and labile carbon to a heath. New Phytologist, 143:523–538.

Miller, O.K. and D.F. Farr, 1975. Index of the Common Fungi of North America, Synonymy and Common Names. Bibliotheca Mycologia, 44:206–230. J. Cramer, Vaduz.

Mirchink, T.G., 1988. Soil Mycology. Moscow University Publishers.

Molau, U., 1996. Climatic impacts on flowering, growth, and vigour in an arctic-alpine cushion plant, Diapensia lapponica, under different snow cover regimes. In: P.S. Karlsson and T.V. Callaghan (eds.). Plant Ecology in the Sub-Arctic Swedish Lapland. Ecological Bulletin 45:210–219.

Molau, U., 2001. Tundra plant responses to experimental and natural temperature changes. Memoirs of National Institute of Polar Research, Special Issue 54:445–466.

Molau, U. and J.M. Alatalo, 1998. Responses of subarctic-alpine plant communities to simulated environmental change: biodiversity of bryophytes, lichens, and vascular plants. Ambio, 27:322–329.

Molau, U. and E.-L. Larsson, 2000. Seed rain and seed bank along an alpine altitudinal gradient in Swedish Lapland. Canadian Journal of Botany, 78:728–747.

Molau, U. and G.R. Shaver, 1997. Controls on seed production and seed germinability in *Eriophorum vaginatum*. Global Change Biology, 3(S1):80–88.

Mølgaard, P., 1982. Temperature observations in high Arctic plants in relation to microclimate in the vegetation of Peary Land, North Greenland. Arctic and Alpine Research, 14:105–115.

Moody, S.A., K.K. Newsham, P.G. Ayres and N.D. Paul, 1999. Variation in the responses of litter and phylloplane fungi to UV-B radiation (290–315 nm). Mycological Research, 103(11):1469–1477.

Moody, S.A., N.D. Paul, L.O. Björn, T.V. Callaghan, J.A. Lee, Y. Manetas, J. Rozema, D. Gwynn-Jones, U. Johanson, A. Kyparissis and A.M.C. Oudejans, 2001. The direct effects of UV-B radiation on *Betula pubescens* litter decomposing at four European field sites. Plant Ecology, 154:27–36.

Mooney, H.A. and W.D. Billings, 1961. Comparative physiological ecology of Arctic and Alpine populations of *Oxyria digyna*. Ecological Monographs, 31:1–29.

Moran, P.A.P., 1953. The statistical analysis of the Canadian lynx cycle. II. Synchronization and meteorology. Australian Journal of Zoology, 1:291–298.

Morneau, C. and S. Payette, 2000. Long-term fluctuations of a caribou population revealed by tree-ring data. Canadian Journal of Zoology, 78:1784–1790.

Morozov, V.V., 1998. Distribution of breeding waders in the north-east European Russian tundras. In: H. Hotker, E. Lebedeva, P.S. Tomkovich, J. Gromadzka, N.C. Davidson, J. Evans, D.A. Stroud and R.B. West (eds.). Migration and International Conservation of Waders: Research and Conservation on North Asian, African and European Flyways. International Wader Studies, 10:186–194.

Morrison, R.I.G., Y. Aubry, R.W. Butler, G.W. Beyersbergen, C. Downes, G.M. Donaldson, C.L. Gratto-Trevor, P.W. Hicklin, V.H. Johnston and R.K. Ross, 2001. Declines in North American shorebird populations. Wader Study Group Bulletin, 94:34–38.

Morrissey, L.A and G.P. Livingston, 1992. Methane emissions from Alaska arctic tundra: An assessment of local spatial variability. Journal of Geophysical Research, 97(D15):16661–16670.

Murray, D.F., 1995. Causes of Arctic plant diversity: origin and evolution. In: F.S. Chapin III and C. Körner (eds.). Arctic and Alpine Biodiversity: Patterns, Causes and Ecosystem Consequences. Ecological Studies, 113:21–32.

Myneni, R.B., C.D. Keeling, C.J. Tucker, G. Asrar and R.R. Nemani, 1997. Increased plant growth in the northern high latitudes from 1981–1991. Nature, 386:698–702.

Myneni, R.B., J. Dong, C.J. Tucker, R.K. Kaufmann, P.E. Kauppi, J. Liski, L. Zhou, V. Alexeyev and M.K. Hughes, 2001. A large carbon sink in the woody biomass of northern forests. Proceedings of the National Academy of Sciences, 98:14784–14789.

Myrberget, S., 1973. Geographic synchronism of cycles of small rodent in Norway. Oikos, 24:220–224.

Nadelhoffer, K.J., A.E. Giblin, G.R. Shaver and J.A. Laundre, 1991. Effects of temperature and substrate quality on element mineralization in six Arctic soils. Ecology, 72:242–253.

Nadelhoffer, K.J., A.E. Giblin, G.R. Shaver and A.E. Linkins, 1992. Microbial processes and plant nutrient availability in arctic soils. In: F.S. Chapin III, R.L. Jefferies, J.F. Reynolds, G.R. Shaver and J. Svoboda (eds.). Arctic Ecosystems in a Changing Climate. An Ecophysiological Perspective, pp. 281–300. Academic Press.

Nadelhoffer, K.J., G. Shaver, B. Fry, A.E. Giblin, L. Johnson and R. McKane, 1996. 15N natural abundances and N use by tundra plants. Oecologia, 107:386–394.

Neilson, P.R. and R.J. Drapek, 1998. Potentially complex biosphere responses to transient global warming. Global Change Biology, 4:505–521.

Nellemann, C., L. Kullerud, I. Vistnes, B.C. Forbes, E. Husby, G.P. Kofinas, B.P. Kaltenborn, J. Rouaud, M. Magomedova, R. Bobiwash, C. Lambrechts, P.J. Schei, S. Tveitdal, O. Gron and T.S. Larsen, 2001. GLOBIO. Global Methodology for Mapping Human Impacts on the Biosphere. UNEP/DEWA/TR.01-3. United Nations Environment Programme, 47pp.

Neuvonen, S., P. Niemela and T. Virtanen, 1999. Climatic change and insect outbreaks in boreal forests: the role of winter temperatures. In: A. Hofgaard, J.P. Ball, K. Danell and T.V. Callaghan (eds.). Animal Responses to Global Change in the North. Ecological Bulletin 47:63–67.

New, M., M. Hulme and P.D. Jones, 1999. Representing twentieth-century space-time climate variability. Part 1: development of a 1961–90 mean monthly terrestrial climatology. Journal of Climate, 12:829–856.

Niemi, R., P.J. Martikainen, J. Silvola, A. Wulff, S. Turtola and T. Holopainen, 2002. Elevated UV-B radiation alters fluxes of methane and carbon dioxide in peatland microcosms. Global Change Biology, 8:361–371.

Nojd, P. and P. Hari, 2001. The effect of temperature on the radial growth of Scots pine in northernmost Fennoscandia. Forest Ecology and Management, 142:65–77.

Nordström, C., H. Soegaard, T.R. Christensen, T. Friborg and B.U. Hansen, 2001. Seasonal carbon dioxide balance and respiration of a high-arctic fen ecosystem in NE-Greenland. Theoretical and Applied Climatology, 70(1–4):149–166.

Oberbauer, S.F., N. Sionit, S.J. Hastings and W.C. Oechel, 1986. Effects of CO_2 enrichment and nutrition on growth, photosynthesis, and nutrient concentration of Alaskan tundra plant species. Canadian Journal of Botany, 64:2993–2998.

Oberbauer, S.F., J.D. Tenhunen and J.F. Reynolds, 1991. Environmental effects on CO_2 efflux from water track and tussock tundra in arctic Alaska, U.S.A. Arctic and Alpine Research, 23(2):162–169.

Oberbauer, S.F., C.T. Gillespie, W. Cheng, A. Sala, R. Gebauer and J.D. Tenhunen, 1996. Diurnal and seasonal patterns of ecosystem CO_2 efflux from upland tundra in the foothills of the Brooks Range, Alaska, U.S.A.. Arctic and Alpine Research, 28(3):328–338.

Odasz, A.M. and O. Savolainen, 1996. Genetic variation in populations of the arctic perennial *Pedicularis dasyantha* (Scrophulariaceae), on Svalbard, Norway. American Journal of Botany, 83:1379–1385.

Oechel, W.C. and W.D. Billings, 1992. Effects of global change on the carbon balance of Arctic plants and ecosystems. In: F.S. Chapin III, R.L. Jefferies, J.F. Reynolds, G.R. Shaver and J. Svoboda (eds.). Arctic Ecosystems in a Changing Climate: An Ecophysiological Perspective, pp. 139–168. Academic Press.

Oechel, W.C., S.J. Hastings, G.L. Vourlitis, M. Jenkins, G. Riechers and N. Grulke, 1993. Recent change of arctic tundra ecosystems from a net carbon dioxide sink to a source. Nature, 361:520–523.

Oechel, W.C., S. Cowles, N. Grulke, S.J. Hastings, B. Lawrence, T. Prudhomme, G. Riechers, B. Strain, D. Tissue and G. Vourlitis, 1994. Transient nature of CO_2 fertilization in Arctic tundra. Nature, 371:500–503.

Oechel, W.C., G.L. Vourlitis, S.J. Hastings and S.A. Bochkarev, 1995. Change in Arctic CO_2 flux-over two decades: Effects of climate change at Barrow, Alaska. Ecological Applications, 5:846–855.

Oechel, W.C., G. Vourlitis and S.J. Hastings, 1997. Cold season CO_2 emission from Arctic soils. Global Biogeochemical Cycles, 11:163–172.

Oechel, W.C., G.L. Vourlitis, S.J. Hastings, R.P. Ault Jr. and P. Bryant, 1998. The effects of water table manipulation and elevated temperature on the net CO_2 flux of wet sedge tundra ecosystems. Global Change Biology, 4:77–90.

Oechel, W.C., G.L. Vourlitis, S.J. Verfaillie Jr., T. Crawford, S. Brooks, E. Dumas, A. Hope, D. Stow, B. Boynton, V. Nosov and R. Zulueta, 2000a. A scaling approach for quantifying the net CO_2 flux of the Kuparuk River Basin, Alaska. Global Change Biology, 6(S1):160–173.

Oechel, W.C., G.L. Vourlitis, S.J. Hastings, R.C. Zulueta, L. Hinzman and D. Kane, 2000b. Acclimation of ecosystem CO_2 exchange in the Alaskan Arctic in response to decadal climate warming. Nature, 406:978–981.

Ogilvie, A.E.J., 1984. The past climate and sea-ice record from Iceland. Part 1: Data to A.D. 1780. Climatic Change, 6:131–152.

Ogilvie, A.E.J., 1991. Climatic changes in Iceland A.D. c.865 to 1598. In: G.F. Bigelow (ed.). The Norse of the North Atlantic. Acta Archaelogica, 61:233–251.

Ogilvie, A.E.J. and I. Jonsdottir, 2000. Sea ice, climate, and Icelandic fisheries in the eighteenth and nineteenth centuries. Arctic, 53:383–394.

Ogilvie, A.E.J. and T. Jonsson, 2001. 'Little Ice Age' research: A perspective from Iceland. Climatic Change, 48:9–52.

Ögren, E., 1996. Premature dehardening in *Vaccinium myrtillus* during a mild winter: a cause for winter dieback? Functional Ecology, 10:724–732.

Ögren, E., 1997. Relationship between temperature, respiratory loss of sugar and premature dehardening in dormant Scots pine seedlings. Tree Physiology, 17:724–732.

Ögren, E., T. Nilsson and L.-G. Sundblad, 1997. Relationship between respiratory depletion of sugars and loss of cold hardiness in coniferous seedlings over-wintering at raised temperatures: indications of different sensitivities of spruce and pine. Plant, Cell and Environment, 20:247–253.

Oksanen, L., S.D. Fretwell, J. Arruda and P. Niemela, 1981. Exploitation ecosystems in a gradient of primary productivity. American Naturalist, 118:240–261.

Oksanen, L., M. Aunapuu, T. Oksanen, M. Schneider, P. Ekerholm, P.A. Lundberg, T. Amulik, V. Aruaja and L. Bondestad, 1997. Outlines of food webs in a low arctic tundra landscape in relation to three theories on trophic dynamics. In: A.C. Gange and V.K. Brown (eds.). Multitrophic Interactions in Terrestrial Ecosystems, pp. 425–437. Blackwell Scientific.

Öquist, M.G. and B.H. Svensson, 2002. Vascular plants as regulators of methane emissions from a subarctic mire ecosystem. Journal of Geophysical Research, 107(D21):4580.

Osterkamp, T.E and V.E. Romanovsky, 1999. Evidence for warming and thawing of discontinuous permafrost in Alaska. Permafrost and Periglacial Processes, 10:17–37.

Ostler, W.K., K.T. Harper, K.B. McKnight and D.C. Anderson, 1982. The effects of increasing snowpack on a subalpine meadow in the Uinta Mountains, Utah, USA. Arctic and Alpine Research, 14:203–214.

Overpeck, J., K. Hughen, D. Hardy, R. Bradley, R. Case, M. Douglas, B. Finney, K. Gajewski, G. Jacoby, A. Jennings, S. Lamoureux, A. Lasca, G. MacDonald, J. Moore, M. Retelle, S. Smith, A. Wolfe and G. Zielinski, 1997. Arctic environmental change of the last four centuries. Science, 278:1251–1256.

Paetkau, D., S.C. Amstrup, E.W. Born, W. Calvert, A.E. Derocher, G.W. Garner, F. Messier, I. Stirling, M.K. Taylor, Ø. Wiig and C. Strobeck, 1999. Genetic structure of the world's polar bear populations. Molecular Ecology, 8(10):1571–1584.

Pahkala, M., A. Laurila and J. Merilä, 2002. Effects of ultraviolet-B radiation on common frog Rana temporaria embryos from along a latitudinal gradient. Oecologia, 133:458–465.

Palmqvist K., L. Dahlman, F. Valladares, A. Tehler, L.G. Sancho and J.-E. Mattsson, 2002. CO_2 exchange and thallus nitrogen across 75 contrasting lichen associations from different climate zones. Oecologia, 133:295–306.

Panikov, N.S., 1994. Response of soil microbial community to global warming: simulation of seasonal dynamics and multiyear succession exemplified by typical tundra. Microbiology (Russia) - English Translation, 63(3):389–404.

Panikov, N.S., 1995. Microbial Growth Kinetics. Chapman & Hall, 378pp.

Panikov, N.S., 1997. A kinetic approach to microbial ecology in arctic and boreal ecosystems in relation to global change. In: W.C. Oechel, T. Callaghan, T. Gilmanov, J.I. Holten, B. Maxwell, U. Molau and B. Sveinbjornsson (eds.). Global Change and Arctic Terrestrial Ecosystems, pp. 171–189. Springer-Verlag.

Panikov, N.S., 1999. Fluxes of CO_2 and CH_4 in high latitude wetlands: measuring, modelling and predicting response to climate change. Polar Research, 18(2):237–244.

Panikov, N.S. and S.N. Dedysh, 2000. Cold season CH_4 and CO_2 emission from boreal peat bogs (West Siberia): Winter fluxes and thaw activation dynamics. Global Biogeochemical Cycles, 14(4):1071–1080.

Panikov, N.S., M.V. Sizova, V.V. Zelenev, G.A. Machov, A.V. Naumov and I.M. Gadzhiev, 1995. Methane and carbon dioxide emission from several Vasyugan wetlands: spatial and temporal flux variations. Ecological Chemistry, 4:13–23.

Panikov, N.S., S.N. Dedysh, O.M. Kolesnikov, A.I. Mardini and M.V. Sizova, 2001. Metabolic and environmental control on methane emission from soils: mechanistic studies of mesotrophic fen in West Siberia. Water, Air, and Soil Pollution, Focus, 1(5–6):415–428.

Papen, H. and K. Butterbach-Bahl, 1999. A 3-year continuous record of nitrogen trace gas fluxes from untreated and limed soil of a N-saturated spruce and beech forest ecosystem in Germany. Part 1: N_2O emissions. Journal of Geophysical Research, 104(D15):18487–18503.

Parmesan, C. and G. Yohe, 2003. A globally coherent fingerprint of climate change impacts across natural systems. Nature, 421:37–42.

Parmesan, C., N. Ryrholm, C. Stefanescu, J.K. Hill, C.D. Thomas, H. Descimon, B. Huntley, L. Kaila, J. Kullberg, T. Tammaru, W.J. Tennent, J.A. Thomas and M. Warren, 1999. Poleward shifts in geographical ranges of butterfly species associated with regional warming. Nature, 399:579–583.

Parsons, A.N., J.M. Welker, P.A. Wookey, M.C. Press, T.V. Callaghan and J.A. Lee, 1994. Growth responses of four sub–Arctic dwarf shrubs to simulated environmental change. Journal of Ecology, 82:307–318.

Paulsson, M., 2003. Temperature Effects on UV-B Induced DNA Damage and Repair in Plants and a Lichen. Ph.D Thesis, University of Lund, 49pp.

Pavlov, P., J.I. Svendsen and S. Indrelid, 2001. Human presence in the European Arctic nearly 40,000 years ago. Nature, 413:64–67.

Payette, S., M. Eronen and J.J.P. Jasinski, 2002. The circumboreal tundra-taiga interface: Late Pleistocene and Holocene changes. Ambio Special Report, 12:15–22.

Perfect, E., R.D. Miller and B. Burton, 1988. Frost upheaval of overwintering plants: a quantitative field study of the displacement process. Arctic and Alpine Research, 20:70–75.

Peteet, D.M. (ed.), 1993. Global Younger Dryas? Quaternary Science Reviews, 12(5):277–356.

Peteet, D.M. (ed.), 1995. Global Younger Dryas Vol. 2, Quaternary Science Reviews, 14(9):811–958.

Peterson, B.J., T.L. Corliss, K. Kriet and J.E. Hobbie, 1992. Nitrogen and phosphorus concentrations and export for the Upper Kuparuk River on the North Slope of Alaska in 1980. Hydrobiologia, 240:61–69.

Philipp, M., 1997. Genetic diversity, breeding system, and population structure in *Silene acaulis* (Caryophyllaceae) in West Greenland. Opera Botanica, 132:89–100.

Philipp, M., 1998. Genetic variation in four species of *Pedicularis* (Scrophulariaceae) within a limited area in West Greenland. Arctic and Alpine Research, 30:396–399.

Phoenix, G.K., 2000. Effects of ultraviolet radiation on sub-Arctic heathland vegetation. Ph.D Thesis, University of Sheffield, 129pp.

Phoenix, G.K., D. Gwynn-Jones, J.A. Lee and T.V. Callaghan, 2000. The impacts of UV-B radiation on the regeneration of a sub-Arctic heath community. Plant Ecology, 146:67–75.

Phoenix, G.K., D. Gwynn-Jones, T.V. Callaghan, D. Sleep and J.A. Lee, 2001. Effects of global change on a sub-Arctic heath: effects of enhanced UV-B radiation and increased summer precipitation. Journal of Ecology, 89:256–267.

Pielke, R.A. and P.L. Vidale, 1995. The boreal forest and the polar front. Journal of Geophysical Research, 100(D12):25755–25758.

Piersma, T., 1997. Do global patterns of habitat use and migration strategies co-evolve with relative investments in immunocompetence due to spatial variation in parasite pressure? Oikos, 80:623–631.

Pop, E.W., S.F. Oberbauer and G. Starr, 2000. Predicting vegetative bud break in two arctic deciduous shrub species, *Salix pulchra* and *Betula nana*. Oecologia, 124:176–184.

Porsild, A.E., 1951. Plant life in the Arctic. Canadian Geographical Journal, 42:120–145.

Post, E., R.O. Peterson, N.C. Stenseth and B.E. McLaren, 1999. Ecosystem consequences of wolf behavioural response to climate. Nature, 401:905–907.

Potter, J.A., M.C. Press, T.V. Callaghan and J.A. Lee, 1995. Growth responses of *Polytrichum commune* (Hedw.) and *Hylocomium splendens* (Hedw.) Br. Eur. to simulated environmental change in the subarctic. New Phytologist, 131:533–541.

Predavec, M., C.J. Krebs, K. Danell and R. Hyndman, 2001. Cycles and synchrony in the collared lemming (*Dicrostonyx groenlandicus*) in Arctic North America. Oecologia, 126:216–224

Prentice, I.C., W.P. Cramer, S.P. Harrisson, R. Leemans, R.A. Monserud and A.M. Solomon, 1992. A global biome model based on plant physiology and dominance, soil properties and climate. Journal of Biogeography, 19:117–134.

Press, M.C., T.V. Callaghan and J.A. Lee, 1998a. How will European Arctic ecosystems respond to projected global environmental change? Ambio, 27:306–311.

Press, M.C., J.A. Potter, M.J.W. Burke, T.V. Callaghan and J.A. Lee, 1998b. Responses of a subarctic dwarf shrub heath community to simulated environmental change. Journal of Ecology, 86:315–327.

Prestrud, P. and K. Nilssen, 1992. Fat deposition and seasonal variation in body composition of arctic foxes in Svalbard. Journal of Wildlife Management, 56:221–233.

Price, P.W., T.P. Craig and H. Roininen, 1995. Working toward a theory on galling sawfly population dynamics. In: N. Cappuccino and P.W. Price (eds.). Population Dynamics: New Approaches and Synthesis, pp. 321–338. Academic Press.

Prock, S. and C. Körner, 1996. A cross-continental comparison of phenology, leaf dynamics and dry matter allocation in arctic and temperate zone herbaceous plants from contrasting altitudes. In: P.S. Karlsson and T.V. Callaghan (eds.). Plant Ecology in the Sub-Arctic Swedish Lapland. Ecological Bulletin 45:93–103.

Pruitt, W.O., 1957. Observations on the bioclimate of some taiga mammals. Arctic, 10:131–138.

Putkonen, J. and G. Roe, 2003. Rain-on-snow events impact on soil temperatures and affect ungulate survival. Geophysical Research Letters, 30(4):1188, doi:10.1029/2002GL016326.

Quested, H.M., M.C. Press, T.V. Callaghan and J.H.C. Cornelissen, 2002. The hemiparasitic angiosperm Bartsia alpina has the potential to accelerate decomposition in sub-Arctic communities. Oecologia, 130:88–95.

Quested, H.M., J.H.C. Cornelissen, M.C. Press, T.V. Callaghan, R. Aerts, F. Trosien, P. Riemann, D. Gwynn-Jones, A. Kondratchuk and S.E. Jonasson, 2003. Decomposition of sub-arctic plants with differing nitrogen economies: A functional role for hemiparasites. Ecology, 84:3209–3221.

Radajewsky, S., P. Ineson, N.R. Parekh and J.C. Murrell, 2000. Stable-isotope probing as a tool in microbial ecology. Nature, 403:646–649.

Rannie, W.F., 1986. Summer air temperature and number of vascular species in Arctic Canada. Arctic, 39:133–137.

Ranta, E., J. Lindstrom, V. Kaitala, H. Kokko, H. Linden and E. Helle, 1997. Solar activity and hare dynamics: A cross-continental comparison. American Naturalist, 149:765–775.

Rapp, A., 1996. Photo documentation of landscape change in Northern Swedish mountains. In: P.S. Karlsson and T.V. Callaghan (eds.). Plant Ecology in the Sub-Arctic Swedish Lapland. Ecological Bulletin 45:170–179.

Rastetter, E.B., R.B. McKane, G.R. Shaver, K.J. Nadelhoffer and A.E. Giblin, 1997. Analysis of CO_2, temperature, and moisture effects on carbon storage in Alaskan arctic tundra using a general ecosystem model. In: W.C. Oechel, T. Callaghan, T. Gilmanov, J.I. Holten, B. Maxwell, U. Molau and B. Sveinbjornsson (eds.). Global Change and Arctic Terrestrial Ecosystems, pp. 437–451. Springer-Verlag.

Read, D.J., J.R. Leake and A.R. Langdale, 1989. The nitrogen nutrition of mycorrhizal fungi and their host plants. In: L. Boddy, R. Marchant and D.J. Read (eds.). Nitrogen, Phosphorus and Sulphur Utilisation by Fungi. British Mycological Society Symposia, 15:181–204. Cambridge University Press.

Réale, D., A.G. McAdam, S. Boutin and D. Berteaux, 2003. Genetic and plastic responses of a northern mammal to climate change. Proceedings of the Royal Society of London: Biological Sciences, 270:591–596.

Reid, D.G., C.J. Krebs and A. Kenney, 1995. Limitation of collared lemming population growth at low densities by predation mortality. Oikos, 73:387–398.

Reierth, E. and K.-A. Stokkan, 1998. Activity rhythm in High Arctic Svalbard ptarmigan (Lagopus mutus hyperboreus). Canadian Journal of Zoology, 76:2031–2039.

Reimers, E., 1982. Winter mortality and population trends of reindeer on Svalbard, Norway. Arctic and Alpine Research, 14:295–300.

Rensen, H., H. Goosse, T. Fichefet and J.-M. Campin, 2001. The 8.2 kyr BP event simulated by a global atmosphere-sea-ice-ocean model. Geophysical Research Letters, 28(8):1567–1570.

Richardson, S.J., 2000. Response of a Sub-arctic Dwarf Shrub Heath Community to Nutrient Addition and Warming. Ph.D Thesis, University of Sheffield, 159pp.

Riedlinger, D., 2001. Community-based assessments of change: contributions of Inuvialuit knowledge to understanding climate change in the Canadian Arctic. MSc. Thesis. University of Manitoba.

Riedlinger, D. and F. Berkes, 2001. Contributions of traditional knowledge to understanding climate change in the Canadian Arctic. Polar Record, 37(203):315–328.

Ries, G., W. Heller, H. Puchta, H. Sandermann, H.K. Seidlitz and B. Hohn, 2000. Elevated UV-B radiation reduces genome stability in plants. Nature, 406:98–101.

Ritchie, J.C., 1987. Postglacial Vegetation of Canada. Cambridge University Press, 178pp.

Ritchie, J.C. and G.M. MacDonald, 1986. The patterns of post-glacial spread of white spruce. Journal of Biogeography, 13:527–540.

Rivkina, E.M., E.I. Friedmann, C.P. McKay and D.A. Gilichinsky, 2000. Metabolic activity of permafrost bacteria below the freezing point. Applied and Environmental Microbiology, 66:3230–3233.

Robberecht, R., M.M. Caldwell and W.D. Billings, 1980. Leaf ultraviolet optical properties along a latitudinal gradient in the Arctic-Alpine life zone. Ecology, 61(3):612–619.

Robinson, C.H., O.B. Borisova, T.V. Callaghan and J.A. Lee, 1996. Fungal hyphal length in litter of Dryas octopetala in a high-Arctic polar semi-desert, Svalbard. Polar Biology, 16:71–74.

Robinson, C.H., A. Michelsen, J.A. Lee, S.J. Whitehead, T.V. Callaghan, M.C Press and S. Jonasson, 1997. Elevated atmospheric CO_2 affects decomposition of Festuca vivipara (L.) Sm. litter and roots in experiments simulating environmental change in two contrasting Arctic ecosystems. Global Change Biology, 3:37–49.

Robinson, C.H., P.A. Wookey, J.A. Lee, T.V. Callaghan and M.C. Press, 1998. Plant community responses to simulated environmental change at a high Arctic polar semi-desert. Ecology, 79:856–866.

Roff, D.A., 1997. Evolutionary Quantitative Genetics. Chapman & Hall, 493pp.

Roulet, N.T., T. Moore, J. Bubier and P. Lafleur, 1992. Northern fens: methane flux and climatic change. Tellus B, 44:100–105.

Roulet, N.T., A. Jano, C.A. Kelly, L.F. Klinger, T.R. Moore, R. Protz, J.A. Ritter and W.R. Rouse, 1994. Role of the Hudson Bay lowland as a source of atmospheric methane. Journal of Geophysical Research, 99(D1):1439–1454.

Rouse, W.R., 2000. The energy and water balance of high-latitude wetlands: controls and extrapolation. Global Change Biology, 6(S1):59–68.

Rovansek, R.J., L.D. Hinzman and D.L. Kane, 1996. Hydrology of a tundra wetland complex on the Alaskan Arctic Coastal Plain. Arctic and Alpine Research, 28(3):311–317.

Rozema, J., B. van Geel, L.O. Björn, J. Lean and S. Madronich, 2002. Toward solving the UV puzzle. Science, 296:1621–1622.

Ruess, L., A. Michelsen, I.K. Schmidt and S. Jonasson, 1999a. Simulated climate change affecting microorganisms, nematode density and biodiversity in subarctic soils. Plant and Soil, 212:63–73.

Ruess, L., A. Michelsen and S. Jonasson, 1999b. Simulated climate change in subarctic soils: responses in nematode species composition and dominance structure. Nematology, 1:513–526.

Rundgren, M. and O. Ingolfsson, 1999. Plant survival in Iceland during periods of glaciation? Journal of Biogeography, 26(2):387–396.

Ruohomäki, K., M. Tanhuanpää, M.P. Ayres, P. Kaitaniemi, T. Tammaru and E. Haukioja, 2000. Causes of cyclicity of Epirrita autumnata (Lepidoptera, Geometridae): grandiose theory and tedious practice. Population Ecology, 42:211–223.

Russell, N.J., 1990. Cold adaptation of microorganisms. Philosophical Transactions of the Royal Society of London Series B, 326:595–611.

Russell, R.S., 1940. Physiological and ecological studies on Arctic vegetation. II. The development of vegetation in relation to nitrogen supply and soil micro-organisms on Jan Mayen Island. Journal of Ecology, 28:269–288.

Sakai, A. and W. Larcher, 1987. Frost Survival of Plants: Responses and Adaptation to Freezing Stress. Ecological Studies 62. Springer-Verlag, 321pp.

Sala, O.E. and T. Chapin, 2000. Scenarios of global biodiversity. Global Change Newsletter, 43:7–11, 19. International Geosphere-Biosphere Programme, Stockholm.

Savile, D.B.O., 1971. Arctic Adaptations in Plants. Agriculture Canada Research Branch, Monograph 6, 81pp.

Schimel, J.P., 1995. Plant transport and methane production as controls on methane flux from arctic wet meadow tundra. Biogeochemistry, 28:183–200.

Schimel, J.P. and F.S. Chapin III, 1996. Tundra plant uptake of amino acid and NH_4^+ nitrogen in situ: plants compete well for amino acid N. Ecology, 77:2142–2147.

Schimel, J.P. and J.S. Clein, 1996. Microbial response to freeze–thaw cycles in tundra and taiga soils. Soil Biology and Biochemistry, 28:1061–1066.

Schipperges, B. and C. Gehrke, 1996. Photosynthetic characteristics of subarctic mosses and lichens. In: P.S. Karlsson and T.V. Callaghan (eds.). Plant Ecology in the Sub-Arctic Swedish Lapland. Ecological Bulletin 45:121–126.

Schmidt, I.K., S. Jonasson and A. Michelsen, 1999. Mineralization and microbial immobilization of N and P in arctic soils in relation to season, temperature and nutrient amendment. Applied Soil Ecology, 11:147–160.

Schmidt, I.K., L. Ruess, E. Bååth, A. Michelsen, F. Ekelund and S. Jonasson, 2000. Long term manipulation of the microbes and microfauna of two subarctic heaths by addition of fungicide, bactericide, carbon and fertilizer. Soil Biology and Biochemistry, 32:707–720.

Schmidt, I.K., S. Jonasson, G.R. Shaver, A. Michelsen and A. Nordin, 2002. Mineralization and distribution of nutrients in plants and microbes in four arctic ecosystems: responses to warming. Plant and Soil, 242:93–106.

Schmidt-Nielsen, K., 1979. Animal Physiology: Adaptation and Environment. Cambridge University Press, 560pp.

Scholander, P.F., V. Walters, R. Hock and L. Irving, 1950. Adaptation to cold in arctic and tropical mammals and birds in relation to body temperature, insulation, and basal metabolic rate. Biological Bulletin, 99:259–271.

Schroeter, B., T.G.A. Green, L. Kappen and R.D. Seppelt, 1994. Carbon dioxide exchange at subzero temperatures: field measurements on *Umbilicaria aprina* in Antarctica. Cryptogamic Botany, 4:233–241.

Schultz, A.M., 1969. A study of an ecosystem: the arctic tundra. In: G.M. Van Dyne (ed.). The Ecosystem Concept in Natural Management, pp. 77–93. Academic Press.

Schulze, E.-D., F.M. Kelliher, C. Körner, J. Lloyd and R. Leuning, 1994. Relationships among maximum stomatal conductance, ecosystem surface conductance, carbon assimilation rate, and plant nitrogen nutrition: a global ecology scaling exercise. Annual Review of Ecology and Systematics, 25:629–662.

Scott, D.A., 1998. Global Overview of the Conservation of Migratory Arctic Breeding Birds outside the Arctic. CAFF Technical Report 4. Conservation of Arctic Flora and Fauna, 134pp.

Searles, P.S., S.D. Flint, S.B. Diaz, M.C. Rousseaux, C.L. Ballaré and M.M. Caldwell, 1999. Solar ultraviolet-B radiation influence on *Sphagnum* bog and *Carex* fen ecosystems: first field season findings in Tierra del Fuego, Argentina. Global Change Biology, 5:225–234.

Searles, P.S., S.D. Flint and M.M. Caldwell, 2001. A meta-analysis of plant field studies simulating stratospheric ozone depletion. Oecologia, 127:1–10.

Sekretareva, N.A., 1999. The Vascular Plants of the Russian Arctic and Adjacent Territories. Sofia, Moscow, 160pp.

Seldal, T., K.J. Andersen and G. Högstedt, 1994. Grazing-induced proteinase inhibitors: a possible cause for lemming population cycles. Oikos, 70:3–11.

Semerdjieva, S.I., E. Sheffield, G.K. Phoenix, D. Gwynn-Jones, T.V. Callaghan and G.N. Johnson, 2003. Contrasting strategies for UV-B screening in sub-Arctic dwarf shrubs. Plant, Cell and Environment, 26:957–964.

Semikhatova, O.A., T.V. Gerasimenko and T.I. Ivanova, 1992. Photosynthesis, respiration, and growth of plants in the Soviet Arctic. In: F.S. Chapin III, R.L. Jefferies, J.F. Reynolds, G.R. Shaver and J. Svoboda (eds.). Arctic Ecosystems in a Changing Climate: An Ecophysiological Perspective, pp. 169–192. Academic Press.

Serreze, M.C., J.E. Walsh, F.S. Chapin III, T. Osterkamp, M. Dyurgerov, V. Romanovsky, W.C. Oechel, J. Morison, T. Zhang and R.G. Barry, 2000. Observational evidence of recent change in the northern high-latitude environment. Climatic Change, 46:159–207.

Shaver, G.R. and W.D. Billings, 1975. Root production and root turnover in a wet tundra ecosystem, Barrow, Alaska. Ecology, 56:401–409.

Shaver, G.R. and F.S. Chapin III, 1991. Production: biomass relationships and element cycling in contrasting Arctic vegetation types. Ecological Monographs, 61:1–31.

Shaver, G.R and J.C. Cutler, 1979. The vertical distribution of live vascular phytomass in cottongrass tussock tundra. Arctic and Alpine Research, 11:335–342.

Shaver, G.R. and S. Jonasson, 2000. Response of Arctic ecosystems to climate change: results of long-term field experiments in Sweden and Alaska. Polar Research, 18:245–252.

Shaver, G.R., F.S. Chapin III and W.D. Billings, 1979. Ecotypic differentiation in *Carex aquatilis* on ice-wedge polygons in the Alaskan coastal tundra. Journal of Ecology, 67:1025–1046.

Shaver, G.R., N. Fetcher and F.S. Chapin III, 1986. Growth and flowering in *Eriophorum vaginatum*: Annual and latitudinal variation. Ecology, 67:1524–1535.

Shaver, G.R., W.D. Billings, F.S. Chapin III, A.E. Giblin, K.J. Nadelhoffer, W.C. Oechel and E.B. Rastetter, 1992. Global change and the carbon balance of arctic ecosystems. Bioscience, 42:433–441.

Shaver, G.R., J.A. Laundre, A.E. Giblin and K.J. Nadelhoffer, 1996. Changes in live plant biomass, primary production, and species composition along a riverside toposequence in arctic Alaska, U.S.A. Arctic and Alpine Research, 28:363–379.

Shaver, G.R., L.C. Johnson, D.H. Cades, G. Murray, J.A. Laundre, E.B. Rastetter, K.J. Nadelhoffer and A.E. Giblin, 1998. Biomass and CO_2 flux in wet sedge tundras: responses to nutrients, temperature, and light. Ecological Monographs, 68:75–97.

Shaver, G.R., J. Canadell, F.S. Chapin III, J. Gurevitch, J. Harte, G. Henry, P. Ineson, S. Jonasson, J. Melillo, L. Pitelka and L. Rustad, 2000. Global warming and terrestrial ecosystems: A conceptual framework for analysis. Bioscience, 50:871–882.

Shaver, G.R., M.S. Bret-Harte, M.H. Jones, J. Johnstone, L. Gough, J. Laundre and F.S. Chapin III, 2001. Species composition interacts with fertilizer to control long-term change in tundra productivity. Ecology, 82:3163–3181.

Sher, A., 1997. Late-Quaternary extinction of large mammals in northern Eurasia: A new look at the Siberian contribution. In: B. Huntley, W. Cramer, A.V. Morgan, H.C. Prentice and J.R.M. Allen (eds.). Past and Future Rapid Environmental Changes: The Spatial and Evolutionary Responses of Terrestrial Biota, pp. 319–339. Springer-Verlag.

Shevtsova, A., A. Ojala, S. Neuvonen, M. Vieno and E. Haukioja, 1995. Growth and reproduction of dwarf shrubs in a subarctic plant community: annual variation and above-ground interactions with neighbours. Journal of Ecology, 83:263–275.

Shevtsova, A., E. Haukioja and A. Ojala, 1997. Growth response of subarctic dwarf shrubs, *Empetrum nigrum* and *Vaccinium vitis-ideae*, to manipulated environmental conditions and species removal. Oikos, 78:440–458.

Siegert, M.J., J.A. Dowdeswell and M. Melles, 1999. Late Weichselian glaciation of the Russian High Arctic. Quaternary Research, 52(3):273–285.

Silapaswan, C.S., D. Verbyla and A.D. McGuire, 2001. Land cover change on the Seward Peninsula: The use of remote sensing to evaluate the potential influences of climate change on historical vegetation dynamics. Canadian Journal of Remote Sensing, 27(5):542–554.

Sinclair, A.R.E., J.M. Gosline, G. Holdsworth, C.J. Krebs, S. Boutin, J.N.M. Smith, R. Boonstra and M. Dale, 1993. Can the solar cycle and climate synchronize the snowshoe hare cycle in Canada? Evidence from tree rings and ice cores. American Naturalist, 141:173–198.

Sitch, S., B. Smith, I.C. Prentice, A. Arneth, A. Bondeau, W. Cramer, J.O. Kaplan, S. Levis, W. Lucht, M.T. Sykes, K. Thonicke and S. Venevsky, 2003. Evaluation of ecosystem dynamics, plant geography and terrestrial carbon cycling in the LPJ dynamic global vegetation model. Global Change Biology, 9:161–185.

Sitch, S., A.D. McGuire, J. Kimball, N. Gedney, J. Gamon, R. Engstrom, A. Wolf, Q. Zhuang, and J. Clein, submitted. Assessing the circumpolar carbon balance of arctic tundra with remote sensing and process-based modeling approaches. Ecological Applications.

Sizova, M.V., N.S. Panikov, T.P. Tourova and P.W. Flanagan, 2003. Isolation and characterization of oligotrophic acido-tolerant methanogenic consortia from a *Sphagnum* peat bog. FEMS Microbiology Ecology, 45:301–315.

Sjögersten, S. and P.A. Wookey, 2002. Spatio-temporal variability and environmental controls of methane fluxes at the forest-tundra ecotone in the Fennoscandian mountains. Global Change Biology, 8:885–895.

Skinner, L.J., 2002. The Role of *Exobasidium* in the Ecology of its Dwarf Shrub Hosts. Ph.D Thesis, University of Sheffield, 174pp.

Skinner, W.R., R.L. Jefferies, T.J. Carleton, R.F. Rockwell and K.F. Abraham, 1998. Prediction of reproductive success and failure in lesser snow geese based on early season climatic variables. Global Change Biology, 4:3–16.

Skotnicki, M.L., P.M. Selkirk and C. Beard, 1998. RAPD profiling of genetic diversity in two populations of the moss *Ceratodon purpureus* in Victoria Island, Antarctica. Polar Biology, 19:172–176.

Skotnicki, M.L., P.M. Selkirk and J.A. Ninham, 1999. RAPD analysis of genetic variation and dispersal of the moss *Bryum argenteum* in Ross Island and Victoria Land, Antarctica. Polar Biology, 21:417–422.

Skre, O., R. Baxter, R.M.M. Crawford, T.V. Callaghan and A. Fedorkov, 2002. How will the tundra-taiga interface respond to climate change? Ambio Special Report, 12:37–46.

Smaglik, P., 2002. A climate of uncertainty. Nature, 415:Naturejobs p. 6.

Smith, P.A., J.A. Schaefer and B.R. Patterson, 2002. Variation at high latitudes: the geography of body size and cranial morphology of the muskox, *Ovibos moschatus*. Journal of Biogeography, 29:1089–1094.

Smith, R.I.L., 1972. Vegetation of the South Orkney Islands with particular reference to Signy Island. British Antarctic Survey Scientific Report 68, 124pp.

Sniegowski, P.D., P.J. Gerrish, and R.E. Lenski, 1997. Evolution of high mutation rates in experimental populations of *E. coli*. Nature, 387:703–705.

Søgaard, H., C. Nordstrøm, T. Friborg, B.U. Hansen, T.R. Christensen and C. Bay, 2000. Trace gas exchange in a high-arctic valley. 3: Integrating and scaling CO_2 fluxes from canopy to landscape using flux data, footprint modeling, and remote sensing. Global Biogeochemical Cycles, 14:725–744.

Solberg, E.J., P. Jorhøy, O. Strand, R. Aanes, A. Loison, B.-E. Sæther and J.D.C. Linnell, 2001. Effects of density-dependence and climate on the dynamics of a Svalbard reindeer population. Ecography, 24:441–451.

Solheim, B., U. Johanson, T.V. Callaghan, J.A. Lee, D. Gwynn-Jones and L.O. Björn, 2002. The nitrogen fixation potential of arctic cryptogam species is influenced by enhanced UV-B radiation. Oecologia, 133:90–93.

Soloviev, M.Y., P.S. Tomkovich and N. Davidson, 1998. An international breeding condition survey of arctic waterfowl: progress report. Wader Study Group Bulletin, 87:43–47.

Sommerfeld, R.A., A.R. Mosier and R.C. Musselman, 1993. CO_2, CH_4 and N_2O flux through a Wyoming snowpack and implications for global budgets. Nature, 361:140–142.

Sonesson, M. and J. Hoogesteger, 1983. Recent tree-line dynamics (*Betula pubescens* Ehrh. ssp. *tortuosa* [Ledeb.] Nyman) in northern Sweden. Nordicana, 47:47–54.

Sonesson, M., C. Gehrke and M. Tjus, 1992. CO_2 environment, microclimate and photosynthesis characteristics of the moss *Hylocomium splendens* in a subarctic habitat. Oecologia, 92:23–29.

Sonesson, M., T.V. Callaghan and L.O. Bjorn, 1995. Short term effects of enhanced UV-B and CO_2 on lichens at different latitudes. The Lichenologist, 27:547–557.

Sonesson, M., T.V. Callaghan and B.A. Carlsson, 1996. Effects of enhanced ultraviolet radiation and carbon dioxide concentration on the moss Hylocomium splendens. Global Change Biology, 2:101–107.

Sonesson, M., B.A. Carlsson, T.V. Callaghan, S. Halling, L.O. Björn, M. Bertgren and U. Johanson, 2002. Growth of two peat-forming mosses in subarctic mires: species interactions and effects of simulated climate change. Oikos, 99:151–160.

Sørensen, T., 1941. Temperature relations and phenology of the northeast Greenland flowering plants. Meddelelser om Grönland, 125:1–305.

Srivastava, D.S. and R.L. Jefferies, 1996. A positive feedback: herbivory, plant growth, salinity, and the desertification of an Arctic salt-marsh. Journal of Ecology, 84:31–42.

Staley, J.T., 1997. Biodiversity: Are microbial species threatened? Current Opinion in Biotechnology, 8:340–345.

Staley, J.T. and J.J. Gosink, 1999. Poles apart: biodiversity and biogeography of sea ice bacteria. Annual Review of Microbiology, 53:189–215.

Staley J.T. and A. Konopka, 1985. Measurement of in situ activities of nonphotosynthetic microorganisms in aquatic and terrestrial habitats. Annual Review of Microbiology, 39:321–346.

Steen, H., N.G. Yoccoz and R.A. Ims, 1990. Predators and small rodent cycles: An analysis of a 79-year time series of small rodent population fluctuations. Oikos, 59:115–120.

Stehlik, I., J.J. Schneller and K. Bachmann, 2001. Resistance or emigration: response of the high-alpine plant *Eritrichium nanum* (L.) Gaudin to the ice age within the Central Alps. Molecular Ecology, 10(2):357–370.

Steiglitz, M., A. Giblin, J. Hobbie, M. Williams and G. Kling, 2000. Simulating the effects of climate change and climate variability on carbon dynamics in Arctic tundra. Global Biogeochemical Cycles, 14:1123–1136.

Stenseth, N.C. and R.A. Ims, 1993. The Biology of Lemmings. Academic Press, UK, 683pp.

Stenström, A., B.O. Jonsson, I.S. Jónsdóttir, T. Fagerström and M. Augner, 2001. Genetic variation and clonal diversity in four clonal sedges (*Carex*) along the Arctic coast of Eurasia. Molecular Ecology, 10:497–513.

Stenström, A., I.S. Jónsdóttir and M. Augner, 2002. Genetic and environmental effects on morphology in clonal sedges in the Eurasian Arctic. American Journal of Botany, 89:1410–1421.

Stetter, K.O., R. Huber, E. Blochl, M. Kurr, R.D. Eden, M. Fielder, H. Cash and I. Vance, 1993. *Hyperthermophilic archaea* are thriving in deep North Sea and Alaskan oil reservoirs. Nature, 365:743–745.

Strann, K.-B., N.G. Yoccoz and R.A. Ims, 2002. Is the heart of the Fennoscandian rodent cycle still beating? A 14-year study of small mammals and Tengmalm's owls in northern Norway. Ecography, 25:81–87.

Strathdee, A.T. and J.S. Bale, 1998. Life on the edge: insect ecology in Arctic environments. Annual Review of Entomology, 43:85–106.

Strathdee, A.T., J.S. Bale, W.C. Block, S.J. Coulson, I.D. Hodkinson and N.R. Webb, 1993. Effects of temperature elevation on a field population of *Acyrthosiphon svalbardicum* (Hemiptera: Aphididae) on Spitsbergen. Oecologia, 96:457–465.

Strathdee, A.T., J.S. Bale, F.C. Strathdee, W. Block, S.J. Coulson, I.D. Hodkinson and N.R. Webb, 1995. Climatic severety and responses to warming of Arctic aphids. Global Change Biology, 1:23–28.

Ström, L., A. Ekberg and T.R. Christensen, 2003. The effect of vascular plants on carbon turnover and methane emissions from a tundra wetland. Global Change Biology, 9:1185–1192.

Stuart, A.J., 1982. Pleistocene Vertebrates in the British Isles. Longman, 212pp.

Stuart, A.J., L.D. Sulerzhitsky, L.A. Orlova, Y.V. Kuzmin and A.M. Lister, 2002. The latest woolly mammoths (*Mammuthus primigenius* Blumenbach) in Europe and Asia: a review of the current evidence. Quaternary Science Reviews, 21:1559–1569.

Stuiver, M., P.M. Grootes and T.F. Braziunas, 1995. The GISP2 ‰18O climate record of the past 16,500 years and the role of the sun, ocean, and volcanoes. Quaternary Research, 44:341–354.

Sturm, M., J.P. McFadden, G.E. Liston, F.S. Chapin III, C.H. Racine and J. Holmgren, 2001a. Snow-shrub interactions in Arctic tundra: A hypothesis with climatic implications. Journal of Climate, 14:336–344.

Sturm, M., C. Racine and K. Tape, 2001b. Climate change: Increasing shrub abundance in the Arctic. Nature, 411:546–547.

Summers, R.W. and L.G. Underhill, 1987. Factors relating to breeding production of Brent Geese *Branta b. bernicla* and waders (Charadrii) on the Taimyr Peninsula. Bird Study, 34:161–171.

Sutherland, W.J., 1988. Predation may link the cycles of lemmings and birds. Trends in Ecology and Evolution, 3:29–30.

Sveinbjarnardóttir, G., 1992. Farm Abandonment in Medieval and Post-Medieval Iceland: An Interdisciplinary Study. Oxbow Monograph 17. Oxbow Books, 192pp.

Sveinbjörnsson, B., A. Hofgaard and A. Lloyd, 2002. Natural causes of the tundra-taiga boundary. Ambio Special Report, 12:23–29.

Svensson, B.H., 1980. Carbon dioxide and methane fluxes from the ombrotrophic parts of a subarctic mire. In: M. Sonesson (ed.). Ecology of a Subarctic Mire. Ecological Bulletin, 30:235–250. Swedish Natural Science Research Council, Stockholm.

Svensson, B.H., T.R. Christensen, E. Johansson and M. Öquist, 1999. Interdecadal changes in CO_2 and CH_4 fluxes of a subarctic mire: Stordalen revisited after 20 years. Oikos, 85:22–30.

Svoboda, J. and B. Freedman (eds.), 1994. Ecology of a Polar Oasis. Alexandra Fiord, Ellesmere Island, Canada. Captus Press, 268pp.

Syroechovski, E.E. and A.G. Kuprianov, 1995. Wild reindeer of the arctic Eurasia: geographical distribution, numbers and population structure. In: E. Grønlund and O. Melander (eds.). Swedish-Russian Tundra Ecology – Expedition 94: A Cruise Report, pp. 175–180. Swedish Polar Research Secretariat, Stockholm.

Taalas, P., J. Kaurola, A. Kylling, D. Shindell, R. Sausen, M. Dameris, V. Grewe, J. Herman, J. Damski and B. Steil, 2000. The impact of greenhouse gases and halogenated species on future solar UV radiation doses. Geophysical Research Letters, 27(8):1127–1130.

Talbot, S.S. and D.F. Murray (eds.), 2001. Proceedings of the First International Conservation of Arctic Flora and Fauna (CAFF) Flora Group Workshop. CAFF Technical Report 10. 65pp.

Talbot, S.S., B.A. Yurtsev, D.F. Murray, G.W. Argus, C. Bay and A. Elvebakk, 1999. Atlas of Rare Endemic Vascular Plants of the Arctic. CAFF Technical Report 3. Conservation of Arctic Flora and Fauna. 73pp.

Tanhuanpää, M., K. Ruohomaki and E. Uusipaikka, 2001. High larval predation rate in non-outbreaking populations of a geometrid moth. Ecology, 82:281–289.

Tannerfeldt, M. and A. Angerbjörn, 1998. Fluctuating resources and the evolution of litter size in the arctic fox. Oikos, 83:545–559.

Tast, J. and O. Kalela, 1971. Comparisons between rodent cycles and plant production in Finnish Lapland. Annales Academiae Scientiarum Fennicae, 186:1–14.

Taulavuori, E., K. Taulavuori, K. Laine, T. Pakonen and E. Saari, 1997. Winter hardening and glutathione status in the bilberry (*Vaccinium myrtillus* L.) in response to trace gases (CO_2, O3) and nitrogen fertilization. Physiologia Plantarum, 101:192–198.

Taulavuori, K., A. Niinimaa, K. Laine, E. Taulavuori and P. Lähdesmäki, 1997a. Modelling frost resistance of Scots pine seedlings using temperature, daylength and pH of cell effusate. Plant Ecology, 133:181–189.

Taulavuori, K., K. Laine, E. Taulavuori, T. Pakonen and E. Saari, 1997b. Accelerated dehardening in bilberry (*Vaccinium myrtillus* L.) induced by a small elevation in air temperature. Environmental Pollution, 98:91–95.

Taulavuori, K., E. Taulavuori, T. Sarjala, E.-M. Savonen, P. Pietiläinen, P. Lähdesmäki and K. Laine, 2000. In vivo chlorophyll fluorescence is not always a good indicator of cold hardiness. Journal of Plant Physiology, 157:227–229.

Taulavuori, K., E. Taulavuori, A. Niinimaa and K. Laine, 2001. Acceleration of frost hardening in *Vaccinium vitis-idaea* (L.) by nitrogen fertilization. Oecologia, 127:321–323.

Tenow, O., 1972. The outbreaks of *Oporinia autumnata* Bkh. and Operophtera spp. (Lep., Geometridae) in the Scandinavian mountain chain and northern Finland 1862–1968. Zoologiska Bidrag från Uppsala, Supplement 2:1–107.

Tenow, O., 1996. Hazards to a mountain birch forest - Abisko in perspective. In: P.S. Karlsson and T.V. Callaghan (eds.). Plant Ecology in the Sub-Arctic Swedish Lapland. Ecological Bulletin 45:104–114.

Tenow, O. and H. Bylund, 2000. Recovery of a *Betula pubescens* forest in northern Sweden after severe defoliation by *Epirrita autumnata*. Journal of Vegetation Science, 11:855–862.

Tenow, O. and A. Nilssen, 1990. Egg cold hardiness and topoclimatic limitations to outbreaks of *Epirrita autumnata* in Northern Fennoscandia. Journal of Applied Ecology, 27:723–734.

Thomas, G. and P.R. Rowntree, 1992. The boreal forests and climate. Quarterly Journal of the Royal Meteorological Society, 118:469–497.

Thommessen, T., 1996. The early settlement of northern Norway. In: L. Larsson (ed.). The Earliest Settlement of Scandinavia and its Relationship with Neighbouring Areas. Acta Archaeologica Lundensia, Series in 8°, 24:235–240. Almquist and Wiksell International.

Thorpe, N., 2000. Contributions of Inuit Ecological Knowledge to Understanding the Impacts of Climate Change to Bathurst Caribou Herd in the Kitikmeot Region, Nunavut. MSc Thesis, Simon Fraser University, Vancouver.

Thorpe, N., N. Hakongak, S. Eyegetok and the Kitikmeot Elders, 2001. Thunder on the Tundra: Inuit Qaujimajatuqangit of the Bathurst Caribou. Tuktuk and Nogak Project, Ikaluktuuthak, Nunavut, 240pp.

Thorpe, N., S. Eyegetok, N. Hakongak and the Kitikmeot Elders, 2002. Nowadays it is not the same: Inuit Qaujimajatuqangit, climate and caribou in the Kitikmeot region of Nunavut, Canada. In: I. Krupnik and D. Jolly (eds.). The Earth is Faster Now: Indigenous Observations of Arctic Environmental Change, pp. 198–239. Arctic Research Consortium of the United States, Fairbanks, Alaska.

Tissue, D.T. and W.C. Oechel, 1987. Response of *Eriophorum vaginatum* to elevated CO$_2$ and temperature in the Alaskan tussock tundra. Ecology, 68:401–410.

Torn, M.S. and F.S. Chapin III, 1993. Environmental and biotic controls over methane flux from Arctic tundra. Chemosphere, 26:357–368.

Torsvik, V., L. Øvreås and T. Frede Thingstad, 2002. Prokaryotic diversity – magnitude, dynamics, and controlling factors. Science, 296:1064–1066.

Tremblay, N.O. and D.J. Schoen, 1999. Molecular phylogeography of *Dryas integrifolia*: glacial refugia and postglacial recolonization. Molecular Ecology, 8(7):1187–1198.

Tsapin, A.I., G.D. McDonald, M. Andrews, R. Bhartia, S. Douglas and D. Gilichinsky, 1999. Microorganisms from permafrost viable and detectable by 16SRNA analysis: a model for Mars. The Fifth International Conference on Mars, July 19–23, 1999, Pasadena, California. Abstract 6104.

Turchin, P. and G.O. Batzli, 2001. Availability of food and the population dynamics of arvicoline rodents. Ecology, 82:1521–1534.

Turchin, P. and I. Hanski, 1997. An empirically based model for latitudinal gradients in vole population dynamics. American Naturalist, 149:842–874.

Turchin, P., L. Oksanen, P. Ekerholm, T. Oksanen and H. Henttonen, 2000. Are lemmings prey or predators? Nature, 405:562–565.

Tyler, N.J.C. and N.A. Øritsland, 1989. Why don't Svalbard reindeer migrate? Holarctic Ecology, 12:369–379.

Väisänen, R.A, E. Lammi and P. Koskimies, 1998. Muuttuva Pesimälinnusto. Otava, Helsinki, 567pp.

van Herk, C.M., A. Aptroot and H.F. van Dobben, 2002. Long-term monitoring in the Netherlands suggests that lichens respond to global warming. The Lichenologist, 34(2):141–154.

Van Wijk, M.T., K.E. Clemmensen, G.R. Shaver, M. Williams, T.V. Callaghan, F.S. Chapin III, J.H.C. Cornelissen, L. Gough, S.E. Hobbie, S. Jonasson, J.A. Lee, A. Michelsen, M.C. Press, S.J. Richardson and H. Rueth, 2004. Long term ecosystem level experiments at Toolik Lake, Alaska and at Abisko, Northern Sweden: generalisations and differences in ecosystem and plant type responses to global change. Global Change Biology, 10:105–123.

Vardy, S.R., B.G. Warner and R. Aravena, 1997. Holocene climate effects on the development of a peatland on the Tuktoyaktuk Peninsula, Northwest Territories. Quaternary Research, 47(1):90–104.

Vartanyan, S.L., V.E. Garutt and A.V. Sher, 1993. Holocene dwarf mammoths from Wrangel Island in the Siberian Arctic. Nature, 362:337–340.

Velichko, A.A., 1995. The Pleistocene termination in Northern Eurasia. Quaternary International, 28:105–111.

Virtanen, T. and S. Neuvonen, 1999. Performance of moth larvae on birch in relation to altitude, climate, host quality and parasitoids. Oecologia, 120:92–101.

Vlassova, T.K., 2002. Human impacts on the tundra-taiga zone dynamics: the case of the Russian lesotundra. Ambio Special Report, 12:30–36.

Vourlitis, G.L. and W.C. Oechel, 1997. Landscape-scale CO$_2$, H2O vapor, and energy flux of moist-wet coastal tundra ecosystems over two growing seasons. Journal of Ecology, 85:575–590.

Vourlitis, G.L. and W.C. Oechel, 1999. Eddy covariance measurements of CO$_2$ and energy fluxes of an Alaskan tussock tundra ecosystem. Ecology, 80:686–701.

Vourlitis, G.L., W.C. Oechel, S.J. Hastings and M.A. Jenkins, 1993. The effect of soil moisture and thaw depth on CH$_4$ flux from wet coastal tundra ecosystems on the North slope of Alaska. Chemosphere, 26(1–4):329–337.

Wager, H.G., 1938. Growth and survival of plants in the Arctic. Journal of Ecology, 26:390–410.

Wagner, D., S. Kobabe, E.-M. Pfeiffer and H.-W. Hubberten, 2003. Microbial controls on methane fluxes from a polygonal tundra of the Lena Delta, Siberia. Permafrost and Periglacial Processes, 14:173–185.

Walbot, V., 1999. UV-B damage amplified by transposons in maize. Nature, 397:398–399.

Walker, D.A., 2000. Hierarchical subdivision of Arctic tundra based on vegetation response to climate, parent material and topography. Global Change Biology, 6(S1):19–34.

Walker, D.A. and M.D. Walker, 1991. History and pattern of disturbance in Alaskan Arctic terrestrial ecosystems: a hierarchical approach to analysing landscape change. Journal of Applied Ecology, 28:244–276.

Walker, D.A., E. Binnian, B.M. Evans, N.D. Lederer, E. Nordstrand and P.J. Webber, 1989. Terrain, vegetation, and landscape evolution of the R4D research site, Brooks Range foothills, Alaska. Holarctic Ecology, 12:238–261.

Walker, D.A., N.A. Auerbach, J.G. Bockheim, F.S. Chapin III, W. Eugster, J.Y. King, J.P. McFadden, G.J. Michaelson, F.E. Nelson, W.C. Oechel, C.-L. Ping, W.S. Reeburgh, S. Regli, N.I. Shiklomanov and G.L. Vourlitis, 1998. Energy and trace-gas fluxes across a soil pH boundary in the Arctic. Nature, 394:469–472.

Walker, D.A., J.G. Bockheim, F.S. Chapin III, W. Eugster, F.E. Nelson and C.L. Ping, 2001. Calcium-rich tundra, wildlife, and the 'Mammoth Steppe'. Quaternary Science Reviews, 20:149–163.

Walker, D.A., W.A. Gould, H.A. Maier and M.K. Raynolds, 2002. The Circumpolar Arctic Vegetation Map: AVHRR-derived base maps, environmental controls, and integrated mapping procedures. International Journal of Remote Sensing, 23:4551–4570.

Walker, M.D., 1995. Patterns of arctic plant community diversity. In: F.S. Chapin III and C. Körner (eds.). Arctic and Alpine Biodiversity: Patterns, Causes and Ecosystem Consequences. Ecological Studies, 113:1–18.

Walker, M.D., D.A. Walker and K.R. Everett, 1989. Wetland Soils and Vegetation, Arctic Foothills, Alaska. U.S. Fish and Wildlife Service Biological Report 89(7), 89pp.

Walker, M.D., D.A. Walker, K.R. Everett and S.K. Short, 1991. Steppe vegetation on south-facing slopes of pingos, central arctic coastal-plain, Alaska, USA. Arctic and Alpine Research 23(2):170–188.

Walker, M.D., D.A. Walker and N.A. Auerbach, 1994. Plant communities of a tussock tundra landscape in the Brooks Range Foothills, Alaska. Journal of Vegetation Science, 5:843–866.

Walker, M.D., D.A. Walker, J.M. Welker, A.M. Arft, T. Bardsley, P.D. Brooks, J.T. Fahnestock, M.H. Jones, M. Losleben, A.N. Parsons, T.R. Seastedt and P.L. Turner, 1999. Long-term experimental manipulation of winter snow regime and summer temperature in Arctic and alpine tundra. Hydrological Processes, 13:2315–2330.

Walker, M.J.C., 1995. Climatic changes in Europe during the last glacial/interglacial transition. Quaternary International, 28:63–76.

Wallenius, T.H., 1999. Yield variations of some common wild berries in Finland in 1956 – 1996. Annales Botanici Fennici, 36(4):299–314.

Walter, B.P. and M. Heimann, 2000. A process-based, climate-sensitive model to derive methane emissions from natural wetlands: Application to five wetland sites, sensitivity to model parameters, and climate. Global Biogeochemical Cycles, 14:745–766.

Warren, M.S., J.K. Hill, J.A. Thomas, J. Asher, R. Fox, B. Huntley, D.B. Roy, M.G. Telfer, S. Jeffcoate, P. Harding, G. Jeffcoate, S.G. Willis, J.N. Greatorex-Davies, D. Moss and C.D. Thomas, 2001. Rapid responses of British butterflies to opposing forces of climate and habitat change. Nature, 414:65–69.

Wayne, L.G., D.J. Brenner, R.R. Colwell, P.A.D. Grimont, O. Kandler, M.I. Krichevsky, L.H. Moore, W.E.C. Moore, R.G.E. Murray, E. Stackebrandt, M.P. Starr and H.G. Truper, 1987. Report of the ad hoc committee on reconcilation of approaches to bacterial systematics. International Journal of Systematic Bacteriology, 37(4):463–464.

Webber, P.J., P.C. Miller, F.S. Chapin III and B.H. McCown, 1980. The vegetation: pattern and succession. In: J. Brown, P.C. Miller, L.L. Tieszen and F.L. Bunnell (eds.). An Arctic Ecosystem: The Coastal Tundra at Barrow, Alaska, pp. 186–218. Dowden, Hutchinson and Ross, Pennsylvania.

Weider, L.J. and A. Hobaek, 2000. Phylogeography and Arctic biodiversity: a review. Annales Zoologici Fennici, 37(4):217–231.

Welker, J.M., U. Molau, A.N. Parsons, C.H. Robinson and P.A. Wookey, 1997. Responses of *Dryas octopetala* to ITEX environmental manipulations: a synthesis with circumpolar comparisons. Global Change Biology, 3(S1):61–73.

Welker, J.M., J.T. Fahnestock and M.H. Jones, 2000. Annual CO_2 flux in dry and moist Arctic tundra: field responses to increases in summer temperatures and winter snow depth. Climatic Change, 44:139–150.

Weller, G., 2000. The weather and climate of the Arctic. In: M. Nuttall and T.V. Callaghan (eds.). The Arctic: Environment, People, Policy, pp. 143–160. Harwood Academic Publishers.

Wennerberg, L., 2001. Breeding origin and migration pattern of dunlin (*Calidris alpina*) revealed by mitochondrial DNA analysis. Molecular Ecology, 10(5):1111–1120.

West, R.G., 2000. Plant Life of the Quaternary Cold Stages: Evidence from the British Isles. Cambridge University Press, 332pp.

Whalen, S.C. and W.S. Reeburgh, 1990. A methane flux transect along the Trans-Alaska Pipeline Haul Road. Tellus B: 42:237–249.

Whalen, S.C. and W.S. Reeburgh, 1992. Interannual variations in tundra methane emission: a 4-year time series at fixed sites. Global Biogeochemical Cycles, 6:139–159.

Whalen, S.C. and W.S. Reeburgh, 2001. Carbon monoxide consumption in upland boreal forest soils. Soil Biology and Biogeochemistry, 33:1329–1338.

White, A., M.G.R. Cannell and A.D. Friend, 2000. The high-latitude terrestrial carbon sink: a model analysis. Global Change Biology, 6:227–245.

Wickström, L.M., V. Haukisalmi, S. Varis, J. Hantula, V.B. Fedorov and H. Henttonen, 2003. Phylogeography of the circumpolar *Paranoplocephala arctica* species complex (Cestoda: Anoplocephalidae) parasitizing collared lemmings (*Dicrostonyx* spp.). Molecular Ecology, 12:3359–3371.

Wickström, L.M., V. Haukisalmi, S. Varis, J. Hantula and H. Henttonen, in press. A molecular phylogeny of anoplocephaline cestodes in rodents and lagomorphs. Systematic Parasitology.

Wielgolaski, F.E., S. Kjelvik and P. Kallio, 1975. Mineral content of tundra and forest tundra plants in Fennoscandia. In: F.E. Wielgolaski, P. Kallio and T. Rosswall (eds.). Fennoscandian Tundra Ecosystems: Plants and Microorganisms. Ecological Studies, 16:316–332.

Wielgolaski, F.E., L.C. Bliss, J. Svoboda and G. Doyle, 1981. Primary production of tundra. In: L.C. Bliss, O.W. Heal and J.J. Moore (eds.). Tundra Ecosystems: A Comparative Analysis, pp. 187–225. Cambridge University Press.

Wijk, S., 1986. Performance of *Salix herbacea* in an alpine snow-bed gradient. Journal of Ecology, 74:675–684.

Wiklund, C.G., A. Angerbjorn, E. Isakson, N. Kjellen and M. Tannerfeldt, 1999. Lemming predators on the Siberian tundra. Ambio, 28:281–286.

Wilke, C.O., J.L. Wang, C. Ofria, R.E. Lenski and C. Adami, 2001. Evolution of digital organisms at high mutation rates leads to survival of the flattest. Nature, 412:331–333.

Williams, M. and E.B. Rastetter, 1999. Vegetation characteristics and primary productivity along an arctic transect: implications for scaling up. Journal of Ecology, 87:885–898.

Williams, M., W. Eugster, E.B. Rastetter, J.P. McFadden and F.S. Chapin III, 2000. The controls on net ecosystem productivity along an Arctic transect: a model comparison with flux measurements. Global Change Biology, 6(S1):116–126.

Williams, M., E.B. Rastetter, G.R. Shaver, J.E. Hobbie, E. Carpino and B.L. Kwiatkowski, 2001. Primary production of an arctic watershed: an uncertainty analysis. Ecological Applications, 11:1800–1816.

Wilson, D.J., C.J. Krebs and T. Sinclair, 1999. Limitation of collared lemming populations during a population cycle. Oikos, 87:382–398.

Wilson, E.O., 1992. The Diversity of Life. Belknap Press.

Wookey, P.A., A.N. Parsons, J.M. Welker, J.A. Potter, T.V. Callaghan, J.A. Lee and M.C. Press, 1993. Comparative responses of phenology and reproductive development to simulated environmental change in sub-Arctic and high Arctic plants. Oikos, 67:490–502.

Wookey, P.A., C.H. Robinson, A.N. Parsons, J.M. Welker, M.C. Press, T.V. Callaghan and J.A. Lee, 1995. Environmental constraints on the growth, photosynthesis and reproductive development of *Dryas octopetala* at a high arctic polar semi-desert, Svalbard. Oecologia, 102:478–489.

Yoccoz, N.G. and R.A. Ims, 1999. Demography of small mammals in cold regions: The importance of environmental variability. In: A. Hofgaard, J.P. Ball, K. Danell and T.V. Callaghan (eds.). Animal Responses to Global Change in the North. Ecological Bulletin 47:137–144.

Young, S.B., 1971. The vascular flora of St. Lawrence Island with special reference to floristic zonation in the Arctic regions. Contributions from the Gray Herbarium, 201:11–115.

Yurtsev, B.A., 2001. The Pleistocene 'Tundra-Steppe' and the productivity paradox: the landscape approach. Quaternary Science Reviews, 20:165–174.

Zackrisson, O. and M.-C. Nilsson, 1992. Allelopathic effects by *Empetrum hermaphroditum* on seed germination of two boreal tree species. Canadian Journal of Forest Research, 22:1310–1319.

Zepp, R.G., T.V. Callaghan and D.J. Erikson III, 2003. Interactive effects of ozone depletion and climate change on biogeochemical cycles. Photochemical and Photobiological Sciences, 2(1):51–61.

Zimov, S.A., G.M. Zimova, S.P. Davidov, A.I. Davidova, Y.V. Voropaev, Z.V. Voropaeva, S.F. Prosianikov, O.V. Prosiannikova, I.V. Semiletova and I.P. Semiletov, 1993. Winter biotic activity and production of CO_2 in Siberian soils: A factor in the greenhouse effect. Journal of Geophysical Research, 98:5017–5023.

Zöckler, C., 1998. Patterns in Biodiversity in Arctic Birds. WCMC Biodiversity Bulletin 3. World Conservation Monitoring Centre, Cambridge, 15pp.

Zöckler, C., 2002. A comparison between tundra and wet grassland breeding waders with special reference to the Ruff (*Philomachus pugnax*). Schriftenreihe für Landschaftspflege und Naturschutz 74. Federal Agency for Nature Conservation, Bonn, 115pp.

Zöckler, C. and I. Lysenko, 2000. Water Birds on the Edge: First Circumpolar Assessment of Climate Change Impact on Arctic Breeding Water Birds. WCMC Biodiversity Series No. 11. World Conservation Monitoring Centre, Cambridge, 20pp. plus Annex.

Zöckler, C., J.H. Mooij, I.O. Kostin, K. Günther and R. Bräsecke, 1997. Notes on the distribution of some bird species on the Taimyr Peninsula. Vogelwelt, 118:329–338.

Zöckler, C., S. Delany and W. Hagemeijer, 2003. Wader populations are declining – how will we elucidate the reasons? Wader Study Group Bulletin, 100:202-211.

Chapter 8

Freshwater Ecosystems and Fisheries

Lead Authors
Frederick J. Wrona, Terry D. Prowse, James D. Reist

Contributing Authors
Richard Beamish, John J. Gibson, John Hobbie, Erik Jeppesen, Jackie King, Guenter Koeck, Atte Korhola, Lucie Lévesque, Robie Macdonald, Michael Power, Vladimir Skvortsov, Warwick Vincent

Consulting Authors
Robert Clark, Brian Dempson, David Lean, Hannu Lehtonen, Sofia Perin, Richard Pienitz, Milla Rautio, John Smol, Ross Tallman, Alexander Zhulidov

Contents

Summary

Changes in climate and ultraviolet radiation levels in the Arctic will have far-reaching impacts, affecting aquatic species at various trophic levels, the physical and chemical environment that makes up their habitat, and the processes that act on and within freshwater ecosystems. Interactions of climatic variables, such as temperature and precipitation, with freshwater ecosystems are highly complex and can propagate through the ecosystem in ways that are difficult to project. This is partly due to a poor understanding of arctic freshwater systems and their basic interrelationships with climate and other environmental variables, and partly due to a paucity of long-term freshwater monitoring sites and integrated hydro-ecological research programs in the Arctic.

This chapter begins with a broad overview of the general hydrological and ecological features of the various freshwater ecosystems in the Arctic, including descriptions of each ACIA region, followed by a review of historical changes in freshwater systems during the Holocene. The chapter continues with a review of the effects of climate change on broad-scale hydro-ecology; aquatic ecosystem structure and function; and arctic fish, fisheries, and aquatic wildlife. Special attention is paid to changes in runoff, water levels, and river- and lake-ice regimes; to biogeochemical processes, including carbon dynamics; to rivers, lakes, ponds, and wetlands; to aquatic biodiversity and adaptive capacities; to fish populations, fish habitat, anadromy, and fisheries resources; and to aquatic mammals and waterfowl. Potential synergistic and cumulative effects are also discussed, as are the roles of ultraviolet radiation and contaminants.

The nature and complexity of many of the effects are illustrated using case studies from around the circumpolar north, together with a discussion of important threshold responses (i.e., those that produce stepwise and/or nonlinear effects). The chapter concludes with a summary of key findings, a list of gaps in scientific understanding, and policy-related recommendations.

8.1. Introduction

The Arctic, which covers a significant area of the Northern Hemisphere, has a number of prominent and unique climatic, geological, and biophysical features. The region is typified by extreme variability in climate and weather, prolonged darkness in the winter and continuous daylight in the summer, the prevalence of vast areas of permafrost, and the dominance of seasonal ice and snow cover. The Arctic also has a diversity of terrains that contain a significant number and diversity of freshwater ecosystems.

The Arctic has some of the largest rivers in the world (e.g., the Lena, Mackenzie, Ob, and Yenisey); numerous permanent and semi-permanent streams and rivers draining mountains, highlands, and glaciated areas; large lakes such as Great Bear, Great Slave, and Taymir; a myr-

iad of smaller permanent and semi-permanent lakes and ponds; vast areas of wetlands and peatlands; and coastal estuarine and river delta habitats. In turn, these freshwater systems contain a wide diversity of organisms that have developed adaptations to cope with the extreme environmental conditions they face. Examples include life-history strategies incorporating resting stages and diapause, unique physiological mechanisms to store energy and nutrients, an ability to grow and reproduce quickly during brief growing seasons, and extended life spans relative to more temperate species.

Thus, given the regional complexity of climate and landscape and the diversity of freshwater ecosystems and their associated biota, projecting the potential impacts of future climate change and ultraviolet (UV) radiation exposure presents significant challenges. What is certain is that the responses are likely to be quite variable and highly specific to particular freshwater ecosystems, their biota, and the ecological and biophysical circumstances in which they occur.

8.1.1. Challenges in projecting freshwater hydrologic and ecosystem responses

The first and most significant challenge in projecting responses of freshwater systems to climate change relates to the limited understanding of how the climate system is coupled to, and influences, key physical and biophysical processes pertinent to aquatic ecosystems, and in turn how these affect ecological structure and function. Figure 8.1 summarizes the complex and often hierarchi-

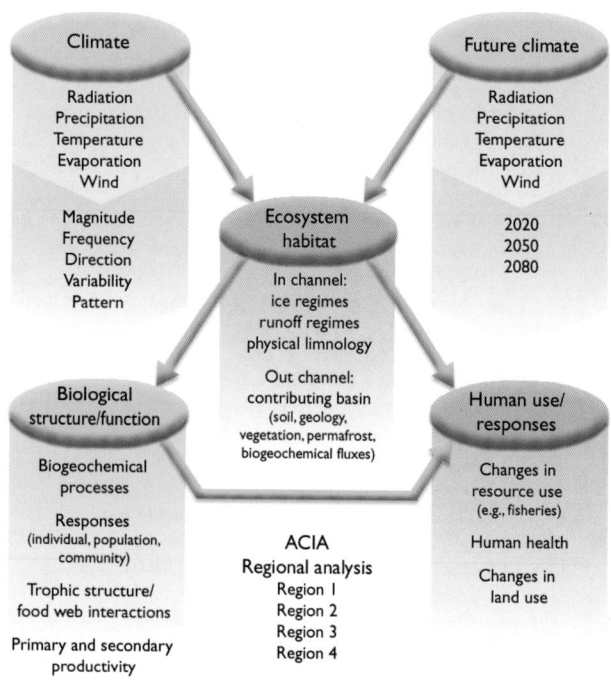

Fig. 8.1. Climate–ecosystem interactions. The interactions among and within components tend to be sequential but complex. However, complex feedbacks also exist both within major classes of components (e.g., trophic structure linkages with biogeochemical cycling), as well as between components (e.g., ice duration and timing feedbacks to the regional climate system), but are not illustrated above for visual clarity.

cal interactions between climatic variables (e.g., radiation, precipitation, and temperature), their influence on the biophysical features of freshwater ecosystem habitat, subsequent effects on biological structure and function, and the interaction of feedbacks within and between components. Freshwater ecosystems are complex entities that consist of groups of species at various trophic levels, the hydrological and physical environment that makes up their habitat, the chemical properties of that environment, and the multiple physical, biogeochemical, and ecological processes that act on and within the system. Hence, any change in these attributes and processes as a result of changes in climate and UV radiation levels will ultimately contribute to variable and dynamic responses within freshwater systems. Even in ecosystems containing only simplified food webs (e.g., those having no predators such as fish or predatory macroinvertebrates), the interactions of environmental parameters such as temperature and precipitation with the system are still complex, and may be propagated in ways that are difficult to project (i.e., nonlinear or stepwise threshold responses in population/community dynamics and stability; see section 8.4.1, Box 8.2). Because freshwater systems receive major inputs from terrestrial systems (Chapter 7) and provide major outputs to marine systems (Chapter 9), altered states and processes within freshwater systems are intimately linked to these arctic ecosystems through feedback and transfer mechanisms.

There are a number of levels within an ecosystem where changes in climate or UV radiation levels may interact with various ecosystem components, including:

- the individual, either within it (e.g., changes in physiological processes affecting thermoregulation, or effects on life processes such as growth and reproductive rates) and/or the whole individual (e.g., behavior);
- the population (e.g., life-history traits, rates of immigration and emigration, migrations, and intra-specific competition);
- the community (e.g., changes in trophic structure and in the levels and magnitudes of food-web interactions such as inter-specific competition, predation, and parasitism); and
- the ecosystem (e.g., changes that affect the nature of the environment that the organisms occupy, such as altered biogeochemical processes and hydrologic regimes).

Hence, there are a number of considerations in assessing the effects of a change in climate or UV radiation levels on freshwater ecosystems. First, changes in the environmental parameters may occur in a variety of ways. Second, these changes may be input to the various aquatic ecosystems in a variety of ways. Third, effects within the ecosystem may manifest at various levels and in various components within the system. Fourth, the effects may propagate through the ecosystem and affect different components or processes differently within the ecosystem. The inherent complexity of such interactions greatly

hampers the ability to make accurate and reasonable projections regarding such effects within arctic freshwater ecosystems with high levels of certainty. Finally, the internal complexity of potential responses makes it difficult to project output effects on key linking ecosystems such as deltas and estuaries that form the interactive zones between terrestrial, marine, and freshwater systems.

General knowledge of how the hydrology, structure, and function of arctic aquatic ecosystems are responding to past (section 8.3) and relatively recent changes in climate and UV radiation levels is gradually improving (e.g., overviews by AMAP, 1997, 1998, 2002; CAFF, 2001; Hessen, 2002; IPCC, 1996, 1998, 2001a; Prowse et al., 2001; Rouse et al., 1997; Vincent and Hobbie, 2000). However, much of the understanding of the processes and mechanics of potential impacts continues to be largely based on studies of aquatic systems outside of the Arctic (e.g., overviews by Antle et al., 2001; Carpenter et al., 1992; Meyer et al., 1999; Scheffer et al., 2001; Schindler D.W., 2001; Schindler D.W. et al., 1996a). Hence, the development of detailed projections of climate change impacts on arctic freshwater ecosystems is limited by a lack of understanding of how these impacts will cascade through arctic ecosystems and create second- and higher-order changes.

With these limitations in mind, using the approach outlined in section 8.1.2, this chapter identifies and discusses projected changes in the hydrology and ecology of

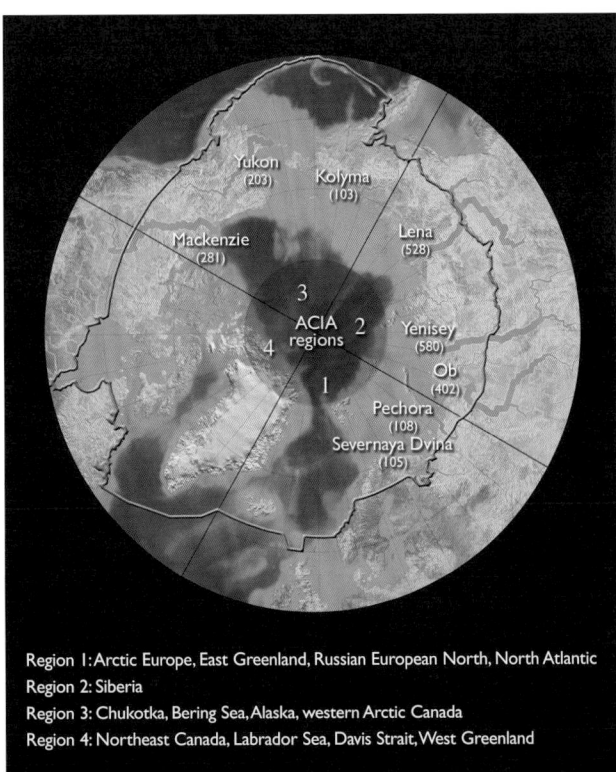

Region 1: Arctic Europe, East Greenland, Russian European North, North Atlantic
Region 2: Siberia
Region 3: Chukotka, Bering Sea, Alaska, western Arctic Canada
Region 4: Northeast Canada, Labrador Sea, Davis Strait, West Greenland

Fig. 8.2. The four ACIA regions, the southern boundary of the Arctic used in this chapter (as defined by AMAP, 1998, modified to include Québec north of the treeline), and the major river systems flowing through these regions to the Arctic Ocean (adapted from AMAP, 1998; discharge data (km³/yr) from R-ArcticNet, 2003).

arctic freshwater ecosystems in response to scenarios of future climate and UV radiation levels for three time slices centered on 2020, 2050, and 2080 generated by the ACIA-designated models (section 4.4). Where possible, similarities and/or differences in projected impacts between the four ACIA regions (Fig. 8.2; see also section 18.3) are identified.

8.1.2. Quantifying impacts and likelihood

The confidence level associated with projecting potential impacts of changes in climate and UV radiation levels is greatly hampered by the rudimentary level of understanding of arctic freshwater hydrology and ecology and their direct and indirect linkages, responses, and feedbacks to present and future climate. Moreover, the coarse spatial resolution of general circulation models (GCMs) and the uncertainty associated with complex, multilayered, and poorly understood interactions between climate variables greatly contribute to uncertainty in projections of future climate. This is exacerbated by other complexities such as inter- and intra-regional variation driven by, for example, latitude or proximity to marine ecosystems. When combined with uncertainties about how individual species and biological communities will respond to changes in climate and UV radiation levels, the ability to forecast hydro-ecological impacts and resulting cascading effects is significantly compromised. This makes precise quantification of climate change impacts difficult and often tenuous.

To address the issue of uncertainty and to recognize the substantial inter-regional and latitudinal differences in understanding and the broad spatial extent of arctic aquatic ecosystems, climate change and UV radiation impacts have been "quantified" using a "weight-of-evidence" approach. This approach uses a hierarchy of classes that represent the range of likelihood of the impact(s)/outcomes(s) occurring based on a compilation of information available from historical data, published literature, model projections, and the expert judgment of the authors. Using the ACIA lexicon (section 1.3.3), projected likelihoods follow a progression from "very unlikely" (i.e., little chance of occurring) through "unlikely", "possible" (some chance), and "likely/probable" to "very likely/very probable".

Although not strictly quantifiable in a numeric sense (e.g., exact probabilities), this approach provides a comparative and relative measure of the likelihood that the impact(s) will occur. Hence, a greater weight-of-evidence indicates a greater confidence in the findings (i.e., an increasing convergence of evidence from a number of independent, comprehensive empirical and/or experimental studies, model projections, etc.) that allows the classification of particular impact(s)/outcome(s) as either "very unlikely" or "very likely". The designation of particular impacts as "possible" or "likely" implies the presence of significant gaps in current knowledge. These gaps must be addressed to achieve a better understanding of impacts at the level of the ecosystem and its components.

This "weight-of-evidence"-based lexicon is directly applied in the conclusions and key findings of the chapter (section 8.8.1), thereby providing a relative "quantification" of the projected responses of freshwater ecosystems to changes in climate and UV radiation levels.

8.1.3. Chapter structure

Section 8.2 provides a broad overview of the general hydrological and ecological features of the various freshwater ecosystems in the Arctic, including descriptions for each ACIA region. Section 8.3 discusses how understanding past climate regimes using paleolimnological and paleoclimatic records helps to better understand present and future responses of freshwater ecosystems. Subsequent sections discuss the climate scenarios generated by the ACIA-designated models and project impacts on the hydrology and ecology of the major types of arctic freshwater ecosystems (section 8.4), impacts on the major components of these ecosystems (section 8.5), impacts of changes in UV radiation levels (section 8.6), and the interactions of these impacts with contaminants (section 8.7). A key feature of arctic freshwater ecosystems is the biota of direct relevance to humans, especially waterfowl, mammals, and fishes that provide the basis for harvests. Species within these groups are of special interest in that they also provide direct biotic linkages between major arctic ecosystems, thus either potentially input or output effects from, or to, terrestrial and marine systems. Fish are of particular relevance since two major ecological groups are present: those wholly associated with freshwaters and those which pass parts of their life history in both fresh and marine waters (i.e., diadromous fishes further divisible into catadromous species such as eels that rear in freshwater and breed in the sea, and anadromous species such as salmon that do the opposite). Anadromous fish provide major nutrient transfers from marine systems back into freshwater systems, thus are of particular significance. A logical extension is to also consider the effects of global change on fisheries for freshwater and diadromous forms; thus, section 8.5.5 parallels the treatment of marine fisheries in Chapter 13. Section 8.8 summarizes key findings and identifies major knowledge gaps and future research needs.

8.2. Freshwater ecosystems in the Arctic

8.2.1. General features of the Arctic relevant to freshwater ecosystems

The nature and severity of climate and weather have a strong influence on the hydrology and ecology of arctic freshwater ecosystems (e.g., Murray et al., 1998a; Pielou, 1994; Prowse and Ommanney, 1990; Prowse et al., 1994; Woo, 1996, 2000). Arctic climate has several prominent features that show extensive variation along strong latitudinal gradients. These include extreme seasonality and severity in temperature extremes (i.e., long, cold winters and relatively short, warm summers, both of which persist long enough to limit biota because of physiological thresholds); high intra- and interannual variability in temperature and precipitation; and strong seasonally driven latitudinal gradients in incident solar and UV radiation levels, to name a few. Extended low temperatures result in extensive ice cover for long periods of the year, significantly affecting physical, chemical, and biological processes in aquatic ecosystems. Extreme seasonality and low levels of incident radiation also have profound effects on aquatic ecosystems: much of this radiation may be reflected owing to the high albedo of ice and snow, especially during the critical early portions of the spring and summer. In addition, the thermal energy of a substantive portion of this incoming energy is used to melt ice, rendering it unavailable to biota. The timing of radiation is also important for some high-latitude aquatic systems that receive a majority of their annual total prior to the melting of their ice cover. Low levels of precipitation generally occur throughout the Arctic and most of this falls as snow, resulting in limited and highly episodic local runoff.

The ecological consequences of these environmental extremes are profound. For instance, overall annual productivity of freshwater systems generally tends to be low because of low levels of nutrient inputs, low temperatures, prolonged periods of ice presence compared to temperate aquatic ecosystems, and short growing seasons (Murray et al., 1998b). In most cases, this results in slower growth and some longer-lived organisms. Seasonal variations in arctic aquatic processes are relatively high, resulting in various adaptations in the organisms that thrive there. In animals, such adaptations include high rates of food consumption when it is available, rapid conversion of food to lipids for energy storage, and later metabolism of stored lipids for over-winter maintenance, growth, and reproduction (Craig, 1989). Additionally, some groups (e.g., fish) exhibit highly migratory behavior to optimize life-history functions, resulting in movements among different habitats triggered by environmental cues (e.g., dramatic temperature decreases) that usually coincide with transitions between particular seasons (Craig, 1989). Migratory organisms such as waterfowl occupy a variety of habitats both seasonally and over their lifetime (CAFF, 2001). Hence, aquatic biota display a wide range of adaptation strategies to cope with the severe environmental conditions to which they are exposed (CAFF, 2001; Pielou, 1994). A critical question is whether future changes in key climatic variables will occur at a rate and magnitude for which current freshwater species have sufficient phenotypic or genetic plasticity to adapt and survive.

8.2.2. Freshwater inputs into arctic aquatic ecosystems

The source, timing, and magnitude of freshwater inputs to arctic freshwater ecosystems has important implications for the physical, chemical and biological properties, as well as the structure, function, and distribution of river, lake, pond, and wetland ecosystems in the Arctic.

Rainfall is a substantial freshwater source for ecosystems at more southerly latitudes, occurring for the most part during the extended summer season. Further north, snowfall dominates the annual freshwater budget. High-latitude polar deserts receive low levels of precipitation and as such have a pronounced moisture deficit. Maritime locations generally receive greater quantities of snow and rain than continental regions.

The most important input of freshwater into aquatic ecosystems is often snowfall. It accumulates over autumn, winter, and spring, and partly determines the magnitude and severity of the spring freshet. Snowpack duration, away from the moderating influences of coastal climates, has been documented to range from ~180 days to more than 260 days (Grigoriev and Sokolov, 1994). In the spring, elevated levels of solar radiation often result in rapid snowmelt. Consequently, this rapid melt of the snowpack translates into spring runoff that can comprise a majority of the total annual flow, and be of very short-term duration – as little as only two to three weeks (Linell and Tedrow, 1981; Marsh, 1990; Rydén, 1981). In addition, at higher latitudes, infiltration of this spring flush of water is constrained by the permafrost. Thus, spring meltwater may flow over land and enter rivers, or accumulate in the many muskegs, ponds, and lakes characteristic of low-lying tundra areas (van Everdingen, 1990). Meltwater can also have major impacts on the quality of water entering lakes and rivers. When highly acidic, it can produce "acid shock" in receiving waters. However, because the incoming meltwater is usually warmer than the lake water, it tends to pass through the lake with little mixing. The potential acidic spring pulse is therefore transient without any marked biological consequences, as documented by paleolimnological investigations (e.g., Korhola et al., 1999; Michelutti et al., 2001).

During the summer, sources of water include not only rain, but also late or perennial snow patches, glaciers, thawing permafrost, and groundwater discharges (Rydén, 1981; van Everdingen, 1990). As temperatures rise in response to climate change, these sources of water are likely to become more pronounced contributors to the annual water budgets of freshwater ecosystems, at least until their ice-based water reserves are depleted.

Groundwater can also have an important influence on the annual water budgets of arctic surface-water ecosystems. Permafrost greatly influences the levels and distribution of groundwater within the Arctic. Groundwater movement through aquifers is restricted by permafrost year-round, and by the frozen active layer for up to ten months of the year (Murray et al., 1998a). Three general types of groundwater systems occur in the Arctic: supra-permafrost, intra-permafrost, and sub-permafrost. Supra-permafrost groundwater lies above the permafrost table in the active layer during summer, and year-round under lakes and rivers that do not totally freeze to the bottom. Intra-permafrost water resides in unfrozen sections within the permafrost, such as tunnels called "taliks", which are located under alluvial flood plains and under drained or shallow lakes and swamps. Sub-permafrost water is located beneath the permafrost table. The thickness of the permafrost determines the availability of sub-permafrost water to freshwater ecosystems, acting as a relatively impermeable upper barrier. These three types of groundwater systems, which may be located in bedrock or in unconsolidated deposits, may interconnect with each other or with surface water (Mackay D. and Løken, 1974; van Everdingen, 1990; Woo, 2000; Woo and Xia, 1995) as outflows via springs, base flow in streams, and icings. Icings (also known as aufeis or naleds) are comprised of groundwater that freezes when it reaches the streambed during winter. Groundwater interactions with surface-water systems greatly influence water quality characteristics such as cation, anion, nutrient, and dissolved organic matter concentrations, and even the fate and behavior of toxic pollutants.

8.2.3. Structure and function of arctic freshwater ecosystems

Arctic freshwater ecosystems are quite varied with respect to their type, physical and chemical characteristics, and their associated biota. Thus, the impacts of climate change and increased UV radiation levels will be variable and highly specific to particular freshwater ecosystems, their biota, and processes. Additionally, in some areas that span a wide latitudinal range (e.g., the arctic regions of Canada and Russia), similar types of freshwater systems exhibit a wide range of characteristics driven in part by latitudinal differences in the environment. These, in turn, will also respond differently to global change. Furthermore, the nature of connections between the various regions of the Arctic and non-arctic areas of the globe differ. Consequently, regional differences between the same types of aquatic systems are likely to exist, despite these being at the same latitude. In addition, historical differences in their development during recent geological time and geomorphic processes that have affected different regions (e.g., extent of Pleistocene glaciations, age, and connectivity to southern areas), will contribute to regional, subregional, and local variability in ecosystem structure and function.

Two major categories of freshwater ecosystems can be defined as lotic (flowing water) and lentic (standing water), but large variation in size, characteristics, and location is exhibited within each. Thus, large differences in response to climate change can be expected. For the purposes of this assessment, lotic ecosystems include rivers, streams, deltas, and estuaries, where flow regimes are a dominant hydrologic feature shaping their ecology. Lentic ecosystems include lakes, ponds, and wetlands (including bogs and peatlands). Although some wetland types may not have standing surface water at all times, they are considered lentic ecosystems for the purposes of this chapter.

Although the Arctic generally contains a relatively low number of aquatic bird and mammal species as compared to more temperate ecozones, it is home to most of the world's geese and calidrid sandpipers (Zöckler, 1998). Migratory birds, including geese, ducks, swans, and gulls,

can be particularly abundant in arctic coastal and inland wetlands, lakes, and deltas (Bellrose, 1980; Godfrey, 1986; Zhadin and Gerd, 1961; for comprehensive review see CAFF, 2001). Most taxonomic groups within the Arctic are generally not very diverse at the species level, although some taxonomic groups (e.g., arctic freshwater fish; see section 8.5.1.1, Box 8.6) have high diversity at and below the species level (e.g., display a large number of ecological morphs). In addition, arctic freshwater systems generally exhibit strong longitudinal gradients in biodiversity, ranging from extremely low biodiversity in high-latitude, low-productivity systems to very diverse and highly productive coastal delta–estuarine habitats (AMAP, 1998; CAFF, 2001; IPCC, 2001a). Very little is known about the biological and functional diversity of taxa such as bacteria/virus, phytoplankton, and zooplankton/macroinvertebrate communities that reside in arctic aquatic ecosystems, despite their undoubted importance as key components of freshwater food webs (Vincent and Hobbie, 2000; Vincent et al., 2000).

8.2.3.1. Rivers and streams, deltas, and estuaries

Rivers and streams

Arctic rivers and streams are most densely distributed in lowlands, including those in Fennoscandia and the Interior Plain of Canada, often in association with lakes and wetlands. Lotic ecosystems include large northward flowing rivers such as the Mackenzie River in Canada (Fig. 8.2), high-gradient mountain rivers, and slow-flowing tundra streams that may be ephemeral and flow only during short periods in the early spring. Flowing-water systems represent a continuum, from the smallest to largest, and although subdividing them at times is arbitrary, river systems of different sizes do vary in terms of their hydrology, water quality, species composition, and direction and magnitude of response to changing climatic conditions. This is particularly relevant in the Arctic, where river catchments may be wholly within the Arctic or extend southward to more temperate locations.

In general, the large rivers of the Arctic have headwaters well south of the Arctic as defined in this chapter (Fig. 8.2; see section 6.8 for a review of major arctic rivers and their historical flow trends), and as such act as conduits of heat, water, nutrients, contaminants, sediment, and biota northward (e.g., Degens et al., 1991). For such systems, not only will local effects of climate change be important, but basin-wide effects, especially those in the south, will also be critical in determining cumulative effects (e.g., see Cohen 1994, 1997). Five of the ten largest rivers in the world fall into this category: the Lena, Ob, and Yenisey Rivers in Russia, the Mackenzie River in Canada, and the Yukon River in Canada and Alaska. These rivers have substantive effects on the entire Arctic, including the freshwater budget of the Arctic Ocean and the hydro-ecology of coastal deltas and related marine shelves. Various portions of these rivers are regulated (Dynesius and Nilsson, 1994), the most affected being the Yenisey River, which is also the largest of the group and the one projected to experience significant further impoundment (an increase of ~50%) over the next few decades (Shiklomanov et al., 2000). For northern aquatic systems, the effects of impoundment on water quantity and quality are wide-ranging, and are expected to be exacerbated by the effects of climate change (Prowse et al., 2001, 2004).

Numerous smaller, but still substantive, rivers also drain much of the Arctic and may arise from headwaters outside of the Arctic. These include the Severnaya Dvina and Pechora Rivers that drain much of the Russian European North, the Khatanga River of Siberia, the Kolyma River of eastern Siberia, and the Churchill and Nelson Rivers that drain much of central Canada and supply water to the Arctic Ocean via Hudson Bay. Although these rivers are much smaller than those in the first group, they are more numerous and in many cases are affected by a similar suite of anthropogenic factors, including agriculture, hydroelectric impoundment, industrialization, mining, and forestry, many of which occur outside of the Arctic and, as climate change progresses, may become more prominent both within and outside of the Arctic.

Still smaller types of lotic systems include medium to small rivers that arise wholly within the Arctic. Examples include the Thelon River in Canada, the Colville River in Alaska, the Anadyr River in Chukotka, many rivers throughout Siberia, and the Tana River of Scandinavia. In many cases, these rivers do not presently have the same degree of local anthropogenic impacts as the previous two types. Despite some level of anthropogenic impacts, many of these arctic rivers harbor some of the largest and most stable populations of important and widely distributed arctic freshwater species. For example, many of the most viable wild populations of Atlantic salmon (*Salmo salar*) are extant in northern systems such as the Tana River of northern Norway, despite widespread declines in southern areas (e.g., Parrish et al., 1998).

Most of the rivers noted above share an important characteristic: their main channels continue flowing throughout the winter, typically beneath ice cover, due to some type of continuous freshwater input from warm southern headwaters, lakes, and/or groundwater inflows. As such, they typically have higher levels of productivity and biodiversity than arctic rivers that do not flow during winter. This latter group consists of numerous rivers that are even smaller and found throughout the Arctic. Fed primarily by snowmelt, they exhibit high vernal flows dropping to low base flows during the summer, with perhaps small and ephemeral flow peaks during summer and autumn precipitation events prior to freeze-up. Glaciers also feed many of these smaller arctic rivers (e.g., in Alaska and Greenland), thus snowmelt feeds initial vernal flows, and glacial melt maintains flows at a relatively high level throughout the summer. Most of these small arctic rivers stop flowing at some point during the winter and freeze to the bottom throughout large reaches. Such is the case for many small rivers in Region 1, those to the

east in Region 2, and the coastal rivers of Chukotka, northern Alaska, and northwestern Canada (Region 3). This hydrology has important implications for the biota present (e.g., habitat and productivity restrictions), and climate change will have important ramifications for such ecosystems (e.g., cascading effects of changes in productivity, migratory routes).

Although the division between rivers and streams is somewhat arbitrary, as a class, local streams are numerous and found throughout the Arctic in association with all types of landforms. Streams feed water and nutrients to lacustrine environments and act as the first-order outflows from many tundra lakes, thus providing connectivity between different aquatic environments and between terrestrial and aquatic systems.

The ecology of arctic rivers and streams is as diverse as are the systems themselves, and is driven in part by size, location, catchment characteristics, nutrient loads, and sources of water. Correspondingly, biotic food webs of arctic rivers (Fig. 8.3) vary with river size, geographic area, and catchment characteristics. For example, benthic algae and mosses, and benthic invertebrate fauna associated with fine sediments, are more common in smaller, slower-flowing rivers and streams, while fish populations are limited in small rivers that freeze over the winter (Hobbie, 1984; Jørgensen and Eie, 1993; Milner and Petts, 1994; Steffen, 1971). Changes to river ecology, whether they are bottom-up (e.g., changes in nutrient loading from catchments will affect primary productivity) or top-down (e.g., predatory fish removal with habitat loss will affect lower-level species productivity and abundance), will affect not only river systems, but also receiving waters. Rivers fed primarily by glaciers are physically dynamic and nutrient-poor, and as such offer challenging environments for primary production and invertebrate communities (Murray et al., 1998a). Spring-fed streams with stable environments of clear water, year-round habitat, and higher winter temperatures exhibit greater diversity in primary producers, including mosses and diatoms, and lower trophic levels such as insects (Hobbie, 1984). Tundra streams tend to be ephemeral and low in pH and nutrients, with correspondingly low productivity. Medium-sized rivers, especially those draining lakes, typically have moderate to high levels of productivity and associated diversity in invertebrate fauna, which in turn are affected by such things as suspended sediment loads. For example, clear-flowing rivers of the Canadian Shield have higher biodiversity at lower trophic levels (e.g., invertebrates) than very turbid rivers of the lowlands of Siberia and the Interior Plain of Canada (Murray et al., 1998a). In general, fish diversity in arctic rivers appears to be related primarily to the size of the river and its associated drainage basin; thus similarly sized rivers differing greatly in suspended sediment loads tend to have a similar overall diversity of fish species. However, the suite of species present differs between clear (e.g., preferred by chars) and sediment-rich (e.g., preferred by whitefishes) rivers. Historical

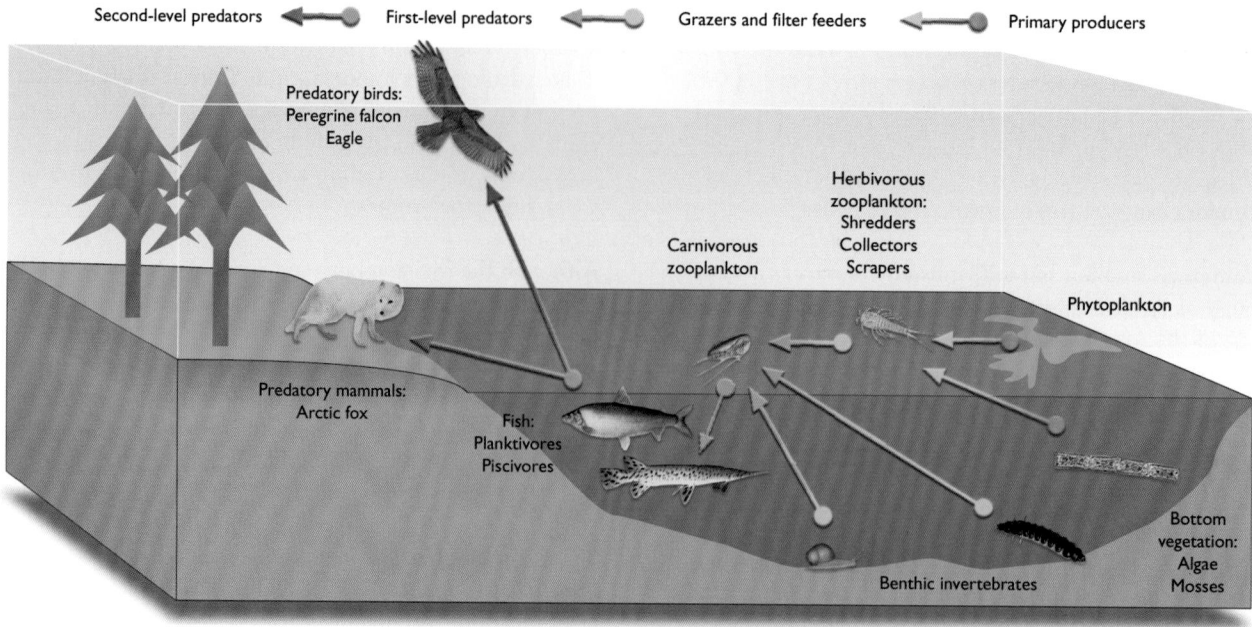

Fig. 8.3. Representative arctic river and stream food web.

factors such as deglaciation events and timing also figure prominently in determining biodiversity at higher trophic levels in these systems (Bodaly et al., 1989).

Another ecological feature of arctic rivers, and one that is likely to be significantly affected by climate change, is that of anadromy or sea-run life histories of many of the fish species present (section 8.5). That is, most of the salmonid fishes found in the Arctic, and several species of other families, use marine environments extensively for summer feeding and, in some instances, for substantial portions of their life history (e.g., much of salmon life history occurs in marine waters). These fish, and to some extent waterfowl, provide a fundamental ecological linkage between freshwater systems, estuarine systems, and marine systems of the Arctic. For such organisms, the effects of changes in climate and UV radiation levels on each environment will be integrated throughout the life of the individual and hence be cumulative in nature.

Deltas and estuaries

Deltas are highly diverse ecosystems that lie at the interface between rivers and lakes or oceans, providing a variety of freshwater habitats that are highly seasonal in nature. The most notable deltas in the Arctic are those of the Lena River in Russia and the Mackenzie River in Canada, where easily eroded sedimentary landscapes contribute to heavy sediment loading in rivers and deltas. Habitats include extensive wetlands, which cover up to 100% of the Mackenzie Delta (Zhulidov et al., 1997), and many ponds and lakes frequented by small mammals, fish, and waterfowl. Arctic deltas are ice-covered for the majority of the year, although flows continue in their major channels throughout the year. A critical hydrologic feature of these systems is the occurrence of ice jams and associated ice-jam floods, both of which are paramount in the maintenance of delta ecosystems

(Prowse, 2001a; Prowse and Gridley, 1993). Spring overland floods are critical to the recharge of delta lakes, such as those of the Yukon, Colville (Dupre and Thompson, 1979; Walker and McCloy, 1969), Mackenzie (Marsh and Hey, 1989, 1991), and Slave Rivers (Peterson E. et al., 1981) in North America, and the Yenisey, Lena, Kolyma, and Indigirka Rivers in Siberia (Antonov, 1969; Burdykina, 1970). Flooding during spring breakup also provides sediments and nutrients to deltas (e.g., Lesack et al., 1991), which in turn help sustain unique and highly productive habitats for plant and animal species, including fish, waterfowl, and small mammals such as muskrats (*Ondatra zibethicus*; e.g., Marsh and Ommanney, 1989). The drastic changes in delta hydrology with seasonal and interannual shifts in flow regimes, and the effect of wind-related disturbance on delta waters, have important implications for delta hydro-ecology. Hence, given the transient and sensitive nature of delta hydro-ecology, climate change is likely to have significant impacts in these areas of the Arctic.

River hydrology not only affects the hydro-ecology of deltas, but also that of estuaries. Examples of large deltas and associated estuaries include the Mackenzie River in Canada, and the Lena, Ob, and Yenisey Rivers in Russia. Arctic estuaries are distinct from those at more southerly latitudes in that their discharge is highly seasonal and ice cover is a key hydrologic variable influencing the ecology of the systems. Winter flows are typically between 5 and 10% of the annual average (Antonov, 1970), and estuarine waters are often vertically stratified beneath the ice cover. This may promote the formation of frazil ice at the freshwater–saltwater boundary. Freshwaters that flow into estuaries during winter typically retain their chemical loads until stratification deteriorates with loss of ice cover. In estuaries that are less than 2 m deep, river discharges in late winter may be impeded by ice and diverted offshore through erosional channels or by tidal inflows (Reimnitz and Kempema, 1987). High-magnitude freshwater discharges in spring carry heavy sediment loads and flow beneath the ice, gradually mixing with saltwater as breakup progresses in the estuary; these discharges dominate estuarine waters when landward fluxes of seawater are less pronounced.

Freshwater inflows from large arctic rivers carry sediment, nutrients, and biota to coastal areas, thereby contributing to the highly productive nature of estuaries and related marine shelves. Furthermore, this production is fostered by the complex nearshore dynamics associated with mixing of water masses differing in density, which in turn, increase the complexity of biological communities (Carmack and Macdonald, 2002). Hence, estuaries provide a significant food source for anadromous species compared to what is available to them from adjacent freshwater streams (Craig, 1989). This productivity typically results in large populations of fish that are important to local fisheries (e.g., Arctic char – *Salvelinus alpinus*, Atlantic salmon) and integral to the food web supporting other arctic organisms such as waterfowl, shorebirds, and marine mammals. The fish populations are keystone components affecting energy transfer (Fig. 8.4). Many anadromous fishes in these systems (e.g., Arctic cisco – *Coregonus autumnalis*, Dolly Varden – *Salvelinus malma*, rainbow smelt – *Osmerus mordax*) overwinter in freshened coastal and estuarine waters that are often used for feeding and rearing during the summer. Fishes migrate upstream in freshwater systems to spawn, and in some cases to overwinter. Given the intimate interaction of anadromous fishes with freshwater and marine environments in these delta/estuary systems, climate-induced changes in freshwater and marine ice and hydrology will significantly affect the life histories of these fishes.

Shorebirds and seabirds that utilize freshwater and/or estuarine habitats, linking freshwater and marine environments, include the red phalarope (*Phalaropus fulicaria*), parasitic jaeger (*Stercorarius parasiticus*), red knot (*Calidris canutus*), dunlin (*C. alpina*), long-tailed jaeger (*S. longicaudus*), northern fulmar (*Fulmarus glacialis*), glaucous gull (*Larus hyperboreus*), white-rumped sandpiper (*C. fuscicollis*), western sandpiper (*C. mauri*), red-necked stint (*C. ruficollis*), Lapland longspur (*Calcarius lapponicus*), black-bellied plover (*Pluvialis squatarola*), semipalmated plover (*Charadrius semipalmatus*), and ruddy turnstone (*Arenaria interpres*). Another important feature of estuarine ecosystems is the potential for transfers (e.g., by waterfowl and anadromous fishes) of significant nutrient loads from marine to freshwater habitats (Bilby et al., 1996). Deltas and estuaries also have high rates of sedimentation and potentially significant rates of sediment suspension, and as such can be important sinks and sources of terrestrial organic carbon (e.g., Macdonald R. et al., 1995) and contaminants (e.g., Milburn and Prowse, 1998), and are thereby capable of producing both positive and negative impacts on the aquatic biota in these systems.

8.2.3.2. Lakes, ponds, and wetlands

Lentic ecosystems of the Arctic are diverse and include an abundance of lakes of varying size, shallow tundra ponds that may contain water only seasonally, and wetlands such as peatlands that are notable stores and sources of carbon. These freshwater systems provide a rich diversity of habitats that are highly seasonal and/or ephemeral.

Lakes and ponds

Arctic lakes are typically prevalent on low-lying landscapes, such as coastal and interior plains (e.g., the

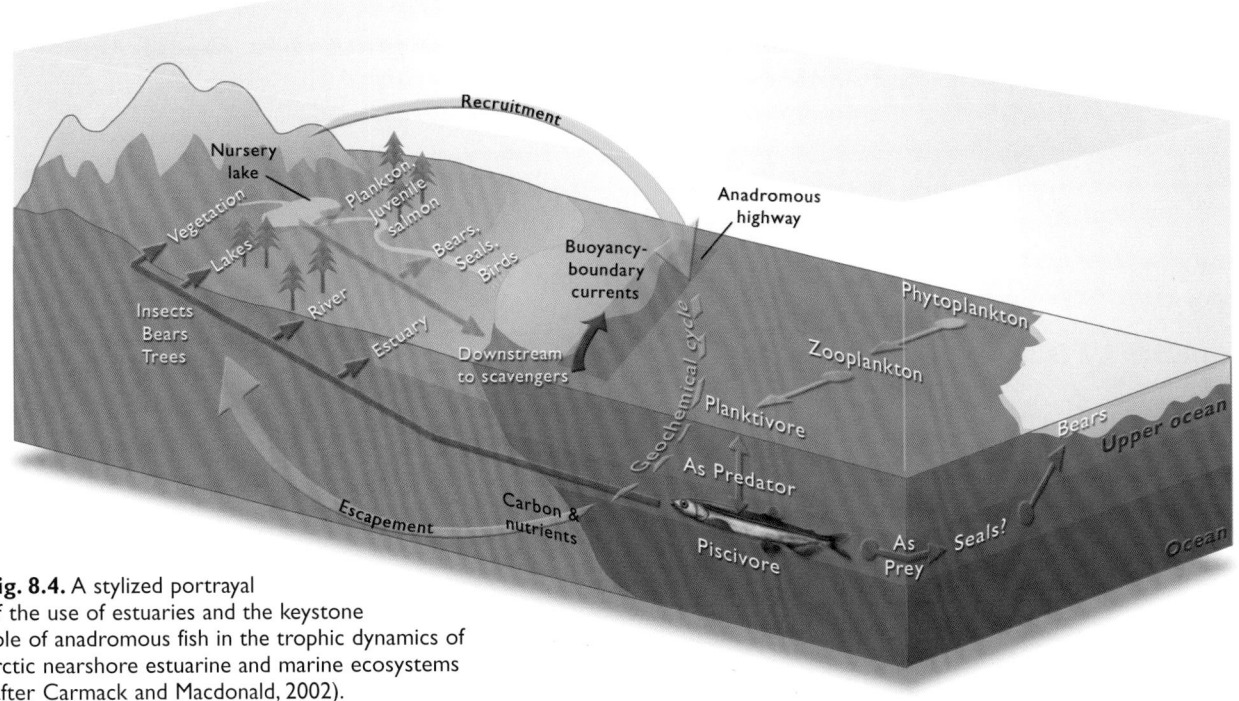

Fig. 8.4. A stylized portrayal of the use of estuaries and the keystone role of anadromous fish in the trophic dynamics of arctic nearshore estuarine and marine ecosystems (after Carmack and Macdonald, 2002).

Canadian Interior Plain and the Finnish Lowlands). There are many kettle (produced by the melting of buried glacial ice), moraine, and ice-scour lakes on the undulating terrain of postglacial arctic landscapes (e.g., the Canadian Shield, Fennoscandia, and the Kola Peninsula; Korhola and Weckström, 2005; Mackay D. and Løken, 1974). Thermokarst lakes are also quite common in the Arctic (e.g., along the Alaskan coast and in Siberia), developing in depressions formed by thawing permafrost. Small ponds also dominate portions of the Arctic landscape (e.g., the low-lying terrain of Fennoscandia); typically less than 2 m deep, these freeze solid over the winter.

Local catchments are typically the primary source of water for arctic lakes (Hartman and Carlson, 1973; Woo and Xia, 1995; Woo et al., 1981). Spring runoff originates from snow accumulation on lake ice, hillslope runoff (Woo et al., 1981), and lateral overflow from wetlands and streams (Marsh and Hey, 1989). Outlets of small lakes may be snow-dammed, and eventually release rapid and large flows downstream (Heginbottom, 1984; Woo, 1980). Arctic lakes also experience considerable evaporative water loss, sometimes resulting in the formation of athalassic (i.e., not of marine origin) saline systems. Water loss may also occur through seepage, which is common in lakes underlain by taliks in the discontinuous permafrost zone (Kane and Slaughter, 1973; Woo, 2000).

The hydro-ecology of the many small arctic lakes is intimately linked with climatic conditions. The timing and speed of lake-ice melt depend on the rate of temperature increase in late spring and early summer, wind, and inflow of basin meltwater and terrestrial heat exchanges (e.g., groundwater inflow, geothermal input, heat loss to maintain any underlying talik; Doran et al., 1996; Welch H. et al., 1987). Some lakes in the high Arctic retain ice

cover throughout the year, while some thermal stratification can occur in arctic lakes where breakup occurs more quickly. In northern Fennoscandia, for example, lakes >10 m deep are usually stratified during the summer and have well-developed thermoclines (Korhola et al., 2002a). In contrast, many high-arctic lakes mix vertically, thereby reducing thermal stratification (Mackay D. and Løken, 1974; Welch H. et al., 1987). Similarly, small shallow lakes do not stratify because they warm quickly and are highly wind-mixed. Heat loss from arctic lakes tends to be rapid in late summer and early autumn and often results in complete mixing. Consequently, shallow lakes and ponds will freeze to the bottom over winter. The duration and thickness of lake-ice cover in larger lakes increases with latitude, reaching thicknesses of up to 2.5 m, and can even be perennial over some years in extreme northern arctic Canada and Greenland (Adams W. et al., 1989; Doran et al., 1996). In addition to air temperature, the insulating effect of snow inversely affects ice thickness. Any shifts in the amounts and timing of snowfall will be important determinants of future ice conditions, which in turn will affect the physical and chemical dynamics of these systems.

The abundance and diversity of biota, productivity, and food web structure in arctic lakes varies regionally with environmental conditions and locally with the physical characteristics of individual lakes (Fig. 8.5). For example, lakes across the Russian European North vary from small, oligotrophic tundra systems (having moderate phytoplankton diversity, low primary productivity and biomass, and relatively high zoobenthos abundance) to larger taiga lakes (displaying greater species diversity and higher primary and secondary productivity and biomass). Mountain lakes of the region tend to have very low phytoplankton diversity, but substantial primary and secondary productivity and biomass, similar to that of taiga lakes. In general, the abundance and diversity of phyto-

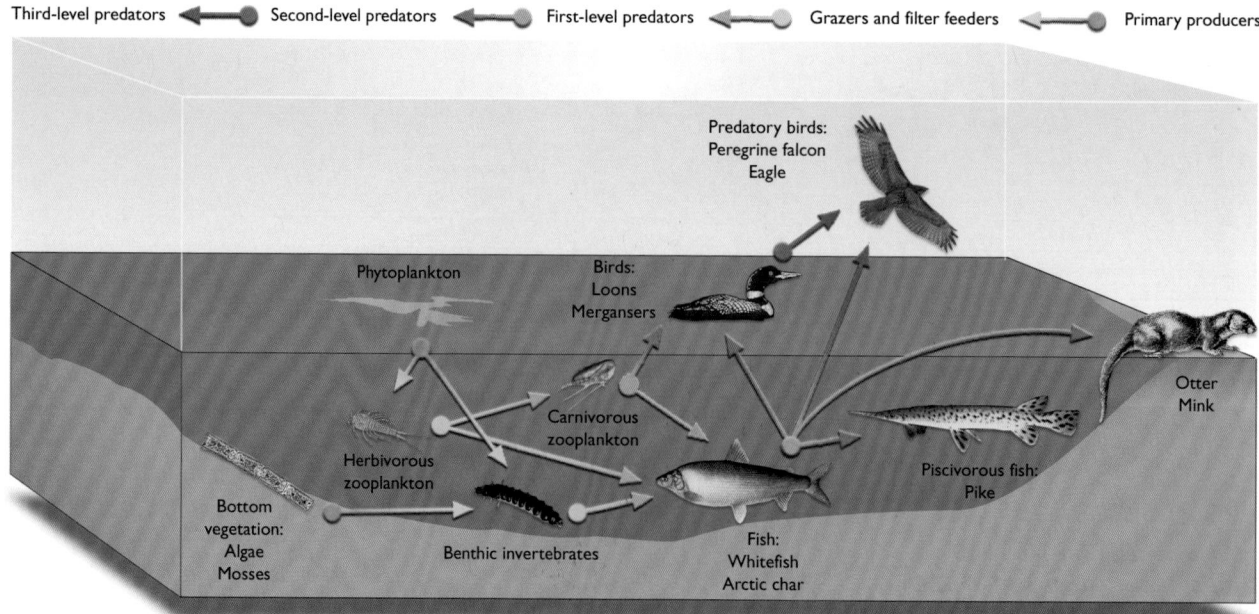

Fig. 8.5. Representative food web in arctic lakes (adapted from AMAP, 1997).

plankton and invertebrates such as rotifers, copepods, and cladocerans increase with lake trophic status (Hobbie, 1984), which is often a function of latitudinal constraints on resources for productivity. For example, some Icelandic lakes have phytoplankton production levels of >100 g C/m²/yr (Jónasson and Adalsteinsson, 1979; Jónasson et al., 1992), contrasting with extremely oligotrophic high-arctic lakes that have phytoplankton production levels of <10 g C/m²/yr (Hobbie, 1984). Although zooplankton are generally limited and at times absent in arctic lakes due to temperature and nutrient limitations, they may be quite abundant in shallow lakes where there is a lack of predators. For example, more than 30 Cladocera species have been documented in certain Finnish Lapland lakes, although generally most of them contain fewer than 10 species (Korhola, 1999; Rautio, 2001). Benthic invertebrate species diversity and abundance also display high latitudinal and inter-lake variability and may be significant in shallow lakes and ponds (Chapin and Körner, 1994; Hansen, 1983; Hobbie, 1984; Jørgensen and Eie, 1993; Vadeboncoeur et al., 2003). For example, in lakes of the Svalbard region, chironomid larvae are often numerically dominant but display low diversity (~10 species; Hirvenoja, 1967; Planas, 1994; Styczynski and Rakusa-Suszczewski, 1963), while more than 49 species have been identified in more southerly Norwegian lakes. Fish in arctic lakes are generally not very diverse, ranging from a few species (one to three) in lakes of Greenland (Riget et al., 2000), Iceland (Sandlund et al., 1992), the Faroe Islands, northwest Scandinavia, and the Kola Peninsula, up to several tens of species near the Pechora River in Russia. These fish may be anadromous or landlocked, depending on life histories and lake–river networks.

In general, tundra ponds tend to have very low annual primary productivity, dominated by macrophytes and benthic bacteria and algae (Hobbie, 1980). The detrital food web is highly important in these systems and phytoplankton growth is limited by nutrients and light. Zooplankton are abundant because fish are mostly absent in these shallow systems; hence, algal turnover is rapid in response to heavy grazing by herbivorous zooplankton (Hobbie, 1980). Pond vegetation typically includes horsetail (*Equisetum* spp.), water smartweed (*Polygonum*

amphibium), duckweed (*Lemna* spp.), and pondweed (*Potamogeton* spp.)(Zhadin and Gerd, 1961), and the resulting plant detritus tends to be mineralized rather than grazed upon. Figure 8.6 illustrates a typical tundra pond food web.

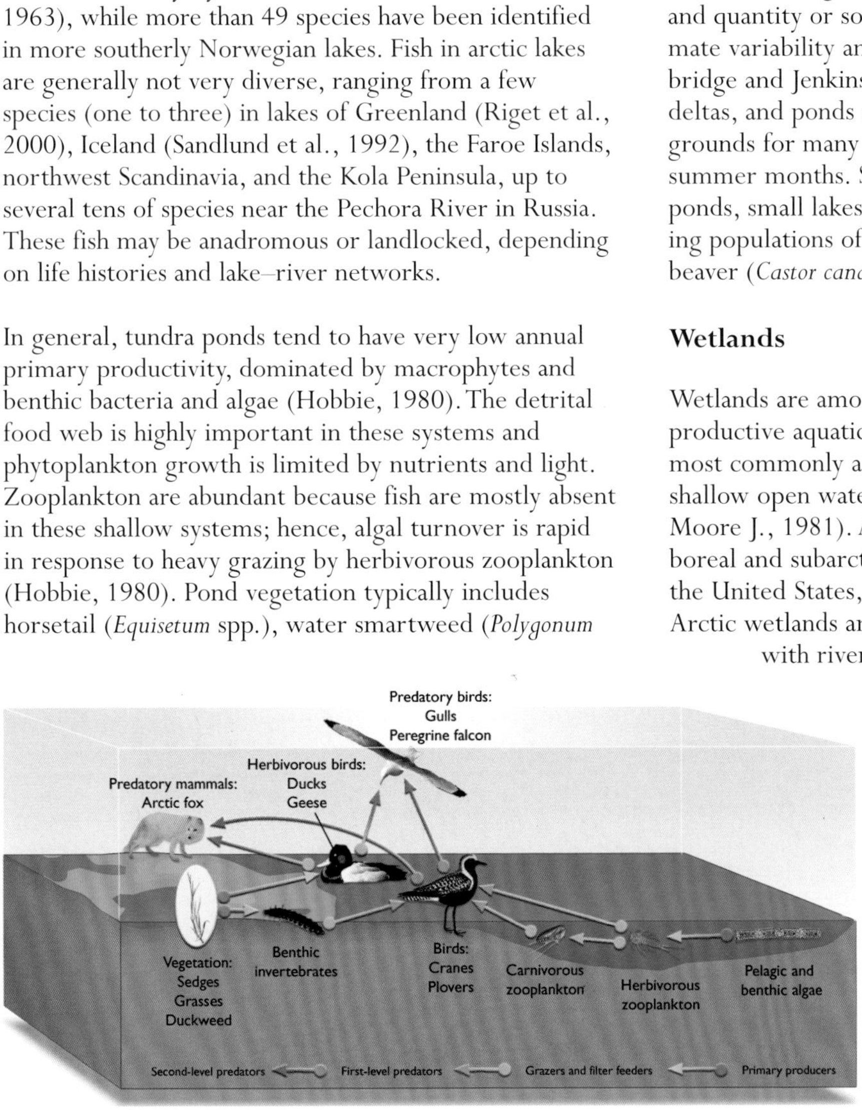

Ponds, as well as lakes and wetlands (discussed below), provide habitat that is critical to a wide variety of waterfowl, as well as small mammals. Typical waterfowl in the Arctic include the Canada goose (*Branta canadensis*), bean goose (*Anser fabalis*), snow goose (*A. caerulescens*), black brent (*B. bernicla*), eider (*Somateria mollissima*), oldsquaw duck (*Clangula hyemalis*), red-throated loon (*Gavia stellata*), yellow-billed loon (*G. adamsii*), Arctic loon (*G. arctica*), tundra swan (*Cygnus columbianus*), ring-necked duck (*Aythya collaris*), canvasback duck (*A. valisineria*), greater scaup (*A. marila*), and king eider (*S. spectabilis*). Some of the most severely endangered species in the world, including the once-abundant Eskimo curlew (*Numenius borealis*), the Steller's eider (*Polysticta stelleri*), and the spectacled eider (*S. fischeri*), are dependent on arctic freshwater systems (Groombridge and Jenkins, 2002). These and other bird species have been affected by a combination of factors such as over-harvesting and changes in terrestrial habitat quality and quantity or some perturbation at sea related to climate variability and/or change (CAFF, 2001; Groombridge and Jenkins, 2002). Coastal and inland wetlands, deltas, and ponds are common feeding and breeding grounds for many species of waterfowl in the spring and summer months. Some more southerly or subarctic ponds, small lakes, and wetlands can also contain thriving populations of aquatic mammals such as muskrat and beaver (*Castor canadensis*).

Wetlands

Wetlands are among the most abundant and biologically productive aquatic ecosystems in the Arctic, and occur most commonly as marshes, bogs, fens, peatlands, and shallow open waters (Mitsch and Gosselink, 1993; Moore J., 1981). Approximately 3.5 million km² of boreal and subarctic peatlands exist in Russia, Canada, the United States, and Fennoscandia (Gorham, 1991). Arctic wetlands are densely distributed in association with river and coastal deltas (e.g., the Lena and Mackenzie Deltas), and low-lying landscapes (e.g., the Finnish and Siberian lowlands and substantive portions of the Canadian Interior Plain). Wetlands are generally less abundant in Region 4 (up to 50% in isolated areas).

Wetlands are a common feature in the Arctic due in large part to the prominence of permafrost and the low rates of evapotranspiration. Aside from precipitation and melt-water, wetlands may also be sustained by groundwater, as is the

Fig. 8.6. Representative food web in arctic tundra ponds (adapted from AMAP, 1997).

case for fens, which are more nutrient-rich, productive wetland systems than bogs, which are fed solely by precipitation. Arctic wetlands may have standing water in the ice-free season or, as in the case of peatlands, may have sporadic and patchy pools. The occurrence of these pools exhibits high seasonal and interannual variability resulting from heat and water fluxes, and high spatial variability resulting from peatland microtopography. As such, arctic wetlands often have a diverse mosaic of microhabitats with different water levels, flow characteristics, and biota. The biogeochemistry of arctic wetlands is also generally distinct from other arctic freshwater systems, with lower dissolved oxygen concentrations, more extreme reducing conditions in sediments, and more favorable conditions for biodegradation (Wetzel, 2001).

Arctic wetlands are highly productive and diverse systems, as they often are important transition zones between uplands and more permanent freshwater and marine water bodies. They are typically dominated by hydrophytic vegetation, with a few species of mosses and sedges, and in some instances terrestrial species such as lichens, shrubs, and trees (e.g., forested bogs in the mountains of Siberia). Insects such as midges (chironomids) and mosquitoes are among the most abundant fauna in arctic wetlands (Marshall et al., 1999). Peatland pools in arctic Finland, for example, host thriving popu-

lations of midges that are more abundant and have greater biomass in areas of standing water than in semi-terrestrial sites, and are an important food source for many peatland bird species (Paasivirta et al., 1988).

Aside from habitat provision, river-flow attenuation, and a number of other ecological functions, wetlands also store and potentially release a notable amount of carbon, with potential positive feedbacks to climate change (e.g., radiative forcing by methane – CH_4 and carbon dioxide – CO_2). It is estimated that northern peatlands store approximately 455 Pg of carbon (Gorham, 1991), which is nearly one-third of the global carbon pool in terrestrial soils. As well, northern wetlands contribute between 5 and 10% of global CH_4 emissions (UNEP, 2003). The role of arctic and subarctic wetlands as net sinks or sources of carbon (Fig. 8.7) is highly dependent on the seasonal water budget and levels; the brief and intense period of summer primary productivity (during which photosynthetic assimilation and respiration of CO_2, and bacterial metabolism and CH_4 generation, may be most active); soil type; active-layer depth; and extent of permafrost. Methane and CO_2 production can occur beneath the snowpack and ice of arctic wetlands. Winter and particularly spring emissions can account for a significant proportion of the annual total efflux of these gases (e.g., West Siberia; Panikov and Dedysh, 2000). Arctic wetlands typically represent net sources of carbon during spring melt and as plants senesce in autumn, shifting to net carbon sinks as leaf-out and growth progress (e.g., Aurela et al., 1998, 2001; Joabsson and Christiensen, 2001; Laurila et al., 2001; Nordstroem et al., 2001). The future status of wetlands as carbon sinks or sources will therefore depend on

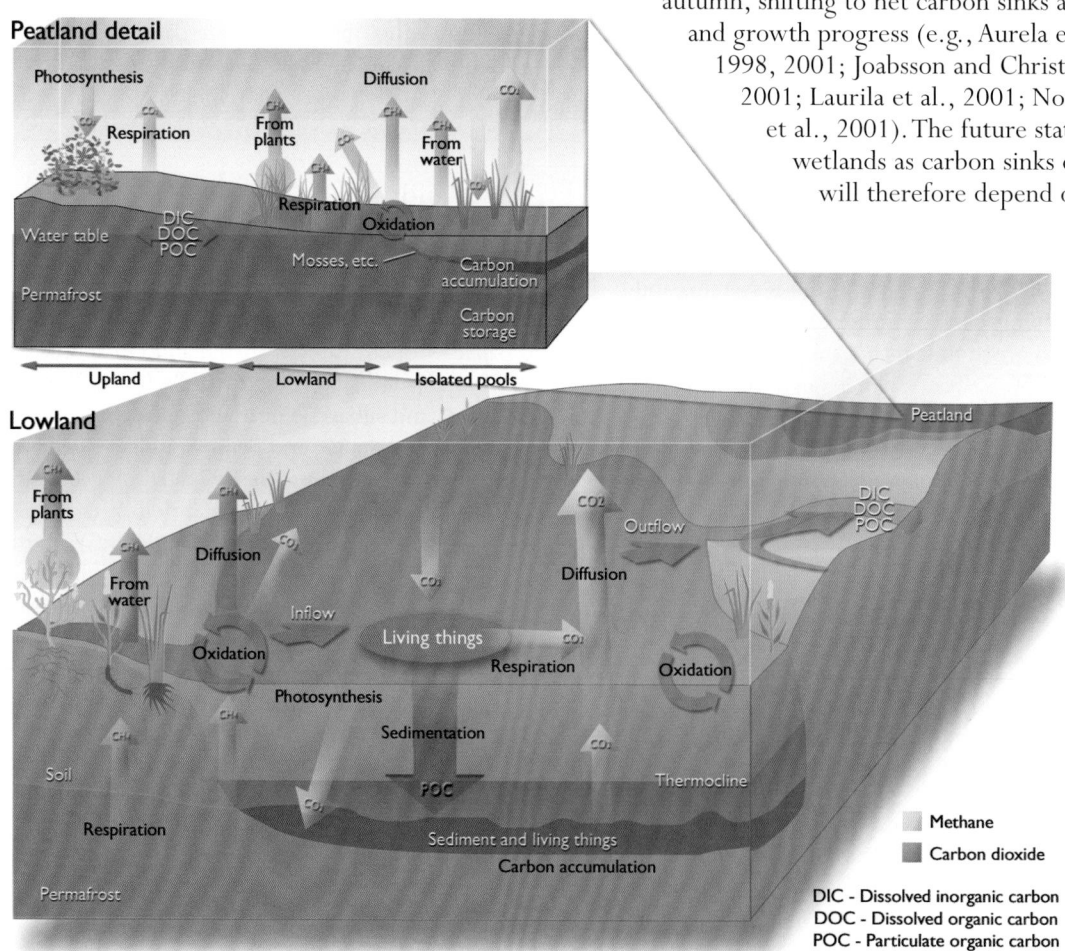

Fig. 8.7. Simplified schematic of carbon cycling in high-latitude aquatic ecosystems.

changes in vegetation, temperature, and soil conditions. Similarly, carbon cycling in lakes, ponds, and rivers will be sensitive to direct (e.g., rising temperatures affecting rates of carbon processing) and indirect (e.g., changes in catchments affecting carbon loading) effects of climate change. Section 7.5 provides a more detailed treatment of carbon cycling and dynamics in Arctic terrestrial and aquatic landscapes.

8.3. Historical changes in freshwater ecosystems

Analysis of the stability, sensitivity, rate, and mode of the response of freshwater ecosystems to past climate change has proven to be a valuable tool for determining the scope of potential responses to future climate changes. Preserved records of ecosystem variations (e.g., trees, fossils, and sedimentary deposits), combined with dating techniques such as carbon-14, lead-210, or ring/varve counting, have been a primary source of information for unraveling past environmental changes that pre-date the age of scientific monitoring and instrumental records. The application of climate change proxies in paleoclimatic analysis has traditionally relied on identification of systematic shifts in ecosystem patterns known from modern analogues or by comparison with independent instrumental or proxy climate records to determine perturbations in climate-driven environmental conditions such as growing-season length, solar insolation, temperature, humidity, ice-cover extent and duration, or hydrologic balance. Such ecosystem-based climate proxies may include the presence, distribution, or diversity of terrestrial, aquatic, or wetland species or assemblages; changes in water or nutrient balances recorded by chemical or isotopic changes; changes in growth rates or characteristics of individual plants and animals; or changes in physical environments (e.g., lake levels, dissolved oxygen content) that are known to be linked to the productivity and health of freshwater ecosystems.

The reliability of and confidence in these ecosystem indicators of climate change has been enhanced through development of spatial networks of paleo-climatic data, by comparison with instrumental climate records where available, and through concurrent examination of abiotic climate change proxies in nearby locations. Such abiotic proxy records include shifts in the isotopic composition of glacial deposits (and to some extent permafrost or pore water), which provide regional information about changes in origin, air-mass evolution, and condensation temperature of precipitation (or recharge); changes in summer melt characteristics of glacial deposits or sedimentary and geomorphological evidence such as the presence of laminated lake sediments (varves), the latter of which are indicative of water depths great enough to produce stratified water columns and meromixis; and variations in varve thicknesses in lakes and fining/coarsening sequences or paleoshoreline mapping that can be used to reconstruct shifts in lake or sea levels.

8.3.1. Ecosystem memory of climate change

The accumulation of ecosystem records of environmental change relies on the preservation of historical signals in ice caps, terrestrial deposits (soils, vegetation, permafrost), and aquatic deposits (wetlands, rivers, lakes, ice), coupled with methods for reconstructing the timing of deposition. As continuity of deposition and preservation potential are not equal in all environments, there is a systematic bias in the paleoclimatic record toward well-preserved lentic environments, and to a lesser extent wetlands, as compared to lotic systems. The following sections describe common archives and the basis of key memory mechanisms.

8.3.1.1. Lentic archives

Biological indicators of environmental change that are preserved in lake sediments include pollen and spores, plant macrofossils, charcoal, cyanobacteria, algae including diatoms, chrysophyte scales and cysts and other siliceous microfossils, biogenic silica content, algal morphological indicators, fossil pigments, bacteria, and invertebrate fossils such as Cladocera, chironomids and related Diptera, ostracods, and fish (Smol, 2002). In general, the best biological indicators are those with good preservation potential, for example, siliceous, chitinized, or (under neutral to alkaline pH conditions) carbonaceous body parts. They also must be readily identifiable in the sedimentary record, and exist within assemblages that have well-defined ecosystem optima or tolerances. Lentic records commonly extend back 6000 to 11 000 years to the time of deglaciation in the circumpolar Arctic.

In general, fossil pollen and spores, plant macrofossils, and charcoal are used to determine temporal shifts in terrestrial ecosystem boundaries, notably past fluctuations in northern treeline and fire history. Pollen and spores from emergent plants may also be useful indicators for the presence and extent of shallow-water environments. Preserved remains of aquatic organisms, such as algae and macrophytes, provide additional information on aquatic ecosystem characteristics and lake-level status. Such indicators, which are used to reconstruct ecological optima and tolerances for past conditions, are normally applied in conjunction with surface-sediment calibration datasets to quantitatively compare present-day ecosystem variables or assemblages with those preserved in the sediment record (Birks, 1995, 1998; Smol, 2002). Douglas and Smol (1999) provide details on the application of diatoms as environmental indicators in the high Arctic, and Smol and Cumming (2000) provided a general treatment of all algal indicators of climate change. Biological indicators useful for lake-level reconstructions include the ratio of planktonic to littoral Cladocera as an index of the relative size of the littoral zone or water depth of northern lakes (Korhola and Rautio, 2002; MacDonald G. et al., 2000a). Chironomids and diatoms may be used in a similar manner. While such information allows for quantitative reconstruction of lake levels, errors in projecting lake water depth from Cladocera, chironomids, and diatoms may be large (Korhola et al., 2000a; Moser et al., 2000; see also MacDonald G. et al., 2000a). Cladoceran remains may also provide evidence of changes in species trophic structure, including fish (Jeppesen et al., 2001a, b, 2003), and chironomids may be used to reconstruct changes in conductivity mediated by variations in runoff and evaporation (Ryves et al., 2002).

Due to their small volume and minimal capacity to buffer climate-driven changes, the shallow lakes and ponds characteristic of large parts of the Arctic may be well suited for hydrological and climate reconstructions. Past shifts in diatom assemblages have been used to track habitat availability for aquatic vegetation, the extent of open-water conditions, shifts in physical and chemical characteristics, and water levels (Moser et al., 2000).

Isotopic analysis (e.g., $\delta^{13}C$, $\delta^{18}O$, $\delta^{15}N$) of fossil material, bulk organic sediments, or components such as cellulose or lignin can provide additional quantitative information. For example, carbon and oxygen isotope analysis of sediment cellulose has been applied in many parts of the circumpolar Arctic (MacDonald G. et al., 2000a). It relies on the key assumptions that fine-grained cellulose in offshore sediments (excluding woody material, etc.) is derived from aquatic plants or algae and that the cellulose–water fractionation is constant (Wolfe B. et al., 2002). Often the source of material (aquatic versus terrestrial) can be confirmed from other tests such as carbon–nitrogen ratios (Wolfe B. et al., 2002), although these two assumptions may not be applicable in all arctic systems (Sauer et al., 2001). Under ideal conditions, the $\delta^{18}O$ signals in aquatic cellulose are exclusively inherited from the lake water and therefore record shifts in the water balance of the lake (i.e., input, through-flow, residency, and catchment runoff characteristics; Gibson J.J. et al., 2002). Studies of ice cores from Greenland and arctic islands support the interpretation of $\delta^{18}O$ signals and other climate proxies across the circumpolar Arctic (Smol and Cumming, 2000). Ice-core records of past precipitation ($\delta^{18}O$, δ^2H) can help to distinguish climatically and hydrologically driven changes observed in lake sediment records.

Stratigraphic reconstructions using $\delta^{13}C$, $\delta^{14}C$, or $\delta^{15}N$ measured in aquatic cellulose and fossil material can likewise be used to examine changes in ecosystem carbon and nitrogen cycles and ecosystem productivity. Trends in chemical parameters such as dissolved inorganic carbon (DIC), dissolved organic carbon (DOC), and total nitrogen can also be reconstructed from fossil diatom assemblages as demonstrated for lakes in the treeline region of the central Canadian Arctic (Rühland and Smol, 2002), Fennoscandia (Seppä and Weckström, 1999), and elsewhere.

While lakes are nearly ideal preservation environments, lake sediment records may not always offer unambiguous evidence of climate-induced ecosystem changes. Other factors not driven by climate, including selective preser-

vation of some organisms (Rautio et al., 2000), erosion or deepening of outlets, damming by peat accumulation, or subsequent permafrost development, can alter lake records (Edwards et al., 2000). Such problems are overcome to some extent by using multi-proxy approaches, by comparing multiple lake records, and by using spatial networks of archives. Further research on modern ecosystems, especially processes controlling the preservation and modification of proxy records, is still required in many cases to reconcile present and past conditions.

8.3.1.2. Lotic archives

Sedimentary deposits in lotic systems are often poorly preserved compared to lentic systems, owing to the relatively greater reworking of most riverine deposits. However, preservation of at least partial sediment records can occur in fluvial lakes, oxbow lakes, estuaries, and artificial reservoirs. Past river discharge can also be studied by tracking the abundance of lotic diatoms in the sediments of lake basins, such as demonstrated for a lake in the high Arctic (Douglas et al., 1996; Ludlam et al., 1996).

8.3.1.3. Terrestrial and wetland archives

Tree rings are a traditional source of climate change information, although there are obvious difficulties in applying the method to tundra environments with sparse vegetation. Conifers are, however, abundant within the circumpolar Arctic (particularly in northwestern Canada, Alaska, and Eurasia), with the northernmost conifers in the world located poleward of 72° N on the Taymir Peninsula, northern Siberia (Jacoby et al., 2000). Tree-ring widths increase in response to warm-season temperatures and precipitation/moisture status and have been used to reconstruct climate changes, in many cases for more than 400 years into the past (Jacoby et al., 2000; Overpeck et al., 1997).

Diatoms, chrysophytes, and other paleolimnological indicators are also preserved in peatlands and may be used to reconstruct peatland development and related water balance and climatic driving forces (Moser et al., 2000). Records of $\delta^{13}C$ and $\delta^{18}O$ from peat cellulose also provide information on climatic variability (Hong et al., 2001), although this method has not been widely applied to date in the Arctic. Selective use of pore water from within peat and permafrost has also been utilized to reconstruct the isotopic composition of past precipitation (Allen et al., 1988; Wolfe B. et al., 2000), although dating control is often imprecise.

8.3.2. Recent warming: climate change and freshwater ecosystem response during the Holocene

The climate of the earth has continuously varied since the maximum extent of ice sheets during the late Pleistocene (e.g., Gajewski et al., 2000). The most recent climate warming trend during the industrial period overprints Holocene climate shifts that have occurred due to orbit-induced variations in solar insolation, as well as oscillations produced by local to regional shifts in sea surface temperatures, atmospheric and oceanic circulation patterns, and the extent of land-ice cover (MacDonald G. et al., 2000a). During the early Holocene (10 000–8000 years BP), orbital variations (the Milankovitch (1941) theory of a 41 000-year cycle of variation in orbital obliquity) resulted in approximately 8% higher summer insolation and 8% lower winter insolation compared to present-day values poleward of 60° N (Kutzbach et al., 1993). This directly altered key factors controlling arctic freshwater systems, including precipitation, hydrology, and surface energy balance. Sea level was also 60 to 80 m below present-day levels, providing an expanded zone (up to several hundred kilometers wide) of nearshore freshwater environments.

During the Holocene, rapidly melting ice sheets presented a shrinking barrier to major airflows, and variations in insolation altered the spatial distribution of atmospheric heating (MacDonald G. et al., 2000a). Several climate heating episodes between 11 000 and 7700 years BP are attributed to the catastrophic drainage of Lake Agassiz and the Laurentide glacial lakes in North America. Paleogeographic data from this interval suggest that the Laurentide Ice Sheet was almost completely gone, with the possible exception of residual ice masses in northern Québec. In general, most of the Arctic experienced summers 1 to 2 °C warmer than today during the early to middle Holocene (Overpeck et al., 1997). A common assumption is that decreases in summer insolation resulted in cooler summers in the late Holocene, which culminated in the Little Ice Age (ca. 1600). This cooling trend ended sometime in the 18th century. Detailed reconstructions of climate and ecosystems in North America at 6000 years BP (Gajewski et al., 2000) confirm that the Holocene was also a time of increased moisture, resulting in the spread of peatlands. In the European Arctic, combined evidence from oxygen isotope and pollen-inferred precipitation records, cladoceran-inferred lake levels, diatom-inferred lake-water ionic strength, and elemental flux records of erosion intensity into lakes, all suggest more oceanic conditions in the region during the early part of the Holocene than today, with a shift towards drier conditions between approximately 6000 and 4500 years BP (Hammarlund et al., 2002; Korhola et al., 2002c). In the late Holocene, there has been a general tendency towards increased moisture, resulting in more effective peat formation (Korhola, 1995).

Despite pervasive orbit-driven forcings, climate changes during the Holocene varied significantly between regions of the Arctic due to differences in moisture sources (Overpeck et al., 1997). In general, arctic Europe, eastern Greenland, the Russian European North, and the North Atlantic were dominated by Atlantic moisture sources; Siberia was dominated by Nordic Seas moisture; Chukotka, the Bering Sea region, Alaska, and the western Canadian Arctic were dominated by Pacific moisture; and northeastern Canada, the

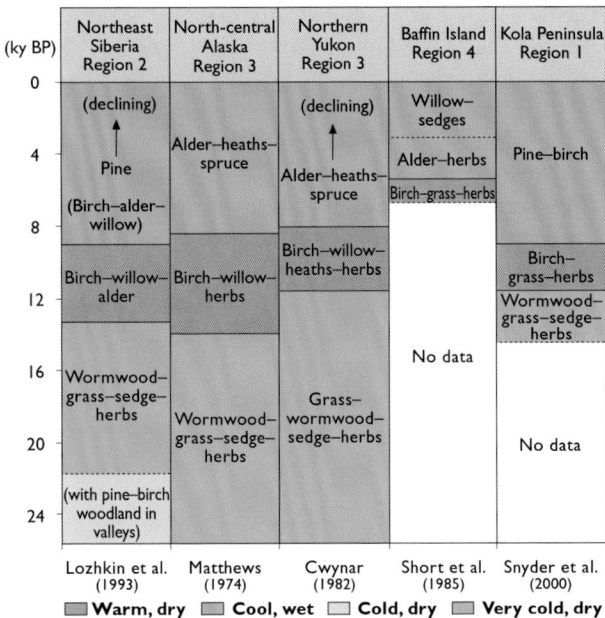

Fig. 8.8. Pollen record of regional arctic climate change (after I. Hutchinson, Simon Fraser University, British Columbia, pers. comm., 2004).

Labrador Sea and Davis Strait regions, and western Greenland were dominated by Labrador Sea and Atlantic moisture. The following sections describe significant regional differences in climate and ecosystem evolution during the Holocene (Fig. 8.8). Much of the subsequent discussion focuses on historical changes in hydroclimatology (e.g., atmospheric moisture sources) and terrestrial landscape features (e.g., vegetation) in the context of their primary control over the water cycle affecting freshwater ecosystems. More details about these changes can be found in sections 2.7 and 7.2.

8.3.2.1. Region 1: Arctic Europe, eastern Greenland, the Russian European North, and the North Atlantic

The present-day climate in northern Fennoscandia is dominated by westerly airflow that brings cyclonic rains to the area, especially during winter (see Seppä and Hammarlund, 2000). The Scandes Mountains of mid-central Sweden mark the boundary between oceanic climate conditions to the west and more continental conditions to the east, especially in northern Finland and Russia, which are strongly influenced by the Siberian high-pressure cell that allows easterly air flow into northern Fennoscandia during winter. Climate and freshwater ecosystem changes during the Holocene have been attributed largely to fluctuations in the prevailing air circulation patterns in the region. Pollen, diatom, chironomid, and oxygen isotope records from lake sediments have been used to reconstruct climate conditions and ecosystem responses during the Holocene (e.g., Korhola et al., 2000b, 2002c; Rosén et al., 2001; Seppä and Birks, 2002; Seppä and Hammarlund, 2000). These studies suggest that northern Fennoscandia was a sparse, treeless tundra environment following final disintegration of the Scandinavian Ice Sheet (10 000–9000 years BP) until birch (*Betula* spp.) forests spread to the shores of the Arctic Ocean and to an altitude of at least 400 m in the mountains between 9600 and 8300 years BP (see section 7.2 for discussion of changes in terrestrial vegetation). Increased moisture during this period has been attributed to strengthening of the Siberian High which may have enhanced sea-level pressure gradients between the continent and the Atlantic Ocean, strengthened the Icelandic Low, and produced greater penetration of westerly winter storms and increased snowfall over western Fennoscandia (see also Hammarlund and Edwards, 1998). Associated strengthening of westerlies and northward shifts in the Atlantic storm tracks may also have produced higher snowfall in Greenland during this period (MacDonald G. et al., 2000a).

The decline of birch forests was accompanied by rapid increases in pine (*Pinus* spp.) forests between 9200 to 8000 years BP in the extreme northeast and 7900 to 5500 years BP in the western and southwestern parts of the region, signaling a shift toward drier summers and increased seasonality (Seppä and Hammarlund, 2000). Pollen evidence suggests that the late-Holocene treeline retreat in northern Norway and Finland started about 5000 years BP, and included southward retreat of treeline species of both pine and birch, which were subsequently replaced by tundra vegetation. This retreat has been attributed to decreased summer insolation during the latter part of the Holocene. It has also been suggested that later snowmelt and cooler summers gradually favored birch at the expense of pine along the boreal treeline. Similar climate changes may explain peatland expansion in the late Holocene within both boreal and tundra ecozones (Seppä and Hammarlund, 2000).

The response of aquatic ecosystems to the climate-induced changes during the Holocene has been inferred from lake-sediment and peat stratigraphic records. The very dry period corresponding to shifts from birch to pine corresponds to increasing frequency of diatom and cladoceran taxa indicating lake-level reduction and vegetation overgrowth of numerous lakes (Seppä and Hammarlund, 2000). Likewise, diatom and cladoceran evidence suggest dry warm summers during the period dominated by pine (~7000–3500 years BP).

There have been a variety of recent quantitative reconstructions of Holocene changes and variability in climatic and environmental variables through analysis of isotopic records and sedimentary remains of pollen, diatom assemblages, and/or chironomid head capsules from arctic and subarctic lakes in northern Sweden (Bigler and Hall, 2002, 2003; Bigler et al., 2002, 2003; Korhola et al., 2000b, 2002c; Rosén et al., 2001; Seppä et al., 2002; Shemesh et al., 2001). Comparative analyses revealed that the timing and scale of development of historical biotic assemblages were attributable to local geology, site-specific processes such as vegetation development, climate, hydrological setting, and in-lake biogeochemical and ecological processes. Several general climate-related trends were deduced for the region: a decrease in the average annual temperature of approximately 2.5 to 4 °C from the early Holocene to the present; summer temperatures during the early Holocene that were 1.7 to 2.3 °C above present-day measurements; winter temperatures that were 1 to 3 °C warmer than at present during the early Holocene; and a decrease in lake-water pH since the early Holocene.

Collectively, proxy records for closed-basin (i.e., a basin that has very little continuous surface outflow so that water-level variations strongly mirror changes in precipitation or moisture status) lakes suggest that water levels were high during the early Holocene, declined during the mid-Holocene dry period (~6000–4000 years BP), and rose again during the latter part of the Holocene. During the culmination of the Holocene dry period, many shallower water bodies in this region decreased greatly in size or may have dried up entirely (Korhola and Weckström, 2005).

In contrast to often quite distinct changes in physical limnology, changes in chemical limnological conditions have been relatively moderate during lake development in the Fennoscandian Arctic and on the Kola Peninsula (Korhola and Weckström, 2005; Solovieva and Jones, 2002). Because of changing climate and successional changes in surrounding vegetation and soils, lakes close to the present treeline are typically characterized by a progressive decline in pH, alkalinity, and base cations, and a corresponding increase in DOC over the Holocene. In contrast, lakes in the barren arctic tundra at higher altitudes manifest remarkable chemical stability throughout the Holocene. Excluding the initial transient alkaline period following deglaciation evident at some sites, the long-term natural rate of pH decline in the arctic lakes of the region is estimated to be approximately 0.005 to 0.01 pH units per 100 yr. This is a generally lower rate than those of more southerly sites in boreal and temperate Fennoscandia, where rates between 0.01 and 0.03 pH units per 100 yr have been observed. No evidence of widespread recent "industrial acidification" is apparent from extensive paleolimnological assessments in arctic Europe (Korhola and Weckström, 2005; Korhola et al., 1999; Sorvari et al., 2002; Weckström et al., 2003). However, fine-resolution studies from a number of remote lakes in the region demonstrate that aquatic bio-assemblages have gone through distinct changes that parallel the post-19th century arctic temperature increase (Sorvari et al., 2002).

8.3.2.2. Region 2: Siberia

Siberian climate was affected by increased summer insolation between 10000 and 8000 years BP, which probably enhanced the seasonal contrast between summer and winter insolation and strengthened the Siberian High in winter and the Siberian Low in summer (Kutzbach et al., 1993). Following final disintegration of the Scandinavian Ice Sheet approximately 10000 to 9000 years BP, cool easterlies were replaced by predominantly westerly flows from the North Atlantic, which now could penetrate western Russia and Siberia (Wohlfarth et al., 2002). Warm, wet summers and cold, dry winters probably dominated the early to mid-Holocene, with more northerly Eurasian summer storm tracks, especially over Siberia (MacDonald G. et al., 2000a). Warm periods were generally characterized by warmer, wetter summer conditions rather than by pronounced changes in winter conditions, which remained cold and dry. Pollen reconstructions from peatlands across arctic Russia suggest that temperatures were 1 to 2 °C higher than at present during the late glacial–Holocene transition, which was the warmest period during the Holocene for sites in coastal and island areas. The warmest period of the Holocene for non-coastal areas (accompanied by significantly greater precipitation) occurred between 6000 and 4500 years BP, with notable secondary warming events occurring at about 3500 and 1000 years BP (Andreev and Klimanov, 2000).

Pollen evidence from permafrost and peat sequences suggests that boreal forest development commenced across northern Russia and Siberia by 10000 years BP, reached the current arctic coastline in most areas between 9000 and 7000 years BP, and retreated south to its present position by 4000 to 3000 years BP (MacDonald G. et al., 2000b). Early forests were dominated by birch, but larch (*Larix* spp.), with some spruce (*Picea* spp.) became prevalent between 8000 and 4000 years BP. The northward expansion of the forest was facilitated by increased solar insolation at the conclusion of the Scandinavian glaciation, and by higher temperatures at the treeline due to enhanced westerly airflow (MacDonald G. et al., 2000b). The eventual southward retreat of the treeline to its present-day position is likewise attributed to declining summer insolation towards the late Holocene, as well as cooler surface waters in the Norwegian, Greenland, and Barents Seas (MacDonald G. et al., 2000b).

Increases in precipitation in some portions of northern Russia occurred during the interval from 9000 to 7000 years BP, followed by gradual drying to 6000 years BP (Andreev and Klimanov, 2000; Wolfe B. et al., 2000). This has been attributed to strengthening of the sea-level pressure gradients that also affected climate and ecosystems in Region 1 at this time (see MacDonald G. et al., 2000a).

Northward migration of the treeline also had a systematic impact on the ecosystem characteristics of some Siberian lakes. For a lake in the Lena River area, diatom assemblages dated prior to treeline advance were found to be dominated by small benthic *Fragilaria* species, and diatom indicators also suggest high alkalinity and low productivity at this time. Following the treeline advance, lakes shifted to stable diatom assemblages dominated by *Achnanthes* species and low alkalinity, attributed to the influence of organic runoff from a forested landscape. Re-establishment of *Fragilaria*-dominated assemblages and higher alkalinity conditions accompanied the subsequent reversion to shrub tundra. Laing et al. (1999) attributed recent declines in alkalinity and minor changes in diatom assemblages to the influx of humic substances from catchment peatlands.

8.3.2.3. Region 3: Chukotka, the Bering Sea, Alaska, and western Arctic Canada

The Laurentide Ice Sheet strongly influenced early Holocene (10000–9000 years BP) climate in northwestern North America, particularly in downwind areas. High albedos, cold surface conditions, and ice-sheet height apparently disrupted westerly airflows (or may possibly have maintained a stationary surface high-pressure cell with anticyclonic circulation), which promoted the penetration of dry, warm air from the southeast (MacDonald G. et al., 2000a). Dry conditions were also prevalent at this time in unglaciated areas such as northwestern Alaska and portions of the Yukon, where a 60 to 80 m reduction in sea level increased distances to marine moisture sources by several hundred kilometers. Biological indicators from Alaskan lakes suggest dry, more productive conditions, with lower lake levels between 11000 and 8000 years BP, followed by a gradual shift to modern moisture levels by 6000 years BP (Barber and Finney, 2000; Edwards et al., 2000).

Terrestrial vegetation (and the northern treeline) clearly indicates a warmer-than-present early Holocene (e.g., Spear, 1993). Vegetation shifts reconstructed mainly from fossil pollen evidence reveal the northward advance and southward retreat of the boreal forest in western North America, which has been attributed mainly to short-term changes in atmospheric circulation and associated storm tracks (i.e., shifts in the mean summer position of the Arctic Frontal Zone; MacDonald G. et al., 1993). Higher temperatures and increased moisture during the mid-Holocene (especially between about 5000 and 3500 years BP) also produced episodes of 250 to 300 km northward advances of the treeline that are recorded in the isotopic, geochemical, diatom, and fossil-pollen records of lakes near present-day treeline in the Yellowknife area of Canada (MacDonald G. et al., 1993). Additional evidence for significant changes in diatom community structure (shifts from planktonic to benthic forms) and increased productivity is recorded in lake sediments during this mid-Holocene warming interval (ca. 6000 to 5000 years BP) in the central Canadian subarctic (Pienitz and Vincent, 2000; Wolfe B. et al., 2002). This period was also accompanied by significant increases in DOC in lakes, lower water transparency, and less exposure to photosynthetically active radiation (PAR) and UV radiation in the water column. On the Tuktoyaktuk Peninsula (near the Mackenzie Delta), forest limits were at least 70 km north of the current treeline between 9500 and 5000 years BP (e.g., Spear, 1993). Permafrost zones were also presumably located north of their present-day distribution during this period. In general, present-day forest types were established in Alaska by 6000 years BP and northwestern Canada by approximately 5000 years BP.

Prolonged development and expansion of peatlands in North America commencing between 8000 and 6000 years BP have been attributed to progressive solar-insolation driven moisture increases towards the late Holocene (MacDonald G. et al., 2000a).

Box 8.1. Northern Québec and Labrador: Long-term climate stability

Northern Québec and Labrador in the eastern Canadian Low Arctic is a landscape dominated by lakes, wetlands, and streams. Few studies have addressed the effects of recent climate change in this region, but a variety of paleolimnological studies have provided insights into long-term change.

This region is comprised of four terrestrial ecozones: "taiga shield" with bands and patches of wetland forest, forest-tundra, and lichen-woodland vegetation; "southern arctic" consisting of shrub tundra; "northern arctic" consisting of true tundra; and "arctic cordillera" (the Torngat Mountains). Major changes in the chemical and biological characteristics of lakes are associated with the transition between these different vegetation types, notably changes in alkalinity and, in particular, colored dissolved organic matter (CDOM; see section 8.6.1, Box 8.10). The darkest water color (greatest CDOM concentrations) is associated with the heavily vegetated wetland-forest catchments, while much lower values occur in lakes completely surrounded by drier lichen woodland and tundra (see figure). As dissolved organic matter has a broad range of effects on high-latitude aquatic ecosystems (Vincent and Pienitz, 1996), this implies that any climate-related shift in catchment vegetation is very likely to have major impacts on the limnology of these eastern Canadian lakes.

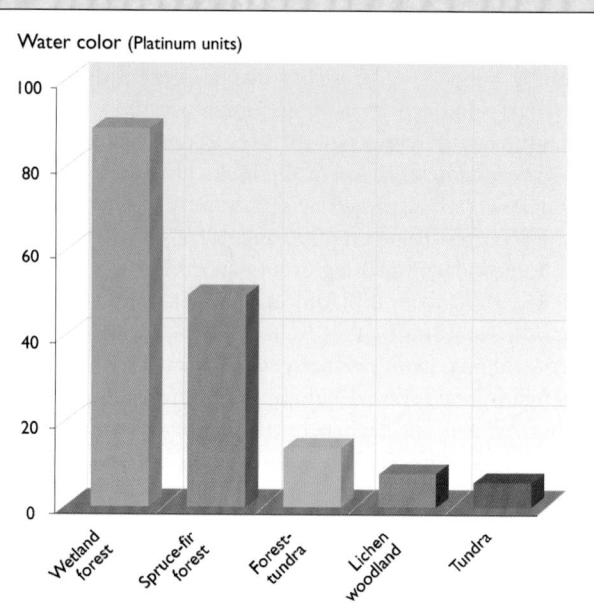

Colored dissolved organic matter (measured by relative water color) in lakes from different vegetation zones in Labrador. Values shown represent the mean of between 7 and 20 lakes (data from Fallu et al., 2002).

In the western Canadian subarctic, variations in climate over the last 5000 years caused large shifts in the position of the northern treeline. These shifts resulted in substantial changes in water color, the underwater light regime (including biologically damaging UV radiation exposure), and in the relative importance of benthic versus pelagic production (as indicated by their diatom communities; MacDonald G. et al., 1993; Pienitz and Vincent, 2000; Pienitz et al., 1999). In contrast, the eastern Canadian subarctic seems to have experienced relatively little change in vegetation structure at the millennial timescale. The stable forest–tundra of northern Québec and Labrador has been referred to as "an ecological museum" (Payette et al., 2001), and is partly a consequence of the extremely slow pace of northward migration and colonization by black spruce (*Picea mariana*) during periods of warming. This region also appears to be less prone to temperature change owing to the strong marine influence of the Hudson Strait and Labrador Current outflows from the Arctic. Like much of the eastern Canadian Arctic and southern Greenland, this region has shown little temperature change or even cooling from the mid-1960s to the mid-1990s, while most other sectors of the circumpolar Arctic have shown strong warming trends over the same period (Capellen and Vraae Jørgensen, 2001; Weller, 1998).

The paleolimnology of the region also reflects this long-term stability in climate and vegetation. Diatom assemblages in the sediment cores taken from Labrador lakes show very little change over the last 200 years, while there were major shifts in community structure elsewhere in the Arctic (Laing et al., 2002). Similarly, in a lake at the northern Québec treeline, CDOM and other inferred limnological variables remained relatively stable over the last 3000 years. For example, the mean inferred DOC concentration over this period is 5 mg/l, with a coefficient of variation for 107 strata of only 8% (Ponader et al., 2002). Coastal lakes in this region exhibited major changes associated with plant succession immediately after deglaciation, and isostatic uplift causing the severing of connections with the sea, but relatively constant conditions over the last 1000 years (Pienitz et al., 1991; Saulnier-Talbot and Pienitz, 2001; Saulnier-Talbot et al., 2003).

This remarkable stability at timescales of decades, centuries, and millennia suggests that northern Québec and Labrador lakes are likely to experience less short-term climate change relative to other regions of the circumpolar Arctic. Projections from the ACIA-designated climate models, however, suggest that these east–west differences in climate change will largely disappear by 2080 (see section 4.4.2), by which time the lakes, rivers, and wetlands of northern Québec and Labrador will begin experiencing the climate impacts that are well advanced in other regions.

8.3.2.4. Region 4: Northeastern Canada, Labrador Sea, Davis Strait, and West Greenland

The climate change and ecosystem response in northeastern Canada is distinguished from northwestern regions by generally colder conditions during the early Holocene because of delayed melting of ice sheet remnants until close to 6000 years BP. Consequently, tundra and taiga with abundant alder (*Alnus* spp.) covered more of Labrador and northern Québec than at present (Gajewski et al., 2000). Shrub tundra and open boreal forest were also denser than at present (Gajewski et al., 2000). Warming in the eastern Arctic reached a maximum shortly after 6000 years BP, with higher sea surface temperatures and decreased sea-ice extent. Peatland expansion was apparently similar to that of northwestern Canada, and is likewise attributed to insolation-driven increases in moisture and cooler conditions in the late Holocene. In contrast with other parts of the circumpolar north, this region has had a relatively stable climate over the last few thousand years but may experience significant temperature increases in the future (see Box 8.1 and projections of future temperature increases in section 4.4.2).

8.3.3. Climate change and freshwater ecosystem response during the Industrial Period

An abrupt shift in the rate of climate and related ecosystem changes occurred around 1840 that distinguish these impacts from those due to climate change observed during the preceding part of the Holocene. A compilation of paleoclimate records from lake sediments, trees, glaciers, and marine sediments (Overpeck et al., 1997) suggests that in the period following the Little Ice Age (~1840 until the mid-20th century), the circumpolar Arctic experienced unprecedented warming to the highest temperatures of the preceding 400 years. The last few decades of the 20th century have seen some of the warmest periods recorded, thus continuing this early industrial trend. The effects of this increase in temperature include glacial retreat, thawing permafrost, melting sea ice, and alteration of terrestrial and aquatic ecosystems. These climate changes are attributed to increased atmospheric concentrations of greenhouse gases, and to a lesser extent shifts in solar irradiance, decreased volcanic activity, and internal climate feedbacks. Examples of profound responses to the recent temperature increases relevant to freshwater systems are numerous. Selected examples are given below.

Sorvari and Korhola (1998) studied the recent (~150 years) environmental history of Lake Saanajärvi, located in the barren tundra at an elevation of 679 m in the northwestern part of Finnish Lapland. They found distinctive changes in diatom community composition with increasing occurrences of small planktonic diatoms starting about 100 years ago. Since no changes in lake water pH were observed, and because both airborne pollution and catchment disturbances are known to be almost nonexistent in the region, they postulated that recent arctic temperature increases are the main reason for the observed ecological change.

To further test the hypothesis that temperature increases drove the system, Korhola et al. (2002b) analyzed additional sedimentary proxy indicators from Lake Saanajärvi. The biological and sedimentological records were contrasted with a 200-year climate record specifically reconstructed for the region using a compilation of measured meteorological data and various proxy sources. They found synchronous changes in lake biota and sedimentological parameters that seemed to occur in parallel with the increasing mean annual and summer temperatures starting around the 1850s, and hypothesized that the rising temperature had increased the metalimnion steepness and thermal stability in the lake, which in turn supported increasing productivity by creating more suitable conditions for the growth of plankton.

Sorvari et al. (2002), using high-resolution (3–10 yr) paleolimnological data from five remote and unpolluted lakes in Finnish Lapland, found a distinct change in diatom assemblages that parallels the post-19th century arctic temperature increase detected by examination of regional long-term instrumental data, historical records of ice cover, and tree-ring measurements. The change was predominantly from benthos to plankton and affected the overall diatom richness. A particularly strong relationship was found between spring temperatures and the compositional structure of diatoms as summarized by principal components analysis. The mechanism behind the change is most probably associated with decreased ice-cover duration, increased thermal stability, and resultant changes in internal nutrient dynamics.

Douglas et al. (1994), using diatom indicators in shallow ponds of the high Arctic (Ellesmere Island), found relatively stable diatom populations over the past 8000 years, but striking successional changes over the past 200 years. These changes probably indicate a temperature increase leading to decreased ice- and snow-cover duration and a longer growing season (Douglas et al., 1994). Although temperature changes are difficult to assess, they were sufficient to change the pond communities. In these ponds, there are no diatoms in the plankton; however, a shift occurred from a low-diversity, perilithic (attached to rock substrates) diatom community to a more diverse periphytic (attached to plants) community living on mosses.

Diatom indicators in lakes also show shifts in assemblages, most likely caused by temperature increases over the past 150 years. Rühland et al. (2003) documented changes in 50 lakes in western Canada between 62° and 67° N, spanning the treeline. Shifts in diatoms from *Fragilaria* forms to a high abundance of the planktonic *Cyclotella* forms are consistent "with a shorter duration of ice cover, a longer growing season, and/or stronger thermal stratification patterns", such as a shift from unstratified to stratified conditions.

Char Lake on Cornwallis Island, Canada (74° N) is the most-studied high-arctic lake. Recent studies (Michelutti et al., 2002) show no change in water quality over time but do show a subtle shift in diatom assemblages as evidenced in the paleolimnological record. These changes are consistent with recent climate changes (1988–1997) and are probably a result of "reduced summer ice cover and a longer growing season". Section 6.7.2 reviews recently documented observations of the general, although not ubiquitous, decline in the duration of lake and river ice cover in the Arctic and subarctic.

Chrysophyte microfossils show changes that parallel diatom changes (Wolfe A. and Perren, 2001) and are also probably related to reduced ice-cover duration. For example, chrysophyte microfossils were absent or rare in Sawtooth Lake (Ellesmere Island, 79° N) over the past 2500 years but suddenly became abundant 80 years ago. Similarly, in Kekerturnak Lake (Baffin Island, 68° N), planktonic chrysophytes increased greatly in the upper sediments dated to the latter part of the 20th century. In contrast, lakes in regions without temperature increases show no change in sediment-based indicators. For example, Paterson et al. (2003) found no change in chrysophyte and diatom indicators over the past 150 years in the sediment of Saglek Lake (northern Labrador, Canada).

This section has provided documented changes in some aspects of freshwater ecosystems driven by climate shifts in the recent past. As such, it provides a baseline against which future effects of climate change can be both projected and measured.

8.4. Climate change effects

8.4.1. Broad-scale effects on freshwater systems

Arctic freshwater systems are particularly sensitive to climate change because numerous hydro-ecological processes respond to even small changes in climate. These processes may adjust gradually to changes in climate, or abruptly as environmental or ecosystem thresholds are exceeded (Box 8.2). This is especially the case for cryo-

Box 8.2. Thresholds of response: step changes in freshwater systems induced by climate change

Some climate change effects projected for arctic freshwaters are likely to result in small, slow responses in the environment; other changes are likely to exceed environmental or ecosystem thresholds and cause a dramatic switch in organisms or a change of state of the system. Thresholds may be physical (e.g., permafrost is likely to begin to slowly thaw when the mean annual air temperature approaches 0 °C); chemical (e.g., the bottom waters of a lake are likely to lose all oxygen when lake productivity increases or allochthonous carbon increases); or biological (e.g., insect larvae frozen in the bottom of tundra ponds will die when their temperature falls below -18 °C; Scholander et al., 1953). While thresholds are only a part of the whole picture of response, they are critical to the understanding and assessment of the full scope of climate change impacts.

One obvious physical threshold is the amount of heat necessary to melt the ice cover of a lake. At the northern limit of the terrestrial Arctic, such as on Ellesmere Island, there are lakes that have only recently begun to have open water during the summer; other lakes now have open water for more summers every decade than in the past. Sediment records of algae, in particular diatom species and chrysophyte abundance, show that in Finland and northern Canada, lakes were ice-bound for thousands of years but conditions began to change about 150 years ago. With open water, the algal community shifts from a predominance of benthic diatoms to planktonic forms, and chrysophytes begin to occur. This is the result of an increase in summer air temperatures caused by climate change and the earlier onset of melt. Some scientists believe that the likely increase in growing-season length may also be important in controlling algal species. Shifts in species composition at the level of primary producers are also likely to have consequences for higher trophic levels through the alteration of food pathways. This could possibly lead to local extirpation of benthic and planktonic animals, as well as overall shifts in productivity.

Another physical threshold is the onset of stratification in lakes. Once lakes begin to have open water, wind-driven water circulation becomes one of the controls of biological processes. Almost all lakes have a period of complete mixing of the water column immediately after the ice cover disappears. Very cold waters may continue to circulate for the entire summer so that each day algae spend a significant amount of time in deep waters where there is not enough light for growth. When a lake stratifies (i.e., when only the uppermost waters mix), algae have better light conditions and primary production increases. The higher temperatures in the upper waters increase the rates of all biotic processes. There is a threshold, probably tied to increased primary production, when entirely new trophic levels appear. For example, the sediment record from a lake in Finland shows that Cladocera, a type of zooplankton, began to appear around 150 years ago. Most lakes in the Arctic already exhibit summer stratification, so this threshold will apply mostly to lakes in the far north.

When air temperatures increase above a mean annual air temperature of -2 °C, permafrost begins to thaw. When the upper layers of ice-rich permafrost thaw, the soil is disturbed; lakes may drain, and ponds form in spheric components that significantly affect the water cycle of lakes, rivers, and ponds; the habitat characteristics of these freshwater systems; and the flora and fauna that occupy them. In the case of large arctic rivers (e.g., the Lena, Mackenzie, Ob, and Yenisey), the effects of climate change must be evaluated for areas outside of as well as within the Arctic. The dynamics of such large systems depend on hydrologic processes prevailing within their water-rich headwaters in more temperate southern latitudes. In addition, many of these headwater areas are regulated in some way, a factor that may interact in some way with downstream arctic climate change impacts.

Prior to considering the specific effects of climate change on arctic freshwater systems, it is useful to place the climate projections generated by the five ACIA-designated atmosphere–ocean general circulation models (AOGCMs) for the Arctic as a whole into a more suitable freshwater context. For the most part, this requires focusing on model projections for the major arctic terrestrial landscapes, including some extra-arctic headwater areas, since these are the domains of freshwater

systems. The following paragraphs review the ACIA-designated model projections (primarily for the final time slice, 2071–2090, to illustrate the most pronounced changes) and, through additional processing of the model projections, provide a perspective on how such changes may be important to broad-scale features of arctic freshwater ecosystems, and a background template for the subsequent discussions of specific effects.

For the area north of 60° N, the five ACIA-designated models project that the mean annual temperature will increase by 3.7 °C (five-model average) between the 1981–2000 baseline and the 2071–2090 time slice, or approximately twice the projected increase in global mean annual temperature (section 4.4.2). At a global scale, AOGCMs used in the Third Assessment Report of the Intergovernmental Panel on Climate Change (IPCC, 2001b) project that it is very likely that nearly all land areas, which include freshwater systems, will warm more rapidly than the global average, particularly during the cold season at northern high latitudes. Within the Arctic, the spatial distribution of the projected tempera-

depressions. In eastern Siberia, newly thawed soils that are rich in organic matter slump into lakes. Microbial action depletes the oxygen in the lake allowing the bacteria to produce so much methane that the lakes and ponds become a significant source of this greenhouse gas, and enhance an important feedback to the climate system. This threshold is likely to affect lakes in the more southerly regions of the Arctic.

It is well known that lakes surrounded by shrubs and trees contain much more colored dissolved organic matter (CDOM; see section 8.6.1, Box 8.10) than lakes in the tundra zone. The CDOM comes from the organic matter produced by plants and modified by soil microbes. It strongly absorbs light, such that the algae of the upper waters become light limited and primary production is reduced. This may also be accompanied by a shift towards increasing primary production by attached algae in the shallow inshore zone relative to offshore planktonic production. The threshold described here is related to the treeline, often demarcated by patches rather than a continuous zone of vegetation. As air temperatures increase over the 21st century these patches are likely to expand, fuse, and move further north, resulting in a slowly moving band of affected lakes. Extreme polar-desert catchments in the high Arctic are very likely to experience their first arrival of higher plants, and a sudden increase in the transfer of organic materials from land to water.

As lakes warm, some species or populations of species will probably reach a temperature threshold for survival. This threshold is linked to increases in the rate of metabolism and growth. For example, a bioenergetic model based on laboratory studies projects that the young-of-the-year lake trout in northern Alaska will not obtain enough food for growth if their metabolic rate rises in response to a temperature increase of a degree or so. Evidence from field studies of a stream fish, the Arctic grayling (*Thymallus arcticus*), also suggests that a population at the northern limit of its distribution is unlikely to survive an increase of only a few degrees in summer water temperatures. Both these examples are of fish species at the northern limit of their distribution. Other types of widely distributed fish, such as whitefish and Arctic char, are less likely to be affected unless new competing species arrive from southerly regions.

A different type of threshold involves a shift in the ecological behavior of migratory fish such as the Arctic char. In many arctic rivers, char migrate to the sea for some months every year; the productive marine food web allows them to grow to a large size. Local fishers harvest many of these sea-run fish in the rivers each time the fish congregate and migrate. When freshwaters become more productive, migrations to the sea are projected to decrease and may eventually cease, thus the char are likely to remain in freshwater rivers and lakes for the entire year. These freshwaters are much less productive than the marine ecosystems; therefore, adult freshwater char are likely to be much smaller than are migrating char. As a result of climate change, a valuable food resource for arctic peoples is likely to change with respect to sizes available, and could possibly be lost.

ture increases in terrestrial areas is associated with even greater projected temperature increases over the central Arctic Ocean. For example, the five-model average projects that autumn (October–December) temperatures over large areas of the Arctic Ocean will increase by up to 9 °C by 2071–2090 compared to the 1981–2000 baseline (section 4.4.2). Adjacent to the Arctic Ocean, the models project substantial temperature increases for extensive terrestrial areas, with the largest projected temperature increases closest to the coastal margins and decreasing to the south.

This pattern of temperature increases is likely to have serious implications for high-latitude coastal areas such as the Russian polar desert and northern tundra, where

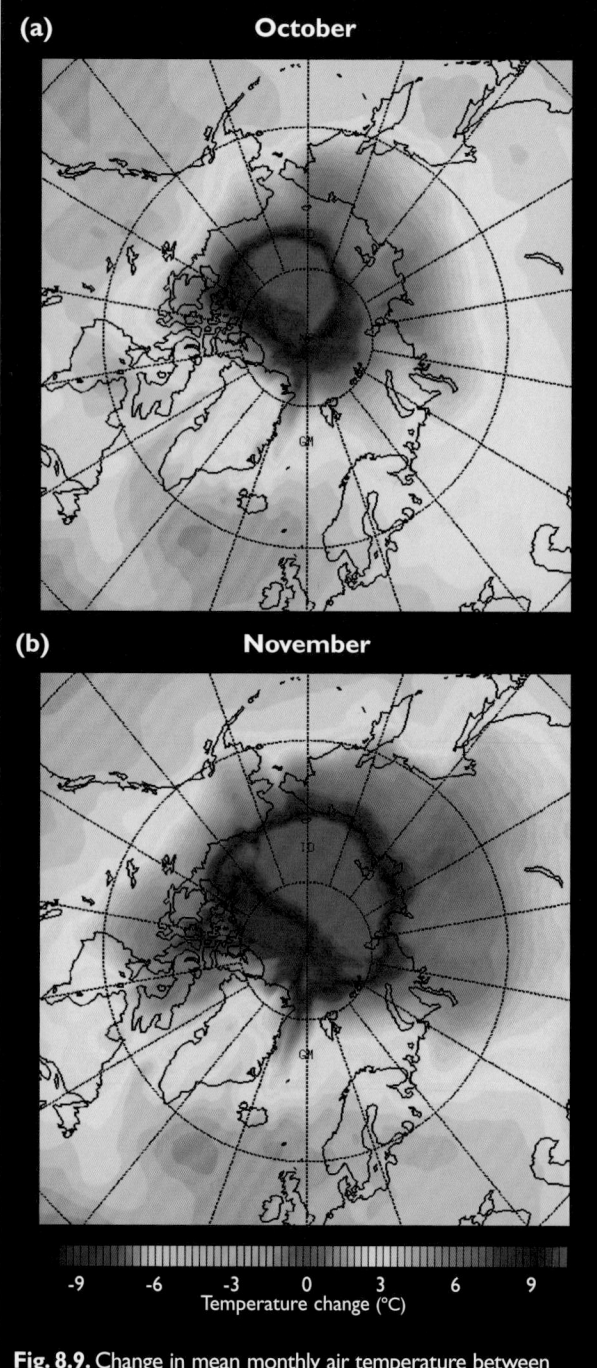

Fig. 8.9. Change in mean monthly air temperature between 1981–2000 and 2071–2090 projected by the ACIA-designated models (five-model average) for (a) October and (b) November.

temperature and associated species distribution gradients are steep (e.g., vascular species abundance increases five-fold from north to south on the Taymir Peninsula; see section 7.3.1.1). Figure 8.9a shows the spatial pattern of October warming projected for the 2071–2090 time slice. (Note the spatial congruence of warming between the ocean and the adjacent arctic coastal zone and the extension to more southerly latitudes.) Areas where projected temperature increases are particularly pronounced include northern Siberia and the western portions of the Canadian Archipelago. Notably, however, the maximum projected air temperature increases in these areas are about 5 °C (greatest near the coasts), compared to the almost two-fold greater projected increases in temperature over the Arctic Ocean. Such pronounced potential temperature increases in freshwater systems in October are particularly important because this is typically the time when freshwater lake and river systems along the coastal margins currently experience freeze-up. Employing a typical rate of change for freeze-up of 1 day per 0.2 °C increase in temperature (Magnuson et al., 2000), the projected temperature increases could cause delays of up to 25 days in freeze-up by 2071–2090. This is likely to have the greatest effect on higher-latitude, near-coastal freshwater systems (see also sections 8.4.3.1 and 6.7.3).

Even more dramatic temperature increases are projected for coastal land areas in November (Fig. 8.9b). Significant temperature increases are projected for most coastal areas in Region 3 and more southerly latitudes in Region 2, including the headwater regions of the major Siberian Arctic rivers below 60° N. Latitudinal gradients of temperature increases are especially important for arctic freshwater systems because of the influence of extra-arctic basins on the timing and magnitude of flow in the major northward-flowing arctic rivers. In the case of Region 2, projected temperature increases in November south of 60° N are significant because this is typically the month that marks the beginning of major snow accumulation. Similar to the delay in freeze-up, such higher temperatures would effectively decrease the length of time available to accumulate a winter snowpack. This would subsequently be reflected in the magnitude of the spring snowmelt that forms the major hydrologic event of the year at northern latitudes and is known to significantly affect downstream arctic river and delta systems. The effect of a reduced period of winter snow accumulation on the freshet magnitude, however, is likely to be offset by the projected increase in winter precipitation. The terrestrial regions of North America and Eurasia are among the areas with the greatest projected precipitation increases; similar to temperature, the largest increases are projected for autumn and winter (section 4.4.3). Although caution must be used in interpreting regional trends from the simulated precipitation patterns because of large variations in model projections, the average of the five ACIA-designated model projections also shows winter increases in precipitation for the extra-arctic headwater regions of the large northern rivers (Fig. 8.10). The degree to which this would compensate for the reduced duration of winter snow accumulation, however, requires detailed regional analysis.

Over the terrestrial regions of the northern latitudes, it is the cold season (defined here as October to May, the current period of dominant snow and ice cover for freshwater systems) that is characterized by the steepest latitudinal gradients in projected temperature increases. Figure 8.11 displays the projected changes in average temperature (from the ACIA 1981–2000 baseline) over terrestrial areas for the four ACIA regions broken into three latitudinal bands of 70°–85°, 60°–70°, and 50°–60° N. The latter represents the zones of higher precipitation that feed the major arctic rivers in Region 2 (Lena, Ob, and Yenisey) and eastern Region 3 plus western Region 4 (Mackenzie River). The steepest latitudinal gradients in projected cold-season temperature increases are evident in Regions 1 and 3, becoming particularly magnified in the latter by 2071–2090, whereas Region 4 shows a slight decrease in the level of warming with latitude. Hence, except for Region 4, it appears that with continued temperature increases the higher-latitude zones will continue to experience the relatively highest degree of warming. This would lead to a reduction in the thermal gradient along the course of some of the major arctic rivers. If such reductions prevail during particular parts of the cold season, they are likely to have major implications for the dynamics of particular hydrologic events such as the spring freshet and ice breakup.

In general, the most severe spring floods on cold-regions rivers are associated with a strong climatic gradient between the headwaters and the downstream reaches –

typically from south to north on most large arctic rivers (e.g., Gray and Prowse, 1993). In such cases, the spring flood wave produced by snowmelt must "push" downstream into colder conditions, and hence towards a relatively competent ice cover that has experienced little thermal decay. Changes in the strength of this climatic gradient would alter the severity of breakup and the associated flooding. Figure 8.12a illustrates the change in average air temperature projected for April 2071–2090. This is currently the month of freshet initiation, with May the primary month of freshet advance, in the southern headwaters of the major arctic rivers. With projected advances in the timing of ice and flow conditions (see also section 6.7.3), April should become the primary month of freshet advance by 2071–2090. Of particular note in Fig. 8.12a are the substantial projected temperature increases in the downstream areas of the major Russian rivers. Such high-latitude temperature increases are likely to lead to less severe ice breakups and flooding as the spring flood wave pushes northward. A comparable degree of high-latitude temperature increase is absent for the Mackenzie River. Of additional note in Fig. 8.12 is the degree of warming projected to occur in the headwater regions of the three large Russian

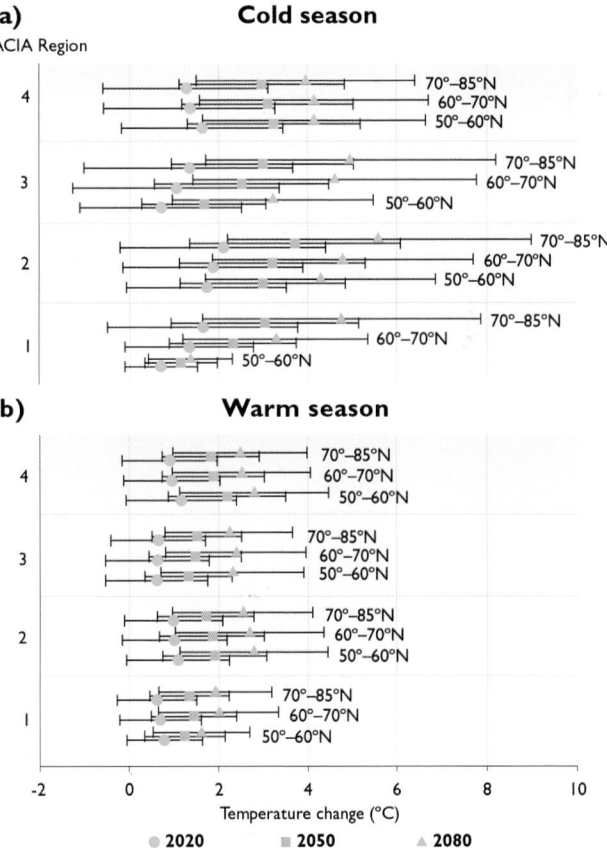

(a)   **Cold season**

ACIA Region

(b)   **Warm season**

Temperature change (°C)

● 2020   ■ 2050   ▲ 2080

Fig. 8.11. Changes in mean air temperature projected by the ACIA-designated models (five-model average) for the land areas of the four ACIA regions at three time slices in three latitudinal bands. Error bars represent standard deviation from the mean. The (a) cold season (October–May) and (b) warm season (June–September) were divided based on approximate ice-covered and open-water conditions prevailing in the current climate for major freshwater systems located in the 60°–70° N band. Longer (shorter) duration ice-covered periods prevail in the more northerly (southerly) latitudinal band.

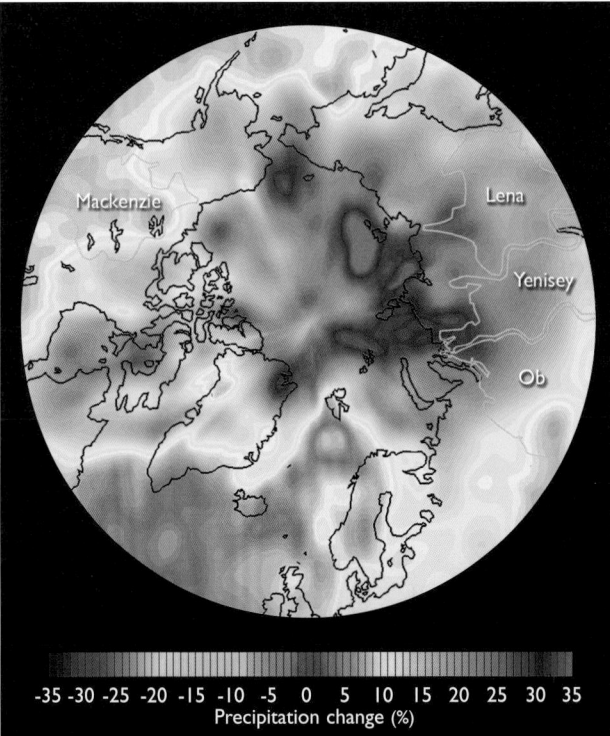

-35 -30 -25 -20 -15 -10 -5 0 5 10 15 20 25 30 35
Precipitation change (%)

Fig. 8.10. Change in November to April precipitation between 1981–2000 and 2071–2090 projected by the ACIA-designated models (five-model average). The basins of four major arctic rivers are also shown.

rivers during April and March. This is very likely to result in an early onset of snowmelt along these rivers. Again, however, a comparable degree of headwater warming and hence snowmelt runoff is not projected for the Mackenzie Basin. This regional dichotomy is likely to produce future differences in the spring timing of lake- and river-ice breakup and associated freshet, including the ultimate export of freshwater to the Arctic Ocean.

Although the smallest temperature increases are projected for the open-water warm season in all regions and at all latitudes (Fig. 8.11b), even the projected ~1 to 3 °C temperature increase is likely to significantly increase evaporative losses from freshwater systems, especially

with a shortened ice season, and via evapotranspiration from the terrestrial landscape that feeds them. Similar to changes in winter snowpack, increases in precipitation could offset such temperature-induced evaporative losses, but the five-model average projects that precipitation increases will be smallest during the summer. More detailed consideration of the changes and effects on cold-regions hydrology that could result from the changes in climate projected by the ACIA-designated models are provided in subsequent sections and in Chapter 6.

8.4.2. Effects on hydro-ecology of contributing basins

The regional patterns of projected changes in temperature and precipitation reviewed in the previous section are useful to understand some of the broad-scale effects that may occur. Specific effects, however, will be much more diverse and complex, even within regions of similar temperature and precipitation changes, because of intra-regional heterogeneity in freshwater systems and the surrounding landscapes that affect them. For example, elevational difference is one physical factor that will produce a complex altered pattern of snow storage and runoff. Although warmer conditions are very likely to reduce the length of winter, snow accumulation could either decrease or increase, with the latter most likely to occur in higher-elevation zones where enhanced storm activity combined with orographic effects will probably increase winter snowfall. Increased accumulation is likely to be most pronounced at very high elevations above the elevated freezing level, where the summer season is likely to remain devoid of major melt events, thereby creating the conditions for the preservation of more semi-permanent snowpacks at high altitudes (Woo, 1996). In contrast, temperature increases at lower elevations, especially in the more temperate maritime zones, are likely to increase rainfall and rain-on-snow runoff events. Snow patterns will be affected by a number of other factors, including vegetation, which is also projected to be altered by climate change (section 7.5.3.2). For example, shifts from tundra vegetation to trees have led to greater snow interception and subsequent losses through sublimation (e.g., Pomeroy et al., 1993), whereas shifts from tundra to shrubs have been shown to reduce snow losses (Liston et al., 2002), thereby affecting the magnitude of the snowpack available for spring melt.

An advance of the spring warming period means that snowmelt will occur during a period of lower insolation, which, other things being equal, will lead to a more protracted melt and less intense runoff. Traditional ecological knowledge indicates that through much of northern Canada, including the western Canadian Arctic and Nunavut, spring melt is already occurring earlier than in the past, and spring air temperatures are higher (Krupnik and Jolly, 2002), although observations near eastern Hudson Bay indicate a delay in the initiation of spring melt (McDonald M.A. et al., 1997). See Chapter 3 for local accounts of such changes in the Arctic. The effects of early and less intense spring melt will be most

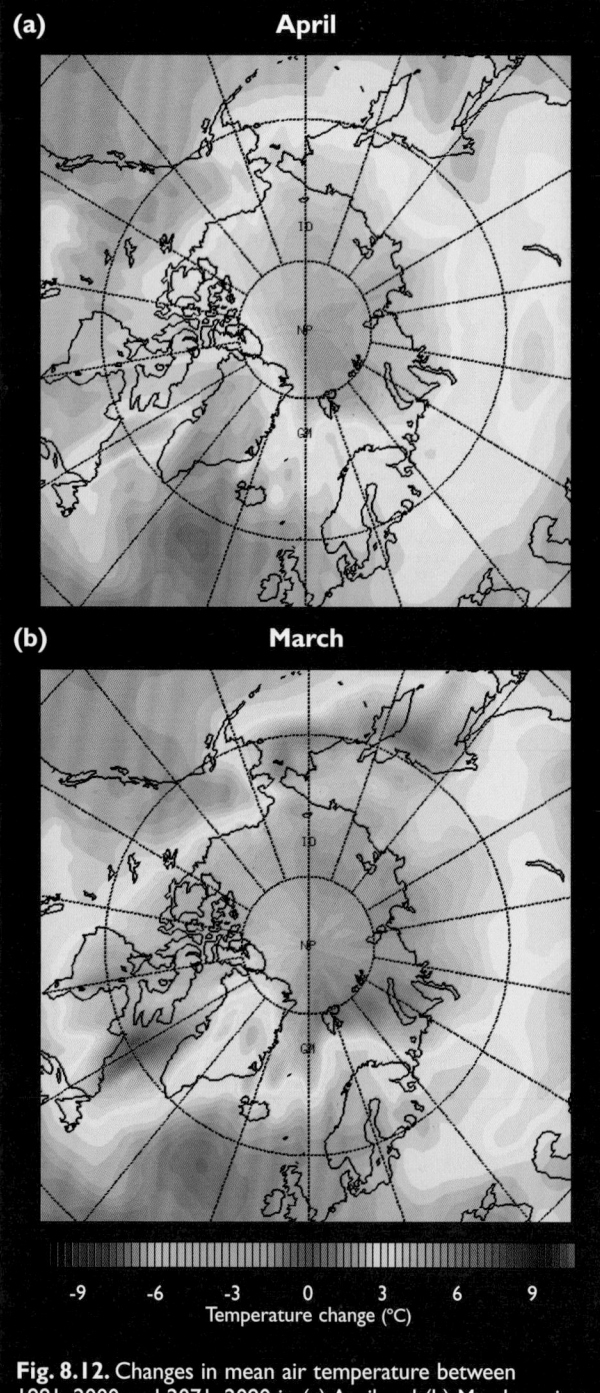

Fig. 8.12. Changes in mean air temperature between 1981–2000 and 2071–2090 in (a) April and (b) March projected by the ACIA-designated models (five-model average).

dramatic for catchments wholly contained within the northern latitudes, where snowmelt forms the major and sometimes only flow event of the year. Reductions in the spring peak will be accentuated where the loss of permafrost through associated warming increases the capacity to store runoff, although there will also be a compensating increase in summer base flow. Overall, the magnitude and frequency of high flows will decline while low flows will increase, thereby flattening the annual hydrograph. This impact is similar to that observed as a result of river regulation, and hence will tend to compound such effects.

Loss of permafrost or deepening of the active layer (seasonal melt depth; see section 6.6.1 for changes in permafrost) will also reduce the peak response to rainfall events in summer, increase infiltration, and promote groundwater flow. This is consistent with the analogue of northern basins where those with less permafrost but receiving comparable amounts of precipitation have a lowered and smaller range of discharge (Rouse et al., 1997). Changes in the rate of evapotranspiration and its seasonal duration will also directly affect stream runoff from permafrost basins. As suggested by the modeling results of Hinzman and Kane (1992) for areas of Alaska, the greatest reduction in summer runoff is likely to occur in years experiencing light, uniformly spaced rainfall events whereas in years characterized by major rainfalls comprising most of the summer precipitation, total runoff volume is likely to be affected least.

Changes in the water balance will vary by regional climate and surface conditions, but particular areas and features are believed to be especially sensitive to such alterations. Such is the case for the unglaciated lowlands of many arctic islands where special ecological niches, such as found at Polar Bear Pass on Bathurst Island or Truelove Lowland on Devon Island, are produced by unique hydro-climatic regimes and are largely dependent on ponded water produced by spring snowmelt. On a broad scale, arctic islands and coastal areas are likely to experience significant changes in local microclimates that will probably affect water balance components, especially evaporation rates. Here, longer open-water seasons in the adjacent marine environments are likely to enhance the formation of fog and low clouds and reduce associated solar radiation. Increased water vapor and lower energy flux would thereby offset any potential increase in evaporation resulting from higher air temperatures (Rouse et al., 1997).

Large regional differences in water balance will also occur because of differences in plant communities (see also section 7.4.1). For example, surface drying of open tundra is restricted when non-transpiring mosses and lichens overlie the tundra. Over the longer term, a longer growing season combined with a northward expansion of more shrubs and trees will very probably increase evapotranspiration. Quite a different situation is very likely to exist over the multitude of wetlands that occupy so much of the northern terrain. Although evap-

oration is inhibited after initial surface drying on those wetlands covered by sphagnum moss or lichen, evapotranspiration continues throughout the summer in wetlands occupied by vascular plants over porous peat soils, and only slows as the water table declines. Higher summer temperatures have the ability to dry such wetlands to greater depths, but their overall storage conditions will depend on changes in other water balance components, particularly snowmelt and rainfall inputs.

As the active layer deepens and more unfrozen flow pathways develop in the permafrost, an enhancement of geochemical weathering and nutrient release is very likely (e.g., phosphorus; Hobbie et al., 1999; see also section 6.6.1.3). Ultimately, this is very likely to affect productivity in arctic freshwater systems such as Toolik Lake, Alaska (Box 8.3). In the short term, the chemical composition of surface runoff and groundwater flows is very likely to change. In addition, suspended sediment loads will very probably increase as a result of thermokarst erosion, particularly in ice-rich locations. Suspended sediment and nutrient loading of northern freshwater systems will probably also increase as land subsidence, slumping, and landslides increase with permafrost degradation, as traditional ecological knowledge has documented in the western Canadian Arctic where the depth of the active layer has increased (Krupnik and Jolly, 2002). Thermokarst erosion is very likely to continue until at least the large near-surface ice deposits are depleted and new surface flow patterns stabilize. Such fluvial-morphological adjustment is likely to be very lengthy, of the order of hundreds of years, considering the time that has been estimated for some northern rivers to reach a new equilibrium after experiencing a major shift in their suspended-sediment regimes (e.g., Church, 1995). A major reason for such a protracted period is the time it takes for new vegetation to colonize and stabilize the channel landforms. The stabilization that will occur in the Arctic under climate change is further complicated by the projected change in vegetation regimes, particularly the northward advance of shrubs and trees (see section 7.5.3.2). Such vegetation shifts will cause further changes in stream water chemistry by altering DOC concentrations. Current data indicate that DOC is negatively correlated with latitude (Fallu and Pienitz, 1999; Rühland and Smol, 1998) and decreases with distance from treeline (Korhola et al., 2002b; Pienitz and Smol, 1994) and along gradients from boreal forest to tundra (Vincent and Hobbie, 2000). Hence, as vegetation shifts from mosses and lichens to grasses and woody species, runoff is very likely to contain increasing concentrations of DOC and particulate detrital material. Verification of enhanced DOC supply associated with northward treeline advance is provided by various paleolimnological and paleoclimatic studies (e.g., Korhola and Weckström, 2005; Seppä and Weckström, 1999; Solovieva and Jones, 2002). Although such increases will be long-term, given the slow rates of major vegetation shifts (see also section 7.5.3.2), earlier increases in DOC and DIC are very likely to result from the earlier thermal and mechanical erosion of the

Box 8.3. Ecological transitions in Toolik Lake, Alaska, in the face of changing climate and catchment characteristics

Toolik Lake (maximum depth 25 m, area 1.5 km²) lies in the foothills north of the Brooks Range, Alaska, at 68° N, 149° W. The river study site is the headwaters of the Kuparuk River. Details of the research project and related publications are available on the Arctic Long-Term Ecological Research site (http://ecosystems.mbl.edu/ARC/). The mean annual temperature of the area is -9 °C, and annual precipitation is approximately 300 to 400 mm. Permafrost is 200 m thick with an active layer up to 46 cm deep. Acidic tussock tundra covers the hillslopes. Sedges dominate a small area of wetlands in the study site, while the dry uplands have a cover of lichens and heaths. Lakes and streams are ultra-oligotrophic, and are ice-free from July to September with strong summer stratification and oxygen saturation. Stream flow is nival, and carries DOC-enriched spring runoff from peaty catchments to Toolik Lake. Primary producers in Toolik Lake consist of 136 species of phytoplankton, dominated by chrysophytes with dinoflagellates and cryptophytes, as well as diatoms. Annual primary productivity averages 12 g C/m² and is co-limited by nitrogen and phosphorus. Zooplankton are sparse.

Fish species are lake trout, Arctic grayling, round whitefish (*Prosopium cylindraceum*), burbot, and slimy sculpin (*Cottus cognatus*), which feed on benthic chironomid larvae and snails, the latter controlling epilithic algae in the lake. Dissolved organic carbon drives microbial productivity (5–8 g C/m²/yr).

The average air temperature of northern Alaska has increased by nearly 2 °C over the past 30 years. Warming of Alaskan waters will possibly have a detrimental effect on adult grayling, which grow best during cool and wet summers and which may actually lose weight during warm and dry summers (Deegan et al., 1999). Approximately 20 km from Toolik (Hobbie et al., 2003), permafrost temperatures at 20 m depth increased from -5.5 to -4.5 °C between 1991 and 2000. This warming of frozen soils probably accounts for recent increases in stream- and lake-water alkalinity.

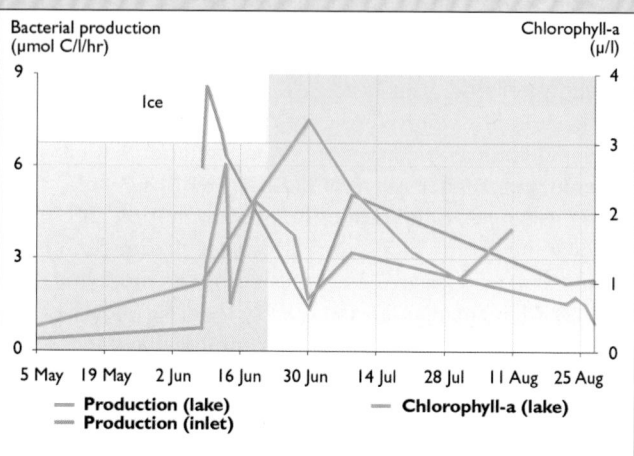

Bacterial productivity and chlorophyll over the spring and summer of 2002 in Toolik Lake (averaged over depths of 3 to 12 m; adapted from Crump et al., 2003).

Climate has been shown to have a significant control on the vegetation of the site, which in turn has affected aquatic resources for productivity. Runoff from thawing soils within the catchment of Toolik Lake has affected lake productivity in a number of ways. Dissolved organic carbon from excretion, leaching, and decomposition of plants in the catchment, along with associated humic materials, has been found to reduce photosynthesis in the lake and absorb 99% of the UV-B radiation in the upper 20 cm (Morris et al., 1995). In spring, meltwater carries terrestrially derived DOC and abundant nutrients. Upon reaching the lake, meltwater flows cause a two-week high in bacterial productivity (~50% of the annual total) beneath the lake-ice cover (see figure). This peak in production, which takes place at 2 °C, illustrates that bacteria are carbon- and energy-limited, not temperature-limited, and as such will be indirectly affected by climate change. Phytoplankton biomass and primary production peak soon after the ice leaves the lake, and as solar radiation peaks. The lake stratifies so rapidly that no spring turnover occurs, causing oxygen-depleted bottom waters to persist over the summer. This effect is very likely to be amplified with higher temperatures, and will probably reduce the habitat available to fish species such as lake trout.

Future increases in average air temperature and precipitation are very likely to further affect freshwater systems at Toolik. Lakes will very probably experience early breakup and higher water temperatures. Stream waters are very likely to warm as well, and runoff is very likely to increase, although evapotranspiration could possibly offset increased precipitation. As waters warm, primary production in lakes and rivers at the site is very likely to increase, although most species of aquatic plants and animals are unlikely to change over the 21st century. Lake and river productivity are also very likely to increase in response to changes in the catchment, in particular, temperature increases in permafrost soils and increased weathering and release of nutrients. Increased precipitation will also affect nutrient supply to freshwater systems, and is likely to result in increased decomposition of organic matter in soils (Clein et al., 2000), formation of inorganic nitrogen compounds, and increased loss of nitrogen from land to water. The shift in terrestrial vegetation to predominantly shrubs is very likely to cause greater loading of DOC and humic materials in streams and lakes, and a reduction in UV-B radiation penetration. However, increases in organic matter are likely to have detrimental effects on the stream population of Arctic grayling at this site, resulting in their disappearance in response to high oxygen depletion. Lake trout, on the other hand, are likely to survive but their habitat will probably be slightly reduced by the combination of reduced deep-water oxygen and warmer surface waters.

permafrost landscape (see also section 6.6.1.3). Zepp et al. (2003) and Häder et al. (2003) provided comprehensive reviews of the projected interactive effects of changes in UV radiation levels and climate on DOC and DIC and related aquatic biogeochemical cycles.

Changes in freshwater catchments with climate change will affect not only loadings of nutrients, sediments, DOC, and DIC to freshwater systems but also the transport and transformation of contaminants. Contaminant transport from surrounding catchments to freshwaters is likely to increase as permafrost degrades and perennial snow melts (Blais et al., 1998, 2001; McNamara et al., 1999). The contaminants released from these frozen stores, and those originating from long-range transport and deposition in contributing basins, can then be stored in sediments or metabolized and biomagnified through the food web. Section 8.7 discusses this topic in more detail.

8.4.3. Effects on general hydro-ecology

8.4.3.1. Streams and rivers, deltas, and estuaries

A number of hydrologic shifts related to climate change will affect lakes and rivers, including seasonal flow patterns, ice-cover thickness and duration, and the frequency and severity of extreme flood events. In the present climate, most streams and rivers originating within the Arctic have a nival regime in which snowmelt produces high flows and negligible flow occurs in winter. In areas of significant glaciers, such as on some Canadian and Russian islands, Greenland, and Svalbard, ice melt from glaciers can sustain flow during the summer, whereas many other streams produce summer flow only from periodic rainstorm events unless they are fed by upstream storage in lakes and ponds.

The subarctic contains a much broader range of hydrologic regimes, which vary from cold interior continental (comparable to those of the Arctic) to maritime regimes fed moisture directly from open seas even during winter. Overall, a warmer climate is very likely to lead to a shift toward a more pluvial runoff regime as a greater proportion of the annual precipitation falls as rain rather than snow; the magnitude of the peak of spring snowmelt declines; thawing permafrost increases near-surface storage and reduces runoff peaks; and a more active groundwater system augments base flows.

Enhancement of winter flow will very probably lead to the development of a floating ice cover in some streams that currently freeze to the bed. This is very likely to be beneficial to the biological productivity of arctic streams and fish survival where winter freshwater habitat is limited to unfrozen pools (Craig, 1989; Prowse, 2001a,b). For other arctic streams and rivers, warming is very likely to result in a shortened ice season and thinner ice cover (section 6.7.3). Since river ice is such a major controller of the ecology of northern streams and rivers, there are likely to be numerous significant impacts.

Under conditions of overall annual temperature increases, a delay in the timing of freeze-up and an earlier breakup will very probably reduce the duration of river-ice cover. Data compiled over the last century or more indicate that changes in timing of these events are likely to be at a rate of approximately one day per 0.2 °C increase in air temperature (Magnuson et al., 2000; see also sections 8.4.1 and 6.7.3). For freeze-up, higher water and air temperatures in the autumn combine to delay the time of first ice formation and eventual freeze-up. If there was also a reduction in the rate of autumn cooling, the interval between these two events would increase. Although all major ice types would continue to form, unless there were also significant changes in the flow regime, the frequency and magnitude of, for example, periods of major frazil ice growth will probably be reduced. This has implications for the types of ice that constitute the freeze-up cover and for the creation of unique under-ice habitats such as air cavities and those influenced by frazil concentrations (Brown et al., 1994; Cunjak et al., 1998; Prowse, 2001b).

Changes in the timing and duration of river ice formation will also alter the dissolved oxygen (DO) regimes of arctic lotic ecosystems. Following freeze-up and the elimination of direct water–atmosphere exchanges, DO concentrations steadily decline, sometimes to near-critical levels for river biota (e.g., Chambers et al., 1997; Power G. et al., 1993; Prowse, 2001a,b). Reductions in ice-cover duration and a related increase in the number of open-water re-aeration zones are very likely to reduce the potential for this biologically damaging oxygen depletion. Such benefits will possibly be offset by the projected enhanced input of DOC and its subsequent oxidation (e.g., Schreier et al., 1980; Whitfield and McNaughton, 1986), the rate and magnitude of which would also be increased as a result of the above-noted higher nutrient loading. Worst-case scenarios would develop on rivers where the flow is already comprised of poorly oxygenated groundwater, such as that supplied from extensive bogs and peatlands. Some rivers in the West Siberian Plain offer the best examples of this situation. Here, the River Irtysh drains large quantities of de-oxygenated water from vast peatlands into the River Ob, resulting in DO levels of only about 5% of saturation (Harper, 1981; Hynes, 1970).

The greatest ice-related ecological impacts of climate change on arctic lotic systems are likely to result from changes in breakup timing and intensity. As well as favoring earlier breakup, higher spring air temperatures can affect breakup severity (Prowse and Beltaos, 2002). While thinner ice produced during a warmer winter would tend to promote a less severe breakup, earlier timing of the event could counteract this to some degree. Breakup severity also depends on the size of the spring flood wave. While greater and more rapid snowmelt runoff would favor an increase in breakup severity, the reverse is true for smaller snowpacks and more protracted melt. Hence, changes in breakup severity will vary regionally according to the variations in winter precipitation and spring melt patterns.

For regions that experience a more "thermal" or less dynamic ice breakup (Gray and Prowse, 1993), the magnitude of the annual spring flood will very probably be reduced. For the many northern communities that historically located near river floodplains for ease of transportation access, reductions in spring ice-jam flooding would be a benefit. In contrast, however, reductions in the frequency and severity of ice-jam flooding would have a serious impact on river ecology since the physical disturbances associated with breakup scouring and flooding are very important to nutrient and organic matter dynamics, spring water chemistry, and the abundance and diversity of river biota (Cunjak et al., 1998; Prowse and Culp, 2003; Scrimgeour et al., 1994). Specifically, ice-induced flooding supplies the flux of sediment, nutrients, and water that is essential to the health of the riparian system; river deltas being particularly dependent on this process (e.g., Lesack et al., 1991; Marsh and Hey, 1989; Prowse and Conly, 2001). More generally, given that the magnitude and recurrence interval of water levels produced by ice jams often exceed those of open-water conditions, breakup is probably the main supplier of allochthonous organic material in cold-regions rivers (Prowse and Culp, 2003; Scrimgeour et al., 1994). In the same manner, breakup serves as an indirect driver of primary and secondary productivity through the supply of nutrients – a common limiting factor for productivity in cold-regions rivers. Even the mesoscale climate of delta ecosystems and spring plant growth depends on the timing and severity of breakup flooding (Gill, 1974; Hirst, 1984; Prowse, 2001a).

River ice is also a key agent of geomorphological change and is responsible for the creation of numerous erosional and depositional features within river channels and on channel floodplains (e.g., Prowse, 2001a; Prowse and Gridley, 1993). Since most geomorphological activity occurs during freeze-up and breakup, changes in the timing of these events are very unlikely to have any significant effect. If, however, climatic conditions alter the severity of such events, this is likely to affect particular geomorphological processes. Furthermore, breakup events affect the general processes of channel enlargement, scour of substrate habitat, and the removal and/or succession of riparian vegetation. All such major river-modifying processes would be altered by any climate-induced shift in breakup intensity.

In summary, if climate change alters the long-term nature of breakup dynamics, the structure and function of rivers and related delta ecosystems are very likely to be significantly altered with direct effects on in-channel and riparian biological productivity. If, for example, significant reductions in dynamic breakups and the related level of disturbance occur, this will reduce overall biological diversity and productivity, with the most pronounced effects on floodplain and delta aquatic systems.

Owing to the reduced ice-cover season and increased air temperatures during the open-water period, summer water temperatures will very probably rise. Combined with greater DOC and nutrient loadings, higher water temperatures are likely to lead to a general increase in total stream productivity, although it is unclear whether temperature will have a significant direct effect on the processing rate of additional particulate detrital material. Irons et al. (1994), for example, found a comparable rate of litter processing by invertebrates in Michigan and Alaska and concluded that temperature was not a main factor. The effect of increased temperature on processing efficiency by "cold-climate" species of invertebrates, however, has not been evaluated. The effect of enhanced nutrient loading to arctic streams is more predictable. The current nutrient limitation of many arctic streams is such that even slight increases in available phosphorus, for example, will produce a significant increase in pri-

Table 8.1. A synthesis of the potential effects of climate change on arctic estuarine systems from both the bottom-up and top-down ecological perspectives (adapted from Carmack and Macdonald, 2002).

Bottom-up: nutrients/production/biota etc.	Top-down: humans/predators/biota etc.
• More open water, more wind mixing, upwelling and greater nutrient availability for primary producers (+) • More open water, more light penetration especially seasonally hence more primary production (+); potential for increased UV radiation levels (-/?) • Decreased ice cover, decreased ice-associated algal production, and subsequent impacts on pelagic and benthic food webs (-) • Increased basin rainfall, increased export of carbon to nearshore (+) • Increased storms and open water, increased coastal erosion (-), increased sediment loads, nutrients and mixing (+), possibly increased productivity especially in late season (?) but offset by decreased light penetration (-) • Potential positive feedback to climate change processes (e.g., permafrost thawing, release of methane, and increased radiative forcing) (-) • Contaminant inputs, mobilization, or increased fluxes driven by temperature changes will increase availability and biomagnification of contaminants in food chains (-)	• Shifting water masses and currents will affect biotic cues for habitat use and migrations of biota such as fish and marine mammals (?) • Redistribution of grazers will affect underlying trophic structure (-/?) • Climate-induced changes in freshwater, estuarine, and marine habitats seasonally used by anadromous fishes will affect distribution and suitability for use, with consequences for the prey communities and possibly fish availability for humans (-/?) • Physical absence or alteration of seasonality or characteristics of ice platforms will affect ice-associated biota (e.g., polar bears, seals, algae) (-), with cascading consequences for fish (+/-) • Increased open water will facilitate whale migrations (+) but increase predator risk to calves (-); shifts in whale populations may cascade through the trophic structure (e.g., shifted predation on fish by belugas; increased predation on plankton by bowhead whales) with unknown trophic consequences for anadromous and marine fish (?)

Cascading consequences from a human perspective as generally: positive (+); negative (-); neutral (0); or unknown (?).

mary productivity (Flanagan et al., 2003). Where productivity responses of stream biota are co-limited by phosphorus and nitrogen (e.g., as suggested by the experimental results of Peterson B. et al., 1993), increased loadings of both nutrients would be required to sustain high levels of enhanced productivity.

Table 8.1 summarizes the potential impacts of climate change on the dynamics of arctic estuaries (Carmack and Macdonald, 2002). The major factor affecting arctic estuarine systems given the degree of climate change projected by the ACIA-designated models will be the increase in freshwater discharge (section 6.8.3). In some arctic basins, such as the Chukchi Sea, there is presently very little freshwater runoff and consequently no estuarine zones. Increased river discharge could possibly create estuarine areas, providing new habitat opportunities for euryhaline species. In established estuarine systems, such as the Mackenzie River system and the Ob and Yenisey Rivers, increased freshwater input in summer (e.g., Peterson B. et al., 2002) is likely to increase stratification, making these habitats more suitable for freshwater species and less suitable for marine species. There are likely to be shifts in species composition to more euryhaline and anadromous species. In addition, increased freshwater input is likely to deposit more organic material, changing estuarine biogeochemistry and perhaps increasing primary productivity, the positive effects of which will possibly be offset in part by increased resuspension of contaminated sediments in these systems.

A secondary impact of increased freshwater discharge that is of serious concern, particularly for Siberian rivers that traverse large industrialized watersheds, is the potential for increased contaminant input. The Ob and Yenisey Rivers, for example, have high levels of organochlorine contamination compared to the Lena River (Zhulidov et al., 1998), which is considered relatively pristine (Guieu et al., 1996). Larsen et al. (1995) noted that arctic fishes have a life strategy that involves intensive feeding in spring and summer, allowing for the buildup of lipid stores and coping with food shortages in winter. The high body-lipid content of arctic fishes may make them more vulnerable to lipid-soluble pollutants such as polycyclic aromatic hydrocarbons (PAHs) or polychlorinated biphenyls (PCBs). In addition, reduced sea-ice coverage that leads to increased marine traffic is likely to have cascading negative consequences (e.g., pollution, risk of oil spills) for estuarine systems.

Arctic deltas provide overwintering habitat for many species that tolerate brackish waters. These areas are maintained as suitable habitat by a combination of continuous under-ice freshwater flow and the formation of the nearshore ice barrier in the stamukhi zone (area of grounded, nearshore ice pressure ridges). As temperatures rise, the seasonal ice zone of estuaries is likely to expand and the ice-free season lengthen (Carmack and Macdonald, 2002). Disruption of either the flow regime or the ice barrier could possibly have profound effects on the availability of suitable overwintering habitat for

desired fish species. Given that such habitat is probably limited and hence limits population abundance, the consequences for local fisheries will probably be significant. In addition, in early winter, subsistence and commercial fisheries target fish that overwinter in deltas. Thinning ice is likely to limit access to these fisheries.

Similar to freshwater systems, ecological control of marine systems can be viewed from bottom-up (i.e., nutrients–production–biota linkages) and/or top-down (i.e., human activities–predators–keystone biota) perspectives (Parsons, 1992). The special role of ice as both a habitat and a major physical force shaping the estuarine and nearshore arctic environment suggests that climate change will work in both modes to affect these systems (Carmack and Macdonald, 2002). One example is the loss of the largest epishelf lake (fresh and brackish water body contained behind the ice shelf) in the Northern Hemisphere with the deterioration and break up of the Ward Hunt ice shelf (Mueller et al., 2003). The loss of this nearshore water body has affected a unique community of marine and freshwater planktonic species, as well as communities of cold-tolerant microscopic algae and animals that inhabited the upper ice shelf.

8.4.3.2. Lakes, ponds, and wetlands

Lentic systems north of the Arctic Circle contain numerous small to medium lakes and a multitude of small ponds and wetland systems. Relatively deep lakes are primarily contained within alpine or foothill regions such as those of the Putorana Plateau in the lower basin of the Yenisey River. One very large and deep lake, Great Bear Lake (Northwest Territories, Canada), is found partly within the Arctic Circle. Variations in its water budget primarily depend on flows from its contributing catchment, comprised largely of interior plains lowlands and exposed bedrock north of 60° N. Its southern counterpart, Great Slave Lake, provides a strong hydrologic contrast to this system. Although also part of the main stem Mackenzie River basin and wholly located north of 60° N, its water budget is primarily determined by inflow that originates from Mackenzie River headwater rivers located much further to the south. Moreover, its seasonality in water levels reflects the effects of flow regulation and climatic variability in one of its major tributaries, the Peace River, located about 2000 km upstream in the Rocky Mountain headwaters of western Canada (Gibson J.J. et al., in press; Peters and Prowse, 2001). As such, the Mackenzie River system offers the best example of a northern lentic system that is unlikely to be significantly affected by changes in hydrologic processes operating within the north (e.g., direct lake evaporation and precipitation) but will be dependent principally on changes in water-balance processes operating well outside the Arctic.

The other major arctic landscape type that contains large, although primarily shallower, lakes is the coastal plains region found around the circumpolar north. As mentioned previously, these shallow systems depend on snowmelt as their primary source of water, with rainfall gains often negated by evapotranspiration during the summer. Evaporation from these shallow water bodies is very likely to increase as the ice-free season lengthens. Hence, the water budget of most lake, pond, and wetland systems is likely to depend more heavily on the supply of spring meltwater to produce a positive annual water balance, and these systems are more likely to dry out during the summer. Another possible outcome of climate change is a shift in vegetation from non-transpiring lichens and mosses to vascular plants as temperatures rise and the growing season extends (Rouse et al., 1997), potentially exacerbating water losses. However, factors such as increasing cloud cover and summer precipitation will possibly mitigate these effects.

Loss of permafrost increases the potential for many northern shallow lotic systems to dry out from a

warmer temperature regime. Ponds are likely to become coupled with the groundwater system and drain if losses due to downward percolation and evaporation are greater than resupply by spring snowmelt and summer precipitation. Patchy arctic wetlands are particularly sensitive to permafrost degradation that can link surficial waters to the supra-permafrost groundwater system. Those along the southern limit of permafrost, where increases in temperature are most likely to eliminate the relatively warm permafrost, are at the highest risk of drainage (Woo et al., 1992). Traditional ecological knowledge from Nunavut and eastern arctic Canada indicates that recently there has been enhanced drying of lakes and rivers, as well as swamps and bogs, enough to impair access to traditional hunting grounds and, in some instances, fish migration (Krupnik and Jolly, 2002; see section 3.4.5 for detailed discussion of a related case study).

Box 8.4. Lake-ice duration and water column stratification: Lake Saanajärvi, Finnish Lapland

Lake Saanajärvi (maximum depth 24 m, area 0.7 km²; 69° N, 20° E) is the key Finnish site in the European research projects Mountain Lake Research and European Mountain Lake Ecosystems: Regionalisation, Diagnostics and Socio-economic Evaluation. Lake Saanajärvi has been intensively monitored since 1996. The data presented here have been published in several papers, including those by Korhola et al. (2002a), Rautio et al. (2000), Sorvari and Korhola (1998), and Sorvari et al. (2000, 2002). The mean annual temperature of the area is -2.6 °C, and annual precipitation is approximately 400 mm. The catchment area is mostly covered by bare rocks and alpine vegetation. Lake Saanajärvi is a dimictic, ultra-oligotrophic, clear-water lake. The lake is ice-free for nearly four months of the year, with highly oxygenated waters, and is strongly stratified for two months after spring overturn. Phytoplankton biomass and densities are low (Forsström, 2000; Forsström et al., in press; Rautio et al., 2000), consisting predominantly of chrysophytes and diatoms. Bacterial biomass is low as well, and zooplankton are not very abundant. Freshwater shrimp (*Gammarus lacustris*) are common and form an important food source for fish, which include Arctic char and brown trout (*Salmo trutta lacustris*).

Changes in water temperature and stratification of Lake Saanajärvi have been associated with climate changes in Finnish Lapland over the past 200 years (Alexandersson and Eriksson, 1989; Sorvari et al., 2002; Tuomenvirta and Heino, 1996). Mean annual air temperatures in Finnish Lapland, as in much of the Arctic, rose 1 to 2 °C following the Little Ice Age. During this period of warming, diatom communities changed from benthic–periphytic to pelagic, Cladocera increased in abundance, and chrysophytes became less numerous. These changes have been shown to be associated with increased rates of organic matter accumulation and increased concentrations of algal pigments during the climatic warming (Korhola et al., 2002a; Sorvari and Korhola, 1998; Sorvari et al., 2002; see figure). After a period of cooling from the 1950s to the 1970s, air temperatures in the Arctic continued to rise. More recently, interannual variability in temperatures has been shown to account for changes in the thermal gradient and mixing of Lake Saanajärvi surface waters. For example, Lake Saanajärvi normally stratifies in early July, two weeks after ice breakup, and retains a distinct, steep thermocline at a depth of 10 to 12 m throughout the summer. In 2001, this summer stratification was broken after a period of slight cooling in early August, after which the lake was only weakly stratified. In 2002, on the other hand, spring and summer temperatures were extremely warm. Spring ice breakup was early and waters warmed quickly, resulting in a very sharp thermocline that was stable during the entire summer stratification period.

Future temperature increases are therefore very likely to affect the thermal structure of lakes in Finnish Lapland and throughout the Arctic, which is likely to have dramatic consequences for lake biota. Rising mean annual temperatures are very likely to influence the duration of summer stratification and the stability and depth of the thermocline in Finnish lakes. As such, many of the presently isothermal lakes are likely to become dimictic as temperatures increase. In addition, the prolonged thermal stratification that is likely to accompany rising temperatures could possibly lead to lower oxygen concentrations and increased phosphorus concentrations in the hypolimnion, benefiting nutrient-limited primary production. As spring temperatures rise and the ice-free period extends, not only is thermal stratification likely to stabilize, but production in many high latitude lakes could possibly peak twice

Warming of surface permafrost, however, will very probably enhance the formation of thermokarst wetlands, ponds, and drainage networks, particularly in areas characterized by concentrations of massive ground ice. Thawing of such ice concentrations, however, is very likely to lead to dramatic increases in terrain slumping and subsequent sediment transport and deposition in rivers, lakes, deltas, and nearshore marine environments. This is likely to produce distinct changes in channel geomorphology in systems where sediment transport capacity is limited, and will probably have a significant impact on the aquatic ecology of the receiving water bodies. Catastrophic drainage of permafrost-based lakes that are now in a state of thermal instability, such as those found along the western arctic coast of Canada, is also very likely (Mackay J., 1992; Marsh and Neumann, 2001, 2003). Losses of thermokarst lakes within low-lying deltaic areas are also likely to result from rising sea

levels. Marine inundation resulting from continually rising sea level commonly drains lakes in the outer portion of the Mackenzie Delta (northern Richards Island: Dallimore et al., 2000). Moreover, Mackay J. (1992) estimated that one lake per year has drained in the Tuktoyaktuk coastlands of northern Canada over the last few thousand years. Future, more pronounced rises in sea level are likely to accelerate this process.

Changes in the water balance of northern wetlands are especially important because most wetlands in permafrost regions are peatlands, which can be sources or sinks of carbon and CH_4 depending on the depth of the water table (see also section 8.4.4.4). An analysis by Rouse et al. (1997) of subarctic sedge fens in a doubled-CO_2 climate suggested that increases in temperature (4 °C) would reduce water storage in northern peatlands even with a small and persistent increase in precipitation. While

rather than once during the open-water season (e.g., Catalan et al., 2002; Hinder et al., 1999; Lepistö, 1999; Lotter and Bigler, 2000; Medina-Sánchez et al., 1999; Rautio et. al., 2000). On a broader scale, changes in lake stratification and water mixing will probably affect species composition (e.g., diatoms; Agbeti et al., 1997).

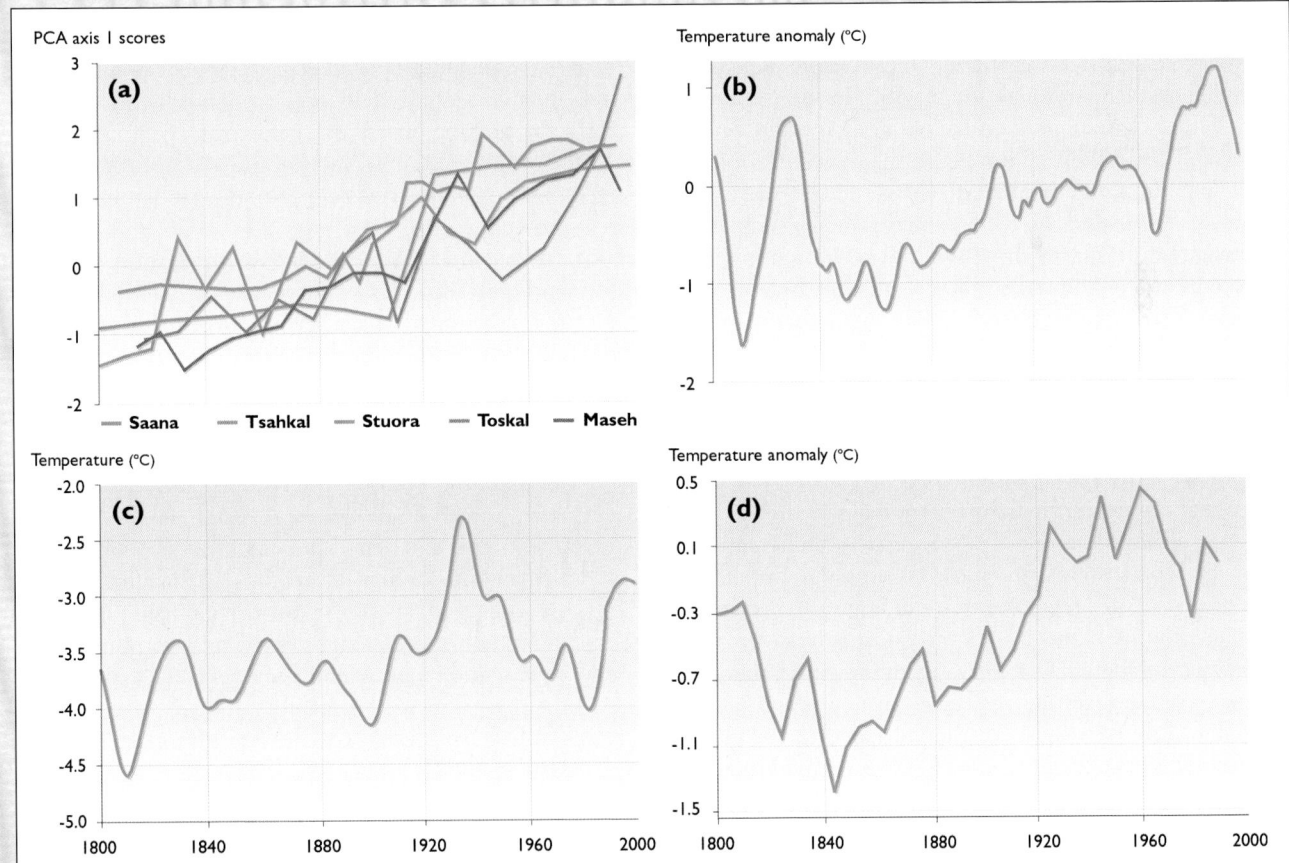

Comparison of diatom assemblage changes with regional and arctic-wide temperature anomalies, showing (a) principal components analysis (PCA) primary axis scores derived from the correlation matrices of the diatom percentage counts from the five study sites; (b) spring (March–May) temperature anomalies for northwestern Finnish Lapland, smoothed using a 10-year low-pass filter; (c) trend in mean annual air temperature in northwestern Finnish Lapland, smoothed using a 10-year low-pass filter; and (d) standardized proxy arctic-wide summer-weighted annual temperature, plotted as departure from the mean (panels a–c from Sorvari et al., 2000; panel d from Overpeck).

acknowledging that storage changes depend on variability in soil moisture and peat properties, projected declines in the water table were 10 to 20 cm over the summer.

As the ice cover of northern lakes and ponds becomes thinner, forms later, and breaks up earlier (section 6.7.3), concomitant limnological changes are very likely. Thinner ice covers with less snow cover will increase the under-ice receipt of solar radiation, thereby increasing under-ice algal production and oxygen (e.g., Prowse and Stephenson, 1986) and reducing the potential for winter anoxia and fish kills. Lower water levels, which reduce under-ice water volumes and increase the likelihood of winterkill, could possibly counteract this effect. Similarly, greater winter precipitation on a thinner ice cover is very likely to promote the formation of more highly reflective snow and white-ice layers. Such layers would reduce radiation penetration well into the spring because they also tend to delay breakup compared to covers comprised of only black ice. Notably, the ACIA-designated models project that incident radiation will decline. Reductions are likely to be relatively small (i.e., 10–12 W/m^2 in May-June between 1981–2000 and 2071–2090, section 4.4.4), however, compared to the major reductions that are likely to result from greater reflective loss from enhanced white-ice formation.

A longer ice-free season will also increase the length of the stratified season and generally increase the depth of mixing (Box 8.4), although the magnitude and duration of the effects will depend on factors such as basin depth and area. This is likely to lower oxygen concentrations in the hypolimnion and increase stress on cold-water organisms (Rouse et al., 1997). Furthermore, such an enhancement of mixing processes and reduction in ice cover will probably increase the potential for many northern lakes and ponds to become contaminant sinks (section 8.7).

With a longer and warmer ice-free season, total primary production is likely to increase in all arctic lakes and ponds, and especially in the oligotrophic high-arctic ponds that are currently frozen for a majority of the year (Douglas and Smol, 1999). Similar to the situation for arctic lotic systems, an enhanced supply of nutrients and organic matter from the more biologically productive contributing basins is likely to boost primary productivity (Hobbie et al., 1999). Again, however, there are likely to be offsetting effects because of reductions in light availability resulting from enhanced turbidity due to higher inputs of DOC and suspended sediment. Hecky and Guildford (1984) noted that analogous factors caused a switch from nutrient limitation, which is a common control of primary production in northern lakes, to light limitation.

8.4.4. Changes in aquatic biota and ecosystem structure and function

Climate change is very likely to have both direct and indirect consequences on the biota and the structure and function of arctic freshwater ecosystems. Changes in key physical and chemical parameters described previously are very likely to affect community and ecosystem attributes such as species richness, biodiversity, range, and distribution, and consequently alter corresponding food web structures and primary and secondary production levels. The magnitude and extent of the ecological consequences of climate change in arctic freshwater ecosystems will depend largely on the rate and magnitude of change in three primary environmental drivers: the timing, magnitude, and duration of the runoff regime; temperature; and alterations in water chemistry such as nutrient levels, DOC, and particulate organic matter loadings (Poff et al., 2002; Rouse et al., 1997; Vincent and Hobbie, 2000).

8.4.4.1. Effects on biological communities, biodiversity, and adaptive responses

Climate change will probably produce significant effects on the biodiversity of freshwater ecosystems throughout the Arctic and possibly initiate varying adaptive responses. The magnitude, extent, and duration of the impacts and responses will be system- and location-dependent, and difficult to separate from other environmental stressors. Biodiversity is related to, or affected by, factors including:

- the variability of regional and local climate;
- the availability of local resources (e.g., water, nutrients, trace elements, energy, substrate) affecting the productivity potential;
- the nature, timing, and duration of disturbance regimes in the area (e.g., floods, catastrophic water loss, fire);
- the original local and regional "stock" of species and their dispersal opportunities or barriers;
- the physiological capacity of individuals and populations to cope with new environmental conditions (e.g., physiological thresholds and tolerances);
- the levels of spatial heterogeneity (habitat fragmentation) and connections among aquatic systems;
- the intensity of biotic interactions such as competition, predation, disease, and parasitism;
- phenotypic and genotypic flexibility in reproductive and life-history strategies (e.g., facultative versus obligatory anadromy for certain fish; plasticity in sexual versus asexual reproductive strategies in aquatic invertebrate and plant species); and
- the overall genetic variability and adaptive capacity of the species (IPCC, 2001a; Pimm, 1991; UNEP, 2003).

Many arctic freshwater systems are exposed to multiple environmental stressors or perturbations including point- and/or nonpoint-source pollution (e.g., long-range aerial transport of contaminants; section 8.7); altered hydrologic regimes related to impoundments and diversions; water quality changes from landscape alterations (e.g., mining, oil and gas exploration); and biological resource exploitation (e.g., subsistence and commercial fisheries and harvesting of waterfowl and mammals;

section 8.5), to name a few. These stressors, along with climate variability, can synergistically contribute to the degradation of biological diversity at the species, genetic, and/or habitat–ecosystem levels (CAFF, 2001; IPCC, 2001a; Pimm et al., 1995; UNEP, 2003). There is little evidence to suggest that climate change will slow species loss. There is growing evidence, however, that climate change will contribute to accelerated species losses at regional and global levels (UNEP, 2003) and that the effects of alterations in the biodiversity of ecosystem structure and function are likely to be more dependent on given levels of functional diversity than on the total number of species (Chapin et al., 2000). Moreover, both the number and type of functional units present in a community largely affect ecosystem resilience and vulnerability to change (UNEP, 2003).

For these reasons, large uncertainties remain in projecting species- and system-specific responses and the impacts of changes in climate and UV radiation levels on biodiversity at local and regional spatial scales. However, several broad projections can be made.

First, locally adapted arctic species are likely to be extirpated from certain areas as environmental conditions begin to exceed their physiological tolerances and/or ecological optima. Hence, species with limited climatic ranges and/or restricted habitat requirements (related to particular physiological or phenological traits) are very likely to be vulnerable to climate change effects. Species with low population numbers and/or that reside in restricted, patchy, and highly specialized environments will be particularly at risk (UNEP, 2003). While wholesale extinctions of entire arctic species are unlikely, some highly valued species (e.g., certain fish species) may possibly become geographically or ecologically marginalized. For example, there are pronounced north–south gradients in the taxonomic composition of stream macroinvertebrate communities in the Arctic, with decreasing species diversity and an increasing importance of taxa such as dipterans with distance northward (Oswood, 1997). Moreover, many of the high-latitude filamentous algal species have temperature optima well above the low ambient water temperatures at which they reside, and are therefore likely to respond positively to moderate increases in temperature (Tang and Vincent, 1999; Tang et al., 1997). Hence, many high-latitude species are currently at their physiological limits and are likely to be very sensitive to future shifts in climate (Danks, 1992). Projected changes in regional runoff patterns and temperature regimes are very likely to affect river and stream environments, possibly reducing the severity of disturbance events that are an integral component of their current hydro-ecology (section 8.4.2). Specifically, Scrimgeour et al. (1994) suggested that if these disturbances play a role in maintaining habitat complexity and associated species richness and diversity, then climate-related changes in the severity of these events will affect macroinvertebrate and aquatic algal species distribution and associated biodiversity patterns (see also Prowse and Culp, 2003).

In estuarine habitats, there are likely to be shifts in species composition to more euryhaline and anadromous species (e.g., fourhorn sculpin – *Myoxocephalus quadricornis*, ninespine stickleback – *Pungitius pungitius*, threespine stickleback – *Gasterosteus aculeatus*, Arctic flounder – *Pleuronectes glacialis*, salmonines, and coregonines). Such shifts in species composition will possibly have cascading effects resulting from competition for food resources with marine species (e.g., Arctic cod – *Boreogadus saida*) that currently inhabit many estuarine zones. The subsequent effects on higher trophic levels (e.g., the impact of potentially decreased Arctic cod abundance on marine mammals and birds) remains unknown (see also section 9.3.4).

For other fish species (e.g., Arctic char), alterations in environmental conditions could possibly shift or reduce the availability of preferred habitats of certain morphs, leading, in the extreme case, to the extirpation of particular morphs from certain locations. For example, pelagic forms of Arctic char in Thingvallavatn, Iceland, occupy portions of the water column that experience summer heating. Should such heating ultimately exceed thermal preferences for this morph, its growth is likely to decrease, with a concomitant reduction in reproduction and productivity. Ultimately, exclusion from the habitat during critical times could possibly occur, permanently extirpating that morph from such areas.

Changes in habitat characteristics driven by climate change are also likely to differentially affect specific populations of fish. For example, some aspects of life-history variation in Dolly Varden on the Yukon north slope appear to be particularly associated with inter-river variation in groundwater thermal properties (e.g., egg size is larger and development time is shorter in rivers that have significant groundwater warming, and reproduction occurs annually in these warmer rivers because sea access allows for earlier feeding, compared to reproduction every two years or less often in colder rivers; Sandstrom, 1995). Thus, climate change effects that mimic this natural local inter-population variability are likely to result in similar shifts in populations presently occupying colder habitats.

A second major effect of climate change will probably be alterations in the geographic range of species, thereby affecting local and regional biodiversity. This is likely to occur through a combination of compression or loss of optimal habitat for "native" arctic species, and the northward expansion of "non-native" southern species. For instance, the large number of northward-flowing arctic rivers provides pathways for colonization of the mainland by freshwater species that, due to climatic limitations, are presently restricted to subarctic or temperate portions of the drainage basins. As climate change effects become more pronounced (e.g., degree-day boundaries or mean temperature isotherms shift northward), the more ecologically vagile species are likely to extend their geographic ranges northward (Oswood et al., 1992). In North America, for example, the distribution of yellow perch (*Perca flavescens*) is projected to expand northward beyond its current, primarily subarctic distribution. Traditional ecological knowledge from the western Canadian Arctic has identified new species of fish (Pacific salmon – *Oncorhynchus* spp. and least cisco – *Coregonus sardinella*) that were not previously present in some aquatic systems of the area (Krupnik and Jolly, 2002; see also Chapter 3, specifically Fig. 3.2). The complete consequences of such new colonizations are unknown, but could include the introduction of new diseases and/or parasites; population reduction or extirpation through competition for critical resources; increased predation; increased hybridization of closely related taxa; and others (see sections 8.5.1.1 and 8.5.2 for detailed discussions of climate-related range extensions in selected fish species and their potential ecological consequences).

Emergent aquatic plants are also expected to expand their distribution northward and thus alter the overall levels of primary production in ponds and small lakes in the Arctic. Alexander et al. (1980) reported total primary production of 300 to 400 g C/m²/yr in ponds of emergent *Carex* (covering one-third of the pond) in Barrow, Alaska, compared to total primary production of 1 g C/m²/yr for phytoplankton and 10 g C/m²/yr for epilithic algae. Traditional ecological observations by trappers on the Peace-Athabasca Delta of the Mackenzie River system, Canada, suggest that muskrat abundance is likely to increase in high-latitude lakes, ponds, and wetlands as emergent aquatic vegetation becomes more prominent (Thorpe, 1986). While the potential northern limit for emergent aquatic macrophytes is not fully known, their projected increased presence will clearly influence the overall productivity and structural complexity of arctic pond and lake habitats.

An overarching issue affecting the responses of arctic aquatic biota and related biodiversity to rapid climate change is "adaptive capacity". The magnitude of change in arctic climate projected for the next 100 years does not exceed that experienced previously, at least at a geological timescale. The future rate of change however, is very likely to be unprecedented. To survive such a challenge, arctic aquatic biota, especially those that are truly arctic in nature, must have the inherent capacity to adapt (i.e., have sufficient genetic capacity at the population level to evolve at the required rate); acclimate (i.e., the phenotypic ability at the population and/or individual level to survive in the new conditions); and/or move (i.e., emigrate to more optimal situations). High levels of diversity that are present below the species level in many arctic organisms imply that some evolutionary compensation for rapid climate change is possible. Taxa with short generation times (e.g., zooplankton) will be able to evolve more rapidly than those with longer generation times (e.g., fish). Furthermore, assessment of genetic variability for some taxa (e.g., mitochondrial DNA in Arctic char; Wilson et al., 1996) suggests that previous events that reduced genetic diversity may have limited their capacity for such rapid evolution. This will probably further hamper responses by

such taxa and, with the projected rapid rate of climate change and other factors (e.g., competition from new colonizers), is likely to result in an increased risk of local extirpation and/or extinction.

Many arctic taxa may already be pre-adapted to acclimate successfully to rapid change. For example, many organisms already have enzymes with different thermal optima to allow them to cope with changing environmental conditions. Such capacity, which is presumed but not demonstrated to exist in most arctic taxa, could possibly counterbalance the increased risk of extinction noted above. Taxa that are capable of emigrating to new areas have additional options to cope with rapid climate change, although access issues are likely to preclude such movements to suitable conditions.

Clearly, significant changes in aquatic biodiversity are very likely to result from climate change, and biota have varying capacities to cope with the rate of this change. Ecologically speaking, any change will have significant ramifications in that adjustments in the ecosystem will follow (sections 8.4.4.2, 8.4.4.3, 8.4.4.4, and 8.5). However, from the human perspective, important questions surround the perceived significance of such changes from economic, cultural, and value perspectives (see Chapters 3, 11, and 12 for discussions of possible socioeconomic implications).

8.4.4.2. Effects on food web structure and dynamics

The impacts of climate change on the structure and dynamics of aquatic food webs remain poorly understood. To date, many of the insights as to how arctic food webs will respond (directly or indirectly) to climate change effects have been obtained from either descriptive studies or a select few manipulative/experimental studies where ecosystem-level or food web manipulations were conducted and response variables measured. Stream processes and biotic populations of the Kuparuk River and Oksrukuyik Creek, Alaska, have been shown to be controlled by the geomorphology of the systems (i.e., input from nutrient-rich springs; Craig and McCart, 1975); climate (i.e., precipitation affects discharge, which affects insect and fish production; Deegan et al., 1999; Hershey et al., 1997); resource fluxes from the surrounding catchments (Peterson B. et al., 1993); and corresponding biotic interactions. For example, nutrient enrichment of the streams resulted in greater primary and fish production, and a corresponding increase in the abundance of benthic macroinvertebrates (Harvey et al., 1998; Peterson B. et al., 1993). In addition, after seven years of artificial enrichment of the Kuparuk River, the dominant primary producers changed from diatoms to mosses (Bowden et al., 1994), which subsequently altered the abundance, distribution, and taxonomic composition of the macroinvertebrate community (Bowden et al., 1999).

Other recent studies of arctic systems have identified the structural and functional importance of the microbial freshwater food web (Fig. 8.13). Work in this area has shown that the microbial food web can comprise a significant fraction of the total community biomass in arctic rivers and lakes, and that energy flow is routed through a diverse trophic network of microbial species displaying a wide array of nutritional modes (heterotrophic bacteria, phototrophic bacteria, phagotrophic protozoa, and mixotrophic flagellates; Vincent and Hobbie, 2000). How climate change will influence the response of the microbial food web is not entirely certain, but studies of temperate systems might help provide insight. Interestingly, research on microbial food webs of more temperate aquatic systems shows that in the absence of heavy grazing pressure on bacteria by macrozooplankton or benthic macroinvertebrates, the principal role of the microbial food web is the degradation (respiration) of organic matter (Kalff, 2002). Hence, the microbial food web is a significant source of energy to plankton, being largely responsible for recycling nutrients in the water column and thereby helping to sustain planktonic and benthic primary production and ultimately higher secondary and tertiary consumers in the food chain (Kalff, 2002). Projected increases in water temperature and inputs of DOC, particulate organic carbon (POC), and DIC arising from climate change are very likely to affect the structural and functional dynamics of the microbial food web, and are likely to increase rates of carbon processing. Pienitz et al. (1995) showed that the same abiotic parameters, along with lake morphometry, explain the greatest percentage of variance in diatom community composition in northwestern Canada. Furthermore, diatom community structure was highly correlated with DOC gradients in Siberian and subarctic Québec lakes (Fallu and Pienitz, 1999; Lotter et al., 1999). Hence, concomitant changes in the phytoplankton component of the food web probably will also cascade through the ecosystem.

Increasing temperature has the potential to alter the physiological rates (e.g., growth, respiration) of individuals, and the vital rates and resulting dynamics of populations (Beisner et al., 1997; McCauley and Murdoch, 1987). Mesocosm studies by Beisner et al. (1996, 1997), which investigated the influence of increasing temperature and food chain length on plankton predator–prey dynamics, showed that the predator–prey system is destabilized at higher temperatures (i.e., the macrozooplankton herbivore *Daphnia pulex* always became extinct), irrespective of the complexity of the food web (i.e., whether a two- or three-level food web was involved). Long-term studies of Toolik Lake, Alaska, project that rising temperatures are likely to eliminate lake trout (*Salvelinus namaycush*) populations in this lake, with concomitant impacts on the food web (see Box 8.3). The bioenergetics model used by McDonald M.E. et al. (1996) projects that a 3 °C rise in July epilimnetic (surface mixed-layer) temperatures could cause young-of-the-year lake trout to require eight times more food than at present just to maintain adequate condition. This requirement greatly exceeds the current food availability in the lake, although it is probable that food availability will increase as temperatures rise. Furthermore, the oxygen concentrations projected by the lake model (Hobbie et al., 1999) show that a future combination of higher temperatures and increased loading of total phosphorus would greatly reduce the hypolimnetic (bottom-water) habitat available for lake trout.

An example of top-down control through size-selective predation was found in ponds and lakes in Barrow, Alaska: lakes with fish had small and transparent *Daphnia longiremis*, while lakes without fish and all ponds had large and pigmented *D. middendorffiana* and *D. pulex* as well as fairy shrimp and the copepod *Heterocope* spp. (Stross et al., 1980). Rouse et al. (1997) concluded that since top predators (fish) in arctic systems tend to be long-lived, population changes owing to recruitment failure may not be reflected in the adult populations for many years. However, the effects of the eventual loss of top predators from these systems are likely to cascade through the food web, affecting the structure and function of both benthic and planktonic communities (Carpenter et al., 1992; Goyke and Hershey, 1992; Hanson et al., 1992; Hershey, 1990; Jeppesen et al., 2003; O'Brien et al., 1992).

Given the information presented in this section, it is very probable that climate change will substantially affect biological interactions, including trophic structure and food chain composition. With top-down and bottom-up processes operating simultaneously in ecosystems (Golden and Deegan, 1998; Hansson, 1992; McQueen et al., 1989; and references therein), the degree to which each process influences producer biomass varies (McQueen et al., 1989). Consequently, the well-established relationship between phosphorus and algal biomass may differ between systems with different levels of productivity. For example, in a two-level trophic system (relatively unproductive), grazing zooplankton may control the algal biomass and the expected positive relationship between chlorophyll-a (Chl-a) and total phosphorus (P) would not be observed. Therefore, differences in productivity and trophic level interactions may explain the discrepancy in the Chl-a–total P relationship between temperate and arctic lakes (Fig. 8.14). The low productivity that has been observed in many arctic lakes may limit the presence of fish predators (i.e.,

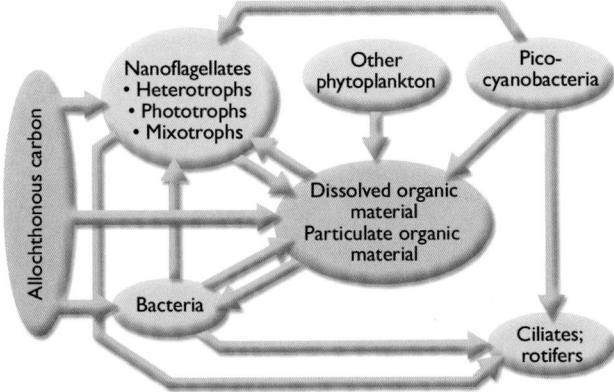

Fig. 8.13. Microbial freshwater food web in arctic lakes (adapted from Vincent and Hobbie, 2000).

more closely represent a two-level trophic system) and may result in systems where algal biomass is controlled by extensive zooplankton grazing (Flanagan et al., 2003).

Top-down control of food web structure in arctic stream and river ecosystems is also important. Golden and Deegan (1998) found that young Arctic grayling (*Thymallus arcticus*) have the potential to produce top-down cascading trophic effects in arctic streams where nutrients are not limited. The grayling were found to affect trophic structure through consumption, nutrient excretion, and the modification of prey behavior. Epilithic Chl-a increased with increasing fish density in both reference P-limited) and fertilized (P-enriched) zones of the Kuparuk River, Alaska, while mayfly density decreased with increasing fish density in the fertilized zone only. These results further illustrate that projecting climate change impacts is not straightforward.

8.4.4.3. Effects on primary and secondary production

Primary and secondary productivity relationships in arctic aquatic ecosystems are highly susceptible to structural and functional alterations resulting from changes in cli-

mate, although the direction and absolute magnitude of the responses are likely to be difficult to project (Hobbie et al., 1999; Laurion et al., 1997; Rouse et al., 1997; Vincent and Hobbie, 2000). For example, while constituents of microbial food webs (e.g., the picocyanobacteria, heterotrophic bacteria, etc.; Fig. 8.13) are likely to respond positively to temperature increases, the photosynthesis rate in the picoplankton fraction (0.2–2 µm) is strongly stimulated by increased temperature to a greater extent than nanoplankton (2–20 µm) and microplankton (20–200 µm) fractions (Rae and Vincent, 1998b).

In general, lake primary productivity will probably increase because higher temperatures correlate with higher primary productivity (a longer ice-free season and more sunlight before the summer solstice are very likely to result in greater primary production by plankton; see Box 8.5). Brylinsky and Mann (1973) analyzed lake productivity in 55 lakes and reservoirs from the tropics to the Arctic, and found the best abiotic variables for estimating productivity to be latitude and air temperature. A closer examination of the relationship between total P, total nitrogen, latitude, and algal biomass (n=433 lake years) also revealed that average algal biomass during the ice-free season is significantly negatively related to the latitude of

Box 8.5. Productivity of northeastern Greenland lakes: species composition and abundance with rising temperatures

The Danish BioBasis monitoring program (part of Zackenberg Ecological Research Operations), initiated in 1995, includes two lakes located in the Zackenberg Valley, northeastern Greenland (74° N). The monitoring at Zackenberg is expected to continue for at least 50 years. The area is situated in a high-arctic permafrost area in North-East Greenland National Park. More information about the area and the monitoring program can be found in Meltofte and Thing (1997) and Christoffersen and Jeppesen (2000), as well as at http://biobasis.dmu.dk. The two lakes, Sommerfuglesø and Langemandssø, have areas of 1.7 and 1.1 ha and maximum depths of 1.8 m and 6.1 m, respectively. Lakes and ponds in the area are ice-covered for most of the year, except from the end of July to the beginning of September. Most water bodies probably freeze solid during the late winter and spring. Primary producers in these nutrient-poor lakes are dinophytes, chlorophytes, and diatoms. Zooplankton grazers are sparse, consisting of *Daphnia*, copepods, and protozoans. Benthic invertebrates include *Lepidurus*. There are no fish in Sommerfuglesø. Dwarf Arctic char in Langemandssø prey on *Daphnia*, therefore, copepods and rotifers dominate zooplankton populations in this lake.

Plants and animals in Sommerfuglesø and Langemandssø are active prior to ice melt, thus phytoplankton biomass becomes substantial as incoming solar radiation increases (Rigler, 1978). However, primary productivity slows as nutrients are consumed, and phytoplankton density varies annually with nutrient abundance. For example, in warmer years, nutrient concentrations and thus productivity are higher due to increased loading of nutrients and humus from catchments as the active layer thaws.

Monitoring of plankton species in the two lakes during 1999 (a year of late ice melt and low water temperatures) and 2001 (a year of early ice melt and high water temperatures) has shown that not only do biomass and abundance change with temperature, but species composition changes as well. In 1999, when water temperatures were lower, chrysophytes and dinophytes represented 93% of total phytoplankton abundance in Sommerfuglesø, while dinophytes dominated phytoplankton (89% of total abundance) in the deeper and colder Langemandssø. In 2001, when ice broke early and water temperatures were higher, chrysophytes completely dominated phytoplankton in both lakes (94–95% of total phytoplankton abundance). Compared to 1999, total phytoplankton abundance was approximately twice as great in 2001, when nutrient levels were higher as well. Zooplankton abundance, in turn, was 2.5 times greater in 2001 than in 1999 in both lakes, likely in response to greater phytoplankton abundance. *Daphnia* and copepods were more abundant in 2001, while rotifers were less abundant than in 1999, perhaps in response to competition for food resources.

the system, independent of the nutrient concentration (Flanagan et al., 2003). This strong latitudinal effect on algal biomass yield suggests that arctic lakes are likely to show a significant increase in productivity if temperature and nutrient loadings in these systems increase as scenarios of future climate change project. While Shortreed and Stockner (1986) found that arctic lakes have lower primary productivity than temperate lakes, Flanagan et al. (2003) showed that at a given level of phosphorus, the productivity of arctic lakes is significantly less than lakes in the temperate zone, with the biomass of lower (trophic level) producers not accounted for simply by lower nutrient concentrations in the Arctic. Further examination of detailed observations of phytoplankton community structure from arctic Long-Term Ecological Research sites indicates that there is no fundamental shift in taxonomic group composition between temperate and arctic phytoplankton communities. This suggests that the difference in the Chl-a–total P relationship between temperate (Watson et al., 1997) and arctic lakes is not an artifact of changes in the Chl-a–biomass ratio resulting from a taxonomic shift in algal communities (Flanagan et al., 2003). Hence, the observed difference in the Chl-a–total P relationship for temperate and arctic lakes may provide insight about the future effects of climate change (Fig. 8.14).

Primary productivity is likely to increase if climate conditions at high latitudes become more suitable for industrial development, and if the associated pollution of currently nutrient-poor aquatic systems increases. For example, mountain lakes in the Kola Peninsula (e.g., Imandra Lake) and lakes and ponds in the Bolshezemelskaya tundra are currently stressed by heavy loadings of anthropogenic organic matter, heavy metals, and crude oil and drilling fluid, as well as thermal pollution. Phytoplankton structure (e.g., species) in these systems has changed, and primary as well as secondary productivity and biomass have increased significantly.

Arctic lakes, although relatively unproductive at present, will probably experience a significant increase in productivity as climate changes. If temperature and nutrient loads increase as projected, it is likely that phytoplankton will no longer experience temperature-induced photosynthetic rate inhibition, and growth rates will probably become more similar to those in the temperate zone, thus allowing for a greater accumulation of algae. If algae are heavily grazed by herbivores at present because of a lack of predation, higher-level predators are likely to invade as the productivity of the system increases. Subsequent increased predation of the grazer community

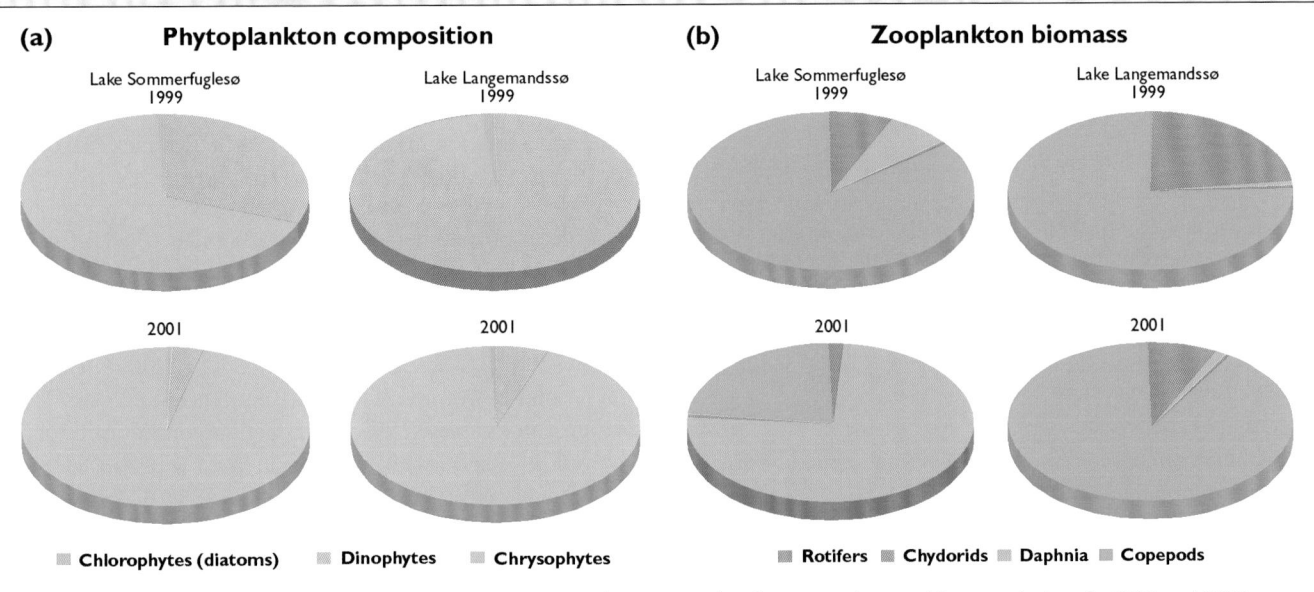

(a) Phytoplankton composition

Lake Sommerfuglesø 1999   Lake Langemandssø 1999

2001   2001

▨ Chlorophytes (diatoms)   ▨ Dinophytes   ▨ Chrysophytes

(b) Zooplankton biomass

Lake Sommerfuglesø 1999   Lake Langemandssø 1999

2001   2001

▨ Rotifers   ▨ Chydorids   ▨ Daphnia   ▨ Copepods

Fig. 1. (a) Phytoplankton composition (average of three samples per year) in Langemandssø and Sommerfuglesø in 1999 and 2001. In both years, total plankton abundance in Langemandssø was twice that in Sommerfuglesø; (b) relative biomass (dry weight) of different zooplankton groups in mid-August 1999 and 2001 in Langemandssø and Sommerfuglesø. In both years, total zooplankton biomass in Sommerfuglesø was approximately twice that in Langemandssø (data provided by BioBasis, http://biobasis.dmu.dk).

The climate of northeastern Greenland is projected to become more maritime in the future. Based on the five years of monitoring at Zackenberg thus far, increasing temperature and precipitation are projected to have major impacts on physicochemical and biological variables in the lakes. If snowfall increases, ice-cover duration is likely to be prolonged, shortening the growing season and reducing productivity, and possibly reducing food availability to the top predator in arctic lakes, the Arctic char. Greater runoff will probably increase nutrient loading and primary productivity, which could possibly result in oxygen depletion and winter fish kill. Thus, one probable outcome of climate change will be the extirpation of local fish populations in shallow lakes in similar ecological situations. Increased particulate loading is likely to limit light penetration for photosynthesis. Increased humus input with snowmelt is also likely to limit light penetration, reducing UV radiation damage to biota.

(a)

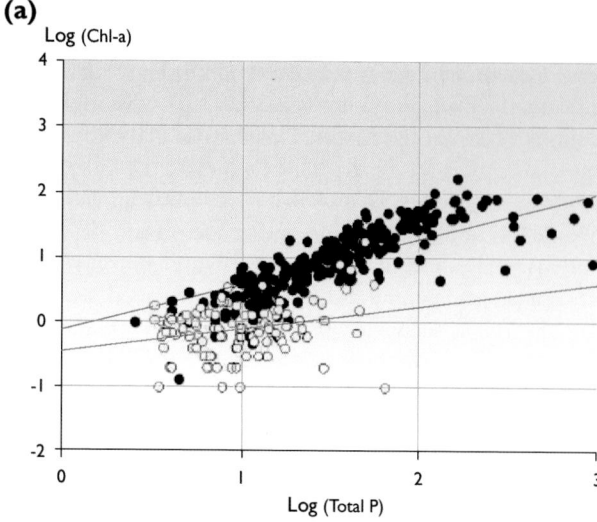

(b)

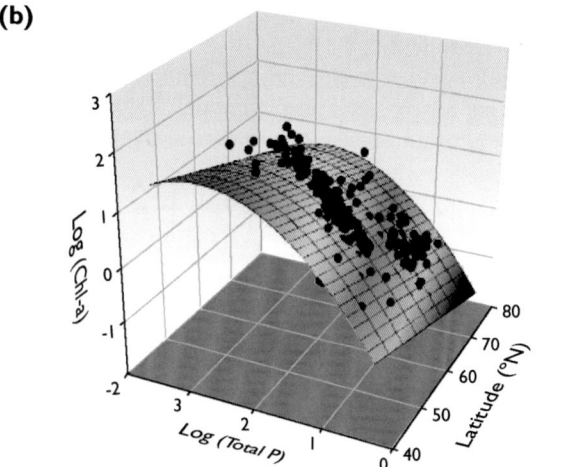

Fig. 8.14. (a) Comparison of the relationship between chlorophyll-a (Chl-a) and total phosphorus (P) in temperate and arctic freshwater systems. The difference between the slopes of the regression lines is statistically significant ($p<0.05$). Solid circles represent temperate lakes ($r^2=0.28$, n=316, $p<0.05$) and open circles represent arctic lakes ($r^2=0.07$, n=113, $p<0.05$). (b) The observed non-linear response of Chl-a to log total P and latitude (from Flanagan et al., 2003).

would permit an increase in algal biomass. In addition, the projected increase in nutrient concentration would augment these changes, making the increase in productivity even more dramatic.

Several empirical studies support this hypothesis. One study compared Swedish lakes (three-level trophic system) to unproductive antarctic lakes (two-level trophic system). The slope of the Chl-a–total P relationship for the antarctic lakes was significantly less than the Swedish lakes. This was hypothesized to be a consequence of the different trophic structures of the lakes, since productive (three-level) Swedish lakes showed a Chl-a–total P relationship similar to temperate lakes, suggesting that climate-related abiotic factors were not causing the differences between the Swedish and unproductive antarctic lakes (Hansson, 1992). Other empirical evidence supporting this hypothesis comes from subarctic lakes in the Yukon, which showed higher levels of zooplankton biomass relative to P concentrations than in temperate

regions, suggesting a two-level trophic system. Shortreed and Stockner (1986) attributed the high abundance of zooplankton to the low abundance of planktivorous fish, which led to an over-consumption of algae.

However, a significant factor further complicating the possible productivity response in arctic systems is the interaction of productivity with DOC. High DOC levels can differentially affect measured primary productivity by influencing light penetration (more DOC leads to darker water), affecting turbidity, and adding carbon for processing. For example, benthic diatoms and total diatom concentrations increased significantly during conditions of high DOC concentrations and low water transparency, whereas planktonic forms decreased (Pienitz and Vincent, 2000). In Southern Indian Lake, northern Manitoba, Hecky and Guildford (1984) found that high DOC concentrations decreased light penetration sufficiently to cause a switch from nutrient to light limitation of primary production. In shallow tundra ponds, over 90% of algal primary production was by benthic algae (Stanley D., 1976), although this level of productivity is very likely to decline if there is appreciable DOC-related light reduction. By contrast, increased DOC is likely to reduce harmful UV-B radiation levels, and thereby have a possible countervailing effect on productivity (Vincent and Hobbie, 2000).

Changes in primary productivity resulting from climate change, whether attributed to increased water temperatures or increased DOC loading, are likely to affect secondary production in arctic freshwaters. Productivity of lake zooplankton is very likely to rise in response to increases in primary production. At Toolik Lake, a 12-fold increase in primary production yielded a less than 2-fold increase in secondary production (O'Brien et al., 1992). This enhanced production is very likely to result in an increase in the abundance of secondary producers, as observed in Alaska, where the abundance of microplankton (rotifers, protozoans) rose with increased primary production (Rublee, 1992). Although larger zooplankton showed little species change with increasing productivity, microzooplankton increased both in number of species (i.e., biodiversity) and trophic levels (i.e., productivity). Observations by Kling et al. (1992a) indicated that zooplankton abundance and diversity are more sensitive to changes in primary productivity with latitude, with species number and types declining to the north, than to changes in lake primary productivity at any given latitude. Therefore, lake productivity and species abundance and diversity will probably shift in favor of zooplankton as primary production increases in a progressively northward direction with climate change.

Climate change is unlikely to affect bacterial species assemblages. Bahr et al. (1996) found that the species of plankton in Toolik Lake, Alaska, were identical to species found in other lakes in temperate regions. The overall productivity of the lake did not appear to be related to the species of bacteria involved, instead, the total bacterial biomass in the plankton was affected by overall primary

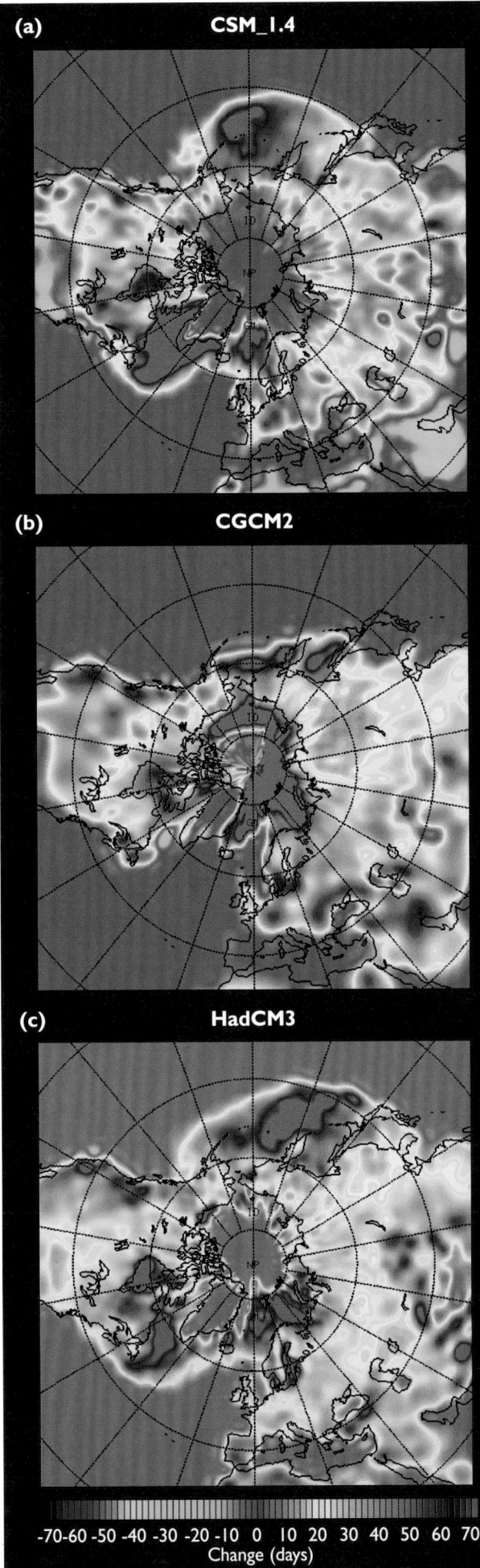

(a) CSM_1.4

(b) CGCM2

(c) HadCM3

-70 -60 -50 -40 -30 -20 -10 0 10 20 30 40 50 60 70
Change (days)

production and by the amount of allochthonous organic matter entering the lake from the drainage basin. Crump et al. (2003) and O'Brien et al. (1997) reported that over half of the bacterial productivity in Toolik Lake was based on terrestrial DOC. As a result, the bacterial numbers, biomass, and productivity in this lake are many times higher than they would be if it contained a plankton-based microbial food web. In contrast, protozoa and rotifer communities are likely to change with the increasing primary productivity that will probably result from climate change. For example, major changes occurred in protozoan, algal, and rotifer assemblages and production owing to significant artificial nutrient additions (four times ambient loading rates; Bettez et al., 2002).

8.4.4.4. Effects on carbon dynamics

The ACIA-designated models project that by 2080, the majority of the Arctic will experience increases in air temperature (section 4.4.2), precipitation (section 4.4.3), evaporation (section 6.2.3), available degree-days for biological growth, and major changes in the extent and nature of permafrost (section 6.6.1.3). Although there are variations between model projections, permafrost degradation is projected to occur most extensively at more southerly latitudes of the circumpolar Arctic, with regional west-to-east gradients across North America and Eurasia (for distribution among ACIA regions see Figs. 6.23 and 8.2). Overall (based on a "median" model projection; section 6.6.1.3), total permafrost area is projected to decrease by 11, 18, and 23% by 2030, 2050, and 2080, respectively. The loss of permafrost and deepening of the active layer is projected to be greatest in the western and southern areas of arctic and subarctic North America and Eurasia because initial permafrost temperatures are closer to 0 °C than in more easterly and northern areas, and these areas are more likely to become snow-free earlier in the spring, permitting enhanced soil warming (e.g., Groisman et al., 1994; section 6.4.4). Growing degree-days are projected to increase across the Arctic (Fig. 8.15), with the greatest increase in Regions 1 and 4, with the exception of Greenland where growing degree-days are projected to remain the same or decrease. On average, a 20- to 30-day increase in growing-season length is projected for areas north of 60° N by the end of the 21st century. As the number of degree-days increases in the Arctic, carbon cycling in arctic wetlands is very likely to not only be affected by changes in the rates and magnitudes of primary and microbial productivity, but also in the quantity and quality of organic material that accumulates in these systems. This in turn will affect carbon loading to, and processing within, arctic lakes and rivers.

Fig. 8.15. Projected changes in the length of the growing season defined as the number of days where the minimum temperature is greater than 0 °C, derived from the ACIA-designated models: (a) CSM_1.4 projected change between 1980–1999 and 2080–2099; (b) CGCM2 projected change between 1976–1994 and 2071–2090; and (c) HadCM3 projected change between 1971–1990 and 2071–2090 (see section 4.2.7 for model descriptions).

Wetlands are a very prominent feature of the Arctic, and are particularly sensitive to climate change. The structure of these systems, and their function as net sources or sinks of carbon, is likely to respond dramatically to changes in permafrost thawing, peatland distribution, and air temperatures and water budgets.

Thawing of perennially frozen wetland soil and ice is likely to result initially in a substantial efflux of carbon, as perennial stores of CO_2 and CH_4 are released to the atmosphere. Such an effect accounted for an estimated 1.6- to 3-fold increase in carbon emissions from degrading permafrost along the 0 °C isotherm in Canada (Turetsky et al., 2002). Permafrost thaw and warming has also accounted for a 100-fold increase in the rate of CO_2 and CH_4 formation in the Ob River basin (Panikov and Dedysh, 2000), and is responsible for drastically increased effluxes of these two gases from a high-latitude mire in Sweden (Friborg et al., 1997; Svensson et al., 1999). This initial increase in CO_2 and CH_4 emissions with permafrost thaw has potential positive climate feedbacks. The effect is likely to decline over time as gas stores are depleted, and as wetland vegetation, hydrology, and carbon sink/source function progressively change with climate.

Permafrost thaw and a greater number of growing degree-days are very likely to result in increased distribution and biomass of wetland vegetation at more northerly latitudes, increasing carbon storage in arctic and subarctic landscapes. Projections based on doubled atmospheric CO_2 concentrations (Gignac and Vitt, 1994; Nicholson and Gignac, 1995) indicate a probable 200 to 300 km northward migration of the southern boundary of peatlands in western Canada, and a significant change in their structure and vegetation all the way to the coast of the Arctic Ocean. Increases in carbon accumulation have been associated with peatland expansion, along with northward movement of the treeline, during Holocene warming, a process that slowed and eventually reversed with the onset of the Little Ice Age (section 8.3.2; Gajewski et al., 2000; Vardy et al., 1997). Similar expansion of peatlands and enhanced biomass accumulation have been recorded in North America in more recent times (Robinson and Moore, 2000; Turetsky et al., 2000; Vitt et al., 2000). Hence, as temperatures rise, wetland/ peatland distribution is likely to increase at high latitudes, and arctic landscapes are likely to become greater carbon sinks. Carbon accumulation at high latitudes is likely to be limited by loss due to disturbance (e.g., increased occurrence of fire as temperatures and evapotranspiration increase in some areas; Robinson and Moore, 2000; Turetsky et al., 2002), which will possibly result in greater carbon loading to lakes and rivers (Laing et al., 1999).

Changes in available growing degree-days, along with changes in the energy and water balances of high-latitude wetlands, will have varying effects on the rates and magnitudes of photosynthetic assimilation of CO_2, and anaer-

obic and aerobic production of CO_2 and CH_4 in existing arctic and subarctic wetlands:

- Rates and magnitudes of primary productivity, and hence carbon sequestration, are very likely to increase in arctic wetlands as air and soil temperatures rise, growing season lengthens (e.g., Greenland: Christensen et al., 2000; Finland: Laurila et al., 2001), and as vegetation changes (as discussed above in the context of permafrost degradation). Carbon fixation in arctic and subarctic wetlands will, however, possibly be limited by UV radiation effects on vegetation (Niemi et al., 2002; see also sections 7.4.2 and 8.6).
- Carbon dioxide accumulation in high-latitude wetlands is likely to be limited by warming and drying of wetland soils, and the associated production and loss of CO_2 through decomposition (e.g., Alaska: three-fold increase, Funk et al., 1994; Finland: Aurela et al., 2001). This effect is likely to lead to substantial losses of CO_2 and potential climate feedbacks.
- Methane production and emissions are likely to decline as high-latitude wetland soils dry with rising temperatures and increased evapotranspiration, and with regional declines in precipitation (e.g., Finland: Minkkinen et al., 2002; Greenland: Joabsson and Christensen, 2001). Moore T. and Roulet (1993) have suggested that only a 10 cm deepening of the water table in northern forested peatlands results in their conversion from a source to a sink of atmospheric CH_4. Methanotrophy is likely to be most pronounced in drier wetlands that tend toward aerobic conditions. The projected shift in vegetation toward woody species will possibly also limit CH_4 release to the atmosphere (section 7.5.3.2; Liblik et al., 1997).
- Methane production in some wetlands is likely to increase as temperatures and rates of methanogenesis increase, and as water tables rise in response to regional increases in water availability (e.g., Finland – projected 84% increase in CH_4 release from wet fen with 4.4 °C temperature increase, Hargreaves et al., 2001; Alaska – 8 to 33-fold increase in CH_4 emissions with high water table, Funk et al., 1994). Methane production will probably increase in those wetlands that have highly saturated soils and standing water, and those that may become wetter with future climate change, with potential climate feedbacks.

Overall, arctic and subarctic wetlands are likely to become greater sources of CO_2 (and in some instances CH_4) initially, as permafrost melts, and over the long term, as wetland soils dry (Gorham, 1991; Moore T. et al., 1998). Although many high-latitude wetlands are likely to experience a net loss of carbon to the atmosphere under future climate change, the expansion of wetland (e.g., peatland) distribution in the Arctic, and the increase in carbon accumulation with permafrost degradation, is likely to offset this loss (see section 7.5 for further treatment of this topic).

In addition to wetlands, wholly aquatic systems (rivers, lakes, and ponds) are also important to carbon cycling in the Arctic. Kling et al. (2000) showed that high-latitude lakes in Alaska were net producers of DOC, whereas streams were typically net consumers. Many arctic lakes and rivers are supersaturated with CO_2 and CH_4, often emitting these gases to the atmosphere via diffusion; increases in productivity (e.g., primary and secondary) deplete carbon in surface waters, resulting in diffusion of CO_2 into the water (see Fig. 8.7; Kling et al., 1992b; Schindler D.E. et al., 1997). Kling et al. (1992b) found that coastal freshwater systems release carbon in amounts equivalent to between 20 and 50% of the net rates of carbon accumulation in tundra environments. Enhanced loadings of carbon to arctic lakes and rivers as permafrost degrades (surface and groundwater flows contribute dissolved CO_2 and CH_4, as well as POC) will affect carbon cycling in these systems in a number of ways.

Dissolved organic carbon loading of lakes and rivers is likely to result in increased primary productivity and associated carbon fixation. This increase in photosynthetic CO_2 consumption by aquatic vegetation (e.g., algae, macrophytes) will possibly reduce emissions of this gas from lake waters to the atmosphere. This effect has been noted in experimental fertilization of both temperate and arctic lakes (Kling et al., 1992b; Schindler D.E. et al., 1997). Nutrient loading of high-latitude rivers, however, is unlikely to have a similar effect, as these waters have a rapid rate of renewal.

Although DOC loading of surface waters will possibly cause a decline in CO_2 emissions from some lakes, increased inputs of DOC and POC will possibly offset this effect and, in some cases, increase CO_2 production. Enhanced DOC and POC loads increase turbidity in some lakes, reducing photosynthesis (section 8.4.4.3). This rise in the availability of organic matter will probably result in a concomitant increase in benthic microbial respiration, which produces CO_2 (Ramlal et al., 1994). These effects are likely to be less pronounced or absent in flowing-water systems.

Increased nutrient loading and water temperature in high-latitude freshwater bodies are also likely to enhance methanogenesis in sediments. Slumping of ice-rich Pleistocene soils has been identified as a major source of CH_4 release from thermokarst lakes, such as in extensive areas of north Siberian lakes (e.g., Zimov et al., 1997), and may explain high winter concentrations of atmospheric CH_4 between 65° and 70° N (Semiletov, 2001; Zimov et al., 2001). Methane produced in such systems is often released to the atmosphere via ebullition, a process that will probably increase as the open-water season lengthens. Emission of CH_4 to the atmosphere is also likely to be enhanced in lakes, ponds, and streams that experience an increase in macrophytic growth, and an associated increase in vascular CH_4 transport.

8.5. Climate change effects on arctic fish, fisheries, and aquatic wildlife

Fish and wildlife intimately associated with arctic freshwater and estuarine systems are of great significance to local human populations (Chapter 3) as well as significant keystone components of the ecosystems (see e.g., section 8.2). Accordingly, interest in understanding the impacts of climate change on these components is very high. However, in addition to the problems outlined in section 8.1.1, detailed understanding of climate change impacts on higher-order biota is complicated by a number of factors:

- Fish and wildlife will experience first-order effects (e.g., increased growth in arctic taxa due to warmer conditions and higher productivity) of climate change as well as large numbers of second-order effects (e.g., increased competition with species extending their distribution northward). The responses of such biota will integrate these sources in complex and not readily discernible ways; further, responses to climate change will be embedded within those resulting from other impacts such as exploitation and habitat alteration, and it may be impossible to differentiate these. These multiple impacts are likely to act cumulatively or synergistically to affect arctic taxa.
- Higher-level ecosystem components affect lower levels in the ecosystem (i.e., top-down control) and in turn are affected by changes in those levels (i.e., bottom-up control). The balance between such controlling influences may shift in indiscernible ways in response to climate change.
- Higher-level ecosystem components typically migrate seasonally between habitats or areas key to their life histories – arctic freshwater fishes and aquatic mammals may do so locally, and aquatic birds tend to do so globally between arctic and non-arctic areas. Thus, the effects of climate change on such organisms will represent the integrated impacts across numerous habitats that indirectly affect the species of interest.

These biotic circumstances increase the uncertainty associated with developing understanding of species-specific responses to climate change, particularly for key fish and other aquatic species that are of economic and ecological importance to arctic freshwater ecosystems and the communities of northern residents that depend on them.

8.5.1. Information required to project responses of arctic fish

Implicit in much of the previous text is the linkage between atmospheric climate parameters and habitat parameters present in aquatic ecosystems, and the linkage of these to effects manifested in organisms and populations. It follows from this logic that changes in climate regimes, however they may manifest, will only indirectly affect aquatic organisms of interest. That is, the aquatic environment itself will be directly affected by changes in climate, but will modify and then transmit the influences in some fashion. Thus, for example, substantive shifts in atmospheric temperature regimes will affect water temperatures, but given the density differences between water and air and the influence of hydrodynamic factors, the effects on aquatic systems will be modified to some degree. In turn, changes in atmospheric parameters will have indirect effects on biota present in aquatic systems and thus may be ameliorated or partially buffered (e.g., thermal extremes or seasonal timing shifted). In some instances, however, climate change effects may be magnified or exacerbated, increasing the multiplicity of possible outcomes resulting from these changes. For example, stream networks amplify many environmental signals that occur at the watershed level, and that are concentrated in the stream channel (Dahm and Molles, 1992). This added level of complexity and uncertainty in the magnitude and direction of climate change manifestations in arctic freshwater ecosystems is not as acute for terrestrial environments. It results in greater uncertainty in projecting potential impacts on aquatic organisms. Figure 8.16 provides an example of the logical associations and direct and indirect effects of climate parameters on anadromous fish and the various aquatic environments used.

8.5.1.1. Fish and climate parameters

The Arctic as defined in this chapter (Fig. 8.2) includes high-, low-, and subarctic areas defined by climate, geography, and physical characteristics. In addition, many areas included in this assessment (e.g., southern

Alaska, the southern Northwest Territories, northern Scandinavia, and Russia) are significantly influenced by nearby southern maritime environments and/or large northward-flowing rivers. This proximity ameliorates local climatic regimes, resulting in more northerly distributions of aquatic taxa than would otherwise occur based strictly on latitudinal position. Moreover, the Arctic includes many different climatological zones. Given that the distribution of many freshwater and anadromous fish species is controlled or significantly influenced either directly or indirectly by climate variables (particularly temperature), it follows that primary associations of fish distribution with climate variables will be important.

Fish are ectotherms, thus, for the most part, their body temperature is governed by that of the surrounding waters. In addition, individual fish species can behaviorally choose specific thermal preferenda (preferred optimal temperatures; Beitinger and Fitzpatrick, 1979) at which physiological processes are optimal (i.e., greatest net benefit is achieved for the individual). This is typically a thermal range that may be fairly narrow; temperatures outside this are suboptimal (i.e., net benefit is still attained but it is not the greatest possible), grading to detrimental (i.e., non-lethal but net energy is expended while in such conditions) and ultimately to lethal conditions (i.e., death ensues after some level of exposure). Furthermore, within a species, local northern populations often have such preferenda set lower than do southern representatives, which presumably represents differential adaptation to local conditions. In addition, individual life stages (e.g., egg, alevin, juve-

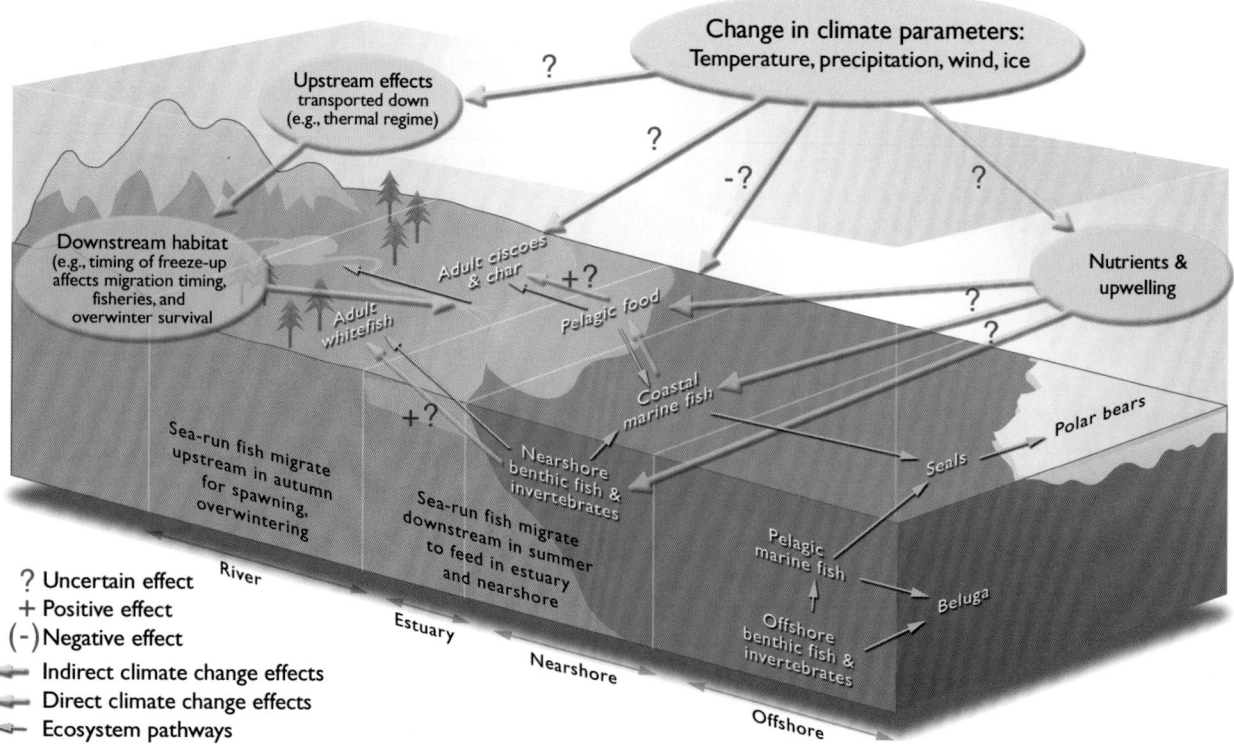

Fig. 8.16. A stylized portrayal of some potential direct effects of climate parameters on arctic aquatic environments and some potential indirect effects on aquatic organisms such as anadromous fish.

nile, adult) differ in their thermal preferenda linked to optimizing criteria specific to their developmental stage. For most species, only limited understanding of such thermal optima is available, and typically only for some life stages of southern species. Fish control body temperatures behaviorally, sensing and moving into appropriate, or from inappropriate, zones (Coutant, 1987). Aquatic thermal regimes are spatially and temporally heterogeneous and availability of water at the preferred temperature may be limited, making it an important resource for which competition may ensue. This may be particularly important in species found in Alaskan and Yukon North Slope rivers (e.g., Dolly Varden and Arctic grayling) during winter, when physical habitat is limited due to rivers freezing to the bottom over long reaches (Craig, 1989). Thus, the thermal niche of individual fish species can be defined.

Magnuson et al. (1979) grouped temperate species into three thermal guilds defined by thermal niches: Warmwater (preferred summer temperatures centered upon 27–31 °C), Coolwater (21–25 °C), and Coldwater (11–15 °C). Following this approach, Reist (1994) defined an Arctic Guild as fish distributed wholly or primarily in northern areas and adapted to relatively colder waters (<10 °C) and related aspects of the habitat such as short growing seasons, extensive ice presence, and long periods of darkness. Freshwater fishes occurring within the geographic definition of the Arctic as used here represent all of these guilds (Box 8.6), however, those of the Warmwater Guild tend to be present only along the southern margins of arctic waters, often associated with local climatic amelioration resulting from inputs from nearby maritime areas or northward-flowing rivers. Some of these guilds can be further subdivided based upon the nature of the fish distribution. Within the generalities discussed below, the impacts of climate change will be species- and ecosystem-specific, thus the following should be viewed as a range of possibilities only. In addition, although thermal regimes are emphasized in this discussion, the influence of other climate parameters may be equally or more important to specific species in particular areas or at particular times during life.

Species of the Arctic Guild have their center of distribution in the Arctic with the southern limits defined by, for example, high temperatures and associated ecological factors including competition from southern fish species. Fish such as broad whitefish (*Coregonus nasus*), Arctic cisco, and many char taxa are examples of Arctic Guild species. The pervasive and ultimate impacts of climate change upon such species are likely to be negative. These impacts generally will appear as range contractions northward driven by thermal warming that exceeds preferences or tolerances; related habitat changes; and/or increased competition, predation, or disease resulting from southern taxa extending their range northward, possibly preceded by local reductions in growth, productivity, and perhaps abundance. Many of these effects will possibly be driven or exacerbated by shifts in the life history of some species (e.g., from

anadromy to freshwater only). Other than conceptual summaries, no detailed research has been conducted to outline such impacts for most fish species of this guild.

Fish that have distributions in the southern arctic are northern members of the Coldwater Guild. This group includes species such as the lake/European/Siberian whitefish complex and lake trout, which have narrow thermal tolerances but usually are widely distributed due to the availability of colder habitats in water bodies (e.g., deeper layers in lakes; higher-elevation reaches in streams; Schelsinger and Regier, 1983). Two distributional subtypes can be differentiated: those exhibiting a wide thermal tolerance (eurythermal) as implied, for example, by a wide latitudinal distribution often extending well outside the Arctic (e.g., lake whitefish – *Coregonus clupeaformis*); and those exhibiting a narrow thermal tolerance (stenothermal) implied by occupation of very narrow microhabitats (e.g., lake trout occupy deep lakes below thermoclines in the south but a much wider variety of coldwater habitats in the north) and/or narrow latitudinal distribution centered in northern areas (e.g., pond smelt – *Hypomesus olidus*). The overall impacts of arctic climate change on these two distributional subtypes are likely to be quite different. Thus, eurythermal species are likely to have the capacity for reasonably quick adaptation to changing climate and, all other things being equal, are likely to exhibit increases in growth, reproduction, and overall productivity. Such species are also likely to extend the northern edge of their distribution further northward where this is at present thermally limited, but this is likely to be a secondary, relatively small response. Conversely, stenothermal coldwater species are likely to experience generally negative impacts. Lake trout in northern lakes, for example, will possibly be forced into smaller volumes of suitable summer habitat below deeper lake thermoclines and will possibly have to enter such areas earlier in the season than at present. Subsequent impacts on such species are very likely to be negative as well. To some degree, northern members of the Coldwater Guild are likely to experience the same general impacts as described for arctic-guild species in the previous paragraph (i.e., reductions in productivity characteristics, increased stress, local extirpation, and/or range contractions). Similar to the arctic-guild species, little or no detailed research assessing impacts on northern coldwater-guild fishes has been conducted to date.

Coolwater Guild species (such as perches) have southern, temperate centers of distribution but range northward to the southern areas of the Arctic as defined herein; these include northern pike (*Esox lucius*), walleye (*Sander vitreus*), and yellow perch (Schlesinger and Regier, 1983). Like those of the Coldwater Guild, these species can also be differentiated into eurythermal and stenothermal species. For example, the perches have a wide latitudinal range and occupy a number of ecological situations extending outside temperate regions, and hence can be described as eurythermal. Northward range extensions of approximately two to eight degrees of latitude are projected for yellow perch in North America under a climate change scenario where annual mean temperatures increase by 4 °C (Fig. 8.17; Shuter and Post, 1990). Shuter and Post (1990) found that the linkage between perch distribution and climate was indirect; that is, the first-order linkage was the direct dependency of overwinter survival (and related size at the end of the first summer of life) on food supply, which limited growth. The food supply, in turn, was dependent upon climate parameters. Alternatively, many northern minnows (e.g., northern redbelly dace – *Chrosomus eos* – in North America) and some coregonines (e.g., vendace – *Coregonus albula* – in Europe) are proba-

bly stenothermal, as implied by their limited latitudinal range and habitat associations. Range contraction along southern boundaries is likely for these species, initially manifested as contraction of distribution within the local landscapes, followed by northward retraction of the southern range limit. Because of their stenothermal tolerances, however, their northward extension is not likely to be as dramatic as that described for perch. To some degree, the presence of many of these species in the large northward-flowing arctic rivers such as the Lena, Mackenzie, Ob, and Yenisey is very likely to promote their northward penetration. The associated effects of heat transfer by such river systems will facilitate north-

Box 8.6. Freshwater and diadromous fishes of the Arctic

There are approximately 99 species in 48 genera of freshwater and diadromous (i.e., anadromous or catadromous forms moving between fresh and marine waters) fish present in the Arctic as defined in this chapter (Fig. 8.2). These represent 17 families (see table). Ninety-nine species is a conservative estimate because some groups (e.g., chars and whitefishes) in fact contain complexes of incompletely resolved species. Many species are also represented by local polymorphic forms that biologically act as species (e.g., four morphs of Arctic char in Thingvallavatn, Iceland). The most species-rich family is the Salmonidae with more than 33 species present, most of which are important in fisheries. The next most species-rich family is the Cyprinidae with 23 species, few of which are fished generally, although some may be fished locally. All remaining families have six or fewer species, and five families are represented in the Arctic by a single species. These generalities hold true for the individual ACIA regions as well. All of the families represented in the Arctic are also present in lower-latitude temperate and subtemperate regions. Most have a southern center of distribution, as do many of their associated species (Berra, 2001). Individual species may be confined to the Arctic, or may penetrate northward from subarctic areas to varying degrees.

Freshwater and diadromous fish present in the Arctic.

Family name	Number of arctic forms in ACIA area		Number of species in ACIA regions				Thermal guild[a]	Exploitation in ACIA regions	Comments
	Genera	Species	1	2	3	4			
Petromyzontidae (lampreys)	2	5	4	2	2	1	Cool	Some in Region 1	
Acipenseridae (sturgeons)	1	5	2	2	1	1	Warm/Cool	Region 1, some in Region 4	
Hiodontidae (goldeyes)	1	1	0	0	1	0	Warm	Region 3 where they occur and are abundant	Goldeye (*Hiodon alosoides*), in North America only
Anguillidae (freshwater eels)	1	2	1	0	0	1	Warm	Region 1, eastern Region 4	Mostly in southern areas only
Clupeidae (shads[b])	1	3	2	0	1	0	Warm/Cool	Region 1, limited in Region 3	Southern areas in interior; also in northern coastal areas influenced by warm currents
Cyprinidae (minnows)	14	23	11	4	13	1	Warm	Some species in Regions 1 and 2; not exploited or limited elsewhere	Most species only occur in southern ACIA area
Catostomidae (suckers)	2	3	0	1	2	3	Warm/Cool/Cold	Limited in Region 2 and western Region 4	
Cobitidae (loaches)	1	1	1	0	0	0	Warm	Not fished	Stone loach (*Noemacheilus barbatulus*) only. Subarctic only; very southern edge of Region 1
Esocidae (pikes)	1	1	1	1	1	1	Cool/Cold	Extensively fished in all four Regions	Northern pike (*Esox lucius*) only, and widely distributed

ward colonization by these species as well as eurythermal species also present in the systems. Knowledge of the association of ecological processes with climate parameters and research quantifying the potential impacts of climate change on coolwater-guild species, although inadequate overall, generally tends to be more comprehensive than for the previous two guilds, but is often focused upon southern populations. Hence, its applicability to arctic populations of the species may be limited.

Warmwater Guild species have their center of distribution well south of the Arctic. Those present in the Arctic as defined in this chapter are few in number (Box 8.6)

and with few exceptions (some cyprinid species) are generally distributed only in the extreme southern portions of the ACIA regions. In many areas of the Arctic, a number of species of this guild are present in southerly areas immediately outside the boundary of the Arctic. Presumably, their northward limit is in most cases determined by present thermal and ecological regimes, especially in the large northward-flowing rivers of Siberia and the western Northwest Territories. As the effects of climate change increasingly ameliorate local limiting factors, species of this guild are very likely to extend their geographic ranges into the Arctic or, if already there, to more northerly locales.

Substantive differences in the number of species present are apparent between the ACIA regions. Region 3 (unglaciated Beringia and the western Canadian Arctic) contains 58 named taxa, followed by Region 1 (Arctic Europe and Russia) with 38, while Regions 2 (Siberia) and 4 (eastern North America) are about equal at 29 and 32, respectively. This probably represents a combination of historical effects (e.g., glacial events, postglacial recolonization routes and access) as well as present-day influences such as local climate, habitat diversity, and ecological processes (e.g., competition and predation). Arctic char is the only species that is truly holarctic, being present on all landmasses in all ACIA regions, occurring the farthest north to the extremes of land distribution (~84° N), and also exhibiting the widest latitudinal range (about 40 degrees) of all true arctic species (i.e., south in suitable lakes to ~45° N). A few additional species are distributed almost completely across the Holarctic but are absent from one or more areas within an ACIA region (e.g., burbot with ~75% of a complete circumpolar distribution; northern pike with ~85%; lake whitefish, European whitefish, and Siberian whitefish (*Coregonus pidschian*) with ~85%; and ninespine stickleback with ~90%). With the exception of the stickleback, all are fished extensively where they occur, representing the mainstays of food fisheries for northern peoples and supporting significant commercial fisheries in most areas. These species are often the only ones present in extremely remote areas, inland areas, and higher-latitude areas, and thus are vital for local fisheries. Where they are regionally present, many other species are exploited to a greater or lesser degree.

Family name	Number of arctic forms in ACIA area		Number of species in ACIA regions				Thermal guild[a]	Exploitation in ACIA regions	Comments
	Genera	Species	1	2	3	4			
Umbridae (blackfish[c])	1	2	0	0	2	0	Cold/Arctic	Limited at most, where they occur	Blackfishes (*Dallia* spp.) only
Osmeridae (smelts)	2	3	0	0	2	0	Cool/Cold	Limited at most, where they occur	
Salmonidae (salmon, char, whitefishes, ciscoes)	10	33+	9	14	22	14	Cool/Cold/Arctic	Most species extensively fished in all four Regions	Salmonids are the most widely distributed and abundant arctic group and fisheries mainstay
Percopsidae (trout-perches)	1	1	0	0	1	1	Cool/Cold	Not fished	Trout perch (*Percopsis omiscomaycus*) only
Gadidae (cods)	1	1	1	1	1	1	Cool/Cold	Extensively fished in all four Regions	Burbot (*Lota lota*) only, and widely distributed
Gasterosteidae (sticklebacks)	3	3	2	1	3	2	Warm/Cool/Cold	Not fished	
Cottidae (sculpins)	2	6	2	2	4	3	Cool/Cold	Not fished	
Percidae (perches)	4	6	2	1	3	3	Warm/Cool	Fished where they occur, especially Region 1	Mostly temperate species, but enter the Arctic via warmer northward-flowing rivers
Totals	48	99	38	29	58	32			
Total families	17	--	12	10	14	12			

[a]Magnuson et al. (1979) and Reist (1994); see section 8.5.1.1 for definitions; [b]only Alosinae (shads) are arctic representatives; [c]only *Dallia* (blackfish) are arctic representatives.

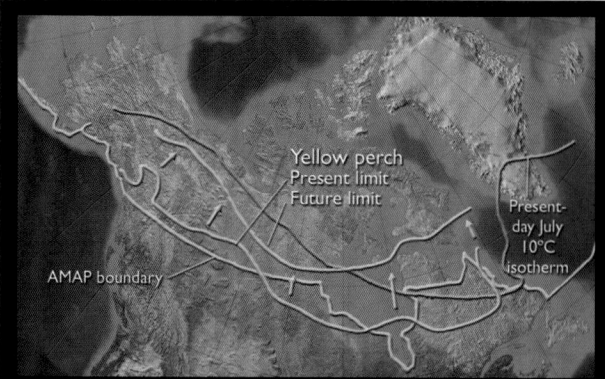

Fig. 8.17. Present and projected future distributional limits of yellow perch in North America. Northward displacements (shown by arrows) are based on overwinter survival assuming a 4 °C increase in mean annual temperature (adapted from Shuter and Post, 1990).

Thermal preferenda presumably optimize all internal physiological processes (i.e., benefits outweigh costs) in individual fish associated with digestion, growth, muscle (hence swimming) efficiency, gas exchange across gills, cellular respiration, reproduction, and so on. The relationship of temperature to such processes is perhaps most easily seen with respect to growth (e.g., increase in size or weight over time; Fig. 8.18). In addition to exhibiting higher growth rates at lower temperatures, arctic fish species also exhibit narrower ranges of temperature preference and tolerance (i.e., stenothermic; Fig. 8.18), which has profound effects on productivity. Stenothermic tolerances also imply that the species may have little capacity to accommodate thermal impacts of climate change. Conversely, species exhibiting eurythermic or wide thermal tolerances or responses are likely to have a much wider capacity to accommodate climate changes (see above and Box 8.6).

Population-level influences of thermal regimes are also apparent. Effects on individuals, such as temperature effects on mortality, feeding, parasitism, and predation, are integrated into consequences for fish populations through the various processes that connect fish populations to their ecosystems (Fig. 8.19). As noted previously, environmental parameters such as temperature may affect various life stages differently and thus can be modeled separately, but it is important to remember that the ultimate effects of all these influences are integrated throughout the fish population of interest. Similarly, environmental changes also have specific effects on other organisms relevant to fish, such as predators, parasites, and food organisms. Therefore, a single environmental parameter may exert both indirect and direct effects at many levels that influence the fish population, but the actual effect of this may be indiscernible from the effects of other natural and anthropogenic influences. Figure 8.19 provides examples of linkages between environmental parameters that affect key processes at the fish population level. Migratory aspects of life history are not shown in the figure, but will also (especially in anadromous fish) be significantly affected by abiotic processes.

Salinity will also be a factor for sea-run phases of adult life history. Climate change and increased variability in climate parameters will drive changes in aquatic abiotic parameters. Such changes will affect the fish directly as well as indirectly via impacts on their prey, predators, and parasites. This cascade of effects, and synergies and antagonisms among effects, greatly complicates the projection of climate change impacts on valued northern fish populations. In addition, other parameters not shown in Fig. 8.19, such as groundwater inflows to spawning beds, will affect the survival of various life-history stages. The ultimate effects of all these interacting factors will in turn affect sustainability of the populations and human uses in a fishery context.

Temperature effects on individual fish and fish populations are perhaps the most easily understood ones, however, other climate parameters such as precipitation (amount and type) will directly affect particular aquatic environmental parameters such as productivity (e.g., see Box 8.7) and flow regimes (amounts and timing). For example, flattening hydrographs and shifts in water sources (sections 8.4.2 and 8.4.3) are very likely to alter the availability of arctic rivers as migratory routes for anadromous fish. Increased and earlier vernal flows are very likely to enhance fish survival during out-migration and lengthen the potential summer feeding period at sea (both positive effects at the levels of the individual fish and the population). However, autumnal flows are required in many smaller rivers to provide access to returning fish (Jonsson, 1991); reduction in amounts and shifts in timing of these flows are very likely to have negative effects. Svenning and Gullestad (2002) examined environmental influences on migrations of Arctic char, particularly local temperature

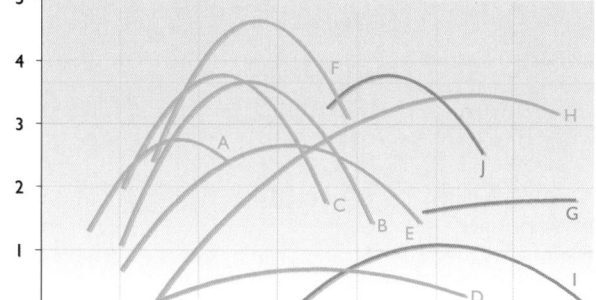

Fig. 8.18. Growth rates of fish species at varying temperatures determined from laboratory studies. Stenothermic northern species (e.g., A, B, C, and F) are grouped towards the lower temperatures on the left, whereas mesothermic southern species (e.g., G, I, and J) are grouped towards the right. Stenothermic species tend to have a more peaked curve indicating only narrow and typically low temperature ranges over which optimal growth is achieved. Wide-ranging eurythermic species (e.g., D, E, and H) probably exhibit the greatest possibilities for adapting rapidly to shifting thermal regimes driven by climate change.

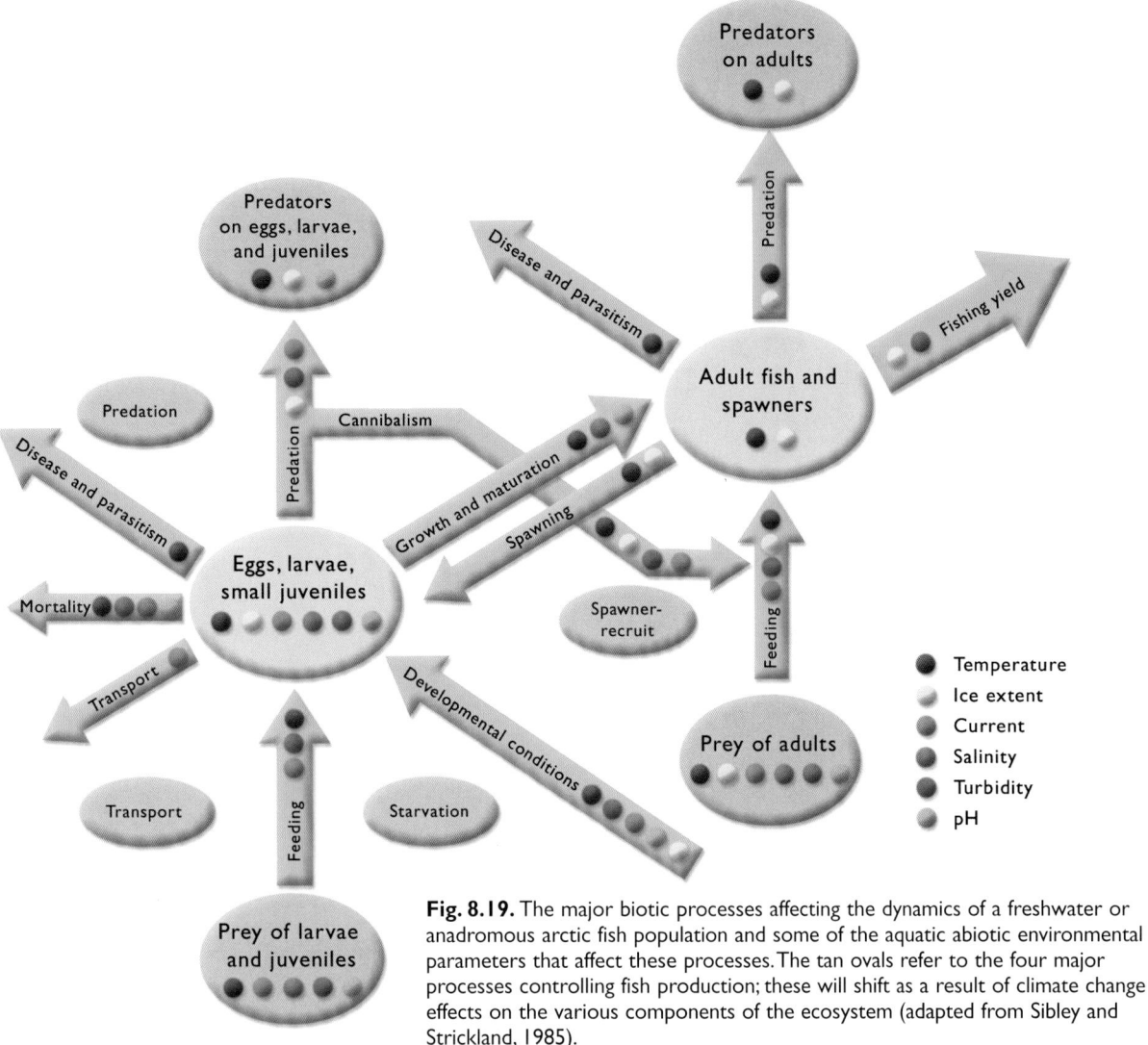

Fig. 8.19. The major biotic processes affecting the dynamics of a freshwater or anadromous arctic fish population and some of the aquatic abiotic environmental parameters that affect these processes. The tan ovals refer to the four major processes controlling fish production; these will shift as a result of climate change effects on the various components of the ecosystem (adapted from Sibley and Strickland, 1985).

effects on flow regimes and the consequences for fish population abundance and structure.

Additional secondary environmental factors that may change in response to direct changes in basic climate parameters will also have important effects on aquatic biota. These include the nature and duration of freeze-up, ice types, ice-cover periods, and breakup, and the nature and penetration of incident radiation into aquatic systems. Similarly, terrestrial impacts of climate change may influence aquatic habitat and indirectly affect its biota (e.g., permafrost alteration and runoff influences on sediment loads, pH and related water chemistry, etc.). Another potential class of indirect effects of climate change includes those affecting the behavior of aquatic biota. For example, fish use thermal regimes and spatiotemporal shifts in these regimes, at least in part, as behavioral cues or thresholds to trigger critical life history functions. Water mass boundaries defined by temperature act as barriers to movement and may define feeding areas (Coutant, 1987). Final gametic maturation in autumn-spawning species is probably triggered by decreasing water temperatures and perhaps also photoperiod in arctic whitefishes. There is anecdotal evidence that decreased sediment loads resulting from freezing of riverbanks trigger final

upstream movements by broad whitefish from holding areas to spawning sites (Reist and Chang-Kue, 1997), an adaptation to ensure eggs are not smothered. Water temperature integrated over time (e.g., as degree-days) affects the rate of egg development. Thus, aquatic thermal regimes affect ectotherms such as fish in two basic ways: by influencing physiology and as cues for behavioral changes. Although typically less understood, similar effects probably result from other physical (e.g., currents, flows, turbidity, ice dynamics) and chemical (e.g., pH, oxygen) parameters in the aquatic habitat (e.g., Sibley and Strickland, 1985). Climate-induced alteration of these habitat characteristics is very likely to significantly affect arctic fish populations, although substantive research is required to quantify such effects.

Freshwater and diadromous fishes of the Arctic exhibit high diversity in the way that climate parameters affect their distributions, physiology, and ecology. These factors, together with the more complex indirect effects that climate may have upon their habitats, implies a wide range of possible responses to climate change. Other than logical extrapolations, most responses to climate change are impossible to quantify due to the absence of basic biological information for most arctic

Box 8.7. Effects of environmental change on life-history and population characteristics of Labrador Arctic char

Present-day relationships between environmental and biological parameters must be understood to provide the foundation for assessing future climate change effects on fish populations. The general lack of such understanding for most arctic fishes currently precludes in-depth development of comprehensive and accurate qualitative scenarios of impacts, and limits quantification of effects under those scenarios. Development of such understanding requires substantive long-term data that are relatively sparse for most arctic fish; a circumstance that demands redressing. A notable exception is the availability of data for Arctic char. The distribution and life-history patterns of Arctic char are complex, and few attempts have been made to relate fluctuations in abundance, catch rates, and stock characteristics to environmental variables such as temperature and precipitation. The table lists associations between biology and variability in environmental parameters for Arctic char from northern Labrador, Canada.

Environmental associations for Nain Arctic char.

Timing	Significant environmental parameters	Probable environmental effect	Observed biological effect on individual fish	Observed biological effect on fish population
Within year	Summer air temperature	Increased marine productivity within limits	Increased weight / Increased length	Better condition
	Sea-surface temperature		Increased growth	
First summer of life	Winter precipitation	Increased snowpack / Decreased seasonal freezing	Increased overwinter survival	More fish
		More overwintering habitat	Decreased energetic costs for maintenance (increased growth)	Earlier recruitment to the fishery (i.e., lower age-at-catch)
Fourth year of life (first year at sea)	Summer air temperature	Increased nutrient loading to nearshore habitat	Increased growth	Increased weight at catch
	Summer precipitation	Increased nearshore productivity	Increased survival	Decreased age-at-catch

Long-term (1977–1997) monitoring of the char fishery at Nain, Labrador (56° 32' N, 61° 41' W) has produced data on both anadromous fish and environmental variables that have been applied in assessing long-term variability in catch biometrics (Power M. et al., 2000). Climate variability, particularly annual and seasonal, was found to have effects at critical life-history stages, and to affect average stock age, weight, and length characteristics, thus determining the dynamics of exploited Arctic char populations several years later and their eventual spawning success (Power M. et al., 2000). The table also summarizes aspects of climate variability and the probable effects on the population.

Mean age-at-catch and weight of Arctic char from the Nain fishery declined significantly, with a lag of four years, in response to high summer precipitation. This precipitation-related change is probably due to fluctuations in river flow and nutrient dynamics during the initial migration of Arctic char to nearshore marine areas. First-time migrants tend to stay in the nearshore areas (Berg, 1995; Bouillon and Dempson, 1989) and are most likely to be immedi-

fish species and the incomplete understanding of the overall associations of ecological processes with present-day climate parameters.

8.5.1.2. Ecosystems, habitat, and fish: climate change in the context of multiple stressors

Aquatic ecosystems are highly structured and complex entities consisting of both abiotic and biotic elements, and functional relationships within and between those elements. Similarly, individual components of ecosystems, such as a single fish species, exhibit a unique but complex structure. From the perspective of an individual or even a population, climate parameters and change in them may be experienced as either a stressor (i.e., that which perturbs homeostatic systems; Adams

S., 1990), or as a promoter (i.e., that which promotes homeostasis). Stressors and promoters directly and indirectly influence underlying physiological processes and their outcomes at both the level of the individual and that of the population. Points of action within individuals range from the molecular to the organismal level; those affecting whole individuals overlap with effects on populations and communities (Adams S., 1990). Events occurring at other levels of the hierarchy also influence the various organizational levels. Levels within the individual tend to have short-term responses and limited overall ecological relevance, whereas those at the population/community/ecosystem level tend to be longer-term responses with higher ecological relevance (Adams S., 1990). Given the structured complexity inherent in a fish population, the effect of any particular ately affected by changes in nutrient inputs resulting from variability in river flow. High-precipitation years increase nutrient and POC exports from river and lake catchments (Allan, 1995; Meyer and Likens, 1979), which increase nutrient inputs to nearshore marine feeding areas and probably increases productivity at all trophic levels.

The significance of increased winter precipitation is related to events occurring in the first critical winter of life for char. Heavier, more frequent snowfalls in Labrador maintain ice cover in an isothermal state and limit ice thickness (Gerard, 1990). Deeper snowpack maintains taliks, or unfrozen areas, in lake and river beds (Allan, 1995; Prowse, 1990), improving winter refugia conducive to fish survival (Allan, 1995; Power G. and Barton, 1987; Power G. et al., 1999).

The possible effects of temperature on Arctic char are complex. Mean fish length increased with rising summer temperatures and the persistence of optimal growth temperatures (12–16 °C) over a longer period of time (Baker, 1983; Johnson, 1980). High spring temperatures and accelerated ice breakup, however, can have negative effects on populations migrating with ice breakup (Nilssen et al., 1997). In the Fraser River (Labrador), breakup typically occurs in late April or early May (Dempson and Green, 1985) and would be well advanced, as would seaward migration, in years experiencing above-normal May temperatures. Although temperature increases can advance preparatory adaptations for marine residency (i.e., smoltification), they also result in a more rapid loss of salinity tolerances and a shortening of the period for successful downstream migration (McCormick et al., 1997). Rapid increases in temperatures are likely to impinge on the development of hypo-osmoregulatory capabilities in migrants and decrease growth due to the increased energetic costs of osmoregulatory stress, increase the probability of death during migrations to the sea, and decrease average growth by reducing the average duration of marine residence.

Several conclusions arise from this study:

- Long-term, comprehensive biological and environmental datasets are critical to assess and monitor climate change impacts on fish populations.
- Climate variables are very important in understanding year-to-year variability in stock characteristics.
- Causative relationships appear to exist between life history and environment but precise roles played, timing of the effect, and limits to the effect need more thorough investigation.
- For long-lived arctic fish, the effects of particular environmental conditions are often lagged by many years, with cascading effects on fishery production and management.
- Environmental effects are manifested in the fish population in the same way that other effects such as exploitation are (e.g., in terms of individual growth that translates into survival, fitness, reproduction, and ultimately into population-dynamic parameters such as abundance), thus distinguishing specific effects of climate change from other proximate drivers may be problematic.
- Particular environmental effects tended to reinforce each other with respect to their effect on the fish; although generally positive in this study, effects from several environmental parameters could presumably act antagonistically resulting in no net effect, or could synergistically act in a negative fashion to substantially impact the population.

stressor or promoter can be manifested at many levels simultaneously; can interact with others in an additive or cumulative fashion; and can typically be observed (if at all) in wild populations only at the more general population or community level.

For example, the effect of a temperature change may induce a short-term physiological response in an individual (e.g., processes occurring outside of the zone of optimal enzyme performance), a medium-term acclimatization response such as expression of new enzyme alleles optimal for the new temperature, and a somewhat longer-term response of changed biological condition. Concomitant expressions at the population level might be a shift in age structure, lower overall abundance, and ultimately local extirpation or adaptation to new conditions.

Another important point is that at least locally, the impacts of climate change on fish populations will be but one of several stressors. Other stressors affecting arctic fish both now and in the future include exploitation, local habitat change due to industrial development or river regulation, contaminant loadings, and changes in incident UV radiation levels. These stressors will result in effects similar to those described for climate change at individual and population levels. However, all these stressors will also interact additively and multiplicatively on individual fish and fish populations; hence, the effects are likely to be cumulative (Reist, 1997b). Perhaps the greatest future challenge associated with climate change will be to effectively recognize and manage in an integrated fashion all potential and realized impacts on arctic fish populations to ensure their conservation and sustainability.

8.5.1.3. Effects of climate-induced changes on physical habitat

Physical changes in aquatic habitats will very probably affect arctic fishes as climate changes in the north. This section provides some examples to illustrate the linkages and various potential effects on biota, but the underlying absence of data precludes quantification of causal linkages in most cases. Rectifying these and similar knowledge gaps is a major future challenge.

Groundwater and fish

Groundwater flows sustain fish habitat and are extremely important during periods of low flow in many arctic rivers (Power G. et al., 1999) and perhaps some lakes (sections 8.2 and 8.4). For stream-dwelling salmonids, inflows along stream bottoms clear fine-grained sediments from spawning areas, supply thermally regulated and oxygenated water to developing eggs and larval fish, and in many cases provide physical living space for juvenile and adult fish. In highly channeled shallow arctic rivers that characterize many areas of the North American Arctic and Chukotka, groundwater inputs are critical to fish migrations and stranding prevention (Power G. et al., 1999). In winter, many Alaskan and western Canadian North Slope rivers cease flowing and freeze to the bottom over large stretches, and groundwater provides refugia that support entire populations of Arctic grayling and Dolly Varden as well as any co-occurring species (Craig, 1989). Overwintering mortality, especially of adults weakened from spawning activities, is suspected as a primary regulator of the populations of Dolly Varden in this area and a major factor in such mortality is the quality and amount of winter habitat maintained by groundwater. Possible increases in groundwater flows resulting from climate change are likely to positively affect overwinter survival, especially if coupled with shorter duration and thinner ice cover. However, increased nutrient loadings in groundwater will possibly have more complex impacts (e.g., increases in in-stream primary and secondary productivity are likely to promote growth and survival of larval fish, but increases in winter oxygen demand associated with vegetation decomposition will possibly decrease overwinter survival of larger fish). How these various effects will balance in specific situations to result in an overall net effect on particular fish populations is unknown.

In summer, ground and surface water inflows ameliorate summer temperatures and provide thermal refugia, especially along southern distributional margins (Power G. et al., 1999). This is probably especially relevant for fish belonging to the arctic and coldwater thermal guilds. However, even the small increases in water temperatures (2 to 4 °C) that are likely to result from climate change (e.g., warmer surface flows) will possibly preclude some species from specific aquatic habitats (e.g., temperature in higher-elevation cold-water stream reaches determines habitat occupancy of bull trout – *Salvelinus confluentus*; Paul and Post, 2001). Increased

ambient conditions above physiological thermal optima are very likely to further stress populations and, combined with other possible effects such as competition from colonizing southern taxa, such impacts are likely to exacerbate range contractions for arctic species.

Ice and fish

The influence of ice on arctic fish and fish habitat is significant, especially in smaller lotic systems important to salmonids (Craig, 1989; Cunjak et al., 1998; Power G. et al., 1999; Prowse, 2001a,b). Effects include possible physical damage (e.g., from frazil ice), limitation of access to habitat (e.g., decreasing water volumes in winter due to ice growth), and annual recharge of habitat structure during dynamic breakup (e.g., cleansing of interstitial spaces in gravel). Shifts in the timing and duration of ice-related events are very likely to affect the survival and success of fish, with some effects being advantageous and others disadvantageous. In the north, these effects will be superimposed upon a poorly known but complex biological and environmental situation. Limited knowledge precludes accurate forecasting of many of these potential effects, and novel approaches are required to redress this (Cunjak et al., 1998).

Decoupling of environmental cues due to differential effects of climate change

A speculative issue, which may present surprises and unanticipated effects, is the potential for decoupling of various types of environmental drivers due to climate change affecting some differentially. Fish and other organisms use progressive and/or cusp-like changes in environmental parameters as cues to trigger key life-history functions such as migration, reproduction, and development. For example, although quantitative linkages are lacking, change in photoperiod (e.g., declining light period) is probably coupled with declining water temperatures in the autumn and together these trigger final gonadal maturation and reproductive activities in many northern fishes (especially salmonids). Environmental cues that drive major life-history events are especially critical for migratory species, and in the Arctic, particularly for anadromous species. This coupling is probably especially strong in the north where both parameters change rapidly on a seasonal basis. Although not explored to date in the context of climate change, as seasonal photoperiod shifts remain unchanged but coincident cues such as declining temperatures occur later in the autumn, such decoupling will possibly have profound impacts on population processes. The initial impact of such decoupling may be quite subtle (e.g., lowered fecundity, fertilization success, and/or egg survival in the previous example), not readily discernible, and almost certainly not directly attributable to climate change. However, a critical threshold is likely to be reached when impacts become significant (e.g., total reproductive failure in one year resulting in a failed year class, ultimately leading to population extirpation if it occurs over successive years approaching the generation time of the population). Investigation of coupling between cues, their influence upon population processes in fish and other aquatic organisms, and their potential for decoupling due to climate change in the Arctic should be a priority.

These are but a few examples of likely influences of physical habitat on fish populations and the potential effects of climate-induced change on them that will have cascading effects on the integrity, sustainability, and future productivity of northern fishes. These serve to illustrate the general lack of knowledge that exists regarding associations between physical habitat and biology in northern aquatic biota, and thus how the impacts of climate change impacts will manifest. Redress of this knowledge gap is required on a community- and/or species-specific basis to account for local and historical influences and filters, which greatly affect the present-day structure and function of these aquatic ecosystems (Tonn, 1990).

8.5.1.4. Issues at the level of fish populations

As implied previously, projecting climate change impacts at the population level for most species is complex and fraught with uncertainty, especially for arctic species for which there is a dearth of fundamental biological information. A variety of approaches to address this problem are available (section 8.5.2) and most have been applied in one way or another to develop some understanding of climate change impacts on northern fish populations.

In North America, much of the research focus on climate change effects on freshwater fish populations and communities has been in the south, for example, in the Great Lakes region and associated fisheries (e.g., Assel, 1991; Hill and Magnuson, 1990; Magnuson et al., 1990; Meisner et al., 1987; Minns and Moore, 1992; Regier et al., 1990, 1996; Shuter and Post, 1990; Smith J., 1991). In that region, climate change is projected to result in effects similar to those projected for the Arctic (e.g., significant reductions in the duration and extent of ice cover, an earlier seasonal disappearance of the 4 °C depth isotherm, measurable declines in DO, and slight hypolimnetic anoxia in shallower basins; Blumberg and Di Toro, 1990; Schertzer and Sawchuk, 1990). Loss of suitable cool-water habitat associated with lake warming is also projected, which will very probably differentially affect species within lacustrine fish communities (e.g., promote growth and survival in lake whitefish but negatively affect these in lake trout; Magnuson et al., 1990). Preliminary consideration of northern areas has occurred for European systems (e.g., Lehtonen, 1996). Relatively less attention has been paid to the possible effects of climate change on resident fish communities in other ecosystems, particularly those in the Arctic. With respect to freshwater fish populations, the IPCC (Arnell et al., 1996) concluded that fish populations in streams and rivers on the margins of their geographic distributions (e.g., arctic and

subarctic species) will be the first to respond to the effects of climate change because these systems have a high rate of heat transfer from the air. Some of these effects include:

- Nutrient level and mean summer discharge explained 56% of the variation in adult Arctic grayling growth over a 12-year period in two Alaskan rivers (Deegan et al., 1999). Summer temperature added to these variables explained 66% of variation in young-of-the-year growth. Correlation with discharge was positive for adults and negative for young, thus grayling life history appears able to respond to variability in the arctic environment by balancing adult growth with year-class strength. How this balance will shift under climate change is uncertain at present.
- Temperature effects on growth appear to be greatest at the extremes of the geographic range of the species (Power M. and van den Heuvel, 1999), and local effects will be species-specific (King et al., 1999). Generally, young-of-the-year fish appear to grow better in warmer summers and reach relatively larger sizes, predisposing them to higher overwinter survival, which determines year-class strength and population abundance (Shuter et al., 1980); potentially a positive result of climate change assuming food is not a limiting factor.
- Northern lake cisco (*Coregonus artedi*) populations along the coast of Hudson Bay exhibited reduced growth and later maturity due to lower temperatures and shorter growing seasons (Morin et al., 1982). Individual fecundity did not change, but the most northerly populations skipped reproduction more frequently (hence overall population productivity was lower). This latitudinal gradient represents responses to temperature stresses whereby further trade-offs in energy allocation between reproduction and growth currently are not possible (Morin et al., 1982); a common circumstance for most arctic fish populations and one that will probably be ameliorated under scenarios of increased temperature, potentially resulting in increased population abundances.
- Counter-gradient variation (Levins, 1969), whereby genetic influences on growth in species such as brown trout (*Salmo trutta*) vary inversely with mean annual water temperatures (Jensen et al., 2000), suggest that trout in the coldest rivers are specifically adapted to low temperatures and short growing seasons. Thus, increased temperatures are likely to negatively affect growth rates, age/size structure, and abundances of northern populations.

8.5.2. Approaches to projecting climate change effects on arctic fish populations.

Uncertainty in projections of future temperature, hydrology, and precipitation, and their associated consequences for vegetation and nutrient patterns in arctic aquatic ecosystems, makes projecting the specific effects of climate change on a fish species difficult. To date, fisheries literature has suggested three approaches to this problem:

1. the use of regionally specific climate projections that can be coupled directly to knowledge of the physiological limits of the species;
2. the use of empirical relationships relating local climate (weather) to measurements of species or stock dynamics (e.g., abundance, size, growth rate, fecundity) and comparison of population success temporally (e.g., from a period of climatically variable years) or spatially (e.g., locales representing the extremes of variation in weather conditions such as latitudinal clines); and
3. the use of current distributional data and known or inferred thermal preferences to shift ecological residency zones into geographic positions that reflect probable future climate regimes.

8.5.2.1. Physiological approaches

Temperature is typically regarded as a factor affecting individual physiological and behavioral processes, but it is also a key characteristic of the habitat of an organism. Hutchinson (1957) defined the niche of an animal as the complete range of environmental variables to which it must be adapted for survival. At the fringes of the distributional range, abiotic parameters associated with particular niche axes are likely to exert a greater influence over the physiological responses (e.g., growth) of the species to its environment than elsewhere. Growth rates and population dynamics of fish living at the limits of their distribution usually differ from those of the same species living in the optimum temperature range (Elliott, 1994). For example, in studies of northern populations of yellow perch, Power M. and van den Heuvel (1999) noted that although heterogeneous thermal environments allow fish opportunities to compensate for temperature fluctuations by selecting for preferred temperatures, such opportunities are limited in the portion of the geographic range where temperatures do not typically exceed those that define the optimum scope for growth. Accordingly, unless future temperatures increase above the point where the maximum scope for growth is realized, northern fish will be limited in their abilities to select for optimal growth temperatures and, consequently, are very likely to more strongly reflect the influence of temperature on growth than southern populations. This also suggests that analogues derived from lower-latitude populations will not be accurate guides to the probable impacts of temperature increases on subarctic and arctic fishes. Nevertheless, the effects of climate change in the north are very likely to include faster, temperature-driven growth and maturation rates, reductions in winter mortality, and expanded habitat availability for many species (Regier et al., 1996). However, somatic gains will possibly be offset by increased maintenance-ration demands to support temperature-induced increases in metabolism. Ration demands for lacustrine fish are likely to be met as temperatures increase, since warm-water lakes are generally more productive than cold-water lakes (Regier et al., 1996). Basic knowledge of temperature–growth relationships and temperature-dependent energy demands is lacking for many key arctic fish species, particularly those exhibiting primarily riverine life histories, thus accurate physiologically based projections of climate change impacts cannot be made.

8.5.2.2. Empirical approaches

Empirical approaches to projecting the possible effects of climate change on fish populations can be subdivided into two groups. The first group examines the integrated responses of a population measured by yield or production over time. The second group examines the population characteristics spatially and uses inherent latitudinal variability to make inferences about how they will change under climate change scenarios.

Temporal Yield/Production Projections: There are numerous models for projecting freshwater fish production in lakes (see Leach et al., 1987). However disagreement exists among researchers as to which lake characteristics most significantly influence productivity. Comparative studies based on lakes covering a wide range of geographic areas and trophic status have suggested that fish production in oligotrophic to hypereutrophic lakes of moderate depth is better correlated with primary production than the morphoedaphic index (Downing et al., 1990). Limitations surrounding such modeling center on the deficiencies in fish distribution data and knowledge of the interactive effects of climate-induced changes in key environmental variables (Minns and Moore, 1992). Together with limited fishery databases of sufficient length, these limitations in most cases preclude this approach for projecting productivity changes in arctic populations.

Latitudinal Projections: Organism life-history characteristics often vary with latitude because of predictable changes in important environmental factors (e.g., Fleming and Gross, 1990; L'Abée-Lund et al., 1989; Leggett and Carscadden, 1978; Rutherford et al., 1999). Among the most important environmental factors which may vary with latitude is temperature, which is known to influence growth rate in fish populations (Elliott, 1994; Wootton, 1990) and thereby indirectly affect life-history attributes that determine population dynamics (e.g., longevity, age-at-maturity, and fecundity). In salmonids, temperature has been shown to influence movement and migration (Jonsson, 1991), habitat occupancy (Paul and Post, 2001), migration timing (Berg and Berg, 1989), smolting (McCormick et al., 1998; Power G., 1981), growth rate (Brett et al., 1969; Jensen et al., 2000), age-at-maturity (L'Abée-Lund et al., 1989; Power G., 1981; Scarnecchia, 1984), fecundity (Fleming and Gross, 1990), and the proportion of repeat spawners (Leggett and Carscadden, 1978). Many studies have demonstrated latitudinally separated disparate populations of the same species with distinctive metabolic rates, thermal tolerances, egg development rates, and spawning temperature requirements consistent with a compensatory adaptation to maximize growth rates at a given temperature (Levinton, 1983). Fish living in low-temperature, high-latitude locales would therefore be expected to compensate by increasing metabolic and growth rates at a given temperature relative to fish in high-temperature, low-latitude locales. There are two generalizations that may be made from studies on latitudinal variation in growth rates: high-latitude fish populations often attain larger maximum body size than conspecifics at lower latitudes; and, although lower temperatures often reduce activity and constrain individuals to grow more slowly, they compensate by accelerating growth rate or larval development time relative to low-latitude conspecifics when raised at identical temperatures. Although adaptation to low temperature probably entails a form of compensation involving relative growth acceleration of high-latitude forms at low temperature, the shift in metabolism increases metabolic costs at higher temperatures, leaving cold-adapted forms with an energetic disadvantage in the higher-temperature environments (Levinton, 1983) that are likely to result from climate change. Accordingly, fish populations are likely to be locally adapted for maximum growth rate and sacrifice metabolic efficiency at rarely experienced temperatures to maximize growth efficiency at commonly experienced temperatures. This suggests that the effects of temperature increases on northern fish will possibly include decreased growth efficiency and associated declines in size-dependent reproductive success. Therefore, particular responses to temperature increases are likely to be population-specific rather than species-specific, which greatly complicates the ability to project future situations for particular species over large areas of the Arctic.

8.5.2.3. Distributional approaches

Many attempts to project biological responses to climate change rely on the climate-envelope approach, whereby present-day species distributions are mapped with respect to key climate variables (e.g., temperature, precipitation) and the distributions shifted in accordance with climate change projections (e.g., Minns and Moore, 1992). For example, Shuter and Post (1990) have argued that weight-specific basal metabolism increases as size decreases with no associated increase in energy storage capacity, resulting in smaller fish being less tolerant of the starvation conditions typically associated with overwintering. Size-dependent starvation endurance requires that young-of-the-year fish complete a minimum amount of growth during their first season of life. Growth opportunity, however, is increasingly restricted on a south–north gradient and the constraint has been demonstrated to effectively explain the northern distributional limit of yellow perch in central and western North America (Fig. 8.17), European perch (*Perca fluviatilis*) in Eurasia, and the smallmouth bass in central North America. If winter starvation does form the basis for the geographic distributions of many

fishes (e.g., 11 families and 25 genera of fish within Canadian waters; Shuter and Post, 1990), climate-induced changes in growing-season length, and consequent reductions in the period of winter starvation, are very likely to be associated with significant range extensions of many species. Species already well established within low-arctic watersheds are likely to show the greatest potential for range extensions. Associated changes in species assemblages are likely to shift patterns of energy flow in many aquatic systems. For example, increasing the number of cyprinids that consume plankton (e.g., emerald shiner – *Notropis atherinoides*, lake chub – *Couesius plumbeus*) in northern waters will possibly divert energy from existing planktivores (e.g., ciscoes) and reduce their population abundances. In turn, top predators (e.g., lake trout) are likely to have altered diets and changes in the ratio of pelagic and benthic sources of carbon in piscivore diets are likely, in turn, to alter tissue mercury concentrations (Power M. et al., 2002; section 8.7), thus linking general climate change impacts with local contaminant loadings.

The dominant result of simulations used to project the impact of climate change on the distribution and thermal habitat of fish in north temperate lakes is an increase in available warmer habitat. Temperature influences on thermal habitat use are strong enough that Christie and Regier (1988) were able to develop measures of thermal habitat volume during the summer period by weighting the amount of lake-bottom area and pelagic volume with water temperatures within species' optimal thermal niches. Thermal habitat volume explained variations in total sustained yield of four commercially important species: lake trout, lake whitefish, walleye, and northern pike.

Although distributional changes provide a convenient and easy means of assessing possible range extensions, the flaw in the approach is that species distribution often reflects the influence of interactions with other species (Davis et al., 1998; Paine, 1966) or historical effects (Tonn, 1990). Projections based on changes in single-species climate envelopes will therefore be misleading if interactions between species are not considered when projections are made. Microcosm experiments on simple assemblages showed that as the spatial distribution of interdependent populations changed as a result of temperature increases, the pattern and intensity of dispersal also changed. Thus, climate change will possibly produce unexpected changes in range and abundance in situations incorporating dispersal and species interaction (e.g., competition and predator–prey dynamics). Feedbacks between species are likely to be even more complex than simple experiments allow (Davis et al., 1998); for example, distributions of stream-resident salmonids are not simple functions of either temperature or altitude (Fausch, 1989). Accordingly, whenever dispersal and interactions operate in natural populations, climate change is likely to provoke similar phenomena and projections based on extrapolation of the climate envelope may lead to serious errors (Davis et al., 1998).

In theory, the temperature signal should be strong enough to project long-term changes in the availability of fish thermal habitat and to use available empirical relationships to project sustainable yields. However, until the results of such research are available for arctic fishes, interannual variability and latitudinal differences in climate will provide the best tests for hypotheses about the importance and effects of climate change on arctic fish species (Magnuson and DeStasio, 1997).

8.5.3. Climate change effects on arctic freshwater fish populations

The ability of fish to adapt to changing environments is species-specific. In the case of rapid temperature increases associated with climate change, there are three possible outcomes for any species: local extinction due to thermal stress, a northward shift in geographic range where dispersive pathways and other biotic and abiotic conditions allow, and genetic change within the limits of heredity through rapid natural selection. All three are likely to occur, depending on the species (Lehtonen, 1996). Local extinctions are typically difficult to project without detailed knowledge of critical population parameters (e.g., fecundity, growth, mortality, population age structure, etc.). Dispersal and subsequent colonization are very likely to occur, but will very probably be constrained by watershed drainage characteristics and ecological and historical filters (Tonn, 1990). In watershed systems draining to the north, increases in temperature are very likely to allow some species to shift their geographic distribution northward (see section 8.5.1.1). In watershed systems draining to the east or west, increases in temperature will possibly be compensated for by altitudinal shifts in riverine populations where barriers to movements into headwaters do not exist. Lake populations needing to avoid temperature extremes are very likely to be confined to the hypolimnion during the warmest months provided anoxic conditions do not develop. Patterns of seasonal occurrence in shallower littoral zones are very likely to change, with consequent effects on trophic dynamics. Changes in species dominance will very probably also occur because species are adapted to specific spatial, thermal, and temporal characteristics that are very likely to alter as a result of climate-induced shifts in precipitation and temperature.

Before successful range extensions can occur, habitat suitability, food supply, predators, and pathogens must be within the limits of the niche boundaries of the species. In addition, routes to dispersal must exist. Physiological barriers to movement such as salinity tolerances or velocity barriers (i.e., currents) will possibly restrict range extensions where physical barriers to migration (e.g., waterfalls, non-connected drainage basins) do not exist. Against this background of dynamic physical and biotic changes in the environment, some regional and species-specific climate change projections have been made.

8.5.3.1. Region 1: European percids

Under scenarios of climate change, spawning and hatching of spring and summer spawning populations are likely to occur earlier in the year. For example, European perch are very likely to advance spring spawning by as much as a month (Lehtonen, 1996) and juveniles will very probably experience longer growth periods and reach larger sizes at the end of the first summer. However, this species may not realize the potential benefits of increased size if higher egg incubation temperatures are associated with smaller larvae having smaller yolk sacs and increased metabolic rates (e.g., Blaxter, 1992; Peterson R. et al., 1977). Small larvae are more susceptible to predation, have higher mortality rates, and have a shorter period during which they must adapt to external feeding to survive (Blaxter, 1992). In addition, increased overwinter survival is very likely to be associated with increased demand for prey resources and will possibly lead directly to population stunting (i.e., smaller fish sizes).

The zander (*Sander lucioperca*) is a eurythermal species distributed widely in Europe whose growth and recruitment success correlates with temperature (Colby and Lehtonen, 1994). The present northern distribution coincides with the July 15 °C isotherm and is likely to shift northward with climate change. Successive year-class strengths and growth rates in northern environments are also likely to increase as temperatures increase. Increases in both abundance and size are very likely to have consequences for the competitiveness of resident coldwater-guild fishes if concomitant increases in lake productivity fail to yield sufficient ration to meet the needs of expanding populations of zander and other percids. Evidence that northward colonization is already occurring comes from the Russian portion of Region 1. Over the last 10 to 15 years, northern pike, ide (*Leuciscus idus*), and roach (*Rutilus rutilus lacustris*) have become much more numerous in the Pechora River Delta and the estuary Sredinnaya Guba (~68° N) of the Barents Sea (A. Kasyanov, Institute of Inland Waters, Russian Academy of Sciences, pers. comm., 2004).

8.5.3.2. Region 2: Fishes in Siberian rivers

Many species of fish in the large northward flowing rivers of Siberia have the potential for significant northward range extensions and/or responses to climate change. Several species in the Yenisey and Lena Rivers that prefer warmer boreal-plain habitats (e.g., roach, ide, common dace – *Leuciscus leuciscus baicalensis*, European perch, and ruffe – *Gymnocephalus cernuus*) are likely to move into the northern mouth areas of these rivers that are currently dominated by whitefishes and chars. Overall, fish species diversity is likely to increase, but this probably will be at the expense of the coldwater salmonids. The speed at which this process might occur is uncertain, however, it may already be occurring and is likely to be within approximately the next ten years. In addition, as environments change, intentional stocking of other species (e.g., carp bream – *Abramis brama* and zander) is likely to occur in the area, which is likely to result in additional pressures upon native arctic fish populations.

8.5.3.3. Region 3: Alaskan game fish

Nutrient availability often determines food availability and lotic productivity, which are believed to be major controlling factors in riverine fish production. Several studies have found that fish density and growth correlate with nutrient status and food availability in streams, with larger standing crops in nutrient-rich streams (Bowlby and Roff, 1986; McFadden and Cooper, 1962; Murphy et al., 1981). In particular, salmonid biomass in nutrient-poor environments varies with nutrient levels, habitat type, and discharge (Gibson R.J. and Haedrich, 1988). The bottom-up propagation of nutrients through algal and invertebrate production to fish has been projected to be a possible result of climate-induced increases in nutrient additions associated with permafrost degradation. However, this premise has rarely been tested, and the relationship between nutrient loading and fish production is poorly understood (Peterson B. et al., 1983). Shifts in stable carbon and nitrogen isotope distributions have demonstrated a coupling between the stimulation of benthic algal photosynthesis and accelerated growth in stream-resident insect and fish populations (Peterson B. et al., 1993). In addition, experimental fertilization of Alaskan tundra rivers has demonstrated increased growth rates for adult and young-of-the-year Arctic grayling, with the strongest response observed in the latter (Deegan and Peterson, 1992).

Temperature increases associated with climate change are also likely to be associated with lower flows, with which growth of adult Arctic grayling is also highly correlated. At low flows, adult growth is low, whereas young-of-the-year continue to grow well (Deegan and Peterson, 1992). As Arctic grayling in many Alaskan systems are already food-limited, the associated increases in metabolic costs are likely to be associated with decreased survival unless nutrient loading associated with permafrost degradation offsets the increased metabolic costs of low-flow conditions (Rouse et al., 1997).

Lake trout are a keystone predator in many Alaskan lakes. Low food supply and temperatures, however, keep the species near physiological limits for survival with the result that lake trout will possibly be particularly sensitive to changes in either temperature or food supply initiated by climate change (McDonald M.E. et al., 1996). Increases in temperature are very likely to increase metabolic demands, which will very probably lead to lower realized growth rates unless met by sufficient increases in ration.

Many populations are already food-limited, which suggests that further increases in temperature are very likely to have significant effects on population abundance. Bio-energetic modeling of juvenile populations of lake trout in the epilimnion of Toolik Lake suggests that they

will not survive a 3 °C increase in mean July epilimnetic temperatures given existing ration, and would require a greater than eight-fold increase in food to achieve historical end-of-year sizes (McDonald M.E. et al., 1996). Documented increases in epilimnetic temperatures, however, have not been associated with increased food availability. If recent changes in the lake foreshadow long-term trends, these modeling results suggest that young lake trout will not overwinter successfully, and the associated changes in mortality patterns may lead to local extinction and the disruption of lake-trout control of the trophic structure in many arctic lakes (McDonald M.E. et al., 1996).

8.5.3.4. Region 4: Northern Québec and Labrador salmonid and pike populations

Among the salmonids of northern Québec and Labrador, the response to temperature changes is very likely to track physiological preferences for warmer waters. Several species, such as native Atlantic salmon and brook trout (*Salvelinus fontinalis*) and introduced brown trout and rainbow trout (*Oncorhynchus mykiss*), are very likely to extend their ranges northward. While the warmer-water percid and cyprinid species are restricted to the southwest and unlikely to extend their range to the north (unless moved by humans) because of dispersal barriers (Power G., 1990b), the euryhaline salmonids are able to move from estuary to estuary as conditions allow. For example, Dumont et al. (1988) documented the successful movement of rainbow and brown trout and exotic salmon species in the estuary of the Gulf of St. Lawrence, and there is some indication that brown trout dispersal in Newfoundland has been temperature-limited (Crossman, 1984). As a result of probable range extensions, Arctic char are very likely to be reduced or replaced by anadromous Atlantic salmon and/or anadromous brook trout throughout much of the southern portion of the region and brook trout are very likely to become a more important component of native subsistence fisheries in rivers now lying within the tundra zone (Power G., 1990b). Lake trout are likely to disappear from rivers and the shallow margins of many northern lakes and behave as currently observed in temperate regions (Martin and Olver, 1980).

Northern pike habitats in much of subarctic North America and Europe are projected to sustain some of the most severe consequences of global climate change. Adult northern pike actively avoid surface temperatures in excess of 25 °C, which are very likely to become more frequent as air temperatures increase throughout much of the distributional range. In shallower lakes, changes in lake chemistry associated with temperature increases will possibly result in cooler bottom waters becoming anoxic and a restriction of suitable habitat (e.g., Schindler D.W. et al., 1990). Studies in Ohio impoundments have shown that although northern pike show summer growth, there is an associated weight loss during the periods of habitat restriction (Headrick and Carline, 1993). Accordingly, northern pike throughout much of their current range are expected to be restricted in both numbers and size as a result of climate change.

Attempts to relate fish yields and mean annual air temperatures have been coupled with geographic information techniques to project shifts in both distribution and yields of important freshwater fishes in this region (Minns and Moore, 1992). In general throughout subarctic Québec, yields for lake whitefish are projected to increase by 0.30 to >1.0 kg/ha/yr. Northern pike yields in southern portions of the Hudson Bay drainage are projected to increase by 0.03 to 0.10 kg/ha/yr, and those in northern portions to increase marginally (0.01–0.03 kg/ha/yr). Walleye yields in the southern drainage basin of Hudson Bay are projected to increase by 0.01 to 0.10 kg/ha/yr. These changes are projected to result from occupancy of new, presently unsuitable areas in the north, and increased overall productivity throughout the entire area. Declining production in southern areas that become unsuitable due to suboptimal thermal regimes for these species or local population extirpation may possibly offset the overall productivity gains.

8.5.4. Effects of climate change on arctic anadromous fish

About 30 species within the arctic regions belonging to the families Petromyzontidae, Acipenseridae, Anguillidae, Clupeidae, Osmeridae, Salmonidae, and Gasterosteidae (Box 8.6) exhibit diadromous behavior (i.e., spend part of their lives in the marine environment and migrate to freshwater to spawn, or the converse). Most arctic diadromous species are actually anadromous (i.e., use estuarine and/or marine environments for feeding and rearing; and freshwater environments for spawning, early life history, and, in the case of most arctic species, overwintering); only freshwater eels (Anguillidae) and some lampreys (Petromyzontidae) are catadromous (i.e., breed at sea and rear in freshwater). Most anadromous species in the Arctic are facultatively anadromous (Craig, 1989) in that many individuals in a population do not necessarily migrate to sea even though it is accessible. Typically, anadromous behavior is most prevalent at northern latitudes (McDowall, 1987) because the ocean is more productive than adjacent freshwater habitats in

Box 8.8. Projecting stock-specific effects of climate change on Atlantic salmon

Differences in stock characteristics, local geography, and interannual variations in spawning escapement of Atlantic salmon confound attempts to apply the results of specific field studies (e.g., Buck and Hay, 1984; Chadwick, 1987; Egglishaw and Shackley, 1977, 1985) in projecting the effects of climate change (Power M. and Power, 1994). Further complications arise from the ongoing debate regarding whether environmental variation and population effects are greatest in fresh or marine waters (Friedland, 1998), and how these act to determine survival of various life stages and population abundance. Knowledge of Atlantic salmon biology, however, is sufficient to describe the range of temperature conditions required for optimal growth and reproductive success, and thus to allow inferences of climate change effects. Atlantic salmon life-history stages all occur within optimal temperature ranges (Dwyer and Piper, 1987; Peterson R. and Martin-Robichaud, 1989; Power G., 1990a; Wankowski and Thorpe, 1979). However, variation in the required range of optimal temperatures for salmon at different life stages makes projecting the effects of climate change difficult. To date, three approaches to tackling the problem have been proposed in the scientific literature (see section 8.5.2).

Results of modeling experiments projecting the possible effects of climate change on different populations of Atlantic salmon (Power M. and Power, 1994).

Population location	Temperature increase		Temperature decrease	
	Smolt production	Parr density	Smolt production	Parr density
47° 01' N, 65° 27' W	decrease	increase	no change	no change
50° 11' N, 61° 49' W	increase	decrease	decrease	increase
53° 42' N, 57° 02' W	increase	decrease	decrease	increase

In the first approach, regional climate scenarios and projections are coupled directly to knowledge of the physiological limits within which salmon operate. For example, winter discharges and associated overwintering habitat will respond to precipitation changes (Power G., 1981). Low summer discharge on the east coast of Newfoundland and in southern Québec, which limits parr (young salmonid with parr-marks before migration to the sea) territory and hampers upstream adult migration, is also very likely to change, affecting population abundances in many rivers (Power G., 1981). Problems with this approach include uncertainty in precipitation and extreme events forecasts, and coupling of regional climate models with ocean circulation models.

A second approach to understanding the possible impacts of climate change on Atlantic salmon is to apply what is known about relationships between weather and salmon population dynamics. For example, historical records from the salmon fisheries in the Ungava region of northern Québec show a correlation between ice conditions, the late arrival of salmon, and poor catches. This relationship suggests that an improvement in salmon abundances will possibly occur in the future associated with a climate-induced reduction in the extent and duration of sea-ice cover (Power G., 1976; Power G. et al., 1987). The correlation between stock characteristics and latitude (Power temperate and arctic zones (Gross et al., 1988). For a number of facultative anadromous species (e.g., Arctic char, Dolly Varden, brook trout, brown trout, and three-spine stickleback), anadromous behavior declines in frequency or ceases toward the southern portion of the distributional range of the species (several references in McDowall, 1987). Anadromy in Arctic char also declines or ceases towards the extreme northern geographic limits, probably because access to and time at sea, hence benefits, are limited. Facultative anadromous species exhibit anadromy in polar regions to take advantage of marine coastal productivity and escape extreme oligotrophic conditions that typify arctic lake systems. Generally, individuals of a population that exhibit anadromous behavior have a larger maximum size and higher maximum age, indicating some benefit to seaward migration and feeding.

Diadromous fishes will integrate climate change effects on freshwater, estuarine, and marine areas, hence the total impact on these fishes is very likely to be significant (e.g., see Fleming and Jensen, 2002; Friedland, 1998).

This will have major resulting impacts since these fishes support important fisheries in all arctic regions (section 8.5.5; Chapter 3). The following paragraphs discuss the consequences of climate change for diadromous fishes.

The projected impacts of climate change on arctic lakes suggest that, overall, productivity of these limited systems will very probably increase due to a longer ice-free growing season and higher nutrient loads. Anadromous fish populations will probably benefit initially with increases in survival, abundance, and size of young freshwater life-history stages, which will possibly cascade to older, normally anadromous stages. Thus, facultatively anadromous species will possibly exhibit progressively less anadromous behavior if the benefits of remaining in freshwater systems outweigh the benefits of migrating to coastal areas for summer feeding over time. Nordeng (1983) reported that when the freshwater food supply was experimentally increased, the incidence of anadromous migration by Arctic char decreased. However, the increased estuarine production discussed previously will possibly offset any tendency to reduce facultative

G., 1981) suggests that mean smolt (young salmonid which has developed silvery coloring on its sides, obscuring the parr marks, and which is about to migrate or has just migrated into the sea) ages are likely to decrease in association with increases in average temperatures and growing-season length. The modeling results of Power M. and Power (1994) projected that temperature increases and decreases will have varying effects on populations at different latitudes (see table). Where present-day temperatures are at the upper end of the optimal temperature range for growth, increases in temperature reduced growth, increased average riverine residency and associated riverine mortalities, decreased smolt production, and increased parr densities. The reverse (increased smolt production and decreased parr densities) occurred when temperatures at the lower end of the temperature range optimal for growth were raised. Modest changes in precipitation, and thus available habitat, had no significant direct effect or interactive effect with changes in temperature on either smolt production or parr density under any of the considered temperature scenarios. Thus, depending upon the exact location and characteristics of the salmon population, the precise impact of a given environmental change under a future climate scenario may be positive or negative relative to present conditions. This makes regional differences in fish biology, present-day local climate, and climate change scenarios extremely important in projecting future situations.

A third approach to projecting the effects of climate change involves attempting to shift ecological zones into more appropriate geographic locations to reflect probable future climate regimes and the known physiology of potentially affected species. The present distribution of many fish is limited by the position of the summer isotherms that limit the fish either directly due to thermal relationships or indirectly through effects on critical resources such as food (Shuter and Post, 1990). Use of this approach suggests that Atlantic salmon will possibly disappear from much of their traditional southern range in both Europe and North America as temperatures rise, and find more suitable habitat in cold rivers that experience warming. In the eastern Atlantic, the overall area occupied by salmon is likely to shrink due to a lack of landmasses to the north with potentially suitable environments. In the western Atlantic, rivers in the Ungava Bay area will possibly become more productive and are likely to experience increases in the numbers of salmon (e.g., the Koroc and Arnaux Rivers). Rivers that currently have large salmon runs are also likely to become more productive (e.g., the George, Koksoak, and Whale Rivers) and experience associated increases in salmon abundances (Power G., 1990a). There are also rivers on Baffin Island and Greenland that will possibly become warm enough for Atlantic salmon to colonize. Such colonization, however, is likely to come at the expense of Arctic char populations that currently inhabit the rivers because of competition between the two species. Constraints on redistribution northward with climate change include reductions in the availability of spawning substrate with increased sediment loading of rivers, changes in stream and river hydrology, and delay in the establishment of more diverse and abundant terrestrial vegetation and trees known to be important for the allochthonous inputs that provide important sources of carbon for salmon (Doucett, 1999).

anadromy in response to increased freshwater production. The exact balance and circumstances of how such scenarios unfold will be ecosystem-specific and will depend on the details of present productivity, accessibility, and ease of migration by fish, as well as the nature and degree of any climate-related effects.

The variability associated with projected changes in productivity is uncertain. The anadromous species listed in Box 8.6, are typically long-lived (15–50 years) compared to other fish species. Longevity benefits species living in variable environments by ensuring a relatively long reproductive cycle, thus minimizing the risk that prolonged environmentally unfavorable periods (5–15 years) will result in the loss of a spawning stock (Leaman and Beamish, 1981). Anadromous forms of arctic fish species are relatively long-lived (>10–15 years) and are probably suited to cope with increased variability that will possibly accompany climate change. Initially, as environmental conditions improve, successful spawning episodes are very likely to increase in frequency. Anadromous fish that are short-lived (<10–15 years) are likely to exhibit more variability in abundance trends with increased variability in environmental conditions.

When in freshwater, anadromous species (Box 8.6) also inhabit streams or rivers in addition to lakes. Projected climate impacts on arctic hydrology (section 8.4) suggest that runoff is very likely to be driven by increased precipitation and will very probably not be as seasonally variable; winter flows are very likely to be enhanced and summer flows reduced. In addition, warmer conditions are projected to reduce the length of winter, shorten the ice season, and reduce ice-cover thickness. Thus, streams that were previously frozen solid will very probably retain water beneath the ice, benefiting anadromous species that utilize streams for winter habitat (e.g., Dolly Varden). Overwintering habitat is critical for arctic species and is typically limited in capacity (Craig, 1989). However, the shortened ice season and thinner ice are very likely to reduce ice-jam severity. This will have implications for productive river deltas that require flooding. There are several anadromous species, such as Arctic cisco, that rely on deltas as feeding areas, particularly in spring (Craig, 1989).

Anadromous fish are by definition highly migratory and tolerant of marine conditions. Thus, as limiting environmental factors ameliorate, a number of sub- or low-arctic anadromous species are likely to extend their northern limits of distribution to include areas within the Arctic. Pacific salmon species are likely to colonize northern areas of Region 3. Sockeye salmon (*Oncorhynchus nerka*) and pink salmon (*O. gorbuscha*) have already been incidentally recorded outside of their normal distribution range on Banks Island, Northwest Territories, Canada (Babaluk et al., 2000). Similarly, anadromous species such as Atlantic salmon, alewife (*Alosa* spp.), brown trout, and brook trout will possibly also extend their northern range of distribution in Regions 1 and 4. New anadromous species invading the Arctic are likely to have negative

impacts on species already present. However, for many of these subarctic species, climate change is likely to have negative impacts on southern populations, offsetting any positive benefits that will possibly accrue in the north (e.g., Welch D. et al., 1998). Catadromous species such as European eel (*Anguilla anguilla*; Region 3) are primarily warm-water species limited by colder arctic temperatures (e.g., Nordkappe, northern Norway is the present limit; Dekker, 2003). Eastward colonization of Russian areas of Region 2, where the species does not now occur, is possible; additionally, increased abundances are likely in some areas where the European eel presently occurs but where populations are insufficient for fisheries (e.g., Iceland).

Two arctic anadromous species are particularly important in northern fisheries: Arctic char (all regions) and Atlantic salmon (Regions 1 and 4). To indicate the range of possible responses of these species to climate change, they are treated separately in Boxes 8.7 and 8.8, respectively.

8.5.5. Impacts on arctic freshwater and anadromous fisheries

The potential and realized impacts of changes in climate and UV radiation parameters on arctic fisheries must be viewed in terms of direct impacts upon the fish and fisheries as well as indirect impacts mediated through the aquatic environment. For fisheries, however, the human context is of great importance and must be considered. Fisheries are managed to have a sustainable harvest. Harvests (i.e., *quantity*) in fisheries affect different species and their life stages differently. Fisheries must also be viewed from the perspective of product *quality*, which affects its suitability for human consumption as well as its economic value. Finally, success of a fishery implies that the fishers themselves have suitable access to and *success* in the fishery, typically the result of experience and local knowledge. This also means that fishers are able to return high-quality catch in good condition to points of consumption or transport to market. All these components of fisheries in arctic freshwaters (quantity, quality, and success) are subject to both direct and indirect impacts of climate change and increased UV radiation levels to a greater or lesser extent. Similarly, climate change is very likely to affect aquaculture operations conducted in arctic freshwaters. The following sections explore the implications for fisheries conducted in freshwaters, estuarine waters, and nearshore coastal waters.

8.5.5.1. Nature of fisheries in arctic freshwaters

Fisheries for arctic freshwater and diadromous fish are conducted in all polar countries including Canada, Denmark (Greenland), the Faroe Islands, Finland, Iceland, Norway, Russia, Sweden, and the United States (Alaska). Freshwater fisheries as described here include those for species that live their entire lives in freshwater, such as lake trout, and those for diadromous species such as Atlantic salmon. Chapter 13 addresses offshore marine fisheries conducted on anadromous species and the relevant impacts of climate change on these species in marine waters.

Arctic freshwater fisheries generally involve mostly local indigenous peoples, although some may also involve non-indigenous local people as well as visitors to the Arctic. (See Chapter 3 for indigenous accounts of changes in fishes and fishing in recent years.) Although details vary locally, at least three types of freshwater fisheries can be distinguished:

- commercial fisheries, where the product is sold commercially either locally or often in markets far removed from the sources;
- recreational fisheries in which non-indigenous people participate primarily for the experience rather than for economic, cultural, or nutritional reasons; and
- domestic or subsistence fisheries conducted by indigenous or local peoples primarily for cultural and sustenance reasons (see also Chapter 12).

Arctic freshwater fisheries can be substantial but generally never achieve the same economic significance that marine fisheries do, in part due to abundances and in part due to the lack of fishery infrastructure (e.g., absence of processing plants in many areas such as the lower Lena River, extremely long distances to markets). For the nine arctic countries, reported commercial catches for northern fishes in 2000 (8 to 350000 tonnes) represented 0.002 to 32% of total commercial catches for all species within those countries (FAO, 2002), although about 10% or less of this catch was truly "arctic" as defined herein. Rather, arctic fisheries are diverse, locally widely dispersed, and target a variety of species that are locally abundant. Such fisheries are extremely important in meeting the needs of the local peoples and contribute significantly to the economy and society of northern peoples (see also Chapter 12), thus their value must be measured in more than simple economic terms and understood in the context of climate impacts.

8.5.5.2. Impacts on quantity and availability of fish

Over the short term, projected productivity increases in arctic freshwater ecosystems, increased summer survival and growth of young fish, and increased overwinter survival of fish will probably result in increased biomass and yields of many fished species. Production shifts will depend upon local conditions such as faunal composition of the fishes and food species, tolerances and reactions of individual species to climate change, and general productivity shifts in aquatic ecosystems (Lehtonen, 1996). However, there will be much regional and local variation and responses are likely to be primarily species- or ecosystem-specific (Tonn, 1990). Thus, for wholly freshwater species, shifts in productivity are more likely to occur in lakes along the southern fringe of the Arctic, and less likely to be observed in flowing-water ecosystems. For anadromous species, increased summer nearshore productivity will possibly enhance

growth rates, hence biomass and potential fishery yields. Furthermore, recent work conducted on Atlantic salmon while in marine waters suggests that warmer sea-surface temperatures (i.e., of 8–10 °C) enhance survival in both winter (Friedland et al., 1993) and early summer (Friedland et al., 2003). Increased growth and survival are very likely to enhance fish returns to freshwaters. Shifts in river flow regimes critical to upstream migrations of anadromous fish, especially in the late summer, will possibly have a negative effect, counterbalancing any positive effects to some degree. Arctic freshwater fisheries production will probably show some increases over the next decade or two. The greatest manifestation of this increase is likely to occur at the southern boundary of the Arctic, and is very likely to involve species that are primarily subarctic (i.e., occurring throughout northern temperate regions and extending into the Arctic). Fisheries yields for such subarctic fish species (i.e., northern pike, lake whitefish, and walleye in eastern North America, and northern pike, European whitefish – *Coregonus lavaretus*, and percids in northern Europe) have been linked with species-specific (and perhaps region-specific) habitat optima (e.g., Christie and Regier, 1988; Lehtonen, 1996; Schlesinger and Regier, 1983). Such yield relationships have been further examined in the context of GCM projections of temperature increase for some areas (e.g., Québec and subarctic Canada; Minns and Moore, 1992; Shuter and Post, 1990). This regional approach suggests that, at least for deeper lakes and perhaps larger rivers, substantial redistribution of fishery potential driven by population productivity as well as by redistribution of species is very likely.

As thermal optima are exceeded locally, and perhaps as ecosystems re-equilibrate and nutrient limitations occur, reductions in biomass and yields are possible. For example, climate change will affect species individually owing to differential colonization, extinction, and productivity rates (Tonn, 1990). This will possibly lead to substantive ecological reorganization (Peterson G. et al., 1997). These effects are likely to be most severe for true arctic species such as broad whitefish and Arctic char, which will possibly also be affected by increased competition from more southerly species extending their geographic distributions northward. Thus, in the longer term, the effects of climate change on the yields of arctic fisheries are likely to be negative for true arctic fish species but positive for subarctic and northern temperate species.

As freshwater productivity increases, the frequency of anadromy will possibly decrease within populations that exhibit facultative anadromy (e.g., Arctic char, Dolly Varden, and broad whitefish). Given that anadromy and feeding at sea results in greater size at a given age and larger populations (Gross et al., 1988), a switch away from anadromy is likely to result in decreased productivity. To ensure sustainability, this may necessitate lower harvests of native anadromous species; shifts in harvesting of alternate species, if available; and/or a change in location or timing of fisheries. The consequences of these changes in fisheries of local indigenous people who rely on the autumn upstream runs of anadromous fish are very likely to be substantial from economic (i.e., protein replacement and increased costs to travel to new fishing areas for smaller catches), social, and cultural perspectives.

As noted previously, one of the hallmarks of climate change in the Arctic is likely to be increased interannual variability in climate parameters. Although it may be partially lost in the background noise of typically high inherent variability in arctic climate, this in turn will possibly increase the variability of good and poor year-classes in arctic fish. The consequence of this for fisheries will probably be increased variability in fishing success and unstable yields of targeted species. Such increased variability is very likely to exacerbate problems discussed previously that affect the biomass and yields of fisheries. Consequences include those associated with domestic sustenance if the local people rely heavily on the fished species, as well as difficulties with developing stable commercial or recreational fisheries that are economically viable and sustainable. As climate change becomes more pronounced, southern fish species are very likely to colonize newly available areas, enhancing the possibility of negative impacts on arctic fish species from competition. However, they may also represent opportunities for new fisheries. Hence, flexible, adaptive management will be key to the success of future fisheries (Peterson G. et al., 1997), particularly in responding to uncertainties associated with available data; an attribute not currently present in many fishery management regimes in the Arctic (e.g., Reist, 1997a; Reist and Treble, 1998).

Availability of fish species to fisheries will probably change as a result of several factors. For example, most fished arctic species are salmonids that tend to prefer cool or cold thermal regimes especially as adults (e.g., lake trout), thus they seek summer refuge in colder waters below thermoclines in lakes. As thermocline depths deepen, the availability of these species to fisheries will possibly change because deeper waters are more difficult to fish. This is very likely to occur in larger, deeper arctic lakes (e.g., Great Slave Lake in Canada) and will possibly necessitate gear changes for fisheries and/or retraining of fishers in new techniques. Questions as to how this might occur and how costs can be covered are currently not being addressed. In addition, the optimal temperature habitats of salmonids (e.g., European whitefish and brown trout) are very likely to change in northern Europe and summer temperatures in shallow arctic lakes will possibly become too high for these species (Lappalainen and Lehtonen, 1997), with consequent effects on local fisheries.

8.5.5.3. Impacts on quality of fish

Quality of fish captured in a fishery refers to its suitability for marketing (e.g., locally or distantly by trade, cultural exchange, and/or sale) and for consumption by humans. This suitability is affected by factors inherent in the fish resulting from environmental conditions experienced prior to capture, as well as factors that affect the fish product after capture. Factors influencing fish quality before capture include "fish condition" (typically an index of weight and length that measures fatness or nutritional state or "well being"; Busacker et al., 1990); flesh firmness, which is typically influenced by water temperatures immediately prior to capture (i.e., warmer waters generally result in poorer-quality flesh); general appearance (e.g., color and lack of imperfections) of both the fish itself and key consumed organs such as livers; parasite loads and disease; and contaminant burdens. Factors influencing fish quality after capture include preservation (e.g., cooling or freezing), and the ease, conditions, and time associated with transport to the consumption site, market, or processing facility. As for all other impacts of change in climate parameters or UV radiation levels, both direct and indirect impacts will influence fish quality.

Indirect and direct impacts on quality of fish before capture are primarily those considered in previous sections. For example, impacts on ecosystem structure and trophic pathways are very likely to affect food availability (both amount and quality) to the fish, influencing fat levels and condition; impacts on migratory patterns or access are very likely to influence growth and condition; and impacts such as higher late-season water temperatures will possibly decrease flesh firmness of cold- and cool-water fishes such as salmonids, reducing either perceived or real quality. There is evidence that the color, size, and firmness of livers and flesh in some species is affected by nutritional state, for example, burbot (*Lota lota*) livers appear to be affected by fat content and presumably nutritional state (Lockhart et al., 1989). This appears to relate in part to seasonal variance in nutrition rather than specifically to environmental impacts such as contamination. Thus, climate change impacts that affect nutrition of fished species are very likely to have consequent effects on fish quality, but these may be difficult to distinguish from ongoing typical seasonal effects.

Some additional potential impacts are worth noting or emphasizing. In general, climate change is very likely to result in increased contaminant burdens in fish flesh, with a concomitant decrease in fish quality and acceptability for human consumption; these contaminant burdens will possibly exceed safe consumption limits. This will possibly be particularly acute for some contaminants such as heavy metals (e.g., mercury) and in some areas of the Arctic. Thus, cautions and caveats associated with arctic contaminants as discussed in the Arctic Monitoring Assessment Programme report (AMAP, 1998) will possibly become more relevant as climate change occurs (see also section 8.7).

Furthermore, the potential impacts of climate change on fish parasites and hence on fish quality have been poorly addressed but appear to represent major higher-order impacts (Marcogliese, 2001). Potential direct impacts on aquatic parasites include many of the same ones noted for fish species, for example, both biological challenges and opportunities associated with parasite physiology either as a direct effect of the environment on the parasite (e.g., higher temperatures and/or shorter durations of low temperatures accelerating development) or as mediated through the host fish (e.g., shifts in fish feeding affecting parasite development). Higher parasite developmental rates suggest increased burdens upon fish hosts, which are very likely to result in decreased productivity of the population and/or poorer condition of individuals (Marcogliese, 2001). A further potential impact of parasites on arctic freshwater fishes is the introduction of new parasites to new host species or new areas (i.e., those not presently colonized) via host colonization of such areas through range extension. This will be complicated by a tendency toward higher levels of eutrophication in arctic water bodies associated with a general increase in temperature, resulting in changes in the species composition of both parasite and fish communities. In addition, disruption of normal developmental synchronicities between parasites and host fish, such as seasonal migrations within a water body, will possibly result in shifts in transmission rates to various hosts necessary to the life cycle of the parasite, but will possibly also result in switching to different hosts. Thus, parasites typically found in temperate fishes will possibly switch to arctic fishes, affecting the latter both biologically and from the perspective of quality. Shifts in thermal regimes that result in increased local densities of hosts, especially intermediate ones such as planktonic or benthic invertebrates, are also very likely to increase parasite species diversity (Marcogliese, 2001 and references therein). Conversely, activities such as fishing (which reduce the density of larger and older fish in relatively pristine fish populations, increasing the density of younger and smaller fish) can result in the "repackaging" of parasites and a net overall increase in parasite density within individual fish (T. Dick, University of Manitoba, Winnipeg, pers. comm., 2001). This reduces fish quality and marketability. The nature and timing of water delivery and potential shifts in overall amounts of precipitation may also affect parasite levels: a general increase in parasites and associated problems is likely to accompany a general decrease in water levels. Although poorly studied at present, the potential impact of fish diseases must also be addressed. Climate change is likely to result in increased incidence and spread of diseases, and perhaps increased intensity locally as fish populations are stressed. Furthermore, effects such as those associated with parasites, disease, and contaminants are part of the cumulative effects on local populations and must be considered when addressing issues of impacts on fish quality.

Many of these effects are most likely to occur, and present major problems for fish quality, in areas of the southern Arctic that presently have both reasonably high levels of exploitation and large southern catchments that flow north to the Arctic Ocean (section 8.2.3). Thus, problems that may be small at present and confined marginally to the southern Arctic will possibly increase in intensity and spatial distribution as climate change becomes more pronounced throughout the Arctic.

The impacts of increased UV radiation levels on some fish parasites will possibly be beneficial for fish by slowing infection rates and/or inhibiting the spread of some parasites (Marcogliese, 2001). However, immunosuppression resulting from increased UV radiation exposure will possibly exacerbate the effects of parasitism, disease, and contaminant loading on individual fish. This will possibly lower population productivity by decreasing survival. In addition, any obvious physical damage such as lesions or growths resulting from increased UV radiation exposure is very likely to decrease fish condition and quality.

8.5.5.4. Impacts on access to and success of fisheries

From the perspective of the fishers, climate change is very likely to have substantive impacts on how, when, and where fisheries may be conducted. Climate change will very probably affect access to and from fishing sites, and local knowledge associated with fish presence, migratory timing, and species composition. The success of the fisheries, especially as measured by transportation of high-quality product to market or point of consumption, is very likely to be similarly affected. Section 16.3 addresses some aspects of the latter impacts, such as transportation and infrastructure issues. Most arctic freshwater fisheries are small in scale, conducted locally and seasonally, and often use limited and relatively simple gear. Climate change impacts that fishers will very probably have to accommodate include increased frequency of extreme events such as high-intensity storms, and increased winter precipitation and stronger water flows that will possibly imperil the fishers, restrict their access to fishing sites, or result in the loss of fishing gear (hence economic burden). Generally, arctic freshwaters have long winter periods during which ice provides a stable platform for transportation across lakes and rivers and for deploying some types of fishing gear such as gill nets. Decreased length of the ice season, concomitant increases in the duration of freeze-up and breakup, and increased ice roughness from storms are very likely to result in substantive changes in timing, duration, and methods by which fisheries are conducted in the future. Chapter 3 documents indigenous observations of changes in ice, including declines in ice duration, thickness, stability, and predictability, which not only alter the timing and safety of ice travel, but also limit and reduce the success of traditional and subsistence activities such as ice fishing (e.g., sections 3.4.1 and 3.4.9).

Success of fisheries often depends upon the experience of the fishers, and for domestic fisheries is intimately connected with traditional knowledge of where and when to fish for particular species. The predictability associated with this will possibly decrease as climate change impacts occur. The timing of migratory runs, the typical keystone event in many northern domestic fisheries, is very likely to exhibit increased variability and decrease the ability of the fishers to know when best to begin fishing. Such circumstances will possibly result in decreased success of fisheries. Furthermore, such variability is likely to become the norm as ecosystems undergo shifts, at least until new equilibriums are established. Because the changes wrought by climate change are likely to be protracted and depend in large part upon local ecological circumstances and the nature of the biota present, new equilibriums are unlikely to be quickly established. Thus, fishers will possibly have to tolerate highly variable and unstable conditions in freshwater and coastal ecosystems. This is very likely to result in highly variable successes in freshwater and anadromous fisheries, at least over longer timeframes.

Table 8.2. Summary of possible, likely, and very likely effects of changes in climate or UV radiation levels on *quantity* of fish in arctic freshwater and anadromous fisheries.

Climate change or UV radiation effect	Potential impact on fisheries	Consequences/comments
Increased productivity at lower trophic levels is very likely to result in increased growth, recruitment, and survival of freshwater species	Biomass and yields increased	Short-term management for increased fishery yields, especially for temperate species in the southern Arctic
Increased productivity at lower trophic levels is likely to result in increased growth in early years for facultatively anadromous species that promotes a shift to wholly freshwater life histories	Shifts in balance of anadromy versus non-anadromy decreases yields overall (i.e., smaller fish and perhaps more fish)	Long-term management for change in type and location of fisheries, and for decreased fishery yields
Local water temperature increases will at some point exceed thermal optima for individuals, possibly decreasing growth	Biomass and yields decreased	Especially true for arctic species and for cool-water species in the southern Arctic; population declines and local extirpation; synergistic effects from other factors such as competition from southern taxa; management issues associated with declining fishery yields
Reduced ice-cover duration on arctic lakes especially in northern arctic areas, increased and more rapid stratification, earlier and increased primary production, and decreased oxygenation at depth will possibly result in a reduction in the quality and quantity of habitat for species such as lake trout	Survival, biomass, and ultimately yields of preferred species generally decreased	Management for decreased fishery yields; potential management for declining fisheries and loss of populations
Improved quality of winter habitat will possibly result in increased survival (but this would also be affected by summer conditions, stratification, and overturns)	Biomass and yields increased	Short-term management for increased fishery yields; long-term implications unknown
Increased water temperatures generally and seasonally, but ultimately a decrease in summer habitat (e.g., deeper thermoclines in lakes, shrunken hypolimnia in lakes, reduced colder waters in rivers) are likely to reduce available habitat and decrease fish productivity, resulting in fish movements to deeper areas and/or fatal stresses on some fish species (e.g., Arctic grayling)	Short-term increase in biomass and yields (several to tens of years) Long-term decrease in biomass and yields (greater than tens of years) Decreased availability of traditionally targeted species and/or loss of key populations	Short-term management for increased fishery yields (e.g., limit growth of fishery) Long-term decrease in traditional fisheries, switch to alternative fisheries if available Long-term relocation of fisheries to new areas such as deeper portions of lakes, possible cost issues to support this relocation
Southern arctic and subarctic fish species very likely to extend distribution ranges northward, which is likely to result in some significant negative effects on native species	Decreased availability or local loss of native species; increased opportunity to fish new species (especially in southern arctic areas)	Management issues for emerging fisheries, i.e., manage to allow increase in populations and successful colonization of arctic areas; retooling and education in new ways of fishing if needed
Northern (wholly arctic) species are very likely to experience range contraction and/or local extirpation	Decreased availability of arctic species to local fisheries, potential for replacement by other species low or uncertain	Management issues for declining fisheries, and ultimately addressing rare or endangered species; in Canada this also has implications under land claim legislation for basic needs provisions
Decreased water flow in summer is likely to decrease habitat availability and possibly deny or shift access for migrating fish	Decreased biomass and yields Decreased availability due to changes in migratory runs	Management issues for declining fisheries Replacement of protein and potential social issues for peoples that heavily rely on traditional fishing; switch to other wildlife
Increased UV radiation levels in surface waters are likely to disrupt development and/or cause damage to young fish consequently decreasing survival, or forcing fish deeper thus slowing growth	Decreased biomass and yields	Management for declining fishery yields
Increased interannual variability in climate, aquatic habitats, productivity, and fish growth and production characteristics are very likely	Unknown: some arctic species are relatively long-lived indicating an ability to withstand prolonged periods of poor year-class success Increased frequency of good and poor year classes	Variability in fishing success; conservative management for median (at best) or low-yield year classes to ensure sustainability; management for highly unpredictable fisheries Instability in yields of targeted species results in uncertainty of product for fisheries
Increased water flows in winter, increased runoff in winter, and decreased spring floods	Changes in migratory runs; possible decreased biomass and yields	Revised management needs for relocated or declining fisheries

Another aspect that deserves consideration is shifts in species composition as new species colonize an area. If the new colonizer is similar to existing species (e.g., Pacific salmon as another salmonid present in an area), the existing experience and interest of fishers is likely to be applicable. Alternatively, if the new species represents an unfamiliar taxon, fishers will possibly have to build the experience base for capture and marketing, assuming the species is desirable. Undesirable species (defined by local needs and wants such as, for example, spiny-rayed species) will possibly prove to be pests by clogging nets and reducing capture efficiency. Such species may also be considered substandard for local use based upon either tradition or physical characteristics. Although highly adaptable, northern peoples will still require time and experience to modify existing practices and develop necessary adaptations for continuing successful fisheries.

Fishing as an industry carries relatively high inherent risks associated with the environment and with the tools employed. These include loss of equipment (i.e., fishing gear and boats) and injury and death of the fishers. Along with the projected changes and increased variability in climate systems, and thus decreased predictability associated with forecasts and environmental conditions, the incidence and severity of catastrophic climatic events such as severe storms are very likely to increase. Such

circumstances will imperil fishers exposed to the elements. For example, protracted breakup or freeze-up periods will make ice conditions more unpredictable. Travel over ice is essential to arctic life and especially to early winter fisheries conducted through the ice; the choice facing fishers will be increased risk or decreased fishing time, hence decreased catch.

This general summary of the impacts of changes in climate and UV radiation levels on arctic freshwater and anadromous fisheries is by no means comprehensive. Tables 8.2, 8.3, and 8.4 summarize numerous additional potential impacts.

Detailed regional and local analyses of particular types of fisheries (e.g., commercial, recreational, or domestic) and of specific arctic freshwater and anadromous fisheries are required to more fully elucidate all impacts, understand their consequences for local fisheries, and stimulate the development of appropriate short- and long-term adaptive responses by fishery managers and related constituents of the fishery infrastructure. Failure to address these issues in a timely fashion will undermine coherent and comprehensive preparedness to meet challenges that changes in climate and UV radiation levels present for arctic freshwater and anadromous fisheries.

Table 8.3. Summary of possible, likely, and very likely effects of changes in climate or UV radiation levels on *quality* of fish in arctic freshwater and anadromous fisheries.

Climate change or UV radiation effect	Potential impact on fisheries	Consequences/comments
If water temperatures increase, thermal optima for individual growth are likely to be exceeded, resulting in negative effects on individuals	Individual fish condition reduced, thus quality is lower Biomass and yields are reduced	Especially true for arctic species and arctic-adapted cool-water species requiring thermal refugia; value of individual fish and total amount landed are reduced
If water temperatures increase, flesh firmness will possibly decrease due to capture in warmer waters	Flesh quality reduced	Value is reduced; preservation compromised
If air temperatures increase, fisheries may occur under warmer conditions, which is very likely to increase problems of preserving and transporting the product	Problems with immediate preservation increased (e.g., on-board refrigerators required) Transportation costs increased (i.e., faster method or more return trips to fish plants) or impossible Quality of product decreased Costs of production increased	Low-value marginal fisheries will not be economically viable; northern fishery development compromised; and some fisheries may be abandoned if transportation is not possible Consumption of lower-quality or poorly preserved product may increase human health risks
Changes in climate and/or UV radiation levels will possibly result in physical disfiguration of fish (e.g., discolorations, lesions, growths, etc.)	Perceived and real quality and value of fish decreased	Increased concern voiced by local peoples requiring appropriate investigation and response from management agencies, e.g., ruling out potential proximate causes other than changes in climate or UV radiation levels Increased inspection and addressing of real and perceived health concerns required
Changes in climate and/or UV radiation levels will possibly result in increased parasitism, and new parasites and/or diseases in traditionally fished arctic species	Decreased interest in fisheries especially those based upon high-quality fish (e.g., recreational fisheries)	Economic development compromised
Persistent contaminants mobilized from natural sources (e.g., mercury liberated by permafrost thawing or flooding), or fluxes from anthropogenic sources to arctic ecosystems increase, which is likely to result in higher body burdens in arctic fish and cascade effects on other higher trophic levels	Real and perceived quality of fish decreased Compromised fish health reduces growth, decreases biomass and fishery yields	Increased inspection and monitoring required Health concerns about fish consumption, especially for domestic fisheries that typically are not routinely monitored

Table 8.4. Summary of possible, likely, and very likely effects of changes in climate or UV radiation levels on the *success* of arctic freshwater and anadromous fisheries.

Climate change or UV radiation effect	Potential impact on fisheries	Consequences/comments
Increased climate variability and frequency of extreme events (e.g., storms affecting fishing, catastrophic winter fish kills) will possibly result in biological consequences for fish populations, consequent synergistic effects on biotic systems (e.g., parasites), and synergistic effects from other impacts (e.g., local industrialization)	Increased unpredictability in places, times, and amounts of fish present in an area, and amounts captured and transported to processing, distribution, or consumption points Increased risk of gear and boat loss Increased personal risk to fishers	Extreme unpredictability in fish volumes has significant consequences for local peoples relying on fish for sustenance, for infrastructure development to support fisheries (e.g., fishing supplies, fish processing plants, transportation), and for development of markets for products from commercial and sport fisheries Loss of gear decreases success, economic viability, and persistence of fishery Need for search and rescue increased; fishing as an occupation falls from favor with a societal cost
Shifted environmental regimes are likely to affect time and difficulty of transportation to fishing sites and of product from sites to distribution or consumption points	Decreased economic value (or increased cost) of many arctic fisheries remote from communities or without permanent access Costs associated with fishing are increased	Marginal fisheries not economically viable, fishery development compromised; increased reliance on local easily accessible domestic fisheries raises the probability of over-exploitation with consequent sustainability and management issues As/if domestic fisheries fail, issues with protein replacement from other sources increase
Changes in the distribution and abundance of traditionally harvested fishes will cause traditional fishing sites to have fewer fish available	Decreased harvests and fewer fish available for communities	Indigenous fishers are tied to location and particular species by tradition and adaptation may be difficult

8.5.5.5. Impacts on specific fishery sectors

In addition to the general impacts discussed previously, climate change will have impacts specific to the various types of fisheries conducted in the Arctic.

Commercial fisheries

Perhaps the most significant challenge facing commercial fisheries will be development of appropriate adaptive management strategies that deal with the complex, synergistic, and cumulative effects of climate change on fish populations and their environment, particularly in the context of sustainable use and long-term conservation.

For example, the conundrums of how to manage both declining and increasing populations of two fishable species in a particular location, how to understand and integrate climate change impacts through functional ecosystem pathways to project future states, and how to balance the needs of local peoples and competing demands all represent real problems for northern fishery management. Clearly, some sort of adaptive or heuristic approach that incorporates elements of both fishery and ecosystem management is required (Reist and Treble, 1998). Generally this is either unavailable or not being applied at present, a situation that must be rectified in order to adapt to climate change as it unfolds in the north.

Furthermore, the research necessary to both underpin management approaches and to elucidate ecosystem linkages to fisheries must be undertaken in the north to fill gaps in understanding. The best approach would be to leave sufficient resilience and compensatory capacity within fished populations and their supporting ecosystems to account for all impacts, and to provide sufficient

buffers for increased variability and surprises associated with climate change. Current management practices incorporate such buffers in a limited way, especially in the Arctic. The development and application of such buffers (e.g., through risk analysis or other techniques) need to be extended. This presents a significant challenge in terms of developing or modifying appropriate tools for use in arctic fisheries.

Domestic fisheries

The subsistence sector in the arctic portion of the Canadian northern economy is estimated to be approximately CAN\$ 15 000 per year per household (Fast and Berkes, 1998). This represents one-quarter to one-half of the total local economy, and this proportion may be growing. Similar values are likely for domestic fisheries in other arctic countries, especially in more remote regions where the proportion of the total economy, hence value, may be even higher. Furthermore, replacement of this sector by wage or industrial economies is generally unlikely. Fisheries, which in the Canadian Arctic include marine mammals in coastal areas, comprise as much as 20% of the overall subsistence harvest in some areas (Fast and Berkes, 1998). Thus, climate-mediated impacts on fish habitat, individual fish, and fish populations are very likely to have significant effects on the availability, use, and sustainability of domestic fisheries. In addition to those discussed previously, the following effects are likely to occur within domestic fisheries.

Climate change will possibly compromise traditional ecological knowledge developed over hundreds or thousands of years of direct environmental contact (Fast and Berkes, 1998), with more pronounced impacts in particular areas where climate change effects are acute. Extreme events, which are unpredictable, are likely to exacerbate these impacts. Thus, increased climate variability and concomitant unpredictability of environmental conditions will possibly be more significant than will change in the trends of such conditions. As noted previously, this will possibly alter access to traditional fishing areas, increase risk associated with travel on the land or ice, and change fishery success. Loss of a significant portion of fish from the household economy is very likely to require replacement with some other means – perhaps increased reliance on food transported from the south and/or on other local northern foods (e.g., terrestrial mammals), further stressing those populations. The former solution (i.e., a dietary shift to southern transported foods) will possibly contribute to dietary problems and increased health costs (Fast and Berkes, 1998). Increased transportation costs for such foods are likely to be covered in some manner by subsidies from southern portions of the national economies, but this feedback from the north would increase the economic impact of climate change in southern areas. The availability of new fish species in some areas will possibly mitigate these problems but may not provide immediate solutions.

Recreational fisheries

Impacts similar to those outlined previously will also affect sport or recreational fisheries. In addition, management demands and economics associated with such fisheries will possibly alter. For example, recreational fishing for Atlantic salmon in eastern North America is regulated in part through river closures driven by higher temperatures and low water levels. The rationale is that such conditions stress fish, and catch-and-release angling (the norm for the area) would further stress individuals and affect populations (Dempson et al., 2001). Between 1975 and 1999, about 28% of 158 rivers on average were closed annually, with up to 70% affected in some years. This resulted in a 35 to 65% loss of potential fishing days with the warmest period (1995–1999) most affected. In part, this stress was the result of increased upstream migratory energy demands associated with lower water levels and higher water temperatures. Although this study was conducted in Newfoundland, it represents a possible future situation for arctic sport fisheries based upon riverine migrating fishes such as Atlantic salmon and Arctic char. Such fishes support significant local tourist economies in many areas of the Arctic, hence climate change impacts on recreational fisheries will possibly result in substantive economic impacts by increasing the frequency and duration of closures (Dempson et al., 2001).

8.5.5.6. Impacts on aquaculture

Aquaculture of fish in northern areas of arctic countries tends to focus upon cold-water species with high economic value such as Atlantic salmon and Arctic char. In general, such culture is presently located in areas south of the Arctic as defined herein, but this is likely to change as demand and opportunity increase. Aquaculture can be conducted wholly in freshwater using locally available or exotic species either indoors or out, or in protected nearshore marine areas primarily using anadromous species (see also Chapter 13). Climate change is very likely to result in a number of possible shifts in this industry, however, similar to those described previously for fisheries dependent upon wild populations, these will possibly be complex and interactive with both positive and negative consequences. The details will be specific to the local situations. Possible changes include production increases, especially in northern locations, due to temperature-driven increased growth rates of cultured fish and also decreased times necessary for culture to marketable sizes (Lehtonen, 1996), but this is likely to increase food requirements. Increased production will depend upon other factors not becoming limiting, especially available volumes of freshwater needed for inland operations. Knowledge of projected shifts in precipitation and evaporation with concomitant impacts on groundwater levels will be important to the viability of such endeavors. Production increases will possibly be offset by increased costs associated with oxygenating warmer waters, especially those for summer use, and increased loss to disease or costs associated with prevention (Lehtonen, 1996).

As warmer conditions extend northward, the areas in which aquaculture is economically viable (i.e., revenue exceeds costs) will possibly increase, opening new areas for this activity. However, increased climate variability and frequency of extreme events will possibly also increase engineering costs. New aquaculture efforts will present economic opportunities, but also have potential negative impacts on local native species, especially if the cultured species is exotic (e.g., a non-native southern species). As climate change effects are realized, the suite of southern species potentially viable for aquaculture will probably increase and present new economic opportunities. This will increase the need for regulatory scrutiny of such development, especially if the risk of escape and naturalization of such species is high. A related issue is very likely to be an increased risk of intentional but unauthorized introductions of exotic species into natural systems already affected to some degree by climate change. Escape and naturalization of Atlantic salmon along the Pacific Coast of North America (Volpe et al., 2000) serves as a valuable model of potential negative effects. Appropriate management and control of such activities will be required; such activities will possibly add significant additional stress to native fish populations already highly stressed by climate change. Strategies to deal with such possibilities are presently lacking or extremely limited, especially for potential transfers within countries.

8.5.6. Impacts on aquatic birds and mammals

Given the increasing understanding of the critical role of climate in driving the population dynamics of waterfowl and aquatic wildlife, it is very likely that progressive, rapid change in climate will trigger substantial fluctuations in endemic fauna and flora. Population- and community-level responses of aquatic birds and mammals will probably result from combinations of direct and indirect impacts. These include changes in winter severity; seasonal snow and ice distribution and depths; timing and peaks of lake, pond, and wetland productivity; predator–prey dynamics; parasite–host interactions; habitat quality and distribution; and fire frequency, intensity, and distribution.

As discussed in section 8.4.1, projections from the five ACIA-designated models suggest that coastal land areas (and associated estuarine and freshwater habitats) are likely to experience dramatic temperature increases and changes in their hydrologic regimes. Such changes are likely to produce significant alterations in the quantity and quality of existing coastal estuarine and delta habitats, thereby affecting associated communities of birds and aquatic mammals.

It is therefore probable that changes in freshwater and estuarine habitat will result in altered routes and timing of migration. Emigration of aquatic mammals and waterfowl is likely to extend northward as more temperate ecosystems and habitats develop at higher latitudes (section 7.3.5). Migration will possibly occur earlier in the spring with the onset of high temperatures, and later in the autumn if high temperatures persist. Breeding-ground suitability and access to food resources are likely to be the primary driving forces in changes in migration patterns. However, many species living in these areas are adapted to, even dependent on, extreme natural fluctuations in climate and associated impacts on water resources. Hence, their responses to such changes are likely to be species-specific and quite varied.

A number of direct and indirect effects are likely to occur in shallow arctic lakes and ponds that lack a thermocline. Summer maximum temperatures are likely to climb above physiological preferences or thresholds of algae, plankton, and benthic invertebrates, which would produce substantial shifts through time in diversity and/or abundance at these lower trophic levels. Such shifts will probably result in earlier or reduced seasonal peaks in abundance of key foods, thereby creating mismatches between resource availability and timing of breeding. This will possibly lead to a lowering of reproductive success in higher-level consumers such as waterfowl.

Changes in water regimes are very likely to dramatically alter the quantity and quality of aquatic and riparian habitat, leading to local changes in the distribution of birds and mammals, and at larger scales, are likely to affect overall habitat availability, carrying capacity, and reproductive success. Aquatic mammals and waterfowl are highly dependent on the availability and quality of aquatic habitats for successful breeding, and in the case of waterfowl, nesting. Northern species will possibly have diminished reproductive success as suitable habitat either shifts northward or declines in availability and access. Northward colonization of southern species will

possibly result in competitive exclusion of "northern" species for habitat and resources. Many of the projected responses are likely to result from changes in temperature and precipitation. For example, Boyce and Miller (1985) showed that water depths have a significant positive effect on the annual production of juvenile whooping cranes (*Grus americana*), and suggested that increased summer temperatures are likely to create drier conditions in whooping crane nesting marshes over the long term, decreasing production of young and slowing the annual population growth rate.

Many shorebirds (e.g., sandpipers, plovers, snipe, godwits, curlews) are also dependent on water levels and the persistence of shallow wetlands. For instance, most North American species of shorebirds breed in the Arctic, with ten species common to the outer Mackenzie Delta (Gratto-Trevor, 1994, 1997). These species are dependent on invertebrate prey during reproduction, and hatchlings are highly dependent on mosquitoes and chironomids, the preferred foods of developing young. Any changes in timing and availability of food at staging sites in the Arctic, let alone the availability of wetland habitat, are likely to have detrimental effects on the success of hatchlings. Therefore, most species are very likely to be adversely affected by loss of shallow wetland habitat as ponded areas dry in response to rising temperatures, a potential decline in precipitation, and permafrost degradation. Conversely, thawing permafrost and precipitation increases are very likely to increase the occurrence and distribution of shallow wetlands, and probably the success of shorebirds in the Arctic.

Long-term survey data are available for a limited number of wetland-dependent migratory birds in Canada that demonstrate some of the possible effects of climate-related change. These data clearly indicate dramatic declines in the abundance of several waterfowl species (e.g., scoters – *Melanitta* spp., lesser scaup – *Aythya affinis*) with core breeding areas located in the northwestern boreal forest of Canada. Several hypotheses have been proposed to explain these patterns, including changes in wetland systems (e.g., food resources for breeding birds or their offspring). It is difficult to identify causes of decline because changes have also occurred simultaneously in the wintering, migration, and breeding areas of each species; however, breeding-ground changes are the probable cause because indices of productivity have decreased during the past 20 years (Afton and Anderson, 2001; Austin et al., 2000).

The dynamics and stability of aquatic mammal populations have also been linked to observed variability and extremes in hydrologic conditions. Thorpe (1986) found that in the Peace-Athabasca Delta, Canada, years with observed spring ice-jam flooding (and associated re-flooding of perched basins) had high success in local trapping of muskrats. A decade with low water levels in the delta resulted in dryer perched basins and fewer muskrats, followed by a decade of higher water levels and high muskrat harvesting. In this case, perched-basin water levels and the extent of emergent vegetation development seemed to be the controlling factors in muskrat occurrence and abundance. Independent traditional ecological knowledge studies of the area also provided corroborative evidence of this trend (Crozier, 1996). Hence, projected decreases in the frequency and intensity of ice-jam flooding under future climate scenarios would probably cause decreases in the re-flooding of perched basins, negatively affecting muskrat populations in years with low water levels.

It is also possible that projected climate change in the Arctic will produce an increased incidence of mortality from disease and/or parasites in bird and aquatic mammal populations. As temperatures rise, southern species of mammals and waterfowl are likely to shift northward. These species will probably carry with them new diseases and/or parasites to which northern species are not adapted.

8.6. Ultraviolet radiation effects on freshwater ecosystems

Ultraviolet radiation is the most photochemically reactive wavelength of solar energy reaching the surface of the earth, and has a broad range of effects on aquatic biogeochemistry, biota, and ecosystems. As a result of anthropogenic impacts on the atmosphere of the earth, UV radiation exposure in arctic environments is changing substantially. Although anthropogenic emissions of ozone-depleting substances have declined since the ratification of the Montreal Protocol and its amendments, future levels of ozone and UV radiation in the Arctic are uncertain, depending not only on continued compliance with the Protocol and changes in legislation, but also on climate change effects on temperatures and trace gases (e.g., sections 5.6.2, 7.5, and 8.4.4.4). This section provides an overview of how underwater UV radiation exposure is linked to climate, followed by a discussion of general principles concerning UV radiation impacts in aquatic ecosystems (including natural protection mechanisms) and a systematic analysis of potential UV radiation impacts on arctic freshwater habitats.

8.6.1. Climate effects on underwater ultraviolet radiation exposure

To understand the overall impact of changes in UV radiation levels, the synergistic and antagonistic processes resulting from climate change have to be considered since they have the potential to modify the underwater UV radiation regime and consequently the stress on aquatic organisms. Climate change is very likely to be accompanied by shifts in biological UV radiation exposure in arctic river, lake, and wetland environments via three mechanisms (Vincent and Belzile, 2003): changes in stratospheric ozone levels, changes in snow- and ice-cover duration, and changes in the colored materials dissolved in natural waters that act as sunscreens against UV radiation.

Although it is projected that the downward trends in ozone levels are likely to reverse in the near future as a consequence of reduced anthropogenic emissions of chlorofluorocarbons (CFCs) and related compounds, some of the longer-lived ozone-depleting substances are still accumulating in the stratosphere and climate change is likely to prolong the effects of depletion. Temperature increases in the troposphere are projected to be accompanied by temperature decreases in the lower stratosphere, and there is already some evidence of this effect in the polar regions. Temperature decreases in the lower stratosphere are very likely to increase the frequency and extent of polar stratospheric clouds (PSCs) that catalyze CFC–ozone reactions, and result in a strengthening of the polar vortex, which in turn is likely to lead to longer-lasting conditions for ozone depletion (Staehelin et al., 2001). The minimum winter temperatures in the arctic stratosphere are very close to the threshold for the formation of PSCs and the chlorine reactions that lead to ozone loss, and the Arctic remains vulnerable to large-scale ozone depletion (Dahlback, 2002). It is also possible that temperature increases could lead to increased zonal flow at mid-latitudes causing the polar vortex to be more stable, again favoring ozone depletion and a delay in the eventual recovery of the ozone layer (Shindell et al., 1998). Furthermore, as greenhouse gas concentrations increase, the tropical tropopause is very likely to become warmer, resulting in the transport of more water vapor into the stratosphere, which in turn is likely to lead to the formation of PSCs at higher temperatures (Kirk-Davidoff et al., 1999). Therefore, the ozone in the arctic stratosphere would be at greater risk of depletion (see section 5.6.2 for further discussion).

The underwater UV radiation environment changes dramatically with a decrease in snow- and ice-cover duration, especially if this occurs during periods of greatest UV radiation flux and ozone depletion. Analyses of the effects of melting arctic sea and lake ice show that this process results in order-of-magnitude increases in biological UV radiation exposure that greatly exceed those caused by moderate ozone depletion (Box 8.9; Vincent and Belzile, 2003). Lake and river ice are relatively transparent to UV radiation because of CDOM exclusion from the ice during freeze-up (Belzile et al., 2002a). Small changes in snow cover and white ice, however, can radically influence the below-ice UV radiation levels in arctic waters (Belzile et al., 2001).

In arctic aquatic environments, variations in suspended particulates, and especially CDOM, affect transmission of UV radiation (see Box 8.10). These variations can be more important than ozone depletion in determining the UV radiation exposure in the water column of freshwater systems.

In some areas of the Arctic, climate change is very likely to be accompanied by increased vegetation, a concomitant increase in CDOM loading (Freeman et al., 2001), and reduced exposure to underwater UV radiation.

These positive effects are likely, however, to be offset by reduced availability of PAR (Arrigo and Brown, 1996; Neale, 2001; Pienitz and Vincent, 2000). Marked south–north gradients in present-day CDOM concentrations in arctic waters are associated with the latitudinal distribution of terrestrial vegetation. Colored dissolved organic matter loading of freshwater systems is less pronounced at higher latitudes. Lakes in the tundra and polar-desert biomes contain low amounts of these materials; small variations in CDOM concentration in these systems can cause major changes in underwater UV radiation exposure (Laurion et al., 1997; Vincent and Pienitz, 1996). Freshwaters in northern Scandinavia are low in dissolved carbon similar to water bodies in North America; the median DOC concentration for 25 lakes above the treeline in Finnish Lapland was 18 mg C/l (Rautio and Korhola, 2002b). Although acid precipitation enhances underwater UV radiation levels by reducing DOC concentrations in the water, increased thawing of permafrost with climate change is very likely to increase soil runoff and levels of DOC (or CDOM) in arctic freshwater systems. This is very likely to be accompanied by an increase in water turbidity, which will probably not only decrease PAR penetration but also increase the relative proportion of UV radiation, thereby hindering repair processes in aquatic organisms that are stimulated by longer wavelengths. Increased physical turbulence is also likely to expose planktonic organisms to unfavorable irradiance conditions (e.g., exposure to high levels of surface UV radiation and PAR), the effects of which are likely to be especially severe for species that cannot migrate.

8.6.2. Ultraviolet radiation effects on aquatic biota and ecosystems

The effects of UV radiation in the aquatic environment range from molecular to whole-ecosystem. Photobiological damage includes the direct effects of UV radiation in which photons are absorbed by biological molecules such as nucleic acids and proteins that then undergo photochemical alteration. An alternative damage pathway is via the interaction of UV radiation and organic compounds or other photosensitizing agents to produce reactive oxygen species such as superoxide and hydroxyl radicals. These can diffuse away from the site of production and cause oxidative damage to enzymes, lipid membranes, and other cellular constituents.

Aquatic biota have four main lines of defense against UV radiation damage: escape, screening, quenching (chemical inactivation), and repair. The net stress imposed by the UV radiation environment reflects the energetic costs of protection and repair in addition to the rate of photochemical degradation or alteration of cellular components (Vincent and Neale, 2000). These defenses are well illustrated by arctic zooplankton (see Box 8.11), but despite this protection these organisms remain vulnerable to ambient UV radiation levels (Zellmer, 1998), particularly in the cold, shallow, CDOM-poor waters that characterize many arctic lakes and ponds.

Changes in underwater UV radiation exposure are likely to directly affect the species composition of aquatic

Box 8.9. Implications of changing snow and ice cover for ultraviolet radiation exposure

The warming northern climate is prolonging open water conditions. The loss of UV radiation-attenuating snow and ice earlier in the season, when water temperatures are still low but UV irradiances are maximal, is likely to be especially stressful for aquatic biota. As shown below, white ice (ice with air inclusions) has a strong attenuating effect on PAR (visible light) and an even greater effect on UV-A and UV-B radiation. This snow-clearing experiment on Hudson Bay showed that only 2 cm of snow reduced the below-ice exposure to UV radiation and to PAR by about a factor of three, with slightly greater effects at the shorter wavelengths.

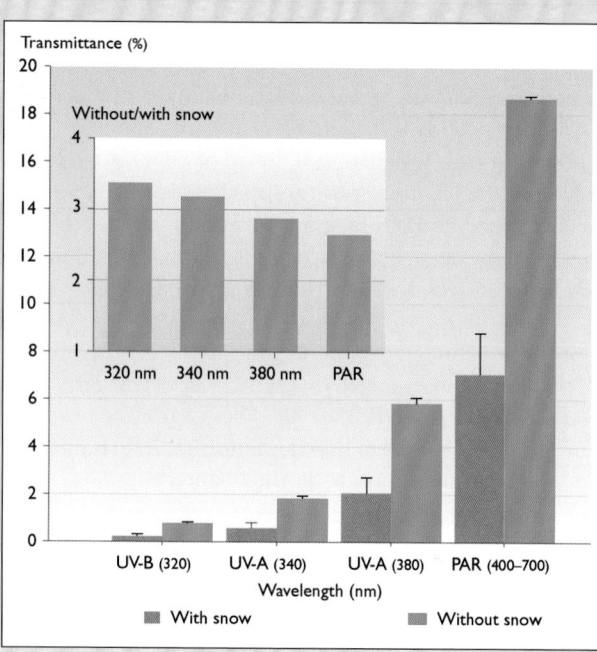

The UV radiation-attenuating effect of ice and snow over the plume of the Great Whale River, in Hudson Bay, Northern Québec, immediately offshore from the river mouth (April 1999). The ice was 1 m thick with white ice at its surface, and the snow was 2 cm thick. The percentage transmittance values are for the penetration of each wavelength into the water beneath before and after the removal of the snow. Inset shows the ratio of transmittance through the ice in each wavelength after (without snow) relative to before (with snow) the 2 cm of snow was cleared from the ice (adapted from Vincent and Belzile, 2003).

biota at each trophic level, as well as cause effects that cascade throughout the benthic (e.g., Bothwell et al., 1994) or pelagic (e.g., Mostajir et al., 1999) food webs and the coupling between them. Some trophic responses are likely to be "bottom-up effects" in that UV radiation exposure reduces the quantity or quality of prey and thereby reduces food supply to the next level of consumer organisms. This could occur, for example, via shifts towards inedible or less digestible algal species (Van Donk and Hessen, 1995; Van Donk et al., 2001) or by reducing the nutritional value of food organisms (Scott J. et al., 1999). The effects of variations in UV-B radiation on the quality of phytoplankton photosynthetic products have received little attention except for studies at the pigment level (e.g., Buma et al., 1996; Zudaire and Roy, 2001). Any alteration in the biochemi-

cal composition of primary producers is likely to change the nutritional value of food consumed by grazers (thus influencing energy flow throughout the food web) as well as restrict the production of photoprotective compounds against UV radiation. Short-term exposure to enhanced UV-B radiation levels in phytoplankton populations of various lakes in the Canadian High Arctic influenced the allocation of newly fixed carbon into the major macromolecular classes (Perin, 2003). Generally, synthesis of both protein and polysaccharides was inhibited by enhanced UV-B radiation levels, and the photosynthate would remain or accumulate in the pool of low molecular weight compounds. Lipid synthesis was insensitive to UV-B radiation levels and represented the most conservative and uniform class, accounting for about 20% of total carbon fixed.

Box 8.10. Colored dissolved organic matter: the natural sunscreen in arctic lakes and rivers

Colored dissolved organic matter (CDOM) is composed of humic and fulvic materials (average to low molecular weight) that are derived from terrestrial soils, vegetation, and microbial activities, and is known to be an effective protective screen against UV radiation for freshwater biota (e.g., plankton, Vincent and Roy, 1993; amphibians, Palen et al., 2002). These compounds absorb UV-A and UV-B radiation and short-wavelength visible light, and in high concentrations such as in arctic rivers they stain the water yellow or brown. Colored dissolved organic matter is now known to be the primary attenuator of underwater UV radiation in subarctic and high-arctic lakes (Laurion et al., 1997); Toolik Lake, Alaska (Morris et al., 1995); arctic ponds (Rautio and Korhola, 2002a); and arctic coastal seas influenced by river inflows (Gibson J.A. et al., 2000; Vincent and Belzile, 2003). The concentrations of CDOM in natural waters are influenced by pH (acidification can cause a severe decline; Schindler D.W. et al., 1996b), catchment morphology, runoff, and the type and extent of terrestrial vegetation. The latter aspects are especially dependent on climate.

The paleoecological record has been helpful in examining past impacts of climate on biological underwater UV radiation exposure, specifically by using fossil diatoms in lake sediments as quantitative indicators of variations in CDOM. This record also underscores the large regional differences in the magnitude and direction of change in underwater UV radiation levels (Ponader et al., 2002; Saulnier-Talbot et al., 2003). Shifts in vegetation and hydrology caused by warming or cooling trends are very likely to affect the quantity of CDOM exported from catchments to their receiving waters, in turn affecting underwater UV radiation levels. For example, an analysis of the past underwater climate (paleo-optics) of subarctic treeline lakes indicated that recent Holocene cooling (from about 3500 years ago to the present) was accompanied by a southward retreat of the treeline and a large decrease in CDOM concentrations in lake waters. This decreasing CDOM resulted in an increase in biological UV radiation exposure that was two orders of magnitude greater than that associated with moderate (30%) ozone depletion (Pienitz and Vincent, 2000). Saulnier-Talbot et al. (2003) reported large decreases in biological UV radiation exposure (starting about 3000 years BP) in a coastal lake in subarctic Québec that were associated with the establishment of terrestrial vegetation in its catchment.

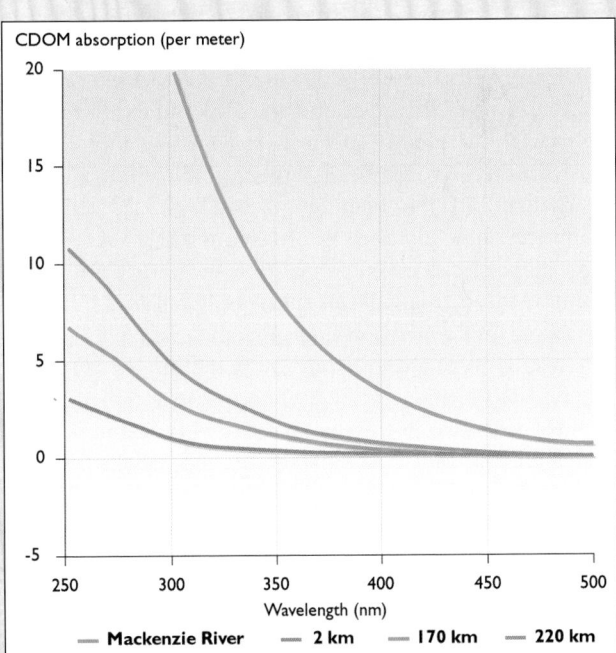

Ultraviolet (<400 nm) and blue-light (400–500 nm) radiation absorption by CDOM in the Mackenzie River (Inuvik, October 2002). The lower curves are for surface samples near the same date in the Beaufort Sea showing the CDOM influence at 2, 170, and 220 km offshore from the mouth of the river (W.F. Vincent and L. Retamal, Laval University, Québec City, unpubl. data, 2004).

Overall, these results were similar to those observed for Lake Ontario (Smith R. et al., 1998). However, the various classes of lipid may respond differently to variations in UV-B radiation levels. For example, exposure to UV radiation influenced fatty acid composition in algal cultures (Goes et al., 1994; Wang and Chai, 1994). Other studies observed that the effect of UV-B radiation on the major lipid classes is species-specific (De Lange and Van Donk, 1997).

Higher trophic levels are dependent on phytoplankton either directly as food or indirectly via trophic cascades. Inhibition of growth and cell division in phytoplankton will most often affect the food quality of these cells by placing stoichiometric constraints on the grazer (Hessen et al., 1997). Hessen and Alstad Rukke (2000) also showed that water hardness could be a major determinant of susceptibility to UV radiation damage among calcium-demanding species such as *Daphnia*. They suggested that calcium, which is an important element for invertebrates with calcified exoskeletons, in low concentrations (low-pH lakes, acidification) could reduce the stress tolerance of organisms. Although several studies of freshwater invertebrate species have reported increased mortality in response to increased UV radiation levels, especially zooplankton (Hurtubise et al., 1998; Leech and Williamson, 2000; Rautio and Korhola, 2002a; Siebeck and Böhm, 1994; Vinebrooke and Leavitt, 1999a; Williamson et al., 1994), the variation in UV radiation tolerance is high among species and life stages (Leech and Williamson, 2000). In general, small zooplankton (small rotifers) are considered to have a high UV radiation tolerance, while large species vary in their tolerance both among and within species. Leech and Williamson (2000) found that cladocerans had the lowest UV radiation tolerance and exhibited high variability among species. *Daphnia* was one of the most sensitive groups of organisms, while adult calanoid and cyclopoid copepods had high UV radiation tolerances. In a comparison of lakes across a successional gradient of catchment vegetation and thus CDOM content, three zooplankton species (*Asplanchna priodonta*, *Ceriodaphnia quadrangula*, and *Bosmina longirostris*) were absent from low-CDOM, UV-transparent waters, and perished when transplanted from a CDOM-rich lake in the series and held at 0.5 m depth under full UV radiation exposure in a clear lake. In contrast, two species that avoided high UV radiation exposure in the near-surface waters (*Daphnia pulicaria* and *Cyclops scutifer*, a highly UV-tolerant species) occurred in even the clearest lakes (Williamson et al., 2001). Morphotypic and biochemical differences among populations of a given species may also play an important role. Pigmented clones of *Daphnia* were more tolerant of UV radiation than transparent clones (Hessen et al., 1999), and pigmentation appears to increase in response to increased UV radiation exposure (Rautio and Korhola, 2002b; Box 8.11). Studies of the effects of natural and enhanced UV radiation levels on fish are rare, but laboratory experiments have shown that high-latitude species of trout have sunburns,

increased fungal infections, and higher mortality when exposed to increased dosages of UV radiation (Little and Fabacher, 1994).

Other trophic responses are likely to be top-down effects, in which some species are released from grazing pressure or predation by UV inhibition of the consumers and thereby achieve higher population densities (Bothwell et al., 1994). This complex combination of direct and indirect effects makes any future shifts in aquatic ecosystem structure extremely difficult to project. In addition, the deleterious effects of UV-B radiation at the community level are difficult to assess since they are generally species-specific. For example, Wickham and Carstens (1998) showed that the responses of the planktonic microbial communities in Greenland ponds to ambient UV-B radiation levels varied greatly between species, especially rotifers and ciliates.

Multiple factors seem to affect amphibians negatively. These factors include both site-specific, local effects (e.g., pesticide deposition, habitat destruction, and disease) as well as global effects (e.g., increased UV-B radiation exposure and climate change; e.g., Häder et al., 2003). Amphibians have been the focus of special interest at temperate latitudes because of the recent widespread decline in many frog populations and the recognized value of these organisms as sensitive indicators of environmental change. Although many amphibians can be relatively resistant to UV-B radiation, it can cause deformities, delays in development, behavioral responses, physiological stress, and death in frogs. The rise in UV-B radiation levels associated with stratospheric ozone depletion has been widely promoted as one of several hypotheses to account for their decline (Collins and Storfer, 2003, and references therein). However, the effects are controversial and in many habitats where the frogs are declining, the animals are well protected by CDOM (Box 8.10). A small number of frog species occur in the subarctic and the Arctic, including the common frog (*Rana temporaria*) and the wood frog (*Rana sylvatica* – North America), with distributions extending north of the Arctic Circle. Contrary to expectation, however, these populations may experience higher UV radiation exposures under natural conditions relative to temperate regions, and therefore be more pre-adapted, because of lower concentrations of UV-screening CDOM in high-latitude waters (e.g., Palen et al., 2002) and life-cycle characteristics (phenology: higher UV-B radiation doses during the breeding season at higher latitudes; Merilä et al., 2000). There is considerable variation in UV radiation tolerance between amphibian strains and species; for example, a latitudinal comparison in northern Sweden found that *R. temporaria* embryos were relatively tolerant of UV-B radiation, with no clear latitudinal differences (Pahkala et al., 2002). The positive and negative effects of climate change on arctic habitats (e.g., duration of open water, extent of wetlands) are likely to have much greater impacts on amphibians than changes in UV radiation exposure.

8.6.3. Impacts on physical and chemical attributes

The large arctic rivers are relatively protected from UV radiation exposure because of their high CDOM content (Gibson J.A. et al., 2000). Conversely, natural and increased UV radiation levels are likely to be important for photochemical loss of carbon from these systems. For example, there is evidence that the duration of ice-free conditions has increased in the Mackenzie River, Canada (Magnuson et al., 2000) and the River Tornio, Finland. The resultant increased exposure to UV radiation is likely to favor increased annual rates of UV degradation of riverine DOC, with possible impacts on the inshore coastal waters that receive these inputs.

Box 8.11. Ultraviolet radiation protection and recovery mechanisms in arctic freshwaters

Aquatic organisms have varying abilities to counter the effects of UV radiation. Photoprotective and repair processes are particularly important in preventing and reversing UV radiation damage to photosynthetic mechanisms. A range of potential repair processes is stimulated by longer wavelengths to counteract the damaging effects of UV radiation. The relative importance of repair versus protection will vary depending on specific conditions and the physiological characteristics of the species assemblage (Banaszak and Neale, 2001). Organisms living in arctic lakes have evolved several strategies to cope with UV radiation, which play an important role in shallow and highly UV-transparent arctic lakes and ponds. Some species of algae and zooplankton have an ability to reduce their exposure to UV radiation by vertical migration, which may be a response to high intensities of UV radiation (Huntsman, 1924). Leech and Williamson (2001) and Rhode et al. (2001) provided further evidence that organisms avoid highly irradiated areas by escaping the brightly lit surface zone.

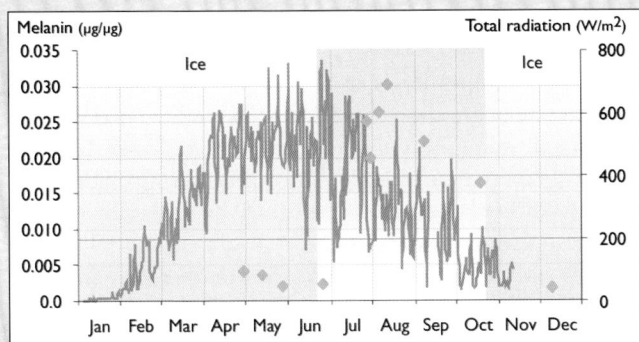

Changes over time in solar radiation (solid line) and the rise in the UV-screening pigment melanin (data points) in the zooplankton *Daphnia umbra* in Lake Saanajärvi, northern Finland, immediately after ice-out (adapted from Rautio and Korhola, 2002b).

In addition to avoidance, aquatic organisms can escape from UV radiation by reducing the effective radiation that penetrates the cell. A number of UV-protecting compounds have been described; the three major types are carotenoids, mycosporine-like amino acids (MAAs), and melanin. The photoprotective properties of carotenoids are mainly associated with antioxidant mechanisms and inhibition of free radicals, as opposed to direct UV radiation screening (Hessen, 1994). Carotenoids absorb wavelengths in the visible light spectrum and do not therefore provide direct protection from UV radiation. Mycosporine-like amino acids have absorption maxima ranging from 310 to 360 nm within the UV wavelength range. They are present in alpine phytoplankton and zooplankton (Sommaruga and Garcia-Pichel, 1999; Tartarotti et al., 2001) and also occur in arctic freshwater organisms although there is no research on this. Animals are unable to synthesize MAAs and carotenoids, and must therefore acquire these compounds from their diet.

Cladocera and fish produce melanin, with absorption maxima between 250 and 350 nm. Melanin acts by absorbing radiation before it enters the body tissues, and its synthesis seems to be a direct response to UV radiation (Hobæk and Wolf, 1991). Melanic zooplankton are typically found in clear arctic waters where the absorbance of UV radiation is low, and in shallow ponds where high DOC levels may not provide enough protection from UV radiation (Hebert and Emery, 1990; Rautio and Korhola, 2002b). It has also been shown that melanin synthesis followed the annual variation in UV radiation levels (i.e., synthesis peaked at the time of maximum underwater UV irradiance, see figure). Aquatic organisms can also repair damage from UV radiation by nucleotide excision repair or by photoreactivation mechanisms, such as photoenzymatic repair (Leech and Williamson, 2000).

Brief exposure to UV radiation triggers only the initial UV radiation stress response. However, responses over long periods show that organisms can acclimate to the UV radiation stress and/or recover growth rates with the development of photoprotective strategies (e.g. the synthesis of photoprotective compounds). A long-term enclosure experiment conducted in a high-arctic lake on Ellesmere Island (Nunavut, Canada) showed an initial decrease in phytoplankton productivity with enhanced UV-B radiation exposure, with recovery after 19 days (Perin, 2003). Long-term acclimation to and recovery from increased levels of UV radiation were also observed in a cultured marine diatom (Zudaire and Roy, 2001). Antecedent light conditions, temperature, nutrient availability, and/or variations between species are all factors that can affect acclimation of organisms to high intensities of UV-B radiation (Zudaire and Roy, 2001).

In addition to having low CDOM concentrations and resultant deep penetration of UV radiation, many arctic lakes, ponds, and wetlands are shallow systems. The mean measured depth for more than 900 lakes in northern Finland was less than 5 m (Blom et al., 1998), and was about 5 m for 31 lakes in the vicinity of Tuktoyaktuk (Northwest Territories; Pienitz and Smol, 1994) and 46 lakes on Ellesmere Island (Nunavut; Hamilton et al., 1994). Consequently, all functional groups, including the benthos, are often exposed to UV radiation throughout the entire water column. In addition, many aquatic species stay in the offshore pelagic zone. Even species that are more benthic or littoral are protected minimally by macrophytes, as arctic waters, especially those in barren catchments, often contain little aquatic vegetation.

Changes in UV radiation exposure in these northern ecosystems are amplified by the low CDOM concentrations. Most have concentrations of DOC below 4 mg DOC/l, the threshold below which there are marked changes in UV radiation penetration through the water column, and in the ratio of wavelengths controlling the damage–repair balance, with only minor changes in CDOM (Laurion et al., 1997).

The initial impacts of climate change are likely to be associated with the loss of permanent ice covers in the far northern lakes: this appears to have already recently taken place in the Canadian High Arctic (Belzile et al., 2002a). These effects are likely to be amplified by prolonged open-water conditions in lakes and ponds. However, other physical changes in these environments (e.g., wind-induced mixing) are likely to have greater perturbation effects than those associated with increased UV radiation exposure. Although increases in CDOM are very likely to mitigate the effects of increased UV radiation levels, decreases in PAR are very likely to hamper photosynthesis. Furthermore, increased turbidity associated with thawing permafrost is likely to further reduce the exposure of organisms to damaging UV radiation (for turbidity effects on UV radiation, see Belzile et al., 2002b).

The photochemical effects of increased UV radiation levels are also likely to influence the toxicity of contaminants (see also section 8.7). Mercury (Hg) is the principal toxic chemical of concern in the Arctic and elsewhere. Methyl mercury (MeHg) is the most toxic form and the only form that biomagnifies in food chains. It was recently shown that:

- ultraviolet radiation exposure photoreduces divalent mercury (Hg^{2+}, the soluble form in lakes) to elemental mercury (Hg^0, the form that can volatilize from lakes; Amyot et al., 1997);
- ultraviolet radiation can also influence photooxidation (the conversion of Hg^0 to Hg^{2+}; Lalonde et al., 2001);
- the formation of MeHg in arctic wetlands is very sensitive to temperature;
- ultraviolet radiation photodegrades MeHg; and

- most of the Hg in recently fallen snow moves back to the atmosphere within a few days of exposure to solar radiation (Lalonde et al., 2002).

Not only is photochemical reduction of Hg important in the Arctic, but microbial reduction and oxidation may also occur as shown previously in temperate lake waters. Microbial oxidation is turned on by a hydrogen peroxide-dependent enzyme likely triggered by photochemical production of hydrogen peroxide (Scully et al., 1997). The interactions between UV radiation, temperature, and pH can alter Hg mobilization and speciation, and regulate the levels of Hg in organisms at the base of the food web. Photochemical events during spring are likely to be especially sensitive to rising UV-B radiation levels. Large quantities of Hg are photochemically oxidized and precipitate out of the arctic atmosphere with the first sunlight each spring, resulting in a rapid rise in Hg concentrations in snow; 24-hour variations in these atmospheric Hg-depletion events correlate with fluctuations in UV-B irradiance (Lindberg et al., 2002). Increasing spring levels of UV radiation due to stratospheric ozone depletion would probably enhance this so-called "mercury sunrise" phenomenon.

Wetlands and peatlands are rich in CDOM and the aquatic biota are therefore well protected from UV radiation exposure. Early loss of snow and ice, however, is likely to increase exposure during a critical growth phase. Photochemical processes may be especially active in these shallow waters, and this mechanism of CDOM loss is very likely to accelerate with temperature increases (snow-cover loss) and ozone depletion.

8.6.4. Impacts on biotic attributes

Mild increases in UV radiation levels are likely to stimulate biological processes via photochemical release of low molecular weight organic carbon substrates and nutrients. More severe increases are likely to cause damage and/or a shift toward UV-tolerant species with a potential loss of diversity or other unique ecosystem attributes (Fig. 8.20).

8.6.4.1. Rivers and streams

Benthic mats and films are a common feature of high-latitude streams as well as many ponds, lakes, and wetlands, and are often dominated by cyanobacteria, especially the nitrogen-fixing genus *Nostoc* and filamentous species of the order Oscillatoriales (Vincent, 2000). These communities commonly occur in shallow water systems where UV radiation exposure is likely to be high. Ultraviolet radiation has a broad range of deleterious effects on benthic cyanobacteria including on their pigment content, nitrogenase activity, photosynthesis, and respiration (Castenholz and Garcia-Pichel, 2000; Vincent and Quesada, 1994). Much of the literature, however, reports experiments conducted under unrealistically high UV radiation dosages provided by artificial lamps, and many of the effects are likely to be much less

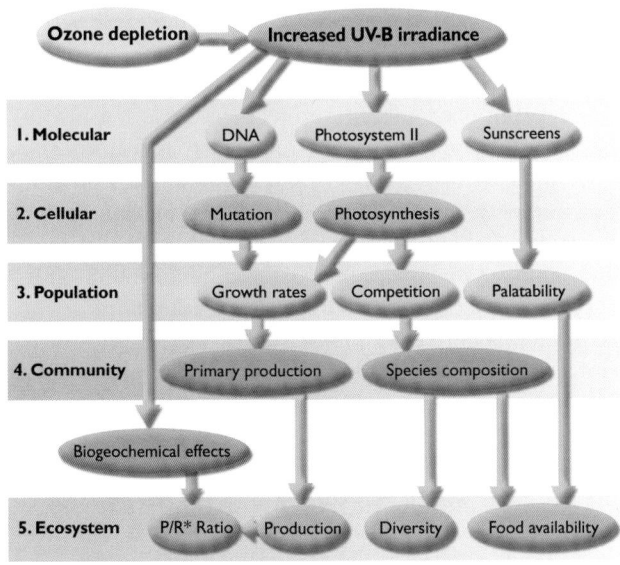

Fig. 8.20. Ultraviolet radiation is the most photochemically reactive wavelength in the solar spectrum and has a wide range of effects, from molecular to whole ecosystem (Vincent and Roy, 1993).

apparent or absent in natural ecosystems, even under conditions of severe ozone depletion (Vincent and Neale, 2000). Periphyton and benthic invertebrates are well protected given their avoidance and sunscreen capacities. Zooplankton and phytoplankton communities are well developed in large arctic rivers (e.g., Rae and Vincent, 1998a), however, they are generally protected by high CDOM concentrations in these waters.

8.6.4.2. Lakes, ponds, and wetlands

In the Arctic, lake organisms have to cope with low nutrient conditions and/or low food availability, low temperatures, and short growing seasons (3–5 months). The UV radiation damage–repair balance may be especially sensitive to features of the arctic freshwater environment. Most of these ecosystems are oligotrophic and phytoplankton are therefore commonly limited by nutrient supply (Bergeron and Vincent, 1997) in addition to low temperatures (Rae and Vincent, 1998a). As a result, the photosynthetic rates per unit biomass tend to be extremely low, even in comparison with other low-temperature systems such as sea ice, polar oceans, and low-temperature cultures (Markager et al., 1999). Because of the low temperatures and low nutrients, phytoplankton photosynthetic rates are extremely low in Canadian high-arctic lakes (usually less than 1.5 µg C/µg Chl-a/hr) and phytoplankton perform very poorly under high-light regimes (Kalff and Welch, 1974; Perin, 2003; Rigler, 1978). There have been few explicit tests of the effect of temperature on UV radiation damage of planktonic systems (Rae and Vincent, 1998a). However, since enzymatic processes are temperature-dependent (whereas damage induction is not), the slow metabolic rates of northern phytoplankton are likely to have a direct effect on the net stress imposed by increased UV radiation exposure by reducing all cellular processes including the rate of repair of photochemical damage. In the antarctic

waters, for example, low temperatures drastically reduce repair to the extent that algal cells failed to show any photosynthetic recovery for at least five hours after UV radiation exposure (Neale et al., 1998). The low nutrient conditions that characterize northern lakes are likely to further compound this effect by reducing the availability of elemental resources for building enzyme systems involved in the functioning of the cell, including the repair of UV radiation damage, and are also likely to limit the investment in photoprotective mechanisms. Moreover, the lower temperatures probably reduce the affinity of the phytoplankton cells for nutrient uptake by membrane transport processes (Nedwell, 2000), thereby increasing nutrient limitation.

The paleoclimatic record has also provided insights into the possible effects of past climate change on UV radiation exposure in aquatic ecosystems. For instance, analyses of fossil diatom assemblages in northern and alpine lake sediments have indicated that variations in underwater UV irradiance during the Holocene had major impacts on algal community structure and productivity (Leavitt et al., 1997; Pienitz and Vincent, 2000). Paleo-optical studies from subarctic lakes have revealed large fluctuations in biologically damaging underwater UV irradiance over the last 6000 years, accompanied by pronounced shifts in algal species composition and changes in the balance between benthic and pelagic primary producers (Pienitz and Vincent, 2000). Vinebrooke and Leavitt (1999b) observed similar effects in low-CDOM mountain lakes.

There are only a few studies of temperature-dependent UV radiation damage to zooplankton (Borgeraas and Hessen, 2000). In general, it is assumed that low temperatures would slow down the UV radiation damage repair mechanisms such as DNA repair and detoxification of reactive oxygen species. However, contrary to expectations, Borgeraas and Hessen (2000) found that reduced temperatures increased survival among UV-irradiated *Daphnia*. They argued that although repair mechanisms are slower in the cold, UV-triggered activation processes (such as reactive oxygen species metabolism and lipid peroxidation) also slow down with decreasing temperature, thereby increasing *Daphnia* survival.

At ambient levels, UV-B radiation can contribute up to 43% of the photoinhibition of photosystem II function in phytoplankton populations of Canadian high-arctic lakes (as measured by both *in vivo* and dichlorophenyl-dimethyl urea-enhanced fluorescence) as well as decreasing (by up to 40%) phytoplankton productivity rates near the water surface (Perin, 1994). The smallest size fraction (i.e., picoplankton: 0.2–2 µm) usually represents more than 50% of total phytoplankton productivity in high-arctic lakes (Perin, 1994). Oligotrophic conditions tend to select for small cells with a high surface-to-volume ratio that favors nutrient transport at low substrate concentrations. Small cells are especially sensitive to UV radiation because they have high illuminated surface-to-volume ratios, little self-shading, and low effectiveness of screening pigments (Karentz et al., 1991; Raven, 1998). Even the production of UV radiation sunscreens is unlikely to confer much protection given the short path length in these cells (Garcia-Pichel, 1994), and studies of a variety of organisms have shown that enhanced UV radiation exposure inhibits the growth of larger cells less than that of smaller cells (Karentz et al., 1991). Laurion and Vincent (1998) evaluated the size dependence of UV radiation effects on photosynthesis in subarctic lakes with a series of short-term photosynthetic experiments, which showed that, in contrast to expectations, the smaller cells were more resistant to UV radiation than larger cells. This smaller-cell fraction was dominated by cyanobacteria, a group known to have a broad range of effective UV-protective mechanisms (Vincent, 2000). Kaczmarska et al. (2000) also found low UV radiation susceptibility in a pico-cyanobacteria-dominated phytoplankton assemblage from a clear lake in southern Canada. On the other hand, short-term experiments in several high-arctic lakes showed that the relative contribution of picoplankton (0.2–2 μm) to phytoplankton production generally decreased with increased UV-B radiation exposure while the larger cells (>20 μm) were more UV-B radiation tolerant and their contribution to productivity usually increased after UV-B radiation exposure. Arctic lake experiments at Spitsbergen also indicated a greater sensitivity of the picocyanobacteria relative to larger colonial species to UV-B radiation exposure (Van Donk et al., 2001). A study by Boelen et al. (2001) showed that for marine tropical plankton, UV-B-induced DNA damage was not significantly different between two size classes (0.2–0.8 μm and 0.8–10 μm). Given the variability in results between studies, other aspects such as species-specific sensitivity, repair capacities, or cell

morphology might be more important than cell size (Boelen et al., 2001).

The level of photoinhibition by UV radiation in phytoplankton can be modified by many factors that influence the extent of exposure within the water column. For instance, vertical mixing can affect the time and duration of phytoplankton exposure to UV radiation and diminish or aggravate projected inhibition of photosynthesis obtained under simulated conditions (continuous UV radiation exposures). In a well-mixed water column, the planktonic community can seek refuge from UV-B radiation, and photo-repair in the deeper portion of the column. However, the formation of near-surface thermoclines caused by high solar irradiance, calm winds, and solar heating of the surface water can retain the phytoplankton under high irradiances for longer periods of times (Milot-Roy and Vincent, 1994) and result in UV radiation damages that can exceed what can be repaired (Xenopoulos et al., 2000). During the spring and summer in the Arctic, climatic conditions (e.g., clouds, rain, snow, fog, and wind) that change tremendously from day-to-day can affect the amount of UV radiation exposure as well as the ratios of UV radiation to longer wavelengths reaching the surface of the earth. In the coastal areas of northern Norway, variation in cloudiness was demonstrated to influence UV radiation levels. The relative amount of UV-A and UV-B radiation to PAR increased during periods of heavy cloud cover (Eilertsen and Holm-Hansen, 2000) because clouds reflect and return radiation (Madronich et al., 1995).

Ultraviolet radiation may impair the transfer of carbon from the microbial food web to higher trophic levels, including zooplankton and fish. However, increased pho-

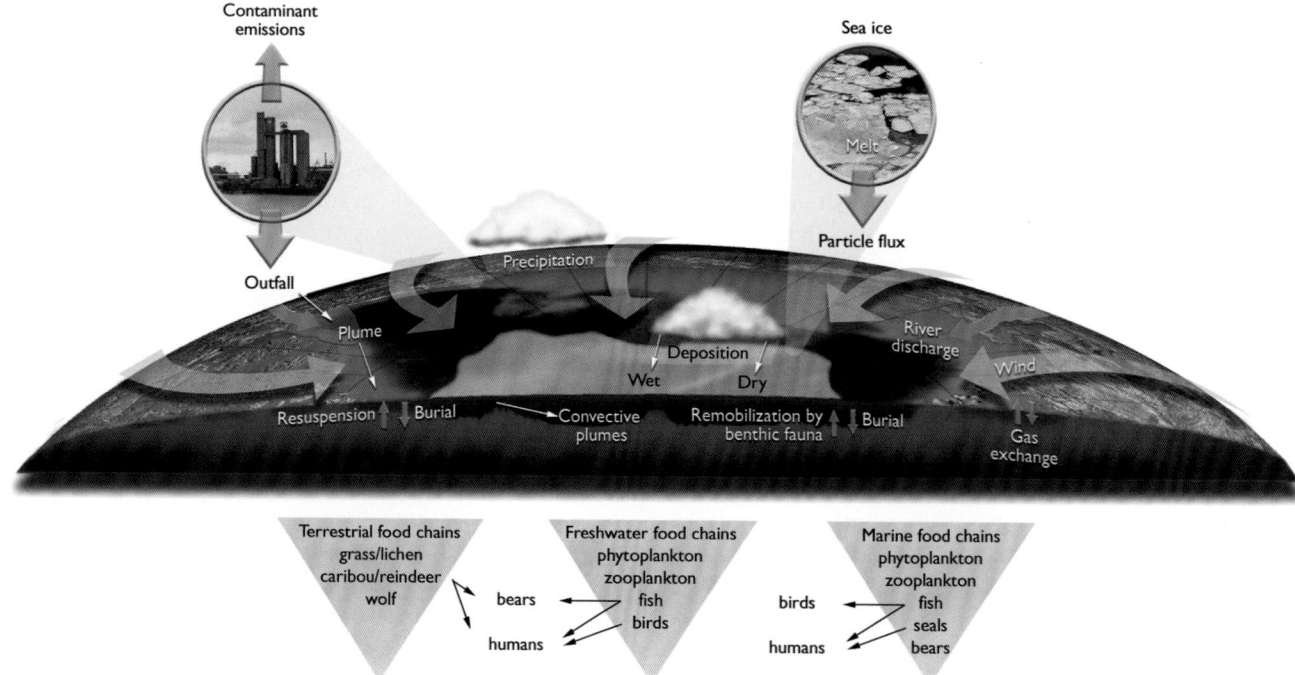

Fig. 8.21. Contaminant pathways, transfers, and exchanges in freshwater systems: land–atmosphere–water interactions (adapted from AMAP, 1998).

tochemical activity associated with UV radiation also has the potential to stimulate some heterotrophic species by causing the breakdown of high molecular weight organic compounds into a more available form (Bertilsson et al., 1999; Lindell et al., 1996; Reche et al., 1998; Wetzel et al., 1995) that can then be taken up for bacterial and protist growth. A study using large enclosures in a high-arctic lake with high levels of dissolved organic compounds showed that after long-term exposure to enhanced UV-B radiation, heterotrophic bacterial production and zooplankton density increased, which may have resulted from an increase in nutrient availability caused by photodegradation of organic compounds and the stimulation of heterotrophic pathways (Perin, 2003).

Some wetland biota such as amphibians are known to be highly sensitive to UV radiation, although sensitivity varies greatly among populations, and other factors such as climate effects on habitat extent are likely to have much greater impacts on northern species (section 8.6.2). A variety of complex responses have been observed to date in wetland plants. For example, UV-B radiation effects on the growth of high-latitude mosses appear to be a function of water supply as well as species. Field irradiation (UV-B radiation levels increased 15%) of *Sphagnum fuscum* caused a 20% reduction in growth, however, growth of the other moss species increased by up to 31% with the enhanced UV-B radiation. This stimulatory effect, however, ceased or was reversed under conditions of reduced water supply (Gehrke et al., 1996).

8.7. Global change and contaminants

During the past 50 years, persistent organic pollutants (POPs), metals, and radionuclides have been widely distributed into northern freshwater ecosystems by long-range atmospheric transport (Barrie et al., 1998; Macdonald R. et al., 2000). Within some catchments, deposition from the atmosphere may be augmented locally by industry or agriculture (AMAP, 1998) or bio-transport (AMAP, 2004b; Ewald et al., 1998; Zhang et al., 2001). Figure 8.21 illustrates contaminant pathways, transfers, and exchanges in freshwater systems. This section briefly discusses how projected global change might alter these pathways, focusing especially on POPs and Hg because they have the greatest potential for change in risks to freshwater ecosystems as a result of climate change (Macdonald R. et al., 2003a,b).

8.7.1. Contaminant pathways and arctic freshwater ecosystems

There are two components of long-range transport pathways: transport to arctic freshwater catchments, and processes within the catchments (Fig. 8.22). Transports and transfers within each of these components are altered by climate change manifested in such things as wind fields, precipitation (amount, timing, form), snow cover, permafrost, extreme events, UV radiation exposure, the hydrological cycle, ice cover, the organic carbon cycle,

and food webs, and are very likely to result in enhanced bioaccumulation of contaminants (e.g., Box 8.12).

Before describing specifically how global change may alter contaminant pathways, it is important to understand how contaminants become concentrated in the environment. Macdonald R. et al. (2002) suggest that there are two distinct concentrating processes, which they term solvent switching and solvent depletion. Solvent switching can, for example, lead spontaneously to concentration amplification of hexachlorocyclohexanes (HCHs) in water because HCH partitions strongly out of air (Li et al., 2002), or high concentrations of PCBs in phytoplankton due to strong partitioning out of water and into lipids (Mackay D., 2001). Solvent depletion involves a reduction in the mass of solvent in which the contaminant is held, a process that can lead to fugacity amplification (i.e., POP concentrations exceeding thermodynamic equilibrium with the surrounding media). Examples include inefficient fat transfers in aquatic food webs (i.e., biomagnification, Kidd et al., 1995c), the loss of organic carbon in settling particles or during sediment diagenesis (Jeremiason et al., 1998; Larsson et al., 1998), the decrease of snow surface area as crystals become more compact during aging or the entire loss of snow surface during melting (Macdonald R. et al., 2002; Wania, 1999), or cryogenic concentration during the formation of ice (Macdonald R. et al., 2003b; Fig. 8.23). Many of the consequences of climate change for the solvent-switching processes are relatively easy to project and model because the effect of temperature on partition coefficients is known. For example, for contaminants that presently are saturated in arctic surface waters, increased temperatures will generally lead to net gas evasion (Harner, 1997) and the ocean is very likely to become a net source of those contaminants to the atmosphere. McKone et al. (1996) concluded that, with temperature increases, the risk hexachlorobenzene (HCB) presents to aquatic biota is likely to decrease slightly because HCB will partition less into water (McKone et al., 1996). The solvent-depleting processes, however, provide a much greater challenge to projection and have not yet been incorporated realistically into models.

8.7.2. Persistent organic pollutants in arctic catchments

The freeze, melt, and hydrological cycles and the organic carbon cycles of arctic lakes are likely to provide sensitive sentinels of change. As discussed in more detail in the previous sections and Chapter 6, probable changes as a result of projected temperature increases include reduced thermal contrast between winter and spring; reduced duration of snowmelt (later freeze-up, earlier melting); reduced ice formation; increased annual precipitation; thawing of permafrost producing a deeper active layer, enhanced soil erosion, mobilization of organic carbon, and reduced pond areas owing to drainage; more frequent extreme weather events; changes in catchment vegetation (i.e., more leaf-bearing plants); changes in nutrient availability; warming of

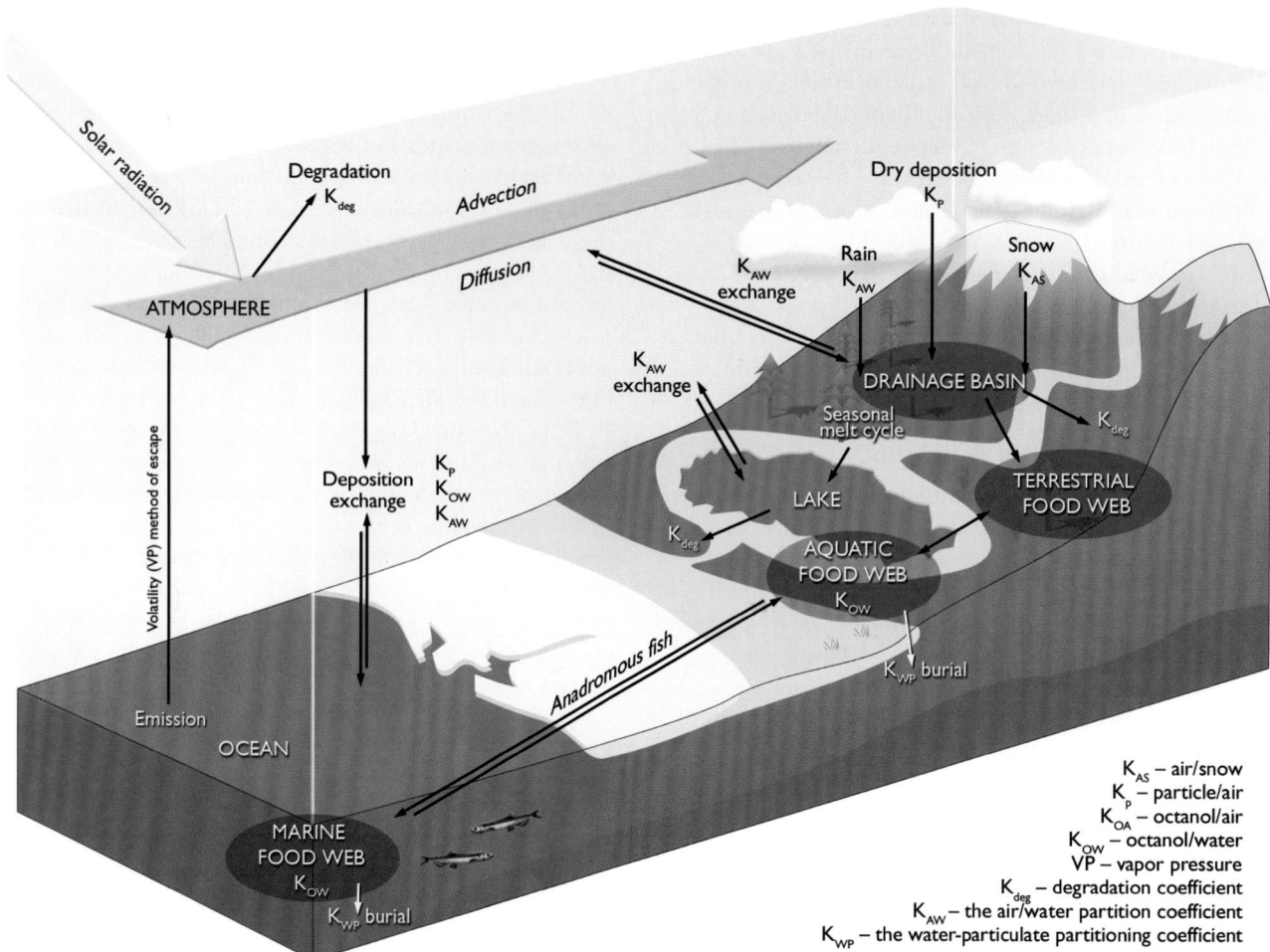

Fig. 8.22. Processes involved in transporting POPs to the Arctic and depositing them into terrestrial ecosystems. Transport, deposition, and exchange can occur anywhere along the transport pathway. Contaminants can also be transported within aquatic and terrestrial food chains. Climate change can alter the physical couplings between the systems (e.g., by changing rain or snowfall patterns), or alter the biological couplings by changing trophic structure or migratory pathways.

lakes; and an increase in the frequency of wildfires (e.g., Hinzman et al., in press; Schindler D.W., 1997; Schindler D.W. et al., 1996a, 1997; Vörösmarty et al., 2001). For some lakes, permafrost degradation together with reduced ice cover is very likely to result in enhanced nutrient and organic carbon loadings and higher productivity. Conversely, if dry summer conditions produce extensive fires, affected lakes are very likely to receive reduced spring melt, fewer nutrients from the catchment, reductions in productivity (Schindler D.W. et al., 1996a), and higher burdens of combustion PAHs.

Most arctic lakes receive their contaminant burdens from the atmosphere, with the catchment acting as a receptor through snow, rain, and dry deposition especially during winter, and a conveyor through snow and ice melt and runoff in spring (e.g., see Larsson et al., 1998; Macdonald R. et al., 2000). This section describes the stages from contaminant release to its final emergence in top freshwater predators (Fig. 8.22), noting in particular those components of the pathway likely to be altered as a result of climate change.

Upon release, contaminants are transported through the atmosphere either as gases or adsorbed onto particles.

During atmospheric transport, washout and air–surface exchange remove some of the contaminant to the surface where it may become permanently sequestered or re-volatilized as a result of, for example, seasonal heating cycles, eventually arriving in the Arctic via a number of "hops" (Barrie et al., 1998). Accordingly, POPs undergo hemispheric-scale chromatography, with surfaces (soil, water, vegetation) providing the stationary phase and the atmosphere providing the moving phase. Global temperature increases will generally accelerate this cycling. Processes that are effective at capturing contaminants in arctic drainage basins (e.g., strong partitioning onto particles, into precipitation, into vegetation) are also effective at removing contaminants during transport. For example, Li et al. (2002) suggested that air–water partitioning alone restricts the entry of beta-hexachlorocyclohexane (β-HCH) into the Arctic by removing it to surfaces by rain and air–sea exchange. In the case of β-HCH, and in the case of contaminants that partition strongly onto particles (e.g., many PAHs, dichloro-diphenyltrichloroethane (DDT), and highly chlorinated PCBs), changes in rainfall patterns (amount and location) are very likely to alter the efficiency of transport to arctic locations and capture within the Arctic. Heavy metals provide an instructive example that will likewise apply to many of the POPs. Presently, less than 20% of particulate metal entering the Arctic is captured there (Akeredolu et al., 1994). Since the five ACIA-designated models project that the Arctic will become a "wetter" place (section 4.4.3, see also Manabe et al., 1992; Serreze et al., 2000), the capture of particulates and contaminants that partition strongly into water is likely to significantly increase by a factor that could more than offset efforts to reduce global emissions.

Because much of the contaminant delivery to the Arctic occurs during late winter as "arctic haze" or as "brown snow" events (Hileman, 1983; Welch H. et al., 1991), it is clear that sequestering by snow is an important

Box 8.12. Temperature-induced metal accumulation and stress responses in fish from Canadian Arctic lakes

High-altitude and high-latitude lakes are very sensitive ecosystems, where even slight environmental changes will possibly substantially affect ecosystem function (Köck and Hofer, 1998; Köck et al., 2001). Environmental changes can alter fish habitat and toxicant accumulation from water and diet. Long-range transport of pollutants also tends to endanger fish populations by leading to highly elevated metal accumulation (Köck et al., 1996). Water temperature has been shown to be the driving force of excessive metal accumulation in these fish. A multi-year project, centered around small sensitive lake ecosystems in the Canadian Arctic Archipelago (Cornwallis Island, Somerset Island, and Devon Island), was designed to explain the interactions between short- or longer-term climatic variation, the bioaccumulation of metals, and various biochemical stress indicators in land-locked populations of Arctic char.

Arctic char were collected at monthly intervals from Resolute Lake (Cornwallis Island, Nunavut, 74° 41' N, 94° 57' W) during summers from 1997 to 2001. Fish were dissected and liver tissue subsampled for analysis of metal content (cadmium – Cd and zinc – Zn) and biochemical stress indicators (glutathione – GSH, glutathione disulfide – GSSG, glycogen, and Vitamin C). Glutathione is an antioxidant, which is reduced to GSSG in the presence of reactive oxygen species. The GSSG/GSH ratio is a sensitive indicator of oxidative stress in cells (Lackner, 1998). Stress response was indicated by a decrease in GSH accompanied by an increase in GSSG.

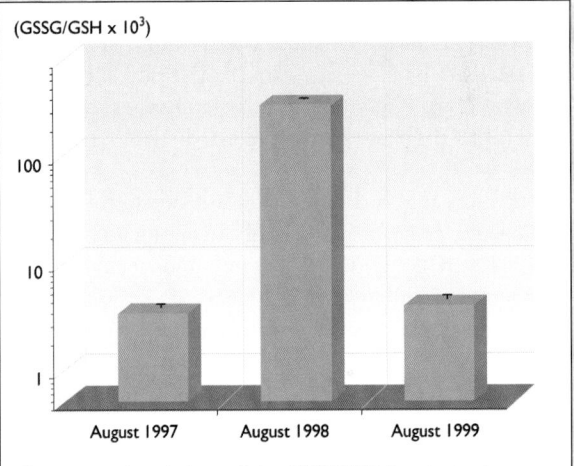

Interannual variation of the GSSG/GSH ratio in the liver of Arctic char (n=20 individuals per group) from Resolute Lake, showing means ± standard deviation (Köck et al., 2002).

Similar to Arctic char from Austrian high-mountain lakes, concentrations of Cd and Zn in the liver of high-latitude char exhibited a marked seasonal change during summer, and were significantly higher at the end of the ice-free period. A similar pattern was found for concentrations of metallothione (an inducible metal-binding protein) in the liver. Concentrations of Cd and Zn in the liver of char collected in August 1998 were significantly higher than those in fish collected during the same period in 1997, 1999, 2000, and 2001, which coincides with much higher lake water temperatures in the El Niño year of 1998 (Köck et al., 2002). Cadmium concentrations in the liver were positively correlated with frequencies of high temperatures (4–12 °C) and negatively with those of lower temperatures (<0–4 °C).

Interannual variation of the GSSG/GSH ratio, glycogen, and Vitamin C levels in the char studied indicate a higher level of stress in 1998 than in 1997 and 1999: the GSSG/GSH ratios in the livers of fish collected in 1998 were significantly higher than in 1997 and 1999 (see figure). Furthermore, concentrations of glycogen and Vitamin C were significantly lower in 1998. The severe depletion of glycogen energy reserves indicates that atypically high lake temperatures enhanced metal bioaccumulation and detoxification responses, diverting energy resources from other important physiological functions.

These results illustrate that Arctic char are extremely susceptible to even slight changes in lake water temperatures. Rising water temperatures lead to increased metabolic rates and thus pumping of higher volumes of water across the gills, which in turn results in increased uptake of dissolved metals from the water. The rapid increase in temperatures projected by various GCMs will possibly be a serious threat to the stability of Arctic char populations in high-latitude lakes.

process. Hence, careful consideration must be given to any changes in arctic snow conditions. Newly formed snow has a large surface area (as much as 0.4 m²/g, Hoff et al., 1998) that scavenges both particulate and gaseous POPs, eventually sequestering them into the snowpack (Gregor, 1990). Precipitation form (snow, rain, fog) is therefore important and, considering the seasonal modulation in atmospheric concentrations of contaminants (Heintzenberg, 1989; Hung et al., 2001; Macdonald R. et al., 2000), so is timing. For example, snowfall during a period of arctic haze would be much more important for transferring contaminants to the ground than at other times of the year.

As snow ages or melts, its surface area to volume ratio decreases, resulting in the removal of the solvent that captured the POPs (e.g., Wania, 1997). Macdonald R. et al. (2002) estimated that this process could lead to fugacity amplification of ~2000 times that of the air – clearly an enormous thermodynamic forcing. Depending on the exact circumstances under which snow loses its surface area, the POPs will be vaporized back to the air or partitioned into particles, soil, vegetation, or meltwater. Changes in the frequency and timing of snowfall, unusual events like freezing rain, or the rate and timing of snowmelt are likely to effect large changes in the proportion of POPs that enter the arctic hydrological cycle.

Terrestrial organic carbon in soils and vegetation has a large capacity to store many POPs (Simonich and Hites, 1994), with PCBs, DDT, HCH, and chlorobenzenes figuring prominently (AMAP, 2004b). Wania and McLachlan (2001) have shown that forests "pump" organochlorines from the atmosphere into foliage and subsequently to long-term soil reservoirs. Accordingly, increased proportions of leaf-bearing plants in arctic catchments will enhance this "pump". Increased metabolism of soil organic carbon owing to temperature increases, changes in soil moisture, or reduced snow cover will force POPs associated with soil organic carbon to redistribute, probably into groundwater or meltwater.

Climate variation results in the storage of contaminants in perennial snow and ice or in soils, vegetation, and delta/estuarine sediments during periods of cold climate (years to decades). These stored contaminants may subsequently be released during a period of warming and, although this process may not be sustainable, it is likely to produce episodes of high contaminant loadings into water (Blais et al., 1998, 2001). During permafrost degradation, a shift toward dendritic drainage patterns (e.g., McNamara et al., 1999) allows a more complete transport of contaminants into ponds and lakes and possibly re-mobilizes contaminants stored in soils. Simultaneously, the reduction of pond areas owing to drainage channels in permafrost (Hinzman et al., in press) is likely to enhance contaminant transport into the remaining surface water.

After POPs enter the hydrological cycle through the mechanisms discussed, a proportion of them will be stored in lakes and lake sediments. Evidence from a limited number of studies (Diamond et al., 2003; Helm et al., 2002; Macdonald R. et al., 2000) suggests that meltwater currently enters arctic lakes when they are thermally stratified beneath an ice cover. Therefore, much of the annual snowmelt traverses under the ice to exit at the outflow carrying its contaminant burden; that is, arctic lakes are not efficient at capturing POPs entering via streamflow. It is probable that many of the previously described alterations in freshwater systems induced by climate change (such as reduced ice cover, increased mixing and primary production, and increased loading of organic carbon and sediment from the contributing catchments) will also enhance contaminant capture in lakes (e.g., see Hinzman et al., in press).

Because most arctic lakes tend to be oligotrophic, only a small proportion of POPs is transported by vertical flux of organic particles and buried in sediments (e.g., Diamond et al., 2003; Muir et al., 1996; Stern and Evans, 2003). A second solvent-depletion process occurs due to organic carbon metabolism during particle settling and within bottom sediments (Fig. 8.23a). The loss of organic carbon can provide exceptionally strong thermodynamic forcing to drive the POPs off solid phases and into sediment pore water, where they may diffuse into bottom waters or partition into benthos (Gobas and MacLean, 2003; Jeremiason et al., 1994; Macdonald R. et al., 2002). An increase in the vigor of the organic carbon cycle (e.g., increased primary production, organic carbon loadings, and microbial activity) will enhance this thermodynamic pump. Cryogenic concentration (Fig. 8.23d) is likely to work together with organic carbon metabolism in sediments during winter to produce exceptionally high concentrations of POPs in bottom water. Although relatively poorly studied, contaminants are believed to be excluded from ice as it forms. For shallow water that freezes nearly to the bottom, dissolved contaminants are likely to be forced into a very small volume of remaining water and the resultant high contaminant concentrations will promote partitioning into remaining organic material including sediment surfaces, benthos, plankton, and larger animals. As noted previously, such under-ice zones are often a critical winter refuge for biota (e.g., Hammar, 1989). It is likely that a general reduction in the depth of ice formed during warmer winters will reduce cryogenic concentration. However, cryogenic concentration interacts with water levels, which are likely to decrease during permafrost degradation (Hinzman et al., in press).

The transfer of lipid-soluble POPs upward in aquatic food webs is one of the most important routes of exposure to apex feeders, including humans. In this solvent-depleting process, much of the lipid is metabolized at each trophic level while the organochlorines are retained. This results in higher trophic levels exhibiting organic carbon biomagnification factors of 3 to 100 in their lipids, and a net bioaccumulation of 10^7 to 10^9 times higher than in the water (Braune et al., 1999; Kidd et al., 1995a,c; Macdonald R. et al., 2003b). Changes in aquatic trophic structure either through alteration of the number of food web steps or the size distribution of predatory fish will likewise change contaminant burdens. With climate change, wide-ranging shifts in zoogeographic distributions have the potential to affect every step in freshwater food chains (Hinzman et al., in press; Schindler D.W., 1997).

There are several other ways that global change can alter contaminant pathways in arctic aquatic ecosystems. As noted in section 8.5.4, recent evidence suggests that salmon migrations undergo large, climate-related variation (Finney et al., 2000, 2002) and that Pacific salmon may respond to change by invading arctic rivers (Babaluk et al., 2000). Given that these salmon biomagnify and bioaccumulate contaminants in the Pacific Ocean, they are an important means of contaminant transport into particular arctic catchments. In specific lakes, fish may supply more POPs than atmospheric deposition (Ewald et al., 1998). Similarly, bird migrations that change in location and intensity have the potential to concentrate, transport, and deposit contaminants in particular catchments (AMAP, 2004b, Braune et al., 1999; Hinzman et al., in press). For example, detailed studies of Lake Ellasjoen, Norway,

found that seabirds can serve as important biological pathways carrying contaminants (in this case POPs) from marine to freshwater environments (AMAP, 2002). Climate change or human intervention is also very likely to lead to the introduction of exotic species to the Arctic. Although probably not a risk to arctic lakes, the invasion of the Great Lakes by the zebra mussel (*Dreissena polymorpha*) provides an instructive example of just how disruptive an exotic species can be to organic carbon and POP cycles (Morrison et al., 1998, 2000; Whittle et al., 2000).

Changes within arctic catchments that cause apex feeders (e.g., humans, bears, mink, birds) to switch their diet from aquatic to terrestrial food sources or vice versa have a large potential to alter contaminant exposure. Whereas arctic aquatic food webs exhibit endemic contamination from biomagnifying chemicals, arctic land-based food webs are among the cleanest in the world (AMAP, 2004b; de March et al., 1998). Dietary changes are forced by fluctuations in the populations of prey species or by changes in access to the species due to early ice melt or permafrost degradation (e.g., Fast and Berkes, 1998; Riedlinger, 2001).

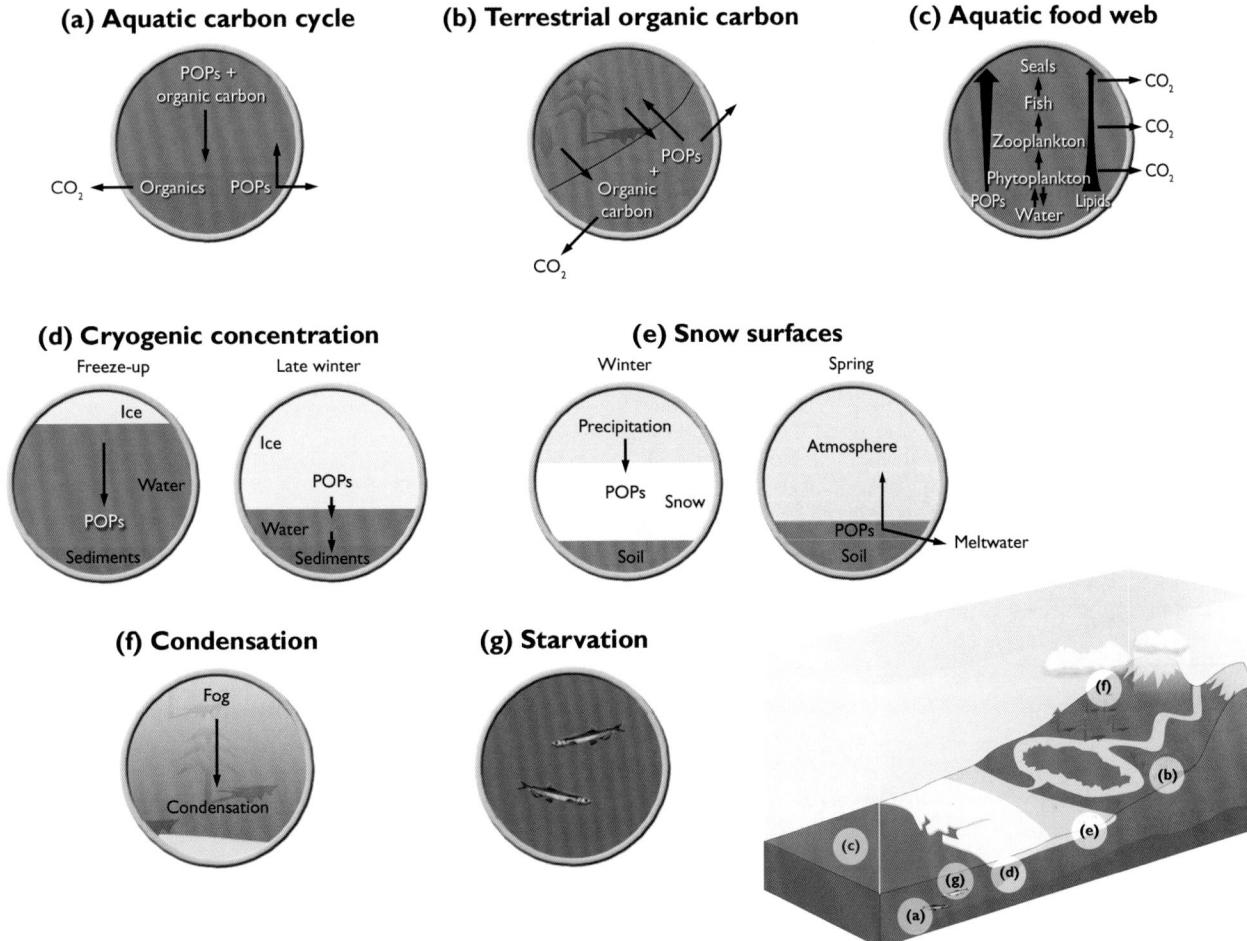

Fig. 8.23. In the illustrated solvent-depleting processes, POPs are concentrated beyond thermodynamic equilibrium through the removal of solvent by organic carbon metabolism in (a) aquatic and (b) terrestrial carbon cycles, by (c) inefficient lipid transfer in aquatic food webs, by (d) exclusion into a dwindling layer of water during the growth of ice, by (e) the loss of snow surfaces during aging or melting, by (f) the loss of surface area through condensation of fog into water droplets or onto surfaces, and (g) through loss of lipid pools during periods of starvation.

As conditions more suitable for domestic crops develop, agriculture or silviculture within arctic drainage basins and associated chemical use is likely to expand. Demographic shifts and population increases in northern regions could possibly lead to increased local release of contaminants. South of the Arctic, global temperature increases and alteration of hydrological cycles will probably result in insect and other pest outbreaks (e.g., West Nile virus or malaria), provoking the re-introduction of banned pesticides (Harner, 1997). Finally, contaminants in dumps or sumps presently contained by permafrost are very likely to be released by permafrost degradation (Williams and Rees, 2001).

Increased fluxes of PAHs are likely to result from the erosion of peat-rich soils (Yunker et al., 1993) or drying trends leading to an increase in wildfires, and are likely to have a greater impact on small rivers that presently receive most of their PAHs from combustion sources (Yunker et al., 2002).

8.7.3. Mercury in arctic catchments

Mercury exhibits a natural global cycle that has been enhanced by human activities such as coal burning, soft and ferrous metal smelting, cement production, and municipal waste with the consequence that two to three times as much Hg is now cycling through the atmosphere and surface waters than was before the rise of industry (Lamborg et al., 2002). Pacyna and Pacyna (2001) estimated that worldwide anthropogenic Hg emissions totaled 2235 x 10³ kg in 1995, with fossil fuel consumption contributing over half of that. This value may be compared to the 2500 x 10³ kg/yr estimate of natural emissions (Nriagu, 1989). The largest emitter of Hg from fossil fuel consumption is China (495 x 10³ kg

in 1995), which is directly upwind from the Bering Sea, Alaska, and the western Arctic. In comparison, Russia released about 54 x 10³ kg in 1995.

Many of the concentrating processes discussed for POPs (Fig. 8.23) apply equally to Hg. However, atmospheric Hg depletion events (MDEs) after polar sunrise provide a unique, climate-sensitive pathway to Hg deposition in arctic catchments (Fig. 8.24, Lindberg et al., 2002; Lu et al., 2001; Macdonald R. et al., 2000; Schroeder et al., 1998; Steffen et al., 2003). The process requires snow surfaces, solar radiation, and the presence of sea salts (bromides and chlorides). Although MDEs are initiated over the ocean, and especially over the marginal seas where halides are more available from frost flowers (crystals of ice that form directly from the vapor phase, often associated with new sea ice, which can become very salty by channeling brine upward from the ice) or first-year ice (Shepson et al., 2003), atmospheric advection can subsequently deposit reactive Hg in arctic catchments (Steffen et al., 2003). As with POPs, Hg can be transferred and concentrated during snow aging and melting, such that a large pulse of Hg is released to terrestrial and freshwater environments during spring melt (e.g., see AMAP, 2004a; Diamond et al., 2003; Stanley J. et al., 2002). Scott K. (2001) showed that Hg deposited through the MDE mechanism is in a form that can readily be taken up by biota. Once Hg enters the hydrological cycle, it can be concentrated and transferred through the carbon cycle and food webs, both of which are vulnerable to change. In addition, the efficiency of arctic lakes in capturing Hg is very likely to be altered by changes in the timing of freshet, ice melt, and productivity (Braune et al., 1999; Diamond et al., 2003; Kidd et al., 1995b; Macdonald R. et al., 2000).

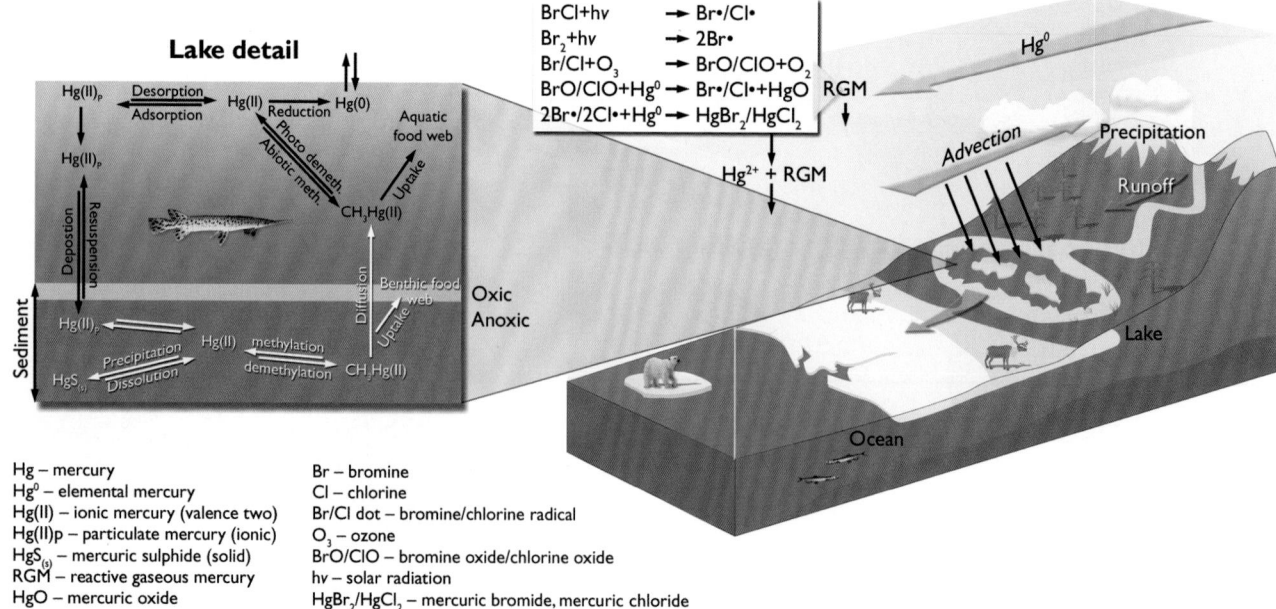

Fig. 8.24. Production of particulate and reactive gaseous mercury over the ocean after polar sunrise (right side) and the advection of reactive and bioavailable forms of mercury into catchments where it is deposited. After deposition, the mercury enters lakes through meltwater and is then subject to reduction and methylation (meth.) processes. Methyl mercury (CH₃Hg(II)) is the most toxic form.

Table 8.5. Environmental factors affecting mercury concentration in aquatic top predators.

Projected change	Effect on mercury concentration in predatory fish	Reference
Flooding of soil	Increase	Bodaly and Johnston, 1992
Increased primary production	Reduce	Pickhardt et al., 2002
Increased number of trophic levels	Increase	Kidd et al., 1995b
Shift toward larger fish	Increase	Sherwood et al., 2001
Reduced lake size	Increase	Bodaly et al., 1993
Increased anadromous fish migration	Increase	Zhang et al., 2001

Once Hg has been deposited into arctic catchments, a number of processes may lead to elevated concentrations in old, predatory fish (Table 8.5). The coupling between Hg deposition on surfaces and its entry into lakes is likely to be enhanced by projected changes in the hydrological and organic carbon cycles (Diamond et al., 2003; Stanley J. et al., 2002). Apex feeders are most vulnerable to any change in the Hg cycle considering that biomagnification factors are 250 to 3000 (Atwell et al., 1998; Folt et al, 2002; Kidd et al., 1995b; Muir et al., 1999). Because MeHg presents a far greater health hazard than inorganic or elemental Hg, methylation is a crucial process upon which climate change operates. Wetlands and wetland sediments are net producers of MeHg (Driscoll et al., 1998; Suchanek et al., 2000), and Hg observed in fish from small lakes appears to correlate with the amount of watershed occupied by wetlands (Greenfield et al., 2001). Flooding of terrestrial landscapes has the well-known consequence of releasing Hg from submerged soils (Bodaly et al., 1984). Therefore, alteration of wetland distribution or area in the Arctic resulting from thawing permafrost (section 8.4.4.4) is very likely to release Hg, which will be more serious if arctic soils contain an inventory of contaminant Hg accumulated during the past century or two.

8.8. Key findings, science gaps, and recommendations

In general, changes in climate and UV radiation levels in the Arctic are very likely to have far-reaching impacts, affecting aquatic species at various trophic levels, the physical and chemical environment that makes up their habitat, and the processes that act on and within freshwater ecosystems. Interactions of climatic variables such as temperature and precipitation with freshwater ecosystems are highly complex and can be propagated through the ecosystem in ways that are not readily predictable. This reduces the ability to accurately project specific effects of climate and UV radiation change on freshwater systems. This is particularly the case when dealing with threshold responses (i.e., those that produce stepwise and/or nonlinear effects). Forecasting ability is further hampered by the poor understanding of arctic freshwater systems and their basic interrelationships with climatic and other environmental variables, as well as by a paucity of long-term freshwater monitoring sites and integrated hydro-ecological research programs in the Arctic.

A significant amount of the understanding of potential impacts is based on historical analogues (i.e., historical evidence from past periods of climate change), as well as from a limited number of more recent studies of ecosystem response to environmental variability. Paleo-reconstructions indicate that during the most recent period of climatic warming, which followed the Little Ice Age, the Arctic reached its highest average annual temperatures observed in the past 400 years, resulting in glacier retreat, permafrost thaw, and major shifts in freshwater ecosystems. Examples of ecosystem effects included altered water chemistry, changes in species assemblages, altered productivity, and an extended growing-season length. Importantly, however, past natural change in the Arctic occurred at a rate much slower than that projected for anthropogenic climate change over the next 100 years. In the past, organisms had considerable time to adapt; their responses may therefore not provide good historical analogues for what will result under much more rapid climate change. In many cases, the adaptability (i.e., adaptation, acclimation, or migration) of organisms under rapidly changing climate conditions is largely unknown. Unfortunately, no large-scale attempts have been made to study the effects of rapid climate change on aquatic ecosystems using controlled experiments, as have been attempted for terrestrial systems (e.g., see International Tundra Experiment studies of tundra plant response in section 7.3.3.1). However, field studies in areas that have recently experienced rapid changes in climate have provided important knowledge. Information about ongoing climate change impacts is provided by results from long-term environmental monitoring and research sites in the Arctic, including the case studies presented in this chapter: Northern Québec and Labrador, Canada (Box 8.1), Toolik Lake, Alaska (Box 8.3), Lake Saanajärvi, Finnish Lapland (Box 8.4), and the Zackenberg Valley in northeastern Greenland (Box 8.5).

8.8.1. Key findings

This section lists a number of broad-scale findings for major components of arctic freshwater ecosystems. Although it was possible in this assessment to evaluate mesoscale regional differences in, for example, the timing and severity of the freshet and/or breakup on large rivers, difficulties in ecological downscaling of most climatic and related hydrological changes precluded regional discrimination of variations in impacts. Hence, most of the following statements are broad-scale and not regionally specific. To indicate the probability of each impact occurring, the ACIA lexicon regarding the range of likelihood of outcome (section 8.1.2) has been applied to the findings (i.e., the bold-text statement(s) in each bullet). Each assigned likelihood is the product of a multi-author scientific judgment based on knowledge synthesized from the scientific literature, including the previously noted case studies, and the interpretation of effects deduced from the ACIA-designated model projections. The same level of likelihood of occurrence is applied to the subsequent, more detailed descriptions following each major finding.

Ecological impacts of changes in runoff, water levels, and river-ice regimes

- **A probable shift to a more pluvial system with smaller and less intense freshet and ice breakup is very likely to decrease the frequency and magnitude of natural disturbances, and reduce the ability of flow systems to replenish riparian ecosystems, particularly river deltas.** As rainfall becomes a more prominent component of high-latitude river flow regimes, and as temperatures rise, freshets will be less intense and ice breakup less dramatic. Furthermore, differential changes in climatic gradients along the length of large arctic rivers will produce varying responses in freshet timing and breakup severity. Decreased frequency and intensity of physical disturbances will result in decreased species richness and biodiversity in riverine, deltaic, and riparian habitats.

- **Reduced climatic gradients along large northern rivers are likely to alter ice-flooding regimes and related ecological processes.** Projected differential rates of temperature increases across the major latitudinal ranges of some large northern rivers and the corresponding reduction in the latitudinal spring temperature gradient are likely to reduce the frequency and magnitude of dynamic river-ice breakups and lead to more placid thermal events. Because such disturbances play a major role in maintaining habitat complexity and associated species richness and diversity, this will have significant implications for ecosystem structure and function.

- **A very probable increase in winter flows and reduced ice-cover growth is very likely to increase the availability of under-ice habitat.** High-latitude rivers that typically freeze to the bottom during the winter will experience increased flow in response to increasing precipitation and winter temperatures, increasing base flow, and declining ice thickness. The subsequent presence of year-round flowing water in these river channels will increase habitat availability, ensuring survival of species previously restricted by the limitation of under-ice habitat. Migration and the geographic distribution of aquatic species (e.g., fish) may also be affected.

- **A probable decrease in summer water levels of lakes and rivers is very likely to affect quality and quantity of, and access to, aquatic habitats.** In areas where combinations of precipitation and evaporation lead to reductions in lake and/or river water levels, pathways for fish movement and migration will be impaired, including access to critical habitat. In addition, declining water levels will affect physical and chemical processes such as stratification, nutrient cycling, and oxygen dynamics.

Changes in biogeochemical inputs from altered terrestrial landscapes

- **Enhanced permafrost thawing is very likely to increase nutrient, sediment, and carbon loadings to aquatic systems.** This is very likely to have a mixture of positive and negative effects on freshwater chemistry. As permafrost and peat warm and active layers deepen with rising temperatures, nutrients, sediment, and organic carbon will be flushed from soil reserves and transported into aquatic systems. Increased nutrient and organic carbon loading will enhance productivity in high-latitude lakes, as well as decrease exposure of biota to UV radiation. Conversely, heavily nutrient-enriched waters (i.e., systems with enhanced sediment and organic matter loads), may result in increased light limitation and reduced productivity.

- **An enhanced and earlier supply of sediment is likely to be detrimental to benthic fauna.** As soils warm in high-latitude permafrost landscapes and become more susceptible to erosion, surface runoff will transport larger sediment loads to lakes and rivers. Aerobes in lake and river bottom sediments will initially be threatened by oxygen deprivation due to higher biological oxygen demand associated with sedimentation. Larger suspended sediment loads will also negatively affect light penetration and consequently primary production levels. Similarly, negative effects such as infilling of fish spawning beds associated with increased sediment loads are also likely in many areas.

- **Increases in DOC loading resulting from thawing permafrost and increased vegetation are very likely to have both positive and negative effects.** The balance will be ecosystem- or site-specific. For example, as DOC increases, there will be a positive effect associated with reduced penetration of damaging UV radiation, but also a negative effect because of the decline in photochemical processing of organic material. An additional negative effect would be a decrease in primary production due to lower light availability (quantity and quality).

Alterations in ponds and wetlands

- **Freshwater biogeochemistry is very likely to be altered by changes in water budgets.** As permafrost soils in pond and wetland catchments warm, nutrient and carbon loading to these freshwater systems will rise. Nutrient and carbon enrichment will enhance nutrient cycling and productivity, and alter the generation and consumption of carbon-based trace gases.

- **The status of ponds and wetlands as carbon sinks or sources is very likely to change.** High-latitude aquatic ecosystems function as sinks or sources of carbon, depending on temperature, nutrient status, and moisture levels. Initially, arctic wetlands (e.g., peatlands) will become sources of carbon as permafrost thaws, soils warm, and accumulated organic matter decomposes. Decomposition rates in aquatic ecosystems will also increase with rising temperatures and increases in rates of microbial activity. Increases in wetland water levels could enhance anaerobic decay and the production and release of methane.

- **Permafrost thaw in ice-rich environments is very likely to lead to catastrophic lake drainage; increased groundwater flux is likely to drain others.** As permafrost soils warm, freshwater bodies will become increasingly coupled to groundwater systems and experience drawdown. Lake drawdown will result in a change in the limnology and the availability and suitability of habitat for aquatic biota. Over the long term, terrestrial habitat will replace aquatic habitat.

- **New wetlands, ponds, and drainage networks are very likely to develop in thermokarst areas.** Thawing permafrost and melting ground ice in thermokarst areas will result in the formation of depressions where wetlands and ponds may form, interconnected by new drainage networks. These new freshwater systems and habitats will allow for the establishment of aquatic species of plants and animals in areas formerly dominated by terrestrial species.

- **Peatlands are likely to dry due to increased evapotranspiration.** As temperatures increase at high latitudes, rates of evapotranspiration in peatlands will rise. Drying of peat soils will promote the establishment of woody vegetation species, and increase rates of peat decomposition and carbon loss.

Effects of changing lake-ice cover

- **Reduced ice thickness and duration, and changes in timing and composition, are very likely to alter thermal and radiative regimes.** Rising temperatures will reduce the maximum ice thickness on lakes and increase the length of the ice-free season. Reduced lake-ice thickness will increase the availability of under-ice habitat, winter productivity, and associated dissolved oxygen concentrations. Extension of the ice-free season will increase water temperatures and lengthen the overall period of productivity.

- **A longer open-water season is very likely to affect lake stratification and circulation patterns.** Earlier breakup will lead to rapid strat-

ification and a reduction in spring circulation. In certain types of lakes this will cause a transfer of under-ice oxygen-depleted water to the deep water of stratified lakes in the summer (i.e., lakes will not get a chance to aerate in the spring). A longer open-water season will result in an increase in primary production over the summer that will lead to increased oxygen consumption in deeper waters as algae decompose. Correspondingly, fish habitat will be substantially reduced by the combination of upper-water warming and the low-oxygen conditions in deeper water. As a result, certain fish species (e.g., lake trout) may become severely stressed.

- **Reduced ice cover is likely to have a much greater effect on underwater UV radiation exposure than the projected levels of stratospheric ozone depletion.** A major increase in UV radiation levels will cause enhanced damage to organisms (biomolecular, cellular, and physiological damage, and alterations in species composition). Allocations of energy and resources by aquatic biota to UV radiation protection will increase, probably decreasing trophic-level productivity. Elemental fluxes will increase via photochemical pathways.

Aquatic biota, habitat, ecosystem properties, and biodiversity

- **Climate change is very likely to affect the biodiversity of freshwater ecosystems across most of the Arctic.** The magnitude, extent, and duration of the impacts and responses will be system- and location-dependent, and will produce varying outcomes, including local and/or regional extinctions or species loss; genetic adaptations to new environments; and alterations in species ranges and distributions, including invasion by southern species.

- **Microbial decomposition rates are likely to increase.** Rates of microbial decomposition will rise in response to increasing temperatures and soil drying and aeration. Enhanced decomposition of organic materials will increase the availability of dissolved organic carbon and emissions of carbon dioxide, with implications for the carbon balance of high-latitude lakes and rivers and, in particular, wetlands, which are significant carbon reserves.

- **Increased production is very likely to result from a greater supply of organic matter and nutrients.** Organic matter and nutrient loading of rivers, lakes, and wetlands will increase as temperatures and precipitation rise. Thawing permafrost and warming of frozen soils with rising temperature will result in the release of organic matter and nutrients from catchments. Rising temperatures will increase the rates and occurrence of weathering and nutrient release. Organic matter contributions may also increase with the establishment of woody species. Primary productivity will rise, the effects of which may translate through the food chains of aquatic systems, increasing freshwater biomass and abundance. In some instances, high loading of organic matter and sediment is very likely to limit light levels and result in a decline in productivity in some lakes and ponds.

- **Shifts in ranges and community composition of invertebrate species are likely to occur.** Temperature-limited species from more southerly latitudes will extend their geographic ranges northward. This will result in new invertebrate species assemblages in arctic freshwater ecosystems.

Fish and fisheries

- **Shifts in species ranges, composition, and trophic relations are very likely to occur.** Southern species will shift northward with warming of river waters, and are likely to compete with northern species for resources. The ranges of anadromous species may shift as oceanic patterns shift. The geographic ranges of northern or arctic species will contract in response to habitat impacts as well as competition. Changes in species composition at northern latitudes are likely to have a top-down effect on the composition and abundance of species at lower trophic levels. The broad whitefish, Arctic char complex, and the Arctic cisco are particularly vulnerable to displacement as they are wholly or mostly northern in their distribution. Other species of fish, such as the Arctic grayling of northern Alaska, thrive under cool and wet summer conditions, and may have less reproductive success in warmer waters, potentially causing elimination of populations.
- **Spawning grounds for cold-water species are likely to diminish.** As water temperatures rise, the geographic distribution of spawning grounds for northern species will shift northward, and is likely to contract. Details will be ecosystem-, species-, and site-specific.
- **An increased incidence of mortality and decreased growth and productivity from disease and/or parasites is likely to occur.** As southern species of fish migrate northward with warming river waters, they could introduce new parasites and/or diseases to which arctic fish species are not adapted, leading to a higher risk of earlier mortality and decreased growth.
- **Subsistence, sport, and commercial fisheries will possibly be negatively affected.** Changes in the range and distribution of fish species in northern lakes and rivers in response to changing habitat and the colonization of southerly species have implications for the operation of commercial fisheries and will possibly have potentially devastating effects on subsistence fishing. Changes in northern species (e.g., range, abundance, health) will diminish opportunities for fisheries on such species, calling for regulatory and managerial

changes that promote sustainable populations. Subsistence fisheries may be at risk in far northern areas where vulnerable species, such as the broad whitefish, the Arctic char complex, and Arctic cisco, are often the only fish species present. Fisheries will have to change to secure access, and to ensure that fishery function and duration of operation are effective, given a change in fish species and habitat. Alternatively, new opportunities to develop fisheries may occur.

Aquatic mammals and waterfowl

- **Probable changes in habitat are likely to result in altered migration routes and timing.** Migration routes of aquatic mammals and waterfowl are likely to extend northward in geographic extent as more southerly ecosystems and habitats develop at higher latitudes with increasing temperature. Migration may occur earlier in the spring with the onset of high temperatures, and later in the autumn if high temperatures persist. Breeding-ground suitability and access to food resources will be the primary drivers of changes in migration patterns. For example, wetlands are important feeding and breeding grounds for waterfowl, such as geese and ducks, in the spring. As permafrost landscapes degrade at high latitudes, the abundance of thermokarst wetlands may increase, promoting the northward migration of southerly wetland species, or increasing the abundance and diversity of current high-latitude species.
- **An increased incidence of mortality and decreased growth and productivity from disease and/or parasites will possibly occur.** As temperatures rise, more southerly species of mammals and waterfowl will shift northward. These species may carry with them new diseases and/or parasites to which northerly species are not adapted, which is likely to result in both an increased susceptibility to disease and parasites, and an increase in mortality.
- **Probable changes in habitat suitability and timing of availability are very likely to alter reproductive success.** Aquatic mammals and waterfowl are highly dependent on the availability and quality of aquatic habitats for successful breeding, and in the case of waterfowl, nesting. Northern species may have diminished reproductive success as suitable habitat either shifts northward or declines in availability and access. Northward migration of southern species may result in competitive exclusion of northern species from habitat and resources.

Climate–contaminant interactions

- **Increases in temperature and precipitation are very likely to increase contaminant capture in the Arctic.** Projected increases in temperature and changes in the timing and magnitude of precipitation will affect the deposition of contaminants at high latitudes. Climate change will accelerate rates of contaminant transfer. Climate scenarios currently project a "wetter" Arctic, increasing the probability of wet deposition of contaminants such as heavy metals and persistent organic pollutants.

- **Episodic releases of high contaminant loadings from perennial snow and ice are very likely to increase.** As temperatures rise at high latitudes, snow and ice accumulated over periods of years to decades will melt, releasing associated stored contaminants in the meltwater. This will increase episodes of high contaminant loadings into water, which may have toxic effects on aquatic organisms. Permafrost degradation may also mobilize contaminants. Lower water levels will amplify the impacts of contamination on high-latitude freshwater bodies.

- **Arctic lakes are very likely to become more prominent contaminant sinks.** Spring melt waters and associated contaminants typically pass through thermally stratified arctic lakes without transferring their contaminant burden. Contaminant capture in lakes will increase with reduced lake-ice cover (decreased stratification), increased mixing and primary production, and greater organic carbon and sediment loading. Contaminants in bottom sediments may dissociate from the solid phase with a rise in the rate of organic carbon metabolism and, along with contaminants originating from cryogenic concentration, may reach increasing levels of toxicity in lake bottom waters.

- **The nature and magnitude of contaminant transfer in the food web are likely to change.** Changes in aquatic trophic structure and zoogeographic distributions will alter biomagnification of contaminants, including persistent organic pollutants and mercury, and potentially affect freshwater food webs, especially top-level predatory fish (e.g., lake trout) that are sought by all types of fisheries.

Cumulative, synergistic, and overarching interactions

- **Decoupling of environmental cues used by biota is likely to occur, but the significance of this to biological populations is uncertain.** Photoperiod, an ultimate biological cue, will not change, whereas water temperature, a proximate biological cue, will change. For arctic species, decoupling of environmental cues will probably have significant impacts on population processes (e.g., the reproductive success of fish, hatching and feeding success of birds, and the migratory timing and success of birds and anadromous fish may be compromised).

- **The rate and magnitude at which climate change takes place and affects aquatic systems are likely to outstrip the capacity of many aquatic biota to adapt or acclimate.**

Evolutionary change in long-lived organisms such as fish cannot occur at the same rate as the projected change in climate. The ability to acclimate or emigrate to more suitable habitats will be limited, thus effects on some native arctic biota will be significant and detrimental. Shorter-lived organisms (e.g., freshwater invertebrates) may have a greater genetic and/or phenotypic capacity to adapt, acclimate, or emigrate.

- **Climate change is likely to act as a multiple stressor, leading to synergistic impacts on aquatic systems.** For example, projected increases in temperature will enhance contaminant influxes to aquatic systems, and independently increase the susceptibility of aquatic organisms to contaminant exposure and effects. The consequences for the biota will in most cases be additive (cumulative) and multiplicative (synergistic). The overall result will be higher contaminant loads and biomagnification in ecosystems.

- **Climate change is very likely to act cumulatively and/or synergistically with other stressors to affect physical, chemical, and biological aspects of aquatic ecosystems.** For example, resource exploitation (e.g., fish or bird harvesting) and climate change impacts will both negatively affect population size and structure.

8.8.2. Key science gaps arising from the assessment

In conducting this assessment, a number of key gaps in scientific understanding became evident. These are noted throughout the chapter, and include:

- the limited records of long-term changes in physical, chemical, and biological attributes throughout the Arctic;
- differences in the circumpolar availability of biophysical and ecological data (e.g., extremely limited information about habitat requirements of arctic species);
- a lack of circumpolar integration of existing data from various countries and disparate programs;
- a general lack of integrated, comprehensive monitoring and research programs, at regional, national, and especially circumpolar scales;
- a lack of standardized and networked international approaches for monitoring and research;
- the paucity of representative sites for comparative analyses, either by freshwater ecosystem type (e.g., small rivers, wetlands, lakes) or by regional geography (ecozone, latitude, elevation);
- the unknown synergistic impacts of contaminants and climate change on aquatic organisms;
- a limited understanding of the cumulative impacts of multiple environmental stressors on freshwater ecosystems (e.g., land use, fisheries, forestry, flow regulation and impoundment, urbanization, mining, agriculture, and poleward transport of contaminants by invasive/replacement species);

- the unknown effects of extra-arctic large-river transport on freshwater systems induced by southern climate change;
- a limited knowledge of the effects of UV radiation–temperature interactions on aquatic biota;
- a deficient knowledge of the linkages between structure (i.e., biodiversity) and function of arctic aquatic biota;
- a poor knowledge of coupling among physical/chemical and biotic processes; and
- a lack of coupled cold-regions hydrological and ecological theories and related projective models.

Filling these gaps, the most outstanding of which include inter-regional differences in the availability of, and access to, circumpolar research (hence the North American and European bias in this assessment), would greatly improve understanding of the effects of climate and UV radiation change on arctic freshwater ecosystems. Furthermore, comprehensive monitoring programs to quantify the nature, regionality, and progress of climate change and related impacts require development and rapid implementation at representative sites across a broad range of the type and size of aquatic ecosystems found within the various regions of the Arctic. Coupling such programs with ongoing and new research will greatly facilitate meeting the challenges sure to result from climate change and increased UV radiation levels in the Arctic.

8.8.3. Science and policy implications and recommendations

A number of the above gaps in scientific understanding could be addressed by the following policy and/or program-related adjustments:

- Establish funding and mechanisms for the creation of a coordinated network of key long-term, representative freshwater sites for comparative monitoring and assessment studies among arctic regions (e.g., creation of a Circumpolar Arctic Aquatic Research and Monitoring Program).
- Based on the results of this assessment, establish a science advisory board (preferably at the international level) for targeted funding of arctic freshwater research.
- Secure long-term funding sources, preferably for an international cooperative program, for integrated arctic freshwater research.
- Adjust current northern fisheries management policies and coordinate with First Nations resource use and consumption.
- Establish post-secondary education programs focused on freshwater arctic climate change issues at both intra- and extra-arctic educational institutions, preferably involving a circumpolar educational consortium.

Acknowledgements

The extensive assistance rendered by Theresa Carmichael and Lucie Lévesque with respect to research activities, editorial actions, and related preparatory activities is gratefully acknowledged. The following individuals provided information on various topics: William Chapman, Brenda Clark, Paul Grabhorn, Chantelle Sawatsky, and John E. Walsh. Special thanks are also given to Dr. Robert Corell for his overall guidance and direction.

References

Adams, S.M., 1990. Status and use of biological indicators for evaluating the effects of stress on fish. In: S.M. Adams (ed.). Biological Indicators of Stress in Fish. American Fisheries Society Symposium, 8:1–8.

Adams, W.P., P.T. Doran, M. Ecclestone, C.M. Kingsbury and C.J. Allan, 1989. A rare second year lake ice cover in the Canadian High Arctic. Arctic, 42:299–306.

Afton, A.D. and M.G. Anderson, 2001. Declining scaup populations: a retrospective analysis of long-term population and harvest survey data. Journal of Wildlife Management, 65:781–796.

Agbeti, M.D., J.C. Kingston, J.P. Smol and C. Watters, 1997. Comparison of phytoplankton succession in two lakes of different mixing regimes. Archiv für Hydrobiologie, 140:37–69.

Akeredolu, F.A., L.A. Barrie, M.P. Olson, K.K. Oikawa, J.M. Pacyna and G.J. Keeler, 1994. The flux of anthropogenic trace metals into the Arctic from the mid-latitudes in 1979/80. Atmospheric Environment, 28:1557–1572.

Alexander, V., D.W. Stanley, R.J. Daley and C.P. McRoy, 1980. Primary producers. In: J.E. Hobbie (ed.). Limnology of Tundra Ponds: Barrow, Alaska, pp. 179–250. Dowden, Hutchinson and Ross.

Alexandersson, H. and B. Eriksson, 1989. Climate fluctuations in Sweden 1860–1987. Reports of Meteorology and Climatology, 58:1–54. Swedish Meteorological and Hydrological Institute, Stockholm.

Allan, J.D., 1995. Stream Ecology: Structure and Function of Running Waters. Chapman & Hall, 400pp.

Allen, D.M., F.A. Michel and A.S. Judge, 1988. The permafrost regime in the Mackenzie Delta, Beaufort Sea region, N.W.T. and its significance to the reconstruction of the paleoclimatic history. Journal of Quaternary Science, 3:3–13.

AMAP, 1997. Arctic Pollution Issues: A State of the Environment Report. Arctic Monitoring and Assessment Programme, Oslo, 188pp.

AMAP, 1998. AMAP Assessment Report: Arctic Pollution Issues. Arctic Monitoring and Assessment Programme, Oslo, Norway, 859pp.

AMAP, 2002. Arctic Pollution 2002 (Persistent Organic Pollutants, Heavy Metals, Radioactivity, Human Health, Changing Pathways). Arctic Monitoring and Assessment Programme, Oslo, 112pp.

AMAP, 2004a. AMAP Assessment 2002: Heavy Metals in the Arctic. Arctic Monitoring and Assessment Programme, Oslo.

AMAP, 2004b. AMAP Assessment 2002: Persistent Organic Pollutants in the Arctic. Arctic Monitoring and Assessment Programme, Oslo. xvi+310pp.

Amyot, M., D.R.S. Lean and G. Mierle, 1997. Photochemical formation of volatile mercury in high Arctic lakes. Environmental Toxicology and Chemistry, 16:2054–2063.

Andreev, A.A. and V.A. Klimanov, 2000. Quantitative Holocene climatic reconstruction from Arctic Russia. Journal of Paleolimnology, 24:81–91.

Antle, J.M., S.M. Capalbo, S. Mooney, E.T. Elliott and K.H. Paustian, 2001. Economic analysis of agricultural soil carbon sequestration: an integrated assessment approach. Journal of Agricultural and Resource Economics, 26:344–367.

Antonov, V.S., 1969. Ice regime of the mouth region of the Lena river in natural and regulated states, translated from Geograficheskogo Obshchestvo SSSR. Izvestiia No. 3:210–209.

Antonov, V.S., 1970. Siberian rivers and Arctic seas. Problemy Arktiki i Antarktiki, 36/37:142–152.

Arnell, N., B. Bates, H. Lang, J.J. Magnuson, P. Mulholland, S. Fisher, C. Liu, D. McKnight, O. Starosolszky and M. Taylor, 1996. Hydrology and freshwater ecology. In: R.T. Watson, M. Zinyowera and R.H. Moss (eds.). Climate Change 1995: Impacts, Adaptations and Mitigation of Climate Change: Scientific-Technical Analyses. Contribution of Working Group II to the Second Assessment Report of the Intergovernmental Panel on Climate Change, pp. 325–363. Cambridge University Press.

Arrigo, K.R. and C.W. Brown, 1996. Impact of chromophoric dissolved organic matter on UV inhibition of primary productivity in the sea. Marine Ecology Progress Series, 140:207–216.

Assel, R.A., 1991. Implications of CO_2 global warming on Great Lakes ice cover. Climatic Change, 18:377–395.

Atwell, L., K.A. Hobson and H.E. Welch, 1998. Biomagnification and bioaccumulation of mercury in an arctic marine food web: insights from stable nitrogen isotope analysis. Canadian Journal of Fisheries and Aquatic Sciences, 55:1114–1121.

Aurela, M., J.-P. Tuovinen and T. Laurila, 1998. Carbon dioxide exchange in a subarctic peatland ecosystem in northern Europe measured by the eddy covariance technique. Journal of Geophysical Research, 103(D10):11289–11301.

Aurela, M., T. Laurila and J.-P. Tuovinen, 2001. Seasonal CO_2 balances of a subarctic mire. Journal of Geophysical Research, 106(D2):1623–1637.

Austin, J.E., A.D. Afton, M.G. Anderson, R.G. Clark, C.M. Custer, J.S. Lawrence, J.B. Pollard and J.K. Ringleman, 2000. Declining scaup populations: issues, hypotheses, and research needs. Wildlife Society Bulletin, 28:254–263.

Babaluk, J.A., J.D. Reist, J.D. Johnson and L. Johnson, 2000. First records of sockeye (Onchorhynchus nerka) and pink salmon (O. gorbuscha) from Banks Island and other records of Pacific salmon in Northwest Territories, Canada. Arctic, 53:161–164.

Bahr, M., J.E. Hobbie and M.L. Sogin, 1996. Bacterial diversity in an Arctic lake – a freshwater SAR11 cluster. Aquatic Microbial Ecology, 11:271–277.

Baker, R., 1983. The effects of temperature, ration and size on the growth of Arctic charr (Salvelinus alpinus L.). M.Sc. Thesis, University of Manitoba, 227pp.

Banaszak, A.T. and P.J. Neale, 2001. Ultraviolet radiation sensitivity of photosynthesis in phytoplankton from an estuarine environment. Limnology and Oceanography, 46:592–603.

Barber, V.A. and B.P. Finney, 2000. Late Quaternary paleoclimatic reconstructions for interior Alaska based on paleolake-level data and hydrologic models. Journal of Paleolimnology, 24:29–41.

Barrie, L., E. Falck, D. Gregor, T. Iverson, H. Loeng, R. Macdonald, S. Pfirman, T. Skotvold and E. Wartena, 1998. The influence of physical and chemical processes on contaminant transport into and within the Arctic. In: D. Gregor, L. Barrie and H. Loeng (eds.). AMAP Assessment Report: Arctic Pollution Issues, pp. 25–116. Arctic Monitoring and Assessment Programme, Oslo.

Beisner, B.E., E. McCauley and F.J. Wrona, 1996. Temperature-mediated dynamics of planktonic food chains: the effect of an invertebrate carnivore. Freshwater Biology, 35:219–231.

Beisner, B.E., E. McCauley and F.J. Wrona, 1997. The influence of temperature and food chain length on plankton predator-prey dynamics. Canadian Journal of Fisheries and Aquatic Sciences, 54:586–595.

Beitinger, T.L. and L.C. Fitzpatrick, 1979. Physiological and ecological correlates of preferred temperature in fish. American Zoologist, 19:319–329.

Bellrose, F.C., 1980. Ducks, Geese and Swans of North America. Third Edition. Stackpole Books, 540pp.

Belzile, C., W.F. Vincent, J.A.E. Gibson and P. Van Hove, 2001. Bio-optical characteristics of the snow, ice and water column of a perennially ice-covered lake in the high Arctic. Canadian Journal of Fisheries and Aquatic Sciences, 58:2405–2418.

Belzile, C., J.A.E. Gibson and W.F. Vincent, 2002a. Colored dissolved organic matter and dissolved organic carbon exclusion from lake ice: implications for irradiance transmission and carbon cycling. Limnology and Oceanography, 47:1283–1293.

Belzile, C., W.F. Vincent and M. Kumagai, 2002b. Contribution of absorption and scattering to the attenuation of UV and photosynthetically available radiation in Lake Biwa. Limnology and Oceanography, 47:95–107.

Berg, O.K., 1995. Downstream migration of anadromous Arctic charr (Salvelinus alpinus (L.)) in the Vardnes River, northern Norway. Nordic Journal of Freshwater Research, 71:157–162.

Berg, O.K. and M. Berg, 1989. Sea growth and time of migration of anadromous Arctic char (Salvelinus alpinus) from the Vardnes river in northern Norway. Canadian Journal of Fisheries and Aquatic Sciences, 46:955–960.

Bergeron, M. and W.F. Vincent, 1997. Microbial food web responses to phosphorus supply and solar UV radiation in a subarctic lake. Aquatic Microbial Ecology, 12:239–249.

Berra, T.M., 2001. Freshwater Fish Distribution. Academic Press, 604pp.

Bertilsson, S., R. Stepanauskas, R. Cuadros-Hansson, W. Granéli, J. Wikner and L. Tranvik, 1999. Photochemically induced changes in bioavailable carbon and nitrogen pools in a boreal watershed. Aquatic Microbial Ecology, 19:47–56.

Bettez, N.D., P.A. Rublee, J. O'Brien and M.C. Miller, 2002. Changes in abundance, composition and controls within the plankton of a fertilised arctic lake. Freshwater Biology, 47:303–311.

Bigler, C. and R.I. Hall, 2002. Diatoms as indicators of climatic and limnological change in Swedish Lapland: a 100-lake calibration set and its validation for paleoecological reconstructions. Journal of Paleolimnology, 27:97–115.

Bigler, C. and R.I. Hall, 2003. Diatoms as quantitative indicators of July temperature: a validation attempt at century-scale with meteorological data from northern Sweden. Palaeogeography, Palaeoclimatology, Palaeoecology, 189:147–160.

Bigler, C., I. Larocque, S.M. Peglar, H.J.B. Birks and R.I. Hall, 2002. Quantitative multiproxy assessment of long-term patterns of Holocene environmental change from a small lake near Abisko, northern Sweden. The Holocene, 12:481–496.

Bigler, C., E. Grahn, I. Larocque, A. Jeziorski and R.I. Hall, 2003. Holocene environmental change at Lake Njulla (999 m a.s.l.), northern Sweden: a comparison with four small nearby lakes along an altitudinal gradient. Journal of Paleolimnology, 29:13–29.

Bilby, R.E., B.R. Fransen and P.A. Bisson, 1996. Incorporation of nitrogen and carbon from spawning coho salmon into the trophic system of small streams: evidence from stable isotopes. Canadian Journal of Fisheries and Aquatic Sciences, 53:164–173.

Birks, H.J.B., 1995. Quantitative paleoenvironmental reconstructions. In: D. Maddy and J.S. Brew (eds.). Statistical Modeling of Quaternary Science Data, Technical Guide 5, pp. 161–254. Quaternary Research Association, Cambridge.

Birks, H.J.B., 1998. D.G. Frey and E.S. Deevey Review 1: Numerical tools in paleolimnology – progress, potentialities and problems. Journal of Paleolimnology, 20:307–332.

Blais, J.M., D.W. Schindler, D.C.G. Muir, L.E. Kimpe, D.B. Donald and B. Rosenberg, 1998. Accumulation of persistent organochlorine compounds in mountains of western Canada. Nature, 395:585–588.

Blais, J.M., D.W. Schindler, D.C.G. Muir, M. Sharp, D.B. Donald, M. Lafreniere, E. Braekevelt and W.M.J. Strachan, 2001. Melting glaciers: a major source of persistent organochlorines to subalpine Bow Lake in Banff National Park, Canada. Ambio, 30:410–415.

Blaxter, J.H.S., 1992. The effect of temperature on larval fishes. Netherlands Journal of Zoology, 42:336–357.

Blom, T., A. Korhola and J. Weckström, 1998. Physical and chemical characterisation of small subarctic lakes in Finnish Lapland with special reference to climate change scenarios. In: R. Lemmelä and N. Helenius (eds.). Proceedings of the Second International Conference on Climate Change and Water, Espoo, Finland, 17–20 August 1998, pp. 576–587. Helsinki University.

Blumberg, A.F. and D.M. Di Toro, 1990. Effects of climate warming on dissolved oxygen concentrations in Lake Erie. Transactions of the American Fisheries Society, 119:210–223.

Bodaly, R.A. and T.A. Johnston, 1992. The mercury problem in hydroelectric reservoirs with predictions of mercury burdens in fish in the proposed Grand Baleine complex, Québec. In: James Bay Publication Series, Hydro-Electric Development: Environmental Impacts, Paper No. 3. North Wind Information Services, Montreal, 15pp.

Bodaly, R.A., R.E. Hecky and R.J.P. Fudge, 1984. Increases in fish mercury levels in lakes flooded by the Churchill River diversion, northern Manitoba. Canadian Journal of Fisheries and Aquatic Sciences, 41:682–691.

Bodaly, R.A., J.D. Reist, D.M. Rosenberg, P.J. McCart and R.E. Hecky, 1989. Fish and fisheries of the Mackenzie and Churchill River basins, northern Canada. In: D.P. Dodge (ed.). Proceedings of the International Large River Symposium. Canadian Special Publication of Fisheries and Aquatic Sciences, 106:128–144.

Bodaly, R.A., J.W.M. Rudd, R.J.P. Fudge and C.A. Kelly, 1993. Mercury concentrations in fish related to size of remote Canadian Shield lakes. Canadian Journal of Fisheries and Aquatic Sciences, 50:980–987.

Boelen, P., M.J.W. Veldhuis and A.G.J. Buma, 2001. Accumulation and removal of UVBR-induced DNA damage in marine tropical plankton subjected to mixed and simulated non-mixed conditions. Aquatic Microbial Ecology, 24:265–274.

Borgeraas, J. and D.O. Hessen, 2000. UV-B induced mortality and antioxidant enzyme activities in Daphnia magna at different oxygen concentrations and temperatures. Journal of Plankton Research, 22:1167–1183.

Bothwell, M.L., D.M.J. Sherbot and C.M. Pollock, 1994. Ecosystem response to solar ultraviolet-B radiation: influence of trophic-level interactions. Science, 265:97–100.

Bouillon, D.R. and J.B. Dempson, 1989. Metazoan parasite infections in landlocked and anadromous Arctic charr (Salvelinus alpinus Linneaus), and their use as indicators of movement to sea in young anadromous charr. Canadian Journal of Zoology, 67:2478–2485.

Bowden, W.B., J.C. Finlay and P.E. Maloney, 1994. Long-term effects of PO_4 fertilization on the distribution of bryophytes in an arctic river. Freshwater Biology, 32:445–454.

Bowden, W.B., D.B. Arscott, D. Pappathanasi, J.C. Finlay, J.M. Glime, J. LeCroix, C.-L. Liao, A.E. Hershey, T. Lampella, B.J. Peterson, W. Wollheim, K. Slavik, B. Shelley, M. Chesterton, J.A. Lachance, R. Le Blanc, A. Steinman and A. Suren, 1999. Roles of bryophytes in stream ecosystems. Journal of the North American Benthological Society, 18:151–184.

Bowlby, J.N. and J.C. Roff, 1986. Trout biomass and habitat relationships in southern Ontario streams. Transactions of the American Fisheries Society, 115:503–514.

Boyce, M.S. and R.S. Miller, 1985. Ten-year periodicity in the whooping crane census. Auk, 102:658–660.

Braune, B., D.C.G. Muir, B. de March, M. Gamberg, K. Poole, R. Currie, M. Dodd, W. Duschenko, J. Eamer, B. Elkin, M. Evans, S. Grundy, C. Hebert, R. Johnstone, K. Kidd, B. Koenig, L. Lockhart, H. Marshall, K. Reimer, J. Sanderson and L. Shutt, 1999. Spatial and temporal trends of contaminants in Canadian Arctic freshwater and terrestrial ecosystems: a review. Science of the Total Environment, 230:145–207.

Brett, J.R., J.E. Shelbourne and C.T. Shoop, 1969. Growth rate and body composition of fingerling sockeye salmon, *Oncorhynchus nerka*, in relation to temperature and ration size. Journal of the Fisheries Research Board of Canada, 26:2363–2394.

Brown, R.S., S.S. Stanislawski and W.C. Mackay, 1994. Effects of frazil ice on fish. In: T.D. Prowse (ed.). Proceedings of the Workshop on Environmental Aspects of River Ice. National Hydrology Research Institute, Saskatoon, Symposium No. 12, pp. 261–278.

Brylinsky, M. and K.H. Mann, 1973. An analysis of factors governing productivity in lakes and reservoirs. Limnology and Oceanography, 18:1–14.

Buck, R.J.G. and D.W. Hay, 1984. The relation between stock size and progeny of Atlantic salmon, *Salmo salar* L., in a Scottish stream. Journal of Fish Biology, 23:1–11.

Buma, A.G.J., H.J. Zemmelink, K. Sjollema and W.W.C. Gieskes, 1996. UVB radiation modifies protein and photosynthetic pigment content, volume and ultrastructure of marine diatoms. Marine Ecology Progress Series, 142:47–54.

Burdykina, A.P., 1970. Breakup characteristics in the mouth and lower reaches of the Yenisey River. Soviet Hydrology: Selected Papers, No. 1:42–56.

Busacker, G.P., I.R. Adelman and E.M. Goolish, 1990. Growth. In: C.B. Shreck and P.B. Moyle (eds.). Methods for Fish Biology, pp. 363–388. American Fisheries Society, Bethesda, Maryland.

CAFF, 2001. Arctic Flora and Fauna: Status and Conservation. Conservation of Arctic Flora and Fauna, Helsinki, 272pp.

Capellen, J. and B. Vraae Jørgensen, 2001. Danmarks Klima 2000 - med tillæg om Færøerne og Grønland. Technical Report No. 01–06. Danish Meteorological Institute, Copenhagen.

Carmack, E.C. and R.W. Macdonald, 2002. Oceanography of the Canadian Shelf of the Beaufort Sea: a setting for marine life. Arctic, 55(S1):29–45.

Carpenter, S.R., S.G. Fisher, N.B. Grimm and J.F. Kitchell, 1992. Global change and freshwater ecosystems. Annual Review of Ecology and Systematics, 23:119–139.

Castenholz, R.W. and F. Garcia-Pichel, 2000. Cyanobacterial responses to UV-radiation. In: B.A. Whitton and M. Potts (eds.). The Ecology of Cyanobacteria: Their Diversity in Time and Space, pp. 591–614. Kluwer Academic Publishers.

Catalan, J., M. Ventura, A. Brancelj, I. Granados, H. Thies, U. Nickus, A. Korhola, A.F. Lotter, A. Barbieri, E. Stuchlik, L. Lien, P. Bitusik, T. Buchaca, L. Camarero, G.H. Goudsmit, J. Kopacek, G. Lemcke, D.M. Livingstone, B. Müller, M. Rautio, M. Sisko, S. Sorvari, F. Sporka, O. Strunecky and M. Toro, 2000. Seasonal ecosystem variability in remote mountain lakes: implications for detecting climatic signals in sediment records. Journal of Paleolimnology, 28(1):25–46.

Chadwick, E.M.P., 1987. Causes of variable recruitment in a small Atlantic salmon stock. American Fisheries Society Symposium, 1:390–401.

Chambers, P.A., G.J. Scrimgeour and A. Pietroniro, 1997. Winter oxygen conditions in ice-covered rivers: the impact of pulp mill and municipal effluents. Canadian Journal of Fisheries and Aquatic Sciences, 54:2796–2806.

Chapin, F.S. III and C. Körner, 1994. Arctic and alpine biodiversity: patterns, causes and ecosystem consequences. Trends in Ecology and Evolution, 9:45–47.

Chapin, F.S. III, E.S. Zavaleta, V.T. Eviner, R.L. Naylor, P.M. Vitousek, H.L. Reynolds, D.U. Hooper, S. Lavorel, O.E. Sala, S.E. Hobbie, M.C. Mack and S. Dìaz, 2000. Consequences of changing biodiversity. Nature, 405:234–242.

Christensen, T.R., T. Friborg, M. Sommerkorn, J. Kaplan, L. Illeris, H. Soegaard, C. Nordstroem and S. Jonasson, 2000. Trace gas exchange in a high-arctic valley. 1. Variations in CO_2 and CH_4 flux between tundra vegetation types. Global Biogeochemical Cycles, 14:701–713.

Christie, G.C. and H.A. Regier, 1988. Measures of optimal thermal habitat and their relationships to yields for four commercial fish species. Canadian Journal of Fisheries and Aquatic Sciences, 45:301–314.

Christoffersen, K. and E. Jeppesen, 2000. Lake monitoring. In: K. Caning and M. Rasch (eds.). Zackenberg Ecological Research Operations, 5th Annual Report, 1999, pp. 43–46. Danish Polar Center, Ministry of Research and Information Technology, Copenhagen.

Church, M., 1995. Geomorphic response to river flow regulation: case studies and time-scales. Regulated Rivers: Research and Management, 11:3–22.

Clein, J.S., B.L. Kwiatkowski, A.D. McGuire, J.E. Hobbie, E.B. Rastetter, J.M. Melillo and D.W. Kicklighter, 2000. Modelling carbon responses of tundra ecosystems to historical and projected climate: a comparison of a plot- and a global-scale ecosystem model to identify process-based uncertainties. Global Change Biology, 6(S1):127–140.

Cohen, S.J. (ed.), 1994. Mackenzie Basin Impact Study (MBIS) Interim Report #2. Environment Canada, Ontario, xvi+484pp.

Cohen, S.J. (ed.), 1997. Mackenzie Basin Impact Study (MBIS) Final Report. Environment Canada, Ontario, viii+372pp.

Colby, P.J. and H. Lehtonen, 1994. Suggested causes for the collapse of zander *Stizostedion lucioperca* (L.) populations in northern and central Finland through comparisons with North American walleye, *Stizostedion vitreum* (Mitchill). Aqua Fennica, 24:9–20.

Collins, J.P. and A. Storfer, 2003. Global amphibian declines: sorting the hypotheses. Diversity and Distributions, 9:89–98.

Coutant, C.C., 1987. Thermal preference: when does an asset become a liability? Environmental Biology of Fishes, 18:161–172.

Craig, P.C., 1989. An introduction to anadromous fishes in the Alaskan Arctic. In: D.W. Norton (ed.). Research Advances on Anadromous Fish in Arctic Alaska and Canada. Biological Papers of the University of Alaska, 24:27–54.

Craig, P.C. and P.J. McCart, 1975. Classification of stream types in Beaufort Sea drainages between Prudhoe Bay, Alaska and the Mackenzie Delta, N.W.T., Canada. Arctic and Alpine Research, 7:183–198.

Crossman, E.J., 1984. Introduction of exotic fishes into Canada. In: W.R. Courtenay Jr. and J.R. Stauffer Jr. (eds.). Distribution, Biology and Management of Exotic Fishes, pp. 78–101. John Hopkins University Press, Baltimore.

Crozier, J., 1996. A Compilation of Archived Writings about Environmental Change in the Peace, Athabasca and Slave River Basins. Report No. 125. Northern River Basins Study, Edmonton.

Crump, B.C., G.W. Kling, M. Bahr and J.E. Hobbie, 2003. Bacterioplankton community shifts in an Arctic lake correlate with seasonal changes in organic matter source. Applied and Environmental Microbiology, 69:2253–2268.

Cunjak, R.A., T.D. Prowse and D.L. Parrish, 1998. Atlantic salmon (*Salmo salar*) in winter: the season of parr discontent? Canadian Journal of Fisheries and Aquatic Sciences, 55(S1):161–180.

Cwynar, L.C., 1982. A Late Quaternary vegetation history from Hanging Lake, Northern Yukon. Ecological Monographs, 52(1):1–24.

Dahlback, A., 2002. Recent changes in surface UV radiation and stratospheric ozone at a high arctic site. In: D.O. Hessen (ed.). UV Radiation and Arctic Ecosystems. Ecological Studies, 153:3–22.

Dahm, C.N. and M.C. Molles Jr., 1992. Streams in semi-arid regions as sensitive indicators of global climate change. In: P. Firth and S.G. Fisher (eds.). Troubled Waters of the Greenhouse Earth, pp. 250–260. Springer-Verlag.

Dallimore, A., C.J. Schröder-Adams and S.R. Dallimore, 2000. Holocene environmental history of thermokarst lakes on Richards Island, Northwest Territories, Canada: Theocamoebians as paleolimnological indicators. Journal of Paleolimnology, 23:261–283.

Danks, H.V., 1992. Arctic insects as indicators of environmental change. Arctic, 45:159–166.

Davis, A.J., L.S. Jenkinson, J.H. Lawton, B. Shorrocks and S. Wood, 1998. Making mistakes when predicting shifts in species range in response to global warming. Nature, 391:783–786.

De Lange, H.J. and E. Van Donk, 1997. Effects of UVB-irradiated algae on life history traits of *Daphnia pulex*. Freshwater Biology, 38:711–720.

de March, B.G.E., C.A. de Wit and D.C.G. Muir, 1998. Persistent organic pollutants. In: AMAP Assessment Report: Arctic Pollution Issues, pp. 183–371. Arctic Monitoring and Assessment Programme, Oslo.

Deegan, L.A. and B.J. Peterson, 1992. Whole-river fertilization stimulates fish production in an Arctic tundra river. Canadian Journal of Fisheries and Aquatic Sciences, 49:1890–1901.

Deegan, L.A., H.E. Golden, C.J. Harvey and B.J. Peterson, 1999. Influence of environmental variability on the growth of age-0 and adult Arctic grayling. Transactions of the American Fisheries Society, 128:1163–1175.

Degens, E.T., S. Kempe and J.E. Richey (eds.), 1991. Biogeochemistry of Major World Rivers. John Wiley and Sons, 356pp.

Dekker, W., 2003. On the distribution of the European eel (*Anguilla anguilla*) and its fisheries. Canadian Journal of Fisheries and Aquatic Sciences, 60:787–799.

Dempson, J.B. and J.M. Green, 1985. Life history of anadromous arctic charr, *Salvelinus alpinus*, in the Fraser River, northern Labrador. Canadian Journal of Zoology, 63:315–324.

Dempson, J.B., M.F. O'Connell and N.M. Cochrane, 2001. Potential impact of climate warming on recreational fishing opportunities for Atlantic salmon, *Salmo salar* L., in Newfoundland, Canada. Fisheries Management and Ecology, 8:69–82.

Diamond, M., P. Helm, R. Semkin and S. Law, 2003. Mass balance and modelling of contaminants in lakes. In: T.F. Bidleman, R. Macdonald and J. Stow (eds.). Canadian Arctic Contaminants Assessment Report II: Sources, Occurrence, Trends and Pathways in the Physical Environment, pp. 187–197. Indian and Northern Affairs Canada, Ottawa.

Doran, P.T., C.P. McKay, W.P. Adams, M.C. English, R.A. Wharton and M.A. Meyer, 1996. Climate forcing and thermal feedback of residual lake-ice covers in the high Arctic. Limnology and Oceanography, 41:839–848.

Doucett, R.R., 1999. Food-web relationships in Catamaran Brook, New Brunswick, as revealed by stable-isotope analysis of carbon and nitrogen. Ph.D Thesis, University of Waterloo, Ontario.

Douglas, M.S.V. and J.P. Smol, 1999. Freshwater diatoms as indicators of environmental change in the High Arctic. In: E.F. Stoermer and J.P. Smol (eds.). The Diatoms: Applications for the Environmental and Earth Sciences, pp. 227–244. Cambridge University Press.

Douglas, M.S.V., J.P. Smol and W. Blake, 1994. Marked post-18th century environmental change in High-Arctic ecosystems. Science, 266:416–419.

Douglas, M.S.V., S. Ludlam and S. Feeney, 1996. Changes in diatom assemblages in Lake C2 (Ellesmere Island, Arctic Canada): response to basin isolation from the sea and to other environmental changes. Journal of Paleolimnology, 16:217–226.

Downing, J.A., C. Plante and S. Lalonde, 1990. Fish production correlated with primary productivity, not the morphoedaphic index. Canadian Journal of Fisheries and Aquatic Sciences, 47:1929–1936.

Driscoll, C.T., J. Holsapple, C.L. Schofield and R. Munson, 1998. The chemistry and transport of mercury in a small wetland in the Adirondack region of New York, USA. Biogeochemistry, 40:137–146.

Dumont, P., J.F. Bergeron, P. Dulude, Y. Mailhot, A. Rouleau, G. Ouellet and J.-P. Lebel, 1988. Introduced salmonids: where are they going in Québec watersheds of the Saint-Laurent River? Fisheries, 13:9–17.

Dupre, W.R. and R. Thompson, 1979. The Yukon Delta: a model for deltaic sedimentation in an ice-dominated environment. In: Proceedings of the 11th Annual Offshore Technology Conference, pp. 657–661.

Dwyer, W.P. and R.G. Piper, 1987. Atlantic salmon growth efficiency as affected by temperature. The Progressive Fish Culturist, 49:57–59.

Dynesius, M. and C. Nilsson, 1994. Fragmentation and flow regulation of river systems in the northern third of the world. Science, 266:753–762.

Edwards, M.E., N.H. Bigelow, B.P. Finney and W.R. Eisner, 2000. Records of aquatic pollen and sediment properties as indicators of late-Quaternary Alaskan lake levels. Journal of Paleolimnology, 24:55–68.

Egglishaw, H.J. and P.E. Shackley, 1977. Growth, survival and production of juvenile salmon and trout in a Scottish stream, 1966–75. Journal of Fish Biology, 11:647–672.

Egglishaw, H.J. and P.E. Shackley, 1985. Factors governing the production of juvenile Atlantic salmon in a Scottish stream. Journal of Fish Biology, 27(Suppl. A):27–33.

Eilertsen, H.C. and O. Holm-Hansen, 2000. Effects of high latitude UV radiation on phytoplankton and nekton modeled from field measurements by simple algorithms. Polar Research, 19:173–182.

Elliott, J.M., 1994. Quantitative Ecology and the Brown Trout. Oxford University Press, 304pp.

Ewald, G., P. Larsson, H. Linge, L. Okla and N. Szarzi, 1998. Biotransport of organic pollutants to an inland Alaska lake by migrating sockeye salmon (*Oncorhynchus nerka*). Arctic, 51:40–47.

Fallu, M.-A. and R. Pienitz, 1999. Diatomées lacustres de Jamésie-Hudsonie (Québec) et modèle de reconstitution des concentrations de carbone organique dissous. Ecoscience, 6:603–620.

Fallu, M.-A., N. Allaire and R. Pienitz, 2002. Distribution of freshwater diatoms in 64 Labrador (Canada) lakes: species-environmental relationships along latitudinal gradients and reconstruction models for water colour and alkalinity. Canadian Journal of Fisheries and Aquatic Sciences, 59:329–349.

FAO, 2002. Capture Production 2000. FAO Yearbook of Fishery Statistics, Volume 90/1. Food and Agricultural Organization of the United Nations, 617pp.

Fast, H. and F. Berkes, 1998. Climate change, northern subsistence and land-based economies. In: N. Mayer and W. Avis (eds.). Canada Country Study: Climate Impacts and Adaptation. Vol. 8: National Cross-Cutting Issues, pp. 205–226. Environmental Adaptation Research Group, Environment Canada.

Fausch, K.D., 1989. Do gradient and temperature affect distributions of, and interactions between, brook charr (*Salvelinus fontinalis*) and other resident salmonids in streams? Physiology and Ecology Japan Special Volume, 1:303–322.

Finney, B.P., I. Gregory-Eaves, J. Sweetman, M.S.V. Douglas and J.P. Smol, 2000. Impacts of climatic change and fishing on Pacific salmon abundance over the past 300 years. Science, 290:795–799.

Finney, B.P., I. Gregory-Eaves, M.S.V. Douglas and J.P. Smol, 2002. Fisheries productivity in the northeastern Pacific Ocean over the past 2,200 years. Nature, 416:729–733.

Flanagan, K., E. McCauley, F. Wrona and T.D. Prowse, 2003. Climate change: the potential for latitudinal effects on algal biomass in aquatic ecosystems. Canadian Journal of Fisheries and Aquatic Sciences, 60:635–639.

Fleming, I.A. and M.R. Gross, 1990. Latitudinal clines – a trade-off between egg number and size in Pacific salmon. Ecology, 71:1–11.

Fleming, I.A. and A.J. Jensen, 2002. Fisheries: effects of climate change on the life cycles of salmon. In: I. Douglas (ed.). Encyclopedia of Global Environmental Change, Vol. 3, Causes and Consequences of Global Environmental Change, pp. 309–312. John Wiley and Sons.

Folt, C.L., C.Y. Chen and P.C. Pickhardt, 2002. Using plankton food web variables as indicators for the accumulation of toxic metals in fish. In: S.H. Wilson and W.A. Suk (eds.). Biomarkers of Environmentally Associated Disease: Technologies, Concepts, and Perspectives, pp. 287–304. CRC Press.

Forsström, L., 2000. Seasonal variability of phytoplankton in Lake Saanajärvi. M.Sc. Thesis, University of Helsinki, 53pp. (In Finnish)

Forsström, L., S. Sorvari, A. Korhola and M. Rautio, in press. Seasonality of phytoplankton in subarctic Lake Saanajärvi in NW Finnish Lapland. Polar Biology.

Freeman, C., C.D. Evans, D.T. Monteith, B. Reynolds and N. Fenner, 2001. Export of organic carbon from peat soils. Nature, 412:785.

Friborg, T., T.R. Christensen and H. Soegaard, 1997. Rapid response of greenhouse gas emission to early spring thaw in a subarctic mire as shown by micrometeorological techniques. Geophysical Research Letters, 24:3061–3064.

Friedland, K.D., 1998. Ocean climate influences on critical Atlantic salmon (Salmo salar) life history events. Canadian Journal of Fisheries and Aquatic Sciences, 55(S1):119–130.

Friedland, K.D., D.G. Reddin and J.F. Kocik, 1993. Marine survival of North American and European Atlantic salmon: effects of growth and environment. ICES Journal of Marine Science, 50:481–492.

Friedland, K.D., D.G. Reddin, J.R. McMenemy and K.F. Drinkwater, 2003. Multidecadal trends in North American Atlantic salmon (*Salmo salar*) stocks and climate trends relevant to juvenile survival. Canadian Journal of Fisheries and Aquatic Sciences, 60:563–583.

Funk, D.W., E.R. Pullman, K.M. Peterson, P.M. Crill and W.D. Billings, 1994. Influence of water table on carbon dioxide, carbon monoxide, and methane fluxes from taiga bog microcosms. Global Biogeochemical Cycles, 8:271–278.

Gajewski, K., R. Vance, M. Sawada, I. Fung, L.D. Gignac, L. Halsey, J. John, P. Maisongrande, P. Mandell, P.J. Mudie, P.J.H. Richard, A.G. Sherin, J. Soroko and D.H. Vitt, 2000. The climate of North America and adjacent ocean waters ca. 6 ka. Canadian Journal of Earth Sciences, 37:661–681.

Garcia-Pichel, F., 1994. A model for internal self-shading in planktonic organisms and its implications for the usefulness of ultraviolet sunscreens. Limnology and Oceanography, 39:1704–1717.

Gehrke, C., U. Johanson, D. Gwynn-Jones, L.O. Björn, T.V. Callaghan and J.A. Lee, 1996. Effects of enhanced ultraviolet-B radiation on terrestrial subarctic ecosystems and implications for interactions with increased atmospheric CO_2. In: P.S. Karlsson and T.V. Callaghan (eds.). Plant Ecology in the Sub-Arctic Swedish Lapland. Ecological Bulletin, 45:192–203.

Gerard, R., 1990. Hydrology of floating ice. In: T.D. Prowse and C.S.L. Ommanney (eds.). Northern Hydrology: Canadian Perspectives. National Hydrology Research Institute, Saskatoon, Scientific Report No. 1, pp. 103–134.

Gibson, J.A.E., W.F. Vincent, B. Nieke and R. Pienitz, 2000. Control of biological exposure to UV radiation in the Arctic Ocean: comparison of the roles of ozone and riverine dissolved organic matter. Arctic, 53:372–382.

Gibson, J.J., E.E. Prepas and P. McEachern, 2002. Quantitative comparison of lake throughflow, residency, and catchment runoff using stable isotopes: modelling and results from a survey of Boreal lakes. Journal of Hydrology, 262:128–144.

Gibson, J.J., T.D. Prowse and D.L. Peters, in press. Hydroclimatic controls on water balance and water level variability in Great Slave Lake. Hydrological Processes.

Gibson, R.J. and R.L. Haedrich, 1988. The exceptional growth of juvenile Atlantic salmon (*Salmo salar*) in the city waters of St. John's, Newfoundland, Canada. Polskie Archiwum Hydrobiologii, 35:385–407.

Gignac, L.D. and D.H. Vitt, 1994. Responses of northern peatlands to climatic change, effects on bryophytes. Journal of the Hattori Botanical Laboratory, 75:119–132.

Gill, D., 1974. Significance of spring breakup to the bioclimate of the Mackenzie River Delta. In: J.C. Reed and J.E. Sater (eds.). The Coast and Shelf of the Beaufort Sea: Proceedings of a Symposium on Beaufort Sea Coast and Shelf Research, pp. 543–544. Arctic Institute of North America.

Gobas, F.A.P.C. and L.G. Maclean, 2003. Sediment-water distribution of organic contaminants in aquatic ecosystems: The role of organic carbon mineralization. Environmental Science and Technology, 37:735–741.

Godfrey, W.E., 1986. The Birds of Canada. National Museums of Canada, Ottawa, 595pp.

Goes, J.I., N. Handa, S. Taguchi and T. Hama, 1994. Effect of UV-B radiation on the fatty acid composition of the marine phytoplankton *Tretraselmis* sp.: relation to cellular pigments. Marine Ecology Progress Series, 114:259–274.

Golden, H.E. and L.A. Deegan, 1998. The trophic interactions of young arctic grayling (*Thymallus arcticus*) in an Arctic tundra stream. Freshwater Biology, 39:637–648.

Gorham, E., 1991. Northern peatlands: role in the carbon cycle and probable responses to climatic warming. Ecological Applications, 1:182–195.

Goyke, A.P. and A.E. Hershey, 1992. Effects of fish predation on larval chironomid (*Diptera, Chironomidae*) communities in an arctic ecosystem. Hydrobiologia, 240:203–212.

Gratto-Trevor, C.L., 1994. Potential effects of global climate change on shorebirds in the Mackenzie Delta lowlands. In: S.J. Cohen (ed.). Mackenzie Basin Impact Study (MBIS) Interim Report #2, pp. 360–371. Environment Canada.

Gratto-Trevor, C.L., 1997. Climate change: proposed effects on shorebird habitat, prey, and numbers in the outer Mackenzie Delta. In: S.J. Cohen (ed.). Mackenzie Basin Impact Study (MBIS) Final Report, pp. 205–210. Environment Canada.

Gray, D.M. and T.D. Prowse, 1993. Snow and floating ice. In: D.R. Maidment (ed.). Handbook of Hydrology, pp. 7.1–7.58. McGraw-Hill.

Greenfield, B.K., T.R. Hrabik, C.J. Harvey and S.R. Carpenter, 2001. Predicting mercury levels in yellow perch: use of water chemistry, trophic ecology, and spatial traits. Canadian Journal of Fisheries and Aquatic Sciences, 58:1419–1429.

Gregor, D., 1990. Deposition and accumulation of selected agricultural pesticides in Canadian Arctic snow. In: D.A. Kurtz (ed.). Long Range Transport of Pesticides, pp. 373–386. Lewis Publishers.

Grigoriev, V.Y. and B.L. Sokolov, 1994. Northern hydrology in the former Soviet Union. In: T.D. Prowse, C.S.L. Ommanney and L.E. Watson (eds.). Northern Hydrology: International Perspectives. National Hydrology Research Institute, Saskatoon, Science Report No. 3, pp. 147–179.

Groisman, P.Y., T.R. Karl and R.W. Knight, 1994. Observed impact of snow cover on the heat balance and the rise of continental spring temperatures. Science, 263:198–200.

Groombridge, B. and M.D. Jenkins, 2002. World Atlas of Biodiversity. Earth's Living Resources in the 21st Century. UNEP World Conservation Monitoring Centre, University of California Press, 340pp.

Gross, M.R., R.M. Coleman and R.M. McDowall, 1988. Aquatic productivity and the evolution of diadromous fish migration. Science, 239:1291–1293.

Guieu, C., W.W. Huang, J.-M. Martin and Y.Y. Yong, 1996. Outflow of trace metals into the Laptev Sea by the Lena River. Marine Chemistry, 53:255–267.

Häder, D.-P., H.D. Kumar, R.C. Smith and R.C. Worrest, 2003. Aquatic ecosystems: effects of solar ultraviolet radiation and interactions with other climatic change factors. In: J.F. Bornman, K. Solomon, and J.C. van der Leun (eds.). Environmental Effects of Ozone Depletion and its Interactions with Climate Change: 2002 Assessment. Photochemical and Photobiological Sciences, 2:39–50.

Hamilton, P.B., D.R.S. Lean and M. Poulin, 1994. The physicochemical characteristics of lakes and ponds from the Northern regions of Ellesmere Island. In: P.B. Hamilton (ed.). Proceedings of the Fourth Arctic-Antarctic Diatom Symposium, pp. 57–63. Canadian Technical Report of Fisheries and Aquatic Sciences No. 1957.

Hammar, J., 1989. Freshwater ecosystems of polar regions: vulnerable resources. Ambio, 18:6–22.

Hammarlund, D. and T.W.D. Edwards, 1998. Evidence of changes in moisture transport efficiency across the Scandes mountains in northern Sweden during the Holocene, inferred from oxygen isotope records of lacustrine carbonates. In: Isotope Techniques in the Study of Environmental Change. Proceedings of a Symposium held in Vienna, 14–18 April 1997, pp. 573–580. STI/PUB/1024. International Atomic Energy Agency, Vienna.

Hammarlund, D., L. Barnekow, H.J.B. Birks, B. Buchardt and T.W.D. Edwards, 2002. Holocene changes in atmospheric circulation recorded in the oxygen-isotope stratigraphy of lacustrine carbonates from northern Sweden. The Holocene, 12:339–351.

Hansen, T., 1983. Bunnfaunastudier i et vassdrag pa Svalbard. Thesis, University of Oslo.

Hanson, K.L., A.E. Hershey and M.E. McDonald, 1992. A comparison of slimy sculpin (*Cottus cognatus*) populations in arctic lakes with and without piscivorous predators. Hydrobiologia, 240:189–202.

Hansson, L.-A., 1992. The role of food chain composition and nutrient availability in shaping algal biomass development. Ecology, 73:241–247.

Hargreaves, K.J., D. Fowler, C.E.R. Pitcairn and M. Aurela, 2001. Annual methane emission from Finnish mires estimated from eddy covariance campaign measurements. Theoretical and Applied Climatology, 70:203–213.

Harner, T., 1997. Organochlorine contamination of the Canadian Arctic, and speculation on future trends. International Journal of Environment and Pollution, 8:51–73.

Harper, P.P., 1981. Ecology of streams at high latitudes. In: M.A. Lock and D.D. Williams (eds.). Perspectives in Running Water Ecology, pp. 313–337. Plenum Press.

Hartman, C.W. and R.F. Carlson, 1973. Water balance of a small lake in a permafrost region. Institute of Water Resources Report IWR-42. University of Alaska, Fairbanks, 23pp.

Harvey, C.J., B.J. Peterson, W.B. Bowden, A.E. Hershey, M.C. Miller, L.A. Deegan and J.C. Finlay, 1998. Biological responses to fertilization of Oksrukuyik Creek, a tundra stream. Journal of the North American Benthological Society, 17:190–209.

Headrick, M.R. and R.F. Carline, 1993. Restricted summer habitat and growth of northern pike in two southern Ohio impoundments. Transactions of the American Fisheries Society, 122:228–236.

Hebert, P.D.N. and C.J. Emery, 1990. The adaptive significance of cuticular pigmentation in *Daphnia*. Functional Ecology, 4:703–710.

Hecky, R.E. and S.J. Guildford, 1984. Primary productivity of Southern Indian Lake before, during and after impoundment and Churchill River Diversion. Canadian Journal of Fisheries and Aquatic Sciences, 41:591–604.

Heginbottom, J.A., 1984. The bursting of a snow dam, Tingmisut Lake, Melville Island, Northwest Territories. In: Current Research, Part B, Geological Survey of Canada Paper 84-01B, pp. 187–192.

Heintzenberg, J., 1989. Arctic haze: air pollution in polar regions. Ambio, 18:50–55.

Helm, P.A., M.L. Diamond, R. Semkin, W.M.J. Strachan, C. Teixeira and D. Gregor, 2002. A mass balance model describing multiyear fate of organochlorine compounds in a high arctic lake. Environmental Science and Technology, 36:996–1003.

Hershey, A.E., 1990. Snail populations in arctic lakes: competition mediated by predation. Oecologia, 82:26–32.

Hershey, A.E., W.B. Bowden, L.A. Deegan, J.E. Hobbie, B.J. Peterson, G.W. Kipphut, G.W. Kling, M.A. Lock, R.W. Merritt, M.C. Miller, J.R. Vestal and J.A. Schuldt, 1997. The Kuparuk River: a long-term study of biological and chemical processes in an arctic river. In: A.M. Milner and M.W. Oswood (eds.). Freshwaters of Alaska: Ecological Syntheses. Ecological Studies 119:107–130.

Hessen, D.O., 1994. *Daphnia* responses to UV-light. In: C.E. Williamson and H.E. Zagarese (eds.). Impact of UV-B Radiation on Pelagic Freshwater Ecosystems. Archiv für Hydrobiologie – Advances in Limnology, 43:185–195.

Hessen, D.O. (ed.), 2002. UV Radiation and Arctic Ecosystems. Ecological Studies 153. Springer-Verlag, 310pp.

Hessen, D.O. and N. Alstad Rukke, 2000. UV radiation and low calcium as mutual stressors for Daphnia. Limnology and Oceanography, 45:1834–1838.

Hessen, D.O., H.J. De Lange and E. Van Donk, 1997. UV-induced changes in phytoplankton cells and its effects on grazers. Freshwater Biology, 38:513–524.

Hessen, D.O., J. Borgeraas, K. Kessler and U.H. Refseth, 1999. UV-B susceptibility and photoprotection of Arctic *Daphnia* morphotypes. Polar Research, 18:345–352.

Hileman, B., 1983. Arctic haze. Environmental Science and Technology, 17:232A–236A.

Hill, D.K. and J.J. Magnuson, 1990. Potential effects of global climate warming on the growth and prey consumption of Great Lakes fish. Transactions of the American Fisheries Society, 119:265–275.

Hinder, B., M. Gabathuler, B. Steiner, K. Hanselmann and H.R. Preisig, 1999. Seasonal dynamics and phytoplankton diversity in high mountain lakes (Jöri lakes, Swiss Alps). Journal of Limnology, 58:152–161.

Hinzman, L.D. and D.L. Kane, 1992. Potential response of an Arctic watershed during a period of global warming. Journal of Geophysical Research, 97(D3):2811–2820.

Hinzman, L.D., N. Bettez, F.S. Chapin III, M.B. Dyurgerov, C.L. Fastie, B. Griffith, R.D. Hollister, A.S. Hope, H.P. Huntington, A. Jensen, D.L. Kane, A.H. Lynch, A. Lloyd, A.D. McGuire, F.E. Nelson, W.C. Oechel, T.E. Osterkamp, C.H. Racine, V.E. Romanovsky, D. Stow, M. Sturm, C.E. Tweedie, G.L. Vourlitis, M.D. Walker, P.J. Webber, J.M. Welker, K. Winker and K. Yoshikawa, in press. Evidence and implications of recent climate change in terrestrial regions of the Arctic. Climatic Change.

Hirst, S.M., 1984. Effects of spring breakup on microscale air temperatures in the Mackenzie River Delta. Arctic, 37:263–269.

Hirvenoja, M., 1967. Chironomidae and Culicidae (Dipt.) from Spitsbergen. Annales Entomologici Fennici, 33:52–61.

Hobæk, A. and H.G. Wolf, 1991. Ecological genetics of Norwegian Daphnia. II. Distribution of *Daphnia longispina* genotypes in relation to short-wave radiation and water colour. Hydrobiologia, 225:229–243.

Hobbie, J.E., 1980. Limnology of Tundra Ponds: Barrow, Alaska. Dowden, Hutchinson and Ross, 514pp.

Hobbie, J.E., 1984. Polar limnology. In: F.B. Taub (ed.). Lakes and Reservoirs. Ecosystems of the World, 23:63–104.

Hobbie, J.E., B.J. Peterson, N. Bettez, L. Deegan, W.J. O'Brien, G.W. Kling, G.W. Kipphut, W.B. Bowden and A.E. Hershey, 1999. Impact of global change on the biogeochemistry and ecosystems of an arctic freshwater system. Polar Research, 18:207–214.

Hobbie, J.E., G. Shaver, J. Laundre, K. Slavik, L.A. Deegan, J. O'Brien, S. Oberbauer and S. MacIntyre, 2003. Climate forcing at the Arctic LTER site. In: D. Greenland, D.G. Goodin and R.C. Smith (eds.). Climate Variability and Ecosystem Response in Long-Term Ecological Research (LTER) Sites, pp. 74–91. Oxford University Press.

Hoff, J.T., D. Gregor, D. Mackay, F. Wania and C.Q. Jia, 1998. Measurement of the specific surface area of snow with the nitrogen adsorption technique. Environmental Science and Technology, 32:58–62.

Hong, Y.T., Z.G. Wang, H.B. Jiang, Q.H. Lin, B. Hong, Y.X. Zhu, Y. Wang, L.S. Xu, X.T. Leng and H.D. Li, 2001. A 6000-year record of changes in drought and precipitation in northeastern China based on a δ13C time series from peat cellulose. Earth and Planetary Science Letters, 185:111–119.

Hung, H., C.J. Halsall, P. Blanchard, H.H. Li, P. Fellin, G. Stern and B. Rosenberg, 2001. Are PCBs in the Canadian Arctic atmosphere declining? Evidence from 5 years of monitoring. Environmental Science and Technology, 35:1303–1311.

Huntsman, A.G., 1924. Limiting factors for marine animals, II: The lethal effect of sunlight. Contributions to Canadian Biology, 2:83–88.

Hurtubise, R.D., J.E. Havel and E.E. Little, 1998. The effects of ultraviolet-B radiation on freshwater invertebrates: experiments with a solar simulator. Limnology and Oceanography, 43:1082–1088.

Hutchinson, G.E., 1957. Concluding remarks. Population studies: animal ecology and demography. Cold Spring Harbor Symposia on Quantitative Biology, 22:415–427.

Hynes, H.B.N., 1970. The Ecology of Running Waters. Liverpool University Press, 555pp.

IPCC, 1996. Climate Change 1995: Impacts, Adaptations and Mitigation of Climate Change: Scientific-Technical Analyses. Contribution of Working Group II to the Second Assessment Report of the Intergovernmental Panel on Climate Change. R.T. Watson, M.C. Zinyowera and R.H. Moss (eds.). Cambridge University Press, 878pp.

IPCC, 1998. The Regional Impacts of Climate Change. An Assessment of Vulnerability. A Special Report of Working Group II of the Intergovernmental Panel on Climate Change. R.T. Watson, M.C. Zinyowera and R.H. Moss (eds.). Cambridge University Press, 527pp.

IPCC, 2001a. Climate Change 2001: Synthesis Report. A Contribution of Working Groups I, II and III to the Third Assessment Report of the Intergovernmental Panel on Climate Change. R.T. Watson and Core Writing Team. Cambridge University Press, 398pp.

IPCC, 2001b. Climate Change 2001: The Scientific Basis. Contribution of Working Group I to the Third Assessment Report of the Intergovernmental Panel on Climate Change. J.T. Houghton, Y. Ding, D.J. Griggs, M. Noguer, P.J. van der Linden, X. Dai, K. Maskell and C.A. Johnson (eds.). Intergovernmental Panel on Climate Change. Cambridge University Press, 881pp.

Irons, J.G., M.W. Oswood, R.J. Stout and C.M. Pringle, 1994. Latitudinal patterns in leaf litter breakdown: is temperature really important? Freshwater Biology, 32:401–411.

Jacoby, G.C., N.V. Lovelius, O.I. Shumilov, O.M. Raspopov, J.M. Karbainov and D.C. Frank, 2000. Long-term temperature trends and tree growth in the Taymir Region of Northern Siberia. Quaternary Research, 53:312–318.

Jensen, A.J., T. Forseth and B.O. Johnsen, 2000. Latitudinal variation in growth of young brown trout *Salmo trutta*. Journal of Animal Ecology, 69:1010–1020.

Jeppesen, E., P.R. Leavitt, L. De Meester and J.P. Jensen, 2001a. Functional ecology and palaeolimnology: using cladoceran remains to reconstruct anthropogenic impact. Trends in Ecology and Evolution, 16:191–198.

Jeppesen, E., K. Christoffersen, F. Landkildehus, T. Lauridsen, S. Amsinck, F. Riget and M. Søndergaard, 2001b. Fish and crustaceans in northeast Greenland lakes with special emphasis on interactions between Arctic charr (*Salvelinus alpinus*), *Lepidurus arcticus* and benthic chydorids. Hydrobiologia, 442:329–337.

Jeppesen, E., J.P. Jensen, C. Jensen, B. Faafeng, D.O. Hessen, M. Søndergaard, T. Lauridsen, P. Brettum and K. Christoffersen, 2003. The impact of nutrient state and lake depth on top-down control in the pelagic zone of lakes: a study of 466 lakes from the temperate zone to the Arctic. Ecosystems, 6:313–325.

Jeremiason, J.D., K.C. Hornbuckle and S.J. Eisenreich, 1994. PCBs in Lake Superior, 1978–1992: decreases in water concentrations reflect loss by volatilization. Environmental Science and Technology, 28:903–914.

Jeremiason, J.D., S.J. Eisenreich, J.E. Baker and B.J. Eadie, 1998. PCB decline in settling particles and benthic recycling of PCBs and PAHs in Lake Superior. Environmental Science and Technology, 32:3249–3256.

Joabsson, A. and T.R. Christensen, 2001. Methane emissions from wetlands and their relationship with vascular plants: an Arctic example. Global Change Biology, 7:919–932.

Johnson, L., 1980. The Arctic charr, *Salvelinus alpinus*. In: E.K. Balon (ed.). Salmonid Fishes of the Genus Salvelinus, pp. 15–98. Dr. W. Junk Publishers.

Jónasson, P.M. and H. Adalsteinsson, 1979. Phytoplankton production in shallow eutrophic Lake M˘vatn, Iceland. Oikos, 32:113–138.

Jónasson, P.M., H. Adalsteinsson and G. St. Jónsson, 1992. Production and nutrient supply of phytoplankton in subarctic, dimictic Thingvallavatn, Iceland. Oikos, 64:162–187.

Jonsson, N., 1991. Influence of water flow, water temperature and light on fish migration in rivers. Nordic Journal of Freshwater Research, 66:20–35.

Jørgensen, I. and J.A. Eie, 1993. The distribution of zooplankton, zoobenthos and fish in lakes and ponds of the Mossel peninsula, Svalbard. Norwegian Institute for Nature Research, Forsknings-rapport 45. Trondheim, 25pp. (In Norewgian)

Kaczmarska, I., T.A. Clair, J.M. Ehrman, S.L. MacDonald, D.R.S. Lean and K.E. Day, 2000. The effect of ultraviolet B on phytoplankton populations in clear and brown temperate Canadian lakes. Limnology and Oceanography, 45:651–663.

Kalff, J., 2002. Limnology: Inland Water Ecosystems. Prentice Hall, 592pp.

Kalff, J. and H.E. Welch, 1974. Phytoplankton production in Char Lake, a natural polar lake, and in Meretta Lake, a polluted polar lake, Cornwallis Island, Northwest Territories. Journal of the Fisheries Research Board of Canada, 31:621–636.

Kane, D.L. and C.W. Slaughter, 1973. Recharge of a central Alaskan lake by subpermafrost groundwater. In: Permafrost: The North American Contribution to the Second International Conference, Yakutsk, pp. 458–462. National Academy of Sciences, Washington, D.C.

Karentz, D., J.E. Cleaver and D.L. Mitchell, 1991. Cell survival characteristics and molecular responses of Antarctic phytoplankton to ultraviolet-B radiation. Journal of Phycology, 27:326–341.

Kidd, K.A., D.W. Schindler, R.H. Hesslein and D.C.G. Muir, 1995a. Correlation between stable nitrogen isotope ratios and concentrations of organochlorines in biota from a freshwater food web. Science of the Total Environment, 160/161:381–390.

Kidd, K.A., R.H. Hesslein, R.J.P. Fudge and K.A. Hallard, 1995b. The influence of trophic level as measured by δ15N on mercury concentrations in freshwater organisms. Water, Air and Soil Pollution, 80:1011-1015.

Kidd, K.A., D.W. Schindler, D.C.G. Muir, H.L. Lockhart and R.H. Hesslein, 1995c. High concentrations of toxaphene in fishes from a subarctic lake. Science, 269:240–242.

King, J.R., B.J. Shuter and A.P. Zimmerman, 1999. Empirical links between thermal habitat, fish growth, and climate change. Transactions of the American Fisheries Society, 128:656–665.

Kirk-Davidoff, D.B., E.J. Hintsa, J.G. Anderson and D.W. Keith, 1999. The effect of climate change on ozone depletion through changes in stratospheric water vapour. Nature, 402:399–401.

Kling, G.W., J. O'Brien, M.C. Miller and A.E. Hershey, 1992a. The biogeochemistry and zoogeography of lakes and rivers in arctic Alaska. Hydrobiologia, 240:1–14.

Kling, G.W., G.W. Kipphut and M.C. Miller, 1992b. The flux of CO_2 and CH_4 from lakes and rivers in arctic Alaska. Hydrobiologia, 240:23–36.

Kling, G.W., G.W. Kipphut, M.M. Miller and W.J. O'Brien, 2000. Integration of lakes and streams in a landscape perspective: the importance of material processing on spatial patterns and temporal coherence. Freshwater Biology, 43:477–497.

Köck, G. and R. Hofer, 1998. Origin of cadmium and lead in clear softwater lakes of high altitude and high latitude, and their bioavailability and toxicity to fish. In: T. Braunbeck, D.E. Hinton and B. Streit (eds.). Fish Ecotoxicology. Experientia Supplementa, 86:225–257.

Köck, G., M. Triendl and R. Hofer, 1996. Seasonal patterns of metal accumulation in Arctic char (*Salvelinus alpinus*) from an oligotrophic Alpine lake related to temperature. Canadian Journal of Fisheries and Aquatic Sciences, 53:780–786.

Köck, G., C. Doblander, H. Niederstätter, B. Berger and D. Bright, 2001. Fish from sensitive ecosystems as bioindicators of global climate change. Report on the Austrian-Canadian research cooperation *High-Arctic 2000* to the Austrian Academy of Science. Vienna, Austria, 71pp. (In German)

Köck, G., C. Doblander, B. Berger, H. Niederstätter, D. Bright, D. Muir, J.D. Reist, J.A. Babaluk and Y. Kalra, 2002. Temperature induced metal accumulation and stress response in fish from Canadian arctic lakes. In: 12th SETAC Europe Annual Meeting, Challenges in Environmental Risk Assessment and Modelling: Linking Basic and Applied Research, 12–16 May, Vienna. Society of Environmental Toxicology and Chemistry.

Korhola, A., 1995. Holocene climatic variations in southern Finland reconstructed from peat initiation data. The Holocene, 5:43–58.

Korhola, A., 1999. Distribution patterns of Cladocera in subarctic Fennoscandian lakes and their potential in environmental reconstruction. Ecography, 22:357–373.

Korhola, A. and M. Rautio, 2002. Cladocera and other branchiopod crustaceans. In: J.P. Smol, H.J.B. Birks and W.M. Last (eds.). Tracking Environmental Change Using Lake Sediments. Vol. 4: Zoological Indicators, pp. 5–41. Kluwer Academic Publishers.

Korhola, A. and J. Weckström, 2005. Paleolimnological studies in arctic Fennoscandia and the Kola Peninsula (Russia). In: R. Pienitz, M.S.V. Douglas and J.P. Smol (eds.). Long-Term Environmental Change in Arctic and Antarctic Lakes, pp. 381–418. Springer.

Korhola, A., J. Weckström, and M. Nyman, 1999. Predicting the long-term acidification trends in small subarctic lakes using diatoms. Journal of Applied Ecology, 36:1021–1034.

Korhola, A., H. Olander and T. Blom, 2000a. Cladoceran and chironomid assemblages as quantitative indicators of water depth in subarctic Fennoscandian lakes. Journal of Paleolimnology, 24:43–54.

Korhola, A., J. Weckström, L. Holmström and P. Erästö, 2000b. A quantitative Holocene climatic record from diatoms in northern Fennoscandia. Quaternary Research, 54:284–294.

Korhola, A., S. Sorvari, M. Rautio, P.G. Appleby, J.A. Dearing, Y. Hu, N. Rose, A. Lami and N.G. Cameron, 2002a. A multi-proxy analysis of climate impacts on the recent development of subarctic Lake Sannajärvi in Finnish Lapland. Journal of Paleolimnology, 28(1):59–77.

Korhola, A., J. Weckström and T. Blom, 2002b. Relationships between lake and land-cover features along latitudinal vegetation ecotones in arctic Fennoscandia. Archiv für Hydrobiologie, Supplementbände (Monograph Studies), 139(2):203–235.

Korhola, A., K. Vasko, H.T.T. Toivonen and H. Olander, 2002c. Holocene temperature changes in northern Fennoscandia reconstructed from chironomids using Bayesian modelling. Quaternary Science Reviews, 21:1841–1860.

Krupnik, I. and D. Jolly (eds.), 2002. The Earth is Faster Now: Indigenous Observations of Arctic Environmental Change. Arctic Research Consortium of the United States, Fairbanks, Alaska, 384pp.

Kutzbach, J.E., P.J. Guetter, P.J. Behling and R. Selin, 1993. Simulated climatic changes: results of the COHMAP Climate-Model experiments. In: H.E. Wright Jr., J.E. Kutzbach, T. Webb III, W.F. Ruddiman, F.A. Street-Perrot and P.J. Bartlein (eds.). Global Climates Since the Last Glacial Maximum, pp. 24–93. University of Minnesota Press.

L'Abée-Lund, J.H., B. Jonsson, A.J. Jensen, L.M Saettem, T.G. Heggberget, B.O. Johnson and T.F. Naesje, 1989. Latitudinal variation in life-history characteristics of sea-run migrant brown trout Salmo trutta. Journal of Animal Ecology, 58:525–542.

Lackner, R., 1998. Oxidative stress in fish by environmental pollutants. In: T. Braunbeck, D.E. Hinton and B. Streit (eds.). Fish Ecotoxicology. Experientia Supplementa, 86:203–224.

Laing, T.E., K.M. Rühland and J.P. Smol, 1999. Past environmental and climatic changes related to tree-line shifts inferred from fossil diatoms from a lake near the Lena River Delta, Siberia. The Holocene, 9:547–557.

Laing, T.E., R. Pienitz and S. Payette, 2002. Evaluation of limnological responses to recent environmental change and caribou activity in the Rivière George region, Northern Québec, Canada. Arctic, Antarctic and Alpine Research, 34:454–464.

Lalonde, J.D., M. Amyot, A.M.L. Kraepiel and F.M.M. Morel, 2001. Photooxidation of Hg(0) in artificial and natural waters. Environmental Science and Technology, 35:1367–1372.

Lalonde, J.D., A.J. Poulain and M. Amyot, 2002. The role of redox reactions in snow on snow-to-air Hg transfer. Environmental Science and Technology, 36:174–178.

Lamborg, C.H., W.F. Fitzgerald, J. O'Donnell and T. Torgerson, 2002. A non-steady-state compartmental model of global-scale mercury biogeochemistry with interhemispheric atmospheric gradients. Geochimica et Cosmochimica Acta, 66:1105–1118.

Lappalainen, J. and H. Lehtonen, 1997. Temperature habitats for freshwater fishes in a warming climate. Boreal Environment Research, 2:69–84.

Larsen, L.H., A. Evenset and B. Sirenko, 1995. Linkages and impact hypotheses concerning valued ecosystem components (VECs) invertebrates, fish, the coastal zone and large river estuaries and deltas. International Northern Sea Route Programme, working paper 12, 39pp. +app.

Larsson, P., L. Okla and G. Cronberg, 1998. Turnover of polychlorinated biphenyls in an oligotrophic and an eutrophic lake in relation to internal lake processes and atmospheric fallout. Canadian Journal of Fisheries and Aquatic Sciences, 55:1926–1937.

Laurila, T., H. Soegaard, C.R. Lloyd, M. Aurela, J.-P. Tuovinen and C. Nordstroem, 2001. Seasonal variations of net CO_2 exchange in European Arctic ecosystems. Theoretical and Applied Climatology, 70:183–201.

Laurion, I. and W.F. Vincent, 1998. Cell size versus taxonomic composition as determinants of UV sensitivity in natural phytoplankton communities. Limnology and Oceanography, 43:1774–1779.

Laurion, I., W.F. Vincent and D.R.S. Lean, 1997. Underwater ultraviolet radiation: development of spectral models for northern high latitude lakes. Photochemistry and Photobiology, 65:107–114.

Leach, J.H., L.M. Dickie, B.J. Shuter, U. Borgmann, J. Hyman and W. Lysack, 1987. A review of methods for prediction of potential fish production with application to the Great Lakes and Lake Winnipeg. Canadian Journal of Fisheries and Aquatic Sciences, 44(Suppl.2):471–485.

Leaman, B.M. and R.J. Beamish, 1981. Ecological and management implication of longevity in some northeast Pacific groundfish. Bulletin of the International North Pacific Fisheries Commission, 42:85–97.

Leavitt, P.R., R.D. Vinebrooke, D.B. Donald, J.P. Smol and D.W. Schindler, 1997. Past ultraviolet radiation environments in lakes derived from fossil pigments. Nature, 388:457–459.

Leech, D.M. and C.E. Williamson, 2000. Is tolerance to UV radiation in zooplankton related to body size, taxon, or lake transparency? Ecological Applications, 10:1530–1540.

Leech, D.M. and C.E. Williamson, 2001. In situ exposure to ultraviolet radiation alters the depth distribution of Daphnia. Limnology and Oceanography, 46:416–420.

Leggett, W.C. and J.E. Carscadden, 1978. Latitudinal variation in reproductive characteristics of American shad (*Alosa sapidissima*): evidence for population specific life history strategies in fish. Journal of the Fisheries Research Board of Canada, 35:1469–1478.

Lehtonen, H., 1996. Potential effects of global warming on northern European freshwater fish and fisheries. Fisheries Management and Ecology, 3:59–71.

Lepistö, L., 1999. Phytoplankton assemblages reflecting the ecological status of lakes in Finland. Monographs of the Boreal Environment Research, 16. 97pp.

Lesack, L., R.E. Hecky and P. Marsh, 1991. The influence of frequency and duration of flooding on the nutrient chemistry of the Mackenzie Delta lakes. In: P. Marsh and C.S.L. Ommanney (eds.). Mackenzie Delta: Environmental Interactions and Implications for Development, National Hydrology Research Institute, Saskatoon, Symposium No. 4, pp. 19–36.

Levins, R., 1969. Thermal acclimation and heat resistance in *Drosophila* species. The American Naturalist, 103:483–499.

Levinton, J.S., 1983. The latitudinal compensation hypothesis: growth data and a model of latitudinal growth differentiation based upon energy budgets. 1. Interspecific comparison of *Ophryotrocha* (Polychaeta: Dorvilleidae). Biological Bulletin, 165:686–698.

Li, Y.-F., R.W. Macdonald, L.M.M. Jantunen, T. Harner, T.F. Bidleman and W.M.J. Strachan, 2002. The transport of ß-hexachlorocyclohexane to the western Arctic Ocean: a contrast to α-HCH. Science of the Total Environment, 291:229–246.

Liblik, L.K., T.R. Moore, J.L. Bubier and S.D. Robinson, 1997. Methane emissions from wetlands in the zone of discontinuous permafrost: Fort Simpson, Northwest Territories, Canada. Global Biogeochemical Cycles, 11:485–494.

Lindberg, S.E., S. Brooks, C.-J. Lin, K.J. Scott, M.S. Landis, R.K. Stevens, M. Goodsite and A. Richter, 2002. Dynamic oxidation of gaseous mercury in the arctic troposphere at polar sunrise. Environmental Science and Technology, 36:1245–1256.

Lindell, M.J., W. Granéli and L.J. Tranvik, 1996. Effects of sunlight on bacterial growth in lakes of different humic content. Aquatic Microbial Ecology, 11:135–141.

Linell, K.A. and J.C.F. Tedrow, 1981. Soil and Permafrost Surveys in the Arctic. Clarendon Press, 279pp.

Liston, G.E., J.P. McFadden, M. Sturm and R.A. Pielke Sr., 2002. Modelled changes in arctic tundra snow, energy, and moisture fluxes due to increased shrubs. Global Change Biology, 8:17–32.

Little, E.E. and D.L. Fabacher, 1994. Comparative sensitivity of rainbow trout and two threatened salmonids, Apache trout and Lahontan cut-throat trout, to ultraviolet-B radiation. In: C.E. Williamson and H.E. Zagarese (eds.). Impact of UV-B Radiation on Pelagic Freshwater Ecosystems. Archiv für Hydrobiologie - Advances in Limnology, 43:217–226.

Lockhart, W.L., D.A. Metner, D.A.J. Murray, R.W. Danell, B.N. Billeck, C.L. Baron, D.C.G. Muir and K. Chang-Kue, 1989. Studies to determine whether the condition of fish from the lower Mackenzie River is related to hydrocarbon exposure. Environmental Studies No. 61. Indian Affairs and Northern Development Canada, Ottawa, Ontario, viii+84pp.

Lotter, A.F. and C. Bigler, 2000. Do diatoms in the Swiss Alps reflect the length of ice-cover? Aquatic Sciences, 62:125–141.

Lotter, A.F., R. Pienitz and R. Schmidt, 1999. Diatoms as indicators of environmental change near Arctic and Alpine treeline. In: E.F. Stoermer and J.P. Smol (eds.). The Diatoms: Applications to the Environmental and Earth Sciences, pp. 205–226, Cambridge University Press.

Lozhkin, A.V., P.M. Anderson, W.R. Eisner, L.G. Ravako, D.M. Hopkins, L.B. Brubaker, P.A. Colinvaux and M.C. Miller, 1993. Late Quaternary lacustrine pollen records from southwestern Beringia. Quaternary Research, 39:314–324.

Lu, J.Y., W.H. Schroeder, L.A. Barrie, A. Steffen, H.E. Welch, K. Martin, L. Lockhart, R.V. Hunt, G. Boila and A. Richter, 2001. Magnification of atmospheric mercury deposition to polar regions in springtime: the link to tropospheric ozone depletion chemistry. Geophysical Research Letters, 28:3219–3222.

Ludlam, S.D., S. Feeney and M.S.V. Douglas, 1996. Changes in the importance of lotic and littoral diatoms in a high arctic lake over the last 191 years. Journal of Paleolimnology, 16:187–204.

MacDonald, G.M., T.W.D. Edwards, K.A. Moser, R. Pienitz and J.P. Smol, 1993. Rapid response of treeline vegetation and lakes to past climate warming. Nature, 361:243–246.

MacDonald, G.M., B. Felzer, B.P. Finney and S.L. Forman, 2000a. Holocene lake sediment records of Arctic hydrology. Journal of Paleolimnology, 24:1–14.

MacDonald, G.M., A.A. Velichko, C.V. Kremenetski, O.K. Borisova, A.A. Goleva, A.A. Andreev, L.C. Cwynar, R.T. Riding, S.L. Forman, T.W.D. Edwards, R. Aravena, D. Hammarlund, J.M. Szeicz and V.N. Gattaulin, 2000b. Holocene treeline history and climate change across northern Eurasia. Quaternary Research, 53:302–311.

Macdonald, R.W., D.W. Paton, E.C. Carmack and A. Omstedt, 1995. The freshwater budget and under-ice spreading Mackenzie River water in the Canadian Beaufort Sea based on salinity and $^{18}O/^{16}O$ measurements in water and ice. Journal of Geophysical Research, 100(C1):895–920.

Macdonald, R.W., L.A. Barrie, T.F. Bidleman, M.L. Diamond, D.J. Gregor, R.G. Semkin, W.M.J. Strachan, Y.-F. Li, F. Wania, M. Alaee, L.B. Alexeeva, S.M. Backus, R. Bailey, J.M. Bewers, C. Gobeil, C.J. Halsall, T. Harner, J.T. Hoff, L.M.M. Jantunen, W.L. Lockhart, D. Mackay, D.C.G. Muir, J. Pudykiewicz, K.J. Reimer, J.N. Smith, G.A. Stern, W.H. Schroeder, R. Wagemann and M.B. Yunker, 2000. Contaminants in the Canadian Arctic: 5 years of progress in understanding sources, occurrence and pathways. Science of the Total Environment, 254:93–234.

Macdonald, R.W., D. Mackay and B. Hickie, 2002. Contaminant amplification in the environment: revealing the fundamental mechanisms. Environmental Science and Technology, 36:457A–462A.

Macdonald, R.W., T. Harner, J. Fyfe, H. Loeng and T. Weingartner, 2003a. AMAP Assessment 2002: The Influence of Global change on Contaminant Pathways to, within, and from the Arctic. Arctic Monitoring and Assessment Programme, Oslo, xii+65pp.

Macdonald, R.W., D. Mackay, Y.-F. Li and B. Hickie, 2003b. How will global climate change affect risks from long-range transport of persistent organic pollutants? Human and Ecological Risk Assessment, 9:643–660.

Mackay, D., 2001. Multimedia Environmental Models: The Fugacity Approach. Second Edition. Lewis Publishers, 272pp.

Mackay, D.K. and O.H. Loken, 1974. Arctic hydrology. In: J.D. Ives and R.G. Barry (eds.). Arctic and Alpine Environments, pp. 111–132. Methuen and Co.

Mackay, J.R., 1992. Lake stability in an ice-rich permafrost environment. Examples from the Western Arctic Coast. In: R.D. Roberts and M.L. Bothwell (eds.). Aquatic Ecosystems in Semi-Arid Regions. Implications for Resource Management. National Hydrology Research Institute, Saskatoon, Symposium Series 7, pp. 1–26.

Madronich, S., R.L. McKenzie, L.O. Björn and M.M. Caldwell, 1995. Changes in ultraviolet radiation reaching the Earth's surface. Ambio, 24:143–152.

Magnuson, J.J. and B.T. DeStasio, 1997. Thermal niche of fishes and global warming. In: C.M. Wood and D.G. McDonald (eds.). Global Warming: Implications for Freshwater and Marine Fish. Society for Experimental Biology Seminar Series, 61:377–408. Cambridge University Press.

Magnuson, J.J., L.B. Crowder and P.A. Medvick, 1979. Temperature as an ecological resource. American Zoologist, 19:331–343.

Magnuson, J.J., J.D. Meisner and D.K. Hill, 1990. Potential changes in the thermal habitat of Great Lakes fish after global climate warming. Transactions of the American Fisheries Society, 119:254–264.

Magnuson, J.J., D.M. Robertson, B.J. Benson, R.H. Wynne, D.M. Livingstone, T. Arai, R.A. Assel, R.G. Barry, V. Card, E. Kuusisto, N.G. Granin, T.D. Prowse, K.M. Stewart and V.S. Vuglinski, 2000. Historical trends in lake and river ice cover in the Northern Hemisphere. Science, 289:1743–1746.

Manabe, S., M.J. Spelman and R.J. Stouffer, 1992. Transient responses of a coupled ocean-atmosphere model to gradual changes of atmospheric CO_2. Part II: Seasonal response. Journal of Climate, 5:105–126.

Marcogliese, D.J., 2001. Implications of climate change for parasitism of animals in the aquatic environment. Canadian Journal of Zoology, 79:1331–1352.

Markager, S., W.F. Vincent and E.P.Y. Tang, 1999. Carbon fixation by phytoplankton in high Arctic lakes: implications of low temperature for photosynthesis. Limnology and Oceanography, 44:597–607.

Marsh, P., 1990. Snow hydrology. In: T.D. Prowse and C.S.L. Ommanney (eds.). Northern Hydrology: Canadian Perspectives. National Hydrology Research Institute, Saskatoon, Scientific Report No. 1, pp. 37–61.

Marsh, P. and M. Hey, 1989. The flooding hydrology of Mackenzie Delta lakes near Inuvik, N.W.T., Canada. Arctic, 42:41–49.

Marsh, P. and M. Hey, 1991. Spatial variations in the spring flooding of Mackenzie Delta lakes. In: P. Marsh and C.S.L. Ommanney (eds.). Mackenzie Delta: Environmental Interactions and Implications for Development. National Hydrology Research Institute, Saskatoon, Symposium No. 4, pp. 9–18.

Marsh, P. and N. Neumann, 2001. Processes controlling the rapid drainage of two ice-rich permafrost-dammed lakes in NW Canada. Hydrological Processes, 15:3433–3446.

Marsh, P. and N. Neumann, 2003. Climate and hydrology of a permafrost dammed lake in NW Canada. In: M. Phillips, S.M. Springman and L.U. Arenson (eds.). Permafrost: Proceedings of the 8th International Conference on Permafrost, Zurich, Switzerland, 21–25 July 2003, Vol. 2, pp. 729:734. International Permafrost Association, Longyearbyen.

Marsh, P. and C.S.L. Ommanney (eds.), 1989. Mackenzie Delta: environmental interactions and implications for development. National Hydrology Research Institute, Saskatoon, Symposium No. 4, 195pp.

Marshall, S.A., A.T. Finnamore and D.C.A. Blades, 1999. Canadian peatlands: the terrestrial arthropod fauna. In: D.P. Batzer, R.B. Rader and S.A. Wissinger (eds.). Invertebrates in Freshwater Wetlands of North America: Ecology and Management, pp. 383–400. John Wiley and Sons.

Martin, N.V. and C.H. Olver, 1980. The lake charr, *Salvelinus namaycush*. In: E.K. Balon (ed.). Charrs: Salmonid Fishes of the Genus Salvelinus, pp. 205–277. Dr. W. Junk Publishers.

Matthews, J.V. Jr., 1974. Wisconsin environment of Interior Alaska: Pollen and macrofossil analysis of a 27 meter core from the Isabella Basin (Fairbanks, Alaska). Canadian Journal of Earth Sciences, 11:828–841.

McCauley, E. and W.W. Murdoch, 1987. Cyclic and stable populations: plankton as paradigm. The American Naturalist, 129:97–121.

McCormick, S.D., J.M. Shrimpton, J.D. Zydlewski, C.M. Wood and D.G. McDonald, 1997. Temperature effects on osmoregulatory physiology of juvenile anadromous fish. In: C.M. Wood and D.G. McDonald (eds.). Global Warming: Implications for Freshwater and Marine Fish. Society for Experimental Biology Seminar Series, 61:279–301.

McCormick, S.D., L.P. Hansen, T.P. Quinn and R.L. Saunders, 1998. Movement, migration, and smolting of Atlantic salmon (Salmo salar). Canadian Journal of Fisheries and Aquatic Sciences, 55(S1):77–92.

McDonald, M.A., L. Arragutainaq and Z. Novalinga, 1997. Voices from the Bay: Traditional Ecological Knowledge of Inuit and Cree in the Hudson Bay Bioregion. Canadian Arctic Resources Committee and Environmental Committee of the Municipality of Sanikiluaq, Ottawa, Canada.

McDonald, M.E., A.E. Hershey and M.C. Miller, 1996. Global warming impacts on lake trout in Arctic lakes. Limnology and Oceanography, 41:1102–1108.

McDowall, R.M., 1987. Evolution and the importance of diadromy: the occurrence and distribution of diadromy among fishes. American Fisheries Society Symposium, 1:1–13.

McFadden, J.T. and E.L. Cooper, 1962. An ecological comparison of six populations of brown trout (Salmo trutta). Transactions of the American Fisheries Society, 91:53–62.

McKone, T.E., J.I. Daniels and M. Goldman, 1996. Uncertainties in the link between global climate change and predicted health risks from pollution: Hexachlorobenzene (HCB) case study using a fugacity model. Risk Analysis, 16:377–393.

McNamara, J.P., D.L. Kane and L.D. Hinzman, 1999. An analysis of an arctic channel network using a digital elevation model. Geomorphology, 29:339–353.

McQueen, D.J., M.R.S. Johannes, J.R. Post, T.J. Stewart and D.R.S. Lean, 1989. Bottom-up and top-down impacts on freshwater pelagic community structure. Ecological Monographs, 59:289–309.

Medina-Sánchez, J.M., M. Villar-Argaiz, P. Sánchez-Castillo, L. Cruz-Pizarro and P. Carrillo, 1999. Structure changes in a planktonic food web: biotic and abiotic controls. Journal of Limnology, 58:213–222.

Meisner, J.D., J.L. Goodier, H.A. Regier, B.J. Shuter and W.J. Christie, 1987. An assessment of the effects of climate warming on Great Lakes basin fishes. Journal of Great Lakes Research, 13:340–352.

Meltofte, H. and H. Thing (eds.), 1997. Zackenberg Ecological Research Operations, 2nd Annual Report, 1996. Danish Polar Center, Ministry of Research and Information Technology, Copenhagen, 80pp.

Merilä, J., M. Pahkala and U. Johanson, 2000. Increased ultraviolet-B radiation, climate change and latitudinal adaptation – a frog perspective. Annales Zoologici Fennici, 37:129–134.

Meyer, J.L. and G.E. Likens, 1979. Transport and transformation of phosphorus in a forest stream ecosystem. Ecology, 60:1255–1269.

Meyer, J.L., M.J. Sale, P.J. Mulholland and N.L. Poff, 1999. Impacts of climate change on aquatic ecosystem functioning and health. Journal of the American Water Resources Association, 35:1373–1386.

Michelutti, N., T.E. Laing and J.P. Smol, 2001. Diatom assessment of past environmental changes in lakes located near the Noril'sk (Siberia) smelters. Water, Air and Soil Pollution, 125:231–241.

Michelutti, N., M.S.V. Douglas and J.P. Smol, 2002. Tracking recent recovery from eutrophication in a high arctic lake (Meretta Lake, Cornwallis Island, Nunavut, Canada) using fossil diatom assemblages. Journal of Paleolimnology, 28:377–381.

Milankovitch, M., 1941. Canon of insolation and the ice age problem. Special Publication 132, Koniglich Serbische Akademie, Belgrade. (English translation by the Israel Program for Scientific Translations, Jerusalem, 1969).

Milburn, D. and T.D. Prowse, 1998. An assessment of a northern delta as a hydrologic sink for sediment-bound contaminants. Nordic Hydrology, 29:64–71.

Milner, A.M. and G.E. Petts, 1994. Glacial rivers: physical habitat and ecology. Freshwater Biology, 32:295–307.

Milot-Roy, V. and W.F. Vincent, 1994. UV radiation effects on photosynthesis: the importance of near-surface thermoclines in a subarctic lake. In: C.E. Williamson and H.E. Zagarese (eds.). Impact of UV-B Radiation on Pelagic Freshwater Ecosystems. Archiv für Hydrobiologie - Advances in Limnology, 43:171–184.

Minkkinen, K., R. Korhonen, I. Savolainen and J. Laine, 2002. Carbon balance and radiative forcing of Finnish peatlands 1900–2100 - the impact of forestry drainage. Global Change Biology, 8:785–799.

Minns, C.K. and J.E. Moore, 1992. Predicting the impact of climate change on the spatial pattern of freshwater fish yield capability in eastern Canadian lakes. Climatic Change, 22:327–346.

Mitsch, W.J. and J.G. Gosselink, 1993. Wetlands. Van Nostrand Reinhold, 722pp.

Moore, J.J., 1981. Mires. In: L.C. Bliss, O.W. Heal and J.J. Moore (eds.). Tundra Ecosystems: A Comparative Analysis, pp. 35–37. Cambridge University Press.

Moore, T.R. and N.T. Roulet, 1993. Methane flux: water table relations in northern wetlands. Geophysical Research Letters, 20:587–590.

Moore, T.R., N.T. Roulet and J.M. Waddington, 1998. Uncertainty in predicting the effect of climatic change on the carbon cycling of Canadian peatlands. Climatic Change, 40:229–245.

Morin, R., J.J. Dodson and G. Power, 1982. Life history variations of anadromous cisco (Coregonus artedii), lake whitefish (C. clupeaformis), and round whitefish (Prosopium cylindraceum) populations of eastern James-Hudson Bay. Canadian Journal of Fisheries and Aquatic Sciences, 39:958–967.

Morris, D.P., H.E. Zagarese, C.E. Williamson, E.G. Balseiro, B.R. Hargreaves, B. Modenutti, R. Moeller and C. Queimalinos, 1995. The attenuation of solar UV radiation in lakes and the role of dissolved organic carbon. Limnology and Oceanography, 40:1381–1391.

Morrison, H.A., F.A.P.C. Gobas, R. Lazar, D.M. Whittle and G.D. Haffner, 1998. Projected changes to the trophodynamics of PCBs in the western Lake Erie ecosystem attributed to the presence of zebra mussels (Dreissena polymorpha). Environmental Science and Technology, 32:3862–3867.

Morrison, H.A., D.M. Whittle and G.D. Haffner, 2000. The relative importance of species invasions and sediment disturbance in regulating chemical dynamics in western Lake Erie. Ecological Modelling, 125:279–294.

Moser, K.A., A. Korhola, J. Weckström, T. Blom, R. Pienitz, J.P. Smol, M.S.V. Douglas and M.B. Hay, 2000. Paleohydrology inferred from diatoms in northern latitude regions. Journal of Paleolimnology, 24:93–107.

Mostajir, B., S. Demers, S.J. de Mora, C. Belzile, J.-P. Chanut, M. Gosselin, S. Roy, P.Z. Villegas, J. Fauchot, J. Bouchard, D.F. Bird, P. Monfort and M. Levasseur, 1999. Experimental test of the effect of ultraviolet-B radiation in a planktonic community. Limnology and Oceanography, 44:586–596.

Mueller, D.R., W.F. Vincent and M.O. Jeffries, 2003. Break-up of the largest Arctic ice shelf and associated loss of an epishelf lake. Geophysical Research Letters, 30: doi:10.1029/2003GL017931.

Muir, D.C.G., A. Omelchenko, N.P. Grift, D.A. Savoie, W.L. Lockhart, P. Wilkinson and G.J. Brunskill, 1996. Spatial trends and historical deposition of polychlorinated biphenyls in Canadian midlatitude and arctic lake sediments. Environmental Science and Technology, 30:3609–3617.

Muir, D.C.G., B. Braune, B. de March, R.J. Norstrom, R. Wagemann, L. Lockhart, B. Hargrave, D. Bright, R. Addison, J. Payne and K. Reimer, 1999. Spatial and temporal trends and effects of contaminants in the Canadian Arctic marine ecosystem: a review. Science of the Total Environment, 230:83–144.

Murphy, M.L., C.P. Hawkins and N.H. Anderson, 1981. Effects of canopy modification and accumulated sediment on stream communities. Transactions of the American Fisheries Society, 110:469–478.

Murray, J.L., 1998a. Physical/geographical characteristics of the Arctic. In: AMAP Assessment Report: Arctic Pollution Issues, pp. 9–24. Arctic Monitoring and Assessment Programme, Oslo.

Murray, J.L., 1998b. Ecological characteristics of the Arctic. In: AMAP Assessment Report: Arctic Pollution Issues, pp. 117–140. Arctic Monitoring and Assessment Programme, Oslo.

Neale, P.J., 2001. Modeling the effects of ultraviolet radiation on estuarine phytoplankton production: impact of variations in exposure and sensitivity to inhibition. Journal of Photochemistry and Photobiology, 62:1–8.

Neale, P.J., R.F. Davis and J.J. Cullen, 1998. Interactive effects of ozone depletion and vertical mixing on photosynthesis of Antarctic phytoplankton. Nature, 392:585–589.

Nedwell, D.B., 2000. Life in the cooler – starvation in the midst of plenty; and implications for microbial polar life. In: C.R. Bell, M. Brylinski and P. Johnson-Green (eds.). Proceedings of the 8th International Symposium on Microbial Ecology, pp. 299–305. Atlantic Canada Society for Microbiology, Halifax.

Nicholson, B.J. and L.D. Gignac, 1995. Ecotype dimensions of peatland bryophyte indicator species along gradients in the Mackenzie River Basin, Canada. The Bryologist, 98:437–451.

Niemi, R., P.J. Martikainen, J. Silvola, A. Wulff, S. Turtola and T. Holopainen, 2002. Elevated UV-B radiation alters fluxes of methane and carbon dioxide in peatland microcosms. Global Change Biology, 8:361–371.

Nilssen, K.J., O.A. Gulseth, M. Iversen and R. Kjol, 1997. Summer osmoregulatory capacity of the world's northernmost living salmonid. American Journal of Physiology: Regulatory, Integrative and Comparative Physiology, 272:R743–R749.

Nordeng, H., 1983. Solution to the 'char problem' based on Arctic char (*Salvelinus alpinus*) in Norway. Canadian Journal of Fisheries and Aquatic Sciences, 40:1372–1387.

Nordstroem, C., H. Soegaard, T.R. Christensen, T. Friborg and B.U. Hansen, 2001. Seasonal carbon dioxide balance and respiration of a high-arctic fen ecosystem in NE-Greenland. Theoretical and Applied Climatology, 70:149–166.

Nriagu, J.O., 1989. A global assessment of natural sources of atmospheric trace metals. Nature, 338:47–49.

O'Brien, W.J., A.E. Hershey, J.E Hobbie, M.A. Hullar, G.W. Kipphut, M.C. Miller, B. Moller and J.R. Vestal, 1992. Control mechanisms of arctic lake ecosystems: a limnocorral experiment. Hydrobiologia, 240:143–188.

O'Brien, W.J., M. Bahr, A.E. Hershey, J.E. Hobbie, G.W. Kipphut, G.W. Kling, H. Kling, M. McDonald, M.C. Miller, P. Rublee and J.R. Vestal, 1997. The limnology of Toolik Lake. In: A.M. Milner and M.W. Oswood (eds.). Freshwaters of Alaska: Ecological Syntheses. Ecological Studies 119:61–106.

Oswood, M.W., 1997. Streams and rivers of Alaska. In: A.M. Milner and M.W. Oswood (eds.). Freshwaters of Alaska: Ecological Syntheses. Ecological Studies 119:331–356.

Oswood, M.W., A.M. Milner and J.G. Irons III, 1992. Climate change and Alaskan rivers and streams. In: P. Firth and S.G. Fisher (eds.). Global Climate Change and Freshwater Ecosystems, pp. 192–210, Springer-Verlag.

Overpeck, J., K. Hughen, D. Hardy, R. Bradley, R. Case, M. Douglas, B.P. Finney, K. Gajewski, G.C. Jacoby, A.E. Jennings, S. Lamoureux, A. Lasca, G.M. MacDonald, J. Moore, M. Retelle, S. Smith, A. Wolfe and G. Zielinski, 1997. Arctic environmental change of the last four centuries. Science, 278:1251–1266.

Paasivirta, L., T. Lahti and T. Perätie, 1988. Emergence, phenology and ecology of aquatic and semi-terrestrial Insects on a boreal raised bog in central Finland. Holarctic Ecology, 11:96–105.

Pacyna, J.M. and E.G. Pacyna, 2001. An assessment of global and regional emissions of trace metals to the atmosphere from anthropogenic sources worldwide. Environmental Reviews, 9:269–298.

Pahkala, M., A. Laurila and J. Merilä, 2002. Effects of ultraviolet-B radiation on common frog *Rana temporaria* embryos from along a latitudinal gradient. Oecologia, 133:458–465.

Paine, R.T., 1966. Food web complexity and species diversity. The American Naturalist, 100:65–75.

Palen, W.J., D.E. Schindler, M.J. Adams, C.A. Pearl, R.B. Bury and S.A. Diamond, 2002. Optical characteristics of natural waters protect amphibians from UV-B in the U.S. Pacific Northwest. Ecology, 83:2951–2957.

Panikov, N.S. and S.N. Dedysh, 2000. Cold season CH_4 and CO_2 emission from boreal peat bogs (West Siberia): winter fluxes and thaw activation dynamics. Global Biogeochemical Cycles, 14:1071–1080.

Parrish, D.L., R.J. Behnke, S.R. Gephard, S.D. McCormick and G.H. Reeves, 1998. Why aren't there more Atlantic salmon (*Salmo salar*)? Canadian Journal of Fisheries and Aquatic Sciences, 55(S1):281–287.

Parsons, T.R., 1992. The removal of marine predators by fisheries and the impact of trophic structure. Marine Pollution Bulletin, 25:51–53.

Paterson, A.M., A.A. Betts-Piper, J.P. Smol and B.A. Zeeb, 2003. Diatom and chrysophyte algal response to long-term PCB contamination from a point-source in northern Labrador, Canada. Water, Air and Soil Pollution, 145:377–393.

Paul, A.J. and J.R. Post, 2001. Spatial distribution of native and non-native salmonids in streams of the eastern slopes of the Canadian Rocky Mountains. Transactions of the American Fisheries Society, 130:417–430.

Payette, S., M.-J. Fortin and I. Gamache, 2001. The subarctic forest-tundra: the structure of a biome in a changing climate. BioScience, 51:709–718.

Perin, S.L., 1994. Short-term influences of ambient UV-B radiation on phytoplankton productivity and chlorophyll fluorescence in two lakes of the High Arctic. M.Sc. Thesis, Trent University, Ontario.

Perin, S.L., 2003. Influences of UVB radiation on lake ecosystems of High Arctic lakes. Ph.D Thesis, University of Ottawa.

Peters, D.L. and T.D. Prowse, 2001. Regulation effects on the lower Peace River, Canada. Hydrological Processes, 15:3181–3194.

Peterson, B.J., J.E. Hobbie, T.L. Corliss and K. Kriet, 1983. A continuous-flow periphyton bioassay: tests of nutrient limitation in a tundra stream. Limnology and Oceanography, 28:583–591.

Peterson, B.J., L. Deegan, J. Helfrich, J.E. Hobbie, M. Hullar, B. Moller, T.E. Ford, A. Hershey, A. Hiltner, G. Kipphut, M.A. Lock, D.M. Fiebig, V. McKinley, M.C. Miller, J.R. Vestal, J. Robie, R. Ventullo and G. Volk, 1993. Biological responses of a tundra river to fertilization. Ecology, 74:653–672.

Peterson, B.J., R.M. Holmes, J.W. McClelland, C.J. Vorosmarty, R.B. Lammers, A.I. Shiklomanov, I.A. Shiklomanov and S. Rahmstorf, 2002. Increasing river discharge to the Arctic Ocean. Science, 298:2171–2173.

Peterson, E.B., L.M. Allison and R.D. Kabzems, 1981. Alluvial ecosystems. Mackenzie River Basin Committee, Mackenzie River Basin Board, Fort Smith, Northwest Territories, Canada, 129pp.

Peterson, G., G.A. De Leo, J.J. Hellmann, M.A. Janssen, A. Kinzig, J.R. Malcolm, K.L. O'Brien, S.E. Pope, D.S. Rothman, E. Shevliakova and R.R.T. Tinch, 1997. Uncertainty, climate change, and adaptive management. Conservation Ecology (online), 1(2), www.bdt.fat.org.br/cons_ecol/vol1/iss2/art4/index.html

Peterson, R.H. and D.J. Martin-Robichaud, 1989. First feeding of Atlantic salmon (*Salmo salar* L.) fry as influenced by temperature regime. Aquaculture, 78:35–53.

Peterson, R.H., H.C.E. Spinney and A. Sreedharan, 1977. Development of Atlantic salmon (*Salmo salar*) eggs and alevins under varied temperature regimes. Journal of the Fisheries Research Board of Canada, 34:31–43.

Pickhardt, P.C., C.L. Folt, C.Y. Chen, B. Klaue and J.D. Blum, 2002. Algal blooms reduce the uptake of toxic methylmercury in freshwater food webs. Proceedings of the National Academy of Sciences, 99:4419–4423.

Pielou, E.C., 1994. A Naturalist's Guide to the Arctic. University of Chicago Press, xvi+328pp.

Pienitz, R. and J.P. Smol, 1994. The ecology and physicochemical characteristics of lakes in the subarctic and arctic regions of the Yukon Territory, Fennoscandia (Finland, Norway), the Northwest Territories and Northern Quebec. In: P.B. Hamilton (ed.). Proceedings of the Fourth Arctic-Antarctic Diatom Symposium, pp. 31–43. Canadian Technical Report of Fisheries and Aquatic Sciences No. 1957.

Pienitz, R. and W.F. Vincent, 2000. Effect of climate change relative to ozone depletion on UV exposure in subarctic lakes. Nature, 404:484–487.

Pienitz, R., G. Lortie and M. Allard, 1991. Isolation of lacustrine basins and marine regression in the Kuujjuaq area (northern Québec), as inferred from diatom analysis. Géographie physique et Quaternaire, 45:155–174.

Pienitz, R., J.P. Smol and H.J.B. Birks, 1995. Assessment of freshwater diatoms as quantitative indicators of past climate change in the Yukon and Northwest Territories, Canada. Journal of Paleolimnology, 13:21–49.

Pienitz, R., J.P. Smol and G.M. MacDonald, 1999. Paleolimnological reconstructions of Holocene climatic trends from two boreal tree-line lakes, Northwest Territories, Canada. Arctic, Antarctic and Alpine Research, 31:82–93.

Pimm, S.L., 1991. The Balance of Nature? Ecological Issues in the Conservation of Species and Communities. The University of Chicago Press, 434pp.

Pimm, S.L., G.J. Russell, J.L. Gittleman and T.M. Brooks, 1995. The future of biodiversity. Science, 269:347–350.

Planas, D., 1994. The high north: present and perspectives. In: R. Margalef (ed.), Limnology Now: A Paradigm of Planetary Problems, pp. 315–351, Elsevier.

Poff, N.L., M.M. Brinson and J.W. Day, 2002. Aquatic Ecosystems and Global Climate Change. Pew Center on Global Climate Change, Arlington, Virginia, 45pp.

Pomeroy, J.W., D.M. Gray and P.G. Landine, 1993. The Prairie Blowing Snow Model: characteristics, validation, operation. Journal of Hydrology, 144:165–192.

Ponader, K., R. Pienitz, W.F. Vincent and K. Gajewski, 2002. Limnological conditions in a subarctic lake (Northern Québec, Canada) during the late Holocene: analyses based on fossil diatoms. Journal of Paleolimnology, 27:353–366.

Power, G., 1976. History of the Hudson's Bay Company salmon fisheries in the Ungava Bay region. Polar Record, 18:151–161.

Power, G., 1981. Stock characteristics and catches of Atlantic salmon (*Salmo salar*) in Québec, and Newfoundland and Labrador in relation to environmental variables. Canadian Journal of Fisheries and Aquatic Sciences, 38:1601–1611.

Power, G., 1990a. Warming rivers (or a changing climate for Atlantic salmon). Atlantic Salmon Journal, 39(4):40–42.

Power, G., 1990b. Salmonid communities in Quebec and Labrador: temperature relations and climate change. Polskie Archiwum Hydrobiologii, 37:13–28.

Power, G. and D.R. Barton, 1987. Some effects of physiographic and biotic factors on the distribution of anadromous Arctic char (*Salvelinus alpinus*) in Ungava Bay, Canada. Arctic, 40:198–203.

Power, G., M. Power, R. Dumas and A. Gordon, 1987. Marine migrations of Atlantic salmon from rivers in Ungava Bay, Québec. American Fisheries Society Symposium, 1:364–376.

Power, G., R. Cunjak, J. Flannagan and C. Katopodis, 1993. Biological effects of river ice. In: T.D. Prowse and N.C. Gridley (eds.). Environmental Aspects of River Ice. National Hydrology Research Institute, Saskatoon, Science Report No. 5, pp. 97–119.

Power, G., R.S. Brown and J.G. Imhof, 1999. Groundwater and fish – insights from northern North America. Hydrological Processes, 13:401–422.

Power, M. and G. Power, 1994. Modeling the dynamics of smolt production in Atlantic salmon. Transactions of the American Fisheries Society, 123:535–548.

Power, M. and M.R. van den Heuvel, 1999. Age-0 yellow perch growth and its relationship to temperature. Transactions of the American Fisheries Society, 128:687–700.

Power, M., J.B. Dempson, G. Power and J.D. Reist, 2000. Environmental influences on an exploited anadromous Arctic charr stock in Labrador. Journal of Fish Biology, 57:82–98.

Power, M., G.M. Klein, K.R.R.A. Guiguer and M.K.H. Kwan, 2002. Mercury accumulation in the fish community of a sub-Arctic lake in relation to trophic position and carbon sources. Journal of Applied Ecology, 39:819–830.

Prowse, T.D., 1990. Northern hydrology: an overview. In: T.D. Prowse and C.S.L. Ommanney (eds.). Northern Hydrology: Canadian Perspectives. National Hydrology Research Institute, Saskatoon, Scientific Report No. 1, pp. 1–36.

Prowse, T.D., 2001a. River-ice ecology. I: hydrology, geomorphic, and water-quality aspects. Journal of Cold Regions Engineering, 15:1–16.

Prowse, T.D., 2001b. River-ice ecology. II: biological aspects. Journal of Cold Regions Engineering, 15:17–33.

Prowse, T.D. and S. Beltaos, 2002. Climatic control of river-ice hydrology: a review. Hydrological Processes, 16:805–822.

Prowse, T.D. and F.M. Conly, 2001. Multiple-hydrologic stressors of a northern delta ecosystem. Journal of Aquatic Ecosystem Stress and Recovery, 8:17–26.

Prowse, T.D. and J.M. Culp, 2003. Ice break-up: a neglected factor in river ecology. Canadian Journal of Civil Engineering, 30:145–155.

Prowse, T.D. and N.C. Gridley (eds.), 1993. Environmental Aspects of River Ice. National Hydrology Research Institute, Saskatoon, Science Report No. 5, 155pp.

Prowse, T.D. and C.S.L. Ommanney, 1990. Northern Hydrology: Canadian Perspectives. National Hydrology Research Institute, Saskatoon, Science Report No.1, 308pp.

Prowse, T.D. and R.L. Stephenson, 1986. The relationship between winter lake cover, radiation receipts and the oxygen deficit in temperate lakes. Atmosphere-Ocean, 24:386–403.

Prowse, T.D., C.S.L. Ommanney and L.E. Watson, 1994. Northern Hydrology: International Perspectives. National Hydrology Research Institute, Saskatoon, Science Report No.3, 215pp.

Prowse, T.D., J.M. Buttle, P.J. Dillon, M.C. English, P. Marsh, J.P. Smol and F.J. Wrona, 2001. Impacts of dams/diversions and climate change. In: Threats to Sources of Drinking Water and Aquatic Ecosystems Health in Canada. National Water Research Institute, Burlington, Ontario. Scientific Assessment Report Series No.1, pp. 69–72.

Prowse, T.D., F.J. Wrona and G. Power, 2004. Dams, reservoirs and flow regulation. In: Threats to Water Availability in Canada. NWRI Scientific Assessment Report Series No. 3 and ACSD Science Assessment Series No. 1, pp. 9–18. National Water Research Institute, Burlington, Ontario.

R-ArcticNET, 2003. A Regional, Electronic, Hydrographic Data Network for the Arctic Region. www.r-arcticnet.sr.unh.edu.

Rae, R. and W.F. Vincent, 1998a. Effects of temperature and ultraviolet radiation on microbial food web structure: potential responses to global change. Freshwater Biology, 40:747–758.

Rae, R. and W.F. Vincent, 1998b. Phytoplankton production in subarctic lake and river ecosystems: development of a photosynthesis-temperature-irradiance model. Journal of Plankton Research, 20:1293–1312.

Ramlal, P.S., R.H. Hesslein, R.E. Hecky, E.J. Fee, J.W.M. Rudd and S.J. Guildford, 1994. The organic carbon budget of a shallow arctic tundra lake on the Tuktoyaktuk Peninsula, NWT, Canada: Arctic lake carbon budget. Biogeochemistry, 24:145–172.

Rautio, M., 2001. Zooplankton assemblages related to environmental characteristics in treeline ponds in Finnish Lapland. Arctic, Antarctic and Alpine Research, 33:289–298.

Rautio, M. and A. Korhola, 2002a. Effects of ultraviolet radiation and dissolved organic carbon on the survival of subarctic zooplankton. Polar Biology, 25:460–468.

Rautio, M. and A. Korhola, 2002b. UV-induced pigmentation in subarctic Daphnia. Limnology and Oceanography, 47:295–299.

Rautio, M., S. Sorvari and A. Korhola, 2000. Diatom and crustacean zooplankton communities, their seasonal variability, and representation in the sediments of subarctic Lake Saanajärvi. Journal of Limnology, 59(Suppl.1):81–96.

Raven, J.A., 1998. The twelfth Tansley Lecture. Small is beautiful: the picophytoplankton. Functional Ecology, 12:503–513.

Reche, I., M.L. Pace and J.J. Cole, 1998. Interactions of photobleaching and inorganic nutrients in determining bacterial growth on colored dissolved organic carbon. Microbial Ecology, 36:270–280.

Regier, H.A., J.J. Magnuson and C.C. Coutant, 1990. Introduction to proceedings: symposium on effects of climate change on fish. Transactions of the American Fisheries Society, 119:173–175.

Regier, H.A., P. Lin, K.K. Ing and G.A. Wichert, 1996. Likely responses to climate change of fish associations in the Laurentian Great Lakes Basin: concepts, methods and findings. Boreal Environment Research, 1:1–15.

Reimnitz, E. and E.W. Kempema, 1987. Field observations of slush ice generated during freeze-up in arctic coastal waters. Marine Geology, 77:219–231.

Reist, J.D., 1994. An overview of the possible effects of climate change on northern freshwater and anadromous fishes. In: S.J. Cohen (ed.). Mackenzie Basin Impact Study (MBIS), Interim Report 2, pp. 377–385. Environment Canada, Ottawa.

Reist, J.D., 1997a. The Canadian perspective on issues in arctic fisheries management and research. In: J.B. Reynolds (ed.). Fish Ecology in Arctic North America. American Fisheries Society Symposium, 19:4–12.

Reist, J.D., 1997b. Potential cumulative effects of human activities on broad whitefish populations in the lower Mackenzie River basin. In: R.F. Tallman and J.D. Reist (eds.). The Proceedings of the Broad Whitefish Workshop: The Biology, Traditional Knowledge and Scientific Management of Broad Whitefish (*Coregonus nasus* (Pallas)) in the Lower Mackenzie River, pp. 179–197. Canadian Technical Report of Fisheries and Aquatic Sciences 2193.

Reist, J.D. and K. Chang-Kue, 1997. The life history and habitat usage of broad whitefish in the lower Mackenzie River basin. In: R.F. Tallman and J.D. Reist (eds.). The Proceedings of the Broad Whitefish Workshop: The Biology, Traditional Knowledge and Scientific Management of Broad Whitefish (*Coregonus nasus* (Pallas)) in the Lower Mackenzie River, pp. 63–84. Canadian Technical Report of Fisheries and Aquatic Sciences 2193.

Reist, J.D. and M.A. Treble, 1998. Challenges facing northern Canadian fisheries and their co-managers. In: J. Oakes and R. Riewe (eds.). Issues in the North, Vol. III. Occasional Publication 44, pp. 155–165. Canadian Circumpolar Institute, University of Alberta.

Rhode, S.C., M. Pawlowski and R. Tollrian, 2001. The impact of ultraviolet radiation on the vertical distribution of zooplankton of the genus Daphnia. Nature, 412:69–72.

Riedlinger, D., 2001. Responding to climate change in northern communities: impacts and adaptations. Arctic, 54:96–98.

Riget, F., E. Jeppesen, F. Landkildehus, T.L. Lauridsen, P. Geertz-Hansen, K. Christoffersen and H. Sparholt, 2000. Landlocked arctic charr (*Salvelinus alpinus*) population structure and lake morphometry in Greenland – is there a connection? Polar Biology, 23:550–558.

Rigler, F.H., 1978. Limnology in the high Arctic: a case study of Char Lake. Verheissungen der Internationale Vereinigung der gesamten Limnologie, 20:127–140.

Robinson, S.D. and T.R. Moore, 2000. The influence of permafrost and fire upon carbon accumulation in high boreal peatlands, Northwest Territories, Canada. Arctic, Antarctic, and Alpine Research, 32:155–166.

Rosén, P., U. Segerström, L. Eriksson, I. Renberg and H.J.B. Birks, 2001. Holocene climatic change reconstructed from diatoms, chironomids, pollen and near-infrared spectroscopy at an alpine lake (Sjuodjijaure) in northern Sweden. The Holocene, 11:551–562.

Rouse, W.R., M.S.V. Douglas, R.E. Hecky, A.E. Hershey, G.W. Kling, L. Lesack, P. Marsh, M. McDonald, B.J. Nicholson, N.T. Roulet and J.P. Smol, 1997. Effects of climate change on the freshwaters of Arctic and subarctic North America. Hydrological Processes, 11:873–902.

Rublee, P., 1992. Community structure and bottom-up regulation of heterotrophic microplankton in arctic LTER lakes. Hydrobiologia, 240:133–142.

Rühland, K.M. and J.P. Smol, 1998. Limnological characteristics of 70 lakes spanning arctic treeline from Coronation Gulf to Great Slave Lake in the central Northwest Territories, Canada. International Review of Hydrobiology, 83:183–203.

Rühland, K.M. and J.P. Smol, 2002. Freshwater diatoms from the Canadian arctic treeline and development of paleolimnological inference models. Journal of Phycology, 38:249–264.

Rühland, K.M., A. Priesnitz and J.P. Smol, 2003. Paleolimnological evidence from diatoms for recent environmental changes in 50 lakes across Canadian Arctic treeline. Arctic, Antarctic and Alpine Research, 35:110–123.

Rutherford, S., S. D'Hondt and W. Prell, 1999. Environmental controls on the geographic distribution of zooplankton diversity. Nature, 400:749–753.

Rydén, B.E., 1981. Hydrology of northern tundra. In: L.C. Bliss, O.W. Heal and J.J. Moore (eds.). Tundra Ecosystems: A Comparative Analysis, pp. 115–137. Cambridge University Press.

Ryves, D.B., S. McGowan and N.J. Anderson, 2002. Development and evaluation of a diatom-conductivity model from lakes in West Greenland. Freshwater Biology, 47:995–1014.

Sandlund, O.T., K. Gunnarsson, P.M. Jónasson, B. Jónsson, T. Lindem, K.P. Magnússon, H.J. Malmquist, H. Sigurjónsdóttir, S. Skúlason and S.S. Snorrason, 1992. The Arctic charr *Salvelinus alpinus* in Thingvallavatn. Oikos, 64:305–351.

Sandstrom, S.J., 1995. The effect of overwintering site temperature on energy allocation and life history characteristics of anadromous female Dolly Varden char (*Salvelinus malma*), from northwestern Canada. M.Sc. Thesis, University of Manitoba, 161pp.

Sauer, P.E., G.H. Miller and J.T. Overpeck, 2001. Oxygen isotope ratios of organic matter in arctic lakes as a paleoclimate proxy: field and laboratory investigations. Journal of Paleolimnology, 25:43–64.

Saulnier-Talbot, E. and R. Pienitz, 2001. Isolation au post-glaciaire d'un bassin côtier près de Kuujjuaraapik-Whapmagoostui, en Hudsonie (Québec): une analyse biostratigraphique diatomifère. Géographie physique et Quaternaire, 55:63–74.

Saulnier-Talbot, É., R. Pienitz and W.F. Vincent, 2003. Holocene lake succession and palaeo-optics of a subarctic lake, northern Québec, Canada. The Holocene, 13:517–526.

Scarnecchia, D.L., 1984. Climatic and oceanic variations affecting yield of Icelandic stocks of Atlantic salmon (*Salmo salar*). Canadian Journal of Fisheries and Aquatic Sciences, 41:917–935.

Scheffer, M., S.R. Carpenter, J.A. Foley, C. Folke and B. Walker, 2001. Catastrophic shifts in ecosystems. Nature, 413:591–596.

Schertzer, W.M. and A.M. Sawchuk, 1990. Thermal structure of the lower Great Lakes in a warm year: implications for the occurrence of hypolimnion anoxia. Transactions of the American Fisheries Society, 119:195–209.

Schindler, D.E., S.R. Carpenter, J.J. Cole, J.F. Kitchell and M.L. Pace, 1997. Influence of food web structure on carbon exchange between lakes and the atmosphere. Science, 277:248–251.

Schindler, D.W., 1997. Widespread effects of climate warming on freshwater ecosystems in North America. Hydrological Processes, 11:1043–1067.

Schindler, D.W., 2001. The cumulative effects of climate warming and other human stresses on Canadian freshwaters in the new millennium. Canadian Journal of Fisheries and Aquatic Sciences, 58:18–29.

Schindler, D.W., K.G. Beaty, E.J. Fee, D.R. Cruikshank, E.R. DeBruyn, D.L. Findlay, G.A. Linsey, J.A. Shearer, M.P. Stainton and M.A. Turner, 1990. Effects of climate warming on lakes of the central boreal forest. Science, 250:967–970.

Schindler, D.W., S.E. Bayley, B.R. Parker, K.G. Beaty, D.R. Cruikshank, E.J. Fee, E.U. Schindler and M.P. Stainton, 1996a. The effects of climate warming on the properties of boreal lakes and streams at the Experimental Lakes Area, northwestern Ontario. Limnology and Oceanography, 41:1004–1017.

Schindler, D.W., P.J. Curtis, B.R. Parker and M.P. Stainton, 1996b. Consequences of climate warming and lake acidification for UV-B penetration in North American boreal lakes. Nature, 379:705–708.

Schindler, D.W., P.J. Curtis, S.E. Bayley, B.R. Parker, K.G. Beaty and M.P. Stainton, 1997. Climate-induced changes in the dissolved organic carbon budgets of boreal lakes. Biogeochemistry, 36:9–28.

Schlesinger, D.A. and H.A. Regier, 1983. Relationship between environmental temperature and yields of subarctic and temperate zone fish species. Canadian Journal of Fisheries and Aquatic Sciences, 40:1829–1837.

Scholander, P.F., W. Flagg, R.J. Hock and L. Irving, 1953. Studies on the physiology of frozen plants and animals in the Arctic. Journal of Cellular and Comparative Physiology, 42:1–56.

Schreier, H., W. Erlebach and L. Albright, 1980. Variations in water quality during winter in two Yukon rivers with emphasis on dissolved oxygen concentration. Water Research, 14:1345–1351.

Schroeder, W.H., K.G. Anlauf, L.A. Barrie, J.Y. Lu, A. Steffen, D.R. Schneeberger and T. Berg, 1998. Arctic springtime depletion of mercury. Nature, 394:331–332.

Scott, J.D., L. Chalker-Scott, A.E. Foreman and M. D'Angelo, 1999. *Daphnia pulex* fed UVB-irradiated *Chlamydomonas reinhardtii* show decreased survival and fecundity. Photochemistry and Photobiology, 70:308–313.

Scott, K.J., 2001. Bioavailable mercury in Arctic snow determined by a light-emitting mer-lux bioreporter. Arctic, 54:92–95.

Scringeour, G.J., T.D. Prowse, J.M. Culp and P.A. Chambers, 1994. Ecological effects of river ice break-up: a review and perspective. Freshwater Biology, 32:261–275.

Scully, N.M., W.F. Vincent, D.R.S. Lean and W.J. Cooper, 1997. Implications of ozone depletion for surface-water photochemistry: sensitivity of clear lakes. Aquatic Sciences, 59:260–274.

Semiletov, I.P., 2001. Atmospheric methane and carbon dioxide in the arctic. In: I.P. Semiletov (ed.). Proceedings of the Arctic Regional Centre/V.I. Il'ichev Pacific Oceanological Institute, Volume 3. Hydrochemistry and Greenhouse Gas, pp. 127–164. Dalnauka, Vladivostok.

Seppä, H. and H.J.B. Birks, 2002. Holocene climate reconstructions from the Fennoscandian tree-line area based on pollen data from Toskaljavri. Quaternary Research, 57:191–199.

Seppä, H. and D. Hammarlund, 2000. Pollen-stratigraphical evidence of Holocene hydrological change in northern Fennoscandia supported by independent isotopic data. Journal of Paleolimnology, 24:69–79.

Seppä, H. and J. Weckström, 1999. Holocene vegetational and limnological changes in the Fennoscandian tree-line area as documented by pollen and diatom records from Lake Tsuolbmajavri, Finland. Ecoscience, 6:621–635.

Seppä, H., M. Nyman, A. Korhola and J. Weckström, 2002. Changes of treelines and alpine vegetation in relation to post-glacial climate dynamics in northern Fennoscandia based on pollen and chironomid records. Journal of Quaternary Science, 17:287–301.

Serreze, M.C., J.E. Walsh, F.S. Chapin III, T. Osterkamp, M. Dyurgerov, V. Romanovsky, W.C. Oechel, J. Morison, T. Zhang and R.G. Barry, 2000. Observational evidence of recent change in the northern high-latitude environment. Climatic Change, 46:159–207.

Shemesh, A., G. Rosqvist, M. Rietti-Shati, L. Rubensdotter, C. Bigler, R. Yam and W. Karlén, 2001. Holocene climatic change in Swedish Lapland inferred from an oxygen-isotope record of lacustrine biogenic silica. The Holocene, 11:447–454.

Shepson, P., P. Matrai, L. Barrie and J. Bottenheim, 2003. Ocean-atmosphere-sea ice-snowpack interactions in the Arctic, and global change. Eos, Transactions of the American Geophysical Union, 84:349–355.

Sherwood, G.D., J. Kovecses, A. Iles, J. Rasmussen, A. Gravel, H. Levesque, A. Hontela, A. Giguère, L. Kraemer and P. Campbell, 2001. 'The bigger the bait...' Metals in the Environment Research Network News, Winter 2001, p. 4, University of Guelph, Ontario.

Shiklomanov, I.A., A.I. Shiklomanov, R.B. Lammers, B.J. Peterson and C.J. Vorosmarty, 2000. The dynamics of river water inflow to the Arctic Ocean. In: E.L. Lewis, E.P. Jones, P. Lemke, T.D. Prowse and P. Wadhams (eds.). The Freshwater Budget of the Arctic Ocean, pp. 281–296. Kluwer Academic Publishers.

Shindell, D.T., D. Rind and P. Lonergan, 1998. Increased polar stratospheric ozone losses and delayed eventual recovery owing to increasing greenhouse-gas concentrations. Nature, 392:589–592.

Short, S.K., W.N. Mode and T.P. Davis, 1985. The Holocene record from Baffin Island; modern and fossil pollen studies. In: J.T. Andrews (ed.). Quaternary Environments: Eastern Canadian Arctic, Baffin Bay and West Greenland, pp. 608–642. Allen & Unwin.

Shortreed, K.S. and J.G. Stockner, 1986. Trophic status of 19 subarctic lakes in the Yukon Territory. Canadian Journal of Fisheries and Aquatic Sciences, 43:797–805.

Shuter, B.J. and J.R. Post, 1990. Climate, population viability, and the zoogeography of temperate fishes. Transactions of the American Fisheries Society, 119:314–336.

Shuter, B.J., J.A. MacLean, F.E.J. Fry and H.A. Regier, 1980. Stochastic simulation of temperature effects on first year survival of smallmouth bass. Transactions of the American Fisheries Society, 109:1–34.

Sibley, T.H. and R.M. Strickland, 1985. Fisheries: some relationships to climate change and marine environmental factors. In: M.R. White (ed.). Characterization of Information Requirements for Studies of CO_2 Effects: Water, Resources, Agriculture, Fisheries, Forests and Human Health, pp. 95–143. DOE/ER-0236. United States Department of Energy, Washington, D.C.

Siebeck, O. and U. Böhm, 1994. Challenges for an appraisal of UV–B effects upon planktonic crustaceans under natural radiation conditions with a non-migrating (*Daphnia pulex obtusa*) and a migrating cladoceran (*Daphnia galeata*). In: C.E. Williamson and H.E. Zagarese (eds.). Impact of UV–B Radiation on Pelagic Freshwater Ecosystems. Archiv für Hydrobiologie - Advances in Limnology, 43:197–206.

Simonich, S.L. and R.A. Hites, 1994. Importance of vegetation in removing polycyclic aromatic hydrocarbons from the atmosphere. Nature, 370:49–51.

Smith, J.B., 1991. Potential impacts of climate change on the Great Lakes. Bulletin of the American Meteorological Society, 72:21–28.

Smith, R.E.H., J.A. Furgal and D.R.S. Lean, 1998. The short-term effects of solar ultraviolet radiation on phytoplankton photosynthesis and photosynthate allocation under contrasting mixing regimes in Lake Ontario. Journal of Great Lakes Research, 24:427–441.

Smol, J.P., 2002. Pollution of Lakes and Rivers: A Paleoenvironmental Perspective. Arnold Publishers, 280pp.

Smol, J.P. and B.F. Cumming, 2000. Tracking long-term changes in climate using algal indicators in lake sediments. Journal of Phycology, 36:986–1011.

Snyder, J.A., G.M. Macdonald, S.L. Forman, G.A. Tarasov and W.N. Mode, 2000. Postglacial climate and vegetation history, north-central Kola Peninsula, Russia: pollen and diatom records from Lake Yarnyshnoe-3. Boreas, 29:261–271.

Solovieva, N. and V.J. Jones, 2002. A multiproxy record of Holocene environmental changes in the central Kola Peninsula, northwest Russia. Journal of Quaternary Science, 17:303–318.

Sommaruga, R. and F. Garcia-Pichel, 1999. UV-absorbing mycosporine-like compounds in planktonic and benthic organisms from a high-mountain lake. Archiv für Hydrobiologie, 144:255–269.

Sorvari, S. and A. Korhola, 1998. Recent diatom assemblage changes in subarctic Lake Saanajärvi, NW Finnish Lapland, and their paleoenvironmental implications. Journal of Paleolimnology, 20:205–215.

Sorvari, S., M. Rautio and A. Korhola, 2000. Seasonal dynamics of subarctic Lake Saanajärvi in Finnish Lapland. In: W.D. Williams (ed.). Internationale Vereinigung für Theoretische und Angewandte Limnologie: Verhandlungen, 27th Congress, Dublin, 1998, 27:507–512.

Sorvari, S., A. Korhola and R. Thompson, 2002. Lake diatom response to recent Arctic warming in Finnish Lapland. Global Change Biology, 8:153–163.

Spear, R.W., 1993. The palynological record of late-Quaternary Arctic tree-line in northwest Canada. Review of Palaeobotany and Palynology, 79:99–111.

Staehelin, J., N.R.P. Harris, C. Appenzeller and J. Eberhard, 2001. Ozone trends: a review. Reviews of Geophysics, 39:231–290.

Stanley, D.W., 1976. Productivity of epipelic algae in tundra ponds and a lake near Barrow, Alaska. Ecology, 57:1015–1024.

Stanley, J.B., P.F. Schuster, M.M. Reddy, D.A. Roth, H.E. Taylor and G.R. Aiken, 2002. Mercury on the move during snowmelt in Vermont. Eos, Transactions of the American Geophysical Union, 83:45–48.

Steffen, A., 1971. Chiromonid (Diptera) biocoenoses in Scandinavian glacier brooks. Canadian Entomologist, 103:477–486.

Steffen, A., W.H. Schroeder, L. Poissant and R. Macdonald, 2003. Mercury in the Arctic atmosphere. In: T.F. Bidleman, R. Macdonald and J. Stow (eds.). Canadian Arctic Contaminants Assessment Report II: Sources, Occurrence, Trends and Pathways in the Physical Environment, pp. 120–138. Indian and Northern Affairs Canada, Ottawa.

Stern, G. and M. Evans, 2003. Persistent organic pollutants in marine and lake sediments. In: T.F. Bidleman, R. Macdonald and D. Stow (eds.). Canadian Arctic Contaminants Assessment Report II: Sources, Occurrence, Trends and Pathways in the Physical Environment, pp. 96–111. Indian and Northern Affairs Canada, Ottawa.

Stross, R.G., M.C. Miller and R.J. Daley, 1980. Zooplankton. In: J.E. Hobbie (ed.). Limnology of Tundra Ponds: Barrow, Alaska, pp. 251–296. Dowden, Hutchinson and Ross.

Styczynski, B. and S. Rakusa-Suszczczewski, 1963. Tendipedidae of selected water habitats of Hornsrund region (Spitsbergen). Polish Archives of Hydrobiology, 11:327–341.

Suchanek, T.H., P.J. Richerson, J.R. Flanders, D.C. Nelson, L.H. Mullen, L.L. Brester and J.C. Becker, 2000. Monitoring inter-annual variability reveals sources of mercury contamination in Clear Lake, California. Environmental Monitoring and Assessment, 64:299–310.

Svenning, M.-A. and N. Gullestad, 2002. Adaptations to stochastic environmental variations: the effects of seasonal temperatures on the migratory window of Svalbard Arctic charr. Environmental Biology of Fishes, 64:165–174.

Svensson, B.H., T.R. Christensen, E. Johansson and M. Öquist, 1999. Interdecadal changes in CO_2 and CH_4 fluxes of a subarctic mire: Stordalen revisited after 20 years. Oikos, 85:22–30.

Tang, E.P.Y. and W.F. Vincent, 1999. Strategies of thermal adaptation by high-latitude cyanobacteria. New Phytologist, 142:315–323.

Tang, E.P.Y., R. Tremblay and W.F. Vincent, 1997. Cyanobacterial dominance of polar freshwater ecosystems: Are high latitude mat-formers adapted to low temperature? Journal of Phycology, 33:171–181.

Tartarotti, B., I. Laurion and R. Sommaruga, 2001. Large variability in the concentration of mycosporine-like amino acids among zooplankton from lakes located across an altitude gradient. Limnology and Oceanography, 46:1546-1552.

Thorpe, W., 1986. A Review of the Literature and Miscellaneous Other Parameters Relating to Water Levels in the Peace-Athabasca Delta Particularly with Respect to the Effect on Muskrat Numbers. Environment Canada.

Tonn, W.M., 1990. Climate change and fish communities: a conceptual framework. Transactions of the American Fisheries Society, 119:337–352.

Tuomenvirta, H. and R. Heino, 1996. Climatic changes in Finland – recent findings. Geophysica, 32:61–75.

Turetsky, M.R., R.K. Wieder, C.J. Williams and D.H. Vitt, 2000. Organic matter accumulation, peat chemistry, and permafrost melting in peatlands of boreal Alberta. Ecoscience, 7:379–392.

Turetsky, M.R., R.K. Wieder and D.H. Vitt, 2002. Boreal peatland C fluxes under varying permafrost regimes. Soil Biology and Biochemistry, 34:907–912.

UNEP, 2003. Review of the interlinkages between biological diversity and climate change, and advice on the integration of biodiversity considerations into the implementation of the United Nations Framework Convention on Climate Change and its Kyoto Protocol. UNEP/CBD/SBSTTA/9/11.

Vadeboncoeur, Y., E. Jeppesen, M.J. Vander Zanden, H.-H. Schierup, K. Christoffersen and D.M. Lodge, 2003. From Greenland to green lakes: cultural eutrophication and the loss of benthic pathways in lakes. Limnology and Oceanography, 48:1408–1418.

Van Donk, E. and D.O. Hessen, 1995. Reduced digestibility of UV-B stressed and nutrient-limited algae by *Daphnia magna*. Hydrobiologia, 307:147–151.

Van Donk, E., B.A. Faafeng, H.J. De Lange and D.O. Hessen, 2001. Differential sensitivity to natural ultraviolet radiation among phytoplankton species in Arctic lakes (Spitsbergen, Norway). Plant Ecology, 154:247–259.

van Everdingen, R.O., 1990. Groundwater hydrology. In: T.D. Prowse and C.S.L. Ommanney (eds.). Northern Hydrology: Canadian Perspectives. National Hydrology Research Institute, Saskatoon, Science Report No.1, pp. 77–101.

Vardy, S.R., B.G. Warner and R. Aravena, 1997. Holocene climate effects on the development of a peatland on the Tuktoyaktuk Peninsula, Northwest Territories. Quaternary Research, 47:90–104.

Vincent, W.F., 2000. Cyanobacterial dominance in the polar regions. In: B.A. Whitton and M. Potts (eds.). The Ecology of Cyanobacteria: Their Diversity in Time and Space, pp. 321–340. Kluwer Academic Publishers.

Vincent, W.F. and C. Belzile, 2003. Biological UV exposure in the polar oceans: Arctic-Antarctic comparisons. In: A.H.L. Huiskes, W.W.C. Gieskes, J. Rozema, R.M.L. Schorno, S.M. van der Vies and W.J. Wolff (eds.). Antarctic Biology in a Global Context. Proceedings of the VIIIth SCAR International Biology Symposium, 27 August –1 September 2001, pp. 176–181.

Vincent, W.F. and J.E. Hobbie, 2000. Ecology of Arctic lakes and rivers. In: M. Nuttall and T.V. Callaghan (eds.). The Arctic: Environment, People, Policy, pp. 197–231. Harwood Academic Press.

Vincent, W.F. and P.J. Neale, 2000. Mechanisms of UV damage to aquatic organisms. In: S.J. de Mora, S. Demers and M. Vernet (eds.). The Effects of UV Radiation in the Marine Environment, pp. 149–176. Cambridge University Press.

Vincent, W.F. and R. Pienitz, 1996. Sensitivity of high latitude freshwater ecosystems to global change: temperature and solar ultraviolet radiation. Geoscience Canada, 23:231–236.

Vincent, W.F. and A. Quesada, 1994. Cyanobacterial responses to UV radiation: implications for Antarctic microbial ecosystems. In: C.S. Weiler and P.A. Penhale (eds.). Ultraviolet Radiation in Antarctica: Measurement and Biological Effects. American Geophysical Union, Antarctic Research Series, Vol. 62, pp. 111–124.

Vincent, W.F. and S. Roy, 1993. Solar ultraviolet-B radiation and aquatic primary production: damage, protection, and recovery. Environmental Reviews, 1:1–12.

Vincent, W.F., J.A.E. Gibson, R. Pienitz, V. Villeneuve, P.A. Broady, P.B. Hamilton and C. Howard-Williams, 2000. Ice shelf microbial ecosystems in the high Arctic and implications for life on Snowball Earth. Naturwissenschaften, 87:137–141.

Vinebrooke, R.D. and P.R. Leavitt, 1999a. Differential responses of littoral communities to ultraviolet radiation in an alpine lake. Ecology, 80:223–237.

Vinebrooke, R.D. and P.R. Leavitt, 1999b. Phytobenthos and phytoplankton as potential indicators of climate change in mountain lakes and ponds: a HPLC-based pigment approach. Journal of the North American Benthological Society, 18:14–33.

Vitt, D.H., L.A. Halsey and S.C. Zoltai, 2000. The changing landscape of Canada's western boreal forest: the current dynamics of permafrost. Canadian Journal of Forest Research, 30:283–287.

Volpe, J.P., E.B. Taylor, D.W. Rimmer and B.W. Glickman, 2000. Evidence of natural reproduction of aquaculture-escaped Atlantic salmon in a coastal British Columbia river. Conservation Biology, 14:899–903.

Vörösmarty, C.J., L.D. Hinzman, B.J. Peterson, D.H. Bromwich, L.C. Hamilton, J. Morison, V.E. Romanovsky, M. Sturm and R.S. Webb, 2001. The Hydrologic Cycle and its Role in Arctic and Global Environmental Change: A Rationale and Strategy for Synthesis Study. Arctic Research Consortium of the U.S., Fairbanks, Alaska, 84pp.

Walker, H.J. and J.M. McCloy, 1969. Morphologic Change in Two Arctic Deltas: Blow River Delta, Yukon Territories and Colville River Delta, Alaska. Research Paper 49. Arctic Institute of North America, 91pp.

Wang, K.S. and T.-J. Chai, 1994. Reduction in omega-3 fatty acids by UV-B radiation in microalgae. Journal of Applied Phycology, 6:415–421.

Wania, F., 1997. Modelling the fate of non-polar organic chemicals in an ageing snow pack. Chemosphere, 35:2345–2363.

Wania, F., 1999. On the origin of elevated levels of persistent chemicals in the environment. Environmental Science and Pollution Research, 6:11–19.

Wania, F. and M.S. McLachlan, 2001. Estimating the influence of forests on the overall fate of semivolatile organic compounds using a multi-media fate model. Environmental Science and Technology, 35:582–590.

Wankowski, J.W.J. and J.E. Thorpe, 1979. The role of food particle size in the growth of juvenile Atlantic salmon (*Salmo salar* L.). Journal of Fish Biology, 14:351–370.

Watson, S.B., E. McCauley and J.A. Downing, 1997. Patterns in phytoplankton taxonomic composition across temperate lakes of differing nutrient status. Limnology and Oceanography, 42:487–495.

Weckström, J., J.A. Snyder, A. Korhola, T.E. Laing and G.M. MacDonald, 2003. Diatom inferred acidity history of 32 lakes on the Kola Peninsula, Russia. Water, Air and Soil Pollution, 149:339–361.

Welch, D.W., Y. Ishida and K. Nagasawa, 1998. Thermal limits and ocean migrations of sockeye salmon (*Onchorynchus nerka*): Long-term consequences of global warming. Canadian Journal of Fisheries and Aquatic Sciences, 55:937–948.

Welch, H.E., J.A. Legault and M.A. Bergmann, 1987. Effects of snow and ice on the annual cycles of heat and light in Saqvaqjuac lakes. Canadian Journal of Fisheries and Aquatic Sciences, 44:1451–1461.

Welch, H.E., D.C.G. Muir, B.N. Billeck, W.L. Lockhart, G.J. Brunskill, H.J. Kling, M.P. Olson and R.M. Lemoine, 1991. Brown snow: a long-range transport event in the Canadian Arctic. Environmental Science and Technology, 25:280–286.

Weller, G., 1998. Regional impacts of climate change in the Arctic and Antarctic. Annals of Glaciology, 27:543–552.

Wetzel, R.G., 2001. Limnology: Lake and River Ecosystems. Third Edition. Academic Press, 1006pp.

Wetzel, R.G., P.G. Hatcher and T.S. Bianchi, 1995. Natural photolysis by ultraviolet irradiance of recalcitrant dissolved organic matter to simple substrates for rapid bacterial metabolism. Limnology and Oceanography, 40:1369–1380.

Whitfield, P.H. and B. McNaughton, 1986. Dissolved-oxygen depressions under ice cover in two Yukon rivers. Water Resources Research, 22:1675–1679.

Whittle, D.M., R.M. Kiriluk, A.A. Carswell, M.J. Keir and D.C. MacEachen, 2000. Toxaphene congeners in the Canadian Great Lakes basin: temporal and spatial food web dynamics. Chemosphere, 40:1221–1226.

Wickham, S. and M. Carstens, 1998. Effects of ultraviolet-B radiation on two Arctic microbial food webs. Aquatic Microbial Ecology, 16:163–171.

Williams, P.J. and W.G. Rees, 2001. Proceedings, Second International Conference on Contaminants in Freezing Ground, Cambridge, 2–5 July 2000. Part 2. Cold Regions Science and Technology, 32(2–3):85–203.

Williamson, C.E., H. Zagarese, P.C. Schulze, B.R. Hargreaves and J. Seva, 1994. The impact of short-term exposure to UV-B radiation on zooplankton communities in north temperate lakes. Journal of Plankton Research, 16:205–218.

Williamson, C.E., O.G. Olson, S.E. Lott, N.D. Walker, D.R. Engstrom and B.R. Hargreaves, 2001. Ultraviolet radiation and zooplankton community structure following deglaciation in Glacier Bay, Alaska. Ecology, 82:1748–1760.

Wilson, C.C., P.D.N. Hebert, J.D. Reist and J.B. Dempson, 1996. Phylogeography and postglacial dispersion of arctic charr (*Salvelinus alpinus* L.) in North America. Molecular Ecology, 5:187–198.

Wohlfarth, B., L. Filimonova, O. Bennike, L. Björkman, L. Brunnberg, N. Lavrova, I. Demidov and G. Possnert, 2002. Late-glacial and early Holocene environmental and climatic change at Lake Tambichozero, southeastern Russian Karelia. Quaternary Research, 58:261–272.

Wolfe, A.P. and B.B. Perren, 2001. Chrysophyte microfossils record marked responses to recent environmental changes in high- and mid-Arctic lakes. Canadian Journal of Botany, 79:747–752.

Wolfe, B.B., T.W.D. Edwards, R. Aravena, S.L. Forman, B.G. Warner, A.A. Velichko and G.M. MacDonald, 2000. Holocene paleohydrology and paleoclimate at treeline, North-Central Russia, inferred from oxygen isotope records in lake sediment cellulose. Quaternary Research, 53:319–329.

Wolfe, B.B., T.W.D. Edwards, K.R.M. Beuning and R.J. Elgood, 2002. Carbon and oxygen isotope analysis of lake sediment cellulose: methods and applications. In: W.M. Last and J.P. Smol (eds.). Tracking Environmental Change Using Lake Sediments, Vol. 2: Physical and Geochemical Methods, pp. 373–400. Kluwer Academic Publishers.

Woo, M.-K., 1980. Hydrology of a small lake in the Canadian High Arctic. Arctic and Alpine Research, 12:227–235.

Woo, M.-K., 1996. Hydrology of northern North America under global warming. In: J.A.A. Jones (ed.). Regional Hydrological Responses to Climate Change, pp. 73–86. Kluwer Academic Publishers.

Woo, M.-K., 2000. Permafrost and hydrology. In: M. Nuttall and T.V. Callaghan (eds.). The Arctic: Environment, People, Policy, pp. 57–96. Harwood Academic Press.

Woo, M.-K. and Z.J. Xia, 1995. Suprapermafrost groundwater seepage in gravelly terrain, Resolute, NWT, Canada. Permafrost and Periglacial Processes, 6:57–72.

Woo, M.-K., R. Heron and P. Steer, 1981. Catchment hydrology of a High Arctic lake. Cold Regions Science and Technology, 5:29–41.

Woo, M.-K., A.G. Lewkowicz and W.R. Rouse, 1992. Response of the Canadian permafrost environment to climatic change. Physical Geography, 134:287–317.

Wootton, R., 1990. Ecology of Teleost Fishes. Chapman and Hall, 404pp.

Xenopoulos, M.A., Y.T. Prairie and D.F. Bird, 2000. Influence of ultraviolet-B radiation, stratospheric ozone variability, and thermal stratification on the phytoplankton biomass dynamics in a mesohumic lake. Canadian Journal of Fisheries and Aquatic Sciences, 57:600–609.

Yunker, M.B., R.W. Macdonald, W.J. Cretney, B.R. Fowler and F.A. McLaughlin, 1993. Alkane, terpene, and polycyclic aromatic hydrocarbon geochemistry of the Mackenzie River and Mackenzie Shelf: riverine contributions to Beaufort Sea coastal sediment. Geochimica et Cosmochimica Acta, 57:3041–3061.

Yunker, M.B., S.M. Backus, E. Graf Pannatier, D.S. Jeffries and R.W. Macdonald, 2002. Sources and significance of alkane and PAH hydrocarbons in Canadian arctic rivers. Estuarine, Coastal and Shelf Science, 55:1–31.

Zellmer, I.D., 1998. The effects of solar UVA and UVB on subarctic *Daphnia pulicaria* in its natural habitat. Hydrobiologia, 379:55–62.

Zepp, R.G., T.V. Callaghan and D.J. Erickson III, 2003. Interactive effects of ozone depletion and climate change on biogeochemical cycles. In: J.F. Bornman, K. Solomon, and J.C. van der Leun (eds.). Environmental Effects of Ozone Depletion and its Interactions with Climate Change: 2002 Assessment. Photochemical and Photobiological Sciences, 2:51–61.

Zhadin, V.I. and S.V. Gerd, 1961. Fauna and Flora of the Rivers, Lakes and Reservoirs of the U.S.S.R. Moskva, 626pp. Smithsonian Institution and National Science Foundation, Washington, D.C., Technical Translation No. 63-1116. Translated by Israel Program of Scientific Translations, Jerusalem, Israel, 1963.

Zhang, X., A.S. Naidu, J.J. Kelley, S.C. Jewett, D. Dasher and L.K. Duffy, 2001. Baseline concentrations of total mercury and methylmercury in salmon returning via the Bering Sea (1999–2000). Marine Pollution Bulletin, 42:993–997.

Zhulidov, A.V., J.V. Headley, R.D. Robarts, A.M. Nikanorov and A.A. Ischenko, 1997. Atlas of Russian Wetlands: Biogeography and Metal Concentrations. National Hydrology Research Institute, Saskatoon, 309pp.

Zhulidov, A.V., J.V. Headley, D.F. Pavlov, R.D. Robarts, L.G. Korotova, V.V. Fadeev, O.V. Zhulidova, Y. Volovik and V. Khlobystov, 1998. Distribution of organochlorine insecticides in rivers of the Russian Federation. Journal of Environmental Quality, 27:1356–1366.

Zimov, S.A., Y.V. Voropaev, I.P. Semiletov, S.P. Davidov, S.F. Prosiannikov, F.S. Chapin III, M.C. Chapin, S. Trumbore and S. Tyler, 1997. North Siberian lakes: A methane source fueled by Pleistocene carbon. Science, 277:800–802.

Zimov, S.A., Y.V. Voropaev, S.P. Davidov, G.M. Zimova, A.I. Davidova, F.S. Chapin III and M.C. Chapin, 2001. Flux of methane from north Siberian aquatic systems: influence on atmospheric methane. In: R. Paepe, V. Melnikov, E. Van Overloop and V.D. Gorokhov (eds.). Permafrost Response on Economic Development, Environmental Security and Natural Resources, pp. 511–524. Kluwer Academic Publishers.

Zöckler, C., 1998. Patterns in Biodiversity in Arctic Birds. World Conservation Monitoring Centre, Cambridge, Biodiversity Bulletin 3, 15pp.

Zudaire, L. and S. Roy, 2001. Photoprotection and long-term acclimation to UV radiation in the marine diatom *Thalassiosira weissflogi*. Journal of Photochemistry and Photobiology B, 62:26–34.

Appendix. Scientific names of arctic fishes alphabetically listed by common name used in the text and boxes

(see Fishbase, http://www.fishbase.org/home.htm for further details on species). See also Box 8.6.

Common Name	Scientific Name or Group	Family
Alewife[c]	*Alosa* spp.	Alosidae
Arctic char[c,b]	*Salvelinus alpinus*	Salmonidae
Arctic cisco[c]	*Coregonus autumnalis*	Salmonidae
Arctic cod[a]	*Boreogadus saida*	Gadidae
Arctic grayling[b]	*Thymallus arcticus*	Salmonidae
Arctic flounder[a]	*Pleuronectes glacialis*	Pleuronectidae
Atlantic salmon[c]	*Salmo salar*	Salmonidae
blackfishes[b]	*Dallia* spp.	Umbridae
bluegill[b]	*Lepomis macrochirus*	Centrarchidae
broad whitefish[c]	*Coregonus nasus*	Salmonidae
brook trout[c]	*Salvelinus fontinalis*	Salmonidae
brown trout[b,c]	*Salmo trutta*	Salmonidae
bull trout[b]	*Salvelinus confluentus*	Salmonidae
burbot[b]	*Lota lota*	Gadidae
carp bream[b]	*Abramis brama*	Cyprinidae
chars	*Salvelinus* spp.	Salmonidae
cisco	*Coregonus (Leucichthys)* subgenus	Salmonidae
common dace[b]	*Leuciscus leuciscus baicalensis*	Cyprinidae
coregonines	whitefishes and ciscoes, Coregoninae	Salmonidae
cyprinids	minnows	Cyprinidae
Dolly Varden[c]	*Salvelinus malma*	Salmonidae
emerald shiner[b]	*Notropis atherinoides*	Cyprinidae
European eel[d]	*Anguilla anguilla*	Anguillidae
European perch[b]	*Perca fluviatilis*	Percidae
European whitefish[b,c]	*Coregonus lavaretus*	Salmonidae
fourhorn sculpin[a]	*Myoxocephalus quadricornis*	Cottidae
goldeye[b]	*Hiodon alosoides*	Hiodontidae
ide[b]	*Leuciscus idus*	Cyprinidae
lake chub[b]	*Couesius plumbeus*	Cyprinidae
lake cisco[b]	*Coregonus artedi*	Coregonidae
lake trout[b]	*Salvelinus namaycush*	Salmonidae
lake whitefish[b,c]	*Coregonus clupeaformis*	Salmonidae

Common Name	Scientific Name or Group	Family
lampreys[c,b]	*Lampetra, Petromyzon* spp.	Petromyzontidae
least cisco[c]	*Coregonus sardinella*	Salmonidae
ninespine stickleback[b,c]	*Pungitius pungitius*	Gasterosteidae
northern pike[b]	*Esox lucius*	Esocidae
northern redbelly dace[b]	*Chrosomus eos*	Cyprinidae
Pacific salmon[c]	*Oncorhynchus* spp.	Salmonidae
percids[b]	perch family	Percidae
pink salmon[c]	*Oncorhynchus gorbuscha*	Salmonidae
pond smelt[c,b]	*Hypomesus olidus*	Osmeridae
rainbow smelt[c]	*Osmerus mordax*	Osmeridae
rainbow trout[b]	*Oncorhynchus mykiss*	Salmonidae
roach[b]	*Rutilus rutilus lacustris*	Cyprinidae
round whitefish[b]	*Prosopium cylindraceum*	Salmonidae
ruffe[b]	*Gymnocephalus cernuus*	Percidae
salmon	*Oncorhynchus* spp.	Salmonidae
salmonines	trouts, chars, salmons	Salmonidae
sculpins	*Cottus* spp.	Cottidae
Siberian whitefish[c]	*Coregonus pidschian*	Salmonidae
slimy sculpin[b]	*Cottus cognatus*	Cottidae
smallmouth bass[b]	*Micropterus dolomieu*	Centrachidae
sockeye salmon[c]	*Oncorhynchus nerka*	Salmonidae
sticklebacks	*Pungitius, Gasterosteus* spp.	Gasterosteidae
stone loach[b]	*Noemacheilus barbatulus*	Cobitidae
suckers[b]	*Catostomus* spp.	Catostomidae
threespine stickleback[c,b]	*Gasterosteus aculeatus*	Gasterosteidae
trout perch[b]	*Percopsis omiscomaycus*	Percopsidae
vendace[c,b]	*Coregonus albula*	Salmonidae
walleye[b]	*Sander vitreus*	Percidae
whitefishes	*Coregonus (Coregonus)* subgenus	Salmonidae
yellow perch[b]	*Perca flavescens*	Percidae
zander[b]	*Sander lucioperca*	Percidae

Dominant life history exhibited in arctic areas; more than one note indicates multiple life history types with order indicating primary mode in the Arctic as defined in this assessment: [a]marine species found in nearshore, brackish, and estuarine areas; [b]freshwater species found in freshwater or freshened brackish water areas only; [c]primarily anadromous species (although freshwater forms are also present); [d]primarily catadromous species (although freshwater forms may be present).

Chapter 9

Marine Systems

Lead Author
Harald Loeng

Contributing Authors
Keith Brander, Eddy Carmack, Stanislav Denisenko, Ken Drinkwater, Bogi Hansen, Kit Kovacs, Pat Livingston, Fiona McLaughlin, Egil Sakshaug

Consulting Authors
Richard Bellerby, Howard Browman, Tore Furevik, Jacqueline M. Grebmeier, Eystein Jansen, Steingrimur Jónsson, Lis Lindal Jorgensen, Svend-Aage Malmberg, Svein Osterhus, Geir Ottersen, Koji Shimada

Contents

9.1. Introduction

Approximately two-thirds of the area addressed by the Arctic Climate Impact Assessment is ocean. This includes the Arctic Ocean and its adjacent shelf seas, as well as the Nordic Seas, the Labrador Sea, and the Bering Sea. These are very important areas from a climate change perspective since processes occurring in the Arctic affect the rate of deep-water formation in the convective regions of the North Atlantic, thereby influencing the global ocean circulation. Also, climate models consistently show the Arctic to be one of the most sensitive regions to climate change.

Many arctic life forms, including humans, are directly or indirectly dependent on productivity from the sea. Several physical factors combine to make arctic marine systems unique including: a very high proportion of continental shelves and shallow water; a dramatic seasonality and overall low level of sunlight; extremely low water temperatures; presence of extensive areas of multi-year and seasonal sea-ice cover; and a strong influence from freshwater, coming from rivers and ice melt. Such factors represent harsh conditions for many types of marine life. In geological terms, the arctic fauna is young; recent glaciations resulted in major losses in biodiversity, and recolonization has been slow owing to the extreme environmental conditions and low productivity of the arctic system. This has resulted in arctic ecosystems, in a global sense, being considered "simple". They largely comprise specialist species that have been able to adapt to the extreme conditions, and overall species diversity is low. The large seasonal pulse of summer production in the Arctic, which occurs during the period of 24 hours light, is particularly pronounced near the ice edge and in shallow seas such as the Barents and Bering Seas. This attracts seasonal migrants that travel long distances to take advantage of the arctic summers and then return south to overwinter.

This assessment has also considered the effects of changes in ultraviolet (UV) radiation. However, although UV-B radiation can result in negative impacts on marine organisms and populations, it is only one of many environmental factors that can result in the types of mortality typically observed. It is thus important to assess the relative importance, and hence potential impact, of ozone depletion-related increases in solar UV-B radiation on arctic marine ecosystems.

The Arctic Ocean has not been considered a significant sink for carbon. This is because its extensive sea-ice cover constrains atmosphere–ocean exchange, and because levels of biological production under multi-year sea ice were believed low. Under warmer climate conditions, however, the amount of carbon sequestered by the Arctic Ocean may increase significantly. In addition, the Arctic's role as a source of carbon (methane and carbon dioxide, CH_4 and CO_2 respectively) is poorly understood owing to frozen reserves in permafrost and gas hydrate layers.

This chapter addresses physical features and processes related to marine climate and their impact on the marine ecosystem. Climate change scenarios for the ocean are very uncertain as most models focus mainly on changes in the atmosphere. Such models are not definitive about changes to ocean circulation, deep-water formation, or the fate of major ocean fronts. Therefore, the conclusions drawn in this chapter regarding likely changes in the marine ecosystem are based on scenarios determined from the projected changes in the atmosphere coupled with the present understanding of how atmospheric forcing influences the ocean, as well as the output from a few ocean models.

9.2. Physical oceanography

Climate changes impact upon the marine ecosystem mainly through their effects on the physical oceanography. This section provides an overview of the physical oceanography of the Arctic sufficient to enable an examination of potential impacts on the biological system. It also addresses the feedback mechanisms between the atmosphere and the ocean through which changes in the oceanography of the Arctic could have global consequences for the atmosphere.

9.2.1. General features

The marine Arctic is defined within this assessment as comprising the Arctic Ocean, including the deep Eurasian and Canadian Basins and the surrounding continental shelf seas (Barents, White, Kara, Laptev, East Siberian, Chukchi, and Beaufort Seas), the Canadian Archipelago, and the transitional regions to the south through which exchanges between temperate and arctic waters occur. The latter includes the Bering Sea in the Pacific Ocean and large parts of the northern North Atlantic Ocean, including the Nordic, Iceland, and Labrador Seas, and Baffin Bay. Also included are the Canadian inland seas of Foxe Basin, Hudson Bay, and Hudson Strait. Those arctic areas that receive most of the heat input from inflowing warm Atlantic water, i.e., the eastern parts of the Nordic Seas and the Arctic Ocean, are collectively referred to as the Arctic Mediterranean. A detailed description of the topography, water properties, and circulation of these areas is given in Chapter 2. The present chapter presents a brief summary of some of the salient features.

Sea ice is one of the dominant physical features for most of these areas, with coverage ranging from year-round cover in the central Arctic Ocean to seasonal cover in most of the remaining areas. Exceptions occur over the deep basins, which are ice-free throughout the year, e.g., the Nordic Seas and the Labrador Sea, and the deep parts of the Bering Sea.

Relatively warm waters from the Atlantic flow through the Nordic Seas into the Arctic Ocean via the Barents Sea and through Fram Strait while the warm Pacific waters flow across the Bering Sea and enter the Arctic

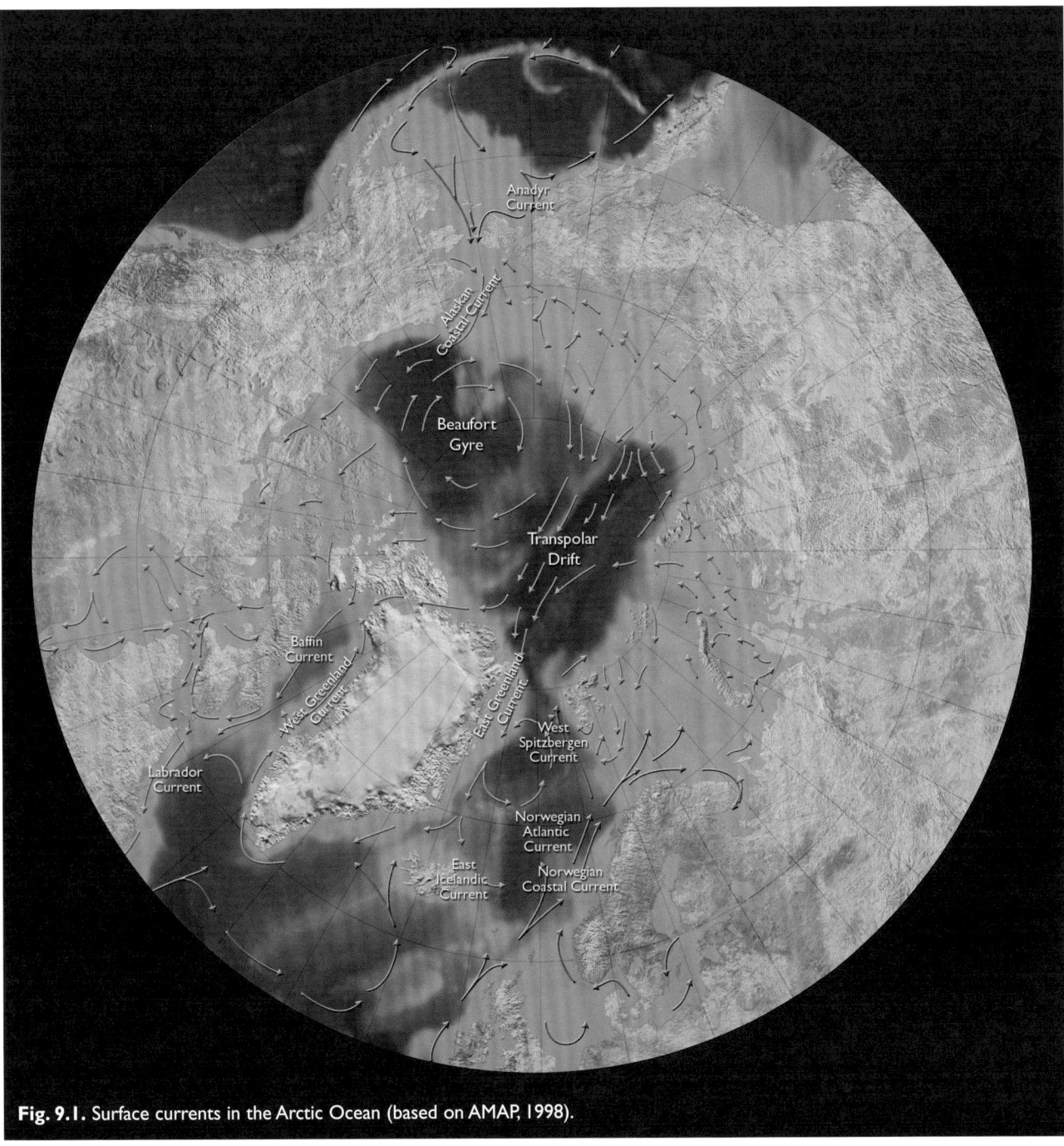

Fig. 9.1. Surface currents in the Arctic Ocean (based on AMAP, 1998).

through the Bering Strait (Fig. 9.1). Approximately ten to twenty times more Atlantic water than Pacific water by volume enters the Arctic Ocean. Within the Arctic Ocean the dominant features of the surface circulation are the clockwise Beaufort Gyre, extending over the Canadian Basin, and the Transpolar Drift that flows from the Siberian coast out through Fram Strait. Both features are strongly influenced by wind forcing. The surface currents along the coast are principally counterclockwise, moving from Atlantic to Pacific on the Eurasian side and from Pacific to Atlantic on the North American side. The subsurface circulation is also counterclockwise and influenced by the inflows from the Atlantic and Pacific Oceans. Waters exit the Arctic Ocean primarily through Fram Strait and the Canadian Archipelago. The arctic waters leaving through Fram Strait are then transported southward along East Greenland, and around the

Labrador Sea and Baffin Bay where they merge with the arctic waters flowing out through the Canadian Archipelago before continuing southward.

The temperature and salinity levels of the various water bodies in the marine Arctic vary considerably, reflecting the extent of the Pacific and Atlantic influence, heat exchange with the atmosphere, direct precipitation, freshwater runoff, and the melting and freezing of sea ice. In the Arctic Ocean, the surface waters are generally near the freezing point owing to the ice cover, whereas the salinity levels exhibit seasonal and spatial fluctuations caused by the freezing and melting of sea ice and river runoff. Density stratification within the Arctic Ocean is principally due to vertical salinity differences. The layer containing the greatest change in salinity is called the halocline. Its characteristics vary across the Arctic Ocean and are largely characterized by the presence or absence of Pacific-origin water. Waters below the halocline are modified Atlantic waters that flowed into the Arctic through Fram Strait and the Barents Sea. The Atlantic and Pacific inflows carry relatively warm and saline waters into the Arctic and their vertical density stratification is usually controlled more by temperature than salinity differences. As these inflows move northward they are cooled by the atmosphere and freshened by river runoff. Mixing with ambient waters also generally leads to cooling and freshening. The waters leaving the Arctic Ocean also mix with ambient waters, in this case becoming warmer and saltier.

9.2.2. Sea ice

Sea ice controls the exchange of heat and other properties between the atmosphere and ocean and, together with snow cover, determines the penetration of light into the sea. Sea ice also provides a surface for particle and snow deposition, a habitat for plankton, and contributes to stratification through ice melt. The zone seaward of the ice edge is important for plankton production and planktivorous fish. For some marine mammals sea ice provides a place for birth and also functions as a nursery area.

This section describes features of sea ice that are important for physical oceanographic processes and the marine ecosystem. More detailed information about sea ice is given in Chapter 6.

9.2.2.1. Seasonal cycle

Sea-ice extent in the Arctic has a clear seasonal cycle and is at its maximum (14–15 million km²) in March and minimum (6–7 million km²) in September (Parkinson et al., 1999). There is considerable interannual variability both in the maximum and minimum coverage. In addition, there are decadal and inter-decadal fluctuations in

the areal sea-ice extent due to changes in atmospheric pressure patterns and their associated winds, continental discharge, and influx of Atlantic and Pacific waters (Gloersen, 1995; Mysak and Manak, 1989; Polyakov et al., 2003; Rigor et al., 2002; Zakharov, 1994).

At the time of maximum advance, sea ice covers the entire Arctic Basin and the Siberian shelf seas (Fig. 9.2). The warm inflow of Atlantic water keeps the southern part of the Barents Sea open, but in cold years even its shallow areas in the southeast are covered by sea ice. Also, the west coast of Spitsbergen generally remains free of ice. It is here that open water is found closest to the Pole in winter, beyond 81° N in some years (Wadhams, 2000). Sea ice from the Arctic Ocean is transported out through Fram Strait and advected southward by the East Greenland Current to cover the entire east coast of Greenland, although in mild winters it does not reach the southern tip of Greenland. In cold years, the sea ice may also extend south to the northern and eastern coasts of Iceland. In most years there is a thin band of sea ice off West Greenland, which is a continuation of the sea ice from East Greenland and is known as "Storis". Only rarely does the Storis meet the dense sea-ice cover of Baffin Bay and Davis Strait to completely surround Greenland. The whole of the Canadian Archipelago, as well as Hudson Bay and Hudson Strait are usually ice-covered (Wadhams, 2000). The Labrador Shelf is also covered by sea ice and the Labrador Current transports this southward to Newfoundland. Further west, a complete sea-ice cover extends across the arctic coasts of northwestern Canada and Alaska and fills the Bering Sea as far south as the shelf break (Wadhams, 2000).

In March or April, the sea ice begins to retreat from its low latitude extremes. By May the coast off northeastern Newfoundland is clear, as is much of the Bering Sea. By June the area south of the Bering Strait is ice-free and open water is found in Hudson Bay and at several arctic coastal locations. August and September are

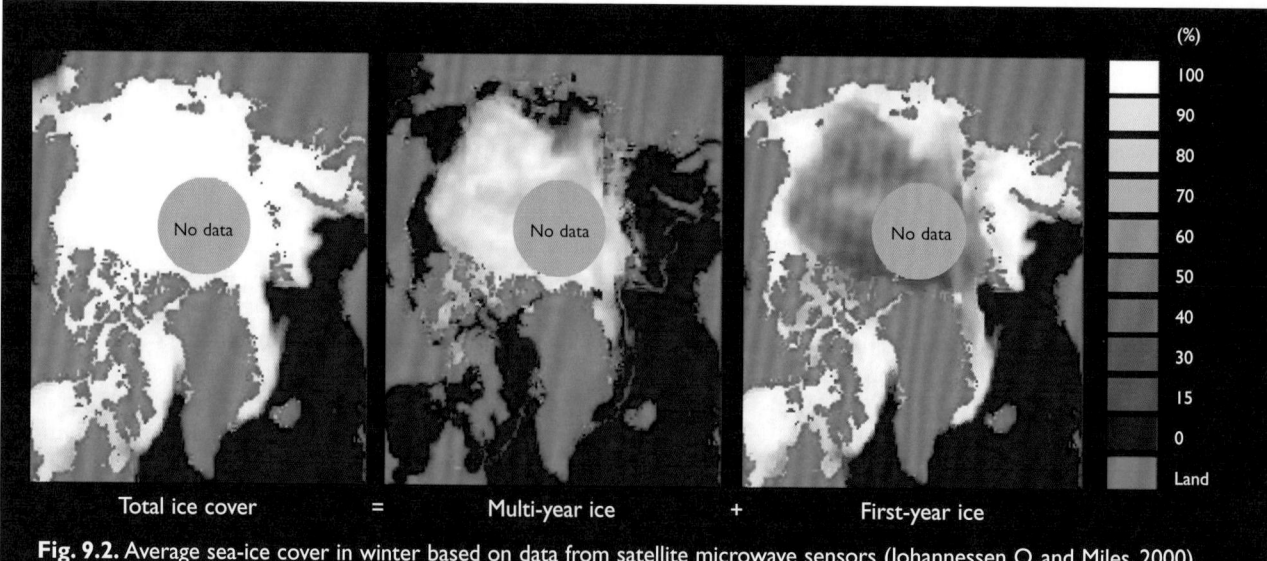

Total ice cover = Multi-year ice + First-year ice

Fig. 9.2. Average sea-ice cover in winter based on data from satellite microwave sensors (Johannessen O. and Miles, 2000). The illustration shows total sea-ice cover, plus the distribution of its two components; multi-year ice and first-year ice. The multi-year ice represents the minimum sea-ice extent in summer.

the months of greatest retreat. At this time most of the Barents and Kara Seas are free of sea ice as far as the northern shelf break. The Laptev Sea and part of the East Siberian Sea have open water along their coastline. In East Greenland, the ice has retreated northward to about 72–73° N, while Baffin Bay, Hudson Bay, and the Labrador Sea become ice-free. In the Canadian Archipelago the winter fast ice usually breaks up. North of Alaska, some open water is typically found along the coast (Wadhams, 2000).

By October, new sea ice has formed in areas that were open in summer, especially around the Arctic Ocean coasts, and in November to January there is a steady advance everywhere toward the winter peak.

9.2.2.2. Fast ice and polynyas

Fast ice grows seaward from a coast and remains in place throughout the winter. Typically, it is stabilized by grounded pressure ridges at its outer edge, and therefore extends to the draft limit of such ridges, usually about 20 to 30 m. Fast ice is found along the whole Siberian coast, the White Sea, north of Greenland, the Canadian Archipelago, Hudson Bay, and north of Alaska.

Polynyas are semi-permanent open water regions ranging in area up to thousands of square kilometers. Flaw leads occur at the border of fast ice when offshore winds separate the drift ice from the fast ice. Polynyas and flaw leads are environmentally important for several reasons (AMAP, 1998):

- they are areas of high heat loss to the atmosphere;
- they typically form the locus of sea-ice breakup in spring;
- they are often locations of intense biological activity; and
- they are regions of deep-water formation.

9.2.2.3. Distribution and thickness

From a combination of satellite observations and historical records, the area covered by sea ice in the Arctic during the summer has been reported to have decreased by about 3% per decade during recent decades (Cavalieri et al., 1997). Multi-year ice is reported to have declined at an even greater rate; 7% per decade during the last 20 years or approximately 600 000 km^2 (Johannessen O. et al., 1999). Combined, these results imply that the area of first-year ice has been increasing. Sea-ice distribution within subregions of the Arctic has also changed dramatically in the past. For example, warming in the Barents Sea in the 1920s and 1930s reduced sea-ice extent there by approximately 15%. This warming was nearly as great as the warming observed over the last 20 years (see section 9.2.4.2, Barents Sea).

In addition to the recent general decrease in sea-ice coverage, submarine observations suggest that the sea ice over the deep Arctic Ocean thinned from an average

thickness of about 3.1 m (1958–1976) to about 1.8 m (1993–1997), or about 15% per decade (Rothrock et al., 1999). In addition, the ice thinned at all 26 sites examined. Overall, the arctic sea ice is estimated to have lost 40% of its volume in less than three decades. However, according to some models (Holloway and Sou, 2002; Polyakov and Johnson, 2000), the submarine observations may have been conducted over part of the ocean that underwent thinning through shifting sea ice in response to changing winds associated with a high Arctic Oscillation (AO) index (see Chapter 2 for descriptions of the AO and the associated North Atlantic Oscillation). Thus, the conclusion of reduced sea-ice thickness, while valid for the domain of submarine measurements, may not necessarily be true for the Arctic Ocean as a whole and an alternative hypothesis that sea-ice thickness *distribution* changed in response to the AO but that sea-ice *volume* may not have changed needs to be carefully evaluated.

Scientific debate continues as to the cause of the areal shrinkage of the arctic sea ice. There is some support for the idea that it is probably part of a natural fluctuation in polar climate (Rothrock et al., 1999), while others claim it is another indication of the response to global warming due to increased levels of greenhouse gases (GHGs; Vinnikov et al., 1999).

9.2.2.4. Length of melt season

Smith D. (1998) used satellite data, predominantly from the Beaufort Sea, to estimate that the melt season increased by about 5.3 days per decade during 1979 to 1996. Rigor et al. (2000) found an increase of about 2.6 days per decade in the length of the melt season in the eastern Arctic but a shortening in the western Arctic of about 0.4 days per decade. These trends parallel general observations of a 1 °C per decade *increase* in air temperature in the eastern Arctic compared to a 1 °C per decade *decrease* in the western Arctic for the same time period (Rigor et al., 2000).

9.2.2.5. Sea-ice drift

General sea-ice motion in the Arctic Ocean is organized by the Transpolar Drift in the Eurasian Basin and by the Beaufort Gyre in Canada Basin (Fig. 9.1). Although it has long been recognized that large-scale ice-drift patterns in the Arctic undergo interannual changes, it was not until the International Arctic Buoy Programme (IABP) that sufficient data became available to map the ice drift in detail and thereby directly link changes in sea-ice trajectories to the AO. The IABP data from 1979 to 1998 suggest two characteristic modes of arctic sea-ice motion (Fig. 9.3), one during a low AO index (AO⁻) and the other during a high AO index (AO⁺) (Macdonald et al., 2003a; Rigor et al., 2002). The ice motion revealed by drifting buoys released onto the ice is reasonably well simulated by models (Maslowski et al., 2000; Polyakov and Johnson, 2000). There are two principal differences between the two modes. First, during pronounced AO⁻ conditions (Fig. 9.3a), sea ice in the Transpolar Drift

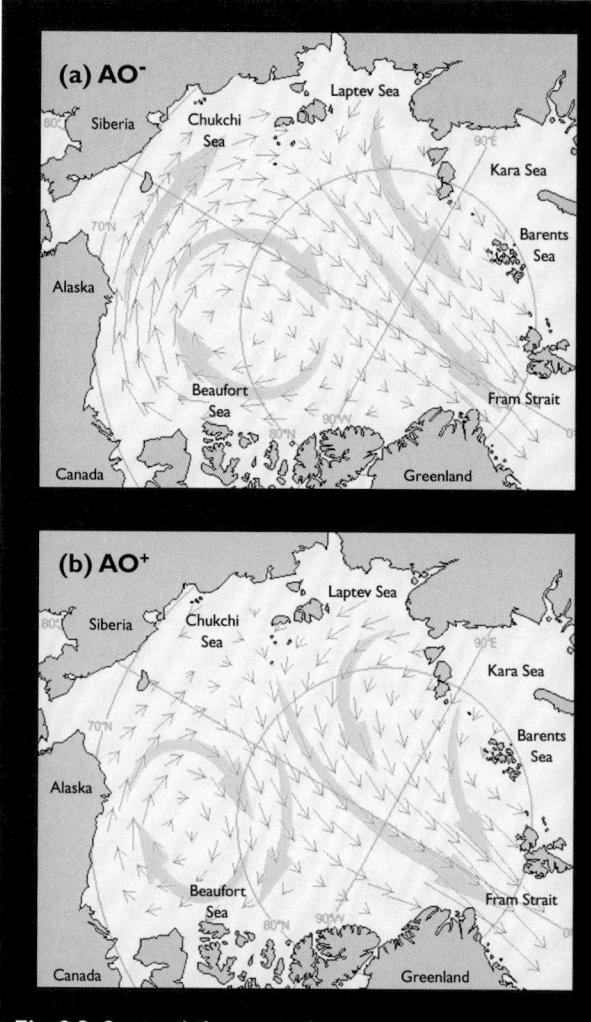

Fig. 9.3. Sea-ice drift patterns for years with (a) pronounced AO⁻ (anticyclonic) conditions and (b) pronounced AO⁺ (cyclonic) conditions (after Maslowski et al., 2000; Polyakov and Johnson, 2000; Rigor et al., 2002). The small arrows show the detailed ice drift trajectories based on an analysis of sea level pressure (Rigor et al., 2002). The large arrows show the general ice drift patterns.

tends to move directly from the Laptev Sea across the Eurasian Basin and out into the Greenland Sea, whereas during pronounced AO⁺ conditions (Fig. 9.3b), ice in the Transpolar Drift takes a cyclonic diversion across the Lomonosov Ridge and into Canada Basin (Mysak, 2001). Second, during pronounced AO⁺ conditions (Fig. 9.3b), the Beaufort Gyre shrinks back into the Beaufort Sea and becomes more disconnected from the rest of the Arctic Ocean, exporting less sea ice to the East Siberian Sea and importing little sea ice from the region to the north of the Canadian Archipelago that contains the Arctic's thickest multi-year ice (Bourke and Garrett, 1987). These changes in sea-ice drift are principally due to the different wind patterns associated with the two AO modes.

During AO⁻ conditions the East Siberian Sea receives much of its ice from the Beaufort Sea and there is an efficient route to carry ice clockwise around the arctic margin of the East Siberian Sea and out toward Fram Strait. Under the strong AO⁺ conditions of the early 1990s, the Beaufort Sea ice became more isolated whereas sea ice from the Kara, Laptev, and East Siberian

Seas was displaced into the central Arctic and toward the Canadian Archipelago. It is not clear from the IABP data how much sea ice from the Russian shelves might be transported into the Canadian Archipelago or the Beaufort Gyre under AO⁺ conditions, but models (Maslowski et al., 2000; Polyakov and Johnson, 2000) suggest that such transport may be important at times.

Fram Strait is the main gateway for arctic ice export. Satellite data, drifting buoys, numerical models, and budgets have been used to construct estimates of the sea-ice flux through Fram Strait (Kwok and Rothrock, 1999; Vinje et al., 1998). Widell et al. (2003) observed a mean sea-ice thickness of 1.8 m and a monthly mean volume flux of 200 km^3 for the period 1990 to 1999. They found no trends in ice thickness and volume flux. The maximum sea-ice volume flux occurred in 1994/95 due to strong winds, combined with relatively thick ice.

9.2.3. Ocean processes of climatic importance

The marine Arctic plays an important role in the global climate system (Box 9.1). A number of physical processes will be affected by the changes anticipated in global climate during the 21st century, but this assessment focuses on those that are expected to have strong impacts on the climate or biology of the Arctic. These include the effects of wind on the transport and mixing of water, and the circulation systems generated by freshwater input and thermohaline ventilation (Fig. 9.4). A key issue is the extent to which each of these processes contributes to driving the inflow of Atlantic water to the Arctic. Models (Seager et al., 2002) have shown that the heat transported by this inflow in some areas elevates the sea surface temperature to a greater extent than the temperature increase projected for the 21st century (see Chapter 4). A weakening of the inflow could therefore significantly reduce warming in these areas and might even induce regional cooling, especially in parts of the Nordic Seas. Thus, special attention is paid to the processes that affect the inflow, especially the thermohaline circulation (see section 9.2.3.4).

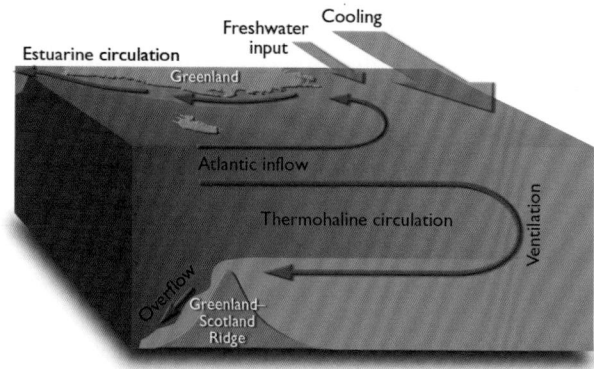

Fig. 9.4. Two types of processes create unique current systems and conditions in the marine Arctic. The input of freshwater, its outflow to the Atlantic, and the en-route entrainment of ambient water create an estuarine type of circulation within the marine Arctic. In addition to this horizontal circulation system, thermohaline ventilation creates a vertical circulation system. Both patterns of circulation are sensitive to climate change.

Box 9.1. Role of the marine Arctic in the global climate system

The marine Arctic is an interconnected component of the global climate system whose primary role is to balance heat gain at low latitudes and heat loss at high latitudes. At low latitudes about half the excess heat is sent poleward as warm (and salty) water in ocean currents (sensible heat, Q_S) and the other half is sent poleward as water vapor in the atmosphere (latent heat, Q_L). At low latitudes the subtropical gyres in the ocean collect excess heat and salt, the western boundary currents carry them poleward, and the Atlantic inflow brings them into the marine Arctic. Heat carried by the atmosphere is released at high latitudes by condensation, thus supplying freshwater to the ocean through precipitation and runoff. Freshwater is stored in the surface and halocline layers of the marine Arctic. To prevent the build-up of salt (by evaporation) at low latitudes, freshwater is exported from the high latitudes, thus completing the hydrological cycle by reuniting the atmospheric water content and the salty ocean water. At high latitudes the return flows include export by ice and transport in low-salinity boundary currents, intermediate water (which forms and sinks along the subpolar fronts), and deep water (which sinks on shelves and in gyres). Export of these low-salinity waters southward couples the Arctic to the world thermohaline circulation (THC) through intermediate and deep-water formation. The role of intermediate water in governing THC is unclear.

The marine Arctic plays an active role in the global climate system with strong feedbacks, both positive and negative.

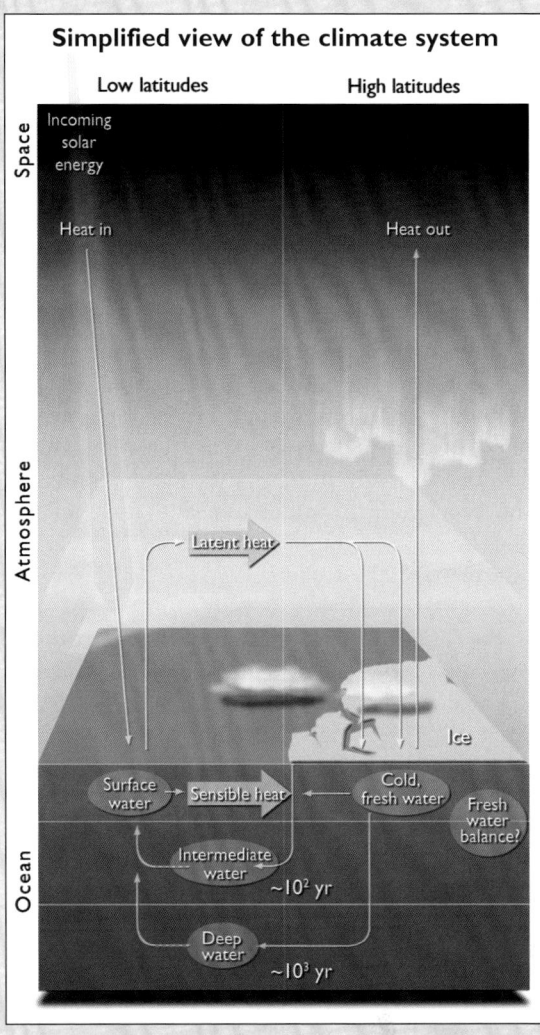

Arctic climate feedbacks

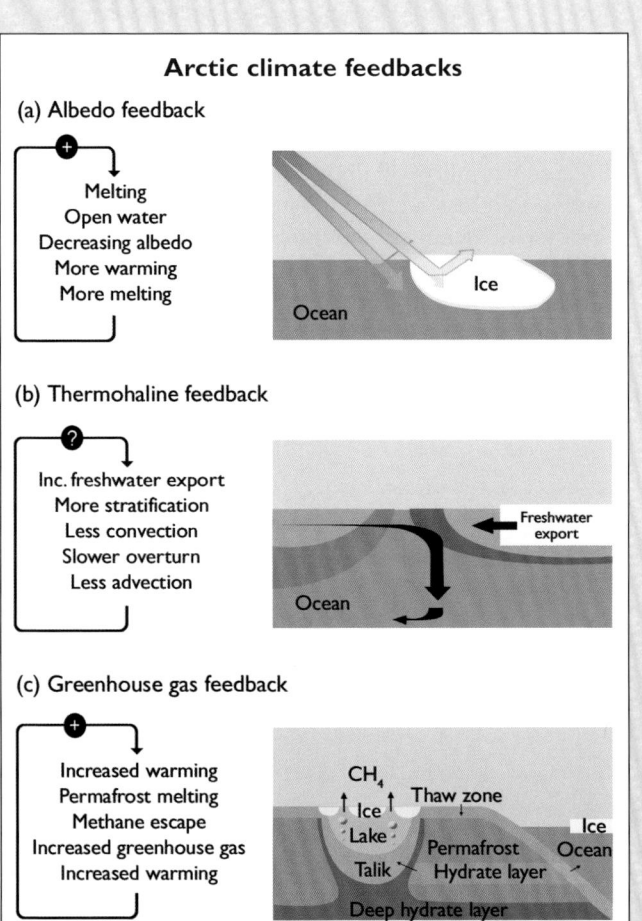

(a) Albedo feedback

Melting
Open water
Decreasing albedo
More warming
More melting

(b) Thermohaline feedback

Inc. freshwater export
More stratification
Less convection
Slower overturn
Less advection

(c) Greenhouse gas feedback

Increased warming
Permafrost melting
Methane escape
Increased greenhouse gas
Increased warming

For example: albedo feedback, thermohaline feedback, and greenhouse gas feedback.

Albedo feedback – Ice and snow reflect most of the solar radiation back into space. With initial warming and sea-ice melting, more heat enters the ocean, thus melting more sea ice and increasing warming.

Thermohaline feedback – If the export of freshwater from the Arctic Ocean should increase, then stratification of the North Atlantic would probably increase, and this could slow the THC. A decrease in the THC would then draw less Atlantic water into high latitudes, leading to a slowdown in the global overturning cell and subsequent localized cooling. (This scenario does not take into account the formation of intermediate water.)

Greenhouse gas feedback – Vast amounts of methane and carbon dioxide are currently trapped in the permafrost and hydrate layers of the arctic margins (Zimov et al., 1997). With warming, arctic coastal lakes will act as a thermal drill to tap this greenhouse gas source and further exacerbate warming.

9.2.3.1. Freshwater and entrainment

Freshwater is delivered to the marine Arctic by atmospheric transport through precipitation and by ocean currents, and to the coastal regions through river inflows (Lewis et al., 2000). Further net distillation of freshwater may occur within the region during the melt/freeze cycle of sea ice, provided that the ice and rejected brine formed by freezing in winter can be separated and exported before they are reunited by melting and mixing the following summer (Aagaard and Carmack, 1989; Carmack, 2000).

The freshwater has decisive influences on stratification and water column stability as well as on ice formation. Without the freshwater input, there would be less freezing, less ice cover, and less brine rejection (Rudels, 1989). This is also illustrated by the difference between the temperature-stratified low latitude oceanic regime and the salinity-stratified high latitude oceanic regime (Carmack, 2000; Rudels, 1993).

In the Arctic Ocean, freshwater is stored within the various layers above and within the halocline, the latter serving as an extremely complex and poorly understood reservoir. This is especially true for the Beaufort Gyre, which represents the largest and most variable reservoir of freshwater storage in the marine Arctic. The ultimate sink for freshwater is its export southward into the North Atlantic to replace the freshwater evaporating from low latitude oceans and to close the global freshwater budget. This southward transport occurs partly through the THC since the overflow from the Nordic Seas into the Atlantic is less saline than the inflowing Atlantic water. The role of the freshwater is illustrated in Fig. 9.5. The figure shows the processes responsible for the development of the horizontal and vertical circulation systems unique to the marine Arctic.

Most of the freshwater in the Arctic Ocean returns southward in the surface outflows of the East Greenland Current and through the Canadian Archipelago. These flows carry low-salinity water as well as sea ice. They include most of the water that enters the Arctic Ocean

through the Bering Strait and water of Atlantic origin entrained into the surface flow. Since the estimated total volume flux of the surface outflows greatly exceeds the combined fluxes of the Bering Strait inflow and the freshwater input, most of the surface outflows must derive from entrained Atlantic water. This process therefore induces an inflow of Atlantic water to the Arctic, which by analogy to the flows in estuaries is usually termed "estuarine circulation". This estuarine-type circulation is sensitive to climate change.

9.2.3.2. Mixed-layer depth

The vertical extent of the surface mixed layer is critical to the primary production and depends on the vertical density stratification and the energy input, especially from the wind. Density stratification is affected by heat and freshwater fluxes from the atmosphere or by advection from surrounding ocean areas. Some areas, for example the Arctic Ocean, are salt-stratified whereas other areas, such as the Nordic Seas and the Bering Sea, are temperature-stratified. In a classic study, Morison and Smith (1981) found that seasonal variations in mixed-layer depth are largely controlled by buoyancy (i.e., heat and salt) fluxes.

Winds blowing over the sea surface transfer energy to the surface mixed layer. In ice-free areas, increased winds would tend to deepen the surface mixed layer, depending upon the strength of the vertical density stratification. In the presence of sea ice, however, the efficiency of energy transfer from wind to water is a complex function of sea-ice roughness and internal ice stress which, in turn, is a function of sea-ice concentration and compactness (see McPhee and Morison, 2001). Because warming will decrease sea-ice concentrations (and so decrease internal ice stress) and increase the duration of "summer" conditions (i.e., earlier breakup and later freeze-up), the efficiency of wind mixing in summer is likely to increase. This is especially true for late summer in the Arctic Ocean when energy input from storms is greatest. However, owing to the poorly understood role of air–ice–ocean coupling and the present level of salt-stratification, this increased exposure

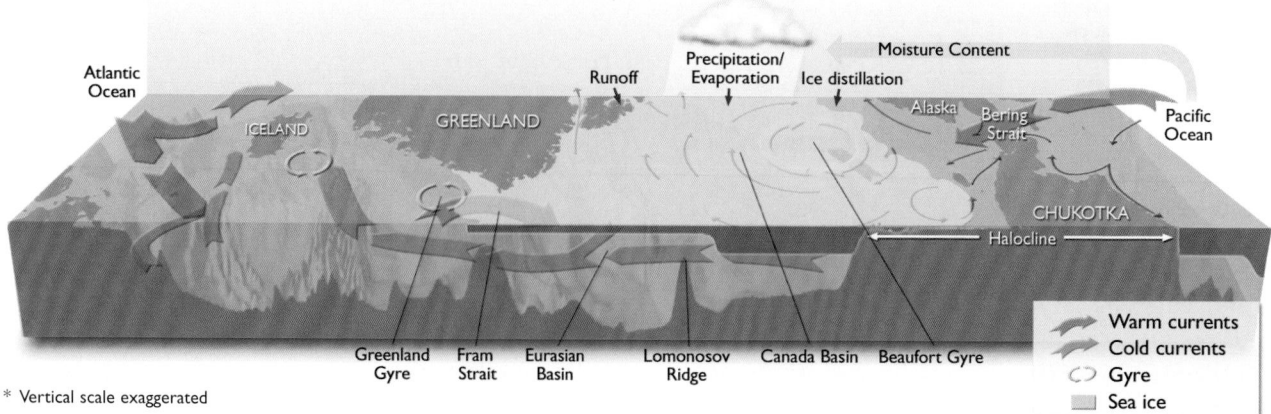

* Vertical scale exaggerated

Fig. 9.5. The freshwater budget of the Arctic Ocean. Low salinity waters are added to the surface and halocline layers via precipitation and runoff, Pacific inflow via the Bering Strait, and the sea-ice distillation process. Low salinity waters and sea ice are subsequently advected through Fram Strait and the Canadian Archipelago into the convective regions of the North Atlantic.

will not necessarily lead to significant increases in mixed-layer depth. Furthermore, the role that lateral advection plays in establishing the underlying halocline structure of the Arctic Ocean must also be considered.

9.2.3.3. Wind-driven transport and upwelling

A number of studies have shown the effect of wind stress on the circulation of particular regions within the marine Arctic (e.g., Aagaard, 1970; Isachsen et al., 2003; Jónsson, 1991). Winds have also been shown to have a strong influence on exchanges between regions (e.g., Ingvaldsen, 2002; Morison, 1991; Orvik and Skagseth, 2003; Roach et al., 1995). If winds were to change significantly, wind-driven currents and exchanges would also change. These wind-induced changes in turn would redistribute the water masses associated with the different currents, thereby affecting the location and strength of the fronts separating the water masses (Maslowski et al., 2000, 2001; Zhang J. et al., 2000).

Retraction of the multi-year ice cover seaward of the shelf break in the Arctic Ocean may lead to wind-induced upwelling at the shelf break, which is currently not happening. This process might substantially increase the rate of exchange between the shelf and deep basin waters, the rate of nutrient upwelling onto the shelves, and the rate of carbon export to the deep basin (Carmack and Chapman, 2003).

9.2.3.4. Thermohaline circulation

Thermohaline circulation is initiated when cooling and freezing of sea water increase the density of surface waters to such an extent that they sink and are exchanged with waters at greater depth. This occurs in the Labrador Sea, in the Nordic Seas, and on the arctic shelves. Together, these regions generate the main source water for the North Atlantic Deep Water; the main ingredient of the global ocean "Great Conveyor Belt" (Broecker et al., 1985). All these arctic areas are therefore important for the global THC. More importantly from the perspective of this assessment is the potential impact of a changing THC on flow and conditions within the marine Arctic. Some areas are more sensitive than others, because the oceanic heat transport induced by the THC varies regionally. The most sensitive areas are those that currently receive most of the heat input from inflowing warm Atlantic water, i.e., the eastern parts of the Nordic Seas and the Arctic Ocean (Seager et al., 2002), namely the Arctic Mediterranean.

The THC in the Arctic Mediterranean is often depicted as more or less identical to open-ocean convection in the Greenland Sea. This is a gross over-simplification since, in reality, there are several different processes contributing to the THC and they occur in different areas. The THC can be subdivided into four steps (Fig. 9.4).

1. Upper layer inflow of warm, saline Atlantic water into the Arctic Ocean and the Nordic Seas.
2. Cooling and brine rejection making the incoming waters denser.
3. Vertical transfer of near-surface waters to deeper layers.
4. The overflow of the dense waters in the deep layers over the Greenland–Scotland Ridge and their return to the Atlantic.

Although these steps are linked by feedback loops that prevent strict causal relations, the primary processes driving the THC seem to be steps 2 and 3, which are termed thermohaline ventilation. By the action of the thermohaline ventilation, density and pressure fields are generated that drive horizontal exchanges between the Arctic Mediterranean and the Atlantic (steps 1 and 4). Box 9.2 illustrates the basic mechanisms of the thermohaline forcing.

Thermohaline ventilation

The waters of the Arctic Ocean and the Nordic Seas are often classified into various layers and a large number of different water masses (Carmack, 1990; Hopkins, 1991). For the present assessment, it is only necessary to distinguish between "surface" (or upper layer) waters and "dense" waters, which ultimately leave the Arctic Mediterranean as overflow into the North Atlantic. The term "dense waters" is used to refer to deep and intermediate waters collectively and the term "thermohaline ventilation" is used as a collective term for the processes that convert surface waters to dense waters. Thermohaline ventilation is a two-step process that first requires cooling and/or brine rejection to increase the surface density and then a variety of processes that involve vertical transfer.

Cooling and brine rejection

Production of dense waters in the arctic Nordic Seas is due initially to atmospheric cooling, and then to brine rejection during sea-ice formation (Aagaard et al., 1985). The waters flowing into the Nordic Seas from the Atlantic exhibit a range of temperatures depending on location and season. On average, their temperature is close to 8 °C, but it decreases rapidly after entering the Nordic Seas. The temperature decrease is especially large in the southern Norwegian Sea. The simultaneous salinity decrease indicates that some of the temperature decrease may be due to admixture of colder and less saline adjacent water masses. Except for relatively small contributions of freshwater from river inflow and the Pacific-origin waters flowing along the east coast of Greenland, the adjacent water masses are predominantly of Atlantic origin. Thus, atmospheric cooling in the Nordic Seas is the main cause of the decreasing temperature of the inflowing Atlantic water.

Attempts have been made to calculate the heat loss to the atmosphere from climatological data, but the sensitivity of the results to different parameterizations of the heat flux makes these estimates fairly uncertain (Simonsen and

Haugan, 1996). Most of the heat loss from the ocean to the atmosphere occurs in ice-free areas of the Nordic and Barents Seas (Simonsen and Haugan, 1996).

Brine rejection, however, is intimately associated with sea-ice formation (Carmack, 1986). When ice forms at the ocean surface, only a small fraction of the salt follows the freezing water into the solid phase, the remainder flowing into the underlying water. Brine also continues to drain from the recently formed ice. Both processes increase the salinity, and therefore density, of the ambient water. In a stationary state, the salinity increase due to brine rejection in cold periods is compensated for

by freshwater input from melting ice in warm periods, but freezing and melting often occur in different regions. For example, on the shallow shelves surrounding the arctic basins rejected brine results in shelf waters sufficiently dense to drain off the shelves, thus becoming separated from the overlying ice (Anderson L. et al., 1999). Winds can also remove newly formed ice from an area while leaving behind the high salinity water.

Vertical transfer of water

The second step in thermohaline ventilation is the vertical descent of the surface waters made denser by cool-

---

Box 9.2. Thermohaline forcing of Atlantic inflow to the Arctic

The processes by which thermohaline ventilation induces Atlantic inflow to the Arctic Mediterranean can be illustrated by a simple model where the Arctic Mediterranean is separated from the Atlantic by a ridge (the Greenland–Scotland Ridge). South of the ridge, Atlantic water (red) with uniform temperature, salinity, and density (ρ) extends to large depths. North of the ridge, the deep layers (blue) are less saline, but they are also much colder than the Atlantic water and therefore denser ($\rho+\Delta\rho$). Above this deep, dense layer is the inflowing Atlantic water, which is modified by cooling and brine rejection to become increasingly similar to the deep layer as it proceeds away from the ridge. The causal links between the processes involved can be broken into three steps.

Thermohaline ventilation – Cooling and brine rejection make the inflowing Atlantic water progressively denser until it has reached the density of the deeper layer. At that stage, the upper-layer water sinks and is transferred to the deeper layer. This is equivalent to raising the interface between the two layers in the ventilation areas, which are far from the ridge.

Overflow – When ventilation has been active for some time, the interface will be lifted in the ventilation areas and will slope down towards the ridge. Other things being equal, this implies that the pressure in deep water will be higher in the ventilation areas than at the same depth close to the ridge. A horizontal internal (so-called baroclinic) pressure gradient will therefore develop which forces the deep water towards and across the ridge. In this simple model, the overflow is assumed to pass through a channel, sufficiently narrow to allow neglect of geostrophic effects. If the rate at which upper-layer water is converted to deeper-layer water is constant, the interface will rise until it can drive an overflow with a volume flux that equals the ventilation rate.

Sea-level drop – When thermohaline ventilation has initiated a steady overflow, there will be a continuous removal of water from the Arctic Mediterranean. Without a compensating inflow, the sea level would drop rapidly north of the ridge. Thus an uncompensated overflow of the present-day magnitude would make the average sea level in the Arctic Mediterranean sink by more than one meter a month. As soon as the water starts sinking north of the ridge, there will, however, develop a sea-level drop across the ridge. This sea-level drop implies that water in the upper layer north of the ridge will experience lower pressure than water at the same level in the Atlantic. A sea surface (so-called barotropic) pressure gradient therefore develops that pushes water northward across the ridge. The amount of Atlantic water transported in this way increases with the magnitude of the sea-level drop. In the steady state, the sea-level drop is just sufficient to drive an Atlantic inflow of the same volume flux as the overflow and the ventilation rate.

When upper-layer water is converted to deeper-layer water at a certain ventilation rate (in m³/s), an overflow and an Atlantic inflow are therefore generated which have the same volume flux on long timescales. In the present state, these fluxes must equal the estimated overflow flux of about 6 Sv. Simple, non-frictional, models indicate that the required interface rise is several hundred meters, as is observed, while the required sea-level drop is only of the order of 1 cm.

ing and brine rejection. Several processes contribute to the transfer. These include the sinking of the boundary current as it flows around the Arctic Mediterranean, open-ocean convection, and shelf convection as well as other ventilation processes (Fig. 9.6).

1. The boundary current enters the Arctic Mediterranean as pure Atlantic water with relatively high temperature (>8 °C) and salinity (>35.2). It enters mainly through the Faroe–Shetland Channel and within the Channel joins with part of the Iceland–Faroe Atlantic inflow. Part of the boundary current continues as an upper-layer flow along the continental slope to Fram Strait. There, one branch moves toward Greenland while the other enters the Arctic Ocean and flows sub-surface along its slope to join the first branch as it exits again through Fram Strait. The flow continues as a subsurface boundary current over the slope off East Greenland all the way to Denmark Strait with the core descending en route (Rudels et al., 2002). While circulating through the Arctic Mediterranean, boundary current waters experience a large temperature decrease, much of it during the initial flow along the Norwegian shelf. While the associated density increase is partly offset by a salinity decrease, there is still a considerable net density increase. After passing Fram Strait, both branches are submerged without direct contact to the atmosphere such that temperature and salinity changes occur mainly through isopycnal mixing with surrounding waters. Isopycnal mixing occurs between waters of the same density but different temperatures and salinities.

2. Open-ocean convection is very different from boundary current deepening, being essentially a vertical process. After a pre-conditioning phase in which the waters are cooled and mixed, further intensive cooling events may trigger localized intense descending plumes or eddies with horizontal scales of the order of a few kilometers or less (Budéus et al., 1998; Gascard et al., 2002; Marshall and Schott, 1999; Watson et al., 1999). They have strong vertical velocities (of the order

of a few hundredths of a meter per second), but do not represent an appreciable net volume flux since they induce upward motion in the surrounding water (Marshall and Schott, 1999). They do, however, exchange various properties (such as CO_2) between the deep and near-surface layers as well as to the atmosphere. They also help maintain a high density at depth. Open-ocean convection is assumed to occur to mid-depths in the Iceland Sea (Swift and Aagaard, 1981). In the Greenland Sea, convective vortices have been observed to reach depths of more than 2000 m (Gascard et al., 2002) and it is assumed that convection in earlier periods penetrated all the way to the bottom layers to produce the very cold Greenland Sea Deep Water, as observed in 1971 (Malmberg, 1983).

3. Shelf convection results from brine rejection and convection, and can lead to the accumulation of high salinity water on the shelf bottom (Jones et al., 1995; Rudels et al., 1994, 1999). Freezing of surface waters limits the temperature decrease, but if winds or other factors remove the sea ice while leaving the brine-enriched water behind, prolonged cooling can produce a high salinity water mass close to the freezing point. Eventually, gravity results in this saline, dense water mass flowing off the shelf and sinking into the arctic abyss. As it sinks, it entrains ambient waters and its characteristics change (Jones et al., 1995; Quadfasel et al., 1988; Rudels, 1986; Rudels et al., 1994). Shelf convection is the only deep-reaching thermohaline ventilation process presumed to enter the Arctic Ocean and hence is responsible for local deep-water formation.

There are at least two additional sinking mechanisms (not included in Fig. 9.6) that may transfer dense water downward; isopycnal sinking and frontal sinking. Overflow water is often defined as water denser than $\sigma_\theta = 27.8$ (Dickson and Brown, 1994) and such water is widely found in the Arctic Ocean and the Nordic Seas, close to the surface. During winter, mixing and cooling result in surface densities up to and above this value. This water can therefore flow over the ridge, sinking below the top of the ridge but without crossing isopycnals. This is termed "isopycnal sinking". A somewhat-related mechanism has been termed "frontal sinking", which indicates that near-surface water from the dense side of a front can sink in the frontal region and flow under the less dense water. In the Nordic Seas, this has been observed in the form of low-salinity plumes sinking at fronts between Arctic and Atlantic waters (Blindheim and Ådlandsvik, 1995).

Horizontal water exchange

The Nordic Seas and the Arctic Ocean are connected to the rest of the World Ocean through the Canadian Archipelago, across the Greenland–Scotland Ridge, and through the Bering Strait, and they exchange water and various properties with the World Ocean through these gaps. Four exchange branches can be distinguished

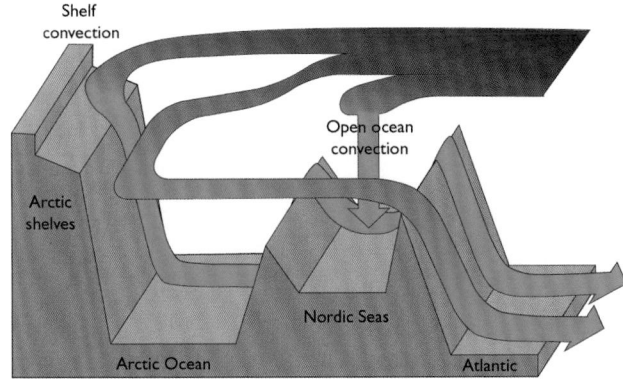

Fig. 9.6. Three of the thermohaline ventilation processes that occur in the Arctic Mediterranean: boundary current deepening, open-ocean convection, and shelf convection.

(Fig. 9.7). The near-surface outflow from the Arctic Ocean through the Canadian Archipelago and Denmark Strait, and the Bering Strait inflow to the Arctic Ocean from the Pacific are important in connection with freshwater flow through the Arctic Ocean and the Nordic Seas. For the THC, the overflow of cold and dense water from the Nordic Seas into the Atlantic and the inflow of Atlantic water to the Nordic Seas and the Arctic Ocean are the most important factors.

Overflow

The term overflow is used here to describe near-bottom flow of cold, dense ($\sigma_\theta > 27.8$; Dickson and Brown, 1994) water from the Arctic Mediterranean across the Greenland–Scotland Ridge into the Atlantic. It occurs in several regions. In terms of volume flux, the most important overflow site is the Denmark Strait, a deep channel between Greenland and Iceland with a sill depth of 620 m. The transport in this branch is estimated at 3 Sv, or about half the total overflow flux (Dickson and Brown, 1994). Mauritzen C. (1996) and Rudels et al. (2002) argue that water from the East Greenland Current forms the major part of this flow. Other sources contribute, however (Strass et al., 1993); some workers suggest the Iceland Sea as the primary source for the Denmark Strait overflow (Jónsson, 1999; Swift and Aagaard, 1981).

The Faroe Bank Channel is the deepest passage across the Greenland–Scotland Ridge and the overflow through the channel is estimated to be the second largest in terms of volume flux, approximately 2 Sv (Saunders, 2001). Owing to the difference in sill depth, the deepest water flowing through the Faroe Bank Channel is usually colder than water flowing through the Denmark Strait and the Faroe Bank Channel is thus the main outlet for the densest water produced in the Arctic Mediterranean.

Overflow has also been observed to cross the Iceland–Faroe Ridge at several sites, as well as the Wyville–Thomson Ridge, but more intermittently. The total overflow across these two ridges has been estimated at slightly above 1 Sv, but this value is fairly uncertain compared to the more reliable estimates for the Denmark Strait and Faroe Bank Channel overflow branches (Hansen and Østerhus, 2000).

As the overflow waters pass over the ridge, their temperature varies from about -0.5 °C upward. A large proportion of the water is significantly colder than the 3 °C value often used as a limit for the overflow (approximately equivalent to $\sigma_\theta > 27.8$). After crossing the ridge, most of the overflow continues in two density-driven bottom currents that are constrained by the effects of the earth's rotation (i.e., the Coriolis force) to follow the topography, although gradually descending. The bottom current waters undergo intensive mixing and entrain ambient waters from the Atlantic Ocean, which increases the water temperature. When the Denmark Strait and Faroe Bank Channel overflow waters join in the region southeast of Greenland, they have been warmed to 2 to 3 °C, typical of the North Atlantic Deep Water. Through entrainment, enough Atlantic water is added to approximately double their volume transport.

Atlantic inflow

Inflow of Atlantic water to the Nordic Seas occurs across the Greenland–Scotland Ridge along its total extent except for the westernmost part of the Denmark Strait. Iceland and the Faroe Islands divide this flow into three branches (Fig. 9.7); the Iceland branch (Jónsson and Briem, 2003), the Faroe branch (Hansen et al., 2003), and the Shetland branch (Turrell et al., 2003). There is a gradual change in water mass characteristics with the most southeastern inflow being the warmest (and most saline). There is also a difference in the volume fluxes, with that for the Iceland branch being much less than for the other two, which are similar in magnitude.

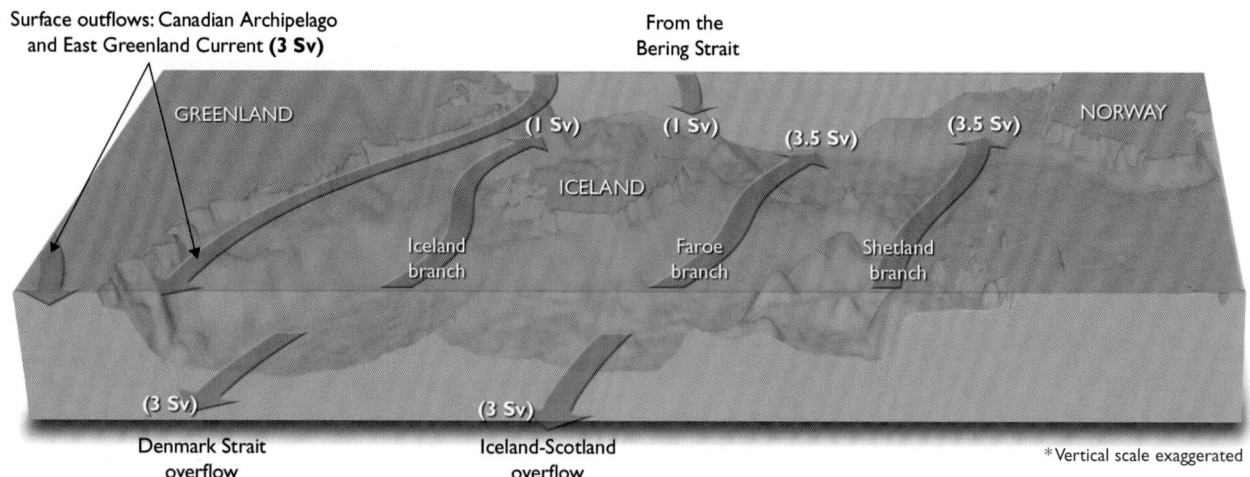

Fig. 9.7. The Arctic Mediterranean has four current branches that import water into the upper layers; three from the Atlantic (the Iceland, Faroe, and Shetland branches), and one from the Pacific. The outflow occurs partly at depth through the overflows and partly as surface (or upper-layer) outflow through the Canadian Archipelago and the East Greenland Current. The numbers indicate volume flux in Sverdrups (10^6 m³/s) rounded to half-integer values and are based on observations, with the exception of the surface outflow, which is adjusted to balance (based on Hansen and Østerhus, 2000).

The Iceland branch flows northward on the eastern side of the Denmark Strait. North of Iceland, it turns east and flows toward the Norwegian Sea, but the heat and salt content of this branch are mixed with ambient water of polar or Arctic Ocean origin and freshwater runoff from land. By the time it reaches the east coast of Iceland it has lost most of its Atlantic character. The Faroe and Shetland branches flow directly into the Norwegian Sea. On their way they exchange water, but still appear as two separate current branches off the coast of northern Norway. Their relative contribution to various regions is not clarified in detail but the Barents Sea is clearly most affected by the inner (Shetland) branch, while the western Norwegian Sea and the Iceland Sea receive most of their Atlantic water from the outer (Faroe) branch.

Budgets

The horizontal exchanges between the Arctic and oceans to the south transfer water, heat, salt, and other properties such as nutrients and CO_2. Since typical temperatures, salinities, and concentrations of various properties are known, quantifying the exchanges is mainly a question of quantifying volume fluxes.

The water budget for the Arctic Ocean and the Nordic Seas as a whole is dominated by the Atlantic inflow and the overflow (Fig. 9.7). The Bering Strait inflow is fairly fresh ($S < 33$) and most of it can be assumed to leave the Arctic Mediterranean in the surface outflow (Rudels, 1989). The deeper overflow is formed from Atlantic water, which means that 75% of the Atlantic inflow is ventilated in the Arctic Ocean and the Nordic Seas. Errors in the flux estimates may alter this ratio somewhat, but are not likely to change the conclusion that most of the Atlantic inflow exits via the deep overflow rather than in the surface outflow.

The question as to how the thermohaline ventilation is split between the Nordic Seas and the Arctic Ocean and its shelves can be addressed in different ways. One method is to measure the fluxes of the various current branches that flow between these two ocean areas; another is to estimate the amount of water produced by shelf convection. Both methods involve large uncertainties, but generally imply that most of the ventilation occurs in the Nordic Seas with perhaps up to 40% of the overflow water produced in the Arctic Ocean (Rudels et al., 1999). That most of the heat loss also appears to occur in the Nordic and Barents Seas (Simonsen and Haugan, 1996) highlights the importance of these areas for the THC.

9.2.3.5. What drives the Atlantic inflow to the Arctic Mediterranean?

The Atlantic inflow is responsible for maintaining high temperatures in parts of the marine Arctic and potential changes in the Atlantic inflow depend on the forces driving the flow. The few contributions to this discussion to be found in the literature (e.g., Hopkins, 1991) generally cite direct forcing by wind stress, estuarine circulation, or thermohaline circulation as being the main driving forces.

The freshwater input combined with entrainment generates southward outflows from the Arctic Mediterranean in the upper layers, which for continuity reasons require an inflow (estuarine circulation). Similarly, thermohaline ventilation generates overflows, which also require inflow (thermohaline circulation). If inflows do not match outflows, sea-level changes are induced, which generate pressure gradients that tend to restore the balance (Box 9.2). To the extent that the water budget (Fig. 9.7) is reliable, it is therefore evident that the processes that generate the estuarine circulation can account for 2 Sv of the Atlantic inflow, whereas thermohaline ventilation is responsible for an additional 6 Sv. This has led some workers to claim thermohaline ventilation as the main driving force for the Atlantic inflow (Hansen and Østerhus, 2000).

Wind affects both the estuarine and the thermohaline circulation systems in many different ways (e.g., through entrainment, cooling, brine rejection, flow paths). Direct forcing by wind stress has also been shown to affect several current branches carrying Atlantic water (Ingvaldsen et al., 2002; Isachsen et al., 2003; Morison, 1991; Orvik and Skagseth, 2003), but there is no observational evidence for a strong direct effect of wind stress on the total Atlantic inflow to the Nordic Seas. On the contrary, Turrell et al. (2003) and Hansen et al. (2003) found that seasonal variation in the volume flux for the two main inflow branches (the Faroe Branch and Shetland Branch on Fig. 9.7) was negligible, in contrast to the strong seasonal variation in the wind stress. Thermohaline ventilation is also seasonal, but its effect is buffered by the large storage of dense water in the Arctic Mediterranean, which explains why the total overflow and hence also thermohaline forcing of the Atlantic inflow has only a small seasonal variation (Dickson and Brown, 1994; Hansen et al., 2001; Jónsson, 1999). In a recent modeling study, Nilsen et al. (2003) found high correlations between the North Atlantic Oscillation (NAO) index and the volume flux of Atlantic inflow branches, but that variations in the total inflow were small in relation to the average value.

These studies indicate that the Atlantic inflow to the Arctic Mediterranean is mainly driven by thermohaline (Box 9.2) and estuarine forcing, but that fluctuations at annual and shorter timescales are strongly affected by wind stress. Variations in wind stress also have a large influence on how the Atlantic water is distributed within the Arctic Mediterranean.

9.2.4. Variability in hydrographic properties and currents

Ocean climate changes on geological time scales in the Arctic are briefly discussed in Box 9.3.

Box 9.3. Arctic climate – a long-term perspective

At the start of large-scale glaciation around 3 million years ago, the Arctic was relatively warm with forests growing along the shores of the Arctic Ocean (Funder et al., 1985; Knies et al. 2002). About 2.75 million years ago a marked phase of global cooling set in, leading to a widespread expansion of ice sheets across northern Eurasia and North America (Jansen et al., 2000). Before this marked cooling, climates were only cold enough to sustain glaciers on Greenland, indicating that the ocean was warmer and the sea-ice cover less than at present (Fronval and Jansen, 1996; Larsen et al., 1994). This cooling is believed due to reduced northward heat transport to the Arctic. After this cooling event, multi-year sea-ice cover and cold conditions probably existed throughout the Arctic, however, less freshwater influx may have reduced surface ocean stratification and open areas and polynyas may have prevailed. Lower sea level also left major portions of the shelf areas exposed.

The next major change occurred approximately 1 million years ago. Glacial episodes became longer, with a distinct 100000 year periodicity and glaciation more severe. Yet between the glacial periods, warmer but short interglacial periods persisted, due to stronger inflow of warm Atlantic waters to the Nordic Seas (Berger and Jansen, 1994; Jansen et al., 2000). The long-term effects of sea-level change through ice sheet erosion affected the ocean exchange with the Arctic. For example, water mass exchange could take place between the Atlantic and the Arctic through the Barents Sea when it changed from a land area to a sea.

After the last glacial period, which ended about 11000 years ago, the marginal ice zone was farther north than at present since the summer insolation was higher in the Northern Hemisphere than now. In the early phase of the postglacial period (Holocene), 8000 to 6000 years ago, mollusks with affinities for ice-free waters were common in Spitsbergen and along the east coast of Greenland. Summer temperatures over Greenland and the Canadian Arctic were at their highest, 3 °C above present values (Dahl-Jensen et al., 1998). The sea-ice cover expanded southward again in the Barents and Greenland Seas 6000 to 4000 years ago, concomitant with the expansion of glaciers in Europe. This expansion was most likely to be a response to the diminishing summer insolation.

Superimposed on these long-term trends, there is evidence of high amplitude millennial- to century-scale climate variability. The millennial-scale events are recorded globally and shifts in temperature and precipitation occurred with startling speed, with changes in annual mean temperature of 5 to 10 °C over one to two decades (Alley et al., 2003; Dansgaard et al., 1993; Haflidason et al., 1995; Koc et al., 1993). These abrupt climate changes occurred repeatedly during glacial periods with a temporal spacing of 2000 to 10000 years. The latest was the Younger Dryas cooling about 12000 years ago, which was followed by two cold phases of lower amplitude, the last 8200 years ago. Cooling periods in the regions surrounding the Arctic were associated with widespread drought over Asia and Africa, as well as changes in the Pacific circulation. Mid-latitude regions were most affected, while the amplitudes of these climate shifts were lower in the high Arctic.

The rapid climate shifts were accompanied by changes in the deep-water formation in the Arctic and the northward protrusion of warm water towards the Arctic (Dokken and Jansen, 1999), yet it would be wrong to say that they shut off entirely during the rapid change events. Instead they were characterized by shifts in the strength and in the depth and location of ocean overturning. The high amplitude climate shifts are hypothesized to be caused by, or at least amplified by, freshwater release from calving and melting of ice sheets in the Arctic.

Bond et al. (2001) identified events when icebergs originating from Greenland were more strongly advected into the North Atlantic and proposed that changes in insolation may have been the cause. Some of these events coincide with known climate periods, such as the Medieval Warm Period and an increase in icebergs during the following cooling period, known as the Little Ice Age. Temperature data from the Greenland Ice Sheet show a general warmer phase (800 to 1200 AD) and a general cold phase (1300 to 1900 AD) during these periods, respectively (Dahl-Jensen et al., 1998). Proxy data with higher temporal resolution from the Nordic Seas suggest similar temperature trends there, but it is clear that neither the Medieval Warm Period nor the Little Ice Age was monotonously warm or cold (Koc and Jansen, 2002).

9.2.4.1. Seasonal variability

Upper-layer waters in the Arctic Ocean that are open or seasonally ice-free experience seasonal fluctuations in temperature due to the annual cycle of atmospheric heating and cooling. The extent of the summer temperature rise depends on the amount of heat used to melt sea ice (and hence not used for heating the water) and the depth of the surface mixed layer. For shallow mixed layers caused by ice melt, surface temperatures can rise substantially during the summer. Seasonal temperature ranges in the near-surface waters generally tend to increase southward. The melting and formation of sea ice leads to seasonal changes in salinity. Salt is rejected from newly formed ice, which increases the salinity of the underlying water. This water sinks as it is denser than its surroundings. Salinity changes in some coastal regions are governed more by the annual cycle of freshwater runoff than by ice, e.g., along the Norwegian coast, in the Bering Sea, and Hudson Bay. Except for areas in which brine rejection from sea-ice formation occurs annually, seasonal changes in temperature and salinity below the mixed layer are usually small.

9.2.4.2. Interannual to decadal variability

Variability observed at interannual to decadal time scales is important as a guide for predicting the possible effect of future climate change scenarios on the physical oceanography of the Arctic.

Arctic Ocean

Long-term oceanographic time series from the Arctic Ocean deep basins are scarce. Data collections have been infrequent, although there was a major increase in shipboard observations during the 1990s (Dickson et al., 2000). These efforts identified an increased presence of Atlantic-derived upper ocean water relative to Pacific-derived water (Carmack et al., 1995; Morison et al., 1998). Temperatures and salinities rose, especially in the Eurasian Basin. The rise in temperature for the Atlantic waters of the arctic basins ranged from 0.5 to 2 °C. The major cause of the warming is attributed to increased transport of Atlantic waters in the early 1990s and to the higher temperatures of the inflowing Atlantic water (Dickson et al., 2000; Grotefendt et al., 1998). At the same time, the front between the Atlantic- and Pacific-character waters moved 600 km closer to the Pacific from the Lomonosov Ridge to the Alpha-Mendeleyev Ridge (Carmack et al., 1995; McLaughlin et al., 1996; Morison et al., 1998). This represented an approximate 20% increase in the extent of the Atlantic-derived surface waters in the Arctic Ocean. In addition, the Atlantic Halocline Layer, which insulates the Atlantic waters from the near-surface polar waters, became thinner (Morison et al., 2000; Steele and Boyd, 1998). As the Atlantic-derived waters increased their dominance in the Arctic Ocean, there was an observed shrinking of the Beaufort Gyre and a weakening and eastward deflection of the Transpolar Drift (Kwok, 2000; Morison et al., 2000). These were shown to be a direct response to changes in the wind forcing over the Arctic associated with variability in the AO (Maslowski et al., 2000, 2001; Zhang et al., 2000).

Barents Sea

Inflow to the Arctic via the Barents Sea undergoes large variability on interannual to decadal time scales (Ingvaldsen et al., 1999, 2003; Loeng et al., 1997). The inflows change in response to varying atmospheric pressure patterns, both local (Ådlandsvik and Loeng, 1991) and large-scale, as represented by the NAO, with a larger transport associated with a higher index (Dickson et al., 2000; Dippner and Ottersen, 2001; Ingvaldsen et al.,

2003). The Shetland Branch of the Atlantic inflow (Fig. 9.7; also known as the Norwegian Atlantic Current) is a major contributor to the inflow to the Barents Sea. It is strongly correlated with the North Atlantic wind stress curl with the current lagging the wind stress curl by 15 months (Orvik and Skagseth, 2003).

Variability in both the volume and temperature of the incoming Atlantic water to the Barents Sea strongly affects sea temperatures. A series of hydrographic stations along a line north of the Kola Peninsula in northwest Russia has been monitored for over 100 years. Annual mean temperatures for this section show relatively warm conditions since the 1990s. It was also warm between 1930 and 1960, but generally cold prior to the 1930s and through much of the period between 1960 and 1990 (Fig. 9.8). Since the mid-1970s there has been a trend of increasing temperature, although the warmest decade during the last century was the 1930s (Ingvaldsen et al., 2003). Also evident are the strong near-decadal oscillations since the 1960s and prior to the 1950s. Annual ocean temperatures in the Barents Sea are correlated with the NAO; higher temperatures are generally associated with the positive phase of the NAO (Ingvaldsen et al., 2003; Ottersen and Stenseth, 2001). The correlation is higher after the early 1970s, which is attributed to an eastward shift in the Icelandic Low (Dickson et al., 2000; Ottersen et al., 2003).

Willem Barentsz was the first to provide information on sea ice conditions in the northern Barents Sea when he discovered Spitsbergen in 1596 (de Veer, 1609). Observations became more frequent when whaling and sealing started early in the 17th century (Vinje, 2001) and since 1740 there have been almost annual observations of sea-ice conditions. Typically, interannual variation in the position of the monthly mean ice edges is about 3 to 4 degrees of latitude. Variations on decadal and centennial scales are also observed. In all probability, the extreme northern position of the ice edge in summer coincides with an increased influx of Atlantic water entering the Arctic Ocean north of Svalbard. Complete disintegration of the sea ice in the Barents Sea proper (south of 80° N) was reported between 1660 and 1750. A similar north-

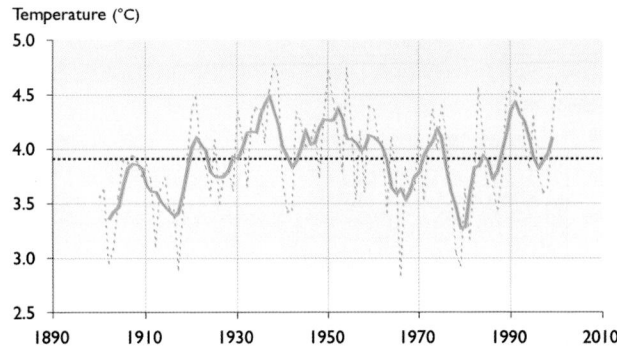

Fig. 9.8. Annual and five-year running means in sea temperature (at 50–200 m) from a series of hydrographic stations along a line north of the Kola Peninsula in northwest Russia (based on data supplied by the Knipovich Polar Research Institute of Marine Fisheries and Oceanography, Russia).

ern retreat of the sea ice was seen again in recent decades (after 1937). In contrast, sea ice completely covered the Barents Sea, as well as the Greenland and Iceland Seas, and the northern part of the Norwegian Sea, during 1881. This coincided with the lowest mean winter air temperature on record.

Northern North Atlantic

In the 1910s and 1920s, a major and rapid atmospheric warming took place over the North Atlantic and Arctic, with the greatest changes occurring north of 60° N (Fig. 9.9; Johannessen O. et al., 2004; Rogers, 1985). Warm conditions generally continued through to the 1950s and 1960s. Sea ice thinned and the maximum extent of the seasonal ice edge retracted northward (Ahlmann, 1949). Increases in surface temperature were reported over the northern North Atlantic (Smed, 1949) and throughout the water column over the shelf off West Greenland (Jensen, 1939). Higher temperatures between the 1930s and 1960s were also observed in the Barents Sea along the Kola Section (Fig. 9.8). The cause of this warming is uncertain although a recent hypothesis suggests that it was due to an increase in the transport of the North Atlantic Current into the Arctic (Johannessen O. et al., 2004).

At the end of this warm period, water temperatures declined rapidly. For example, at a monitoring site off northern Iceland, temperatures (at 50 m) suddenly declined in 1964 by 1 to 2 °C (Malmberg and Blindheim, 1994). This was caused by the replacement of the warm Atlantic inflow by the cold waters of the East Greenland Current. Also, the front to the east of Iceland between the warm Atlantic waters and the cold arctic water moved southward. These observations signified that the cooling had coincided with large-scale changes in circulation.

In the Labrador Sea, temperatures reached maximum values in the 1960s and did not decline substantially until the early 1970s. Shelf temperatures on the western Grand Banks at a site 10 km off St. John's, Newfoundland have been monitored since the late 1940s. Low-frequency subsurface temperature trends at this site are representative of the Grand Banks to southern Labrador (Petrie et al., 1992). Temperatures

continued a general decline superimposed upon by quasi-decadal oscillations until the mid-1990s. Temperature minima were observed near the mid-1970s, mid-1980s, and mid-1990s that correspond to peaks in the NAO index (Colbourne and Anderson, 2003). After the mid-1990s, temperatures rose. Winter temperatures off Newfoundland are negatively correlated with those in the Barents Sea (Fig. 9.10) and linked through their opposite responses to the NAO. The Barents Sea and Newfoundland temperatures however have only been closely linked to the NAO since the 1960s (Ottersen et al., 2003).

During the 1970s, an upper-layer surface salinity minimum was observed in different regions of the North Atlantic (e.g., Dickson and Blindheim, 1984; Dooley et al., 1984; Malmberg, 1984). The generally accepted explanation for this observation was given by Dickson et al. (1988). During the 1960s, an intense and persistent high-pressure anomaly became established over Greenland. As the northerly winds increased through to a peak in the late 1960s, there was a pulse of sea ice and freshwater out of the Arctic via Fram Strait with the result that the waters in the East Greenland Current and the East Icelandic Current became colder and fresher. In addition, convective overturning north of Iceland and in the Labrador Sea was minimal, preserving the fresh characteristics of the upper layer. Beginning in the Greenland Sea in 1968, significant quantities of freshwater were advected via Denmark Strait into the Subpolar Gyre. The low salinity waters (called the Great Salinity Anomaly) were tracked around the Labrador Sea, across the Atlantic, and around the Nordic Sea before returning to the Greenland Sea by 1981–1982. Similar transport of low salinity features around the Subpolar Gyre was suggested to have occurred in the early 1900s (Dickson et al., 1988) and in the mid-1980s (Belkin et al., 1998). Belkin et al. (1998) proposed that the source of the mid-1980s salinity anomaly originated in Baffin Bay.

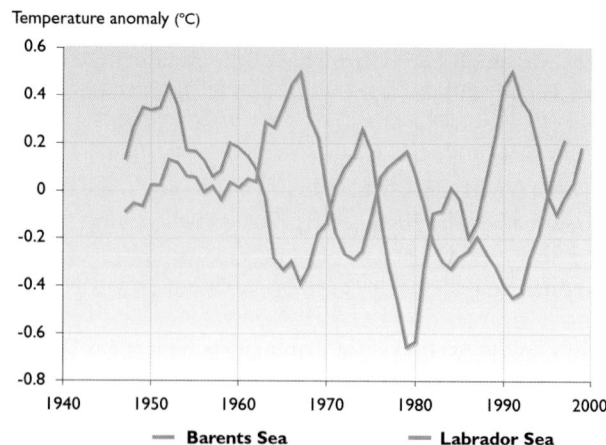

Fig. 9.10. Five-year average winter temperature anomalies (relative to the mean for 1971 to 2000) for the Barents Sea (the Kola Section off northwestern Russia, 0–200 m mean) and the Labrador Sea (Station 27 on the western Grand Bank off Newfoundland, near bottom at 175 m).

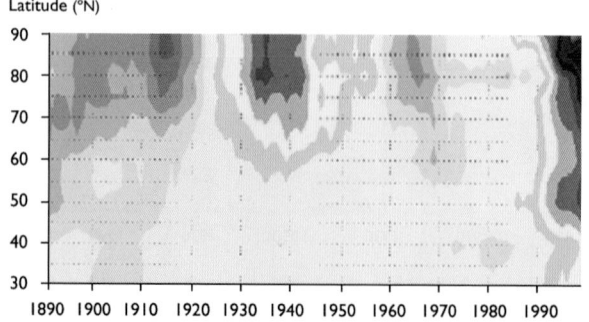

Fig. 9.9. Observed time–latitude variability in surface air temperature anomalies north of 30° N (Johannessen O. et al., 2004).

The deep water of the Norwegian Sea has for a long time been considered to have a relatively stable temperature. However, since the mid-1980s there has been a steady increase of more than 0.05 °C for the waters between 1200 m and 2000 m, and even the deepest water has shown a small temperature increase (Østerhus and Gammelsrød, 1999). In the surface layer there has been a steady decrease in salinity. In the deep, south-flowing waters of the Greenland Sea there has been a 40-year trend toward decreasing salinity and this trend toward decreasing salinity has spread throughout much of the northern North Atlantic (Dickson et al., 2002). Dickson et al. (2003) suggest this may correspond to a general freshening of the whole Atlantic.

Interannual variability in the depth of convection in the Greenland Sea (Budéus et al., 1998; Meincke et al., 1992) and Labrador Sea (Lazier, 1980, 1995) depends upon wind, air temperature, upper layer salinity and temperature, and the pre-winter density structure. Dickson et al. (1996) and Dickson (1997) found the convective activity in the two areas to be of opposite phase, linked to shifting atmospheric circulation as reflected in the NAO index. In the late 1960s when the NAO index was low, there was intense convection in the Greenland Sea and little convection in the Labrador Sea owing to reduced winds and freshwater accumulation at the surface. In contrast, in the late 1990s when the NAO index was high, the reverse occurred with deep convection in the Labrador Sea and minimal convection in the Greenland Sea. Deep-reaching convection in the Greenland Sea contributes to overflow waters but Hansen et al. (2001) did not observe any NAO-like variations in their 50-year time series of Faroe Bank Channel overflow. However, deep convection is only one of several ventilation processes affecting the overflow (see section 9.2.3.4, vertical transfer of water). Hansen et al. (2001) did however find a general decreasing trend in the overflow, as was observed for the overflows across the southern part of the Iceland–Faroe Ridge and the Wyville–Thompson Ridge (Hansen et al., 2003). In the 1990s, higher temperatures offset the corresponding reduced Atlantic inflow to the Nordic Seas such that there was no net change in the heat flux but Turrell et al. (2003) suggested that a reduced salt flux may account for some of the freshening observed in large parts of the Nordic Seas.

Hudson Bay and Hudson Strait

The timing of the sea-ice advance and retreat in Hudson Bay and Hudson Strait varies between years by up to a month from their long-term means. This sea-ice variability has been linked to dominant large-scale atmospheric modes, in particular the NAO and the El Niño–Southern Oscillation (ENSO; Mysak et al., 1996; Wang J. et al., 1994). In years of high positive NAO and ENSO indices, heavy ice conditions occur in Hudson Bay as well as in Baffin Bay and the Labrador Sea. This increase in sea ice is attributed to cold air masses and stronger northwesterly winds over the region.

Between 1981 and the late 1990s air temperatures over Hudson Bay and Hudson Strait increased. This led to an earlier breakup of sea ice (Stirling et al., 1999) and an earlier spring runoff of river discharge into Hudson Bay (Gagnon and Gough, 2001).

Bering Sea

At decadal and longer timescales, the Bering Sea responds to two dominant climate patterns: the Pacific Decadal Oscillation (PDO) and the AO (see Chapter 2 for a detailed discussion). The PDO is strongly coupled to the sea level pressure pattern with stronger winds in the Aleutian low-pressure system during its positive phase (Mantua et al., 1997). It has a major impact on the southern Bering Sea. Thus the 40- to 50-year oscillation in the PDO led to higher sea surface temperatures in the North Pacific from 1925 to 1947 and 1977 to 1998, and cold conditions in 1899 to 1924 and 1948 to 1976. The AO had major shifts around 1977 and 1989 and there has been a long-term strengthening from the 1960s through the 1990s. Heavy sea-ice years in the Bering Sea generally coincide with negative values of the PDO, such as occurred in the early 1970s. The late 1970s and 1980 were warm years with reduced sea-ice cover. Heavy sea ice was again observed in the 1990s, but was not as extensive as in the early 1970s. In the 1990s, there was a shift toward warmer spring temperatures that resulted in sea ice in the Bering Sea melting one week earlier than in the 1980s, and the snow melting up to two weeks earlier (Stabeno and Overland, 2001).

9.2.5. Anticipated changes in physical conditions

During the 1990s it became apparent that global warming would occur more rapidly and with greater impact in the high latitudes (Morison et al., 2000; SEARCH, 2001). Observations showed substantial variability in the arctic water column, atmosphere, ice cover, and export to the North Atlantic (e.g., Belkin et al., 1998; Carmack et al., 1995; Morison et al., 1998; Rothrock et al., 1999; Walsh J.E. et al., 1996). This variability spans temporal scales that include interannual fluctuations, inter-decadal patterns, and long-term trends. The first challenge is to define the temporal scales and magnitudes of arctic variability, for example to distinguish recurrent modes from trends and to separate natural from anthropogenic climate forcing. The second challenge is to understand and predict the impact of changes in the physical environment on the biota.

This section links the various sub-components of the physical system (e.g., land/ocean exchanges, shelf/basin interactions, inter-basin fronts, and the transport of ice and water properties) to climate-scale forcing at seasonal and decadal timescales. The assessment is based on the outcome of Chapter 4, plus the most recent results from the Intergovernmental Panel on Climate Change (IPCC, 2001) and information from the peer-reviewed literature.

9.2.5.1. Atmospheric circulation

General features of projected changes in the arctic atmosphere relevant to marine processes are summarized in Table 9.1. Air temperatures are very likely to increase by 4 to 5 °C over most of the Arctic by 2080. As air temperatures are very likely to increase more in winter than in summer there is very likely to be an associated decrease in the amplitude of the seasonal cycle. The IPCC (2001) reported that some studies have shown increasingly positive trends in the indices of the NAO/AO in simulations with increased concentrations of GHGs. The magnitude and character of the changes vary for the different models. In general, the intensity of winter storms and the zonal temperature gradient are likely to decrease. However, in some regions (e.g., the Labrador, Nordic, Bering, and Beaufort Seas) an increase in storm activity is likely. Storm tracks are likely to shift northward under stronger AO and NAO conditions. Christensen and Christensen (2003) projected that the atmosphere will contain more water under a warmer climate, making more water available for precipitation. Model scenarios project an increase in precipitation of 10% by 2080 and an increase in cloud cover of 8%.

Paeth et al. (1999) assessed changes in the mean and variance of the NAO at decadal scales. They predicted that the mean value will increase, while the variance will decrease, suggesting that the NAO will stabilize in the positive phase. The consequences of such a scenario are likely to be more westerly winds and milder weather over Europe during winter, while the Labrador Sea would be likely to experience more northwesterly winds and colder conditions. Shindell et al. (1999) and Fyfe et al. (1999) also predicted a positive trend in the NAO index. Ulbrich and Chrisroph (1999) concluded that there will be a northeastward shift of the NAO's northern variability center from a position close to the east coast of Greenland to the Norwegian Sea while Shindell (2003) stated that if the dynamic strengthening of the arctic vortex continues the Northern Hemisphere is likely to continue to warm up rapidly during winter.

Despite present uncertainties, it can be concluded that if the NAO increased, it would be likely to lead to increased westerly winds over the North Atlantic and more frequent storm patterns. Any trend toward positive AO conditions would be very likely to result in a weakening of the Beaufort High and increased cyclonicity over Canada Basin, as noted by Proshutinsky and Johnson (1997). Winds over the Bering/Chukchi Seas would probably also weaken. Changes in atmospheric forcing will impact upon most of the features discussed in the following sections; sea-ice conditions, ocean cir-

Table 9.1. Changes in surface and boundary forcing based on model projections and/or extrapolation of observed trends. Unless otherwise specified these projected changes are very likely to happen.

	2020	2050	2080
Air temperature			
annual mean[a]	1–1.5 °C increase	2–3 °C increase	4–5 °C increase
winter	2.5 °C increase	4 °C increase	6 °C increase in the central Arctic
summer	0.5 °C increase	0.5–1.0 °C increase	1 °C increase
seasonality	Reduced seasonality (warmer winters compared to summer)		
interannual variability	No change	No change	No change
Wind			
means	While changes in winds are expected, there is at present no consistent agreement from general circulation models as to the magnitude of the changes in either speed or direction		
storm frequency	Possible increase in storm intensity regionally (Labrador, Beaufort, Nordic Seas); in general, winter storms will decrease slightly in intensity because the pole to equator temperature gradient decreases		
storm tracks	Probable northward shift in storm tracks		
regional issues	In areas of sea-ice retreat, there will be an increase in wind-driven effects (currents, waves) because of longer fetch and higher air–sea exchange		
Precipitation/runoff			
mean[b]	2% increase	6% increase	10% increase
seasonality	Decreased seasonality in runoff related to earlier snow melt. Seasonality in precipitation unclear		
snow on ice	1–2% increase	3–5% increase	6–8% increase
Sea level	5 cm rise	15 cm rise	25 cm rise
Cloud cover			
general	3% increase	5% increase	8% increase
spring, autumn	4–5% increase	5–7% increase	8–12% increase
winter, summer	1–2% increase	3–5% increase	4–8% increase
Cloud albedo	Not available	Not available	Not available

[a]These numbers are averages and should be higher in the central Arctic and lower over southern regions; [b]based on the estimates of precipitation minus evaporation in Chapter 6.

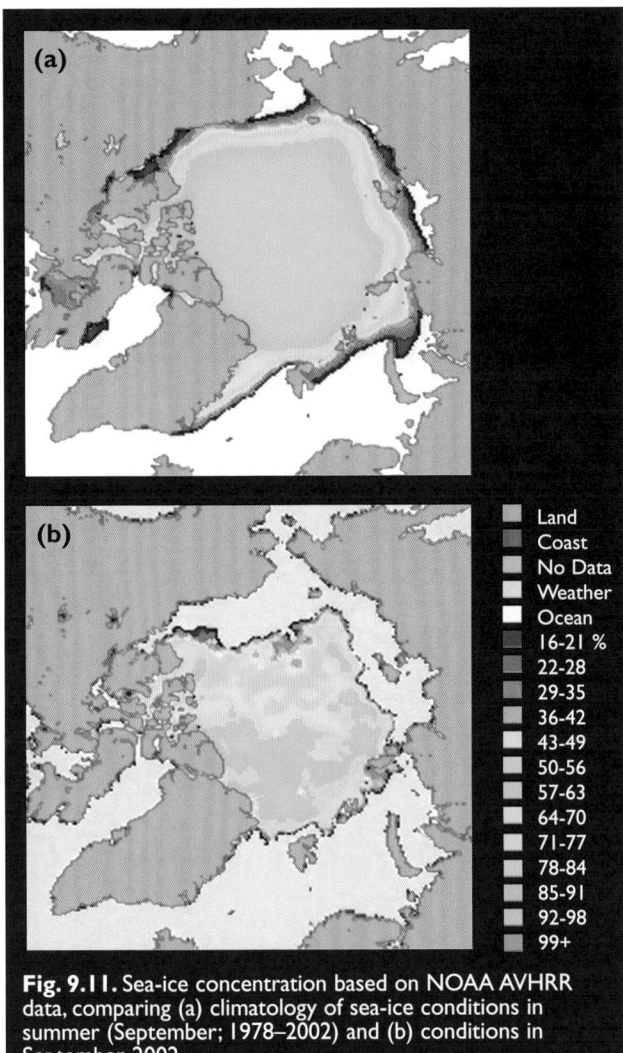

Fig. 9.11. Sea-ice concentration based on NOAA AVHRR data, comparing (a) climatology of sea-ice conditions in summer (September; 1978–2002) and (b) conditions in September 2002.

Legend:
- Land
- Coast
- No Data
- Weather
- Ocean
- 16-21 %
- 22-28
- 29-35
- 36-42
- 43-49
- 50-56
- 57-63
- 64-70
- 71-77
- 78-84
- 85-91
- 92-98
- 99+

for the period 1978 to 2002 (Fig. 9.11, see also Serreze et al., 2003). Substantial changes in sea-ice conditions are evident in Canada Basin north of the Chukchi Sea, and in the sea-ice extent in Fram Strait and north of Svalbard.

Projected changes in sea-ice conditions for the 21st century are summarized in Chapter 6 based on output from the five ACIA-designated global climate models. Tables 9.2 and 9.3 show the maximum and minimum values for sea-ice extent projected by these five models, respectively. The values shown are the adjusted model values, meaning that the data have been "normalized" by forcing a fit to the 1981–2000 baseline observations. The projections vary widely, especially for the summer. The CSM_1.4 (National Center for Atmospheric Research) model consistently projects the greatest sea-ice extent and the least amount of change, while the CGCM2 (Canadian Centre for Climate Modelling and Analysis) model consistently projects the least sea ice and the greatest amount of change. However, all five ACIA-designated models agree in projecting that sea-ice coverage will decrease both in summer and winter.

Areal ice extent

Increases in the AO index are likely to result in the Transpolar Drift taking a strongly cyclonic diversion across the Lomonosov Ridge and into Canada Basin and the Beaufort Gyre shrinking back into the Beaufort Sea (section 9.2.2.5; Fig. 9.3). This is very likely to alter the advective pathways and basin residence times of sea ice formed in winter on the Eurasian shelves. Furthermore, the ice extent in early autumn is also likely to be reduced, due to expected changes in wind forcing and winter air temperature in the eastern Russian Arctic (Rigor et al., 2002). By 2050, the CGCM2 model, which results in the greatest rate of sea-ice melt, projects that the entire marine Arctic may be sea-ice free in summer (Table 9.3). The other four models agree in projecting the presence of summer sea ice, at least until the end of the 21st century, but disagree in their projections of the extent of areal coverage. While the changes in winter sea-ice coverage are generally projected to be much smaller than in summer (Table 9.2), it is likely that the Barents Sea and most of the Bering Sea may be totally ice free by 2050 (see Chapter 6).

culation and water properties, ocean fronts, and thermohaline circulation.

9.2.5.2. Sea-ice conditions

Under scenarios of climate warming, sea-ice cover is expected to "retreat" further into the Arctic Basin, to breakup earlier and freeze-up later, and to become thinner and more mobile. For example, substantial differences in sea-ice conditions were observed in summer 2002 compared to the climatology of sea-ice conditions in summer

Table 9.2. Sea-ice extent in March (10^6 km^2) as projected by the five ACIA-designated models.

Model	1981–2000	2011–2030	2041–2060	2071–2090
CGCM2	16.14	15.14	13.94	13.26
CSM_1.4	16.32	15.00	14.16	14.01
ECHAM4/ OPYC3	16.19	15.62	14.97	14.38
GFDL-R30_c	16.17	15.60	14.86	14.52
HadCM3	16.32	15.53	14.87	13.74

CGCM2: Canadian Centre for Climate Modelling and Analysis; CSM_1.4: National Center for Atmospheric Research; ECHAM4/OPYC3: Max-Planck Institute for Meteorology; GFDL-R30_c: Geophysical Fluid Dynamics Laboratory; HadCM3: Hadley Centre for Climate Prediction and Research.

Table 9.3. Sea-ice extent in September (10^6 km^2) as projected by the five ACIA-designated models.

Model	1981–2000	2011–2030	2041–2060	2071–2090
CGCM2	7.28	3.33	0.55	0.05
CSM_1.4	7.22	7.00	6.72	6.59
ECHAM4/ OPYC3	7.02	6.03	4.06	2.68
GFDL-R30_c	7.28	5.91	4.33	2.91
HadCM3	7.41	6.22	5.12	3.22

CGCM2: Canadian Centre for Climate Modelling and Analysis; CSM_1.4: National Center for Atmospheric Research; ECHAM4/OPYC3: Max-Planck Institute for Meteorology; GFDL-R30_c: Geophysical Fluid Dynamics Laboratory; HadCM3: Hadley Centre for Climate Prediction and Research.

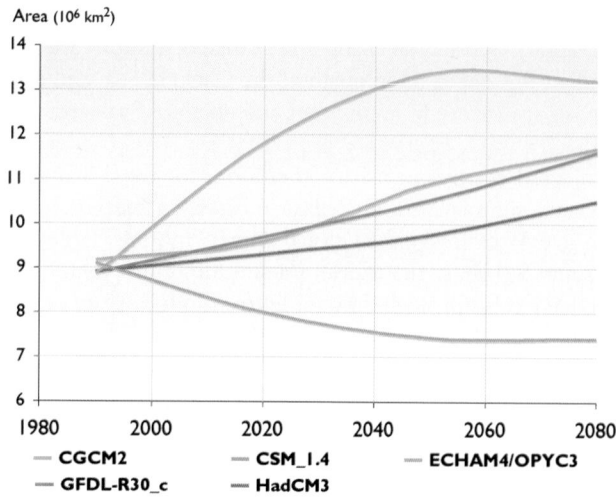

Fig. 9.12. Changes in the areal coverage of the seasonal sea-ice zone over the 21st century as projected by the five ACIA-designated models. The illustration is based on the values in Tables 9.2 and 9.3.

Seasonal sea-ice zone

Every year around 7 to 9 million km² of sea ice freezes and melts in the Arctic (Parkinson et al., 1999). Four of the five ACIA-designated models project that the seasonal sea-ice zone is likely to increase in the future because sea-ice coverage will decrease more during summer than winter (Fig. 9.12). This suggests that sea-ice thickness is also likely to decrease because a single winter of sea-ice growth is an insufficient period to reach equilibrium thickness. There is very likely to be a shorter period of sea-ice cover due to earlier breakup and later freeze-up. Longer ice-free periods will significantly increase sub-surface light availability. (At present, sea ice lingers in the Arctic Ocean through May and June, months of high levels of insolation.) A delayed freeze-up will also expose more open water to forcing by autumn storms. Retreat of the seasonal sea-ice zone northward into the central arctic basins will affect nutrient and light availability on the continental shelves during summer and autumn by increasing the areas of open water, wind mixing, and upwelling.

Fast ice is not explicitly included in climate model scenarios. Although reductions in the extent, thickness, and stability of fast ice are likely to occur, the implications of climate change for fast ice is recognized as a gap in knowledge.

9.2.5.3. Ocean circulation and water properties

Changes in the surface and boundary forcing (Table 9.1) will probably result in changes in ocean circulation, water mass properties, and ocean processes (section 9.2.3). Sea surface temperatures are likely to increase by approximately the same amount as air temperatures in areas that are sea-ice free, but are very likely to remain the same (i.e., near freezing) in ice-covered waters. By 2020 the upper water layer of all arctic shelves is very likely to exhibit stronger seasonality in

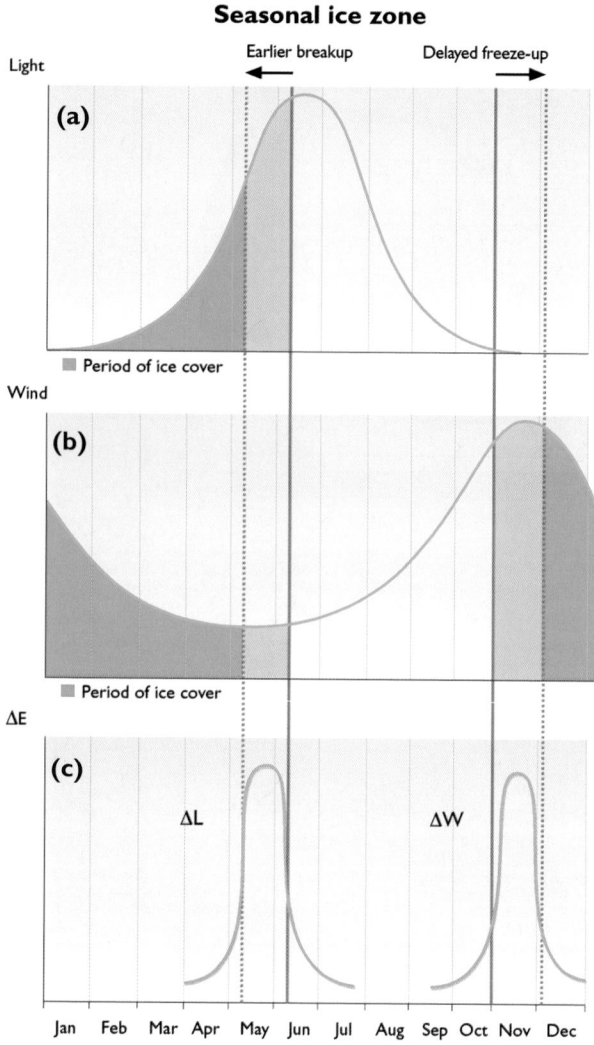

Fig. 9.13. The annual pattern of (a) incoming solar radiation, (b) wind speed, and (c) incremental changes in incoming solar radiation (ΔL) and wind energy (ΔW) to the water column from earlier breakup and delayed freeze-up. Earlier breakup allows a disproportionate amount of solar energy into the water column, while delayed freeze-up exposes the water column to autumn storms.

terms of sea-ice cover, and by 2080, 50 to 100% of the Arctic Ocean is likely to undergo such variability. Whether or not there is a concurrent increase in mixed-layer depth in summer depends on the relative coupling of wind and ocean in the presence or absence of sea ice, which in turn depends on the magnitude of internal ice stress (Wadhams, 2000). McPhee and Morison (2001) argue that, away from coastal areas, in summer most of the wind momentum transferred to the sea ice is subsequently passed on to the water and that sea ice may even serve to enhance the coupling of wind to water. Mixed-layer deepening is also very likely to be influenced by increased river discharge. Sub-sea light levels will increase in areas where sea ice is absent, but are very likely to decrease where sea ice remains due to increased snow.

General circulation models project a strengthening of the AO leading to increased atmospheric cyclonicity over the Arctic Ocean during the 21st century (Fyfe et al.,

1999). This, in turn, is very likely to affect the sea-ice drift and surface currents in the Arctic Basin. With increased cyclonicity the Transpolar Drift is very likely to shift eastward to favor drift directly toward Fram Strait (Rigor et al., 2002). The Beaufort Gyre is very likely to weaken and retreat into Canada Basin, following the "cyclonic mode" discussed by Proshutinsky and Johnson (1997) and Polyakov and Johnson (2000). In turn, changes in the Beaufort Gyre are very likely to affect the storage and release of freshwater in Canada Basin (Proshutinsky et al., 2002). Under this scenario the position of the Atlantic/Pacific Front will tend to align with the Alpha-Mendeleyev Ridge rather than the Lomonosov Ridge (McLaughlin et al., 1996; Morison et al., 1998) and perhaps retreat further into Canada Basin. Modeling studies by Zhang and Hunke (2001) and Maslowski et al. (2000) are in general agreement.

Relatively small changes in the timing of sea-ice breakup and freeze-up (of the order of a few weeks) are very likely to have a disproportionate effect on the physical forcing of arctic waters (Fig. 9.13). For example, under present-day conditions, much of the incoming solar radiation during the long summer days is reflected back by the ice and snow and so does not reach the water column to warm it. A later freeze-up will mean that the ocean surface is exposed to wind forcing by autumn storms for a longer period of time. Combined with prolonged exposure of the shelf break to wind forcing, this is very likely to enhance vertical mixing and the shelf–basin exchange of heat, salt, nutrients, and carbon.

Thresholds for change in the Arctic Ocean

Three potential thresholds for substantial changes in ocean circulation and water mass properties are described in this section (also see Box 9.4).

1. If and when the seasonal sea-ice zone retreats annually beyond the shelf break.
2. If and when the Arctic Ocean becomes sea-ice free in summer.
3. If and when parts of the deep arctic basins (e.g., the western Nansen Basin and western Canada Basin) remain sea-ice free in winter.

The seasonal retreat of sea ice from shelf domains to the deep Arctic Basin, anticipated as soon as 2020 (Tables 9.2 and 9.3, and Chapter 6), will expose the shelf-break region to upwelling- and downwelling-favorable winds, both for longer and more often. The coupling of wind and water in the presence of sea ice is not straightforward and can be of greater significance than in sea-ice free waters if internal ice stress (a function of ice concentration and compactness) is sufficiently small (McPhee and Morison, 2001). It is thus likely that a zone of maximum coupling exists in the transition from full sea-ice cover to open water, and that if this zone were located over the shelf break, then shelf–basin exchange would also increase (Carmack and Chapman, 2003). Such exchange would draw more Pacific- and Atlantic-origin waters onto the shelves, with an associated increase in the delivery of salt, heat, and nutrients.

Box 9.4. Effects of climate change in the Arctic on global ocean circulation and climate

The Arctic plays a key role in the global climate through its production of North Atlantic Deep Water (NADW). North Atlantic Deep Water is formed by the mixture of waters produced by thermohaline ventilation in the Arctic Mediterranean, entrained Atlantic water, and water convected in the Labrador Sea. Once formed the NADW flows southward through the Atlantic Ocean and, together with the denser Antarctic Bottom Water (AABW), forms the source of all the deep and bottom waters of the World Ocean. NADW is of considerable significance for the global thermohaline circulation (THC).

If climate change should result in reduced thermohaline ventilation in the Arctic, there is considerable – although not unambiguous – evidence for a reduced NADW–THC through the Atlantic (Ganachaud and Wunsch, 2000; Munk and Wunsch, 1998; Rahmstorf and England, 1997; Toggweiler and Samuels, 1995). A proper understanding of this scenario requires an understanding of the relative magnitudes of the NADW and the AABW contributions to the global THC. Traditional estimates of NADW production (e.g., Schmitz and McCartney, 1993) are of the order of 15 Sv and this is supported by modern estimates based on the WOCE (World Ocean Circulation Experiment) data set (Ganachaud and Wunsch, 2000). Estimates of AABW production are less consistent, but even the highest estimates (Broecker et al., 1999) indicate that AABW production is currently significantly less than NADW production. The NADW is therefore considered to account for more than half the deep-water production of the World Ocean at present. The latest IPCC assessment (IPCC, 2001) concludes that most models show a weakening of the Northern Hemisphere THC, which contributes to a reduction in surface warming in the northern North Atlantic. The more extreme scenario of a complete shutdown of the THC would have a dramatic impact on the climate of the North Atlantic region, on the north–south distribution of warming and precipitation, on sea-level rise, and on biogeochemical cycles (IPCC, 2001). The IPCC concluded this to be a less likely, but not impossible, scenario. More reliable estimates of its likelihood and consequences require more reliable coupled ocean–atmosphere models than are presently available (see section 9.7).

Box 9.5. The Chukchi albedo feedback loop: An Achilles Heel in the sea-ice cover of the western Arctic?

Attention has long focused on the role of Atlantic inflow waters in the transport of heat within the Arctic Basin, and its potential to impact upon the overlying sea ice should the arctic halocline weaken or break down. This box highlights the potential for Pacific inflow waters to impact upon the overlying sea ice, and the potential for this inflow to amplify locally the well-known albedo feedback mechanism.

Pacific inflow waters are warmed in summer as they travel northward across the seasonally ice-free parts of the Bering and Chukchi Seas. On reaching the shelf break, these waters subduct below the polar mixed layer and enter the arctic halocline, forming Pacific Summer Water (PSW), identified by a shallow temperature maximum at depths 40 to 60 m and salinities near 31.5. The water at the temperature maximum may be higher in years with extensive open waters over the Bering and Chukchi Seas. Summer climatological data (see panel, Timokhov and Tanis, 1998; Shimada et al., 2001) demonstrates the accumulation of such water within the Beaufort Gyre over the eastern flank of the Northwind Ridge.

One possibility for the fate of this stored heat within the PSW in the southwestern Canada Basin is that it acts to retard the growth of sea ice during the subsequent winter. Therefore, the amount of summer melting (freshwater addition) and winter freezing (freshwater removal) are not balanced, and ice floes drifting over the "warm patch" will be thinner than in surrounding waters. The thin sea ice observed east of the Northwind Ridge, also noted by Bourke and Garrett (1987), is evidence of local PSW influence. When winter sea-ice growth falls below a critical (and unknown) value at the start of the melt season, the thickness and concentration of sea ice over the region would be sufficient to reduce albedo and initiate further sea-ice reduction, thus initiating a feedback. An alternate explanation for the record low sea-ice concentrations in summer 2002 is given by Serreze et al. (2003).

The disappearance of sea-ice cover in the Arctic in summer, as projected by the CGCM2 model by 2050, will have far reaching effects on upper-layer circulation and water properties. Direct exposure of surface waters to wind will enhance wind-driven circulation. Also, it is probable that wind-driven vertical mixing will increase the depth of the surface mixed layer, depending upon the strength of local stratification. For example, wind-driven deepening of the mixed layer is very likely to be more pronounced in the more weakly stratified Nansen Basin than in Canada Basin with its strong Pacific influence. Concurrently, the seasonal sea-ice zone is very likely to increase (perhaps by 10 million km^2) owing to projections that the rate of decrease in sea-ice cover for summer will be greater than for winter (Fig. 9.12; Tables 9.2 and 9.3).

Some model scenarios project that by 2080 the formation of sea ice in winter will no longer completely cover the Arctic Basin. If this does occur, two parts of the Arctic Basin are potential sites for sea-ice free or at least decreased ice concentrations in winter: the western Nansen Basin and western Canada Basin.

The first site is the weakly stratified western Nansen Basin adjacent to the inflow and subduction of Atlantic waters (Martinson and Steele, 2001). Here the incoming waters are warmest and the overlying halocline is weakest. At present, this region has the deepest winter mixed layer in the central Arctic Ocean. Under the extreme climate change scenario in which sea ice in winter no longer completely covers the Arctic Basin, the Nansen Basin is likely to become a region of strong convection and deepwater formation. However, the dynamics are more likely to resemble the present day Nordic Sea system, i.e., with deeper mixed-layer ventilation or convection (see Muench et al., 1992 and Rudels et al., 2000 for a discussion of water masses). However, this argument supposes an increased transport of warmer Atlantic water into the Arctic whereas some models (e.g., Rahmsdorf, 1999) suggest a weakening or southward shift of the THC.

The second site is located in the western Canada Basin immediately north of the Chukchi Sea and above the Northwind Ridge (Box 9.5). This area is adjacent to the inflow of shallow and relatively warm summer water through the Bering Strait and across the Chukchi Sea. The spread of this relatively warm water takes place within the Beaufort Gyre at depths of 40 to 60 m, and is thus within the limits of winter haline convection (Shimada et al., 2001).

At both sites, it is their proximity to warm inflows from the Atlantic and Pacific that establishes conditions that may reduce winter sea-ice cover. It is not clear, however, if release of heat from subsurface sources would serve to melt the sea ice, or merely keep new ice from forming. In either case ecosystems currently located in the Nordic and Bering Seas are very likely to shift northward.

Changes in the Nordic and the Barents Seas

An 80-year CMIP2 integration (1% per year increase in the atmospheric CO_2 concentration) with the Bergen Climate Model (BCM) was used to estimate changes in the Nordic and Barents Seas (Furevik et al., 2003). This model has a relatively high spatial resolution in these areas and is believed to give as reliable projections for these areas as can be obtained at present. However, in common with other such models, its predictive capability is limited and the results presented should be seen as possible, rather than likely outcomes.

The evolution of the winter sea surface temperature field is shown in Fig. 9.14. From the present to 2020 a minor cooling is projected over most of the area. The greatest decrease is projected to occur along the marginal ice zone in the Barents Sea and off the East Greenland coast, with a maximum decrease of more than 1 °C projected in Denmark Strait. Some of this cooling is likely to be associated with the weaker westerlies projected for this period (Furevik et al., 2002). In the central Nordic Seas a warming of 0.5 °C is projected. By 2050, the entire Nordic Seas are projected to become warmer with the exception of a small area in Denmark Strait. The largest warming is projected to occur in the northeastern Barents Sea and to the south of Iceland. With the doubling of the atmospheric CO_2 concentration assumed by 2070, surface temperatures in the Nordic Seas are projected to increase by 1 to 2 °C, with the highest values in the Barents Sea. Minimum warming (<0.5 °C) is projected in the Denmark Strait.

(a) **(b)**

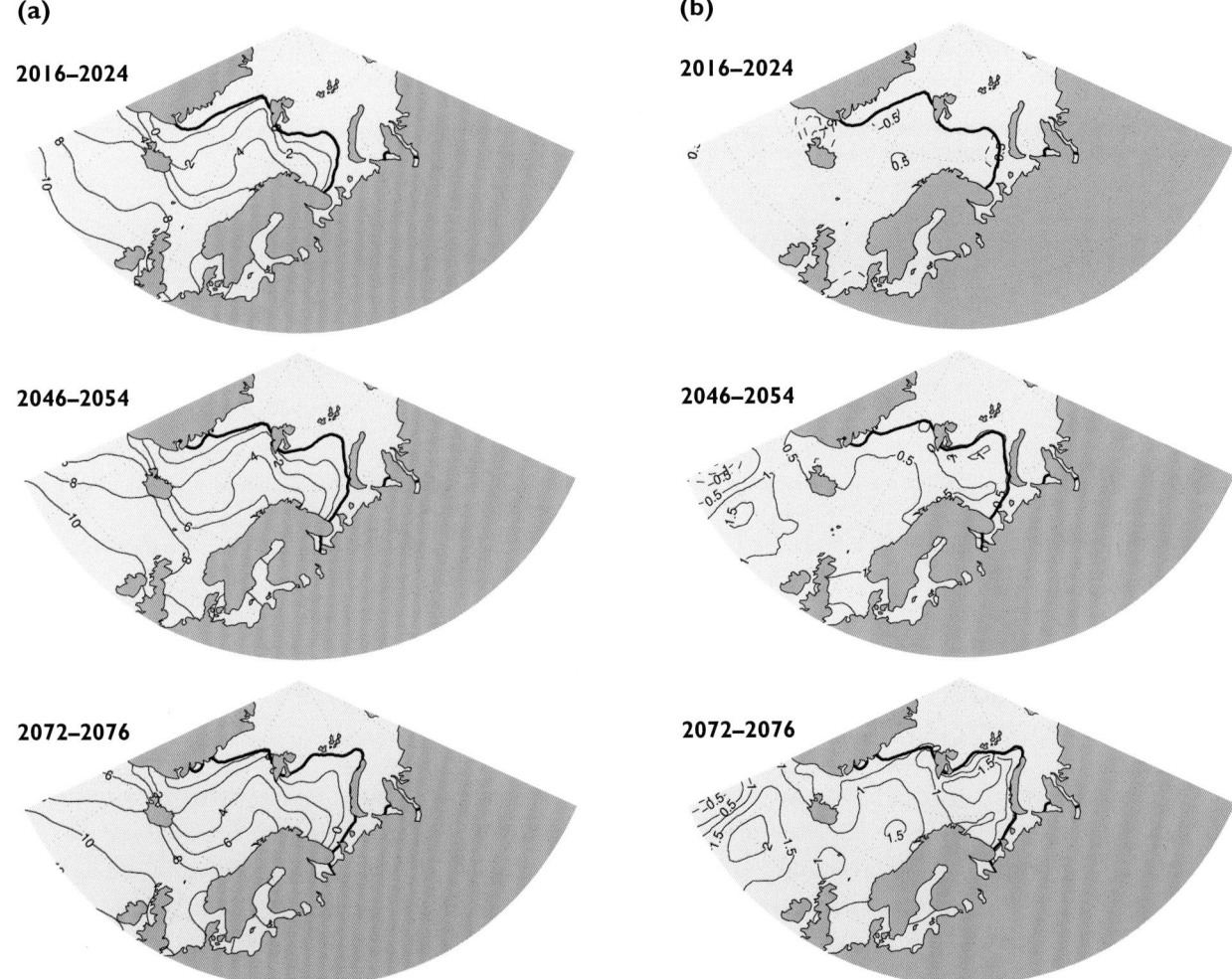

Fig. 9.14. Evolution of the sea surface temperatures and the sea-ice edge (heavy black line) in the BCM CMIP2 integration; (a) shows the March sea surface temperatures and sea-ice distribution around the years 2020, 2050, and 2075, (b) shows projected changes from 2000 to 2020, 2050, and 2075, respectively (Furevik et al., 2002).

Projected salinity changes in the Nordic Seas are generally small, except for areas influenced by coastal runoff and the melting of sea ice. By 2020, there is projected to be a freshening (a salinity decrease of 0.1 to 0.3) in the southeast Barents Sea and the Kara Sea, and a weak freshening along the East Greenland coast. The freshening continues to the 2050s, with salinity reductions north of Siberia in the range 0.1 to 0.5. A significant freshening is also projected in the Arctic Ocean (a salinity decrease of 0.3 to 0.5), which is advected southward with the East Greenland Current into the Denmark Strait and East Icelandic Current. The arctic waters are projected to become slightly more saline, but not exceeding a salinity increase of 0.1. By the 2070s, the model output suggests 0.1 to 0.2 more saline water south of the inflow area, and less than 0.1 more saline arctic waters in the Nordic Seas. North of Siberia and in the Arctic Ocean, salinities are projected to decrease by 0.5 to 1.0, and a tongue of fresher water is projected along the East Greenland Coast.

In terms of volume flow, from 2000 to 2020 the Bergen Climate Model projects a small (<10%) increase in the net Atlantic inflow through the Iceland–Scotland Gap, mainly near Iceland, and a corresponding increase in the Denmark Strait outflow. There is generally a weakening by a few percent of the cyclonic gyre in the Nordic Seas. By 2050, the Nordic Seas gyre is projected to have weakened by a further 10%. A greater inflow of arctic waters is projected via the eastern branch (east of the Faroe Islands), and less via the western. No significant changes are projected for the Barents Sea. Toward 2070 a further reduction in the internal cyclonic flow in the Nordic Seas is projected. There is also a strengthening (~0.25 Sv, ~12%) in the transport of arctic waters through the Barents Sea with a compensating reduction through Fram Strait (Furevik et al., 2002).

Seas of the North American Arctic

Projections of change in the Bering, Chukchi, and Beaufort Seas, the Canadian Archipelago, Baffin and Hudson Bays, and the Labrador Sea are highly uncertain as many important aspects of these regions (e.g., the presence of fast ice, strong seasonality, complex water mass structure, through flow) are not included in the current global climate models. The following discussion is thus highly speculative.

These seas are expected to experience the general changes in sea ice, sea surface temperature, mixed-layer depth, currents, fronts, nutrient and light levels, air temperature, winds, precipitation and runoff, sea level, and cloud cover summarized in Tables 9.1 and 9.4, but owing to their more southerly latitude and contact with terrestrial systems, the changes may be greater and perhaps faster. Because the Bering/Chukchi shelf is very shallow the effects of the albedo feedback mechanism

Table 9.4. Summary of changes projected in ocean conditions according to the five ACIA-designated models relative to baseline conditions. Unless otherwise specified these projected changes are very likely to happen.

	2020	2050	2080
Sea ice			
duration	Shorter by 10 days	Shorter by 15–20 days	Shorter by 20–30 days
winter extent	6–10% reduction	15–20% reduction	Probable open areas in high Arctic (Barents Sea and possibly Nansen Basin)
summer extent	Shelves likely to be ice free	30–50% reduction from present	50–100% reduction from present
export to North Atlantic	No change	Reduction beginning	Strongly reduced
type	Some reduction in multi-year ice, especially on shelves	Significant loss of multi-year ice, with no multi-year ice on shelves	Little or no multi-year ice
landfast ice	Possible thinning and a retreat in southern regions	Probable thinning and further retreat in southern regions	Possible thinning and reduction in extent in all arctic marine areas
Sea surface temperature			
winter/summer (outside THC regions and depending upon stratification and advection)	An increase by about the same amount as the air temperatures in ice-free regions. No change in ice-covered regions		
seasonality	All shelf seas to undergo seasonal changes	30–50% of Arctic Ocean to undergo seasonal changes	50–100% of Arctic Ocean to undergo seasonal changes
Mixed-layer depth	Increase during summer in areas with reduced ice cover and increased wind		
Currents	In regions affected by THC, modifications to the THC will change the strength of the currents		
Ocean fronts	Fronts are often tied to topography but with altered current flows, may rapidly shift their position		
Light exposure	With decreasing ice duration and areal extent, more areas to be exposed to direct sunlight		
Nutrient levels	Substantial increases over the shelf regions due to retreat of the sea ice beyond the shelf break	High levels on shelves and in deep arctic basins; higher levels due to deeper mixed layer in areas of reduced ice cover	

are likely to be amplified as water moves across ice-free parts of the shelves. Preconditioning of Pacific inflow waters during their transport across the shelf supplies a reservoir of heat at shallow depths within the offshore halocline, which may affect conditions to the east on the Beaufort Shelf (Box 9.5). Such heat could potentially retard sea-ice growth the following winter.

The Canadian Archipelago is a large (~2.9 million km², including Foxe Basin and Hudson Strait) and complex shelf domain for which it is particularly difficult to draw conclusions regarding global warming (Melling, 2000). Sea ice remains landfast for more than half the year there, but the presence of fast ice is not included in the global climate models. The general trends projected by the ACIA-designated models and summarized in Table 9.1 are likely to be representative for this region. Two additional features of the Canadian Archipelago are (1) that it serves as a passageway for water masses moving from the Arctic Ocean to the North Atlantic via Baffin Bay and the Labrador Sea, and (2) that its sea-ice domain is a variable mixture of local growth and floes imported from the Arctic Basin, and that transport through the Canadian Archipelago is governed in the present climate by ice bridges across connecting channels (Melling, 2002).

Large uncertainties exist in the changes projected for the Labrador Sea. If the NAO increases as some models project, then there are likely to be stronger northwesterly winds and colder air masses over this region. This would lead to increased sea-ice cover, colder water temperatures, and increased deep convection. Conversely, general atmospheric warming would lead to warmer water temperatures, decreased sea-ice cover, and decreased convection. The slight increase in precipitation may possibly lower salinities over the Labrador Sea, with the largest decline occurring over the shelves due to the accumulation of river discharges. Temperatures in the region are also likely to be greatly influenced by the relatively warm Irminger Current inflow but given the poor understanding of future wind fields, changes in its strength are highly uncertain. Polynyas, such as the North Water Polynya in northern Baffin Bay, owe their existence, at least in part, to winds that move sea ice from the area of its formation southward, so maintaining the area as open water even in the middle of winter. If the winds change, the number and size of polynyas are also likely to change.

9.2.5.4. Ocean fronts

Open ocean fronts generally separate water masses, are associated with strong current flows, and act as barriers for marine organisms. It is difficult, however, to provide reliable estimates of how fronts will respond to climate change since few models provide such information. Most of the deep ocean fronts are linked to bottom topography and so it is likely that these will maintain their present positions, e.g., along the Mohn Ridge in the Greenland Sea and around the Svalbard Bank in the

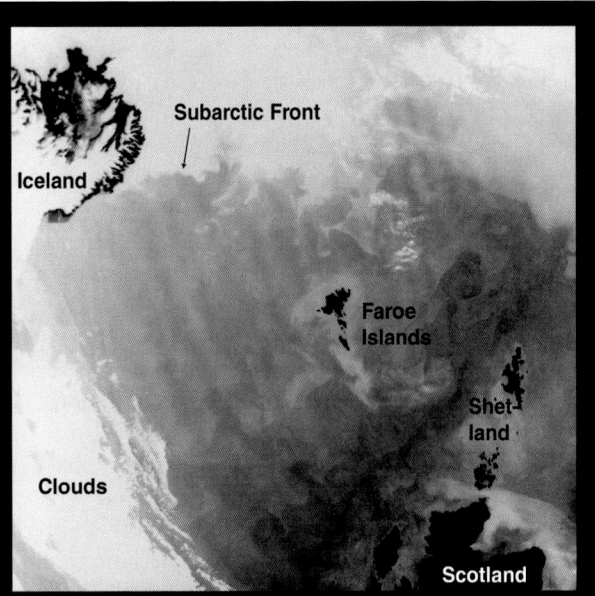

Fig. 9.15. Infrared satellite image showing the position of the Subarctic Front (also called the Iceland–Faroe Front in this region) between Iceland and Scotland on 18 May 1980. Dark areas indicate warm water, and light areas indicate cold water, except where cloud cover occurs (satellite image supplied by University of Dundee, UK).

Barents Sea. However, where topographic steering is weak, fronts may disappear or be displaced. The eastern part of the Polar Front in the Barents Sea is very likely to disappear as a result of climate change (Loeng, 2001). In the Norwegian Sea, the front to the east of Iceland is likely to move northeastward to the position it occupied during the warm period at Iceland between 1920 and 1964.

The reduced inflow of Atlantic water projected by some models would be very likely to shift ocean fronts toward the continental slope region. For example, in the Norwegian Sea the Subarctic Front separates Atlantic and arctic waters, typically lies a few hundred kilometers north of the Faroe Islands (Fig. 9.15), and reduced inflow would be likely to move the front closer to or even onto the Faroe Shelf. If such a shift takes place a cooling of the order of 5 °C would possibly occur in the areas affected. Assessing the likelihood of its occurrence is, however, far beyond the capability of present-day models.

9.2.5.5. Possibility and consequences of altered thermohaline circulation

A major uncertainty in projecting the extent of climate change in the Arctic concerns the response of the THC to altered freshwater flux. In turn, the THC of the Arctic is an integral part of the global THC (see Boxes 9.2 and 9.4). At present, climate models do not generate unambiguous results. Some project a significant weakening, or even collapse, of the THC, while others project a stable THC. An alternative view is that the THC will not weaken or shut down, but that the sites of ventilation will relocate north or south within the system (Aagaard and Carmack, 1994; Ganopolski and Rahmstorf, 2001).

Several coupled atmosphere–ocean general circulation models have been used to simulate the effects of increased GHG emissions on the North Atlantic THC. Rahmstorf (1999) summarized the outcome of six such simulations, all of which projected a weakening of the Atlantic overturning. Latif et al. (2000), however, did not find weakening of the overturning in their model. The models tend to agree that global climate change is very likely to include increased freshwater input to the Arctic Mediterranean, but tend to disagree on the associated consequences. Much of the uncertainty involves the response of the Atlantic inflow and the positive feedback mechanism that it can induce through salt advection. In the simulation by Latif et al. (2000), the feedback mechanism was counteracted by the increased salinity of the Atlantic inflow to the Arctic Mediterranean. The salinity increase in their model was explained by increased freshwater transport from the Atlantic to the Pacific in the tropical atmosphere. Latif (2001) used the observed salinity increase at Bermuda to support their conclusions. In the Nordic Seas, observations indicate the opposite with a general freshening in the upper layers (Blindheim et al., 1999; Verduin and Quadfasel, 1999).

Most of the general circulation models that project a weakening of the THC project a reduction of no more than 50% for the 21st century (IPCC, 2001). Some, however, project instabilities and the possibility of a more or less total collapse of the THC when the intensity of the circulation falls below a certain threshold (Tziperman, 2000). Although such results may explain the instabilities reported for the glacial climate state, their applicability to a GHG-warming scenario cannot be assessed objectively at present.

Observations of the salinity of overflow water in the Atlantic confirm a long-term decrease (Dickson et al., 2002). However, observational evidence for or against a reduction in the THC itself is uncertain. While many observations indicate a reduction in deep convection in the Greenland Sea since 1970, deep convection to depths below the sill level of the Greenland–Scotland Ridge is still occurring (Budéus et al., 1998; Gascard et al., 2002; Meincke et al., 1997) and there are also

other sources of dense water. Observations of the Faroe Bank Channel overflow (Fig. 9.16) indicate significant decreases in volume flux during the latter half of the 20th century, especially since 1970 (Hansen et al., 2001). A lack of similar data for the Denmark Strait overflow leaves open the question as to whether the change in the Faroe Bank Channel overflow is representative of the total overflow flux.

A significantly weakened THC in the Nordic Seas is thus a possible scenario. Reduced ventilation implies reduced renewal rates for deep water in some of the basins, and this seems to be happening in the Norwegian Basin (Østerhus and Gammelsrød, 1999). These changes are slow, however. The magnitude of the inflow weakening and its spatial extent will possibly be influenced by changes in the wind field (Blindheim et al., 1999). The waters most likely to be affected in this scenario are those to the north of Iceland and the Faroe Islands, those in the southern Norwegian Sea, and those in the Barents Sea.

The situation in the waters to the north of Iceland, where the present-day climate is associated with a highly variable Atlantic inflow, can be used to illustrate a potential impact of climate change. Hydrographic investigations show clear seasonal variation in this inflow, with a maximum inflow in summer. There are, however, pronounced interannual differences in the variability of the inflow that affect the temperature, salinity, and stability of the water column (Ástthórsson and Vilhjálmsson, 2002; Thordardóttir, 1977). Most of the Atlantic inflow (80–90%, see Fig. 9.7) enters the Arctic Mediterranean through the Norwegian Sea. This region is characterized by abnormally high sea surface temperatures (up to almost 10 °C) compared to zonal averages (Rahmstorf and Ganopolski, 1999). Much of this elevation is due to the heat flux from the inflowing Atlantic water. The temperature decrease in some areas, especially in winter, resulting from a severely weakened Atlantic inflow would thus be much larger than the projected warming (Chapter 4) by the end of the 21st century according to certain models (Seager et al., 2002). Thus, there is the possibility that some areas of the Arctic Ocean will experience significant _regional_ cooling rather than warming, but present models can assess neither the probability of this occurring, nor its extent and magnitude. According to Rahmstorf (2003) the extent to which Europe's mild winters depend on the transport of heat by the North Atlantic Current is presently unknown.

9.3. Biota

Following a general introduction to the biota of the marine Arctic, this section reviews the dominant species and, where possible, presents relevant life history and ecological information. The section then addresses the influence of physical factors on the biota and discusses variations in abundance and distribution observed in response to past climate fluctuations. The section concludes by presenting possible future changes in the arctic biota induced by the projected

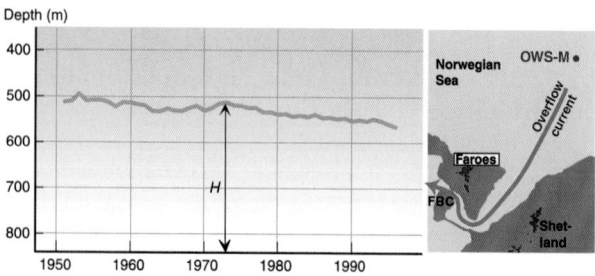

Fig. 9.16. Temporal variability in the intensity of the overflow through the Faroe Bank Channel. The lavender line shows the five-year running mean for the depth of the 28.0 density surface ($\gamma_\theta = 28.0$ kg/m^3) at Ocean Weather Ship M (OWS-M). This surface is considered the upper limit of the dense overflow water and its height (H) above the sill level of the Faroe Bank Channel (FBC) is used as an indicator for overflow intensity. A deepening trend in the density surface implies a decreasing overflow intensity through this channel (Hansen et al., 2001).

changes in the atmospheric forcing functions and potential future sea-ice conditions discussed in Chapter 6. Salmon ecology and response to climate change are addressed in Chapters 8 and 13.

9.3.1. General description of the community

Biological production in the oceans is based primarily on phytoplankton or planktonic algae. These are microscopic unicellular plants that mostly reside within the water column but in the Arctic are also found in and on the sea ice. Through photosynthesis, they reduce CO_2 while releasing oxygen and producing carbohydrates. The carbohydrates are converted, according to the needs of the algae, into essential compounds such as proteins and nucleic acids by incorporating nitrogen, phosphorus, sulfur, and other elements.

The organic matter produced by the algae is primarily consumed by herbivorous (i.e., plant-eating) animals, mainly zooplankton, which in turn may be eaten by fish. The fish are then consumed by seabirds and mammals, including humans. Each segment of the food web within which organisms take in food in the same manner is called a trophic level. Thus phytoplankton are considered the first tropic level, zooplankton the second, etc. The loss of organic matter between one trophic level and the next is about 75 to 80%. The main losses are associated with respiration (i.e., the burning of food) within the organisms themselves, consumption by bacteria (i.e., microbial degradation) of dissolved organic matter, and sinking cellular remains and fecal pellets (i.e., the body's waste). These processes all result in the release of CO_2 or nutrients. Only a small fraction of the organic matter reaches the seabed – the deeper the water column, the smaller this fraction (Box 9.6).

Table 9.5. Average carbon biomass and annual carbon productivity for different trophic levels within the Barents Sea, compared with that for human populations in Norway and Japan. Data recalculated from Sakshaug et al. (1994).

	Biomass (mg C/m²)	Productivity (mg C/m²/yr)
Bacteria	400	60000
Phytoplankton	2000	90000
Zooplankton (copepods and krill)	>3000	9500
Zoobenthos[a]	5160	1550
Capelin[b]	600	300
Cod[c]	300	100
Minke whales	110	2.6
Seals	30	0.5
Seabirds	2.5	0.4
Polar bears	0.25	0.027
People, Norway	107	1.5
People, Japan	2200	22

[a]Interannual biomass variation, 3000–7350 mg C/m² (Denisenko and Titov, 2003); [b]Interannual biomass variation, 30–700 mg C/m²; [c]Interannual biomass variation, 150–700 mg C/m².

Pelagic ecosystems are those which occur within the water column of the open ocean away from the ocean floor. Arctic pelagic ecosystems, like pelagic ecosystems elsewhere and in contrast to terrestrial ecosystems, are dominated by animal biomass. In the Barents Sea, for example, the mean annual plant biomass is 2 g C/m² whereas the mean annual animal biomass is at least four times more. Globally, annual marine primary production is about 40 Pg C (i.e., 10^{15} grams) or 40% of the

Box 9.6. Organisms in the food web

Population abundance, whether for algae, fish, or polar bears (*Ursus maritimus*), is dependent on the population growth and death rates. Given a growth rate higher than the death rate, the population size will increase, and *vice versa*. If the two rates are equal, the population is in steady state. Another variable is population migration; stocks may arrive in or leave a given ecosystem. Essentially, a change in any environmental variable, including those affected by climate change, has a direct impact on one or more processes by changing their rate, which in turn causes a change in population biomass. Thus, while the population growth rate is determined by light or nutrient levels (algae), or food availability (animals), the loss rate represents the sum of losses due to natural death, pollution, sedimentation, and being eaten, fished, or hunted.

Populations can be arranged hierarchically within a food web on the basis of what they eat, with the lowest trophic level comprising photosynthetic organisms. Animals can move up the food-web hierarchy as they grow, by becoming able to eat larger prey. Because most of the food intake is spent on maintaining life, reproduction, movements, etc., only 15 to 25% contributes to population growth, which during steady state represents food for the next trophic level. Consequently, marine food chains are short, with a maximum of five trophic levels.

In models, population growth is described by exponential functions in which growth and mortality rates themselves are functions of environmental change, including changes caused by the evolution of the ecosystem itself (feedback). Ideally, ecosystem models should include all trophic levels, including major species as separate entities; however, coupling plankton and fish is difficult, as is the coupling of fish and higher animals.

Box 9.7. Sea-ice communities

Sympagic organisms are those that live in close association with sea ice, either within channels in the ice itself, on the underside of the ice, or at the interface with the water immediately below the ice. The organisms that inhabit this environment are highly specialized, but cover a wide taxonomic range, from bacteria and simple algae, to vertebrate fauna. Some species, particularly microorganisms, become incorporated into the sea ice as the ice crystals grow. While it may seem an inhospitable environment, the sea ice is actually quite a stable and organically-enriched environment for those organisms that can tolerate its extreme conditions. While some organisms occupy the sea ice as it forms, others actively or passively migrate into the ice ecosystems. Organisms that live within the interstitial spaces of sea ice include microfauna such as protists, and larger organisms such as ciliates, nematodes, rotatorians, turbellarians, and copepods. Multi-year sea ice has the most complex communities and often serves as a platform for colonizers to young ice. In addition, the abundance and biomass of the multi-year sea ice organisms can be very high. For example, copepods may easily exceed densities of 150 individuals per square meter. Given the correct conditions vast algal mats can form on the under-surface of sea ice, including both microalgae and macroalgae such as sea-tangle (*Fucus distichus*), and their associated epiphytic organisms such as *Pylaiella littoralis*. In comparison, seasonal sea ice normally has lower densities and lower biomasses and tends to support more simple communities.

The spatial distribution of sea-ice fauna is generally patchy, even within single ice fields, because the origin, history, size, snow thickness, and the thickness of the sea ice itself can vary dramatically. Interannual variability within a community is also high. In areas such as the Barents Sea, the size of the Atlantic Water inflow varies from year to year, causing dramatic changes in the sympagic community. Nevertheless, sympagic communities are characterized by fauna that can withstand high levels of variation in food availability and low temperatures. Generally, the older the sea ice, the more complex and established the sympagic community. Ice-living invertebrates tend to have low basal metabolic rates, concomitant slow growth, and long life cycles. For example, some arctic ice amphipods live for five or six years, while their more temperate counterparts, or amphipods in other arctic habitats, have average life spans of two to three years. Large, lipid-rich ice-dwelling amphipods are prime prey for the circumpolar polar cod (*Boreogadus saida*). This small, arctic fish is an opportunistic feeder that can live pelagically, or in association with sea-ice communities. In ice-filled waters its diet largely comprises *Themisto libellula* and *Apherusa glacialis*. The fish capture these small "lipid packages" and convert them into prey that is substantial enough to support higher vertebrates such as seabirds and marine mammals. Some seabirds and marine mammals also eat large invertebrate ice-dwellers directly. Black guillemots (*Cepphus grylle*) and thick-billed murres (*Uria lomvia*) feed on the amphipod *Gammarus wilkitzkii*. Little auks (*Alle alle*) and ivory gulls (*Pagophila eburnea*) also eat sympagic amphipods. First-year harp seal (*Phoca groenlandica*) feed extensively on sympagic amphipods when they start to self-feed. However, the preferred prey within the sympagic community is polar cod for most marine mammals.

Thinning, and reduced coverage of arctic sea ice will have dramatic impacts on the entire sympagic ecosystem, particularly on interstitial organisms as these do not have alternate habitats in which to live. Also, given that the sympagic community is important in providing pelagic and benthic communities with food, particularly during the summer when the sea ice melts, changes in this highly specialized environment are likely to have repercussions throughout the arctic marine community as a whole.

---

global total (marine plus terrestrial) production. Macroalgal biomass (i.e., large plants such as kelp and sea-tangle) in the Arctic is believed to be small due to habitat restrictions caused by freezing, ice scouring by small icebergs, and local freshwater input. In some areas, however, macroalgal biomass can be large as kelp forests do occur in the Arctic.

Generally, the higher the trophic level, the smaller the production. In the Barents Sea, major pelagic fish species represent a few hundred milligrams of carbon biomass per square meter and seabirds and polar bears, only 2.5 and 0.25 mg C/m^2, respectively (Table 9.5). The table shows the inverted biomass pyramid which is typical for phytoplankton and zooplankton in the marine pelagic food web. On the Bering Shelf, the annual primary pro-

duction is higher than in the Barents Sea. In the shallow areas of the Bering Sea (40 to 100 m depth), the "rain" of organic particles from the upper layers to the benthic (bottom-dwelling) animals can be higher than the fraction grazed by pelagic animals (Walsh J.J. et al., 1989). This input is also much higher than in the Barents Sea, which has an average depth of 230 m. Benthic biomass and production are lowest in the deep Arctic Ocean.

Box 9.7 reviews the highly specialized communities associated with seasonal and multi-year sea-ice.

Although phytoplankton generally grow more slowly in the Arctic than in warmer areas, near-freezing temperatures would not delay the onset of the initial phytoplankton bloom (i.e., period of very high production) by more than two to three days compared to that at 5 to 10 °C. Light- and nutrient-limitation is more important than temperature. Arctic zooplankton, certainly the predominant copepods, have adapted to cold conditions by having life cycles that are two to ten times longer than corresponding species in temperate conditions.

The Arctic Ocean as a whole is not particularly productive yet seasonal productivity in patches of the Barents and Chukchi Seas, and on the Bering Shelf, is among the highest of anywhere in the world (Rysgaard et al., 2001; Sakshaug, 2003; Sakshaug et al., 1994; Springer et al., 1996; Walsh, 1989). In these areas, the primary production supports large populations of migratory seabirds, a large community of various mammals, and some of the world's richest fisheries.

Many seabirds and some marine mammal species either migrate into the Arctic during the summer pulse of productivity or can cope with the long periods when food supplies are limited. Many of the permanent residents store large quantities of reserve energy in the form of lipids (oils) during periods of abundant food supply while others survive winter in a dormant stage.

9.3.1.1. Phytoplankton, microalgae, and macroalgae

Phytoplankton are often classified according to size. Nanoplankton (2–20 μm) are the most abundant yet several microplankton species (>20 μm; some reaching 500–750 μm) can produce intense blooms given sufficient light, nutrients, and stratification. Microalgae can join together and then sink to form thick mats on the bottom in shallow coastal waters (Glud et al., 2002). Among the approximately 300 species of marine phytoplankton known in high northern latitudes, diatoms and dinoflagellates comprise around 160 and 35 species, respectively (Sukhanova et al., 1999). Diatoms have non-growing siliceous shells and thus need silicate for growth while dinoflagellates move by the action of tail-like projections called flagella. Diatoms are responsible for most of the primary production in arctic pelagic ecosystems. Within the Arctic, the Arctic Ocean has the lowest number of different species and the western Barents Sea the most (Horner, 1984; Loughlin et al., 1999; Melnikov, 1997).

Prymnesiophytes (another group of swimming flagellates) include the two bloom-forming species; *Phaeocystis pouchetii* and *Emiliania huxleyi* (the latter being an exception among prymnesiophytes by lacking flagella and having a cover of calcite platelets, and as such are highly relevant to the carbon cycle). *Phaeocystis pouchetii* is common throughout the Arctic except in the deep Arctic Ocean (Hasle and Heimdal, 1998; Sukhanova et al., 1999). *Emiliania* blooms have been observed south of Iceland (Holligan et al., 1993), in the Norwegian and Bering seas (Paasche, 1960; Sakshaug et al., 1981; Sukhanova et al., 1999), and in Norwegian fjords (Berge, 1962; Johnsen and Sakshaug, 2000). *Emiliania*

huxleyi blooms were first recorded on the southeastern Bering Shelf in 1997 during an extremely bright summer (Napp and Hunt, 2001) and in the Barents Sea in 2000 (Fossum et al., 2002). *Emiliania huxleyi* continues to bloom in both areas.

Dinoflagellates, chrysophytes, cryptophytes, and green flagellates are common in arctic waters. Cyanobacteria (formerly called blue-green algae), common in temperate and tropical waters, are abundant in the deep reaches of the Bering Sea (Sukhanova et al., 1999). They are also transported into the Barents Sea by the Atlantic inflow. Dinoflagellates are particularly important in multi-year ice, and a variety of flagellates thrive in melt ponds on top of the sea ice in summer (Braarud, 1935; Gosselin et al., 1997).

The major species of diatom and prymnesiophyte possess the water-soluble reserve carbohydrate ß-1,3 glucan (chrysolaminarin), which is by far the most important carbon source for marine bacteria. Although in most phytoplankton species lipids comprise <10% of dry weight, a large proportion comprises essential polyunsaturated fatty acids that are distributed throughout the ecosystem (Falk-Petersen et al., 1998; Henderson et al., 1998). Healthy phytoplankton cells are protein-rich, with proteins comprising up to 50% of dry weight (Myklestad and Haug, 1972; Sakshaug et al., 1983).

Locally, the hard-bottom intertidal zone in the Arctic Ocean supports beds of sea-tangle (*Fucus distichus*) and in the littoral and sublittoral regions (down to about 40 m in clear water) are kelp forests of *Alaria esculenta*, *Laminaria saccharina*, *L. digitata*, and *L. solidungula* (Borum et al., 2002; Hop et al., 2002; Zenkevich, 1963). *Laminaria saccharina*, *L. digitata*, and the red alga *Ahnfeltia plicata* are commercially important in the northern coastal areas of Russia (Korennikov and Shoshina, 1980).

9.3.1.2. Microheterotrophs

Microheterotrophs are non-photosynthetic microorganisms. Their role is not well documented in the Arctic, but bacterial production is generally thought to be high, albeit somewhat reduced due to the low temperatures (Pomeroy et al., 1990). Rates of bacterial production are mainly determined by the amount of decaying organic matter available, although limitation by mineral nutrients cannot be excluded in some cases (Rich et al., 1997).

There are upward of 10^{11} to 10^{12} bacteria cells per cubic meter in the water column (Steward et al., 1996). Phages, a group of highly species-specific viruses, which are even more abundant than bacteria, attack and kill bacteria and phytoplankton, thus regulating their abundance (Bratbak et al., 1995). A well-developed community of heterotrophic flagellates grazes on the bacteria. These in turn are eaten by a variety of protozoans such as ciliates, which are in turn eaten by copepods. Thus the ciliates form an important link between the microbial (i.e., bacteria-based) and grazing food webs.

Excluding bacteria, the microheterotrophs in sea ice and ice-filled waters comprise 60 to 80 species of flagellate and about 30 species of protozoan, especially ciliates (Ikävalko and Gradinger, 1997). In contrast to first-year ice, multi-year ice has a well-developed microbial community. The abundance of microheterotrophs is particularly high during and immediately after phytoplankton maxima (Booth and Horner, 1997).

9.3.1.3. Zooplankton

Mesozooplankton play a major role in pelagic ecosystems including those of the Arctic, where a diverse array of planktonic animals comprise, on average more than 50% of the total pelagic biomass (Sakshaug et al., 1994). Marine mesozooplankton comprises ~260 species in the Arctic, ranging from less than 40 species in the East Siberian Sea to more than 130 species in the Barents Sea (Zenkevich, 1963).

Herbivorous mesozooplankton belonging to the family Calanoidae in the crustacean order Copepoda are predominant in terms of species richness, abundance, and biomass. Large herbivorous copepods (2–5 mm adult size) can make up 70 to 90% of the mesozooplankton biomass in the arctic seas. The most important are *Calanus finmarchicus*, *C. hyperboreus*, and *C. glacialis* in Atlantic and Arctic Water, and *C. marshallae*, *Eucalanus bungii*, *Neocalanus* spp., *Metridia longa*, and *M. pacifica* in the North Pacific and the Bering Sea. *Calanus finmarchicus* predominates in Atlantic Water, *C. hyperboreus* is found in both Atlantic and Arctic Water, and *C. glacialis* is found almost exclusively in Arctic Water. Variations in the distribution and abundance of *Calanus* species are considered early indicators of climate-induced change in the North Atlantic system (Beaugrand et al., 2002) with major consequences for the recruitment of fish species such as cod, which depend on them (Beaugrand et al., 2003).

The large copepods in the Arctic represent, as elsewhere, important links between primary production and the upper levels of the food web because they store large amounts of lipid for overwintering and reproduction (e.g., Scott et al., 2000). Calanoid copepods overwinter at depths of several hundred meters and then ascend to surface waters in spring to reproduce. Adults and the late copepodite stage V feed on phytoplankton in the surface waters storing lipids through the spring and summer (e.g., Dawson, 1978; Hargrave et al., 1989). Daily vertical migrations, common in most seas, have not been observed in the Arctic, not even under sea ice (Fortier et al., 2002). Many small copepods, <2 mm adult size, are known to be herbivorous while some are carnivorous (Loughlin et al., 1999; Smith and Schnack-Schiel, 1990; Stockwell et al., 2001).

Krill (euphausiids) are swarming shrimp-like crustaceans that are common on the Atlantic side of the Arctic Ocean and in the Bering Sea but are not common in the central Arctic Ocean. They can make up to

45% of mesozooplankton catches by weight (Dalpadado and Skjoldal, 1991, 1996) but are generally less abundant in the Arctic than in some areas of the Southern Ocean (Dalpadado and Skjoldal, 1996; Loughlin et al., 1999; Smith, 1991). Some species, for example *Thysanoessa inermis*, are herbivorous whereas others are omnivorous or even carnivorous, for example *T. raschii*, *T. longipes*, *T. longicauda*, and *Euphausia pacifica*. Most graze diatoms and *Phaeocystis pouchetii* efficiently (Båmstedt and Karlson, 1998; Falk-Petersen et al., 2000; Hamm et al., 2001; Loughlin et al., 1999; Mackas and Tsuda, 1999; Smith, 1991).

Amphipods, another crustacean group, are represented in the Arctic by *Apherusa glacialis*, *Onisimus* spp., *Gammarus wilkitzkii*, and *Themisto libellula*, all of which are associated with sea ice or ice-influenced waters. Except for the latter, they live in the interstitial cavities (brine channels) in the ice and on the underside of the pack ice, where *G. wilkitzkii* constitutes >90% of the amphipod biomass at times. *Apherusa* is common in first-year ice, and *Onisimus* in fast ice (Hop et al., 2000).

Themisto libellula lives in ice-filled waters but is not dependent on sea ice. It is an important food source for the upper trophic levels and is itself carnivorous, feeding on herbivorous copepods and other ice-associated zooplankton. It appears to fill the same niche as krill where these are absent (Dunbar, 1957). The largest of the ice amphipods, *Gammarus wilkitzkii*, can reach 3 to 4 cm in length. *Apherusa glacialis* and *G. wilkitzkii*, which are closely associated with multi-year ice, have a high fecundity (Melnikov, 1997; Poltermann et al., 2000).

Although copepods, amphipods, and euphausiids are predominant in terms of mesozooplankton biomass in the arctic seas, virtually all major marine zooplankton groups are represented, namely, hydrozoans, ctenophores, polychaetes, decapods, mysids, cumaceans, appendicularians, chaetognaths, and gastropods (Hop et al., 2002; Murray, 1998). Pteropods (planktonic snails) such as *Limacina helicina* occur in vast swarms some years (Grainger, 1989; Kobayashi, 1974).

9.3.1.4. Benthos

The benthic fauna differs substantially between the continental shelves and the abyssal areas of the Arctic due to differences in hydrography, with warmer and more saline water in the deeper areas (Curtis, 1975). The benthos of the Bering Sea and the Canadian Archipelago between the New Siberian Islands and Bathurst Island is primarily Pacific (Dunton, 1992). The Atlantic fauna are carried into the Barents Sea by the Atlantic inflow and into the central Arctic by strong boundary currents. The fauna of the shallow Kara, Laptev, and Pechora Seas has to contend with large seasonally fluctuating physical conditions and massive amounts of freshwater from the Russian rivers. The littoral (i.e., near-coastal) zone varies from the rocky shore of exposed coasts, to sand and mud in sheltered areas of fjords and bays, and is influenced to varying degrees by ice cover and scouring. Despite the formative studies by Russian workers in the first decades of the twentieth century (summarized by Zenkevich, 1963) detailed quantitative information on the distribution of the benthos and the structure of benthic communities in the Eurasian Arctic (especially in coastal and estuarine areas) is limited. Since around 1980, extensive regions of the North American arctic shelf and fjord areas have been sampled and their communities described and related to environmental influences, see for example studies by Stewart et al. (1985), Aitken and Fournier (1993), Grebmeier et al. (1989), and Feder et al. (1994). The greatest numbers of benthic species are found in areas of mixing between cold polar waters and temperate waters, for example between the Barents Sea and the Bering Sea, and off West Greenland and Iceland. The total number of benthic invertebrate species in the Barents Sea has been estimated at around 1600, but in the western parts of the Bering Sea alone the total number may exceed 2000 (Zenkevich, 1963). In the shallow waters of the Laptev Sea there are 365 benthic species (Zenkevich, 1963) and even fewer in the Beaufort Sea owing to the cold, unproductive arctic water masses, and to the brackish conditions (Curtis, 1975). In the deep Arctic Ocean, the number of benthic macrofauna species varies from 0 to 11 (Kröncke, 1994). The number of species in the intertidal zone of Svalbard (Weslawski et al., 1993), Bjørnøya (Weslawski et al., 1997), Baffin Island (Ellis, 1955), and Greenland (Madsen, 1936) varies between 30 and 50. The low number of benthic macrofauna species in the arctic intertidal zone is usually attributed to ice scouring (Ellis, 1955), a combination of tidal height and ice thickness (Ellis and Wilce, 1961), or heavy wave action (Weslawski et al., 1997).

Most recent benthic research has focused on specific patterns and processes resulting in biological hot spots such as below predictable leads in the sea ice, polynyas, oceanographic fronts, areas of intense mixing, and the marginal ice zone (Dayton et al., 1994).

Because a relatively large proportion of the primary production in highly productive water columns can potentially reach the bottom, primary and benthic production tends to be coupled. The fraction of sinking matter that reaches the bottom is related to bottom depth; the shallower the water body, the greater the amount of material reaching the bottom. In shallow arctic waters, the benthic food web plays a greater role than in the deep seas or at lower latitudes (Cooper et al., 2002; Grebmeier and Barry, 1991). The Bering Shelf and the southern Chukchi Sea exhibit some of the highest levels of faunal biomass in the world's oceans (Fig. 9.17), supporting a rich fauna of bottom-feeding fish, whales, seals, walruses, and sea ducks (Grebmeier et al., 1995; Hood and Calder, 1981; Joiris et al., 1996; Welch et al., 1992). Other rich benthic communities in the Arctic occur in Lancaster Sound and the shallow parts of the Barents Sea.

The benthic fauna varies with depth and habitat. For example, off Svalbard the most common species in the steep rocky littoral zone include the macroalgae *Fucus* spp., sessile (i.e., non-mobile) barnacles (*Balanus balanoides*), and motile (i.e., mobile) gastropods (*Littorina saxatilis*) and amphipods (*Gammarus setosus* and *G. oceanicus*). The tidal flats are inhabited by a rich and diverse non-permanent fauna due to sediment freezing for six to eight months each year (Weslawski et al., 1999). The sediment fauna is dominated by small polychaetes (*Scoloplos armiger, Spio filicornis, Chaetozone setosa*) and oligochaetes (Weslawski et al., 1993). Sublittoral organisms include the barnacle *Balanus balanus* that contributes a large proportion of the biomass of sessile species (Jørgensen and Gulliksen, 2001). Other conspicuous, sessile species are the bivalve *Hiatella arctica*, actinarians *Urticina eques* and *Hormathia nodosa*, bryozoans, and *Ophiopholis aculeata*. Many, small, motile amphipods (*Calliopidae* sp.), isopods (*Munna* sp. and *Janira maculosa*), snails (*Alvania* sp.) and barnacles (*Tonicella* sp.), are observed together with infaunal polychaetes, nematodes, bivalves (*Thyasira* sp.), and amphipods (*Harpinia* spp.). The infauna occur in pockets of sediment on the rocky wall. At depths between 100 and 300 m in soft bottom areas of the northern Barents Sea (Cochrane et al., 1998), the polychaetes *Maldane sarsi, Spiochaetopterus typicus*, and *Chone paucibranchiata* are among the dominant species.

Some crustaceans occur or have occurred in the arctic regions at densities sufficient for commercial interest. These include the deepwater prawns *Pandalus borealis* (Aschan and Sunnanå, 1997) and *Pandalopsis dispar*, and several crab species: red king crab (*Paralithodes camtschatica*, Hjelset et al., 2002; Jewett and Feder, 1982), *Lithodes aequispina*, Tanner and snow crab (*Chionoecetes* spp.), and Dungeness crab (*Cancer magister*, Orensanz et al., 1998). Commercially harvested arctic mollusks include clams (*Mya truncata, M. arenaria*), blue mussel (*Mytilus edulis*), and Iceland scallop (*Chlamys islandica*). Commercial fisheries and aquaculture are addressed in detail in Chapter 13.

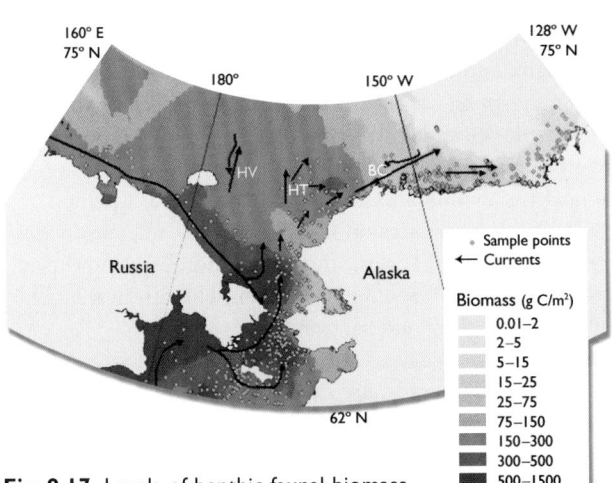

Fig. 9.17. Levels of benthic faunal biomass in the northern regions of the Bering, Chukchi, East Siberian, and Beaufort Seas (Dunton et al., 2003).

9.3.1.5. Fish

Arctic or Arctic-influenced waters are inhabited by more than 150 species of fish (Murray, 1998). Few are endemic to the Arctic, unlike the situation in the Southern Ocean where endemic species predominate. Most fish species found in the Arctic also live in boreal (northern) and even temperate regions. Arctic fish communities are dominated by a small number of species. The most abundant being Greenland halibut (*Reinhardtius hippoglossoides*), polar cod, Atlantic and Pacific cod (*Gadus morhua* and *G. macrocephalus*), Greenland cod (*G. ogac*), walleye pollock (*Theragra chalcogramma*), capelin (*Mallotus villosus*), long rough dab, also known as American plaice (*Hippoglossoides platesoides*), yellowfin sole (*Pleuronectes asper*), Atlantic and Pacific herring (*Clupea harengus* and *C. pallasi*), and redfish (*Sebastes* spp. e.g., *S. mentella*, *S. marinus*).

Greenland halibut, polar cod, and capelin have a circumpolar distribution. Greenland cod is a predominantly arctic species that is restricted to Greenland waters. The other species principally occur in waters to the south of the Arctic Ocean, except for parts of the Barents and Chukchi Seas.

Capelin

Capelin is a small circumpolar pelagic fish (Fig. 9.18). It is planktivorous (i.e., eats plankton), feeding mainly on copepods, followed by krill and amphipods. It is particularly abundant in the North Atlantic and the Barents Sea (Gjøsæter, 1995, 1998), and around Iceland (Vilhjálmsson, 1994). In the eastern Bering Sea, capelin tend to occur in cooler or more northerly areas. Capelin populations are subject to extreme fluctuations (e.g., Gjøsæter and Loeng, 1987; Sakshaug et al., 1994) in their distribution and abundance. Capelin is heavily exploited in the Atlantic but not the Pacific sector of the Arctic.

Table 9.6. Annual productivity and food requirement of higher trophic levels: average for the whole Barents Sea over several years. Data recalculated from Sakshaug et al. (1994) by Sakshaug and Walsh (2000).

	Annual production (mg C/m²)	Food requirement (mg C/m²)
Capelin	280	
Cod	90	550
Whales	3.6	360
Seals	0.8	95
Seabirds	0.5	78
Capelin fishery		200
Total	375	1280

Capelin is important in the diet of other fishes, marine mammals, and seabirds (e.g., Haug et al., 1995; Lawson and Stensen, 1997; Mehlum and Gabrielsen, 1995) and is thus regarded as a key prey species. Capelin can provide more than 20% of the food required by seabirds, higher predators, and the capelin fishery collectively in an average year (Table 9.6).

Fluctuations in the abundance of capelin have a big impact on their predators, particularly cod, seals, and seabirds. The growth rate of cod and their somatic and liver condition, for example, are correlated with capelin population abundance (Carscadden and Vilhjálmsson, 2002; Vilhjálmsson, 1994, 2002; Yaragina and Marshall, 2000).

Herring

Atlantic herring is generally restricted to waters south of the Polar Front, for example in the Nordic Seas and the Barents Sea (Vilhjálmsson, 1994). Like capelin, Atlantic herring is planktivorous, feeding in highly productive frontal areas of the open sea. Larval herring are important prey for seabirds. Adult herring is an important food item for larger fish and marine mammals.

The principal population of Atlantic herring in the Arctic is the Norwegian spring-spawning stock; one of the largest fish stocks in the world and with a spawning biomass that exceeded ten million tonnes for much of the 20th century. This population, together with the Icelandic spring- and summer-spawning herring make up the Atlanto-Scandian herring group. The migration route from nursery areas to feeding areas to overwintering areas to spawning areas takes the Norwegian spring-spawning herring around the Norwegian Sea (Box 9.8), over a distance of several thousand kilometers, and even into the Icelandic Sea during certain climatic warm periods (see Fig. 9.19 and section 9.3.3.3).

Spawning occurs at many sites along the Norwegian coast between 58° and 70° N. The spawning grounds comprise five main areas, but their relative importance, the time of arrival on the spawning grounds, and the spawning time have often changed (Slotte, 1998). These changes are not solely due to varying environmen-

Fig. 9.18. Distribution of capelin (green) (based on Vilhjálmsson, 1994).

tal conditions, but are also affected by population structure, and the optimum life history strategy for individual fish under varying levels of food supply. Flexibility in spawning behavior offers an adaptive advantage to the population during changing climates.

Pacific herring are common in the Bering Sea shelf regions (NRC, 1996). This species is, however, of relatively minor importance for seabirds and marine mammals in that region (Livingston, 1993).

Polar cod

Polar cod is a key species in many arctic food chains and forms a major link in the transfer of energy from zooplankton to top carnivores (Fig. 9.20). Large polar cod (23–27 cm) consume mainly fish and are themselves eaten by a variety

of large fish as well as by many seabird species and most arctic marine mammals (e.g., Dahl T. et al., 2000; Hobson and Welch, 1992; Holst M. et al., 2001; Lawson and Stenson, 1997; Lowry and Frost, 1981; Mehlum et al., 1999; Nilssen et al., 1995; Orr and Bowering, 1997; Rowe et al., 2000; Wathne et al., 2000). Polar cod spend much of their time associated with sea ice and stay in arctic waters throughout their life cycle. This species is broadly distributed, from inshore surface waters to very deep waters (Falk-Petersen et al., 1986; Jarvela and Thorsteinson, 1999; Pedersen and Kanneworff, 1995; Walters V., 1955). Polar cod occur in large schools (Crawford and Jorgenson, 1996; Welch et al., 1993)

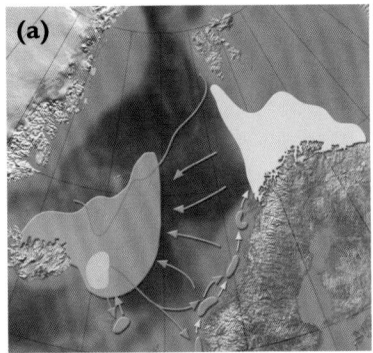
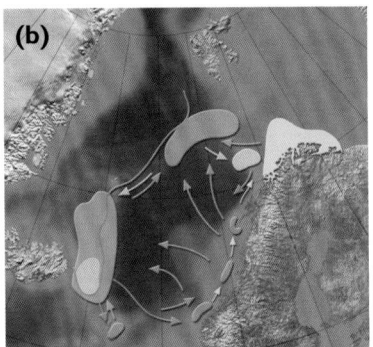
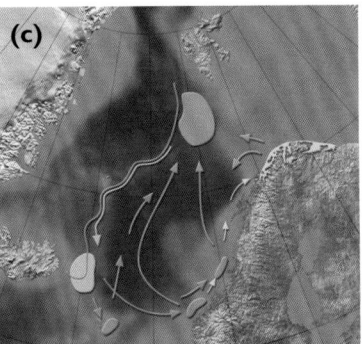
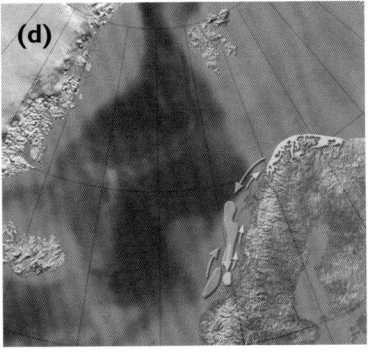
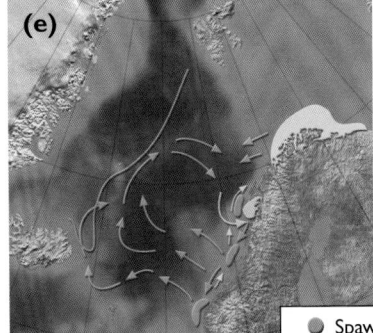

Box 9.8. Effects of climate on Norwegian spring-spawning herring

In the Icelandic area, herring was the fish species most affected by the environmental adversities of the 1960s (Dragesund et al., 1980; Jakobsson, 1980; Jakobsson and Østvedt, 1999). This is not surprising since herring are plankton feeders and in Icelandic waters are near their northern limit of distribution. Thus, the traditional feeding migrations of the Norwegian spring-spawning herring stock to the waters off northern Iceland (Fig. 9.19a) stopped completely when the Atlantic plankton community collapsed. In 1965–1966, the oldest herring were instead forced to search for food in the Norwegian Sea near the eastern boundary of the East Icelandic Current, i.e., around 150 to 200 nautical miles farther east than previously (Fig. 9.19b). In 1967–1968, the stock migrated north to feed west of Svalbard during summer (Fig. 9.19c). This was also the case in 1969 when the overwintering grounds also shifted from 50 to 80 nautical miles east of Iceland to the west coast of Norway (Fig. 9.19d). The Norwegian spring-spawning herring stock collapsed in the latter half of the 1960s (Dragesund et al., 1980) and the feeding migrations to the west into the Norwegian Sea ceased altogether (Fig. 9.19e).

The abundance of the Norwegian spring-spawning herring stock increased dramatically in the 1990s. This process has, however, taken about twenty-five years despite a ban on commercial fishing in the period 1973 to 1983. It was not until the mid-1990s that these herring resumed some semblance of their previous feeding pattern. The Norwegian spring-spawning herring still overwinter in fjords in the Lofoten area on the northwest coast of Norway. When and if they will revert completely to the traditional distribution and migration pattern cannot be predicted.

Fig. 9.19. Changes in the migration routes, and feeding and wintering areas of Norwegian spring-spawning herring during the latter half of the twentieth century. The plots show (a) the normal migration pattern during the warm period before 1965, (b and c) the pattern following the Great Salinity Anomaly until the stock collapsed in 1968, (d) during years of low stock abundance, and (e) the present migration pattern (based on Vilhjálmsson, 1997).

Legend:
- ● Spawning areas
- ◖ Juvenile areas
- ● Main feeding areas
- ⇒ Spawning migrations
- ← Feeding migrations
- ⇒ Over-wintering area

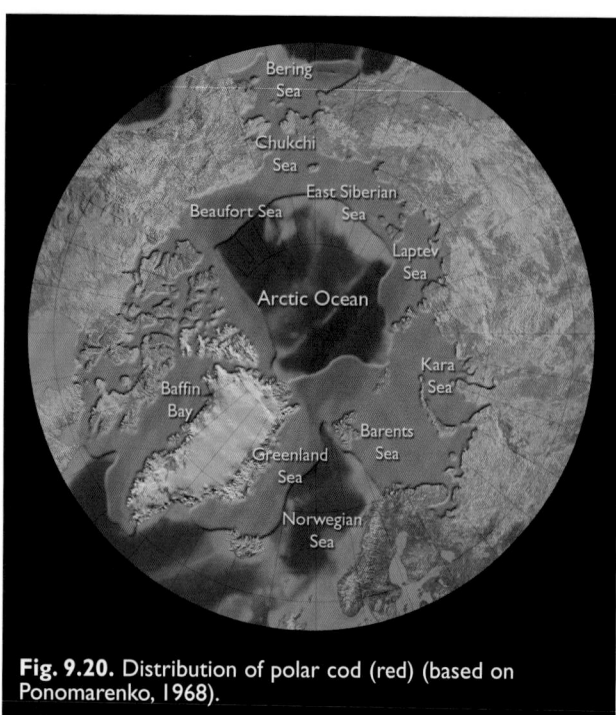

Fig. 9.20. Distribution of polar cod (red) (based on Ponomarenko, 1968).

mostly north of 60° N in the eastern Bering Sea or in the cold pool of the mid-shelf region (Wyllie-Echeverria and Wooster, 1998). The importance of this species is likely to have been underestimated in the past, in part owing to its patchy distribution. Polar cod displays a variety of physiological and biochemical adaptations to life in cold waters, including bioenergetic adjustment to low temperature (Hop et al., 1997; Ingebrigtsen et al., 2000; Steffensen et al., 1994).

Cod

Cod species found in the Arctic include Atlantic cod, Pacific cod (Bergstad et al., 1987), Pacific tomcod (*Microgadus proximus*), which occurs as far north as the Bering Sea, Greenland cod, and Arctic cod (*Arctogadus glacialis*) that resides in the Arctic Ocean, but about which little is known (Mikhail and Welch, 1989; Morin et al., 1991; Sufke et al., 1998). The majority of these species appear regularly in the diet of marine mammals (e.g., Holst M. et al., 2001; Welch et al., 1992).

Atlantic cod is the most abundant gadoid species in the northern North Atlantic. Like Atlantic herring it occurs mainly to the south of the Polar Front, yet can live in temperatures below 0 °C by producing antifreeze proteins.

Four large cod populations occurred in the arctic areas of the North Atlantic during the 20th century. The Northeast Arctic cod spawns along the Norwegian coast, with more than 50% of this occurring in the Lofoten area. Cod from Iceland spawn around the coast with more than 50% of this occurring off the southwest corner. The cod off Greenland have inshore and offshore spawning components, and an immigrant contribution from Icelandic waters. The history of the increase and

collapse of this cod population during the 20th century is described in section 9.3.3.3. The cod population off Newfoundland and Labrador also collapsed during the 1990s, owing to high fishing mortality combined with adverse environmental changes (Drinkwater, 2002).

Pacific cod is a mixed feeder that consumes a wide variety of fish (primarily walleye pollock), shellfish, and invertebrates in the eastern Bering Sea (Livingston et al., 1986).

Walleye pollock

Walleye pollock is the single most abundant fish species in the Bering Sea, comprising the bulk of the commercial catch in this area (Akira et al., 2001; Livingston and Jurado-Molina, 2000; Wespestad et al., 2000). It is mainly semi-pelagic, dominating the outer shelf regions. Walleye pollock is primarily planktivorous, feeding on copepods and euphausiids but adults become cannibalistic, feeding on juveniles seasonally (Dwyer et al., 1987). Juvenile pollock is an important prey item for other fish species, marine mammals, and seabirds (Springer, 1992).

Redfish

Several redfish species are broadly distributed and common in arctic deep waters (100 to >500 m). They are slow-growing and long-lived species. The three common species which are exploited in the northern North Atlantic are *Sebastes marinus*, *S. mentella*, and *S. viviparus*, but the latter, which is the smallest of the three, is not caught in significant amounts (Frimodt and Dore, 1995; Hureau and Litvinenko, 1986; Muus and Nielsen, 1999). There are two distinct populations of *S. mentella*. These vary in their habitat and fishery and are commonly known as deep-sea redfish and oceanic redfish, respectively. The relationship between the two forms and the extent to which the populations are separated spatially is not clear (ICES, 2003).

Oceanic redfish are caught in the Irminger Sea during the summer at depths of 100 to 200 m and water temperatures of 5 to 6 °C. Mature fish feed on krill and small fish such as capelin and herring and undertake extensive feeding migrations. They mate in early winter and the female carries the sperm and eggs, and later larvae, which are born in April/May (Wourms, 1991). The juveniles stay near the bottom, along the edge of the continental shelf.

Greenland halibut

Greenland halibut is commercially important in the North Atlantic and the Pacific, and is an important food item for deep-feeding marine mammals (e.g., narwhal and hooded seals) and sharks feeding on benthos such as the Greenland shark (*Somniosus macrocephalus*). During their first four to five years as immature fish in the eastern Bering Sea, the Greenland halibut inhabit depths to

200 m. On the Atlantic side, immature fish occur mainly between 200 and 400 m depth. Adults mainly occupy slope waters between 200 and 1000 m or more (Alton et al., 1988). Walleye pollock and squid are the main prey items for Greenland halibut in the eastern Bering Sea (Yang and Livingston, 1988).

Other flatfish

Other arctic flatfish include the long rough dab, which is an abundant bottom-dweller in some parts of the Arctic seas, including the Barents Sea (Albert et al., 1994). On the Pacific side in the eastern Bering Sea, yellowfin sole, flathead sole (*Hippoglossoides elassodon*), rock sole (*Pleuronectes bilineatus*), Alaska plaice (*Pleuronectes quadrituberculatus*), and arrowtooth flounder (*Atheresthes stomias*) are important members of the groundfish community (Livingston, 1993). Yellowfin sole, Alaska plaice, and rock sole consume mostly infaunal prey such as polychaetes, clams, and echiuran worms. These fish are distributed at depths generally less than 50 m. The highly piscivorous (i.e., fish-eating) arrowtooth flounder is found mostly on the outer shelf area, as is flathead sole, which mainly consumes brittle stars.

9.3.1.6. Marine mammals and seabirds

Arctic marine mammals to a large extent escaped the mass extinctions that affected their terrestrial counterparts at the end of the Pleistocene (Anderson, 2001). Like fish, mammals and birds have the advantage of having great mobility and hence are good colonizers. Thus, it is not surprising that these groups dominate the arctic marine megafauna, represented both by resident and migratory species. Their high abundance was a major attractant for people to this region historically, becoming the mainstay of the diet of coastal communities throughout the Arctic (Chapter 12) and later the subject of extreme levels of commercial exploitation. The massive harvests of marine mammals and seabirds that began in the 1600s and lasted for several hundred years decimated many arctic populations. Bowhead whales (*Balaena mysticetus*) and sea otters (*Enhydra lutris*) were almost driven to extinction throughout the Arctic (Burns et al., 1993; Kenyon, 1982), while the great auk (*Pinguinus impennis*) and Steller sea cow (*Hydrodamalis gigas*) did become extinct. Walruses (*Odobenus rosmarus*) were all but extirpated in some arctic regions (Gjertz and Wiig, 1994, 1995). Polar bears, all the great whales, white whales (*Dephinapterus leucas*), and many species of colonially nesting seabird were dramatically reduced. Harvesting of marine mammals and seabirds is now undertaken in accordance with management schemes based on sustainability in most Arctic countries, although overexploitation of some species is still occurring (CAFF, 2001).

Marine mammals are the top predators in the Arctic other than humans. Virtually all large-scale taxonomic groupings of marine mammals have arctic representatives (Perrin et al., 2002).

Polar bear

The polar bear, the pinnacle predator, has a circumpolar distribution and is dependent on sea ice to provide for most of its needs (Ferguson et al., 2000a,b; Mauritzen M. et al., 2001; Stirling et al., 1993). Polar bears feed almost exclusively on ice-associated seals (e.g., Lønø, 1970; Stirling and Archibald, 1977; Smith T., 1980). Adult bears can swim quite long distances if required, but mothers with cubs depend on ice corridors to move young cubs from terrestrial denning areas to prime hunting areas on the sea ice (Larsen T., 1985, 1986). Pregnant females dig snow dens in the early winter and give birth several months later. This requires a significant depth of snow, thus females return year after year to land sites that accumulate sufficient snow early in the season. A mother that emerges from the den with her young has not eaten for five to seven months (Ramsay and Stirling, 1988). Therefore, successful spring hunting is essential for the family's survival and largely dictates condition, reproductive success, and survival for all polar bears (e.g., Stirling and Archibald, 1977). Factors that influence the distribution, movement, duration, and structure of sea ice profoundly affect the population ecology of polar bears, not least due to their influence on the principal prey species, ringed seal (*Phoca hispida*) (Stirling and Øritsland, 1995; Stirling et al., 1999). The global polar bear population is estimated at 22000 to 27000 (IUCN, 1998).

Walrus

Walruses, like polar bears, are circumpolar, but with a more disjointed distribution. Two sub-species are recognized: the Pacific walrus (*Odobenus rosmarus divergens*) and the Atlantic walrus (*O. r. rosmarus*) (Fay, 1981, 1982). The global walrus population is estimated at about 250000, of which 200000 belong to the Pacific sub-species. The Atlantic walrus is distributed from the central and eastern Canadian Arctic eastward to the Kara Sea (Fay, 1981; Zyryanov and Vorontsov, 1999), including several more or less well-defined sub-

Fig. 9.21. Walrus routinely use sea ice as a haul-out platform in shallow areas where they feed on benthic fauna (photo supplied by Kit Kovacs & Christian Lydersen, Norwegian Polar Institute).

populations (Andersen L. et al., 1998; Buchanan et al., 1998; Outridge and Stewart, 1999). Walruses haul-out on pack ice most months of the year (Fig. 9.21), using land-based sites only during summer when sufficient sea ice is unavailable. Walruses have a narrow ecological niche, depending on the availability of shallow water (<80 m) with bottom substrates that support a high production of bivalves (e.g., Born et al., 2003; Fisher and Stewart, 1997; Wiig et al., 1993).

Seals

Ringed seals represent the "classical" arctic ice seal, being uniquely able to maintain breathing holes in thick sea ice. Thus, they can occupy areas far from sea-ice edges, unreachable by other seal species. They are distributed throughout the Arctic, even at the North Pole (Reeves, 1998). They number in the millions and this is by far the most abundant seal species in the Arctic. This species exclusively uses the sea ice for breeding, molting, and resting (haul-out), and rarely, if ever, moves onto land. Although quite small, ringed seals survive the thermal challenges posed by the arctic winter by building lairs in the snow on top of sea ice, where they rest in inclement weather and where they house their new-born pups (e.g., Lydersen and Kovacs, 1999; Smith T. and Stirling, 1975). Ice amphipods and fish constitute much of their diet (e.g., Gjertz and Lydersen, 1986; Weslawski et al., 1994).

The bearded seal (*Erignathus barbatus*) has a patchy circumpolar Arctic distribution (Burns, 1981a). This species breeds on drifting sea ice (Kovacs et al., 1996) but occasionally hauls out on land during the summer. These animals are mostly benthic feeders, eating a wide variety of fish, mollusks, and other invertebrates in shallow areas. Some bearded seal populations are thought to be resident throughout the year, while others follow the retreating pack ice in summer, and then move southward again in the late autumn and winter (Burns, 1967; Gjertz et al., 2000). The global population has not been assessed but is thought to number in the hundreds of thousands in the Arctic (Kovacs, 2002a).

Harbour seals (*Phoca vitulina*) have one of the broadest distributions of the pinnipeds, from temperate areas as far south as southern California to arctic waters of the North Atlantic and into the Bering Sea in the Pacific (Bigg, 1969; Rice, 1998). They are coastal, non-migratory, and aggregate in small numbers on rocky outcrops, beaches, or inter-tidal areas (Grellier et al., 1996; Pitcher and McAllister, 1981). They are opportunistic feeders that eat a wide variety of fish species and some cephalopods and crustaceans (Bowen and Harrison, 1996). Harbour seals are not numerous in the Arctic and several of the populations that live north of the Arctic Circle are very small (Boveng et al., 2003; Henriksen et al., 1996).

In the Atlantic sector of the Arctic, there are three additional phocid (i.e., true) seal species: harp seals, hooded seals (*Cystophora cristata*), and grey seals (*Halichoerus grypus*). Harp seals are highly gregarious and migratory, moving southward to three traditional breeding sites (off the east coast of Canada, in the White Sea, and between Jan Mayen and Svalbard) for the birthing period on pack ice in March (Lavigne and Kovacs, 1988). Following the breeding season, harp seals from each population move northward into molting sites before dispersing into the Arctic for the rest of the year (e.g., Folkow and Blix, 1992). Adult harp seals feed mainly on small marine fish such as capelin, herring, sculpins (*Cottidae*), sand lance (*Ammodytes americanus*), and polar cod (e.g., Lawson and Stenson, 1997; Lawsen et al., 1995; Nilssen, 1995), and then on krill and amphipods. The global population is thought to exceed seven million animals (Lavigne, 2002).

The hooded seal is a large, pack-ice breeding northern phocid that ranges through a large sector of the North Atlantic. In spring the adults gather to breed in two main groups; one off the east coast of Canada and the other either in Davis Strait or off East Greenland depending on conditions (Lavigne and Kovacs, 1988). Some weeks after breeding, the animals move northward into traditional molting areas before dispersing for the summer and autumn, preferring the outer edges of pack ice (Folkow and Blix, 1995). They feed on a variety of deep-water fishes including Greenland halibut and a range of redfish species, as well as squid (Folkow and Blix, 1999). The global population is very difficult to estimate because hooded seals are difficult to survey, but is certainly in excess of half a million animals (Kovacs, 2002b).

Grey seals were historically abundant in Icelandic waters and along the coastal regions of northern Norway and northeastern Russia (Collett, 1912). They have been depleted through hunting and government culling programs (Wiig, 1987) and in some areas have been extirpated (Haug et al., 1994). A crude estimate of the population of grey seals inhabiting northern Norway and the Murman coast of Russia is 4500 (Haug et al., 1994).

Two additional ice-breeding seals that occur in the Bering Sea are the spotted seal (*Phoca largha*) and the ribbon seal (*Phoca fasciata*). The spotted seal breeds in eight largely discrete birthing areas (Rice, 1998). They have a coastal distribution during the summer and early autumn, but migrate offshore to the edge of the ice pack for the rest of the year (Lowry et al., 1998). Spotted seals eat a wide variety of prey, including fish, crustaceans, and cephalopods (Bukhtiyarov et al., 1984; Lowry and Frost, 1981; Lowry et al., 1982). There are no recent, reliable population estimates for this species (Burns J., 2002). Ribbon seals are poorly known, pack-ice breeders that congregate loosely in suitable areas of thick pack ice in the North Pacific during the breeding season (Rice, 1998). They do not haul out on land and are assumed to be either pelagic or northern pack-ice dwellers in summer (Burns J., 1981b). They are reported to eat crustaceans, fish, and cephalopods (Frost K.

and Lowry, 1980; Shustov, 1965). Current data on population size are not available, but counts in the 1970s revealed 100 000 to 200 000 animals (Burns J., 1981b).

Northern fur seals (*Callorhinus ursinus*), Steller sea lions (*Eumetopias jubatus*), and sea otters all breed terrestrially on the Pribilof, Aleutian, Commander, and Kurile Islands in the North Pacific. The latter two species breed as far south as the Californian coast.

Whales

White whales, narwhal (*Monodon monoceros*), and bowhead whales live only in the high Arctic (see Perrin et al., 2002) and are commonly found in ice-covered waters where they use edges, leads, and polynyas to surface for breathing. Narwhal mainly occur within the Atlantic region, while the others have patchy circumpolar ranges (Rice, 1998). All three migrate seasonally, largely in relation to the northward retraction and southward expansion of the seasonal sea ice. They prey on small fishes, especially polar cod, although narwhal also eat large quantities of cephalopods, and bowhead whales consume a greater proportion of planktonic crustaceans than either of the other two species.

Other cetaceans also frequent arctic waters in summer, but these remain in relatively ice-free waters and spend most of the year elsewhere. These include white-beaked dolphin (*Lagenorhynchus albirostris*) in the North Atlantic/Barents/Greenland Sea and Dahl's porpoise (*Phocoenoides dalli*), right whales (*Eubalaena glacialis*), and grey whales (*Eschrichtius robustus*) in the North Pacific/Bering Sea. Harbour porpoise (*Phocoena phocoena*) and killer whales (*Orcinus orca*) are among the toothed whales, and blue whales (*Balaenoptera musculus*), fin whales (*B. physalus*), minke whales (*B. acutorostrata*), humpback whales (*Megaptera novaeangliae*), and sei whales (*B. borealis*) are some of the baleen whales that are regular summer residents in arctic waters. Many of the great whales inhabit the Bering Sea in summer.

Seabirds

Some of the largest seabird populations in the world occur in the Arctic (e.g., Anker-Nilssen et al., 2000; Boertmann et al., 1996; Gaston and Jones, 1998; Norderhaug et al., 1977). Over 60 seabird species frequent the Arctic, and over 40 breed there (Murray, 1998). Many species take advantage of the summer peak in productivity and then overwinter elsewhere. In the extreme, the red phalarope (*Phalaropus fulicarius*), the northern phalarope (*P. lobatus*), and the Arctic tern (*Sterna paradisaea*) spend the summer in the high Arctic and overwinter in the southern hemisphere off Peru or West Africa. In contrast, the spectacled eider (*Somateria fisheri*), black guillemot (*Cepphus grylle*), ivory gull, and northern fulmar (*Fulmaris glacialis*) stay in the Arctic all year round, using the southern edges of the sea ice or open water areas for feeding in winter. Polynyas are extremely important winter habitats for these species

(Brown and Nettleship, 1981; Stirling, 1997). Most of the global population of the threatened spectacled eider overwinters in single-species flocks in a few polynyas in a restricted area of the Bering Sea (Petersen et al., 1999). In the rare instance that such polynyas freeze for longer than a few days, mass mortalities can occur, altering population growth and affecting the species for decades (Ainley and Tynan, 2003).

Most arctic seabird species nest in large colonies on cliffs, which offers some protection from terrestrial predators such as the Arctic fox (*Alopex lagopus*). Other species, such as Sabine's gull (*Xema sabini*), nest on the ground on isolated islands, while others use burrows either on sloping ground (e.g., little auk) or in rock crevices (e.g., black guillemot). Several of the auk species are among the most abundant nesting arctic seabirds, including the little auk, thick-billed murre, common murre (*Uria aalge*), and the Atlantic puffin (*Fratercula arctica*). The blacked-legged kittiwake (*Rissa tridactyla*) is the most numerous Arctic gull, but glaucous gulls (*Larus hyperboreus*) are also common. Arctic terns are abundant in some regions, as are common eider (*Somateria mollissima*). The Pribilof Islands, in the eastern Bering Sea, are breeding sites for large numbers of piscivorous seabirds including black-legged and red-legged kittiwake (*R. brevirostris*), and common and thick-billed murre.

The foraging ecology and energetics of seabirds have been studied quite extensively in many arctic areas (e.g., Barrett et al., 2002; Bogstad et al., 2000; Croxall, 1987; Montevecchi, 1993) and despite species differences, some basic patterns are evident. Most arctic seabirds forage on small fish and large copepods, primarily in the upper and mid-water column (e.g., Garthe, 1997; Montevecchi and Myers, 1996). Foraging is often concentrated in frontal areas or at ice edges, where convergences can concentrate marine zooplankton (Hunt et al., 1999). Eiders are the exception, foraging in shallow water for benthic animals, particularly echinoderms and mollusks. Polar cod is an extremely important prey item for most arctic seabirds, but other small school-forming species (such as capelin and herring in the Barents Sea) are extremely important regionally. Surface feeders (e.g., kittiwakes and fulmars) forage on-the-wing, dipping into the water to capture prey, or feed while sitting on the water surface when prey concentrations are high and available within the top few centimeters. The alcids and related species dive to considerable depths (Schreer and Kovacs, 1997) in search of prey. They also travel considerable distances and can stay underwater for relatively extended periods, allowing them to take advantage of fish and invertebrates that reside under the sea ice, e.g., euphausiids, amphipods, and polar cod (Bradstreet, 1980). As foragers, most seabird species are generalists responding to changing spatial and temporal prey availability (e.g., Montevecchi and Myers, 1995, 1996). However, the little auk and the Bering Sea least auklet (*Aethia pusilla*), which specialize on calanoid copepods, have a narrow foraging niche (Karnovsky et al., 2003). Ivory gulls are one of the most specialized of the arctic seabirds, living in association with pack ice for most of their lives and breeding on exposed mountain peaks in glaciated areas of the high Arctic. One of their favorite foods is the blubber of marine mammals, acquired by scavenging on carcasses. Yet, similar to many arctic seabirds, a large part of their diet comprises polar cod and other small fish and invertebrates. The small fish and invertebrates are usually taken after being washed onto the surface of ice floes and edges (Haney and MacDonald, 1995; Hunt et al., 2002). Ross' gulls (*Rhodostethia rosea*) also perform this type of foraging behavior.

In addition to seabirds that are strictly marine feeders, skuas, a host of arctic shorebirds, and some ducks (beside the marine feeding eiders), geese, and divers also spend time at sea.

9.3.2. Physical factors mediating ecological change

There are a variety of means by which climate can affect marine biota. These can be direct or indirect. Examples of the former include temperature, which affects the metabolism and distribution of organisms; wind-driven currents, which transport planktonic organisms; sea ice, which provides higher predators with a platform for birthing or foraging; and snow, which allows for the construction of overwintering lairs. An indirect means by which climate can affect biota is through those climate processes that affect nutrient levels and surface mixed layer depth, which in turn influence primary and secondary productivity, and ultimately food availability to the upper trophic levels. Figure 9.22 illustrates those

climatic factors that can influence the Barents Sea ecosystem, both directly and indirectly. Similar interactions are also valid for other marine areas. The timing of sea-ice formation and melt-back, as well as temperature, can influence the timing, location, and intensity of biological production.

Of the main factors mediating ecological change in the Arctic, the distribution of sea ice is most important. Sea ice, together with its snow cover, can reduce light levels at the water surface to those observed at 40 m or more in an ice-free water column. Primary production in the water column below the sea ice is thus severely light-limited. However, the sea ice is of major importance as a habitat for marine mammals and the location of ice edges is extremely important to seabirds. Moreover, the melting of sea ice in spring results in a stratification of the upper water column that promotes primary production.

The flow of warm water into the Arctic and the mixing and stratification of the water column are also important. The flow of warm water into the Arctic is important for the northward transport of zooplankton populations, such as the transport of *Calanus finmarchicus* from the Norwegian Sea to the Barents Sea. The mixing and stratification of the water column is determined by the opposing forces of wind and freshwater supply (Sakshaug and Slagstad, 1992).

Generally, sea surface temperatures in the Arctic are low, but true ectotherms (previously called "cold-blooded organisms", i.e., their body temperatures vary with the temperature of their surroundings) can grow at the freezing point of seawater. In principle, organisms

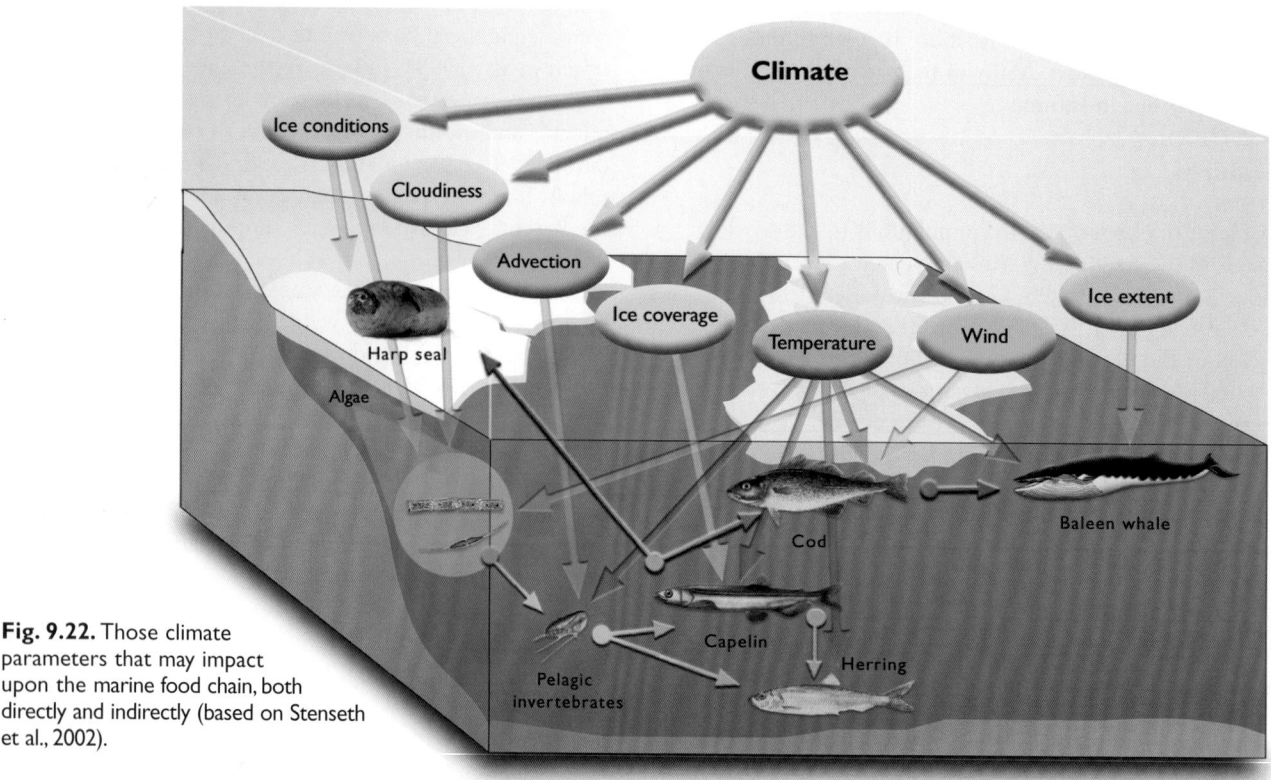

Fig. 9.22. Those climate parameters that may impact upon the marine food chain, both directly and indirectly (based on Stenseth et al., 2002).

grow faster the higher the temperature up to an optimum range, which can be from 8 to 15 °C for species living in the Arctic. A temperature increase of 10 °C would roughly double the biochemical rates, and thus the growth rate.

9.3.2.1. Primary production

The effect of temperature on primary production is largely indirect, through its effect on sea-ice cover and the mixing characteristics of the water column. The direct effect of rising temperature, through its effect on growth rate, would primarily shorten the spring bloom by two to five days, and perhaps slightly increase regenerative production. New production would be likely to increase because it is primarily regulated by the vertical nutrient supply.

Limiting factors

Potentially limiting nutrients in the Arctic are nitrogen or phosphorus, and for diatoms, also silicate. Iron controls primary production by retarding nitrate uptake in the Northeast Pacific and the deep regions of the eastern Bering Sea (Frost and Kishi, 1999). It has also been observed to limit temporarily spring bloom production in the Trondheimsfjord (Õzturk et al., 2002). Silicate, which like nitrogen is also affected by iron control, limits diatom growth in some areas of the Barents Sea (Nielsen and Hansen, 1995; Wassmann et al., 1999). Because arctic rivers are rich in nitrogen and silicate but poor in phosphate, phosphorus limitation is likely in and around some estuaries.

Most microalgae are probably not limited by CO_2 because they contain the enzyme carbonic acid anhydrase, which can furnish CO_2 from bicarbonate (Anning et al., 1996; Goldman, 1999; Reinfelder et al., 2000; Sültemeyer, 1998). Production of the coccoliths that cover coccolithophorids also furnishes CO_2.

In nature, an increase in the supply of the limiting nutrient typically causes a predominance of large-celled species. A sufficient supply of iron and silicate favors large bloom-forming diatoms that enhance the sedimentation rate.

Nutrient status in winter differs strongly between arctic regions, reflecting the nutrient concentration of the deep or intermediate waters that supply nutrients to the upper layers. This is related to the increasing age of the intermediate and deep waters along their THC route. Thus, Atlantic water (which is relatively young) exhibits the lowest concentrations and the deep Bering Sea water (which is older) the highest (Table 9.7). However, because mixing between surface and intermediate water in the Bering Sea is low owing to the high stability of the water column, surface water concentrations in the Bering Sea are actually lower than in the Southern Ocean.

Owing to the high winter nutrient concentrations on the Bering Shelf and in the southern Chukchi Sea, productivity in these regions can be two to four times higher than in the Barents Sea (Coachman et al., 1999; Grahl et al., 1999; Olsen et al., 2003; Schlosser et al., 2001; Shiomoto, 1999; Walsh J.J. and Dieterle, 1994). Because of its distance from shelf-break upwelling, however, the northeast coastal Alaskan Shelf exhibits low nutrient levels, on a par with those of the Atlantic sector (Coachman and Walsh, 1981).

North of 85° N, severe light limitation restricts primary production in the water column to a six-week growth season, which is initiated by the melting of the snow on top of the sea ice in July (English, 1961; Kawamura, 1967; Usachev, 1961). In multi-year ice, the dense biomass on the underside of the sea ice is also strongly light-limited, but in melt ponds, intense small-scale production can occur (Booth and Horner, 1997; Gosselin et al., 1997; Sherr et al., 1997). Productivity in the multi-year ice in the shelf seas is an order of magnitude greater than in first-year ice, presumably because of a greater nutrient supply, however, the latter generally has very low levels of primary production (Andersen O., 1989; Gradinger, 1996; Juterzenka and Knickmeier (1999). In polynyas, early melting of sea ice can prolong the growth season by three months (Smith et al., 1997; Suzuki et al., 1997).

Timing

In seasonally ice-covered areas, the onset of the phytoplankton bloom is usually determined by the timing of the breakup of the sea ice (Alexander and Niebauer, 1981; Braarud, 1935; Gran, 1931; Head et al., 2000; Stabeno et al., 2001; Wassmann et al., 1999). Typically, an ice-edge bloom unfolds in a 20 to 100 km wide belt south of the northward-retreating ice edge. The bloom develops rapidly because water from the melting sea ice establishes a shallow wind-mixed layer of 15 to 35 m depth. The ice-edge bloom generally begins in mid-

Table 9.7. Winter nutrient levels (mmol/m³) in the Barents Sea, the Bering Sea (surface and at depths >300 m), and the Southern Ocean (the Ross and Scotia Seas) (Sakshaug, 2003).

	Barents Sea (Atlantic Water)	Bering Sea (surface water)	Bering Sea (deep water)	Ross Sea (surface water)	Scotia Sea (surface water)
Nitrate	10–12	10–30	45	25	30
Phosphate	0.85	1.0–2.0	3.5	2	2
Silicate	6–8	25–60	100–300	50–60	100

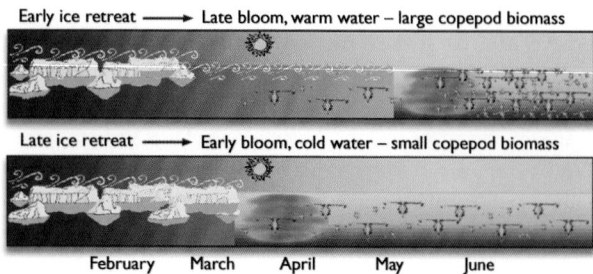

Fig. 9.23. The relative timing of the sea-ice retreat and the spring bloom in the Bering Sea (Hunt et al., 2002).

April to early May at the southernmost fringes of the first-year ice, both in the Barents and Bering Seas and in the Labrador/Newfoundland region (Alexander and Niebauer, 1981). In the Bering Sea, years with early sea-ice retreat (i.e., starting in winter) have delayed blooms as the blooms cannot begin until light levels and stratification are sufficient to support them. Thus, in the Bering Sea, early ice retreat implies a late bloom, while late ice retreat implies an early bloom (Fig. 9.23). In the Barents Sea, however, very cold winters that result in a more southern distribution of the ice edge (with sea ice forming over Atlantic water to the south of the Polar Front) can have very early blooms because once melting starts the sea ice melts rapidly from below. Near multi-year ice in the Arctic Ocean, melting is delayed until July, resulting in a short growing season (Strass and Nöthig, 1996), and in the ice-filled regions of the Greenland Sea, late melting can delay the ice-edge bloom until late May as far south as the Denmark Strait (Braarud, 1935).

Impact of physical and chemical forcing

After the ice-edge bloom, primary production becomes very low in the strongly stratified waters, with nutrients near the limit of detection (Fujishima et al., 2001; Taniguchi, 1999; Whitledge and Luchin, 1999). In iron-controlled waters, however, there are still high nitrate concentrations in the water column. Near the pycno-cline (i.e., the region of strongest vertical density gradient) in arctic waters, a restricted vertical supply of nutrients enables the development of a 3 to 10 m thick chlorophyll maximum layer that is strongly light-limited (Heiskanen and Keck, 1996; Luchetta et al., 2000; Nielsen and Hansen, 1995).

In ice-free waters, it is the onset of thermally-derived stratification that determines the timing of the spring bloom. The blooms deplete the upper layer nutrient concentrations. In the Norwegian Sea and the Atlantic (southwest) part of the Barents Sea, thermally-derived water-column stability is established in late May to early June (Halldal, 1953; Olsen et al., 2003; Paasche, 1960; Steemann-Nielsen, 1935). In ice-free estuaries and fjords, and waters surrounding Iceland, freshwater-induced stability triggers a bloom in late March to late April (Gislason and Astthórsson, 1998; Braarud 1935; Sakshaug, 1972). On continental shelves, nutrient sup-

plies from upwelling or strong tidal mixing can maintain high levels of production, as observed in both the Barents and Bering Seas.

Pulsed (wind-driven) nutrient supplies associated with passing atmospheric low pressure systems often result in small blooms, however, in arctic waters, the pycno-cline is usually too strong to allow a temporary deepening of the surface mixed layer and so bring in nutrients from sub-pycnocline waters (Overland et al., 1999b; Sakshaug and Slagstad, 1992). In the Bering Sea, storms, especially those in mid- to late May, lead to a large nutrient supply and prolonged primary production, whereas a weakening of the summer winds lowers the nutrient supply for continuing summer blooms (Stabeno et al., 2001).

Wind-driven nutrient supply supports about 50% of the annual primary production in the southern Barents Sea influenced by the Atlantic inflow and this supply exhibits no clear temporal trend. In the northern Barents Sea, however, primary production clearly follows variations in the NAO index, being high following NAO$^+$ years – which correspond to years with relatively warm winters and little sea ice (Slagstad and Støle Hansen, 1991). The higher production is a result of the reduced sea-ice cover allowing a larger area of the northern Barents Sea exposure to the strong light levels.

For the outer and mid-shelf domains of the Bering Shelf, the wind-driven nutrient supply supports 30 to 50% of the annual primary production, depending on the frequency and intensity of summer storms. Interdecadal trends in chlorophyll-a (Chl-a) concentration were observed by Sugimoto and Tadokoro (1997) in eastern Bering Sea regions deeper than 150 m but it is not known if these resulted in changes in either the spring or overall annual primary production levels. The few available data suggest that the summer contribution to annual new production may have decreased in recent years with the advent of calmer, sunnier summers. Coastal domain production is not thought to vary much between years. On the northern shelf, variability in phytoplankton biomass and production has been linked to variability in the transport of the Bering Slope Current that leads to the Anadyr Stream (Springer et al., 1996).

Distribution of primary production

The distribution of primary production in the Arctic provides a good illustration of the effects of physical and chemical forcing (Table 9.8). Annual primary production in the deep Arctic Ocean, the lowest known for any sea, reflects the high incidence of multi-year sea ice with snow, and thus the short growing season (Cota et al., 1996; Gosselin et al., 1997). Nevertheless, present estimates are far higher than the pre-1990 estimates, which ignored production within the multi-year ice.

Due to the inflow of Atlantic and Bering Sea water, the Barents Sea and a patch of the Chukchi Sea, respectively, have enhanced annual production (Hegseth, 1998; Noji et al., 2000; Sakshaug and Slagstad, 1992; Smith et al., 1997; Walsh J.J. and Dieterle, 1994). In the other Siberian shelf areas, annual production is low due to multi-year ice hindering wind-driven upwelling of nutrient-rich deep water along the shelf break, leaving re-mixing of nutrient-poor shelf water and phosphorus-poor river water as the main nutrient sources.

In Atlantic water, annual primary production is high, in part due to wind-driven episodic upwelling in summer (Fig. 9.24) (Olsen et al., 2003; Sakshaug and Slagstad, 1992). The most productive area is the Bering Shelf where a highly productive "greenbelt" is associated with the upwelling of extremely nutrient-rich water along the shelf break and the Anadyr Current (Hansell et al., 1993; Nihoul et al., 1993; Springer et al.1996; Walsh J.J. et al., 1989). In the deep eastern Bering Sea, annual primary productivity is similar to, or slightly higher than that in the Barents Sea (Maita et al., 1999; Springer et al., 1996).

9.3.2.2. Secondary production

Although the zooplankton database is small, it suggests that growth rates of calanoid copepods and other crustaceans are dependent on temperature such that the time from hatching to the next adult generation is shorter in warmer water. The growth rate, however, is also very

Table 9.8. Estimated levels of primary production, defined as the integrated net photosynthesis (corrected for respiration) over at least 24 hours, plus the grazing rate of mesozooplankton (compiled by Sakshaug (2003) on the basis of data from several authors).

	Area (10^3 km^2)	Total primary production (g C/m^2)	New primary production (g C/m^2)	Grazing rate of zooplankton (g C/m^2)	Total primary production (Tg C)
Central Deep Arctic	4489	>11	<1	-	>50
Arctic shelves	5052	32	8	10	279
Barents Sea	1512	<20–200[a]	<8–100	15–50	136
Barents north slope	-	35	16	-	-
White Sea	90	25	6	-	2
Kara Sea	926	30–50	7–12	-	37
Laptev Sea	498	25–40	6–10	-	16
East Siberian Sea	987	25–40	6–10	-	30
Chukchi Sea	620	20–>400	5–>160	7–>90	42
Beaufort Sea	178	30–70	7–17	-	8
Lincoln Sea	64	20–40	5–10	-	3
Other (Canadian Arctic)	182	20–40	5–10	-	5
Northeast Water Polynya	<50	20–50	13–32	-	-
North Water Polynya	-	150	70	-	-
Total Arctic Ocean	9541	>26	<5	-	>329
Atlantic sector	5000	97	50	-	483
Baffin Bay	690	60–120	25–50	-	62
Hudson Bay	820	50–70	25–35	-	49
Greenland Sea	600	70	40	-	42
Labrador Sea	1090	100	45	-	110
Norwegian Sea	1400	80–150	35–65	-	160
Icelandic Sea	400	100–200[b]	45–90	-	60
West Spitsbergen	-	120	55	-	-
Bering Shelf	1300	>230	-	-	>300
Alaskan coastal	-	50–75	<20	32–50	-
Siberian coastal	-	>400	>160	>90	-
Middle, outer shelf	-	150–175	30–50	35–70	-
Shelf Break	-	450–900	170–360	-	-
Bering oceanic	1000	60–180	-	-	155
Okhotsk Sea	1600	100–200	-	-	240
Global, ocean	362000	110	-	-	40000[c]
Global, land	148000	405	-	-	60000

[a]Highest values occur where topography and currents cause continuous nutrient supply in Atlantic sector, lowest values in northernmost part; [b]production to the south and east of Iceland (i.e., in Atlantic water) is four times that to the north and east; [c]plus 5000 Tg benthic (seaweed) carbon production and 4000 to 7000 Tg of dissolved organic carbon.

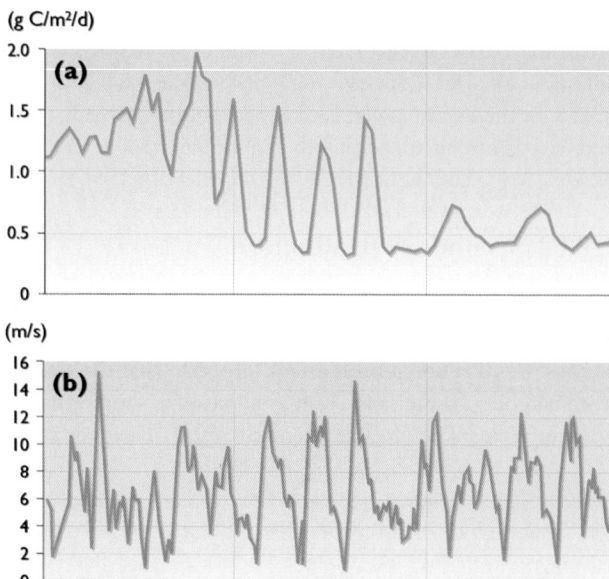

Fig. 9.24. Estimated (a) primary production and (b) wind speed for the Atlantic water of the Barents Sea in summer 1998 (wind data and hind-cast model data from the Norwegian Meteorological Institute; production model by D. Slagstad, Norwegian University of Science and Technology).

dependent on food supply. More specifically, the growth rate depends on the extent to which the fat-storage organs of the zooplankton are filled to capacity, which in turn is highly dependent on phytoplankton availability (Hygum et al., 2000). Nauplii (early-stage larvae) and early-stage copepodite stages can be food-limited at <0.5 to 0.7 mg Chl-a/m³ (Campbell et al., 2001). This level of concentration is common in waters which receive a low supply of new nutrients due to strong stratification and are therefore dominated by low levels of regenerative primary production (Båmstedt et al., 1991; Booth et al., 1993; Hirche and Kwasniewski, 1997; Irigoien et al., 1998).

Match versus mismatch

The concept of match and mismatch is very important in food-web energy transfer. A match implies that the predators are located in the same space and time as their prey and a mismatch when they are not. In principle, grazing by zooplankton is efficient when a large and growing population of zooplankton coincides with a phytoplankton bloom. Production of mesozooplankton is small in areas characterized by a mismatch. This is a highly non-linear event because phytoplankton blooms and zooplankton swarms are episodic. To ensure a match higher in the food web, fish and zooplankton populations also need to coincide in time and space. Physical oceanographic conditions, such as temperature, salinity, stratification, mixing, and currents can influence the timing and location of the plankton production and biomass as well as the eggs and larvae of fish and invertebrates. In this sense, oceanographic conditions play a large role in determining the extent of a match or mismatch between trophic levels.

Non-grazed phytoplankton sink except for most of the (nanoplankton) fraction that is based on regenerated

nutrients. Thus, sedimentation rates are lower when there is a match between phytoplankton and zooplankton. Grazing and sedimentation are thus competing processes and both are strongly dependent on large-celled new production.

In Atlantic water, late development of copepodite stages of *Calanus finmarchicus* is a good match with the late and relatively long-lasting phytoplankton blooms that occur in mid-May to June (Dalpadado and Skjoldal, 1991; Skjoldal et al., 1987). But it is mismatched with the timing of the initial blooms, which is presumably one of the main reasons why *C. finmarchicus* is allochthonous in the Barents Sea (Melle and Skjoldal, 1998). The mismatch is greatest in very cold winters when sea ice covers Atlantic water and the blooms are typically four to six weeks earlier than usual (Olsen et al., 2003). The reason that the blooms are earlier than usual in such winters is because once melting starts the sea ice over the Atlantic water melts rapidly from below. This can result in the phytoplankton bloom being too early for the zooplankton, thus causing a mismatch in timing with the peak in zooplankton (Olsen et al., 2003; Skjoldal and Rey, 1989). Such years can have very low levels of secondary production. Although not strongly correlated, a match seems likely to occur in Atlantic water with mixing depths greater than 40 m, while a mismatch seems likely with mixing depths less than 40 m.

In the generally ice-free Norwegian fjords, the major phytoplankton blooms occur from February to early April, depending on latitude and the extent of fresh-water-induced stability. As the major zooplankton peak does not occur until April or May, the zooplankton must feed on the secondary summer and autumn blooms (Wassmann, 1991). Owing to the extreme mismatch, almost all of the early spring bloom sinks to the bottom of the fjord.

9.3.2.3. Fish

Climate fluctuations affect fish directly, as well as by causing changes in their biological environment (i.e., in relation to predators, prey, species interactions, and disease). Direct physiological effects include changes in metabolic and reproductive processes. Climate variability may influence fish population abundance, principally through effects on recruitment. Variability in the physical environment may also affect feeding rates and competition by favoring one species relative to another, as well as by causing changes in the abundance, quality, size, timing, spatial distribution, and concentration of prey. Variability in the physical environment also affects predation through influences on the abundance and distribution of predators. Fish diseases leading to a weakened state or even death may also be environmentally triggered. Particular temperature ranges may, for instance, be more conducive to allowing disease outbreaks. While water temperature is typically the main source of environmental impact on fish, salinity and oxygen conditions, and ocean mixing and transport processes are also important.

Reproduction, recruitment, and growth

The physical environment affects the reproductive cycle of fish. For example, ambient temperatures may determine the age at sexual maturity. Atlantic cod off Labrador and the northern Grand Banks mature at 7 yr, while in the warmer waters off southwest Nova Scotia and on Georges Bank they mature at 3.5 and 2 yr, respectively (Drinkwater, 1999). Reproduction is typically temperature-dependent with gonad development occurring more quickly under warm conditions. Thus, temperature determines the time of spawning. Examples of low temperatures resulting in delayed spawning have been observed off Newfoundland, both in capelin (Nakashima, 1996) and Atlantic cod (Hutchings and Myers, 1994).

Understanding variability in recruitment (the number of young surviving long enough to potentially enter the fishery) has long been a prime issue in fisheries science. Evidence of changes in fish abundance in the absence of fishing suggests environmental causes. Following spawning, cod eggs and later young stages are generally distributed within the upper water column before they settle toward the bottom as half-year olds. The strength of a year-class is to a large degree determined during the first six months of life (Helle et al., 2000; Hjort, 1914; Myers and Cadigan, 1993; Sundby et al., 1989); life stages during which ocean climate may have a decisive effect (Cushing, 1966; De Young and Rose, 1993; Dickson and Brander, 1993; Ellertsen et al., 1989; Ottersen and Sundby, 1995; Sætersdal and Loeng, 1987). The effects of temperature on recruitment of Atlantic cod across its entire distribution range were examined by Ottersen (1996) and Planque and Fredou (1999). Populations inhabiting areas at the lower end of the overall temperature range of the species (i.e., West Greenland, Labrador, Newfoundland, and the Barents Sea) had higher than average recruitment when temperature anomalies were positive, while recruitment to populations occupying the warmer areas (e.g., the Irish and North Seas) seemed better with negative temperature anomalies. For populations inhabiting regions with mid-range temperatures the results were inconclusive. The recruitment of Norwegian spring-spawning herring is also linked to variability in water temperature (Toresen and Østvedt, 2000; see section 9.3.3.3).

The pelagic ecosystem in the southeastern Bering Sea may, according to the recently published Oscillating Control Hypothesis, alternate between primarily bottom-up control in cold regimes and primarily top-down control in warm regimes (Hunt and Stabeno, 2002; Hunt et al., 2002). The timing of spring primary production in the southeastern Bering Sea is determined predominately by the timing of sea-ice retreat. Late retreat leads to an early, ice-associated bloom in cold water, whereas no ice, or early retreat, leads to an open-water bloom in warm water. In years when the spring bloom occurs in cold water, low temperatures limit the production of zooplankton, and the survival of larval and juvenile fish, and their recruitment into the populations of large piscivorous fish, such as walleye pollock, Pacific cod, and arrowtooth flounder. Continued over decadal scales, this will lead to bottom-up limitation and a decreased biomass of piscivorous fish. Alternatively, in periods when the bloom occurs in warm water, zooplankton populations should grow rapidly, providing plentiful prey for larval and juvenile fish. Abundant zooplankton will support strong recruitment of fish and will lead to abundant predatory fish that control forage fish, including in the case of walleye pollock, their own juveniles (Hunt and Stabeno, 2002; Hunt et al., 2002).

Because fish are ectothermic, temperature is the key environmental factor. Individual growth is the result of a series of physiological processes (i.e., feeding, assimilation, metabolism, transformation, and excretion) whose rates are all controlled by temperature (Brett, 1979; Michalsen et al., 1998). Brander (1994, 1995) examined 17 North Atlantic cod populations and showed that mean bottom temperature accounted for 90% of the observed (ten-fold) difference in growth rates between populations. Higher temperatures led to faster growth rates over the temperature range experienced by these populations. Growth rate decreases at higher temperatures and the temperature for maximum growth decreases as a function of size (Björnsson, 2001).

The biomass of zooplankton, the main food for larval and juvenile fish, is generally greater when temperature is high in the Norwegian and Barents Seas (Nesterova, 1990). High food availability for the young fish results in higher growth rates and greater survival through the vulnerable stages that determine year-class strength. Temperature also affects the development rate of fish larvae directly and, thus, the duration of the high-mortality and vulnerable stages decreases with higher temperature (Blood, 2002; Coyle and Pinchuk, 2002; Ottersen and Loeng, 2000; Ottersen and Sundby, 1995). Also, in the Barents Sea, mean body size as half-year olds fluctuates in synchrony for herring, haddock, and Northeast Arctic cod and the length of all three is positively correlated with water temperature. This indicates that these species, having similar spawning and nursery grounds, respond in a similar manner to large-scale climate fluctuations (Loeng et al., 1995; Ottersen and Loeng, 2000). For Barents Sea cod, mean lengths-at-age for ages 1 to 7 are greater in warm periods (Dementyeva and Mankevich, 1965; Michalsen et al., 1998; Nakken and Raknes, 1987).

For 2- and 3-year old Barents Sea capelin, Gjøsæter and Loeng (1987) found positive correlations between temperature and growth for different geographical regions and for different years. Changes in water temperature through altered climate patterns may also affect predator–prey interactions. In the Barents Sea, the increase in basic metabolic rates of Northeast Arctic cod, associated with higher temperatures, can result in a rise in the consumption of capelin by 100 000 tonnes per degree centigrade (Bogstad and Gjøsæter, 1994).

Distribution and migration

Temperature is one of the main factors, together with food availability and suitable spawning grounds, which determines the large-scale distribution pattern of fish. Because most fish species (and stocks) tend to prefer a specific temperature range (Coutant, 1977; Scott J., 1982), long-term changes in temperature can lead to expansion or contraction of the distribution range of a species. These changes are generally most evident near the northern or southern boundaries of the species range; warming results in a northward shift and cooling draws species southward. For example, in the Barents Sea, temperature-related displacement of Northeast Arctic cod has been reported on interannual time scales as well as at both small and large spatial scales. In warm periods, cod distribution is extended eastward and northward compared to colder periods when the fish tend to concentrate in the southwestern part of the Barents Sea (Ottersen et al., 1998). Capelin distribution also responds to changes in water temperature both in the Barents Sea (Sakshaug et al., 1992) and off Newfoundland and Labrador.

The relatively high interannual stability of residual currents, which prevail in most regions, maintains the main features of larval drift patterns from spawning area to bottom settlement area for each population, and consolidates differences between populations. Interannual variation is introduced through changes in large- and regional-scale atmospheric pressure conditions. These affect winds and upper ocean currents, which in turn modify drift patterns of fish larvae and introduce variability in water temperature and the availability of prey items. While a long and unrestricted larval drift is important for some cod populations, such as those in the Barents Sea and the Icelandic component at West Greenland, recruitment to populations residing in small and open systems depends on larval retainment and the avoidance of massive advective losses (Ottersen, 1996; Sinclair M., 1988).

Many species that undertake seasonal migrations appear to use environmental conditions as cues. For example, April sea surface temperatures and sea-ice conditions in the southern Gulf of St. Lawrence determine the average arrival time of Atlantic herring on their spawning grounds (Lauzier and Tibbo, 1965; Messieh, 1986). Sea-ice conditions also appear to control the arrival time in spring of Atlantic cod onto the Magdalen Shallows into the Gulf of St. Lawrence (Sinclair A. and Currie, 1994). Atlantic salmon arrive earlier along the Newfoundland and Labrador coasts during warmer years (Narayanan et al., 1995).

The Norwegian spring-spawning herring stock, inhabiting the Norwegian and Icelandic Seas, is highly migratory. Larvae and fry drift into the Barents Sea, while adults undergo substantial feeding and spawning migrations (Holst J. et al., 2002). Since around 1950, biomass and migration patterns have fluctuated dramatically. While

these shifting migration patterns may be dominated by density-dependence, environmental conditions are also likely to have been important (Holst J. et al., 2002).

In the Bering Sea, warmer bottom temperatures lead to the distribution of adult walleye pollock, Greenland turbot, yellow Irish lord (*Hemilepidotus jordani*), and thorny sculpin (*Icelus spiniger*) being more widespread on the shelf, while Arctic cod are restricted to the cold pool (Wyllie-Echeverria and Wooster, 1998).

The combination of environmentally influenced distribution patterns and politically restricted fisheries patterns can have pronounced impacts on the availability of fish to fishers. For instance, most of the Barents Sea is under either Norwegian or Russian jurisdiction, but there is a small, disputed region of international waters in the center. This area is aptly named the "Loophole" and at times it is the site of extensive fishing activity by the international fishing fleets. Most fishing occurs in the southern part of the Loophole, where in warmer years several species of all sizes are found throughout the year. However, in colder years there may be hardly any fish in the area for prolonged periods. The reason for this pattern is that the southern part of the Loophole lies to the south of the Polar Front so that even relatively small east–west movements of the water masses may result in large temperature changes. In cold years, the Polar Front is displaced farther south and west than the Loophole. The fish move in order to remain within the warmer water, thereby making them unavailable to the international fishing fleets (Aure, 1998). The movement of the Polar Front is most pronounced between warm and cold years in the Barents Sea as a whole, but movements may also occur on time scales of weeks.

9.3.2.4. Marine mammals and seabirds

Some important predator–prey match–mismatch issues also occur with higher predators. The timing of reproduction in many seal species is thought to match the availability of large zooplankton and small fishes at the time when pups are weaned and when polar bear den emergence occurs during the peak reproductive period of their favorite prey, ringed seal. Likewise, invertebrate or fish species must be available in the upper parts of the water column when seabird young commence self-feeding. Higher predators might not easily track shifts in the production of zooplankton and fish, which are more directly influenced by temperature.

Factors that influence the distribution and annual duration of sea ice or snow availability in the spring can potentially have profound influences on the population ecology of some arctic marine mammals. Sea ice is the breeding habitat for all pagophilic (i.e., ice-loving) seal species and it is the primary hunting platform for polar bear. Changes in the time of formation or disappearance of seasonal sea ice, in the quality of the sea ice, and in the extent of total coverage of both seasonal and multi-year ice could all affect ice-dependent species. Snow cover is very important for polar bears and ringed seals and changes in average snow depth or duration of the snow season could affect their breeding success.

Walruses appear to follow an annual migratory pattern, moving with the advance and retreat of the sea ice in most parts of their range (Fay, 1981, 1982; Wiig et al., 1996). However, this may be due to the sea ice blocking access to shallow-water feeding areas, rather than to it serving as an essential habitat element.

The primary requirement for seabirds in the Arctic is suitable breeding cliffs near abundant prey sources. If ice edges or frontal regions shift such that the distance between these highly productive areas and the nesting areas becomes too great, the mismatch would have serious consequences for seabirds.

9.3.3. Past variability – interannual to decadal

Previous data collections combined with present-day models shows that climate variability is very likely to have influenced population parameters of marine organisms, especially fish (section 9.3.2.3). Water temperature undoubtedly affects species composition in different areas, as well as the recruitment, growth, distribution, and migration of different fish species. However, most of the relationships between water temperature and population variables are qualitative and few of those discussed here can be quantified.

9.3.3.1. Plankton

There are few long time series for phytoplankton in the Arctic. Exceptions include (1) datasets covering 20 years or more for Icelandic waters (Thordardóttir, 1984) and Norway (Oslofjord, Trondheimsfjord; Johnsen et al., 1997); (2) a program undertaken during the 1990s to monitor harmful algae along the Norwegian coast (Dahl E. et al., 1998, 2001); and (3) zooplankton data provided by the Continuous Plankton Recorder, which has been used in much of the North Atlantic between 50° and 65° N for over fifty years (Johns et al., 2001). This has generated one of the most detailed records of seasonal, interannual, and decadal variability in zooplankton to date. Sampling in the Northwest Atlantic is less complete but extends across the Labrador Sea to the Grand Banks, the Scotia Shelf, and the Gulf of Maine.

The copepod *Calanus finmarchicus* contributes >50% of the biomass of sampled plankton in the North Atlantic. Its population has declined substantially in the Northeast Atlantic since the early 1960s (Fig. 9.25), apparently as a function of variation in the NAO (Planque and Batten, 2000). Also, recent and persistent declines seem to be related to a low-frequency change in the volume of Norwegian Sea Deep Water, where *Calanus finmarchicus* overwinters (Heath et al., 1999). Figure 9.26 shows that, in contrast, the arctic species *Calanus glacialis* extended its range in the Northwest Atlantic during the

1990s as a consequence of the extension of cold Labrador Slope Water (Johns et al., 2001).

Historical time series of zooplankton biomass suggest a decrease in biomass between 1954 and 1995 in the oceanic and outer shelf regions of the eastern Bering Sea (Sugimoto and Tadokoro, 1998). However, when the data are separated by shelf region, such a trend is not apparent (Napp et al., 2002). Inshore sampling of *Calanus marshallae* indicates a much higher biomass in the late 1990s compared to the early 1980s. Water temperature is the most important factor influencing zooplankton growth rates and may be responsible for the observed interannual variability in mid-shelf zooplankton biomass (Coyle and Pinchuk, 2002; Napp et al., 2000). During cold springs when the spring bloom is dominated by ice-edge blooms, reduced coupling between the mesozooplankton and phytoplankton means more phytoplankton will be ungrazed and sink to the bottom, so enhancing the benthic food web. Stronger coupling between mesozooplankton and phytoplankton in warmer springs may result in a stronger pelagic production.

The population of the jellyfish *Chrysaora melamaster* increased at least ten-fold during the 1990s (Brodeur et al., 1999b). These large jellyfish compete for food with young walleye pollock (consuming an estimated 5% of the annual crop of zooplankton) and also feed upon them (consuming an estimated 3% of newborn walleye pollock). Jellyfish have very low energy requirements compared with fish (20 times less) and mammals (200 times less than whales) on a per unit weight basis. Their increased abundance may be due to reduced nutrients and a lower-energy plankton regime.

9.3.3.2. Benthos

Data are available for sedentary and long-lived macrozoobenthos, which are relevant indicators of multi-year environmental fluctuations between the late 1700s and the present. Biogeographical boundaries in the Barents Sea have shifted as a result of temperature fluctuations (Blacker, 1965). Based on analyses using temperature paleo-reconstructions, it appears that high arctic species tend to survive only when temperatures remain between -1.8 and 6 °C, whereas adults of boreal species can survive temperatures of -1 to 25 °C. Also, biogeographical changes in the bottom fauna appear to occur faster and are more easily detected during warm periods than cold periods.

The zoobenthos of the Russian Arctic seas has been most intensively studied in the Barents Sea. Deryugin (1924) detected several unusual species in Kola Bay in 1908 and 1909 and related this to fluctuations in water temperature. Some boreal species in the Barents Sea have responded to environmental change by shifting their biogeographical borders (Fig. 9.27; Chemerisina, 1948; Nesis, 1960). This reflects variations in population size at habitat boundaries, not changes in the size and shape of the habitats themselves (Galkin, 1998).

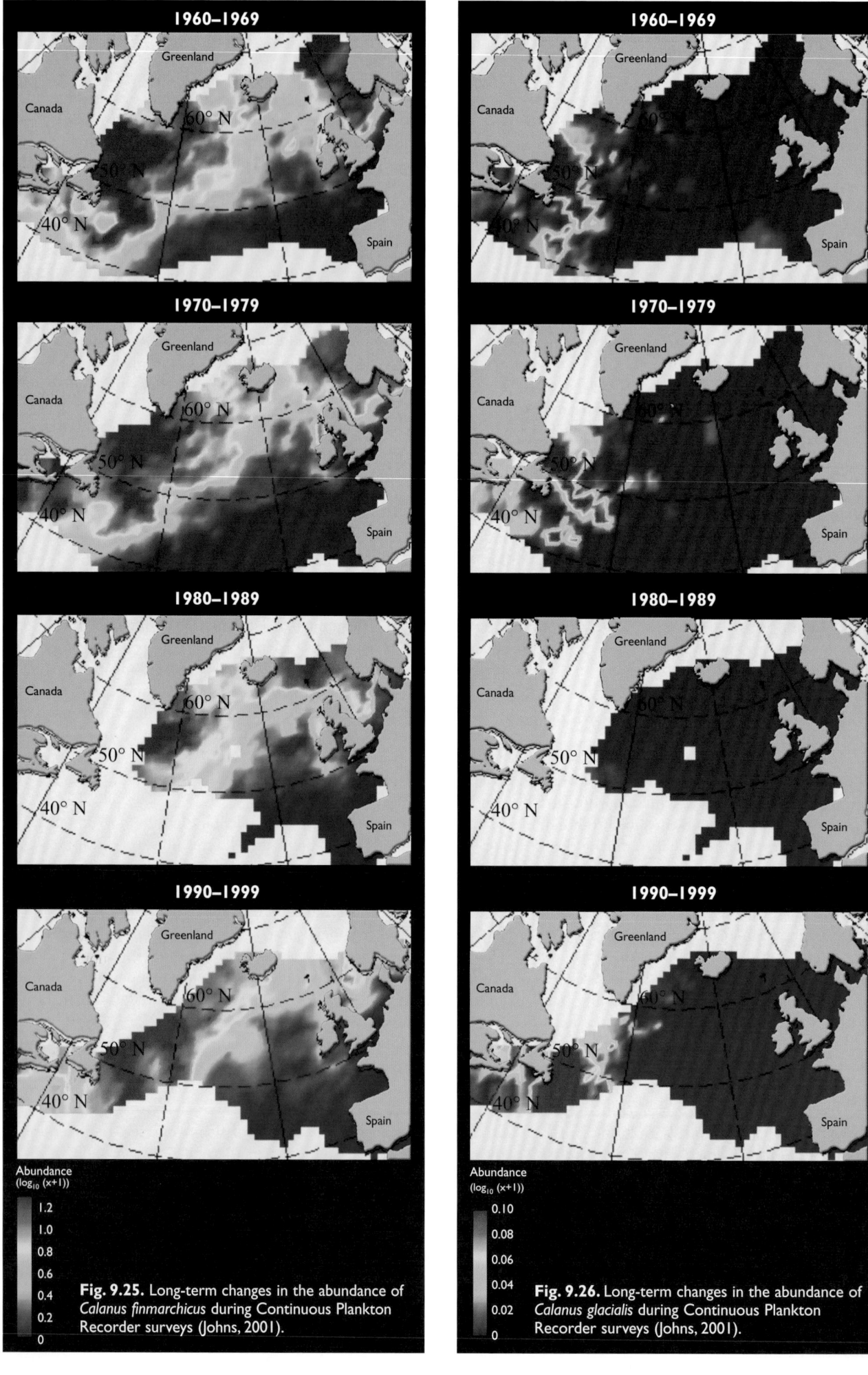

Fig. 9.25. Long-term changes in the abundance of *Calanus finmarchicus* during Continuous Plankton Recorder surveys (Johns, 2001).

Fig. 9.26. Long-term changes in the abundance of *Calanus glacialis* during Continuous Plankton Recorder surveys (Johns, 2001).

In years following warming, the polychaete *Spiochaetopterus typicus* predominates along the Kola Section in the Barents Sea. Following cold years, the polychaete *Maldane sarsi* predominates. *Spiochaetopterus typicus* is thus an indicator of warming or warm conditions.

Estimating natural fluctuations in zoobenthos biomass in the Barents Sea is difficult owing to the impact of commercial bottom trawling (Denisenko, 2001). In the Pechora Sea, where there is no traditional demersal fishery, changes in zoobenthos biomass in 1924, 1958 to 1959, 1968 to 1970, and 1992 to 1995 show a negative correlation between zoobenthos biomass and temperature (Denisenko, pers. comm., Zoological Institute RAS, St. Petersburg, 2003).

In the Bering Sea, long-term change in zoobenthos communities is known for the eastern regions as a result of Soviet and American investigations in 1958 to 1959 and 1975 to 1976. In the 1950s, maximum biomass occurred in the northwestern part of the eastern Bering Sea in the mid-shelf region at bottom depths between 50 and 150 m. In the early 1970s, the highest biomasses occurred in the mid-shelf area, southeast of the Pribilof Islands. Because the early 1970s were cold compared to the late 1950s, it may be that the difference in zoobenthos biomass related to changes in the southern limit of the ice edge and thus to the amount of ice-edge primary production that fell to the benthos, ungrazed by pelagic zooplankton.

Recent studies indicate ongoing change in the benthic communities of the Bering and Chukchi Seas (Francis et al., 1998; Grebmeier and Cooper, 1995; Sirenko and Koltun, 1992). The region just north of the Bering Strait is a settling basin for organic carbon, which results in a high benthic standing stock and high oxygen uptake rates (Grebmeier, 1993; Grebmeier et al., 1988, 1989).

Benthic productivity in this region near 67°30' N, 169° W has historically maintained the highest benthic faunal biomass of the entire Bering/Chukchi system (Grebmeier, 1993; Grebmeier and Cooper, 1994; Grebmeier et al., 1995; Stoker, 1978, 1981). Although benthic biomass remains high in the area, regional changes in the dominant benthic species have occurred. This is likely to indicate changing hydrographic conditions (Grebmeier, 1993; Grebmeier et al., 1995).

In the St. Lawrence Island polynya region, changes in regional oceanography due to the position or size of the Gulf of Anadyr gyre are ultimately related to the northward transport of water through the Bering Strait, and to the geostrophic balance within the Arctic Ocean basin. The latter, which is related to variations in the NAO/AO index, drives the northward current system in the northern Bering Sea (Walsh et al., 1989). Roach et al. (1995) found little flow through the Bering Strait into the Arctic Ocean during the NAO-positive period of the early 1990s, and a large increase in flow when the NAO became negative in 1996. Small flow into the Arctic Ocean is coincident with reduced northward transport of water south of St. Lawrence Island.

The Gulf of Anadyr "cold pool" is maintained by sea-ice production and brine formation in the St. Lawrence Island polynya. Reduced sea-ice production to the south of the polynya resulting in a decreased supply of nutrients for early-season primary production would limit benthic populations (Grebmeier and Cooper, 1995). However, it is possible that an enhanced and more energetic polynya could result from warming. This could maintain a chemostat-type bloom system, as to the north of St. Lawrence Island (Walsh J.J. et al., 1989), allowing a longer growing season and greater production and thus transport.

The three species of crab that inhabit the eastern Bering Sea shelf (red king crab, Tanner crab, and snow crab) exhibit highly periodic patterns in abundance. Rosenkranz et al. (2001) found that anomalously cold bottom temperatures in Bristol Bay may adversely affect the reproductive cycle of Tanner crab and that northeasterly winds may promote coastal upwelling, which advects larvae to regions of fine sediments favorable for survival upon settling. Incze and Paul (1983) linked low densities of copepods within the 70 m isobath in Bristol Bay with low abundance of Tanner crab larvae. Recruitment patterns for red king crab in Bristol Bay show the populations to be negatively correlated with the deepening of the Aleutian Low and warmer water temperatures (Zheng and Kruse, 2000). Red king crabs were commercially exploited during the late 1970s, which has also contributed to the population decline.

9.3.3.3. Fish

There are few records of marine biota showing interannual and longer-term variability in the Arctic Ocean, but records of the abundance of commercial fish species for

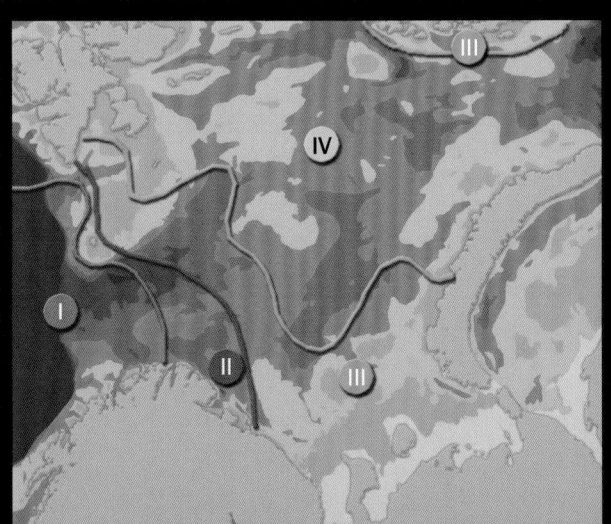

Fig. 9.27. Biogeographical boundaries in the Barents Sea during the 20th century. I maximal western extent of arctic species in cold periods; II line of 50% average relation between boreal and arctic species; III maximal eastern extent of boreal species in warm periods; IV transition zone.

the Labrador, Greenland, Iceland, Norwegian, and Barents Sea go back to the start of the twentieth century and even earlier in some cases. Within these areas capelin, cod, and herring populations have undergone very large fluctuations in biomass and distribution.

The period of warming from the mid-1920s to the mid-1960s, which affected Greenland and Iceland in particular (see section 9.2.4.2), had a profound effect on the major commercial fish species and also on most other marine life. A number of species, which had been rare in offshore areas west of Greenland, became abundant at this time and population biomass increased by several orders of magnitude. These changes were not related to fishing and are clearly due to climate variability.

Boreal species such as cod are likely to respond strongly to temperature variability and so show greater variability in recruitment at the extremes of their range (Brander, 2000; Ottersen and Stenseth, 2001). However, for the period over which records are available most populations have been reduced to low levels as a result of fishing pressure and may therefore show high variability throughout their distribution.

The warming period of the 1920s

The warming period of the 1920s caused a poleward extension in the range of distribution for many fish and other marine and terrestrial species from Greenland to Iceland and eastward to the Kara Sea. Records of changes in species distribution during the 1920s provide some of the most convincing evidence of the pervasive effects of a change in climate on the marine ecosystem as a whole. Jensen (1939) published a comprehensive review of the effects of the climate change in the Arctic and subarctic regions during this period, which presents much of this information. Some of the salient points concerning fish species are summarized in Table 9.9.

The marine shelf ecosystem off West Greenland is affected by cold polar water masses and temperate Atlantic water. Changes in the distribution of these water masses, under the influence of the NAO, affect the distribution and abundance of fish species and hence fisheries yields (Pedersen and Rice, 2002; Pedersen and Smidt, 2000; Schopka, 1994).

The distribution of cod extended poleward by about 1000 km between 1920 and 1930 and can be followed in some detail, because fishing stations were established progressively further north as directed coastal fisheries were established by the Greenland Administration.

The international offshore fishery for cod off West Greenland reached a peak of over 400000 tonnes in the early 1960s before collapsing. The decline was due to a combination of fishing pressure and reduced water temperature, and probably to a lack of recruitment from Iceland. The relationship between water temperature and recruitment level is clear for this area (Brander, 2000; Buch et al., 1994), with poor recruitment occurring at temperatures below about 1.5 °C (measured on Fylla Bank, 64° N, in June). The warming of the North Atlantic that has taken place since about the early 1990s has also affected Greenland but temperatures remained below 1.5 °C until 1996. Thus it is too early to expect a recovery of the cod population; cod take about seven years to reach maturity and may be adversely affected by the trawl fishery for shrimps which is now the mainstay of the Greenland fisheries.

One of the principal changes which took place off Iceland during the 1920s warming period occurred within the major pelagic populations – those of herring and capelin (Table 9.9). Prior to 1920, the capelin spawned regularly on the south and southwest coasts of Iceland, but from 1928 to 1935 very few capelin were taken in these areas. In contrast, herring extended their spawning areas from the south and southwest coast to the east, northwest, and north coasts in this period (Saemundsson, 1937).

Similar changes are also recorded for Jan Mayen, the Barents Sea, the Murman Coast, the White Sea, Novaya Zemlya, and the Kara Sea, where cod and herring extended their ranges and became more abundant.

Table 9.9. Changes in the distribution and abundance of fish species off West Greenland and Iceland during the period of warming from 1920 onwards. Prepared by Brander (2003) based on Saemundsson (1937) and Jensen (1939).

	West Greenland	Iceland
Species previously absent, but which appeared from 1920 onwards	Haddock (*Melanogrammus aeglefinus*), tusk (*Brosme brosme*), ling (*Molva molva*)	Bluntnose sixgill shark (*Notidanus griseus*), swordfish (*Xiphias gladius*), horse mackerel (*Trachurus trachurus*)
Rare species which became more common and extended their ranges	Coal fish (*Pollachius virens*; new records of spawning fish), Atlantic salmon (*Salmo salar*), spurdog (*Squalus acanthias*)	Witch (*Glyptocephalus cynoglossus*), turbot (*Psetta maxima*), basking shark (*Cetorhinus maximus*), northern bluefin tuna (*Thunnus thynnus*), mackerel (*Scomber scombrus*), Atlantic saury (*Scomberesox saurus*), ocean sunfish (*Mola mola*)
Species which became abundant and extended their ranges poleward	Atlantic cod, Atlantic herring (new records of spawning fish)	Atlantic cod, Atlantic herring (both extended their spawning distribution)
Arctic species which no longer occurred in southern areas, and extended their northern limits	Capelin, Greenland cod, Greenland halibut (became much less common)	Capelin

Climate effects on fish in the Barents Sea

Understanding of the processes underlying major fluctuations in the fish ecosystem of the Barents Sea is considerably better than for most other areas of the northern North Atlantic (Rødseth, 1998). The main species involved are Atlantic cod, capelin, and Atlantic herring. These species are closely linked through the food web. Cod is highly dependent on capelin as its main prey. One- to two-year old herring prey heavily on the larvae of capelin, whose mortality increases greatly in years with a large biomass of young herring. Interactions between these species are strongly affected by the highly variable oceanographic conditions of the Barents Sea (Hamre, 1994).

The early years of the twentieth century, particularly 1902, were extremely cold in the Barents Sea, with extensive sea-ice cover. This resulted in a crisis for the Norwegian fisheries, with low catches of small Northeast Arctic cod in very poor condition. Large numbers of seals, primarily harp seals, moved down the Norwegian coast from the Barents Sea. A similar sequence of events occurred during the cold period in the 1980s, when the capelin population collapsed; the cod were small and in poor condition and harp seals again invaded the northern coast of Norway.

For cod in particular, the consequences of variability in water temperature, transport, and food during early life stages have been studied closely (Michalsen et al., 1998; Ottersen and Loeng, 2000; Ottersen et al., 1998; Sætersdal and Loeng, 1987). Growth and survival rates of larvae and juveniles are higher in warm years and the large year-classes of cod spread further east into the Barents Sea, where they encounter cooler water and their growth rate slows as a result (Ottersen et al., 2002).

Norwegian spring-spawning herring

The biomass of Norwegian spring-spawning herring increased almost ten-fold between 1920 and 1930, when the Norwegian Sea and much of the North Atlantic went through a period of rapid warming (Toresen and Østvedt, 2000). The herring population declined rapidly from the late 1950s and by 1970 had decreased by more than four orders of magnitude. The decline was coincident with a period of cooling (Fig. 9.28). Although this cooling may have been a contributing factor, it is likely that heavy fishing pressure was the primary cause of the collapse of the population.

The collapse of the Norwegian spring-spawning herring population coincided with a retraction of the summer feeding distribution due to the southward and eastward shift in the location of the Polar Front. The Polar Front was to the north of Iceland prior to 1965 but has since stayed west of Iceland (see Box 9.8). Despite a complete recovery of the herring spawning stock and a rise in water temperature north of Iceland in 2000 to levels similar to those of the mid-1960s (Malmberg and Valdimarsson, 2003), the herring have not returned to their earlier feeding areas.

The rapid cooling during the mid- and late 1960s also resulted in reduced growth and recruitment of the Norwegian spring-spawning herring (Toresen and Østvedt, 2000). The same happened with the Icelandic summer spawning herring (Jakobsson et al., 1993).

Temperature-mediated habitat changes in Canadian capelin

Capelin off Newfoundland and Labrador spread southward as far as the Bay of Fundy when water temperatures declined south of Newfoundland in the mid-1960s and retracted northward as water temperatures rose in the 1970s (Colton, 1972; Frank et al., 1996; Tibbo and Humphreys, 1966). During cooling in the latter half of the 1980s and into the 1990s, capelin again extended their range, eastward to Flemish Cap and southward onto the northeastern Scotia Shelf off Nova Scotia (Frank et al., 1996; Nakashima, 1996). For example, small quantities of capelin began to appear in the groundfish trawl surveys on the Scotia Shelf in the mid-1980s and since then numbers have increased dramatically (Frank et al., 1996). Initially, only adult capelin were caught, but juveniles later appeared, suggesting capelin were successfully spawning.

This shift appears to have been part of a larger-scale ecosystem change. While capelin were spreading onto the Scotia Shelf, polar cod, whose primary grounds have traditionally been the Labrador Shelf stretching southward to northern Newfoundland, were moving southward. In the late 1980s and early 1990s, as water temperatures decreased, polar cod pushed southward onto the Grand Banks and into the Gulf of St. Lawrence in large numbers (Gomes et al., 1995; Lilly et al., 1994).

Historical climate and fish in the Bering Sea

The direct effects of atmospheric forcing resulting from climate variations are very important to the physical

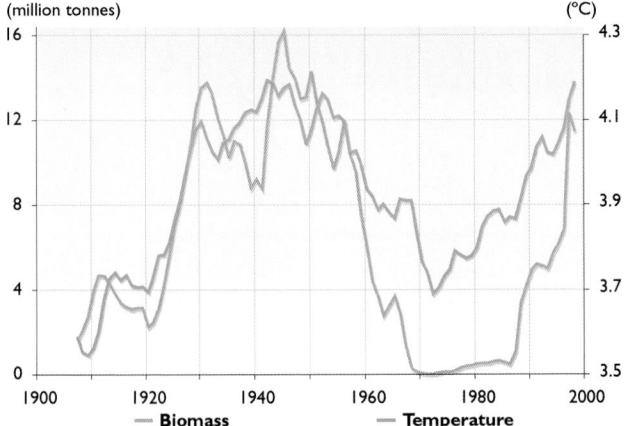

Fig. 9.28. Relationship between water temperature and the biomass of the Norwegian spring-spawning herring stock (Toresen and Østvedt, 2000).

oceanographic conditions of the Bering Sea. Since the eastern Bering Sea shelf has a characteristically sluggish mean flow and is separated from any direct oceanographic connection to the North Pacific Ocean by the Alaska Peninsula, linkages between the eastern Bering Sea and the climate system are primarily a result of the ocean–atmosphere interaction (Stabeno et al., 2001). Climate variations in this region are directly linked to the location and intensity of the Aleutian Low pressure center, which affects winds, surface heat fluxes, and the formation of sea ice (Hollowed and Wooster, 1995). The pressure index has experienced eight statistically significant shifts on roughly decadal time scales that alternated between cool and warm periods (Overland et al., 1999a). A well-documented shift (Trenberth, 1990 among others) from a cool to a warm period occurred between 1977 and 1989, which coincided with the start of fishery-independent sampling programs and fishery catch monitoring of major groundfish

species. Information from the contrast between this period and the previous and subsequent cool periods (1960 to 1976 and 1989 to 2000) forms the basis of the following discussion.

Changes in atmospheric climate are primarily transmitted through the Bering Sea to the biota via the mechanisms of wind stress (Francis et al., 1998) and the annual variation in sea ice extent (Stabeno et al., 2001). These mechanisms directly alter the timing and abundance of primary and secondary production by changing the salinity, mixed-layer depth, nutrient supply, and vertical mixing in the ocean system.

The extent and timing of the sea ice also determines the area where cold bottom water temperatures will persist throughout the following spring and summer. This eastern Bering Sea area of cold water, known as the cold pool, varies with the annual extent and dura-

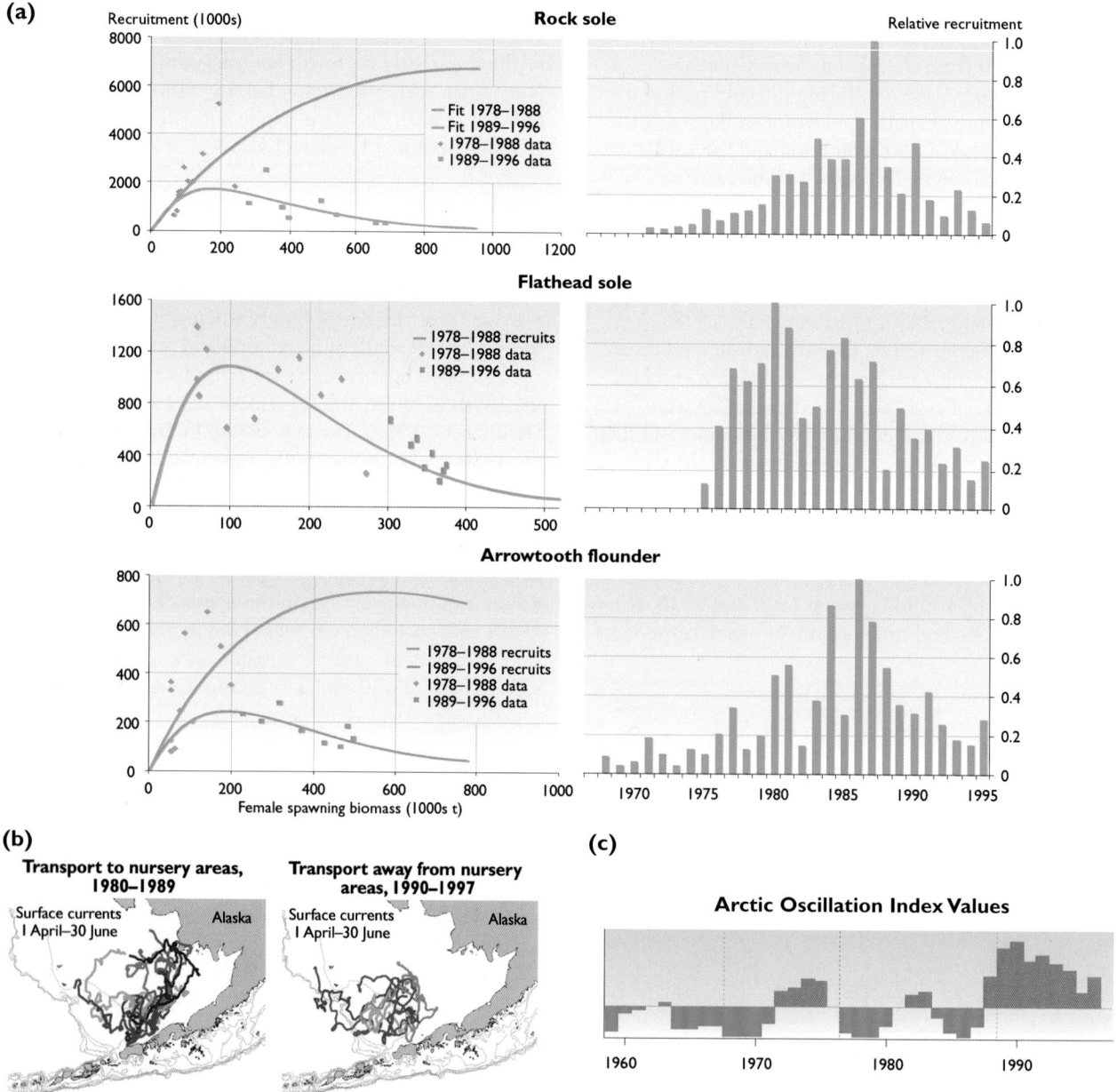

Fig. 9.29. Winter spawning flatfish (a) recruitment and (b) predicted wind-driven larval drift patterns relative to (c) decadal-scale atmospheric forcing in the eastern Bering Sea (Wilderbuer et al., 2002).

Box 9.9. Effect of atmospheric forcing in the Bering Sea

Recruitment responses of many Bering Sea fish and crab species are linked to decadal scale patterns of climate variability (Francis et al., 1998; Hare and Mantua, 2000; Hollowed et al., 2001; Wilderbuer et al., 2002; Zheng and Kruse, 2000). Decadal changes in the recruitment of some flatfish species in the eastern Bering Sea appear to be related to patterns in atmospheric forcing (see Fig. 9.29). The AO index, which tracks the variability in atmospheric pressure at polar and mid-latitudes, tends to vary between negative and positive phases on a decadal scale. The negative phase brings higher-than-normal pressure over the polar region and the positive phase does the opposite, steering ocean storms farther north. These patterns in atmospheric forcing in winter may influence surface wind patterns that advect fish larvae onto or off the shelf. When the index was in its negative phase in the 1980s, southwesterly winds tended to dominate, which is likely to have resulted in the transport of flatfish larvae to favorable nursery grounds. The positive phase in the 1990s showed winds to be more southeasterly, which would tend to advect larvae off the shelf. The relative recruitment of three species of winter spawning flatfish in the Bering Sea – arrowtooth flounder, rock sole, and flathead sole – was high in 1977 to 1988 and low in 1988 to 1998, indicating a link between surface wind advection patterns during the larval stage and flatfish survival.

However, periods of strong Aleutian Lows are associated with weak recruitment for some Bering Sea crab species and are unrelated to others (Zheng and Kruse, 2000) depending on species-specific life history traits. Winds from the northeast favor retention of crab larvae in offshore mud habitats that serve as suitable nursery areas for young Tanner crabs since they bury themselves for protection (Rosenkranz et al., 2001). However, southwesterly winds promote inshore advection of crab larvae to coarse, shallow water habitats in inner Bristol Bay that serve as nursery areas for red king crabs who find refuge among biogenic structures (Tyler and Kruse, 1998). Timing and composition of the plankton blooms may also be important, as red king crab larvae prefer to consume *Thalassiosira* spp. diatoms, whereas Tanner crab larvae prefer copepod nauplii. Some species, such as Bering Sea herring, walleye pollock, and Pacific cod, show interannual variability in recruitment that appears more related to ENSO-driven climate variability (Hollowed et al., 2001; Williams and Quinn, 2000). Years of strong onshore transport, typical of warm years in the Bering Sea, correspond to strong recruitment of walleye pollock, possibly due to separation of young fish from cannibalistic adults (Wespestad et al., 2000). Alaskan salmon also exhibit decadal scale patterns of production, which are inversely related to the pattern of west coast salmon production (Hare and Mantua, 2000). Including environmental variables such as sea surface temperature and air temperature significantly improved the results of productivity models for Bristol Bay sockeye salmon (*Oncorhynchus nerka*) compared to models containing density-dependent effects only (Adkison et al., 1996).

tion of the ice pack and can influence fish distributions. Pollock have shown a preference for warmer water, and exhibit an avoidance of the cold pool (Wyllie-Echeverria, 1995). In cold years they utilize a smaller portion of the shelf waters, in contrast to warm years when they have been observed as far north as the Bering Strait and the Chukchi Sea. Strong year-classes of pollock have been found to occur synchronously throughout the Bering Sea (Bulatov, 1995) and to coincide with above-normal air and bottom water temperatures and reduced sea-ice cover (Decker et al., 1995; Quinn and Niebauer, 1995). These favorable years of production are the result of good juvenile survival and are related to how much cold water habitat is present (Ohtani and Azumaya, 1995), the distribution of juveniles relative to the adult population which influences the level of predation (Wespestad et al., 2000), and enhanced rates of embryonic development in warmer water (Blood, 2002; Haynes and Ignell, 1983).

The distributions of forage fishes including Pacific herring, capelin, eulachon (*Thaleichthys pacificus*), and juvenile Pacific cod and pollock indicate temperature-related differences (Brodeur et al., 1999a). Capelin exhibits an expanded range in years with a larger cold pool and contracts in years of reduced sea-ice cover. Although the

productivity of capelin populations in relation to water temperature is not known, Bering Sea herring populations exhibited improved recruitment during warm years (Williams and Quinn, 2000), similar to other herring populations where the timing of spawning is also temperature-related (Zebdi and Collie, 1995).

Recruitment and stock biomass have been examined for evidence that climatic shifts induce responses in the production of groundfish species in the Bering Sea and North Pacific Ocean (Hollowed and Wooster, 1995; Hollowed et al., 2001). Even though results from these studies are highly variable, strong autocorrelation in recruitment associated with the significant change in climate in 1977 was observed for many salmonids and some winter-spawning flatfish species such as eastern Bering Sea arrowtooth flounder and Greenland halibut. The two latter species showed opposite changes post-1977 (increasing biomass for arrowtooth flounder and decreasing biomass for Greenland halibut). Substantial increases in the abundance of Pacific cod, skates, flatfish such as rock sole, and non-crab benthic invertebrates also took place on the Bering Shelf in the 1980s (Conners et al., 2002).

The decadal-scale patterns in recruitment success for winter-spawning flatfish (Fig. 9.29) may be associated with decadal shifts in the Aleutian low pressure system that affects cross-shelf advection patterns of larvae to favorable nursery areas rather than with water temperature (Wilderbuer et al., 2002). Box 9.9 describes the effects of atmospheric forcing in the Bering Sea in more detail.

9.3.3.4. Marine mammals and seabirds

Although fragmented, there is a lot of evidence to suggest that climate variations have profound effects on marine mammals and seabirds. The capelin collapse in the Barents Sea in 1987 had a devastating effect on seabirds breeding on Bjørnøya. Repeated years (1967, 1981, 2000, 2001, and 2002) with little or no sea ice in the Gulf of St. Lawrence resulted in years with almost zero production of seal pups, compared to hundreds of thousands in good sea-ice years.

Vibe (1967) explored the relationship between climate fluctuations and the abundance and distribution of animals, including marine mammals and seabirds, in Greenland. During the cold, dry, and stable "drift-ice stagnation" phase in West Greenland (approximately 1810 to 1860), marine mammals and seabirds concentrated at central West Greenland because the sea ice did not advance far north into Davis Strait. During the "drift-ice pulsation stage" (1860 to 1910), when the sea ice of the Arctic Ocean drifted into the Atlantic Ocean

in larger amounts than before, marine mammal and seabird populations decreased in the unstable and wet climate of West Greenland because the East Greenland Current and the East Greenland sea ice advanced far north into Davis Strait in summer. In the same period, the Greenland right whale (*Balaena mysticetus*) population "stagnated" in the Atlantic region. During the "drift-ice melting stage" (1910 to 1960) the East Greenland sea ice decreased in Davis Strait and populations of marine mammals and seabirds increased in northern West Greenland. Cod were abundant along the coast of West Greenland and multiplied in Greenland waters.

The condition of adult male and female polar bears has declined in Hudson Bay since the early 1980s, as have birth rates and the proportion of first-year cubs in the population. Stirling et al. (1999) suggest that the proximate cause of these changes in physical and reproductive parameters is a trend toward earlier breakup of the sea ice, which has resulted in the bears coming ashore in poorer condition.

9.3.4. Future change – processes and impacts on biota

Table 9.4 summarizes the potential physical oceanographic changes in the Arctic based on the projected changes in the atmospheric forcing functions (Table 9.1)

Table 9.10. Potential long-term ecological trends due to climate warming. Unless otherwise specified these projected changes are very likely to happen.

	Phytoplankton	Zooplankton	Benthos	Fish	Marine mammals and seabirds
Distribution	Increased spatial extent of areas of high primary production in the central Arctic Ocean.	Southern limit of distribution for colder water species to move northward. Distribution of more southerly species to move northward.	Southern limit of distribution for colder water species to move northward. Distribution of more southerly species to move northward.	Southern limit of distribution for colder water species to move northward. Distribution of more southerly species to move northward. Timing and location of spawning and feeding migrations to alter.	Poleward shift in species distributions.
Production	Increased production in central Arctic Ocean, and Barents and Bering Sea shelves.	Difficult to predict, will depend on the timing of phytoplankton production and seawater temperatures.	Difficult to predict, will partly depend on the degree of match/mismatch between phytoplankton/zooplankton production and on water temperature. Production by shrimp and crab species may decline.	Wind-driven advection patterns of larvae may be critical as well as a match/mismatch in the timing of zooplankton production and fish larval production.	Dramatic declines in production by ice-associated marine mammals and increases by more temperate species. Seabird production likely to be mediated through forage availability, which is unpredictable.
Species composition/ diversity	Dependent on mixing depth: shallow mixing favors diatoms, intermediate depth mixing favors *Phaeocystis*, deep mixing may favor nanoflagellates.	Adaptable arctic copepods, such as *Calanus glacialis*, may be favored.	Cold-water species may decline in abundance along with some clams and crustaceans, while warm water polychaetes, blue mussel (*Mytilus edulis*), and other types of benthos may increase.	Cod, herring, walleye pollock, and some flatfish are likely to move northward and become more abundant, while capelin, polar cod, and Greenland halibut will have a restricted range and decline in abundance.	Declines in polar bear, and in ringed, harp, hooded, spotted, ribbon, and possibly bearded seals. Increased distribution of harbour seals and grey seals. Possible declines in bowhead, narwhal, grey, and beluga whales. Ivory gulls and several small auk species are likely to decline while other changes in bird populations are unpredictable.

and potential future sea-ice conditions discussed in Chapter 6. Table 9.10 summarizes the potential long-term ecological changes in the marine system that are considered likely to arise as a result of these physical changes. The time frames for these changes to the biological system are addressed in this section by trophic level and by region where appropriate. The most pronounced physical changes are likely to include a substantial loss of sea ice, an increase in air and sea surface temperature, and changes in the patterns of wind and moisture transport.

Changes in the distribution of many species, ranging from phytoplankton to whales, are very likely to occur. The main habitat changes affecting marine mammals and seabirds include a reduction in sea ice, changes in snow cover, and a rise in sea level. Phenological changes, species replacements, and changes at lower trophic levels are also likely to have a strong influence on upper trophic level species.

9.3.4.1. Primary production

Changes in sea ice, water temperature, freshwater input, and wind stress will affect the rate of nutrient supply through their effect on vertical mixing and upwelling. Changes in vertical mixing and upwelling will affect the timing, location, and species composition of phytoplankton blooms, which will in turn affect the zooplankton community and the productivity of fish.

Changes in the timing of the primary production will determine whether this production is utilized by the pelagic community or is exported and utilized by the benthos (Box 9.10). The retention to export ratio also depends upon the advection and temperature preferences of grazing zooplankton, which together determine the degree of match or mismatch between primary and secondary production. The projected disappearance of seasonal sea ice from the Barents and Bering Seas (and thus elimination of ice-edge blooms) implies that these areas would have blooms resembling those of more southerly seas. The timing of these open ocean blooms in the Barents and Bering Seas will then be determined by the onset of seasonal stratification, again with consequences for a match/mismatch in timing with zooplankton.

Removal of light limitation in areas presently covered by multi-year sea ice is likely to result in a two- to five-fold increase in primary production, provided wind mixing is sufficient to ensure adequate nutrient supply. Moreover, earlier melting in the seasonal sea-ice zone is likely to enhance annual primary production by extending the growing season. The actual outcome in terms of annual production, however, is highly dependent upon regional and local changes in upwelling, wind-driven vertical mixing, and freshwater supply from sea ice and rivers. Note, for example, that it takes only a small increase in salt stratification (i.e., a decrease in surface salinity) to offset the effect of increased winds on vertical mixing. Regional cooling, as projected by some of the ACIA-designated models, would result in the opposite effects to those of the warming scenarios described in the rest of this section.

The disappearance of sea ice from the Barents Sea is likely to result in a more than doubling of the present levels of primary production, especially in the northernmost part. This is a consequence of a deeper wind-mixed layer and an increased vertical supply of nutrients from the underlying Atlantic water. Predicting changes in the timing of the spring bloom requires a better understanding of, and capability of modeling, the combined effects of

---

Box 9.10. Effects of a variable ice edge on key biological processes affecting carbon flux on an arctic shelf

Primary production (PP) occurs in the euphotic zone when light and nutrient conditions allow. This primary production may be retained by recycling within the euphotic zone or exported to deeper waters and be available for the benthos. The efficiency of retention is strongly determined by the occurrence of a match (where zooplankton are available to graze and recycle the primary production) or mismatch (where zooplankton are not present in sufficient numbers and primary production sinks out of the euphotic zone to be grazed by the benthos). Zooplankton densities may be affected by

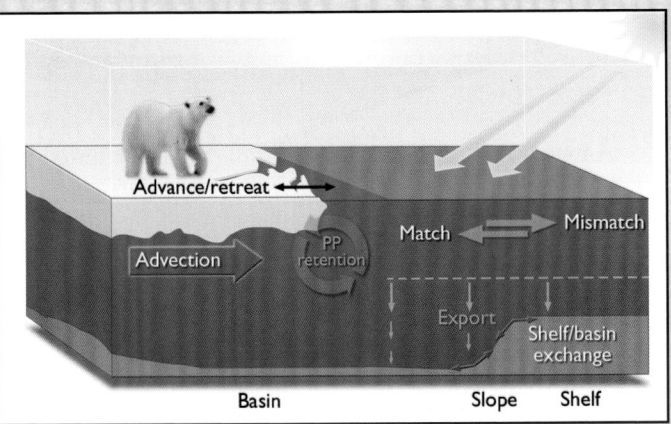

advection in certain shelf locations such as the Barents and Chukchi Seas. Additional concerns involve sequestration of carbon in shelf, slope, and basin sediments, and exchange processes that act to move carbon from one regime to another (red arrows). The location of the ice edge, where much primary production occurs, relative to topography (e.g., the shelf break and slope) strongly impact upon all of these processes. Under climate change scenarios, the ice edge will retreat further and faster into the basin, thus increasing the export of PP first to the slope and then to the abyssal ocean (E.C. Carmack, pers. comm. 2004).

ice-edge retreat and stability in the position of the Polar Front. To the south of the Polar Front, the absence of sea ice will reduce stratification thereby delaying the spring bloom until the onset of thermal stratification and the development of the seasonal surface mixed layer. North of the Polar Front, however, the timing of the spring bloom is strictly tied to light availability. At present, the spring bloom in the northern Barents Sea must await the retreat of the marginal ice zone for adequate light levels. In the absence of sea ice, the spring bloom is likely to occur earlier, and is very likely to occur earlier than in the region to the south of the Polar Front.

Primary production on the Bering Shelf is also likely to be enhanced if it becomes permanently ice-free, primarily due to an extended growing season and continuous upwelling of nutrient-rich water along the highly productive zone associated with the Bering Shelf break. More intense wind and more arid conditions at and near the Gobi and Takla Makan deserts in northeast Asia will possibly lessen the impact of iron control in the Northeast Pacific and the eastern Bering Sea.

In the shelf seas of the Arctic Ocean (e.g., the Kara, Laptev, East Siberian, and Beaufort Seas), a significant increase in nutrient supply is very likely to happen when the edge of the permanent ice pack retreats beyond the shelf break. This is very likely to trigger the onset of shelf-break upwelling and the delivery of nutrient-rich offshore waters to shallow shelf regions, perhaps more than doubling present levels of productivity.

In the central Arctic Ocean, two additional conditions of sea-ice retreat are important to primary production: the disappearance of sea-ice cover in summer and the regional appearance of open water areas in winter (e.g., north of Svalbard and northeast of the Chukchi Sea). In open water areas during summer, productivity is likely to increase due to increased wind mixing and nutrient re-supply. Within areas regionally open in winter, additional nutrients are likely to be supplied through the combined effects of wind stress and convective mixed layer deepening. It is possible that these two types of area will be as productive as is currently the case in their southern counterparts (the Greenland and deep Bering Seas, respectively). Before the development of these two distinctive conditions, areal primary production is likely to increase as the number and size of leads in the multi-year ice increase.

Surface mixed-layer depth is likely to have a strong impact on phytoplankton community structure, particularly in the Nordic Seas. Regions where the seabed or the depth of mixing (due to ice melt or river inflow) is less than about 40 m are likely to favor diatom blooms. Deeper mixing, to about 80 m, is likely to favor *Phaeocystis*. Thus, unless there is an increase in freshwater input, stronger winds are likely to result in *Phaeocystis* becoming more common than at present. This is possible in Atlantic water to the south of the Polar Front. If the surface mixed layer in the Atlantic water extends beyond

about 80 m, it is possible that a low-productive community dominated by nanoflagellates would be favored, as currently occurs in the off-shelf parts of the Southern Ocean (Hewes et al., 1990). This implies little transfer of carbon to herbivores and sediments because the grazers would be largely ciliates (Sakshaug and Walsh, 2000).

9.3.4.2. Zooplankton production

Any northward extension of warm water inflows is likely to carry with it temperate zooplankton, for instance into the Siberian Shelf Seas and the Bering Shelf (Brodeur and Ware, 1992; Overland et al., 1994; Skjoldal et al., 1987). Such inflows are likely to include gelatinous plankton in summer and autumn (Brodeur et al., 2002). Ice fauna such as the large amphipods will suffer massive loss of habitat if multi-year ice disappears. The possibility of increased transport of cold water on the western side of the North Atlantic could bring cold-loving zooplankton species farther south. Correspondingly, the southern limit of distribution of northern species may shift northward on the eastern side of the North Atlantic and southward on the western side, as indicated by zooplankton studies over the last 40 years (Beaugrand et al., 2002).

If the Siberian Shelf Seas become warmer in the future, it is possible that *Calanus finmarchicus* will thrive and multiply throughout the area as a whole, rather than being restricted to the Siberian Shelf water as currently occurs. There is, however, risk of a mismatch with phytoplankton blooms in that earlier melting will cause earlier stratification and, thus, an earlier bloom. However, if sea ice is absent during summer and autumn, there will be deeper vertical mixing, making the system more like that of the southern Barents Sea, with later blooms, albeit dependent on stratification caused by freshwater inputs from rivers. If water temperatures in the Siberian Shelf Seas stay lower than presently occur in the southern Barents Sea, the development of *C. finmarchicus* is likely to be retarded.

Grazing versus sedimentation

If a mismatch occurs in the timing of phytoplankton and zooplankton production due to early phytoplankton blooms, the food web will be highly inefficient in terms of food supply to fish and export production (Hansen et al., 1996). Export production and protozoan biomass are likely to increase.

A match with phytoplankton blooms can be achieved by arctic copepods, such as *C. glacialis*, which can adjust its egg production to the development of the phytoplankton bloom whether early or late in the season. This may also pertain to other important copepods in arctic waters. If so, actively grazing zooplankton "for all seasons" are very likely to exist for any realistic climate change and thus future ratios of grazed to exported phytoplankton biomass in the Arctic Ocean are unlikely to be much different to those at present.

Fish versus zooplankton

The crucial issue concerning the effects of climate change on zooplankton production is likely to be related to the match versus mismatch between herbivorous zooplankton and fish. The extent to which commercially valuable fish will migrate northward and the extent to which they will be able to utilize early developing populations of *C. glacialis* along the Siberian Shelf are unknown. A worst-case scenario would be a mismatch resulting in starving and, ultimately, dying fish in a summer ecosystem characterized by protozoans and unsuccessful, inflexible copepods such as *C. finmarchicus*.

9.3.4.3. Benthos

Future fluctuations in zoobenthic communities are very likely to be related to the temperature tolerance of the animals and to future water temperatures. While the majority of boreal forms have planktonic larvae that require a fairly long period to develop to maturity, arctic species do not (Thorson, 1950). Thus, boreal species should be quick to spread in warm currents in periods of warming, while the more stenothermal arctic species (i.e., those able to live within a narrow temperature range only) will perish quickly. In periods of cooling, the arctic species, with their absence of pelagic stages are very likely to slowly follow the warmer waters as they recede. Boreal species that can survive in near-freezing water could remain within the cooler areas.

From the prevailing direction of warm currents in the Barents Sea, shifts in the geographical distribution of the fauna should be quicker and more noticeable during periods of warming than periods of cooling. Any change in the abundance or biomass of benthic communities is most likely to result from the impact of temperature on the life cycle and growth rate of the various species. If warming occurs within the Barents Sea over the next hundred years, thermophilic species (i.e., those capable of living within a wide temperature range) will become more frequent. This is likely to force changes in the zoobenthic community structure and, to a lesser extent, in its functional characteristics, especially in coastal areas.

The highly productive region to the north of the Bering Strait is likely to undergo changing hydrographic conditions, which in turn are likely to result in changes to the dominant species (Grebmeier, 1993; Grebmeier et al., 1995). The hydrography of the St. Lawrence Island polynya region and the Anadyr region is ultimately related to the northward transport of water through the Bering Strait. Because the latter is related to variations in the AO, the future of the northern Bering Shelf is very likely to be closely related to variations in these oscillations (Walsh J.J. et al., 1989). If AO$^+$ conditions predominate in the future, it is likely that the flow of Bering Water into the Arctic Ocean will be small, resulting in a reduction in northward transport of water south of St Lawrence Island.

Because the Gulf of Anadyr "cold pool" is maintained by sea-ice production/brine formation in the St Lawrence Island polynya, an enhanced and more energetic polynya resulting from warming is likely to maintain a chemostat-type bloom system (Walsh J.J. et al., 1989), allowing a longer growing season and higher levels of production.

9.3.4.4. Fish production

Understanding how climate variability affects individual fish populations and fisheries and how the effects differ between species is extremely important when projecting the potential impacts of climate change. Projections of the response of local marine organisms to climate change scenarios have a high level of uncertainty. However, by using observations of changes in fish populations due to past climate variability it is possible to predict some general responses.

Climate change can affect fish production through a variety of means. Direct effects of temperature on the metabolism, growth, and distribution of fish could occur. Food web effects could also occur, through changes in lower trophic level production or in the abundance of top-level predators, but such effects are difficult to predict. However, it is expected that generalist predators are more adaptable to change than specialists. Fish recruitment patterns are strongly influenced by oceanographic processes such as local wind patterns and mixing and by prey availability during early life stages, which are also difficult to predict. Recruitment success could be affected by changes in the time of spawning, fecundity rates, survival rate of larvae, and food availability.

General trends in distribution and production

Poleward extensions of the distribution range for many fish species are very likely under the projected climate change scenarios (see Fig. 9.30 and Box 9.11). Some of the more abundant fish species that would be very likely to move northward under the projected warming include Atlantic and Pacific herring and cod, walleye pollock in the Bering Sea, and some of the flatfishes that might presently be limited by bottom temperatures in the northern areas of the marginal arctic seas. The southern limit of colder-water fishes such as polar cod and capelin would be very likely to move northward. The Greenland halibut is also likely either to shift its southern boundary northward or restrict its distribution more to continental slope regions. Salmon, which show high fidelity of return to natal streams, might possibly be affected in unknown ways that relate more to conditions in natal streams, early marine life, or feeding areas that might be outside the Arctic.

Fish production patterns are also very likely to be affected, although there are large uncertainties regarding the timing and location of zooplankton and benthic production that serve as prey resources for fish growth, and the wind advection patterns and direction that favor

Box 9.11. Climate impact on the distribution of fish in the Norwegian and Barents Seas

An increase in water temperature of 1 to 2 °C in the Atlantic part of the Norwegian and Barents Sea is very likely to result in a change in distribution for several species of fish. However, in both seas there are fronts between the warm Atlantic water and the cold arctic water masses, whose position is partly determined by bottom topography. How these fronts may move in future is addressed in section 9.2.5.4.

Previous experience of how fish react to changes in water temperature in the Barents Sea may be used to speculate about future changes. The most likely impact of an increase in water temperature on some commercial fish species in shown in Fig. 9.30. Capelin is very likely to extend its feeding area north and northeastward. During summer it might feed in the Arctic Basin and migrate to the Kara Sea. Whether the capelin maintain their spawning ground along the coast of northern Norway and the Kola Peninsula is unknown. They may possibly move eastward, and may even spawn along the west coast of Novaya Zemlya. Cod is also likely to expand its feeding area eastward, especially as capelin is its main food source. As cod is demersal (i.e., a near-bottom fish), it is not likely to migrate north of the Barents Sea and into the deep Arctic Basin. Haddock will probably follow the same track as cod, but as at present is likely to remain further south than cod.

In the Norwegian Sea, herring is likely to return to the feeding and overwintering area used before 1964 (see Box 9.8), but is likely to maintain the same spawning areas along the Norwegian coast. Mackerel (*Scomber scombrus*) and blue whiting (*Micromesistius poutassou*) are likely to migrate northeast to the Barents Sea. The mackerel and blue whiting will then compete with the other pelagic species in the Barents Sea for a limited supply of food. It is also likely that new species may enter the Norwegian Sea.

survival of some fish species relative to others. This is an active area of research, presently being addressed by GLOBEC (Global Ocean Ecosystem Dynamics) research programs around the world. Given historical recruitment patterns, it seems likely that herring, cod, and walleye pollock recruitment would be increased under future climate warming scenarios. Benthic-feeding flatfish, such as rock sole in the eastern Bering Sea, would be likely to have higher average recruitment in a warmer Bering Sea. Greenland halibut, capelin, and polar cod would be likely to decline in abundance. The

greatest variability in recruitment would occur for all species at the extremes of their ranges.

Migration patterns are very likely to shift, causing changes in arrival times along the migration route. The timing of the spring migration of cod into the Gulf of St. Lawrence appears to be related to the timing of ice melt. In winter, cod appear to congregate at the edge of the sea ice but do not pass beneath it (Fréchet, 1990). The spring migration appeared to be delayed by as much as 20 days in 1992, when ice melt was particularly late in the southern region of the Gulf. Change in sea ice distribution is one of the expected effects of climate change that is likely to have pronounced impacts on many fish species. Growth rates are very likely to vary, with the amplitude and direction being species dependent. While cod growth rates in the southern areas of the Arctic are very likely to increase with a rise in water temperature (Brander, 1995; Michalsen et al., 1998), this may not be the case for Arctic Ocean species.

Qualitative predictions of the consequences of climate change on fish resources require good regional atmospheric and oceanic models of the response of the ocean to climate change. Dynamically or statistically downscaled output from global circulation models, which are only recently becoming available, could be very useful. Greater understanding is needed concerning the life histories for those species for which predictions are required, and concerning the role of the environment, species interactions, and fishing in determining the variability of growth, reproduction, distribution, and abundance of fish populations. The multi-forcing and numerous past examples of "failed" predictions of environment–fish relationships indicate the difficulties faced by fisheries scientists in providing reliable predictions of the response to climate change.

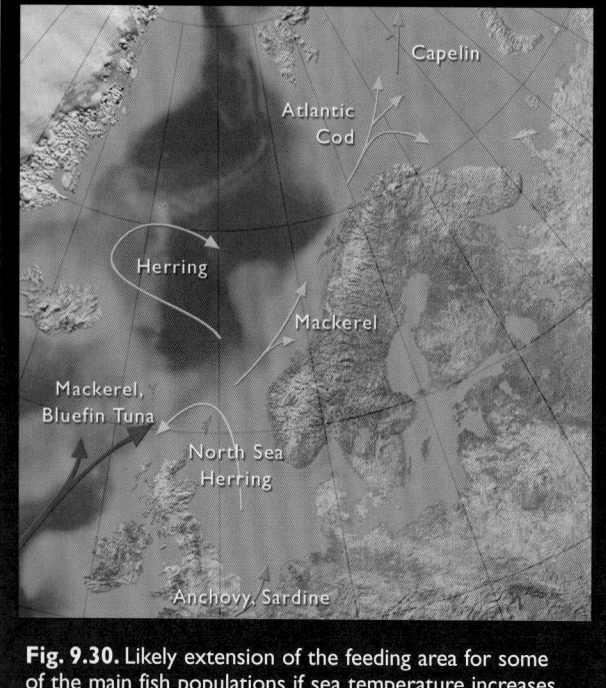

Fig. 9.30. Likely extension of the feeding area for some of the main fish populations if sea temperature increases. For herring, see also Box 9.8 (modified after Blindheim et al., 2001).

9.3.4.5. Marine mammals and seabirds

The impacts of climate change scenarios on marine mammals and seabirds in the Arctic are likely to be profound, but the precise form these impacts will take is not easy to determine (Jarvis, 1993; Shugart, 1990). Patterns of change are non-uniform (Parkinson, 1992) and highly complex. Oscillations occurring at a variety of scales (e.g., Mysak et al., 1990) complicate regional predictions of long-term trends. Also, species responses will vary dramatically (e.g., Kitaysky and Golubova, 2000). Meso-scale environmental features, e.g., frontal zones and eddies, that are associated with enhanced productivity are important to apex predators, but future changes in these features are not represented well at the present spatial resolution of circulation models (Tynan and DeMaster, 1997). Regional, small-scale coupled air–sea–ice models are needed in order to make reliable projections of change in mesoscale environmental features.

Given the most likely scenarios for changes in oceanographic conditions within the ACIA region by 2020 (Table 9.4), changes in seabird and marine mammal communities are very likely to be within the range(s) observed over the last 100 years. If, however, the increase in water temperature and the sea-ice retreat continue as projected until 2050 and 2080, marine ecosystems will change in ways not seen in recent history. One of the first changes expected is a poleward shift in species (and broader assemblages). However, there is a limit to how far north arctic species can shift following the sea ice. Once seasonal sea-ice cover retreats beyond the shelf regions, the oceanographic conditions will change dramatically and become unsuitable for many species. If the loss of sea ice is as dramatic temporally and spatially as has been projected by the ACIA-designated models, negative consequences are very likely within the next few decades for arctic animals that depend on sea ice for breeding or foraging (Brown, 1991; Burns, 2001; Stirling and Derocher, 1993; Tynan and DeMaster, 1997). The worst-case scenarios in terms of reduced sea-ice extent, duration, thickness, and concentration by 2080 are very likely to threaten the existence of whole populations and, depending on their ability to adapt to change, are very likely to result in the extinction of some species. Prospects for long-term abundance projections for populations of large marine predators are not good (e.g., Jenkins, 2003).

Climate change also poses risks to marine mammals and seabirds in the Arctic in terms of increased risk of disease for arctic-adapted vertebrates owing to improved growing conditions for the disease vectors and from introductions via contact with non-indigenous species (Harvell et al., 1999); increased pollution loads via increased precipitation bringing more river borne pollution northward (Macdonald et al., 2003b); increased competition from northward temperate species expansion; and impacts via increased human traffic and development in previously inaccessible, ice-covered areas. Alterations to the density, distribution, or abundance of

keystone species at various trophic levels, such as polar bears and polar cod, are very likely to have significant and rapid effects on the structure of the ecosystems they currently occupy.

Although many climate change scenarios focus on negative consequences for ecosystems, climate change will provide opportunities for some species. The ability to adapt to new climate regimes is often vast, and this potential should not be underestimated; many higher marine vertebrates in the Arctic are adapted to dealing with patchy food resources and high variability in the abundance of food resources.

Marine mammals

Changes in the extent and type of sea ice will affect the distribution and foraging success of polar bears. The earliest impact of warming had been considered most likely to occur at the southern limits of their distribution, such as James and Hudson Bays (Stirling and Derocher, 1993), and this has now been documented (Stirling et al., 1999). Late sea-ice formation and early breakup means a longer period of annual fasting for polar bears. Reproductive success is strongly linked to their fat stores; females in poor condition have smaller litters and smaller cubs, which are less likely to survive, than females in good condition. There are also concerns that direct mortality rates are likely to increase with the climate change scenarios projected by the ACIA-designated models. For example, increased frequency or intensity of spring rain could cause dens to collapse resulting in the death of the female as well as the cubs. Earlier spring breakup of ice could separate traditional den sites from spring feeding areas, and young cubs forced to swim long distances from breeding areas to feeding areas would probably have a lower survival rate. It is difficult to envisage the survival of polar bears as a species given a zero summer sea-ice scenario. Their only option would be a terrestrial summer lifestyle similar to that of brown bears, from which they evolved. In such a case, competition, risk of hybridization with brown bears and grizzly bears, and increased interactions with people would then number among the threats to polar bears.

Ice-living seals are particularly vulnerable to the changes in the extent and character of arctic sea ice projected by the ACIA-designated models because they depend on the sea ice as a pupping, molting, and resting platform, and some species forage on many ice-associated prey species (DeMaster and Davis, 1995). Of the high arctic pinnipeds ringed seals are likely to be most affected because many aspects of their life history and distribution are linked to sea ice (Finley et al., 1983; Smith T. et al., 1991; Wiig et al., 1999). They are the only arctic seal species that can create and maintain holes in thick sea ice and hence their distribution extends further north than that of all other pinnipeds. Ringed seals require sufficient snow cover to construct their lairs and the sea ice must be sufficiently stable in spring to rear young successfully (Fig. 9.31) (Lydersen

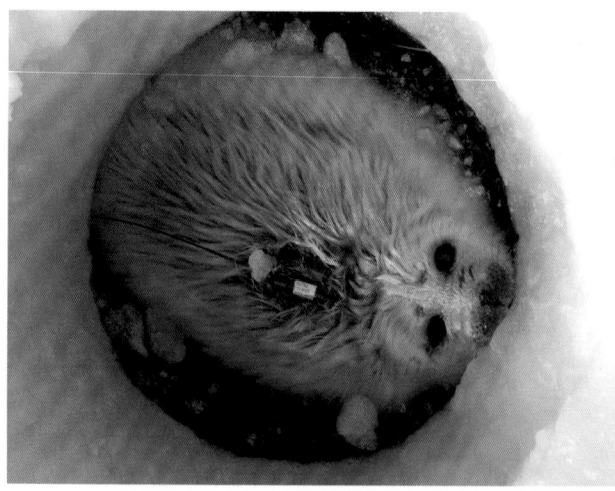

Fig. 9.31. A ringed seal pup outfitted with a radio-transmitter that was deployed as part of a haul-out behavior study in Svalbard, spring 2003 (photo by Kit Kovacs & Christian Lydersen, Norwegian Polar Institute).

and Kovacs, 1999). Premature breakup of the sea ice could result in premature separation of mother–pup pairs and hence high neonatal mortality. Ringed seals do not normally haul out on land and to do this would represent a dramatic change in behavior. Land breeding would expose the pups to much higher predation rates, even in a best-case scenario.

Bearded seals use regions of thin, broken sea ice over shallow areas with appropriate benthic prey communities (Burns J., 1981a). Their distribution, density, and reproductive success are dependent on the maintenance of suitable sea-ice conditions in these shallow, often coastal, areas. Walruses, another predominantly benthic feeder, also have quite specific sea-ice requirements. They overwinter in areas of pack ice where the ice is sufficiently thin that they can break through and maintain breathing holes (Stirling et al., 1981), but is sufficiently thick to support the weight of groups of these highly gregarious animals. Ice retreat may result in much of the remaining arctic sea ice being located over water that is too deep for these benthic foragers. Also, there is a more general concern that the likely decline in the community of plants, invertebrates, and fishes that live in close association with the underside of sea ice are very likely to result in a dramatic decrease in the flux of carbon to the benthic community, upon which bearded seals, walruses, and other animals such as grey whales depend (Tynan and DeMaster, 1997).

Harp seals are flexible about the nature of their summer sea-ice habitat, but during breeding travel to traditional sites in southern waters where they form large herds on extensive areas of pack ice. Massive pup mortality occurs during poor ice years. Hooded seals also breed in traditional areas, but select thicker sea ice than harp seals, and prefer areas where individual floes are large. Females move away from ice edges, presumably to reduce harassment from males (Kovacs, 1990). Pup mortality is also high for hooded seals during poor ice years. The situation which occurred during three years

in the last two decades when ice did not form in the Canadian Gulf of St. Lawrence breeding area implies severe consequences for harp and hooded seals if spring sea-ice conditions continue to follow current and projected trends. The range and relative abundance of these species is linked to sea-ice cover and climatic conditions (Vibe, 1967) and it is not known whether natal site fidelity is maintained for life, regardless of reproductive outcome. Thus, it is difficult to predict whether harp and hooded seals will adjust the location of their breeding and molting activities if spring sea-ice distribution changes dramatically over a relatively short period.

Spotted seals require sea ice over waters of specific depth and so, like bearded seals in the Atlantic, are very likely to be strongly affected by reduced sea-ice extent. The ecological requirements of ribbon seals are so poorly known that the effects of changes in sea-ice conditions are impossible to predict. Their flexibility in shifting from traditional breeding and foraging sites is unknown. Poor seasonal sea-ice conditions will result in a decimation of year-classes in the short term, but in the longer term, herds may form at more northerly sites that meet their needs. Those species that haul out on land when sea ice is not available, such as walrus and spotted seal, may be less affected by changes in sea-ice conditions than the other ice-associated seals.

In contrast, harbour seals and grey seals are likely to expand their distribution in an Arctic with less sea ice. They are for the most part temperate species that have a broad enough niche that they can occupy warm spots in the current Arctic. Other pinnipeds that breed on land in the Arctic are otariid seals. These are likely to be profoundly affected by changes in their food base, as is thought to be happening in the present regime shift in the North Pacific. They could also be affected by heat stress, but Steller sea lions have a present distribution that includes the Californian coast, implying a considerable tolerance for warm conditions given access to the ocean. Sea otters, like Steller sea lion, have a broad distribution at present and are likely to be most affected by changes at lower trophic levels which affect their food availability.

The impact of climate-induced perturbations on cetaceans is less certain than for ice-breeding pinnipeds and polar bears (Tynan and DeMaster, 1997), although Burns (2001) suggests grave implications for cetaceans in the Arctic. The uncertainty arises because the link between arctic cetaceans and sea ice is largely via prey availability rather than the sea ice itself (Moore, 2000; Moore et al., 2000). All the northern whales exhibit habitat selection, with sea-ice cover, depth, bathymetric structure, for example, of varying importance (Moore, 2000; Moore et al., 2000). Bowhead whales, beluga whales (*Delphinapterus leucas*), narwhals, and minke whales can all break young sea ice with their backs in order to breathe in ice-covered areas, but their distribution is generally restricted to areas containing leads or polynyas and open-water areas at the periphery of the pack-ice zone. Bowhead whales are considered the most ice-adapted cetacean. They feed largely on high arctic copepods and euphausiids (Lowry, 1993), and the distribution of these prey species determines their movements and distribution. Bowhead whales have evolved as ice whales, with elevated rostrums (i.e., beaks) and blow holes that allow them to breathe more easily in sea ice; it is not known whether they could adjust to ice-free waters (Tynan and DeMaster, 1997). Bowhead whales are presently an endangered species despite decades of protection from commercial hunting. They consume *Calanus* spp. and euphausiids and changes in sea-ice conditions are likely to have a major impact on their foraging (Finley, 2001).

Narwhal and beluga are known to forage at ice edges and cracks (Bradstreet, 1982; Crawford and Jorgenson, 1990), but are highly migratory and range well south of summer edges in the arctic pack ice (Rice, 1998), foraging along the fronts of glaciers (Lydersen et al., 2001) or even in areas of open water (Reeves, 1990). A small, threatened, population of belugas is resident in the Canadian Gulf of St. Lawrence, well south of the Arctic Circle, which has been affected by industrial pollution and habitat disturbance. Tynan and DeMaster (1997) predicted that arctic belugas might alter the timing and spatial patterns of seasonal migration given a retreat of the southern ice edge, particularly in the Canadian Archipelago. Vibe (1967) reported that the historical beluga distributions are linked to sea ice, wind, and current conditions along the Greenland coast (see section 9.3.3.4). The changes projected for arctic sea ice over the coming decades may promote genetic exchange between populations that are currently isolated due to the barrier formed by the southern ice edge. Narwhal utilize coastal habitats in summer, but in winter move offshore to deep-water areas with complex bathymetry. These areas are completely ice-covered except for shifting leads and cracks. Narwhal are thought to feed on cephalopods at this time (Dietz et al., 2001), thus the effects of climate change on narwhal are likely to be via sea-ice distribution patterns and effects on key prey species.

All other cetacean species that frequent the Arctic avoid ice-covered areas. Their distributions are predominantly determined by prey availability (Ridgway and Harrison, 1981-1999) and so the impact of climate change will occur indirectly via changes to their potential prey base. Grey whales are unusual in that they are benthic feeders, and so are very likely to be affected by climate change in ways more similar to walruses and bearded seals than other cetaceans.

Seabirds

The effects of climate change on seabird populations, both direct and indirect through effects on the oceans, are likely to be detected first near the limits of the species range and near the margins of their oceanographic range (Barrett and Krasnov, 1996; Montevecchi and Myers, 1997). Brown (1991) suggests that the southern limits for many arctic seabird species will move north-ward, as will their breeding ranges. Changes in patterns of distribution, breeding phenology, and periods of residency in the Arctic are likely to be some of the first responses to climate change observed in arctic seabird populations. This is partly because these are more easily detected than subtle or complex changes such as changes in population size and ecosystem function (Furness et al., 1993; Montevecchi, 1993). Because arctic seabirds are long-lived, have generally low fertility, and live in a highly variable environment, effects of climate change on population size, even if quite significant, may take several years to show (Thompson and Ollason, 2001).

Seabirds are likely to be influenced most by indirect changes in prey availability (Brown, 1991; IPCC, 1998; Schreiber E., 2001). Seabirds respond to anything that affects food availability and so are often good indicators of a system's productivity (Bailey et al., 1991; Hunt et al., 1991; Montevecchi, 1993). Several studies have shown that climate-induced changes in oceanographic conditions can have large-scale and pervasive effects on vertebrate trophic interactions, affecting seabird population size and reproductive success (Duffy, 1990; Montevecchi and Myers, 1997; Schreiber R. and Schreiber, 1984). Species with narrow food or habitat requirements are likely to be the most sensitive (Jarvis, 1993; Vader et al., 1990). As warmer (or colder) water would affect the distribution of prey, the distribution of individual seabird species is likely to change in accordance with changes in the distribution of macrozooplankton and fish populations. Brown (1991) suggests that improved foraging conditions will result in range expansions northward for many species. This is because the retreating pack ice will open up more feeding areas in spring and will provide phytoplankton with earlier exposure to daylight, thereby increasing productivity throughout the Arctic. However, from analyses of probable changes in food availability in subantarctic waters, Croxall (1992) concluded that it was not possible to be certain whether a change in the amount of sea ice would mean more or less prey for seabirds. Many of these uncertainties are also relevant to arctic areas.

Changes in water temperature are very likely to have significant consequences for pelagic fish species (see section 9.3.4.4 and Chapter 13). Most fish species are sensitive to changes in water temperature (e.g., Gjøsæter, 1998), and only slight changes in the thermal regime can induce changes in their temporal and spatial (both vertical and horizontal) distributions (Blindheim et al., 2001; Loeng, 2001; Methven and Piatt, 1991; Shackell et al., 1994). For example, increases in air temperature will probably lead to a greater inflow of warm Atlantic water into the Barents Sea, caused by complex interactions between different water masses, ocean currents, and wind systems in the north Atlantic. This inflow is very likely to displace the Polar Front north- and eastward, especially in the eastern Barents Sea. The ice edge would then be located further north in winter, with a consequent reduction in the phytoplankton bloom which normally follows the receding ice edge during spring and summer. It is likely that the distribution of the Barents Sea capelin would be displaced northeastward, from the central to the northeastern Barents Sea. Important life-cycle changes are likely to include changes in the timing of spawning, with a consequent shift in the timing of migration and a displacement of migration routes (Loeng, 2001). Such changes to capelin alone could have profound consequences for many arctic seabirds in the Barents Sea.

Extreme changes in the spatial and temporal availability of food can have dramatic effects on the survival of adult seabirds (Baduini et al., 2001; Piatt and van Pelt, 1997). However, seabirds are able to travel great distances and so are insulated to some extent from environmental variability. They are able to exploit locally and ephemerally favorable conditions during much of the year quite freely. However, during the breeding season when they are constrained to return to a land-based breeding site but are dependant on marine resources for foraging, less extreme reductions in prey availability can affect reproductive success. Most Northern Hemisphere seabirds forage within 200 km of their colonies (Hunt et al., 1999). Because seabirds generally lay only one egg they cannot alter clutch size to compensate for low food availability in a given season. Instead, they reduce the extent of their parental care contribution when resources are in short supply in order to protect their own long-term survival (Øyan and Anker-Nilssen, 1996; Weimerskirch, 2001). Because they are long-lived, have delayed sexual maturity, and have conservative reproductive output, even dramatic reductions in fledgling survival may not be apparent in terms of overall population size for several years.

If climate change induces long-term shifts in the spatial distribution of macrozooplankton (predominantly crustaceans) and small schooling pelagic fish, seabird breeding distribution patterns are likely to alter. These prey species are usually concentrated in frontal or upwelling areas, which provide a spatially and temporally predictable food supply for seabirds (Hunt, 1990; Hunt et al., 1999; Mehlum et al., 1998a,b; Schneider, 1990; Watanuki et al., 2001). If changing environmental conditions cause these oceanographic features to relocate, then prey distributions are very likely to change. If new breeding sites become available in close proximity to the new feeding areas, little change is likely. However, if suitable breeding areas are not available near the relocated fronts or upwelling, the seabirds may not be able to take advantage of available food at its new location during the reproductive season, resulting in reproductive failure. The impacts of future climate change on seabirds are likely to be extremely variable in a spatial context.

Temporal changes in prey availability can also change the timing of breeding in seabirds (Schreiber E., 2001), and potentially result in a mismatch between the timing of reproduction and the time of food abundance (Visser et al., 1998). Such a mismatch may have profound impacts on reproductive success (Brinkhof et al., 1997). The timing of breeding is especially critical for birds breeding in

arctic areas; low temperatures and a restricted period of prey availability create a narrow temporal window in which the nesting period sits (Hamer et al., 2001).

The ivory gull is an exception to many of these general patterns. This species is closely associated with sea ice throughout most of its life cycle. Changes in sea-ice extent and concomitant changes in the distribution of ice-associated seals and polar bears are very likely to result in changes in ivory gull distribution and potentially negative effects on abundance. There is concern that major reductions in ivory gull populations have already occurred (Krajick, 2001; Mallory et al., 2003). There is also concern that little auks, specialist feeders on arctic copepods during the summer, would be negatively affected by the changes predicted in the "*Calanus* complex" in the Barents Sea and other parts of the North Atlantic.

Changes in sea level may restrict breeding at existing sites, but may increase the suitability of other sites that are not currently usable owing to, for example, predator access.

Direct evidence of negative effects of environmental conditions (weather) for seabirds is rare, although wind is thought to be important for foraging energetics. Healthy arctic seabirds have little difficulty coping with extreme cold; they are insulated by feathers and subcutaneous fat. However, owing to these adaptations they may have difficulty keeping cool. Warmer temperatures in the Arctic are very likely therefore to set southern limits to seabird distributions that are unrelated to the availability of prey or breeding sites (Brown, 1991). Extreme weather can result in direct mortality of chicks or even adults, but it is most likely that the greatest effect of inclement weather would be to restrict the opportunity for seabirds to forage (Harris and Wanless, 1984). Heavy rain could flood the nests of burrowing species such as little auks or puffins (Rodway et al., 1998; Schreiber E., 2001) and freezing rain could affect the thermal balance of exposed chicks leading to mortality (Burger and Gochfeld, 1990). Changes to the normal patterns of wind speed and direction could alter the cost of flight, particularly during migration (Furness and Bryant, 1996; Gabrielsen et al., 1987), but it is the nature and extent of the change that determine whether the consequences are negative (or positive) for individual seabird species.

9.4. Effects of changes in ultraviolet radiation

This section assesses the potential impacts of ozone depletion-related increases in solar ultraviolet-B radiation (280–315 nm = UV-B) on arctic marine ecosystems. For a comprehensive review of the extensive and rapidly growing technical literature on this subject, readers are referred to several recent books (DeMora et al., 2000; Häder, 1997; Helbling and Zagarese, 2003), and particularly to Hessen (2001) with its focus on the Arctic. UV-B optics in marine waters and ozone layer depletion and solar ultraviolet radiation are described in Chapter 5.

The exponential relationship between the capacity of ozone to filter ultraviolet light – lower wavelengths are much more strongly filtered – means that small reductions in stratospheric ozone levels result in large increases in UV-B radiation at the earth's surface (e.g., Kerr and McElroy, 1993; Madronich et al., 1995). Since ozone layer depletion is expected to continue for many more years, albeit at a slower rate (Shindell et al., 1998; Staehelin et al., 2001; Taalas et al., 2000), the possible impacts of solar UV-B radiation on marine organisms and ecosystems are currently being investigated (Browman, 2003; Browman et al., 2000; De Mora et al., 2000; Häder, 1997; Häder et al., 2003; Helbling and Zagarese, 2003; Hessen, 2001). A growing number of studies have found that current levels of UV-B radiation are harmful to aquatic organisms and may, in some extreme instances, reduce the productivity of marine ecosystems (De Mora et al., 2000; Häder, 1997; Häder et al., 2003; Helbling and Zagarese, 2003; Hessen, 2001). Reductions in productivity induced by UV-B radiation have been reported for phytoplankton, heterotrophic organisms, and zooplankton; the key intermediary levels of marine food chains (De Mora et al., 2000; Häder, 1997; Häder et al., 2003; Helbling and Zagarese, 2003; Hessen, 2001). Similar studies on planktonic fish eggs and larvae indicated that exposure to levels of UV-B radiation currently incident at the earth's surface results in higher mortality and may lead to reduced recruitment success (Hunter et al., 1981, 1982; Lesser et al., 2001; Pommeranz, 1974; Walters C. and Ward, 1998; Williamson et al., 1997; Zagarese and Williamson, 2000, 2001).

Ultraviolet radiation also appears to affect biogeochemical cycling within the marine environment and in a manner that could affect overall ecosystem productivity and dynamics (Zepp et al., 2003).

9.4.1. Direct effects on marine organisms

The majority of UV-B radiation research examines direct effects on specific organisms. Some marine copepods are negatively affected by current levels of UV-B radiation (Häder et al., 2003). UV-B-induced mortality in the early life stages, reduced survival and fecundity in females, and changes in sex ratios have all been reported (Chalker-Scott, 1995; Karanas et al., 1979, 1981; Lacuna and Uye, 2001; Naganuma et al., 1997; Tartarotti et al., 2000). UV-B-induced damage to the DNA of crustacean zooplankton has also been detected in samples collected up to 20 m deep (Malloy et al., 1997). Eggs of *Calanus finmarchicus* – a prominent member of the mesozooplankton community throughout the North Atlantic – incubated under UV-B radiation exhibited a lower percentage hatch rate than those protected from UV-B radiation (Alonso Rodriguez et al., 2000). This indicates that *Calanus finmarchicus* may be sensitive to variation in incident UV-B radiation. Results for the few other species that have been studied are highly variable with some showing strong negative impacts, while others are resistant (Damkaer, 1982; Dey et al., 1988;

Thomson, 1986; Zagarese and Williamson, 2000). The factors determining this susceptibility are many and complex, but include seasonality and location of spawning, vertical distribution, presence of UV-B-screening compounds, and the ability to repair UV-B-induced damage to tissues and DNA (Williamson et al., 2001).

The work of Marinaro and Bernard (1966), Pommeranz (1974), and Hunter et al. (1979, 1981, 1982) provided clear evidence of the detrimental effect of UV-B radiation on the planktonic early life stages of marine fish. Hunter et al. (1979), working with northern anchovy (*Engraulis mordax*) and Pacific mackerel (*Scomber japonicus*) embryos and larvae, reported that exposure to surface levels of UV-B radiation could be lethal. Significant sub-lethal effects were also reported: lesions in the brain and retina, and reduced growth rate. The study concluded that, under some conditions, 13% of the annual production of northern anchovy larvae could be lost as a result of UV-B-related mortality (Hunter et al., 1981, 1982). Atlantic cod eggs were negatively affected by exposure to UV-B radiation in very shallow water; 50 cm deep or less (Béland et al., 1999; Browman and Vetter, 2001). With the exception of a small (but rapidly growing) number of recent studies, little additional information is available on the effects of UV-B radiation on the early life stages of fish. However, as for copepods, the early life stages of fish will vary in their susceptibility to UV-B radiation and for the same reasons. Thus, some studies conclude that the effects of UV-B radiation will be significant (e.g., Battini et al., 2000; Lesser et al., 2001; Williamson et al., 1997), while others conclude that they will not (e.g., Dethlefsen et al., 2001; Kuhn et al., 2000; Steeger et al., 2001).

9.4.2. Indirect effects on marine organisms

Exposure to UV radiation, especially UV-B radiation, has many harmful effects on health. These may result in poorer performance, or even death, despite not being *directly* induced by exposure to UV-B radiation. UV-B radiation suppresses systemic and local immune responses to a variety of antigens, including microorganisms (Garssen et al., 1998; Hurks et al., 1994). In addition to suppressing T-cell-mediated immune reactions, UV-B radiation also affects nonspecific cellular immune defenses. Recent studies demonstrate disturbed immunological responses in UV-B-irradiated roach (*Rutilus rutilus*): the function of isolated head kidney neutrophils and macrophages (immuno-responsive cells) were significantly altered after a single dose of UV-B radiation (Salo et al., 1998). Natural cytotoxicity, assumed to be an important defense mechanism in viral, neoplastic, and parasitic diseases, was also reduced. A single dose of UV-B radiation exposure decreased the ability of fish lymphocytes to respond to activators, and this was still apparent 14 days later (Jokinen et al., 2001). This indicates altered regulation of lymphocyte-dependent immune functions. Finally, exposure to UV-B radiation induces a strong systemic stress response which is manifested in fish blood by an increased number of circulating phagocytes and elevated plasma cortisol levels (Salo et al., 2000a). Exposure to UV-A (315–400 nm) radiation induced some of the same negative effects on the immune system (Salo et al., 2000b). Since high cortisol levels induce immuno-suppression in fish (Bonga, 1997) the effect of exposure to UV-B radiation on the immune system clearly has both direct and indirect components. Taken together, these findings indicate that the immune system of fish is significantly affected by exposure to a single, moderate-level dose of UV-B radiation. At the population level, a reduction in immune response might be manifested as lowered resistance to pathogens and increased suscepti-bility to disease. The ability of the fish immune system to accommodate increases in solar UV-B radiation is not known. Also, the immune system of young fish is likely to be highly vulnerable to UV-B radiation because lym-phoid organs are rapidly developing and because critical phases of cell proliferation, differentiation, and matura-tion are occurring (Botham and Manning, 1981; Chilmonczyk, 1992; Grace and Manning, 1980). It is also possible that exposure to ambient UV-B radiation impedes the development of the thymus or other lym-phoid organs resulting in compromised immune defense later in life. The effect of UV-B radiation on the immune function of fish embryos and larvae, and on the development of the immune system, is unknown.

Other indirect effects of UV-B radiation are also possi-ble. For example, UV-B radiation may affect sperm qual-ity for species that spawn in the surface layer (Don and Avtalion, 1993; Valcarcel et al., 1994) and so affect fer-tilization rate and/or genome transfer.

Studies on the impact of UV-B radiation have almost all examined the effects of short-term exposure on biological end-points such as skin injury (sunburn), DNA damage, development and growth rates, immune function, or out-right mortality. Few have examined the potential effects of longer-term (low-level) exposure (but see Fidhiany and Winckler, 1999). All these indirect (and/or longer-term) effects of UV-B radiation have yet to be investigated.

9.4.3. Ecosystem effects

9.4.3.1. Food chains

Although the effects of UV-B radiation are strongly species-specific, marine bacterioplankton and phytoplank-ton can be negatively affected (De Mora et al., 2000; Hessen, 2001). Severe exposure to UV-B radiation can, therefore, decrease productivity at the base of marine food chains. The importance of this decrease is highly speculative, but decreases in carbon fixation of 20 to 30% have been proposed (Helbling and Villafañe, 2001). Arctic phytoplankton appear more susceptible than antarctic species, possibly owing to deeper surface mixed layers in the Arctic (Helbling and Villafañe, 2001). Also, if UV-B radiation reduces the productivity of protozoans and crustacean zooplankton there will be less prey available for fish larvae and other organisms that feed upon them.

The few studies that have investigated the indirect effects of UV-B radiation on specific organisms con-clude that UV-B-induced changes in food-chain interac-tions can be far more significant than direct effects on individual organisms at any single trophic level (Bothwell et al., 1994; Hessen et al., 1997; Williamson et al., 1999). Recent investigations indicate the possi-bility of food-chain effects in both the marine and freshwater environment: exposure to UV-B radiation (even at low dose rates) reduces the total lipid content of some microalgae (Arts and Rai, 1997; Arts et al., 2000; Plante and Arts, 1998) and this includes the polyunsaturated fatty acids (PUFAs) (Goes et al., 1994; Hessen et al., 1997; Wang K. and Chai, 1994). For zooplankton and fish larvae, the only source of PUFAs is the diet – they cannot be synthesized and so must be obtained from prey organisms (Goulden and Place, 1990; Rainuzzo et al., 1997; Reitan et al., 1997; Sargent et al., 1997). Dietary deficiencies are manifest-ed in many ways. For example, in the freshwater clado-ceran *Daphnia* spp., growth rates are correlated with the concentration of eicosapentaenoic acid in the water column (De Lange and Van Donk, 1997; Müller-Navarra, 1995a,b; Scott C. et al., 1999). In Atlantic herring, dietary deficiencies of essential fatty acids, in particular docosahexaenoic acid, reduce the number of rods in the eyes (Bell M. and Dick, 1993) and negative-ly affect feeding at low light levels (Bell M. et al., 1995; Masuda et al., 1998). Other negative conse-quences of essential fatty acid deficits have also been reported (Bell J. et al., 1998; Kanazawa, 1997; Rainuzzo et al., 1997). A UV-B-induced reduction in the PUFA content of microalgae will be transferred to the herbivorous zooplankton that graze on them, there-by decreasing the availability of this essential fatty acid to fish larvae. Since fish larvae (and their prey) require these essential fatty acids for proper development and growth, a reduction in the nutritional quality of the food base has potentially widespread and significant implications for the overall productivity and health of aquatic ecosystems.

9.4.3.2. Quantitative assessments

Quantitative assessments of the effects of UV-B radia-tion on marine organisms at the population level are scarce. However, several studies are currently underway using mathematical simulation models. Neale et al. (1998, 2001) estimated that a 50% seasonal reduction in stratospheric ozone levels could reduce total levels of primary production – integrated throughout the water column – by up to 8.5%. Kuhn et al. (2000) developed a model that incorporates physical and biological infor-mation and were able to generate an absolute estimate of mortality under different meteorological and hydro-graphic conditions. As a result, they were able to evalu-ate the relative impacts of different combinations of environmental conditions – for example, a typical clear sky versus a typical overcast sky; a typical clear water column versus a typical opaque coastal water column; current ambient ozone levels versus a realistically thinned ozone layer. For *Calanus finmarchicus* eggs in the estuary and Gulf of St. Lawrence, UV-B-induced mortality for all model scenarios ranged from <1% to 51%, with a mean of 10.05% and an uncertainty of ±11.9% (based on 1 standard deviation and 48 modeled scenarios). For Atlantic cod, none of the scenarios gave a UV-B-induced mortality greater than 1.2%, and the mean was 1.0±0.63% (72 modeled scenarios).

In both assessments (Kuhn et al., 2000; Neale et al., 1998, 2001), the most important determinant of UV-B-related effects was water column transparency (see Fig. 9.32): even when ozone layer depletions of 50% were modeled, the effect on mortality remained far lower than that resulting from either thick cloud cover or opacity of the water column. This demonstrates that variability in cloud cover, water quality, and vertical distribution and displacement within the surface mixed layer have a greater effect on the flux of UV-B radiation to which planktonic marine organisms are exposed than ozone layer depletion. In contrast, Huot et al. (2000) showed that ozone thickness could in some instances be the single most important determinant of DNA damage in bacterioplankton.

Since the concentrations of dissolved organic carbon (DOC) and Chl-a are strongly correlated with the transparency of the water column to UV-B radiation, it follows that their concentrations are an overriding factor affecting UV-B-induced mortality. The Kuhn et al. (2000) model supports this contention. DOC levels in eutrophic coastal zones are often greater than 3 to 4 mg/L; the diffuse attenuation coefficients for UV-B radiation at such levels essentially protect *Calanus finmarchicus* and cod eggs from UV-B-induced mortality (Fig. 9.33). Thus, DOC can be considered as a sunscreen for organisms inhabiting eutrophic coastal zone waters. DOC concentrations in arctic waters are typically <1 mg/L (Aas et al., 2001). At these levels, DOC is not as effective at protecting planktonic marine organisms from UV-B-related damage (Fig. 9.33).

Although these model-based predictions are useful, there are limited data to parameterize the models, and it will be some time before similar predictions can be made for the many species inhabiting the full range of conditions within the world's ocean, including those of the Arctic.

9.4.4. General perspectives

Although UV-B radiation *can* have negative impacts (direct effects) on marine organisms and populations, it is only one of many environmental factors (e.g., bacterial and/or viral pathogens, predation, toxic algae) that result in the mortality typically observed in these organisms. Recent assessments indicate that UV-B radiation is generally only a minor source of direct mortality (or decreases in productivity) for populations, particularly in "DOC-protected" coastal zones. However, for those species whose early life stages occur near the surface, there may be circumstances (albeit rare) –

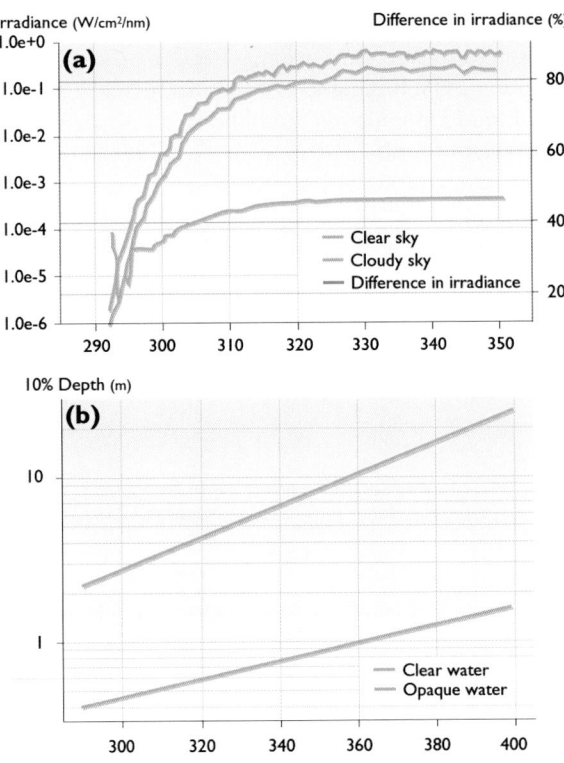

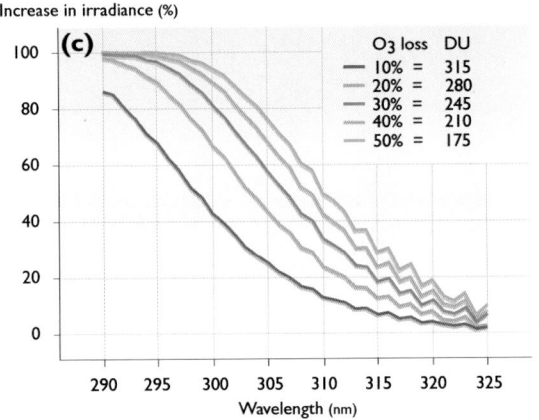

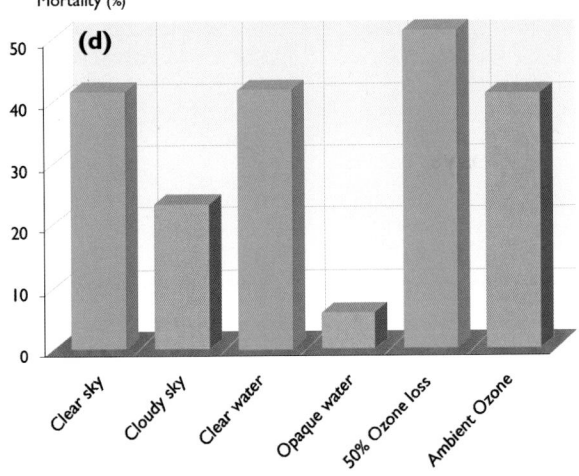

Fig. 9.32. Output of a mathematical simulation model (Kuhn et al., 2000) illustrating the relative effects of selected variables on UV-induced mortality in *Calanus finmarchicus* embryos (modified from Browman et al., 2000). The plots illustrate the effects on irradiance of (a) clear versus cloudy sky, (b) clear versus an opaque water column, (c) 50% thinning of ozone versus ambient ozone, while (d) compares the relative impacts of all three on mortality. This graphic illustrates that water column transparency is the single most important determinant of UV exposure – of considerably more importance than ozone layer depletion.

such as a cloudless sky, thin ozone layer, lack of wind, calm seas, low nutrient loading – under which the contribution of UV-B radiation to the productivity and/or mortality of a population could be far more significant. The impact of indirect effects has not as yet been adequately evaluated.

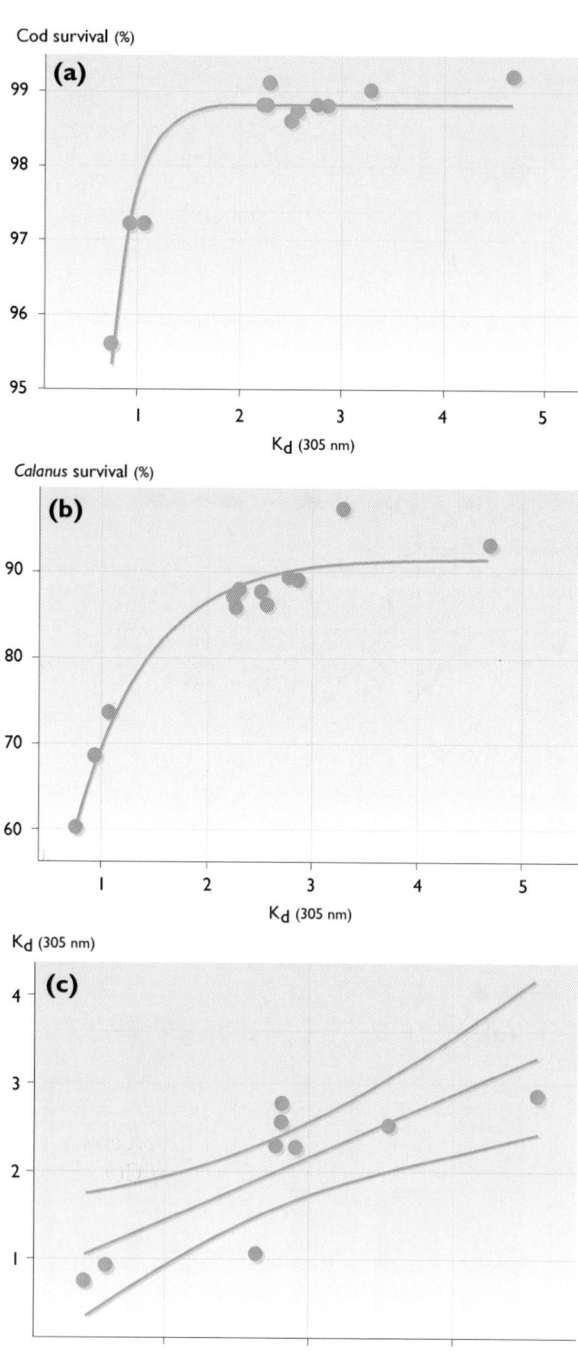

Fig. 9.33. Level of protection from UV damage afforded by the organic matter content of the water column. (a) diffuse attenuation coefficient (K_d) at 305 nm versus modeled survival of Atlantic cod embryos exposed to UV radiation in a mixed water column; (b) K_d at 305 nm versus modeled survival of *Calanus finmarchicus* embryos exposed to UV radiation in a mixed water column; (c) dissolved organic carbon (DOC) versus K_d at 305 nm from field measurements in temperate marine coastal waters (the estuary and Gulf of St. Lawrence, Canada). The straight line is the regression; the curved lines the 95% confidence intervals (modified from Browman, 2002).

9.5. The carbon cycle and climate change

The Arctic Ocean has not been considered a significant carbon sink; first, because extensive sea-ice cover constrains atmosphere–ocean exchange, and second, because levels of biological production under perennial sea ice were considered low (English, 1961). Under warmer conditions, however, the amount of carbon sequestered by the Arctic Ocean is very likely to increase significantly. The role of the Arctic as a potential carbon source, in the form of CH_4 and CO_2, is unclear owing to limited information on the likely impact of climate change on the substantial frozen reserves in permafrost and gas hydrate layers.

The ocean carbon cycle comprises a physical pump, a biological pump, and an alkalinity or anion pump. The physical pump is driven by physical and chemical processes, which affect the solubility of CO_2 and the transport of water from the surface mixed layer to depth. The biological pump is driven by primary production, consuming dissolved CO_2 through photosynthesis and producing particulate organic carbon (POC) and DOC. The alkalinity pump concerns the removal of carbon by calcification in the upper waters and the release of carbon when calcium carbonate is dissolved at depth. The alkalinity pump is not affected by temperature itself, but is affected indirectly through shifts in biological speciation.

9.5.1. Physical pump

The presence of sea ice strongly affects the physical pump, which regulates the exchange of CO_2 between the atmosphere and the ocean. This exchange is primarily determined by the difference in partial pressure of CO_2 (pCO_2) over the air–sea interface. Physical factors, such as wind mixing, temperature, and salinity, are also important in this exchange. Dissolved inorganic carbon (DIC) is the largest component of the marine carbon pool.

Multi-year ice restricts air–sea exchange over the central Arctic Ocean and seasonal sea ice restricts air–sea exchange over shelf regions to ice-free periods. Because the solubility of CO_2 in seawater increases with decreasing temperature, the largest uptake of atmospheric CO_2 occurs primarily in the ice-free Nordic Seas (~86 x 10^{12} g C/yr; Anderson L. and Katlin, 2001) where northward flowing Atlantic waters are rapidly cooled. Similarly, the Barents and Bering/Chukchi Seas, where inflowing Atlantic and Pacific waters undergo cooling, are also important uptake regions: uptake in the Barents Sea is ~9 x 10^{12} g C/yr (Fransson et al., 2001) and in the Bering/Chukchi Seas is ~22 x 10^{12} g C/yr (Katlin and Anderson, 2005). Uptake in the Bering/Chukchi Seas is higher than in the Barents Sea for reasons discussed in greater detail in section 9.5.2; namely, a higher potential for new production owing to a greater supply of nutrients, and a larger area of retreating ice edge along which much of the primary production occurs. Carbon uptake in the ice-covered Arctic Ocean and interior shelf seas is ~31 x 10^{12} g C/yr (Katlin and Anderson, 2005).

Although these fluxes are not large on a global scale ($\sim$2000 x 10^{12} g C/yr), the air–sea CO_2 flux is very likely to increase regionally under scenarios of climate warming. For example, the ACIA-designated models project the Barents Sea and the northern Bering Sea to be totally ice-free by 2050 (see section 9.2.5.2). Such changes in ice cover and longer periods of open water will result in more regions that resemble the Greenland Sea, where the physical pump is strong due to low surface water temperatures and high wind speeds (Johannessen T. et al., 2002). Atmospheric exchange will also increase as the areal coverage of the permanent ice pack is reduced and more leads and polynyas are formed. Here, the combination of increased atmospheric exchange (driven by winds) and ventilation (driven by sea-ice formation and convection) transport CO_2 from the atmosphere into the halocline and potentially deeper, eventually entering the deep North Atlantic Ocean and the THC. Ventilation of Arctic Ocean intermediate waters has been estimated to sequester $\sim$0.026 Gt C/yr, nearly an order of magnitude more than the sink due to convection in the Greenland Sea (Anderson L. et al., 1998) and this is very likely to increase, possibly significantly.

Seasonally ice-covered shelf regions are also important dense water formation areas. Brine release during sea-ice formation increases the density of surface waters which then sink and are advected from the shelf to basin interiors, transporting CO_2 into the halocline and deeper waters. Under warming conditions, ice formation on shelves will occur later and ice melt will occur earlier, thereby increasing the time available for air–sea

interaction/equilibration and CO_2 uptake. The coincidence of open water with late summer storms will also increase air–sea exchange and CO_2 uptake.

Changes in dense water production and the THC will affect the ocean carbon reservoir (Hopkins, 2001). The global ocean stores approximately fifty times more carbon than the atmosphere, mostly in the deep waters of the Pacific Ocean owing to their volume and long residence time. Slowing or stopping the THC would make the Atlantic circulation more like that of the Pacific, increasing its carbon storage and thus weakening the greenhouse effect and cooling the atmosphere – a negative feedback. In contrast however, if sites of deep ventilation were to move northward into the Arctic Basin (Aagaard and Carmack, 1994), the resulting overturn may result in a positive feedback due to CO_2 release to the atmosphere.

Changes in ice cover extent also affect the uptake of atmospheric CO_2 by altering the equilibrium concentrations in the water column. Anderson L. and Katlin (2001), using the Roy et al. (1993) solubility equations, calculated that melting 2 to 3 m of sea ice and mixing the resulting freshwater into the top 100 m of the water column would increase CO_2 uptake and could remove $\sim$3 g C/m^2. But, where warming is sufficient to increase surface water temperatures by 1 °C, $\sim$8 g C/m^2 could be released due to the decrease in solubility. At high latitudes, surface waters are often undersaturated because heat is lost to the atmosphere more quickly than CO_2 can dissolve. If ice cover retreated and the contact period with the atmosphere increased, this undersaturation would result in atmospheric CO_2 uptake. Anderson L. and Katlin (2001), using data for the Eurasian Basin where Atlantic waters dominate the upper water column, calculated that surface waters in the St. Anna Trough, the Eurasian Basin, and the Makarov Shelf slope have a potential carbon uptake of 35, 48, and 7 g C/m^2, respectively, when ice cover conditions allow saturation.

Regionally, the effects of upwelling of halocline waters onto the shelf must also be considered. For example, a profile of the fugacity (partial pressure corrected for the fact that the gas is not ideal) of CO_2 ($f\,CO_2$) shows that Pacific-origin waters below 50 m in Canada Basin are oversaturated due to their origin in the productive Bering/Chukchi Seas (see Fig. 9.34). If upwelling brought these oversaturated waters onto the shelf and they mixed with surface waters CO_2 would be released. Upwelling of waters with salinity $\sim$33 (near 150 m in the Canada Basin) has been observed on the Alaskan and Beaufort shelves (Aagaard et al., 1981; Melling, 1993; Melling and Moore, 1995). Upwelling is also expected to increase when the ice edge retreats beyond the shelf break (Carmack and Chapman, 2003). In contrast, the $f\,CO_2$ profile of Atlantic-origin waters shows that waters below 50 m in the Eurasian Basin are undersaturated and will take up atmospheric CO_2 if moved onto the shelf by upwelling (Anderson L. and Katlin, 2001). Hence, the recent shift in the Makarov Basin from a

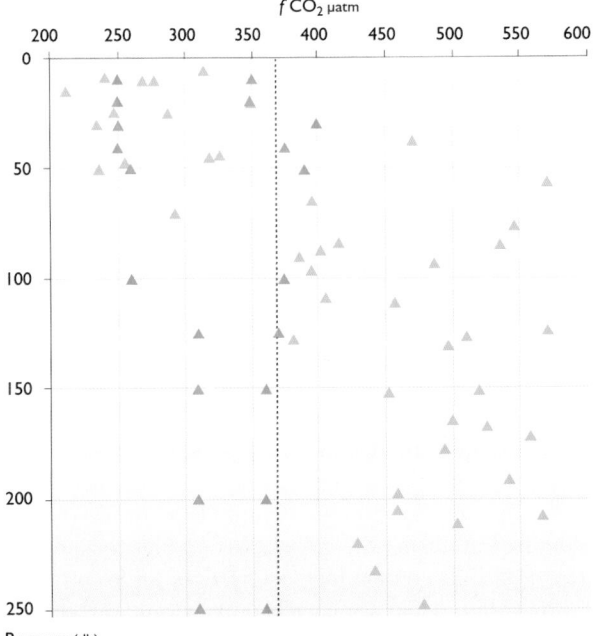

Pressure (db)

▲ Eurasian Basin - 1996 (Includes data from St. Anna Trough and Laptev Sea)
▲ Canada Basin - 1997 (from shelf to deep basin)

Fig. 9.34. Profiles of the fugacity (partial pressure corrected for the fact that the gas is not ideal) of CO_2 in Canada Basin and the Eurasian Basin. Data to the left of the dotted line are undersaturated and to the right are over-saturated.

Pacific- to an Atlantic-origin halocline has modified shelves on the perimeter from a potential source to a potential sink of atmospheric CO_2.

9.5.2. Biological pump

The DOC concentrations in the deep arctic regions are comparable to those in the rest of the world's oceans (Agatova et al., 1999; Borsheim et al., 1999; Bussmann and Kattner, 2000; Gradinger, 1999; Wheeler et al., 1997). Within the Arctic Ocean, shelves are regions of high biological production, especially those within the Bering, Chukchi, and Barents Seas. Here, CO_2 uptake is increased because CO_2 fixation during photosynthesis affects the physical pump by reducing pCO_2.

Levels of primary production are high on shelves due to increased light levels during ice-free periods and the supply of new nutrients by advection or vertical mixing. Although phytoplankton blooms are patchy, they are strongly associated with the retreating ice edge and the position of the ice edge in relation to the shelf break. In the northern Bering and southern Chukchi Seas, primary production occurs over a shallow shelf (50 to 200 m) and as the zooplankton and bacterioplankton cannot fully deplete this carbon source, it is either transferred to the benthos or advected downstream (Shuert and Walsh, 1993). On the southeast Bering Sea shelf, which is deeper at ~200 m, there is potential for a match/mismatch of primary production and zooplankton grazing due to water temperature (Box 9.10). An early bloom in cold melt water means most of the primary production goes to the benthos. A shift from an ice-associated bloom to a water-column bloom in the central and northern Bering Sea shelf as a result of ice retreat provides the potential for development of the plankton community at the expense of the benthic community (Hunt et al., 2002). Under climate warming, the benthic community is very likely to be most affected if this carbon is transferred to the deep basin instead of the shelf. Under these circumstances, carbon is disconnected from the food web and can be buried. In contrast, the Barents Sea shelf is much deeper (300 m) and primary production supports a large pelagic community that is unlikely to be affected. Nevertheless, a larger quantity of carbon is likely to be buried in future as deposition shifts from the shelf region to the deeper slope and basin region due to the northward movement of the ice edge.

Projections that the Arctic Ocean will be ice-free in summer (see section 9.2.5.2) imply that production will increase in waters where it was previously limited by ice cover. Based on nutrient availability, Anderson L. et al. (2003) estimated that the biological carbon sink would increase by 20 x 10^{12} g C/yr under ice-free conditions. However, mesocosm studies on the effect of high initial ambient CO_2 (750 µatm) on coccolithophore assemblages have shown an increase in POC production (Zondervan et al., 2002). This would be a negative response to atmospheric CO_2 increase.

9.5.3. Alkalinity pump

Removal of carbonate ions during the formation of calcareous shells and the subsequent sinking of these shells is important in the transfer of inorganic carbon to deeper waters and eventually the sediments. Carbonate shell sinking is also an efficient means of removing organic carbon from the euphotic zone (see section 9.5.2). Together, these processes will provide a negative feedback. However, calcification results in an increase in oceanic pCO_2 through the redistribution of carbonate species, which represents a positive feedback. Partial equilibrium with the atmospheric CO_2 will result in an increase in pH that may reduce calcification (Riebesell et al., 2000).

9.5.4. Terrestrial and coastal sources

The Arctic Ocean accounts for 20% of the world's continental shelves and these receive, transport, and store terrestrial organic carbon (primarily from rivers and coastal erosion sources) to an extent significant at the global scale (Rachold et al., 2004). Olsson and Anderson (1997) estimated that 33 to 39 x 10^{12} g of inorganic carbon are delivered to the Arctic Ocean each year by rivers. Although the amount of total organic carbon is more difficult to estimate because more than 90% is deposited in deltas (Rachold et al., 1996), it may be similar. An increase in precipitation due to climate warming will not necessarily increase carbon burial, however, as the geological composition of the drainage basin and the amount of flow are both controlling factors. For example, the Mackenzie and Yukon are both erosional rivers, while the Siberian rivers are depositional, especially the Ob for which the drainage basin includes marsh lowlands (Pocklington, 1987). Thus, increased precipitation is likely to lead to increased DIC delivery in the first case but not the second, and depends on the timing and intensity of the freshwater flow into the sea. Burial will occur on the shelf, and in adjacent ocean basins if transported offshore by sea ice, ocean currents, or turbidity currents.

Regional transport of terrestrial organic carbon to the marine system also results from coastal erosion. For example, the near-shore zone of the Laptev and East Siberian Seas is the most climatically sensitive area in the Arctic and has the highest rates of coastal retreat (Are, 1999; Grigoriev and Kunitsky, 2000). Biodegradation of this coastal material is a regional source of high pCO_2 in surface waters of the Laptev and East Siberian Seas (Semiletov, 1999a). Longer ice-free conditions and late-summer storms may accelerate the release of terrestrial carbon frozen during the last glaciation. Pleistocene permafrost soils contain huge ice wedges (up to 60 to 70% by volume) and are enriched by organic carbon (~1 to 20% by weight; Are, 1999, Romanovsky et al., 2000). The amount of organic carbon stored in permafrost is large (~450 Gt C), similar to the quantity of dissolved carbon stored in the Arctic Ocean (Semiletov, 1999b), and its release to the atmosphere depends on sediment burial rates and competing consumption by biota.

The rate of coastal erosion in the Arctic appears to have increased from a few meters per year to tens of meters per year (Are, 1999; Tomirdiaro, 1990). The highest rates of coastal retreat have been observed at capes; regions important as hunting locations. Bottom erosion is also evident. The bottom depth in the near-shore zone of the Northern Sea Route has increased by ~0.8 m over the past 14 years (Tomirdiaro, 1990). Many climate-related factors affect coastal retreat in the Arctic: permafrost ice content, air temperature, wind speed and direction, duration of open water, hydrology, and sea-ice conditions. In addition to the direct effects of climate change, rates of coastal retreat might also increase indirectly due to wave fetch and storm surge activity. Sea-level rise (~15 cm per 100 years, Proshutinsky et al., 2001) will further accelerate coastal erosion.

9.5.5. Gas hydrates

The release of CH_4 and CO_2 trapped in vast gas-hydrate reservoirs in permafrost is very likely to play a key but largely overlooked role in global climate, particularly as CH_4 is 60 times more efficient as a GHG (on a molar basis) than CO_2. For example, Semiletov (1999a) estimated that the upper 100 m layer of permafrost contains at least 100 000 Gt of organic carbon in the form of CH_4 and CO_2. Although CH_4 is one of the most important GHGs, there are currently only ~4 Gt of CH_4 carbon in the atmosphere. If a small percentage of CH_4 from the gas-hydrate reservoir were released to the atmosphere, it could result in an abrupt and significant increase in global temperature through positive feedback effects (Bell P., 1982; Nisbet, 1990; Paull et al., 1991; Revelle, 1983).

The marine Arctic is a particularly important source region for CH_4. Following glacial melting and sea-level rise during the Holocene, relatively warm (0 °C) Arctic Ocean waters flooded the relatively cold (-12 °C) Arctic permafrost domain (Denton and Hughes, 1981). As a result, permafrost sediments underlying the arctic shelf regions are still undergoing a dramatic thermal regime change as this heat is conducted downward as a thermal pulse. Subsurface temperatures within the sediment may have risen to the point that both gas hydrate and permafrost may have begun to thaw. In this case CH_4 would undergo a phase change, from a stable gas hydrate to a gas, and therefore rise through the sediment. Little is known about the fate of CH_4 released in this manner. Depending on the structure and ice matrix of surrounding sediments, CH_4 can be either consumed by anaerobic CH_4 oxidation or released upward through conduits into the overlying seawater. Evidence of elevated CH_4 concentrations in seawater has been observed in the Beaufort Sea (Macdonald, 1976) and along the North Slope of Alaska (Kvenvolden, 1991; Kvenvolden et al., 1981). Kvenvolden et al. (1993) noted that CH_4 concentrations under sea ice in the Beaufort Sea were 3 to 28 times higher in winter than summer, suggesting that CH_4 accumulates under the sea ice in winter and is rapidly released into the atmosphere when the sea ice

retreats. The timing and release of these under-ice accumulations will change with changes in ice cover.

9.6. Key findings

This section summarizes the conclusions from sections 9.2.5, 9.3.4, 9.4, and 9.5.

The Arctic is a major component of the global climate system; it both impacts and is impacted upon by the larger global system. This interaction is illustrated in the bulleted list of key findings, respectively labeled A>G and G>A. There are also forcing mechanisms and responses that remain internal to the Arctic (A>A). Any change in atmospheric forcing (wind, temperature, and precipitation) is of great importance for the ocean circulation and ocean processes (G>A).

- Large uncertainties in the response of the arctic climate system to climate change arise through poorly quantified feedbacks and thresholds associated with the albedo, the THC, and the uptake of GHGs by the ocean. Since climate models differ in their projections of future change in the pressure fields and hence their associated winds, much uncertainty remains in terms of potential changes in stratification, mixing, and ocean circulation.
- The Arctic THC is a critical component of the Atlantic THC. The latest assessment by the Intergovernmental Panel on Climate Change (IPCC, 2001) considered a reduction in the Atlantic THC likely, while a complete shutdown is considered unlikely but not impossible. If the Arctic THC is reduced, it will affect the global THC and thus the long-term development of the global climate system (A>G). Reduction in the global THC may also result in a lower oceanic heat flux to the Arctic (G>A). If the THC is reduced, local regions of the Arctic are likely to undergo cooling rather than warming, and the location of ocean fronts may change (A>A). The five ACIA-designated models cannot assess the likelihood of these occurrences.
- Most of the present ice-covered arctic areas are very likely to experience reductions in sea-ice extent and thickness, especially in summer. Equally important, it is very likely that there will be earlier sea-ice melt and later freeze-up (G>A). This is likely to lead to an opening of navigation routes through the Northwest and Northeast Passages for greater periods of the year and thus to increased exploration for reserves of oil and gas, and minerals.
- Decreased sea-ice cover will reduce the overall albedo of the region, which is very likely to result in a positive feedback for global warming (A>G).
- Upper water column temperatures are very likely to increase, especially in areas with reduced sea-ice cover.
- The amount of carbon that can be sequestered in the Arctic Ocean is likely to increase significantly under

scenarios of decreased sea-ice cover, through surface uptake and increased biological production (A>G).

- Greenhouse gases (CO_2 and CH_4) stored in permafrost may be released from marine sediments to the atmosphere subsequent to warming, thus initiating a strong positive feedback (A>G).

- In areas of reduced sea-ice cover, primary production is very likely to increase, which in turn is likely to increase zooplankton and possibly fish production. Increased cloud cover is likely to have the opposite effect on primary production in areas that are currently ice free (G>A).

- The area occupied by benthic communities of Atlantic and Pacific origin is very likely to increase, while areas occupied by colder-water species are very likely to decrease. Arctic species with a narrow range of temperature preferences, especially long-lived species with late reproduction, are very likely to be the first to disappear. A northward retreat for the arctic benthic fauna may be delayed for the benthic brooders (the reproductive strategy for many dominant polar species), while species producing pelagic larvae are likely to be the first to colonize new areas in the Arctic (G>A).

- A reduction in sea-ice extent is very likely to decrease the natural habitat for polar bears, ringed seals, and other ice-dependent species, which is very likely to lead to reductions in the survival of these species. However, increased areas and periods of open water are likely to be favorable for some whale species and the distribution of these species is very likely to move northward (G>A).

- Some species of seabird such as little auk and ivory gull are very likely to be negatively affected by the changes predicted to occur within the arctic communities upon which they depend under climate warming, while it is possible that other species will prosper in a warmer Arctic, as long as the populations of small fish and large zooplankton are abundant (G>A).

- Increased water temperatures are very likely to lead to a northward shift in the distribution of many species of fish, to changes in the timing of their migration, to a possible extension of their feeding areas, and to increased growth rates. Increased water temperatures are also likely to lead to the introduction of new species to the Arctic but are unlikely to lead to the extinction of any of the present arctic fish species. Changes in the timing of biological processes are likely to affect the overlap of spawning for predators and their prey (match/mismatch; Box 9.10) (G>A).

- Stratification in the upper water column is likely to increase the extent of the present ice-free areas of the Arctic, assuming no marked increase in wind strength (G>A).

- There are strong correlations between DOC, Chl-a, and the attenuation of UV radiation in marine waters. This is particularly significant within the context of possible UV-B attenuation in marine coastal systems, since DOC and Chl-a are usually more highly concentrated in ice-free waters than ice-covered waters.

- Present assessments indicate that UV-B radiation generally represents only a minor source of direct mortality (or decreased productivity) for populations, particularly in DOC-protected coastal zones. However, for those species whose early life stages occur near the surface, it is possible that under some circumstances – a cloudless sky, thin ozone layer, lack of wind, calm seas, low nutrient loading – the contribution of UV-B radiation to the productivity and/or mortality of a population could be far more significant. Thus, it is likely that UV-B radiation can have negative impacts (direct and/or indirect effects) on marine organisms and populations. However, UV-B radiation is only one of many environmental factors responsible for the mortality typically observed in these organisms.

9.7. Gaps in knowledge and research needs

Many aspects of the interaction between the atmosphere and the ocean, and between climate and the marine ecosystem require a better understanding before the high levels of uncertainty associated with the predicted responses to climate change can be reduced. This can only be achieved through monitoring and research, some areas requiring long-term effort. For some processes, the ocean responds more or less passively to atmospheric change, while for others, changes in the ocean themselves drive atmospheric change. The ocean clearly has a very important role in climate change and variability. Large, long-lived arctic species are generally conservative in their life-history strategies, so changes, even dramatic changes, in juvenile survival may not be detected for long periods. Zooplankton, on the other hand, can respond within a year, while microorganisms generally exhibit large and rapid (within days or weeks) variations in population size, which can make it difficult to detect long-term trends in abundance. Long data series are thus essential for monitoring climate-induced change in arctic populations.

Although the ACIA-designated models all project that global climate change will occur, they are highly variable in their projections. This illustrates the great uncertainty underlying attempts to predict the impact of climate change on ecosystems. The models do not agree in terms of changes projected to wind fields, upon which ocean circulation and mixing processes depend. Thus, conclusions drawn in this chapter regarding future changes to marine systems are to a large extent based on extrapolations from the response of the ocean to past changes in atmospheric circulation. This is also the case for predictions regarding the effects of climate change on marine ecosystems. The present assessment has been able to provide some qualitative answers to questions raised regarding climate change, but has rarely been able to account for non-linear effects or multi-species interactions. Consequentially, reliable quantitative information on the response of the marine ecosystem to climate change is lacking.

9.7.1. Gaps in knowledge

This section highlights some of the most important gaps in knowledge. These require urgent attention in order to make significant progress toward predicting and understanding the impacts of climate change on the marine environment. Each item includes an explanation as to why it is considered important.

Thermohaline circulation

Global circulation models provide an ambiguous assessment of potential changes to the THC. Most project a decrease in the strength of the THC; however, some recent models project little or no change. The THC is extremely important for the thermal budget of the Arctic Ocean and the North Atlantic.

Vertical stratification

Present climate models are unable to project future wind conditions, or to project how increased air temperatures, ice melt, and freshwater runoff will influence the vertical stability of the water column. The amount of vertical mixing that will occur is thus uncertain. Such information is required in order to project the effects of climate change on vertical heat and nutrient fluxes.

Ocean currents and transport pathways

It is necessary to understand the forces driving ocean circulation (wind, freshwater runoff, sea-ice freezing/melting) and their variability. Ocean circulation is fundamental to the distribution of water masses and thus the distribution and mixture of species within the marine ecosystem.

Fronts

Open ocean fronts act as barriers to many marine organisms and are important feeding areas for higher trophic organisms. The relative importance of production at frontal regions compared to that at non-frontal regions has not been assessed for the Arctic, nor has the importance of fronts in terms of recruitment success for fish. Few climate models provide information on fronts and their variability, and even less have an adequate spatial resolution with which to address this issue.

Release of greenhouse gases and sequestration of carbon

Changes in the balance of GHGs (i.e., sources relative to sinks) are known to impact upon climate yet little is known about the arctic reservoir. This is made all the more important through positive feedback mechanisms. Carbon can be sequestered by physical and biological processes, and can be released during ocean mixing events and the thawing of permafrost; estimates of the rates and reservoir sizes need refining before they can be used in global circulation models. Changes in the extent and timing of sea-ice cover may affect trophic structure and thus the delivery of carbon to the sediment.

Species sensitivity to climate change

Little is known about the response times of species to climate change. For example, the rapid disappearance of sea ice may not allow for adaptive change by many arctic specialists and may possibly result in the disappearance of ice-dependent species. Microorganisms, zooplankton, and fish are all expected to exhibit shifts in distribution but the rates at which this will occur cannot be predicted at present.

Match/mismatch between predators and prey

The timing of reproduction for many species is related to that of their prey. How the timing and location of the production or spawning of most species might alter in response to climate change is unclear and so therefore is the extent of a potential match/mismatch between predators and their prey. Potentially, this could impact upon the whole arctic ecosystem.

Indirect and non-linear effects on biological processes

Biota are indirectly affected by atmospheric climate change through effects on their surrounding environment and on the food web. While the response of a species to change in one particular variable can often be surmised, although generally not quantified, its response to a collection of direct and indirect effects occurring simultaneously is considerably more difficult to address. This is further complicated by the non-linearity of many processes.

Competition when/if new species are introduced into the ecosystem

Many arctic specialists have relatively narrow habitat and other niche requirements. Their likely response to a possible increase in competition from more opportunistic/generalist species in a warmer Arctic is unclear.

Gelatinous zooplankton

The abundance and variability of gelatinous zooplankton such as jellyfish has not been determined for most arctic regions. Although gelatinous zooplankton are known to be important as both predators and prey, and that they can represent a significant component of the biomass at times, their actual role within the ecosystem is unclear.

UV-B radiation exposure

Almost all existing evaluations of the effects of UV radiation are based on short-term studies. Studies are lacking on longer-term sub-lethal exposure to UV radiation, on both individual species and the overall productivity of marine ecosystems. UV-induced reductions in the nutritional quality of the food base could possibly pass through the food chain to fish, potentially reducing their growth rates as well as their nutritional condition.

9.7.2. Suggested research actions

This section lists possible actions that could be undertaken to improve the knowledge and understanding of important processes related to climate change. To reduce the uncertainties in the predicted responses to climate change it is necessary for work to proceed on several fronts simultaneously. Research actions that are considered to be of highest priority are identified by an **H**.

Observational technologies

- *Increase the application of recently developed technologies.* Recent developments range from current meters, to satellite sensors, to monitors for marine mammals.
- *Develop Remote Underwater Vehicles (RUVs) capable of working reliably under the sea ice for extended periods.* This will reduce sampling costs and enable data collection in regions difficult to access using conventional sampling methods. Instrumentation on the RUVs should include means to sample the biota.

Surveying and monitoring

- *Undertake surveys in those areas of the marine Arctic that are poorly mapped and whose resident biota have not been surveyed* **(H)**. These include surveys under the permanent ice cap in winter (perhaps using RUVs), and surveys to quantify the CH_4 and carbon reserves in the arctic marine sediments.
- *Continue and expand existing monitoring programs* **(H)**, both spatially and in breadth of measurement. New monitoring activities should be established in areas where they are presently lacking and these should be designed to address the effects of climate change. Issues to be addressed include the timing and amount of primary and secondary production, larval fish community composition, and reproductive success in marine mammals and seabirds. Key ecosystem components, including non-commercial species, must be included.
- *Evaluate monitoring data through data analysis and modeling* to determine their representativeness in space and time.

Data analysis and reconstruction

- *Reconstruct the twentieth-century forcing fields over the arctic regions.* Present reconstructions only extend back to around 1950. These reconstructions would help to model past climates.
- *Establish an arctic database that contains all available physical and biological data.* There should be open access to the database.
- *Recover past physical and biological data from the Arctic.* There are many data that are not presently available but could be recovered.

- *Undertake analysis of past climate events* to better understand the physical and biological responses to climate forcing. An example is the dramatic air temperature warming that took place from the 1920s to 1960 in the Arctic.

Field programs

- *Undertake field studies to quantify climate-related processes* **(H)**. Examples of particular processes that require attention are: open ocean and shelf convection; forces driving the THC; physical and biological processes related to oceanic fronts; sequestrating of carbon in the ocean, including a quantification of air–ice–ocean exchange; long-term effects of UV-B radiation on biota; and, interactions between benthic, ice, and pelagic fauna.

Modelling

- *Improved modeling of the ocean and sea ice in global circulation models* **(H)**. For example, how will the THC change? What are the consequences of change in the THC for the position and strength of ocean fronts, ocean current patterns, and vertical stratification?
- *Development of reliable regional models for the Arctic* **(H)**. These are essential for determining impacts on the physics and biology of the marine Arctic.
- *Strengthen the bio-physical modeling of the Arctic.* Increased emphasis is required on coupling biological models with physical models in order to improve predictive capabilities.

Approaches

- *Prioritize ecosystem-based research* **(H)**. Previous biological research programs were often targeted on single species. While these data are essential for input to larger-scale programs, the approach must be complemented by a more holistic ecosystem-based approach. Alternative concepts, methods, and modeling approaches should be explored. More effort should be placed on integrating multiple ecosystem components into modeling efforts concerning climate effects. Research on microbial communities, which may play a future role in a warmer Arctic, must be included.

References

Aagaard, K., 1970. Wind-driven transports in the Greenland and Norwegian seas. Deep-Sea Research and Oceanographic Abstracts, 17:281–291.

Aagaard, K. and E.C. Carmack, 1989. The role of sea ice and other fresh-water in the Arctic circulation. Journal of Geophysical Research - Oceans, 94(C10):14485–14498.

Aagaard, K. and E.C. Carmack, 1994. The Arctic Ocean and climate: a perspective. In: O.M. Johannessen, R.D. Muench and J.E. Overland (eds.). The Polar Oceans and their Role in Shaping the Global Environment. Geophysical Monograph Series, 85:5–20. American Geophysical Union.

Aagaard, K., L.K. Coachman and E.C. Carmack, 1981. On the halocline of the Arctic Ocean. Deep-Sea Research Part A, 28:529–545.

Aagaard, K., J.H. Swift and E.C. Carmack, 1985. Thermohaline circulation in the Arctic Mediterranean seas. Journal of Geophysical Research - Oceans, 90:4833–4846.

Aas, E., J. Høkedal, N.K. Højerslev, R. Sandvik and E. Sakshaug, 2001. Spectral properties and UV attenuation in Arctic marine waters. In: D.O. Hessen (ed.). UV Radiation and Arctic Ecosystems, pp. 23–56. Springer-Verlag.

Adkison, M.D., R.M. Peterman, M.F. Lapointe, D.M. Gillis and J. Korman, 1996. Alternative models of climatic effects on sockeye salmon (Oncorhynchus nerka) productivity in Bristol Bay, Alaska, and the Fraser River, British Columbia. Fisheries Oceanography, 5:137–152.

Ådlandsvik, B. and H. Loeng, 1991. A study of the climatic system in the Barents Sea. Polar Research, 10(1):45–49.

Agatova, A.I., N.V. Arzhanova and N.I. Torgunova, 1999. Organic matter in the Bering Sea. In: T.R. Loughlin and K. Ohtani (eds.). Dynamics of the Bering Sea, pp. 261–283. University of Alaska Fairbanks.

Ahlmann, H.W., 1949. Introductory address. ICES Rapports et Procès-Verbaux des Réunions, 125:9–16.

Ainley, D.G. and C.T. Tynan, 2003. Sea ice: a critical habitat for polar marine mammals and birds. In: D.N. Thomas and G.S. Dieckmann (eds.). Sea Ice - An Introduction to its Physics, Chemistry, Biology and Geology, Chapter 8. Blackwell Science.

Aitken, A.E. and J. Fournier, 1993. Macrobenthos communities of Cambridge, McBeth and Itirbilung fiords, Baffin Island, Northwest Territories, Canada. Arctic, 46(1):60–71.

Akira, N., T. Yanagimoto., K. Mito and S. Katakura, 2001. Interannual variability in growth of walleye Pollock, Theragra chalcogramma, in the central Bering Sea. Fisheries Oceanography, 10(4):367–375.

Albert, O.T., N. Mokeeva and K. Sunnanå, 1994. Long rough dab (Hippoglossoides platessoides) of the Barents Sea and Svalbard area: ecology and resource evaluation. ICES CM 1994/O:8. International Council for the Exploration of the Sea, Copenhagen, 9pp.

Alexander, V. and H.J. Niebauer, 1981. Oceanography of the eastern Bering Sea ice-edge zone in spring. Limnology and Oceanography, 26:1111–1125.

Alley, R.B., J. Marotzke, W.D. Nordhaus, J.T. Overpeck, D.M. Peteet, R.A. Pielke Jr., R.T. Pierrehumbert, P.B. Rhines, T.F. Stocker, L.D. Talley and J.M. Wallace, 2003. Abrupt climate change. Science, 299:2005–2010.

Alonso Rodriguez, C., H.I. Browman, J.A. Runge and J.-F. St-Pierre, 2000. Impact of solar ultraviolet radiation on hatching of a marine copepod, Calanus finmarchicus. Marine Ecology Progress Series, 193:85–93.

Alton, M.S., R.G. Bakkala, G.E. Walters and P.T. Munro, 1988. Greenland turbot Reinhardtius hippoglossoides of the eastern Bering Sea and Aleutian Islands Region. NOAA Technical Report 71.

AMAP, 1998. The AMAP Assessment Report: Arctic Pollution Issues. Arctic Monitoring and Assessment Programme, Oslo, xii + 859pp.

Andersen, L.W., E.W. Born, I. Gjertz, Ø. Wiig, L.-E. Holm and C. Bendixen, 1998. Population structure and gene flow of the Atlantic walrus (Odobenus rosmarus rosmarus) in the eastern Atlantic Arctic based on mitochondrial DNA and microsatellite variation. Molecular Ecology, 7:1323–1336.

Andersen, O.G.N., 1989. Primary production, chlorophyll, light, and nutrients beneath the Arctic sea ice. In: Y. Herman (ed.). The Arctic Seas: Climatology, Oceanography, Geology and Biology, pp. 147–191. van Nostrand Reinhold, New York.

Anderson, L.G. and S. Katlin, 2001. Carbon fluxes in the Arctic Ocean - potential impact by climate change. Polar Research, 20:225–232.

Anderson, L.G., K. Olsson, E.P. Jones, M. Chierici and A. Fransson, 1998. Anthropogenic carbon dioxide in the Arctic Ocean: Inventory and sinks. Journal of Geophysical Research - Oceans, 103:27707–27716.

Anderson, L.G., E.P. Jones and B. Rudels, 1999. Ventilation of the Arctic Ocean estimated by a plume entrainment model constrained by CFCs. Journal of Geophysical Research - Oceans, 104:13423–13429.

Anderson, L.G., E.P. Jones and J.H. Swift, 2003. Export production in the central Arctic Ocean evaluated from phosphate deficits. Journal of Geophysical Research, 108:3199, doi:10.1029/2001JC001057.

Anderson, P.K., 2001. Marine mammals in the next one hundred years: twilight for a Pleistocene megafauna? Journal of Mammalogy, 82(3):623–629.

Anker-Nilssen, T., V. Bakken, H. Strom, A.N. Golovkin, V.V. Bianki and I.P. Tatarinkova (eds.), 2000. The Status of Marine Birds Breeding in the Barents Sea Region. Norsk Polarinstitutt rapportser. No. 113. Norwegian Polar Institute, Tromsø.

Anning, T., N. Nimer, M.J. Merrett and C. Brownlee, 1996. Costs and benefits of calcification in coccolithophorids. Journal of Marine Systems, 9:45–56.

Are, F.E., 1999. The role of coastal retreat for sedimentation in the Laptev Sea. In: H. Kassens, H.A. Bauch, I.A. Dmitrienko, H. Eicken, H.-W. Hubberten, M. Melles, J. Thiede and L.A. Timokhov (eds.). Land-Ocean Systems in the Siberian Arctic, pp. 287–298. Springer-Verlag.

Arts, M.T. and H. Rai, 1997. Effects of enhanced ultraviolet-B radiation on the production of lipid, polysaccharide and protein in three freshwater algal species. Freshwater Biology, 38:597–610.

Arts, M.T., H. Rai and V.P. Tumber, 2000. Effects of artificial UV-A and UV-B radiation on carbon allocation in Synechococcus elongatus (cyanobacterium) and Nitzschia palea (diatom). Verhandlungen der Internationalen Vereinigung fur Theoretische und Angewandte Limnologie, 27:2000–2007.

Aschan, M. and K. Sunnanå, 1997. Evaluation of the Norwegian shrimp surveys conducted in the Barents Sea and the Svalbard area 1980–1997. ICES CM 1997/Y:07. International Council for the Exploration of the Sea, Copenhagen.

Ástthórsson, Ó.S. and H. Vilhjálmsson, 2002. Iceland shelf large marine ecosystem: decadal assessment and resource sustainability. In: K. Sherman and H.R. Skjoldal (eds.). Large Marine Ecosystems of the North Atlantic, pp. 219–243. Elsevier Science.

Aure, J., 1998. Havets Miljø 1998. Fisken og Havet, 2:1–90.

Baduini, C.L., K.D. Hyrenbach, K.O. Coyle, A. Pinchuk, V. Mendenhall and G.L. Hunt Jr., 2001. Mass mortality of short-tailed shearwaters in the southeastern Bering Sea during summer 1997. Fisheries Oceanography, 10:117–130.

Bailey, R.S., R.W. Furness, J.A. Gauld and P.A. Kunzlik, 1991. Recent changes in the population of the sandeel (Ammodytes marinus Raitt) at Shetland in relation to estimates of seabird predation. ICES Marine Science Symposia, 193:209–216.

Båmstedt, U. and K. Karlson, 1998. Euphausiid predation on copepods in coastal waters of the Northeast Atlantic. Marine Ecology Progress Series, 172:149–168.

Båmstedt, U., H.C. Eilertsen, K.S. Tande, D. Slagstad and H.R. Skjoldal, 1991. Copepod grazing and its potential impact on the phytoplankton development in the Barents Sea. Polar Research, 10:339–353.

Barrett, R.T. and Y.V. Krasnov, 1996. Recent responses to changes in stocks of prey species by seabirds breeding in the southern Barents Sea. ICES Journal of Marine Science, 53:713–722.

Barrett, R.T., T. Anker-Nilssen, G.W. Gabrielsen and G. Chapdelaine, 2002. Food consumption by seabirds in Norwegian waters. ICES Journal of Marine Science, 59(1):43–57.

Battini, M., V. Rocco, M. Lozada, B. Tartarotti and H.E. Zagarese, 2000. Effects of ultraviolet radiation on the eggs of landlocked Galaxias maculatus (Galaxiidae, Pisces) in northwestern Patagonia. Freshwater Biology, 44:547–552.

Beaugrand, G., P.C. Reid, F. Ibañez, J.A. Lindley and M. Edwards, 2002. Reorganization of North Atlantic marine copepod diversity and climate. Science, 296:1692–1694.

Beaugrand, G., K.M. Brander, J.A. Lindley, S. Souissi and P.C. Reid, 2003. Plankton effect on cod recruitment in the North Sea. Nature, 427:661–664.

Béland, F., H.I. Browman, C. Alonso Rodriguez and J.-F. St-Pierre, 1999. Effect of solar ultraviolet radiation (280–400 nm) on the eggs and larvae of Atlantic cod (Gadus morhua). Canadian Journal of Fisheries and Aquatic Sciences, 56:1058–1067.

Belkin, I.M., S. Levitus, J. Antonov and S.-A. Malmberg, 1998. "Great Salinity Anomalies" in the North Atlantic. Progress in Oceanography, 41:1–68.

Bell, J.G., D.R. Tocher, B.M. Farndale and J.R. Sargent, 1998. Growth, mortality, tissue histopathology and fatty acid compositions, eicosanoid production and response to stress, in juvenile turbot fed diets rich in gamma-linolenic acid in combination with eicosapentaenoic acid or docosahexaenoic acid. Prostaglandins, Leukotrienes and Essential Fatty Acids, 58:353–364.

Bell, M.V. and J.R. Dick, 1993. The appearance of rods in the eyes of herring and increased didocosahexaenoyl molecular species of phospholipids. Journal of the Marine Biological Association of the UK, 73:679–688.

Bell, M.V., R.S. Batty, J.R. Dick, K. Fretwell, J.C. Navarro and J.R. Sargent, 1995. Dietary deficiency of docosahexaenoic acid impairs vision at low light intensities in juvenile herring (Clupea harengus L.). Lipids, 30:443–449.

Bell, P.R., 1982. Methane hydrate and the carbon dioxide question. In: W.C. Clark (ed.). Carbon Dioxide Review, pp. 401–406. Oxford University Press.

Berge, G., 1962. Discolouration of the sea due to Coccolithus huxleyi 'bloom'. Sarsia 6:27–40.

Berger, W.H. and E. Jansen, 1994. Mid-Pleistocene climate shift: The Nansen connection. In: O.M. Johannessen, R.D. Muench and J.E. Overland (eds.). The Polar Oceans and Their Role in Shaping the Global Environment. Geophysical Monograph Series, 85:295–311.

Bergstad, O.A., T. Jørgensen and O. Dragesund, 1987. Life history and ecology of the gadoid resources of the Barents Sea. Fisheries Research, 5:119–161.

Bigg, M.A., 1969. Clines in the pupping season of the harbour seal, *Phoca vitulina*. Journal of the Fisheries Research Board of Canada, 26:449–455.

Björnsson, B., A. Steinarsson and M. Oddgeirsson, 2001. Optimal temperature for growth and feed conversion of immature cod (*Gadus morhua* L.). ICES Journal of Marine Science, 58(1):29–38.

Blacker, R.W., 1965. Recent changes in the benthos of the West Spitzbergen fishing grounds. International Commission for the Northwest Atlantic Fisheries, Special Publication, 6:791–794.

Blindheim, J. and B. Ådlandsvik, 1995. Episodic formation of intermediate water along the Greenland Sea Arctic Front. ICES CM 1995/Mini:6. International Council for the Exploration of the Sea, Copenhagen, 11pp.

Blindheim, J., V. Borovkov, B. Hansen, S.-Aa. Malmberg, W.R. Turrell and S. Østerhus, 1999. Upper layer cooling and freshening in the Norwegian Sea in relation to atmospheric forcing. Deep-Sea Research I, 47:655–680.

Blindheim, J., R. Toresen and H. Loeng, 2001. Fremtidige klimatiske endringer og betydningen for fiskeressursene. Havets miljø 2001. Fisken og Havet, 2:73–78.

Blood, D.M., 2002. Low-temperature incubation of walleye pollock (*Theragra chalcogramma*) eggs from the southeast Bering Sea shelf and Shelikof Strait, Gulf of Alaska. Deep-Sea Research II, 49:6095–6108.

Boertmann, D., A. Mosbech, K. Falk and K. Kampp, 1996. Seabird Colonies in Western Greenland (60°–79° 30' N lat.). National Environmental Research Institute of Denmark. Technical Report 170, 148pp.

Bogstad, B. and H. Gjøsæter, 1994. A method for estimating the consumption of capelin by cod in the Barents Sea. ICES Journal of Marine Science, 51:273–280.

Bogstad, B., T. Haug and S. Mehl, 2000. Who eats whom in the Barents Sea? In: G.A. Vikingsson and F.O. Kapel (eds.). Minke Whales, Harp and Hooded Seals: Major Predators in the North Atlantic Ecosystem. The North Atlantic Marine Mammal Commission Scientific Publications, 2:98–118.

Bond, G.C., B. Kromer, J. Beer, R. Muscheler, M.N. Evans, W. Showers, S. Hoffmann, R. Lotti-Bond, I. Hajdas and G. Bonani, 2001. Persistent solar influence on North Atlantic climate during the Holocene. Science, 294:2130–2136.

Bonga, S.E.W., 1997. The stress response in fish. Physiological Reviews, 77:591–625.

Booth, B.C. and R.A. Horner, 1997. Microalgae on the Arctic Ocean Section, 1994: species abundance and biomass. Deep-Sea Research II, 44:1607–1622.

Booth, B.C., J. Lewin and J.R. Postel, 1993. Temporal variation in the structure of autotrophic and heterotrophic communities in the subarctic Pacific. Progress in Oceanography, 32:57–99.

Born, E.W., S. Rysgaard, G. Ehlme, M. Sejr, M. Acquarone and N. Levermann, 2003. Underwater observations of foraging free-living Atlantic walruses (*Odobenus rosmarus rosmarus*) and estimates of their food consumption. Polar Biology, 26(5):348–357.

Børsheim, K.Y., S.M. Myklestad and J.-A. Sneli, 1999. Monthly profiles of DOC, mono- and polysaccharides at two locations in the Trondheimsfjord (Norway) during two years. Marine Chemistry, 63:255–272.

Borum, J., M.F. Pedersen, D. Krause-Jensen, P.B. Christensen and K. Nielsen, 2002. Biomass, photosynthesis and growth of *Laminaria saccharina* in a high-arctic fjord, NE Greenland. Marine Biology, 141:11–19.

Botham, J.W. and M.J. Manning, 1981. The histogenesis of the lymphoid organs in the carp *Cyprinus carpio* L. and the ontogenetic development of allograft reactivity. Journal of Fish Biology, 19:403–414.

Bothwell, M.L., D.M.J. Sherbot and C.M. Pollock, 1994. Ecosystem response to solar ultraviolet-B radiation: influence of trophic-level interactions. Science, 265:97–100.

Bourke, R.H. and R.P. Garrett, 1987. Sea ice thickness distribution in the Arctic Ocean. Cold Regions Science and Technology, 13:259–280.

Boveng, P.L., J.L. Bengtson, D.E. Withrow, J.C. Cesarone, M.A. Simpkins, K.J. Frost and J.J. Burns, 2003. The abundance of harbor seals in the Gulf of Alaska. Marine Mammal Science, 19(1):111–127.

Bowen, W.D. and G.K. Harrison, 1996. Comparison of harbour seal diets in two inshore habitats of Atlantic Canada. Canadian Journal of Zoology, 74:125–135.

Braarud, T., 1935. The Øst Expedition to the Denmark Strait in 1929. II. The phytoplankton and its conditions of growth. Hvalradets Skrifter, 10:1–173.

Bradstreet, M.S.W., 1980. Thick-billed murres and black guillemots in the Barrow Strait area, N.W.T., during spring: diets and food availability along ice edges. Canadian Journal of Zoology, 58:2120–2140.

Bradstreet, M.S.W., 1982. Occurrence, habitat use, and behavior of seabirds, marine mammals, and arctic cod at the Pond Inlet ice edge. Arctic, 35:28–40.

Brander, K.M., 1994. Patterns of distribution, spawning and growth in North Atlantic cod: the utility of inter-regional comparisons. ICES Marine Science Symposia, 198:406–413.

Brander, K.M., 1995. The effect of temperature on growth of Atlantic cod (*Gadus morhua* L.). ICES Journal of Marine Science, 52:1–10.

Brander, K., 2000. Effects of environmental variability on growth and recruitment in cod (*Gadus morhua*) using a comparative approach. Oceanologica Acta, 23(4):485–496.

Brander, K., 2003. Fisheries and climate. In: G. Wefer, F. Lamy and F. Mantoura (eds.). Marine Science Frontiers for Europe, pp 29–38. Springer-Verlag.

Bratbak, G., M. Levasseur, S. Michaud, G. Cantin, E. Fernández, B.R. Heimdal and M. Heldal, 1995. Viral activity in relation to *Emiliania huxleyi* blooms: a mechanism of DMSP release? Marine Ecology Progress Series, 128:133–142.

Brett, J.R., 1979. Environmental factors and growth. In: W.S. Hoar, D.J. Randal and J.R. Brett (eds.). Fish Physiology, vol. 8, pp. 599–675. Academic Press.

Brinkhof, M.W.G., A.J. Cavé and A.C. Perdeck, 1997. The seasonal decline in the first-year survival of juvenile coots: an experimental approach. Journal of Animal Ecology, 66:73–82.

Brodeur, R.D. and D.M. Ware, 1992. Long-term variability in zooplankton biomass in the subarctic Pacific Ocean. Fisheries Oceanography, 1(1):32–38.

Brodeur, R.D., M.T. Wilson, G.E. Walters and I.V. Melnikov, 1999a. Forage fishes in the Bering Sea: distribution, species associations, and biomass trends. In: T.R. Loughlin and K. Ohtani (eds.). Dynamics of the Bering Sea, pp. 509–536. University of Alaska Fairbanks.

Brodeur, R.D., C.E. Mills, J.E. Overland, G.E. Walters and J.D. Schumacher, 1999b. Evidence for a substantial increase in gelatinous zooplankton in the Bering Sea, with possible links to climate change. Fisheries Oceanography, 8:296–306.

Brodeur, R.D., H. Sugisaki and G.L. Hunt Jr., 2002. Increases in jellyfish biomass in the Bering Sea: implications for the ecosystem. Marine Ecology Progress Series, 233:89–103.

Broecker, W.S., D.M. Peteet and D. Rind, 1985. Does the ocean-atmosphere system have more than one stable mode of operation? Nature, 315:21–26.

Broecker, W.S., S. Sutherland and T.-H. Peng, 1999. A possible 20th-century slowdown of Southern Ocean deep water formation. Science, 286:1132–1135.

Browman, H.I., 2003. Assessing the impacts of solar ultraviolet radiation on the early life stages of crustacean zooplankton and ichthyoplankton in marine coastal systems. Estuaries, 26:30–39.

Browman, H.I. and R.D. Vetter, 2001. Impact of UV radiation on crustacean zooplankton and ichthyoplankton: case studies from sub-arctic marine ecosystems. In: D.O. Hessen (ed.). UV Radiation and Arctic Ecosystems, pp. 261–304. Springer-Verlag.

Browman, H.I., C. Alonso Rodriguez, F. Beland, J.J. Cullen, R.F. Davis, J.H.M. Kouwenberg, P.S. Kuhn, B. McArthur, J.A. Runge, J.-F. St-Pierre and R.D. Vetter, 2000. Impact of ultraviolet radiation on marine crustacean zooplankton and ichthyoplankton: a synthesis of results from the estuary and Gulf of St. Lawrence, Canada. Marine Ecology Progress Series, 199:293–311.

Brown, R.G.B., 1991. Marine birds and climatic warming in the northwest Atlantic. In: W.A. Montevecchi and A.J. Gaston (eds.). Studies of High-Latitude Seabirds. 1. Behavioural, Energetic, and Oceanographic Aspects of Seabird Feeding Ecology. Canadian Wildlife Service Occasional Paper 68, pp. 49–54. Environment Canada, Ottawa.

Brown, R.G.B. and D.N. Nettleship, 1981. The biological significance of polynyas to arctic colonial seabirds. In: I. Stirling and H. Cleator (eds.). Polynyas in the Canadian Arctic. Canadian Wildlife Service Occasional Paper Number 45, pp. 59–66. Environment Canada, Ottawa.

Buch, E., S.A. Horsted and H. Hovgård, 1994. Fluctuations in the occurrence of cod in the Greenland waters and their possible causes. ICES Marine Science Symposia, 198:158–174.

Buchanan, F.C., L.D. Maiers, T.D. Thue, B.G.E. de March and R.E.A. Stewart, 1998. Microsatellites from the Atlantic walrus *Odobenus rosmarus rosmarus*. Molecular Ecology, 7:1083–1085.

Budéus, G., W. Schneider and G. Krause, 1998. Winter convective events and bottom water warming in the Greenland Sea. Journal of Geophysical Research - Oceans, 103:18513–18527.

Bukhtiyarov, Y.A., K.J. Frost and L.F. Lowry, 1984. New information on foods of the spotted seal, *Phoca largha*, in the Bering Sea in spring. In: F.H. Fay and G.A. Fedoseev (eds.). Soviet-American Cooperative Research on Marine Mammals. Vol. 1. Pinnipeds, pp. 55–59. NOAA Technical Report 12.

Bulatov, O.A., 1995. Biomass variation of walleye pollock of the Bering Sea in relation to oceanological conditions. In: R.J. Beamish (ed.). Climate Change and Northern Fish Populations. Canadian Special Publication of Fisheries and Aquatic Sciences, 121:631–640.

Burger, J. and M. Gochfeld, 1990. The Black Skimmer: Social Dynamics of a Colonial Species. Columbia University Press, New York. 355pp.

Burns, J.J., 1967. The Pacific Bearded Seal. Alaska Department of Fish and Game, Juneau, 66pp.

Burns, J.J., 1981a. Bearded seal *Erignathus barbatus* Erxleben, 1777. In: S.H. Ridgway and R.J. Harrison (eds.). Handbook of Marine Mammals, Vol. 2. Seals, pp. 145–170. Academic Press.

Burns, J.J., 1981b. Ribbon seal *Phoca fasciata* Zimmermann, 1783. In: S.H. Ridgway and R.J. Harrison (eds.). Handbook of Marine Mammals, Vol. 2. Seals, pp. 89–109. Academic Press.

Burns, J.J., 2002. Harbor seal and spotted seal *Phoca vitulina* and *P. largha*. In: W.F. Perrin, B. Wursig and J.G.M. Thewissen (eds.). Encyclopedia of Marine Mammals, pp. 552–560. Academic Press.

Burns, J.J., J.J. Montague and C.J. Cowles (eds.), 1993. The Bowhead Whale. Society for Marine Mammalogy Special Publication Number 2. Allen Press, Lawrence, Kansas. 787pp.

Burns, W.C.G., 2001. From the harpoon to the heat: climate change and the International Whaling Commission in the 21st century. Georgetown International Environmental Law Review, 13:335–359.

Bussmann, I. and G. Kattner, 2000. Distribution of dissolved organic carbon in the central Arctic Ocean: the influence of physical and biological properties. Journal of Marine Systems, 27:209–219.

CAFF, 2001. Arctic Flora and Fauna: Status and Conservation. Conservation of Arctic Flora and Fauna, Edita, Helsinki, 272pp.

Campbell, R.G., J.A. Runge and E.G. Durbin, 2001. Evidence for food limitation of *Calanus finmarchicus* production rates on the southern flank of Georges Bank during April 1997. Deep-Sea Research II, 48:531–549.

Carmack, E.C., 1986. Circulation and mixing in ice-covered waters. In: N. Untersteiner (ed.). Geophysics of Sea Ice, NATO ASI Series B, 146:641–712.

Carmack, E.C., 1990. Large-scale physical oceanography of polar oceans. In: W.O. Smith Jr. (ed.). Polar Oceanography, Part A, Physical Science, pp. 171–222. Academic Press.

Carmack, E.C., 2000. The Arctic Ocean's freshwater budget: sources, storage and export. In E.L. Lewis, E.P. Jones, P. Lemke, T.D. Prowse and P. Wadhams (eds.). The Freshwater Budget of the Arctic Ocean, pp. 91–126. Kluwer Academic Press.

Carmack, E.C. and D. Chapman, 2003. Wind-driven shelf/basin exchange on an Arctic shelf: The joint roles of ice cover extent and shelf-break bathymetry. Geophysical Research Letters, 30:1778, doi:10.1029/2003GL017526.

Carmack, E.C., R.W. Macdonald, R.G. Perkin, F.A. McLaughlin and R.J. Pearson, 1995. Evidence for warming of Atlantic water in the southern Canadian Basin of the Arctic Ocean: Results from the Larsen-93 expedition. Geophysical Research Letters, 22:1061–1064.

Carscadden, J.E. and H. Vilhjálmsson, 2002. Capelin – What are they good for? Introduction. ICES Journal of Marine Science, 59(5):863–869.

Cavalieri, D.J., P. Gloersen, C.L. Parkinson, J.C. Comiso and H.J. Zwally, 1997. Observed hemispheric asymmetry in global sea ice changes. Science, 278:1104–1106.

Chalker-Scott, L., 1995. Survival and sex ratios of the intertidal copepod, *Tigriopus californicus*, following ultraviolet-B (290–320 nm) radiation exposure. Marine Biology, 123:799–804.

Chemerisina, V.T., 1948. On zoogeography of the Barents Sea. Trudy Murmanskoy Biologicheskoy Stantzii AN SSSR, 1:293–298. (In Russian)

Chilmonczyk, S., 1992. The thymus in fish: Development and possible function in the immune response. Annual Review of Fish Diseases, 2:181–200.

Christensen, J.H. and O.B. Christensen, 2003. Climate modelling: severe summertime flooding in Europe. Nature, 421:805–806.

Coachman, L.K. and J.J. Walsh, 1981. A diffusion model of cross-shelf exchange of nutrients in the southeastern Bering Sea. Deep-Sea Research A, 28:819–846.

Coachman, L.K., T.E. Whitledge and J.J. Goering, 1999. Silica in Bering Sea deep and bottom water. In: T.R. Loughlin and K. Ohtani (eds.). Dynamics of the Bering Sea, pp. 285–310. University of Alaska Fairbanks.

Cochrane, S., S. Dahle, E. Oug, B. Gulliksen and S.G. Denisenko, 1998. Benthic fauna in the northern Barents Sea. Akvaplan-niva report 434-97-1286. 33pp +app.

Colbourne, E.B. and J.T. Anderson, 2003. Biological response in a changing ocean environment in Newfoundland waters during the latter decades of the 1900s. ICES Marine Science Symposia 219:169–181.

Collett, R., 1912. Norway's Wildlife. Vol. 1. Norway's Mammalian Fauna. H. Aschehoug and Co., Kristiania. (In Norwegian)

Colton, J.B. Jr., 1972. Temperature trends and the distribution of groundfish in continental shelf waters, Nova Scotia to Long Island. Fishery Bulletin, 70:637–657.

Conners, M.E., A.B. Hollowed and E. Brown, 2002. Retrospective analysis of Bering Sea bottom trawl surveys: regime shift and ecosystem reorganization. Progress in Oceanography, 55:209–222.

Cooper, L.W., J.M. Grebmeier, I.L. Larsen, V.G. Egorov, C. Theodorakis, H.P. Kelly and J.R. Lovvorn, 2002. Seasonal variation in sedimentation of organic materials in the St. Lawrence Island polynya region, Bering Sea. Marine Ecology Progress Series, 226:13–26.

Cota, G.F., L.R. Pomeroy, W.G. Harrison, E.P. Jones, F. Peters, W.M. Sheldon Jr. and T.R. Weingartner, 1996. Nutrients, primary production and microbial heterotrophy in the southeastern Chukchi Sea: Arctic summer nutrient depletion and heterotrophy. Marine Ecology Progress Series, 135:247–258.

Coutant, C.C., 1977. Compilation of temperature preference data. Journal of the Fisheries Research Board of Canada, 34:739–745.

Coyle, K.O. and A.I. Pinchuk, 2002. Climate-related differences in zooplankton density and growth on the inner shelf of the southeastern Bering Sea. Progress in Oceanography, 55:177–194.

Crawford, R. and J. Jorgenson, 1990. Density distribution of fish in the presence of whales at the Admiralty inlet landfast ice edge. Arctic, 43(3):215–222.

Crawford, R.E. and J.K. Jorgenson, 1996. Quantitative studies of Arctic cod (*Boreogadus saida*) schools: important energy stores in the Arctic food web. Arctic, 49(2):181–193.

Croxall, J.P. (ed.), 1987. Seabirds: Feeding Ecology and Role in Marine Ecosystems. Cambridge University Press.

Croxall, J.P., 1992. Southern Ocean environmental changes: effects on seabird, seal and whale populations. Philosophical Transactions of the Royal Society, London, B, 338(1285):319–328.

Curtis, M.A., 1975. The marine benthos of arctic and sub-arctic continental shelves; a review and regional studies and their general results. Polar Record, 17(111):595–626.

Cushing, D.H., 1966. Biological and hydrographic changes in British Seas during the last thirty years. Biological Reviews, 41:221–258.

Dahl, E., B. Edvardsen and W. Eikrem, 1998. *Chrysochromulina* blooms in the Skagerrak after 1988. In: B. Reguera, J. Blanco, M.L. Fernández and T. Wyatt (eds.). Harmful Microalgae, pp. 104–105. Intergovernmental Oceanographic Commission.

Dahl, E., T. Aune and K. Tangen, 2001. Shellfish toxicity in Norway – experiences from regular monitoring, 1992–1999. In: G.M. Hallegraeff, S.I. Blackburn, C.J. Bolch and R.J. Lewis (eds.). Harmful Algal Blooms 2000, pp. 425–428. Intergovernmental Oceanographic Commission

Dahl, T.M., C. Lydersen, K.M. Kovacs, S. Falk-Petersen, J. Sargent, I. Gjertz and B. Gulliksen, 2000. Fatty acid composition of the blubber in white whales (*Delphinapterus leucas*). Polar Biology, 23(6):401–409.

Dahl-Jensen, D., K. Mosegaard, N. Gundestrup, G.D. Clow, S.J. Johnsen, A.W. Hansen and N. Balling, 1998. Past temperatures directly from the Greenland ice sheet. Science, 282:268–271.

Dalpadado, P. and H.R. Skjoldal, 1991. Distribution and life history of krill from the Barents Sea. Polar Research, 10:443–460.

Dalpadado, P. and H.R. Skjoldal, 1996. Abundance, maturity and growth of the krill species *Thysanoessa inermis* and *T. longicaudata* in the Barents Sea. Marine Ecology Progress Series, 144:175–183.

Damkaer, D.M., 1982. Possible influences of solar UV radiation in the evolution of marine zooplankton. In: J. Calkins (ed.). The Role of Solar Ultraviolet Radiation in Marine Ecosystems, pp. 701–706. Plenum Press.

Dansgaard, W., S.J. Johnsen, H.B. Clausen, D. Dahl-Jensen, N.S. Gundestrup, C.U. Hammer, C.S. Hvidbjerg, J.P. Steffensen, A.E. Sveinbjörnsdottir, J. Jouzel and G. Bond, 1993. Evidence for general instability of past climate from a 250-kyr ice-core record. Nature, 364:218–220.

Dawson, J.K., 1978. Vertical distribution of *Calanus hyperboreus* in the central Arctic ocean. Limnology and Oceanography, 23(5):950–957.

Dayton, P.K., B.J. Mordida and F. Bacon, 1994. Polar marine communities. American Zoologist, 34:90–99.

Decker, M.B., G.L. Hunt Jr. and G.V. Byrd Jr., 1995. The relationships among sea surface temperature, the abundance of juvenile walleye pollock, and the reproductive performance and diets of seabirds at the Pribilof Islands, southeastern Bering Sea. In: R.J. Beamish (ed.). Climate Change and Northern Fish Populations. Canadian Special Publication of Fisheries and Aquatic Sciences, 121:425–43

De Lange, H.J. and E. Van Donk, 1997. Effects of UVB-irradiated algae on life history traits of *Daphnia pulex*. Freshwater Biology, 38:711–720.

DeMaster, D.P. and R. Davis, 1995. Workshop on the Use of Ice-Associated Seals in the Bering and Chukchi Seas as Indicators of Environmental Change. Report of the workshop on Ice-Associated Seals held 29–31 March 1994, National Marine Mammal Laboratory, NOAA, Seattle, 10pp.

Dementyeva, T.F. and E.M. Mankevich, 1965. Changes in growth rate of Barents Sea cod as affected by environmental factors. International Commission for the Northwest Atlantic Fisheries, Special Publication, 6:571–577.

De Mora, S.J., S. Demers and M. Vernet (eds.), 2000. The Effects of UV Radiation in the Marine Environment. Cambridge University Press.

Denisenko, S.G., 2001. Long-term changes of zoobenthos biomass in the Barents Sea. Proceedings of the Zoological Institute of the Russian Academy of Sciences, 289:59–66.

Denisenko, S.G. and O.V. Titov, 2003. Distribution of zoobenthos and primary production of plankton in the Barents Sea. Oceanology, 43(1):72–82.

Denton, G.H. and T.J. Hughes, 1981. The Last Great Ice Sheets. Wiley Interscience, 484pp.

Deryugin, K.M., 1924. The Barents Sea at the Kola Meridian (33°30E). Trudy Severnoi Naucho-Promyslovoi Ekspeditsii, 19:3–103. (In Russian)

Dethlefsen, V., H. von Westernhagen, H. Tug, P.D. Hansen and H. Dizer, 2001. Influence of solar ultraviolet-B on pelagic fish embryos: osmolality, mortality and viable hatch. Helgoland Marine Research, 55:45–55.

Dey, D.B., D.M. Damkaer and G.A. Heron, 1988. UV-B dose/dose-rate responses of seasonally abundant copepods of Puget Sound. Oecologia, 76:321–329.

DeYoung, B. and G.A. Rose, 1993. On recruitment and distribution of Atlantic cod (*Gadus morhua*) off Newfoundland. Canadian Journal of Fisheries and Aquatic Sciences, 50:2729–2741.

de Veer, G., 1609. The True and Perfect Description of Three Voyages. T. Pauier, London.

Dickson, R.R., 1997. From the Labrador Sea to global change. Nature, 386:649–650.

Dickson, R.R. and J. Blindheim, 1984. On the abnormal hydrographic conditions in the European Arctic during the 1970s. ICES Rapports et Procès-Verbaux des Réunions, 185:201–213.

Dickson, R.R. and K.M. Brander, 1993. Effects of a changing windfield on cod stocks of the North Atlantic. Fisheries Oceanography, 2:124–153.

Dickson, R.R. and J. Brown, 1994. The production of North Atlantic Deep Water: sources, rates and pathways. Journal of Geophysical Research, 99(C6):12,319–12,342.

Dickson, R.R., J. Meincke, S.-A. Malmberg and A.J. Lee, 1988. The "great salinity anomaly" in the northern North Atlantic 1968–1982. Progress in Oceanography, 20:103–151.

Dickson, R.R., J. Lazier, J. Meincke, P. Rhines and J. Swift, 1996. Long-term coordinated changes in the convective activity of the North Atlantic. Progress in Oceanography, 38:241–295.

Dickson, R.R., T.J. Osborn, J.W. Hurrell, J. Meincke, J. Blindheim, B. Aadlandsvik, T. Vinje, G. Alekseev and W. Maslowski, 2000. The Arctic Ocean response to the North Atlantic Oscillation. Journal of Climate, 13:2671–2696.

Dickson, R., I. Yashayaev, J. Meincke, B. Turrell, S. Dye and J. Holfort, 2002. Rapid freshening of the deep North Atlantic Ocean over the past four decades. Nature, 416:832–837.

Dickson R.R., R. Curry and I. Yashayaev, 2003. Recent changes in the North Atlantic. Philosophical Transactions of the Royal Society, London, A, 361:1917–1934.

Dietz, R., M.P. Heide-Jørgensen, P.R. Richard and M. Acquarone, 2001. Summer and fall movements of narwhals (*Monodon monoceros*) from northeastern Baffin Island towards northern Davis Strait. Arctic, 54:244–261.

Dippner, J.W. and G. Ottersen, 2001. Cod and climate variability in the Barents Sea. Climate Research, 17(1):73–82.

Dokken, T.M. and E. Jansen, 1999. Rapid changes in the mechanism of ocean convection during the last glacial period. Nature, 401:458–461.

Don, J. and R.R. Avtalion, 1993. Ultraviolet irradiation of tilapia spermatozoa and the Hertwig effect: electron microscopic analysis. Journal of Fish Biology, 42:1–14.

Dooley, H.D., J.H.A. Martin and D.J. Ellett, 1984. Abnormal hydrographic conditions in the Northeast Atlantic during the 1970s. ICES Rapports et Procès-Verbaux des Réunions, 185:179–181.

Dragesund, O., J. Hamre and Ø. Ulltang, 1980. Biology and population dynamics of the Norwegian spring spawning herring. ICES Rapports et Procès-Verbaux des Réunions, 177:43–71.

Drinkwater, K.F., 1999. Changes in ocean climate and its general effect on fisheries: examples from the North-west Atlantic. In: D. Mills (ed.). The Ocean Life of Atlantic Salmon - Environmental and Biological Factors Influencing Survival, pp. 116–136. Fishing News Books, Oxford.

Drinkwater, K.F., 2002. A review of the role of climate variability in the decline of northern cod. In: N.A. McGinn (ed.). Fisheries in a Changing Climate. American Fisheries Society Symposium, 32:113–130.

Duffy, D.C., 1990. Seabirds and the 1982–84 El Niño-Southern Oscillation. In: P.W. Glynn (ed.). Global Ecological Consequences of the 1982–83 El Niño-Southern Oscillation, pp. 395–415. Elsevier.

Dunbar, M.J., 1957. The determinants of production in northern seas: a study of the biology of *Themisto libellula*. Canadian Journal of Zoology, 35:797–819.

Dunton, K., 1992. Arctic biogeography: the paradox of the marine benthic fauna and flora. Trends in Ecology Evolution, 7:183–189.

Dwyer, D.A., K.M. Bailey and P.A. Livingston, 1987. Feeding habits and daily ration of walleye pollock (*Theragra chalcogramma*) in the eastern Bering Sea, with special reference to cannibalism. Canadian Journal of Fisheries and Aquatic Sciences, 44:1972–1984.

Ellertsen, B., P. Fossum, P. Solemdal and S. Sundby, 1989. Relation between temperature and survival of eggs and first-feeding larvae of northeast Arctic cod (*Gadus morhua* L.). ICES Rapports et Procès-Verbaux des Réunions, 191:209–219.

Ellis, D.V., 1955. Some observations on the shore fauna of Baffin Island. Arctic, 8:224–236.

Ellis, D.V. and R.T. Wilce, 1961. Arctic and subarctic exsamples of intertidal zonation. Arctic, 14:224–235.

English, T.S., 1961. Some biological observations in the central North Polar Sea. Drift Station Alpha 1957–1958. Arctic Institute of North America Research Paper, 13:8–80.

Falk-Petersen, I.-B., V. Frivoll, B. Gulliksen and T. Haug. 1986. Occurrence and age/size relations of polar cod, *Boreogadus saida* (Lepechin), in Spitsbergen coastal waters. Sarsia, 71(3–4):235–245.

Falk-Petersen, S., J.R. Sargent, J. Henderson, E.N. Hegseth, H. Hop and Y.B. Okolodkov, 1998. Lipids and fatty acids in ice algae and phytoplankton from the Marginal Ice zone in the Barents Sea. Polar Biology, 20:41–47.

Falk-Petersen, S., W. Hagen, G. Kattner, A. Clarke and J. Sargent, 2000. Lipids, trophic relationships, and biodiversity in Arctic and Antarctic krill. Canadian Journal of Fisheries and Aquatic Sciences, 57(suppl.3):178–191.

Fay, F.H., 1981. Walrus *Odobenus rosmarus* (Linnaeus, 1758). In: S.H. Ridgway and R.J. Harrison (eds.). Handbook of Marine Mammals. Vol. 1. The Walrus, Sea Lions, Fur Seals and Sea Otter, pp. 1–23. Academic Press.

Fay, F.H., 1982. Ecology and biology of the pacific walrus, *Odobenus rosmarus divergens* Illiger. North American Fauna, 74:1–279.

Feder, H.M., A.S. Naidu, S.C. Jewett, J.M. Hameedi, W.R. Johnson and T.E. Whitledge, 1994. The north-eastern Chukchi Sea: benthos environmental interactions. Marine Ecology Progress Series, 111:171–190.

Ferguson, S.H., M.K. Taylor and F. Messier, 2000a. Influence of sea ice dynamics on habitat selection by polar bears. Ecology, 81:761–772.

Ferguson, S.H., M.K. Taylor, A. Rosing-Asvid, E.W. Born and F. Messier, 2000b. Relationships between denning of polar bears and conditions of sea ice. Journal of Mammalogy, 81:1118–1127.

Fidhiany, L. and K. Winckler, 1999. Long term observation on several growth parameters of convict cichlid under an enhanced ultraviolet-A (320–400 nm) irradiation. Aquaculture International, 7:1–12.

Finley, K.J., 2001. Natural history and conservation of the Greenland whale, or bowhead, in the Northwest Atlantic. Arctic, 54:55–76.

Finley, K.J., G.W. Miller, R.A. Davis and W.R. Koski, 1983. A distinctive large breeding population of ringed seals (*Phoca hispida*) inhabiting the Baffin Bay pack ice. Arctic, 36:162–173.

Fisher, K.I. and R.E.A. Stewart, 1997. Summer foods of Atlantic walrus, *Odobenus rosmarus rosmarus*, in northern Foxe Basin, Northwest Territories. Canadian Journal of Zoology, 75:1166–1175.

Folkow, L.P. and A.S. Blix, 1992. Satellite tracking of harp and hooded seals. In: I.G. Priede and S.M. Swift (eds.). Wildlife Telemetry: Remote Monitoring and Tracking of Animals, pp. 214–218. Ellis Horwood Ltd.

Folkow, L.P. and A.S. Blix, 1995. Distribution and diving behaviour of hooded seals. In: A.S. Blix, L. Walloe and O. Ulltang (eds.). Whales, Seals, Fish and Man, pp. 193–202. Elsevier.

Folkow, L.P. and A.S. Blix, 1999. Diving behaviour of hooded seals (*Cystophora cristata*) in the Greenland and Norwegian Seas. Polar Biology, 22(1):61–74.

Fortier, M., L. Fortier, C. Michel and L. Legendre, 2002. Climatic and biological forcing of the vertical flux of biogenic particles under seasonal Arctic sea ice. Marine Ecology Progress Series, 225:1–16.

Fossum, P., H. Gjøsæter and R. Ingvaldsen, 2002. Spesielle økologiske forhold i Barentshavet høsten 2001 – hva hendte? Havets miljø 2002. Fisken og Havet, 2:67–70.

Francis, R.C., S.R. Hare, A.B. Hollowed and W.S. Wooster, 1998. Effects of interdecadal climate variability on the oceanic ecosystems of the NE Pacific. Fisheries Oceanography, 7:1–21.

Frank, K.T., J. Carscadden and J.E. Simon, 1996. Recent excursions of capelin (*Mallotus villosus*) to the Scotian Shelf and Flemish Cap during anomalous hydrographic conditions. Canadian Journal of Fisheries and Aquatic Sciences, 53:1473–1486.

Fransson, A., M. Chierici, L.G. Anderson, I. Bussmann, G. Kattner, E.P. Jones and J.H. Swift, 2001. The importance of shelf processes for the modification of chemical constituents in the waters of the Eurasian Arctic Ocean: implication for carbon fluxes. Continental Shelf Research, 21(3):225–242.

Fréchet, A., 1990. Catchability variations of cod in the marginal ice zone. Canadian Journal of Fisheries and Aquatic Sciences, 47(9):1678–1683.

Frimodt, C. and I. Dore, 1995. Multilingual Illustrated Guide to the World's Commercial Coldwater Fish. Fishing News Books, Oxford. 264pp.

Fronval, T. and E. Jansen, 1996. Late Neogene paleoclimates and paleo-ceanography in the Iceland-Norwegian sea: evidence from the Iceland and Voring Plateaus. In: J. Thiede, A.M. Myhre, J.V. Firth, G.L. Johnson and W.F. Ruddiman (eds.). Proceedings of the Ocean Drilling Program, Scientific Results, 151:455–468. Ocean Drilling Program, College Station, Texas.

Frost, B.W. and M.J. Kishi, 1999. Ecosystem dynamics in the eastern and western gyres of the subarctic Pacific – a review of lower trophic level modelling. Progress in Oceanography, 43:317–333.

Frost, K.J. and L.F. Lowry, 1980. Feeding of ribbon seals (*Phoca fasciata*) in the Bering Sea in spring. Canadian Journal of Zoology, 58:1601–1607.

Fujishima, Y., K. Ueda, M. Maruo, E. Nakayama, C. Tokutome, H. Hasegawa, M. Matsui and Y. Sohrin, 2001. Distribution of trace bioelements in the Subarctic North Pacific Ocean and the Bering Sea (the R/V Hakuho Maru cruise KH-97-2). Journal of Oceanography, 57:261–273.

Funder, S., N. Abrahamsen, O. Bennike and R.W. Feyling-Hanssen, 1985. Forested Arctic: evidence from North Greenland. Geology, 13:542–546.

Furevik, T., H. Drange and A. Sorteberg, 2002. Anticipated changes in the Nordic Seas marine climate. Fisken og Havet, 4:1–13.

Furevik, T., M. Bentsen, H. Drange, I.K.T. Kindem, N.G. Kvamstø and A. Sorteberg, 2003. Description and validation of the Bergen Climate Model: ARPEGE coupled with MICOM. Climate Dynamics, 21:27–51.

Furness, R.W. and D.M. Bryant, 1996. Effect of wind on field metabolic rates of breeding Northern Fulmars. Ecology, 77(4):1181–1188.

Furness, R.W., J.J.D. Greenwood and P.J. Jarvis, 1993. Can birds be used to monitor the environment? In: R.W. Furness and J.J.D. Greenwood (eds.). Birds as Monitors of Environmental Change, pp. 1–41. Chapman & Hall.

Fyfe, J.C., G.J. Boer and G.M. Flato, 1999. The Arctic and Antarctic oscillations and their projected changes under global warming. Geophysical Research Letters, 26(11):1601–1604.

Gabrielsen, G.W., F. Mehlum and K.A. Nagy, 1987. Daily energy expenditure and energy utilization of free-ranging black-legged kitti-wakes. Condor, 89:126–132.

Gagnon, A. and W.P. Gough, 2001. Influence of climate variability on the river discharge of the Hudson Bay region, Canada. Canadian Association of Geographers, 2001 Annual General Meeting, Montreal.

Galkin, Yu.I., 1998. Long-term changes in the distribution of molluscs in the Barents Sea related to the climate. Berichte zur Polarforschung, 287:100–143.

Ganachaud, A. and C. Wunsch, 2000. Improved estimates of global ocean circulation, heat transport and mixing from hydrographic data. Nature, 408:453–457.

Ganopolski, A. and S. Rahmstorf, 2001. Rapid changes of glacial climate simulated in a coupled climate model. Nature, 409:153–158.

Garssen, J., M. Norval, A. El-Ghorr, N.K. Gibbs, C.D. Jones, D. Cerimele, C. De Simone, S. Caffieri, F. Dall'Acqua, F.R. De Gruijl, Y. Sontag and H. Van Loveren, 1998. Estimation of the effect of increasing UVB exposure on the human immune system and related resistance to infectious diseases and tumours. Journal of Photochemistry and Photobiology B, 42:167–179.

Garthe, S., 1997. Influence of hydrography, fishing activity, and colony location on summer seabird distribution in the south-eastern North Sea. ICES Journal of Marine Science, 54(4):566–577.

Gascard, J.-C., A.J. Watson, M.-J. Messias, K.A. Olsson, T. Johannessen and K. Simonsen, 2002. Long-lived vortices as a mode of deep venti-lation in the Greenland Sea. Nature, 416:525–527.

Gaston, A.J. and I.L. Jones, 1998. The Auks. Oxford University Press, 349pp.

Gislason, A. and O.S. Ástthórsson, 1998. Seasonal variations in biomass, abundance and composition of zooplankton in the subarctic waters north of Iceland. Polar Biology, 20:85–94.

Gjertz, I. and C. Lydersen, 1986. The ringed seal (*Phoca hispida*) spring diet in northwestern Spitsbergen, Svalbard. Polar Research, 4:53–56.

Gjertz, I. and Ø. Wiig, 1994. Past and present distribution of walruses in Svalbard. Arctic, 47(1):34–42.

Gjertz, I. and Ø. Wiig, 1995. The number of walruses (*Odobenus rosmarus*) in Svalbard in summer. Polar Biology, 15(7):527–530.

Gjertz, I., K.M. Kovacs, C. Lydersen and Ø. Wiig, 2000. Movements and diving of adult ringed seals (*Phoca hispida*) in Svalbard. Polar Biology, 23:651–656.

Gjøsæter, H., 1995. Pelagic fish and the ecological impact of the modern fishing industry in the Barents Sea. Arctic, 48(3):267–278.

Gjøsæter, H., 1998. The population biology and exploitation of capelin (*Mallotus villosus*) in the Barents Sea. Sarsia, 83:453–496.

Gjøsæter, H. and H. Loeng, 1987. Growth of the Barents Sea capelin, *Mallotus villosus*, in relation to climate. Environmental Biology of Fishes, 20:293–300.

Gloersen, P., 1995. Modulation of hemispheric sea-ice cover by ENSO events. Nature, 373:503–506.

Glud, R.N., M. Kühl, F. Wenzhöfer and S. Rysgaard, 2002. Benthic diatoms of a high Arctic fjord (Young Sound, NE Greenland): impor-tance for ecosystem primary production. Marine Ecology Progress Series, 238:15–29.

Goes, J.I., N. Handa, S. Taguchi and T. Hama, 1994. Effect of UV-B radiation on the fatty acid composition of the marine phytoplankton *Tetraselmis* sp.: relationship to cellular pigments. Marine Ecology Progress Series, 114:259–274.

Goldman, J.C., 1999. Inorganic carbon availability and the growth of large marine diatoms. Marine Ecology Progress Series, 180:81–91.

Gomes, M.C., R.L. Haedrich and M.G. Villagarcia, 1995. Spatial and temporal changes in the groundfish assemblages on the northeast Newfoundland/Labrador Shelf, northwest Atlantic, 1978–91. Fisheries Oceanography, 4:85–101.

Gosselin, M., M. Levasseur, P.A. Wheeler, R.A. Horner and B.C. Booth, 1997. New measurements of phytoplankton and ice algal production in the Arctic Ocean. Deep-Sea Research II, 44:1623–1644.

Goulden, C.E. and A.R. Place, 1990. Fatty acid synthesis and accumu-lation rates in daphnids. Journal of Experimental Zoology, 256:168–178.

Grace, M.F. and M.J. Manning, 1980. Histogenesis of the lymphoid organs in rainbow trout, *Salmo gairdneri*. Developmental and Comparative Immunology, 4:255–264.

Gradinger, R., 1996. Occurrence of an algal bloom under Arctic pack ice. Marine Ecology Progress Series, 131:301–305.

Gradinger, R., 1999. Vertical fine structure of the biomass and composi-tion of algal communities in Arctic pack ice. Marine Biology, 133:745–754.

Grahl, C., A. Boetius and E.M. Nöthig, 1999. Pelagic-benthic coupling in the Laptev Sea affected by ice cover. In: H. Kassens, H.A. Bauch, I.A. Dmitrenko, H. Eicken, H.-W. Hubberten, M. Melles, J. Thiede and L.A. Timokhov (eds.). Land-Ocean Systems in the Siberian Arctic: Dynamics and History, pp. 143–152. Springer, Berlin.

Grainger, E.H., 1989. Vertical distribution of zooplankton in the central Arctic Ocean. In: L. Rey and V. Alexander (eds.). Proceedings of the Sixth Conference of the Comite Arctique International, 13–15 May 1985, pp. 48–60. E.J. Brill, The Netherlands.

Gran, H.H., 1931. On the conditions for the production of the plankton in the sea. ICES Rapports et Procès-Verbaux des Réunions, 75:37–46.

Grebmeier, J.M., 1993. Studies of pelagic-benthic coupling extended onto the Soviet continental shelf in the northern Bering and Chukchi Seas. Continental Shelf Research, 13(5–6):653–668.

Grebmeier, J.M. and J.P. Barry, 1991. The influence of oceanographic processes on pelagic-benthic coupling in polar regions: A benthic perspective. Journal of Marine Systems, 2(3–4):495–518.

Grebmeier, J.M. and L.W. Cooper, 1994. A decade of benthic research on the continental shelves of the northern Bering and Chukchi seas: Lessons learned. In: R.H. Meehan, V. Sergienko and G. Weller (eds.). Bridges of Science Between North America and the Russian Far East, pp. 87–98. American Association for the Advancement of Science, Arctic Division, Fairbanks, Alaska.

Grebmeier, J.M. and L.W. Cooper, 1995. Influence of the St Lawrence Island polynya upon the Bering Sea benthos. Journal of Geophysical Research – Oceans, 100(C3):4439–4460.

Grebmeier, J.M. and K.H. Dunton, 2000. Benthic processes in the northern Bering/Chukchi Seas: status and global change. In: H.P. Huntington (ed.). Impacts of Changes in Sea Ice and Other Environmental Parameters in the Arctic, pp. 18–93. Marine Mammal Commission Workshop, Girdwood, Alaska, 15–17 February 2000.

Grebmeier, J.M., C.P. McRoy and H.M. Feder, 1988. Pelagic-benthic coupling on the shelf of the northern Bering and Chukchi Seas. 1. Food supply source and benthic biomass. Marine Ecology Progress Series, 48(1):57-67.

Grebmeier, J.M., H.M. Feder and C.P. McRoy, 1989. Pelagic-benthic coupling on the shelf of the northern Bering and Chukchi Seas. 2. Benthic community structure. Marine Ecology Progress Series, 51(3):253–268.

Grebmeier, J.M., W.O. Smith and R.J. Conover, 1995. Biological processes on Arctic continental shelves: ice-ocean-biotic interactions. In: W.O. Smith and J.M. Grebmeier (eds.). Arctic Oceanography: Marginal Zones and Continental Shelves. Coastal and Estuarine Studies, 49:231–261.

Grellier, K., P.M. Thompson and H.M. Corpe, 1996. The effect of weather conditions on harbour seal (*Phoca vitulina*) haulout behaviour in the Moray Firth, Northeast Scotland. Canadian Journal of Zoology, 74:1806–1811.

Grigoriev, M.N. and V.V. Kunitsky, 2000. Destruction of the sea coastal ice-complex in Yakutia. In: I.P. Semiletov (ed.). Hydrometeorological and Biogeochemical Research in the Arctic (in Russian). Trudy Arctic Regional Center, 2, pp.109–116. Vladivostok.

Grotefendt, K., K. Logemann, D. Quadfasel and S. Ronski, 1998. Is the Arctic Ocean warming? Journal of Geophysical Research, 103:27679–27687.

Häder, D.-P. (ed.), 1997. The Effects of Ozone Depletion on Aquatic Ecosystems. R.G. Landes Company, Austin, Texas.

Häder, D.-P., H.D. Kumar, R.C. Smith and R.C. Worrest, 2003. Aquatic ecosystems: effects of solar ultraviolet radiation and interactions with other climatic change factors. Photochemical and Photobiological Sciences, 2:39–50.

Haflidason, H., H.P. Sejrup, D.K. Kristensen and S. Johnsen, 1995. Coupled response of the late glacial climatic shifts of northwest Europe reflected in Greenland ice cores: Evidence from the northern North Sea. Geology, 23:1059–1062.

Halldal, P., 1953. Phytoplankton investigations from weather ship M in the Norwegian Sea, 1948–1949. Hvalradets Skrifter, 38:1–91.

Hamer, K.C., E.A. Schreiber and J. Burger, 2001. Breeding biology, life histories, and life history –environment interactions in seabirds. In: E.A. Schreiber and J. Burger (eds.). Biology of Marine Birds. CRC Marine Biology Series, 1:215–259. CRC Press.

Hamm, C., M. Reigstad, C. Wexels Riser, A. Mühlebach and P. Wassmann, 2001. On the trophic fate of *Phaeocystis pouchetii*. VII. Sterols and fatty acids reveal sedimentation of P. pouchetii-derived organic matter via krill fecal strings. Marine Ecology Progress Series, 209:55–69.

Hamre, J., 1994. Biodiversity and exploitation of the main fish stocks in the Norwegian-Barents Sea ecosystem. Biodiversity and Conservation, 3:473–492.

Haney, J.C. and S.D. MacDonald, 1995. Ivory Gull (*Pagophila eburnean*). In: A. Poole and F. Gill (eds.). The Birds of North America, No. 175. Academy of Natural Sciences, Philadelphia and the American Ornithologists' Union, Washington, D.C. 24pp.

Hansell, D.A., T.E. Whitledge and J.J. Goering, 1993. Patterns of nitrate utilization and new production over the Bering-Chukchi shelf. Continental Shelf Research, 13:601–627.

Hansen, B. and S. Østerhus, 2000. North Atlantic–Nordic Seas exchanges. Progress in Oceanography, 45:109–208.

Hansen, B., S. Christiansen and G. Pedersen, 1996. Plankton dynamics in the marginal ice zone of the central Barents Sea during spring: carbon flow and structure of the grazer food chain. Polar Biology, 16:115–218.

Hansen, B., W.R. Turrell and S. Østerhus, 2001. Decreasing overflow from the Nordic seas into the Atlantic Ocean through the Faroe Bank channel since 1950. Nature, 411:927–930.

Hansen, B., S. Østerhus, H. Hátún, R. Kristiansen and K.M.H. Larsen, 2003. The Iceland-Faroe inflow of Atlantic Water to the Nordic Seas. Progress in Oceanography, 59:443–474.

Hare, S.R. and N.J. Mantua, 2000. Empirical evidence for North Pacific regime shifts in 1977 and 1989. Progress in Oceanography, 47:103–145.

Hargrave, B.T., B. Von Bodungen, R.J. Conover, A.J. Fraser, G. Phillips and W.P. Vass, 1989. Seasonal changes in sedimentation of particulate matter and lipid content of zooplankton collected by sediment trap in the Arctic Ocean off Axel Heiberg island. Polar Biology, 9(7):467–475.

Harris, M. and S. Wanless, 1984. The effect of the wreck of seabirds in February 1983 on auk populations on the Isle of May (Fife). Bird Study, 31:103–110.

Harvell, C.D., K. Kim, J.M. Burkholder, R.R. Colwell, P.R. Epstein, D.J. Grimes, E.E. Hofmann, E.K. Lipp, A.D.M.E. Osterhaus, R.M. Overstreet, J.W. Porter, G.W. Smith and G.R. Vasta, 1999. Emerging marine diseases - climate links and anthropogenic factors. Science, 285:1505–1510.

Hasle, G.R. and B.R. Heimdal, 1998. The net phytoplankton in Kongsfjorden, Svalbard, July 1998, with general remarks on species composition of Arctic phytoplankton. Polar Research, 17:31–52.

Haug, T., G. Henriksen, A. Kondakov, V. Mishin, K.T. Nilssen and N. Rov, 1994. The status of grey seals *Halichoerus grypus* in North Norway and on the Murman coast, Russia. Biological Conservation, 70:59–67.

Haug, T., H. Gjøsæter., U. Lindstrom and K.T. Nilssen, 1995. Diet and food availability for north-east Atlantic minke whales (*Balaenoptera acutorostrata*), during the summer of 1992. ICES Journal of Marine Science, 52(1):77–86.

Haynes, E.G. and G. Ingell, 1983. Effect of temperature on rate of embryonic development of walleye pollock (*Theragra chalcogramma*). Fishery Bulletin, 81:890–894.

Head, E.J.H., L.R. Harris and R.W. Campbell, 2000. Investigations on the ecology of *Calanus* spp. in the Labrador Sea. I. Relationship between the phytoplankton bloom and reproduction and development of *Calanus finmarchicus* in spring. Marine Ecology Progress Series, 193:53–73.

Heath, M.R., J.O. Backhaus, K. Richardson, E. McKenzie, D. Slagstad, D. Beare, J. Dunn, J.G. Fraser, A. Gallego, D. Hainbucher, S. Hay, S. Jonasdottir, H. Madden, J. Mardaljevic and A. Schacht, 1999. Climate fluctuations and the spring invasion of the North Sea by *Calanus finmarchicus*. Fisheries Oceanography 8 (suppl.1):163–176.

Hegseth, E.N., 1998. Primary production of the northern Barents Sea. Polar Research, 17(2):113–123.

Heiskanen, A.-S. and A. Keck, 1996. Distribution and sinking rates of phytoplankton, detritus, and particulate biogenic silica in the Laptev Sea and Lena River (Arctic Siberia). Marine Chemistry, 53:229–245.

Helbling, E.W. and V.E. Villafañe, 2001. UV radiation effects on phytoplankton primary production: a comparison between Arctic and Antarctic marine ecosystems. In: D.O. Hessen (ed.). UV Radiation and Arctic Ecosystems, pp. 203–226. Springer-Verlag.

Helbling, E.W. and H. Zagarese, 2003. UV Effects in Aquatic Organisms and Ecosystems. Springer-Verlag. 574pp.

Helle, K., B. Bogstad, C.T. Marshall, K. Michalsen, G. Ottersen and M. Pennington, 2000. An evaluation of recruitment indices for Arcto-Norwegian cod (*Gadus morhua* L.). Fisheries Research, 48(1):55–67.

Henderson, R.J., E.N. Hegseth and M.T. Park, 1998. Seasonal variation in lipid and fatty acid composition of ice algae from the Barents Sea. Polar Biology, 20:48–55.

Henriksen, G., I. Gjertz and A. Kondakov, 1997. A review of the distribution and abundance of harbour seals, *Phoca vitulina*, on Svalbard, Norway, and in the Barents Sea. Marine Mammal Science, 13:157–163.

Hessen, D.O. (ed.), 2001. UV Radiation and Arctic Ecosystems. Springer-Verlag. 310pp.

Hessen, D.O., H.J. De Lange and E. Van Donk, 1997. UV-induced changes in phytoplankton cells and its effects on grazers. Freshwater Biology, 38:513–524.

Hewes, C.D., E. Sakshaug, F.M.H. Reid and O. Holm-Hansen, 1990. Microbial autotrophic and heterotrophic eucaryotes in Antarctic waters: relationships between biomass and chlorophyll, adenosine triphosphate and particulate organic carbon. Marine Ecology Progress Series, 63:27–35.

Hirche, H.-J. and S. Kwasniewski, 1997. Distribution, reproduction and development of *Calanus* species in the Northeast water in relation to environmental conditions. Journal of Marine Systems, 10:299–317.

Hjelset, A.M., M.A. Pinchukov and J.H. Sundet, 2002. Joint report for 2002 on the red king crab (*Paralithodes camtschaticus*) investigations in the Barents Report to the 31st session for the mixed Russian-Norwegian Fisheries Commission, 18 pp.

Hjort, J., 1914. Fluctuations in the great fisheries of northern Europe viewed in the light of biological research. ICES Rapports et Procès-Verbaux des Réunions, 20:1–228.

Hobson, K.A. and H.E. Welch, 1992. Observations of foraging Northern Fulmars (*Fulmarus glacialis*) in the Canadian High Arctic. Arctic, 45(2):150–153.

Holligan, P.M., S.B. Groom and D.S. Harbour, 1993. What controls the distribution of the coccolithophore, *Emiliania huxleyi*, in the North Sea? Fisheries Oceanography, 2:175–183.

Holloway, G. and T. Sou, 2002. Has Arctic sea ice rapidly thinned? Journal of Climate, 15:1691–1701.

Hollowed, A.B. and W.S. Wooster, 1995. Decadal-scale variations in the eastern subarctic Pacific: B. Response of northeast Pacific fish stocks. In: R.J. Beamish (ed.). Climate Change and Northern Fish Populations. Canadian Special Publication of Fisheries and Aquatic Sciences, 121:373–385.

Hollowed, A.B., S.R. Hare and W.S. Wooster, 2001. Pacific Basin climate variability and patterns of Northeast Pacific marine fish production. Progress in Oceanography, 49:257–282.

Holst, J.C., O. Dragesund, J. Hamre, O.A. Misund and O.J. Østevedt, 2002. Fifty years of herring migrations in the Norwegian Sea. ICES Marine Science Symposia, 215:352–360.

Holst, M., I. Stirling and K.A. Hobson, 2001. Diet of ringed seals (*Phoca hispida*) on the east and west sides of the North Water Polynya, northern Baffin Bay. Marine Mammal Science, 17(4):888–908.

Hood, D.W. and J.A. Calder, 1981. The Eastern Bering Sea Shelf: Oceanography and Resources. University of Washington Press, Seattle, 1339pp. (2 volumes)

Hop, H., W.M. Tonn and H.E. Welch, 1997. Bioenergetics of Arctic cod (*Boreogadus saida*) at low temperatures. Canadian Journal of Fisheries and Aquatic Sciences, 54(8):1772–1784.

Hop, H., M. Poltermann, O.J. Lønne, S. Falk-Petersen, R. Korsnes and W.P. Budgell, 2000. Ice amphipod distribution relative to ice density and under-ice topography in the northern Barents Sea. Polar Biology, 23:357–367.

Hop, H., T. Pearson, E.N. Hegseth, K.M. Kovacs, C. Wiencke, S. Kwasniewski, K. Eiane, F. Mehlum, B. Gulliksen, M. Wlodarska-Kowalczuk, C. Lydersen, J.M. Weslawski, S. Cochrane, G.W. Gabrielsen, R.J.G. Leakey, O.J. Lønne, M. Zajaczkowski, S. Falk-Petersen, M. Kendall, S.-Å. Wängberg, K. Bischof, A.Y. Voronkov, N.A. Kovaltchouk, J. Wiktor, M. Poltermann, G. di Prisco, C. Papucci and S. Gerland, 2002. The marine ecosystem of Kongsfjorden, Svalbard. Polar Research, 21:167–208.

Hopkins, T.S., 1991. The GIN Sea – A synthesis of its physical oceanography and literature review 1972–1985. Earth-Science Reviews, 30:175–318.

Hopkins, T.S., 2001. Thermohaline feedback loops and Natural Capital. Scientia Marina, 65(suppl.2):231–256.

Horner, R., 1984. Phytoplankton abundance, chlorophyll *a*, and primary productivity in the western Beaufort Sea. In: P.W. Barnes, D.M. Schell and E. Reimnitz (eds.). The Alaskan Beaufort Sea: Ecosystems and Environments, pp. 295–310. Academic Press.

Hunt, G.L. Jr., 1990. The pelagic distribution of marine birds in a heterogeneous environment. Polar Research, 8:43–54.

Hunt, G.L. Jr. and P.J. Stabeno, 2002. Climate change and the control of energy flow in the southeastern Bering Sea. Progress in Oceanography, 55:5–22.

Hunt, G.L. Jr., J.F. Piatt and K.E. Erikstad, 1991. How do foraging seabirds sample their environment? In: Proceedings of the 20th International Ornithological Congress, pp. 2272–2279. New Zealand Ornithological Congress Trust Board, Wellington, New Zealand.

Hunt, G.L. Jr., F. Mehlum, R.W. Russell, D. Irons, M.B. Decker and P.H. Becker, 1999. Physical processes, prey abundance, and the foraging ecology of seabirds. In: N.J. Adams and R. Slotow (eds.). Proceedings of the 22nd International Ornithological Congress, Durban, pp. 2040–2056. BirdLife South Africa, Johannesburg.

Hunt, G.L. Jr., P. Stabeno, G. Walters, E. Sinclair, R.D. Brodeur, J.M. Napp and N.A. Bond, 2002. Climate change and control of the southeastern Bering Sea pelagic ecosystem. Deep-Sea Research II, 49:5821–5853.

Hunter, J.R., J.H. Taylor and H.G. Moser, 1979. Effect of ultraviolet irradiation on eggs and larvae of the northern anchovy, *Engraulis mordax*, and the Pacific mackerel, *Scomber japonicus*, during the embryonic stage. Photochemistry and Photobiology, 29:325–338.

Hunter, J.R., S.E. Kaupp and J.H. Taylor, 1981. Effects of solar and artificial ultraviolet-B radiation on larval northern anchovy, *Engraulis mordax*. Photochemistry and Photobiology, 34:477–486.

Hunter, J.R., S.E. Kaupp and J.H. Taylor, 1982. Assessment of effects of UV radiation on marine fish larvae. In: J. Calkins (ed.). The Role of Solar Ultraviolet Radiation in Marine Ecosystems, pp. 459–493. Plenum Press.

Huot, Y., W.H. Jeffrey, R.F. Davis and J.J. Cullen, 2000. Damage to DNA in bacterioplankton: A model of damage by ultraviolet radiation and its repair as influenced by vertical mixing. Photochemistry and Photobiology, 72:62–74.

Hureau, J.-C. and N.I. Litvinenko, 1986. Scorpaenidae. In: P.J.P. Whitehead, M.-L. Bauchot, J.-C. Hureau, J. Nielsen and E. Tortonese (eds.). Fishes of the North-eastern Atlantic and the Mediterranean, Vol 3, pp. 1211–1229. UNESCO, Paris.

Hurks, M., J. Garssen, H. van Loveren and B.-J. Vermeer, 1994. General aspects of UV-irradiation on the immune system. In: G. Jori, R.H. Pottier, M.A.J. Rodgers and T.G. Truscott (eds.). Photobiology in Medicine, pp. 161–176. Plenum Press.

Hutchings, J.A. and R.A. Myers, 1994. Timing of cod reproduction: interannual variability and the influence of temperature. Marine Ecology Progress Series, 108:21–31.

Hygum, B.H., C. Rey and B.W. Hansen, 2000. Growth and development rates of *Calanus finmarchicus* nauplii during a diatom spring bloom. Marine Biology, 136:1075–1085.

ICES, 2003. Report of the North-Western Working Group. ICES CM 2003/ACFM:24. International Council for the Exploration of the Sea, Copenhagen, 325pp.

Ikävalko, J. and R. Gradinger, 1997. Flagellates and heliozoans in the Greenland Sea ice studied alive using light microscopy. Polar Biology, 17:473–481.

Incze, L.S. and A.J. Paul, 1983. Grazing and predation as related to energy needs of stage I zoeae of the Tanner crab *Chinoectes bairdi* (Brachyura, Majidae). Biological Bulletin, 165:197–208.

Ingebrigtsen, K., J.S. Christiansen, O. Lindhe and I. Brandt, 2000. Disposition and cellular binding of 3H-benzo[a]pyrene at subzero temperatures: studies in an aglomerular arctic teleost fish – the polar cod (*Boreogadus saida*). Polar Biology, 23(7):503–509.

Ingvaldsen, R., L. Asplin and H. Loeng, 1999. Short time variability in the Atlantic inflow to the Barents Sea. ICES CM 1999/L:05. International Council for the Exploration of the Sea, Copenhagen, 12pp.

Ingvaldsen, R., H. Loeng and L. Asplin, 2002. Variability in the Atlantic inflow to the Barents Sea based on a one-year time series from moored current meters. Continental Shelf Research, 22:505–519.

Ingvaldsen, R., H. Loeng, G. Ottersen and B. Ådlandsvik, 2003. The Barents Sea climate during the 1990s. ICES Marine Science Symposia, 219:160–168.

IPCC, 1998. The Regional Impacts of Climate Change: An Assessment of Vulnerability. A Special Report of Working Group II of the Intergovernmental Panel on Climate Change. R.T. Watson, M.C. Zinyowera and R.H. Moss (eds.). Cambridge University Press, 527pp.

IPCC, 2001. Climate Change 2001: The Scientific Basis. Contribution of Working Group I to the Third Assessment Report of the Intergovernmental Panel on Climate Change. J.T. Houghton, Y. Ding, D.J. Griggs, M. Noguer, P.J. van der Linden, X. Dai, K. Maskell and C.A. Johnson (eds.). Cambridge University Press, 881pp.

Irigoien, X., R. Head, U. Klenke, B. Meyer-Harms, D. Harbour, B. Niehoff, H.-J. Hirche and R. Harris, 1998. A high frequency time series at Weathership M, Norwegian Sea, during the 1997 spring bloom: feeding of adult female *Calanus finmarchicus*. Marine Ecology Progress Series, 172:127–137.

Isachsen, P.E., J.H. LaCasce, C. Mauritzen and S. Häkkinen, 2003. Wind-driven variability of the large-scale recirculating flow in the Nordic Seas and Arctic Ocean. Journal of Physical Oceanography, 33:2534–2550.

IUCN, 1998. Status of the polar bear. In: A.E. Derocher, G.W. Garner, N.J. Lunn and O. Wiig (eds.). Polar Bears: Proceedings of the 12th Working Meeting of the IUCN/SSC Polar Bear Specialist Group, pp. 23–44. World Conservation Union (IUCN), Gland, Switzerland and Cambridge, U.K.

Jakobsson, J., 1980. The north Icelandic herring fishery and environmental conditions 1960–1968. ICES Rapports et Procès-Verbaux des Réunions, 177:460–465.

Jakobsson, J. and O.J. Østvedt, 1999. A review of joint investigations on the distribution of herring in the Norwegian and Iceland Seas 1950–1970. Rit Fiskideildar, 16:208–238.

Jakobsson, J., A. Gudmundsdottir and G. Stefansson, 1993. Stock-related changes in biological parameters of the Icelandic summer-spawning herring. Fisheries Oceanography, 2(3–4):260–277.

Jansen, E., T. Fronval, F. Rack and J.E.T. Channell, 2000. Pliocene-Pleistocene ice rafting history and cyclicity in the Nordic Seas during the last 3.5 Myr. Paleoceanography, 15:709–721.

Jarvela, L.E. and L.K. Thorsteinson, 1999. The epipelagic fish community of Beaufort Sea coastal waters, Alaska. Arctic, 52(1):80–94.

Jarvis, P.J., 1993. Environmental changes. In: R.W. Furness and J.J.D. Greenwood (eds.). Birds as Monitors of Environmental Change, pp. 42–77. Chapman & Hall, London.

Jenkins, M., 2003. Prospects for biodiversity. Science, 302:1175–1177.

Jensen, A.S., 1939. Concerning a change of climate during recent decades in the Arctic and subarctic regions, from Greenland in the west to Eurasia in the east, and contemporary biological and geophysical changes. Biologiske Meddelelser 14, 75pp.

Jewett, S.C. and H.M. Feder, 1982. Food and feeding habits of the king crab *Paralithodes camtschatica* near Kodiak Island, Alaska. Marine Biology, 66(3):243–250.

Johannessen, O.M. and M. Miles, 2000. Arctic sea ice and climate change – will the ice disappear in this century? Science Progress, 83(3):209–222.

Johannessen, O.M., E.V. Shalina and M.W. Miles, 1999. Satellite evidence for an Arctic sea ice cover in transformation. Science, 286:1937–1939.

Johannessen, O.M., L. Bengtsson, M.W. Miles, S.I. Kuzmina, V.A. Semenov, G.V. Alekseev, A.P. Nagurnyi, V.F. Zakharov, L. Bobylev, L.H. Pettersson, K. Hasselmann and H.P. Cattle, 2004. Arctic climate change – observed and modeled temperature and sea ice variability. Tellus, 56A:328–341.

Johannessen, T., L.G. Anderson, R. Bellerby, M.-J. Messias, A. Olsen, A. Olsson, A. Omar, I. Skejelvan and A. Watson, 2002. The role of convection and seasonal to interannual variability on carbon uptake in the Nordic seas. EOS, Transactions of the American Geophysical Union, 83(2002 Ocean Sciences Meeting Supplement):Abstract #OS11H-10.

Johns, D.G., 2001. *Calanus* abundance in the North Atlantic as determined from CPR-surveys. In: Proceedings of the workshop on The Northwest Atlantic Ecosystem - a Basin Scale Approach, pp. 44–47. Canadian Science Advisory Secretariat, Proceedings Series 2001/23.

Johns, D.G., M. Edwards and S.D. Batten, 2001. Arctic boreal plankton species in the Northwest Atlantic. Canadian Journal of Fisheries and Aquatic Sciences, 58(11):2121–2124.

Johnsen, G. and E. Sakshaug, 2000. Monitoring of harmful algal blooms along the Norwegian coast using bio-optical methods. South African Journal of Marine Science, 22:309–321.

Johnsen, G., Z. Volent, K. Tangen and E. Sakshaug, 1997. Time series of harmful and benign phytoplankton blooms in northwest European waters using the Seawatch buoy system. In: M. Kahru and C.W. Brown, (eds.). Monitoring Algal Blooms: New Techniques for Detecting Large-scale Environmental Change, pp. 115–143. Landes Bioscience.

Joiris, C.R., J. Tahon, L. Holsbeek and M. Vancauwenberghe, 1996. Seabirds and marine mammals in the eastern Barents Sea: late summer at-sea distribution and calculated food intake. Polar Biology, 16(4):245–256.

Jokinen, E.I., H.M. Salo, S.E. Markkula, A.K. Immonen and T.M. Aaltonen, 2001. Ultraviolet B irradiation modulates the immune system of fish (*Rutilus rutilus*, Cyprinidae). Part III. Lymphocytes. Photochemistry and Photobiology, 73:505–512.

Jones, E.P., B. Rudels, and L.G. Anderson, 1995. Deep waters of the Arctic Ocean: origins and circulation. Deep-Sea Research I, 42:737–760.

Jónsson, S. 1991, Seasonal and interannual variability of wind stress curl over the Nordic Seas. Journal of Geophysical Research, 96(C2):2649–2659.

Jónsson, S., 1999. The circulation in the northern part of the Denmark Strait and its variability. ICES CM 1999/L:06. International Council for the Exploration of the Sea, Copenhagen, 9pp.

Jónsson, S. and Briem, J., 2003. Flow of Atlantic Water west of Iceland and onto the North Icelandic Shelf. ICES Marine Science Symposia, 219:326–328.

Jørgensen, L.L. and B. Gulliksen, 2001. Rocky bottom fauna in arctic Kongsfjord (Svalbard) studied by means of suction sampling and photography. Polar Biology, 24:113–121.

Juterzenka, K.V. and K. Knickmeier, 1999. chlorophyll *a* in water column and sea ice during the Laptev Sea freeze-up study in autumn 1995. In: H. Kassens, H.A. Bauch, I.A. Dmitrenko, H. Eicken, H.-W. Hubberten, M. Melles, J. Thiede and L.A. Timokhov (eds.). Land-Ocean Systems in the Siberian Arctic: Dynamics and History, pp. 153–160. Springer.

Kanazawa, A., 1997. Effects of docosahexaenoic acid and phospholipids on stress tolerance of fish. Aquaculture, 155:129–134.

Karanas, J.J., H. Van Dyke and R.C. Worrest, 1979. Midultraviolet (UV-B) sensitivity of *Acartia clausii* Giesbrecht (Copepoda). Limnology and Oceanography, 24:1104–1116.

Karanas, J.J., R.C. Worrest and H. Van Dyke, 1981. Impact of UV-B radiation on the fecundity of the copepod *Acartia clausii*. Marine Biology, 65:125–133.

Karnovsky, N.J., S. Kwasniewski, J.M. Weslawski, W. Walkusz and A. Beszczynska-Moller, 2003. Foraging behavior of little auks in a heterogeneous environment. Marine Ecology Progress Series, 253:289–303.

Katlin, S. and L.G. Anderson, 2005. Uptake of atmospheric carbon dioxide in Arctic Shelf seas: evaluation of the relative importance of processes that influence pCO_2 in water transported over the Bering-Chukchi Sea shelf. Marine Chemistry, 94:67–79.

Kawamura, A., 1967. Observations of phytoplankton in the Arctic Ocean in 1964. Bulletin of the Plankton Society of Japan, 1967:71–89.

Kenyon, K.W., 1982. Sea otter (*Enhydra lutris*). In: J.A. Chapman and G.A. Feldhamer (eds.). Wild Mammals of North America: Biology, Management, Economics, pp. 704–710. Johns Hopkins University Press, Baltimore.

Kerr, J.B. and C.T. McElroy, 1993. Evidence for large upward trends of ultraviolet-B radiation linked to ozone depletion. Science, 262:1032–1034.

Kitaysky, A.S. and E.G. Golubova, 2000. Climate change causes contrasting trends in reproductive performance of planktivorous and piscivorous alcids. Journal of Animal Ecology, 69:248–262.

Knies, J., J. Mathhiessen, C. Vogt and R. Stein, 2002. Evidence of 'Mid-Pliocene (~3 Ma) global warmth' in the eastern Arctic Ocean and implications for the Svalbard/Barents Sea ice sheet during the late Pliocene and early Pleistocene (~3-1.7 Ma). Boreas, 31:82–93.

Kobayashi, H.A., 1974. Growth cycle and related vertical distribution of the cosomatous pteropod *Spiratella* ('Limacina') *helicina* in central Arctic Ocean. Marine Biology, 26(4):295–301.

Koc, N. and E. Jansen, 2002. Holocene climate evolution of the North Atlantic Ocean and the Nordic Seas - a synthesis of new results. In: G. Wefer, W.H. Berger, K.-E. Behre and E. Jansen (eds.). Climate Development and History of the North Atlantic Realm, pp. 165–173. Springer Verlag.

Koc, N., E. Jansen and H. Haflidason, 1993. Paleoceanographic reconstructions of surface ocean conditions in the Greenland, Iceland and Norwegian seas through the last 14 ka, based on diatoms. Quaternary Science Reviews, 12:115–140.

Korennikov, S.P. and E.V. Shoshina, 1980. Composition and distribution of algae in the south-west part of the Barents Sea from Mikulkin to the Cape Russkii Zavorot. Botanichyeskii Zhurnal (Leningrad), 65(6):855–859.

Kovacs, K.M., 1990. Mating strategies of male hooded seals (*Crystophora cristata*). Canadian Journal of Zoology, 68:2499–2505.

Kovacs, K.M., 2002a. Bearded seal. In: W.F. Perrin, B. Wursig and J.G.M. Thewissen (eds.). Encyclopedia of Marine Mammals, pp. 84–87. Academic Press.

Kovacs, K.M., 2002b. Hooded seal. In: W.F. Perrin, B. Wursig and J.G.M. Thewissen (eds.). Encyclopedia of Marine Mammals, pp. 580–582. Academic Press.

Kovacs, K.M., C. Lydersen and I. Gjertz, 1996. Birth site characteristics and prenatal molting in bearded seals (*Erignathus barbatus*). Journal of Mammalogy, 77(4):1085–1091.

Krajick, K., 2001. Arctic life, on thin ice. Science, 291:424–425.

Kristiansen, S. and B.Aa. Lund, 1989. Nitrogen cycling in the Barents Sea – I. Uptake of nitrogen in the water column. Deep-Sea Research A, 36:255–268.

Kristiansen, S., T. Farbrot and P.A. Wheeler, 1994. Nitrogen cycling in the Barents Sea – Seasonal dynamics of new and regenerated production in the marginal ice zone. Limnology and Oceanography, 39:1630–1642.

Kröncke, I., 1994. Macrobenthos composition, abundance and biomass in the Arctic ocean along a transect between Svalbard and the Makarov basin. Polar Biology, 14(8):519–529.

Kuhn, P.S., H.I. Browman, R.F. Davis, J.J. Cullen and B.L. McArthur, 2000. Modeling the effects of ultraviolet radiation on embryos of *Calanus finmarchicus* and Atlantic cod (*Gadus morhua*) in a mixing environment. Limnology and Oceanography, 45:1797–1806.

Kvenvolden, K.A., 1991. A review of Arctic gas hydrates as a source of methane in global change. In: G. Weller, C.L. Wilson and B.A.B. Severin (eds.). Proceedings of the International Conference on the Role of the Polar Regions in Global Change, 11–15 June 1990, University of Alaska Fairbanks, vol. 2, pp. 696–701.

Kvenvolden, K.A., T.M. Vogel and J.V. Gardner, 1981. Geochemical prospecting for hydrocarbons in the outer continental shelf, Southern Bering Sea, Alaska. Journal of Geochemical Exploration, 14:209–219.

Kvenvolden, K.A., M.D. Lilley, T.D. Lorenson, P.W. Barnes and E. McLaughlin, 1993. The Beaufort Sea continental shelf as a seasonal source of atmospheric methane. Geophysical Research Letters, 20:2459–2462.

Kwok, R., 2000. Recent changes in Arctic Ocean sea ice motion associated with the North Atlantic Oscillation. Geophysical Research Letters, 27:775–778.

Kwok, R., and D.A. Rothrock, 1999. Variability of Fram Strait ice flux and North Atlantic Oscillation, Journal of Geophysical Research, 104(C3):5177–5189.

Lacuna, D.G. and S.-I. Uye, 2001. Influence of mid-ultraviolet (UVB) radiation on the physiology of the marine planktonic copepod *Acartia omorii* and the potential role of photoreactivation. Journal of Plankton Research, 23:143–156.

Larsen, H.C., A.D. Saunders, P.D. Clift, J. Beget, W. Wei, S. Spezzaferri and the ODP Leg 152 Scientific Party, 1994. Seven million years of glaciation in Greenland. Science, 264:952–955.

Larsen, T., 1985. Polar bear denning and cub production in Svalbard, Norway. Journal of Wildlife Management, 49:320–326.

Larsen, T., 1986. Population biology of the polar bear (*Ursus maritimus*) in the Svalbard area. Norsk Polarinstitutt Skrifter, 184:1–55.

Latif, M., 2001. Tropical Pacific/Atlantic Ocean interactions at multidecadal time scales. Geophysical Research Letters, 28(3):539–542.

Latif, M., E. Roeckner, U. Mikolajewicz and R. Voss, 2000. Tropical stabilization of the thermohaline circulation in a greenhouse warming simulation. Journal of Climate, 13:1809–1813.

Lauzier, L.M. and S.N. Tibbo, 1965. Water temperature and the herring fishery of Magdalen Islands, Quebec. International Commission for the Northwest Atlantic Fisheries, Special Publication, 6:591–596.

Lavigne, D.M., 2002. Harp seal. In: W.F. Perrin, B. Wursig and J.G.M. Thewissen (eds.). Encyclopedia of Marine Mammals, pp. 560–562. Academic Press.

Lavigne, D.M and K.M. Kovacs, 1988. Harps & Hoods: Ice-breeding Seals of the Northwest Atlantic. University of Waterloo Press, Ontario, 174pp.

Lawson, J.W. and G.B. Stenson, 1997. Diet of northwest Atlantic harp seals (Phoca groenlandica) in offshore areas. Canadian Journal of Zoology, 75:2095–2106.

Lawson, J.W., G.B. Stenson and D.G. McKinnon, 1995. Diet of harp seals (Phoca groenlandica) in nearshore waters of the northwest Atlantic during 1990–1993. Canadian Journal of Zoology, 73:1805–1818.

Lazier, J.R.N., 1980. Oceanographic conditions at Ocean Weather Ship Bravo, 1964–1974. Atmosphere-Ocean, 18: 227–238.

Lazier, J.R.N., 1995. The salinity decrease in the Labrador Sea over the past thirty years. In: D.G. Martinson, K. Bryan, M. Ghil, M.M. Hall, T.R. Karl, E.S. Sarachik, S. Sorooshian and L.D. Talley (eds.). Natural Climate Variability on Decade-to-Century Time Scales, pp. 295–302. National Academy Press, Washington, D.C.

Lesser, M.P., J.H. Farrell and C.W. Walker, 2001. Oxidative stress, DNA damage and p53 expression in the larvae of Atlantic cod (Gadus morhua) exposed to ultraviolet (290–400 nm) radiation. Journal of Experimental Biology, 204:157–164.

Lewis, E.L., E.P. Jones, P. Lemke, T.D. Prowse and P. Wadhams (eds.), 2000. The Freshwater Budget of the Arctic Ocean. Kluwer Academic Press, 623 pp.

Lilly, G.R., H. Hop, D.E. Stansbury and C.A. Bishop, 1994. Distribution and abundance of polar cod (Boreogadus saida) off southern Labrador and eastern Newfoundland. ICES CM 1994/O:6. International Council for the Exploration of the Sea, Copenhagen, 21pp.

Livingston, P.A., 1993. Importance of predation by groundfish, marine mammals and birds on walleye pollock Theragra chalcogramma and Pacific herring Clupea pallasi in the eastern Bering Sea. Marine Ecology Progress Series, 102:205–215.

Livingston, P.A. and J. Jurado-Molina, 2000. A multispecies virtual population analysis of the eastern Bering Sea. ICES Journal of Marine Science, 57(2):294–299.

Livingston, P.A., D.A. Dwyer, D.L. Wencker, M.S. Yang and G.M. Lang, 1986. Trophic interactions of key fish species in the eastern Bering Sea. International North Pacific Fisheries Commission Bulletin, 47:49–65.

Loeng, H., 2001. Klima og fisk – hva vet vi og hva tror vi. Naturen, 125(3):132–140.

Loeng, H., H. Bjørke and G. Ottersen, 1995. Larval fish growth in the Barents Sea. In: R.J. Beamish (ed.). Climate Change and Northern Fish Populations. Canadian Special Publication of Fisheries and Aquatic Sciences, 121:691–698.

Loeng, H., V. Ozhigin and B. Ådlandsvik, 1997. Water fluxes through the Barents Sea. ICES Journal of Marine Science, 54:310–317.

Lønø, O., 1970. The polar bear (Ursus maritimus Phipps) in the Svalbard area. Norsk Polarinstitutt Skrifter, 149:1–103.

Loughlin, T.R., I.N. Sukhanova, E.H. Sinclair and R.C. Ferrero, 1999. Summary of biology and ecosystem dynamics in the Bering Sea. In: T.R. Loughlin and K. Ohtani (eds.). Dynamics of the Bering Sea, pp. 387–407. University of Alaska Fairbanks.

Lowry, L.F., 1993. Foods and feeding ecology. In: J.J. Burns, J.J. Montague and C.J. Cowles (eds.). The Bowhead Whale, pp. 201–238. Society for Marine Mammalogy Special Publication No. 2. Allen Press, Kansas.

Lowry, L.F. and K.J. Frost, 1981. Feeding and trophic relationships of phocid seals and walruses in the eastern Bering Sea. In: D.W. Hood and J.A. Calder (eds.). The Eastern Bering Sea Shelf: Oceanography and Resources. Vol. 2, pp. 813–824. University of Washington Press, Seattle.

Lowry, L.F., K.J. Frost, D.G. Calkins, G.L. Swartzman and S. Hills, 1982. Feeding habits, food requirements, and status of Bering Sea marine mammals. Council Document 19. North Pacific Fishery Management Council, Anchorage, Alaska, 292pp.

Lowry, L.F., K.J. Frost, R. Davis, D.P. DeMaster and R.S. Suydam, 1998. Movements and behavior of satellite-tagged spotted seals (Phoca largha) in the Bering and Chukchi Seas. Polar Biology, 19:221–230.

Luchetta, A., M. Lipizer and G. Socal, 2000. Temporal evolution of primary production in the central Barents Sea. Journal of Marine Systems, 27:177–193.

Lydersen, C. and K.M. Kovacs, 1999. Behaviour and energetics of ice-breeding, North Atlantic phocid seals during the lactation period. Marine Ecology Progress Series, 187:265–281.

Lydersen, C., A.R. Martin, K.M. Kovacs and I. Gjertz, 2001. Summer and autumn movements of white whales Delphinapterus leucas in Svalbard, Norway. Marine Ecology Progress Series, 219:265–274.

Macdonald, R.W., 1976. Distribution of low-molecular-weight hydrocarbons in the southern Beaufort Sea. Environmental Science and Technology, 10:1241–1246.

Macdonald, R.W., T. Harner, J. Fyfe, H. Loeng and T. Weingartner, 2003a. The Influence of Global Change on Contaminant Pathways to, within, and from the Arctic. Arctic Monitoring and Assessment Programme, Oslo, xii+65pp.

Macdonald, R.W., E. Sakshaug and R. Stein, 2003b. The Arctic Ocean: modern status and recent climate change. In: R. Stein and R.W. Macdonald (eds.). The Organic Carbon Cycle in the Arctic Ocean, Chapter 1.2, pp. 6–21. Springer.

Mackas, D.L. and A. Tsuda, 1999. Mesozooplankton in the eastern and western subarctic Pacific: community structure, seasonal life histories, and interannual variability. Progress in Oceanography, 43(2–4):335–363.

Madronich, S., R.L. McKenzie, M.M. Caldwell and L.O. Bjorn, 1995. Changes in ultraviolet radiation reaching the Earth's surface. Ambio, 24:143–152.

Maita, Y., M. Yanada and T. Takahashi, 1999. Seasonal variation in the process of marine organism production based on downward fluxes of organic substances in the Bering Sea. In: T.R. Loughlin and K. Ohtani (eds.). Dynamics of the Bering Sea, pp. 341–352. University of Alaska Fairbanks.

Malloy, K.D., M.A. Holman, D. Mitchell and H.W. Detrich III, 1997. Solar UVB-induced DNA damage and photoenzymatic DNA repair in Antarctic zooplankton. Proceedings of the National Academy of Sciences, 94:1258–1263.

Mallory, M.L., H.G. Gilchrist, A.J. Fontaine and J.A. Akearok, 2003. Local ecological knowledge of ivory gull declines in Arctic Canada. Arctic, 56(3):293–298.

Malmberg, S.-A., 1983. Hydrographic investigations in the Iceland and Greenland Seas in late winter 1971- 'Deep Water Project.' Jokull, 33:133–140.

Malmberg, S.-A., 1984. Hydrographic conditions in the east Icelandic Current and sea ice in North Icelandic waters 1970–1980. ICES CM 1984/C:20. International Council for the Exploration of the Sea, Copenhagen, 9pp.

Malmberg, S.-A. and J. Blindheim, 1994. Climate, cod, and capelin in northern waters. ICES Marine Science Symposia, 198:297–310.

Malmberg, S.-A. and H. Valdimarsson, 2003. Hydrographic conditions in Icelandic waters, 1990–1999. ICES Marine Science Symposia, 219:50–60.

Mantua, N.J., S.R. Hare, Y. Zhang, J.M. Wallace and R.C. Francis, 1997. A Pacific interdecadal climate oscillation with impacts on salmon production. Bulletin of the American Meteorological Society, 78:1069–1079.

Marinaro, J. and M. Bernard, 1966. Contribution à l'étude des oeufs et larves pélagiques de poissons méditerranés. I. Notes preliminaires sur l'influence léthale du rayonnement solaire sur les oeufs. Pelagos, 6:49–55.

Marshall, J. and F. Schott, 1999. Open-ocean convection: Observations, theory, and models. Reviews of Geophysics, 37(1):1–64.

Martinson, D.G. and M. Steele, 2001. Future of the Arctic sea ice cover: implications of an Antarctic analog. Geophysical Research Letters, 20:307–310.

Madsen, H., 1936. Investigations on the shore fauna of Eastern Greenland with a survey of the shore of other Arctic regions. Meddelelser om Gronland, 100(8):1–112.

Maslowski, W., B. Newton, P. Schlosser, A. Semtner and D. Martinson, 2000. Modelling recent climate variability in the Arctic Ocean. Geophysical Research Letters, 27(22):3743–3746.

Maslowski, W., D.C. Marble, W. Walczowski and A.J. Semtner, 2001. On large scale shifts in the Arctic Ocean and sea-ice conditions during 1979–98. Annals of Glaciology, 33:545–550.

Masuda, R., T. Takeuchi, K. Tsukamoto, Y. Ishizaki, M. Kanematsu and K. Imaizumi, 1998. Critical involvement of dietary docosahexaenoic acid in the ontogeny of schooling behaviour in the yellowtail. Journal of Fish Biology, 53:471–484.

Mauritzen, C., 1996. Production of dense overflow waters feeding the North Atlantic across the Greenland-Scotland Ridge. Deep-Sea Research I, 43(6):769–835.

Mauritzen, M., A.E. Derocher and Ø. Wiig, 2001. Space-use strategies of female polar bears in a dynamic sea ice habitat. Canadian Journal of Zoology, 79:1704–1713.

McLaughlin, F.A., E.C. Carmack, R.W. Macdonald and J.K.B. Bishop, 1996. Physical and geochemical properties across the Atlantic/Pacific water mass front in the southern Canadian Basin. Journal of Geophysical Research, 101(C1):1183–1198.

McPhee, M.G. and J.H. Morison, 2001. Turbulence and diffusion: Under-ice boundary layer. In: J. Steele, S. Thorpe and K. Turekian (eds.). Encyclopedia of Ocean Sciences, pp 3071–3078. Academic Press.

Mehlum, F. and G.W. Gabrielsen, 1995. Energy expenditure and food consumption by seabird populations in the Barents Sea region. In: H.R. Skjoldal, C. Hopkins, K.E. Erikstad and H.P. Leinaas (eds.). Ecology of Fjords and Coastal Waters, pp. 457–470. Elsevier.

Mehlum, F., G.L. Hunt, M.B. Decker and N. Nordlund, 1998a. Hydrographic features, cetaceans and the foraging of thick-billed murres and other marine birds in the northwestern Barents Sea. Arctic, 51(3):243–252.

Mehlum, F., N. Nordlund and K. Isaksen, 1998b. The importance of the 'Polar Front' as a foraging habitat for guillemots *Uria* spp. breeding at Bjørnøya., Barents Sea. Journal of Marine Systems, 14:27–43.

Mehlum, F., G.L. Hunt Jr., Z. Klusek and M.B. Decker, 1999. Scale-dependent correlations between the abundance of Brunnich's guillemots and their prey. Journal of Animal Ecology, 68(1):60–72.

Meincke, J., S. Jonsson and J.H. Swift, 1992. Variability of convective conditions in the Greenland Sea. ICES Marine Science Symposia, 195:32–39.

Meincke, J., B. Rudels and H.J. Friedrich, 1997. The Arctic Ocean-Nordic Seas thermohaline system. ICES Journal of Marine Science, 54:283–299.

Melle, W. and H.R. Skjoldal, 1998. Reproduction and development of *Calanus finmarchicus*, *C. glacialis* and *C. hyperboreus* in the Barents Sea. Marine Ecology Progress Series, 169:211–228.

Melling, H., 1993. The formation of a haline shelf front in wintertime in an ice-covered arctic sea. Continental Shelf Research, 13:1123–1147.

Melling, H., 2000. Exchanges of freshwater through the shallow straits of the North American Arctic. In: E.L. Lewis, E.P. Jones, P. Lemke, T.D. Prowse and P. Wadhams (eds.). The Freshwater Budget of the Arctic Ocean, pp. 479–502. Kluwer Academic Press.

Melling, H., 2002. Sea ice of the northern Canadian Arctic Archipelago. Journal of Geophysical Research, 107(C11), doi: 10.1029/2001JC001102.

Melling, H. and R.M. Moore, 1995. Modification of halocline source waters during freezing on the Beaufort Sea shelf: evidence from oxygen isotopes and dissolved nutrients. Continental Shelf Research, 15:89–113.

Melnikov, I.A., 1997. The Arctic Ice Ecosystem. Gordon and Breach Science Publishers, 204pp.

Messieh, S.N., 1986. The Enigma of Gulf Herring Recruitment. NAFO Scientific Council Research Document, 86/103, Serial No. N1230. Northwest Atlantic Fisheries Organization, Nova Scotia.

Methven, D.A. and J.F. Piatt, 1991. Seasonal abundance and vertical distribution of capelin (*Mallotus villosus*) in relation to water temperature at a coastal site off eastern Newfoundland. ICES Journal of Marine Science, 48:187–193.

Michalsen, K., G. Ottersen and O. Nakken, 1998. Growth of North-east Arctic cod (*Gadus morhua* L.) in relation to ambient temperature. ICES Journal of Marine Science, 55:863–877.

Mikhail, M.Y. and H.E. Welch, 1989. Biology of Greenland cod, *Gadus ogac*, at Saqvaqjuac, northwest coast of Hudson Bay. Environmental Biology of Fishes, 26(1):49–62.

Montevecchi, W.A., 1993. Birds as indicators of change in marine prey stocks. In: R.W. Furness and J.J.D. Greenwood (eds.). Birds as Monitors of Environmental Change, pp. 217–266. Chapman & Hall.

Montevecchi, W.A. and R.A. Myers, 1995. Prey harvests of seabirds reflect pelagic fish and squid abundance on multiple spatial and temporal scales. Marine Ecology Progress Series, 117:1–9.

Montevecchi, W.A. and R.A. Myers, 1996. Dietary changes of seabirds indicate shifts in pelagic food webs. Sarsia, 80:313–322.

Montevecchi, W.A. and R.A. Myers, 1997. Centurial and decadal oceanographic influences on changes in northern gannet populations and diets in the north-west Atlantic: implications for climate change. ICES Journal of Marine Science, 54:608–614.

Moore, S.E., 2000. Variability of cetacean distribution and habitat selection in the Alaskan Arctic, autumn 1982–91. Arctic, 53:448–460.

Moore, S.E., D.P. DeMaster and P.K. Dayton, 2000. Cetacean habitat selection in the Alaskan Arctic during summer and autumn. Arctic, 53:432–447.

Morin, B., C. Hudon and F. Whoriskey, 1991. Seasonal distribution, abundance, and life-history traits of Greenland cod, *Gadus ogac*, at Wemindji, eastern James Bay. Canadian Journal of Zoology, 69(12):3061–3070.

Morison, J.H., 1991. Seasonal fluctuations in the West Spitsbergen Current estimated from bottom pressure measurements. Journal of Geophysical Research, 96 (C10):18381–18395.

Morison, J.H. and J.D. Smith, 1981. Seasonal variations in the upper Arctic Ocean as observed at T-3. Geophysical Research Letters, 8:753–756.

Morison, J., M. Steele and R. Anderson, 1998. Hydrography of the upper Arctic Ocean measured from the nuclear submarine U.S.S. Pargo. Deep-Sea Research I, 45:15–38.

Morison, J., K. Aagaard and M. Steele, 2000. Recent environmental changes in the Arctic: a review. Arctic, 53:359–371.

Muench, R.D., M.G. McPhee, C.A. Paulson and J.H. Morison, 1992. Winter oceanographic conditions in the Fram Strait – Yermak Plateau region. Journal of Geophysical Research, 97:3469–3483.

Müller-Navarra, D., 1995a. Evidence that a highly unsaturated fatty acid limits *Daphnia* growth in nature. Archives of Hydrobiology, 132:297–307.

Müller-Navarra, D., 1995b. Biochemical versus mineral limitation in *Daphnia*. Limnology and Oceanography, 40:1209–1214.

Munk, W. and C. Wunsch, 1998. Abyssal recipes II: energetics of tidal and wind mixing. Deep-Sea Research I, 45:1977–2010.

Murray, J.L., 1998. Ecological characteristics of the Arctic. In: AMAP Assessment Report: Arctic Pollution Issues, pp.117–140. Arctic Monitoring and Assessment Programme, Oslo.

Muus, B.J. and J.G. Nielsen, 1999. Sea Fish. Scandinavian Fishing Year Book. Hedehusene, Denmark, 340 pp.

Myers, R.A. and N.G. Cadigan, 1993. Density-dependent juvenile mortality in marine demersal fish. Canadian Journal of Fisheries and Aquatic Sciences, 50:1576–1590.

Myklestad, S. and A. Haug, 1972. Production of carbohydrates by the marine diatom *Chaetoceros affinis* var. *willei* (Gran)Hustedt. I. Effect of the concentration of nutrients in the culture medium. Journal of Experimental Marine Biology and Ecology, 9:125–136.

Mysak, L.A., 2001. Patterns of Arctic circulation. Science, 293:1269–1270.

Mysak, L.A. and D.K. Manak, 1989. Arctic sea-ice extent and anomalies, 1953–1984. Atmosphere-Ocean, 27(2):376–405.

Mysak, L.A., D.K. Manak and R.F. Marsden, 1990. Sea-ice anomalies observed in the Greenland and Labrador Seas during 1901–1984 and their relation to an interdecadal Arctic climate cycle. Climate Dynamics, 5:111–133.

Mysak, L.A., R.G. Ingram, J. Wang and A. van der Baaren, 1996. The anomalous sea-ice extent in Hudson Bay, Baffin Bay and the Labrador Sea during three simultaneous NAO and ENSO episodes. Atmosphere-Ocean, 34:313–343.

Naganuma, T., T. Inoue and S. Uye, 1997. Photoreactivation of UV-induced damage to embryos of a planktonic copepod. Journal of Plankton Research, 19:783–787.

Nakashima, B.S., 1996. The relationship between oceanographic conditions in the 1990s and changes in spawning behaviour, growth and early life history of capelin (*Mallotus villosus*). NAFO Scientific Council Studies, 24:55–68.

Nakken, O. and A. Raknes, 1987. The distribution and growth of Northeast Arctic cod in relation to bottom temperatures in the Barents Sea, 1978–1984. Fisheries Research, 5:243–252.

Napp, J.M. and G.L. Hunt Jr., 2001. Anomalous conditions in the south-eastern Bering Sea, 1997: Linkages among climate, weather, ocean, and biology. Fisheries Oceanography, 10:61–68.

Napp, J.M., A.W. Kendall Jr. and J.D. Schumacher, 2000. A synthesis of biological and physical processes affecting the feeding environment of larval walleye pollock (*Theragra chalcogramma*) in the eastern Bering Sea. Fisheries Oceanography, 9:147–162.

Napp, J.M., C.T. Baier, R.D. Brodeur, K.O. Coyle, N. Shiga, and K. Mier, 2002. Interannual and decadal variability in zooplankton communities of the southeastern Bering Sea shelf. Deep-Sea Research II, 49:5991–6008.

Narayanan, S., J. Carscadden, J.B. Dempson, M.F. O'Connell, S. Prinsenberg, D.G. Reddin and N. Shackell, 1995. Marine climate off Newfoundland and its influence on Atlantic salmon (*Salmo salar*) and capelin (*Mallotus villosus*). In: R.J. Beamish (ed.). Climate Change and Northern Fish Populations. Canadian Special Publication of Fisheries and Aquatic Sciences, 121:461–474.

Neale, P.J., R.F. Davis and J.J. Cullen, 1998. Interactive effects of ozone depletion and vertical mixing on photosynthesis of Antarctic phytoplankton. Nature, 392:585–589.

Neale, P.J., J.J. Fritz and R.F. Davis, 2001. Effects of UV on photosynthesis of Antarctic phytoplankton: models and their application to coastal and pelagic assemblages. Revista Chilena de Historia Natural, 74:283–292.

Nesis, K.N., 1960. Fluctuations of the Barents Sea bottom fauna under the effect of the hydrobiological regime variation (at the Kola Meridian transect). Sovetskie rybokhozyaystvennye issledovania v moryakh Evropeyskogo Severa. Moscow, pp. 129–137. (In Russian)

Nesterova, V.N., 1990. Plankton biomass along the drift route of cod larvae. Pinro, Murmansk, 64pp. (In Russian)

Niebauer, H.J., V. Alexander and S.M. Henrichs, 1995. A time-series study of the spring bloom at the Bering Sea ice edge. I. Physical processes, chlorophyll and nutrient chemistry. Continental Shelf Research, 15:1859–1877.

Nielsen, T.G. and B. Hansen, 1995. Plankton community structure and carbon cycling on the western coast of Greenland during and after the sedimentation of a diatom bloom. Marine Ecology Progress Series, 125:239–257.

Nihoul, J.C.J., P. Adam, P. Brasseur, E. Deleersnijder, S. Djenidi and J. Haus, 1993. Three-dimensional general circulation model of the northern Bering Sea's summer ecohydrodynamics. Continental Shelf Research, 13:509–542.

Nilsen, J.E.O., Y. Gao, H. Drange, T. Furevik and M. Bentsen, 2003. Simulated North Atlantic-Nordic Seas water mass exchanges in an isopycnic coordinate OGCM. Geophysical Research Letters, 30(10):1536, doi: 10.1029/2002GL016597.

Nilssen, K.T., 1995. Seasonal distribution, condition and feeding habits of Barents Sea harp seals (Phoca groenlandica). In: A.S. Blix, L. Walloe and O. Ulltang (eds.). Whales, Seals, Fish and Man, pp. 241–254. Elsevier Science.

Nilssen, K.T., T. Haug., V. Potelov and Y.K. Timoshenko, 1995. Feeding habits of harp seals (Phoca groenlandica) during early summer and autumn in the northern Barents Sea. Polar Biology, 15(7):485–493.

Nisbet, E.G., 1990. The end of the ice age. Canadian Journal of Earth Science, 27:148–157.

Noji, T., L.A. Miller, I. Skjelvan, E. Falck, K.Y. Børsheim, F. Rey, J. Urban-Rich and T. Johannessen, 2000. Constraints on carbon drawdown and export in the Greenland Sea. In: P. Schäfer, W. Ritzrau, M. Schlüter and J. Thiede (eds.). The Northern North Atlantic: A Changing Environment, pp. 39–52. Springer, Berlin.

Norderhaug, M., E. Bruun and G.U. Mollen, 1977. Barentshavet sjofuglressurser. Norsk Polarinstitutt Meddelelser, 104:1–119. (In Norwegian with English summary)

NRC, 1996. The Bering Sea Ecosystem. National Research Council. National Academy Press, Washington, D.C., 320pp.

Ohtani, K. and T. Azumaya, 1995. Influence of interannual changes in ocean conditions on the abundance of walleye pollock in the eastern Bering Sea. In: R.J. Beamish (ed.). Climate Change and Northern Fish Populations. Canadian Special Publication of Fisheries and Aquatic Sciences, 121: 87–95.

Olsen, A., T. Johannessen and F. Rey, 2003. On the nature of the factors that control spring bloom development at the entrance to the Barents Sea and their interannual variability. Sarsia, 88(6):379–393.

Olsson, K. and L.G. Anderson, 1997. Input and biogeochemical transformation of dissolved carbon in the Siberian shelf seas. Continental Shelf Research, 17:819–833.

Orensanz, J.M.L., J. Armstrong, D. Armstrong and R. Hilborn, 1998. Crustacean resources are vulnerable to serial depletion – the multifaceted decline of crab and shrimp fisheries in the Greater Gulf of Alaska. Reviews in Fish Biology and Fisheries. 8(2):117–176.

Orr, D.C. and W.R. Bowering, 1997. A multivariate analysis of food and feeding trends among Greenland halibut (Reinhardtius hippoglossoides) sampled in Davis Strait, during 1986. ICES Journal of Marine Science, 54(5):819–829.

Østerhus, S. and T. Gammelsrød, 1999. The abyss of the Nordic Seas is warming. Journal of Climate, 12(11):3297–3304.

Orvik, K.A. and Ø. Skagseth, 2003. The impact of the wind stress curl in the North Atlantic on the Atlantic inflow to the Norwegian Sea toward the Arctic. Geophysical Research Letters, 30(17), doi: 10.1029/2003GL017932.

Ottersen, G. 1996. Environmental Impact on Variability in Recruitment, Larval Growth and Distribution of Arcto–Norwegian Cod. Geophysical Institute, University of Bergen, 136pp.

Ottersen, G. and H. Loeng, 2000. Covariability in early growth and year-class strength of Barents Sea cod, haddock and herring: The environmental link. ICES Journal of Marine Science, 57:339–348.

Ottersen, G. and N.C. Stenseth, 2001. Atlantic climate governs oceanographic and ecological variability in the Barents Sea. Limnology and Oceanography, 46:1774–1780.

Ottersen, G. and S. Sundby, 1995. Effects of temperature, wind and spawning stock biomass on recruitment of Arcto-Norwegian cod. Fisheries Oceanography, 4:278–292.

Ottersen, G., K. Michalsen and O. Nakken, 1998. Ambient temperature and distribution of north-east Arctic cod. ICES Journal of Marine Science, 55:67–85.

Ottersen, G., K. Helle and B. Bogstad, 2002. Do abiotic mechanisms determine interannual variability in length-at-age of juvenile Arcto-Norwegian cod? Canadian Journal of Fisheries and Aquatic Sciences, 59:57–65.

Ottersen, G., H. Loeng, B. Adlandsvik and R. Ingvaldsen, 2003. Temperature variability in the Northeast Atlantic. ICES Marine Science Symposia, 219:86–94.

Outridge, P.M. and R.E.A. Stewart, 1999. Stock discrimination of Atlantic walrus (Odobenus rosmarus rosmarus) in the eastern Canadian Arctic using lead isotope and element signatures in teeth. Canadian Journal of Fisheries and Aquatic Sciences, 56:105–112.

Overland, J.E., M.C. Spillane, H.E. Hurlburt and A.J. Wallcraft, 1994. A numerical study of the circulation of the Bering Sea Basin and exchange with the North Pacific Ocean. Journal of Physical Oceanography, 24:736–758.

Overland, J.E., J.M. Adams and N.A. Bond, 1999a. Decadal variability of the Aleutian Low and its relation to high latitude circulation. Journal of Climate, 12:1542–1548.

Overland, J.E., S.A. Salo, L.H. Kantha and A.C. Clayson, 1999b. Thermal stratification and mixing on the Bering Sea shelf. In: T.R. Loughlin and K. Ohtani (eds.). Dynamics of the Bering Sea, pp. 129–146. University of Alaska Fairbanks.

Øyan, H. and T. Anker-Nilssen, 1996. Allocation of growth in food-stressed Atlantic Puffin chicks. The Auk, 113(4):830–841.

Öztürk, M., E. Steinnes and E. Sakshaug, 2002. Iron speciation in the Trondheim Fjord from the perspective of iron limitation for phytoplankton. Estuarine Coastal and Shelf Science, 55:197–212.

Paasche, E., 1960. Phytoplankton distribution in the Norwegian Sea in June, 1954, related to hydrography and compared with primary production data. Fiskeridirektoratets Skrifter, Serie Havundersokelser, 12(2):1–77.

Paeth, H., A. Hense, R. Glowienka-Hense, R. Voss and U. Cubasch, 1999. The North Atlantic Oscillation as an indicator for greenhouse-gas induced regional climate change. Climate Dynamics, 15:953–960.

Parkinson, C.L., 1992. Spatial patterns of increases and decreases in the length of the sea ice season in the north polar region, 1979–1986. Journal of Geophysical Research, 97(C9):14377–14388.

Parkinson, C.L., D.J. Cavalieri, P. Gloersen, H.J. Zwally and J.C. Comiso, 1999. Arctic sea ice extents, areas, and trends, 1978–1996. Journal of Geophysical Research, 104(C9):20837–20856.

Paull, C.K., W. Ussler III and W.P. Dillon, 1991. Is the extent of glaciation limited by marine gas hydrates? Geophysical Research Letters, 18:432–434.

Pedersen, S.A. and P. Kanneworff, 1995. Fish on the West Greenland shrimp grounds, 1988–1992. ICES Journal of Marine Science, 52(2):165–182.

Pedersen, S.A. and J.C. Rice, 2002. Dynamics of fish larvae, zooplankton, and hydrographical characteristics in the West Greenland large marine ecosystem 1950–1984. In: K. Sherman and H.R. Skjoldal (eds.). Large Marine Ecosystems of the North Atlantic, pp. 151–193. Elsevier.

Pedersen, S.A. and E.L.B. Smidt, 2000. Zooplankton distribution and abundance in West Greenland waters, 1950–1984. Journal of Northwest Atlantic Fishery Science, 26:45–102.

Perrin, W.F., B. Wursig and J.G.M. Thewissen, 2002. Encyclopedia of Marine Mammals. Academic Press, 1414pp.

Petersen, M.R., W.W. Larned and D.C. Douglas, 1999. At-sea distribution of spectacled eiders: a 120-year-old mystery solved. The Auk, 116:1009–1020.

Petrie, B., J.W. Loder, S. Akenhead and J. Lazier, 1992. Temperature and salinity variability on the eastern Newfoundland shelf: the residual field, Atmosphere-Ocean, 30:120–139.

Piatt, J.F. and T.I. van Pelt, 1997. Mass-mortality of Guillemots (Uria aalge) in the Gulf of Alaska in 1993. Marine Pollution Bulletin, 34:656–662.

Pitcher, K.W. and D.C. McAllister, 1981. Movements and haulout behaviour of radio-tagged harbor seals, Phoca vitulina. The Canadian Field-Naturalist, 95:292–297.

Planque, B. and S.D. Batten, 2000. Calanus finmarchicus in the North Atlantic: the year of Calanus in the context of interdecadal change. ICES Journal of Marine Science, 57(6):1528–1535.

Planque, B. and T. Fredou, 1999. Temperature and the recruitment of Atlantic cod (Gadus morhua). Canadian Journal of Fisheries and Aquatic Sciences, 56:2069–2077.

Plante, A.J. and M.T. Arts, 1998. Photosynthate production in laboratory cultures (UV conditioned and unconditioned) of Cryptomonas erosa under simulated doses of UV radiation. Aquatic Ecology, 32:297–312.

Pocklington, R., 1987. Arctic rivers and their discharges. Geologisch-Paläontologisches Institut, University of Hamburg, SCOPE/UNEP Sonderband, Heft 64:261–268.

Poltermann, M., H. Hop and S. Falk-Petersen, 2000. Life under Arctic sea ice – reproduction strategies of two sympagic (ice-associated) amphipod species, Gammarus wilkitzskii and Apherusa glacialis. Marine Biology, 136:913–920.

Polyakov, I.V. and M.A. Johnson, 2000. Arctic decadal and interdecadal variability. Geophysical Research Letters, 27(24):4097–4100.

Polyakov, I.V., G.V. Alekseev, R.V. Bekryaev, U.S. Bhatt, R. Colony, M.A. Johnson, V.P. Karklin, D. Walsh and A.V. Yulin, 2003. Long-term ice variability in Arctic marginal seas, Journal of Climate, 16 (12):2078–2085.

Pomeroy, L.R., S.A. Macko, P.H. Ostrom and J. Dunphy, 1990. The microbial food web in the Arctic seawater: concentration of dissolved free amino acids and bacterial abundance and activity in the Arctic Ocean and in Resolute Passage. Marine Ecology Progress Series, 61(1–2):31–40.

Pommeranz, T., 1974. Resistance of plaice eggs to mechanical stress and light. In: J.H.S. Blaxter (ed.). The Early Life History of Fish, pp. 397–416. Springer-Verlag.

Ponomarenko, V.P., 1968. Some data on the distribution and migrations of Polar cod in the seas of the Soviet Arctic. Rapports et Process-verbaux des Réuniuns, Conseil Permanent International pour l'Exploration de la Mer, 158:131–135.

Proshutinsky, A.Y. and M.A. Johnson, 1997. Two circulation regimes of the wind-driven Arctic Ocean. Journal of Geophysical Research, 102(C6):12493–12514.

Proshutinsky, A., V. Pavlov and R.H. Bourke, 2001. Sea level rise in the Arctic Ocean. Geophysical Research Letters, 28:2237–2240.

Proshutinsky, A., R.H. Bourke and F.A. McLaughlin, 2002. The role of the Beaufort Gyre in Arctic climate variability: Seasonal to decadal climate scales. Geophysical Research Letters, 29, doi:10.1029/2002GL015847.

Quadfasel, D., B. Rudels and K. Kurz, 1988. Outflow of dense water from a Svalbard fjord into the Fram Strait. Deep-Sea Research A, 35:1143–1150.

Quinn, T.J. II and H.J. Niebauer, 1995. Relation of eastern Bering Sea walleye pollock (Theragra chalcogramma) recruitment to environmental and oceanographic variables. In: R.J. Beamish (ed.). Climate Change and Northern Fish Populations. Canadian Special Publication of Fisheries and Aquatic Sciences, 121: 497–507.

Rachold, V., A. Alabyan, H.-W. Hubberton, V.N. Korotaev and A.A. Zaitsev, 1996. Sediment transport to the Laptev Sea - hydrology and geochemistry of the Lena River. Polar Research, 15:183–196.

Rachold, V., H. Eicken, V.V. Gordeev, M.N. Grigoriev, H.-W. Hubberten, A.P. Lisitzin, V.P. Shevchenko, L. Schirrmeister, 2004. Modern terrigenous organic carbon input to the Arctic Ocean. In: R. Stein and R.W. Macdonald (eds.). The Organic Carbon Cycle in the Arctic Ocean, pp. 33–54. Springer-Verlag.

Rahmstorf, S., 1999. Shifting seas in the greenhouse? Nature, 399:523–524.

Rahmstorf, S., 2003. Thermohaline circulation: The current climate. Nature, 421(6924):699.

Rahmstorf, S. and M.H. England, 1997. Influence of southern hemisphere winds on North Atlantic Deep Water flow. Journal of Physical Oceanography, 27:2040–2054.

Rahmstorf, S. and A. Ganopolski, 1999. Long-term global warming scenarios computed with an efficient coupled climate model. Climatic Change, 43:353–367.

Rainuzzo, J.R., K.I. Reitan and Y. Olsen, 1997. The significance of lipids at early stages of marine fish: a review. Aquaculture, 155:103–115.

Ramsay, M.A. and I. Stirling, 1988. Reproductive biology and ecology of female polar bears (Ursus maritimus). Journal of Zoology, London, 214:601–634.

Reeves, R.R., 1990. An overview of the distribution, exploitation and conservation status of belugas, worldwide. In: J. Prescott and M. Gauquelin (eds.). For the Future of the Beluga: Proceedings of the International Forum for the Future of the Beluga, pp. 47–58. University of Quebec Press.

Reeves, R.R., 1998. Distribution, abundance and biology of ringed seals (Phoca hispida): an overview. In: M.P. Heide-Jørgensen and C. Lydersen (eds.). Ringed Seals in the North Atlantic. The North Atlantic Marine Mammal Commission, Scientific Publications 2:9–45.

Reinfelder, J.R., A.M.L. Kraepiel and F.M.M. Morel, 2000. Unicellular C_4 photosynthesis in a marine diatom. Nature, 407:996–999.

Reitan, K.I., J.R. Rainuzzo, G. Øie and Y. Olsen, 1997. A review of the nutritional effects of algae in marine fish larvae. Aquaculture, 155:207–221.

Revelle, R., 1983. Methane hydrates in continental slope sediment and increasing CO_2. In: Changing Climate, Report of the Carbon Dioxide Assessment Committee, National Research Council, pp. 252–261. National Academy Press, Washington, D.C.

Rice, D.W., 1998. Marine Mammals of the World - Systematics and Distribution. Society for Marine Mammalogy Special Publication No. 4. Allen Press, Kansas, 231pp.

Rich, J., M. Gosselin, E. Sherr, B. Sherr and D.L. Kirchman, 1997. High bacterial production, uptake and concentrations of dissolved organic matter in the Central Arctic Ocean. Deep-Sea Research II, 44:1645–1663.

Ridgway, S.H. and R.J. Harrison, 1981–1999. Handbook of Marine Mammals. Vols. 1–6. Academic Press.

Riebesell, U., I. Zondervan, B. Rost, P.D. Tortell, R.E. Zeebe and F.M.M. Morel, 2000. Reduced calcification of marine plankton in response to increased atmospheric CO_2. Nature, 407:364–367.

Rigor, I.G., R.L. Colony and S. Martin, 2000. Variations in surface air temperature observations in the Arctic, 1979–97. Journal of Climate, 13:896–914.

Rigor, I.G., J.M. Wallace and R.L. Colony, 2002. Response of sea ice to the Arctic Oscillation. Journal of Climate, 15:2648–2663.

Roach, A.T., K. Aagaard, C.H. Pease, S.A. Salo, T. Weingartner, V. Pavlov and M. Kulakov, 1995. Direct measurements of transport and water properties through the Bering Strait. Journal of Geophysical Research, 100(C9):18443–18458.

Rødseth, T., 1998. Models for Multispecies Management. Springer-Verlag, 246pp.

Rodway, M.S., J.W. Chardine and W.A. Montevecchi, 1998. Intra-colony variation in breeding performance of Atlantic puffins. Colonial Waterbirds, 21(2):171–184.

Rogers, J.C., 1985. Atmospheric circulation changes associated with the warming over the northern North Atlantic in the 1920s. Journal of Applied Meteorology, 24:1303–1310.

Romanovsky, N.N., H.-W. Hubberten, A.V. Gavrilov, V.E. Tumskoy, G.S. Tipenko, M.N. Grigoriev and C. Siegert, 2000. Thermokarst and land–ocean interactions, Laptev Sea region, Russia. Permafrost and Periglacial Processes, 11:137–152.

Rosenkranz, G.E., A.V. Tyler and G.H. Kruse, 2001. Effects of water temperature and wind on year-class success of Tanner crabs in Bristol Bay, Alaska. Fisheries Oceanography, 10:1–12.

Rothrock, D.A., Y. Yu and G.A. Maykut, 1999. Thinning of the Arctic sea-ice cover. Geophysical Research Letters, 26(23):3469–3472.

Rowe, S., I.L. Jones, J.W. Chardine, R.D. Elliot and B.G. Veitch, 2000. Recent changes in the winter diet of murres (Uria spp.) in coastal Newfoundland waters. Canadian Journal of Zoology, 78(3):495–500.

Roy, R.N., L.N. Roy, K.M. Vogel, C. Porter-Moore, T. Pearson, C.E. Good, F.J. Millero and D.M. Campbell, 1993. The dissociation constants of carbonic acid in seawater at salinities 5 to 45 and temperatures 0 to 45 °C. Marine Chemistry, 44:249–267.

Rudels, B., 1986. The theta-S relations in the northern seas: Implications for the deep circulation. Polar Research, 4:133–159.

Rudels, B., 1989. The formation of polar surface water, the ice export and the exchanges through the Fram Strait. Progress in Oceanography, 22:205–248.

Rudels, B., 1993. High latitude ocean convection. In: D.B. Stone and S.K. Runcorn (eds.). Flow and Creep in the Solar System: Observations, Modeling and Theory, pp. 323–356. Kluwer Academic Press.

Rudels, B., E.P. Jones, L.G. Anderson and G. Kattner, 1994. On the intermediate depth waters of the Arctic Ocean. In: O.M. Johannessen, R.D. Muench, and J.E. Overland (eds.). The Polar Oceans and their Role in Shaping the Global Environment. Geophysical Monograph Series, 85:33–46. American Geophysical Union, Washington D.C.

Rudels, B., H.J. Friedrich and D. Quadfasel, 1999. The Arctic Circumpolar Boundary Current. Deep-Sea Research II, 46:1023–1062.

Rudels, B., R. Meyer, E. Fahrbach, V.V. Ivanov, S. Østerhus, D. Quadfasel, U. Schauer, V. Tverberg and R.A. Woodgate, 2000. Water mass distribution in Fram Strait and over the Yermak Plateau in summer 1997. Annales Geophysicae, 18:687–705.

Rudels, B., E. Fahrbach, J. Meincke, G. Budéus and P. Eriksson, 2002. The East Greenland Current and its contribution to the Denmark Strait overflow. ICES Journal of Marine Science, 59:1133–1154.

Rysgaard, S., M. Kuhl, R.N. Glud and J.W. Hansen, 2001. Biomass, production and horizontal patchiness of sea ice algae in a high-Arctic fjord (Young Sound, NE Greenland). Marine Ecology Progress Series, 223:15–26.

Saemundsson B., 1937. Fiskirannsoknir. Andvari, vol. 62 Reykjavik. (In Icelandic)

Sætersdal, G. and H. Loeng, 1987. Ecological adaption of reproduction in Northeast Arctic cod. Fisheries Research, 5:253–270.

Sakshaug, E., 1972. Phytoplankton investigations in Trondheimsfjord, 1963–1966. Det Kongelige Norske Videnskabernes Selskabs Skrifter, 1972(1):1–56.

Sakshaug, E., 2003. Primary and secondary production in Arctic Seas. In: R. Stein and R.W. Macdonald (eds.). The Organic Carbon Cycle in the Arctic Ocean, pp. 57–81. Springer, Berlin.

Sakshaug, E. and D. Slagstad, 1992. Sea ice and wind: Effects on primary productivity in the Barents Sea. Atmosphere-Ocean, 30:579–591.

Sakshaug, E. and J.J. Walsh, 2000. Marine biology: Biomass, productivity distributions and their variability in the Barents and Bering Seas. In: M. Nuttall and T.V. Callaghan (eds.). The Arctic: Environment, People, Policy, pp. 163–196. Harwood, Amsterdam.

Sakshaug, E., S. Myklestad, K. Andresen, E.N. Hegseth and L. Jorgensen, 1981. Phytoplankton off the More coast in 1975–1976: distribution, species composition, chemical composition and conditions for growth. In: R. Saetre and M. Mork (eds.). The Norwegian Coastal Current, pp. 688–711. University of Bergen.

Sakshaug, E., K. Andresen, S. Myklestad and Y. Olsen, 1983. Nutrient status of phytoplankton communities in Norwegian waters (marine, brackish, and fresh) as revealed by their chemical composition. Journal of Plankton Research, 5:175–196.

Sakshaug, E., A. Bjørge, B. Gulliksen, H. Loeng and F. Mehlum (eds.), 1992. Okosystem Barentshavet. Universitetsforlaget (2. utgivelse), Oslo, 304pp.

Sakshaug, E., A. Bjørge, B. Gulliksen, H. Loeng and F. Mehlum, 1994. Structure, biomass distribution, and energetics of the pelagic ecosystem in the Barents Sea: a synopsis. Polar Biology, 14:405–411.

Sakshaug, E., F. Rey and D. Slagstad, 1995. Wind forcing of marine primary production in the northern atmospheric low-pressure belt. In: H.R. Skjoldal, C. Hopkins, K.E. Erikstad and H.P. Leinaas (eds.). Ecology of Fjords and Coastal Waters, pp. 15–25. Elsevier Science.

Salo, H.M., T.M. Aaltonen, S.E. Markkula and E.I. Jokinen, 1998. Ultraviolet B irradiation modulates the immune system of fish (*Rutilus rutilus*, Cyprinidae). I. Phagocytes. Journal of Photochemistry and Photobiology, 67:433–437.

Salo, H.M., E.I. Jokinen, S.E. Markkula and T.M. Aaltonen, 2000a. Ultraviolet B irradiation modulates the immune system of fish (*Rutilus rutilus*, Cyprinidae). II. Blood. Journal of Photochemistry and Photobiology, 71:65–70.

Salo, H.M., E.I. Jokinen, S.E. Markkula ,T.M. Aaltonen and H.T. Penttila, 2000b. Comparative effects of UVA and UVB irradiation on the immune system of fish. Journal of Photochemistry and Photobiology B, 56:154–162.

Sambrotto, R.N. and J.J. Goering, 1983. Interannual variability of phytoplankton and zooplankton production on the southeast Bering Shelf. In: W.S. Wooster (ed.). From Year to Year: Interannual Variability of the Environment and Fisheries of the Gulf of Alaska and the Eastern Bering Sea, pp. 161–177. Washington Sea Grant Program, University of Washington, Seattle.

Sargent, J.R., L.A. McEvoy and J.G. Bell, 1997. Requirements, presentation and sources of polyunsaturated fatty acids in marine fish larval feeds. Aquaculture, 155:117–127.

Saunders, P.M., 2001. The dense northern overflows. In: G. Siedler, J. Church and J. Gould (eds.). Ocean Circulation and Climate: Observing and Modelling the Global Ocean, pp. 401–417. Academic Press.

Schlosser, P., J.L. Bullister, R. Fine, W.J. Jenkins, R. Key, J. Lupton, W. Roether and W.M. Smethie Jr., 2001. Transformation and age of water masses. In: G. Siedler, J. Church and J. Gould (eds.). Ocean Circulation and Climate; Observing and Modelling the Global Ocean, pp. 431–452. Academic Press.

Schmitz, W.J. Jr. and M.S. McCartney, 1993. On the North Atlantic circulation. Reviews of Geophysics, 31:29–50.

Schneider, D.C., 1990. Seabirds and fronts: a brief overview. Polar Research, 8:17–21.

Schopka, S.A., 1994. Fluctuations in the cod stock off Iceland during the twentieth century in relation to changes in the fisheries and environment. ICES Marine Science Symposia, 198:175–193.

Schreer, J.F. and K.M. Kovacs, 1997. Allometry of diving capacity in air-breathing vertebrates. Canadian Journal of Zoology, 75(3):339–358.

Schreiber, E.A., 2001. Climate and weather effects on seabirds. In: E.A. Schreiber and J. Burger (eds.). Biology of Marine Birds. CRC Marine Biology Series, Vol. 1, pp. 179–215. CRC Press.

Schreiber, R.W. and E.A. Schreiber, 1984. Central Pacific seabirds and the El Niño Southern Oscillation: 1982–1983 perspectives. Science, 225:713–716.

Scott, C.L., S. Falk-Petersen, J.R. Sargent, H. Hop, O.J. Lønne and M. Poltermann, 1999. Lipids and trophic interactions of ice fauna and pelagic zooplankton in the marginal ice zone of the Barents Sea. Polar Biology, 21:65–70.

Scott, C.L., S. Kwasniewski, S. Falk-Petersen, and J.R. Sargent, 2000. Lipids and life strategies of *Calanus finmarchicus*, *Calanus glacialis* and *Calanus hyperboreus* in late autumn, Kongsfjorden, Svalbard. Polar Biology, 23(7):510–516.

Scott, J.S., 1982. Depth, temperature and salinity preferences of common fishes of the Scotian Shelf. Journal of Northwest Atlantic Fishery Science, 3:29–39.

Seager, R., D.S. Battisti, J. Yin, N. Gordon, N. Naik, A.C. Clement and M.A. Cane, 2002. Is the Gulf Stream responsible for Europe's mild winters? Quarterly Journal of the Royal Meteorological Society, 128:2563–2586.

SEARCH, 2001. SEARCH: Study of Environmental Arctic Change, Science Plan. Polar Science Center, Applied Physics Laboratory, University of Washington, Seattle, 89pp.

Semiletov, I.P., 1999a. Aquatic sources and sinks of CO_2 and CH_4 in the polar regions. Journal of Atmospheric Sciences, 56:286–306.

Semiletov, I.P. 1999b. Destruction of the coastal permafrost ground as an important factor in biogeochemistry of the Arctic Shelf waters. Doklady Earth Sciences, 368: 679–682.

Serreze, M.C., J.A. Maslanik, T.A. Scambos, F. Fetterer, J. Stroeve, K. Knowles, C. Fowler, S. Drobot, R.G. Barry and T.M. Haran, 2003. A record minimum in Arctic sea ice extent and area in 2002. Geophysical Research Letters, 30(3), doi: 10.1029/2002GL016406.

Shackell, N.L., J.E. Carscadden and D.S. Miller, 1994. Migration of pre-spawning capelin (*Mallotus villosus*) as related to temperature on the northern Grand Bank, Newfoundland. ICES Journal of Marine Science, 51:107–114.

Sherr, E.B., B.F. Sherr and L. Fessenden, 1997. Heterotrophic protists in the Central Arctic Ocean. Deep-Sea Research II, 44:1665–1673.

Shimada, K., E.C. Carmack, K. Hatakeyama and T. Takizawa, 2001. Varieties of shallow temperature maximum waters in the western Canadian Basin of the Arctic Ocean. Geophysical Research Letters, 28:3441–3444.

Shindell, D., 2003. Whither Arctic climate? Science, 299:215–216.

Shindell, D.T., D. Rind and P. Lonergan, 1998. Increased polar stratospheric ozone losses and delayed eventual recovery owing to increasing greenhouse-gas concentrations. Nature, 392:589–592.

Shindell, D.T., R.L. Miller, G.A. Schmidt and L. Pandolfo, 1999. Simulation of recent northern climate trends by greenhouse-gas forcing. Nature, 399:452–455.

Shiomoto, A., 1999. Effect of nutrients on phytoplankton size in the Bering Sea Basin. In: T.R. Loughlin and K. Ohtani (eds.). Dynamics of the Bering Sea, pp. 323–340. University of Alaska Fairbanks.

Shuert, P.G. and J.J. Walsh, 1993. A coupled physical-biological model of the Bering-Chukchi Seas. Continental Shelf Research, 13:543–573.

Shugart, H.H., 1990. Using ecosystem models to assess potential consequences of global climatic change. Trends in Ecology and Evolution, 5:303–307.

Shustov, A.P., 1965. The food of ribbon seals in the Bering Sea. Transactions of the Pacific Research Institute of Fisheries and Oceanography, 59:178–183.

Simonsen, K. and P.M. Haugan, 1996. Heat budgets of the Arctic Mediterranean and sea surface heat flux parameterizations for the Nordic Seas. Journal of Geophysical Research, 101(C3):6553–6576.

Sinclair, A. and L. Currie, 1994. Timing of cod migration into and out of the Gulf of St. Lawrence based on commercial fisheries, 1986–1993. Department of Fisheries and Oceans Canadian Science Advisory Secretariat Research Document, 1994/047, 18pp.

Sinclair, M., 1988. Marine Populations: An Essay on Population Regulation and Speciation. University of Washington Press, Seattle, 252pp.

Sirenko, B.I. and V.M. Koltun, 1992. Characteristics of benthic biocenoses of the Chukchi and Bering Seas. In: P.A. Nagel (ed.). Results of the Third Joint US-USSR Bering and Chukchi Seas Expedition (BERPAC), Summer 1988, pp 251–258. U.S. Fish and Wildlife Service, Washington, D.C.

Skjoldal, H.R. and F. Rey, 1989. Pelagic production and variability of the Barents Sea ecosystem. In: K. Sherman and L.M. Alexander (eds.). Biomass Yields and Geography of Large Marine Ecosystems, pp. 241–286. AAAS Selected Symposium 111. Westview Press, Colorado.

Skjoldal, H.R., A. Hassel, F. Rey and H. Loeng, 1987. Spring phytoplankton development and zooplankton reproduction in the central Barents Sea in the period 1979–84. In: H. Loeng (ed.). Proceedings of the Third Soviet-Norwegian Symposium, Murmansk, 1986, pp. 59–89. Institute of Marine Research, Bergen, Norway.

Slagstad, D. and K. Støle-Hansen, 1991. Dynamics of plankton growth in the Barents Sea: model studies. Polar Research, 10:173–186.

Slotte, A., 1998. Spawning migration of Norwegian spring spawning herring (*Clupea harengus* L.) in relation to population structure. Ph.D Thesis, University of Bergen.

Smed, J., 1949. The increase in the sea temperature in northern waters during recent years. ICES Rapports et Procès-Verbaux des Réunions, 125:21–25.

Smith, D.M., 1998. Recent increase in the length of the melt season of perennial Arctic sea ice. Geophysical Research Letters, 25(5):655–658.

Smith, S.L., 1991. Growth, development and distribution of the euphausiids *Thysanoessa raschi* (M. Sars) and *Thysanoessa inermis* (Krøyer) in the south-eastern Bering Sea. Polar Research, 10:461–478.

Smith, S.L. and S.B. Schnack-Schiel, 1990. Polar zooplankton. In: W.O. Smith Jr. (ed.). Polar Oceanography Part B, pp. 527–598. Academic Press.

Smith, T.G. 1980. Polar bear predation of ringed and bearded seals in the land-fast sea ice habitat. Canadian Journal of Zoology, 58(12):2201–2209.

Smith, T.G. and I. Stirling, 1975. The breeding habitat of the ringed seal (*Phoca hispida*): The birth lair and associated structures. Canadian Journal of Zoology, 53:1297–1305.

Smith, T.G., M.O. Hammill and G. Taugbøl, 1991. Review of the developmental, behavioural and physiological adaptations of the ringed seal, *Phoca hispida*, to life in the arctic winter. Arctic, 44:124–131.

Smith, W.O. Jr., L.A. Codispoti, D.M. Nelson, T. Manley, E.J. Buskey, H.J. Niebauer and G.F. Cota, 1991. Importance of *Phaeocystis* blooms in the high-latitude carbon cycle. Nature, 352:514–516.

Smith, W.O. Jr., M. Gosselin, L. Legendre, D. Wallace, K. Daly and G. Kattner, 1997. New production in the Northeast Water Polynya: 1993. Journal of Marine Systems, 10:199–209.

Springer, A.M., 1992. A review: walleye pollock in the North Pacific – how much difference do they really make? Fisheries Oceanography, 1:80–96.

Springer, A.M., C. McRoy and M.V. Flint, 1996. The Bering Sea green belt: shelf edge processes and ecosystem production. Fisheries Oceanography, 5:205–223.

Stabeno, P.J. and J.E. Overland, 2001. The Bering Sea shifts toward an earlier spring transition. EOS, Transactions, American Geophysical Union, 82(29):317–321.

Stabeno, P.J., N.A. Bond, N.B. Kachel, S.A. Salo and J.D. Schumacher, 2001. On the temporal variability of the physical environment over the south-eastern Bering Sea. Fisheries Oceanography, 10(1):81–98.

Staehelin, J., N.R.P. Harris, C. Appenzeller and J. Eberhard, 2001. Ozone trends: a review. Reviews of Geophysics, 39:231–290.

Steeger, H.U., J.F. Freitag, S. Michl, M. Wiemer and R.J. Paul, 2001. Effects of UV-B radiation on embryonic, larval and juvenile stages of North Sea plaice (*Pleuronectes platessa*) under simulated ozone-hole conditions. Helgoland Marine Research, 55:56–66.

Steele, M. and T. Boyd, 1998. Retreat of the cold halocline layer in the Arctic Ocean. Journal of Geophysical Research, 103(C5):10419–10435.

Steemann-Nielsen, E., 1935. The production of phytoplankton at the Faroe Isles, Iceland, East Greenland and in the waters around. Meddelelser fra Kommissionen for Danmarks Fiskeri - Og Havunder Soegelser, 3(1):1–93.

Steffensen, J.F., P.G. Bushnell and H. Schurmann, 1994. Oxygen consumption in four species of teleosts from Greenland: no evidence of metabolic cold adaptation. Polar Biology, 14(1):49–54.

Stenseth, N.C., A. Mysterud, G. Ottersen, J.W. Hurrell, K.-S. Chan and M. Lima, 2002. Ecological effects of climate fluctuations. Science, 297:1292–1296.

Steward, G.F., D.C. Smith and F. Azam, 1996. Abundance and production of bacteria and viruses in the Bering and Chukchi Seas. Marine Ecology Progress Series, 131(1–3):287–300.

Stewart, P.L., P. Pocklington and R.A. Cunjak, 1985. Distribution, abundance and diversity of benthic macroinvertebrates on the Canadian continental shelf and slope of Southern Davis Strait and Ungava Bay. Arctic, 38:281–291.

Stirling, I., 1997. The importance of polynyas, ice edges and leads to marine mammals and birds. Journal of Marine Systems, 10:9–21.

Stirling, I. and W.R. Archibald, 1977. Aspects of predation of seals by polar bears. Journal of the Fisheries Research Board of Canada, 34:1126–1129.

Stirling, I. and A.E. Derocher, 1993. Possible impacts of climatic warming on polar bears. Arctic, 46(3):240–245.

Stirling, I. and N.A. Øritsland, 1995. Relationships between estimates of ringed seal (*Phoca hispida*) and polar bear (*Ursus maritimus*) populations in the Canadian Arctic. Canadian Journal of Fisheries and Aquatic Sciences, 52:2594–2612.

Stirling, I., H. Cleator and T.G. Smith, 1981. Marine mammals. In: I. Stirling and H. Cleator (eds.). Polynyas in the Canadian Arctic, pp. 45–58. Canadian Wildlife Service Occasional Paper No. 45.

Stirling, I., D. Andriashek and W. Calvert, 1993. Habitat preferences of polar bears in the western Canadian Arctic in late winter and spring. Polar Record, 29:13–24.

Stirling, I., N.J. Lunn and J. Iacozza, 1999. Long-term trends in the population ecology of polar bears in western Hudson Bay in relation to climate change. Arctic, 52:294–306.

Stockwell, D.A., T.E. Whitledge, S.I. Zeeman, K.O. Coyle, J.M. Napp, R.D. Brodeur, A.I. Pinchuk and G.L. Hunt Jr., 2001. Anomalous conditions in the south-eastern Bering Sea, 1997: nutrients, phytoplankton and zooplankton. Fisheries Oceanography, 10:99–116.

Stoker, S.W., 1978. Benthic Invertebrate Macrofauna of the Eastern Continental Shelf of the Bering and Chukchi Seas. Ph.D. Thesis, University of Alaska, Fairbanks, 155pp.+ app.

Stoker, S.W., 1981. Benthic invertebrate macrofauna of the eastern Bering/Chukchi continental shelf. In: W. Hood and J.A. Calder (eds.). The Eastern Bering Sea Shelf: Oceanography and Resources. Vol. 2, pp. 1069–1090. University of Washington Press, Seattle.

Strass, V.H. and E.-M. Nöthig, 1996. Seasonal shifts in ice edge phytoplankton blooms in the Barents Sea related to the water column stability. Polar Biology, 16:409–422.

Strass, V.H., E. Fahrbach, U. Schauer and L. Sellmann, 1993. Formation of Denmark Strait Overflow Water by mixing in the East Greenland Current. Journal of Geophysical Research, 98(C4):6907–6919.

Sufke, L., D. Piepenburg and C.F. von Dorrien, 1998. Body size, sex ratio and diet composition of *Arctogadus glacialis* (Peters, 1874) (Pisces: Gadidae) in the Northeast Water Polynya (Greenland). Polar Biology, 20(5):357–363.

Sugimoto, T. and K. Tadokoro, 1997. Interannual-interdecadal variations in zooplankton biomass, chlorophyll concentration and physical environment in the subarctic Pacific and Bering Sea. Fisheries Oceanography, 6(2):74–93.

Sugimoto, T. and K. Tadokoro, 1998. Interdecadal variations of plankton biomass and physical environment in the North Pacific. Fisheries Oceanography, 7(3–4):289–299.

Sukhanova, I.N., H.J. Semina and M.V. Venttsel, 1999. Spatial distribution and temporal variability of phytoplankton in the Bering Sea. In: T.R. Loughlin and K. Ohtani (eds.). Dynamics of the Bering Sea, pp. 453–483. University of Alaska Fairbanks.

Sültemeyer, D., 1998. Carbonic anhydrase in eukaryotic algae: characterization, regulation, and possible function during photosynthesis. Canadian Journal of Botany, 76:962–972.

Sundby, S., H. Bjørke, A.V. Soldal and S. Olsen, 1989. Mortality rates during the early life stages and year class strength of northeast Arctic cod (*Gadus morhua* L.). ICES Rapports et Procès-Verbaux des Réunions, 191:351–358.

Suzuki, Y., S. Kudoh and M. Takahashi, 1997. Photosynthetic and respiratory characteristics of an Arctic ice algal community living in low light and low temperature conditions. Journal of Marine Systems, 11:111–121.

Swift, J.H. and K. Aagaard, 1981. Seasonal transitions and water mass formation in the Iceland and Greenland seas. Deep-Sea Research A, 28(10):1107–1129.

Taalas, P., J. Kaurola, A. Kylling, D. Shindell, R. Sausen, M. Dameris, V. Grewe, J. Herman, J. Damski and B. Steil, 2000. The impact of greenhouse gases and halogenated species on future solar UV radiation doses. Geophysical Research Letters, 27:1127–1130.

Taniguchi, A., 1999. Differences in the structure of the lower trophic levels of pelagic ecosystems in the eastern and western subarctic Pacific. Progress in Oceanography, 43:289–315.

Tartarotti, B., W. Cravero and H.E. Zagarese, 2000. Biological weighting function for the mortality of *Boeckella gracilipes* (Copepoda, Crustacea) derived from experiments with natural solar radiation. Photochemistry and Photobiology, 72:314–319.

Taylor, A.E., 1991. Marine transgression, shoreline emergence: evidence in seabed and terrestrial ground temperatures of changing relative sea levels, Arctic Canada. Journal of Geophysical Research, 96:6893–6909.

Thompson, P.M. and J. Ollason, 2001. Lagged effects of ocean climate change on fulmar population dynamics. Nature, 413:417–420.

Thomson, B.E., 1986. Is the impact of UV-B radiation on marine zooplankton of any significance? In: J.G. Titus (ed.). Effects of Changes in Stratospheric Ozone and Global Climate, Vol. 2, pp. 203–209. US Environmental Protection Agency and United Nations Environment Programme.

Thordardóttir, T., 1977. Primary production in North Icelandic waters in relation to recent climate change. In: Polar Oceans. Proceedings of the Oceanographic Congress, pp. 655–665.

Thordardóttir, T., 1984. Primary production north of Iceland in relation to water masses in May-June 1970–1989. ICES CM 1984/L:20. International Council for the Exploration of the Sea, Copenhagen, 17pp.

Thorson, G., 1950. Reproductive and larval ecology of marine bottom invertebrates. Biological Reviews, 25:1–45.

Tibbo, S.N. and R.D. Humphreys, 1966. An occurrence of capelin (*Mallotus villosus*) in the Bay of Fundy. Journal of the Fisheries Research Board of Canada, 23:463–467.

Timokhov, L. and F. Tanis (eds.), 1998. Arctic Climatology Project, 1998. Environmental Working Group Joint U.S.-Russian Atlas of the Arctic Ocean - Summer Period. Environmental Research Institute of Michigan in association with the National Snow and Ice Data Center, Ann Arbor, Michigan. CD-ROM.

Toggweiler, J.R. and B. Samuels, 1995. Effect of Drake Passage on the global thermohaline circulation. Deep-Sea Research I, 42:477–500.

Tomirdiaro, S.V., 1990. Lessovo-ledovaya formatsia Vostochnoi Sibiri v pozdnem pleistocene i golocene. Nauka, Moscow.

Toresen, R. and O.J. Østvedt, 2000. Variation in abundance of Norwegian spring-spawning herring (*Clupea harengus*, Clupeidae) throughout the 20th century and the influence of climatic fluctuations. Fish and Fisheries, 1:231–256.

Trenberth, K.E., 1990. Recent observed interdecadal climate changes in the Northern Hemisphere. Bulletin of the American Meteorological Society, 71:988–993.

Turrell, W.R., B. Hansen, S. Hughes and S. Østerhus, 2003. Hydrographic variability during the decade of the 1990s in the Northeast Atlantic and southern Norwegian Sea. ICES Marine Science Symposia, 219:111–120.

Tyler, A.V. and G.H. Kruse, 1998. A comparison of year-class variability of red king crabs and Tanner crabs in the eastern Bering Sea. Memoirs of the Faculty of Fisheries, Hokkaido University, 45(1):90–95, Japan.

Tynan, C.T. and D.P. DeMaster, 1997. Observations and predictions of Arctic climatic change: potential effects on marine mammals. Arctic, 50(4):308–322.

Tziperman, E., 2000. Proximity of the present-day thermohaline circulation to an instability threshold. Journal of Physical Oceanography, 30:90–104.

Ulbrich, U. and M. Christoph, 1999. A shift of the NAO and increasing storm track activity over Europe due to anthropogenic greenhouse gas forcing. Climate Dynamics, 15:551–559.

Usachev, P.I., 1961. Phytoplankton of the North Pole. Fisheries Research Board of Canada Translation Series, 1285.

Vader, W., R.T. Barrett, K.E. Erikstad and K.B. Strann, 1990. Differential responses of common and thick-billed murres to a crash in the capelin stock in the southern Barents Sea. Studies in Avian Biology, 14:175–180.

Valcarcel, A., G. Guerrero and M.C. Maggese, 1994. Hertwig effect caused by UV-irradiation of sperm of the catfish, *Rhamdia sapo* (Pisces, Pimelodidae), and its photoreactivation. Aquaculture, 128:21–28.

Verduin, J. and D. Quadfasel, 1999. Long-term temperature and salinity trends in the central Greenland Sea. In: E. Jansen (ed.). European Subpolar Ocean Programme (ESOP) II, Final Scientific Report, A1:1–11. University of Bergen.

Vibe, C., 1967. Arctic Animals in Relation to Climate Fluctuations. Danish Zoogeographical Investigations in Greenland. Meddelelser om Gronland, 170(5):227pp.

Vilhjálmsson, H., 1994. The Icelandic capelin stock. Capelin (*Mallotus villosus* Muller) in the Iceland-Greenland-Jan Mayen area. Journal of the Marine Research Institute, Reykjavik, 13:1–281.

Vilhjálmsson, H., 2002. Capelin (*Mallotus villosus*) in the Iceland–East Greenland–Jan Mayen ecosystem. ICES Journal of Marine Science, 59:870–883.

Vinje, T., 2001. Anomalies and trends of sea-ice extent and atmospheric circulation in the Nordic Seas during the period 1864–1998. Journal of Climate, 14:255–267.

Vinje, T., N. Nordlund and Å. Kvambekk, 1998. Monitoring ice thickness in Fram Strait, Journal of Geophysical Research, 103(C5):10437–10449.

Vinnikov, K.Y., A. Robock, R.J. Stouffer, J.E. Walsh, C.L. Parkinson, D.J. Cavalieri, J.F.B. Mitchell, D. Garrett and V.F. Zakharov, 1999. Global warming and Northern Hemisphere sea ice extent. Science, 286(5446):1934–1937.

Visser, M.E., A.J. van Noordwijk, J.M. Tinbergen and C.M. Lessells, 1998. Warmer springs lead to mistimed reproduction in great tits (*Parus major*). Proceedings of the Royal Society of London: Biological Sciences, 265:1867–1870.

Wadhams, P., 2000. Ice in the Ocean. Gordon and Breach Science Publishers, Amsterdam, 351pp.

Walsh, J.E., W.L. Chapman and T.L. Shy, 1996. Recent decreases of sea level pressure in the central Arctic. Journal of Climate, 9:480–488.

Walsh, J.J., 1989. Arctic carbon sinks: present and future. Global Biogeochemical Cycles, 3:393–411.

Walsh, J.J. and D.A. Dieterle, 1994. CO_2 cycling in the coastal ocean. I. A numerical analysis of the southeastern Bering Sea with applications to the Chukchi Sea and the northern Gulf of Mexico. Progress in Oceanography, 34:335–392.

Walsh, J.J., C.P. McRoy, L.K. Coachman, J.J. Goering, J.J. Nihoul, T.E. Whitledge, T.H. Blackburn, P.L. Parker, C.D. Wirick, P.G. Stuert, J.M. Grebmeier, A.M. Springer, R.D. Tripp, D.A. Hansell, S. Djenidi, E. Deleersnijder, K. Henriksen, B.A. Lund, P. Andersen, F.E. Muller-Karger and K. Dean, 1989. Carbon and nitrogen cycling within the Bering/Chukchi Seas: Source regions for organic matter effecting AOU demands of the Arctic Ocean. Progress in Oceanography, 22:277–359.

Walters, C. and B. Ward, 1998. Is solar radiation responsible for declines in marine survival rates of anadromous salmonids that rear in small streams? Canadian Journal of Fisheries and Aquatic Sciences, 55:2533–2538.

Walters, V., 1955. Fishes of western Arctic America and eastern Arctic Siberia: Taxonomy and zoogeography. Bulletin of the American Museum of Natural History, 106:255–368.

Wang, J., L.A. Mysak and R.G. Ingram, 1994. Interannual variability of sea-ice cover in Hudson Bay, Baffin Bay and the Labrador Sea. Atmosphere-Ocean, 32:421–447.

Wang, K.S. and T. Chai, 1994. Reduction in omega-3 fatty acids by UV-B irradiation in microalgae. Journal of Applied Phycology, 6:415–421.

Wassmann, P., 1991. Dynamics of primary production and sedimentation in shallow fjords and polls of western Norway. In: H. Barnes, A.D. Ansell and R.N. Gibson (eds.). Oceanography and Marine Biology: An Annual Review, 29:87–154.

Wassmann, P., T. Ratkova, I. Andreassen, M. Vernet, C. Pedersen and F. Rey, 1999. Spring bloom development in the marginal ice zone and the Central Barents Sea. Pubblicazioni della Stazione Zoologica di Napoli I: Marine Ecology, 20:321–346.

Watanuki, Y., F. Mehlum and A. Takahashi, 2001. Water temperature sampling by foraging Brunnich's Guillemots with bird-borne data loggers. Journal of Avian Biology, 32:189–193.

Wathne, J.A., T. Haug and C. Lydersen, 2000. Prey preference and niche overlap of ringed seals *Phoca hispida* and harp seals *P. groenlandica* in the Barents Sea. Marine Ecology Progress Series, 194:233–239.

Watson, A.J., M.-J. Messias, E. Fogelqvist, K.A. Van Scoy, T. Johannessen, K.I.C. Oliver, D.P. Stevens, F. Rey, T. Tanhua, K.A. Olsson, F. Carse, K. Simonsen, J.R. Ledwell, E. Jansen, D.J. Cooper, J.A. Kruepke and E. Guilyardi, 1999. Mixing and convection in the Greenland Sea from a tracer release experiment. Nature, 401:902–905.

Weimerskirch, H., 2001. Seabird demography and its relationship with the marine environment. In: E.A. Schreiber and J. Burger (eds.). Biology of Marine Birds. CRC Marine Biology Series Volume 1, pp. 115–135. CRC Press.

Welch, H.E., M.A. Bergmann, T.D. Siferd, K.A. Martin, M.F. Curtis, R.E. Crawford, R.J. Conover and H. Hop, 1992. Energy flow through the marine ecosystem of the Lancaster Sound region, Arctic Canada. Arctic, 45(4):343–357.

Welch, H.E., R.E. Crawford and H. Hop, 1993. Occurrence of Arctic cod (*Boreogadus saida*) schools and their vulnerability to predation in the Canadian High Arctic. Arctic, 46(4):331–339.

Weslawski, J.M., J. Wiktor, M. Zajaczkowski and S. Swerpel, 1993. Intertidal zone of Svalbard 1. Macroorganism distributution and biomass. Polar Biology, 13:73–79.

Weslawski, J.M., M. Ryg, T.G. Smith and N.A. Øritsland, 1994. Diet of ringed seals (*Phoca hispida*) in a fjord of west Svalbard. Arctic, 47:109–114.

Weslawski, J.M., M. Zajaczkowski, J. Wiktor and M. Szymelfenig, 1997. Intertidal zone of Svalbard. 3. Litoral of a subarctic oceanic island: Bjornoya. Polar Biology, 18:45–52.

Weslawski, J.M., M. Szymelfenig, M. Zajaczkowski and A. Keck, 1999. Influence of salinity and suspended matter on benthos of an Arctic tidal flat. ICES Journal of Marine Science, 56(S):194–202.

Wespestad, V.G., L.W. Fritz, W.J. Ingraham and B.A. Megrey, 2000. On relationships between cannibalism, climate variability, physical transport, and recruitment success of Bering Sea walleye pollock (*Theragra chalcogramma*). ICES Journal of Marine Science, 57(2):272–278.

Wheeler, P.A., J.M. Watkins and R.L. Hansing, 1997. Nutrients, organic carbon and organic nitrogen in the upper water column of the Arctic Ocean: implications for the sources of dissolved organic carbon. Deep-Sea Research II, 44:1571–1592.

Whitledge, T.E. and V.A. Luchin, 1999. Summary of chemical distributions and dynamics in the Bering Sea. In: T.R. Loughlin and K. Ohtani (eds.). Dynamics of the Bering Sea, pp. 217–249. University of Alaska Sea.

Widell, K., S. Østerhus and T. Gammelsrød. 2003. Sea ice velocity in the Fram Strait monitored by moored instruments. Geophysical Research Letters, 30(19):1982, doi:10.1029/2003GL018119.

Wiig, Ø., 1987. The grey seal *Halichoerus grypus* (Fabricius) in Finmark, Norway. Fiskeridirektoratets Skrifter, Serie Havundersokelser, 18:241–246.

Wiig, Ø., I. Gjertz, D. Griffiths and C. Lydersen, 1993. Diving patterns of an Atlantic walrus *Odobenus rosmarus rosmarus* near Svalbard. Polar Biology, 13:71–72.

Wiig, Ø., I. Gjertz and D. Griffiths, 1996. Migration of walrus (*Odobenus rosmarus*) in the Svalbard and Franz Josef Land area. Journal of Zoology, London, 238:769–784.

Wiig, Ø., A.E. Derocher and S.E. Belikov, 1999. Ringed seal (*Phoca hispida*) breeding in the drifting pack ice of the Barents Sea. Marine Mammal Science, 15:595–598.

Wilderbuer, T.K., A.B. Hollowed, W.J. Ingraham Jr., P.D. Spencer, M.E. Conners, N.A. Bond and G.E. Walters. 2002. Flatfish recruitment response to decadal climate variability and ocean conditions in the eastern Bering Sea. Progress in Oceanography, 55:235–247.

Williams, E.H. and T.J. Quinn II, 2000. Pacific herring, *Clupea pallasi*, recruitment in the Bering Sea and north-east Pacific Ocean, II: relationships to environmental variables and implications for forecasting. Fisheries Oceanography, 9:300–315.

Williamson, C.E., S.L. Metzgar, P.A. Lovera and R.E. Moeller, 1997. Solar ultraviolet radiation and the spawning habitat of yellow perch, *Perca flavescens*. Ecological Applications, 7:1017–1023.

Williamson, C.E., B.R. Hargreaves, P.S. Orr and P.A. Lovera, 1999. Does UV play a role in changes in predation and zooplankton community structure in acidified lakes? Limnology and Oceanography, 44:774–783.

Williamson, C.E., P.J. Neale, G. Grad, H.J. De Lange and B.R. Hargreaves, 2001. Beneficial and detrimental effects of UV on aquatic organisms: Implications of spectral variation. Ecological Applications, 11:1843–1857.

Wourms, J.P., 1991. Reproduction and development of *Sebastes* in the context of the evolution of piscine viviparity. Environmental Biology of Fishes, 30:111–126.

Wyllie-Echeverria, T., 1995. Sea-ice conditions and the distribution of walleye pollock (*Theragra chalcogramma*) on the Bering and Chukchi shelf. In: R.J. Beamish (ed.). Climate Change and Northern Fish Populations. Canadian Special Publication of Fisheries and Aquatic Sciences, 121:131–136.

Wyllie-Echeverria, T. and W.S. Wooster, 1998. Year-to-year variations in Bering Sea ice cover and some consequences for fish distributions. Fisheries Oceanography, 7:159–170.

Yang, M.S. and P.A. Livingston, 1988. Diet and daily ration of Greenland halibut, *Reinhardtius hippoglossoides*, in the eastern Bering Sea. Fishery Bulletin, 86:675–690.

Yaragina, N.A. and C.T. Marshall, 2000. Trophic influences on interannual and seasonal variation in the liver condition index of Northeast Arctic cod (*Gadus morhua*). ICES Journal of Marine Science, 57:42–55.

Zagarese, H.E. and C.E. Williamson, 2000. Impact of solar UV radiation on zooplankton and fish. In: S.J. de Mora, S. Demers and M. Vernet (eds.). The Effects of UV Radiation in the Marine Environment, pp. 279–309. Cambridge University Press.

Zagarese, H.E. and C.E. Williamson, 2001. The implications of solar UV radiation exposure for fish and fisheries. Fish and Fisheries, 2:250–260.

Zakharov, V.F., 1994. On the character of cause and effect relationships between sea ice and thermal conditions of the atmosphere. Berichte zur Polarforschung, 144:33–43.

Zebdi, A. and J.S. Collie, 1995. Effect of climate on herring (*Clupea pallasi*) population dynamics in the Northeast Pacific Ocean. In: R.J. Beamish (ed.). Climate Change and Northern Fish Populations. Canadian Special Publication of Fisheries and Aquatic Sciences 121:277–290.

Zepp, R.G., T.V. Callaghan and D.J. Erickson III, 2003. Interactive effects of ozone depletion and climate change on biogeochemical cycles. Photochemical and Photobiological Sciences, 2:51–61.

Zenkevich, L., 1963. Biology of the Seas of the USSR. George Allen and Unwin, London, 955pp.

Zhang, J., D. Rothrock and M. Steele, 2000. Recent changes in arctic sea ice: The interplay between ice dynamics and thermodynamics. Journal of Climate, 13(17):3099–3114.

Zhang, Y.X. and E.C. Hunke, 2001. Recent Arctic change simulated with a coupled ice-ocean model. Journal of Geophysical Research, 106(C3):4369–4390.

Zheng, J. and G.H. Kruse, 2000. Recruitment patterns of Alaskan crabs in relation to decadal shifts in climate and physical oceanography. ICES Journal of Marine Science, 57:438–451.

Zimov, S.A., Y.V. Voropaev, I.P. Semiletov, S.P. Davidov, S.F. Prosiannikov, F.S. Chapin III, M.C. Chapin, S. Trumbore and S. Tyler, 1997. North Siberian lakes: A methane source fueled by Pleistocene carbon. Science, 277:800–802.

Zondervan I., B. Rost and U. Riebesell, 2002. Effect of CO_2 concentration on the PIC/POC ratio in the coccolithophore *Emiliania huxleyi* grown under light-limiting conditions and different day lengths. Journal of Experimental Marine Biology and Ecology, 272:55–70.

Zyryanov, S.V. and A.V. Vorontsov, 1999. Observations on the Atlantic walrus, *Odobenus rosmarus rosmarus*, in spring of 1997 in the Kara Sea and southeastern part of the Barents Sea. Zoologichesky Zhurnal, 78:1254–1256.

Chapter 10

Principles of Conserving the Arctic's Biodiversity

Lead Author
Michael B. Usher

Contributing Authors
Terry V. Callaghan, Grant Gilchrist, Bill Heal, Glenn P. Juday, Harald Loeng, Magdalena A. K. Muir, Pål Prestrud

Contents

Summary

Biodiversity is fundamental to the livelihoods of arctic people. The Convention on Biological Diversity defines biodiversity as "the variability among living organisms from all sources including, *inter alia*, terrestrial, marine and other aquatic ecosystems and the ecological complexes of which they are a part: this includes diversity within species, between species and of ecosystems". A changing climate can affect all three levels of biodiversity. There are many predicted influences of climate change on the Arctic's biodiversity. These include (1) changes in the distribution ranges of species and habitats; (2) changes in the extent of many habitats; (3) changes in the abundance of species; (4) changes in genetic diversity; (5) changes in the behavior of migratory species; (6) some non-native species becoming problematic; and (7) the need for protected areas to be managed in different ways.

What should be done now before the anticipated changes occur? First, it is important to document the current state of the Arctic's biodiversity. Local inventories of biodiversity have generally not been carried out, although the inventory for Svalbard is a striking exception, recording both native and non-native species in both terrestrial and marine environments. Such work requires trained ecologists, trained taxonomists, circumpolar knowledge, and a focus on all three levels of biodiversity (genes, species, and ecosystems). Second, the changes that take place in the Arctic's biodiversity need to be identified. Management of the Arctic's biodiversity, in the sea, in freshwater, or on land, must work with ecological succession and not against it. Considerably more effort needs to be invested in developing predictive models that can explore changes in biodiversity under the various scenarios of climate change. Third, changes in the Arctic's biodiversity need to be recorded and the data shared. In a situation where so much uncertainty surrounds the conservation of biodiversity, knowledge of what has changed, where it has changed, and how quickly it has changed becomes critically important. Monitoring biodiversity, especially on a circumpolar basis, must be a goal, and a circumpolar monitoring network needs to be fully implemented so as to determine how the state of biodiversity is changing, what the drivers of change are, and how other species and people respond. Finally, new approaches to managing the Arctic's biodiversity need to be explored. Best practice guidelines should be available on a circumpolar basis. The Circumpolar Protected Area Network needs to be completed and reviewed so as to ensure that it does actually cover the full range of the Arctic's present biodiversity. An assessment needs to be made, for each protected area, of the likely effects of climate change, and in the light of this assessment the methods of management for the future. This poses questions of resources and priorities, but it is essential that the Arctic's ecosystems continue to exist and function in a way that such services as photosynthesis, decomposition, and purification of pollutants continue in a sustained manner.

10.1. Introduction

Arctic peoples obtain their primary source of food and many of the materials used in clothing and building from the plant and animal species indigenous to the Arctic. These species range from mammals, fish, and birds, to berries and trees. However, the relationship between arctic people and those arctic species upon which they depend is not simple since each of these species is in turn dependent on a range of other arctic species and on the ecological processes operating within the arctic ecosystems. The biological diversity of the arctic environment is thus fundamental to the livelihoods of arctic peoples. Relevant information from indigenous peoples on arctic biodiversity is given in Chapter 3.

The two major processes operating within ecosystems are photosynthesis and decomposition. Photosynthesis is the biochemical process whereby radiant energy from the sun is used to synthesize carbohydrates from carbon dioxide (CO_2) and water in the presence of chlorophyll. The energy fixed during photosynthesis is transferred from the primary producers through successive trophic levels by feeding and thus starts the food chains and food webs upon which all animal life depends. The organisms responsible are green plants – predominantly vascular plants in the terrestrial environment and algae in the freshwater and marine environments. The vascular plants, which include all flowering plants and ferns, are relatively well-known taxonomically and feature in most books on the terrestrial environment of the Arctic (e.g., CAFF, 2001; Sage, 1986). The non-vascular plants such as the mosses, liverworts, and lichens are less well-known taxonomically. The algae are taxonomically the least well-known plants of the Arctic; most are single-celled and many have a wide distribution range within the northern hemisphere (John et al., 2002).

Decomposition is the process whereby dead plant and animal material is broken down into simple organic and inorganic compounds, with a consequent release of energy. The carbon is released back into the atmosphere as CO_2, and nutrients such as nitrogen, phosphorus, and potassium are available for recycling. Decomposition processes are undertaken by an enormous range of organisms in soils and in aquatic sediments. These organisms include bacteria, actinomycetes, fungi, protozoa, nematodes, worms (especially enchytraeid worms), mollusks, insects (especially collembolans – springtails, and dipteran larvae – flies), crustaceans, and arachnids (especially mites). Species richness can be outstanding, with up to 2000 species within a square meter of grassland soil (Usher, 1996), which has led to soil being considered "the poor man's tropical rain forest". However, many of the species in soils and sediments are unknown and undescribed, and their roles in the soil or sediment ecosystem, and in the processes of decomposition, are very poorly understood. This means that, within a changing climate, there are many questions about the decomposition process that need addressing (Heal, 1999).

In addition to photosynthesis and decomposition, there are many other important ecological processes operating within arctic ecosystems, for example: pollutant breakdown and detoxification, the purification of water, the release of oxygen, and nutrient recycling.

The major ecosystems of the Arctic, and their biological diversity, are addressed in detail in other chapters: Chapter 7 addresses the terrestrial environment, focusing on the tundra and polar desert ecosystems; Chapter 8 addresses freshwater ecosystems; and Chapter 9 addresses marine systems. This chapter focuses on the principles of conserving biodiversity, exploring the ecosystems, species, and genes in the Arctic, and the threats faced in a changing environment. The starting point for this discussion is the Convention on Biological Diversity (SCBD, 2000), which states that its objectives are "... the conservation of biological diversity, the sustainable use of its components and the fair and equitable sharing of the benefits arising out of the utilization of genetic resources..." (Article 1).

The Convention on Biological Diversity defines "biological diversity" (often shortened to "biodiversity") as "the variability among living organisms from all sources including, *inter alia*, terrestrial, marine and other aquatic ecosystems and the ecological complexes of which they are a part; this includes diversity within species, between species and of ecosystems" (Article 2). This definition clearly implies that biodiversity, and both its conservation and utilization, must be viewed at three levels – the level of the gene, the species, and the ecosystem (or habitat).

A changing climate can affect all three levels of biodiversity, and Chapters 7, 8, and 9 address such issues. What the human population wishes to conserve, and the way that biodiversity conservation is practiced, will also be affected by a changing climate. The exploitation of the Arctic's biodiversity resources, and the potential for their exploitation in the future, will equally be affected, and these topics are considered in greater detail in Chapter 11 (wildlife conservation and management), 12 (hunting, herding, fishing, and gathering by indigenous peoples), 13 (marine fisheries and aquaculture), and 14 (forests and agriculture). The present chapter deals primarily with the first two tenets of the Convention on Biological Diversity, namely the conservation of biodiversity and its sustainable use by the peoples of the Arctic. The first involves all aspects of the Arctic's wildlife, from the smallest organisms (viruses, bacteria, and protozoa) to the largest plants and animals. The latter invokes the concept of stewardship: stewardship implies a sustainable form of management rather than the preservation of species and ecosystems without change. Climate change will result in changes in the productivity of ecosystems through photosynthesis and changes in the rates of decomposition. The balance between these two major processes will, to a large extent, determine the future nature of the arctic environment, the resources upon which arctic peoples (and

visitors) depend, and whether the Arctic exacerbates climate change by releasing greater quantities of CO_2 to the atmosphere or helps to control climate change by acting as a sink for atmospheric CO_2. Biodiversity is therefore both affected by and affects climate change.

The first two lines of approach to biodiversity conservation are often the development of lists of species and habitats to be given special protection (usually through legislation, and often on the basis of "Red Lists"), and the designation of protected areas where biodiversity conservation takes primacy over other forms of water and land use. By 1990, there had been significant achievements (IUCN, 1991) in establishing protected areas in the Arctic. Norway, Sweden, and Finland, for example, all had strict nature reserves (IUCN management category I), national parks (IUCN category II), and/or other nature reserves (IUCN category IV) within their arctic territories. In fact, the extent of these protected arctic areas is often considerably greater than the extent of equivalent protected areas further south. In Sweden, four of the seven national parks located within the Arctic are each larger than the total area of the 18 national parks south of the Arctic (Table 10.1). One of these, Abisko, has as its aim "to preserve the high Nordic mountain landscape in its natural state" (Naturvårdverket, 1988), while others have similar aims to preserve landscapes and, by implication, the biodiversity that those landscapes contain.

In 1996, Conservation of Arctic Flora and Fauna (CAFF) developed a strategy, with an associated action plan, for a Circumpolar Protected Area Network. CAFF's efforts, jointly with other international governmental and non-governmental organizations, and a range of local, regional, and national bodies, led to the establishment of nearly 400 protected areas (each greater than 10 km^2) by 2000 (CAFF, 2001). The selection process for potential protected areas has been studied in many parts of the world and tends to be a blend of science (what is most desirable to protect?) and pragmatism (what is possible to

Table 10.1. Details of the 25 national parks in Sweden (Hanneberg and Löfgren, 1998).

	Extent (ha)
National parks in the Arctic	
Abisko	7700
Muddus	50350
Padjelanta	198400
Pieljekaise	15340
Sarek	197000
Stora Sjöfallet	127800
Vadvetjåkka	2630
Average extent of the seven national parks in the Arctic	85603
Average extent of the 18 national parks south of the Arctic (range: 27 to 10440 ha)	2446

protect?), and is not always easy even with a broad measure of agreement between the public and government.

Internationally, many criteria have been proposed as a basis for selecting sites for protection and designation as nature reserves and national parks. These were reviewed by Margules and Usher (1981) and further developed by Usher (1986) into a "popularity poll" reflecting frequency of use (Table 10.2). Whereas some of these may be inappropriate in the Arctic (being better suited to the more fragmented environments of industrialized regions), the criteria ranked highest are all relevant to northern ecosystems. However, one of the difficulties of applying such criteria is that comprehensive habitat and species inventories may not exist, and so it is impossible to make meaningful comparisons or to determine the areas of greatest priority (see also section 10.5.1).

Table 10.2 essentially contains "scientific" criteria, without the socio-economic criteria necessary for assessing existing and proposed land and water use plans. So although it might be possible to establish a

Table 10.2. Criteria used for selecting areas of land or water for protection and designation as nature reserves and national parks (Usher, 1986). The 26 criteria are ranked from those most frequently used (1) to those used only once in the review of 17 published sets of criteria (19=).

Rank	Criterion or criteria
1=	• Diversity of species
	• Diversity of habitats
3=	• Naturalness
	• Rarity of species
	• Rarity of habitats
6	• Extent of habitat
7	• Threat of human interference or disturbance
8=	• Educational value
	• Representativeness
	• Amenity value for local human population
11	• Scientific value
12	• Recorded history
13=	• Size of population of species of conservation concern
	• Typicalness
15=	• Uniqueness
	• Potential value
	• Ecological fragility
	• Position in an ecological or geographical unit
19=	• Archaeological interest
	• Availability
	• Importance for migratory wildfowl
	• Ease of management
	• Replaceability
	• Silvicultural gene bank
	• Successional stage
	• Wildlife reservoir potential

range of assessments based on the scientific criteria listed in Table 10.2, to gain a balanced perspective it is also important to establish plans for land and water use and the aspirations of people living in the area. Local economies depend on the biodiversity resources, and in balancing the various criteria it is essential to include long-term views and to ensure that demands for short-term gains do not predominate. The possible effects of climate change on biodiversity also need to be included in assessments, especially effects that will be experienced over the longer term.

Thus, there are many competing pressures on the ability of an individual, group, organization, or nation to conserve the biodiversity of the Arctic. These can be summarized in six points:

- all species native to the Arctic need to be conserved (i.e., neither allowed to become extinct nor driven to extinction by human activity);
- the genetic variation within these species needs to be conserved because this ensures the greatest chance of species' adaptation to a changing environment and hence their long-term survival under a changing climate;
- the habitats of these species need to be conserved because each species is an integral part of a food web, being itself dependent on a set of other species and with a different set of species dependent upon it;
- human populations living in the Arctic are themselves an integral part of the Arctic's biodiversity and food webs;
- non-native species and external human pressures may present challenges to arctic genes, species, and ecosystems, and hence risk assessments are a vital factor in managing new pressures on the arctic environment; and
- protected areas are not a universal panacea for the conservation of the Arctic's biodiversity, but should be viewed as land and water managed for the primacy of nature in a broader geographical area where other land- and water-uses may have primacy.

CAFF (2002a) summarized these points by stating that "The overall goal of Arctic nature conservation is to ensure that Arctic ecosystems and their biodiversity remain viable and vigorous for generations to come and, therefore, able to sustain human socio-economic and cultural needs". Balancing this duality of biodiversity conservation and sustainable use, CAFF developed five strategic issues (see Table 10.3) and these are further developed throughout this chapter.

This chapter comprises four main sections. Section 10.2 provides a brief introduction to the special features of arctic ecosystems and arctic species that justify conservation attention; possible threats to the Arctic's biodiversity are considered in section 10.3. Eight issues are then addressed in relation to the management and conservation of the Arctic's biodiversity (section 10.4). The chap-

Table 10.3. The five key strategic issues facing nature conservation in the Arctic (as quoted from CAFF, 2002a).

Strategic issue	Overall goal
Conserving arctic species	... to maintain vigorous populations of Arctic plant and animal species
Conserving arctic ecosystems and habitats	... to maintain and enhance ecosystem integrity in the Arctic and to avoid habitat fragmentation and degradation
Assessing and monitoring arctic biodiversity	... to monitor status and trends in Arctic biodiversity as an integral part of assessing the overall state of the Arctic environment
Global issues	... to understand and minimize the impacts of global changes and activities on Arctic biodiversity
Engaging society	... to promote circumpolar and global awareness of Arctic biodiversity issues

ter concludes with an exploration of some general principles concerning the conservation of the Arctic's biodiversity, some of the implications, and a series of recommendations (section 10.5).

10.2. Conservation of arctic ecosystems and species

Earlier chapters focused on the terrestrial, freshwater, and marine environments of the Arctic, and their component species. Several physical characteristics distinguish polar environments from the environments of other regions: limited daylight for much of the year, low temperatures, and low levels of precipitation. Collectively, these limit biological productivity over a large part of the year because photosynthesis and decomposition are severely constrained. In contrast, the brief arctic summer, which experiences continuous daylight and warmer temperatures, generates a large pulse of primary productivity. These dramatic seasonal changes strongly influence the Arctic's biodiversity. For example, productivity in summer is sufficient to attract migratory species of birds and mammals to the region.

Recent glaciations have resulted in major losses of the resident arctic fauna and recolonization has been slow (particularly in the terrestrial and freshwater environments), owing to both the extreme environmental conditions and the low overall productivity of arctic ecosystems. This has resulted in the arctic ecosystems, in a global sense, being considered "simple", i.e., having relatively few species. The species that they do contain are mainly "specialists" in the sense that they have been able to adapt to the extreme conditions. Thus, there are few species at any particular trophic level, and overall species diversity in terrestrial, freshwater, and marine habitats is low.

The seasonal constraints result in similar life-history traits in many arctic plant and animal species. Compared to species living in temperate regions, species living in the Arctic throughout the year are typically long-lived, slow-growing, and have low rates of annual reproduction. These factors appear to be adaptive to environments that can vary greatly from year to year, and where productivity is constrained to a short period of time, even in a favorable year (MacArthur and Wilson, 1967; Pianka, 1970). Specifically, these life-history traits are suitable for plant and animal species living in environments where reproductive attempts within a single year

may need to be abandoned to ensure adult survival (Trathan et al., 1996; Weimerskirch, 2002).

Several of these traits may limit the capacity of species to respond to rapid environmental change. High adult survival rates, coupled with low rates of reproduction, make populations slow to recover from catastrophic events (Danchin et al., 1995; Jenouvrier et al., 2003). Also, the adaptations unique to species living in polar environments also limit their ability to respond to warming conditions or to the greater environmental variability projected to result from climate change scenarios for the Arctic (Laxon et al., 2003; Parkinson, 2000; Parkinson et al., 1999; Vinnikov et al., 1999).

The rest of section 10.2 considers the special features of arctic habitats that make their biological diversity vulnerable to climate change. In their analysis of the European Arctic, Hallanaro and Pylvänäinen (2002) recognized nine broad habitat types. Six of these have not been significantly affected by human activities: habitats above and beyond (i.e., north of) the treeline; forests; wetlands; lakes and rivers; coasts and shores; and the sea. The other three have been strongly affected: farmland; urban areas; and mosaic landscapes.

In this chapter the Arctic is considered in terms of five broad habitat groupings, including marine environments; freshwater environments; environments north of the treeline; boreal forests; and habitats intensively modified by people. The term *wildlife* was defined in Anon (2001a) as "in a more scientific sense...wildlife refers to all non-domesticated organisms. It includes mammals, birds, fish, amphibians, and reptiles, as well as vascular plants, algae, fungi, bacteria, and all other wild living organisms". Anon (2001a) defined *habitats* as "all the elements of the Earth that are used by wildlife species to sustain themselves throughout their life cycles. This includes the spaces (i.e., terrestrial and aquatic) that they require as well as the properties of those places (e.g., biota, climate, soils, ecological processes and relationships). Habitats function in providing such needs as food, shelter, and a home place. Habitats can be thought of as distinctive places or ecosystems...". These broad definitions are used in this chapter.

Although it might seem simple to identify terrestrial, freshwater, and marine habitats, as well as the wildlife that occurs in each, in practice it is not because each habitat merges into another. For example, catchments or watersheds on land are terrestrially defined, but water percolating through the soil or running off the soil surface eventually enters streams and rivers. So where do terrestrial habitats end and freshwater habitats begin? Similarly, rivers enter estuaries where they are subject to tides, and species characteristic of rivers meet species characteristic of the sea. Where do freshwater habitats end and marine habitats begin? Along the shore the sea and the land interact, and there may be no clear demarcation between terrestrial and marine habitats. The situation is further complicated by anadromous species, such as Atlantic salmon (*Salmo salar*). These spawn in rivers, and the young pass through the estuaries on their way to the sea where they mature before returning several years later to their natal rivers to begin the cycle again. The reverse occurs with catadromous species, such as the eel (*Anguilla anguilla*), which spawns at sea. There are thus gradients, rather than clear boundaries between the wildlife of terrestrial, freshwater, and marine environments, and a pragmatic approach to allocating species and habitats to these broad groupings is taken within sections 10.2.1 to 10.2.4.

10.2.1. Marine environments

The arctic marine environment covers about 13 million km^2 (CAFF et al., 2000), of which about 45% is a permanent ice cap that covers part of the Arctic Ocean. Seasonal sea ice forms during winter, and recedes during the short arctic summer, exposing large areas of open water. The marine environment is thus dominated by sea ice (CAFF, 2001) and by the dynamics of that ice and especially the location of the ice edge. The transition zone between the sea ice and the open water has intense algal growth in spring and summer, and it is the primary production by these phytoplankton that supports the arctic marine food webs. Only in exceptional cases can the energy that drives the marine food webs be obtained from other sources. CAFF (2001) recorded the recent discoveries of "hot vents" and "cold seeps" in the Arctic. At these sites, bacteria are capable of deriving energy from methane (CH_4) or hydrogen sulfide (H_2S) gases that emerge as bubbles or in solution from the vents and seeps. These bacteria are then fed on by other organisms and so form the basis of some very specialized and localized food webs. Research on marine biodiversity is usually expensive, which is probably why comparatively less is known about marine biodiversity than terrestrial biodiversity (Anon, 2001a).

Projected changes in sea ice, temperature, freshwater, and wind will affect nutrient supply rates through their effects on vertical mixing and upwelling. These will in turn result in changes in the timing, location, and species composition of phytoplankton blooms and, subsequently, in the zooplankton community and the productivity of fishes. Changes in the timing of primary production can affect its input to the pelagic community as well as the amount exported to and taken up by the benthic community. The retention: export ratio also depends on the advection of plankton and nutrients within the water body (Shuert and Walsh, 1993) and on the temperature preferences of the grazing zooplankton; these both determine the degree of match or mismatch between primary and secondary production (see Chapter 9).

The projected disappearance of seasonal sea ice from the Barents and Bering Seas, and so the elimination of ice-edge blooms, would result in these areas having blooms resembling those presently occurring in more southerly seas (Alexander and Niebauer, 1981). The timing of such blooms will be determined by the onset of seasonal stratification, again with consequences for a match or mismatch between phytoplankton and zooplankton production. If a mismatch occurs, due to early phytoplankton blooms, the food webs will be highly inefficient in terms of food supply to fish (Hansen B. and Østerhus, 2000). Both export production and protozoan biomass is likely to increase. However, both the areal extent of export production and grazing by copepods are projected to increase slightly because of the larger ice-free area (see Chapter 9).

Future fluctuations in zoobenthic communities will be related to the temperature tolerance of the animals and to the future temperature of the seawater. Whereas most boreal species have planktonic larvae that need a fairly long period to develop to maturity, arctic species do not (Thorson, 1950). Consequently, boreal species should be quick to spread with warm currents during periods of warming, while the more stenothermal arctic species (i.e., those only able to tolerate a small temperature range) will quickly perish. Shifts in the distribution of the fauna are likely to be quicker and more noticeable during periods of warming than periods of cooling. Change in the abundance or biomass of benthic communities is most likely to result primarily from the impact of temperature on the life cycles and growth rates of the species concerned. If warming occurs, thermophilic species (i.e., those tolerating a wide temperature range) will become more frequent (see Chapter 9). This will force changes to the zoobenthic community structure and, to a lesser extent, to its functional characteristics, especially in coastal areas.

Climate change affects fish production through direct and indirect pathways. Direct effects include the effects of temperature on metabolism, growth, and distribution. Food web effects could also occur, through changes in lower trophic level production or in the abundance of top-level predators, but the effects of these changes on fish are difficult to predict. However, generalist predators are likely to be more adaptable to changed conditions than specialist predators (see Chapter 9). Fish recruitment patterns are strongly influenced by oceanographic processes such as local wind patterns, mixing, and prey availability during early life stages; these are also difficult to predict. Recruitment success could be affected by changes in the timing of spawning, fecundity rates, larval survival rates, and food availability.

Poleward extensions of the range of many fish species are very likely under the projected climate change scenarios discussed in Chapter 4. Some of the more abundant species that are likely to move northward under the projected warming include Atlantic and Pacific herring (*Clupea harengus* and *C. pallasi* respectively), Atlantic and Pacific cod (*Gadus morhua* and *G. macrocephalus* respectively), walleye pollock (*Theragra chalcogramma*) in the Bering Sea (Blindheim et al., 2001), and some of the flatfishes that might presently be limited by bottom temperatures in the northern areas of the marginal arctic seas. The southern limit of colder-water fish species, such as polar cod (*Boreogadus saida*) and capelin (*Mallotus villosus*), are likely to move northward. Greenland halibut (*Reinhardtius hippoglossoides*) might possibly shift its southern boundary northward or restrict its distribution more to continental slope regions (see Chapter 9). Migration patterns are very likely to shift, causing changes in arrival times along the migration route (Holst et al., 2002). Qualitative predictions of the consequences of climate change on fish resources require good regional atmospheric and ocean models of the response of the ocean to climate change. There is considerable uncertainty about the effects of non-native species moving into a region in terms of their effects on the "balance" within an ecosystem.

The impacts of the projected climate change scenarios on marine mammals and seabirds in the Arctic are likely to be profound (Vibe, 1967), but are difficult to predict in precise terms. Patterns of change are non-uniform and highly complex. The worst-case scenarios for reductions in sea-ice extent, duration, thickness, and concentration by 2080 threaten the existence of entire populations of marine mammals and, depending on their ability to adapt, could result in the extinction of some species (Jenkins, 2003).

Climate change also poses risks to marine mammals and seabirds in the Arctic beyond the loss of habitat and forage bases. These include increased risk of disease for arctic-adapted vertebrates owing to improved growing conditions for the disease vectors and to contact with non-native species moving into the Arctic (Harvell et al., 1999); increased pollution loads resulting from an increase in precipitation bringing more river borne pollution northward (Macdonald R. et al., 2003); increased competition from the northward expansion of temperate species; and impacts via increased human traffic and development in previously inaccessible, ice-covered areas. Complexity arising from alterations to the density, distribution, or abundance of keystone species at various trophic levels, such as polar bears (*Ursus maritimus*) and polar cod, could have significant and rapid consequences for the structure of the ecosystems in which they currently occur.

Although many climate change scenarios focus on the potentially negative consequences for ecosystems, environmental change can also bring opportunities. The ability of some species to adapt to new climate

regimes is often considerable, and should not be underestimated. Many marine vertebrates in the Arctic, especially mammals and birds, are adapted to dealing with patchy food resources and to a high degree of variability in its abundance.

Ice-living seals are particularly vulnerable to changes in the extent and character of the sea ice because they use it as a pupping, molting, and resting platform, and some species also forage on ice-associated prey. Of the arctic pinnipeds, ringed seals (*Phoca hispida*) are likely to be the most affected because so many aspects of their life history and distribution are tied to sea ice (Smith and Stirling, 1975). They require sufficient snow cover to construct lairs and the ice must be sufficiently stable in spring for them to rear young successfully. Early break-up of the sea ice could result in premature separation of mother–pup pairs and hence increased neonatal mortality. Ringed seals do not normally haul out on land and to do this would be a very dramatic change in their behavior. Land breeding would expose ringed seal pups to much higher predation rates.

Changes in the extent and type of sea ice affect the distribution and foraging success of polar bears (Ferguson et al., 2000a,b; Mauritzen et al., 2001; Stirling et al., 1993). The earliest impacts of warming will occur at their southern limits of distribution, such as at James and Hudson Bays; and this has already been documented by Stirling et al. (1999). Late sea-ice formation and early break-up also mean a longer period of annual fasting. Reproductive success in polar bears is closely linked to their fat stores. Females in poor condition have smaller litters, as well as smaller cubs that are less likely to survive. There are also concerns that direct mortality rates might increase. For example, increased frequency or intensity of spring rains could cause dens to collapse, resulting in the death of the female as well as the cubs. Earlier spring break-up of sea ice could separate traditional den sites from spring feeding areas, and if young cubs were forced to swim long distances between breeding areas and feeding areas this could decrease their survival rate. The survival of polar bears as a species is difficult to envisage under conditions of zero summer sea-ice cover. Their only option would be to adopt a terrestrial summer lifestyle similar to brown bears (*Ursus major*), from which they evolved. But competition, risk of hybridization with brown and grizzly bears (both *U. major*), and an increase in human interactions, would also pose a threat to their long-term survival.

The effects of climate change on seabird populations, both direct and indirect, are very likely to be detected first near the limits of the species range and the margins of their oceanographic range (Barrett and Krasnov, 1996; Montevecchi and Myers, 1997). The southern limits of many arctic seabirds are likely to retract northward, also causing breeding ranges to shift northward (Brown, 1991). Changes in patterns of distribution, breeding phenology, and periods of residency in the Arctic are likely to be some of the first observed responses to climate change. Seabirds will also be affected by changes in prey availability and so can serve as indicators of ecosystem productivity. Since warmer (or colder) water would affect the distribution of prey species, the distribution of individual seabird species is likely to reflect changes in the distribution of macrozooplankton and fish populations. Changes in sea level may restrict the use of current breeding sites, but may increase the suitability of other sites that are not currently used owing to predator access or for other reasons.

With climate change already underway, planning for the conservation of marine biodiversity is an imperative. Series of actions are being proposed (CAFF et al., 2000; Anon, 2001a). These can be grouped into five key issues, namely:

- the implementation of an inventory of the Arctic's biodiversity and of schemes for monitoring trends in the biodiversity resource, including appropriate indicators;
- the completion of a circumpolar network of marine and maritime protected areas;
- the development of circumpolar guidelines for managing arctic biodiversity in a sensitive manner, bearing in mind the needs of local communities and the fact that "controlled neglect" may be an appropriate means of management;
- the establishment of fora for developing integrated management schemes for coasts and seas; and
- the review of marine regulatory instruments, with recommendations for further actions where necessary.

Conservation is unlikely to be easy (CAFF, 2001), but as many as possible of these five key issues should be developed on a circumpolar basis. This is particularly the case for the marine environment because many of the species tend not to be localized, but to be widely distributed throughout the Arctic Ocean as a whole. Indeed, some species have regular, seasonal patterns of migration. Satellite tracking has shown that walrus (*Odobenus rosmarus*) and narwhal (*Monodon monoceros*) can move great distances within the Arctic Ocean in relatively short periods of time (Anon, 2001b). Similarly, polar bears, ringed seals, and beluga whales (*Delphinapterus leucas*) have been shown to exhibit extensive and rapid circumpolar movements.

The main requirement for the conservation of marine biodiversity is the need to take a holistic approach. The majority of national parks and reserves are predicated primarily upon the protection of coastal birds and mammals (Bernes, 1993). This needs to be expanded to include the ecosystems upon which these birds and mammals depend, and upon which the commercially-exploited fish populations also depend. It is not just the vertebrate animals that are important, but the whole range of biodiversity, and especially those small and often unknown organisms that are either trapping solar energy by photosynthesis or decomposing organic mat-

ter to enable the recycling of nutrients. It is the totality of the biodiversity of the marine habitats and ecosystems of the Arctic that support the sustainable production of the biological resources upon which the indigenous peoples, and others, depend. This holistic approach is underlined in the final sections of Chapter 9 which discuss the effects of climate change on phytoplankton; zooplankton production; benthic organisms; fish production; marine mammal distribution, especially in relation to sea-ice cover; and seabird distribution and prey availability.

Although there are many unknowns, it is likely that many of the vertebrate animals will move northward, with many of these species likely to become less abundant. However, for the phytoplankton, it is the extent of the mixing of the ocean layers that will determine the increases and decreases for the various taxonomic groups.

10.2.2. Freshwater environments

The Arctic has many types of freshwater habitat. There is a wide range of wetlands, including mires, marshes, sedge and reed beds, floodplain "grasslands", salt marshes, and coastal lagoons, as well as a large number of rivers, streams, and lakes. In fact, excluding the freshwater locked up in permanent ice in the Antarctic, a large proportion of the earth's liquid freshwater resources occur in the Arctic.

There is no universally accepted definition of a "wetland". Hallanaro and Pylvänäinen (2002) described a wetland as "areas where the water table lies near the surface for much of the year. Shallow water bodies can also be considered as wetlands if they are mainly covered by vegetation. In wetlands at least half of all of the plants should be hydrophytes, which can withstand or may even depend on high water levels". With such a loose definition, there can be many gradients from a wetland to some other sort of habitat. For example, as wetlands border onto colder areas, permafrost could become common, whereas near the coast the influence of sea ice will be greater, and toward the taiga there will be an assortment of wet woodland habitats.

Lakes and rivers are abundant in the Arctic. Norway is estimated to have in excess of 200 000 lakes with a surface area greater than 0.01 km^2 but less than 1 km^2, and 2457 lakes larger than this. Sweden is estimated to have 2908 rivers and the Republic of Karelia 1210 rivers. The 18 largest lakes in Europe are all in northern Europe, although some are located outside the Arctic (located between 60° and 66° N). Such statistics demonstrate the extent of the liquid freshwater resource in the Arctic.

Thus, there is a great range in the type and extent of arctic freshwater environments (see Chapter 8 for further details), and this extent is perhaps proportionally greater than in other geographical areas. For example, the rivers, lakes, and wetlands of Siberia are mainly fed by thaw and summer rains, which account for up to 80% of total annual flow (Zhulidov et al., 1997) and which do not usually penetrate the impermeable permafrost barrier. Rivers in eastern Siberia typically freeze over in winter, flowing mainly, if not solely, in summer. The larger rivers in western Siberia have greater flows, controlled by discharges from their substantial catchments that extend into more southerly latitudes. The Rivers Ob and Yenisey provide significant contributions to the total freshwater discharge from Asia to the Arctic Ocean. Another example, is the Mackenzie Delta in North America, which is the second largest delta in the Arctic and subarctic (Lewis, 1991), being 200 km long and 65 km wide (Prowse, 1990). The delta has about 50% lake coverage (Mackay, 1963) and extensive wetlands. The small coastal rivers in the western Mackenzie Delta freeze over in winter. The spring break-up in the upstream parts of the Mackenzie River catchment causes rapid increases in water and suspended sediment discharges into the delta. These flood low-lying land and can recharge delta lakes.

These examples illustrate two of the special features of arctic freshwater environments. First, that the ecosystems can be frozen for much of the year, meaning water is available for relatively short periods of time. Second, that there is considerable variability, both within and between years, in terms of flooding, drying out, freezing, freeze–thaw cycles, and the periods of time over which these occur.

The dynamics of many of the lotic (river) and lentic (lake) environments in the Arctic are related to permafrost, and freezing can reduce or even halt the flow of rivers. The relationships between river flow, lake depth, and the onset or cessation of freezing conditions are also features of the arctic environment. Sources of water during the summer include, in addition to rain, late or perennial snow patches, glaciers, thawing of permafrost, and groundwater discharges (Rydén, 1981; van Everdingen, 1990). The projected increases in temperature are likely to result in these water sources becoming greater contributors to the annual water budgets of freshwater ecosystems. Many of the lentic environments are relatively shallow, and so the species within them have to be able to withstand considerable environmental variability, especially when the water bodies freeze.

Arctic freshwater ecosystems are species-poor compared to similar ecosystems in temperate and tropical areas (Bazely and Jefferies, 1997). This makes them particularly suitable for trophic studies, as for example the research by Kling et al. (1992) using isotopes of nitrogen and carbon. As Bazely and Jefferies (1997) reported, aquatic food chains in the Arctic are long, which is unusual given the low overall productivity per unit area. This paradox may reflect the pulse-regulated nature of the ecosystems, whereby seasonal resource acquisition and population growth are restricted to short periods. During unfavorable periods for growth and reproduction, low maintenance costs (or migration) enable populations to survive. It is postulated that this "idling" survival strategy allows extended food

chains to occur because high-energy demands by organisms do not occur year-round.

A crucial feature of the biodiversity of the Arctic's freshwater environment is the fish, generally occurring at high trophic levels and providing an important resource for the human population. Given the slow growth rates and low overall productivity, these fish populations can easily be over-exploited. Chapter 8 outlines the possible effects of climate change on a number of fish stocks, both those resident in freshwater and those that are diadromous (migrating between freshwater and sea water). Anadromous behavior (migrating from salt to freshwater, as in the case of a fish moving from the sea into a river to spawn) is most prevalent in northern latitudes (McDowall, 1987) because the ocean is more productive than the freshwater environments.

Climate change will affect arctic freshwater habitats by causing local extinctions and by changing the distribution ranges of species (see Chapter 8). Changes in the amount of precipitation and the length of snow lie will be important. The effects of increased precipitation for freshwater habitats will be primarily geomorphological, especially in the increased sediment loads in rivers and the increased deposition of sediments in lakes, at hydroelectric dams, and in estuaries. Such changes will affect habitats and the species they support, and so are likely to impact adversely on the biodiversity of the Arctic. The effects of decreased precipitation could be even more severe, resulting in the drying of wetlands, oxidation of organic compounds in sediments, and so a further release of CO_2 to the atmosphere. Changes in temperature are likely to affect the physiology of individuals, altering population dynamics and interactions between species. Temperature effects are very likely to be most pronounced in relation to fish, potentially opening up arctic freshwater ecosystems to fish species that currently have a more southern distribution.

Conservation of the biodiversity of freshwater habitats in the Arctic has been hampered by the lack of a common classification of habitats, especially for the wetlands. With each country using different definitions, it is difficult to determine trans-Arctic trends and to compare differences between regions. Classification schemes can be contentious, but it is vital that schemes are adopted as soon as possible (Naiman et al., 1992). For conservation, classification of habitats or species provides a framework for communication, management and, where necessary, legislation or regulation. This is important because of the many threats to arctic freshwater biodiversity. An analysis of environmental trends in the Nordic countries viewed threats to the freshwater environment from a two-dimensional perspective (Fig. 10.1). The vertical axis shows the area over which the threat operates and the horizontal axis represents the perceived seriousness of the threat. The illustration includes 14 current threats to biodiversity and ten long-term threats to the natural resources of the Nordic countries. The position of the ellipses on each diagram is therefore analogous to a risk assessment for that particular threat. The diagram does not show how these threats will change as the climate changes, but it is likely that many of the ellipses will move to the right.

Such predictions contain many uncertainties. Nevertheless, Chapter 8 concludes with a series of nine predictions about the effects of climate change on freshwater environments and their biodiversity:

- microbial decomposition rates are likely to increase;
- increased production is very likely to result from a greater supply of organic matter and nutrients;
- shifts in invertebrate species' ranges and community compositions are likely to occur;
- shifts in fish species' ranges, composition, and trophic relations will very probably occur;

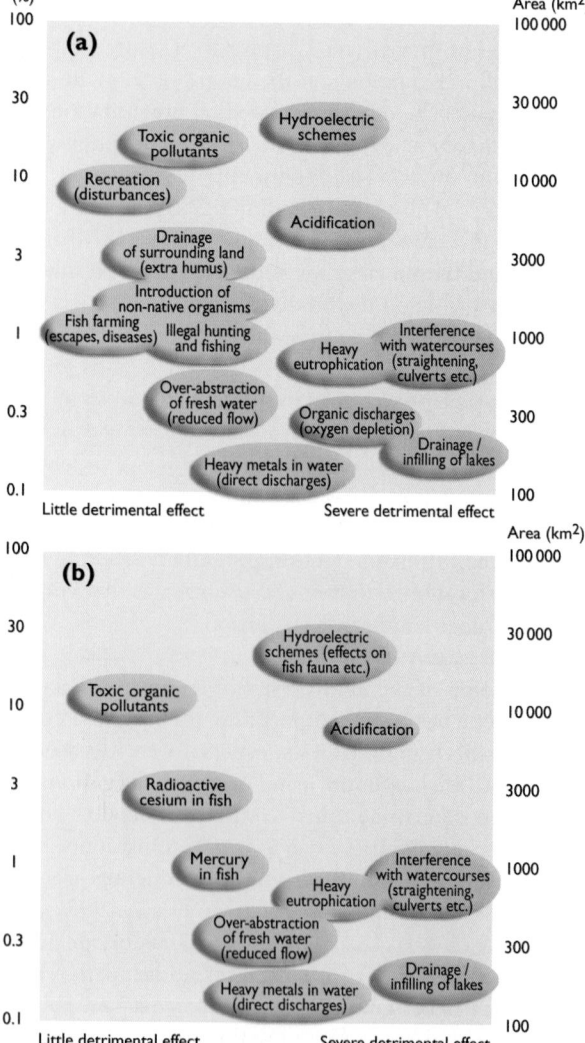

Fig. 10.1. A representation of the impacts of various threats to the freshwater environment of the Nordic nations. The vertical axis is a logarithmic representation of the extent, ranging from 100 to 100 000 km². The horizontal axis represents the perceived severity of the threat. Thus in each diagram threats to the lower left are of least concern, while those to the upper right are of greatest concern. (a) current threats to biodiversity, (b) long-term threats to natural resources. (Based on Bernes, 1993; reproduced with permission from The Nordic Council of Ministers, Denmark).

- spawning grounds for cold-water fish species are likely to diminish;
- an increased incidence of mortality and decreased growth and productivity from disease/parasites are likely to occur in fish species, and will possibly occur in aquatic mammals and waterfowl;
- subsistence, sport, and commercial fisheries will possibly be negatively affected;
- probable changes in habitat are likely to result in altered migration routes and timing of migration for aquatic mammals and waterfowl; and
- probable changes in timing of habitat availability, quality, and suitability are very likely to alter reproductive success in aquatic mammals and waterfowl.

These issues pose many challenges, and neither traditional knowledge nor scientific knowledge are able to meet these challenges completely. In addition to the need for more research, the development of generic models is essential if research in one area, on one species, or on one habitat, is to be applied to other areas, to other species, or to other habitats.

10.2.3. Environments north of the treeline

Arctic organisms must either survive or avoid the long, cold winters. Adaptations range from avoidance behavior (long-distance migration, migration from tundra to forest, migration down the soil profile) to specific physiological, morphological, and life history traits in both plants and animals. Species with specific adaptations to cold conditions often lack the flexibility to adapt to new conditions, particularly interactions with immigrant, competitive species from the south. For example the displacement of Arctic fox (*Alopex lagopus*) by red fox (*Vulpes vulpes*), and many arctic plant species that are shade intolerant (see Chapter 7).

In addition to the constraints of low temperatures on biodiversity, the contrast between summer and winter conditions is also important. The photoperiod is likely to constrain budburst, frost hardening, and reproduction in some potentially immigrant shrubs and trees. It is also likely to affect the endocrinology of mammals leading to constraints on reproduction and the onset of appetite. Short growing seasons select for plants that are perennials and have long development periods, for example three to four years from flower bud initiation to seed set. Marked temperature differences between summer and winter conditions currently select for plants that accumulate and store resources: up to 98% of biomass can be below ground. Such storage organs are likely to become a respiratory burden with warmer winters, and slow-growing plant species with multi-year development are eventually likely to be displaced by faster growing species, including annuals.

Overall, species richness in the Arctic north of the treeline is low (see Chapter 7). About 3% of the species making up the global flora occur in the Arctic. However, lower taxonomic groups are better represented than higher orders: only 0.7% of the flowering plant species occur in the Arctic compared with 1.6% of the cone-bearing plants. At a scale of 100 m², however, the diversity of the flora of some arctic communities can equal that of temperate or boreal latitudes owing to the generally small size of arctic plants. Within the Arctic, the diversity of animals (about 6000 species) is twice that of plants. Again, with lower taxonomic groups better represented. Springtails, at 6% of the global total, are better represented than advanced invertebrate groups such as beetles with 0.1% of the global total. Climatic warming is very likely to increase the total number of species in the Arctic as species with more southern ranges shift northward, but the overall composition of the flora and fauna is vulnerable to the loss of arctic species at lower taxonomic orders (Cornelissen et al., 2001). Some taxonomic groups are particularly species rich in a global context: any impact of climate warming on such species, for example, willows (*Salix* spp.), sawflies, stoneflies, wading birds, and salmonid fish, is likely to affect their diversity at the global level.

An important consequence of the decline in numbers of species with increasing latitude is a corresponding increase in dominance. For example, one species of collembolan, *Folsomia regularis*, may constitute 60% of the total collembolan density in polar deserts (Babenko and Bulavintsev, 1997). Examples for plants include the cotton-grass *Eriophorum vaginatum*, and *Dryas* species. These "super-dominants" are generally highly adaptable, occupy a wide range of habitats, and have significant effects on ecosystem processes. Lemmings (*Lemmus* spp. and *Dicrostonyx* spp.) are super-dominant species during peak years in their population cycles (Stenseth and Ims, 1993).

Trophic structure is less complex in the Arctic than further south. In all taxonomic groups, the Arctic has an unusually large proportion of carnivorous species and a low proportion of herbivores (Chernov, 1995). As herbivores are strongly dependent on the response of vegetation to climate variability, warming is likely to alter the trophic structure substantially as well as the dynamics of arctic ecosystems. The herbivore-based system in most tundra habitats is dominated by one or two lemming species (Batzli et al., 1980; Oksanen et al., 1997; Wiklund et al., 1999), while the abundance of phytophagous (plant-eating) insects relative to plant biomass is small on arctic tundra (Strathdee and Bale, 1998). Large predators such as wolves, wolverines, and bears are less numerous in the tundra than the boreal forest (Chernov and Matveyeva, 1997) and predation impacts on tundra ungulates are usually low. Thus, the dynamics and assemblages of vertebrate predators in arctic tundra are almost entirely based on lemmings and other small rodent species (*Microtus* spp. and *Clethrionomys* spp.) (Batzli, 1975; Wiklund et al., 1999), while lemmings and small rodents consume more plant biomass than other herbivores. Climate has direct and indirect impacts on the interactions among trophic levels, but there is greater

uncertainty about the responses to climate change of animals at higher trophic levels.

Mechanical disturbance to plants and soils (animals can avoid or respond to such problems) occurs at various scales. Large-scale slope failures, such as active layer detachment, destroy plant communities but open niches for colonization by new generations of existing species or immigrant species with ruderal characteristics (fast growth, short life span, large reproductive capacity, and widespread dispersal of seeds). Such disturbances can also lead to recruitment of old genotypes of species producing long-lived seed that has been buried for hundreds of years (Vavrek et al., 1991). Sorting of stones and sediments in the active layer from daily to seasonal freeze–thaw cycles causes patterning of the ground and the creation of a mosaic of habitats at the landscape scale and a range of niches at the centimeter to meter scale (Matveyeva and Chernov, 2000). Such sorting, together with longer term permafrost degradation, movement of soils on slopes, and displacement by moving compacted snow and ice, exerts strong forces on plant roots. Above ground, wind-blasted ice crystals can erode plant tissues that extend above the protective snow cover. Mechanical impacts in the soil select for species without roots (mosses, lichens, algae), species with very shallow and simple root systems (e.g., *Pinguicula* spp.), and species with mechanically elastic roots (e.g., *Phippsia algida* and *Tofieldia pusilla*) (Jonasson and Callaghan, 1992). Amelioration of the mechanical impacts is likely to lead to displacement of specialized species by more competitive neighboring species.

Super-dominant species such as lemmings have large effects on ecosystem processes (Batzli et al., 1980; Laine and Henttonen, 1983; Stenseth and Ims, 1993). Lemming peak densities exceed 200 individuals per hectare in the most productive *Lemmus* habitats of Siberia and North America (Batzli, 1981) and the standing crop may approach 2.6 kg dry weight per hectare. Lemmings have a high metabolic rate and *Lemmus* spp. in particular has low digestive efficiency (about 30%, compared to 50% in other small rodents). Consequently, their consumption rate and impact on the vegetation exceeds that of all other herbivores combined (with the exception of the local effects of geese near breeding colonies). Also, lemmings destroy more vegetation than they ingest and after population peaks typically 50% of the above-ground biomass has been removed by the time of snow melt (Turchin and Batzli, 2001). In unproductive snowbeds, which are favored winter habitats of the lemming *Lemmus lemmus* (Kalela, 1961), between 90 and 100% of the moss and graminoids present during winter may have been removed (Koskina, 1961).

In forest near the treeline, insect defoliators can have devastating impacts on the ecosystem. The autumnal moth (*Epirrita autumnata*) shows cyclicity in its populations and outbreak proportions occur approximately every 10 to 11 years (Tenow, 1972, 1996). Many thousands of hectares of forests are defoliated in outbreak years and defoliated forests require about 70 years to attain their former leaf area. However, insect outbreaks in sub-arctic Finland, followed by heavy reindeer browsing of regenerating birch shoots, have led to more or less permanent tundra (Kallio and Lehtonen, 1973; Lehtonen and Heikkinen, 1995).

These outbreaks are important for predators, such as snowy owl (*Nyctea scandiaca*) and arctic fox, which both prey on lemmings, and parasitoids such as the wasp *Cotesia* sp., which lays its eggs in caterpillars of the autumn moth. Changes to the populations and population trends of species such as lemmings and forest insect pests are very likely to have far reaching consequences for the biodiversity of the vegetation they consume, and for their predators and parasitoids, as well as for ecosystem processes like nutrient cycling.

The geography of the Arctic forces a range of constraints on the ability of vegetation zones and species to shift northward. In mainland Fennoscandia and many parts of the Russian Arctic, apart from Taymir and the western Siberian lowland, the strip of tundra between the boreal forest and the ocean is relatively narrow. Trees already occur close to the Arctic Ocean at Prudhoe Bay and Khatanga. Any northward movement of the forest will completely displace the tundra zone, and hence its biodiversity, from these areas. On the western Siberian plain, extensive bog ecosystems limit the northward expansion of forest and in arctic Canada, the high Arctic archipelago presents a natural barrier to dispersal of plants and range extensions of animals, while the barrens (polar desert and prostrate dwarf shrub tundra with less than 50% of the ground covered by vegetation) consist of soils that will constrain forest development for perhaps hundreds of years.

Continuous and discontinuous permafrost are characteristic of the Arctic. Permafrost, particularly its effect on the thickness of the active layer, limits the depth and volume of biologically available soil and reduces summer soil temperatures. These constraints limit plant rooting, the activity of soil flora, fauna, and microbes, and ecosystem process such as decomposition. Soil movements associated with permafrost dynamics are discussed in Chapter 7. Thawing of permafrost can have dramatic effects on biodiversity, depending upon drainage, precipitation changes, and, consequently, soil moisture. Permafrost thawing associated with waterlogging can prevent the northward advance of the treeline and even initiate a southward retreat (Crawford et al., 2003). In other areas, such as the North Slope of Alaska, where precipitation is only about 125 mm/yr, permafrost thawing is likely to lead to drying and in some areas novel communities, reminiscent of the tundra-steppe, could form.

In addition to the effects of permafrost on biodiversity, biodiversity can also affect permafrost. A complete cover of vegetation, particularly highly insulative mosses, buffers soil temperatures from climate warming. In extreme cases, vegetation can lead to permafrost growth and a thinning of the active layer.

Arctic terrestrial ecosystems have the same types of feedback to the climate system as many other ecosystems, but the magnitude of these feedbacks is greater than most others. Per square meter, the tundra stores about half as much carbon as the boreal forests (about 9750 g/m^2 and 20500 g/m^2, respectively, 15900 g/m^2 at the interface between tundra and boreal forest according to McGuire et al., 1997). However, most of the carbon in the tundra occurs in the soil (about 94%), whereas about half (46%) of the carbon in the boreal forest occurs in the vegetation. The carbon stored in the tundra (about 102 Pg) is about 40% of that stored in the boreal forests (excluding the boreal woodlands). The tundra, boreal forest, and boreal woodlands together store 461 Pg of carbon; this is equivalent to about 71 years of annual global carbon emissions (based on emission data for the 1960s) of CO_2 from fossil fuels (about 6.5 Pg of carbon per year). In contrast to the boreal forest, tundra has a high albedo and reflects about 80% of incoming radiation and this can lead to local cooling. Displacement of tundra vegetation by shrubs increases winter soil temperatures by 2 °C (Sturm et al., 2001).

Feedbacks that change the rate of climate change (although probably not the direction) will affect the rates of changes in biodiversity. For example, the effect of shrubs on soil temperatures is expected to increase decomposition rates and nutrient cycling, and so further shrub expansion. Also, it is possible that glacial dynamics (as well as more generally the dynamics of frozen ground) will have an effect (Chernov, 1985). Glaciers have expanded and contracted in response to climatic variations. For example, in Iceland the maximum extent of the glaciers in historical times occurred in 1890. The majority of the glaciers contracted during the first half of the 20th century, particularly during the warm 1930s. Then from about 1940 the climate cooled, slowing the retreat of the glaciers, and some even started to advance again (Jóhannesson and Sigurðsson, 1998). This dynamic behavior of glaciers can have a marked effect on the biodiversity of nunataks (hills or mountains completely surrounded by glacial ice), which often contain a large proportion of the regional biodiversity. For example, there are over 100 species of vascular plants growing on Esjufjöll, a 9 km long nunatak within the glacier Vatnajökull, which is more than 20% of Iceland's total vascular plant flora (Einarsson, 1968).

Glacial dynamics are not entirely related to temperature. In Norway, there is some evidence that inland glaciers are currently retreating while coastal glaciers are advancing in response to greater quantities of snowfall. This indicates the difficulties of predicting the effects of climate change on glaciers. The different rates of warming at different seasons of the year, as well as changes in seasonal precipitation patterns, especially for snow, will

Fig. 10.2. Pine (*Pinus sylvestris*) forest in the Arctic. This area of almost natural forest is on an island in Inarijärvi, Europe's eighth largest lake, near Inari in Finland (68° 55' N). (Photo: M.B. Usher, July 1999).

Fig. 10.3. The mosaic structure of northern boreal forest; pine and birch forest associated with mires and small areas of open water north of Inari, Finland (69° 12' N). (Photo: M.B. Usher, July 1999).

all determine the future dynamics of glaciers. These in turn influence the nunataks, the extent of areas of new ground available for primary ecological succession after glacial retreat, and the loss of ecosystems covered by advancing glaciers.

10.2.4. Boreal forest environments

The Arctic encompasses the northern edge of the boreal forest and the woody communities, often containing shrubby trees, that are associated with the northern treeline. These northern forests are often dominated by four coniferous genera: the pines (*Pinus* spp.), spruces (*Picea* spp.), larches (*Larix* spp.), and firs (*Abies* spp.), as well as by two broadleaved genera, the birches (*Betula* spp.) and the aspens (*Populus* spp.), most of which have transcontinental distributions across Eurasia or North America (Nikolov and Helmisaari, 1992). An example of a pine-dominated forest near Inari, Finland (about 69° N) is shown in Fig. 10.2. This is typical of the near-natural forest, with slow-growing trees, dead wood, and natural regeneration in gaps where the dead and moribund trees allow sufficient light to penetrate to the forest floor. The forests frequently give way to mires and small lakes leading to a mosaic structure of forest and wetland. Figure 10.3, also near Inari in Finland, shows this transition, with both pine trees and birch woodland in the distance. The boreal forest region has a distinctive set of biodiversity characteristics at each of the three levels of biodiversity – genetic diversity, species diversity, and ecological communities. These are the key to assessing vulnerability of the boreal forest biodiversity to climate change.

When two or more distinct ecological communities or habitats are adjacent, there is a unique opportunity for organisms to live and reproduce in a diverse landscape. Landscape diversity is controlled by the physical arrangement of ecological communities. Climate change, by influencing the distribution of forest species, communities, and conditions, is a major factor controlling landscape diversity.

The extensive ecotone between boreal forest and tundra (a treeline 13 500 km long) is a prominent feature of the northern boreal region (some of the major climate-related fluctuations of the treeline are discussed in Chapter 14). The juxtaposition of trees and tundra increases the diversity of species that can exploit or inhabit the tundra. For example, insectivorous ground-dwelling birds that feed in the tundra but nest in trees are able to survive because of the mixture of habitats. Local human inhabitants can obtain shelter and make useful items for outdoor activities at this interface. The probability of climate warming causing the development of new treeline communities is described in Chapter 14. During recent decades of warming, the white spruce (*Picea glauca*) limit in Alaska (and almost certainly in western Canada) has developed two populations with opposite growth responses to the warming. Under extreme levels of projected warming, white spruce with negative growth responses would be likely to disappear from the dry central part of the northern boreal forest. In moister habitats, white spruce with positive growth responses to warming would expand in distribution. It is possible that part of the southern tundra boundary in North America would no longer border spruce forest but would border aspen (*Populus tremuloides*) parkland instead (Hogg and Hurdle, 1995).

The changes in boreal forests caused by fire and insect disturbance produce higher order effects due to the patterns and timing of the habitat conditions that they create at larger scales. Microtine rodents, birds, and hares (*Lepus timidus*) in the Fennoscandian boreal region undergo cyclic population fluctuations, generally on a three- to four-year cycle (Angelstam et al., 1985). Many factors contribute to these population cycles, including predator numbers, food plant quantity and/or quality, pathogens, parasites, and habitat heterogeneity. Some weather and climatic factors, such as snow depth, also directly influence animal numbers. In the future, population cycles of boreal animals are likely to remain primarily under the control of predators, although overall numbers of animals will respond to the overall amount of suitable habitat produced by events, such as forest fires, that are in turn related to climate warming. A ten-year study of trophic structure in the boreal forest in the Kluane area of southwest Yukon Territory, Canada, examined the ten-year animal population cycle. In this region the boreal community is a top-down system driven by the predators, and snowshoe hare (*Lepus americanus*) is a keystone species without which much of the community would collapse (Krebs et al., 2001). Hares influence all other cycles, and hare cycles are themselves controlled by the interaction of predator effect and food supply with little or no climate or fire effect detected. However, by the end of the study, 30% of the white spruce forest in the study area had been killed by spruce bark beetle (*Dendroctonus rufipennis*), which was probably related to climate warming (see Chapter 14). The change in habitat condition in the Kluane study area is one of the largest disturbances resulting from climate warming in the region over the last few centuries.

Specific areas of the boreal region are more species-rich than others (Komonen, 2003). Areas that have not been glaciated or which were deglaciated earliest are generally more species rich than more recently deglaciated areas (Komonen et al., 2003), suggesting that risks of major migrations of the boreal forest increase the probability of species loss. Boreal regions with a diversity of geological and soil substrates, such as Far East Russia, the Scandes Mountains, and the northern Rocky Mountains of North America, are relatively species-rich compared to more uniform areas such as the Canadian Shield or the Ob Basin. Boreal areas that have experienced interchange between the ecosystems (Asian Steppes, North American Plains) or continents (Beringia) are relatively species-rich.

Total species richness in the boreal region is greater than in the tundra to the north and less than in the temperate deciduous forest to the south, in line with levels of total ecosystem productivity (Waide et al., 1999). The southern boreal region contains more species than the northern boreal region, and one effect of climate warming is likely to be the addition of species to what is now the northern boreal region. A global summary of changes in phenology (the distribution and timing of events) across a number of organism groups already indicates the existence of a coherent signal of warming (i.e., poleward and upward migration, earlier activity in spring) (Root et al., 2003). However, the processes that eliminate boreal species (fire, insects, and drought) operate quickly, while those that add species (migration) operate more slowly. This raises the possibility that climate warming, in certain areas, could result in reduced species richness in the short term followed later by species gains as long as migratory barriers were not limiting. However, intensive forest management in Fennoscandia is one of the main causes of decline in the most rare or endangered boreal forest species there (Nilsson and Ericson, 1992) and managed forest landscapes do pose movement and connection barriers to the species in them (Hanski and Ovaskainen, 2000).

The conservation of certain boreal forest habitats is particularly important for maintaining species diversity, and climate change can bring serious challenges in this respect. Of the major ecological regions of the earth, boreal forest is distinctive for being conifer dominated (Juday, 1997). Older conifer forests on productive sites are the focal habitats of biodiversity conservation across the boreal region for several reasons. They are particularly rich in canopy lichens, mosses, and bryophytes; in the fungi responsible for decomposing wood; and in specialized insects, for woodpeckers and other cavity-nesting animals, and for insectivorous songbirds (Berg et al., 1994; Essen et al., 1992).

The reason that old-growth (or natural) forests are so important for the conservation of biodiversity lies in the holistic approach to nature conservation. Natural forests, with their J-shaped stem-number curve (a few old, large trees and many small, young trees) provide a range of habitats that support a range of different species of plants and animals. Old trees provide nesting holes for some bird species, diseased and moribund trees provide a substrate for many species of fungi, dead wood provides a resource for saproxylic (wood-feeding) insects, and some moth species will only lay their eggs on the foliage of young trees, etc. Wood-feeding arthropods form a diverse taxonomic group that is under pressure throughout Europe (Pavan, 1986; Speight, 1986) and elsewhere. In contrast, managed forests of younger trees tend to have little dead wood, few nesting holes for birds, and less light reaching the forest floor and thus a less well developed dwarf shrub, herbaceous, moss, and lichen flora, which in turn supports fewer invertebrates. A focus on the beetles of the northern forests (Martikainen and Kouki, 2003) has demonstrated both that these semi-natural forests contain a relatively large number of rare species and that there are difficulties in making accurate inventories.

Owing to the natural rate of stand-replacing disturbances (fire and insects) in the boreal forest, old-growth conifer stands are not necessarily abundant even in landscapes with little direct human impact. Human modification of the boreal forest landscape typically makes these old forests rarer because management for wood products is usually based on the good returns from cutting large conifers. In parts of the boreal region, where commercial forest management is established or expanding, productive stands of mature and old conifers are already rare (eastern Canada, northern Fennoscandia; Linder and Ostlund, 1992) or the target for early harvest (Siberia; Rosencranz and Scott, 1992). One of the major effects of climate warming on boreal forests is to increase tree death from fire and insects, and conifer stands are more flammable and often more susceptible to insect-caused tree death than broadleaved forests. Thus the ecosystem of greatest conservation interest, old conifer forest, is the one at most risk of decline due to climate warming.

Fire is a natural and recurrent feature of boreal forests, aiding the maintenance of biodiversity in these northern forests. Fire is expected to pass through a forest every 100 to 200 years (Korhonen et al., 1998). Some species are adapted to using the resources of burnt forests – charred trees which are still standing, trees which have started to decay, and the early stages of ecological succession following fire. Because fires in managed forests are usually extinguished quickly, burnt forest habitats have become rare and the species that depend on them are increasingly threatened and even locally extinct. In Finland, 14 species, mostly beetles (Coleoptera) and bugs (Hemiptera), associated with burnt areas in forests are threatened with extinction (Korhonen et al., 1998).

However, can extensive fires be tolerated in managed forests when the trees are required for extraction and as the raw material for the timber industry? Growth rates of trees near the transition from forest to tundra are extremely slow, which makes management of these far northern forests uneconomic (except for the initial exploitation of the few trees large enough to be used in timber mills, etc.). However, with climate change (and eutrophication by nitrogen deposition) productivity is likely to increase, and so the management of these northern forests becomes a potentially more viable economic activity, with consequent effects on forest biodiversity.

Fire itself is not the risk factor for the maintenance of boreal forest species diversity, but rather the altered characteristics of fire that can result from climate warming, especially amount, frequency, and severity. Conifer dominance itself promotes the occurrence of large, landscape-scale fires through characteristics such as flammable foliage and ladder fuels (defined by Helms (1998) as "combustible material that provides vertical continuity between vegetation strata and allows fire to climb into the crowns of trees or shrubs with relative ease"). Many boreal trees and other plants show adaptations to fire such as seed dormancy until fire, serotinous cones, fire-resistant bark, and sprouting habit. Many understory plant species of the boreal forest have means of persistence from underground structures following fire or are

effective re-colonizers (Gorshkov and Bakkal, 1996; Grime, 1979; Grubb, 1977; Rees and Juday, 2002). Fire in the boreal forest sustains a set of species in early post-fire communities that are distinct from later successional species. These include species from a range of groups, including birds, beetles, spiders, and vascular and non-vascular plants (Essen et al., 1992; Haeussler and Kneeshaw, 2003; Rees and Juday, 2002). Changes in natural fire regimes by human management interacting with climate warming can disrupt the specific fire regimes that sustain these species. For example, in some circumstances climate warming combined with human fire suppression results in less frequent but more intense fire. This change can kill species adapted to periodic light ground fires.

The boreal landscape also includes areas that never burn. These fire-free areas are important for the persistence of fire-sensitive species. Fire-free refuges occur across most of Fennoscandia (Essen et al., 1992); in the southeast Yukon Territory such an area contains an exceptionally rich flora (Haeussler and Kneeshaw, 2003). With the more frequent, more extensive, and more intense fires projected to result from climate warming, current fire refuges are likely to burn for the first time in recent history, thus reducing or locally eliminating fire-sensitive species.

After a sustained period of enhanced burning caused by climate warming, some boreal forests are likely to undergo type conversion from conifer to broadleaf tree dominance as a result of the depletion of fuels (see Chapter 14). An abrupt shift in forest composition of that type would significantly decrease the amount of old conifer habitat present at a given time from the large landscape perspective, possibly decreasing populations of some dependent organisms to critically low levels.

The boreal forest is characterized by large numbers of individuals of the few tree species with wide ecological amplitude, in contrast to tropical forests that sustain a small number of individuals of many species. Genetic diversity in any species is in part the result of opportunity for gene recombinations and so follows the laws of probability. In the boreal forest, probability favors the survival of large numbers of different gene combinations because of the characteristically large populations of each species (Widen and Svensson, 1992). To the degree that these genotypes reflect specific adaptations to local environments, they promote the survival and success of the species (Li et al., 1997). For example, foresters have developed seed transfer guidelines in order to define areas in which it is safe to collect seed for planting in a given site, based on their practical experience of failures in tree plantations from seed collected outside the local environment; boreal Alaska includes several hundred seed transfer zones (Alden, 1991), suggesting that a high degree of local adaptation may be typical.

The optimum growth and survival of the major boreal tree species across their large and varied natural distributions requires the survival of a large proportion of current genes, including genes that are rare today but would help survival of the species under future environmental conditions. One of the main risks for boreal forest from climate change is that major areas of the current distribution of boreal tree species might become climatically unsuitable for their survival faster than populations of the species could migrate, resulting in the loss of many adaptive genes. Fire and insect outbreaks are known to be triggered by warm weather (see Chapter 14), and gene loss would be likely to result from larger areas of more complete tree death. Gene survival in a changing climate becomes even more difficult if the native gene diversity is already diminished, as is usually the case in a managed forest and where human activities have reduced forests to remnants (Lieffers et al., 2003). In human-dominated landscapes the appropriate genes for an adaptive response of boreal forest plants to some aspects of climate change may already be rare if the trait was not associated with traits selected for in the forest management program. In addition, when the landscape is fragmented by human activities (for example by roads, pipelines, power lines, industrial and agricultural development, and excessive grazing), even the plant species with adaptive genes are very unlikely to migrate effectively under future climate change.

Nearly all the boreal forest tree species are open wind pollinated, which facilitates a wide distribution of genes (Widen and Svensson, 1992). The present boreal forest is the product of major periods of global warming and cooling that forced the boreal organisms to migrate far to the south of current limits and back several times. These climatic displacements imply that today's plants have considerable adaptive abilities as they have survived past climate changes. Even so, some loss of genes is almost inevitable in populations of trees and other plants coping with the major and rapid environmental changes that have been projected (see Chapter 4).

From the geological record, Spicer and Chapman (1990) considered that climate change is most strongly expressed at the poles. There is a dynamic equilibrium between the climate, the soils, and the vegetation. Arctic soils are crucial to the functioning of the terrestrial ecosystems (Fitzpatrick, 1997). Heal (1999) considered that "soil biology has changed dramatically since…the 1970s" and "the emphasis and approach has changed from descriptive to predictive, structure to function, organism to process, local to global". Much of the descriptive data collected in the 1970s were summarized by Swift et al. (1979), where the soils of the tundra and taiga were compared with those of temperate and tropical areas. However, these shifts in emphasis highlight that scientific knowledge of arctic soils is out of date, and is particularly weak because the information gained during the International Biological Programme (the first international collaborative research program of the International Council of Scientific Unions, running from 1964 to 1974, with a focus on "the biological basis of productivity and human welfare" – see Clapham, 1980 and Bliss et al., 1981) in the 1970s lacks experimental evidence relevant to the current issues of climate change. Evidence for the change in ecological thinking is evident in the studies by Robinson and Wookey (1997) on Svalbard, in which the emphasis was on decomposition and nutrient cycling.

Soils have frequently been neglected when biodiversity and its conservation are considered (Usher, in press). However, soils often contain the most species-rich communities in the Arctic, and so need to be considered in any planning or action for conserving biodiversity. However, many fundamental questions remain (Heal, 1999). What are the physical drivers of change? How will the ecological processes that occur within soil respond to climate change? How will the populations and communities of soil organisms adapt to climate change? It is known that environmental perturbations can change the dominance and trophic structure of the nematode community (Ruess et al., 1999a) in the subarctic soils of northern Sweden, and that such changes can have a large impact on microbial biomass and microbial turnover rates (Ruess et al., 1999b). In the boreal forest, there appears to be little correlation between taxonomic diversity and the process rates within the soils (Huhta et al., 1998), but it is not known whether this is typical of other arctic soils

It is widely held that diversity promotes ecosystem function, and so that biodiversity loss threatens to disrupt the functioning of ecosystems (Luck et al., 2003). More research is needed on arctic soils to determine whether the many species in these soils are all required, or whether there is some "redundancy" whereby the ecosystem could function efficiently with far fewer species. Also, with climate change, it becomes increasingly important to understand the carbon fluxes through arctic and subarctic soils – will there be net accumulations of soil carbon or net losses of carbon in the form of CO_2 or CH_4 to the atmosphere? Such knowledge is critical for the development of conservation policies and for the management of arctic ecosystems and their biodiversity.

10.2.5. Human-modified habitats

The concept of the Arctic as a pristine environment is a widespread fallacy. Humans have long been involved in the Arctic, both directly and indirectly, with little effect on its biodiversity, although hunting and gathering activities, and grazing of domesticated stock, must have had some effect. Damming of rivers to create fish traps is one of the few examples of early intensive environmental modification by people, as is the effects of over-grazing in Iceland. It is only since about 1800 that people have had significant impacts on arctic biodiversity through intensive intentional, or unintentional, modification of terrestrial, freshwater, or marine environments. The main environmental modifications have been through:

- expansion of land management for agriculture (including herding) and forestry, both of which have been very limited;

- expansion of marine and, to a lesser extent, freshwater commercial fisheries, especially with the advent of recent technologies;
- aquaculture as an emerging marine industry; and
- industrial, urban, and recreational developments, which have expanded considerably in recent decades, resulting in modifications to most types of habitat, regional production and dispersal of contaminants, and associated expansion of communication networks.

The actual proportions of terrestrial, freshwater, and marine habitats that are directly managed for human use in the Arctic are still very small, in contrast with the situation in other areas of the world (except the Antarctic), where agricultural habitats growing crop plants abound, and where derelict land, left over from activities such as mining, quarrying, or municipal development, is not uncommon. Agriculture within the Arctic is very limited; forestry is slightly more frequent. Around settlements and industrial developments there have been substantial changes to the natural environment, and non-native (weed) species have been able to establish in these disturbed habitats. However, the projected changes in climate are very likely to result in significant expansion and intensification of these human activities across the region, particularly where climate warming is most marked. The greatest potential impacts on biodiversity are likely to be through fragmentation of terrestrial ecosystems and the expansion of marine traffic as sea-ice conditions become less severe in the Northeast and Northwest Passages. There are at least four fundamental characteristics of arctic biodiversity that make it sensitive to these developments.

1. Many arctic plants and animals have slow growth rates and are long-lived as adaptations to the short summer season. These characteristics limit their capacity to respond to relatively rapid changes in their environment, especially when these recur over relatively short time periods. Recurrent disturbance tends to select for species with ruderal characteristics, some of which are found in species living in sites where freeze–thaw cycles predominate.
2. The low productivity of most habitats forces fauna to forage or hunt over large areas. Finding suitable habitats for breeding and shelter further extends the range requirements. Thus fragmentation of habitats and limitations to movement could potentially affect many species.
3. The flora and fauna have been selected to survive under extreme climatic conditions. This has given them a competitive advantage in the Arctic over species from warmer climates. Climate warming is very likely to result in a gradual northward shift in arctic species as a result of a natural northward shift in the ranges of more southerly species. However, the projected increase in human activities will also result in the introduction of non-native species, some of which are

expected to compete successfully with the native species. This is analogous to the experience of species introductions on isolated islands.
4. Some species that breed in the Arctic migrate to lower latitudes to avoid the extreme winter conditions. Migration places significant energetic stress on the animals; this means that the animals have evolved specific routes which provide access to transit feeding areas. The modification of habitats by people, both within and outside the Arctic, can have significant impacts on particular migratory species or populations.

These four characteristics of the flora and fauna of the Arctic make them particularly sensitive to the expansion of human activities in the region. For example, the effects of over-grazing by domestic livestock are clearly evident in Iceland where the vegetation cover has been lost and soil erosion is severe (Arnalds et al., 2001). This has led to desertification, with more than 50% of Iceland's land area (excluding that under permanent ice) being classified as either in "poor condition" or "bad condition". The history of desertification in Iceland was outlined by Arnalds (2000), and stands as a reminder of what can happen when the land's vegetative cover is damaged. The vegetation in other areas of the Arctic has evolved in the presence of large herbivorous mammals, unlike Iceland's vegetation, a factor which was thought by Arnalds (2000) to be significant.

Climate change is likely to cause gradual expansion at the northern boundary and contraction at the southern boundary of the range of arctic species. In contrast, the expansion of human activities in response to climate change is very likely to cause more rapid northward movement and the introduction of non-native species. The latter will occur mainly through accidental transport and release of individual organisms and propagules beyond their current, natural distribution limits. Such introductions, although having a very low probability of survival (the 10%:10% rule, resulting in only 1% becoming problematic (Williamson, 1996)), will occasionally result in the establishment of populations that expand rapidly, causing invasions which are highly predictable in general but highly unpredictable in detail. Thus, a key lesson is "to expect the unexpected".

Conservation action needs to both prevent serious loss of biodiversity and hence ecosystem function, and to restore past damage. The work of the Soil Conservation Service in Iceland demonstrates the difficulty of restoring grossly damaged ecosystems, how long the process is likely to take, and the potential problems that can be caused by non-native, invasive species. In a changing environment it is also necessary to recognize that a few of the wild relatives of cultivated plants occur in the Arctic (Heywood and Zohary, 1995). Being on the northern edge of their ranges, these might have particular genetic traits that prove valuable in breeding new varieties of crop plants for use under different climatic conditions.

10.2.6. Conservation of arctic species

The Arctic is generally species-poor compared with other large geographical areas of the world. There are, however, a number of charismatic species that capture people's imagination; including the polar bear, the reindeer or caribou (*Rangifer tarandus*), the gyrfalcon (*Falco rusticolus*), and the apparently frail Arctic poppy (*Papaver polare*). Terrestrial mammals number only 48 species, although some might be more properly considered as subarctic species, straying into the Arctic by a short distance only. Of these 48 species, 9 occur in Greenland, 29 in Alaska, 31 in the Canadian Arctic, and 33 in the Russian Arctic. Sage (1986) lists these species, but noted some taxonomic uncertainties which could result in these numbers changing slightly following further taxonomic research. Corresponding figures for breeding birds, noting the caveat that some species breed only very occasionally in the Arctic, are 183 for the Arctic as a whole, and 61, 113, 105, and 136 for Greenland, Alaska, Canada, and Russia respectively.

Arctic species, especially mammals and birds, feature strongly in books on wildlife (e.g., CAFF, 2001; Sage, 1986) and ecology (e.g., Chernov, 1985; Stonehouse, 1989). The purpose of this section is not to list the species of the Arctic, but to reinforce the ecological characteristics of the species that live in the Arctic. An understanding of these characteristics is essential for the conservation management of the Arctic's biodiversity.

The main characteristic essential for a species to survive in the Arctic is the ability to cope with cold temperatures. Most species have evolved strategies for surviving the arctic winter, i.e., cold tolerance, with the remainder developing strategies for cold avoidance. There are many ways of developing cold tolerance. For mammals that spend the whole year in the Arctic, this often involves depositing a layer of fatty tissue under the skin, as occurs in species of whales and seals. These species provide a valuable resource for the local human populations that harvest them for meat and for the oil that can be extracted from the blubber. A similar physiological system is used in some seabirds, such as the Atlantic puffin (*Fratercula arctica*), a vital oily food in the diet of the former inhabitants of the North Atlantic island of St. Kilda (Quine, 1989).

Invertebrate animals have a different system of cold tolerance. They accumulate glycerol in their tissues and, although they are usually susceptible to freezing, are able to "supercool" whereby the body fluids remain liquid at temperatures well below the freezing point (Sømme and Conradi-Larsen, 1977a). The majority of the alpine, arctic, and antarctic insects and mites are able to supercool, developing glycerol concentrations of up to 42 µg/mg of fresh weight and being able to survive temperatures below -15 °C (Sømme, 1981). This has an effect on the life cycles of these invertebrates in that they cannot reach the reproductive state until they are two to three years old, largely because they have to empty their guts before they supercool and have relatively limited opportunities for growth during the short arctic summer (Birkemoe and Sømme, 1998; Sømme and Birkemoe, 1999). However, it is known that some species enter a reproductive diapause when reared at constant temperature in the laboratory (e.g., the collembolan *Hypogastrura tullbergi*), and that this diapause can only be terminated by exposure to cold (Birkemoe and Leinaas, 1999). This poses the question as to whether, with the warming of the terrestrial environment, some invertebrate species may be unable to breed. Hodkinson et al. (1998) have reviewed the whole subject in relation to invertebrates that live in arctic soils.

Cold avoidance is a strategy adopted by a number of species of vertebrate animals. Arctic rodents, such as the insular vole (*Microtus abbreviatus*) of the Alaskan and

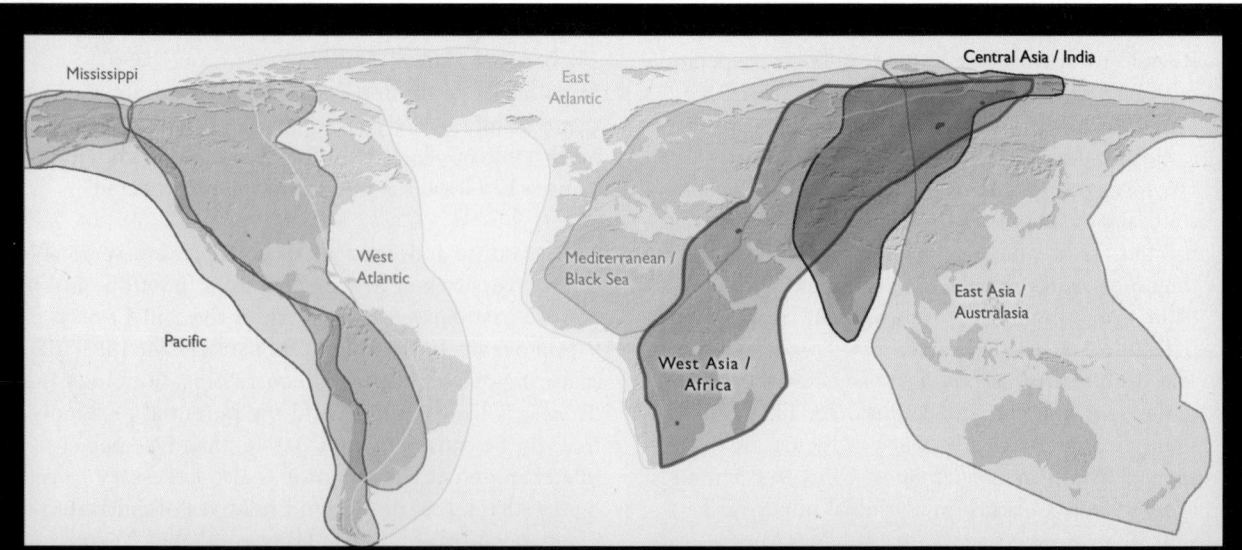

Fig. 10.4. The eight main international flyways used by shorebirds (waders) on migration. Within each flyway reasonably constant routes are used between the breeding grounds and the wintering grounds, although the southbound and northbound routes might differ. Each flyway comprises many different individual routes used by the different species and by different populations within a species. All arctic areas used by breeding shorebirds are included in these eight flyways. (Based on Thompson D. and Byrkjedal, 2001).

Canadian Arctic, avoid the coldest conditions by living within or under the snow (Stonehouse, 1989). Reindeer and caribou migrate to the forest on the southern edge of the Arctic, to over-winter in the more sheltered conditions of the boreal forest, before migrating north in the spring to the arctic tundra grazing grounds. Many of the fish species of the Arctic Ocean follow the edge of the sea ice in its seasonal movements southward during the autumn/winter and northward in the spring/summer.

Some species of bird have perfected the cold avoidance strategy by undergoing long-distance migrations. BirdLife (2002) featured the movement of the buff-breasted sandpiper (*Tryngites subruficollis*) which nests predominantly in the Canadian Arctic (with a small population in the Alaskan Arctic), but over-winters in South America in an area stretching from southern Brazil, through the northeast corner of Argentina, and into Paraguay. This is an example of one of the eight recognized flyways, known as the Mississippi Flyway, for shorebirds that breed in the Arctic (Thompson D. and Byrkjedal, 2001). Figure 10.4 shows the routes between the arctic breeding grounds, the staging areas which allow the birds to feed while they are *en route*, and the wintering grounds (which are often in the Southern Hemisphere). Conservation efforts for these migratory species must be international so that the species gain protection along the whole of flyway as well as in the arctic breeding grounds.

It is more difficult to characterize the strategies of plants in terms of cold tolerance or cold avoidance. Virtually all arctic plants are perennial, and so are able to reproduce over several years or remain in a vegetative state until climatic conditions in a particular year favor reproduction. Perennial plants have overwintering organs, such as roots and buds, which are protected by snow or soil from the coldest temperatures. One of the very few annual species is the snow gentian (*Gentiana nivalis*), which occurs in the north American Arctic and Greenland; in Europe it is predominantly a mountain species (Fig. 10.5). The snow gentian flowers and sets seed rapidly in the summer, and is said to have a seed bank so that it can survive climatically adverse years without flowering or with very restricted flowering, and hence demonstrates extreme year-to-year variability in population size (Raven and Walters, 1956).

Anoxia is a potential problem for species that overwinter in the Arctic. Marine mammals surface in order to obtain fresh air, and use a number of ways to maintain breathing holes in sea ice. The migration of fish in relation to the extent of the sea ice may also be related to the oxygen content of the seawater as well as to temperature. Terrestrial invertebrates have also developed mechanisms to cope with anoxia: for example, the two mite species studied by Sømme and Conradi-Larsen (1977b) survived for at least three months at 0 °C under anoxic conditions, whereas a species from further south in Norway died within six to eight days under similar conditions. Arthropods form lactate under anoxic conditions, with concentrations rising to nearly 2 µg/mg fresh weight, indicating this as a possible mechanism for coping with the anaerobic conditions that might prevail in arctic soils during winter.

As well as developing strategies for cold tolerance and cold avoidance, arctic species need to cope with freeze–thaw cycles in spring and autumn, and warm conditions in summer when there might be excess water due to the ice melt or desiccation due to low precipitation (Hodkinson et al., 1998). Over the year, each species has to be able to survive many ecological conditions. This is particularly evident in two features of arctic populations: extended life cycles and extreme year-to-year variability in population size.

It has already been mentioned that very few arctic plant species are annuals, and that the soil arthropods are generally not reproductive until two or three years old (whereas in temperate Europe and North America such species would have at least one generation per year). An example of the extended life cycle was given by CAFF (2001) where the life cycle of "woolly bear" larva of the moth *Gynaephora groenlandica* can vary from 7 to 14 years. In much of northern Europe and America such "woolly bears" (of other moth species) have an annual life cycle.

There is often extreme year-to-year variability in the sizes of arctic populations. This is particularly evident in relation to the occasional outbreaks of the autumnal moth, *Epirrita autumnata*. The larvae of this moth can cause widespread defoliation of downy birch (*Betula pubescens*) trees, for example in Arctic Finland, and in the most severe cases the trees subsequently die. These two

Fig. 10.5. The snow gentian is one of the very few species of vascular plants in the Arctic that have an annual life history; germinating, flowering, and setting seed within the short growing season of the arctic summer. (Photo: M.B. Usher, July 1997).

features of arctic populations – the extended life cycles and the extreme fluctuations in size – both make conservation management, and particularly the monitoring of species, more difficult.

Although the Arctic might be species-poor compared to other regions of the world, there are very few arctic species that are currently threatened with extinction. BirdLife (2002) produced a world map, shaded from white (no species of bird known to be threatened with extinction), through shades of yellow and orange, to red (where at least 25 species are threatened). The majority of the Arctic is white, although there are some areas of pale yellow in the Russian Arctic. How this map might change with climatic warming is not known, but the situation in the Arctic at the start of the 21st century is healthier than in virtually any other major geographical region. If the arctic environment is conserved, with particular attention given to arctic ecosystems (Muir et al., 2003), it is possible that a smaller proportion of the Arctic's species will be threatened with extinction than in other geographical areas.

This ecosystem approach to conservation has been defined as "the comprehensive integrated management of human activities based on best available scientific knowledge about the ecosystem and its dynamics, in order to identify and take action on influences which are critical to the health of the ecosystems, thereby achieving sustainable use of ecosystem goods and services and maintenance of ecosystem integrity" (as quoted by Muir et al., 2003). The ecosystem approach can thus be applied either to the marine environment or to the terrestrial and freshwater environments of the Arctic, and is discussed further in section 10.5. It is fundamental to the conservation of any species that its ecosystem is conserved, with its variety of species and the genetic variability of those species. As relatively few arctic species are currently threatened with extinction, the Arctic must be one of the places where an ecosystem approach can most readily be adopted, bringing together the human, plant, animal, microbial, marine, freshwater, and terrestrial perspectives.

10.2.7. Incorporating traditional knowledge

Other chapters within this assessment address the impacts of climate change on indigenous peoples and local communities, as well as on their traditional lifestyles, cultures, and economies. Other chapters also report on the value of traditional knowledge, and the observations of indigenous peoples and local communities in understanding past and future impacts of climate change. This section focuses on the relationship between biodiversity and climate change, impacts on indigenous peoples, and the incorporation of traditional knowledge.

There has been increasing interest in recent years in understanding traditional knowledge. Analyses often link traditional knowledge with what is held sacred by local peoples. Ramakrishnan et al. (1998) explored these links

with a large number of case studies, largely drawn from areas of India, but also including studies based in other parts of Asia, Africa, the Middle East, and southern Europe. A focus on northern America, again with a number of case studies, was reported by Maynard (2002). The many case studies demonstrate that traditional knowledge is held by peoples worldwide, except perhaps in the most developed societies where the link between people and nature has largely been broken. A recognition of this breakdown is the first step toward restoring biodiversity and its conservation in a changing world using knowledge that has been built up over centuries or millennia. As Ramakrishnan et al. (2000) reported "although the links between traditional ecological knowledge on the one hand, and biodiversity conservation and sustainable development on the other, are globally recognized, there is a paucity of models which demonstrate the specificity of such links within a given ecological, economic, socio-cultural and institutional context". They state that "we need to understand how traditional societies…have been able to cope up with uncertainties in the environment and the relevance of this about their future responses to global change". These concepts point the way to a greater integration of the knowledge of indigenous peoples into the present and future management of the Arctic's biodiversity.

A recent report by the Secretariat for the Convention on Biodiversity on interlinkages between biological diversity and climate change (SCBD, 2003) specifically addresses projected impacts on indigenous and traditional peoples. The term "traditional peoples" is used by the Intergovernmental Panel on Climate Change in its report on climate change and biodiversity (IPCC, 2002) to refer to local populations who practice traditional lifestyles that are often rural, and which may, or may not, be indigenous to the location. This definition thus includes indigenous peoples, as used in the present assessment. The SCBD report began by noting that indigenous and traditional peoples depend directly on diverse resources from ecosystems for many goods and services. These ecosystems are already stressed by current human activities and are projected to be adversely affected by climate change (SCBD, 2003). In addition to incorporating the main findings of the IPCC report (IPCC, 2002), the SCBD report concluded as follows:

1. The effects of climate change on indigenous and local peoples are likely to be felt earlier than the general impacts. The livelihood of indigenous peoples will be adversely affected if climate and land-use change lead to losses in biodiversity, especially mammals, birds, medicinal plants, and plants or animals with restricted distribution (but have importance in terms of food, fiber, or other uses for these peoples) and losses of terrestrial, coastal, and marine ecosystems that these peoples depend on.
2. Climate change will affect traditional practices of indigenous peoples in the Arctic, particularly fisheries, hunting, and reindeer husbandry. The ongoing interest among indigenous groups relating

to the collection of traditional knowledge and their observations of climate change and its impact on their communities could provide future adaptation options.

3. Cultural and spiritual sites and practices could be affected by sea-level rise and climate change. Shifts in the timing and range of wildlife species due to climate change could impact the cultural and religious lives of some indigenous peoples. Sea-level rise and climate change, coupled with other environmental changes, will affect some, but not all, unique cultural and spiritual sites in coastal areas and thus the people that reside there.

4. The projected climate change impacts on biodiversity, including disease vectors, at the ecosystem and species level could impact human health. Many indigenous and local peoples live in isolated rural living conditions and are more likely to be exposed to vector- and water-borne diseases and climatic extremes and would therefore be adversely affected by climate change. The loss of staple food and medicinal species could have an indirect impact and can also mean potential loss of future discoveries of pharmaceutical products and sources of food, fiber, and medicinal plants for these peoples.

The SCBD report commented directly on the incorporation of traditional knowledge and biodiversity by noting that the collection of traditional knowledge, and the peoples' observations of climate change and its impact on their communities, could provide future adaptation options. Traditional knowledge can thus be of help in understanding the effects of climate change on biodiversity and in managing biodiversity conservation in a changing environment, including (but not limited to) genetic diversity, migratory species, and protected areas. The report also noted the links between biodiversity conservation, climate change, and cultural and spiritual sites and practices of indigenous people, emphasizing that shifts in the timing and range of wildlife species could impact on the cultural and religious lives of some indigenous peoples. A detailed consideration of the links between cultural and spiritual sites and practices on the one hand and indigenous peoples on the other has been published recently (CAFF, 2002b). Although this report focused on sacred sites of indigenous peoples in the Yamal-Nenets Autonomous Okrug and the Koryak Autonomous Okrug in northern Russia, it also examined wider arctic and international aspects with some consideration given to the conservation value of sacred sites for indigenous peoples in Alaska and northern Canada.

Local people have knowledge about biodiversity, although it might neither be recognized as such nor formulated using the terminology of scientific biodiversity, that can be of great assistance in the management of arctic biodiversity. Muir (2002b) discussed the models and decision frameworks for indigenous participation in coastal zone management using Canadian experience, and pointed out that commercial harvesting of fish and marine mammals, as well as the effects of tourism, can conflict with local peoples' subsistence harvesting rights for fish and marine mammals. Traditional knowledge is multi-faceted (Burgess, 1999) and very often traditional methods of harvesting and managing wildlife have been sustainable (Jonsson et al., 1993). It is these models of sustainability that need to be explored more fully as the biodiversity resource changes, and the potential for its sustainable harvesting changes with a changing climate.

10.2.8. Implications for biodiversity conservation

In terms of conserving arctic ecosystems and habitats, CAFF (2002a) stated that "the overall goal is to maintain and enhance ecosystem integrity in the Arctic and to avoid habitat fragmentation and degradation". This goal is elaborated by recognizing the holistic nature of biodiversity conservation, including not just the flora and fauna, but also the physical environment and the socio-economic environment of people living within the area. It is the socio-economic factors that particularly affect arctic ecosystems, exerting pressures that have the potential to degrade habitats, to force declines in population sizes and numbers of species, and to reduce the functioning of ecosystems. Habitat fragmentation is probably the greatest threat to arctic ecosystems, which seem particularly ill-equipped to deal with it.

Although an important means of conserving the natural and cultural heritage is through protected areas, it is not a panacea. The arctic countries, through CAFF, have promoted the establishment of the Circumpolar Protected Area Network (CPAN), which aims to link protected areas throughout the Arctic; to ensure adequate representation of the various biomes; and to increase the public's understanding of the benefits and values of protected areas throughout the Arctic.

This is a useful start to the conservation of the arctic biodiversity, but many productive areas, such as coastal zones and marine ecosystems, are currently very underrepresented in the CPAN (CAFF, 2002a). At best, protected areas will only cover a relatively small proportion of the total land and sea area of the Arctic, and so conservation thinking is required beyond the established protected areas. This means that conservation of biodiversity must be integral to all aspects of social policy, including health and education of local people, planning for visitors and the associated developments, control and regulation of developments, and all aspects of the use of land, water, and air. Biodiversity conservation must be an important aspect of thinking, or as CAFF (2002a) stated, there needs to be a principle of "conservation first".

CAFF recommended that "the Arctic States in collaboration with indigenous people and communities, other Arctic residents, and stakeholders (1) identify important freshwater, marine and terrestrial habitats in the Arctic and ensure their protection through the establishment of protected areas and other appropriate conservation measures, and (2) promote an ecosystems approach to resource use and management in the circumpolar Arctic, through, *inter alia*, the development of common guidelines and best practices". This provides a way forward, but the generalities need to be expanded into the detail needed for the practical application of biodiversity conservation alongside the sustainable development of the Arctic, and the sustainable use of its resources, for the benefit of local people and visitors alike. A consensus approach, as fostered at an Arctic Council meeting on freshwater, coastal, and marine environments (Muir et al., 2003), needs to be promoted and developed on a circumpolar basis.

10.3. Human impacts on the biodiversity of the Arctic

The projected climatic changes in the Arctic, particularly the projected decrease in sea-ice extent and thickness, will result in increased accessibility to the open ocean and surrounding coastal areas. This is very likely to make it easier to exploit marine and coastal species, over a larger area and for a greater proportion of the year. Decreased extent and thickness of sea ice and increased seawater temperatures will, however, also result in changes in the distribution, diversity, and productivity of marine species in the Arctic and so will change the environment for hunters and indigenous peoples. However, increased traffic and physical disturbance caused by increased access to the marine areas is likely to pose a more significant threat to biodiversity than increased hunting pressure. On land, snow and ice cover in winter enable access into remote areas by snowmobile and the establishment of ice roads; however, in summer, transportation and movement become more difficult. A shorter winter season and increased thawing of permafrost in summer, potentially resulting from a warming climate, could reduce hunting pressure in remote areas.

There are at least four types of pressure acting on marine, coastal, freshwater, and terrestrial habitats that affect both their conservation and biodiversity: (1) issues relating to the exploitation of species, especially stocks of fish, birds, and mammals, and to forests; (2) the means by which land and water are managed, including the use of terrestrial ecosystems for grazing domesticated stock and aquatic ecosystems for aquaculture; (3) issues relating to pollutants and their long-range transport to the Arctic; and (4) development issues relating to industrial development and to the opening up of the Arctic for recreational purposes. These factors were discussed by Hallanaro and Pylvänäinen (2002) and Bernes (1993), who included hydroelectricity generation as a major impact on freshwater systems.

10.3.1. Exploitation of populations

Exploitation and harvest of living resources have been shown to pose a threat to arctic biodiversity. Species like the Steller sea cow (*Hydrodamalis gigas*), in the Bering Sea, and the great auk (*Pinguinus impennis*), in the North Atlantic, were hunted for food by early western explorers and whalers, and became extinct in the 18th and 19th centuries, respectively. Increasing demands for whale products in Europe, and improvements to the ships and harvesting methods intensified the exploitation of several arctic baleen whale species from the 17th century onward. Over-exploitation resulted in severely depleted populations of almost all the northern baleen whale species, and few have recovered their pre-17th century population sizes. For example, even though a few individuals have been observed in recent years, the bowhead whale (*Balaena mysticetus*) is still considered extinct in the North Atlantic. The Pacific population is bigger, but still considered endangered. Both subpopulations used to number in the tens of thousands. Many baleen whales, feeding on zooplankton, were a natural part of the arctic ecosystems 400 years ago. Their large biomass implies that they may have been a "keystone" species in shaping the biodiversity of the Arctic Ocean.

Many populations of charismatic arctic species have been over-exploited over the last few hundred years. The history of the slaughter of walruses (*Odobenus rosmarus*) in the North Atlantic and Pacific is well documented (Gjertz and Wiig, 1994, 1995). The walrus survived because its range of distribution included inaccessible areas, and the species is now expanding back into its previous distributional range due to its protection and to a ban on harvesting the animals in many areas. The International Polar Bear Treaty (1973) protected the polar bear (*Ursus maritimus*) after several sub-populations became severely depleted due to hunting (Prestrud and Stirling, 1994). Some subspecies of reindeer/caribou have also been close to extinction due to hunting pressure both in the European and North American Arctic (Kelsall, 1968). Similarly, several goose populations have approached extinction due to hunting on the breeding and wintering grounds (Madsen et al., 1999).

There have also been effects on a number of tree species. Wood has always been a valued commodity and since the first human populations were able to fell trees and process the felled trunks, forests have been cut for their timber. During the last few centuries, systems of forest management have developed to enable the forest to be regenerated more rapidly, either naturally or artificially by planting young trees. The need to exploit these

Table 10.4. Percentage distribution of age classes of coniferous forests in countries with arctic territory (Hallanaro and Pylvänäinen, 2002). The index, *I*, is the ratio of the percentage of trees over 80 years old to the percentage less than 40 years old, and so indicates the naturalness of the forests.

	0–40 yr	41–80 yr	81–100 yr	>100 yr	Index (*I*)
Murmansk (Russia)	31	19	5	45	1.61
Norway	33	21	13	33	1.39
Finland	32	33	13	22	1.09
Karelia (Russia)	40	19	7	34	1.02
Sweden	52	22	10	16	0.50

Fig. 10.6. The reef forming deep-sea coral, *Lophelia pertusa* (white coral, upper left hand corner), occurs on the continental shelf and shelf break off the northwest European coast. The red gorgonian, *Paragorgia arborea*, occurs on these reefs. The brittle star, *Gorgonocephalus caputmedusae* (yellow, center), frequently occurs on top of the gorgonians to take advantage of stronger currents. (Photo: CAFF, 2001; reproduced with permission from CAFF, Iceland).

Fig. 10.7. Fragments and larger pieces of dead coral, *Lophelia pertusa*, from a trawling ground on the Norwegian continental shelf at a depth of about 190 m. The benthic communities have been severely disturbed and are virtually devoid of larger animals. (Photo: CAFF, 2001; reproduced with permission from CAFF, Iceland).

forests for wood is demonstrated by the age structure of the trees in national forest estates (Table 10.4). Natural (unmanaged) forests have a large proportion of old trees compared to young trees, whereas managed forests have a large proportion of younger trees (often managed on rotations of 40 to 80 years). Table 10.4 appears to indicate a positive correlation between northerliness and naturalness (indicated by the index, *I*).

Since around the 1970s, modern management systems, improved control, and changed attitudes have largely diminished threats from sports hunting and harvesting for subsistence purposes. Most of the previously over-exploited populations are recovering or showing signs of recovery. However, there are still examples where hunting is a problem. In accordance with the International Polar Bear Treaty, local and indigenous peoples are allowed to hunt polar bears. In Canada, populations in some of the 14 management areas were over-exploited in the 1990s, and hunting was stopped periodically in some of these areas (Lunn et al., 2002). Similarly, in Greenland, uncertainties about the number of polar bears taken, and about their sex and age composition, have created concerns about the sustainability of the current harvest (Lunn et al., 2002). In southwestern Greenland, seabird populations have been over-exploited for a number of years by local peoples and the populations of guillemots (*Uria* spp.) have decreased by more than 90% in this area (CAFF, 2001).

Arctic and subarctic oceans, like the Barents, Bering, and Labrador Seas, are among the most productive in the world, and so have been, and are being, heavily exploited. For example, (1) commercial fish landings in Canada decreased from 1.61 million tonnes in 1989 to 1.00 million tonnes in 1998 (Anon, 2001a); (2) the five-fold decline in the cod (*Gadus morhua*) stock in the Arctic Ocean between about 1945 and the early 1990s; and (3) the huge decline (more than 20-fold) in the herring

(*Clupea harengus*) stock in the Norwegian Sea (Bernes, 1993). A report on the status of wildlife habitats in Canada stated that "Canadian fisheries are the most dramatic example of an industry that has had significant effects on the ocean's habitats and ecosystems" (Anon, 2001a).

Considerable natural annual variability in productivity, mainly due to variations in the influx of cold and warm waters to the Arctic, is a considerable challenge for fisheries management in the Arctic. Collapses in fish populations caused by over-exploitation in years of low productivity have occurred frequently and have resulted in negative impacts on other marine species. The stocks of almost all the commercially exploitable species in the Arctic have declined, and Bernes (1993) went as far as to state that several fish stocks are just about eliminated. Hamre (1994) suggested that the relative occurrence of species at some trophic levels has been displaced. Such changes in the few commercially-valuable fish species can have tremendous impacts on the coastal communities which are dependent upon the fishing industry for their livelihoods (CAFF, 2001). Even though supporting information is scarce, it is likely that the disappearance of the big baleen whales and the heavy exploitation (or over-exploitation) of fish stocks over many years have changed the original biodiversity and ecosystem processes of the subarctic oceans.

Heavy exploitation of benthic species, such as shrimps and scallops, also affects other species in the benthic communities. Bottom trawls damage species composition and so affect the food web. An example is the damage that can be caused to the cold water coral community. This coral reef habitat, often in deep water near the edge of the continental shelf, supports many other species such as gorgonians and brittle stars (Fig. 10.6). Passes over this community with a trawl leave only fragments of dead coral that can support no other species (Fig. 10.7). It has been estimated that, within commercial fishing grounds, all points on the sea floor are trawled at least twice per year.

10.3.2. Management of land and water

Changes in both land and water use influence biodiversity in the Arctic. This is different to the situation in most of the more southern biomes where changes in land use predominate (Sala and Chapin, 2000). In the Arctic, the limited expansion of forestry and agriculture is likely to be restricted to particularly productive environments, although there is greater potential for aquaculture in the Arctic.

In the Arctic, the original change in land use might not be obvious and impacts may be progressive and long-lasting. Thus the gradual increase in grazing pressure, particularly by sheep, has resulted in the loss of sward diversity and eventual soil erosion. This was probably a contributory factor in the extinction of agricultural colonies in Greenland between AD 1350 and 1450. In Iceland, "desert" with unstable and eroding soils resulted from a combination of removal of the 25% forest cover and the introduction of sheep since settlement in the 9th century. Soil rehabilitation is now a priority, but is a long, slow process. Establishment of long-term grass swards has had some success, and planting birch (*Betula pubescens*) and native willows (*Salix lanata* and *S. phylicifolia*) is proving a successful conservation measure, using mycorrhizal inocula, for re-establishing species and habitat diversity of grasslands, shrublands, and woodlands that were lost through overgrazing (A. Aradottir, Icelandic Soil Conservation Service, pers. comm., 2004; Enkhtuya et al., 2003) although non-native species can cause problems.

Draining of peatlands, and other wetlands including marshes and salt marshes, has been widely undertaken to bring the land into productive use, mainly for forestry but to a limited extent also for agriculture. In general there is an inverse correlation between the extent of drainage and northerliness. Data for relatively small areas are not available, but national data are presented in Table 10.5. The index, *P*, gives an indication of how much of the national peatland has been drained, which in the most northerly areas is relatively small. Drainage has a major impact on biodiversity. Invariably

Fig. 10.8. In Norwegian Finnmark the number of reindeer trebled between 1950 and 1989 resulting in extensive overgrazing of the vegetation. The ground to the left and above the fence had been overgrazed, while that to the right and in the foreground had been protected from grazing. Note the presence of shrubs and the green nature of the herbaceous ground cover. (Source: Hallanaro and Pylvänäinen, 2002; reproduced with permission from Georg Bangjord, Statens Naturoppsyn, Norway).

most of the species characteristic of the wetland are lost, except where small populations survive in drainage ditches. The newly created habitats are more prone to invasion by non-native species, and soil erosion may become more problematic. Migratory bird species may lose nesting places, and the land cannot retain as much water as before and so runoff increases during and immediately after storms. Drainage therefore has a major effect on the functioning of ecosystems, as well as encouraging biodiversity loss, usually for very limited economic gains at a time when climate change is likely to increase both the risk and rate of desertification in the Arctic. Biodiversity conservation in the Arctic should recognize the importance of wetlands as functional ecosystems with their full biodiversity complement.

Overgrazing on the tundra can be severe; the subject has been reviewed by Hallanaro and Usher (in press). In Finland, there were around 120000 reindeer at the start of the 20th century. This increased to around 420000 animals by 1990, but subsequently declined to around 290000 animals by 2000. The effects of overgrazing are clearly shown wherever areas of countryside are fenced off. Figure 10.8 shows an area of Norwegian Finnmark where the density of reindeer trebled between 1950 and 1989. Overgrazing eliminates ground cover by shrubs and dwarf shrubs, as well as reducing the cover of herbs,

Table 10.5. Extent of peatland (Data: Hallanaro and Pylvänäinen, 2002). The index, *P*, is the proportion of the total peatland not drained (the figure in the second column minus the sum of the figures in the third and fourth columns) to the total peatland area. Because different countries use different definitions for peatland, the data are not comparable between countries, although the values of *P* are comparable between countries.

Country	Total area of peatland (million hectares)	Area drained for forestry	Area drained for agriculture	*P*
Iceland	1.00	Small	0.13	0.86
Karelia (Russia)	5.40	0.64	0.09	0.86
Norway	3.00	0.41	0.19	0.80
Sweden	10.70	1.50	0.60	0.80
Finland	10.40	5.70	0.60	0.39

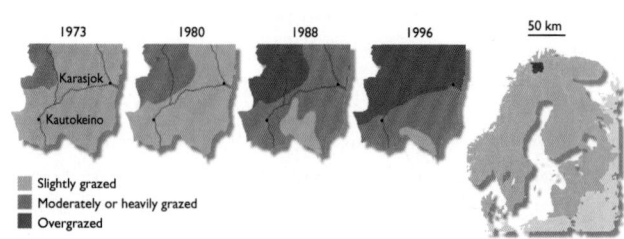

Fig. 10.9. Changes in grazing pressure in Finnmarksvidda, northern Norway, between 1973 and 1996. The increase in areas of lichen communities assessed as being overgrazed rises from none in 1973 to approximately two-thirds of the area in 1996. (Source: Hallanaro and Pylvänäinen, 2002; reproduced with permission from The Nordic Council of Ministers, Denmark).

grasses, and lichens. A more detailed analysis of the area where this photograph was taken is shown in Fig. 10.9. Over the 23 years from 1973 to 1996, the area changed from one having around a sixth of the land being moderately to heavily grazed (with the remainder being slightly grazed), to one having around two-thirds being overgrazed, a little under a third being moderately to heavily grazed, and only a small proportion (probably less than 5%) being slightly grazed.

The long-term effects of overgrazing are unknown, but if it results in the elimination of key species, such as shrubs, the recovery of the overgrazed ecosystems will be very slow. If all the key plant species remain in the community, even at very low densities, and are able to re-grow and set seed after the grazing pressure is lifted, then recovery could be faster. Two factors are important – the intensity of the grazing pressure and the period of time over which it occurs. Experimental exclosures have shown that, once grazing pressure by large herbivores is lifted, the regrowth of shrubs and tree species can be remarkable. Outside the fence, willows are reduced to small plants, of no more than a couple of centimeters high and with a few horizontal branches of up to 20 cm. These plants have few leaves and generally do not flower. Inside the fence the willows grow to at least 40 cm high, and are full of flowers with abundant seed set (Fig. 10.10). It is unknown how long these dwarf, overgrazed plants can both survive and retain the ability to re-grow after the grazing pressure is reduced. There have been no studies on the associated invertebrate fauna of these willows. So, it also unknown whether the phytophagous insects and mites are able to survive such a "bottleneck" in the willow population, or for how long they can survive these restricted conditions.

Although the vascular plants are the most obvious, it is the lichen component of arctic habitats that can be most affected by overgrazing. In areas with reindeer husbandry, the lichen cover has generally thinned on the winter grazing grounds. In the most severely impacted

Fig. 10.10. Whortle-leaved willow (*Salix myrsinites*) fruiting and growing in a grazing exclosure on limestone grassland that had been heavily overgrazed. After about 20 years without grazing by sheep or deer, this willow forms an understorey with other shrubs to a sparse woodland of birch (*Betula pubescens*) and rowan (*Sorbus aucuparia*) trees. (Photo: M.B. Usher, June 1998).

areas the lichens have been almost completely grazed out of the plant communities, or have been trampled, exposing bare ground which is then subject to erosion. Lichens, which are capable of surviving the harshest of environmental conditions, are frequently the most important photosynthetically active organisms in tundra ecosystems. Albeit slow-growing, many lichen species only thrive at low temperatures, and there is concern that if climate change results in a reduction in the number of lichen species or individuals, there could be a massive release of CO_2 to the atmosphere (Dobson, 2003). The combination of very low growth rates, overgrazing by domesticated or wild mammals and birds, and climate change indicates that large areas of the Arctic are susceptible to huge habitat changes in the future. Potentially, the lichen cover could be replaced by bare ground, with the risk of erosion by wind and running water, or by species that are currently not native to the Arctic.

Forests provide shelter during the coldest months of the year, and some of the mammals that feed on the tundra in summer migrate to the forests in winter. Pressure on herbaceous ground vegetation, especially on the lichens, can be severe. This is likely to be more of a problem in managed forests where the trees are grown closer together, less light reaches the forest floor, and the herbaceous and lichen layer is thus sparser. Overgrazing of the forest floor vegetation, including the young regeneration of tree species, is a problem in some areas and a potential problem in all other areas. Overgrazing, however, may not just result from agricultural and forestry land use; it may also result from successful conservation practices. For example, the population of the lesser snow goose (*Chen caerulescens*) in northern Canada rose from 2.6 million in 1990 to 6 million in 2000 as a result of protection. In summer, the geese feed intensively on the extensive coastal salt marshes (of western Hudson Bay), but large areas are now overgrazed, the salinity of the marshes is increasing, and vegetation has deteriorated. These examples demonstrate the potential fragility of ecosystems in which the food web is dominated by a few key species – a situation not uncommon in the Arctic.

The introduction of species into species-poor northern ecosystems is a disturbance which can have major impacts on the existing flora and fauna. The impact of introduced foxes and rats on seabird populations on arctic islands is particularly strong. A similar situation also occurs when new species are introduced into isolated freshwater ecosystems or when conditions change within a lake. For example, opossum shrimps (*Mysis relicta*) were introduced into dammed lakes in the mountains of Sweden and Norway by electric companies to enhance prey for burbot (*Lota lota*) and brown trout (*Salmo trutta*). Unexpectedly, the shrimps ate the zooplankton that was a food source for Arctic char (*Salvelinus alpinus*) and whitefish (*Coregonus lavaretus*), leading to an overall decline in fish production. Arctic char provide many interesting insights into arctic species. The resident population in Thingvallavatn, Iceland, was isolated from the sea 9600 years ago by a volcanic eruption, and became trapped within the lake. There are now four distinct forms that, although closely related genetically, are very different with respect to morphology, habitat, and diet. The Arctic has been described as a "theatre of evolution" as the few resident species capitalize on those resources that are not contested by other species. This encourages genetic diversification, a feature that is strongly shown by the Arctic char, a genetically diverse species and the only freshwater fish inhabiting high-arctic waters (Hammer, 1989, 1998).

The subtle and sensitive interactions within food webs are illustrated by an experiment at Toolik Lake LTER (Long Term Ecological Research) site in Alaska. Lake trout (*Salvelinus namaycush*) play a key role controlling populations of zooplankton (*Daphnia* spp.), snails (*Lymnaea elodes*), and slimy sculpin (*Cottus cognatus*). To test the hypothesis that predation by lake trout controls populations of slimy sculpin, all large trout were removed from the lake. Instead of freeing slimy sculpin from predation, the population of burbot rapidly expanded and burbot became an effective predator, restricting slimy sculpin to rocky littoral habitats, and allowing the density of its prey, chironomid larvae, to remain high. This is an example of changes in "top-down" control of populations by predators, contrasting with "bottom-up" control in which lower trophic levels are affected by changes in nutrient or contaminant loading (Vincent and Hobbie, 2000; see also Chapter 8).

Disturbance resulting from management in marine ecosystems has not been widely studied, other than by observing the impacts of trawling on seabed fauna and habitats (Figs. 10.6 and 10.7) and preliminary consideration of the potential impacts of invasive species through aquaculture, ballast water, and warming (Muir et al., 2003). Impacts of trawling are not particularly apparent in shallow waters where sediments are soft and organisms are adapted to living in habitats that are repeatedly disturbed by wave action. In deeper waters, undisturbed by storms and tides, large structural biota have developed, such as corals and sponges, and which provide habitats for other organisms. These relatively long-lived, physically fragile communities are particularly vulnerable to disturbance and are not adapted to cope with mechanical damage or the deposition of sediment disturbed by trawls.

Fish farming also affects marine ecosystems. This can be local due to the deposition of unused food and fish feces on the seabed or lake floor near the cages in which the fish are farmed. Such deposits are poor substrates for many marine organisms, and bacterial mats frequently develop. There can also be polluting effects over wider areas due to the use of veterinary products. Over a wider area still, escaped fish can interbreed with native fish stocks, thereby having a genetic effect. Thus, commercial fishing and fish farming can have adverse effects on arctic biodiversity. Sustainable management practices may be difficult to develop, but their introduction and implementation are essential if the fishery industries are to persist into the future.

There is a particular need to assess the potential problems faced by migratory fauna. The challenges met by migratory species are illustrated by the incredible dispersion of shorebirds to wintering grounds in all continents (Fig. 10.4). Recent evidence on waders from the East Atlantic flyway compares the population trends in seven long-distance migrant species that breed in the high Arctic with 14 species that have relatively short migrations from their breeding grounds in the sub-arctic. The long-distance migrants all show recent population declines and are very dependent on the Wadden Sea on the Netherlands coast as a stopover feeding ground. The waders with shorter migrations are much less dependent on the Wadden Sea and show stable or increasing populations. The emerging hypothesis is that waders with long migrations are critically dependent on key stopover sites for rapid refueling. For the Wadden Sea, although the extent available has not changed, the quality of resources available has declined through expansion of shellfish fisheries (Davidson, 2003).

There is evidence of a similar impact on migratory waders at two other sites. In Delaware Bay, a critical spring staging area in eastern North America, the impact is again due to over-exploitation of food resources by people. Similarly, the requirements of people and waders are in conflict in South Korea where a 33 km seawall at Saemangeum has resulted in the loss of 40 000 hectares of estuarine tidal flats and shallows. This site is the most important staging area on the East Asian Australasian Flyway, hosting at least 2 million waders of 36 species during their northward migration. At least 25 000 people are also dependent on this wetland system.

Thus, there are many forms of physical and biological disturbance in the Arctic (as well as in southern regions used by arctic species during migration). Such disturbances arise directly or indirectly from human intervention and the management of land and water. Although deliberate intervention can generate unexpected consequences, there is no doubt that conservation management is essential if the biodiversity of the Arctic is to be protected. In particular, implementation of international agreements, such as the Convention on the Conservation of Migratory Species of Wild Animals (also known as the Bonn Convention) and the Ramsar Convention on Wetlands, is increasingly urgent as a means to protect wetland and coastal areas.

10.3.3. Pollution

Pollution levels in the Arctic are generally lower than in temperate regions (AMAP, 1998, 2002). Locally, however, pollution from mining, industrial smelters, military activities, and oil and gas development has caused serious harm or posed potential threats to plant and animal life. Long-range transport of pollutants from sources outside the Arctic, in the atmosphere, rivers, or ocean currents, is also of concern (Anon, 2001a; Bernes, 1993). Particular problems include nitrogen and phosphorus causing eutrophication (especially in the

Baltic Sea), organic wastes from pulp mills creating an oxygen demand in the benthos, the effects of toxic metals (especially mercury), and bioaccumulation of organic compounds such as polychlorinated biphenyls (PCBs).

A recent report on the status of wildlife habitats in the Canadian Arctic (Anon, 2001a) listed four major classes of pollutant in the Arctic: mercury, PCBs, toxaphene, and chlorinated dioxins and furans (Table 10.6). Two main points are evident from Table 10.6: that pollutants are carried over long distances in the atmosphere and that pollutants accumulate in arctic food chains. Pollution is an international issue that needs to be resolved in a multi-national manner. However, wildlife is possibly more tolerant than might first appear because no arctic species are known to have become globally extinct due to pollution. However, the trends in pollutant uptake (see Table 10.6) are of concern.

Emissions of sulfur from industrial smelters and mining in the Russian Arctic have caused environmental disasters, killing vegetation and damaging freshwater ecosystems (AMAP, 1998). These impacts have, however, been restricted to relatively small areas surrounding the

Table 10.6. Major groups of pollutants in freshwater ecosystems and species in the Canadian Arctic (Anon, 2001a).

Mercury
- mercury is the most important metal in arctic lakes from a toxicological viewpoint
- observations show, and models confirm, that about a third of the total mercury that enters a high-arctic lake is retained in the sediments, around half is exported downstream, and the rest is lost to the atmosphere
- mercury concentrations consistently exceed guideline limits in fish for subsistence consumption or commercial sale
- mercury concentrations in fish tend to increase with increasing fish size

PCBs
- subarctic lakes first show PCB concentrations in the 1940s (±10 years)
- high-arctic lakes show no significant PCB concentrations until the 1960s (±10 years)
- PCB concentrations in fish tend to increase with increasing fish size

Toxaphene
- toxaphene is the major organochlorine contaminant in all fish analyzed
- highest toxaphene levels are generally seen in fish that are strictly piscivorous
- toxaphene concentrations in fish tend to increase with increasing fish size

Chlorinated dioxins and furans
- chlorinated dioxins and furans are found in fishes from some Yukon lakes
- levels of chlorinated dioxins and furans in fish throughout the Canadian Arctic are low compared to levels in fish obtained either near bleached Kraft mills or in the lower Great Lakes

sources. Long-range transport of sulfur and acid rain to the Arctic has reduced in recent years. The problems of acidification due to sulfur deposition are well known and ameliorative procedures have been established (Bernes, 1991). Acidification results in lakes becoming clear and devoid of much of their characteristic wildlife, so causing considerable local loss of biodiversity. Data from well water in Sweden (Bernes, 1991) showed a north–south gradient in acidification, with fewest effects in the north. Liming the inflow waters of some lakes has seen a recovery or partial recovery in pH, the aquatic plant and animal communities, and recolonization and recovery of the fish populations. An analysis of Scandinavian rivers (Bernes, 1993) also showed a north–south gradient, with relatively few acidified rivers in the arctic areas.

Pollution is also a threat to the boreal forests. The problems of increased aerial deposition of nitrogen have been well documented (e.g., Bell, 1994), and result in both eutrophication and acidification. The acidifying effects of sulfur deposition tend to be least severe in the Arctic, owing to its distance from areas where sulfur oxide (SO_x) gases are emitted. However, there are areas of the Arctic where the degree of acid deposition exceeds the soil's capacity to deal with it, i.e., the critical load (Bernes, 1993).

Levels of anthropogenic radionuclides in the Arctic are declining (AMAP, 2002). Radionuclides in arctic food chains are derived from fallout from atmospheric nuclear tests, the Chernobyl accident in 1986, and from European reprocessing plants. Radiocesium is easily taken up by many plants, and in short food chains is transferred quickly to the top consumers and people, where it is concentrated. Radiocesium has been a problem in arctic food chains, but after atmospheric nuclear tests were stopped 40 years ago, and the effects of the Chernobyl accident have declined, the problem is diminishing. Hallanaro and Pylvänäinen (2002) discussed the effects of the nuclear tests in Novaya Zemla, Russia and the Chernobyl accident, and concluded that neither had "resulted in any evident changes in biodiversity".

Oil pollution in the Arctic has locally caused acute mortality of wildlife and loss of biodiversity. Long-term ecological effects are also substantial: even 15 years after the Exxon Valdez accident in Alaska, toxic effects are still evident in the wildlife (Peterson et al., 2003). A more acute form of pollution is due to major oil spills, although minor discharges are relatively common. Devastation of wildlife following an oil spill is obvious, with dead and dying oiled birds and the smothering of intertidal algae and invertebrate animals. The type of oil spilled, whether heavy or light fuel oil, determines the effects on the fish. Light oils that are partially miscible with seawater can kill many fish, even those that generally occur only at depth (Ritchie and O'Sullivan, 1994). Less sea ice resulting from a warming climate is likely to increase accessibility to oil, gas, and mineral resources, and to open the Arctic Ocean to transport between the Pacific and Atlantic Oceans. Such activities will increase the likelihood of accidental oil spills in the Arctic, increasing the risk of harm to biodiversity. A warmer climate may, however, make combating oil spills easier and increase the speed at which spilled oil decomposes.

With the possible exception of mercury, heavy metals are not considered a major contamination problem in the Arctic or to threaten biodiversity (AMAP, 2002). The Arctic may, however, be an important sink in the global mercury cycle (AMAP, 2002). Mercury is mainly transported into the Arctic by air and deposited on snow during spring; the recently discovered process involves ozone and is initiated by the returning sunlight (AMAP, 2002). Mercury deposited on snow may become bioavailable and enter food chains, and in some areas of the Arctic levels of mercury in seabirds and marine mammals are increasing.

Persistent organic pollutants (POPs) are mainly trans-ported to the Arctic by winds. Even though levels in the Arctic are generally lower than in temperate regions, several biological and physical processes, such as short food chains and rapid transfer and storage of lipids along the food chain, concentrate POPs in some species at some locations. AMAP (2002) conclud-ed that "adverse effects have been observed in some of the most highly exposed or sensitive species in some areas of the Arctic". Persistent organic pollutants have negative effects on the immune system of polar bears, glaucous gulls (*Larus hyperboreus*), and northern fur seals (*Callorhinus ursinus*), and peregrine falcons (*Falco peregrinus*) have suffered eggshell thinning. The ecologi-cal effects of POPs are unknown.

The direct effects of pollutants on trees are compound-ed by the effects of diseases and defoliating arthropods, and by interactions between all three. Across Europe, these have been codified into the assessment of crown defoliation and hence crown density (e.g., Innes, 1990). Each country prepares an annual report to allow the international situation to be assessed and trends deter-mined. These assessments provide a measure of forest condition and changes in condition. These assessments are currently made in the main timber producing areas of Europe, but it would be of benefit to establish an international forest condition monitoring network across the boreal forests of the subarctic.

A warmer Arctic will probably increase the long-range transport of contaminants to the Arctic. Flow rates in the big Siberian rivers have increased by 15 to 20% since the mid-1980s (see Chapter 6) due to increased precipitation. Northerly winds are likely to increase in intensity with climatic warming, bringing more volatile compounds such as some POPs and mercury into the Arctic. Conservation action must aim to reduce the amounts of the pollutants resulting in chronic effects from entering arctic ecosystems, and to reduce the risk of accidents for pollutants resulting in acute effects.

10.3.4. Development pressures

Biodiversity in the Arctic is affected by pervasive, small-scale, and long-lasting physical disturbance and habitat fragmentation as a side-effect of industrial and urban developments and recreation. Such disturbances, often caused by buildings, vehicles, or pedestrians, can alter vegetation, fauna, and soil conditions in localized areas. A combination of these "patches" can result in a landscape-level mosaic, in effect a series of "new" ecosystems with distinctive, long-term, biodiversity characteristics. These are becoming more widespread in the Arctic and in some cases can, through enhanced productivity and vegetation quality, act as "polar oases" having a wide influence on local food webs.

Forbes et al. (2000) reviewed patch dynamics generated by anthropogenic disturbance, based on re-examination of more than 3000 plots at 19 sites in the high and low arctic regions of Alaska, Canada, Greenland, and Russia. These plots were established from 1928 onward and resurveyed at varying intervals, often with detailed soil as well as vegetation observations. Although these patches have mostly experienced low-intensity and small-scale disturbances, "none but the smallest and wettest patches on level ground recovered unassisted to something approaching their original state in the medi-um term (20–75 years)". Forbes et al. (2000) conclud-ed that "in terms of conservation, anthropogenic patch dynamics appear as a force to be reckoned with when plans are made for even highly circumscribed and ostensibly mitigative land use in the more productive landscapes of the increasingly accessible Arctic".

Development in the marine environment of the Arctic is currently very limited. However, a recent report on the status of wildlife habitats in the Canadian Arctic (Anon, 2001a) stated that "the Arctic landscapes and seascapes are subject to…oil and gas and mining devel-opments [which] continue to expand". Muir's (2002a) analysis of coastal and offshore development concluded that pressures on the marine environment are bound to increase. There will be further exploration for oil and gas. If substantial finds are made under the arctic seas then development is likely to take place. While most known oil reserves are currently on land, offshore exploration, such as that west of the Fylla Banks 150 km northwest of Nuuk in Greenland (Anon, 2001b), will continue to have local impacts on the seabed. Muir (2002a) also predicted that marine navigation and trans-port are likely to increase in response to both economic development and as the ice-free season extends as a result of climate change, with the consequent infra-structure developments.

Recreational use of arctic land by people, largely from outside the Arctic, is increasing. Although hikers and their associated trails potentially present few problems, this is not the case for the infrastructure associated with development and for off-road vehicles. Potential prob-lems with trails are associated with vegetation loss along and beside the trail. This leads to erosion of the skeletal soils by wind, frost, or water. There is current discussion about the use of trekking poles (Marion and Reid, 2001) and whether, by making small holes in the ground that can fill with water, followed by freeze–thaw cycles, they increase the potential for erosion.

Use of off-road vehicles has increased with their greater accessibility. They can also exert greater environmental pressures than trampling by people. As a result various laws and regulations have been introduced to reduce or eliminate the damage that they cause. In Russia, off-road vehicles are frequently heavy, such as caterpillar tractors. Although it is forbidden to use these in treeless areas in summer, violations are thought to be common. Norway has prohibited off-road driving throughout the year, although different rules apply to snowmobiles. Use of the latter is becoming more frequent, with 10–11 per thousand of the population owning them in Iceland and Norway by the late 1990s; this increases to 17 in Finland, 22 in Sweden, and 366 in Svalbard. The Fennoscandian countries have established special snowmobile routes to concentrate this traffic and so prevent more widespread damage and disturbance to snow-covered habitats.

Implications of infrastructure development and habitat fragmentation, especially the construction of linear features such as roads and pipelines, are less clearly understood. However, Nellemann et al.'s (2003) research gave some indications about effects on reindeer. Reindeer generally retreat to more than 4 km from new roads, power lines, dams, and cabins. The population density dropped to 36% of its pre-development density in summer and 8% in winter. In areas further than 4 km from developments, population density increased by more than 200%, which could result in overgrazing of these increasingly small "isolated" areas. If reindeer, easily able to walk across a road, behaviorally prefer to avoid roads, what are the effects of such developments on smaller animals, vertebrates and invertebrates, that are less capable of crossing such obstacles? This indicates that arctic habitats must be of large extent if they are to preserve the range of species associated with such habitats. How large should habitats be? Two developments 8 km apart, on the basis of Nellemann et al.'s (2003) research, can only accommodate 8% of the wild reindeer density (using winter data), and so developments will have to be more distant from each other if there is not to be undue pressure on the reindeer population and the habitats into which they move. Nellemann et al.'s (2001) conclusion was that the impacts of development in the Arctic extend for 4 to 10 km from the infrastructure. So, two developments separated by 20 km may leave no land unimpacted. Developments must therefore be carefully planned, widely separated, and without the fragmentation of habitats by roads, trails, power lines, or holiday cabins.

As well as potential impacts from development, habitats will change with a changing climate. An example of where this is important for tourism is in the Denali National Park, the most visited national park in Alaska. Bus tours provide the main visitor experience by providing viewing of wildlife and scenery along the park road. The Denali park road begins in boreal forest at the park headquarters and extends through treeline into broad expanses of tundra offering long vistas. Climate-driven changes in the position of forest versus tundra would have significant effects on the park by changing the suitability of certain areas for these experiences. A tree-growth model for the park has been developed based on landscape characteristics most likely to support trees with positive growth responses to warming versus landscapes most likely to support trees with negative responses (M.W. Wilmking, Columbia University, pers. comm., 2004). The results were projected into the 21st century using data from the five general circulation models climate scenarios used in the ACIA analysis. The scenarios project climates that will cause dieback of white spruce at low elevations and treeline advance and infilling at high elevations. The net effect of tree changes is projected to be a forest increase of about 50% along the road corridor, thus decreasing the possibility for viewing scenery and wildlife at one of the most important tourist sites in Alaska. The maps of potential forest dieback and expansion should be useful for future planning.

Developments have two important implications for conservation, and both can potentially be implemented *a priori*. First, what regulations are needed to reduce environmental risks? A study for the Hudson Bay area of Canada (Muir, 2000) provided possible mechanisms for safeguarding local communities, biodiversity, and the environment, while not totally restricting development. Second, how can competing interests be reconciled? Muir (2002a) advocated forms of integrated management, although stating that such "approaches to integrated management which reconcile economic and conservation values will be complex and consultative". There is a need for biodiversity conservation interests to form an integral part of any consultations over the use of the marine, coastal, freshwater, and terrestrial resources of the Arctic.

10.4. Effects of climate change on the biodiversity of the Arctic

This section examines how climate change might affect the biodiversity of the Arctic. The effects are grouped into six categories: potential changes in the ranges of species and habitats (section 10.4.1); changes in their amounts, i.e., the extent of habitats and population sizes (sections 10.4.2 and 10.4.3); possible genetic effects (section 10.4.4); changes in migratory habits (section 10.4.5); likely problems from non-native species (section 10.4.6); and implications for the designation and management of protected areas (section 10.4.7).

The discussions should be read alongside the appropriate sections of Chapters 7 (tundra and polar desert ecosystems), 8 (freshwater ecosystems), and 9 (marine systems), which also include analyses of the effects of climate change. This section should also be read alongside the appropriate sections of Chapters 11 (wildlife conservation and management) and 14 (forests and agriculture). In this chapter analyses are oriented toward the conservation of arctic genes, arctic species, and arctic ecosystems.

10.4.1. Changes in distribution ranges

In a warming environment it is generally assumed that the distribution range of a species or habitat will move northward, and that locally it will move uphill. Although such generalizations may be true, they hide large differences between species and habitats, in terms of how far they will move and whether they are actually able to move.

Some of the earlier studies were undertaken in Norway and investigated the "climate-space" then occupied by a few communities and plant species. The "climate-space" comprised two factors – altitude and distance inland (Holten and Carey, 1992). Figure 10.11 shows the effect of a probable climate change scenario on the distribution of blueberry (*Vaccinium myrtillus*) heaths. The heath is predicted to move uphill, with its mean altitude changing from about 760 m to about 1160 m. The questions for the conservation of this type of heathland are whether all heaths below 700 m will cease to exist (and how quickly this will happen) and whether the heaths can actually establish at altitudes of between about 1300 and 1600 m. Similar studies for other plant species generally predict that they will move to occupy a climate-space that is at a higher altitude and further inland (Holten, 1990).

Norway spruce (*Picea abies*) presently occurs throughout Fennoscandia and Russia, more or less as far north as the shore of the Arctic Ocean. If winter temperatures rise by 4 °C, the distribution range projected for Norway

spruce virtually halves, with the majority of the southern and southwestern populations disappearing (Holten and Carey, 1992). Owing to the barrier caused by the Arctic Ocean, Norway spruce cannot expand its distribution northward, and so is squeezed into a smaller area. Holten and Carey (1992) also projected the distribution of beech (*Fagus sylvatica*), a tree whose present distribution is more southern. They forecast that this species will spread northward into the Arctic, and may potentially replace the spruce in some of the more coastal areas. The distribution range of the beech thus expands as it shifts north and moves into the Arctic, there being apparently no barriers to its expansion (except perhaps for the size of its seed which makes dispersal more difficult).

In modeling changes in distribution ranges, attempts are made to identify the "climate-space" which a species or habitat currently occupies, and then to identify where that climate-space will occur under scenarios of climate change, for example in 2050 or 2100. Such models assume that the species or habitat currently occupies its optimal climate-space, and that the species or habitat will be able to move as the climate changes. This brings up a range of questions about the suitability of areas for moving through and of barriers, such as mountains for terrestrial species and habitats, or the difficulty of moving from lake to lake, or river to river, for freshwater species. Such models have been used to project what might happen to species on nature reserves (Dockerty and Lovett, 2003), in mountain environments (Beniston, 2003), and to the species of the major biomes isolated on nature reserves (Dockerty et al., 2003). Dockerty et al. (2003) predicted that the relict arctic and boreo-arctic montane species in temperate regions are all likely to have a decreased probability of occurrence in the future.

Arctic species and habitats are thus likely to be squeezed into smaller areas as a result of climate change. However, there are some caveats. Cannell et al. (1997), exploring interactions with pollutant impacts (the CO_2 fertilization effect and nitrogen deposition), concluded that the movement of plant species may be less than expected, but that the stress-tolerant species, including those characteristic of the Arctic, are likely to be lost. Oswald et al. (2003) also explored possible changes in

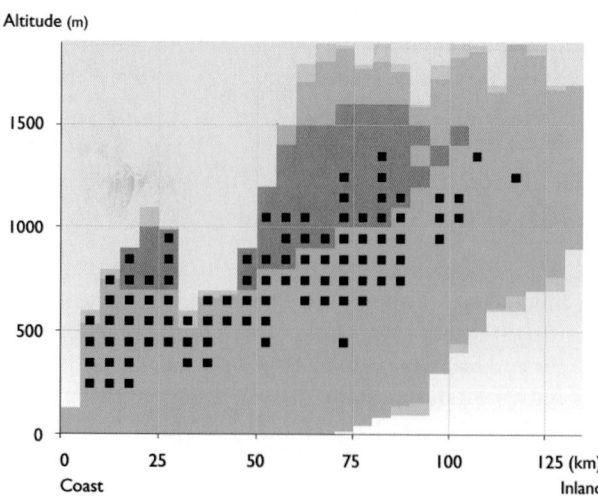

Fig. 10.11. A correlative model showing the current (black squares) and predicted (shaded purple) range of *Vaccinium* heaths in Norway. The grid cells represent steps of 100 m in altitude on the vertical axis and 5 km distance from the sea on the horizontal axis. The model is derived from the then most probable scenario of climate change in Norway, i.e., a 2 °C increase in July temperatures and a 4 °C increase in January temperatures (Holten and Carey, 1992).

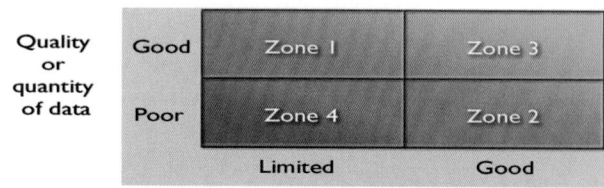

Fig. 10.12. A representation of extent of understanding and the quality or quantity of data when applied to modeling problems. For the majority of potential applications in conservation the level of understanding of the system is low and the quantity of data small, and so the modeling would fall in the lower left corner of Zone 4 (Usher, 2002a).

plant species in northern Alaska, and concluded that the responses of species and habitats are likely to be heterogeneous. The continued northward push of the more southern species and habitats has been outlined by Pellerin and Lavoie (2003) in relation to changes in ombrotrophic bogs due to forest expansion. It is these individualistic responses to climate change (Graham and Grimm, 1990), by species and habitats, which make prediction difficult. Individualistic responses appear to be the norm rather than the exception for plants and invertebrate animals (Niemelä et al., 1990).

The individualistic responses of species may produce novel effects. This is illustrated using the example of a simple and hypothetical community with a broadly similar abundance of three species: A, B, and C (community A+B+C). Under a climate change scenario with species moving northward, if species A moved rapidly, species B moved more slowly, and species C hardly moved at all, this could result in a community dominated by species A with species B as a sub-dominant (community A+b) in the north and a community dominated by species C with species B as a sub-dominant (community C+b) more or less where A+B+C used to occur. It is possible that neither A+b nor C+b would be recognized as communities, and so, in the geographical contraction of A+B+C, two new communities – A+b and C+b – had arisen, both of which were novel. Climate change could thus give rise to some new habitat types, and although this might not change the overall biodiversity of the Arctic at the species level, there could be changes to biodiversity at the habitat level.

Current distribution ranges of plants and animals in the marine environment depend upon the ocean currents as well as on the extent of the sea-ice cover at different times of the year. With the projected decrease in sea-ice cover and the more northerly position of the ice edge, the distribution of the algae, phytoplankton, invertebrates, and fish will also change. An analysis of the effects of climate change on marine resources in the Arctic (Criddle et al., 1998) left much in doubt, stating that "the effects of climate variation on some Bering Sea fish populations are fairly well known in terms of empirical relationships but generally poorly known in terms of mechanisms". The authors proposed a program of research to help predict the effects of climate change on the commercially-exploited fish stocks and more widely on marine biodiversity as a whole.

The lack of knowledge on this topic was addressed by Starfield and Bleloch (1986). They presented a simple model of the context within which most conservation work could be undertaken (Fig. 10.12). Conservation generally has little understanding of the system to be conserved, and managers have poor data upon which to build models. The conservation of biodiversity falls in zone 4. This is the zone where statistical models are most helpful, indicating expectations with some probability attached and often very wide confidence limits.

What are the implications for conservation? The most detailed assessment of changes in distribution ranges of species and ecosystems in relation to conservation are probably the studies on national parks and other conser-

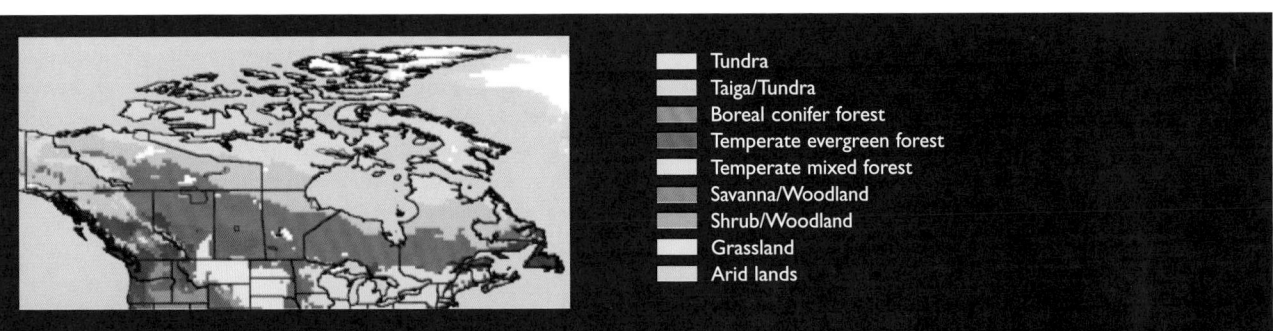

	Tundra
	Taiga/Tundra
	Boreal conifer forest
	Temperate evergreen forest
	Temperate mixed forest
	Savanna/Woodland
	Shrub/Woodland
	Grassland
	Arid lands

Fig. 10.13. The present MAPSS vegetation distribution in Canada's national parks. Nine vegetation zones are shown, excluding the permanent ice in the north (reproduced with permission from Daniel Scott, University of Waterloo, Canada).

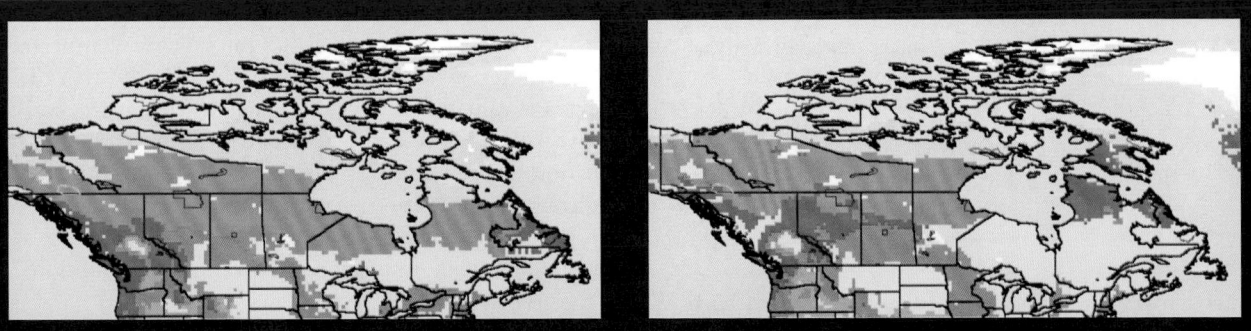

Fig. 10.14. The projected MAPSS vegetation distribution in Canada's national parks using two scenarios of climate change. Although the details of these two projections differ, they both demonstrate the northward movement of vegetation zones relative to current conditions (reproduced with permission from Daniel Scott, University of Waterloo, Canada).

vation areas in Canada (Scott and Lemieux, 2003; Scott and Suffling, 2000; Scott et al., 2002). The large scale of biomes and environmental conditions in Canada facilitate the definition of spatial patterns by models with a grid resolution of 0.5° latitude by 0.5° longitude. The studies of 36 national parks and other designated conservation areas involved the application of two global vegetation models (BIOME3 and MAPSS) which represent the effects of enhanced CO_2 on nine or ten biome types consistent with IPCC analysis. The different number of biomes is because BIOME3 combined boreal and taiga/tundra biomes which were separated in MAPSS. Five general circulation models (three equilibrium models: UKMO, GFDL-R30, and GISS; two transient models: HadCM2 and MPI-T106) were applied, providing some direct cross-reference to the present assessment.

A northward movement of the major biomes was projected in all five scenarios, changes in the dominant biomes of tundra, taiga/tundra, and boreal conifer forest were particularly clear (compare Fig. 10.13, which shows present conditions, with Fig. 10.14, which shows two projections for the northerly movement of the Canadian vegetation zones). As is the case for the ACIA-designated climate models (see Chapter 4), although the trends were similar between models, the actual values and local spatial patterns showed considerable variation. Regardless of the vegetation and climate change scenarios used, the potential for substantial changes in biome representation within the national parks was shown repeatedly. At least one non pre-existing biome type appeared in 55 to 61% of parks in the MAPSS-based scenarios and 39 to 50% in the BIOME3-based scenar-

ios. Representation of northern biomes (tundra, taiga/tundra, and boreal conifer forest) in protected areas was projected to decrease due to the overall contraction of these biomes in Canada. Projections for the southern biomes were more variable but their representation in protected areas generally increased.

The seven arctic national parks range in size from Vuntut in Yukon Territory at 4345 km² to Quttinirpaaq (formerly Ellesmere Island) at 37775 km² in Nunavut. The parks cover a range of conditions from high arctic polar desert and glaciers to taiga, extensive wetlands, coastal areas, lakes, and rivers. They also contain, and were often designated to conserve, a variety of species and populations; for example, they contain one of the greatest known musk oxen (*Ovibos moschatus*) concentrations, calving grounds for Peary caribou (*Rangifer tarandus pearyi*), migration corridors and staging areas, one of the largest polar bear denning areas, spawning and over-wintering sites for Arctic char, considerable species richness with over 300 plant species in one area, plus important historical, cultural, and archaeological sites and unique fossils from Beringia. Some of the significant impacts of climate change within the arctic national parks are outlined in Table 10.7.

10.4.2. Changes in the extent of arctic habitats

The previous section showed that distribution ranges of many arctic habitats are likely to decrease with climate change and that this generally implies a reduction in the overall extent of the habitat. The response of each habitat is likely to be individualistic (Oswald et al., 2003), and to depend upon the dynamics of the populations and communities, as well as on a range of species interactions such as competition, predation, parasitism, hyperparasitism, and mutualism. Habitat extent will depend upon the individualistic responses of the component species, and these in turn will depend upon the physiological responses of the individuals that form those species populations (see section 10.4.3).

In the marine environment far less is known about the potential effects of warmer temperatures, increased atmospheric CO_2 concentrations, and increased irradiance by ultraviolet-B (UV-B) on the species populations and habitats. A review of marine nature reserves by Halpern and Warner (2002) showed that changes in population sizes and characteristics can be fast. Compared with undesignated areas, their study indicated that the average values of density, biomass, organism size, and diversity all increased within one to three years of designation. These rapid responses, the result of protection through conservation designation, indicate that marine organisms and marine habitats have the potential to respond quickly to changed environmental conditions.

Change will occur, and in general it appears that arctic habitats are likely to have smaller population sizes within smaller distribution ranges. What will replace them? Habitats that currently occur in the sub-Arctic or in the

Table 10.7. Potential impacts of climate change on the arctic national parks and other protected areas (H.G. Gilchrist, Canadian Wildlife Service, pers. comm., 2004).

Impact	Effects of impact
Northward treeline extension	Up to 200–300 km movement in the next 100 years (where movement is not impeded by soil condition)
Increased active layer and permafrost thawing	May extend northward by 500 km, causing altered drainage patterns
Sea-level rise	Variable, either moderated by isostatic rebound or exacerbated by subsidence
Reduced sea- and lake-ice seasons	Altered sea mammal distributions (especially for polar bears and ringed seals), as well as more northerly distribution of ice-edge phytoplankton blooms, zooplankton, and fish
Increased snow pack and ice layers	Reduced access to browse for ungulates
Greater severity and length of insect seasons	Increased harassment of ungulates and potential for pest outbreaks in boreal forests
Altered migration patterns	Diminished genetic exchange among arctic islands
Altered predator–prey and host–parasite relationships	Changes in species abundance, and potentially the establishment of novel interactions between pairs of species

northern boreal zone are likely to move northward, and their responses to climate change are likely to be individualistic. So it is possible that habitats currently south of the Arctic might migrate northward and occur "naturally" within the Arctic, as for example with the northward movement of beech forest (section 10.4.1).

This will make it difficult to establish, if indeed there is a distinction, whether species and habitats of the Arctic in the future are native or non-native (see section 10.4.6). Owing to the different responses of habitats and species, it is likely that novel species assemblages will occur in the future, being habitat types that are currently unknown or not envisaged. Thus, the current habitat classifications are likely to have to change as novel habitat types evolve in response to rapid climate change. This has considerable implications for species and habitat conservation and for management today, and may lead to alterations in the priorities for biodiversity conservation in the future. While the name of a species is more or less stable, and so easily incorporated into legislative frameworks (i.e., appended lists of protected species), a habitat's name and description is less stable, implying a need for periodic reviews of legislative frameworks.

10.4.3. Changes in the abundance of arctic species

As sections 10.4.1 and 10.4.2 imply, it is the species composition of an area that will change, forcing changes to the communities in which they occur. The individualistic responses of the species (Oswald et al., 2003) will depend upon the dynamics of the species populations, the competitive or mutualistic interactions between species, and the biochemical and physiological responses of the individuals.

Biochemistry and physiology are fundamental to how an individual responds to its environment and to changes in that environment. Rey and Jarvis (1997) showed that young birch (*Betula pendula*) trees grown in an atmosphere with elevated CO_2 levels had 58% more biomass than trees grown in ambient CO_2 concentrations. They also found that the mycorrhizal fungi associated with the roots of the experimental trees differed; those grown in elevated CO_2 levels were late successional species, while those grown in ambient CO_2 levels were the early successional species. This showed the complexity of understanding the effects of climate change on the conservation of biodiversity. Normally, with regenerating birch trees, the whole successional suite of fungi would be expected to occur on the young trees' roots as they emerge from the seed, establish themselves, grow, and then mature. Does the work of Rey and Jarvis' (1997) imply that more attention needs to be given to protecting the early successional mycorrhizal species? They will clearly be needed in the ecosystem if climate cools or CO_2 levels fall in the future.

Other physiological studies have detected a 4 to 9% thickening of the leaves of lingonberry (*Vaccinium*

vitis-idaea) under enhanced UV-B radiation, whereas the deciduous blueberry and bog blueberry (*V. uliginosum*) both had 4 to 10% thinner leaves under similar conditions (Björn et al., 1997). Growth of the moss *Hylocomium splendens* was strongly stimulated by enhanced UV-B radiation, as long as there was additional water, whereas the longitudinal growth of the moss *Sphagnum fuscum* was reduced by about 20%. Björn et al.'s (1997) results for lichen growth under enhanced UV-B radiation were variable, leading them to conclude that "it is currently impossible to generalize from these data". They also investigated the decomposition of litter from *Vaccinium* plants grown under normal conditions and under conditions of enhanced UV-B radiation. Litter from the *V. uliginosum* plants treated with UV-B radiation had a decreased α-cellulose content, a reduced cellulose/lignin ratio, and increased tannins compared to the control litter, and so was more resistant to decomposition. Slower decomposition was also observed for *V. myrtillus* litter. Björn et al. (1997) did not investigate the palatability of the leaves to invertebrate animals. Moth larvae, particularly those in the family Geometridae (the "loopers" or "spanworms"), are a large component of the diet of many passerine birds in the boreal forest and near the forest/tundra margin. If the larval population densities are reduced due to a lack of palatability of the leaves on which they feed, the effects of UV-B radiation could be far-reaching on the below- and above-ground food webs of the terrestrial Arctic.

Changes in phenology, the time of year when events happen, will also affect the size of populations. A number of studies have already shown that vascular plants are flowering earlier, insects (especially butterflies) are appearing earlier in the year, some birds are starting to nest earlier in spring, amphibians are spawning earlier, and migratory birds are arriving earlier (see a review by Usher, 2002b). Some of these phenological observations are beginning to be used as indicators of the effects of climate change on biodiversity, although most studies are just recording data on the changes in species populations in the earlier part of the year (usually spring) and do not record data for the end-of-summer changes that could be affecting plant growth rates in the autumn or autumnal flight periods for species of insect. The important ecological impact of phenology concerns how changes will affect interactions between pairs of species. If one species changes its phenology more than another, will this then increase or decrease the effects of competition, herbivory, predation, parasitism, etc.? If synchrony occurs, and the organisms become less synchronous, this could have considerable effects on population sizes and biodiversity.

In the marine environment, seabirds show strong preferences for regions of particular sea surface temperatures (SSTs) (Schreiber, 2002). Some seabird populations have been found to respond to long-term climatic changes in the North Atlantic Ocean (Aebischer et al., 1990; Thompson P. and Ollason 2001), the North Pacific Ocean (Anderson and Piatt, 1999; Bertram et al., 2001; Jones I. et al., 2002; Sydeman et al., 2001; Veit et al.,

1997), and Antarctica. Although global SSTs are generally increasing, this long-term trend is superimposed on cyclical patterns created by climatic oscillations, such as the North Pacific, North Atlantic, and Arctic Oscillations (Francis et al., 1998; Hare and Mantua, 2000; Hurrell et al., 2003; Wilby et al., 1997). These oscillations cause periodic reversals in SST trends, two of which have occurred since 1970 in the Northern Hemisphere; from 1970 much information has been accumulated on seabird population trends in the circumpolar Arctic (Dragoo et al., 2001; Gaston and Hipfner, 2000).

To examine the effect of SST changes on seabird populations at a global scale, data on population changes throughout the distribution ranges of the common guillemot or murre (*Uria aalge*) and Brünnich's guillemot or thick-billed murre (*U. lomvia*) were examined to document how they changed in response to climate shifts, and potential relationships with SSTs (D.B. Irons, U.S. Fish and Wildlife Service, pers. comm., 2003). Both species breed throughout the circumpolar north from the high Arctic to temperate regions, although Brünnich's guillemots tend to be associated with colder water than common guillemots and are the dominant species in the Arctic (Gaston and Jones, 1998).

The analysis showed that positive population trends occurred at guillemot colonies where SST changes were small, while negative trends occurred where large increases or large decreases in SST occurred. Highest rates of increase for the southerly species, the common guillemot, occurred where SST changes were slightly negative, while increases for the arctic-adapted Brünnich's guillemot were most rapid where SST changes were slightly positive. These results demonstrate that most guillemot colonies perform best when temperatures are approximately stable, suggesting that each colony is adapted to local conditions (D.B. Irons, U.S. Fish and Wildlife Service, pers. comm., 2003). This study also demonstrates how seabirds respond to changes in climatic conditions in the Arctic over large temporal and spatial scales.

A study on the Atlantic puffin in the Lofoten Islands, northern Norway, has shown that sea temperatures from March through July (which is the first growth period for newly hatched Atlantic herring) and the size of herring in the food intake of adult puffin together explain about 84% of the annual variation in fledging success of puffin chicks (T. Anker-Nilssen, Norwegian Institute for Nature Research, pers. comm., 2003). Although there are relatively few data for the marine environment, what there are (especially for seabirds) indicate reduced population sizes for many of the marine wildlife species of the Arctic, and so conservation activity must aim to ameliorate such declines. Protected areas are an important aspect of such activity and are discussed further in section 10.4.7.

10.4.4. Changes in genetic diversity

Little attention had been paid to genetic diversity, despite it being one of the major themes in the Convention on Biological Diversity. For example, Groombridge's (1992) book on biological diversity had 241 pages on species diversity, 80 pages on the diversity of habitats, but only 6 pages on genetic diversity. Similarly, Heywood's (1995) Global Biodiversity Assessment had only 32 pages on the subject of "genetic diversity as a component of biodiversity" of its total of 1140 pages.

The reason for this discrepancy is because species tend to be tangible entities and many are easily recognizable. The species concept does not work well, however, for the single-celled forms of life, which often live in soils or sediments under freshwater or the sea, where the genetic variability is often more important than the species itself. Habitats are also recognizable, often on the basis of their species, but present complications because they tend to merge into one another. Compared with these tangible entities, genetic variability is often not recognizable and can only be detected by sophisticated methods of analysis using molecular techniques. Of the millions of species that exist, very little is known about their genetic diversity except for a few species of economic importance, a few species that are parasites of people or their domestic stock, and a few other species that geneticists have favored for research (e.g., the *Drosophila* flies). As in all other parts of the world, relatively little is known about the genetic variability of species that occur in the Arctic.

What then can be done to conserve the Arctic's genetic diversity? On the basis that natural selection requires a genetic diversity to operate, conservation practice should aim to find a surrogate for the unknown, or almost unknown, genetic diversity. This is best done by conserving each species over as wide a distribution range as possible and in as many habitats as possible. This ensures maximum geographical and ecological variability, assuming that local adaptation of species represents different genotypes. Attempting to map population genetics to landscape processes is relatively new (Manel et al., 2003) and has been termed "landscape genetics". Manel et al. (2003) stated that it "promises to facilitate our understanding of how geographical and environmental features structure genetic variation at both the population and individual levels, and has implications for...conservation biology". At the moment it must be assumed that the geographical and environmental features have structured the genetic variation, and this assumption must be made before the links can be proved. How this variability has actually arisen is unclear.

Throughout continental Europe, a continuous postglacial range expansion is assumed for many terrestrial plant and animal species. This has often led to a population structure in which genetic diversity decreases with distance from the ancestral refugium population (Hewitt, 2000), and so northern populations are often genetically less diverse than their southern counterparts (Hewitt, 1999).

Among discontinuously distributed species, such as those living on remote islands, this pattern can be obscured by differences in local effective population sizes. For example, considerable genetic diversity exists among populations of common eider ducks (*Somateria mollissima*) nesting throughout the circumpolar Arctic. Historical and current processes determining phylogeographic structure of common eiders have recently been reconstructed, based on maximum parsimony and nested clade analysis (A. Grapputo, Royal Ontario Museum, pers. comm., 2004; Tiedemann et al., 2004). Five major groups (or "clades") have been identified; the three most different include common eiders from Alaska, Svalbard, and Iceland. The remaining two include eider populations from the eastern Canadian Arctic and West Greenland, and from northwest Europe.

Nested clade analysis also suggests that the phylogeographic patterns observed have a strong historical pattern indicating past fragmentation of eider populations due to glacial events. Following the retreat of the glaciers, eiders surviving in refugia expanded to re-colonize their range, and populations apparently remixed. These refugial populations occurred across Arctic Canada and Greenland (A. Grapputo, Royal Ontario Museum, pers. comm., 2004), and apparently in a single refugium in northwest Europe (Tiedemann et al., 2004). The oldest population split was estimated between Pacific eiders and birds that colonized the western Canadian Arctic islands about 120 000 years ago after the retreat of ice sheets in the previous glacial maximum. In North America, this was likely to have been followed by a second expansion that began in warmer periods about 80 000 years ago from Alaska eastward across the Palearctic to establish populations in the eastern Canadian Arctic and West Greenland. In Europe, genetic analyses suggest that common eiders underwent a postglacial range expansion from a refugium in Finland, north and west to the Faroe Islands and subsequently to Iceland. Despite this relatively recent mixing of haplotypes, extant populations of common eider ducks are strongly structured matrilineally in the circumpolar Arctic. These results reflect the fact that current long-distance dispersal is limited and that there is considerable philopatry of female eiders to nesting and wintering areas (Tiedemann et al., 2004).

In contrast to common eider ducks, king eider ducks (*Somateria spectabilis*) show a distinct lack of spatial genetic structure across arctic North America (Pearce et al., 2004). In the western Palearctic, the king eider has been delineated into two broadly distributed breeding populations in North America, in the western and eastern Arctic, on the basis of banding (ringing) data (Lyngs, 2003) and of isotopic signatures of their diet while on wintering grounds (Mehl et al., 2004, in press). These studies indicated the use of widely separated Pacific and Atlantic wintering areas. Despite this, recent studies of microsatellite DNA loci and cytochrome *b* mitochondrial DNA show small and non-significant genetic differences based on samples from three wintering and four nesting areas in arctic North America, Russia, and Greenland (Pearce et al., 2004). Results from nested clade analysis and coalescent-based analyses suggest his-

torical population growth and gene flow that collectively may have homogenized gene frequencies. However, the presence of several unique mtDNA haplotypes among birds wintering in West Greenland suggested that gene flow may now be more limited between the western and eastern arctic populations than in the past (Pearce et al., 2004); this would be consistent with recent banding data from eastern Canada and West Greenland (Lyngs, 2003).

Collectively, these two examples of closely related duck species illustrate how climatic events can influence the genetic structure of arctic species over time. They also show how historical periods of isolation, combined with little gene flow currently (matrilineally, at least), have contributed to maintain genetic diversity. However, the fact that the common and king eider differ so markedly in their degree of genetic diversity throughout the circumpolar Arctic, despite sharing many ecological traits, suggests that the effects of more rapid climate change on genetic diversity may be difficult to predict.

There are at least three features of this genetic variability that need to be considered in the conservation of the Arctic's biodiversity. First, the genetic structure of a species at the edge of its range, where it is often fragmented into a number of small and relatively isolated populations, is often different from that at the center of the range, where populations can be more contiguous and gene flow is likely to be greater. It is these isolated, edge-of-range populations that are possibly undergoing speciation, and which might form the basis of an evolution toward different species with different ecologies in the future.

Second, hybridization can be both a threat and an opportunity. Although arctic examples are rare, it can be a threat where two species lose their distinctive identities, as is happening with the introduction of Sika deer (*Cervus nippon*) into areas where red deer (*C. elaphus*) naturally occur. This is one of the potential problems with the introduction into the Arctic of non-native species (section 10.4.6). Hybridization can also be an opportunity. The hybrid between the European and American *Spartina* grasses doubled its number of chromosomes and acts as a newly evolved species in its own right.

Third, there are suggestions (Luck et al., 2003) that the genetic variability of populations is important in maintaining the full range of ecosystem services. Although this concept is little understood, it is intuitively plausible because, as factors in the environment change, individuals of differing genetic structure may be more or less able to fulfill the functional role of that species in the ecosystem. Thus, with a variable environment, the ecosystem needs species whose individuals have a variable genetic makeup.

Although little is known about genetic variability, a geographically spread suite of protected areas, encompassing the full range of habitat types, is probably the best conservation prescription for the Arctic's biodiversity that can currently be made. It should be appropriate for conserving the biodiversity of habitats and species, and is probably also appropriate for conserving genetic biodiversity.

10.4.5. Effects on migratory species and their management

Migration was briefly addressed in sections 10.2.6 and 10.3.2, and the eight major international flyways for shorebirds breeding in the Arctic are shown in Figure 10.4. Migration is a cold and ice avoidance strategy used by birds, marine mammals, and fish. Although some species of insect also migrate, it is uncommon for the milkweed butterfly (*Danaus plexippus*), well known for its migrations through North America, to migrate in the spring and early summer as far north as the Canadian Arctic.

The goose species of the western Palearctic region provide good examples of migratory species that have been the subject of considerable research and conservation action (Madsen et al., 1999). Of the 23 populations, five populations of greylag goose (*Anser anser anser* and *A. a. rubirostris*) do not nest in the Arctic; neither do the two populations of Canada goose (*Branta canadensis*) which are not native to the region. The remaining 16 populations of seven species (11 subspecies) are listed in Table 10.8. There are a variety of flyways, some moving southeast from the breeding grounds in northeast Canada, Greenland, and Iceland, and others moving southwest from the breeding grounds in the Russian Arctic, both into Western Europe. The three populations of barnacle goose (*Branta leucopsis*) can be used as an example (see Box 10.1).

The examples demonstrate a number of features of migratory populations and their conservation. The geese require sufficient food resources to make two long journeys each year. The summer feeding grounds in the Arctic and the winter feeding grounds in temperate Europe provide the majority of the food requirements. However, while on migration, the geese need to stage and replenish their energy reserves. In years when winter comes early and Bjørnøya is iced over before the geese arrive, it is known that many are unable to gain sufficient energy to fly on to Scotland and there can be very heavy mortality, especially of that year's young. Although the three populations appear from the brief descriptions in Box 10.1 to be geographically isolated from each other, there is a very small amount of mixing between these populations, and so gene flow is probably sufficient for this one species not to have sub-speciated.

The examples also demonstrate that conservation efforts need to be international. For each of the three populations, protection is required for parts of the year in the breeding grounds, in the wintering grounds, and in the staging areas. Conservation action needs to be taken wherever the geese land. The fact that there is some straying from the main flight paths implies that conservation is required all along these migration routes. In Europe, the Bonn Convention aims to provide such an instrument for the conservation of migratory species; this could form a model for all migratory species, including those that use the Arctic for part of their life cycle.

Climate change could affect these species through changes in their habitats. For the Greenland nesting population it would be possible for their breeding grounds to move northward because there is land north of the current breeding range. This could hardly happen for the populations breeding on Svalbard and in Russia because there is very little ground north of the current breeding areas (just the north coast of Svalbard and the north of Novaya Zemlya). Because many of the wintering

Table 10.8. The sixteen goose populations that nest in the Arctic and overwinter in the western Palearctic. The data were extracted from Madsen et al. (1999).

		Breeding area	Wintering area
Taiga bean goose	*Anser fabalis fabalis*	Scandinavia and Russia	Baltic
Tundra bean goose	*Anser fabalis rossicus*	Russia	Central and Western Europe
Pink-footed goose	*Anser brachyrhynchus*	Iceland and Greenland	Great Britain
Pink-footed goose	*Anser brachyrhynchus*	Svalbard	Northwest Europe
White-fronted goose	*Anser albifrons albifrons*	Russia	Western Europe
Greenland white-fronted goose	*Anser albifrons flavirostris*	West Greenland	British Isles
Lesser white-fronted goose	*Anser erythropus*	Scandinavia and Russia	Central and southeast Europe
Greylag goose	*Anser anser anser*	Iceland	Scotland
Greylag goose	*Anser anser anser*	Northwest Europe	Northwest and southwest Europe
Barnacle goose	*Branta leucopsis*	East Greenland	British Isles
Barnacle goose	*Branta leucopsis*	Svalbard	Scotland and northern England
Barnacle goose	*Branta leucopsis*	Russia and the Baltic	Northwest Europe
Dark-bellied brent goose	*Branta bernicla bernicla*	Russia	Western Europe
Light-bellied brent goose	*Branta bernicla hrota*	Northeast Canada	Ireland
Light-bellied brent goose	*Branta bernicla hrota*	Svalbard	Northwest Europe
Red-breasted goose	*Branta ruficollis*	Russia	Black Sea

Box 10.1. The three populations of barnacle goose in the western Palearctic

The western population of barnacle goose (*Branta leucopsis*) in the western Palearctic breeds near the coast along northeast Greenland from about 70° to 78° N. On the autumn migration the geese stage in Iceland, near the south coast, where they spend about a month feeding before they fly on to the wintering grounds along the west coast of Ireland and the west and north coasts of Scotland. In the spring the geese leave the British Isles in April and stage on the northwest coast of Iceland for three or four weeks before flying back to Greenland to recommence the annual cycle. These geese are legally protected in Greenland from 1 June to 31 August, although a few are legally hunted by local people. In Iceland the geese are protected in the spring, although it is considered that some are illegally killed, but few are thought to be killed in autumn. In the United Kingdom the geese are fully protected as a result of domestic legislation and of being listed in Annex 1 to Council Directive 79/409/EEC on the conservation of wild birds (also known as the Birds Directive).

A second (or central) population of about 25000 birds breeds in Svalbard between about 77° and 80° N. After breeding, the geese leave Svalbard in August, and many arrive on Bjørnøya at the end of August staying until late September or early October when they fly on to the Solway Firth in southwest Scotland. They return north in the spring, staging in the Helgeland Archipelago off the coast of Norway (between 65° and 66° N) for two to three weeks before flying on to Svalbard. The geese are legally protected in Svalbard, Norway, and the United Kingdom, and it is thought that very few are illegally shot.

The eastern population breeds in northern Russia, from the Kola Peninsula in the west to Novaya Zemlya and the Yugor Peninsula in the east. In the autumn the birds fly southwest, along the Gulf of Bothnia and the southern part of the Baltic Sea, staging on the Estonian and Swedish Baltic islands. The majority of the birds winter on the North Sea coast of Denmark, Germany, and the Netherlands. The species is legally protected in Russia, although Madsen et al. (1999) reported that it appears that many are shot and that both the adults and the eggs are used as an important part of the diet of local people. Within the countries of the European Union, the geese are fully protected by the Birds Directive.

Barnacle geese from the Greenland population overwintering on the island of Islay, western Scotland

grounds are managed as grasslands for cattle and sheep grazing, it is possible that these may change less than the breeding grounds. The staging areas are also likely to change, and it is possible that the distance between breeding and wintering grounds might become longer, requiring more energy expenditure by the migrating birds. This leaves a series of unknowns, but at present these goose populations are increasing in size, are having an economic impact on the wintering grounds, and have raised what Usher (1998) has termed "the dilemma of conservation success". This is the problem of reconciling the interests of the local people with the need to conserve species that the people either depend upon harvesting or that damage their livelihoods.

10.4.6. Effects caused by non-native species and their management

Biological invasions have fascinated ecologists for well over 50 years (Elton, 1958). The many problems caused by non-native species are becoming more apparent, and the World Conservation Union (IUCN) identifies them as the second most important cause of loss in global biodiversity (the primary reason being loss and fragmen-

tation of habitats). A word of caution is, however, needed with language. Why a species is geographically where it is currently found cannot always be determined; if it is known to be there naturally, it is generally referred to as "native". If it has been brought in from another geographical area by human agency, either intentionally or unintentionally, it is referred to as "non-native" (Usher, 2000, discussed these distinctions and the gradations between them). The term "non-native" is essentially synonymous with "alien", "exotic", and "introduced", all of which occur in the literature. Williamson (1996) described the "10:10 rule", suggesting that 10% of species introduced to an area would establish themselves (i.e. they do not die out within a few years of introduction, and start to reproduce) and that 10% of these established species would become "pests". While this rule seems reasonably true for plants, it seems to underestimate the numbers of vertebrate animals that become problematic (Usher, 2002b). It is this 1% (10% of 10%) of species that are introduced, or rather more for vertebrate animal species, which can be termed "invasive".

To date, the Arctic has escaped the major problems that invasive species have caused in many other parts of the world. During the 1980s there was a major international program on the ecology of biological invasions. The synthesis volume (Drake et al., 1989) does not mention the Arctic (or the Antarctic), although global patterns of invasion into protected areas indicated that the problems diminished with latitude north or south of the regions with a Mediterranean climate (Macdonald I. et al., 1989).

In terrestrial ecosystems, climate change is very likely to mean that more species will be able to survive in the Arctic. It is arguable whether new species arriving in the Arctic can be classified as "native" or "non-native" when the rapidly changing climate is anthropogenically driven. However, with a changing climate new species will very probably arrive in the Arctic, some of which will establish and form reproducing populations. Although there is no obvious candidate for a non-native species to be invasive in the Arctic, it needs to be remembered that at least 1% of species introduced into the Arctic are likely to become invasive. At present there are no means of determining the major risks, but the introduction of disease organisms, for wildlife and people, is a distinct possibility.

In the boreal forests, the insects, as a group, pose the most serious challenge because of their ability to increase rapidly in numbers and because of the scarcity of effective management tools. From past experience, it is probable that many forest-damaging insects have the potential to appear at outbreak levels under a warmer climate and increased tree stress levels, but this has not been observed to date. Two examples demonstrate the risks. First, the bronze birch borer (*Agrilus anxius*) has been identified as a species that can cause severe damage to paper birch (*Betula papyrifera*), and may be effective in limiting the birch along the southern margin of its distribution (Haak, 1996). It is currently present at relatively low levels in the middle and northern boreal region of North America. Second, an outbreak of the Siberian silkworm (*Dendrolimus sibiricus*) in west Siberia from 1954 to 1957 caused extensive tree death on three million hectares of forest. Movement of outbreak levels northward would considerably alter the dynamics of Siberian forests.

There are similar concerns in the freshwater environment. In much of northern Europe and northern America, it is the introduction of fish species that cause most problems. For example, in Loch Lomond in Scotland the invasive ruffe (*Gymnocephalus cernuus*) eats the eggs of an arctic relict species, the powan (*Coregonus lavaretus*), thereby threatening this species in one of its only British habitats (Doughty et al., 2002). Similarly, in North America the invasion of the Great Lakes by the lamprey (*Petromyzon marinus*), first seen in Lake Erie in 1921, led to the collapse of a number of fisheries following its establishment and first known breeding in the 1930s. For example, the trout fishery in Lake Michigan was landing about 2600 tonnes of fish each year between 1935 and 1945, but this dropped to 155 tonnes by 1949 when the fishery essentially ended (Watt, 1968). Although these examples are outside the Arctic, they highlight potential problems with non-native fish species

as arctic rivers and lakes become warmer. There are also potential problems with fish that escape from fish farms and enter the natural environment and breed with native fish stock. The genetic effects of this interbreeding can be profound, altering the behavior of the resulting fish stock, as has been found with Atlantic salmon (*Salmo salar*) in Norway.

In the marine environment one of the major potential problems is the discharge of ballast water. With thinning of the sea ice and the opening up of the Arctic Ocean to more shipping for more of the year, the possibility of the introduction of non-native species is greater and the environmental risks are increased. Analyses of ballast water have shown that it can contain a large number of different species of marine organisms, including marine algae and mollusks that are potentially invasive. Also, ballast water has occasionally been found to contain organisms that could be pathogenic to people. Regulating discharges of ballast water in not easy, nor is its enforcement always possible, but to prevent the threat of invasive marine organisms it is essential that international agreements regulate such discharges in coastal waters and on the high seas of the Arctic.

The effects of introduced Arctic foxes on seabird populations is an example that links the marine and terrestrial environments. Seabirds commonly nest on offshore islands, in part to avoid terrestrial predators to which they are vulnerable, both to the loss of eggs and chicks and to direct predation on adults. Several seabird populations have declined when mammalian predators were accidentally or intentionally introduced to nesting islands (Burger and Gochfeld, 1994). Arctic foxes were intentionally introduced for fur farming in the late 1800s and early 1900s on several of the Aleutian Islands of Alaska. Before these introductions, the islands supported large populations of breeding seabirds and had no terrestrial predators. Although most fox farming ended prior to the Second World War, the introduced animals persisted on many islands, preying on breeding seabirds at rates affecting their population sizes (Bailey, 1993). Evidence from southwestern Alaska (Jones R. and Byrd, 1979), and comparisons of islands with and without foxes in the Shumagin Islands (Bailey, 1993), suggest it is likely that foxes are responsible for the reduced seabird population sizes on islands supporting foxes. Those species nesting underground, in burrows or in rock crevices, were less affected (Byrd et al., 1997).

Foxes have recently been eradicated from several islands (Bailey, 1993) and the responses of seabird populations have been dramatic. Pigeon guillemot (*Cepphus columba*) populations began to increase within three to four years following fox removal at Kiska Island and 20-fold increases occurred in guillemot numbers at Niski-Alaid Island within 15 years of fox removal (Byrd et al., 1994). The introduction of Arctic foxes to the Aleutian Islands, and their influence on native seabird species, provides a dramatic example of how the intentional introduction or movement of species can influence arctic biodiversity.

The report by Rosentrater and Ogden (2003) contained the cautionary note "presently, the magnitude of the threat of invasive species on Arctic environments is unclear: however, the potential impacts of this threat warrant further investigation and precautionary action on species introductions, especially since climate change is expected to result in the migration of new species into the region". The risk to the environment and to bio-diversity of intentionally introducing any non-native species into the Arctic must be established before the species is introduced. Experience worldwide indicates that it is often too late if the risk is assessed after the introduction; it might then also be too late to control the spread and effects of the invasive species. The pre-cautionary action is to stop the arrival of the invasive species in the first place because its later eradication may be impossible, and even if possible worldwide experi-ence shows that it is likely to be extremely expensive.

10.4.7. Effects on the management of protected areas

Establishment of protected areas has been a core activi-ty of conservation legislation throughout the world. The concept is implemented in different ways by differ-ent national governments, with differing degrees of suc-cess, as is clear from reviews of international activities (e.g., IUCN, 1991). This section reviews the underlying ecological concepts related to the conservation of bio-diversity and the potential effects of climate change.

Reviews by CAFF (2001, 2002a) showed that much progress has been made in designating protected areas in the Arctic, but that further progress is needed, espe-cially in the marine environment. Halpern and Warner (2002) indicated that marine reserves are very effective at conserving biodiversity, and Halpern (2003) consid-ered that marine protected areas need to be large in extent. In the terrestrial and freshwater environments, some of the largest protected areas worldwide occur in the Arctic. Few studies explore whether such protection is achieving its stated aims.

In general the establishment of protected areas has a scientific foundation. As Kingsland (2002) stated "its goal is to apply scientific ideas and methods to the selection and design of nature reserves and to related problems, such as deciding what kinds of buffer zones should surround reserves or how to establish corridors to link reserves and allow organisms to move from one area to another. As in other areas of conservation biolo-gy, designing nature reserves is a 'crisis' science, whose practitioners are driven by an acute sense of urgency over the need to stem the loss of species caused by human population growth". This to some extent misses a vital point: the social sciences are also involved with conservation. Why is it important to conserve bio-diversity, why are particular species favored over others, or how do people fit into the conservation framework? Such questions are not addressed here, despite their importance to the local communities of the Arctic (section 10.2.7); this section focuses on the scientific bases of conservation.

Three main facets of ecological thinking have affected the design of potential protected areas. The concepts of island biogeography, of habitat fragmentation, and the establishment of metapopulations (and of corridors) are not unrelated and can all impact upon protected areas in a changing climate.

The concept of island biogeography (MacArthur and Wilson, 1967) includes the idea that the number of species on an island is dynamic, representing the equi-librium between the arrival of new species and the extinction of existing species. Larger islands would have greater immigration rates, and possibly smaller stochas-tic extinction rates, than small islands, and hence the equilibrium number of species would be greater. Similarly, distant islands would have smaller immigra-tion rates than similarly sized islands nearer the source of immigrants, but would probably have similar extinc-tion rates, and so would have fewer species. Using many sets of data for island biota, these concepts are formu-lated into the empirical relationship:

$$S = CA^z$$

where S is the number of species on the island, A is the area of the island, and C and z are constants (C represents the number of species per unit area, and z generally takes a value of about 0.3. This relationship implies that if the island area is increased ten-fold, the number of species will about double). Although there have been few island biogeographical studies in the Arctic, Deshaye and Morisset (1988, 1989) confirmed that larger islands in the subarctic (in the Richmond Gulf, northern Québec, Canada) contain more species than smaller islands.

Island biogeography has thus been used to justify larger rather than smaller protected areas. With climate change, and with arctic wildlife populations and their distribution ranges likely to diminish (sections 10.4.1 to 10.4.3), use of the precautionary principle would also suggest that larger rather than smaller protected areas should be established.

Fragmentation of ecosystems has been viewed as the "islandization" of habitats. Although fragments cannot be thought of as real islands, the use of island biogeograph-ical concepts tends to apply relatively well (Harris, 1984). This has led to the formulation of "rules" for the design of protected areas, starting with Diamond (1975), but leading to more sophisticated designs as in Fig. 10.15. Size and shape are the key factors in the design of protected areas, but the inclusion of fragments of natural ecosystems is helpful for biodiversity conser-vation. Under a changing climate, fragmentation of arc-tic ecosystems should be avoided. Fragmentation always causes problems (Saunders et al., 1987), even if at some scales it might appear to increase biodiversity (Olff and Ritchie, 2002).

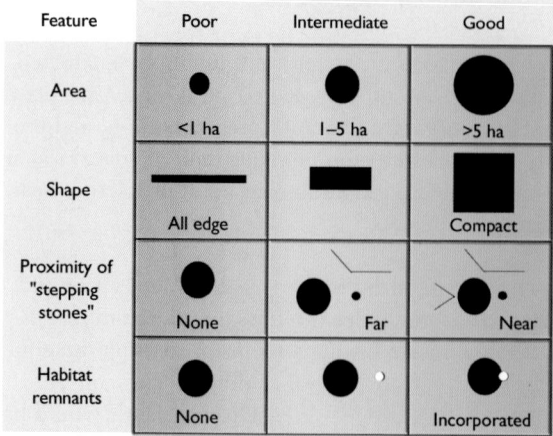

Value for species richness

Fig. 10.15. A representation of the biodiversity conservation value of potential protected areas, based on a study of insects in farm woodlands but also applicable to other habitats and other taxonomic groups (Usher, 2002a). The scaling should change to reflect the larger areas prescribed for the Arctic. Habitats are in black and habitat fragments are small white circles. Linear features, such as small rivers, are represented by straight lines.

With fragmentation an integral part of modern development, corridors appear to be a useful concept. How does the landscape fit together such that individuals can move from habitat patch to habitat patch? As pointed out by Weber et al. (2002), land managers and wildlife biologists must collaborate to determine the patterns of protected areas within the landscape that will be of most benefit to wildlife. Some scientists advocate corridors: Saunders and Hobbs (1991) gave a number of examples where corridors appear to work. Others have argued that corridors allow invasive species entry into protected areas, while more recent research calls into question the whole value of corridors. Albeit a beguilingly simple concept, at present neither the value of corridors, nor their lack of value, has been proven. With climate change underway, it is thus best to avoid the necessity for corridors by focusing on larger protected areas and a reduction in the processes leading to habitat fragmentation. This will promote real connectivity, rather than an apparent connectivity, for species and habitats.

However, will the protected areas that exist today, even if they have been located in the best possible place to conserve biodiversity, still be effective in the future with climate change? The answer is probably "no". Designations have been widely used, but are based on assumptions of climatic and biogeographic stability and usually designated to ensure the maintenance of the *status quo*. Available evidence indicates that these assumptions will not be sustainable during the 21st century. So what can be done to make the network of protected areas more appropriate to the needs of the Arctic and its people?

First, today's protected areas should encompass land or water that will potentially be useful for biodiversity conservation in the future. This is where models of the changing distribution of species and habitats are useful and where their outputs should be included in the design

of protected areas (see the example of the Canadian national parks in section 10.4.1). This means that designation should reflect both the present value of the areas for biodiversity as well as the projected future value (the potential value).

Second, boundaries may need to be more flexible. In general, boundaries are lines on maps, and enshrined in legislation, and so are difficult to change. The present practices could be described as having "hard boundaries". An alternative could be that the boundaries change with changes in the distribution of the flora or fauna being protected. That is, over time (probably decades rather than years) the location of the protected areas would shift geographically (this could be described as the protected areas having "soft boundaries"). However, it is important that sociological and developmental pressures do not destroy the value of the protected areas in safeguarding the biodiversity that is their *raison d'etre* – nothing would be worse than in 50 years time having a network of sites that were protecting very little. More flexible systems of designation, adding areas which are or will become important, and dropping areas that are no longer important, would appear to be one way forward to conserve biodiversity within the Arctic. A system of designations with "soft boundaries" has not yet been tried anywhere in the world, but could become a policy option that is pioneered in the Arctic.

Protected area designations are a major policy and management system for the conservation of biodiversity, as well as for historical and cultural artifacts. Climate change might result in designated communities and species moving out of the designated area; communities and species new to the area will tend to colonize or visit, especially from the south; and assemblages of species without current analogues will form as individual species respond to climate change at different rates and in different ways. It will therefore be necessary to adjust such concepts as "representative communities" and "acceptable limits of change" that are part of the mandate of many national and international designations. The expected changes will include many surprises resulting from the complex interactions that characterize ecosystems and the non-linearity of many responses.

The scientific basis of biodiversity conservation planning in the era of climate change argues against procedures designed to maintain a steady state. There are four general policy options to respond to climate change that have been used in the Canadian national parks (summarized by Scott and Lemieux, 2003).

1. Static management. Continuing to manage and protect current ecological communities and species within current protected area boundaries, using current goals.
2. Passive management. Accepting the ecological response to climate change and allowing evolutionary processes to take place unhindered.
3. Adaptive management. Maximizing the capacity of species and ecological communities to adapt to cli-

mate change through active management (for example, by fire suppression, species translocation, or suppression of invasive species), either to slow the pace of ecological change or to facilitate ecological change to a new climate adapted state.

4. Hybrid management. A combination of the three previous policy options.

It is likely that adaptive management will be the most widely applied. This is likely to include actions to maintain, for as long as possible, the key features for which the original designation was made, for example by adjusting boundaries. Past experience indicates that intervention strategies tend to be species-specific, and to be strongly advocated, but this must not detract from the more scientific goal of conserving the Arctic's biodiversity in a holistic manner.

10.4.8. Conserving the Arctic's changing biodiversity

Preceding sections have addressed issues such as the effects of climate change on the size and spatial extent of species populations and the communities in which the species occur, the need to conserve genetic diversity, potential problems resulting from the arrival of nonnative species, and problems faced by migrant species. This section addresses a few topics that cut across those already discussed. The two main topics discussed here are taxonomy and monitoring.

Biodiversity depends upon taxonomy. It is necessary to be able to name species and habitats, or to understand variation in DNA, to be able to start to think about biodiversity and its conservation, and to communicate thoughts. Taxonomy is therefore fundamental to the work on biodiversity (Blackmore, 2002). It is necessary to know the species being considered – knowledge of birds, mammals, and fish is certainly satisfactory, but is this true for all the insects in the Arctic and their roles in the arctic freshwater and terrestrial ecosystems? Knowledge of vascular plants (flowering plants and ferns) is probably satisfactory, but is this true for the mosses, liverworts, lichens, and algae that are responsible for much of the photosynthesis, in the sea, freshwaters, and on land? As in almost all parts of the world, is there knowledge about the species of protozoa or bacteria that are associated with the processes of decomposition in arctic soils and in the sediments under lakes or on the sea floor? There are many areas of arctic taxonomy that require exploration and research, and it is vital to the conservation of the Arctic's biodiversity that these taxonomic subjects are addressed.

Monitoring is important for understanding how the Arctic's biodiversity is changing and whether actions to conserve this are being successful. As Cairns (2002) pointed out, monitoring needs to occur at both the system level and the species level. Monitoring will help now, and in the future, to determine if current predictions are correct and to modify and improve the systems

of management. From a scientific perspective, monitoring will allow more data to be collected and, if coupled with research, will also allow a greater understanding of the mechanisms involved with change. In time, therefore, with increasing data and increasing understanding, the conservation of biodiversity would move in the plane shown in Figure 10.12 from the bottom left hand corner and, perhaps only slightly, toward the top right hand corner. With data and understanding it should be possible in the future to build better models and hence make better predictions.

Conservation of the Arctic's biodiversity at present relies upon two approaches. One is through the establishment of protected areas, and this was discussed in section 10.4.7. Greater knowledge of taxonomy and monitoring of what is happening within those protected areas are both important for their future management. The other approach is more educational, bringing biodiversity thinking into all aspects of life in the Arctic. Considerations of biodiversity need to be explicit in planning for developments at sea or on land. Biodiversity needs to be considered explicitly in the management of land, freshwater, and the sea. Links between biodiversity and the health of the local people need to be established. Biodiversity forms the basis of most tourism into the Arctic, but facilities for tourists need particular care so as not to damage the very reason for their existence (Rosentrater and Ogden, 2003). Biodiversity conservation as a concept therefore needs to permeate all aspects of life in the Arctic.

If it is accepted that protected areas are only ever going to cover a relatively small percentage of the land and sea area of the Arctic (possibly between 10 and 20%), then it is the land and sea outside the protected areas that will hold the majority of the Arctic's biodiversity. Just as within protected areas it is vital to have knowledge of taxonomy and programs of monitoring, there must also be taxonomic knowledge and monitoring throughout the Arctic. The majority of the biodiversity resource in the nonprotected areas must not be sacrificed because a minority of that resource is within protected areas. Apart from the Antarctic, it is probably easier to achieve this balance between protected areas and the rest of the land and sea area in the Arctic than in other areas of the world, but it will require international effort if the Arctic's biodiversity is to be conserved for future generations to use and enjoy. All this, in the face of climate change, will need "building resilience" (the expression used by Rosentrater and Ogden, 2003) into all arctic ecosystems, whether or not they lie within protected areas.

10.5. Managing biodiversity conservation in a changing environment

To conclude this chapter on conserving the Arctic's biodiversity, it is appropriate to explore a number of topics that have been implicit in the various descriptions and discussions of sections 10.1 to 10.4. Four topics are addressed in this final section: documenting the current biodiversity; predicting changes in that biodiversity resource over the next 50 or 100 years; determining how that biodiversity resource is actually changing; and managing the Arctic's biodiversity resource in a sustainable manner.

Each topic generates a number of questions, and their answers involve many concepts, most of which have already been introduced in this chapter. Sixteen recommendations are made in relation to the various discussions and conclusions in this section.

10.5.1. Documenting the current biodiversity

The Arctic nations have very good inventories of their mammals and birds (listed by Sage, 1986). Although it is possible that a few more species might have been recorded in the Arctic since the mid-1980s, it is unlikely that the numbers of 183 species of bird and 48 species of terrestrial mammal will have changed significantly.

It is notable that Sage (1986) was unable to provide similar lists for any other taxa of wildlife in the Arctic. From the literature on the Arctic it would probably now be possible to prepare reasonably good inventories of the marine mammals, freshwater and marine fish, and vascular plants. Although this is as much as most nations in the world can compile for national inventories, such lists omit the most species-rich taxa. Large numbers of species of bryophyte (mosses and liverworts), lichen (or lichenized fungi), fungi, and algae occur, as well as many species of invertebrate animals. Terrestrially, it is likely that the insects and arachnids (mites and spiders) will be the most species-rich, whereas in the sea it is likely to be the crustaceans and mollusks that are most species-rich. However, there are many other taxonomic groups, especially the nematodes and many marine taxa of worms, sponges, and hydroids, as well as single-celled organisms in which the "species" concept is more difficult to apply.

Inventories are important. They form the building blocks for biodiversity conservation because, unless the biodiversity is known, it is not possible to begin to conserve it or to recognize when it is changing. Documentation of the numbers and types of species living in the Arctic has focused mainly on terrestrial systems and is detailed in Chapter 7. The Arctic has around 1735 species of vascular plants, 600 bryophytes, 2000 lichens, 2500 fungi, 75 mammals, 240 birds, 3300 insects dominated by the Diptera (two-winged flies), 300 spiders, 5 earthworms, 70 enchytraeid worms, and 500 nematodes. This species diversity represents a small but variable percentage of the world's species, with some groups relatively strongly represented. Thus, there are about 0.4% of the world's insects but 6.0% of the Collembola; as well as 0.6% of the world's ferns but 11.0% of the lichens. There is currently no comparable documentation of numbers of species in the freshwater and marine environments of the Arctic, although there is significant environmental overlap for some taxa, for example, the birds.

An excellent example of an arctic inventory is the work done on Svalbard (Elvebakk and Prestrud, 1996; Prestrud et al., 2004). An overview is given in Table 10.9, giving Svalbard a species richness of about 5700 (terrestrial, freshwater, and marine environments combined). However, this total does not include many of the single-celled organisms, such as the protozoa, and so a full inventory would be substantially longer.

Many species, particularly vascular plants, are endemic to the Arctic. However, there are few endemic genera. This has been attributed to the youthfulness of the arctic flora and fauna, with insufficient time undisturbed to allow the evolution of endemic genera. The proportions in many taxa that are endemic to the Arctic, especially for the lower plants and invertebrates, is unknown, a feature that deserves more attention. The level of information varies widely between taxonomic groups, especially for the soil invertebrates and lower plants that have been examined at few sites. In contrast, information on vascular plants, birds, and mammals is detailed, both in terms of species identification, and in terms of population size and distribution.

In documenting current arctic biodiversity as a basis for conservation, a key feature is that many of the vertebrate

Table 10.9. Species richness in the terrestrial, freshwater, and marine environments of Svalbard (summarized from Elvebakk and Prestrud, 1996, and Prestrud et al., 2004). Detailed species lists are contained in the references quoted.

	Number of species
Plants	
Cyanobacteria[a]	73
Algae[a,b]	1049
Fungi and lichenised fungi[c]	1217
Mosses and liverworts[d]	373
Vascular plants[e]	173
Animals	
Marine crustacea[f]	467
Marine mollusks[f]	252
Other marine invertebrates[f]	924
Marine vertebrates (fish)[f]	70
Terrestrial and freshwater arachnids[g]	134
Terrestrial and freshwater insects[g]	289
Other terrestrial and freshwater invertebrates[g]	617
Birds[h,i]	53
Mammals[h,j]	9
Total	**5700**

[a]Skulberg (1996); Hansen J. and Jenneborg (1996); [b]Hasle and Hellum von Quillfeldt (1996) [c]Alstrup and Elvebakk (1996); Elvebakk and Hertel (1996); Elvebakk et al. (1996); Gulden and Torkelsen (1996); [d]Frisvoll and Elvebakk (1996); [e]Elven and Elvebakk (1996); [f]Palerud et al. (2004); [g]Coulson and Refseth (2004); [h]Strøm and Bangjord (2004); [i]202 species recorded, of which 53 are known to be breeding, to have bred in the past, or are probably breeding; [j]23 species recorded (plus another 8 species which are known to have been introduced), of which 9 are known to be breeding or to have bred in the past.

species spend only a small proportion of their time in the Arctic. This adaptive behavior is found in most birds, some marine mammals, and some freshwater and marine fish. As a result, documentation of their status and conservation action for them is dependent on international cooperation. It is also probable that the main threats to these migratory species occur during their migrations or during their winter period outside the Arctic. Current threats include changes in land- and water-use, human exploitation of resources upon which the animals depend, direct cropping of the animals for food or sport, accidental killing (as in the by-catch resulting from other fisheries), or pollution. A particular benefit of detailed and long-term observations, particularly for migratory birds that cover all continents (Figure 10.4), is that they provide a highly sensitive indicator of global environmental change.

After drawing up biodiversity inventories, individual items (species or habitats) can be assessed for their ability to survive into the future. For example, the IUCN has established criteria for assessing the degree of threat to the continued existence of species (IUCN, 1994). Many nations have used these IUCN criteria as the basis for compiling their national "Red Lists". Species are allocated to the various threat groups on the basis of criteria (Table 10.10). These criteria are grouped into four sets, which are briefly outlined here (see IUCN, 1994 for the various nuances).

First, there is a criterion of the known or suspected reduction in a species' population size. If this is known to have declined by at least 80% over the last ten years or three generations, then the species might be categorized as "critically endangered". Similarly, if the reduction in population size is more than 50% or more than 20% over the last ten years or three generations, then the species could be categorized as "endangered" or "vul-

nerable" respectively. Good data are necessary for such changes in population size to be known or estimated.

Second, there is a criterion relating to the known or estimated decline in the range of the species. Again somewhat arbitrary thresholds are set where the extent of occurrence is estimated to be less than 100 km^2, 5000 km^2, and 20000 km^2, or the area of occupancy is estimated to be less than 10 km^2, 500 km^2, and 2000 km^2, for the "critically endangered", "endangered", and "vulnerable" categories respectively. For these, the populations must be severely fragmented or located in a single place and either declining or demonstrating extreme fluctuations, in order to be categorized as "critically endangered". There are similar weaker criteria for the "endangered" and "vulnerable" categories (for example, populations must be at no more than 5 or 10 places respectively).

Third, the total population size can be used. The thresholds are less than 250 mature individuals and declining, or less than 50 mature individuals, for the "critically endangered" category. These thresholds are raised to 2500 and 250 for the "endangered" category and 10000 and 1000 for the "vulnerable" category. At these small total population sizes it is feared that inbreeding could occur, thus reducing the genetic variability within the species. Consequently, conservation action is needed, encouraging all of the mature individuals to contribute to future generations so that the present genetic diversity is not lost.

Finally, assessments can be on the basis of quantitative analyses estimating the risk of extinction in the wild over a period of either a number of years or over a number of generations, whichever is the longer. For the "critically endangered" category, the risk of extinction in the wild would have to be greater than 50% over 10 years or three generations. For the "endangered" category, the risk would have to be at least 20% within 20 years or

Table 10.10. The categories proposed by the IUCN for assessing the vulnerability, and hence the conservation priority, of species (abstracted from IUCN, 1994).

Species evaluated	Data adequate	IUCN category and code	Notes
Yes	Yes	Extinct (EX)	There is no reasonable doubt that the last individual of the species has died
Yes	Yes	Extinct in the wild (EW)	As above, but the species survives in cultivation, in captivity, or in at least one naturalized population outside its native distribution range
Yes	Yes	Critically endangered (CR)	The species is facing an extremely large risk of extinction in the wild in the immediate future
Yes	Yes	Endangered (EN)	The species is facing a large risk of extinction (but not as large as the category above) in the wild in the near future
Yes	Yes	Vulnerable (VU)	The species is facing a large risk of extinction in the wild in the medium-term future
Yes	Yes	Lower risk (LRcd, LRnt, LRlc)	The species does not fit into the above categories, but this category can be divided into three. *Conservation dependent* taxa are those that have a conservation program, cessation of which is likely to result in the species being moved into one of the above categories within five years. *Near threatened* taxa are those that are close to being vulnerable. *Least concern* taxa are those that do not fit into either of the above categories
Yes	No	Data deficient (DD)	There are insufficient data for a decision to be made about allocating the species to any of the above categories
No	No	Not evaluated (NE)	The species has not been assessed for sufficiency of data and hence does not fit into any of the above categories

Box 10.2. Five examples of the causes and possible consequences of genetic variability

1. Low levels of genetic variation in arctic plants, especially in the high Arctic, have been considered to result from widespread vegetative propagation and low sexual recruitment. The Swedish-Russian Tundra Ecology Expedition in 1994 provided the opportunity to sample 16 sites in a coastal transect from the Kola Peninsula to eastern Russia and up to 77° N. Four sedge species, *Carex bigelowii*, *C. ensifolia*, *C. lugens*, and *C. stans*, all showed a relatively high degree of genetic variation within most populations. Those populations with the lowest variation were associated with sites that were recently glaciated (10000 years ago) rather than populations from refugia which were already deglaciated 60000 to 70000 years ago (Stenstrom et al., 2001). Thus, although individual species may be geographically widespread, their genetic makeup and ecotypic variation, and hence their capacity to react to change, can be variable.

2. In Sweden, the rare wood-inhabiting polyporous fungus, *Fomitopsis rosea*, illustrates the limitation of genetic variability resulting from isolation of populations. Populations in isolated forest stands in Sweden had much narrower genetic structure than populations within the continuous taiga forests of Russia (Seppola, 2001). This suggests that habitat fragmentation can restrict genetic differentiation and potentially limit responses to environmental change.

3. Survival of reciprocal transplants of *Dryas octopetala* between snowbed and fellfield sites was followed for 15 years. Non-native genotypes have shown variable mortality rates after experiencing the rapid environmental change of transplanting. Some non-native transplants have survived, with variable rates between sources, but were far fewer than native transplants within their own environment. McGraw (1995) concluded that the existence of ecotypes adapted to different environments improves the probability that the species as a whole will survive rapid environmental change.

4. Musk oxen (*Ovibos moschatus*), despite a circumpolar distribution, have extremely low genetic variability and it is uncertain how they will respond to environmental change or to new parasites and diseases. However, since 1930, reintroduction following local extinction has proved successful from Greenland to Alaska, from Alaska to Wrangel Island, and from Alaska to the Taymir Peninsula. Reintroductions in Norway have been less successful (Gunn, 2001).

5. The genetic composition of plant populations, for example the purple saxifrage (*Saxifraga oppositifolia*) and the moss campion (*Silene acaulis*), determines their capacity to respond to short- or long-term environmental change. Species and populations also respond to the contrasting wet and dry micro-environments within high-arctic habitats. Evidence indicates that current populations in the high Arctic are derived from survivors in refugia during the last glaciation and from migrants that colonized more recently. It is likely that heterogeneity of sites and populations, combined with the history of climate variation, has provided the present flora with the resilience to accommodate substantial and even rapid changes in climate without loss of species (Crawford 1995; Crawford and Abbott 1994).

five generations, whereas for the "vulnerable" category it would have to be at least 10% within 100 years. Such an assessment depends on good data as well as on a suitable model that can be used to assess the risks.

The IUCN criteria are predicated upon species conservation. However, genetic diversity is also a part of the Convention on Biological Diversity. Many species have widespread distributions within the Arctic and occur in different habitats, landforms, and communities. This is a feature of the low species diversity, providing the opportunity for species to exploit resources and environments with little or no competition. Under the conditions of low species diversity, it is thought that the width of the ecological niche of the remaining species is wide. Measures of species richness underestimate the genetic diversity and there is a need to increase documentation of genetic variation within species, especially for those of conservation concern. Ecotypic differentiation is likely to be an important attribute in species response to climate change and is recognized as a key characteristic of arctic biodiversity. Five examples that illustrate genetic variability, its causes, and possible consequences emphasize the importance of both understanding and maintaining genetic variation within species by conserving diverse populations as a basis for conservation – an application of the precautionary principle (see Box 10.2).

This poses a number of questions for nations with arctic territory and for nations interested in the Arctic's biodiversity. Can inventories be prepared for more taxa than just the mammals and birds, which already exist? Are there data of sufficient quality and quantity to allocate the species to the IUCN categories? Are the data good enough and are there suitable models that can be used to estimate the risks of extinction? Are there sufficient taxonomists to be able to recognize, identify, and list the Arctic's species? Although the work of the IUCN is aimed at species, it is also important to have an inventory of habitats. Initially, however, on a circumpolar basis there needs to be agreement on the classification of habitats in the marine environment, the freshwater environment, and the terrestrial environment. This will require ecological expertise and international agreement, but is a requisite first step in drawing up an inventory of the Arctic's habitats, and then assessing which habitats are priorities for conservation action.

These considerations lead to the first four recommendations. These are made without attempting to allocate responsibility for undertaking the work involved.

1. There needs to be a supply of trained ecologists who can devise appropriate circumpolar classifications of habitats and then survey these so as to measure their extent and quality and to establish their dynamics.
2. There needs to be a supply of trained taxonomists who can draw up inventories of the Arctic's species. There are already good data on which species of vertebrate animals and vascular plants are to be found in the Arctic, so particular attention needs to be given to the training of taxonomists who can work with non-vascular plants, invertebrate animals, fungi, and microorganisms (protozoa, bacteria, etc.).
3. Inventories need to be generated for the Arctic's biodiversity (both species and habitats), indicating for each entry in the inventory where it occurs and either the size of the overall species population or the extent of the habitat. Such inventories need to be on a circumpolar basis rather than on a national basis as nations with arctic territory also have territory south of the Arctic.
4. The genetic diversity of many of the Arctic's species is presently poorly known or unknown. Much research is needed to explore this aspect of the Arctic's biodiversity and conservation management will need to ensure that genetic diversity is not lost.

10.5.2. Identifying changes in the Arctic's biodiversity

In section 10.4, seven series of changes were explored, focusing on the distribution range of species and habitats, on the total size of species populations and the extent of habitats, and on genetic variability within populations. Each of these interacts with the success and failure of non-native species to establish themselves in the Arctic, with the migration routes and timing of migration of migratory species, and with the selection and management of protected areas. Change is expected, and each species is likely to respond in an individualistic way so that novel assemblages of species are very likely to occur in the future. Sources of information on changes to biodiversity are many and varied and analyses of past changes can provide insights into the future (Box 10.3).

Change in ecological communities is often referred to as "ecological succession". A distinction is drawn between "primary succession", which occurs on new substrates such as when a glacier recedes (Miles and Walton, 1993), and "secondary succession", which occurs following a disturbance or perturbation. A preservationist attitude might be to maintain what occurs today and so manage a habitat in such a way as to oppose ecological succession. A conservationist attitude would be to work with ecological succession. This dichotomy of thinking is highlighted by Rhind (2003), who said "we have become fixated with the idea of preventing natural succession and, in most cases, would not dream of allowing a grassland or heathland to develop into woodland". In the Arctic, climate change will drive primary and secondary successional changes and, in the interests of conserving the Arctic's biodiversity, management should work with these changes rather than opposing them.

Species might adapt to new environmental conditions if they have a sufficient genetic diversity and sufficient time. This is outlined in Chapter 7 where it is stated that a key role of biodiversity is to provide the adaptive basis for accommodating the extreme levels of environmental variability that characterize much of the Arctic. The genetic level of biodiversity allows populations to meet the challenges of an extremely variable arctic environment and this ensures persistence of the populations, at least in the short to medium term. Over the longer term, such genetic diversity is the basis for evolutionary change leading to the emergence of new subspecies and species. With projections of a rapidly changing climate, genetic diversity is important as a kind of insurance that the species will be able to successfully meet the environmental challenges that they will face.

As stated by Walls and Vieno (1999) in their review of Finnish biodiversity "...mere biological information is not enough for successful biodiversity conservation. Conservation decisions and the design of biodiversity management are primarily questions of social and economic policy...Biodiversity conservation requires, in fact, the whole spectrum of sociological, economic and policy analyses to complement the basic biological information". Traditional knowledge was addressed in section 10.2.7, but the implications of Walls and Vieno's (1999) comment are that the knowledge gained in the past is insufficient since the aspirations of today's people for the future also need to be considered. This highlights one of the central divisions of thought about biodiversity conservation. Is it "nature-centric", because it is believed that nature has an inherent right to exist? Or, is biodiversity conservation "human-centric", because it is believed that the biological world must be molded to suit the needs of people, now and in the future? The problem with the former approach is that it can neglect the fact that humans (*Homo sapiens*) are an integral part of the ecosystem and the food web. The problem with the latter is that it places *H. sapiens* as the only species that really matters, and hence it is of limited concern if other species become extinct. A middle way needs to be found.

In the Arctic, people have been part of the food web more or less since the end of the last ice age when ecological succession began with the northward movement of plants and animals, in the sea and on land, as the ice retreated. As well as the obvious changes in distribution,

Box 10.3. Some sources of information on changes in the Arctic's biodiversity

Paleo-ecological evidence

Probably the most dramatic ecological event in arctic prehistory was the conversion of a vegetation mosaic dominated by semi-arid grass–steppe with dry soils and a well developed grazing megafauna to a mosaic dominated by wet-moss tundra without a large grazing fauna. There are three main hypotheses to explain the changes.

- The "pleistocene overkill hypothesis". This suggests that Beringia was colonized by people with hunting skills who developed spears with stone micro-blades which enabled them to drive the megafauna to extinction and that it was this loss of grazing that caused the vegetation change. Corroborative evidence for intensive killing comes from paleolithic sites where large quantities of bones have been unearthed. At Mezhirich in the Ukraine, bones of 95 individual mammoths (*Mammuthis primigenius*) were found.
- The "climate hypothesis". This assumes that an arid, continental climate prevailed in Beringia during the Pleistocene giving low summer precipitation and dry soils, promoting productive steppe vegetation which supported the populations of large grazers (mammoths, bison, and horses). As the climate became wetter during the Holocene, snow depth increased, the moss–lichen cover developed, and herbaceous vegetation reduced. This vegetation change is shown in the Pleistocene pollen and plant macrofossil record and it is hypothesized that the vegetation change resulted in the decline and eventual extinction of the megafauna.
- The "keystone-herbivore hypothesis". This hypothesis combines the overkill and climate hypotheses with a more detailed understanding of vegetation changes that results from current knowledge of changes in both grazing and climate (Zimov et al., 1995).

Evidence from refugia such as Beringia, which remained without ice cover during past glaciations as a result of local climate conditions, and changes in sea level have been important in documenting long-term development of species and genetic diversity. Documentation of past ecological changes through analyses of plant and animal remains in stratified terrestrial, freshwater, and marine sediments has contributed much to the analysis of climate change.

Historical documentation

Historical records show that Greenland was first colonized by Norsemen around AD 986. The population rose to about 3000 based on up to 280 farms and enhanced by fishing and trading in walrus skins and ivory. The colony became extinct in the 15th century, probably due to climatic deterioration and possibly disease. Analysis of the vegetation in the vicinity of the farms and habitations indicates that about 50 vascular plants were probably introduced by the Norsemen and have survived to the present day – an ecological footprint detected and quantified through historical documentation (Fogg, 1998). It is the historical records of fishing, whaling, and sealing in the arctic seas that provide some of the most detailed documentation of the distribution and population changes of marine fauna. These are extensively detailed in Chapters 11 and 13. The data reflect the impacts of variation in climate and exploitation often over the past 50 to 100 years or more.

---

number, extent, etc., there are likely to be many more subtle changes in the functions of ecosystems and in the physiology of individuals, but prediction of what these changes might be is largely elusive. Predictions are based on models. The concept of modeling biodiversity conservation has already been addressed (see Fig. 10.12) and has been shown to be within the domain of statistical models rather than precise models that give a definitive result. However, despite such limitations, models are useful in attempting to explore the likely changes to the Arctic's biodiversity and their effects on the human population.

For example, in Finland models have been used to project the likely changes in the distribution of the major forest trees – pine (*Pinus sylvestris*), spruce (*Picea abies*) and birch (*Betula* spp.) – predicting the movement north of the two coniferous species (Kuusisto et al., 1996). At the same time, the models have projected that whereas at present only the southern fifth of Finland is

thermally suitable for cultivating spring wheat, by 2050 it is likely that this crop could be grown throughout the southern half of Finland. Herein lies the social problems. Finland currently is a country with an economy largely based on forestry and it has a biodiversity rich in forest species. If the economy were to change to one more agriculturally based, how would this affect the social structure of the human population? Would the loss of the forest biodiversity and the loss of the social aspects of its use (e.g., collecting berries and mushrooms, hiking, and other leisure activities in the forest) be acceptable?

These considerations of change lead to two further recommendations.

5. Management of the Arctic's biodiversity must work with ecological succession and not against it. This thinking needs to be incorporated into all aspects

Indigenous knowledge

Insights into environmental and ecological change that are based on indigenous knowledge are now fully recognized and increasingly documented (see Chapters 3 and 12). The documentation includes insights into changes in biodiversity over recent decades, particularly regarding species of importance to hunters. The knowledge is specific to local areas but can be accumulated and compared across regions. For example, maps of migration routes indicate species-specific changes around Hudson Bay (McDonald et al., 1997), whereas recent changes in fish and wildlife, described by Inuvialuit hunters in Sachs Harbour, illustrate specific evidence of other responses to climate changes (Krupnik and Jolly, 2002):

Two species of Pacific salmon caught near the community.
Increased numbers of Coregonus sardinella (least cisco).
Fewer polar bears in area because of less ice.
Increasing occurrence of "skinny" seal pups at spring break-up.
Observation of robins; previously unknown small birds.
Increased forage availability for caribou and muskox.
Changes to timing of intra-island caribou migration
Identification of current and future changes

Documentation of changes in many mammals, birds, and fish is already well developed in national programs of individual arctic nations and internationally for migratory species. Monitoring is particularly strong where international agreements and commercial interests are involved and where individual species are classified as "endangered" on the national or international "Red Lists" drawn up using IUCN criteria (see section 10.5.1). There are, however, other aspects of biodiversity where documentation of change is seriously lacking. Documentation of changes in various aspects of plant diversity is very weak. There are only two programs that approximate to systematic, circumpolar observations of plants. (1) One is the International Tundra Experiment (ITEX), which has routinely recorded changes in vegetation cover and plant performance at about 30 sites (including some alpine and antarctic sites). Experimental passive warming of about 1 to 2 °C is achieved by installing replicated open-topped chambers, with adjacent plots without experimental warming as controls. ITEX has been in operation for a decade, but initial data synthesis has already begun (Arft et al., 1999). The serious limitation in ITEX as a monitoring program is that individual sites are largely dependent on short-term research funding. (2) The other, on a totally different spatial scale and level of resolution, is the use of satellite measurements to detect changes in vegetation greenness (Myneni et al., 1997). This assessment of change in greenness between 1981 and 1991 cannot be validated owing to the total lack of systematic ground observations at a compatible spatial scale.

of the management of biodiversity in the sea, in freshwater, and on the land.

6. Models need to be further developed to explore changes in biodiversity under the various scenarios of climate change. Again, these models will need to explore biodiversity change in the sea, in freshwater, and on land.

10.5.3. Recording the Arctic's changing biodiversity

There are two aspects to recording the Arctic's changing biodiversity that need to be addressed: monitoring (or surveillance) and indicators. Monitoring involves the periodic recording of data so that trends can be detected. Usually, it also involves assessing progress toward some target, but often it only involves determining if the resource being monitored still exists and how the

amount of that resource is changing (and this is often referred to as surveillance). Indicators are regularly monitored measures of the current state of the environment, the pressures on the environment, and the human responses to changes in that state. This three-point set of indicators is often referred to as the "pressure-state-response model" (Wilson et al., 2003). It is often easier to find indicators of state than indicators of either pressures or responses.

Monitoring of wildlife has a long history. There have been attempts to coordinate monitoring, as outlined for the Nordic Nations by From and Söderman (1997). The aim in these nations was "to monitor the biodiversity and its change over time with appropriate and applicable mechanisms, and to monitor the cause-effect relationship between pressure and response on biodiversity by using specific biological indicators". There were five implications of these objectives: (1) the program would

Box 10.4. The seven long-term objectives for CAFF's biodiversity monitoring (CAFF, 2002c)

Overall objective

To provide an information basis for sound decision-making regarding conservation and sustainable use of arctic flora and fauna.

Detailed objectives

1. To detect change and its causes amongst flora and fauna of the circumpolar Arctic.
2. To strengthen the infrastructure for and harmonization of long-term monitoring of arctic flora and fauna.
3. To provide an early warning system and strengthen the capacity of arctic countries to respond to environmental events.
4. To ensure the participation of arctic residents, including indigenous peoples, and to incorporate their knowledge into monitoring.
5. To establish a circumpolar database of biodiversity monitoring information and contribute to existing European and global database systems.
6. To contribute to national, circumpolar, European, and global policies concerned with conservation of biodiversity and related environmental change.
7. To integrate circumpolar biodiversity monitoring information with physical and chemical monitoring information of the Arctic Monitoring and Assessment Programme and others.

exclude chemical and physical aspects of environmental monitoring; (2) the focus would be on ecosystems and species and the data would be analyzed in the simplest manner to provide appropriate, qualitative, and quantitative information; (3) another focus would be anthropogenic changes, although the analyses would need to distinguish these from natural changes; (4) monitoring would include, among others, threatened habitats and species, and hence their disappearance or extinction would become known; and (5) the monitoring would not directly focus on administrative performance indicators, although it might provide important information for understanding these. The main problem with this Nordic monitoring program is that it relates only to the terrestrial environment, although this does include wetland and coastal habitats. More attention needs to be paid to the marine environment.

Progress is being made in relation to monitoring biodiversity in the Arctic (CAFF, 2002c) with the Circumpolar Biodiversity Monitoring Program. Its goal is "to improve understanding of biodiversity through harmonization and/or expansion of existing programs and networks. The proposed approach focuses on three large ecosystems (terrestrial, freshwater, marine) and selected criteria include ecological importance, socio-economic importance, and feasibility". CAFF (2002c) then continued with accounts of a number of monitoring programs, covering Arctic char, caribou and reindeer, polar bear, ringed seal, shorebirds (also known as waders), seabirds, geese, and work in relation to the International Tundra Experiment. The strengths of this proposal are that the connections between the marine, freshwater, and terrestrial environments are recognized and that the monitoring would be on a circumpolar basis; the weakness is that so few actual species are being monitored, although the aspirations are more ambitious. At present there is no explicit botanical monitoring, and the invertebrate

animals have been overlooked. For example, a program focused on the many species of fritillary butterfly of the genus *Clossiana* (although taxonomically this has now been divided into a number of genera), which occur in northern Asia, northern Europe, and North America, would indicate much about the effects of climate change on insects and their food plants, and on the interrelationships between plants and specialized herbivores. For the future, the Circumpolar Arctic Biodiversity Monitoring Network project is challenging, having the twin goals to "develop the infrastructure, strengthen ecological representation, and create data management systems for circumpolar Arctic species biodiversity monitoring networks", and to "establish functional links between these arctic networks and European and global biodiversity observation systems and programs". The long-term objectives of CAFF's biodiversity monitoring are listed in Box 10.4.

Monitoring is widely advocated. For example, BirdLife (2000) indicated that it wished to "monitor and report on progress in conserving the world's birds, sites and habitats", but also that it wished to monitor the effectiveness of its work in achieving the objectives set out in its strategy. Usher (1991) posed five questions about monitoring. These related to the *purpose* (what are the objectives?), the *methods* to be used (how can the objectives be achieved?), the form of *analysis* (how are the data, which will be collected periodically, to be analyzed statistically and stored for future use?), the *interpretation* (what might the data mean and can they be interpreted in an unbiased manner?), and *fulfillment* (when will the objectives have been achieved?). It is vital that all five questions are asked and answered before a monitoring scheme begins. All too frequently *ad hoc* monitoring programs provide data that cannot be analyzed statistically and so the confidence that can be placed in resulting trends is minimal.

The basic need is for the establishment of a circumpolar network of sites where large-scale (hectares or square kilometers) replicated plots can be distributed where vegetation cover and composition can be documented. Following scientific principles, the network could be spatially located to test the hypotheses of vegetation change that have been generated during the ACIA process. Establishment of some sites within the CPAN could further test the performance of this approach to conservation. Further, fine-scale observations, for example of species performance, could be nested within the landscape-scale plots. Such a hierarchy of spatial scales would be similar to that defined in the Global Terrestrial Observing System (GTOS) led by the FAO. 171 arctic sites and a number of arctic site networks are currently registered on the Terrestrial Ecosystem Monitoring Sites of the GTOS, and they could provide the basis for an appropriate monitoring network. The GTOS has developed a Biodiversity Module with seven core variables to guide development in the program (threatened species, species richness, pollinator species, indicator species, habitat fragmentation, habitat conversion, and colonization by invasive species). The relationship with the sister programs, the Global Ocean Observing System (GOOS) and the Global Climate Observing System (GCOS), needs to be clarified. This would correspond with the recommendations in Chapters 7, 8, and 9. Each chapter identifies the need for improved systematic, long-term observation and monitoring programs.

Based on the aspects of the conservation of biodiversity identified in this chapter, further attention should be given to the five subsidiary aspects of monitoring outlined in Box 10.5. It would be too resource intensive to

Box 10.5. Five other aspects of monitoring that relate to the principles of biodiversity conservation outlined in this chapter

Phenology monitoring

This has a long tradition, especially in Russia, but has not been developed to meet future needs. Observation of the timing of specific phenomena, for example leaf and flower emergence, arrival and departure of migratory birds, and timing of emergence and feeding of specific insects, can be directly related to climatic conditions if repeated annually. Such observations are particularly suited to remote rural communities where other monitoring is not feasible. It also has a strong educational potential.

Genetic diversity

This is generally poorly and unsystematically documented. The establishment of a baseline for future detection of change is a priority. Selection of a limited number of distinct taxonomic and functional groups, with particular conservation concern, should allow establishment of an initial circumpolar baseline, including storage of appropriate material.

Invertebrate fauna

Both the diversity and distribution of invertebrates, especially in soils and freshwater sediments, are poorly documented, despite their importance as a basis for food webs and in the decomposition of organic matter and nutrient cycling. Establishment of basic survey information is best developed through a short-term targeted program at a limited number of existing research bases and field sites, supplemented where necessary so as to obtain a representative coverage of broad habitat types.

Integrated monitoring

Potential cause and effect variables would be recorded; this is seen to be increasingly important as the complexity of the systems is recognized. The ACIA has provided the best available understanding of the complex system responses to climate change. The next critical step is to express these as system models and test these through existing and expanded data at a limited number of selected field sites, so as to test and refine the hypotheses and to assess the potential establishment of long-term integrated monitoring.

A rapid response network

The ACIA has highlighted the probability of increased frequency and intensity of climatic events, increased outbreaks of pests and diseases, increased pollution, and other environmental accidents. The timing and location of such events are currently unpredictable. Yet the need for rapid initial documentation of impacts on biodiversity as a basis for longer-term observations is regularly required. The use of existing distributed field stations to provide an initial, international rapid response network is a logical development that would benefit from a feasibility study.

(a) Global level

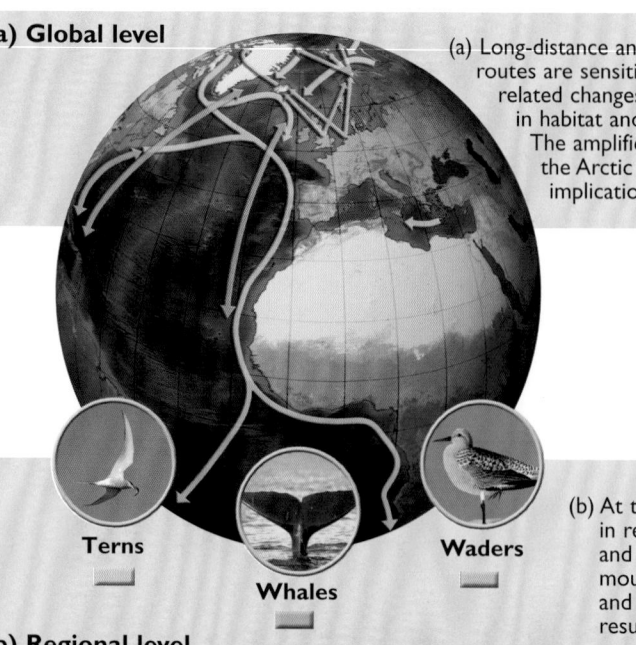

(a) Long-distance animal migration routes are sensitive to climate-related changes such as alterations in habitat and food availability. The amplification of warming in the Arctic thus has global implications for wildlife.

Terns

Whales

Waders

(b) Regional level

attempt to monitor all aspects of the Arctic's biodiversity. So in order to reduce the amount of work required indicators are often advocated. For indicators to be valuable they should ideally fulfill the following four criteria (modified from Wilson et al., 2003). First, they should reflect the state of the wider ecosystems of which they are a part. Second, indicators should have the potential to be responsive to the implementation of biodiversity conservation policies. Third, indicators should be capa-

(b) At the regional level, vegetation and the animals associated with it will shift in response to warming, thawing permafrost, and changes in soil moisture and land use. Range shifts will be limited by geographical barriers such as mountains and bodies of water. Shifts in plankton, fish, and marine mammals and seabirds, particularly those associated with the retreating ice edge, will result from changes in air and ocean temperatures and winds.

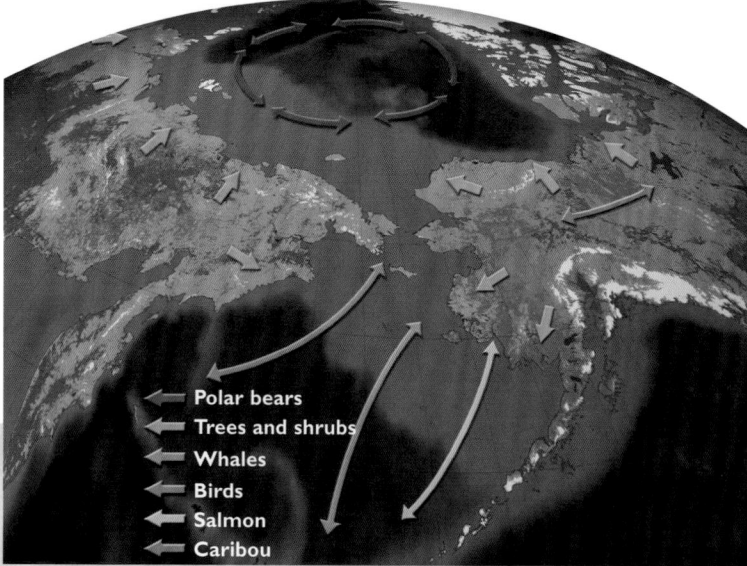

Polar bears
Trees and shrubs
Whales
Birds
Salmon
Caribou

(c) Landscape level

ble of being measured reliably on a regular (not necessarily annual) basis, and should be comparable with similar measures at greater spatial scales. Fourth, they should have, or have the potential for, strong public resonance. Such a set of criteria for indicators fits well with the set of seven long-term objectives of CAFF's Circumpolar Arctic Biodiversity Monitoring Network proposal, outlined in Box 10.4.

(c) At the landscape level, shifts in the mosaic of soils and related plant and animal communities will be associated with warming-driven drying of shallow ponds, creation of new wet areas, land use change, habitat fragmentation, and pests and diseases. These changes will affect animals' success in reproduction, dispersal, and survival, leading to losses of northern species and range extensions of southern species.

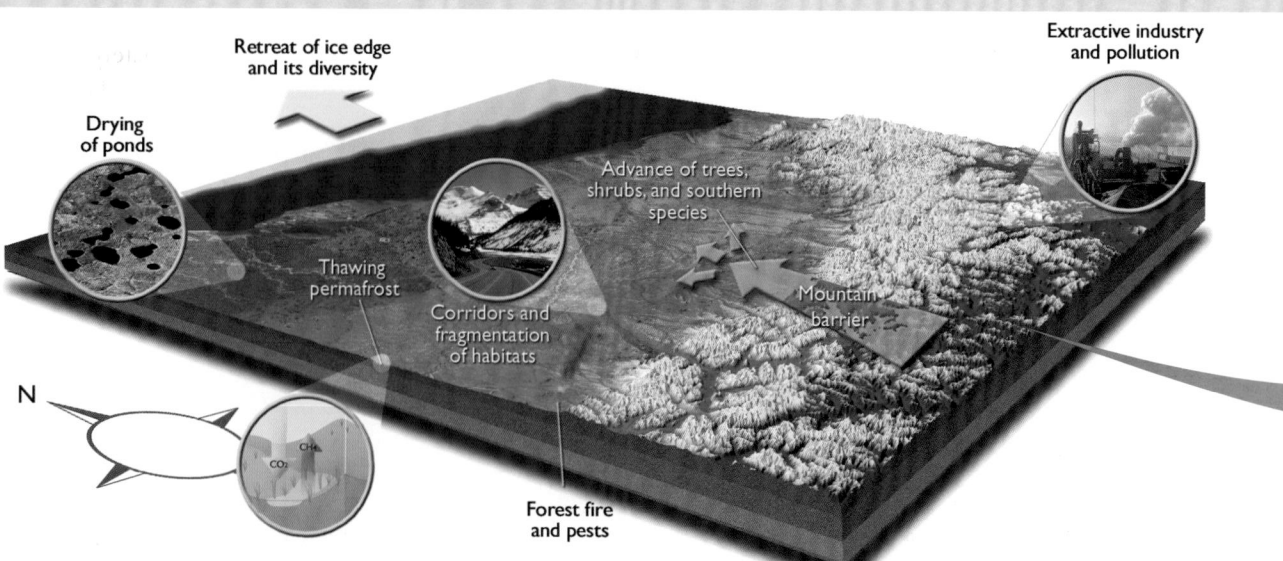

Retreat of ice edge and its diversity

Extractive industry and pollution

Drying of ponds

Thawing permafrost

Advance of trees, shrubs, and southern species

Mountain barrier

Corridors and fragmentation of habitats

Forest fire and pests

N

Fig. 10.16. A representation of the effects of climate change on biodiversity at different spatial scales. The text focuses on species diversity and to some extent on habitat diversity, but genetic diversity is not included.

These discussions lead to three further recommendations.

7. Circumpolar monitoring networks need to be fully implemented throughout the Arctic. The proposals are challenging, but data on the state of the Arctic's biodiversity, on the drivers of change in that biodiversity, and on the effectiveness of responses to those changes, needs to be collected, analyzed, and used in the development of future arctic biodiversity policy.

8. Attention needs to be given to establishing the kinds of subsidiary aspects of monitoring, examples of which are outlined in Box 10.5. These are vital if a holistic view is to be taken of the Arctic's biodiversity, its conservation in the face of a changing climate, and the management of the biodiversity resource for future generations of people to use and enjoy.

9. A suite of indicators needs to be devised and agreed, monitoring for them undertaken, and the results made publicly available in a format (or formats) so as to inform public opinion, educators, decision-makers, and policy-makers.

10.5.4. Managing the Arctic's biodiversity

"The Arctic is a distinct and significant component of the diversity of life on Earth" was a statement made at a meeting in 2001 to celebrate ten years of arctic environmental cooperation (Vanamo, 2001). This probably encapsulates why the conservation of the Arctic's biodiversity is not only essential to the peoples of the Arctic but also why the Arctic is important globally. It sets the imperative to do something to conserve the biodiversity of one of the more pristine geographical parts of the

(d) PLOT LEVEL

(d) Changes in snow conditions, ice layers, the cavity beneath the snow, summer temperatures, and nutrient cycling act on individual plants, animals, and soil microorganisms leading to changes in populations. It is at the level of the individual animal and plant where responses to the climate take place leading to global-scale vegetation shifts.

world, but nevertheless a geographical area that is threatened with a series of human-induced changes due to developments and over-exploitation within the Arctic, and to long-range pollution and climate change, which are both global problems.

One of the first requirements is to collate information about the best way to manage the Arctic's biodiversity in a changing climate. This will be based on knowledge held by local people together with knowledge that has been gained by scientists, either through observation or experiment. There have been a number of attempts to bring together guidelines for best practice, usually either in a nation or for a particular area. An example would be the proposals developed in Finland for practical forest management (Korhonen et al., 1998). These guidelines integrate concern for the environment with the needs of production forestry, and the use of forests for recreation, protection of the quality of soil and water, and the management of game species. They provide an example of what can be done when all the interest groups work together for a common goal. Such an approach would also be useful on a circumpolar basis for the conservation and sustainable use of the Arctic's biodiversity. This leads to a further recommendation.

10. Best practice guidelines need to be prepared for managing all aspects of the Arctic's biodiversity. These need to be prepared on a circumpolar basis and with the involvement of all interested parties.

The value of protected areas has been discussed (section 10.4.7), as well as the plans for developing a comprehensive network of these areas throughout the Arctic. Such a start is excellent, setting aside areas of land, freshwater, and sea where nature has primacy over any other forms of land- and water-use. The three questions that need to be asked are how quickly can this network of protected areas be completed, how will they need to change as the climate is changing, and are they doing what they were designed to do? First, the reviews by CAFF (2001, 2002a) indicated that there were some of the Arctic's habitats, especially in the marine environment, that were not adequately covered by the CPAN. It is important that work on establishing a comprehensive CPAN is undertaken so that protection can be afforded to the breadth of the Arctic's biodiversity before any is lost. Second, work on understanding how climate change will affect each protected area will allow management to have a greater chance of protecting the biodiversity in that area, or of adopting the "soft boundary" approach outlined in section 10.4.7. Work needs to be undertaken, and made widely available in management guidelines, on the management of these protected areas; an example for the protected areas in Finland is as in Anon (1999). Work also needs to analyze how climate change is likely to affect each of the protected areas. Such work has been carried out for the Canadian national parks (Scott and Suffling, 2000), stressing the importance of sea-level rise for the many national parks that are located on the coast. These considerations give rise to two further recommendations.

11. The CPAN needs to be completed and then reviewed so as to ensure that it does actually cover the full range of the Arctic's present biodiversity.
12. An assessment needs to be made for each protected area of the likely effects of climate change, and in the light of this assessment the management methods and any revisions of the area's boundary need to be reviewed.

In undertaking these reviews, one of the important questions is whether or not the protected area is conserving what it was designed to conserve. This is not always a simple task, especially with year-to-year variation in population sizes and with longer term changes in habitat quality, but such assessments are becoming more commonplace (e.g., Parrish et al., 2003).

Protected areas are just one method for attempting to conserve the Arctic's biodiversity. Although biodiversity conservation is the primary focus of management within the protected areas, they will only ever cover a relatively small proportion of the land and water area of the Arctic, and thus will only contain a small proportion of the Arctic's biodiversity resource. Hence, it is imperative that biodiversity is also considered in the land and water outside protected areas. Forms of integrated management need to be adopted whereby biodiversity is not forgotten among all the other competing claims for space on land or at sea. The kind of approach proposed for the Canadian Arctic, with forms of integrated management of coastal and marine areas (M.A.K. Muir, Arctic Institute of North America and CAFF, pers. comm., 2003), is just one example of practical applications of a biodiversity approach to the wider environment. The need is to incorporate biodiversity thinking into all forms of policy development, not just environmental policies, but also policies on education, health, development, tourism, and transport. This is clearly a part of this wider environmental approach for biodiversity conservation. In this way more of the Arctic's biodiversity is likely to be protected in the face of a changing climate than by relying solely on the protected areas. These considerations give rise to two further recommendations.

13. Integrated forms of management, incorporating the requirement for biodiversity conservation, need to be explored for all uses of the land, freshwater, and sea in the Arctic.
14. Biodiversity conservation needs to be incorporated into all policy development, whether regional, national, or circumpolar.

In order to assist in these processes, the "ecosystem approach", sometimes also referred to as the "ecosystem-based approach", has been advocated (Hadley, 2000). This sets out a series of 12 principles, some of which are science-oriented, but all of which form an essentially socio-economic context for conservation. In relation to climate change in the Arctic, two of the 12 principles are particularly relevant. Principle 5 focuses on ecosystems services, and is that "conservation of ecosystem structure and function, in order to maintain ecosystem services, should be a priority target for the ecosystem approach". Principle 10 states that "the ecosystem approach should seek the appropriate balance between, and integration of, conservation and use of biological diversity". An example of the possible application of this approach for the marine environment in the Arctic is as reported by CAFF et al. (2000, the summary of the presentation by K. Sherman) and Muir et al. (2003). Since this approach is still comparatively new, its details have as yet been worked out in very few situations. Hence, a further recommendation.

15. The ecosystem approach (or ecosystem-based approach) should be trialed for a number of situations in the Arctic, so as to assess its ability to harmonize the management of land and water both for the benefit of the local people and for the benefit of wildlife.

In all this work, it should be remembered that the conservation of the Arctic's biodiversity is necessary for itself, for the peoples of the Arctic, and more generally for this planet. These concepts were implicitly enshrined in the Convention on Biological Diversity, the final text of which was agreed at a conference in Nairobi, Kenya, in May 1992. Within a year, the Convention had received 168 signatures. As a result, the Convention entered into force on 29 December 1993, and there is now considerable international activity to implement the Convention in the majority of nations globally. This gives rise to a final recommendation.

16. All nations with arctic territory should be working toward full implementation of the Convention on Biological Diversity, coordinating their work on a circumpolar basis, and reporting both individually and jointly to the regular Conferences of the Parties.

10.5.5. Concluding remarks

Biodiversity is not the easiest of concepts to grasp. On the biological side, biodiversity needs to be considered at three scales – variation within species (genetic diversity), variation between species (species diversity), and variation among assemblages of species (habitat diversity). Whereas habitat diversity in the Arctic's land, freshwater, and sea would probably be measured in hundreds of habitats, species diversity would be measured in thousands or tens of thousands of species, and genetic diversity in millions of genes. These are all influenced by a changing climate. On the geographical side, biodiversity can be considered at many different scales, from the individual plant or animal and its immediate surroundings, to the whole world. Again, a changing climate can affect each of these scales, and indeed the effects at one scale may be different to the effects at another.

This chapter has shown that the Arctic's biodiversity is important in relation to the biodiversity of the world at the largest extreme and to local people at the smallest extreme. The types of impacts that climate change might have are illustrated in Fig. 10.16, which endeavors to highlight the importance of four of the spatial scales. Each of the ecological processes is affected by climate change, whether the migrations at the global scale or decomposition of dead plant and animal material at the plot level. A small shift in a climatic variable can have very different effects at these scales, and a small change at one scale can cause other changes in scales both above and below. Cause and effect are often difficult to determine, and so models to project changes as a result of climate change are still problematic.

Herein lies the difficulty in conserving the Arctic's biodiversity. Among this multitude of scales, what are the priorities? Should the primary focus be on habitats, species, or genes? Which of the many spatial scales is the most important? It is clear that not every aspect of the Arctic's biodiversity can be conserved, so priorities have to be attached to actions that can conserve the greatest amount of biodiversity or, in some situations, the greatest amount of useful biodiversity. But to set these priorities, information is required about the present state of biodiversity and about how it is changing. With such information, models of a more or less sophisticated type can be used to project what might happen in the future. It is within this context that the 16 recommendations have been made, and their acceptance should assist the peoples of the Arctic in conserving their biodiversity into the future.

Acknowledgements

Michael Usher would like to thank the contributing authors for their inputs to this chapter, but especially Magdalena Muir for the provision of much literature and Pål Prestrud for arranging a meeting in Oslo in May 2004. Michael Usher's participation in the ACIA has been funded by the Universities of Alaska, USA, and Oslo, Norway, and by the Joint Nature Conservation Committee (UK) through CAFF, Iceland; for all of this funding he is grateful.

References

Aebischer, N.J., J.C. Coulson and J.M. Colebrook, 1990. Parallel long-term trends across four marine trophic levels and weather. Nature, 347:753–755.

Alden, J., 1991. Provisional tree seed zones and transfer guidelines for Alaska. USDA Forest Service, General Technical Report PNW 270.

Alexander, V. and H.J. Niebauer, 1981. Oceanography of the eastern Bering Sea ice-edge zone in spring. Limnology and Oceanography, 26:1111–1125.

Alstrup, V. and A. Elvebakk, 1996. Fungi III. Lichenicolous fungi. In: A. Elvebakk and P. Prestrud (eds.). A Catalogue of Svalbard Plants, Fungi, Algae and Cyanobacteria, pp. 261–270. Norsk Polarinstitutt.

AMAP, 1998. Arctic Pollution Issues: a State of the Arctic Environment Report. Arctic Monitoring and Assessment Programme, Oslo.

AMAP, 2002. Arctic Pollution 2002. Persistent Organic Pollutants, Heavy Metals, Radioactivity, Human Health, Changing Pathways. Arctic Monitoring and Assessment Programme, Oslo.

Anderson, P.J. and J.F. Piatt, 1999. Community reorganization in the Gulf of Alaska following ocean climate regime shift. Marine Ecology Programme Series, 189:117–123.

Angelstam, P., E. Lindstrom and P. Widen, 1985. Synchronous short-term population fluctuations of some birds and mammals in Fennoscandia – occurrence and distribution. Holarctic Ecology, 8:285–298.

Anon, 1999. The Principles of Protected Area Management in Finland: Guidelines on the Aims, Function and Management of State-owned Protected Areas. Metsähallitus, Helsinki.

Anon, 2001a. The Status of Wildlife Habitats in Canada 2001. Wildlife Habitat Canada/Canada Habitat Faunique.

Anon, 2001b. NERI Report and Activities 2000–2001. National Environment Research Institute, Roskilde.

Arft, A.M. and 28 other authors, 1999. Responses of tundra plants to experimental warming: meta-analysis of the International Tundra Experiment. Ecological Monographs, 69:491–511.

Arnalds, O., 2000. Desertification: an appeal for a broader perspective. In: O. Arnalds and S. Archer (eds.). Rangeland Desertification, pp. 5–15. Kluwer Academic Press.

Arnalds, O., E.F. Þorarinsdottir, S. Metusalemsson, A. Jonsson, E. Gretarsson and A. Arnason, 2001. Soil Erosion in Iceland. Iceland Soil Conservation Service, Hella.

Babenko, A.B. and V.I. Bulavintsev, 1997. Springtails (Collembola) of Eurasian polar deserts. Russian Journal of Zoology, 1:177–184.

Bailey, E.P., 1993. Introduction of Foxes to the Alaskan Islands – History, Effects on Avifauna, and Eradication. United States Department of the Interior, Fish and Wildlife Service Resource Publication.

Barrett, R.T. and Y.V. Krasnov, 1996. Recent responses to changes in stocks of prey species by seabirds breeding in the southern Barents Sea. ICES Journal of Marine Science, 53:713–722.

Batzli, G.O., 1975. The role of small mammals in arctic ecosystems. In: F.B. Golley, K. Petrusewicz and L. Ryszkowski (eds.). Small Mammals: their Productivity and Population Dynamics, pp. 243–267. Cambridge University Press.

Batzli, G.O., 1981. Population and energetics of small mammals in the tundra ecosystem. In: L.C. Bliss, O.W. Heal and J.J. Moore (eds.). Tundra Ecosystems: a Comparative Analysis, pp. 377–396. Cambridge University Press.

Batzli, G.O., R.G. White, S.F. MacLean, F.A. Pitelka and B.D. Collier, 1980. The herbivore-based trophic system. In: J. Brown, P.C. Miller, L.L. Tieszen and F.L. Bunnell (eds.). An Arctic Ecosystem: the Coastal Tundra at Barrow, Alaska, pp. 335–410. Hutchinson and Ross.

Bazely, D.R. and R.L. Jefferies, 1997. Trophic interactions in Arctic ecosystems and the occurrence of a terrestrial trophic cascade. In: S.J. Woodin and M. Marquiss (eds.). Ecology of Arctic Environments, pp. 183–207. Blackwell Science.

Bell, N. (ed.), 1994. The Ecological Effects of Increased Aerial Deposition of Nitrogen. British Ecological Society.

Beniston, M., 2003. Climatic change in mountain regions: a review of possible impacts. Climate Change, 59:5–31.

Berg, A., B. Ehnstrom, L. Gustafsson, T. Hallingback, M. Jonsell and J. Weslien, 1994. Threatened plant, animal, and fungus species in Swedish forests: distributions and habitat associations. Conservation Biology, 8, 718–731.

Bernes, C., 1991. Acidification and Liming of Swedish Freshwaters: Monitor 12. Swedish Environmental Protection Agency.

Bernes, C., 1993. The Nordic Environment – Present State, Trends and Threats. Nordic Council of Ministers, Copenhagen.

Bertram, D.F., D.L. Mackas and S.M. McKinell, 2001. The seasonal cycle revisited: interannual variation and ecosystem consequences. Progress in Oceanography, 49:283–307.

BirdLife, 2000. BirdLife 2000: the Strategy of BirdLife International, 2000–2004. BirdLife International, Cambridge.

BirdLife, 2002. Globally Threatened Birds Indicating Priorities for Action. BirdLife International, Cambridge.

Birkemoe, T. and Leinaas, H.P., 1999. Reproductive biology of the Arctic collembolan Hypogastrura tullbergi. Ecography, 22:31–39.

Birkemoe, T. and L. Sømme, 1998. Population dynamics of two collembolan species in an Arctic tundra. Pedobiologia, 42:131–145.

Björn, L.O., T.V. Callaghan, C. Gehrke, D. Gwynn-Jones, B. Holmgren, U. Johanson and M. Sonesson, 1997. Effects of enhanced UV-B radiation on subarctic vegetation. In: S.J. Woodin and M. Marquiss (eds.). Ecology of Arctic Environments, pp. 241–253. Blackwell.

Blackmore, S., 2002. Biodiversity update – progress in taxonomy. Science, 298:365.

Blindheim, J., R. Toresen and H. Loeng, 2001. Fremtidige klimatiske endringer og betydningen for fiskeressursene. Havets miljø. Fisken og Havet, 2:73–78.

Bliss, L.C., O.W. Heal and J.J. Moore, 1981. Tundra Ecosystems: a Comparative Analysis. Cambridge University Press.

Brown, R.G.B., 1991. Marine birds and climate warming in the northwest Atlantic. In: W.A. Montevecchi and A.J. Gaston. Studies of High-latitude Seabirds. 1. Behavioural, Energetic, and Oceanographic Aspects of Seabird Feeding Ecology, pp. 49–54. Canadian Wildlife Service.

Burger, J. and M. Gochfeld, 1994. Predation and effects of humans on island-nesting seabirds. In: D.N. Nettleship, J. Burger and M. Gochfeld (eds.). Seabirds on Islands: Threats, Cases Studies, and Action Plans, pp. 39–67. Birdlife International, Cambridge.

Burgess, P., 1999. Traditional Knowledge. Arctic Council Indigenous Peoples' Secretariat, Copenhagen.

Byrd, G.V., J.L. Trapp and C.F. Zeillemaker, 1994. Removal of introduced foxes: a case study in restoration of native birds. Transactions of the North American Natural Resources Conference, 59:317–321.

Byrd, G.V., E.P. Bailey and W. Stahl, 1997. Restoration of island populations of black oystercatchers and pigeon guillemots by removing introduced foxes. Colonial Waterbirds, 20:253–260.

CAFF, 2001. Arctic Flora and Fauna: Status and Conservation. Conservation of Arctic Flora and Fauna, Edita, Helsinki.

CAFF, 2002a. Arctic Flora and Fauna: Recommendations for Conservation. Conservation of Arctic Flora and Fauna, International Secretariat, Akureyri.

CAFF, 2002b. The conservation value of sacred sites of indigenous peoples of the Arctic: a case study in northern Russia – report on the state of sacred sites and sanctuaries. Conservation of Arctic Flora and Fauna, Technical Report, 10.

CAFF, 2002c. Circumpolar Biodiversity Monitoring Program. Coordination Meeting, Akureyri, Iceland, April 11–12, 2002. Conservation of Arctic Flora and Fauna, Technical Report, 12.

CAFF, PAME and IUCN, 2000. Circumpolar Marine Workshop, 28 November – 2 December 1999: Report and Recommendations. Conservation of Arctic Flora and Fauna, Akureyri; Protection of the Arctic Marine Environment, Akureyri; and The World Conservation Union, Gland.

Cairns, J., 2002. Environmental monitoring for the preservation of global biodiversity: the role in sustainable use of the planet. International Journal for Sustainable Development and World Ecology, 9:135–150.

Cannell, M.G.R., D. Fowler and C.E.R. Pitcairn, 1997. Climate change and pollutant impacts on Scottish vegetation. Botanical Journal of Scotland, 49:301–313.

Chernov, Y.I., 1985. The Living Tundra. Cambridge University Press.

Chernov, Y.I., 1995. Diversity of the Arctic terrestrial fauna. In: F.S. Chapin and C. Körner (eds.). Arctic and Alpine Biodiversity: Patterns, Causes and Ecosystem Consequences, pp. 81–95. Springer-Verlag.

Chernov, Y.I. and N.V. Matveyeva, 1997. Arctic Ecosystems in Russia. In: F.E. Wielgolaski (ed.). Ecosystems of the World, pp. 361–507. Elsevier.

Clapham, A.R. (ed.), 1980. The IBP Survey of Conservation Sites: an Experimental Study. Cambridge University Press.

Cornelissen, J.H.C., T.V. Callaghan, J.M. Alatalo, A.E. Hartley, D.S. Hik, S.E. Hobbie, M.C. Press, C.H. Robinson, G.R. Shaver, G.R. Phoenix, D. Gwynn-Jones, S. Jonasson, M. Sonesson, F.S. Chapin, U. Molau and J.A. Lee, 2001. Global change and Arctic ecosystems: is lichen decline a function of increases in vascular plant biomass? Journal of Ecology, 89:984–994.

Coulson, S.J. and D. Refseth, 2004. The terrestrial and freshwater invertebrate fauna of Svalbard (and Jan Mayen). In: P. Prestrud, H. Strøm and H.V. Goldman (eds.). A Catalogue of the Marine and Terrestrial Animals of Svalbard, pp. 57–122. Norwegian Polar Institute, Tromsø.

Crawford, R.M.M., 1995. Plant survival in the High Arctic. Biologist, 42:101–105.

Crawford, R.M.M. and R.J. Abbott, 1994. Pre-adaptation of Arctic plants to climate change. Botanica Acta, 107:271–278.

Crawford, R.M.M, C.E. Jeffree and W.G. Rees, 2003. Paludification and forest retreat in northern oceanic environments. Annals of Botany, 91:213–226.

Criddle, K.R., H.J. Niebauer, T.J. Quinn, E. Shea and A. Tyler, 1998. Marine biological resources. In: G. Weller and P.A. Anderson. Implications of Global Change in Alaska and the Bering Sea Region, pp. 75–94. Center for Global Change and Arctic System Research, University of Alaska, Fairbanks.

Danchin, E., G. Gonzalez-Davila and J.D. Lebreton, 1995. Estimating bird fitness correctly by using demographic models. Journal of Avian Biology, 26:67–75.

Davidson, N., 2003. Declines in East Atlantic wader populations: is the Wadden Sea the problem? Waders Study Group Bulletin, 101/102, Abstracts of Declining Waders Workshop, Cadiz.

Deshaye, J. and P. Morisset, 1988. Floristic richness, area, and habitat diversity in a hemiarctic archipelago. Journal of Biogeography, 15:747–757.

Deshaye, J. and P. Morisset, 1989. Species-area relationships and the SLOSS effect in a subarctic archipelago. Biological Conservation, 48:265–276.

Diamond, J.M., 1975. The island dilemma: lessons of modern biogeographic studies for the design of nature reserves. Biological Conservation, 7:129–146.

Dobson, F., 2003. Getting a liking for lichens. The Biologist, 50:263–267.

Dockerty, T. and A. Lovett, 2003. A location-centred, GIS-based methodology for estimating the potential impacts of climate change on nature reserves. Transactions in GIS, 7:345–370.

Dockerty, T., A. Lovett and A. Watkinson, 2003. Climate change and nature reserves: examining the potential impacts, with examples from Great Britain. Global Environmental Change, 13:125–135.

Doughty, C.R., P.J. Boon and P.S. Maitland, 2002. The state of Scotland's fresh waters. In: M.B. Usher, E.C. Mackey and J.C. Curran (eds.). The State of Scotland's Environment and Natural Heritage, pp. 117–144. The Stationery Office, Edinburgh.

Dragoo, D.E., G.V. Byrd and D.B. Irons, 2001. Breeding status, population trends and diets of seabirds in Alaska, 2000. United States Fish and Wildlife Service Report, AMNWR 01/07.

Drake, J.A., H.A. Mooney, F. di Castri, R.H. Groves, F.J. Kruger, M. Rejmánek and M. Williamson, 1989. Biological Invasions: a Global Perspective. Wiley.

Einarsson, E., 1968. Vegetationen på nogle nunatakker i Vatnajökull, p. 106. Naturens Verden, April.

Elton, C.S., 1958. The Ecology of Invasions by Plants and Animals. Methuen.

Elvebakk, A. and H. Hertel, 1996. Lichens. In: A. Elvebakk and P. Prestrud. A Catalogue of Svalbard Plants, Fungi, Algae and Cyanobacteria, pp. 271–359. Norsk Polarinstitutt, Oslo.

Elvebakk, A. and P. Prestrud, (eds.), 1996. A Catalogue of Svalbard Plants, Fungi, Algae and Cyanobacteria. Norsk Polarinstitutt, Oslo.

Elvebakk, A., H.B. Gjærum and S. Sivertsen, 1996. Fungi II. Myxomycota, Oomycota, Chytrodiomycota, Zygomycota, Ascomycota, Deuteromycota, Basidiomycota: Uredinales and Ustilaginales. In A. Elvebakk and P. Prestrud (eds.). A Catalogue of Svalbard Plants, Fungi, Algae and Cyanobacteria, pp. 207–259. Norsk Polarinstitutt, Oslo.

Elven, R. and A. Elvebakk, 1996. Vascular plants. In: A. Elvebakk and P. Prestrud (eds.). A Catalogue of Svalbard Plants, Fungi, Algae and Cyanobacteria, pp. 9–55. Norsk Polarinstitutt, Oslo.

Enkhtuya, B., U. Oskarsson, J.C. Dodd and M. Vosatka, 2003. Inoculation of grass and tree seedlings used for reclaiming eroded areas in Iceland with mycorrhizal fungi. Folia Geobotanica, 38:209–222.

Essen, P.A., B. Ehnstrom, L. Ericson and K. Sjoberg, 1992. Boreal forests – the focal habitats of Fennoscandia. In: L. Hansson (ed.). Ecological Principles of Nature Conservation. Applications in Temperate and Boreal Environments, pp 252–325. Elsevier Applied Science.

Ferguson, S.H., M.K. Taylor and F. Messier, 2000a. Influence of sea ice dynamics on habitat selection by polar bears. Ecology, 81:761–772.

Ferguson, S.H., M.K. Taylor, A. Rosing Asvid, E.W. Born and F. Messier, 2000b. Relationships between denning of polar bears and conditions of sea ice. Journal of Mammalogy, 81:1118–1127.

Fitzpatrick, E.A., 1997. Arctic soils and permafrost. In: S.J. Woodin and M. Marquiss. Ecology of Arctic Environments, pp. 1–39. Blackwell.

Fogg, G.E., 1998. The Biology of Polar Habitats. Oxford University Press.

Forbes, B.C., J.J. Ebersole and B. Strandberg, 2000. Anthropogenic disturbance and patch dynamics in circumpolar Arctic ecosystems. Conservation Biology, 15:954–969.

Francis, R.C., S.R. Hare, A.B. Hollowed and W.S. Wooster, 1998. Effects of interdecadal variability on the NE Pacific. Fisheries Oceanography, 7:1–21.

Frisvoll, A.A. and A. Elvebakk, 1996. Bryophytes. In: A. Elvebakk and P. Prestrud (eds.). A Catalogue of Svalbard Plants, Fungi, Algae and Cyanobacteria, pp. 57–172. Norsk Polarinstitutt, Oslo.

From, S. and G. Söderman, 1997. Nature Monitoring Scheme: Guidelines to Monitor Terrestrial Biodiversity in the Nordic Countries. Nordic Council of Ministers, Copenhagen.

Gaston, A.J. and J.M. Hipfner, 2000. Brünnich's Guillemot (*Uria lomvia*). In: A. Poole and F. Gill (eds.). The Birds of North America, p.32. The Birds of North America Inc., Philadelphia.

Gaston, A.J. and I.L. Jones, 1998. The Auks. Oxford University Press.

Gjertz, I. and Ø. Wiig, 1994. Past and present distribution of walruses in Svalbard. Arctic, 47:34–42.

Gjertz, I. and Ø. Wiig, 1995. The number of walruses (*Odobenus rosmarus*) in Svalbard in summer. Polar Biology, 15:527–530.

Gorshkov, V.V. and I.J. Bakkal, 1996. Species richness and structure variations of Scots pine forest communities during the period from 5 to 210 years after fire. Silva Fennica, 30:329–340.

Graham, R.W. and E.C. Grimm, 1990. Effects of global climate change on the patterns of terrestrial biological communities. Trends in Ecology and Evolution, 5:289–292.

Grime, J.P., 1979. Plant Strategies and Vegetation Processes. Wiley.

Groombridge, B. (ed.), 1992. Global Biodiversity: Status of the Earth's Living Resources. Chapman and Hall.

Grubb, P.J., 1977. The maintenance of species-richness in plant communities: the importance of the regeneration niche. Biological Reviews, 52:107–145.

Gulden, G. and A.-E. Torkelsen, 1996. Fungi I. Basidiomycota: Agaricales, Gasteromycetales, Aphyllophorales, Exobasidiales, Dacrimycetales and Tremellales. In: A. Elvebakk and P. Prestrud (eds.). A Catalogue of Svalbard Plants, Fungi, Algae and Cyanobacteria, pp. 173–206. Norsk Polarinstitutt, Oslo..

Gunn, A., 2001. Muskoxen. In: Arctic Flora and Fauna: Status and Conservation, pp. 240–241. Conservation of Arctic Flora and Fauna, Edita.

Haak, R.A., 1996. Will global warming alter paper birch susceptibility to bronze birch borer attack? In: W.J. Mattson, P. Niemila and M. Rossi (eds.). Dynamics of Forest Herbivory: Quest for Pattern and Principle, pp. 234– 247. USDA Forest Service, North Central Forest Experiment Station, St. Paul.

Hadley, M. (ed.), 2000. Solving the Puzzle: the Ecosystem Approach and Biosphere Reserves. UNSCO, Paris.

Haeussler, S. and Kneeshaw, D., 2003. Comparing forest management to natural processes. In: P.J. Burton, C. Messier, D.W. Smith and W.L. Adamowicz (eds.). Towards Sustainable Management of the Boreal Forest, pp. 307–368. NRC Research Press.

Hallanaro, E.-L. and M. Pylvänäinen, 2002. Nature in Northern Europe: Biodiversity in a Changing Environment. Nordic Council of Ministers, Copenhagen.

Hallanaro, E.-L. and M.B. Usher, in press. Natural heritage trends: an upland saga. In: D.B.A. Thompson and M.F. Price (eds.). People and Nature: Conservation and Management in the Mountains of Northern Europe. The Stationery Office, Edinburgh.

Halpern, B.S., 2003. The impact of marine reserves: do reserves work and does reserve size matter? Ecological Applications, 13:S117–S137.

Halpern, B.S. and R.R. Warner, 2002. Marine reserves have rapid and lasting effects. Ecology Letters, 5:361–366.

Hammer, J., 1989. Freshwater ecosystems of polar regions: vulnerable resources. Ambio, 18:6–22.

Hammer, J., 1998. Evolutionary Ecology of Arctic Char (*Salvelinus alpinus* (L)). Intra- and Interspecific Interactions in Circumpolar Populations. Dissertations from the Faculty of Science and Technology, 408. Acta Universitatis Upsaliensis, Uppsala.

Hamre, J., 1994. Biodiversity and exploitation of the main fish stocks in the Norwegian – Barents Sea ecosystem. Biodiversity and Conservation, 3:473–492.

Hanneberg, P. and Löfgren, R., 1998. Sweden's National Parks. Swedish Environmental Protection Agency, Stockholm.

Hansen, B. and Osterhus, S., 2000. North Atlantic – North Sea exchanges. Progress in Oceanography, 45:109–208.

Hansen, J.R. and L.H. Jenneborg, 1996. Benthic marine algae and cyanobacteria. In: A. Elvebakk and P. Prestrud (eds.). A Catalogue of Svalbard Plants, Fungi, Algae and Cyanobacteria, pp. 361–374. Norsk Polarinstitutt, Oslo.

Hanski, I. and O. Ovaskainen, 2000. The metapopulation capacity of a fragmented landscape. Nature, 404:755–758.

Hare, S.R., and N.J. Mantua, 2000. Empirical evidence for North Pacific regime shifts in 1977 and 1989. Progress in Oceanography, 47:103–145.

Harris, L.D., 1984. The Fragmented Forest: Island Biogeography Theory and the Preservation of Biotic Diversity. University of Chicago Press.

Harvell, C.D., K. Kim, J.M. Burkholder, R.R. Colwell, P.R. Epstein, D.J. Grimes, E.E. Hofmann, E.K. Lipp, A.D.M.E. Oterhaus, R.M. Overstreet, J.W. Porter, G.W. Smith and G.R. Vasta, 1999. Emerging marine diseases – climate links and anthropogenic factors. Science, 285:1505–1510.

Hasle, G.R. and C. Hellum von Quillfeldt, 1996. Marine microalgae. In: A. Elvebakk and P. Prestrud (eds.). A Catalogue of Svalbard Plants, Fungi, Algae and Cyanobacteria, pp. 375–382. Norsk Polarinstitutt, Oslo.

Heal, O.W., 1999. Looking north: current issues in Arctic soil ecology. Applied Soil Ecology, 11:107–109.

Helms, J.A. (ed.), 1998. The Dictionary of Forestry. The Society of American Foresters, Bethesda.

Hewitt, G., 1999. Post-glacial re-colonization of European biota. Biological Journal of the Linnean Society, 68:87–112.

Hewitt, G., 2000. The genetic legacy of the Quaternary ice ages. Nature, 405:907–913.

Heywood, V.H. (ed.), 1995. Global Biodiversity Assessment. Cambridge University Press.

Heywood, V.H. and D. Zohary, 1995. A catalogue of the wild relatives of cultivated plants native to Europe. Flora Mediterranea, 5:375–415.

Hodkinson, I.D., N.R. Webb, J.S. Bale, W. Block, S.J. Coulson and A.T. Strathdee, 1998. Global change and Arctic ecosystems: conclusions and predictions from experiments with terrestrial invertebrates on Spitsbergen. Arctic and Alpine Research, 30:306–313.

Hogg, E.H. and P.A. Hurdle, 1995. The aspen parkland in western Canada: a dry-climate analogue for the future boreal forest? Water, Air and Soil Pollution, 82:391–400.

Holst, J.C., O. Dragesund, J. Hamre, O.A. Misund and O.J. Østevedt, 2002. Fifty years of herring migrations in the Norwegian Sea. ICES Marine Science Symposia, 215:352–360.

Holten, J.I., 1990. Predicted floristic change and shift of vegetation zones in a coast-inland transect in central Norway. In: J.I. Holten (ed.). Effects of Climate Change on Terrestrial Ecosystems, pp. 61–77. Norsk Institutt for Naturforskning, Trondheim.

Holten, J.I. and P.D. Carey, 1992. Responses of Climate Change on Natural Terrestrial Ecosystems in Norway. Norsk Institutt for Naturforskning, Trondheim.

Huhta, V., T. Persson and H. Setälä, 1998. Functional implications of soil faunal diversity in boreal forests. Applied Soil Ecology, 10:277–288.

Hurrell, J.W., Y. Kushnir, G. Ottersen and M. Visbeck, 2003. An overview of the North Atlantic Oscillation. In: J.W. Hurrell, Y. Kushnir, G. Ottersen and M. Visbeck (eds.). The North Atlantic Oscillation: Climate Significance and Environmental Impact. Geophysical Monograph Series, 134:1–35.

Innes, J.L., 1990. Assessment of Tree Condition: Forestry Commission Field Book 12. HMSO, London.

IPCC, 2002. Climate Change and Biodiversity. Gitay, H., A. Suarez, R.T. Watson and D.J. Dokken (eds.). World Meteorological Organization and Intergovernmental Panel on Climate Change.

IUCN, 1991. Protected Areas of the World: a Review of National Systems. Vol. 2, Palaearctic. World Conservation Union, Gland.

IUCN, 1994. IUCN Red List Categories. World Conservation Union, Gland.

Jenkins, M., 2003. Prospects for biodiversity. Science, 302:1175–1177.

Jenouvrier, S., C. Barbraud and H. Weimerskirch, 2003. Effects of climate variability on the temporal population dynamics of southern fulmars. Journal of Animal Ecology, 72:576–587.

Jóhannesson, T. and O. Sigurðsson, 1998. Interpretation of glacier variations in Iceland 1930–1995. Jökull, 45:27.

John, D.M., B.A. Whitton and A.J. Brook (eds.), 2002. The Freshwater Algal Flora of the British Isles: an Identification Guide to Freshwater and Terrestrial Algae. Cambridge University Press.

Jonasson, S. and T.V. Callaghan, 1992. Mechanical properties of roots in relation to frost heave in the Arctic. New Phytologist, 122:179–186.

Jones, I.L., F.M. Hunter and G.J. Robertson, 2002. Annual adult survival of least auklets (Aves, Alcidea) varies with large scale climatic conditions in the North Pacific Ocean. Oecologia, 133:38–44.

Jones, R.D. and G.V. Byrd, 1979. Interrelations between seabirds and introduced animals. In: J.C. Bartonek and D.N. Nettleship (eds.). Conservation of Marine Birds of Northern North America, pp. 221–226. United States Fish and Wildlife Service, Wildlife Research Report No. 11.

Jonsson, B., R. Andersen, L.P. Hansen, I.A. Fleming and A. Bjørge, 1993. Sustainable Use of Biodiversity. Norsk Institutt for Naturforskning, Trondheimones.

Juday, G.P., 1997. Boreal forests (taiga). In: The Biosphere and Concepts of Ecology, pp. 1210–1216. Volume 14 Encyclopedia Britannica, 15th edition.

Kalela, O., 1961. Seasonal change of the habitat in the Norwegian lemming, *Lemmus lemmus*. Annales Academiae Scientarium Fennicae, Series A, IV, Biologica, 55:1–72.

Kallio, P. and J. Lehtonen, 1973. Birch forest damage caused by *Oporinia autumnata* (Bkh.) in 1965–66, in Utsjoki, N Finland. Reports of the Kevo Subarctic Research Station, 10:55–69.

Kelsall, J., 1968. The Migratory Barren-ground Caribou of Canada. Canadian Wildlife Service, Ottawa.

Kingsland, S., 2002. Designing nature reserves: adapting ecology to real-world problems. Endeavour, 26:9–14.

Kling, G.W., B. Fry and W.J. O'Brien, 1992. Stable isotopes and planktonic trophic structure in Arctic lakes. Ecology, 73:561–566.

Komonen, A., 2003. Hotspots of insect diversity in boreal forests. Conservation Biology, 17:976–981.

Komonen, A., J. Ikävalko and W. Weiyung, 2003. Diversity patterns of fungivorous insects: comparison between glaciated vs. refugial boreal forests. Journal of Biogeography, 30:1873–1881.

Korhonen, K.-M., R. Laamanen and S. Savonmäki (eds), 1998. Environmental Guidelines to Practical Forest Management. Metsähallitus, Helsinki.

Koskina, T.V., 1961. New data on the nutrition of Norwegian lemming (*Lemmus lemmus*). Bulletin of the Moscow Society of Naturalist, 66:15–32.

Krebs, C.J., R. Boonstra, S. Boutin and A.R.E. Sinclair, 2001. Conclusions and future directions. In: C.J. Krebs, S. Boutin and R. Boonstra (eds.). Ecosystem Dynamics of the Boreal Forest. The Kluane Project, pp. 492–501. Oxford University Press.

Krupnik, I. and D. Jolly (eds.), 2002. The Earth is Faster Now: Indigenous Observations of Arctic Environmental Change. Arctic Research Consortium of the United States, Fairbanks.

Kuusisto, E., L. Kauppi and P. Heikinheimo (eds.), 1996. Climate Change and Finland: Summary of the Finnish Research Programme on Climate Change (SILMU). The Academy of Finland, Helsinki.

Laine, K. and H. Henttonen, 1983. The role of plant production in microtine cycles in northern Fennoscandia. Oikos, 40:407–418.

Laxon, S., N. Peacock and D. Smith, 2003. High interannual variability of sea ice thickness in the Arctic region. Nature, 425:947–950.

Lehtonen, J. and R.K. Heikkinen, 1995. On the recovery of mountain birch after Epirrita damage in Finnish Lapland, with a particular emphasis on reindeer grazing. Ecoscience, 2:349–356.

Lewis, P., 1991. Sedimentation in the Mackenzie Delta. In: P. Marsh and C.S.L. Ommaney (eds.). Mackenzie Delta: Environmental Interactions and Implications of Development, pp. 37–38. Environment Canada, Saskatoon.

Li, P., J. Beaulieu and J. Bousquet, 1997. Genetic structure and patterns of genetic variation among populations in eastern white spruce (*Picea glauca*). Canadian Journal of Forest Research, 27:189–198.

Lieffers, V.J., C. Messier, P.J. Burton, J.-C. Ruel and B.E. Grover, 2003. Nature-based silviculture for sustaining a variety of boreal forest values. In: P.J. Burton, C. Messier, D.W. Smith and W.L. Adamowicz (eds.). Towards Sustainable Management of the Boreal Forest, pp. 481–530. NRC Research Press.

Linder, P. and Ostlund, L., 1992. Changes in the boreal forests of Sweden 1870–1991. Svensk Bot. Tidskr., 86:199–215. (In Swedish)

Luck, G.W., G.C. Daily and P.R. Ehrlich, 2003. Population diversity and ecosystem services. Trends in Ecology and Evolution, 18:331–336.

Lunn, N.J., S. Schliebe and E. Born, 2002. Polar bears. Proceedings of the 13th working meeting of the IUCN/SSC Polar Bear Specialist Group, 23–28 June 2001, Nuuk, Greenland. Occasional paper of the IUCN Species Survival Commission, No. 26.

Lyngs, P., 2003. Migration and winter ranges of birds in Greenland. Dansk Ornitologisk Forenings Tidsskrift, 97:1–167.

MacArthur, R.H. and E.O. Wilson, 1967. The Theory of Island Biogeography. Princeton University Press.

Macdonald, I.A.W., L.L. Loope, M.B. Usher and O. Hamann, 1989. Wildlife conservation and the invasion of nature reserves by introduced species: a global perspective. In: J.A. Drake, H.A. Mooney, F. di Castri, R.H. Groves, F.J. Kruger, M. Rejmánek and M. Williamson (eds.). Biological Invasions: a Global Perspective, pp. 215–255. Wiley.

Macdonald, R.W., T. Harner, J. Fyfe, H. Loeng and T. Weingartner, 2003. AMAP Assessment 2002: the Influence of Global Change on Contaminant Pathways to, within, and from the Arctic. Arctic Monitoring and Assessment Programme, Oslo.

Mackay, J.R., 1963. The Mackenzie Delta Area, N.W.T. Geographical Branch Memoir No. 8, Department of Mines and Technical Surveys, Ottawa.

Madsen, J., G. Cracknell and T. Fox, (eds.), 1999. Goose Populations of the Western Palearctic: a Review of Status and Distribution. Wetlands International, Wageningen, and National Environmental Research Institute, Rónde.

Manel, S., M.K. Schwartz, G. Luikart and P. Taberlet, 2003. Landscape genetics: combining landscape ecology and population genetics. Trends in Ecology and Evolution, 18:189–197.

Margules, C.R. and M.B. Usher, 1981. Criteria used in assessing wildlife conservation potential: a review. Biological Conservation, 21:79–109.

Marion, J.L. and S.E. Reid, 2001. Development of the United States 'Leave No Trace' programme: a historical perspective. In: M.B. Usher (ed.). Enjoyment and Understanding of the Natural Heritage, pp. 81–92. The Stationery Office, Edinburgh..

Martikainen, P. and J. Kouki, 2003. Sampling the rarest: threatened beetles in boreal forest biodiversity inventories. Biodiversity and Conservation, 12:1815–1831.

Matveyeva, N. and Y. Chernov, 2000. Biodiversity of terrestrial ecosystems. In: M. Nuttall and T.V. Callaghan (eds.). The Arctic Environment: People, Policy, pp. 233–274. Harwood Academic Publishers.

Mauritzen, M., A.E. Derocher and Ø. Wiig, 2001. Female polar bear space use strategies in a dynamic sea ice habitat. Canadian Journal of Zoology, 79:1704–1713.

Maynard, N.G. (ed.), 2002. Native Peoples – Native Homelands: Climate Change Workshop. Final Report: Circles of Wisdom. National Aeronautics and Space Administration, Albuquerque.

McDonald, M., L. Arragutainaq and Z. Novalinga, (eds.), 1997. Voices from the Bay: Traditional Ecological Knowledge of Inuit and Cree in the Hudson Bay Bioregion. Canadian Arctic Resources Committee, Ottawa.

McDowall, R.M., 1987. Evolution and the importance of diadromy. American Fisheries Society Symposium, 1:1–13.

McGraw, J.B., 1995. Patterns and causes of genetic diversity in Arctic plants. In: F.S. Chapin and C. Korner. Arctic and Alpine Biodiversity, pp. 33–43. Springer Verlag.

McGuire, A.D., J.M. Melillo, D.W. Kicklighter, Y. Pan, X. Xiao, J. Helfrich, B.M. Moore, C.J. Vorosmarty and A.L. Schloss, 1997. Equilibrium responses of global net primary production and carbon storage to doubled atmospheric carbon dioxide: sensitivity to changes in vegetation nitrogen concentration. Global Biochemistry Cycles, 11:173–189.

Mehl, K.R., R.T. Alisauskas, K.A. Hobson and D.K. Kellett, 2004. To winter east or west? Heterogeneity in winter site philopatry in a central Arctic population of king eiders. Condor, 106:241–247.

Mehl, K.R., R.T. Alisauskas, K.A. Hobson and F.R. Merkel, in press. Linking breeding and wintering grounds of king eiders: making use of polar isotopic gradients. Journal of Wildlife Management.

Miles, J. and D.W.H. Walton (eds.), 1993. Primary Succession on Land. Blackwell.

Montevecchi, W.A. and R.A. Myers, 1997. Centurial and decadal oceanographic influences on changes in northern gannet populations and diets in the north-west Atlantic: implications for climate change. ICES Journal of Marine Science, 54:608–614.

Muir, M.A.K., 2000. Regulation of Marine Transportation and Implications for Ocean Management in Hudson Bay. Report for Fisheries and Oceans Canada, underpinning the October 2000 Western Hudson Bay Workshops and supporting information for the Hudson Bay Oceans Working Group (www.umanitoba.ca/academic/institutes/natural_resources/im-node/hudson_bay).

Muir, M.A.K., 2002a. Integrated coastal and marine management in northern regions: reconciling economic development and conservation. Journal of Coastal Research, special issue 36:522–530.

Muir, M.A.K., 2002b. Models and decision frameworks for indigenous participation in coastal zone management in Queensland, based on Canadian experience. Coat to Coast 2002, Australia's National Coastal Conference, pp.303–306.

Muir, M.A.K., T. van Pelt and K. Wohl, 2003. Ecosystem-based approaches for conserving Arctic biodiversity. Discussion paper for the Arctic Council's October 2003 Arctic Marine Strategic Plan Workshop (www.pame.is).

Myneni, R.B., C.D. Keeling, C.J. Tucker, G. Asrar and R.R. Nemani, 1997. Increased plant growth in the northern high latitudes from 1981–1991. Nature, 386:698–702.

Naiman, R.J., D.G. Lonzarich, T.J. Beechie and S.C. Ralph, 1992. General principles of classification and the assessment of conservation potential in rivers. In: P.J. Boon, P. Calow and G.E. Petts (eds.). River Conservation and Management, pp.93–123. Wiley.

Naturvårdverket, 1988. Sveriges Nationalparker. Naturvårdverket, Stockholm.

Nellemann, C., L. Kullerud, I. Vistnes, B.C. Forges, G.P. Kofinas, B.P. Kaltenborn, O. Gron, D. Henry, M. Magomedova, C. Lambrechts, R. Bobiwash, P.J. Schei and T.S. Larsen, 2001. GLOBIO – Global Methodology for Mapping Human Impacts on the Biosphere. United Nations Environment Programme.

Nellemann, C., I. Vistnes, P. Jordhøy, O. Strand and A. Newton, 2003. Progressive impact of piecemeal infrastructure development on wild reindeer. Biological Conservation, 113:307–317.

Niemelä, J., Y. Haila, E. Halme, T. Pajunen and P. Punttila, 1990. Diversity variation in carabid beetle assemblages in the southern Finnish taiga. Pedobiologia, 34:1–10.

Nikolov, N. and H. Helmisaari, 1992. Silvics of the circumpolar boreal forest tree species. In: H.H. Shugart, R. Leemans and G.B. Bonan (eds.). A Systems Analysis of the Global Boreal Forest, pp 13–84. Cambridge University Press.

Nilsson, S.G. and L. Ericson, 1992. Conservation of plant and animal populations in theory and practice. In: L. Hansson (ed.). Ecological Principles of Nature Conservation. Applications in temperate and Boreal Environments, pp. 71–112. Elsevier Applied Science.

Oksanen, L., M. Aunapuu, T. Oksanen, M. Schneider, P. Ekerholm, P.A. Lundberg, T. Amulik, V. Aruaja and L. Bondestad, 1997. Outlines of food webs in a low arctic tundra landscape in relation to three theories on trophic dynamics. In: A.C. Gange and V.K. Brown (eds.). Multitrophic Interactions in Terrestrial Ecosystems, pp. 351–373. Blackwell Scientific Publications.

Olff, H. and M.E. Ritchie, 2002. Fragmented nature: consequences for biodiversity. Landscape and Urban Planning, 58:83–92.

Oswald, W.W., L.B. Brubaker, F.S. Hu and G.W. Kling, 2003. Holocene pollen records from the central Arctic Foothills, northern Alaska: testing the role of substrate in the response of tundra to climate change. Journal of Ecology, 91:1034–1048.

Palerud, R., B. Gulliksen, T. Brattegard, J.-A. Sneli and W. Vader, 2004. The marine macro-organisms in Svalbard waters. In: P. Prestrud, H. Strøm and H.V. Goldman (eds.). A Catalogue of the Marine and Terrestrial Animals of Svalbard, pp. 5–56. Norwegian Polar Institute, Tromsø.

Parkinson, C.L., 2000. Variability of Arctic sea-ice: the view from space, an 18-year record. Arctic, 53:341–358.

Parkinson, C.L., D.J. Cavalieri, P. Gloersen, H.J. Zwally and J.C. Comiso, 1999. Arctic sea ice extents, areas and trends, 1978–1996. Journal of Geophysical Research, 104:20837–20856.

Parrish, J.D., D.P. Braun and R.S. Unnasch, 2003. Are we conserving what we say we are? Measuring ecological integrity within protected areas. BioScience, 53:851–860.

Pavan, M., 1986. A European Cultural Revolution: The Council of Europe's «Charter on Invertebrates». Council of Europe, Strasbourg.

Pearce, J.M., S.L. Talbot, B.J. Pierson, M.R. Petersen, K.T. Scribner, D.L. Dickson and A. Mosbech, 2004. Lack of special genetic structure among nesting and wintering king eiders. Condor, 106:229–240.

Pellerin, S. and C. Lavoie, 2003. Reconstructing the recent dynamics of mires using a multitechnique approach. Journal of Ecology, 91:1008–1021.

Peterson, C.H., S.D. Rice, J.W. Short, D. Esler, J.L. Bodkin, B.E. Ballachey and D.B. Irons, 2003. Long-term ecosystem response to the Exxon Valdez oil spill. Science, 302:2082–2086.

Pianka, E.R., 1970. On r- and k-selection. American Naturalist, 104:592–597.

Prestrud, P. and I. Stirling, 1994. The international polar bear agreement and the current status of polar bear conservation. Aquatic Mammals, 20:113–124.

Prestrud, P., H. Strøm and H.V. Goldman (eds.), 2004. A Catalogue of the Terrestrial and Marine Animals of Svalbard. Norwegian Polar Institute, Tromso.

Prowse, T.D., 1990. Northern hydrology: an overview. In: T.D. Prowse and C.S.L. Ommaney (eds.). Northern Hydrology: Canadian Perspectives, pp. 1–36. Environment Canada, Saskatoon.

Quine, D.A., 1989. St. Kilda Revisited, 3rd edition. Dowland Press.

Ramakrishnan, P.S., K.G. Saxena and U.M. Chandrashekara (eds.), 1998. Conserving the Sacred for Biodiversity Management. Oxford and IBH Publishing.

Ramakrishnan, P.S., U.M. Chandrashekara, C. Elouard, C.Z. Guilmoto, R.K. Maikhuri, K.S. Rao, S. Sankar and K.G. Saxena (eds.), 2000. Mountain Biodiversity, Land Use Dynamics, and Traditional Ecological Knowledge. Oxford and IBH Publishing.

Raven, J. and M. Walters, 1956. Mountain Flowers. Collins.

Rees, D.C. and G.P. Juday, 2002. Plant species diversity and forest structure on logged and burned sites in central Alaska. Forest Ecology and Management, 155:291–302.

Rey, A. and P.G. Jarvis, 1997. An overview of long-term effects of elevated atmospheric CO_2 concentration on the growth and physiology of birch (*Betula pendula* Roth.). Botanical Journal of Scotland, 49:325–340.

Rhind, P., 2003. Britain's contribution to global conservation and our coastal temperate rainforest. British Wildlife, 15:97–102.

Ritchie, W. and M. O'Sullivan, 1994. The Environmental Impact of the Wreck of the *Braer*; the Ecological Steering Group on the Oil Spill in Shetland. The Scottish Office, Edinburgh.

Robinson, C.H. and P.A. Wookey, 1997. Microbial ecology, decomposition and nutrient cycling. In: S.J. Woodin and M. Marquiss (eds.). Ecology of Arctic Environments, pp. 41–68. Blackwell.

Root, T.L., J.T. Price, K.R. Hall, S.H. Schneider, C. Rosenzweig and J.A. Pounds, 2003. Fingerprints of global warming on wild animals and plants. Nature, 421:57–60.

Rosencranz, A. and A. Scott, 1992. Siberia's threatened forests. Nature, 335:293–294.

Rosentrater, L. and A.E. Ogden, 2003. Building resilience in Arctic ecosystems. In: L.J. Hansen, J.L. Biringer and J.R. Holfman (eds.). Buying Time: a User's Manual for Building Resistance and Resilience to Climate Change in Natural Systems, pp. 95–121. World Wide Fund For Nature.

Ruess, L., A. Michelsen and S. Jonasson, 1999a. Simulated climate change in subarctic soils: responses in nematode species composition and dominance structure. Nematology, 1:513–526.

Ruess, L., A. Michelsen, I.K. Schmidt and S. Jonasson, 1999b. Simulated climate change affecting microorganisms, nematode density and biodiversity in subarctic soils. Plant and Soil, 212:63–73.

Rydén, B.E., 1981. Hydrology of northern tundra. In: L.C. Bliss, O.W. Heal and J.J. Moore (eds.). Tundra Ecosystems: a Comparative Analysis, pp. 115–137. Cambridge University Press.

Sage, B., 1986. The Arctic and its Wildlife. Croom Helm.

Sala, O.E. and T. Chapin, 2000. Scenarios of global biodiversity. International Geosphere-Biosphere Programme, Newsletter 43:9–11.

Saunders, D.A. and R.J. Hobbs, 1991. Nature Conservation 2: the Role of Corridors. Surrey Beatty.

Saunders, D.A., G.W. Arnold, A.A. Burbidge and A.J.M. Hopkins, 1987. Nature Conservation: the Role of Remnants of Native Vegetation. Surrey Beatty.

SCBD, 2000. Convention on Biological Diversity. Text and Annexes. Secretariat of the Convention on Biological Diversity, Montreal.

SCBD, 2003. Interlinkages between biological diversity and climate change. Advice on the integration of biodiversity considerations into the implementation of the United Nations Framework Convention on Climate Change and the Kyoto Protocol. Secretariat of the Convention on Biological Diversity, Technical Series, 10.

Schreiber, E.A., 2002. Climate and weather effects on seabirds. In: E.A. Schreiber and J. Burger (eds.). Biology of Marine Birds, pp. 179–216. CRC Press.

Scott, D. and C.J. Lemieux, 2003. Vegetation Response to Climate Change: Implications for Canada's Conservation Lands. Environment Canada.

Scott, D. and R. Suffling, 2000. Climate Change and Canada's National Park System: a Screening Level Assessment. Adaptation and Impacts Research Group, Environment Canada, Hull and University of Waterloo.

Scott, D., J.R. Malcolm and C. Lemieux, 2002. Climate change and modelled biome representation in Canada's national park system: implication for system planning and park mandates. Global Ecology and Biogeography, 11:475–484.

Seppola, A-L., 2001. Protected areas in Northern Fennoscandia: an important corridor for taiga species. In: Arctic Flora and Fauna: Status and Conservation, p. 82. Conservation of Arctic Flora and Fauna, Edita, Helsinki.

Shuert, P.G. and J.J. Walsh, 1993. A coupled physical-biological model of the Bering/Chukchi Seas. Continental Shelf Research, 13:19–93.

Skulberg, O.M., 1996. Terrestrial and limnic algae and cyanobacteria. In: A. Elvebakk and P. Prestrud (eds.). A Catalogue of Svalbard Plants, Fungi, Algae and Cyanobacteria, pp. 383–395. Norsk Polarinstitutt, Oslo.

Smith, T.G. and I. Stirling, 1975. The breeding habitat of the ringed seal (*Phoca hispida*). The birth lair and associated structures. Canadian Journal of Zoology, 53:1297–1305.

Sømme, L., 1981. Cold tolerance of alpine, Arctic, and Antarctic Collembola and mites. Cryobiology, 18:212–220.

Sømme, L. and T. Birkemoe, 1999. Demography and population densities of *Folsomia quadrioculata* (Collembola, Isotomidae) on Spitsbergen. Norwegian Journal of Entomology, 46:35–45.

Sømme, L. and E.-M. Conradi-Larsen, 1977a. Cold-hardiness of collembolans and oribatid mites from windswept mountain ridges. Oikos, 29:118–126.

Sømme, L. and E.-M. Conradi-Larsen, 1977b. Anaerobiosis in overwintering collembolans and oribatid mites from windswept mountain ridges. Oikos, 29:127–132.

Speight, M.C.D., 1986. Saproxylic Invertebrates and their Conservation. Council of Europe.

Spicer, R.A. and J.L. Chapman, 1990. Climate change and the evolution of high-latitude terrestrial vegetation and floras. Trends in Ecology and Evolution, 5:279–284.

Starfield, A.M. and A.L. Bleloch, 1986. Building Models for Conservation and Wildlife Management. Macmillan.

Stenseth, N.C. and R.A. Ims, 1993. The Biology of Lemmings. Academic Press.

Stenstrom, A., B.O. Jonsson, I.S. Jonsdottir, T. Fagerstrom and M. Augner, 2001. Genetic variation and clonal diversity in four clonal sedges (Carex) along the Arctic coast of Eurasia. Molecular Ecology, 10:497–513.

Stirling, I., D. Andriashek, and W. Calvert, 1993. Habitat preferences of polar bears in the western Canadian Arctic in late winter and spring. Polar Record, 29:13–24.

Stirling, I., N.J. Lunn, and J. Iacozza, 1999. Long-term trends in the population ecology of polar bears in western Hudson Bay in relation to climate change. Arctic, 52:294–306.

Stonehouse, B., 1989. Polar Ecology. Blackie.

Strathdee, A.T. and J.S. Bale, 1998. Life on the edge: insect ecology in Arctic environments. Annual Reviews of Entomology, 43, 85–106.

Strøm, H. and G. Bangjord, 2004. The bird and mammal fauna of Svalbard. In: P. Prestrud, H. Strøm and H.V. Goldman (eds.). A Catalogue of the Marine and Terrestrial Animals of Svalbard, pp. 123–137. Norwegian Polar Institute, Tromsø.

Sturm, M., J.P. McFadden, G.E. Liston, F.S. Chapin, J. Holmgren and M. Walker, 2001. Snow-shrub interactions in Arctic tundra: a feedback loop with climate implications. Journal of Climatology, 14:336–344.

Swift, M.J., O.W. Heal and J.M. Anderson, 1979. Decomposition in Terrestrial Ecosystems. Blackwell.

Sydeman, W.J., M.M. Hester, J.A. Thayer, F. Gress, P. Martin and J. Buffa, 2001. Climate change, reproductive performance and diet composition of marine birds in the southern California Current system 1969–1997. Progress in Oceanography, 49:309–329.

Tenow, O., 1972. The outbreaks of *Oporinia autumnata* Bkh. and *Operophtera* spp. (Lep., Geometridae) in the Scandinavian mountain chain and northern Finland 1862–1968. Zoologiska Bidrag från Uppsala, Supplement 2, 1–107.

Tenow, O., 1996. Hazards to a mountain birch forest - Abisko in perspective. Ecological Bulletins, 45:104–114.

Thompson, D. and I. Byrkjedal, 2001. Shorebirds. Colin Baxter Photography.

Thompson P.M. and J.C. Ollason, 2001. Lagged effects of ocean climate change on fulmar population dynamics. Nature, 413:417–420.

Thorson, G., 1950. Reproductive and larval ecology of marine bottom invertebrates. Biological Reviews, 25:1–45.

Tiedemann, R., K.B. Paulus, M. Scheer, K.G. VonKistowski, K. Sirnisson, D. Bloch and M. Dam, 2004. Mitochondrial DNA and microsatellite variation in the eider duck (*Somateria mollissima*) indicate stepwise postglacial colonization of Europe and limited current long-distance dispersal. Molecular Ecology, 13:1481–1494.

Trathan, P.N., J.P. Croxall and E.J. Murphy, 1996. Dynamics of Antarctic penguin populations in relation to inter-annual variation in sea ice distribution. Polar Biology, 16:321–330.

Turchin, P. and G.O. Batzli, 2001. Availability of food and population dynamics of arvicoline rodents. Ecology, 82:1521–1534.

Usher, M.B., 1986. Wildlife conservation evaluation: attributes, criteria and values. In: M.B. Usher (ed.). Wildlife Conservation Evaluation, pp. 3–44. Chapman and Hall..

Usher, M.B., 1991. Scientific requirements of a monitoring programme. In: F.B. Goldsmith (ed.). Monitoring for Conservation and Ecology, pp. 15–32. Chapman and Hall.

Usher, M.B., 1996. The soil ecosystem and sustainability. In: A.G. Taylor, J.E. Gordon and M.B. Usher (eds.). Soils, Sustainability and the Natural Heritage, pp. 22–43. HMSO.

Usher, M.B., 1998. Minimum viable population size, maximum tolerable population size, or the dilemma of conservation success. In: B. Gopal, P.S. Pathak and K.G. Saxena (eds.). Ecology Today: an Anthology of Contemporary Ecological Research, pp. 135–144. International Scientific Publications.

Usher, M.B., 2000. The nativeness and non-nativeness of species. Watsonia, 23:323–326.

Usher, M.B., 2002a. An Archipelago of Islands: the Science of Nature Conservation. The Royal Society of Edinburgh, Edinburgh and Scottish Natural Heritage.

Usher, M.B., 2002b. Scotland's biodiversity: trends, changing perceptions and planning for action. In: M.B. Usher, E.C. Mackey and J.C. Curran (eds.). The State of Scotland's Environment and Natural Heritage, pp. 257–269. The Stationery Office, Edinburgh.

Usher, M.B., in press. Soil biodiversity, nature conservation and sustainability. In: R.D. Bardgett, M.B. Usher and D.W. Hopkins (eds.). Biological Diversity and Function in Soils. Cambridge University Press.

van Everdingen, R.O., 1990. Groundwater hydrology. In: T.D. Prowse and C.S.L. Ommaney (eds.). Northern Hydrology: Canadian Perspectives, pp. 77–101. Environment Canada.

Vanamo, S., 2001. Ten Years of Arctic Environmental Cooperation: a Compilation of Speeches. Unit for the Northern Dimension, Ministry for Foreign Affairs of Finland, Helsinki.

Vavrek, M.C., J.B. McGraw and C.C. Bennington, 1991. Ecological genetic variation in seed banks, III. Phenotypic and genetic differences between young and old seed populations of *Carex bigelowii*. Journal of Ecology, 79:645–662.

Veit, R.R., J.A. McGowan, D.G. Ainley, T.R. Wahls and P. Pyle, 1997. Apex marine predator declines ninety percent in association with changing oceanic climate. Global Change Biology, 3:23–28.

Vibe, C., 1967. Arctic Animals in Relation to Climate Fluctuations: Meddelelser on Grønland 170(5).

Vincent, W.F. and J.E. Hobbie, 2000. Ecology of Arctic lakes and rivers. In: M. Nuttall and T.V. Callaghan. The Arctic: Environment, People, Policy, pp. 197–232. Harwood Academic Publishers.

Vinnikov, K.Y., A. Robock, R.J. Stouffer, J.E. Walsh, C.L. Parkinson, D.J. Cavalieri, J.F.B. Mitchell, D. Garrett and V.F. Zakharov, 1999. Global warming and northern hemisphere sea ice extent. Science, 286:1934–1937.

Waide R.B., M.R. Willig, G. Mittelbach, C. Steiner, L. Gough, S.I. Dodson, G.P. Juday and R. Parmenter, 1999. The relationship between productivity and species richness. Annual Review of Ecology and Systematics, 30:257–300.

Walls, M. and M. Vieno (eds.), 1999. Natural Resources and Social Institutions: Workshop Proceedings. Academy of Finland, Helsinki.

Watt, K.E.F., 1968. Ecology and Resource Management: a Quantitative Approach. McGraw-Hill.

Weber, W.L., J.L. Roseberry and A. Woolf, 2002. Influence of the Conservation Reserve Programme on landscape structure and potential upland wildlife habitat. Wildlife Society Bulletin, 30:888–898.

Weimerskirch, H., 2002. Seabird demography and its relationship with the marine environment. In: E.A. Schreiber and J. Burger. Biology of Marine Birds, pp. 115–135. CRC Press.

Widen, B. and L. Svensson, 1992. Conservation of genetic variation in plants – the importance of population size and gene flow. In: L. Hansson (ed.). Ecological Principles of Nature Conservation. Applications in Temperate and Boreal Environments, pp. 71–112. Elsevier Applied Science.

Wiklund, C.G., A. Angerbjörn, E. Isakson, N. Kjellén and M. Tannerfeldt, 1999. Lemming predators on the Siberian tundra. Ambio, 28:281–286.

Wilby, R.L., G. O'Hare and N. Barnsley, 1997. The North Atlantic Oscillation and British Isles climate variability, 1865–1996. Weather, 52:266–276.

Williamson, M., 1996. Biological Invasions. Chapman and Hall.

Wilson, J., E. Mackey, S. Mathieson, G. Saunders, P. Shaw, I. Walker, A. Watt and V. West, 2003. Towards a Strategy for Scotland's Biodiversity: Developing Candidate Indicators of the State of Scotland's Biodiversity. Scottish Executive Environment and Rural Affairs Department Paper 2003/6.

Zhulidov, A.V., J.V. Headley, R.D. Roberts, A.M. Nikanorov and A.A. Ischenko, 1997. Atlas of Russian Wetlands. Environment Canada.

Zimov, S.A., V.I. Chuprynin, A.P. Oreshenko, F.S. Chapin, M.C. Chapin and J.F. Reynolds, 1995. Effects of mammals on ecosystem change at the Pleistocene-Holocene boundary. Ecological Studies, 113:128–135.

Chapter 11

Management and Conservation of Wildlife in a Changing Arctic Environment

Lead Author
David R. Klein

Contributing Authors
Leonid M. Baskin, Lyudmila S. Bogoslovskaya, Kjell Danell, Anne Gunn, David B. Irons, Gary P. Kofinas, Kit M. Kovacs, Margarita Magomedova, Rosa H. Meehan, Don E. Russell, Patrick Valkenburg

Contents

Summary

Climate changes in the Arctic in the past have had major influences on the ebb and flow in availability of wildlife to indigenous peoples and thus have influenced their distribution and the development of their cultures. Trade in animal parts, especially skins and ivory of marine mammals, and trapping and sale of fur-bearing animals go far back in time. Responsibility for management and conservation of wildlife in the Arctic falls heavily on the residents of the Arctic, but also on the global community that shares in the use of arctic resources. A sense of global stewardship toward the Arctic is critical for the future of arctic wildlife and its peoples.

This chapter, drawing on Chapters 7 to 9, emphasizes that throughout most of the Arctic, natural ecosystems are still functionally intact and that threats to wildlife typical for elsewhere in the world – extensive habitat loss through agriculture, industry, and urbanization – are absent or localized. There is increasing evidence that contaminants from the industrialized world to the south are entering arctic food chains, threatening the health and reproduction of some marine mammals and birds and the humans who include them in their diets. Protection of critical wildlife habitats in the Arctic is becoming recognized by those living inside as well as outside the Arctic as essential for both the conservation of arctic wildlife and its sustainable harvest by residents of the Arctic.

Management of wildlife and its conservation, as practiced in most of the Arctic, is conceptually different to that at lower latitudes where management efforts often focus on manipulation of habitats to benefit wildlife. The history of over-exploitation of marine mammals and birds for oil and skins to serve interests outside the Arctic is now being balanced by international efforts toward conservation of the flora and fauna of the Arctic, focusing on maintaining the Arctic's biodiversity and valuing its ecosystem components and relationships. Case studies from Russia and Canada focusing on harvest strategies and management of caribou (wild reindeer) highlight the complex nature of this species. One reports the development of a co-management system, involving shared responsibility between users of the wildlife and the government entities with legal authority over wildlife, giving local residents a greater role in wildlife management.

Throughout much of the Arctic, harvesting of wildlife for food and furs through hunting and trapping has been the most conspicuous influence that residents of the Arctic have had on arctic wildlife in recent decades. It was the overexploitation of wildlife during the period of arctic exploration and whaling, largely in the 18th and 19th centuries, that led to the extinction of the Steller sea cow in the Bering Sea and the great auk in the North Atlantic, and drastic stock reductions and local extirpation of several other terrestrial and marine mammals and birds. In regions of the Eurasian Arctic, the adoption of reindeer herding by indigenous hunting cultures led to the extirpation or marked reduction of wild reindeer (caribou) and drastic reductions of wolves, lynx, wolverines, and other potential predators of reindeer. Heavy grazing pressure by semi-domestic reindeer along with encroachment of timber harvest, agriculture, hydroelectric development, and oil and gas exploration have altered plant community structure in parts of the Fennoscandian and Russian Arctic. Large-scale extraction of nonrenewable resources accelerated in the Arctic during the latter half of the 20th century with impacts on some wildlife species and their habitats, especially in Alaska from oil production, in Canada from mining for diamonds and other minerals, and in Russia primarily from extraction of nickel, apatite, phosphates, oil, and natural gas. Among the factors that influence arctic wildlife, harvest of wildlife through hunting and trapping is potentially the most manageable, at least at the local level. Indigenous peoples throughout much of the North are asserting their views and rights in management of wildlife, in part through gains in political autonomy over their homelands. Arctic residents are now starting to influence when, where, and how industrial activity may take place in the Arctic. Part of this process has been the consolidation of the efforts of indigenous peoples across national boundaries to achieve a greater voice in management of wildlife and other resources through international groups such as the Inuit Circumpolar Conference and the Indigenous Peoples Secretariat of the Arctic Council. The stage appears to be set for indigenous peoples of the Arctic to become major participants in the management and conservation of arctic wildlife. The legal institutions, however, encompassing treaty and land rights and other governmental agreements vary regionally and nationally throughout the Arctic, posing differing opportunities and constraints on how structures for wildlife management and conservation can be developed.

This chapter provides examples from throughout the Arctic which show that conservation of wildlife requires sound management and protection of wildlife habitats at the local, regional, and national levels if the productivity of those wildlife populations upon which arctic peoples depend is to be sustained. Wildlife populations and their movements in both the marine and terrestrial environments transcend local, regional, and national boundaries, thus successful management and conservation of arctic wildlife requires international agreements and treaties. The chapter concludes that responsibility for maintaining the biodiversity that characterizes the Arctic, the quality of its natural environment, and the productivity of its wildlife populations must be exercised through global stewardship. Guidelines are provided for effective management and associated conservation of wildlife in a changing Arctic with emphasis on the complexity and limitations of managing wildlife in marine systems. The guidelines also stress the need for development of regional land and water use plans as a basis for protection of critical wildlife habitats in relation to existing and proposed human activities on the lands and waters of the Arctic.

11.1. Introduction

What can be learned from present wildlife management systems in the Arctic that can be drawn upon to alter existing systems or to design new ones to more effectively deal with climate-induced changes, and other changes that may occur in the future? Climate is the driver of change that has been the primary focus of the Arctic Climate Impact Assessment, however, it is important to remember that changes from other causes are also underway within the Arctic and that these are also affecting arctic ecosystems, as well as the economies, lifestyles, and dependency on wildlife of people in the Arctic. Many of these changes will continue along similar trajectories into the future, influenced by changing climate. The effects of climate change on wildlife populations, their productivity, and their distributions, will increasingly threaten arctic wildlife at the species, population, and ecosystem levels. Systems for management and conservation of wildlife in the Arctic will face new challenges and must become adaptable to the changes taking place in the natural environment accelerated by climate change. However, management and conservation of wildlife serve human interests, therefore in addition to becoming adaptable to those changes taking place in the natural environment, efforts toward management and conservation of wildlife in the Arctic must also be adaptable to those changes taking place among human societies, both within the Arctic and within the global community as a whole.

The objectives of this chapter are:

- To present an overview of structures for management and associated conservation of wildlife of land and sea in the Arctic, emphasizing current functioning structures.
- To assess the effectiveness of existing structures for management and conservation of wildlife in the Arctic in view of wide variation in regional social, economic, and cultural conditions.
- To emphasize the role of indigenous people in management of wildlife and its conservation in the Arctic.
- To explain how the distinctive regional and cultural perspectives of arctic residents affect management and conservation of wildlife in the Arctic within the context of the broader perspectives of the Arctic by the global community.
- To assess the adaptability of existing structures for management and conservation of wildlife in the Arctic within the context of expected climate change, and in association with resource extraction, other industrial development, the local economy, and community life.

11.2. Management and conservation of wildlife in the Arctic

11.2.1. Background

The term "wildlife" is used in this chapter in the modern sense inclusive, relevant to the Arctic, of non-domesticated birds and mammals living primarily in natural habitats in both terrestrial and marine environments. Wildlife management is an applied science that had its main development in continental Europe and North America. Aldo Leopold pioneered the development of modern, science-based wildlife management in the United States early in the 20th century, publishing in 1933 the first college-level text on wildlife management (Leopold, 1933). The initial focus of wildlife management was on species hunted or harvested by humans and has been parallel to, but distinct from, fishery management. Where practiced in most countries of the world today, however, it encompasses all aspects of conservation of wildlife species (including amphibians and reptiles) whether hunted or not, and encompasses harvest regulation, habitat protection and enhancement, wildlife population inventory and monitoring, and related ecosystem dynamics and research. Aldo Leopold's writings on environmental ethics and philosophy (Leopold, 1938, 1949, 1953) have also played a major role in the developing conservation and environmental movements following the Second World War.

Wildlife provided the foundation for the establishment of people and the development of their cultures in the Arctic. Wildlife was the primary source of food for humans living in the Arctic, and provided materials for clothing, shelter, fuel, tools, and other cultural items. Arctic-adapted cultures show similarity but also diversity in their dependency on specific species of wildlife. Caribou and reindeer, both the wild and semi-domesticated forms (all are the same species, *Rangifer tarandus*, reindeer being the term used for the Eurasian forms, and caribou for those native to North America), are of primary importance to most inland dwelling peoples throughout the Arctic. Marine mammals support indigenous peoples in coastal areas of the Arctic. Birds are also important in the annual cycle of subsistence harvest of wildlife in most arctic environments. Many wildlife species of the Arctic that are migratory, especially birds, but also marine mammals and some caribou and wild reindeer herds, are dependent during part of their annual life cycles on ecosystems outside the Arctic. As a consequence, efforts to ensure the conservation and sustainable human harvests of migratory species require management and conservation efforts that extend beyond the Arctic. The indigenous peoples of the Arctic include the marine mammal hunting Iñupiaq and Inuit of Alaska, Canada, and Greenland; the Dene who hunt the caribou herds of arctic Canada; the hunting, fishing, and reindeer herding Saami of the arctic regions of Fennoscandia and adjacent Russia; the reindeer herding and woodland hunting Dolgans of the central Siberian Arctic; and nearly twenty other cultural groups present throughout the circumpolar region (see Chapter 12).

Past climate changes have had major influences on the ebb and flow in availability of wildlife to indigenous peoples and thus have influenced the distribution of indigenous peoples in the Arctic and the development of their cultures. The accelerated climate warming observed in recent decades (Chapters 2 and 4), however, is resulting in major and more rapid changes in the ecology of arctic wildlife (Chapters 7, 8, 9), necessitating reassessment of structures for the management and conservation of arctic wildlife. As northern cultures developed, including those of indigenous and non-indigenous arctic residents, their relationships to wildlife were also influenced beyond strictly subsistence dependency through trade or other economic relationships, both internal to their own cultures and with other cultures. Trade in animal parts, especially skins and ivory of marine mammals; the semi-domestication of reindeer; and trapping and sale of fur-bearing animals go far back in time. Over the last two to three centuries cash income has become important for indigenous and non-indigenous residents from selling meat and hides and as well as through home industries producing saleable craft items from animal parts (see Chapters 3 and 12). Arctic wildlife is valued by many living outside the Arctic for its attraction for viewing and photographing, especially whales, seabirds, polar bears (*Ursus maritimus*), and caribou; for incorporation in art depicting the arctic environment; and for associated tourism. Sport and trophy hunting of wildlife bring many to the Arctic, with associated economic benefits to local residents through services provided. Others value the Arctic through virtual recognition of and fascination for the role of wildlife species in the dynamics of arctic ecosystems, many of whom may never visit the Arctic but learn about arctic wildlife through the printed and visual media. Responsibility for management and conservation of wildlife in the Arctic clearly falls heavily on the residents of the Arctic, now especially through empowerment of indigenous people, but also on the global community that benefits from the exploitation of arctic resources and shares in the appreciation of the wildlife and other values of the arctic environment. A consequence of conservation efforts affecting wildlife and their habitats, generated largely outside the Arctic, has been the many "protected areas" (UNESCO Biosphere Reserves, national parks, wildlife refuges, nature preserves, and sanctuaries) established by arctic countries, often with the encouragement and support of international conservation organizations such as the Conservation of Arctic Flora and Fauna (CAFF), the World Conservation Union (IUCN), and the World Wide Fund for Nature (WWF). A sense of global stewardship toward the Arctic is critical for the future of arctic wildlife and its peoples.

11.2.2. Present practices

Throughout most of the Arctic, natural ecosystems are still functionally intact (see Chapters 7, 8, 9). Most

threats to wildlife typical for elsewhere in the world – extensive habitat loss through agriculture, industry, and urbanization – are absent in much of the Arctic or are localized. Similarly, introduced and invading wildlife species are few throughout most of the Arctic and tend to be localized at the interface between forest and tundra. Changes, however, are accelerating. Contaminants from the industrialized world to the south have reached arctic food chains, threatening the health and reproduction of some wildlife, especially marine mammals and birds, and the humans who include them in their diet (AMAP, 1998a,b, 2002). Energy and mineral extraction developments in the Arctic, although localized and widely scattered, tend to be of large scale, for example the Prudhoe Bay oil field complex in Alaska, the mining and associated metallurgical developments in the Taymir and Kola regions of Russia, and the hydroelectric development in northern Quebec. These contribute to the pollution and contamination of the arctic waters, atmosphere, and lands and result in local loss of wildlife through habitat destruction, excessive hunting, and other cumulative impacts. Protection of critical wildlife habitats in the Arctic is becoming increasingly recognized as essential for both the conservation of arctic wildlife and management of its harvest by arctic residents as pressures from outside the Arctic for exploitation of its resources increase (CAFF, 2001a; NRC, 2003).

Management of wildlife and its conservation, as practiced in most of the Arctic, is conceptually different in the minds of arctic dwellers in contrast to most people living at lower latitudes where management efforts often focus on manipulation of habitats to benefit wildlife (Fig. 11.1). Thus, "management of wildlife" in the Arctic may seem to some inappropriate terminology that has

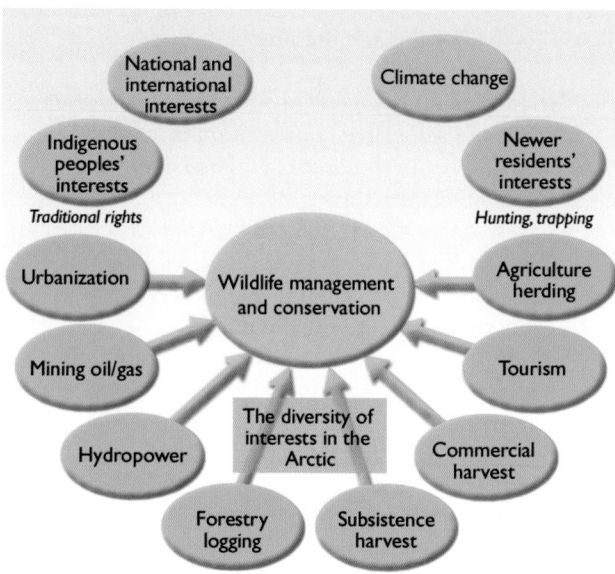

Fig. 11.1. Management and conservation of wildlife in the Arctic is driven by internal and external forces that involve wide-ranging interests and uses of wildlife. These include traditional harvest and dependency by indigenous peoples, the effects of resource extraction and associated industrial development, tourism, and valuation of wildlife at national and international levels through legal structures and conservation efforts.

developed through its application outside the Arctic. Arctic residents have often seen little justification for conventional wildlife management throughout much of the Arctic in the past, and have questioned the need for science-based wildlife management when harvest levels have posed little threat to sustained viability of the species harvested (e.g., Huntington, 1992). To the contrary, many arctic peoples see the current health of arctic ecosystems as evidence of their effectiveness as conservationists over the centuries and their often aggressive resistance in the past to commercial overexploitation of marine mammals and birds for oil and skins (Burch, 1998). Prior to the presence of Europeans in the Arctic, the archeological evidence indicates that communities and entire cultures either moved or died out as a consequence of changing climate and associated unsustainable levels of wildlife harvest (Knuth, 1967; Schledermann, 1996), as was also the case at lower latitudes (Grayson, 2001). As well, these perceptions grow from historical conditions of "internal colonialism" in which southern populations viewed the arctic resources as open to access and available for exploitation, contrasting to indigenous views of territoriality with soft borders and property held in common by groups (Osherenko and Young, 1989). In recent years, many indigenous residents have resisted systems for wildlife management and conservation imposed from outside the Arctic, particularly when these rely heavily on new and strange technologies and are based on tenets that are unfamiliar or inappropriate to arctic cultures (Klein, 2002).

Increased emphasis by those living outside the Arctic on conservation of the flora and fauna of the Arctic and associated emphasis on maintaining its biodiversity, and valuing all its ecosystem components and relationships, has understandably appeared hypocritical to many arctic indigenous peoples dependent on sustainable harvest of arctic wildlife (e.g., Freeman and Kreuter, 1994). Thus, some indigenous peoples have questioned the justification for wildlife management in the Arctic as a discrete aspect of ecosystem or land use management, when in much of the Arctic the need is for integrated land, coastal, and oceanic plans for management.

The legacy of relations and emergent conditions require the development of wildlife management approaches in the Arctic that foster collective action among a highly diverse set of stakeholders and also assume high ecological uncertainty (Jentoft, 1998; Young and Osherenko, 1993). Research on the sustainability of common property resources of the past two decades, which has questioned conventional approaches of "state control" as reflected in Hardin's (1968) *Tragedy of the Commons*, points to social institutions as key determinants of human behavior and ecological change (Berkes and Folke, 1998; Hanna et al., 1996; Ostrom, 1990; Ostrom et al., 2002; Young, 2001). The findings of institutional analysis identify design principles that are critical for effective institutional performance, and note how effective institutions of wildlife management can reduce transaction costs among actors and build trust

among players. In some regions of the Arctic, the settlement of indigenous land claims has provided opportunities to create new institutional arrangements with these principles in mind, and thus giving local communities a greater role in the practice of wildlife management if not in determining the premises on which it is based (e.g., Adams et al., 1993; Berkes, 1989; Caulfield, 1997; Freeman, 1989; Huntington, 1992; Osherenko, 1988; Usher, 1995).

Throughout much of the Arctic, harvesting of wildlife for food and furs through hunting and trapping has, nevertheless, been the most conspicuous influence that residents of the Arctic have had on arctic wildlife in recent decades. It was the overexploitation of wildlife during the period of arctic exploration and whaling in the 18th and 19th centuries that led to the extinction of the Steller sea cow (*Hydrodamalis gigas*) in the Bering Sea and the great auk (*Pinguinus impennis*) in the North Atlantic, and drastic stock reductions and local extirpation of several other terrestrial and marine mammals and birds. In many regions of the Eurasian Arctic, the adoption of reindeer herding by indigenous hunting cultures led to the extirpation or marked reduction of wild reindeer and drastic reductions of wolves (*Canis lupus*), lynx (*Lynx lynx*), wolverines (*Gulo gulo*), and other potential predators of reindeer (Chapter 12). In recent decades heavy grazing pressure by semi-domestic reindeer has altered plant communities in parts of the Fennoscandian and Russian Arctic, that has in some areas been exacerbated by encroachment into traditional grazing areas of timber harvest, agriculture, hydroelectric development, and oil and gas exploration (e.g., Forbes, 1999). Large-scale extraction of nonrenewable resources accelerated in the Arctic during the latter half of the 20th century with consequences for some wildlife species and their habitats, especially in Alaska from oil production, in Canada from mining for diamonds and other minerals, and in Russia primarily from extraction of nickel, apatite, phosphates, oil, and natural gas (CAFF, 2001a).

Among the factors that influence arctic wildlife, harvest of wildlife through hunting and trapping is potentially the most manageable, at least at the local level. At a more regional level, these influences come through decisions on wildlife habitat as a land use issue. Indigenous peoples throughout much of the North are asserting their views and rights in wildlife management, in part through increased political autonomy over their homelands or involvement in cooperative management regimes (Caulfield, 1997; Huntington, 1992; Klein, 2002; Nuttall, 1992, 2000). However, people still feel largely limited in controlling the influences on wildlife and wildlife habitats brought about through climate change, or large-scale resource extraction in both the marine and terrestrial environments, changes largely resulting from the effects of, and pressures generated by, people living outside the Arctic. Similarly, arctic residents are generally poorly informed about conditions and management of migratory species in their wintering environments far from the Arctic, especially waterfowl

Box 11.1. The Inuit Circumpolar Conference

The Inuit Circumpolar Conference (ICC) defends the rights and furthers the interests of Inuit in Greenland, Canada, Alaska, and Chukotka – in the far east of the Federation of Russia. Established in 1977, the ICC maintains national offices in each of the four countries and has official observer status in the United Nations Economic and Social Council. Noted for its efforts to conserve and protect the environment and to promote sustainable development, the ICC also defends and promotes the human rights of Inuit, the Arctic's original inhabitants.

and some whale species, and seek greater involvement in management of migratory species governed by international treaties. The influence that Canadian arctic peoples had, however, in the negotiations leading to the 2001 Stockholm Convention on Persistent Organic Pollutants has shown the potential for concerted action by arctic peoples at the global level (Downie and Fenge, 2003).

Throughout most of the Arctic where efforts have been directed at conservation and management of wildlife, the primary focus has been on regulation of the harvest of wildlife to ensure the long-term sustainability of the wildlife populations and the associated human harvest from them. Secondly, protection of wildlife habitats from loss or degradation has been acknowledged as essential for the sustainability of wildlife populations, however, where large-scale development activity has occurred local interests in wildlife have often been poorly represented in land use decisions. Although there are similarities throughout much of the Arctic in the distribution of wildlife species and their use by humans, there are major local and regional differences in the importance of specific wildlife species in the local subsistence and cash economies. These differences relate to past traditions of use of wildlife, relative availability of wildlife for harvest, and the role that wildlife play in the local economy. For example, in Eurasia, commercial harvest of wildlife is generally supported by legal structures that assign wildlife ownership to the land owner, in contrast to North America where wildlife remains the property of the state and commercial harvest of wildlife is prohibited or discouraged.

Along with the increasing political autonomy of indigenous peoples in recent decades, these arctic residents are developing their capacity to influence when, where, and how industrial activity may take place in the Arctic. Part of this process has been the consolidation of the efforts of indigenous peoples across national boundaries to achieve a greater voice in management of wildlife and other resources through international groups such as the Inuit Circumpolar Conference (see Box 11.1) and the Indigenous Peoples Secretariat of the Arctic Council. In addition to the eight arctic countries that make up membership of the Arctic Council, indigenous organizations have representation as Permanent Participants of the Council and include the Russian Association of Indigenous Peoples of the North, the Inuit Circumpolar Conference, the Saami Council, the Aleutian International Association, the Arctic Athabaskan Council, and the Gwich'in Council International.

Through the resulting increased political voice and sharing of interests, the stage appears set for indigenous peoples of the Arctic to become major participants in the management and conservation of arctic wildlife. The legal institutions, however, encompassing treaty and land rights and other governmental agreements vary regionally and nationally throughout the Arctic, posing differing opportunities and constraints on how structures for wildlife management and conservation can be developed.

Conservation of wildlife in the Arctic requires sound management and protection of habitats at the local, regional, national, and international levels if the productivity of those wildlife populations that arctic peoples are dependent upon is to be sustained. Wildlife populations and their movements in both the marine and terrestrial environments often transcend local, regional, and national boundaries, thus successful management and conservation of arctic wildlife, requiring scientific investigation, monitoring, and management action, must also transcend political boundaries through international agreements and treaties (CAFF, 2001a). Many of the pressures on arctic wildlife originate outside the Arctic, such as contaminants in marine wildlife, habitat alteration through petroleum and mining developments, and climate changes exacerbated by increased concentrations of greenhouse gases. It seems clear that responsibility for maintaining the biodiversity that characterizes the Arctic, the quality of its natural environment, and the productivity of its wildlife populations must be supported through a sense of stewardship at both the local and global levels.

11.2.3. The role of protected areas

A goal of ecosystem conservation in the Arctic as elsewhere is maintenance of the health of the unique complex of ecosystems that characterize the Arctic, and in doing so, to attempt to ensure the protection and sustainability of the unique biodiversity for which the Arctic is valued both by arctic residents and the rest of the world community. An important process in the efforts to achieve this goal has been the identification of natural habitats of critical importance in the life cycles of wildlife species, and their subsequent protection through legal processes at local, regional, national, and international levels of government. Although "protected areas" are often established with the well-being of a single species or a group of related species being the primary focus (e.g., Ramsar sites for waterfowl, Round Island in Alaska for walrus (*Odobenus rosmarus*); see Fig. 11.2), all

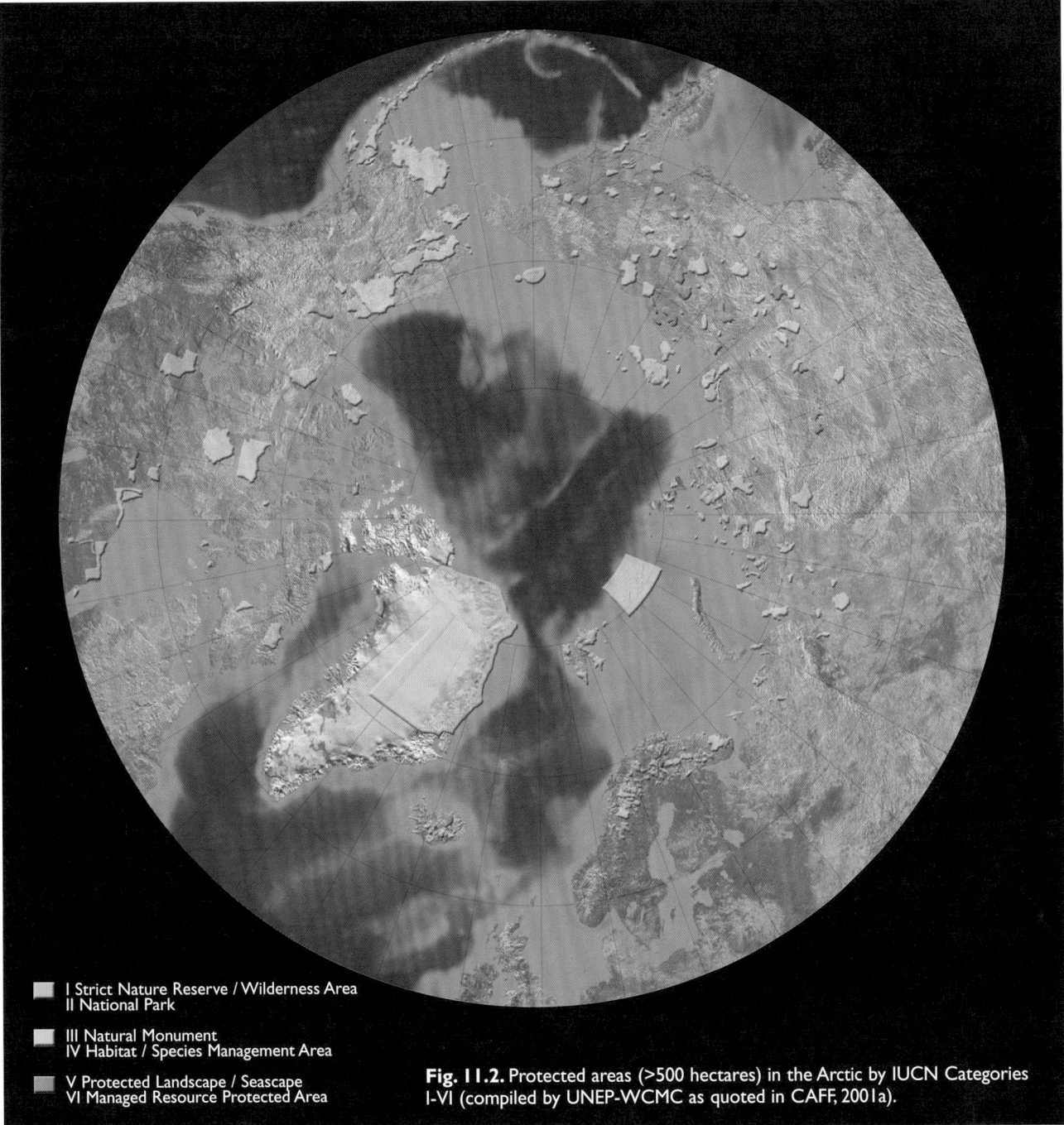

I Strict Nature Reserve / Wilderness Area
II National Park

III Natural Monument
IV Habitat / Species Management Area

V Protected Landscape / Seascape
VI Managed Resource Protected Area

Fig. 11.2. Protected areas (>500 hectares) in the Arctic by IUCN Categories I-VI (compiled by UNEP-WCMC as quoted in CAFF, 2001a).

forms of life that are encompassed within these units generally benefit. Conversely, other areas may be protected primarily in recognition of the unique biodiversity that they encompass. In 1996, CAFF developed a Strategy and Action Plan for a Circumpolar Protected Area Network. Execution of the plan was designed to perpetuate the dynamic biodiversity of the arctic region through habitat conservation in the form of protected areas to represent arctic ecosystems, and to improve physical, informational, and managerial ties among circumpolar protected areas. As a result of CAFF's efforts, jointly with other international governmental and non-governmental organizations, and local, regional, and national governments and interests, nearly 400 protected areas (greater than 10 km²) were established throughout the Arctic in 2000, totaling over 2.5 million km² (CAFF, 2001a).

Selection of areas needed for protection in the interest of wildlife conservation is not a task easily accomplished even when there is broad public and governmental support for the process. Identifying those areas of critical habitat needing protection for the effective conservation of wildlife in the Arctic requires comprehensive habitat inventories and assessment of all existing and proposed land uses within areas under consideration. Part of these assessments is the weighing up of consequences of the present and proposed uses of the areas under consideration for protection (e.g., subsistence, commercial, and sport hunting; reindeer grazing; transportation corridor construction; and other resource extraction uses). Establishment of protected areas critical to effective conservation of wildlife, and acceptance and respect for their legal protection, generally requires advance involvement, open discussion, and often compromise among all poten-

Box 11.2. Balancing nature conservation and industrial development in Canada

There should be no new or expanded large-scale industrial development in Canada until a network of protected areas is reserved which adequately represents the natural region(s) affected by that development. The Conservation First Principle (WWF Canada, 2001).

An essential element of conserving Canada's natural heritage is to permanently protect an ecologically viable, representative sample of each of the country's terrestrial and aquatic natural regions. These protected areas conserve a basic level of natural habitat for Canadian wildlife and the ecological processes that provide freshwater, fertile soils, clean air, and healthy animals and plants. In many places, these natural areas are crucial to the continued livelihoods and cultural integrity of Canada's indigenous peoples.

Protecting representative samples of every natural region in Canada should be accomplished in a way that fully respects the constitutional rights of indigenous peoples, and provides genuine economic opportunities for local residents. This goal can with careful planning be accomplished without sacrificing jobs or economic development.

Canada signed and ratified the international Convention on Biological Diversity in 1992. The same year, all Canadian Ministers responsible for wildlife, parks, the environment, and forestry (federally, provincially, and territorially) agreed in the Tri-Council Commitment to take a critical first step in conserving biodiversity by completing a network of ecologically representative protected areas in land-based natural regions by 2000, and by accelerating the protection of representative protected areas in Canada's marine natural regions.

The area of representative protected areas in Canada doubled in the 1990s, but the Tri-Council Commitment has not yet been met. Not all natural terrestrial regions have been moderately or adequately represented in protected areas, and marine regions remained largely unrepresented. Canadian government bodies have continued to approve new oil and gas leases, forest allocations, mining projects, hydro dams, and other large-scale development projects in Canada's natural habitats. WWF Canada (November 2001) stated that: "Every time a development project is proposed in a natural region that is not yet adequately represented by protected areas, we erode the options to establish these natural and cultural safeguards".

tial users of the areas and representatives of the governments with legal responsibility for their establishment. An example of the complex process for justification and establishment of protected areas for wildlife conservation was initiated in northern Yukon Territory of Canada and adjacent Alaska through an agreement between Canada and the United States establishing the International Porcupine Caribou Board. Through these international efforts a report on the sensitive habitats of the Porcupine Caribou Herd was prepared (IPCB, 1993) and is being used in an ongoing process of providing justification and protection of critical habitats within existing protected areas in Alaska and Yukon Territory, and in the regional planning process and establishment of additional protected areas in northern Yukon Territory. Non-governmental organizations can and have played an important role in the establishment of protected areas for wildlife conservation in the Arctic. Another example is the "Conservation First Principle" concept under development for the Canadian North through shared governmental and non-governmental efforts (see Box 11.2).

Protected areas set aside by governmental action, merely through establishment of their boundaries, do help to bring about public recognition of the importance of their role in wildlife conservation. Unless their establishment is accompanied by enforceable laws that govern their use, however, the areas remain protected in name only and remain vulnerable to overexploitation of the

wildlife, and habitat alteration and destruction through competing land uses. Political pressures generated by large and often multinational industries interested in protected areas as loci for energy or mineral extraction, mass tourism, or other developments destructive to wildlife and their habitats, may be successful in persuading governments to allow them into these areas. Examples of where the protection offered to arctic areas set aside for wildlife conservation has been violated are widespread throughout the Arctic (e.g., seismic exploration for oil in the Arctic National Wildlife Refuge and atomic bomb testing in the Alaska Maritime National Wildlife Refuge, both in Alaska; illegal harassment of walrus in the Wrangel Island Reserve and uncontrolled poaching of wildlife in Kola Peninsula reserves by military personnel, both in Russia).

Although the importance of existing protected areas and the need for establishment of additional protected areas for effective conservation of wildlife in the Arctic are internationally recognized, climate change adds an additional layer of complexity in use of protected areas as a tool in wildlife conservation. If plants and animals change their distribution in response to a changing climate as is expected (Chapters 7, 8, 9), critical habitats of wildlife (seabird nesting colony sites, reindeer/caribou calving grounds, waterfowl and shore-bird nesting and staging areas, marine mammal haul-out areas) will also change in their distribution over time. Consequently, anticipating the needs for new protected areas important for conservation as wildlife and their habitats change in their distributions on the landscape will be an extremely difficult process. The process will necessarily need to be dynamic, with ongoing assessment of wildlife habitat use and dependency. This should enable recognition of the continued importance of some existing protected areas, and conversely, recognition that others that become abandoned by wildlife may no longer be needed, though they may retain value for protection of plant species or other ecosystem components. Wildlife management and conservation in an Arctic under the influence of climate change must be adaptive to ecosystem level changes that are not feasibly reversible within the human timescale, such as the northward movement of boreal ecotones into the Arctic along with the associated wildlife. Thus, protected areas will have value as areas where climate-induced or other externally influenced changes within ecosystems can be observed and monitored, free of major direct human impacts.

The establishment and use of protected areas is an essential component of conservation of wildlife and their habitats in the Arctic and in the protection of the biodiversity that characterizes arctic ecosystems. However, protected areas alone cannot ensure the sustained integrity of arctic ecosystems under the influences of a changing climate and accelerating pressures from resource extraction, tourism, and associated construction of roads, pipelines, and other transportation corridors. Of major concern is the fracturing of habitats through development activities, especially transportation corridors that may restrict the free movement and exchange of plants and animals between habitats even though significant parts of these habitats may have protected status. Ecological requirements for subpopulations of both plants and animals may be encompassed within protected areas, but the long-term integrity and sustainability of arctic ecosystems and the wildlife and other organisms within them requires opportunity for genetic exchange between components. Although critical habitat units may merit rigid protection, the intervening natural environment must be managed so that movement of species within entire ecosystems remains possible. Establishment of protected areas should be consistent with subsistence harvesting activities and not designed to exclude them. Management of the harvest of wildlife must be adaptable to changes that may take place in the population status of wildlife species.

Transportation corridors, especially roads and their associated vehicle traffic, may fracture habitats and limit free movement of species within ecosystems, however, they also provide corridors for the movement of invasive plant and animal species, with often detrimental consequences for native species with which they may compete, prey upon, parasitize, or infect. "Invasive species" is an all-inclusive generic term. It includes plants and animal species truly exotic to most regions of the Arctic and subarctic, such as the dandelion (*Taraxacum officinale*), house mouse (*Mus musculus*), and Norway rat

(*Rattus norvegicus*) that have inadvertently been introduced by humans. There are, however, invasive species native to adjacent biomes, such as the moose (*Alces alces*) and snowshoe hare (*Lepus americanus*), that have expanded into parts of the North American Arctic from the boreal forest with consequences for arctic species and ecosystems. Humans have also been responsible for the deliberate introduction of plant and animal species into the Arctic. Examples are the introduction of lupine (*Lupinus* spp.) and coniferous trees to Iceland associated with erosion control and forest reestablishment, which through their subsequent dispersal have become nuisance species in areas where they crowd out native or introduced forage species for domestic livestock, and threaten preservation of the natural biodiversity. Among animals, the deliberate introduction of Arctic foxes (*Alopex lagopus*) to the Aleutian and Commander Islands in the 18th century for harvest of their pelts led to the marked reduction or extirpation of populations of marine birds, waterfowl, and other ground nesting birds. The intensive, decades-long efforts of the U.S. Fish and Wildlife Service to eliminate the Arctic foxes on many of the Aleutian Islands has resulted in rapid reestablishment of successful bird nesting on islands from which the foxes have been removed, but this has involved a great expenditure of effort and money. It can be expected that the appearance of invasive species in the Arctic will increase through deliberate and accidental human activities, as well as by natural dispersal assisted by transportation corridors and parameters of climate change that may favor the new species over native plants and animals.

It is important to remember that the decrease in biodiversity with increasing latitude that is a characteristic of arctic ecosystems is partly a consequence of the slow rate of dispersal of species into the Arctic following deglaciation. It is very likely that climate change, especially the climate warming projected to occur throughout much of the Arctic (see Chapter 4), and other forces will accelerate the "natural" movement of plant and animal species into the Arctic. It remains for human judgment to determine whether invading plant and animal species are to be considered part of the natural ongoing process of ecosystem change in the Arctic, whether they pose threats to the natural biodiversity of arctic ecosystems, or whether they are detrimental to human efforts to manage arctic ecosystems for human exploitation. Important tasks facing managers of wildlife in a changing Arctic will be assessing consequences for native species and ecosystems of the effects of invasive species within the constraints of a changing climate. It may also be necessary, where regionally appropriate, to develop procedures that restrict invasion of species that may have undesirable consequences for native species.

11.2.4. Change in human relationships with wildlife and managing human uses of wildlife

On the basis of early archeological evidence, human cultures with the technologies that allowed them to live under the climatic extremes of the Arctic while exploiting its marine resources did not appear until the mid-Holocene Epoch ~7000 years ago (Giddings, 1967). The entrance of humans to the Americas from Asia via Beringia 7000 to 8000 years earlier, however, occurred near the end of the Pleistocene Epoch when sea levels were lower, land areas greater, and the environment markedly different to how it later became in the Holocene (Meltzer, 1997). During much of the Holocene, following the first major wave of human movement into North America, as the Pleistocene ice retreated from the land, changes in human distribution, demography, culture, and movements were predominantly tied to changes in availability of wildlife. Humans located where species that were essential components of their diets, and provided materials for their clothing, shelter, tools, and weapons, were available. This pattern of human use of the land and adjacent sea prevailed as the Arctic was settled and cultures evolved in adaptation to the wildlife and other resources available for their exploitation (Schledermann, 1996; Syroechkovskii, 1995).

Wildlife species in both marine and terrestrial systems have undergone changes in their abundance and distribution in the past, and therefore in their availability for use by people in the Arctic. Some of these changes have resulted from heavy commercial exploitation of marine wildlife for their skins and oil and of terrestrial mammals largely for their pelts. Longer-term changes in distribution and abundance of wildlife in the Arctic are thought to have been largely the result of changes in climate affecting temperature, precipitation, snow characteristics, and sea-ice conditions and their influence on food chain relationships (see Chapters 7, 8, 9). All the peoples of the Arctic and the animals they hunt and use are subject to the vagaries of arctic climate. The global warming observed in the latter half of the 20th century, consistent with projections by general circulation models, has advanced most rapidly in certain parts of the Arctic, however, there have been regional inconsistencies (see Chapters 2, 4, 6). The western Canadian Arctic and the Alaskan Arctic have shown decadal temperature increases of 1.5 °C, whereas a nearly opposite cooling trend has been recorded in Labrador, northern Quebec, Baffin Island, and adjacent southwest Greenland (Serreze et al., 2000). Nevertheless, although some regions of the Arctic may not have experienced the pronounced warming in recent decades that has characterized most of the Arctic, changes in other climate-related parameters such as precipitation, frequency and severity of storm events, and thinning and reduced seasonal extent of sea ice are being observed in all regions of the Arctic (Chapter 2). Increases in ultraviolet-B (UV-B) radiation levels in the Arctic associated with thinning of the atmospheric ozone layer may have consequences for life processes of both plants and animals, however little is known of possible effects on wildlife (Chapter 7). However, to the extent that increased UV-B radiation levels may result in differential changes in tissue structure and survival of plant species, resulting in changes in their quality and abundance as food for herbivores, wildlife and their food chain relationships will be affected.

As a general rule the numbers of plant and animal species decline with increasing latitude from the equator to the poles, as does the complexity of species interrelationships and associated ecosystem processes. Since external influences tend to be buffered by the complexity of biological processes within ecosystems, the less complex arctic ecosystems can be expected to respond more dynamically to climate change than the more complex systems that exist at lower latitudes, and this seems to have been the case during past periods of climate change (Chapter 7). An additional compounding factor is that rates of climate-related change in much of the Arctic, reflected in climate warming and decrease in sea-ice thickness and extent, exceed those at lower latitudes.

11.3. Climate change and terrestrial wildlife management

11.3.1. Russian Arctic and subarctic

Hunting is an important part of the Russian economy, both through harvest of wildlife products and through pursuit of traditional sport and subsistence hunting. Fur production has been an essential part of the economy of the Russian North throughout history. Management of wildlife also has a long history in Russia, from early commercial and sport hunting to the creation of a complicated multifunctional state system under the Soviet government. Early attempts at regulation of hunting are known from the 11th century, and these attempts at wildlife management were connected with protection of species or groups of species. The first national law regarding hunting was imposed in 1892 as a reaction to widespread sport hunting, the establishment of hunter's unions, and the efforts of naturalists and others with interests in wildlife. These early efforts toward managing wildlife were based on wildlife as a component of private property.

Under the Soviet system, wildlife management developed on the basis of state ownership of all resources of the land, including wildlife, and a state monopoly over foreign trade and fur purchasing. Commercial hunting was developed as an important branch of production within the national economy. The state-controlled wildlife management system resulted in an elaborate complex of laws as the basis for governing commercial and sport hunting, for investigation of resources and wildlife habitats, for organization of hunting farms or collectives, for establishment of special scientific institutes and laboratories, for incorporation of scientific findings in wildlife management, and for the development of a system of protected natural areas. Justification for identifying natural areas deserving protection in the Russian Arctic became apparent as major segments of the Russian economy increased their dependence on exploitation of arctic resources during the Soviet period, stimulated by the knowledge that 70 to 90% of the known mineral resources of the country were concentrated in the Russian North (Shapalin, 1990). More than 300 protected natural areas of varying status were established for restoration and conservation of wildlife resources in the Russian Far North (Baskin, 1998).

Wildlife management was concentrated in a special Department of Commercial Hunting and Protected Areas within the Ministry of Agriculture. Local departments were organized in all regions of the Russian Federation for organization, regulation, and control of hunting with the intent to make them appropriate for actual conditions. Hunting seasons were established for commercial and sport hunting by species, regulation of numbers harvested, and designation of types of hunting and trapping equipment to be used. The major hunting activity was concentrated in specialized hunting farms. Their organization was initially associated with designated areas. The main tasks of the state hunting farms were planning, practical organization of hunting, and management for sustained production of the wildlife resources. At the same time, the system of unions of sport hunters and fishers was organized for regulation of sport hunting and fishing under the control of the Department of Commercial Hunting and Protected Areas (Ammosov et al., 1973; Dezhkin, 1978).

Commercial hunting has been primarily concentrated in the Russian Far North (tundra, forest–tundra, northern taiga), which makes up 64% of the total hunting area of the Russian Federation. During the latter decades of the Soviet system the Russian Far North produced 52% of the fur and 58% of the meat of ungulates and other wildlife harvested. The proportional economic value of the three types of resident wildlife harvested was 41% for fur (sable (*Martes zibellina*) – 50%, arctic fox – 9%, ermine (*Mustela erminea*) – 18%), 40% for ungulates (moose – 41%, wild reindeer – 58%), and 19% for small game (ptarmigan (*Lagopus* spp.) – 68%, hazel grouse (*Tetrastes bonasia*) – 15%, wood grouse (*Tetrao urogallus*) – 11%)

(Zabrodin et al., 1989). Variation by region in characteristics of the harvest of wildlife in the Russian Arctic and subarctic is compared in Table 11.1. Participation in commercial hunting by the able-bodied local population was 25 to 30%. Profit from hunting constituted 52 to 58% of the income of the indigenous population. Of the meat of wild ungulates harvested, the amount obtained per hunter per year was 233 kg for professional hunters, 143 kg for semi-professional hunters, and 16 kg for novice hunters. The proportion of total wild meat harvested that was purchased by the state was 60%. Of that purchased by the state, 73% was for consumption by the local population. Fish has also been an important food resource for local populations, as well as for the professional hunters/ fishers. A professional hunter's family would use about 250 kg of fish per year, and 2000 kg of fish were required per year to feed a single dog team (eight dogs). By the end of the 1980s state purchase of wildlife and fish was 34% of potential resources, and local consumption was 27% (Zabrodin et al., 1989).

Indigenous residents of the Russian Arctic and subarctic have not had limitations on hunting for their subsistence use. However, all those engaged in professional, semi-professional, and sport hunting have been required to purchase licenses. Indigenous people involved in the state-organized hunting system were also provided with tools and consumer goods. The main problems that have confronted effective wildlife management in the Russian Arctic are widespread poaching, uneven harvest of wildlife, and loss of wildlife habitats and harvestable populations in connection with industrial development.

The wildlife management system in the Russian Arctic was not destroyed by the transformation of the political and economic systems that took place at the end of the

Table 11.1. Regional variation in wildlife harvest in the Russian Arctic and subarctic under the Soviet system (Zabrodin et al., 1989).

	European Russia	Western Siberia	Eastern Siberia	Northern Far East Russia
Share of area (%)	7	14	25	54
Ranking of relative biological productivity	4	2	1	3
Proportion of available resource harvested (%)	23	48	76	63
Expenditure (%)	9	15	34	42
Breakdown of value by species within region				
Fur				
Sable (%)	–	14	24	23
Polar fox (%)	5	7	3	4
Ungulate				
Moose (%)	15	18	12	20
Wild reindeer (%)	4	8	42	15
Game				
Partridge (%)	51	26	4	8
Distribution of the harvest				
Purchased by the state (%)	33	37	61	58
Local consumption (%)	67	63	39	42

20th century, but it was weakened. Partly as a consequence of this weakening, but also due to expansion of industrial development in the Russian Arctic and the effects of climate change, there has been the development of several major threats to effective wildlife conservation.

- Transformation of habitats in connection with industrial development. From an ecological standpoint the consequences of industrial development affect biological diversity, productivity, and natural dynamics of ecosystems. As far as environmental conditions are concerned it is important to note that apart from air and water pollution there is a possibility of food pollution. In terms of reindeer breeding, hunting, and fishing, industrial development has resulted in loss of habitats and resources, a decrease in their quality and biodiversity, and destruction of grazing systems (Dobrinsky, 1995, 1997; Yablokov, 1996; Yurpalov et al., 2001). A considerable portion of the biological resources presently exploited is from populations outside regions under industrial development (Yurpalov et al., 2001).
- Reduction in wildlife populations as a result of unsystematic and uncontrolled exploitation through commercial hunting.
- Curtailment of wildlife inventory and scientific research, resulting in loss of information on population dynamics, health, and harvest of wildlife.
- Changes in habitat use by wildlife, in migration routes, and in structure and composition of plant and animal communities as a consequence of climate change. Such changes include increased frequency and extent of fires in the northern taiga, displacement northward of active breeding dens of the Arctic fox on the Yamal Peninsula (Dobrinsky, 1997), as well as in other areas (Yablokov, 1996), and replacement of arctic species by boreal species as has occurred in the northern part of the Ob Basin (Yurpalov et al., 2001).

Both commercial and sport hunting are permitted throughout the Russian North. Commercial hunting for wild reindeer for harvest of velvet antlers is permitted for 20 days in the latter part of June. Commercial hunting of reindeer for meat can take place from the beginning of August through February. Sport hunting is permitted from 1 September to 28 February. A license is required to hunt reindeer (cost for sportsmen about US$4, for commercial enterprise about US$3). There are no restrictions on numbers of reindeer to be hunted. Hunting is permitted everywhere, with the exception of nature reserves. Regional wildlife harvest systems are compared in Table 11.2, together with associated wildlife population trends, threats to wildlife and their habitats, and conservation efforts.

In recent years in the Russian North, marketing of venison experienced an economic revival. In mining settlements in 2001 the cost of venison commonly approached US$2.5 per kilogram, making commercial hunting of reindeer potentially profitable. A significant demand has

also existed for velvet antlers. However, under existing conditions in most of the Russian North where there are no roads and settlements are few, hunting of wild reindeer at river crossings remains the most reliable and productive method of harvest (see the case study on river crossings as focal points for wild reindeer management in the Russian Arctic in Box 11.3). Additionally, concentration of hunting effort at specific river-crossing sites provides an opportunity to influence hunting methods and for monitoring the number of animals killed. A proposal has been made to protect the traditional rights of indigenous hunters by granting them community ownership of some of the reindeer river crossings. This would presumably allow them to limit increasing competition from urban hunters for the reindeer. At present, indigenous people hunt reindeer only for their personal or community needs, but as owners of reindeer harvest sites at river crossings they would have a basis for developing a commercial harvest. Some large industrial companies have indicated a readiness to support commercial harvest of reindeer by indigenous people by assisting in the transportation of harvested reindeer to cities and mining settlements. Already, there are plans to open some of the more accessible river crossings for hunting by people from nearby towns and this will include personal use as well as commercial sale of the harvested reindeer. However, there is a need for development of regulations to prevent excessive harvesting of the reindeer and associated alteration of their migration routes. The inability in the past to predict the availability over extended periods of time of wild reindeer for human harvest because of their natural long-term population fluctuations led many indigenous peoples in the Arctic to include more than one ecologically distinct resource (e.g., reindeer and fish)

Fig. 11.3. Harvesting by indigenous people of wild reindeer in the Russian North and caribou in North America was traditionally done at river crossings on migration routes. This continues to be an efficient method of hunting reindeer and caribou in some regions, a hunting system that lends itself to managed control of the harvest.

as their primary food base. Similarly, a balance between harvest of reindeer for local consumption and commercial sale in communities in the Russian North would appear to offer greater flexibility for management of the reindeer and sustainability of local economies than large-scale commercial harvesting of reindeer. Flexibility in options for management of wild reindeer will be essential in the Arctic of the future that is expected to experience unpredictable and regionally variable ecological consequences of climate change. Increased adaptability of the arctic residents to climate change will be best achieved through dependence on a diverse resource base. This applies to the monetary and subsistence economies of arctic residents, as well as to the species of wildlife tar-

geted for management, if wildlife is to remain an essential base for community sustainability.

Changes have occurred over time in methods and patterns of harvesting wild reindeer in the Russian North and these changes provide perspective on wildlife management in a changing climate. Since prehistoric times indigenous peoples throughout Eurasia and North America have hunted wild reindeer and caribou during their autumn migration at traditional river crossings. Boats were used to intercept the swimming animals where they were killed with spears (Fig. 11.3). This method of harvesting wild reindeer may offer potential for management of wild reindeer under the

Table 11.2. Comparison of wildlife harvest systems in the Russian North.

Harvest system	Wildlife population trends	Threats to wildlife and their habitats	Conservation efforts
Kola Peninsula			
Hunting for subsistence and for local market sales	Over-harvest of ungulates, drastic decline in wild reindeer	Over-harvest of ungulates by military and for subsistence, fracturing of habitats by roads and railroads, habitat degradation from industrial pollution	Laplandsky Reserve (1930) 2784 km². Pasvik Reserve (1992) 146 km² (International, with Norway's Oevre Pasvik Park 66.6 km²)
Nenetsky Okrug, Yamal, Gydan			
Intensive reindeer husbandry, control of large predators, incidental subsistence hunting, Arctic fox trapping	Decline in wolves, wolverines, and foxes	Over-grazing by reindeer, habitat damage by massive petroleum development with roads and pipelines, hunting by workers, control of predators	Nenetsky Reserve (1997) 3134 km² (near Pechora delta – waterfowl and marine mammals)
Khanty-Mansiysky Okrug			
Hunting focus on wild reindeer, moose, and fur-bearers; indigenous hunting culture in decline	Low hunting pressure, populations stable	Industrial development, forest and habitat destruction, fragmentation by roads and pipelines, pollution from pipeline leaks	Reserves: Malaya Sosva 2256 km², Gydansky 8782 km², Yugansky 6487 km², Verkhne-Tazovsky 6133 km²
Taymir			
Hunting focus on wild reindeer and waterfowl, mostly subsistence, commercial harvest of velvet antlers at river crossings, restrictions limiting commercial antler harvest being enforced	Decline or extirpation of wild reindeer subpopulations near Norilsk, inadequate survey methods	Wild reindeer total counts are basis for management; lack of knowledge of identity and status of discrete herds; extensive habitat loss from industrial pollution; habitat fracturing and obstructed movements by roads, railroad, pipelines, and year-round ship traffic in Yenisey River for metallurgical and diamond mining, and oil and gas production	Reserves: Putoransky 18873 km², Taimyrsky 17819 km², Bolshoy Arctichesky 41692 km²; region-wide ecosystem/community sustainability plan being developed
Evenkiya			
Hunting for subsistence and local markets, primarily moose, wild reindeer, and bear, little trapping effort	Little information, assumed stable	Low human (Evenki) density and poor economy result in little threat at present to wildlife and habitats	Need is low due to remoteness and low population density. No nature reserves
Yakutia (Sakha)			
Hunting primarily for wild reindeer, moose, snow sheep, and fur bearers, heavy commercial harvest as well as for subsistence, decline of reindeer herding increases dependency on subsistence hunting	Heavy harvest of reindeer and snow sheep for market results in population declines, introduced muskox increasing	Diamond mining provides markets for meat leading to over-harvest and non-selective culling, decrease in sea ice restricts seasonal migrations of reindeer on Novosiberski Islands to and from mainland	Ust Lensky Reserve 14330 km². Muskox introduction adds new species to regional biodiversity and ecosystem level adjustments
Chukotka			
Wild reindeer, snow sheep, and marine mammals hunted for subsistence by Chukchi and Yupik people	Increases in wild reindeer, snow sheep, and large predators with decline in reindeer herding, muskoxen on Wrangel Island increasing	Major decline in reindeer herding, movement of Chukchi to the coasts, poor economy, and low extractive resource potential results in greatly reduced threats to wildlife inland from the coasts, increased pressure on marine mammals for subsistence	Reserves: Wrangel Island 22256 km², Magadansky 8838 km², Beringia International Park – proposed but little political support

Box 11.3. River crossings as focal points for wild reindeer management in the Russian Arctic

Harvesting wild reindeer at river crossing sites (see Fig. 11.3) has played a significant role in regional economies and the associated hunting cultures in the Russian North (Khlobystin, 1996). Many crossing sites were the private possession of families (Popov, 1948). When reindeer changed crossing points it sometimes led to severe famine, and entire settlements vanished (Argentov, 1857; Vdovin, 1965). Such changes in use of migration routes are thought to result from fluctuations in herd size and interannual climate variability. Under the Soviet government, large-scale commercial hunting at river crossings displaced indigenous hunters.

Importance of river crossings for wild reindeer harvest

On the Kola Peninsula and in western Siberia there are few known locations for hunting reindeer at river crossings. In Chukotka, a well-known place for hunting reindeer was located on the Anadyr River at the confluence with Tahnarurer River. In autumn, reindeer migrated from the tundra to the mountain taiga and hunters waited for them on the southern bank of the Anadyr River. Reindeer often select different routes when migrating from the summering grounds. Indigenous communities traditionally arranged for reconnaissance to try to predict the migration routes. In Chukotka, mass killing sites at river crossings were known only in the tundra and forest–tundra, not in the taiga (Argentov, 1857). In Yakutia, reindeer spend summers on the Lena Delta where forage is abundant and cool winds, and the associated absence of harassment by insects, provide favorable conditions for reindeer. In August–September, as the reindeer migrate southwestward, hunters wait and watch for them on the slightly elevated western bank of the Olenekskaya Protoka channel of the Lena Delta where the reindeer traditionally swim across the channel. In the Taymir, 24 sites for hunting reindeer by indigenous people were located along the Pyasina River and its tributaries (Popov, 1948). The killing sites at river crossings occupy fairly long sections of the river. In more recent times when commercial slaughtering occurred, hunter teams occupied sections 10 to 20 kilometers long along the river and used observers to signal one another by radio about approaching reindeer; motor boats carrying the hunters then moved to points on the river where hunting could take place (Sarkin, 1977). In the more distant past, hunters used canoes and needed to be more precise in determining sites and times of the reindeer crossing. Reindeer are very vulnerable in water, and although their speed in water is about 5.5 km/hr (Michurin, 1965) humans in light boats could overtake the animals. In modern times, using motorboats and rifles, hunters were able to kill up to 70% of the animals attempting to cross the rivers at specific sites. A special effort was made to avoid killing the first reindeer entering the water among groups approaching the river crossings. Experience showed that if the leading animals were shot or disturbed those following would be deflected from the crossing. Conversely, if the leading animals were allowed to cross, following animals continued to cross despite disturbance by hunting activities (Savel'ev, 1977).

recent drastic changes that have taken place in social and economic conditions among the indigenous peoples of the Russian North resulting from the dissolution of the Soviet Union. Can management of wild reindeer through harvesting primarily at river crossings ensure sustainable harvests from the large migratory herds under conditions of human social and economic change compounded by the effects of climate change on the reindeer and their habitats? Addressing this question may be possible by comparing the population dynamics of reindeer and caribou herds in regions of the Arctic with differing climate change trends (Post and Forchhammer, 2002; Human Role in Reindeer/Caribou Systems project, see www.rangifer.net).

11.3.2. The Canadian North

11.3.2.1. Historical conditions and present status

In comparison to ecosystems at lower latitudes in Canada most ecosystems in the Canadian Arctic are considered functionally intact, although the consequences for marine ecosystems of contaminants introduced from industrial activity to the south and climate-induced thawing are not

known. Most threats typical for elsewhere in the world – such as habitat loss through agriculture, industry, and urbanization – are localized. Introduced species primarily associated with agriculture at lower latitudes are scarce, or largely confined to areas near communities. Invasive wildlife species from the south, such as moose and snowshoe hares, are primarily restricted to the tundra–forest interface. Within most arctic ecosystems, resource use through hunting is the most conspicuous influence that people have on wildlife with the exception of localized resource extraction and expanding tourism. Among the factors that can influence arctic wildlife, hunting is potentially the most manageable and its quantitative assessment needed for management is feasible. Although hunting is not currently considered a threat to terrestrial wildlife in the Canadian Arctic, it has recently interacted with other factors such as weather to locally reduce caribou abundance on, for example, some arctic islands (Gunn et al., 2000). Managed hunting is considered an important part of wildlife conservation through its emphasis on sustainability of harvest. Hunting, however, poses a threat when it causes or contributes to undesired declines or through interaction with other species with detrimental consequences. The latter is especially rele-

Commercial harvest at river crossings

During the Soviet period, large-scale commercial harvest of reindeer at river crossings displaced indigenous hunters from these traditional hunting sites (Sarkin, 1977; Zabrodin and Pavlov, 1983). In Yakutia, after commercial hunting began in the 1970s, hunting techniques included the use of electric shocks to kill reindeer as they came out of the water. In recent years these commercially harvested reindeer populations in Yakutia declined precipitously (Safronov et al., 1999). In the Taymir, indigenous people practiced subsistence hunting at river crossings until the 1960s. However, by 1970, hunting regulations had banned hunting at river crossings by indigenous people and other local residents because of concern that over-harvest of the reindeer would occur. The Taymir reindeer increased greatly in the following years. Biologists working with the reindeer proposed reinstatement of the traditional method of killing animals at river crossings in order to establish a commercial harvest from the large Taymir population and to stabilize the population in line with the carrying capacity of the available habitat. The Taymir state game husbandry system was established by 1970. Up to 500 hunters participated in the annual harvests. All appropriate river hunting locations on the Pyasina River and the Dudypta, Agapa, and Pura tributaries were taken over for the commercial harvests. Large helicopters and in some cases refrigerated river barges were used to transport reindeer carcasses to markets in communities associated with the Norilsk industrial complex. Over a period of 25 years about 1.5 million reindeer were harvested by this system (Pavlov et al., 1993). After 1992, there was a decrease in the number of reindeer arriving at most of these river crossings, resulting in an abrupt decline in the harvest from about 90000 per year in peak years to about 15000 per year in subsequent years. This was associated with the disproportionate harvest of female reindeer (Klein and Kolpashchikov, 1991).

Consequences of climate change

Climate change may affect river crossings as sites for controlled harvest of reindeer in several ways. If patterns of use of summering areas change in relation to climate-induced changes in plant community structure and plant phenology then migratory routes between summer and winter ranges may also change. Thus, some traditional crossings may be abandoned and new crossings established. Changes in the timing of freeze-up of the rivers in autumn at crossing sites may interfere with successful crossings by the reindeer if the ice that is forming will not support the reindeer attempting to cross. These conditions have occurred infrequently in the past in association with aberrant weather patterns; however timing of migratory movements would also be expected to change with a consistent directional trend mirroring seasonal events.

vant in marine systems where knowledge of ecosystem relationships and processes are less well understood than they are for terrestrial systems. Hunting remains inextricably part of the long relationship between indigenous people of the Arctic and their environment, and they see themselves as part of the arctic ecosystems within which they dwell (Berkes and Folke, 1998).

Fluctuations in caribou numbers over decades in the Canadian Arctic have been a frequently reiterated observation in indigenous knowledge (e.g., Ferguson and Messier, 1997), and this parallels archaeological evidence from western Greenland (Meldgaard, 1986). The increased hunting that followed European colonization, with the introduction of firearms and commercial hunting, accentuated or over-rode natural fluctuations in caribou numbers and contributed to the so-called caribou crisis of low numbers between 1949 and 1955 (Kelsall, 1968). Subsequently, the herds of barren-ground caribou increased five-fold. The number of caribou on the mainland tundra in four of the largest herds (Bathurst, Beverly, Qamanirjuaq, and Bluenose) was estimated at 1.4 million in the mid-1990s and numbers are believed to be remaining relatively stable.

Historically, muskoxen (*Ovibos moschatus*) were sufficiently numerous to be an important part of the indigenous culture on the mid-arctic islands, but were less so on the mainland until a brief pulse in commercial hunting for hides in the late 1800s and early 1900s (Barr, 1991). However, sharp declines in muskox numbers on the Northwest Territories (NWT) mainland followed unregulated commercial trade in muskox hides. Muskox numbers quickly collapsed and within 30 years only a handful of scattered herds remained on the mainland. Muskox hunting was banned between 1917 and 1967, after which populations had started to recover by the 1970s when subsistence hunting was resumed under quotas. Numbers of muskoxen in the NWT and Nunavut have been recently estimated at about 100000 on the arctic islands and about 20000 on the mainland (Gunn and Fournier, 1998).

Hunting was not the cause of all known historic wildlife declines – muskoxen virtually disappeared from Banks and western Victoria Islands in the late 1800s, before European influences. Inuvialuit elders have memory from their youth of an icing storm that encased vegetation in ice and many muskoxen died on Banks Island (Gunn et al., 1991). Muskox numbers rebounded on Banks Island from a few hundred to 3000 by 1972 and to 64000 by 2001 (Nagy et al., 1996; J. Nagy pers. comm., 2001).

The number of polar bears killed by hunters increased with European exploration and trading in the Canadian Arctic. Hunting for hides was not significant until the 1950s when prices climbed in response to market demands. Snow-machines were becoming available in the 1960s, leading to increased hunting and stimulating international concern over sustainability of the polar bear harvest. In 1968, regulations imposed quotas to reduce hunting of polar bears. Canada has about 14800 polar bears of the entire arctic population of 25000 to 30000 bears (IUCN Polar Bear Specialist Group, 1998).

11.3.2.2. Present wildlife management arrangements and co-management

The federal and territorial governments responded to the wildlife declines in the NWT during the first half of the 20th century with well-meaning but mostly poorly explained regulations that restricted hunting. These regulations largely ignored local knowledge and emphasized hunting as a threat, which alienated indigenous hunters and left them feeling bitter. Those feelings still influence discussions about hunting, although changes in management practices as a result of establishing new management regimes in recent years may be reducing mistrust (Kruse et al., 2004; Richard and Pyke, 1993; Usher, 1995).

Co-management is a type of regime that has emerged in response to such conditions of conflict and mistrust to shift power and responsibility to boards comprising wildlife users, as well as government representatives. Co-management agreements establish boards of user representatives and agency managers, and typically have authority for wildlife management subject to conservation, public safety, and public health interests. Although overall authority for management is vested in the appropriate government ministry and/or indigenous governing organization, co-management boards make day-to-day decisions on wildlife and are valuable in assessing problems, achieving regional consensus, and making recommendations to user communities, management agencies, and government policy-makers. Co-management potentially helps to ensure that indigenous ecological knowledge is included in wildlife management, although there is debate over its effectiveness in this regard (Usher, 1995). Under land claims legislation, the territorial government determines a total allowable harvest using species-specific methods and recommends to the boards the allowable harvest for species that are regulated. If the total allowable harvest exceeds the basic needs levels, then the surplus can be allocated to non-beneficiaries or for commercial wildlife harvest, including sale of meat and guided hunts for non-resident sport/trophy hunters.

The NWT and Nunavut territorial governments use a variety of methods for determining allowable harvest. Differences in methodology are a complex of practicali-

ty, species life history, and management history. For caribou and muskox harvest management, pragmatic flexibility often takes precedence over application of theory (Caughley, 1977; Milner-Gulland and Mace, 1998). Aerial surveys are used to track caribou and muskox population trends. For barren-ground caribou, the survey findings have not been used to limit subsistence hunting, although they have been used to set quotas for commercial use. In a few instances, communities voluntarily took action to reduce hunting on some arctic islands, based on hunter reports of decline in caribou numbers. In contrast to caribou, muskoxen are hunted under an annual quota based on a 3 to 5% harvest of the total muskoxen estimated within the management unit. The local community decides whether the quota is for subsistence or commercial use.

Managing polar bears has taken a different direction from managing caribou and muskoxen, at least partly because tracking polar bear abundance is logistically difficult and prohibitively expensive. The total allowable harvest is based on modeling the maximum number of female bears that can be taken without causing a population decline (Taylor et al., 1987). The flexible quota system, allowing sex-selective hunting, assumes that the sustainable annual harvest of adult females (greater than two years of age) is 1.6% of the estimated population, and that males can be harvested at twice that rate. Within the total annual quota, each community is allocated a maximum number of males and females. If the quota of females killed is exceeded, the total quota for the subsequent year is reduced by the exceeded amount. During the period 1995–1996 to 1999–2000 the average annual harvest of polar bears in Canadian territories, combined with harvest statistics reported in Alaska and Greenland, was 623 animals while the sustainable harvest estimate was 608 (Lunn et al., 2002). Communities and territorial governments developed and jointly signed Local Management Agreements in the mid-1990s that provide background, provide for use of both scientific and traditional knowledge, and provide the procedure for estimating population size and establishing the annual harvest quota.

Progress has also been made in developing co-management for other marine mammals, notably the small whales in the eastern and western Canadian Arctic. Conservation and management of the beluga whale (*Delphinapterus leucas*) in Alaska and the NWT is through the Alaskan and Inuvialuit Beluga Whale Committee, which includes representatives from communities and governments as well as technical advisors (Adams et al., 1993). However, only representatives from beluga hunting communities vote on hunting issues. In the eastern Arctic less progress has been made toward co-management for narwhal (*Monodon monoceros*) partly because of a failure to involve fully the Inuit hunters (Richard and Pike, 1993). Advisory and co-management boards and agreements are not necessarily a guarantee of widespread hunter support (Usher, 1995). Klein et al. (1999) compared caribou management under the Beverly–Qamanirjuaq Caribou

Management Board with management of the Western Arctic Caribou Herd in Alaska through a statewide Board of Game. They concluded that information was not flowing effectively from user representatives on the co-management board to the user communities, thus the users did not feel as involved in management of the caribou as in Alaska where regionally based biologists collecting data for management had more interaction with the users.

How do co-management arrangements help to meet the goals of sustainability in conditions of climate change? Experience with Canadian co-management arrangements demonstrates that these systems can be critical tools for tracking the trends in climate change, reducing human vulnerabilities, and facilitating optimal human adaptation to impacts in single-species management. Trust relations growing from formal co-management arrangements also provide conditions from which innovative ecological monitoring and research involving local/traditional knowledge and science add to the system's capacity to cope with change. In short, a focus on biological aspects of wildlife management should be complemented with institutional considerations to understand their full effectiveness in addressing the possible impacts of climate change.

Co-management is defined both with respect to institutional features of an arrangement (Osherenko, 1988) as well as by outcome of sharing of decision-making authority by local communities of resource users and agencies in the management of common pool resources (Pinkerton, 1989). Power-sharing arrangements can emerge through informal relations between parties (e.g., regional biologists and local hunters), as a result of formal agreements, or, as is most common, from a combination of *de jure* and *de facto* relations. Structures for co-management of wildlife therefore differ from conventional state resource management systems in which decision-making is bureaucratically organized and driven primarily by the principles of scientific management. As well, co-management differs from local control in which a resource user community pursues self-determination, largely independent of external parties. In practice, these arrangements result in considerable latitude in the range of authority and responsibility exercised by resource users (Berkes, 1989).

In the Canadian Arctic, formal co-management has become a common feature of the political landscape either through constitutionally entrenched land-claims agreements or as stand-alone arrangements. Implementation is typically directed through boards of users and agency representatives that are advisory to government ministers, agencies, local communities, and various indigenous governance bodies. In most cases, co-management agreements have been struck to specify community rights to hunting and provide a meaningful role for indigenous subsistence users in management decision-making. In several cases they have proven critical in achieving compliance when facing scarcity of resource stocks (e.g., Peary Caribou (*Rangifer tarandus*

pearyi) of Banks Island and co-management system of the Inuvialuit Final Agreement).

What is the significance of co-management to sustainability? Meeting the goals of sustainability requires that resource managers, local communities, and other parties cooperate in resource management. These management functions typically include ecological monitoring and impact assessment, research, communication between parties, policy-making, and enforcement. As a part of this process, there is a need for adequate and integrated knowledge at multiple scales of population regulators, habitat relationships, and potential impacts of human activity, including harvesting, on the population (Berkes, 2002; Berkes and Folke, 1998).

A case study of the Canadian co-management of the Porcupine Caribou Herd, toward sustainability under conditions of climate change, is given in the Appendix.

11.3.2.3. Hunting as a threat to wildlife conservation

Hunting can become a threat to wildlife conservation if population size changes unpredictably in response to environmental perturbations or density dependent changes (unless the population size is closely monitored and hunting is adjusted quickly). Most large mammals in the Arctic are relatively long-lived and thus somewhat resilient to interannual environmental variability that may result in loss of a single age class through breeding failure or heavy mortality of young animals. However, extreme conditions such as icing of vegetation or deep snows restricting access to forage may result in near total mortality across age classes (Miller, 1990) or rarely, regional extirpation of populations or subspecies (Vibe, 1967). Muskoxen are large-bodied grazers capable of using low quality forage during winter and with a predominantly conservative lifestyle. Thus, they are adapted to buffering some of the consequences of variable weather and forage supplies (Adamczewski, 1995; Klein, 1992; Klein and Bay, 1994). Caribou, in their much greater range of latitudinal distribution (muskoxen are rarely found in the boreal forests) are less strongly coupled as a species by feedback loops to their forage (Jefferies et al., 1992). However, their more energetic life style, associated with their morphology and behavior, predisposes them to feeding selectively for high quality forage, necessitating extensive movements and often long seasonal migrations between the barren grounds and the boreal forests (Klein, 1992). Long migrations may be an evolutionary strategy that buffers localized variables in forage quality and availability, which may be weather-related. Icing of vegetation in winter and fires on winter ranges in summer are examples of these weather-related influences on winter forage availability. Caribou are vulnerable to other aspects of weather that affect quality and availability of forage on calving grounds, the level of insect harassment and parasitism, and in the Canadian Arctic Archipelago, freedom of inter-island movement. In the northernmost arctic islands, environmental variability becomes more significant as many processes are near their limits of variability, such as plant growth, which plays a large role in determining herbivore reproduction and survival. Consequently, annual variation in population attributes such as pregnancy rates and calf survival is high. For example, Thomas (1982) documented annual pregnancy rates of between 0 and 80% for Peary caribou and the range in calf production and survival between 1982 and 1998 was 23 to 76 calves per 100 cows for caribou on Banks Island (Larter and Nagy, 1999). The amount of environmental variability may exceed the capability of large mammals to buffer changes and lead to unexpected surges in recruitment or mortality. Rate of population change and size will be more unpredictable and thus hunting will be at more risk of being out of phase with the population trend. Changes in caribou numbers on Banks Island is an example of hunting accelerating a decline likely to have already been underway in response to an environmental change (severe snow winters). Caribou declined from 11 000 in 1972 to perhaps less than 1000 (Nagy et al., 1996; J. Nagy pers. comm., 2001).

North of Banks Island is the range of the Peary caribou, which are only found on Canada's high-arctic islands. Trends in Peary caribou numbers are only available from

Fig. 11.4. Throughout the Arctic, traditional modes of transport (a) have been largely replaced by mechanized all-terrain vehicles (b) that permit people in many regions of the Arctic to range more widely for subsistence hunting. While this spreads wildlife harvest over greater areas it also requires more extensive survey of the status of wildlife populations as a basis for wildlife management (photo: D.R. Klein).

the western high-arctic islands where numbers have fluctuated within a long-term decline from 26 000 in 1961 to 1000 by 1997 (Gunn et al., 2000). In 1991, the Committee on the Status of Endangered Wildlife in Canada classified caribou on the high-arctic and Banks islands as Endangered based on the steep population declines during the 1970s and 1980s. This was believed to have been caused by climatic extremes – warmer than usual autumn storms causing dense snow and icing, which limit access to forage (Miller, 1990).

Institutional circumstances that may lead to wildlife vulnerability to hunting start with limitations in the ability to detect population declines. Detecting declines in caribou or muskox numbers partly depends on recognizing trends in population size (Graf and Case, 1989; Heard, 1985). The aim is to conduct regular surveys, but high costs and large survey areas have increased survey intervals to the extent that population changes have been missed. For example, the inter-island caribou population of Prince of Wales and Somerset Islands was considered to be relatively stable between 1974 and 1980 (estimated at 5000 caribou in 1980). In the early 1990s, Inuit hunters reported seeing fewer caribou on those two islands, which triggered a survey, but not until 1995. The survey revealed that caribou had declined to less than 100 (Gunn et al., 2000).

Problems with detecting population declines are not just technical. Hunters frequently distrust survey techniques and disbelieve the results, especially when declines in caribou are reported (Klein et al., 1999), but the same may be true for muskoxen and hunted whales (Richard and Pike, 1993). Disbelief stems from historical relationships that have involved poor communication, as well as cultural differences in relying on abstract concepts and numbers as opposed to personal observation. Further differences arise over interpretation of factors causing declines – for example, whether caribou have moved away from the survey area or whether numbers declined because deaths exceeded births (Freeman, 1975; Miller and Gunn, 1978). However, merging information derived from scientific investigation and existing weather records with information gleaned from indigenous hunters is increasingly employed as a tool in monitoring wildlife population response to climate change (Ferguson and Messier, 1997; Kofinas, 2002).

Socio-economic factors can affect the vulnerability of wildlife to hunting. The two territories of NWT and Nunavut have been described as having a "Fourth World" economy (Weissling, 1989) with the indigenous population often forming enclaves within the larger communities that are economically dominated by the North American society. The growing human population in the north, nevertheless, remains heavily dependent on hunting and fishing (Bureau of Statistics, 1996). At present, wage earning provides the cash needed for the purchase and operation of equipment and supplies necessary for hunting and fishing, which have become highly dependent on mechanized transport (Wenzel, 1995) (Fig. 11.4), which in turn creates the need for at least part-time work. However, wage-earning opportunities are relatively limited, shifting the emphasis to commercial use of wildlife and fisheries, but the distinction between subsistence and commercial use is by no means simple. In West Greenland, for example, small-scale sales of minke whales (*Balaenoptera acutorostrata*) and fin whales (*B. physalus*) were considered necessary to maintain cash flow to purchase supplies for subsistence hunting (Caulfield, 1993). But managing for commercial use that is not focused on maximizing profits is inconsistent with systems for management of commercial harvest. Clark (1976) explained the economic rationale for the ease with which commercial harvesting can lead to over-harvesting, especially for long-lived species with low rates of reproduction.

Finally, a mixture of concern and defensiveness exists in response to "outside" (i.e., southern Canada and elsewhere) views or opinions about wildlife harvest and management. In a workshop on future action over the endangered Peary caribou, this was recognized as a serious issue (Gunn et al., 1998), especially in the context of allowing caribou hunting while considering reduction of wolf predation through translocations or other predator control methods. Response to "outside" opinions stems partly from previous experience with some organized animal rights activists and some who see hunting as a threat to animal welfare or conservation. Indigenous hunters, who view their dependency on local resources as sustainable in contrast to the heavy dependency by southern urban dwellers on nonrenewable resources, perceive such urban-based organizations as a threat to their way of life. This view has proven to be the case, for example in the movement against seal hunting that led to the European Common Market's ban on seal skins, which resulted in a substantial loss of income from seal-skins in some Inuit communities (Wenzel, 1995).

11.3.2.4. Additional threats to wildlife conservation

The risk that hunting can become unsustainable and cause or contribute to population declines may lie in the unexpected (Holling, 1986). The unexpected ranges from shortcomings in data collection or predictive models, to environmental changes accumulating in unanticipated ways not encompassed by traditional knowledge. Within this context, this includes threats to wildlife from outside the Arctic, such as atmospheric transfer of contaminants and climate change, even if there is uncertainty as to how those threats may unfold in practice. However, management of use of wildlife and associated conservation of wildlife is most difficult in the absence of available methods to monitor both the harvest levels and the status of the populations that are harvested.

Global climate change and the atmospheric transport of contaminants are factors that are already affecting some arctic populations. Global warming in the near future is projected to trigger a cascade of effects (Oechel et al., 1997). Evidence consistent with projections of global climate change in the western Arctic includes Inuvialuit reports of ecological changes such as the appearance of previously unknown birds and insects following trends of warmer weather (IISD, 1999). Along the mainland central arctic coast, Inuit are expressing concerns for the deaths of caribou crossing sea ice as freeze-up is later and break-up earlier than before (Thorpe, 2000).

Sustainability of wildlife for hunting can be affected by influences of climate change on the hunted populations. For example, an increased difficulty in finding winter forage is likely for caribou on the western arctic islands if warmer temperatures bring a greater frequency of freezing rain and deeper snow. Annual snowfall for the western high Arctic increased during the 1990s and the three heaviest snowfall winters coincided with Peary caribou numbers on Bathurst Island dropping from 3000 to an estimated 75 caribou between 1994 and 1997. Muskoxen declined by 80% during the same three winters (Gunn et al., 2000).

Atmospheric and aquatic transport of contaminants has resulted in contaminants reaching detectable levels in arctic wildlife (AMAP, 1997, 2002; Elkin and Bethke, 1995), although effects on population ecology are poorly understood. Although many contaminants that may be detrimental to living organisms are of anthropogenic origin, many derive from natural sources. Persistent organochlorine compounds are carried in the atmosphere, but cadmium is almost entirely from natural sources and mercury is from ocean degassing, natural breakdown, and atmospheric and anthropogenic sources (AMAP, 1997). Bioaccumulation of contaminants can reach levels in marine mammals that pose threats to humans who consume them, especially pregnant and lactating women and their infants (see Chapter 15).

If global warming imposes increased environmental stress on wildlife it is likely to interact with contaminants. For example polar bears, at the top of the marine food chain, accumulate contaminants by eating ringed seals (*Phoca hispida*) and other marine mammals. Relatively high levels of organochlorine compounds and metals are found in polar bears, with relatively strong regional patterns (AMAP, 1997). In female polar bears, although the existing body levels of organochlorine compounds may be sequestered effectively when fat reserves are high, the sequestration away from physiological pathways may be inadequate during a poor feeding season (AMAP, 1997; Polischuk et al., 1994). On western Hudson Bay, there is a trend for female bears to have less fat reserves as sea ice break-up occurs progressively earlier, forcing them ashore where they are required to fast for increasingly longer periods (Stirling et al., 1999). How contaminants in marine systems may change with a changing climate, and what may be the consequences for wildlife and the humans who consume wildlife is not understood, yet an understanding of the nature of the threats posed by contaminants in arctic systems and the processes and pathways involved is critical for the management and conservation of arctic wildlife.

11.3.3. The Fennoscandian North

11.3.3.1. Management and conservation of wildlife under change

In the boreal forest and mountainous areas of northern Fennoscandia the major hunted wildlife species are moose, grouse, dabbling ducks and some diving ducks, and bean geese (*Anser fabalis*). There is increased interest, largely among urban dwellers, to conserve large carnivores. These predatory species are now recovering from high hunting pressures during past decades by farmers and reindeer herders in defense of their livestock. Nevertheless, there have been centuries-long habitat changes in the Fennoscandian Arctic brought about by human activities, including community development and expansion, road and other transportation corridor construction, hydropower development, mining, tourism development, forest clearing, and establishment of military training or test sites (Fig. 11.5). This has resulted in substantial reduction of available habitat for wildlife as well as fragmentation of existing habitats. The consequences for wildlife have been limitations on the freedom of seasonal movements of wildlife, as well as restricted dispersal, and associated genetic exchange, fragmentation of wildlife populations, and lowered overall productivity of the land and waters of northern Fennoscandia for wildlife.

In Norway and Sweden, wolves were completely exterminated during the mid-20th century. Animals from Finland/Russia have recently recolonized the southern, forested part of the peninsula. Bears (*Ursus arctos*) were exterminated in Norway, except for a small population on the border with Russia and Finland. Recovery of bears by dispersing animals from Sweden has occurred in some border areas farther to the south. Decisions have been made that determine areas in which these predators will be tolerated and areas where they will be excluded, largely on the basis of the presence of freely ranging domestic livestock and Saami reindeer. In the exclusion zones in Norway, targeted hunts are held to

kill individual large carnivores or groups of them regardless of the status of the species. No wolves have been permitted to reestablish in the Saami reindeer herding areas, which lie north of approximately 63° N.

The climate record and outputs from climate models (Chapter 2 and 4) indicate little change in temperature patterns in northern Fennoscandia in recent decades, in contrast to other parts of the Arctic. Similarly, models projecting future climate trends in the Arctic suggest slow rates of warming in Fennoscandia. An exception is the north coastal region of Norway where models project substantial increases in winter temperature and precipitation. The effects of global warming in the region include ablation of mountain glaciers, altitudinal advances in the treeline, increases in magnitude of defoliating insect outbreaks, and, possibly, a decline in the frequency and magnitude of small mammal population cycles (see Chapter 7). Thus far, there has been little serious research effort focused directly on how changing temperature and precipitation will influence wildlife populations in Fennoscandia.

11.3.3.2. Hunting systems

In general, the moose hunt is based on licenses issued by the regional governments to hunting teams. Each license allocates the number of moose to be harvested from the specific land area for which the license is issued, whether it is private or government owned land. The hunting quota is based on population estimates derived from hunter observations and aerial surveys, including assessment of sex and age composition, but consideration is given to the number of traffic accidents and damage done by moose to forest stands. The timing and length of moose hunting seasons vary within and between countries.

Large carnivore populations are estimated through observations incidental to surveys of other wildlife, local or regional field studies of carnivore species and their prey relationships, and other techniques. Hunting quotas and conservation measures are based on population esti-

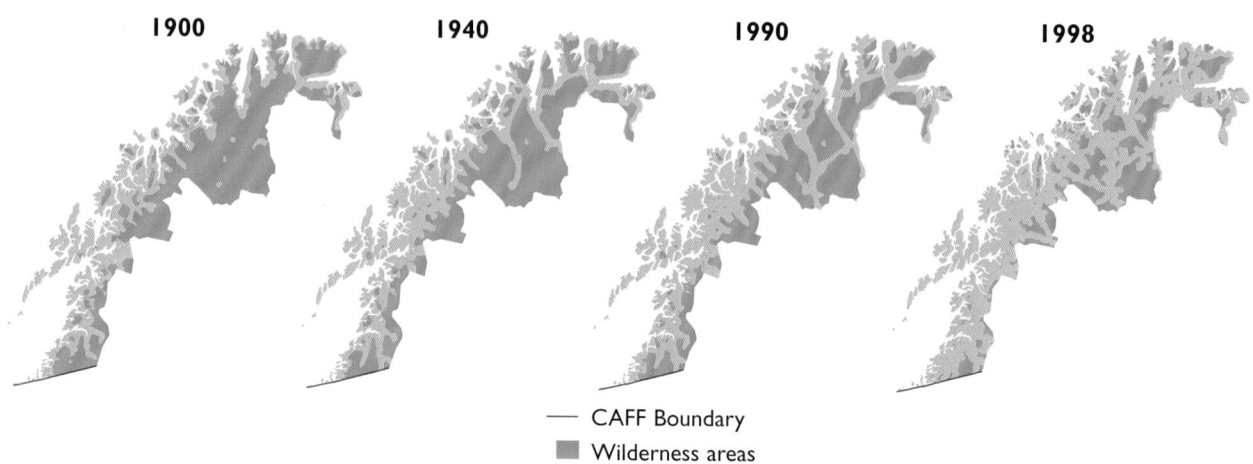

Fig. 11.5. Natural habitat fragmentation in northern Norway is exemplified by the decrease in wilderness areas in Norway north of the CAFF boundary since 1900. Wilderness is defined as an area lying more than five kilometers from roads, railways, and regulated water-courses (Norwegian Mapping Authority as quoted in CAFF, 2001a).

mates, reproductive rates, and levels of predation on reindeer, sheep, and other domestic animals.

The hunting system for ptarmigan and grouse rests primarily on setting of the hunting season dates, which traditionally fall between late August and mid-February. In some areas there is a bag limit, often based on local monitoring programs. Grouse hunting in mountain areas is currently undergoing discussion and the different hunting systems are under evaluation from both the biological and hunters' perspectives.

Wildlife management for hunter harvest of ducks is based primarily on setting the start and duration of the hunting season within the period from late August through late November. Some areas are closed to hunting, including areas around villages.

11.3.3.3. Monitoring systems

In the Fennoscandian countries there is a strong tradition for hunters to report the number of animals killed, and hunters voluntarily assist in wildlife surveys. This is a valuable aid to wildlife management in Finland, Sweden, and Norway and efforts continue to improve the hunter reporting system to ensure greater reliability of the information obtained. Systems for monitoring the population status of moose and large carnivores are among the most highly developed, whereas the least developed system is for ducks, with systems for monitoring ptarmigan and grouse populations intermediate. There is a concern in some areas of the Arctic that these hunter-based systems will be less effective because many young hunters who were born and raised in the rural areas of the North, and having familiarity with the specific wildlife habitats and wildlife of their region, are moving to urban areas to seek employment. Consequently, the number of hunters living close to the land in the Fennoscandian Arctic is decreasing while those from urban centers outside the region are increasing.

11.3.3.4. Flexibility of hunting systems under climate change

With increasing temperatures, in concert with other long-term changes, such as wetland eutrophication, populations of some waterfowl species, for example whistling swans (*Cygnus columbianus*), eider ducks (*Somateria* spp.), and greylag geese (*Anser anser*), are expected to increase in size and to expand their distribution. Consequently, there will be demand for hunting opportunities on these species in areas where today there is no hunting. The procedure for establishing hunting regulations under the present system should be adaptable to allow changes in hunter harvest levels to ensure optimal sustainable harvest through hunting of these waterfowl species. Restrictions on hunting have also allowed recovery of species such as common eider (*Somateria mollissima*) and barnacle goose (*Branta leucopsis*) that nest in the high Arctic, to the point where it may be justified to reconsider opening hunting seasons on them.

Adjustments in moose hunting in response to moose population changes can be achieved through flexibility in establishment of hunting quotas. However, some difficulties can be foreseen. For example, if temperatures during the early part of the hunting season are high there may be difficulties preserving the meat in the field without access to cold storage rooms. This may limit hunting to periods of suitable weather before snow accumulation. This might make it difficult for small hunting teams to fill their quotas. If snow arrives early in the autumn/ early winter, access to the hunting grounds may be limited due to difficulties for vehicle travel on logging roads. For the large carnivores, there is similar flexibility in the establishment of hunting quotas.

For grouse and ducks, discussions on hunting regulations mainly concern timing of the hunting season. If the season starts too early the birds are still unfledged and considered too small to hunt. If the hunting season starts too late in the North migratory birds may have already moved south.

Possibilities exist to adjust hunting and the associated management systems in the Fennoscandian North to changes in wildlife populations that may result from the effects of climate change. However, social and economic factors that relate to the various interests in wildlife by local residents and those who come from outside the region also need to be considered in developing wildlife management plans. Management of wildlife in the Fennoscandian Arctic under conditions of a changing climate must be "adaptive" and thus capable of responding to changes in ecosystem dynamics that at times may be unpredictable and therefore unanticipated.

There is a need to establish a comprehensive monitoring program for all wildlife species (moose and some of the large carnivores are currently monitored within each country), with monitoring stations spread out over the Fennoscandian countries, and with coordination of these efforts. There is an urgent need for long-term data as a basis for identifying trends, and a similar need to secure information from remote areas. It is important to develop systems that give "early warning". Such procedures have been in development by the Finnish Game and Fisheries Research Institute that stimulate discussions on changes in hunting systems among and between hunters, wildlife biologists, and regional government wildlife consultants/managers. The resulting adjustment of hunting regulations based on a melding of the interests, concerns, experience, and observations of hunters with the expertise and investigative findings of trained wildlife biologists should provide relatively effective tracking of changes in wildlife populations as a consequence of possible changes in climate.

11.3.4. The Alaskan Arctic

The management system for terrestrial wildlife in Alaska that developed following its admission to statehood in 1959 initially followed the institutional structures adopted by most other states. In both the United States and

Canada, wildlife has been considered by law common property of the people and therefore control of its use, management, and conservation has fallen within the jurisdiction of the state or province within which it occurred. This is in contrast to the system throughout most of Europe where wildlife is the property of the landowner. With the settlement of claims of indigenous peoples in the Canadian North, however, varying levels of responsibility for management of wildlife have been granted to regional indigenous governing authorities. The federal governments of the United States and Canada hold jurisdiction over migratory birds and inter-state or inter-province traffic in harvested wildlife.

In Alaska, a Board of Game, comprising residents of the state appointed by the Governor, establishes regulations governing wildlife harvesting. Regulations established by the board are based on recommendations from professional biologists and managers employed by the Alaska Department of Fish and Game in collaboration with biologists of federal land management agencies, as well as on recommendations from regional citizens advisory groups, and the general public, within the constraints of laws passed by the State Legislature governing wildlife conservation and use. Administrative structure for wildlife management in the State of Alaska mirrors that of other states. Actual wildlife management in Alaska, however, now differs markedly from the other states with similar involvement of the public in resource management decision-making. In Alaska, the federal government, primarily through the U.S. Fish and Wildlife Service, assumes a much greater role in regulation of the harvest of wildlife than in other states. This federal participation in the wildlife regulatory process came about through legislation resulting from settlement of the land claims of the indigenous peoples of Alaska and related legislation by the U.S. Congress (the Alaska Native Claims Settlement Act of 1971 and the Alaska National Interest Lands Conservation Act of 1980). These federal laws mandate that rural residents of Alaska, comprised mostly of indigenous peoples, should receive priority over urban and non-resident hunters in harvesting for subsistence use of the annual surplus of fish and wildlife from federal lands. The state's failure to pass similar subsistence priority legislation, consistent with the state constitution, resulted in the loss of management authority by the state for fish and wildlife on federal lands in Alaska. Since federal lands in national forests, wildlife refuges, national parks, military and other federal reserves, and federal public domain lands constitute 60% of the total land area of Alaska (1.48 million km²), the federal role in management and conservation of wildlife in Alaska is unique among the states. This federal–state partnership in management of Alaska's fish and wildlife resources has been both controversial and complex and has contributed to political polarization between urban and rural users of fish and wildlife resources (Klein, 2002). However, in most regions of the Alaskan Arctic sufficiently remote from urban centers there is little competition in the harvest of wildlife between the mainly indigenous, rural population and

urban hunters, although hunting methods and especially means of transport have changed markedly in recent decades (Fig. 11.4).

In spite of the legal complexities involved in managing Alaska's wildlife, state and federal wildlife biologists and managers increasingly are working together with the users toward maintaining sustainable harvests of wildlife, achieving equitable allocation of the harvest among wildlife users, and improving efficiency of the management process. Biologists and managers with the Alaska Department of Fish and Game, involved with management of caribou of the Western Arctic Herd, Alaska's largest caribou herd estimated to contain 490000 animals in 2003, have been instrumental in establishing a Western Arctic Caribou Herd Working Group whose members represent indigenous and non-indigenous hunters, federal land management agencies, state resource management agencies, and environmental organizations. This working group is viewed as a preliminary step in the process of

Fig. 11.6. Management for sustained harvest and conservation of the large herds of caribou and wild reindeer in North America, Greenland, and Eurasia requires periodic aerial monitoring of the populations. Shown here (a) is a survey flight over the Porcupine Caribou Herd shared by the United States and Canada, involving a photo census over summer concentrations, in conjunction with lower-level flights to obtain sex and age composition counts. Placing collars equipped with radio transmitters on some of the animals (b) enables tracking and locating the herds in their seasonal movements, and assessing mortality rates, fidelity to calving grounds, and other indicators of population status (photos: K. Whitten).

Table 11.3. Status and trends in major land-based wildlife species in the Alaskan Arctic (based on data for 2003 from the Alaska Department of Fish and Game).

	Population status[a] (number per estimate)	Trend	Harvest level[b] (number per year)	Threats
Caribou (by herd)				
Western Arctic	430000	down	22000	weather, coal mining[c]
Porcupine Herd	125000	down	4500	oil development
Central Arctic	27000	stable	1100	oil development
Teshekpuk	27000	stable	3000	oil development
Mulchatna	130000	down	6000	disease[d]
Nushagak	1500	stable	300	no immediate threats
Northern Peninsula	7000	stable	500	disease[d]
Southern Peninsula	3500	up	100	no immediate threats
Adak Island	1500	up	200	no immediate threats
Muskox				
North Slope	1000	stable	40	illegal harvest
Seward Peninsula	800	up	50	no immediate threats
Nunivak Island	500	stable	75	no immediate threats
Nelson Island	230	stable	25	illegal harvest
Moose				
North Slope	750	up	30	disease[e]
Selawik/Kobuk/Noatak	10000	stable	400	no immediate threats
Seward Peninsula	5000	stable	350	no immediate threats
Yukon/Kuskokwim	3000	up	200	illegal harvest[f]
Northern Bristol Bay	3500	up	600	no immediate threats
Alaska Peninsula	5000	stable	300	no immediate threats
Brown bear				
North Slope	2000	up	40	no immediate threats
Selawik/Kobuk/Noatak	3000	up	40	no immediate threats
Seward Peninsula	1250	up	75	illegal harvest[g]
Yukon/Kuskokwim	750	up	25	no immediate threats
Northern Bristol Bay	1500	up	75	no immediate threats
Alaska Peninsula	8500	up	300	no immediate threats

Wolf

Wolf numbers in coastal areas of Alaska vary widely from year to year because wolves are susceptible to rabies that is periodically enzootic in Arctic foxes. Wolves are probably more common now than at any time over the last 100 years because of the relatively high numbers of moose and caribou that now occur. Wolf densities are higher in the more forested areas where they also can prey on moose. In some local areas (e.g., the North Slope and Seward Peninsula) wolf numbers are below natural levels due to legal and illegal harvest. There are no foreseeable human-related threats to wolves, except on the Seward Peninsula where reindeer herders attempt to exclude them from reindeer grazing areas

Black bear

Black bears are abundant in the Kobuk Valley, Yukon Flats, and in most other forested areas, but the Alaskan Arctic is the periphery of black bear range, so they are absent or rare from most arctic areas. Numbers of black bears will probably increase as forest cover expands in northwest Alaska. There are no foreseeable human-related threats to black bears

Wolverine

Wolverines are common throughout the Alaskan Arctic, and with the worldwide decline in fur prices, interest in harvesting them has decreased. Wolves commonly kill wolverines and wolverine densities appear to be higher in areas where wolf numbers are low. There are no foreseeable human-related threats to wolverines

Lynx

Lynx are cyclic or irruptive in the Alaskan Arctic, and in areas where snowshoe hares become periodically abundant, lynx can become abundant. Lynx are virtually absent from most areas in most years

Other fur-bearers

Other common fur-bearers in the Alaskan Arctic are mink, river otter, marten, red fox, and Arctic fox, although all of these except red and Arctic foxes are uncommon on Alaska's North Slope

[a]Population estimates are based on the most recent census or survey. For some species (e.g., brown bears) data are extrapolated from intensively surveyed areas to larger areas; [b]Estimates adjusted annually based on subsistence harvest surveys. About 85% of the caribou harvest is by local residents. About 50% of moose and brown bears harvested is by local residents. Almost 100% of the harvest of black bears and fur-bearers is by local residents; [c]Most Western Arctic caribou winter in a relatively small area where food could become inaccessible due to unusual and extreme coastal snowstorms and icing. The calving area contains about 50% of the known U.S. coal reserves, but these reserves are unlikely to be developed within the next 50 years; [d]Pneumonia has been prevalent as these caribou herds have declined from high population levels; [e]Moose are recovering from a long-term decline that was possibly related to Brucellosis; [f]Moose, especially in the Kuskokwim River drainage area, are illegally harvested at a rate that prevents population expansion into suitable habitat on the lower reaches of the river; [g]Brown bears are heavily hunted (legally and illegally) in areas with reindeer herding.

establishing a multi-stakeholder system for the Western Arctic Herd in which users play an important role in the management process similar to the Canadian co-management boards for the Beverly and Qamanirjuaq and Porcupine caribou herds (Klein et al., 1999; Thomas and Schaefer, 1991; see also Appendix).

Management of wildlife in the arctic regions of Alaska also differs from wildlife management as traditionally practiced in most of the United States in that natural plant communities that constitute wildlife habitat have undergone little alteration through conversion of the land for agriculture, intensive forest management, industrial development, and urban sprawl. As a consequence, the focus of wildlife management in the Alaskan Arctic by state and federal wildlife biologists has largely been on monitoring structure and harvest levels of the most important hunted and trapped wildlife species rather than on aspects of habitat manipulation or restoration. Population estimates and condition and trend information are collected on caribou, moose, muskoxen, and mountain sheep largely through aerial surveys, facilitated through limited use of radio transmitters placed on some animals in more intensively monitored populations (Fig. 11.6). Similar, but less intensive survey work is also focused on wolves and brown bears as a basis for assessing their potential influence on ungulate populations through predation. This survey information is increasingly being supplemented by harvest information obtained from hunters for development of annual recommendations of harvest levels that are made to the Alaska Board of Game and the Federal Subsistence Division. The status and trends in populations of wild mammals in the arctic regions of Alaska that are hunted and trapped, their harvest levels, and possible threats to their populations are shown in Table 11.3.

Fig. 11.7. The increasing frequency of fires and total area burned in the northern forest zones and in the ecotone between forest and tundra (see Chapter 14), a consequence of climate warming, poses difficult decisions for wildlife managers. Although fire has been a natural feature of the ecology of these plant communities, a reduction in the ratio of older plant communities with high lichen biomass to post-fire early succession stages can be detrimental to caribou and reindeer that feed on the lichens in winter. The shrubs that are characteristic of the post-fire vegetation are also favored by recent climate warming, and provide suitable forage for moose. More intensive efforts at fire suppression may benefit caribou and reindeer, at least in the short term, to the detriment of moose.

Focus on the population dynamics of wildlife is a relatively efficient and cost-effective approach to management of wildlife. However, without inventory and monitoring of vegetation, its quality, and availability as forage within the habitats of large herbivores, knowledge of vegetation changes brought about through climate change, wildfire, or other factors cannot be integrated into the management and conservation of wildlife. The recent and expected continuing increase in area burned by wildfires in the ecotone between the boreal forest and the arctic tundra is of special relevance. This is because much of the lichen-dominated winter range of the large migratory herds of caribou in Alaska lies within this ecotone (Weladji et al., 2002) (Fig. 11.7). Thus the adaptability of management to respond to the effects of climate change is substantially limited.

11.3.4.1. Minimizing impacts of industrial development on wildlife and their habitats

Increasingly, industrial development activities associated with energy and mineral exploration and extraction in the Alaskan Arctic are encroaching on wildlife habitats and threatening wildlife populations through habitat loss, expanded legal and illegal wildlife harvest, and environmental contamination from industrial pollutants entering wildlife food chains. Assessing the magnitude and importance of impacts from existing and proposed industrial development activities in the Arctic is a time-consuming and difficult process under the best of circumstances. This task is rendered even more difficult when the ongoing effects on the environment of accelerated climate change in the Arctic must be factored into the assessment. Although an environmental impact assessment is required under the National Environmental Policy Act of 1969 as a basis for seeking approval for any large-scale federal project "significantly affecting the quality of the human environment", there has been relatively little effort made to undertake follow-up assessments of the actual impacts of projects once they have been approved. The assessments that have been made of the magnitude and ecological significance of threats resulting from specific development projects have been simple, general overviews (Klein, 1973, 1979), or have focused on specific wildlife species (Cameron et al., 2002), or have been limited to a few specific impacts or types of projects (Douglas et al., 2002). Assessment of the consequences of cumulative impacts from multiple interrelated projects taking place over extended periods has only recently been attempted through analysis and synthesis of past studies. The most recent and comprehensive effort in this regard was the *Assessment of the cumulative effects of petroleum development on Alaska's North Slope* that was compiled by a panel of experts appointed through the National Research Council, with a primary focus on the giant Prudhoe Bay and related oil fields (NRC, 2003). Investigative assessment of the environmental consequences of development projects can provide valuable information for the government bodies responsible for weighing the potential consequences of proposed new development projects.

In most of the Alaskan Arctic there is insufficient knowledge of plant and animal distributions on the lands and in the waters, and the ecological relationships existing there, as a basis for carrying out environmental impact assessments in advance of proposed development projects. Short-term studies specifically designed to address postulated impacts on wildlife and their habitats in the absence of an understanding of the complexity of the ecosystem relationships that may be affected are usually inadequate to enable a comprehensive assessment of the environmental impacts that may result from a project. An exception was the proposal to drill for oil in the coastal plain of the Arctic National Wildlife Refuge in northeastern Alaska resulting in the debate before the U.S. Congress in spring 2002. In that case, the back-ground of 20 years of detailed environmental studies of the proposed development area, including mapping of the vegetation and multi-year investigation of the population dynamics and ecosystem relationships of wildlife species, enabled a comprehensive assessment of the expected impacts of the proposed oil development (Douglas et al., 2002). As a consequence, information about the wildlife and other environmental values and the magnitude of the risks to which they would be exposed should oil development be allowed there played a major role in Congress' unwillingness to open the Arctic Refuge to oil development. Assessment of the impacts of proposed industrial development on the ecosystems of the Arctic Refuge was compounded by the difficulty of distinguishing between ecosystem-level effects resulting from climate change influences versus those resulting from the proposed development (Fig. 11.8).

A major obstacle to effective wildlife management in the Arctic in the face of increasing national and global pressures for large-scale energy and mineral extraction is the lack of specific information at the landscape level of wildlife distribution, habitat types and their seasonal use patterns, definition and mapping of critical habitats, and mapping of human land use and related wildlife harvests. An ultimate goal for effective management and conservation of wildlife and wildlife habitats in the Alaskan Arctic, as well as for all regions of the Arctic, is the accumulation of sufficient knowledge of the wildlife and other resources of the lands and waters to enable development of detailed regional land and water use plans. Such plans should employ the use of technology for remote sensing of landscape characteristics and Geographical Information System maps, and include analysis of plant community and soil characteristics, determination of wildlife and fish distribution, identification of critical fish and wildlife habitats, and designation of existing and proposed protected areas. An important part of regional land and water use plans, as the name implies, is mapping of existing patterns of land and water use for subsistence and other human activities, and other physical and biological features of the environment. This documentation of the physical and biological characteristics of the lands and waters of arctic regions would provide a basis for identifying and contrasting changes that may occur in the environment as a consequence of climate change. Its primary value, however, would be in assisting industrial interests in advance planning of development activities in the Arctic to minimize their potential impact on fish and wildlife resources and the users of these resources, and in the evaluation and assessment of proposed industrial developments by local, regional, and national governing bodies prior to their decisions over approval. Details of development of regional land and water use plans and use of environmental impact statements and environmental impact assessments as the basis for land and water use decisions in relation to wildlife management and conservation in northern ecosystems were described by Klein and Magomedova (2003) from which the text in Box 11.4 is abstracted.

Fig. 11.8. Oil fields in the Alaskan Arctic where displacement of caribou from calving grounds, obstruction of their movements, and herd fracturing has occurred (photos: D.R. Klein). Assessment of the impacts of oil, gas, and mining developments on arctic wildlife is rendered more complex because of the difficulty of differentiating the influences of the changing climate, thus the task of planning to minimize effects of proposed new developments on wildlife and their habitats has become equally complex.

Box 11.4. The potential role of regional land and water use planning in wildlife management and conservation in the Arctic

Components of ecosystem planning

- Regional land and water area (aquatic, estuarine, and marine coastal) use plans should be developed by the responsible government units prior to consideration of possible resource extraction developments in all regions of the Arctic.
- The plans should define and map habitat characteristics for wildlife, including identification of critical habitats that may need special protection.
- Traditional and existing patterns of human use of wildlife that are basic to the social, economic, and cultural well-being of the residents of the region should be inventoried, mapped, and included in the plans.
- People, and their use of the land and water resources, should be recognized in the plans as integral components influencing processes of arctic and subarctic ecosystems.

Value and use of the land and water use plan

- Regional land and water use plans available to the responsible governing units, the people residing within the region, and those proposing developments within the region (industry, politicians, and others) clarify limits of acceptability of proposed development activities and structures that may affect wildlife prior to their approval.
- Designation and mapping of critical and sensitive wildlife habitat units that need protection in advance of development proposals simplifies planning and minimizes costly and time-consuming conflicts.

Climate change as a factor in assessing industrial impacts

- Changes in global climate, with pronounced effects in the Arctic and subarctic (see Chapters 7, 8, and 9), add complexity to the task of assigning wildlife-related values and anticipating uses of the land and waters, as well as assessing consequences of development in northern ecosystems.
- Changes in climate globally, and locally within the Arctic (see Chapters 2, 4, and 6), are accelerating social, economic, and cultural changes among human societies within the Arctic, rendering assessment of the consequences of existing and proposed industrial development in the North on human use of wildlife more complex and difficult than in the past.

11.4. Management and conservation of marine mammals and seabirds in the Arctic

Coastal people of the Arctic have, throughout history, depended on marine mammals and seabirds as principal subsistence resources. Seabirds have provided eggs and meat and in some cases skins, and various marine mammal products have been used for meat, clothing, heat, light, tools, toys, and a host of other essential components of day-to-day living (e.g., Donovan, 1982; Kinloch et al., 1992; Pars et al., 2001; Riewe and Gamble, 1988). The great abundance of these animals in the Arctic also attracted attention from the south as early as the 1500s, and large-scale commercial harvests of these animals have been undertaken by a variety of nations within arctic regions – particularly harvests focused on whales and seals. Subsistence harvesting of marine mammals and seabirds currently occurs in most arctic nations. However, hunting intensities differ markedly with community size and density, and the wildlife species present regionally. National and local management regimes are highly varied. Also, the line between commercial and subsistence hunting is not

clear, given that some meat as well as skins and tusks from "subsistence" hunts are sold commercially, and sport hunting is conducted on some species within quotas assigned to indigenous communities.

Large-scale commercial harvests of arctic marine mammals are restricted to harp (*Phoca groenlandica*) and hooded (*Cystophora cristata*) seals. But non-indigenous people also commercially harvest a variety of species at smaller scales, such as minke whales in Norwegian waters, belugas in the White Sea, and pilot whales (*Globicephala melaena*) in the Faroe Islands. Sport hunting by non-indigenous peoples is also conducted on grey (*Halichoerus grypus*) and harbour (*Phoca vitulina*) seals and to a lesser extent, ringed and bearded (*Erignathus barbatus*) seals, as well as on a variety of seabird species.

The changes that will occur in hunting patterns due to climate change and the management initiatives that will be necessary to achieve sustainable harvests under new environmental conditions are highly speculative at the moment. Analyses are currently becoming available, such as this assessment, which will help to predict change in the next decades in the Arctic due to climate change (e.g., Newton, 2001; Riedlinger, 1999, 2002; Weller et al., 1999). Some of the most likely changes are:

- modifying the timing and location of harvest activities;
- adjusting the species and quantities harvested; and
- minimizing risk and uncertainty while harvesting in less stable climatic and ice conditions.

The analyses presented in the rest of this section largely serve to document current management regimes in the arctic countries with respect to marine mammals and seabirds. Hopefully, this will serve to highlight where future climate-related impacts might be dealt with via international measures or within the administration of the various arctic countries. It is important to recognize that the marine and terrestrial environments are not distinct from one another. Marine birds and many marine mammals require a land base for some of their life activities, be it nesting sites for birds, maternal dens for polar bears, or haul-out areas used by many marine mammals for resting, breeding, or giving birth. Also, most arctic residents who harvest marine wildlife live in coastal communities at the interface of land and sea.

Several marine mammal and seabird species are managed in part via international agreements or conventions and management issues are also discussed in international fora such as CAFF working groups. For example, polar bear research and management is coordinated internationally via the IUCN Polar Bear Specialist Group. This group was formed following the first international meeting on polar bear conservation, held in Fairbanks, Alaska in 1965, and subsequently led to the development and negotiation of the International Agreement for the Conservation of Polar Bears and their Habitat, which was signed in Oslo, Norway in 1973. The agreement came into effect for a five-year trial period in 1976. It was unanimously confirmed for an indefinite period in January 1981. This agreement stipulates that the contracting parties will conduct national research programs on polar bears related to the conservation and management of the species, will coordinate such research with research carried out by the other parties, will consult with the other parties regarding management of migrating polar bear populations, and will exchange information on research and management and data on bears taken (Wiig et al., 1995). A treaty between the United States and Russia defines a Bilateral Agreement for the Conservation of Polar Bears in the Chukchi/Bering Seas that deals with the management of this specific polar bear stock (USFWS, 1997, 2002a). The North Atlantic Fisheries Organization's Harp and Hooded Seal Working Group performs a similar role regarding coordination of the management of stocks of these two commercially harvested seal species. The North Atlantic Marine Mammal Commission is another international body that promotes cooperation on the conservation, management, study, and sustainable use of marine mammals in the North Atlantic. The International Whaling Commission sets quotas for the commercial harvest of all large

cetacean species (currently operating with a total moratorium on commercial harvesting), and also provides a format for discussions regarding small cetaceans. The North Pacific Fur Seal Treaty regulated harvesting of the northern fur seal (*Callorhinus ursinus*) between Japan, Russia, United States, and Great Britain (for Canada) from early in the 1900s until 1985, when the commercial hunt was terminated. A Joint Commission on Conservation and Management of Narwhal and Beluga was established in 1989 to address conservation and management of stocks that migrate between Canadian and Greenland waters. Organizations operating within the Arctic Council, such as CAFF, are playing an increasing role as advisory bodies in conservation and management of sea mammals and seabirds, largely via international working groups. For example, the CAFF Circumpolar Seabird Working Group recently produced the International Murre Conservation Strategy and Action Plan (CAFF, 1996) that identifies management issues related to common (*Uria aalge*) and thick-billed (*U. lomvia*) murres, which experienced significant declines in several circumpolar countries throughout the twentieth century. This group has also developed the International Eider Conservation Strategy (CAFF, 1997). Not all international agreements are legally binding, however, and most legislation regarding wildlife management is undertaken at the national level within the various arctic countries.

The following sections discuss the basic characteristics of management and conservation of marine mammals and seabirds for the main arctic regions. Further information on these regions may be found in Chapter 13.

11.4.1. Russian Arctic

Along with the continental shelf and exclusive economic zone adjacent to its boundaries, the Russian Arctic region accounts for over 30% of the area of the Russian Federation. The Russian continental shelf in the Arctic extends to the greatest distance and has the largest area of any country in the world. The associated shoreline and the area of the basins drained by the Russian rivers flowing into the Arctic Ocean are both huge. The region comprises the Central Arctic zone (roughly north of 80° N) and the Atlantic, Siberian, and Pacific sectors. The Russian Arctic, in particular the Atlantic and Pacific sectors, is characterized by a great diversity of marine ecosystems. Sea ice has an exceptionally important role in the life of marine mammals and birds of the Arctic. The nature of the sea-ice cover and the system of stationary polynyas and ice leads essentially determine the intra-specific structure, dynamics of number of species and populations, and the dates and pathways of their seasonal migrations. Of the marine mammals, the walrus alone is capable of successfully breaking gray ice 10 to 15 cm thick, and adult bowhead whales (*Balaena mysticetus*) break gray-white ice up to 15 to 30 cm thick with their backs. But similar to other marine mammals and birds, walruses and bowhead whales completely depend on the sustainable system of clear water space between pack-ice fields for their northward progress. A number of arctic marine mammal and bird species are circum-polar, and are represented by several populations and even subspecies (e.g., the bowhead whale, walrus, bearded seal, ringed seal, herring gull (*Larus argentatus*), glaucous gull (*L. hyperboreus*), and kittiwake (*Rissa tridactyla*)). They may be fairly isolated geographically as are the Svalbard and Chukchi–Bering sea stocks of the bowhead whale, Atlantic and Pacific subspecies of the walrus, and populations of the same species of seabirds of the Atlantic and Pacific sectors, but occasionally are only separated by massifs of heavy pack ice (the Laptev and Pacific subspecies of the walrus). There are some species that dwell in contiguous regions of Norway and the United States. The Russian Arctic also provides feeding grounds for some southern species, for instance, the Californian gray whales (*Eschrichtius robustus*) and short-tailed shearwaters (*Puffinus tenuirostris*) that nest in southern Australia. A number of species and populations have been classified as rare and protected and accordingly are listed in the Red Data Books of the IUCN and the Russian Federation.

Climate changes have occurred repeatedly in the history of the Arctic. The entire arctic zone exhibited a warming trend during 1961 to 1990; the region from 60° to 140° E (i.e., a substantial part of the Russian Siberian Arctic) showed the greatest warming. Over the last 30 years changes in sea ice have been in conformity with that of the warming trend. The climate, however, in different sectors of the Russian Arctic, both in the past and today, has been variable (AMAP, 1998b; Yablokov, 1996). The causal connection between global and regional changes in climate, on the one hand, and the number and distribution of arctic marine mammals, on the other, is far more complex than commonly believed. Despite switches between warming and cooling periods, the ranges for the majority of higher vertebrates of the Russian Arctic have been fairly stable over the last millennia. This is confirmed by dating the remains of whales, pinnipeds, and birds (1500 to 2680 years) from ancient coastal villages on the Chukchi Peninsula, located on the main migration routes of the animals, and in their breeding and feeding areas (Dinesman et al., 1996).

Animals of the marine environment are capable of maintaining and even expanding their ranges owing to physiological, biochemical, and behavioral mechanisms for adaptation to changing environmental conditions. An example is found in the Californian gray whale. Migrating between the feeding areas in the Bering Strait, Chukchi Sea, and East Siberian Sea and the breeding areas in the subtropical lagoons of Mexican California, gray whales annually cover 18 000 to 20 000 km. This huge migration route covers over 50° of latitude and exposes whales to the effects of constantly changing environmental factors, in particular the strong fluctuations in temperature and photoperiod.

In marine ecosystems, it is primarily the higher vertebrates that have been the most threatened by the rapidly developing direct consequences of human activities, now aggravated by climate change. The increased rate of these impacts frequently exceeds the adaptive capacities of living organisms. Over-harvest of the fish resources of the Barents Sea, primarily capelin (*Mallotus villosus*), resulted in profound rearrangements of the trophic relationships of the entire marine ecosystem, causing massive mortality of marine colonial birds and harp seals in the late 1980s. In 1988, on the southern island of Novaya Zemlya, fish-eating marine birds switched to a zooplankton diet (L.S. Bogoslovskaya pers. obs., 2003).

Throughout the 1990s the economic development activities that caused these detrimental processes increased many times due primarily to sharp increases in oil and gas production in the coastal regions and increased ship transit through polynyas and stationary ice leads, which are vital for marine mammals and birds in the high latitudes. This was most pronounced in the Atlantic sector. Pollutants associated with these activities and those from industrial activities on land that reach the sea through the major river systems flowing into the arctic seas have been found at all trophic levels of the biota, frequently causing morbidity and mortality of marine animals. In the early 1990s maps were compiled indicating levels of pollution by heavy metals, organochlorine compounds, petroleum products, and phenols in surface waters and bottom sediments of the seas of the Russian Arctic (Melnikov et al., 1994).

In the former Soviet Union, the Arctic was never legally defined geographically. Depending on current needs of the state, the southern border of the Arctic was delineated to serve immediate and short-term interests. The Soviet government apparently intended to extend the region's northern boundary to the North Pole but never made a full-scale claim over such a Soviet Arctic sector. The Soviet Arctic was always classified as a closed frontier zone and administrated accordingly. All services, including those purely civil, were to a large extent included in the classified status. This also applied to environmental monitoring of terrestrial and marine areas, and particularly to plant and wildlife species. In present-day Russia this situation has, nevertheless, deteriorated. The limited system of arctic environment monitoring, developed in Soviet times, was virtually discontinued. For lack of funds, no new national parks or coastal and marine reserves and sanctuaries were established through federal, regional, or local jurisdictions. Existing protected areas, for lack of financing, have had funding reduced or eliminated for research as well as for protection from detrimental human activities (Yablokov, 1996). The network of specialized marine sanctuaries, reserves, and parks considered necessary for protection of arctic cetaceans and pinnipeds has not had any significant development.

The Parliament (Federal Assembly) of the Russian Federation has so far enacted no law, amendment, or supplement to the current laws on the protection of the arctic environment. Moreover, the term "Arctic" is absent from the federal legislation. In some instances the term "Extreme North" is used, but this term is not used in international documents. No national arctic doctrine has been elaborated to reflect the many diverse interests of the Russian society in the Arctic, including protection of polar marine ecosystems. Thus, the Russian situation is unique. Federal governing bodies have signed a number of important international acts and bilateral agreements on the environment and sustainable development of the Arctic, but national legislation or statutory framework for management and protection of the arctic ecosystems has not been developed. At present no adequate legal framework exists for management and protection of the marine ecosystems of the Arctic and the associated species, subspecies, and populations of birds and mammals. There are, however, international documents, including ratified conventions and agreement on a number of species.

Russia is a member of the IUCN Polar Bear Specialist Group and operates under a 1973 International Polar Bear Agreement. In fact, polar bear hunting was banned in the former Soviet Union in 1956 and until very recently only problem polar bears could be killed (Belikov, 1993). However, the level of protection has diminished recently due to economic and political changes that make nature conservation and control of the use of the environment ineffective, and an increased interest by Russian people in using polar bears as a resource has been expressed. An agreement signed by the Government of the United States and the Russian Federation on October 16, 2000, recognized the need of indigenous people to harvest polar bears for subsistence purposes. It includes provisions for developing sustainable harvest limits, allocation of the harvest between jurisdictions, and the need for compliance and enforcement. Half the harvest limit, which is yet to be decided, will be allotted to each country. The agreement reiterates requirements of the multi-lateral polar bear agreement and restricts harvesting of denning females, females with cubs, or cubs less than one year old, and prohibits use of aircraft, large motorized vessels, snares, or poison. The agreement does not allow hunting for commercial purposes or commercial uses of polar bears

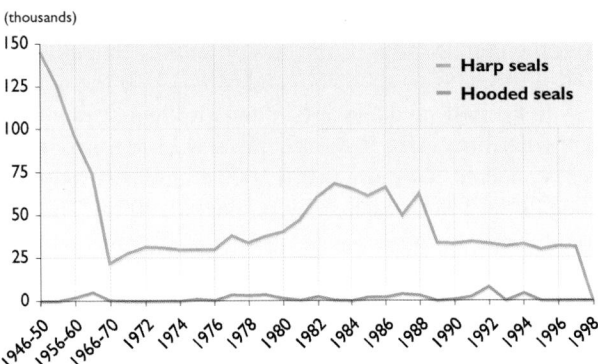

Fig. 11.9. Commercial harvest of harp and hooded seals by Russian vessels since the mid-1940s (East and West Ice combined).

or their parts. It commits the partners to the conservation of ecosystems and important habitats, with a focus on conserving specific polar bear habitats such as feeding, congregating, and denning areas. Mechanisms to coordinate management programs with the Chukotka government and with the Chukotka indigenous organizations are currently being determined. The agreement is currently undergoing procedural handling by the U.S. Congress and required legislative steps in Russia are being determined (USFWS, 1997, 2002a).

Other marine mammal harvests within Russia are managed on the basis of Total Allowable Catches (TACs) that are assigned by species and geographical region (Table 11.4). Catches of commercial species such as harp and hooded seals have remained constant over the last few decades (Fig. 11.9). However, reporting of harvest statistics and enforcement of TACs is difficult to manage in outlying areas given Russia's current economic and administrative difficulties and the status of populations and their harvests is in reality largely unknown.

Indigenous people in Russia have collected seabirds and their eggs since ancient times. Non-indigenous people have also harvested seabirds in coastal areas since the colonization of northwest and northeast Russia more than two centuries ago (Golovkin, 2001). In the Barents Sea

Table 11.4. Total allowable catches of marine mammals in Russia for 2002 (Government of the Russian Federation, 2001).

	Western Bering Sea	Eastern Kamchatka		Sea of Okhotsk			Caspian Sea	Barents Sea	White Sea
		Karaginskaya	Petropavlovsk-Komandorskaya	Northern Sea of Okhotsk	Western Kamchatskaya	Eastern Sakhalinskaya			
White whale	300			400	100	200		500	50
Killer whale						5			
Northern fur seal			3400			1800			
Walrus	3000								
Ringed seal	5900	600		18500	6000	3500		1500	1100
Ribbon seal	5800	200		9000	500	5500			
Bearded seal	4000			4800	1900	700		250	100
Caspian seal							500		

Note: Total Allowable Catch of white whales, killer whales and walruses are given for subsistence needs of small peoples of the North and Far East and for scientific and cultural-educational purposes.

region tens of thousands of eggs were collected annually from the middle of the 19th century until the beginning of the 20th. During the 1920s and early 1930s the number of eggs collected increased dramatically. For example, at Besymmyannaya Bar, Novaya Zemlya 342500 murre eggs were collected and more than 12000 adult birds were killed in the 1933 season alone (Golovkin, 2001). The need for conservation was recognized at the time, and several state reserves were established in the late 1930s where egg collecting and bird harvesting were prohibited. In the Commander Islands, near Kamchatka in the southern Bering Sea, seabird exploitation began with the first Russian expeditions to the area. Pallas's cormorant (*Phalacrocorax perspicillatus*) was harvested heavily and this is thought to have contributed to the extinction of this species. In the 19th century the Commander Islands were settled by Russians and Aleuts. These established residents began to harvest eggs and birds in the tens of thousands annually. Their preferred species were northern fulmars (*Fulmarus glacialis*), pelagic cormorants (*P. pelagicus*), thick-billed murres, horned puffins (*Fratercula corniculata*), tufted puffins (*F. cirrhata*), and glaucous-winged gulls (*Larus glaucescens*). In Kamchatka, local people collected 4000 to 5000 glaucous-winged and black-headed (*L. ridibundus*) gull eggs annually in the past, but the collection is thought to be negligible currently.

Traditional patterns of harvesting seabird eggs continued despite national hunting regulations prohibiting harvest of eggs of all bird species everywhere in Russia. In the Murmansk region however, local hunting regulations permit hunting of alcids (auks, puffins, guillemots, etc.) in autumn and winter (Golovkin, 2001). All four eider species are protected along the entire coast of Russia. It is known that some illegal harvesting takes place due to a general lack of enforcement. In the Barents Sea region it is thought that thousands of eggs are collected annually (Golovkin, 2001). It is known that 2000 glaucous-winged gull eggs were collected in 1999 and again in 2001 from Toporkov Island, where the largest colony of the species exists among the Commander Islands (CAFF, 2001b). Illegal egg collecting is also known to be a common activity among inhabitants of villages and crews of visiting vessels in the northern Sea of Okhotsk and human influences on easily accessible colonies of common eiders has increasingly been evident on the northern coast of the Koryak Highlands, Chukotka. The need to improve seabird management plans, conservation laws, and hunting regulations is recognized (Golovkin, 2001).

The scientific community of Russia, the indigenous minorities of the North, and the non-governmental environmental organizations have been campaigning for a refinement of the legislative framework regarding the Arctic. There are, however, few examples of fruitful cooperation between governmental bodies and indigenous and local organizations for management and protection of the natural environment of the Arctic. One positive example, however, concerns the 25-year monitoring of marine mammals and their harvest by the indigenous Inuit and Chukchi peoples of the Chukchi

Peninsula, associated with Russian participation in the International Whaling Commission. These activities have been possible through the active role and support of agencies of the US government responsible for marine mammal and bird conservation and management, and indigenous peoples' corporations of Alaska.

11.4.2. Canadian Arctic

Polar bear harvesting in Canada is undertaken in accordance with the 1973 International Polar Bear Agreement. Between 500 and 600 polar bears are taken annually in Canada by Inuit and Amerindian hunters under a system of annual quotas that is reviewed annually in Nunavut, the NWT, Yukon Territory, Ontario, Manitoba, Quebec, and Newfoundland/Labrador. Within the quota assigned to each coastal village in the NWT and Nunavut, hunters are allowed to allocate a number of hunting tags to non-resident sport hunters, who are guided by local Inuit hunters. Sport hunting and the sale of skins are important sources of cash income for small settlements in northern Canada. The annual economic value of the polar bear hunt is about one million Canadian dollars (CWS/CWF, 2002). The Canadian Wildlife Service represents Canada in the International Polar Bear Working Group.

Seal and whale management falls within the jurisdiction of the Department of Fisheries and Oceans (DFO) at the federal level. Harp and hooded seals are commercially hunted using a quota system in Canadian waters and shared stocks with Greenland involve some co-management. For 2002, the TAC for harp seals was 275000 and for the hooded seal 10000 (DFO, 2002a). Sale, trade, or barter of harp seal white-coat pups or hooded seal blue-backs (pups) is prohibited under Canada's Marine Mammal Regulations. The use of vessels over 65 ft (19.8 m) in length is also prohibited. The actual number of animals harvested varies from year to year depending on sea-ice conditions, market prices, or subsidy systems (Fig. 11.10), although the actual harvest quota has remained constant in recent years. Some subsistence hunting of harp and hooded seals takes place in northern regions, but this hunt only numbers a few thousand animals. Grey seals are harvested in a small, traditional commercial hunt in an area off the

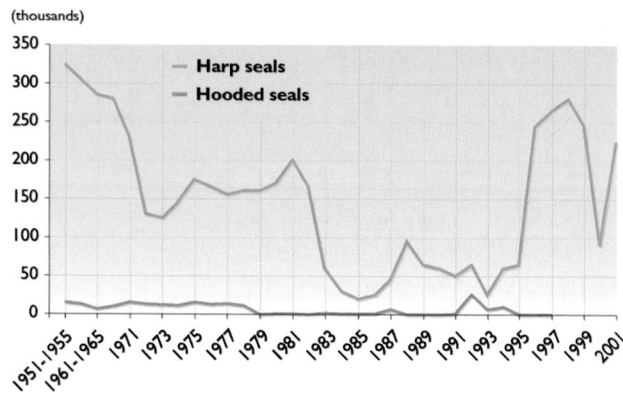

Fig. 11.10. Commercial catches of harp and hooded seals in Canadian waters since 1950 (from seal catch data for Canada, Norway, and Russia collated at the North Atlantic Marine Mammal Commission, Polar Environmental Center, Tromsø, Norway).

Magdalen Islands and at a few locations in the Maritimes. The numbers taken are small and thus a TAC has not been established. Ringed seals and bearded seals are taken in subsistence harvests in Labrador and throughout the Canadian Arctic, but figures are not available regarding harvest levels. Ringed seals are by far the most important arctic seal for human consumption and utilization in the Canadian Arctic. The Nunavut Wildlife Management Board has conducted a five-year harvest study on all species of seals and the resulting report is available via their website (www.nwmb.com/english). Subsistence hunting of arctic seals is not regulated.

Commercial harvesting of walrus was banned in Canada in 1931. All hunting currently conducted is indigenous subsistence hunting. Walrus harvest regulations are undergoing changes with the establishment of Nunavut but currently, residents of Coral Harbour, Sanikilqaq, Arctic Bay, and Clyde River have DFO-established quotas, and all other Inuk residents are permitted to hunt up to four walrus per year. Similar to the situation for polar bears, communities can set some of their quota aside for sport hunting by non-indigenous people (McCluskey, 1999). Four Atlantic walrus stocks occur in the eastern Canadian Arctic: Foxe Basin, Southern and Eastern Hudson Bay, Northern Hudson Bay–Hudson Strait– Southern Baffin Island–Northern Labrador, and the North Water (Baffin Bay–Eastern Canadian Arctic). The status of three of the four is classified as poorly known and the fourth is "fair". In the latter case, the stock is currently being harvested at a removal rate of 300 animals, which may exceed sustainable yield (Born et al., 1995). There is a similar concern regarding the North Water stock and the Southern and Eastern Hudson Bay stocks. The final stock is so poorly known that it is not reasonable to attempt to determine whether current harvest levels are sustainable or not.

Canada discontinued commercial whaling in 1972. However, whaling has been important to Inuit in the Arctic since prehistoric times and Arctic Inuit currently hunt about 700 beluga and about 300 narwhal annually in Canada. There is concern for the conservation of several beluga stocks in eastern Canada, while those in the west are harvested well within sustainable yields (DFO, 2000, 2002b). The St. Lawrence River population is endangered, although it has been completely protected from hunting since 1979. Also, populations in Southeast Baffin Island–Cumberland Sound and Ungava Bay are endangered, and the Eastern Hudson Bay population is threatened. The Eastern High Arctic/Baffin Bay population is classified as a special concern. Subsistence hunting of belugas in some parts of the Arctic is a concern because of its potential to cause continued decline or lack of recovery of depleted populations (DFO, 2002b). Narwhal are hunted in Hudson Bay and Baffin Bay under a quota system in 19 communities (DFO, 1998a,b). Baffin Bay narwhals summer in waters that include areas in northwestern Greenland and thus are a shared stock. Recent reviews of these stocks have been performed in consideration of new management options for this species.

Hunting of bowhead whales has recently been resumed in both the eastern and western Canadian Arctic following the settlement of land claim agreements, based on the traditional cultural value of these hunts (DFO, 1999). Both bowhead whale populations are classified as endangered (COSEWIC, 2002; DFO, 1999). Harvesting of these populations violates the intent of the International Whaling Commission (Finley, 2001), although Canada is not a member of this whaling regime. Subsistence whaling is currently managed under three separate land claim agreements – the James Bay Northern Quebec agreement, the Inuvialuit Final Agreement, and the Nunavut Land Claims Agreement – in the Canadian North.

Seabird harvesting in Canada dates back thousands of years to early colonization by indigenous peoples. Historically, seabirds were an important component of the subsistence lifestyle for coastal peoples, but today seabird harvesting for birds and eggs is much less widespread, although improved hunting technologies have tended to increase harvests on species such as murres (Chardine, 2001). Regulation of seabird harvesting (with the exception of cormorants and eiders) is done under the Migratory Bird Convention Act of 1917 that protects them year-round from hunting. Indigenous people in Canada are exempt from this restriction and can at any time take various auk species and scoters (*Melanitta* spp.) for human food and clothing. Eiders are hunted as game birds by both indigenous and nonindigenous people in a controlled annual hunt. Seabird egg collecting is not permitted under the general terms of the convention, but indigenous people are allowed to take auk eggs (Chardine, 2001).

Common eiders, thick-billed murre, and black guillemot (*Cepphus grylle*) are the most commonly harvested seabird species in arctic Canada, and are utilized by indigenous people wherever they are available (Chardine, 2001). There is no comprehensive monitoring of seabird harvests in Canada, but the total annual seabird take in the Arctic is thought to number about 25 000 individuals, about half of which are common eiders. Egg collecting is not as widespread as bird hunting, and has usually involved ground nesting species such as common eiders, Arctic terns (*Sterna paradisaea*), and gulls, which is technically illegal, as well as little auks (*Alle alle*). It is thought that some few thousand eggs are collected annually (Chardine, 2001). The most intense consumptive use of seabirds in Canada occurs in Newfoundland and Labrador, where thick-billed and common murres are harvested based on a set hunting season and bag limits. In the past, hunting levels were extreme, and recently enacted legislation is attempting to bring the harvest to sustainable levels. Currently, 200 000 to 300 000 murres are shot in the Newfoundland/Labrador hunt and approximately 20 000 common eiders are taken in Atlantic Canada. Atlantic puffin (*Fratercula arctica*), dovekie (little auk), razorbills (*Alca torda*), and black guillemots are legally hunted in Labrador. Illegal harvesting of other species such as shearwaters (*Puffinus* spp.), gulls, and terns is also known to occur (Chardine, 2001). Seabird harvests on the north shore of the St. Lawrence River in Quebec were considered large enough to have reduced seabird populations, but a recent education program is thought to have reduced local hunting pressure to a level where population recovery is expected (Blanchard, 1994). One of the primary needs for improving the management of seabird harvests in Canada is to improve knowledge regarding the current level of seabird harvesting, particularly in regions where harvest is thought to be substantial but little information exists (Chardine, 2001). The Canadian Wildlife Service, in cooperation with various indigenous wildlife management boards, co-manages seabirds in the Arctic.

11.4.3. Fennoscandian North

Marine mammal harvesting has been a tradition in Norway, Iceland, the Faroe Islands, and Greenland for centuries. Norway and Greenland, the only Fennoscandian countries that have polar bears, are both signing members of the International Agreement on the Conservation of Polar Bears (IACPB). Denmark signed the original agreement, but Greenland Home Rule took over legal responsibility for management of renewable resources, including polar bears, in 1979. Polar bears are completely protected in Svalbard (Wiig, 1995). Only bears causing undue risk to human property or life have been shot since the closure of the harvest some decades ago; these cases are dealt with under the authority of the Governor of Svalbard, which acts under the Norwegian Ministry of the Environment and the Ministry of Justice. Polar bears are legally harvested in Greenland. Full-time, licensed hunters have taken an average of 150 bears per year in Greenland in recent decades in accordance with most international recommendations for harvesting, although some local rules in some regions do not entirely conform to the IACPB (Born, 1993). The Greenland Institute of Natural Resources, which has been operating since 1995, is concerned that polar bears may require increased protection in Greenland.

In Norway, harp seals and hooded seals are commercially harvested, based on government-set quotas. The harp and hooded seal harvests are managed in agreement with the North Atlantic Fisheries Organisation. Current harvest levels are low compared to takes early in the 20th century (Fig. 11.11), and are set within sustainable limits. Ringed seals and bearded seals can be freely harvested in Svalbard outside their respective breeding seasons, but actual takes are very low. Harbour seals on Svalbard are Red Listed, and are completely protected. Coastal seals along the northern coast of Norway, which include grey and harbour seals as well as small numbers of ringed and bearded seals, are hunted through species-based quotas and licensing of individual hunters. Grey seals and harbour (common) seals are harvested in Iceland; catches of these two species have dropped gradually over recent decades and currently about 1000 harbour seals and a few hundred grey seals are caught annually. In Greenland, about 170 000 seals are taken annually, mainly harp and ringed seals. They are an important source of traditional food, and about 100 000 skins are sold annually to the tannery in Nuuk, Greenland (Jessen, 2001). There are few national regulations in Greenland regarding seal hunting; there are four Executive Orders, two related to catch reporting, one banning exportation of skins from pups, and the fourth is a regulation on harbour seal hunting in spring. There is concern that harbour seals may be in threatened status in Greenland (Jessen, 2001). With the exception of harbour seals, Greenland's seal stocks are plentiful.

Walruses were commercially harvested in Svalbard historically, to the brink of extirpation, but are now totally protected and are recovering (Born et al., 1995). The walrus population that winters off central West Greenland is harvested at a level that is thought to exceed sustainable yield (Born et al., 1995). Approximately 65 walruses are taken annually from this area where only about 500 animals remain. The same is true of the North Water stock that winters along the west coast of Greenland as far south as Disko Island. At this location about 375 walruses are taken annually from a group that only numbers a few thousand. In East Greenland, the small harvests focus mainly on adult males, and are thought to be within the replacement yield.

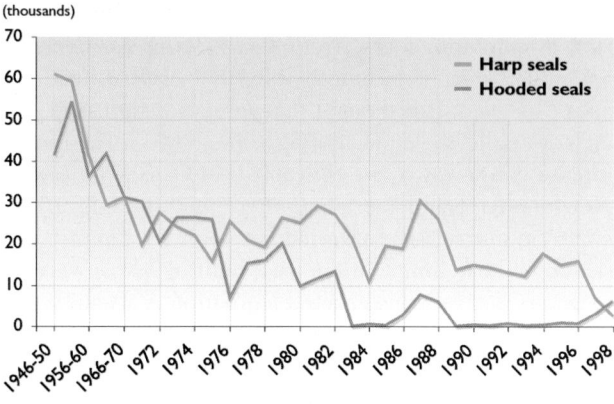

Fig. 11.11. Commercial catches of harp and hooded seals in the West Ice by Norwegian vessels since the mid-1940s (from seal catch data for Canada, Norway, and Russia collated at the North Atlantic Marine Mammal Commission, Polar Environmental Center, Tromsø, Norway).

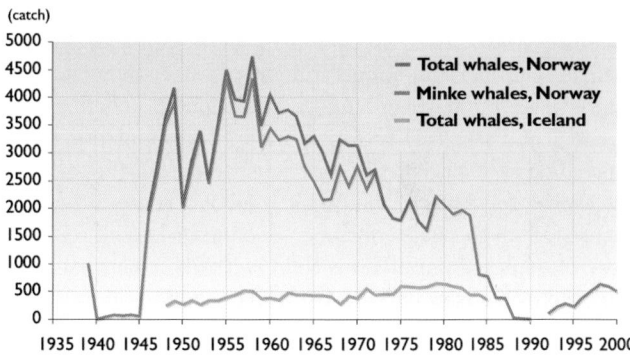

Fig. 11.12. Commercial catches of cetaceans in Norway and Iceland since 1939 (whale harvest data for Iceland from the Marine Research Institute, Reykjavik, Iceland; and for Norway from Statistisk Sentralbyrå, Oslo, Norway).

Whaling has been a traditional undertaking in Norway for centuries. However, cetaceans, large and small, with the exception of minke whales, are completely protected in Norwegian waters currently. Approximately 600 minke whales have been taken annually in recent years in the commercial hunt in Norwegian waters (Fig. 11.12). Management of this harvest is the responsibility of the Ministry of Fisheries, as is the case for all commercial marine mammal hunting in Norway. The harvest quota for minke whales is set by the Norwegian Government. This harvest is considered sustainable, and was sanctioned by the Scientific Advisory Board of the International Whaling Commission, but is in violation of the International Whaling Commission's total ban on commercial whaling. Norway, however, has entered a reservation against the moratorium, so its harvest is not strictly speaking a violation of International Whaling Commission decisions. Some poaching of harbour porpoise (*Phocoena phocoena*) is thought to take place along the Finnmark coast, but the level of this harvest is unknown. Although Iceland is technically not currently whaling commercially (Fig. 11.12), it was announced in August 2003 that Iceland would begin culling minke whales for "scientific purposes"; Iceland has been importing whale meat from the Norwegian minke whale harvest. In the Faroe Islands whales are harvested for local meat consumption. The majority of the harvest is pilot whales. The hunt has ranged from a few hundred animals to a few thousand in recent years. Other species are also taken, although less frequently, including humpback whales (*Megaptera novaeangliae*), bottlenose dolphins (*Tursiops truncatus*), harbour porpoises, and white-sided dolphins (*Lagenorhynchus acutus*). Indigenous people in Greenland continue their long tradition of subsistence whaling. The harvests currently focus mainly on white whales and narwhal coastally, but fin and minke whales are also taken, as well as pilot whales sporadically, and killer (*Orcinus orca*) and humpback whales have been taken on occasion. The fin and minke whale catches are sanctioned by the Interna-

tional Whaling Commission (2002), within the agreements for indigenous subsistence whaling. West Greenland is permitted an annual catch of 19 fin whales. West Greenland has an annual quota of 175 minke whales and East Greenland can take up to 12 of this species annually (until 2006). The Institute of Natural Resources, of the Home Rule Government, has documented that beluga have declined due to overexploitation in Greenland, and suggests that this species needs increased protection along with the narwhal and harbour porpoise (Fig. 11.13). The Greenland Home Rule Government is currently revising the management plan and hunting regulations for small cetaceans (K. Mathiasen pers. comm., 2004).

The cultural traditions for seabird harvesting in the Fennoscandian North are varied. In Finland there is no tradition for hunting alcids. Species such as eiders, oldsquaw (*Clangula hyemalis*), common merganser (*Mergus merganser*), and red-breasted mergansers (*M. serrator*), are hunted by set seasonal open and closing dates. Egging has been forbidden since 1962, with the exception of the autonomous region of the Åland Islands in the southwest archipelago (Hario, 2001). This region has its own hunting act that regulates the take of seabirds. Present harvests in Finland are thought to be sustainable; selling harvested birds is not allowed. In Iceland, there is a long tradition of harvesting seabirds, including northern fulmars, Arctic terns, black-headed gulls, great (*Larus marinus*) and lesser (*L. fuscus*) black-backed gulls, herring gulls, glaucous gulls, eiders, Atlantic puffins, common and thick-billed murres, razorbills, and black-legged kittiwakes (Petersen, 2001). Great cormorants (*Phalacrocorax carbo*), shags (*P. aristotelis*), black guillemots, and northern gannets (*Sula bassanus*) are also harvested but to a lesser degree, and eggs of gulls, terns, and sometimes eiders are also collected, although there are no records of egg numbers (Petersen, 2001). Eider down is also utilized. Seabird meat is sold in Iceland, and there has been increasing market demand for this during the last 10 to 15 years as a specialty item for tourists. The Ministry of the Environment supervises the Act on Conservation, Protection, and Hunting of Wild Birds and Land Mammals in Iceland. Seasons are set for shooting individual bird species, but the periods for egg collecting and catching of young are not specified. Three gull species that are classified as pests can be killed year-round. Information on current population sizes and the effects of harvesting, as well as more information on egg collecting, is needed to improve managements of seabird harvests in Iceland (Petersen, 2001).

The Faroe Islands have a long tradition of seabird harvesting that continues today. The two dominant target species are northern fulmars and puffins. Norway also has a long tradition of harvesting marine birds. Down collecting and harvesting eggs, adult birds, and chicks have been important subsistence and commercial activities for rural residents of coastal northern Norway (Bakken and Anker-Nilsson, 2001). Significant hunting

Fig. 11.13. Harvests of some seabird and marine mammal species for sale in country food markets (as shown here at Nuuk, Greenland in 1991) may exceed the sustainability of their populations, justifying setting of harvest quotas and establishment of protected areas (photo: D.R. Klein).

and collecting have also taken place at Bjørnøya and Svalbard until recent decades. Currently, hunting is only permitted on a small number of marine birds (Svalbard – northern fulmar, thick-billed murre, black guillemot, and glaucous gull; mainland – great cormorant and shag, greylag goose, oldsquaw and red-breasted merganser, black-headed gull, common gull, herring gull, great black-backed gull, and black-legged kittiwake) during set

Table 11.5. The status of marine birds breeding in the Barents Sea region (Anker-Nilssen et al., 2000).

	National Red List[a]		Bern Convention[b]	Bonn Convention[c]
	Norway	Russia		
Great northern diver (*Gavia immer*)	R		II	II
Northern fulmar (*Fulmarus glacialis*)			III	
European storm petrel (*Hydrobates pelagicus*)			II	
Leach's storm-petrel (*Oceanodroma leucorhoa*)			II	
Northern gannet (*Morus bassanus*)			III	
Great cormorant (*Phalacrocorax carbo*)			III	
European shag (*Phalacrocorax aristotelis*)		R	III	
Greylag goose (*Anser anser*)			III	II
Barnacle goose (*Branta leucopsis*)			II	II
Brent goose (*Branta bernicla*)	V	R	III	II
Common eider (*Somateria mollissima*)			III	II
King eider (*Somateria spectabilis*)			II	II
Steller eider (*Polysticta stelleri*)			II	I/II
Long-tailed duck (*Clangula hyemalis*)	DM		III	II
Black scoter (*Melanitta nigra*)	DM		III	II
Velvet scoter (*Melanitta fusca*)	DM		III	II
Red-breasted merganser (*Mergus serrator*)			III	II
Eurasian oystercatcher (*Haematopus ostralegus*)			III	
Purple sandpiper (*Calidris maritima*)			II	II
Ruddy turnstone (*Arenaria interpres*)	R		II	II
Red-necked phalarope (*Phalaropus lobatus*)			II	II
Grey phalarope (*Phalaropus fulicarius*)	V		II	II
Arctic skua (*Stercorarius parasiticus*)			III	
Great skua (*Catharacta skua*)			III	
Sabine's gull (*Xema sabini*)	R		II	
Black-headed gull (*Larus ridibundus*)			III	
Mew gull (*Larus canus*)			III	
Lesser black-backed gull (*Larus fuscus*)	E[d]			
Herring gull (*Larus argentatus*)				
Glaucous gull (*Larus hyperboreus*)			III	
Great black-backed gull (*Larus marinus*)				
Black-legged kittiwake (*Rissa tridactyla*)			III	
Ivory gull (*Pagophila eburnea*)	DM	R	II	
Common tern (*Sterna hirundo*)			II	II
Arctic tern (*Sterna paradisaea*)			III	II
Common murre (*Uria aalge*)	V		III	
Thick billed murre (*Uria lomvia*)			III	
Razorbill (*Alca torda*)	R		III	
Black guillemot (*Cepphus grylle*)	DM		III	
Little auk (*Alle alle*)			III	
Atlantic puffin (*Fratercula arctica*)	DC		III	

[a]Categories: E (Endangered), V (Vulnerable), R (Rare), DC (Declining, care demanding), DM (Declining, monitor species); [b]II should be protected against all harvesting, III should not be exploited in a way that may threaten their populations; [c]I includes species that are considered threatened by extinction. II are not threatened by extinction, but international co-operation is needed to ensure protection; [d]Red List for Svalbard only.

seasons. Egg collecting is permitted from herring gulls, great black-backed gulls, common gulls, and black-legged kittiwakes early in the laying season. Eider eggs are collected only in areas where construction of nest shelters for eiders is traditional. Harvests within Norway are considered within sustainable limits, but there is concern that some seabird stocks shared with Russia and Greenland may currently be excessively harvested (Table 11.5; Bakken and Anker-Nilssen, 2001).

Seabird harvesting in Greenland has a long history and continues to have a key role in Greenland's subsistence hunting. Murres and eiders are the most heavily harvested species, but others such as dovekies and kittiwakes are also harvested frequently in some regions of the country (Fig. 11.13). There are acknowledged management problems and murres, eiders, and Arctic terns have all recently declined due to overexploitation. For example, the number of thick-billed murre breeding colonies has been reduced from 48 to 23 during the last 30 years on the west coast of Greenland (Nordic Ministers Advisory Board, 1999). This is the result of over-harvesting eggs and adult birds. The colonies closest to human settlements have been the most impacted. Prior to the 20th century, communities in Greenland were small and hunting was done from kayaks, which resulted in little impact on seabirds. However, the human population has increased substantially, motorboats and shotguns are now common hunting tools, and the resulting increased harvest has brought about a drastic decrease in the number of formerly large colonies – particularly for murres (Christensen, 2001). Commercial harvests of tens of thousands of birds have been conducted annually since 1990 in southern Greenland municipalities. Most of this hunting pressure takes place in autumn and winter. In northwest and East Greenland seabirds have always been exploited during the breeding period; the only time that they are available in the region. In an attempt to prevent further reductions in the murre breeding population, a closed season was introduced north from Kangatsiaq Municipality in the late 1980s. Subsequent interviews and meetings with hunters showed that illegal hunting continued to be intensive through much of the breeding period (Christensen, 2001). This illegal harvesting, particularly in the Upernavik District, is thought to be a serious threat to breeding colonies. In the small settlements of Avanersuaq and Ittoqqortoormiit, murre shooting is permitted throughout the year. By-catch in fishing nets and increased disturbance near colonies by boat and helicopter traffic are thought to be factors additional to hunting contributing to the reduction in seabird populations. A complete ban was put in force in 1998 on collecting murre eggs, but harvesting continued illegally in some regions (Christensen, 2001). More restrictive legislation on seabird harvesting was put in place on 1 January 2002, but was later retracted due to complaints from hunters. This was followed by attempts to revise existing legislation. Enforced hunting bans will be necessary in some important areas (Christensen, 2001) to bring about population recovery.

A major obstacle for the management and conservation of marine mammals in the North Atlantic, as elsewhere in the Arctic, has been the limited information available on the general biology of marine mammals, their distribution and seasonal movements, use of marine habitats, food chain relationships, and general ecology. This is understandable in view of the difficulty of carrying out research in the marine environment of the Arctic and studying animals that spend most or all of their lives at sea, much of which is below the sea surface. Recently developed technology, however, enables monitoring of movements, feeding behavior, and aspects of the general ecology of marine mammals. These techniques can also provide essential information on the relationship of marine mammals to commercial fisheries, needed to base conservation efforts and to develop management plans (Fig. 11.14)

Fig. 11.14. Recent advances in electronic technology and methodology for handling arctic marine mammals allows for collection of data on movements, seasonal habitat preferences, food chain relationships, and other aspects of their social behavior and ecosystem relationships that were previously unavailable to those responsible for their management and conservation. Shown here (a) on the sea ice adjacent to Svalbard, an anesthetized polar bear is being weighed and other biological data collected prior to its release by a team of scientists from the Norwegian Polar Institute. In (b) a similar team is releasing a beluga whale in the waters adjacent to Svalbard after having glued a package to its back containing environmental sensing instrumentation, a data logger, and a radio transmitter capable of sending data to receivers in aircraft, ships, and polar-orbiting satellites (photos: Kit Kovacs and Christian Lydersen, Norwegian Polar Institute).

11.4.4. Alaskan Arctic

Physical changes in the marine system have the capability to dramatically affect marine species. Marine mammals that depend upon sea ice, such as walrus, polar bears, and the several species of ice seals, use ice as a platform for resting, breeding, and rearing young. While sea ice is a dynamic environment, general seasonal patterns exist and subsistence harvest practices have developed in concert with these seasonal rhythms. Hunters have reported changes in winds, sea-ice distribution, and sea-ice formation that particularly affect hunting (Krupnik and Jolly, 2002). Winds are reported to be stronger now compared with the recent past, and there are fewer calm days. For hunters out in small boats, even a 10 to 12 mph wind creates waves of sufficient size to swamp boats (see Chapter 3). Winds also affect distribution of sea ice. Early season strong winds move sea ice northward and the marine mammals on the retreating ice are quickly out of range of some villages (notably those on St. Lawrence and Diomede Islands, and Shishmaref). Winds can also pack sea ice so tightly against shorelines that hunters are unable to get their boats out. These changes are not predictable, which affects both hunting opportunity and safety. For example, in spring 2001 Barrow whalers made trails at least 50 miles long through the shore ice to reach open leads for hunting. In spring 2002, hunters out on the ice edge became stranded as a large lead unexpectedly developed between their hunting camp and the shore, necessitating a major rescue effort.

Marine mammals are an integral part of the culture and economy of indigenous communities in Alaska, as they have been for centuries. Indigenous people depend on marine mammal species for food and other subsistence needs and utilize all species that are available within Alaskan waters to some degree. The United States is a participant in the 1973 Agreement on the Conservation of Polar Bears and the 2000 U.S./Russia Bilateral Agreement on the Conservation and Management of the Alaska–Chukotka Polar Bear Population. The Alaskan Department of Fish and Game is the state authority dealing with management issues related to polar bears. However, national responsibility for polar bears remains under legislation of the Marine Mammal Protection Act of 1972. Polar bears can be harvested for subsistence purposes or for creating items of handicraft or clothing by coastal dwelling indigenous people provided that the populations are not depleted and the taking is not wasteful. There are no limits on quotas, seasons, or other aspects of the hunt. No commercial hunting or sale of polar bears or their parts are permitted (USFWS, 1994, 2002a,b). Polar bear stocks in Alaska are linked to the east (Southern Beaufort Sea stock) with Canada and to the west (Chukchi/Bering Sea stock) with Russia. A joint agreement exists between the Inuvialuit Game Council, NWT and the Iñupiaq of the North Slope Borough, Alaska for the management of the Southern Beaufort Sea group and negotiations are near completion with Russia for the western areas. Polar bear catches in Alaska vary annual-

ly, depending largely on how many bears approach areas near settlements, because there is little targeted hunting effort on this species. The number of bears shot annually in the 1990s varied between approximately 60 and 300. There is no indication that the current level of hunting is not sustainable, although information is lacking for the Chukchi/Bering Sea stock (USFWS, 2002g).

In 1994, an amendment to the Marine Mammal Protection Act of 1972 included provisions for the development of cooperative management agreements between the U.S. Fish and Wildlife Service and Alaska Native organizations to conserve marine mammals and provide for the co-management of subsistence use by Alaskan indigenous people. A mandatory marking, tagging, and reporting program implemented by the U.S. Fish and Wildlife Service in 1988 for some species has provided considerable data for subsistence harvests in recent years. The Indigenous People's Council for Marine Mammals, the U.S. Geological Survey's Biological Resources Division, and the National Marine Fisheries Service and the U.S. Fish and Wildlife Service jointly administer co-management funds provided to the State of Alaska under the Marine Mammal Protection Act of 1972. The U.S. Fish and Wildlife Service works with a number of groups to manage marine mammals in Alaska such as the Alaska Sea Otter and Steller Sea Lion Commission, the Alaska Nanuuq Commission, and the Eskimo Walrus Commission. For example, the Cooperative Agreement developed in 1997 between the U.S. Fish and Wildlife Service and the Eskimo Walrus Commission has served to facilitate the participation of subsistence hunters in activities related to the conservation and management of walrus stocks in Alaska. The agreement has resulted in the strengthening and expansion of harvest monitoring programs in Alaska and Chukotka, as well as efforts to develop locally based subsistence harvest regulations. The mean annual harvest of Pacific walrus over the period 1996 to 2000 was about 5800 animals. However, the hunt has varied quite dramatically from year to year depending primarily on ice conditions and hunting effort, and has varied between 4000 and 16000 animals per year over the 1980s and 1990s (USFWS, 2002c). Sustainable level of harvest cannot be prescribed because of a lack of information on population size and trend, but the population is thought to number in excess of 200000 animals, having recovered dramatically from heavy exploitation early in the 20th century. Other seals, such as ringed seals, bearded seals, harbour seals, and spotted seals (*Phoca largha*) are important in the diet of indigenous people in Alaska and are harvested in significant numbers.

Sea otters (*Enhydra lutris*) were heavily depleted by commercial harvests during the 1700s, and probably numbered only a few thousand animals in 13 remnant colonies when they became protected by the International Fur Seal Treaty in 1911 (USFWS, 2002d,e,f). Following protection and translocation of animals, they recovered and re-colonized much of their historic range in Alaska. Sea otter populations in southcentral Alaska and those reintroduced into southeast Alaska are growing and each of the two stocks is subject to a subsistence harvest of about 300 animals. The southwestern Alaskan stock in the Aleutian Islands is undergoing a population decline that is not explained by the level of human-induced mortality. Heavy predation by killer whales, previously not known to be a significant predator on sea otters, has been reported and postulated as a cause for the decline (Estes et al., 1998). This apparent shift in trophic level relationships is also thought to be tied to other changes brought about through heavy commercial fishing pressure and warming of these marine waters through strong El Niño events and climate warming (Benson and Trites, 2002).

The northern fur seal historically underwent population reductions through heavy commercial harvests both at breeding colonies and at sea. It then was managed by international treaty through the North Pacific Fur Seal Commission. Commercial hunting of this species was terminated in 1985. However, like the sea otter in the west and several other marine mammal populations including Steller sea lions (*Eumetopias jubatus*) (Loughlin, 2002) and harbour seals (Boveng et al., 2003), the fur seal has been declining since about 1990 at a rate of 2% per year (Gentry, 2002) despite small subsistence harvests. The current marine mammal population declines in the Bering Sea and North Pacific appear to be part of a complex regime shift that is thought to be the result of temperature shifts that caused several major fish stocks to collapse and the impacts are cascading through the system (e.g., Benson and Trites, 2002). The collapse of these fish stocks, however, may be tied to the intense commercial exploitation of the Bering and North Pacific fisheries. Management responses to the population declines have been undertaken through a host of plans and agreements, such as the Co-management Agreement between the Aleut community of St. George Island and the National Marine Fisheries Service that was signed in 2001 to address management of the northern fur seal and Steller sea lion at St. George Island (NMFS, 2001).

Fig. 11.15. The bowhead whale harvest at Barrow, Alaska is carried out under a regional harvest quota established by the Alaska Eskimo Whaling Commission (photo: Department of Wildlife Management, North Slope Borough).

Subsistence hunting of bowhead, gray, beluga, and minke whales takes place in Alaska (Fig. 11.15). At the local level the Alaska Eskimo Whaling Commission regulates whaling activities. Eskimo whaling is conducted from nine traditional whaling communities. The current quota of 51 bowhead whales is hunted from St. Lawrence Island and Little Diomede Island in the Bering Sea and from coastal villages along the northern Alaskan coast. This hunt was not sanctioned by the International Whaling Commission in 2002 (IWC, 2002), however, an emergency session of the commission in 2003 agreed on a new quota for the Alaskan subsistence harvest. The bowhead is classified as an endangered species. The beluga is the second most important cetacean species harvested for subsistence in Alaska and it is hunted in significant numbers. The Alaska Beluga Whale Committee oversees this hunt. The gray whale quota, which is sanctioned by the International Whaling Commission, is 140 animals per year in the eastern North Pacific (620 animals from 2003 to 2006). This species was removed from the endangered species list in 1995 following a dramatic recovery in the eastern Pacific; western Pacific stocks (off Korea) have not recovered and remain listed. Minke whales are opportunistically taken in Alaska.

Seabird harvests in Alaska are managed through a co-management council that includes indigenous, federal, and State of Alaska representatives that provide recommendations to the U.S. Fish and Wildlife Service and North American Flyway Councils. The latter bodies are included because most harvested species fall under the North American Migratory Bird Treaty Act that prohibits hunting from 10 March until 1 September, but provisions for Alaska provide that indigenous inhabitants of the State of Alaska may harvest migratory birds and their eggs for subsistence uses at any time as long as there is no wasteful taking of birds or eggs. Seabirds and their eggs cannot be bought or sold in Alaska. Subsistence harvest information is only available for the last decade, and these statistics are thought to represent minimal harvest estimates. The two most harvested species are crested auklet (*Aethia cristatella*) (about 12000) and common murre (about 10000) (CAFF, 2001b; Denlinger and Wohl, 2002). Smaller numbers of other seabirds taken include: cormorants, gulls, common loons (*Gavia immer*), red-legged kittiwakes (*Rissa brevirostris*), black-legged kittiwakes, yellow-billed loons (*G. adamsii*), thick-billed murres, least auklets (*Aethia pusilla*), parakeet auklets (*A. psittacula*), Pacific loons (*G. pacifica*), Arctic loons (*G. arctica*), red-throated loons (*G. stellata*), ancient murrelets (*Synthliboramphus antiquus*), tufted puffins, Arctic terns, and horned puffins. Harvests from St. Lawrence Island communities dominate the overall harvest statistics. The ten-year average for eider harvests are common eider 2000, king eider (*Somateria spectabilis*) 5500, spectacled eider (*S. fischeri*) 200, and Steller's eider (*Polysticta stelleri*) 50. Seabird egg collecting is more evenly spread geographically than the hunting of birds, with eggs of gulls, murres, and terns being the most commonly harvested (Denlinger and Wohl, 2002). More information is needed on population trends and the harvests themselves as a basis for establishing sustainable harvest levels. Harvest by humans, however, is not recognized by the U.S. Fish and Wildlife Service as a threat to seabirds in Alaska (USFWS, 2001). Recommendations for regulations governing harvest of game and non-game birds for each season are adopted by the Co-management Council and then forwarded to the U.S. Fish and Wildlife Service for action. These include seasons, bag limits, restrictions on methods for taking birds, law enforcement policies, and recommendations for programs to monitor populations, provide education for the public, assist integration of traditional knowledge, and instigate habitat protection (CAFF, 2001a).

A new conservation tool for seabirds and their habitats is Important Bird Areas (IBAs). The program was started in Europe in 1989 by Birdlife International and has grown into a worldwide wildlife conservation initiative. The goal of the IBA program is to get indigenous people, landowners, scientists, government agencies, non-governmental organizations, and land trusts to work together and set priorities for bird conservation. There are criteria regarding high concentration areas and rare species that are considered for inclusion into the IBA program. In Alaska, Audubon Alaska has recently completed a draft list of IBAs for the Bering and Chukchi Seas. While IBAs are for all birds, the ones identified in Alaska were mostly set up because of high concentra-

tions of seabirds. There are 138 sites, the majority of which are in the Bering Sea (Fig. 11.16).

11.4.5. Future strategies

A changing environment will result in changes in subsistence hunting patterns. Harvest levels may decrease for some species as their seasonal availability decreases, while for others, harvest levels may increase. Close documentation of harvest levels and patterns will be needed to track these changes and to contribute to site-specific information on wide-ranging species. Hunter participation in collection of population and other biological information is essential for effective marine mammal management. Changes in health of walrus were reported in 2000, when hunters reported that adults appeared skinny and that few calves were present, potentially reflecting poor access to food resources. For ice-dependent species that are difficult to study directly, information from subsistence-harvested animals can be of considerable value for their management. In addition, hunters are interested in and concerned about changes they are observing. Should harvest restrictions become necessary, direct involvement of the subsistence community in developing the restrictions will facilitate such changes.

Marine mammals, throughout most of the Arctic, are the primary subsistence food base for coastal residents of the Arctic. Seabirds, including eiders and other sea ducks,

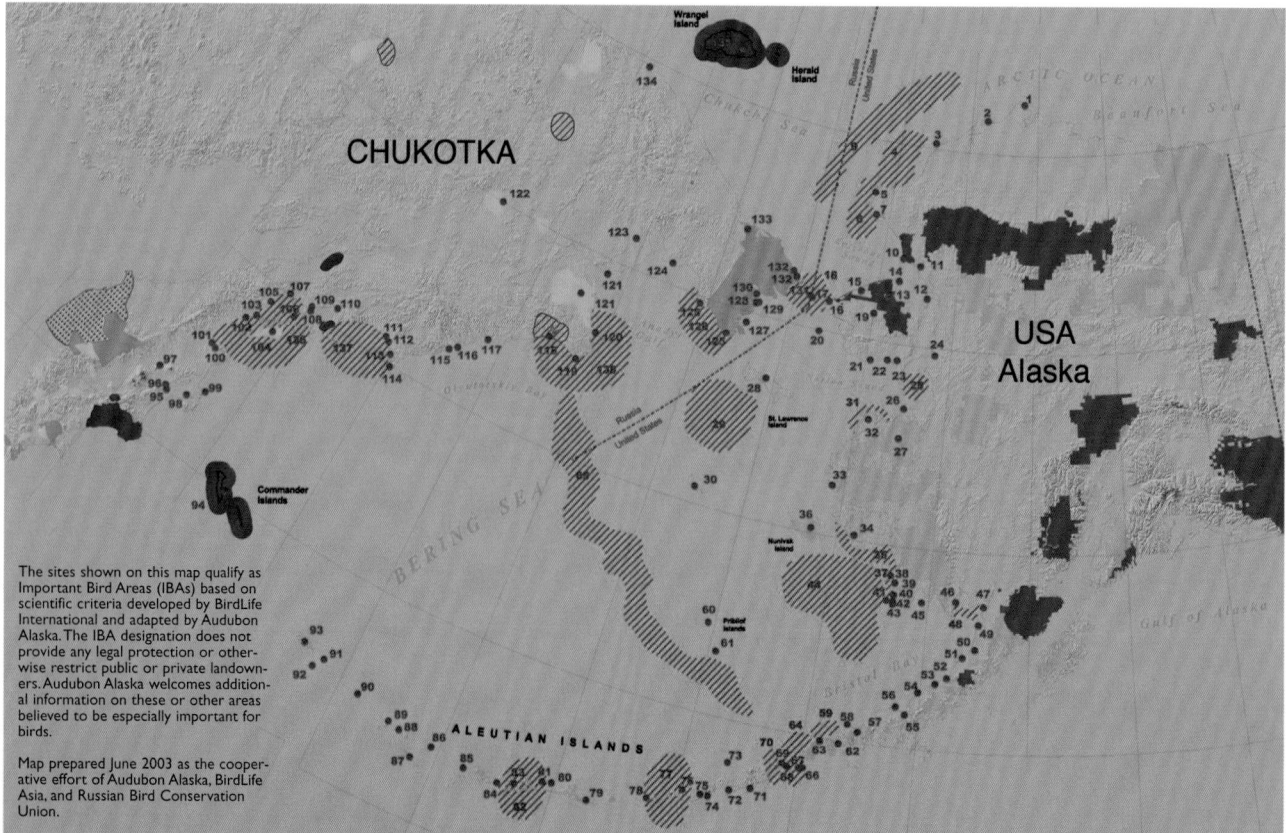

Fig. 11.16. Important Bird Areas in the Bering and Chukchi Sea regions. These have been generated through cooperative efforts of scientists in government agencies and non-governmental organizations, working with indigenous people and other coastal residents in Russia and the United States. The map is an essential step in the planning for a network of protected areas critical for the conservation of seabirds and their habitats in the Bering–Chukchi region (map supplied by the National Audubon Society).

alcids, and gulls, are also important to many coastal communities as a source of food. In some areas seabirds are also harvested commercially. The most productive regions for seabirds in the Northern Hemisphere are between approximately 50° and 70° N. Well over 100 million seabirds live in these arctic and subarctic regions, an order of magnitude more than seabirds living in the temperate regions (Croxall et al., 1984). The management implications of climate change are complicated and largely unknown, but increasing temperatures, thawing of the sea ice with associated movement of the pack ice edge northward, and rising sea levels will cer-

Table 11.6. Population trends, management, and threats to marine birds in the Bering, Chukchi, and Beaufort Seas.

	Population trends: Bering Sea	Population trends: Beaufort and Chukchi Seas	Management regulations[a]	Harvest birds	Harvest eggs	Threats[b]	Status[c]
Common loon	Unknown	Unknown	Yes	Yes	Yes	1,2,3,4,5,6,7	3
Yellow-billed loon	Unknown	Stable	Yes	Yes	Yes	1,2,3,4,5,6,7	2
Pacific loon	Stable	Stable	Yes	Yes	Yes	1,2,3,4,5,6,7	3
Arctic loon	Unknown	Unknown	Yes	Yes	Yes	1,2,3,4,5,6,7	3
Red-throated loon	Decrease	Stable	Yes	Yes	Yes	1,2,3,4,5,6,7	2
Short-tailed albatross	Increase	N/A	Yes	No	No	3,5,6,7	1
Black-footed albatross	Decrease	N/A	Yes	No	No	3,5,6,7	2
Laysan albatross	Decrease	N/A	Yes	No	No	3,5,6,7	2
Northern fulmar	Increase	N/A	Yes	No	No	3,7	3
Fork-tailed storm petrel	Increase	N/A	Yes	No	No	3,7	3
Double-crested cormorant	Unknown	N/A	Yes	No	No	1,3,7	3
Pelagic cormorant	Decrease	Unknown	Yes	Yes	No	1,3,7	3
Red-faced cormorant	Unknown	N/A	Yes	Yes	Yes	1,3,7	2
Common eider	Stable	Decrease	Yes	Yes	Yes	1,3,6,7	3
King eider	N/A	Decrease	Yes	Yes	Yes	1,3,6,7	2
Spectacled eider	Stable	Stable	Yes	Yes	Yes	1,2,3,6,7	1
Steller's eider	Unknown	Unknown	Yes	Yes	Yes	1,2,3,6,7	1
Herring gull	Unknown	N/A	Yes	No	Probable	7	3
Glaucous-winged gull	Decrease	N/A	Yes	Probable	Yes	7	3
Glaucous gull	Unknown	Unknown	Yes	Probable	Yes	7	3
Red-legged kittiwake	Decrease	N/A	Yes	Yes	Yes	7	2
Black-legged kittiwake	Decrease	Increase	Yes	Yes	Yes	7	3
Arctic tern	Unknown	Unknown	Yes	Yes	Yes	4,7	2
Aleutian tern	Unknown	Unknown	Yes	No	Yes	4,7	2
Common murre	Decrease	Stable	Yes	Yes	Yes	1,3,7	3
Thick-billed murre	Stable	Stable	Yes	Yes	Yes	1,3,7	3
Black guillemot	N/A	Decrease	Yes	No	No	1,7	3
Pigeon guillemot	Unknown	N/A	Yes	Yes	No	1,3,6,7	3
Marbled murrelet	Unknown	N/A	Yes	No	No	1,3,4,5,7	2
Kittlitz's murrelet	Unknown	N/A	Yes	No	No	1,3,4,7	2
Ancient murrelet	Unknown	N/A	Yes	No	No	1,7	2
Cassin's auklet	Unknown	N/A	Yes	No	No	1,7	3
Parakeet auklet	Unknown	N/A	Yes	Yes	Yes	1,7	3
Crested auklet	Unknown	N/A	Yes	Yes	No	1,7	3
Whiskered auklet	Unknown	N/A	Yes	No	No	1,7	2
Least auklet	Unknown	N/A	Yes	Yes	No	1,7	3
Horned puffin	Unknown	Unknown	Yes	Yes	Probable	1,7	3
Tufted puffin	Increase	N/A	Yes	Yes	Probable	1,3,7	3

N/A not applicable; [a]Regulated within the 3 nautical mile territorial waters zone by the U.S. Migratory Bird Treaty Act; [b]1:oil pollution, 2:over-harvest, 3: fisheries by-catch, 4:human disturbance, 5:habitat alteration, 6:contaminants, 7:climate change; [c]1:Threatened or Endangered (U.S.), 2:Birds of Conservation Concern (U.S.), 3:Low or moderate concern.

tainly reduce the availability of seabirds as food to many arctic communities. This will complicate the role of management to ensure the health of seabird populations as components of ecosystems undergoing change, while providing for the sustainable use of the seabirds by the people that depend upon them. A further complication in assessing how seabirds may move northward and possibly establish new nesting colonies within the context of a warmer climate is the difficulty of predicting how the marine food species upon which seabirds are dependent may change their distribution and productivity in relation to climate change and other human impacts such as commercial fisheries.

11.4.5.1. North Pacific, Bering, Chukchi, and Beaufort Seas

If temperatures increase for sustained periods, with associated melting of Arctic Ocean ice, and the band of high seabird productivity shifts northward, there is likely to be a dramatic overall decline in the number of seabirds living in the arctic and subarctic regions of the North Pacific and adjacent Arctic Ocean where high latitude nesting islands are extremely limited. This is particularly apparent when contrasting the rugged island and coastal topography of the southern Bering Sea with the low-lying coastal plains that border much of the northern Bering, Chukchi, and Beaufort Seas. A different situation exists in the North Atlantic and Canadian High Arctic because of more high latitude islands with rugged coastal topography that might serve as new nesting sites.
In addition, if the sea level rises as projected as a consequence of climate warming, many low-elevation nesting islands used by eiders, terns, and gulls will be inundated, resulting in decreased numbers of these species.

Estimates of population trends and status, current management, and threats for arctic seabirds of the North Pacific and associated Arctic Ocean, including the Bering, Okhotsk, Chukchi, and Beaufort Seas, are summarized in Table 11.6. Presently there is little or no information on population trends of many seabird species nesting in the Arctic. Better data on population trends are critical for effective management and conservation of these species, especially in areas where they are harvested for human use.

11.5. Critical elements of wildlife management in an Arctic undergoing change

The expected effects of climate change on arctic wildlife have been addressed in other chapters, particularly Chapters 7 (tundra and polar desert ecosystems), 8 (freshwater ecosystems and fisheries), and 9 (marine systems). Chapters 3 (indigenous perspectives), 12 (hunting, herding, fishing, and gathering), 13 (marine fisheries and aquaculture), and 14 (forests and agriculture) assess human relationships to climate change in the Arctic via commercial and subsistence harvest of resources, land use practices, and socio-cultural change. The latter chap-

ters assess the interface between people and the natural biological systems of the Arctic, recognizing that people of the Arctic are both components of arctic ecosystems as well as major drivers of these systems. Humans living outside the Arctic have become a major driving influence on arctic systems as a consequence of their industrialization and associated urbanization, accelerated pressures for exploitation of the world's non-renewable mineral and energy resources, globalization of the economy, and exportation of their cultural, social, and economic values and aspirations. These pressures, largely generated at temperate latitudes, reach into the Arctic through their effects on climate, atmospheric and marine pollution, and their social, economic, and cultural influences on the human and nonhuman residents of the Arctic.

11.5.1. User participation

This chapter deals primarily with assessment of the effectiveness of existing structures for management and conservation of wildlife in the Arctic and the adaptability of these structures to changes that are expected to continue and to accelerate in the Arctic in the future. A comparative analysis of the existing arrangements and their processes of evolution would serve as the basis for assessing the capacity of management to meet the challenges that may come with various climate change scenarios. While it is not possible to determine with a high level of specificity the nature of these challenges, it can be assumed that managers and users of arctic wildlife resources will be confronted with increased variability, a greater likelihood of surprise, and rapid change which may stress even the most robust wildlife institutions. It is, however, important to recognize that climate change, although of major consequence for arctic systems, is one of several driving forces influencing the broad spectrum of accelerated changes that are occurring in the Arctic. These forces of change, the climatic, the economic, the social, the cultural, and the political, operate through influences both internal to the Arctic as well as those of a global nature. It therefore follows that management and conservation of wildlife in the Arctic should serve the interests of all those, both within and outside the Arctic, who would use and value the wildlife of the Arctic. Responsibility for management and conservation of arctic wildlife, therefore, extends to the entire global community.

A major political change has taken place in the Arctic in recent decades bringing increased regional autonomy and a stronger voice for the residents of the Arctic in managing their own affairs. This has major consequences for wildlife conservation and management in the Arctic. The increased interest in, and broader participation by, residents of the Arctic in management of the species of importance to them should receive major emphasis in the design of systems for conservation and management of wildlife in all regions of the Arctic where indigenous peoples reside. Existing systems that have incorporated the concept of participation in, and shared responsibility for, wildlife management by residents of the Arctic who are the users of the wildlife are often referred to as co-management. These management systems have proved preferable to the wildlife users, have improved the collection of biological and harvest information on the target species, served as a means for integrating traditional knowledge and science, and have increased efficiency in managing wildlife for sustained harvest and conservation. These regimes vary in the degree to which they are based on formal legal standing and reflect the cultural, ecological, and economic conditions in which they emerge. Examples are: 1) the Canadian Beverly-Qamanirjuaq Caribou Management Board, a highly complex management system spanning several jurisdictions and involving numerous groups, some of whom have settled land claims and others that have not; 2) the Canadian Porcupine Caribou Management Board, which is relatively simple in composition compared to the Beverly-Qamanirjuaq arrangement and interfaces with the United States–Canada caribou management system that provides limited authority to Alaskan caribou user communities; 3) the Alaska Eskimo Whaling Commission, which is homogenous in composition and highly effective when interacting at the international level; and 4) the Inuvialuit–Inupiat Beluga Commission, a strong bilateral arrangement where there are few third-party interests and local resource users have significant influence. While the range of conditions for joint management differ, the conditions of sharing the responsibility for the conservation and management of wildlife between users of the wildlife and the governmental units that have legal jurisdiction over the lands, waters, and their resources in the Arctic has generally proved workable and effective. Legal jurisdiction over wildlife in large regions of the Arctic is often shared between governments and indigenous peoples through treaties, land claim settlements, and other governmental agreements that influence how co-management systems can be developed and how authority over wildlife is partitioned.

How, if at all, might principles of co-management be applied to regions like Russia, where local resource users have limited legal rights and non-local interests commonly influence policy making? To what extent is co-management possible in regions where traditionally semi-nomadic reindeer herders hold no title to land? What are the limits to co-management for addressing the problems of climate change in Alaska under the existing system of state–federal dual management? These questions highlight the need for more in-depth comparative research in this area of institutional analysis.

Wildlife management has always been a source of contention among wildlife users, and the adoption of co-management systems must be accompanied by trial periods to ensure that both government managers and wildlife user representatives have time to learn the process and accept their relative responsibilities prior to passing judgment on the effectiveness of the system. A major question raised is, can co-management systems manage wildlife populations, assuring their sustained production and conservation, if they become alarmingly depressed as a consequence of climate change or other causes? It seems reasonable to expect that effective management under such difficult conditions, whatever the management system, would require the ability to investigate causes of the population change as a basis for prescribing management action. If the users of wildlife are acknowledged to be a source of information about wildlife ecology, as well as participants in wildlife surveys and scientific investigations, then achieving an understanding of the relative importance of population regulatory mechanisms seems more likely than in management systems in which the users play no active role and managers live remote from the system (Klein et al., 1999). In a similar context, when management decisions are made within a true co-management system, the users, through their representation on the management board, are participants in a democratic process and are therefore more inclined to accept and comply with restrictive regulations than if management decisions are made by a remote governmental authority. Although regulations established through the co-management process may be more acceptable and complied with by the majority of resource users than regulations imposed from outside the region, total agreement is unlikely, and enforcement of harvest regulations is as important as in other management systems.

11.5.1.1. Lateral collaboration and cooperation

In addition to the hierarchical structure of management systems that are vertically structured within national or international jurisdictions, there is need for increasing lateral connections that result in sharing of knowledge, experience, and responsibility for wildlife management and conservation. Lateral connections can include increased interaction between communities sharing a common wildlife resource, between a community and an industrial development activity that both affect a wildlife resource but in differing ways, and region-to-region communication regarding experiences and knowledge about management of similar species. An example of the latter is the "Profile of Herds" concept being developed through an International Arctic Science Committee project. It provides a basis for inter-herd comparison of the management and conservation of caribou and wild and domestic reindeer. The project has as its goal the collation and organization of data on population status and dynamics, management practices, human interactions (herding, hunting, subsistence and commercial uses, and cultural relationships), and range size and characteristics of caribou and reindeer in a circumpolar context. The data are being archived through Environment Canada and the Institute of Arctic Biology at the University of Alaska Fairbanks, with access via www.rangifer.net. These files on caribou and reindeer herds throughout the Arctic and subarctic will enable ongoing comparison of harvest methods and levels, predator relationships, range conditions, and carrying capacity under varying climatic, environmental, and human influences, and under differing management regimes. Caribou and reindeer share common ecological relationships with their environment that are characteristic of the species, however, the relative importance of the driving variables within their environment may vary widely over the total range of distribution of the species. The capability to compare the relative effectiveness of a given herd management system with others throughout the North should assist in adapting management systems and practices in response to changes brought about by climate, industrial impacts on herd ranges and habitats, trends in subsistence and economic needs, and evolving indigenous cultures.

11.5.2. A regional land use perspective

In order to effectively manage wildlife within an environment of change in the Arctic, basic inventories of wildlife populations and their dynamics, and investigation of ecosystem relationships of wildlife on a regional basis are a prerequisite, as well as providing early warning indicators (Fig. 11.17). This information is critical to meet proximal needs of management for prescribing methods, means, and seasons of harvest and for setting harvest quotas. Inventory information is also critical for longer-term monitoring of animal populations and ecosystem relationships as a basis for assessing changes in distribution, movements, and population trajectories that may be the consequence of climate change or other human-induced changes in the natural environment. Basic inventory data on wildlife, wildlife habitats and movements, and patterns of human use of wildlife are also of critical importance in assessment of impacts of proposed development projects.

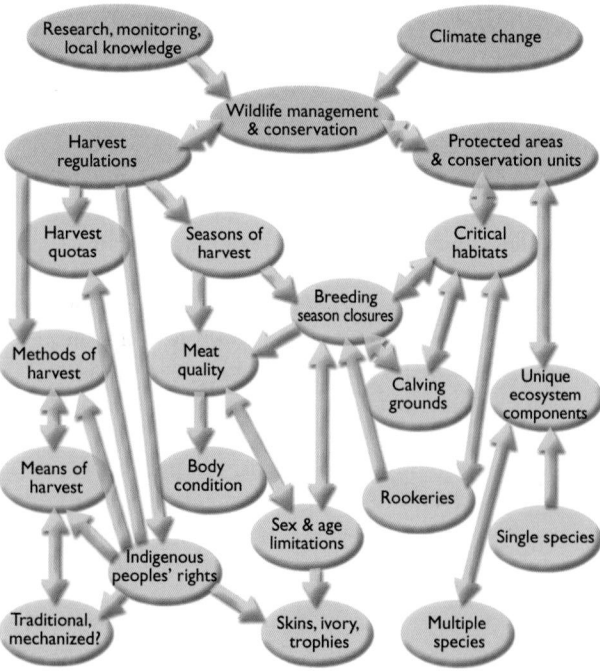

Fig. 11.17. Effective management of wildlife and its conservation involves accumulation of knowledge of animal population biology and ecological relationships through research, monitoring, and accessing local knowledge. This then provides a basis for defining critical habitats and providing for their legal protection, and for establishing wildlife harvest regulations with local involvement to ensure sustainability of wildlife populations with continuing opportunity for their harvest.

Needs for effective management and conservation of wildlife in a changing Arctic vary regionally. For example, to deal with threats to management and conservation of wildlife in the Russian North and to return effective wildlife management to the Russian Arctic and subarctic, the following changes in existing structures for management and their application are widely acknowledged as needed:

- adaptation of existing wildlife management systems consistent with existing social and economic conditions, constraints, and opportunities;
- elaboration of legal and economic mechanisms for protection of wildlife resources and habitats to ensure sustainability of wildlife populations and their production in conjunction with industrial resource development;
- elaboration of legal and economic mechanisms for protection of traditional hunting cultures in conjunction with industrial resource development;
- increasing the effectiveness and the technological level of commercial hunting, the processing of wildlife products, and their marketing consistent with resource conservation; and
- systematically organized inventory and monitoring of wildlife resources based on both scientific and traditional knowledge and methodology.

These needed changes, however, are not unique to Russia, and are basic to effective wildlife management and conservation throughout the Arctic. It is the needed focus on these structural components of management that is particularly timely in Russia in the current post-Soviet transition period.

The process of development of regional land use plans, with adequate wildlife inventory data available, enables layout of proposed human activities on the land, such as roads, communities, and other structures, in consideration of protection of wildlife habitat values, movement routes, and patterns of human use of the wildlife. Development of regional land use plans based on adequate wildlife inventory data should enable designation of protected areas to encompass critical wildlife habitats, such as caribou calving grounds, wetland bird nesting habitats, and coastal haul-out sites and nesting colonies of critical importance for marine mammals and birds. However, regional land use plans must be subject to periodic revision, based on continuing wildlife inventory and monitoring data, and therefore be adaptable to environmental change brought about through changes in climate, and the continuing and cumulative consequences on the land of all human activities. Thus, areas designated to protect critical wildlife habitat units may at times need to be altered through expansion, relocation, or removal of protection in response to major changes in wildlife distribution and habitat use brought about through climate-induced or other changes in the environment.

If land use plans are in place in arctic regions prior to proposals for large-scale industrial development projects, such as energy or mineral extraction, hydropower development, construction of roads, railroads, pipelines, and power-lines, initial decisions on the feasibility of proposed projects will be simplified. Project planning can proceed with knowledge of regional wildlife values that need to be protected, critical habitats that need to be avoided, and provisions necessary for the sustainable harvest and other uses of wildlife. The controversy, associated political polarization, and animosity that often develops among interest groups over proposed development projects in the Arctic can be minimized if comprehensive land use plans have been prepared. Efforts to develop comprehensive regional land use plans involving local residents and government are currently underway in a few regions of the Arctic. Examples include the Swedish MISTRA project, *Sustainable Management of the Mountain Region*, that includes assessment of the natural resources in the mountains of northern Sweden, their levels of use, and their economic and socio-cultural relationships within and outside the region in development of a land use plan aimed at long-term community and resource sustainability; and the Canadian *Yukon North Slope Wildlife Conservation and Management Plan* of a similar nature that evolved from the joint Alaska and Yukon *Community Sustainability Project*. Reindeer and caribou, and their ecosystem relationships and associated human dependency on them, have provided initial stimuli for development of these land use plans. Wild reindeer and the indigenous cultures that evolved in association with them are also the focus of recently initiated land use investigations in the Taymir of the western Siberian Arctic through the *Taymir Reindeer Project*, which is a first stage in development of a regional land use plan.

The concept of regional land use planning as a basis for management and conservation of wildlife in an environment of change also has application in the marine environment. In most of the marine environment of the Arctic, offshore petroleum exploration and production and permanent infrastructure development has not been at all comparable to that on land. Nevertheless, the need for protection of critical wildlife habitats and associated ecosystem relationships is as important in the marine environment as it is on land. The international or binational nature of many species of marine wildlife clearly requires international efforts in the development of marine area use agreements to ensure protection of critical habitats for marine wildlife. Planning processes for where to place major shipping routes, where bottom trawling can take place without irreversible damage to benthic systems, where ship-based tourist traffic can be focused to provide good experiences while minimizing effects on wildlife, and where restrictions on ice-breaking activity might be essential to protect breeding habitats of seal species all require information that is very similar to that needed to assess land-based human development activities in the Arctic. Marine ecosystems in the Arctic, and worldwide, are less well known than terrestrial ecosystems, largely because humans are land-dwelling creatures with limited capabilities for operating below the surface of the sea. The task of carrying out

needed research to understand ecosystem relationships of marine wildlife that spend major parts of their life cycles beneath the sea surface is, therefore, more complex. However, inventory and monitoring methods for assessing marine wildlife abundance are developing rapidly, as is tracking technology needed to record movement patterns and habitat use. So, although marine research tends to be more costly than terrestrial studies, a great deal is now possible that is highly relevant to developing good, responsive management practices.

Integral to the effective use of regional land use planning as a basis for management and conservation of wildlife in the Arctic is assessment of the cumulative impacts of development projects that have taken place within the region. Although environmental impact assessments of proposed major projects are now prescribed by government policy in most arctic countries, these assessments have been restricted to the project under consideration and have rarely considered the cumulative impacts on wildlife to which the proposed project would contribute. A recent assessment of the cumulative impacts of petroleum development in the Alaskan Arctic requested and financed by the U.S. Congress has pointed out major consequences for wildlife that have affected their management and conservation, and that were not anticipated through environmental assessments required for the individual projects (NRC, 2003).

11.5.3. Concluding recommendations

Shared responsibility for management and conservation of wildlife in the Arctic requires involvement, cooperation, and collaboration among all interest groups. Indigenous peoples of the Arctic, the majority of whom are dependent on annual harvests of wildlife for the subsistence component of their economy, are gaining increased, and often primary, responsibility for management of local harvests of wildlife. In most of the Arctic it is the indigenous peoples who will play the key role in management and conservation of wildlife (Klein, 2002). Many of the non-indigenous residents of the Arctic are also consumptive users of wildlife and depend upon wildlife as a supplement to their economy. Direct involvement of the users of wildlife in its management at the local level has the potential for rapid management response to changes in wildlife populations and their availability for harvest. Rapid response to changes in numbers and distribution of wildlife is a prerequisite for effective management of many resident species of arctic wildlife and their conservation under present conditions of limited predictability of ecosystem response to climate change, and is an important component in management of migratory species often requiring international collaboration.

Non-consumptive use of wildlife through viewing and photography, as part of tourism in the Arctic, can affect wildlife through disturbance and stress during sensitive periods in their annual cycle, by displacement from habitats, or through attraction to food wastes. Management of the relationship of tourism to wildlife in the Arctic requires collaboration between management regimes at regional, national, and international levels. Since marine species of wildlife are among the most accessible, and therefore attractive to tourists visiting the Arctic, ship-based tourism poses a major threat to arctic wildlife. Establishment of specific areas to ensure protection from excessive disturbance at breeding sites will continue to be an important part of wildlife conservation in the Arctic in an environment of change. Most tourism companies operating in the Arctic are based outside the Arctic, thus guidelines and regulations for management of tourism impacts on wildlife must include bi-national or international processes and cooperation. Assuring compliance with guidelines and enforcement of regulations also requires cooperation among countries that share wildlife resources that are the focus of the tourism industry.

Many wildlife species of importance as food and other components of the economy of arctic residents are migratory and therefore spend parts of their annual life cycles in different ecosystems, some of which may be at great distances from the Arctic. Migratory wildlife are necessarily subject to management responsibilities that transcend local interests, whether they move with the annual advance and recession of sea ice as many marine wildlife do, whether they travel overland seasonally to track food quality and availability characteristic of caribou and reindeer, or whether they journey through many thousands of kilometers from the Arctic to wintering areas as do most arctic nesting birds and many whale species that feed in arctic waters during the summer months. Management and conservation of migratory or wide-ranging species requires broad participation by all those with interests and responsibilities for arctic wildlife. This requires that management be expanded from local jurisdiction to include regional, national, and international collaboration and shared responsibility in management of migratory and wide-ranging wildlife. Spreading responsibility for management and conservation of wildlife over broader geographical interests is clearly important where it is not possible for those responsible at the local level to be aware of the status and ecosystem relationships of wildlife species after they have left the local area. Sharing the responsibility for management also generally results in greater total effort expended for the collection of biological and harvest information needed to ensure the well-being of wildlife populations. This may improve the chances for early detection of responses of migratory wildlife to the effects of climate change. Conversely, achieving action deemed necessary for management of migratory wildlife to compensate, correct, or adapt to climate-related changes may be difficult and drawn out because of bureaucratic complexity inherent in international governing bodies. Where international overseeing may be justified and needed for aspects of arctic wildlife management and conservation at the policy level, efficient and more timely execution of policy through management actions may be possible through a reduction in bureau-

cratic layering by delegation of authority to bi-regional or multi-regional councils or committees whose membership is representative of the specific national interests involved. Such management bodies would be most effective if their focus and responsibility were restricted to a single species (e.g., beluga whales) or a group of ecologically similar species (e.g., seabirds) and their membership included local users of the wildlife.

The role of international agencies and organizations in wildlife management and conservation is particularly important with regard to the insidious consequences of pollutants and contaminants entering arctic food chains largely from sources outside the Arctic. Inventories and monitoring of the pollutants and contaminants entering arctic ecosystems, and research on their consequences for the health of arctic wildlife as well as the health of the arctic residents who consume the wildlife, are critical to management of arctic wildlife. An understanding of the role of pollutants and contaminants in wildlife food chains, wildlife health, and associated human health, and the influence of climate change on these relationships, underlies interpretation of the consequences of other environmental variables on wildlife, which is basic to management and conservation of wildlife in the Arctic. Clearly, international oversight, coordination, reporting, collating of information, and associated stimulation of national efforts are needed to better understand the importance of pollutants and contaminants entering the Arctic. The reduction of levels of pollutants and contaminants entering the Arctic, and management of their impacts on arctic wildlife, will require action at the national level through joint international efforts.

Achieving effective conservation and management of wildlife in a changing Arctic will require a team-building approach among governments at all levels that relate to the environment and human well-being, and with all other groups with an interest in the Arctic. This effort should include the indigenous peoples and other residents of the Arctic, and scientists undertaking research in the Arctic, representatives of industry and business seeking development of arctic resources or other economic opportunities in the Arctic, those who travel to the Arctic for recreation or tourism, and the non-governmental organizations seeking to protect or sustain environmental, aesthetic, and other less tangible values of the Arctic in the broader interest of society. Interests in the Arctic by these diverse groups are often overlapping and sometimes conflicting, but the successful management and conservation of arctic wildlife requires that these groups be represented in the management process and that adequate information is available for equitable consideration of the diverse interests that relate to arctic wildlife. The role of international, non-governmental environmental organizations is particularly important in maintaining focus of the public on the broad spectrum of environmental values existing in the Arctic when proposals for large-scale industry- or government-sponsored projects become politicized at the regional or national levels.

Acknowledgements

We appreciate the helpful assistance and comments provided by the following people during the preparation of this chapter: Patricia Anderson, Andrew Balser, Terry Callaghan, Eric Cline, Dorothy Cooley, Barb Hameister, Olav Hjeljord, Igor Krupnik, Magdalena Muir, Pål Prestrud, Lynn Rosentrater, Michael Usher, and Kent Wohl. The Alaska office of the National Audubon Society, Yukon Renewable Resources, Canadian Porcupine Caribou Management Board, Sustainability of Arctic Communities Project, the Arctic Programme of the World Wildlife Fund, and the Institute of Arctic Biology of the University of Alaska Fairbanks provided information useful in preparation of the chapter.

Personal communications

Bogoslovskaya, L.S., 2003. Center of Traditional Subsistence Studies, Russian Research Institute of Cultural and Natural Heritage, Moscow.

Kasianov, M., 2002. Prime Minister of Russian Federation, Moscow.

Kofinas, G., 2004.

Mathiasen, K., 2004. Greenland Home Rule Government, Nuuk.

Nagy, J., 2001. Government of the Northwest Territories, Inuvik, Canada

Valkenburg, P., 2004.

Whitten, K., 2004.

References

Adams, M., K.J. Frost and L.A. Harwood, 1993. Alaska and Inuvialuit Beluga Whale Committee (AIBWC) - an initiative in 'at home management'. Community-based whaling in the North. Arctic, 46:134–137.

Adamczewski, J., 1995. Digestion and body composition in the muskox. Ph.D. Thesis, University of Saskatchewan.

AMAP, 1997. Arctic Pollution Issues: A State of the Arctic Environment Report. Arctic Monitoring and Assessment Programme, Oslo, 188pp.

AMAP, 1998a. The AMAP Assessment Report: Arctic pollution Issues. Arctic Monitoring and Assessment Programme, Oslo, xii + 859pp.

AMAP, 1998b. Arctic Monitoring and Assessment Programme. Hydrometeoizdat. St. Petersburg, 188pp.

AMAP, 2002. Arctic Pollution 2002. Arctic Monitoring and Assessment Programme, Oslo. xi + 111pp.

Ammosov, V.A., N.N. Bakeev, A.T. Voilochnikov, M.P. Vorobjova, S.N. Grakov, V.N. Derjagin, I.P. Karpuhin, G.V. Korsakov, S.A. Korytin, S.A. Larin, S.V. Marakov, B.A. Mihailovskii, A.P. Nikulcev, M.P. Pavlov, E.V. Stahrovskii and J.P. Jazan, 1973. Commercial Hunting in USSR. Lesnaja promyshlennost, Moscow. 408pp. (In Russian)

Anker-Nilssen, T., V. Bakken, H. Strøom, A.N. Golovkin, V.V. Bianki and I.P. Tatarinkova (eds.), 2000. The Status of Marine Birds Breeding in the Barents Sea Region. Norsk Polarinstitutt Rapport Nr. 113.

Argentov, A., 1857. A description of the Nikolaevskiy Chaunskiy parish. -Zapiski Sibirskogo otdeleniya imperatorskogo Rossiskogo Geograficheskogo obshchestva, 111(1):70–101. (In Russian)

Bakken, V. and T. Anker-Nilssen, 2001. Harvest of seabirds in North Norway and Svalbard. In: L. Denlinger and K. Wohl (eds.). Seabird Harvest Regimes in the Circumpolar Nations, pp.42–45. Conservation of Arctic Flora and Fauna, Technical Report No. 9.

Barr, W., 1991. Back from the brink: the road to muskox conservation in the Northwest Territories. Komatik Series 3. The Arctic Institute of North America, University of Calgary, 127pp.

Baskin, L.M., 1998. Hunting of game animals in the Soviet Union. In: E.J. Milner-Gulland and R. Mace (eds.). The Conservation of Biological Resources, pp. 331–345. Blackwell Science.

Belikov, S.E., 1993. Status of polar bear populations in the Russian Arctic 1993. In: Ø. Wiig, E.W. Born and G.W. Garner (eds.). Polar Bears: Proceedings of the Eleventh Working Meeting of the IUCN/SC Polar Bear Specialist Group, pp. 115–120. The World Conservation Union

Berman, M. and G.P. Kofinas, 2004. Hunting for models: rational choice and grounded approaches to analyzing climate effects on subsistence hunting in an arctic community. Ecological Economics, 49:31–46.

Benson, A. and A.W. Trites, 2002. Ecological effects of regime shifts in the Bering Sea and eastern North Pacific Ocean. Fish and Fisheries, 3(2):95–113.

Berkes, F. 1989. Co-management and the James Bay Agreement. In: E. Pinkerton (ed.). Co-operative Management of Local Fisheries: New Directions for Improved Management and Community Development, pp. 181–182. University of British Columbia Press.

Berkes, F. 2002. Cross-scale institutional linkages: perspectives form the bottom up. In: E. Ostrom, T. Dietz, N. Dol_ak, P.C. Stern, S. Stovich and E.U. Weber (eds.). The Drama of the Commons, pp. 293–322. National Academy Press.

Berkes, F. and C. Folke (eds.), 1998. Linking Social and Ecological Systems. Cambridge University Press, 459pp.

Blanchard, K.A., 1994. Culture and seabird conservation: the north shore of the Gulf of St. Lawrence, Canada. In: D.N. Nettleship, J. Burger and M. Gochfeld (eds.). Seabirds on Islands: Threats, Case Studies and Action Plans. Birdlife Conservation Series No. 1.

Born, E.W., 1993. Status of polar bears in Greenland (1993). In: Ø. Wiig, E.W. Born and G.W. Garner (eds.). Polar Bears: Proceedings of the Eleventh Working Meeting of the IUCN/SC Polar Bear Specialist Group, pp 81–104. IUCN/SC.

Born, E.W., I. Gjert and R.R Reeves, 1995. Population Assessment of Atlantic Walrus. Meddelelser om Gronland No. 138.

Boveng, P.L., J.L. Bengtson, D.E. Withrow, J.C. Cesarone, M.A. Simpkins, K.J. Frost and J.J. Burns, 2003. The abundance of harbor seals in the Gulf of Alaska. Marine Mammal Science, 19(1):111–127.

Burch, E.S. Jr., 1998. The Iñupiaq Eskimo nations of Northwest Alaska. University of Alaska Press, 473pp.

Bureau of Statistics, 1996. Statistics Quarterly. Department of Public Works and Services, Government of the Northwest Territories, Yellowknife, Northwest Territories. Vol. 18.

CAFF, 1996. International Murre Conservation Strategy and Action Plan. Conservation of Arctic Flora and Fauna, Reykjavik.

CAFF, 1997. Circumpolar Eider Conservation Strategy and Action Plan. Conservation of Arctic Flora and Fauna, Reykjavik.

CAFF, 2001a. Arctic Flora and Fauna: Status and Conservation. Conservation of Arctic Flora and Fauna, Helsinki, 272pp.

CAFF, 2001b. Seabird Harvest Regimes in the Circumpolar Nations. Conservation of Arctic Flora and Fauna, Technical Report No. 9, 56pp.

Cameron, R.D., W.T. Smith, R.G. White and B. Griffith, 2002. The Central Arctic Caribou Herd. In: D.C. Douglas, P.E. Reynolds and E.B. Rhode (eds.). Arctic Refuge Coastal Plain Terrestrial Wildlife Research Summaries, pp. 38–45. U.S. Geological Survey, Biological Resources Division, Biological Science Report USGS/BRD/BSR-2002–0001.

Caulfield, R.A., 1993. Aboriginal subsistence whaling in Greenland: the case of Qeqertarsuaq municipality in west Greenland. Community-based whaling in the North. Arctic, 46:144–155.

Caulfield, R.A., 1997. Greenlanders, Whales, and Whaling: Sustainability and Self-Determination in the Arctic. University Press of New England, 203pp.

Caughley, G., 1977. Analysis of Vertebrate Populations. John Wiley and Sons.

Chardine, J.W., 2001. Seabird harvests in Canada. In: L. Denlinger and K. Wohl (eds.). Seabird Harvest Regimes in the Circumpolar Nations, pp. 11–20. Conservation of Arctic Flora and Fauna, Technical Report No. 9.

Christensen, T., 2001. Seabird harvests in Greenland. In: L. Denlinger and K. Wohl (eds.). Seabird Harvest Regimes in the Circumpolar Nations. Conservation of Arctic Flora and Fauna, pp. 21–36. Technical Report No. 9.

Clark, C.W., 1976. Mathematical Bioeconomics: The Optimum Management of Renewable Resources. Wiley Interscience.

COSEWIC, 2002. Canadian Species at Risk, May 2002. Committee on the Status of Endangered Wildlife in Canada, 34pp.

Croxall, J.P., P.G.H. Evans and R.W. Schreiber (eds.), 1984. Status and Conservation of the World's Seabirds. Proceedings, International Council for Bird Preservation Symposium, Cambridge. ICBP Technical Publication No. 2.

CWS/CWF, 2002. Polar bear. Hinterland Who's who. Canadian Wildlife Service, Canadian Wildlife Federation. www.hww.ca/index [last accessed 18-Feb-04]

Denlinger, L.M. and K.D. Wohl, 2001. Harvest of seabirds in Alaska. In: L. Denlinger and K. Wohl (eds.). Seabird Harvest Regimes in the Circumpolar Nations, pp. 3–10. Conservation of Arctic Flora and Fauna, Technical Report No. 9.

Dezhkin, V.V. (ed.), 1978. Commercial Hunting in RSFSR. Lesnaja promyshlennost, Moscow, 256pp. (In Russian)

DFO, 1998a. Baffin Bay Narwhal. Department of Fisheries and Oceans, Canada. Science Stock Status Report E5-43.

DFO, 1998b. Hudson Bay Narwhal. Department of Fisheries and Oceans, Canada, Science Stock Status Report E5-44.

DFO, 1999. Hudson Bay-Foxe Basin Bowhead whales. Department of Fisheries and Oceans, Canada, Science Stock Status Report E5-52.

DFO, 2000. Eastern Beaufort Sea Beluga Whales. Department of Fisheries and Oceans, Canada, Science Stock Status Report E5-38.

DFO, 2002a. Atlantic seal hunt - 2002 Management Plan. Department of Fisheries and Oceans, Canada.

DFO, 2002b. Beluga. Underwater World, Department of Fisheries and Oceans, Canada, 10pp.

Dinesman, L.G., N.K. Kiseleva, A.B. Savinetsky and B.F. Khassanov, 1996. Secular Dynamics of Coastal One of Northeastern Chukotka. Argus, Moscow, 189pp.

Dobrinsky, L.N. (ed.), 1995. Nature of Yamal. Nauka, Ekaterinburg, 435pp. (In Russian)

Dobrinsky, L.N. (ed.), 1997. Monitoring of the Biota at Yamal Peninsula in Relation to the Development of Facilities for Gas Extraction and Transportation. Ekaterinburg, 191pp. (In Russian)

Donovan, G.P. (ed.), 1982. Aboriginal Subsistence Whaling. Reports of the International Whaling Commission, Special Issue 4.

Douglas, D.C., P.E. Reynolds and E.B. Rhode (eds.), 2002. Arctic Refuge Coastal Plain Terrestrial Wildlife Research Summaries. U.S. Geological Survey, Biological Resources Division, Biological Science Report USGS/BRD/BSR-2002-0001.

Downie, D.L. and T. Fenge (ed.), 2003. Northern Lights against POPs: Combating Toxic Threats in the Arctic. McGill University Press, xxv + 347pp.

Elkin, B.T. and R.W. Bethke, 1995. Environmental contaminants in the Northwest Territories, Canada. Science of the Total Environment, 160/161:307–321.

Estes, J.A., M.T. Tinker, T.M. Williams and D.F. Doak, 1998. Killer whale predation on sea otters linking ocean and nearshore systems. Science, 282:473–476.

Ferguson, M.A.D. and F. Messier, 1997. Collection and analysis of traditional ecological knowledge about a population of Arctic tundra caribou. Arctic, 50:17–28.

Finley, K.J., 2001. Natural history and conservation of the Greenland whale, or bowhead in the Northwest Atlantic. Arctic, 54(1):55–76.

Forbes, B.C., 1999. Reindeer herding and petroleum development on Poluostrov Yamal: sustainable or mutually incompatible uses? Polar Record, 35(195):317–322.

Freeman, M.M.R., 1975. Assessing movement in an Arctic caribou population. Journal of Environmental Management, 3:251–257.

Freeman, M.M.R. 1989. The Alaska Eskimo Whaling Commission: successful co-management under extreme conditions. In: E. Pinkerton, (ed). Co-operative Management of Local Fisheries: New Directions for Improved Management and Community Development, pp. 137–154. University of British Columbia Press.

Freeman, M.M.R. and U.P. Kreuter (ed.), 1994. Elephants and whales: resources for whom? Gordon and Breach, Basel, xiii + 321pp.

Gentry, R.L., 2002. Northern fur seal. In: W.F. Perrin, B. Wursig and J.G.M. Thewissen (eds.). Encyclopedia of Marine Mammals, pp. 813. Academic Press.

Giddings, J.L., 1967. Ancient Man of the Arctic. Alfred J. Knopf, New York, 391pp.

Golovkin, A., 2001. Seabird harvest in Russia. In: L. Denlinger and K. Wohl (eds.). Seabird Harvest Regimes in the Circumpolar Nations, pp. 44–46. Conservation of Arctic Flora and Fauna, Technical Report No. 9.

Government of the Russian Federation, 2001. Government of the Russian Federation Decree 20 November 2001 #1551-p. To approve enclosed Limits of Total Allowed Catches of aquatic biological resources in internal marine waters, in territorial waters, on continental shelf and in exclusive economic zone of the Russian Federation in 2002. Prime Minister of the Russian Federation, M. Kasianov.

Graf, R. and R. Case, 1989. Counting muskoxen in the Northwest Territories. Canadian Journal of Zoology, 67:1112–1115.

Grayson, D.K., 2001. The archaeological record of human impacts on animal populations. Journal of World Prehistory, 15:1–68.

Griffith, B., D.C. Douglas, N.E. Walsh, D.D. Young, T.R. McCabe, D.E. Russell, R.G. White, R.D. Cameron and K.R. Whitten, 2002. The Porcupine Caribou Herd. In: D.C. Douglas, P.E. Reynolds and E.B. Rhode (eds). Arctic Refuge Coastal Plain Terrestrial Wildlife Research Summaries, pp. 8–37. U.S. Geological Survey, Biological Resources Division.

Gunn, A. and B. Fournier, 1998. Muskox numbers and distribution in the Northwest Territories, 1997. Northwest Territories Department of Resources, Wildlife and Economic Development. File Rep. No. 121. 55pp.

Gunn, A., C.C. Shank and B. McLean, 1991. The status and management of muskoxen on Banks Island. Arctic, 44:188–195.

Gunn, A., U.S. Seal and P.S. Miller (eds.), 1998. Population and Habitat Viability Assessment Workshop for Peary Caribou and Arctic-Island Caribou (Rangifer tarandus). Conservation Breeding Specialist Group, Apple Valley, Minnesota.

Gunn, A., F.L. Miller and J. Nishi, 2000. Status of endangered and threatened caribou on Canada's Arctic islands. Rangifer Special Issue 12.

Hanna, S.S., C. Folke and K-G. Maler, (eds.), 1996. Rights to nature: ecological, economic, cultural, and political principles of institutions for the environment. Island Press, Washington, DC.

Hario, M., 2001. Review of the hunting regime of seabirds in Finland. In: L. Denlinger and K. Wohl (eds.). Seabird Harvest Regimes in the Circumpolar Nations, pp. 17–20. Conservation of Arctic Flora and Fauna, Technical Report No. 9.

Hardin, G., 1968. Tragedy of the commons. Science, 162:1243–1248.

Heard, D.C., 1985. Caribou census methods used in the Northwest Territories. McGill Subarctic Research Papers, 40:229–238.

Holling, C.S., 1986. The resilience of terrestrial ecosystems: local surprise and global change. In: W.C. Clark and R.E. Munn (eds.). Sustainable Development of the Biosphere, pp. 292–317. Cambridge University Press.

Huntington, H.P., 1992. Wildlife Management and Subsistence Hunting in Alaska. Belhaven Press, xvii + 177pp.

IISD, 1999. Inuit observations on climate change. Trip report 1, International Institute for Sustainable Development, Winnipeg, 20pp.

IPCB, 1993. Sensitive Habitats of the Porcupine Caribou Herd. Report of the International Porcupine Caribou Board by the Porcupine Technical Committee, United States and Canada.

IUCN Polar Bear Specialist Group, 1998. Worldwide status of the polar bear. In: Proceedings of the 12th Working Meeting of the IUCN Polar Bear Specialists, January 1997, Oslo.

IWC, 2002. Catch Limits for Aboriginal Subsistence Whaling. International Whaling Commission. www.iwcoffice.org/2002PressRelease [last accessed 18-Feb-04]

Jefferies, R.L., J. Svoboda, G. Henry, M. Raillard and R. Ruess, 1992. Tundra grazing systems and climatic change. In: F.S. Chapin, R.L. Jefferies, J.F. Reynolds, G.R. Shaver, J. Svoboda and E.W. Chu (eds.). Arctic Ecosystems in a Changing Climate: An Ecophysiological Perspective, pp. 391–412. Academic Press.

Jentoft, S. (ed.), 1998. Commons in a Cold Climate: Coastal Fisheries and Reindeer Pastoralism in North Norway: The Co-management Approach. Parthenon Publishing Group, Tromsø. 372pp.

Jessen, A., 2001. Seals in the Marine Ecosystem. Report of the Seal Seminar, March 20 and 21, Nuuk, Greenland. Nordic Council of Ministers.

Kelsall, J., 1968. The Migratory Barren-ground Caribou of Canada. Canadian Wildlife Service Monograph No. 3.

Khlobystin, L., 1996. Eastern Siberia and Far East, Neolithic of Northern Eurasia. Arkheologiya, 3:270–329. (In Russian)

Kinloch, D., H. Kuhnlein and D.C.G. Muir, 1992. Inuit foods and diet: a preliminary assessment of benefits and risks. Science of the Total Environment, 122:247–278.

Klein, D.R., 1973. The impact of oil development in the northern environment. Proceedings of the 3rd Interpetrol Congress, Rome, Petrolio e ambiente, pp. 109–121.

Klein, D.R., 1979. The Alaska Oil Pipeline in retrospect. Transactions of the North American Wildlife and Natural Resources Conference, 44:235–246.

Klein, D.R., 1992. Comparative ecological and behavioral adaptations of Ovibos moschatus and Rangifer tarandus. Rangifer, 12:47–55.

Klein, D.R., 2002. Perspectives on wilderness in the Arctic. In: A. Watson, L. Alessa and J. Sproull (eds.). Wilderness in the Circumpolar North: Searching for Compatibility in Ecological, Traditional, and Ecotourism Values, pp. 1–6. 2001 May 15–16; Anchorage.

Klein, D.R. and C. Bay, 1994. Resource partitioning by mammalian herbivores in the high Arctic. Oecologia, 97:439–450.

Klein, D.R. and L.S. Kolpashchikov, 1991. Current status of the Soviet Union's largest caribou herd. In: C. Cutler and S.P. Mahoney (eds.). Proceedings of the 4th North American Caribou Workshop, pp. 251–255. St. John's, Newfoundland.

Klein, D.R., L. Moorhead, J. Kruse and S.R. Braund, 1999. Contrasts in use and perceptions of biological data for caribou management. Wildlife Society Bulletin, 27:488–498.

Klein, D.R. and M. Magomedova, 2003. Industrial development and wildlife in arctic ecosystems: can learning from the past lead to a brighter future? In: R.O. Rasmussen and N.E. Koroleva (eds.). Social and Economic Impacts in the North, pp. 35–56. Kluwer Academic.

Knuth, E., 1967. Archaeology of the Musk-ox Way. Contributions du Centre d' Études Arctiques et Finno-Scandinaves, No. 5. École Pratigue des Hautes Études, Paris, 78pp.

Kofinas, G. P., 1998. The cost of power sharing: Community involvement in Canadian Porcupine caribou co-management. Ph.D. Thesis, University of British Columbia, 467pp.

Kofinas, G. (with contributions from the communities of Aklavik, Arctic Village, Old Crow, and Fort McPherson), 2002. Community contributions to ecological monitoring: knowledge co-production in the U.S.-Canada Arctic borderlands. In: I. Krupnik and D. Jolly (eds.). The Earth is Faster Now: Indigenous Observations of Arctic Environmental Change, pp. 54–91. Arctic Research Consortium of the United States, Fairbanks, Alaska.

Kofinas, G., C. Nicolson, M. Berman and P. McNeil, 2002. Caribou Harvesting Strategies and Sustainability Workshop Proceedings. Inuvik, Northwest Territories, April 15–16, 2002. NSF Sustainability of Arctic Communities Project (Phase II).

Krupnik, I. and D. Jolly (eds.), 2002. The Earth is Faster Now: Indigenous Observations of Arctic Environmental Change. Arctic Research Consortium of the United States, Fairbanks, Alaska, 384pp.

Kruse, J., D. R. Klein, L. Moorehead, B. Simeone and S. Braund, 1998. Co-management of natural resources: a comparison of two caribou management systems. Human Organization, 57:447–458.

Kruse, J.A., R.G. White, H.E. Epstein, B. Archie, M.D. Berman, S.R. Braund, F.S. Chapin III, J. Charlie Sr., C.J. Daniel, J. Eamer, N. Flanders, B. Griffith, S. Haley, L. Huskey, B. Joseph, D.R. Klein, G.P. Kofinas, S.M. Martin, S.M. Murphy, W. Nebesky, C. Nicolson, D.E. Russell, J. Tetlichi, A. Tussing, M.D. Walker and O.R.Young, 2004. Sustainability of arctic communities: an interdisciplinary collaboration of researchers and local knowledge holders. Ecosystems, 7:1–14.

Larter, N. and J. Nagy, 1999. Sex and age classification surveys of Peary caribou on Banks Island, 1982–1998: a review. Northwest Territories Department of Resources, Wildlife and Economic Development. Manuscript Rep. No. 114, 33pp.

Leopold, A., 1933. Game Management. Macmillan.

Leopold, A., 1938. Conservation esthetic. Bird Lore, 40:101–109.

Leopold, A., 1949. A Sand County almanac. Oxford University Press.

Leopold, A., 1953. Round River. Random House.

Loughlin, T.R., 2002. Steller's sea lion. In: W.F. Perrin, B. Wursig and J.G.M. Thewissen (eds.). Encyclopedia of Marine Mammals, pp. 1181–1185. Academic Press.

Lunn, N.J., S. Schliebe and E.W. Born, 2002. Polar Bears, Proceedings of the 13th Working Meeting of the IUCN/SSC Polar Bear Specialist Group, 23–28 June 2001, Nuuk, Greenland. Occasional Paper of the IUCN Species Survival Commission No. 26, 155pp.

McCluskey, K., 1999. Managing walrus: NWMB beginning to look at a new system. Northern News Service, October 04/99.

Meldgaard, M., 1986. The Greenland caribou – zoogeography, taxonomy, and population dynamics. Meddelelser om Grønland, Bioscience, 20:1–88.

Melnikov, S.A., C.V. Vlasov, O.V. Rishov, A.N. Gorshkov and A.I. Kuzin, 1994. Zones of relatively enhanced contamination levels in the Russian Arctic Seas. Arctic Research of the United States, 18:277–283.

Meltzer, D.L., 1997. Monte Verde and the Pleistocene peopling of the Americas. Science, 276:754–755.

Michurin, L.N., 1965. Wild reindeer of the Taimyr Peninsula and rational utilization of its resources [Dikiy severnyi olen' Taymyrskogo polu-ostrova/ratsional'naya utilizatsiya ego resursov]. Thesis. Vsesoyuznyi sel'skokhozyaistvennyi institut zaochnogo obucheniya publ, Moscow. 24pp. (In Russian)

Miller, F.L., 1990. Peary Caribou Status Report. Environment Canada, Canadian Wildlife Service Western and Northern Region, 64pp.

Miller, F.L. and A. Gunn, 1978. Inter-island movements of Peary caribou south of Viscount Melville Sound, Northwest Territories. Canadian Field Naturalist, 92(4):331–333.

Milner-Gulland, E.J. and R. Mace (eds.), 1998. Conservation of Biological Resources. E.J. Blackwell Science.

Nagy, J.A., N.C. Larter and V.P. Fraser, 1996. Population demography of Peary caribou and muskox on Banks Island, N.W.T., 1982–1992. Rangifer Special Issue No. 9:213–222.

Newton, J., 2001. Background document to Climate Change Policy Options in Northern Canada. John Newton Associates.

NMFS, 2001. Co-management agreement between the Aleut community of St. George Island and the National Marine Fisheries Service. National Marine Fisheries Service, Seattle.

Nordic Ministers Advisory Board, 1999. Nature protection in the Arctic - Nordic strategy for the protection of nature and cultural heritage in the Arctic - Greenland, Iceland and Svalbard, 95pp.

NRC, 2003. Assessment of the cumulative effects of petroleum development on Alaska's North Slope. National Academy Press, xiii + 452pp.

Nuttall, M., 1992. Arctic Homeland: Kinship, Community and Development in Northwest Greenland. Belhaven Press, 194pp.

Nuttall, M., 2000. Indigenous peoples, self-determination, and the Arctic environment. In: M. Nuttall and T.V. Callaghan (eds.). The Arctic: Environment, People, Policy, pp. 377–409. Harwood Academic Publishers.

Oechel, W.C., T. Callaghan, J.I. Holten, B. Maxwell, U. Molau and B. Sveinbjornsson (eds.), 1997. Global Change and Arctic Terrestrial Ecosystems. Springer-Verlag. 493pp.

Osherenko, G., 1988. Can co-management save arctic wildlife? Environment, 30:6–13,29–34.

Osherenko, G., and O. R. Young, 1989. The age of the Arctic: Hot conflicts and cold realities. Cambridge University Press.

Ostrom, E., 1990. Governing the commons: The evolution of institutions for collective action. Cambridge University Press.

Ostrom, E., T. Dietz, N. Dolsak, P.C. Stern, S Stovich and E.U. Weber (eds.), 2002. The Drama of the Commons. The Committee on the Human Dimensions of Global Change. Division of Behavioral and Social Sciences and Education. National Academy Press.

Pavlov, B.M., L.A. Kolpashchikov and V.A. Zyryanov, 1993. Taimyr wild reindeer populations: management experiment. Rangifer, Special Issue 9:381–384.

Pars, T., M. Osler and P. Bjerregaard, 2001. Contemporary use of traditional and imported food among Greenlandic Inuit. Arctic, 54(1):22–31.

Petersen, A., 2001. Review of the hunting and harvest regimes for seabirds in Iceland. In: L. Denlinger and K. Wohl (eds.). Seabird Harvest Regimes in the Circumpolar Nations, pp. 37–40. Conservation of Arctic Flora and Fauna, Technical Report No. 9.

Pinkerton, E. (ed.), 1989. Co-operative Management of Local Fisheries: New Directions for Improved Management and Community Development. University of British Columbia Press. 312pp.

Polischuk, S.C., R.J. Letcher, R.J. Norstrom, S.A. Atkinson and M.A. Ramsay, 1994. Relationship between PCB concentration, body burden and percent body fat in female polar bears while fasting. Organohalogen Compounds, 20:535–539.

Popov, A.A., 1948. Nganasany. Material culture. Akademiya Nauka Publ. 128pp. (In Russian)

Post, E. and M.C. Forchhammer, 2002. Synchronization of animal population dynamics by large scale climate fluctuations. Nature, 420:168–171.

Richard, P.R. and D.G. Pike, 1993. Small whale co-management in the eastern Canadian Arctic: a case history and analysis. Arctic, 46:138–155.

Riedlinger, D., 1999. Climate change and the Inuvialuit of Banks Island, NWT: using traditional environmental knowledge to complement western science. Arctic, 52(4):430–432.

Riedlinger, D., 2002. Responding to climate change in northern communities: impacts and adaptations. Arctic, 54(1):96–98.

Riewe, R.R. and L. Gamble, 1988. The Inuit and wildlife management today. In: M.M.R. Freeman and L.N. Carbyn (eds.). Traditional Knowledge and Renewable Resources Management in Northern Regions, pp. 31–37. Boreal Institute for Northern Studies.

Safronov, V.M., I.S. Reshetnikov and A.K. Akhremenko, 1999. Reindeer of Yakutiya. Ecology, morphology and use. Nauka Publ., 222pp. (In Russian)

Sarkin, A.V., 1977. Establishing and economy of hunting and restoration of hunting resources in the state hunting husbandry 'Taymyrskiy'. In: G.A. Sokolov (ed.). Ekologiya/ispol'zovanie okhotnich'ikh zhivotnykh Krasnoyarskogo kraya, pp. 84–88. Institut Lesa Publ. (In Russian)

Savel'ev, V.D., 1977. Behavior of reindeer in river-crossings. In: G.A. Sokolov (ed.). Ekologiya/ispol'zovanie okhotnich'ikh zhivotnykh Krasnoyarskogo kraya, pp. 17–20. Institut Lesa Publ., (In Russian)

Schledermann, P., 1996. Voices in stone. Komatik Series No. 5. Arctic Institute of North America, University of Calgary, 221pp.

Serreze, M.C., J.E. Walsh, F.S. Chapin III, T. Osterkamp. M. Romanovsky, W.C. Oechel, J. Morison, T. Zhang and R.G. Barry, 2000. Observational evidence of recent change in the northern high-latitude environment. Climate Change, 46:159–207.

Shapalin, B.F., 1990. The problems of economic and social development of the Soviet North. In: V.M. Kotlyakov and V.E Sokolov (eds.): Arctic Research. Advances and Prospects, pp. 415–419. Proceedings of the Conference of Arctic and Nordic Countries on Coordination of Research in the Arctic, Leningrad, December 1988. Part 2. Moscow: Nauka..

Stirling, I., N.J. Lunn and J. Iacozza, 1999. Long-term trends in the population ecology of polar bears in western Hudson Bay in relation to climatic change. Arctic, 52:294–306.

Syroechkovskii, E.E., 1995. Wild Reindeer. Smithsonian Institution Libraries, Washington, D.C., 290pp. (translated from the 1986 Russian edition).

Taylor, M., D. DeMaster, F.L. Bunnell and R.E. Schweinsburg, 1987. Modelling the sustainable harvest of female polar bears. Journal Wildlife Management, 51:811–820.

Thomas, D.C., 1982. The relationship between fertility and fat reserves of Peary caribou. Canadian Journal Zoology, 60:597–602.

Thomas, D.C. and J. Schaefer, 1991. Wildlife co-management defined: The Beverley and Kamanuruak Caribou Management Board. Proceedings of the Fifth North American Caribou Workshop. Rangifer, Special Issue 7:73–89.

Thorpe, N., 2000. Contributions of Inuit Ecological knowledge to
   Understanding the Impacts of Climate Change on the Bathurst
   Caribou Herd in the Kitikmeot Region, Nunavut. M.R.M. Thesis,
   Simon Fraser University, British Columbia, 279pp.
USFWS, 1994. Conservation Plan for the Polar Bear in Alaska. United
   States Fish and Wildlife Service.
USFWS, 1997. Final Environmental Assessment - Development of
   Proposed Treaty US/Russian Bilateral Agreement for the
   Conservation of Polar Bears in the Chukchi/Bering Seas. United
   States Fish and Wildlife Service.
USFWS, 2001. Subsistence Migratory Bird Harvest Survey. United
   States Fish and Wildlife Service.
USFWS, 2002a. Polar Bear (*Ursus maritimus*): Chukchi/Bering Sea Stock.
   United States Fish and Wildlife Service.
USFWS, 2002b. Polar Bear (*Ursus maritimus*): Southern Beaufort Sea
   Stock. United States Fish and Wildlife Service.
USFWS, 2002c. Pacific Walrus (*Odobenus rosmarus divergens*): Alaska
   Stock. United States Fish and Wildlife Service.
USFWS, 2002d. Sea Otter (*Enhydra lutris*): Southwest Alaska Stock.
   United States Fish and Wildlife Service.
USFWS, 2002e. Sea Otter (*Enhydra lutris*): Southeast Alaska Stock.
   United States Fish and Wildlife Service.
USFWS, 2002f. Sea Otter (*Enhydra lutris*): Southcentral Alaska Stock.
   United States Fish and Wildlife Service.
USFWS. 2002g. Stock assessment for polar bear (*Ursus maritimus*) -
   Chukchi/Bering Sea Stock (revision 14 March). United States Fish
   and Wildlife Service, 6pp.
Usher, P., 1995. Co-management of natural resources: some aspects of
   the Canadian experience. In: D.L. Peterson and D.R. Johnson (eds.).
   Human Ecology and Climate Change, pp. 197–206. Taylor and
   Francis.
Vdovin, I.S., 1965 Historical and ethnographic overview of Chukchi.
   Nauka Publ., 403pp. (In Russian)
Vibe, C., 1967. Arctic animals in relation to climatic fluctuations.
   Meddelelser om Grønland, Bd., 170(5):1–227.
Weissling, L.E., 1989. Arctic Canada and Zambia: a comparison of the
   development processes in the Fourth and Third Worlds. Arctic,
   42:206–216.
Weladji, R., D.R. Klein, Ø. Holand and A. Mysterud, 2002.
   Comparative response of *Rangifer tarandus* and other northern
   ungulates to climatic variability. Rangifer, 22:33–50.
Weller, G., P. Anderson and B. Wang (eds.), 1999. Preparing for a
   Changing Climate: The Potential Consequences of Climate Variability
   and Change. A report of the Alaska Regional Assessment group pre-
   pared for the U.S. Global Change Research Program. Center for
   Global Change and Arctic System Research, Fairbanks, Alaska.
Wenzel, G.W., 1995. Animal Rights, Human Rights. Ecology,
   Economy and Ideology in the Canadian Arctic. University of
   Toronto Press, 206pp.
Wiig, Ø., 1995. Status of polar bears in Norway (1993). In: Ø. Wiig,
   E.W. Born and G.W. Garner (eds.). Polar Bears: Proceedings of the
   Eleventh Working Meeting of the IUCN/SSC Polar Bear Specialist
   Group, January 25–29 1993. pp. 109–114.
Wiig, Ø, E.W. Born and G.W. Garner (eds.), 1995. Polar Bears:
   Proceedings of the Eleventh Working Meeting of the IUCN/SSC
   Polar Bear Specialist Group, January 25–29 1993. v + 192pp.
WWF, 2001. The Conservation First Principle. World Wildlife Fund,
   Ottawa, Canada.
Yablokov, A.A. (ed.), 1996. Russian Arctic: On the Edge of
   Catastrophe. The Centre of Ecological Politics of Russia, Moscow,
   207pp. (In Russian)
Young, O.R., 2001. The Institutional Dimensions of Environmental
   Change: Fit, Interplay, and Scale. MIT Press.
Young, O.R. and G. Osherenko (eds), 1993. Polar Politics: Creating
   International Environmental Regimes. Cornell University Press.
Yurpalov, S.Y., V.G. Loginov, M.A. Magomedova and V.D. Bogdanov,
   2001. Traditional economy in conditions of industrial expansion
   (as an example of Yamal-Nenets autonomous Okrug). UD RAN,
   Ekaterinburg, 36pp. (In Russian)
Zabrodin, V.A. and B.M. Pavlov, 1983. Status and rational use of Taimyr
   population of wild reindeer. In: V.E. Razmakhnin (ed.). Dikiy sev-
   ernyi olen'. Ekologiya, voprosy okhrany/ratsional'nogo ispol'zo-
   vaniya, pp. 60–75. TSNIL Glavokhota Publ., Moscow. (In Russian)
Zabrodin, V.A., A.M. Karelov and A.V. Dragan, 1989. Commercial
   Hunting in the Far North. Agropromizdat, Moscow, 204pp.
   (In Russian)

Appendix. Canadian co-management of the Porcupine Caribou Herd, toward sustainability under conditions of climate change

Climate change projections are an additional factor that must be incorporated into co-management considerations. As it is unclear if and how humans can affect the trajectories of climate change, the ultimate effects of climate change on indigenous caribou hunting societies are likely to depend on the capacity of their management systems to detect change, decipher its implications, and facilitate human adaptation in ways that meet societal needs. Also, it will be difficult to differentiate possible effects of climate change on the ecology of caribou from other human-induced influences, such as habitat fragmentation or disturbance from industrial development, construction of transportation corridors, or expanding tourism. Thus, there is a need for caribou management arrangements to be highly adaptive in their approach, and thus more resilient.

Porcupine Caribou and their environment

The Porcupine Caribou Herd is one of approximately 184 wild *Rangifer tarandus* herds (102 in North America), is the eighth largest herd in North America, and the largest shared migratory herd of mammals of the United States and Canada. The Porcupine Herd has been monitored and the subject of intensive research since the early 1970s. The population has grown at about 4% per year since the early 1970s to a high of 178 000 animals in 1989. During this period all major herds increased throughout North America. The synchrony in the population trends of these herds suggests that they have been responding to continental-scale events, presumably weather-related. Since 1989 the herd has declined at 3.5% per year to a low of 123 000 in 2001. Compared to other migratory herds across North America, the Porcupine Herd has the lowest growth rate and one of the highest adult cow mortality rates (Griffith et al., 2002).

Institutional and organizational features of the Porcupine Caribou Herd management system

From an institutional and organizational viewpoint, the Porcupine Caribou Herd system is complex, including two nation states, seven indigenous claimant groups, three territorial/state-level governments, and approximately 17 local communities (see Kofinas, 1998). This complexity contains important contrasts, including legal and cultural differences between U.S. and Canadian governance systems, and highlights the need for coordination of activities in uses of the herd and its habitat.

Two agreements specifically contain language for this type of coordination and provide for Canadian local community involvement. These agreements are the Canadian Porcupine Caribou Management Agreement and the Agreement between the Government of Canada and the Government of the United States of America on the Conservation of Porcupine Caribou. This case study deals primarily with the Canadian agreement. The Canadian Porcupine Caribou Management Agreement was signed in 1985 by federal and territorial governments and indigenous organizations of the region. The agreement is implemented by the Porcupine Caribou Management Board, which includes an equal number of indigenous and other representatives. The Porcupine Caribou Management Agreement states objectives that its signatories cooperatively manage the Porcupine Caribou Herd and its habitat within Canada so as to ensure the conservation of the herd with a view to providing for the ongoing subsistence needs of indigenous users; to provide for participation of indigenous users in Porcupine Caribou Herd management; to protect certain priority harvesting rights in the Porcupine Caribou Herd for indigenous users, while acknowledging that other users may also share the harvest; acknowledge the rights of indigenous users; and to improve communications between governments, indigenous users, and others with regard to the management of the Porcupine Caribou Herd within Canada. The Porcupine Caribou Management Agreement states that the Porcupine Caribou Management Board is an advisory body to the Canadian federal and territorial governments, and is directed to assume responsibility for harvest allocations in the event that it determines that they are needed. With respect to the imposition of harvest quotas, a burden of proof rests with government management agencies that such actions are warranted by conservation needs.

The Porcupine Caribou Management Agreement is a single population co-management arrangement, with its jurisdictional authority limited, by the terms of the agreement, to activities in Canada. Although it has no jurisdictional authority over activities in the United States, by virtue of the subsequently signed bi-lateral U.S.–Canada Porcupine Caribou Herd agreement it is linked to the International Porcupine Caribou Board for such activities. The table (p. 646) lists various functions of the Canadian co-management, problem areas, and provisions in the Canadian Porcupine Caribou Herd agreement providing for community involvement.

Structure of the Canadian Porcupine Caribou Management Board, showing proportional representation on the board of user groups, biologists, and government agency managers.

Implications of climate change for management

While there is considerable uncertainty regarding the specific impacts of climate change on caribou and caribou hunting, it is clear that climate change is likely to affect caribou body condition, herd movements and distribution, and abundance, as well as hunters' access to hunting grounds (Berman and Kofinas, 2004; see also section 12.3.5). Given the central role of caribou in the socio-cultural systems of indigenous people of the region, the negative impacts of climate change could result in significant social costs.

The capacity of a co-management regime to limit vulnerabilities and facilitate human adaptation in conditions of climate change is critical to the long-term sustainability of the system. More specifically, climate change suggests that certain functions of wildlife management may be critical in coping with possible climate change scenarios. They include:

- creating a regional forum for deliberations on caribou management issues;
- maintaining collaborative and systematic ecological monitoring;
- focusing research that draws on local and scientific knowledge;
- evaluating sensitive habitat, protecting important habitat, and participating in impact assessments;
- developing a strategic harvest management plan;
- overseeing appropriate policies for traditional barter and trade;
- guiding effective forest fire management polices and practice;
- developing climate-related communication tools; and
- achieving regional consensus and good compliance with co-management endorsed policies.

Creating a regional forum for deliberations on caribou management issues

For 17 years Porcupine Caribou Management Board members have met on caribou related issues and concerns. Meetings occur in local communities and regional centers on a rotating basis to ensure broad community input. Board-level transactions provide the basis for building relations among members who collectively represent various organizations as well as differing perspectives. Achieving consensus among the many Porcupine Caribou Herd stakeholders is rarely simple, yet the terms of the Porcupine Caribou Management Agreement and its board forum provide mechanisms for linking local, regional, and international decision-making.

The linkages in this process, however, have included problems. Western notions of efficient, representative, democratic process differ from traditional indigenous notions of consensus. Moreover, meeting agendas are often overwhelmed by discussion on policy issues, leaving little time to explore the broader implications of climate change to the region and the Porcupine Caribou Herd.

Overview of key Porcupine Caribou Management Agreement terms for community involvement (Kofinas, 1998).

Function	Problem area	Provisions guiding community involvement in various activity areas
Communication	Linking community with management	Agreement explicitly states that communities will participate and sit on board with government members. Authority for community membership is held by signatory organizations and the co-management board chair is selected by the board's membership.
Research and data collection	Adequate knowledge of caribou resource and its habitat	Agreement gives co-management board role in reviewing research and methods and encourages community members to participate in the collection of data.
Impact assessment and habitat protection	Providing a role for community to participate in the assessment of impacts and protection of habitat	Agreement includes directive to conserve resource and habitat. Also directs co-management board to participate in land management planning and impact reviews.
Policy on caribou as an exchangeable good	Exploitation of caribou resource by unchecked market forces and maintenance of traditional systems of exchange	Agreement allows for traditional systems of exchange and barter, and trade guidelines to be established to regulate those transactions. Agreement also prohibits the commercial sale of Porcupine Caribou Herd meat, but allows for the commercial sale of non-edible parts.
Enforcement	Regulations of general laws of application (e.g., safety) and hunting using traditional methods	Agreement says little about enforcement directly, but charges the board to recommend the establishment of quotas if necessary and make other recommendations to the Minister. Agreement also states that indigenous hunters can continue to harvest caribou using traditional and new methods of hunting.

In crisis conditions, however, community members and governments alike look to the Porcupine Caribou Management Board as the forum to voice concerns and engage in deliberations on management. Plus, the board's long experience with communication problems has provided time to experiment and improve its communications strategies.

Maintaining collaborative and systematic ecological monitoring

In 1996, the Porcupine Caribou Management Board in collaboration with Environment Canada, other co-management bodies and agencies of the region, various First Nations, Inuvialuit governing organizations, and local caribou users' communities recognized that climate change required a more intensive monitoring program, and created the Arctic Borderlands Ecological Knowledge Co-op. The objective of the "Knowledge Co-op" is to draw on local knowledge and science-based indicators to understand what is changing in the region and why (Kofinas et al., 2002). The focus of the monitoring is climate change, regional development, and contaminants. Interviews are conducted by local research associates each year to document local observations of unusual sightings, caribou body condition, difficulties with access to hunting grounds, caribou distribution, and movements. Findings are spatially referenced and entered into a Geographical Information System database. The findings of science and community-based monitoring are discussed at annual gatherings (see for example the Arctic Borderlands Ecological Knowledge Co-op at www.taiga.net/coop).

Focusing research that draws on local and scientific knowledge

The special problems of climate change require research by agencies and universities to link global phenomena to regional and local conditions. As a result of proposals for oil and gas development in the region, the Porcupine

Caribou Herd has been the subject of intensive research and is considered the most studied caribou herd in the Arctic. Interest in climate change provides an opportunity to make comparisons with other herds, and improve overall understanding. Building on the Porcupine Caribou co-management experience, several unique research endeavors have been established involving user communities, university researchers, and agency scientists. Among them is the Sustainability of Arctic Communities project (www.taiga.net/sustain), a seven-year integrated assessment research project that has involved 22 university researchers and Porcupine Caribou Herd communities from both sides of the border (Kruse et al., 2004). This research has led to new findings on the relationships between spring green-up and calf survival, and exploration of decadal trends in timing of green-up correlating with the Arctic Oscillation (Griffith et al., 2002).

Evaluating and protecting important habitat, and participating in impact assessments

Assessment of distribution patterns of the herd during variable climatic conditions will identify sensitive habitats used by the herd. Protection of sensitive habitat maintains the resilience of the herd to endure periods of climate hardship.

The Porcupine Caribou co-management process prompted publication of the *Sensitive Habitats of the Porcupine Caribou Herd* (IPCB, 1993). Ongoing research by agencies has continued to assess at a smaller scale habitat questions and questions relating to the possible effects of changes in caribou distribution and movement due to climate change.

Planning a strategic harvest management

The potential for negative effects of climate change on the population of the Porcupine Caribou Herd suggests the need for a clear and comprehensive harvest management plan and its implications for caribou conservation and community subsistence needs. Ideally, such planning is undertaken well before there is a dramatic decrease in caribou numbers and a crisis situation occurs.

With the recent decline in population of the herd since 1989, the Porcupine Caribou Management Board has facilitated international gatherings of hunters and managers to identify thresholds at which harvest policies are necessary, and the elaborate details of those policies. To help this process, the board is using gaming scenarios with a caribou population simulation model that projects climate change conditions and serves as a discussion tool among local residents, agency managers, and managers (Kofinas et al., 2002).

Overseeing policies for traditional barter and trade

Projected population declines due to climate change may restrict harvest opportunities and create the need for greater exchange of caribou between households and between communities. Sharing and reciprocity through exchange of caribou is a traditional adaptation of subsistence hunting economies which ensures survival through periods of resource scarcity. Terms of the Canadian co-management agreement acknowledge the traditional barter and trade practices of Porcupine Caribou user communities, and direct the Porcupine Caribou Management Board to establish guidelines. Through the International Porcupine Caribou Board, interagency discussions have addressed federal government food and drug administration policies that restrict the transportation of caribou across international borders, and have come to agreement allowing the free exchange of Porcupine caribou meat between user communities across borders.

The co-management process for the Porcupine Caribou Herd brings parties of diverse interests together to discuss difficult wildlife management issues. A discussion at a special workshop on harvest management policy, held April 2002 in Inuvik, NWT, included Gwitchin hunters from Old Crow, Yukon Territory, Fort McPherson and Aklavik, NWT, Inuvialuit hunters from Aklavik and Inuvik, NWT, Iñupiaq hunters from Kaktovik, Alaska, wildlife managers from the Canadian Wildlife Service and Yukon Renewable Resources, and representatives of a Yukon sport hunters organization (photo: G. Kofinas).

Guiding effective forest fire management polices and practice

The Porcupine Caribou Management Board has worked with the Canadian Department of Indian Affairs and Northern Development (one of Canada's federal land management agencies) to map historic fire patterns across the range of the herd and to assess how the department's burn–let burn policies are sensitive to caribou habitat concerns. The board has concluded that terrain typical of the herd's winter range is so diverse, offering numerous natural firebreaks, that fires serve to maintain a healthy, uneven aged mix of habitats.

Developing climate-related communication tools

The increased uncertainty over the consequences of climate change create a demand for meaningful information exchanges between government, indigenous, private, and academic sectors. The co-management arrangement for the Porcupine Caribou Herd in Canada has pioneered several approaches to communication exchange to discuss the state of knowledge about the possible effects of climate change. These include a web-based discussion tool called "The Possible Futures Model" that simulates the combined effects of climate change, development, changes in tourism, and government spending on the Sustainability of Arctic Communities project. A more detailed explanation of the Sustainability of Arctic Communities project is provided in Chapter 12.

Achieving regional consensus and compliance with co-management endorsed policies

It can be argued that the success of co-management is best measured by the compliance of resource users and agency personnel with co-management board recommendations (Kruse et al., 1998). This measure of success recognizes that while the co-management process may facilitate new relationships among those directly involved, it is possible that miscommunications and limited support among the greater set of organizations and individuals can remain. The level of compliance of hunters to a quota due to a climate-driven decline in population is not yet known, yet there is evidence of the system's potential to enlist the support of local hunters and managers. One example is a board recommendation for a prohibition on the sale of caribou antlers, which resulted when antler buyers representing oriental medicinal markets offered to buy antlers from hunters. To date, hunter compliance with the prohibition on antler sales is high. In response to the recent decline in the herd's population, there has been a community call by Old Crow, Yukon's hunters to voluntarily (without formal policy) restrict all cow harvests.

The Porcupine Caribou Herd co-management system and climate change

Climate change is expected to be more gradual than the institutional and development-related changes of the co-management system. If the Porcupine Caribou Management

Board and its partner organizations can play the central role in monitoring, anticipating, evaluating, and responding to climate change, the board will have to maintain its legitimacy with key players that utilize the range of the herd. To be effective, the full set of groups has to feel a sense of ownership in its decision-making process. That the Porcupine Caribou Management Board has jurisdiction only in Canada may lead to a significant challenge in ensuring that a coordinated approach is taken in both countries. For that reason, there is a need for the International Porcupine Caribou Board, recently inactive because of political positions on oil development, to rejuvenate its structure and mission, and to provide the coordinated link that was intended by the international Porcupine Caribou Herd agreement.

Chapter 12

Hunting, Herding, Fishing, and Gathering: Indigenous Peoples and Renewable Resource Use in the Arctic

Lead Author

Mark Nuttall

Contributing Authors

Fikret Berkes, Bruce Forbes, Gary Kofinas, Tatiana Vlassova, George Wenzel

Contents

Summary

This chapter discusses the present economic, social, and cultural importance of harvesting renewable resources for indigenous peoples, provides an assessment of how climate change has affected, and is affecting, harvesting activities in the past and in the present, and considers what some of the future impacts may be. Key to this chapter are several detailed case studies based on extensive research with indigenous communities in a number of arctic settings. These case studies discuss past, present, and potential impacts of climate change on specific activities and livelihoods. It is not possible to provide circumpolar coverage of the situation for all indigenous peoples, as detailed descriptions are not available for all regions of the Arctic. The material presented in this chapter, however, does illustrate some of the common challenges faced by indigenous peoples in a changing Arctic.

One aim of this chapter is to assess the adaptive strategies that have enabled communities to respond to and cope with climate change in the past, and to assess to what extent these options, if any, remain open to them. While there are few data available on this topic, research shows that while indigenous peoples have generally adapted well to past climate change, the scale, nature, and extent of current and projected climate change brings a very different sense of uncertainty, presenting different kinds of risks and threats to their livelihoods and cultures. The chapter also points to pressing research needs. Compared to the extensive scientific literature on climate change considered in most other chapters in this assessment, data on the impacts of climate change on the livelihoods of the Arctic's indigenous peoples are limited, particularly in the case of the Russian North.

This chapter illustrates the complexity of problems faced by indigenous peoples today and underscores the reality that climate change is but one of several, often interrelating problems affecting their livelihoods and cultures. This chapter is, therefore, as much a scoping exercise, the beginning of a process, as it is an assessment of current knowledge. It emphasizes the urgency of extensive, regionally-focused research on the impacts of climate change on hunting, herding, fishing, and gathering activities, research that will contribute to a much greater understanding of climate change impacts, as well as to place these impacts within the much broader context of rapid social, economic, and environmental change.

12.1. Introduction

Indigenous peoples throughout the Arctic maintain a strong connection to the environment through hunting, herding, fishing, and gathering renewable resources. These practices provide the basis for food production and have endured over thousands of years, with cultural adaptations and the ability to utilize resources often associated with or affected by seasonal variation and changing ecological conditions.

Climatic variability and weather events often greatly affect the abundance and availability of animals and thus the abilities and opportunities to harvest and process animals for food, clothing, and other purposes. Many species are only available seasonally and in localized areas and indigenous cultures have developed the capacity and flexibility to harvest a diversity of animal and plant species. Indigenous cultures have, in many cases, also shown resilience in the face of severe social, cultural, and economic change, particularly in the last 100 years.

The longstanding dependence of present indigenous societies on hunting, herding, fishing, and gathering continues for several critically important reasons. One is the economic and dietary importance of being able to access customary, local foods. Many of these local foods – fish, and meat from marine mammals or caribou and birds, for instance, as well as berries and edible plants – are nutritionally superior to the foodstuffs which are presently imported (and which are often expensive to buy). Another reason is the cultural and social importance of hunting, herding, and gathering animals, fish, and plants, as well as processing, distributing, consuming, and celebrating them (Freeman, 2000).

These activities remain important for maintaining social relationships and cultural identity in indigenous societies. They define a sense of family and community and reinforce and celebrate the relationships between indigenous peoples and the animals and environment upon which they depend (Callaway, 1995; Nuttall, 1992). Hunting, herding, fishing, and gathering activities are based on continuing social relationships between people, animals, and the environment (Brody, 1983; Callaway, 1995; Freeman et al., 1998; Nuttall, 1992; Wenzel, 1991). As such, they link people inextricably to their histories, their present cultural settings, and provide a way forward for thinking about sustainable livelihoods in the future.

The significance of hunting, herding, fishing, and gathering has wide cultural ramifications. Seal hunting, for example, is not only an occupation and a way of life, but also a symbolic part of Inuit cultures (Nuttall, 1992; Wenzel, 1991). The cultural role of activities relating to the use of living marine and terrestrial resources is not only of concern to those who depend economically on these activities, but also to those who live in towns and are involved in occupations with no direct attachment to hunting, fishing, and herding (e.g., Caulfield, 1997). Yet whatever the importance for social identity and cultural life, the primary need for, and use of animals is based purely on a need for survival.

Arctic communities have experienced, and are experiencing, stress from a number of different forces that threaten to restrict harvesting activities and sever these relationships. The arctic regions are tightly tied politically, economically, and socially to the national mainstream and are inextricably linked to the global economy (Caulfield, 2000; Nuttall, 1998; Osherenko and Young, 1989; Young, 1992). Rapid social, economic, and demographic change, resource development, trade barriers, and animal-rights campaigns have all had impacts on hunting, herding, fishing, and gathering activities. The material in this chapter on the Russian North, for example, illustrates how poaching, oil development, and clear-cutting of forests undermine the subsistence base for indigenous peoples. Hunting, herding, fishing, and gathering are also being challenged by environmental changes such as climate variability. Despite this, indigenous peoples have reasserted cultural rights and identities, have called for the recognition of self-determination, and are achieving significant levels of regional government (Nuttall, 1998).

For many arctic residents, consuming food from animals is fundamentally important for personal and cultural well-being. Indigenous peoples have reported a loss of vitality, a decline in health, and a decrease in personal well-being when they are unable to eat traditional/country foods (Wein and Freeman, 1992). These problems do not just emerge when climate change denies people access to traditional/country foods, but are very much linked to problems associated with the undermining of local modes of production. The erosion of a person's position as a provider of welfare to family and community also has serious ramifications. A recent study of the importance of whaling for Inuit societies illustrates the negative social, cultural, economic, and nutritional consequences of not being able to gain access to, and to eat, traditional/country foods (Freeman et al., 1998) and points to the kinds of problems that indigenous peoples may experience if climate change denies them access to wild food resources.

The conservation of arctic wildlife and ecosystems depends in part on maintaining the strength of the relationship between indigenous peoples, animals, and the environment, and in part on securing the rights of indigenous peoples to continue customary harvesting activities. As this assessment shows, these activities and relationships appear to be threatened by severe climate change. The potential impacts of climate change on harvesting wildlife resources are of fundamental concern for the social and economic well-being, the health, and the cultural survival of indigenous peoples throughout the Arctic, who live within institutional, legal, economic, and political situations that are often quite different from non-indigenous residents. Furthermore, indigenous peoples rely on different forms of social organization for their livelihoods and well-being (Freeman, 2000).

Many of the concerns about climate change arise from what indigenous peoples are already experiencing in some areas, where climate change is an immediate and pressing problem, rather than something that may happen, or may or may not have an impact in the future. For example, Furgal et al. (2002) discussed local anxieties over environmental changes experienced by communities in northern Quebec and Labrador, and argued that the impacts on human health and the availability of important traditional/country foods from plants and ani-

mals can already be observed. Indigenous accounts of current environmental change say that such changes in climate and local ecosystems are not just evident in animals such as caribou shifting their migration routes and altering their behavior, but in the very *taste* of animals.

As the various chapters of this assessment show, scientific projections and scenarios suggest that there will be significant changes in the climate of the Arctic, in the character of the environment, and in its resources. For example, latitudinal shifts in the location of the taiga–tundra ecotone will have significant effects on ecosystem function and biodiversity at the regional scale. One dramatic anticipated change, taking place over several decades to hundreds of years, is the gradual forestation of tundra patches in the present forest–tundra mosaic and a northward shift of the treeline by hundreds of kilometers. These changes will affect vegetation structure and the composition of the flora and fauna and will have implications for indigenous livelihoods, particularly reindeer herding and hunting and gathering (see Chapter 7).

The aims of this chapter are:

- to discuss the present economic, social, and cultural importance of harvesting renewable resources for indigenous peoples;
- to provide an assessment of how climate change has affected, and is affecting, harvesting activities in the past and in the present; and
- through a selection of detailed case studies based on extensive research with indigenous communities in several arctic settings, to discuss some of the past, present, and potential impacts of climate change on specific activities and livelihoods.

The case studies were selected to provide a sense of the impacts that climate change is having in the present, or could have in the near future, on the livelihoods of indigenous peoples. It is not possible to provide circumpolar coverage of the situation for all indigenous peoples. Apart from space constraints, detailed descriptions are not available for all regions of the Arctic. The material presented in this chapter, especially through the case studies, illustrates the common challenges faced by indigenous peoples in a changing Arctic.

Another purpose of this chapter is to assess the adaptive strategies that have enabled communities to respond to and cope with climate change in the past and to establish the extent to which these options remain open to them. There are few data published on this topic, but based on those that are available the chapter shows that while indigenous peoples have often adapted well to past climate change, the scale and nature of current and projected climate change brings a very different sense of uncertainty for indigenous peoples, presenting different kinds of risks and threats to their livelihoods.

It should be noted that, compared to the scientific chapters in this assessment, data on the impacts of climate change on the livelihoods of indigenous peoples are limited, particularly in the case of the indigenous peoples of Russian North. The case studies in this chapter illustrate the complexity of problems faced by indigenous peoples today and underscore the reality that climate change is but one of several, often interrelating problems affecting their livelihoods.

This chapter is, therefore, as much a scoping exercise as it is an assessment of current knowledge. It emphasizes the urgency for extensive, regionally-focused research on the impacts of climate change on hunting, herding, fishing, and gathering activities, research that will not just contribute to a greater understanding of climate impacts, but will place these impacts within the broader context of rapid social, economic, and environmental change.

12.2. Present uses of living marine and terrestrial resources

12.2.1. Indigenous peoples, animals, and climate

12.2.1.1. Animals, food, and survival

The indigenous peoples of the Arctic include the Iñupiat, Yup'ik, Alutiiq, Aleuts, and Athapaskans of Alaska; the Inuit, Inuvialuit, Dene, and Athapaskans of northern Canada; the Kalaallit and Inughuit of Greenland; the Saami of Fennoscandia and Russia's Kola Peninsula; and the Chukchi, Even, Evenk, Nenets, and Yukaghir of the Russian Far North and Siberia (see Chapters 1 and 3 for an extended discussion). These peoples have subsisted for thousands of years on the resources of land and sea, as hunters, gatherers, fishers, and reindeer herders. Today, many indigenous communities across the Arctic continue to depend on the harvesting and use of living terrestrial, marine, and freshwater resources. In recent decades indigenous peoples have demanded the right to be involved in the policy-making processes that affect their lives, lands, and communities. Responding to rapid social change and threats to the arctic environment, demands for land claims and self-government have been based on historical and cultural rights to lands and resources.

The species most commonly harvested by the indigenous peoples of the Arctic are marine mammals such as seals; walrus (*Odobenus rosmarus*); narwhal (*Monodon monoceros*); beluga (*Delphinapterus leucas*), fin (*Balaenoptera physalus*), and minke (*Balaenoptera acutorostrata*) whales; polar bear (*Ursus maritimus*) and land mammals such as caribou (*Rangifer tarandus*), reindeer (*Rangifer tarandus*)[4], and muskox (*Ovibos moschatus*); and fish such as salmon (*Oncorhynchus* spp.), Arctic char (*Salvelinus alpinus*), northern pike (*Esox lucius*), and other species, such as whitefishes (*Coregonus* spp.). Many of these species are used as food, and for clothing and other products, as well as figuring prominently in the cash economy of local households and communities (Caulfield, 2000; Dahl, 2000; Huntington, 1992; Nuttall, 1992).

Ringed seals (*Phoca hispida*), bearded seals (*Erignathus barbatus*), and hooded seals (*Cystophora cristata*) are widely hunted in Greenland and Canada. Harp seals (*Phoca groenlandica*) and harbour seals (*Phoca vitulina*) are also used locally. Smaller toothed whales like the beluga and the narwhal are hunted in many areas of Canada and Greenland and are prized for their *mattak* (skin) and meat. Baleen whales like bowhead (*Balaena mysticetus*), minke, fin, grey (*Eschrichtius robustus*), pilot (*Globicephala melaena*), and other larger whales are also a valued source of food. Walrus are also commonly taken in Inuit areas, especially in the Bering Strait region and in the Canadian Arctic (Caulfield, 2000).

Fish species used by arctic communities include those that move seasonally from marine to freshwater environments, such as salmon and Arctic char, which are particularly important for indigenous peoples of Alaska (including Inuit communities around Kotzebue Sound, Norton Sound, and the Yukon and Kuskokwim Deltas). The five species of Pacific salmon are also an important food source and a major source of cash income for many households (Caulfield, 2000). Other arctic species used locally include Atlantic salmon (*Salmo salar*), lake trout (*Salvelinus namaycush*), several species of whitefish, pike, and grayling (*Thymallus arcticus*).

Marine fish are an important source of food and a cornerstone of economic life in the Arctic. Arctic cod (*Boreogadus saida*) is used for domestic consumption but also has a long history of use for commercial purposes, especially in Greenland. While its numbers today are reduced, it remains an important part of northern economies in Canada, Greenland, Iceland, and Norway. Greenlandic-owned (and largely Greenlandic-crewed) fishing vessels also fish in waters beyond Greenland, such as in the Barents Sea. In the Bering Sea, the large fishery for pollock (*Theragra chalcogramma*) is undertaken mainly by vessels coming from outside the Arctic, but indigenous peoples are increasingly participating in this and other Bering Sea fisheries. Several flatfish, including halibut (*Hippoglossus hippoglossus*), Greenland halibut (*Reinhardtius hippoglossoides*), and flounder (*Pleuronectes ferrugineus*) are important locally for food and for cash. In Greenland, deep-water shrimp (*Pandalus borealis*) is the major source of export income; indeed, Greenland is the world's largest exporter of shrimp, while the economies of small communities along the west coast are increasingly based on fishing for local stocks of Greenland halibut and cod. Capelin (*Mallotus villosus*), which spawns in large numbers on rocky beaches, is a particularly important coastal fish used locally in Canada and Greenland for human and sled dog food.

Several terrestrial species – especially caribou, reindeer, muskox, and moose (*Alces alces*) – are extremely impor-

---

[4]Caribou are wild animals in North America. Reindeer are domesticated animals in or originally from Eurasia. There are also "wild reindeer", meaning the wild relatives of the animals that were domesticated in Eurasia.

tant in local economies. Caribou, in particular, are hunted widely in Alaska and Canada and in some parts of Greenland, and are used both for food and for other products. Caribou populations are known to vary dramatically over time, and hunters are attuned to the near predictability of their seasonal abundance and migratory routes. Reindeer underpin the culture and economy of herding societies in Fennoscandia and Siberia. Moose are common in the subarctic boreal forest, but their range is expanding into more northerly environments. Other terrestrial species of economic importance to arctic residents include muskox, grizzly bear (*Ursus arctos*), wolf (*Canis lupus*), arctic fox (*Alopex lagopus*), muskrat (*Ondatra zibethicus*), and ground squirrel (*Spermophilus parryii*).

Indigenous peoples have also collected eggs and hunted birds among coastal colonies of auks and other seabirds. For example, Greenlanders hunt, among others, Brünnich's guillemot (*Uria lomvia*), common eider (*Somateria mollissima*), king eider (*Somateria spectabilis*), and kittiwakes (*Rissa* spp.), and take the eggs of all these species. They also collect the eggs of birds not hunted for food, such as the Arctic tern (*Sterna paradisaea*).

Literally hundreds of harvest studies have been carried out in the Arctic and subarctic, particularly in Alaska and Canada. The wide range and diversity of plant and animal species used for food by indigenous peoples is illustrated by data from recent studies and surveys from the Canadian Arctic summarized in reports by the Arctic Monitoring and Assessment Programme (e.g., AMAP, 1998). Figure 12.1 shows harvest levels in the different Inuit regions and in the Yukon.

In 1989, the total harvest in the Northwest Territories was estimated to be about 5000 tonnes, or 232 kilograms per person per year, excluding commercial fish catches. There is very little information about the harvesting activities of most Dene and Métis communities, except for fur-bearing species and commercially significant fish. Employment figures indicate that subsistence activities are important, as almost 40% of the indigenous population in Dene communities were not part of the labor force according to a survey in 1991 (AMAP, 1998). Almost 38% of people over 15 years old said that they used non-cash activities to provide for their families. A slightly larger percentage said that they had lived on the land in the previous twelve months. An estimate of the per-capita harvest suggests that the communities are self-sufficient in their protein requirements. Yukon First Nations also rely heavily on subsistence activities. About one third of the people in the 1991 Aboriginal People's Survey said that they had lived on the land in the previous year and 30% supported their families with activities that were not part of the cash economy (AMAP, 1998).

The AMAP assessment shows that studies support the picture of a high reliance on subsistence production throughout northern Canada (AMAP, 1998). Even if

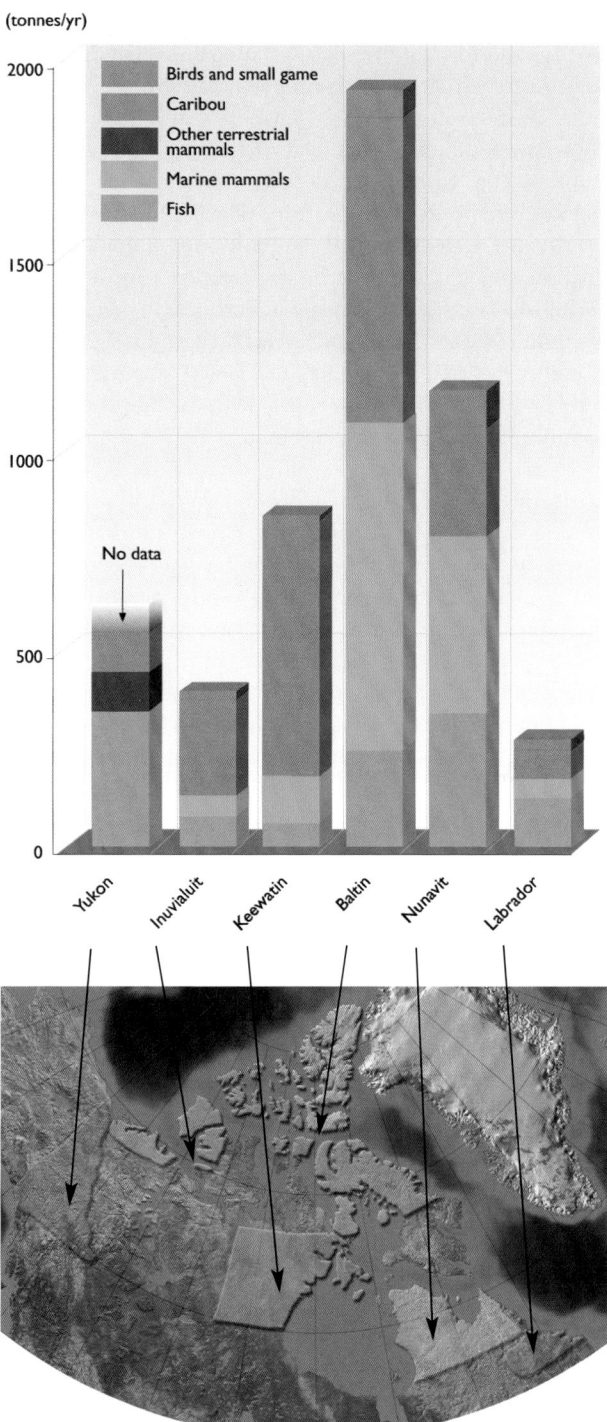

Fig. 12.1. Average annual indigenous subsistence production in arctic Canada (based on AMAP, 1998).

store-bought foods are also common, traditional/country foods contribute a significant proportion of the daily nutritional intake.

The traditional diets of indigenous peoples in northern Canada are more balanced than a diet of foods imported from southern Canada, which have higher levels of sugar and more saturated fats. Using traditional/country foods is regarded by indigenous peoples as more economical than purchasing food in the store. This becomes especially important in communities where many people are not employed or where many have incomes below the poverty line. Traditional/country foods are also important for reinforcing the social relationships that are central to the culture and subsistence economy.

Diets and food preferences vary between communities and between families, but detailed studies provide some examples of what people eat. In Aklavik, Northwest Territories, more than half the Inuit households consume caribou, beluga, Arctic hare (*Lepus arcticus*), muskrat, whitefish, cisco (*Coregonus* spp.), burbot (*Lota lota*), inconnu (*Stenodus leucichthys*), Arctic char, ducks, geese, cloudberries (*Rubus* spp.), cranberries (*Oxycoccux* spp.), and blueberries (*Vaccinium* spp.), with caribou the most common food (Wein and Freeman, 1992).

The types of food eaten also depend on the time of year. In Aklavik, autumn is caribou and moose hunting season, as well as being the hunting season for Dall sheep (*Ovis canadensis dalli*), ducks, and geese. Winter activities are trapping small fur-bearing animals and fishing. When the ice breaks up in April, muskrat are harvested for their pelts and meat. The waterfowl return, and are used as food until they begin to nest. Fishing resumes after ice break-up. Spring is the time for gathering roots. Summer is whaling time, and people travel out to the Yukon coast to hunt beluga. Willow tops, bird eggs, and wild rhubarb supplement the diet. As autumn approaches again, it is time to dry fish and caribou meat and to pick berries. Among the Dene, a few dietary studies have been carried out specifically to estimate the amount of contaminants in traditional/country foods. These surveys show, for example, that moose are eaten in summer, barrenland caribou in winter, and ducks in spring. Other important foods are inconnu, whitefish, cisco, and blueberries. In the winter, moose, rabbit, whitefish, and loche are part of the diet, and in the spring woodland caribou (Wein and Freeman, 1992). The influence of the fear of contaminants on food harvesting is an important issue that needs development (Weinstein, 1990).

A survey of dietary preferences in the communities of Fort Smith, Northwest Territories, and Fort Chipewyan, Alberta, showed that people ate traditional/country foods six times per week and that animals from the land made up one-third of the diet. In a survey of Yukon First Nations (Haines Junction in the traditional territory of the Champagne–Aishihik First Nation, Old Crow, which is a remote community on the Porcupine River relying heavily on the Porcupine Caribou Herd that migrates through their land, Teslin at Teslin Lake, and Whitehorse, which is the territorial capital with a more diverse population), virtually all households used moose and salmon, as well as berries and other plant foods. Many also used caribou, hare, ground squirrel, beaver, ducks, grouse, chinook salmon (*Oncorhynchus tshawytscha*), sockeye salmon (*O. nerka*), coho salmon (*O. kisutch*), whitefish, lake trout, and Labrador tea (*Ledum* spp.). In total, mammals accounted for about half the traditional food, fish for a fifth, berries for a fifth, other plants for a tenth, and birds for a twentieth. People got most of their food from hunting and fishing (AMAP, 1998).

As the dietary surveys carried out in Yukon First Nation communities show, traditional/country food harvested from the local environment has a central role in the daily lives of individuals, families, households, and communities. Traditional/country foods improve the quality of the diet as shown by the lower fat and saturated fat content of the diet when traditional food is consumed. Traditional/country foods also represent important sources of dietary energy, protein, iron, and zinc. The increased physical activity associated with traditional food harvest, and the role of the traditional/country food system in cultural and social support systems is also likely to contribute to health (Receveur et al., 1998; Wein, 1994).

12.2.1.2. Animals and cultural identity

Successful harvesting of all the species used by indigenous peoples requires specialized knowledge of animal and fish behavior, sea ice and terrestrial conditions, and arctic weather. The detailed knowledge of the Arctic's indigenous peoples about these factors is widely recognized. Indigenous peoples have detailed and complex systems of classification and knowledge about the natural world which is developed and enhanced through long-term experience and generational transmission (Nuttall, 1998). This knowledge has enabled indigenous societies to exploit highly productive ecosystems effectively in the region for thousands of years (Caulfield, 2000) and provides a foundation for economic, cultural, spiritual, and ethical concerns that guide the use and management of natural resources (Nuttall, 1998).

The living resources of the Arctic do not just sustain indigenous peoples in an economic and nutritional sense, but provide a fundamental basis for social identity, cultural survival, and spiritual life. As such they are as much important cultural resources as they are economic ones. This dependence on animals for food and social, cultural, and economic well-being is reflected in rules for community hunting, in herding traditions, and in patterns of sharing and gift-giving based on kinship ties and other forms of close social relatedness. Participation in family and community hunting, herding, and fishing activities contributes to defining and establishing a sense of social relatedness and is important for community and cultural identity, as well as for providing a moral framework for relationships between people and between people and animals.

Across the Arctic, the sharing and distribution of meat and fish is central to daily social life and expresses and sustains social relationships, and the case study from Nunavut (section 12.3.2) illustrates vividly the sharing practices and networks in one particular region. Harvesting and its associated processing and sharing activities reaffirm fundamental values and attitudes towards animals and the environment and provide a moral foundation for continuity between generations (Callaway et al., 1999; Nuttall, 1992; Wenzel, 1991).

In seal hunting households in Greenland and Canada, for example, the meat, fat, and skin of the seal is utilized. There is rarely much wasted. Complex and precise local rules determine the sharing and distribution of the catch, and seal meat is commonly shared out to people beyond the household, whether those people are related to the hunter or not (Petersen, 2003). For arctic hunting peoples, sharing can only be understood with reference to the sense of social relatedness that people feel they have with each other and with animals and the environment. This has been well documented by recent research on the consumption of traditional/country foods in Greenland (Caulfield, 1997; Nuttall, 1992; Pars et al., 2001).

The cultural expression of the relationships between humans and animals is evident in first-catch celebrations. At an early age, boys are taken on hunting trips with their fathers, who begin to teach them the skills and impart the knowledge necessary to be a successful hunter. In small Greenlandic hunting villages, for example, when a boy catches his first seal, he will give gifts of meat to every household in his community and people are invited to his parents' home for coffee or tea and cake. A first catch celebration is not only a recognition by the community of the boy's development as a hunter, but is a statement of the vitality and cultural importance of the hunting way of life (Nuttall, 1992). For arctic hunting peoples such as the Inuit, sharing the products of the hunt is a social event that demonstrates relatedness, affection, and concern. Obligations to share underlie customary ideologies of subsistence and contribute to the reproduction of kinship ties and other close social relationships (Nuttall, 1992; Wenzel, 1991). Climate change not only disrupts hunting activities, it has an impact on such social relationships, as the case study from Nunavut (section 12.3.2) shows.

Rich mythologies, vivid oral histories, festivals, and animal ceremonialism also illustrate the social, economic, and spiritual relationships that indigenous peoples have with the arctic environment. Animals have a spiritual essence as well as a cultural and economic value, and land and water are not just seen as commodities. For indigenous peoples, many features in the landscape are sacred places, especially along migration routes, where animals reveal themselves to hunters in dreams, or where people encounter animal spirits while traveling (Brody, 1983).

In Alaska and Canada, Athapaskan oral histories describe how features of the landscape, or the elements, such as the moon, sun, wind, stars, and so on, were originally human beings and whose spirits are now embodied in aspects of the natural world. In Greenland, Canada, and Alaska, Inuit stories about the origin of the elements, the sun and the moon, and other celestial bodies, are often related to myths about the balance between daylight and darkness, time and space, and between the human and natural worlds. In Siberia and Sapmi, one can find reindeer antlers that have been placed at sacred sites and adorned with gifts, and sacred stones placed on the tops of mountains and near lakes and rivers.

12.2.1.3. Place, environment, and climate

Although the Arctic is often labeled one of the last remaining wilderness areas on earth, this ignores the fact that the Arctic is a homeland for indigenous peoples. The indigenous names for features of the landscape – for streams, lakes, mountains, valleys, plains, and tundra meadows – as well as the icescape and features of the sea are not merely geographically descriptive. The names that indigenous peoples have given to the arctic landscape are multidimensional, in that they contain information about physical features, the availability and movement of animals, community history, and historical and mythological events (Nuttall, 1992, 2001). This differs sharply from the practice of naming places by explorers, colonialists, and settlers in order to control, own, and dominate the landscape.

Often, place names provide information about climate change and significant weather-related events. For indigenous peoples, stories and discussions about the weather and climate are interwoven with stories and experiences of particular tasks, like hunting, herding, fishing, berry-picking, or traveling (see Chapter 3). Much of this is bound up with memories of past events, of local family histories, and of a strong sense of attachment to place and locality (Nuttall, 2001). The weather connects people to the environment and to animals.

One example of this is the understanding of *sila* in Greenland. In Kalaallisut (Greenlandic) the word for weather and climate is *sila*. *Sila* is also used to mean "the elements" or "the air". But *sila* is also the word for "intelligence/consciousness", or "mind," and is understood to be the fundamental principle underlying the natural world. *Sila* is manifest in each and every person. It is an all-pervading, life-giving force – the natural order, a universal consciousness, and a breath soul (Nuttall, 1992). *Sila* connects a person with the rhythms of the universe, integrating the self with the natural world. As *sila* links the individual and the environment, a person who lacks *sila* is said to be separated from an essential relationship with the environment that is necessary for human well-being. When people in Greenland experience a change in the weather, this change is experienced in a deeply personal way. And when they talk about their concerns about climate change, they articulate this not only in terms of how their own sense of self, personhood, and well-being is changing in relation to external climatic fluctuations, but in their concerns for their own sense of self and well-being in terms of climate change (Nuttall, in press).

Memories and knowledge of how the weather and climate has changed are also found in oral histories as well as in contemporary observations. For Athapaskan people of Canada's Yukon Territory and southeast Alaska, memories of the Little Ice Age play a significant role in indigenous oral traditions. Cruikshank (2001) shows how these stories are "sedimented" on land just like geological processes. Athapaskan clan histories document travel across glaciers from several directions. Eyak, Athapaskan, and Tlingit place names encapsulate information and local ecology and climate now rendered invisible by English names. Cruikshank (2001) shows that surging glaciers present navigational, spiritual, and intellectual challenges of a sentient "land that listens". Stories about changes in the weather, to the landscape, and to glaciers persist with a richness, range, and variety because of ongoing risks they posed to everyday life well into the 20th century.

Today, as Athapaskan people demonstrate concern with climate change, there is a contemporary validity to these stories. They not only record the consequences of climate change, and enrich scientific understandings of past climatic conditions, but also provide information on the responses that helped indigenous communities cope with and adapt to climate change. Observations and understandings of change are invaluable to scientists working on the impacts of climate change and increased levels of ultraviolet-B radiation by providing long-term records of observed changes with which to compare and contrast their results (De Fabo and Bjorn, 2000).

12.2.2. Mixed economies

In indigenous communities in the Arctic today, households are economic units within villages, settlements, and small towns characterized by a blend of formal economies (e.g., commercial harvesting of fish and other animals, oil and mineral extraction, forestry, and tourism) and informal economies (e.g., harvesting renewable resources from land and sea). The ability to carry out harvesting activities is not just dependent on the availability of animals, but on the availability of cash, as the technologies of modern harvesting activities are extremely expensive in remote and distant arctic communities. Throughout the Arctic, many indigenous communities (whether they are predominantly seal hunting communities in northern Greenland or Canada, fishing communities in Norway, or reindeer herding societies in Siberia) are increasingly characterized by pluri-activity in that cash is generated through full-time or part-time paid work, seasonal labor, craft-making, commercial fishing, or other pursuits such as involvement in tourism that support and supplement renewable resource harvesting activities.

In mixed economies, half or more of household incomes may come from wage employment, simple commodity production, or from government transfer payments (Caulfield, 2000; Langdon, 1986; Weinstein, 1996). Increasing reliance on other economic activities does not mean that production of food for the household has declined in importance. Hunting, herding, gathering, and fishing activities are mainly aimed at satisfying the important social, cultural, and nutritional needs, as well as the economic needs, of families, households, and communities (see Bodenhorn (2000) for northern Alaska, Hovelsrud-Broda (2000) for East Greenland, and Wenzel (2000) for eastern Baffin Island).

Research points to the continued importance of harvesting activities despite a growing proportion of the population of indigenous communities not being directly involved in harvesting (e.g., Usher, 2002). Purchased foodstuffs supplement diets composed mainly of wildlife resources (Callaway, 1995; Nuttall, 1992) and individuals and households that do not have the means or ability to hunt often have regular access to country foods through local distribution channels and networks of sharing (see the case study of Inuit sharing patterns in Nunavut in section 12.3.2).

Nor has money diminished subsistence-oriented production as a central feature of life in the Arctic – indeed cash has made the continuation of hunting, herding, and gathering possible in some cases, rather than contributing to its decline (Kruse, 1991; Nuttall, 1992; Wenzel, 1991; Wolfe and Walker, 1987). In parts of the Arctic commercial and subsistence uses of country foods are intrinsically linked. In Alaskan villages fish for the household are often taken during commercial fishing trips, the profits from which are often invested in new equipment for subsistence pursuits (Callaway et al., 1999).

Cash is often used to buy equipment for procuring food from harvesting activities (e.g., boats, rifles, snow-machines). Cash also meets demands for a rising standard of living: to purchase oil to heat homes, to buy consumer goods, or to travel beyond the community. While food procured from renewable resource harvesting continues to provide arctic peoples with important nutritional, socio-economic, and cultural benefits, finding ways to earn money is a major concern in many arctic communities (Caulfield, 2000). The interdependence between formal and informal economic sectors, as well as the seasonal and irregular nature of wage generating activities (such as tourism) means that families and households are often faced with a major problem in ensuring a regular cash flow. For example, Callaway et al. (1999) demonstrated that the ability to carry out harvesting activities in Alaska – and thus the quality of life in rural communities – is linked to the state's economic and political environments.

The impacts of climate change on formal economic activities will also have implications for renewable resource harvesting activities. In Alaska, recent climate change has increased the cost and risk of subsistence pursuits. On the coast of northern Alaska, where the ice pack has retreated a significantly greater distance from land, North Slope hunters have to cross a greater expanse of open water to reach hunting grounds. The increased time and distance added to a hunting trip adds to the cost and risk of accessing marine mammal resources. Fuel and maintenance costs are greater because of the longer distance to travel, which also decreases the use and expectancy of the technology used (boats, engines, rifles). For safety reasons, boats with larger engines are required, adding strain to limited budgets (Callaway et al., 1999).

The economic value of traditional/country food is emphasized by the level of food insecurity common among indigenous peoples. In a major dietary survey in Yukon First Nation communities, 39% of respondents reported having insufficient resources to purchase all the food they would need from the store if traditional food was not available; the average weekly "northern food basket" was priced at Can\$ 164 in communities, compared to Can\$ 128 in Whitehorse (Receveur et al., 1998). The Nunavut case study (section 12.3.2) illustrates the problems hunters face in gaining access to money. While hunting produces large amounts of high quality food – the Government of Nunavut estimates that it would cost approximately Can\$ 35 000 000 to replace this harvest production – as the case study illustrates, virtually none of this traditional wealth can be converted into the money needed to purchase, operate, and maintain the equipment hunters use. Abandoning hunting for imported food would be less healthy and immensely costly.

12.2.3. Renewable resource use, resource development, and global processes

Despite variations in economic and cultural practices, many indigenous communities throughout the Arctic share one important characteristic – their economies are vulnerable to changes caused by the global processes affecting markets, technologies, and public policies in addition to the environmental impacts of climate change. Residents of arctic communities are increasingly tied to world markets and the growth of the mixed economies of arctic communities points to widening interaction of arctic societies with the global economy (Caulfield, 1997; Nuttall, 1998). Greenland's largest single source of export income, for example, is deep-water shrimp, marketed in Europe, North America, and Japan. Oil from Alaska's North Slope meets 25% of total US demand, and provides healthy tax revenues for the North Slope Borough's Iñupiat residents. Development of hydropower has sparked major conflict between Saami in northern Norway and industry and governments to the south (Caulfield, 2000).

> Renewable resources are a part of this global dynamic: salmon from Alaska's Bering Sea is found in fashionable restaurants of Boston and Los Angeles within hours of being caught; Japanese technicians advise Greenlanders about how to produce specialized shrimp products ["fantails"] for Tokyo markets; wealthy European and North American hunters pursue polar bear in northern Canada for trophies; wilderness enthusiasts in places like Alaska's Denali National Park seek wildlife experiences where subsistence hunting by indigenous peoples is banned; and animal rights activists lobby to keep Inuit hunters from selling seal skins on the European market, no matter how justifiable the practice on biological grounds.
> Caulfield, 2000

Arctic fisheries are a good example of how the effects and influences of global processes are increasingly felt in all aspects of social, economic, and cultural life in the Arctic today. Many problems experienced by North Atlantic coastal communities in the Arctic, for instance, can be attributed in part to the global restructuring of fisheries, the balance of competition between different species and different fishing areas, the globalization of the sourcing of supplies for processing plants and retail markets, and the redistribution of wealth from traditional actors, such as local fishers and local processors, to powerful global players in the form of transnational corporations. Fisheries are being transformed from industries or ways of life subject to the control and regulation of local, regional, and national authorities to a global enterprise dominated by a handful of transnational corporations (Nuttall, 2000).

Industrial development, deforestation, and pollution are also significant. In northern Russia, domesticated reindeer populations are decreasing due to the degradation of winter reindeer pastures by deforestation, industrial pollution, and overgrazing. Fewer winter pastures are available for reindeer as large territories are being occupied by the mining, oil, and gas industries, leading to greater pressure from grazing on increasingly fragile tundra and lesotundra ecosystems (Callaghan et al., 2002; Vlassova, 2002). Several ecosystems in northern Russia are already overgrazed by reindeer. The reindeer population of the Yamal Peninsula, for example, exceeds the carrying capacity of pastures by 1.5 times, with 70% of pasture registered as low quality (Vlassova, 2002).

In Yukon Territory, concern over contaminants and dietary risks include: (1) the risks associated with increased market food consumption (for example, fewer of the protective factors associated with traditional food use, lower nutritional intake, and higher saturated fat intake), and (2) risks associated with exposure to chemical contaminants from the consumption of traditional food. Concentrations of organochlorine compounds and heavy metals are known to be very low in most market food, but are of potential concern in traditional food. Standard government analyses assume therefore that market food does not contain chemical contaminants, and that risk from contaminant intake via traditional food will be related to the level of exposure; the higher the level of exposure, the higher the supposed risk (Receveur et al., 1998).

Thus, for some indigenous communities climate change may not be the most immediate issue of local concern. Yet the interrelations between industrial development, pollution and contaminants, international trade, sustainable development, and climate change (and their cumulative impacts) are poorly understood and further research is needed. With an increased focus on sustainable development of both renewable and non-renewable resources in the Arctic, future research on how local, regional, and national economies throughout the circumpolar North are being affected by climate change will need to contextualize arctic case studies with reference to the internationalization of production and exchange, the globalization of economic and industrial activity, and the activities and influences of transnational corporations and transnational practices.

There is scientific difficulty in stating how far climate change alone has affected arctic marine ecosystems in the past fifty years, for instance, as the impacts of over-fishing and over-hunting may be far greater (Sakshaug and Walsh, 2000). However, Finney et al. (2002) presented results that support a strong role for climate forcing in regulating abundance of northeastern Pacific fisheries over the last two millennia. Sockeye salmon return to spawn and die in the lakes in which they were born, releasing nutrients into the lake which accumulate in the sediments. By analyzing sediment cores from nursery lakes in Alaska, their research revealed the existence of multi-century regimes in salmon abundance. The two noticeable multi-century shifts in salmon abundance at ~100 BC (the beginning of a sustained period of low abundance) and ~AD 800–1200 (the beginning of a sustained period of high abundance) correspond to periods of major change in ocean–atmosphere circulation in the northeastern Pacific. Historical catch records, being of short duration, provide only a limited understanding of fish population dynamics and their response to climate change. This 2200-year record demonstrates that very low productivity regimes, lasting for centuries, can occur even without the influence of commercial fisheries, in response to climate changes and associated oceanic changes (Finney et al., 2002).

Nor is climate change the only cause of changes to the treeline and tundra. The overgrazing of reindeer pastures in northern Russia leads to the overgrazing of leso-tundra, damaging shrubs, and has an impact on the treeline, pushing it further south in some areas (Vlassova, 2002). In Fennoscandia the development of reindeer husbandry over the last 100 years has also increased the risk of overgrazing. The shift from intensive to extensive reindeer husbandry probably reduced pressure on vegetation in some places; however it also meant that larger numbers of reindeer could be kept. In Finland, for example, the number of reindeer rose dramatically in the 1950s, with herds growing rapidly throughout Fennoscandia during the 1970s. The result was increasing grazing pressure over very wide areas (Bernes, 1996).

In Norway, the growing numbers of reindeer and herds, together with the reduction of available pasture, have strongly reduced the most important asset of Saami pastoralists, namely flexibility. As a result, it is increasingly difficult for Saami herders to cope with variations in climate and pasture conditions (Bjørklund, 2004). Herders have strategies for dealing with climatic variability or changes in pasture which are becoming harder to utilize for a number of reasons. For example, if pasture becomes too scarce in summer owing to growing herd sizes, or if conditions become difficult because of climatic fluctuations one year, herders might leave the area early and keep their reindeer longer on autumn and winter pastures, or move their herds to temporarily

vacant neighboring pasture. This flexibility is becoming increasingly problematic as fences, pasture regulations, a growing number of herds, and changing management systems combine to reduce the possibility of using such strategies (Bjørklund, 2004).

Human activities, industrial development, resource use regulations, and global economic processes have far-reaching consequences for the environment and so magnify the likely impacts on indigenous livelihoods of variations in weather and climate. Indigenous economies are not self-reliant closed systems and although their involvement in global networks of production and consumption may provide means of strengthening and extending possibilities for arctic communities, they also introduce greater elements of risk and could make people and their livelihoods less resilient to coping with and adapting to climate impacts.

12.2.4. Renewable resource use and climate change

12.2.4.1. Climate change impacts: some key facts

Renewable resources will continue to be central to the sustainable development strategies of numerous arctic communities. However, renewable resources and the harvesting of renewable resources by indigenous peoples in the Arctic could be affected by global climate change and increased levels of ultraviolet radiation caused by ozone depletion. Climate change scenarios suggest that climate change will have impacts on marine and terrestrial animal populations, affecting population size and structure, reproduction rates, and migration routes (IPCC, 2001). Arctic residents, particularly indigenous peoples who depend on renewable resources for their livelihoods and cultural survival, will feel these climate change impacts first and most intensely.

However, because of the interdependence between arctic economies and global markets, indigenous peoples are multiply exposed – to climate change, to changes caused by the global processes affecting markets, technologies, and public policies, and to local and regional political and economic situations. It is important to contextualize climate change impacts with reference to other changes experienced by arctic residents. Being able to access traditional food resources and ensuring food security will be a major challenge in an Arctic affected increasingly by climate change and global processes.

This assessment shows that the results of scientific research and evidence from indigenous peoples (see Chapter 3) have increasingly documented climatic changes that are more pronounced in the Arctic than in any other region of the world. Yet although this indicates that the physical environment, as well as the flora and fauna, has been undergoing noticeable change, the impacts felt throughout the Arctic will be unique and will vary from region to region. Different climatic trends have been observed in different parts of the Arctic – while average temperatures in the North American western Arctic and Siberia have been increasing over the last 30 years (e.g., annual temperatures in the Canadian western Arctic have climbed by 1.5 °C and those over the central Arctic have warmed by 0.5 °C), temperatures in Canada's Hudson Bay and in Greenland, particularly in the Davis Strait area, have decreased (Chapman and Walsh, 1993), suggesting that climate change involves regional cooling as well as global warming.

If the scientific projections and scenarios are realized, climate change could have potentially devastating impacts on the Arctic and on the peoples who live there, particularly those indigenous peoples whose livelihoods and cultures are inextricably linked to the arctic environment and its wildlife. Some scenarios suggest that the most direct changes will be noticeable in a reduction in the extent of sea ice and permafrost, less ice in lakes and rivers, pronounced reductions in seasonal snow, and the disappearance of the existing glacier mass, leading to a corresponding shift in landscape processes (Lange, 2000; Siegert and Dowdeswell, 2000; Weller, 2000).

Scientific research shows that over the last 100 years there has already been a significant reduction in the extent and thickness of arctic sea ice. Since 1979 alone, the extent of sea ice throughout the Arctic has decreased by 0.35%, and record reductions were observed in the Beaufort and Chukchi Seas in 1998 (Johannessen et al., 1999; Maslanik et al., 1999). Sea ice is highly dependent on the temperature gradient between ocean and atmosphere and on near-surface oceanic heat flow and will react swiftly to changes in atmospheric conditions (Lange, 2000). Atmosphere–ocean climate models project a reduction in sea ice of around 60% in the next 50 to 100 years under a scenario in which atmospheric CO_2 concentrations double. Models also project that permafrost will thaw more quickly in spring, but take longer to refreeze in autumn, and that the active layer boundaries will gradually move poleward, with most of the ice-rich discontinuous permafrost disappearing by the end of the 21st century.

Climate variability appears to have caused relatively rapid shifts in the organization of arctic marine ecosystems. In the Bering Sea ecosystem and the Barents Sea ecosystem climate-driven variability is significant (Sakshaug and Slagstad, 1992). There are difficulties, however, in establishing which of these changes result from natural environmental fluctuations and which result from human activities. In the eastern Bering Sea upper trophic levels have undergone significant changes in the past 100 to 150 years, largely due to commercial exploitation of mammals, fish, and invertebrates. Climatic changes may have contributed in part to the changes in animal populations. Higher ocean temperatures and lower salinities, changes in seasonal sea-ice extent, rising sea levels, and many other (as yet undefined) effects are certain to have significant impacts on marine species, with implications for arctic coastal communities dependent on hunting and fishing (Weller and Lange, 1999).

Most arctic species of marine mammal and fish depend on the presence of sea ice and many indigenous coastal communities depend on the harvesting of these species. The ice edge is unique among the world's ecosystems in that it moves thousands of kilometers each year, north in spring and south in autumn. Walrus, numerous species of seal, and cetaceans such as beluga and narwhal all follow the ice edge as it moves, taking advantage of the ready access to food and (for walrus and seals) the availability of ice to haul out on for sunning, mating, and raising pups in late winter and spring (an important time for Inuit hunting communities).

The almost complete elimination of multi-year ice projected for the Arctic Ocean is likely to be immensely disruptive to ice-dependent microorganisms, which will lack a permanent habitat. Preliminary results from research in the Beaufort Sea suggest that ice algae and other microorganisms may have already been profoundly affected by warming over the last 20 years. Research indicates that most of the larger marine algae have died out, and been replaced by a much less productive community of microorganisms more usually associated with freshwater ecosystems (see Chapters 8 and 9).

Walrus, seals, and whales are likely to undergo shifts in range and abundance in response to the projected changes in multi-year sea ice, while the migration routes of caribou will alter. These changes could impact upon the hunting, trapping, and fishing economies of many small, remote arctic settlements. Although warming may increase biological production in some wildlife species, the distribution of many species crucial to the livelihoods and well-being of indigenous peoples could change. Important wetlands may disappear, or drainage patterns and tundra landscapes will be altered significantly, which could affect ducks and other waterfowl. Changes in terrestrial vegetation will have consequences for reindeer herding, subsistence lifestyles, and agriculture (see Chapters 7, 11 and 14).

Terrestrial animals such as caribou and reindeer are important for indigenous communities throughout the Arctic and would be affected by climate change directly through changes in thermal stress in animals, and indirectly by difficulties gaining access to food and water. Arctic communities located on coastlines may be affected by rising sea levels, increased coastal erosion, and severe storms. The fortunes of subsistence fisheries will depend on marine fish stocks and their climate-related variations (Lange, 2000). As the amount of sea ice decreases, seals, walrus, polar bears, and other ice-dependent species will suffer drastically.

Recent observations have demonstrated that there has been a distinct warming trend in lowland permafrost of 2 to 4 °C over the last 100 years (Fitzharris et al., 1996; Lange, 2000), leading to disturbances of animal and human activities due to thawing, thermokarst formation, and severe erosion. Further warming is likely to continue this trend and increase the likelihood of natural hazards for people (particularly affecting hunting and herding), buildings, communication links, and pipelines. The documentation of widespread thawing of discontinuous permafrost in Alaska illustrates some of these hazards and the implications for habitat change and the physical infrastructure of communities. In western Alaska several communities in low-lying areas, including Shishmaref, Kivalina, and Little Diomede, are affected by recent climate changes and face severe problems as a result of erosion and thawing of the discontinuous permafrost (Callaway et al., 1999).

Unstable sea ice could make ice-edge hunting more difficult and dangerous. Temperature and precipitation changes could affect migration patterns of terrestrial mammals like caribou and alter breeding and molting areas for birds. Salmon, herring, walrus, seals, whales, caribou, moose, and various species of waterfowl are expected to undergo shifts in range and abundance (IPCC, 2001). Changes in snow cover could affect the growth and distribution of plants essential for survival of caribou and reindeer. Changes in snow cover could also make accessing hunting, fishing, and herding areas more difficult by dogsled, snow-machine, or other vehicles, making local adjustments in hunting practices and harvesting strategies necessary.

12.2.4.2. Indigenous observations of climate change

In many parts of the Arctic, indigenous peoples are reporting that they are already experiencing the effects of climate change. In Canada's Nunavut Territory, Inuit hunters have noticed the thinning of sea ice and the appearance of birds not usually found in their region; Iñupiat hunters in Alaska report that ice cellars are too warm to keep food frozen; Inuvialuit in the western Canadian Arctic report thunderstorms and lightning (a rare occurrence in the region); Gwich'in Athapaskan people in Alaska have witnessed dramatic changes in weather, vegetation, and animal distribution patterns over the last 50 years or so; Saami reindeer herders in Norway have observed that prevailing winds relied on for navigation have shifted and that snow cannot be relied on for traveling over on trails that people have always used and considered safe (see Chapter 3).

Indigenous peoples in Alaska, for example, have already reported that there has been little snow in autumn and early winter, but substantial snowfall in late winter and early spring (Chapter 3). According to local hunters, the lack of snow makes it difficult for polar bears and ringed seals to make dens for giving birth or, in the case of male polar bears, to seek protection from the weather. The lack of ringed seal dens may affect the numbers and condition of polar bears, which prey on ringed seals and often seek out the dens. People in northern coastal Alaska are concerned that hungry polar bears may be more likely to approach villages and encounter people.

Inuit observations of climate change have been recently documented for the Kitikmeot region of Nunavut

(Thorpe et al., 2002). People have spoken of a changed climate in the 1990s compared with previous decades: increasing temperatures with earlier spring melts and later freeze-ups in autumn have meant periods of longer summer-like conditions, while weather has become variable and unpredictable. This change and variability has had many impacts on caribou. Migration routes and the location of calving grounds have shifted and food sources have sometimes become inaccessible. Inuit have recently noticed more frequent short-term changes in temperature, especially in freeze–thaw cycles, which, because these cycles help form an icy layer on the top of snow or tundra, prevent caribou accessing vegetation (Thorpe et al., 2002).

12.2.4.3. Consequences of climate change for the livelihoods of indigenous peoples: caribou hunting and reindeer herding

The case studies in section 12.3 provide detailed analyses of the current and potential consequences of climate change for the livelihoods of indigenous peoples. The case study on caribou hunting in Arctic North America (section 12.3.5), for example, shows how the location of modern-day human settlements relative to caribou migration routes has consequences for the success of community caribou hunting. Communities like Old Crow in Yukon Territory, located in the center of the range of large migratory herds, have opportunities to intercept caribou during autumn and spring migrations, whereas communities situated on the margin of a herd's range may have access to animals only during winter or briefly during the summer calving and post-calving periods. The range of a large herd can contract at low population levels and expand at high population levels, and the implications for local communities situated at a distance from a herd's range can mean a decline in successful hunting and even the abandonment of caribou hunting for several decades. Shifting migration routes because of climate change will have consequences for hunting success.

All caribou and reindeer herds depend on the availability of abundant tundra vegetation such as lichen for forage, especially during the calving season. Climate-induced changes to arctic tundra may cause major vegetation zones to shift significantly northward, as well as having an effect on freeze–thaw cycles. The timing and occurrence of ice crusts due to refreezing of molten snow layers, which might be affected by changes in climate, will be a major factor for the sustenance of caribou and reindeer herds (Lange, 2000). This will have significant implications for reindeer populations in relation to their ability to find food and raise calves. Future variations in weather and climate could mean a potential decline in caribou and reindeer populations and have an adverse effect on hunting and herding practices. This could threaten human nutrition for indigenous households and threaten a whole way of life for arctic communities.

Russian historical records from the 1800s and early 1900s provide startling documented evidence of devastating losses of reindeer stocks of Siberian indigenous herders due to occasional and dramatic weather events and environmental changes (Krupnik, 2000). Such changes also had severe social impacts, pushing wealthy pastoralists below the poverty line. Declines and increases in caribou and reindeer populations are cyclical. Reindeer populations display consistent instability, indicating that herds and grazing systems are strongly influenced by climatic variation (Chapter 17). Severe weather conditions in spring, or a late snow melt, can have significant effects on reindeer populations, resulting in the death of young or weak animals during winter periods of starvation (Lee et al., 2000). Research suggests that climate change may already be contributing to the decline of caribou and reindeer herds. For example, the caribou disappeared from northern East Greenland in 1900 through migration to West Greenland in search of an adequate food supply as a result of climatic changes; this, in turn, caused the arctic wolf to disappear by 1934 owing to the loss of its main source of food (WCMC, 1990).

The disappearance of some caribou on Canada's Banks Island may be linked to climate change according to recent research (see Riedlinger and Berkes, 2001; and the case study in section 12.3.1) and also the observations of Inuvialuit, as discussed by Nagy M. (2004):

> In the '70s I guess, that's when they really started noticing it, muskox taking over. But [regarding] caribou, sometimes [...] in the fall, we get freeze-up on the whole island. Then, before the snow is really deep, we get our mild weather and rain. Then it's cold enough for the rain to freeze on top the snow and that's when the caribou try to leave the island, even go out into the ocean. 'Cause they were eating mostly ice.

> We were still here when one year it happened. When dogs started seeing the caribou, they'd be running. Nothing wrong with them but they just stop and start kicking. They have too much water in their stomach, their heads are spinning. So a lot of big bulls died off by spring [...]. There was even one year, that worst year that time, the cows didn't have any calves, they didn't. That hit them just before the rutting season.

> I don't think [the muskox] really pushed the caribou away. Like right now the caribou are just dying, now. [...] in the fall time, [...] when the weather is not good, the ones that are born, they just freeze when the weather is not good.

Using the results of Wilkinson et al. (1976), Gunn et al. (1991) dismissed forage competition between muskox and caribou and linked the disappearance of caribou on Banks Island to changing climate conditions associated with earlier spring snow disappearance, warmer winters that are snowier (hence more difficult for accessing forage), and with higher incidence of freezing rain.

Although annual die-offs of 60 to 300 caribou occurred during the winters of 1987–1988, 1988–1989, and 1990–1991 when freezing rains occurred (Nagy J. et al., 1996), Larter and Nagy (2000, 2001) concluded that the drop in number of Banks Island caribou in 1994 and in 1998 happened despite high calf production, high overwinter survival rates of calves, and less severe winter snow conditions. Thus, severe winter weather might not be the major cause of caribou decline. According to Nagy M. (2004), some Inuvialuit think that caribou do not like the strong smell of muskox and prefer to be away from them. Accordingly, some Inuvialuit say that caribou have moved out of the island to avoid muskox. Lent (1999) noted that reindeer herders in Alaska believed that "caribou and reindeer will avoid muskox, moving away when muskox enter their vicinity" but added "there is no quantitative evidence to support this contention, nor has a controlled study been undertaken". Hence expressing some of the distrust wildlife scientists might have towards local knowledge.

As Chapter 17 discusses in more detail, recent modeling studies indicate that the mean annual temperature over northern Fennoscandia is likely to increase by 0.3 to 0.5 °C per decade during the next 20 to 30 years, with the annual amount of precipitation increasing by 1 to 4% per decade. Such changes are likely to affect snow conditions and foraging conditions for reindeer. In Finland there is increasing concern about the effect of a changing climate on the winter snowpack and on the distribution of lichens, the main winter food for reindeer. Climate change is expected to mean that fast-growing vascular plants may out-compete slower growing lichens, which will affect the eating habits of reindeer. In Finland, Saami reindeer herders are aware of when reindeer numbers fall due to adverse weather and attempt to preserve their herds by adjusting the number of animals they slaughter (Lee et al., 2000).

12.2.4.4. Concerns over irreversible impacts

Indigenous peoples live with fluctuations in weather and climatic conditions. Experiencing year-to-year changes in weather, ice and snow patterns, animal behavior and movement, and in hunting conditions is part of life in the Arctic. Yet the trends currently being observed give concern over major, irreversible impacts on indigenous communities and livelihoods. For example, since the late 1970s Alaska Natives in communities along the coast of the northern Bering and Chukchi Seas have noticed substantial changes in the ocean and the animals that live there, particularly in the patterns of wind, temperature, ice, and currents (see also Chapter 3).

A significant collection of indigenous environmental observations was recorded during a study of environmental changes in Canada's Hudson Bay region. The results are published in *Voices from the Bay* by the Canadian Arctic Resources Committee and the Municipality of Sanikiluaq, a small Inuit community on the Belcher Islands in the midst of Hudson Bay. Completed in 1996 and published in 1997, the study brought together 78 Inuit and Cree hunters and elders from 28 communities on the shores of Hudson and James Bays in a series of workshops held over three years to describe, record, and verify ecological changes in the region, including but not limited to climate change (McDonald et al., 1997). Observations include wholesale changes in location, number, and duration of polynyas – open water areas in winter – in eastern Hudson Bay, and changing routes of Canada geese (*Branta canadensis*) and snow geese (*Anser caerulescens*), but the study indicates that alterations in weather and climate are by no means uniform within the region. *Voices from the Bay* and other observations by indigenous peoples (see Chapter 3) illustrate an important and inescapable fact: that much of the impact of climate change on northern indigenous peoples will be channeled through ecological changes to which they will have to respond, cope, and adapt.

As indigenous peoples perceive and experience it, the Arctic is becoming an environment at risk (Nuttall, 1998) in the sense that sea ice is now unstable where hunters previously knew it to be safe, more dramatic weather-related events such as floods are occurring, vegetation cover is changing, and particular animals are no longer found in traditional hunting areas during specific seasons. The weather is becoming increasingly unpredictable and local landscapes, seascapes, and icescapes are becoming unfamiliar.

Hunters and herders in some places are already altering their hunting patterns to accommodate changes to ice, tundra vegetation, and the distribution of marine and terrestrial harvested species (Callaway et al., 1999). As the case study from Sachs Harbour shows (see section 12.3.1), physical environmental change is immediately observable in terms of reduced sea-ice cover and lack of old (or multi-year) ice around the community in summer, and the thawing of permafrost. These changes challenge Inuvialuit knowledge and understanding of the environment, and make prediction, travel safety, and resource access more difficult. The Inuvialuit, like most indigenous groups who live off the land, rely on their ability to predict environmental phenomena such as snow and ice conditions, the weather, and the timing of wildlife migrations. For the Inuvialuit, as is increasingly reported throughout the Arctic by many other peoples, seasons have become less consistent, and weather events have become less predictable in the last few years (Krupnik and Jolly, 2002).

12.2.5. Responding to climate change

12.2.5.1. Flexibility and adaptation

The Arctic has experienced significant climate change in the past, just as the global climate has changed historically in response to natural variations. What may seem to be relatively minor variations in temperature have produced large positive feedbacks in the environment that have often had dramatic impacts on physical and biologi-

cal systems (e.g., Vibe, 1967). The successful long-term occupation of the Arctic by indigenous peoples has been possible, in part, owing to their adaptive capacity (in social, economic, and cultural practices) to adjust to climate variation and change. Hundreds and even thousands of years ago, arctic populations adapted to gradual or even rapid environmental change by settling amid favorable climate conditions and along the paths of animal migration routes.

The study of the origins, migration patterns, and socioeconomic development of arctic cultures is significant to any assessment of climate change in that it offers insight into long-term environmental adaptations, the impact of environmental change on humans, and in turn how humans have utilized resources and impacted upon the environment (e.g., Sabo, 1991). Historical, archaeological, and anthropological evidence suggests that indigenous peoples had elaborate ecological knowledge that was crucial to their successful adaptation to changing environmental conditions, as well as to seizing the opportunities presented by climate change. The archaeological and ethno-historical record reveals that, in dealing with climate change, resource availability, social and economic change, and the introduction of new technology, indigenous populations have developed significant flexibility in resource procurement techniques and in social structure.

Climate change or the overexploitation of animal and fish populations meant that arctic hunting bands would have been forced to move to other areas in search of game, or to have adapted and diversified their range of subsistence techniques. Odner (1992), for example, has argued that Saami populations in northern Norway coped with the periodic scarcity of wild reindeer in the middle ages by diversifying their subsistence activities, intensifying the exploitation of other species, moving on to other hunting grounds, developing techniques of animal husbandry, or by storing meat.

In the Canadian Arctic, Sabo (1991) showed how Inuit in the eastern Canadian Arctic coped with the effects of climatic change on the population dynamics, distribution, and availability of terrestrial and marine resources by rescheduling their hunting activities and adapting their hunting techniques, and by maintaining flexibility in settlement patterns and social organization. Developing an ecosystem model and reviewing evidence for climate change over a 1000-year period for southern Baffin Island, Sabo (1991) demonstrated that the rescheduling of resource procurement systems and the continuation of a flexible arrangement in Inuit settlement patterns and demographic organization ensured both the availability and production of food and acted as regulatory social mechanisms which were able to respond to environmental change. Sabo (1991) argued that, while there is paleoenvironmental evidence to suggest climate change did affect Inuit subsistence activities on Baffin Island during this period, climate change is only one of several factors contributing to adaptive responses. Rather than resulting in environmental determinism, the ecology and climate of southern Baffin Island enabled successive human populations to develop long-term strategies of environmental diversification. By using a variety of resources and habitats the prehistoric population and historic Inuit retained a resilient human/ecosystem relationship during a long period of continuity and change.

The expansion of the Thule tradition across the North American Arctic, from western Alaska eastward to the central Canadian Arctic and beyond to Hudson Bay, Labrador, and Greenland offers another example of how indigenous peoples have adapted and migrated as the climate has changed. During the Neo-Atlantic Optimum (ca. AD 1000), the Canadian Arctic passed through a period of 400 to 500 years (the Scandic Period) during which mean summer temperatures were 1 to 2 °C below the current average to a warmer period with summer temperatures around 2 °C higher than at present. This warming period resulted in the Canadian eastern Arctic experiencing less summer sea ice, longer periods of open water, and ice-free summers. For Inuit groups, access was opened up to maritime habitats with a variety of marine mammals, mainly narwhal, beluga, harp seal and, significantly, the bowhead whale (Wenzel, 1995a). While this climatic shift changed the ecology of the Canadian eastern Arctic, the cultural effects of the Neo-Atlantic Optimum on coastal Inuit groups were also far-reaching. The major shift was perhaps the replacement of the paleo-eskimo Dorset culture by Thule migrants from the Beaufort/Chukchi Seas region, whose subsistence culture was underpinned by their dependence on the bowhead whale (Wenzel, 1995a). The eastward movement of these migrants was facilitated by the changing ecological conditions and the movement of the bowhead whale into previously ice-closed areas of the eastern Arctic (Wenzel, 1995a).

The Thule tradition bore the hallmark of what is the essence of successful indigenous resource use systems throughout the Arctic – flexibility in technology and social organization and an ability to cope with climate change, responding both to its associated risks and seizing its opportunities. The archaeological record, ethnohistorical accounts, and the memories of elders provide detailed accounts of how human life in the Arctic has always been dominated and influenced by periodic, irregular, and often dramatic ecosystem changes, triggered by periods of warming and cooling, extreme weather events, and fluctuations in animal populations (Krupnik, 2000).

12.2.5.2. Barriers to adaptation

Change is a fact of life for arctic peoples, and they have a rich heritage of cultural adaptations to deal with it. Many of the short-term (or coping) responses appear to be based on this tradition of flexibility and innovation. The transition from sedentary to nomadic subsistence livelihoods and vice versa was the key to the survival and sustainability of arctic indigenous cultures. Cultural and

ecological diversity required flexibility and resilient coping strategies during periods of extreme change and subsistence diversity was the outcome of a successful cultural and social response to climate variation and the resource instability of the Arctic (Krupnik, 1993).

Yet, a word of caution must be added: while there are success stories in terms of adaptation to climate change, it would be wrong to assume that adaptation is simple and not fraught with difficulties. There are losers as well as winners when climate change challenges indigenous peoples to respond in ways that can mitigate the negative impacts. In the Canadian eastern Arctic, the Dorset people lost out while the Thule migration was facilitated by climatic change, and as research on the social consequences of climate change in Greenland shows, people living in towns with similar social and economic settings and political and institutional structures showed a marked difference in their abilities and readiness to adapt to changing conditions (see Rasmussen and Hamilton, 2001).

Environmental changes, particularly in climate and ocean currents, that have affected fisheries in West Greenland are well documented, as are the associated social and economic changes, especially at the beginning of the 20th century (Hamilton et al., 2000). As the waters of southern and west Greenland warmed, seals moved further north, making seal hunting harder for the Inuit population. Cod and other fish (halibut and shrimp) moved into the now warmer waters and made the development of a cod fishery possible. The development of fishing in West Greenland shows how climate change can provide opportunities for some people, some local communities, and some local regions. As Thuesen (1999) argued, the political and economic changes taking place in West Greenland at the beginning of the 20th century meant that Greenlanders were now involved in and participating in the new political structures of local municipal councils and two provincial councils established in 1908. In 1910, experimental fisheries were taking place in West Greenland and Greenlandic fishers were learning new skills in fisheries training programs. The west coast town of Sisimiut was able to take advantage of these new developments, advantageously situated as it is at the northernmost limit of the ice-free waters on the west coast.

For those Greenlanders who embraced change and the opportunities now arising, some were able to benefit more than others because they played crucial roles as local entrepreneurs and took advantage of the opportunities to diversify local economies. Thuesen (1999) argued that the development of Sisimiut as an important fishing centre was due in part to a strong sense of local identity and strong dynamism in the community – in short, people had a willingness to embrace change, to diversify the economic base, and to work to develop new industries. This stands in contrast to the development of the southwest Greenlandic town of Paamiut around the same time. Paamiut's development was based largely on plentiful resources of cod. With few other resources available in commercially viable quantities, there was little incentive to diversify the local economy (Rasmussen and Hamilton, 2001). The concentration on a single resource demonstrated the vulnerability of Paamiut in the face of environmental change. The cod population began to fall, due to a combination of climatic change and overfishing, and the economy and population of Paamiut declined as a result (Rasmussen and Hamilton, 2001). This highlights the importance of recognizing that, in any adaptive strategy, local conditions and social differences are considerable factors in the success of a region affected by change, be it from climate, social, economic, or political factors. The development of cod fishing in Greenland also shows, however, how climate change and social change go hand in hand. Cod fishing developed at a time when climate change was having an adverse effect on seal hunting, yet the population of Greenland was also growing, making it necessary to find alternative ways for the majority of the population to make a living.

Arctic hunters and herders have always lived with and adapted to shifts and changes in the size, distribution, range, and availability of animal populations. They have dealt with flux and change by developing significant flexibility in resource procurement techniques and in social organization. Yet the ecological and social relations between indigenous peoples and animal species are not just affected by climate-induced disruption, changing habitats and migration routes, or new technology. The livelihoods of the indigenous peoples of the Arctic are subject both to the influences of the market economy and to the implementation of government policy that either contributes to a redefinition of hunting, herding, and fishing, or threatens to subvert subsistence lifestyles and indigenous ideologies of human–animal relationships.

Today, arctic peoples cannot adapt, relocate, or change resource use activities as easily as they may have been able to do in the past, because most now live in permanent communities and have to negotiate greatly circumscribed social and economic situations. The majority of indigenous peoples live in planned settlements with elaborate infrastructures, and their hunting and herding activities are determined to a large extent by resource management regimes, by land use and land ownership regulations, and by local and global markets. As the case study on Inuit in Nunavut shows (section 12.3.2), the mobility that Inuit once possessed to move in response to shifts in the pattern and state of their resource base is no longer possible. Inuit in Nunavut now live in communities that are a direct result of Canadian government policy and which represent hundreds of millions of dollars of infrastructure and other investment. Clyde River, for instance, which is home to about 800 people and more or less representative of the kind of infrastructure and services found across Nunavut, is the result of some Can$ 50 million of government investment. In today's social, political, and economic climate,

migration to remain in contact with animals and, more broadly, to maintain traditional Inuit hunting livelihoods would seem to be virtually impossible.

Changes to settlement patterns and to the ecological relations between humans and animals often arise from government attempts to introduce new economic activities or to "sedentarize" indigenous peoples. In northern Russia and Siberia, for example, the Soviet authorities "industrialized" reindeer herding as a way of facilitating the development of the Soviet North. The new settlements and industries in Siberia came to depend on reindeer herders to supply them with meat. Today, in post-Soviet Russia, privatization and the transition to a market economy bring new challenges to reindeer herding peoples in Siberia and the Russian Far East, highlighting the dependence of arctic reindeer systems on the complex interlinkages between local, regional, and global economies.

In a similar vein, caribou management on the Canadian Barrens became an integral part of a broad program of social engineering – federal, provincial, and territorial authorities imposed management strategies based on their own (rather than Inuit and Dene) ideas about conservation and hunting (Usher, 2004). There are similar stories from other parts of the Arctic. For example, the introduction of reindeer to the Seward Peninsula in western Alaska during 1892 to 1902 was done to provide meat for Iñupiat communities, yet was also intended as a way of transforming Iñupiat from being subsistence marine mammal hunters to reindeer herders and thus to play an active role in the wider cash economy of the United States (Anderson and Nuttall, 2004).

Strict regulatory regimes and management practices imposed by states and federal and provincial agencies increasingly affect hunting and herding (Anderson and Nuttall, 2004). Some, while aiming, in principle, to protect and conserve wildlife also restrict access to resources. In Alaska, for example, state and federal policies make subsistence issues extremely complex. State and federal law define subsistence as the customary and traditional non-commercial use of wild resources and regulations limit the prospects of finding markets for caribou meat. Earning money through more commercial channels is not an option for Alaskan subsistence hunters. In northern Fennoscandia, Saami reindeer herders have traditionally ranged far and wide, crossing national borders as they follow their reindeer herds between winter and summer pastures. In modern times, political developments have restricted migration routes over the last 100 years or so. Economic development in the 19th and 20th centuries, such as mining, forestry, railways, roads, hydro-electric power, and tourism have all had an impact on traditional Saami livelihoods.

In Greenland, threats to the cultural and economic viability of hunting livelihoods in small communities come from transformations in resource management regimes and Home Rule government regulations, which conflict with local customary practices and knowledge systems (Dahl, 2000; Nuttall, 2001). Caribou, whales, seals, and fish, which have traditionally been subject to common use rights vested in members of a local community, are becoming national and privately-owned divisible commodities subject to rational management regimes defined by the state and the interest groups of hunters and fishers, rather than to locally understood and worked out rights, obligations, and practices. As is still evident in some parts of Greenland today, it has traditionally been the case that no-one owns animals – everyone has the right to hunt and fish as a member of a local community. A caribou, fish, or marine mammal does not become a commodity until it has been caught and transformed into private property. Even then, complex local rules, beliefs, and cultural practices counter the exclusive sense of individual ownership (Nuttall, 2001). However, trends in caribou hunting since the 1980s are illustrative of general wildlife management policies in Greenland, where membership of a territorial, or place-based, community no longer gives hunters exclusive rights to harvest caribou. In West Greenland, caribou hunting was largely a family event until the 1970s. Kinship, locality, and territory were the mechanisms for regulating harvesting activities. Today, hunting rights are vested in people as members of social and economic associations irrespective of a local focus. Discussing the situation in central West Greenland, Dahl (2000) showed how the traditional hunting territories of various communities are not the same as the administrative boundaries that surround villages, towns, districts, and municipalities. The relevant territorial unit for hunting caribou (and other animals such as beluga and narwhal) is Greenland, rather than a place-based community.

Hunters and herders are thus constrained by institutional frameworks and management structures, as well as by the legal recognition to resource use rights. They are commonly experiencing a transition from herding and hunting, from what may be called a "way of life", to an occupation and industry. The similarities with commercial fisheries management in the circumpolar North are notable, especially the effects of the implementation of individual transferable quotas (ITQs). The ITQ system is a management response to overfishing and to declining catches of major fish species, particularly demersal species. Although designed to ensure the viability of fish stocks, sustainable catch levels, and economic efficiency, ITQ management results in the transformation of traditional common use rights in fish stocks into privately owned, divisible commodities. As Helgason and Palsson (1997) argued, ITQs represent the idea that both the human and natural worlds can be organized, controlled, and managed in a rational way. Nature is not only "presented as an inherently technical and logical domain, the project of the resource economist and manager is sometimes likened to that of the engineer or the technician". Helgason and Palsson (1997) described the public discontent in Iceland with the commoditization of fishing rights as a consequence of the ITQ system and

which has resulted in fishing rights being concentrated in the hands of a few large operators – a discontent articulated in feudal metaphors such as "tenancy" and "lordships of the sea". The ITQ system, although ostensibly seen by economists and resource managers as a way of achieving the sustainable use of fish stocks, has in reality a social impact in terms of changing power relations within local communities and regional fisheries, by contributing to the concentration of wealth into the hands of a few large fishing vessel owners. The ITQ system has effectively meant the enclosure of the commons and the privatization of resources, which allows parallels to be drawn between fisheries and rural land use debates throughout the Arctic.

12.2.5.3. Opportunities for adaptation and response

Commercial, political, economic, legal, and conservation interests have reduced the ability of indigenous peoples to adapt and be flexible in coping with climatic variability. The contemporary reality for many hunters and herders is that they are placed in very inflexible situations. Faced with climate change they are not necessarily in a position to respond appropriately. However, indigenous peoples have demonstrated resilience and adaptability in the face of change. In the climate-changed Arctic that this assessment considers, how indigenous peoples can take advantage of the opportunities that may arise, as well as how they can modify or change their mode of production in response to climatic variability, for example by switching hunting and fishing activities, is a critical research need.

For some arctic peoples, the political and management systems are already in place that could assess the impacts of climate change, allow local and regional governments to act on policy recommendations to deal with the consequences, and improve the chances for indigenous peoples to deal successfully with climate change. Although complex, solutions to environmental problems are potentially realistic.

Significant political changes since the 1970s have included land claims in Alaska and Canada and the formation of regional governments in Greenland and Nunavut. Settlements include the Alaska Native Claims Settlement Act (1971), Greenland Home Rule (1979), the James Bay and Northern Quebec Agreement (1975–1977), the Inuvialuit Final Agreement (1984), and the Nunavut Agreement of 1992 (the Territory of Nunavut was inaugurated in 1999). These political changes often include changes in the ways that living and non-living resources are managed. A greater degree of local involvement in resource use management decisions has been introduced, including in some cases the actual transfer of decision-making authority to the local or regional level (CAFF, 2001).

In addition, significant steps have been taken with innovative co-management regimes that allow for the sharing of responsibility for resource management between indigenous and other uses and the state (Huntington, 1992; Osherenko, 1988; Roberts, 1996). Examples include the Alaska Eskimo Whaling Commission, the Kola Saami Reindeer Breeding Project, the Inuvialuit Game Council, and the North Atlantic Marine Mammal Commission. Self-government is about being able to practice autonomy. The devolution of authority and the introduction of co-management allow indigenous peoples opportunities to improve the degree to which management and the regulation of resource use considers and incorporates indigenous views and traditional resource use systems (Huntington, 1992).

Co-management projects involve greater recognition of indigenous rights to resource use and emphasize the importance of decentralized, non-hierarchical institutions, and consensus decision-making. This presents tremendous opportunities for collaboration between indigenous peoples, scientists, and policy-makers concerned with the sustainable use of living resources (Caulfield, 2000). And it is within this new political and scientific environment of power sharing and dialogue that indigenous communities, scientists, and policy-makers can work together to find solutions (such as building flexibility into otherwise constraining wildlife management regimes) to the pressing problems climate change may bring to the Arctic. Although knowledge integration in co-management systems remains fraught with technical, methodological, and political difficulties (Nadasdy, 2003), some of the case studies presented in this chapter show how evolving forms of co-management institutions create opportunities to increase local resilience and the ability to cope with, respond to, and deal with change. For example, new governance mechanisms through the Inuvialuit Final Agreement of 1984 are helping Inuvialuit to negotiate and manage the impacts of change. For instance, the five co-management bodies established by the Agreement provide an effective means for Inuvialuit communities to communicate with regional, territorial, and federal governments and, indeed, to the Arctic Council.

The detailed case studies that follow show how climate change is having an impact on hunting, herding, gathering, and fishing activities. However, they also show that some of the impacts have been absorbed through the flexibility of the seasonal cycle and local ways of life. For the Inuvialuit of Sachs Harbour, for example, coping strategies relate to adjusting subsistence activity patterns: modifying timing of harvest activity; modifying location of harvest activity; modifying method of harvest activity; adjusting the species harvested; and minimizing risk and uncertainty. Yet, for indigenous peoples, dependence on animals and involvement in complex global processes, combined with the natural vulnerability of the Arctic and the concern with the accelerated nature of climate change, magnify the potential effects of global climate change on their cultures and livelihoods.

12.3. Understanding climate change impacts through case studies

12.3.1. Canadian Western Arctic: the Inuvialuit of Sachs Harbour

Sachs Harbour has been studied and reported on intensively through the Inuit Observations of Climate Change project, undertaken jointly by the Community of Sachs Harbour and the International Institute for Sustainable Development. The Inuvialuit (the Inuit of the Canadian western Arctic) themselves initiated the study because they wanted the documentation of the severe and disturbing environmental changes that they were witnessing. The project was undertaken with several objectives (Ford, 1999; IISD, 2000; Riedlinger and Berkes, 2001):

- to produce a video on how climate change is affecting the people;
- to disseminate Inuit observations to the world;
- to document local knowledge of climate change; and
- to explore the potential contributions of traditional knowledge to climate change research.

The project was planned and carried out using participatory research methods. Results are based on a 12-month study of Sachs Harbour covering all four seasons in 1999/ 2000, with follow-up visits for verification and project evaluation (Jolly et al., 2002). Inuvialuit perceptions shaped the study from the very beginning; the project started with a planning workshop which asked the people of Sachs Harbour *their* objectives and what *they* considered important for the project to focus on. Video documentation plans, research questions, and the overall process were all defined jointly by the study team and the community (Berkes and Jolly, 2001; Jolly et al., 2002).

The community of Sachs Harbour is located on Banks Island in the Canadian western Arctic. It is a tiny community of some 30 households, and the smallest of the six Inuvialuit communities in the region covered by the comprehensive native land claims agreement; the Inuvialuit Final Agreement of 1984 (Fast and Mathias, 2000). Sachs Harbour, a permanent settlement only since 1956, is an outgrowth of the white fox trade beginning in the 1920s (Usher, 1970). Many of the current residents have relations in the Mackenzie Delta area. Some are descendants of the Copper Eskimo of Victoria Island to the east; many are related to the Iñupiat (Alaska Inuit) who had earlier moved to the Delta.

There are no previous studies about how climate change may have affected resource use in the past on Banks Island. Major changes in resource use concern the development of the white fox trade and its subsequent collapse with the disappearance of the European fur market in the 1980s, and the dramatic changes in muskox and caribou numbers on the island. Muskox were present in extremely low numbers in the early 1900s, but populations increased in the latter half of the 20th century, giving Banks Island the largest muskox population in the world. In the meantime, however, caribou numbers have declined. There is no consensus on the question of whether the caribou decline is related to muskox increase. Nor is there agreement regarding the impact of climate change on these two species, but a number of potential negative impacts are possible, including those related to extreme weather events (Gunn, 1995).

Although Sachs Harbour, as the permanent village, only dates from the 1950s, local observations, as captured by the Inuit Observations of Climate Change project, go back to the 1930s (Jolly et al., 2002). Perceptions of Sachs Harbour hunters and fishers are consistent in indicating that changes observed in the 1990s are without precedent and outside the range of variation that the Inuvialuit consider normal. Before addressing the observations of change and how the people have coped with them, it is necessary to review patterns of subsistence.

12.3.1.1. Patterns of subsistence and the impact of climate change

Some 20 species of terrestrial and marine mammals, fish, and birds were taken in 1999/2000 at Sachs Harbour. During the winter, people hunted muskox and, to a lesser extent, caribou, Arctic fox, wolf, polar bear, and ringed seals. Small game included ptarmigan (*Lagopus* spp.) and Arctic hare. As the weather began to warm in March and April, people headed out to numerous inland lakes to ice-fish for lake trout and Arctic char.

In May, fishing slowed down as the snow goose season approached. Banks Island supports a large breeding colony of snow geese. Goose hunting and egg-collecting were important community activities. Family groups camped at rivers and inland lakes, and the entire community harvested and processed geese, some of it for inter-community trade. The goose hunt was over by mid-June, as people returned to lakes to fish where there still was ice. They also fished for Arctic cod on sea ice and went sealing. With ice break-up in June and July, people hunted mainly for ringed seals, and some bearded seals, off the ice floes and from boats in open water. July through early September, people set gillnets for char, Arctic cod, and least cisco (*Coregonus sardinella*), and some did rod-and-reel fishing in lakes. In September, people turned to muskox and caribou.

In some years, including 1999/2000, the muskox hunt is a commercial harvest that employs almost the entire community throughout November. Guiding and outfitting for sport hunting for polar bears and muskox also provide employment and cash income. These commercial activities complement the subsistence harvest, and are a major source of cash income for the community.

The cycle of hunting and fishing varies from year to year, but the usual pattern has been affected by environmental changes being observed by the people of Sachs Harbour. These changes, as documented by IISD (2000), Riedlinger and Berkes (2001), and Jolly et al. (2002), may be sum-

marized under five headings: physical environmental change; predictability of the environment; travel safety on land and ice; access to resources; and changes in animal distributions and condition (see Table 12.1).

Physical environmental change is most readily observable in terms of reduced sea-ice cover and lack of old (or multi-year) sea ice around the community in summer, and the thawing of permafrost. These changes challenge Inuvialuit knowledge and understanding of the environment, and make prediction, travel safety, and resource access more difficult. The Inuvialuit, like most indigenous groups who live off the land, rely on their

Table 12.1. Examples of environmental changes impacting upon subsistence activities (adapted from Riedlinger and Berkes, 2001; Jolly et al., 2002).

Physical environmental change
- Multi-year sea ice no longer comes close to Sachs Harbour in summer
- Less sea ice in summer means that water is rougher
- Open water is now closer to land in winter
- More rain in summer and autumn makes travel difficult
- Permafrost is no longer solid in places
- Lakes draining into the sea from ground thawing and slumping
- Loose, soft snow (as opposed to hard-packed snow) makes it harder to travel

Predictability of the environment
- It has become difficult to tell when ice is going to break-up on rivers
- Arrival of spring has become unpredictable
- Difficult to predict weather and storms
- There are "wrong" winds sometimes
- More snow, blowing snow, and whiteouts

Travel safety on the land and ice
- Too much broken ice in winter makes travel dangerous
- Unpredictable sea-ice conditions make travel dangerous
- Less multi-year ice means traveling on first-year ice all winter, which is less safe
- Less sea-ice cover in summer means rougher, more dangerous storms at sea

Access to resources
- It is more difficult to hunt seals because of a lack of mult-year sea ice
- In winter, cannot go out as far when hunting because of a lack of firm sea-ice cover
- Harder to hunt geese because the spring melt occurs so fast
- Warmer summers and more rain mean more vegetation and food for animals

Changes in animal distributions and condition
- Less fat on seals
- Observe fish and bird species never before seen
- Increase in biting flies; never had mosquitoes before
- Seeing fewer polar bears in the autumn because of a lack of sea ice
- More least cisco caught now

ability to predict environmental phenomena such as snow and ice conditions, the weather, and the timing of wildlife migrations. Seasons have become less consistent, and weather events have now become less predictable.

Travel safety is closely related to physical environmental change and loss of ability to predict the environment. For example, sea ice near the community is used for travel, ice-fishing, and seal and polar bear hunting. Sound knowledge of the sea ice and the ability to monitor and predict changes are critical to hunting success and safety. In the 1990s, people in Sachs Harbour observed increased ice movement in winter and spring, changes in the distribution of leads, cracks, and pressure ridges, as well as overall thinning of the ice. People say that in the past they rarely had to worry about the ice the way they do now; one has to be more cautious than ever before when traveling on ice.

Access to resources is often related to travel access and safety. For example, changes in the rate of spring melt and increased variability associated with spring weather conditions have affected access to hunting and fishing camps. When families go out to camps at lakes for ice fishing and goose hunting in May, they travel by snowmobile, pulling a *qamutik* (sled), staying on snow-covered areas or using coastal sea ice and frozen rivers. However, warmer springs have resulted in earlier, faster snow melt and river break-up, making access difficult. The availability of some species has changed due to the inability of people to hunt them under changing environmental conditions. For example, less summer ice means that ringed seals are harder to spot and hunt.

However, not all changes in species availability are related to access. Changes in animal distributions have also occurred, with respect to birds (many new mainland species never before seen on Banks Island), fish (two species of Pacific salmon), and insects. Some of the changes may operate through ecological mechanisms. Sachs Harbour hunters discuss and speculate on the impacts of environmental change on species distributions and availability. For example, warmer temperatures and higher rainfall may have increased summer forage for caribou and muskox. But these changes may also increase the risk of extreme weather events such as freezing rain in autumn that may cover the ground with a layer of ice, making forage unavailable.

12.3.1.2. Short-term and long-term responses to change

The Inuvialuit of Sachs Harbour draw on accumulated knowledge and experience in dealing with change. They recognize that they have always adapted to change – social, political, and economic change, as well as environmental change. When asked about the impact of environmental change on subsistence activities, most people are quick to point out that they always find some way to deal with changes. Change is a fact of life for indigenous peoples, and they have a rich heritage of cultural adapta-

tions to deal with change. Many of the short-term (or coping) responses appear to be based on this tradition of flexibility and innovation.

Environmental changes observed in Sachs Harbour are not trivial, and these are having an impact on subsistence activities. However, many of the impacts have been absorbed through the flexibility of the seasonal cycle and the Inuvialuit way of life. Inuvialuit coping strategies mostly relate to adjusting subsistence activity patterns: modifying timing of harvest activity; modifying location of harvest activity; modifying method of harvest activity; adjusting the mix of species harvested; and minimizing risk and uncertainty. Table 12.2 provides examples of each.

Modifying the timing of harvest activity is often related to increased seasonal variability. Hunters adjust their seasonal calendars to deal with change. Since change is unpredictable, hunters also use waiting as a coping strategy; people wait for the geese to arrive, for the weather to improve, and so on. Modifying the location of harvest activity is often necessitated by physical changes. Changes related to sea ice require hunters to stay close to the community because of safety concerns. The thawing of permafrost in many areas has left travelers to make new trails to avoid slumps and mudslides. Also, hunters have had to use different modes of transport to adjust how they harvest animals.

Table 12.2. Short-term or coping responses to environmental change in Sachs Harbour: changing when, where, or how hunting and fishing takes place (adapted from Berkes and Jolly, 2001).

Modifying the timing of harvest activity
• Warmer temperatures and unpredictable sea-ice conditions mean hunters go out earlier for polar bear
• Shorter springs and an increased rate of snow melt have reduced the time spent on the land; people return to the community after the goose hunt, instead of proceeding to lakes for ice fishing
Modifying the location of harvest activity
• Erosion and slumping at one fishing lake near the community has necessitated fishing at other lakes instead
• More bare ground and unreliable snow conditions mean families travel over coastal sea ice rather than along inland routes
Adjusting how harvesting is done
• Use of all terrain vehicles instead of snowmobiles to travel to spring camps when there is not enough snow
• Hunters take seals from boats in open water, necessitated by the lack of summer sea ice on which seals normally haul out
Adjusting the mix of species harvested
• More *qaaqtaq* (least cisco) caught in nets at the mouth of the Sachs River
• Hunters are taking different kinds of mainland ducks previously rare in the area
Minimizing risk and uncertainty
• River- and sea-ice conditions monitored more closely
• Only the more experienced hunters travel on certain types of sea ice

A major coping strategy is switching species. Reduced fishing opportunity in one area (e.g., spring ice-fishing in lakes) may be compensated for by an increase in another (least cisco). Climate change has brought new potential resources through range extensions. Pintail (*Anas acuta*) and mallard ducks (*A. platyrhynchos*), both mainland species, and white-fronted goose or "yellow legs" (*Anser albifrons*) and tundra swans (*Cygnus columbianus*), both historically rare on Banks Island, have been observed in increasingly larger numbers.

Hunters have adopted a number of strategies to minimize risk and uncertainty. In response to increased variability and unpredictability associated with the weather and other environmental phenomena, they monitor ice conditions more closely and take fewer chances. Hunters say that "you really need to have experience to travel on the sea ice now", and describe being more careful when they travel.

The short-term coping strategies summarized in Table 12.2 are ultimately based on cultural adaptations. Berkes and Jolly (2001) compiled from various sources a list of cultural practices which are considered to be adaptive responses to arctic ecosystems: (1) mobility and group size flexibility; (2) flexibility of seasonal cycles of harvest; (3) detailed local environmental knowledge; (4) sharing mechanisms and social networks; and (5) inter-community trade.

Table 12.3 provides a summary of these adaptive mechanisms and evidence from Sachs Harbour as to whether they are still viable. The first of these adaptive mechanisms is no longer operative owing to settlement of people into permanent villages, but the other four seem viable.

The flexibility of seasonal cycles and the creativity with which hunters take advantage of harvesting potentials are backed up by oral traditions and by Inuvialuit cultural values that emphasize the appropriateness of harvesting what is available and acting opportunistically.

Regarding local environmental knowledge (traditional or indigenous knowledge) and related skill sets, some have obviously been lost, and some are being transmitted incompletely. Certain kinds of skills that were once universal in Inuvialuit society have become restricted to relatively few families who are active on the land. For example, almost all teenage boys in Sachs Harbour can use guns, but not many can build snow houses. The nature of people's practical engagement with the environment has changed; skill sets and land-based knowledge have also changed. For example, hunters use GPS units for navigation and safety, a very recent skill. The use of snowmobiles since the 1970s, also a new skill, has necessitated a greater knowledge of ice conditions because sled dogs can sense dangerous ice while snowmobiles cannot.

Sharing mechanisms for food and social networks for mutual support are still very much in evidence in Sachs Harbour, especially within extended family units and in providing for elders. A relatively small number of hunters account for most of the harvest; thus, relatively few people are providing for the families of occasional hunters and non-hunters. The imbalance is addressed by new forms of reciprocity whereby food-rich members of extended families share with cash-rich members, thus bringing wage income into the realm of sharing relationships. Inter-community trade is extensive. Sharing between communities does not seem to have declined but rather increased in importance. Sachs Harbour has an abundance of snow geese and muskox, and these are exported to other communities, in return for caribou and beluga whale products. These exchanges use the norms of generosity (giving without asking), sharing and generalized reciprocity, and not the Western rules of economic exchange involving cash exchange.

In sum, the Inuvialuit of Sachs Harbour have coped with the effects of recent climatic changes by modifying when, where, and how hunting and fishing are carried out. These coping strategies borrow from traditions of

Table 12.3. Cultural practices which may be considered adaptive responses to changes in the arctic environment, and evidence of their viability in Sachs Harbour (adapted from Berkes and Jolly, 2001).

Cultural practice	Viability–evidence from Sachs Harbour
Mobility of hunting groups; seasonal settlements; group size flexibility with grouping and regrouping of self-supporting economic units	No longer operative owing to permanent settlements; compensated for by the use of mechanized transport to increase mobility of family groups and all-male hunting groups
Flexibility of seasonal cycles of harvest and resource use, backed up by oral traditions to provide group memory	Source of major short-term coping strategies, aided by rapid transport and communication technology to monitor animal population movements
Detailed local environmental knowledge (traditional knowledge) and related skill sets for harvesting, navigating, and food processing	Underpins ability to change when, where, or how subsistence harvesting occurs; loss of universality of some skills; loss of some knowledge and skills compensated for by new knowledge and skills
Sharing mechanisms and social networks for mutual support and risk minimization; high social value attached to sharing and generosity	Sharing of food and associated social values still important, especially within extended family units; special considerations for elders; new forms of reciprocity involving cash
Inter-community trade along networks and trading partnerships, to deal with regional differences in resource availability	Active inter-community networks, especially within Inuvialuit region; more extensive than practiced by previous generations; norms of generosity and generalized reciprocity still alive

flexible resource use, and dynamic traditional environmental knowledge and skills. Also important among adaptive strategies is food sharing through intra-community social networks and inter-community trade. All these cultural practices are still largely intact in Sachs Harbour and the Canadian western Arctic in general. All these strategies provide considerable buffering capacity to deal with climate change, or with any other kind of social or environmental perturbation.

12.3.1.3. Climate change and social and ecological relations

There is no evidence that climate change, as observed in the 1990s, has altered the ecological relations between the people of Sachs Harbour and their resources, or altered social relations within the community. It has not resulted in increased or decreased pressures on any of the major resources. However, it has had some consequences for the local perceptions of the environment and local cultural understandings of resources. For example, the Inuvialuit of Sachs Harbour are concerned about the impact of the lack of sea ice in the summer on ringed seal pups. Some of them are also concerned about the risk of extreme events to animal populations, such as the potential impact of freezing rain on caribou forage.

One major impact of climate on the local perception of the environment concerns the issue of loss of predictability. Land-based livelihoods in the Arctic depend on the peoples' ability to predict the weather (is the storm breaking so I can get out?), read the ice (should I cross the river?), judge the snow conditions (could I get back to the community before nightfall?), and predict animal movements and distributions. A hunter who cannot predict the weather or read the ice would be limited in mobility; one who cannot decide what to hunt and where cannot bring back much food.

Climate change has the potential to impact on indigenous environmental knowledge and predictive ability, thus damaging the self-confidence of local populations in making a living from their resources. Such changes may ultimately leave them as strangers on their own land. Arctic peoples are experts at adapting to conditions that outsiders consider difficult. However, climate change impacts raise the issues of speed and magnitude of change, as compared to how fast people can learn and adapt. The evidence from Sachs Harbour hunters indicates that current environmental change is beginning to stress their ability to adapt. Rapid change requires rapid learning, and unpredictability superimposed on change interferes with the ability to learn. Predictability is affected by extreme weather events and higher variability, and appears to be an area of climate change research that deserves consideration in its own right.

Even though this case study focuses on impacts and adaptations associated with harvests and subsistence, climate change also has other economic and cultural consequences. For example, in addition to harvesting implica-

tions the lack of sea ice also makes some people "lonely for the ice", as the ice is a central feature of Inuvialuit life (Riedlinger and Berkes, 2001). Other environmental changes that are permafrost-related (e.g., thaw slumps, soil erosion) may not be a major threat to subsistence, but may have direct impacts on other aspects of community life, such as the maintenance of buildings and roads.

12.3.1.4. Climate change impacts in context

Inuvialuit society in Sachs Harbour has been affected by many social and environmental changes over recent decades. Major changes in subsistence and other resource use patterns have been caused by changes in global fur markets (white fox), commercialization of muskox (early 1900s depletions), and their subsequent protection followed by population recovery. These changes, plus government policies, have resulted in major social and economic transformations in Inuvialuit society, turning these migratory hunting peoples into village-dwellers who use mechanized transport to go out on the land. Further changes in recent years have seen the introduction of commercial muskox hunts, and sport hunting based on muskox and polar bears.

Compared to these major changes, the impact of climate change is relatively minor, at least so far, and it is not beyond the ability of the community to adapt. However, climate change is a relatively recent event, and the ability of Sachs Harbour Inuvialuit to respond to and cope with it, mainly by adjusting subsistence activities, may not be a reliable indication of the community's ability to adapt in the future. How much change can be accommodated by the Inuvialuit and their resource use systems? Elsewhere, recent publications have focused on the *resilience*, or the amount of perturbation that the Sachs Harbour hunting system can absorb and adapt to by learning and self-organization (e.g., Berkes and Jolly, 2001). The question of resilience is important because little is known about building adaptive capacity in the face of climate change.

Evolving co-management institutions in the area create additional opportunities to increase resilience and the ability to deal with change. New governance mechanisms through the Inuvialuit Final Agreement of 1984 seem to be helping the people of Sachs Harbour to negotiate and manage the impacts of change. There are five co-management bodies established through the Agreement that make it possible for the Inuvialuit communities in the area to communicate with the regional, territorial, and federal governments, and eventually with the Arctic Council.

Co-management has created linkages that were not possible only a few years ago. For example, indigenous hunters have been interacting with scientists in meetings such as the Beaufort Sea 2000 Conference, organized by one of the co-management agencies, the Fisheries Joint Management Committee (FJMC, 2000). Co-management bodies, connecting local-level institutions with government agencies, provide vertical linkages across levels of organization and horizontal linkages across geographic areas. Berkes and Jolly (2001) hypothesized that such governance mechanisms have the potential to contribute to learning and to self-organization, and hence to build adaptive capacity to deal with change.

12.3.2. Canadian Inuit in Nunavut

The impact of climate on Inuit has been a dominant, if not the predominant, theme in Eskimo anthropology since Franz Boas (Boas, 1888) undertook research on Baffin Island. At a time when the study of hunter-gatherers has become a virtual sub-discipline within anthropology, the "attribute" that still sets Inuit apart from the Kalahari San and other hunting peoples is the same one that European visitors to Nunavut, from Martin Frobisher (an early explorer) to today's tourist, remark upon. That is, how can any people adapt to the arctic environment, and to most people the arctic environment is epitomized by climate, especially the cold and the long, dark winters.

This case study focuses on the adaptability (or adaptiveness) of the traditional Inuit economy in Nunavut in a (presumed) time of climate-induced ecological instability. The relationship between Inuit ecology and Inuit economy is almost too obvious. Inuit are hunters and the most referenced passage in Boas's seminal *The Central Eskimo* (Boas, 1888) is about the relationship between sea ice, ringed seal distribution, and Inuit hunting and settlement. So a part of this case study is necessarily about Inuit hunting and wildlife harvesting. In other words, it will speak to the production component of the traditional economy, particularly Inuit hunting and the production of *niqituinnaq* (real food) including what at Clyde River (the community from which much of the material in this case study is derived) is called *ningiqtuq* – the sharing or, put formally, the Inuit system of resource allocation and redistribution.

Gaining an understanding of how environmental change due to a warming (or cooling) climate may affect the material aspects of Inuit resource production (i.e., the

economics) is important. And so are the possible effects of climate-induced ecological instability on the traditional economy because it is the socio-cultural rules that order who gets what when that make the economy Inuit.

With regard to generating hypotheses, or at least envisioning scenarios, modern workers have the benefit of the archaeology and paleoclimatology undertaken over the past 40 years in the North American Arctic. Much of this was to answer questions about how climate has influenced the economics of Inuit life. There is less information about the Inuit economy as it is impossible to know exactly how a seal or caribou was shared within communities, let alone who received what piece, 500 or 1000 years ago. However, as Inuit economics and economy are linked, there is at least the possibility, using data about past changes and about how the system currently functions, to model the socio-economic impact among Nunavummiut (the Inuit of Nunavut) because of a large-scale change in climate.

12.3.2.1. Inuit subsistence and climate: the long-term record

The relations between climate and Inuit material subsistence and cultural adaptation can be examined through what is known from climatology, physical oceanography, and biology about two long-term climate trends. These are the Little Climatic Optimum–Medieval Warm Epoch (also known as the Neo-Atlantic Period, ca. AD 1000–1250), and the Second Climatic Optimum/Neo-Boreal Period/Little Ice Age, which lasted from ca. AD 1550 to 1900 (Andrews and Andrews, 1979; Grove, 1988; Lamb, 1982; Vasari et al., 1972).

Data from northern Europe, Iceland, and the eastern Arctic indicate that during the Neo-Atlantic Period temperatures across the high latitudes of the North Atlantic region were as much as 2 to 2.5 °C above the annual average that prevailed in the eastern Arctic through most of the 20th century. Conversely, the Little Ice Age involved a significant cooling of this region, with the most pronounced thermal effect in summer. Data from northwestern and mid-Europe suggest that summers

Box 12.1. Inuktitut terms

Akpallugitt	form of sharing between individuals ("inviting in")
Ilagiit	extended family
Isumataq	head of an *ilaqiit* (lit. "one who thinks")
Katujiyuk	apportioning of meat within a cooperating task group
Minaqtuq	commensal sharing/distribution of stored food
Nalaqtuk	behavioral terms meaning respect or obedience
Ningiq	a share of a hunted animal
Ningiqtuq	to share a portion of a hunted animal
Niqiliriiq	those who share; "neighbors"
Niqisutaiyuq	a form of commensal sharing
Niqitatianaq	transfers of food between two unrelated hunters
Niqituinnaq	meat from a hunted animal ("real food")

Nirriyaktuqtuq	commensal meal
Nunavummiut	people of Nunavut
Paiyuktuq	a gift of food (related forms: *quaktuaktuq*, *niqisutaiyuq*)
Quaktuaktuq	a form of commensal sharing; food gifts to close affines
Sila	weather, climate; also: mind, consciousness
Tigutuinnaq	transfers (usually food) from an *isumataq* to a subordinate
Tugagauyuk	transfers (usually food) from a subordinate hunter to superior kin
Uummajusiutiit	unrelated cooperating hunters
Ungayuk	behavioral term meaning affection or solidarity
Umiaqa	traditional woman's boat
Qayaq	kayak

averaged 0.5 to 0.8 °C less then those in the preceding moderating Pacific Period. Further north, in Scandinavia, the first half of the 17th century saw 13 summers at least 1 °C colder then the estimated average for the 16th century (Briffa et al., 1990; Pfister, 1988).

These episodes also produced large-scale positive feedback in the North American arctic ecosystem. The impact of each episode on northern physical and biological systems in turn correlates with climate-related adaptive adjustments by Inuit (see Barry et al., 1977; Dekin, 1969; Maxwell, 1985).

The most discussed episode is the Second Climatic Optimum, which warmed the North American polar stage from the Chukchi Sea to West Greenland. This warming, beginning around AD 1000, saw the central and eastern Canadian Arctic experience a spatial and temporal reduction in the amount of seasonal sea ice present.

This change in the physical environment created extensive new range for bowhead whales and for longer periods. And the expansion of bowhead whales from north Alaskan waters eastward (while North Atlantic bowhead whales were able to penetrate farther west) enabled Thule Culture people, the direct ancestors of modern Inuit, with their whale hunting experience, to follow. With a technology adapted to exploit a resource that the indigenous population of this part of the Arctic could not, the migrants rapidly displaced the late Paleo-Eskimo population that had developed *in situ* over the previous two millennia from the whole of Nunavut, Ungava–Labrador, and Greenland. Thus, while Thule Culture lasted only the few centuries that this extreme warm period allowed large whales passage into most of the Canadian Arctic, many of the technologies that the *Qallunaat* (non-Inuit) world associates with Inuit – dog traction, the *umiaq* and *qayaq*, and large marine mammal hunting – are Thule legacies.

The Little Ice Age, the deep cold that set in following a transitional cooling from the Medieval Warm Epoch, is the reason why the Inuit culture that Europeans met as they quested for a northern route to Cathay looked as it did (and still looked until about 1970). The long summers with almost ice-free open water were gone and, except on the western and easternmost fringes of the Inuit area, so were bowhead whales. The whole tenor of Inuit life had changed.

The winter security that came with the harvesting of a 20- or 30-tonne whale was gone and so was the large supply of fuel and building material that came with capturing a bowhead whale. Instead, Inuit developed what McGhee (1972) somewhat over-generally called a "*Netsilik* adaptation" based on the exploitation of a variety of seasonally available smaller prey species, chiefly caribou in summer, ringed seals through the winter, and anadromous Arctic char during their passage to and from the ocean.

In addition, the Inuit pattern of winter settlement across much of Nunavut changed from the land to the sea ice and the Thule Culture Classic Stage semi-subterranean whalebone and boulder house was abandoned in many areas for the snow *igliuk* or iglu. Overall, Inuit became less sedentary because large supplies of food could no longer be rapidly developed and the new primary resource suite comprised species that were highly mobile and/or elusive.

12.3.2.2. *Ningiqtuq*: the traditional/contemporary economy

An economy is the orderly movement of goods and services from producers to consumers (Langdon, 1986).

> ...*a subsistence economy is a highly specialized mode of production and distribution of not only goods and services, but of social forms*... Lonner, 1980

An extensive discussion on the economy of Nunavut is beyond the scope of this case study. However, other than in the territorial capital and main regional government centers, the term "subsistence", as it is used by Lonner (1980), describes the situation for the rest of Nunavut.

Table 12.4. Clyde Inuit *ningiqtuq* interaction sets. *Ningiqtuq* is generally seen as a multi-layered strategy by which participants achieve the widest possible intra-community distribution of resources. However, while Damas used *ningiq* to refer only to the social movement of *niqituinnaq*, *ningiqtuq* is conceptualized here as a set of socio-economic operations that also encompass labor and non-traditional resources.

Interaction set	Flow direction	Reference
Traditional	1a. *isumataq* << *ilagiit* subordinates	*Tugagaujuq*[a]
	1b. *isumataq* >> *ilagiit* subordinates	*Tigutuinnaq*[a]
	2. father-in-law << son-in-law	*tugagaujuq* (?)
	3. *isumataq* >> community	*Nirriyaktuqtuq/minatuq* (?)
Modern	4. between unrelated hunters	*Uummajusiutiit*
	5. *angijukak* << unrelated hunters	*Taliqtuq*
	6. *angijukak* >> community	*Nirriyaktuqtuq* (?)
Other	7. between unrelated young and elders	*nalaktuq* related
	8. between same generation non-kin; generally among the elderly	inviting in and "gifting"

[a]*Tugagaujuq* and *tigutuinnaq* are complementary and participants are generally seen as being *niqiliriiq* (sharers of food).

Put another way, it is a mixed economy (sometimes described wrongly as a dual economy) in which traditional and non-traditional resources – represented by wild foods and money, respectively – interact, although "optimal economy" is probably a more accurate description. The reality for most Nunavummiut is that the best return for one dollar comes from hunting, but without a dollar hunting is not possible. What is optimal (i.e., how much of each resource type is best) differs from household to household, but few households can manage reasonably without some mix of country and imported food. As Fienup-Riordan (1986) observed, "…[monetary income] is perceived as the means to accomplish and facilitate the harvest, and not an end in itself".

With respect to the traditional economy, this case study concentrates on the social form(s) that organize the material flow of food once it has been captured. To some extent, these rules also apply when money is "captured" (Wenzel and White, 2001). However, this is more uneven owing to the scarcity of money and the costs that are almost always associated with its acquisition.

12.3.2.3. The system in outline

Ningiqtuq is not a single defined process by which seal meat or *maktaaq* are distributed. It is generally translated as meaning "to share", but it is in fact a web of social mechanisms for distributing and redistributing food and other resources. How allocation is accomplished differs across Nunavut (see Collings et al., 1998; Damas, 1972), but the term is used in almost all regions of the territory to describe the overall process of transferring food between individuals, households, and across entire communities. Table 12.4 outlines the array of distributional mechanisms in Clyde River. Not all the processes included in the table are "traditional", there are several that the older generation of Clyde River people consider the result of modern village circumstances. However, each form shown was referenced by at least three informants to a traditional type or behavioral precept (see Table 12.5).

As Table 12.5 shows, food sharing at Clyde River is a multi-level system that encompasses social relations ranging from the action that occurs between paired isolates (as in *akpallugiit*) to means that span the entire community (*minaqtuq*). And while *ningiqtuq*, as practiced today by Clyde River Inuit, includes aspects related to the changed pattern of settlement that came about through Canadian government centralization policies in the 1950s and 1960s, organization of the system based on traditional principles of, foremost, kinship and, second, intra-generational solidarity, remains.

In functional terms, almost every form of sharing encompassed by the concept of *ningiqtuq* has as its basis a social, rather than an economic, referent. The greatest sharing activity in terms of social focus occurs within the context of the restricted extended family. Within the *ilagiit* essentially all members are in a *niqiliriiq* (literally, "those who share food") relationship. And it is within the *ilagiit* that the *nalaqtuk* (Damas' (1963) respect–obedience dyad, but which may be conceptualized as responsibility–obligation (Wenzel, 1981)) directive that structures intergenerational/interpersonal behavior is most apparent.

Whereas *tugagaujuk–tigutuinnaq* activities function almost wholly within the social context of the extended family, as Tables 12.4 and 12.5 indicate, mechanisms for the more generalized distribution of food resources are also present. The main one being *nirriyaktuqtuq*, or communal meal. Such commensalism may be restricted to the *ilagiit*, particularly when resources are scarce, or may include a large segment of the community. In either circumstance, communal meals are always held in, or immediately adjacent to, the dwelling of the hosting extended family head.

12.3.2.4. Generalized reciprocity

A major reason for presenting an exhaustive review of the Inuit economy in Nunavut is to dispel the commonly held view that the Inuit traditional economy can be summed up by the term generalized reciprocity. It cannot, and this is as inappropriate as saying that catching a seal sums up the traditional economy (Collings et al., 1998; Damas, 1972; Wenzel, 1981, 1989, 1995b, 2000).

Table 12.5. Aspects of Clyde Inuit *ningiqtuq*.

Social context	Behavioral directive	Form	Description
1a. Individual	*Ungayuk* (solidarity–affection)	*akpallugiit*	inviting in guests (typically same generation non-kin)
1b.	*Ungayuk*	*quaktuaktuq/niqisutaiyuq/ paiyuktuq*	food gifts to close affines and non-kin (generally restricted to elders)
1c.	*Ungayuk*	*niqitatianaq*	*Uummajusiutiit* ("partnered" hunters)
2a. Intra-*Ilagiit*	*Nalaqtuk* (respect–obedience)	*niqiliriiq*	*tugagauyuk-tigutuinnaq* complementary
2b.	*Nalaqtuk*	*nirriyaktuqtuq*	restricted commensalism
3a. Inter-*Ilagiit*/community	*Ungayuk*	*nirriyaktuqtuq*	open commensalism
3b.	*Ungayuk*	*minaqtuq*	distribution of stored food
3c.	*Nalaqtuk*	*Katujiyuk*	within task group

The *ningiqtuq* economy is socially complex. Although some of its forms are general in scope – commensalism being an example – most of its operations are founded in balanced reciprocal relations, with reciprocity enforced by social precepts that provide for inclusion as well as sanction.

12.3.2.5. Climate change and the economy

Warming versus cooling

Based on what is known of the impacts of the two most recent major climatic events to have affected the Inuit (section 12.3.2.1), warming would appear to be a good thing for the Inuit economy; the Second Climatic Optimum spurred an amazing cultural expansion, with Inuit traveling nearly 8000 km in barely 200 years, in the process displacing a cultural tradition nearly 2000 years old.

However, the *Netsilik* hunting adaptation (section 12.3.2.1) was a response to a cold environment and the *ningiqtuq* economy differs markedly from the economy practiced around bowhead whaling in North Alaska since at least the 19th century (Burch, 1985; Spencer, 1959). (This is not to say that a *ningiqtuq*-type of sharing is absent among Iñupiat (Bodenhorn, 2000), but rather that it is overlain by a more corporately-oriented mechanism.) This suggests that the present warming, should it continue to increase, may not be good for either the traditional economy or the subsistence economy.

The best evidence for testing this theory comes from the West Greenland work of Vibe (1967) on the effect of climate change on northern biota and Inuit resource use. Using a 150-year database (1800–1950) drawn from Danish colonial meteorological, ice, and trading records, Vibe (1967) correlated the episodes of warming and cooling over this period with the rise and fall in the capture of ringed seals and polar bears. By comparing the official trading records with sea-ice conditions during this period it was apparent that when the local climate ameliorated, which reduced the duration of the seasonal sea ice, the capture of both species declined. Vibe (1967) also pointed out that ringed seals are the main prey item for polar bears and that a stable sea-ice environment is critical to ringed seal ecology, especially for successful spring pupping.

Vibe's study, which drew on the rich scientific and commercial records available from Greenland, is unique in those terms. However, the conclusion that ringed seal pup production suffers when increased temperatures seasonally destabilize the sea ice and that this affects the polar bear harvest supports statements by Inuit based on their long empirical experience with both species.

Ringed seals

Ringed seals and polar bears are as important now as at any time in the past to the economic well-being of small Nunavut communities. The ringed seal, or *natsiq*, is one of the principal items in the traditional Inuit diet. Besides being the most abundant marine mammal in circumpolar waters, ringed seals are present throughout the year along the entire Nunavut coastline. Their presence through winter offsets the absence of most other important food species at this time. Finally, *natsiq* provide high quality nutrition when few alternatives, except for the most costly imported foods, are available.

To Inuit, *natsiq* are an all-season, all-year food. At Clyde River, where it is one of eighteen species of mammal, fish, and bird that are regularly harvested, ringed seals comprised 54% of the edible biomass captured by Clyde hunters between 1979 and 1983 (Wenzel, 1991). In 1979, of the 169 tonnes of country food that came into the community, 109 tonnes (64.9%) were ringed seal (Wenzel, 1991). Thus, a substantial reduction in the seal harvest would have profound implications for the ecological economics of Inuit life. This is even more apparent in terms of the seasonal dietary contribution of ringed seal. Ringed seals represent 58% of the winter food supply, but 66, 81, and 64% in spring, summer, and autumn, respectively (Wenzel, 1991). Caribou, the next most important food species by edible weight, comprises 39% of the winter food capture, and 30, 13, and 18.5% for the other seasons, respectively.

A substantial reduction in ringed seals would also affect the overall economy of Inuit subsistence. This is mainly because there is no other species on the land or in the waters of Nunavut that is as abundant or as available as *natsiq*. In simple terms, no other species could sustain the subsistence requirements of Inuit. But, more importantly, the cultural meaning of *ningiqtuq* would suffer. This is because *niqituinnaq* (real food) is quite literally the stuff of sharing. To hunt, catch, and share this kind of food is to an Inuk the essence of living *Inuktitut* (Wenzel et al., 2000). Ringed seal is as much a cultural commodity in Inuit subsistence culture as it is an item of diet.

Polar bears

Polar bears also play an important role in the contemporary subsistence system. Like ringed seals, they are also *niqituinnaq*. And, if climate change affects the ecology and distribution of ringed seals, it will thus affect polar bears. However, in food terms, polar bears, especially when compared to ringed seals, are of minor importance. Nevertheless, they represent one of the few sources of money that Inuit can access through traditional activities. While polar bear hides have long had a market outside the traditional uses to which Inuit put them, a polar bear hunt sold today to an American, Swiss, Mexican, or Japanese sport hunter may bring as much as US$15 000 per bear to a community. Rifles, snowmobiles, and gasoline are now as effective a part of Inuit subsistence as dog teams, seal oil lamps, and fishing leisters were sixty years ago. (Why this is requires looking at Canadian internal colonial policy from 1945 to 1985.)

The quandary that confronts every Inuk hunter is how to gain access to money at a minimum cost in time. While hunting produces large amounts of high quality food – the Government of Nunavut estimates that it would cost approximately Can$ 35 000 000 to replace this harvest production – virtually none of this traditional wealth can be converted into the money needed to purchase, operate, and maintain the equipment hunters use. Yet abandoning hunting for imported food would not only be less healthy but would also be immensely costly. But this is not in fact a viable alternative as approximately 30 to 35% of adult Nunavummiut are unemployed and another 15 to 20% are underemployed or only able to work seasonally.

Polar bear sport hunting helps meet the cash resource needs of many hunters while imposing a minimal cost in time. In 2001, ten sport polar bear hunts at Clyde River brought approximately Can$ 212 000 into the community, with half going directly to the Inuit – more income than entered Clyde River from four years of hiking, kayaking, and other forms of ecotourism. And these hunt revenues directly capitalized the purchase of five snowmobiles, a 7 m inboard-engine equipped boat, a large outboard engine, and two all-terrain 4-wheel drive vehicles (some Can$ 75 000–90 000 of equipment) by sport hunt workers for use in sealing and other subsistence activities. (Note: hunt workers purchased one all-terrain vehicle and one snowmobile for relatives not involved in sport hunting; money does enter the *ningiqtuq* sharing system.)

Projected climate change

If the projected climate change scenarios are correct (see Chapter 4), some Nunavut communities, possibly even Clyde River, may find that the traditional and contemporary aspects of their subsistence systems are affected as described by Vibe (1967). In which case, if access to ringed seals and polar bears decreased, could Clyde River hunters shift to other subsistence sources, like narwhal, caribou, and harp seal that are at present of only minor importance?

The answer is probably yes, but not easily for a variety of reasons. Firstly, because it is highly likely that at least some potential "fallback" species will also be affected by a continued warming. For instance caribou, now the principal terrestrial resource for Inuit, are highly sensitive to the kinds of wet/cool conditions that may occur in autumn when rain, rather than snow, may lead to the icing over of vegetation and so limit their ability to obtain winter food. This occurred in autumn 1972 (see Kemp et al., 1978) on several islands in the Canadian High Arctic with the result that caribou disappeared for nearly six years from Bathurst and Cornwallis Islands.

The present reduced state of the Peary Caribou Herd, sufficiently serious for a number of central arctic communities to limit and even ban their subsistence harvests of this species, may have been triggered by autumn rains that iced the winter food supply and crusted the snow

cover. In most areas, muskox, which are better adapted to these conditions, have replaced the caribou, but are themselves vulnerable to exploitation.

Narwhal and harp seal may provide some replacement for any reduction in ringed seals. Neither is an arctic winter species, but if summers come earlier and stay ice-free for longer, the harvest of both may be increased significantly. Narwhal, because its *maktaaq* (skin) is a favored food and the ivory tusk of males has commercial value, would draw increased subsistence attention. And the northwest Atlantic harp seal herd, which summers between Baffin Island and West Greenland, has grown geometrically since southern Canadian commercial exploitation was limited in the mid-1980s.

However, there are serious issues concerning both species. Narwhal probably do not possess the population size to sustain any significant increase in their harvest. Moreover, there are serious Canadian and international regulatory issues that would need to be addressed even if an expanded harvest were solely for food. Similarly, any increase in the use of harp seals, which at present draw minor attention from Nunavummiut, would re-ignite the political activity that caused the collapse of markets for seal products (Wenzel, 1991).

One thing is certain. The mobility that Inuit once possessed to move in response to shifts in the pattern and state of their resource base is no longer possible. Inuit in Nunavut now live in communities that are a direct result of Canadian government policy and which represent hundreds of millions of dollars of infrastructure and other investment. Clyde River, for instance, which is home to about 800 people and more or less representative of the kind of infrastructure and services found across Nunavut, is the result of some $50 million of government investment. In today's political-economic climate, migration to remain in contact with *natsiq*, polar bear, or more broadly, to maintain traditional Inuit subsistence culture is virtually impossible.

12.3.2.6. Conclusions

Inuit, whether Nunavummiut, Alaskan, or Kalaallit, have shown adaptiveness in the face of the incredibly rapid change in their cultural environment as they have passed through successive stages of colonization in just six or seven decades. Having been able to adapt to that kind of environmental change, global warming will be far less formidable.

12.3.3. The Yamal Nenets of northwest Siberia

Given the lack of data on historic and prehistoric patterns of indigenous wildlife harvests and subsistence hunting in relation to climate or "weather" change, this case study focuses on the potential interactions between climate, land use, and reindeer management in the Yamal Nenets Autonomous Okrug of northwest Siberia. This is a region of ice-rich permafrost that has been subject to large-scale petroleum development over the past few decades, while at the same time giving indications of its sensitivity to decadal and even interannual variations in climate.

For at least a millennium (Fedorova, 1998), this region has also served as the homeland of the Yamal Nenets, nomads who have either hunted or herded reindeer as their main livelihood, supplemented by fishing, hunting, and gathering. Nenets have recently expressed great concern in a number of fora regarding their future in reindeer husbandry because of forces largely beyond their control (Forbes and Kofinas, 2000; Jernsletten and Klokov, 2002; Khorolya, 2002). These concerns are discussed here within the context of climate change.

Arctic peoples are experts in adapting to changing conditions (environmental, social, and economic) and recognize their abilities in this regard. Nonetheless, Yamal Nenets are currently showing signs of stress adapting to the recent barrage of simultaneous changes in their homeland – from health and demography (Pika and Bogoyavlensky, 1995) to questions of land tenure (Golovnev and Osherenko, 1999) and increasingly severe "overgrazing", predation, and poaching on reindeer pastures (Jernsletten and Klokov, 2002). There is a risk that a rapidly changing climate may accelerate ecosystem degradation in ways that Nenets are unable to cope with, given the constellation of other factors impinging upon their ability to maintain herding as a viable livelihood.

Krupnik (1993) argued that indigenous reindeer pastoralism expanded rapidly throughout the Russian Arctic during the 18th century as a result of two interwoven factors – socio-economic transformation and environmental change, in particular climate change resulting in "ecologically favorable conditions". The biological factors Krupnik (1993) cited which positively affected semi-domestic herd development were improvements in summer pastures and a concomitant increase in reproductive rates, coupled with a drop in summer and winter mortality. Proxy climate data for the past millennium for Yamal are summarized by Shiyatov and Mazepa (1995) and indicate a summer warming trend throughout the 1700s, as described by Krupnik (1993) for Eurasia in general.

Shiyatov and Mazepa (1995) also reported late 19th and 20th century climate trends for Yamal and these indicate a warming trend during the early and mid-summer periods between about 1940 and 1960. The total reindeer population began a period of rapid growth around 1950, following the near decimation of the herds during the Second World War, and this continued through the 1990s (Golovnev and Osherenko, 1999). At present there are around 180000 animals on the Yamal Peninsula and more than 600000 animals in the okrug managed by 2618 mostly family-based units with a nomadic lifestyle (WRH, 1999).

Since the collapse of the Soviet Union, various collaborative teams of scientists have made available a great deal of data on the relationship between climate and permafrost across the Russian Arctic. The Circumpolar Active Layer Monitoring (CALM) program is paramount among these efforts to observe changes in the seasonally thawed active layer and near-surface permafrost, including thermokarst erosion. Although the annual depth of substrate thaw (the "active layer") poorly reflects contemporary climatic warming on Yamal (Pavlov, 1998), inter-seasonal variability is strongly correlated with summer thawing degree-days. At the same time, the frozen ground beneath Yamal is characterized as "warm" permafrost with its temperature amplitude not far below 0 °C and these substrates have been warming in recent years (Pavlov, 1994).

With regard to air temperature, combined regional data from the mid-1970s onward show relatively small magnitude, positive trends in thawing degree-day totals, and a rise in mean annual air temperature. There is evidence that this is not the case in other parts of the Russian Arctic, such as neighboring Taymir and Chukotka in the Russian Far East (Kozhevnikov, 2000). Some recent modeling efforts project the onset of a climatic regime that is not conducive to the maintenance of permafrost over extensive areas of northwestern Siberia, with warmer spring and summer temperatures and additional precipitation. The authors concluded that such a development would have serious ramifications for engineered works in the region, owing to the extensive area underlain by massive ground ice (Anisimov and Nelson, 1997).

In general, the ecological impact of large-scale climate variability and recent climate change on northern ungulates is well documented. Variations in growth, body size, survival, fecundity, and rates of population increase have all been correlated with major atmospheric phenomena including the North Atlantic and Arctic Oscillations (Griffith et al., 2002). There is evidence from northern Fennoscandia, for example, that both extremely high and low oscillation indices have adverse effects on reindeer (Helle and Timonen, 2001). The mechanisms underlying these correlations derive from direct and indirect impacts on grazing conditions for the animals, such as the phenological development and nutritional quality of forage plant species, late-lying snow cover in spring and early winter icing events, and the animals' immediate thermal environment (Mysterud et al., 2001; Post and Stenseth, 1999).

Regardless of these historical trends in climate impacts and future scenarios emphasizing risk, the overriding concerns for contemporary Nenets herders of Yamal revolve around what is collectively referred to as "the pasture problem" (Jernsletten and Klokov, 2002; see also Podkoritov, 1995) and related issues pertaining to land tenure (Osherenko, 2001). Nenets have constantly adapted to change prior to, during, and since the development of intensive reindeer management that became the dominant management regime in the early 1900s. They have survived first Tsarist and later Soviet dreams of establishing state and religious authority over even the most remote human populations. Yet nothing has challenged them like the ongoing search for petroleum beneath their ancestral lands.

Oil and gas development began in the 1960s and intensified steadily through the 1970s and 1980s, quickly followed by the collapse of the Soviet Union and, almost simultaneously, the overnight disappearance of the largely artificial market for reindeer meat, and the replacement of barter with a cash economy. In the confusion of sorting out ownership of animals and title to land in a newly capitalist society, herd sizes continued to increase to historic highs as land withdrawals for industry pushed the animals onto progressively smaller parcels of land and restrictive migration routes, resulting in extensive pasture degradation (Forbes, 1999; Golovnev and Osherenko, 1999).

The so-called pasture problem is multifaceted and has developed over a long period of time. The collectivization of the herds, which took place under Stalin, is partly to blame, as it instituted the restrictive "brigade" system of management and sought to maximize meat production for the Soviet "market". This took away Nenets' ability to adjust to changing environmental conditions resulting from changes in weather, climate, social relations, and forage conditions, including grazing/trampling impacts. At present there are no fallow or "reserve" pastures on Yamal, as there previously were under traditional Nenets management. However, there were already reports of heavy grazing in some areas even before the onslaught of Soviet-style management (Golovnev and Osherenko, 1999; WRH, 1999).

The continued presence of the Nenets after the fall of communism, with their culture, livelihood, and ecosystems more or less intact, shows how successfully they had adapted to the period of collectivization. However, petroleum exploration developed rapidly and relatively unchecked, with a virtual lack of meaningful protocols and lax enforcement of the few new rules (Forbes, 1999). The problem now, from the herders' perspective, principally concerns land withdrawals for petroleum exploration, infrastructure development, and related degradation processes such as quarrying for sand and gravel, blowing sand and dust, and off-road vehicle traffic in summer (Forbes, 1995; Jernsletten and Klokov, 2002; Khorolya, 2002; WRH, 1999).

Alongside herding, Nenets have always supplemented their diet, clothing, and other needs by fishing and hunting. Nenets observe that the massive influx of industrial workers to Yamal and the concomitant increase in hunting and fishing pressure has meant the decimation of many freshwater ecosystems and some preferred game species (e.g., polar fox) in areas around the main gas fields and transport corridors (Okotetto and Forbes, 1999).

In attempting to adapt to the heavy grazing pressure on the pastures the herders now see themselves as "racing" along their migration routes (Golovnev and Osherenko,

1999). During field research in late summer 1991, the same week as the coup in Moscow took place, Bruce Forbes, one of the authors of this case study, met with herders near the main gas field of Bovanenkovo on north-central Yamal. The number of fully loaded sledges scattered around the camp surpassed the number of empty sledges. The head of the brigade explained that the reason was that they were breaking camp every 24 to 48 hours. He explained that as the herds have become larger they must now have the animals on the move almost constantly. One reason is to avoid rupturing the vegetation mat and exposing the fine-grained sand and loess beneath, which are prone to aeolian erosion, another is reduced forage quality.

Assessing the consequences of climate change and petroleum development, either individually or in combination, is particularly difficult for *Rangifer* spp. compared to other ungulates due to the complexity of their ecological relations (Gunn and Skogland, 1997; Klein, 1991). These involve traditional patterns of migratory movements, resulting in transitory dependence on different ecosystems and special physiological and morphological adaptations that enable them to use a unique food resource. Also, their complex social structure varies seasonally (Klein, 1991).

In the west, reindeer herders and caribou hunters display an acute awareness of the need for coupling indigenous knowledge about wildlife and environment with scientists' efforts to understand climate change and have clearly expressed their concerns as they pertain to traditional livelihoods (Krupnik and Jolly, 2002; Turi, 2000). Among reindeer herders in northern Russia, impacts other than those arising from changes in climate appear to be of more immediate concern and the overall situation has been described as a crisis (Krupnik, 2000). This has led to what has been described as passive rather than active adaptation (Klokov, 2000) to the many and drastic changes.

Dmitri Khorolya is himself Nenets and is both president of the Reindeer Herders' Union of Russia and director of Yarsalinski sovkhoz, the largest collective management unit on Yamal. In his address to the Second World Reindeer Herders' Congress in June 2001 he observed that:

> the vulnerable ethnic-economical systems of [Russian] reindeer peoples are frequently exposed to hard market conditions, particularly where oil and gas mining has become the principal factor in the development of arctic areas. Industrial activity in the [Russian] north has resulted in the destruction of many thousands of hectares of reindeer pasture. The process is continuing. In some regions pasture degradation threatens preservation of reindeer husbandry and the anxiety of reindeer herders for their future should be heard by the world community.

Although permafrost-related changes may not be a major threat to subsistence in Inuvialuit (see section

12.3.1), the situation in Yamal has the potential to be different owing to the long, restricted migrations involved, and the loss of the Nenets' traditional capacity for flexibility. As studies have shown, permafrost in the form of massive ground ice is common and the landscapes range from moderately to highly unstable even in the absence of industrial development or intensive reindeer management (Nelson and Anisimov, 1993). For Yamal, "what if" scenarios pertaining to climate change must include:

- the prospect of early melting/late freezing sea ice in the Ob River delta, as this would remove access between winter and summer pastures for the main herds (e.g., Yarsalinski sovkhoz); and
- increased traffic from the Northern Sea Route, perhaps inevitable but certainly benefiting from early melting/late freezing sea ice in the Kara Sea. This could accelerate the pace of regional development.

Either scenario risks additional stress on the adaptive abilities of the Yamal Nenets. Yet even in the absence of climate change, within the next two to three decades there are critical and immediate threats from questions of title to land and accelerating changes in land use. The latter includes local and widespread damage from industry, and the ecosystem-level effects of reindeer grazing and trampling, as well as poaching. Throughout this period the various parties must strive to minimize conflict (Klein, 2000). In the longer term, if the current climate warming continues, extensive changes to existing tundra communities can be expected as permafrost begins to thaw and large areas are either denuded by landslide events (Leibman and Egorov, 1996) or subject to paludification by thawing ground ice via thermokarst. Adaptation to such changes will require: (1) greater efforts on the part of industry to prevent or mitigate additional disturbance; (2) a flexible system of land use, emphasizing property rights, that is acceptable to both the Nenets and the State; and (3) additional practitioners' and scientific knowledge on the composition and potential forage utility of emergent plant communities which will necessarily be exploited by the reindeer.

12.3.4. Indigenous peoples of the Russian North

This case study is based on the ongoing work of the Russian Association of Indigenous Peoples of the North (RAIPON), together with the NorthSet project of the Institute of Geography at the Russian Academy of Sciences. This work concerns the assessment of climate change impacts on the indigenous peoples of the Russian North within the context of broader social, economic, and political changes. This case study is based on the preliminary results of initial research, but is included here because it illustrates the tremendous challenges faced by indigenous peoples throughout the Russian North.

The indigenous peoples of the Russian North have depended on traditional hunting, fishing, and gathering for thousands of years and, for several hundred years,

many groups have practiced nomadic reindeer breeding. Human impacts and environmental transformation in the Russian Arctic have intensified over the last few decades. Significant climate change is also becoming evident, as is the destructive impact of industry. The biggest sources of pollution are the oil and gas industries, as well as mineral extraction and processing, aggravated by poor purification facilities. The main negative impacts of industrial development threatening the livelihoods of indigenous peoples include:

- the destruction of reindeer pastures and widespread degradation of ecosystems, especially due to the construction of industrial infrastructures and industrial pollution;
- massive toxic pollution of marine and freshwater environments, affecting the habitats and spawning grounds of fish and causing the destruction of fisheries;
- deforestation due to the timber industry using concentrated methods of clear-cutting, leading to the destruction of the non-timber forest resources of high cultural and economic importance;
- large-scale landscape and soil destruction, erosion (especially thermokarst erosion), and the degradation of tundra and taiga vegetation as a result of air pollution from industrial emissions (especially emissions from the non-ferrous metal industry);
- flooding of valuable subsistence areas due to the construction of hydroelectric power dams; and
- forest fires, partly associated with poaching and partly with increased recreational pressure around the regions of industrial development.

These impacts have added to the tremendous problems faced by Russia's northern indigenous peoples, which can only be understood by reference to Soviet and post-Soviet transformations. During Soviet times, public policies resulted in the resettlement of the inhabitants of small settlements into large villages. This coercive resettlement of indigenous peoples signaled the beginning of the destruction of the social and ecological relationships that characterized their subsistence lifestyles. Resettlement, the separation of children from their parents in favor of education at boarding schools, preservation orders on vital grasslands and reindeer pastures, and the reduced possibilities for engaging in traditional activities, together with many other changes, led to a spiritual and social crisis among the indigenous peoples (Vlassova, 2002). Since the 1970s, unemployment and alcoholism have become widespread, family structures are breaking down, and traditional culture is being destroyed.

In recent years, the destruction of traditional subsistence activities, especially reindeer herding – the most important activity for many indigenous groups – has continued apace. The difficult period of transition to a market economy in post-Soviet Russia has brought sharp changes to the economic and social conditions of the indigenous peoples of the Russian North, to which they have had to adapt quickly in order to survive. In the

1990s, when the formation of the market economy and democratization of society in the Russian Federation began, the situation in reindeer husbandry changed dramatically. This period of transition has seen a rapid decay of collective reindeer husbandry and a partial return to the private ownership of reindeer herds. This has occurred without the introduction of sufficient legal reforms, particularly affecting agricultural and traditional lands. One major trend has been a significant reduction in the population of domesticated reindeer. Combined with a lack of approaches for the development of an alternative program for sustainable development, and faced with increasing climate variability and change, the situation for the indigenous peoples of the Russian North is increasingly bleak.

The indigenous peoples of the Russian North comprise a mere 2% of the entire northern Russian population and number approximately 200 000 individuals belonging to forty different peoples (Haruchi, 2001). The most numerous are the Nenets, who comprise around 35 000 persons; the least numerous are the Enets with about 209 and the Orok with 109. The subsistence area of the indigenous peoples is roughly 60% of the overall territory of the Russian Federation and their traditional subsistence activities include reindeer herding, hunting (including marine mammals), fishing, gathering wild plants and, to a certain degree, craft-making and traditional art. The specific activities of the different peoples vary significantly from region to region.

The indigenous communities of Russia are the most endangered social group in the current period of transition to a market economy. Between 1990 and 2000, the number of indigenous people employed on northern livestock farms, as well as in hunting and fishing, fell by 37%. In these years of market reforms, the actual rate of unemployment in the indigenous settlements of the Russian North is, on average, not less than 40 to 50% of the economically active population. The situation is worse for young people in remote areas. Some small villages of autonomous okrugs (e.g., in the Koryak Autonomous Okrug) face an unemployment rate of 75 to 80%; in some districts of Habarov Kray the unemployment rate among the indigenous peoples increased six-fold during the 1990s.

Social ills associated with unemployment – poverty, disease, family breakdown, crime, suicide, and alcoholism – are increasing in indigenous settlements. Mortality among indigenous peoples increased by 35.5% in the 1990s (Abdulatipov, 1999). The nature of mortality has changed over the last few decades: the main risk group is no longer children, but young adults, and the main cause of death is no longer sickness, but death as a result of injuries, accidents, and suicide. The main cause of this situation is the destruction of traditional lifestyles (Vlassova, 2002). For Saami living in Lovozero on the Kola Peninsula, whose health and livelihoods have been affected by pollution and ecological degradation, environmental improvement is an even greater priority than the

improvement of housing conditions, which are extremely poor (Afanasieva, 2002). According to the Saami, the climate is becoming less comfortable, and they articulate this in terms of their livelihoods and health. The environment is one in which they dwell and comfortable housing will not improve their health if the climate is changing. During a workshop organized by RAIPON in April 2003 many Saami participants expressed concerns about a link between rapid and frequent climate/weather changes and the increase in cases of high blood pressure. As they spoke, they connected health and illness directly to climate variability and change.

Significant shifts have occurred in unemployment structure. The indigenous share in municipal positions, and in the service, educational, and cultural sectors has increased considerably while the participation of indigenous peoples in the traditional economy has decreased sharply. The highest levels of unemployment are observed in the areas where indigenous peoples retain traditional livelihoods. In larger settlements with a developed service sector, employment within the indigenous population is slightly higher. Yet, an increasing reliance on service sector activities does not always mean that harvesting renewable resources and production of traditional food for the household has declined in importance. As in other arctic states, hunting, herding, gathering, and fishing still satisfy important cultural, social, and nutritional needs, as well as the economic needs of families, households, and communities.

In this changing social and economic climate, indigenous systems of traditional resource use are under threat. Traditional land use areas are mainly located within zones of political and economic interest, particularly those concerning oil, mineral, and timber production, and military complexes with nuclear test sites. From the initial results of the research being conducted by RAIPON, a majority of indigenous people consider poaching, forest fires caused by humans, industrial logging, and clearing of forests for firewood to be some of the most significant issues that affect the physical environments and well-being of their communities.

- Decreasing populations of animal and plant species are a serious concern and it may be that this is not due to climate and ecological changes alone, but is aggravated by poaching, which is a serious problem in several regions.
- Fires, the frequency and scale of which have recently increased, are either natural or man-made. In the Tyumen region alone, which is now being intensively explored for natural resources, over 1.5 million hectares of reindeer pasture have been destroyed by fire. One of the causes of escalation of fires in the tundra, the taiga–tundra zone, and the taiga might also be climate warming, especially summer droughts.
- In recent decades, commercial logging operations have advanced closer to the taiga–tundra zone across much of the boreal forest region. The transforma-

tion of the northern parts of the taiga zone into a taiga–tundra, or even tundra, as a result of human activity is occurring in Russia (Vlassova, 2002).

- The fuel deficit in remote communities is one reason for illegal logging. Serious ecological problems arise with cutting of forests for fuel in Kovran, Loveozero, and Kuumba.

One of the causes of the decrease in reindeer numbers is the degradation of the treeline (taiga–tundra) winter reindeer pastures caused by industrial forestry, clearing of forests for firewood, and industrial pollution. The traditional ways of life of indigenous peoples are characterized by high adaptability to seasonal as well as to spatial differences in the physical environment. Climate changes may interfere with the human–nature cycle of reindeer herding, where herders follow the paths of reindeer between summer grazing lands in the tundra and mountains and winter grazing lands in the treeline. Winter pastures are of great importance for reindeer herding. During the long arctic winter, reindeer depend upon access to pasture rich in ground lichens, which are their basic food. In the autumn, reindeer start to move to forested areas that provide layers of soft snow that they can dig through to find the ground lichens. Epiphytic lichens on old trees are important reserve fodder when the ground lichens can not be reached due to ice layers on or within the snow. The lichens almost exclusively provide these animals with the carbohydrates required to maintain their body temperature in winter (Vlassova and Volkov, 2001).

Another cause of the decrease in reindeer numbers is the overgrazing of tundra and taiga–tundra pastures. Increasingly fewer winter pastures are available for reindeer herding as large territories are being occupied by mining and petroleum industries. This increases the pressure by domesticated reindeer on the tundra and taiga–tundra ecosystems, and thus leads to further degradation. Ecosystems are completely overgrazed by reindeer in many areas. The overgrazing of reindeer pastures leads to deforestation of the taiga–tundra winter pastures, especially owing to the damage to trees and shrubs. This has the effect of pushing the treeline southward in many areas (Vlassova and Volkov, 2001).

Fires are contributing to the degradation of reindeer pastures and to the decline in reindeer herding. Although their frequency and scale have increased, the interaction of fires with pastures and forest is complicated. For example, fires may play an important role in forest regeneration as they provide important minerals and free soils from leaf litter and ground vegetation cover, which under some conditions inhibit forest growth. Such interactions should be included in ecosystem management schemes. A decline in reindeer herding could also have a negative impact on reforestation as reindeer promote the removal of leaf litter and thereby the ability of new trees to become established.

It is within this extremely complex socio-economic and changing ecological situation that indigenous peoples in the Russian North must deal with climate change issues. RAIPON's initial work on climate change impacts suggests an important way forward: indigenous observations of climate change must be examined together with greater emphasis given to the concerns of indigenous peoples in terms of environmental degradation and habitat loss due to other factors. A broader understanding of change and discussions on how to deal with this must be included in environmental impact assessments, in environmental policy, and in the elaboration of local programs for sustainable development.

12.3.5. Indigenous caribou systems of North America

Subsistence caribou hunting in North America is practiced by Dogrib, Koyukon, Gwich'in, Dene, Cree, Chipewyan, Innu, Naskapi, Yupiit, Iñupiat, Inuvialuit, Inuit, and other indigenous peoples from the Ungava Peninsula of Labrador (Canada) to the western Arctic of Alaska (Fig. 12.2). While the cultural role of caribou differs among these groups, caribou is arguably the most important terrestrial subsistence resource for indigenous hunters in arctic North America (Hudson et al., 1989; Klein, 1989; Kofinas et al., 2000). The total annual harvest by North American hunters is estimated to be more than 160 000 animals, with its replacement value as store-bought meat roughly equivalent to US$ 30 000 000.

While this monetary value illustrates the enormous contribution caribou make to the northern economy, it does not capture the social, psychological, and spiritual value of caribou to its users. For many indigenous culture groups who identify themselves as "caribou people", like the Gwich'in, Naskapi, and Nunamiut, caribou–human relations represent a bond that blurs the distinction between people, land, and resources, and links First Peoples of the North with their history. This intimate relationship between people and caribou suggests that negative impacts from climate change on caribou and caribou hunting would have significant implications for the well-being of many indigenous communities, their sense of security and tradition, and their ability to meet their basic nutritional needs.

The caribou production system of the Vuntut Gwitchin, people whose traditional territories and settlement are centered on Old Crow, Yukon, may be used to illustrate variables that must be considered in climate change assessments of northern caribou user communities.

12.3.5.1. The enduring relationship of people and caribou

Caribou have been of critical importance to northern peoples of North America for millennia (Burch, 1972; Lynch, 1997; Hall, 1989). Archaeological evidence suggests that during the Wisconsin Glaciation, the distribution of *Rangifer* extended across much of the western hemisphere (Banfield, 1961, 1974; Kelsall, 1968; Spiess, 1979), from as far south as New Jersey to New Mexico

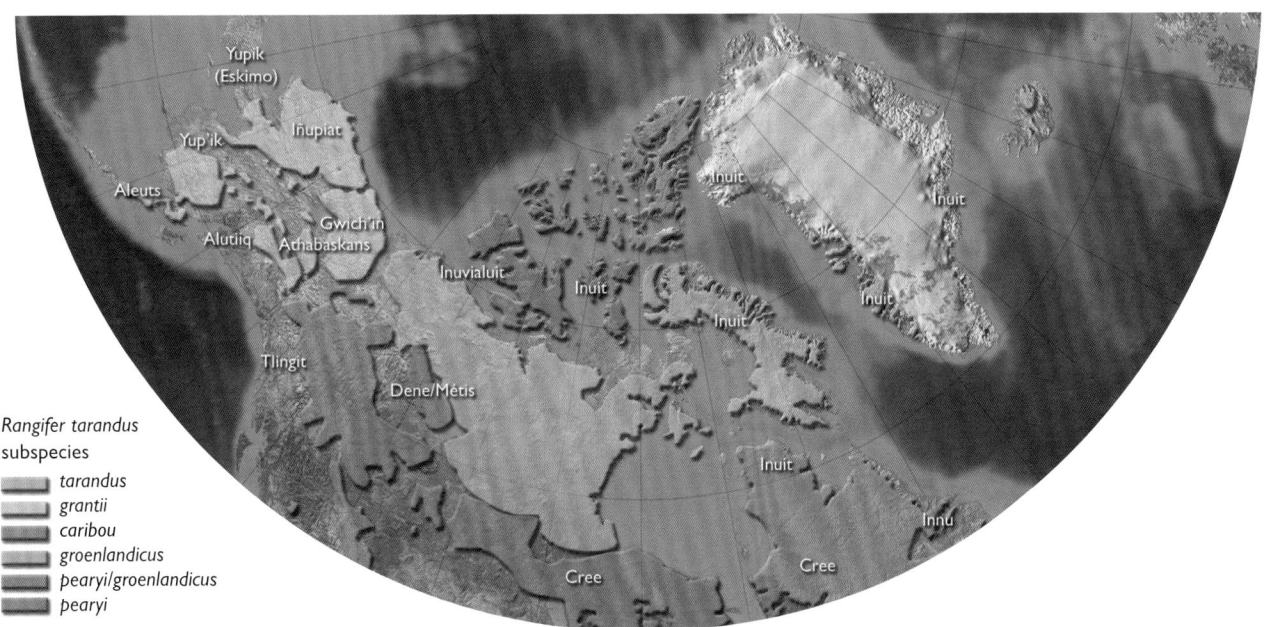

Fig. 12.2. North American distribution of *Rangifer* subspecies and selected indigenous peoples of North America (Kofinas and Russell, 2004).

and Nevada (Jackson and Thacker, 1997; Lynch, 1997). Caribou were available to paleo-indigenous hunters seasonally, with variation in availability related to a herd's ecological rhythms, human territoriality and mobility, and access to other living resources. Shifts in climate regimes that precipitated glacial epochs had dramatic consequences for caribou and the peoples that depended upon them. Recent evidence from the southern Yukon shows how shifts in climate have resulted in dramatic changes in the distribution of caribou, while remaining a part of the oral traditions and identity of indigenous peoples well after the disappearance of large herds (Farnell et al., 2004; Hare et al., in press).

The traditional caribou hunting grounds of the Vuntut Gwitchin are located within the caribou range of the Porcupine Caribou Herd, a region referred to as the Yukon–Alaska Refugium (see Chapter 11), and considered to have been unglaciated throughout the four glacial epochs. Paleontological evidence suggests that caribou have continually inhabited the Alaska–Yukon Refugium for over 400 000 years, through the Wisconsin Glaciation. Archaeological evidence of human habitation in this region is among the oldest excavated in North America. While questionable artifacts have been used to suggest the presence of humans in the area 25 000 to 29 000 years ago, confirmed findings at the Bluefish Caves, located on the Bluefish River southeast of Old Crow, Yukon have been dated 17 000 to 12 000 years old, including the bones of caribou.

Archaeological research linking proto Gwich'in to the present-day hunters identifies a complex of sites on the Porcupine and Crow Rivers, and indicates continual human inhabitation of the region and use of Porcupine Caribou for around 2000 years. Many of the sites are situated at present-day caribou river crossings, with material culture and subsistence patterns closely related to the caribou resource. Other noteworthy sites include

more than 40 caribou fences, strategically located across the southern range of the Porcupine Caribou Herd, and used by the Gwich'in families until the beginning of the 19th century (Greer and Le Blanc, 1992; McFee, no date given; Warbelow et al., 1975).

Social organization of caribou production traditionally reflects the seasonal cycle of caribou movements, overall changes in herd population, and access to other important subsistence resources. In winter, when caribou herds are mostly sedentary, traditional hunting involved small-group hunts and stalking; autumn migration brought large numbers of caribou and was undertaken by larger parties and family groups, intercepting caribou at traditional river crossings and/or directing movements of wild caribou into corrals. High demand for caribou meat in preparation for winter required large-scale harvests involving considerable effort by family groups. Limited summer hunts of young caribou provided lighter hides that were important for clothing. While traditional caribou hunting is often described as cooperative in behavior and egalitarian in social structure, recognition of exceptional hunting abilities was critical to survival. Caribou fences of the Vuntut Gwitchin are reported to have been "owned" by skilled hunting leaders, with a fence complex capable of harvesting as many as 150 animals in a single round-up, and managed by as many as 12 families. Cooperation among groups situated at different fences was necessary for managing the annual variability in migration patterns and uneven hunting success of family groups.

Ethnological studies of Porcupine Caribou users document the central role of caribou in community life (Balicki, 1963; Slobodin, 1962, 1981). Oral histories are replete with accounts of human migration, exceptional hardship, and starvation, due to the unavailability of caribou. While some have argued that over-hunting has been a key factor driving the declines observed in many northern wildlife populations (Martin, 1978), there is little evidence that over-hunting of caribou by indigenous peoples was the sole cause of population decline in large herds. Given the population estimates of indigenous hunters in the pre-contact period, it is more likely that changes in caribou populations of large herds and shifts in their distributions were driven primarily by climate (Peterson and Johnson, 1995), with hunting contributing to these changes at low population levels.

12.3.5.2. Modern-day subsistence systems

Caribou is still a vibrant component of many caribou user communities' dual cash–subsistence economies. For example, harvest data for the community of Old Crow (population ~275) show that the annual per capita caribou harvest has been as high as five animals (Kofinas, 1998). Modern-day harvesting in the community of Old Crow generally occurs during autumn, winter, and spring, with the autumn harvest the most important. In autumn, bull caribou are in prime condition (i.e., fat) and the cooler temperatures allow open-air production of drying meat, with use of boats to hunt at crossings. Winter harvesting does occur, but is generally limited because the herd's winter distribution is too far from the community to allow affordable access. The spring hunt by the Vuntut Gwitchin provides a supply of fresh meat after the long winter, but is limited by the warmer temperatures which constrain caribou production and storage of meat in caches. Governing the activities of harvesters is a strong local ethic against waste.

The location of modern-day settlements has consequences for the success of community caribou hunting. Communities, like Old Crow, sited at the center of the range of large migratory herds are able to intercept caribou during autumn and spring migrations, whereas communities sited on the margin of a herd's range may have access to animals only during winter or briefly during the summer calving and post-calving periods. History shows that the

range of a large herd can contract at low population levels and expand at high levels. The consequence for local communities situated away from the heart of a herd's range can be a decline in hunting success and in some cases abandonment of caribou hunting for several decades until the herd returns to a higher population level.

An important mechanism for adaptation and survival of traditional indigenous subsistence economies is a system of reciprocity through the sharing of harvested animals. Data from the Alaska Department of Fish and Game show the extent to which household sharing occurs in fifteen Western Arctic Herd user communities (Fig. 12.3). Networks of exchange are internal to communities and usually kinship-based. These networks also extend to residents of neighboring communities and regional centers (Magdanz et al., 2002). Central to this exchange process and in hunting success for many traditional hunters is the concept of luck. Like many hunting peoples, luck in hunting is regarded by many Vuntut Gwitchin not just as a matter of hit-or-miss probability, but also as the consequence of human deference and respect for animals, and generosity in sharing harvests with fellow community members (Feit, 1986).

Caribou subsistence hunting in indigenous communities of the north is today practiced as part of a dual cash–subsistence economy (Langdon, 1986). Cash inputs (e.g., jobs, transfer payments, investments by Native Corporations) supply essential resources for the acquisition of modern-day hunting tools. The transition to improved hunting technologies (e.g., bigger and faster snowmobiles and boats, outboard motors with greater horsepower, high-powered rifles, and access to caribou radio-collar distribution and movement data via the internet), allows greater access to caribou than in previous years, and a more consistent availability of fresh meat, which thus changes the level and type of uncertainty that has traditionally been associated with indigenous caribou hunting.

Government policies and agreements dictate if and how caribou harvesting enters into the realm of monetary exchange. For example, the State of Alaska and the U.S.-Canada International Agreement for the Conservation of the Porcupine Caribou Herd (1987) prohibits the commercial harvest and sale of caribou, whereas commercial tags for caribou of other herds are permitted for herds of the Northwest Territories, Nunavut, and Quebec, where several for-profit native and non-native corporations participate. Outfitter caribou hunting is also practiced as a component of local mixed economies in some regions. In some indigenous communities, there is a resistance to engage in guided caribou hunting, a policy that is defended as a need to retain traditional values and avoid commercialization of a sacred resource.

Many have speculated that engagement of subsistence hunters with the cash economy and the effects of modernization would ultimately lead to a decease in participation in the subsistence way of life (Murphy, 1986).

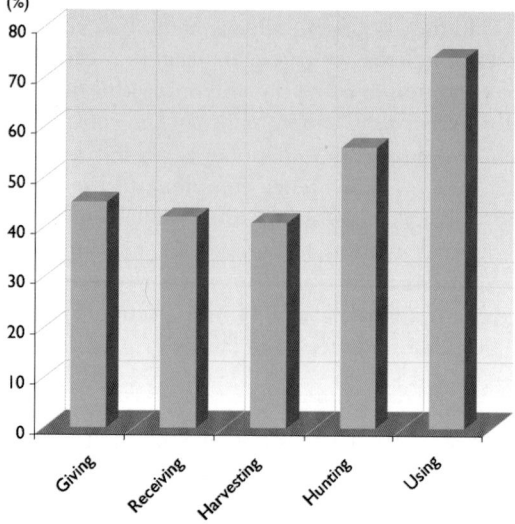

Fig. 12.3. Subsistence systems of reciprocity (data source: Alaska Department of Fish and Game/Division of Subsistence Community Profiles).

Yet, evidence demonstrates that under some conditions subsistence hunting can thrive in a modern context (Kruse, 1992; Langdon, 1986, 1991). The changes have, however, affected the allocation of time resources, including the time community members spend on the land pursuing a subsistence way of life (Kruse, 1991). Whereas before 1960 there was great flexibility in the allocation of time for subsistence harvesting and trapping, today's pursuit of employment and educational opportunities and its attendant shift to "clock time" is noted by many people at the local level as constraining opportunities for harvesting and affecting the transmission of cultural hunting traditions to younger generations. Relevant in the discussion about climate change and subsistence caribou hunting is the process by which financial resources compensate for the more constrained schedules of today's hunters, by improving technologies for time-efficient travel to hunting grounds (Berman and Kofinas, 2004). The shift to improved harvesting technologies also suggests that climate change impacts on community caribou hunting be considered within the context of a cultural system that is highly dynamic and with some (but not infinite) capacity for adaptation.

Critically relevant to a community's adaptive capacity is its collective knowledge of caribou and caribou hunting (Berkes, 1999), which includes an understanding of the distribution and movement of animals in response to different weather conditions (Kofinas et al., 2002). Community ecological knowledge of caribou is local in scale, and provides an important basis for hunters' decisions about the allocation of hunting resources (time, gas, and wear and tear on snow machines and boats) and the quantity of caribou to be harvested. This knowledge is sustained through the practice of caribou hunting traditions (i.e., time spent hunting and being on the land), and the transmission of knowledge and its cultural traditions to younger generations.

12.3.5.3. Conditions affecting caribou availability

Maintaining conditions for successful caribou hunting is not just a question of sustaining caribou herds at healthy population levels, but includes consideration of a complex and interacting set of social, cultural, political, and ecological factors. While environmental conditions (e.g., autumn storms, snow depth, rate of spring snow melt) may affect the Porcupine Caribou Herd's seasonal and annual distribution and movements (Eastland, 1991; Fancy et al., 1986; Russell et al., 1992, 1993), associated factors (e.g., timing of freeze-up and break-up, shallow snow cover, and the presence of "candle ice" on lakes) may affect hunters' access to hunting grounds. Individual and community economic conditions affecting hunters' access to equipment and free time for hunting are also key elements. Consequently, assessing caribou availability in conditions of climate change requires an approach that is more multifaceted than standard subsistence use documentation or "traditional ecological

Table 12.6. Key variables and their implications for assessments of climate change effects on caribou availability (based on Berman and Kofinas, 2004; Kruse et al., 2004).

Caribou population level	A decrease in total stock of animals has implications for the total range occupied by the herd; the likelihood hunters will see caribou while hunting, and management policy affecting the allocation and possible restrictions of harvests.
Distribution and movement of herd	Climate conditions are critical in the caribous' selection of autumn and spring migratory patterns and winter grounds. Calving locations affect community hunters' proximity to caribou.
Time for hunting; time on the land	Time for hunting emerges as an important variable as more community members engage in full-time participation in the wage economy. It is also important functionally in the maintenance and transmission of knowledge of caribou hunting.
Community demographics	Community demographics determine present and future demand for caribou. Out migration of people to distant cities may also affect the knowledge base if residence outside the community is for an extended period.
Household structure	Household structure affects resources (people and gear) that can be pooled for hunting. For example, households comprising adult bachelors often serve as important providers for households with non-hunters (e.g., elders, women, full-time working members who have limited time or no skill).
Cash income	Cash income provides for acquisition of gear needed in harvesting and compensation when time restrictions limit hunting opportunities. Where barter and trade allow for monetary exchange, it permits direct acquisition of meat.
Technology for harvesting caribou	Faster boats and snow machines, improved GPS systems, and lighter outdoor gear can bring hunting areas previously inaccessible due to high travel costs, within reach. Increased use of high technology gear can increase the demand for cash to support subsistence harvesting.
Cultural value	Cultural value affects interest in caribou hunting rates of consumption, and ethics of hunting practice.
Sharing	Inter- and intra-community sharing buffers against household caribou shortfall. Some indigenous belief that sharing also ensures future hunting success.
Social organization of the hunt	Hunting as individuals or in "community hunts" are both successful strategies.
Formal state institutions for management of caribou	Formally recognized institutions, such as a rural or native hunting priority, may prove critical when a herd is determined by management boards to be at low levels (see Chapter 11 for more details).

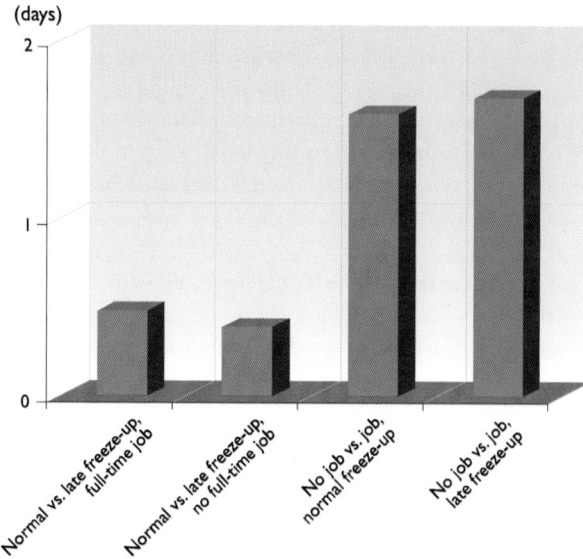

Fig. 12.4. Change in compensating variation from caribou hunting (Berman and Kofinas, 2004).

knowledge" documentation. Table 12.6 lists some of the key variables important in climate change assessments for caribou subsistence systems.

A simulation model based on Old Crow caribou hunting and the cash economy was constructed based on local knowledge and quantitative socio-economic data (Berman and Kofinas, 2004). The model drew on local knowledge and science-based research to assess the implications of a climate change scenario that assumed later break-up of ice on the Porcupine River, an important watercourse for intercepting caribou during the autumn migration. The model results revealed variation in compensatory levels for households with different types of employment. Figure 12.4 shows the estimated compensating variation for the possible changes in work and climate patterns based on 1993 hunting conditions for a household with three adults, two children, and no autumn harvest. The model suggests that late freeze-up costs the example household the equivalent of about half a day in lost leisure or family time. The loss for this sce-

nario is modest because caribou were present near the community during early winter in 1993, which meant that the access restriction still left hunters with some harvest opportunities even though these were not plentiful for the season for which data were available, and relatively few households would have hunted even under normal climate conditions. The loss in leisure or family time is slightly less if no-one in the household has full-time work. The compensating variation for having a job is negative, and about three times as large as the cost of the late freeze-up. This suggests that having a full-time job under these conditions reduces the household's welfare because it leaves insufficient time to hunt. If the data were for a year with more plentiful caribou, the cost would have been greater. The model does not include the increased risk of exposure for hunters attempting to intercept caribou during late freeze-up conditions, which typically include traveling up river by boat through moving ice (Berman and Kofinas, 2004).

The Sustainability of Arctic Communities Synthesis Model (Kruse et al., 2004), based on the integrated assessment of 22 scientists and four indigenous Porcupine Caribou user communities, projected the effects of a 40-year climate change scenario. Scenario assumptions included warmer and longer summers and greater variability in snow conditions, including deeper snow in winter, shallower snow in winter, and fewer "average" snow years. The results show that the combined effects of these conditions result in a significant decrease in the herd's population (see Chapter 11 for details of the impacts on caribou populations). The model assumes that no harvest restriction is implemented, and that intra-community sharing of caribou and community hunts are organized in years when most of the community households do not meet their target needs. In this climate change scenario, the model projects that within seven years of the final decade, less than half the households will meet half their caribou needs (Fig. 12.5).

12.3.5.4. Keeping climate assessment models in perspective

Community involvement in the Sustainability of Arctic Communities Synthesis Model project and documentation of local knowledge on climate change through the Arctic Borderlands Ecological Knowledge Cooperative (Kofinas, 2002) provide insights into the challenges associated with trying to assess the impacts of climate change on subsistence caribou hunting. Despite the effort of researchers to capture the key drivers and stochastic characteristics of the systems, a local review of the model pointed to problems because of the complexity of the system. For example, the Sustainability of Arctic Communities Synthesis Model assumes that warmer summer temperatures under climate change will result in an increase in insect harassment for caribou and an associated cost to the caribou energy budget. Caribou hunters of Aklavik, Northwest Territories, who have observed a recent increase in summer temperature, have also observed an increase in summer winds, and thus an

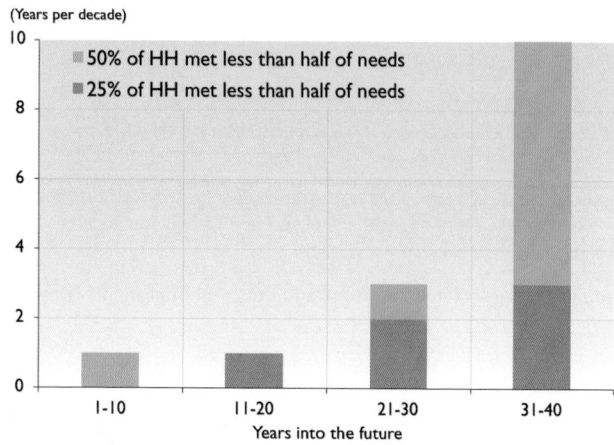

Fig. 12.5. Projected number of years in each decade that households (HH) have caribou harvest shortfalls (Kruse et al., 2004).

overall *decrease* in insect harassment to caribou. Community members from Old Crow questioned the use of a climate change scenario that assumes a regional increase in snow depth, since the model does not include the mosaic of landscape variation in their region. While these interactions illustrate some of the limitations of models in the assessment of climate change effects on local community use of subsistence resources, they do highlight that the involvement of communities as full partners in an integrated assessment can be of value in identifying data gaps, directing the work of future research, and portraying assessment results in ways that reflect the uncertainty and complexity of changes in the relationships between people and their environment.

12.3.5.5. Conclusions

Local community hunters of Old Crow and many others across North America have noted an overall increase in the variability of weather conditions (Jolly et al., 2002; Kofinas, 2002; Thorpe et al., 2002). These observations are coupled with an awareness of social and cultural changes in communities. While it will be difficult to make predictions about the trajectories of future climate conditions and their anticipated impacts on caribou and caribou subsistence systems, it is clear that the associated variability and overall uncertainty pose special problems for caribou people like the Vuntut Gwitchin of Old Crow. Indeed, local observations of later autumn freeze-up of rivers, drying of lakes and lowering of water levels in rivers, an increase in willow and birch in some areas, shifts in migrations and distribution patterns, and restrictions on access suggest that the problems of climate change are already being experienced. While the challenges of climate change and climate change assessments for local hunters and researchers are considerable, there are clear opportunities for collaboration between groups to ensure the sustainability of the subsistence way of life.

12.4. Summary and further research needs

This chapter has considered existing research on climate change impacts and has illustrated, through detailed case studies for several arctic settings, some of the most pressing issues currently faced by indigenous peoples. Although the material is diverse and extensive, some common themes emerge.

- Indigenous peoples around the Arctic maintain a strong and vibrant connection to the environment through hunting, herding, fishing, and gathering renewable resources.
- Hunting, herding, fishing, and gathering activities provide the primary means for obtaining and producing food in indigenous communities. These practices have endured over thousands of years, with cultural adaptations and the ability to utilize resources often associated with or affected by seasonal variation and changing ecological conditions.
- Hunting, herding, fishing, and gathering remain important for maintaining social relationships and cultural identity in indigenous societies. Hunting, herding, fishing, and gathering activities link people inextricably to their histories, their contemporary cultural settings, and provide a way forward for thinking about sustainable livelihoods in the future.
- As the climate changes, the indigenous peoples of the Arctic are facing special challenges and their abilities to harvest wildlife and food resources are already being tested. Although this chapter shows that climatic variability and weather events often greatly affect the abundance and availability of animals and thus the abilities and opportunities to harvest and process animals for food, the rate and extent of current and projected change give cause for alarm.
- Adaptation refers to the potential to react in a way that mitigates the impacts of negative change. Becoming resilient to climate change, and preparing to respond, cope with, adapt to, and negotiate climate change and its impacts, risks, and opportunities will require urgent and special attention.
- Climate change cannot be understood in isolation from other environmental changes, rapid social and cultural change, and globalization. Arctic communities have experienced, and are experiencing, stress from a number of different forces that threaten to restrict harvesting activities and sever these relationships.
- Rapid social and economic change, resource development, trade barriers, and animal-rights campaigns have all impacted upon hunting, herding, fishing, and gathering activities. Section 12.3.4 on the Russian North, for example, illustrates how poaching, oil development, and clear-cutting of forests undermine the subsistence base for indigenous peoples.
- Arctic peoples cannot adapt, relocate, or change resource use activities as easily as in the past, because most now live in permanent communities and must negotiate greatly circumscribed social and economic situations.
- Hunting, herding, and fishing activities are determined to a large extent by resource management regimes, land use and land ownership regulations, and local and global markets. The mobility and flexibility that indigenous peoples once possessed to move in response to shifts in the pattern and state of their resource base is no longer possible.
- Commercial, political, economic, legal, and conservation interests have reduced the abilities of indigenous peoples to adapt and be flexible in coping with climatic variability. However, for some peoples of the Arctic, the political and management systems are already in place that could assess the impacts of climate change, allow local and regional governments to act on policy recommendations to deal with the consequences, and improve the chances for indigenous peoples to deal successfully with climate change. Although complex, solutions to environmental problems are potentially realistic.

This chapter demonstrates an urgent need for a greater understanding of the scope of these environmental, social, political, and economic issues and challenges in a rapidly changing milieu. The chapter is intended as a scoping exercise as much as it is an assessment of current knowledge. The case studies are based on extensive work in partnership with indigenous communities, and the chapter as a whole has developed with significant advice, guidance, and input from the Permanent Participants to the Arctic Council.

Communities across the Arctic are culturally and economically diverse and are affected by environmental change in different ways. Such diversity also means that local experiences of climate impacts and responses to climate variability and change may not be universal. How do communities, therefore, differ in how they utilize strategies for mitigating negative change and in the effectiveness of their adaptive capacity? Given this question, the case studies illustrate the importance of research on the localized, regional, and circumpolar studies of socio-economic impacts of recent climate change.

The emphasis of scientific research on climate change is to assess the impacts on the environment, ecosystem processes, and wildlife. One gap in knowledge is how climate change affects social relations in indigenous communities. This chapter highlights this as a critical aspect of climate change research, arguing that a change in the ability of indigenous peoples to access traditional/country food resources can have a corresponding impact on the social fabric of their communities. In a very real sense, therefore, the discussion of climate impacts on hunting, herding, fishing, and gathering by indigenous peoples is about sustaining human/food resource relationships and activities in indigenous societies, as well as being aware that climate change impacts pose a threat of severe and irreversible social changes. The case studies illustrate the complexity of problems faced by indigenous peoples today and underscore the reality that climate change is but one of several problems affecting their livelihoods. Clearly, research should place emphasis on understanding climate change impacts within the broader context of rapid social and economic change and seek to determine ways of distinguishing between changes that occur as a result of societal, cultural, and economic events, and changes that result from physical processes.

Future research on climate change should result in a deeper understanding of what exactly forms the basis for the social, cultural, political, and economic viability of arctic communities, and attempt to explore the research priorities highlighted by communities themselves. Significant and promising new research initiatives are currently underway that promise to break new ground in contributing the knowledge needed to formulate climate change impact assessments, national policies, and adaptation strategies, including the major U.S. interagency-led Study of Environmental Arctic Change (SEARCH, 2001) and the Canadian ArcticNet (http://www.arcticnet.ulaval.ca/index_en.asp) programs.

The case studies in section 12.3 were selected to provide a sense of climate change impacts on indigenous communities and their livelihoods. It was not possible to provide circumpolar coverage of the situation for all indigenous peoples. Nor was it possible to include detailed information on plant resources and freshwater fish. Indeed, there is a lack of good material on this, especially for the Russian Arctic, where industrialization and nonrenewable resource development is happening much faster, and with a greater discernible impact than climate change, even though this is a region of recent and substantial warming. There are many kinds of side effects from industrialization in the Russian North, especially from accelerated road and railway building, such as the destruction of huge areas of pasture for gravel and sand mining, and the rapid increase in poaching, which give cause for immediate concern. The situation in the Russian North clearly demonstrates the importance of assessing and understanding the impacts of climate change within the broader context of other rapid changes.

One of the aims of this chapter was to assess the adaptations that have enabled communities to succeed in the past and to establish the extent to which these options remain open to them. There are few published data on this topic, but based on what is available this chapter shows that while indigenous peoples have often adapted well to past climate change, the scale and nature of current and projected climate change brings a very different sense of uncertainty and presents different kinds of risks and threats to the livelihoods of indigenous peoples.

Is an ability to respond and cope with climate change, mainly by adjusting subsistence activities, a reliable indication of an ability to adapt in the future? Research is needed on understanding how much change can be accommodated by the existing ways of life of indigenous peoples. Case studies in this chapter have pointed to the *resilience*, or the amount of perturbation that the resource use systems of indigenous peoples can absorb, and how they can adapt by learning and self-organization. How are indigenous communities adapting to and coping with change? How will reindeer, marine mammals, and fish themselves adapt to the changes in their habitats? Can the return of subsidies for meat distribution and marketing help in making coping and adaptation responses possible? These are questions of pressing concern for hunters, fishers, and herders active across the entire Arctic. The question of resilience and limits to adaptability is important and further research is needed since little is known about building adaptive capacity in the face of climate change. Above all, there is need to match or coordinate physical climate data on extreme weather events and the impact of these events on actual hunting, fishing, and herding practices.

Further research is also needed on co-management and governance institutions and whether they can create additional opportunities to increase resilience, flexibility, and the ability to deal with change. How can, for example, new governance mechanisms help indigenous peoples to negotiate and manage the impacts of climate change? With a capacity-building strategy now being a key objective for the Arctic Council, tremendous opportunities exist for cooperation and constructive dialogue on dealing with climate change between communities, organizations, institutions, and governments at circumpolar and wider international levels.

Acknowledgements

Mark Nuttall wishes to thank the Polar Regions Section of the UK Foreign and Commonwealth Office for providing generous support to cover his participation in this assessment. The case studies are based on extensive work in partnership with indigenous communities, and the contributing authors express their thanks to all those involved. The chapter as a whole has developed with significant advice, guidance, and input from the Permanent Participants to the Arctic Council, and special thanks also goes to the large number of international reviewers who provided detailed, thoughtful and constructive comments and criticism.

References

Abdulatipov, R., 1999. The Aboriginal Face of Russia. The Independent Newspaper – Scenario N5, p. 7.

Afanasieva, N., 2002. The socio-economic and legal situation of the Saami in the Murmansk oblast – an example to take urgent measures. In: T. Kohler and K. Wessendorf (eds.). Towards a New Millennium. Ten Years of the Indigenous Movement in Russia, pp. 139–144. International Working Group on Indigenous Affairs, Copenhagen.

AMAP, 1998. The AMAP Assessment Report: Arctic Pollution Issues. Arctic Monitoring and Assessment Programme, Oslo, xii + 859pp.

Anderson, D.G. and M. Nuttall (eds.), 2004. Cultivating Arctic Landscapes: Knowing and Managing Animals in the Circumpolar North. Oxford Berghahn Books Ltd.

Andrews, M. and J.T. Andrews, 1979. Bibliography of Baffin Island environments over the last 1000 years. In: A. McCartney (ed.). Thule Eskimo Culture: An Anthropological Retrospective. ASC Mercury Paper No. 88. National Museum of Man, Ottawa.

Anisimov, O.A. and F.E. Nelson, 1997. Permafrost zonation and climate change in the Northern Hemisphere: results from transient general circulation models. Climatic Change, 35:241–258.

Balikci, A., 1963. Family organization of the Vunta Kutchin. Arctic Anthropology, 1(2):62–69.

Banfield, A.W.F., 1961. Revision of the Reindeer and Caribou, genus *Rangifer*. National Museum of Canada Bulletin, 177, vi+137pp.

Banfield, A.W.F., 1974. The Mammals of Canada. University of Toronto Press. xxv + 438pp.

Barry, R., W. Arundale, J.T. Andrews, R. Bradley and H. Nichols, 1977. Environmental and cultural change in the eastern Arctic during the last five thousand years. Arctic and Alpine Research, 9:193–210.

Berkes, F., 1999. Sacred Ecology: Traditional Ecological Knowledge and Resource Management. Taylor and Francis, 209pp.

Berkes, F. and D. Jolly, 2001. Adapting to climate change: social-ecological resilience in a Canadian western Arctic community. Conservation Ecology, 5(2):18 [online] www.consecol.org/vol5/iss2/art18

Berman, M. and G. Kofinas, 2004. Hunting for models: rational choice and grounded approaches to analyzing climate effects on subsistence hunting in an Arctic community. Ecological Economics, 49(1):31–46.

Bernes, C., 1996. The Nordic Arctic Environment: unspoilt, exploited, polluted? Nordic Council of Ministers, Copenhagen, 240pp.

Bjørklund, I., 2004. Saami pastoral society in northern Norway: the national integration of an indigenous management system. In: D.G. Anderson and M. Nuttall (eds.). Cultivating Arctic Landscapes: Knowing and Managing Animals in the Circumpolar North. Oxford Berghahn Books.

Boas, F., 1888. The Central Eskimo. Sixth Annual Report of the Bureau of American Ethnology for the Years 1884–1885, pp. 399–699. Smithsonian Institution, Washington, D.C.

Bodenhorn, B., 2000. It's good to know who your relatives are but we are taught to share with everybody: shares and sharing among Inupiaq households. In: G.W. Wenzel, G. Hovelsrud-Broda and N. Kishigami (eds.). The Social Economy of Sharing: Resource Allocation and Modern Hunter-Gatherers, pp. 27–60. Senri Ethnological Series. National Museum of Ethnology, Osaka.

Briffa, K., T. Bartholin, D. Eckstein, P. Jones, W. Karlen, F. Schweingruber and P. Zetterberg, 1990. A 1400-year tree-ring record of summer temperatures in Fennoscandia. Nature, 346:434–439.

Brody, H., 1983. Maps and Dreams: Indians and the British Columbia Frontier. Penguin, Harmondsworth.

Burch, E.S. Jr., 1972. The caribou/wild reindeer as a human resource. American Antiquity, 37(3):337–368.

Burch, E.S. Jr., 1985. Subsistence Production in Kivalina, Alaska: A Twenty-Year Perspective. Technical Paper No. 128. Division of Subsistence, Alaska Department of Fish and Game.

CAFF, 2001 Arctic Flora and Fauna. Status and Conservation. Conservation of Arctic Flora and Fauna, Helsinki, 272pp.

Callaghan, T.V., R.M.M. Crawford, M. Eronen, A. Hofgaard, S. Payette, W.G. Rees, O. Skre, B. Sveinbjörnsson, T.K. Vlassova and B.R. Werkman, 2002. The dynamics of the tundra-taiga boundary: an overview and suggested coordinated and integrated approach to research. Ambio Special Report, 12:3–5.

Callaway, D., 1995. Resource use in rural Alaskan communities. In: D.L. Peterson and D.R. Johnson (eds.). Human Ecology and Climate Change: People and Resources in the Far North. Taylor and Francis.

Callaway, D., J. Eamer, E. Edwardsen, C. Jack, S. Marcy, A. Olrun, M. Patkotak, D. Rexford and A. Whiting, 1999. Effects of climate change on subsistence communities in Alaska. In: G. Weller, P. Anderson (eds.). Assessing the Consequences of Climate Change for Alaska and the Bering Sea Region, pp. 59–73. Center for Global Change and Arctic System Research, University of Alaska Fairbanks.

Caulfield, R.A., 1997. Greenlanders, Whales and Whaling: Sustainability and Self-determination in the Arctic Hanover. University Press of New England, 219pp.

Caulfield, R., 2000. The political economy of renewable resource harvesting in the Arctic. In: M. Nuttall and T.V. Callaghan (eds.). The Arctic: Environment, People, Policy. Harwood Academic Press.

Chapman, W.L. and J.E. Walsh, 1993. Recent variations of sea ice and air temperatures in high latitudes. Bulletin of the Meteorological Society of America, 73:34–47.

Collings, P., G.W. Wenzel and R. Condon, 1998. Modern food sharing networks and community integration in the central Canadian Arctic. Arctic, 51(4):301–314.

Cruikshank, J., 2001. Glaciers and climate change: perspectives from oral tradition. Arctic, 54(4):377–393.

Dahl, J., 2000. Saqqaq: an Inuit hunting community in the modern world. University of Toronto Press, 336pp.

Damas, D., 1963. Igluligmiut Kinship and Local Groupings: A Structural Approach. National Museum of Canada.

Damas, D., 1972. Central Eskimo systems of food sharing. Ethnology, 11(3):220–240.

De Fabo, E. and L.O. Bjorn, 2000. Ozone depletion and UV-B radiation. In: M. Nuttall and T.V. Callaghan (eds.). The Arctic: Environment, People, Policy, pp. 555–574. Harwood Academic Press.

Dekin, A., 1969. Climate change and cultural change: a correlative study from Eastern Arctic prehistory. Polar Notes, 12:11–31.

Eastland, W.G., 1991. Influence of Weather on Movements and Migration of Caribou. M.S. Thesis. University of Alaska Fairbanks, 111pp.

Fancy, S.G., L.F. Pank, K.R. Whitten and L.W. Regelin, 1986. Seasonal movement of caribou in Arctic Alaska as determined by satellite. Canadian Journal of Zoology, 67:644–650.

Farnell, R., P.G. Hare, E. Blake, V. Bowyer, C. Schweger, S. Greer and R. Gotthardt, 2004. Multidisciplinary investigations of alpine ice patches in southwest Yukon, Canada: paleoenvironmental and paleobiological investigations. Arctic, 57(3):247–259.

Fast, H. and J. Mathias, 2000. Directions towards marine conservation in Canada's western Arctic. Ocean and Coastal Management, 34:183–205.

Fedorova, N. (ed.), 1998. Ushedshie v kholmy (Gone to the hills: culture of the coastal residents of the Yamal Peninsula during the Iron Age). History and Archaeology Institute, Ekaterinburg.

Feit, H., 1986. Hunting and the quest for power: The James Bay Cree and white men in the twentieth century. In: R.B. Morrison and C.R. Wilson (eds.). Native Peoples: The Canadian Experience, pp. 171–207. McClelland and Stewart.

Fienup-Riordan, A., 1986. When Our Bad Season Comes. Alaska Anthropological Association Monograph No. 1. Alaska Anthropological Association, Anchorage.

Finney, B.P., I. Gregory-Eaves, M.S.V. Douglas and J.P. Smol, 2002. Fisheries productivity in the northeastern Pacific Ocean over the past 2,200 years. Nature, 416(6882):729–733.

Fitzharris, B.B., I. Allison, R.J. Braithwaite, J. Brown, P.M.B. Foehn, W. Haeberli, K. Higuchi, V.M. Kotlyakov, T.D. Prowse, C.A. Rinaldi, P. Wadhams, M-K Woo, Y. Xie, O. Anisimov, A. Aristarain, R.A. Assel, R.G. Barry, R.D. Brown, F. Dramis, S. Hastenrath, A.G. Lewkovicz, E.C. Malagnino, S. Neale, F.E. Nelson, D.A. Robinson, P. Skvarca, A.E. Taylor and A. Weidick, 1996. The cryosphere: changes and their impacts. In: R.T. Watson, M.C. Zinyovera, R.H. Moss and D.J. Dokken (eds.). Climate Change 1995: The Science of Climate Change, pp. 241–265. Contribution of Working Group II to the Second Assessment Report of the Intergovernmental Panel on Climate Change. Cambridge University Press.

FJMC, 2000. Beaufort Sea 2000. Renewable Resources for Our Children. Conference Summary Report. Fisheries Joint Management Committee, Inuvik Northwest Territories. Online at www.fjmc.ca

Forbes, B.C., 1995. Tundra disturbance studies. III. Short-term effects of aeolian sand and dust, Yamal Region, northwest Siberia, Russia. Environmental Conservation, 22:335–344.

Forbes, B.C., 1999. Land use and climate change in the Yamal-Nenets region of northwest Siberia: some ecological and socio-economic implications. Polar Research, 18:1–7.

Forbes, B.C. and G. Kofinas (eds.), 2000. The human role in reindeer/caribou grazing systems. Polar Research, 19:1–142.

Ford, N., 1999. Communicating climate change from the perspective of local people: A case study from Arctic Canada. Journal of Development Communication, 1(11):93–108.

Freeman, M.M.R. (ed.), 2000. Endangered Peoples of the Arctic. Greenwood Press, Westport, 304pp.

Freeman, M.M.R., L. Gogolovskaya, R.A. Caulfield, I. Egede, I. Krupnik and M.G. Stevenson, 1998. Inuit, Whaling and Sustainability. Altamira Press.

Furgal, C., D. Maritn and P. Gosselin, 2002. Climate change and health in Nunavik and Labrador: lessons from Inuit knowledge. In: I. Krupnik and D. Jolly (eds.). The Earth is Faster Now: Indigenous Observations of Arctic Environmental Change, pp 266–300. ARCUS.

Golovnev, A.V. and G. Osherenko, 1999. Siberian Survival: the Nenets and their Story. Cornell University Press.

Greer, S.C. and R.J. Le Blanc, 1992. Background Heritage Studies – Proposed Vuntut National Park. Northern Parks Establishment Office, Canadian Parks Service, 116pp.

Griffith, B., D.C. Douglas, N.E. Walsh, D.D. Young, T.R. McCabe, D.E. Russell, R.G. White, R.D. Cameron and K.R. Whitten, 2002. The Porcupine caribou herd. In: D.C. Douglas, P.E. Reynolds and E.B. Rhode (eds.). Arctic Refuge Coastal Plain Terrestrial Wildlife Research Summaries, 8–37: U.S. Geological Survey, Biological Resources Division.

Grove, J., 1988. The Little Ice Age. Methuen.

Gunn, A., 1995. Responses of Arctic ungulates to climate change. In: D.L. Peterson and D.R. Johnson (eds.). Human Ecology and Climate Change: People and Resources in the Far North, pp. 89–104. Taylor and Francis.

Gunn, A. and T. Skogland, 1997. Responses of caribou and reindeer to global warming. In: W.C. Oechel et al. (eds.). Global Change and Arctic Terrestrial Ecosystems, pp. 189–200. Springer-Verlag.

Gunn, A., C. Shank and B. McLean, 1991. The history, status and management of muskoxen on Banks Island. Arctic, 44(3):188–195.

Hall, E., (ed.), 1989. People and Caribou in the Northwestern Territories. Department of Renewable Resources, Government of Northwest Territories, Yellowknife.

Hamilton, L.C., P. Lyster and O. Otterstad, 2000. Social change, ecology and climate in 20th century Greenland. Climatic Change, 47(1/2):193–211.

Hare et al., in press. Multidisciplinary investigations of Alpine ice patches in southwest Yukon, Canada: ethnographic and archaeological Investigations. Arctic Ms#03–151.

Helgason, A. and G. Palsson, 1997. Contested commodities: the moral landscape of modernist regimes. Journal of the Royal Anthropological Institute, 3:451–471.

Helle, T. and M. Timonen, 2001. The North Atlantic Oscillation and reindeer husbandry. In: P. Kankaanpää et al. (eds.). Arctic Flora and Fauna: Status and Conservation. Helsinki, p. 25

Hovelsrud-Broda, G., 2000. Sharing, transfers transactions and the concept of generalized reciprocity. In: G.W. Wenzel, G. Hovelsrud-Broda, N. Kishigami (eds.). The Social Economy of Sharing: Resource Allocation and Modern Hunter-Gatherers, pp. 61–85. Senri Ethnological Series. National Museum of Ethnology, Osaka.

Hudson, R.J., K.R. Drew and L.M. Baskin (eds.), 1989. Subsistence Hunting. Wildlife Production Systems: Economic Utilisation of Wild Ungulates. Cambridge University Press.

Huntington, H.P., 1992. Wildlife Management and Subsistence Hunting in Alaska. University of Washington Press.

IISD, 2000. Sila Alangotok. Inuit Observations of Climate Change. International Institute for Sustainable Development, Winnipeg. Video available online: www.iisd.org/casl/projects/inuitobs.htm

IPCC, 2001. Climate Change 2001: The Scientific Basis. Contribution of Working Group I to the Third Assessment Report of the Intergovernmental Panel on Climate Change. J.T. Houghton, Y. Ding, D.J. Griggs, M. Noguer, P.J. van der Linden, X. Dai, K. Maskell and C.A. Johnson (eds.). Cambridge University Press, 881pp.

Jackson, L.J. and P.T. Thacker (eds.), 1997. Caribou and Reindeer Hunters of the Northern Hemisphere. Ashgate Publishing, 272pp.

Jernsletten, J.-L. and K. Klokov, 2002. Sustainable Reindeer Husbandry. Arctic Council/Centre for Saami Studies, Tromsø.

Johannessen, O.M., E.V. Shalina and M.V. Miles, 1999. Satellite evidence for an arctic sea ice cover in transformation. Science, 286:1937–1939.

Jolly, D., F. Berkes, J. Castleden, T. Nichols and the Community of Sachs Harbour, 2002. We can't predict the weather like we used to. Inuvialuit observations of climate change, Sachs Harbour, western Canadian Arctic. In: I. Krupnik and D. Jolly (eds.). The Earth Is Faster Now. Arctic Research Consortium of the United States, Fairbanks, Alaska.

Kelsall, J.P., 1968. The Migratory Barren-Ground Caribou of Canada. Department of Indian Affairs and Northern Development. Canadian Wildlife Service, Ottawa.

Kemp, W., G.W. Wenzel, E. Val and N. Jensen, 1978. A Socioeconomic Baseline Study of Resolute Bay and Kuvinaluk. Polargas Project, Toronto, 354pp.

Khorolya, D., 2002. Reindeer husbandry in Russia. In: S. Kankaanpää et al. (eds.). The Second World Reindeer Herders' Congress. Anár 2001, pp. 40–42. Arctic Centre Reports 36, University of Lapland.

Klein, D., 1989. Subsistence hunting. In: R.J. Hudson, K.R. Drew and L.M. Baskin (eds.). Wildlife Production Systems: Economic Utilisation of Wild Ungulates, pp. 96–111. Cambridge University Press.

Klein, D.R., 1991. Caribou in the changing North. Applied Animal Behaviour Science, 29:279–291.

Klein, D.R., 2000. Arctic grazing systems and industrial development: can we minimize conflicts? Polar Research, 19:91–98.

Klokov, K., 2000. Nenets reindeer herders on the lower Yenisei River: traditional economy under current conditions and responses to economic change. Polar Research, 19:39–47.

Kofinas, G.P., 1998. The Cost of Power Sharing: Community Involvement in Canadian Porcupine Caribou Co-management. Ph.D. Thesis. University of British Columbia, 471pp.

Kofinas, G., 2002. Community contributions to ecological monitoring: knowledge co-production in the US-Canada Arctic borderlands. In: I. Krupnik, D. Jolly (eds.). The Earth is Faster Now: Indigenous Observations of Arctic Environmental Change, pp. 54–91. Arctic Research Consortium of the United States, Fairbanks, Alaska.

Kofinas, G. and D. Russell, 2004. North America. In: family-based Reindeer Herding and Hunting Economies and the Status and Management of Wild Reindeer/Caribou Populations. Centre for Saami Studies, University of Tromsø.

Kofinas, G., G. Osherenko, D. Klein and B. Forbes, 2000. Research planning in the face of change: the human role in reindeer/caribou systems. Polar Research, 19(1):3–22.

Kofinas, G., Aklavik, Arctic Village, Old Crow and Fort McPherson, 2002. Community contributions to ecological monitoring: knowledge co-production in the U.S.-Canada Arctic borderlands. In: I. Krupnik and D. Jolly (eds.). The Earth is Faster Now: Indigenous Observations of Arctic Environmental Change, pp. 54–91. Arctic Research Consortium of the United States, Fairbanks, Alaska.

Kozhevnikov, Yu.P., 2000. Is the Arctic getting warmer or cooler? In: B.S. Ebbinge, Yu.I. Mazourov and P.S. Tomkovich (eds.). Heritage of the Russian Arctic: Research, Conservation and International Cooperation, pp. 145–157. Ecopros Publishers, Moscow.

Krupnik, I., 1993. Arctic Adaptations: Native Whalers and Reindeer Herders of Northern Eurasia. University Press of New England.

Krupnik, I., 2000. Reindeer pastoralism in modern Siberia: research and survival in the time of crash. Polar Research, 19:49–56.

Krupnik, I. and D. Jolly (eds.), 2002. The Earth is Faster Now: Indigenous Observations of Arctic Environmental Change. Arctic Research Consortium of the United States, Fairbanks, Alaska.

Kruse, J., 1991. Alaska Inupiat subsistence and wage employment patterns: understanding individual choice. Human Organization, 50(4):317–326.

Kruse, J.A., 1992. Alaska North Slope Inupiat Eskimo and resource development: why the apparent success? Prepared for presentation to the American Association for the Advancement of Science, Annual Meeting, January 1992, Chicago.

Kruse, J.A., R.G. White, H.E. Epstein, B. Archie, M.D. Berman, S.R. Braund, F.S. Chapin III, J. Charlie Sr., C.J. Daniel, J. Eamer, N. Flanders, B. Griffith, S. Haley, L. Huskey, B. Joseph, D.R. Klein, G.P. Kofinas, S.M. Martin, S.M. Murphy, W. Nebesky, C. Nicolson, K. Peter, D.E. Russell, J. Tetlichi, A. Tussing, M.D. Walker and O.R. Young, 2004. Sustainability of Arctic communities: an interdisciplinary collaboration of researchers and local knowledge holders. Ecosystems, 7:1–14.

Lamb, H., 1982. Climate, History, and the Modern World. Methuen.

Langdon, S.J., 1986. Contradictions in Alaskan Native economy and society. In: S. Langdon (ed.). Contemporary Alaskan Native Economics, pp. 29–46. University Press of America.

Langdon, S.J., 1991. The integration of cash and subsistence in Southwest Alaskan Yup'ik Eskimo communities. In: T. Matsuyama and N. Peterson (eds.). Cash, Commoditisation and Changing Foragers, pp. 269–291. Senri Publication No. 30. Osaka, Japan, National Museum of Ethnology.

Lange, M., 2000. Integrated global change impact assessments. In: M. Nuttall and T.V. Callaghan (eds.). The Arctic: Environment, People, Policy. Harwood Academic Press.

Larter, N.C. and J.A. Nagy, 2000. Calf production and overwinter survival estimates for Peary caribou, *Rangifer tarandus pearyi*, on Banks Island, Northwest Territories. The Canadian Field-Naturalist, 114:661–670.

Larter, N.C. and J.A. Nagy 2001. Variation between snow conditions at Peary caribou and muskox feeding sites and elsewhere in foraging habitats on Banks Island in the Canadian High Arctic. Arctic, Antarctic and Alpine Research, 33(2):123–130.

Lee, S.E., M.C. Press, J.A. Lee, T. Ingold and T. Kurttila, 2000. Regional effects of climate change on reindeer: a case study of the Muotkatunturi region in Finnish Lapland. Polar Research, 19(1):99–105.

Leibman, M.O. and I.P. Egorov, 1996. Climatic and environmental controls of cryogenic landslides, Yamal, Russia. In: K. Senneset (ed.). Landslides, pp. 1941–1946. A.A. Balkema.

Lent, P.C. 1999. Muskoxen and their Hunters: A History. Animal Natural History Series, vol. 5. University of Oklahoma Press.

Lonner, T., 1980. Subsistence as an Economic System in Alaska: Theoretical and Policy Implications. Technical Paper No. 67. Division of Subsistence, Alaska Department of Fish and Game.

Lynch, T.F., 1997. Introduction. In: L.J. Jackson and P.T. Thacker (eds.). Caribou and Reindeer Hunters of the Northern Hemisphere. Ashgate Publishing.

Magdanz, J.S., C.J. Utermohle and R.J. Wolfe, 2002. The production and distribution of wild food in Wales and Deering, Alaska. Division of Subsistence, Alaska Department of Fish and Game. Technical Paper 259.

Martin, C., 1978. Keepers of the Game. University of California Press.

Maslanik, J.A., M.C. Serreze and T. Agnew, 1999. On the record reduction in 1998 western Arctic sea ice cover. Geophysical Research Letters, 26(13):1905–1908.

Maxwell, M., 1985. Eastern Arctic Prehistory. Academic Press.

McDonald, M., L. Arragutainaq and Z. Novalinga, 1997. Voices from the Bay. Ottawa Canadian Arctic Resources Committee and Municipality of Sanikiluaq.

McFee, R.D. (no date). Caribou fence facilities of the historic Yukon Kutchin. In: Megaliths to Medicine Wheels: Boulder Structures in Archaeology, pp. 159–170. Proceedings of the Eleventh Annual Chacmool Conference, University of Toronto.

McGhee, R., 1972. Climatic change and the development of Canadian Arctic cultural traditions. In: Y. Vasari, H. Hyvarinen and S. Hicks (eds.). Climatic Change in Arctic Areas during the Last Ten Thousand Years, pp. 39–57. Acta Universitatis Ouluensis Series A: Scientiae Rerum Naturalium 3, Geologica 1. University of Oulu, Finland.

Murphy, S.C., 1986. Valuing Traditional Activities in the Northern Native Economy: the Case of Old Crow, Yukon Territory. M.A. Thesis, University of British Columbia, 189pp.

Mysterud, A., N.C. Stenseth, N.G. Yoccoz, R. Langvatn and G. Steinhelm, 2001. Nonlinear effects of large-scale climatic variability on wild and domestic herbivores. Nature, 410:1096–1099.

Nadasdy, P. 2003. Reevaluating the co-management success story. Arctic, 56(4):367–380.

Nagy, M., 2004. We did not want the muskox to increase. Inuvialuit knowledge about muskox and caribou populations on Banks Island, Canada. In: D.G. Anderson and M. Nuttall (eds.). Cultivating Arctic Landscapes: Knowing and Managing Animals in the Circumpolar North. Oxford Berghahn Books.

Nagy, J.A., N.C. Larter and V.P. Fraser, 1996. Population demography of Peary caribou and muskox on Banks Island, N.W.T., 1982–1992. Rangifer Special Issue, 9:213–222.

Nelson, F.E. and O.A. Anisimov, 1993. Permafrost zonation in Russia under anthropogenic climatic change. Permafrost and Periglacial Processes, 4:137–148.

Nuttall, M., 1992. Arctic Homeland: Kinship, Community and Development in Northwest Greenland. University of Toronto Press.

Nuttall, M., 1998. Protecting the Arctic: Indigenous Peoples and Cultural Survival. Routledge Harwood.

Nuttall, M. 2000. Barriers to the sustainable uses of living marine resources in Vestnorden. In: J. Allansson and I. Edvardsson (eds.). Community Viability, Rapid Change and Socio-Ecological Futures, pp. 22–38. University of Akureyri and Stefansson Arctic Institute, Akureyri.

Nuttall, M., 2001. Locality, identity and memory in South Greenland. Etudes/Inuit/Studies, 25:53–72.

Nuttall, M. in press. Inuit, climate and marine resources: risk and resilience in a changing Arctic. In: N. Kishigami, J. Savelle (eds.). Marine Resources Conservation. National Museum of Ethnology, Osaka.

Odner, K., 1992. The Varanger Saami: habitation and economy AD1200–1900. Scandinavian University Press.

Okotetto, M.N. and B.C. Forbes, 1999. Conflicts between Yamal-Nenets reindeer husbandry and petroleum development in the forest tundra and tundra region of northwest Siberia. In: S. Kankaanpää, T. Tasanen and M.-L. Sutinen (eds.). Sustainable Development in Timberline Forests, pp. 95–99. Finnish Forest Research Administration, Helsinki.

Osherenko, G., 1988. Wildlife Management in the North American Arctic: The Case for Co-Management. In: M.M.R. Freeman and L.N. Carbyn (eds.). Traditional Knowledge and Renewable Resource Management, pp. 92–104. Boreal Institute for Northern Studies, University of Alberta.

Osherenko, G., 2001. Indigenous rights in Russia: is title to land essential for cultural survival? Georgetown International Environmental Law Review, 13:695–734.

Osherenko, G. and O R. Young, 1989. The Age of the Arctic: Hot Conflicts and Cold Realities. Cambridge University Press.

Pars, T., M. Osler and P. Bjerregaard, 2001. Contemporary use of traditional and imported food among Greenlandic Inuit. Arctic, 54:22–31.

Pavlov, A.V., 1994. Current changes of climate and permafrost in the Arctic and Sub-Arctic of Russia. Permafrost and Periglacial Processes, 5:101–110.

Pavlov, A.V., 1998. Active layer monitoring in northern West Siberia. Nordicana, 57:875–881.

Petersen, R., 2003. Settlements, Kinship and Hunting Grounds in Traditional Greenland. MoG Man and Society, vol. 27, 324pp.

Peterson, D.L. and D.R. Johnson (eds.), 1995. Human Ecology and Climate Change: People and Resources in the Far North. Taylor and Francis.

Pfister, C., 1988. Variations in the spring-summer climate of Central Europe from the High Middle Ages to 1850. In: H. Wanner and U. Siegenthaler (eds.). Long and Short Term Variability of Climate. Springer.

Pika, A. and D. Bogoyavlensky, 1995. Yamal Peninsula: oil and gas development and problems of demography and health among indigenous populations. Arctic Anthropology, 32:61–74.

Podkoritov, F.M., 1995. Reindeer herding on Yamal. Leningrad Atomic Electrical Station, Sosnovyi Bor. (In Russian)

Post, E. and N.C. Stenseth, 1999. Climatic variability, plant phenology, and northern ungulates. Ecology, 80:1322–1339.

Rasmussen, R.O. and L.C. Hamilton, 2001. The Development of Fisheries in Greenland Roskilde. North Atlantic Regional Studies Res. Rep. 53, Roskilde University, 124pp.

Receveur, O., N. Kassi, H.M. Chan, P.R. Berti and H.V. Kuhnlein, 1998. Yukon First Nations' Assessment Dietary Benefit/Risk. Report to communities. Centre for Indigenous Peoples' Nutrition and Environment.

Riedlinger, D. and F. Berkes, 2001. Contributions of traditional knowledge to understanding climate change in the Canadian Arctic. Polar Record, 37:315–328.

Roberts, K., 1996. Circumpolar Aboriginal People and Co-management Practice: Current Issues in Co-Management and Environmental Assessment. Joint Secretariat, Inuvialuit Renewable Resources Committee, Calgary, Alberta.

Russell, D.E., K.R. Whitten, R. Farnell and D. van de Wetering, 1992. Movements and Distribution of the Porcupine Caribou Herd, 1970–1990. Tech. Rep. 139. Canadian Wildlife Service, Pacific and Yukon Region.

Russell, D.E., A.M. Martell and W.A.C. Nixon, 1993. The range ecology of the Porcupine Caribou Herd in Canada. Rangifer Special Issue, 8: 168.

Sabo III G., 1991. Long Term Adaptations among Arctic Hunter Gatherers. Garland Publishing.

Sakshaug, E. and D. Slagstad, 1992. Sea ice and wind: effects on primary productivity in the Barents Sea. Atmosphere Ocean, 30:579–591.

Sakshaug, E. and J. Walsh, 2000. Marine biology: biomass productivity distributions and their variability in the Barents and Bering Seas. In: M. Nuttall and T.V. Callaghan (eds.). The Arctic: Environment, People, Policy. Harwood Academic Press.

SEARCH, 2001. Study of Environmental Arctic Change. Science Plan. Polar Science Center, University of Washington, 89pp.

Siegert, M. and J. Dowdeswell, 2000. Glaciology. In: M. Nuttall and T.V. Callaghan (eds.). The Arctic: Environment, People, Policy, p. 27–55. Harwood Academic Press.

Shiyatov, S.G. and V.S. Mazepa, 1995. Climate. In: L.N. Dobrinskii (ed.). The Nature of Yamal, pp. 32–68. Nauka, Ekaterinburg. (In Russian)

Slobodin, R., 1962. Band Organization of the Peel River Kutchin. National Museum of Canada.

Slobodin, R., 1981. Kutchin. In: J. Helm (ed.). Handbook of North American Indians. Vol. 6: Subarctic, pp. 514–532. Smithsonian Institution, Washington.

Spencer, R., 1959. The North Alaskan Eskimo: A Study in Ecology and Society. Bureau of American Ethnology Bulletin 171. Smithsonian Institution, Washington, D.C.

Spiess, A.E., 1979. The Reindeer and Caribou Hunters: An Archeological Study. Academic Press, 312pp.

Thorpe, T., S. Eyegetok, N. Hakongak and the Kitikmeot Elders, 2002. Nowadays it is not the same: Inuit Qaumjimajatuqangit, climate and caribou in the Kitikmeot Region of Nunavut, Canada. In: I. Krupnik and D. Jolly (eds.). The Earth is Faster Now: Indigenous Observations of Arctic Environmental Change, pp. 198–239. Arctic Research Consortium of the United States, Fairbanks, Alaska.

Thuesen, S.T., 1999. Local identity and history of a Greenlandic town: The making of the town of Sisimiut (Holsteinsborg) from the 18th to the 20th century. Etudes/Inuit/Studies, 23(1–2):55–67.

Turi, J.M., 2000. Native reindeer herders' priorities for research. Polar Research, 19:131–133.

Usher, P.J. 1970. The Bankslanders: Economy and Ecology of a Frontier Trapping Community. 2 vols. Department of Indian Affairs and Northern Development, Ottawa.

Usher, P. 2002. Inuvialuit use of the Beaufort Sea and its resources, 1960–2000. Arctic, 55(1):18–28.

Usher, P.J., 2004. Caribou crisis or administrative crisis? Wildlife and Aboriginal policies on the Barren Grounds of Canada, 1947–60. In: D.G. Anderson, M. Nuttall (eds.). Cultivating Arctic Landscapes: Knowing and Managing Animals in the Circumpolar North. Berghahn.

Vasari, Y., H. Hyvärinen and S. Hicks (eds.), 1972. Climatic Change in Arctic Areas during the Last Ten Thousand Years. Acta Universitatis Ouluensis Series A: Scientiae Rerum Naturalium 3, Geologica 1. University of Oulu, Finland.

Vibe, C., 1967. Arctic Animals in Relation to Climatic Fluctuations. Meddelelser øm Grønland, Bd 170, Nr 5. C.A. Reitzels, Copenhagen.

Vlassova, T.K., 2002. Human impacts on the tundra - taiga zone dynamics: the case of the Russian Lesotundra. Ambio Special Report 12:30–36.

Vlassova, T.K. and S.G. Volkov, 2001. Ecological condition of the reindeer pastures in Russia and abroad. Russian Academy of Agricultural Sciences Review, 2:101–117. (In Russian)

Warbelow, C., D. Roseneau and P. Stern, 1975. The Kutchin Caribou Fences of Northeastern Alaska and the Northern Yukon. Studies of Large Mammals along the Proposed Mackenzie Valley Gas Pipeline Route from Alaska to British Columbia. Biological Report Series. Volume 32. J.R.D., Renewable Resources Consulting Services Ltd, 129pp.

WCMC, 1990. North-east Greenland National Park. World Conservation Monitoring centre. www.unep-wcmc.org/protected_areas/data/sample/0352v.htm [accessed December 2001].

Wein, E.E., 1994. The high cost of a nutritionally adequate diet in four Yukon communities. Canadian Journal of Public Health, 85:310–312.

Wein, E.E. and M.M.R. Freeman, 1992. Inuvialuit food use and food preferences in Aklavik, Northwest Territories, Canada. Arctic Medical Science, 51:159–172.

Weinstein, M.S., 1990. Notes for understanding the impacts of contaminants on Native subsistence economies. Collection Nordicana, 56:51–52.

Weinstein, M.S., 1996. The Ross River Dena: a Yukon Aboriginal Economy Ottawa. Royal Commission on Aboriginal Peoples.

Weller, G., 2000. The weather and climate of the Arctic. In: M. Nuttall and T.V. Callaghan (eds.). The Arctic: Environment, People, Policy. Harwood Academic Publishers.

Weller, G. and M. Lange (eds.), 1999. Impacts of Global Climate Change in the Arctic Regions. International Arctic Science Committee, Oslo.

Wenzel, G.W., 1981. Clyde Inuit Ecology and Adaptation: The Organization of Subsistence. Canadian Ethnology Service Mercury Paper No. 77. National Museums of Canada, Ottawa.

Wenzel, G.W., 1989. Sealing at Clyde River, N.W.T.: A discussion of Inuit economy. Etudes/Inuit/Studies, 13(1):3–23.

Wenzel, G.W., 1991. Animal Rights, Human Rights: Ecology, Economy and Ideology in the Canadian Arctic. University of Toronto Press.

Wenzel, G.W., 1995a. Warming the Arctic: environmentalism and Canadian Inuit. In: D.L. Peterson and D.R. Johnson (eds.). Human Ecology and Climate Change: People and Resources in the Far North. Taylor and Francis.

Wenzel, G.W., 1995b. Ningiqtuq: Inuit resource sharing and generalized reciprocity in Clyde River, Nunavut. Arctic Anthropology, 32(2):43–60.

Wenzel, G.W., 2000. Sharing, money, and modern Inuit subsistence: obligation and reciprocity at Clyde River, Nunavut. In: G.W. Wenzel, G. Hovelsrud-Broda and N. Kishigami (eds.). The Social Economy of Sharing: Resource Allocation and Modern Hunter-Gatherers, pp. 61–85. Senri Ethnological Series. National Museum of Ethnology, Osaka.

Wenzel, G.W. and L.-A. White, 2001. Chaos and Irrationality(!): Money and Inuit Subsistence. Paper presented at International Congress of Arctic Social Sciences, Québec City.

Wenzel, G.W., G. Hovelsrud-Broda and N. Kishigami, 2000. Introduction. In: G.W. Wenzel, G. Hovelsrud-Broda and N. Kishigami (eds.). The Social Economy of Sharing: Resource Allocation and Modern Hunter-Gatherers, pp. 1–6. Senri Ethnological Series. National Museum of Ethnology, Osaka.

Wilkinson, P.F., C.C. Shank and D.F. Penner, 1976. Muskox-caribou summer range relations on Banks Island, N.W.T. Journal of Wildlife Management, 40(1):151–162.

Wolfe, R.J., and R.J. Walker, 1987. Subsistence economies in Alaska: productivity, geography and development. Arctic Anthropology, 24(2):56–81.

WRH, 1999. Reindeer husbandry in Yamal Nenets Autonomous Okrug. Working paper No. 4/99. Association of World Reindeer Herders, Tromsø.

Young, O., 1992. Arctic Politics: Conflicts and Cooperation in the Circumpolar North. Hanover, Dartmouth College.

Chapter 13

Fisheries and Aquaculture

Lead Authors
Hjálmar Vilhjálmsson, Alf Håkon Hoel

Contributing Authors
Sveinn Agnarsson, Ragnar Arnason, James E. Carscadden, Arne Eide, David Fluharty, Geir Hønneland, Carsten Hvingel, Jakob Jakobsson, George Lilly, Odd Nakken, Vladimir Radchenko, Susanne Ramstad, William Schrank, Niels Vestergaard, Thomas Wilderbuer

Contents

Summary

This chapter addresses fisheries and aquaculture in four large marine ecosystems, three in the northern North Atlantic and one in the North Pacific. The ecosystems around Greenland and off northeast Canada (east of Newfoundland and Labrador) are of a true arctic type. Owing to a greater influence of warm Atlantic or Pacific water, the other systems are of a cold-temperate type. Historical data are used to project the effects of a warming climate on commercial and other marine stocks native to these ecosystems.

Modeling studies show that it is difficult to simulate and project changes in climate resulting from the response to forces that can and have been measured and even monitored on a regular basis for considerable periods and on which the models are built. Furthermore, current climate models do not include scenarios for ocean temperatures, watermass mixing, upwelling, or other relevant ocean variables such as primary and secondary production, on either a global or regional basis. As fisheries typically depend on such variables, any predictions concerning fisheries in a changing climate can only be of a very tentative nature.

Commercial fisheries in arctic regions are based on a number of species belonging to physically different ecosystems. The dynamics of many of these ecosystems are not well understood and therefore it is often difficult to identify the relative importance of fishing and the environment on changes in fish populations and biology. Moreover, current fish populations differ in abundance and biology from those in the past due to anthropogenic effects (i.e., exploitation rates). As a result it is unclear whether current populations will respond to climate change as they may have done in the past. Thus the effects of climate change on marine fish stocks and the eventual socio-economic consequences of those effects for arctic fisheries cannot be accurately predicted.

In general, it is likely that a moderate warming will improve conditions for some of the most important commercial fish stocks, e.g., Atlantic cod, herring, and walleye pollock. This is most likely to be due to enhanced levels of primary and secondary production resulting from reduced sea-ice cover. Reduced sea ice would automatically improve recruitment to Atlantic cod, herring, and walleye pollock stocks, as well as to a number of other smaller stocks.

Such changes could also lead to extensive expansions of habitat areas for species such as cod and herring. The most spectacular examples are cod at Greenland and the Norwegian spring-spawning herring. Atlantic cod appear to be unable to propagate off West Greenland except under warm conditions when a very large self-sustaining cod stock has been observed. At the same time, there has sometimes been a large-scale drift of juvenile cod from Iceland to Greenland. Many of these cod have returned to Iceland to spawn as adults, thus expanding the distribution range of Icelandic cod. In warm periods, the Norwegian spring-spawning herring forages for food westward across the Norwegian Sea to the north of Iceland, but is excluded from the western half of the Norwegian Sea and northern Icelandic waters during cold periods. This results in a loss of about a third of the summer feeding grounds for the largest single herring stock in the world.

Global warming is also likely to induce an ecosystem regime shift in some areas, resulting in a very different species composition. In such cases, relative population sizes, fish growth rates, and spatial distributions of fish stocks are likely to change. This will result in the need for adjustments in the commercial fisheries. However, unless there is a major climatic change, such adjustments are likely to be relatively minor and, although they may call for fresh negotiations of fishing rights and total allowable catches, such changes are unlikely to entail significant economic and social costs.

The total effect of a moderate warming of climate on fish stocks is likely to be of less importance than the effects of fisheries policies and their enforcement. The significant factor in determining the future of fisheries is sound resource management practices, which in large part depend upon the properties and effectiveness of resource management regimes and the underlying research. Examples supporting this statement are the collapse of the "northern cod" off Newfoundland and Labrador, the fall and rise of the Norwegian spring-spawning herring, and the stable condition of the Alaska pollock of the Bering Sea. However, all arctic countries are currently making efforts to implement management strategies based on precautionary approaches, with increasing emphasis on the inclusion of risk and uncertainty in all decision-making.

The economic and social impacts of altered environmental conditions depend on the ability of the social structures involved, including the fisheries management system, to generate the necessary adaptations to the changes. These impacts will be very different to those experienced in earlier times, when the concept of fisheries management was almost unknown. Furthermore, in previous times general poverty, weak infrastructure, and lack of alternative job opportunities meant that the ability of societies to adapt to change, whether at a national or local level, was far less than today. Thus, it is unlikely that the impact of the climate change projected for the 21st century (see Chapter 4) on arctic fisheries will have significant long-term economic or social impacts at a national level. Some arctic regions, especially those very dependent on fisheries may, however, be greatly affected.

13.1. Introduction

This chapter identifies the possible effects of climate change on selected fish stocks and their fisheries in the

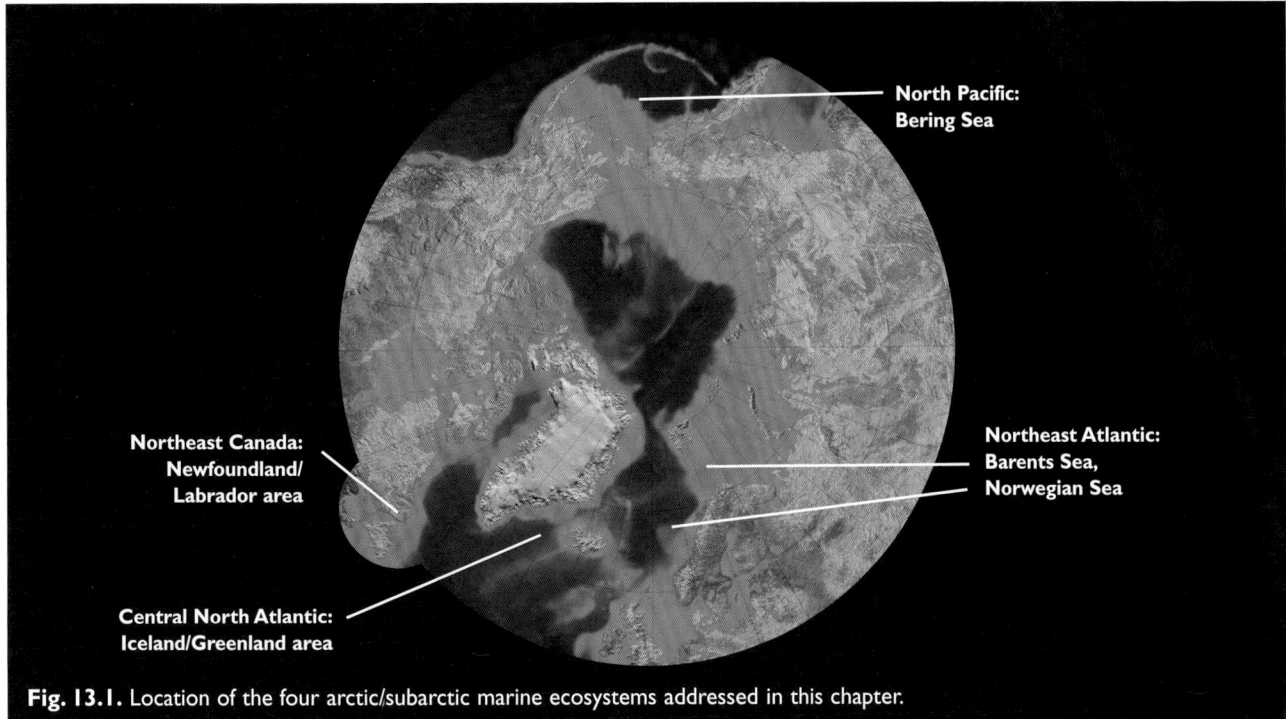

North Pacific: Bering Sea

Northeast Canada: Newfoundland/ Labrador area

Central North Atlantic: Iceland/Greenland area

Northeast Atlantic: Barents Sea, Norwegian Sea

Fig. 13.1. Location of the four arctic/subarctic marine ecosystems addressed in this chapter.

Arctic. Arctic fisheries of selected species are described in the northeast Atlantic (i.e., the Barents and the Norwegian Seas), the waters around Iceland and Greenland, the waters off northeastern Canada, and the Bering Sea (Fig. 13.1). The species discussed are those few circumpolar species (capelin (*Mallotus villosus*), Greenland halibut (*Reinhardtius hippoglossoides*), northern shrimp (*Pandalus borealis*), and polar cod (*Boreogadus saida*)) and those of commercial importance in specific regions. The latter include Atlantic cod (*Gadus morhua*), haddock (*Melanogrammus aeglefinus*), Alaska pollock (*Theragra chalcogramma*), Pacific cod (*Gadus macrocephalus*), snow crab (*Chionoecetes opilio*), plus a number of others. Marine mammals are also considered in this chapter as they form an important component of northern marine ecosystems and several are of commercial importance.

This chapter focuses on the effects of climate change on commercial fisheries and the impacts on society as a whole. Chapters 9, 10, and 12 address the implications of fisheries and aquaculture for indigenous peoples.

This chapter is organized such that for each of the four regions the discussion follows a standard format: introduction; ecosystem essentials; fish stocks and fisheries; past climatic variations and their impact on commercial stocks; possible impacts of global warming on fish stocks; the economic and social importance of fisheries; past economic and social impacts of climate change on fisheries; economic and social impacts of global warming: possible scenarios; and ability to cope with change. The chapter concludes with a synthesis of the regional assessments of the impacts of climate change on arctic fisheries and societies, and with research recommendations.

13.1.1. Biological and model uncertainties/certainties

Precise forecasts of changes in fish stocks and fisheries and their effects on society are not possible. The sources of uncertainty can be grouped into three categories: (1) uncertainties in identifying the reasons for past changes in fish biology, (2) uncertainties in the projections of potential changes in the ocean climate under climate change scenarios, and (3) uncertainties relating to the socio-economic effects of changes in fish stocks.

There are many biological characteristics of fish that change in response to natural variability in the physical environment. However, when fish stocks are heavily exploited, as many arctic stocks have been, it has proven difficult to identify the relative importance of fishing and environment on observed changes in biology. Also, many fish stocks are currently much less abundant than in the past and are showing extreme changes in population characteristics. Thus, even if historical observations of variability in fish biology could be associated with past changes in ocean climate, it is not known whether the present populations would respond in a manner similar to the historical response.

Some of the uncertainties surrounding the response of the ocean to the projected changes in global climate discussed in Chapter 4 were addressed in Chapter 9. One of the most important components of the arctic environment is the thermohaline circulation. Possible changes in the thermohaline circulation and their consequences are described in section 9.2.5.5. Present climate models are considered to generate reasonably reliable projections of climate change at a global scale but are considered to generate less reliable results at the regional level. This results in uncertainty in evaluations of potential effects of climate change on the large marine ecosystems considered in this chapter.

Some key findings in Chapter 9 reflect a high degree of certainty about changes in the arctic seas. Although regional changes were not identified in Chapter 9, the chapter concludes that in most arctic areas *upper water column temperatures are very likely to increase, especially in areas with reduced sea-ice cover* and that *increased water temperatures are very likely to lead to a northward shift in the distribution of many species of fish, to changes in the timing of their migration, to a possible extension of their feeding areas, and to increased growth rates.* Chapter 9 also concludes that *most of the present ice-covered arctic areas are very likely to experience reductions in sea-ice extent and thickness, especially in summer* and that *in areas of reduced sea-ice cover, primary production is very likely to increase, which in turn is likely to increase zooplankton and possibly fish production.* In addition, Chapter 9 concludes that *increased areas and periods of open water are likely to be favorable for some whale species and the distribution of these species is very likely to move northward.* An expansion of their feeding grounds would presumably lead to an increase in their abundance. Thus, although the Chapter 9 conclusions are global in scale and do not identify specific changes in the four marine ecosystems considered here, they do provide, with a high degree of probability, a basis for considering these conclusions within the context of the fish stocks, fisheries, and possible effects on human societies resulting from the projected changes in the four areas.

13.1.2. Societal uncertainties

Once fish population changes have been evaluated, it becomes necessary to relate those changes to changes in society. This raises new difficulties. Even when changes in fish populations are predictable to a high degree of accuracy, there is no deterministic relationship between these changes and those in society. Social change is driven by a number of different forces; with climate change only one of a number of natural factors. Also, humans are important drivers of change, through economic and political activities. It is extremely difficult to isolate the relative impact of the various drivers of change. In addition, societies have the capacity to adapt to change. Changes in fish stocks, for example, are met by adjustments in fisheries management practices and the way fisheries are performed.

The result of these uncertainties is that there are few firm predictions in this chapter. Instead, changes in potential effects and likely outcomes are considered.

13.1.3. The global framework for managing living marine resources

A global framework for the management of living marine resources has been developed over recent decades, providing coastal states with extended jurisdiction over natural resources. The Third United Nations Law of the Sea Conference (UNCLOS) was convened in

1973 and ended nine years later with the adoption in 1982 of the United Nations Law of the Sea Convention, which lays down the rules and principles for the use and management of the natural resources in the ocean. The most important elements are the provisions that enable coastal states to establish exclusive economic zones (EEZs) up to 200 nautical miles (360 kilometers) from their coastal baselines. Coastal states have sovereign rights over the natural resources in their EEZs. The Convention also mandates that coastal states manage resources in a sustainable manner and that they be used optimally. Where fish stocks are shared among countries, they shall seek to cooperate on their management.

A country's authority to manage fish stocks is defined by its 200 mile EEZ. Within its EEZ, a coastal state has sovereign rights over the natural resources, and therefore the authority to manage the living marine resources there. During the 1980s it became evident that the framework provided by the Convention was inadequate to cope with two major developments in fisheries worldwide: the dramatic increase in fishing in the high seas beyond the EEZs and a corresponding increase in catches within the EEZs. Both developments were driven by rapidly growing fishing capacity. The consequence was that many stocks were overfished. A treaty was therefore negotiated under the auspices of the United Nations to supplement the Convention, seeking to provide a legal basis for restricting fisheries on the high seas and introducing more restrictive management principles, enhanced international cooperation in management, and improved enforcement of management measures. The Agreement for the Implementation of the Provisions of the United Nations Convention on the Law of the Sea of 10 December 1982 Relating to the Conservation and Management of Straddling Fish Stocks and Highly Migratory Fish Stocks (The UN Fish Stocks Agreement) was thus adopted in 1995 and mandates the application of a precautionary approach to fisheries management. It also emphasizes the need for cooperation between countries at a regional level in this respect. These two elements have proved crucial in the development of international fisheries conservation and management policies since the mid-1990s, not least in arctic areas. Existing regional arrangements have been improved upon in order to implement the agreements. This applies to the Northwest Atlantic Fisheries Organization (NAFO), which covers the Northwest Atlantic, and the North East Atlantic Fisheries Commission (NEAFC), which covers the international waters in the Northeast Atlantic. An agreement placing a moratorium on fishing on the high seas in the Bering Sea has been in force since 1994.

The development of this global framework for fisheries management has been accompanied by a corresponding development of fisheries management regimes in individual countries. The design and performance of such regimes are crucial to the fate of fish stocks. At the global level, the major challenges to fisheries management are related to the need to reduce a substantial overcapacity in the world's fishing fleets, and the need to introduce more sustainable management practices. To achieve the latter, countries are introducing precautionary approaches to fisheries management – a crucial requirement of the 1995 UN Fish Stocks Agreement. In addition, ecosystem-based approaches to the management of living marine resources, where natural factors such as climate change are taken into account in decision-making, are under development. The 2002 World Summit on Sustainable Development stated in its implementation plan that ecosystem-based approaches to management are to be in place by 2010.

All arctic countries with significant fisheries have well established resource management regimes with comprehensive systems for producing the knowledge base required for management, the promulgation of regulations to govern fishing activities, and arrangements to ensure compliance with regulations. While the various regimes vary considerably with regard to the design of management policies, the challenges they confront in attempting to reduce overcapacity and in introducing precautionary approaches to fisheries are similar.

For marine mammals there is a single international body at the global scale, and several regional bodies. At the global scale the 1946 International Convention for the Regulation of Whaling mandates an International Whaling Commission (IWC) to regulate the harvest of great whales. A moratorium on commercial whaling was adopted in 1982. A number of countries, among them Norway and Russia, availed themselves of their right

under the convention not to be bound by this decision. Canada and Iceland left the Commission due to the preservationist developments there. Iceland rejoined the Commission in 2003. The North Atlantic Marine Mammal Commission (NAMMCO) is tasked with the management of marine mammals in the North Atlantic.

13.2. Northeast Atlantic – Barents and Norwegian Seas

This section addresses the potential impacts of climate change on the fisheries in the arctic area of the Northeast Atlantic. The area comprises the northern and eastern parts of the Norwegian Sea to the south, and the north Norwegian and northwest Russian coasts and the Barents Sea to the east and north. The fisheries take place in areas under Norwegian and Russian jurisdictions as well as in international waters. The total fisheries in the area were around 2.1 million t in 2001 (based on data in Michalsen, 2003). Aquaculture is dominated by salmon and trout and produced 86000 t in 2001 (Fiskeridirektoratet, 2002a).

The legal and political setting of the fisheries in the Northeast Atlantic is complex. Norway and Russia established 200 nm EEZs in 1977, as a consequence of developments in international ocean law at the time. The waters around Svalbard come under a Fisheries Protection Zone set up by Norway, which according to the 1920 Svalbard Treaty holds sovereignty over the Svalbard archipelago. The waters around the Norwegian island of Jan Mayen, north of Iceland, are covered by a Fisheries Zone. Two areas occur on the high seas beyond the EEZs: in the Barents Sea the so-called "Loophole" and in the Norwegian Sea the so-called "Herring hole" (Fig. 13.2). Norway and Russia have long traditions of cooperation both in trade and management issues. In the 18th century, Norwegian fishermen in the north traded cod for commodities from Russian vessels – the so-called "Pomor-trade" (Berg, 1995). Joint management of the Barents Sea fish stocks has been negotiated since 1975. Since then, a comprehensive framework for managing the living marine resources in the area has been developed, including the high seas. The resources in the area are exploited with vessels from Norway and Russia, as well as from other countries.

Northern Norway includes three counties: Finnmark, Troms, and Nordland, and covers an area of 110000 km^2 – about the same size as Great Britain. The total population is 460000. Owing to the influence of the North Atlantic Current, the climate in this region is several degrees warmer than the average in other areas at the same latitude. While the Norwegian fishing industry occurs in many communities along the northern coast, the northwest Russian fishing fleet is concentrated in large cities, primarily Murmansk. In addition to the Murmansk Oblast, Russia's "northern fishery basin" comprises Arkhangelsk Oblast, the Republic of Karelia, and Nenets Autonomous Okrug (see Fig. 13.2). There is no significant commercial fishing activity east of these

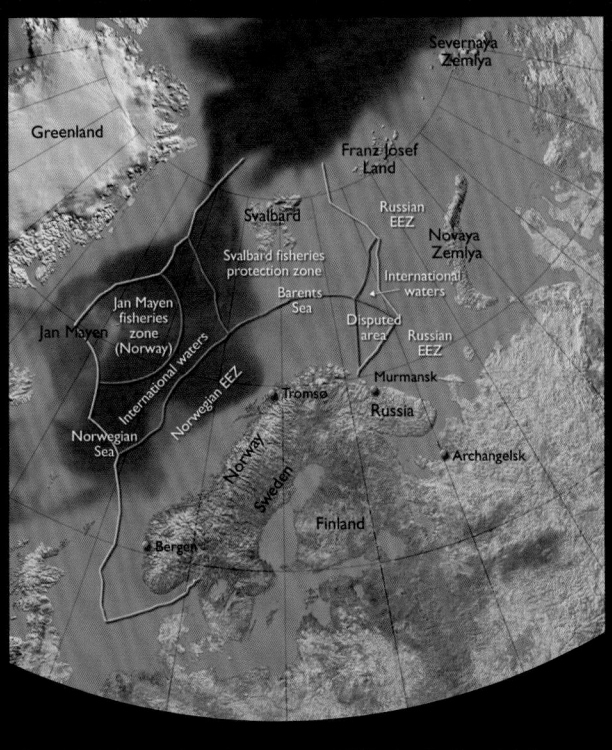

Fig. 13.2. Map of the Norwegian EEZ, the Svalbard fisheries protection zone, and the Russian EEZ in the Barents and Norwegian Seas. The international areas in the central Norwegian and Barents Seas are often referred to as the "herring hole" and "loophole", respectively.

regions until the far eastern fishery basin in the North Pacific. Since 1 January 2002, the population in the four federal subjects constituting Russia's northern fishery basin was 3.2 million people.

13.2.1. Ecosystem essentials

There are large seasonal variations in the upper water layers of the Barents Sea (see section 9.2.4.1). The spring bloom starts in the southwestern areas and spreads north- and eastward following the retreat of the sea ice. Fish and marine mammals also exhibit directed migrations: spawning migrations south- and westward in late autumn and winter, and feeding migrations north- and eastward in late spring and summer.

Relatively few species and stocks make up the bulk of the biomass at the various trophic levels. Fifteen to twenty species of whales and seals forage regularly in the area. Harp seals (*Phoca groenlandica*) and minke whales (*Balaenoptera acutorostrata*) are the two most important predators in the pelagic ecosystem. The harp seals breed in the southeastern parts of the Barents Sea, i.e., in the White Sea, and feed close to the ice edge, mainly on amphipods and capelin. In periods of low capelin abundance, harp seals feed on other fish, such as cod, haddock, and saithe (*Pollachius virens*), and migrate southward along the Norwegian coast (Nilssen K., 1995). Minke whales feed on various species of fish and over most of the area from May to September (Nordøy et al., 1995). During the winter the whales occur further south in the Atlantic Ocean.

The spawning grounds of most species are situated along the coast of Norway and Russia. Spawning normally occurs in winter and spring (February to May) and egg and larval drift routes are toward the north and east. Juveniles and adults feed in the area; polar cod in the north- and northeasternmost parts, saithe and herring (*Clupea harengus*) in the southwest, as well as the easternmost Norwegian Sea and off the Norwegian coast. Capelin reside mainly on the Atlantic side of the Polar Front during winter, but feed on the zooplankton production in the large ice-free areas north of the Polar Front in summer and autumn. Cod has the most extensive distribution. Adult cod spawn in Atlantic water far south along the coast of Norway in March to April, and then feed along the Polar Front and even far into arctic water masses during summer and autumn. All species exhibit seasonal migrations, which coincide with the formation and melting of sea ice: north- and eastward during spring and summer, south- and westward during autumn and winter.

Cod, saithe, haddock, and redfish (*Sebastes marinus* and *S. mentella*) have their main spawning grounds on the coastal banks and off the shelf edge (redfish only) of Norway between 62° and 70° N and return to the Barents Sea after spawning. Herring migrate out of the Barents Sea before maturing, feed as adults in the Norwegian Sea, and have their main spawning grounds far-

ther south along the Norwegian coast, between about 59° and 68° N. Capelin spawn in the northern coastal waters mainly between 20° and 35° E, while polar cod has two main spawning areas; one in Russian waters in the southeastern part of the Barents Sea and another in the northwest, close to the Svalbard archipelago. The capelin spawning schools are followed by predating immature cod, four to six years old. Adult Greenland halibut inhabit the slope waters at depths between 400 and 1000 m over the entire area. Northern shrimp occur over most of the area in regions with bottom depths of between 100 and 700 m on the "warm" side of the Polar Front. Individuals are four to seven years old when they change sex from male to female and spawning (hatching of eggs) occurs in summer and autumn over most of the area.

From simulations of interactions between capelin, herring, cod, harp seals, and minke whales, Bogstad et al. (1997) found the herring stock to be sensitive to changes in minke whale abundance because whale predation in the Barents Sea affects the number of recruits to the mature herring stock. They also found that an increasing harp seal stock will reduce the capelin and cod stocks, implying that an unexploited seal population would lead to a substantial loss of catch in the cod fishery.

Cod, capelin, and herring are considered key fish species in the ecosystem and interactions between them generate changes which also affect other fish stocks as well as marine mammals and birds (Bogstad et al., 1997). Recruitment of cod and herring is enhanced by inflows of Atlantic water carrying large amounts of suitable food (especially the "redfeed" copepod *Calanus finmarchicus*) for larvae and fry of these species. Consequently, survival increases, so that juvenile cod and herring become abundant in the area. However, since young and juvenile herring prey on capelin larvae in addition to zooplankton, capelin recruitment might be negatively affected and thus cause a temporal decline in the capelin stock, an occurrence that would affect most species in the area (fish, birds, and marine mammals) since capelin is their main forage fish. Predators would then prey on other small fish and shrimps. In particular, cod cannibalism may increase and thus affect future recruitment of cod to the fishery (Hamre, 2003).

In periods of low abundance or absence of capelin and/or herring, the top predators will have to feed somewhere else or shift to prey on the zooplankton group. For cod, such shifts have been observed twice in the past 15 years and were related to the collapses of the capelin stock in 1986–1988 and 1993–1994.

13.2.2. Fish stocks and fisheries

For the past thousand years, fishing for cod and herring has been important for coastal communities in Norway and northern Russia (Solhaug, 1983). Throughout the centuries, fishing was purely coastal and seasonal and based on the large amounts of adult cod and herring

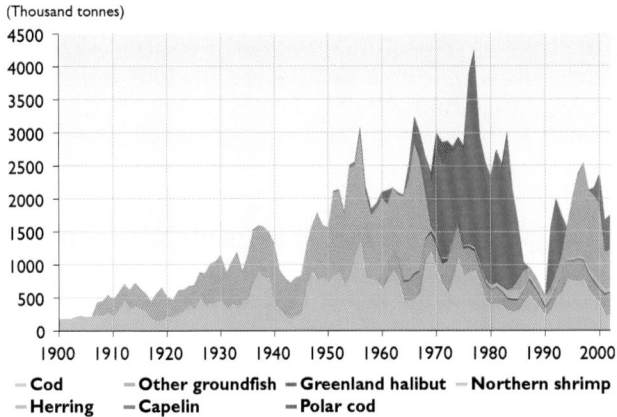

(Thousand tonnes)

Cod — Other groundfish — Greenland halibut — Northern shrimp
Herring — Capelin — Polar cod

Fig. 13.3. Landings in Norway from the most important commercial catches taken in the Arctic (data from the Ministry of Fisheries and Coastal Affairs, Oslo).

migrating into near-shore waters for spawning during winter–spring and on the schools of immature cod feeding on spawning capelin along the northern coasts in April to June. A certain development toward offshore fishing took place at the end of the 19th century when cod were caught on the Svalbard banks and driftnetting of herring began off northern Iceland. However, the quantities caught in these "offshore" fisheries were small compared to the near-shore catches in the traditional fisheries for both species. Estimates of annual yields of cod and herring prior to 1900 were given by Øiestad (1994). For both species large fluctuations were experienced. The dominant feature is the 5- to 10-fold increases between 1820 and 1880 as compared to yields in previous centuries. For fish species other than cod and herring reliable estimates of yield prior to the 20th century are not available.

Landings for herring, capelin, polar cod, Greenland halibut, northern shrimp, and northeast Atlantic cod in the 20th century are shown in Fig. 13.3. Total fish landings from the area increased from about 0.5 million t at the beginning of the century to about 3 million t in the 1970s. This increase was mainly due to a series of major technological improvements of fishing vessels and gear, including electronic instruments for fish finding and positioning, which took place during the 20th century and dramatically increased the effectiveness of the fishing fleet. Furthermore, there was a growing market demand for fish products.

13.2.2.1. Capelin

When herring became scarce in the late 1960s the purse seine fleet targeted capelin and catches increased rapidly in the 1970s. Management measures such as minimum allowable catch size and closing of areas where undersized fish occurred, as well as limited fishing seasons, were introduced in the early 1970s, first by Norway and later jointly by Norway and Russia. Total allowable catches (TACs) have been enforced since 1978. Landings have fluctuated widely. In 2002, the total catch of capelin was 628 000 t (Fig. 13.3). During the 1980s, the

importance of capelin and juvenile herring as food sources for cod and other predators was fully realized (see Nakken, 1994 for references). As a consequence, there was increased research effort on species interactions and since 1990 the cod stock's need for capelin as food has been taken into account in the scientific advice on management measures.

13.2.2.2. Polar cod

Russia and Norway started regular fisheries with bottom and pelagic trawls for polar cod in the late 1960s. The catches increased to approximately 350 000 t in 1971. The Norwegian fleet was active until 1973, when fishers lost interest because of declining catches. Since then landings have been exclusively Russian. Catches in 2001 were about 40 000 t.

13.2.2.3. Greenland halibut

Until the early 1960s, the Greenland halibut fishery (Fig. 13.3) was mainly pursued by coastal longliners off the coast of northern Norway. Annual landings were about 3000 t. An international trawl fishery developed in the area between 72° and 79° N and catches increased to about 80 000 t in the early 1970s. Landings decreased throughout the 1970s; the spawning stock biomass declined from more than 200 000 t in 1970 to about 40 000 t in the early 1990s and has since remained at this low level. Since 1992, only vessels less than 28 m in length using long lines or gillnets have been permitted to carry out a directed fishery. The rest of the fishing fleet has been restricted by by-catch rules. The total catch in 2002 was 13 000 t.

13.2.2.4. Northern shrimp

Prior to 1970, trawling for northern shrimp took place in the fjords of northern Norway and catches were low. During the 1970s offshore grounds were exploited. Catches increased until 1984 when 128 000 t were landed. Since then, catch levels have fluctuated (Fig. 13.3). Fisheries have been regulated by bycatch rules and closed areas since the mid-1980s. Areas are closed to fishing when the catch rates of young cod, haddock, and Greenland halibut exceed a certain limit. In later years, young redfish has also been included in the bycatch quota. Areas are also closed when the proportion of minimum-size shrimp (15 mm carapace length) is too high. In the Russian EEZ an annual TAC is also enforced. Estimated cod consumption of shrimp has since 1992 been approximately ten times higher than the landings, which were about 58 000 t in 2001.

13.2.2.5. Herring

Until the 1950s, herring fisheries remained largely seasonal and near shore. The bulk of the landings came from Norwegian vessels. In the 1950s Russian fishers developed a gillnet fishery in offshore waters in the Norwegian Sea, and in the early 1960s purse seiners started using echo sounding equipment to locate herring. These technological developments resulted in a large increase in the total catches until 1966 (2 million t). Thereafter, catches decreased rapidly and the stock collapsed (Fig. 13.3, and see Box 13.1). Although individual scientists expressed concern about the stock, effective management measures were neither advised nor implemented until after the stock had collapsed completely. Minor catches in the early 1970s (between 7000 and 20000 t) removed most of the remaining spawning stock as well as juveniles and it was not until 1975 that the fishing pressure was brought to a level which permitted the stock to start recovering. For 25 years the stock was very small and remained in Norwegian coastal waters throughout the year. Norway introduced management measures including minimum allowable landing size and annual TACs. Furthermore, a complete ban on fishing herring was enforced for some

years. During the 1990s the stock recovered, started to make feeding migrations into the Norwegian Sea, and catch quotas and landings increased. In 2002 the total landings were 830000 t.

13.2.2.6. Northeast Atlantic cod

Prior to 1920, the bulk of the northeast Atlantic cod (*Gadus morhua*) catch was from two large seasonal and coastal fisheries: the fishery for immature cod feeding on spawning capelin along the northern coast of Norway and Russia and the fishery for spawning cod ("skrei") further south off northern Norway (the Lofoten fishery). In the 1920s and 1930s an international bottom trawl fishery targeting cod as well as other species (haddock, redfish) developed in offshore areas of the Barents Sea and off Svalbard. Annual catches increased from about 400000 t in 1930 to 700000 to

Box 13.1. The fall and rise of the Norwegian spring-spawning herring

In the early 1950s, the spawning stock of Norwegian spring-spawning herring was estimated at 14 million t – one of the largest fish stocks in the world. Most of the adult stock migrated between Norwegian and Icelandic coastal waters to spawn in winter and feed in summer, respectively. The herring fishery was important for several countries, especially Norway, Iceland, Russia, and the Faroe Islands. However, after 15 years of over-exploitation and a decreasing spawning stock, the stock collapsed in the late 1960s.

Deteriorating climatic conditions north of Iceland and in the western Norwegian Sea are crucial in explaining changes of feeding areas and migration routes of these herring in the late 1960s. High fishing intensity was, however, the major factor behind the actual stock collapse. The breakdown had large social and economic consequences for those depending on the fishery. Nevertheless, the industry managed to redirect its effort to other pelagic species – primarily capelin.

Over the following decades, the remaining herring kept close to the Norwegian coast. The stock was strictly regulated and fishing was prohibited for several years. These regulations, probably in combination with favorable climatic conditions, contributed to a considerable increase in stock size from the mid-1980s, making it possible to resume fishing. By the late 1980s the spawning stock had reached a level of 3 to 4 million t, mainly due to above average recruitment by the 1983 year class.

By 1995, the spawning stock had reached 5 million t. As a consequence, the stock extended its feeding grounds by resuming its old migration pattern westward into the Norwegian Sea. It therefore became available for fishing beyond areas under Norwegian jurisdiction. The unilateral Norwegian management regime was no longer adequate to regulate fishing of the stock. Meanwhile, there was no arrangement to oversee the international management of the fishery. Negotiations between Norway, Russia, Iceland, and the Faroe Islands failed, and the total catch quota recommended by ICES was exceeded in the following year.

High economic values were at stake for all actors. Fishers and fisheries managers in all involved countries and in the EU were very engaged in the conflict. A first agreement was reached between Norway, Russia, Iceland, and the Faroe Islands in May 1996. In December 1996, the EU was included in the arrangement, where the five parties set and distribute TACs of Norwegian spring-spawning herring, based on ICES advice. The responsibility to manage the share of the stock in international waters is vested with the NEAFC, of which the aforementioned parties are members. Negotiations are held every year, but the percentage allocation key has not changed since the 1996 agreement. However, changes in the migration pattern may upset the present arrangement. The arrangement is, however, not currently functional due to disagreement over quota distribution.

This example shows that not only negative, but also positive changes in stock abundance may create management problems. If the parties had not reached agreement, there would have been devastating consequences for the exploitation and development of the Norwegian spring-spawning herring stock, almost certainly resulting in significant economic losses. This example shows the importance of political efforts to solve such conflicts.

800 000 t at the end of the decade. Landings also remained high after the Second World War until the end of the 1970s when catches declined sharply due to reduced stock size and the introduction of EEZs. Management advice was given by the International Council for the Exploration of the Sea (ICES) from the early 1960s. Increases in trawl mesh sizes were recommended in 1961 and in 1965 a variety of further conservation measures were recommended in order to increase yield per recruit and to limit the overall fishing mortality. From 1969 onward, ICES has expressed concern about the future size of the spawning stock, considering that at low levels of spawning stock biomass there would be an increased risk of poor recruitment to the stock. The first TAC for cod was set in 1975, but was far too high. Although minimum mesh size regulations had been in force for some years at that time, it is fair to conclude that no effective management measures were in operation for demersal fish in the area prior to the establishment of 200 nm EEZs in 1977.

The estimated average fishing mortality for the five-year period 1997 to 2001 is a record high (0.90) and about twice the fishing mortality corresponding to the precautionary approach (0.42). In the period 1998 to 2000 the spawning stock biomass was well below the recommended precautionary level of 500 000 t. However, despite relatively low recruitment in most recent years, the spawning stock has increased since 2000 and is now considered to be above precautionary levels. Landings have varied considerably over time and in 2002 were 430 000 t (Fig. 13.3).

13.2.2.7. Marine mammals

Three species of marine mammals are commercially exploited in the Northeast Atlantic by Norwegian and Russian fishers, i.e., minke whales, hooded seals (*Cystophora cristata*), and harp seals. In addition, grey seals (*Halichoerus grypus*) and harbour seals (*Phoca vitulina*) are exploited along the Norwegian coast by local hunters. Offshore exploitation of marine mammals in the area began in the 16th century. Basque and later Dutch and British vessels hunted Greenland right whales (*Balaena mysticetus*) and seals. Processing plants were established at shore stations as far north as northwestern Spitzbergen (Arlov, 1996). Russian and Norwegian hunters have caught walrus (*Odobenus rosmarus*), polar bear (*Ursus maritimus*), and seals at the Svalbard archipelago since the 16th century. By the first decades of the 19th century the stocks of right whales had almost disappeared, and the walrus was so depleted that the hunt became unprofitable. A new era of offshore exploitation began around 1860 to 1870 when the use of smaller ice-going vessels ("sealers") permitted Norwegian hunters to penetrate into the drift ice. At about the same time the invention of the grenade harpoon made hunting of great whales profitable. Catches of great whales increased between 1870 and 1900, but leveled off and decreased rapidly during the first decade of the 20th century.

Minke whales

Minke whales have been hunted in landlocked bays ("whaling bays") along the coast of Norway since olden times. Offshore hunting, using small motorized vessels, developed prior to the Second World War, essentially as an extension of fishing activities. Catches increased until the 1950s, the mean annual take at that time being about 2300 animals. Since 1960, catches have decreased due to reductions in annual TACs. Between 1987 and 1992 no commercial hunting was allowed. In recent years annual catches have been 400 to 600 animals and the quota for 2002 is 674 minke whales. The stock in the area is estimated at 112 000 animals (Michalsen, 2003).

Harp seals and hooded seals

Two stocks of harp seal, in the West Ice (Greenland Sea) and the East Ice (White Sea – Barents Sea), and one stock of hooded seal in the West Ice are subject to offshore sealing; since about 1880 mainly by Norwegian and Russian hunters. The total annual catch from these stocks increased from about 120 000 animals around 1900 to an average of about 350 000 per year in the 1920s. Since then catches have declined, mainly because of catch regulations (i.e., TACs). In recent years the loss of markets has been the main limiting factor. In the 1990s, catches of harp seal in the West Ice were 8000 to 10 000 animals each year and 8000 to 9000 for hooded seal, while catches of harp seal in the East Ice ranged from 14 000 to 42 000 per year. Russian catches, which constitute about 82% of the total, are taken in the East Ice, while the Norwegian catches (about 18%) are taken in both the West Ice and East Ice.

Hooded seals are found in the North Atlantic between Novaya Zemlya, Svalbard, Jan Mayen, Greenland, and Labrador. All the Norwegian catch of hooded seal takes place in the West Ice (Greenland Sea). Russia has not caught hooded seals since 1995. The total catch in 2001 was 3820 animals. All seal stocks are assessed every second year by a joint ICES/NAFO working group, which provides ICES with sufficient information to give advice on stock status and catch potential. All three stocks are well within safe biological limits, and harvesting rates are sustainable.

13.2.3. Past climatic variations and their impact on commercial stocks

The relationship between the physical effects of climate change and effects on the ecosystem is complex. It is not possible to isolate, let alone quantify, the effects of climate change on biological resources. The following discussion is therefore of a tentative and qualitative nature.

A number of climate-related events have been observed in the Northeast Atlantic fisheries (see section 9.3.3.3). During the warming of the Nordic Seas between 1900 and 1940, there were substantial northward shifts in the geographical boundaries for a range of marine species from plankton to commercial fish, as well as for terrestrial mammals and birds (Dickson, 1992). Recruitment of both cod and herring is positively related to inflows of Atlantic waters to the area and thus to temperature changes. Both stocks increased significantly between 1920 and 1940 when water temperatures increased (Hylen, 2002; Toresen and Østvedt, 2000). The increase in stock size was probably an effect of enhanced recruitment, because catches increased in the same period. A similar development may have occurred between 1800 and 1870 (Øiestad, 1994). Øiestad (1994) also provided evidence that cod abundance was low during the cold period between 1650 and 1750.

Since the Second World War both cod and herring have been subject to overfishing. This resulted in a collapse of the herring stock in the 1960s, with serious consequences for other inhabitants of the ecosystem as well as man (see Box 13.1). For cod, the most likely result of the overfishing has been a far lower average annual yield since 1980 than the stock has potential to produce. Recruitment of cod depends heavily on parent stock size in addition to environmental factors (Ottersen and Sundby, 1995; Pope et al., 2001). For several decades heavy fishing pressure has prevented maintenance of the cod spawning stock at a level which optimizes recruitment levels in the long run. Therefore, management of these stocks is the key issue in assessing the effects of potential climate variations (Eide and Heen, 2002).

13.2.4. Possible impacts of climate change on fish stocks

Global models project an increase in surface temperature in the Northeast Atlantic area of 3 to 5 °C by 2070 (see Chapter 4). Regional models however, project that for surface temperatures in this area there will be "a cooling of between 0 and -1 °C" by 2020 (Furevik et al., 2002). By 2050 the area is projected to have become warmer and by 2070 surface temperatures are projected to have increased by 1 to 2 °C (Furevik et al., 2002).

Research over the last few decades shows that cod production increases with increasing water temperature for stocks inhabiting areas of mean annual temperature below 6 to 7 °C, while cod stocks in warmer waters exhibit reduced recruitment when the temperature increases (Sundby, 2000). The mean annual ambient temperature for northeast Atlantic cod is 2 to 4 °C (depending on age group) and the stock has experienced greatly improved recruitment during periods of higher temperature in the past (Sundby, 2000). A rise in mean annual temperature in the Barents Sea over the period to 2070 is therefore likely to favor cod recruitment and production, and result in an extended distribution area (i.e., spawning and feeding areas) to the north and east. A similar statement may be made for herring (see Chapter 9). This statement is based on the assumption that the production and distribution of animals at lower trophic levels (particularly copepods – the food for larvae) remain unchanged. The projection is also based on

the assumption that harvest rates are kept at levels that maintain spawning stock biomass above the level at which recruitment is adversely affected.

Experience indicates that it is likely that a rise in water temperature, as projected for the area, will result in large displacements to the north and east of the distribution ranges of resident marine organisms, including fish, shrimps, and marine mammals. Their boundaries are very likely to be extended as waters get warmer and sea-ice cover decreases. "Warm water" pelagic species, such as blue whiting (*Micromesistius poutassou*) and mackerel (*Scomber scombrus*), are likely to occur in the area in higher concentrations and more regularly than in the past. Eventually, these species will possibly inhabit the southwestern parts of the present "arctic area" on a permanent basis.

The effects of a temperature rise on the production by the stocks of fish and marine mammals presently inhabiting the area are more uncertain. These depend on how a temperature increase is accompanied by changes in ocean circulation patterns and thus plankton transport and production. In the past, recruitment to several fish stocks in the area, cod and herring in particular, has shown a positive correlation with increasing temperature. This was due to higher survival rates of larvae and fry, which in turn resulted from increased food availability. Food is transported into the area via inflows of Atlantic water, which have also caused the ocean temperature to increase. Hence, high recruitment in fish is associated with higher water temperature but is not caused by the higher water temperature itself (Sundby, 2000).

Provided that the fluctuations in Atlantic inflows to the area are maintained along with a general warming of the North Atlantic waters, it is likely that annual average recruitment of herring and cod will be at about the long-term average until around 2020 to 2030. This projection is also based on the assumption that harvest rates are kept at levels that maintain spawning stocks well above the level at which recruitment is impaired. How production will change further into the future is impossible to guess, since the projected temperatures, particularly for some of the global models, are so high that species composition and thus the interactions in the ecosystem may change completely.

13.2.5. The economic and social importance of fisheries

The fishery sector is of considerable economic significance in Norway, being among the country's main export earners. Data used in this section are based on statistics from "Fisken og Havet" and the Norwegian Directorate of Fisheries, and include landings from catches taken in ICES statistical areas I, IIa, and IIb. In 2001, the export of fish products accounted for 14% of the total exports from mainland Norway (based on data from the Statistical Yearbook of Norway and information from the Norwegian Seafood Exports Council). The fisheries constituted 1.5% of the Norwegian Gross National Product in 1999, excluding petroleum. In northwest Russia, fisheries are of less economic importance nationally. A substantial share of the catches taken in Russian fisheries in the north is landed abroad.

Most northern coastal communities are heavily dependent on the fisheries in economic terms, as well as being culturally and historically attached to fisheries. As early as AD 1000 an extensive trade in dried cod had developed in northern Norway, through the Hanseatic trade (Solhaug, 1983). The coastal fishery and trade made up the economic foundation for the communities along the northern coast. Since the early 1980s, aquaculture has become increasingly important, accounting for a significant part of the economic value of the fisheries sector (Ervik et al., 2003).

The total fishery in the arctic Northeast Atlantic yields about 2.1 million t and has a total annual value of around US$ 2 billion. The resources occurring in the Arctic are also significant to fishery communities elsewhere. A substantial component of the catches in the Arctic is taken by fishers from outside the region, such as those from southern Norway and elsewhere in Europe.

13.2.5.1. Fish stocks and fisheries

Most of the Norwegian fish harvest is taken in the Norwegian EEZ (Fig. 13.2). Altogether, the waters under Norwegian jurisdiction cover around 2 million km^2 – more than six times the area of mainland Norway. The arctic fisheries occur in three main areas: the Barents Sea/Svalbard area, the north Norwegian coast, and around Jan Mayen.

In the Norwegian fisheries, northeast Atlantic cod is by far the most important stock in economic terms. The landed value was approximately US$ 350 million in 2000, but had declined to just below US$ 209 million

in 2002 (Fig. 13.4). The landed value of herring also increased considerably throughout the 1990s, to about US$ 205 million in 2002. The third most valuable species is northern shrimp, of which the landed value was approximately US$ 100 million in 2000, but had declined to about US$ 85 million by 2002. Other important fisheries include those for capelin, Greenland halibut, king crab (*Paralithodes camtschaticus*), haddock, and saithe. These fisheries are important to the processing plants along the coast, and so to the viability of coastal communities.

For the northwest Russian fishing fleet, northeast Atlantic cod is also the most important fish stock. Catches are taken in Russian as well as Norwegian waters. Since the early 1990s, most of the cod caught by Russian fishers in the Barents Sea has been landed abroad, primarily in Norway. Only small quantities of mainly pelagic fish have been landed in Russia from the Barents Sea in recent years. The share of the total catch from the Northeast Atlantic has however increased. The northwest Russian fishing fleet, previously engaged mainly in distant water fishing, now works in the immediate northern vicinity. While only 234000 t were taken in the Northeast Atlantic in 1990, catches have been over 500000 t in all years since.

The economic value of the commercial exploitation of marine mammals in Norway and Russia is of minor direct significance nationally and regionally. But since marine mammals are major consumers of commercial fish species, their harvest is seen as an important contribution to maintaining a balance in the ecosystem. The marine mammal fishery also has a long tradition. Archeological excavations and early historical records clearly show that whaling has been conducted since ancient times and that whales were exploited before AD 1000 (Haug et al., 1998). In the 17th century, British and Dutch whalers killed an annual average of 250 Greenland right whales in the arctic and subarctic regions. These whales were processed at shore stations along the west coast of Spitsbergen (Arlov, 1996; Hacquebord, 2001).

13.2.5.2. Fishing fleets and fishers

The fishing fleet in northern Norway consists of around 1250 vessels operating on a year-round basis (Fiskeridirektoratet, 2002b). More than half are small vessels of 13 m or less. The fleet has been considerably reduced since the early 1970s. Small vessels fishing with conventional gear such as nets, lines, and jigs dominate. A large part of the fishery therefore occurs close to shore and in the fjords. Larger coastal vessels are ocean going. Trawlers and purse-seiners dominate the offshore fisheries. The vessels are required to carry a license to fish, and also need a fish quota to be admitted to a particular fishery. There are almost no open access fisheries in Norwegian waters. Most coastal communities have a number of vessels attached to them.

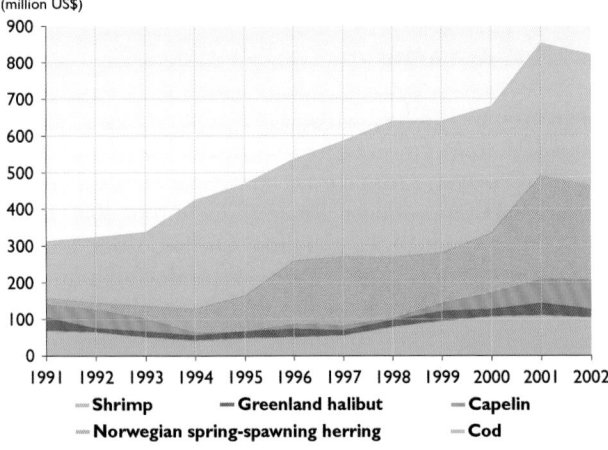

Fig. 13.4. Nominal value of the landings in Norway from the arctic fisheries, 1991–2002 (data from the Fisheries Directorate, Bergen, Norway).

Northwest Russian fisheries include a variety of fishery-related activities and participants. They are based in Murmansk and Arkhangelsk Oblasts, and in the Republic of Karelia (Hønneland and Nilssen, 2000; Nilssen F. and Hønneland, 2001). Most of the activity is located in the city of Murmansk, where most vessel owners, fish processing plants, and management authorities have their premises. The association of fishing companies in "the northern basin" of the Soviet Union, *Sevryba* ("North Fish"), was founded in 1965 and given the status of General Directorate of the Soviet Ministry of Fisheries in Northwestern Russia. Sevryba was made a private joint-stock company in 1992. The majority of the approximately 450 fishing vessels located in northwestern Russia are controlled by a handful of fishing companies (referred to henceforth as the "traditional" companies). The rest are distributed between *kolkhozy* (fishing collectives) and private fishing companies (referred to henceforth as the "new" companies). The total number of vessels has been stable since the early 1990s: few old vessels have been taken out of service and few new vessels have been purchased (Hønneland, 2004).

The "traditional" fishing companies are a legacy from the Soviet period. This fleet mainly consists of medium-sized (50 to 70 m) and large (over 70 m) vessels, and has around 250 to 300 ships. Before the dissolution of the Soviet Union, their main activity was the exploitation of pelagic species in distant waters and fisheries in the northern Atlantic Ocean. These companies now deploy a fleet of mid-sized factory trawlers for fishing and processing codfish. The collective fleet is significantly smaller in number, with some 80 to 100 vessels. Nearly all are of medium size (50 to 70 m). The fishing collectives are more diversified than other companies. Like the traditional companies, the collectives also aim at upgrading their fleet. The "new" companies (including the so-called coastal fishing fleet) have the smallest fleet, both in number and vessel size, limiting the range of the vessels and so the markets for the sale of the fish. The fleet comprises around 100 vessels, including around 30 coastal fishing vessels of less than 50 m in length.

The Russian perception of "coastal fishing" differs from that in neighboring countries. While a Norwegian "coastal" fishing vessel normally has a small crew and goes to port for daily delivery of catches, a northwest Russian "coastal" fishing vessel has a crew of more than a dozen and stays at sea for weeks before landing the catch. The reasons for this are two-fold. The fishing industry that was developed during the Soviet period was based on large-scale fishing and processing. Traditions, skills, and infrastructure for small-scale coastal fisheries are therefore non-existent in the main fishing regions of the Russian Federation. In addition, fish stocks for developing a viable coastal fishery are not available. Also, the financial status of the fishing companies is an obstacle to the development of coastal fisheries (Hønneland, 2004).

13.2.5.3. The land side of the fishing industry

More than 90% of the fish landed in Norway – by Norwegian, Russian, and other countries' vessels – is exported. Changes in the international market for fish and fish products may thus have substantial effects on the processing plants as well as on the rest of the industry. Many fish processing plants are heavily dependent on landings by Russian vessels. In 2001, around 70% of the Russian cod quota was landed in Norway. This percentage has since decreased, with the increase in landings in other countries and trans-shipments in the open ocean. The fishing industry, especially the fillet-producing plants, has experienced low profitability and an increasing number of bankruptcies in recent years (Bendiksen and Isaksen, 2000). Increased competition for raw materials and high production costs in Norway help to explain the problems. In addition, the advantage of the Norwegian industry has been its location near the resources. New freezing and defrosting technologies, and infrastructure developments that make frozen products more valuable (Dreyer, 2000), reduce the advantage of proximity to the resource.

There are around 170 fish processing plants in northern Norway (Roger Richardsen, Fiskeriforskning, pers. comm., 2002 data). The size of the plants varies substantially. Most are engaged in producing traditional white-fish products, for example dried cod, salted fish, and stockfish. In Finnmark, a relatively large proportion of the plants concentrate on fillet production, while the shrimp industry is more important in Troms (NORUT, 2002). In Nordland, both fillet and traditional production is important.

Before the dissolution of the Soviet Union, Murmansk had the largest fish processing plant of the entire Union. Since fishing in distant waters has been reduced and catches from northern waters landed abroad, activities at the fish processing plants in Murmansk have been drastically reduced. The production of consumer products fell from 83 300 t in 1990, to 10 100 t in 1998 (Nilssen F. and Hønneland, 2001). Processing of fish outside Murmansk is insignificant.

13.2.5.4. Aquaculture

Since around 1980, Atlantic salmon (*Salmo salar*) and trout (*Oncorhynchus mykiss*)-based aquaculture has developed in Norway, making this country the world's biggest farmed salmon producer. Total production in 2000 was 485 000 t, worth US\$ 1.6 billion. Of this, around 145 000 t of salmon and trout were produced in northern Norway, at a production (i.e., before sales) value of approximately US\$ 470 million. This makes salmon the single most important species in terms of economic value, both in northern Norway and in the Norwegian fishing industry as a whole.

In 2000, there were 854 licenses for salmon and trout production in Norway, of which some 30% were for sites located in the three northern counties (Fiskeri-direktoratet, 2001). The number of plants and sites in northern Norway is expected to increase considerably in the future (Hartvigsen et al., 2003). In addition to salmon, this development will also involve other fish species such as Atlantic halibut (*Hippoglossus hippoglossus*) and cod. Over time, aquaculture is expected to become more important to the north Norwegian economy than the combined marine fisheries.

An important aspect of the aquaculture industry is that it is dependent on a huge supply of pelagic fish species. Fishmeal and oils are important components of the diet of many species of farmed fish, including salmon and trout. The quantity needed is so high that the industry at a global level is sensitive to rapid fluctuations in important pelagic stocks. El Niño–Southern Oscillation (ENSO) events in the Pacific have already affected the industry through impacts on anchovy (*Engraulis* spp.) stocks. From 1997 to 1998, the global marine fishery was reduced by nearly 8 million t, mainly due to ENSO events (FAO, 2000). Reduced supply on the international market led to increased prices of fishmeal in this period. The latest assessment by the Intergovernmental Panel on Climate Change (IPCC, 2001) states that unless alternative sources of protein are found, aquaculture could in the future be limited by the supply of fishmeal and oils.

Aquaculture is in its infancy in northwest Russia and the total production is negligible. It is however likely to increase in the future.

13.2.5.5. Employment in the fisheries sector and the fisheries communities

There are approximately 17 000 fishers in Norway, of which almost half live in the three northern counties. In northern Norway it is common to combine fishing with other trades to make a living, particularly in remote areas. Part-time fishers make up about a third of the total number of people in the profession. The number of fishers has been sharply reduced over recent decades. This reflects broader societal changes with a shift in the workforce from primary to secondary and tertiary occupations, as well as technological development in the industry. A total of 12 420 persons worked in fish processing in Norway in 2000 (Ministry of Fisheries, 2002). About half of these worked in the northernmost counties.

In 2001, around 3600 people worked in aquaculture in Norway (Ministry of Fisheries, 2002). Of these about a third worked in the three northernmost counties. The combined direct employment in the fisheries sector in northern Norway is 16 000 to 17 000 people. The fisheries also generate substantial employment in related activities, such as shipbuilding, ship repairs, and gear production, as well as sales and exports. The number of people employed in the related industries has increased substantially over recent decades. The employment generated in related industries by the fish-

eries sector is 0.75 man-years per year in the fisheries (KPMG and SINTEF, 2003), amounting to some 12 000 people in northern Norway. The total employment generated is therefore close to 30 000 people. With a total population in northern Norway of 460 000, this implies that the fisheries are crucial to employment and income in the region.

Corresponding data on employment in the fisheries sector for northwest Russia were not available.

According to Lindkvist (2000) there are 96 communities in Norway that can be characterized as fishing communities. Of these, 42 occur in the three northern counties. Of these, 31 may be defined as fisheries-dependent in the sense that more than 5% of the working population is employed in fisheries and fish processing (Lindkvist, 2000). These communities are typically small and located in remote areas. Most face depopulation and problems such as lack of qualified personnel to maintain public services, but at the same time have few alternative trades to fishing. In Finnmark county, about 10% of the total employment is in the fisheries sector (Hartvigsen et al., 2003). Remote, fisheries-dependent communities in northern Norway have the highest depopulation rates in the country. Since the 1980s, none of its municipalities have increased in population. On average the coastal municipalities have experienced a population reduction of around 30% (Hartvigsen et al., 2003).

Demographic pressure towards urbanization, which is expected to continue (IPCC, 2001), may be said to be one of the major driving forces behind this development. Other factors, such as lack of employment opportunities and inferior public services, may be seen both as a cause of the problem as well as a consequence. There is also the trend of fishing boats being sold out of the communities. These trends indicate that the small fishery-dependent societies are under continuous pressure. These societies are subject to a "double exposure" (O'Brien and Leichenko, 2000), where climate change occurs simultaneously with economic marginalization. The Norwegian government has for a long period run programs aimed at strengthening the viability of fishery-dependent societies in the north. In recent years these efforts have been directed towards market orientation, flexibility, and a more robust industrial structure, rather than towards subsidies to the industry. Some regional development programs are aimed at diversification of the economic activity in remote areas by supporting, among other things, female-run enterprises (Lotherington and Ellingsen, 2002).

Among the Russian Federation subjects in the northwest, the Murmansk Oblast is most important from the point of view of fisheries. This region is one of the most urbanized in Russia, with around 92% of the population living in cities and towns. Most of the northwest Russian fishing fleet is concentrated in the city of Murmansk. Some companies are located in the three other Russian Federation subjects: Arkhangelsk (Arkhangelsk Oblast),

Petrozavodsk (Republic of Karelia), and Narjan-Mar (Nenets Autonomous Okrug).

The fishing industry is important for several major cities in northwestern Russia, but these cannot be characterized as "fishing communities" in the sense that this concept is understood in the West. Their viability is not dependent on fisheries. Also, the significance of the fishing industry has been severely reduced in the post-Soviet period as the catches of Russian vessels are mainly delivered to the West. The redirection of landings to the home market has been one of the main ambitions of Russian fishery authorities at both the federal and regional level since the early 1990s. That this has not been achieved points to the relative impotency of these bodies. At the federal level, the State Committee for Fisheries has twice lost its status as an independent body of governance (subsumed into the Ministry of Agriculture in 1992–1993 and 1997–1998) and seen its traditional all-embracing influence over fisheries management significantly reduced. In 2000, the Ministry of Trade and Economic Development succeeded in introducing a system for quota auctions, against the will of the State Committee for Fisheries. Regional authorities increased their influence during the 1990s. This development has now been reversed owing to the re-centralization that began around 2000, commensurate with wider developments in Russia since President Putin came to power. Hence, while regional authorities in northwestern Russia have a declared aim of developing coastal fisheries, actual development in this sphere can only be considered minimal.

13.2.5.6. Markets

All data in this section are from the Norwegian Seafood Export Council (http://www.seafood.no).

Norway is one of the worlds biggest fish exporters – more than 90% of the landings are exported (in 2001 Norway was the world's second largest fish exporter, after Thailand). There are two aspects to this. First, the income generated by fish exports is substantial – around US$ 4 billion in 2001. As the production in aquaculture will increase, and the production of petroleum will decrease, exports of fish products can be expected to become more important in the future. The Ministry of Fisheries envisages that aquaculture will become a mainstay of the Norwegian economy in the years to come, and that the sales value in northern Norway will be nearly five times higher in 2020 than today. Second, Norway is a major supplier to many markets. The Norwegian imports are important to, for example, the EU market for seafood, which is therefore vulnerable to fluctuations in Norwegian fisheries.

The single most important species in terms of export value is salmon, which had an export value of US$ 1.8 billion in 2000. The second most important category is whitefish, the exports of which (consisting mainly of cod, haddock, and saithe) are worth in the range of

US$ 1.2 billion annually. Pelagic species, of which herring is the most important, had an export value of US$ 920 million in 2001. The fourth most important species in terms of export value is northern shrimp.

Landings of Russian-caught cod in Norway have increased since 1990. During 1995 to 1997, landings were around 250000 to 300000 t per year. Since then, there has been a reduction in Russian landings of cod as well as other fish in Norway. Trans-shipments of fish at sea and landings in other countries are increasing while landings in Norway are decreasing. Catches landed in Russia mostly go to the Russian consumer market. Imports of fish to Russia from Norway are rapidly increasing.

13.2.5.7. The management regime

In addition to the EEZ, Norway also manages the resources in the Fishery Zone around Jan Mayen and in the Fishery Protection Zone around Svalbard. The Norwegian EEZ borders the EU zone to the south, the Faroe Islands to the southwest, and Russia to the east. A large area beyond the EEZ boundary in the Norwegian Sea and a smaller area in the Barents Sea are international waters. Most of the economically important stocks move between the zones of two or more states.

Cooperation between the owner countries in the management of these stocks is essential to ensure their sustainable use. A series of agreements has been negotiated among the countries in the Northeast Atlantic that establish bilateral and multilateral arrangements for cooperation on fisheries management. The most extensive management regime on arctic stocks in the Northeast Atlantic is that between Norway and Russia. A joint fisheries commission meets annually to agree on TACs and the allocation for the major fisheries in the Barents Sea: i.e., those for cod, haddock, and capelin (since 2001 a total quota has also been set for the king crab fishery). The total quotas set are shared between the two countries – the allocation key is 50-50 for cod and haddock, and 60-40 for capelin. A fixed additional quantity is traded to third countries. There are also agreements on mutual access to the EEZs and exchange of quotas through this arrangement (Hoel, 1994). An important aspect of the cooperation with Russia is that a substantial part of the Russian harvest in the Barents Sea is taken in the Norwegian zone and landed in Norway. The cooperation also entails joint efforts in fisheries research and in enforcement of fisheries regulations.

Despite disagreement between Norway and Russia on the delimitation of the boundary between their EEZ and the shelf in the Barents Sea, the cooperation on resource management between the two countries may generally be characterized as well functioning (Hønneland, 1993). However, agreed TACs by Norway and Russia have, in some years, exceeded those recommended by fisheries scientists. In addition, the actual catches have sometimes been larger than those agreed. Since the late 1990s, a precautionary approach has been gradually implemented in the management of the most important fisheries. However, retrospective analyses have shown that ICES estimates of stock sizes have often been too high, thereby incorrectly estimating the effect of a proposed regulatory measure on the stock. This has had the unfortunate effect that stock sizes for a given year are adjusted downward in subsequent assessments, rendering adopted management strategies ineffective (Korsbrekke et al., 2001; Nakken, 1998). However, the Joint Norwegian–Russian Fisheries Commission has decided that from 2004 onward multi-annual quotas based on a precautionary approach will be applied. A new management strategy adopted in 2003 shall ensure that TACs for any three-year period shall be in line with the precautionary reference values provided by ICES.

A number of other agreements are also in effect in the area, notably a five-party agreement among the coastal states in the Northeast Atlantic to manage Atlanto-Scandian herring (Ramstad, 2001). Total quotas for the following year's herring fishery are set, and divided among the parties. A separate quota is set for the area on the high-seas in the Norwegian Sea. The high seas quota, most of which is given to the same coastal states, is formally managed by the NEAFC, which is mandated to manage the fishing on the high seas in the Northeast Atlantic. Norway also has an extensive cooperation with the EU on the management of shared stocks in the North Sea, as well as on the exchange of fish quotas, which entails access for EU vessels to north Norwegian waters. The EU is given a major share of the third country quota of cod in the Norwegian waters north of 62° N.

Management measures for marine mammals harvested in the area are decided by the IWC, NAMMCO, and the Joint Norwegian–Russian Fisheries Commission. The IWC has not been able to adopt a Revised Management Scheme and so does not set quotas. Since 1993, Norway has set unilateral quotas for the take of minke whales, on the basis of the work of the IWC Scientific Committee (Hoel, 1998). NAMMCO adopts management measures for cetaceans and seals in the northern Northeast Atlantic (Hoel, 1993).

A precondition for sound management of living marine resources is that sufficient knowledge about the resources is available. In Norway, the Institute of Marine Research is the main governmental research institution, while the Northern Institute of Marine Research (PINRO) plays the same role on the Russian side. ICES is the international institution for formulating scientific advice to the fisheries authorities in the North Atlantic countries. Its work is generally based on inputs from the research institutions in the member countries. The ICES advice is now based on a precautionary approach, which seeks to introduce a greater sensitivity to risk and uncertainty into management. Three of the challenges for fisheries management in the future are: a better understanding of species interac-

tions (multi-species management), more reliable data from scientific surveys, and a better understanding of the impact of physical factors – such as changing climatic conditions – on stocks. A major challenge is the development and implementation of an ecosystem-based approach to the management of living marine resources, where the effects of climate change are also considered when establishing management measures.

The management measures essentially fall into three categories:

- input regulations in the form of licensing schemes restricting access to a fishery;
- output regulations, consisting of the fish quotas given to various groups of fishers which limit the amount of fish they are entitled to in any given season; and
- technical measures specifying for example the type of fishing gear to be used in a particular fishery.

The objectives of fisheries management in Norway are related to conservation, efficiency, and regional considerations (Report to Parliament, 1998). Conservation of resources is seen as a precondition for the development of an efficient industry and maintenance of viable fishing communities. An important objective of the fisheries policy is to improve the economic efficiency of the industry. An important issue is therefore to reduce the capacity of the fishing fleet, which is much larger than needed to take the quotas available and therefore makes the costs of fishing too high. Attempts to remove excess capacity include scrapping of vessels, regulatory mechanisms, and vessel construction regulations. A quota arrangement allowing for merging two vessels' quotas while removing one of the vessels from the fishery gives vessel owners an incentive to remove excess fishing capacity, and can contribute to a more efficient fleet. However, this can result in coastal communities seeing their local fleet reduced or even disappearing, threatening the viability of that community.

The enforcement of the fisheries regulations in Norway is carried out both at sea and when the fish is landed. At sea, the Coast Guard is responsible for inspecting fishing vessels and checking their catch against vessel logbooks. Foreign vessels fishing in Norwegian waters are also inspected. The activity of the Coast Guard is vital for the functioning of the management regime as a whole. Ocean-going vessels are required to install and use a satellite-based vessel-monitoring system enabling the authorities to continually monitor their activities. The Directorate of Fisheries also inspects activities on the fishing grounds, as well as at the landing sites. When fish is landed, the sales organization buying the fish reports the landed quantity to the Fisheries Directorate, which is responsible for maintaining the fisheries statistics.

The regulation of Soviet fisheries in the Northeast Atlantic used to be the responsibility of the Sevryba association. As this organization lost its status in fisheries regulation in the mid-1990s, the regulatory tasks were partly taken over by the enforcement body Murmanrybvod, partly by the fisheries departments of regional authorities in each federal subject in the area, and since 2000 to an increasing extent the regulatory tasks have been the remit of federal authorities. During the 1990s, the Russian share of the Barents Sea quotas was first divided among the four federal subjects of the region by the so-called Scientific Catch Council (formerly headed by Sevryba, since 2001 by the federal State Committee for Fisheries). Within each federal subject, a Fisheries Council (led by regional authorities) distributed quota shares among individual ship owners. The influence of both the Scientific Catch Council and the regional Fisheries Councils was reduced after the introduction of quota auctions in 2000/2001. Since then, an increasing share of the quotas has been sold at auctions, administered by the federal Ministry of Trade and Economic Development. In November 2003, the Russian Government decided to abolish the auctions and instead introduce a resource rent (a fee on quota shares). The quotas will from 2004 be distributed by an inter-ministerial commission at the federal level, so the regional authorities will also lose the influence of inter-regional quota allocation (Hønneland, 2004).

Apart from quotas, the Russians have fishery regulations similar to those in the Norwegian system: regulations pertaining to fishing gear, size of the fish, and composition of individual catches. In addition, the Russians have a more fine-meshed system than the Norwegians for closing and opening of fishing grounds. Individual inspectors from the enforcement body Murmanrybvod or researchers from the scientific institute PINRO can close a "rectangle" (a square nautical mile) on site for a period of three days. After three days, the "rectangle" is reopened if scientists make no objections, i.e., if the proportion of undersized fish in catches does not continue to exceed legal limits.

Traditionally, the civilian fishery inspection service Murmanrybvod, subordinate to the Russian State Committee for Fisheries, has been responsible for enforcing Russian fishery regulations in the Barents Sea. In 1998, responsibility for fisheries enforcement at sea in the Russian Federation was transferred to the Federal Border Service. In the northern fishery basin, the Murmansk State Inspection of the Arctic Regional Command of the Federal Border Service was established to take care of fisheries enforcement. However, this body is only responsible for physical inspections at sea, while inspection of landed catches has been transferred to the Border Guard. Murmanrybvod is still in charge of keeping track of how much of the quotas has been caught by individual ship owners at any one time. It has also retained its responsibility for the closing of fishing grounds in areas with excessive intermingling of undersized fish, a very important regulatory measure in both the Russian and Norwegian part of the Barents Sea. Finally, Murmanrybvod is still responsible for

enforcement in international convention areas. In practice, Murmanrybvod places its inspectors on board northwest Russian fishing vessels that fish in the NEAFC or NAFO areas.

The reorganization of the Russian enforcement system is generally believed to have led to a reduction in the system's effectiveness, at least from a short-term perspective. For example, officers in the Murmansk State Inspection of the Federal Border Service generally lack experience in fisheries management and enforcement. This has partly been compensated for by the transfer of some of Murmanrybvod's inspectors. More apparent is the lack of material resources to maintain a presence at sea. Contrary to the intentions of the reorganization of the enforcement system, the presence at sea by monitoring vessels has declined since the Border Guard took over this duty in 1998. Precise data for presence at sea and inspection frequency are not available, but Jørgensen (1999) estimated that the Border Guard performed around 160 inspections at sea in 1998, which represents a significant reduction compared to an estimated 700 to 1000 annual inspections at sea by Murmanrybvod prior to the reorganization. For periods of several months during 1998, not a single enforcement vessel was present on the fishing grounds in the Russian part of the Barents Sea. Officials of the Border Service explain this by a lack of funds to purchase fuel. Critics question the genuineness of the Border Service's will to play a role in fisheries management. The result of the reorganization has, in any event, so far led to a tangible reduction in the effectiveness of Russian enforcement in the Barents Sea.

13.2.6. Economic and social impacts of climate change on fisheries in the Northeast Atlantic

The economic importance of fisheries to northern Norway is substantial, cod being the most significant species. Problems related to profitability in the fishing industry have been evident for a long time, and have contributed to depopulation problems in remote, fishery-dependent areas. Aquaculture is, however, a growing industry and is expected to be important to the future viability of local communities in northern Norway. In northwest Russia, the fishing industry is based in big cities, Murmansk in particular, and is therefore not as significant to local communities as it is in Norway.

A study by Furevik et al. (2002) developing regional ocean surface temperature scenarios for the Northeast Atlantic concluded that for the 2020 scenario, no substantial change is likely in the physical parameters. The authors concluded that a slight cooling in ocean surface temperature is likely by 2020 with warming likely in the longer-term scenarios. For the near-term future, climate change is therefore not likely to have a major impact on the fisheries in the region. Uncertainties surrounding these scenarios are however considerable. These are amplified when the physical effects on biota are included, and amplified again when the effects of climate change on society are added. In addition, social change is driven by a vast number of factors, of which climate change is only one. The rest of this section is therefore tentative and should be read more as discussions of likely patterns of change than predictions of future developments.

The effects of climate change are closely related to the vulnerability of industries and communities, and to their capability to adapt to change and mitigate the effects of change. Within this context vulnerability is defined as "the extent to which a natural or social system is susceptible to sustaining damage from climate change" (IPCC, 2001). It depends on the ability and capacity of society at the international, national, and regional level to cope with change and to remedy its negative effects. Climate change may also result in positive changes.

The fisheries sector is one in which the industry has always had to adapt to and cope with environmental change: the abundance of various species of fish and marine mammals has varied throughout history, often dramatically and also within short periods of time. Adapting to changing circumstances is therefore second nature to the fishing industry as well as to the communities that depend upon it. An important issue is thus whether climate change brings about changes at scales and rates that are unknown, and whether adaptation can be achieved within the existing institutional structures.

13.2.6.1. Resource management

Resource management is the key factor in deciding the biological and economic sustainability of the fisheries. The fishing opportunities are decided by the management regime. There are virtually no remaining fisheries where the economic result is decided by the industry itself. The design and operation of both the domestic and international management regimes are crucial to the sustainability and economic efficiency of the fisheries, and hence to the economic viability of the communities that depend upon them. The development and implementation of a precautionary approach, as well as the emergence of ecosystem-based management, may enhance the resilience of the stocks and therefore make the industry and communities more robust to future external shocks. As discussed in section 13.2.5.7, the main arrangements for managing living marine resources in the Northeast Atlantic are being modified in this direction, with the implementation of a precautionary approach and the development of an ecosystem-based approach to management.

A major challenge for the management regime is that of adjusting to the possible changes in migration patterns of stocks resulting from climate change. This finding is in conformity with that of the IPCC (2001) and Everett et al. (1996). Changes in migration patterns of fish stocks have previously upset established arrangements for resource management, and can trigger conflicts between countries. One example is that of

northeast Atlantic cod: in the early 1990s, the stock extended its range northward in the Barents Sea, into the high seas in the area (the so-called "loophole"). Vessels from a number of countries without fishing rights in the cod fishery took the opportunity to initiate an unregulated fishery in the area, thereby undermining the Norwegian–Russian management regime. This triggered a conflict between Norway and Russia on the one hand, and Iceland on the other. The conflict was later resolved through a trilateral agreement (Stokke, 2001). Another example is that of the Norwegian spring-spawning herring (Box 13.1): following more than two decades of effort at rebuilding the stock on the part of Norwegian authorities, in the mid-1990s the stock began to migrate from the Norwegian EEZ and into international waters for parts of the year. By doing so the stock became accessible to vessels from other countries, and in the absence of an effective management regime for the stock in the high seas, efforts at rebuilding the stock could prove futile. A regime securing a management scheme for the stock eventually came into place, but took several years to negotiate (Box 13.1). Thus, changes in migration patterns, which are likely to be triggered by changes in water temperatures, tend to result in unregulated fishing and conflicts among countries. The outcome of such conflicts may be conflicting management strategies, new distribution formulas, or even new management regimes.

Another important factor is that negative events tend to be a liability to the management regime. The so-called "cod crisis" in the late 1980s, for example, led to several modifications of the existing regime. The management regime is likely to be held responsible for social and economic consequences of climate change. This may in turn affect the legitimacy and authority of the regime, and its effectiveness in regulating the industry. An important aspect in that regard is the way decisions about resource management and allocation of resources are made. A regime that involves those interests that are affected by decisions in the decision-making processes tends to produce regulations that are considered more legitimate than regimes that do not involve stakeholders (Mikalsen and Jentoft, 2003).

Current fisheries management models are mainly based on general assumptions of constant environmental factors. The current methods applied in fisheries management can not accommodate environmental changes. A study by Eide and Heen (2002) investigated the economic output from the fisheries under different environmental scenarios and under different management regimes for the cod and capelin fisheries in the Barents Sea. Using the ECONMULT fleet model (Eide and Flaaten, 1998) and a regional impact model for the north Norwegian economy (Heen and Aanesen, 1993), they concluded that even a narrow range of management regimes has a variety of possible economic outcomes. Even though climate change may result in significant potential effects on catches, profitability, employment, and income, changes in the management regimes seem to have an even larger impact. This conclusion sets the discussion of effects of global climate change in perspective. It implies that a large number of factors influence the economic activities and their output and, furthermore, that the operation of the management regime seems to be the most significant of these factors.

The crucial factor for resource management under conditions of climate change is therefore the development of robust and precautionary approaches and institutions for managing the resources. The decisive factor for the health of fish stocks, and therefore the fate of the fishing industry and its dependent communities, appears to be the resource management regime.

13.2.6.2. The fishing fleet

The ability to adapt to changes in migration patterns or stock size of commercially exploited species will vary between different vessel groups in the fishing fleet. The ocean-going fleet is capable of adjusting to changes in migration patterns, as it has a wide operating range. Small coastal vessels are more limited in that regard. Thus, northern communities with a strong dependency on small coastal vessels are likely to be more affected if migration patterns and availability of important fish stocks change significantly. If fish stocks move closer to the coast it is an advantage to the coastal fleet, while it is a disadvantage for this fleet if the stocks move more seaward. Such a development may be confounded by changing weather patterns with severe weather events becoming more prevalent. All vessel groups will be affected if changes lead to stocks crossing jurisdictional borders. That may imply a change in distribution of resources among countries.

Increased production and larger stocks of cod and herring are possible outcomes of climate change in the Northeast Atlantic. A question arises as to which fleet groups are most capable of making the best of such positive changes in the resource. Such changes may result in different availability of the resources between groups of fisheries (e.g., coastal versus ocean-going vessels), affecting the domestic allocation of resources. It may also lead to a greater political pressure to change the allocation of resources between the main groups of resource users.

Changes in stock abundance and migration patterns are not new to the industry. The availability of fish stocks and their accessibility to the coastal fleet has changed throughout recorded history, and the industry as well as the management regime is used to adapting to changing circumstances. The key question is whether climate change would amplify such variations and aggravate their effects beyond the scale with which the industry and the regulating authorities are familiar.

Changes in oceanic conditions may also affect the migrating ranges of marine mammals, and hence marine mammal–fisheries interactions. Such interactions could include marine mammals preying on fish, thus increasing competition with fishers, or marine mammals interacting directly with the fishery, for example by interfering with fishing gear. Marine mammals are also vectors of parasites that may affect fish and fisheries.

13.2.6.3. Aquaculture

Higher water temperature generally has positive effects on aquaculture in terms of fish growth. The IPCC reported that warming and consequent lengthening of the growing season could have beneficial effects with respect to growth rates and feed conversion efficiency (IPCC, 2001). Warmer waters may also have negative effects on aquaculture since the presence of lice and diseases may be related to water temperature. In recent years high water temperatures in late summer have caused high mortality at farms rearing halibut and cod, the production of which is still at a pre-commercial stage. Salmon is also affected by high temperatures and farms may expect higher mortalities of salmon. A rise in sea temperatures may therefore favor a northward movement of production, to sites where the peak water temperatures are unlikely to be above levels at which fish become negatively affected.

An increase in severe weather events can be a cause of escapes from fish pens and consequent loss of production. Escapes are also a potential problem in terms of the spread of disease. However, technological developments may compensate for this.

The aquaculture industry is dependent on capture fish for salmon feed. Climate change may cause a lack of and/or variability in the market for such products, but this is also an area where research may lead to the development of other feed sources.

13.2.6.4. The processing industry, communities, and markets

The fish processing industry in the north faces challenges in the structural changes both in the first-hand market (from fisher to buyer) and in the export market. Increased international competition for scarce resources has left the processing side of the industry increasingly vulnerable to globalization pressures. At the same time many of the communities, depending on fisheries for their existence, experience economic marginalization- and depopulation-related problems. The vulnerability of the fishing industry and fishing communities can therefore be considered as relatively high at the outset, rendering them particularly susceptible to any negative influences resulting from climate change. Such impacts may however be minor compared to that of other drivers of change. Furthermore, the fish processing industry is very varied. The size of fish processing plants is one aspect of this, their versatility and ability to vary production and adapt to changing circumstances is another. The ability of the particular type of industry to adapt to various earlier "crises", whether in terms of demand or supply failures, could be an indicator of their future "coping-capacity" for effects resulting from climate change. Another issue is that climate-induced changes elsewhere in the world may affect the situation for the north Norwegian fishing industry and fishing communities. Experience from, for example, the fisheries crisis in Canada in the 1990s indicates that such situations tend to intensify competition for further processing of the raw material. To the industry in Norway, with high labor costs, such a scenario is negative.

13.2.7. Ability to cope with change

Many factors contribute to a community's "coping capacity" in relation to depopulation and to structural changes in the fisheries sector (Baerenholdt and Aarsaether, 2001). The future of these settlements may depend on their ability to adapt to increased competition, efficiency, deregulation, and liberalization of the markets, as much as on the accessibility of fishing resources for their local production systems (Lindkvist, 2000).

While the management regime can be seen as an instrument to ease negative effects of climate change, it is however also important to consider public measures beyond the fisheries management regime that affect the conditions of the fishing industry more broadly, as for example regional policies and the development of alternative means of employment. Measures for building infrastructure such as roads or to develop harbor facilities are but one example. Government support for fisheries in the form of direct subsidies is now effectively prohibited by international agreements. But in Norway in particular there is a strong tradition for supporting regional development in a broader sense, and programs to this end may enhance the resilience of northern communities.

In addition to adapting to possible changes in the resource resulting from climate change, the fishing communities will also need to adapt to possible other climate-related changes in their vicinity (e.g., weather events) and their effects on terrestrial biota and infrastructure. These may have indirect effects on the fishery sector, related economic activities, or on other aspects of life, valued by the people in the respective communities.

13.2.8. Concluding comments

The Northeast Atlantic area comprises the northern and eastern parts of the Norwegian Sea to the south, and the north Norwegian coast and the Barents Sea to the east and north. The total fisheries in the area amounted to 2.1 million t in 2001. Aquaculture production is dominated by salmon and trout and amounted to 86 000 t in 2001. Norway and Russia have long traditions for co-operating both in trade and management issues. Since 1975, a comprehensive framework for managing the living marine resources in the area has been developed, covering also the areas on the high seas. While the Norwegian fishing industry is located in numerous communities all along the northern coast, the northwest

Russian fishing fleet is concentrated in large cities, primarily Murmansk.

Owing to the influence of the North Atlantic Current, the climate in this region is several degrees warmer than the average in other areas at the same latitude. Historically, a number of climate-related events have been observed in the Northeast Atlantic fisheries. Since the Second World War both cod and herring, the two major fish stocks in the area, have been subject to overfishing. This has resulted in a far lower average annual yield than these stocks have the potential to produce. Therefore, the management of stocks is the key issue in assessing the effects of potential climate variations on fish stocks.

Provided that the fluctuations in Atlantic water inflows to the area are maintained along with a general warming of the North Atlantic waters, it is likely that the annual average recruitment in herring and cod will be at about the long-term average during the first two to three decades of the 21st century. This projection is based on the assumption that harvest rates are kept at levels that maintain spawning stocks well above the level at which recruitment is impaired. How production will change further in the future is impossible to guess, since the projected temperatures, particularly for some global models, are so high that species composition and thus the interactions in the ecosystem may change completely.

Resource management is the key factor in deciding the biological and economic sustainability of the fisheries. The design and operation of both the domestic and international management regimes are therefore crucial in determining sustainability and economic efficiency. The development and implementation of a precautionary approach, as well as the emergence of ecosystem-based management, may enhance the resilience of the stocks and thus lessen the vulnerability of the industry to future external shocks. A large number of factors influence economic activities and their output, and an effective rational management regime seems to be the most significant of these. The crucial factor for resource management under conditions of climate change is therefore the development of robust and precautionary approaches and institutions for resource management.

13.3. Central North Atlantic – Iceland and Greenland

This section deals with the marine ecosystems of Iceland and Greenland. Although there are large differences, both physical and biological, between these two ecosystems there are also many similarities. Seafood exports represent a major source of revenue for both countries. Figure 13.5 shows the locations of the sites referred to most frequently in the text.

The waters around Iceland are warmer than those around Greenland due to a greater Atlantic influence and are generally ice free under normal circumstances. Exceptions are infrequent and usually last for relatively short periods in late winter and spring when drift ice may come close inshore and or even become landlocked off the north and east coasts. However, drift ice has been known to surround Iceland during cold periods, such as during the winter of 1918. Greenlandic waters are colder, sea-ice conditions more severe, and ports on the coastline commonly close for long periods due to the presence of winter sea ice and icebergs.

The reason for treating these apparently dissimilar ecosystems together is the link between the stocks of Atlantic cod at Iceland and Greenland. There is a documented drift of larval and 0-group cod (in its first year of life) from Iceland to Greenland with the western branch of the warm Irminger Current (Jensen, 1926). Spawning migrations in the reverse direction have been confirmed by tagging experiments (e.g., Hansen et al., 1935; Jónsson, 1996; Tåning, 1934, 1937). There are, however, large variations in the numbers of cod and other fish species, which drift from Iceland to Greenland and not all these fish return to Iceland as adults.

The history of fishing the waters around Iceland and Greenland dates back hundreds of years but is mainly centered on Atlantic cod, the preferred species in northern waters in olden times. Icelandic waters are usually of a cold/temperate nature and are therefore relatively species-rich. Consequently, with the diversification of fishing gear and vessel types in the late 19th century and the beginning of the 20th century, numerous other fish species, both demersal and pelagic, began to appear in catches from Icelandic waters. The Greenlandic marine environment is much colder and commercially exploitable species are therefore fewer. Present-day catches only comprise nine demersal fish species, two pelagic fish species, and three species of invertebrates. There is currently almost no catch of cod at Greenland.

Whale products feature in Icelandic export records from 1948 until the whaling ban (zero quotas) was implemented in 1986, but their value was never a significant component of exported seafood. Iceland has a long history of hunting porpoises, seals, and seabirds, and gathering seabird eggs for domestic use. Although this hunting and gathering gradually decreased with time, it is still a traditional activity in some coastal communities. For Greenland, several species of marine mammals (at least five different whale species, five species of seals, plus walrus) and six species of seabird are listed in catch statistics. Catches of marine mammals and seabirds are still important in Greenland, culturally and socially, as well as in terms of the local economy.

13.3.1. Ecosystem essentials

The marine ecosystem around Iceland is located south of the Polar Front in the northern North Atlantic (Fig. 13.5). The area to the south and west of Iceland is dominated by the warm and saline Atlantic water of the North Atlantic Current, the most important component being its westernmost branch, the Irminger Current (Fig. 13.5). The Irminger Current bifurcates off the northern west coast of Iceland. The larger branch flows west across the northern Irminger Sea towards Greenland. The smaller branch is advected eastward onto the North Icelandic shelf where the Atlantic water mixes with the colder waters of the East Icelandic Current, an offshoot from the cold East Greenland Current. On the shelf north and east of Iceland the

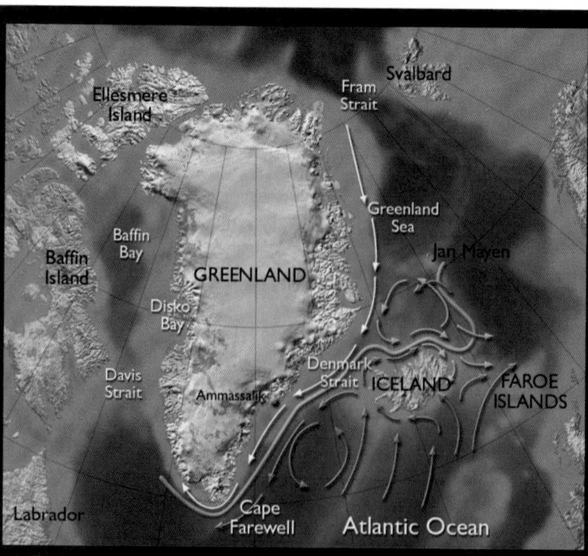

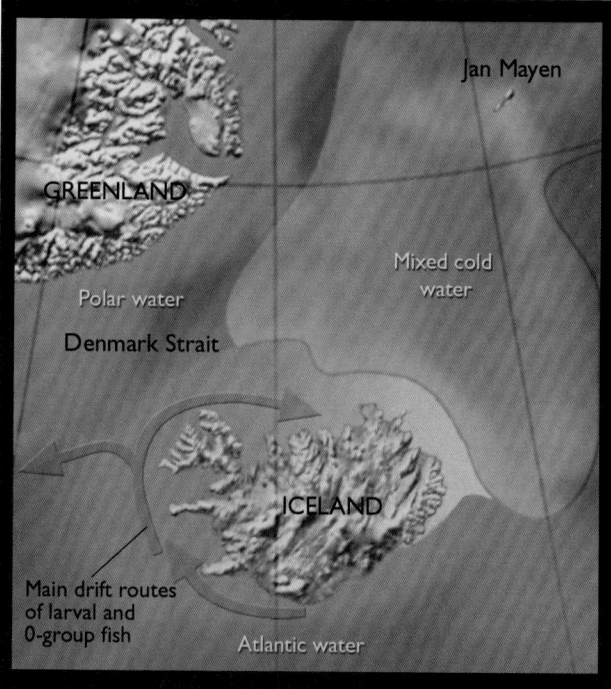

Fig. 13.6. The main water masses in the Iceland–East Greenland–Jan Mayen areas. The larval drift is driven by the two branches of the Irminger Current, which splits to the west of northwest Iceland (based on Stefánsson, 1999; Vilhjálmsson, 1994, 2002).

Polar Front
— Cold polar water
— Mixed cold water
— Warm water of the Irminger Current and the North Atlantic Drift
— Mixed cool water

Fig. 13.5. Location map for the Iceland/Greenland area. The arrows show the main surface ocean currents (based on Blindheim, 2004; Stefánsson, 1999).

degree of mixing increases in the direction of flow and the influence of Atlantic water is therefore lowest on the east Icelandic shelf as shown in Fig. 13.6. Hydrobiological conditions are relatively stable within the domain of the Atlantic water to the south and west of Iceland, while there may be large seasonal as well as interannual variations in the hydrography and levels of biological production in the mixed waters on the north and east Icelandic shelf (Anon, 2004b; Astthorsson and Gislason, 1995), depending on the intensity of the flow of Atlantic water and the proximity of the Polar Front. Large variations in the flow of Atlantic water onto the shelf area north of Iceland on longer timescales have also been demonstrated (Malmberg, 1988; Malmberg and Kristmannsson, 1992; Malmberg et al., 1999; Vilhjálmsson, 1997).

The East Greenland Current carries polar water south over the continental shelf off the east coast of Greenland and after rounding Cape Farewell (about 60° N; 43° W) continues north along the west coast. Off the east coast, the temperature of these cold polar waters may be ameliorated by the warmer Atlantic waters of the Irminger Current, especially near the shelf break and on the outer parts of the shelf (see Fig. 13.5). Off West Greenland, the surface layer is dominated by cold polar water, while relatively warm mixed water of Atlantic origin is found at depths between 150 and 800 m, north to about 64° N. Mixing and diffusion of heat between these two layers, as well as changes in the relative strength of their flow, are fundamental in determining the marine climatic conditions and the levels of primary and secondary production off West Greenland (e.g., Buch, 1993; Buch and Hansen, 1988; Buch et al., 1994, 2002).

The Irminger Current is also important as a transport mechanism for juvenile stages of various species of fish (Fig. 13.6). Thus, its eastern branch plays a dominant

role in transporting fish fry and larvae from the southern spawning grounds to nursing areas on the shelf off northwest, north, and east Iceland, while the western branch may carry large numbers of larval and 0-group fish across the northern Irminger Sea to East Greenland and from there to nursery areas in southern West Greenland waters. The main ocean currents in the Iceland/Greenland area are shown in Fig. 13.5.

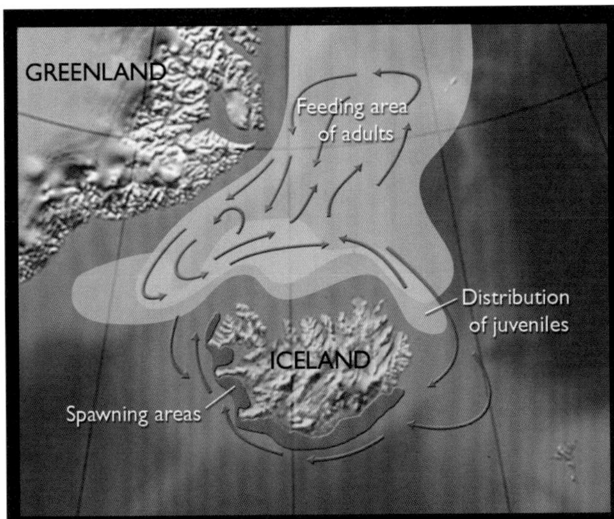

The Icelandic marine ecosystem contains large stocks of zooplankton such as calanoid copepods and krill, which are eaten by adult herring and capelin, adolescents of numerous other fish species, as well as by baleen whales. The larvae and juveniles of both pelagic and demersal fish also feed on eggs and juvenile stages of the zooplankton. Benthic animals are also important in the diet of many fish species, especially haddock, wolffish (*Anarhichas lupus lupus*), various species of flatfish, and cod.

Owing to the influence of warm Atlantic water, the fauna of Icelandic waters is relatively species-rich and contains over 25 commercially exploited stocks of fish and marine invertebrates. In contrast, there are only a few commercial fish and invertebrate species in Greenlandic waters (Muus et al., 1990) and these are characterized by cold water species such as Greenland halibut, northern shrimp, capelin, and snow crab. Redfish are also found, but mainly in Atlantic waters outside the cold waters of the East Greenland continental shelf and cod can be plentiful at West Greenland in warm periods.

Around Iceland, most fish species spawn in the warm Atlantic water off the south and southwest coasts. Larvae and 0-group fish drift westward and then northward from the spawning grounds to nursery areas on the shelf off northwest, north, and east Iceland, where they grow in a mixture of Atlantic and Arctic water (e.g., Schmidt, 1909). Larval and 0-group cod and capelin, as well as species such as haddock, wolffish, tusk (*Brosme brosme*), and ling (*Molva molva*) may also be carried by the western branch of the Irminger Current across to East Greenland and onward to West Greenland (e.g., Jensen, 1926, 1939; Tåning, 1937; see also Fig. 13.6). The drift of larval and 0-group cod to Greenland was especially extensive during the 1920s and 1940s.

Capelin is the largest fish stock in the Icelandic marine ecosystem. Unlike other commercial stocks, adult capelin undertake extensive feeding migrations northward into the cold waters of the Denmark Strait and the Iceland Sea during summer. The capelin return to the outer reaches of the north Iceland shelf in October/November from where they migrate to the spawning grounds south and west of Iceland in late December/early January (Fig. 13.7). Spawning is usually over by the end of March. Capelin are especially important in the diet of small and medium-sized cod (Pálsson, 1997). Most juvenile capelin aged 0, 1, and 2 years reside on or near the shelf off northern Iceland and on

Fig. 13.7. Distribution and migration of capelin in the Iceland–Greenland–Jan Mayen area (Vilhjálmsson, 2002).

the East Greenland plateau west of the Denmark Strait (Fig. 13.7). These components of the stock are therefore accessible to fish, marine mammals, and seabirds throughout the year. On the other hand, the summer feeding migrations of maturing capelin into the colder waters of the Denmark Strait and the Iceland Sea place the larger part of the adult stock out of reach of most fish, except Greenland halibut, for about five to six months. However, these capelin are then available to whales, seals, and seabirds. During the feeding migrations, adult capelin increase 3- to 4-fold in weight and their fat content increases from a few percentage points up to 15 to 20%. When the adult capelin return to the north Icelandic shelf in autumn they are preyed on intensively by a number of predators, apart from cod, until the end of spawning in the near-shore waters to the south and west of Iceland. Thus, adult capelin represent an enormous energy transfer from arctic regions to important commercial fish stocks in Icelandic waters proper (Vilhjálmsson, 1994, 2002).

Off West Greenland, northern shrimp and Greenland halibut spawn at the shelf edge off the west coast. This is also the case for the northern shrimp stock, which is found in the general area of the Dohrn Bank, about mid-way between East Greenland and northwest Iceland. Greenlandic waters also contain capelin populations that spawn at the heads of numerous fjords on the west and east coasts. These capelin populations appear to be self-sustaining and local, feeding at the mouths of their respective fjord systems and over the shallower parts of the shelf area outside these fjords (Friis-Rødel and Kanneworff, 2002). During the warm period from the early 1930s until the late 1960s there was also an extensive spawning of cod to the southeast, southwest, and west of Greenland (e.g., Buch et al., 1994).

In the pelagic ecosystem off Greenland the population dynamics of calanoid copepods and to some extent krill play a key role in the food web, being a direct link to fish stocks, baleen whales, and some important seabirds, such as little auk (*Alle alle*) and Brünnich's guillemot (*Uria lomvia*). But polar cod, capelin, sand eel (*Ammodytes* spp.), and squid (*Illex illecebrosus*) are probably the most important pelagic/semi-pelagic macrofauna acting as forage for fish such as Greenland halibut and cod, marine mammals, and seabirds. Benthic animals are also important. Northern shrimp is a major food item for Atlantic cod and many other species of fish and marine mammals (e.g., Jarre, 2002).

13.3.2. Fish stocks and fisheries

13.3.2.1. Atlantic cod

Historically, demersal fisheries at Iceland and Greenland fall into two categories: land-based fisheries conducted by local inhabitants and those of distant water foreign fleets. For centuries the main target species was cod. Until the late 19th century, the local fisheries were primarily conducted with open rowboats, while the distant

water fishing fleets consisted of much larger, decked ocean-going sailing vessels. Until the end of the 19th century, almost all fishing for demersal species, whether from small open rowboats or larger ocean going sailing vessels, was by hand lines.

Jónsson (1994) estimated that the combined landings by Icelandic, Dutch, and French fishing vessels were around 35 000 t per year for the period 1766 to 1777. One hundred years later, the combined French and Icelandic catches averaged about 55 000 t per year. From the subsequent development of fishing effort and knowledge of stock sizes and exploitation rates, it is obvious that even large fleets of several hundred sailing vessels and open rowboats, fishing with primitive hand lines, can not have had a serious effect on the abundant cod stock and other demersal species at Iceland.

This situation changed dramatically with the introduction of steam and combustion engines to the fishing fleet, and the adoption of active fishing gear at the turn of the 19th century. By the beginning of the 20th century the otter trawl had been adopted by the foreign fleet (e.g., Thor, 1992), while the smaller motor powered Icelandic boats began to use gill nets, long lines, and Danish seines. Landings from the Icelandic area were no longer almost exclusively cod, but species such as haddock, halibut, plaice (*Pleuronectes platessa*), and redfish (*Sebastes marinus*) also became common items of the catch. The demersal catch at Iceland is estimated to have increased from about 50 000 t in the 1880s to about 160 000 t in 1905, reaching 250 000 t just before the First World War. Although cod was still the most important species, the proportion of other demersal species landed had increased to about 30% (Fig. 13.8).

With the increasing effort and efficiency of the international distant water and local fishing fleets, cod catches in Icelandic waters increased to peak at 520 000 t in 1933, while the catch of other demersal species increased to about 200 000 t (Fig. 13.8).

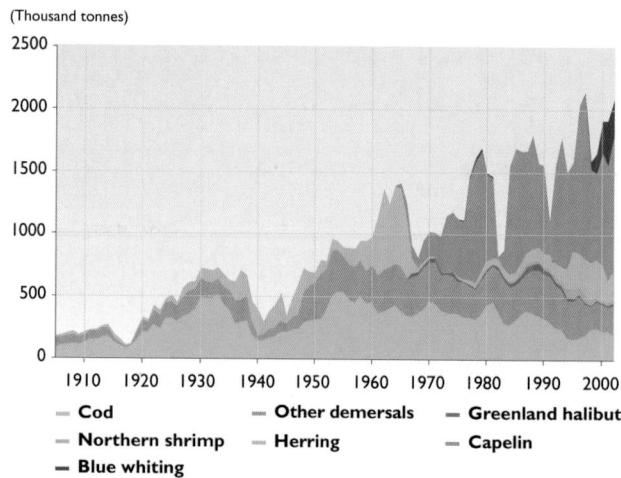

(Thousand tonnes)

Fig. 13.8. Total catch from Icelandic fishing grounds, 1905–2002 (data from the Icelandic Directorate of Fisheries and the Marine Research Institute).

Catches declined during the late 1930s, while the exploitation rate increased until the fishing effort fell drastically due to the Second World War. Nevertheless, the exploitation rate of cod remained at a moderate level due to recruitment from the superabundant 1922 and 1924 year classes (Schopka, 1994). After the Second World War, catches of demersal fish from Icelandic grounds increased again. Landings peaked at about 860 000 t in 1954, with cod accounting for about 550 000 t (Fig. 13.8). Because of the very strong 1945 cod year class and good recruitment to other demersal stocks, the exploitation rate of cod and other demersal species remained at a low level, although almost 50% higher than during the late 1920s and early 1930s. From 1955, the exploitation rate of all demersal stocks at Iceland, but especially that of cod, increased rapidly and with few exceptions has since been far too high. Until 1976, this was due to the combined effort of Icelandic and foreign distant water fleets. However, since the extension of the Icelandic EEZ to 200 nautical miles in 1977, the high rate of fishing has continued due to the enhanced efficiency of Iceland's fishing fleet.

Although cod has been fished intermittently off West Greenland for centuries, the success of the cod fishery at Greenland has been variable. Despite patchy data from the 17th and 18th centuries, there is little doubt that cod abundance at West Greenland fluctuated widely (e.g., Buch et al., 1994). Information from the 19th century suggests that cod were plentiful in Greenlandic waters until about 1850. After that there seems to have been very few cod on the banks and in inshore waters off Greenland until the late 1910s to early 1920s, when a small increase in the occurrence of cod in inshore areas was noted (Hansen, 1949; Jensen, 1926, 1939). Cod were also registered in offshore regions off West Greenland in the late 1920s, where fisheries by foreign vessels expanded quickly and catches increased from about 5000 t in 1926 to 100 000 t in 1930. From then until the end of the Second World War in 1945, this fishery yielded annual catches between about 60 000 and 115 000 t (Fig. 13.9). The total cod catch reached

about 200 000 t by 1950 and then fluctuated around 300 000 t between 1952 and 1961. After that the cod catch increased dramatically and landings varied from about 380 000 to 480 000 t between 1962 and 1968. By 1970, the catch had fallen to 140 000 t and was, with large variations, within the range 10 000 to 150 000 t until the early 1990s (Fig. 13.9). Since 1993, almost no Atlantic cod has been caught in Greenlandic waters. Before the introduction of the 200 nm EEZ around Greenland in 1978 the cod fishery was mostly conducted by foreign fleets, but since then the Greenlandic fleet has dominated the fishery.

13.3.2.2. Greenland halibut

An Icelandic Greenland halibut fishery began in the early 1960s (Fig. 13.8). Initially, long line was the main fishing gear but this method was abandoned because killer whales (*Orcinus orca*) removed more than half the catch from the hooks. Since the early 1970s this fishery has been conducted using otter trawls.

At Greenland, a fishery for Greenland halibut began in a very modest way around 1915 and had by 1970 only reached an annual catch of about 2700 t, most of which was taken by Greenland. From 1970 to 1980 other countries participated in the Greenland halibut fishery, which peaked in 1976 at about 26 000 t. By 1980 the catch had fallen to about 7000 t. During the 1990s, the catch increased rapidly to about 25 000 t in 1992 and was in the range of 30 000 to 35 000 during 1998 to 2002. Since 1980, foreign vessels have not played a significant role in the Greenland halibut fishery off West Greenland. The total catch of Greenland halibut in West Greenland waters is shown in Fig. 13.9.

13.3.2.3. Northern shrimp

A small inshore fishery for northern shrimp began in Icelandic waters in the mid-1950s. Initially, this was a fjordic fishery of high value to local communities. An offshore shrimp fishery, which began in the mid-1970s on the outer shelf off the western north coast, soon expanded to more eastern areas. Annual landings from this fishery increased to between 25 000 and 35 000 t in the late 1980s and to between 45 000 and 75 000 t in the 1990s. Recently, catches have declined drastically, both in offshore and coastal areas (Fig. 13.8).

The catch of northern shrimp off West Greenland has increased steadily since its beginning in 1960. At the outset, this species was fished only by the Greenlandic fleet, but from 1972 large vessels from other countries joined this fishery. This led to a large increase in the total catch of northern shrimp, which peaked at about 61 000 t in 1976. Between 1976 and the early 1980s, the catch by other countries decreased and has been insignificant since. On the other hand, the Greenlandic catch increased steadily, from a total catch in 1960 of about 1800 t to 132 000 t in 2002 as shown in Fig. 13.9.

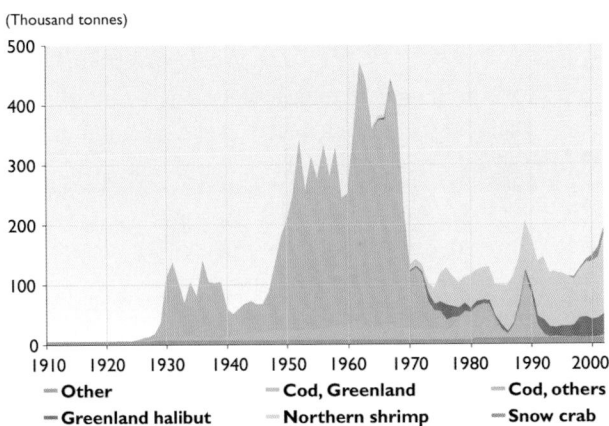

Fig. 13.9. Total catch off West Greenland, 1900–2002 (data from the Greenland Statistical Office and Directorate of Hunting and Fishing).

13.3.2.4. Herring

Commercial fishing for herring started at Iceland in the 1860s when Norwegian fishermen initiated a land-based fishery on the north and east coasts using traditional Scandinavian beach seines. This fishery proved very unstable and was abandoned in the late 1880s. Drift netting was introduced at the turn of the 19th century and purse seining in the early 20th century (1904). The latter proved very successful off the north coast, where the herring schools used to surface regularly, while drift nets had to be used off the south and west coasts where the herring rarely surfaced. The north coast herring fishery increased gradually during the 1920s and 1930s and had reached 150000 to 200000 t by the beginning of the 1940s (Fig. 13.8). During this period, the fishery was limited mainly by lack of processing facilities. Around 1945 the herring behavior pattern changed and as a result purse seining for surfacing schools north of Iceland became ineffective and catches declined. The reasons for this change in behavior have never been identified.

Horizontally ranging sonar, synthetic net fibers, and hydraulic power blocks for hauling the large seine nets were introduced to the herring fishery during the late 1950s and early 1960s (Jakobsson, 1964; see also Box 13.1). These technical innovations, as well as better knowledge of the migration routes of the great Atlanto-Scandian herring complex (i.e., Norwegian spring-spawning herring and much smaller stocks of Icelandic and Faroese spring-spawning herring), lead to an international herring boom in which Icelandic, Norwegian, Russian (USSR), and Faroese fishermen were the main participants (for Icelandic catches see Fig. 13.8). This extraordinary herring fishery ended with a collapse of the Atlanto-Scandian herring complex during the late 1960s due to overexploitation of both adults and juveniles (Box 13.1). Catches of Atlanto-Scandian herring (now called Norwegian spring-spawning herring since the Icelandic and Faroese components have not recovered) in the Icelandic area have been negligible since the late 1960s and Iceland's share of the TAC of this herring stock since the mid-1990s has mainly been taken outside Icelandic waters. There is no fishery for herring at Greenland.

It took the Norwegian spring-spawning stock about two and a half decades to recover despite severe catch restrictions (Box 13.1). Both the Icelandic spring- and summer-spawning herring suffered the same fate. Retrospective analysis of historical data shows that there were no more than 10000 to 20000 t left of the Icelandic summer-spawning herring stock in the late 1960s/early 1970s (Jakobsson, 1980). A fishing ban was introduced and since 1975 the fishery has been regulated, both by area closures and minimum landing size, as well as by having a catch rule corresponding to a TAC of roughly 20% of the estimated adult stock abundance in any given year. The stock recovered gradually, is at a historical high at present, and the annual yield over the 1980s and 1990s was on average about 100000 t.

13.3.2.5. Capelin

An Icelandic capelin fishery began in the mid-1960s and within a few years replaced the rapidly dwindling herring fishery, as was also witnessed in the Barents Sea (Vilhjálmsson, 1994, 2002; Vilhjálmsson and Carscadden, 2002). The capelin fishery is conducted by the same high-technology fleet as used for catching herring. During the first eight to ten years, the fishery only pursued capelin spawning runs in near-shore waters off the southwest and south coasts of Iceland in February and March and annual yields increased to 275000 t. In 1972, the fishery was extended to deep waters east of Iceland in January, resulting in an increase in the annual catch by about 200000 t. In 1976, an oceanic summer fishery began north of Iceland and in the Denmark Strait. In 1978, the summer fishery became international as it extended north and northeast into the EEZs of Greenland and Jan Mayen (Norway). Within two years the total seasonal (July to March) capelin catch increased to more than one million t. Total annual international landings of capelin from this stock during 1964 to 2002 are shown in Fig. 13.8.

Historically, capelin have been caught at Greenland for domestic use and animal fodder. A small commercial fishery for roe-bearing females began at West Greenland in 1964 with a catch of 4000 t, which is also the largest catch on record. There were relatively large fluctuations in the capelin catch from 1964 to 1975, but since then the catch has been insignificant. This fishery is conducted by Greenlanders.

13.3.2.6. Blue whiting

The most recent addition to Icelandic fisheries is that of the semi-pelagic blue whiting. This is a straddling species commonly encountered in that part of the Icelandic ecosystem dominated by Atlantic water, i.e., off the west, south, and southern east coast. A small blue whiting fishery began in the early 1970s, increased to about 35000 t in 1978 and then dwindled to 105 t in

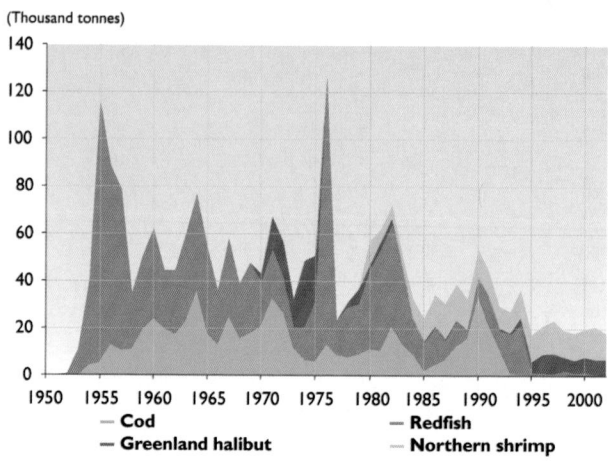

Fig. 13.10. Total catch off East Greenland, 1950–2002 (data from the Greenland Statistical Office and Directorate of Hunting and Fishing).

1984. There was renewed interest in this fishery in the mid-1990s and from 1997 to 2002 the blue whiting catch increased from 10 000 to 285 000 t (Fig. 13.8).

13.3.2.7. Fisheries off East Greenland

East Greenland waters have been fished commercially only since the Second World War (Fig. 13.10). The main reason for this is the rough bottom topography as well as the speed and irregularity of the ocean currents, especially near the edge of the continental shelf. These conditions render it difficult to fish East Greenland waters except with large powerful vessels and robust fishing gear. The main species that have been fished commercially off East Greenland are Greenland halibut, northern shrimp, cod, and redfish. With the exception of northern shrimp since the 1980s, the fisheries off East Greenland have almost exclusively been conducted by foreign fleets.

13.3.2.8. Marine mammals and seabirds

The Icelandic marine ecosystem contains a number of species of large and small whales, most of which are migratory. Commercial whaling has been conducted intermittently in Iceland for almost a century. Initially, large Norwegian whaling stations were operated from the mid-1880s until the First World War, first on the Vestfirdir peninsula (northwest Iceland) and later on the east coast. By about 1912, stocks had become depleted to the extent that whaling was no longer profitable and in 1916 the Icelandic Parliament passed an act prohibiting all whaling. In the following decades whale stocks gradually recovered and from 1948 until zero quotas on whaling were set in 1986, a small Icelandic company operated with four boats from a station on the west coast, just north of Reykjavík. The main target species were fin (*Balaenoptera physalus*), sei (*B. borealis*), and sperm (*Physeter catodon*) whales and the average yearly catches were 234, 68, and 76 animals respectively. In addition, 100 to 200 (average 183) minke whales were taken annually by small operators between 1974 and 1985. Although never commercially important at a national level, whaling was very profitable for those

engaged in the industry. Icelandic whale catches by species are shown in Fig. 13.11.

The numbers of seals in Icelandic waters are comparatively small. The populations of the two main species, harbour seals and grey seals, are estimated at 15 000 and 6000 animals, respectively (Anon, 2004c). Harbour seal abundance is stable while the numbers of grey seals have decreased. Sealing has never reached industrial proportions in Iceland, the total number of skins varying between 1000 and 7000 annually since the 1960s.

Although foreign fleets have pursued large-scale whaling in Greenlandic waters, native Greenlanders have hunted whales for domestic use only. Harvest of the main species has been modest and is unlikely to have had any effect on stocks. Five seal species are exploited in Greenland, with harp and ringed (*Phoca hispida*) seals by far the most important. Ringed seal catches increased from the mid-1940s until the late 1970s and then dropped until the mid-1980s after which they increased. The harp seal catches increased until the 1960s at which point they began to decrease and were very low during the 1970s. Since then, harp seal catches have increased continuously and at the time of writing were higher than ever.

Greenlandic catches of whales, seals, walrus, and seabirds between 1993 and 2000 are shown in Fig. 13.12. Sealskin prices were subsidized in Greenland when prices started to decline on the world market and sealskin campaigns are thought unlikely to have influenced hunting effort for seals in Greenland. There have, however, been indirect positive effects, in that Canadian catches (Labrador plus Newfoundland) of both species fell dramatically and the harp seal population increased to double its size within a relatively few years. The decrease in ringed seal catches during the early 1980s coincided with the sealskin campaign, but the underlying cause was probably population dynamics, triggered by climatic fluctuations (Rosing-Asvid, 2005).

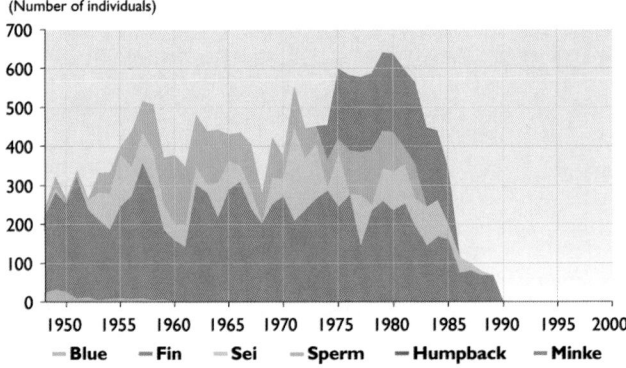

Fig. 13.11. Catch of large whales at Iceland, 1948–2000 (data from the Icelandic Directorate of Fisheries and the Marine Research Institute).

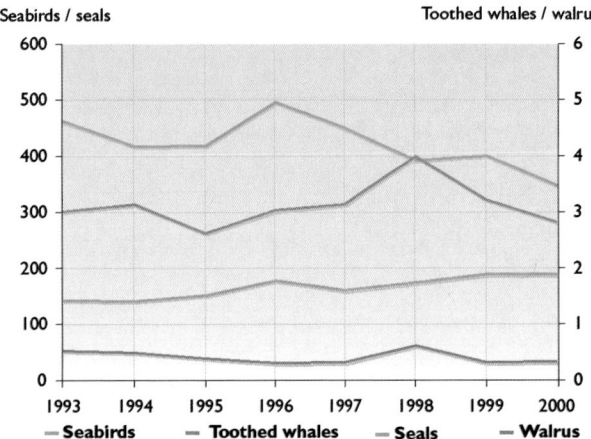

Fig. 13.12. Greenland catch of marine mammals and seabirds, 1993–2000 (data from the Greenland Statistical Office and Directorate of Hunting and Fishing).

13.3.2.9. Aquaculture

In the late 1970s and 1980s there was much interest in aquaculture in Iceland. A number of facilities were developed for the cultivation of salmon, rainbow trout (*Salmo gairdneri*), and Arctic char (*Salvelinus alpinus*) at various sites on the coast. Practically all failed, either for financial reasons or lack of expertise, or both. The few that survived, or were rebuilt on the ruins of others, have until recently not produced much more than necessary for the domestic market.

In comparative terms, aquaculture has therefore been of little economic importance for Iceland in the past. However, renewed interest began in the 1990s. Iceland is once again investing heavily in fish farming – but this time it is private capital rather than short-term loans or state funding which governs the progress. The largest quantitative increase will almost certainly be in salmon. Total production in 2001 was around 4000 t of salmon and related species. It is expected that by 2010 the production of these species will have increased to around 25 000 to 30 000 t. In addition, there is increased interest and success in the farming of Atlantic halibut, sea bass (*Dicentrachus labrax*), turbot (*Psetta maxima*), cod, and some other marine fish, and recently there has been a considerable increase in the production of abalone (*Haliotis rufuscens*) and blue mussel (*Mytilus edulis*).

Despite fish farmers working closely with the industry and with researchers to accelerate growth in production of both salmonids and whitefish species, it is expected to be a few more years before the industry is operating smoothly. Area conflicts with wild salmon have not been resolved, cod farming is still at the fry stage, and char – a high price product – has a limited market. Nevertheless, aquaculture is being developed to become more than an extra source of income and as a consequence, major fisheries companies are investing in development projects in this sector.

Aquaculture was attempted in Greenland in the 1980s. The experiment failed and aquaculture is not conducted in Greenland at the present time.

13.3.3. Past climatic variations and their impact on commercial stocks

The main climate change over the Nordic Seas and in the northwest North Atlantic over the 20th century was a rise in air temperature during the 1920s and 1930s with a concurrent increase in sea temperature and a decrease in drift ice. There was distinct cooling in the 1940s and early 1950s followed by reversal to conditions similar to those of the 1920s and 1930s. These changes and their apparent effect on marine biota and commercial stocks in Icelandic and Greenlandic waters were studied and reported on by a number of contemporary researchers (e.g., Fridriksson, 1948; Jensen, 1926, 1939; Sæmundsson, 1934; Tåning, 1934, 1948). Summaries have been given by, for example, Buch et al. (1994) and Vilhjálmsson (1997).

Figure 13.13 shows five-year running averages of sea surface temperature anomalies off the central north coast of Iceland and illustrates trends in the physical marine environment of Icelandic waters over the 20th century. The main features are an increased flow of Atlantic water onto the shelf north of Iceland between 1920 and 1964 followed by a sudden cooling in 1965 to 1971 and more variable conditions since then. A strong presence of Atlantic water on the north and east Icelandic shelf promotes vertical mixing and thus favors both primary and secondary production, i.e., prolongs algal blooms and increases zooplankton biomass. Greenland also experienced a climatic warming in the 1920s probably with similar effects on the lowest levels of the food chain (Fig. 13.14).

At Iceland, one of the most striking examples of the effects of the climatic warming during the 1920s was a mass spawning of cod off the north and east coasts in addition to the usual spawning off south and west Iceland (Sæmundsson, 1934). Furthermore, there was large-scale drift of larval and 0-group cod across the northern Irminger Sea to Greenland in 1922 and 1924

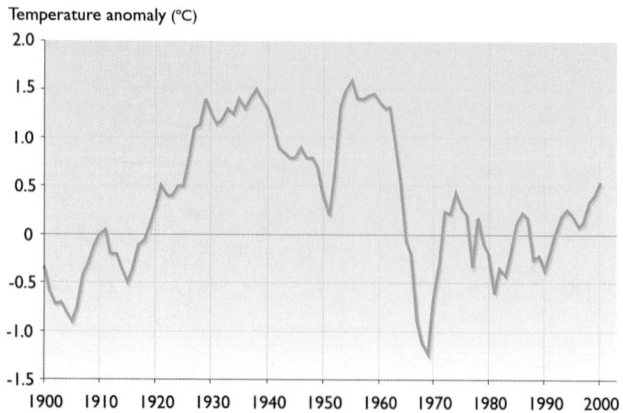

Fig. 13.13. Sea surface temperature anomalies north of Iceland (based on Anon, 2004b; Stefánsson, 1999). Five-year running means, 1900–2001.

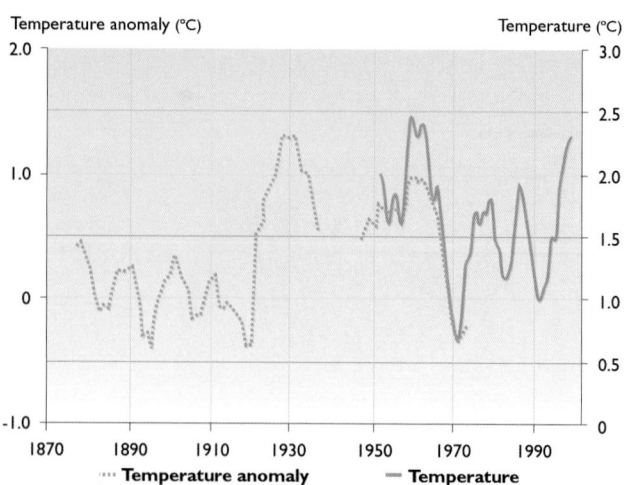

Fig. 13.14. Variations in sea temperature and temperature anomalies on the Fylla Bank off southwest Greenland (adapted from Buch et al., 1994, 2002). Five-year running means, 1875–2000.

(Jensen, 1926; Schopka, 1994). This is described in detail in Box 13.2.

Changes in the marine fish fauna off West Greenland were even more spectacular than those off Iceland. There was a large increase in cod abundance and catches in the 1920s (Fig. 13.15), and other gadids, such as saithe, haddock, tusk, and ling, previously rare or absent at Greenland, also appeared there in the 1920s and 1930s. Furthermore, herring appeared in large numbers off West Greenland in the 1930s and began to spawn there in the period July through September, mainly south of 65° N (Jensen, 1939). These herring spawned near beaches, similar to capelin in these waters. Like capelin, herring are bottom spawners with their eggs adhering to the substrate or even, as in this case, the fronds of seaweed. In 1937, the northernmost distribution of adult herring reached 72° N (Jensen, 1939). However, a herring fishery of commercial scale has never been pursued at Greenland.

In the early 1900s capelin were very common at West Greenland between Cape Farewell and Disko Bay (Fig. 13.5), but unknown further north (Jensen, 1939). In the 1920s and 1930s, the center of the West Greenland capelin populations gradually shifted north and capelin became rare in their former southern area of distribution. By the 1930s, the main spawning had shifted north by 400 nm to the Disko Bay region (Fig. 13.5). Off East Greenland capelin have gradually extended their distribution northward along the coast to Ammassalik (Jensen, 1939). However, capelin are an arctic species and have probably been common in that area for centuries since Ammassalik means "the place of capelin".

During the latter half of the 1960s there was a sudden and severe climatic cooling with an associated drop in sea temperature, salinity, and plankton production (Fig. 13.16), and an increase in sea ice to the north and east of Iceland (e.g., Astthorsson and Gislason, 1995; Malmberg, 1988; Thórdardóttir, 1977, 1984). Temperatures increased again in the 1970s, but were

then more variable during the previous warm period. The low sea temperatures were also recorded in West Greenland waters (Fig. 13.14). This low temperature, low salinity water (the "Great Salinity Anomaly") drifted around the North Atlantic and had noticeable, and in some cases serious, effects on marine ecosystems (reviewed e.g., by Jakobsson, 1992).

In the Icelandic area, herring was the fish species most affected by the cold conditions of the 1960s (Dragesund et al., 1980; Jakobsson, 1969, 1978, 1980; Jakobsson and Østvedt, 1999). This is not surprising as herring are plankton feeders and in north Icelandic waters are near their limit of distribution. This was manifested in large-scale changes in migrations and distribution (see Fig. 9.19) and a sudden and steep drop in abundance (which however was mostly brought about by overfishing – see Box 13.1). The abundance of the Norwegian spring-spawning herring stock increased dramatically in the 1990s (see section 13.2.2.5 and Box 13.1) and regained some semblance of its previous feeding pattern (for an overview of these changes see Chapter 9). Presently, Norwegian spring-spawning herring still overwinter in the Lofoten area on the northwest coast of Norway. Whether and when they revert completely to the "traditional" distribution and migration pattern cannot be predicted.

The two Icelandic herring stocks, i.e., the spring- and summer-spawning herring stocks, suffered the same

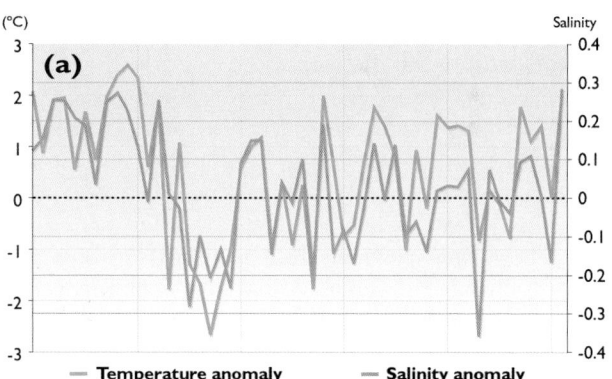

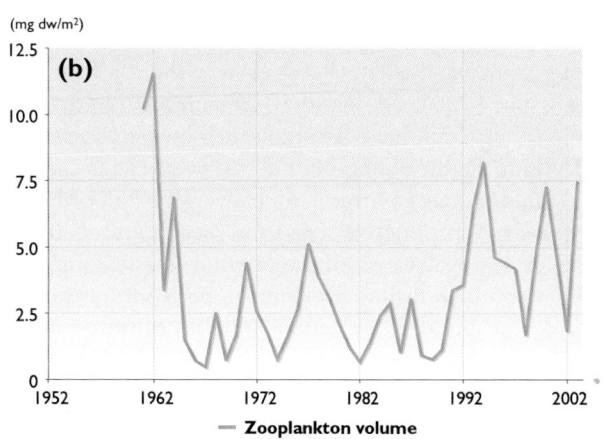

Fig. 13.16. Deviations of (a) temperature and salinity, and (b) zooplankton volume north of Iceland, spring 1952–2003 (Anon, 2004b).

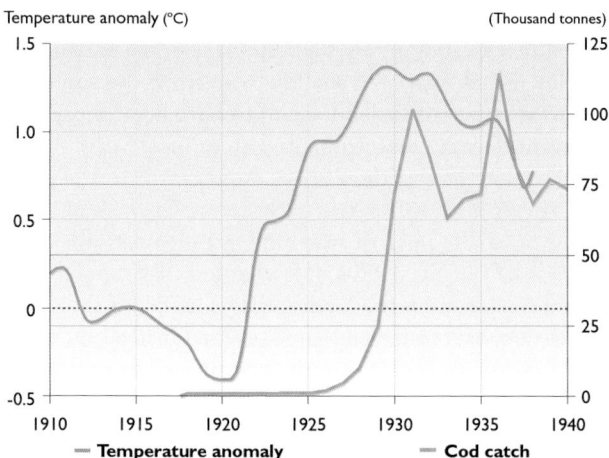

Fig. 13.15. Temperature anomaly and the catch of cod off West Greenland, 1910–1940 (Vilhjálmsson, 1997).

Box 13.2. The Iceland/Greenland cod and climate variability

Although the abundance of the Icelandic cod stock prior to 1920 is not known, it was unquestionably large (e.g., Schmidt, 1909). Furthermore, the climatic warming of the 1920s and 1930s appears to have greatly increased reproductive success of Icelandic cod through extended spawning areas and increased primary and secondary production in the mixed waters north and east of Iceland compared to previous decades. In addition, huge amounts of larval and 0-group cod drifted west across the northern Irminger Sea in 1922 and 1924, grew off West Greenland, and returned to Iceland in large numbers to spawn (Schopka, 1994; Vilhjálmsson, 1997). Tagging experiments indicate that the majority of these fish then remained within the Icelandic marine ecosystem (Hansen, 1949; Hansen et al., 1935; Jakobsson, 2002; Jónsson, 1996; Tåning, 1934, 1937). Thus, the distribution area and biomass of cod in the Icelandic marine ecosystem can be enormously enlarged through larval drift and returning adults during warm periods.

The climatic warming in the 1920s (Fig. 13.14) resulted in far greater changes in the distribution and abundance of cod at Greenland than Iceland. Until the 1920s, cod occurred in scattered numbers in inshore waters near Cape Farewell, the southernmost promontory of Greenland (Jensen, 1926, 1939; Tåning, 1948). In the 1920s, cod appeared over wider areas and in increasing numbers. This is shown in the rapid rise in the international catch of cod at West Greenland in the late 1920s, which coincides with the time needed for the 1922 and 1924 year classes to grow to marketable size. Furthermore, cod extended their distribution northward along the west coast of Greenland by 600 to 800 nm in the 1920s and 1930s (Tåning, 1948). At East Greenland, cod appeared in small schools in the Ammassalik area around 1920 and became common around 1930 along the east coast south from Ammassalik (Schmidt, 1931). The drift of 0-group cod from Iceland to Greenland continued on and off from the 1930s to the mid-1960s, although on a smaller scale than for the superabundant year classes of 1922, 1924, and 1945 (Schopka, 1994).

By the early 1930s, West Greenland waters were warm enough for successful spawning of cod (Buch et al., 1994; Hansen, 1949; Hansen et al., 1935; Jensen, 1939; Tåning, 1937). Some members of the 1922 and 1924 year classes took advantage of this, spawned off West Greenland and, with the small inshore cod population, were instrumental in giving rise to a local self-sustaining component. The West Greenland cod stock became very large and sustained annual catches of 300000 to 470000 t throughout the 1950s and 1960s. From 1973 to 1993 the average annual catch off West Greenland was about 55000 t. Peak catches in this period are associated with year classes which drifted as 0-group from Iceland to Greenland. At present, there are few cod at East and West Greenland and no local recruitment to the cod stock (Buch et al., 1994, 2002).

Although fishing mortalities at Greenland increased in the 1950s and 1960s and accelerated the crash of the Greenland cod in the 1970s, the spawning stock remained above 500000 t until 1970 and produced large year

fate. The spring-spawning stock still shows no sign of recovery, while the summer-spawning stock recovered a few years after a fishing ban was imposed in the early 1970s (Jakobsson and Stefánsson 1999). It seems that, like the West Greenland cod, the Icelandic spring-spawning herring had difficulties in self propagation in cold periods and would probably have collapsed in the late 1960s and early 1970s, even without a fishery (Jakobsson, 1980). The summer-spawning herring, on the other hand, have adapted much better to variability in Icelandic waters. For all three stocks it can be concluded that environmental adversities placed them under reproductive stress and disrupted feeding and migration patterns. Environmental stress, coupled with far too high fishing pressure on both adults and juveniles, resulted in the actual collapses of these herring populations.

While the growth rate of Icelandic capelin has shown a significant positive correlation with temperature and salinity variations in the north Icelandic area since the mid-1970s, this relationship probably describes feeding conditions in the Iceland Sea rather than a direct effect of temperature (Astthorsson and Vilhjálmsson, 2002; Vilhjálmsson, 1994, 2002). Results of attempts to relate recruitment of the Icelandic capelin stock to physical and biological variables, such as temperature, salinity, and zooplankton abundance, have been ambiguous. Nevertheless, judging by their stock size, the Icelandic capelin, which spawn in shallow waters off the south and west coasts of Iceland, seem to have been successful in recent decades and probably also in most years during the latter half of the 20th century.

However, at the peak of warming in the late 1920s and the first half of the 1930s, it was noted that capelin had ceased to spawn on the traditional grounds off the south and west coasts of Iceland and spawned instead off the easternmost part of the south coast as well as in fjords and inlets on the southeast and north coasts (Sæmunds-son, 1934). Sæmundsson also noted that the cod had become unusually lean and attributed this to lower capelin abundance. Although there can be other causes of reduced growth of cod, e.g., competition due to a classes until 1964. Like at Iceland, there was a severe cooling of the Greenlandic marine environment in the latter half of the 1960s and since then the only year classes of commercial significance at Greenland are those of 1973 and 1984, both of which drifted to Greenland as 0-group from Iceland. Despite warmer Greenlandic waters after the cooling of the late 1960s, no year classes of Greenlandic origin have appeared (Vilhjálmsson and Friðgeirsson, 1976; Vilhjálmsson and Magnússon, 1984; Schopka, 1994). This indicates that cod cannot reproduce efficiently at Greenland except under hydrographic conditions that are warmer than "normal".

The fishable part of the Icelandic cod stock (age 4+) declined from almost 2.5 million t in the early 1950s to below 600000 t in 1986. The spawning stock decreased from about 1260 to below 200000 t over this period. The initial large stock size was due to low fishing pressure in and immediately after the Second World War and to the recruitment of the superabundant 1945 year class. A large part of this year class drifted across to Greenland as 0-group and grew in Greenlandic waters. Later, around 500 million members of this year class migrated back to Iceland for spawning and appear not to have left (Schopka, 1994). Despite the cold period of 1965 to 1971 and warmer but more variable conditions since then, recruitment remained at a normal level until 1985, with occasional boosts by immigrants from Greenland, although on a much smaller scale than in 1922, 1924, and 1945 (Schopka, 1994).

Compared to other cod stocks in arctic/subarctic areas, recruitment variability of cod which grow within the Icelandic ecosystem is low or about 1:4 in the period 1920 to 1984. Although it seems that the Icelandic ecosystem cannot support juvenile year classes much beyond sizes corresponding to 300 million recruits at age 3, it has easily accommodated very large numbers of adult cod migrating back from Greenland to their natal spawning grounds. Even the very cold period from 1965 to 1971, and the variable conditions since then, do not appear to have had much detrimental effect on recruitment to the cod stock by fish that grew locally. Average recruitment during 1920 to 1985 was 210 million age 3 cod per annum.

However, since 1985 there has been a large and protracted decline in recruitment, from 210 million to about 135 million age 3 cod per annum. A very small and young spawning stock in the range of 120000 to 210000 t is the only common denominator over this period. This is very likely to have resulted in lower quality eggs, shorter spawning time, smaller spawning grounds, and possibly different drift routes, and seems to be the most plausible explanation for the reduced recruitment (Marteinsdottir and Begg, 2002; Marteinsdottir and Steinarsson, 1998). The most likely explanation for the large year classes of 1983 and 1984, which derived from small spawning stocks, is that old fish from the abundant year classes of 1970 and 1973 were still present in the spawning stock in sufficient numbers to enhance recruitment.

large stock size, Sæmundsson's conclusion may have been correct. The change in capelin spawning areas he described is probably disadvantageous for this capelin stock. The reason being that suitable spawning areas would be much reduced compared to those previously and presently occupied by the stock. Furthermore, larval drift routes could be quite different and a proportion of the larvae would probably end up in the western Norwegian Sea and be spread to regions where their survival rate might be much lower.

The catch history and series of stock assessments of northern shrimp in deep waters northwest, north, and east of Iceland, as well as at Greenland are too short for establishing links with environmental variability. Being a frequent item in the diet of small and medium-sized cod, stocks of northern shrimp are likely to be larger when cod abundance is low. However, in general terms, the stock probably benefits from cooler sea temperatures, possibly through both enhanced recruitment and a reduced overlap of shrimp and cod distribution.

13.3.4. Possible impacts of climate change on fish stocks

To project the effects of climate change on marine ecosystems is a very difficult task, despite knowing the effects of previous climatic change. Previous sections described how the marine climate around Iceland changed over the 20th century, from a cold to a warm state in the 1920s, lasting with some deviations for about 45 years, with a sudden cooling in 1965 which lasted until 1971. Since then, conditions have been warmer but variable and temperatures have not risen to the 1925 to 1964 levels. Available evidence suggests that, as a general rule, primary and secondary production and thereby the carrying capacity of the Icelandic marine ecosystem is enhanced in warm periods, while lower temperatures have the reverse effect. Within limits, this is a reasonable assumption since the northern and eastern parts of the Icelandic marine ecosystem border the Polar Front, which may be located close to the coast in cold years but occurs far offshore in warm periods when levels of biological production are enhanced through nutrient renewal and associated mixing processes, resulting from an increased flow of Atlantic water onto the north and east Icelandic plateau.

Over the last few years the salinity and temperature levels of Atlantic water off south and west Iceland have increased and approached those of the pre-1965 period. At the same time, there have been indications of increased flow of Atlantic water onto the mixed water areas over the shelf north and east of Iceland in spring and, in particular, in late summer and autumn. This may be the start of a period of increased presence of Atlantic water, resulting in higher temperatures and increased vertical mixing over the north Icelandic plateau, but the time series is still too short to enable firm conclusions.

However, there are many other parameters which can affect how an ecosystem and its components, especially those at the upper trophic levels, will react to changes in temperature, salinity, and levels of primary and secondary production. Two of the most important are stock sizes and fisheries, which are themselves connected. Owing to high fishing pressure since the early 1970s, most of the important commercial fish stocks in Icelandic waters are smaller than they used to be, and much smaller than at the onset of the warming period in the 1920s. Associated with this are changes in age and size distributions of spawning stocks; spawners are now fewer, younger, and smaller. These changes can affect reproductive success through decreased spawning areas and duration of spawning, smaller eggs of lower quality, and changes in larval drift routes and survival rates (Marteinsdottir and Begg, 2002; Marteinsdottir and Steinarsson, 1998). It is unlikely that the response of commercial fish stocks to a warming of the marine environment at Iceland, similar to that of the 1920s and 1930s, will be the same in scope, magnitude, and speed as occurred then. Nevertheless, a moderate warming is likely to improve survival of larvae and juveniles of most species and thereby contribute to increased abundance of commercial stocks in general. The magnitude of these changes will, however, be no less dependent on the success of future fishing policies in enlarging stock sizes in general and spawning stock biomasses in particular, since the carrying capacity of Icelandic waters is probably about two to three times greater than that needed by the biomass of commercial species in the area at present.

The following sections describe three possible scenarios of warming for the marine ecosystems of Iceland and Greenland and attempt to project the associated biological and socio-economic changes.

13.3.4.1. No climate change

Although the marine climate may dictate year-class success in some instances, there is little if any evidence to suggest that year-class failure and thereby stock propagation is primarily due to climate-related factors. Therefore, assuming no change from the ACIA baseline climate conditions of 1981–2000, the development and

potential yield in biomass of commercial stocks will in most cases depend on effective rational management, i.e., a management policy aimed at increasing the abundance of stocks through reduced fishing mortalities and protection of juveniles. This is the present Icelandic policy. Although it has not yet resulted in much tangible success, it should eventually do so and with a speed that largely depends on how well incoming year classes of better than average size can be protected from being fished as adolescents.

A successful fishing policy of this kind should ensure an increase in the abundance of many demersal fish stocks by around 2030. This would considerably increase the sustainable yield from these stocks compared to the present. This could also apply for the Icelandic summer-spawning herring, although that stock is already exceeding its historical maximum abundance. The increase in yield in tonnes is, however, not directly proportional to increase in stock abundance. Thus, a doubling of the fishable biomass of the Icelandic cod stock would probably increase its long-term sustainable yield in tonnes by about 20 to 30% compared to the present annual catch of about 200 000 t. Furthermore, due to natural variability in the size of recruiting year classes, increases in stock biomasses of the various species are most likely to occur in a stepwise fashion and the value of the catch would not necessarily increase proportionally.

However, on the negative side, it is likely that the northern shrimp catch would decrease due to increased predation by cod and that the capelin summer/autumn fishery would have to be reduced or stopped altogether, in order for the needs of their more valuable fish predators to be met and those of large whales, if whales remain subject to a moratorium on commercial whaling. Increases in abundance, but especially extended migrations of the Norwegian spring-spawning herring to feed in north Icelandic waters, will determine the value of the yield from that stock for Iceland. For this to occur on a long-term basis, the intensity of the cold East Icelandic Current must weaken and temperatures north of Iceland must increase. Such conditions are not envisaged under this scenario.

At Greenland, the no-change scenario will have little effect on the present situation, given that stocks are presently managed in a rational manner and that this is expected to continue.

13.3.4.2. Moderate warming

Most criteria in the no-change scenario are probably also valid for a moderate warming of 1 to 3 °C. However, due to greater primary and secondary production and a direct temperature effect *per se*, stock-rebuilding processes are likely to be accelerated in most cases. Nevertheless, as for the no-change scenario, a rational fishing policy must be maintained. Indeed, it is very likely that harvesting strategies can be used which would give higher returns from most of the major demersal stocks in the Icelandic area. As under the no-change scenario, a side effect of such a policy would be a rise in the mean age and number of older fish in the spawning stock of cod, which would further enhance larval production and survival.

Drift of larval and 0-group cod across the northern Irminger Sea to East Greenland and onward to West Greenland waters is likely to become more frequent and the number of individuals transported to increase compared to the latter half of the 20th century. Since sea temperature off West Greenland will also increase under this scenario, it is very likely that the drift of cod larvae and juveniles from Iceland will lead to the establishment of a self-sustaining Greenlandic cod stock. With a successful management strategy and in the light of past events, that cod stock could become very large and have enormous positive economic benefits for Greenland (see section 13.3.6.2). However, it is unlikely that this will contribute much to cod abundance at Iceland. This is because present fish finding and catch technologies are so effective that these cod can, and very likely will, be easily fished in Greenlandic waters before they could return to Iceland for spawning at the age of seven to eight years.

An increase in temperature of 1 to 3 °C in the north Icelandic area is large in comparative terms and will, among other things, be associated with a weakening of the East Icelandic Current and a considerable reduction in its domain. The degree of reduction is very likely to be sufficient to enable the Norwegian spring-spawning stock to again take advantage of the rich supply of *Calanus finmarchicus* over the north Icelandic shelf. This scenario would make it easier and cheaper for Iceland to take its share of this stock, and would also make the stock more valuable. The reason for this is a large increase in the proportion of the catch which could be processed for human consumption compared to the current situation where a large proportion must be reduced to the comparatively cheaper fishmeal and oil. It is also very likely that more southern species such as mackerel and tuna will enter Icelandic waters in sufficient concentrations for commercial fishing in late summer and autumn.

13.3.4.3. Considerable warming

According to the B2 emissions scenario, model results indicate that a rise in temperature beyond 2 to 3 °C in the Icelandic area in the 21st century is unlikely. However, should that happen, the high temperature is likely to lead to dramatic changes to the Icelandic marine ecosystem. Section 13.3.1 described the key role of capelin for the well-being of many demersal stocks, and highlighted the large reduction in weight-at-age of Icelandic cod during the two capelin stock collapses. Capelin spawning also ceased on their traditional grounds off the south and west coasts of Iceland in the late 1920s and early 1930s, occurring instead in fjords and inlets on the southeast and north coasts (Sæmunds-

son, 1934). Under such conditions the extent of capelin spawning grounds would reduce considerably. Should the rise in sea temperature increase beyond that of the 1920 to 1940 period, it is likely that capelin spawning might be even further reduced and limited to the north and east coasts of Iceland. This would result in major changes in larval drift routes and survival and, eventually, to a large reduction in, or even a complete collapse of, the Icelandic capelin stock.

Owing to the key role of capelin as forage fish in the Icelandic marine ecosystem this scenario would be very likely to have a considerable negative impact on most commercial stocks of fish, whales, and seabirds which are dominant in this ecosystem at present. Such a scenario is also very likely to result in species from more temperate areas moving into the area and at least partially replacing those most affected by a lack of capelin.

13.3.5. The economic and social importance of fisheries

13.3.5.1. The fishing industry and past economic fluctuations

Iceland

During the 20th century, the Icelandic gross domestic product (GDP) had an average annual growth of about 4% per year. This was largely driven by expansion in the fisheries and fish processing industries. Furthermore, fluctuations in aggregate economic output were highly correlated with variations in the fishing industry. Good catches and high export prices resulted in economic growth, while poor catches and adverse foreign market conditions led to economic slowdown and even depression. All five major economic depressions in the 20th century can be directly related to changes in the fortunes of the fishing sector, either wholly or partially (Agnarsson and Arnason, 2003).

The first of these major depressions covers the period of the First World War, which had catastrophic effects on Iceland, as it did on many other European countries. The first two years of the war were favorable for the fishing sector however, as increased demand pushed up foreign prices, but in 1916 the international trade structure broke down and Iceland had to accept harsh terms of trade with the Allies. In 1917, Iceland was forced to sell half its trawler fleet to France. This led to substantially reduced demersal fish and herring catches in 1917 and 1918. The result was a sharp drop in GDP and a depressed economy until 1920 (Fig. 13.17).

The effects of the "Great Depression" were first felt in Iceland in autumn 1930, and in the following two years GDP fell by 0.5% and 5% respectively as demand for maritime exports declined sharply. Following a brief recovery, the economy was hit again when the Spanish Civil War broke out in 1936 and closed Iceland's most important market for fish products. Despite these

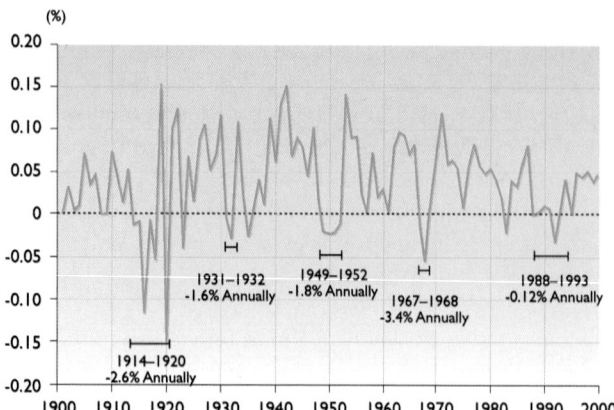

Fig. 13.17. GDP growth in Iceland, 1901–2000: showing major depressions (Agnarsson and Arnason, 2003).

events, economic growth still averaged 3% in the 1930s, mostly because of strong rebound in the fisheries, especially the herring fisheries, in 1933 to 1939. The strong performance of the fisheries in the 1930s appears to be the reason that the "Great Depression" was felt less in Iceland than most other countries of Western Europe.

The Second World War was a boom period for Iceland led by good catches and very favorable export prices. But in 1947 and subsequent years, herring catches fell considerably and real export prices subsided from the high wartime levels. The result was a prolonged economic recession from 1949 to 1952.

During the 1960s, the economy grew at an average rate of 4.8%. This was largely due to very good herring fisheries. When the herring stocks collapsed toward the end of the decade the result was a severe economic depression in 1968 and 1969, when the GDP declined by 1.3% and 5.5% respectively. Unemployment reached over 2% – a great shock for an economy used to excess demand for labor since the 1930s – and many households moved abroad in search of jobs. Net emigration amounted to 0.6% of the total population in 1969 and 0.8% in 1970.

High economic growth resumed between 1971 and 1980 with annual rates averaging 6.4%. However, just as during the 1960s, this growth was to a significant extent based on overexploitation of the most important fish stocks. Reduced fishing quotas and weak export prices reduced fishing profitability in the late 1980s. And, partly as a consequence of this, the Icelandic economy was stagnant between 1988 and 1993, with an average annual decline in GDP of 0.12%.

Since 1993, the Icelandic economy has shown steady and impressive annual growth rates. One reason for this is a recovery of some fish stocks. More important, however, are more favorable fish export prices and the impact of the individual transferable quota (ITQ) system. The ITQ system has enabled the fishing industry to increase and stabilize profits and more easily adjust to changing quotas and fish availability.

Thus, over the 20th century as a whole, it appears that major fluctuations in the Icelandic economy largely reflect changes in the fortunes of the fishing industry both in terms of harvest quantity and output prices. This implies that possible changes in fish stocks due to climate change may have similar macro-economic effects. However, it is very likely the macro-economic impact of any given change in fish availability will be smaller in the future than in the past. First, because the importance of the fishing industry for the Icelandic economy has declined substantially, and second, because the ITQ system has probably made the fishing industry more capable of adapting to changes in fish stocks. However, it must be noted that if the current depressed state of some of the most important fish stocks persists, adverse environmental changes may actually translate into larger biological shocks than those experienced in the past.

Greenland

Greenland does not offer the same overwhelming evidence of the national economic importance of the fishing industry as Iceland. This, however, does not mean that the economic importance of the Greenland fishing industry is any less than in Iceland. In fact it is probably much greater.

First, the Greenland fishing industry developed much later than that in Iceland. Thus, the Greenland fishing activity was relatively insignificant over the first half of the 20th century (see Fig. 13.9) even when compared to the rest of the Greenland economy. Second, being based on underexploited fish stocks, the Greenland fishing industry expanded relatively smoothly until the 1980s, resulting in far fewer of the dramatic fluctuations in fisheries output experienced in Iceland. Third, the Greenland economic statistics are less comprehensive than in Iceland, meaning fewer data.

Since 1970, there have been two major cycles in the Greenland economy (Fig. 13.18) both associated with changes in the fishing industry, more precisely the cod fishery.

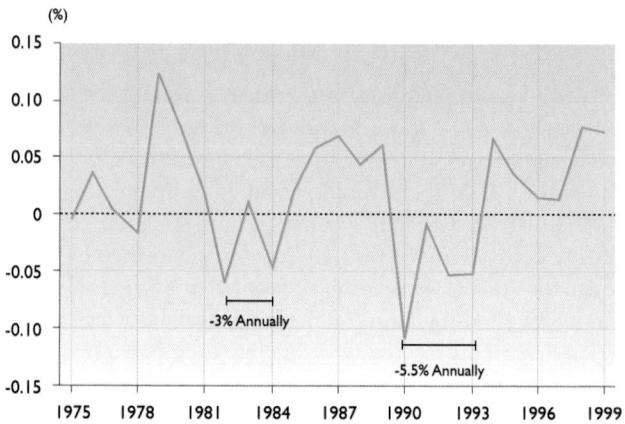

Fig. 13.18. GDP growth in Greenland, 1975–1999: showing major depressions (Anon, 2000).

Historically, the cod fishery has been Greenland's most important fishery (although this has now been superceded by the shrimp fishery). The cod fishery underwent a major expansion in the latter half of the 1970s due to reduction in foreign fishing following the extension of the Greenland fisheries jurisdiction to 200 nm and a greatly expanded Greenland fishing effort. This led to a period of good economic growth that reversed abruptly in 1981 with a major contraction of the cod fishery due to a combination of overfishing and low export prices. The subsequent period of economic depression lasted for three years during which the GDP decreased by 9% per year. Another short-lived boom in the cod fishery from about 1985 led to a corresponding boom and bust cycle in the economy with a five-year growth period followed by a sharp depression lasting four years during which GDP decreased by over 20%. Economic growth resumed in Greenland in 1995, not on the basis of cod, which has not reappeared, but shrimp fishing which expanded very rapidly during the latter half of the 1990s.

As in Iceland, historical evidence indicates a close connection between fluctuations in GDP and variations in the Greenland fishing industry.

13.3.5.2. The economic and social role of fisheries

Iceland

The relative importance of the fishing industry in the Icelandic economy seems to have peaked before the middle of the 20th century. Since then, both the share of fish products in merchandise exports and the fraction of the total labor force engaged in fishing have declined significantly. In 2000, the fishing industry employed 8% of the labor force, accounted for 63% of merchandise exports, and generated 42% of export earnings. Total export value of fish products in 2000 was about US$ 1220 million.

National accounts estimates of the contribution of the fishing industry to GDP – available since 1980 – confirm this trend. Thus, in 1980 the direct contribution of the fishing industry to GDP was over 16%. In 2000, this had dropped to just over 11%, which corresponds to an added US$ 900 million.

These aggregate statistics will understate the real contribution of the fishing industry to the Icelandic economy. There are two fundamental reasons for this. The first is that there are a number of economic activities closely linked to the fishing industry but not part of it. These comprise the production of inputs to the fishing industry, the so-called "backward linkages", and the various secondary uses of fish products, the so-called "forward linkages" (Arnason, 1994). The backward linkages include activities such as shipbuilding and maintenance, fishing gear production, the production of fishing industry equipment and machinery, the fish packaging industry, fisheries research, and education. The forward linkages comprise the transport of fish products, the production of animal feed from fish products, the marketing of fish products, and retailing of fish products. According to Arnason (1994), these backward and forward linkages may add at least a quarter to the direct GDP contribution of the fishing industry.

The other reason why the national accounts may underestimate the contribution of the fishing industry to GDP is the role of the fishing industry as a disproportionately strong exchange earner. To the extent that the availability of foreign currency constrains economic output, the economic contribution of a disproportionately strong export earner may be greater than is apparent from the national accounts. While the size of this "multiplier effect" is not easy to measure, some studies suggest it may be quite significant (Agnarsson and Arnason, 2003; Arnason, 1994). If this is the case, the total contribution of the fishing industry to GDP may be much higher than estimates suggest, in the sense that removing the fishing industry would, with all other things remaining the same, lead to this reduction in GDP.

There are also economic reasons as to why a change in the conditions of the fishing industry due, for example, to climate change, might have a lesser economic impact than suggested by the direct (and indirect) contribution of the fishing industry to GDP. Most economies exhibit some resilience to exogenous shocks. This means that the initial impact of such shocks is at least partly counteracted by the movement of labor and capital to economic activities made comparatively more productive by the shock. Thus, a negative shock in the fishing industry would to a certain extent be offset by labor and capital moving from the fishing industry to alternative industries and vice versa. Thus, the long-term impact of such a shock may be much less than the initial impact. The extent to which this happens depends on the availability of alternative industries. However, with increased labor mobility, communication technology, and human capital this type of flexibility is probably much greater than in the past.

Regional importance

Analysis in terms of macro-economic aggregates does not take into account that the economic importance of the fishing industry varies from one region of the country to another. In 2000, when the fishing industry (harvesting and processing) employed only about 8% of the

Table 13.1. The importance of the fishing sectors to Icelandic communities in 1997.

Labor share of the fishing sectors (%)	Number of communities	Number of inhabitants	Percentage of total population
>40	24	12812	7.7
25–40	16	23063	8.6
10–25	14	36959	13.7
5–10	16	26832	10.0
<5	54	161922	60.1

national workforce, it provided jobs for over 35% of the working population in the western fjords and almost 30% of the working population in the eastern fjords. Both regions are sparsely populated and account for only a small proportion of the total Icelandic population. Near the capital, Reykjavík, where most of the alternative industries such as manufacturing and services are located, the fishing industry employed only about 3% of the working population.

The local importance of the fishing industry is even more apparent at the community level. In 1997, the fishing industry accounted for over 40% of the local employment in 24 out of a total of 124 municipalities in Iceland (Table 13.1). A typical example of a community totally dependent on fishing is Raufarhöfn, a small community of 400 inhabitants in northeast Iceland. Almost 70% of the adult population worked in the fishing industry in 1997. In four other communities the fishing industry accounted for over 60% of total employment in 1997.

By contrast, in 54 communities the fishing industry accounted for less than 5% of total employment. Most of the largest municipalities in Iceland belong to this group. It is mainly the smaller, economically less developed communities that depend heavily on fisheries (see Table 13.1).

Thus, the effects of a significant reduction in fish availability around Iceland, or the benefits of fisheries expansion would be differently felt in the various regions and communities of Iceland. In general terms, a significant reduction in fish availability is liable to be economically and socially disastrous for the western and eastern fjord regions and for certain other smaller regions of Iceland, while in the more densely populated southwest of Iceland such a reduction would be felt mainly as an increased influx of labor from the outlying regions and the corresponding realignment of economic activity.

Although labor mobility is high in Iceland, it may not be easy for inhabitants of fishing villages to find jobs elsewhere following a decline in fisheries due to climate change, especially if the economy is already depressed. Also, as many of the employment opportunities in and around the capital require particular education and training, individuals transferring from the fishing industry may have to accept relatively inferior jobs. At the same time, reduced employment and movement out of the fisheries-dependent regions and communities of Iceland will decrease real estate values in these areas, meaning that these migrants may have to suffer a significant decrease in the value of their assets at the same time as moving to seek new employment.

Thus, a significant reduction in the Icelandic fishing industry would lead to noticeable social disruption. However, given the nature of Icelandic society, it would probably be resolved within five to ten years of the initial shock, although the disruption would impose a certain stress on the social and political system during this period of adjustment.

Greenland

The fishing industry is by far Greenland's most important production sector. In the 1960s and early 1970s fish and fish products accounted for between 80 and 90% of Greenland's total export value. In 1974, there was a very large increase in the export of lead and zinc, which increased GDP by about 50% and caused fish and fish products to fall to between 60 and 70% of total export value. The export of lead and zinc ceased in 1990. Since then, export of fish and fish products has accounted for about 90% of Greenland's total export value. In 2000, the export value of fish and fish products was about US$ 270 million and the total export value about US$ 285 million.

Exact statistics about the direct contribution of the fishing industry to the Greenland GDP are not available. However, the contribution to the gross national income (GNI) may be as high as 20%. This, however, does not tell the complete story. Greenland is part of Denmark with a "Home Rule" government. This means that Greenlanders can decide their own policies, except for foreign and defense policy. Every year, the Home Rule government receives economic support from the Danish State. In 2000 this amounted to about US$ 350 million or almost 25% of GNI. Correcting for this indicates a direct contribution of the fishing industry to the Greenland GDP of 25 to 30%.

As for Iceland, however, the fishing industry also has an indirect contribution to the Greenland economy via forward and backward linkages as well as multiplier effects. Adding these may bring the total contribution of the fishing industry to the Greenland economy as a whole to over 50%.

Regional importance

Greenland as a whole is highly dependent on the fishing industry. This is even more the case in less populated communities along the coast. About 20% of Greenland's population lives in small villages and settlements with an average population of about 150 inhabitants. Many more live in small towns with less than a thousand inhabitants. The economic activity in these communities is almost exclusively based on the exploitation of living marine resources, i.e., through fishing and hunting. Also, the geographical isolation of many of these communities means alternative employment opportunities are few if any.

Thus, a significant drop in the fish stocks and other living marine resources would have a devastating impact on these communities. Most would decrease significantly and many would disappear altogether, causing those inhabitants that left to become economically and socially dispossessed. A secondary effect would be the substantial influx of these people to the more urban areas of Greenland and the problems that this would cause.

A significant increase in the stocks of fish and other living marine resources would cause the reverse effect and would strengthen the economic basis of Greenland's smaller communities. While larger towns may benefit disproportionately from such a change, the net effect would probably be to increase population in the smaller communities and to expand the geographical extent of habitation in Greenland.

13.3.6. Economic and social impacts of climate change: possible scenarios

From an economic point of view, climate change may impact on fisheries in at least two ways: by altering the availability of fish to fishers and by changing the price of fish products and fisheries inputs. Although both types of impact may be initiated by climate change, the former is a more direct consequence of climate change than the latter.

The possible impact of climate change on fish availability may occur through changes in the size of commercial fish stocks, changes in their geographical distribution, and changes in their catchability. These changes, if they occur, will affect the availability of fish for commercial harvesting. The direction of this impact is uncertain. It may be negative, and so reduce the maximum sustainable economic yield from the fish stocks, or positive, and so increase the maximum sustainable economic yield from the fish stocks. Also, the impact may vary for different fish stocks and for different regions. Irrespective of the direction of the impact, however, it is very likely that climate change will, at least temporarily, cause instability or fluctuations in harvesting possibilities while ecosystems adjusts to new conditions. The adjustment period may be long, and may even continue after the period of climate change has ended.

The same applies to changes in economic value in that relative prices may continue to adjust after an exogenous shift, such as climate change, has been resolved. In fact, economic adjustments following climate change, being dependent on biological/ecological adjustments, will by necessity continue after the latter are complete.

This section speculates on the possible economic and social impacts in Iceland and Greenland of changes in fish availability. The possible impacts of relative price changes are not discussed. However, the economic and social impacts of price changes will be similar to those of changes in fish stock availability. In terms of drawing inferences from historical evidence, it is not important whether expansions and contractions in the fishing industry result from changes in prices or fish availability.

Empirical evidence of possible economic impacts of changes in fish stock availability is either qualitative historical evidence or quantitative evidence. Qualitative evidence (discussed in section 13.3.5.1) relates economic fluctuations to qualitative evidence of expansions and contractions in the fishing industry. Quantitative evidence, in the form of time series for fisheries production and production values, provides a basis for statistical estimates of the relationship between the production value of the Icelandic and Greenland fishing industries and their respective GDP and Gross National Product (GNP) growth.

13.3.6.1. Iceland

Reliable time series data for the output and output value of the fishing industry are available since 1963. These data have been used to estimate the form and parameters of a relationship between economic growth rates and the output value of the fishing industry as well as other relevant economic variables such as capital and labor (Agnarsson and Arnason, 2003). The equation exhibits good statistical properties and actual and fitted GDP growth rates are illustrated in Fig. 13.19.

This equation can, with certain modifications, be used to predict the short- and long-term impact of a change in fish stock availability due to climate change. It is important to realize, however, that to use this equation it is necessary to project (1) the extent and timing of climate change, (2) the impact of global climate change on fish stock availability, and (3) the impact of changed fish stock availability on the value of fish production (which involves both the volume and price of fish production).

Impact on GDP

This section presents the outcome of calculations to estimate the possible impact on GDP of changed fish stock availability as a result of climate change. The impact of other variables on the value of fish production is ignored. The calculations are based on two key factors: the impact of future climate change on the value of fish production in Iceland and the estimated relationship between economic growth and the value of fish production (see Agnarsson and Arnason, 2003). Both are highly uncertain. Thus, the following calculations must not be regarded as predictions. They are intended to serve as

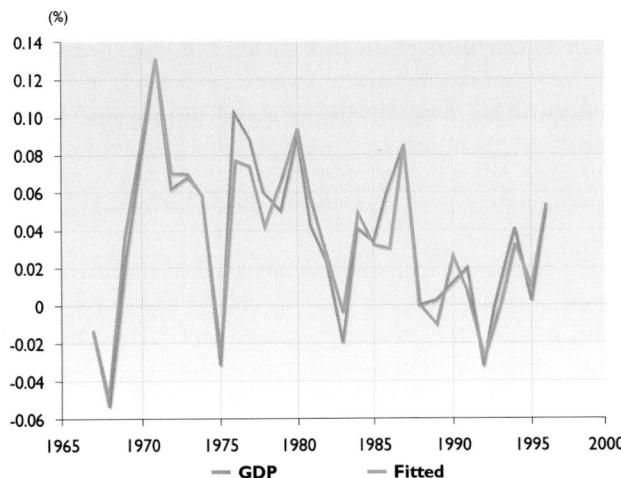

Fig. 13.19. GDP growth in Iceland, 1966–1997: actual and fitted values.

indications of the likely magnitudes of the impact on GDP in Iceland resulting from certain stated premises regarding changes in fish stock availability.

Available projections (see section 9.3.4.4) suggest that climate change over the next 50 to 100 years is (1) unlikely to have a great impact on fish stock availability in Icelandic waters and (2) is very likely to benefit the most valuable fish stocks. As a result, the overall effects of climate change on the Icelandic fisheries are likely to be positive. As these expectations are very uncertain, the rest of this section illustrates this point using three scenarios.

The first scenario assumes a gradual increase in fish stock availability of 20% over a period of 50 years. This is known as the "optimistic" scenario and corresponds to a 0.4% increase in the value of fish production annually. The second scenario assumes a gradual reduction in fish stock availability of 10% over 50 years. This is known as the "pessimistic" scenario and corresponds to an annual reduction in the value of fish production by 0.2%. The third scenario assumes a 25% reduction in fish stock availability over a relatively short period of five years. This corresponds to a collapse in the stock size of one major species or a group of important commercial species. In fact, there are some indications that the response of fish stocks to climatic change may be sudden and discontinuous rather than gradual. Owing to the magnitude and suddenness of this reduction it is known as the "dramatic" scenario.

These scenarios illustrate the likely range of economic impacts of climate change around Iceland. In interpreting their outcomes it is important to remember that these scenarios are restricted to the impact of climate change assuming all other variables affecting fish stocks and their economic contribution are unchanged. These outcomes do not incorporate the possibly simultaneous impact of improved fisheries management or other variables affecting the size of fish stocks and the value of the fisheries. In fact, given the currently depressed state of many of the most valuable fish stocks in Iceland, a better harvesting policy may easily contribute at least as much to the overall economic yield of the fisheries as the most optimistic climate scenario. However, such a policy will also improve the outcomes of the more pessimistic climate scenarios.

Optimistic scenario

In the optimistic scenario, fish stock availability is assumed to increase in equal steps by 20% over the next 50 years. The impact of this scenario on GDP relative to a benchmark GDP of unity is illustrated in Fig. 13.20. The figure illustrates that this quite considerable increase in fish stock availability has only a relatively minor impact on GDP. The maximum impact occurs in year 50, when increased fish stock availability has fully materialized. At this point GDP has increased (com-

pared to the initial level) by less than 4%. The long-term impact, after economic adjustment processes are complete, is even less at around 2.5%. The largest annual increase in GDP is small at under 0.2%, and is substantially less than the GDP measurement error. Thus it would be hardly noticeable. In the years following the end of the increase in fish production, growth rates decline as production factors (which move to the fishing industry) reduce economic production elsewhere. Long-term GDP growth rates are, of course, unchanged. The main conclusion to be drawn is that a 20% increase in the output of the fishing industry equally spread over 50 years has a very small, hardly noticeable, impact on the short-term economic growth rates in Iceland as well as on long-term GDP.

Pessimistic scenario

In the pessimistic scenario, fish stock availability is assumed to decrease in equal steps by 10% over the next 50 years. The impact of this scenario on GDP relative to a benchmark GDP of unity is also illustrated in Fig. 13.20. As in the optimistic scenario, it is apparent that this considerable decrease in fish stock availability has a relatively minor impact on long-term GDP. The maximum impact also occurs in year 50, at which point GDP has been reduced by less than 2%. The long-term impact, after economic adjustment processes are complete, is even less or just over 1%. The largest annual decrease in GDP is well over -0.1%. This occurs a few years after the decrease in fish stock availability begins. For most of the period, however, the impact on annual economic growth rates is much less. In the years following the end of the decrease in fish production, growth rates improve, as production factors (which move from the fishing industry) find productive employment elsewhere. All these deviations in annual GDP growth rates are well within GDP measurement errors. Long-term GDP growth rates are, of course, unchanged. As under the optimistic scenario, the main conclusion to be drawn is that a 10% decrease in output from the fishing industry equally spread over 50 years has a negligible impact on the short-term economic growth rates in Iceland as well as on long-term GDP.

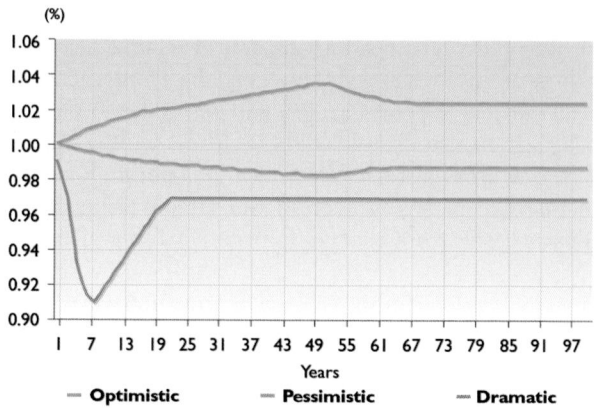

Fig. 13.20. Iceland: impact of different scenarios on GDP (benchmark GDP = 1.0).

Dramatic scenario

The dramatic scenario assumes a fairly substantial drop in fish stock availability and, hence, fish production of 25% over the next five years. The impact of the dramatic scenario on GDP relative to a benchmark GDP of unity is illustrated in the lowest curve in Fig. 13.20. This sudden drop in fish stock production has a significant negative impact on GDP in the short term. At its lowest point, in year eight, GDP is reduced by over 9% (compared to the initial level). The long-term negative impact, after economic adjustment processes are complete, is only about a 3% reduction in GDP. The decrease in annual GDP growth rates for the first seven years following the start of reduced fish stock production are significant or -1 to -2%. The maximum decline occurs toward the end of the reduction process in years four and five. However, only four years later, the contraction ends, and the deviation in annual GDP growth rates is reversed as production factors released from the fishing industry find productive employment elsewhere. According to these calculations, this adjustment process is complete by year 25 when growth rates have reverted to the underlying economic growth rate.

Social and political impacts

If the change in fishing industry output is gradual and the economic impact comparatively small (as in the "optimistic" and "pessimistic" scenarios), it is unlikely that the accompanying social and political impacts will be noticeable at a national level. Although over the long term, social and political impacts will undoubtedly occur, whether these will be large enough to be distinguished from the impact of other changes is uncertain. Regionally, however, the situation may be very different. In some parts of Iceland (see section 13.5.2) the economic and social role of the fishing industry is far above the national average. In these areas, the economic, social, and political impact of an expansion or contraction in the fishing industry will be much greater than for Iceland as a whole and in some areas undoubtedly quite dramatic.

If, on the other hand, the change is fairly sudden (as in the "dramatic" scenario) the short-term social and political impact may be quite drastic. In the long term, however, after the initial impact, social and political conditions will revert to the long-term scenario described by the "pessimistic" scenario. Whether the economic and social adjustments will then also revert to their initial state is not clear.

Impacts on fish markets

Reductions or increases in fish production in Iceland alone will not have a significant impact on global fish markets. Neither are they likely to have a large impact on the marketing of Icelandic fish products, provided the changes are gradual. If there is an overall decline in the global supply of species of fish that Iceland currently

exploits, the impact on marketing of these species is uncertain. Almost certainly the marketing of the species in reduced supply will become easier. Thus, prices will rise counteracting the decrease in volume. However, the marketing impact might actually be the opposite. For some species, a large and steady supply is required to maintain marketing channels. If this is threatened, these channels may close and alternative outlets will have to be found.

Discussion

The main conclusion to be drawn is that the changes in fish stock availability that seem most likely to be induced by climate change over the next 50 to 100 years are unlikely to have a significant long-term impact on GDP in Iceland and, consequently, on social and political conditions in Iceland. Also, it appears that any impact, small as it may be, is more likely to be positive than negative. However, if on the other hand, climate change results in sudden rather than gradual changes in fish stock availability, the short-term impact on GDP and economic growth rates may be quite significant. The impact seems very unlikely to be dramatic (i.e., over 5% change in GDP between years) however. Over the long term, the impact on GDP of a sudden change in fish stocks will be indistinguishable from the effects of more gradual change. Long-term social and political impacts may differ although there is no clear evidence to support this.

13.3.6.2. Greenland

Reliable time series data for the export value of the Greenland fishing industry are available since 1966. These data have been used to estimate the form and parameters of a relationship between GDP and the real export value of fish products (Vestergaard and Arnason, 2004). The equation exhibits reasonable statistical properties. Actual and fitted GDP growth rates according to this equation are illustrated in Fig. 13.21. It is projected that a 1% increase in the export value of fish products will lead to a 0.29% increase in the Greenland GDP. Subject to the same qualifications as for Iceland, this equation can be

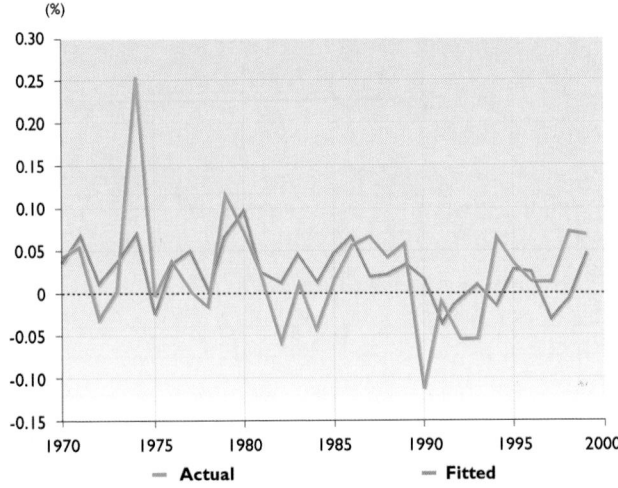

Fig. 13.21. GDP in Greenland, 1970–1999: actual and fitted values.

used to predict the economic impact of a change in fish stock availability resulting from climate change.

Impact on GDP

Available projections (see section 9.3.4.4 and sections 13.3.3.3, 13.3.3.4, and 13.3.4.1 to 13.3.4.3) suggest that climate change over the next 100 years is very likely to benefit the most valuable fish stocks at Greenland. This is particularly likely to be the case for the cod stock, which could experience a revival from its current extremely depressed state to a level, seen during warm periods of the 20th century, where it could yield up to 300000 t on a sustainable basis. However, climate change and increased predation by cod could lead to a dramatic fall in the sustainable harvest of shrimp by up to 70000 t (Fig. 13.22). The value of the increased cod harvest would, however, greatly exceed losses due to a possibly reduced harvest of shrimp. In fact, this change could lead to doubling or even tripling of the total production value of the Greenland fishing industry. Thus, the projected climate change could have a major positive impact on the Greenland fishing industry. However, this is highly uncertain. As was the case for Iceland this section continues on the basis of three scenarios.

The first scenario, termed the "pessimistic" scenario, assumes that despite more favorable habitat conditions, cod will not reestablish permanently in Greenland waters. Instead, there will be periodic bursts of cod availability accompanied by a corresponding drop in shrimp, based on occasional large-scale larval drift from Iceland similar to that seen in warmer periods in the past. The overall impact will be a slight average increase in fish harvests with some peaks and troughs. The second scenario, termed the "moderate" scenario, assumes a modest and gradual return of cod to Greenland which in 20 years would be capable of yielding 100000 t per year on average. This would be accompanied by a corresponding decline in the shrimp stock. The third scenario, termed the "optimistic" scenario, assumes a return of the Greenland cod stock, initially generated by Icelandic cod larval drift, to the levels of the 1950s

and 1960s. A full revival, however, would take some decades and would occur in a fluctuating manner. Ultimately, in about 30 years, the cod stock would be capable of producing an average yield of 300000 t per year, compared to almost nothing at present. The average shrimp harvest, however, would be reduced from a current level of almost 100000 t per year to about 20000 t per year. Nevertheless, the overall value of the Greenland fish harvest would almost double. These scenarios illustrate the likely range of economic impacts of climate change around Greenland. The harvest projections are all based on a two-species fisheries model developed for the Greenland fisheries (Hvingel, 2003).

Pessimistic scenario

The pessimistic scenario assumes an insubstantial change in overall average fish stock availability. However, due to the occasional large-scale influx and survival of Icelandic cod larvae, periodic bursts in cod availability occur. This situation results in fluctuating fish production rates and GDP impacts over time (see lowest curve in Fig. 13.23) with a small average increase. The average increase in GDP after 50 years is about 2% higher than would otherwise have been the case.

Moderate scenario

In the moderate scenario, fish stock availability is assumed to increase gradually by about 20% over the next 100 years. The impact of this scenario on GDP relative to a benchmark GDP of unity is shown by the middle curve in Fig. 13.23. This increase in the availability of fish leads to a moderate long-term increase in GDP of 6% (compared to the initial level). However, as most of this increase is projected to occur over the first ten years (in fact the initial impact is projected to be greater than the long-term impact) there would be a significant addition to GDP of 1% per year during this initial period.

Optimistic scenario

In the optimistic scenario, fish stock availability is assumed to increase gradually, but in a fluctuating man-

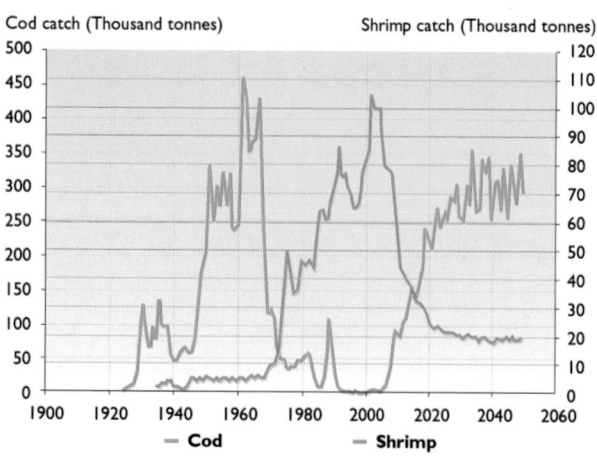

Fig. 13.22. The history and possible future development of cod and shrimp harvests in Greenland under global warming (Hvingel, in prep).

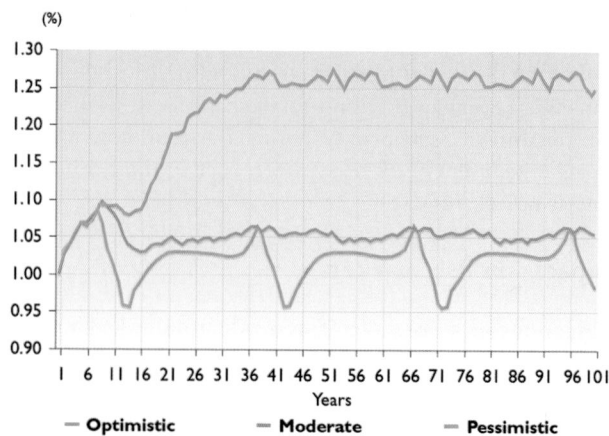

Fig. 13.23. Greenland: impact of different scenarios on GDP (benchmark GDP=1.0).

ner, by about 100% over the next 50 years. The projected relationship between the value of fish exports and GDP suggests that this will lead to an ultimate increase in GDP of 28% compared what would otherwise have been the case. This impact relative to a benchmark GDP of unity is illustrated in Fig. 13.23. Such an impact would be very noticeable. The addition to economic growth would be close to 0.8% per year over the first 30 years but would then decrease, stopping after 40 years.

Social and political impacts

If the optimistic scenario were to occur and the increased cod harvest was mainly caught and processed by Greenlanders there would be a dramatic improvement in the Greenland unemployment/underemployment situation and in the income of the large group of self-employed small-boat fishers/hunters. Everything else being the same, the current high level of unemployment would disappear and the total income of many families would increase markedly. This would have a major social impact in Greenland. Cod fishing of the magnitude projected under the optimistic scenario might easily lead to the establishment of large-scale fish processing factories in the more densely populated regions of Greenland.

Under the pessimistic and moderate scenarios employment and consequently social impact would be far more moderate, especially if the changes appeared gradually over a long period. If, however, the change is sudden (as in the "dramatic" scenario for Iceland), short-term social and political impacts may be drastic.

Discussion

The likely impacts of the possible changes in fish stock availability in Greenland waters resulting from climate change cover a wider range than for Iceland. At one extreme, they could lead to a 30% increase in GDP, while at the other extreme the impact on GDP could be negligible. This range in outcomes reflects (1) the greater importance of the fisheries sector in Greenland compared to Iceland and (2) that during warm periods the very large marine habitat around Greenland can accommodate a biomass of commercial species which is many times greater than that at present. Since the arctic influence is more pronounced in the Greenland marine environment than in that around Iceland, it follows that impacts resulting from a moderate climate change, if it occurs, are very likely to be more dramatic in the marine environment around Greenland.

13.3.7. Ability to cope with change

Climate change will almost certainly lead to changes in the relative sizes, biological productivity, and spatial distribution of commercial fish stocks. These changes may be predominantly advantageous or predominantly disadvantageous. They may be sudden or may emerge gradually.

The economic and social impacts of changes in fish stock availability depend on the direction, magnitude, and rapidity of these changes. The economic and social impacts also depend, possibly even more so, on the ability of the relevant social structures to adapt to altered conditions. Good social structures facilitate fast adjustments to new conditions and thus mitigate negative impacts. Weak or inappropriate social structures exhibit sluggish and possibly inappropriate responses and thus may exacerbate problems resulting from adverse environmental changes.

One of the most crucial social structures in this respect is the fisheries management system. This determines the extent to which the fisheries can adapt in an optimal manner to new conditions. Other important social structures relate to (1) the adaptability of the economic system – especially price flexibility, labor education and mobility, and the extent of economic entrepreneurship; (2) the ability to adjust macro-economic policies; and (3) the extent and nature of the social welfare system. These structures influence the form of the necessary adjustments to new conditions, for example whether they are smooth and quick or difficult and long-term.

The Icelandic fisheries management system is based on permanent harvest shares in the form of ITQs and so is inherently forward-looking and thus probably well-suited to adjust optimally to changes in the availability of fish, especially if these changes are to an extent foreseen. For any time path of fish stock productivity, ITQ-holders have a strong incentive to maximize the expected present value of the fishery as this will also maximize their expected wealth. As a result, there is a high likelihood that TACs and other stock size determinants will be adjusted optimally to altered conditions. In other words, with the Icelandic fisheries management system, there is little chance that the fishing activity will exacerbate a negative biological impact arising from climate change. Also, the opportunities generated by a positive biological impact will probably be close to fully exploited by the fishing sector. In Greenland, however, fishers' rights to harvest shares are considerably weaker. As a result, Greenland seems less prepared than Iceland to adapt in an economically efficient manner to changed fish stock availability.

For the other relevant social structures, it also appears that those in Iceland are well-suited to facilitating adjustment to adverse environmental changes. First, Icelandic social structures are used to having to adapt to quite drastic fluctuations in the economy (see section 13.3.5.1), which implies that these institutions have evolved to cope with such fluctuations. Second, labor education and mobility is high in Iceland. High labor mobility refers to labor movements within Iceland as well as to between Iceland and other countries. This means that negative regional impacts resulting from climate change are unlikely to lead to permanently depressed unemployment areas or even persistent national unemployment. Third, the level of private and commercial entrepreneurship is high in Iceland. This suggests that in the case of negative impacts, economic substitute activities are likely to be quickly spotted and exploited. Fourth, there is an extensive social welfare system in Iceland that would provide at least temporary compensation to individuals adversely affected by environmental change. Although this social "safety net" will tend to delay adjustments to new conditions, it will also ensure that the burden of adverse changes is shared among the total population and so prevents individual hardship.

Thus, broadly speaking, it appears that Icelandic social structures are well suited to cope with sudden changes in fish stock availability resulting from climate change. This means that adjustments to changes are likely to be fast and smooth. Adverse environmental changes will nevertheless be economically and socially costly. However, they are unlikely to be exacerbated by social and economic responses. On the other hand, although the Icelandic economy seems well prepared to deal with changes in fish stock availability, an adverse biological impact may be felt more strongly than in the past. This is because some of the most valuable commercial stocks are currently close to their historical minimum. Hence, if there is an adverse environmental change, the initial reduction in harvests may be more dramatic than previously experienced and, also, the risk of a long-tem stock depression greater.

The ability of social structures in Greenland to adjust to new conditions is similar to the situation in Iceland. Greenlanders are used to variable environmental and economic conditions. Although perhaps not at the level of Iceland, general education and labor training is also high in Greenland. However, partly due to the size of Greenland and the isolation of many communities, labor mobility is considerably less than in Iceland. Most importantly, however, the degree of private and commercial entrepreneurship in Greenland seems much less than in Iceland.

Thus, while the Greenland social structure and institutions seem reasonably well placed to adjust to the changes in fish stock availability that might result from climate change, they presently appear less suited in this respect than those in Iceland.

13.3.8. Concluding comments

The ecosystems of Iceland and Greenland are very different. Icelandic waters occur to the south of the Polar Front under normal circumstances and are therefore under the influence of warm Atlantic water. Greenland waters are dominated by the cold East Greenland Current. This difference is reflected in higher numbers of exploited species at Iceland than at Greenland where there is also a dominance of arctic species of plankton, commercial invertebrates, and fish.

Both ecosystems were subject to large climatic changes in the 20th century. After a prolonged period of cold

conditions (lasting several decades), a warm period started around 1920 and peaked in the 1930s. Conditions were cooler in the 1940s, but warmed again and stayed warm for several years. There was a sudden cooling associated with decreased salinity and severe ice conditions in the mid-1960s. These changes reverberated around the North Atlantic at least twice during the next two decades. Conditions improved at Iceland in 1972 but remained variable for the next two and a half decades. Since the late 1990s, conditions have been persistently warmer at Iceland and even more so off West Greenland.

The warming of the 1920s and the early 1930s was followed by spectacular changes in the fish fauna of Icelandic and, in particular, Greenlandic waters. Cod and herring began to spawn in large numbers off the north coast of Iceland in addition to the traditional spawning areas in the Atlantic water off the south and west coasts. For some years, capelin were absent from their usual spawning areas to the south and west of Iceland, but spawned instead off southeast Iceland as well as in fjords and bays on the north and east coasts. Despite searches by Danish biologists and fishers, almost no cod were found to the south and west of Greenland between 1900 and 1920. However, after 1920 cod began to appear in increasing numbers in these areas and in the 1930s a cod fishery off West Greenland yielded on average about 100 000 t per year. This change is attributed to a massive drift of 0-group cod from Iceland across the northern Irminger Sea to the east and then west of Greenland.

Other changes associated with the warm period were manifested in a regular appearance of more southerly species such as mackerel and tuna at Iceland, while cold-temperate species such as haddock, saithe, and herring became fairly common at Greenland. In the latter case, cod extended their distribution northward to the west of Greenland for hundreds of kilometers to Disco Bay. Likewise, the center of capelin distribution shifted from off southwest Greenland to the Disco Bay area and the northern limit of their distribution reached as far as Thule.

Although many of the cod which had drifted across to Greenland as juveniles returned to spawn in their native areas off south and west Iceland, many did not, but spawned instead off West Greenland and eventually gave rise to a self-sustaining Greenlandic cod component. The Greenland cod were very successful for a number of decades and eventually formed a large local stock that supported catches in the hundreds of thousands of tonnes.

The cooling of the mid-1960s had a devastating effect, both at Iceland and Greenland. The spawning of cod at Greenland seems to have ceased completely and the stock crashed. At Iceland, the zooplankton community in the north and east changed from an atlantic to an arctic type. The most drastic effect was the disappearance of the Norwegian spring-spawning herring (the largest known herring stock in the world) from its traditional feeding areas north of Iceland. In the following years the herring fed further east in the Norwegian Sea until the stock crashed in the late 1960s. The two local and much smaller Icelandic herring stocks suffered the same fate. These stocks were all subject to a large fishing pressure in the 1960s. While the collapses in the herring stocks can be traced to exploitation beyond sustainable levels, this is not the case for the Greenland cod stock. Although the stock was also heavily fished, and its downfall accelerated by the fishery, evidence suggests that cod cannot reproduce effectively at Greenland except under warm climatic conditions. On the other hand, the low temperatures of the latter 1960s and relatively cold but variable conditions of Icelandic waters since the early 1970s do not seem to have adversely affected the Icelandic cod stock. The relatively low abundance of cod at Iceland at present appears to be the result of overfishing.

From a socio-economic point of view, climate-driven changes in fish abundance at Iceland over the 20th century had very large effects. In particular, the disappearance of the Atlanto-Scandian herring had severe consequences at all levels of society. Iceland lost about half its foreign revenue from fish products almost instantly, resulting in severe economic depression. However, the depressed state did not last long. The herring sector of the fishery quickly targeted other local species and also shifted to fishing herring elsewhere, mainly in the North Sea and adjacent waters. The Icelandic fishery sector also adapted quickly to various fishing restrictions imposed during the last quarter of the 20th century.

The lucrative cod fishery, which started off West Greenland in the late 1920s and lasted until the stock collapse in the early 1970s, did not have much effect on the Greenland economy. This was because the lack of suitable vessels and gear, as well as the necessary infrastructure, meant Greenland was unable to benefit from these conditions except on a very small scale.

Three scenarios of possible future climate change (no change, moderate warming (1–3°C), and considerable warming (4°C or more)) were used to examine likely outcomes for Iceland and Greenland. Changes in the size and distribution of commercial stocks are very unlikely under the no-change scenario. Thus, the Greenlandic fishing sector would mainly depend on cold water fish such as Greenland halibut and invertebrates such as northern shrimp and snow crab. At Iceland, fish species such as Atlantic cod, haddock, saithe, and redfish would dominate demersal fisheries, while capelin, local herring, and possibly blue whiting would dominate the pelagic fisheries. Catches of some species under the no-change scenario could be increased considerably through effective fisheries management, particularly in Iceland.

Moderate warming is likely to result in quite large positive changes in the catch of many species. Through larval drift from Iceland, a self sustaining cod stock is likely to be established off West Greenland which could yield annual catches of around 300000 t. If that happens, catches of northern shrimp are likely to decrease to about 30% of the present level, while catches of snow crab and Greenland halibut are not likely to alter much. Such changes would probably approximately double the export earnings of the Greenland fishing industry, which roughly translates into the sum presently paid by Denmark to subsidize the Greenland economy. Such dramatic changes are not likely in the Icelandic marine ecosystem. Nevertheless, it is likely that there will be an overall gain through larger catches of demersal species such as cod, and pelagic species such as herring, and new fisheries of more southern species such as mackerel. On the other hand, capelin catches are likely to decrease, both through diminished stock size and the necessity of conserving this important forage fish for other species. Effective fisheries management is very likely to continue to play a key role for Greenland and Iceland.

Little can be said about potential changes under the scenario of considerable warming. This is because such a situation is outside any recorded experience.

13.4. Newfoundland and Labrador Seas, Northeastern Canada

Fisheries in ACIA Region 4 may be subdivided into those near the coast of Greenland, those near the coast of Canada, and those in the deep waters of Baffin Bay and Davis Strait between Greenland and Canada. The whole area is within the fisheries convention area of NAFO (Fig. 13.24) and the stocks are currently managed by the coastal state or by NAFO.

Along the northeast coast of Canada the study area extends southward to the central Grand Bank (46° N) in

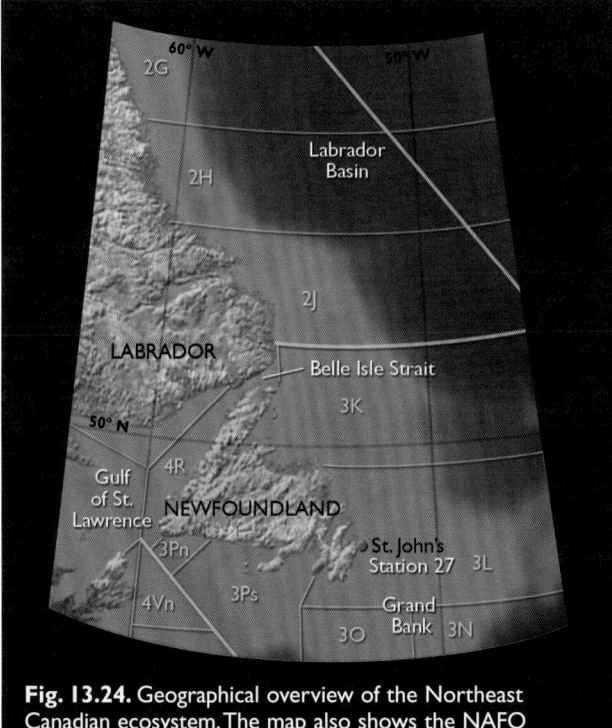

Fig. 13.24. Geographical overview of the Northeast Canadian ecosystem. The map also shows the NAFO statistical areas referred to in the text.

order to assess climate-driven impacts on marine ecosystems that are comparable to those considered for the northeast Atlantic (section 13.2) and around Iceland (section 13.3). This southward extension reflects the presence of the Labrador Current, which transports cold water southward from Davis Strait, the Canadian Archipelago, and Hudson Bay. The median southerly extent of sea ice is on the northern Grand Bank at approximately 47° N (Anon, 2001) and bottom water temperatures on the northern Grand Bank are below 0 °C for long periods. The southerly extent of cold conditions is also indicated by the regular presence of polar cod along the northeast coast of Newfoundland and their occasional occurrence on the northern Grand Bank (Lilly and Simpson, 2000; Lilly et al., 1994).

Fish has dominated the history of Newfoundland since the time of British colonization. The British interest in Newfoundland after its "discovery" during the Cabot voyage of 1497 was due to the incredibly large amounts of codfish. Exploitation of this fishery by the British reduced its dependence on Iceland for fish, a dependence that was creating difficulties. The French also saw the value of Newfoundland's fishery, and possession of the island became an important part of the colonial wars of the 18th century (for the historical background of Newfoundland, see Chadwick, 1967; Innis, 1954; Lounsbury, 1934). As an inducement for France to enter the revolutionary war on the side of the American colonies, Benjamin Franklin offered a share of the Newfoundland fishery to the French as bounty once the war was won (Burnett, 1941). Indeed, until the late 1800s, when a cross-island railroad was built, fishing was Newfoundland's only industry. There was then a series of diversification programs, which have continued in one form or another until the present day. Although in the early 1970s Newfoundland had the world's largest hydroelectric plant (in Labrador), and despite many attempts to diversify the economy with both small-scale industries (e.g., cement production, knitting mills, a shoe factory, a chocolate factory) and numerous large-scale industries (e.g., the Churchill Falls hydroelectric station, a petroleum refinery, a third paper mill, iron mines in Labrador) in the twenty years following Confederation with Canada in 1949, none of these made any difference to the dominance of the fishery in Newfoundland (Letto, 1998). However, what did change in Newfoundland with Confederation, and after revisions in Canadian federal/provincial intergovernmental arrangements, was the emergence of extremely large government, health, and education sectors which, as shares of GDP, eclipsed the fishery. By 1971, the fish and fish processing sectors accounted for less than 5% of Newfoundland GDP. By 2001, their contribution was 3.5 %.

13.4.1. Ecosystem essentials

The ecosystem off northeastern Canada has been characterized by a relatively small number of species, a few of which have historically occurred in high abundance

(Bundy et al., 2000; Carscadden et al., 2001; Livingston and Tjelmeland, 2000). The dominant fodder fish has been capelin, with polar cod more prominent to the north and sand lance (*Ammodytes dubius*) more prominent to the south on the plateau of Grand Bank. Herring is found only in the bays and adjacent waters. These four species of planktivorous fish feed mainly on calanoid copepods and larger crustaceans, the latter predominantly hyperiid amphipods to the north and euphausiids to the south. The dominant piscivorous fish has been Atlantic cod, but Greenland halibut and American plaice (*Hippoglossoides platessoides*) have also been important. Snow crab and northern shrimp have been the dominant benthic crustaceans. The top predators are harp seals and hooded seals which migrate into the area from the north during late autumn and leave in spring. Other important predators include whales, most of which migrate into the area from the south during late spring and leave during autumn. The most important are humpback (*Megaptera novaeangliae*), fin, minke, sei, sperm, and pilot whales (*Globicephala melaena*). Additional immigrants from the north during the winter include many birds, such as thick-billed murre, northern fulmar (*Fulmarus glacialis*), and little auk. Additional immigrants from the south during summer include short-finned squid, fish such as mackerel and bluefin tuna (*Thunnus thynnus*), and birds such as greater shearwater (*Puffinus gravis*) and sooty shearwater (*P. griseus*).

The Labrador/Newfoundland ecosystem has experienced major changes since 1980. Atlantic cod and most other demersal fish, including species that were not targeted by commercial fishing, had declined to very low levels by the early 1990s (Atkinson, 1994; Gomes et al., 1995). In contrast, snow crab (DFO, 2002a) and especially northern shrimp (DFO, 2002b) increased considerably in abundance during the 1980s and 1990s and now support the most important fisheries in the area. Harp seals increased in abundance between the early 1970s and the mid-1990s (DFO, 2000c). Capelin have been found in much reduced quantities in offshore acoustic surveys since the early 1990s, but indices of capelin abundance in the inshore surveys have not shown similar declines, leaving the status of capelin uncertain and controversial (DFO, 2000b, 2001). Atlantic salmon, the major anadromous fish in the area, has declined in abundance, due in part to lower survival at sea (DFO, 2003b; Narayanan et al., 1995).

The waters of eastern Newfoundland have been fished for centuries, primarily for Atlantic cod but with an increasing emphasis on other species during the latter half of the 20th century. These fisheries have undoubtedly had an influence on both the absolute abundance of some species and the abundance of species relative to one another. However, the role of the fisheries in structuring the ecosystem is often difficult to distinguish from the role of changes in the physical environment. The area cooled during the last three decades of the 20th century, with particularly cold periods in the early 1970s, early to mid-1980s, and early 1990s. This cooling, which was

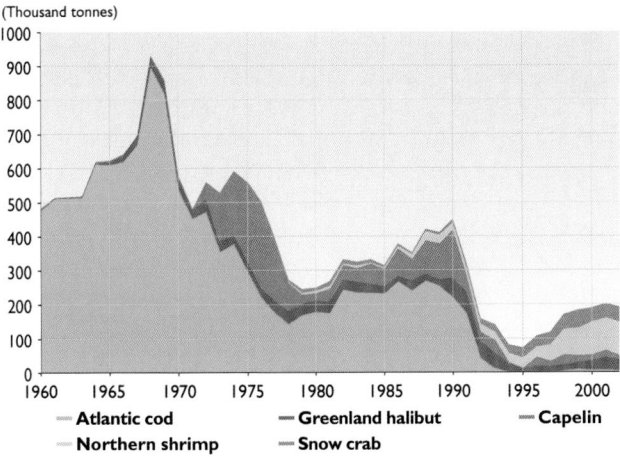

(Thousand tonnes)

— Atlantic cod — Greenland halibut — Capelin
— Northern shrimp — Snow crab

Fig. 13.25. Total catch of selected species off northeastern Canada, 1960–2002 (data from NAFO STATLANT 21A online database www.nafo.int/activities/FRAMES/AcFrFish.html; Anon, 2004a; Darby et al., 2004; Dawe et al., 2004; Hvingel, 2004; Lilly et al., 2003; Murphy and Bishop, 1995; Orr et al., 2003, 2004; Stenson et al., 2000).

associated with an intensification of the positive phase of the North Atlantic Oscillation (Colbourne and Anderson, 2003; Colbourne et al., 1994; Mann and Drinkwater, 1994; Narayanan et al., 1995), may have played an important role in the dramatic decline in Atlantic cod and other demersal fish, and the increase in crustaceans, especially northern shrimp.

13.4.2. Fish stocks and fisheries

Catches are from official NAFO statistics (as of February 2004) or from relevant assessment documents if there is a difference between the two (e.g., NAFO 2001a,b). Figure 13.25 provides an overview of developments in the main fisheries off Newfoundland and Labrador since 1960.

13.4.2.1. Atlantic cod

The distribution of Atlantic cod off Canada has historically been from the northern Labrador Shelf southward to beyond the limit of this study, although during the 1990s there were few cod off Labrador. Atlantic cod tends to occur on the continental shelf, but has been

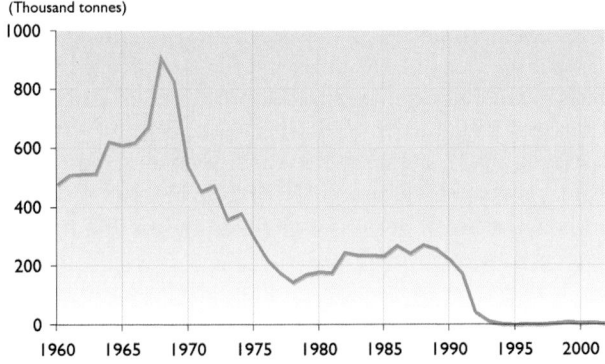

(Thousand tonnes)

Fig. 13.26. Catch of Atlantic cod, 1960–2002 (data from Murphy and Bishop, 1995; Lilly et al., 2003).

found at depths of at least 850 m on the upper slope off eastern Newfoundland (Baird et al., 1992).

The European fishery for Atlantic cod off eastern Newfoundland began in the late 15th century. For the first few centuries fishing was by hook and line, so the cod were exploited only from late spring to early autumn and only in shallow water along the coast and on the plateau of Grand Bank to the southeast of the island. There is evidence that local inshore over-exploitation was occurring in the 19th century (Cadigan, 1999), but improvements in gear and an increase in the area fished tended to compensate for local reductions in catch rate. Annual landings increased through the 18th and 19th centuries to about 300 000 t in the early 20th century. The deep waters were refugia until the 1950s, when larger vessels with powered gur-dys were introduced to exploit cod in deep coastal waters and European trawlers started to fish the deeper water on the banks. Landings increased dramatically in the 1960s as large numbers of trawlers located and exploited the overwintering aggregations on the edge of the Labrador Shelf and the Northeast Newfoundland Shelf. At the same time, the numbers of large cod in deep water near the coast of Newfoundland are thought to have declined quickly as the longliner fleet switched to synthetic gillnets. Catches peaked at 894 000 t in 1968, and then declined steadily to only 143 000 t in 1978. Following Canada's declaration of a 200 nm EEZ in 1977, the stock recovered slightly and catches were between 230 000 and 270 000 t for most of the 1980s. However, catches fell rapidly in the early 1990s as the stock declined to very low levels. A moratorium on directed fishing was declared in 1992 (Fig. 13.26). A small cod-directed inshore fishery was opened in 1998 but closed in 2003. Additional details on the history of the Atlantic cod fishery of Newfoundland and Labrador, including changes in technology and temporal variability in the spatial distribution of fishing effort, may be found in Templeman (1966), Lear and Parsons (1993), Hutchings and Myers (1995), Lear (1998), Neis et al. (1999), and Hutchings and Ferguson (2000).

13.4.2.2. Greenland halibut

Greenland halibut (also called Greenland turbot) is distributed off West Greenland from Cape Farewell northward to about 78° N and then southward off eastern Canada to beyond the limit of this study. It is a deep-water species, occurring at depths from about 200 m to at least 2200 m off West Greenland (Bowering and Brodie, 1995). The history of the fishery is complicated by temporal and spatial variation in effort and catch by different fleets and by alleged underreporting of landings. For details of the fisheries, refer to Bowering and Brodie (1995), Bowering and Nedreaas (2000), and NAFO (2001b).

The fishery off eastern Newfoundland dates back to the mid-19th century (Bowering and Brodie, 1995; Bowering and Nedreaas, 2000). Annual catches from

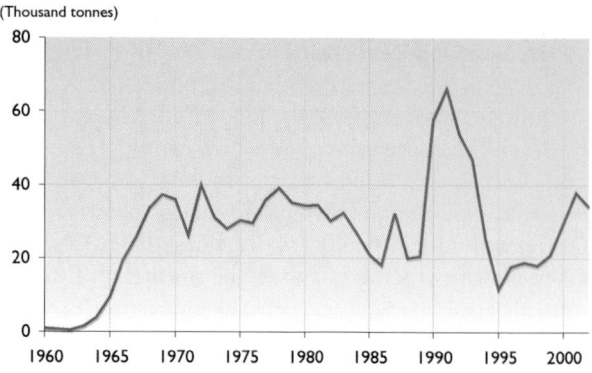

Fig. 13.27. Catch of Greenland halibut, 1960–2002 (data from NAFO STATLANT 21A online database www.nafo.int/activities/FRAMES/AcFrFish.html; Darby et al., 2004).

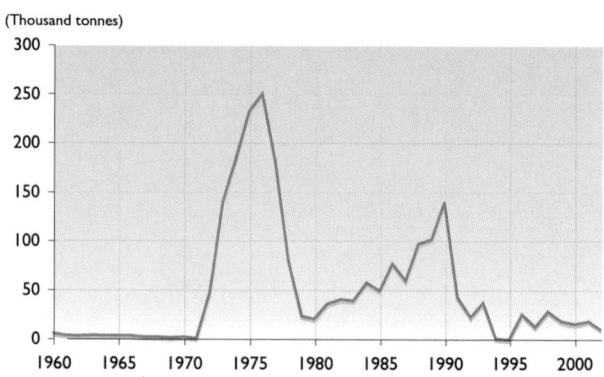

Fig. 13.28. Catch of capelin, 1960–2002 (data from NAFO STATLANT 21A online database www.nafo.int/activities/FRAMES/AcFrFish.html).

longlines were less than 1000 t until the early 1960s, when catches began to increase substantially. Landings from offshore trawlers, mainly from European countries, also increased after the mid-1960s. Catches in SA 2 + Div. 3KL fluctuated around 25000 to 35000 t from the late 1960s to the early 1980s, after which there was a gradual decline to about 15000 t in 1986. Landings increased dramatically in 1990 with the arrival of many non-Canadian trawlers that fished deep waters on the northern Grand Bank (see Fig. 13.24 for location). Catches over the next four years were high (estimated at between 55000 and 75000 t in 1991; NAFO, 2001b), declined substantially in 1995 due to an international dispute, and increased again in the late 1990s under NAFO quotas that maintained catches well below those of the early 1990s (Fig. 13.27).

The fishery to the north (NAFO SA 0), which has been conducted primarily with otter trawlers in the second half of the year (Bowering and Brodie, 1995), reported an average annual catch of 2100 t between 1968 and 1989 (including a high of 10000 t in 1972). Catches increased dramatically to 14500 t in 1990 with increased effort by Canada, but declined to about 4000 t from 1994 onward. These landings came mainly from off southeastern Baffin Island. The fishery expanded even further north into Baffin Bay in the mid- to late 1990s (Treble and Bowering, 2002). This fishery, which extended to 73° N in 2002 (M.A. Treble, Fisheries and Oceans Canada, pers. comm., 2003), has been limited by sea-ice cover in September through November.

13.4.2.3. Capelin

Before the start of a commercial offshore fishery in the early 1970s, capelin were fished on or near the spawning beaches. Annual catches, used for local consumption, may have reached 20000 to 25000 t (Templeman, 1968). Offshore catches by foreign fleets increased rapidly, peaking in 1976 at about 250000 t, and then declined rapidly. This offshore fishery continued at a low level until 1992. Catches in the offshore fishery were taken at different times of the year in different areas. The spring fishery was dominated by large midwater trawlers operating in Div. 3L. During the autumn, the offshore fishery first

occurred in Div. 2J, off the coast of Labrador, and gradually moved south into Div. 3K as the capelin migrated toward their overwintering area (see Fig. 13.24 for NAFO statistical areas). This fishery was also dominated by large midwater trawlers, which mostly took feeding capelin that would spawn the following year. During the late 1970s, as the foreign fishery declined, Canadian fishers began fishing mature capelin near the spawning beaches to supply the Japanese market for roe-bearing females. This fishery expanded rapidly, exhibited highest catches during the 1980s, and declined over the 1990s. Catches in the inshore fishery have generally been lower than from the offshore fishery. The total international catch of capelin off Newfoundland and Labrador from 1960 to 2002 is shown in Fig. 13.28.

13.4.2.4. Herring

Herring in the Newfoundland and Labrador area are at the northern extent of their distribution. Stocks are coastal in distribution and stock abundance is low compared to other stocks in the Atlantic. A peak catch of 30000 t occurred in 1979, supported by strong year classes from the 1960s. Recruitment since the 1960s has been lower. Stock sizes in the late 1990s were less than 90000 t and annual catches less than 10000 t (DFO, 2000a).

13.4.2.5. Polar cod

Polar cod is broadly distributed through the Arctic and in cold waters of adjacent seas. It occurs on the shelf from northern Labrador to eastern Newfoundland, with the average size of individuals and the size of aggregations decreasing from north to south (Lear, 1979). There has been no directed fishery for polar cod off eastern Canada, but a small bycatch was reported in the Romanian capelin fishery in 1979 (Maxim, 1980), and it is likely that small quantities were also taken in other years and by other countries.

13.4.2.6. Northern shrimp

Northern shrimp is distributed off West Greenland from Cape Farewell northward to about 74° N and then

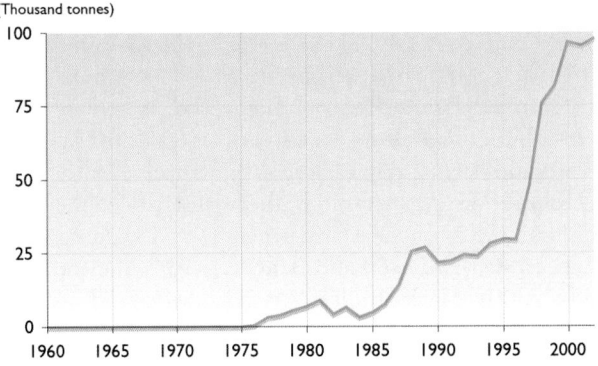

Fig. 13.29. Catch of northern shrimp, 1960–2002 (data from NAFO STATLANT 21A online database www.nafo.int/activities/FRAMES/AcFrFish.html; Hvingel, 2004; Orr et al., 2003, 2004).

southward off eastern Canada to beyond the limit of this study. The depth of highest concentration tends to vary from area to area but is generally between 200 to 600 m. A fishery with large trawlers began off northeastern Canada in the late 1970s (Orr et al., 2001a). For the first decade most of the catch was taken from two channels in the central and southern Labrador Shelf, but in the late 1980s there was an increase in effort and landings both to the south on the Northeast Newfoundland Shelf and to the north off northern Labrador. Catches increased above 25 000 t by the mid-1990s. New survey technology introduced in 1995 indicated that commercial catches were very small relative to survey biomass, and quotas were increased considerably in the late 1990s. Total landings rose to more than 90 000 t by 2000 (Fig. 13.29). Much of the increase in catch from 1997 onward was from a new fleet of small (<100 feet) vessels that fished with bottom trawls mainly on the mid-shelf. In the 1990s fishing also expanded to Div. 3L (Orr et al., 2001b).

13.4.2.7. Snow crab

Snow crab is distributed from the central Labrador Shelf at approximately 55° N southward off eastern Canada to beyond the limit of this study. The depth distribution extends from approximately 50 to 1400 m, but most of the fishery occurs at 100 to 500 m. The fishery off eastern Newfoundland began in the late

1960s as a small bycatch fishery, but soon expanded into a directed fishery with crab traps (pots) along most of the inshore areas of eastern Newfoundland (Div. 3KL) (Taylor and O'Keefe, 1999). During the late 1970s and early 1980s there was an increase in effort and an expansion of fishing grounds. Catches in Div. 3KL reached almost 14 000 t in 1981, but then declined. In the mid-1980s there was expansion of the fishery to the area off southern Labrador (Div. 2J) and new entrants gained access to supplement declining incomes from the groundfish fisheries. The number of participants and the area fished expanded further during the 1990s, and total catches rose quickly, reaching almost 55 000 t in 1999. Quotas and landings were reduced for the next two years following concerns that the resource may have declined.

Commercial catch rates in Div. 3KL increased during the late 1970s to a peak in about 1981, declined to their lowest point by 1987, and then increased in the late 1980s and early 1990s to a level comparable to that in the early 1980s (DFO, 2002a). Catch rates remained high to the end of the 1990s, despite the substantial increase in fishing effort and landings (Fig. 13.30). This partly reflects an increase in the area fished, although there must also have been an increase in productivity.

13.4.2.8. Marine mammals

Harp seals summer in the Canadian Arctic or Greenland but winter and breed in Canadian Atlantic waters. There are two major breeding groups: the first breeding in the Gulf of St. Lawrence and the second breeding off southern Labrador and northeast Newfoundland (Bundy et al., 2000). The total population increased from less than 2 million in the early 1970s to more than 5 million in the mid-1990s (Healey and Stenson, 2000; Stenson et al., 2002). The increase was largely due to a reduction in the hunt after 1982 (Stenson et al., 2002). The population stabilized when the hunt was increased in the mid-1990s. Reported Canadian catches of harp seals include

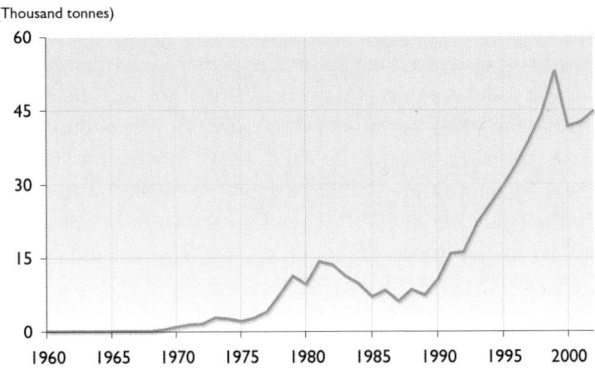

Fig. 13.30. Catch of snow crab, 1960–2002 (data from NAFO STATLANT 21A online database www.nafo.int/activities/FRAMES/AcFrFish.html; Hvingel, 2004; Orr et al., 2003, 2004).

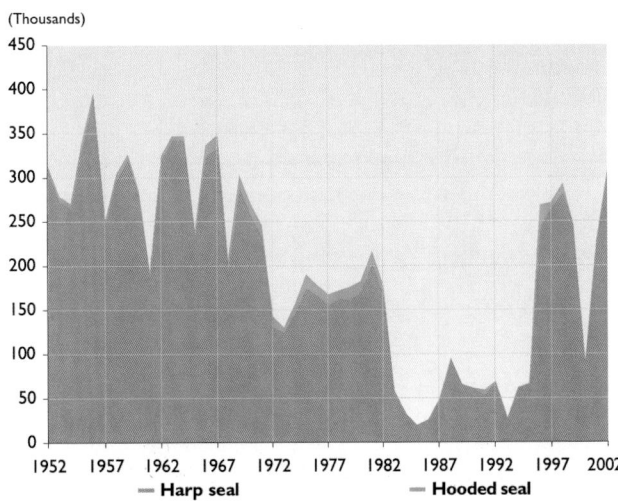

Fig. 13.31. Catch of harp and hooded seals, 1952–2002. The catches also include seals taken in the Gulf of St. Lawrence (data from Anon, 2004a; Stenson et al., 2000).

harvests off the coast of Newfoundland/Labrador (the "Front") and in the Gulf of St. Lawrence. Seals caught in both areas belong to the same population: the Northwest Atlantic Harp Seals. The proportion of the population that occurs in the two areas varies among years, as does the relative number of seals caught in each area. Catches from both areas are combined in official statistics and so those presented here are combined "Front" and Gulf of St. Lawrence catches (Fig. 13.31).

Hooded seals are less abundant than harp seals. Whelping occurs on pack ice off northeast Newfoundland, in Davis Strait, and in the Gulf of St. Lawrence. Pups migrate into arctic waters and remain there as juveniles. Adults migrate south in the autumn and return to the Arctic in April (Bundy et al., 2000). The harvest of hooded seals ("Front" and Gulf of St. Lawrence combined) is shown in Fig. 13.31.

There has been no commercial whaling in the area since the late 1970s. Using north Atlantic population estimates, assumed growth rates, and an assumed proportion of the total population in the Newfoundland and Labrador area, Bundy et al. (2000) estimated population abundances of 33000 for humpback whales, 1000 for fin whales, 5000 for minke whales, 1000 for sperm whales, 1000 for sei whales, and 9000 for pilot whales.

13.4.3. Past climatic variations and their impact on commercial stocks

13.4.3.1. Atlantic cod

The severe decline in Atlantic cod in the Newfoundland and Labrador area seems to have occurred from north to south. On the northern and central Labrador shelf (Div. 2GH), catches of 60000 to 90000 t were reported for the period 1965 to 1969, but catches declined to less than 5000 t for most of the 1970s and early 1980s, and to less than 1000 t in the latter half of the 1980s (Fig. 13.32). There are no analyses of factors that contributed to the decline in this area.

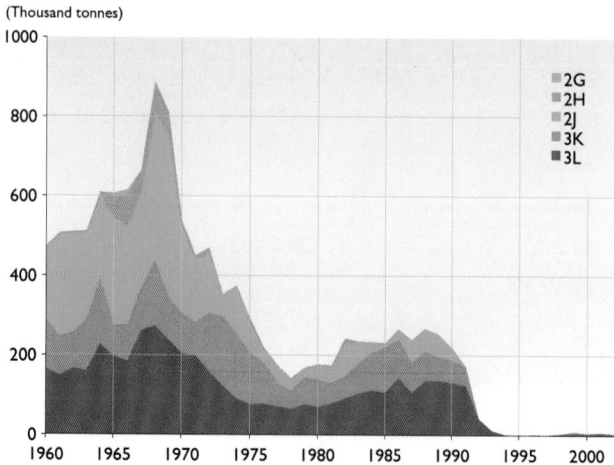

Fig. 13.32. Catch of Atlantic cod by NAFO statistical division, 1960–2000 (data from Murphy and Bishop, 1995; Lilly et al., 2003).

In the area from southern Labrador to the northern Grand Bank, the Div. 2J+3KL stock (the so-called "northern cod") collapsed in the 1970s in response to severe overfishing. The stock recovered slightly in the 1980s but collapsed to even lower levels in the late 1980s and early 1990s. There is controversy as to whether there was a rapid but progressive decline from the mid-1980s onward or a precipitous decline in the early 1990s (Atkinson and Bennett, 1994; Shelton and Lilly, 2000). Many studies (e.g., Haedrich et al., 1997; Hutchings, 1996; Hutchings and Myers, 1994; Myers and Cadigan, 1995; Myers et al., 1996a,b, 1997a,b) have concluded that the final stock collapse was entirely due to fishing activity (landed catch plus discards). However, several authors have pointed to ways in which the decline in water temperature and increase in sea-ice cover might have contributed to the collapse, either directly by reducing productivity (Drinkwater, 2000, 2002; Mann and Drinkwater, 1994; Parsons L. and Lear, 2001) or indirectly by affecting distribution (Rose et al., 2000).

Despite many studies on this cod stock, there are few uncontested demonstrations of the influence of climate variability on stock dynamics. There is an expectation that recruitment might be positively influenced by warm temperatures, because the stock is at the northern limit of the species' range in North America (Planque and Frédou, 1999). However, there have been conflicting reports of whether such a relationship can be detected (deYoung and Rose, 1993; Hutchings and Myers, 1994; Planque and Frédou, 1999; Taggart et al., 1994). Part of the problem is that recruitment is also positively influenced by the number and size of spawners in the population (the spawning stock biomass or SSB; Hutchings and Myers, 1994; Morgan et al., 2000; Myers et al., 1993; Rice and Evans, 1988; but see Drinkwater, 2002). Both temperature and SSB declined from the 1960s to the 1990s, increasing the difficulty of demonstrating a temperature effect. A reported positive relationship between recruitment and salinity (Sutcliffe et al., 1983) was subsequently supported (Myers et al., 1993) and later rejected (Hutchings and Myers, 1994; Shelton and Atkinson, 1994) as data for additional years became available. The negative effect of temperature on individual growth has been well documented (Krohn et al., 1997; Shelton et al., 1999). Additional aspects of cod biology that changed during the early 1990s, possibly in response to changes in the physical environment, include a delay in arrival on traditional inshore fishing grounds in early summer (Davis, 1992), a concentration of distribution toward the shelf break in autumn (Lilly, 1994; Taggart et al., 1994), a move to deeper water in winter (Baird et al., 1992), and an apparent southward shift in distribution (Kulka et al., 1995; Rose and Kulka, 1999; Rose et al., 1994).

Of much interest is the possibility that an increase in natural mortality contributed to the rapid disappearance of cod in the early 1990s. The sharp decline in survey abundance indices occurred during a period of severe cold and extensive sea-ice cover. A considerable decline in the condition of the cod occurred at the same time, especially in the north (Bishop and Baird, 1994; Lilly, 2001). Steep declines in abundance also occurred among other groundfish in the 1980s and 1990s (Atkinson, 1994; Gomes et al., 1995), and while there have been some suggestions that these declines were caused by captures during fishing for cod and other species (Haedrich and Barnes, 1997; Haedrich and Fischer, 1996; Haedrich et al., 1997; Hutchings, 1996), there is no direct evidence of large removals. In the case of American plaice, a species studied in detail, Morgan et al. (2002) demonstrated that the declines were too large to have resulted from fishing alone. The contribution of increased natural mortality to the decline in cod and other demersal fish in this area during the last two decades of the 20th century, and particularly during the early 1990s, remains unresolved (Lilly, 2002; Rice, 2002).

The northern cod stock was still at a very low level a decade after the moratorium on directed fishing (DFO, 2003a; Lilly et al., 2001). Recruitment to ages 0 to 2 remained very low, possibly due in part to a very small spawning stock biomass; juveniles in the offshore areas appeared to show very high mortality, possibly due in part to predation by harp and hooded seals; and a directed fishery during 1998 to 2002 targeted the inshore aggregations, resulting in increased mortality on the larger fish. The unquantified impacts of low spawning stock biomass, high predation, and fishing make it difficult to establish whether some aspect of ocean climate has had a role in impeding recovery.

13.4.3.2. Greenland halibut

The status of Greenland halibut in the northwest Atlantic has been uncertain because the stock structure is still unclear, the fish have extensive ontogenetic migrations, there appear to have been shifts in distribution, the fisheries have undergone many changes in fleet composition and in areas and depths fished, and individual research surveys have only covered part of the distribution range. Nevertheless, evidence suggests that the biomass of Greenland halibut on the western side of the Labrador Sea declined substantially during the 1980s, with the decline off Baffin Island and northern Labrador (Div. 0B and 2GH) most pronounced in the first half of the decade and the decline off southern Labrador and eastern Newfoundland (Div. 2J3K) to the south most pronounced in the latter half of the 1980s and the early 1990s (Bowering and Brodie, 1995). Evidence for a decline in biomass in Div. 2J3K is also seen in the declining success of the gillnet fishery in the 1980s. The history of the fish exploited during the 1990s by the new deep-water trawler fishery to the south on the northern Grand Bank (Div. 3L) is less clear. At least some of these fish may have migrated into the area from the shelf to the north (Bowering and Brodie, 1995), in which case the decline in Div. 2J3K was partly due to a southward shift in distribution.

Reasons for the declines in biomass and shift in distribution remain unclear. Bowering and Brodie (1995) drew attention to the decline in water temperatures on the shelf in the early 1990s, but thought it unlikely that such a change would in itself have affected the distribution and abundance of Greenland halibut because this species occupies relatively deep water. Also, much of the shift in distribution must have occurred in the latter half of the 1980s, a period during which water temperatures were low but not as low as during the early 1990s.

Variability in the physical environment had no observed effect on either size at age (Bowering and Nedreaas, 2001) or maturity at size or age (Morgan and Bowering, 1997) between the late 1970s and mid-1990s.

13.4.3.3. Capelin

The relationship between capelin biology and the physical environment has been extensively studied in the Newfoundland and Labrador area. Of particular relevance to this assessment is the observation that many aspects of capelin biology changed during the 1990s and, initially, it appeared that these were the result of changes in water temperature. However, water temperatures in the latter half of the 1990s returned to normal while the biological changes exhibited by capelin did not revert to earlier patterns. There are many environmental variables that are linked to capelin biology which may be relevant in the event of global climate change and these are briefly described in the rest of this section.

Mean fish length of the mature population was smaller during the 1990s (Carscadden et al., 2002). These small sizes have been attributed to smaller fish sizes at age with fewer older and more younger fish in the population. Condition (calculated as a relationship between length and weight and regarded as a measure of "well-being") of capelin was generally higher in the 1980s than the 1990s. Condition was not related to temperature (Carscadden and Frank, 2002).

Spawning occurs most often on fine gravel and grain size and beach orientation have been shown to explain 61% of the variation in egg concentration among beaches (Nakashima and Taggart, 2002; Vilhjálmsson, 1994). Water temperature is also a determinant of capelin spawning. The lowest and highest recorded temperatures for beach spawning in Newfoundland are 3.5 and 11.9 °C, with beach spawning ceasing when temperatures exceed 12.0 °C (Nakashima and Wheeler, 2002). Capelin eggs are very cold- and salinity-tolerant, surviving down to -5 °C and in salinities from 3.4 to 34 (Davenport, 1989; Davenport and Stene, 1986). The rate of egg development in the beach gravel is directly related to average incubation temperatures, which in turn are determined by water temperature, maximum and minimum air temperature, and hours of sunlight (Frank and Leggett, 1981).

Some capelin that move close to spawning beaches eventually spawn in deeper water adjacent to beaches. This demersal spawning can occur simultaneously with intertidal spawning when temperatures are suitable as well as when water temperatures at the beach–water interface become too warm. Egg mortality among these demersal eggs has been observed to be higher. Reproductive success may have been lower during the 1990s because the water temperatures encountered when the capelin reached the spawning beaches would have increased the incidence of demersal spawning (Nakashima and Wheeler, 2002). Historically, the spawning of capelin off Newfoundland beaches in June and July was a predictable event. In the early 1990s, spawning was later, and 80% of the variation in spawning time (1978–1994) was significantly and negatively related to mean fish size and sea temperatures experienced during gonadal maturation (Carscadden et al., 1997). Spawning on Newfoundland beaches continued to be delayed through 2000 despite sea temperatures having returned to normal. However, mean lengths of capelin continued to be small.

There are historical reports of capelin occurring outside their normal distribution range. Unusual appearances in the Bay of Fundy and on the Flemish Cap were attributed to cooler water temperatures while occurrences in Ungava Bay coincided with warming trends (summarized by Frank et al., 1996).

In the early 1990s, capelin distribution occurred more to the south, centered on the northern Grand Banks. Originally attributed to the colder water temperatures (Frank et al., 1996), this shift within the normal distribution area continued through 2000. Because capelin did not return to their usual pattern of seasonal distribution as water temperatures increased, this suggests that factors other than water temperature were also operating. Outside their normal distribution area, capelin occurred on the Flemish Cap and eastern Scotian Shelf in the early 1990s and occasionally during earlier cold periods. Capelin continued to appear on the Flemish Cap and on the eastern Scotian Shelf through 2000. In this case, capelin appear to be gradually declining in abundance as the waters warm. For mature capelin offshore during spring, Shackell et al. (1994) concluded that temperature was not a proximate cue during migration but that seasonal temperatures moderated offshore capelin migration patterns through the regulation of growth, maturation, food abundance, and distribution.

Capelin typically move up and disperse throughout the water column at night, descending and aggregating at greater depths during the day. However, during spring surveys throughout the 1990s they remained deeper in the water column and exhibited reduced vertical migration (Mowbray, 2002; Shackell et al., 1994). This change in vertical distribution was not related to the several factors tested, including temperature and predation, but may have been linked to feeding success (Mowbray, 2002).

Recruitment of beach-spawning capelin is partly determined by the frequency of onshore winds during larval residence in the beach gravel (Carscadden et al., 2000; Leggett et al., 1984). Capelin assessments have been especially problematic since the early 1990s, resulting in considerable uncertainty in the status of the stock. However, there is no evidence to indicate that exploitation has had a direct effect on population abundance (Carscadden et al., 2001), suggesting that any variations in abundance are due to environmental factors. It is not known whether some changes in biology such as condition and distribution have affected abundance, however, spawning time and increased demersal spawning may be contributing to poor survival.

Thus, exploitation has not been shown to affect any aspect of capelin biology in this area. Although there have been several changes in capelin biology beginning in the early 1990s, there is no clear indication of what external factor(s) has (have) influenced the changes. Earlier studies concluded that temperature was an important factor for some changes, but it now seems unlikely that temperature is the sole factor, given that water temperatures have returned to normal. There are suggestions that changes in food supply (zooplankton) may be affecting capelin biology but the exact mechanisms have not been identified.

13.4.3.4. Herring

Recruitment is positively related to warm overwintering water temperatures and high salinities (Winters and Wheeler, 1987); these conditions seldom exist in this region and so, large year classes rarely occur.

13.4.3.5. Polar cod

The distribution of polar cod off eastern Newfoundland expanded to the south and east during the cold period of the early 1990s (Lilly and Simpson, 2000; Lilly et al., 1994).

13.4.3.6. Northern shrimp

The shrimp resource off northeastern Canada has increased in density and expanded in distribution since the mid-1980s. There is no indication that increased catches have negatively affected the resource (DFO, 2002b).

There is much support for the hypothesis that the increase in northern shrimp off northeastern Canada was, at least in part, a consequence of a reduced predation pressure by Atlantic cod and other groundfish (Bundy, 2001; Lilly et al., 2000; Worm and Myers, 2003). Nevertheless, there is evidence that other factors were involved. For example, Lilly et al. (2000) noted that the increase in shrimp density on the Northeast Newfoundland Shelf might have started in the early 1980s, a time when the biomass of Atlantic cod was increasing following its first collapse in the 1970s. Parsons D. and Colbourne (2000) found that catch per unit effort in the shrimp fishery on the central Labrador Shelf was positively correlated with sea-ice cover six years earlier. They suggested that cold water or sea-ice cover itself was beneficial to the early life history stages of shrimp in that area.

13.4.3.7. Snow crab

The increased productivity of snow crab in the 1990s may have been caused, at least in part, by the release in predation pressure from Atlantic cod and other demersal fish (Bundy, 2001). However, the relationships between Atlantic cod and snow crab have not yet been explored to the same extent as for Atlantic cod and northern shrimp. A preliminary examination of the influence of oceanographic conditions on snow crab productivity has shown a negative relationship between ocean temperature and lagged catch rates (DFO, 2002a). This has been interpreted to indicate that cold conditions early in the life cycle are associated with the production of strong year classes of snow crab in this area.

13.4.3.8. Marine mammals

Trends in populations of marine mammals over recent decades appear to be influenced mainly by the commercial harvest. As populations of harp seals have increased in abundance, changes in biological characteristics indicate that density-dependence may be operating (Stenson et al., 2002). Density-independent influences may also regulate harp seal populations. Harp seals whelp on sea ice and mortalities may vary according to sea-ice conditions in this critical period. Mortalities of newly whelped pups may also occur during winter storms.

Concerns regarding the impact of predation by seals on commercial fish species increased as seal populations increased. It has been estimated that 74% (about 3 million t) of the total annual consumption by four species of seals in eastern Canada occurred off southern Labrador and Newfoundland (Hammill and Stenson, 2000). Predation by harp seals has been implicated in the lack of recovery of the northern cod stock (DFO, 2003a), and predation on cod by hooded seals may be large (DFO, 2003a).

13.4.3.9. Aquaculture

Salmonid aquaculture does not occur in the ACIA part of Newfoundland because the water is too cold in winter. The main species cultured is blue mussel. Production of this species has grown over the last twenty years such that, in 2002, around 1700 t were raised in the whole of Newfoundland.

13.4.4. Possible impacts of climate change on fish stocks

Two recent papers (Frank et al., 1990; Shuter et al., 1999) discussed the possible influence of climate change on ecosystems and fisheries off eastern Canada. Frank et al. (1990) predicted shifts in the ranges of several groundfish stocks because of redistribution of populations and changing recruitment patterns. Stocks at the southern limit of a species' distribution should retract northward, whereas those near the northern limit should expand northward. Frank et al. (1990) did not make predictions specifically for Labrador and eastern Newfoundland, but events during the decade following publication of their paper were in many respects opposite to these general predictions. The changes off Labrador and eastern Newfoundland were unprecedented and not predicted, and illustrate the uncertainty of predictions, even on a regional scale and in the relatively short term. Shuter et al. (1999) had the advantage of witnessing the dramatic changes that occurred in the physical and biotic environment during the 1990s. They concluded that greenhouse gas accumulation will lead to a warmer, drier climate and, for the fisheries of Atlantic Canada, this will result in a "decrease in overall sustainable harvests for coastal and estuarine populations due to decreases in freshwater discharge and consequent declines in ecosystem productivity". For fisheries in the Arctic, they predicted "increases in sustainable harvests for most fish populations due to increased ecosystem productivity, as shrinkage of ice cover permits greater nutrient recycling".

As the relative importance of fishing and environment is difficult to determine for any species or group of species, it is not surprising that the importance attributed to each has varied for different studies. It is also not surprising, given the differences among species in the magnitude of fishery removals relative to stock size, that opinion favors fishing as the dominant factor for some species and environment for others. For demersal fish, there are many statements to the effect that declines were caused entirely by overfishing, but there is evidence that changes in oceanographic properties contributed to changes in distribution and declines in productivity. For crab and especially shrimp, it has been suggested that increases in biomass were simply a consequence of a release in predation pressure from Atlantic cod and perhaps other demersal fish, but again there is evidence that changes in oceanographic factors contributed to an increase in reproductive success. For capelin, most information supports the hypothesis that fishing had little impact on population dynamics, and that environmental factors were the primary determinant of stock size, well-being (growth and condition), distribution, and timing of migrations. For polar cod, fishing may be dismissed as a contributor to changes in distribution and biomass.

An important constraint on predicting changes in fish stocks and the fisheries that exploit them off Labrador and eastern Newfoundland is uncertainty about the direction and magnitude of change in important oceanographic variables. For surface air temperature, some model outputs project a cooling over the central North Atlantic, and it is not clear where the Labrador/Newfoundland region lies within the gradation from significant warming in the high Arctic to cooling over the central North Atlantic. In addition, there is no model for downscaling the output of general circulation models to specifics of the Labrador/Newfoundland area. As an example of the importance of regional models, many large-scale models project an increase in air temperature over the Norwegian and Barents Seas, while simulations with one specific regional model (Furevik et al., 2002) indicate that sea surface temperature in that area may decline in the next 20 years before increasing later in the century. Another concern is that natural variability in a specific region, such as the Labrador Shelf, may be greater than variability in the global mean (Furevik et al., 2002). Thus, a warming trend in shelf waters off Labrador and Newfoundland might be accompanied by substantial annual variability, such as was witnessed during the last three decades of the 20th century, and it is even possible that the amplitude of that variability could increase. For biota, extreme events associated with this variability might be at least as influential as any long-term trend. For the Labrador Shelf and Northeast Newfoundland Shelf, it is probably at least as important to know how the North Atlantic Oscillation will behave (especially the intensity and location of the Icelandic Low) as it is to know that global temperature will rise.

In the absence of region-specific information on likely future developments of climate, all predictions of climate-driven changes in the marine ecosystem off Newfoundland and Labrador can only be highly tentative. The following subsections describe the changes that seem most likely under three different scenarios: no change or even cooling of climate, moderate warming, and considerable warming.

13.4.4.1. No change

As temperatures were generally below the long-term average during the ACIA baseline period (1981–2000), no change from present conditions or even a cooling are likely to favor the current balance of species in the system. This implies a predominance of commercial invertebrates like northern shrimp and snow crab and cold water species of fish such as Greenland halibut, polar cod, and capelin.

13.4.4.2. Moderate warming

The moderate warming scenario (an increase of 1 to 3 °C) assumes that there will be a gradual warming of the shelf waters off Labrador and Newfoundland. Using the events in West Greenland during the first half of the 20th century (Vilhjálmsson, 1997) as a spatial/temporal analogue, there is likely to be better recruitment success and northward expansion of Atlantic cod and some other demersal fish that live mainly on the shelf. Capelin is also likely to shift northward. If zooplankton abundance is enhanced by warmer water, capelin growth is likely to improve. It is possible that many existing capelin spawning

beaches will disappear with the projected rise in sea level (Shaw et al., 1998). Depending on the increase in sea level, storm events, and the availability of glacial deposits, some beaches may move and new beaches be formed, while others may disappear completely. While beach-spawning capelin can adapt to spawning on suitable sediment in deeper water, survival of eggs and larvae appears to be adversely affected (Nakashima and Wheeler, 2002), suggesting that a rise in sea level is likely to result in reduced survival and recruitment for capelin. A warming of sea temperatures is likely to retard recruitment to snow crab and northern shrimp, so these species might experience gradual reductions in productivity. Thus, a gradual, moderate warming of sea temperature is likely to promote a change back to a cod–capelin system from the present system where snow crabs and northern shrimp are the major commercial species. In addition, both cod and capelin are also likely to become more prominent off central Labrador than they were during the 1980s.

A gradual warming of shelf waters is also likely to promote a shift of more southerly species into the area. For example, haddock is likely to become more abundant on the southern part of Grand Bank, and expand into the study area. Migrants from the south, such as short-finned squid, mackerel, and bluefin tuna, are likely to occur more regularly and in greater quantities than in the 1980s and early 1990s.

The simple scenario of a gradual change back to a cod–capelin system under moderate warming conditions is uncertain. This is because the influence of oceanographic variability in the past is still not clear, and because it is likely that the dynamics of some species are now dominated by a different suite of factors than was the case in the past. It is highly likely that the ecosystem off northeastern Canada changed substantially as a consequence of fishing during the first four centuries after the arrival of European fishers, changed even further with the increasingly intensive fishing of the 20th century, and has changed dramatically from the 1960s onward. The magnitude of these changes is such that it would be difficult to predict accurately the future state of this ecosystem even without the added complications of climate change. Thus, the system could remain in its current state, could revert to some semblance of an historic state (or at least the state of the early 1980s), or could evolve toward something previously unseen.

Changes in sea ice (see Tables 9.2 and 9.3) are likely to have a negative impact on harp seals, the most important marine mammal predator in the area. Sea-ice duration is projected to shorten and it is not known whether harp seals would be able to adjust their breeding time to accommodate this change. A decrease in sea-ice extent is unlikely to affect harp seals because they would probably shift their distribution with the sea ice. However, thinner sea ice may be deleterious, resulting in increased pup mortality. Increases in regional storm intensities (see Table 9.1) are likely to result in higher pup mortalities if such storms occur during the critical period shortly after birth (G. Stenson, Fisheries and Oceans Canada, St. John's, pers. comm., 2003). Changes in seal abundance are likely to cascade through the ecosystem, since seals are important predators on many fodder fish and commercially important groundfish (Bundy et al., 2000; Hammill and Stenson, 2000), and are thought by some to be important in impeding the recovery of cod (DFO, 2003a) and thus maintaining the present balance within the ecosystem.

In addition to uncertainty regarding the response of individual species and the ecosystem as a whole, there is uncertainty regarding the influence of changing sea-ice cover on the fisheries themselves. A reduction in the extent and duration of sea ice may permit fishing further to the north and would increase the period during which ships would have access to certain fishing grounds. In particular, these changes in sea-ice cover would affect the Greenland halibut and shrimp fisheries in Baffin Bay and Davis Strait. For example, an increased open water season and extended fishing period is thought to have the potential to increase the harvest of Greenland halibut at the time of spawning (late winter/spring).

A reduction in sea-ice cover (see Tables 9.2 and 9.3) is also likely to negatively impact upon Greenland halibut fisheries that are conducted through fast ice. For example, a fishery that was developed in Cumberland Sound on Baffin Island in the late 1980s has developed into a locally important enterprise (Crawford, 1992; Pike, 1994). The fishery is conducted with longlines set through ice over deep (600–1125 m) water, with the season extending in some years from mid-January to June. Since the mid-1990s, the season has been shorter, typically from early February to May (M.A. Treble, Fisheries and Oceans Canada, Winnipeg, pers. comm., 2003). To date, attempts at fishing during the open water season have not proved successful. The catches have been small and the fish appear dispersed. It is unclear whether the fish would be present in commercial concentrations in the winter/spring if sea ice were not present. Even if they were, the absence of sea ice would certainly affect the conduct of the fishery.

13.4.4.3. Considerable warming

Since a warming of 4 to 7 °C is beyond any recorded experience in the Newfoundland–Labrador area, a meaningful discussion of the considerable warming scenario is not practicable. In very general terms, such a shift could favor cold-temperate species such as cod, improve conditions for more southern species such as haddock and herring, and even lead to the formation of demersally spawning stocks of capelin in addition to beach spawning stocks. However, there are likely to be other changes, such as a freshening of the surface layer due to freshwater from melting sea ice further north, which would be likely to reduce primary production in the area.

13.4.5. The economic and social importance of fisheries

From an economy based primarily on the fishery, Newfoundland has, along with most of North America, moved to a service economy. By 1971, for instance, the fishing and fish processing sectors accounted for less than 5% of Newfoundland's GDP, whereas the service sector accounted for more than half. Mining accounted for 11% and construction 18% (although that included construction of some of the large diversification projects). Nearly twenty years later, in 1989, shortly before the groundfish collapse, the fishery harvesting and processing sectors together accounted for slightly more than 5%, service industries had grown to 68%, mining had fallen to less than 6%, and construction to 8% (for a more extensive discussion see Schrank et al., 1992). By 2001, the fishery harvesting and processing sectors accounted for only 3% of GDP, the service industries remained constant at 68%, construction had slipped to 4.7%, and conventional mining to 3%. Oil production, a new industry in Newfoundland, already accounted for 8.4% of the provincial GDP, with every prospect of growing (the 2001 data were from the Newfoundland Statistics Agency; www.nfstats.gov.nf.ca/statistics/GDP/GDP_Industry.asp). Mining was also expected to see resurgence with the potential opening of a large nickel mine in Labrador.

While the fishery may not be of great importance to the overall Newfoundland economy, it continues to dominate completely the economy in rural areas, and perhaps even more importantly, its culture. After fifty years, there is still a daily *Fisheries Broadcast* on radio and when the Canadian Broadcasting Company decided to cancel the weekly television program *Land and Sea*, which often focuses on the fishery, public pressure forced the crown corporation to continue the program. With the fishery in deep trouble in 1989, the dominant newspaper in the province, *The Evening Telegram*, commented in an editorial entitled "Too Many Fishermen?" on 1 June 1989, that "Newfoundland's fishery must eventually be expanded and diversified so that it can employ more people, not fewer…"

With the spectacular change in fisheries employment that accompanied the collapse of the northern cod stock, there has been a sharp reemphasis on economic diversification. The two areas paraded as holding the hope of the future are tourism and information technology (Government of Newfoundland and Labrador, 2001). Progress has been made in both areas (e.g., fishing vessels converted to tour boats for whale watching, and many bed and breakfast establishments), but has been uneven, and some government policies have been inconsistent with the promotion of these industries. For instance, despite its interest in developing tourism, the Newfoundland government decommissioned or privatized a substantial number of the parks in the province's extensive parks system (Overton, 2001). How many of those sold remain as parks is unknown.

The real problem with the emphasis on tourism and information technology is that it is happening at the time as it is happening throughout the world. Why should Newfoundland have an advantage over the rest of the world in either field? It is too early to tell whether this diversification will be successful.

The story of the Newfoundland fishery does not end with the collapse of the groundfishery, its catastrophic consequences for many families, and the serious pressures it placed on the government. Two critical sets of changes have occurred since 1992: (1) the fishery management process has evolved and (2) shellfish have replaced groundfish as the main components of fishery landings in Newfoundland.

Fishery management by restricting total allowable catches began in Newfoundland in the mid-1970s. Following a particularly dramatic race to the fish in 1981, the government imposed, at the industry's request, enterprise allocations on the offshore groundfish fleet in 1982. These allocations were divisible and transferable in the year in which they were assigned. Government emphasized that these were rights to fish as opposed to property rights, which could be permanently sold. By this time, gear and geographic restrictions had been imposed, as had limited entry to non-groundfish inshore fisheries. With the expansion of the crab fishery, enterprise allocations have also been assigned to this inshore sector. These allocations are not transferable. The federal government has relinquished the licensing of fishers in favor of the provincial Professional Fish Harvester's Certification Board. This board was established as part of a professionalization program and it licenses harvesters either as apprentices or in one of two professional classes. The federal government, in turn, in 1996 established "core" fishing enterprises for the inshore fisheries. Senior level professional fishers who met certain conditions were declared the heads of core fishing enterprises. Approximately 5500 of these were established and the government claimed that no additional core licenses would be issued; the only way that fishers could obtain such status would be to buy out the core license of an existing core fisher. In an attempt to reduce the number of fishers, the federal government bought approximately 1500 core licenses, claiming that these will not be reissued. The final major changes in the management system are that (1) species license fees (access fees) are no longer nominal, for most important species they are based on the anticipated gross income from the fishery; and (2) there is now an extensive system of public consultations before recommendations concerning total allowable catches are made to the Minister of Fisheries and Oceans. For a discussion of the current fishery management system, see Schrank and Skoda (2003).

The value of landings in all Newfoundland fisheries in 1991 was Can$ 282838000. This fell during the first year of the moratorium to Can$194745000 and then doubled to Can$ 388700000 by 2000. Viewed alternatively, the period of the northern cod moratorium saw an increase of one-third in the real value of Newfoundland fish and shellfish landings. But, while in 1991 43% of the value of landings was for cod, in 2000 46% was accounted for by crab and a further 30% by shrimp. With some cod stocks reopened on a limited basis for commercial fishing, cod accounted for 9% of the total landings in 2000, although this had earlier (in 1996) fallen to less than one-half of one percent (www.dfo-mpo.gc.ca/communic/statistics/landings).

For environmental reasons, whether the lack of predation or favorable climatic conditions, the shellfish population has surged and there has been a nearly complete conversion of Newfoundland's fishing industry from groundfish to shellfish. The conversion took years to occur. Labor requirements for shellfish are lower than for groundfish and, having higher unit prices, the shellfish quantities that yield these landings figures are much smaller than for groundfish.

Since all major species are under quota, total allowable catches for shrimp and crab have been increasing rapidly, along with the number of harvesting and processing licenses for shrimp and crab. The number of shrimp harvesting licenses rose from 19 in 1986 and 57 in 1991 to 438 in 2000. The numbers of crab licenses for those years were 274, 721, and 3333 (Corbett, 2002). There has been much controversy as to whether the old error of issuing too many licenses has occurred for crab. Even with the exodus from the province, in February 2003 there was still a 17.5% seasonally adjusted unemployment rate in the province (where the national figure was 7.4%) with continued pressure to open closed plants and increase licenses for crab fishers. While the number of crab fishing licenses has increased substantially, the increase in the number of crab processing plants has been modest.

13.4.6. Past variations in the fishing industry and their economic and social impacts

The *Evening Telegram* editorial, referred to in section 13.4.5, appeared within the context of a fishery that had long been in trouble. In 1967, a provincially financed report supported the trend away from a seasonal inshore fishery in Newfoundland toward a capital-intensive year-round offshore fishery. The report also noted that the "number of people dependent on the fishery should be reduced" (Pushie, 1967, 185). This has been a recurring theme. In 1970, the federal fisheries department appealed to the Canadian cabinet to permit the department to establish regulations that would lead to a 50% reduction in the number of Atlantic Canadian (meaning mainly Newfoundland) fishers.

The authority to effect these changes was denied (Schrank, 1995). The fishery faced repeated crises, was repeatedly studied, and the conclusion was repeatedly drawn that too many people were dependent on it. One study estimated that of the then 35000 licensed fishers, only 6000 could be supported unsubsidized by the fishery at a better than poverty-level income. The same study concluded that for every dollar of fish landed, there was a dollar of subsidy (Schrank et al., 1986). In 1976, with the extension of coastal states' fishery jurisdiction to 200 nm from shore, the two Canadian departments concerned (fisheries and regional economic expansion) both published reports stating that there was sufficient extra capacity in the industry that no significant employment benefits could be expected from the expanded jurisdiction (Government of Canada, 1976a,b). Two years later, a provincial government report made a similar point by stating that the Canadianized fishery, when fully developed, could employ only 9000 inshore fishers (Government of Newfoundland and Labrador, 1978). Yet, in response to popular pressure, both federal departments, as well as the provincial government, licensed and subsidized a tremendous expansion in the physical capacity of the industry: from 13 636 registered fishers in 1975 to 33 640 in 1980; from 9517 registered inshore vessels in 1976 to 19 594 in 1980; from a fish freezing capacity of 181 000 t in 1974 to 467 000 t in 1980; from net fishers's unemployment insurance benefits of US$ 30 724 000 in 1976/77 to US$ 66 060 000 in 1980/81; and from outstanding loans of the provincial Fisheries Loan Board (to finance inshore vessels) from US$ 36 869 000 in 1976/77 to US$ 78 558 000 in 1980/81 (Schrank, 1995). By 1981 the expansion had stopped and the two federal departments agreed that no further expansion of fish processing facilities would be built with federal financing (LeBlanc and De Bané, 1981). However, the damage had already occurred and, in the face of the anti-inflation recession of the early 1980s in the United States (where most of the Newfoundland fish production was sold), the market for Newfoundland fish products shrank dramatically and most Newfoundland fish processing companies faced bankruptcy.

As a result, the industry was financially, but not structurally, reorganized and the massive industrial closures implicit in bankruptcy were averted (Kirby, 1982). Yet, starting in 1982, inshore groundfish catches fell and after 1986 offshore catches followed. By 1992, the situation was so bad that a moratorium on the commercial catching of the formerly massive northern cod stock was put into place. Shortly after, nearly all Newfoundland groundfisheries were closed and the moratorium was extended to non-commercial fishing (Schrank, 1995). The closure of most of these groundfisheries continues, in whole or in part, to the present. The closure of the Newfoundland groundfisheries is reputed to have involved the largest mass layoff of labor in Canadian history. In social terms (due to the mass layoff), in biological terms (due to the decimation of the fish stock), and in governmental financial terms (due to billions of dollars spent on income maintenance for fishers and fish plant workers) the moratoria were disasters.

The response of government, industry, and the public to the moratoria indicates what might happen with climate change. Although the cause of the stock destruction in Newfoundland waters may be debated, the dramatic effect on the fish population is incontrovertible. Should significant changes in environmental conditions occur, and should these changes have substantial effects on commercial fish stocks, then the Newfoundland experience may provide a template for what might be expected to happen elsewhere. Moreover, the Newfoundland experience may also indicate the need for alternative policies.

The decline in the cod fisheries was better understood after the reports by Alverson (1987) and Harris L. (1989, 1990). That major problems were developing in the groundfishery was no longer debatable. In 1990, in response to the decline of the fishery, the federal government introduced the Atlantic Fisheries Adjustment Program (AFAP; see Schrank, 1997). The emphasis was on the word "adjustment". People were to be retired from the fishery, rural communities were to receive money to help them diversify their economies away from the fishery, and steps were to be taken to increase scientific understanding of the declining fish stocks. But only a few hundred fishers left the industry. With the shock of the total closure of the commercial northern cod fishery in July 1992, the federal government, anticipating that the fishery would revive in two years, created the Northern Cod Adjustment and Recovery Program (NCARP; see Schrank, 1997). Again there was an emphasis on people adjusting out of the fishery. This program called for early retirement of fishers, buybacks of fishing licenses, training of fishers for other trades, and income maintenance payments to fishers and fish plant workers. A third of the 9000 northern cod fishers and half the 10 000 plant workers affected by the northern cod shutdown were expected to leave the fishery. In fact, only 1436 took early retirement and 876 fishers sold their licenses. Fishers were not convinced that the shutdown would continue for long, and believed that the government would support them until the fishery reopened; the relatively uneducated, potential low end laborers did not see a need to leave an industry in which they were skilled and for which they were trained from an early age. One reason for low educational levels (until 1991, less than half Newfoundland's adult population had completed high school) was the fishing tradition. Boys started fishing with their fathers at a young age and looked forward to leaving school as soon as possible to join the family fishing enterprise. Boys and girls with little interest in fish harvesting could work for life in the local fish plant, in jobs which mostly required little formal education.

However, the fish did not return after two years, and have still not returned a decade after the start of the moratorium. As NCARP was ending, a new adjustment program began. The Atlantic Groundfish Strategy (TAGS; see Schrank, 1997) was to be a five-year program of income maintenance and adjustment (license buybacks and retirements) in which a 50% reduction in fishing capacity was anticipated. Again, there was very little movement of people out of the fishery. Their reluctance to abandon the fishery was for the same reasons as under the NCARP program.

As TAGS drew to a close at the end of the 1990s, the government took a harder line. The post-TAGS program did not resemble its predecessors: income maintenance was severely cut and many people were removed from the program. With government financial support gone, or going, and the fish still not returned, an exodus from the fishery finally occurred.

Between 1986 and 1991, the Newfoundland population stagnated, at least partially from a dramatic drop in family size. From the highest birth rate of Canadian provinces, by 1991 Newfoundland had the lowest. Also, there had always been modest migration out of the province. But in the five years between 1991 and 1996 (from the year after the start of AFAP to halfway through TAGS) the population actually fell by 3%. With the continuing moratorium and the change in government policy, the exodus increased significantly and between 1996 and 2001 there was a further drop of 7%. Census figures for 2001 are from Statistics Canada (2002a), while those for 1991 and 1996 are from Statistics Canada (1999).

Even though such a population drop in a province over a decade is dramatic, this value of 10% actually hides the severity of the impact of the fishery collapse on Newfoundland's rural communities. Trepassey, on the southern shore of the Avalon Peninsula, was the location of a major groundfish processing plant. Newfoundland groundfish operators had been operating under an Enterprise Allocation scheme since 1982. With the drop in fish stocks toward the end of the 1980s, enterprise allocations were cut and, in response, a number of fish plants closed. One of the first to close, in 1990, was that in Trepassey. The result was that a town with 1375 inhabitants in 1991 had shrunk by more than 35% to 889 in 2001. Many rural communities in Newfoundland have seen population declines since 1990 of 15 to 30% (Statistics Canada, 2002b).

13.4.7. Economic and social impacts of climate change: possible scenarios

Climate change is likely to cause changes in the size of fish stocks. The effects are unlikely to be greater than the historical changes described in section 13.4.6. With human society, responses to impulses are not "natural" in the sense that a climatic change "causes" a human response. The human response is determined by the magnitude of the stimulus plus the political response of the society. In this sense, societal responses to climate change will not be qualitatively different from society's responses to past changes. The political system will respond, and the details of that response are impossible to predict. But models exist from past experience. However, many of the federal and provincial interventions in the fishery since 1992 have appeared ill thought out, often unfair, and have raised controversy.

A recent case provides an illustration. The groundfish plant in Twillingate was once owned by the largest fish company in Newfoundland, Fishery Products International, Ltd., but had been sold to another operator in the mid-1980s (Schrank et al., 1995). With the northern cod moratorium in 1992, the plant was shut and remained shut until 2002 when it opened as a shellfish plant with more technologically sophisticated equipment than had been used for groundfish. The Marine Institute, a branch of Memorial University, introduced a course to teach fish plant workers to use the new equipment, charging more than CAN$ 400 per person. For unemployed people receiving (un)employment insurance the fee was paid by the federal government. Others had to pay for themselves. Most unemployed people without employment insurance could not afford the fee. Most working people, wanting higher paying jobs in the fish plant, or former fish plant workers wanting to return to the industry, would need to quit their jobs to take the course, unless they were granted time off, which is unlikely in unskilled trades. But if they quit their jobs and completed the course, there was no guarantee of a job in the Twillingate, or any other technologically advanced, fish plant (CBC, 2002). Thus, every aspect of the long adjustment process which started with the decline of the groundfishery in the 1980s has been characterized by a deep sense of unfairness.

The Newfoundland experience shows that a "catastrophic" event concerning the fishery leads to severe adjustment problems, and that the adjustment period may be very long, but that it also raises new potential for a successful industry. The issues seem to be:

- how to convince participants in the industry that there is a crisis;
- adjustments that need to be made;
- the role of government; and
- how to protect the new fishery from the mistakes of the failed fishery.

In terms of predicting the socio-economic effects of long-term climate change, this is one case where it is easier to prescribe than to predict. The Newfoundland experience has shown reactions to expanding fish populations and to shrinking fish populations. In neither case was the reaction, in terms of government action or political pressure, appropriate. During the expansion period of the late 1970s, the fishery expanded too much, with excessive and ultimately largely immobile labor and capital entering the industry. Whereas a properly managed fishery would have restricted the expansion of production factors, the expansion was almost without letup until stopped by a general economic crisis. It was understood at the time that employment expansion was an incorrect response. Should fish stocks off Newfoundland increase over the next 20 or 50 years, care should be taken (1) to restrict by government regulation the magnitude of any expansion of capital and labor in the fishery; or (2) to ensure that such economic incentives are in place that excessive growth does not occur; or (3) to combine the two.

Should fish stocks decrease over the next 20 or 50 years, then it should be clear from the start that endless subsidies will not be forthcoming. License buyouts, even generous license buyouts of core enterprises for instance, would help. While these payments are subsidies, they are limited in scope and time and would have the effect of permanently shrinking the factor base of the industry. In the 1990s, such a policy would have been much cheaper and much less stressful for the fishing families affected, than the offering of income maintenance payments.

A gradual warming of shelf waters is likely to lead to increased opportunity for aquaculture. Warmer temperatures and shorter periods of sea-ice cover are likely to enable mussel farming to be more productive. Warmer waters are also likely to promote the development of Atlantic cod farming. If inshore waters become sufficiently warm, it is likely to be possible to farm Atlantic salmon along the east coast of Newfoundland. This is presently impossible because water temperature in winter falls below the lethal temperature for salmon.

13.4.8. Ability to cope with change

Climate change will affect all aspects of the fishery: the range of existing species, the relative populations of different species, and the economic circumstances of people who depend on the fishery for a living. How ready the economic and social systems of Newfoundland are to cope with these changes is not clear. When the Newfoundland fishery was revitalized after the declaration of the 200 nm limit, its economic structure overexpanded with largely immobile capital and labor and resulted in disaster. When the cod fishery started to decline in the late 1980s, several years passed before many of the necessary adjustments occurred. Whether the situation will be any different in response to climate change depends on whether lessons have been learned, and whether the social and political systems are prepared to adjust. Both the expansion of the ground-fishery in the late 1970s and the failure of the fishery to adjust to decimated stocks in the early and mid-1990s were largely due to the subsidies. During the 1970s the expansion was mainly financed by the federal and provincial governments. Despite the efforts of the federal government to adjust fishers out of the industry in the 1990s, the adjustment programs became income maintenance programs, which in effect encouraged fishers to remain in the industry in the hope that the fish would return. It was only when the subsidies were substantially reduced after 1998 that a significant number of fishers left the industry (based on the assumption that departures from the fishery are reflected in the census figures).

Subsidies are not the results of whimsical acts of governments or politics but are responses to real social and economic concerns. As long as the government considers the survival of small rural communities a major priority, subsidies to the fishery (the primary industry in these communities) will continue. While subsidies exist, the response of people to changes in the industry will be slow. Without subsidies, economic forces will require change, probably rapid change if the fishery is declining. If the biological base of the fishery is expanding, there is always the possibility that the industry will overexpand without government help. Government financial assistance would virtually ensure that overexpansion would occur.

To the extent that adjustments induced by climate change cause human suffering, the government can be expected to ameliorate the situation and ease the necessary transitions. But there is strong precedent for transition programs being transformed into short-run income maintenance programs. If that were to happen again, the process of adjustment is likely to be as long, painful, and wasteful as before.

Thus, it is impossible to predict how ready society is to cope with the effects of climate change. The response mechanisms are not automatic and political reactions will play a major role.

13.4.9. Concluding comments

The ecosystem off the northeast coast of Canada is under the influence of the Labrador Current, which carries cold water south from Davis Strait, the Canadian Archipelago, and Hudson Bay. As a result, climate impacts in this ecosystem can be compared to impacts on comparable ecosystems in the Northeast Atlantic and Iceland. Historically, the dominant demersal species were cod, Greenland halibut, and American plaice, the dominant invertebrates were northern shrimp and snow crabs, the dominant pelagic fish was capelin, and the dominant top predators were harp seals and whales.

The Labrador/Newfoundland ecosystem experienced major changes in the 1980s and 1990s. Atlantic cod and most other demersal fish, including species that were not targeted by commercial fishing, had declined to very low levels by the early 1990s. In contrast, snow crab and especially northern shrimp surged during the 1980s and 1990s and now support the most important fisheries in the area. Harp seals increased in abundance between the early 1970s and the mid-1990s. Capelin have been found in much reduced quantities in offshore acoustic surveys since the early 1990s, but indices of capelin abundance in the inshore surveys have not experienced similar declines, leaving the status of capelin uncertain and controversial.

The relative importance of overfishing and the environment on changes in cod and Greenland halibut has not been determined, although fishing is generally accepted as the most important factor affecting cod abundance. Ocean climate is thought to have had an impact on the lack of cod recovery, although this has not been quantified. Exploitation has not been shown to have affected any aspect of capelin biology in this area. Although there have been several changes in capelin biology since the early 1990s, there is no clear indication of what external factor(s) has (have) influenced the changes. A combination of reduced predation and favorable environmental conditions probably contributed to the success of northern shrimp and snow crab. Harp seals increased because of reduced commercial harvesting.

Changes of the magnitude that have occurred in the biological components of the ecosystem since the early 1980s are unprecedented and together with the lack of regional predictions of changes in the ocean due to climate change, make predictions of biological responses to climate change highly speculative.

If there is no change from the present state or even a cooling, it is likely that the current balance of species will persist.

With a moderate, gradual warming, there is likely to be a change back to a cod–capelin system with a gradual decline in northern shrimp and snow crab. Cod and other demersal, shelf-dwelling species and capelin are likely to move north. Many existing capelin spawning beaches are likely to disappear as sea levels rise. If there is an increase in demersal spawning by capelin in the absence of new spawning beaches, capelin survival is likely to decline. Seals are likely to experience higher pup mortality as sea ice thins. Increases in regional storm intensities are also likely to result in higher pup mortality. A reduction in the extent and duration of sea ice is likely to permit fishing further to the north. A reduction in sea-ice cover is likely to shorten Greenland halibut fisheries that are conducted through fast ice.

If a more intense regional warming occurs as a consequence of extensive climatic warming, then predicting the responses of the biological community to these changes must occur in the absence of historic precedence and be completely speculative. Such an event is likely to improve conditions for cold-temperate species such as cod, improve conditions for more southern species such as haddock and herring, and even result in the formation of demersally spawning stocks of capelin.

Although the fishery in Newfoundland has accounted for 5% or less of provincial GDP since 1971, it dominates the economy and culture in rural areas. The cod fishery expanded rapidly in the 1980s and then contracted rapidly in the 1990s, the latter in response to the fishing moratorium. The social and economic effects of changes in fish stocks due to climate change are likely to be less than the historical changes experienced in the latter part of the 20th century in Newfoundland and Labrador.

Past experience suggests that the political system will respond but that the details of the response are impossible to predict. It is, however, possible to prescribe directions that governments should follow in the event of expansions or contractions of fish stocks resulting from climate change. If fish stocks off Newfoundland increase over the next 20 or 50 years, care should be taken (1) to restrict by government regulation the magnitude of any expansion of capital and labor in the fishery; (2) to ensure that such economic incentives are in place that excessive growth does not occur; or (3) to combine the two. If fish stocks decrease over the next 20 or 50 years, then it should be clear from the start that endless subsidies will not be forthcoming. License buyouts, even generous license buyouts of core enterprises for instance, would help. While these payments are subsidies, they are limited in scope and time and would have the effect of permanently shrinking the factor base of the industry.

Aquaculture in Newfoundland and Labrador is relatively small but there is interest in expansion, especially with the lack of recovery of cod stocks. A gradual warming of shelf waters is likely to lead to increased opportunity for aquaculture. Warmer temperatures and shorter periods of sea-ice cover are likely to enable mussel farming to be more productive. Warmer waters are also likely to promote the development of Atlantic cod farming and the farming of Atlantic salmon along the east coast of Newfoundland.

13.5. North Pacific – Bering Sea

The continental shelves of the eastern and western Bering Sea together produce one of the world's largest and most productive fishing areas (Fig. 13.33). They contain some of the largest populations of marine mammals, birds, crabs, and groundfish in the world (Overland, 1981). A quarter of the total global yield of fish came from here in the 1970s. The central Bering Sea contains a deep basin that separates the shelves on the Russian and American sides and falls partly outside the 200 nm EEZs of the two countries. Prior to extended fishing zones, a complex set of bi- and multilateral fisheries agreements was established for the area. These range from agreements on northern fur seal (*Callorhinus ursinus*) harvests and Canada/US fisheries for Pacific salmon (*Oncorhynchus* spp.) and Pacific halibut (*Hippoglossus stenolepis*), to the multilateral International

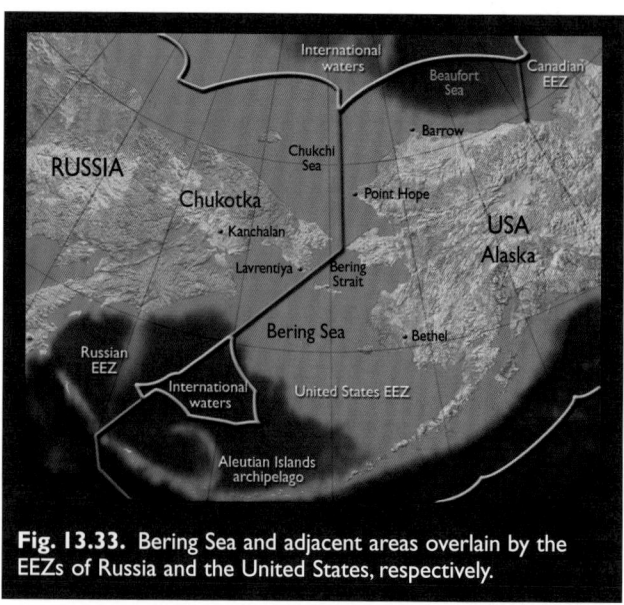

Fig. 13.33. Bering Sea and adjacent areas overlain by the EEZs of Russia and the United States, respectively.

North Pacific Fisheries Convention for the development and use of scientific information for managing fisheries on the high seas (Miles et al., 1982a,b). (Various post-EEZ license agreements have permitted fishing by non-Russian and US fleets. At present such fishing is precluded in the US EEZ and is much reduced in waters of the Russian Federation.) In the so-called "Donut Hole", a pocket of high seas area surrounded by US and Russian EEZs, scientific research and commercial fishing are carried out in accordance with the Convention on the Conservation and Management of Pollock Resources in the Central Bering Sea by the two coastal states and Japan, Korea, Poland, and China. The North Pacific Science Organization and the North Pacific Anadromous Fish Commission were established to facilitate fisheries and ecosystem research in the North Pacific region, including the Bering Sea.

Commercial fisheries in the Bering Sea are generally large-scale trawl fisheries for groundfish of which about 30% of the total catch is processed at sea and the rest delivered to shoreside processing plants in Russia and the United States. Home port for many of the Bering Sea vessels is outside the ACIA region reflecting the comparative advantage of supply and service available in lower cost regions. Small coastal communities have a strong complement of indigenous peoples with subsistence fishing interests. They depend on coastal species, especially salmon, herring, and halibut, but the overlap with commercial activities is generally small. Anadromous species extend far inland via the complex river systems and are critical resources for indigenous peoples. The chief indigenous involvement in the marine commercial sector is the Community Development Program in the Northeast Pacific where 10% of TACs are allocated to coastal communities and their chosen partners (Ginter, 1995). Because the eastern Bering Sea is within the EEZ of the United States, harvest levels of commercially important species of fish and invertebrates are regulated through federal laws. Management plans exist for the major target species that specify target fishing mortality levels calculated to maintain the long-term female spawning stock levels at 40% of the unfished equilibrium level for fully exploited species. In the western Bering Sea, within the Russian EEZ, fishery management is executed on the basis of an annual TAC established for all commercial stocks of fish, invertebrates, and marine mammals. Allowable catch is calculated as a percentage of the fishable stock. Percentages for individual stocks and species were based on early scientific studies and do not exhibit annual change. However, since 1997, these harvest percentages have been revised by government research institutes, using new modeling applications and adaptive management approaches. The recommended TACs are approved by the special federal agency and issued as a governmental decree.

Annual catches of all commercial groundfish species between 1990 and 2001 in the US eastern Bering Sea EEZ ranged from 1.3 to 1.8 million t and averaged 1.6 million t. The walleye pollock (*Theragra chalcogramma*)

catch averaged 1.2 million t and ranged from 0.99 to 1.45 million t (Hiatt et al., 2002). In the western Bering Sea, the total groundfish catch reached 1.45 million t in 1988 of which walleye pollock contributed 1.29 million t. The annual catch of walleye pollock between 1990 and 2001 averaged 0.73 million t ranging from 0.45 to 1.06 million t. Walleye pollock comprised 89% of the catch, on average, over the 11-year period.

Aquaculture is not a particularly important activity in the Bering Sea region. In the eastern Bering Sea region, Alaska has adopted policies that prohibit aquaculture but enable some land-based hatcheries that produce salmon for release into the sea to supplement at times of low escapement. Some of these salmon pass through the eastern Bering Sea and may have some effect on larvae, for example red king crab (*Paralithodes camtschaticus*) larvae, but this has not been demonstrated. None of the hatcheries operate in the western Bering Sea region (NPAFC, 2001).

13.5.1. Ecosystem essentials

The Bering Sea is a subpolar sea bounded by the Bering Strait to the north and the Aleutian Islands archipelago to the south (Fig. 13.33). Geographically, the Bering Sea lies between 52° and 66° N, and 162° E and 157° W. The narrow (85 km long) and shallow (<42 m deep) passage of the Bering Strait connects the Bering Sea to the more northern Chukchi Sea and the Arctic Ocean to the north. The sea area covers almost 3 million km^2 and is divided almost equally between a deep basin in the southwest and a large, extensive continental shelf in the east and north. The eastern continental shelf is 1200 km in length, exceeds 500 km in width at its narrowest point, and is the widest continental shelf outside the Arctic Ocean (Coachman, 1986). The shelf is a featureless plain that deepens gradually from its extensive shoreline to the shelf break at about 170 m depth. There are very limited commercial fisheries in the Chukchi Sea or the Arctic Ocean north of the Bering Strait due to a known lack of resources, operating difficulties, and distance from markets. Marine mammal populations are locally important for subsistence use.

13.5.2. Fish stocks and fisheries

This section describes the life history characteristics, distribution, and trends in abundance and fisheries for the main species which are or have been the subject of important fisheries or which are important as forage fish for such species. Catch records for the major groundfish species of the eastern and western Bering Sea are shown by species in Figs. 13.34 and 13.35 respectively.

13.5.2.1. Capelin

In the Bering Sea, adult capelin only occur near shore during the month surrounding the spawning run. In other months they occur far offshore. In the eastern Bering Sea capelin occur in the vicinity of the Pribilof

Islands and the continental shelf break; in the western Bering Sea they occur in the northern Anadyr Gulf and near the northwestern Kamchatka coast. The seasonal migration may be associated with the advancing and retreating sea-ice edge. In the eastern Bering Sea, sea ice retreats during summer. As a coldwater species, capelin may migrate in close association with the retreating ice edge resulting in the summer capelin biomass located in the northern Bering Sea, an area not covered by surveys and with very little commercial fishing. Capelin aggregations near the northwestern Kamchatka coast have a stable distribution over the warm season. It is reported that the biomass of capelin and smelt grows in periods of climatic transition, when the abundance of other common pelagic fish (walleye pollock and herring) are low in the western Bering Sea (Naumenko et al., 1990). Capelin biomass was estimated at 200000 t on the western Bering Sea shelf between 1986 and 1990. Their biomass may be much larger on the expanded eastern shelf. Nevertheless, capelin are not commercially exploited in the Bering Sea. In Russia, some attempts were made to include capelin and polar cod in a commercial fishery in the mid-1990s. Capelin are a major component of the diets of marine mammals feeding along the ice edge in winter (Wespestad, 1987) and of seabirds in spring.

13.5.2.2. Greenland halibut

In the Bering Sea, Greenland halibut (commonly known as Greenland turbot) spend the first three or four years of life on the continental shelf after which they migrate to deep waters of the continental slope where they live as adults (Alton et al., 1988; Shuntov, 1970; Templeman, 1973). Although tagging studies show that they undergo feeding and spawning migrations in the North Atlantic Ocean, it is unknown to what extent this happens in the Bering Sea. A slow-growing and long-lived species, Greenland halibut reach over 100 cm in length and 20 years of age in the Bering Sea. Greenland halibut are a valuable commercial product and have been caught in trawling operations and by longlines. Catches of Greenland halibut and arrowtooth flounder were reported together in

the 1960s; combined catches ranged from 10000 to 58000 t per year with an average annual catch of 33700 t. The Greenland halibut fishery intensified in the 1970s with catches of this species peaking between 1972 and 1976 at 63000 to 78000 t per year, primarily taken by distant-water trawl fleets from Japan. Catches declined after implementation of the Magnuson Fishery Conservation and Management Act (FCMA) in 1977, where the US fisheries jurisdiction was extended to 200 nm from the coast. However, catches were still relatively high in 1980 to 1983 with an annual range of 48000 to 57000 t. After that, trawl harvest declined steadily and averaged 8000 t between 1989 and 2000. This decline is mainly due to catch restrictions placed on the fishery because of declining recruitment and market conditions. In the western Bering Sea, Greenland halibut were lightly exploited due to low stock abundance before the FCMA took effect in the eastern Bering Sea. In 1978, a Greenland halibut fishery began on the northwestern continental slope, mostly by longlines. Annual harvest varied from 2010 to 6589 t between 1978 and 1990 with part of the harvest resulting from bycatch in the Pacific cod longline fishery. Since the early 1990s, Greenland halibut stock abundance and catches have declined. Resource assessment surveys on the continental shelf in 1975 and between 1979 and 2002 showed that intermediate size Greenland halibut (40–55 cm) were present throughout the region from 50 to 200 m depth during the late 1970s and early 1980s (Alton et al., 1988). By 1985 and 1986 the distribution range had decreased such that Greenland halibut were only encountered in the area to the west and south of St. Matthew Island and at much reduced densities. Since then, fish of this size range have only been caught in small quantities in the northern part of the survey area. It is unknown whether environmental conditions in the late 1970s and early 1980s were favorable for strong recruitment of Greenland halibut and levels have since returned to more normal recruitment levels, or whether there has been reduced recruitment to

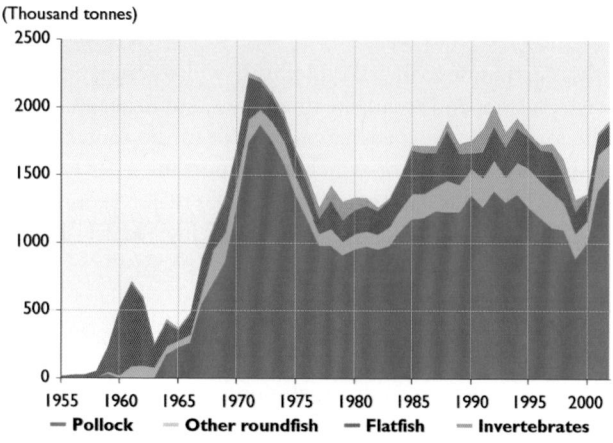

Fig. 13.34. Catch by species from the eastern Bering Sea, 1955–2002 (NPFMC, 2004).

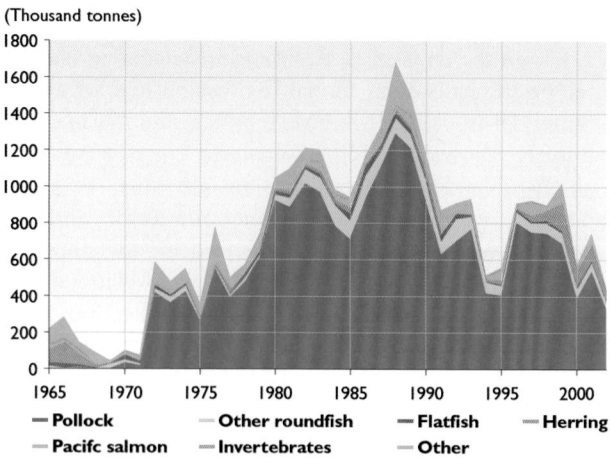

Fig. 13.35. Catch by species from the western Bering Sea, 1965–2002 (1965–1993 data from Committee on the Bering Sea Ecosystem, 1996; 1994–2003 data from the TINRO-Center archive, Vladivostok, Russia).

this stock since the mid-1980s. However, stock assessment models suggest a declining population since 1985 (Ianelli at el., 2001). Greenland halibut are widely distributed in the western Bering Sea but are not abundant there. The most significant Greenland halibut aggregations occur on the outer continental shelf and slope along the Korjak coast (Borets, 1997; Novikov, 1974). Survey results indicate that Greenland halibut abundance was higher in the northern Bering Sea in the 1990s than in the 1980s. However, the total biomass and overall distribution of this flatfish decreased in the Bering Sea region as a whole.

13.5.2.3. Shrimp

Pandalid shrimp (primarily *Pandalus jordani*) are widely distributed along the outer third of the eastern continental shelf where they are consistently caught in resource assessment trawl catches in small numbers. Humpy shrimp (*P. goniurus*) are distributed throughout the northern Bering Sea shelf and the Anadyr Gulf, in contrast to northern shrimp (*P. borealis*), which are much less abundant. Northern shrimp were the first commercially exploited shrimp in the Bering Sea after aggregations were discovered on the outer shelf north of the Pribilof Islands in 1960 (Ivanov, 1970). This fishery was conducted by Japanese vessels and peaked at 31 600 t in 1963. After that the northern shrimp stock declined sharply and commercial fishing ceased after 1967. Since then there has been no fishery for pandalid shrimp in the Bering Sea. Humpy shrimp aggregations were discovered in the Anadyr Gulf in 1967. A large-scale Russian trawl fishery harvested humpy shrimp in the northern Bering Sea in late 1960s to 1970s until they too became less abundant. Individual trawl catches of Humpy shrimp reached 10 t per 15 minute haul in the Anadyr Gulf, which became the catch value record in the world shrimp fishery. Humpy shrimp biomass was estimated at 350 000 t in the Anadyr Gulf in 1975. The annual Russian harvest of humpy shrimp exceeded 11 200 t in 1978 (Ivanov, 2001) but then declined due to the lack of a market for small-sized shrimp. Other pandalid shrimp species were also caught as bycatch in the pursuit of other target species.

13.5.2.4. Polar cod

Polar cod are caught in small amounts in resource assessment surveys at the northernmost survey stations on the eastern Bering Sea shelf. The southern extent of their summer distribution is related to bottom water temperature where they have been found to range from 59° N in 1999 (coldest year) to 62° N in 1996 (warmest survey year on record, except 2003). Since polar cod are found at such high latitudes, little information is available on their life history characteristics in the eastern Bering Sea and they are not pursued as a commercial species due to their low abundance. In the northwestern Bering Sea and the Chukchi Sea, polar cod are distributed at depths from 15 to 251 m (Tuponogov, 2001). A local fishery on polar cod existed there during years of high abundance (1967–1970; see Tuponogov, 2001).

13.5.2.5. Crabs

Snow crab and Tanner crab (*Chionoecetes bairdi*) are distributed throughout the eastern Bering Sea shelf with the exception of the shallow waters of Bristol Bay (Otto, 1998). The abundance of commercial size males was estimated at 183.5 million crabs in 1988 (Stevens et al., 1993). The distribution extends beyond the study area to the north and west, and to a small extent into the Gulf of Alaska. Owing to the relatively narrow shelf area of the western Bering Sea, snow crab abundance is notably less there. In 1969 the number of commercial size males was estimated at 25 million crabs (Slizkin and Fedoseev, 1989). An intensive directed fishery began for snow crab in the Bering Sea in the 1980s. They were initially caught incidental to the pursuit of red king crab until 1964 when both Japan and Russia increased their effort for this species due to a bilateral agreement with the United States to limit king crab catches (Davis, 1982). The combined Japanese–Russian catch of snow and Tanner crab increased until 1970 to 22 844 t (ADF&G, 2002), after which quotas were established for these nations' fishing fleets and the catch was sharply reduced. The American pot fishery (non trawl) began shortly after and catches increased during the 1980s to a peak in 1991 at 172 588 t. Catches rapidly declined with stock decrease but increased again in the mid-1990s as the snow crab stock condition improved. Since 2000, the stock has again declined and the commercial fishery is presently operating under reduced quotas. The Tanner crab fishery has been closed since 1997 in the eastern Bering Sea (NPFMC, 2002). In the western Bering Sea, there was no commercial snow crab or Tanner crab fishery in 2000 and 2001. Only insignificant catches (250 t) were allowed during research surveys. The results indicated some improvement in stock condition and a small commercial fishery was allowed in 2002.

13.5.2.6. Pollock

Walleye pollock (hereafter referred to as pollock) is the most abundant species within the Bering Sea and is widely distributed throughout the North Pacific Ocean in temperate and subarctic waters (Shuntov et al., 1993; Wolotira et al., 1993). Pollock are a semi-demersal schooling fish, which become increasingly demersal with age. They are a relatively short-lived (natural mortality estimated at 0.3) and fast-growing fish, females usually become sexually mature at four years of age. The maximum recorded age is about 22 years. The stock structure of Bering Sea pollock is not well defined. In the US part of the Bering Sea, pollock are considered to form three stocks for management purposes: the eastern Bering Sea stock (which comprises pollock occurring on the eastern Bering Sea shelf from Unimak Pass and to the US–Russian Convention line), the Aleutian Islands Region stock (which occurs within the Aleutian Islands shelf region from 170° W to the US–Russian Convention line), and the Central Bering Sea stock (known as the Bogoslof Island pollock, and which are thought be a mixture of pollock that migrate from the US and Russian shelves to the Aleutian Basin around the time of maturity). There are only two stocks in the Russian EEZ. Pollock currently support the largest fishery in US waters and comprise 75 to 80% of the annual catch in the eastern Bering Sea and around the Aleutian Islands. From 1954 to 1963, pollock were only harvested at low levels in the eastern Bering Sea. Directed foreign fisheries first began in 1964 after which catches increased rapidly during the late 1960s, and peaked in 1970 to 1975 when they ranged from 1.3 to 1.9 million t per year. Following a peak catch of 1.9 million t in 1972, catches were reduced through bilateral agreements with Japan and Russia. Since the US claim to extended jurisdiction in 1977, the annual average eastern Bering Sea pollock catch has been 1.2 million t, ranging from 0.9 million t in 1987 to nearly 1.5 million t (including the Bogoslof Islands area catch in 1990), while stock biomass has ranged from a low of 4 to 5 million t to highs of 10 to 12 million t (NPFMC, 2002). In 1980, US vessels began fishing for pollock and by 1987 were able to take 99% of the quota. Since 1988, only US vessels have been operating in this fishery and by 1991, the current domestic observer program for this fishery was fully operational. In the southwestern Bering Sea, the pollock fishery developed slowly during the mid-1960s stabilizing at 200000 to 300000 t in the latter half of the 1970s and the 1980s. After 1995, there was a reduction in harvest due to a decline in pollock stocks in the western Bering Sea. After that, the total pollock catch in the Russian EEZ was maintained by increasing fishing activity in the Navarin region between 1996 and 1999, and ranged from 596000 to 753000 t. The pollock catch subsequently declined in the northern region due to poor stock condition as well as to the application of precautionary approaches in pollock fishery management. The total pollock catch in the Russian EEZ declined from 1327000 t in 1988 to 393180 t in 2000. Vessels of "third countries" began fishing in the mid-1980s in the international zone of the Bering Sea (commonly referred to as the "Donut Hole"). The Donut Hole is located in the deep water of the Aleutian Basin and is distinct from the customary areas of pollock fisheries, namely the continental shelves and slopes. Japanese scientists began reporting the presence of large quantities of pollock in the Aleutian Basin in the mid- to late 1970s, but large-scale fisheries did not begin until the mid-1980s. Thus, the Donut Hole catch was only 181000 t in 1984, but grew rapidly and by 1987 exceeded the catch within the US Bering Sea EEZ. The outside-of-EEZ catch peaked in 1989 at 1.45 million t and then declined sharply. By 1991, the Donut Hole catch was 80% less than the peak value, with subsequent low catches in 1992 and 1993. A moratorium was enforced in 1993 and since then only minimal pollock catches have been harvested from the Aleutian Basin by resource assessment fisheries.

In response to continuing concerns over the possible impacts groundfish fisheries may have on rebuilding populations of Steller sea lions (listed as an endangered species after four decades of decline), changes have been made in regulations of the pollock fisheries in the eastern Bering Sea and at the Aleutian Islands. Pollock are important prey items for Steller sea lions and these changes were designed to reduce the possibility of competitive interaction of the fishery with Steller sea lions. For the fisheries, comparisons of seasonal fishery catch and pollock biomass distributions in the eastern Bering Sea led to the conclusion that the fishery had disproportionately high seasonal harvest rates within critical sea lion habitat which could lead to reduced sea lion prey densities. Consequently, management measures were designed to redistribute the fishery both temporally and spatially according to pollock biomass distributions (the underlying assumption being that the independently derived area-wide and annual exploitation rate for pollock would not reduce local prey densities for sea lions).

13.5.2.7. Pacific cod

Pacific cod are widely distributed from southern California to the Bering Sea, although the Bering Sea is the center of greatest abundance for this species. Tagging studies have shown that they migrate seasonally over large areas. In late winter, Pacific cod converge in large spawning concentrations over relatively small areas. Spawning takes place over a wide depth range (40–290 m) near the bottom. Eggs are demersal and adhesive. Estimates of natural mortality range from 0.29 to 0.99, while a value of 0.37 is used in the stock assessment model. Pacific cod have been found aged up to 19 years and females are estimated to reach 50% maturity at 5.7 years, corresponding to an average length of 67 cm. Pacific cod are the second largest Bering Sea groundfish fishery. Beginning in 1964, the Japanese trawl fishery for pollock expanded and cod became an important bycatch species and an occasional target species during pollock operations (in the early 1960s, a Japanese longline fishery harvested Bering Sea Pacific cod for the frozen fish market). By 1977, foreign catches of Pacific cod had consistently been in the 30000 to 70000 t range for a full decade (Thompson and Dorn, 2001). The foreign and joint venture sectors dominated catches through 1988, when a US domestic trawl fishery and several joint venture fisheries began operations. By 1989, the domestic sector was dominant and by 1991 the foreign and joint venture operations had been displaced entirely. Catches of Pacific cod since 1978 have ranged from 33000 t in 1979 to 232600 t in 1997 with an average of about 141900 t. At present, the Pacific cod stock is exploited by a multiple-gear fishery, including trawl, longline, pot, and jig components (with the exception of 1992, the trawl catch was the largest component of the fishery (in terms of catch weight) between 1978 and 1996. Since 1997, the longline fleet has taken the greatest proportion of Pacific cod). Pacific cod were estimated to be at low abundance levels in 1978 but experienced strong recruitment (age 3) in the early 1980s, which built the stock to high levels. The population biomass peaked at 2.5 million t in 1987 and then declined gradually to about half the peak value in 2001. In the western Bering Sea, the Russian cod fishery developed slowly and was mostly unsuccessful until the late 1960s. Several attempts were undertaken by Japanese and local fishermen in longline and trawl fisheries development in the 1920s and 1930s. Meanwhile, commercially significant Pacific cod concentrations were described by scientific expeditions. In particular, dense aggregations were found in the northwestern area in 1950 to 1952 near the Navarin Cape (Gordeev, 1954). This led to the organization of a special cod fishery expedition in 1968 (Vinnikov, 1996). Pacific cod harvest from this area ranged from 6500 to 24 500 t in the first years, and peaked at 117 650 t in 1986. In the 1990s, catches declined due to a restructuring of the fishery and, in recent years, from decreases in cod abundance in the North Pacific. Pacific cod biomass was estimated at 766 000 t in 1989 (Vinnikov, 1996) and had declined to 172 000 t by 2000.

13.5.2.8. Flatfish

The flatfish harvest and resource is much smaller in the southwestern Bering Sea with its relatively narrow shelf than in the eastern Bering Sea. A directed flatfish fishery began in the mid-1950s in the southwestern Bering Sea. This is a small-scale land-based fishery using Danish seines and, to a lesser extent, trawls. Yellowfin sole (*Limanda aspera*) comprise the main part of the flatfish harvest in the southwestern Bering Sea (72.7% of the predicted flatfish TAC for 2002 and about 74% in Danish seine catches in recent years) and its biomass is estimated at 78 000 t on the southwestern shelf compared to 1.6 million t in the eastern Bering Sea. Maximum catches in the southwestern Bering Sea, 32 000 and 20 000 t respectively, were registered in 1958 and 1959. The situation changed dramatically in 1960 and 1961 when the flatfish harvest fell to its lowest ever values (100 to 160 t). The stock condition improved over the following decades. By the mid-1960s catches had stabilized at approximately 6000 t per year which continued through 1974, after which they declined until the early 1980s. During 1996 to 2002, the flatfish catch in the southwestern Bering Sea varied from 6000 to 13 500 t. In terms of other flatfish species, Alaska plaice (*Pleuronectes quadrituberculatus*), rock sole (*Lepidopsetta bilineata*), and northern flathead sole (*Hippoglossoides robustus*) are the most important in the southwestern Bering Sea.

The abundance of yellowfin sole is low in the northwestern Bering Sea. The most important flatfish species is northern flathead sole, which accounts for about two-thirds of the total flatfish biomass, followed by Alaska plaice, and rock sole. A directed flatfish fishery did not begin in the northwestern region until the 1990s and never developed extensively. However, the flatfish bycatch sometimes reached significant levels and between 1965 and 1984 ranged from 2440 to 29 140 t in the northwestern Bering Sea. The flatfish bycatch increased to 33 460 t in 1985 and 39 900 t in 1986, leveling off at 24 000 to 29 000 t over the next six years, and then declining to an average of 9700 t after 1993. A target flatfish fishery did not develop extensively, and the target catches remained less than the bycatch in the large cod and pollock fisheries.

In the eastern Bering Sea, yellowfin sole is distributed from British Columbia to the Chukchi Sea, into the western Bering Sea, and south along the Asian coast to about 35° N off the South Korean coast (Hart, 1973). In the Bering Sea, it is the most abundant flatfish species and is the target of the largest flatfish fishery in the United States. While also found in the Aleutian Islands region and the Gulf of Alaska, the center of abundance for this stock is on the eastern Bering Sea shelf. Adults are benthic and occupy separate winter and spring/summer spawning and feeding grounds. They overwinter near the shelf break at approximately 200 m depth and move into nearshore spawning areas as the shelf ice recedes (Nichol, 1997). Spawning is protracted and variable, beginning as early as May and continuing through August, occurring primarily in shallow water at depths less than 30 m (Wilderbuer et al., 1992). Eggs, larvae, and juveniles are pelagic and usually found in shallow areas. The estimated age at 50% maturity is 10.5 years with a length of about 29 cm (Nichol, 1994). The natural mortality rate is likely to be within the range 0.12 to 0.16, with a maximum recorded age of 33 years (Wilderbuer, 1997). Yellowfin sole have been caught with bottom trawls on the Bering Sea shelf every year since the fishery began in 1954. Between 1959 and 1962 yellowfin sole was overexploited by Japanese and Russian trawl fisheries when catches averaged 404 000 t annually. As a result stock abundance declined. Catches also declined to an annual average of 117 800 t between 1963 and 1971, declining further to an annual average of 50 700 t between 1972 and 1977. The yield in this latter period was partially due to the discontinuation of the Russian fishery. In the early 1980s, catches increased peaking at over 227 000 t in 1985. In the 1980s, there was a major transition in the characteristics of the fishery in the eastern Bering Sea. Before this, yellowfin sole were taken exclusively by non-US fisheries and these fisheries continued to dominate through 1984. However, US fisheries developed rapidly in the 1980s, and foreign fisheries were phased out. Since 1990, only domestic harvesting and processing has occurred. The 1997 catch of 181 389 t was the largest since the fishery became completely domestic, but decreased to 101 201 t in 1998. The 2000 catch totaled 83 850 t and the 2001 catch was 63 400 t. For many years in the 1990s the yellowfin sole fishery was constrained by closures in order to attain the bycatch limit of Pacific halibut allowed in the yellowfin sole directed fishery. Stock biomass has declined by 1 million t from the peak biomass observed in 1985 and was estimated at 1.6 million t in 2002.

13.5.2.9. Salmon

The Bering Sea is important habitat for many stocks comprising the five species of Pacific salmon during the ocean phase of their life history. Here, the various stocks intermingle from origins in Siberia, Alaska, the Aleutian Islands, Japan, Canada, and the US west coast. The earliest fisheries for salmon were probably indigenous subsistence fisheries in which salmon were captured returning to their native streams to spawn. During the 20th century there were three main fisheries for salmon in the Bering Sea: the Russian and Alaskan domestic fisheries, the Japanese high-seas gillnet and longline fishery, and the bycatch of salmon in the groundfish fisheries.

Salmon canneries first appeared on the Alaskan side of the Bering Sea in the late 1890s to process fish returning to Bristol Bay. It is reported that between 1894 and 1917 the Kvichak and Nushagak rivers flowing into Bristol Bay produced 10 million sockeye salmon (*Oncorhynchus nerka*) annually (Netboy, 1974). Purse seines and gill nets were the primary fishing gear in the early days of the fishery. Gill nets were hauled from the beach using horses, which were later replaced by engines, whereas the purse seine fishery started around 1915 with the advent of powered fishing craft. Purse seining continues to the present as the primary gear in a highly mobile fleet fishing near-shore, which assures the targeting of specific salmon stocks. Although all five species of Pacific salmon are present in Bristol Bay, sockeye salmon are the most abundant and have dominated the salmon catch for years. The Bristol Bay salmon catch for all species totaled 42 million fish in 1993, of which 41 million were sockeye salmon, the largest catch on record (fishery statistics from the Pacific salmon fishery on the western Bering Sea coast (eastern Kamchatka region) are available since 1906). On average, pink salmon (*O. gorbuscha*) contributed 73.8% of the Russian salmon catch in the western Bering Sea between 1952 and 1993, chum salmon (*O. keta*) 24.2%, sockeye salmon 1.3%, chinook salmon (*O. tshawytscha*) 0.6%, and coho salmon (*O. kisutch*) only 0.1% (Chigirinsky, 1994). Since 1989, the runs of pink salmon to the eastern Kamchatka coast have been in good condition in odd years. The historical highest catch totaled 83 640 t in 1999. The average pink salmon catch (38 390 t) for 1989 to 2001 is more than twice the average level of 15 996 t for 1952 to 1993 (Chigirinsky, 1994). Similarly, chum salmon catches were stable at 11 000 to 12 000 t in 2000 to 2001 compared to 5250 t for 1952 to 1993. The recent improved stock conditions coincide with new fishery regulations, which limit the chum salmon bycatch during the pink salmon fishery. The main sockeye salmon fishery in eastern Kamchatka results from the productive Kamchatka River, slightly south of the Bering Sea.

The Japanese high-seas gillnet and longline salmon fishery expanded into the Bering Sea in 1952 with three motherships and 57 catcher boats, which increased to 14 motherships and 407 catcher boats by 1956 (Netboy, 1974). (Motherships are large vessels to which catcher boats deliver their catches and where the fish are processed for human consumption or reduced to meal and oil, they also carry fuel and other provisions for the catcher fleet.) The peak catch of 116 200 t occurred in 1955 and annual catches ranged from 71 000 to 87 000 t between 1957 and 1977 (Harris C., 1989). Sockeye, chum, and pink salmon comprised 95% of the catch in this fishery, which ceased operations in 1983. The bycatch of salmon in the commercial groundfish fisheries is of less importance than for the directed fisheries, but still accounts for fishing mortality important to resource managers. Observer sampling of the groundfish fishery indicates that chinook salmon are more frequent in bottom trawls and the other species more frequent in the pelagic trawls (Queirolo et al., 1995). In the western Bering Sea, primarily chinook and chum salmon were present in the bottom trawl catches during research surveys in 1974 to 1991 (Radchenko and Glebov, 1998).

13.5.2.10. Marine mammals

The Bering Sea contains a rich and diverse assemblage of marine mammals, including north temperate, arctic and subarctic species. Twenty-six species from the orders Pinnipedia (sea lions, walrus, and seals), Cetacea (whales, dolphins, and porpoises), and Carnivora (sea otter), and polar bears are present at varying times of the year (Lowry and Frost, 1985). Some species are resident throughout the year (e.g., harbour seal, Steller sea lion, sea otter (*Enhydra lutris*), beluga whale (*Delphinapterus leucas*), and Dall's porpoise (*Phocoenoides dalli*)) while others migrate into the Bering Sea during the summer on feeding excursions. Arctic species including polar bears, walrus, ringed and bearded seals (*Erignathus barbatus*), and bowhead whales (*Balaena mysticetus*) mostly occur in the Bering Sea during autumn and winter and are associated with the presence of seasonal sea ice. Most of the marine mammal species are found over the continental shelf and in coastal areas, although five whale species reside in the deep/oceanic waters of the Bering Sea basin (Lowry et al., 1982).

Harvesting of marine mammals has occurred since at least 1790, the first year when northern fur seal harvests were recorded (Langer, 1980). The harvest peaked in the 1870s at over 100 000 animals and was at levels exceeding 40 000 males annually until 1985 when the northern fur seal commercial harvest was stopped and only subsistence hunting by Aleuts was allowed in the Pribilof Islands. In the Russian EEZ, fur seal hunting has seen many changes since the mid-1980s. Since 1987, the experimental hunting of "silver" fur seals (aged 3–4 months) has been conducted on the Commander Islands (Boltnev, 1996). The harvest rate was established at 60% from the average annual male abundance for 1987 to 1989. Actually, significantly less than 50% were killed, which has further decreased to less than 30% since 1989. The number of animals killed decreased from 6700 in 1995 to 3000 in 1999 and 2180 in 2000. The declining harvest is related to the decline in the fur seal population and the negative effect of disturbance by hunters on seal reproduction. All fur seal hunting is presently restricted to Bering Island. Bachelor males aged from two to five years were hunted on Medny Island until the mid-1990s (2134 animals were killed in 1994) and this area was then closed to harvesting in 1995. Whaling spread to the Bering Sea in the mid-19th century when large numbers (2500 in 1853) of bowhead whales were taken (NMFS, 1999). This harvest continued for 50 years until the bowhead whale population became depleted. The current subsistence harvest totals 60 to 70 whales annually. Some species, such as humpback and grey whales (*Eschrictius robustus*), which are present in the Bering Sea in summer, were historically harvested during the winter near Hawaii and California and in waters off the Chukotski Peninsula (about 130 to 135 whales). Kenyon (1962) reported that Steller sea lions were very abundant in the Pribilof Islands when discovered in 1786, but were soon overhunted. After protective measures were taken, numbers grew from a few hundred in 1914 to about 6000 in 1960. The population has since declined to low numbers and has been the subject of extensive research to find the cause of the decline.

In the United States, stock assessment information on the 39 stocks of the 24 species of marine mammals in the Bering Sea are used to classify each stock as either strategic, non-strategic, or not available (Angliss et al., 2001). Strategic stocks are those considered threatened, endangered, or depleted under US law. The strategic stocks include: northern fur seal, sperm whale, humpback whale, fin whale, the North Pacific right whale (*Eubalaena japonica*), and the bowhead whale. Three Bering Sea stocks also have further designations: northern fur seals are designated as depleted under the Marine Mammal Protection Act, and the western stock of Steller sea lion is listed as endangered under the Endangered Species Act, as is the bowhead whale. Nine of the 39 marine mammal stocks are estimated to be increasing, five are stable, three are declining, and the status of the others is unknown. Subsistence harvest is allowed for three species: northern fur seals, beluga, and bowhead whales. In Russia, marine mammal populations are classified as commercial, non-commercial, or protected. Protected species include all whales and dolphins (with the exception of grey whales, whaled by indigenous people for subsistence), sea otter, and polar bear. Some commercial quota is established for beluga whales, but is not taken. Walrus, spotted seal (*Phoca largha*), ringed seal, and ribbon seal (*P. fasciata*) are hunted in the northwestern Bering Sea. However, their harvest has been relatively low since the cessation of ship-based hunting operations. In 1998 to 2000, the harvest was less than 60% of the established TAC on different seal species and averaged 32.8%.

13.5.3. Past climatic variations and their impact on commercial stocks

Climate change primarily influences ocean water temperatures through the regulation of synoptic atmospheric processes and water exchange between the western Bering Sea and the Pacific Ocean. Four physical processes determine the change in ocean climate regimes in the North Pacific (Schumacher, 2000): the lunar tidal cycle, variations in solar radiation (Davydov, 1972; Van Loon and Shea, 1999), changes in the North Pacific circulation that affect air–sea exchange of heat and, finally, changes in the momentum of the Aleutian Low atmospheric pressure pattern. These processes generate a subset of basin-scale factors, each of which contributes to the oceanographic conditions of the Bering Sea. The Aleutian Low is an example of an atmospheric activity center in the northern-hemisphere (Beamish and Bouillon, 1993; Francis et al., 1998; Hollowed and Wooster, 1992; Latif and Barnet, 1994; Luchin et al., 2002; Wooster and Hollowed, 1995). Water inflow and atmospheric forcing appear to serve as links in the signal transfer chain for the Bering Sea region. Their functioning reflects the direct effect of the atmosphere on the marine environment through the temperature regime of shelf waters, and the undirected oceanographic phenomena offshore. The signal propagates through changes in the general current pattern and tidal wave parameters, which determine the intensity of the water exchange between the shelf and open sea regions.

The direct effects of atmospheric forcing resulting from climate variations are very important to the physical oceanographic dynamics of the eastern Bering Sea shelf, which has a characteristically sluggish mean flow and is separated from any direct oceanographic connection to the North Pacific Ocean by the Alaska Peninsula. Therefore, linkages between the eastern Bering Sea shelf and the climate system are mainly a result of the ocean–atmosphere interaction (Stabeno et al., 2001). Climate variations in this region are directly linked to the location and intensity of the Aleutian Low pressure center which affects winds, surface heat fluxes, and the formation of sea ice (Hollowed and Wooster 1995). The pressure index shows eight statistically significant

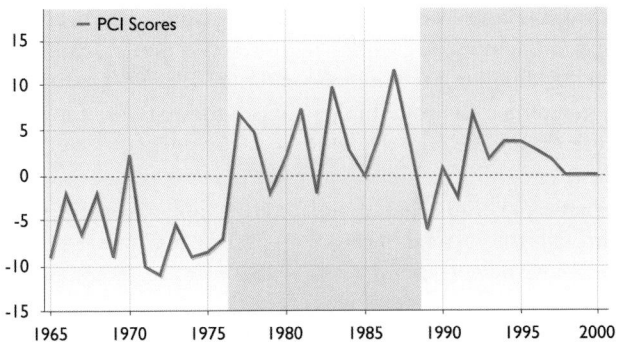

Fig. 13.36. Regime shifts in the Bering Sea, 1968–2000 (composite of 31 physical indicators: Hare and Mantua, 2000).

shifts, alternating between cool and warm periods, over the 20th century, which occurred on roughly decadal time scales (Overland et al., 1999). A well-documented shift (Trenberth 1990; Hare and Mantua, 2000) from a cool to a warm period occurred between 1977 and 1989, which coincided with the commencement of fishery-independent sampling programs and fishery catch monitoring of major groundfish species. Information from the contrast between this period and the prior and subsequent cool periods (1960–1976 and 1989–2000) forms the basis of the following discussion of the response of eastern Bering Sea species to climate-induced system changes (Fig. 13.36).

13.5.3.1. Effects on primary productivity

The influx of Pacific waters northward into the western Bering Sea results in a warming effect. The dynamics of the environmental conditions of the Bering Sea offshore zone and the relatively narrow western shelf are largely determined by the periodic behavior of current patterns (Shuntov and Radchenko, 1999). The direction and velocity of these currents coincide with changes in the atmospheric circulation pattern, effects which are manifested through the change in intensity of the inflow of North Pacific Ocean water. From 1977 to 1989, a period of enhanced atmospheric transport, an intensification of currents into the Bering Sea resulted in enhanced fluctuations in the thermal properties of the system towards a warmer state. During those years, the effect of horizontal water movement and mixing on primary production was almost as important as vertical mixing due to the renewed supply of nutrients necessary for phytoplankton blooms. According to long-term data series, the highest concentrations of spring-time nutrients in the upper mixed layer were observed in the Aleutian straits, over the continental slope, and in areas where the influx of North Pacific water was present. The enhanced rate of primary production may be as much as 10 to 13 g C/m² per day (Sapozhnikov et al., 1993), which is more than can be used by the zooplankton and microheterotrophs (especially in the western Bering Sea shelf). The unutilized primary production accumulates at the upper boundary of subsurface waters, which is relatively cold for microheterotrophs, and the organic matter gradually rises into the upper layers in divergence zones and cyclonic eddies during the warm season. Therefore, favorable conditions for plankton development during spring, both from heating and nutrient supply from Pacific waters, may cascade through higher trophic levels and play a large role in determining the total biological productivity for the year (Radchenko et al., 2001).

Changes in atmospheric climate are mainly transmitted through the eastern Bering Sea physical environment to the biota through wind stress (Francis et al., 1998) and annual variation in sea-ice extent (Niebauer et al., 1999; Stabeno et al., 2001). These mechanisms directly alter the timing and abundance of primary and secondary production through changes in salinity, mixed-layer depth,

upwelling, nutrient supply, and vertical mixing. These environmental changes vary at a decadal scale and resulted in higher levels of primary and secondary production during the warm period of 1977 to 1988 than in the earlier cool period (Brodeur and Ware, 1992; Hollowed et al., 2001; Luchin et al., 2002; Minobe, 1999; Polovina et al., 1995; Sugimoto and Tadokoro, 1997). During periods of low summer storm activity in the Bering Sea region, as in 1993 to 1998, water column stratification increases. Heating of a thin surface layer above the seasonal thermocline prevents vertical nutrient transport from the underlying, stratified layers, which reduces levels of primary production and biological productivity in the Bering Sea (Shuntov et al., 1997), despite warmer surface water temperature. This is consistent with the total heat budget of the upper layer of the Bering Sea, which was lower in 2002 than in the warmer period of the previous decade (Fig. 13.37).

In the relatively warm years of 1997 to 1998, there was significant growth in euphausiid biomass in the western Bering Sea (Radchenko et al., 2001) suggesting that warmer waters provide favorable conditions for the survival and growth of some subarctic zooplankton species. Crustacean growth rates have also been found to be above average in warm conditions (Vinogradov and Shushkina, 1987; Zaika, 1983). This enhanced growth rate allows for a longer maturation period and spawning season. A meta-analysis of marine copepod species indicates that growth rate is positively correlated with increasing temperature and that generation time decreases, allowing more productivity in warmer climates (Huntley and Lopez, 1992). The oceanographic conditions in the epipelagic layer are not considered crucial for copepod reproduction in the Bering Sea, since copepod species reproduce in relatively stable deeper layers below 500 m. However, calanoid copepod biomass was much higher in the eastern Bering Sea

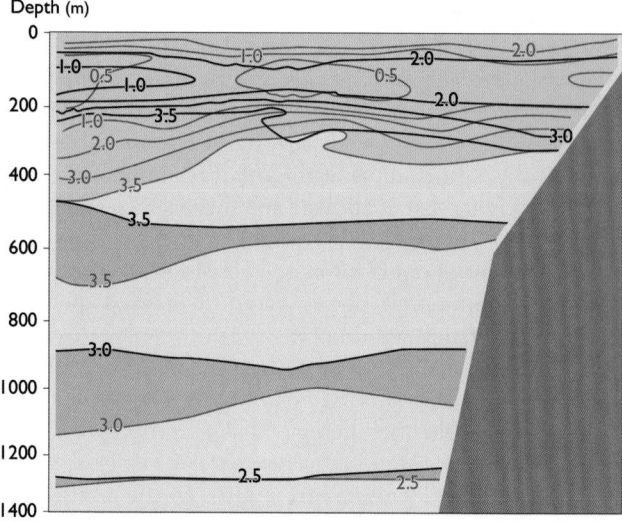

Fig. 13.37. Comparison of vertical temperature (°C) sections across the Kamchatka Strait on August 13–14, 1991 (black lines, after Stabeno et al., 2001) and June 27, 2002 (blue lines, TINRO-Center data, Vladivostok). Pink area indicates the layers where the water temperature was higher in 2002, green area – in 1991.

middle shelf in warm years (Smith and Vidal, 1986), probably due to higher growth rates. These findings suggest that climate change to a warm period enhances ecosystem productivity from the lower trophic levels, particularly for planktonic crustaceans.

13.5.3.2. Effects on sea-ice formation, distribution, and longevity

If it is assumed that any future climate change maintains the scale and periodicity of recent climate change events in the Bering Sea, then the period of meridional-type predominance in the wind transport above the Bering Sea, which began in the early 1990s, may last for 10 to 12 years before changing to a period of enhanced zonal transport. During the warm 1980s the zonal pattern of atmospheric circulation predominated (Luchin et al., 1998), as was the case in the 1920s and 1930s (Shuntov and Vasilkov, 1982). Periods of decreased zonal atmospheric circulation index (Girs, 1974) are characterized by colder arctic air masses over the Bering Sea region and a decrease in air temperature. The transitional 1989 to 1990 years were also characterized by a decrease in the zonal atmospheric circulation pattern above the far-eastern seas (Glebova, 2001; Overland, 2004).

Sea-ice distribution and residence time are frequently regarded as integral with the thermal regime of the Bering Sea pelagic zone (Ikeda, 1991; Khen, 1997; Luchin et al., 2002; Niebauer et al., 1999; Overland, 1991; Wyllie-Echeverria and Ohtani, 1999). The dynamics of sea-ice conditions directly depend on the intensity of the shelf water cooling in winter, wind direction, and water exchange between the shelf and the open sea. Similarly, ice conditions determine the intensity and degree of winter convection, the formation of cold near-bottom shelf waters, and the temperature distribution of surface and intermediate layers. The extent and timing of the sea ice also determine the area where cold bottom water temperatures will persist throughout the following spring and summer. This area of cold water, known as the "cold pool", varies with the annual extent and duration of the ice pack, and can influence fish distributions. For example, adult pollock have shown a preference for warmer water and exhibit an avoidance of the cold pool (Wyllie-Echeverria, 1995) such that in colder years they utilize a smaller proportion of the shelf waters and in warm years have been observed as far north as the Bering Strait and the Chukchi Sea.

13.5.3.3. Oscillating control hypothesis

During warm periods, favorable environmental conditions after the seasonal sea-ice retreat can result in a significant increase in the Bering Sea biological productivity. In contrast, physical factors during cold periods adversely affect zooplankton growth and biomass, and thus the viability of the pelagic fish juveniles feeding on this production. The "oscillating control hypothesis"

proposes that the southeastern Bering Sea pelagic ecosystem alternates between primarily bottom-up control in cold regimes and primarily top-down control in warm regimes (Hunt et al., 2002). Late ice retreat (late March or later) leads to an early, ice-associated bloom in cold water (as occurred in 1995, 1997, and 1999), whereas no sea ice, or early ice retreat before mid-March, leads to an open-water bloom in May or June in warm water (as occurred in 1996, 1998, and 2000). Zooplankton, particularly crustaceans, are sensitive to water temperature. In years when the spring bloom occurs in cold water, low temperatures limit the production of zooplankton, the survival of larval and juvenile fish, and their recruitment. Such a phenomenon may be important for large piscivorous fish, such as walleye pollock, Pacific cod, and arrowtooth flounder. When continued over decadal scales, this situation leads to bottom-up limitation and a decreased biomass of piscivorous fish. Alternatively, in periods when the bloom occurs in warm water, zooplankton populations should grow rapidly, providing plentiful prey for larval and juvenile fish. In the southeastern Bering Sea, important changes in the biota since the mid-1970s include a marked increase in the biomass of large piscivorous fish and a concurrent decline (due to predation) in the biomass of forage fish, including age-1 walleye pollock, particularly over the southern part of the shelf (Hunt et al., 2002).

13.5.3.4. Effects on forage fish

Spatial distributions of forage fishes including herring, capelin, eulachon (*Thaleichtys pacificus*), and juvenile cod and pollock indicate temperature-related differences (Brodeur et al., 1999; Wyllie-Echeverria and Ohtani, 1999). Annual capelin distributions exhibit an expanded range in years with a larger cold pool and contract in years of reduced sea-ice cover. Although the productivity of capelin stocks in relation to temperature is not known, population growth of this relatively cold-water dwelling fish is not expected under the conditions of a warm regime. As discussed, capelin biomass increased when the abundance of walleye pollock and Pacific herring were low in the western Bering Sea (Naumenko et al., 1990), possibly due to a reduction in predation pressure of these species on capelin larvae. The eastern Bering Sea herring stocks showed improved recruitment in warm years (Williams and Quinn, 2000), similar to herring stocks on the Pacific coast of the United States where the timing of spawning is also temperature related (Zebdi and Collie, 1995). In the western Bering Sea, Pacific herring have also demonstrated a dependence on reproductive success related to the thermal conditions of coastal waters. However, herring stock increase and large-scale fishery restoration are related to the "historically most abundant" (Naumenko, 2001) year class, which appeared in the anomalously cold year of 1993. Generally, strong herring year classes have appeared in the western Bering Sea in years with high sea surface temperatures in May but the lowest sea surface temperatures in June

(Naumenko, 2001). After 2000, herring biomass decreased in the western Bering Sea but still exceeds the average level for the last warm period (1977–1989). In general, the distributions of all forage species from trawl surveys in a cold year (1986) were more widespread and with greater overlap among species than in a warm year (1987) (Brodeur et al., 1999).

13.5.3.5. Effects on pollock stocks

Pollock larvae concentrate in the water mass under the seasonal thermocline (Nishiyama et al., 1986). More productive year classes of pollock coincided with better nursery conditions for their larvae, which were related to a well-developed thermocline (pycnocline), large biomass of copepod nauplii, and low abundance of predators (Bailey et al., 1986; Nishiyama et al., 1986; Shuntov et al., 1993). The first two factors are related to warm conditions in the Bering Sea epipelagic layer. Age-1 pollock may also be distributed throughout the cold pool and move between water masses. During cold conditions, predation pressure on age-1 pollock is intense by their major piscine predators (adult pollock, arrowtooth flounder, and Pacific cod). As the cold pool reduced, predation on age-1 pollock increased due to overlapping distributions of Greenland halibut, yellow Irish lords (*Hemilepidotus jordani*), and thorny sculpins (*Icelus spiniger*) (Wyllie-Echeverria and Ohtani, 1999). The total biomass of the first group of predators was much higher in the 1980s than the second group (Aydin et al., 2002) and has remained higher until the present, despite some declines in western Bering Sea walleye pollock and cod stocks. In addition, the second group of predators comprises relatively small-sized fish (except Greenland halibut) and age-1 pollock could avoid predation through higher growth rates during warm conditions. In the relatively warm 1980s, strong year classes of pollock occurred synchronously throughout the Bering Sea (Bulatov, 1995) and coincided with above-normal air and bottom temperatures and reduced sea-ice cover (Decker et al., 1995; Quinn and Niebauer, 1995). These favorable years of production were due to high juvenile survival and are related to how much cold water habitat is present (Ohtani and Azumaya, 1995), the distribution of juveniles relative to the adult population to avoid predation (Wespestad et al., 2000), and enhanced rates of embryonic development in warmer water (Haynes and Ingell, 1983). Strong year classes of pollock were also observed in the eastern Bering Sea in the 1990s (Stepanenko, 2001), which may be related to the higher frequency of ENSO events, which contributed to heat transport throughout the region (Hollowed et al., 2001). However, there were no strong year classes of pollock in the western Bering Sea in the 1990s. This could be due to a general cooling of the Bering Sea climate and the oceanographic regime in a period of less intensive Pacific water inflow in the 1990s. The pelagic layer heat budget may need to be similar to that of the late 1970s and 1980s for the pollock reproduction conditions to improve in the Bering Sea as a whole.

13.5.3.6. Effects on other groundfish

Time series of recruitment and stock biomass have been examined for evidence that climate shifts induce responses in the production of groundfish species in the Bering Sea and North Pacific Ocean (Hollowed and Wooster, 1995; Hollowed et al., 2001). Even though results from these studies can be highly variable, strong autocorrelation in recruitment, associated with the significant change in climate in 1977, was observed for salmonids and some winter-spawning flatfish species. Substantial increases in the abundance of Pacific cod, skates, flatfish, and non-crab benthic invertebrates also occurred on the Bering Sea shelf in the 1980s as evidenced from trawl survey CPUE (Conners et al., 2002). This warm period was characterized by larger research catches and a change in the benthic invertebrate species composition from a system largely dominated by crabs to a more diverse mix of starfish, ascidians, and sponges.

In the southwestern Bering Sea, transition from the relatively warm period of 1977 to 1989 to the subsequent cool period was also evident in the groundfish community. The proportion of Pacific cod decreased from 80% in 1985 to 12 to 26.3% in the 1990s, while sculpin (8.2% in 1985) and flatfish (9.3% in 1985) proportions increased by 15.1 to 31.5% and 24.2 to 39.6%, respectively (Gavrilov and Glebov, 2002). Anthropogenic factors can also affect the state and dynamics of benthic communities. For example, large fishery removals of red king crab occurred in the 1970s and may have contributed to the reorganization of the benthos in the eastern Bering Sea. The climatic change related to recruitment success for winter-spawning flatfish may be associated with cross-shelf advection of larvae to favorable nursery areas, instead of with water temperature (Wilderbuer et al., 2002).

Sea-ice conditions and water temperatures can influence fishery effectiveness in addition to fish stock distributions and abundance. Coldwater effects have been observed in the behavior of flatfish species that may cause changes in the annual operation of the fishery. Because cold water causes slower metabolism in high latitude fish stocks, spawning migrations of yellowfin sole may be delayed in cooler years (Wilderbuer and Nichol, 2001), which can alter the temporal and spatial characteristics of the fishery. In addition, high catch rates have been obtained by targeting yellowfin sole close to the retreating ice edge, which has a high temporal variability and in warm years only occurs in areas north of the spring distribution of yellowfin sole. The catch process can also be affected as it is believed that flatfish bury themselves in muddy substrate during cold years and so become less vulnerable to herding by the sweep lines of bottom trawls (Somerton and Munro, 2001). This would result in lower catch rates in cold years for shelf flatfish species. These temperature-related behavior effects may also occur in other commercial species, particularly in pelagic fish, which react to avoid capture (Sogard and Olla, 1998).

13.5.3.7. Effects on salmon

Throughout their century-long exploitation, Alaskan salmon stocks have had periods of high and low production which persist for many consecutive years before abruptly reversing to the opposite production state. These production regimes coincide with low frequency climate changes in the North Pacific Ocean and the sub-arctic Bering Sea (i.e., the Pacific Decadal Oscillation and the Aleutian Low Pressure Index). In the 1930s and early 1940s, and then again in the late 1970s, Bering Sea salmon catches reached high levels during warm temperature regimes in their oceanic habitat. It is hypothesized that improved feeding conditions may prevail during warm oceanic regimes (Hare and Francis, 1995). There is also evidence of an upper thermal tolerance for salmon species that has set limits on their distributions (Welch et al., 1995), but it is doubtful that this effect would occur in the Bering Sea because the historical temperature range there is much lower.

13.5.3.8. Effects on crab stocks

The three species of crab that inhabit the eastern Bering Sea shelf (red king crab, Tanner crab, and snow crab) exhibit highly periodic patterns of increased abundance. Rosenkranz et al. (2001) investigated five hypotheses on factors affecting year class strength of Tanner crab in Bristol Bay in order to understand these patterns. They determined that anomalously cold bottom temperatures may adversely affect the Tanner crab reproductive cycle and that northeast winds may promote coastal upwelling, which advects larvae to regions of fine sediments favorable for survival upon settling. Incze et al. (1987) linked low densities of copepods inside the 70 m isobath of Bristol Bay with low abundance of Tanner crab larvae. An examination of recruitment patterns of red king crab in relation to decadal shifts in climate indicates that the Bristol Bay stocks are negatively correlated with the deepening of the Aleutian Low and warmer water temperatures (Zheng and Kruse, 2000). Red king crabs were also moderately exploited during the late 1970s, which contributed to the population decline.

13.5.4. Possible impacts of climate change on fish stocks

Given the present state of knowledge of complex marine ecosystems such as in the Bering Sea, it is not possible to predict with any certainty the effects of future atmospheric forcing, in this case increased sea surface temperature, on commercial fish and invertebrate species. Evaluation of a future state of nature would require knowledge of the future values of many ocean–atmosphere parameters to describe how these changes would be manifested in upper trophic level commercial stocks. These parameters include storm activity and frequency, wind direction and intensity, shelf stratification characteristics, effects on circulation and transport activity, sea level pressure (location and intensity of the Aleutian Low pressure system), and precipitation as well as projections of sea surface temperature.

Three future climate scenarios are considered for the Bering Sea: no change from present conditions; moderate warming; and considerable warming.

13.5.4.1. No change

Under the no-change scenario the Bering Sea climate will continue to exhibit decadal-scale shifts alternating between warm and cool periods. These shifts in temperature regime have been shown to favor some species while their effect on others is unclear (section 13.5.3.4 to 13.5.3.8).

Under the present US and Russian management systems, it is expected that fish and invertebrate populations would be at or rebuilt to target spawning biomass levels as dictated by the management plans. This should result in an increase in total catches from the Bering Sea. Over the long term, however, a large total average increase is unlikely, but could nevertheless be considerable in individual cases.

13.5.4.2. Moderate warming

A moderate warming scenario can be developed by extrapolating trends characterizing the decadal-scale variability in the key physical factors influencing the Bering Sea ecosystem. On the basis of a moderate increase in air temperature (of 1 to 3 °C) and a general warming of the upper pelagic zone, several changes are likely:

- an increase in the zonal type repetition of atmospheric circulation for the early 2000s and for the period of the next 11-year cycle of solar radiation;
- an increase in storm activity and wind-induced turbulence for the same period;
- a gradual increase in water exchange with the Pacific Ocean, reaching a maximum in 2015 to 2020; and
- reduced sea ice, accelerated by an increase in air temperature, for the next 10 to 20 years after which time sea ice might increase again.

Variability in solar radiation correlates with many phenomena (Schumacher, 2000). It is a potential forcing mechanism for decadal-period oscillations of the coupled air–ice–sea system in the northern-hemisphere (Ikeda, 1990). Changes in solar fluxes correlate with change in the height of an atmospheric pressure surface in the troposphere of the northern-hemisphere (correlation coefficient = 0.72; Van Loon and Shea, 1999). Spectral maxima occur roughly every 7–17 (with an average of 11) and 22 years. The North Pacific Index also has phases similar to those noted for changes in solar radiation (Minobe, 1999). Storm activity and wind-induced turbulence of the sea surface layer are determined by the tracks and strength of cyclones, which are in turn determined by the nature of the pressure field. In the Bering Sea, a strong Aleutian Low is the source of most of the storm energy, and results in intense mixing of the sea surface layer in winter. Strengthening of the Aleutian Low occurs in years of zonal air transfer predominance (Shuntov, 2001). Such interrelations enable a prediction of high storm activity and wind-induced turbulence for all of the next 11-year cycle of solar activity.

Available information for the recent warm period in the Bering Sea suggests that primary productivity, and thus carrying capacity, would be enhanced under the warming scenario. However, because mixed-layer depth and water movements are not available for this scenario, the extent of this increase cannot be predicted owing to uncertainties concerning the renewal of the nutrient supply essential for sustaining the phytoplankton and zooplankton blooms. Also, as spring blooms are associated with the ice edge, a decrease in sea-ice extent associated with climate warming could

delay the onset of primary production in spring (Hunt et al., 2002). High water-column stability, which occurs at the ice edge during ice retreat, also supports intense phytoplankton blooms.

Recent studies on phytoplankton sinking velocities show that diatom cells sink more quickly than flagellates, which are lighter (Huisman and Sommeijer, 2002). Thus, it is possible that iceless winters could create unfavorable conditions for diatom blooms. This implies that climate warming could result in decreased biological production in the Bering Sea until the start of the projected increase in sea-ice cover after 2010. The dynamics of the Bering Sea sea-ice conditions are characterized by several periods of cyclic recurrence, ranging from 2–3 to 50 years (Plotnikov, 1996; Plotnikov and Yurasov, 1994; Ustinova and Sorokin, 1999). Obviously, this series is short for an exact tracking of the 50-year cycle. However, dramatic shifts in ice-cover anomalies were noted in the Bering and Chukchi Seas between 1976 and 1979, which divide

Table 13.2. Changes to stocks in the western Bering Sea and projected stock dynamics in response to a moderate warming (+ positive effect evident, - negative effect evident, 0 no effect evident or unclear effect).

Group	Increase in water temperature in upper pelagic layer	Increase in wind stress, zonal transport oscillation	Increase in water exchange with Pacific Ocean	Mild sea-ice conditions	Prevalent biological effects related to the physical environment changes	Key reference
Adult pollock	+	0	+	+	+ (food supply)	Shuntov et al. 1993
Juvenile pollock	+	-	+	+	- (predation)	Nishiyama et al. 1986
Pacific cod	+	0	+	0	+ (food supply)	Bakkala 1993
Pacific herring, WBS	-	0	-	+	- (competition)	Naumenko 2001
Pacific herring, EBS	+	0	-	+	- (competition)	Wespestad 1987
Pacific salmon	+	-	+	0	+ (food supply)	Hollowed et al. 2001
Cephalopods	-	0	+	0	- (predation)	Sinclair et al. 1999
Capelin	-	-	-	-	- (predation)	Wespestad 1987
Arctic cod	-	+	-	-	- (competition)	Wyllie-Echeverria and Ohtani 1999
Pacific halibut	-	0	0	0	-	Clark et al. 1999
Greenland halibut	-	0	-	0	- (competition)	Livingston et al. 1999
Arrowtooth flounder	0	+	0	0	+ (food supply)	Wilderbuer et al. 2002
Small flatfish	+	+	0	-	+ (food supply)	Wilderbuer et al. 2002
Skates	0	0	0	0	+ (food supply)	Borets 1997
Sculpins	+	0	0	-	- (competition)	Borets 1997
Atka mackerel	+	-	+	+	+ (food supply)	Shuntov et al. 1994
Mesopelagic fish	0	0	+	0	- (predation)	Radchenko 1994
Tanner crab	+	+	0	-	+ (food supply)	Rosenkranz et al. 2001
King crab	+	+	0	-	- (predation by flatfishes)	Haflinger and McRoy 1983
Shrimp	+	-	0	-	- (predation)	Ivanov 2001
Benthic epifauna	+	0	+	-	- (predation)	Conners et al. 2002
Benthic infauna	+	0	+	-	- (predation)	Livingston et al. 1999
Jellyfish	+	-	+	0	- (competition)	Brodeur et al. 1999
Euphausiids	+	0	+	+	- (predation)	Shuntov 2001
Copepods	+	0	+	+	- (predation)	Shuntov 2001
Phytoplankton	+	+	+	-	- (grazing)	Shuntov 2001

positive trend in stock abundance   negative trend in stock abundance   no trend expected or uncertain trend the 1952–1994 data series into two distinct periods, which differ by 5.4% in average values (Niebauer et al., 1999). Alternatively, a warmer period could increase thermal stratification such that the bloom, which is not ice-dependent, would start sooner. However, it could also be that the nutrient supply is quickly depleted during a short and intense bloom and that photosynthesis is slowed. In addition to upwelling and nutrient recycling in the pelagic layer, Pacific water inflow is also a source of nitrogen and phosphorus. Intensification of water exchange with the Pacific Ocean under climate warming is thus likely to result in increased primary production in the Bering Sea (Shuntov et al., 2002, 2003; Ustinova, and Sorokin, 1999). Increased levels of primary production are usually associated with improved survival for juveniles of most fish species (Cushing, 1969) and subsequent contribution to the adult spawning stock.

Predictions of the relationship between climate change and commercial species distribution, abundance, and harvest patterns are based on the assumption that future management policies will be the same as at present. Namely, that target fishing mortality values will be designed to maintain the female spawning stock at a minimum of 40 to 60% of the unfished level (depending on species). Also, that when stocks are assessed to be below this level, harvest is reduced proportionally to rebuild the spawning stock to the target level. This approach is likely to result in fisheries for species which respond favorably to warmer conditions realizing greater catches and possibly shifting to areas of increased abundance or expanded habitat, while fisheries for species which are negatively affected by a warmer climate are likely to have smaller quotas, reduced areas of operation, and even vastly different areas.

Literature documenting changes in the Bering Sea ecosystem under the previous warm period suggests that increases in the abundance of many groundfish species are very likely under future warming. Pollock, Pacific cod, Pacific halibut (Clark and Hare, 2002), skates, some flatfish species, salmon, eastern Bering Sea herring, and Tanner crab are all likely to benefit under warmer conditions (Table 13.2), although the mechanisms underlying the increase are not clear in most cases. Strickland and Sibley (1984) proposed a possible northward expansion of pollock feeding and spawning habitat under a warming scenario due to reduced sea-ice cover, water column stratification, and increased food supply.

There are very likely to be many changes in the Bering Sea ecosystem following a change to warmer conditions from those at present. Centers of capelin distribution are likely to move northward to colder waters and forego the large areas of spawning habitat that would be available in colder years. Polar cod and five species of marine mammals that are associated with the ice edge (harbour seal, Steller sea lion, sea otter, beluga whale, and Dall's porpoise) are likely to be restricted to the Chukchi Sea for large parts of the year. Red king crab, which decreased in abundance during the past warming

period, is unlikely to do well under future warming. Deep-water species such as Greenland halibut are unlikely to be affected during their larval and juvenile stages, except when they are present in shallow water. Greenland halibut recruitment is likely to be enhanced in colder years, perhaps due to a decrease in overlap with adult pollock (Livingston et al., 1999). Similarly, predation by large piscivorous fish is likely to affect the pandalid shrimp stocks; these shrimp species are among the preferred prey of Pacific cod during summer feeding in the northern Bering Sea, especially for new recruits (Napazakov et al., 2001). Predation on shrimp stocks is likely to increase with increased recruitment of Pacific cod. Annual consumption of gonatid squid species was estimated at 4.2 million t in the Bering Sea for the late 1980s (Radchenko, 1992). A growth in squid stocks is unlikely under a warming regime due to predicted increases in the abundance of their main predators (except for Greenland halibut). Increased abundance of sculpins and western Bering Sea herring in the 1990s was attributed to the weakening of interspecific competition in the pelagic and benthic fish communities by Naumenko et al. (2001). These stocks are likely to experience future decreases due to increases in pollock stock size.

Although some flatfish species on the Bering Sea shelf have shown increased productivity under warmer conditions, their distributions are unlikely to shift in response to ocean temperature. These species migrate to deeper, warmer waters in winter (where some spawn) and then migrate to the mid-shelf area in spring and early summer where they feed on the benthic infauna. Site fidelity is important for feeding purposes and these species are likely to tolerate moderate thermal changes. Habitat specialists, which have successfully developed niche preferences, are much less likely to be affected by climate change than "colonizing" species which have a more diverse diet so might be inclined to shift with the changing conditions. Such characteristics are expected to influence the extent to which fisheries would change under a warmer climate. However, the total fishery catch occurring under a climate change scenario would only increase to the extent allowable under current management practices. This also corresponds to historical data for the western Bering Sea fishery (Fig. 13.35). Attempts to forecast this increase, as described in the rest of this section, are based on the previously achieved maximum fishery harvest and assume that current management philosophies continue.

The maximum pollock catch in the Bering Sea was 4.07 million t in 1988, and averaged 3.55 million t between 1986 and 1990 (Fadeev and Wespestad, 2001). The total walleye pollock biomass in the Bering Sea over that period was about 20 million t (Shuntov et al., 1997). Conventional wisdom assumes that the 1990s stock reduction was due to a decrease in productivity in response to environmental conditions rather than to overfishing. This means that the pollock harvest in the favorable period of the late 1980s can be used as the reference point for predicting future catch for the project-ed warm period. As a rule, an increase in the pollock fishery stock is due to several average and strong year classes, as in the latter 1960s, or one super-strong year class. A super-strong generation appeared in the Bering Sea in 1978 and ensured a stabilization of stock abun-dance and the development of the large-scale pollock fishery at the end of the 1980s (Stepanenko, 2001). The eastern Bering Sea population of pollock almost doubled between 1995 and 2001 and supports an annual catch of more than 1 million t with strict regulations. The stable condition of this population provides for the likely increase in future abundance associated with a moderate climatic warming. However, the 1978 year class occurred 13 years after the first strong year class in 1965 (Wespestad, 1993) and no cohorts of this strength have been observed since. Thus, a swift increase in stock size and catch of pollock in the near future is unlikely.

The annual Pacific cod harvest ranged from 33 100 to 117 650 t and averaged 65 210 t in the western Bering Sea during the period of Dutch seine trawls and, to a lesser degree, longline fisheries between 1981 and 2001. The catch for the whole Bering Sea for that period totaled 207 110 t. This is a relatively low catch compared to an estimated cod biomass of 3.27 million t. Adult Pacific cod are the main predator for some commercially important fish (e.g., pollock and herring) and crus-taceans, particularly Tanner crab and shrimp. Relative to their weight, one unit of Pacific cod biomass consumes about 1.11 biomass units of Tanner crab juveniles, 1.12 of shrimp, 0.8 of walleye pollock, 0.39 of squid, and 0.31 of herring on the western Kamchatka shelf during the six months of the warm season (Chuchukalo et al., 1999). Whether an assumed increase in fishing pressure is justifiable for the purpose of decreasing predation by Pacific cod on other species in the ecosystem is under investigation. If the Pacific cod stock attains the same abundance in the Bering Sea as in the mid-1980s, it is likely that the total harvest could be increased (in both the western and eastern Bering Sea) to the level experi-enced in the 1980s and 1990s, i.e., around 350 000 t.

Between 1981 and 1991, herring fisheries in the south-eastern Bering Sea, in the vicinity of the Alaskan coast, harvested around 30 000 t, while harvests in the south-western part of the Bering Sea (Fig. 13.38) were around 17 000 t over this period, relative to a total biomass level of nearly 500 000 t. The same level of harvest is likely for the next warm period. The western Bering Sea "fat herring" fishery is very likely to decline during the next decade, but the Alaskan roe-sac herring fishery is likely to increase.

The Pacific salmon fishery recently surpassed its top harvest level in the northeastern Kamchatka area due to the record pink salmon catches in 1997 (82 300 t) and 1999 (83 600 t). However, some decline is likely to occur there since the local stocks of other Pacific salmon species are not as abundant. The chum salmon coastal catch did not exceed 12 200 t, while the sockeye salmon catch did not exceed 7000 t for these years. However, these relatively high catches were made in the latter half of the 1990s. On the basis of the 5-years rate of increase in the 1990s, the total chum and sockeye salmon harvest could reach a surplus of 20 and 12%, respectively, by the end of the 2020s. Eastern Bering Sea salmon production is dominated by sockeye salmon which contributed 41 million fish from a total catch of 42 million Pacific salmon in Bristol Bay in 1993. This was a historical record and followed the previous record of 37 million fish (101 550 t) in 1983 (National Research Council, 1996). The Bristol Bay sockeye stock has since declined and is presently in a period of low production. Stock dynamics observed over the 1960s to 1990s (Chigirinsky, 1994) suggest that periods of low productivity can last for 15 to 20 years and with an average annual sockeye catch of about 20 000 t. However, if moderate climatic warming is favorable for Pacific salmon growth and survival during the marine part of its life cycle, it is likely that the annual catch will approach that of the previous warm period (1977 to 1993), i.e., about 110 000 t. The proportion of sockeye salmon is very likely to increase from the mid-2000s to the 2020s from 25 to 30%, to 50 to 55% of the total and the proportion of pink salmon will decrease accordingly. Chum salmon are very likely to take third place, chinook salmon fourth, and coho salmon fifth.

Although flatfish biomass will possibly increase in future warm periods, the catch is likely to remain low due to bycatch and market constraints. The Atka mackerel (*Pleurogrammus monoptergius*) of the Aleutian Islands, skates, smelt, and saffron cod (*Eleginus gracilis*) of the southeastern shelf are other potential stocks in good condition. Development of new markets for these fish-ery products could increase future harvests. Comments concerning a future crab fishery in the Bering Sea cannot be made as it is not well understood which environmen-tal conditions would enable better survival and growth of crab larvae and juvenile stages. Also, the reasons for the sharp crab stock decrease in the Bering Sea in the 1980s are not known and there is debate as to whether the decline was due to overfishing or environmental

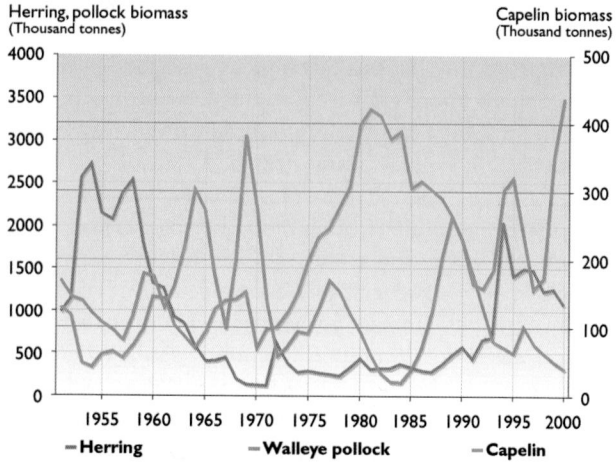

Fig. 13.38. Long-term changes in pelagic fish biomass in the western Bering Sea, 1951–2000 (after Naumenko et al., 2001).

change. Polar cod, capelin, sculpin, mesopelagic fish, shrimp, and squid fisheries are presently undeveloped in the Bering Sea, and no precondition exists to develop these fisheries under a warmer climate regime. Some bycatch of commander squid (*Berryteuthis magister magister*) and sculpins in the trawl fishery for pollock in the Dutch seine fishery on groundfish is used, but the total value of this harvest is insignificant.

13.5.4.3. Considerable warming

Since a warming of >4 °C has not previously been observed, it is not possible to comment on changes which might occur in the marine ecosystem based on past cause and effects. It is likely that the distributions of many species would shift poleward and that there would be significant changes in the arctic ecosystem. Ice-associated species would encounter a shrinking habitat and there would be greater potential for stock collapse for species forced to forego past areas of desirable spawning and nursery habitats due to thermal intolerance. The species succession likely under a scenario of considerable warming is not known, but a sudden reduction in the economic potential of Bering Sea fisheries is possible.

13.5.5. The economic and social importance of fisheries

In comparison to other areas of the Arctic, the commercial fisheries of the North Pacific, including the Sea of Okhostk, and the Bering Sea, are relative newcomers. Near-shore artesanal fisheries by indigenous peoples have occurred for centuries in the Bering Sea (Frost, 2003; Ray and McCormick-Ray, 2004; see also Chapters 3 and 12). The first documented commercial exploitation of groundfish dates back to 1864, when a single schooner fished for Pacific cod in the Bering Sea (Cobb, 1927), although salmon were part of commerce during earlier times. In 1882, American sailing schooners began a regular handline cod fishery. As recorded in Russian literature, the California-based fishers ceased to sail to fish in the Sea of Okhostk after the cod shoals near the Shumagin Islands in the Gulf of Alaska were discovered. In the western Bering Sea, the early Russian fisheries were poorly developed and limited to near shore subsistence fishing by indigenous peoples and settlers (Ray and McCormick-Ray, 2004). However, even at this early date the Bering Sea was known to contain a rich resource of fish. The herring fishery area expanded northward to the Bering Strait and operated during two weeks in May when herring migrated near the coasts. The Pacific salmon fishery yielded 12 million fish each year of which 2 to 4 million fish were from the Yukon Delta area, while the remainder were caught by Russian development companies and Japanese corporations operating concessions on Russian rivers (Netboy, 1974).

In contrast to the slow development of the early fisheries, the hunting of marine mammals developed rapidly. In the western Bering Sea, the fur seal harvest ranged from 20 000 to 50 000 animals on the Commander Islands. A Russian–American Company was mainly responsible for the hunting and fur purchase operations in the eastern Bering Sea, Gulf of Alaska, and Aleutian Archipelago regions between 1786 and 1862. The sea otter harvest totaled 201 403 animals during the time of the Russian–American Company of which nearly a third was purchased by merchants from the indigenous peoples. Other marine mammal harvests included sea lion hunting on St. George Island (on the Pribilof Archipelago), which yielded 2000 animals per year, and walrus hunting, which yielded 300 to 2000 animals per year until the harvest was reduced in the 1830s due to a declining population. Owing to overexploitation, the fur seal breeding grounds disappeared from the Pribilof Islands, Unalashka Island, and adjacent areas in 1830 to 1840. From 1743 to 1823, 2 324 364 fur seals, 200 839 sea otters, about 44.2 t of walrus tusk, and 47.8 t of baleen were harvested from the Aleutian Arc, other islands, and the Alaskan coast. The first protective measures on fur seal populations from Japanese and American illegal sealers were set by Russia in 1893. There is an illustration of this in the Rudyard Kipling ballad *The Rhyme of the Three Sealers*:

> *Now this is the Law of the Muscovite, that he proves with shot and steel*

> *When ye come by his isles in the Smoky Sea ye must not take the seal.*

In 1911, a three-sided treaty was concluded between Russia, the United States, and Japan, which established a sealing prohibition on the high seas in exchange for compensation paid from harvests in the rookeries (Miles et al., 1982a,b).

Large-scale commercial exploitation of the Bering Sea fish stocks developed slowly. Between 1915 and 1920, as many as 24 US vessels fished Pacific cod. Annual harvests ranged from 12 000 to 14 000 t (Pereyra et al., 1976). Small and infrequent halibut landings were made by US and Canadian fishers between 1928 and 1950, which increased sharply and exceeded 3300 t between 1958 and 1962 (Dunlop et al., 1964). In the early 1970s, the halibut catch fell to a low of 130 t before recovering to a high in 1987, and then slowly declined. The International Pacific Halibut Commission, established by Canada and the United States in 1923 to manage the halibut resource, determined that factors such as over-exploitation by the setline fishery, juvenile halibut bycatch, and adverse environmental conditions led to the decline in abundance (National Research Council, 1996). In the western Bering Sea, the exploitation of groundfish resources was mainly by small-scale coastal operations. Information on groundfish abundance was lacking until the first Soviet Pacific Integrated Expedi-tion in 1932 to 1933. This covered the entire Bering Sea and found the eastern shelf and continental slope to be more productive fishing grounds than the narrower western ones. As a result, Soviet fisheries concentrated their efforts in the eastern Bering Sea after 1959. By the mid-1960s, newly organized Soviet fishing on the eastern Bering Sea shelf and in the Gulf of Alaska yielded about 600 000 t of Pacific ocean perch, yellowfin sole, herring, cod, crabs, and shrimps (Zilanov et al., 1989).

The Japanese and Russian fleets expanded rapidly between 1959 and 1965, with vessels from the Republic of Korea and other nations also participating in later years. These fishery efforts were added to the solely Japanese fishery efforts, which have actively operated in the Bering Sea since the 1930s, especially after the Second World War. By 1960, 169 vessels from Japan were present on the Bering Sea fishing grounds along with 50 to 200 vessels from the Soviet Union (Alverson et al., 1964). Significant growth in fishing effort led to overfishing of several stocks. The Soviet walleye pollock fishery began in the early 1970s after the decline of some commercially valuable fish stocks. Before that, walleye pollock was not regarded in the Soviet fishery as a target species. The Japanese mothership operations had three to five conventional catcher/trawlers and as many as eight pairs of trawlers associated with each mothership (Alverson et al., 1964). The catch was processed at sea with the frozen products transported ashore for food. Japanese catches were mostly processed aboard motherships into fishmeal, with livers extracted for vitamin oil. Female walleye pollock in spawning condition soon became an important source of roe-bearing fish, which were processed into valuable products such as different kinds of fish roe and surimi. The increase in product value, combined with an increase in pollock abundance after the latter half of the 1960s led to the gradual increase in catch: up to 550 000 t in 1967 and 1 307 000 t in 1970 (Fadeev and Wespestad, 2001). Groundfish catches were mainly by vessels from Japan and the Soviet Union until 1986, when US fishing vessels participating

in joint ventures with foreign processing vessels took a larger proportion of the catch. By 1990, the distant water fleets were phased out of the eastern Bering Sea (the US EEZ) and US fishing vessels became the sole participants in the fishery. Some fishing occurs under license from the Russian Federation in its EEZ.

13.5.5.1. Fisheries

United States fisheries off Alaska constitute more than half of landings and about half the value of national landings of fish and shellfish from federal waters (NMFS, 2003a). Depending on species, approximately 90% of the landings in Alaska are from the Bering Sea/Aleutian Islands area. All the groundfish, crab, and salmon in the US EEZ of the Bering Sea are caught by domestic fishing bodies (Hiatt et al., 2002). In the Russian EEZ the majority of the harvests are taken by domestic fleets with a decreasing amount harvested under agreements with neighboring states. In 1997 it is estimated that the Russian Far East fisheries accounted for 70% of the Russian Federation total fisheries production (Conover, 1999; Zilanov, 1999), however this proportion may be decreasing due to the declines in pollock, crab, herring, and other species not being offset by the increases in Pacific salmon.

In the Bering Sea, walleye pollock is the major harvest by volume and value, with Pacific cod, flatfish, salmon, and crabs constituting most of the rest (Table 13.3). The total wholesale (raw fish landings) value for groundfish harvests in the eastern Bering Sea was approximately US$ 426 million in 2001. The total primary processed value was approximately US$ 1.4 billion. Crab harvests, mainly from the Bering Sea/Aleutian Islands area, amounted to US$ 124 million even at the low population abundances noted earlier (Hiatt et al., 2002). Pacific salmon, a large amount of which comes from the Bristol

Table 13.3. Trends in abundance and value of major Alaskan fisheries (inflation-adjusted US dollars) (Alaska Department of Fish and Game, as cited by Pacific Fishing, January 2002).

Species	Stock 1977	Value 1977	Stock 2001	Value 2001	Discussion
Salmon	200 million fish	US$ 500 million with peak value in 1988 of US$ 1.18 billion	175 million fish	US$ 205 million	A small decrease in total catch but a large decrease in price due to competition with farmed fish
Groundfish	Very small US harvest	US$ 2–3 million but rapidly increasing to US$ 1.0 billion in 1988 as a result of Americanization	1.65 million t harvested	US$ 400 million	Whitefish markets strong yet price weak but US dollar also weak
Shellfish (primarily crab species but some shrimp in early years)	Red king crab strong, other species small harvests	US$ 440 million. Drops when red king crab bubble bursts but Opilio crab takes over	Most species at low levels	US$ 125 million	Strong competition in Opilio fishery from Eastern Canada but weak competition from Russia
Pacific halibut	Low catch most likely due to foreign fleet bycatch	Less than US$ 30 million	High abundance	US$ 150 million	Strong stocks and good price vis a vis other white fish
Herring	Low abundance	Less than US$ 30 million although value increased in mid-1980s/mid-1990s to US$ >50 million	Low abundance	Less than US$ 30 million	Herring in same situation in 1977 and 2001

Bay and Yukon River areas, had a Bering Sea catch value of between US$ 122 million (2001) and US$ 179 million (2000) (Link et al., 2003). The Community Development Quota (CDQ) Program, which allocates 10% of the total Bering Sea TAC to 65 coastal communities organized into six CDQ corporations, earns more than US$ 40 million annually (NPFMC, 2003a). A separate value is not assigned in this study to recreation or subsistence harvests in the Bering Sea due to lack of adequate analyses, despite their local and cultural significance. Economic value data for the Russian Far East are difficult to locate (Pautzke, 1997). Press reports for product value estimate the total 2001 production to account for US$ 3.0 billion (Pacific Rim Fisheries Update, May 2002). Since the transition to a market economy began in the early 1990s and the Soviet style management of fisheries has changed, it appears that there are significant tracking and reporting difficulties with less fish being landed to avoid taxation and fees. Instead, harvests may be transferred at sea or transported directly to foreign markets by fishing vessels (Velegjanin, 1999). Thus, production and value data must be treated with caution until a more robust accounting system is developed.

13.5.5.2. Fishing fleet and fishers

Almost every fishing vessel in the Bering Sea fleet is registered outside the region. Vessels must be of requisite size to weather the environmental conditions and to have adequate scale efficiencies to operate in the area. These factors plus the lack of deepwater moorings and other support services make the eastern North Pacific a largely "distant water" fishery. Overall, the number of vessels eligible to fish for the increasing stocks of groundfish in the federal waters of the Bering Sea has decreased since the mid-1990s from 464 vessels in 1995 to 398 in 2001. This is the case for all groundfish vessel classes and types. In 2001, there were 163 hook and line (longline) vessels, 81 pot vessels, and 162 trawl vessels fishing, of which around 20 were at-sea capture/processors for pollock. The overall decrease in number results from rationalization programs for pollock under the American Fisheries Act 1998 and the North Pacific Fishery Management Council's license limitation program for all species (although this figure does not include halibut/sablefish vessels which have Individual Fishing Quota qualification) (Hiatt et al., 2002). For other sectors, there were around 274 eligible Bering Sea/Aleutian Islands crab fishing vessels, 2500 catcher longliners (including Alaska state-water vessels) mostly involved in halibut/sablefish and Pacific cod fisheries, and some 5200 salmon fishing vessels of various types (Natural Resources Consultants, 1999).

Employment in the groundfish harvesting sector (at-sea catching and processing on land as well as motherships) in 2001 amounted to 4000 full-time equivalent jobs including skippers, fishing crew, processing crew, and home office staff (NMFS, 2003b). With few exceptions, most of this employment is in relatively small corporations. North Pacific Fishery Management Council license limitation regulations limit the size and ability to grow of existing catching bodies. Thus, few large integrated harvesting and processing companies exist. Still, even the smaller organizations deal in multi-million dollar investments with substantial annual operating expenses, e.g., a typical catcher vessel of about 35 to 40 m in length would require a family owner or small business to have a fair market value of US$ 2.5 million to 3.5 million (Natural Resources Consultants, 1999).

In the western Bering Sea, the situation is similar to that in the Alaskan EEZ. A large part of the harvesting capacity is located in the southern parts of the area, as are the financial and supply and repair services. The number of fishing vessels has declined drastically since the end of the Soviet era distant water fishing, owing to other nations extending their EEZs and to efforts to renew the fishing fleet and to reorganize it on market economy terms (Zilanov, 1999). Between 1990 and 1999 the Russian fishing fleet decreased by nearly 44% in number. Most of the fleet was privatized in the form of joint stock companies (56.7%), or transferred to cooperatives (*kolkhozes*; 23.7%), private companies (12.5%), or joint Russian–foreign ventures (2.4%) (Zilanov, 1999). In the Russian Far East, this has enabled small and mid-scale fisheries to develop while some large entities under Soviet style fisheries have changed and remained dominant forces. Likewise, total employment in the fisheries sector fell from 550000 in 1990 to 398000 in 1998. Contributing to the decline in employment in the Russian Far East was an exodus of people assigned to duties there returning to families and friends in their home regions.

13.5.5.3. The land side of the fishing industry

Approximately 70% of the Bering Sea harvests are processed on shore in a relatively small number (8) of groundfish processing plants near Dutch Harbor/Onalaska (NMFS, 2003b). Recent efforts have been made to locate processing facilities on Adak Island in the western Aleutians. Crabs are processed on the Pribilof Islands during periods of high abundance of red king crabs and snow crab in the Bering Sea. Salmon tendering and processing is focused around Bristol Bay although not exclusively. Sites where processing occurs require significant infrastructure for processing as well as for providing services to the fishing fleet. Given the remote nature of the Bering Sea fish processing activities, the communities in which these occur are highly dependent on the fishing industry for economic activity, with government services and tourism distant rivals. Most of the groundfish processing occurs adjacent to the densest aggregations of groundfish and where catcher vessels with refrigerated sea-water holds can make relatively rapid trips to maintain product quality. However, for some species and products (e.g., high grade surimi) it is difficult for shoreside processors to compete.

Employment in Alaskan shoreside processing for groundfish is estimated at 3525 full-time equivalents (NMFS,

2003b). The number of processing jobs onshore in the Bering Sea has increased by as much as 50% between the early 1990s and the present because of policies decreasing the amount allocated to at-sea processing versus onshore processing. Much of the work force is an ethnically diverse group of work permit holders from other parts of the world, mostly west coast United States, Mexico, and the Philippines. Over time, Alaskan communities in the Bering Sea region are being transformed as workers stay on and climb the corporate ladder.

In the Russian Far East a significant proportion of the catches have been processed at sea with the rest processed on shore or kept in cold storage, etc. With domestic demand low in terms of the ability to compete with global market price and other tax and regulatory issues onshore, there is a substantial incentive to process offshore and export directly (Velegjanin, 1999). This has contributed to a sizeable decrease in domestic consumption and employment in shoreside processing and other services to the fishing industry. The transition to a market economy has been difficult but the learning curve is trending upward with new management institutions and experience. However, without the full cooperation of the fishing industry and management, and tensions over the allocation of revenue between the Far East and Moscow, it will be some time before the industry stabilizes.

13.5.5.4. Fisheries communities

The North Pacific fishing communities surrounding the Bering Sea are different to those of the North Atlantic. There is no history of small coastal fishing communities developing commercial fishing on the currently harvested large stocks of pollock, Pacific cod, etc. In the eastern Bering Sea some 65 communities exist with a total population of around 27 500. They are frequently inhabited by a large percentage of indigenous Alaskans, but not exclusively (NPFMC, 2003a). Until they became participants in the CDQ program, they had limited coastal subsistence fisheries as well as some small-scale commercial fisheries for salmon and halibut. Involvement in the groundfish and crab fisheries has provided valuable income and employment as well as a role in management of the offshore fisheries. The main location of the fish processing on Akutan and Dutch Harbor/Onalaska had been important for crab, halibut, and some salmon fisheries. It was not until foreign and domestic investment was encouraged in shoreside processing of groundfish in the late 1980s that these communities were transformed. Loss of access to fishing in the US EEZ prompted Japanese investment in processing so that raw fish could be purchased at low prices and benefits gained in value-added processing from shore-based plants.

The history of the purchase of Alaska from Russia in 1867 and its status as a territory until Statehood in 1959 was that of a domestic colony. In particular, fishing interests in western Washington and Oregon were some of the prime early investors in Alaskan fisheries. Ownership of the highly seasonal Alaskan canneries was mostly outside Alaska. Salmon fishing brought labor from the south. Halibut fisheries were developed as soon as ice-making and refrigeration technologies permitted catching and transport of fish to southern markets. Early crab fishing interests were based out of Seattle. Thus, the fisheries of Alaska have strong personal, financial, and service connections to Seattle due to the laws of comparative advantage. Alaska is a high cost area for living and carrying out a business (Natural Resources Consultants, 1999). In the federal water fisheries, residents of other states must not be discriminated against in management regulations, which further enforces the long, mostly cooperative, relationship between fishing interests in Washington and Oregon and those in Alaska. Overall dependence on fisheries varies by community but in Alaska as a whole, fisheries is a distant second to oil production in terms of revenue from resource extraction and for some cities with onshore processing, fisheries are the prime source of local landing tax revenue.

Similar to Alaska, small indigenous Russian settlements existed around the western Bering Sea. With the colonization by Russians, larger towns developed and during the Soviet era these grew as bases for resource development and national defense. Population in the seven administrative regions of the Far East is concentrated in coastal cities and declined slowly throughout the 1990s (Zilanov, 1999). Several large cities account for the majority of the population such that much of the Russian coastline is undeveloped. Fisheries are dominated by fishing interests in Vladivostok and Nakhodka. Increasingly stronger demands are being made by other regional fishing bases for more autonomy in management and greater allocations to proximate users.

13.5.5.5. Markets

The relatively low populations of the Bering Sea region do not constitute a very large local market for the large-scale fisheries. Thus, both Alaska and the Russian Far East look to distant markets at home and abroad. For Alaska, the prime markets are Japan, Korea, and China with Europe providing entry for some products. Over 90% of Alaskan fish is exported. Korea and now China with their relatively low wage labor have served as processing centers for some products that are re-exported, i.e., imported back in some value-added form. For the Russian Far East, exports have started to play an increasing role in the fisheries economy. During the Soviet era up to 80% of the Far Eastern fish products were processed and sent on to domestic markets in the western more populous parts of the country. The rest was exported or taken under fisheries agreements with neighboring states to obtain hard currencies needed by the central government. While low effective demand (i.e., domestic consumers) is not able to pay international prices for seafood products, many of the higher value species are exported and low value species and products are imported so that around 50% of the seafood harvested is destined for export (Zilanov, 1999). Thus, the remote Bering Sea is a major player in terms of global seafood markets where declines in abundance of Atlantic cod, for example, open markets for fillets of pollock at the same time demand for pollock surimi products seems to be slackening as a result of weak Japanese and Korean markets. Similarly, high abundance of snow crab in Canada causes market erosion for this species in the eastern Bering Sea.

Owing to the significant price competition from farmed Atlantic salmon, the wild salmon dependent fishing enterprises and communities are facing major adjustments. Even though North Pacific wild salmon stocks are abundant at present, the large quantity of farm raised salmon and its method of sale and delivery reduce the price that can be obtained. There is some consideration in Alaska and Russia about starting aquaculture but it is recognized that the investment, organization, and technology may be significant hurdles (Link et al., 2003). Given the experience with salmon, there is also concern over the farming of halibut, sablefish, and cod becoming competitive with wild stock harvests.

13.5.5.6. Management regime

The US and Russian EEZs are the major management jurisdictions in the Bering Sea although the multilateral conventions for management of the "Donut Hole" fishery outside these boundaries also has an important role in fisheries management. Similarly, the Convention for a North Pacific Marine Science Organization and the Convention for the Conservation of Anadromous Stocks in the North Pacific Ocean provide frameworks for scientific exchange and cooperation. Even though the major activities covered by these conventions occur to the south of the ACIA boundary, the Wellington Convention for the Prohibition of Fishing with Large Driftnets constrains fisheries on the high seas with potential to intercept salmon of Russian and US origin as well to have negative bycatch effects on Dall's porpoise and some seabird species. Bilateral agreements, such as between Canada and the United States for salmon and halibut management and between Russia and Japan for salmon, also exist.

At the national level, the Magnuson-Stevens Fishery Conservation and Management Act is the prime legislation guiding fisheries management in federal waters. In Alaska, this means that all waters between 3 nm from the state's baselines and 200 nm is under federal jurisdiction. Other relationships exist, such as federal management for halibut in all waters due to the Convention between Canada and the United States for the Preservation of the Halibut Fishery of the Northern Pacific Ocean and Bering Sea, and Alaskan state jurisdiction (with federal oversight) over crabs as creatures of the continental shelf and salmon that are harvested within state waters (Miles et al., 1982a). The waters off Alaska constitute one of the nation's eight fishery management regions. This is administered by the regional office of

the National Marine Fisheries Service, with management decision-making taking place in the North Pacific Fishery Management Council – an advisory body to the regional director and thereby to the Secretary of Commerce. The federal regulations aim to develop a decision process that is comprehensive, transparent, and open to participation by all interested parties (NMFS, 2003b).

The main tools for fishery management are Fishery Management Plans that set out the rules and regulations for management of each species or species complexes. Under the current management approach, TAC is set on an annual basis in the Stock Assessment Fishery Evaluation process (e.g., NPFMC, 2002). As part of this process, ecosystem considerations are made explicit in the form of a chapter of the Stock Assessment Fishery Evaluation document that addresses ecosystem trends and relationships to fishing, as well as in the environmental assessments required in accordance with the National Environmental Policy Act. All meetings of the Council and its Advisory Committee and Scientific and Statistical Committee are open to the public. Thus, any interested party can observe and participate in deliberations of Plan Development Teams setting TACs.

The North Pacific Fishery Management Council has developed innovative approaches to management. Scientific advice is rigorously adhered to in the setting of TACs and conservative harvest limits are applied. A cap of two million tonnes has been set on total removals in the fishery even when allowable catches might be considerably higher. Bycatch is counted against TAC and target fisheries can be closed if the bycatch limit is reached before the target fishery TAC. Larger boats are required to carry and pay for one or more observers to gather scientific information about harvests. Species such as halibut, salmon, and herring are considered prohibited species in the groundfish and other non-target species fisheries. Finally, significant areas of the fishing grounds are closed to trawling to protect habitat necessary for other species, e.g., red king crab savings area (Witherell et al., 2000). In addition, much of the present work of the North Pacific Fishery Management Council is on developing spatially explicit relationships between fisheries and fish habitats under the Essential Fish Habitat Provisions of the Magnuson-Stevens Fishery Conservation and Management Act (NPFMC, 2003b). There is also a Council emphasis on rationalization of fisheries through share-based management systems such as the Individual Fishing Quota program for halibut and sablefish (and as proposed for Bering Sea and Aleutian Islands crabs) or through using a cooperative approach as for pollock under the American Fisheries Act, 1999.

In the EEZ of the Russian Far East, the issues and basic management system are similar to those in the Northeast Atlantic (see section 13.2.5) with the exception of the reciprocal fishing agreements. The regional administration is subject to central control for setting allocations and for Border Guard enforcement. From comments about the implementation of enforcement in western Russia, it seems the US Coast Guard and the Border Guards have developed a more effective cooperation on enforcement in the Bering Sea, particularly with respect to the fishing zone boundary and high seas driftnet fishing. The scientific basis for setting allocations in Russia is similar to that of Alaska. Significant concerns have been expressed about how well such allocations are being followed and enforced (Velegjanin, 1999). Similarly, the role of the central as opposed to the regional fishery administrations in the setting and allocation of quotas is being challenged. For several years, significant proportions of the total allowable harvest are being auctioned to the highest bidders. This innovative effort has been controversial.

13.5.6. Variations in Bering Sea fisheries and socio-economic impacts: possible scenarios

The major changes in the commercial fisheries of the Bering Sea have been in the distribution of the harvests among nations and sub-nation user groups. Changes in the species composition of the catch due to changes in environmental conditions and fishing pressures have also affected those employed in the fishing industry and their communities. However, while the latter are of considerable interest in the present assessment, it is important to note that the adjustments to changing claims to jurisdiction in the Bering Sea have been extensive (Miles et al., 1982a). The enormous dislocation of fishing fleets from Japan and then the Soviet Union post-EEZ extension, shows that major adjustments can be made but with considerable hardship. Similarly, the response of the US fishing industry assisted by favorable government incentives shows how quickly it can respond to opportunity. The question thus is how fully occupied fisheries can respond to sustainable and precautionary management.

Fisheries in the Bering Sea are largely a post Second World War phenomenon in terms of the technology and scale of enterprise necessary to fish the inhospitable and enormous expanses of the remote Bering Sea shelf. With the developments in mothership operations and food processing technology came the development of new markets for species such as walleye pollock that despite being available in large quantities had not previously been considered a target species. Little is known about the fisheries ecosystem of the Bering Sea prior to the development of the intensive industrial-scale fisheries. Attention has been given to the early whaling activity in the North Pacific as this affected the more valuable and easier to harvest species. The effects of removing this biomass of whales on controls in the Bering Sea ecosystem is not clear (National Research Council, 1996) but cannot have been insignificant. The decline in the North Pacific whaling was offset by effort directed toward other areas, including the Southern Ocean. For communities where rending and processing occurred onshore, the displacement of effort meant the end of whaling as a source of employment and income.

Fisheries development after the Second World War tended to target the highest value species first. Despite efforts to develop sufficient scientific information and international management under the International Convention for High Sea Fisheries of the North Pacific Ocean, some stocks such as Pacific ocean perch and other longer-lived, high value species were overfished. The opportunity to fish on previously unfished stocks of very large size and extent resulted in significant employment and income benefits. With the development of coastal state management came the need to manage these large-scale fisheries properly. Most observers do not consider the harvests reported for the early period to be an accurate representation of catches. The valuable but limited joint scientific survey of stocks, performed by Canada, Japan, and the United States under the Convention, provides some information. This results in the period of record being extremely limited. Another factor is that the Bering Sea is a large, remote, and difficult area to characterize and monitor. Thus, the linking of scientific advice to fisheries management objectives has been a process of successive refinement. The ability to assess the range in natural variability in stock sizes is very imprecise and how the ecosystems function is only now being modeled with a significant degree of sophistication to begin to understand some of the issues involved (National Research Council, 2003).

Eight periods of alternating cold/warm sea temperatures are evident in the instrumental record. The extent to which these have altered population sizes and concentrations is difficult to establish for the reasons mentioned. Furthermore, population sizes may have been affected by high levels of fishing for some high value species, and low levels of fishing for species with low market value or with high levels of bycatch. Fishery management is generally thought to mediate for overfishing and to manage to maintain abundance of desired species. Since the mid- to late 1970s warmer temperatures and the associated patterns of atmospheric and sea surface circulation may have favored salmonids, winter-spawning flatfish, walleye pollock, Pacific cod, and Pacific halibut, and have been detrimental to capelin, Pacific herring, shrimps, and several species of large crab. Fisheries have developed on those species that are at high levels of abundance and left those whose abundance is low (NMFS, 2003b).

The US fishing industry in the Bering Sea survived changes in the relative abundance of particular species during the growth phase by, for example, shifting from crab fishing to walleye pollock fishing and Pacific cod fisheries. This has altered conditions for traditional crab processing ports in the Pribilof Islands but has contributed to the growth of groundfish processing in Dutch Harbor/Onalaska and Akutan. The question is what would happen to the industry under a pronounced shift to a coldwater period. Fisheries management is attempting to rationalize effort in these fisheries to increase efficiency, to reduce bycatch of prohibited species, and to increase capture value through higher quality products and utilization rates. This tends to reduce flexibility of movement, as occurred when the domestic fisheries developed. There is little planning in place for how fishery management could operate in a transition between cold and warm regimes. For most of the groundfish species management under quotas, the expectation is that small or large year classes would be detected in the assessments and that quotas would rise and fall to prevent overfishing. For species with short lifespans this approach may be less effective, although high natural variability is considered by managers. For exceptionally long-lived species such as rockfish (*Sebastes* spp.), experience shows that very conservative harvest rates may need to be used and no-take marine reserves have been suggested as a tool to insure against loss of older highly productive fish.

This is an important issue, as is evident from the massive buildup of the red king crab fleet in the late 1970s to harvest anomalously large quantities of a Bristol Bay stock that subsequently crashed – probably due to recruitment failure following changes in environmental conditions. Additional effort entered the crab fleet with the strong stocks of snow crabs in the 1980s and 1990s. A sharp decline in these fisheries, again associated with changes in environmental conditions, caused severe problems for operators with high debt service and relatively few assets. These problems in the eastern Bering Sea crab fisheries provided an incentive to find other pot gear fishing opportunities and so other fixed gear operators in Pacific cod are now being squeezed by the entry of crab vessels into their traditional fisheries. This domino effect is highly predictable even if the underlying phenomena driving the process are not.

Warmer conditions are less favorable for pinnipeds. This appears to be an indirect food web effect rather than a direct effect through predation, although there may be interacting effects. This complex interaction between climate and pinniped survival has a pronounced effect on major commercial fisheries in the eastern Bering Sea under US jurisdiction. The spatial extent and timing of walleye pollock, Atka mackerel, and Pacific cod fisheries have been modified as a precautionary measure to protect Steller sea lions (National Research Council, 2003). In this way, changes in environmental conditions that result in effects on non-target species can be sufficiently significant in terms of the management of endangered and threatened species that they result in increased fishing costs and thus reduced profits.

The many subsistence fishing villages on the shores of the Bering Sea experience climate variability directly. The 65 CDQ communities in the eastern Bering Sea region have direct connections with climate variability through subsistence fishery activities and participation in the industrial fisheries through their partners. Industrial fisheries in the Bering Sea are dependent on large-scale shore-based processing plants that can operate, like the

fishery itself, under difficult conditions. This is because catcher vessels that deliver to the shoreside plants must now operate further offshore because of the closed areas to buffer sea lion competition for prey. At-sea processors are more adaptable to changing environmental conditions because they can follow the fish and fishing conditions and can deliver to various ports.

Salmonids have well-documented aggregate north/south shifts in production under warm and cold periods (Beamish and Bouillon, 1993; Hare and Francis, 1995; Mantua et al., 1997). Although this does not explain all sources of variability it has been used successfully to gain a better management understanding. These trends are now being exacerbated by the decrease in market price following the decline in the Asian market and competition from farmed sources of Atlantic salmon. Even at high levels of abundance fishing for wild salmonids in the Bering Sea is at best marginal. This may force fundamental change in the structure and practices of salmon fishing. Also, extremely low returns to the Yukon River make survival of the Alaskan and Canadian indigenous peoples dependent on the abundance of migrating salmonids precarious. This has brought disaster relief in the form of federal and state loans and welfare programs. Recent studies (Kocan et al., 2001) suggest that the decline in Yukon stocks may be due to warmer environmental conditions and so beyond the control of fishery managers. The low levels of salmon have already resulted in renewed calls for reducing the salmon bycatch in Bering Sea trawl fisheries. Even though salmon bycatch rates have been reduced, more salmon are wanted by Yukon and other peoples. The trawl industry that has been pushed from low to higher bycatch areas due to measures for Steller sea lion protection has taken proactive real time measures to avoid salmon bycatch.

The location of the sea-ice edge and of the extent and timing of the melting of the sea ice as well as the development of the "cold pool" can have positive and negative effects on fisheries through their tendency to concentrate or disperse certain species or to contribute to increased levels of primary and secondary production within the Bering Sea ecosystem. Direct impacts on crab pot loss resulting from shifts in the position of the ice edge have been noted in the opilio fisheries in some cold years. The economic consequences of these types of variability are considered part of the risks of fishing in the Bering Sea. At present, it is possible to make only general comments about the effects of climate variability on fisheries in the Bering Sea from a socio-economic perspective. Better analyses require a better scientific understanding of ecosystem dynamics within the Bering Sea and a better ability to predict. A complicating factor is the difficulty of understanding the dynamics within the fisheries due to the very short period of record. Also, external market forces are currently affecting the value of the fisheries to a very significant extent and this may be more important than variability in landings or overall fish abundance.

At the industrial scale of fishing and processing that is characteristic of groundfish and crab fisheries in the Bering Sea, the social effects reflect broader economic trends. Lower prices and quantities generate fewer and less well paid jobs. However, high world market prices for species such as red king crab may offset declines in stocks when other sources of supply decrease (e.g., in Russian waters), or increase (e.g., red king crab in northern Norway). Rationalization through the economic system or fishery management systems may allow greater long-term stability with less overall investment in harvesting and processing. Fewer operators earning a greater return on investment are more likely to absorb swings in abundance due to changes in environmental or other conditions. It is difficult to assess impacts on consumers as the world trade in fisheries tends to find ways to satisfy market demands. However, impacts on fishery dependent communities and small family-owned enterprises can be devastating as the high costs of fishing may exceed the price available (Link et al., 2003). Having most assets tied up in ownership of a fishing vessel and gear, a limited entry area permit, and nowhere to sell is a formula for disaster. Many operations face bankruptcy and in communities with many such entities, there are few alternatives.

13.5.7. Ability to cope with change

Over the past few decades Bering Sea fisheries have been built around fairly consistent warm water species although there are some differences between the western and eastern Bering Sea. Coastal states have benefited more in recent years than distant water fishing nations. However, the management response to a transition to a cold phase has not been adequately considered nor has the response to continued warm periods. Changes to stocks in the western Bering Sea and projected stock dynamics in response to a moderate warming are explored in Table 13.2. Assuming a shift between a cold and warm regime in the mid-1970s, which for the Bering Sea is only ±1 °C (see section 13.5.4.2), could result in many effects and other coincident changes. For example: salmon increase in number but the world market price declines; groundfish abundance increases but the Asian market is weak owing to other economic factors; US snow crab stocks decline but Canadian stocks increase due to possible unfavorable or favorable environmental conditions.

A very small difference in ocean conditions can be detected as a cold or warm phase in the Bering Sea. Although a global climate change scenario for the Bering Sea *per se* does not exist, this shift between cold and warm periods provides some working hypotheses about what could be expected. At a minimum, it is likely that the conditions that have prevailed over the past few decades might constitute a baseline for slightly warmer conditions. Which means there is unlikely to be a resurgence of crab or shrimp populations or herring and capelin and other small pelagic species. The ecosystem would continue to be dominated by walleye

pollock, Pacific cod, and flatfish. Walleye pollock juveniles may continue to occupy the role of coldwater forage fish. Salmonids would probably remain abundant in the aggregate in northern waters but in the south off British Columbia and Washington and Oregon stocks would decrease.

Socio-economically this baseline case would replicate the current system in terms of production of fish commodities. Through improvements in fishery management, it may be possible to increase the harvests of certain stocks by managing for recovery to levels of former abundance. However, it is just as likely that unforeseen events or interactions may result in management mistakes that offset such gains. Exploitation of underutilized species may be feasible to some degree. There may be some gains in catching the whole TAC due to changes in gear and fishing practices to generate lower bycatch rates. To attain increases in value added and utilization rates, it may be necessary to further rationalize the industry.

Additional factors to be included in the scenario of a continuation of prevailing conditions are declines in marine mammal and seabird populations. In some cases, fishery interactions, while modest and indirect, may justify further efforts to protect the numbers of seabirds and marine mammals under an adverse environmental regime, and such requirements may constrain fisheries more than would be the case if the stock was the sole interest of management. Similarly, environmental groups may change the level of performance that they expect fishery management to attain, i.e., no detectable impact standard or negligible effect standard and this would alter the management "field of play".

With continued warming, there is likely to be a range of sea temperatures that would continue to generate positive recruitment and growth scenarios for some of the warm water species (Table 13.2). This is likely to result in unfavorable conditions (i.e., increased predation) for pandalid shrimp and most crab species. If walleye pollock stocks increase, their impact as a predator on fish may also increase with unpredictable outcomes. Migration paths, timing of spawning, timing of the start of primary production, and species composition are very unlikely to remain the same. Similarly, reduced sea ice is likely to change the early spring ecosystem processes but greater surface exposure to winter storm conditions is likely to increase nutrient cycling and resuspension from shallower waters. To date, there are no credible published predictions of changes to fisheries north of the Bering Strait under a no or low sea ice scenario.

13.5.8. Concluding comments

In comparison to fisheries in other areas of the Arctic, commercial fisheries of the North Pacific, including the Sea of Okhostk and the Bering Sea, are relative newcomers. Commercial fishing for groundfish stocks other than Pacific halibut began in the Bering Sea in the 1950s by fleets from Japan and Russia and soon developed into large-scale operations involving many nations. These fleets primarily harvested walleye pollock, Pacific cod, flatfish, sablefish, Atka mackerel, crab, herring, and salmon stocks. In the late 1970s, EEZs were established 200 nm seaward from the coast by Russia and the United States and fisheries management plans were established. By 1990, the distant water fleets were phased out of the eastern Bering Sea (i.e., the US EEZ). US fisheries off Alaska constitute more than half the landings and about half the value of national landings of fish and shellfish from federal waters. In the Russian EEZ, most catches are taken by domestic fleets with a decreasing proportion harvested under agreements with neighboring states.

Well-documented climate regime shifts occurred in the Bering Sea over the 20th century at roughly decadal time scales, alternating between warm and cool periods. A climate regime shift in the Bering Sea in 1977 changed the marine environment from a cool to a warm state. The warming-induced ecosystem shifts favored recruitment to herring stocks and enhanced productivity for Pacific cod, skates, flatfish, and non-crustacean invertebrates. The species composition of the benthic community changed from a crab-dominated assemblage to a more diverse mix of starfish, ascidians, and sponges. Pacific salmon production was found to be positively correlated with warmer temperatures. Consecutive strong year classes were established and historically high commercial catches were taken. Levels of walleye pollock biomass were low in the 1960s and 1970s (2 to 6 million t) but subsequently increased to levels greater than 10 million t and have remained large in most years since 1980.

Information from the contrast between the 1977 to 1989 warm period and the prior and subsequent cool periods (1960–1976 and 1989–2000) form the basis of the predicted response of the Bering Sea ecosystem to scenarios of future warming. Predictions include increased primary and secondary productivity with a greater carrying capacity, increased catches for species favored by a warm regime, poleward shifts in the distributions of some cold-water species, and possible negative effects on ice-associated species.

Walleye pollock is the major harvest species by volume and value, with Pacific cod, flatfish, salmon, and crabs constituting most of the rest. Total wholesale value for groundfish harvests in the eastern Bering Sea is approximately US$ 426 million, while the total primary processed value is approximately US$ 1.4 billion. The North Pacific fishing communities surrounding the Bering Sea are different from those of the North Atlantic. On the coast of the eastern Bering Sea there are some 65 communities with a total population of around 27 500 inhabitants, but these do not have a long history of fishing.

Fishery Management Plans are the main tool for fishery management in US waters. These set forth the rules and regulations for the management of each species or species complexes. Under the current management approach, TACs are set on an annual basis in the Stock Assessment Fishery Evaluation process. Ecosystem considerations are explicitly made available at the time of the TAC setting process. The North Pacific Fishery Management Council has developed some fairly innovative approaches to management. In the EEZ of the Russian Far East, the regional administration is subject to central control for setting TAC allocations and for Border Guard enforcement.

The main changes over the years in the commercial fisheries of the Bering Sea have been in the distribution of the harvests among nations and sub-national user groups. There have been extensive adjustments to changing claims to jurisdiction in the Bering Sea. The tremendous dislocation of fishing fleets from Japan and Russia (then the Soviet Union) after the EEZ extension to 200 nm shows that major adjustments can be made but with considerable hardship. Similarly, the response of the US fishing industry, assisted by favorable government incentives, shows how quickly the fishery can respond to changed opportunities.

Eight periods of alternating cold/warm sea temperatures are evident in the instrumental record. Population sizes may have been affected by both high levels of fishing for some high-value species and low levels of fishing for species with low market value or with high levels of bycatch. Fishery management is generally intended to prevent overfishing and maintain the abundance of desired species. Since the mid- to late 1970s, warmer temperatures and associated atmospheric and sea surface circulation may have favored salmonids, winter-spawning flatfish, walleye pollock, Pacific cod, and Pacific halibut but have been detrimental to capelin, Pacific herring, shrimps, and several species of large crab. Fisheries are fully developed on those species that are at high levels of abundance, but have essentially ceased on those whose abundance is low. The last few decades of Bering Sea fisheries have been built around species that consistently favor warm water. However, there are some contradictions between the western and eastern Bering Sea. There is no question that coastal states have benefited more in recent years than distant water fishing nations. However, the management response to a transition to a cold phase has not been adequately considered, nor has it for the opposite, i.e., continued prevailing warm conditions.

Previous sections of this chapter have demonstrated that a very small difference in ocean environmental conditions can be detected as a cold or warm phase in the Bering Sea. While there is not a global climate change scenario for the Bering Sea per se, this shift between cold and warm periods does provide a basis for some working hypotheses about what to expect in the area in future. At a minimum, it is likely that the conditions which have prevailed over the last few decades might constitute a baseline for slightly warmer conditions. Therefore, there is not likely to be a resurgence of crab or shrimp populations, or herring and capelin and other small pelagic fish species. The ecosystem is likely to continue to be dominated by walleye pollock, Pacific cod, and flatfish. Walleye pollock juveniles are likely to continue their role as cold-water forage fish. Salmonids are likely to remain abundant in the aggregate in northern waters, but south off British Columbia and Washington and Oregon stock abundance would be depressed.

Socio-economically this baseline case would replicate the current system in terms of production of fish commodities. Through improvements in fisheries management, it may be possible to increase harvests of certain stocks by managing for recovery to levels of former abundance. However, it is probably just as likely that unforeseen events or interactions may produce management mistakes that offset such gains in a dynamic ocean system. Exploitation of underutilized species may be feasible to some degree. Some gains in catching the whole TAC might be possible due to improvements in gear and fishing practices to lower bycatch rates. In order to attain increases in value added and in utilization rates, the industry may need to be further rationalized.

Under a continued warming scenario, it is very likely that there could be a range of temperatures that would continue to generate positive recruitment and growth scenarios for some of the warm advantaged species. These conditions would be negative for pandalid shrimp and most crab species. If walleye pollock stocks increase, their impact as a predator on fish may also increase with unpredictable outcomes. It is very unlikely that migration paths, timing of spawning, timing of start of primary production, and composition of species would remain the same. Similarly, loss of sea ice may result in changes to the early spring bloom and associated ecosystem processes, however greater surface exposure to winter storm conditions might increase nutrient circulation and resuspension in shallower waters. To date, there are no credible published data on what could happen in the waters north of the Bering Strait with respect to fisheries under a change to a significantly warmer climate.

13.6. Synthesis and key findings

Modeling experiments show that it is not easy to project changes in climate due to forces, which can and have been measured and even monitored on a regular basis for considerable periods of time and are the data upon which such models are built. The main reason being that major natural events occur over time scales greater than decades or even centuries and the period of regular monitoring of potentially important forcing events is relatively short. Also, current climate models do not include scenarios for ocean temperatures, watermass mixing, upwelling, and other relevant ocean variables such as primary and secondary production, neither globally nor regionally. Thus, it is not possible to predict the effects of climate change on marine fish stocks with any degree of certainty and so the eventual socio-economic consequences of these effects for arctic fisheries.

Nevertheless, and despite these difficulties, the scientific community should still rise to the challenge of predicting reactions of marine stocks in or near the Arctic to climate change, basing initial studies on past records of apparent interactions, however imperfect and inconclusive. It is on such bases — and such bases only — that effective future research can and should be planned and undertaken.

Commercial fisheries in arctic regions are based on a number of species belonging to physically different ecosystems. The dynamics of many of these ecosystems are not well understood. This adds a significant degree of uncertainty to attempts to predict the response of individual species and stocks to climate change. Indeed, to date it has been difficult to identify the relative importance of fishing and the environment on changes in fish populations and biology. Moreover, current fish populations differ in abundance and biology from past populations due to anthropogenic effects (i.e., exploitation rates). As a result it is unclear whether current populations will respond to climate change as they may have done in the past.

Nevertheless, it does appear likely that a moderate warming will improve the conditions for some of the most important commercial fish stocks, as well as for aquaculture. This is most likely to be due to enhanced levels of primary and secondary production resulting from reduced sea-ice cover and more extensive habitat areas for subarctic species such as cod and herring. Global warming is also likely to induce an ecosystem regime shift in some areas, resulting in a very different species composition. Changing environmental conditions are likely to be deleterious for some species and beneficial for others. Thus, relative population sizes, fish growth rates, and spatial distributions of fish stocks are likely to change (see Table 9.11). This will result in the need for adjustments in the commercial fisheries. However, unless there is a major climatic change over a very short period, these adjustments are likely to be relatively minor and are unlikely to entail significant economic and social costs.

The total effect of climate change on fish stocks is probably going to be of less importance than the effects of fisheries policies and their enforcement. The significant factor in determining the future of fisheries is sound resource management practices, which in large part depend upon the properties and effectiveness of resource management regimes. All arctic countries are currently making efforts to implement management strategies based on precautionary approaches, with increasing emphasis on ecosystem characteristics, effects of climate changes, and including risk and uncertainty analyses in decision-making. Ongoing adjustments to management regimes are likely to enhance the ability of societies to adapt to the effects of climate change.

The economic and social impacts of altered environmental conditions depend on the ability of the social structures involved, including the fisheries management system, to generate the necessary adaptations to the changes. It is unlikely that the impact of the climate change projected for the 21st century (see Chapter 4) on arctic fisheries will have significant long-term economic or social impacts at a national level. Some arctic regions, especially those very dependent on fisheries or marine mammals and birds in direct competition with a fishery may, however, be greatly affected. Local communities in the north are exposed to a number of forces of change. Economic marginalization, depopulation, globalization-related factors, and public policies in the different countries are very likely to have a stronger impact on the future development of northern communities than climate change, at least over the next few decades.

This chapter considers the possible effects of projected climate change on four major ecosystems: the Northeast Atlantic (Barents Sea), the central North Atlantic (Iceland/Greenland), Northeast Canada (Newfoundland/Labrador), and the North Pacific (Bering Sea). There are substantial differences between these regions in that the Barents Sea and Icelandic waters are of a subarctic/temperate type, while the arctic influence is much greater in Greenland waters, the waters off northeast Canada, and the Bering Sea. It follows, therefore, that climate change need not affect these areas in the same or a similar manner. Also, the length of useful time series on past environmental variability and associated changes in hydrobiological conditions, fish abundance, and migrations varies greatly among regions. Finally, there are differences in species interactions and variable fishing pressure, which must also be considered.

Owing to heavy fishing pressure and stock depletions, the Barents Sea, Icelandic waters, and possibly also the Bering Sea could, through more efficient management, yield larger catches of many fish species. For that to happen research must increase, and more cautious management strategies must be developed and enforced. However, a moderate warming could enhance the rebuilding of stocks and could also result in higher sustainable yields of most stocks, among others, through enlarged distribution areas and increased availability of food in general. On the other hand, warming could also cause fish stocks to change their migratory range and area of distribution. This could (as history has shown) trigger conflict among nations over distribution of fishing opportunities and would require tough negotiations to generate viable solutions regarding international cooperation in fisheries management.

The waters around Greenland and off northeast Canada are very different from the above. These regions are more arctic in nature. Greenland appears unable to support subarctic species such as cod and herring except during warm periods. Examples from the 20th century prove this point. For example, there were no cod in the first two and a half decades, but a large local self-sustaining cod stock from 1930 until the late 1960s, apparently initiated by larval and 0-group drift from Iceland. If current climate conditions remain unchanged little change is likely around Greenland. On the other hand, a "moderate warming" such as that between 1920 and the late 1960s is likely to result in dramatic changes in species composition – a scenario where cod would play the major role. The northeast Canadian case is an extreme example of a situation where a stock of Atlantic cod (the so-called "northern" cod), which had sustained a large fishery for at least two centuries, is suddenly gone. Opinion differs as to how this has happened; most people believe that the decline was due entirely to overfishing, whereas others think that adverse environmental factors were significant contributors. In the present situation, however, the northern cod stock is so depleted that it is very likely to take decades to rebuild – even under the conditions of a warming climate.

An evaluation of what could happen to marine fisheries and aquaculture in the Arctic should the climate warm by more than 1 to 3 °C is not attempted in the present assessment. This is beyond the range of available data and would be of limited value. In general terms, however, it is likely that at least some of the ecosystems would experience reductions in present-day commercial stocks which might be replaced partially or in full by species from warmer waters.

13.7. Research recommendations

Past experience shows that marine living resources are not unlimited and must be harvested with caution. Although management practices have improved in recent decades, the present situation still leaves much room for improvement. More and better research is required to fill this gap.

1. Present monitoring of the physical and biological marine environment must be continued and in many cases increased. Basic research is often considered a burden, but is a prerequisite for understanding biological processes. Modern technology enables the automation of many of the time consuming tasks previously conducted from expensive research vessels. For example, buoys can now be deployed in strategic locations on land and at sea for continuous measurement of many variables required in marine biological studies. The monitoring of commercial stocks must also continue, applying new technologies as these become available. There is a general shortage of ship time for sea-based work. Administrators (governments) are often unaware of this, also that despite computers enabling more extensive and deeper analyses of existing datasets, people are still required to operate and program the computers.

2. Although the modeling of marine processes, particularly the modeling of climate variability, is still in its infancy, such work is the key to increasing understanding of the effects of the projected climate change scenarios (see Chapter 4). The devel-

opment of regional applications is particularly
important. Regional effects might differ substantially from those considered average global effects. In
order to relate physical changes in the atmosphere
and oceans to changes within specific ecosystems,
the modeling of regional effects is essential.
Current fisheries management models are based on
general assumptions of constant environmental factors. The use of ecosystem-based approaches for
fisheries management will require that physical and
biological factors that do not directly affect the target species are also taken into account.

3. It is extremely difficult to estimate the economic
consequences of climate change on the world fisheries or the fisheries for any given region. It is
important to invest in the development of better
methods for examining the economic and social consequences of climate change, at both the global and
regional level, and at the national and local level.

References

ADF&G, 2002. Alaska Commercial Harvests of King, Tanner, and Snow
Crab, 1953–2000. Alask department of Fish and game,
www.cf.adfg.state.ak.us/region4/shellfish/crabs/1953–00.htm

Agnarsson, S. and R. Arnason, 2003. The Role of the Fishing Industry in
the Icelandic Economy. An Historical Examination. Working paper
W03:08. Institute of Economic Studies, University of Iceland.

Alton, M.S., R.G. Bakkala, G.E. Walters and P.T. Munro, 1988.
Greenland turbot, *Rheinhardtius hippoglossoides*, of the eastern Bering
Sea and Aleutian Islands. NOAA Tech. Rep. NMFS 71, 31 p.

Alverson, D.L., 1987. A Study of Trends of Cod Stocks off Newfoundland
and Factors Influencing their Abundance and Availability to the
Inshore Fishery: A Report to the Honorable Tom Siddon, Minister of
Fisheries of Canada. Task Group on Newfoundland Inshore Fisheries.

Alverson, D.L., A.T. Pruter and L.L. Ronholt, 1964. A study of demer-
sal fishes and fisheries of the northeastern Pacific Ocean. H.R.
MacMillan Lectures in Fisheries. Institute of Fisheries, University of
British Columbia.

Angliss, R.P., D.P. Demaster and A.L. Lopez, 2001. Alaska marine
mammal stock assessments, 2001. NOAA Tech. Mem. NMFS-AFSC-
124. 203p.

Anon, 2000. Udvikling i nogle centrale økonomiske konjunkturindikator-
er. Konjunkturstatistik 200:2. Statistics Greenland.
http://www.statgreen.gl/dk/publ/konjunk/00-1konj.pdf

Anon, 2001. Sea Ice Climatic Atlas, East Coast of Canada, 1971–2000.
Canadian Ice Service, Ministery of Public Works.

Anon, 2004a. Report of the joint ICES/NAFO Working Group on Harp
and Hooded Seals. ICES CM 2004/ACFM:6.

Anon, 2004b. Environmental Conditions in Icelandic Waters 2004.
Technical Report 101. Marine Research Institute, Reykjavík, 37p.
(In Icelandic – English summary, table and figure legends)

Anon, 2004c. State of Marine Stocks in Icelandic Waters 2003/2004.
Prospects for the Quota Year 2004/2005. Technical Report 202.
Marine Research Institute, Reykjavík, 181p. (In Icelandic – English
summary, table and figure legends)

Arlov, T.B., 1996. Svalbards historie. H. Aschehoug & Co.

Arnason, R., 1994. The Icelandic Fisheries: Evolution and Management
of a Fishing Industry. Fishing News Books, Oxford.

Astthorsson, O.S. and Á. Gislason, 1995. Long-term changes in zooplankton biomass in Icelandic waters in spring. ICES Journal of
Marine Science, 52:657–668.

Astthorsson, O.S. and H. Vilhjálmsson, 2002. Iceland Shelf LME: decadal
assessment and resource sustainability. In: K. Sherman and H.R.
Skjoldal (eds.). Large Marine Ecosystems of the North Atlantic,
pp. 219–243. Elsevier Science.

Atkinson, D.B., 1994. Some observations on the biomass and abundance
of fish captured during stratified-random bottom trawl surveys in
NAFO Divisions 2J and 3KL, autumn 1981–1991. NAFO Scientific
Council Studies, 21:43–66.

Atkinson, D.B. and B. Bennett, 1994. Proceedings of a Northern Cod
Workshop Held in St. John's, Newfoundland, Canada, January
27–29, 1993. Canadian Technical Report of Fisheries and Aquatic
Sciences No. 1999, 64p.

Aydin, K.Y., V.V. Lapko, V.I. Radchenko and P.A. Livingston, 2002.
A comparison of the eastern Bering and western Bering Sea shelf
and slope ecosystems through the use of mass-balance food web
models. NOAA Technical Memorandum, NMFS-AFSC-130. 78p.

Baerenholdt, J.O and N. Aarsaether, 2001. Transforming the Local.
Coping Strategies and Regional Policies. Nord 2001:25, 208p.

Bakkala, R.G., 1993. Structure and Historical changes in the groundfish complex of the eastern Bering Sea. NOAA Tech. Rep NMFS
114. 91p.

Bailey, K.M., R.C. Francis and J.D. Schumacher, 1986. Recent information on the causes of variability in recruitment of Alaska pollock
in the eastern Bering Sea: physical conditions and biological interactions. Bulletin of the International North Pacific Fisheries
Commission, 47:155–165.

Baird, J.W., C.R. Stevens and E.F. Murphy, 1992. A review of hydroa-
coustic surveys conducted during winter for 2J3KL cod,
1987–1992. Canadian Atlantic Fisheries Scientific Advisory
Committee, Res. Doc. 92/107, 14p.

Beamish, R.J. and D.R. Bouillon, 1993. Pacific salmon production
trends in relation to climate. Canadian Journal of Fisheries and
Aquatic Sciences, 50(5):1002–1016.

Bendiksen, B.I and J.R. Isaksen, 2000. Fiskerinæringen i Finnmark -
Analyse av verdiskapning og råstoffomsetning. Rapport nr.
12/2000, Tromsø: Fiskeriforskning.

Berg, R., 1995. Norge på egen hånd. Norsk utenrikspolitisk historie
bd 2. Universitetsforlaget, Oslo.

Bishop, C.A. and J.W. Baird, 1994. Spatial and temporal variability in
condition factors of Divisions 2J and 3KL cod (*Gadus morhua*).
Northwest Atlantic Fisheries Organization, Scientific Council
Studies, 21:105–113.

Blindheim, J., 2004. Oceanography and climate. In: H.R. Skjoldal,
R. Sætre, A. Færnø, O.A. Misund, I. Røttingen (eds.). The
Norwegian Sea Ecosystem, pp. 65–96. Tapir Academic Press.

Boltnev, A.I., 1996. Status of the northern fur seal (*Callorhinus ursinus*)
population on the Commander Islands. In: O.A. Mathisen and
K.O. Coyle (eds.). Ecology of the Bering Sea: A Review of Russian
Literature. Alaska Sea Grant College Program Report No. 96–01,
pp. 277–288.

Borets, L.A., 1997. Bottom ichthyocoenoses of the Russian shelf of
the far-eastern seas: composition, structure, functioning elements,
and commercial significance. TINRO-Center. 217 p. (In Russian)

Bowering, W.R. and W.B. Brodie, 1995. Greenland halibut
(*Reinhardtius hippoglossoides*). A review of the dynamics of its distri-
bution and fisheries off Eastern Canada and Greenland. In: A.G.
Hopper (ed.). Deep-water Fisheries of the North Atlantic Oceanic
Slope, pp. 113–160. Kluwer Academic.

Bowering, W.R. and K.H. Nedreaas, 2000. A comparison of Greenland
halibut (*Reinhardtius hippoglossoides* (Walbaum)) fisheries and distribution in the Northwest and Northeast Atlantic. Sarsia, 85:61–76.

Bowering, W.R. and K.H. Nedreaas, 2001. Age validation and growth
of Greenland halibut (*Reinhardtius hippoglossoides* (Walbaum)): a comparison of populations in the Northwest and Northeast Atlantic.
Sarsia, 86:53–68.

Brodeur, R.D. and D.M. Ware, 1992. Long term variability in zooplankton biomass in the subarctic Pacific Ocean. Fisheries
Oceanography, 1(1):32–38.

Brodeur, R.D., M.T. Wilson, G.E. Walters and I.V. Melnikov, 1999.
Forage fishes in the Bering Sea: distribution, species associations,
and biomass trends. In: T. Loughlin and K. Ohtani (eds.). Dynamics
of the Bering Sea, pp. 509–536. University of Alaska Fairbanks.

Buch, E., 1993. The North Atlantic water component of the West
Greenland Current. ICES CM 1993/C: 20.

Buch, E. and H.H. Hansen, 1988. Climate and cod fishery at West
Greenland. In: T. Wyatt and M.G. Larraneta (eds.). Long Term
Changes in Marine Fish Populations. A symposium held in Vigo,
Spain, 18–21 November 1986, pp. 345–364.

Buch, E., S.A. Horsted and H. Hovgaard, 1994. Fluctuations in the
occurrence of cod in Greenland waters and their possible causes.
ICES Marine Science Symposia, 198:158–174.

Buch, E., M.H. Nielsen and S.A. Pedersen, 2002. Ecosystem variability and regime shift in West Greenland waters. Northwest Atlantic
Fisheries Organization, Standing Committee on Research Doc.
02/16, Ser. No. N4617.

Bulatov, O.A., 1995. Biomass variation of walleye pollock of the
Bering Sea in relation to oceanological conditions. In: R.J. Beamish
(ed.). Climate Change and Northern Fish Populations, pp.
631–640. Canadian Special Publication of Fisheries and Aquatic
Sciences, 121.

Bundy, A., 2001. Fishing on ecosystems: the interplay of fishing and
predation in Newfoundland-Labrador. Canadian Journal of
Fisheries and Aquatic Sciences, 58:1153–1167.

Bundy, A., G.R. Lilly and P.A. Shelton, 2000. A Mass Balance Model of the Newfoundland-Labrador Shelf. Canadian Technical Report of Fisheries and Aquatic Sciences 2310, xiv + 157p.

Burnett, E.C., 1941. The Continental Congress. Macmillan.

Cadigan, S.T., 1999. Failed proposals for fisheries management and conservation in Newfoundland, 1855–1880. In: D. Newell and R.E. Ommer (eds.). Fishing Places, Fishing People: Traditions and Issues in Canadian Small-scale Fisheries, pp. 147–169. University of Toronto Press.

Carscadden, J.E. and K.T. Frank, 2002. Temporal variability in the condition factors of Newfoundland capelin (*Mallotus villosus*) during the past two decades. ICES Journal of Marine Science, 59:950–958.

Carscadden, J.E., B.S. Nakashima and K.T. Frank, 1997. Effects of fish length and temperature on the timing of peak spawning in capelin (*Mallotus villosus*). Canadian Journal of Fisheries and Aquatic Sciences, 54:781–787.

Carscadden, J.E., K.T. Frank and W.C. Leggett, 2000. Evaluation of an environment-recruitment model for capelin (*Mallotus villosus*). ICES Journal of Marine Science, 57:412–418.

Carscadden, J.E., K.T. Frank and W.C. Leggett, 2001. Ecosystem changes and the effects on capelin (*Mallotus villosus*), a major forage species. Canadian Journal of Fisheries and Aquatic Sciences, 58:73–85.

Carscadden, J.E., W.A. Montevecchi, G.K. Davoren and B.S. Nakashima, 2002. Trophic relationships among capelin (*Mallotus villosus*) and seabirds in a changing ecosystem. ICES Journal of Marine Science, 59:1027–1033.

CBC, 2002. Item broadcast on *The Morning Show* in St. John's, Newfoundland on Canadian Broadcasting Company Radio 1 in the half hour starting at 7:00 a.m. Wednesday, April 24, 2002.

Chadwick, St. J., 1967 Newfoundland: Island into Province. Cambridge University Press.

Chigirinsky, A.I., 1994. Commercial fishing of Pacific salmon in the Bering Sea. TINRO Transactions, 116:142–151. (In Russian)

Chuchukalo, V.I., V.I. Radchenko, V.A. Nadtochii, V.N. Koblikov, A.M. Slabinskii and D.A. Terent'ev, 1999. Feeding and some features of ecology of Gadidae of the western Kamchatka shelf in summer 1996. Journal of Ichthyology, 39(4):309–321. (Translated from Voprosy Ikhthiologii, 1999, 39(3):362–374)

Clark, W.G. and S.R. Hare, 2002. Effects of climate and stock size on recruitment and growth of Pacific halibut. North American Journal of Fisheries Management, 22:852–862.

Clark, W.G., S.R. Hare, A.M. Parma, P.J. Sullivan and R.J. Trumble, 1999. Decadal changes in growth and recruitment of Pacific halibut (*Hippoglossus stenolepis*). Canadian Journal of Fisheries and Aquatic Sciences, 56:242–252.

Coachman, L.K. 1986. Circulation, water masses, and fluxes on the southeastern Bering Sea shelf. Continental Shelf Research, 5:23–108.

Cobb, J.N., 1927. Pacific cod fisheries. US Bureau of Fisheries, Document 1014, Appendix VII to the Report of the US Commissioner of Fisheries for 1926. US Government Printing Office.

Colbourne, E.B. and J.T. Anderson, 2003. Biological response in a changing ocean environment in Newfoundland waters during the latter decades of the 1900s. ICES Marine Science Symposia, 219:169–181.

Colbourne, E.B., S. Narayanan and S. Prinsenberg, 1994. Climate changes and environmental conditions in the Northwest Atlantic, 1970–1993. ICES Marine Science Symposia, 198:311–322.

Committee on the Bering Sea Ecosystem, 1996. R.C. Francis (Chair), L.G. Anderson, W.D. Bowen, S.K. Davis, J.M. Grebmeier, L.F. Lowry, I. Merculieff, C.H. Peterson, C. Pungowiyi, T.C. Royer, A.M. Springer and W.S. Wooster, 1996. The Bering Sea Ecosystem. Commission on Geosciences, Environment and Resources (CGER), Polar Research Board (PRB). 320p.

Conners, M.E., A.B. Hollowed and E. Brown, 2002. Retrospective analysis of Bering Sea bottom trawl surveys: regime shift and ecosystem reorganization. Progress in Oceanography, 55:209–222.

Conover, S., 1999. The 1997/98 Directory of Russian Far East Fishing Companies. Alaska Center for International Business, Anchorage, 99p.

Corbett, F., 2002. Fax message from F. Corbett, Department of Fisheries and Oceans to A. St. Croix, May 27, 2002.

Crawford, R.E., 1992. Life history of the Davis Strait Greenland halibut, with reference to the Cumberland Sound fishery. Canadian Manuscript Report of Fisheries and Aquatic Sciences 2130.

Cushing, D.H., 1969. The fluctuation of year-classes and the regulation of fisheries. Fiskeridirektoratets skrifter, 15:368–379.

Darby, C., B. Healey, J.-C. Mahé and W.R. Bowering, 2004. Greenland halibut (*Reinhardtius hippoglossoides*) in Subarea 2 and Divisions 3KLMNO: an assessment of stock status based on Extended Survivors Analysis, ADAPT, and ASPIC analyses, with stochastic projections of potential stock dynamics. Northwest Atlantic Fisheries Organization, Scientific Council Research Doc. 04/55, Serial No. N5008.

Davenport, J., 1989. The effects of salinity and low temperature on eggs of the Icelandic capelin *Mallotus villosus*. Journal of the Marine Biological Association of the United Kingdom, 69:1–9.

Davenport, J. and A. Stene, 1986. Freezing resistance, temperature and salinity tolerance in eggs, larvae and adults of the capelin, *Mallotus villosus*, from Balsfjord. Journal of the Marine Biological Association of the United Kingdom, 66:145–157.

Davis, M.B., 1992. Description of the inshore fishery during 1991 for 2J3KL cod as reported by inshore fisherpersons. Canadian Atlantic Fisheries Scientific Advisory Committee, Research Document 93/37.

Davydov, I.V., 1972. On oceanological fundamentals of yield formation of the separate herring generations in the western Bering Sea. TINRO Transactions, 82:281–307. (In Russian)

Dawe, E., D. Taylor, D. Orr, D. Parsons, E. Colbourne, D. Stansbury, J. Drew, P. Beck, P. O'Keefe, P. Veitch, E. Seward, D. Ings and A. Pardy, 2004. An assessment of Newfoundland and Labrador snow crab in 2003. Department of Fisheries and Oceans, Canadian Science Advisory Secretariat Res. Doc. 2004/024.

Decker, M.B., G.L. Hunt Jr. and G.V. Byrd Jr., 1995. The relationships among sea surface temperature, the abundance of juvenile walleye pollock, and the reproductive performance and diets of seabirds at the Pribilof Islands, southeastern Bering Sea. In: R.J. Beamish (ed.). Climate Change and Northern Fish Populations, pp. 425–437. Canadian Special Publication of Fisheries and Aquatic Sciences 121.

deYoung, B. and G.A. Rose, 1993. On recruitment and distribution of Atlantic cod (*Gadus morhua*) off Newfoundland. Canadian Journal of Fisheries and Aquatic Sciences, 50:2729–2741.

DFO, 2000a. East and Southeast Newfoundland Atlantic Herring. Department of Fisheries and Oceans, Canada, Science Stock Status Report B2-01.

DFO, 2000b. Capelin in Subarea 2 + Div. 3KL. Department of Fisheries and Oceans, Canada, Science Stock Status Report B2-02.

DFO, 2000c. Northwest Atlantic harp seals. Department of Fisheries and Oceans, Canada, Science Stock Status Report E1-01.

DFO, 2001. Capelin in Subarea 2 + Div. 3KL - Update. Department of Fisheries and Oceans, Canada, Science Stock Status Report B2-02.

DFO, 2002a. Newfoundland and Labrador snow crab. Department of Fisheries and Oceans, Canada, Science Stock Status Report C2-01.

DFO, 2002b. Northern shrimp (*Pandalus borealis*) - Div. 0B to 3K. Department of Fisheries and Oceans, Canada, Science Stock Status Report C2-05.

DFO, 2003a. Northern (2J+3KL) Cod. Department of Fisheries and Oceans, Canada, Stock Status Report 2003/018.

DFO, 2003b. Newfoundland and Labrador Atlantic salmon 2003 stock status update. Department of Fisheries and Oceans, Canada, Stock Status Update Report 2003/048.

Dickson, R., 1992. Hydrobiological variability in the ICES area, 1980–1990. Introduction. ICES Marine Science Symposia, 195:1–10.

Dragesund, O., J. Hamre and Ø. Ulltang, 1980. Biology and population dynamics of the Norwegian spring spawning herring. ICES Rapports et Procès-Verbaux des Réunions, 177:43–71.

Dreyer, B., 2000. Globalisering av råvaremarkedet - strategiske utfordringer for lokal fiskeindustri. Økonomisk Fiskeriforskning Report no. 10:2.

Drinkwater, K.F., 2000. Changes in ocean climate and its general effect on fisheries: examples from the north-west Atlantic. In: D. Mills (ed.). The Ocean Life of Atlantic Salmon: Environmental and Biological Factors Influencing Survival, pp. 116–136. Fishing News Books, Oxford.

Drinkwater, K.F., 2002. A review of the role of climate variability in the decline of northern cod. American Fisheries Society Symposium, 32:113–130.

Dunlop, H.A., Bell, F.H., Myhre, R.J., Hardman, W.H., and Southward, G.M. 1964. Investigation, Utilization and Regulation of Halibut in Southeastern Bering Sea. International Pacific Halibut Commission Report 35. Seattle, Washington: International Pacific Halibut Commission.

Eide, A. and O. Flaaten, 1998. Bioeconomic Multispecies Models of the Barents Sea Fisheries. In T. Rødseth (ed.). Models for Multispecies Management, pp. 141–172. Physica-Verlag.

Eide, A. and K. Heen, 2002. Economic impacts of global warming. A study of the fishing industry in North Norway. Fisheries Research, 56(3):261–274.

Ervik, A., A. Kiessling, O. Skilbrei and T. van der Meeren (eds.), 2003. Havbruksrapport. Fisken og havet, særnummer 3.

Everett, J.T., A. Krovnin, D. Luch-Belda, E. Okemwa. H.A. Regier and J.P. Troadec, 1996. Fisheries. In: Climate Change 1996. Impacts, Adaptations, and Mitigation of Climate Change: Scientific-technical Analyses. Contribution of Working Group II to the Second Assessment Report of the Intergovernmental Panel of Climate Change, pp. 511-537.

Fadeev, N.S. and V. Wespestad, 2001. Review of walleye pollock fishery. TINRO Transactions, 128:75–91. (In Russian)

FAO, 2000. The State of World Fisheries and Aquaculture 2000. U.N. Food and Agriculture Organization.

Fiskeridirektoratet, 2001. Nøkkeltall fra norsk havbruksnæring. Rapport, Fiskeridirektoratet, Bergen.

Fiskeridirektoratet, 2002a. Beskrivelse av havbruksnæringen i området Lofoten til den norsk-russiske grense. Fiskeridirektoratet, Bergen.

Fiskeridirektoratet, 2002b. Fiskeriaktiviteten i området Lofoten – Barentshavet. Fiskeridirektoratet, Bergen.

Francis, R.C., S.R. Hare, A.B. Hollowed and W.S. Wooster, 1998. Effects of interdecadal variability on the oceanic ecosystems of the NE pacific. Fisheries Oceanography, 7:1–21.

Frank, K.T. and W.C. Leggett, 1981. Prediction of egg development and mortality rates in capelin (*Mallotus villosus*) from meteorological, hydrographic, and biological factors. Canadian Journal of Fisheries and Aquatic Sciences, 38:1327–1338.

Frank, K.T., R.I. Perry and K.F. Drinkwater, 1990. Predicted response of Northwest Atlantic invertebrate and fish stocks to CO_2-induced climate change. Transactions of the American Fisheries Society, 119:353–365.

Frank, K.T., J.E. Carscadden and J.E. Simon, 1996. Recent excursions of capelin (*Mallotus villosus*) to the Scotian Shelf and Flemish Cap during anomalous hydrographic conditions. Canadian Journal of Fisheries and Aquatic Sciences, 53:1473–1486.

Fridriksson, Á., 1948. Boreo-tended changes in the marine vertebrate fauna of Iceland during the last 25 years. ICES Rapports et Procès-Verbaux des Réunions, 125:30–32.

Friis-Rødel, E. and P. Kanneworff, 2002. A review of capelin (*Mallotus villosus*) in Greenland waters. ICES Marine Science Symposia, 216:890–896.

Frost, O., 2003. Bering: The Russian Discovery of America. Yale University Press, 330p.

Furevik, T., H. Drange and A. Sorteberg, 2002. Anticipated changes in the Nordic Seas marine climate: scenarios for 2020, 2050 and 2080. Institute of Marine Research, Bergen. Fisken og Havet No. 4.

Gavrilov, G.M. and I.I. Glebov, 2002. The composition of demersal ichthyocenosis in the western part of Bering Sea in November, 2002. TINRO Transactions, 130:1027–1037. (In Russian)

Ginter, J., 1995. The Alaska community development quota fisheries management program. Ocean and Coastal Management, 28(1–3):147–163.

Glebova, S.Yu., 2001. Coordination of types of atmospheric processes above the far-eastern seas. TINRO Transactions, 128:58–74. (In Russian)

Gomes, M.C., R.L. Haedrich and M.G. Villagarcia, 1995. Spatial and temporal changes in the groundfish assemblages on the north-east Newfoundland/Labrador Shelf, north-west Atlantic, 1978–1991. Fisheries Oceanography, 4:85–101.

Gordeev, V.D., 1954. Results of the Bering Sea trawl expedition of 1950–52. TINRO Transactions, 41:253–269. (In Russian)

Government of Canada, 1976a. Policy for Canada's Commercial Fisheries. Fisheries and Marine Service of the Department of the Environment, Ottawa.

Government of Canada, 1976b. Community and Employment Implications of Restructuring the Atlantic Fisheries. Analysis and Liaison Branch of the Department of Regional Economic Expansion, Ottawa.

Government of Newfoundland and Labrador, 1978. Setting a Course: A Regional Strategy for Development of the Newfoundland Fishing Industry to 1985. Department of Fisheries, St. John's, Newfoundland.

Government of Newfoundland and Labrador, 2001. Securing Our Future Together: Final Report on the Renewal Strategy for Jobs and Growth. Government of Newfoundland and Labrador, St. John's, Newfoundland.

Hacquebord, L., 2001. Three centuries of whaling and walrus hunting in Svalbard and its impact on the Arctic ecosystem. Environment and History, 7:169–185.

Haedrich, R.L. and S.M. Barnes, 1997. Changes over time of the size structure in an exploited shelf fish community. Fisheries Research, 31:229–239.

Haedrich, R.L. and J. Fischer, 1996. Stability and change of exploited fish communities in a cold ocean continental shelf ecosystem. Senckenbergiana Maritima, 27:237–243.

Haedrich, R.L., J. Fischer and N.V. Chernova, 1997. Ocean temperatures and demersal fish abundance on the northeast Newfoundland continental shelf. In: L.E. Hawkins and S. Hutchinson (eds.). The responses of marine organisms to their environments - Proceedings of the 30th European Marine Biology Symposium, pp. 211–221. University of Southampton.

Haflinger, K.E. and C.P. McRoy, 1983. Final Report: Yellowfin sole (*Limanda aspera*) predation on three commercial crab species (*Chionoecetes opilio*, C. Bairdi, and Paralithodes camtschatica) in the southeastern Bering Sea. National Marine Fisheries Service Final Report, Contract #82-ABC-00202.

Hammill, M.O. and G.B. Stenson, 2000. Estimated prey consumption by harp seals (*Phoca groenlandica*), hooded seals (*Cystophora cristata*), grey seals (*Halichoerus grypus*) and harbour seals (*Phoca vitulina*) in Atlantic Canada. Journal of Northwest Atlantic Fishery Science, 26:1–23.

Hamre, J., 2003. Capelin and herring as key species for the yield of north-east Arctic cod. Results from multispecies model runs. Scientia Marina, 67(1):315–323.

Hansen, P.M., 1949. Studies of the biology of the cod in Greenland waters. ICES Rapports et Procès-Verbaux des Réunions, 123:1–83.

Hansen, P.M., A.S. Jensen and Å.V. Tåning, 1935. Cod marking experiments in the waters of Greenland 1924–1933. Meddelelser fra Kommissionen for Danmarks Fiskeri- og Havundersøgelser. Serie: Fiskeri, 10(1):3–119.

Hare, S.R. and R.C. Francis, 1995. Climate change and salmon production in the northeast Pacific Ocean. In: R.J. Beamish (ed.). Climate Change and Northern Fish Populations, pp. 357–372. Canadian Special Publication of Fisheries and Aquatic Sciences 121.

Hare, S.R. and N. Mantua, 2000. Empirical evidence for North Pacific (climatic) regime shifts in 1977 and 1989. Progress in Oceanography, 47(2–4):103–145.

Harris, C.K., 1989. The effect of international treaty changes on Japan's high seas salmon fisheries, with emphasis on their catches of North American sockeye salmon, 1972–1984. University of Washington, Phd. Dissertation, 231p.

Harris, L. (Chairman), 1989. Independent Review of the State of the Northern Cod Stock. Department of Fisheries and Oceans, Ottawa.

Harris, I. (Chairman), 1990. Independent Review of the State of the Northern Cod Stock: Final Report of the Northern Cod Review Panel. Department of Supply and Services, Ottawa.

Hart, J.L. 1973. Pacific Fishes of Canada. Fisheries Research Board of Canada Bulletin 180, 740p.

Hartvigsen, R., A. Olsen, P.K. Hansen, O. Bjerk, E. Bolle and S. Jensen, 2003. Beskrivelse av havbruksnæringen I området Lofoten til den norsk-russiske grense. Fiskeridirektoratet, Bergen.

Haug, T., L. Walløe, S. Grønvik, N. Hedlund, M. Indregaard, H. Lorentzen, D. Oppen-berntsen and N. Øien, 1998. Sjøpattedyr - om hval og sel i norske farvann. Universitetsforlaget, Oslo.

Haynes, E.G. and G. Ingell, 1983. Effect of temperature on rate of embryonic development of walleye pollock (*Theragra chalcogramma*). Fishery Bulletin, 81:890–894.

Healey, B.P. and G.B. Stenson, 2000. Estimating pup production and population size of the Northwest Atlantic harp seal (*Phoca groenlandica*). Department of Fisheries and Oceans, Canadian Stock Assessment Secretariat Res. Doc. 2000/081.

Heen, K. and M. Aanesen, 1993. Kryssløpsmodell for nordnorsk økonomi 1987. Rapport NORUT samfunn forskning, Tromsø. (In Norwegian)

Hiatt, T., R. Felthoven and J. Terry, 2002. Stock assessment and fishery evaluation report for the groundfish fisheries of the Gulf of Alaska and Bering Sea/Aleutian Island Areas. Appendix D economic status of the groundfish fisheries of Alaska, 2001. Alaska Fisheries Science Center, National Marine Fisheries Service, National Oceanic and Atmospheric Administration, Seattle, 119p.

Hoel, A.H., 1993. Regionalisation of international whale management: the case of the Northeast Atlantic Marine Mammals Commission. Arctic, 46, 2:116–123.

Hoel, A.H., 1994. The Barents Sea: fisheries resources for Europe and Russia. In: O.S. Stokke and O. Tunander (eds.). The Barents Region. Cooperation in Arctic Europe. International Peace Research Institute, Oslo and the Fridtjof Nansen Institute.

Hoel, A.H., 1998. Political uncertainty in international fisheries management. Fisheries Research, 37:239–250.

Hollowed, A.B. and W.S. Wooster, 1992. Variability of winter ocean conditions and strong year classes of Northeast Pacific groundfish. ICES Marine Science Symposia, 195:433–444.

Hollowed, A.B. and W.S. Wooster, 1995. Decadal-scale variations in the eastern subarctic Pacific: II. Response of Northeast Pacific fish stocks. In: R.J. Beamish (ed.). Climate Change and Northern Fish Populations, pp. 373–385. Canadian Journal of Fisheries and Aquatic Sciences 121.

Hollowed, A.B., S.R. Hare and W.S. Wooster, 2001. Pacific Basin climate variability and patterns of Northeast Pacific marine fish production. Progress in Oceanography, 49:257–282.

Hønneland, 1993. Fiskeren og allmenningen; forvaltning og kontroll: Makt og kommunikasjon i kontrollen med fisket i Barentshavet, University of Tromsø.

Hønneland, G., 2004. Russian Fisheries Management: The Precautionary Approach in Theory and Practice. Martinus Nijhoff Publishers/Brill Academic Publishers, 210p.

Hønneland, G. and F. Nilssen, 2000. Comanagement in Northwest Russian fisheries. Society and Natural Resources, 13:635–648.

Huisman, J. and B. Sommeijer, 2002. Maximal sustainable sinking velocity of phytoplankton. Marine Ecology Progress Series, 244:39–48.

Hunt, G.L., P. Stabeno, G. Walters, E. Sinclair, R.D. Brodeur, J.M. Napp and N.A. Bond, 2002. Climate change and control of the southeastern Bering Sea pelagic ecosystem. Deep Sea Research II, 49(26):5821–5853.

Huntley, M.E. and M.D.G. Lopez, 1992. Temperature-dependent production of marine copepods: a global analysis. American Naturalist, 140:201–242.

Hutchings, J.A., 1996. Spatial and temporal variation in the density of northern cod and a review of hypotheses for the stock's collapse. Canadian Journal of Fisheries and Aquatic Sciences, 53:943–962.

Hutchings, J.A. and M. Ferguson, 2000. Temporal changes in harvesting dynamics of Canadian inshore fisheries for northern Atlantic cod, *Gadus morhua*. Canadian Journal of Fisheries and Aquatic Sciences, 57:805–814.

Hutchings, J.A. and R.A. Myers, 1994. What can be learned from the collapse of a renewable resource? Atlantic cod, *Gadus morhua*, of Newfoundland and Labrador. Canadian Journal of Fisheries and Aquatic Sciences, 51:2126–2146.

Hutchings, J.A. and R.A. Myers, 1995. The biological collapse of Atlantic cod off Newfoundland and Labrador: an exploration of historical changes in exploitation, harvesting technology, and management. In: R. Arnason and L. Felt (eds.). The North Atlantic Fisheries: Successes, Failures, and Challenges, pp. 39–93. The Institute of Island Studies, Charlottetown, Prince Edward Island.

Hvingel, C., 2003. A Predictive Fisheries Model for Greenland. Discussion paper.

Hvingel, C., 2004. The fishery for northern shrimp (*Pandalus borealis*) off West Greenland, 1970–2004. Northwest Atlantic Fisheries Organization, Scientific Council Research Doc. 04/75, Serial No. N5045.

Hvingel, C., in prep. Shrimp (*P. borealis*)-cod (*G. morhua*) interactions and their implications for the management of the West Greenland fisheries.

Hylen, A., 2002. Variation in abundance of Arcto-Norwegian cod during the 20th century. ICES Marine Science Symposia, 215:543–550.

Ianelli, J.N., T.K. Wilderbuer and T.M. Sample, 2001. Greenland turbot. In: Stock Assessment and Fishery Evaluation Report for the Groundfish Resources of the Bering Sea/Aleutian Islands regions. Chapter 4.

Ikeda, M., 1990. Decadal oscillations of the air–ice–sea system in the northern hemisphere. Atmosphere-Ocean, 28:106–139.

Incze L.E., D.A. Armstrong and S.L. Smith, 1987. Abundance of larval Tanner crabs (*Chionoecetes* spp.) in relation to adult females and regional oceanography of the southeastern Bering Sea. Canadian Journal of Fisheries and Aquatic Sciences, 44:1143–1156.

Innis, H.A., 1954. The Cod Fisheries. University of Toronto Press, revised edition.

IPCC, 2001. Climate Change 2001 – Impacts, Adaptation, and Vulnerability. Contribution of Working Group II to the Third Assessment Report of the Intergovernmental Panel on Climate Change. Cambridge University Press.

Ivanov, B.G., 1970. Distribution of northern shrimp (*Pandalus borealis* Kr.) in the Bering Sea and the Gulf of Alaska. TINRO Transactions vol. 72. VNIRO Transactions, Mutual issue, 70:131–148. (In Russian)

Ivanov, B.G., 2001. Studies and fisheries of pandalid shrimps (Crustacea, Decapoda, Pandalidae) in the Northern Hemisphere. A review in the XXI century eve (with special reference to Russia), pp. 9–31. In: Ivanov, B.G. (ed.). Study of biology of commercial crustaceans and algae of Russian seas. (In Russian)

Jakobsson, J., 1964. Recent developments in the Icelandic purse-seine herring fishery. In: H. Kristjonsson (ed.). Modern Fishing Gear of the World 2, pp. 294–395. Fishing News Books, Oxford.

Jakobsson, J., 1969. On herring migration in relation to changes in sea temperature. Jökull, 19:134–145.

Jakobsson, J., 1978. The north Icelandic herring fishery and environmental conditions 1960–1968. ICES Symposium on the Biological Basis of Pelagic Fish Stock Management. Paper no. 30, 101p.

Jakobsson, J., 1980. Exploitation of the Icelandic spring- and summer-spawning herring in relation to fisheries management, 1947–1977. ICES Rapports et Procès-Verbaux des Réunions, 177:23–42.

Jakobsson, J., 1992. Recent variability in the fisheries of the North Atlantic. ICES Marine Science Symposia, 195:291–315.

Jakobsson, J., 2002. Ocean Travellers. ICES Marine Science Symposia, 215: 330–342.

Jakobsson, J. and O.J. Østvedt, 1999. A review of joint investigations of the distribution of herring in the Norwegian and Iceland Seas 1950–1970. In: H. Vilhjalmsson, J. Olafsson and O.S. Astthorsson (eds.). Unnsteinn Stefánsson Festschrift. Rit Fiskideildar, XVI:209–238.

Jakobsson, J. and G. Stefánsson, 1999. Management of Icelandic summer spawning herring off Iceland. ICES Journal of Marine Science, 56:827–833.

Jarre, A. (ed.), 2002. Ecosystem West Greenland. Workshop held at the Greenland Institute of Natural Resources, Nuuk 29 November–03 December 2001. Arctic Research Journal 1.

Jensen, A.S., 1926. Indberetning av S/S Dana's praktisk videnskabelige fiskeriundersøgelser ved Vestgrønland i 1925. Beretninger og Kundgørelser vedr. Styrelsen af Grønland, 2:291–315.

Jensen, A.S., 1939. Concerning a change of climate during recent decades in the Arctic and Subarctic regions, from Greenland in the west to Eurasia in the east, and contemporary biological and physical changes. The Royal Danish Science Society. Biological Reports, 14, 8:1–77.

Jónsson, J., 1994. Fisheries off Iceland, 1600–1900. ICES Marine Science Symposia, 198:3–16.

Jónsson, J., 1996. Tagging of cod (*Gadus morhua*) in Icelandic waters 1948–1986. Rit Fiskideildar, 14(1):5–82.

Jørgensen, A.K., 1999. Norwegian and Russian Fisheries Enforcement in the Barents Sea – a Comparison with Regard to Effectiveness. MA Thesis, University of Oslo.

Kenyon, K.W., 1962. History of the Steller sea lion at the Pribilof Islands, Alaska. Journal of Mammalogy, 43:68–75.

Khen, G.V., 1997. Main regularities of multi-year changes in ice cover of the Bering Sea and Sea of Okhotsk. Complex investigations of the Sea of Okhotsk Ecosystem. Moscow: VNIRO. pp. 64–67. (In Russian)

Kirby, M.J.L. (Chairman), 1982. Navigating Troubled Waters – A New Policy for the Atlantic Fisheries. Department of Supply and Services of the Government of Canada, Ottawa.

Kocan, R., P. Hershberger and J. Winton, 2001. Effects of Ichthyophonus on survival and reproductive success of Yukon River Chinook salmon: Final Report for Study 01-200 for the US Fish and Wildlife Service, Subsistence Management, Fisheries Resource Monitoring Program. University of Washington, Seattle.

Korsbrekke, K., S. Mehl, O. Nakken and M. Pennington, 2001. A survey-based assessment of the Northeast Arctic cod stock. ICES Journal of Marine Science, 58:763–769.

KPMG and SINTEF, 2003. Betydningen av fiskeri- og havbruksnæringen for Norge – en ringvirkningsanalyse. SINTEF (Foundation for Scientific and Industrial Research at the Norwegian Institute of Technology) and KPMG, Trondheim, Norway. (In Norwegian)

Krohn, M., S. Reidy and S. Kerr, 1997. Bioenergetic analysis of the effects of temperature and prey availability on growth and condition of northern cod (*Gadus morhua*). Canadian Journal of Fisheries and Aquatic Sciences, 54(1):113–121.

Kulka, D.W., J.S. Wroblewski and S. Narayanan, 1995. Recent changes in the winter distribution and movements of northern Atlantic cod (*Gadus morhua* Linnaeus, 1758) on the Newfoundland-Labrador Shelf. ICES Journal of Marine Science, 52:889–902.

Langer, R.H., 1980. Summary of Northern Fur Seal data and collection procedures. NOAA Tech. Mem. NMFS F/NWC-3.

Latif, M. and T. Barnett, 1994. Causes of decadal climate variability over the North Pacific and North America. Science, 266:634–637.

Lear, W.H., 1979. Distribution, size, and sexual maturity of Arctic cod (*Boreogadus saida*) in the northwest Atlantic during 1959–1978. Canadian Atlantic Fisheries Scientific Advisory Committee, Res. Doc. 79/17.

Lear, W.H., 1998. History of fisheries in the Northwest Atlantic: the 500-year perspective. Journal of Northwest Atlantic Fishery Science, 23:41–73.

Lear, W.H. and L.S. Parsons, 1993. History and management of the fishery for northern cod in NAFO Divisions 2J, 3K and 3L. In: L.S. Parsons and W.H. Lear (eds.). Perspectives on Canadian Marine Fisheries Management. pp. 55–89. Canadian Bulletin of Fisheries and Aquatic Sciences 226.

LeBlanc, R. and P. de Bané, 1981. A Proposed Memorandum of Understanding between DFO and DREE Respecting Joint Action for Development of the Fisheries of Eastern Canada. Signed by Roméo LeBlanc and Pierre De Bané, March 20, 1981. National Archives of Canada, RG124, Accession 87–88/35, File #4560-F-0 Part 1.

Leggett, W.C., K.T. Frank and J.E. Carscadden, 1984. Meteorological and hydrographic regulation of year-class strength in capelin (*Mallotus villosus*). Canadian Journal of Fisheries and Aquatic Sciences, 41:1193–1201.

Letto, D.M., 1998. Chocolate Bars and Rubber Boots: the Smallwood Industrialization Plan. Blue Hill Publishers.

Lilly, G.R., 1994. Predation by Atlantic cod on capelin on the southern Labrador and Northeast Newfoundland shelves during a period of changing spatial distributions. ICES Marine Science Symposia, 198:600–611.

Lilly, G.R., 2001. Changes in size at age and condition of cod (*Gadus morhua*) off Labrador and eastern Newfoundland during 1978–2000. ICES CM 2001/V:15, 34p.

Lilly, G.R., 2002. Ecopath modelling of the Newfoundland Shelf: observations on data availability within the Canadian Department of Fisheries and Oceans. In: T. Pitcher, M. Vasconcellos, S. Heymans, C. Brignall and N. Haggan (eds). Information Supporting Past and Present Ecosystem Models of Northern British Columbia and the Newfoundland Shelf. Fisheries Centre, University of British Columbia. Fisheries Centre Research Reports, 10(1):22–27.

Lilly, G.R. and M. Simpson, 2000. Distribution and Biomass of Capelin, Arctic Cod and Sand Lance on the Northeast Newfoundland Shelf and Grand Bank as Deduced from Bottom-trawl Surveys. Department of Fisheries and Oceans, Canadian Stock Assessment Secretariat Res. Doc. 2000/091, 40p.

Lilly, G.R., H. Hop, D.E. Stansbury and C.A. Bishop, 1994. Distribution and abundance of polar cod (*Boreogadus saida*) off southern Labrador and eastern Newfoundland. ICES C.M.1994/O:6, 21p.

Lilly, G.R., D.G. Parsons and D.W. Kulka, 2000. Was the increase in shrimp biomass on the Northeast Newfoundland Shelf a consequence of a release in predation pressure from cod? Journal of Northwest Atlantic Fishery Science, 27:45–61.

Lilly, G.R., P.A. Shelton, J. Brattey, N.G. Cadigan, B.P. Healey, E.F. Murphy and D.E. Stansbury, 2001. An Assessment of the Cod Stock in NAFO Divisions 2J+3KL. Department of Fisheries and Oceans, Canadian Stock Assessment Secretariat Res. Doc. 2001/044, 148p.

Lilly, G.R., P.A. Shelton, J. Brattey, N.G. Cadigan, B.P. Healey, E.F. Murphy, D.E. Stansbury and N. Chen, 2003. An assessment of the cod stock in NAFO Divisions 2J+3KL in February 2003. Department of Fisheries and Oceans, Canadian Science Advisory Secretariat Res. Doc. 2003/023, 157p.

Lindkvist, K.B., 2000. Dependent and independent fishing communities in Norway. In: D. Symes (ed.). Fisheries Dependent Regions. Fishing News Books, Blackwell Science.

Link, M.R., M.L. Hartley, S.A. Miller, B. Waldrop, J. Wilen and J. Barnett. 2003. An analysis of options to restructure the Bristol Bay salmon fishery. Prepared for Bristol Bay Economic Development Corporation and the Joint Legislative Salmon Industry Task Force, Anchorage. 104p.

Livingston, P.A. and S. Tjelmeland, 2000. Fisheries in boreal ecosystems. ICES Journal of Marine Science, 57:619–627.

Livingston, P.A., L-L. Low and R.J. Marasco, 1999. Eastern Bering Sea ecosystem trends. In: Tang, Q. and K. Sherman (eds.). Large Marine Ecosystems of the Pacific Rim – Assessment, Sustainability, and Management, pp. 140–162. Blackwell Science.

Lotherington, A.-T. and M.B. Ellingsen, 2002. Små penger og Store Forventninger. Evaluering av Nettverkskreditt. NORUT, rapport nr. 8.

Lounsbury, R.G., 1934. The British Fishery in Newfoundland, 1634–1763. Yale University Press.

Lowry, L.F. and K. J. Frost, 1985. Biological interactions between marine mammals and commercial fisheries in the Bering Sea. In: J.R. Beddington, R.J.H. Beverton and D.M. Lavigne (eds.). Marine Mammals and Fisheries, pp. 42–61. George Allen & Unwin.

Lowry, L.F., K.J. Frost, D.G. Calkins, G.L. Swartzman and S. Hills, 1982. Feeding habits, food requirements, and status of Bering Sea marine mammals. North Pacific Fishery Management Council, Anchorage, Alaska. Document Nos. 19 and 19A. 574p.

Luchin, V.A., A.V. Saveliyev and V.I. Radchenko, 1998. Long-period climatic waves in the western Bering Sea ecosystem. In: V.F. Kozlov (ed.). Climate and Interannual Variability in the Atmosphere-Land-Sea system in the American-Asian sector of the Arctic, pp. 31–42. Proceedings of the Arctic regional center. V.1. Vladivostok: Far-Eastern State University Press. (In Russian)

Luchin, V.A., I.P. Semiletov and G.E. Weller, 2002. Changes in the Bering Sea region: atmosphere-ice-water system in the second half of the twentieth century. Progress in Oceanography, 55(1–2):23–44.

Malmberg, S.A., 1988. Ecological impact of hydrographic conditions in Icelandic waters. In: T. Wyatt and M.G. Larraneta (eds.). Long-term changes in marine fish populations, pp. 95-123. Symposium held in Vigo, Spain, 18–21 November 1986.

Malmberg, S.A. and S. Kristmannsson, 1992. Hydrographic conditions in Icelandic waters, 1980–1989. ICES Marine Science Symposia, 195:76–92.

Malmberg, S.A., J. Mortensen and H. Valdemarsson, 1999. Decadal scale climate and hydrobiological variability in Icelandic waters in relation to large scale atmospheric conditions in the North Atlantic. International Council for the Exploration of the Sea. ICES CM 1999/L:13, 9p.

Mann, K.H. and K.F. Drinkwater, 1994. Environmental influences on fish and shellfish production in the Northwest Atlantic. Environmental Reviews, 2:16–32.

Mantua, N.J., S.R. Hare, Y. Zhang, J.M. Wallace and R.C. Francis, 1997. A Pacific interdecadal climate oscillation with impacts on salmon production. Bulletin of the American Meteorological Society, 78(6):1069–1080.

Marteinsdottir, G. and G. Begg, 2002. Essential relationships incorporating the influence of age, size and condition on variables required for estimation of reproductive potential in Atlantic cod *Gadus morhua* stocks. Marine Ecology Progress Series, 235:235–256.

Marteinsdottir, G. and A. Steinarsson, 1998. Maternal influence on the size and viability of Iceland cod (*Gadus morhua* L.) eggs and larvae. Journal of Fish Biology, 52(6):1241–1258.

Maxim, C., 1980. Capelin (*Mallotus villosus*) Catch, Effort, and Biological Characteristics in the Romanian Fishery in Division 2J, September–October 1979. Northwest Atlantic Fisheries Organization Scientific Council Research Doc. 80/4, Serial No. N030.

Michalsen, K. (ed.), 2003. Fisken og havet, særnummer 1. Institute of Marine Research, Bergen.

Mikalsen, K. and S. Jentoft, 2003. Limits to participation? On the history, structure, and reform of Norwegian fisheries management. Marine Policy, 27(5):397–407.

Miles, E., S. Gibbs, D. Fluharty, C. Dawson and D. Teeter, 1982a. The Management of Marine Regions: The North Pacific. California University Press, 656p.

Miles, E., J. Sherman, D. Fluharty and S. Gibbs, 1982b. Atlas of Marine Use: The North Pacific. California University Press, 121p.

Ministry of Fisheries, 2002. Fiskeri og havbruk 2002. Statistical Report, Ministry of Fisheries, Oslo.

Minobe, S, 1999. Resonance in bidecadal and pentadecadal climate oscillations over the North Pacific: Role in climate regime shifts. Geophysical Research Letters, 26(7):853–858.

Morgan, M.J. and W.R. Bowering, 1997. Temporal and geographic variation in maturity at length and age of Greenland halibut (*Reinhardtius hippoglossoides*) from the Canadian north-west Atlantic with implications for fisheries management. ICES Journal of Marine Science, 54:875–885.

Morgan, M.J., P.A. Shelton, D.P. Stansbury, J. Brattey and G.R. Lilly, 2000. An Examination of the Possible Effect of Spawning Stock Characteristics on Recruitment in four Newfoundland Groundfish Stocks. Canadian Stock Assessment Secretariat Res. Doc. 2000/028.

Morgan, M.J., W.B. Brodie and D.W. Kulka, 2002. Was over-exploitation the cause of the decline of the American plaice stock off Labrador and northeast Newfoundland? Fisheries Research, 57:39–49.

Mowbray, F.K., 2002. Changes in the vertical distribution of capelin (*Mallotus villosus*) off Newfoundland. ICES Journal of Marine Science, 59:942–949.

Murphy, E.F. and C.A. Bishop, 1995. The status of 2GH cod, 3LNO haddock, 3Ps haddock and 3Ps pollock. Canadian Department of Fisheries and Oceans, Atlantic Fisheries Research Doc. 95/33.

Muus, B.J., F. Salomonsen and F.C. Vibe, 1990. Grønlands fauna: fisk, fugle, pattedyr. The Greenland Fauna: fish, birds and mammals, 463p. (In Danish)

Myers, R.A. and N.G. Cadigan, 1995. Was an increase in natural mortality responsible for the collapse of northern cod? Canadian Journal of Fisheries and Aquatic Sciences, 52:1274–1285.

Myers, R.A., K.F. Drinkwater, N.J. Barrowman and J.W. Baird, 1993. Salinity and recruitment of Atlantic cod (*Gadus morhua*) in the Newfoundland region. Canadian Journal of Fisheries and Aquatic Sciences, 50:1599–1609.

Myers, R.A., N.J. Barrowman, J.M. Hoenig and Z. Qu, 1996a. The collapse of cod in Eastern Canada: the evidence from tagging data. ICES Journal of Marine Science, 53:629–640.

Myers, R.A., J.A. Hutchings and N.J. Barrowman, 1996b. Hypotheses for the decline of cod in the North Atlantic. Marine Ecology Progress Series, 138:293–308.

Myers, R.A., N.J. Barrowman and J.A. Hutchings, 1997a. Inshore exploitation of Newfoundland Atlantic cod (*Gadus morhua*) since 1948 as estimated from mark-recapture data. Canadian Journal of Fisheries and Aquatic Sciences, 54(1):224–235.

Myers, R.A., J.A. Hutchings and N.J. Barrowman, 1997b. Why do fish stocks collapse? The example of cod in Atlantic Canada. Ecological Applications, 7:91–106.

NAFO, 2001a. STATLANT 21 reported catches by stock tabulated against STACFIS estimates, 1985–1999. Northwest Atlantic Fisheries Organization, Scientific Council Summary Document 01/5, 23p.

NAFO, 2001b. Scientific Council Report, Northwest Atlantic Fisheries Organization. June 2001, 339p.

Nakashima, B.S. and C.T. Taggart, 2002. Why are some beaches better for spawning than others? ICES Journal of Marine Science, 59:897–908.

Nakashima, B.S. and J.P. Wheeler, 2002. Capelin (*Mallotus villosus*) spawning behaviour in Newfoundland waters – the interaction between beach and demersal spawning. ICES Journal of Marine Science, 59:909–916.

Nakken, O., 1994. Causes of trends and fluctuations in the Arcto-Norwegian cod stock. ICES Marine Science Symposia, 198:212–228.

Nakken, O., 1998. Past, present and future exploitation and management of marine resources in the Barents Sea and adjacent areas. Fisheries Research, 37:23–35.

Napazakov, V.V., V.I. Chuchukalo, N.A. Kuznetsova, V.I. Radchenko, A.M. Slabinskii and V.A. Nadtochii, 2001. Feeding and some features of ecology of Gadidae fish in the western part of Bering Sea in the summer – autumn season. TINRO Transactions, 128:907–928. (In Russian)

Narayanan, S., J. Carscadden, J.B. Dempson, M.F. O'Connell, S. Prinsenberg, D.G. Reddin and N. Shackell, 1995. Marine climate off Newfoundland and its influence on Atlantic salmon (*Salmo salar*) and capelin (*Mallotus villosus*). In: R.J. Beamish (ed.). Climate Change and Northern Fish Populations, pp. 461–474. Canadian Special Publication of Fisheries and Aquatic Sciences 121.

National Research Council, 1996. The Bering Sea Ecosystem. Committee on the Bering Sea Ecosystem, Polar Research Board, Commission on Geosciences, Environment and Resources. National Academy Press.

National Research Council, 2003. Decline of the Steller Sea Lion in Alaskan Waters: Untangling Food Webs and Fishing Nets. National Academies Press.

Natural Resources Consultants, 1999. Status of Washington-based commercial fisheries and the fleets' future utilization of Fishermen's Terminal. Port of Seattle, Seattle. 195p.

Naumenko, N.I., 2001. Biology and fisheries on marine herring of the far-eastern seas. Kamchatsky Publishers. 330p. (In Russian)

Naumenko, N.I., P.A. Balykin, E.A. Naumenko and E.R. Shaginyan, 1990. Long-term changes in pelagic fish community in the western Bering Sea. TINRO Transactions, 111:49–57. (In Russian)

Naumenko, N.I., P.A. Balykin and E.A. Naumenko, 2001. Long-term fluctuations in the pelagic community in the western Bering Sea. In: Abstracts of PICES Tenth Annual Meeting, Victoria, BC, Canada. pp. 125.

Neis, B., D.C. Schneider, L. Felt, R.L. Haedrich, J. Fischer and J.A. Hutchings, 1999. Fisheries assessment: what can be learned from interviewing resource users? Canadian Journal of Fisheries and Aquatic Sciences, 56:1949–1963.

Netboy, A., 1974. The salmon: their fight for survival. Houghton Mifflin, Boston.

Nichol, D.G., 1994. Maturation and spawning of female yellowfin sole in the eastern Bering Sea. Proceedings of the International Pacific Flatfish Symposium, Anchorage, Alaska.

Nichol, D.G., 1997. Effects of geography and bathymetry on growth and maturity of yellowfin sole, *Pleuronectes asper*, in the eastern Bering Sea. Fishery Bulletin, 95:494–503.

Niebauer, H.J., N.A. Bond, L.P. Yakunin and V.V. Plotnikov, 1999. An update on the climatology and sea ice of the Bering Sea. In: T.R. Loughlin and K. Ohtani (eds.). Dynamics of the Bering Sea, pp. 29–59. Alaska Sea Grant College Program. AK-SG-99-03.

Nilssen, F. and G. Hønneland, 2001. Institutional change and the problems of restructuring the Russian fishing industry. Post-Communist Economies, 13:313–330.

Nilssen, K.T., 1995. Seasonal distribution, condition, and feeding habits of Barents Sea harp seals. In: A.S. Blix, L. Walloe and O. Ulltang (eds.). Whales, Seals, Fish and Man. Proceedings of the International Symposium on the Biology of Marine Mammals in the Northeast Atlantic, pp. 241–254. Tromsø, Norway, 29 November – 1 December 1994. Developments in Marine Biology 4. Elsevier.

Nishiyama, T., K. Hirano and T. Haryu, 1986. The early life history and feeding habits of larval walleye pollock in the southeast Bering Sea. Bulletin of the International North Pacific Fisheries Commission, 45:177–227.

NMFS, 1999. Our Living Oceans. Report on the Status of US Living Marine Resources, 1999. NOAA Tech. Memo. NMFS-F/SPO-41, 301p. National Marine Fisheries Service, Washington, D.C.

NMFS, 2003a. Fisheries of the United States, 2002. National Marine Fisheries Service, Washington, D.C.

NMFS, 2003b. Alaska Groundfish Fisheries: Draft Programmatic Supplemental Environmental Impact Statement. Vol. V. App. A, 484p. National Marine Fisheries Service.

Nordøy, E., P.E. Mortensson, A.R. Lager, L.P. Folkow and A.S. Blix, 1995. Food consumption of the Northeast Atlantic stock of harp seals. In: A.S. Blix, L. Walloe and O. Ulltang (eds.). Whales, Seals, Fish and Man. Proceedings of the International Symposium on the Biology of Marine Mammals in the Northeast Atlantic, pp. 255–260. Tromsø, Norway, 29 November – 1 December 1994. Developments in Marine Biology 4, Elsevier

NORUT Samfunnsforskning, 2002. Konjunkturbarometer for Nord-Norge, NORUT, Tromsø, 63p.

Novikov, N.P., 1974. Commercial fishes of continental slope of the northern part of Pacific Ocean. Moscow: Food Industry. 308 p. (In Russian)

NPAFC, 2001. Annual Report, 2001. North Pacific Anadromous Fish Commission, Vancouver.

NPFMC, 2002. Stock Assessment Fishery Evaluation. North Pacific Fishery Management Council, Anchorage.

NPFMC, 2003a. Discussion paper on Western Alaska Community Development Quota (CDQ) Program. North Pacific Fishery Management Council, Alaska.

NPFMC, 2003b. Preliminary draft Environmental Impact Statement for Essential Fish Habitat identification and conservation in Alaska. North Pacific Fishery Management Council Anchorage. Vol. I,II.

NPFMC, 2004. Stock Assessment and Resource Evaluation Report. The groundfish resources of the Bering Sea. North Pacific Fisheries Management Council.

O'Brien, K. and R. Leichenko, 2000. Double exposure: assessing the impacts of climate change within the context of globalization. Global Environmental Change, 10:221–232.

Ohtani, K. and T. Azumaya, 1995. Influence of interannual changes in ocean conditions on the abundance of walleye pollock in the eastern Bering Sea. In: R.J. Beamish (ed.). Climate Change and Northern Fish Populations, pp. 87–95. Canadian Special Publication of Fisheries and Aquatic Sciences 121.

Øiestad, V., 1994. Historic changes in cod stocks and cod fisheries: Northeast Arctic cod. ICES Marine Science Symposia, 198:17–30.

Orr, D., D.G. Parsons, P.J. Veitch and D.J. Sullivan, 2001a. Northern Shrimp (*Pandalus borealis*) off Baffin Island, Labrador and Northeastern Newfoundland – First Interim Review. Department of Fisheries and Oceans, Canadian Stock Assessment Secretariat Res. Doc. 2001/043.

Orr, D.C., D.G. Parsons, P. Veitch and D. Sullivan, 2001b. An Update of Information Pertaining to Northern Shrimp (*Pandalus borealis*) and Groundfish in NAFO Divisions 3LNO. Northwest Atlantic Fisheries Organization, Scientific Council Research Doc. 01/186, Serial No. N4576.

Orr, D.C., P.J. Veitch and D.J. Sullivan, 2003. Northern shrimp (Pandalus borealis) off Baffin Island, Labrador and Northeastern Newfoundland. Department of Fisheries and Oceans, Canadian Science Advisory Secretariat Res. Doc. 2003/050.

Orr, D.C., P.J. Veitch and D.J. Sullivan, 2004. Divisions 3LNO northern shrimp (Pandalus borealis) – interim monitoring update. Northwest Atlantic Fisheries Organization, Scientific Council Research Doc. 04/65, Serial No. N5026.

Ottersen, G. and S. Sundby, 1995. Effects of temperature, wind and spawning stock biomass on recruitment of Arcto-Norwegian cod. Fisheries Oceanography, 4:278–292.

Otto, R.S., 1998. Assessment of the eastern Bering Sea snow crab, *Chionoecetes opilio*, stock under the terminal molting hypothesis. In: G.S. Jamieson and A. Campbell (eds.). Proceedings of the North Pacific Symposium on Invertebrate Stock Assessment and Management. Can. Spec. Publ. Fish. Aquat. Sci. 125:109–124.

Overland, J.E. 1981. Marine Climatology of the Bering Sea. In: D.W. Hood and J.A. Calder (eds.). The eastern Bering Sea shelf: oceanography and resources, pp. 15–22. University of Washington Press.

Overland, J.E., 1990: Prediction of vessel icing at near-freezing sea temperatures. Weather and Forecasting, 5:62–77.

Overland, J.E., 2004. Ocean and climate changes. In: V. Alexander, R.I. Perry (eds.). PICES Report on Marine Ecosystems of the North Pacific. Special Publication of the North Pacific Marine Science Organization, 1:39–57.

Overland, J.E., J.M. Adams and N.A. Bond, 1999. Decadal variability and the Aleutian Low and its relation to high-latitude circulation. Journal of Climate, 12:1542–1548.

Overton, J., 2001. Official acts of vandalism: privatizing Newfoundland's Provincial Parks in the 1990s. In: D. McGrath (ed.). From Red Ochre to Black Gold. Flanker Press, Newfoundland.

Pálsson, Ó.K., 1997. Food and feeding of cod (*Gadus morhua*). In: J. Jakobsson and Ó. K. Pálsson (eds.). Multispecies Research 1992–1995, pp. 177–191. Marine Research Institute, Reykjavík, Technical Report, 57. (In Icelandic)

Parsons, D.G. and E.B. Colbourne, 2000. Forecasting fishery performance for northern shrimp (*Pandalus borealis*) on the Labrador Shelf (NAFO Divisions 2HJ). Journal of Northwest Atlantic Fishery Science, 27:11–20.

Parsons, L.S. and W.H. Lear, 2001. Climate variability and marine ecosystem impacts: a North Atlantic perspective. Progress in Oceanography, 49:167–188.

Pautzke, C., 1997. Russian Far East Fisheries Management – North Pacific Fishery Management Council Report to Congress. Anchorage, 68p.

Pereyra, W.T., J.E. Reeves and R.A. Bakala, 1976. Demersal Fish and Shellfish Resources of the Eastern Bering Sea in the Baseline Year 1975. Northwest Fisheries Center, National Marine Fisheries Service. Seattle.

Pike, D.G., 1994. The Fishery for Greenland Halibut (*Reinhardtius hippoglossoides*) in Cumberland Sound, Baffin Island, 1987–1992. Canadian Technical Report of Fisheries and Aquatic Sciences 1924.

Planque, B. and T. Frédou, 1999. Temperature and the recruitment of Atlantic cod (*Gadus morhua*). Canadian Journal of Fisheries and Aquatic Sciences, 56:2069–2077.

Plotnikov, V.V., 1996. Variability of ice conditions of the far-eastern (the Bering, Okhotsk and Japan) seas and their forecast. Ph.D. thesis. Vladivostok: POI DVO of Russian Academy of Sciences. In Russian

Plotnikov, V.V. and G.I. Yurasov, 1994. Seasonal and interannual variability of ice cover in the North Pacific marginal seas, pp. 13–24. PICES-STA workshop on monitoring subarct. Pacific Ocean, Nemuro, Japan.

Polovina, J.J., G.T. Mitchum and G.T. Evans, 1995. Decadal and basin-scale variation in mixed layer depth and the impact on biological production in the Central and North Pacific, 1960–88. Deep-Sea Research, 42(10):1701–1716.

Pope, J., P. Large and T. Jakobsen, 2001. Revisiting the influences of parent stock, temperature and predation on the recruitment of the Northeast Arctic cod stock, 1930–1990. ICES Journal of Marine Science, 58:967–972.

Pushie, G.F. (Chairman), 1967. Report of the Royal Commission on the Economic Prospects of Newfoundland and Labrador. Office of the Queen's Printer, St. John's, Newfoundland.

Queirolo, L.E., L.W. Fritz, P.A. Livingston, M.R. Loefflad, D.A. Colpo and Y.L. deReynier, 1995. Bycatch, utilization, and discards in the commercial groundfish fisheries of the Gulf of Alaska, Eastern Bering Sea and Aleutian Islands. NOAA Tech. Mem. NMFS-AFSC-58. 148p.

Quinn, T.J. II and H.J. Niebauer, 1995. Relation of eastern Bering Sea walleye pollock (*Theragra chalcogramma*) recruitment to environmental and oceanographic variables. In: R.J. Beamish (ed.). Climate Change and Northern Fish Populations, pp. 497–507. Canadian Special Publication of Fisheries and Aquatic Sciences 121.

Radchenko, V.I., 1992. The role of squids in the Bering Sea pelagic ecosystem. Oceanology, 32(6): 1093–1101. (In Russian)

Radchenko V.I., 1994. Composition, structure and dynamics of the epipelagic nekton communities. MSc. Thesis. Vladivostok. 24p. (in Russian)

Radchenko, V.I. and I.I. Glebov, 1998. Incidental by-catch of Pacific salmon during Russian bottom trawl surveys in the Bering Sea and some remarks on its ecology. North Pacific Anadromous Fish Commission Bulletin, 1:367–374.

Radchenko, V.I., G.V. Khen and A.M. Slabinsky, 2001. Cooling in the western Bering Sea in 1999: quick propagation of La Niña signal or compensatory processes effect? Progress in Oceanography, 49(1–4):407–422.

Ramstad, S., 2001. Etableringen av et Internasjonalt Forvaltningsregime for Norsk Vårgytende Sild. En kamp om ressurser mellom Norge, Russland, Island, Færøyene og EU. Hovedoppgave i Statsvitenskap, Universitetet i Tromsø.

Ray, G.C. and J. McCormick-Ray, 2004. Coastal Marine Conservation: Science and Policy. Blackwell.

Rice, J., 2002. Changes to the large marine ecosystem of the Newfoundland-Labrador Shelf. In: K. Sherman and H.R. Skjoldal (eds.). Large Marine Ecosystems of the North Atlantic, pp. 51–103. Elsevier Science.

Rice, J.C. and G.T. Evans, 1988. Tools for embracing uncertainty in the management of the cod fishery of NAFO divisions 2J+3KL. Journal du Conseil International pour l'Exploration de la Mer, 45:73–81.

Report to Parliament, 1998. Perspectives on the Development of the Norwegian Fishing Industry. Ministry of Fisheries, Oslo. (In Norwegian)

Rose, G.A. and D.W. Kulka, 1999. Hyperaggregation of fish and fisheries: how catch-per-unit-effort increased as the northern cod (*Gadus morhua*) declined. Canadian Journal of Fisheries and Aquatic Sciences, 56(1):118–127.

Rose, G.A., B.A. Atkinson, J. Baird, C.A. Bishop and D.W. Kulka, 1994. Changes in distribution of Atlantic cod and thermal variations in Newfoundland waters, 1980–1992. ICES Marine Science Symposia, 198:542–552.

Rose, G.A., B. deYoung, D.W. Kulka, S.V. Goddard and G.L. Fletcher, 2000. Distribution shifts and overfishing the northern cod (*Gadus morhua*): a view from the ocean. Canadian Journal of Fisheries and Aquatic Sciences, 57:644–663.

Rosenkranz, G.E., A.V. Tyler and G.H. Kruse, 2001. Effects of water temperature and wind on year-class success of Tanner crabs in Bristol Bay, Alaska. Fisheries Oceanography, 10(1):1–12.

Rosing-Asvid, A., 2005. Climate variability around Greenland and its influence on ringed seals and polar bears. Ph.D. Thesis, University of Copenhagen.

Sæmundsson, B., 1934. Probable influence of changes in temperature on the marine fauna of Iceland. ICES Rapports et Procès-Verbaux des Réunions, 86:3–6.

Sapozhnikov, V.V., N.V. Arzhanova and V.V. Zubarevitch, 1993. Estimation of primary production in the ecosystem of western Bering Sea. Russian Journal of Aquatic Ecology, 2(1):23–34.

Schmidt., J., 1909. The distribution of the pelagic fry and spawning regions of the Gadoids in the North-N-Atlantic from Iceland to Spain. ICES Rapports et Procès-Verbaux des Réunions, 10:1–229.

Schmidt, J., 1931. Summary of the Danish marking experiments on cod, 1904–1929, at the Faroes, Iceland and Greenland. Rapp. P-v. Reun.Cons.perm.int..Explor.Mer, 72. 13 pp.

Schopka, S.A., 1994. Fluctuations in the cod stock off Iceland during the twentieth century in relation to changes in the fisheries and environment. In: J. Jakobsson, O.S. Asthorsson, R.J.H. Beverton, B. Björnsson, N.Daan, K.T. Frank, J.B. Rothschild, S.Sundby and S.Tilseth (eds.). Cod and Climate. ICES Marine Science Symposia, 198:175–193.

Schrank, W.E., 1995. Extended fisheries jurisdiction: origins of the current crisis in Atlantic Canada's fisheries. Marine Policy, 19:285–299.

Schrank, W.E., 1997. The Newfoundland fishery: past, present and future. In S. Burns (ed.). Subsidies and Depletion of World Fisheries: Case Studies. World Wildlife Fund, Washington, D.C.

Schrank, W.E. and B. Skoda, 2003. The cost of marine fishery management in eastern Canada: Newfoundland, 1989/1990 to 1999/2000. In: W.E. Schrank, R. Arnason and R. Hannesson (eds.). The Cost of Fisheries Management, pp, 91–128. Ashgate Publishers.

Schrank, W.E., N. Roy and E. Tsoa, 1986. Employment prospects in a commercially viable Newfoundland fishery: An application of 'An Econometric Model of the Newfoundland Groundfishery'. Marine Resource Economics, 3:237–263.

Schrank, W.E., N. Roy, R. Ommer and B. Skoda, 1992. An inshore fishery: A commercially viable industry or an employer of last resort. Ocean Development and International Law, 23:335–367.

Schrank, W.E., B. Skoda, P. Parsons and N. Roy, 1995. The cost to government of maintaining a commercially unviable fishery: the case of Newfoundland 1981/82 to 1990/91. Ocean Development and International Law, 26:357–390.

Schumacher, J.D., 2000. Regime shift theory: A review of changing environmental conditions in the Bering Sea and the Eastern North Pacific. In: S. Connover (ed.). Proceedings of the Fifth North Pacific Rim Fisheries Conference, pp. 127–142. Cooperation, coordination and communication, AK: Pacific Rim Fisheries Program, Institute of the North, Alaska Pacific University.

Shackell, N.L., J.E. Carscadden and D.S. Miller, 1994. Migration of pre-spawning capelin (*Mallotus villosus*) as related to temperature on the northern Grand Bank, Newfoundland. ICES Journal of Marine Science, 51:107–114.

Shaw, J., R.B. Taylor, D.L. Forbes, M.-H. Ruz and S. Solomon, 1998. Sensitivity of the Coasts of Canada to Sea-level Rise. Geological Survey of Canada Bulletin 505, 79p.

Shelton, P.A. and D.B. Atkinson, 1994. Failure of the Div. 2J3KL Cod Recruitment Prediction using Salinity. Department of Fisheries and Oceans, Canadian Stock Assessment Secretariat Res. Doc. 94/66.14p.

Shelton, P.A. and G.R. Lilly, 2000. Interpreting the collapse of the northern cod stock from survey and catch data. Canadian Journal of Fisheries and Aquatic Sciences, 57:2230–2239.

Shelton, P.A., G.R. Lilly and E. Colbourne, 1999. Patterns in the annual weight increment for Div. 2J+3KL cod and possible prediction for stock projection. Journal of Northwest Atlantic Fishery Science, 25:151–159.

Shuntov, V.P. 1970. Seasonal distribution of black and arrow-toothed halibuts in the Bering Sea. TINRO Transactions, vol. 72. VNIRO Transactions, Mutual issue, 70:391–401. (In Russian)

Shuntov, V.P., 2001. Biology of far-eastern seas of Russia. TINRO-Center. V.1. 580p. (In Russian)

Shuntov, V.P. and V.I. Radchenko, 1999. Summary of TINRO Ecosystem Investigations in the Bering Sea. In: T.R. Loughlin and K. Otani (eds.). Dynamics of the Bering Sea. Alaska Sea Grant Program. AK-SG-99-03. pp. 771–776.

Shuntov, V.P. and V.P. Vasilkov, 1982. Epochs of atmospheric circulation and cyclic reoccurrence of abundance dynamics of Far-Eastern and Californian sardine. Journal of Ichthyology, 22(1):187–199. (In Russian)

Shuntov, V.P., A.F. Volkov, O.S. Temnykh and E.P. Dulepova, 1993. Walleye pollock in the far-eastern seas ecosystems. TINRO, 426p. (In Russian)

Shuntov, V.P., V.V. Lapko, V.V. Nadtochiy and E.V. Samko, 1994. Interannual changes in fish communities in upper epipelagic layer in the western Bering Sea and Pacific waters off Kamchatka. Journal of Ichthyology, 34(5):642–648. (In Russian)

Shuntov, V.P., V.I. Radchenko, E.P. Dulepova and O.S. Temnykh, 1997. Biological resources of the Far Eastern Russian economic zone: structure of pelagic and bottom communities, up-to-date status, tendencies of long-term dynamics. TINRO Transactions, 122:3–15. (In Russian)

Shuntov, V.P., E.P. Dulepova and I.V. Volvenko, 2002. Up-to-date status and long-term dynamics of biological resources in the Far Eastern economic zone of Russia. TINRO Transactions, 130:3–11. In Russian

Shuntov, V.P., L.N. Bocharov, E.P. Dulepova, A.F. Volkov, O.S. Temnykh, I.V. Volvenko, I.V. Melnikov and V.A. Nadtochiy, 2003. Results of monitoring and ecosystem research of biological resources of the Far East seas of Russia. TINRO Transactions, 132:3–26. In Russian

Shuter, B.J., C.K. Minns, H.A. Regier and J.D. Reist, 1999. Canada country study: Climate impacts and adaptations: fishery sector. Vol. VII, National Sectoral Volume, pp 258–317.

Sinclair E.H., A.A. Balanov, T. Kubodera, V.I. Radchenko and Yu. A. Fedorets, 1999. Distribution and ecology of mesopelagic fishes and cephalopods. In: T.R. Loughlin and K. Ohtani (eds.). Dynamics of the Bering Sea, pp. 485–508. Fairbanks: Alaska Sea Grant College Program.

Slizkin, A.G. and V.Ya. Fedoseev, 1989. Distribution, biology, population structure and abundance of Tanner crabs in the Bering Sea. In: Proceedings of the International Scientific Symposium on Bering Sea Fisheries. Sitka, Alaska. Seattle, pp. 316–347.

Smith, S.L., and J. Vidal, 1986. Variations in the distribution, abundance, and development of copepods in the southeastern Bering Sea in 1980 and 1981. Continental Shelf Research, 5:215–239.

Sogard, S.M. and B.L. Olla, 1998. Contrasting behavioral responses to cold temperatures by two marine fish species during their pelagic juvenile interval. Environmental Biology of Fishes, 53:405–412.

Solhaug, T., 1983. De norske fiskeriers historie 1815–1880. Universitetsforlaget, Bergen.

Somerton, D.A. and P. Munro, 2001. Bridle efficiency of a survey trawl for flatfish. Fishery Bulletin, 99:641–652.

Stabeno, P.J., N.A. Bond, N.B. Kachel, S.A. Salo and J.D. Schumacher, 2001. On the temporal variability of the physical environment over the south-eastern Bering Sea. Fisheries Oceanography, 10(1):81–98.

Statistics Canada, 2002a. www12.statcan.ca/english/census01/products/standard/popdwell/Table-PR.cfr (April 18, 2005).

Statistics Canada, 2002b. 2001 Community Profiles. ww12.statcan.ca/english/profil01/PlaceSearchForm1.cfm (March 18, 2003).

Stefánsson, U., 1999. Hafi? (The Ocean). Háskólaútgáfan, Reykjavík. 480p. (In Icelandic)

Stenson, G.B., B.P. Healey, B. Sjare and D. Wakeham, 2000. Catch-at-age of northwest Atlantic harp seals, 1952–1999. Department of Fisheries and Oceans, Canadian Stock Assessment Secretariat Res. Doc. 2000/079.

Stenson, G.B., M.O. Hammill, M.C.S. Kingsley, B. Sjare, W.G. Warren and R.A. Myers, 2002. Is there evidence of increased pup production in northwest Atlantic harp seals, *Pagophilus groenlandicus*? ICES Journal of Marine Science, 59:81–92.

Stepanenko, M.A., 2001. The state of stock, interannual variability of recruitment, and fisheries of the eastern Bering Sea pollock in 1980s – 1990s. TINRO Transactions, 128:145–152. (In Russian)

Stevens, B.G. R.A. MacIntosh, G.A. Haaga and J.H. Bowerman, 1993. Report to the industry on the 1992 eastern Bering Sea crab survey. Alaska Fisheries Science Center. Processed Rep. 93–14. 53 p.

Stokke, O.S. (ed.) 2001. Governing High Seas Fisheries: The Interplay of Global and Regional Regimes. Oxford University Press.

Strickland, R.M. and T. Sibley, 1984. Projected Effects of CO2 Induced Climate Change on the Alaska Walleye Pollock (*Theragra chalcogramma*) Fishery in the Eastern Bering Sea and the Gulf of Alaska. University of Washington, FRI-UW-8411. 112p.

Sugimoto, T. and K. Tadokoro, 1997. Interannual-interdecadal variations in zooplankton biomass, chlorophyll concentration and physical environment in the subarctic Pacific and Bering Sea. Fisheries Oceanography, 6(2):74–93.

Sundby, S., 2000. Recruitment of Atlantic cod stocks in relation to temperature and advection of copepod populations. Sarsia, 85:277–289.

Sutcliffe, W.H., R.H. Loucks, K.F. Drinkwater and A.R. Coote, 1983. Nutrient flux onto the Labrador Shelf from Hudson Strait and its biological consequences. Canadian Journal of Fisheries and Aquatic Sciences, 40:1692–1701.

Taggart, C.T., J. Anderson, C. Bishop, E. Colbourne, J. Hutchings, G. Lilly, J. Morgan, E. Murphy, R. Myers, G. Rose and P. Shelton, 1994. Overview of cod stocks, biology, and environment in the Northwest Atlantic region of Newfoundland, with emphasis on northern cod. ICES Marine Science Symposia, 198:140–157.

Tåning, Å.V., 1934. Survey of long distance migrations of cod in the North Western Atlantic according to marking experiments. ICES Rapports et Procès-Verbaux des Réunions, 89(3):5–11.

Tåning, Å.V., 1937. Some features in the migration of cod. Journal du Conseil International pour l'Exploration de la Mer, 12:1–35.

Tåning, Å.V., 1948. On changes in the marine fauna of the North-Western Atlantic Area, with special reference to Greenland. ICES Rapports et Procès-Verbaux des Réunions, 25:26–29.

Taylor, D.M. and P.G. O'Keefe, 1999. Assessment of the 1998 Newfoundland and Labrador Snow Crab Fishery. Department of Fisheries and Oceans, Canadian Stock Assessment Secretariat Res. Doc. 99/143.

Templeman, W., 1966. Marine resources of Newfoundland. Fisheries Research Board of Canada Bulletin 154.

Templeman, W., 1968. Review of some aspects of capelin biology in the Canadian area of the northwest Atlantic. ICES Rapports et Procès-Verbaux des Réunions, 158:41–53.

Templeman, W., 1973. Distribution and abundance of the Greenland halibut, *Rheinhardtius hippoglossoides* (Walbaum) in the Northwest Atlantic. International Commission Northwest Atlantic Fisheries Research Bulletin, 10:82–98.

Thompson, G.G. and M. Dorn, 2001. Pacific cod. In: Stock Assessment and Fishery Evaluation Report for the Groundfish Resources of the Bering Sea/Aleutian Islands regions. Chapter 2.

Thor, J.Th., 1992. British trawlers and fishing grounds at Iceland 1889–1916. Hi? íslenska bókmenntafélag, Reykjavik. 237p. (In Icelandic)

Thórdardóttir, Th., 1977. Primary production in North Icelandic waters in relation to recent climatic changes. In: M.J. Dunbar (ed.). Polar Oceans. Proceedings of the Polar Oceans Conference held at McGill University, Montreal, 1974, pp. 655–665. Arctic Institute of North America, Canada.

Thórdardóttir, Th., 1984. Primary production north of Iceland in relation to water masses in May–June 1970–1980. ICES CM 1984/L:20.

Toresen, R. and O.J. Østvedt, 2000. Variations in abundance of Norwegian spring-spawning herring throughout the 20th century and the influence of climatic fluctuations. Fish and Fisheries, 1:231–256.

Treble, M.A. and R. Bowering, 2002. The Greenland Halibut (*Reinhardtius hippoglossoides*) Fishery in NAFO Division 0A. NAFO SCR Doc. 02/46. Serial No. N4658. 9p.

Trenberth, K.E., 1990. Recent observed interdecadal climate changes in the Northern Hemisphere. Bulletin of the American Meteorological Society, 71:988–993.

Tuponogov, V.N., 2001. Polar cod. In: Hydrometeorology and Hydrochemistry of Seas, pp. 199–204. Vol. 10. The Bering Sea. Issue 2. Hydrochemical Conditions and Oceanological Fundamentals of Formation of Biological Productivity. Hydrometeoizdat. (In Russian)

Ustinova, E.I. and Yu.I. Sorokin, 1999. Some regularities of climate variability of the far-eastern seas, pp. 50–51. Proc. of 11th all Russian conference on fish. Oceanogr. Moscow: VNIRO. In Russian

Van Loon, H. and D.J. Shea, 1999. A probable signal of the 11-year solar cycle in the troposphere of the northern hemisphere. Geophysical Research Letters, 26:2893–2896.

Velegjanin, A.N., 1999. International legal aspects of the Russian Far East seafood trade with the North Pacific Rim Countries. In: V.M. Kaczynski and D.L. Fluharty (eds.). Impacts of Populations and Markets on the Sustainability of Ocean and Coastal Resources: Perspectives of Developing and Transition Economies of the North Pacific, pp 189–198. Collection of papers from the International Conference June 3–4, 1999. School of Marine Affairs, University of Washington.

Vestergaard, N. and R. Arnason, 2004. On the relationship between Greenland's Gross Domestic Product and her fish exports: empirical estimates. Working paper W04:07. Institute of Economic Studies, University of Iceland.

Vilhjálmsson, H., 1994. The Icelandic capelin stock. Capelin *Mallotus villosus* (Muller) in the Iceland, Greenland, Jan Mayen area. Rit Fiskideildar, 13(1), 281p.

Vilhjálmsson, H., 1997. Climatic variations and some examples of their effects on the marine ecology of Icelandic and Greenland waters, in particular during the present century. Rit Fiskideildar, 15(1):7–29.

Vilhjálmsson, H., 2002. Capelin (*Mallotus villosus*) in the Iceland–East Greenland–Jan Mayen ecosystem. ICES Marine Science Symposia, 216:870–883.

Vilhjálmsson, H. and J.E. Carscadden, 2002. Assessment surveys for capelin in the Iceland-East Greenland-Jan Mayen area, 1987–2001. ICES Marine Science Symposia, 216:1096–1104.

Vilhjálmsson, H. and E. Fri?geirsson, 1976. A review of 0-group surveys in the Iceland-East Greenland area in the years 1970–1975. ICES Cooperative Research Report, 54:1–34.

Vilhjálmsson, H. and J.V. Magnússon, 1984. Report on the 0-group fish survey in Icelandic and East-Greenland waters, August 1984. ICES C.M. 1984/H:66, 26p.

Vinnikov, A.V., 1996. Pacific cod (*Gadus macrocephalus*) of the western Bering Sea. In: O.A. Mathisen and K.O. Coyle (eds.). Ecology of the Bering Sea: A review of Russian literature, pp. 183–202. Alaska Sea Grant College Program Report No. 96–01.

Vinogradov, M.E. and E.A. Shushkina, 1987. Functioning of pelagic communities in the oceanic epipelagic layer. Nauka Publishers, 240p. (In Russian)

Welch, D.W., A.I. Chigirinsky and Y. Ishida, 1995. Upper thermal limits on the oceanic distribution of Pacific salmon in the spring. Canadian Journal of Fisheries and Aquatic Sciences, 52:489–503.

Wespestad, V.G., 1987. Population dynamics of Pacific herring (*Clupea pallasii*), capelin (*Mallotus villosus*) and other coastal pelagic fishes in the eastern Bering Sea. In: Forage fishes of the Southeastern Bering Sea. US Dep. Int., Min. Mgt. Serv., Alaska OCS Region, MMS87-0017:55–60.

Wespestad, V.G., 1993. The status of the Bering Sea pollock and the effect of 'donut hole' fishery. Fisheries, 18(3):18–24.

Wespestad, V.G., L.W. Fritz, W. Ingraham and B.A. Megrey, 2000. On relationships between cannibalism, climate variability, physical transport, and recruitment success of Bering Sea walleye pollock (*Theragra chalcogramma*). ICES Journal of Marine Science, 57(2):272–278.

Wilderbuer, T.K., 1997. Yellowfin sole. In: Stock Assessment and Fishery Evaluation Report for the Groundfish Resources of the Bering Sea and Aleutian Islands Regions as projected for 1998, pp. 160–187.

Wilderbuer, T.K. and D.G. Nichol, 2001. Yellowfin sole. In: Stock Assessment and Fishery Evaluation Report for the Groundfish Resources of the Bering Sea/Aleutian Islands. Chapter 3. North Pacific Fisheries Management Council.

Wilderbuer, T.K., A.B. Hollowed, W.J. Ingraham, P.D. Spencer, M.W. Conners, N.A. Bond and G.E. Walters, 2002. Flatfish recruitment response to decadal climate variability and ocean conditions in the eastern Bering Sea. Progress in Oceanography, 55(1–2):235–248.

Williams, E.H. and T.J. Quinn II, 2000. Pacific herring, *Clupea pallasi*, recruitment in the Bering Sea and north-east Pacific Ocean, II: relationships to environmental variables and implications for forecasting. Fisheries Oceanography, 9(4):300–315.

Winters, G.H. and J.P. Wheeler, 1987. Recruitment dynamics of spring-spawning herring in the Northwest Atlantic. Canadian Journal of Fisheries and Aquatic Sciences, 44:882–900.

Witherell, D., C. Pautzke and D. Fluharty, 2000. An ecosystem-based approach for Alaska groundfish management. ICES Journal of Marine Science, 57:771–777.

Wolotira Jr., R.J., T.M. Sample, S.F. Noel and C.R. Iten, 1993. Geographic and bathymetric distributions for many commercially important fishes and shellfishes off the west coast of North America based on research survey and commercial catch data, 1912–1984. NOAA Tech. Memorandum NMFS-AFSC-6, 184p.

Wooster, W.S. and A.B. Hollowed, 1995. Decadal-scale variations in the eastern subarctic Pacific: I. Winter ocean conditions. In: R.J. Beamish (ed.). Climate Change and Northern Fish Populations, pp. 81–85. Canadian Special Publication in Fisheries and Aquatic Science 121.

Worm, B. and R.A. Myers, 2003. Meta-analysis of cod-shrimp interactions reveals top-down control in oceanic food webs. Ecology, 84:162–173.

Wyllie-Echeverria, T., 1995. Sea-ice conditions and the distribution of walleye pollock (*Theragra chalcogramma*) on the Bering and Chukchi Shelf. In: R.J. Beamish (ed.). Climate Change and Northern Fish Populations, pp. 131–136. Canadian Special Publication in Fisheries and Aquatic Science 121.

Wyllie-Echeverria, T. and K. Ohtani, 1999. Seasonal sea ice variability and the Bering Sea ecosystem. In: T.R. Loughlin and K. Ohtani (eds.). Dynamics of the Bering Sea. Alaska Sea Grant College Program. AK-SG-99-03. pp. 435–451.

Zaika, V.E., 1983. Comparative hydrobiont productivity. Naukova Dumka Publishers, 203p. (In Russian)

Zebdi, A. and J.S. Collie, 1995. Effect of climate on herring (*Clupea pallasi*) population dynamics in the Northeast Pacific Ocean. In: R.J. Beamish (ed.). Climate Change and Northern Fish Populations. Canadian Special Publication of Fisheries and Aquatic Sciences 121: 277–290.

Zheng, J. and G.H. Kruse, 2000. Recruitment patterns of Alaska crabs in relation to decadal shifts in climate and physical oceanography. ICES Journal of Marine Science, 57:438–451.

Zilanov, V.K., 1999. Marine living resources of the Russian Far East in the time of change: impacts of population, markets, reforms, and climate. In: V.M. Kaczynski and D.L. Fluharty (eds.). Impacts of Populations and Markets on the Sustainability of Ocean and Coastal Resources: Perspectives of Developing and Transition Economies of the North Pacific, pp. 97–111. Collection of Papers from the International Conference June 3–4, 1999. School of Marine Affairs. University of Washington. 277p.

Zilanov, V.K., L.I. Shepel, A.N. Vylegjanin, M.A. Stepanenko, T.I. Spivakova and N.V. Yanovskaya, 1989. International problems of conservation and exploitation of Bering sea living resources. World Fisheries, 1:1–102. (In Russian)

Chapter 14

Forests, Land Management, and Agriculture

Lead Author
Glenn P. Juday

Contributing Authors
Valerie Barber, Paul Duffy, Hans Linderholm, Scott Rupp, Steve Sparrow, Eugene Vaganov, John Yarie

Consulting Authors
Edward Berg, Rosanne D'Arrigo, Olafur Eggertsson, V.V. Furyaev, Edward H. Hogg, Satu Huttunen, Gordon Jacoby, V.Ya. Kaplunov, Seppo Kellomaki, A.V. Kirdyanov, Carol E. Lewis, Sune Linder, M.M. Naurzbaev, F.I. Pleshikov, Ulf T. Runesson, Yu.V. Savva, O.V. Sidorova, V.D. Stakanov, N.M. Tchebakova, E.N. Valendik, E.F. Vedrova, Martin Wilmking

Contents

Summary

The boreal region covers about 17% of global land area, and the arctic nations together contain about 31% of the global forest (non-boreal and boreal). The boreal forest is affected by and also contributes to climate change through its influence on the carbon cycle and albedo. Boreal forests influence global levels of atmospheric carbon dioxide and other greenhouse gases by taking up carbon dioxide in growth, storing carbon in live and dead plant matter, and releasing carbon through decomposition of dead organic matter, live plant and animal respiration, and combustion during fire. Human management influences on carbon uptake and storage include the rearrangement of forest age classes through timber harvest or wildfire suppression, selection of tree species, fertilization, and thinning regimes. The combined effect of all management actions can either enhance or reduce carbon uptake and storage.

Agriculture has existed in the Arctic as defined in this chapter for well over a millennium, and today consists of a mixture of commercial agriculture on several thousand farms and widespread subsistence agriculture. Potatoes and forage are characteristic crops of the cooler areas, and grains and oilseed crops are restricted to areas with the warmest growing seasons. The main livestock are dairy cattle and sheep, which have been declining, and diversified livestock such as bison or other native animals, which have generally been increasing in commercial operations. The five ACIA-designated models all project rising temperatures that are very likely to enable crop production to advance northward throughout the century, with some crops now suitable only for the warmer parts of the boreal region becoming suitable as far north as the Arctic Circle. The average annual yield of farms is likely to increase at the lower levels of warming due to climate suitability for higher-yielding crop varieties and lower probabilities of low temperatures limiting growth. However, in the warmest areas, increased heat units during the growing season are very likely to cause a slight decrease in yields since warmer temperatures can speed crop development and thereby reduce the amount of time organic matter accumulates. Under the ACIA-designated model projections, water deficits are very likely to increase or appear in most of the boreal region. By the end of the 21st century, unless irrigation is practiced, water stress is very likely to reduce crop yields. Water limitation is very likely to become more important than temperature limitations for many crops in much of the region. Overall, negative effects are unlikely to be stronger than positive effects. Lack of infrastructure is likely to remain a major limiting factor for commercial agricultural development in the boreal region in the near future. Even under model-projected levels of climate change, government policies regarding agriculture and trade will still have a very large, and perhaps decisive, influence on the occurrence and rate of agricultural development in the north.

Understanding the condition or character of the forest resource system that climate change affects is crucial in assessing forests and land management. Russia has made commitments to management of carbon stocks that are of global interest because of the amounts involved. Fire and insect disturbance at very large scales have generated resource management challenges in Canada. A large proportion of Alaska is managed as strict nature reserves and as resource lands for biodiversity and ecosystem services. Large forest disturbances associated with climate change have occurred in Alaska, disrupting ecosystems and imposing direct costs, but the large area of reserves improves the ultimate prospects of species surviving potential future climate change. In highly managed forests of Finland, Sweden, and Norway, forests are generally managed effectively and are increasing in volume, but the prospect of climate change puts at risk human expectations of specific future resource returns. In Iceland, temperature increases have improved tree growth at a time of a large afforestation program designed to increase forest land cover and sequester carbon.

About 6000 years BP (the end of the postglacial thermal maximum), radial growth of larch trees on the Taymir Peninsula of Russia surpassed the average of the last two millennia by 1.5 to 1.6 times. Tree growth and warm-season temperature have irregularly decreased in northernmost Eurasia and North America from the end of the postglacial thermal maximum through the end of the 20th century. Long-term tree-ring chronologies from Russia, Scandinavia, and North America record the widespread occurrence of a Medieval Warm Period about 1000 years BP, a colder Little Ice Age ending about 150 years ago, and more recent warming. Recent decades were the warmest in a millennium or more at some locations. Temperature and tree growth records generally change at the same time and in the same direction across much of the Arctic and subarctic. However, intensified air-mass circulation associated with a warmer climate has introduced a stronger flow of warm air into specific regions of the Arctic and enhanced the return flow of cold air out of the Arctic in other regions. Temperature and tree-growth trends are correlated but opposite in sign in these contrasting regions.

Between 9000 and 7000 years BP, trees occurred in at least small groups in what is now treeless tundra nearly to the arctic coastline throughout northern Russia. Around 6000 years BP, the northern treeline on the Taymir Peninsula (currently the farthest north in the world) was at least 150 km further north than at present. During the period of maximum forest advance, mean July temperature in northern Russia is estimated to have been 2.5 to 7.0 °C higher than the modern mean. This record of past forest advance suggests that there is a solid basis for projecting similar treeline change under climate change producing similar temperature increases. It also suggests that the components of ecosystems present today have the capacity to respond and adjust to such climate fluctuations. The greatest retreat of forest and expansion of tundra occurred between 4000 and 3000 years BP. In northeast Canada, the black spruce forest limit has remained stable for the past 2000 to 3000 years. In recent decades, milder winters have permitted stems that were restricted to snow height by cold and snow abrasion to emerge in upright form, and future climate projected by the ACIA-designated models would permit viable seed production, which is likely to result in infilling of the patchy forest–tundra border and possibly begin seed rain onto the tundra. In the Polar Ural Mountains, larch reproduction is associated with warm weather, and newly established trees have measurably expanded forest cover during the 20th century, although there is a time lag between climate warming and upslope treeline movement.

Across the boreal forest, warmer temperatures in the last several decades have either improved or decreased tree growth, depending on species, site type, and region. Some tree-growth declines are large in magnitude and have been detected at different points across a wide area, although the total extent has not been delineated. Temperature-induced drought stress has been identified as the cause of reduced growth in some areas, but other declines are not currently explained. Reduced growth in years with high temperatures is common in treeline white spruce in western North America, suggesting reduced potential for treeline movement under a warming climate. Tree growth is increasing in some locations, generally where moisture and nutrients are not limiting, such as in the boreal regions of Europe and eastern North America. The five ACIA-designated models project climates that empirical relationships suggest are very unlikely to allow the growth of commercially valuable white spruce types and widespread black spruce types in major parts of Alaska and probably western boreal Canada. The models project climates that are very likely to increase forest growth significantly on the Taymir Peninsula. The upper range of the model projections represents climates that may cross ecological thresholds, and it is possible that novel ecosystems could result, as during major periods of global climate change in the past.

Large-scale forest fires and outbreaks of tree-killing insects are characteristic of the boreal forest, are triggered by warm weather, and promote many important ecological processes. On a global basis, atmospheric carbon equal to 15 to 30% of annual emissions from fossil fuels and industrial activities is taken up annually and stored in the terrestrial carbon sink. Between 1981 and 1999, it is estimated that the three major factors affecting the terrestrial carbon sink were biomass carbon gains in the Eurasian boreal region and North American temperate forests, and losses in areas of the Canadian boreal forest. Particular characteristics of forest disturbance by fire and insects, such as rate, timing, and pattern of disturbance, are crucial factors in determining the net uptake or release of carbon by forests. The evidence necessary to establish a specific climate change effect on disturbance includes a greater frequency of fire or insect outbreaks, more extensive areas of tree mortality, and more intense disturbance resulting in higher average levels of tree death or severity of burning. Some elements of the record of recent boreal forest disturbance are consistent with this profile of climate change influence, especially forest fires in some parts of Russia, Canada, and Alaska and insect disturbances in North America.

Carbon uptake and release at the stand level in boreal forests is strongly influenced by the interaction of nitrogen, water, and temperature influences on forest litter quality and decomposition. Warmer forest-soil temperatures that occur following the death of a forest canopy due to disturbance increase the rate of organic litter breakdown, and thus the release of elements for new plant growth (carbon uptake). The most likely mechanism for significant short-term change in boreal carbon cycling as a result of climate change is the control of species composition caused by disturbance regimes. Successional outcomes from disturbance have different effects on carbon cycling especially because of the higher level and availability of nutrient elements (and thus decomposition) in organic litter from broadleaf trees compared to conifers. Net global land-use and land-cover change, especially aggregate increases or decreases in the area of forest land, may be the most important factor influencing the terrestrial sink of carbon. When water and nitrogen remain available at the higher growth rates typical of enhanced carbon dioxide environments, further carbon uptake occurs. Broadleaf litter produced under elevated carbon dioxide conditions is lower in quality (less easily decomposed) than regular litter because of lower nitrogen concentration, but quality of conifer litter in elevated carbon dioxide environments may not be as affected.

Different crop species and even varieties of the same species can exhibit substantial variability in sensitivity to ultraviolet-B (UV-B) radiation. In susceptible plants, UV-B radiation causes gross disruption of photosynthesis, and may inhibit plant cell division. Determining the magnitude of the effect of elevated UV-B radiation levels is difficult, because interactions with other environmental factors, such as temperature and water supply, affect crop reactions and overall growth. Damage by UV-B radiation is likely to accumulate over the years in trees. Evergreens receive a uniquely high UV radiation dose in the late winter, early spring, and at the beginning of the growing season because they retain vulnerable leaf structures during this period of maximum seasonal UV-B radiation exposure, which is amplified by reflectance from snow cover. Exposure to enhanced levels of UV-B radiation induces changes in the anatomy of needles on mature Scots pine similar to characteristics that enhance drought resistance. UV-B radiation plays an important role in the formation of secondary chemicals in birch trees at higher latitudes. Secondary plant chemicals released by birch exposed to increased UV-B radiation levels might stimulate its herbivore resistance.

14.1. Introduction

The Arctic has been defined in somewhat different ways in various studies, reports, and assessments, based primarily on the purpose of the project. While the most restrictive definitions limit the Arctic to treeless tundra, snow, and ice in the high latitudes, most definitions of the Arctic encompass some elements of the boreal forest. The definition used by the Arctic Monitoring and Assessment Programme (AMAP, 1997; see section 8.1.1, Fig. 8.2), for example, includes the productive boreal forests of northwest Canada and Alaska, but includes mostly marginal treeline forest and woodland in eastern Canada. Permafrost-free forests in the northern portion of the Nordic countries are within the AMAP-defined Arctic, but across central and eastern Siberia, the boundary follows the margin of sparse northern taiga and forest–tundra. This chapter focuses on the northernmost portion of the boreal forest region, but broadens consideration of the subject for two important reasons. First, many elements of the boreal forest are best understood as a whole (e.g., the gradients of changing tree responses to the environment from south to north), and this chapter includes an extensive and well studied Siberian transect that uses such an approach. Second, the five scenarios of climate change generated by the ACIA-designated models (section 4.4) project temperatures within the Arctic that today only occur in the boreal forest far to the south. If temperature increases of a magnitude similar to those projected by these models actually occur, the nearest analogues of climate (and eventually ecosystems) that would exist in the Arctic are those of more southerly boreal forest regions.

Sections 14.2 and 14.3 describe forest characteristics across the northern boreal forest to provide the context for understanding the importance of recent climate-related changes in the region and potential future change. Section 14.4 provides an overview of the climate scenarios generated by the ACIA-designated models and describes how the scenarios were used in different aspects of the assessment.

While many factors affect agriculture in the far north (e.g., changing markets, social trends, and national and international policies), section 14.5 focuses on the climate-sensitive aspects of crop production systems that would be affected under the scenarios of future climate, focusing on climate stations representative of areas with agricultural production or potential. Section 14.3 also considers the challenges that climate change poses for land management. Tree rings are one of the most important sources of information about past climates, especially in the sparsely populated far north, and section 14.6 reviews the record of climate and tree rings across the Arctic and northern boreal region.

Section 14.7 presents new information about the direct effects of climate on tree growth in the northern boreal forest, in both the distant and more recent past, and uses scenarios generated by the ACIA-designated models to project how climate change may affect the growth of selected tree species during the 21st century. Section 14.8 identifies key climate controls on large-scale population increases in insects that damage trees, and provides some recent evidence of these effects.

Forest fire is another major indirect effect of climate on the status of forests in the far north. Section 14.9 examines some of the climate-sensitive aspects of fire and possible future fire conditions and effects. The climate-related changes in growth, insect-caused tree death or reductions in tree growth, and fire are major factors that control the uptake and storage of atmospheric carbon (section 14.10). The implications of future climate change for forest distribution are briefly considered in section 14.11. Finally, section 14.12 summarizes some recently published information on the effects of increased UV-B radiation levels on boreal forest species, and section 14.13 reviews critical research needs.

14.2. The boreal forest: importance and relationship to climate

14.2.1. Global importance

The boreal region covers about 17% of the terrestrial area of the earth (Bonan et al., 1992), with a broad zone of forest in a continuous distribution across the Eurasian and North American landmasses. The boreal forest is defined as a belt of forest south of the tundra characterized by a small number of coniferous species including spruce, larch, pine, and fir and a limited number of broad-leaved species, primarily birch and poplar (see Appendix D for common and scientific names of tree and other woody species mentioned in this chapter). At the landscape scale, conifers dominate the boreal forest, although broad-leaved trees can be locally dominant. Forest and woodland in the arctic nations (excluding

Table 14.1. Total forest area in arctic nations and percentage of global forest area (Smith W. et al., 2001; US Forest Service, www.fs.fed.us/r10/spf/facts/spffact.htm).

	Total forest area (10^6 ha)	% of global forest area
Russia	851.4	22.4[a]
Canada	44.6	6.4[a]
United States (all)	302.4	8.0[a]
United States (Alaska boreal only)	35	0.9[b]
Finland	21.9	0.6[a]
Sweden	27.1	0.7[a]
Norway	8.8	0.2[a]
Iceland	0.034	<0.001[a]
Arctic nation total	1456.2	38.3
World total	3800	100

[a]"forest" category (FAO, 2002); [b]"boreal forest" category (Labau and van Hees, 1990)

Denmark), most of which is boreal forest, represent about 1.5 billion hectares (ha) of the total global forest area of 3.8 billion ha (in 2000), or about 38% of global forest area (FAO, 2001; Table 14.1). Russian forests (excluding woodlands), the vast majority of which are boreal, represent 22.4% of the global total, by far the largest proportion of any nation in the world (Table 14.1). Two of the remaining three countries with the largest percentage of global forest area (Canada and the United States) are also arctic nations.

This chapter focuses on the northern portion of the boreal forest, but many aspects of the topic must be considered from a broader perspective. The boreal forest contains trees growing at the highest latitude on earth, and along its northern margin it merges into the circumpolar tundra. The boreal region is the northernmost part of the world where agricultural crops are produced regularly on a significant scale and where a settled agricultural way of life has historical continuity.

The boreal forests of North America and Eurasia share some plant and animal species and display a number of other similarities. The boreal forests of Russia and especially Siberia are often referred to as the taiga, an indigenous term meaning "little sticks". The term taiga is equivalent to the true boreal forest.

The boreal forest is both affected by and contributes to climate change: both topics are examined in this chapter. Globally, the existence of large areas of boreal forest cover has a significant effect on the radiative balance of the planet (Bonan et al., 1992). The rough-textured, dark surface of land covered with boreal forest canopy intercepts and absorbs a high proportion of solar radiation, converting it to heat (Bonan et al., 1992). In contrast, the smooth, snow-covered surface of the tundra is highly reflective. In high-latitude regions where snow covers the ground for half of the year or more, the albedo effect of tundra versus boreal forest cover is magnified. Future expansion of the forest into present-day tundra regions resulting from a warming climate would thus amplify the warming further.

Another important influence of the global boreal forest on climate is its influence on levels of atmospheric carbon dioxide (CO_2) and other greenhouse gases (GHGs). Boreal forests take up CO_2 through photosynthesis, and store carbon in live and dead plant matter, including substantial long-term accumulations in large tree boles and in soil. Forests release CO_2 to the atmosphere through decomposition of dead organic matter, live plant and animal respiration, and combustion that takes place during fires. Both natural (e.g., fire) and anthropogenic (e.g., timber removal) disturbances are important influences on the boreal forest. An ecosystem disturbance is defined as a change in state or condition that disrupts the way in which the system has been functioning (photosynthesis, water regulation, etc.), causing it to reinitiate successional development. Disturbances vary by cause, rate, intensity, extent, timing, frequency, and duration.

Management-related factors influence carbon uptake and storage in the form of tree mass. These management practices include the rearrangement of forest age classes by timber harvest or suppression of wildfires, selection of tree species, fertilization, and thinning regimes. The combined effect of all management actions can either enhance or reduce carbon uptake and storage. For example, across the Russian boreal region, for many years after logging the forests that regrow take up less atmospheric CO_2 than nearby old-growth forests (Schulze et al., 1999). In the Boreal Cordilleran ecozone of Canada (see Wiken, 1986 for definition), it is estimated that total suppression of natural disturbances and their complete replacement by harvesting for maximum sustainable yield would increase carbon storage in soils and wood products over a period of a century or two (Price et al., 1997). Direct climate effects that increase or decrease tree growth in unmanaged natural forests also influence short-term uptake of atmospheric carbon.

The boreal forest and northern tundra together contain 40% of global reactive (readily decomposable to CO_2, methane, water, and mineral nutrients) soil carbon, an amount similar to the amount of carbon held in the atmosphere (McGuire et al., 1995b; Melillo et al., 1993). The extensive boreal forest plains of northeast Europe, western Siberia, and central and eastern North America that are within or immediately south of the discontinuous permafrost region occupy the zone of maximum carbon storage in soil organic matter on the earth. Climate change, interacting with human use and management of boreal forest, northern agricultural, and tundra ecosystems, would enhance the decomposition of carbon stored in soil organic matter and its subsequent release into the atmosphere, thus compounding climate change caused by anthropogenic GHG emissions.

The boreal forest is one of the most intact major vegetation regions of the earth, but in some areas boreal forest has been extensively converted to other land uses or severely damaged by air pollution (e.g., in Iceland and particular areas of Russia, respectively). Boreal forests in Finland, Sweden, Norway, and parts of Canada are generally intensively managed for timber production, and in such intensively managed stands, tree age structure, tree species, and spacing are controlled (section 14.3.4). However, huge areas of central and eastern Siberia and northwestern North America represent the most extensive remaining areas of natural forest on the planet (Bryant et al., 1997). Not all natural boreal forests consist of older trees: large areas are burned or subject to insect-caused tree mortality annually. Climatic factors, especially prolonged periods of warm weather, often create the conditions that result in fire and insect disturbances in boreal forests. The boreal forest is subject to rapid changes causing long-term consequences as a result of these climate-related effects.

The boreal forest is the breeding zone for a huge influx of migratory forest birds that perform many important roles (e.g., insect consumption, seed dispersal) in the boreal region and in other forests of the world during migration and winter residence in the south. Climate affects the population level of these migratory birds and their food resources. Climate-associated processes also determine the amount and quality of forest habitat available to migratory birds.

From 1981 to 1999, three regions of the world primarily affected the terrestrial carbon sink, which takes up and stores atmospheric carbon equivalent to 15 to 30% of annual global emissions from fossil fuel combustion and industrial activities (Myneni et al., 2001). The majority of the net terrestrial carbon change came from biomass carbon gains in the Eurasian boreal region and North American temperate forests, and carbon losses from some Canadian boreal forests. Some of the terrestrial biomass change was a response to direct and indirect climate effects. However, human use and management of the boreal forest was an important factor as well, and could be a significant future contributor to human management of the carbon cycle. Certain forest biomass carbon sinks can be used to meet national commitments to reduce GHG emissions under the Kyoto Protocol of the Framework Convention on Climate Change. Land and resource managers in the Arctic and boreal regions are interested in potential "carbon cropping", which might involve payments from organizations wishing to sustain or enhance carbon storage (Bader, 2004). Mechanisms to place values on the various carbon transfers are not fully in place. However, if effective, exchangeable systems of placing values on transfers of carbon are adopted at an international level, boreal forests could potentially generate a flow of wealth into arctic and subarctic regions from other parts of the world for boreal forest and land management treatments, offsets for emissions elsewhere, or policies designed to store or retain carbon.

Uncertainties remain about the influence of the boreal forest on each of the key processes that determine global carbon balance. For example, the uptake of atmospheric CO_2 by tree and other plant growth may either increase or decrease with increasing temperature, depending on the species, the geographic region where the growth occurs, the range of the temperature increase, and other climate factors such as precipitation that are likely to change in a changing climate. However, there has been substantial recent progress in understanding the response of elements of the boreal system to temperature. Across the boreal regions, a first generation of studies, models, databases, and measurements have provided a significantly better understanding of one of the most extensive and important vegetation types on earth. Continued and expanded data collection from research and management activities will provide a reasonable basis for determining the net contribution of the boreal forest to GHG balance and climate, the further changes a warming climate would induce in the boreal forest, and the agricultural and forest management opportunities available to the region in the future.

14.2.2. Arctic importance

Trees occur on only a small proportion of the land surface within the Arctic as defined in this chapter. Even so, forests and woodlands are important on a regional basis within the Arctic for several reasons. Where trees do occur, they serve as indicators of more productive terrestrial ecosystems with longer growing seasons than treeless tundra. Trees, even when present in small numbers on the arctic landscape, offer resources to arctic residents for a variety of uses. Finally, some areas of full-canopy forest within the Arctic are generally the most productive natural systems within the political jurisdiction where they occur (e.g., the galley forests along rivers that extend into the tundra in the northern Yukon and Northwest Territories in Canada). Specific reasons for the importance of boreal forest and agriculture in the Arctic include the following:

- Portions of the boreal forest devoted to forest products production are major contributors to the national economies of some arctic nations. Although the current zone of optimum climate for boreal forest growth is in the middle or southern boreal region, nearly all scenarios of climate change place the climatically optimum growth region within the present-day Arctic within a century or so.
- Residents of the boreal region depend on the products and resources of the forest for a variety of ways of life, including traditional ways of life.
- The major rivers of the boreal region transport large volumes of wood into the Arctic (Eggertsson, 1994; Ott et al., 2001), and this wood resource supports ecosystems that decompose the wood and feed organisms in rivers, oceans, and beaches. Climate change is very likely to affect all the processes in this system, including tree growth, erosion, river transport, and wood decay.
- Wood transported into the Arctic was an important resource for people in a naturally treeless environment during prehistoric times (Alix, 2001) and is still a useful and valued resource for many arctic residents today.
- The boreal forest collects, modifies, and distributes much of the freshwater that enters the Arctic Basin (see sections 6.8 and 8.2), and changes in boreal forests resulting from climate change would certainly affect many of these important functions.
- Portions of the boreal forest region have experienced some of the greatest temperature increases reported during the 20th century, and the responses of the forest system and the societal consequences in the region provide lessons that may be useful to other regions that could eventually experience similar change.
- Recent temperature increases in the boreal region have increased the frequency of occurrence of critical temperature thresholds for the production of agricultural crops currently grown in the region. Possible future temperature increases almost cer-

tainly would increase the land area on which crops could be produced successfully, and are very likely to increase the variety of agricultural crops that could be grown.

14.2.3. Climatic features

The boreal region is often assumed to be a zone of homogenous climate, but climate across the region is actually surprisingly diverse. During the long summer days, interior continental locations under persistent high-pressure systems experience hot weather that facilitates extensive forest fires frequently exceeding 100 000 ha. In maritime portions of the boreal region affected by air masses that originate over the North Atlantic, North Pacific, or Arctic Oceans, summer daily maximum temperatures are on average cooler than interior locations and seldom reach the high temperatures experienced at locations further inland.

Precipitation is abundant in the boreal zone of most of the Nordic countries, western Russia, and certain coastal and mountain regions of western North America. By contrast, in the topographically complex

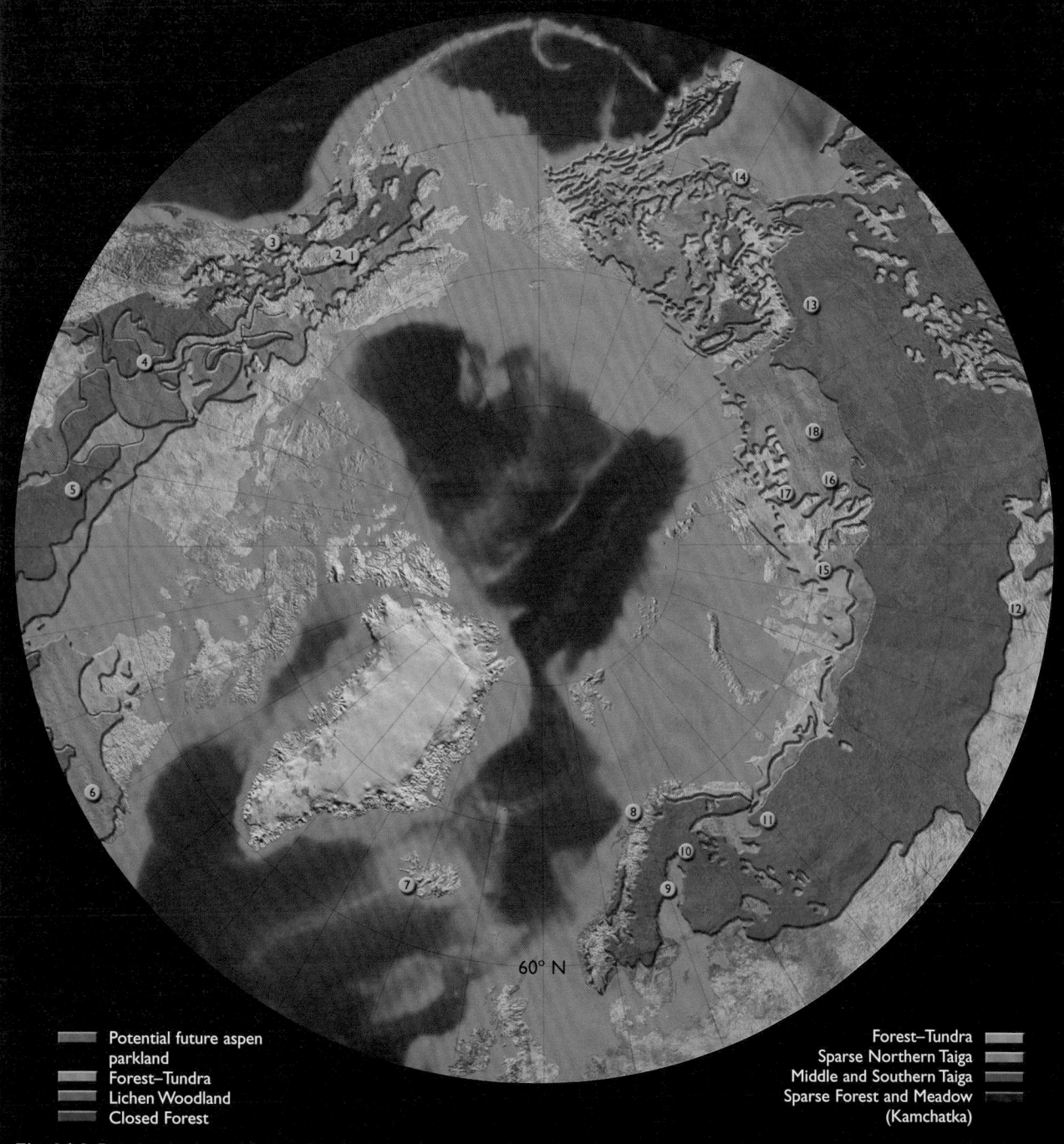

Potential future aspen parkland
Forest–Tundra
Lichen Woodland
Closed Forest

Forest–Tundra
Sparse Northern Taiga
Middle and Southern Taiga
Sparse Forest and Meadow (Kamchatka)

60° N

Fig. 14.1. Present-day boreal forest distribution, using a simplified formation system representing a gradient of decreasing productivity and species diversity from south to north (compiled using data from Anon., 1983; Anuchin and Pisarenko, 1989; Elliott-Fisk, 1988; Kuchler, 1970; Kurnayev, no date; Oswald and Senyk, 1977; Rowe, 1972; Viereck and Little, 1972). The area depicted in orange represents a broad zone where the ratio of precipitation to evapotranspiration is nearly one: under a scenario of doubled atmospheric CO_2, this area is projected to become too dry to support closed-canopy boreal forest, shifting instead to aspen "parkland", a woodland formation (Hogg and Hurdle, 1995). Numbered locations are sites analyzed in this chapter (see section 14.4.1, Table 14.2).

landscapes of Alaska, northwestern Canada, and central and northeastern Siberia, precipitation sometimes limits forest growth so that natural grasslands are part of the landscape.

Precipitation in the boreal region of western North America is influenced by storms in the southern Bering Sea and North Pacific Ocean and reaches a distinct maximum in late summer. In other parts of the boreal forest region, precipitation is more evenly distributed throughout the year or exhibits a winter maximum. East-central Siberia experiences low winter snow depths because the strong Siberian High suppresses precipitation. The boreal landscapes of far eastern Siberia and western North America are mountainous, whereas the topography of most of central and western Siberia and eastern Canada is characterized by low, smooth hills and level terrain. The mountainous boreal regions are characterized by sharply varying local climates (Pojar, 1996) and aspect-controlled differences in forest types (Viereck et al., 1986). As a result, the forest–tundra boundary is much more irregular there than on the plains of the central portions of the continents (Fig. 14.1). All of this regional climatic variation must be taken into account as a fundamental backdrop when considering climate change and ecological response in boreal forests.

In Eurasia and North America, both the northern and the southern boundaries of the boreal zone are not aligned at the same latitude east to west. The Icelandic and Aleutian Lows deflect storm tracks and advect relatively mild air masses northward as they approach the western margins of Eurasia and North America, respectively. As a result the boreal forest belt is located considerably farther north in both the Nordic countries and western North America than in the center of the continents (Fig. 14.1). In contrast, cold polar air flowing southward follows a persistent path along the eastern portion of both continents, and consequently the boreal forest belt reaches its southernmost limits there. In the center of the Eurasian landmass, the latitudinal position of the boreal forest depends mainly on the degree of continentality of the climate. The most northerly forests in the world occur along the Lower Khatanga River in central Siberia (72° 32' N), where the climate is more continental than either western or eastern Siberia. The warmer summer temperature in northernmost central Siberia is the critical factor that allows trees to survive there despite extreme winter cold.

14.2.4. Climate variability

Essentially all of the boreal forest in Alaska is north of 60° N, and practically all the boreal forest of eastern Canada is south of 60° N (Fig. 14.1). The boreal forest region is particularly prone to climatic variability because minor variations in key features of the atmospheric circulation can either intensify the advection of warm air into this naturally cold region, or enhance the distribution of cold air southward through the region. There is some evidence that the climate system in the far

north operates in a way that positive (western continent) and negative (eastern continent) anomalies operate in synchrony with each other. It appears that the intensification of meridional air mass movement is especially effective in warming the western margin of the North American Arctic while cooling the eastern margin of the continent (Fig. 14.2), leading to east–west temperature anomalies in the arctic and boreal regions (see also section 2.6.2.1, e.g., Fig. 2.7, and section 6.7.2).

Periods of major climate change, including alternating glacial and interglacial conditions, have repeatedly and drastically affected the northern regions of the planet (section 2.7). During the late Pleistocene and several previous glaciations, the present-day boreal region was mostly covered with glacial ice and forest organisms were largely displaced south of the current limits of the region (Wright, 1983; Wright and Barnosky, 1984; section 2.7). The relatively small unglaciated portion of the present-day boreal region was almost entirely treeless and contained assemblages of species unlike any found today (Anderson and Brubaker, 1994).

Present-day boreal forest vegetation characteristically has a large ecological amplitude (i.e., the ability to survive across a wide range of environmental conditions). The paleoenvironmental record and modern instrumental measurements demonstrate major shifts in temperature regimes in this region even during the time that forest has been present (sections 2.7 and 14.6), and the presence of this large ecological amplitude in the trees indicates that this wide range of possible temperatures is a consistent enough feature of the environment that it has required an adaptive response. Rapid and large changes in weather over the short term send signals through boreal systems that initiate vital processes such as infrequent periodic tree reproduction (Juday et al., 2003). On the one hand, this high level of natural climate variability suggests that during periods of climate

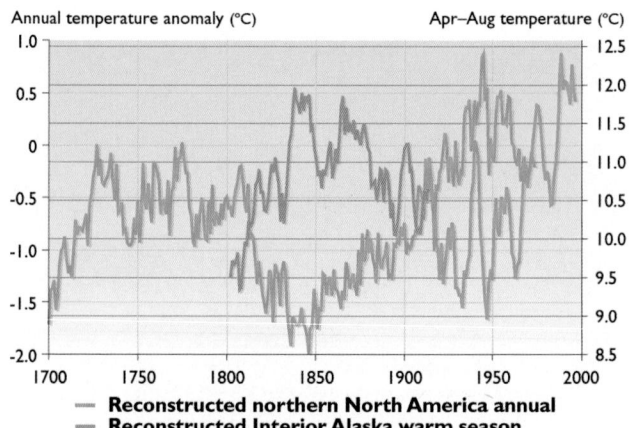

Fig. 14.2. Opposing high-latitude temperature trends from tree-ring reconstructions of mean annual temperature with a strong signal from eastern and central North America (data adapted from D'Arrigo and Jacoby, 1992, 1993; Jacoby and D'Arrigo, 1989) versus warm-season (Apr–Aug) temperature in western North America (observed and reconstructed from tree rings; Barber et al., 2004). Anomalies are calculated from the 1671–1973 mean.

change, the effects are more likely to be detectable at an earlier time in the boreal region than in many other parts of the earth. On the other hand, a long and persistent history of climate variability in the region suggests that organisms in the boreal forests of today may be among those better adapted to climate change because they have been filtered by many climate fluctuations in the past.

Because ecosystems quite unlike those of today have existed in the region in the past (Anderson and Brubaker, 1994; Wright, 1983; Wright and Barnosky, 1984), major climate change in the future is likely to produce ecosystems unknown today. The emergence of novel ecosystems (from the human perspective) is partly the result of individualistic species responses to changes in the environment. Each species has its own environmental requirements, tolerances, and thresholds, so that some species that co-occur today may not in the future, or existing sets of species may be joined by additional species (sections 7.3 and 7.6). Because of this property of individualistic responses to change in the environment, conservation efforts must be informed by monitoring the status of key species on a continuing basis. Conservation measures, such as modified harvest limits or fire management, must account for rapidly changing environmental conditions or changes in species populations not anticipated in management planning assumptions or outside historical experience in order to meet a goal of sustainability (Chapter 11).

The simultaneous pattern of temperature and precipitation anomalies can have important ecological impacts in forests of the north. Wildfire in the boreal forest is the product of short- to medium-term warm and dry conditions, usually associated with high-pressure dominance during the long days of summer (Johnson, 1992). Alternating periods of warm and dry versus cool and moist summer climate in central Alaska regulate the growth and reproduction of white spruce (*Picea glauca*), ultimately providing a mechanism to synchronize the production of seed crops to periods immediately following major forest fires (Barber et al., 2000; Juday et al., 2003). If future climate change alters not just the mean of climate parameters, but also the pattern of alternating warm/dry and cool/moist conditions, the resulting climate pattern could interfere with the reproductive success of one of the most widely distributed and dominant North American conifers. This example indicates the potential for subtle influences to be major factors in climate change effects.

14.2.5. Unique influences on climate

The Boreal Ecosystem–Atmosphere Study (BOREAS) was a large-scale, international interdisciplinary experiment in the northern boreal forests of Canada that began in 1993. Its goal was to understand how boreal forests interact with the atmosphere, how much CO_2 they were capable of storing, and how climate change will affect them.

Albedo measurements from BOREAS are among the lowest ever measured over vegetated regions, and indicate that the boreal forest (especially forest dominated by black spruce – *Picea mariana*) absorbs nearly 91% of incident solar radiation (Hall et al., 1996). In terms of water and energy balance, BOREAS found that the boreal ecosystem often behaves like an arid landscape, particularly early in the growing season. Even though the moss layer is moist for most of the summer, nutrient-poor soils and limiting climatic conditions result in low photosynthetic rates, leading to low evapotranspiration. As a result, relatively little of the available moisture is transferred to the atmosphere. Much of the precipitation penetrates through the moss layer into the soils, which are permeable, then encounters the underlying semi-impermeable layer and runs off. Most of the incoming solar radiation is intercepted by the vegetation canopy, which exerts strong control over transpiration water losses, rather than by the moist underlying moss/soil surface. As a result, much of the available surface energy is dissipated as sensible heat.

The BOREAS experiment also found that coniferous vegetation in particular follows a very conservative water-use strategy. Stomatal closure drastically reduces transpiration when the foliage is exposed to dry air, even if soil moisture is freely available. This feedback mechanism acts to keep the surface evapotranspiration rate at a steady and surprisingly low level (less than 2 mm/d over the season). The low evapotranspiration rates coupled with high available energy during the growing season can lead to high sensible heat fluxes and the development of deep (3000 m) planetary boundary layers, particularly during the spring and early summer. These planetary boundary layers are often characterized by intense mechanical and sensible heat-driven turbulence.

14.3. Land tenure and management in the boreal region

The influence of climate change on forest values and forest users depends on the amount and initial condition of the forest resource and the uses or intangible values of the forest for people, cultures, and economies. This section reviews forest extent, the overall allocation of forest land to different uses, the main patterns of forest use, the management systems, and the values generated by the boreal forest. Where these characteristics can be singled out by political jurisdiction or other means, the discussion is focused on the northern boreal forest. This discussion forms the basis for considering climate change impacts.

14.3.1. Russia

Russia contains the largest forested area of any nation, amounting to an estimated 763.5 million ha (FAO, 1999). The boreal forest of Russia can be thought of as three roughly parallel belts of southern and middle taiga, sparse northern taiga, and in the farthest north a forest–tundra region extending to the completely treeless tun-

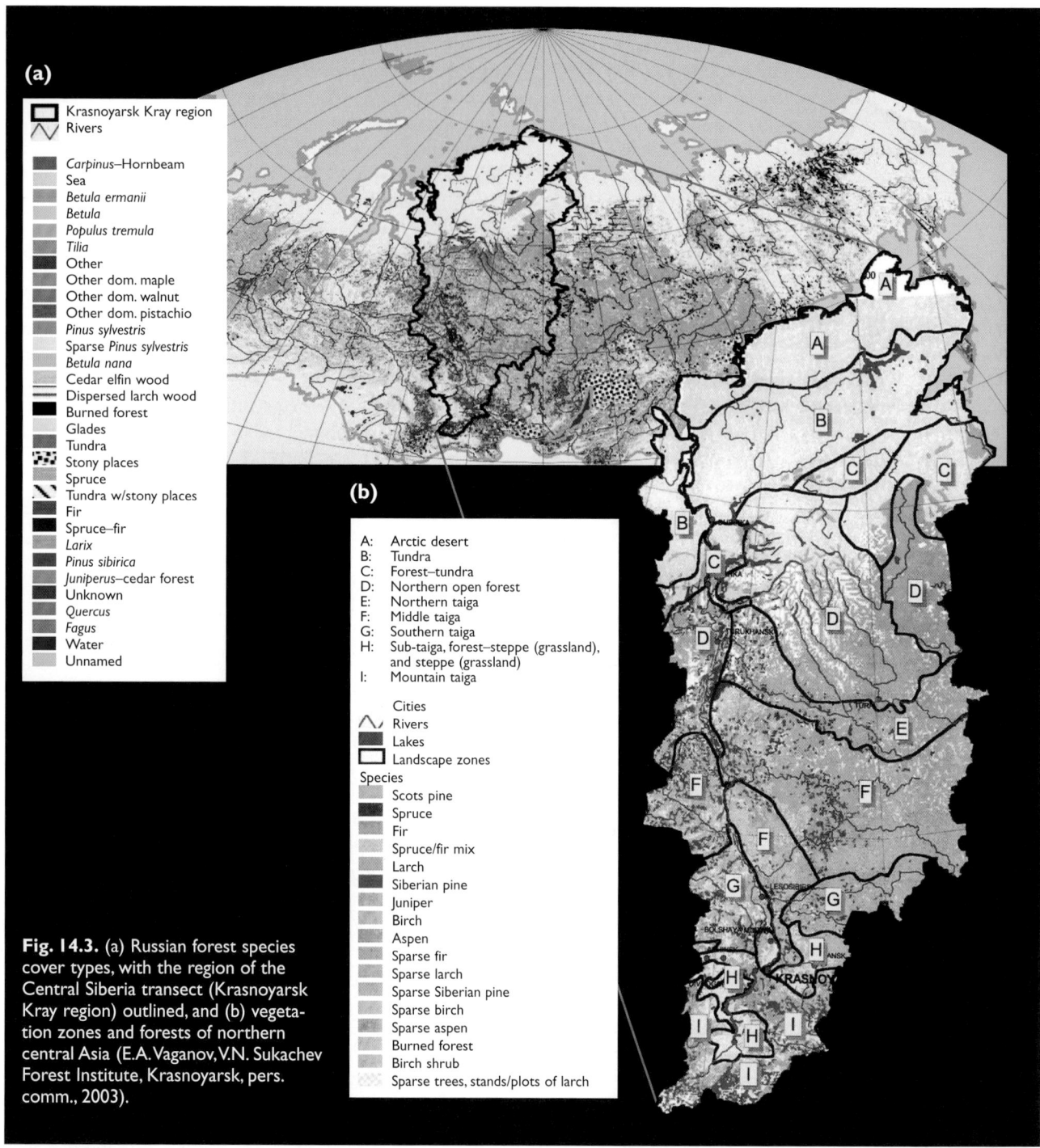

Fig. 14.3. (a) Russian forest species cover types, with the region of the Central Siberia transect (Krasnoyarsk Kray region) outlined, and (b) vegetation zones and forests of northern central Asia (E.A. Vaganov, V.N. Sukachev Forest Institute, Krasnoyarsk, pers. comm., 2003).

dra (Fig. 14.1). Dominant components of the Russian boreal forest change from spruce in the west (European Russia), to larch in the center and east, to pine at various locations (Fig. 14.3a). In Northern Eurasia, especially in central Siberia (Fig. 14.3b) and the Far East, as much as 70 to 75% of taiga forests appear to be close to a natural state. The remaining forests consist of fragments and other human-influenced forest, and areas dominated by marsh–bog complexes.

The central and especially the southern taiga zones have a long history of quite intensive land use. Although large untouched areas are rare or absent even in northern European Russia, a much greater proportion of these forests exists in a natural state than in similar vegetation zones in the Nordic countries. The Karelian Isthmus

region in the St. Petersburg Oblast is a good example. The forests of the Green Belt along the Finnish–Russian (Karelian) border are one of the most important centers of boreal biodiversity in Europe (Silfverberg and Alhojärvi, 2004). Many species in the Red Data Book (endangered) lists for Sweden and Finland still exist in relatively healthy populations in northwest Russia. Disruption or decline in the ecological health of these remnant natural forests in Russia resulting from climate change is very likely to have severe implications for the survival of the listed species, at least within Europe.

Economically exploitable forests total about half of the forested areas under state forest management in Russia, and are generally located in the southern taiga. The northern boreal forest of Russia is generally too distant from transportation infrastructure and consumption centers to repay the costs of timber removal. The Russian forest sector is a major employer, with a work force estimated at 1.8 million people in the mid- and late 1990s. However, during the 1990s, public-sector forest management organizations often did not have enough money to retain employees, fight forest fires, enforce logging regulations, or make periodic inventories. A slow recovery of public-sector capacity has begun in more recent years.

A 1993 reform, the Basic Forest Law, started a movement toward market transactions in this sector. The Basic Forest Law allows forest leasing and auctions of standing timber, and forest leasing is the main market mechanism currently in use. In the early 1990s, rapid and unsustainable cutting of old-growth and mature forests began. After 1996, changes in forest management policy and better information resulted in a slowing of the pace of timber cutting in old-growth forests. The new Forest Code of the Russian Federation was issued in 1997. The Russian forest management system uses the term "Forest Fund" to refer to all forest and related lands under governmental jurisdiction, which in practice is nearly all the forest. The dominant part of the Forest Fund was (and still is) under the management of the Russian Federal Forest Service (Rosleskhoz), which manages about 94% of the total forest area in Russia, with another 4% belonging to agricultural organizations, 1% to the Committee of Environmental Protection, and 1% to other state bodies.

For management purposes, Russian forests have been divided into three categories based on economic and ecological characteristics. The first category comprises forests with a protective function, for example, watershed forests (20% of forested land), which are available for partial tree removal, sanitary tree felling, and small (maximum 10 ha) clear cuts. The second category consists of forests in inhabited areas and forests with low productivity (5.5% of forest area). The vast majority of the forest (74.5%) is included in the third category, industrially exploitable forests, where clear-cutting (up to 50 ha) is the main management practice. In addition to managing nearly all Russian forests, Rosleskhoz is also responsible for about 20% of the logging, in the form of partial and sanitary logging. Climate change impacts that disturb forest cover in category one and two forest are very likely to impose extra costs on managers, local governments, and forest users to stabilize or recover forests. In category three forest, actual or potential forest product values are at risk from climate change. However, because of the economic inaccessibility of the majority of this forest in the northern taiga, direct losses are likely to be relatively small overall.

As of 2000, the Russian Federation had 99 state zapovedniks, or strict scientific nature reserves, totaling 31 million ha or 1.82% of Russia's territory (Ostergren and Shvarts, 2000). Zapovedniks generally meet category I criteria of the World Conservation Union classification of protected areas (IUCN, 2000). During the 1990s, Russia established 35 national parks totaling 6.8 million ha (0.40% of Russia's territory). There are plans to establish additional zapovedniks and parks (Colwell et al., 1997). Practically all the national parks are located in Forest Fund areas and are managed by the state forestry authorities. These areas are managed for a range of scientific and biodiversity values, and the main concerns are climate changes that could reduce the chances for survival of the protected species or ecosystems.

The forests of Russia are an important component of the global carbon cycle because of the extensive area of forest land and the high storage of carbon in cold soils. In consideration of its extensive and significant forest resources, Russia has obtained substantial carbon emission credits as part of its participation in the Kyoto Protocol (Webster, 2002). Fulfilling the potential of Russian forests to offset carbon emissions will require sustaining, and to some degree rebuilding, a land management capability over a vast area with certain fundamental aspects: forest inventory and measurement, surveillance and detection of forest health problems, trained and deployable fire control and management forces, and various resource management specialists. Many of the benefits of increasing carbon sequestration in Russian forests can be obtained as a direct result of implementing policies that are widely agreed to be rational and beneficial (Shvidenko et al., 1997), including measures such as harvest levels in line with actual growth, effective fire control forces in regions of high-value timber, and adequate regeneration efforts.

14.3.2. Canada

The Canadian boreal forest represents nearly 6.4% of global forest area according to the United Nations Food and Agriculture Organization (FAO) definition of forest (Table 14.1). Forests play a large role in the Canadian environment, economy, culture, and history. Forest (tree-covered land with a full canopy) and woodland (tree-covered land with less than a complete forest canopy) cover nearly half (44%) of the Canadian landscape, totaling about 401.5 million ha, according to Canada's Forest Inventory 2001 (Natural Resources Canada, 2004). Canadian forest land totals about 309.8 million ha, according to the 2001 inventory, and about 294.7 million ha are not reserved and therefore potentially available for commercial forest activities, although much of the land has not been definitively allocated as to use. The large majority of Canadian forest is crown land held for the public, with 71% controlled by the provinces and 23% under federal control (Natural Resources Canada, 2003). A few percent of Canadian forest land is managed by territorial governments and the balance is in private hands. Of the 401.5 million ha of tree-covered land in Canada, 22.8 million (5.7%) are by law managed to remain in a natural state. On another 27.5 million ha (6.8%), timber harvesting is excluded by administrative policy (e.g., on unstable soils or as habitat buffers along important lakes or rivers). The most accessible forest land, and therefore the most likely to experience forest management activities, covers 144.6 million ha, or 36.0% of the total tree-covered area.

In the Canadian land tenure system, provincial governments are responsible for managing most of the land within their boundaries held for the benefit of the public (crown land). Until recently, the federal government held and managed land north of 60° N (an area not organized into provinces). However, in this area, by progressive steps, ownership and decision-making responsibility are passing to indigenous peoples and territorial governments. Simultaneously, lands of major conservation interest are being established as new national parks and wildlife refuges managed by the federal government. The indigenous peoples of Canada, who meet their cultural, spiritual, and material needs from their forest homeland, have a unique perspective and set of goals in forest management. Canadian land and forest management has changed significantly and is likely to change further as aboriginal title, treaty rights, and governmental responsibility to protect these rights are all more specifically defined.

The large majority of Canadian forest is boreal, with species such as white spruce, black spruce, aspen (*Populus tremuloides*), and paper birch (*Betula papyrifera*) having essentially transcontinental distributions. The forests of Canada are naturally dynamic, with large-scale disturbances quite typical. Across all Canadian forest types, insect defoliation affected 18.6 million ha in 2002 (Natural Resources Canada, 2003). Section 14.10.2 details the role of disturbance in the boreal forest with respect to carbon, and section 14.8.1 describes the role of forest insect disturbances in the boreal forest.

In the late 1990s, more than 300 communities in Canada depended largely on jobs in the forestry sector. During that period, the wood and paper industries and associated organizations employed more than 830 000 people, and paid more than Can$ 11.8 billion in wages annually. In 2002, 361 400 people were directly employed in the forest industry (Natural Resources Canada, 2003). Historically, Canada has been one of the largest suppliers of wood and paper products in the world, with 1995 shipments of manufactured forest products valued at Can$ 71.4 billion. Forest products exports from Canada contributed Can$ 39.6 billion to its net balance of trade in 2003 – almost as much as energy, fishing, mining, and agriculture combined. Canadian forests also contribute to uses and support industries providing billions of dollars in sales, including recreation, tourism, natural foods, furs, Christmas trees, and maple syrup. Much of the rapidly increasing recreation activity is forest-based. The number of visitor-days to forested national parks was 29.7 million in 1994.

Timber is harvested from about one million ha in Canada annually, or 0.7% of the total accessible, managed forest land. Allocations of timber resources are based on long-term goals for land use and forest man-

agement established in forest plans, and regional analyses and estimates of wood supply. On public (crown) lands, tenure arrangements with forest companies or communities to harvest timber are usually issued through contracts or licenses. Recent changes to legislation and tenure arrangements include provisions to license the harvesting of other forest resources such as blueberries or mushrooms. All harvest activities must also complement or integrate management objectives for wildlife, water, subsurface resources, hydroelectric energy, and transportation. The northwestern Canadian boreal forests of northern Alberta and British Columbia and the southern Yukon and Northwest Territories are the last regions of Canada to experience large-scale forest products harvest, beginning primarily in the 1980s. The installation of wood products processing facilities stimulated the expanded harvest in the northwestern Canadian boreal forest, and under the leasing system, large areas that are currently primary forest are now committed to eventual harvest.

The new emphasis in Canadian forest management typically includes the identification of objectives for the conservation of forests as a source of economic wealth, of habitat for wildlife and fish, of gene pools for biological diversity, and of water and carbon. Climate change calls into question the ability to adequately forecast future forest condition and growth and thus conduct meaningful planning. However, Canadian forest land managers are considering how to deal with climate change effects with specifically adapted silvicultural techniques for maintaining forest health, managing declining stands, regenerating disturbed areas with desired genotypes and species, and assisting in species migration (Parker et al., 2000).

Because of the large share of productive forest resources under Canadian provincial jurisdiction, an important source of leadership in developing coordinated forest policy has been the Canadian Council of Forest Ministers (CCFM), made up of the principal forestry officials of the provinces. Faced with public concerns about the extent of timber harvesting, and in response to the 1992 National Forest Strategy and the United Nations Conference on Environment and Development (UNCED), the CCFM developed a framework of criteria and indicators to define and measure progress toward sustainable forest management, in consultation with the entire Canadian forest community. The framework reflects the values of Canadians and identifies the forest features and uses they want to sustain or enhance, including indicators of environmental, social, and economic health.

Canada and 11 other countries have collaborated in the development of criteria and indicators for the conservation and sustainable management of boreal and temperate forests outside Europe (known as the "Montreal Process"). Climate change adds major uncertainty to basic assumptions about future forest condition, growth, and uses that are critical in making decisions in the present. For example, current forest harvest levels developed to meet the test of sustainability must be based on projections of future forest growth and mortality. A major climate shift would alter these factors in ways not fully understood but very likely to be disruptive. The challenge is to decide what forest activities should be allowed today based on an assumed future in which climate change outside the range experienced in previous planning horizons may be having an effect.

14.3.3. United States (Alaska)

Alaska is by far the largest state in the United States, occupying about 20% of the area covered by the remainder of the nation or an area greater than the Nordic countries combined. The two different types of forest found in Alaska are coastal rainforest in southeast and south-central Alaska and boreal forest in northern and Interior Alaska. The coastal forest in Alaska covers about five million ha, but the most productive areas of this forest type are south of the Arctic as defined in this chapter. However, much of south-central Alaska is either coastal forest or a boreal–coastal forest transition and is within the Arctic as defined here. The amount of land in the boreal region of Alaska that supports at least 10% forest cover is about 46 million ha, or 41% of the state. Statewide, about 6.4 million ha or 16.3% of total Alaska forest land is classified as "productive" forest, that is, land capable of an average growth rate of 1.4 m^3/ha/yr (Labau and van Hees, 1990). Even less of the Alaska boreal forest (12% or 5.5 million ha) is considered productive commercial timberland (Labau and van Hees, 1990).

Of the 114 million ha that make up Alaska, the federal government owned over 95% until Alaska became a state in 1959. The state government was granted the right to eventual ownership of 32 million ha (28% of Alaska) as a condition of statehood. To date the state of Alaska has received 27.4 million ha (85%) of its entitlement. Under terms of the Alaska Native Claims Settlement Act, Alaska Native corporations are entitled to receive 13.4 million ha (11.7% of Alaska), and most of that land has been conveyed. Individuals own only about 0.6 million ha (0.5% of Alaska). The private individual ownership category is expected to slowly increase as a result of government land sales and transfers. The federal government retains nearly 68 million ha in Alaska (60% of the state), including about 20.6 million ha in national parks and 31.1 million ha in national wildlife refuges.

National parks and wildlife refuges generally preclude resource development, but there are a few exceptions. In some circumstances, petroleum development can take place in wildlife refuges, and rural residents with a history of local use may obtain resources such as house logs, fuel wood, and poles for fish traps from national parks. Taking all federal land designations together, and including other protected land such as state parks, Alaska has probably the highest percentage (about 40%) of its area devoted to strict protection of natural habitats in the world. At least 25% of the productive boreal timberland in Alaska is reserved by law from forest harvest, and a similar amount is estimated to be reserved by administrative policy (Labau and van Hees, 1990). Climate change effects on this strictly preserved land base are likely to involve primarily the temporary reduction or increase in the populations of certain species resulting from land-cover change. The intactness and extent of these ecosystems enhance the prospect of species survival, even following large-scale climate change. The prescription offered by conservation biologists that best equips species to withstand major movement of optimum climate zones is to maintain large-scale, topographically diverse landscapes with naturally functioning ecosystems (Markham and Malcom, 1996). Such a strategy preserves complete gene pools and specially adapted ecotypes, and provides maximum opportunity for natural migration and disturbance recovery. The current boreal forest is largely the result of such adjustment by the biota to the many cycles of glacial and interglacial climate changes during the Pleistocene. The current land allocation situation matches the conservation biology prescription for climate change resilience better in Alaska than in almost any other major forest region of the world.

The Alaskan boreal forest is currently used for a variety of economic, subsistence, recreational, scientific, and other purposes. Local-scale logging has been a traditional use for much of the 20th century. The boreal region has only small-scale wood products facilities: mainly small sawmills and facilities to manufacture specialty products such as house logs and birchwood items. Employment in forest products manufacturing industries, mostly in the coastal region, peaked in 1990 at just under 4000 people, constituting 1.4% of total Alaska employment in that year (Goldsmith and Hull, 1994), but was only about 600 in 2002 (Gilbertson, 2002). In the 1990s, the two major wood products manufacturing facilities (pulp mills) in the state, which were supplied by long-term (50-year) contracts, permanently stopped operations. Much of the current economic activity associated with the Alaska boreal forest is generated by the basic activities of exercising the rights and responsibilities of ownership. These activities include forest inventory, monitoring conditions and trends, wildland fire management, access administration, and permits for use. Those administrative activities will occur under any scenario for the future, although they might need to be intensified under certain conditions that could be caused by climate change.

In Alaska, there are so few roads that timber removal generally must meet the costs of building or extending surface transportation routes. Much of the productive forest is distributed in scattered small stands across large landscapes. As a result, productive timberland is, with a few exceptions, not economically accessible. If low-cost forms of access (e.g., winter roads on frozen ground) can be used, the area of forest with positive stumpage value increases. Climate change will decrease the amount of time when winter access is safe on ice bridges and frozen winter roads (section 16.3.6, Fig. 16.21). Even a small amount of additional warming would initiate permafrost thawing across a sizeable portion of the interior boreal region (sections 6.6.1.3 and 16.2.2.3), severely disrupting the ground surface and causing widespread death of existing forest cover. Over the long term, however, sustained temperature increases are likely to expand the productive forest area significantly as long as available moisture does not become limiting (Juday et al., 1998).

The contributions of Alaska forests in providing subsistence food, fuel, building materials, and indirect ecosystem services, generally not measured by dollar flows, are very important in Alaska and probably exceed values from commercial timber operations in most of its boreal region. Alaska forests contribute ecosystem services especially important to the cash economy by providing commercial fisheries, sport hunting and fishing, and values of non-consumptive uses of the forest involving tourism, recreation, and enhancement of the quality of life. Scientific research is one of the most important current uses of the Alaska boreal forest, with climate change effects and carbon-cycle investigations major topics of continuing interest.

Over one million nonresident tourists visit Alaska annually, and the number of tourist visits has increased steadily over the last few decades (Northern Economics Inc., 2004). Forests create specific scenic resources for major segments of the tourism industry, including cruise ships and state ferry routes in Prince William Sound, and near the rights-of-way of the Alaska state highway system and the Alaska railroad. Beginning in the 1990s, large-scale insect-caused tree death and injury related to temperature increases was observed along several of the most popular and heavily traveled tourist routes. Recreation continues in those areas, but the quality of the experience for some visitors has been reduced.

If the assumption is made that demand and prices will rise and the public will allow expanded timber cutting under certain circumstances, the size and economic value of the Alaska forest products sector are likely to increase, and thus the risk of harm to this sector from climate change is likely to be greater. If the assumption is made that access will continue to be expensive (prohibitive), that manufacturing cost disadvantages in Alaska will persist, and that public attitudes about expanded timber cutting on the public lands are negative, the size and economic value of the Alaska forest products sector are likely to remain at current (historically depressed) levels or even decline further, limiting the magnitude of exposure to risk from climate change (Berman et al., 1999).

Much of recent Alaska forest management has been described as "opportunistic" (i.e., taking advantage of other events or projects to build programs and capabilities). Nevertheless, as inventory data accumulated, publicly built access systems expanded, and new scientific insights and data handling tools became available in Alaska, professional forest managers anticipated they would be able to set and accomplish goals more systematically. However, the increased uncertainty associated with climate change that has already been experienced in Alaska makes long-term forest planning and management considerably more difficult. It is already unclear whether there is a reasonable probability that the cost of planting or regenerating certain stands in certain regions can be recovered in future forest harvests because of higher risk from climate-triggered insect outbreaks (section 14.8.2) among other climate change effects (drying lakes, shrinking glaciers, and large burned areas) already obvious on the landscape.

14.3.4. Fennoscandia

Finland, Sweden, and Norway have certain features in common. These nations extend from at least 60° N (or further south) to north of the Arctic Circle. Across that distance, these nations encompass north-temperate deciduous or transition forest and an entire gradient of the boreal forest to treeline and tundra in the north. Despite its northerly location, the climate of the region is the warmest of equivalent latitudes in the circumpolar world, due to the strong influence of the Atlantic Gulf Stream. Annual precipitation varies from 1500 mm or greater in western Norway, which receives the strongest Gulf Stream influence, to 300 mm in the northeastern portions of Finland. The growing season lasts 240 days in the south and 100 days in the north. Mean temperatures range from 14 to 17 °C in July to between -14 and 1 °C in January and February. The combination of relatively mild temperatures and a deep, dependable snowpack that insulates the ground even in the coldest northern areas means that permafrost is practically absent in forest regions.

Most of Sweden and Finland is characterized by relatively even topography and is less than 300 m above sea level. In the west, mainly in Norway, the Scandinavian mountain range reaches elevations of 1000 to 2000 m above sea level. These mountain peaks are not forest-covered. The treeline varies from 700 m above sea level in the southern part of the mountains to 400 m in the northern part.

In northern areas of Fennoscandia (and the adjacent Kola region of Russia), the Saami (Lapp) people pursue reindeer husbandry in forest lands on the basis of ancient rights. The Saami are legally entitled to use lands that belong to others in order to feed and protect their reindeer herds. Saami earn their living from reindeer breeding, and reindeer move through different areas throughout the year. The forest owners affected by the herds are under a mandate to cooperate with Saami communities to ensure that reindeer can obtain their life-cycle requirements. However, because these rights are exercised across four national jurisdictions, legal systems, and boundary controls, traditional flexibility is hindered (section 12.2.5.2). This jurisdictional complexity works against adaptability of the herding system that would be desirable in response to climate change.

Due to the generally low buffering capacity of forest soils in the Fennoscandian region, a high level of air pollution, mainly originating abroad, has resulted in widespread soil acidification. Leaching of mineral nutrients has reduced soil-buffering levels by as much as half in recent decades. High levels of acid deposition (both sulfur oxides and nitrogen oxides) represent future risks to the growth and health of forest ecosystems. Overall, however, the measured rates of growth of managed and tended stands have increased considerably, which might be due at least partially to the nitrogen input.

14.3.4.1. Finland

Finland is the most heavily forested country in Europe, with 16 times more forest per capita than the European average. Forests as defined by the FAO cover 23 million ha or 74.2% of the land area. Finnish forests have been intensively harvested over the last few decades. Despite an active harvest program and the reduction of national territory after the Second World War (territory Finland ceded to the Soviet Union contained over 12% of its forest area and about 20% of the best saw-timber stands), Finnish reserves of wood volume are now greater than during the 20th century, and continue to increase. Today the annual aggregate wood growth in Finland is about 75 million cubic meters, while around 60 million cubic meters or less are harvested or die of natural causes. Of the total logged area, regeneration felling accounts for roughly one-third and thinning two-thirds.

In the boreal region, the most common trees are Scots pine (*Pinus sylvestris*), Norway spruce (*Picea abies*), silver birch (*Betula pendula*), and downy birch (*B. pubescens*). Usually two or three tree species dominate a given stand. Pure pine stands occur in rocky terrain, on top of arid eskers, and in pine swamps. Natural spruce stands occur on richer soil. Birch is commonly found as an admixture, but can occasionally form pure stands. About half of the forest area consists of mixed stands. Various kinds of peatlands are also an important part of the Finnish landscape. Originally, peatlands covered about one-third of Finland. Many peat areas have been drained for farming, forestry, and peat extraction purposes. Bog drainage reached a peak in the 1970s, when nearly 1% of the total land area of the country was being drained each year. About half of the original peatland area has been preserved in its virgin state.

Nearly all of the Finnish forest is intensively managed. In the 20th century, foresters favored conifers, especially pine, at the expense of other species. As the oldest generations of trees have been felled, the average tree age in the forest has become younger. Forest stand treatments and forest roads have fragmented large contiguous wilderness areas, and forest fires are largely prevented. Managed commercial forests of this kind now cover more than 90% of the productive forest land in Finland.

About 40% of all endangered species in Finland are dependent on older age classes of natural forest (Parviainen, 1994). Any climate change effects (intensified physiological stress, fire) that disproportionately affect the remnant of older forest are virtually certain to have large negative impacts on the survival of the rare elements of biodiversity.

With the implementation of new forest management and protected area programs, the contribution of forest harvest and management to species endangerment will diminish. For the moment, however, forestry is still a greater threat to the preservation of species than any other human activity (Essen et al., 1992). One reason for this is that about half of the plant, animal, and fungus species found in the country live in forests. Of the various factors attributable to forestry that reduce biodiversity, the most important is the rarity of larger, old trees – both live and dead (Essen et al., 1992). Specialist organisms such as arboreal lichens, wood-rotting fungi, and mosses and invertebrates have declined with the decrease in large old trees, snags, and logs.

Finland has 35 national parks that cover 8150 km^2. Together with other nature reserves, the total protected area amounts to approximately 29 000 km^2 or about 9% of the total land area of Finland. The Finnish government adopted a special protection program for old-growth forests in 1996 and for broadleaf woodland in the late 1980s. There are ten programs designed to protect various types of natural features and areas. The goal is to extend protection to over 10% of the national territory. As of March 2003, about 3.57 million ha of land to be protected had been identified. About 2.7 million ha had been included in the programs by the beginning of 1999. Protecting habitats either totally or partially from human activities is the main strategy to recover endangered species and encourage biodiversity. In the case of forests, this mainly means protecting the remaining old-growth stands of both conifer and broadleaf forest growing on rich soil as public reserves, as these habitats have been declining most rapidly. There are elaborate arrangements for consultation with landowners during the planning stages of protection programs. This process of consultation was followed when the areas for inclusion in the Finnish Natura 2000 scheme were being designated. The largest category of protected areas is wilderness in Lapland. Wilderness areas, established by the Finnish Parliament, cover about 15 000 km^2 of forest, treeline, and tundra habitats. Although strictly limited, some forest cutting is permitted in some parts of these wilderness areas.

Most of the highly productive forests in Finland, especially those in the southern and central parts of the country, are privately owned. Nearly 60% of the Finnish forest is privately owned and one in five people belongs to a forest-owning family, including a broad spectrum of the population. The average size of a private forest holding is 30 ha. Most of the forest in

Lapland is publicly owned, and as a result, the northern forests have been managed more uniformly and on a larger scale than those elsewhere.

Until the mid- to late 1990s, one-third of Finnish export earnings came from forests, and the forest products industry as a whole was second only to metal products as an export sector. The economic activity (turnover) of the forest sector was roughly Eur$ 23.5 billion: management activities contributed Eur$ 1.7 billion, the forest industry Eur$ 17 billion, and machines and other equipment EurR$ 5 billion. The Finnish share of the total value of global export trade in forest products was about 7.6%. Forest management and the forest industry employed about 100 000 people while the rest of the sector employed about 50 000 people. About 100 000 timber sales deals are made every year between forest owners and forest industry companies.

An environmental program for forestry was adopted in 1994 based on forest-related principles approved at UNCED in 1992 and the 1993 general principles for sustainable forestry adopted by European forest ministers in Helsinki. The goal of Finnish forestry has become ensuring a sustainable economic return as well as preserving biodiversity and facilitating multiple use of the forests. Management practices that strongly alter the environment, such as bog drainage, deep plowing of forest soil, and use of herbicides to kill undergrowth, have been almost totally abandoned. Habitats important to biodiversity preservation have been excluded from timber operations, and both living and dead trees are now routinely left in cutting areas to support biodiversity. Natural regeneration of trees has increased in importance relative to planting nursery-grown seedlings and more attention is paid to preserving forest landscapes.

The Government of Finland adopted a National Forest Program (NFP) in March 1999. The goals of the NFP are to increase the industry's annual consumption of domestic wood by 5 to 10 million cubic meters by 2010, double the wood processing industry's export value, and increase the annual use of wood for energy to 5 million cubic meters. The Finnish government, in collaboration with the forest industry, will also ensure competitive conditions for the forest industry (e.g., supplying energy at a competitive price) and launch the technology and development programs needed for promoting the wood processing industry and wood-based energy production. In addition to wood product utilization, the NFP recognizes and promotes other forest uses, including hunting, reindeer husbandry, wild mushroom and berry picking, scenic and cultural values, recreation, and tourism.

The forest sector in Finland also makes a unique contribution toward the goal of zero net emissions of CO_2. At present, about 20% of the total energy production in Finland is based on wood, which is substantial compared to the global average. The forest industry produces about 80% of the wood energy by burning black liquor, a by-product of pulp mills, and sawdust and chips from the wood processing industry. Pulp mills are completely self-sustaining in terms of energy and are even able to supply other plants with energy. Households and small heating plants produce about 20% of the wood energy. They use primarily small dimension wood from thinning, chips made out of logging waste, and building waste.

14.3.4.2. Sweden

Sweden is a heavily forested nation. Closed forest and forest plantations cover slightly more than 60% of the nation or about 27.1 million ha. Forested shrubland, which is especially common in the transition between closed forest and tundra, represents another 3 million ha or 6.6% of Sweden. Inland waters make up almost 10% of the country, and the non-forest land base is about 23% (FAOSTAT, 2000). Sweden has 25 national parks that together cover more than 0.6 million ha, or 1.5% of the area of the country. Sweden also has 1563 nature reserves covering 2.6 million ha, or around 5.5% of the area of the country. There are also other nature reserves and protected areas that are increasingly provided by private forest owners, including individuals and large forest companies.

One striking feature in Sweden compared to other timber-producing countries is that the national government owns only 5% of the productive forest lands. In 1993, most government-owned timberland was transferred to a forest product corporation – AssiDomän – of which the national government owns 51% of the shares and the remaining 49% are traded on the stock exchange. Private individuals (families), owning approximately half of Swedish forests, are the largest single category of forest owners. Forest companies own approximately 3% and other owners account for 12%. Forest industry land holdings are concentrated in central Sweden and some portions of Norrland (the northern three-fifths of Sweden), where the industry also operates many large, modern production facilities. Swedish forest companies have globalized their operations and established themselves firmly in the European countries and on other continents. At present, some 8 million ha of forest land have been certified in Sweden. Certification is a process of formal evaluation and recognition by third-party evaluators that forest products have been produced using sustainable management practices, and can be advertised to consumers as such. Forest companies account for most of the certified forest land.

Sweden has the highest population (8.8 million) of the Nordic countries, but is still largely a sparsely populated country, especially in the north. Despite the relatively low population density, Swedish forests show the effects of many centuries of human use. Only in the northern interior are the forests less affected by humans. Today much of the northern Swedish forest is protected, either as nature reserves or by other means.

The average size of a private forest is about 50 ha. Until the Second World War, most private forest owners were farmers who lived on their property and were engaged in crop cultivation as well as harvesting wood products. Since then, the area of forest land devoted to this type of combined agriculture and forestry enterprise has dropped from more than 9 million to less than 4 million ha. Today many individual forest owners do not live on their forest land, but often in communities close to it or in more distant cities and towns. The bulk of forestry work on their properties is now performed by employees of forest owners associations or by other contractors.

Early in the 19th century, Sweden began the process of industrialization by rapidly expanding its sawmill industry. The major Norrland rivers were suitable for floating timber to the Baltic Sea, thereby opening up previously untouched inland forests to large-scale logging. The sawmill companies, which purchased very large tracts of land from farmers until 1905, soon gained a strong position in European timber export markets. Northern Sweden was very sparsely populated until the beginning of the 19th century. At that time, an accelerating agrarian colonization took place, and later in the century, large-scale exploitation of the virgin forest began. The first forest resources to be exploited were those close to the Bothnian coast, and subsequent exploitation moved further inland. This exploitation affected almost all forest land in northern Sweden between 1850 and 1950.

The boreal forest landscape in Sweden has changed dramatically during the last 150 years. Owing to the vigorous pursuit of forest management goals over the last 100 years, widespread stands of commercially tended Scots pine of generally similar age and structure make up most of the northern Swedish landscape (Essen et al., 1992). Less than 3% of the forest area of Sweden supports trees older than 160 years, and less than 2% of the conifer volume is in trees with diameters greater than 45 cm (Essen, 1992). The volume of standing dead trees in Sweden has decreased by more than 90% in the last century and the number of large-diameter conifers (>30 cm diameter at breast height) has decreased by more than 80% (Linder and Ostland, 1992). Many old-growth specialist species, including arboreal lichens, mosses, insects, wood-rotting fungi, and cavity-nesting insectivorous birds, have become rare in boreal Sweden (Essen, 1992). All the taxonomic groups that are represented in the 1487 threatened forest species in Sweden include elements that are largely dependent on old forests or the habitat elements of old forests (Berg A. et al., 1994). Many of these species belong to functional groups, such as insect predators, that perform vital functions for the health and productivity of the forest system as a whole. As a result, the northern Swedish forest is particularly vulnerable to any climate change effects that would accelerate tree mortality in older stands.

About 20% of the energy consumption in the country originates from forest biomass, and there is presently a significant annual accumulation of carbon in standing biomass, equal to approximately 30% of Swedish fossil fuel emissions. The high level of air pollution originating outside the country and the accelerated leaching of mineral nutrients in Sweden has resulted in widespread soil acidification. Forest-damaging amounts of both sulfur dioxide and nitrogen oxides are being deposited. The accelerated leaching of mineral nutrients caused by acid deposition has reduced nutrient levels by half on some sites in recent decades, although nitrogen deposition increases supplies of a critical nutrient. Air pollution poses a serious threat to the health and growth of Swedish forest ecosystems. In general, researchers believe that multiple stresses may act synergistically, resulting in acceleration of change due to other stresses (Oppenheimer, 1989). Climate change, therefore, must be understood to be acting in this context in Sweden, as one among many factors.

The present-day timber stock is 50% larger than it was when detailed measurements began in the 1920s. However, for much of the early and mid-20th century, forests in Sweden were exploited in ways that today, with increased knowledge and insights, are understood to be detrimental to the environment. Clear-cutting was extensive, and little effort was made to protect biodiversity. Since the 1992 UNCED meeting, the concept of sustainability has broadened. This is reflected in the revised Swedish forestry policy that went into effect in 1994. The 1994 Forestry Law states: "The forest is a national resource which should be managed so that it provides a good return on a sustainable basis and ensures the preservation of biodiversity". These two goals have equal priority (Lamas and Fries, 1995). Underlying this policy is the conviction that there will continue to be a demand for renewable products in the future and that Swedish forests can remain an important source of natural raw material produced using principles based on ecological cycles. Swedish forest legislation protects key woodland habitats, forests located near high mountains, and wetland forests. Special regulations govern four million ha of low-productivity woodlands that are not included in the productive forest land, allowing only careful low-intensity utilization and ensuring that their basic character remains unchanged. Hydrologic and other possible changes resulting from climate change represent potential risks for some of these protected features.

Swedish forest policy also states that forests should be able to sustain hunting and the gathering of wild mushrooms and berries as well as active silviculture. The "right of common access" is traditional in Sweden. People are entitled to hike through the natural landscape and to pick mushrooms and berries regardless of who owns the land. This tradition broadens the pool of resources and users with specific concerns about potential climate change impacts beyond those of timber and a comparatively small number of forest-land owners. Hunting and fishing remain widespread and significant activities in Sweden, and are considered in forest management since they are connected to key management practices. Sports such as orienteering, cross-country skiing, and other popular outdoor activities that take place in the forests involve a relatively large proportion of people in Sweden. Participants in these sports have a strong interest in the health of the forests and are an important forest and climate change constituency even though they do not remove products from the forest.

14.3.4.3. Norway

Forests and other wooded land cover about 37%, or 11.9 million ha, of the Norwegian mainland. Nearly 23%, or 7.2 million ha, is regarded as productive forest. The productive forest is distributed among 125 000 forest properties, and about 79% is owned by private individuals.

The total area of wilderness territory in Norway has been greatly reduced over the past 100 years. Wilderness territory is defined as areas more than 5 km from roads, railways, regulated watercourses, power lines, and vehicle tracks. Although the largest 20th-century reduction in wilderness occurred in the lowlands of southern Norway, where wilderness is now virtually eliminated, mountainous areas and northern Norway have experienced major wilderness reductions and fragmentation as well. Today, wilderness represents 12% of the total land area of Norway (not including Svalbard and Jan Mayen). Approximately 6.4% of mainland Norway has official protected area status. New protection plans, especially for additional national parks, are being developed, and about 15% of the land base will be within protected areas by the year 2010. Both public and private lands that lie within or adjacent to designated national parks are protected from construction, pollution, or any other encroachment.

Norwegian forests have been exploited intensively for the export of roundwood, sawn timber, and wood tar for hundreds of years. In addition, there is a long tradition of using the forests for domestic animal grazing and game hunting. For many years, the amount of wood removed from Norwegian forests has been less than the amount of growth. The biomass of standing trees has almost doubled since 1925, and the volume of standing forest increased by more than 95% from 1925 to 1994, to about 616 million cubic meters. In 1994, the net increase (increment or growth minus removal) in forest mass was 9.5 million cubic meters, or 1.5% of the total volume of standing forest. The 1994 volume of Norwegian forest consisted of 46% spruce, 33% pine, and 22% deciduous trees. The net volume increase was greatest for deciduous trees and pine. Because of their increase in total biomass, Norwegian forests have contributed to a net transfer of CO_2 from the atmosphere into storage during the 20th century. It is estimated that in 1994, the net amount of CO_2 assimilated by forest was 15 Tg, amounting to about 40% of Norway's annual anthropogenic emissions.

The increase in wood volume can be attributed to many factors. A sustained effort at intensive forest management has systematically removed older stands that have low rates of net growth, and increased the proportion

of stands in the early and most rapid stages of growth. An extensive afforestation effort has established forests on sites such as formerly open wetlands that previously supported no (or only minor) forest cover. Natural forest vegetation has returned to uncultivated land that is no longer farmed, producing a gradual buildup of vegetation mass. Long-range transport of nitrogen in precipitation has had a fertilizing effect on Norwegian forests. Finally, the introduction of new species that grow faster than native species on some sites has also led to greater forest growth. It should also be noted that the particularly large increase reflected in the inventory data for the most recent years is probably due in part to new methods of calculation.

Today the forest is used first and foremost as a source of raw materials for sawmills and the pulp and paper industries. Government grants or subsidies for forest planting began as early as 1863 in Norway. In recent years, forest planting has been somewhat reduced to between 200 and 300 km^2 of forest planted or sown annually. Norwegian plantations are often spruce monocultures. The diversity and abundance of the fauna and flora in planted spruce forest is much lower than that in naturally regenerated forest.

Although the volume of forest has increased considerably since the beginning of the 20th century, the present-day forest has been transformed from its original condition. Clear-cutting, plantation establishment, introduction of alien species, ditching, and forest road construction are some of the intensive forest management measures that have been applied. Acidification and pollution have also affected forests. As in the other Nordic countries, management practices applied in order to produce increased tree biomass have caused a reduction in biological diversity. Much of the present forest growth is concentrated in large monocultures. Large stands of the same age class contribute to a reduction in the number of species of flora and fauna compared to what would occur in a more natural, mixed type of vegetation.

Forestry and the forest industry are important sectors of the Norwegian economy. Wood and forest products represent about 11% of Norwegian non-oil exports. This is slightly less than the export value of the Norwegian fishing industry, somewhat higher than both aluminum and natural gas export values, and twice the value of Norwegian high-technology exports. About 30 000 people receive their income from primary forestry and the forest industry. Most of the economic activity in the forestry sector takes place in rural districts and is particularly important in these areas that have fewer economic alternatives than urban zones. Norway began exporting large quantities of sawn lumber in the 16th century and continued this practice for several centuries. However, since the Second World War, most of the sawn lumber has been utilized nationally. Most homes in Norway are constructed of wood, with wooden interior fittings, and wood is an important part of the everyday life of Norwegians. In 1995, the wood processing industry used 5.1 million cubic meters of wood and employed about 16 000 people.

Sawmills use about 50% of the Norwegian roundwood harvested. There are 225 sawmills in Norway operating on an industrial scale. Paper products have the highest export values of all the forest-based products. Paper and board products are currently produced by 36 different machines in Norway. Wood pulp and chemical pulp are produced by 17 production units. Every year, Norway exports about one million tonnes of newsprint. The pulp and paper industry employs 9000 people.

Until a century ago, wood was the dominant energy resource in Norway. Oil and hydropower are currently the major energy sources, and fuelwood use is only 7% of the volume used 100 years ago. However, the forest may play an increasing role as an energy supply. Shortages of electricity and CO_2 taxes on the use of oil have increased the interest in bio-energy. Today, the pulp and paper industry is by far the largest producer of bio-energy in Norway.

14.3.4.4. Iceland and Greenland

As a geologically young landform situated along the mid-Atlantic Ridge, isolated from both Europe and North America, Iceland has always supported ecosystems with a restricted set of species (only about 483 native and naturalized vascular species; Icelandic Institute of Natural History, 2001). Iceland is also the area of the Arctic with the most substantial human impact on ecosystems. At the time of Viking settlement in the year 874, vegetation was estimated to cover about 65% of Iceland, while vegetation covers only about 25% of the island today (Blöndal, 1993). This reduction in vegetative cover is most likely the result of intensive land and resource utilization by a farming and agrarian society over 11 centuries, although recent investigations also suggest that specific years of low temperature and low precipitation possibly initiated devegetated patches under a traditional winter grazing system that was abandoned in the early 20th century (Hellden and Olafsdottir, 1999). Estimates vary as to the percentage of the island originally covered with forest and woody vegetation at settlement, but a range of 25 to 30% is plausible (Blöndal, 1993). The native forest of Iceland was principally comprised of downy birch and a few tall willow species with a relatively large proportion of the woody stems in a dwarfed and somewhat contorted growth form. Rowan (*Sorbus aucuparia*) and European aspen (*Populus tremula*) occasionally occur among the birches. Tea-leaved willow (*Salix phylicifolia*), hairy willow (*Salix lanata*), and creeping juniper (*Juniperus communis*) are common shrubs in birch woodlands. Ancient native birch woodlands and scrub are key areas of terrestrial biodiversity in Iceland. The birch woodlands still retain most of their original biodiversity and are a key habitat for a number of threatened and endangered species, while the deforested land has suffered degradation and in some cases has been reduced to deserts.

When Norse farmers, traders, warriors (Vikings), and seafarers arrived in Iceland from western and/or northern Scandinavia, Ireland, and Britain in the 870s and 880s, the land was uninhabited and vegetated from the seashore to all but the glaciated mountains. At higher altitudes, the woodlands gave way to dwarf birch (*Betula nana*), willow, grasses, and moss. Birch woodlands separated by wetlands dominated the lowland vegetation. The lowlands were the only habitable land areas and were occupied in the first decade or two of settlement. Most of the first settlers were farmers that brought livestock to the island. Forests were cleared rapidly by burning and harvest for building materials and charcoal manufacture, generally within 50 years of the arrival of the settlers. Nearly all of the habitable woodlands were converted to pastures and only the mountainsides and highland margins retained forests, which survived to the early 19th century.

In southern and western Iceland, the land was cleared and barley was grown. Domestic livestock were sustained at numbers that caused landscape degradation, soil loss, and reduced long-term productivity of pastures. A pollen record from Iceland (Hallsdóttir, 1987) confirms the rapid decline of birch and the expansion of grasses between AD 870 and 900, a trend that continued to the present. As early as AD 1100, more than 90% of the original Icelandic forest was gone and by 1700, about 40% of the soils had been washed or blown away. Vast gravel-covered plains were created where there was once vegetated land.

The exploration of Greenland in 984 by Eric the Red gave the more adventurous Icelanders new farm sites and freedom from limits on establishing themselves in an occupied landscape. The small areas of woodlands in Greenland, dominated by birch and willow in the south and alder (*Alnus* spp.) in the valleys of the Western Settlement, were highly valued. The Greenland Norse were also farmers, although they supplemented their diets with seal, birds, and reindeer/caribou. For several centuries the Norse remained committed to their farming way of life but gradually experienced various stresses (land degradation, trade and genetic isolation, social and economic challenges), many of which were aggravated by a cooling climate. The Norse settlements in Greenland and Iceland were established shortly after the beginning of a period of relatively warm and stable climate that persisted from the 8th through the 12th centuries (Ogilvie and McGovern, 2000). With the gradual degradation of the landscape that reduced its capacity to support livestock and crops, and during a severe cooling period between 1343 and 1362, the Norse presence in the Western Settlement of Greenland ended and the Norse population disappeared. The mid-14th century cold interval in Greenland was probably the period of the lowest temperature in the last 700 years, and the effect of this climate change on the Norse community is highly likely to have been an important factor in its terminal decline (Ogilvie and McGovern, 2000). Simultaneously, arctic-adapted

Thule people arrived for the first time as the culmination of their historic expansion eastward and may have played a role in displacing the indigenous Norse people who first inhabited the area.

Until quite recently, efforts to conserve and increase the area covered by native woodlands in Iceland were hindered because the majority of woodlands were subject to grazing, mainly by sheep: the total area covered by birch woodlands does not appear to have changed significantly in recent decades. However, the establishment of the Icelandic Soil Conservation Service in 1907 was instrumental in fencing off and halting the most severe erosion processes in lowland areas of southern and northeastern Iceland. This is a unique context in which to consider climate change effects in the Arctic – humans have damaged soils and eliminated so much of the native forest ecosystems in Iceland that climate change represents an opportunity for forest recovery and expansion.

Starting in 1907 but especially since the mid-20th century, efforts were made to establish forest plantations in Iceland. Plantations now total at least 20000 ha (Gunnarsson K., 1999). Until the 1980s, the lands most suitable (from the standpoint of soils and climate) for re-establishing native birch forests remained committed to farming and winter fodder production for sheep and horses. However, since Iceland joined the European Economic Area and the World Trade Organization, Icelandic agriculture has been integrated into a trading and market zone in which many of its products fall under production quotas or price disadvantages. Farms are going out of traditional agricultural production, while other forms of land use, such as afforestation and horse farming, are increasing.

About 130 tree species from most of the cold regions of the world have been evaluated for their growth and survival characteristics (Blöndal, 1993). Different types of forest are being established for particular purposes. Native birch forest is often favored in recreation areas near urban centers and private land surrounding weekend recreation cottages in the country. Usually, land that is being converted to native birch forest must be fenced to exclude sheep. Protection forests are designed to prevent erosion of exposed soil, which is still quite common. Species commonly used in protection forests include downy birch, Sukachev larch (*Larix sukaczewii*), Siberian larch (*L. sibirica*), lodgepole pine (*Pinus contorta*), willows, and alder. A program to afforest denuded or severely eroded land has resulted in the planting of about one million trees in 70 areas subject to soil erosion. Production forests are designed to produce timber, fuelwood, and Christmas trees. Planted species include Sitka spruce (*Picea sitchensis*), Sukachev and Siberian larch, lodgepole pine, and black cottonwood (*Populus trichocarpa*). State-supported afforestation projects were initiated in recent years in all parts of Iceland, starting in eastern Iceland in 1991. The aims of the projects are to create jobs in the rural areas

(e.g., in nurseries) and to support jobs on those farmsteads taking part in the project. Farmers who participate in the project can receive a state grant of 97% of the total cost of afforestation on their farmlands (Gunnarsson K., 1999).

Until the 1970s, most plantations were established within the existing birch woodland, but that is no longer the case. Approximately 2000 ha of mainly non-native tree species have been planted annually in Iceland since 2003. In the late 1980s, forest and woodland vegetation covered only about 1.33% of Iceland, amounting to about 117000 ha of native birch woodland and 15000 to 20000 ha of plantations. More than 80% of the woodland remaining in Iceland was less than 5 m in height in the 1980s, and 14.2% of the birch woods have trees taller than 5 m (Sigurdsson and Snorrason, 2000). In 2001, Iceland had 3 national parks totaling 189000 ha and 35 nature reserves totaling 225000 ha (Icelandic Institute of Natural History, 2001).

The main challenge to successful introduction of tree species to Iceland has been finding varieties that can survive and reproduce in an environment of strong, steady winds and growing seasons at the cool margin of tree tolerance. Warmer climate conditions generally increase the planting success and particularly the growth of trees in Iceland. The restructuring of the rural economy to include less production agriculture and more ecotourism, recreation, and amenity-based uses is supporting the trend toward the conversion of the Icelandic landscape from a treeless farming region into one with expanded forest area. The government program to sequester CO_2 has increased afforestation efforts. Most of the tree planting for carbon sequestration is contracted to farmers. A national goal is to expand forest cover on 1% of the national area per decade until forests cover about 5% of Iceland, which would represent the greatest forest extent in a millennium. In 1995, the estimated annual carbon sequestration by Icelandic plantation forests was 65 Gg CO_2, or about 2.9% of the annual national emissions at the time. By 1999, the annual sequestration totaled 127 Gg, representing 4.7% of annual emissions (Sigurdsson and Snorrason, 2000).

Approximately 70% of forests and woodlands in Iceland are privately owned, with the remainder owned by the national government or local municipalities. The Iceland Forest Service, an agency under the Ministry of Agriculture, is primarily responsible for implementing and monitoring government forest policy. The Environment and Food Agency of Iceland, part of the Ministry for the Environment, plays a role in the conservation of protected native birch forests. The main forestry laws in Iceland include the Forestry Law (1955), the Farm Afforestation Law (1991), and the Southland Afforestation Law (1997). In response to a request by the Icelandic parliament, the Iceland Forest Service prepared a comprehensive afforestation plan for all of Iceland as a part of its sustainable development strategy, resulting in the Regional Afforestation Law (1999). The Icelandic Soil Conservation Service, also under the Ministry of Agriculture, is responsible for implementing the Soil Conservation Law (1965), which also contains provisions for vegetation conservation and restoration.

14.4. Use and evaluation of the ACIA scenarios

14.4.1. Method of analysis

This chapter uses the projections from the five ACIA-designated models (section 4.2.7) primarily in the form of numerical output for key climate variables in specific grid cells across the area of analysis. Criteria for selecting the models and assumptions used in generating the scenarios are described in Chapter 4. For the purposes of this chapter, 14 sites were chosen that are broadly distributed across the northern part of the boreal region or are located somewhat south of the Arctic but currently experience climate conditions projected to occur in the Arctic and boreal region by the ACIA-designated models. Most of the locations were chosen because they have a medium- to long-term record of climate conditions. Many are major population, trade, or transportation centers within their respective regions, and the sites are representative of most regions in the far north where agriculture and

Table 14.2. Sites used for evaluation of the ACIA-designated model projections. Numbers in first column are those used in Fig. 14.1. Analyses of the Taymir Peninsula used the mean of the four stations on the peninsula.

	Location	Latitude	Longitude
1	Fairbanks, Alaska, United States	64.82° N	147.52° W
2	Big Delta, Alaska, United States[a]	63.92° N	145.33° W
3	Whitehorse, Yukon Territory, Canada	62.72° N	135.10° W
4	Fort Vermillion, Alberta, Canada	58.40° N	116.00° W
5	Thompson, Manitoba, Canada	55.80° N	97.90° W
6	Goose Bay, Newfoundland, Canada	53.30° N	60.4° W
7	Reykjavik, Iceland	64.13° N	21.56° W
8	Tromsø, Norway	69.83° N	18.55° E
9	Umeå, Sweden	63.8° N	20.30° E
10	Rovaniemi, Finland	66.55° N	25.00° E
11	Archangelsk, Russia	64.5° N	40.42° E
12	Novosibirsk, Russia	55.00° N	82.90° E
13	Yakutsk, Russia	62.00° N	130.00° E
14	Magadan, Russia	59.60° N	150.80° E
Stations on the Taymir Peninsula, Siberia			
15	Dudinka	69.67° N	86.28° E
16	Essej	68.78° N	102.62° E
17	Khatanga	71.97° N	86.78° E
18	Olenek	68.83° N	112.72° E

[a]Analysis of growing degree-days only

forest management are currently (or potentially could be) practiced. The selected locations include areas in the eastern, central, and western portions of both Eurasia and North America (Table 14.2 and Fig. 14.1).

The primary variables obtained from the climate model output were mean monthly temperature, total monthly precipitation, and growing degree-days (GDDs). Mean monthly temperature is the mean of all mean daily temperatures for the month. Mean daily temperature is the mean of the high and low temperature on a given day. Mean monthly temperature integrates much of the short-term variability in weather and yet preserves a specific seasonal signal of warmth or coolness that has proven useful in many applications such as tree-ring studies. Total monthly precipitation reflects the amount and timing of moisture, and is widely used in forestry, agriculture, and hydrological applications. Growing degree-days are a measure of accumulated heat energy, calculated by taking the sum of daily mean temperatures above some defined threshold. For example, if the threshold value for accumulation is 5 °C, then a series of daily means of 2, 6, 7, and 5 °C would accumulate respectively 0, 1, 2, and 0 degrees for a total of 3 GDD.

Output climate variables from the ACIA-designated models were obtained generally for the period 1990 to 2099, which allowed an 8- to 10-year overlap with recorded data (typically 1990 or 1991 through 1998 or 1999). In some cases, recorded data from stations were truncated in 1989 or 1990, in which case model output from 1980 to 2099 was used in order to generate a 9- or 10-year overlap. For each station, the differences between monthly or annual values of the model output of temperature or precipitation representing the entire grid cell and the values measured at the specific recording station were calculated for the overlap period. From these differences, an overall mean difference was calculated and applied as a correction factor to the model output for all years beyond the overlap period. The variance was not adjusted, although some features of scenario versus recorded variance were examined (see section 14.4.4). The adjustment of general circulation model (GCM) output for use at a specific location is an active area of ongoing research (e.g., Hewitson, 2003).

14.4.2. Size and placement of grids

Grid cells in GCMs are large polygons that represent the surface of the earth. In the output from the ACIA-designated models, a specific climate variable is represented by a single value for an entire grid cell. Climate at a specific station location within the grid cell can be different from the value that expresses the entire grid cell for several reasons. It is important to understand what causes some of these differences in order to properly evaluate the scenario output and the ways it is used in this chapter.

The different GCMs use grid cells of different sizes and non-uniform placement. A grid cell that contains a given station location may be represented as a polygon largely north of the station in one model but largely south of the station location in another model, and the temperature values generated by the two models will have a different relationship to the location of the climate station. The topography of the earth is another factor that can cause a difference between a climate station and a grid cell value. In one model, values for high- and low-elevation terrain may be integrated to calculate a single number for the climate parameter, whereas in another model the grid cell may be placed in a way that the surface of the earth is more uniformly high or low.

Mountain terrain often produces sharp gradients in precipitation. As a result, grid cells that include mountains and smaller valleys exhibit particularly large differences in precipitation compared to recording stations that are typically located in low-elevation valleys.

A small grid cell occupying an area along a mean frontal position in one model will show greater climate variability than a larger cell in the same model even if both contain the same recording station. In addition, different models produce different patterns of air-mass mixing, and as a result different patterns of mean frontal positions.

14.4.3. Range of scenarios

The ACIA-designated models project a range of annual and seasonal mean temperatures by the end of the model runs in 2099. The grid cell containing Ari Mas (on the Taymir Peninsula, Russia, and site of the mostly northerly treeline in the world; Jacoby et al., 2000) provides an example of these different temperature projections (Fig. 14.4). The Ari Mas grid-cell data was adjusted as described in section 14.4.1 using the mean annual temperature from four stations (Dudinka, Essej,

Khatanga, and Olenek), which averages about -11.5 °C for the available 20th-century record. The long-term mean (regression line fit) for the CGCM2 output of mean annual temperature adjusted to the stations reaches about -7.2 °C at the end of the scenario period, while the CSM_1.4 output increases to about -2.7 °C by the end of the period (Fig. 14.4). The CGCM2 projects a climate in this grid cell at the end of the century that is within the range of forest–tundra transition conditions, whereas the CSM_1.4 projects mean annual temperatures typical of regions where today some of the more productive examples of the northern boreal forest occur.

It was sometimes possible in the analyses to calibrate the response function of a key outcome, such as tree growth, to climate. Often the different scenarios produced similar results, with differences mainly in short-term (interannual) variability. The graphic display of such results overlaps to such an extent that clear trends are difficult to see. As a result, when the outputs of a climate variable of interest were broadly similar among the five scenarios, the high- and low-end members were used to examine the range of consequences produced by the scenarios. Using the Ari Mas region again as an example, warm-season (May–September) temperatures projected by the ECHAM4/OPYC3 model increase by about 5.7 °C and those projected by CSM_1.4 by about 3.0 °C (Fig. 14.5). The warm-season mean is closely related to the mean radial growth of trees at this location, so the scenarios provide some basis for understanding what the effects of temperature increases at this location might be on tree growth and vigor (sustained rate of growth over time). Note the generally close relationship in Fig. 14.5 between the regional tree-ring growth signal, and the mean May through

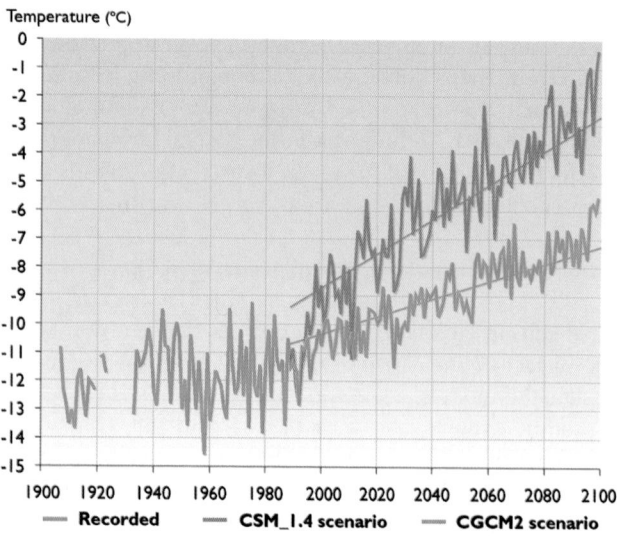

Fig. 14.4. Observations and projections of mean annual temperature for Ari Mas, Taymir Peninsula, Russia. Observations represent the average of four climate stations (Dudinka, Essej, Khatanga, and Olenek) on the Taymir Peninsula; projections from the CSM_1.4 and CGCM2 models have been adjusted using the average mean annual temperatures from these four stations (see section 14.4.1).

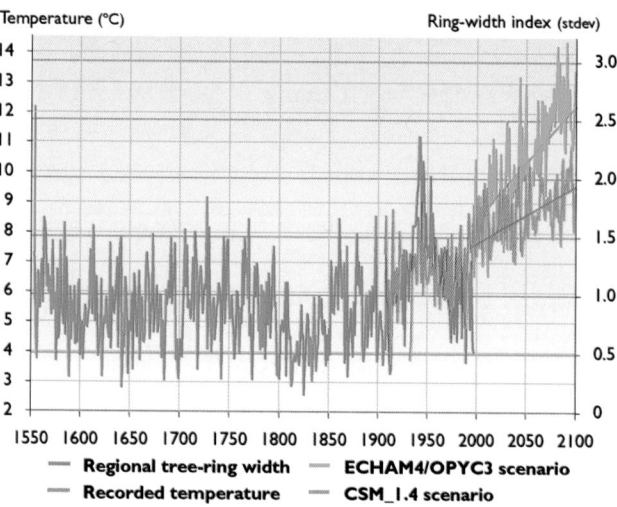

Fig. 14.5. Historic relationship between growth of Siberian larch and warm-season (May–Sep) temperature, and projections of future warm-season temperature increases in the Taymir Peninsula, Russia. Observed temperatures are the average of four stations (Dudinka, Essej, Khatanga, and Olenek); model-projected temperatures are for the entire grid cell and have been adjusted (see section 14.4.1). Tree-ring widths are shown as de-trended and normalized index values (standard deviation from the long-term mean of each measured series; data from Jacoby et al., 2000).

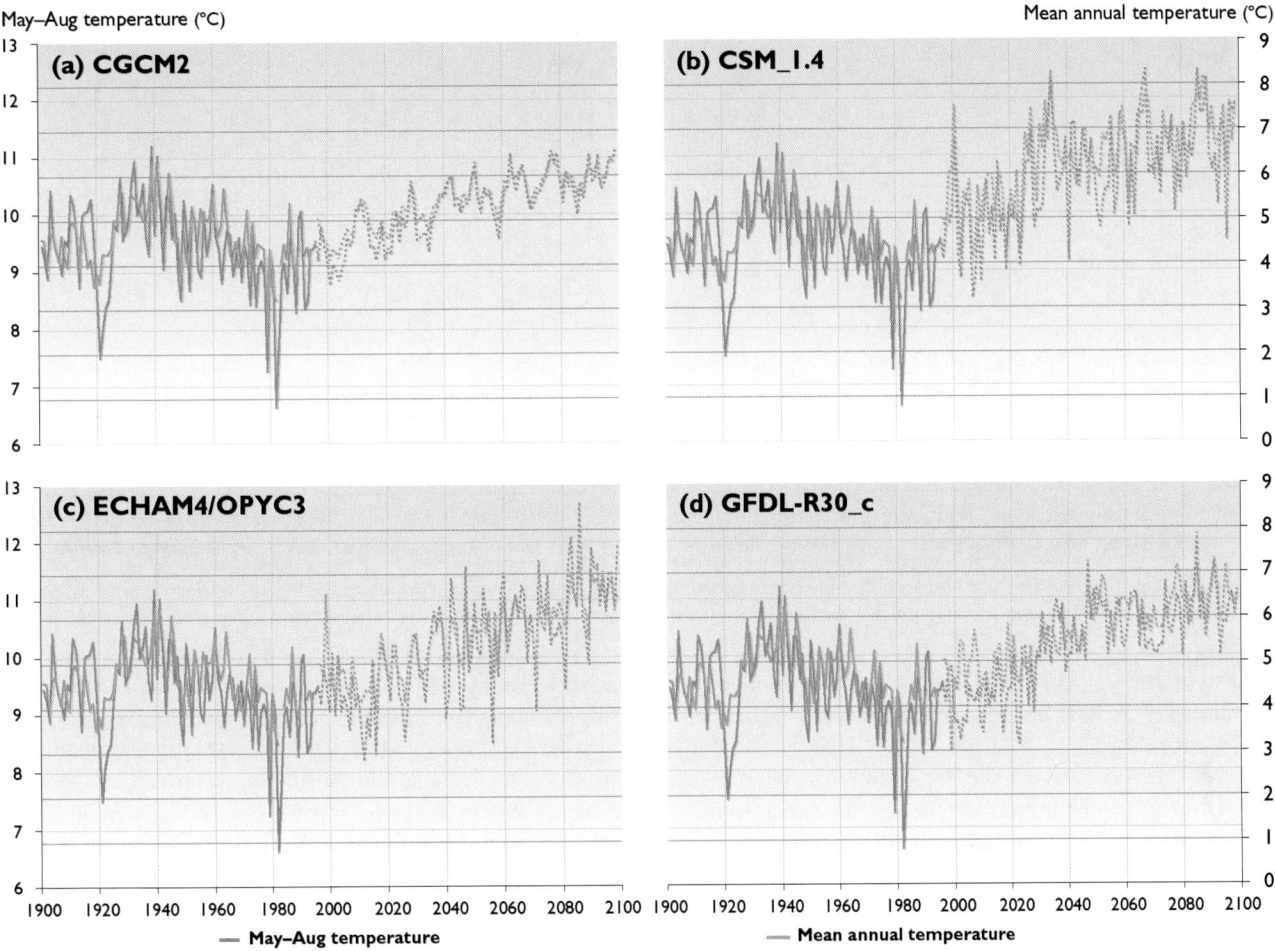

Fig. 14.6. Relationship between summer (May–Aug) and mean annual temperature at Reykjavik, Iceland, for the period of the instrumental record and the ACIA scenario period.

September temperature. Siberian larch is heat-limited and displays a positive growth relationship with temperature. The ECHAM4/OPYC3 scenario for the 21st century projects temperatures that would approximately double the rate of growth and make this marginal site a productive forest. The CSM_1.4 scenario does not project the same degree of warming, but would eliminate periods of severe growth limitation. It should be remembered that the relationship of tree growth to climate may change under an altered climate.

Very few of the examined model outputs showed cooling or no change. The exception in the analysis is Reykjavik, Iceland. The CGCM2 and GFDL-R30_C scenarios for the grid cell that contains Reykjavik project mean annual and summer (May–August) temperatures that generally are not warmer than 20th-century observations, and temperatures at the end of the 21st century that are near the long-term means of the recorded data (Fig. 14.6a,d). Other model scenarios for the Reykjavik grid cell generally show a warming trend (Figs. 14.6b,c), but the magnitude is notably less than most of the other grid cells and stations examined. The four scenarios in Fig. 14.6 produce higher values of both summer and mean annual temperature than were recorded during the mid-20th century temperature peak, but in this highly maritime area none of the scenarios greatly exceed the mid-20th

century temperatures. Note the variability in the degree to which the different scenarios reproduce the historical relationship between the two variables depicted.

14.4.4. Variability and seasonality

In addition to long-term trends and magnitudes of temperature increase, the models produce different features of shorter-term climate variability. Often these short-term climate events or minor trends can be decisive ecological influences. For example, even one or two years of high temperatures can be enough to produce extreme risk of forest fire (Johnson, 1992) or optimal conditions for major outbreaks of insects that kill trees or reduce their growth (section 14.8). Conversely, cold spells during the growth season may harm or kill the trees. A short-term period of favorable weather may trigger production of especially large tree seed crops, with long-lasting ecological consequences (Juday et al., 2003). The stations used for scenario analysis exhibit the opposite temperature relationships mentioned in section 14.2.4 (Fig. 14.7). When summer (June–August) temperatures are high at Fairbanks, Alaska (western interior of North America) they tend to be low at Yakutsk, Russia (eastern interior of Eurasia) (Fig. 14.7a). When mean annual temperatures at Goose Bay, Labrador, Canada (eastern edge of North America) are high, they

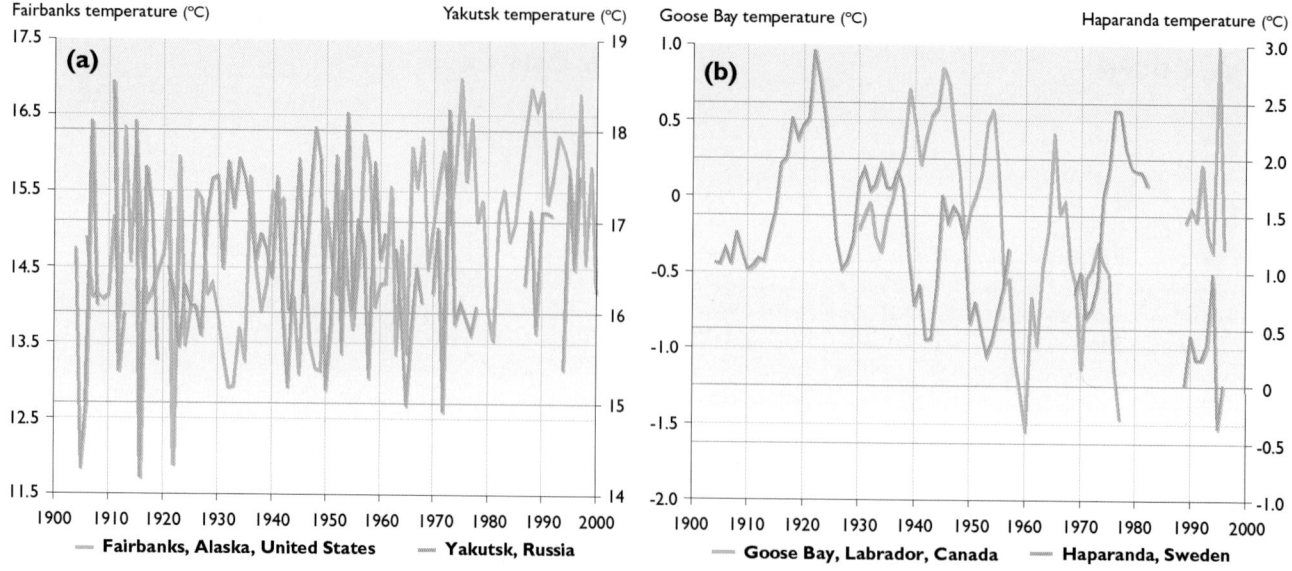

Fig. 14.7. Opposing temperature relationship between stations located in eastern and western portions of the continents in (a) summer (Jun–Aug) and (b) annual (5-yr running mean) (data from the Climatic Research Unit, University of East Anglia, United Kingdom, 2003).

tend to be low at Haparanda, Sweden (western edge of Eurasia) (Fig. 14.7b).

Because these features of climate variability are so important in producing ecological effects, it is important to understand the behavior of the models with respect to key features of short- and medium-term variability and seasonality.

Most of the ACIA-designated model outputs for the grid cells examined reproduce something close to the historical record of the difference between summer and mean annual temperature (Figs. 14.6 and 14.8a,c,d,e). However, sometimes scenario outputs show a different relationship between summer (May–August) and mean annual temperature than do the recorded data (Fig. 14.8b). All the model scenarios shown in Fig. 14.8 project temperature increases that are likely to substantially alter the moderately temperature-limited forest climate typical of the 20th century. The models project temperature ranges that would surpass thresholds for key factors such as outbreaks of insects that attack trees, new species, and altered ecosystem functions such as growth, fire, and decomposition.

The CGCM2 projections for the grid cell containing Fairbanks, Alaska, do not reproduce the same relationship between temperature trends in the coldest winter months and mean annual temperature as the recorded data (Fig. 14.9), and this disparity widens with time. It is possible that climate change would actually produce such an effect, but not all the scenarios agree on such features and realistic projections of effects on forests and agriculture require the resolution of such discrepancies. Different models have different strengths and weaknesses and all should be applied with an awareness of their limitations, circumstances in which they perform well, and novelties that could affect assessments of key climate-dependent processes. Differences in the relationship between seasonal and

annual climate variables can have particular ecological importance. For example, a higher growing-season temperature may increase the growth of trees, leading to one set of effects, but an increasing mean annual temperature may promote thawing of permafrost, which could undermine or destroy the soil rooting zone for trees growing on such sites. Thus, different rates of temperature increase (annual versus seasonal) may produce quite different ecological effects.

14.4.5. "Surprises" in climate change effects

As with any use of models to project the future, a note of caution must be introduced. The projections of climate change and impacts on forests, land management, and agriculture are only as good as the climate models upon which they are based. It is also important to understand that the consequences of change introduced into ecosystems often include elements of chance and contingency that have very real and long-lasting consequences. Even though ecological events *could* unfold in a variety of ways under a given climate scenario, when events *do* take a certain pathway of cause and effect, this closes off another set of outcomes from that point onward. These elements of chance and contingency make it difficult to project ecosystem responses. However, other elements can assist in the projection of ecosystem responses. Including large areas in the analysis allows processes that have a variety of potential outcomes (e.g., the number of trees of a certain species reproducing successfully after a fire) to occur repeatedly. Any process that expands the pool of possible outcomes (either simulated outcomes or those that can be measured in ecosystems) of the climate change "experiment" will then cause the results or outcomes to begin to approach some distribution that may be described. Repeated outcomes can be produced not only as the result of multiple examples across space, but as the result of multiple outcomes across time in the same place. As a result, when larger areas and longer periods

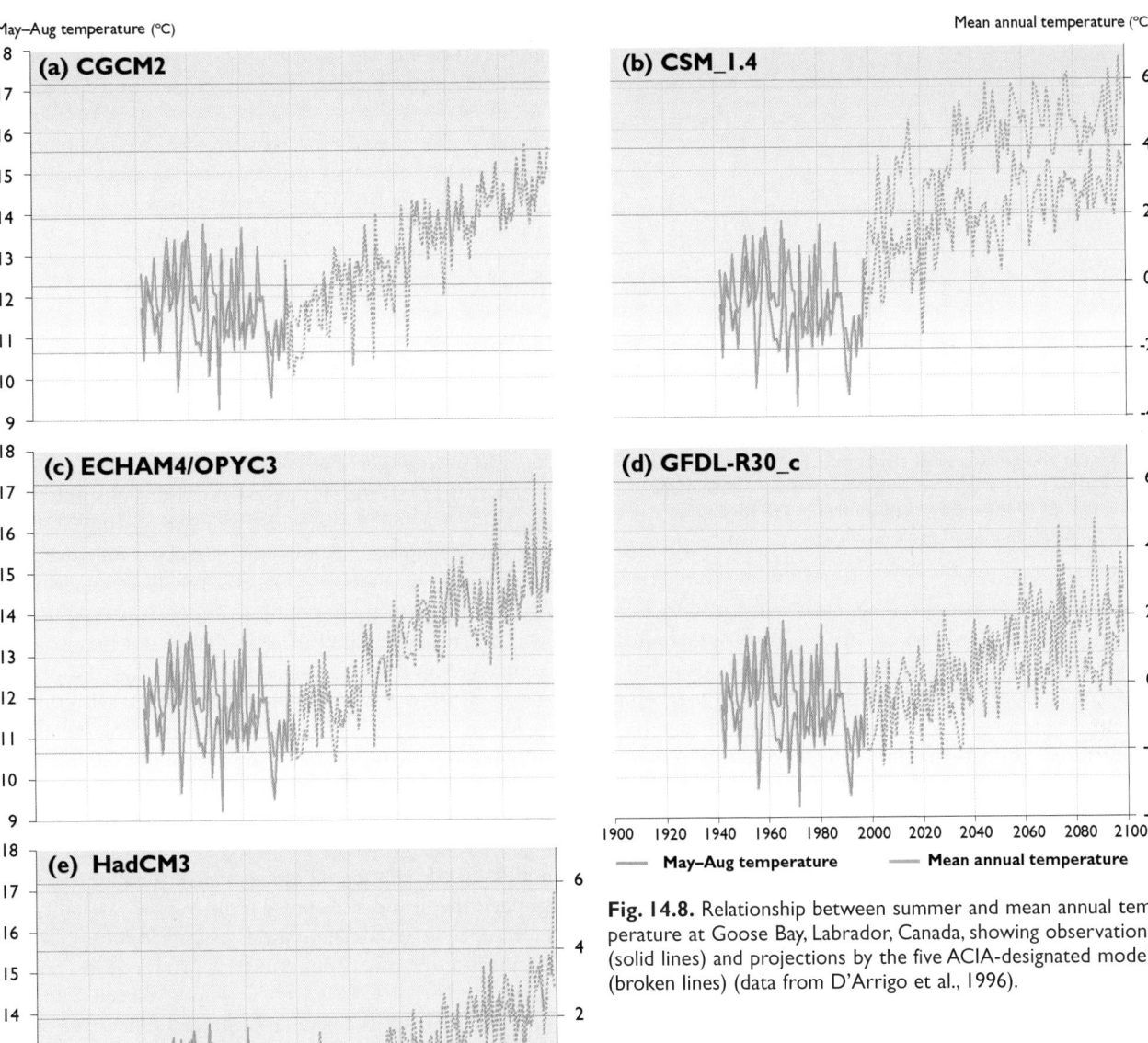

Fig. 14.8. Relationship between summer and mean annual temperature at Goose Bay, Labrador, Canada, showing observations (solid lines) and projections by the five ACIA-designated models (broken lines) (data from D'Arrigo et al., 1996).

of analysis are involved, the outcomes in ecosystems can be better described.

The climate changes that are the most difficult to project are low-probability, high-impact events. By definition, there are few examples of such events to learn from; in fact, analogous events may not have occurred during the period of the observational record. An example in forest ecosystems is the introduction and spread of tree pathogens such as insects or fungi. Newly introduced pathogens are capable of radically reducing the abundance of susceptible trees, with long-lasting consequences, yet there is no direct human experience of such events and no evidence of the influence of such outbreaks in the past. Temperatures maintain the current distribution limits of many of these insect and disease pathogens, so if temperatures increase, then movement

of new pathogens into the northern boreal forest is highly likely. Temperature increases in the boreal region can also increase the risk of freezing damage. Boreal forest vegetation is generally well adapted to survive or recover from brief periods of below-freezing temperatures after the initiation of growth in the spring. However, some of these freeze-protection mechanisms provide protection from temperatures only slightly below freezing or are not fully effective at the earliest stage of growth. Earlier initiation of growth in the spring or even in the late winter in the boreal forest as the result of increasing temperatures is very likely to be followed by a return to seasonal cold temperatures well below freezing, resulting in damage that would not occur with later emergence from winter dormancy (Kellomaki et al., 1995, Prozherina et al., 2003). The upper levels of temperature increases projected by the five ACIA-designated models are within the range of climate change that is likely to include "ecosystem surprises" for which no historic analogue exists and which ecological modeling is not likely to project.

However, humans are not passive spectators of change in forest ecosystems or forest resources that are important to them. Humans actively intervene, plan, and

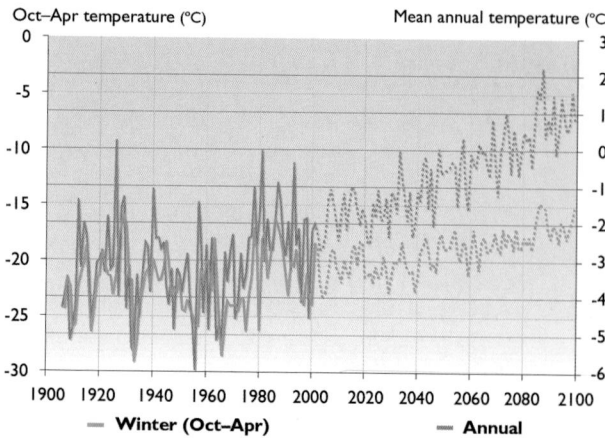

Fig. 14.9. Observed (solid line) and projected (CGCM2; broken line) winter and mean annual temperature at Fairbanks, Alaska. Observed data are a combination of University Experiment Station (1906–1948) and Fairbanks Airport (1948–2001) measurements (data from NOAA, 2002).

manage forest resources and agents of change in forests through their influence on factors such as fire that affect forests and forest values. Humans completely plan and create agricultural crops, livestock herds, and marketing systems. Therefore, it is possible for climate change to initially create effects that humans can respond to in a way that makes humans the most important agent of change. In some ways, these management responses are part of the standard set of land and resource management treatments that forestry and agriculture have developed over time. Reforestation, new species plantations, salvage of dead trees, sanitation treatments of a stand to reduce its vulnerability to pathogens, planting new crops, and altering soil or site conditions have a long history. Crops can be moved to suitable temperature zones, fields irrigated, and tillage systems adjusted to manipulate soil temperature within certain limits. What is new, and only now being developed, is the application of these techniques and the development of additional techniques to achieve specific goals associated with climate change in the far north, including both adaptation and mitigation approaches, and some appreciation of the associated costs.

14.4.6. Differences between the B2 and A2 emissions scenarios

For all of the model analysis presented in this chapter, projections based on the B2 emissions scenario (section 4.4.1) were used. This section evaluates the possible differences between model projections based on the A2 emissions scenario and those based on the B2 emissions scenario.

Broadly speaking, all the ACIA-designated models forced with either emissions scenario project that temperatures within the boreal region will reach levels higher than shown by reconstructions of climate for nearly the last 1000 years or essentially since the Medieval Warm Period. In every case, the models forced with the A2 emissions scenario project higher temperatures at the end of the 21st century than they project when forced

with the B2 emissions scenario (section 4.4.2). The temperature increase over the 21st century projected by the model with the least warming when forced with the B2 emissions scenario (CSM_1.4) is about 3.5 °C, but the same model when forced with the A2 emissions scenario projects a temperature increase of about 4.5 °C over the same period. The greatest temperature increase over the 21st century – slightly more than 7 °C – is projected by the ECHAM4/OPYC3 model forced with the A2 emissions scenario.

Considered from the broadest perspective, the difference between projections based on the A2 emissions scenario versus those based on the B2 emissions scenario is largely a matter of timing of effects on forests rather than different kinds of effects. In practical terms, this means that the thresholds described in this chapter would be reached sooner under climate conditions projected using the A2 emissions scenario compared to conditions projected using the B2 emissions scenario. The difference in timing of these effects is projected to be as much as 40 years earlier in the case of the projections based on the A2 emissions scenario. The temperature projections of each model when forced with the A2 emissions scenario begin to diverge from projections using the B2 emissions scenario in amounts significant for forest processes after about 2040.

In general, the projections of the ECHAM4/OPYC3 model forced with the B2 emissions scenario best matched the features of annual temperature variability in the recorded data for the stations examined in this chapter; most of the other four scenarios produced less annual variability. The ECHAM4/OPYC3 model forced with the A2 scenario projects a pattern of steep temperature rise early in the 21st century, then a period of little change, and finally a steep rise again in the last decades of the century. This suggests that the effects of climate change under this scenario, with large variability possible from year to year, could be especially sudden. Under such a scenario the risk of extreme events, such as widespread forest fires and insect outbreaks, would probably be greatest.

14.5. Agriculture

14.5.1. Arctic agriculture in a global context

The population of the earth relies on two basic agricultural systems for its food supply. The first is the _large-scale commercial production_ and trade of commodities. The commodities that dominate this system are cattle, sheep, and hogs for meat or wool, dairy animals for milk products, grains for animal feed and direct human consumption, and oilseeds. These products move through a complex, and in most cases efficient, network of trading, transportation, and processing before reaching the final consumer. Smaller volumes of products that are important for human dietary variety also move through this network, including potatoes, fruits and vegetables, and raw products for beverages and spices. As a result, a diverse array of foods is available to those consumers who have both physical and economic access to the network. Within the Arctic, this agricultural and trading system serves major population centers, areas served by regular surface transportation, and remote communities. The potential impacts of climate change on this system are global and complex: the system is vulnerable to direct climate influences on crops in multiple climate zones across the earth, to direct climate effects on shipping activities, and to subtle climate influences on prices and relative price differences (Rosenzweig and Parry, 1994).

The second type of agricultural system that supplies food for the population of the world is *subsistence* agricultural production. This system can be generally characterized as regional and largely self-reliant. It does not rely on a complex infrastructure to move either raw or manufactured products to the final consumer. Products are characteristic of the production region, and often obtained as part of a mixture of annual activities (section 12.2.2). These regions are categorized as those having low population density and/or a low economic profile. This subsistence agricultural system serves the majority of the area of the Arctic, but a minority of its population. Generally, regions in which subsistence agriculture predominates have either limited or no physical and/or economic access to the primary global food supply network. Potential climate change impacts on the subsistence system may be acute and local (e.g., a mid-season frost kills a crop) or subtle and longer term (e.g., climate effects on price advantages or barter opportunities).

In the Arctic, many examples can be found of a mixture of the two food systems or a cultural preference for the subsistence agricultural system despite economic and infrastructure advantages that would permit participation in the commercial production system. People in the Arctic can and often do move between participation in one system and the other. It is not clear that the commercial production system will completely displace the subsistence agricultural system in the foreseeable future.

14.5.2. Existing agriculture in the Arctic

Agriculture is a relatively small industry in high-latitude regions and consists mostly of cropping cool-season forage crops, cool-season vegetables, and small grains; raising traditional livestock (cattle, sheep, goats, pigs, poultry); and herding reindeer. This chapter focuses on crop production rather than livestock. Production of inexpensive reliable feeds is often a major constraint to animal agriculture and livestock production is often limited more by non-climate factors such as availability of processing facilities and appropriate waste handling rather than directly by climate. While agriculture is limited by climate in the Arctic, especially in the colder regions, it is also limited by lack of infrastructure, a small population base, remoteness from markets, and land ownership issues. Major climate limitations include short growing seasons (not enough time to mature or to produce high yields of harvestable crop), lack of heat energy (too few

GDDs during the season), long and/or unfavorable winter weather that can limit survival of many perennial crops, and high moisture stress in some areas.

Most of the climate stations selected for examination have climates that are currently at least marginally suitable for growing barley, oats, green peas, and potatoes (although not always to full tuber maturity). Wheat has been grown in experimental plantings as far north as Rovaniemi, Finland, the southern Northwest Territories, Canada, and Fairbanks, Alaska, but is considered too marginal for commercial production in those areas and is a major commercial crop only in southern Siberia (represented by Novosibirsk). Canola (rapeseed) is also produced only in the southernmost part of the Arctic and is considered marginal even in many areas now considered suitable for barley (such as Interior Alaska). Sunflowers currently are produced in generally the same locations as wheat. The magnitude of temperature increases projected by the ACIA-designated models would remove these heat limitations in areas that are presently climatically marginal by the mid- or late 21st century.

All of the climate stations examined in this chapter have summers with enough heat to produce at least one successful harvest of forage crops, including legumes such as alfalfa or clover and grass such as timothy. Slight increases in growing-season temperature are likely to increase the probability of two or more successful harvests and thus increased yields in some areas. However, winter survival is a problem in most areas, thus alfalfa is produced only in the warmer parts of the Arctic, and clovers are successfully grown in areas with relatively mild winters. Perennial grasses, especially smooth bromegrass and timothy, are grown in much of the region.

Agricultural production statistics match the boundaries of the Arctic as defined in this chapter at the national level for Iceland. Relevant statistics are also available for the three northern counties of Norway (Troms, Finnmark, and Norland), Alaska (United States), and the Yukon and Northwest Territories (Canada). These areas are highlighted in this section. Data from the Norrbottens region of Sweden match this chapter's analysis area, and Finnish agricultural statistics are readily available at the national level and for jurisdictions with boundaries that are not as relevant for the purposes of this assessment. Complete and reliable statistics for agriculture in arctic Russia are not available.

Yukon Territory is one of the most mountainous parts of Canada, but agriculture occurs in valleys, especially in the southern interior, and in suitable landforms in the Northwest Territories. The data in this section on Canadian agriculture are from Statistics Canada (2001) and Hill et al. (2002). There are at least 170 farms in the Yukon and 30 in the Northwest Territories, representing a total capital value of over Can$ 67 million, primarily in land and buildings. Most of these farms are operated as part of a yearly cycle of other activities, and only 21 have gross receipts greater than Can$ 25000.

Farmers in the Canadian north take a diversified approach to agriculture. They produce traditional livestock and engage in sled-dog breeding and non-traditional livestock production primarily for local and specialty markets. Along with east-central Alaska, the western Yukon is the driest part of the North American subarctic. Because portions of these regions are already natural grasslands too dry for tree growth, irrigation would certainly be required to exploit improved temperature regimes for crop agriculture in such areas. In the central and eastern Canadian Arctic, temperatures currently fall well below minimums necessary for traditional agriculture, although warming trends have been detected. Across much of the Canadian far north, further temperature increases as projected by the ACIA-designated models would move the zone of temperature suitability northward into a region largely dominated by soils with relatively poor suitability for agriculture. Broad areas of Turbic and Organic Cryosols cover these northern landscapes (Lacelle et al., 1998), and in the glaciated north, soils are often thin, stony, and derived from nutrient-poor bedrock parent materials. Areas with pockets of suitable climates and soils that are close to population centers could possibly become locally important for food production for established populations with a goal of greater self-sufficiency.

Alaska has several climatically distinct agricultural regions, including interior valleys and lowlands with warm and dry summer climates, coastal regions with cooler summers and moderate to high precipitation, and a broad transition region in the major population zone of south-central Alaska. Most of the volume and value of agricultural production takes place in the transition and interior locations. In the dry interior valleys, heat sums necessary to mature grains such as barley are generally achieved, and recent warming trends have increased the probability of achieving this critical factor to a very high level (Sharratt, 1992, 1999). In interior valleys and the south-central transition region, there is a high correlation between years of excellent crop production and adequate and (especially) well-timed precipitation in the early and middle parts of the growing season. However, persistent and heavy precipitation in the late summer can delay, reduce, or even ruin crop harvest and recovery.

In 2001, Alaska had about 580 farms and ranches (sales greater than US$ 1000) covering 370 000 ha, of which the great majority was unimproved pasture (Alaska Agricultural Statistics Service, 2002). Alaskan crop production in the same year included 10 500 t of potatoes (including high-value, certified virus-free seed potatoes), 27 000 t of hay, 800 t of carrots, and over 8800 m^3 of oats and barley. Alaska produced 11 500 cattle, 1200 sheep, 1000 hogs and 15 000 domestic reindeer in 2001. A few farm and ranch operations are producing species such as elk (wapiti), bison, yak, and musk oxen. Some Alaskan crops and pastures are already moisture-limited, especially in the warmer summer climate of recent years (Sharratt, 1994). If temperatures were to increase across

the northern boreal region without a significant increase in precipitation in the early and middle parts of the growing season, the change would be unfavorable for agricultural production, except in the case of irrigated crop production that could take advantage of the greater growing-season heat units. Temperature increases projected for coastal areas generally are very likely to be favorable to agriculture, although much of the area consists of steeply mountainous terrain with little soil development. In the south-central transition region, Alaskan agriculture faces strong economic challenges from competing urban and suburban land uses in the region where historically it has been most well-established. Large new areas are available to be opened to agriculture, especially in central Alaska, but no recent public policy initiative to do so has emerged. National-level agricultural subsidy payments to Alaskan farmers have nearly doubled in the last decade to over US$ 2 million, or about 20% of net farm income.

In 2000, northern Finland contained about 10 000 farms, of which about 50% were dairy farms. The average farm size is about 11.5 ha. While most northern Finnish farm properties remain in their traditional size and configuration (they have not been subdivided), the number that are producing significant amounts of dairy products has decreased rapidly in recent years (Häkkilä, 2002). Reasons for the decline include surpluses of dairy products, small quantities produced, and especially social change. These changes include migration of farm owners to towns and cities with professional opportunities that are more attractive and have higher income potential, and the conversion of inactive farms into country residences for town dwellers.

The three northern counties in Norway had 95 000 ha of land in agricultural use in 1999, distributed among about 6500 owners (Statistics Norway, no date). The area receives nearly the maximum warming influence of the Gulf Stream in the Arctic, so there is a strong maritime influence especially in the south and west of the region. Climate factors limiting to agriculture are maximum temperatures and total warmth over the summer,

Fig. 14.10. Garden in northern Norway (photo: P. Grabhorn, 2003, Grabhorn Studios, Cerrillos, New Mexico).

which are strongly influenced by cool and wet maritime weather. Prolonged cool and wet maritime conditions can result in slow plant growth and failure of crops to complete development, wet fields and crops at harvest time, and, occasionally, outbreaks of plant diseases. Agricultural operations are generally small, with only 570 holdings larger than 30 ha and only 50 holdings larger than 50 ha. Employment in agriculture, horticulture, and forestry in the three-county region in 1999 totaled about 6500 full-time equivalents. Many of these farming operations fit into a diversified set of annual activities that include wage employment, fishing, cultural landscape maintenance, and personal consumption and non-cash trade of crops (Fig. 14.10). Nearly all farming operations include cultivated meadows for hay and pasture, and about one-third produce potatoes. Only 45 farms, all in the southernmost county of Nordland, produce grain or oilseed crops.

Temperature increases are likely to move the grain production boundary northward. Any climate changes that enhance the maritime effect, especially in the summer, would not be favorable to agriculture in this coastal region, although the effects are very likely to be lessened inland. Climate changes that increase growing-season length and daily maximum temperatures while maintaining or slightly decreasing persistence of growing-season clouds and rain are very likely to be favorable to agricultural production in this area. Agriculture in the far north of Norway will be strongly influenced by national and European policies to subsidize rural populations and landscapes, perhaps more than any other factor. The government of Norway is likely to have sufficient revenues available from petroleum to intervene strongly in the economics of northern community infrastructure and agriculture to achieve its social and environmental goals for some time to come.

Iceland is surrounded during the growing season by a cool ocean surface; locations farthest inland from the cool maritime influence are covered with glacial ice. As a result, lack of summer warmth has been the chronic limitation to Icelandic agriculture, especially in years when sea ice reaches the Icelandic shoreline. Icelandic agriculture traditionally has been limited to the crops most tolerant of cool season conditions. Even small temperature increases historically have removed this limitation and permitted the better harvests in the record. Iceland produces over 2 million m^3 of hay, more than 10 000 t of potatoes, and over 4000 t of cereal grains (barley; Statistics Iceland, 2003). Glasshouse agriculture using abundant geothermal heat supports the production of about 1000 t of tomatoes and cucumbers. In 2001, the 3000 farms in Iceland produced 473 000 sheep, 70 000 cattle, 73 000 horses, and 106 000 liters of processed milk (Statistics Iceland, 2003). During the last decade of the 20th century, sheep and hay production declined at double-digit rates, largely for non-climatic market, policy, and social reasons. Numbers of cattle and horses (the famous Icelandic pony) have remained level. Icelandic agriculture employs about 4400 people, which

is 3.1% of the total employment. Greenland supports about 20 000 sheep and more than 3000 reindeer (Statistics Greenland, 2005).

The entrance of Iceland into the European Community Common Agricultural Programme (CAP) has been the major force in Icelandic agriculture in the last decade. In general, CAP subsidies have sustained smaller and higher-cost producers and rural economies, introduced limitations for certain products, reduced risk to producers, and transferred funds from consumers to producers. Temperature increases would almost certainly permit accelerated expansion of cereal grain production and possibly new oilseed crops, reduce the cost of winter forage for livestock, and allow increased per-unit crop yields where nutrients or water are not limiting. There are complex multiple goals in Icelandic agriculture, including food production, employment, rural stability, maintaining an attractive landscape, and environmental protection. Because the agricultural system exists in a highly interventionist public policy environment, a traditional economic analysis is likely to show that national and CAP policies will be more influential than temperature increases in the future of Icelandic agriculture (see section 14.3.4.4 for a discussion of the conversion of agricultural land to forests).

14.5.3. Approach to scenario analysis

The 14 sites that were chosen for examination of the ACIA-designated model projections were selected to represent areas in the northern part of the boreal region currently supporting commercial agriculture and areas north of current potential (section 14.4.1). The analysis did not consider soil effects, although soil can be a major limitation to agriculture, and changes in soil processes following climate change can have significant impacts on agriculture (Gitay et al., 2001). In much of the circumpolar north, large tracts of soils already exist that are suitable for agriculture if other constraints are removed, with the exception of much of northern Canada (composed of the Canadian Shield) and northern Fennoscandia, where the granitic soils are shallow and nutrient-poor.

This analysis used GDDs as a primary determinant of climate suitability for producing annual crops. Growing degree-days (a measure of accumulated heat energy) are considered a good predictor of plant phenology and are often used to project the timing of different plant growth stages. For example, farmers and extension agents often use GDDs to determine the optimum harvest time for forage crops. There are some major limitations to the use of GDDs. They may vary for different cultivars, locations (Sanderson et al., 1994), and growth stages (Bourgeois et al., 2000; Kleemola, 1991). Heat energy requirements for plants interact with other environmental factors, such as moisture stress and photoperiod (Bootsma, 1984; Nuttonson, 1955, 1957), and thus GDDs may not be a good predictor of plant development under stress. Heat requirements for plant development generally decrease with increasing photoperiod, so fewer GDDs are needed to reach a given growth stage at high latitudes than at low latitudes.

In areas with humid autumns, GDDs may equal or exceed the threshold for a given crop to reach maturity, but wet conditions may not allow the crop to dry enough to allow cost-effective mechanical harvesting. Thus, even though projections of future temperature may indicate that certain areas will be suitable for some crops, wet conditions may offset the temperature effects, especially in more humid zones within the northern boreal region.

Growing degree-day requirements for various crops from the highest-latitude areas possible were obtained from the literature, and similar GDD requirements were assumed for a given crop throughout the region. One problem with using GDD data from the literature for determining climate suitability for a given crop is that different reports use different base temperatures for calculating GDDs. For example, many agriculturists often use a base of 5 °C (GDD_5) for cool-season crops, while others use a base of 0 °C (GDD_0). Others use some experimentally determined growth threshold as the base temperature. If no information existed for GDD requirements using a base of 5 or 0 °C, the GDD requirements for the base temperatures of interest were estimated from published GDDs based on similar base temperatures. In some cases, insufficient information was available to do this for both 5 and 0 °C base temperatures. Despite its limitations, use of GDDs is a widely reported and accepted method to project approximate crop phenology. GDD analysis is generally accepted as a good way to estimate climate suitability for producing annual crops.

For projecting potential water stress, the analysis used model-projected potential evapotranspiration minus growing-season precipitation. Positive values indicate a potential water deficit, interpreted to mean potential water stress, which could limit crop yields or quality. However, many crops can produce acceptable yields even under water-deficit conditions, and water stress is also affected by soil water-holding capacity, rooting depth, precipitation timing (e.g., spring versus summer), and type of storms (e.g., infrequent large rainfall events may result in more water loss due to runoff than small, frequent storms).

This assessment ignored the effects of atmospheric CO_2 enrichment and changes in crop pests, other than to estimate the potential general effects. Few field studies have been done on CO_2 enrichment effects on crop plants at high latitudes and results from more temperate zones may not be transferable to northern areas, thus there is little data on which to judge the effects of CO_2 enrichment. Climate effects on pests are complex and difficult to forecast. Such an analysis is obviously needed to understand the complete range of climate change effects, but it is not yet feasible because of a lack of detailed knowledge of insect and other pathogens.

For this analysis, a few annual crops were selected that are currently well-adapted or marginally adapted to parts of the region and which have potential to become economically important crops for animal feed (cereal grains), human food, (beans, peas, potatoes, wheat), or oilseeds (canola, sunflowers).

14.5.4. Climate limitations and influences

If the ACIA-designated model projections of GDDs at a given site equaled or exceeded the minimum GDD requirements for a given crop, but were below the mid-point of the range given in Table 14.3, the crop was considered to be marginal. The analysis was based on 20-year averages, to smooth out much of the variability in the data and because it is assumed that farmers would require about 20 years under a changing climate regime to accept the change and adopt farming practices that included new crops.

The GDD requirements for some of the crops presented in Table 14.3 appear inconsistent between the two base temperatures (for example, with a base of 5 °C, seed peas appear to require fewer GDDs at the low end of the range than canola but require more GDDs at the upper end of the range). This may be due to errors in the literature, errors in the estimation of GDD requirements, or it may mean that either the 5 or 0 °C base temperature is not appropriate for some crops. This may cause errors in determining climate suitability for some crops. Table 14.3 also provides estimates of the number of accumulated GDDs required for forage crops to reach optimum time of harvest. These may be underestimates for high-latitude

Table 14.3. Growing degree-day requirements for various annual crops to reach maturity (Anon., 1996-2000; Ash et al., 1999; Dofing, 1992; Dofing and Knight, 1994; Juskiw et al., 2001; Miller et al., 2001; Nuttonson, 1955, 1957; Sharratt, 1999) and for forage crops to reach optimum harvest stage (Bootsma, 1984; Breazeale et al., 1999).

	Growing degree-days (5 °C base)	Growing degree-days (0 °C base)
Annual crops		
Peas (green for processing)	700–800	1000
Spring barley	700–900	1200–1500
Peas (for seed)	800–1150	1500–1700
Oats		1300–1700
Canola	950–1050	1350–1550
Potatoes	1000–1100	
Spring wheat	1000–1200	1400–1650
Dry beans	1100–1500	
Sunflowers (for seed)		1800–2000
Forage crops		
Alfalfa	350–450	
Red clover	450	
Timothy	350–450	

regions, as low soil temperatures may delay initiation of growth, especially in spring.

The models use large grid cells, which often contain large tracts of terrain not likely to become suitable for agriculture (such as high mountains), and some of the model projections are obviously not realistic even for present conditions. Therefore, this analysis uses proportional changes from the present, compared to observed weather data, rather than using actual temperature projections provided by the models. For each parameter analyzed (GDD_0, GDD_5, and growing-season water deficit), the highest and lowest projections for each location were selected (Table 14.4).

The success of perennial crops is governed more by winter survival success than by growing-season weather conditions. Factors such as the timing of killing frosts, warm spells during winter, length of the dormant season, snow

Table 14.4. Projected growing degree-days calculated from the highest and lowest model-projected temperatures for each site.

	2011–2030		2041–2060		2071–2090	
	Highest	Lowest	Highest	Lowest	Highest	Lowest
Growing degree-days (0 °C base)						
Fairbanks	2175	1800	2525	1850	2625	2025
Big Delta	1925	1600	2225	1650	2350	1800
Whitehorse	1750	1500	2050	1525	2200	1700
Fort Vermillion	1875	1675	2000	1750	2050	1875
Thompson	2025	1825	2125	1900	2375	2050
Goose Bay	1550	1375	1600	1525	1775	1575
Reykjavik	1275	1125	1425	1050	1450	1150
Tromsø	1325	1100	1425	1125	1525	1300
Umeå	1700	1425	1875	1600	1900	1675
Rovaniemi	1650	1400	1825	1575	1850	1675
Archangelsk	1700	1450	1875	1600	2000	1625
Novosibirsk	2025	1900	2125	1850	2250	2000
Yakutsk	2025	1750	2150	1800	2325	1975
Magadan	1625	1400	1825	1425	2050	1775
Growing degree-days (5 °C base)						
Fairbanks	1600	1125	2000	1150	2150	1325
Big Delta	1275	900	1600	875	1725	1050
Whitehorse	1275	900	1600	925	1725	1050
Fort Vermillion	1400	1200	1550	1275	1575	1400
Thompson	1350	1175	1450	1225	1650	1375
Goose Bay	1300	1100	1500	1275	1700	1350
Reykjavik	500	475	625	425	625	500
Tromsø	600	450	675	475	725	625
Umeå	1025	775	1200	975	1200	1075
Rovaniemi	900	700	1050	800	1050	875
Archangelsk	1000	825	1150	975	1250	1025
Novosibirsk	1325	1250	1425	1200	1500	1375
Yakutsk	1350	1075	1450	1100	1625	1325
Magadan	925	725	1075	625	1250	1050

depth, and winter temperatures all interact to affect winter survival. Warm periods during winter, especially prolonged temperatures above 0 °C, can be detrimental to perennial plants by reducing winter dormancy and depleting carbohydrate reserves (Crawford R., 1997). The complexity of these interactions makes projection of climate change effects on these crops difficult.

14.5.5. Growing degree-day analysis

When GDD_0 was used for analysis, all of the ACIA-designated models projected that all the examined locations would be suitable for green pea production early in this century, and the high-extreme models (GFDL-R30_c and CGCM2) projected climates suitable for barley at all locations by 2030. When GDD_5 was used, the models projecting the most GDDs suggested that Reykjavik and Tromsø would be unsuitable for green peas by 2030 because of too many GDDs. The GFDL-R30_c projections indicate that potatoes could be grown in all locations except Reykjavik, Tromsø, Rovaniemi, and Magadan (marginal at Umeå and Archangelsk) by 2030, while the low-extreme models (HadCM3 and CSM_1.4) projected suitable climates only in Fairbanks, Fort Vermillion, Thompson, Goose Bay, Novosibirsk, and Yakutsk. Thus, projections for potatoes may be too low, as potatoes are already produced, at least marginally, in many areas not projected to be suitable by 2030. This may be because use of GDDs may not be a good way to project phenological development in potatoes (Shaykewich and Blatta, 2001). The models generally projected that all except the coolest sites in Scandinavia would have climates at least marginally suitable for potato production by the end of the century.

All of the models project a climate suitable for dry pea production at all locations except Reykjavik and Tromsø by 2030 when GDD_5 was used, although the low-extreme models (HadCM3 and CSM_1.4) projected that Goose Bay, all of the Scandinavian sites, Archangelsk, and Magadan would be unsuitable for dry peas using GDD_0. All the models project that all sites except Reykjavik and Tromsø would have climates suitable (marginally so at Goose Bay) for seed pea production by the end of the century. The high-extreme models (GFDL-R30_c and CGCM2) project that all of the North American sites (marginal at Big Delta, Whitehorse, and Goose Bay), Novosibirsk, and Yakutsk would have climates suitable for dry bean production by 2030, but the low-extreme models (HadCM3 and CSM_1.4) project that climate would only be marginal for such production even at the warmest sites (Fairbanks, Fort Vermillion, Thompson, Novosibirsk) by 2030. Models projecting the highest number of heat units at a given location indicated that all sites except Reykjavik, Tromsø, and Rovaniemi would be at least marginally suitable for dry beans by 2060, whereas the low extreme models (HadCM3 and CSM_1.4) indicated that only Fairbanks, Fort Vermillion, Thompson, Goose Bay, Novosibirsk, and Yakutsk would have climates suitable for dry bean production by 2090.

Most of the models projected that most locations would be at least marginally suitable for all of the cereal grains by the end of the century, although the low-extreme models, using GDD_5, projected that only Fairbanks, Fort Vermillion, Thompson, Goose Bay, Novosibirsk, and Yakutsk would be well-suited for wheat by 2090. The high-extreme model (GDFL-R30_c or CGCM2) for each location projected that climate would be suitable for canola at all locations except Tromsø by 2030 using GDD_0. All of the models projected that climate will be at least marginally suited for oilseed sunflowers by 2030 at Fairbanks, Thompson, and Novosibirsk, and the high-extreme models projected climates suitable for sunflowers at all sites except Goose Bay, Reykjavik, and Tromsø (marginal at Rovaniemi) by 2090. Conversely, the low-extreme models projected that only Fairbanks, Thompson, Novosibirsk, and Yakutsk would be well-suited, and Big Delta and Fort Vermillion would be marginally suited, for this crop by near the end of the century.

Assuming that the amount of heat is the main factor limiting yields and number of harvests of perennial forage crops, model projections indicate that warmer growing seasons are likely to increase the potential number of harvests and hence seasonal yields for perennial forage crops at all locations. Uncertainty about winter conditions make forecasts about survival potential for crops difficult. Warmer winters could actually decrease survival of some perennial crops if winter thaws followed by cold weather become more frequent. Crops adapted to certain types of winter stress are very likely to experience different types of stresses if winter temperatures rise and winter thaws become more frequent. However, lengthened growing seasons, especially in autumn, are very likely to result in northward extension of climate suitable for alfalfa production.

Table 14.5. Projected water deficits calculated from highest and lowest model projections at each site. Negative values indicate a water surplus.

| | Water deficit (mm) | | | | | |
| | 2011–2030 | | 2041–2060 | | 2071–2090 | |
	Highest	Lowest	Highest	Lowest	Highest	Lowest
Fairbanks	250	166	282	207	311	253
Whitehorse	150	116	287	165	321	265
Fort Vermillion	324	148	355	168	371	218
Thompson	335	132	394	157	413	169
Goose Bay	22	-122	62	-151	58	-125
Reykjavik	184	-20	216	-15	241	68
Tromsø	84	-31	117	14	134	44
Umeå	217	139	250	163	262	173
Rovaniemi	185	51	233	111	226	145
Archangelsk	191	-7	232	26	231	31
Novosibirsk	389	126	414	127	469	149
Yakutsk	289	152	315	161	328	165
Magadan	224	117	238	164	269	180

All of the models project high interannual variability in GDD accumulation. Thus, even though 20-year averages may indicate climates well suited for crops for a given area, the frequency of growing seasons cool enough to limit production may convince farmers that the risk for producing these crops is too high. Thus, the perception of risk by farmers could be a limiting factor for the northward advancement of many of the analyzed crops.

14.5.6. Precipitation and potential evapotranspiration analysis

Most of the models project increases in precipitation at most locations, with some forecasting up to 20% increases from values typical of the 1981–2000 baseline period by 2090. There was a great deal of variability among models, with no single model consistently projecting the greatest or smallest change in precipitation at all locations. A few models projected decreases in precipitation at a few sites. For example, CSM_1.4 projects a decrease of 5% at Whitehorse (compared to the present) during the 2011–2030 time slice, and HadCM3 projects a decrease of about 1% at Umeå and about 4% at Tromsø by 2090. All models projected a general increase in potential evapotranspiration over time at all locations, with much variability among models and locations.

Some locations in the subarctic already experience water deficits large enough to reduce crop yields. For example, Sharratt (1994) found that barley yields in Interior Alaska are depressed in about five of every nine years due to insufficient water supply. Water deficits were usually projected to increase over time, with GFDL-R30_c usually projecting the smallest deficits and HadCM3 and CSM_1.4 usually projecting the largest (Table 14.5). Most models project rather large water deficits, especially by the end of the analysis period. However, the HadCM3 model projects rather large water surpluses throughout the 21st century at Goose Bay, and slight surpluses at Reykjavik until about mid-century (Table 14.5). Most models, especially CSM_1.4, project fairly small deficits at Tromsø. These results indicate that water is likely to become a major limiting factor for production of most crops at all but the maritime sites in the boreal zone. Unless irrigation is supplied, which may not be economical for most of the crops analyzed, production is likely to be limited in many areas to drought resistant crops, such as cereal grains.

14.5.7. Indirect effects of climate change

This analysis has focused on the direct effects of changes in growing-season temperatures (using GDDs as indicators of crop growth potential) and moisture relationships on potential agricultural crop production in the next century in high-latitude regions. However, the indirect effects of climate change may have similar or greater impacts on agricultural development. Several of them are outlined here without in-depth analysis, to acknowledge their potential importance and alert those involved in future assessments that these factors need to be considered.

The development of crop disease requires suitable host, pathogen, and environmental conditions. Temperature and moisture are critical for the spread of many plant diseases (Gitay et al., 2001), especially in a region as severely heat-limited as the Arctic. Under conditions of increasing temperatures, the risk of crop damage increases in all regions of the northern half of Europe (Beniston et al., 1998). The severity and number of species capable of reaching infestation levels for North American agriculture are likely to increase with less severe winters (Shriner et al., 1998). These findings suggest that disease and weed pests are likely to increase throughout the Arctic under the ACIA-designated model scenarios. These problems are not likely to offset potential yield increases or eliminate the potential for new crops in most cases. However, it is possible that severe outbreaks could have that effect in specific cases. For example, temperature increases in Finland are very likely to increase the incidence of potato late blight to the point that it will significantly decrease potato yield in that country (Carter et al., 1996).

While the effects of atmospheric CO_2 enrichment were not considered in this analysis, higher CO_2 concentrations are very likely to result in yield increases for most crops (Warrick, 1988). Lack of experimental data on the effects of CO_2 fertilization on crops in high-latitude regions make in-depth analysis of this effect impossible at present.

Lack of infrastructure development, small population sizes (thus limiting local markets), and distance to large markets are likely to continue to be major factors limiting agricultural development in most of the northern part of the boreal region during this century. One possible indirect effect of climate change on agriculture in the Arctic and subarctic is the effect on transportation, including land transport and the prospect of regular ocean shipping across the Arctic Ocean with reduced or absent sea-ice cover (section 16.3). Thawing of permafrost that is susceptible to ground subsidence is disruptive to existing roads and railroads. Ultimately, once the permafrost thaws, in nearly all cases a more stable foundation will be available for permanent transportation routes and facilities. It is possible that shipping across the Arctic Ocean will greatly enhance economical trade in agricultural products, especially products unique to these regions, and thus enhance development of commercial agriculture in the far north.

If the net effect of climate change is greater global agricultural production, then food prices are likely to be lower for the world as a whole as long as the rather large subsidies that influence prices do not markedly increase or decrease. Lower world food prices would reduce the incentive to rely on local commercial or subsistence production in the Arctic. The opposite would be the case if climate change reduces global agricultural production.

A key component of the ability of global food production to cause a fundamental change in the supply/demand relationship for local agricultural production in the Arctic (other factors being equal) is shipping cost. Regular shipping across an ice-free Arctic Ocean (section 16.3.7) is very likely to lower transport costs for nonperishable bulk commodities such as grains or fertilizers, and thereby lower prices for the basic inputs to the agricultural system in the Arctic. In addition, it is possible that the availability of arctic shipping will stimulate the export of bulk finished products from the Arctic. The volume of agricultural imports to or exports from the Arctic is unlikely to generate the economies of scale that would support regular shipping at a major cost advantage for agriculture within the Arctic, at least not within the 21st century. Mining and petroleum industries would be more likely to generate higher volumes of cargo, although there is only a limited ability to change the kind of cargo handled by specialized vessels. However, if arctic ports of call were integrated into a much higher volume of general trade between the northern continents, then price advantages for agriculturally related arctic cargoes would be much more likely.

The impacts of lower transport cost on arctic commercial agricultural production would be complex. Inexpensive, regular arctic shipping is very likely to provide a price advantage that could possibly stimulate production of unique arctic commodities such as reindeer and caribou meat or muskox milk or cheese, as long as other factors affecting demand were favorable. Conversely, lower shipping costs are likely to discourage arctic production of agricultural commodities widely produced elsewhere, unless overall global supply or demand structure shifted descisively. However, basic input data that would permit in-depth analysis, including trade levels and patterns in an ice-free Arctic, economics and practicalities of Arctic Ocean shipping operations, and the overall regional costs of arctic agricultural production compared to global supply and demand, are somewhat speculative or lacking.

In Europe, the United States, and Canada, public policy and interventions are major factors that agricultural producers confront in determining their activities. In many cases, national or trading-area agricultural transfer payments and rules of production correlate especially well with the annual profile of agriculture in the Arctic, as noted in section 14.5.2. There is no obvious trend to suggest that the significant role played by government policy will change during the 21st century. Russia is experiencing a period of population and economic consolidation in its northern regions, an adjustment required after investments made during Soviet times could not be sustained under new conditions. Given the strong dominance of state land ownership, government policy will be a decisive factor in the future of agriculture in most of the arctic nations. The current agricultural profiles also suggest that local markets are vital to the agricultural sector in the Arctic. If climate change results in increases in the human population of the

Arctic, then the entire agricultural sector is likely to expand given the historic positive relationship between regional population and agricultural production.

14.6. Tree rings and past climate

The most recent historical period of Northern Hemisphere warming of similar magnitude (but possibly different in its cause) to that of recent decades is the Medieval Warm Period (MWP) from about AD 900 to 1300 (see also section 2.7.5). However, interpretations of the climate during the MWP do not always agree. Some evidence suggests that the MWP was not general across the planet and did not exceed the current warming in its fluctuations (Hughes and Diaz, 1994; Mann M. et al., 1998). Other evidence suggests temperatures 1 to 1.5 °C higher than at present across the Northern Hemisphere during the MWP (Dahl-Jensen et al., 1998; Esper et al., 2002; Naurzbaev and Vaganov, 2000). The main tool to compare recent warming with temperature levels during the MWP and earlier periods in most terrestrial regions of the far north is long-term climate reconstructions based on tree-ring data, because few or no other historical records exist and marine proxies and ice cores cannot provide the required geographic coverage or detail.

Tree-ring chronologies serve as a useful basis for reconstructing natural temperature fluctuations in the high latitudes over millennial intervals, although the degree of reliability needs to be assessed carefully in each application. An important potential limitation is that the relationship between tree rings and climate may vary with time. However, compared to other indirect sources of climatic information, tree-ring chronologies have certain important advantages. First, tree rings record a complete annual sequence of climatic information. Second, in northern Eurasia, where trees reach a maximum age of 1100 years, there is a dense dendroclimatic network allowing spatially detailed quantitative temperature reconstruction for the last 500 to 600 years, and in some regions for more than two millennia (Briffa et al., 2001; Hughes et al., 1999; Vaganov et al., 1996, 2000).

The rate and magnitude of recent Northern Hemisphere temperature increases are unique within the last several centuries (Briffa et al., 1996, 2001; Mann M. et al., 1998). Some climate models that include anthropogenic effects calculate that the greatest temperature increase, in the range of 3 to 4 °C, should have occurred over the last several centuries in the high latitudes of the Northern Hemisphere (Budyko and Izrael, 1987; Kondrat'ev, 2002). However, temperature reconstructions (generally of the warm season or even a specific portion of the warm season) based on tree-ring chronologies from subarctic Eurasia, a region that makes up a large part of the projected zone of maximum warming, do not show temperature increases of the projected magnitude (Briffa et al., 1998; Naurzbaev and Vaganov, 2000). This may be partly due to the disproportionate influence of warm-season temperatures

on tree growth, genuine local spatial climate variability, issues in calibrating the tree rings to estimate temperature with uniform reliability throughout the whole period of analysis, or errors or missing factors in climate model scenarios.

14.6.1. Past climate change in central Eurasia

This section examines several aspects of high-resolution proxy records based on tree-ring chronologies. Subarctic temperature reconstructions for Asia are compared with the main climatic forcing mechanisms during the last several centuries; reconstructed temperature fluctuations during the MWP are compared with recent temperature changes in the high latitudes of Eurasia using the millennial tree-ring chronologies; and recent temperature fluctuations are compared to temperature fluctuations during most of the Holocene in order to reveal warmest and coolest periods as well as rapid temperature changes in the central Asian sector of the subarctic.

The three main sources of information used for this analysis include local chronologies from the central Asian subarctic dendroclimatic network (Fig. 14.11) based on the analysis of living old trees (Briffa et al., 1998; Vaganov et al., 1996); super-long (at least two millennia) tree-ring chronologies constructed from cross-dating the abundant dead wood material of northern Siberia; and subfossil wood material excavated from alluvial deposits in terraces of small rivers in the Taymir and Lower Indigirka regions and even from sites north of the modern tree limit.

The radiocarbon dates of subfossil wood were used to define preliminary calendar time intervals and then the cross-dating method was used to identify absolute dates of tree-ring formation. Unfortunately, not enough material was available to build an absolute chronology for the entire Holocene in Taymir, so the "floating" chronologies (chronologies that cannot be tied with certainty to absolute calendar dates) with numerous radiocarbon dates were also used to analyze past temperature deviations (Naurzbaev et al., 2001).

A climatic signal was derived from raw tree-ring measurements using the regional (age) curve standardization

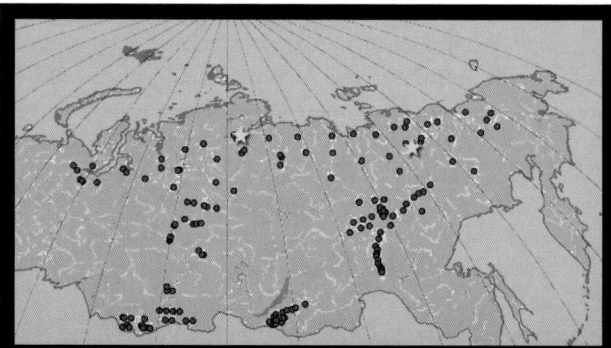

Fig. 14.11. Sites in the dendroclimatic network of the Asian subarctic (circles) and locations of millennial-length chronologies (stars) (Vaganov et al., 1996).

(RCS) approach (Briffa et al., 1996; Esper et al., 2002). This approach is applied to remove the age-dependent variations from single tree-ring series and to retain low-frequency climatic deviations (positive or negative trends) as well as high-frequency variations (year-to-year change) when averaging tree-ring index series. More details about this method of standardization can be found in Briffa et al. (2001), Esper et al. (2002), and Naurzbaev and Vaganov (2000). To verify results, tree-ring chronologies were compared to the instrumental climatic data averaged over a large sector of the subarctic. Finally, the longer-term tree-ring temperature reconstructions were compared to other proxy data including long-term variations in solar radiation (Overpeck et al., 1997), long-term variations in volcanic activity derived from ice-core measurements (Zielinski et al., 1994), and variations in CO_2 concentrations in air trapped in the GISP2 ice core (central Greenland; Wahlen et al., 1991).

14.6.1.1. Climate change in the central Asian subarctic during the last 400 years

Local tree-ring width series for each of 11 sites were obtained by averaging standardized series of individual trees. Between 60 and 70% of the variation in tree-ring width indices in the Eurasian subarctic is caused by changes in summer temperature (Briffa et al., 1998; Vaganov et al., 1996, 1998). A high correlation of local chronologies with temperature (r=0.69 to 0.84) allows the use of simple regression equations for reconstructing temperature based on tree rings. In effect, tree-ring width is used as a simple predictor of summer temperature. In order to avoid an additional procedure of statistical transformation of local tree-ring index series into normalized values of summer temperature variation, each of the local chronologies was normalized to the mean-square deviation in the generalized curve for the Asian subarctic region (Vaganov et al., 1996). To highlight trends and reduce short-term variability, the transformed local series was smoothed with a five-year run-

ning mean. Long-term temperature changes dominate the variability of the resulting generalized series. The generalized curve was compared with other temperature reconstructions for the circumpolar Northern Hemisphere (Overpeck et al., 1997) as well as with the main climatic forcing mechanisms.

Temperature variations in the Asian subarctic over the past 400 years correspond well to those observed across the circumpolar north (Fig. 14.12). Both curves clearly illustrate the temperature rise from the beginning of the 19th century to the middle of the 20th century. The main discrepancies between the two curves occur in the second decade of the 19th century and after the 1950s. The correlation of the two reconstructed temperature curves with each other for the preindustrial period is significant, but low (r=0.38, p<0.05), and markedly increases (r=0.65, p<0.001) for the industrial period (1800–1990) due to a distinct temperature increase that is shown in both curves.

More interesting is the correlation analysis of both curves (Fig. 14.12) with the main climatic forcing factors. The Asian subarctic generalized curve shows a significant correlation with all main climatic forcing factors: with solar radiation (r=0.32 for the entire period and 0.68 for the industrial period from 1800 to 1990); with volcanic activity (r=-0.41 for the entire period and -0.59 for the industrial period); and with atmospheric CO_2 concentration (r=0.65 for the period since 1850). The circumpolar reconstructed temperature curve (Fig. 14.12) is weakly correlated with solar radiation and atmospheric CO_2 concentration, and is not significantly correlated with volcanic activity. This is because more homogenous proxy data (only the tree-ring chronologies) were used for the Asian subarctic reconstruction, while different sources of proxy records (i.e. tree rings, lake sediments, isotopes in ice cores) were used in the circumpolar curve. The correlations with climatic forcing factors further indicate that natural factors (solar irradiance and volcanic activity) explain more of temperature variability that is common to the Asian subarctic and the circumpolar north than does the CO_2 concentration. Spatio-temporal analysis of reconstructed summer temperature variations in the Asian subarctic revealed that recent warming is characterized by an increased frequency of years with anomalously warm summers over the entire Siberian subarctic (the latitudinal dendroclimatic network seen in Fig. 14.11 that is replicated across a distance of 5000 km from east to west; Vaganov et al., 1996).

14.6.1.2. Medieval and current warming in northeastern Eurasia

Millennium length tree-ring chronologies were constructed from samples at two sites close to the northern treeline: east Taymir (71° 00' N, 102° 00' E) and northeast Yakutia (69° 24' N, 148° 25' E). These stands are made up of Gmelin larch (*Larix gmelinii*) and Cajander larch (*L. cajanderi*) in which the oldest living trees are up

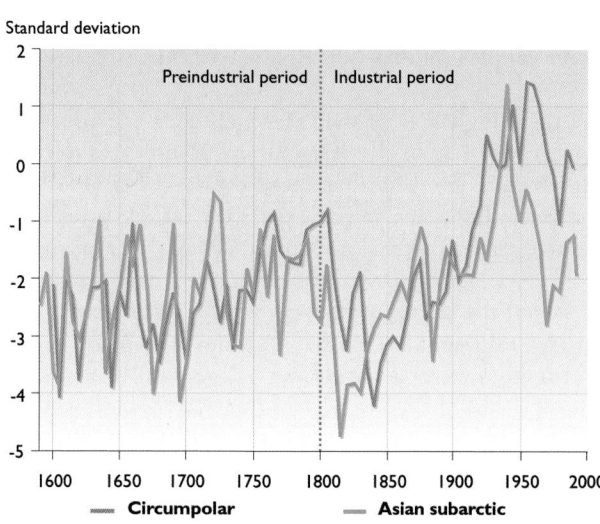

Fig. 14.12. Temperature variations from proxy records for the circumpolar Northern Hemisphere (data from Overpeck et al., 1997) and the Asian subarctic (see Fig. 14.11 for locations of sites).

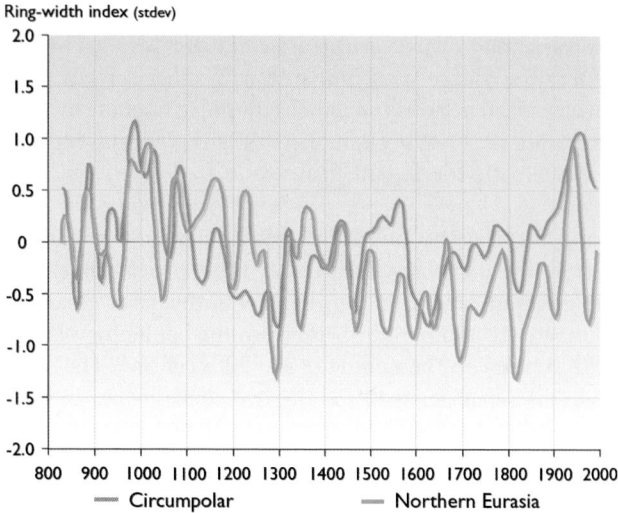

Fig. 14.13. Long-term changes in tree-ring growth in the circumpolar north (data from Esper et al., 2002) and northern Eurasia (combined chronology for east Taymir and northeast Yakutia).

to 1100 years old (Vaganov et al., 2000). Well-preserved dead tree trunks allowed extension of the record farther back than the maximum age of the living samples, allowing the construction of absolutely dated tree-ring chronologies for the periods from 431 BC to AD 1999 (Taymir) and from 359 BC to AD 1998 (Yakutia). The RCS approach was applied to distinguish variation caused by climatic change. These representative tree samples were highly responsive to recorded temperature, allowing the reconstruction of temperatures over the last two millennia with annual resolution. With this long-term record, the instrumental record of 20th-century temperatures can be compared with temperatures during the MWP. Earlier results (Hughes et al., 1999; Vaganov et al., 1996) showed that tree-ring chronologies were highly correlated across distances up to 200 km (up to 500 km in northern regions), so these millennial chronologies represent temperature variations over a large sector of the Siberian subarctic.

The combined chronology for east Taymir and northeast Yakutia was compared with a generalized tree-ring chronology developed for the circumpolar region by Esper et al. (2002; Fig. 14.13). The two chronologies are significantly correlated during last 1200 years (r=0.47, p<0.001), suggesting that they represent long-term temperature trends in the entire Northern Hemisphere subarctic. Both curves show the warming characteristic of the MWP (10th to 13th centuries), decreasing temperature during the Little Ice Age (LIA, 14th to 19th centuries), and a 20th-century temperature increase.

The Siberian summer temperature reconstruction indicates that the warmest centuries were AD 1000 to 1200, with anomalies (from the mean over the last millennium) of 0.70 °C and 0.57 °C for those centuries, and the coolest centuries in were in the LIA period (AD 1600 to 1900), with anomalies for those centuries of -0.42 °C, -0.39 °C and -0.56 °C. The analysis of

long-term trends of summer temperature leads to several conclusions. Present-day warming is estimated to represent an increase in summer temperature of approximately 0.6 °C above the coolest period of the LIA. The reconstruction clearly reveals the timing of the MWP, which occurs in the 10th to 13th centuries. The reconstructed data indicate a greater warming (by about 1.3 °C) above the long-term mean during the MWP compared to the amount of cooling below the mean during the LIA. The results agree well with previous assessments of medieval warming in the Northern Hemisphere (Esper et al., 2002) but show less warming than that projected by climate models or historical analogues that use natural forcing factors such as solar variability and volcanic activity (Budyko and Izrael, 1987).

14.6.1.3. Climate change in the eastern Taymir Peninsula over the past 6000 years

The dendrochronological material discussed in this section was gathered from the Kheta-Khatanga plain region and the Moyero-Kotui plateau in the eastern Taymir Peninsula, near the northernmost present-day limit of tree growth in the world. The wood samples were collected from three areas: the modern treeline in the north forest of Ari-Mas; the modern altitudinal treeline (200–300 m above sea level) in the Kotui River valley; and alluvial deposits in terraces of large tributaries of the Khatanga River (one sample location is 170–180 km north of the modern treeline). The total number of wood samples exceeds 400. The RCS approach was used to standardize individual series (Naurzbaev and Vaganov, 2000). Approximate absolute dating (in contrast to the relative dating developed from the ring series) was established from radiocarbon dates of 45 samples of subfossil wood collected throughout the soil organic layer.

The result of this cross-dating is an absolute tree-ring chronology (see section 14.6.1.2) as well as a series of "floating" chronologies up to 1500 years in length, evenly spaced within the 6000-year interval of the mid- to late Holocene. The relative dating of the "floating" chronologies is based on the calibrated radiocarbon age of the samples (Stuiver and Reimer, 1993).

The resulting curve of the tree-ring index (relative growth) extends over 6000 years, and indicates favorable climatic conditions at about 6000 years BP. This period of warmth represents the latter stages of the postglacial thermal maximum (Lamb, 1977a; section 2.7.4.2). The growth of larch trees at that time surpassed the average radial growth of trees during the last two millennia by 1.5 to 1.6 times. Tree growth (and temperature, accordingly) has generally decreased from the end of the postglacial thermal maximum through the end of the 20th century. Several samples of subfossil wood collected in the flood plain of the Balakhnya River were accurately dated in accordance with the "floating" chronology to the period from 4140 to 2700 BC. Due to the geography of the watershed, these samples could not have originated from the more southern regions, which means that during the postglacial thermal maximum the northern treeline was situated at least 150 km further north than at present. The postglacial thermal maximum can also be clearly identified by the relative levels of the stable isotope carbon-13 (^{13}C) in the annual tree rings (Naurzbaev et al., 2001). Increased ^{13}C concentration in annual layers of wood of this species (Cajander larch) is highly correlated with warm summer temperatures. High ^{13}C content is found in wood dated to the period that ring-width techniques reconstruct as warm, confirming high temperatures during this period.

Quantitative evaluation of mean deviations of average summer and annual temperatures demonstrates higher temperature variability during the postglacial thermal maximum compared to the 3.5 °C variability typical of the 20th-century instrumental temperature record. This corresponds well to the earlier published data (Naurzbaev and Vaganov, 2000) and to findings of subfossil wood in alluvial deposits of the Balakhnya River 150 km north of the present-day treeline. Reliable dates for the postglacial thermal maximum, in agreement with the "floating" chronologies, indicate that during this period sparse larch forest extended at least 1 to 1.5 degrees of latitude further north than the northernmost present-day forest limits in the Ari-Mas massif.

A comparison of the long-term temperature reconstruction for the Taymir Peninsula with other indicators of long-term temperature change in the high latitudes of the Northern Hemisphere during the Holocene (including summer melting on the Agassiz Ice Cap, northern Ellesmere Island, Canada; summer temperature anomalies estimated from the elevation of carbon-14 dated subfossil pine wood samples in the Scandes mountains, central Sweden; and temperature reconstruction from oxygen isotopes in calcite sampled along the growth axis of a stalagmite from a cave at Mo i Rana, northern Norway; Bradley, 2000) reveals several noteworthy features. During much of the Holocene, and especially starting between 9000 and 8000 years BP, the overall high-latitude Northern Hemisphere temperature steadily decreased, although there were shorter fluctuations with significant amplitude. This distinct trend of general temperature decrease agrees with the Taymir chronology. The concurrence of characteristic temperature fluctuations can be seen, for example, in significant decreases at about 6000 years BP and 4000 years BP, and increases at about 3000 years BP and 1000 years BP (the MWP). This coincidence suggests that some regions of the Arctic experienced long-term temperature changes in common with the high-latitude mean, which has been reconstructed using various proxy data. These interpretations of the data are supported, for instance, by the results of radiocarbon and dendrochronological dating of wood remains from the MWP collected north of the modern treeline in the Polar Urals (Shiyatov, 1993; section 14.11.1.3). However, some aspects of a reconstruction that infers higher summer temperatures during the postglacial thermal maximum than have been recorded in the late 20th century remain uncertain and subject to confirmation from additional research. The range in estimates obtained from different sources of the degree to which reconstructed temperature at its postglacial maximum exceeded the maximum warmth of the 20th or early 21st centuries is significant: from 0.6 °C (glacier and stalagmite layers and bottom deposits) to 3 to 3.5 °C (indicated by Taymir tree-ring chronology and the greatest extension of forest in the Scandinavian mountains; Kullman and Kjallgren, 2000). These deviations in reconstructed temperature may have been influenced by local conditions, different sensitivities of proxy sources to temperature change, or inadequate calibration models. Unfortunately, at present it is impossible to determine the cause of these deviations.

To detect the anthropogenic component of climate variations at high latitudes it is important to know whether temperature is *already* affected by increasing GHG concentrations and whether the rate of temperature increase is unprecedented in the period of instrument-based temperature records. A series of synthesizing studies, as well as simulations with GCMs, have established that anthropogenic emissions have had a significant influence on the rate of temperature rise in the Northern Hemisphere (Mann M. et al., 1998). However, in contrast to global trends, in the long-term northern Siberia tree-ring chronologies the present high-latitude summer temperature increase is less than that experienced during the postglacial thermal maximum. In this region, natural climatic forcing factors appear to have been more significant than the combination of human and natural factors producing the current summer warming thus far. In the Siberian study area, the amplitude of the current summer warming is not thus far greater than the warming during the MWP. This may be partly explained by the difference between regional and global trends, and partly by the difference between the trends in summer and annual temperature.

To determine with confidence whether the *rate* of temperature increase is unprecedented, it is necessary to obtain good quantitative data with high temporal resolution for the Holocene that will help to identify periods of drastic natural temperature increases in the past, and then to examine the amplitude of such drastic increases to see if there are natural limits during such periods of change. Therefore, one of the urgent tasks at present is the construction and analysis of super-long-term tree-ring chronologies for Eurasia with an adequate amount of subfossil wood. Such studies are being intensively conducted in Europe (Baillie, 2000; Leuschner and Delorme, 1988; Leuschner et al., 2000) and in north Asia (Hantemirov, 1999; Shiyatov, 1986; Vaganov et al., 1998). They should soon provide high-resolution tree-ring chronologies for Eurasia that can be used for quantitative reconstruction of temperature and for calibration of data obtained from other indirect sources of climatic information with lower temporal resolution.

Temperature anomaly (°C)

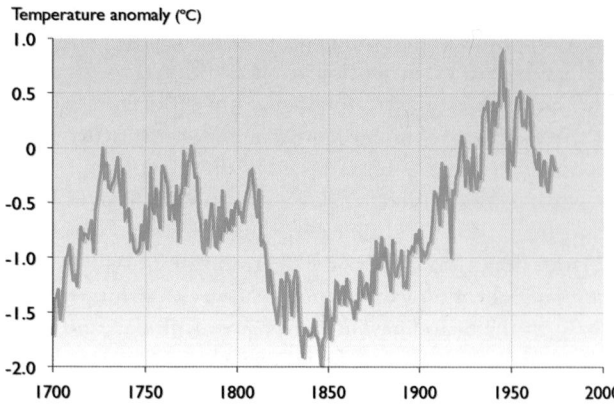

Fig. 14.14. Annual temperature anomaly (from 1671–1973 average) reconstructed from white spruce at the North American treeline (see D'Arrigo et al. (1996, 2001) and Jacoby and D'Arrigo (1989) for details of the reconstruction technique).

14.6.2. Past climate change in Alaska and Canada

One of the first large-scale climate reconstructions based on boreal tree-ring data was a study of treeline white spruce across northern North America (primarily Canada) covering about 90 degrees of longitude or about one-third of the circumpolar northern treeline extent (Jacoby and D'Arrigo, 1989). This chronology was based on the positive response of tree growth at treeline to temperature, and allows the reconstruction of mean annual temperature anomalies back to AD 1700 (Fig. 14.14). Key features of the reconstruction are intermediate temperatures during most of the 18th century, sharp cooling during the first half of the 19th century, gradual warming from the mid-19th century to a mid-20th century peak, and a slight cooling from about 1950 to the 1970s. If recent proxy data or climate records are available in a given locality to compare to the overall long-term record, unusual warming during the last decades of the 20th century is often noted (Jacoby and D'Arrigo, 1989; Jacoby et al., 1988). Some of the recent increase in annual temperatures in this reconstruction can be attributed to recovery from the last stages of the LIA.

Since the studies of the late 1980s, other analyses have added more spatial representation and longer temporal coverage (e.g., Mann M. et al., 1999) leading to an essentially complete coverage of the 20th century (Esper et al., 2002). The most recent reconstructed annual temperature curves confirm the major anomalies in annual temperature of the North American treeline curve of Jacoby and D'Arrigo (1989; see Fig. 14.14) that has served as the basis for standard Arctic temperature reconstructions (e.g., Overpeck et al. 1997).

The need to splice tree-ring records from overlapping generations of trees introduces some questions about whether the successively earlier generations of tree-ring records have been calibrated correctly and adequately preserve low-frequency variations (longer-term trends) in temperature. Methods of processing tree-ring data to

preserve the low-frequency variations correctly have improved, allowing such trends and their possible causes to be identified (Esper et al., 2002). The Esper et al. (2002) reconstruction shows more prominent low-frequency trends, including the MWP and the LIA, than previous reconstructions or global or Northern Hemisphere-wide averages.

New technology, such as x-ray density (Jacoby et al., 1988) and stable isotope techniques, allow measurements of tree-ring properties in addition to ring width. Maximum latewood density of northern conifers increases when mid- to late growing season moisture stress is great (D'Arrigo et al., 1992). Maximum latewood density of boreal conifers also may represent an index of canopy growth where productivity is temperature-related, as indicated by satellite-sensed normalized difference of vegetation index (NDVI; representing "greenness" of the land surface) values (D'Arrigo et al., 2000). Carbon-13 isotope content is generally measured as "discrimination", which represents the difference in the amount of the isotope in sampled plant tissue compared to a reference standard. Less ^{13}C discrimination (greater ^{13}C content in sample) indicates production of the sampled plant tissue under a condition of restricted stomatal exchange, generally as a result of moisture stress (Livingston and Spittlehouse, 1996).

Maximum latewood density, ^{13}C isotope discrimination, and ring width of upland white spruce stands in central Alaska are well correlated with each other (Barber et al., 2000). Latewood density and ^{13}C isotope discrimination contain information specific to the climatic conditions of the year of ring formation, in contrast to ring width (which is influenced by two or more years of temperature), making them ideal for reconstructing past climates (Barber et al., 2000). No continuous instrument-based

Temperature (°C)

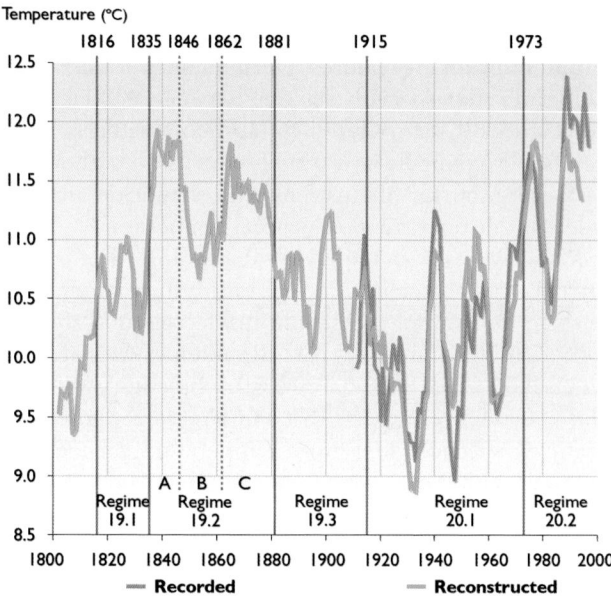

Fig. 14.15. Warm-season (Apr–Aug) temperature regimes and regime shifts in central Alaska from 1800 to 1996 from observations and reconstructed from tree-ring density and ^{13}C isotope discrimination (Barber et al., 2000; Juday et al., 2003).

temperature records exist for the western North American boreal region until the early years of the 20th century, and few records exist until the mid-20th century.

A 200-year reconstruction of warm-season (April to August) temperature at Fairbanks, Alaska, based on tree-ring density and ^{13}C isotope discrimination has been constructed (Barber et al., 2004; Juday et al., 2003; Fig. 14.15). The warm-season temperature reconstruction for Interior Alaska has been divided into multi-decade segments or warm-season temperature regimes. Regimes represent multi-decadal periods of characteristic temperatures that persist between periods of rapid climate change (Fig. 14.15). Note that the first half of the 20th century experienced extended periods of cool summers, which relieved moisture stress of low-elevation white spruce. The reconstruction of warm temperatures in the mid-19th century is out of phase with overall Northern Hemisphere means, but is strongly established by the proxies (^{13}C isotope discrimination and maximum latewood density) used in the reconstruction. A reconstruction of the annual Pacific Decadal Oscillation (section 2.2.2.2) index using western North American tree-ring records, accounting for up to 53% of the variance in instrumental records and extending back to 1700, also indicates that decadal-scale climatic shifts occurred in the northeast Pacific region prior to the period of instrumental record (D'Arrigo et al., 2001). These results suggest that rapid temperature shifts followed by semi-stable periods are a fundamental feature of climate change in that region.

From the perspective of two centuries, the recent very high rate of temperature increase in the second half of the 20th century in Interior Alaska is partially explained by a change from some of the lowest warm-season temperatures to some of the highest in the entire period (Fig. 14.15). Unlike the annual temperature reconstruction based on North America-wide treeline white spruce (Fig. 14.14), the Interior Alaska reconstruction (Fig. 14.15) indicates that the mid-19th century (Regimes 19.2A and 19.2C) was one of the warmest periods and the mid-20th century was a period of unusually cool summers. Because many of the climate records available in this part of the world begin only in the late 1940s or early 1950s (during the one of the coldest periods of the 20th century) and continue to the present (the warmest period of the last millennium), the instrument-based record indicates a higher rate of temperature increase than the longer-term reconstructions that incorporate several cycles of temperature increases and decreases. This suggests that the strong late 20th-century warming (during the warm season) in western North America may have a considerable component of natural climate variability in the signal.

There are still regions of the north that are not represented in large-scale temperature reconstructions, especially on the timescale of the past millennium. Recorded data and climate models strongly indicate that there are very likely to be important regional dif-ferences in temperature trends across the high-latitude north. In fact, a comparison of the large-scale reconstruction of northern North American annual temperature anomalies with the Interior Alaska reconstruction shows opposite trends (Fig. 14.2) that appear to be a consistent part of the climate system.

Across the Northern Hemisphere and beginning in different regions at different times, northern treeline trees display a reduced sensitivity to growing-season warmth (Briffa et al., 1998; Jacoby and D'Arrigo, 1995; Vaganov et al., 1999). The reduction in the positive response of some trees to warm-season temperature seems to have occurred around 1970, when warming resumed after a cool interval in the mid-20th century. When late 20th-century warming resumed, some trees continued to increase in radial growth in response to the warmer conditions. However the growth increase per unit of temperature increase was not nearly as great as previously, and in some trees temperature no longer had a reliable predictive relationship to tree growth at all (Briffa et al., 1998). In Alaska, much of the change can be attributed to greater moisture deficits associated with higher warm-season temperatures (Jacoby and D'Arrigo, 1995). In northern Siberia, the effect is attributed to shorter periods of thawed soil because of increased depth of snow cover, an indirect effect of warmer winters (Vaganov et al., 1999). For other areas, there are hypotheses about air pollution, UV radiation damage, and other factors (Briffa et al., 1998).

14.6.3. Past climate change in northwestern Europe

In the past few decades, Scandinavian tree-ring data have provided an increasing amount of information about past climate variability. Natural reserves preserve a number of old and mature forests virtually untouched by humans, especially in central and northern Scandinavia. Scots pine growing close to altitudinal or latitudinal distribution limits in the Scandinavian Mountains or in northernmost Sweden mainly respond to summer temperatures (with increased growth) and data from such sites have been used to interpret past climate variability (Briffa et al., 1990; Grudd et al., 2002; Gunnarson B. and Linderholm, 2002; Linderholm, 2002). Furthermore, due to the proximity to the Norwegian Sea, and hence the influence of maritime air masses brought in with westerly and south-westerly winds, precipitation may be a growth-limiting factor in moist areas, such as the western slopes of the Scandinavian Mountains or in peatlands (Linderholm et al., 2002, 2003; Solberg et al., 2002). High-frequency North Atlantic Oscillation signals (section 2.2.2.1) have been found in tree-ring data from east-central Scandinavia (Lindholm et al., 2001).

The relationship of climate to tree-ring variability in Scots pine in Scandinavia has been studied in a wide range of growth environments. A comprehensive study was made of pine growing on peatlands along a north–

south profile through Sweden to see if the trees contained high-resolution climate information (Linderholm et al., 2002). Peatland pines were also compared to pines growing on dry sites. Pines growing on peatlands are dependent on growing-season temperature and precipitation, as well as on local water-table variations, which are influenced by longer-term trends in both temperature and precipitation. There is a lag of up to several decades in the response of the pines to water levels, such that trees are integrating the immediate effects of growing-season climate as well as a delayed effect from the water table, making them unsuitable for high-frequency climate reconstruction. The sensitivity of pines growing on peatland also changes depending on climate. When the growing season is wet and cold, temperature is more important and trees respond positively to temperature, in particular to July temperature. Precipitation response increases to the south but is never as important as for pines growing on dry soils. Precipitation is important mainly in controlling water-table levels (Linderholm et al., 2002).

A 1091-year record of tree growth from AD 909 to 1998, developed from living and subfossil Scots pine in the central Scandinavian Mountains, provides evidence of low-frequency climate variation (Gunnarson B. and Linderholm, 2002). July temperatures had the largest effect on the growth of these trees, but growth was also positively and significantly correlated with October to December temperatures in the previous year. The response to precipitation during the vegetative period was negative although not significantly. The authors inferred that the chronology represents summer temperatures for the central Scandinavian Mountains, although it is suggested that care should be taken when interpreting the record. The chronology indicates prolonged excursions below the mean (cool conditions unfavorable for tree growth) in the mid-12th and the 13th centuries, and in the mid-16th and late 17th centuries (corresponding to the early and late LIA, respectively). Below-mean conditions in the late 18th century correlate with a "recent cold period" (Fisher et al., 1998; Grove, 1988; Jones and Bradley, 1992; Lamb 1977b). The chronology also provides evidence for the MWP in the 10th and early part of the 11th centuries as well as warmer periods during the mid-14th, mid-17th, and 20th centuries (Gunnarson B. and Linderholm, 2002).

14.7. Direct climate effects on tree growth

14.7.1. The Flakaliden direct warming experiment

14.7.1.1. Background

In 1994, a soil warming experiment began at Flakaliden near Vindeln, 65 km northwest of Umeå, Sweden (64° 07' N, 19° 27' E). The experiment is in a planted Norway spruce forest established in 1963. The environment of the area is representative of the northern portion of the European boreal forest. Mean annual temperature is 2.3 °C and mean annual precipitation about 590 mm. The goals of the Flakaliden warming experiment are to:

- quantify the effect of soil warming on the seasonal course of plant respiration and phenology of trees at low (irrigated) and high (irrigated and fertilized) availability of soil nutrients;
- test and improve available mechanistic models used to project impacts of climate change on respiratory dynamics in plants and forest soils; and
- to estimate net carbon budgets for boreal Norway spruce at the tree, stand, and regional scale, in present and future climates.

The soil warming treatment was installed in late 1994 in the buffer zone of one irrigated and one irrigated and fertilized stand; air temperature was not directly modified. Each heated subplot has a corresponding unheated control plot. The reason for using treatments including irrigation was to reduce the risk of drying the soil as an effect of the soil warming. The experiment was not designed to produce the effects of climate warming *per se*, but to isolate the effect of one of the most distinctive features of the boreal forest that is thought to contribute to its great carbon storage – soil temperature. While aboveground production of plant material is relatively great in boreal forests, cold soils limit the rate of decomposition that releases the fixed carbon back into the atmosphere. Therefore, soil warming, if it were to increase decomposition of stored soil carbon more than it affected production, could have a disproportionate effect on carbon balance.

In the Flakaliden experiment, soil warming starts in April each year, about five weeks before the soil thaws in the unheated plots. The soil temperature is increased by 1 °C per week, until a 5 °C difference between heated and control plots is reached. In late autumn, when the soil temperature in the control plots approaches 0 °C, the soil temperature of the heated plot is decreased by 1 °C per week. If the control plots do not freeze before 1 November, the temperature reduction is still initiated (CarbonSweden, 2003).

14.7.1.2. Questions, hypotheses, and results

Based on the assumptions of doubled atmospheric CO_2 concentration and a 4 to 6 °C increase in annual mean temperature, the following responses were hypothesized:

- Increased CO_2 and temperature will have a small positive effect on biomass production in boreal forest growing on nutrient-poor sites. The stimulating effect will mainly be due to a shorter period with frozen soils and increased nitrogen mineralization.
- Increased CO_2 and temperature will have a positive effect on photosynthesis in boreal forest growing on sites with good nutrient availability. The net effect on biomass production will, however, be reduced as an effect of increased plant respiration (mainly foliage).

- In boreal forests growing on poor sites, an increase in temperatures will stimulate soil respiration more than biomass production and therefore the net carbon balance will be negative until a new equilibrium is reached. The strength of the carbon source will depend on site index and the size of the soil carbon pool.
- The boreal forest ecosystems will be a major sink for atmospheric carbon, once new equilibriums between carbon fixation and decomposition of soil organic matter have been reached. The time to reach a new equilibrium will depend on site index and the size of the soil carbon pool.

Soil moisture is not normally limiting to growth at Flakaliden (Bergh et al., 1999). The earlier spring soil thawing and later autumn freezing in heated plots increased mineralization of soil organic matter, which increased the concentration of most nutrients in the needles (Bergh and Linder, 1999). The effect was most pronounced during the first years of warming, but was still apparent after the fourth season, by which time stemwood production had increased by approximately 50% compared to the control plots (Jarvis and Linder, 2000). Earlier access to water in spring results in an earlier start of photosynthesis (Bergh and Linder, 1999). After six seasons of warming at Flakaliden, stem volume production (m³/ha/yr) was 115% higher on heated and irrigated plots than on unheated control plots; on heated, irrigated, and fertilized plots production was 57% higher than on unheated plots. The results indicate that in a future warmer climate, with increased nitrogen availability and a longer growing season, biomass production is very likely to increase substantially on both low- and high-fertility sites in the more humid parts of the boreal forest (Strömgren and Linder, 2002). However, the Flakaliden results are specific to soil warming without air-temperature change incorporated and it is too early to determine whether the observed responses are transitory or will be long-lasting.

In recent years, the large contribution of fine root turnover (growth and death within the growing season) as a factor in the annual production and storage of carbon in boreal forests has been recognized. Interactive effects of soil warming and fertilization on root production, mortality, and longevity at Flakaliden demonstrated that that nitrogen addition combined with warmer soil temperatures decreases the risk of root mortality, and annual fine root production is a function of the length of the growing season (Majdi and Öhrvik, 2004). Under scenarios of climate change that increase soil temperature, and maintain adequate soil moisture and sufficient nitrogen, root production (and carbon stored in roots at a given time) in boreal forests is very likely to increase, especially at low-fertility sites.

The influence of soil temperature on boreal forest growth and carbon storage in natural field situations does not appear to be as great as the potential demonstrated in the Flakaliden results. After four years of warming, a major temperature acclimation had occurred and there was only a small difference in soil CO_2 flux between heated and non-heated plots (Jarvis, 2000). The timing of soil thawing (date of near-surface soil temperature rapidly increasing above 0 °C) was not a good predictor for the start of spring photosynthesis in boreal coniferous forest at five field stations in northern and southern Finland, northern and southern Sweden, and central Siberia. The best predictor of the start of spring photosynthesis was air temperature (Tanja et al., 2003). In one case, photosynthesis commenced 1.5 months before soil thawing. At most sites a threshold value for air-temperature indices projected the beginning of photosynthesis in the spring, which varied among the sites by 30 to 60 days. The threshold values varied from site to site, probably reflecting genetic differences among the species and/or differences in the physiological state of trees in late winter and early spring induced by climate. A single physiological temperature threshold for the start of photosynthesis may not exist.

14.7.2. Climate effects on tree growth along the Central Siberia IGBP transect

The International Geosphere-Biosphere Programme (IGBP) is an international, interdisciplinary scientific research program built on networking and integration. It addresses scientific questions requiring an international approach, and undertakes analysis, synthesis, and integration activities on broad earth-system themes. The goals of the IGBP are to develop common frameworks for collaborative research, form research networks, promote standardized methodologies, facilitate construction of global databases, undertake model and data comparisons, and facilitate efficient patterns of resource allocation.

In the early 1990s, the IGBP developed the Global Change and Terrestrial Ecosystems project to establish terrestrial transects for global change research as one way to study ecosystem and climate change across large spatial scales (Canadell et al., 2002). The IGBP terrestrial transects run for more than 1000 km along specific environmental gradients such as temperature or precipitation, and along more conceptual gradients of land-use intensity. They often cross ecotones such as tundra–taiga that are believed to be highly sensitive regions with strong feedbacks to global change. The Central Siberia IGBP transect was one of five high-latitude IGBP transects.

14.7.2.1. Climate response functions of trees along a latitudinal gradient

The geography of central Eurasia offers an excellent opportunity to examine how climate influences tree growth along an uninterrupted transect from the cold tundra margin in the north to the semi-arid steppe grassland of central Asia. This transect approach can provide some ideas about how climate change might affect growth by comparing the climate factors that historically and currently control tree growth as one proceeds southward. However, several limitations to this approach should be considered. For example, the genetics of individual trees of even the same species change from north to south. However, the comprehensive view of how tree growth responds differently to climate from cool to warm regions is still quite useful.

To define the main climatic factors that influence tree-ring growth in various regions along the Central Siberia IGBP transect, correlation coefficients of tree-ring structure chronologies with monthly temperature and precipitation were calculated (Fritts, 1976; Schweingruber, 1988, 1996). Tree-ring data from Siberian larch, Gmelin larch, Siberian spruce (*Picea obovata*), and Scots pine from 46 sites located in regions from the forest–steppe zone in the south to the forest–tundra zone in the north were used (Fig. 14.16). The relationship of tree-ring width to climate was investigated at all sites.

To strengthen the climatic signal common to each region, tree-ring data from sites were averaged as regional chronologies when sufficiently high correlation of master chronologies from the same vegetation zone permitted. Averaged regional chronologies were obtained for the following vegetation zones along the transect: forest–tundra and the northern part of the northern taiga; northern taiga; middle taiga; and southern taiga. These chronologies were compared to regional climatic data averaged for several meteorological stations. In the south, in the forest–steppe zone, the similarity of master chronologies was lower and three regional chronologies were obtained. They were correlated with the data from the nearest meteorological stations.

Comparison of the climatic response functions obtained for trees growing at different regions along the transect show that there is a change in the climatic factor that

defines tree-ring growth at sites located along the temperature gradient in central Eurasia (Fig. 14.17). Summer temperature is one of the most important external factors that define tree-ring growth at the northern treeline (Vaganov et al., 1996, 1999). It positively influences tracheid production and explains up to 70% of the variability in tree-ring width. In the middle taiga region, the effect of summer temperature on tree-ring growth decreases and the influence of precipitation increases (Kirdyanov, 2000). In this region, winter precipitation has a strong negative effect on tree-ring growth, while June temperature has a positive influence. In the forest–steppe zone, tree-ring climatic response functions are typical of regions with limited moisture (Fritts, 1976), where variability in tree-ring width is mainly explained by moisture variability (directly influenced by precipitation and indirectly influenced by temperature). The negative influence of spring and early summer temperature is explained by water loss from the soil at the beginning of growing season. High temperatures in the previous August and September affect soil water content, which is important for tree-ring growth activation during the next growing season (Magda and Zelenova, 2002; Schweingruber, 1996).

From north to south along the transect, the limiting positive effect of summer temperature in the forest–tundra zone is replaced by the limiting effect of spring precipitation (positive) and early summer temperature (negative) in the forest–steppe zone (Fig. 14.17). Climate conditions (soil moisture and temperature) at the beginning of and during the first part of the growing season play the key role in determining annual radial tree growth and wood production at various latitudes from the northern treeline to the forest–steppe zone. Moving from north to south, the start of the tree-ring growth season shifts to earlier dates. June and July conditions are important for trees growing at the northern treeline, whereas April through June precipitation and temperature influence tree-ring formation in the forest–steppe zone (Fig. 14.17).

Under climate scenarios projected by the ACIA-designated models, trees currently growing in a given central Eurasian forest zone begin to experience temperatures near the end of the 21st century that are typical today of the next zone to the south. The simplest response (linear) would be for the propagules of the vegetation of any zone under consideration to migrate northward and eventually reconstitute the zone further north – in effect a "migration" of the zone northward through regeneration over time, so that present-day zones are replaced in the same sequence by the zones found to the south (Fig. 14.17). However, novel features of the ACIA-designated climate projections and their effects through time could bring about a nonlinear forest response (Fig. 14.17). It is difficult to make specific projections of these outcomes, but it is possible that some processes resulting from the ACIA-designated climate projections will produce unique effects not seen within the range of temperature variability experienced during the last millennium.

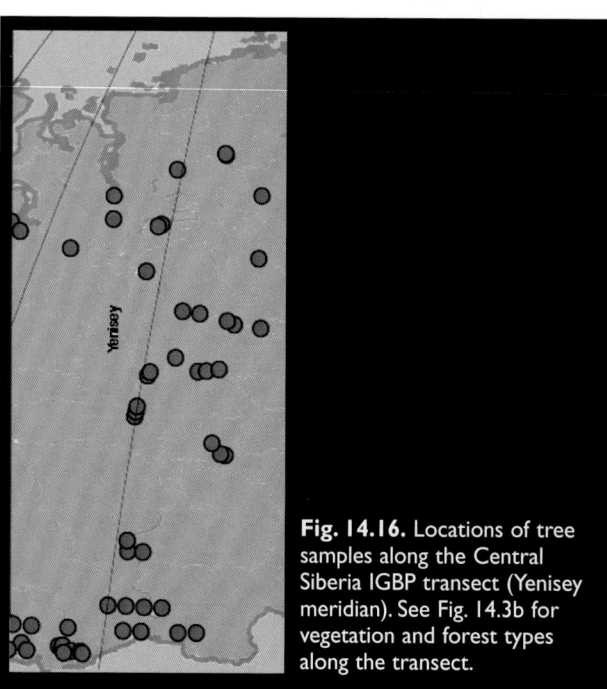

Fig. 14.16. Locations of tree samples along the Central Siberia IGBP transect (Yenisey meridian). See Fig. 14.3b for vegetation and forest types along the transect.

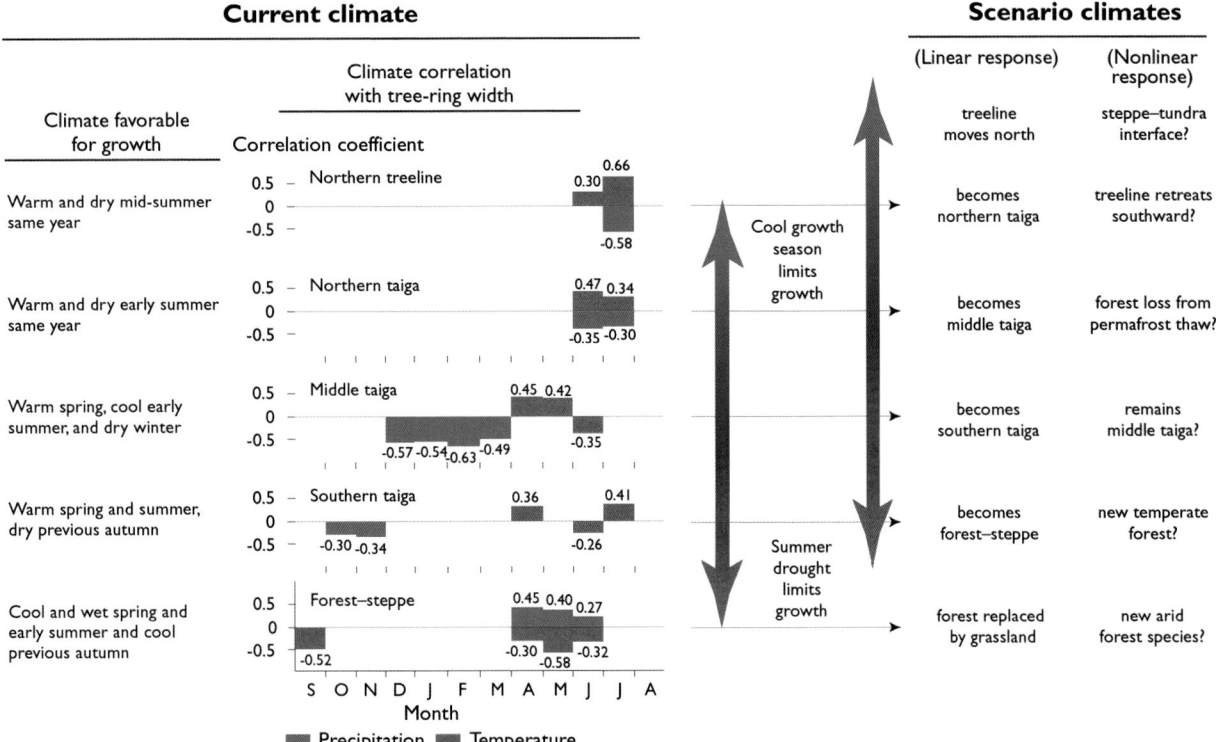

Fig. 14.17. Statistically significant correlations between tree-ring increments and climatic variables (typical climatic response functions) for different vegetation zones of central Eurasia along the Central Siberia IGBP transect. Potential alternative outcomes of climate change are also depicted. A warmer climate could result in the replacement of existing forest vegetation zones in sequence (linear response) or novel ecosystems could appear (nonlinear response) (data from Panyushkina et al., 1997; Vaganov, 1989; Vaganov et al., 1985, 1996).

For example, all the ACIA-designated models project an increase in temperature that is very likely to result in thawing of permafrost along the southern limit of its present-day distribution (section 6.6.1.3) and in low-elevation basins in the southern Yukon, central Alaska, and southwestern portions of the Northwest Territories. Permafrost thawing would transform forest soils and create site conditions that have few or no current analogues. The ACIA-designated models project annual temperatures in the southern boreal forest of central Eurasia at the end of the 21st century that are typical of present-day temperate forest, but the actual amount of moisture supply and species migration could either permit or hinder development of temperate forest in this region. The current territory of southern taiga is projected to be replaced by forest–steppe (Tchebakova et al., 1995) with a 15% phytomass decrease (Monserud et al., 1996) in a warmer climate, and forest degradation and decline during the transition period are very likely to result in a major increase in forest debris and forest flammability. The difference between the ACIA-designated model projecting the largest increase in central Eurasian growing season temperature, generally the CGCM2, and the model projecting the smallest increase, the CSM_1.4, was generally between 15 and 20%. At the extreme upper range of temperature increases in the CGCM2 and the ECHAM4/OPYC3 scenarios, it is possible that warming and drying effects will bring tundra into contact with semi-arid steppe (Fig. 14.17). Empirical relationships between evapotranspiration and vegetation in central North America project that aspen parkland will

extend into the Arctic (Fig. 14.1, Hogg and Hurdle, 1995) under a scenario based on doubled atmospheric CO_2 concentrations.

14.7.2.2. Variability in the strength of climate influence on tree growth

Tree-ring variability in Gmelin larch, Siberian larch, and Scots pine was compared from sites located at different latitudes along the Central Siberian IGBP transect. Tree-ring width chronologies were calculated for trees from the forest–tundra zone (71° N); northern (64°–69° N), central (61° N), and southern (58° N) taiga regions; the forest–steppe zone (53° N); and high-elevation forest (51° N) (Table 14.6). To obtain comparable values of mean tree-ring width for different regions, only growth of mature trees (i.e., the period when the age trend in individual tree-ring width curves is not pronounced) was analyzed. The average age of larch trees growing in the south of the study area is less than that of northern larches. Hence, mean tree-ring width for larch from the north (61°–71° N) was calculated for the most recent 50 years for trees older than 200 years, and at the sites located at 57° to 51° N, for larch after 150 years of growth. Because pine trees are generally younger than larch growing at the same latitude, mean tree-ring width of pine was calculated for the most recent 50 years for trees older than 150 years.

Correlation coefficients of individual chronologies with the master time series and coefficients of sensitivity indicate the strength of environmental influences that synchronize tree-ring growth at the same site. The higher these two parameters, the greater the role of the environment in tree-ring growth. These statistics were averaged for the master chronologies obtained for sites from the same latitudinal belt.

Tree-ring width in larch trees increases from north to south up to the region of southern taiga (57° N; Table 14.6), then decreases in the high-elevation forest zone of Tuva (51° N). Correlation of individual chronologies with master time series and tree-ring sensitivity values indicate a decrease in environmental influence on tree-ring growth from north to south along the transect. The highest correlation coefficients were obtained for the northern treeline region and sites 200 km to the south, while the lowest were in the southern taiga region. Tree growth in the southern taiga region is less sensitive to environmental influences than growth of northern trees and trees at high elevations (Table 14.6). Similar changes in tree-ring variability along the transect were found for pine. Tree-ring width increases from the northern border of the pine area to the south taiga region. At the same time, the correlation of individual series with the master chronology and the sensitivity both decrease (Table 14.6). By contrast, in the forest–steppe zone (53° N) the year-to-year variability in pine growth is more closely synchronized to various local climate factors rather than an overall regional signal.

These trends in tree-ring width variability along the north–south Central Siberian IGBP transects repeated longitudinally are typical of tree growth in the entire Siberian boreal zone. Shashkin and Vaganov (2000) reported similar results (increasing larch tree-ring width with decreasing site latitude) from Yakutia, East Siberia, caused by the gradient of environmental factors that

most influence tree-ring growth. At the northern treeline, lack of summer warmth is the main climatic factor that limits ring growth (Kirdyanov and Zarharjewski, 1996; Vaganov et al., 1996). As a result, at such high latitudes, large-scale patterns of summer temperature synchronize the growth of trees not only at the same site but also at sites located up to 800 km apart (Vaganov et al., 1999). Temperature increases projected by the ACIA-designated models are very likely to have positive effects on larch growth across this area.

The influence of summer temperature decreases moving southward from the northern end of the transect. This gradient of environmental control leads to the higher growth rate and lower sensitivity of tree growth to climate in the middle of the transect. However, toward the southern portion of the transect, there is once again an increase in the correlation between individual tree-ring growth and sensitivity of master chronologies. Summer warmth in these dry regions is a strongly unfavorable factor for pine ring formation at its southern limit of distribution and larch at its lower elevation limit. Therefore, at the southern end of the transect, the strong influence of environmental factors on larch growth in high-elevation forest and on pine in the forest–steppe zone synchronize ring formation in trees on similar types of sites. The broadly shared environmental controls caused better registration of environmental changes in tree-ring structure (higher sensitivity) among sites. Temperature increases projected by the ACIA-designated models are very likely to have negative effects on larch and pine growth across the southern part of the transect.

Table 14.6. Statistical characteristics of larch and pine ring-width chronologies from different locations along the Central Siberia IGBP transect.

Latitude (° N)	Number of sites	Number of trees	Mean tree-ring width (mm)	Correlation with master chronology	Coefficient of sensitivity
Larch					
71	4	48	0.19	0.8	0.42
69	3	55	0.21	0.82	0.4
64	7	71	0.24	0.67	0.34
61	5	78	0.37	0.66	0.28
57	3	28	0.53	0.58	0.21
51	4 and 3[a]	33	0.46	0.68	0.37
Pine					
66	1	19	0.32	0.65	0.3
61	3	24	0.36	0.6	0.22
58	2	24	0.52	0.55	0.21
53	4 and 2[a]	16	0.62	0.65	0.32

[a]Two samples at this latitude; the number of trees is the total for both sites.

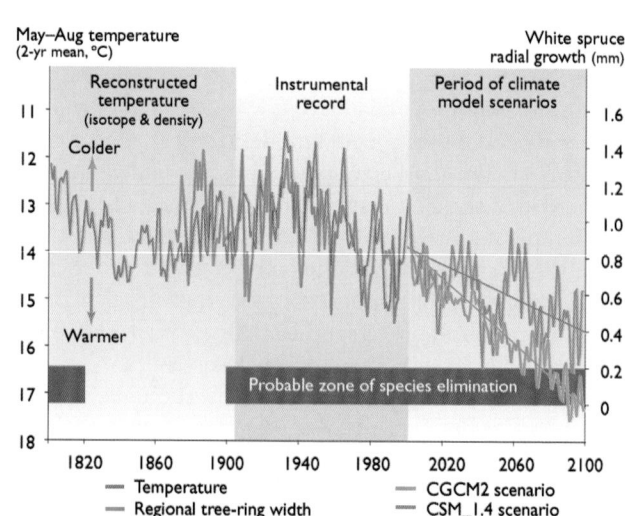

Fig. 14.18. Historic and reconstructed relationship between white spruce growth and summer temperature at Fairbanks, Alaska, and projections based on climate scenarios. The tree growth sample includes 10 stands across central Alaska. Summer temperature is an excellent predictor of white spruce growth. Because higher temperatures are associated with reduced growth and growth is the dependent variable, the temperature scale (left axis) has been inverted. Given the historical relationship between the variables and the scaling of these axes, the temperatures projected by the ACIA-designated models can be used to infer the approximate level of growth (right axis) possible in the future (data from Barber et al., 2004).

14.7.3. Response of high-latitude conifers to climate and climate change scenarios

14.7.3.1. White spruce in Alaska and Canada

The scientific literature on the relationship of tree-ring width to climate in northern North America is dominated by studies of white spruce carefully selected in order to allow the reconstruction of past climates (section 14.6.2) from trees that achieve greater growth with warmer temperatures and lower growth with cooler temperatures. However, radial growth of white spruce on upland sites across a broad area of central Alaska exhibits a strong negative response to summer temperature (Juday et al., 1998, 1999; Fig. 14.18). The negative relationship of radial growth to summer temperature is consistent throughout the 20th century, and occurs in a broad range of dominant and co-dominant trees in mature and old stands (Barber et al., 2000). The growth of white spruce on these sites is best projected by the mean of May through August temperature in the year of ring growth and the year prior (Barber et al., 2000). This relationship between climate and tree growth is sustained during the period of the instrumental record as well as the period for which summer temperatures were reconstructed from ¹³C isotope content and latewood density.

Based on this strong relationship, the two-year mean of May through August temperature for the Fairbanks grid cell was calculated from the ACIA-designated model and used to project future growth of this species on similar sites (Fig. 14.18). Because the relationship between temperature and growth is negative (less growth in warmer conditions) and the response variable of interest is tree growth, the temperature axis is inverted (increasing temperature downward) in Fig. 14.18.

The CGCM2 scenario projects the highest temperatures for the Fairbanks grid cell, although with a reduced range of annual variability. The CSM_1.4 scenario projects the least warming in the grid cell of the five ACIA-designated models, with variability similar to the recorded data (Fig. 14.18). If white spruce growth maintains the same relationship to temperature in the scenario period as during the calibration period, under the CGCM2 scenario growth is very likely to cease (the empirical relationship reaches zero growth) by the end of the scenario period (Fig. 14.18). Under the CSM_1.4 scenario, white spruce growth is very likely to decline to about 20% of the long-term mean. The zone of temperature and tree growth in Fig. 14.18 that corresponds to 20% or less of long-term mean growth has been highlighted as a "zone of probable species elimination". The stressed condition of trees in such a climate is likely to predispose them to other agents of tree mortality such as insect outbreaks (section 14.8) and diseases. While the CSM_1.4 scenario does not produce warming by the end of the scenario period that is empirically associated with zero growth, this white spruce population would be growing in a climate that is very likely to greatly reduce its growth. This climate and reduced level of growth almost certainly would place the trees that occur there now at an elevated risk of mortality, primarily from fire (section 14.9.2.3) and insects (section 14.8.2). White spruce in this region that demonstrate this climatic response are among the largest, most rapidly growing, and commercially valuable in boreal Alaska.

At high latitudes and altitudes where moisture is not limiting, white spruce has a positive growth response to summer temperature. The growth of near-treeline white spruce north of Goose Bay, Labrador in eastern Canada is positively correlated with the mean of monthly temperature in June, July, and September of the growth year and April of the previous year (Fig. 14.19; D'Arrigo et al., 1996). Some of the trees in the sample may have been responding to non-climatic factors in the early 1950s and especially in the late 1990s when a significant short-term growth decline occurred (Fig. 14.19). The long-term tree-ring chronology available from the Labrador near-treeline sample reconstructs past climates in general agreement with those of several instrument-based and modeling studies of this sector of the North Atlantic (D'Arrigo et al., 1996).

If the relationship between temperature and tree growth in the Labrador sample is maintained in the future, the ACIA-designated model projections again provide a basis for evaluating possible tree growth responses. In the grid cell containing Goose Bay, the CSM_1.4 model projects the greatest amount of warming by the end of the scenario period and the GFDL-R30_c scenario projects the least warming. The long-term mean of ring-width growth in the sample is set to 1.0 and the variation is expressed in units of standard deviation (Fig. 14.19). As projected using the regression lines of the scenarios, the growth of trees that retained the historical relationship of temperature and growth

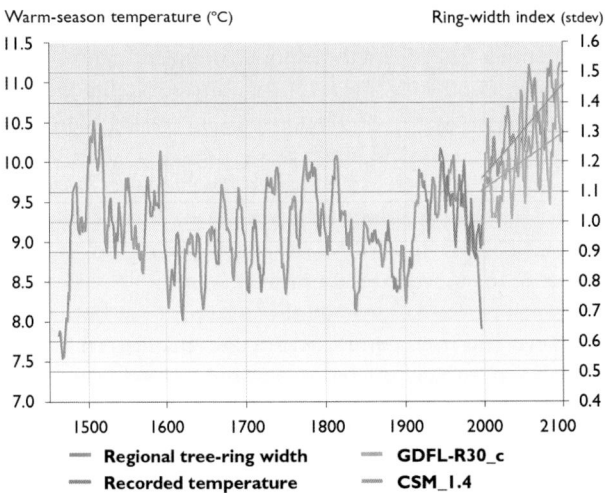

Fig. 14.19. Historic relationship between white spruce growth and mean warm-season temperature (June, July, and September of the growth year and April of the previous year, smoothed with a 5-year running mean), and climate scenarios for central and northern Labrador, Canada (data from D'Arrigo et al., 1996).

would increase to about 0.3 standard deviations greater than the long-term mean under the GFDL-R30_c scenario and nearly 0.5 standard deviations greater under the CSM_1.4 scenario. Both scenarios produce warming and inferred levels of tree growth that this area has not experienced since the beginning of the proxy record in the late 1400s. These populations of white spruce are relatively small, slow growing, and generally not commercially valuable.

The response of treeline white spruce populations to climate change is of particular interest because this species delimits much of the North American treeline (Sirois, 1999) and the performance under warmer conditions of present-day populations is an indication of the possible future effects of climate change. In a large (>1500 trees) sample of treeline white spruce across the mountains of the Brooks and Alaska Ranges, over 40% displayed a statistically significant negative growth response to summer warmth while slightly fewer than 40% had a positive response to late spring warmth (Wilmking et al., 2004). In the negatively responding (warmer = less growth) white spruce population, July temperature explained most of the variability in growth. Growth was strongly reduced at temperatures above a July threshold of about 16 °C at the Fairbanks International Airport climate station, which served as a representative common reference (temperatures were estimated to be 3 to 4 °C cooler at the various treeline sites). Growth of positive responders is correlated to March or April temperature, and the relationship is generally not significant until the second half of the 20th century (Wilmking et al., 2004). D'Arrigo and Jacoby (Lamont-Doherty Earth Observatory, Columbia University, pers. comm., 2003) found a similar dual response in white spruce in the Wrangell St. Elias Mountains of Alaska. A study of eight sites at and near alpine and arctic treeline in three regions of Alaska found mixed populations of temperature response types as well, and growth decreased in response to increasing temperatures at all but the wettest sites after 1950 (Lloyd and Fastie, 2002). The negative growth response to temperature is more common in contiguous stands and tree islands (clusters of trees) than in isolated individual trees (Lloyd and Fastie, 2002; Wilmking et al., 2004). The intensity of the negative effect of July warmth increased after 1950, directly reflecting July temperatures above the threshold in a greater number of years in the second half of the 20th century compared to the first half (Wilmking et al., 2004).

The explanation for reduced growth in treeline white spruce with warming is that negative responders are experiencing temperature-induced drought stress, while the positively responding trees are not (Jacoby and D'Arrigo, 1995; Juday et al., 2003; Wilmking et al., 2004). At some treeline sites in northern Alaska, individual trees have shifted their response to climate during their lifetimes, following the regime shift to warmer conditions that took place in the last decades of the 20th century (Jacoby and D'Arrigo, 1995). Whereas previously the growth of the tree responded positively

to summer temperature, it either became insensitive or began responding negatively to summer temperature after the shift to the warmer regime.

These results establish that from an ecological perspective, recent climate warming has been a major event that has strongly affected the growth performance of the majority of white spruce at and near treeline in Alaska and almost certainly in similar climate zones in Canada. The five ACIA-designated models project July temperatures by the mid- to late 21st century that the empirical relationship associates with very low or no growth in negatively responding white spruce. This suggests that under the ACIA-designated model projections, it is possible that the northern white spruce tree limit in Alaska and adjacent Canada would not readily advance. At the least, treeline is likely to become much more complex, with negative responders disappearing and positive responders expanding. It is possible that novel conditions would emerge, such as a portion of the southern tundra boundary in North America that is separated from the boreal forest by aspen parkland (see Fig. 14.1; Hogg and Hurdle, 1995). The actual response of treeline white spruce in this part of the world also adds support to the suggestion that aridification could be a major issue under climate scenarios projected by the ACIA-designated models.

14.7.3.2. Black spruce in Alaska and Canada

Stands dominated by black spruce represent about 55% of the boreal forest cover of Alaska and a large fraction of the northern boreal region of Canada. The response of black spruce to climate has not been studied as extensively or for as long as white spruce, so the literature on climate/growth relationships is more limited. This section draws on some recent dendrochronology work in Interior Alaska.

Different temperature factors are associated with black spruce growth on different permafrost-dominated sites (Fig. 14.20), a more variable response than the consistent temperature/growth response in upland white spruce (section 14.7.3.1). Of four sites examined, three show a negative growth response to increasing temperature, while one shows a positive response (Fig. 14.20). Both the Toghotthele site (Fig. 14.20a) and the Zasada Road 10 site in the Bonanza Creek Long Term Ecological Research site (Fig. 14.20c) show a negative growth response to summer temperatures (year of growth and 2 years prior). Radial growth of black spruce at both the northern and southern BOREAS sites in western Canada was also negatively related to summer temperature, or conversely, favored by cooler and wetter conditions (Brooks et al., 1998).

Black spruce at the Caribou-Poker Creeks Research Watershed (CPCRW) show both a negative correlation to April and May temperatures during the growth year and a positive correlation to February temperature two years prior to growth (Fig. 14.20b). A simple two-

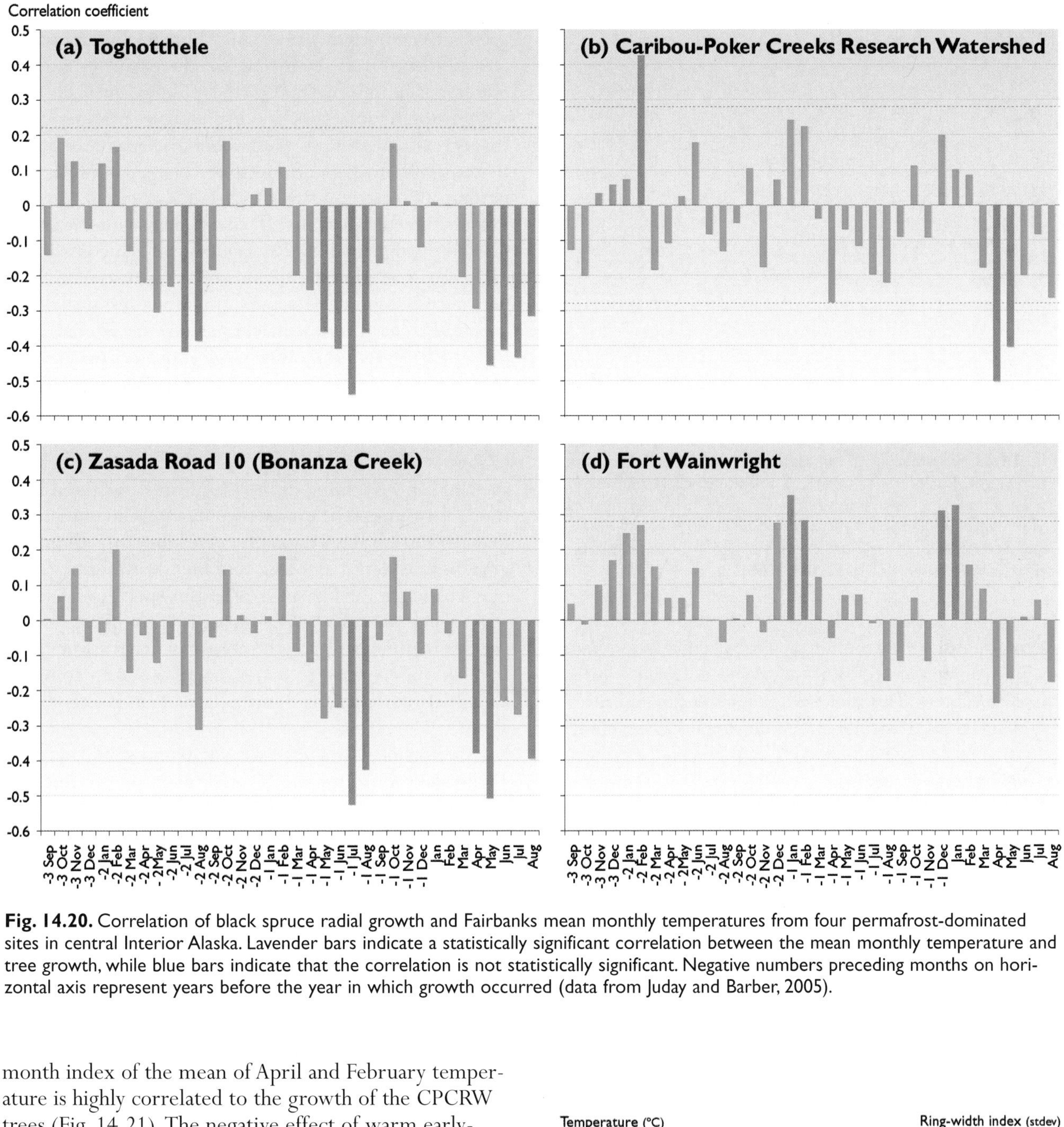

Correlation coefficient

(a) Toghotthele

(b) Caribou-Poker Creeks Research Watershed

(c) Zasada Road 10 (Bonanza Creek)

(d) Fort Wainwright

Fig. 14.20. Correlation of black spruce radial growth and Fairbanks mean monthly temperatures from four permafrost-dominated sites in central Interior Alaska. Lavender bars indicate a statistically significant correlation between the mean monthly temperature and tree growth, while blue bars indicate that the correlation is not statistically significant. Negative numbers preceding months on horizontal axis represent years before the year in which growth occurred (data from Juday and Barber, 2005).

month index of the mean of April and February temperature is highly correlated to the growth of the CPCRW trees (Fig. 14.21). The negative effect of warm early-spring temperatures on growth at CPCRW can be attributed to the onset of photosynthesis in spring when the ground is still frozen, causing desiccation and damage to the needles (Berg E. and Chapin, 1994) early in the growing season. The smoothed (5-year running mean) values are highly correlated during the 20th century (r=0.86), suggesting that tree growth of this species at sites similar to these could be projected using the ACIA-designated scenarios. While warm February temperatures favor growth, warmth in April depresses growth. With substantial temperature increases in the late 20th century, growth of this species has declined because the negative influence of April is stronger.

Growth of black spruce trees at Fort Wainwright is positively correlated with winter temperatures (Fig. 14.20d). When January in the year of growth and December, February, and January in the year prior to growth were all

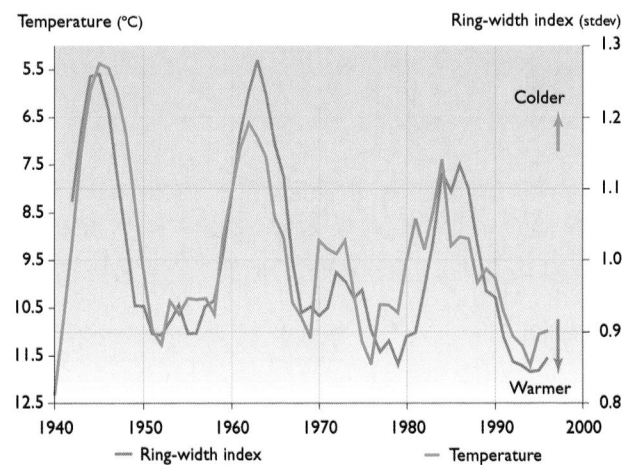

Fig. 14.21. Relationship of radial growth of black spruce at Caribou-Poker Creeks Research Watershed to the mean of April (growth year) and February (two years prior) temperature, smoothed with a 5-year running mean. Temperature axis is inverted as in Fig. 4.18 (data from Juday and Barber, 2005).

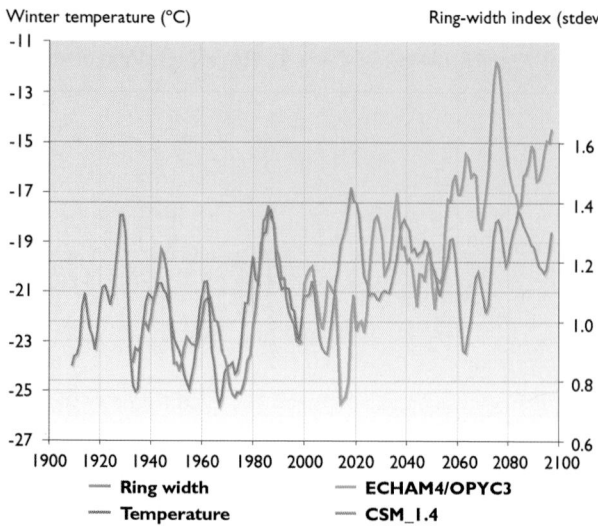

Fig. 14.22. Relationship of radial growth of black spruce at Fort Wainwright, Alaska (n=20 trees), to a 4-month climate index (mean of monthly temperature at Fairbanks, Alaska, in January of growth year and January, February, and December of previous year). Scenario lines show projections of the 4-month climate index (data from Juday and Barber, 2005).

warm, the trees grew better (Fig. 14.22). The smoothed (5-year running mean) values are highly correlated during the 20th century. This is one of the few species and site types in central Alaska for which the empirically calibrated growth rate can be inferred to improve under projected higher temperatures.

The positive relationship between monthly temperatures and black spruce growth at the Fort Wainwright site suggests that growth of tree on sites similar to it is very likely to increase (Fig. 14.22). The ECHAM4/OPYC3 model projects especially strong warming for the Fairbanks grid cell in the winter months that best predict black spruce growth at the site, so the empirical relationship, if it were maintained, suggests a major

increase in growth rate (Fig. 14.22). Black spruce at the Fort Wainwright site occupy a low productivity system, so the overall significance of the projected growth increase is not clear. The CSM_1.4 model projects winter temperatures that do not increase as rapidly or to as great a degree, so the modeled relationship of these positively responding spruce suggests only a modest increase in growth under that scenario. Presumably, the positive effect of warm winter temperatures on growth is experienced though control of active-layer rooting depth and soil temperature in this permafrost-dominated ecosystem. However, well before the end of the scenario period, the CGCM2 (Fig. 14.9) and all the other models project mean annual temperatures above freezing in this grid cell. If temperatures increase to that extent, the temperature/growth relationship depicted in Fig. 14.22 is not likely to persist. However, this permafrost site is very near thawing, and warming of the magnitude projected by the ACIA-designated models would probably initiate thawing during the 21st century, leading to widespread ground subsidence and tree toppling, representing a new challenge for the survival of this species on such sites. Once the soil thawing process is complete, species with higher growth rates than black spruce, such as white spruce or paper birch, are likely to have a competitive advantage on the transformed site.

For the Toghotthele site (Alaska Native-owned land; Toghotthele Corporation), model projections of future temperature were compared to the empirical record of black spruce growth (Fig. 14.23) similar to the approach used for white spruce (section 14.7.3.1). The mean of four summer months is an excellent predictor of tree growth, with warm years resulting in strongly reduced growth. As for white spruce, the results suggest that if climates similar to those projected by the ACIA-designated models actually occur, by the end of the 21st century black spruce would experience climates that are very unlikely to permit the species to survive on similar types of sites. In this case, the CSM_1.4 model projects the highest levels of the particular set of monthly temperatures that drives the relationship while the CGCM2 model projects slightly lower levels. However, in both scenarios, warmth in individual years produces an empirical relationship very near zero growth before the end of the scenario period. Allowing for some differences in calibration and degree of climate control, the generally similar growth of black spruce in the BOREAS study areas suggests that elimination of black spruce is very likely to be widespread across the western North American boreal forest.

14.7.3.3. Scots pine in Scandinavia

Growth responses of Scots pine at the tree limit in the central Scandinavian Mountains varied throughout the 20th century (Linderholm, 2002). Long, hot, and dry summers have traditionally been thought to be optimal for pine growth in that environment. The greatest pine growth of the past three centuries occurred during the

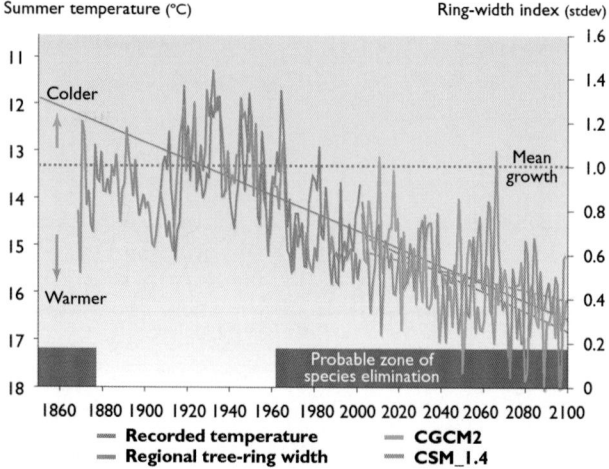

Fig. 14.23. Relationship between summer temperatures (mean of May and June in growth year, and June and July of the previous year) at Fairbanks and relative growth of black spruce at the Toghotthele site in central Alaska, and projections based on climate scenarios. Temperature axis is inverted as in Fig. 14.18 (data from Juday and Barber, 2005).

decade from 1945 to 1954. Although summer temperature was not particularly high during the mid-20th century, higher than average spring and autumn temperatures that extended the growing season were inferred to be the cause of this decade of high Scots pine growth (Linderholm, 2002). Despite similar apparently favorable temperature conditions in the latter part of the 20th century, growth of this species on similar sites declined during the most recent warmth. This suggests the occurrence of some additional unique component of

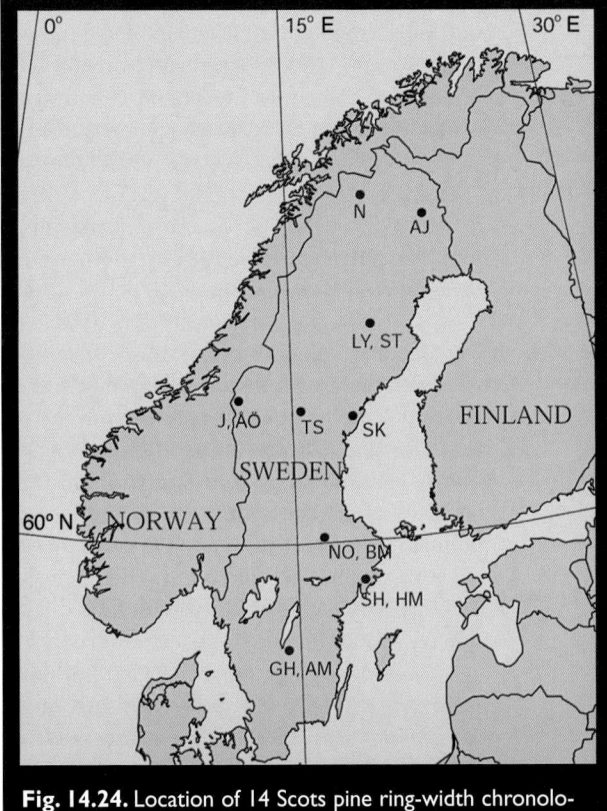

Fig. 14.24. Location of 14 Scots pine ring-width chronologies (see Table 14.7 for key to locations).

Table 14.7. Locations of 14 Scots pine ring-width chronologies. First column shows abbreviations used in Figs. 14.24 and 14.25.

	Location	Site type
N	Nikkaluokta	dry
AJ	Angerusjärvi	wet
LY	Lycksele	dry
ST	Stortjäderbergsmyran	wet
J	Jämtland	dry
ÅÖ	Årsön	wet
TS	Tannsjö	dry
SK	Skuleskogen	dry
NO	Norberg	dry
BM	Bredmossen	wet
SH	Stockholm	dry
HM	Hanvedsmossen	wet
GH	Gullhult	dry
AM	Anebymossen	wet

the recent warming that negated the previously positive influence of extended growing seasons.

In addition to north–south transects of tree-ring chronologies, an east–west transect has been developed in Fennoscandia (Linderholm et al., 2003). Nine tree-ring width chronologies for Scots pine were compared for growth variability and response to climate along a gradient of maritime to continental conditions in central Fennoscandia. The study revealed higher growth variance and stronger response to climate in the oceanic area west of the Scandinavian Mountains, compared to the more continental areas further east. Pine growth responded positively to elevated summer temperatures in the western areas, and positively to high summer precipitation in the east. Generally, pine growth showed a weaker relationship with the North Atlantic Oscillation (section 2.2.2.1) than with temperature and precipitation. During the last half of the 20th century, pine growth in western Fennoscandia displayed reduced sensitivity to climate, while in the east, growth sensitivity increased. Indications of growth stress were found in one site east of the Scandinavian Mountains. Increasing temperatures have been accompanied by increasing precipitation in Fennoscandia throughout the 20th century, and it was suggested that a change in climate regime from subcontinental to sub-maritime caused those trees to experience climatic stress (Linderholm et al., 2003).

A selection of Scots pine tree-ring width chronologies was collected from across Sweden (Fig. 14.24). The standardized chronologies include wet (peatland) and dry (mineral soil) growth environments (Table 14.7). Despite large differences in climate responses among the sites, the collective growth trend for all the sampled Scots pine in the last few decades of the 20th century is negative (Fig. 14.25). Furthermore, a common feature is distinct changes in growth sensitivity to climate at all sites during that period, a feature that has also been observed at several other Fennoscandian sites. The pattern is one of widespread and simultaneous decreases in growth, and a decreased ability of previously established climate factors to project growth as accurately.

Because precipitation, in addition to temperature, has increased over the past decades in Scandinavia, it is unlikely that drought stress is the main reason for decreasing Scots pine growth throughout Sweden. The one exception is northernmost Sweden, where growth responses to summer temperatures are decreasing and responses to autumn precipitation increasing. However, in central Sweden, the evidence suggests a different cause of reduced growth and changing climate responses. On the eastern slopes of the Scandinavian Mountains, the climate could be regarded as subcontinental, with low annual precipitation and high temperature amplitude over the year. Tree-ring data from sites east of the mountains implies that over the past decades, the climate of this area has become more maritime, and consequently the trees may have suffered stress from surplus moisture rather than drought

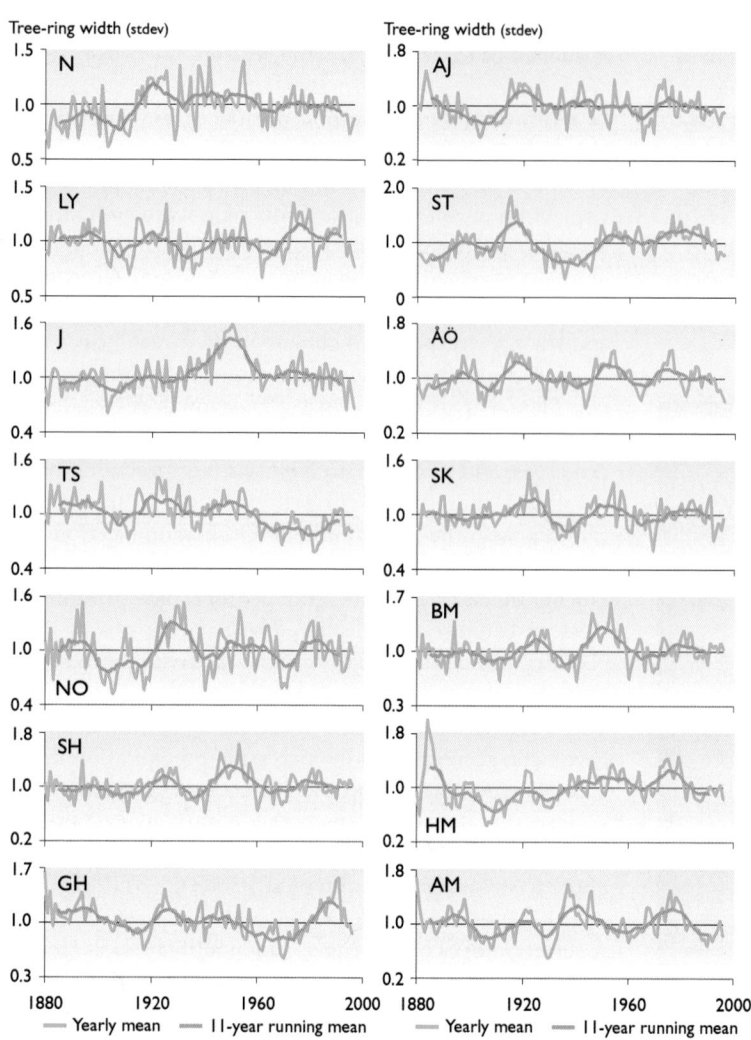

Fig. 14.25. Standard chronologies of Scots pine ring width (age de-trended and normalized with long-term mean set to 1.0 and standard deviation on 1.0) in Sweden. Curves express the degree of growth at any time relative to the long-term mean of the sample (see Table 14.7 for key to locations; data from Linderholm, 2002; Linderholm et al., 2002, 2003).

include bark and wood-boring beetles, defoliating insects (often Lepidoptera – moths and butterflies), and insects that attack roots and cones. Major population increases of these insects (irruptions), along with wildland fires, are the major short-term agents of change in the boreal forest. Insect-caused tree mortality can appear suddenly and within 1 to 10 years affect hundreds of thousands to millions of hectares of boreal forest in an outbreak or related series of outbreaks. Moderate to severe defoliation of trees by insects has affected nearly 16 million ha/yr in the province of Ontario since 1975 (Parker et al., 2000). Nearly all boreal forest regions are subject to this phenomenon, and it may represent an adaptive response in which simultaneous mass tree death helps create conditions for the renewal of site productivity through processes that warm soils, mobilize nutrient elements, and promote simultaneous tree regeneration.

The peculiar vulnerability of boreal forests to large-scale insect disturbance may also be related to the low biodiversity of both the host tree populations (limited number of species) and the limited diversity within the complex of tree-attacking insects and the populations of predators and parasites that work to stabilize their numbers. Ecological theory predicts that systems with few species should be less stable or buffered against major population swings, and of the major forest regions of the world, the boreal forest is the least species-rich in most taxonomic groups. Finally, the occurrence of tree hybrids may be a non-climatic risk factor for large-scale insect outbreaks. Where the distributions of two wind-pollinated tree species of the same genus (e.g., spruce, pine, larches) approach each other or overlap, the trees can interbreed and form hybrid offspring. Because the hybrid trees contain a random assortment of the genes of both parents, they seldom inherit highly integrated and complete defense traits against insect attacks, such as specific chemical defenses or timing of events to avoid vulnerability. As a result, tree defenses against insect attack may be lowest in hybrid zones (Whitham, 1989; Whitham et al., 1994).

The level of tree mortality caused by insect outbreaks can vary from selective removal of a few percent of dominant trees to the death of nearly the entire forest canopy. Although vulnerable tree species and age classes must be present for insect outbreaks to occur, climatic events are often the trigger or proximate cause of the insect population increases and associated widespread tree mortality. In general, many organisms in the boreal region, and especially the northern boreal forest region, are heat-limited, that is, they could perform (survive, grow, and reproduce) better with more

(Linderholm et al., 2003). In conclusion, the observed change in tree growth patterns and growth/climate relationships is most likely an effect of contemporary climate change. How trees respond to increased temperatures and/or precipitation is dependent on the local climate, and it seems that trees in oceanic areas are less affected than are those where the local climate regime has shifted from subcontinental to sub-maritime.

14.8. Climate change and insects as a forest disturbance

14.8.1. Role of insects in the boreal forest

The boreal forest is naturally subject to periodic large-scale tree mortality from insect outbreaks. The death of attacked trees has a considerable influence on the balance of carbon sequestered by forests. In boreal forests of Canada, insect-caused timber losses (tree death) may be up to 1.3 to 2.0 times greater than the mean annual losses due to fires (Volney and Fleming, 2000). The particular insects involved in large-scale boreal outbreaks warmth if other factors did not become limiting. Many insect agents of tree mortality in the northern boreal region are heat-limited, so that sustained periods of abnormally warm weather are often associated with irruptions of insects that attack boreal tree species. Given that this is a natural climate-driven system, it is important to distinguish climate change effects from natural operation of this system.

Generally, a climate change effect imposed on the natural insect/tree-death system is very likely to result in a greater frequency of insect outbreaks; more extensive areas of tree mortality during outbreaks; and greater intensity of insect attack resulting in higher average levels of tree death within outbreak areas. The sustained operation of these factors as a result of temperature increases, if they occur, is very likely to begin to change regional tree species composition within decades. A necessary confirmation of the occurrence of a specific climate change effect on the insect/tree-death process would include the detection of differences between the composition of the new, altered forest and historical and recent paleo-ecological records of previous generations of forest. An alternative possibility that needs to be considered is that temperature increases (along with increased precipitation) will possibly produce a more favorable growth environment for certain tree species on certain site types. The increased vigor of the trees is very likely to increase their ability to produce defensive compounds and successfully resist the similarly likely increased levels of insect attacks.

It is possible that the influences of climate warming on forests and on tree-hosted insects (positive and negative for forest health) could occur simultaneously in different locations of the boreal forest, or could occur sequentially in one place at different magnitudes and rates of temperature increase. As a result, temperature increases could be associated with improved forest health in some locations and forest health problems elsewhere, and the climate change effect at any given time would be the sum of these two outcomes. In addition, the initial stage of temperature increases would produce one profile of the balance of forest health outcomes across the area of analysis, but later stages representing greater temperature increases would shift the previous balance of forest health outcomes. The empirical record strongly suggests overall negative effects on boreal forest health from sustained temperature increases.

14.8.2. Spruce bark beetle in Alaska

During the 1990s, the Kenai Peninsula in south-central Alaska experienced the largest outbreak of spruce bark beetles (*Dendroctonus rufipennis*) in the world (Werner, 1996). The forests of the Kenai lowlands are transitional between coastal Sitka spruce (*Picea sitchensis*) rainforests and the white spruce–paper birch boreal forests of semi-arid central Alaska. From 1989 to 1999, more than 1.6 million ha of mature forest made up of white spruce and hybrids of Sitka and white spruce (called Lutz

spruce) in south-central Alaska experienced at least 10 to 20% tree mortality, the threshold level for aerial mapping detection. In much of the insect outbreak area, virtually all mature canopy trees were killed. Extensive logging was initiated to salvage beetle-killed timber on private lands, and the US federal government has found it necessary to appropriate funds to reduce fuel hazards in stands with numerous dead trees after infestation.

The first documented spruce bark beetle infestation on the Kenai Peninsula occurred in 1950. Local infestations occurred throughout the 1950s and 1960s on the northern Peninsula (Holsten, 1990). A major outbreak in the northern Peninsula that started in 1970 prompted the US Forest Service to initiate annual surveys in 1971, which have continued with few interruptions. Annual survey maps show that spruce bark beetles are endemic to Kenai Peninsula forests, at least under the present climatic regime. In any given year, there is always a background level of infestation somewhere in south-central Alaska, with periodic major outbreaks in some years.

The relationship of spruce bark beetle to climate involves two direct temperature controls over populations of the insect, and an indirect control through host tree resistance. Two successive cold winters depress the survival rate of spruce bark beetle to such a low level that there is little outbreak potential for the following season (Holsten, 1990). The spruce bark beetle is a heat-limited organism on the Kenai Peninsula and normally requires two summers to complete its life cycle. During abnormally warm summers on the Kenai Peninsula, the spruce bark beetle can complete its life cycle in one year, dramatically increasing the potential for population buildup (Werner and Holsten, 1985). Tree health increases host resistance to beetle attack: healthy spruce trees can successfully resist moderate numbers of beetle attacks by opposing the wood-boring activity of females entering the tree to lay eggs with pitch under high turgor pressure (Holsten, 1990). The beetles are unable to overcome the flow of pitch and either are expelled or succumb in their pitch tubes. Host trees that are under stress, including either climate stress or stress from mechanical breakage, have reduced growth reserves, less pitch, and lower turgor pressure, and so are less able to resist spruce bark beetle attacks. When regional events stress entire populations of trees, spruce bark beetle reproductive success is greatly increased.

The US Forest Service conducts annual aerial surveys for "red needles" (i.e., conifer needles dead for a year or less) that allow areas of spruce killed by spruce bark beetle within the previous year to be mapped. These annual surveys describe the areal extent of active infestation; densities of recently killed trees can range from a few stems to hundreds of stems per hectare (USDA Forest Service, 1950-1999). There is a one- to three-year lag between beetles entering a stand and a stand-wide flush of red needles (tree death). Typically, a sequence of red-needle acreage maps on the Kenai Peninsula will show a rise and fall of active infestation over a 5 to 10 year period on a scale of 100 to 1000 ha. After that time, nearly all white/Sitka/Lutz spruce greater than 12 cm in diameter will generally be dead in the affected stand, and subsequent surveys will record no more red-needle acreage for that stand.

Following the 1976–1977 increase in North Pacific sea surface temperatures (Mantua et al., 1997), mean annual and summer temperatures in the Kenai Peninsula lowlands rose 0.5 to 1.5 °C. This temperature increase is the latest acceleration of post-LIA climate warming in south-central Alaska (Jacoby and D'Arrigo, 1995; Wiles et al., 1996). Until recently, the most visible manifestation of the late 20th-century warming was a drying landscape, with dried-up ponds, falling water levels (as much as 1 m) in closed-basin lakes, and black spruce invasion of peat wetlands across the landscape.

Annual red-needle acreages for the southern Kenai Peninsula can be compared with standardized climate

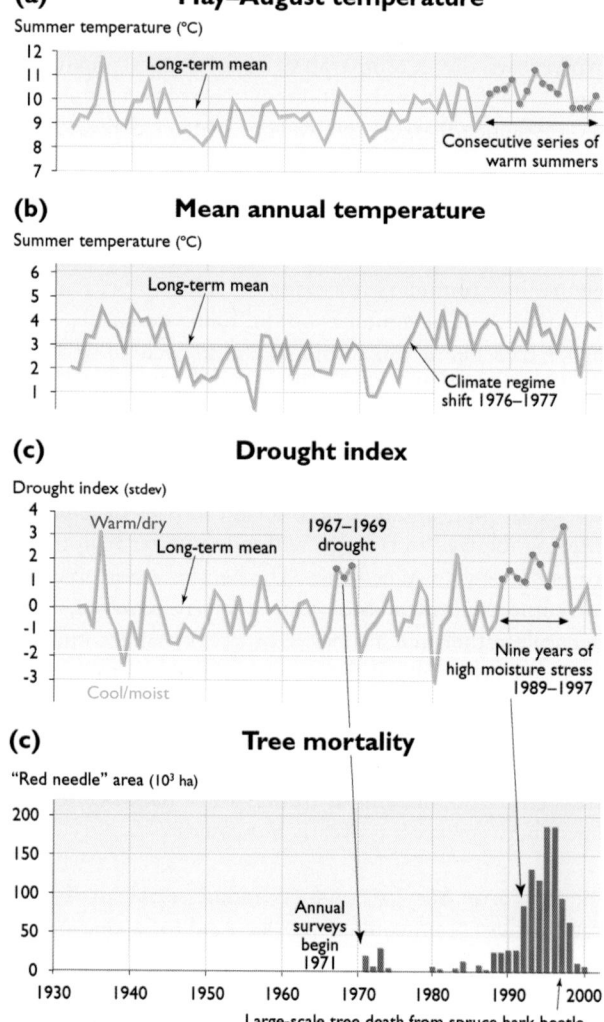

(a)   May–August temperature

Summer temperature (°C)

(b)   Mean annual temperature

Summer temperature (°C)

(c)   Drought index

Drought index (stdev)

(c)   Tree mortality

"Red needle" area (10³ ha)

Large-scale tree death from spruce bark beetle

Fig. 14.26. Timing of climate events that release spruce bark beetles from population limits showing (a) mean summer (May–Aug) temperature; (b) mean annual temperature; and (c) drought index, compared to (d) actual outbreaks of bark beetles (represented by "red-needle" area) during the 20th-century period of record (data from E. Berg, Kenai National Wildlife Refuge, Soldotna, Alaska, pers. comm., 2002; NOAA, 2002 (Homer Airport station); Wittwer, 2004).

parameters recorded at the Homer Airport climate station (Fig. 14.26). In Fig. 14.26c, the drought index is calculated by subtracting the normalized past October to present September total precipitation from the normalized May through August mean temperature of the named year. The drought index is designed to combine the moisture contribution of the previous winter with the growing season moisture and temperature of the identified year. The first major spruce bark beetle outbreak of the post-1950 period occurred in the early 1970s, following the extremely warm and dry period of 1968 to 1969 on the northern Peninsula (Fig. 14.26). Red-needle mortality dropped to nearly zero by 1975, following three cool summers (1973–1975). On both the northern and southern Peninsula, the sustained onset of warm summers beginning in 1987 was followed by substantial increase in red-needle mortality beginning in 1990 and reaching a maximum in 1996. Annually mapped red-needle area declined after 1996 on the southern Peninsula because beetles had killed most of the available mature spruce forest, not because climatic factors became less favorable for insect population increases.

The association of red-needle mortality with drought is particularly striking in the southern Peninsula forests between 1989 and 1997 (Fig. 14.26c,d). Spruce bark beetles first attack the large, slowly growing trees in a stand (Hard, 1985, 1987). The survival of smaller trees (both young saplings and released understory pole-size trees) is attributed to their ability to produce enough pitch to swamp the beetle galleries and kill the larvae. An additional factor may be that large trees are more prone to drought stress. Due to cold soils, spruce trees in Alaska usually root within the upper 30 to 40 cm of the soil and extend their roots laterally for 1 to 2 m. This shallow, extended rooting can bring trees into competition with neighboring trees and surface vegetation, especially in times of drought. Ring widths in many large trees generally decline for 5 to 10 years prior to the tree succumbing to bark beetles, clearly signifying a stressed condition. Drought is the most likely source of this stress (Fig. 14.26, Hard, 1987).

While warm summers are the proximate cause of outbreaks, spruce stands must reach a certain level of maturity before serious spruce bark beetle infestation can develop, regardless of summer temperatures. For example, relatively light levels of tree mortality from bark beetles occurred in an area near the headquarters of the Kenai National Wildlife Refuge that burned in 1926, while a nearby stand of spruce 125 years old or older experienced heavy mortality. Many stands on the southern Kenai Peninsula were composed of trees that regenerated following a moderate spruce bark beetle outbreak in the 1870s and 1880s, and those trees only matured to a condition that would support substantial spruce bark beetle infestations in the mid-20th century. By the time that warm summers began in 1987, these trees had once again become prime bark beetle habitat.

Moisture appears to be another important factor in spruce bark beetle outbreaks on the Kenai Peninsula. Precipitation at the Kenai Airport station reached a record low from 1967 to 1969, and was at a near-record low (to that date) at the Homer Airport station during that period. This dry period was followed by an intense but short-lived spruce bark beetle outbreak from 1970 to 1974 (Fig. 14.26c,d). It appears that at least two warm summers are required to initiate a spruce bark beetle outbreak. This is consistent with the life cycle of the beetles, which normally takes two years to complete, except in the warmest summers when it can be completed in one year (Werner and Holsten, 1985). In 1987, summer temperatures shifted above the mean (Fig. 14.26a), beginning a sustained series of warm summers that is unique in both the instrumental and local reconstructed temperature records. After 1987, the spruce bark beetle outbreak accelerated, and no period of several cool summers, which could thermally arrest the outbreak, has occurred to the present. Red-needle acreage, however, has dropped steadily since 1996 because most of the mature spruce forest has been killed. Food supply, rather than climate, appears to have been the ultimate limiting factor in this outbreak, unlike outbreaks during the past 200 to 250 years studied by staff of the Kenai National Wildlife Refuge.

These results suggest that there is a continuing high risk from climate change to the management of forest land in south-central Alaska for spruce forest crops. Under recent climate conditions, and especially under scenarios of further temperature increases, spruce bark beetle irruption potential is very likely to remain high. As the small surviving (understory) spruce trees in the region mature to commercial forest dimensions, they will move into the prime size and age classes to serve as hosts for spruce bark beetle. Under these circumstances, the regional environment is very likely to remain effectively saturated with spruce bark beetles because climate limitations on beetles have been removed. Investments in regeneration and early tending of new commercial stands of spruce, should that be desired, would carry considerable risk because bark beetles would become effective agents of tree mortality at about the time that stands of spruce became large enough to generate commercial value.

14.8.3. Spruce budworm in North America

The eastern spruce budworm (*Choristoneura fumiferana*), is a defoliating insect affecting conifer trees, generally in the southern and central boreal region of the United States and Canada. The absolute distribution limits of spruce budworm generally follow that of white spruce throughout the boreal forest, and there are records of severe defoliation within 150 km of the Arctic Circle (Harvey, 1985). Outbreaks have at times extended over 72 million ha and lasted for up to 15 years (Fleming and Volney, 1995). The budworm is generally present in the forest at low background numbers. During irruptions, there can be more than 22 million budworm lar-

vae per hectare in suitable forest stands (Crawford H. and Jennings, 1989).

Weather appears to be a critical factor in determining budworm distribution. Irruptions of budworm generally follow drought, and outbreaks also start after hot, dry summers (Fleming and Volney, 1995). Drought stresses the host tree population, reducing host resistance (Mattson and Haack, 1987). Elevated summer temperatures increase budworm reproductive output. Female budworms lay 50% more eggs at 25 °C than at 15 °C (Jardine, 1994). Finally, higher temperatures and drought can shift the timing of budworm reproduction so that natural parasitoid predators are no longer effective in limiting budworm numbers (Mattson and Haack, 1987). Conversely, cold weather can stop a budworm outbreak. Budworms starve if a late spring frost kills the new shoot growth of the host trees on which the larvae feed.

Given this weather/climate sensitivity of budworm, it was inferred that a warming climate would be associated with northward movement of spruce budworm outbreaks (Gitay et al., 2001). Such a northward movement appears to have happened. Before about 1990, spruce budworm had not appeared able to reproduce in the northern boreal forest of central Alaska. In that year, after a series of warm summers, a spruce budworm irruption occurred in the Bonanza Creek Long-Term Ecological Research site and visible canopy damage spread over tens of thousands of hectares of nearby white spruce forest. Populations of budworm have persisted in this area near the Arctic Circle, including a minor outbreak in 2002–2003.

14.8.4. Other forest-damaging insects in North America

Aspen is the most important deciduous tree species in the Canadian boreal forest, with more than 1000 Tg of carbon stored in the aboveground biomass. Aspen dieback has become conspicuous over parts of the southern boreal forest and aspen parkland in western Canada. In 18 aspen stands near Grande Prairie, Alberta, defoliation histories were reconstructed based on tree rings and records of past insect outbreaks. Several factors contributed to the observed dieback. Defoliation by the forest tent caterpillar (*Malacosoma disstria*) and drought in the 1960s and 1980s led to reduced growth and predisposed some stands to secondary damage by wood-boring insects and fungal pathogens. Thaw–freeze events during a period of unusually light late winter snow cover (1984–1993) also contributed to the observed dieback. Under climate warming, the severity of these stressors is very likely to increase, which would pose a serious concern for the future health, productivity, and carbon sequestration of aspen forests in the region (Hogg et al., 2002).

During the last two decades of the 20th century, warmer summers and winters in central Alaska were associated with a noticeable increase in the area of trees killed or damaged by insects. The mapped area of forest affected by spruce budworm, spruce coneworm (*Dioryctria reniculelloides*), and larch sawfly (*Pristiphora erichsonii*) defoliation mapped throughout Interior Alaska increased, totaling over 300 000 ha of combined infestations during the period 1991 to 1996 (Holsten and Burnside, 1997).

Various insects are known have the potential to become serious agents of tree mortality in a much warmer boreal and arctic region. Insect species not currently present are almost certain to be able to disperse readily into the region if higher temperatures allow, and some are likely to develop outbreak potential. In Ontario, Canada, forest land managers estimate that a range of forest-damaging insects that are currently prominent in the central and southern boreal area would move northward or, if already present, develop greater outbreak potential (Parker et al., 2000). These include spruce budworm, forest tent caterpillar, jack pine budworm (*Choristoneura pinus pinus*), and gypsy moth (*Lymantria dispar*). The bronze birch borer (*Agrilus anxius*) is a North American species that can cause severe damage to paper birch and may be effective in limiting the survival of birch along the southern (warm) margin of its distribution (Haak, 1996). The bronze birch borer is present today in small numbers as far north as Alaska. The ACIA-designated models project temperatures in northern boreal North America near the temperatures of areas where the bronze birch borer effectively limits birch today. In boreal Alaska, a larch sawfly outbreak killed most of the larger and older tamarack (*Larix laricina*) trees during a warm period in the decade of the 1990s, and aspen leaf miner (*Phyllocnistis populiella*) appeared at outbreak levels (142 000 ha) by 2003 (Wittwer, 2004).

14.8.5. Tree-damaging insects in northern Europe

Several insect species with tree hosts in northern European boreal forests regularly undergo population outbreaks, and most have some connection to direct or indirect effects of warm weather anomalies and temperature limitation. In Europe, temperature increases, especially in the form of warmer winters, are associated with increased outbreaks of various species of bark beetles and aphids (Beniston et al., 1998). Bark beetles (Coleoptera, Scolytidae) are present in the boreal region of northern Europe. A key feature of temperature sensitivity is the timing of the first flight of mature adults in the spring as they seek new host trees to attack. Many beetles do not begin spring flight until air temperature reaches a threshold level. The bark beetle *Ips typographus* (spruce engraver beetle) built up in numbers in trees damaged by heavy storm and snow damage in the late 1960s, and then broke out in large numbers during a series of warm years in the early 1970s (Heliövaara and Peltonen, 1999). A similar combination of widespread tree host susceptibility and warm summers in the future is very likely to result in similar outbreaks and tree death. There are no examples in Europe of bark beetles simultaneously expanding their distribution in one direction and retracting in another (Heliövaara and Peltonen, 1999).

Tree-defoliating insects that are responsive to temperature increases are capable of causing large-scale tree death and injury in the northernmost forests of Europe (Bylund, 1999; Danell et al., 1999). Geometrid moths of two species occur in the mountain birch (*Betula pubescens* ssp. *czerepanovii*) forest that makes up the ecotone between boreal forest and the tundra of Fennoscandia. The autumnal moth (*Epirrita autumnata*) and the winter moth (*Operophtera brumata*) regularly reach outbreak levels. Although both species lay eggs on a variety of trees and shrubs, the main host is mountain birch. Outbreaks of the autumnal moth mainly occur in mature forests in inland locations, whereas winter moth outbreaks are restricted to warm south-facing slopes or warmer locations along the Norwegian coast (Neuvonen et al., 1999).

Populations of both the winter and autumnal moth fluctuate from low to high levels with 7 to 12 years between peak densities (Altenkirch, 1990). Severe defoliation causes death of older stems, which usually triggers vigorous sprouting of new stems from the base of the tree. However, the recovery process in these forests is slow, and may take about a century (Bylund, 1999). Particularly severe defoliation may kill the underground portions of the tree as well. The natural cyclic fluctuation of geometrid moth numbers is driven by density-dependent factors that lag outbreaks in time, including buildup of natural enemies, disease, and reduced food quality, and by weather conditions at all life stages. As do many of the macrolepidoptera, autumnal moths overwinter as eggs, making them vulnerable to extreme cold temperatures that reduce survival, particularly egg masses placed in the tree canopy above the snow limit. From 1960 to 1990, minimum winter temperatures below -36 °C, along with fine-scale patches of trees on dry and nutrient poor soils, efficiently predicted the distribution of autumnal moth outbreaks in birch forests of Finland (Neuvonen et al., 1999). Climate change involving an increase in minimum winter temperatures is likely to increase the frequency, and possibly the severity, of outbreaks of this species and others that overwinter as eggs (Neuvonen et al., 1999). Warmer summer temperatures with no change in winter temperatures are likely to reduce the geometrid moth outbreak potential, because higher densities of moth predators associated with warm summers would partially protect birch, while the winter limitation on the moths would still operate. Finally, if climate change consists of warmer summers and winters, as projected by the ACIA-designated models, current insect population/ecological models are not adequate to project overall effects on outbreaks (Neuvonen et al., 1999), although new insect distributions and novel climate limitations and outbreak patterns are very likely under such circumstances (see also section 7.4.1.4).

A diprionid sawfly (*Neodiprion sertifer*) is the most serious defoliator of pine forests in northern Europe. Outbreaks mostly occur on dry and infertile sites. Minimum winter temperatures below -36 °C also limit this species. Outbreaks in Finland have been most frequent in southern and central inland areas, and temperature increases are likely to make outbreaks more common in eastern and northern areas (Neuvonen et al., 1999).

14.9. Climate change and fire

Fire is a major climate-related disturbance in the boreal forest, with pervasive ecological effects (Payette, 1992; Van Cleve et al., 1991; Zackrisson, 1977). Climate, disturbance, and vegetation interact and affect each other, and together they influence the rate and pattern of changes in vegetation (Neilson, 1993; Noble, 1993), the rate of future disturbance (Gardner et al., 1996; Rupp et al., 2002; Turner M. et al., 2003), and the pattern of new forest development (Turner D. et al., 1995; Rupp et al., 2000a,b). Understanding these interactions and feedbacks is critical in order to understand how scenarios of climate change will affect future fire regimes and the consequences these ecological changes will have for both boreal forests and forest management.

The total area burned in North America has been increasing concurrently with recent temperature increases and other climatic changes (Stocks et al., 2000). The annual area burned in western North America doubled in the last 20 years of the 20th century (Murphy et al., 2000). Based upon less precise statistics there appears to be a similar trend in the Russian Federation (Kasischke et al., in press).

Three factors are important when examining the impacts of climate change on the fire regime. First, boreal forests become increasingly flammable during succession as the forest floor and understory builds up. Conifer-dominated stands are usually more flammable than broad-leaved and other vegetation types. Early post-fire vegetation is generally less flammable than vegetation that is older and more structurally complex. Vegetation also influences fire probability indirectly through its effects on regional climate (Chapin et al., 2003), with early successional stands absorbing and transferring only half the incoming solar radiation compared to late successional spruce stands (Baldocchi and Vogel, 1995). A second factor is the need to understand the direct effects of climate on the fire regime. Finally, humans and their land-use changes affect the fire regime, and these changes must be considered to understand future fire effects in a warmer climate.

14.9.1. The role of fire in subarctic and boreal forest

Disturbance is the driving force behind vegetation dynamics in the boreal biome. Wildfires, alluvial processes (i.e., erosion and flood deposition), tree-fall events caused by wind, and tree-killing insect outbreaks all play major roles that can affect large portions of the landscape. However, fire is particularly significant because of its pervasive presence, its strong link to climate, and the direct feedbacks it has to permafrost dynamics, regional climate, and the storage and release of carbon. Fire often follows large-scale tree death caused by other factors.

In the mosaic of old burns that covers the boreal forest landscape, there are four tendencies related to ecosystem development, expressed over timescales of decades to millennia and over differing landscape areas. There is a tendency towards paludification, which is the water-logging and cooling of soil caused by the buildup of organic soil layers (Viereck, 1970). Paludification can be a significant factor in treeline retreat in maritime climates (Crawford R. et al., 2003). A second tendency is for the longer-lived conifer species to usurp the canopy from shorter-lived hardwood species (Pastor et al., 1999; Van Cleve et al., 1991) with an accompanying decline in living biomass (Paré and Bergeron, 1995). The third tendency is towards the formation of a shifting-mosaic steady state in which burned areas at different stages of secondary succession form the pieces of the mosaic (Wright, 1974). Finally, there is a tendency toward increased risk of burning with age. Fire opposes paludification by oxidizing the organic matter buildup on the ground surface (Mann D. and Plug, 1999). Fire also interrupts conifer take-over by restarting successional development with the earliest broadleaf-dominated stages. In addition, because the time since the last fire is an important determinant of vegetation composition and soil conditions, fire frequency has a large influence on carbon storage and release within the boreal forest (Kasischke et al., 1995; Kurz and Apps, 1999).

Fire is strongly controlled by both temporal and spatial patterns of weather and climate (Flannigan and Harrington, 1988; Flannigan et al., 2001; Johnson, 1992). Specific fire behavior responds to hourly, daily, and weekly weather conditions. Solar radiation, continentality, topography, and specific terrain features influence these fire-generating conditions. The general fire regime of an area responds to long-term, landscape-level climatic patterns. As climate varies, the controlling weather variables can vary in magnitude and direction (Flannigan et al., 2001).

Case studies from northern coniferous forests have documented weather variables prior to and during specific fire events (Flannigan and Harrington, 1988) and have identified good predictors of conditions that promote the rapid spread of fire. These conditions include warm temperatures, little or no precipitation, low relative humidity, and high winds. Synoptic weather conditions that produce cold frontal systems, drought, and low relative humidity have been found to be good predictors of area-burned activity (Flannigan et al., 2001). Upper-air circulation patterns have been related to area burned (Newark, 1975). Catastrophic burning events are related to the breakdown of the 500 mb long-wave ridge

(Nimchuk, 1983). Breakdown of these ridges generally occur at the same time as documented increases in lightning strikes and strong surface winds.

Regional and global links (teleconnections) between meteorological variables and weather anomalies have been identified. The most documented sources of teleconnections are the El Niño–Southern Oscillation and the Pacific Decadal Oscillation (section 2.2.2.2). Several North American studies have linked teleconnections to area-burned anomalies (Flannigan et al., 2000; Johnson and Wowchuk, 1993; Swetnam and Betancourt, 1990). Teleconnections may offer long-range forecasting techniques for temperature and precipitation anomalies that fire managers could use in estimating fuel moisture and fire potential (Flannigan et al., 2001). Winter sea surface temperatures in the Pacific are significantly correlated with warm-season temperature and seasonal area burned in Canada (Flannigan et al., 2000). Significant correlations between these factors vary by provincial region and phase of the North Pacific Oscillation (1953–1976 versus 1977–1995). Interestingly, the sign of the correlation changed from strongly negative (1953–1976) to strongly positive (1977–1995) for four regions (Alberta, western Ontario, eastern Ontario, and Quebec) at the same time they changed from strongly positive to strongly negative for two regions (British Columbia and Saskatchewan). During the same period, two regions did not change sign (Yukon/Northwest Territories and Manitoba). These shifts occurred at previously identified changes in climate regimes (e.g., section 14.6.2, Fig. 14.15), and emphasize the control the overall atmospheric circulation has on fire at a regional scale in understandable, but varying ways.

There are two different scenarios of the relay floristics (rate of spread of vegetation; *sensu* Egler, 1954) pathway for secondary succession after fire that operate in the boreal forest. The first is the process in which deciduous shrubs and trees initially colonize burned sites, but are replaced in the overstory in approximately 150 to 200 years by coniferous trees that may be limited in movement by the time necessary for successive generations to reach reproductive maturity or the presence of individual surviving trees dispersed throughout the burn. The second process is self-replacement, which occurs after fires by root sprouting (i.e., aspen, birch, balsam poplar (*Populus balsamifera*), and numerous shrub species) or by fire-stimulated seed release (i.e., black spruce) (Greene et al., 1999; Mann D. and Plug, 1999). Both of these successional processes occur in the boreal forest. However, the relative importance of each pathway in determining the structure of the boreal forest at a landscape scale is still unknown.

The fire regime describes the general characteristics of fire and its effects on ecosystems over time. It can be defined by specific components such as frequency, intensity, severity, size, and timing. These characteristics have been used to develop classification systems that aim to describe the principal types of fire regimes associated with different ecosystems. These fire/ecosystem regime categories are general and broad due to the large spatial and temporal variability exhibited by specific ecosystems (Whelan, 1995), but they do provide a conceptual model useful for understanding both fire behavior and fire effects in a particular system. Documentation of the components of fire regimes (i.e., ignition sources, frequency, extent, and severity) for specific ecosystems and geographic regions is also highly variable, and in many cases not well quantified.

Lightning strikes and humans are the two sources of ignitions in the boreal forest. Lightning is the most significant cause of fires (defined by total area burned), although this trend varies among regions. The number of lightning-caused fires generally declines as latitude increases because of decreased heating at the ground surface necessary to produce convective storms, and as climate becomes more maritime (cool layer at ground surface) (Johnson, 1992). Humans are responsible for high numbers of ignitions, but the fires started consume much less area because usually they are actively suppressed. Indigenous peoples ignited fires for specific purposes throughout history, but their impact on past overall fire regimes remains uncertain (Johnson, 1992; Swetnam et al., 1999). Specific regions of the boreal forest in North America experienced substantial anthropogenic fire impacts during the "gold rush" era of the late 19th and early 20th centuries.

14.9.2. Regional fire regimes

14.9.2.1. Russia

Fire statistics for the Russian Federation in general, and the Russian boreal region specifically, are incomplete at best (FAO, 2001; Stocks, 1991). Official fire statistics have been reported only for protected regions of Forest Fund land (section 14.3.1). Furthermore, only 60% of the Forest Fund land is identified as protected (FAO, 2001). Approximately 430 million ha of forested tundra and middle taiga in Siberia and the Far East receive no fire protection. The paucity of reliable statistics is a result of numerous issues including remoteness, lack of detection and mapping technology, lack of fire-management funding, and deliberate falsification of past records for political reasons. Humans were identified as the major source of ignitions at 65%, followed by lightning (16%), prescribed agricultural burning (7%), and other/unknown activities (12%) (Shetinsky, 1994). Intensive prevention and education programs have had little success in decreasing anthropogenic fires. Many of these fires grow to large sizes due to overstretched suppression resources.

Keeping these limitations in mind, the statistics for the protected Forest Fund area provide some perspective on fire across the Russian Federation. Between 18 000 and 37000 forest fires were detected annually from 1950 to 1999. The average annual area burned within the zone of detection for the decades 1950 to 1959, 1960 to 1969, 1970 to 1979, 1980 to 1989, and 1990 to 1999 was 1.54, 0.68, 0.48, 0.54, and 1.2 million ha, respectively (FAO, 2001). Indirect estimates of annual area burned in both protected and unprotected Forest Fund areas have been developed through modeling techniques. Shvidenko and Goldammer (FAO, 2001) used a modified expert system model (Shvidenko and Nilsson, 2000), available fire statistics, and forest inventory data on age and stand structure to calculate an estimated total of the annual average area burned in all of Russia over the past 30 years. The estimate applies to the Forest Fund and State Land Reserve area. An estimated 5.10 million ha burned annually, of which 3.94 million ha were in the boreal bioclimatic zones of forest tundra–northern taiga, middle taiga, and southern taiga (see Fig. 14.1).

Satellite remote-sensing techniques provide insight into the potential total extent of area burned across major regions of the Russian Federation. In 1987, an estimated 14.4 million ha burned in the Russian Far East and eastern Siberia (Cahoon et al., 1994), and in 1992, an estimated 1.5 million ha burned in all of the Russian territories (Cahoon et al., 1996). In 1998, following a very strong El Niño, an estimated 9.4 million ha burned in the Asian regions of Russia (FAO, 2001).

Only a small fraction of total forested area falls under any organized fire-suppression management. Fire suppression is headed by the Ministry of Natural Resources of Russia, which manages suppression efforts through regional offices. Fire-suppression resources include both ground and aerial operations. Aerial operations provide detection and monitoring services, and direct suppression resources (i.e., water and retardant drops, transport of ground personnel and smokejumpers). The State Forest Guard (ground suppression) and the Avialesookhrana (aerial detection and suppression) coordinate fire suppression with local and regional authorities. Under severe fire conditions, military detachments and local populations are recruited.

Operational policy and logistical allocation follow fire danger predictions, which are based on weather and climatic conditions. The Nesterov fire index, similar to the Canadian Forest Fire Danger Rating System or the US National Fire Danger Rating System, is used for prediction. Aircraft patrols and resource deployments are based on the fire index predictions. A severe limitation to successful fire management has been a lack of funding – in 1998, only US$ 0.06 per hectare was allocated to fire suppression (FAO, 2001). In addition, a lack of advanced technology equipment (i.e., satellite monitoring, radios, etc.), aircraft, and State Forest Guard personnel, as well as unfavorable land-use practices, have been identified as major weaknesses in the current system.

14.9.2.2. Canada

Canadian fire management agencies developed a large-fire database for all fires larger than 200 ha for the period 1959 to the present (Stocks et al., 2000). Fires larger than 200 ha represent only 3.5% of the total number of fires, but 97% of the total area burned over this period (Stocks et al., 2000). The fire perimeters were digitized and mapped in a geographic information system, and the database includes ancillary information such as ignition location and date, size, cause, and suppression action(s) taken. This database has been expanded into the past where data exist (as far back as 1918), and is continuously updated with each passing fire season. Total area burned in Canada has more than doubled since the 1970s, and the upward trend is well explained by warmer temperatures (Gillett et al., 2004).

Lightning-caused fires predominate throughout the fire record and account for almost all large fires in the northern portions of Canada. From 1959 to 1997, lightning ignited approximately 68% of all fires, and those fires accounted for approximately 79% of the total area burned (Stocks et al., 2000). Temporal trends show a steady increase in the number of lightning fires and their contribution to total area burned from the 1960s through the 1990s, which fire managers attribute to technology improvements in fire detection and monitoring. Anthropogenic fires are a significant contributor to the total number of fires almost exclusively in populated areas and along the road network, and suppression of these fires is effective in limiting the total area burned. Temporal trends show a steady decrease in total area burned due to anthropogenic ignitions, attributed to aggressive fire suppression tactics.

The average annual area burned between 1959 and 1997 was approximately 1.9 million ha, with interannual variability ranging from 270000 to 7.5 million ha burned (Stocks et al., 2000). Spatial trends identified a few ecozones (taiga plains, taiga shield, boreal shield, and boreal plains; see Wiken, 1986 for definitions) that accounted for the majority (88%) of total area burned between 1959 and 1997, primarily because these ecozones have a continental climate that is conducive to extreme fire danger conditions and have large uninhabited areas with low values-at-risk, so fires are allowed to burn unimpeded (Stocks et al., 2000).

Organized fire suppression and management has operated in Canada since the early 1900s, initiated as a response to large catastrophic fires, much like those experienced in the United States. Canada is recognized worldwide for its advanced technologies and operational efficiencies in both fire prevention and suppression. The Canadian Forest Fire Danger Rating System (CFFDRS) was first developed in 1968 and consists of two subcomponents: the Canadian Forest Fire Weather Index (FWI) system, which provides a quantitative rating of relative fire potential for standard fuel types based upon weather observations; and the Canadian

Forest Fire Behavior Prediction system, which accounts for fire behavior variability of fuel types based on topography and components of the FWI. The CFFDRS is used for training, prevention, operational planning, prescribed burning, and suppression tactics. In addition, researchers use CFFDRS for investigating fire growth modeling, fire regimes, and potential climate change impacts. The system is used extensively in Alaska in place of the US Fire Danger Rating System – a system developed primarily for the lower 48 states.

Aerial reconnaissance in conjunction with lightning detection systems has been employed heavily since the 1970s throughout the North American boreal forest (including Alaska). The lightning detection system identifies areas of high lightning strike density and allows for focused aerial detection operations. This has greatly increased detection efficiencies over remote regions.

14.9.2.3. United States (Alaska)

The fire record for the Alaskan boreal forest has evolved over time and consists of three datasets (Kasischke et al., in press). The first is a tabular summary of total annual area burned since 1940. The second is a tabular database that contains the location, ignition source, size, management option, and initiation and extinguishment date for all fires since 1956. The third is a geographic information system database of the boundary of fires since 1950. The spatial database includes all fires larger than 400 ha occurring from 1950 to 1987, and all fires larger than 25 ha occurring since 1988.

Early studies minimized the importance of lightning as an ignition source in the Alaskan boreal forest because anthropogenic ignitions were thought to be more important (Lutz, 1956). However, the implementation of digital electronic lightning detection systems along with a more thorough review of fire statistics led to the realization that lightning is not only widespread throughout the growing season in Interior Alaska, but is responsible for igniting the fires that burn most of the area (Barney and Stocks, 1983; Gabriel and Tande, 1983). Analysis of fire statistics from the Alaska Fire Service shows that while humans start more than 61% of all fires, these fires are responsible for only 10% of the total area burned (Kasischke et al., in press). The remaining fires are the result of lightning ignitions.

Convective storms and associated lightning can range in size from individual clouds to synoptic thunderstorms covering thousands of square kilometers. A single thunderstorm may produce most of the annual lightning strikes in an area (Nash and Johnson, 1996). In Interior Alaska, the lightning density gradient generally runs from high in the east to low in the west, parallel with the warmer summer climate in the interior continental areas and cooler maritime summer climate near the coast (Kasischke et al., in press). However, mapped fires that originate from lightning strikes are well distributed throughout the interior of the state. In contrast to light-

ning ignitions that are uniformly distributed between the Brooks and Alaska Ranges, anthropogenic fire ignitions are centered around major population centers (Fairbanks and Anchorage), as well as along the major road networks (Kasischke et al., in press).

The presence of the boreal forest landscape may promote convective thunderstorms. In central Alaska, the density of lightning strikes is consistently highest within boreal forest compared to tundra and shrub zones across a climatic gradient (Dissing and Verbyla, 2003). Within the tundra, the number of lightning strikes increases closer to the boreal forest edge. The paleo fire record indicates that wildfires increased once black spruce became established in Interior Alaska (despite a cooler, moister climate). This may have been due to two factors: 1) increased landscape heterogeneity and higher sensible heat fluxes (see section 14.2.5 BOREAS results) leading to increased convective thunderstorms; and 2) increased fuel flammability associated with the black spruce vegetation type (Chambers and Chapin, 2003; Dissing and Verblya, 2003; Kasischke et al., in press; Lynch et al., 2003). These results suggest that climate changes that promote the expansion or increase in density of conifer forest are likely to increase the incidence of fire, but that climate changes that decrease conifers (too-frequent burning, aridification) are likely to decrease subsequent burning.

The annual area burned in Alaska exhibits a bimodal distribution, with years of high fire activity (ignitions and area burned) punctuating a greater number of years of low activity. For the Alaskan boreal forest region, 55% of the total area burned between 1961 and 2000 occurred in just 6 years. The average annual area burned during these episodic fire years was seven times greater than the area burned in the low fire years.

With the observed increase in air temperature and lengthening of the growing season over the past several decades in the North American boreal forest region, a corresponding increase in fire activity between the 1960s and 1990s might be expected (Stocks, 2001). For the Alaskan boreal forest region (which was less affected by fire suppression efforts than southern boreal areas), such an increase in fire activity was not apparent until the summer of 2004, in which a record 2.71 million ha burned. That summer (May–August) also had the highest mean monthly temperatures recorded at Fairbanks since observations began in 1906. A record 1.84 million ha also burned in 2004 in the adjacent Yukon Territory, amounting to about 12.4% of all forest land in the Yukon. Between 1981 and 2000 there was only a slight (7%) rise in the annual area burned (297 624 ha/yr) compared to the period between 1961 and 1980 (276 624 ha/yr) (Kasischke et al., in press). These extensive 2004 fires have clearly established a significant upward trend in the area burned during the period of record. The frequency of large fire years has been greater since 1980 (five large fire years) than before 1980 (two large fire years), and the increase in average annual area burned between these periods is due to the increase in frequency of large fire years and the decrease in fire-fighting activities in remote areas. The fire data record for Alaska is consistent with an increase in large fires in response to recent climate warming, but not sufficient to determine definitively whether the increase is outside the range of natural variability. The official fire statistics for Alaska show that there were two large fire years per decade in the 1940s and 1950s, indicating that the frequency of large fire years has been relatively constant over the past 60 years, although this entire period is distinctly warmer than preceding centuries.

The number and size distribution of fires in the Alaskan boreal forest region is different during low and severe fire years. During low fire years between 1950 and 1999, an average of 17 fires greater than 400 ha occurred per year, with an average size of 7800 ha. In contrast, during high fire years an average of 66 fires greater than 400 ha occurred, with an average size of 20 300 ha (Kasischke et al., in press). In low fire years, 73% of the total area burned occurred in fires larger than 50 000 ha. In high fire years, 65% of the total area burned occurred in fires larger than 50 000 ha. In low fire years, 9% of the total area burned in fires larger than 100 000 ha, with no fires larger than 200 000 ha. In contrast, during high fire years, 33% of the total area burned in fires larger than 100 000 ha.

Fire managers in Alaska typically recognize two fire seasons (Kasischke et al., in press). The major fire season typically occurs after mid-June when the surface of the earth becomes warm enough to drive convective thunderstorm activity. However, there is an earlier fire season throughout the state associated with the extremely dry fuel conditions that occur immediately after snowmelt and before leaf-out in late spring. At this time, precipitation levels are low, and fuel moisture conditions are extremely low because of the curing of dead vegetation during winter and the lack of green vegetation. During this period, human activities result in the majority of fire ignitions.

Fire management and suppression is a relatively new concept in Alaska and did not become an organized effort until 1939. In the early 1980s, fire-management efforts in Alaska were coordinated under the direction of an interagency team. Beginning in 1984, fire-management plans were developed for thirteen planning areas. The fire plans provide for five separate suppression options:

- *Critical protection* is provided for areas of human habitation or development. These areas receive immediate initial attack and continuing suppression action until the fire is extinguished.
- *Full protection* is provided for areas of high resource values. Fires receive immediate initial attack and aggressive suppression to minimize area burned.
- *Modified action* provides for initial attack of new fires during the severe part of the fire season (May 1–July 10). Escaped fires are evaluated by

the land manager and suppression agency for appropriate suppression strategies. On specified dates (generally after July 10), the modified action areas convert to limited action.

- *Limited action* is provided for areas of low resource values or where fire may actually serve to further land management objectives. Suppression responses include the monitoring of fire behavior and those actions necessary to ensure it does not move into areas of higher values.
- *Unplanned areas* exist where land managers declined participation in the planning process. These few unplanned areas receive the equivalent of full protection.

The formation of these plans radically changed wildland fire management in Alaska. The adoption of these cooperative plans created areas where fires were no longer aggressively attacked due to the low economic resource value of those lands. The primary goals of the plans are to restore the natural fire regime to the boreal forest ecosystem and to reduce suppression costs.

14.9.2.4. Fennoscandia

Finland, Sweden, and Norway account for a very small percentage of total area burned in the boreal forest and generally have not experienced large fires in modern times. This can be attributed to the ease of access throughout these relatively small countries and their highly managed forest systems. In addition, lightning fires are not a major factor, accounting for less than 10% of all fires (Stocks, 1991). Fire statistics for this region of the boreal forest are limited and discontinuous. The lack of natural fire is one of the principal causes of endangerment of a set of fire-dependent species in northern Fennoscandia (Essen et al., 1992).

Finland reported an annual average of 800 ha burned during the 1990s (FAO, 2001). Since record keeping began in 1952, there has been a steady decrease in average annual area burned from a high of 5760 ha in the 1950s to 1355 ha in the 1960s to less than 1000 ha from 1971 to 1997 (FAO, 2001). An average of 2224 fires per year occurred during the period 1995 to 1999. Anthropogenic fires accounted for 61% of all fires, followed by 29% of unknown cause, and 10% caused by lightning (FAO, 2001).

Boreal forest covers most of Sweden and includes a mix of flammable conifer trees, shrubs, and mosses. Sweden reported an annual average of 1600 ha burned during the 1990s (FAO, 2001). Humans are the major cause of fires accounting for approximately 65% (FAO, 2001) of the annual average of 3280 fires during the 1990s (FAO, 2001). Norway reported an average of 564 ha burned annually between 1986 and 1996 (FAO, 2001). The highest frequency of fires occurs in the boreal forest region of the eastern lowlands (Mysterud et al., 1998). On average, there were only 513 fires per year in Norway during the 1990s (FAO, 2001).

Forest fires are virtually absent from the sparsely forested regions of Iceland and Greenland. No fire statistics were reported to the FAO for the period 1990 to 1999.

14.9.3. Possible impacts of climate change on fire

Rupp et al. (2000a,b) used a spatially explicit model to simulate the transient response of subarctic vegetation to climatic warming in northwestern Alaska near treeline. In the model simulations, a warming climate led to more and larger fires. Vegetation and fire regime continued to change for centuries in direct response to a 2 to 4 °C increase in mean growing-season temperature. Flammability increased rapidly in direct response to temperature increases and more gradually in response to simulated climate-induced vegetation change. In the simulations, warming caused as much as a 228% increase in the total area burned per decade, which led to a landscape dominated by an increasingly early successional and more homogenous deciduous forest.

Turner M. et al. (2003) used the same model (Rupp et al., 2000b) to simulate fire-regime sensitivity to precipitation trends. Precipitation projections from global climate models have the largest associated errors and the highest variability between models. A simulated instantaneous 2 °C increase in average growing-season temperature was applied with two different precipitation regimes (a 20% increase and a 20% decrease from current precipitation levels) to explore the possible influence of climate change on long-term boreal forest ecosystem dynamics. Both scenarios projected an increase in the total number of fires compared to current climate. However, the distribution of fire sizes was surprising. As expected, the warmer and drier scenario resulted in fewer small fires and an associated shift toward larger fires. Also as expected, the warmer and wetter scenario resulted in fewer large fires and an associated shift toward smaller fires. However, the distribution of very large fires (burning >25% of the total landscape area) was unexpected. The warmer and wetter scenario produced more very large fires compared to the warmer and dryer scenario. The warmer and dryer climate scenario experienced frequent medium-sized fires, which prevented fuels from building up across the landscape and limited the number of large fires. In contrast, the warmer and wetter climate scenario led to frequent small fires, which allowed the development of well-connected, highly flammable late-successional stands across the landscape.

Additional model simulations suggest that vegetation effects are likely to cause significant changes in the fire regime in Interior Alaska (Rupp et al., 2002). Landscapes with a black spruce component were projected to have more fires and more area burned than landscapes with no black spruce component. Black spruce landscapes were also projected to experience numerous fires consuming extensive portions (more than 40%) of the landscape. These results agree with observations in the Canadian boreal forest where 2% of the fires account for

98% of the total area burned (Stocks, 1991). Large-scale fire events need to be realistically represented in ecosystem models because they strongly influence ecosystem processes at landscape and regional scales. These results have strong implications for global-scale models of terrestrial ecosystems. Currently, these models consider only plant functional types distinguished by their physiological (C_3 versus C_4 photosynthesis), phenological (deciduous versus evergreen), and physiognomic (grass versus tree) attributes. The fire-modeling results suggest the need for a finer resolution of vegetation structure and composition related to flammability in order to simulate accurately the dynamics of the fire regime, and to understand how different climate changes are likely to change ecosystems at several different scales.

Modeling results show that fire regime plays an important role in determining the overall relative abundance of ecosystem types at any given time, and that different fire regimes create qualitatively different patterns of ecosystem placement across the landscape (Rupp et al., 2000b, 2002). One process is "contagion" or the spread of effects into one type largely due to its juxtaposition with another. Simulation results in a typical Alaska black spruce landscape projected increases in the total amount of deciduous forest that burned in two landscapes that contained a black spruce component. The results indicated a much shorter fire return interval for deciduous forest in a landscape dominated by black spruce, and the fire return times were similar to actual fire intervals calculated by Yarie (1981) for deciduous forest in Interior Alaska (Rupp et al., 2000b, 2002). Landscape-level changes in the fire return interval of specific fuel types is an important effect of spatial contagion that currently cannot be addressed by statistical formulation within a global vegetation model. Although there is an excellent quantitative understanding of fire behavior as a function of climate/weather and vegetation at the scale of hours and meters (Johnson, 1992), the dynamic simulation of fire effects at landscape or regional scales remains rudimentary (Gardner et al., 1999). A long-term potential consequence of intensified fire regimes under increasing temperatures is that fire-induced changes in vegetation are likely to lead to a more homogenous landscape dominated by early-successional deciduous forest (Rupp et al., 2000b, 2002). This has significant implications for the regional carbon budget and feedbacks to climate (section 14.10.2). A shift from coniferous to deciduous forest dominance is very likely to have a negative feedback to temperature increases due to changes in albedo and energy partitioning (Chapin et al., 2000).

Stocks et al. (1998) used outputs from four GCMs to project forest-fire danger in Canada and Russia in a warmer climate. Temperature and precipitation anomalies between runs forced with current atmospheric CO_2 concentrations and those forced with doubled atmospheric CO_2 concentrations were combined with observed weather data for 1980 to 1989 in both Canada and Russia. All four models projected large increases in the areal extent of extreme fire danger in both countries under the doubled CO_2 scenarios. A monthly analysis identified an earlier start date to the fire seasons in both countries and significant increases in total area experiencing extreme fire danger throughout the warmest months of June and July. Scenarios are still of limited use, however, in projecting changes in ignitions.

Model projections of the spatial and temporal dynamics of a boreal forest under climate change were made for the Kas-Yenisey erosion plain in the southern taiga of western Siberia (Ter-Mikaelian et al., 1991). This study projected that the number of years in which there are severe fires would more than double under a summer temperature increase from 9.8 to 15.3 °C, the area of forests burned annually would increase by 146%, and average stored wood mass would decrease by 10%.

Flannigan and Van Wagner (1991) also investigated the impact of climate change on the severity of the forest fire season in Canada. They used projections from three GCMs forced with doubled atmospheric CO_2 concentrations to calculate the seasonal severity rating across Canada. Their results suggest a 46% increase in seasonal severity rating, with a possible similar increase in area burned, in a doubled CO_2 climate.

Flannigan et al. (1998) looked at future projections of wildfire in circumpolar boreal forests in response to scenarios of climate change. Simulations were based on GCM outputs for doubled atmospheric CO_2 concentrations. The simulation and fire history results suggested that the impact of climate change on northern forests as a result of forest fires may not be severe as previously suggested, and that there may be large regions of the Northern Hemisphere with a reduced fire frequency. These simulation results are attributed to a switch from using monthly to using daily GCM output. The scenarios still produced areas where the interval between fires is likely to decrease (i.e., more frequent fires), but they also produced regions of no change or with greater probability of an increasing interval (i.e., less frequent fires).

14.10. Climate change in relation to carbon uptake and carbon storage

14.10.1. The role of the boreal forest in the global carbon cycle

Within the terrestrial biosphere, forests cover 43% of the land area but are potentially responsible for 72% of the annual net primary productivity (McGuire et al., 1995b). The boreal forest covers roughly 1.37 billion ha and by itself (not including high-latitude tundra) contains approximately 20% of global reactive soil carbon, an amount similar to that held in the atmosphere (IPCC, 2001; Schlesinger, 1997). Climate change can affect high-latitude carbon cycling at multiple timescales. The most likely mechanism for significant short-term change

in boreal carbon cycling resulting from climate change is a change in rates of organic matter decomposition in the forest floor and mineral soil resulting from major changes in species composition caused by alteration of disturbance regimes. Climate change can also strongly affect rates of carbon cycling through its control of the disturbance regime and the subsequent successional development of ecosystems (Barr et al., 2002; Gower et al., 2001; Jiang et al., 2002; Trumbore and Harden, 1997; Wang et al., 2001; Zimov et al., 1999). Climate-induced shifts in dominant tree species composition within the present boreal forest (Carcaillet et al., 2001; Hogg and Hurdle, 1995; Smith T. et al., 1992) are likely to have profound impacts on the global carbon budget (Gower et al., 2001; Kasischke et al., 2002).

Boreal and subarctic peatlands contain approximately 455 Pg of carbon accumulating at an average rate of 0.096 Pg/yr, and constitute a significant proportion of the total boreal carbon pool (Gorham, 1991). The majority of peat consists of molecules that are highly resistant to degradation (e.g., lignin and cellulose). Species composition in sphagnum bogs is highly resilient (likely to remain the same or recover after change) because the mosses modify the local environment to produce highly acidic conditions, and their resiliency increases with age (Kuhry, 1994).

14.10.2. The role of disturbance in the carbon cycle of the boreal forest

Four processes largely control the storage and release of carbon in boreal forests: the rate of plant growth; the rate of decomposition of dead organic matter; the rate of permafrost accretion and degradation; and the frequency and severity of fires (Kasischke et al., 1995). All four processes are affected by landscape-scale disturbance. Differences in carbon cycling between mature and recently disturbed forest ecosystems have been observed in both experimental studies and modeling experiments. Some studies (Arain et al., 2002; Valentini et al., 2000) suggest that the annual carbon budget of the mature northern forest is at equilibrium and in some cases losing carbon to the atmosphere. In addition, model results and field experiments show that when ecosystems are disturbed, significant losses of soil carbon and nutrients can occur (Schimel et al., 1994) for a number of years after the disturbance.

The effects of temperature and disturbance (i.e., fire and grazing) on carbon exchange over three years in five undisturbed sites and five disturbed sites in forests of northeast Siberia were measured by Zimov et al. (1999) and results show that disturbance increased the seasonal amplitude of net carbon exchange. Disturbance had a larger effect on seasonal amplitude than either interannual or geographic differences in growing season temperature.

Fire affects the storage of carbon in the boreal forest in at least five ways: it releases carbon to the atmosphere;

converts relatively decomposable plant material into stable charcoal; re-initiates succession and changes the ratio of forest-stand age classes and age distribution; alters the thermal and moisture regime of the mineral soil and remaining organic matter, which strongly affects rates of decomposition; and increases the availability of soil nutrients through conversion of plant biomass to ash (Kasischke et al., 1995). Each of these effects exerts an influence at different timescales. As a result, the effect of a given climate scenario on carbon storage in the boreal region will be greatly influenced by fire regime (section 14.9), and represents a complex calculation requiring a great deal of specific spatial and temporal information.

Relatively few studies have directly examined carbon emissions in the boreal forest resulting from fire. Nonetheless, valuable contributions have been made by studies that used remotely sensed data to estimate direct carbon emissions from combustion as fires are occurring (Amiro et al., 2001; Isaev et al., 2002). These studies show that wildland fires have the potential to release a significant amount of carbon directly into the atmosphere. The effects of wildfire on this initial carbon loss are highly variable and strongly influenced by forest type and fire severity (Wang et al., 2001). Soil drainage (defined by the water table, moss cover, and permafrost dynamics) is the dominant control of direct fire emissions (Harden et al., 2000).

Fires cause the release of carbon not only during but also for a short time after fires. Tree mortality after surface fires can be extensive, leading to a pulse of carbon released from heterotrophic respiration as fine roots die and aboveground fine fuels (i.e., needles) fall to the ground and decompose rapidly (Conard and Ivanova, 1997). The non-combustion post-fire release of CO_2 has the potential to affect global levels of atmospheric CO_2 over the short term, representing another mechanism by which the boreal forest can play a significant role in the global carbon cycle. Together, these direct and indirect fire-generated carbon emissions from boreal forests worldwide may exceed 20% of the estimated global emissions from all biomass burning (Conard and Ivanova, 1997).

Wildland fires change the distribution of soil organic carbon pools with respect to their turnover (release to CO_2) times. Soil carbon in the form of forest litter that has fallen recently has a relatively rapid turnover rate, but carbon in the form of charcoal is stable (very long turnover rates). If rates of burning increase in the boreal forest, an increasing proportion of soil organic carbon is likely to be converted to stable charcoal. This change in the soil organic carbon pool allocation is difficult to estimate, because forest type and fire severity strongly influence the effects of wildfire on carbon redistribution (Wang et al., 2001). Harden et al. (2000) show the importance of this shifting distribution of turnover times in soil organic carbon. They developed a system of ordinary differential equations to explore constraints on carbon losses to fire,

using modern estimates of carbon production, decomposition, and storage; a model of fire dynamics developed for millennial timescales; and an assessment of the long-term carbon balance of a variety of boreal landscapes in North America. A sensitivity analysis found that their model results were responsive to the rate at which charred plant remains decomposed. Unfortunately, the specific characteristics of fires that result in maximum charcoal production are not well studied. However, it seems reasonable to infer that a moderate-intensity fire, in which combustion is enough to kill trees but not intense enough to consume them completely, would produce the greatest amount of charcoal. In complex mountainous or hilly terrain at high latitudes, north-facing slopes have higher fuel moisture content and as a result generally experience less complete combustion of fuels during fire (Van Cleve et al., 1991). This suggests that slope aspect might be an important factor in the conversion of plant biomass to charcoal.

Much of the difference in carbon cycling after disturbance can be linked to shifts in species composition and ecosystem age structure that enhance both peak summer CO_2 uptake and winter CO_2 efflux. The seasonal amplitude of net ecosystem carbon exchange in northern Siberian ecosystems is greater in disturbed than undisturbed sites, due to increased summer influx and winter efflux (Zimov et al., 1999). Disturbed sites differ from undisturbed sites during the summer, having 2.1 to 2.5 times the daytime CO_2 influx and 1.8 to 2.6 times the nighttime CO_2 efflux. Winter respiration in disturbed sites is 1.7 to 4.9 times that in undisturbed sites. Carbon cycling within disturbed ecosystems is more sensitive to interannual temperature variability than older forests, and disturbed sites also experience a greater difference in annual carbon exchange with the atmosphere in warm versus cold years than older forests (Zimov et al., 1999).

Two hypotheses have been advanced to explain why these differences in carbon cycling caused by enhanced CO_2 summer uptake and winter efflux occur in disturbed areas. One hypothesis is that the recent increase in March and April temperatures in high-latitude continental regions of North America and Siberia has advanced snowmelt and lengthened the growing season (section 6.4). The second hypothesis is that temperature-driven increases in summer carbon gain (greater CO_2 uptake from greater growth) balanced by increased winter respiration (greater CO_2 release from enhanced decomposition and live respiration) could enhance the seasonal amplitude of atmospheric CO_2 concentrations without a change in net annual carbon accumulation. Although the mechanism remains somewhat uncertain, experimental studies confirm that these responses do occur under appropriate conditions.

14.10.3. Climate and carbon allocation in the boreal forest

Changes in species composition modify the way carbon is allocated and stored. Deciduous forests experience greater carbon cycling (production and decomposition) than coniferous stands (Gower et al., 2001). Both aboveground net primary production and overall (including belowground) net primary production were roughly two times greater in a boreal deciduous forest than a coniferous forest. The fraction of net primary production allocated to coarse- and fine-root primary production is roughly two times greater for evergreen conifers than deciduous trees (Gower et al., 2001). Because of these differential allocation patterns between deciduous and evergreen stands, the amount of carbon in the soil of mature black spruce stands is approximately three times the amount of carbon in the biomass of the trees (Kasischke et al., 1995). Since the rate of decomposition is higher in deciduous than in coniferous forests (because of both litter quality and site conditions), nitrogen is probably more available in deciduous forests, further increasing production (Makipaa et al., 1999). These results make sense physiologically, since deciduous species have higher maximum rates of photosynthesis and productivity than evergreens and produce litter that quickly decomposes.

Aboveground carbon pools are directly related to stand age. Gower et al. (2001) found that the effect of stand age (young versus old stands) on net primary production is roughly equivalent to the effect of soils (fertile versus infertile) and annual variation in the environment (favorable versus unfavorable weather). Similarly, Wang et al. (2001) found that wildfire exerts a lingering effect on carbon exchange between the boreal forest and the atmosphere via its effect on the age structure of forests and leaf-area index (LAI) during succession. They also found a strong inverse linear relationship between aboveground net primary productivity and age that was largely explained by a decline in LAI. Modeling results from Kurz and Apps (1999) suggested that forest ecosystems in Canada were a carbon sink from 1920 to 1980 and a source from 1980 to 1989. They suggested that this was a result of a change in the disturbance regime, and this finding is consistent with recent fire statistics (see section 14.9.2). Sometime around 1977, a regime shift in the climate of the North Pacific Ocean occurred that has been suggested to be part of a low-frequency oscillation (Niebauer, 1998). One of the consequences of this shift was that the position of the Aleutian Low associated with El Niño moved even farther eastward than it did in previous El Niño years. This shift is consistent with a more easterly (less southerly) flow component across Interior Alaska, which could exert teleconnective influences on the fire-dominated disturbance regime of Canada (Bonsal et al., 1993). If the key control over fire occurrence was a one-time climate regime shift caused by a change in sea surface temperature in the northeast Pacific Ocean, it suggests that climate changes occurring at low temporal frequencies exert a strong influence on the rate of carbon cycling in regions where the disturbance regime is climatically driven. Kurz and Apps (1999) suggested that as stand age increases, the ability to sequester carbon decreases and the susceptibility to disturbance increases.

Modification of soil thermal regime and permafrost degradation as a result of fire have been documented. Warmer and drier (due to reduced cover of saturated mosses) conditions following a forest fire increase decomposition and decrease carbon storage (Kasischke et al., 1995). Simulation results suggest that a 5 °C increase in average annual air temperature results in a 6 to 20% decrease in the total carbon stored in the soil over a 25-year period (Bonan and Van Cleve, 1992). In China, Wang et al. (2001) found that soil-surface CO_2 flux decreases immediately after wildfire because of a lack of root respiration, which accounts for about 50% of total soil-surface flux. In the time after a fire during which appreciable tree mortality occurred, the majority of respiration shifts from autotrophic to heterotrophic. As a consequence of the increased heterotrophic respiration and low net primary production in the early stages of succession, areas that have recently burned in the boreal forest tend to act as a carbon source for a brief period of time. Rapalee et al. (1998) found evidence that fire scars on the landscape are a net carbon source for about 30 years after burning, after which systems become a net carbon sink. Experiments in the Alaskan boreal forest showed that about 20% of the carbon in the soil surface layer is lost through decomposition during the first 20 to 30 years after a fire due to increased soil temperature (Van Cleve and Viereck, 1983). Rapalee et al. (1998) asserted that ecosystem changes in net ecosystem production are driven by changes in decomposition and species composition. They further suggested that changes in species composition are driven by fire-induced modification of the active layer. The relative importance of active-layer modification depends to a large extent on aspect and fire severity.

Fire frequencies in the sphagnum peatlands have decreased over the past 7000 years due to cooler and wetter conditions (Kuhry, 1994). Carbon emissions from peat combustion are still an order of magnitude greater than warming-induced oxidation (Gorham, 1991). Fire affects not only the amount of carbon in the forested peat systems but also the subsequent rate of accumulation. Kuhry (1994) found that peat accumulation in sphagnum-dominated peatlands of western boreal Canada decreased significantly with increasing fire frequencies. It is estimated that warming and drying would result in relatively rapid decomposition of peat soil organic carbon (Schimel et al., 1994). Valentini et al. (2000) investigated two contrasting land cover types, a regenerating forest and a bog, in the central Siberian region during July 1996, and found that net CO_2 uptake was limited by the decreasing soil water content in the regenerating forest. Their results showed substantial differences in both transpiration and carbon assimilation. The bog used the incoming solar energy principally for transpiration and, because of the constant availability of water, transpiration was not sensitive to seasonal changes in moisture conditions. The bog system also maintained high carbon assimilation potential compared to the regenerating forest. This trend was maintained and amplified in dry conditions, when the carbon uptake of regenerating forest decreased significantly. Bogs and peatlands represent a very different land cover type than forests and must be considered separately when assessing the role of disturbance in the boreal forest (see section 8.4.4.4).

14.10.4. Forest cover type, disturbance, and climate change

Vegetation response to climate change could feed back to cause large changes in regional and global climate through effects on terrestrial carbon storage (Smith T. and Shugart, 1993) and on water and energy exchange (Bonan et al., 1992; Chapin and Starfield, 1997). The rate and magnitude of this feedback is influenced by transient changes in the distribution of terrestrial ecosystems in response to changes in climate, disturbance regime, and recruitment rates. The long-term direction of ecosystem change is also sensitive to spatial patterns and processes operating at the landscape scale (Turner D. et al., 1995).

Current projections of vegetation response to climate change either assume that the disturbance regime does not change (Pastor and Post, 1986) or use globally averaged disturbance rates (Smith T. and Shugart, 1993). Conversely, projections of disturbance regimes in a warmer climate (Flannigan and Van Wagner, 1991; Flannigan et al., 1998; Kasischke et al., 1995) generally neglect rates and patterns of vegetation response to climate and disturbance. However, the climate–disturbance–vegetation interactions clearly influence the rate and pattern of changes in vegetation (Neilson, 1993; Noble, 1993) and disturbance (Gardner et al., 1996) through effects on fire probability and spread and pattern of colonization (Turner M. et al., 1997).

Landscape-scale interactions between vegetation and disturbance are particularly important in the forest–tundra ecotone (Chapin and Starfield, 1997; Noble, 1993; Starfield and Chapin, 1996) where vegetation change is very likely to have large feedbacks to climate (Pielke and Vidale, 1995). The potential colonization of tundra by forest is very likely to increase terrestrial carbon storage (Smith T. and Shugart, 1993), thereby reducing atmospheric carbon, but increase absorption of solar radiation (Bonan et al., 1992) thereby creating a positive feedback to regional temperature increases (Chapin and Starfield, 1997).

Modeling the transient dynamics of vegetation change allows investigation of both the short- and long-term responses of ecosystems to landscape-level disturbance and recruitment, and the subsequent feedbacks between climate and the biosphere. These spatial processes are responsible for long-term changes in vegetation distribution (Dale, 1997) in response to changing climate, and must eventually be incorporated into hemispheric- or global-scale spatio-temporal models of gradual climate change.

14.10.5. Land-use change

Atmospheric CO_2 and oxygen data confirm that the terrestrial biosphere was largely neutral with respect to net carbon exchange during the 1980s but became a net carbon sink in the 1990s (Schimel et al., 2001). However, the cause of this shift remains unclear. Several studies have indicated that land-use change is responsible for the majority of the terrestrial sink (Birdsey and Heath, 1995; Fang et al., 2001; Goodale et al., 2002; Houghton and Hackler, 2000; Houghton et al., 1983). Calculations of land-use changes can be used to calculate associated carbon fluxes (Houghton and Hackler, 2000). Since 1850, there has been a 20% decrease in global forest area, and during this period deforestation has been responsible for approximately 90% of the estimated emissions from land-use change (Houghton et al., 1999).

The majority of land-use change studies have focused on areas outside the boreal region, and a review of the critical issues provides perspective on the impact of land-use change on carbon cycling in the boreal forest. In the tropics, forests contain 20 to 50 times more carbon per unit area than agricultural land, and as a result, tropical deforestation during the early 1990s released 1.2 to 2.3 Pg of carbon annually (Melillo et al., 1996). Once tropical vegetation is cleared, soil mass is quickly lost through erosion and oxidation. When tropical forest soils are cleared of vegetation and cultivated, surface horizons experience exponential mass loss resulting in roughly a 25% decrease in carbon (Melillo et al., 1996). Despite the large atmospheric source of CO_2 from tropical deforestation, the terrestrial system is acting as a net sink for carbon (IPCC, 2001), and in a spatially explicit inversion analysis, Rayner et al. (1999) found no evidence of a large net source from the tropics. The spatial inversion analysis allows a more focused examination of carbon fluxes between discrete regions. For example, the lack of a large tropical net source suggests a tropical terrestrial sink of roughly the same magnitude. The exact nature of this tropical terrestrial sink remains a source of debate. Melillo et al. (1996) used gas flux studies to show that undisturbed tropical forests in the Brazilian Amazon are responsible for a net carbon uptake, but more work needs to be done to examine and quantify carbon flux from tropical forests experiencing and recovering from deforestation.

Like tropical forests, temperate and boreal regions in the Northern hemisphere have experienced substantial land-use changes in the past several hundred years. Siberian forests account for 20% of global forest area and net primary production; Valentini et al. (2000) estimated that approximately 800 000 ha are harvested there annually. Fang et al. (2001) found that Chinese forests acted as a carbon source from 1948 to 1980, and as a sink from 1981 to 1998. Subsequent works focused on constraining spatial and temporal aspects of carbon fluxes and therefore, several atmospheric inversion analyses have indicated a large terrestrial carbon sink in the Northern Hemisphere (Ciais et al., 1995; Fan et al., 1998; Keeling et al., 1996; Rayner et al., 1999). Estimates of carbon flux in the United States derived from independent forest inventory methods (Birdsey and Heath, 1995; Caspersen et al., 2000; Turner D. et al., 1995) and ecosystem models (Hurtt et al., 2002) provide supporting evidence for the presence of a North American sink, although of a lower magnitude than that estimated by Fan et al. (1998). Goodale et al. (2002) found that growth rates in unmanaged forests of the eastern United States have changed little over the past several decades, suggesting that nearly all of the carbon accumulation in the region is due to forest regrowth from past disturbance rather than growth stimulated by increased atmospheric CO_2, nitrogen deposition, or climate change.

14.10.6. Nitrogen deposition and carbon dioxide fertilization

A process-based model that simulates the biomass production of Norway spruce in southeastern Norway under both current climate and climate change scenarios was used to project biomass production responses to three climate change scenarios (Zheng et al., 2002). Net primary production (dry mass) was projected to increase by 7% over the current 10.1 t/ha/yr under a mean annual air temperature elevated by 4 °C over present-day levels. Doubling current ambient CO_2 concentration was projected to increase net primary production by 36%. The scenario of both elevated temperature and elevated CO_2 concentration led to an increase in net primary production of nearly 50%, which was higher than the sum of the two effects alone.

Nitrogen availability is often the limiting factor in net primary productivity. The majority of anthropogenic nitrogen inputs come from combustion (both biomass and fossil fuel) and agricultural fertilizer application (IPCC, 2001). Photosynthetic rate is correlated with the nitrogen content of leaves, since carbon assimilation is driven by the nitrogen-rich enzyme rubisco. Hence, reduced nitrogen availability decreases both leaf nitrogen content and photosynthesis. As a result, the carbon and nitrogen cycles are fundamentally coupled. Due to this coupling, increased anthropogenic nitrogen deposition must be considered in conjunction with elevated CO_2 levels. Elevated CO_2 levels have multiple direct effects on fundamental biochemical processes such as photosynthesis and respiration, which collectively determine net primary production.

Among studies that manipulated both CO_2 and nitrogen availability, the mean enhancement of photosynthesis by elevated CO_2 levels at the lowest level of nitrogen availability was 40%, while the mean enhancement at higher levels of nitrogen availability was 59% (McGuire et al., 1995a). These results indicate that for a fixed increase in CO_2 concentration, biomass increases proportionally to increased nitrogen availability. In a review of the literature, McGuire et al. (1995a) also found that compared to low nitrogen availability and baseline CO_2 levels, increased nitrogen availability and elevated CO_2 concentrations significantly increased biomass accumulation, sometimes by a multiple of more than two. Differential increases in biomass in response to elevated CO_2 concentrations are found in different species: in general, deciduous species exhibit twice the growth response of conifer species to elevated CO_2 levels (Ceulemans and Mousseau, 1994).

The carbon and nitrogen cycles are also strongly correlated with evapotranspiration. Arain et al. (2002) examined the response of net ecosystem productivity and evaporation to elevated atmospheric CO_2 concentrations and found that modeled and measured results showed a linear relationship between CO_2 uptake and evaporation. This coupling implies that as nitrogen deposition increases and plant tissue carbon to nitrogen ratios decrease, nitrogen cycling increases at a fixed level of evapotranspiration (Schimel et al., 1997). This is essentially an example of biological supply and demand. It becomes less efficient for plants to exert energy on translocation before senescence; hence, the quality of litter increases. The scaling of this response from tissue to plant level is seen in the results of modeling and field studies showing that nitrogen fertilization results in increased net primary productivity (Bergh et al., 1999; Chapin et al., 1988; McGuire et al., 1992; Vitousek and Howarth, 1991).

At the ecosystem level, studies indicate that conifers and deciduous species differ in their response to elevated CO_2 levels. Arain et al. (2002) examined the response of boreal net ecosystem productivity to elevated CO_2 levels in both a 70-year-old aspen stand and a 115-year-old black spruce stand. They found that the aspen stand was a weak to moderate carbon sink while the black spruce stand was a weak carbon sink in cool years and a weak carbon source in warm years (consistent with Fig. 14.23 and the BOREAS results described in section 14.7.3.2). These results emphasize the practical importance of the strong coupling between water flux and both the carbon and nitrogen cycles. When midsummer temperatures were high, the net ecosystem production of the black spruce stand decreased significantly due to increased respiration. At longer timescales, reduced litter quality resulting from elevated CO_2 levels has the potential to cause long-term negative feedbacks that constrain the response of net primary productivity. Litter nitrogen concentration is generally positively correlated with decomposition rates, and Cotrufo et al. (1994) found that in deciduous stands, cumulative respiration rates were lower for litter derived from elevated CO_2 conditions while rates for Sitka spruce remained relatively unaffected. These results have implications for the response of forests not only to elevated atmospheric CO_2 concentrations but also to the warmer temperatures that are very likely to ultimately accompany elevated atmospheric CO_2 levels.

Schimel et al. (1994) found that the amount of nitrogen lost from an ecosystem is an increasing function of the rate at which nitrogen cycles through the system. For example, nitrogen mineralization is a key index of soil inorganic nitrogen turnover, and is strongly correlated with evapotranspiration. Schimel et al. (1997) found evidence that losses of nitrogen trace gases are linked to the rate of mineralization of ammonium (NH_4) and nitrate (NO_3) from organic matter, a rate that increases as temperature and soil moisture increase. Similarly, the product of water flux and the ratio of NO_3 to dissolved organic nitrogen (DON) concentration directly control leaching losses of NO_3 and DON (Schimel et al., 1997). Hence, nitrogen losses are controlled by soil moisture and water flux. This linkage between hydrological and nutrient cycles is of critical importance in assessing the relevance of nitrogen mineralization enhanced by increasing temperature to net primary productivity. Schimel et al. (1997) noted that water, energy, and nutrient limitation of net primary productivity and carbon storage tend to equilibrate in near "steady-state" ecosystems. This implies that the greatest potential for discrepancies between carbon, nitrogen, and water cycling exist in recently disturbed ecosystems. In support of this, Schimel et al. (2001) noted that in general, carbon accumulation in recovering ecosystems is high and chronosequence studies show lower accumulation in undisturbed landscapes. The decreased net primary productivity observed in "mature" ecosystems is a consequence of the equilibration of the nitrogen and water fluxes. Hence, the impacts of temperature increases on enhanced nitrogen mineralization are likely to be greatest in recently disturbed ecosystems.

Finally, although it is not always the case that nitrogen limits growth, a review of studies exposing plants to both elevated CO_2 levels and increased soil nitrogen concentrations showed significant increases in net primary productivity (McGuire et al., 1995a). Under the assumption that "mature" ecosystems exhibit decreased net primary productivity because of equilibration of the nitrogen and water fluxes, elevated atmospheric CO_2 concentrations are likely to stimulate growth. The relatively rapid increase in atmospheric CO_2 concentrations makes this scenario even more plausible. Elevated atmospheric CO_2 concentrations have been shown to increase the amount of photosynthesis per unit of water transpired, also known as water-use efficiency. Schimel et al. (1994) suggested that if CO_2 or fertilizer and pollutant nitrogen increase global net primary production over the coming decades, it is possible for soil carbon increases to occur on a commensurate timescale. Despite this prospect of greater carbon storage, the estimated effect of increasing temperatures is sensitive to the feedback between primary production and decomposition via the nitrogen cycle. As soil organic matter is lost through enhanced nitrogen mineralization caused by temperature increases, more nitrogen becomes available for plant growth, which results in the formation of more soil organic matter, thus acting as a negative feedback (Schimel et al., 1994).

Another possible mechanism causing equilibration of the carbon, nitrogen, and water fluxes comes from examina-

tion of the tissue-level response to elevated atmospheric CO_2 concentrations. Acclimation to elevated CO_2 levels can occur through one or more of three processes of leaf-level carbon assimilation: carboxylation, light harvest, and carbohydrate synthesis (McGuire et al., 1995a). Under saturating light conditions at low levels of intercellular CO_2, assimilation is limited by the quantity and activity of rubisco, the enzyme that is primarily responsible for capturing atmospheric carbon in the production of sugars. At high levels of intercellular CO_2, the enzymatically controlled rate of carbohydrate synthesis, which affects the phosphate regeneration that is necessary for harvesting light energy, may regulate the fixation of carbon (McGuire et al., 1995a). Hence, mechanisms acting at scales from tissue-level biochemistry to ecosystem-level nutrient cycling exert influences to equilibrate carbon, nitrogen, and water fluxes in mature ecosystems. This equilibration has implications for recent patterns of carbon flux observed in terrestrial ecosystems. Schimel et al. (2001) noted that the terrestrial carbon sink must eventually become saturated because photosynthesis follows a saturating function with respect to CO_2. As the rate of photosynthesis slows, plant and microbial respiration must catch up eventually, reducing incremental carbon storage to zero. Simulations with the Terrestrial Ecosystem Model (Marine Biological Laboratory, Woods Hole) concur, suggesting that soon after atmospheric CO_2 concentration stabilizes, heterotrophic respiration comes into balance with net primary production and the CO_2-stimulated terrestrial carbon sink disappears (Melillo et al., 1996).

In addition to the effect that anthropogenic modification of carbon and nitrogen cycles has on atmospheric CO_2, temperature in boreal and tundra regions affects the intra-annual variability in atmospheric CO_2 concentrations. Surface temperature in the north is positively correlated with seasonal amplitude of atmospheric CO_2 the following year. Zimov et al. (1999) found that 75% of the annual increases in mean annual air temperature between 1974 and 1989 coincided with decreases in CO_2 amplitude at the Barrow monitoring station, consistent with observations of net summer CO_2 efflux from tundra and boreal forest during warm years.

14.11. Climate change and forest distribution

14.11.1. Historic examples of treeline movement

The contrast between the realm of life where trees dominate the environment and the very different realm where much shorter herbaceous and low woody vegetation is the exclusive plant cover is one of the most visually striking landscape characteristics of the circumpolar Arctic. A number of questions about arctic and associated alpine treelines have attracted scientific interest for well over a century. A simple catalog and summary of this body of research is too large for this chapter. Treeline questions have been studied from a number of different perspectives, but nearly all are at least implicitly connected with the question of how cold temperatures limit tree growth and survival. Because of this overriding interest in temperature limits, and because of the ubiquitous occurrence of treelines across the Arctic and the high state of preservation of dead trees in the cold environment of the Arctic, treeline studies have much to offer on the specific question of climate warming and cooling. This section focuses on some recent studies that shed light on the long-term history of dynamics and movement of tree limits as related to specific temperature-controlled processes. These studies provide a perspective of long-term continuity, and insight into the mechanisms that recent temperature increases have affected and that warming projected by the ACIA-designated models could affect.

14.11.1.1. Northern Eurasia

Between 9000 and 7000 years BP, forest occupied what is now treeless tundra nearly to the arctic coastline throughout northern Russia (MacDonald et al., 2000). These results are based on earlier subfossil wood collections carried out by Russian scientists as well as modern collections of subfossil wood from the Kola, Yamal, and Taymir Peninsulas, and at the mouth of the Lena River. Subsequently, the greatest retreat of forest and expansion of tundra (compared to the modern position of the treeline) took place between 4000 and 3000 years BP. During the period of maximum forest advance, the mean July temperature in northern Russia (at the coast) was 2.5 to 7.0 °C higher than the present-day mean, based on modern tree growth relationships to temperature. The northward advance of the treeline reflected a series of other environmental changes in the Arctic, including increased solar insolation, reduced sea-ice cover in the Arctic Ocean, increased climate continentality, and a significant intrusion of warm North Atlantic water into the Arctic (MacDonald et al., 2000). This documented record of past forest advance suggests that there is a solid basis for projecting similar treeline change from scenarios that project similar temperature increases. It also suggests that the components of ecosystems present today have the capacity to respond and adjust to such climate fluctuations.

A reconstruction of treeline movement in the Swedish Scandes Mountains using 173 dated remains of pine wood reveals a that there has been a gradual decrease in the elevation of the treeline (with small fluctuations), generally consistent with climate cooling. Elevational decline in treeline as the result of decreasing temperatures began sometime after 10700 calendar years BP. Although the rate of treeline recession was greater before about 8000 years BP than after, generally the treeline evidence here indicates a smooth long-term temperature decrease with only a few minor, brief warming and cooling episodes. The total elevational retreat of treeline was estimated to be 500 m, corresponding to a temperature decrease of 6 to 8 °C. Since the beginning of the 20th century, however, a local

warming of about 1 °C has been associated with an upward movement of more than 100 m, which represents the largest adjustment of treeline position in the last 3000 years (Kullman, 2001).

14.11.1.2. Yamal Peninsula

Detailed analyses of the dynamics of polar treeline on the Yamal Peninsula are presented in Hantemirov (1999), Hantemirov and Shiyatov (1999, 2002), and Shiyatov et al. (1996). Holocene wood deposits in the southern Yamal Peninsula include a large quantity of subfossil tree remains, including stems, roots, and branches. This is the result of intensive accumulation and conservation of buried wood in permafrost. The existence of extensive frozen subfossil wood in the southern Yamal Peninsula in what is now the southern tundra zone was first noted and described by B.M. Zhitkov in 1912 during an ornithological expedition through the area. Tikhomirov (1941) showed that tree remains conserved in peat sediments were direct evidence that the northern treeline in the warmest period of the Holocene reached the central regions of the Yamal Peninsula (up to 70° N). Today the polar treeline is considerably further south on the Peninsula (67° 30' N).

Systematic collection of subfossil wood samples started in 1982 in the watersheds of the Khadyta, the Yadayak-hodiyakha, and the Tanlova Rivers in the southern Yamal Peninsula, located between 67° 00' and 67° 50' N and 68° 30' and 71° 00' E. River flow in this area is from north to south, excluding the possibility that wood was transported to the collection site from more southerly locations; thus, there is very high confidence that the region experienced a considerably warmer summer climate in the relatively recent past.

Radiocarbon dating of the subfossil wood (53 dates) was cross dated with ring series from the samples, allowing the construction of a continuous tree-ring chronology 7314 years in length. Absolute dates can

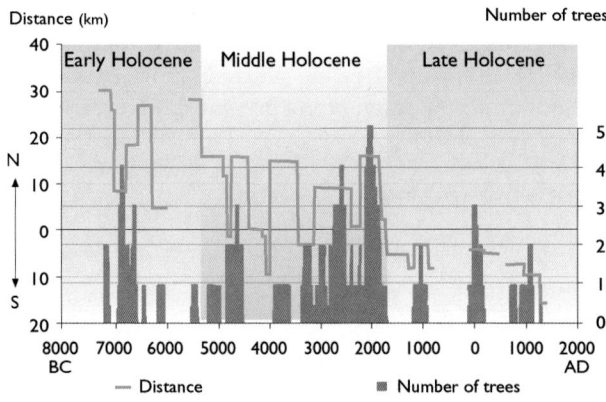

Fig. 14.27. Dynamics of the polar treeline on the Yamal Peninsula in Siberia, showing the relative distance of sampled tree remains from the present position of northernmost open stands of larch in river valleys and the number of tree samples for each radiocarbon date. A proposed division of the Holocene into three stages based on climatic shifts is indicated (data from Hantemirov and Shiyatov, 1999).

generally be assigned to the recovered wood remains. The result makes it possible to reconstruct the dynamics of tree limits on the Yamal Peninsula during the Holocene (Fig. 14.27). From at least 10 000 to 9000 years BP, trees grew across most of the Peninsula. The most favorable conditions for tree growth (almost certainly warm summers) occurred from 7200 to 6000 BC. From about 6000 to 5600 BC, climatic conditions became less favorable for tree growth (almost certainly cooler summers), but trees persisted and did not retreat to the south. Beginning about 5400 BC, trees were displaced southward. The forest stand density also greatly decreased during this period, and it can be considered as a transition to the next stage of the Holocene. From 5400 until 1700 BC, the polar treeline was still located at 69° N, well north of the present-day position. In unfavorable periods (4500–3900 BC and 3600–3400 BC), tree survival was mainly restricted to the river corridors, but in climatically more favorable periods (5200–4500 BC, 3900–3600 BC, and 3400–1800 BC), forests grew on hills and raised surfaces beyond the rivers.

One of the most important periods of displacement of polar treeline to the south and major reduction in the density and productivity of forest stands occurred about 1700 BC. This stage can be regarded as the end of the Middle Holocene and the beginning of the modern stage of treeline evolution on the Yamal Peninsula. During the last 1700 years, forest–tundra and forest associations have been primarily restricted to river valleys in the southern part of the Peninsula. Somewhat more favorable conditions occurred from 1200 to 900 BC, from 100 BC to AD 200 and during the MWP (AD 700–1400).

Treeline dynamics for the last 4000 years of the Holocene were reconstructed with even greater precision using more than 500 cross-dated tree stems with known coordinates of their burial places in valleys of different rivers on the southern Yamal Peninsula (Fig. 14.28). Treeline displacements northward and southward were relatively small and less important during the last 3600 years than those that occurred in the previous few millennia. Treeline generally moved at the most 5 km to the south of the present-day treeline, and

subsequently northward only to the present boundary of open woodland in the river valleys. However, one particularly noteworthy major displacement of treeline to the south occurred during the second half of the 17th century BC. In this relatively short period (not exceeding 100 years), the boundary of larch open woodland moved southward nearly 15 to 20 km, and the treeline retreated a further 8 to 10 km during next 700 years. This major displacement of treeline in the 17th century BC appears to have been driven by strongly inclement climatic conditions (cold summers), representing the lowest reconstructed summer temperatures in the entire series. It was in the years immediately after 1657 BC that the temperature decreased sharply. Fourteen years during the interval 1630 to 1611 BC appear to have been extremely cold, reaching a nadir in two specific years, 1626 and 1625 BC. No other period during the reconstruction is even close. Moreover, it is clear that in 1625 BC, a severe freeze occurred in the middle of the summer (as indicated by characteristic anatomical structures of freeze injury in the tree rings). It is very probable that this short-term extreme climate event represented climate cooling following one of the largest volcanic eruptions of the last few millennia, which happened in about 1628 BC (possibly the eruption of the Santorini volcano in the eastern Mediterranean). The cooling appears to have reinforced another closely spaced cooling event that preceded it. The earlier of the two periods of extreme cold temperatures began sometime after 1657 BC, but in this case, it is difficult to determine the cause (Grudd et al., 2000). These events were the final circumstances that resulted in the most significant southward retreat of treeline during at least the last 4000 years. In spite of extremely favorable summer warmth that returned afterward and even persisted at various intervals, the treeline never returned to its previous boundary.

Another important result of dendrochronological dating of large samples of subfossil wood is the ability to calculate the relative abundance of Siberian spruce in forest stands of the area, which is an index or proxy for the degree of continentality of the climate. Figure 14.29 shows the change in the proportion of spruce in forest stands (the remaining part is all larch). In the first six centuries, from AD 900 to 1500, the proportion of spruce decreased from 22% to 3–5%. After that, the

Latitude of northernmost trees

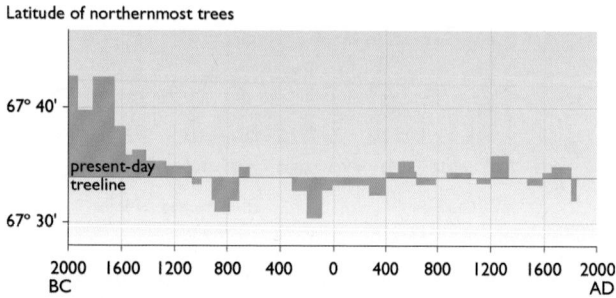

Fig. 14.28. Reconstruction of polar treeline dynamics on the Yamal Peninsula from 2000 BC, showing the latitude of recovered samples by year (reanalysis from the data of Hantemirov and Shiyatov, 2002).

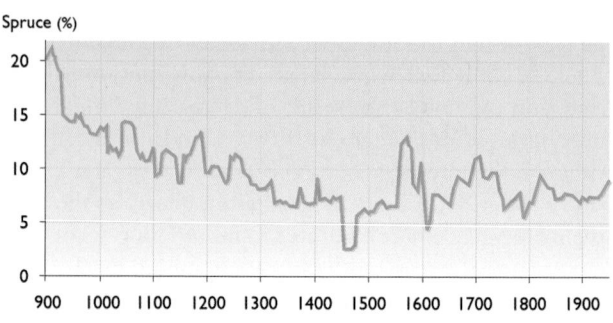

Fig. 14.29. Change in the proportion of subfossil remains made up of spruce compared to larch in Yamal Peninsula samples (data from Shiyatov, 2003).

Table 14.8. Altitudinal displacement of the upper treeline in the Polar Ural Mountains during the last 1150 years (Shiyatov, 2003).

Period (AD)	Number of years	Change in treeline	
		Direction	Rate (m/decade)
850–1280	430	rise	1–2
1280–1580	300	retreat	2–3
1580–1790	210	rise	0.5–1.0
1790–1910	120	retreat	2–4
1910–2000	90	rise	4–6

percentage of spruce stabilized in the range of 7 to 10%. The 20th century is characterized by an increasing percentage of spruce in forest stands in the valley of the River Khadytayakha, and a weak northward advance of the polar treeline.

14.11.1.3. Polar Ural Mountains

Significant spatial and temporal changes took place in the upper treeline ecotone in the Polar Ural Mountains (66°–67° N, 65°–66° E) during the last millennium (Shiyatov, 1993, 1995, 2003). Within the treeline ecotone, located between 100 and 350 m above sea level, open forests of Siberian larch dominate. Patches of closed forests of mixed larch and Siberian spruce grow at lower altitudes in the ecotone. Up to the present, these forests have largely developed under conditions of very little human influence. A large number of wood remnants persist on the ground at elevations up to 60 to 80 m above the present treeline, and within the ecotone between forest and tundra. These wood remains have been preserved for up to 1300 years because of the low rate of wood decomposition in the severe cold. The wood remnants provide material to extend a ring-width chronology back to AD 645, and to date the lifespan of a large quantity of dead trees.

In order to estimate the downward displacement of the altitudinal treeline (the highest altitudinal position of open forest over the last millennium), a transect 430 m long and 20 m wide was set up on the southeastern slope of Rai-Iz Massif, from the highest location of larch wood remnants (340 m above sea level) down to the present upper treeline (280 m). The transect was divided into 10 x 10 m quadrats. All of the wood remnants were

mapped and wood sections cut from the base of the trunk and roots were collected from each individual tree. The calendar year of establishment and death for each tree was determined by cross dating. Altogether, the life history of 270 dead trees was defined. In addition, 16 young living trees and seedlings were also mapped and their age determined.

Using these data, stages in the overall downward displacement of the upper treeline over the last 1150 years were reconstructed (Table 14.8). This time interval were divided into 5 periods, distinguished by differing directions of treeline shifts (rise or retreat in elevation) and differing rates of displacement. During the earliest period (430 years, AD 850–1280), the upper treeline rose from 305 to 340 m above sea level, an average of 1 to 2 m per decade. The highest altitudinal position reached by the treeline and the densest and most productive larch stands of the last millennium occurred in the 13th century. The second period (300 years, AD 1280–1580) was characterized by a substantial retreat of the upper treeline (from 340 m down to 295 m) at a mean rate of 2 to 3 m per decade. During the third period (210 years, AD 1580–1790), the treeline retreat stopped. The calculated rise (from 295 to 305 m, 0.5–1.0 m/decade) was not significant. The most extreme retreat was seen in the fourth period (120 years, AD 1790–1910), during which the upper treeline receded from 305 to down to 270 m, an average of 2 to 4 m per decade. During this fourth period, the upper treeline was at its lowest altitudinal position of the millennium. In the last period, from 1910 to the present, vigorous forest establishment took place on sites that were forested during the Middle Ages. The rate of change was the highest of the millennium (from 270 to 308 m, 4–6 m/decade).

The last period of expansion of forest vegetation is being studied using both direct and indirect evidence (old terrestrial, aerial, and satellite photographs; repeated stand descriptions of permanent plots and transects; morphology and age structure of stands; large-scale mapping within the ecotone; and meteorological and dendroclimatic data). Table 14.9 shows the change in area of different types of forest–tundra ecosystems within the ecotone during the 20th century. These data were obtained at the time of large-scale (1:10 000) mapping of the key study area (3085 ha) located at the bot-

Table 14.9. Area change in different types of forest–tundra ecosystems from 1910 to 2000 in the Polar Ural Mountains (Shiyatov, 2003).

	Tundra with individual trees		Sparse growth of trees		Open forest		Closed forest	
	Area (ha)	Change (ha/decade)	Area (ha)	Change (ha/decade)	Area (ha)	Change (ha/decade)	Area (ha)	Change (ha/decade)
1910	2403		349		328		5	
		-76		37		24		15
1960	2021		535		450		79	
		-26		-52		36		42
2000	1917		327		593		258	

tom of Tchernaya Mountain. Three maps were produced
that show the spatial distribution of these types of forest–tundra ecosystems in the beginning, middle, and
end of the 20th century. Change in area was estimated
for two periods: the 50 years from 1910 to 1960, and
the 40 years from 1960 to 2000.

Between 1910 and 2000, the area of tundra with only
scattered individual trees decreased significantly from
2403 to 1917 ha, or from 78 to 62% of the main study
area. The greatest rate of change (-76 ha/decade) was
observed during the first 50 years, when isolated
seedling establishment above the treeline was the most
intensive. Data obtained on the change in area of the
sparse trees mapping unit is very interesting since during
the first period the change was positive (+37 ha/decade)
but during the second period it was negative (-52 ha/
decade). The decrease occurred when young established
trees began producing seeds, resulting in an increase in
stand density (many stands changed from tundra with
isolated trees into open forest). This led to the great
increase in open-forest area between 1960 and 2000
(from 450 to 593 ha, +36 ha/decade). However, the
most impressive changes were seen in the case of closed
forests. The area of closed forest increased from 5 to
258 ha over the 90 years with the transformation of
open forests into closed forests. To date, more than 550
repeated terrestrial photographs have been taken which
can be used to reconstruct stand parameters for the
middle of the 20th century.

Temperature increases during the 20th century were the
major cause of the expansion of trees in the treeline ecotone. The mean June temperature at the Salekhard
weather station (50 km east of the study area) increased
from 7.2 °C (1883–1919) to 8.5 °C (1920–1998).
The July mean increased from 13.8 to 14.3 °C. Mean
temperatures in the winter months (November–March)
increased from -20.8 to -19.6 °C. The average June–July
temperature, which is critical for tree growth, increased
from 10.5 to 11.4 °C (0.9 °C). This means that the
June–July isotherm rose 120 to 130 m in altitude (in
this area the elevational temperature gradient is 0.7 °C/
100 m). However, the mean rise of the upper treeline
was only 20 to 40 m owing to a deficiency of viable
seeds on sites remote from fertile trees and stands.

In this area, larch cones open and seeds are disseminated
only on days with elevated temperatures, typically from
the end of June to the end of July (Shiyatov, 1966).
Wind is the primary method of seed dissemination;
other means of dispersal, such as animals and birds, are
insignificant. Since seed dissemination takes place in
summer when snow cover is absent, it is difficult for
seeds to be transported upwards over long distances.
Heavy larch seeds are carried no more then 40 to 60 m
from the tree and most of these seeds become lodged in
lichen–moss and shrub layers that prevent them from
germinating successfully. That is why abundant seedling
establishment took place only close to individual trees
and within existing stands.

Impressive changes have occurred in the structure of
existing stands during the last 90 years. Most sites with
trees have become much denser and more productive
(up to 4 to 5 times) and many tundra sites located within the treeline ecotone have been afforested. The degree
of afforestation increased from 22 to 38% (based on data
from Table 14.9). Thus, many factors affecting the forest–tundra ecotone position reacted to summer temperature changes. However, for the purpose of climatic
reconstruction, the best proxies for reconstructing summer temperature are those obtained from existing stands
(tree rings, biomass, stand and canopy density, degree of
afforestation) rather than treeline movement, which
experiences lag effects. Although the displacement rate
of the upper treeline is a good reflection of long-term
climatic fluctuations (Table 14.8), reconstruction of
actual temperature changes is complicated because of
the response lag caused by the slow growth of seedlings
and the lack of seeds on remote sites. For example, the
recent warming observed in summer months is of the
same degree as inferred for the Middle Ages (about
1 °C) but the upper treeline has not yet reached the altitudinal position where forests grew in the 13th century.
To overcome such discrepancies, it is necessary to use
corrective factors. These will be different for each period and study area. For example, in the Polar Ural
Mountains over the 90 years of study, the upper treeline
rose 20 to 40 m in altitude but the June–July temperature isotherm rose 120 to 130 m. Therefore, reconstructed temperatures based on recent treeline movement should be increased by a factor of four.

Recent treeline movement in the Polar Ural Mountains
can be confirmed as a widespread phenomenon by comparison of high-resolution satellite images. The satellite
and *in situ* data analysis shows an increasing proportion
of the area with higher stand crown closure. Specifically,
in 1968 about 23.5% of the area supported stands with
crown closure ≥30%; in 1998 this area had increased to
50.0%. The border of stands with closure ≤10% moved
between 100 and 400 m horizontally, depending on the
site. Some areas that were tree-covered before the LIA
still do not support any trees.

14.11.1.4. Northeast Canada

Treeline does not always fluctuate in a straightforward
way in response to the control of direct climate warming
and cooling on the establishment and death of trees from
seed, as illustrated in northeastern Canada. Treeline populations in northern Quebec are primarily single-species
stands of black spruce, surviving in the upright growth
form in protected sites, or in a damaged, low growth
form where spruce are exposed to wind and snow abrasion above the snowpack. Spruce macrofossils are generally absent in the northern Quebec tundra zone, and the
forest limit and soil charcoal limit occur together,
strongly suggesting that the forest limit has remained
stable during the last 2000 to 3000 years. Apparently,
climate change does not easily trigger treeline advance
or retreat in this location (Lavoie and Payette, 1996).

However, the character of the forest near the treeline has changed over the last few millennia. The Hustich (1966) hypothesis suggests that the regenerative capacity of forests decreases from south to north. At the treeline of northeastern Canada, summer climate is too cold for frequent seed production, although black spruce produce cones that fail to complete only the last stages of development (Sirois, 1999). The black spruce present today were established thousands of years ago in northern Quebec and have persisted by layering since then. Sporadic and patchy fires kill black spruce in this environment and, in the absence of seed production from nearby survivors, trees are eliminated in the local fire patches. By this process, the treeline of today, which is out of equilibrium with the environment that created it, has developed a more irregular and patchy boundary with time (Payette and Gagnon, 1985).

Since the 1960s and 1970s, small-tree stands have emerged above the snowpack, triggered by less harsh conditions associated with snowier winters in the last few decades (Pereg and Payette, 1998). This process decreases the numbers of low, damaged growth-form individuals in the population, and increases the numbers of upright, tree-form stems. The temperature increases projected by the ACIA-designated models are likely to be sufficient to provide the final requirement of summer warmth that would result in viable seed production at treeline in northeastern Canada (see Fig. 14.8 for summer temperature scenarios at Goose Bay, Labrador). The initial effect of warming is very likely to induce viable seed production within the forest–tundra zone, resulting in infilling of the patchy forest–tundra border. It is possible that production of viable seeds at the absolute tree limit would begin seed rain onto the tundra, a process that probably has not occurred in appreciable amounts for thousands of years.

14.11.2. Scenarios of future treeline movement

White et al. (2000) developed a global vegetation model to project how increasing atmospheric CO_2 concentrations and other parameters, such as air temperature and nitrogen deposition, affect vegetation composition and distribution and carbon sequestration in biomass and soils north of 50° N. The model, Hybrid v4.1, is a non-equilibrium, dynamic global vegetation model, with a sub-daily time step, driven by increasing CO_2 and transient climate output from the Hadley Centre GCM (HadCM2) with simulated daily and interannual variability. Three emissions scenarios (IPCC, 1995) were used: IS92a, which results in an atmospheric CO_2 concentration of 790 ppm by 2100; CO_2 stabilization at 750 ppm by 2225; and CO_2 stabilization at 550 ppm by 2150. Land use and future nitrogen deposition were not included. The model projects an expansion of the area of coniferous boreal forest and mixed/temperate forests of the Northern Hemisphere by about 50%. This forest expansion mainly displaces tundra, driven by the direct effects of rising CO_2 levels and temperatures on tree photosynthesis and growth. The forest expansion and an

associated increased carbon sink also depend on the indirect effects of increased nitrogen deposition and improved water-use efficiency. However, the model operates with only positive growth responses to increased temperatures, and new findings, such as negative growth responses in treeline trees to increased temperatures (section 14.7.3.1) and human activities forcing the treeline southward (Vlassova, 2002), suggest that the modeled increase in carbon uptake driven by forest expansion is unlikely to be fully realized.

The central Canadian Arctic is a region of low topographic relief and few barriers to tree migration, and is often depicted as an area of major northward tree movement in a straightforward response to increasing temperatures (e.g., Nuttall and Callaghan, 2000). The BIOME 1.1 model of potential natural vegetation under doubled atmospheric CO_2 conditions projects that about 60% of global tundra will be displaced, largely through the expansion of boreal forest (Skree et al., 2002). When the BIOME 1.1 model is forced with tripled CO_2 concentrations (see section 4.4.1, Fig. 4.12b), it projects that boreal forest will replace 70% of the tundra of northern Europe and virtually all the European Arctic coast will be occupied by what is today recognized as southern taiga or boreal forest (Cramer, 1997). These vegetation models of boreal forest response to climate change generally produce outputs based on the final adjustment of trees to occupy all climatically suitable areas. However, a variety of processes can reduce the potential for tree expansion or create environments that preclude tree establishment or survival, such as paludification (Skree et al., 2002). There is increasing evidence in the Russian Arctic of southward displacement of trees that are subject to strong maritime climate influences or intensified human land uses such as reindeer production systems (Vlassova, 2002). At the large scale, future northward movement of the treeline in response to climate change of the magnitude projected by the ACIA-designated models is very likely. However, a considerable lag period is possible, for example because of challenges to tree establishment (e.g., poor seedbed conditions in tundra) that must be overcome by repeated occurrences of low-probability events. Finally, environmental complexity and novelty in an altered, warmer world is likely to produce unexpected vegetation patterns at a local or regional scale.

14.12. Effects of ultraviolet-B on forest vegetation

During summer in the Arctic, levels of UV-B radiation are generally not much different than at mid-latitudes. However, the Arctic and Antarctic are uniquely susceptible to short-term, intense stratospheric ozone depletion, especially in their respective spring seasons. For general background on ozone depletion and UV-B radiation in the Arctic, see Chapter 5.

Exposure to increased UV-B (280–315 nm) radiation is known to inhibit plant growth, development, and physiological processes. Long-lived, slow-growing plants such as trees may show cumulative effects of increasing UV-B radiation levels (Sullivan and Teramura, 1992). Effects of elevated UV-B radiation levels on plant processes vary in severity and direction and among species, varieties, and clones, as well as among plant parts and developmental stages. To avoid UV-B radiation, plants have developed screening mechanisms including increased leaf thickness or cuticle (Johanson et al., 1995; Newsham et al., 1996), optical structures to scatter and reflect UV-B radiation (Kinnunen at al., 2001), and the accumulation of UV-B screening phenolics and flavonoids in the epidermal cells of the leaves (Alenius et al., 1995; Cen and Bornman, 1993; Day, 1993).

Until the 1980s, relatively few plant species had been screened to determine the effect of enhanced levels of UV-B radiation. Some species show sensitivity to present levels of UV-B radiation while others are apparently unaffected by very large enhancements in UV radiation levels. Dicotyledonous crop plants such as peas (Day and Vogelmann, 1995) and canola seem to be more susceptible to increased UV radiation levels than cereals such as wheat (Beyschlag et al., 1989) and barley, although many other factors play an important role in sensitivity. Even varieties of the same species can exhibit substantial variability in UV-B radiation sensitivity (Teramura and Sullivan, 1994). About two-thirds of the few hundred species and cultivars tested appear to be susceptible to damage from increased UV-B radiation levels. Crop damage caused by UV-B radiation under laboratory conditions generally has been attributed to impairment of the photosynthetic process (Day and Vogelmann, 1995). In addition to gross disruption of photosynthesis, UV-B radiation may inhibit plant cell division as a physiological change, causing reduced growth and yields (Teramura and Sullivan, 1994). Accurately determining the magnitude of the effect of elevated UV-B radiation levels in the field is difficult, because interactions with other environmental factors, such as temperature and water supply, affect the reaction and overall growth of the crop (Balakumar et al., 1993, Mepsted et al., 1996).

Sensitivity to UV-B radiation differs among forest species and populations and is influenced by environmental conditions (Laakso and Huttunen, 1998; Lavola, 1998). The composition of flavonoids varies according to altitude and latitude. High-latitude, low-altitude species and populations are more sensitive (fewer protective mechanisms) to enhanced levels of UV-B radiation than low-latitude, high-altitude species and populations (more developed defense mechanisms), reflecting natural levels of exposure to UV-B radiation (Sullivan and Teramura, 1992). Scots pine populations growing at high latitudes are rich in prodelphinidin (Laracine-Pittet and Lebreton, 1988) and significant differences in the characteristics of UV-absorbing compounds occur among species of pine (Kaundun et al., 1998) and birch (Lavola, 1998).

Ultraviolet-B radiation has many direct and indirect effects on plant growth and development (Caldwell M. et al., 1998). The direct effects are most damaging because the photons of UV-B radiation cause lesions in important UV-B absorbing biomolecules such as nucleic acids and proteins (Caldwell C., 1993; Greenberg et al., 1989; Taylor et al., 1997). Photoproducts of DNA formed by UV-B radiation are all toxic and mutagenic (Taylor et al., 1997) and altered DNA or RNA structures may interfere with transcription and replication causing slower protein synthesis as a result of UV-B radiation stress (Jordan et al., 1994; Taylor et al., 1997). The indirect effects of UV-B radiation on plants can also cause damage by the formation of free radicals and peroxides (Takahama and Oniki, 1997; Yamasaki et al., 1997). Excessive UV-B radiation levels can also alter patterns of gene activity (Caldwell M. et al., 1998).

Field studies in both crop plants and trees suggest that the primary effects of increased UV-B irradiance are subtle, light-induced morphological responses that alter carbon allocation (Bassman et al., 2003; Day, 2001). Recent UV radiation acclimation studies of subarctic and arctic plants have emphasized a multitude of responses ranging from avoidance and protective mechanisms to inhibition and accumulation of effects (Gwynn-Jones et al., 1999; Kinnunen et al., 2001; Laakso, 1999; Latola et al., 2001; Turunen et al., 2002). Enhanced UV-B radiation levels affect litter decomposition directly (photodegradation and mineral nutrient cycling) and indirectly (chemical changes in plant tissues) and also affect the biochemical cycling of carbon (Rozema et al., 1997). Both direct and indirect effects include physical, chemical, and biological components. The indirect effects of UV-B radiation are likely to be more significant than direct effects in subarctic and arctic forest ecosystems.

Subarctic conifers are long-lived with long generation times; damage from UV-B radiation is likely to accumulate over the years. Winter injuries to evergreens are caused by the interaction between freezing, desiccation, and photo-oxidation (Sutinen et al., 2001). In subarctic and arctic conditions, late-winter cold temperatures, enhanced UV radiation levels from intense solar radiation, and water deficiency are the major environmental risks. Xeromorphic leaves and small leaf size of subarctic and arctic trees reduce evapotranspiration. However, late spring in high-latitude subarctic and arctic ecosystems is characterized by high levels of solar radiation and fluctuations between freezing and thawing temperatures. The UV radiation dose received by evergreens in the late winter, early spring, and at the beginning of the short growing season is high due to reflectance from persistent snow cover. Some earlier studies considered evergreens to be tolerant, but other studies have revealed evergreens to be sensitive to enhanced UV-B radiation levels (Kinnunen et al., 2001; Laakso et al., 2000; McLeod and Newsham, 1997).

Experimental field evidence indicates that enhanced UV-B radiation levels mainly increase the amount of soluble UV-B-absorbing compounds in summergreen (deciduous) plants (Searles et al., 2001), but the protective functions of wall-bound phenolic compounds and epicuticular waxes in evergreens are more complicated. The responses are species-specific among pine species (Laakso et al., 2000).

In needles from mature subarctic Scots pine, enhanced UV-B radiation levels induced xeromorphic (change in plant anatomy to enhance drought resistance) characteristics, including smaller epidermal area and enhanced development of the cuticle layer (Latola et al., 2001). Ultraviolet responses increased the concentration of UV-B radiation absorbing compounds in the epidermal cells and induced high and accumulating proportions of oxidized glutathione (Kinnunen et al., 2001; Laakso et al., 2001; Latola et al., 2001; Turunen et al., 1998, 1999). The cumulative stress has been measured as gradually decreased total glutathione and an increased proportion of oxidized glutathione levels in one- to three-year-old needles (Laakso et al., 2001; Laakso and Huttunen, 1998).

The most consistent field response to enhanced UV-B radiation levels (Day, 2001; Searles et al., 2001) is an increase in concentrations of soluble or wall-bound UV-B radiation absorbing compounds in leaves, complicated by great seasonal, daily, and developmental variation, both in epicuticular and internal compounds. It is important to remember that not all soluble flavonoids are UV-inducible (Lavola, 1998) or acid-methanol extracted (e.g., cell-wall bound UV-B absorbing phenyl-propanoids; Jungblut et al., 1995).

Increased amounts of epicuticular waxes and UV-absorbing compounds, such as flavonoids, and smaller leaf/needle surface area are plant defense mechanisms against UV-B radiation. In young Scots pine seedlings, UV-B radiation induces flavonoid biosynthesis (Schnitzler et al., 1997). However, an increase in UV-B absorbing compounds may result in a decrease in cell expansion and cell-wall growth (Day, 2001). Diacylated flavonol glucosides provide protection from UV-B radiation (Turunen et al., 1999) and biosynthesis of these compounds and the development of waxy cuticles appears to provide effective UV radiation protection in young needles (Kinnunen et al., 2001; Laakso et al., 2000). The protective effects come about through rapid development of epicuticular waxes, an increase in cutinization, and an initial increase followed by inhibition of UV-screening compounds (Kinnunen et al., 2001; Laakso et al., 2000).

Plant-surface wax morphology, chemistry, and quantity respond to environmental changes. Many diols (Turunen et al., 1997) are important as reflectors to avoid harmful UV radiation, but they also absorb in the UV spectrum. In many tree species, phenolics and other UV-protective substances are situated in the cuticle. The role of epicuticular waxes has been considered mainly to provide reflectance of UV radiation, but some wax components, for example, secondary alcohols (e.g., nonacosan-10-ol) and β-diketones, absorb UV radiation (Hamilton, 1995). Naturally established treeline Scots pines in a study site at Pallas-Ounastunturi National Park, Finland, were exposed to enhanced UV-B radiation levels during the full period of arctic summer daytime. The average needle dry weight increased and the wax content decreased in the UV-treated trees. The responses were observed both in the previous and current needle year (Huttunen and Manninen, in press).

A long-term study of silver birch conducted over three growing seasons showed changes in growth (i.e., shorter and thinner stems), biomass allocation, and chemical protection, while the effects on secondary metabolites in the bark were minor (Tegelberg and Julkunen-Tiitto, 2002). The changes did not occur until the third growing season, demonstrating the importance of long-term studies and the cumulative effects of UV-B radiation. A three-year study of Scots pine showed how varying defense mechanisms within the season, needle age, and developmental stage protected the Scots pine needles against increased UV-B radiation levels (Kinnunen et al., 2001). However, protective pigment decreased during the third year of exposure, suggesting that cumulative UV-B radiation exposure affects defense mechanisms and possibly makes these defenses insufficient for long-term exposure.

Quantitative changes were detected in secondary metabolites (plant chemicals) in leaves of dark-leaved willow (Salix myrsinifolia) exposed to enhanced UV-B radiation levels (Tegelberg and Julkunen-Tiitto, 2001). The changes in amount of secondary compounds are likely to have indirect effects at the ecosystem level on willow-eating insects and their predators, and on the decomposition process. Both dark-leaved and tea-leaved willow (S. phylicifolia) showed chemical adaptations to increasing UV radiation levels. However, the chemical adaptations were based more on clone-specific than on species-specific responses (Tegelberg and Julkunen-Tiitto, 2002). Several types of phenolic compounds in seedlings of nutrient-limited silver birch respond to UV-B irradiance, and seedlings are less susceptible to UV radiation when grown in nutrient-limiting conditions (Lavola et al., 1997). However, changes in secondary metabolites of birch exposed to increased UV-B radiation levels might increase its herbivore resistance (Lavola et al., 1997). A lower level of animal browsing on birch because of this chemical change induced by increased UV-B radiation levels could possibly improve the performance of birch over its woody plant competitors. Silver birch exposed to UV-B radiation with nutrient addition also displays an efficient defense mechanism through production of secondary metabolites, and demonstrates the additive effects of nutrient addition (de la Rosa et al., 2001). These results clearly establish that UV-B radiation plays an important role in the formation of secondary chemical characteristics in birch trees at higher latitudes (Lavola, 1998).

14.13. Critical research needs

14.13.1. Agriculture

This chapter focused mainly on annual crops and used GDD changes to estimate the likely effects of climate change on northern agricultural potential. Future assessments should include in-depth analyses of perennial crops and livestock, including scenarios of non-climate factors such as markets and public-sector policies. Better climate/crop models specific to agriculture in high-latitude regions (i.e., special crop varieties, long day-length effects) are also needed for future assessments. The effects of temperature increases and moisture changes on soil processes should be an integral part of the modeling systems. Future assessments should include many more climate stations in the boreal and arctic regions. The insect, disease, and weed issues that might accompany climate change and expanded agriculture should be addressed, particularly those with a unique high-latitude dimension unlikely to be part of ongoing studies in more southerly locations.

Given the critical role of public policy in agriculture in the region, an important issue is to identify national and international policies that might accommodate expansion of specific crops in the Arctic and subarctic. The infrastructure change needed to accommodate climate change should be identified along with the economic impacts of constructing new infrastructure and the economic impacts once such projects are constructed.

14.13.2. Boreal forests and climate change

Given the large contribution of the boreal forest to the terrestrial carbon sink, and the large stocks of carbon stored in it, the health and vigor of the forests of this region are a particularly significant input to climate change policy, science, and management. Climatic models, recorded data, and proxies all indicate that there are significant spatial differences in temperature anomalies and associated tree growth changes around the globe. Not all species and relatively few forest site types have been examined to determine the strength and components of climate controls on tree growth. In order to determine the current and probable future carbon uptake of the boreal forest, these spatial differences in climate change and tree growth need to be systematically identified, both in the contemporary environment and in the recent historical past. Particular attention should be given to ways of effectively combining ground-level studies, which clarify mechanisms of control and provide a historical perspective, with remote-sensing approaches, which provide comprehensive spatial coverage that can be easily repeated and updated.

Trees and forest types that change their level of sensitivity to climate or even switch their growth responses to temperature are of particular interest. If climate controlled growth more at some times and less at other times in the past, tree-ring data used to reconstruct past climates will contain biases from the recorded climate data they are calibrated to, especially if the calibrating climate data cover a short time period. It is important to establish whether this change or reduction in tree growth–climate relationships is unique to the 20th century, whether the climate change of recent decades is largely responsible for growth declines, and finally, whether the mechanism of climate control is direct or indirect.

In order to better understand, measure, and project the rapid transformations that are possible in the boreal forest, better information is needed about the role of climate and forest condition in triggering major fire and insect disturbances. The specific characteristics of fires that result in maximum production of charcoal, which represents an important form of long-term carbon sequestration, are not well documented. Further insight is needed into the interaction of climate, forest disturbance, forest succession (and its influences on carbon dynamics), water and nitrogen dynamics, and changes in carbon stocks.

Forest advance into tundra has the potential to generate a large positive temperature feedback. Unfortunately, the understanding of change at this crucial ecological boundary comes from a small number of widely separated studies undertaken to achieve many different objectives. A coordinated, circumpolar treeline study and monitoring initiative will be necessary to address definitively the question of how and why this boundary is changing at the scale required to address its potential global importance.

Improvements in scaling output from GCM grid data to applications at a given location will improve confidence in scenario projections.

14.13.3. Boreal forests and ultraviolet-B radiation

The UV-radiation survival strategies of evergreens appear to include simultaneous inhibition and avoidance. Late-winter and early-spring conditions are critical, and their importance should be studied further and effects quantified. More studies, longer-term studies, and well-designed ecosystem studies that incorporate the previously established effects from individual plot studies and consider the multiple influences of UV-B radiation on plant chemistry in settings of ecological communities will be needed to understand the cumulative effect of increased UV-B radiation levels on forests at high latitudes.

References

Alenius, C.M., T.C. Vogelmann and J.F. Bornman, 1995. A three-dimensional representation of the relationship between penetration of UV-B radiation and UV-screening pigments in leaves of *Brassica napus*. New Phytologist, 131:297–302.

Alix, C., 2001. Exploitation du bois par les populations néo-eskimo entre le nord de l'Alaska et le Haut-Arctique canadien. Ph.D. Thesis. Paris University I ? Panthéon Sorbonne, 2 vols.

Altenkirch, W., 1990. Zyklische Fluktuation beim Kleinen Frostspanner (*Operophtera brumata*). Allgemeine Forst- und Jagdzeitung, 162:2–7.

AMAP, 1997. Arctic Pollution Issues: A State of the Environment Report. Arctic Monitoring and Assessment Programme, Oslo, 188p.

Amiro, B.D., B.J. Stocks, M.E. Alexander, M.D. Flannigan and B.M. Wotton, 2001. Fire, climate change, carbon and fuel management in the Canadian boreal forest. International Journal of Wildland Fire, 10(4):405–413.

Anderson, P.M. and L.B. Brubaker, 1994. Vegetation history of Northcentral Alaska: a mapped summary of late-quaternary pollen data. Quaternary Science Reviews, 13:71–92.

Anon., 1983. Vegetation map of the USSR. 1:16,000,000. In: Geographical Atlas of the USSR, pp. 108–109. Moscow.

Anon., 1996–2000. Climate Applications: Newfoundland and Labrador Heritage. www.heritage.nf.ca/environment/climate_applications.html.

Anuchin, V. and A. Pisarenko, 1989. Forest Encyclopedia, Volume I and II, p. 563. State Forest Committee, Moscow.

Arain, M., T. Black, A. Barr, P. Jarvis, J. Massheder, D. Verseghy and Z. Nesic, 2002. Effects of seasonal and interannual climate variability on net ecosystem productivity of boreal deciduous and conifer forests. Canadian Journal of Forest Research, 32:878–891.

Ash, G.H.B., D.A. Blatta, B.A. Mitchell, B. Davies, C.F. Shaykewich, J.L. Wilson and R.L. Raddatz, 1999. Agricultural Climate of Manitoba. Agriculture, Food and Rural Initiatives, Government of Manitoba, Winnipeg. Available at: www.gov.mb.ca/agriculture/climate/waa50s00.html.

Bader, H.G., 2004. Memorandum to section heads. Alaska Division of Lands and Minerals, Department of Natural Resources. Northern Regional Office, Fairbanks, Alaska. January 6, 2004.

Baillie, M.G.L., 2000. A Slice Through Time: Dendrochronology and Precision Dating. Routledge. London, 286p.

Balakumar, T., V.H.B. Vincent and K. Paliwal, 993. On the interaction of UV-B radiation (280–315 nm) with water stress in crop plants. Physiologia Plantarum, 87:217–222.

Baldocchi, D.D. and C.A. Vogel, 1995. Energy and CO_2 flux densities above and below a temperate broad-leaved forest and boreal pine forest. Tree Physiology, 16:5–16.

Barber, V., G. Juday and B. Finney, 2000. Reduced growth of Alaska white spruce in the twentieth century from temperature-induced drought stress. Nature, 405:668–672.

Barber, V.A., G.P. Juday, B.P. Finney and M. Wilmking, 2004. Reconstruction of summer temperatures in interior Alaska from tree-ring proxies: evidence for changing synoptic climate regimes. Climatic Change, 63:91–120.

Barney, R.J. and B.J. Stocks, 1983. Fire frequencies during the suppression period. In: R.W. Wein and D.A. MacLean (eds.). The Role of Fire in Northern Circumpolar Ecosystems, pp. 45–62. John Wiley and Sons.

Barr, A.G., T.J. Griffis, T.A. Black, X. Lee, R.M. Staebler, J.D. Fuentes, Z. Chen and K. Morgenstern, 2002. Comparing the carbon budgets of boreal and temperate deciduous forest stands. Canadian Journal of Forest Research, 32:813–822.

Bassman, J.H., G.E. Edwards and R. Robberecht, 2003. Photosynthesis and growth in seedlings of five forest tree species with contrasting leaf anatomy subjected to supplemental UV-B radiation. Forest Science, 49(2):176–187.

Beniston, M., R.S.J. Tol, R. Delecolle, G. Hoermann, A. Inglesias, J. Innes, A.J. McMichael, W.J.M. Martens, I. Nemesova, R. Nicholls and F.L. Toth, 1998. Europe. In: R.T. Watson, M.C. Zinyowera, R.H. Moss and D.J. Dokken (eds.). The Regional Impacts of Climate Change: An Assessment of Vulnerability, pp. 151–185. A Special Report of Working Group II of the Intergovernmental Panel on Climate Change. Cambridge University Press.

Berg, A., B. Ehnstrom, L. Gustafsson, T. Hallingback, M. Jonsell and J. Weslien, 1994. Threatened plant, animal, and fungus species in Swedish forests: distributions and habitat associations. Conservation Biology, 8(3):718–731.

Berg, E.E. and F.S. Chapin III, 1994. Needle loss as a mechanism of winter drought avoidance in boreal conifers. Canadian Journal of Forest Research, 24:1144–1148.

Bergh, J. and S. Linder, 1999. Effects of soil warming during spring on photosynthetic recovery in boreal Norway spruce stands. Global Change Biology, 5:245–253.

Bergh, J.S.L., T. Lundmark and B. Elfving, 1999. The effect of water and nutrient availability on the productivity of Norway Spruce in northern and southern Sweden. Forest Ecology and Management, 119:51–62.

Berman, M., G.P. Juday and R. Burnside, 1999. Climate change and Alaska's forests: people, problems, and policies. In: G. Weller and P. Anderson (eds.). Implications of Global Change in Alaska and the Bering Sea Region. Proceedings of a workshop at the University of Alaska Fairbanks, 29–30 October 1998, pp. 21–42.

Beyschlag, W., P.W. Barnes, S.D. Flint and M.M. Caldwell, 1989. Enhanced UV-B irradiation has no effect on photosynthetic characteristics of wheat (*Triticum aestivum* L.) and wild oat (*Avena fatua* L.) under greenhouse and field conditions. Photosynthetica, 22:516–525.

Birdsey, R.A. and L.S. Heath, 1995. Carbon changes in US forests. In: L.A. Joyce (ed.) Climate Change and the Productivity of America's Forests. General Technical Report RM-GTR-271, pp. 56–70. Forest Service, United States Department of Agriculture.

Blöndal, S., 1993. Socioeconomic importance of forests in Iceland. In: J. Alden, J.L. Mastrantonio and S. Odum (eds.). Forest Development in Cold Climates. NATO ASI Series A, Life Sciences, 244:1–13. Plenum Press, New York.

Bonan, G.B. and K. Van Cleve, 1992. Soil temperature, nitrogen mineralization and carbon source-sink relationships in boreal forests. Canadian Journal of Forest Research, 22:629–639.

Bonan, G.B., D. Pollard and S.L. Thompson, 1992. Effects of boreal forest vegetation on global climate. Nature, 359:716–718.

Bonsal, B.R., A.K. Chakravarti and R.G. Lawford, 1993. Teleconnections between North Pacific SST anomalies and growing season extended period dry spells on the Canadian prairies. International Journal of Climatology, 13:865–878.

Bootsma, A., 1984. Forage crop maturity zonation in the Atlantic Region using growing degree days. Canadian Journal of Plant Science, 64:329–338.

Bourgeois, G., S. Jenni, H. Laurence and N. Tremblay, 2000. Improving the prediction of processing pea maturity based on the growing-degree day approach. HortScience, 35:611–614.

Bradley, R.S., 2000. Past global changes and their significance for the future. Quaternary Science Reviews, 19:391–402.

Breazeale, D., R. Kettle and G. Munk, 1999. Using growing degree days for alfalfa production. Fact Sheet 99-71. University of Nevada Cooperative Extension Service.

Briffa, K.R., T.S. Bartholin, D. Eckstein, P.D. Jones, W. Karlén, F.H. Schweingruber and P. Zetterberg, 1990. A 1,400-year tree-ring record of summer temperatures in Fennoscandia. Nature, 346:434–439.

Briffa, K.R., P.D. Jones, F.H. Schweingruber, W. Karlen and S.G. Shiyatov, 1996. Tree-ring variables as proxy-climate indicators: problems with low-frequency signals. In: P.D. Jones, R.S. Bradley and J. Jouzel (eds.). Climate Change and Forcing Mechanisms of the Last 2000 Years. NATO ASI Series I, Global Environmental Change, 41:9–41.

Briffa, K.R., F.H. Schweingruber, P.D. Jones, T.J. Osborn, S.G. Shiyatov and E.A. Vaganov, 1998. Reduced sensitivity of recent tree-growth to temperature at high northern latitudes. Nature, 391:678–682.

Briffa, K.R., T.J. Osborn, F.H. Schweingruber, I.C. Harris, P.D. Jones, S.G. Shiyatov and E.A. Vaganov, 2001. Low-frequency temperature variations from a northern tree ring density network. Journal of Geophysical Research, 106(D3):2929–2941.

Brooks, J.R., L.B. Flanagan and J.R. Ehleringer, 1998. Responses of boreal conifers to climate fluctuations: indications from tree-ring widths and carbon isotope analyses. Canadian Journal of Forest Research, 28:524–533.

Bryant, D., D. Nielsen and L. Tangley, 1997. The Last Frontier Forests: Ecosystems and Economies on the Edge. World Resources Institute, Washington, DC. 42pp.

Budyko, M.I. and Yu.A. Izrael (eds.), 1987. Anthropogenic Forcing of Climate. Gidrometeoizdat, Leningrad, 476p. (In Russian)

Bylund, H., 1999. Climate and the population dynamics of two insect outbreak species in the north. Ecological Bulletins, 47:54–62.

Cahoon, D.R. Jr., B.J. Stocks, J.S. Levine, W.R. Cofer III and J.M. Pierson, 1994. Satellite analysis of the severe 1987 forest fires in northern China and southeastern Siberia. Journal of Geophysical Research, 99(D9):18627–18638.

Cahoon, D.R. Jr., B.J. Stocks, J.S. Levine, W.R. Cofer III and J.M. Pierson, 1996. Monitoring the 1992 forest fires in the boreal ecosystem using NOAA AVHRR satellite imagery. In: J.L. Levine (ed.). Biomass Burning and Climate Change - Vol 2: Biomass Burning in South America, Southeast Asia, and Temperate and Boreal Ecosystems, and the Oil Fires of Kuwait, pp. 795–802. MIT Press.

Caldwell, C.R., 1993. Ultraviolet induced photodegradation of cucumber (*Cucumis sativus* L.) microsomal and soluble protein tryptophanyl residues in vitro. Plant Physiology, 101:947–953.

Caldwell, M.M., L.O. Björn, J.F. Bornman, S.D. Flint, G. Kulandaivelu, A.H. Teramura and M. Tevini, 1998. Effects of increased solar UV radiation on terrestrial ecosystems. Journal of Photochemistry and Photobiology B, 46:40–52.

Canadell, J.G., W.L. Steffen and P.S. White, 2002. IGBP/GCTE terrestrial transects: Dynamics of terrestrial ecosystems under environmental change – Introduction. Journal of Vegetation Science, 13:297–450.

CarbonSweden, 2003. www-carbonsweden.slu.se/database/measurements/flakasoil.html

Carcaillet, C., Y. Bergeron, P.J.H. Richard, B. Frechette, S. Gauthier and Y.T. Prairie, 2001. Change of fire frequency in the Eastern Canadian boreal forest during the Holocene: does vegetation composition or climate trigger the fire regime? Journal of Ecology, 89:930–946.

Carter, T.R., R.A. Saarikko and K.J. Niemi, 1996. Assessing the risks and uncertainties of regional crop potential under a changing climate in Finland. Agricultural and Food Science in Finland, 5:329–350.

Caspersen, J.P., S.W. Pacala, J.C. Jenkins, G.C. Hurtt, P.R. Moorcroft and R.A. Birdsey, 2000. Contributions of land-use history to carbon accumulation in U.S. forests. Science, 290:1148–1151.

Cen, Y.-P. and J.F. Bornman, 1993. The effect of exposure to enhanced UV-B radiation in the penetration of monochromatic and polychromatic UV-B radiation in leaves of *Brassica napus*. Physiologia Plantarum, 87:249–255.

Ceulemans, R. and M. Mousseau, 1994. Effect of elevated atmospheric CO_2 on woody plants. New Phytologist, 127:425–446.

Chambers, S.D. and F.S. Chapin III, 2003. Fire effects on surface-atmosphere energy exchange in Alaskan black spruce ecosystems: Implications for feedbacks to regional climate. Journal of Geophysical Research, 108(D1):doi:10.1029/2001JD000530.

Chapin, F.S. III and A.M. Starfield, 1997. Time lags and novel ecosystems in response to transient climatic change in arctic Alaska. Climatic Change, 35:449–461.

Chapin, F.S. III, C.S.H. Walter and D.T. Clarkson, 1988. Growth response of barley and tomato to nitrogen stress and its control by abscisic acid, water relations and photosynthesis. Planta, 173:352–366.

Chapin, F.S. III, D. Baldocchi, W. Eugster, S.E. Hobbie, E. Kasischke, A.D. McGuire, R. Pielke Sr., J. Randerson, E.B. Rastetter, N. Roulet, S.W. Running and S.A. Zimov, 2000. Arctic and boreal ecosystems of western North America as components of the climate system. Global Change Biology, 6(Suppl. 1):211–223.

Chapin, F.S. III, T.S. Rupp, A.M. Starfield, L. Dewilde, E.S. Zavaleta, N. Fresco, J. Henkelman and A.D. McGuire, 2003. Planning for resilience: modeling change in human-fire interactions in the Alaskan boreal forest. Frontiers in Ecology and the Environment, 1:255–261.

Ciais, P., P.P. Tans, M. Trolier, J.W.C. White and R.J. Francey, 1995. A large northern hemisphere terrestrial CO_2 sink indicated by the $^{13}C/^{12}C$ ratio of atmospheric CO_2. Science, 269:1098–1101.

Colwell, M.A., A.V. Dubynin, A.Y. Koroliuk, and N.A. Sobolev, 1997. Russian nature reserves and conservation of biological diversity. Natural Areas Journal, 17(1):56–68.

Conard, S.G. and G.A. Ivanova, 1997. Wildfire in Russian boreal forests – potential impacts of fire regime characteristics on emissions and global carbon balance estimates. Environmental Pollution, 98(3):305–313.

Cotrufo, M.F., P. Ineson and A.P. Rowland, 1994. Decomposition of tree leaf litters grown under elevated CO_2: effect of litter quality. Plant and Soil, 163:121–130.

Cramer, W., 1997. Modeling the possible impact of climate change on broad-scale vegetation structure: examples from northern Europe. Ecological Studies, No. 124, pp. 1312–1329. Springer.

Crawford, H.S. and D.T. Jennings, 1989. Predation by birds on spruce budworm *Choristoneura fumiferana*: functional, numerical, and total responses. Ecology, 70:152–163.

Crawford, R.M.M., 1997. Oceanity and the ecological disadvantages of warm winters. Botanical Journal of Scotland, 49:205–221.

Crawford, R.M.M., C.E. Jeffree and W.G. Rees, 2003. Paludification and forest retreat in northern oceanic environments. Annals of Botany, 91:213–226.

Dahl-Jensen, D., K. Mosegaard, N. Gundestrup, G.D. Clow, S.J. Johnsen, A.W. Hansen and N. Balling, 1998. Past temperatures directly from the Greenland ice sheet. Science, 282:268–271.

Dale, V.H., 1997. The relationship between land-use change and climate change. Ecological Applications, 7(3):753–769.

Danell, Kj., A. Hofgaard, T.V. Callaghan and J.B. Ball, 1999. Scenarios for animal responses to global change in Europe's cold regions: an introduction. Ecological Bulletins, 47:8–15.

D'Arrigo, R.D. and G.C. Jacoby, 1992. Dendroclimatic evidence from northern North America. In: R.S. Bradley and P.D. Jones (eds.). Climate since AD 1500, pp. 296–311. Routledge, London.

D'Arrigo, R.D. and G.C. Jacoby, 1993. Secular trends in high northern-latitude temperature reconstructions based on tree rings. Climatic Change, 25:163–177.

D'Arrigo, R.D., G.C. Jacoby and R.M. Free, 1992. Tree-ring width and maximum latewood density at the North American tree line: parameters of climate change. Canadian Journal of Forest Research, 22:1290–1296.

D'Arrigo, R.D., E.R. Cook and G.C. Jacoby, 1996. Annual to decadal-scale variations in northwest Atlantic sector temperatures inferred from Labrador tree rings. Canadian Journal of Forest Research, 26:143–148.

D'Arrigo, R.D., C.M. Malmstrom, G.C. Jacoby, S. Los and D.E. Bunker, 2000. Tree-ring indices of interannual biospheric variability: testing satellite-based model estimates of vegetation activity. International Journal of Remote Sensing, 21:2329–2336.

D'Arrigo, R.D., R. Villalba and G. Wiles, 2001. Tree-ring estimates of Pacific decadal climate variability. Climate Dynamics, 18:219–224.

Day, T.A., 1993. Relating UV-B radiation screening effectiveness of foliage to absorbing-compound concentration and anatomical characteristics in a diverse group of plants. Oecologia, 95:542–550.

Day, T.A., 2001. Ultraviolet radiation and plant ecosystems. In: C.S. Cockell and A.R. Blaustein (eds.). Ecosystems, Evolution, and Ultraviolet Radiation, pp. 80–117. Springer Verlag.

Day, T.A. and T.C. Vogelmann, 1995. Alterations in photosynthesis and pigment distributions in pea leaves following UV-B exposure. Physiologia Plantarum, 94:433–440.

de la Rosa, T.M., R. Julkunen-Tiitto, T. Lehto and P.J. Aphalo, 2001. Secondary metabolites and nutrient concentrations in silver birch seedlings under five levels of daily UV-B exposure and two relative nutrient addition rates. New Phytologist, 150:121–131.

Dissing, D. and D.L. Verbyla, 2003. Spatial patterns of lightning strikes in interior Alaska and their relations to elevation and vegetation. Canadian Journal of Forest Research, 33(5):770–782.

Dofing, S.M., 1992. Growth, phenology, and yield components of barley and wheat grown in Alaska. Canadian Journal of Plant Science, 72:1227–1230.

Dofing, S.M. and C.W. Knight, 1994. Variation for grain fill characteristics in northern-adapted spring barley cultivars. Acta Agriculture Scandinavia, B, 44:88–93.

Eggertsson, O., 1994. Origin of the Arctic Driftwood – A Dendrochronological Study. Lundqua Thesis 32. Lund University, 13p+4app.

Egler, F.E., 1954. Vegetation science concepts. I. Initial floristic composition, a factor in old-field vegetation development. Vegetatio, 4:412–417.

Elliott-Fisk, D.L., 1988. The boreal forest. In: M.G. Barbour and W.D. Billings (eds.). North American Terrestrial Vegetation, pp. 33–62. Cambridge University Press.

Esper, J., E.R. Cook and F.H. Schweingruber, 2002. Low-frequency signals in long tree-ring chronologies for reconstructing past temperature variability. Science, 295:2250–2253.

Essen, P.A., B. Ehnstrom, L. Ericson and K. Sjoberg, 1992. Boreal forests – the focal habitats of Fennoscandia. In: L. Hansson (ed.). Ecological Principles of Nature Conservation. Applications in Temperate and Boreal Environments, pp. 252–325. Elsevier Applied Science.

Fan, S., M. Gloor, J. Mahlman, S. Pacala, J. Sarmiento, T. Takahashi and P. Tans, 1998. A large terrestrial carbon sink in North America implied by atmospheric and oceanic carbon dioxide data and models. Science, 282:442–446.

Fang, J., A. Chen, C. Peng, S. Zhao and L. Ci, 2001. Changes in forest biomass carbon storage in China between 1949 and 1998. Science, 292:2320–2322.

FAO, 1999. State of the World's Forests 1999. U.N. Food and Agriculture Organization Documentation Group.

FAO, 2001. Global Forest Resources Assessment 2000. Main Report. U.N. Food and Agriculture Organization, Forestry Paper 140, 512p.

FAO, 2002. Global Forest Resources Assessment 2001. U.N. Food and Agriculture Organization, 181p.

FAOSTAT, 2000. http://apps.fao.org/page/collections

Fisher, H., M. Werner, D. Wagenbach, M. Schwager, T. Thorsteinnson, F. Wilhelms, J. Kipfstuhl and S. Sommer, 1998. Little Ice Age clearly recorded in northern Greenland ice cores. Geophysical Research Letters, 25:1749–1752.

Flannigan, M.D. and J.B. Harrington, 1988. A study of the relation of meteorological variables to monthly provincial area burned by wildfire in Canada (1935–80). Journal of Applied Meteorology, 27:441–452.

Flannigan, M.D. and C.E. Van Wagner, 1991. Climate change and wildfire in Canada. Canadian Journal of Forest Research, 21(1):66–72.

Flannigan, M.D., Y. Bergeron, O. Engelmark and B.M. Wotton, 1998. Future wildfire in circumboreal forests in relation to global warming. Journal of Vegetation Science, 9(4):469–476.

Flannigan, M.D., B.J. Stocks and B.M. Wotton, 2000. Climate change and forest fires. Science of the Total Environment, 262(3):221–229.

Flannigan, M., I. Campbell, M. Wotton, C. Carcaillet, P. Richard and Y. Bergeron, 2001. Future fire in Canada's boreal forest: Paleoecology results and general circulation model – regional climate model simulations. Canadian Journal of Forest Research, 31(5):854–864.

Fleming, R.A. and W.J.A. Volney, 1995. Effects of climate change on insect defoliator population processes in Canada's boreal forest: some plausible scenarios. Water, Soil, and Air Pollution, 82:445–454.

Fritts, H.C., 1976. Tree–Rings and Climate. Academic Press, 567p.

Gabriel, H.W. and G.F. Tande, 1983. A Regional Approach to Fire History in Alaska. U.S. Bureau of Land Management, Technical Report 9, 34p.

Gardner, R.H., W.W. Hargrove, M.G. Turner and W.H. Romme, 1996. Global change, disturbances and landscape dynamics. In: B. Walker and W. Steffen (eds.). Global Change and Terrestrial Ecosystems, pp. 149–172. Cambridge University Press.

Gardner, R.H., W.H. Romme and M.G. Turner, 1999. Predicting forest fire effects at landscape scales. In: D.J. Mladenoff and W.L. Baker (eds.). Spatial Modeling of Forest Landscape Change: Approaches and Applications, pp. 163–185. Cambridge University Press.

Gilbertson, N., 2002. Alaska employment scene. Alaska Economic Trends, 22(5):26–30.

Gillett, N.P., A.J. Weaver, F.W. Zwiers and M.D. Flannigan, 2004. Detecting the effect of climate change on Canadian forest fires. Geophysical Research Letters, 31:L18211, doi:10.1029/2004GL020876.

Gitay, H., S. Brown, W. Easterling and B. Jallow, 2001. Ecosystems and their goods and services. In: J.J. McCarthy, O.F. Canziani, N.A. Leary, D.J. Dokken and K.S. White, (eds.). Climate Change 2001: Impacts, Adaptation, and Vulnerability. Contribution of Working Group II to the Third Assessment Report of the Intergovernmental Panel on Climate Change, pp. 235–342. Cambridge University Press.

Goldsmith, S. and X. Hull, 1994. Tracking the Structure of the Alaska Economy: The 1994 ISER MAP Economic Database. Working Paper 94.1. Institute of Social and Economic Research, University of Alaska, Anchorage, 30p.

Goodale, C.L., M.J. Apps, R.A. Birdsey, C.B. Field, L.S. Heath, R.A. Houghton, J.C. Jenkins, G.H. Kuhlmaier, W. Kurz, S. Liu, G.-J. Nabuurs, S. Nilsson and A.Z. Shvidenko, 2002. Forest carbon sinks in the northern hemisphere. Ecological Applications, 12(3):891–899.

Gorham, E., 1991. Northern peatlands: role in the carbon cycle and probable responses to climatic warming. Ecological Applications, 1(2):182–195.

Gower, S.T., O. Krankina, R.J. Olson, M. Apps, S. Linder and C. Wang, 2001. Net primary production and carbon allocation patterns of boreal forest ecosystems. Ecological Applications, 11(5):1395–1411.

Greenberg, B.M., V. Gaba, O. Canaani, S. Malkin and A.K. Maattoo, 1989. Separate photosensitizers mediate degradation in the 32-kda photosystem II reaction center protein in the visible and UV spectral region. Proceedings of the National Academy of Sciences, 86:6617–6620.

Greene, D.F., J.C. Zasada, L. Sirois, D. Kneeshaw, H. Morin, I. Charron and M.-J. Simard, 1999. A review of the regeneration dynamics of North American boreal forest tree species. Canadian Journal of Forest Research, 29:824–839.

Grove, J.M., 1988. The Little Ice Age. Methuen and Co., 498p.

Grudd, H., K.R. Briffa, B.E. Gunnarson and H.W. Linderholm, 2000. Swedish tree rings provide new evidence in support of a major, widespread environmental disruption in 1628 BC. Geophysical Research Letters, 27:2957–2960.

Grudd, H., K.R. Briffa, W. Karlén, T.S. Bartholin, P.D. Jones and B. Kromer, 2002. A 7400-year tree-ring chronology in northern Swedish Lapland: natural climatic variability expressed on annual to millennial timescales. The Holocene, 12:657–666.

Gunnarson, B. and H.W. Linderholm, 2002. Low frequency climate variation in Scandinavia since the 10th century inferred from tree rings. The Holocene, 12:667–671.

Gunnarsson, K., 1999. Afforestation projects and rural development in Iceland. In: A. Niskanen and J. Väyrynen (eds.). Regional Forest Programmes: A Participatory Approach to Support Forest Based Regional Development. European Forest Institute Proceedings, 32:198–204.

Gwynn-Jones, D., J. Lee, U. Johanson, G. Phoenix, T. Callaghan and M. Sonesson, 1999. The responses of plant functional types to enhanced UV-B radiation. In: J. Rozema (ed.). Stratospheric Ozone Depletion: The Effects of Enhanced UV-B Radiation on Terrestrial Ecosystems, pp. 173–186. Backhuys Publishers.

Haak, R.A., 1996. Will global warming alter paper birch susceptibility to bronze birch borer attack? In: W.J. Mattson, P. Niemila and M. Rossi (eds.). Dynamics of Forest Herbivory: Quest for Pattern and Principle, pp. 234–247. United States Department of Agriculture Forest Service, General Technical Report NC-183.

Häkkilä, M., 2002. Farms of northern Finland. Fennia, 180(1–2):199–211.

Hall, F.G., P.J. Sellers and D.L. Williams, 1996. Initial results from the boreal ecosystem-atmosphere experiment, BOREAS. Silva Fennica, 30(2–3):109–121.

Hallsdóttir, M., 1987. Pollen analytical studies of human influences on vegetation in relation to the Landnám Tephra layer in southwest Iceland. Lundqua Thesis 18, Lund University, 45p.

Hamilton, R.J., 1995. Analysis of waxes. In: R.J. Hamilton (ed.). Waxes: Chemistry, Molecular Biology and Functions, pp. 311–342. The Oily Press, Dundee.

Hantemirov, R.M., 1999. Tree-ring reconstruction of summer temperatures on a north of Western Siberia for the last 3248 years. Siberian Ecological Journal, 6(2):185–191.

Hantemirov, R.M. and S.G. Shiyatov, 1999. Main stages of woody vegetation development on the Yamal Peninsula in the Holocene. Russian Journal of Ecology, 30(3):141–147.

Hantemirov, R.M. and S.G. Shiyatov, 2002. A continuous multimillennial ring-width chronology in Yamal, northwestern Siberia. The Holocene, 12(6):717–726.

Hard, J.S., 1985. Spruce beetles attack slowly growing spruce. Forest Science, 31:839–850.

Hard, J.S., 1987. Vulnerability of white spruce with slowly expanding lower boles on dry, cold sites to early seasonal attack by spruce beetles in south central Alaska. Canadian Journal of Forest Research, 17:428–435.

Harden, J.W., S.E. Trumbore, B.J. Stocks, A. Hirsch, S.T. Gower, K.P. O'Neill and E.S. Kasischke, 2000. The role of fire in the boreal carbon budget. Global Change Biology, 6(1):174–184.

Harvey, G.T., 1985. The taxonomy of the coniferophagous Choristoneura (Lepidoptera Tortricidae): a review. In: C.J. Sanders, R.W. Stark, E.J. Mullins and J. Murphy (eds.). Recent Advances in Spruce Budworms Research, pp. 16–48. Canadian Forestry Service.

Heliövaara, K. and M. Peltonen, 1999. Bark beetles in a changing environment. Ecological Bulletins, 47:48–53.

Hellden, U. and R. Olafsdottir, 1999. Land Degradation in NE Iceland: An assessment of extent, causes and consequences. Lund Electronic Reports in Physical Geography No. 3. Department of Physical Geography, Lund University, Sweden, 30p. www.natgeo.lu.se/ELibrary/LERPG/LERPG/3/3Article.pdf

Hewitson, B., 2003. Developing perturbations for climate change impact assessments. EOS, Transactions of the American Geophysical Union, 84(35):337–341.

Hill, T., D. Beckman, M. Ball, P. Smith and V. Whelan, 2002. Yukon Agriculture: State of the Industry, 2000–2001. Department of Energy, Mines and Resources, Government of Yukon, 48p.

Hogg, E.H. and P.A. Hurdle, 1995. The aspen parkland in western Canada: a dry-climate analogue for the future boreal forest? Water, Air and Soil Pollution, 82:391–400.

Hogg, E.H., J.P. Brandt and B. Kochtubajda, 2002. Growth and dieback of aspen forests in northwestern Alberta, Canada in relation to climate and insects. Canadian Journal of Forest Research, 32:823–832.

Holsten, E.H., 1990. Spruce Beetle Activity in Alaska: 1920–1989. USDA Forest Service Technical Report R10-90-18, 28p.

Holsten, E. and R. Burnside, 1997. Forest health in Alaska: an update. Western Forester, 42(4):8–9.

Houghton, R.A. and J.L. Hackler, 2000. Changes in terrestrial carbon storage in the United States. 1: The roles of agriculture and forestry. Global Ecology and Biogeography, 9(2):125–144.

Houghton, R.A., J.E. Hobbie, J.M. Melillo, B. Moore, B.J. Peterson, G.R. Shaver and G.M. Woodwell, 1983. Changes in the carbon content of terrestrial biota and soils between 1860 and 1980: A net release of CO_2 to the atmosphere. Ecological Monographs, 53(3):235–262.

Houghton, R.A., J.L. Hackler and K.T. Lawrence, 1999. The U.S. carbon budget: contributions from land-use change. Science, 285:574–577.

Hughes, M.K. and H.F. Diaz, 1994. Was there a 'Medieval Warm Period' and if so, where and when? Climatic Change, 26:109–142.

Hughes, M.K., E.A. Vaganov, S.G. Shiyatov, R. Touchan and G. Funkhouser, 1999. Twentieth-century summer warmth in northern Yakutia in a 600-year context. The Holocene, 9(5):603–608.

Hurtt, G.C., S.W. Pacala, P.R. Moorcroft, J. Caspersen, E. Shevliakova, R.A. Houghton and B. Moore III, 2002. Projecting the future of the U.S. carbon sink. Proceedings of the National Academy of Sciences, 99(3):1389–1394.

Hustich, I., 1966. On the forest-tundra and the northern tree lines. Annales Universitatis Turkuensis Series A II, 36:7–47.

Huttunen, S. and S. Manninen, in press. Scots pine and changing environment - Needle responses. Polish Botanical Studies.

Icelandic Institute of Natural History, 2001. Biological Diversity in Iceland. National Report to the Convention on Biological Diversity. Ministry for the Environment, 56 p.

IPCC, 1995. Climate Change 1994: Radiative Forcing of Climate Change and an Evaluation of the IPCC 1992 Emission Scenarios. J.T. Houghton, L.G. Meira Filho, J. Bruce, H. Lee, B.A. Callander, E. Haites, N. Harris and K. Maskell (eds.). Cambridge University Press, 339p.

IPCC, 2001. Climate Change 2001: The Scientific Basis. Contribution of Working Group I to the Third Assessment Report of the Intergovernmental Panel on Climate Change. J.T. Houghton, Y. Ding, D.J. Griggs, M. Noguer, P.J. van der Linden, X. Dai, K. Maskell and C.A. Johnson (eds.). Cambridge University Press, 881p.

Isaev, A.S., G.N. Korovin, S.A. Bartalev, D.V. Ershov, A. Janetos, E.S. Kasischke, H.H. Shugart, N.H.F. French, B.E. Orlick, and T.L. Murphy, 2002. Using remote sensing to assess Russian forest fire carbon emissions. Climatic Change, 55(1–2):235–249.

IUCN, 2000. Guidelines for Protected Area Management Categories. World Conservation Union, Gland, Switzerland.

Jacoby, G.C. and R.D. D'Arrigo, 1989. Reconstructed Northern Hemisphere annual temperature since 1671 based on high latitude tree-ring data from North America. Climatic Change, 14:39–59.

Jacoby, G.C. and R.D. D'Arrigo, 1995. Tree ring width and density evidence of climatic and potential forest change in Alaska. Global Biogeochemical Cycles, 9:227–234.

Jacoby, G.C., I.S. Ivanciu and L.D. Ulan, 1988. A 263-year record of summer temperature for northern Quebec reconstructed from tree-ring data and evidence of a major climatic shift in the early 1800s. Palaeogeography, Palaeoclimatology, Palaeocology, 64:69–78.

Jacoby, G.C., N.V. Lovelius, O.I. Shumilov, O.M. Raspopov, J.M. Karbainov and D.C. Frank, 2000. Long-term temperature trends and tree growth in the Taymir region of northern Siberia. Quaternary Research , 53:312–318.

Jardine, K., 1994. Finger on the carbon pulse. The Ecologist, 24:220–224.

Jarvis, P.G. (coordinator), 2000. Predicted Impacts of Rising Carbon Dioxide and Temperature on Forests in Europe at Stand Scale. ECOCRAFT Environment R&D ENV4-CT95-0077 IC20-CT96-0028.

Jarvis, P. and S. Linder, 2000. Botany: constraints to growth of boreal forests. Nature, 405:904–905.

Jiang, H., M.J. Apps, C. Peng, Y. Zhang and J. Liu, 2002. Modelling the influence of harvesting on Chinese boreal forest carbon dynamics. Forest Ecology and Management, 169(1):65–82.

Johanson, U., C. Gehrke, L.O. Bjorn and T.V. Callaghan, 1995. The effects of enhanced UV-B radiation on the growth of dwarf shrubs in a subarctic heathland. Functional Ecology, 9:713–719.

Johnson, E.A., 1992. Fire and Vegetation Dynamics: Studies from the North American Boreal Forest. Cambridge University Press, 129p.

Johnson, E.A. and D.R. Wowchuk, 1993. Wildfires in the southern Canadian Rocky Mountains and their relationship to mid-tropospheric anomalies. Canadian Journal of Forest Research, 23:1213–1222.

Jones, P.D. and R.S. Bradley, 1992. Climate variations over the last 500 years. In: R.S. Bradley and P.D. Jones (eds.). Climate Since A.D. 1500, pp. 649–665. Routledge Press.

Jordan, B.R., P. James, A. Strid and R. Anthony, 1994. The effect of ultraviolet-radiation on gene expression and pigment composition in etiolated and green pea leaf tissue: UV-B induced changes are gene-specific and dependent on the developmental stage. Plant, Cell and Environment, 17:45–54.

Juday, G.P. and V.A. Barber, 2005. Alaska Tree Ring Data: Long-Term Ecological Research. Bonanza Creek LTER Database, Fairbanks, Alaska. BNZD: 214. http://www.lter.uaf.edu/ data_catalog_detail.cfm?dataset_id=214

Juday, G.P., R.A. Ott, D.W. Valentine and V.A. Barber, 1998. Forests, climate stress, insects and fire. In: G. Weller and P. Anderson (eds.). Implications of Global Climate Change in Alaska and the Bering Sea Region, pp. 23–49. The Center for Global Change and Arctic System Research, Fairbanks, Alaska.

Juday, G.P., V.A. Barber, E. Berg and D. Valentine, 1999. Recent dynamics of white spruce treeline forests across Alaska in relation to climate. In: S. Kankaanpaa, T. Tasanen and M.-L. Sutinen (eds.). Sustainable Development in the Northern Timberline Forests. Proceedings of the Timberline Workshop, May 10–11, 1998, Whitehorse, Canada, pp. 165–187. The Finnish Forest Research Institute, Helsinki.

Juday, G.P., V. Barber, S. Rupp, J. Zasada and M.W. Wilmking, 2003. A 200-year perspective of climate variability and the response of white spruce in Interior Alaska. In: D. Greenland, D. Goodin and R. Smith (eds.). Climate Variability and Ecosystem Response at Long-Term Ecological Research (LTER) Sites, pp. 226–250. Oxford University Press.

Jungblut, T.P., J.P. Schnitzler, W. Heller, N. Hertkorn, J.W. Metzger, W. Szymczak and H. Sandermann Jr., 1995. Structures of UV-B induced sunscreen pigments of the Scots pine (Pinus sylvestris L.). Angewandte Chemie – International Edition in English, 34(3):312–314.

Juskiw, P.E., Y.-W. Jame and L. Kryzanowski, 2001. Phenological development of spring barley in a short-season growing area. Agronomy Journal, 93: 370–379.

Kasischke, E.S., N.L. Christensen and B.J. Stocks, 1995. Fire, global warming, and the carbon balance of boreal forests. Ecological Applications, 5:437–451.

Kasischke, E.S., D. Williams and B. Donald, 2002. Analysis of the patterns of large fires in the boreal forest region of Alaska. International Journal of Wildland Fire, 11(2):131.

Kasischke, E.S., T.S. Rupp and D.L. Verbyla, in press. Fire trends in the Alaskan boreal forest. In: F.S. Chapin III, J. Yarie, K. Van Cleve, L.A. Viereck, M.W. Oswood and D.L. Verbyla (eds.). Alaska's Changing Boreal Forest. Oxford University Press.

Kaundun, S.S., P. Lebreton and B. Fady, 1998. Geographical variability of Pinus halepensis Mill. as revealed by foliar flavonoids. Biochemical Systematics and Ecology, 26:83–96.

Keeling, R.F., S.C. Piper and M. Heimann, 1996. Global and hemispheric CO_2 sinks deduced from changes in atmospheric O_2 concentration. Nature, 381:218–220.

Kellomaki, S., H. Hanninen and M. Kolstrom, 1995. Computations on frost damage to Scots pine under climatic warming in boreal conditions. Ecological Applications, 5(1):42–52.

Kinnunen, H., S. Huttunen and K. Laakso, 2001. UV-absorbing compounds and waxes of Scots pine needles during a third growing season of supplemental UV-B. Environmental Pollution, 112:215–220.

Kirdyanov, A.V., 2000. Dendroclimatic analysis of tree-ring width, tracheid dimension and wood density variations in conifers along the regional temperature gradient (the Yenisey Meridian). In: Abstracts of the XXI IUFRO World Congress 'Forests and Society: The Role of Research', Kuala Lumpur, Vol. 3, pp. 536–537. International Union of Forest Research Organizations.

Kirdyanov, A.V. and D.V. Zarharjewski, 1996. Dendroclimatological study on tree-ring width and maximum density chronologies from Picea obovata and Larix sibirica from the North of Krasnoyarsk region (Russia). Dendrochronologia, 14:227–236.

Kleemola, J., 1991. Effect of temperature on phasic development of spring wheat in northern conditions. Acta Agriculturae Scandinavica, 41:275–283.

Kondrat'ev, K.Ya., 2002. Global climate change: reality, assumptions and fantasies. Earth Research From Space, 1:3–23. (In Russian)

Kuchler, A.W., 1970. Potential natural vegetation of Alaska. In: National Atlas of the United States, p. 92. U.S. Geological Survey, Washington, DC.

Kuhry, P., 1994. The role of fire in the development of Sphagnum-dominated peatlands in western boreal Canada. Journal of Ecology, 82:899–910.

Kullman, L., 2001. 20th century climate warming and tree-limit rise in the southern Scandes of Sweden. Ambio, 30(2):72–80.

Kullman, L. and L. Kjällgren, 2000. A coherent postglacial tree-limit chronology (Pinus sylvestris L.) for the Swedish Scandes: aspect of paleoclimate and 'recent warming', based on megafossil evidence. Arctic, Antarctic, and Alpine Research, 32(4):419–428.

Kurnayev, S.F., no date. Forest-vegetational regionalization of the USSR (1:16,000,000). http://grida.no/prog/polar/ecoreg/ecoap1c.htm.

Kurz, W.A. and M.J. Apps, 1999. A 70-year retrospective analysis of carbon fluxes in the Canadian forest sector. Ecological Applications, 9(2):526–547.

Laakso, K., 1999. Effects of ultraviolet-B radiation (UV-B) on needle anatomy and glutathione status of field-grown pines. Acta Universitatis Ouluensis A, 338, 47p.

Laakso, K. and S. Huttunen, 1998. Effects of the ultraviolet-B radiation (UV-B) on conifers: a review. Environmental Pollution, 99:319–328.

Laakso, K., J.H. Sullivan and S. Huttunen, 2000. The effects of UV-B radiation on epidermal anatomy in loblolly pine (Pinus taeda L.) and Scots pine (Pinus sylvestris L.). Plant, Cell and Environment, 23(5):461–472.

Laakso, K., H. Kinnunen and S. Huttunen, 2001. The glutathione status of mature Scots pines during the third season of UV-B radiation exposure. Environmental Pollution, 111:349–354.

Labau, V.J. and W. van Hees, 1990. An inventory of Alaska's boreal forests: their extent, condition, and potential use. In: Proceedings of the International Symposium on Boreal Forests: Condition, Dynamics, Anthropogenic Effects, Archangelsk, Russia, July 16–26. State Committee of USSR on Forests, Moscow.

Lacelle, B., C. Tarnocai, S. Waltman, J. Kimble, D. Swanson, Ye.M. Naumov, B. Jacobsen, S. Goryachkin and G. Broll, 1998. Northern Circumpolar Soil Map. Agriculture and Agri-Food Canada; USDA; V.V. Dokuchayev Soils Institute; Institutes of Geography – University of Copenhagen and Moscow; Institute of Landscape Ecology – University of Muenster.

Lamas, T. and C. Fries, 1995. Emergence of a biodiversity concept in Swedish forestry. Water, Air, and Soil Pollution, 82(1/2):57–66.

Lamb, H.H., 1977a. Climate: Present, Past and Future. Methuen, 835p.

Lamb, H.H., 1977b. Climate: Present, Past and Future. Vol.2. Climatic History and the Future. Methuen, 603p.

Laracine-Pittet, C. and P. Lebreton, 1988. Flavonoid variability within Pinus sylvestris. Phytochemistry, 27:2663–2666.

Latola, K., H. Kinnunen and S. Huttunen, 2001. Needle ontogeny of mature Scots pines under enhanced UV-B radiation. Trees – Structure and Function, 15:346–352.

Lavoie, C. and S. Payette, 1996. The long-term stability of the boreal forest limit in subarctic Québec. Ecology, 77(4):1226–1233.

Lavola, A., 1998. Accumulation of flavonoids and related compounds in birch induced by UV-B irradiance. Tree Physiology, 18:53–58.

Lavola, A., R. Julkunen-Tiitto, T. de la Rosa and T. Lehto, 1997. The effect of UV-B radiation on UV-absorbing secondary metabolites in birch seedlings grown under simulated forest soil conditions. New Phytologist, 137:617–621.

Lavola, A., P.J. Aphalo, M. Lahti and R. Julkunen-Tiitto, 2003. Nutrient availability and the effect of increasing UV-B radiation on secondary plant compounds in Scots pine. Environmental and Experimental Botany, 49(1):49–60.

Leuschner, H.H. and A. Delorme, 1988. Tree ring work in Goettingen – absolute oak chronologies back to 6255 BC. In: T. Hackens, A.V. Munaut and Claudine Till (eds.). Wood and Archaeology. PACT, 22, II.5: 123–131.

Leuschner, H.H., M. Spurk, M. Baillie and E. Jansma, 2000. Stand dynamics of prehistoric oak forests derived from dendrochronologically dated subfossil trunks from bogs and riverine sediments in Europe. Geolines, 11:118–121.

Linder, P. and L. Ostlund, 1992. Changes in the boreal forests of Sweden 1870–1991. Svensk Botanisk Tidskrift, 86:199–215. (In Swedish)

Linderholm, H.W., 2002. 20th century Scots pine growth variations in the central Scandinavian Mountains related to climate change. Arctic, Antarctic, and Alpine Research, 34:440–449.

Linderholm, H.W., A. Moberg and H. Grudd, 2002. Peatland pines as climate indicators? A regional comparison of the climatic influence on Scots pine growth in Sweden. Canadian Journal of Forest Research, 32:1400–1410.

Linderholm, H.W., B.Ø. Solberg and M. Lindholm, 2003. Tree-ring records from central Fennoscandia: the relationship between tree growth and climate along a west-east transect. The Holocene, 13:887–895.

Lindholm, M., O. Eggertson, N. Lovelius, O. Raspopov, O. Shumilov and A. Laanelaid, 2001. Growth indices of North European Scots pine record the seasonal North Atlantic Oscillation. Boreal Environment Research, 6:275–284.

Livingston, N.J. and D.L. Spittlehouse, 1996. Carbon isotope fractionation in tree ring early and late wood in relation to intra-growing season water balance. Plant, Cell and Environment, 19:768–774.

Lloyd, A.H. and C.L. Fastie, 2002. Spatial and temporal variability in the growth and climate response of treeline trees in Alaska. Climatic Change, 52:481–509.

Lutz, H.J., 1956. Ecological Effects of Forest Fires in the Interior of Alaska. United States Department of Agriculture Technical Bulletin 1133, United States Government Printing Office, 121p.

Lynch, J.A., J.S. Clark, N.H. Bigelow, M.E. Edwards and B.P. Finney, 2003. Geographic and temporal variations in fire history in boreal ecosystems of Alaska. Journal of Geophysical Research, 108(D1):doi:10.1029/2001JD000332.

MacDonald, G.M., A.A. Velichko, C.V. Kremenetski, O.K. Borisova, A.A. Goleva, A.A. Andreev, L.C. Cwynar, R.T. Riding, S.L. Forman, T.W.D. Edwards, R. Aravena, D. Hammarlund, J.M. Szeicz and V.N. Gattaulin, 2000. Holocene treeline history and climate change across Northern Eurasia. Quaternary Research, 53:302–311.

Magda, V.N. and A.V. Zelenova, 2002. Radial growth of pine as an indicator of atmosphere humidity in Minusinsk hollow. Izvestiya Russkogo Geograficheskogo Obshchestva, 134(1):73–78. (In Russian)

Majdi H. and J. Öhrvik, 2004. Interactive effects of soil warming and fertilization on root production, mortality, and longevity in a Norway spruce stand in Northern Sweden. Global Change Biology, 10:182–188(7).

Makipaa, R., T. Karjalainen, A. Pussinen and S. Kellomaki, 1999. Effects of climate change and nitrogen deposition on the carbon sequestration of a forest ecosystem in the boreal zone. Canadian Journal of Forest Research, 29:1490–1501.

Mann, D.H. and L.J. Plug, 1999. Vegetation and soil development at an upland taiga site, Alaska. Ecoscience, 6(2):272–285.

Mann, M.E., R.S. Bradley and M.K. Hughes, 1998. Global scale temperature patterns and climate forcing over the six centuries. Nature, 392:779–782.

Mann, M., R. Bradley and M. Hughes, 1999. Northern Hemisphere temperatures during the past millennium: inferences, uncertainties, and limitations. Geophysical Research Letters, 26:759–762.

Mantua, N.J., S.R. Hare, Y. Zhang, J.M. Wallace and R.C. Francis, 1997. A Pacific interdecadal climate oscillation with impacts on salmon production. Bulletin of the American Meteorological Society, 78:1069–1079.

Markham, A. and J. Malcom, 1996. Biodiversity and wildlife conservation: adaptation to climate change. In: J. Smith, N. Bhatti, G. Menzhulin, R. Benioff, M. Campos, B. Jallow and F. Rijsberman (eds.). Adaptation to Climate Change: Assessment and Issues, pp. 384–401. Springer-Verlag.

Mattson, W.J. and R.A. Haack, 1987. The role of drought in outbreaks of plant-eating insects. BioScience, 37:110–118.

McGuire, A.D., J.M. Melillo, L.A. Joyce, D.W. Kicklighter, A.L. Grace, B. Moore and C.J. Vorosmarty, 1992. Interactions between carbon and nitrogen dynamics in estimating net primary productivity for potential vegetation in North America. Global Biogeochemical Cycles, 6(2):101–124.

McGuire, A.D., J.M. Melillo and L.A. Joyce, 1995a. The role of nitrogen in the response of forest net primary production to elevated atmospheric carbon dioxide. Annual Review of Ecology and Systematics, 26:472–503.

McGuire, A.D., J.M. Melillo, D.W. Kicklighter and L.A. Joyce, 1995b. Equilibrium responses of soil carbon to climate change: Empirical and process-based estimates. Journal of Biogeography, 22:785–796.

McLeod, A.R. and K.K. Newsham, 1997. Impact of elevated UV-B on forest ecosystems. In: P. Lumsden (ed.). Plants and UV-B: Responses to Environmental Change. Society for Experimental Biology Seminar Series, 64:247–281.

Melillo, J.M., A.D. McGuire, D.W. Kicklighter, B. Moore III, C.J. Vorosmarty and A.L. Schloss, 1993. Global climate change and terrestrial net primary production. Nature, 63:234–240.

Melillo, J.M., R.A. Houghton, D.W. Kicklighter and A.D. McGuire, 1996. Tropical deforestation and the global carbon budget. Annual Review of Energy and the Environment, 21:293–310.

Mepsted, R., N.D. Paul, J. Stephen, J.E. Corlett, S. Nogues, N.R. Baker, H.G. Jones and P.G. Ayers, 1996. Effects of enhanced UV-B radiation on pea (Pisum sativum L.) grown under field conditions in the UK. Global Change Biology, 2:325–334.

Miller, P., W. Lenier and S. Brandt, 2001. Using growing degree days to predict plant stages. Montana Guide Fact Sheet MT200103 AG 7/2001. Montana State University Extension Service, 8p.

Monserud, R.A., N.M. Tchebakova, T.P. Kolchugina and O.V. Denisenko, 1996. Change in Siberian phytomass predicted for global warming. Silva Fennica, 30(2–3):185–200.

Murphy, P.J., J.P. Mudd, B.J. Stocks, E.S. Kasischke, D. Barry, M.E. Alexander and N.H.F. French, 2000. Historical fire records in the North American boreal forest. In: E.S. Kasischke and B.J. Stocks (eds.). Fire, Climate Change, and Carbon Cycling in the Boreal Forest, pp. 274–288. Springer-Verlag.

Myneni, R.B., J. Dong, C.J. Tucker, R.K. Kaufmann, P.E. Kauppi, J. Liski, L. Zhou, V. Alexeyev and M.K. Hughes, 2001. A large carbon sink in the woody biomass of Northern forests. Proceedings of the National Academy of Sciences, 98(26):14784–14789.

Mysterud, I., I. Mysterud and E. Bleken, 1998. Forest fires and environmental management in Norway. International Forest Fire News, 18:72–75.

Nash, C.H. and E.A. Johnson, 1996. Synaptic climatology of lightning-caused forest fires in subalpine and boreal forests. Canadian Journal of Forest Research, 26:1859–1874.

Natural Resources Canada, 2003. The State of Canada's Forests 2002–2003. Canadian Forest Service, Natural Resources Canada, Ottawa, 95p.

Natural Resources Canada. 2004. The State of Canada's Forests 2003–2004. Canadian Forest Service, Natural Resources Canada, Ottawa, 93p.

Naurzbaev, M.M. and E.A. Vaganov, 2000. Variation of summer and annual temperature in the east Taymir and Putoran (Siberia) over the last two millennia inferred from tree-rings. Journal of Geophysical Research, 105(D6):7317–7327.

Naurzbaev, M.M., O.V. Sidorova and E.A. Vaganov, 2001. History of the late Holocene climate on the eastern Taymir according to long-term tree-ring chronology. Archaeology, Ethnology and Anthropology of Eurasia, 3(7):17–25.

Neilson, R.P., 1993. Transient ecotone response to climatic change: some conceptual and modeling approaches. Ecological Applications, 3:385–395.

Neuvonen, S., P. Niemalä and T. Virtanen, 1999. Climatic change and insect outbreaks in boreal forests: the role of winter temperatures. Ecological Bulletins, 47:63–67.

Newark, M.J., 1975. The relationship between forest fire occurrence and 500 mb longwave ridging. Atmosphere, 13:26–33.

Newsham, K.K., A.R. McLeod, P.D. Greenslade and B.A. Emmett, 1996. Appropriate controls in outdoor UV-B supplementation experiments. Global Change Biology, 2:319–324.

Niebauer, H.J., 1998. Variability in Bering Sea ice cover as affected by a regime shift in the North Pacific in the period 1947–1996. Journal of Geophysical Research, 103(C12):27717-27737.

Nimchuk, N., 1983. Wildfire Behavior Associated with Upper Ridge Breakdown. Alberta Energy and Natural Resources Report Number T/50, Forest Service, Alberta.

Noble, I.R., 1993. A model of the responses of ecotones to climate change. Ecological Applications, 3:396–403.

NOAA, 2002. Daily Climatological Data, Summary of the Day (TD-3200). U.S. Department of Commerce, National Oceanic and Atmospheric Administration, National Climatic Data Center, Asheville.

Northern Economics Inc., 2004. Alaska Visitor Arrivals. Prepared for the Alaska Department of Community and Economic Development, 16p.

Nuttall, M. and T.V. Callaghan, 2000. The Arctic: Environment, People, Policy. Plate 14. Harwood Academic Publishers, 647p.

Nuttonson, M.Y., 1955. Wheat-Climate Relationships and the Use of Phenology in Ascertaining the Thermal and Photo-thermal Requirements for Wheat. American Institute of Crop Ecology, Washington, D.C., 388p.

Nuttonson, M.Y., 1957. Barley-Climate Relationships and the Use of Phenology in Ascertaining the Thermal and Photo-thermal Requirements for Barley. American Institute of Crop Ecology. Washington, D.C., 200p.

Ogilvie, A.E.J. and T.H. McGovern, 2000. Sagas and science: climate and human impacts in the North Atlantic. In: W.W. Fitzhugh and E.I. Ward (eds.). Vikings: the North Atlantic Saga, pp. 385–393. Smithsonian Institution Press.

Oppenheimer, M., 1989. Climate change and environmental pollution: physiological and biological interactions. Climatic Change, 15(1–2):255–270.

Ostergren, D.M. and E. Shvarts, 2000. Russian zapovedniki in 1998, recent progress and new challenges for Russia's strict nature preserves. In: A.E. Watson, G.H. Aplet and J.C. Hendee (compilers). Personal, Societal, and Ecological Values of Wilderness. Sixth World Wilderness Congress Proceedings on Research, Management and Allocation, Vol. II, pp. 209–213. Proceedings RMRS-P-14.

Oswald, E.T. and J.P. Senyk, 1977. Ecoregions of Yukon Territory. Fisheries and Environment Canada, Canadian Forest Service. 115pp.

Ott, R.A., M.A. Lee, W.E. Putman, O.K. Mason, G.T. Worum and D.N. Burns, 2001. Bank Erosion and Large Woody Debris Recruitment Along the Tanana River, Interior Alaska. Report to the Alaska Department of Environmental Conservation. Prepared by the Alaska Department of Natural Resources. Project No. NP-01-R9.

Overpeck, J., K. Hughen, D. Hardy, R. Bradley, R. Case, M. Douglas, B. Finney, K. Gajewski, G. Jacoby, A. Jennings, S. Lamoureux, A. Lasca, G. MacDonald, J. Moore, M. Retelle, S. Smith, A. Wolfe and G. Zielinski, 1997. Arctic environmental change of the last four centuries. Science, 278:1251–1256.

Panyushkina, I.P., E.A. Vaganov and V.V. Shishov, 1997. Dendroclimatic analysis of larch growth variability in the North of Central Siberia. Geography and Natural Resources, 2:80–90.

Paré, D. and Y. Bergeron, 1995. Above-ground biomass accumulation along a 230-year chronosequence in the southern portion of the Canadian boreal forest. Journal of Ecology, 83:1001–1007.

Parker, W.C., S.J. Colombo, M.L. Cherry, M.D. Flannigan, S. Greifenhagen, R.S. McAlpine, C. Papadopol and T. Scarr, 2000. Third millennium forestry: what climate change might mean to forests and forest management in Ontario. The Forestry Chronicle, 76(3):445–461.

Parviainen, J., 1994. Finnish silviculture, managing for timber production and conservation. Journal of Forestry, 92(9):33–36.

Pastor, J. and W.M. Post, 1986. Influence of climate, soil moisture, and succession of forest carbon and nitrogen cycles. Biogeochemistry, 2:3–27.

Pastor, J., Y. Cohen and R. Moen, 1999. Generation of spatial patterns in boreal forest landscapes. Ecosystems, 2:439–450.

Payette, S., 1992. Fire as a controlling process in the North American boreal forest. In: H.H. Shugart, R. Leemans and G.B. Bonan (eds.). A Systems Analysis of the Global Boreal Forest, pp. 145–169. Cambridge University Press.

Payette, S. and R. Gagnon, 1985. Late Holocene deforestation and tree regeneration in the forest-tundra of Quebec. Nature, 313:570–572.

Pereg, D. and S. Payette, 1998. Development of black spruce growth forms at treeline. Plant Ecology, 138:137–147.

Pielke, R.A. and P.L. Vidale, 1995. The boreal forest and the polar front. Journal of Geophysical Research, 100(D12):25755–25758.

Pojar, J., 1996. Environment and biogeography of the western boreal forest. The Forestry Chronicle, 72(1):51–58.

Price, D.T., D.H. Halliwell, M.J. Apps, W.A. Kurz and S.R. Curry, 1997. Comprehensive assessment of carbon stocks and fluxes in a Boreal-Cordilleran forest management unit. Canadian Journal of Forest Research, 27:2005–2016.

Prozherina, N., V. Freiwald, M. Rousi and E. Oksanen, 2003. Interactive effect of springtime frost and elevated ozone on early growth, foliar injuries and leaf structure of birch (*Betula pendula*). New Phytologist, 159(3):623–636.

Rapalee, G., S.E. Trumbore, E.A. Davidson, J.W. Harden and H. Veldhuis, 1998. Soil carbon stocks and their rates of accumulation and loss in a boreal forest landscape. Global Biogeochemical Cycles, 12(4):687–702.

Rayner, P.J., I.G. Enting, R.J. Francey and R. Langenfelds, 1999. Reconstructing the recent carbon cycle from atmospheric CO_2, $\delta^{13}C$ and O_2/N_2 observations. Tellus B, 51(2):213–231.

Rosenzweig, C. and M. Parry, 1994. Potential impacts of climate change on world food supply. Nature, 367:133–138.

Rowe, J.S., 1972. Forest Regions of Canada. Publication No. 1300. Canadian Forestry Service, Department of the Environment, Ottawa, 172p.

Rozema, J., J. van de Staaij, L.O. Björn and M. Caldwell, 1997. UV-B as an environmental factor in plant life: stress and regulation. Trends in Ecology and Evolution, 12(1):22–28.

Rupp, T.S., A.M. Starfield and F.S. Chapin III, 2000a. A frame-based spatially explicit model of subarctic vegetation response to climatic change: comparison with a point model. Landscape Ecology, 15:383–400.

Rupp, T.S., F.S. Chapin III and A.M. Starfield, 2000b. Response of subarctic vegetation to transient climatic change on the Seward Peninsula in northwest Alaska. Global Change Biology, 6:541–555.

Rupp, T.S., F.S. Chapin III, A.M. Starfield and P. Duffy, 2002. Modeling boreal forest dynamics in interior Alaska. Climatic Change, 55:213–233.

Sanderson, M.A., T.P. Larnezos and A.G. Matches, 1994. Morphological development of alfalfa as a function of growing degree days. Journal of Production Agriculture, 7:239–242.

Schimel, D.S., B.H. Braswell, E.A. Holland, R. McKeown, D.S. Ojima, T.H. Painter, W.J. Parton and A.R. Townsend, 1994. Climatic, edaphic, and biotic controls over storage and turnover of carbon in soils. Global Biogeochemical Cycles, 8(3):279–293.

Schimel, D.S., VEMAP participants and B.H. Braswell, 1997. Continental scale variability in ecosystem processes: models, data, and the role of disturbance. Ecological Monographs, 67(2):251–271.

Schimel, D.S., J.I. House, K.A. Hibbard, P. Bousquet, P. Ciais, P. Peylin, B.H. Braswell, M.J. Apps, D. Baker, A. Bondeau, J. Canadell, G. Churkina, W. Cramer, A.S. Denning, C.B. Field, P. Friedlingstein, C. Goodale, M. Heimann, R.A. Houghton and J. Melillo, 2001. Recent patterns and mechanisms of carbon exchange by terrestrial ecosystems. Nature, 414:169–173.

Schlesinger, W.H., 1997. Biogeochemistry: An Analysis of Global Change. Academic Press, 588p.

Schnitzler, J.P., T.P. Jungblut, C. Feicht, M. Kofferlein, C. Langebartels, W. Heller and H. Sanderman Jr., 1997. UV-B induction of flavonoid biosynthesis in Scots pine (*Pinus sylvestris* L.) seedlings. Trees – Structure and Function, 11(3):162–168.

Schulze, E.-D., J. Lloyd, F.M. Kelliher, C. Wirth, C. Rebmann, B. Luhker, M. Mund, A. Knohl, I.M. Milyukova, W. Schulze, W. Ziegler, A.B. Varlagin, A.F. Sogachev, R. Valentini, S. Dore, S. Grigoriev, O. Kolle, M.I. Panfyorov, N. Tchebakova and N.N. Vygodskaya, 1999. Productivity of forests in the Eurosiberian boreal region and their potential to act as a carbon sink – a synthesis. Global Change Biology, 5:703–722.

Schweingruber, F.H., 1988. Tree Rings: Basics and Applications of Dendrochronology. Reidel Publishers, 276p.

Schweingruber, F.H., 1996. Tree Rings and Environment. Dendroecology. Paul Haupt, Berne, 609p.

Searles, P.S., S.D. Flint and M.M. Caldwell, 2001. A meta-analysis of plant field studies simulating stratospheric ozone depletion. Oecologia, 127:1–10.

Sharratt, B., 1992. Growing season trends in the Alaskan climate record. Arctic, 45(2):124–127.

Sharratt, B., 1994. Observations and modeling of interactions between barley yield and evapotranspiration in the subarctic. Agricultural Water Management, 25:109–119.

harratt, B., 1999. Thermal requirements for barley maturation and leaf development in interior Alaska. Field Crops Research, 63:179–184.

Shashkin, E.A. and E.A. Vaganov, 2000. Dynamics of stem basal area increment in trees from various regions of Siberia related to global temperature changes. Russian Journal of Forestry, 3:3–11. (In Russian)

Shaykewich, C.F and D.A. Blatta, 2001. Heat Units for Potato Production in Manitoba. Manitoba Agriculture, Soils and Crops Branch, 6p. Available at: www.gov.mb.ca/agriculture/climate/wac01s01.html.

Shetinsky, E.A., 1994. Protection of forests and forest pyrology. Ecology, Moscow, 209 p. (In Russian)

Shiyatov, S.G., 1966. The time of dispersion of Siberian larch seeds in north-western part of its areal space and role of that factor in mutual relation between forest and tundra. In: Problems of Physiology and Geobotany. Publication of the Sverdlovsk Branch of the All-Union Botanical Society, 4: 109–113. (In Russian)

Shiyatov, S.G., 1986. Dendrochronology of Upper Timberline in Polar Ural Mountains. Nauka, Moscow, 186p. (In Russian)

Shiyatov, S.G., 1993. The upper timberline dynamics during the last 1100 years in the Polar Ural Mountains. In: B. Frenzel (ed.). Oscillations of the Alpine and Polar Tree Limit in the Holocene, Palaeoclimate Research, 9:195–203.

Shiyatov, S.G., 1995. Reconstruction of climate and the upper timberline dynamics since AD 745 by tree-ring data in the Polar Ural Mountains. In: P. Heikinheimo (ed.). Proceedings of the SILMU International Conference on Past, Present and Future Climate, Helsinki, August 1995. Publications of the Academy of Finland, 6/95:144–147.

Shiyatov, S.G., 2003. Rates of change in the upper treeline ecotone in the Polar Ural Mountains. PAGES News, 11(1):8–10.

Shiyatov, S.G., R.M. Hantemirov, F.H. Schweingruber, K.R. Briffa and M. Moell, 1996. Potential long-chronology development on the Northwest Siberian Plain: Early results. Dendrochronologia, 14:13–29.

Shriner, D.S., R.B. Street, R. Ball, D. D'Amours, K. Duncan, D. Kaiser, A. Maarouf, L.R. Mortsch, P.J. Mulholland, R. Neilson, J.A. Patz, J.D. Scheraga, J.G. Titus, H. Vaughan and M. Weltz, 1998. North America. In: R.T. Watson, M.C. Zinyowera, R.H. Moss, and D.J. Dokken (eds.). The Regional Impacts of Climate Change: An Assessment of Vulnerability. A Special Report of Working Group II of the Intergovernmental Panel on Climate Change, pp. 255–330. Cambridge University Press.

Shvidenko, A.Z. and S. Nilsson, 2000. Extent, distribution and ecological role of fire in Russian forests. In: E.S. Kasischke and B.J. Stocks (eds.). Fire, Climate Change, and Carbon Cycling in the Boreal Forest, pp. 132–150. Springer-Verlag.

Shvidenko, A., S. Nilsson and V. Roshkov, 1997. Possibilities for increased carbon sequestration through the implementation of rational forest management in Russia. Water, Air and Soil Pollution, 94(1/2):137–162.

Sigurdsson, B. and A. Snorrason, 2000. Carbon sequestration by afforestation and revegetation as means of limiting net-CO_2 emissions in Iceland. Biotechnologie, Agronomie Société et Environnement, 4(4):303–307.

Silfverberg, P. and P. Alhojärvi, 2004. Finnish-Russian Development Programme on Sustainable Forest Management and Conservation of Biodiversity in Northwest Russia, Evaluation of the Nature Conservation Component. Evaluation Report 2004. Ministry of the Environment, Finland, 61p.

Sirois, L., 1999. The sustainability of development in northern Quebec forests: social opportunities and ecological challenges. In: S. Kankaanpaa, T. Tasanen and M.-L. Sutinen (eds.). Sustainable Development in the Northern Timberline Forests. Proceedings of the Timberline Workshop, May 10–11, 1998, Whitehorse, Canada, pp. 17–27. Finnish Forest Research Institute, Helsinki.

Skree, O., R. Baxter, R.M.M. Crawford, T.V. Callaghan and A. Fedorkov, 2002. How will the tundra-taiga interface respond to climate change? Ambio Special Report, 12:37–45.

Smith, T.M. and H.H. Shugart, 1993. The transient response of carbon storage to a perturbed climate. Nature, 361:523–526.

Smith, T.M., J.F. Weishampel and H.H. Shugart, 1992. The response of terrestrial C storage to climate change modeling C dynamics at varying spatial scales. Water, Air and Soil Pollution, 64:307–326.

Smith, W.B., J.S. Vissage, D.R. Darr and R.M. Sheffield, 2001. Forest Resources of the United States, 1997. North Central Research Station, U.S. Department of Agriculture, Forest Service, 198p.

Solberg, B.O., A. Hofgaard and H. Hytteborn, 2002. Shifts in radial growth responses of coastal *Picea abies* induced by climatic change during the 20th century, central Norway. Ecoscience, 9:79–88.

Starfield, A.M. and F.S. Chapin III, 1996. Model of transient changes in arctic and boreal vegetation in response to climate and land use change. Ecological Applications, 6:842–864.

Statistics Canada, 2001. 2001 Census of Agriculture. www.statcan.ca/english/freepub/95F0301XIE/tables/territories.htm.

Statistics Greenland, 2005. Greenland in Figures 2005. 3rd Edition. Greenland Home Rule Government, 32p.

Statistics Iceland, 2003. Iceland in figures 2002–2003, Vol. 8. Hagstofa Islands. Statistics Iceland, Reykjavik, 32p.

Statistics Norway, no date. www.ssb.no/english/subjects/10/04/10/

Stocks, B.J., 1991. The extent and impact of fires in northern circumpolar countries. In: J.S. Levine (ed.). Global Biomass Burning: Atmospheric, Climatic, and Biospheric Implications, pp. 197–202. MIT Press.

Stocks, B.J., 2001. Fire Management in Canada. Section 6.1.2. In: Global Forest Fire Assessment 1990–2000. Forest Resources Assessment - WP 55. U.N. Food and Agriculture Organization.

Stocks, B.J., M.A. Fosberg, T.J. Lynham, L. Mearns, B.M. Wotton, Q. Yang, J.-Z. Jin, K. Lawrence, G.R. Hartley, J.A. Mason and D.W. McKenny, 1998. Climate change and forest fire potential in Russian and Canadian boreal forests. Climatic Change, 38(1):1–13.

Stocks, B.J., M.A. Fosberg, M.B. Wotton, T.J. Lynham and K.C. Ryan, 2000. Climate change and forest fire activity in North American boreal forests. In: E.S. Kasischke and B.J. Stocks (eds.). Fire, Climate Change, and Carbon Cycling in the Boreal Forest, pp. 368–376. Springer-Verlag.

Strömgren, M. and S. Linder, 2002. Effects of nutrition and soil warming on stemwood production in a boreal Norway spruce stand. Global Change Biology, 8(12):1194–1204.

Stuiver, M. and P.J. Reimer, 1993. Extended ^{14}C data base and revised CALIB 3.0 ^{14}C age calibration program. Radiocarbon, 35:215–230.

Sullivan, J.H. and A.H. Teramura, 1992. The effects of ultraviolet-B radiation on loblolly pine. 2. Growth of field-grown seedlings. Trees – Structure and Function, 6:115–120.

Sutinen M.-L., R. Arora, M. Wisniewski, E. Ashworth, R. Strimbeck and J. Palta, 2001. Mechanisms of frost survival and freeze-damage in nature. In: F.J. Bigras and S.J. Colombo (eds.). Conifer Cold Hardiness. Tree Physiology, 1:89–120. Kluwer Academic Press.

Swetnam, T.W. and J.L. Betancourt, 1990. Fire-Southern Oscillation relations in the Southwestern United States. Science, 249:1017–1020.

Swetnam, T. W., C. D. Allen and J. L. Betancourt, 1999. Applied historical ecology: Using the past to manage for the future. Ecological Applications, 9(4):1189–1206.

Takahama, U. and T. Oniki, 1997. A peroxidase/phenolics/ascorbate system can scavenge hydrogen peroxide in plants. Physiologia Plantarum, 101:845–852.

Tanja, S., F. Berninger, T. Vesala, T. Markkanen, P. Hari, A. Mäkelä, H. Ilvesniemi, H. Hänninen, E. Nikinmaa, T. Huttula, T. Laurila, M. Aurela, A. Grelle, A. Lindroth, A. Arneth, O. Shibistova and J. Lloyd, 2003. Air temperature triggers the recovery of evergreen boreal forest photosynthesis in spring. Global Change Biology, 9(10):1410–1426.

Taylor, R.M., A.K. Tobin and C.M. Bray, 1997. DNA damage and repair in plants. In: P.J. Lumsden (ed.). Plants and UV-B: Responses to Environmental Change. Society for Experimental Biology Seminar Series, 64:53–76.

Tchebakova, N.M., R.A. Monserud, R. Leemans and D.I. Nazimova, 1995. Possible vegetation shifts in Siberia under climate change. In: J. Pernetta , R. Leemans, O. Elder and S. Humphrey (eds.). Impacts of Climatic Change on Ecosystems and Species: Terrestrial Ecosystems, pp. 67–82. World Conservation Union.

Tegelberg, R. and R. Julkunen-Tiitto, 2001. Quantitative changes in secondary metabolites of dark-leaved willow (*Salix myrsinifolia*) exposed to enhanced ultraviolet-B radiation. Physiologia Plantarum, 113:541–547.

Tegelberg, R., P.J. Aphalo and R. Julkunen-Tiitto, 2002. Effects of long-term, elevated ultraviolet-B radiation on phytochemicals in the bark of silver birch (*Betula pendula*). Tree Physiology, 22:1257–1263.

Ter-Mikaelian, M., V.V. Furyaev and M.I. Antonovsky, 1991. A spatial forest dynamics model regarding fires and climate change. In: Problems of Ecological Monitoring and Ecosystem Modeling, V. 13. Gidrometeoizdat, Leningrad. (In Russian)

Teramura, A.H. and J.H. Sullivan, 1994. Effects of UV-B radiation on photosynthesis and growth of terrestrial plants. Photosynthesis Research, 39:463–473.

Tikhomirov, B.A., 1941. About the forest phase in the post-glacial history of vegetation in a north of Siberia and its relics in a modern tundra. Materials on Flora and Vegetation History of USSR, V.1:315–374.

Trumbore, S.E. and J.W. Harden, 1997. Accumulation and turnover of carbon in organic and mineral soils of the BOREAS northern study area. Journal of Geophysical Research, 102(D24):28,817–28,830.

Turner, D.P., G.J. Koerper, M.E. Harmon and J.J. Lee, 1995. A carbon budget for forests of the conterminous United States. Ecological Applications, 5(2):421–436.

Turner, M. G., W. H. Romme, R. H. Gardner and W. W. Hargrove, 1997. Effects of fire size and pattern on early succession in Yellowstone National Park. Ecological Monographs, 67:411–433.

Turner, M.G., S.L. Collins, A.L. Lugo, J.J. Magnuson, T.S. Rupp and F.J. Swanson, 2003. Disturbance dynamics and ecological response: the contribution of long-term ecological research. BioScience, 53(1):46–56.

Turunen, M., S. Huttunen, K.E. Percy, C.K. McLaughlin and J. Lamppu, 1997. Epicuticular wax of subarctic Scots pine needles: response to sulphur and heavy metal deposition. New Phytologist, 135:501–515.

Turunen, M., T. Vogelmann and W.K. Smith, 1998. Effective UV-screening in lodgepole pine needles. In: L.J. De Kok and I. Stulen (eds.). Responses of Plant Metabolism to Air Pollution, pp. 463–464. Backhuys Publishers.

Turunen, M., W. Heller, S. Stich, H. Sandermann and M.-L. Sutinen, 1999. Effects of UV exclusion on phenolic compounds of young Scots pine seedlings in the subarctic. Environmental Pollution, 106(2):225–234.

Turunen, M., M.-L. Sutinen, K. Derome, Y. Norokorpi and K. Lakkala, 2002. Effects of solar UV radiation on birch and pine seedlings in the subarctic. Polar Record, 38(206):233–240.

USDA Forest Service, 1950–1999. Forest Insect and Disease Conditions in Alaska. Annual Reports. U.S. Department of Agriculture, Forest Health Protection.

Vaganov, E. A., 1989. On the methods of forecasting crop yield using dendroclimatological data. Ecologiya (Russian Journal of Ecology.), 3:15–23. (In Russian)

Vaganov, E.A. and A.V. Shashkin, 2000. Growth and Tree Ring Structure of Conifers. Nauka, Novosibirsk, 232p. (In Russian)

Vaganov, E. A., A.V. Shashkin, I.V. Sviderskaya and L.G. Vysotskaya, 1985. Histometric analysis of woody plant growth. Nauka, Novosibirsk, 108p. (In Russian)

Vaganov, E.A., S.G. Shiyatov and V.S. Mazepa, 1996. Dendroclimatic Studies in Ural-Siberian Subarctic. Nauka, Novosibirsk, 246p. (In Russian)

Vaganov, E.A., S.G. Shiyatov, R.M. Hantemirov and M.M. Naurzbaev, 1998. Summer temperature variations in high latitudes of the Northern Hemisphere during last 1.5 millennium: comparative analysis tree-ring and ice core data. Doklady Akademii Nauk, 338(5):681–684.

Vaganov, E.A., M.K. Hughes, A.V. Kirdyanov, F.H. Schweingruber and P.P. Silkin, 1999. Influence of snowfall and melt timing on tree growth in subarctic Eurasia. Nature, 400:149–151.

Vaganov E.A., K.R. Briffa, M.M. Naurzbaev, F.H. Schweingruber, S.G. Shiyatov and V.V. Shishov, 2000. Long-term climatic changes in the Arctic region of the Northern Hemisphere. Doklady Earth Sciences, 375(8):1314–1317.

Valentini, R., S. Dore, G. Marchi, D. Mollicone, M. Panfyorov, C. Rebmann, O. Kolle and E.-D. Schulze, 2000. Carbon and water exchanges of two contrasting central Siberia landscape types: regenerating forest and bog. Functional Ecology, 14(1):87–96.

Van Cleve, K. and L.A. Viereck, 1983. A comparison of successional sequences following fire on permafrost-dominated and permafrost-free sites in interior Alaska. In: Permafrost: Fourth International Conference, Proceedings, pp. 1286–1291. National Academy Press.

Van Cleve, K., F.S. Chapin III, C.T. Dyrness and L.A. Viereck, 1991. Element cycling in Taiga forests: State-factor control. Bioscience, 41(2):78–88.

Viereck, L.A., 1970. Forest succession and soil development adjacent to the Chena River in interior Alaska. Arctic and Alpine Research, 2:1–26.

Viereck, L.A. and E.L. Little, Jr., 1972. Alaska Trees and Shrubs. Agriculture Handbook No. 410. U.S. Department of Agriculture, Washington, DC, 265p.

Viereck, L.A., K. Van Cleve and C.T. Dyrness, 1986. Forest ecosystem distribution in the taiga environment. In: K. Van Cleve, F.S. Chapin III, P.W. Flanagan, L.A. Viereck and C.T. Dyrness (eds.). Forest Ecosystems in the Alaskan Taiga, pp. 22–43. Springer-Verlag.

Vitousek, P.M. and R.W. Howarth, 1991. Nitrogen limitation on land and in the sea: How can it occur? Biogeochemistry, 13(2):87–116.

Vlassova, T.K., 2002. Human impacts on the tundra-taiga zone dynamics: the case of the Russian lesotundra. Ambio Special Report, 12:30–36.

Volney, W.J.A. and R.A. Fleming, 2000. Climate change and impacts of boreal forest insects. Agriculture, Ecosystems and Environment, 82(1–3):283–294.

Wahlen, M., D. Allen, B. Deck and A. Herchenroder, 1991. Initial measurements of CO_2 concentrations (1530–1940 AD) in air occluded in the GISP2 ice core from central Greenland. Geophysical Research Letters, 18:1457–1460.

Wang, C., S.T. Gower, Y. Wang, H. Zhao, P. Yan and B.P. Bond-Lamberty, 2001. The influence of fire on carbon distribution and net primary production of boreal *Larix gmelinii* forests in north-eastern China. Global Change Biology, 7(6):719–730.

Warrick, R.A., 1988. Carbon dioxide, climate change and agriculture. The Geographical Journal, 154:221–233.

Webster , P., 2002. Climate change: Russia can save Kyoto, if it can do the math. Science, 296:2129–2130.

Werner, R.A., 1996. Forest health in boreal ecosystems of Alaska. The Forestry Chronicle, 72(1):43–46.

Werner, R.A. and E.H. Holsten, 1985. Factors influencing generation times of spruce beetles in Alaska. Canadian Journal of Forest Research, 15:438–443.

Whelan, R.J., 1995. The Ecology of Fire. Cambridge University Press, 346p.

White, A., M.R.G. Cannell and A.D. Friend, 2000. The high latitude terrestrial carbon sink: a model analysis. Global Change Biology, 6:227–246.

Whitham, T.G., 1989. Plant hybrid zones as sinks for pests. Science, 244:1490–1493.

Whitham, T.G., P.A. Morrow and B.M. Potts, 1994. Plant hybrid zones as centers of biodiversity: The herbivore community of two endemic Tasmanian eucalypts. Oecologia, 97:481–490.

Wiken, E.B., 1986. Terrestrial Ecozones of Canada. Ecological Land Classification, Series No. 19. Environment Canada, Hull, Quebec. 26p.+map.

Wiles, G.C., R.D. D'Arrigo and G.C. Jacoby, 1996. Summer temperature changes along the Gulf of Alaska and the Pacific Northwestern coast modeled from coastal tree rings. Canadian Journal of Forest Research, 26:474–481.

Wilmking, M., G.P. Juday, V. Barber and H. Zald, 2004. Recent climate warming forces contrasting growth responses of white spruce at treeline in Alaska through temperature thresholds. Global Change Biology, 10:1–13.

Wittwer, D., 2004. Forest Health Conditions in Alaska, 2003. USDA Forest Service General Technical Report R10-TP-123, 82p.

Wright, H.E. Jr., 1974. Landscape development, forest fires, and wilderness management. Science, 186:487–495.

Wright, H.E. Jr., 1983. Introduction. In: H.E. Wright Jr. (ed.). Late-Quaternary Environments of the United States, Vol. 2, The Holocene, pp. xi-xvii. University of Minnesota Press.

Wright, H.E. Jr. and C.W. Barnosky, 1984. Introduction to the English Edition. In: A.A. Velichko (ed.). Late-Quaternary Environments of the Soviet Union, pp. xiii-xxii. University of Minnesota Press.

Yamasaki, H., Y. Sakihama and N. Ikehara, 1997. Flavonoid-peroxidase reaction as a detoxification mechanism of plants against H_2O_2. Plant Physiology, 115:1405–1412.

Yarie, J., 1981. Forest fire cycles and life tables: a case study from interior Alaska. Canadian Journal of Forest Research, 11(3):554–562.

Zackrisson, O., 1977. Influence of forest fires on the North Swedish boreal forest. Oikos, 29:22–32.

Zheng, D., M. Freeman, J. Bergh, I. Røsberg and P. Nilsen, 2002. Production of *Picea abies* in south-east Norway in response to climate change: a case study using process-based model simulation with field validation. Scandinavian Journal of Forest Research, 17(1):35–46.

Zielinski, G.A., P.A. Mayewski, L.D. Meeker, S. Whitlow, M.S. Twickler, M. Morrison, D.A. Meese, A.G. Gow and R.B. Alley, 1994. Record of volcanism since 7000 BC from the GISP2 Greenland ice core and implications for the volcano-climate system. Science, 264:948–951.

Zimov, S.A., S.P. Davidov, G.M. Zimova, A.I. Davidova, F.S. Chapin, M.C. Chapin and J.F. Reynolds, 1999. Contribution of disturbance to increasing seasonal amplitude of atmospheric CO_2. Science, 284:1973–1976.

Chapter 15

Human Health

Lead Authors

Jim Berner, Christopher Furgal

Contributing Authors

Peter Bjerregaard, Mike Bradley, Tine Curtis, Edward De Fabo, Juhani Hassi, William Keatinge, Siv Kvernmo, Simo Nayha, Hannu Rintamaki, John Warren

Contents

Summary

The nature of projected climate-related changes and variability, and the characteristics of arctic populations, means that impacts of climate change on the health of arctic residents will vary considerably depending on such factors as age, gender, socio-economic status, lifestyle, culture, location, and the capacity of local health infrastructure and systems to adapt. It is more likely that populations living in close association with the land, in remote communities, and those that already face a variety of health-related challenges will be most vulnerable to future climate changes. Health status in many arctic regions has changed significantly over the past decades and the climate, weather, and environment have played, and will continue to play a significant role in the health of residents in these regions.

Direct health impacts may result from changes in the incidence of extreme events (avalanches, storms, floods, rockslides) which have the potential to increase the numbers of deaths and injuries each year. Direct impacts of winter warming in some regions may include a reduction in cold-induced injuries such as frostbite and hypothermia and a reduction in cold stress. As death rates are higher in winter than summer, milder winters in some regions could reduce the number of deaths. Direct negative impacts of warming could include increased heat stress in summer and accidents associated with unpredictable ice and weather conditions. Indirect impacts may include increased mental and social stress related to changes in the environment and lifestyle, potential changes in bacterial and viral proliferation, vector-borne disease outbreaks, and changes in access to good quality drinking water sources. Also, some regions may experience a change in the rates of illnesses resulting from impacts on sanitation infrastructure. Impacts on food security through changes in animal distribution and accessibility have the potential for significant impacts on health as shifts from a traditional diet to a more "western" diet are known to be associated with increased risk of cancers, diabetes, and cardiovascular disease.

Increased exposure to ultraviolet (UV) radiation among arctic residents has the potential to affect the response of the immune system to disease, and to influence the development of skin cancer and non-Hodgkin's lymphoma, as well as the development of cataracts. However, as the current incidence rates for many of these ailments are low in small arctic communities it is difficult to detect, let alone predict, any trends in their future incidence. The presence of environmental contaminants threatens the safety of traditional food systems, which are often the central fabric of communities. The projected warming scenarios will affect the transport, distribution, and behavior of environmental contaminants and thus human exposure to these substances in northern regions.

These changes are all taking place within the context of cultural and socio-economic change and evolution. They therefore represent another of many sources of stress on societies and cultures as they affect the relationship between people and their environment, which is a defining element of many northern cultures. Through potential increases in factors influencing acculturative stress and mental health, climate-related changes may further stress communities and individual psychosocial health. Communities must be prepared to identify, document, and monitor changes in their area in order to adapt to shifts in their local environment. The basis of this understanding is the ability to collect, organize, and understand information indicative of the changes taking place and their potential impacts. A series of community indicators are proposed to support this development of monitoring and decision-making ability within northern regions and communities.

15.1. Introduction

15.1.1. Background and rationale

The arctic regions share common characteristics such as sparse population, harsh climate, similar geographic features, high latitude, and characteristic seasonal extremes of daylight hours and temperatures. The modern climate record shows that regions of the same high latitude often have very different mean annual temperatures, precipitation regimes, and ecosystems. It is important to incorporate this diversity in any assessment of the current and future impacts of climate change on the health of northern peoples.

In general, arctic regions of the United States, Canada, and Nordic Europe have more economic support than those of the former Soviet Union. As a result, the availability and quality of data on human health status vary widely and are not available for some regions. The evaluation, both current and future, of the impact of climate change on human health is entirely data dependent.

As a result, this chapter (unlike those addressing specific climate issues) cannot address the potential health impact of climate change using a regional or time-specific approach.

The ACIA-designated models project climate change relative to baseline conditions (1981–2000) for three 20-year "time slices" (2011–2030, 2041–2060, and 2071–2090) for four arctic regions. The time-slice regional scenario is a useful construct for those ecological components which are not able to quickly relocate or utilize technology to mitigate climate-related impacts, but is not as useful for human populations, where non-climate factors can cause mass relocations over periods of days to weeks. The reactor accident at Chernobyl in 1986 is one such example. Also, economic decisions such as those to develop natural resources (e.g., petroleum, minerals, timber) can bring food, education, and health resources to a region, and may accelerate or mitigate a decrease in numbers of a traditional food species. These changes might result in population growth and an improvement in health status, or the erosion of a community's cultural base, bringing cultural stress and a deterioration in health status.

This chapter does not attempt to predict health impacts for any specific region, or any specific time frame, for the following reasons.

- Climate models, over prolonged intervals (decades), are uncertain, and thus impacts on people and their communities are also uncertain.
- Humans can adopt strategies to mitigate almost any possible health impact, given sufficient support.
- Levels of governmental, public health, social, and cultural support vary dramatically among circumpolar communities and will continue to fluctuate in the future.
- Over the past four decades, many regions of the circumpolar arctic have shown a warming trend, however significant areas have also shown a cooling trend, such that uniform temperature assumptions can not yet be justified.

This chapter presents a discussion of mechanisms by which climate can influence human health in arctic communities such that these communities might plan appropriate monitoring strategies to support the development of adaptive or mitigation actions. In this way, potential negative impacts might be recognized and mitigated, and potential opportunities might be recognized and exploited.

15.1.2. Health in the circumpolar Arctic

According to the World Health Organization (WHO, 1967), health includes aspects of physical, mental, and social well-being and is not simply the absence of disease. In this holistic vision of health, which is very similar to that supported by the many indigenous groups throughout the world, the well-being of individuals and communities is tied to that of their environment. Human health status is a result of the complex interaction of genetic, nutritional, and environmental factors. "Environment" in this context includes the socio-economic, cultural, and physical infrastructure and ecosystem factors. Many of these groups of factors can improve or degrade health status, by enhancing the resilience of a population, or by causing stress. In these instances, the stress can be a direct physical change such as temperature, can take the form of increased prevalence of a disease-causing organism, or be represented by a perceived threat or sense of loss, engendering psychological stress.

Previous reports have concluded that, for the residents of the contiguous United States, climate change will have a small overall health impact, due to the ability of the existing public health system to respond to new threats (Patz et al., 2000). This conclusion is unlikely to be true for the North American Arctic, or for arctic residents in many other countries, for the following reasons.

- Many arctic residents live in very small, isolated communities, with a fragile system of economic

support, dependence on subsistence hunting and fishing, and little or no economic infrastructure.
- Rural arctic public health and acute care systems are often marginal, sometimes poorly supported, and in some cases, non-existent.
- Culture is often critical to community and individual health, and may be affected by climate change via mechanisms such as the loss of a traditional subsistence food source, which can result in a grief response and severe stress. Climate changes can become a source of illness, injury, and mortality for arctic communities.

The combined result of these factors is that rural arctic residents are often uniquely vulnerable to health impacts from climate change, mediated by a variety of mechanisms. Also, owing to their close relationship and dependence on the land, sea, and natural resources for cultural, traditional, social, economic, and physical well-being, indigenous peoples are also uniquely vulnerable to these environmental changes. It is for these reasons that this chapter emphasizes potential climate impacts on health in small arctic communities, among whose residents many are indigenous.

This chapter presents a brief overview of available data on the current health status of arctic residents, followed by a series of sections describing the potential impact mechanisms of climate change on socio-cultural and socio-economic environments and physical infrastructures as they relate to human health. Sets of indicators are then proposed to prospectively monitor potential climate change impacts on human health. The chapter concludes with recommendations for research and action.

15.2. Socio-cultural conditions, health status, and demography

Social conditions and lifestyles among indigenous populations vary widely throughout the Arctic. Many indigenous peoples rely on the food that they hunt and harvest from the land and sea, as it provides for much of their nutritional intake as well as being a critical component of their cultural identity, and in many cases, their local informal economy (Duhaime, 2002). Members of the urban population, of whom many are of European descent, have lives that are to some extent indistinguishable from those of their kinfolk in Europe or North America, although the arctic climate still determines much of their daily life and is an underlying condition for infrastructure and transport.

15.2.1. Socio-cultural conditions and health status

Living conditions are changing throughout the Arctic for indigenous as well as non-indigenous residents as a consequence of the change from an economy based on hunting to modern wage earning (AMAP, 2003). The following description of social change in Greenland (Bjerregaard and Curtis, 2002) is similar to that in many other circumpolar regions. The shift from a traditional Inuit community to a modern society started at the beginning of the 20th century when fishing began to replace the hunting of marine mammals as the main source of livelihood. This was accompanied by population movement from a large number of small villages to larger – although still small by many standards – population centers, and by the gradual supplement of the traditional subsistence economy by a cash economy. By the end of the Second World War, however, Greenland still had a relatively isolated and traditional society where most people lived in small villages and subsisted on small-scale hunting and fishing activities. During the latter half of the 20th century unprecedented changes occurred in Greenland resulting in a very modern society thoroughly integrated in global political and economic systems. Fishing and the associated processing industry are the basis of the present economy but at a very advanced level with ocean-going fishing vessels existing alongside smaller crafts and some fishing from the ice. Subsistence hunting and fishing are still widespread but are increasingly becoming a leisure activity. Daily connections by air to Denmark now exist, and even small villages have telephone service and internet access. Supermarkets contain products such as fresh mangos and papayas, as well as a range of European meats, dairy, and vegetable products, and frozen Greenlandic fish and seal meat.

The influence of such changes on physical health and everyday life are obvious; positive changes include vastly improved housing conditions, a stable supply of food and increased access to western goods, and decreased mortality and morbidity from infectious diseases including tuberculosis. However, societal change and modernization have also brought a number of social and mental health problems as well as increasing prevalence of chronic diseases such as cardiovascular diseases and dia-

betes (Bjerregaard and Young, 1998). Children have been brought up with values that were useful for hunters and hunters' wives living in small communities: independence, self-reliance, non-interference with other people's lives, and physical strength. As adolescents and adults they have had to cope with life in much larger and more densely populated communities, in a world that rewards formal education, language skills, and discussion instead of action. The majority of individuals have adjusted to the new situation but for some the burden has been a significant challenge. In some cases, these changes have been associated with historical changes in climatic conditions. The relationship between climate and settlement in Greenland illustrates the complexity of these changes in arctic communities (see Box 15.1).

Over the last 50 years, the population of most arctic regions has dramatically increased. Much of this increase is due to a reduction in infant mortality and mortality from infectious diseases, particularly tuberculosis and the vaccine preventable diseases of childhood. Also, safe water supplies, sewage disposal, development of rural hospitals, and in some regions, community-based medical providers, have contributed to improved care and access to care for injuries and illness. All regions have greatly improved transportation infrastructure, resulting in the availability of western food items, tobacco, and alcohol on a scale not previously possible. Plus, in many arctic regions communications technology has made western culture visible in even the most remote settlements. Arctic indigenous residents are, in most regions, encouraged to become permanent residents in fixed locations, to facilitate the provision of services and economic opportunities. This has eliminated the historic practice of families and groups of families to move, intact, when climate or other environmental change made a region

Box 15.1. Climate change impacts on settlement in Greenland

In the early 20th century, climate warming resulted in Atlantic cod (*Gadus morhua*) appearing in great shoals off the west coast of Greenland. Cod fishing became a source of cash income for the Greenlanders and the traditional society based on hunting of seals and whales began to make way for a modern fishing society and cash economy. The population of Greenland concentrated in fewer and larger towns and the number of villages decreased. This development was intensified after the Second World War due to deliberate concentration of the population in towns with schools, health care, and shops. Fishing was planned to be the major source of revenue for the Greenland society. In the 1960s, however, climate cooling together with overfishing resulted in the disappearance of cod from the west coast of Greenland (Hamilton et al., 2003) (see also section 15.5.5.3).

In the 1960s, large numbers of shrimp were detected in Disko Bay. Over the course of a few years, the village of Qasigiannguit with a population of only 343 in 1955 developed into a lively town centered on the shrimp factory. During the 1990s, the shrimp disappeared from the coastal waters. Large sea-going vessels now fish and process the shrimp far from the coast and the factory has closed. People have started moving from the town. In 1982 when the population was at its maximum there were 1800 inhabitants; in 2000 the population of Qasigiannguit was only 1400. The unemployment rate is among the highest in Greenland: 14.4% compared to 7.1% in the towns in general.

The examples show how climate change can influence the occurrence of commercially important species and how the disappearance of a species can have negative impacts on socio-economic conditions within a local community. Many Inuit communities are particularly vulnerable to changes in species availability because they often rely on the availability of only one or a few species.

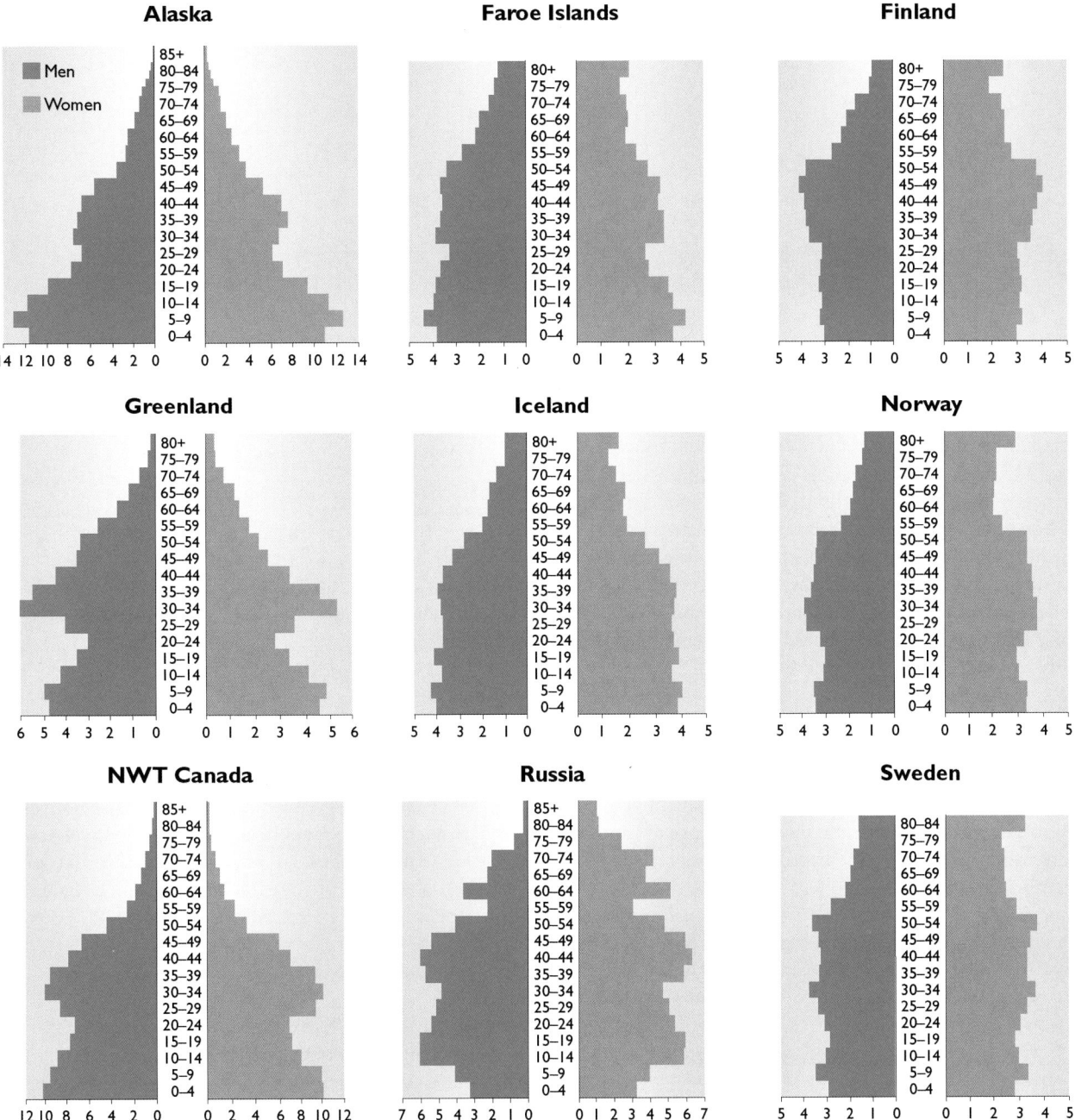

Fig. 15.1. Age structure of arctic populations (WHO, 2000).

unable to sustain them. Also, there are often few products or services that many small communities can contribute to the overall economy, such that survival often depends on a complex web of government-funded economic support, combined with primarily service-based employment in schools, sanitation facilities, and transportation infrastructure. The establishment of fixed village locations has also affected subsistence activities.

Indigenous culture is under stress from competing western culture, and subsistence activities are affected by climate change and concerns about the contamination of traditional food resources by contaminants, both from local sources and from long-range sources transported to the Arctic via ocean and atmospheric currents. Zoonotic diseases (animal diseases that can be passed to humans) and parasitic diseases are also associated with some tradi-

tional food species and traditional food preparation methods (e.g., trichinosis or botulism). Assessments of food safety have resulted in the collection of information on micronutrients and anthropogenic chemicals and have often resulted in the release of confusing or conflicting messages to rural residents (AMAP, 2003; CACAR II, 2003). As a result, erosion of cultural support, a decrease in traditional activities, and substitution of western foods for traditional foods are becoming more important as the causes of morbidity and mortality among arctic populations such that, in some respects, they now more closely resemble western populations.

Historically, there was little heart disease, cancer, obesity, or diabetes in circumpolar populations. Major causes of mortality were infectious diseases, especially tuberculosis, pneumonia, and injury. Life expectancy was short,

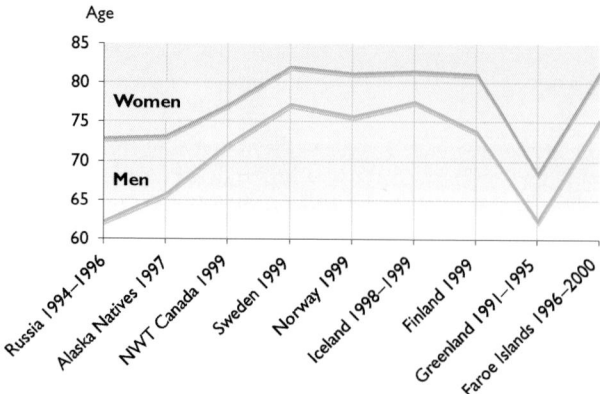

Fig. 15.2. Life expectancy in arctic populations (IHS, 1999b; WHO, 2000).

and infant mortality was high (Bjerregaard and Young, 1998). However, for the reasons discussed in this section much has now changed. Section 15.2.2 presents the current population structure, birth rates, infant mortality, and causes of adult mortality for the countries with arctic residents. Arctic regions of the Russian Federation have few comparable data but there is no reason to believe that this region does not also have similar serious health problems as seen elsewhere.

Many technological advances have made subsistence hunting safer, such as modern protective clothing, Global Positioning System (GPS) devices, radio, cellular telephones, and weather forecasts. Hunting efficiency has also improved dramatically through the use of modern firearms and improved transportation, such as boats, snowmobiles, all terrain vehicles, and aircraft. These advances have the potential to erode traditional knowledge and skills, which could increase risk. For instance, loss of traditional knowledge of short-term weather changes and ice thickness could result in injury or death.

15.2.2. Population structure and health statistics

Some regions of the Arctic have a different population structure compared to that of more temperate regions of the same country. This is true for the indigenous rural arctic populations; including Alaska Natives, Canadian Arctic indigenous groups, Inuit, and Greenland Inuit.

The populations of the Northwest Territories (NWT, including Nunavut) and the Yukon (which are predominantly indigenous), as well as Alaska Natives and Greenland residents (who are 90% Inuit), have a greater percentage of children and a smaller percentage of older people than in the Nordic countries (Fig. 15.1; AMAP, 2002). These three groups represent the majority of rural arctic residents for whom comparable health data exists. All these groups typically reside in very small communities (of around 50 to 5000 inhabitants), in remote regions, and with traditional foods comprising a significant part of the diet.

The health of arctic populations can be determined from a range of health status indicators, including life

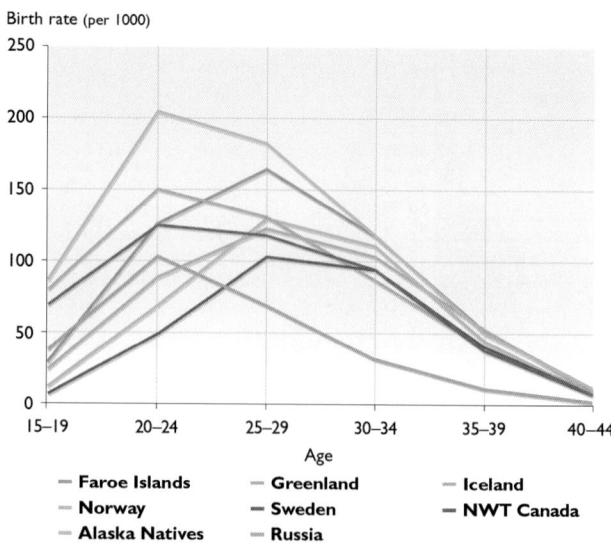

Fig. 15.3. Birth rates by age group in arctic populations (IHS, 1999b; WHO, 2000).

expectancy at birth, birth rate, infant mortality, population mortality, and age-adjusted causes of death.

15.2.2.1. Life expectancy

Life expectancy in arctic populations has improved owing to a wide variety of changes in social conditions and lifestyles. A significant contributor to improved life expectancy is decreased infant mortality. Alaska Natives, NWT residents, and Greenland residents generally have lower life expectancies than residents in the Nordic countries and on average, life expectancy is lower for indigenous populations (Fig. 15.2; Statistics Canada, 2003).

15.2.2.2. Birth rate

Alaska Natives, NWT residents, and Greenland residents have higher birth rates than residents in the Nordic countries (Fig. 15.3). This reflects the greater proportion of children in these populations.

15.2.2.3. Infant mortality

Infant mortality has decreased considerably since around 1950 for Alaska Natives (Fig. 15.4). Despite the improvement the overall infant mortality rates for indigenous arctic residents in Alaska and Greenland remain higher than for all races infant mortality rates in the United States and Canada. Infant mortality rates are lowest in the Nordic countries.

15.2.2.4. Common causes of death

Differences also exist in the most common causes of death based on death certificate data. To account for differences in population age structure (see Fig. 15.1) the mortality rates were adjusted to those for a standardized population structure; although the standard population structure used for the Nordic countries is slightly different to that for Canada and the United States. Figure 15.5 compares death

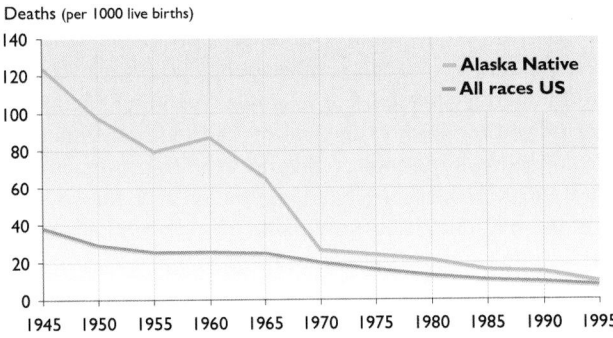

Fig. 15.4. Rates of infant mortality in Alaskan Natives and all races U.S. infants (IHS, 1999b; WHO, 2000).

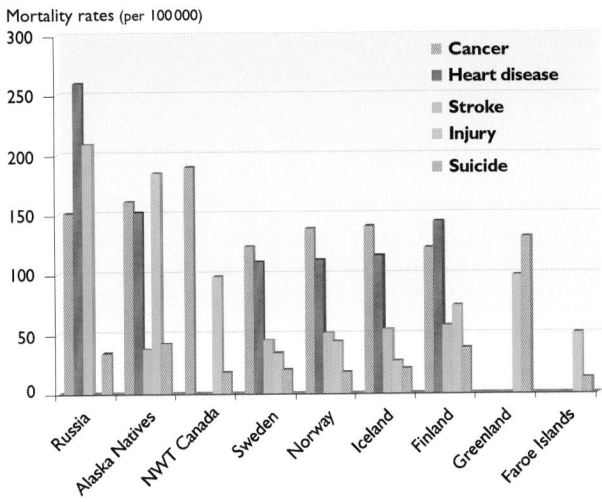

Fig. 15.5. Common causes of death in arctic populations (IHS, 1999b; WHO, 2000).

rates for a range of conditions in indigenous populations (or largely indigenous in the case of the NWT) and the American and European populations. It is evident that Alaska Natives and Greenland residents have much higher mortality rates for injury and suicide. Mortality rates for heart disease and cancer are now similar in arctic indigenous populations and in relation to overall rates for the United States, Canada, and Northern European countries.

15.3. Potential impacts of direct mechanisms of climate change on human health

Human health in northern communities is affected via a number of direct and indirect impacts of climate-related changes. "Direct impacts" refers to those health consequences resulting from direct interactions with aspects of the environment that have changed or are changing with local climate (i.e., resulting from direct interactions with physical characteristics of the environment: air, water, ice, land; and for example exposure to thermal extremes). They include such things as difficulties in dealing with heat and cold stress; alleviation of cold stress due to warmer winters; dangers associated with travel and activities on the land resulting from unpredictable weather patterns and ice conditions; and increased incidences of sunburn and rashes as a result of increased sun intensity and exposure to UV-B radiation.

The direct impacts of climate and UV radiation on human health are primarily related to extreme events, temperature, and changes induced by exposure to UV-B radiation. Much of the discussion in this section on the mechanisms involved in these potential impacts involves associations of health events with observed climate change, without assigning causality. Where the effects are understood, the mechanisms are described. However, in many instances, the exact mechanisms are not known, or the relationships between human health and climate variables are multifaceted.

15.3.1. Extreme events

Some reports indicate that extreme weather events such as droughts, floods, and storms may become more frequent and intense in the future (Haines and McMichael,

1997) and there is some evidence that this is already occurring (see Krupnik and Jolly, 2002). Injury and death are the direct health impacts most often associated with natural disasters. Precipitation regimes are expected to affect the frequency and magnitude of natural processes which can potentially lead to death and injury, such as debris flow, avalanches, and rock falls (Koshida and Avis, 1998).

Thunderstorms and high humidity have been associated with short-term increases in hospital admissions for respiratory and cardiovascular diseases (Kovats et al., 2000). According to Mayer and Avis (1997), there is controversy concerning the incidence and continuation of significant mental health problems, such as post traumatic stress disorders following natural disasters. An increase in the number of mental health disorders has been observed in the United States after natural disasters. Longer periods of extreme weather and storm events could have social impacts on communities that are isolated from regional centers and if major modes of transport are no longer available. The impacts of extreme events on everyday subsistence activities could also affect community and individual well-being.

Indigenous people throughout the Arctic have reported that the weather has become more "unpredictable" and in some cases that extreme or storm events progress more quickly today than in the past (Fox S., 2002; Furgal et al., 2002; Krupnik and Jolly, 2002). Some northern residents report that this unpredictability limits subsistence activities and travel and increases the risks of people being trapped by weather while outside the community (Fox S., 2002; Furgal et al., 2002; Krupnik and Jolly, 2002; see Chapter 3).

Yeah, it changes so quick now you find. Much faster than it used to... Last winter when the teacher was caught out it was perfect in the morning, then it went down flat and they couldn't see a thing. It was like you were traveling and floating in the air, you couldn't see

the ground. Eighteen people were caught out then, and they almost froze, it was bitterly cold. Labrador hunter, as quoted in Furgal et al., 2002

15.3.2. Temperature-related stress

Warming is projected for some regions of the Arctic (see Chapter 4), and this may result in an increase in the number and magnitude of extreme warm days. Exposure to extreme and prolonged heat is associated with heat cramps, heat exhaustion, and heatstroke. However, because of the low mean temperature in many arctic regions, the likelihood of such events having large impacts on public health for the general population is low. Death rates are higher in winter than in summer and milder winters in some regions could actually reduce the number of deaths during winter months. However, the relationship between increased mortality and winter weather is difficult to interpret and more complex than the association between mortality and morbidity and exposure to high temperatures (Haines and McMichael, 1997; Patz et al., 2000). For example, many winter deaths are due to respiratory infections such as influenza and it is unclear how influenza transmission would be affected by warmer winter temperatures. Some studies indicate an association between extreme temperature-related events and mortality. For these associated impacts, groups such as the elderly and people affected by cardio-respiratory problems are more vulnerable (Patz et al., 2000).

In North America, summer heat waves affect more urban populations than northern people, especially because of the urban heat-island effect (Kovats et al., 2000). The impact is greater when the high temperatures (>25 °C) are irregular and occur at the beginning of summer (Thouez et al., 1998). Indigenous people in some regions of the Arctic are reporting incidences of stress related to temperature extremes not previously experienced. For example, shortness of breath and reduced physical activity (e.g., fishing), and an increase in respiratory discomfort (Furgal et al., 2002).

Fewer cold days, associated with a general warming trend in some regions during winter, are reported to have the positive effect of allowing people to get out more in winter and so alleviate stress related to extreme cold (Furgal et al., 2002). However, in Nunavik for example, approximately one to two heat waves occur every 30 years while extreme cold is much more common. In regions where heat waves do not represent a real risk for northern populations, an increase in extreme cold events could have more serious implications. According to Dufour (1991), respiratory problems were responsible for one in seven deaths among the Inuit population of Nunavik. Muir (1991) reported that respiratory problems were the primary reason for visits to the nurse or doctor. Chronic respiratory illnesses are highly prevalent in some northern regions. For example, in Labrador, breathing problems are among the most common long-term medical conditions in adults and children (LIA, 1997). In these two northern Canadian regions,

chronic respiratory illnesses could be amplified by prolonged cold events. Indirect effects of prolonged cold events could also occur as other public health problems are further aggravated. For example, spending a longer period of time in crowded and overheated houses during prolonged cold periods could affect the transmission of viral infections, especially among the elderly, the young, and the physiologically vulnerable (e.g., individuals who are immunosuppressed due to the presence of other diseases or medication). Other factors such as smoking can also modify the incidence of respiratory illnesses.

In the 1970s, scientific research focused for the first time on dramatic rises in mortality every winter, and on smaller rises in unusually hot weather. Heat-related deaths often result from severe dehydration (causing hemoconcentration) resulting from the loss of electrolytes and water in sweat and the inability to regulate body temperature. In northern Sweden, a clear association between atmospheric pressure, changing temperature, and increasing rates of cardiac events was documented (Messner et al., 2003). Exposure to low ambient temperatures for long periods brings specific physiological stresses. Cold exposure is part of daily working life in the Arctic. It affects human outdoor activity significantly because the arctic winter is long and cold conditions are severe. Winter, with mean temperatures of less than 0 °C, lasts for more than seven months in some regions. The interactions between temperature (in this case cold) and health, and the various health consequences are summarized in Fig. 15.6. Responses to cold may be normal, exaggerated (hyperreactions), or damped (hyporeactions). These result in eventual body cooling and associated impacts. In some instances, hyperreactions may occur which themselves result in disease. Climate models project that cool winter temperatures will persist in many circumpolar regions (see Chapter 4). Cold is likely to remain an environmental cause of illness and death.

15.3.2.1. Limits of human survival in the thermal environment

Human body heat balance depends on: the thermal environment (air temperature, air velocity, air moisture, and radiative heat gain from sun or artificial sources); the thermal insulation of clothing; and the rate of physical work producing heat via metabolic pathways (e.g., Parsons K., 1993). For a naked human at rest, the thermoneutral air temperature is 27 °C. In temperatures above the thermoneutral zone, heat loss is increased by sweating, and in lower temperatures, heat production is increased by muscular work (up to about 1200 W) or by shivering (up to about 500 W). By doing heavy physical work, a naked human can survive at an air temperature of about -5 °C for several hours. The extreme limits of behavioral temperature regulation depend on available technology, but working at extreme low or high temperatures is possible with special clothing. The removal of body heat by air movement, and its practical application in designing appropriate clothing, is known as the "windchill" effect (Quayle and Steadman, 1998).

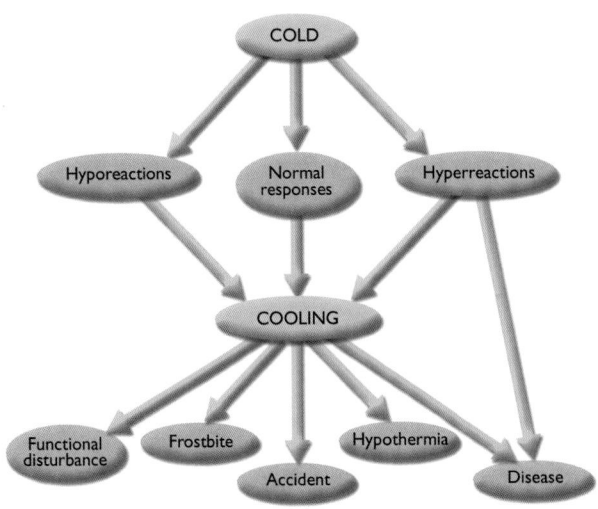

Fig. 15.6. Interactions between temperature and health.

The effects of heat balance are usually classified in terms of five levels: comfort, discomfort, performance degradation, health effects, and tolerance (Lotens, 1988). For an adequately clothed person initial cold problems start to appear at an ambient temperature of 10 °C when fingers start to cool during light manual work. Even with heavy work, cold problems appear at between -20 and -25 °C. For optimal manual performance, skin temperature is 32 to 36 °C. Below a skin temperature of 13 °C manual performance rapidly deteriorates. Marked changes in ambient temperature can increase or decrease cognitive performance or remain without effect. When effects are seen, cold particularly appears to affect the performance of complex cognitive tasks involving short-term or working memory (Palinkas, 2001).

Psychological, whole body, and local physiological acclimatization develops when the thermal environment is changed. Marked acclimatization can be developed within about ten days. In cold, the usual signs of acclimatization are blunted responses of the cardiovascular system (heart rate and blood pressure) and heat production (shivering). Cold-induced vasoconstriction in hands is also attenuated (e.g., Rintamäki, 2001). Heat acclimatization involves increased sweating and earlier onset of sweating.

For healthy active people a 5 to 10 °C decrease in temperature is not expected to result in serious effects on the maintenance of body heat balance during outdoor work. Humans can compensate for a 10 °C decrease by wearing additional clothing or by increasing metabolic heat production by 30 to 40 W/min. If the temperature of arctic or subarctic climates increases by 5 to 10 °C, the climate would still be cool or cold, with cold temperatures still having more impact on human physiology than heat. More serious problems could occur under extreme conditions such as during the coldest winter months in arctic or subarctic climates, if ambient temperatures decrease or the cold season increases markedly in length. There is an upper limit to the thermal insulation of winter clothing. A decrease or increase in ambient temperature, especially if the change is rapid, is a more serious threat for sick and/or elderly people than for healthy individuals capable of a physically active lifestyle.

15.3.2.2. Cold injuries

Cold-related injuries are immediate pathological consequences of cold exposure. As a consequence of direct or indirect effects of cold, the total injury rate may increase in relation to environmental cold exposure. The rate of slip and fall injuries, for example, increases with decreasing temperature. Increasing rates of slip and fall injuries are seen at temperatures of 0 °C and below. Low temperature is often a secondary source of injury and may not be reflected at true frequencies in statistical records. Risk of unintentional injury is least at a temperature of about 20 °C and increases with lower and higher ambient temperatures.

Injuries such as frostbite, hypothermia, and others are linked to body cooling. Cooling injuries occur most often during winter months, especially during the few coldest winter days and are also increased by wind speed. Cooling injuries show a strong relationship with temperature, i.e., the lower the temperature the more injuries occur. The majority of cooling injuries are freezing injuries (e.g., frostbite) (Taylor M., 1992).

Frostbite generally occurs on the most peripheral parts of the body (head, hands, feet). For the head, frostbite of the ears is almost twice as common as frostbite of the nose and cheek. Several areas of the body may be injured simultaneously. Mild frostbite most commonly occurs in the head region. Frostbite of the feet and hands frequently causes severe tissue damage and requires medical treatment or hospitalization. Young Finnish men reported a 2% annual incidence in frostbite over their lifetimes. Twenty-five percent was blister grade or more severe. In general, the incidence of frostbite varies annually from 0 to 27% among different outdoor occupations. Also, urban people experience more frostbite than rural people for the same thermal environments (Ramsey et al., 1983). Frostbites are comparable to burns in their immediate consequence. The immediate effect of frostbite can be a mild or more severe functional limitation of the injured area, requiring medical attention, and in some cases, hospitalization. The most common latent symptoms of frostbite are local hypersensitivity to cold and pain in the injured area, cold-induced sensations and disturbances in muscular function, and potentially excessive sweating. These latent symptoms may have negative impacts on occupational activities in 13 to 43% of cases. Permanent post-symptoms or invalidity commonly develop as a result of severe frostbite requiring hospitalization (Miller and Chasmar, 1980). Factors known to cause a predisposition to frostbite include cold-provoked white finger phenomenon, sensitivity to cold, diabetic vascular disease, psychiatric disorders, prior frostbite, older age, and tobacco smoking (Hassi and Mäkinen, 2000). Use of certain drugs or alcohol, "cold protective ointment" on the face, and inadequate clothing increase risk of frostbite during cold exposure

(Lehmuskallio, 1999). Accidents, fatigue, and poor nutrition are also associated with increased frostbite risk.

15.3.2.3. Cold-related diseases

Cold-related diseases are either caused by cold or are affected by cold exposure. The rate of cooling in different sites of the human body is also modified by individual factors like cardiovascular diseases, diseases of peripheral circulation, respiratory diseases, musculoskeletal diseases, and skin diseases.

Cardiovascular diseases

The higher incidence of cardiovascular events in colder regions and during winter is well known, and several mechanisms have been suggested based on increased blood pressure, hematological changes, and respiratory infections (Keatinge, 1991). Most investigations have used ecological data such as daily temperatures recorded at weather stations and mortality in the general population. Cause-specific mortality is the outcome measure most commonly used. Hospital discharge records, linked with out-of-hospital deaths, provide a powerful tool for detecting even weak effects of temperature. The association of coronary heart disease mortality and temperature is usually U-shaped, mortality being lowest within the range 10 to 20 °C and higher either side. However, the temperature at which mortality reaches a minimum is lower in colder countries (Fig. 15.7). For example, in Yakutsk, Siberia, temperatures as low as -48 °C had no effect on coronary mortality rates (Donaldson et al., 1998a; Näyhä, 2002).

The increase in mortality on the colder side is about 1% per 1 °C decrease in temperature, but the increase on the warmer side may be very steep. The exact point of the minimum temperature and the magnitude of the effect vary between countries. In Finland, the winter excess mortality from coronary heart disease has leveled off over recent decades. The share of annual mortality from cardiovascular diseases due to cold is estimated at 5 to 20%. The detailed mechanisms by which cold is related to cardiovascular mortality, either directly or by respiratory infections or indirect effects of winter behaviors such as shoveling snow, have not been clarified. Cold exposure causes an increase in blood pressure and hemoconcentration resulting from fluid shifts, leading to coronary thromboses one to two days after cold exposure. Following the recent decline in influenza mortality, around half the excess winter deaths are now due to coronary thrombosis. These peak about two days after the coldest part of a long period of very cold weather. Around half the remaining winter deaths are due to respiratory disease, and these peak about 12 days after maximum cold days.

Cerebral vascular diseases

The association of temperature and cerebral vascular accidents is similar to that for coronary heart diseases with morbidity and mortality increasing with a decline

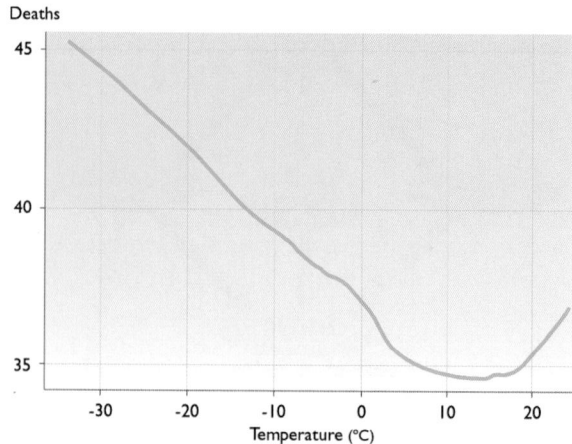

Fig. 15.7. Deaths from coronary heart disease and mean daily temperature in Finland, 1971–1995 (Donaldson et al., 2003).

in temperature. The pattern is often U-shaped, with some increase in numbers at warmer temperatures. The morbidity and mortality of stroke is usually lowest at temperatures of 15 to 20 °C, however some variations exist. In northeastern Russia stroke mortality only increases at temperatures below 0 °C (Donaldson et al., 1998b; Näyhä, 2002).

The gradient of cerebral vascular accidents against temperature is around 1% per 1 °C decrease in temperature, as for coronary heart diseases. In Japan, the dose–response relationship was similar for intracerebral hemorrhages and cerebral infarctions, whereas in Finland a greater winter excess was observed in the incidence of intracerebral hemorrhage than for other forms of stroke, but no gradient relative to temperature has been reported. A change in temperature of at least a two-day duration is needed for stroke mortality to rise, and the time lag between the temperature change and the maximal increase in mortality is estimated at one to four days (Donaldson and Keatinge, 1999).

The long-term trends in the effect of temperature on stroke have not been determined, but the seasonal amplitude of stroke deaths in Finland has diminished since the 1920s. The proportion of stroke-related deaths attributable to the cold season was estimated at 13% in the 1960s, but had diminished to 9% by the 1990s. A British investigation which reported a decline of 57% in the stroke-temperature gradient between 1977 and 1994 also suggested that the effect of environmental temperature on stroke is being modified by other external factors (Donaldson and Keatinge, 1999).

Respiratory diseases

Common respiratory cold-related symptoms are watery rhinitis, and as a consequence of constriction of the bronchi, asthma-like symptoms which include wheezing, coughing, and breathing difficulties. Deaths related to respiratory diseases, primarily pneumonia, increase significantly during the winter months. Watery rhinitis is a physiological irritation response to cold air inhalation and is harmless.

The prevalence of breathing problems provoked by exercise and/or cold weather is high among asthmatic subjects (81.6%) and significantly elevated among allergic subjects (45.1%) and people with chronic obstructive pulmonary disease (74.6%). For people with no known respiratory disease, the prevalence is 10.0%. The risk of chronic bronchitis and bronchitic symptoms at the population scale is elevated in outdoor workers in some populations, but is not elevated in regular recreational cross-country skiers, and the risk of developing asthma is not significantly elevated by regular exercise or work in cold climates. Constriction of the laryngeal area is a momentary reflex in response to cold air and is usually harmless. In very exceptional cases of the disease, known as cold urticaria, this phenomenon may be life-threatening. Air quality and behavioral choices such as smoking are also major influences on the incidence of respiratory diseases.

Peripheral circulatory diseases

The normal responses of the peripheral circulatory system to cold stress can be affected in individuals with vascular diseases. Thermal comfort and physical performance may be decreased and risk of cold injury may be increased. In advanced stages of peripheral arteriolosclerosis, blood vessels are narrowed. Further constriction caused by cold exposure may increase risk of frostbite. A reversible episodic constriction of the blood vessels in fingers and toes is a fairly common pathological response to cold exposure and is known as the Raynaud's phenomenon. Owing to the constriction of the blood vessels, the blood flow in fingers and toes is markedly reduced at temperatures colder than 10 °C. Originally, Raynaud's phenomenon was described as episodic white fingers provoked by cold or other stress factors, together or alone. The population prevalence is 5 to 30% and is related to gender, age, and region of residence (Maricq et al., 1993). As a clinically significant condition, it has a reported prevalence of 2 to 6%. Cold exposure in a patient with the condition may result in a cluster of different symptoms caused by transient constrictions occurring in the circulation of the heart, lung, kidneys, or brains. The symptoms may vary widely and can include migraine headaches, chest pain, and possible visual effects.

Cold urticaria

The most familiar and common abnormal skin reaction related to cold exposure is cold urticaria. It is usually a chronic condition and is often provoked by some other physical agent. Symptoms usually occur locally on exposed areas of skin. They sometimes appear during cold exposure but more frequently appear when the skin re-warms after cooling and then disappear again after 20 to 30 minutes. Fifteen percent of the population is subject to symptoms at some stage and the annual average prevalence in Finland is 2 to 4%. Cold urticaria lasts from months to several years. Prevalence of hospitalization for severely affected patients is only around one in 4000. In cold urticaria, skin reaction to cold exposure is characterized by erythema, swelling, wheals, or papules.

Other symptoms on cold exposure can be more severe, such as vertigo, headache, nausea, vomiting, tachycardia, dyspnea, flushing, faintness, or rarely, life-threatening anaphylactic shock.

Musculoskeletal diseases and symptoms

There is limited scientific understanding of the relationship between musculoskeletal diseases and cold. Extensor tenosynovitis has been described with windy cold exposure in temperatures from 0 to -25 °C (Georgitis, 1978). The increased incidence of tenosynovitis in female food industry workers was attributed to the low ambient temperature (Chen et al., 1991; Chiang et al., 1993). Local cold exposure in a frozen food factory was associated with a ten times higher incidence of carpal tunnel syndrome than in warm environments (Chiang et al., 1990). Symptoms of musculoskeletal diseases can vary, and include local or generalized feelings of pain and fatigue of muscles and joints. Low back pain, knee pain, and shoulder pain were significantly more common in cold storage workers than in a thermoneutral environment and were dependent on the duration of the work in the cold environment.

Cold-related immune effects

Cold temperatures and isolation can be immunosuppressive and, in humans who have over-wintered in the Antarctic, suppression of cell-mediated immunity is well documented (Ando, 1990; Muller et al., 1988; Tingate et al., 1997). The effect of sunlight-induced immunosuppression above (or concomitant with) this temperature/isolation induced immunosuppression remains to be determined.

15.3.2.4. Summary

Changes in the frequency or intensity of natural disasters or extreme weather events can have direct and indirect impacts on human health in the Arctic. In remote locations this is accentuated by a reduced capacity to respond to these events because of the isolated nature of the communities and the often limited health infrastructure present. The variability of such events is likely to increase with future climate changes. Changes in temperature have the potential to influence health in arctic communities in both negative and positive ways. With the low mean annual temperature in many arctic regions, the likelihood of heat events having large health impacts on the general population is low. However, the impacts of these events on individuals with respiratory problems and other conditions can be serious. Fewer colder days associated with winter warming in some regions may actually have several positive health impacts. Impacts of cold temperature are well known and increases in the length or magnitude of extreme cold periods in some regions may have significant negative impacts on the general population, especially for individuals with conditions making them more susceptible to such exposure. Under any climate change scenario, tem-

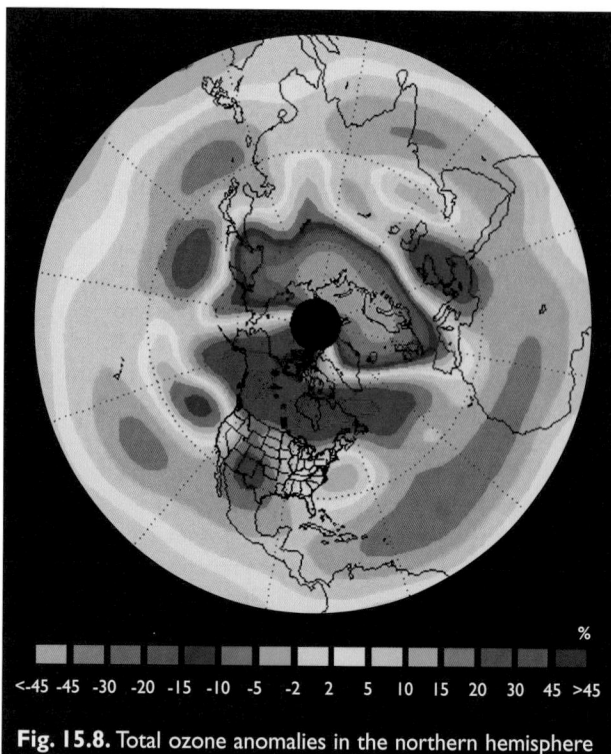

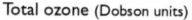

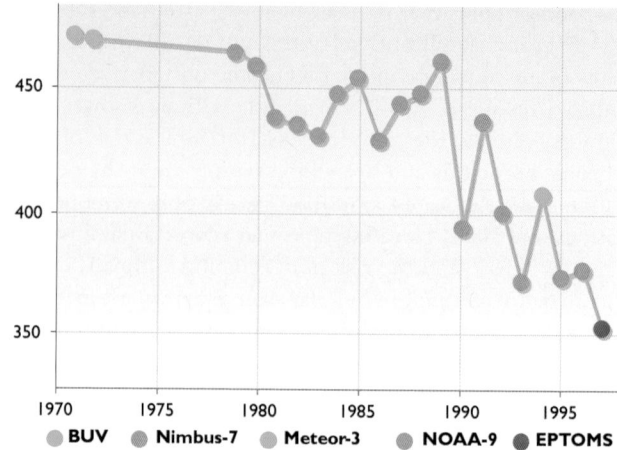

Fig. 15.8. Total ozone anomalies in the northern hemisphere for March 2003 relative to the mean March value for 1979 to 1986. Areas where the March 2003 value is within ±2% of the long-term mean are shown in light gray (data provided by NOAA, National Weather Service, 2003).

Fig. 15.9. Average total ozone over the Arctic (63° to 90° N) in March. 1971 and 1972 data are from the BUV instrument on the Nimbus-4 satellite; 1979–1993 data are from the Total Ozone Mapping Spectrometer (TOMS) on Nimbus-7; 1994 data are from the TOMS on Meteor-3; 1996 data are from the SBUV/2 on NOAA-9; 1997 data are from the TOMS on the Earth Probe satellite. (http://www.cmdl.noaa.gov/star/arcticuv2.html).

perature will continue to influence the health of arctic populations both directly and indirectly.

15.3.3. UV-B radiation and arctic human health

Stratospheric ozone loss has been observed during the winter/spring months over most of the Arctic since the early 1990s. Losses of up to 40% have been recorded in Scandinavia and Siberia, and in Canada, sporadic losses of 10 to 20% or more have been reported. The daily total ozone level in March 1997 at Point Barrow (71.3° N) in Alaska was about 6% below the previous ten-year average, and on 17–18 March 1999, Barrow experienced record low ozone levels for that location in March.

During winter and spring 2001/02, the mean March stratospheric ozone levels show a 5 to 15% loss of ozone compared to the average March value for 1979 to 1986. The 2002/03 winter also had low total ozone values over parts of the Arctic. The decrease was greater than for the previous two winters, but not as great as in the 1990s. During the winter months of 2002/03 (December, January, February, and March) parts of the Arctic, mainly but not limited to Siberia and Scandinavia, had levels up to 45% lower than comparable values for the same area in the early 1980s (NOAA National Weather Service, unpubl. data, 2003). Recent data indicate widely diverse ozone losses continuing throughout the year. Figure 15.8 shows the March 2003 anomalies for stratospheric ozone relative to the average March values for 1979 to 1986.

Figure 15.9 shows the large decline in average total ozone values in March over the Arctic (63°–90° N) during the 1990s. McKenzie et al. (1999) presented some of the strongest data to date regarding the relationship between ozone loss and increased levels of UV-B radiation. Although their data are for the southern hemisphere the same relationship is highly likely to occur in the Arctic. The data, which reflect ozone levels for the austral summers between 1978/79 and 1998/99 and UV-B radiation levels for the austral summers between 1989/90 and 1998/99 at 45° S (Fig. 15.10), provide strong evidence for increases in UV-B radiation levels in areas where baseline levels were already high, suggesting that man-made perturbations to the ozone layer are occurring as predicted (see also Chapter 5).

UV-B related human health effects include increases in the incidence of skin cancer, potential effects associated with increased suppression of the immune system including weakened resistance to some types of infectious disease (Sleijffers et al., 2002), and increased incidence of cataracts as well as changes in Vitamin D_3 production in the skin (IASC, 1995). In the Arctic, increases in UV-B radiation may also interact with other environmental stressors such as chemical pollutants, cold temperature, isolation, and viral illnesses in some populations (IASC, 1995). The rest of this section describes potential UV-B related health effects.

15.3.3.1. Immunosuppression

Ultraviolet-B radiation can initiate a selective down regulation of cell-mediated immunity in mammals, including humans. It is speculated that this may be a natural regulatory mechanism, selected through evolutionary pressure, to prevent autoimmune attack on sunlight-damaged skin (De Fabo and Noonan, 1983). The unusu-

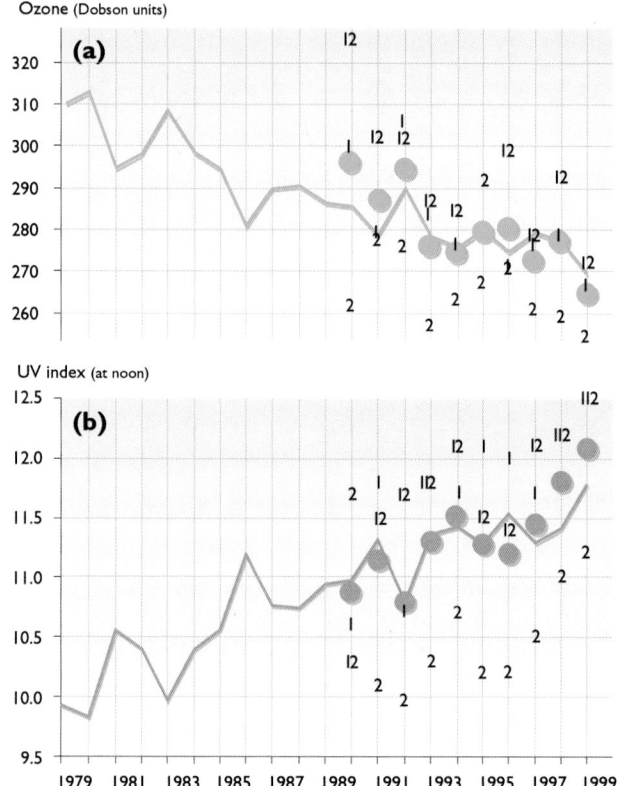

Fig. 15.10. Association between (a) ozone loss and (b) increased levels of UV-B radiation (Mackenzie et al., 1999).

al feature of UV-B-induced immunosuppression is that it redirects cell-mediated immunity to sensitizing antigens from an up or "effector" type response to a down or "suppressor" type response. These antigens can include chemical, viral, and tumor antigens (Noonan and De Fabo, 1992). Significantly, skin pigmentation is not an efficient protection factor against UV-B induced immune suppression. Immunosuppressive effects of UV-B radiation play an important role in UV-B induced skin cancer by preventing the destruction of highly antigenic skin cancers by the immune system (De Fabo and Kripke, 1979, 1980).

An in vivo wavelength dependence study by De Fabo and Noonan, (1983) for immune suppression of contact hypersensitivity in mice has identified a light-absorbing substance, residing on skin, as *trans*-urocanic acid (UCA). The *cis* structure of UCA (*cis*-UCA) formed upon absorption of light by *trans*-UCA, is known to cause immunosuppression in humans similar to that in mice caused by UV-B radiation (van Strien and Korstanje, 1995). In addition to sunlight modulation of cell-mediated immunity, which may involve susceptibility to certain infectious diseases (Sleijffers et al., 2002), *cis*-UCA may be important for arctic populations for another reason. A recent report indicated that a binding receptor for *cis*-UCA has been identified as the neurotransmitter 5-hydroxytrypta-mine, or 5HT (Serotonin) (Nghiem et al., 2002; Walterscheid et al., 2002). Lack of sunlight is known to play a role in mood disorders (Nilssen et al., 1997),

among other factors (Näyhä et al., 1994), in arctic populations (Nilssen et al., 1999). Future studies are needed on the role of UCA in human immunity, as well as on mood disorders linked to sunlight deprivation. Box 15.2 describes the action spectra and biological amplification of UV radiation.

Genetically determined susceptibilities to UV-induced immunosuppression have been shown to exist and appear to be controlled by several interacting *Uvs* genes involving autosomal and X-linked genes. Such an interaction for UV-immunosuppression had not been described previously and may be unique for this mechanism (Noonan and Hoffman, 1994). A genetically determined high susceptibility to UV-induced immunosuppression may be an important risk factor for UV-related human diseases not just in arctic populations but in other populations as well.

15.3.3.2. Skin cancer

Immunosuppressive effects of UV-B radiation play an important role in UV-induced skin cancer by preventing the destruction of highly antigenic skin cancers by the immune system (De Fabo and Kripke, 1979, 1980). There are three main types of skin cancer. Two tend not to metastasize and are known as basal and squamous cell carcinoma, and are often referred to collectively as non-melanoma skin cancer. The third type, which shows a higher mortality, and which can metastasize aggressively, is malignant melanoma of which several subtypes exist (Fears et al., 1976; McGovern et al., 1973). It should be noted however that any potential increase in skin cancer incidence related to reflectance from snow is likely to be mitigated by the projected decrease in snow cover.

There is much experimental evidence of a clear connection between sunlight exposure and non-melanoma skin cancer, and which implicates UV-B radiation as a carcinogen (Armstrong et al., 1997; Fears et al., 1976; Parsons P. and Musk, 1982). A relationship between sunlight and malignant melanoma, while less clear, is considered a near certainty (Armstrong and Kricker, 1993; Berwick, 2000; Bulliard, 2000; Fears et al., 1976; Jemal et al., 2001; Mack and Floderus, 1991). Epidemiological evidence indicates that sporadic or intermittent sunlight exposure can be a very important factor in malignant melanoma development, especially in childhood (Autier et al., 1997). But not all sunburn leads to melanoma, as other predisposing factors are needed. The molecular mechanisms underlying the relationship between malignant melanoma and exposure to UV radiation, particularly wavelength specific mechanisms, such as the importance of UV-B radiation, as opposed to UV-A radiation are, at present, unclear. To help clarify these mechanistic pathways, recent developments include, among others, the genetic engineering of a transgenic mouse capable of producing melanoma tumors following UV radiation of neonatal animals. These tumors show a striking similarity to human melanoma (Noonan et al., 2001). Once the active waveband for melanoma induction is identified, an

Box 15.2. Action spectra and biological amplification of UV radiation

Photobiological responses are by definition wavelength dependent. However, to compare the biologically-inducing activity of the many spectrally different sources available, from sunlight to sun tanning lamps, it is necessary to consider differences in wavelength efficiency in initiating the biological response, whether it is skin cancer, sunburn, photosynthesis, or immune suppression. In order to make such comparisons, it is necessary to calculate, and then deliver "biologically effective" doses from the optical source. Differences in wavelength efficiency can be accounted for by using an appropriate wavelength-dependence or "action spectrum". An action spectrum describes the relative efficiency of radiation at different wavelengths to produce a given effect. Health effects experts, for example, rely upon action spectra to provide information regarding which wavelengths in the full spectrum of sunlight or the full spectrum of artificial sunlamps cause sunburn, or DNA damage (Sutherland, 1995; Young et al., 1998), or immune suppression (De Fabo and Noonan, 1983). Once experimentally derived, the action spectrum can be multiplied by the spectral output of any given source. In the case of sunburn, the International Commission on Illumination action spectrum for erythema (McKinlay and Diffey, 1987) is used to calculate the UV Index, a measure of sun burning effectiveness used worldwide (Long, 2003).

Action spectra are also useful in determining increases in biologically effective UV radiation doses due to ozone depletion, known as the "radiation amplification factor" (RAF), and how these increases in UV-B radiation result in "biological amplification" for a given response. For example, to predict changes in skin cancer incidence as a function of stratospheric ozone depletion, two processes are necessary. First, the increase in biologically effective UV-B radiation that results from an ozone loss of 1% must be determined, i.e., the RAF. Second, the ratio of the percentage change in biological effect to the proportional change in biologically effective irradiance – the BAF – needs to be determined. Thus, the total amplification factor for the biological impact is a product of the two: amplification factor = RAF x BAF. More detailed information on ozone depletion, skin cancer, and RAF/BAF determinations is reported by Moan et al. (1989), Jones (1992), and Strzhizhovskii (1998).

In addition to providing a weighting function to determine biologically-effective doses, action spectra are useful for helping to identify the initial light-absorbing photoreceptor responsible for triggering a light-driven biological response (De Fabo, 1980; Noonan and De Fabo, 1993). Such information can help direct further research on a given photobiological response (De Fabo and Noonan, 1983).

action spectrum can be constructed. A skin cancer action spectrum has been used to predict increases in non-melanoma skin cancer by increased UV-B radiation resulting from ozone destruction between 1979 and 1994 (Slaper et al., 1996).

In the Arctic, skin cancer rates are in general low. This is due primarily to the low UV-B radiation levels relative to equatorial regions. Also, skin cancer is rare in arctic indigenous populations consistent with findings elsewhere that skin pigmentation is protective against skin cancer. A recent study, however, involving Danes working in Greenland and cancer risk indicated an elevated risk of melanoma in females. A role for excessive UV radiation exposure in this regard has been suggested (Nielsen L. et al., 1997). With increasing numbers of non-indigenous people living in the Arctic, the incidence of melanoma and non-melanoma skin cancer must be carefully monitored in both groups. Some indigenous groups in the Arctic are reporting evidence of increased UV-B radiation exposure and are experiencing skin rashes and burns for the first time. They report a sense that the "sun is hotter" (Fox S., 2002; Furgal et al., 2002; see also Chapter 3).

The sun burns us easily, it was not very hot in the past. Kuujjuaq, man aged 62 as quoted in Furgal et al., 2002

The sun was not that hot in the past. Nowadays, it's really hot. My skin burns when I'm out for a while. Sometimes, we stay indoors in a shack. Kuujjuaq, man aged 70 as quoted in Furgal et al., 2002

15.3.3.3. Non-Hodgkin's lymphoma

Certain epidemiological evidence suggests a link between non-Hodgkin's lymphoma and sunlight exposure (Langford et al., 1998; Zheng and Owens, 2000). This is suggested to be via the immunosuppressive effects of UV-B radiation (Langford et al., 1998; McKenna et al., 2000; McMichael and Giles, 1996; Zheng and Owens, 2000). A correlation between the occurrence of skin cancer and the occurrence of non-Hodgkin's lymphoma has also been described (Cliff and Mortimer, 1999). However, in contrast to non-melanoma skin cancer, non-Hodgkin's lymphoma does not show a latitudinal gradient in the United States, suggesting that UV-B radiation may be a co-factor rather than a primary causative agent of this disease. Danish women working in Greenland are reported to show an excess of lymphatic malignancies, which raises the question of a role for excess UV-B radiation (Nielsen L. et al., 1997). Autoimmune diseases such as Type-I diabetes and multiple sclerosis may also have an immunosuppressive connection with UV-B radiation

(Staples et al., 2003). The continued reports regarding correlations between health problems and UV radiation indicate the need for further investigation in relation to ozone loss over the Arctic (McKenna et al., 2003; Okada et al., 2003).

15.3.3.4. Cataract

Cataract is a major cause of blindness. In 2002, there were an estimated 180 million people worldwide who were visually disabled. Between 40 and 45 million people are blind (http://www.who.int/pbd/en). Epidemiological reports and experimental studies indicate that cataract formation is a complicated process with many associated risk factors. The precise mechanism of action is not known although UV-B radiation is very strongly implicated and associations with latitude and climatically different countries have been reported (Hockwin et al., 1999; Katoh et al., 2001; Sasaki H. et al., 2000; Sasaki K. et al., 1999; Taylor A. et al., 2002; Taylor H., 1989; West, 1999). Furthermore, a recent action spectrum indicated that after correcting for corneal transmittance, the biological sensitivity of the rat lens to UV-B is at least as great at 295 nm as at 300 nm. After correcting for transmittance by the atmosphere, UV-B at 305 nm is suggested to be the most likely wave band to damage the rat lens (Merriam et al., 2000).

Several types of cataract exist with a varying degree of association with sunlight (Hockwin et al., 1999; Katoh et al., 2001; Sasaki H. et al., 2000; Sasaki K. et al., 1999; West, 1999). Published evidence tends to support the concept that cortical cataract is more likely to be related to UV-B radiation (Hockwin et al., 1999; Katoh et al., 2001; Sasaki H. et al., 2000; Sasaki K. et al., 1999; West, 1999). Another study, however, suggests the opposite when lifetime cumulative UV-B radiation exposure and exposure after teenage years are considered (Hayashi et al., 2003), underscoring the need for further study including detailed wavelength or action spectrum studies on cataract. The importance of dietary factors and cataract also requires further research (Taylor A. et al., 2002; Valero et al., 2002; Wegener et al., 2002).

15.3.3.5. Vitamin D

Vitamin D (calciferol) is a fat-soluble vitamin. It is present in food, but can also be made in human skin after exposure to UV-B radiation from the sun. In general, skin synthesis provides most of the vitamin D to the body (80 to 100%; Glerup et al., 2000) and with adequate sunlight exposure, dietary vitamin D may be unnecessary (Holick, 2001).

Many factors can affect vitamin D production such as season, latitude, age, skin color, time spent outdoors, and sun angle. Few foods contain significant amounts of vitamin D that can act as a substitute for sunshine exposure. Fish with a high fat content, such as sardines, salmon, herring, and mackerel are excellent sources of vitamin D. Other important sources are meat, milk, eggs, and fortified foodstuffs. Fortified foodstuffs however, may not be sufficient to preclude the need for sunlight exposure. In one study it was reported that dark-skinned, veiled, pregnant women, their infants, and elderly in residential care had the highest vitamin D deficiency of subjects studied (Nowson and Margerison, 2002). This suggests that vitamin D deficiency may be a bigger risk factor in populations worldwide when other factors reducing exposure to sunlight are considered e.g., arctic populations during long, dark winters. A balance is thus needed between sunshine exposure and risk of excessive sunlight leading to skin cancer or other UV-B related health effects, or insufficient sunlight exposure and hence vitamin D deficiency for diseases such as rickets and certain non-skeletal diseases (Fahrleitner et al., 2002; Pasco et al., 2001).

Vitamin D exists in several forms, each with a different activity, and is involved in a large variety of biological functions, including regulatory functions (for a comprehensive review of vitamin D see Feldman, 2003). Briefly, the liver and kidney help convert vitamin D to its active hormone form (calcitriol). For example, in skin vitamin D is produced from UV-B-induced photoconversion of 7-dehydrocholesterol (7-DHC; provitamin D). Vitamin D then undergoes hydroxylation to calcidiol (25-OHD_3) in the liver and becomes the major circulating form, and then to calcitriol (vitamin D_3 hormone; 1alpha 25-(OH)_2D_3) in the kidney. 1alpha, 25-(OH)_2D_3 is carried systemically to distal target organs where it binds to a vast array of nuclear receptors to generate its appropriate biological response. Vitamin D appears, therefore, to be involved in various aspects of fundamental cell regulation.

The major function of vitamin D is to maintain normal blood levels of calcium and phosphorus. Vitamin D aids calcium absorption, helping to form and maintain a strong skeletal structure. Without vitamin D, bones can become thin, brittle, soft, or deformed. Vitamin D prevents rickets in children and osteomalacia in adults, which are skeletal diseases that result in defects that weaken bones. A recent study reports for the first time a link between UV-B radiation exposure and calcitriol synthesis in human skin (Lehmann et al., 2003).

Given the six months of darkness in winter followed in spring/summer by potential excess exposure to UV-B radiation due to ozone depletion, studies in arctic communities on UV-B induction of vitamin D in liver and kidney and vitamin D hormone in skin are of high priority.

15.3.3.6. Other factors

Pollutants

Many types of pollutant have been identified in arctic biota and the arctic environment, including polyaromatic hydrocarbons (PAHs), polychlorinated biphenyls (PCBs), and heavy metals (AMAP, 2002). There is some evidence of interaction between UV-B radiation and chemical pollutants. For example, aquatic organisms that have ingested UV-B-absorbing PAHs have been shown to exhibit phototoxic effects following exposure to UV-B radiation (Huovinen, 2000). A number of persistent organic pollutants, including PCBs, are immunologically damaging even without exposure to UV-B radiation. At present there is little information on the potential for the combined effects of these agents or on immunosuppression from exposure to UV-B radiation alone. In view of such findings and the high levels of persistent organic pollutants in areas of the Arctic, research is required on the combined effects of UV-B radiation, PAHs, and PCBs.

Dietary factors

Diet may also be important in UV-B effects in arctic populations. Experimental evidence shows that UV-related immunosuppression in mice can be increased by increasing dietary histidine (Reilly and De Fabo, 1991) and correlates with a modulating increase in *trans*-urocanic acid (De Fabo et al., 1997). High levels of dietary fat have also been demonstrated to enhance UV-induced immunosuppression in experimental systems (Black et al., 1995) and may be a factor in arctic populations as well as high histidine levels in some species of fish (Suzuki et al., 1987) and seal (Hoppner et al., 1978).

Viral interactions

Activation of viruses by UV radiation is well-documented (Ando, 1990; Coetzee and Pollard, 1974; Tingate et al., 1997; Zhang et al., 1999; Zmudzka et al., 1996). Indigenous populations appear predisposed to cancers such as nasopharyngeal and salivary gland cancer (Nielsen L. et al., 1997). These cancers are thought to be associated with the high viral load in these populations, and genetic factors also appear to be involved. Salivary gland cancer has been linked to UV radiation exposure in several studies on non-arctic populations (Nielsen N. et al., 1996) although a latitude gradient has not been demonstrated. UV-B radiation may be a cofactor. Herpes simplex virus (HSV) is found in all races, cultures, and continents. The virus affects 80 to 90% of the world's population. Most people have herpes simplex as cold sores or genital herpes. In experimental animal models UV-B radiation was shown to release immune inhibition on HSV expression and lead to HSV manifestation. A similar effect may occur in the human population (Taylor J. et al., 1994). Whether enhanced levels of UV-B radiation due to ozone depletion will exacerbate HSV or other viral infections in the arctic population remains to be determined.

Infectious diseases

The emergence of new infectious diseases and the re-emergence of old infectious diseases is also an issue in the Arctic (Butler et al., 1999). Given an association

between the lowering of resistance to some infectious diseases by UV-B induced immunosuppression in experimental animal systems (Sleijffers et al., 2002), plus a lowering of the cellular immune response against hepatitis B vaccination in humans by urocanic acid (Sleijffers et al., 2003; section 15.3.3.1), attention to increases in infectious diseases as a consequence of increased exposure to UV-B due to ozone depletion is of high priority.

15.3.3.7. Summary

A lack of detailed information on human health effects due to increased exposure to UV-B radiation in the Arctic precludes an evaluation of risk assessment at present. Skin cancer appears to be a low risk phenomenon in the Arctic particularly in indigenous populations. However, given the long-term likelihood of ozone loss and increased levels of UV-B radiation, it is not clear how long this low risk will be maintained. Indigenous and non-indigenous populations should both be monitored routinely for skin cancers, cataracts, and precursors to such conditions. The effect of UV-B induced immune suppression, alone or in combination with arctic stressors and pollutants, on human or animal populations is also unknown. The implications of increased UV-B exposure on the incidence of viral and infectious diseases in arctic populations need to be addressed. Also, understanding of the interactions between UV-B radiation, pollutants, and traditional diets requires significant effort in the future. Population effects related to UV-B health effects following stratospheric ozone depletion is a major research topic for the Arctic as the stratospheric ozone layer will remain vulnerable for the next decade or so even with full compliance with the Montreal Protocol (UNEP, 2003).

15.4. Potential impacts of indirect mechanisms of climate change on human health

In addition to direct impacts of climate-related changes on human health, human health in northern communities is affected by a number of indirect impacts. These refer to those health consequences resulting from indirect interactions mediated via human behavior and components of the environment that have changed or are changing with local climate. These indirect effects include critical impact mechanisms such as social and mental stress related to changes in the environment or lifestyle which are related to changes in local climatic regimes. They include such things as effects on diet (decreased subsistence/cultural food species abundance and/or availability) as a result of climate-related impacts on wildlife species or environmental factors influencing indigenous peoples' access to these resources (e.g., ice distribution, land stability, weather predictability); effects on health as a result of changed access to good quality drinking water sources; effects on rates of diseases resulting from climate effects on sanitation infrastructure; and changes in human disease incidence associated with climate impact on zoonotic diseases.

15.4.1. Changes in animal and plant populations

Climate change can have dramatic effects on the numbers and distribution of species in an ecosystem and these changes may have significant social, cultural, and health effects on indigenous human populations. All ecosystems are a dynamic equilibrium of species and climate is one of the most important factors in this delicate balance. Climate changes, even subtle changes, may shift this balance to favor some species, stress other species sometimes to the point of extinction, and allow new species to enter the system. Other chapters of this report address the significant and sometimes dramatic affects of climate change on species. Infectious disease agents of plants, animals, and humans are integral components of all ecosystems and may increase, decrease, or spread into new regions as a result of climate change. The dynamics of these changes and how they may affect human health are described in this section.

15.4.1.1. Species responses

Since the late 19th century, global temperatures have risen by an average of 0.6 °C with the increases more pronounced at higher latitudes and higher elevations (IPCC, 2001). A review of 143 studies involving range distributions of nearly 1500 species showed that over this period 80% had shifted their ranges toward the poles (Root et al., 2003). These shifts are likely to be greater in polar regions where the temperature increase has been greater.

Increased warming from the mid-1970s to the mid-1990s is thought to have been the major reason for a rise in salmon numbers in the Bering Sea and North Pacific (Weller et al., 1999). Salmon is an important traditional food for many indigenous people and an important economic asset in the Bering Sea and North Pacific. However, the increased temperatures also had some adverse health effects. The warming caused increased productivity at all levels of the food chain including increased numbers of those algae which cause toxic red tides and paralytic shellfish poisoning (Patz et al., 2000; Weller et al., 1999). Extreme environmental events such as flooding which washes nutrients into coastal waters could also increase the occurrence of paralytic shellfish poisoning.

The warming trend that caused high salmon numbers in the North Pacific and Bering Sea may also have been the major factor responsible for the dramatic reduction in Alaska salmon since 1997 (Mathews-Amos and Berntson, 2002; Weller et al., 1999). Higher water temperatures may have increased salmon metabolism to levels that could not be sustained by existing food sources and spawning salmon were often smaller than normal. A reduction or disappearance of traditional food species may result in indigenous populations switching from a traditional diet to less healthy diets; such dietary shifts are associated with an increased prevalence of chronic diseases such as diabetes, heart disease, and cancer among northern populations (Bjerregaard and Young, 1998). Health effects related to extreme economic hardship could also follow a decline in traditional food species.

Climate change often creates opportunities for species to move into new regions. Species may colonize a region from an adjacent area, through migration or through accidental or intentional transport by humans. In Alaska, recent warming has increased average growing degree-days by about 20% (USGCRP, 2000). This has had a significant impact on the ecosystems there. Sustained warming on the Kenai Peninsula in south-central Alaska encouraged a spruce bark beetle infestation and the largest tree death in North America. Northern expansion of the boreal forest has favored the steady advance of beaver in northern and western Alaska. Beaver make dramatic alterations in surface water and fish habitat. They may also expand the range of Giardia, a parasitic disease of beaver that can infect many other animals and humans. An expansion in the range occupied by beaver has also been reported by indigenous people in other regions of the Arctic (Furgal, 2002; Krupnik and Jolly, 2002).

> *We see moose now, and never did before. We see them more and more each year. We've seen them up as far as the bay north of here (Nain).* Nain, hunter aged 43; in Furgal et al., 2002

> *It was maybe twenty years ago that we saw the first beaver track ever up in the bay (Webb's Bay, north of Nain)…we didn't know what it was.* Nain, hunter aged 46; in Furgal et al., 2002

It is not unusual for a few individuals of a migratory species to deviate from normal pathways and arrive in a new region. Every year Alaskan bird watchers are thrilled at "exotic" Asian species that visit local bird-feeders. If the new ecosystem is favorable the species can become permanently established. Climate change may create favorable habitats in regions that could not previously be inhabited. Successful establishment of a new species alters the ecosystem and impacts upon other species (Root et al., 2003). Accidental introductions arising from human activities may also result in colonization. There is concern in Alaska about species transported into the region when ships discharge ballast waters. Climate change may create conditions in the marine environment that are favorable to species introduced in this manner. The marine ecosystem is vital to subsistence food users and the economies of Alaska and other northern regions. The introduction of competing species or diseases of existing species could be catastrophic for the fisheries and indigenous communities of the Arctic.

15.4.1.2. Infectious diseases

Infectious disease agents are part of any ecosystem and are also affected by climate change. Stresses posed by climate change may increase the susceptibility of plants and animals to disease agents causing both the rates of infection and severity of infection to increase. High incidences of diseases and die-offs were reported during recent El Niño–Southern Oscillation events (Mathews-Amos and Berntson, 2002; Weller et al., 1999). Climate warming during these events has been associated with sickness in marine mammals, birds, fish, and shellfish. Disease agents associated with these illnesses have included botulism, Newcastle disease, duck plague, influenza in seabirds, and a herpes-like virus epidemic in oysters. Diseases that attack species forming habitat for other species, such as seagrass, are also affected and can have devastating environmental impacts. If such effects arise following the temporary temperature shifts associated with El Niño–Southern Oscillation events, then it is likely that temperature changes arising from longer duration climate change will be associated with an increased occurrence of diseases and epidemics.

Species expanding into new regions will expose resident species to any disease agents they take with them. This is the case for beaver and Giardia. Giardia is a waterborne disease, and beaver dams increase the surface water habitat that promotes the spread of the parasite to other animals such as caribou and humans. Infected caribou may also increase the spread of the parasite as they migrate to other parts of their range. Other waterborne diseases are also likely to become a greater risk due to beaver-engineered surface water changes especially in areas where people use untreated surface water.

Species that benefit from climate changes are likely to increase in numbers and create more hosts for disease agents. Expansion of host species into new regions and the reformulation of ecosystems increase the probability of spread to new host species. Species that have not been exposed to a new infectious disease agent are often extremely susceptible to the infection. The catastrophic epidemics in indigenous North Americans following the introduction of smallpox, tuberculosis, influenza, and many other infectious diseases after European contact are such examples.

The West Nile virus is a recent example of how far and fast a disease agent can spread after colonizing a new region. The virus was first identified on the east coast of North America in 1999 and by the end of 2003 had spread to all lower 48 states except Washington and Oregon in the United States and seven of 13 Canadian provinces and territories (CDC, 2003; Health Canada, 2002). Nearly 9000 human cases and over 200 deaths occurred in the United States in 2003. West Nile virus is primarily a disease of birds that is transmitted by mosquitoes.

Infected migratory birds are responsible for the spread of the West Nile virus into a region, with mosquitoes responsible for the spread of the virus to other birds (and other animals and humans). Even though the virus originated in tropical Africa it has adapted to many North American mosquitoes and so far to over 130 species of North American bird, some of which migrate to the Arctic (CDC, 2003). Mosquito species known to transmit the virus are also found in the American Arctic.

Climate has always posed a barrier for insect-borne diseases but climate change and the extremely adaptive nature of the West Nile virus may favor continued northerly expansion. The extent of future expansion and the possibility of transmission to the Arctic remain to be seen. Some northern regions, such as Alaska, have initiated a West Nile virus surveillance program (ADHSS, 2000). If the virus does reach Alaska, migratory birds could carry it to the massive population centers of China and South Asia.

Zoonotic diseases are diseases of animals that can also be transmitted to humans. Rabies is a classic example. Rabies infects many wild and domestic animals which may then spread the disease to humans usually from saliva through a bite. In the Arctic, rabies is most often carried by fox and conditions which favor increased fox numbers can lead to rabies epidemics and to human cases. Other arctic zoonotic diseases that could be influenced by climate change include botulism, paralytic shellfish poisoning, tularemia, brucellosis, echinococcus, trichinosis, and cryptosporidium. Botulism spores occur in marine sediments and the intestinal tracts of animals and fish (Chin, 2000). In Canada and Alaska outbreaks have been associated with seal meat, salmon, and salmon eggs. Also, warmer temperatures are reported to influence traditional fermentation processes (for the preparation of fermented meat – *igunaq*) which are related to increased cases of botulism in Nunavik (Furgal et al., 2002). Alaska has established a paralytic shellfish poisoning surveillance program in areas where the algae responsible for the production of the toxin occur. Climate warming may favor the spread of these species to new regions.

The mechanisms by which climate change affects disease agents and their hosts vary. Warmer temperatures may allow species with a low rate of infection, such as brucellosis in caribou or echinococcus in voles, to survive in larger numbers increasing the number of susceptible hosts and infected animals. Disease agents that survive in soil and water may benefit by warmer temperatures. Climate change may stress host species reducing their resistance and increasing rates of infection. Host species may enter new regions bringing disease agents with them. Continental cross-over from the Americas to Eurasia and from Eurasia to the Americas, as with the West Nile virus and influenza, is also possible and perhaps even more likely with climate change.

Diseases that infect important traditional and economic species can also affect human health. Indigenous groups throughout the circumpolar north have reported a variety of abnormal conditions and diseases in salmon over recent years (Krupnik and Jolly, 2002). Of increasing concern are diseases of farmed salmon or other fish that enter Alaskan waters via ballast waters and then spread to wild salmon. Reduced salmon consumption by indigenous peoples in favor of less healthy foods could lead to increased levels of diabetes, cardiovascular disease, and cancer.

Diseases that could threaten arctic ecosystems and populations following climate change include those that already exist in the Arctic and those moving in from more southerly areas (see Box 15.3). The Arctic could also be a source of disease agents. For example, if West Nile virus reaches Alaska it could easily jump to Asia via established bird migration routes. Diseases of livestock and poultry could also expand through arctic transmission routes. Expansion of disease agents into new environments, in new hosts, and via new vectors encourages new adaptations through natural selection. These could include changes in virulence and resistance and so result in more dangerous strains.

Influenza is an example of a disease that could be disseminated by arctic bird species to populations in other parts of the world. Bird influenza viruses serve as a genetic reservoir for other animal influenza strains including those that infect humans (Webster, 2002). Migration can spread these viruses to other species and humans. Influenza is not the stuffy nose, headache, and upset digestive tract that people equate with flu. In each of the 11 epidemics of the last three decades influenza killed 20 000 to 40 000 Americans. The Spanish flu strain of 1918 to 1919 killed more than 20 million people. In the 1998 influenza epidemic in Hong Kong a third of all diagnosed cases died (Tam, 2002). Fortunately the outbreak was small and did not spread (see Box 15.4).

Extreme events, such as droughts, floods, and storms (see section 15.3.1), are also linked to changes in ecological systems resulting in bacterial proliferation and impacts on the availability of safe drinking water (Mayer and Avis, 1997). Abnormal rainfall events can trigger mosquito-borne disease outbreaks, flood-related disasters, and depending on existing water infrastructure and systems used in northern communities, contamination of the water supply with human and animal waste. Human health depends on an adequate supply of potable water. If climate change affects the availability of freshwater supply, sanitation systems and the efficiency of local sewage systems will also be affected. Changes in rainfall patterns may also force people to use poorer quality sources of drinking water, potentially increasing the risk of bacterial and other contamination. This is particularly important for those communities in which significant numbers of individuals still rely on traditional sources of drinking water. All these factors could result in an increased incidence of diarrheal diseases.

15.4.1.3. Summary

Arctic climate change is likely to have profound effects on living things and thus human health, both in northern communities and throughout the world. When traditional food species are affected, dietary patterns may shift to less healthy food choices and diseases such as diabetes, cardiovascular disease, and cancer are likely to increase.

Box 15.3. Zoonotic diseases, climate change, and human health

Zoonotic diseases are infectious diseases of animals that can be transmitted to humans. Humans can be infected by these diseases a number of ways. Rabies, tularemia, and many others are transmitted by direct contact with infected animals. Others such as Giardia are shed into the environment and humans are infected following environmental exposure such as consuming contaminated water. The transmission of other diseases may be more complex. Humans are exposed to the West Nile virus from mosquitoes that were infected from birds harboring the virus. Many zoonotic diseases exist in arctic host species. Climate change may influence the spread, proliferation, and transmission of these diseases to humans through a variety of means.

- Tularemia is a bacterial disease of many mammals including rabbits, muskrats, and beaver, as well as humans and could become an increased threat in stressed animals or by animals expanding into new ranges. Tularemia may cause a variety of symptoms in humans. The pneumonic form has a 30 to 60% fatality rate.
- Rabies epidemics in the Arctic are linked to the fox and the cyclical increases in fox populations (Dietrich, 1981). Climate changes which increase rodent and rabbit populations could be a factor in rabies epidemics.
- Brucellosis is a bacterial disease of many hoofed animals and carnivores. It can cause a wide variety of symptoms in humans and the fatality rate is around 2% in untreated cases (Chin, 2000). Bison, caribou, reindeer, foxes, and bears can carry the disease (Dietrich, 1981). Climate changes that affect the distribution of these species could increase or decrease the risk from brucellosis.
- Echinococcus is a tapeworm parasite of animals and humans. The natural cycle involves foxes or dogs and rodents. Humans are an incidental host when they ingest eggs passed by dogs or foxes. The parasite develops invasive destructive cysts in the abdominal cavity and the infection is often fatal (Chin, 2000).
- An arctic strain of trichinella occurs in marine mammals such as walrus (Dietrich, 1981) and accounts for trichinosis outbreaks in some northern regions (e.g., Nunavik). Climate warming that reduces numbers of marine mammals or eliminates them from their current range may decrease the threat and occurrence of human cases of trichinella.
- Cryptosporidium is a protozoan parasite of many animals and humans (Chin, 2000). Some recent epidemics involved contamination of community water sources. Climate changes could favor the spread of the parasite to arctic communities that consume untreated surface water.

Changes in the traditional food lifestyle are also likely to affect human health through changes in social and cultural activities. Changes that affect commercially important species and so increase or decrease local income can also affect human health. Infectious diseases of plants, animals, and humans are also affected by climatic changes. Owing to the indirect nature of these influences, predictions of their likelihood are not possible; however, the potential impacts on human health related to these changes clearly warrant further attention.

15.4.2. Changes in the physical environment

15.4.2.1. Ice and snow

Climate warming scenarios project changes in sea-ice distribution and ice thickness in the Arctic (see Chapter 4). The direct human health effects of a reduction in ice thickness include injuries and death. Travel over increasingly thin ice for fishing, hunting, or recreation activities becomes increasingly dangerous. Mortality statistics show that accidents cause a significant number of deaths in some Inuit populations (e.g., Hodgins, 1997). Inuit in northern Canada report a decrease in ice extent and thickness during key traveling and hunting times in some communities (Furgal et al., 2002; Nickels et al., 2002).

The indirect health effects related to these changes are associated with marine productivity. Sea ice has a major influence on primary production and the ecology of species such as seals, walrus, and polar bear (see Chapter 9). Ice algae are a major source of food for a wide range of zooplankton and crustaceans. A reduction in sea-ice extent would reduce the substrate available for the ice algae and so reduce the food source for the ice-associated zooplankton and crustaceans. Greater melting of sea ice would decrease the salinity of the water column and the rate of the vertical flow which brings nutrients up from

deeper waters (Conover et al., 1986), further reducing the productivity of the phytoplankton. A decrease in primary productivity would affect the crustacean and fish populations upon which seal populations rely (Welch et al., 1992). Also, populations of seals and walrus, which require sea ice for breeding and pupping, may decline as ice-covered areas recede (Maxwell, 1997). This would affect polar bear populations as seals are their major food supply. Polar bears in Hudson and James Bay are particularly vulnerable in this respect. Some species could be extirpated or become extinct if the Arctic Ocean becomes ice-free for much of the year (Maxwell, 1997). Arctic foxes, normally scavenging polar bear kills, may be forced to increase predation on nesting birds in the summer (Welch et al., 1992).

Cooling, as is projected for some arctic regions, could have both positive and negative impacts on indigenous food security. During the major cooling of the Arctic around AD 1400 to 1700, the Inuit are thought to have adapted by hunting ringed seals (*Phoca hispida*) which became more plentiful as the sea ice extended (McGhee, 1987). Similarly, changes in the timing, amount, and composition of snow can affect the health of arctic residents as it influences their abilities to hunt, travel, and access traditional foods at certain times of the year. The changes projected in ice and snow under the various climate change scenarios (see Chapter 4) could have potential impacts on individual and community social health and well-being (see Box 15.5).

> *…and all of a sudden we had a storm and everyone was lost, years ago you'd get the good snow for snow houses but now, you wouldn't be able to make one [snow house] if you had to.* Nain, hunter aged 49; in Furgal et al., 2002

Many villages in the Arctic are connected to other settlements only by sea or air. Because air service to the vil-

Box 15.4. Climate change, arctic birds, and influenza

Wild birds host many species of virus including influenza. Birds may serve as the source of genetic material or of new influenza strains that could infect humans.

The role and mechanism of wild birds in this process is now better understood (Snacken, 1999; Tam, 2002; Webster, 2002). Numerous influenza strains are found in wild water birds. The bird–flu relationship is very stable and most infections even with multiple strains cause no symptoms. Many millions of these birds with their associated influenza viruses inhabit the Arctic. Peak numbers in birds and viruses occur in autumn and move south during migration. As the huge flocks move southward influenza strains circulate and recombine within the migrating birds and infect domestic birds and swine along the flyways. Influenza in domestic species brings the viruses closer to humans. The right combination of strains in domestic animals could be the source of a new human strain and a global epidemic. Climate change that favors nesting success, expansion to new ranges, and intermingling of bird species could contribute to this process.

The Spanish flu strain of 1918 to 1919 that killed more than 20 million people was thought to have emerged from a swine or bird strain (Snacken, 1999). The Asian flu strain of 1957 and the Hong Kong strain of 1968 may have been the product of human and bird strains. The 1997 to 1998 influenza epidemic in Hong Kong originated in poultry (Tam, 2002). That epidemic, even though it was small with only 18 confirmed cases, was especially alarming because six of these cases resulted in fatalities.

lages in many regions is irregular, villages are otherwise isolated for two to five weeks every autumn and spring when there is too much ice in the water to go by boat but not enough ice to go by dog sledge or snowmobile. During this period, hunting, fishing, and travel between villages is limited by means other than plane and there is often reduced provision of goods to local stores, and reduced availability of fresh meat or vegetables in some communities. For short periods, the weather can become so stormy that normal everyday activities within the village are difficult. This increased sense of isolation is reported to be associated with increased incidences of interpersonal conflict, depression, and other forms of social stress (Furgal et al., 2002).

15.4.2.2. Permafrost

Permafrost is very sensitive to temperature fluctuations. The top layer of permafrost, known as the active layer,

thaws in summer and freezes again in winter. A warming of 4 to 5 °C would cause more than half of the discontinuous permafrost zone in Canada to disappear. Under such a scenario, the boundary between the continuous and discontinuous permafrost zones is expected to shift northward by hundreds of kilometers and the active layer in the discontinuous zone is projected to increase to twice its current depth (Maxwell, 1997). The uneven pitted terrain which results can severely affect animal activities and can damage or destroy the ecosystems based on the permafrost (Osterkamp, 1982, 1994). Several impacts of thawing permafrost have already been observed in Alaska (IASC, 1997).

- Destruction of trees and reduction in areas of boreal forest.
- Expansion of thawed lakes, grasslands, and wetlands.
- Destruction of habitat for caribou and terrestrial birds and mammals.

Box 15.5. Climate change and traditional food security

Traditional foods collected from the land, sea, lakes, and rivers are important sources of health and well-being to many indigenous communities. Such foods continue to contribute significant amounts of protein to the total diet and help individuals to meet or exceed daily requirements for several vitamins and essential elements (AMAP, 2003; Blanchet et al., 2000; Kuhnlein et al., 2000; Van Oostdam et al., 1999). Historically, by eating all animal parts, northern indigenous diets provided the nutrients and essential elements required to sustain life in this harsh climate. Such items are still important today as they contribute, for example, nearly 50% of the weekly protein, iron, and vitamin A intake in Nunavik women under 45 years old (Jetté, 1992) and nearly 50% in Labrador Inuit (Lawn and Langer, 1994). Components of the Inuit diet, particularly omega-3 fatty acids in fish oils, have been shown to provide protection against arteriosclerosis and ischemic heart disease (Bjerregaard et al., 1997) and some cancers. Also, marine species are the main source of selenium in northern diets; an antioxidant and a known anticarcinogen that may also help protect individuals from mercury toxicity (Blanchet et al., 2000). In addition to the substantial nutritional benefits, traditional foods provide many cultural, social, and economic benefits to individuals and communities.

Climate changes may affect the consumption of traditional foods by northern people through a variety of means. Impacts may occur via changes in access to food sources, for example by:

- a change in the distribution of important food species;
- the unpredictable nature of weather, as this can influence the possibilities for hunting or fishing;
- low water levels in lakes and streams, the timing of snow, and ice extent and stability, as these can influence access to hunting locations and key species; and by
- a shorter winter season and increased snowfall (two effects of a warmer climate), as these may decrease the ability of northern people to hunt and trap (Maxwell, 1997).

Climate changes may influence the availability and health of traditional food species via:

- impacts on critical components of their diet (e.g., climate impacts on vegetation may influence caribou health and abundance);
- impacts on their ability to forage and survive critical seasons (e.g., deeper snow and changes in freezing rain incidents can negatively affect the ability of caribou and reindeer to forage in winter);
- warming, as this may increase the exposure of some species to insects, pests, and parasites; and via
- temperature changes as these may influence migration and breeding patterns.

The impacts of a decline in the proportion of traditional foods consumed by northern peoples are significant. Shifts away from a traditional diet toward a more western diet, higher in carbohydrates and sugars, are associated with increased levels of cardiovascular diseases, diabetes, vitamin-deficiency disorders, dental cavities, anemia, obesity, and lower resistance to infections. Both climate warming and cooling are as likely to impact on aspects of indigenous food security in the future as they have in the past.

• Clogging of salmon spawning streams with sediment and debris.

Such impacts could further affect traditional food sources (e.g., caribou, salmon) and the related activities (e.g., hunting, gathering, fishing) (AMAP, 2003).

15.4.2.3. Summary

Through changes in the timing and conditions of ice formation, stability, and break up, the amount and timing of snow, and the stability of critical land (e.g., permafrost) in the regions used by indigenous communities, climate change can result in significant negative impacts on the health of community residents. According to some indigenous communities these changes and their effects are already occurring. Such changes are likely to continue to affect the safety of land- and water-based travel, and availability and access to traditional food species by arctic residents and will thus continue to challenge the health of indigenous communities in the future.

15.4.3. Built environments in the north

Infrastructure promotes safe and healthy community environments and provides access to health-related services. In northern regions, housing provides protection against harsh environmental conditions and, with adequate ventilation, a healthy living environment. Sanitation facilities are needed to prevent the spread of disease and are increasingly important when population densities are high, or when even small populations are fixed in one location. In small remote communities transportation is often necessary to gain access to healthcare or emergency services. It is likely that climate change will adversely affect infrastructure and housing throughout the Arctic.

15.4.3.1. Sanitation infrastructure

Unsafe drinking water combined with inadequate sanitation and hygiene is listed sixth in the top ten health risk factors leading to disease, disability, and death worldwide (WHO, 2002). The provision of high quality water can protect against chemical constituents and waterborne diseases such as hepatitis, gastroenteritis, typhoid, cryptosporidiosis, and giardiasis (American Public Health Association, 2001; Smith D. et al., 1996). Sufficient quantities of water are required for personal hygiene, cleaning, and laundry. Epidemics of otherwise commonly preventable diseases such as hepatitis A, hepatitis B, bronchitis, otitis media (a serious ear infection), impetigo, and meningitis in remote Alaskan communities are often attributed to poor sanitation. Alaskan Native villages with inadequate sanitation systems accounted for more than 72% of 596 reported cases of hepatitis A in Alaska in 1988 (US OTA, 1994). The spread of diseases caused by contaminated drinking water or inadequate sanitation is a concern for communities throughout the world.

Sanitation facilities can consist of individual facilities such as septic systems or pit privies (outhouses) or community facilities such as organized haul systems or pipeline networks. The level of health in a community depends on the type of sanitation facilities (US OTA, 1994). Water hauled by individual residents may be safe at the point of collection, but might become contaminated in the containers used for transport and storage. Closed haul systems (sealed containers) or piped utilities can reduce the potential for contaminating water supplies or human contact with sewage (US OTA, 1994). Sanitation facilities can include different levels of service. In the most basic form, water and wastes are hauled to and from the residence by hand. Providing a community water and wastewater haul system can raise the level of service. Such systems provide greater amounts of water for sanitation purposes and therefore improve the level of community health. Piped utility systems provide the highest level of service and commensurate health benefits.

The level of sanitation service provided to arctic communities varies. For example, few Russian arctic communities are served by piped systems, while virtually 100% of Norwegian arctic communities have piped systems or adequate individual facilities. In Canada, Greenland, and the United States, the level of service provided to communities varies substantially from region to region. In some cases, communities must support utility operation via local user fees. In other cases, the cost of utility operation is supplemented by sources outside the community. In the Arctic, where residents may rely heavily upon subsistence and where economic conditions are often strained, even minor increases in costs may negatively affect utility operation and maintenance.

15.4.3.2. Water supply systems

Water supply systems include a water source, storage facility, and distribution system. Water sources contaminated by biological, chemical, or mineral constituents may require treatment to render the water supply safe for human consumption. In the United States it is estimated that contaminated drinking water causes more than 900 000 people to become ill and up to 900 to die each year (American Public Health Association, 2001). In 1993, inadequate water treatment in one city caused an outbreak of approximately 403 000 illnesses, 440 hospitalizations, and 50 deaths (Craun et al., 2002).

Water supplies are required for personal hygiene, cleaning, drinking, and cooking. Owing to the labor involved, when water is hauled individually it is used for drinking and cooking and tends to be used sparingly for hygiene and cleaning (Smith D. et al., 1996). In more sophisticated sanitation systems, water is used to transmit human waste from residences through pipelines or via haul tanks to the point of treatment and/or disposal.

Source

Water sources exist as surface supplies or groundwater wells. The highest quality source with adequate quantity available to the local community is typically used. Sources with significant levels of contaminants require more complex treatment and are undesirable due to increased treatment costs and complexity.

Arctic surface water sources include streams, rivers, lakes, tundra ponds, or man-made impoundments that capture snow and rain. Surface water sources require some form of treatment to ensure that the water is safe for drinking owing to the potential for contamination by pathogens (US EPA, 1992). Groundwater sources generally have less risk of contamination by pathogens. Naturally occurring organic or inorganic substances can exist in surface water and groundwater supplies. These contaminants are often removed early in the treatment process to avoid the possibility of transforming them – through chemical reactions with disinfectants – into potentially dangerous byproducts.

Although high quality arctic groundwater sources exist, permafrost often restricts the volume these aquifers can produce. Alternatively, there might be a sufficient volume of groundwater to meet a community's needs but it may be so highly mineralized that it requires sophisticated and costly treatment.

Climate change can cause water sources to become inadequate in volume or unfit in quality in a number of ways. For example:

- Groundwater supplies can be reduced by less frequent precipitation. Intense but less frequent rainstorms limit aquifer recharge by the majority of the water being lost to runoff.
- Drought or short intense storms can affect surface water impoundments. The water supply in small impoundments or lakes can be depleted during long dry periods. Intense storms may cause watersheds to release water too rapidly creating high but short-duration flows and with most of the precipitation being lost to runoff.
- Coastal communities can experience increased levels of salinity, dissolved solids, or other contaminants in groundwater due to a climate change induced sea-level rise (Linsley et al., 1992). The groundwater may become brackish and unfit for human consumption.
- Flooding of coastal areas by storm surges may become more frequent and severe. Tundra ponds or lakes, located near the coast, can become contaminated by seawater with their water becoming brackish and thus unfit for human consumption.
- Levels of salinity and bromide may increase in river intakes due to rising sea levels. The saline wedge can penetrate farther upstream potentially contaminating river intakes with seawater (Smith O., 2001).
- Intense storms can create high runoff rates that may exceed the design capacity of a water diversion or dam overflow structure. This may result in damage or the complete loss of the facility.

- Thawing permafrost may also damage water diversion or dam structures. As permafrost thaws, structures founded on frozen soil can become unstable. This may compromise the ability of the structure to impound or divert the necessary volumes of water.
- Flooding caused by ice jams in northern rivers often occurs in the spring or early summer when riverbanks are frozen. Flooding of rivers in late summer or autumn is rare. Intense rainstorms that cause flooding when soils are thawed will accelerate riverbank erosion and increase the potential for damage to adjacent structures.
- Northern communities often have limited economies – many based on subsistence. Contaminated water sources or damaged intake structures will require repair, modification, or replacement. If resources are not available for repairs or facility replacement, residents may be forced to use an unsafe water supply.

Treatment

Water treatment systems are designed to remove contaminants and inactivate pathogens. Designs differ, and are based on the properties of a water source. Climate change can result in a decrease in the quality of a water source which can then overwhelm the treatment system. The treated water may be safe for consumption but unpalatable due to taste, smell, or color. This can result in residents seeking alternative sources that are untreated and potentially unsafe. Climate change can adversely affect water treatment systems in a number of ways. For example:

- Rising sea levels can contaminate groundwater or surface water sources. A consequent rise in bromide concentrations may increase the formation of dangerous byproducts during disinfection (Singer, 1999). The process required to treat a water source contaminated by bromide is complex and costly; operation and maintenance of such systems may be prohibitively difficult and expensive for small, remote communities.
- Intense rainstorms can increase turbidity, pathogen, and organic contaminants in a water source. A substantial increase in these contaminants can exceed the ability of a water treatment system to produce safe and palatable water. High levels of suspended material can overwhelm a filtration process and reduce the effectiveness of a disinfectant to inactivate pathogens. Increased levels of organic contaminants can overwhelm a treatment process and increase the formation of dangerous disinfection byproducts (Singer, 1999).
- Warming weather and longer dry periods can cause more frequent and severe algal blooms in lakes or ponds used as a surface water supply. Algae may clog water treatment filters and so reduce the ability of the system to meet demand. The presence of algae can also increase the formation of disinfec-

tion byproducts or cause foul tastes and odors making the water unpalatable (Linsley et al., 1992; Singer, 1999).

Distribution

Water distribution in northern communities consists of self-haul, community-haul, or piped utility systems. Self-haul systems require minimal infrastructure because water can be hauled by foot, sled, or small all-terrain vehicle (Fig. 15.11). Community-haul systems use larger haul containers, which require larger vehicles and therefore substantial all-weather access ways. Self-haul and community-haul systems both require convenient access to the bulk treated water storage tank to be viable.

Access roads and boardwalks must be maintained in passable condition for haul systems to operate (Fig. 15.12). The road, bridge, or boardwalk must be structurally sound and capable of supporting relatively heavily loaded vehicles under repetitive daily cycles of operation.

Piped utilities rest on aboveground supports or are buried below ground. The more desirable and conventional belowground installation requires thaw-stable soils. When thaw-stable soils do not exist, piped utilities are usually constructed aboveground to minimize the potential for thawing of the permafrost and subsequent loss of foundation support for the structure (Fig. 15.13).

When a piped distribution system is used, pipelines must remain sound to ensure the water supply remains safe for human consumption. A breech in a pipeline can allow contamination of the water to occur within the distribution system (Geldreich, 1992). In 1989, contamination of the water supply caused by a pipeline breech in Cabool, Missouri, resulted in 243 people becoming ill and four dying (Fox K., 1993).

Often, arctic piped distribution systems continuously circulate water for freeze protection. Loss of water in a distribution system during cold weather can result in loss of circulation and complete freeze failure of the system (Smith D. et al., 1996).

Large structures such as water storage tanks cannot accommodate significant movement of the foundation. Movement of the foundation or loss of foundation support can cause a breech in the shell of a water storage tank, potentially rendering the facility unusable.

Climate change, can adversely affect water distribution systems in many ways. For example:

* Flooding caused by storm surges or heavy rainstorms can damage roads, boardwalks (Fig. 15.14), water storage facilities, and aboveground pipelines. In some areas, floodwaters can include ice, which may substantially increase floodwater damage.
* Roads, boardwalks, pipelines, and water storage facilities can be adversely affected by erosion. Riverbank erosion may accelerate during late season flooding. Coastal communities may experience accelerated erosion along shorelines due to thawing permafrost, severe storms, rising sea levels, or reduced periods of sea-ice cover.
* Frozen seas protect shorelines and reduce the generation of waves created by severe winter storms. Indigenous people report that the extent and duration of sea-ice cover is changing (e.g., Furgal et al., 2002; Krupnik and Jolly, 2002; Nickels et al., 2002; see Chapters 3 and 6).
* Thawing permafrost can result in the loss of foundation support for aboveground or belowground pipelines, water storage facilities, access roads, or boardwalks. Loss of foundation support for a pipeline can damage the facility and allow contamination of the water supply (Geldreich, 1992). Damage to storage facilities, access roads, or boardwalks can render a water distribution system inoperable.

15.4.3.3. Wastewater systems

Wastewater systems transport human waste from residences, provide treatment, and dispose of effluent. Improper methods of collecting, treating, or disposing of human waste have been attributed to numerous

Fig. 15.11. All-terrain vehicle water haul system, Mekoryuk, Alaska (photo by Mark Baron).

Fig. 15.12. New boardwalk access, Chefornak, Alaska (photo by John Warren).

Fig. 15.13. Aboveground water and sewer utilidor, Selawik, Alaska (photo by John Warren).

Fig. 15.14. Boardwalk damaged by storm surge flooding, Kipnuk, Alaska (photo by Mike Marcaurele).

outbreaks of infectious disease (US OTA, 1994). In Sweden, 3600 people became ill at a ski resort through a cross connection between a drinking water reservoir and a sewage pipeline (Fewtrell and Bartram, 2001). In Alaska, between 1972 and 1995 more than 7000 cases of hepatitis A were reported to the Epidemiology Section of the Health and Social Services Department. The method of transmission was via the fecal–oral route. Inadequate sewage disposal in many remote communities was cited as the cause.

The level of service provided by wastewater collection, treatment, and disposal systems varies throughout the Arctic. In some remote Alaskan villages, residents use small buckets to collect human waste. Buckets are then carried by hand to central disposal points where wastes are dumped into receptacles (Fig. 15.15), or carried directly by the resident to sewage disposal facilities. This is often referred to as a "honey bucket" haul system. A plastic liner is often used to line the bucket and contain the waste. Hauling wastewater from residences by hand is the most unsanitary form of collection and represents the lowest level of wastewater service.

Improved levels of service include pit privies and holding tanks. Pit privies are frequently used when homes are scattered and soil and groundwater conditions are favorable. Holding tanks are used when homes are located in close proximity, or when soil conditions or high groundwater makes septic systems infeasible. Holding tanks are sited at residences and emptied by a community-owned or commercial pumping service. The holding tank size in a particular community is determined by economics and access. Because many villages have narrow roads and boardwalks, large vehicle access is limited and holding tank volumes are typically small (Fig. 15.16). Thus, the amount of water available for personal hygiene and cleaning is minimal. Piped utilities provide the highest level of service. Flush toilets are normally used in piped systems, and water supplies and wastewater removal systems can provide ample water for personal hygiene, cleaning, laundry, or other sanitation needs.

Collection

Wastewater collection systems are designed to minimize the potential for human contact with sewage. Disease transmission can occur in populations where collection systems are inadequate and contact with wastewater is not controlled (IHS, 1999a; Schliessmann et al., 1958). Failed collection systems can discharge human waste to the environment, contaminate water supplies, and transmit disease via human contact. Many of the effects of climate change on water distribution systems also apply to wastewater collection infrastructure. Roads must remain in passable condition throughout the year for haul systems to operate; pipeline integrity must be maintained for piped wastewater collection systems to function properly.

Treatment and disposal

Wastewater treatment for small remote arctic communities is generally limited to simple systems. Mechanical treatment methods, such as aeration, are not typically used due to cost and complexity of operation (Smith D. et al., 1996).

In the Arctic, individual wastewater treatment facilities include pit privies and septic systems. Community facili-

Fig. 15.15. Honey bucket disposal container, Shishmaref, Alaska (photo by John Warren).

Fig. 15.16. Small sewage haul system, Mekoryuk, Alaska (photo by Mark Baron).

Fig. 15.17. Lagoon waste washed through fence by storm surge flooding, Kipnuk, Alaska (photo by Brian Aklin).

ties typically include earthen lagoons, tundra ponds, septic tanks with ocean outfalls, and septic tanks with drainfields. When communities are located near water and favorable soils exist, drainfields are placed within the thaw bulbs of rivers or along seashores.

Wastewater treatment and disposal systems are designed to ensure wastewater remains separate from the water supply, human contact with waste does not occur, and the potential for vector transport is limited.

Climate change can adversely affect wastewater treatment and disposal systems in a number of ways. For example:

- Flooding caused by storm surges or swollen rivers can adversely affect wastewater lagoons, tundra ponds, bunkers, pit privies, or septic systems. Floodwaters may enter these facilities and spread partially treated waste throughout communities or into water supplies (Fig. 15.17).
- Riverbank or shoreline erosion can damage wastewater facilities located along seashores or within the thaw bulb of a river. Erosion can also intercept wastewater lagoons and tundra ponds.
- The warming of ice-rich permafrost beneath lagoon dikes can cause the loss of structural support. As dikes settle, a breech may occur resulting in the discharge of human waste into the environment. In such circumstances, increased maintenance – at a minimum – is required to sustain operational wastewater volumes and treatment efficiencies.
- However, warming weather and longer summers can also increase biological activity in wastewater lagoons and natural tundra ponds used for treatment. This increase in biological activity can improve treatment efficiencies resulting in increased treatment capacity and potentially delaying the need to expand or replace facilities due to community growth.

15.4.3.4. Solid waste systems

Solid waste collection and disposal in the Arctic is performed with relatively conventional methods. Recycling, incineration, and baling facilities are rare and generally limited to larger communities. Collection in very small communities is typically by self-haul. Larger communities often use community-haul systems, which are preferred because wastes are more likely to be discarded in proper locations.

Solid waste disposal sites in the Arctic are generally frozen. Landfill wastes are often inadvertently mixed with snow due to winter operations and later covered with soil. Therefore wastes remain stable as long as materials remain frozen.

Many of the effects of climate change on water and wastewater systems also apply to solid waste collection infrastructure. Landfill access routes must remain passable for collection and disposal to occur. Flooding and erosion of solid waste landfills can spread waste, contaminate water supplies, and transmit disease through human contact. As frozen solid waste materials thaw, they can release contaminants into the environment through runoff.

15.4.3.5. Building structures

Critical requirements for housing in the Arctic include efficient and dependable heating and adequate ventila-

Fig. 15.18. Access road damage by coastal erosion, Shishmaref, Alaska (photo by Tony Weyionanna).

tion. Northern economies are often limited and houses are typically small, and so less costly to construct and heat. Such conditions may result in overcrowding. Also, building envelopes are usually tightly constructed to reduce heat loss and minimize heating costs. This combination of overcrowding and poor ventilation can lead to poor indoor air quality and potentially unhealthy living conditions.

Indoor air pollutants dangerous to health include mold, radon, tobacco smoke, carbon monoxide, and chemical emissions from household products and furnishings. Adverse short-term health effects of poor indoor air include asthma, hypersensitivity pneumonitis, and humidifier fever. Long-term health effects include respiratory disease, heart disease, and cancer.

Climate change can result in the destruction of housing via flooding and erosion, through loss of foundational support due to thawing permafrost, or by severe storm damage. Lost housing can cause further overcrowding, respiratory illness, and mental stress in small communities.

Health service buildings within remote communities offer quick access to basic health care and emergency services. When these structures are damaged by climate change, community health care is disrupted.

Transportation infrastructure

Transportation infrastructure provides access to health services located outside remote communities. Emergency transport of individuals from northern communities, usually by medical evacuation, to sites where comprehensive health services exist is critical. Infrastructure such as airstrips, roads (Fig. 15.18), and docks can be damaged by climate change. This damage can limit access to critical health care and emergency services.

15.4.3.6. Summary

The potential effects of climate change include increased variability in precipitation, reductions in the extent of sea ice, and climate warming and cooling. These changes can increase the frequency and severity of river and coastal flooding and erosion, drought, and degradation of permafrost. Such changes are very likely to impact on arctic infrastructure and housing.

Water sources may be subject to saltwater intrusion and increased contaminant levels, which may overwhelm treatment processes and jeopardize the safety of drinking water supplies. The quantities of water available for basic hygiene can become limited due to drought and damaged infrastructure. The incidence of

---

Box 15.6. Infrastructure, climate and public health: Shishmaref, Alaska

Recent studies indicate reductions in sea-ice thickness and shorter periods of sea-ice cover in Alaska. These climatic changes have increased shoreline erosion of Shishmaref, an Iñupiat village of 560 people located on Sarichef Island in the Chukchi Sea.

Archaeological finds indicate centuries of human habitation in Shishmaref. Villagers state that erosion has been accelerated by soils no longer frozen, limited sea-ice cover, and increasingly violent weather. During the winters of 2000/01 and 2002/03, erosion caused the bluff line to recede more than 12 meters, destroying structures and forcing other homes to be moved. Several other communities in the region, including Kivalina, Wainwright, and Barrow, face similar challenges related to coastal erosion.

Shishmaref uses two different haul systems for water and wastewater transport. One system consists of small water and sewage holding tanks within homes that are filled and emptied by small motorized haul vehicles operated by the community. Remaining residents haul their own water and use a community-operated honey bucket haul system for waste disposal. The community's water source is a rain/snow catchment impoundment located near sea level. Water is treated by a simple unaided sand filtration system with activated carbon enhancement.

Recent storms in Shishmaref have prompted evacuations, washed away a number of boats used for subsistence, and destroyed buildings and roads. Erosion is threatening the sewage lagoon and the village sanitation system, and local officials are concerned that seawater has contaminated the community's water source (Anchorage Daily News, November 23, 2003, Severe Storms Pound Shishmaref).

The destruction, property loss, threatened sanitation systems, and realization that the community will ultimately have to be moved is creating significant stress within the community. In summary, climate-related changes in Shishmaref will have both short-term and long-term effects on human health.

Coastal erosion damage, Shishmaref, Alaska (photo by Curtis Nayokpuk).

disease caused by contact with human waste can increase when sewage is spread by flooding, damaged infrastructure, or inadequate hygiene. Damaged infrastructure increases repair costs and further stresses fragile arctic economies.

The positive effects of climate change include reduced heating costs for buildings and pipelines. Treatment efficiencies in wastewater lagoons may also improve due to warmer water temperatures resulting from longer periods of warm weather. This increased efficiency may delay the need to expand natural wastewater treatment systems as local populations grow. As an example, Box 15.6 summarizes the links between infrastructure, climate, and public health in Shishmaref, Alaska.

15.4.4. Contaminants

Human and ecosystem health in the Arctic is affected by the accumulation of heavy metals and biologically persistent man-made compounds of industrial and agricultural origin, known as persistent organic pollutants (POPs). These contaminants are mostly generated at lower latitudes and transported by various natural mechanisms (termed contaminant pathways) to the Arctic. They then enter the arctic food chain, and are ultimately consumed by human residents who are often highly dependent on wildlife for food.

15.4.4.1. Human health effects

Health effects of chronic low-level exposure to POPs and heavy metals, such as lead, mercury, and cadmium, are, in general, incompletely understood. An assessment on this topic was published recently by AMAP (2003). Several points require emphasizing:

- in terms of low-level chronic food-borne exposure, pregnant women, the developing fetus, and the developing infant are the most sensitive stages of human life;
- the exposure is to a mixture, never a single compound, making assignment of cause and effect very difficult;
- toxicological models and wildlife studies suggest that neurodevelopment, growth, immunological development, and endocrine function are the most likely targets for effects from exposure;
- sensitivity to these compounds varies widely in wildlife and laboratory species, and is not always useful in predicting the toxicity of tissue levels in humans;
- the developmental effects potentially attributable in human infants exposed to these compounds can also be caused by many other exposures, confounding study results;
- arctic communities are often small, making statistically significant sampling difficult;
- long-term studies are further complicated by difficult access to remote communities for follow-up; and

- the arctic marine subsistence diet is rich in antioxidants such as selenium, omega-3 fatty acids, and other micronutrients. Selenium has the potential to mitigate the toxicity of mercury.

15.4.4.2. Major transport pathways

Major contaminant transport pathways include winds, ocean currents, and river outflow, all of which are affected by climate. Important mechanisms within the Arctic also affect contaminant transfer, such as surface ice movement, thawing of permafrost and glaciers, season length, changes in freshwater lakes, and the partitioning of chemical compounds between gas, liquid, and solid phases. Migratory species, which spend significant parts of their life cycle at lower latitudes, can accumulate these contaminants and bring them into the Arctic. Also, increased levels of human activity in the Arctic, including maritime transport, represent a potential mechanism for contaminant transport and release. Each of these mechanisms is reviewed in this section within the context of potential impacts on human health. It must be understood that all these pathways, and their impact on transport, have the same limitations as climate models: (1) it is difficult to detect trends, due to short instrumental records, and (2) linking change in the various pathways to climate change, and predicting their effect on each other is poorly understood in many cases. A more complete discussion on this topic was published recently by AMAP (Macdonald et al., 2003).

Season length

The length of time air masses remain at, or above, critical temperatures influences the extent to which an organic contaminant can volatilize and remain easily within the gas phase, or attached to airborne particles. Air mass movements can be rapid and cover long distances. With the projected increase in season length, contaminant movement could significantly increase, both into and out of the Arctic. Season length, by its effect on ice melt, the active layer of permafrost, periods of ice-free river flow, and longer growing periods has the major mediating influence on contaminant movement into, out of, and within the Arctic (Macdonald et al., 2003).

Atmospheric transport

Atmospheric circulation in the Northern Hemisphere is influenced by atmospheric pressure. Established patterns of pressure variation, including the Northern Hemisphere Annual mode, often referred to as the Arctic Oscillation (AO) have major influence on surface wind (Wallace and Thompson, 2002).

The AO, while a major influence on arctic climate, is thought to account for only 20% of variance in atmospheric pressure (see Chapter 2). Short (5- to 7-year) and longer term (50- to 80-year) variations in the sea-level atmospheric pressure fields, possible contributions from greenhouse gas warming, and the lack of long-term instrumental records make interpretation of AO variations since the 1960s difficult (Fyfe, 2003; Wang and Ikeda, 2001). Major winds into the Arctic fluctuate in intensity, duration, and to some extent, direction based on atmospheric pressure fields.

Various forms of precipitation, including snow and rain, act to remove contaminants from transporting air masses, and add them to land, surface ice, snow, and surface water. Increasing precipitation as a result of climate change could result in increased levels of contaminant deposition (Li et al., 2002).

Winds are the major source of mercury to the Arctic, mostly in its metallic gaseous form (AMAP, 2004). Loss of ozone in the upper layers of the atmosphere allows increased levels of UV-B radiation to reach the earth's surface in the Arctic. This initiates a complex mercury–halide interaction changing mercury to reactive gaseous mercury, which is easily removed from the atmosphere and taken up by the ecosystem in the early spring during the intense growth period. This atmospheric removal of mercury, termed a mercury depletion event, is most prominent during the period of "polar sunrise" in spring. The combination of possibly increased transport of gaseous mercury, less atmospheric ozone to block UV-B radiation, earlier ice-free periods along the Arctic Ocean shoreline, and abundant supplies of bromine and chlorine salts, could impact upon food web and human mercury accumulation. Extreme events might also increase rates of transport and contaminant deposition (see Fig. 8.22).

Ocean currents

Contaminant transport by ocean currents occurs primarily at the surface (Morison et al., 1998) and the largest input is from the North Atlantic between Greenland and Norway, with a smaller input from the North Pacific via the Bering Sea. A significant input from freshwater occurs via rivers entering the Arctic Ocean. The AO and season length influence these inflows.

A large volume of water and ice exits the Arctic Ocean via the Canadian Archipelago, carrying with it suspended and dissolved contaminants brought into the Arctic by wind, rivers, and ocean currents. Climatic conditions favoring increased ocean current transport to the Arctic Ocean would potentially expose all human residents, but might expose those of the Canadian Archipelago, as well as those of West Greenland and eastern Labrador, to higher levels.

Sea ice and glaciers

Ice, in the form of sea ice or land-based glaciers, is capable of storing contaminants deposited by wind. As climate warming causes melting, the ice releases contaminants, either by volatilization or by release of partic-ulate-associated contaminants into surface water, where further transport or entry into the food chain occurs.

Exposure of Arctic Ocean surface water, as a result of decreasing ice cover, could increase the movement of contaminants such as hexachlorocyclohexanes (HCHs) and toxaphene into the atmosphere, speeding movement within or out of the Arctic (Macdonald et al., 2000a,b).

Sea ice can also transport contaminants by other means, particularly by accumulating contaminants in sediment, from grounding on shallow coastal shelves, particularly where rivers enter the Arctic basin (Barrie et al., 1998). Sediment-rich shallow water can also be incorporated into sea ice by freezing as surface ice is formed. The contaminant-bearing ice can then follow established circulation patterns in the Arctic Ocean, influenced by wind and surface air temperature, releasing contaminant-containing sediment as it melts. While the components of this cycle are well described, climate change could affect any particular step, regionally or throughout the Arctic Ocean, with unpredictable effects on contaminant transport and human health.

Rivers and lakes

River flow into the Arctic Ocean will increase if, as projected by the ACIA-designated climate models, there is an increase in precipitation and the length of the ice-free period for much of the Arctic and subarctic. This will promote increased river transport of contaminants from industrial and agricultural sources further south.

Arctic lakes are thought to have functioned as temporary storage for contaminants deposited in snow during the winter, with rapid runoff removal in spring (Macdonald et al., 2000a). It is possible that an earlier melt, increased contaminant inputs to the lake water, and an earlier onset of spring growth in the lake ecosystem could result in greater amounts of contaminants being incorporated into the ecosystem, and thus into organisms consumed by humans (Macdonald et al., 2002).

Permafrost

Permafrost underlies much of the Arctic. In regions of discontinuous permafrost, contaminants deposited onto the surface by wind, rain, and snow are released during thawing and mobilized into active biological systems or into runoff, eventually draining into lakes, rivers, and the ocean. Permafrost also acts as a containment mechanism for man-made waste sites, such as landfills, sewage lagoons, mine tailings, and dumpsites. Increasing season length and surface air temperatures could allow contaminants to migrate through active permafrost layers to surface water sources used by humans and wildlife. Entry into runoff during periods of increased rainfall is also a possibility. The overall effect of permafrost thawing is likely to be increased contaminant exposure by a variety of mechanisms, and represents a "new" transport pathway.

Wildlife

Pacific salmon (*Oncorhynchus* spp.) spawn and die in freshwater streams and lakes. Contaminants accumulated by a salmon during that part of its lifecycle spent in the North Pacific are deposited into the local freshwater ecosystem where it dies. Over the past decades, as warming has occurred, Pacific salmon species have gradually extended their range north into the Arctic, adding contaminant loads to the predators and local biomass where they spawn and die (Babaluk et al., 2000). The magnitude of this input, for some freshwater systems, has been shown to exceed the input from atmospheric pathways (Ewald et al., 1998). Some contaminants, which are present at higher concentrations in the North Pacific than the Arctic, notably β-HCH, could thus increase in local arctic freshwater systems to levels higher than at present, so increasing the potential for human exposure and possible health effects.

Humans

Potentially, the greatest impact from human activities is from increased maritime traffic during ice-free periods with the sudden catastrophic release of hazardous materials into the local, regional, and eventually circumpolar environment. The possible increase in extreme weather events could increase the likelihood of such an event.

15.4.4.3. Summary

The transport mechanisms and pathways for contaminants in the Arctic are incompletely understood. Time trends are not available for the concentrations of most contaminants in most media, and instrumental records for major forcing factors are equally sparse. Models that link contaminant movement to climate events do not yet exist. All components of all contaminant pathways are affected by climate. The degree of uncertainty in the climate processes precludes prediction based on current climate models, but makes a powerful case for further research, as well as environmental, wildlife, and human health monitoring.

Monitoring and research on contaminant health effects in arctic residents and ecosystems is ongoing, and must now be linked to systematic research on contaminant movements into, and out of the Arctic. A warming climate in the Arctic could offer a variety of new economic opportunities to arctic residents, including increased maritime activity. International efforts should continue at preventing massive contaminant spills in the Arctic: such spills could be the most damaging contaminant event for local and regional ecosystems and residents.

15.5. Environmental change and social, cultural, and mental health

The potential impacts resulting from direct (section 15.3) and indirect (section 15.4) mechanisms of climate change on human health have been presented in previous sections of this chapter. This section addresses climate change and its link with social, cultural, and mental health.

The association between health and social, cultural, and economic factors has been reported extensively. A gradient has been consistently demonstrated across different socio-economic classes, regardless of how they are defined, for various measures of mortality and morbidity, both for individual diseases and for all causes combined (Marmot and Wilkinson, 1999). This gradient exists in many countries around the world, and has persisted despite major improvements in the overall health and wealth of populations. Also, numerous studies have shown correlations between unemployment and morbidity and mortality (e.g., Morris et al., 1994; Moser et al., 1984). The disparity in health status between Inuit and the larger national populations to which they belong (Canada, Denmark, Russia, United States) has often been attributed to their relatively poorer socio-economic status. Furthermore, the link between socio-economic conditions and health has been demonstrated within arctic populations. In a study of 49 Inuit and Dene communities in the Northwest Territories, based on community-level data from the 1992 NWT Housing Survey and routinely reported health and social service agency data, Young and Mollins (1996) found a correlation between most indicators for housing and socio-economic status with the rate of health center visits, used as a proxy measurement of morbidity. Among Canadian Inuit the proportion of respondents reporting excellent/very good health increased with the level of formal education (Bjerregaard and Young, 1998; Statistics Canada, 1991). Those in the highest income category are also more likely to report excellent health than those in lower income categories. The association between education and positive self-reported health was also found in the 1993-94 Greenland Health Interview Survey (controlled for age and sex) (Bjerregaard and Young, 1998). Among those with the most formal schooling, self-rated health was better in all age groups and with fewer reported longstanding illnesses.

Figure 15.19 represents one view of the interrelationship between health and social, cultural, and environmental factors. At the center of the model is the individual with his or her genetic endowment, physiological adaptation, and personal life experience. Everyday life, for example, work and lifestyle, determine the extent to which individuals are exposed to factors within their immediate surroundings that directly affect health. Such factors include tobacco, alcohol, diet, contaminants, microorganisms, and psychological factors. Everyday life is in turn determined by social, economic, and cultural factors within the wider society. All the individual, social, cultural, and socio-economic factors are influenced by the environment within which they are embedded and by possible changes in this environment (e.g., see Box 15.1).

The effect of change on communities resulting from climate related impacts, for example, by flooding and ero-

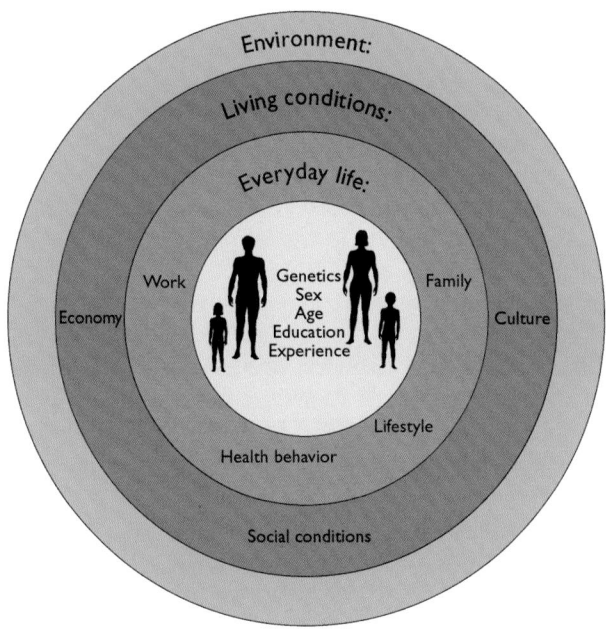

Fig. 15.19. Interrelationship between social, cultural, and environmental factors known to affect health.

sion, or warming temperatures and shifts in local resource access and availability, and the links between these changes in the physical environment and individual health are clear. Research using an ecological approach has also shown that the social environment of the community affects population health (Macintyre and Ellaway, 2000). Studies have focused on the extent to which social relations, social capital, and the social vitality of communities influence individual health status (Berkman, 1995; Kawachi et al., 1997; Veenstra, 2002). Although the relationship between the social environment of the community and the health of the individuals is still not clear, studies have shown associations between, for example, social isolation and mortality (Berkman and Syme, 1979) and social capital and poor self-rated health (Kawachi et al., 1999). In both cases the association was present even after adjusting for individual risk factors. Even though the health advantages are not fully understood, research has shown that a healthy community can be characterized as a safe environment that provides opportunities for social integration, and is neither conflictual, abusive, nor violent (Taylor J. et al., 1994).

Research and practice in health promotion have focused on the community level rather than on disease prevention in individuals. The role of the social environment for individual health argues this inclusion. Also, it is widely believed that the positive effects of health interventions result from a high degree of community ownership and participation in defining objectives, planning, and implementation of the initiative (Bracht, 1999; Green and Kreuter, 1999).

Research on health promotion and community health shows that community members' involvement in the social life and their shared pursuit of broader social goals

through psychosocial processes can positively affect health, and that participation is a condition for positive outcome of health promotion efforts.

The potential for future health promotion in combining these two perspectives on community health is obvious as increasing communities' capacity to address problems and to work together to solve them will strengthen the effect of health promotion projects and help to improve the health of the population. Arctic communities have recently undergone rapid socio-cultural change and these changes will be further accentuated by the projected climate changes addressed in Chapter 4. Building communities' capacity to meet these changes is an area that needs further attention throughout the Arctic.

15.5.5.1. Acculturative stress and mental health

The concept of *acculturation* describes the cultural and psychological changes that result from continuous contact between people belonging to different cultural or ethnic groups (Berry et al., 1986; Redfield et al., 1936). At the population level, changes in social structure, economic base, and political organization frequently occur as a result of this contact, whereas at the individual level the changes are in such areas as behavior, identity, values, and attitudes. Berry et al. (1986) described this process in indigenous cultures in relation to the impact of modernization on indigenous people and traditional groups (Fig. 15.20). Climate change can also impact on the traditional way of living via influences on reindeer herding, fishing, and hunting. Alterations of the physical environment can lead to a rapid and a long-term cultural change and loss of traditional culture which can, in turn, create psychological distress and mental health challenges (see Box 15.7).

Group strategies for dealing with changes in ways of living can vary from readily and easily adapting to the changes, to resisting them, or to collapsing under their weight (Berry, 1985). Individuals adopt different acculturation strategies and their mental health is expected to vary both as a function of the strategy itself and in relation to the balance between the strategy which is possible or preferred by the individual and that of the majority group (see Fig. 15.20). From the mental health perspective, integration seems the most successful strategy and marginalization the most problematic, with separation and assimilation being intermediate in relation to mental health (Berry, 1990a; Berry and Sam,

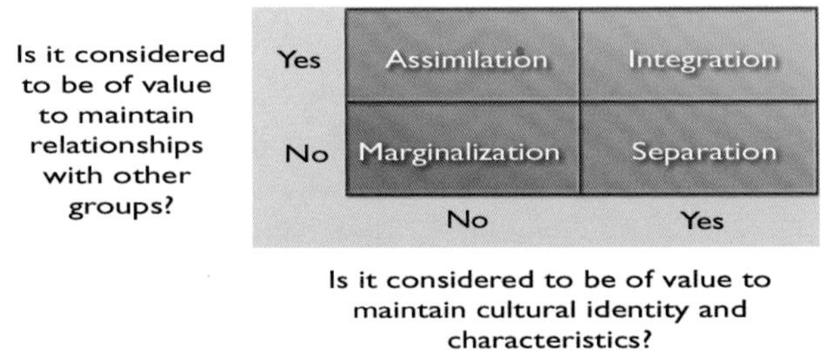

Is it considered to be of value to maintain relationships with other groups?		No	Yes
	Yes	Assimilation	Integration
	No	Marginalization	Separation

Is it considered to be of value to maintain cultural identity and characteristics?

Fig. 15.20. Four modes of acculturation (Berry, 1990b).

Box 15.7. Climate, reindeer herding, and Saami culture

The Saami are indigenous reindeer herding people originally inhabiting arctic regions of Norway, Sweden, Finland, and western Russia. The Saami live in communities which have undergone rapid modernization and cultural revitalization. Some communities are more traditional while others are more modern with regard to the average lifestyle of their residents. In terms of mental health and social well-being, Saami people living in areas strong in culture and traditional ways of life appear just as healthy as the average population in these regions.

However, climate changes could have significant impacts on Saami communities and culture. Increasing temperatures may influence the seasonal migration of reindeer (see Chapters 11 and 12). The consequences of the climate changes are likely to be varied and include causing reindeer to stay for extended periods on less productive fields. Throughout the 1990s, overuse of the pasture lands by reindeer contributed to a crisis within the reindeer herding industry leading to a forced reduction in numbers of reindeer. This in turn forced several reindeer herders to stop herding. According to social workers in the Saami highlands, this led to higher rates of unemployment and dependence on social welfare in these communities. As reindeer herding is seen as a Saami activity, and reindeer herders as the carrier of the Saami culture, those individuals who left herding lost not just an important occupation but also status as an important protector of the Saami culture. Exclusion from this group has been a considerable stress for families and has had significant and far reaching impacts on the Saami culture as a whole.

1997). Marginalization reflects alienation and anomie and is a risk factor for mental health (Berry and Kim, 1988). In integration, relationships to both cultures may provide the strongest socio-cultural foundation for good mental health including bicultural competence and coping strategies necessary for adaptation to both cultures (Berry and Kim, 1988).

The concept of *acculturative stress* refers to stress in which the stressors are identified as being rooted in the process of cultural change or acculturation. For arctic peoples these stressors can be loss of traditional food resources and habitats, unemployment, loss of cultural practices, and migration. Climate change can be an external factor that indirectly initiates the acculturation process by forcing people to behave in new ways, to change their ways of living, and to replace or drop old traditions (see Box 15.7).

Acculturative stress may be associated with psychological changes such as psychosomatic symptoms, feelings of alienation and marginality, and identity confusion (Berry, 1997). If the acculturation experience overwhelms the individual with a feeling of loss of control, psychopathology may occur such as depression and anxiety, substance abuse, and suicide (Fig. 15.21). For instance, if climate change leads to a loss of herding, hunting, or fishing opportunities for people who are closely connected to and dependent upon such activities, this can result in feelings of loss and grief. This might result in longer-term feelings of marginalization which can contribute to substance abuse, depression, and suicide. A study of Saami males living in remote areas found a close relationship between marginalization and depression/anxiety (Kvernmo and Heyerdahl, 2003), while a study conducted in Greenland found that growing up in a town and being fully bilingual, as compared to growing up in a small village and speaking only Greenlandic, was associated with better mental health status (Bjerregaard and Curtis, 2002). This disagrees with the assumptions previ-

ously held that the social and psychological outcomes of acculturation are inevitably negative (Berry and Kim, 1988; Malzberg and Lee, 1956). It appears that successful integration depends on having and taking advantage of the opportunities necessary to meet changing conditions.

15.5.5.2. Examples of the influence of rapid change on psychosocial health

Several examples are used here to illustrate the relationship between the rapid socio-cultural and economic change witnessed in the Arctic over the past 20 to 50 years and the psychosocial health of the populations living there. A good understanding of such links is important in relation to the climate changes projected for the Arctic over the next hundred years (see Chapter 4). Further changes may result from climate-related changes in economic opportunities, traditional lifestyles, and subsistence activities.

A change in occupational patterns from hunting and small-scale fishing to an increase in wage-earning employment is seen across the Arctic. In Greenland villages, wage earning has been introduced along with institutions and service deliveries such as schools, stores, and sanitation infrastructure. In towns, there are factories and enterprises, banks, shops, and administrative services. The changes in occupational patterns are associated with decreased physical activity and a change from a traditional marine diet to a more western diet in many coastal indigenous groups and this is of particular importance in relation to cardiovascular diseases (Bjerregaard and Young, 1998). In many studies, cardiovascular risk increases with modernization and urbanization, but although mortality from ischemic heart disease is slightly lower among Inuit than in the general population of Denmark, Canada, and the United States, it has decreased over recent years while marine food has been increasingly replaced by store bought food. In contrast, mortality from other heart

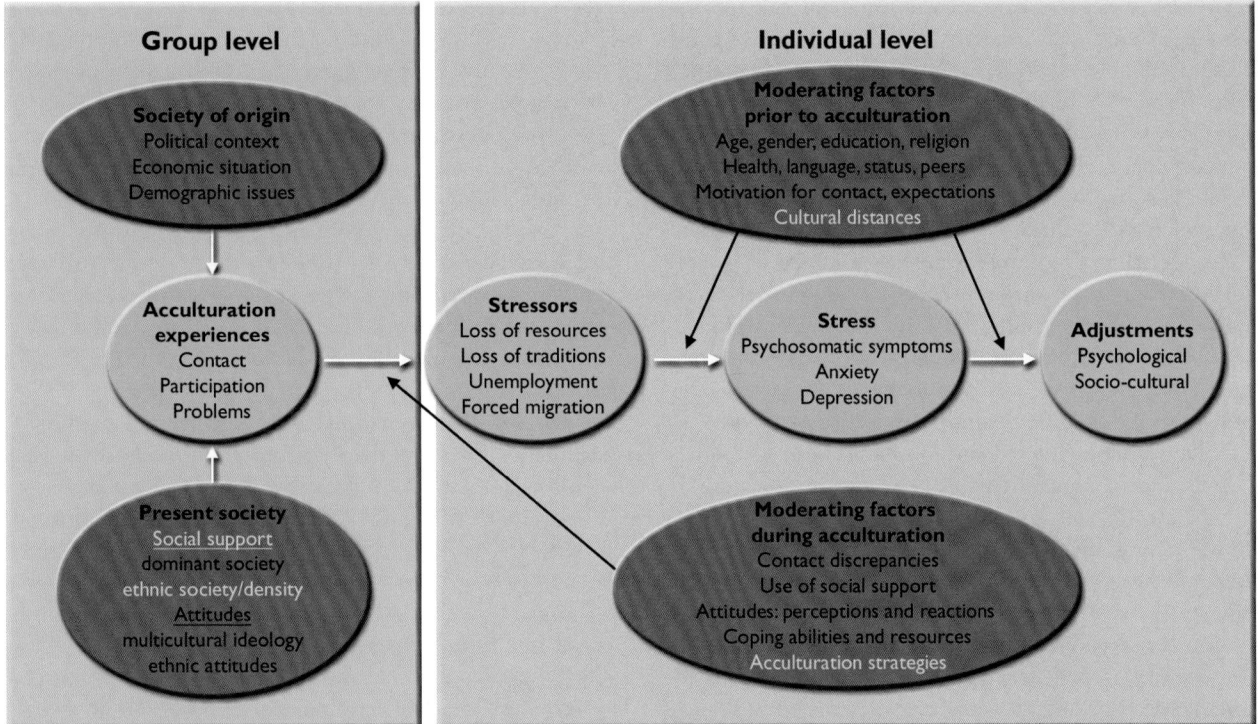

Fig. 15.21. Acculturation stress and moderating factors (based on Sam and Berry, 1995).

diseases and stroke is considerably higher in Inuit and shows an uncertain time trend (Bjerregaard et al., 2003). Type 2 diabetes increased considerably over the last 40 years among the Inuit in parallel with processes of modernization (Ebbesson et al., 1998; Murphy et al., 1992). In Greenland, type 2 diabetes is now more prevalent than in Denmark (9.8 and 7.9% in the 40 to 64 year old age range, respectively; Bjerregaard et al., 2003), which leads to speculation about increased genetic sensitivity to environmental pressure in the Inuit. The change in dietary patterns from traditional foods to a more western diet with a higher consumption of store bought foods is important in relation to both the connection with obesity and other cardiovascular risk factors and with the loss of the socio-cultural values related to eating and sharing traditional foods and their significance for health.

The amount of traditional food and the specific species consumed vary considerably among regions and population groups (Bjerregaard and Young, 1998; Blanchet et al., 2000; Kuhnlein et al., 2000). Older people and people in villages who fish and hunt themselves eat more traditional food, whereas the young wage earners consume more store bought foods often related to their convenience. However, regardless of the levels of consumption, traditional foods are highly valued by all population groups. They are considered filling and healthy and to provide strength, warmth, and energy in ways that store bought food does not (S. Bernier, Public Health Research Unit, Université Laval, Canada, pers. comm., 2002; Borré, 1994; Furgal et al., 2001). Traditional food is also reported to be a significant contributor to cultural identity, tradition, and social cohesion in Inuit communities. To eat and like tradi-

tional food is perceived as a marker of identity in the same way as is speaking the Inuit language (Kleivan, 1996; Searles, 2002).

> *Inuit food has sustained us, nourished us, brought us together, and given us a sense of who we are.*
> Egede 1995

> *... in order to have a Greenlandic identity the person must eat ... and like dried fish, raw mattak, etc.*
> Petersen 1985

> *I can't do without kalaalimernit [traditional Greenlandic food]. I eat it a lot. I was brought up with people who eat kalaalimernit. It tastes good and it feels good for you. For instance when I do sports I feel kind of stronger when I have eaten kalaalimernit.*
> Focus group study: Ilulissat, wage earners

The rapid changes that have taken place within arctic communities and the migration from villages to larger towns has lead to psychosocial stress among populations that within little more than one generation have had to adapt to significantly different ways of life. Those who move may experience a lack of social relations in towns, where still more people live in single-family households and social relations are chosen and individualized. Traditional values may appear irrelevant, as language skills and formal education become the means to succeed and avoid unemployment. However, traditional knowledge and expertise in traditional activities is still very relevant, if not increasingly so, when aspects of the environment become more unpredictable or "risky" in which to practice subsistence activities and travel on land or sea.

Psychosocial stress is reflected in the incidence of social problems seen in many arctic communities today. Whereas alcohol abuse in western societies is usually characterized by an increasing daily consumption over many years, it is the occasional, sometimes regular, drinking spree or binge drinking which creates many problems in arctic communities (Bjerregaard and Young, 1998). The most important health implications are accidents and violence resulting in intentional and non-intentional traumas (cuts, bruises, fractures, head injuries). Drowning, falls, frostbite, burns, and pneumonia also result from intoxication and a direct association between alcohol use and incidence of suicides has been shown. Alcohol consumption is also associated with economic problems and job loss due to instability at work and to domestic abuse.

Studies have shown a high occurrence of violence and sexual abuse in some arctic populations (Bjerregaard and Young, 1998). A survey in Greenland showed that women and men have equally often been victims of violence (47% for women and 48% for men) but that women had more often been sexually abused (25% and 6%, respectively) (Curtis et al., 2002). Having been the victim of violence or sexual abuse was significantly associated with a number of health problems: chronic disease, recent illness, poor self-rated health, and mental health problems. The association between having been the victim of violence or sexual abuse and current health status was stronger for women than for men (Curtis et al., 2002).

Similarly, suicide has played a significant role in many arctic communities and individuals' lives. It has been argued, based on the much higher suicide rate in men than women that women in the Arctic have been more successful than men in adapting to social change. While women have been able to continue their traditional roles as caregivers, both in the family and in the labor market, the transition from hunter and sole bread-winner to wage-earner in a subordinate position or even unemployment has been a difficult transition for many arctic males. On the other hand, studies have shown that women more often than men have suicidal

Table 15.1. Prevalence of serious suicidal thoughts in Greenland Inuit according to childhood experience of alcohol misuse and sexual violence (Curtis et al., 2002).

	No sexual violence during childhood (n=1150)	Sexual violence experienced during childhood (n=65)
	%	%
No alcohol problems in parental home (N=760)	10.3	21.7
Occasional alcohol problems in parental home (N=355)	18.2	48
Often alcohol problems in parental home (N=100)	39.8	82.4

thoughts and mental health problems (Bjerregaard and Curtis, 2002). The finding that men suffer more from socio-cultural change related stress than women may be in part based on the more visible and more commonly reported manifestations of this stress or frustration common among men (i.e., they are more likely to become violent or to commit suicide). A population survey of adults in Greenland showed significant associations between suicidal thoughts and a number of social and cultural factors, the most important determinants being the occurrence of alcohol problems in the parental home and the experience of sexual violence during childhood (Bjerregaard, 2001). Among Greenlanders who had neither experienced alcohol problems nor sexual violence in childhood, 10% reported suicidal thoughts, while for those who had experienced both, 82% reported suicidal thoughts (Table 15.1).

Alteration in subsistence species, changes in habitability of buildings, erosion of village sites (see Box 15.6), and disappearance of commercially critical species (see Box 15.1) may all cause dislocation of residents from smaller to larger communities, with a subsequent overload of scant resources in the receiving community. This combination of factors and interaction with the capacity of individuals and communities to adapt to or change with these climate and environmentally influenced shifts may have a significant bearing on the social, cultural, and mental health of individuals in arctic communities (for a discussion of community capacity and resilience see Chapter 17).

15.5.5.3. Example scenario of interactions between climate warming, ocean temperature, and health

Climate change can influence health in the Arctic in many ways. These fall into two groups: direct (section 15.3) and indirect (section 15.4) influences. Two indirect pathways are of significance to individual and community health issues related to socio-cultural and economic transition. The first results in reduced opportunities for subsistence hunting and fishing as a result of changes in animal or plant populations, an increase in extreme events, and changes in sea-ice distribution and thickness, while the second forces populations to move. These factors include the movement or loss of crucial species for subsistence hunting, coastal erosion, breakdown of sanitation infrastructure, and increasing difficulties in maintaining transportation systems for goods and people. Both pathways, of which the latter is also a consequence of the former, may be related to climate changes and may be seen as mediated through acculturation and cultural loss. However, the modernization process in arctic communities over the last 50 years has been accompanied by a growth in the cash economy, making it increasingly difficult to maintain a livelihood from hunting and fishing and therefore encouraging people to move to more regional economic centers.

Figure 15.22 shows two possible pathways between a cooling of ocean temperatures and decreased health status in indigenous communities. The first is via a reduction in the catch of commercially important species. This happened in Greenland in the 1960s when Atlantic cod disappeared from coastal waters off West Greenland (see Box 15.1). Unless new sources of income are generated, reduced catches will result in deterioration in economic conditions and unemployment. Unemployment is associated with social and mental health problems in communities, increased use of alcohol and violence, and suicide.

The second pathway of impacts is via a change in the availability of subsistence species. Historically, indigenous groups have responded to such a change by moving with the animals to places with better catches, however, human migrations are now far less easy with today's settled villages. Decreased opportunities for catching seals, for example, result in both dietary changes and a loss of the knowledge required to recognize, harvest, and prepare traditional foods (Kuhnlein and Receveur, 1996), and in a sense of cultural loss as hunting and food sharing is important for the identity and maintenance of the social well-being of the community (Borré, 1994). The lack of subsistence species also drives an eventual migration of individuals to towns where some will face unemployment. The shift from a traditional marine diet to a more western diet, along with other behavioral changes that follow the transition from a lifestyle based on hunt-

ing and fishing to a more sedentary wage-based economy lifestyle, is of major importance in relation to the occurrence of cardiovascular disease in Inuit populations.

In spite of of the present cooling trend in parts of Greenland, the climate scenarios described in Chapter 4 project a warming trend for Greenland over the next 100 years. The health impacts of this warming are likely to be mediated by the same indirect mechanisms discussed in section 15.4. Changes in the distribution of commercially important fish species, as well as in subsistence species could affect health in the same way as ocean cooling. Increased risk to hunters and travelers due to ice changes, and political and economic policy decisions which result in the relocation of populations are also likely to have major health impacts.

It is unlikely that periodic warmer weather could result in the same detectable changes in mortality from cardiovascular disease and stroke that are seen in large populations in the northern temperate zones of Europe and North America. The scattered, sparse population of the present cooling region of Greenland is too small for a significant number of well-documented mortality events.

Significant warming in Alaska since the 1970s has resulted in significant impacts. Permafrost thawing and lack of sea-ice protection have resulted in the imminent threat of forced village dislocation, with severe stress on families. River and coastal erosion with continued warming will make this more widespread. In addition, the northward spread of beaver populations has resulted in the obstruction of many streams traditionally used by villages for surface water supply. Contamination of these sources by the zoonotic parasite *Giardia lamblia* makes outbreaks of human disease much more likely. At the same time, moose have extended their range into new regions and waterfowl habitat is increasing, making new subsistence foods available. Thus, in Alaska, new risks and new opportunities have resulted from warming. This makes the development of community monitoring a more urgent task, as only data will enable policy and public health responses to mitigate risks and maximize opportunity.

Despite the arguments made here for the connections between environmental change, socio-cultural transition, and health in the Arctic, many negative health outcomes are attributable to causes other than climate change. The modernization process in indigenous communities in the Arctic since the 1950s has been accompanied by many such health outcomes (e.g., increased use of alcohol, more violence, and higher mortality from suicides, heart disease, and diabetes), without any yet-reported association with climate change.

15.5.5.4. Summary

Health is a multifactorial concept, influenced by a variety of determinants, climate change being one of many environmental factors. Many of these determinants are still poorly understood. The rapid social, cultural, and

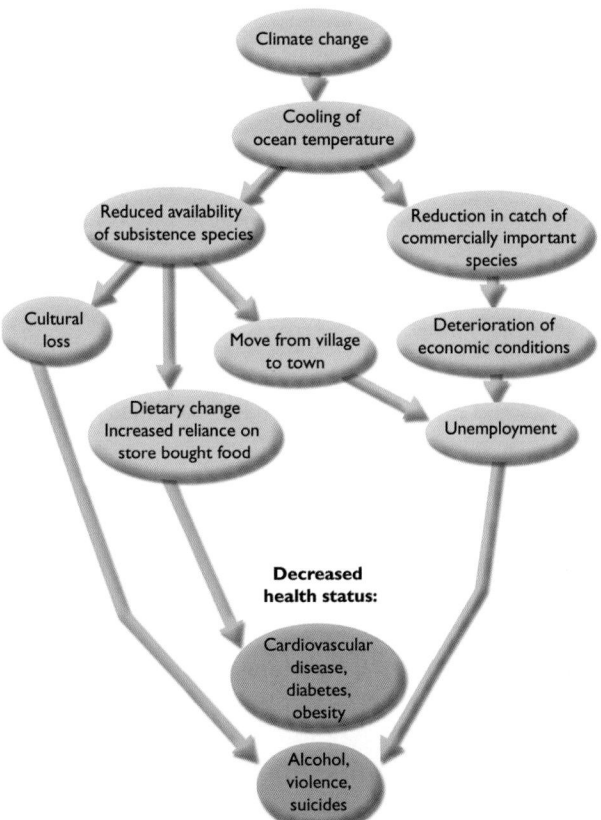

Fig. 15.22. Example scenario of the link between climate-related changes, and community and individual health in indigenous communities in Greenland.

economic transition that arctic communities have seen over the past 50 years has influenced lifestyles and individual and community health. These changes are very likely to be affected and even accentuated by climate change in the future. The influence of adaptive behaviors resulting in changes to lifestyle and cultural loss further influence acculturative stress in arctic communities. Climate change has the potential to influence the rapid changes ongoing in communities today by challenging individuals' and community's relationship with their local environment, which has for thousands of years, been the basis of their identity, culture, and well-being.

15.6. Developing a community response to climate change and health

Health impacts related to climate change in the Arctic are likely to vary across communities and regions, with some changes being positive and others adversely affecting the health of individuals. In response to these impacts, communities need to develop strategies to take advantage of opportunities and to minimize risks. In some cases, communities have already started to adapt to climate related changes with potential impact on aspects of health. For example, in Inuit communities of the western Canadian Arctic, individuals report now taking bottled water on trips due to the lack of fresh water sources while on the land and hunters have adapted their hunting and fishing times to compensate for the changes in species availability and access to continue to procure fresh traditional foods (Nickels et al., 2002). A key component in this ability to adapt and respond is the development of a better understanding of the relationship between climate and the health of northern peoples and access to locally relevant information on the changes taking place. The identification, selection, and monitoring of some basic indicators for climate and health is one tool communities can use to help in the development of their response to these changes. This information can support the community's capacity to know what changes are occurring, what changes are likely to take place in the future, and what impacts these changes may have. Linking these indicators to the projected scenarios of change reported in other chapters of this assessment allow communities to monitor, and where required develop strategies to minimize negative impacts in the future. This section proposes some candidate indicators for this purpose based on the scientific review of direct and indirect climate–human health interactions presented in this chapter.

15.6.1. Goals of community indicators

The identification, selection, and monitoring of indicators is one way in which communities can gather the information to support their diagnostic needs and to support the development of potential response strategies. Within this context, indicators for climate change impacts on human health are measurements or observations of a parameter (e.g., snow, ice, water, temperature, UV-B radiation, permafrost, a component of human health) that link climate, the environment, and an aspect of community or

individual health. The indicators selected by the community need to be issue-specific and must be presented in a way that makes the information they generate easy to understand and useful in making decisions about climate change and health impacts at the community level.

Indicators can potentially serve many purposes. For example:

- To confirm changes or trends in a condition over time (e.g., date of freeze-up of the local bay).
- To assess the current condition of the environment to judge its adequacy with reference to a standard (e.g., safety of ice for travel).
- To anticipate hazardous conditions before negative impacts occur (e.g., to know when a severe storm event is coming or to predict a shortage of a fresh source of a specific traditional food in the community related to difficulties with access and availability).
- To identify causes of effects and to identify appropriate action (e.g., windchill factor to warn people about cold injury).

The type(s) of indicators chosen by a community or region for a monitoring program must be determined by the specific goals of the community. For example, whether it is the intent to be able to warn individuals of future dangerous conditions, whether it is to determine if an increase in accidents experienced by residents while on the land or water is associated with changes in local climate conditions, or whether it is to determine if changes in local food security (access and availability of traditional food species) are associated with changes in climate and whether this situation will become sufficiently bad in the future that they must be prepared to take early action. In many cases, regional and community public health and other authorities (e.g., Meteorological Service/Weather Bureau, Wildlife Dept.) can be helpful in selecting and gathering such indicator-based data. A good definition of goals for the monitoring activity will provide direction as to what indicators are best suited to their needs and are most appropriate to the community or region. The scale at which the indicators are gathered is also critical as some changes can be detected and are best managed at a regional scale (e.g. changes in access to market foods to supplement local diets at critical times of the year) while others require a more local-scale approach (e.g., monitoring permafrost stability at the airstrip of a community). In all cases, for indicators to be appropriate, meaningful, and useful to local communities, community individuals must be directly involved in their identification and selection as well as in the design and implementation of the data gathering monitoring programs (Eyles and Furgal, 2002).

15.6.2. Characteristics of useful indicators

Criteria can be used to guide indicator identification and selection to ensure that appropriate indicators are chosen to help meet community objectives. It is essential that each community or region develops its own criteria, although many may be general and useful to all regions. Two general types of criteria that are often used are scientific and use-based criteria.

Scientific criteria are intrinsic to the issue of scientific quality (sensitivity, reliability, and statistical validity) and are often addressed by using indicators from existing and recognized lists. Indicators based on traditional or indigenous knowledge are very useful and should include a description, where possible, of the understanding of the link between the indicator and the specific health impact. It is also important to consider other "use-based" criteria, such as:

- Feasibility (are they already available and if not, what is the feasibility of collecting new information, taking account of cost, ease and time for collection, capacity to gather data, etc.).
- Perceived importance of the indicator to those affected (community representatives).
- Number of indicators (a manageable number is needed to attain specified goals, but this number must not be too cumbersome for community monitoring system managers).
- Balance (a rough balance among the various aspects of the issue).
- Catalyst for action (those that act as a catalyst to drive action are also useful, e.g., ice thickness on travel routes).
- Understandability by the media, local decision-makers, and policy-makers.
- Minimal environmental impact to collect.
- Relevance to all members of the population.

The candidate indicators listed in the following section are proposed on the basis of the review presented in this chapter and all meet the following basic criteria:

- The indicator relates climate changes, directly or indirectly, to potential human health impacts.
- The data are already being gathered by regional or national governments or are readily available from other sources.
- Where the indicator data are not already gathered, they could be easily collected by communities using standard methods.
- Time trend data for the indicator exist or can be easily gathered.

15.6.3. Proposed candidate indicators

This section lists indicators proposed as potential candidates for community selection. They are not an exhaustive or complete list but are proposed as tools to assist communities in discussing and identifying their own indicators. They are derived from the review of direct and indirect climate–human health interactions presented in this chapter and assume the collection of general environmental data related to climate changes at the local and regional scale (e.g., temperature, precipitation, ice cover).

15.6.3.1. Direct impact mechanism indicators

The collection of some basic indicator data can help communities monitor the direct impacts of climate on health in northern communities. Table 15.2 identifies indicators of health impacts related to direct interactions with climate variables such as accidents while on the land or ice (unintentional trauma) related to bad weather conditions, deaths or injuries related to extreme weather events, and the health impacts of increased UV-B exposure.

15.6.3.2. Indirect impact mechanism indicators

Similarly, the monitoring of basic indicators, such as those presented in Table 15.3 can be used to help

Table 15.2. Direct impact mechanism indicators.

Useful health indicators
- General health statistics (see AMAP, 2003)
- Rates of cold injuries (e.g., frostbite)
- Rates of coronary heart disease
- Rates of unintentional injury
- Rates of intentional injury

Extreme weather events, thermal stress, and health
- Extreme event-related use of regional and community rescue services
- Unintentional injury mortality associated with extreme weather events
- Highest and lowest seasonal temperature
- Number of days in winter with extreme low temperature (where extreme is defined as deviation of more than 20% below the average monthly temperature in winter)
- Number of days in summer with extreme high temperature (where extreme is defined as deviation of more than 20% above the monthly average temperature in summer)
- Reports of respiratory trouble (hospitalization)
- Deaths due to exposure in winter

UV-B radiation and health
- Incidence of skin cancers in arctic regions
- UV-B radiation measurement/ozone depletion measurement in arctic regions

Indicators for annual monitoring[a]
- Measurements of UV-B radiation at ground level and at the personal level by integrating or spectral radiometers and personal dosimeters
- Incidence of sunburns especially the time at which increases in sunburn begin to be noted throughout the Arctic
- Number of cases of snow blindness and frequency of reports
- Increases in cold sore occurrences which may indicate suppression of immunity against Herpes simplex virus (Type 1) and other aspects of cell-mediated immunity by UV-B radiation
- Increases in cataract on an annual basis throughout the Arctic

[a] with emphasis on ozone depletion episodes as sunlight increases throughout the arctic spring and summer

Table 15.3. Indirect impact mechanism indicators.

Wildlife populations and health

- Government harvest data by species of interest (key country food species, sentinel species)
- Stock assessments of species of interest and importance to local economies and diet
- Local arrival/departure dates of migratory species
- Frequency of reports of new species to a region
- Important animal disease frequency (e.g., rabies, brucellosis)
- Appearance of new zoonotic diseases (e.g., West Nile virus)
- Local hunter/fisher reports of animal/fish abnormalities
- Registry of reportable infectious diseases spread from animals to humans and through contaminated water
- Incidence of human cases of zoonotic diseases

Ice, snow, and health

- Rates of cold injuries (e.g., frostbite)
- Mortality rates from coronary heart disease
- Rates of unintentional injury

Infrastructure and health

- Movement measurements of structures on permafrost
- Measurements of key shoreline and river bank erosion rates
- Measurement of trends in flood depth and frequency
- Increased repair costs for sanitation infrastructure, boardwalks, and roads
- Increased operational costs for water treatment systems
- Increase in regulatory noncompliance events for sanitation systems
- Increases in pollution of waterways caused by human waste or solid waste leachate
- Increased incidence of waterborne disease outbreaks
- Increased incidence of certain cancers caused by waterborne contaminants
- Increased incidence of diseases associated with poor personal hygiene
- Increased incidence of diseases caused by contact with wastewater contaminants
- Increased reports of damage to sanitation infrastructure caused by erosion, flooding, or foundation failures
- Increased reports of water rationing caused by drought

Society, culture, and health

- Number of in-migrants to regional hub communities per year, and by number of families
- Number of out-migrants from a region's villages, per year
- Incidence of legal encounters for child abuse, assault, and alcohol-related offences
- Community and regional trends in unemployment
- Community and regional rates of completion of 12 years formal education
- Incidences of treatment for depression and post-traumatic stress disorder
- Consumption of traditional food species (frequency, total amount, and percentage composition of total diet)
- Self-reported health status from regional health surveys

communities identify, understand, and track the indirect impacts climate variables may be having on community and individual health in the circumpolar North. Impacts related to exposure to zoonotic diseases, indirect injuries related to environmental conditions, changes in the stability and safety of community infrastructure, and the combined impacts of climate-related changes on social and mental health and well-being can be monitored via the collection of such data.

15.7. Conclusions and recommendations

This chapter discusses direct and indirect relationships and impacts, both positive and negative, between climate-related changes and human health in northern communities. The likelihood of these impacts occurring in any specific community or region is difficult to determine. The risk of impact depends on many factors, both current and future. The most obvious conclusion that can be drawn is that much research remains to be done on the relationship between climate change and individual and community health in the Arctic. Climate will continue to influence public health in the small remote communities of the Arctic. The recent record warming, and the scenario of future warming, combined with the multiple mechanisms by which climate impacts on health indicate urgent need for adopting community-based monitoring strategies. A network of such communities, within and across regions, reporting a common set of similarly measured climate, health status, infrastructure, and ecosystem observations would serve to identify both emerging threats, and opportunities. In addition, this would allow utilization of public health data to quantify impact, and provide the basis for resource advocacy, and to suggest mitigation strategies, as well as jurisdictional policy. Communities must monitor changes that may represent threats or adverse impacts. Where possible, they must proactively develop coping or adaptation strategies. Similarly, they must take advantage of potential opportunities that these environmental changes present. Thus, regional and national governments need to assist in the design, and provide support for community based monitoring and mitigation strategies to cope with climate change.

15.7.1. Principal conclusions and recommendations

1. **There is a lack of comparability in health status data between countries.**

   A core group of health status indicators, gathered and defined identically, should be a high priority for Arctic Council member nations.

2. **There is a need for a carefully planned strategy, at the community and regional level, to monitor and document environmental change.**

   Arctic Council members and program workgroups should provide technical assistance regarding monitoring strategies, climate impact mitigation and pilot studies, data analysis, and evaluation.

3. **There is a lack of an organized effort to collect and utilize indigenous knowledge regarding climate and climate changes.**

   Indigenous knowledge, and its preservation, should be encouraged among Arctic Council member nations.

4. **There are few data on the impact of changing UV-B exposure in the Arctic on the biota and human residents. There is little systematic monitoring of ground-level UV-B radiation.**

   Academic and United Nations organizations have created UV-B research strategies. With regional and community collaboration, research and monitoring strategies relevant to circumpolar populations should be created.

5. **There are few data on climate change impacts on regional biota. A critical need exists for the monitoring of wildlife diseases, and human–wildlife disease interaction. There are few data on climate-induced changes in the diet of subsistence species, which affects their nutritional value in traditional diets.**

   Arctic Council programs have the expertise to design effective regional and international monitoring programs in cooperation with communities. This critical activity should be given a high priority.

6. **There is no systematic monitoring in all regions for safety of snow and ice conditions for local/regional travel and subsistence activities.**

   Regional governments should collaborate with communities to establish appropriate monitoring and communication networks; dissemination of appropriate traditional and modern survival skills should be systematically taught to children and young people.

7. **Monitoring is critical in regions of the Arctic where physical infrastructure depends on permafrost or where a village site depends on sea ice protection from storm erosion.**

   Regional governments should assist in developing community-based monitoring.

8. **Data on contaminant transport into and out of the Arctic is critical for projecting impact and risk for arctic wildlife and resi-** **dents. Changing climate makes monitoring essential.**

   International coordinated research and monitoring of changing contaminant transport pathways should continue and expand where needed.

In all areas, there is a need for local and regional integrated analyses of the associations between arctic health status and climate variables. This can be accomplished by the establishment of monitoring and data collection mechanisms where they do not already exist. Also, public health education programs should incorporate, where possible, information on risks associated with the environmental changes most relevant to their region.

15.7.2. Recommendations for monitoring and research

Specific recommendations for monitoring and research are as follows.

Thermal stress and arctic human health

Establish organized monitoring and data collection programs (inclusive of local perspectives and indigenous knowledge) involving, but not limited to, the indicators identified in this chapter to support community understanding of changes in arctic health owing to thermal stress.

UV-B radiation and arctic human health

Measure incident UV-B radiation at ground and individual levels using personal dosimeters and ground-based integrating and spectral radiometers.

Carry out cross-sectional population based surveys and follow-up studies in the Arctic and other areas to investigate the causal relationship between UV-B irradiance and cataract and other UV-B induced health effects to the eye such as climatic droplet keratopathy.

Compile local residents' knowledge and perceptions of UV-B radiation and its effects on health in order to supplement scientific data and obtain new knowledge about existing UV-B related habits, particularly with regard to skin, and immune system and eye impacts.

Establish an international network of UV radiation research centers and monitoring programs.

Expand collection of epidemiological data on key UV-B and health issues (e.g., arctic cataract data are difficult to obtain as are sunburn and cold-sore data).

Wildlife, diet, and health

Establish community based and regional scale monitoring programs for the indicators identified in this chapter.

Where problems are identified (e.g., increasing incidences of exposure to zoonotic diseases), establish surveillance programs to protect the health, social, economic, and cultural benefits of harvesting subsistence species while maintaining confidence in and the benefits of consuming these foods.

Snow, ice, and arctic community health

Establish surveillance and communication networks at the community level to support early warning of dangerous conditions for travel and land-based activities (weather, ice conditions, etc.).

Provide support for community freezers and other food supplement and food safety programs to ensure access to safe and healthy traditional foods in arctic communities.

Develop education programs for young people, based on traditional survival skills, to enhance individual capacity to continue aspects of a traditional lifestyle.

Infrastructure and arctic human health

Establish local level monitoring programs for data collection on permafrost and infrastructure stability.

Monitor ground temperatures and compare to historic measurements.

Monitor basal depth of permafrost and compare to historic measurements.

Monitor coastal and river bank erosion and compare to historic measurements.

Monitor incidence of flooding caused by storm surges or heavy precipitation.

Monitor emergency projects for repair of failing infrastructure.

Society, culture, climate, and arctic health

Research the role of the environment, specifically climate-related variables, in community change for small remote locations.

Research to support the understanding of the cultural context and variables influencing individual and community vulnerability, capacity, and resilience in relation to the impacts these changes represent.

Acknowledgements

I would like to acknowledge the support of the Alaska Native Tribal Health Consortium during the preparation of the assessment. For John Warren, Mike Bradley, and myself, it would not have been possible to participate without that support.

References

ADHSS, 2000. The Threat of West Nile Virus in Alaska. State of Alaska Epidemiology Bulletin No. 20. Alaska Department of Health and Social Services, Anchorage.

AMAP, 2002. Arctic Pollution 2002: Persistent Organic Pollutants, Heavy Metals, Radioactivity, Human Health, Changing Pathways. Arctic Monitoring and Assessment Programme, Oslo. xii+112pp.

AMAP, 2003. AMAP Assessment 2002: Human Health in the Arctic. Arctic Monitoring and Assessment Programme, Oslo.

AMAP, 2004. AMAP Assessment 2002: Heavy Metals in the Arctic. Arctic Monitoring and Assessment Programme, Oslo.

American Public Health Association, 2001. Drinking Water Quality and Public Health (Position Paper). American Journal of Public Health, 91(3):499–500.

Ando, M., 1990. Risk evaluation of stratospheric ozone depletion resulting from chlorofluorocarbons (CFC) on human health. Nippon Eiseigaku Zasshi, 45:947–953. (Abstract in English)

Armstrong, B.K. and A. Kricker, 1993. How much melanoma is caused by sun exposure? Melanoma Research, 3(6):395–401.

Armstrong, B.K., A. Kricker and D.R. English, 1997. Sun exposure and skin cancer. Australasian Journal of Dermatology, 38:S1–S6.

Autier, P., J.F. Dore, O. Gefeller, J.P. Cesarini, F. Lejeune, K.F. Koelmel, D. Lienard and U.R. Kleeberg, 1997. Melanoma risk and residence in sunny areas. British Journal of Cancer, 76(11):1521–1524.

Babaluk, J.A., J.D. Reist, J.D. Johnson and L. Johnson, 2000. First records of sockeye (*Oncorhynchus nerka*) and pink salmon (*O. gorbuscha*) from Banks Island and other records of Pacific salmon in Northwest Territories, Canada. Arctic, 53(2):161–164.

Barrie, L., E. Falck, D. Gregor, T. Iverson, H. Loeng, R. Macdonald, S. Pfirman, T. Skotvold and E. Wartena, 1998. The influence of physical and chemical processes on contaminant transport into and within the Arctic. In: D. Gregor, L. Barrie and H. Loeng (eds.). AMAP Assessment Report: Arctic Pollution Issues, pp. 25–116. Arctic Monitoring and Assessment Programme.

Berkman, L.F., 1995. The role of social relations in health promotion. Psychosomatic Medicine, 57(3):245–254.

Berkman, L.F. and S.L. Syme, 1979. Social networks, host resistance, and mortality: A nine-year follow-up study of Alameda County residents. American Journal of Epidemiology, 109:186–204.

Berry, J.W., 1985. Acculturation and mental health among circumpolar peoples. In: R. Fortuine (ed.). Circumpolar Health, pp. 305–311. University of Washington Press.

Berry, J.W., 1990a Psychology of acculturation. In: J. Berman (ed.). Cross-Cultural Perspectives: Nebraska Symposium on Motivation, Vol. 37, pp. 201–234. University of Nebraska Press.

Berry, J.W., 1990b. Four modes of acculturation. Arctic Medical Research, 9:142–150.

Berry, J.W., 1997. Immigration, acculturation and adaptation. Applied Psychology: An International Review, 46:5–34.

Berry, J.W. and U. Kim, 1988. Acculturation and mental health. In: P. Dasen, J.W. Berry and N. Sartorius (eds.). Health and Cross-cultural Psychology: Towards Applications, pp. 207–236. Sage Publications.

Berry, J.W. and D. Sam, 1997. Acculturation and adaptation. In: Handbook of Cross-Cultural Psychology, vol. 3, Social Behavior and Applications, 2nd Edition. Allyn and Bacon, Boston.

Berry, J.W., J.E. Trimble and E.L. Olmedo, 1986. Assessment of acculturation. In: W.L. Lonner and J.W. Berry (eds.). Field Methods in Cross-Cultural Research, pp. 291–324. Sage Publications.

Berwick, M., 2000. Gene-environment interaction in melanoma. Forum (Genova), 10:191–200.

Bjerregaard, P., 2001. Rapid socio-cultural change and health in the Arctic. International Journal of Circumpolar Health, 60:102–111.

Bjerregaard, P. and T. Curtis, 2002. The Greenland Population Study. Cultural change and mental health in Greenland: The association of childhood conditions, language and urbanization with vulnerability and suicidal thoughts among the Inuit of Greenland. Social Science and Medicine, 54(1):33–48.

Bjerregaard, P. and K.T. Young, 1998. The Circumpolar Inuit – Health of a Population in Transition. Munksgaard, Copenhagen.

Bjerregaard, P., C. Curtis, F. Senderovitz, U. Christensen and T. Pars, 1995. Living conditions, life style and health in Greenland. Danish Institute for Clinical Epidemiology, Copenhagen (In Danish)

Bjerregaard, P, G. Mulvad and H.S. Pederson, 1997. Cardiovascular risk factors in Inuit of Greenland. International Journal of Epidemiology, 26:1182–1190.

Bjerregaard, P., T.K. Young and R. Hegele, 2003. Low incidence of cardiovascular disease among the Inuit. What is the evidence? Atherosclerosis, 166:351–357.

Black, H.S., J.I. Thornby, J.E. Wolf, Jr., L.H. Goldberg, J.A. Herd, T. Rosen, S. Bruce, J.A. Tschen, L.W. Scott and S. Jaax, 1995. Evidence that a low-fat diet reduces the occurrence of non-melanoma skin cancer. International Journal of Cancer, 62:165–169.

Blanchet, C., É. Dewailly, P. Ayotte, S. Bruneau, O. Receveur and B.J. Holub, 2000. Contribution of selected traditional and market foods to the diet of Nunavik Inuit women. Canadian Journal of Dietetic Practice and Research, 61:50–59.

Borré, K., 1994. The healing power of the seal: the meaning of Inuit health practice and belief. Arctic Anthropology, 31(1):1–15.

Bracht, N. (ed), 1999. Health Promotion at the Community Level. New Advances. Sage Publications.

Bulliard, J.L., 2000. Site-specific risk of cutaneous malignant melanoma and pattern of sun exposure in New Zealand. International Journal of Cancer, 85:627–632.

Butler, J.C., A.J. Parkinson, E. Funk, M. Beller, G. Hayes and J.M. Hughes, 1999. Emerging infectious diseases in Alaska and the Arctic: a review and a strategy for the 21st century. Alaska Medicine, 41:35–43.

CACAR II, 2003. Canadian Arctic Contaminants Assessment Report II, vols: Human Health, Biotic Environment, Abiotic Environment, Knowledge in Action, Highlights. Indian and Northern Affairs, Canada.

CDC, 2003. West Nile Virus Activity – United States November 20–25. Morbidity Mortality Weekly Report, 5(47):1160. U.S. Centers for Disease Control and Prevention.

Chen, F.Li.T., H. Huang and I. Holmer, 1991. A field study of cold effects among cold store workers in China. Arctic Medical Research, 50(suppl6):99–103.

Chiang H-C., S-S. Chen, H-S. Yu and Y-C. Ko, 1990. The occurrence of carpal tunnel syndrome in frozen food factory employees. Kae Hsiung I Hsueh Tsa Chih, 6(2):73–89.

Chiang, H-C., Y-C. Ko, S-S Chen, T-N. Wu and P-Y. Chang, 1993. Prevalence of shoulder and upper limb disorders among workers in the fish-processing industry. Scandinavian Journal of Work Environment and Health, 19:126–131.

Chin, J. (ed.), 2000. Control of Communicable Diseases Manual. American Public Health Association, Washington, D.C.

Cliff, S. and P.S. Mortimer, 1999. Skin cancer and non-Hodgkins lymphoproliferative diseases: is sunlight to blame? Clinical and Experimental Dermatology, 24(1):40–41.

Coetzee, W.F. and E.C. Pollard, 1974. Near-UV effects on the induction of prophage. Radiation Research, 57(2):319–331.

Conover, R.J., A.W. Herman, S.J. Prinsenberg and L.R. Harris, 1986. Distribution of and feeding by the copepod *Pseudocalamus* under fast ice during the arctic spring. Science, 232:1245–1247.

Craun, G.F., N. Nwachuku, R.L. Calderon and M.F. Craun, 2002. Outbreaks in drinking water systems, 1991–1998. Journal of Environmental Health, 65(1):16–23.

Curtis, T., F.B. Larsen, K. Helweg-Larsen and P. Bjerregaard, 2002. Violence, sexual abuse and health in Greenland. International Journal of Circumpolar Health, 61:110–122.

De Fabo, E.C., 1980. On the nature of the blue-light photoreceptor: still an open question. In: H. Senger (ed.). The Blue-Light Syndrome, pp. 187–197. Springer-Verlag.

De Fabo, E.C. and M.L. Kripke, 1979. Dose-response characteristics of immunologic unresponsiveness to UV-induced tumors produced by UV irradiation of mice. Photochemistry and Photobiology, 30:385–390.

De Fabo, E. and M. Kripke, 1980. Wavelength dependence and dose-rate independence of UV radiation-induced immunologic unresponsiveness of mice to a UV-induced fibrosarcoma. Photochemistry and Photobiology, 32:183–188.

De Fabo, E.C. and F.P. Noonan, 1983. Mechanism of immune suppression by ultraviolet irradiation in vivo. I. Evidence for the existence of a unique photoreceptor in skin and its role in photoimmunology. Journal of Experimental Medicine, 158:84–98.

De Fabo, E.C., L.J. Webber, E.A. Ulman and L.D. Broemeling, 1997. Dietary L-histidine modulates murine skin levels of *trans*-urocanic acid, an immunoregulating photoreceptor, with an unanticipated periodicity: potential relevance to skin cancer. Journal of Nutrition, 127:2158–2164.

Dietrich, R.A. (ed.), 1981. Alaskan Wildlife Diseases. Institute of Arctic Biology, University of Alaska Fairbanks.

Donaldson, G.C. and W.R. Keatinge, 1999. Mortality related to cold weather in elderly people in southeast England, 1979–94. British Medical Journal, 315:1055–1056.

Donaldson, G.C., V.E. Tchernjavskii, S.P. Ermakov, K. Bucher and W.R. Keatinge, 1998a. Winter mortality and cold stress in Yekaterinburg, Russia: interview study. British Medical Journal, 316:514–518.

Donaldson, G.C., S.P. Ermakov, Y.M. Komarov, C.P. McDonald and W.R. Keatinge, 1998b. Cold related mortalities and protection against cold in Yakutsk, eastern Siberia: observation and interview study British Medical Journal, 317:978–982.

Donaldson, G.C., W.R. Keatinge and S. Nayha, 2003. Changes in summer temperature and heat-related mortality since 1971 in North Carolina, South Finland, and Southeast England. Environmental Research, 91(1):1–7.

Dufour, R., 1991. La documentation du cancer au Nunavik. DSC du Centre hospitalier de l'Université Laval, 43pp.

Duhaime, G. (ed), 2002. Sustainable Food Security in the Arctic: State of the Knowledge. Canadian Circumpolar Institute, University of Alberta, Occasional Publication Series No. 52.

Ebbesson, S.E.O., C.D. Schraer, P.M. Risica, A.I. Adler, L. Ebbesson, A.M. Mayer, E.V. Shubnikof, J. Yeh, O.T. Go and D.C. Robbins, 1998. Diabetes and impaired glucose tolerance in three Alaskan Eskimo populations. Diabetes Care, 21:563–569.

Egede, I., 1995. Inuit food and Inuit health: contaminants in perspective. Avativut/Ilusivut Newsletter 2(1).

Ewald, G., P. Larsson, H. Linge, L. Okla and N. Szarzi, 1998. Bio-transport of organic pollutants to an inland Alaska lake by migrating Sockeye salmon (*Oncorhynchus nerka*). Arctic, 51:40–47.

Eyles, J. and C. Furgal, 2002. Indicators in environmental health: identifying and selecting common sets. Canadian Journal of Public Health, 93(1): 62–67.

Fahrleitner, A., H. Dobnig, A. Obernosterer, E. Pilger, G. Leb, K. Weber, S. Kudlacek and B.M. Obermayer-Pietsch, 2002. Vitamin D deficiency and secondary hyperparathyroidism are common complications in patients with peripheral arterial disease. Journal of General Internal Medicine, 17:663–669.

Fears, T., J. Scotto and M. Schneiderman, 1976. Skin cancer, melanoma, and sunlight. American Journal of Public Health, 66:461–464.

Feldman, D.G.F.H., 2003. Vitamin D. Academic Press.

Fewtrell, L. and Bartram, J. (ed.), 2001. Water Quality-Guidelines, Standards and Health: Assessment of Risk and Risk Management for Water-Related Infectious Disease. World Health Organization.

Fox, K.R., 1993. Engineering aspects of waterborne disease: outbreak investigations. Proceedings of the Annual Conference on Water Research, pp. 85–93. June 6–10 1993, San Antonio, Texas. American Water Works Association.

Fox, S., 2002. These are things that are really happening. In: I. Krupnik and D. Jolly (eds.). The Earth is Faster Now: Indigenous Observations of Arctic Environmental Change, pp. 12–53. Arctic Research Consortium of the United States, Fairbanks, Alaska.

Furgal, C., S. Bernier, G. Godin and E. Dewailly, 2001. Decision-making and diet: balancing the physical, economic, and social components (Year 2). In: S. Kalhok (ed.). Synopsis of Research Conducted under the 2001–2002 Northern Contaminants Program, pp.37–42. Department of Indian Affairs and Northern Development, Ottawa.

Furgal, C., D. Martin and P. Gosselin, 2002. Climate change and health in Nunavik and Labrador: lessons from Inuit knowledge. In: I. Krupnik and D. Jolly (eds.). The Earth is Faster Now: Indigenous Observations of Arctic Environmental Change, pp. 266–300. Arctic Research Consortium of the United States, Fairbanks, Alaska.

Fyfe, J., 2003. Separating extratropical zonal wind variability and mean change. Journal of Climate, 16:863–874.

Geldreich, E.E., 1992. Waterborne pathogen invasions: a case study for water quality protection in distribution. American Water Works Asoocation, Water Quality Technology Conference Proceedings.

Georgitis, J., 1978. Extensor tenosynovitis of the hand from cold exposure. Journal of the Maine Medical Association, 69(4):129–131.

Glerup, H., K. Mikkelsen, L. Poulsen, E. Hass, S. Overbeck, J. Thomsen, P. Charles and E.F. Eriksen, 2000. Commonly recommended daily intake of vitamin D is not sufficient if sunlight exposure is limited. Journal of Internal Medicine, 247:260–268.

Green, L.W. and M.W. Kreuter, 1999. Health Promotion Planning. An Educational and Ecological Approach. Mayfield Publishing Company, Mountain View, California, 506pp.

Haines, A. and A.J. McMichael, 1997. Climate change and health: implications for research, monitoring, and policy. British Medical Journal, 315(7112):870–874.

Hamilton, L.C., B.C. Brown and R.O. Rasmussen, 2003. West Greenland's cod-to-shrimp transition: local dimensions of climatic change. Arctic, 56(3P):271–282.

Hassi, J and T.M. Mäkinen, 2000. Frostbite: occurrence, risk factors and consequences. International Journal of Circumpolar Health, 59(2):92–98.

Hayashi, L.C., S. Hayashi, K. Yamaoka, N. Tamiya, M. Chikuda and E. Yano, 2003. Ultraviolet B exposure and type of lens opacity in ophthalmic patients in Japan. Science of the Total Environment, 302:53–62.

Health Canada, 2002. West Nile Virus Surveillance Information. Population and Public Health Branch: www.hc-sc.gc.ca/pphb-dgspsp/wnv-vwn (Dec 2002).

Hockwin, O., M. Kojima, Y. Sakamoto, A. Wegener, Y. Bo Shui and K. Sasaki, 1999. UV damage to the eye lens: further results from animal model studies: a review. Journal of Epidemiology, 9:S39-S47.

Hodgins, H., 1997. Health and what affects it in Nunavik: how is the situation changing. Department of Public Health, Nunavik Regional Board of Health and Social Services, Nunavik. 321pp.

Holick, M.F., 2001. Sunlight 'D'ilemma: risk of skin cancer or bone disease and muscle weakness. Lancet, 357:4–6.

Hoppner, K., J.M. McLaughlan, B.G. Shah, J.N. Thompson, J. Beare-Rogers, J. Ellestad-Sayed and O. Schaefer, 1978. Nutrient levels of some foods of Eskimos from Arctic Bay, N.W.T., Canada. Journal of the American Dietetic Association, 73(3):257–260.

Huovinen, P., 2000. Ultraviolet radiation in aquatic environments: underwater UV penetration and responses in algae and zooplankton. Jyväskylä Studies in Biological and Environmental Science 86.

IASC, 1995. Effects of increased ultraviolet radiation in the Arctic: an interdisciplinary report on the state of knowledge and research needed. 2, 1–56. International Arctic Science Committee, Washington, D.C.

IASC, 1997. Ultraviolet International Research Centers (UVIRC's): A proposal for interdisciplinary UV-B research in the Arctic. 7, 1–36. International Arctic Science Committee.

IHS, 1999a. Criteria for the Sanitation Facilities Construction Program. Washington, D.C.: U.S Indian Health Service, Division of Environmental Engineering, Environmental Engineering Branch

IHS, 1999b. Regional differences in Indian health, 1998–99. Indian Health Service, Rockville, Maryland.

IPCC, 2001. Climate Change 2001: The Scientific Basis. Contribution of Working Group I to the Third Assessment Report of the Intergovernmental Panel on Climate Change. J.T. Houghton, Y. Ding, D.J. Griggs, M. Noguer, P.J. van der Linden, X. Dai, K. Maskell and C.A. Johnson (eds.). Cambridge University Press, 881pp.

Jemal, A., S.S. Devesa, P. Hartge and M.A. Tucker, 2001. Recent trends in cutaneous melanoma incidence among whites in the United States. Journal of the National Cancer Institute, 93(9):678–683.

Jetté, V. (ed.), 1992. Santé Québec, 1992. A health profile of the Inuit, Report of the SANTÉ QUÉBEC health survey among the Inuit of Nunavik, vols. 1 and 2.

Jones, R.R., 1992. Ozone depletion and its effects on human populations. British Journal of Dermatology, 127:2–6.

Katoh, N., F. Jónasson, H. Sasaki, M. Kojima, M. Ono and N. Takahashi, 2001. Cortical lens opacification in Iceland. Risk factor analysis – Reykjavik Eye Study. Acta Ophthalmologica Scandinavica, 79:154–159.

Kawachi, I., B.P. Kennedy, K. Lochner and D. Prothrow-Stith, 1997. Social capital, income inequality, and mortality. American Journal of Public Health, 87:1491–1498.

Kawachi, I., B.P. Kennedy and R. Glass, 1999. Social capital and self-rated health: a contextual analysis. In: I. Kawachi, B.P. Kennedy and R.G. Wilkinson (eds.). The Society and Population Health Reader: Income Inequality and Health, pp. 236–248. New Press.

Keatinge, W.R., 1991. Global warming and health. British Medical Journal, 302:965–966.

Kleivan, I., 1996. An ethnic perspective on Greenlandic food. In: B. Jacobsen, C. Andreasen and J. Rygaard (eds.). Cultural and Social Research in Greenland 95/96. Ilisimatusarfik/Atuakkiorfik, Nuuk.

Koshida, G. and W. Avis, 1998. Étude pan-canadienne sur les impacts et l'adaptation à la variabilité et au changement climatiques. Tome VII: Questions sectorielles.

Kovats, R.S., B. Menne, A.J. McMichael, C. Corvalan and R. Bertollini, 2000. Climate Change and Human Health: Impact and Adaptation. World Health Organization, 48pp.

Krupnik, I. and D. Jolly (eds.), 2002. The Earth is Faster Now: Indigenous Observations of Arctic Environmental Change. Arctic Research Consortium of the United States, Fairbanks, Alaska, 384pp.

Kuhnlein, H.V. and O. Receveur, 1996. Dietary change and traditional food systems of indigenous peoples. Annual Review of Nutrition, 16:417–442.

Kuhnlein, H.V., O. Receveur, H.M. Chan and E. Loring, 2000. Assessment of Dietary Benefit/Risk in Inuit Communities. Centre for Indigenous Peoples' Nutrition and Environment (CINE), McGill University.

Kvernmo, S. and S. Heyerdahl, 2003. Acculturation strategies and ethnic identity as predictors of behaviour problems in arctic minority adolescents. Journal of the American Academy of Child Adolescent Psychiatry, 42(1):57–65.

Langford, I.H., G. Bentham and A.L. McDonald, 1998. Mortality from non-Hodgkin lymphoma and UV exposure in the European Community. Health and Place, 4:355–364.

Lawn, J. and N. Langer, 1994. Air stage subsidiary monitoring program. Department of Indian Affairs and Northern Development, Ottawa. Final Report, vol. 2: Food Consumption Survey.

Lehmann, B., W. Sauter, P. Knuschke, S. Dressler and M. Meurer, 2003. Demonstration of UV-B-induced synthesis of 1alpha, 25-dihydroxyvitamin D(3) (calcitriol) in human skin by microdialysis. Archives of Dermatological Research, 295:24–28.

Lehmuskallio, E., 1999. Cold protecting ointments and frostbite. A questionnaire study of 830 conscripts in Finland. Acta Dermato-Venereologica, 79(1):67–70.

Li, Y.-F., R. W. Macdonald, L.M.M. Jantunen, T. Harner, T.F. Bidleman and W.M.J. Strachan, 2002. The transport of β-hexa-chlorocyclohexane to the western Arctic Ocean: contrast to α-HCH. Science of the Total Environment, 291(1–3):229–246.

LIA, 1997. Environmental Health Study. Final Report. Labrador Inuit Association, Nain, Labrador.

Linsley, R.K., J.B. Franzini, D.L. Freyberg and G. Tchobanoglous, 1992. Water-Resources Engineering. 4th Edition. McGraw Hill.

Long, C., 2003. UV Index forecasting practices around the world. Scholarly Publishing and Academic Resources.

Lotens, W.A., 1988. Comparison of predictive models for clothed humans. ASHRAE Transactions, 94:1321–1340.

Macdonald, R.W., L.A. Barrie, T.F. Bidleman, M.L. Diamond, D.J. Gregor, R.G. Semkin, W.M.J. Strachan, Y.F. Li, F. Wania, M. Alaee, L.B. Alexeeva, S.M. Backus, R. Bailey, J.M. Bewers, C. Gobeil, C.J. Halsall, T. Harner, J.T. Hoff, L.M.M. Jantunen, W.L. Lockhart, D. Mackay, D.C.G. Muir, J. Pudykiewicz, K.J. Reimer, J.N. Smith, G.A. Stern, W.H. Schroeder, R. Wagemann and M.B. Yunker, 2000a. Contaminants in the Canadian Arctic: 5 years of progress in understanding sources occurrence and pathways. Science of the Total Environment, 254:93–234.

Macdonald, R.W., S.J. Eisenreich, T.F. Bidleman, J. Dachs, J. Pacyna, K. Jones, B. Bailey, D. Swackhamer and D.C.G. Muir, 2000b. Case studies on persistence and long range transport of persistent organic pollutants. In: G. Klecka and D. Mackay (eds.). Evaluation of Persistence and Long-range Transport of Organic Chemicals in the Environment, pp. 245–314. SETAC Press.

Macdonald, R.W., D. Mackay and B. Hickie, 2002. Contaminant amplification in the environment: revealing the fundamental mechanisms. Environmental Science and Technology, 36:457A–462A.

Macdonald, R.W., T. Harner, J. Fyfe, H. Loeng and T. Weingartner, 2003. AMAP Assessment 2002: The Influence of Global Change on Contaminant Pathways to, within, and from the Arctic. Arctic Monitoring and Assessment Programme, Oslo, xi+65pp.

Macintyre, S. and A. Ellaway, 2000. Ecological approaches: rediscovering the role of the physical and social environment. In: L.F. Berkman and I. Kawachi (eds.). Social Epidemiology, pp. 332–358. Oxford University Press.

Mack, T.M. and B. Floderus, 1991. Malignant melanoma risk by nativity, place of residence at diagnosis, and age at migration. Cancer Causes and Control, 2:401–411.

Malzberg, B. and E. Lee, 1956. Migration and Mental Disease. Social Science Research Council.

Maricq, H.R., P.H. Carpenter, M.C. Weinrich, J.E. Keil, A. Franco, P. Dronet, O.C.M. Poncot and M.V. Maines, 1993. Geographic Variation in the Prevalence of Raynaud's Phenomenon: Charleston, SC, USA, vs Taentaise, Savoie, France. Journal of Rheumatology, 20:70–76.

Marmot, M. and R. Wilkinson, 1999. Social Determinants of Health. Oxford University Press.

Mathews-Amos, A. and E.A. Berntson, 2002. Turning up the Heat: How Global Warming Threatens Life in the Sea. World Wildlife Fund and the Marine Conservation Biology Institute.

Maxwell, B., 1997. Responding to Global Climate Change in Canada's Arctic, Volume II of the Canada Country Study: Climate Impacts and Adaptation, Cat. No. Eng56-119/5-197E, 82pp.

Mayer, N. and W. Avis (eds.), 1997. Canada Country Study: Climate Impacts and Adaptations, National Cross Cutting Issues, Volume VIII, Cat. No En56-119/7-1997E.

McGhee, R., 1987. Climate and people in the prehistoric arctic. Northern Perspectives, 15(5):13–15.

McGovern, V., M. Mihm, C. Bailly, J. Booth, W. Clark, A. Cochran, E. Hardy, J. Hicks, A. Levene, M. Lewis, J. Little and G. Milton, 1973. The classification of malignant melanoma and its histologic reporting. Cancer, 32(6):1446–1457.

McKenna, D.B., V.R. Doherty, K.M. McLaren and J.A. Hunter, 2000. Malignant melanoma and lymphoproliferative malignancy: is there a shared aetiology? British Journal of Dermatology, 143:171–173.

McKenna, D.B., D. Stockton, D.H. Brewster and V.R. Doherty, 2003. Evidence for an association between cutaneous malignant melanoma and lymphoid malignancy: a population-based retrospective cohort study in Scotland. British Journal of Cancer, 88:74–78.

McKenzie, R., B. Connor and G. Bodeker, 1999. Increased summer-time UV radiation in New Zealand in response to ozone loss. Science, 285:1709–1711.

McKinlay, A.F. and B.L. Diffey, 1987. A reference action spectrum for ultraviolet induced erythema in human skin. CIE Journal, 6(1):17–22.

McMichael, A.J. and G.G. Giles, 1996. Have increases in solar ultra-violet exposure contributed to the rise in incidence of non-Hodgkin's lymphoma? British Journal of Cancer, 73:945–950.

Merriam, J.C, L. Stefan, R. Michael, P. Söderberg, J. Dillon, L. Zheng and A. Marcelo, 2000. An action spectrum for UV-B radiation and the rat lens. Investigative Ophthalmology and Visual Science, 41:2642–2647.

Messner, T., V. Lundberg and B. Wikstrom, 2003. The Arctic Oscillation and incidence of acute myocardial infarction. Journal of Internal Medicine, 253(6):666–670.

Miller, B.J. and L.R. Chasmar, 1980. Frostbite in Saskatoon: a review of 10 winters. Canadian Journal of Surgery, 23(5):423–426.

Moan, J., A. Dahlback, T. Henriksen and K. Magnus, 1989. Biological amplification factor for sunlight-induced nonmelanoma skin cancer at high latitudes. Cancer Research, 49:5207–5212.

Morison, J., M. Steele and R. Andersen, 1998. Hydrography of the upper Arctic Ocean measured from the nuclear submarine U.S.S. Pargo. Deep-Sea Research I, 45:15–38.

Morris, J.K., D.G. Cook and A.G. Shaper, 1994. Loss of employment and mortality. British Medical Journal, 308:1135–1139.

Moser, K.A., A.J. Fox, D.R. Jones and P.O. Goldblatt, 1984. Unemployment and mortality in OPCS longitudinal study. Lancet, 337:1324–1328.

Muir, B.L., 1991. L'état de santé des Indiens et des Inuit du Canada Santé et Bien-être social Canada, Santé Canada, Ottawa, 64pp.

Muller, H.K., D.J. Lugg and D.L. Williams, 1988. Cutaneous immune responses in Antarctica. A reflection of immune status? Arctic Medical Research, 47(suppl 1):249–251.

Murphy, E., A.L. Kinmonth and T. Marteau, 1992. General practice based diabetes surveillance: the views of patients. British Journal of General Practice, 42:279–283.

Näyhä, S., 2002. Cold and the risk cardiovascular diseases. A review. International Journal of Circumpolar Health, 61:373–380.

Näyhä, S., E. Vaisanen and J. Hassi, 1994. Season and mental illness in an Arctic area of northern Finland. Acta Psychiatrica Scandinavica, 377:46–49.

Nghiem, D.X., J.P. Walterscheid, N. Kazimi and S.E. Ullrich, 2002. Ultraviolet radiation-induced immunosuppression of delayed-type hypersensitivity in mice. Methods, 28(1):25–33.

Nickels, S., C. Furgal, J. Castelden, P. Moss-Davies, M. Buell, B. Armstrong, D. Dillon and R. Fongerm, 2002. Putting the human face on climate change through community workshops. In: I. Krupnik and D. Jolly (eds.). The Earth is Faster Now: Indigenous Observations of Arctic Environmental Change, pp. 300–344. Arctic Research Consortium of the United States, Fairbanks, Alaska.

Nielsen, L.G., M. Frisch and F. Melchers, 1997. Cancer risk in a cohort of Danes working in Greenland. Scandinavian Journal of Social Medicine, 25:44–49.

Nielsen, N.H., H.H. Storm, L.A. Gaudette and A.P. Lanier, 1996. Cancer in circumpolar Inuit 1969–1988. A summary. Acta Oncologica, 35(5):621–628.

Nilssen, O., R. Lipton, T. Brenn, G. Hoyer, E. Boiko and A. Tkatchev, 1997. Sleeping problems at 78 degrees north: the Svalbard Study. Acta Psychiatrica Scandinavica, 95:44–48.

Nilssen, O., T. Brenn, G. Hoyer, R. Lipton, J. Boiko and A. Tkatchev, 1999. Self-reported seasonal variation in depression at 78 degree north. The Svalbard Study. International Journal of Circumpolar Health, 58:14–23.

NOAA National Weather Service, 2003. http://www.cpc.ncep.noaa.gov/products/stratosphere/sbuv2to/sbuv2to_anom.html.

Noonan, F.P. and E.C. De Fabo, 1992. Immunosuppression by ultra-violet B radiation: initiation by urocanic acid. Immunology Today, 13:250–254.

Noonan, F.P. and E.C. De Fabo, 1993. UV-induced immunosuppression. In: A. Young, L.O. Björn, J. Moan and W. Nultsch (eds.). Environmental UV Photobiology, pp. 113–148. Plenum Press.

Noonan, F.P. and H.A. Hoffman, 1994. Control of UV-B immuno-suppression in the mouse by autosomal and sex-linked factors. Immunogenetics, 40:247–256.

Noonan, F.P., J.A. Recio, H. Takayama, P. Duray and M.R. Anaver, 2001. Neonatal sunburn and melanoma in mice. Nature, 413(6853):271–2.

Nowson, C.A. and C. Margerison, 2002. Vitamin D intake and vitamin D status of Australians. Medical Journal of Australia, 177:149–152.

Okada, S., E. Weatherhead, I.N. Targoff, R. Wesley and F.W. Miller, 2003. Global surface ultraviolet radiation intensity may modulate the clinical and immunologic expression of autoimmune muscle disease. Arthritis and Rheumatism, 48:2285–2293.

Osterkamp, T.E., 1982. Potential impacts of a warmer climate on permafrost in Alaska. In: Proceedings of a conference on the Potential Effects of Carbon Dioxide-Induced Climatic Changes in Alaska, April 1982, Misc. Pub. 83-1, University of Alaska, Fairbanks.

Osterkamp, T.E., 1994. Evidence for warming and thawing of discontin-uous permafrost in Alaska. Eos, Transactions, American Geophysical Union, 75(44):85.

Palinkas, L.A., 2001. Mental and cognitive performance in the cold. International Journal of Circumpolar Health, 60(3):430–439.

Parsons, K.C., 1993. Human Thermal Environments. The Effects of Hot, Moderate and Cold Environments on Human Health, Comfort and Performance. Taylor and Francis.

Parsons, P. and P. Musk, 1982. Toxicity, DNA damage and inhibition of DNA repair synthesis in human melanoma cells by concentrated sun-light. Photochemistry and Photobiology, 36:439–445.

Pasco, J.A., M.J. Henry, G.C. Nicholson, K.M. Sanders and M.A. Kotowicz, 2001. Vitamin D status of women in the Geelong Osteoporosis Study: association with diet and casual exposure to sunlight. Medical Journal of Australia, 175:401–405.

Patz, J.A., M.A. McGiihin, S.M. Bernhard, K.L. Ebi, Epstein, P.R. A. Grambsch, D.J. Gubler, P. Rieter, I. Romieu, J.B. Rose, J.M. Samet and J. Trtanj, 2000. The potential health impacts of climate variability and change for the United States: Executive Summary of the report of the health sector of the U.S. national assessment. Environmental Health Perspectives online, http://ehp.niehs.nih.gov/topic/global/patz-full.html.

Petersen, R., 1985. The use of certain symbols in connection with Greenlandic identity. In: J. Brøsted, A. Dahl, H.C. Gray, H.C. Gulløv, G. Henriksen, J.B. Jørgensen and I. Kleivan (eds). Native power: The quest for autonomy and nationhood of indigenous peo-ples. Universitetsforlaget, Bergen.

Quayle, R.G. and Steadman, R.G., 1998. The Steadman windchill: an improvement over present scaling. Weather and Forecasting 13:1187–1193.

Ramsey, J.G., C.L. Burford, M.Y. Beshir and R.C. Jensen, 1983. Effects of workplace thermal conditions on safe work behavior. Journal of Safety Research, 14:105–114.

Redfield, R., R. Linton and M.J. Herskovitz, 1936. Memorandum for the study of acculturation. American Anthropologist, 38:149–152.

Reilly, S.K. and E.C. De Fabo, 1991. Dietary histidine increases mouse skin urocanic levels and enhances UV-B-induced immune suppression of contact hypersensitivity. Photochemistry and Photobiology, 53:431–438.

Rintamäki, H., 2001. Human cold acclimatization and acclimation. International Journal of Circumpolar Health, 60(3):422–429.

Root, T.L., J.T. Price, K.R. Hall, S.H. Schneider, C. Rosenzweig and A.J. Pounds, 2003. Fingerprints of global warning on wild animals and plants. Nature, 421:57–60.

Sam, D.L. and J.W. Berry, 1995. Acculturative stress among young immi-grants in Norway. Scandinavian Journal of Psychology, 36(1):10–24.

Sasaki, H., F. Jonasson, M. Kojima, N. Katoh, M. Ono, N. Takahashi and K. Sasaki, 2000. The Reykjavik Eye Study – prevalence of lens opaci-fication with reference to identical Japanese studies. Ophthalmologica, 214:412–420.

Sasaki, K., H. Sasaki, M. Kojima, Y.B. Shui, O. Hockwin, F. Jonasson, H.M. Cheng, M. Ono, N. Katoh, 1999. Epidemiological studies on UV-related cataract in climatically different countries. Journal of Epidemiology, 9:S33–S38.

Schliessmann, D.J., F.O. Atchley, M.J. Wilcomb and S.F. Welch, 1958. Relation of Environmental Factors to the Occurrence of Enteric Diseases in Areas of Eastern Kentucky. U.S. Public Health Service Publication No. 591. U.S. Government Printing Office, Washington, D.C.

Searles, E., 2002. Food and the making of modern Inuit identities. Food & Foodways, 10:55–78.

Singer, P.C. (ed.), 1999. Formation and Control of Disinfection By-Products in Drinking Water. American Water Works Association, Denver, Colorado.

Slaper, H., G.J. Velders, J.S. Daniel, F.R. de Gruijl and J.C. Van der Leun, 1996. Estimates of ozone depletion and skin cancer incidence to examine the Vienna Convention achievements. Nature, 384:256–258.

Sleijffers, A., J. Garssen and H. Van Loveren, 2002. Ultraviolet radiation, resistance to infectious diseases, and vaccination responses. Methods, 28:111–121.

Sleijffers, A., A. Kammeyer, F.R. De Gruijl, G.J. Boland, J. Van Hattum, W.A. Van Vloten, H. Van Loveren, M.B. Teunissen and J. Garssen. 2003. Epidermal cis-urocanic acid levels correlate with lower specific cellular immune responses after hepatitis B vaccination of ultraviolet B-exposed humans. Photochemistry and Photobiology, 77: 271–275.

Smith, D.W., W.L. Ryan, V. Christensen, J. Crum and G.W. Heinke, 1996. Cold Regions Utilities Monograph. American Society of Civil Engineers, New York, 840pp.

Smith, O. P., 2001. Global Warming Impacts on Alaska Coastal Resources and Infrastructure. Testimony at Fairbanks Congressional Hearing, Fairbanks, Alaska

Snacken, R., A.P. Kendal, L.R. Haaheim and J.M. Wood, 1999. The Next Influenza Pandemic: Lessons from Hong Kong, 1997. Emerging Infectious Diseases, 5:1–11.

Staples, J.A., A.L. Ponsonby, L.L. Lim and A.J. McMichael, 2003. Ecologic analysis of some immune-related disorders, including type 1 diabetes, in Australia: latitude, regional ultraviolet radiation, and disease prevalence. Environmental Health Perspectives, 111:518–523.

Statistics Canada. 1991. Aboriginal Peoples Survey, 1991. http://www.statcan.ca/english/Dli/Data/Ftp/aps.htm.

Statistics Canada, 2003. Health Statistics for the Yukon Territory. Statistics Canada, Ottawa.

Strzhizhovskii, A.D., 1998. Biomedical and economic consequences of stratosphere ozone depletion. Radiatsionnaia Biologiia, Radioecologiia, 38:238–247.

Sutherland, B.M. 1995. Action spectroscopy in complex organisms: potentials and pitfalls in predicting the impact of increased environmental UVB. Journal of Photochemistry and Photobiology B 31:29–34.

Suzuki, T., T. Hirano and M. Suyama, 1987. Free imidazole compounds in white and dark muscles of migratory marine fish. Comparative Biochemistry and Physiology B, 87:615–619.

Tam, J., 2002. Influenza A (H5N1) in Hong Kong: an overview. Vaccine, 20:S77–S81.

Taylor, A., P.F. Jacques, L.T. Chylack Jr, S.E. Hankinson, P.M. Khu, G. Rogers, J. Friend, W. Tung, J. K. Wolfe, N. Padhye and W.C. Willett, 2002. Long-term intake of vitamins and carotenoids and odds of early age-related cortical and posterior subcapsular lens opacities. American Journal of Clinical Nutrition, 75:540–549.

Taylor, H., 1989. Ultraviolet radiation and the eye: an epidemiologic study. Transactions of the American Ophthalmological Society, 87:802–853.

Taylor, J.R., G.J. Schmieder, T. Shimizu, C. Tie and J.W. Streilein, 1994. Interrelationship between ultraviolet light and recurrent herpes simplex infections in man. Journal of Dermatological Science, 8:224–232.

Taylor, M.S., 1992. Cold weather injuries during peacetime military training. Military Medicine, 157(11):602-60-4.

Thouez, J.P.M., B. Singh, P. André and C. Bryant, 1998. Le réchauffement du climat terrestre et les impacts potentiels en géographie des maladies. The Canadian Geographer, 42(1):78–85.

Tingate, T.R., D.J. Lugg, H.K. Muller, R.P. Stowe and D.L. Pierson, 1997. Antarctic isolation: immune and viral studies. Immunology and Cell Biology, 75:275–283.

UNEP, 2003. http://www.unep.org/ozone/pdf/Press-Backgrounder.pdf.

US EPA, 1992. Drinking Water Handbook for Public Officials. U.S. Environmental Protection Agency, EPA-810-B-92-016, Washington, D.C.

USGCRP, 2000. Climate Change Impacts on the United States: The Potential Consequences of Climate Variability and Change. Overview: Alaska. Global Change Research Program, Washington, D.C.

US OTA, 1994. An Alaska Challenge: Native Village Sanitation. Office of Technology Assessment, OTA-ENV-591. U.S. Government Printing Office.

Valero, M.P., A.E. Fletcher, B.L. De Stavola, J. Vioque and V.C. Alepuz, 2002. Vitamin C is associated with reduced risk of cataract in a Mediterranean population. Journal of Nutrition, 132:1299–1306.

Van Oostdam, J., A. Gilman, E. Dewailly, P. Usher, B. Wheatley, H. Kuhnlein, S. Neve, J. Walker, B. Tracy, M. Feeley, V. Jerome and B. Kwavnick, 1999. Human Health Implications of Environmental Contaminants in Arctic Canada: a Review. Elsevier Science, 82pp.

van Strien, G.A. and M.J. Korstanje, 1995. Treatment of contact hypersensitivity with urocanic acid. Archives of Dermatological Research, 287:564–566.

Veenstra, G., 2002. Social capital and health (plus wealth, income inequality and regional health governance). Social Science and Medicine, 54:849–868.

Wallace, J.M. and D.W.J. Thompson, 2002. Annular modes and climate prediction. Physics Today, 55:28–33.

Walterscheid, J.P., D.X. Nghiem and S.E. Ullrich, 2002. Determining the role of cytokines in UV-induced immunomodulation. Methods, 28(1):71–78.

Wang, J. and M. Ikeda, 2001. Arctic sea-ice oscillation: regional and seasonal perspectives. Annals of Glaciology, 33:481–492.

Webster, R., 2002. The importance of animal influenza for human disease. Vaccine, 20:S16–S20.

Wegener, A., M. Heinitz and M. Dwinger, 2002. Experimental evidence for interactive effects of chronic UV irradiation and nutritional deficiencies in the lens. Developments in Ophthalmology, 35:113–124.

Welch, H.E., M.A. Bergmann, T.D. Siferd, K.A. Martin, M.F. Curtis, R.E. Crawford, R.J. Conover and H. Hop, 1992. Energy flow through the marine ecosystem of the Lancaster Sound Region, Arctic Canada. Arctic, 45:343–357.

Weller, G., P. Anderson and B. Wang (eds.), 1999. Preparing for a Changing Climate. The Potential Consequences of Climate Variability and Change: Alaska. A report of the Alaska Regional Assessment Group for the U.S. Global Change Research Program. Center for Global Change and Arctic System Research, University of Alaska, Fairbanks.

West, S., 1999. Ocular ultraviolet B exposure and lens opacities: a review. Journal of Epidemiology, 9:S97–S101.

WHO, 1967. The Constitution of the World Health Organization. WHO Chronicle, 1:29

WHO, 2000. World Health Organization International Statistical Database. http://www.who.int/whosis

WHO, 2002. The World Health Report 2002. World Health Organization, Geneva.

Young, A.R., C.A. Chadwick, G.I. Harrison, O. Nikaido, J. Ramsden and C.S. Potten, 1998. The similarity of action spectra for thymine dimers in human epidermis and erythema suggests that DNA is the chromophore for erythema. Journal of Investigative Dermatology, 111:982–988.

Young, T.K. and C.J. Mollins, 1996. The impact of housing on health: an ecologic study from the Canadian Arctic. Arctic Medical Research, 55(2):52–61.

Zhang, P., M. Nouri, J.L. Brandsma, T. Iftner and B.M. Steinberg, 1999. Induction of E6/E7 expression in cottontail rabbit papillomavirus latency following UV activation. Virology, 263:388–394.

Zheng, T. and P.H. Owens, 2000. Sunlight and non-Hodgkin's lymphoma [letter; comment]. International Journal of Cancer, 87(6):884–886.

Zmudzka, B.Z., S.A. Miller, M.E. Jacobs and J.Z. Beer, 1996. Medical UV exposures and HIV activation. Photochemistry and Photobiology, 64:246–253.

Chapter 16

Infrastructure: Buildings, Support Systems, and Industrial Facilities

Lead Author
Arne Instanes

Contributing Authors
Oleg Anisimov, Lawson Brigham, Douglas Goering, Lev N. Khrustalev, Branko Ladanyi, Jan Otto Larsen

Consulting Authors
Orson Smith, Amy Stevermer, Betsy Weatherhead, Gunter Weller

Contents

Summary

This chapter discusses the potential impacts of climate change on arctic infrastructure. Particular concerns are associated with permafrost warming and degradation, coastal erosion, the stability and maintenance of transportation routes, and industrial development. Adaptation, mitigation, and monitoring techniques will be necessary to minimize the potentially serious detrimental impacts.

Infrastructure is defined as facilities with permanent foundations or the essential elements of a community. It includes schools; hospitals; various types of buildings and structures; and facilities such as roads, railways, airports, harbors, power stations, and power, water, and sewage lines. Infrastructure forms the basis for regional and national economic growth.

Climate change is likely to have significant impacts on existing arctic infrastructure and on all future development in the region. In most cases, engineering solutions are available to address climate change impacts, thus the issue is more economic than technological. It is possible that the uncertainty associated with projections of future climate change will increase the cost of new projects in the Arctic.

Permafrost engineers must address the problem of preserving infrastructure under projected future climate conditions. One solution is to construct new buildings as existing ones are damaged and abandoned. It is possible that this method will be inadequate, since the required rate of new construction rises exponentially using the climate projections presented in this assessment. In areas of warm, discontinuous permafrost, it is very difficult to find economic solutions to address the impacts of climate change on foundations or structures. These areas, together with the coastal zone where the combined problems of increased wave action, sea-level rise, and thermal erosion have no simple engineering solutions, present the greatest challenges in a changing climate.

Projected increases in temperature, precipitation, and storm magnitude and frequency are very likely to increase the frequency of avalanches and landslides. In some areas, the probability of severe impacts on settlements, roads, and railways from these events is very likely to increase. Structures located on sites prone to slope failure are very likely to be more exposed to slide activity as groundwater amounts and pore water pressures increase.

An increasing probability of slides coupled with increasing traffic and population concentrations is very likely to require expensive mitigation measures to maintain a defined risk level. The best way to address these problems is to incorporate the potential for increasing risk in the planning process for new settlements and transportation routes.

16.1. Introduction

There are increased concerns related to the impact of projected climate change on arctic infrastructure, particularly how future climate change may:

- increase the environmental stresses structures are exposed to, particularly in comparison to design specifications, and cause increased risk and damage to infrastructure and threat to human lives;
- affect geohazards and the impacts of extreme events;
- affect natural resource development scenarios in the Arctic; and
- affect socioeconomic development in the Arctic.

Figure 16.1 presents a flow diagram of the questions that need to be answered in order to complete an impact study. Relevant information from indigenous peoples on infrastructure is given in Chapter 3.

16.2. Physical environment and processes related to infrastructure

Chapter 6 has a detailed presentation of the physical environment and processes in the Arctic related to permafrost (section 6.6), snow cover (section 6.4), precipitation (section 6.2), and sea-ice cover and extent (section 6.3), and can be used as a reference for the discussions presented in this chapter.

16.2.1. Observed changes in air temperature

Changes in arctic climate over the past century can be determined by using data from standard climate stations on land and measurements taken on drifting ice floes in the Arctic Ocean. These data show a consistent trend of increasing air temperatures in the Northern Hemisphere during the 20th century, although the

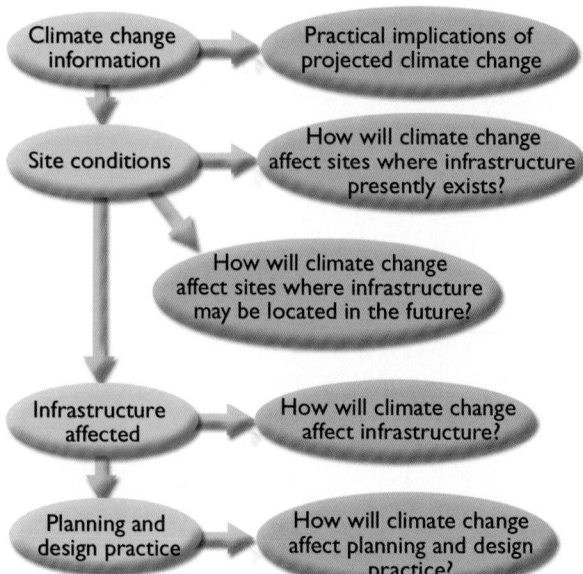

Fig. 16.1. Flow diagram of chapter structure and questions to be answered.

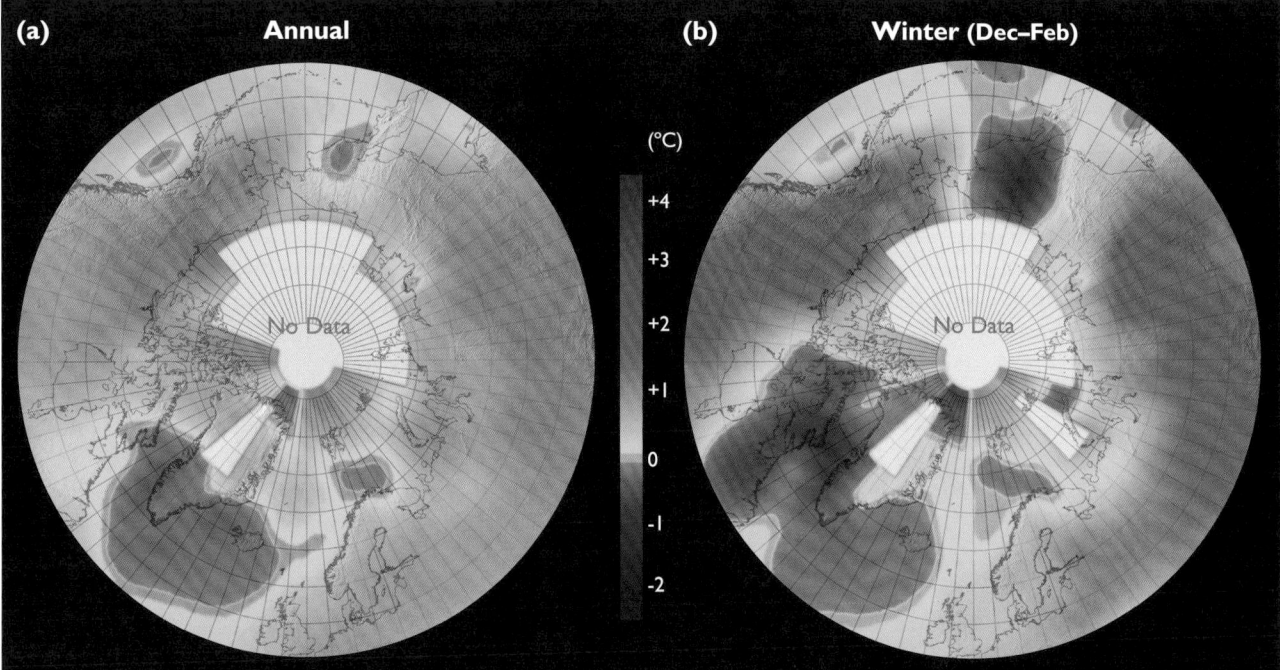

Fig. 16.2. Change in observed surface air temperature between 1954 and 2003: (a) annual mean; (b) winter (Chapman and Walsh, 2003, using data from the Climatic Research Unit, University of East Anglia, www.cru.uea.ac.uk/temperature).

observed changes are not spatially uniform (Anisimov, 2001). While in some regions of the Arctic the warming trend was as great as 5 °C per century, areas of decreasing temperatures were observed in eastern Canada, the North Atlantic, and Greenland (Anisimov and Fitzharris, 2001; Borzenkova, 1999a,b; Jones et al., 1999; Serreze et al., 2000).

Figure 16.2 shows the change in observed surface air temperature between 1954 and 2003 (see also section 2.6.2). Patterns of annual air temperature change indicate that the recent warming has been greatest in Alaska, northwestern Canada, and Siberia (Fig. 16.2a). Temperature increases in winter were much greater than increases in the annual mean temperature: up to 3 to 4 °C over Alaska, northwestern Canada, and Siberia (Fig. 16.2b). In southern Greenland and Iceland, annual mean temperatures decreased by approximately 1 °C, while winter temperatures decreased by 1 to 2 °C. A winter temperature decrease of 1 to 2 °C was also observed in Chukotka.

On the North Slope of Alaska and in northern Siberia, air temperatures increased by 2 to 4 °C, while the global mean air temperature increase over the 20th century was only about 0.6 °C. This pattern is consistent with the hypothesis that the contemporary warming is largely caused by anthropogenic greenhouse gas emissions. Section 2.6.2 discusses observed arctic temperature changes in detail, while section 4.4.2 provides projections of future arctic temperature change.

16.2.2. Permafrost

Permafrost underlies most of the surfaces in the terrestrial Arctic. Permafrost depths vary from a few to

many hundreds of meters (Brown et al., 1997). At selected locations in Yakutia with a cold continental climate, permafrost occurs to depths of 1500 m. Most biogeochemical and hydrological processes in permafrost are confined to the active (seasonally thawed) layer, which varies from several tens of centimeters to several meters in depth. Seasonal thaw depth and the temperature of the frozen ground are two important parameters that must be accounted for in the design of infrastructure built on permafrost. These parameters control key cryogenic processes, such as creep, thaw settlement, adfreeze bond (bond between frozen soil and the material embedded in it), frost heave, and frost jacking (annually repeated foundation uplift caused by frost heave; see Andersland and Ladanyi (1994) for further discussion of these processes). Seasonal thaw depth and frozen-ground temperature both depend on ground-surface temperature, heat flow from the interior of the earth, snow cover, vegetation, and soil properties.

Owing to their low thermal conductivity, snow cover and vegetation (with the underlying organic layer) attenuate annual variations in air temperature and are important regulators of permafrost temperature and depth of seasonal thaw at the local scale. The temperature of permafrost under a thick layer of snow may be several degrees higher than in nearby permafrost that lacks snow cover. In summer, the thermal conductivity of the vegetation and underlying organic layer is typically much smaller than in winter. This reduces summer heat fluxes and keeps permafrost temperatures lower than they would be in the absence of vegetation. A controlled experiment near Fairbanks, Alaska, produced permafrost degradation to a depth of 6.7 m over a 26-year period, simply by removing the insulating layer of vegetation (Linell, 1973). Finally, thermal conductivity is typically 20 to 35% lower in thawed mineral soils than in frozen mineral soils. Consequently, the mean annual temperature below the level of seasonal thawing may be 0.5 to 1.5 °C lower than on the ground surface.

The extreme arctic environment requires unique cold-regions engineering and infrastructure solutions that account for severe climate conditions, the presence of permafrost, and various cryogenic processes that may have destructive effects on structures. Since infrastructure is designed to withstand variations in environmental parameters within a prescribed range, information about past changes in arctic climate and environmental conditions is crucial for developing optimum engineering solutions for future infrastructure and safe management of existing structures.

16.2.2.1. Observed changes in permafrost

Changes in permafrost temperature due to increasing air temperatures were observed in Russia as early as 1970. Pavlov (1997) presented data indicating that the mean annual permafrost temperature increased by 2.0 to 2.5 °C at a depth of 3 m and by 1.0 °C at a depth of 10 m between 1979 and 1995. Observations of soil temperature changes at the Marre-Sale geocryological station in the southwestern region of the Yamal Peninsula (Table 16.1) are especially illustrative. Similar changes have been observed in Alaska (Osterkamp and Romanovsky,

1999) and elsewhere (section 6.6.1.2). Changes in active-layer thickness have also been observed.

The Global Terrestrial Network for Permafrost (GTN-P) and the Circumpolar Active Layer Monitoring (CALM) program were established to monitor such changes. The GTN-P was initiated by the International Permafrost Association (IPA) to organize and manage a global network of permafrost observatories for detecting, monitoring, and projecting climate change (Burgess et al., 2000b). The network, authorized under the Global Climate Observing System and its associated organizations, consists of two observational components: the active layer and the thermal state of the underlying permafrost. CALM, established in 1990, provides the active-layer monitoring component (Brown et al., 2000), while GTN-P provides monitoring of the thermal state of the permafrost. The European Community project, Permafrost and Climate in Europe, contributes to the GTN-P and monitors nine boreholes in mountain permafrost (see also IPCC, 2001).

16.2.2.2. Observed changes in freezing and thawing indices

The strength and deformation characteristics of frozen soils are dependent on soil type, temperature, density, ice content, unfrozen water content, salinity, stress state, and strain rate (section 16.2.2.5). Thawing of frozen soil, or even an increase in the temperature of frozen soil, may lead to deteriorating strength and deformation characteristics, accelerated settlement, and possible foundation failure.

The design of foundations in permafrost regions must, therefore, always include an evaluation of the maximum active-layer thickness and permafrost temperature that may occur in the foundation soils during the lifetime of the structure. The initial and long-term bearing capacity of the foundation can then be determined.

Table 16.1. Soil temperatures measured between 1979 and 1995 at the Marre-Sale station, southwestern Yamal Peninsula, Russia (Pavlov, 1997).

Description of ground surface and subsurface soil type	Depth (m)	Soil temperature (°C) Mean	Soil temperature (°C) Increase
Slope with willow–green moss cover; sand to 0.7 m, loam	3	-5.4	2.2
	6	-5.3	1.2
	10	-5.2	0.8
Horizontal hilly peatland with grass–shrub–moss–lichen cover; peat to 0.75 m, ice, sand	3	-5.6	1.1
	6	-5.6	1.0
	10	-5.6	0.7
Polygonal tundra with moss–lichen–grass–shrub cover; sand	3	-6.5	1.3
	6	-6.5	1.1
	10	-6.4	0.6
Runoff zone on gentle southern slope, cloudberry–sedge–sphagnum–hypnum bog; sand, loam	3	-3.6	0.5
	6	-3.8	0.2
	10	-3.9	0.1
Bottom of dried lakes, meadow bottom with sedge–hypnum bog; peat-enriched sand	3	-3.4	1.1
	6	-3.8	0.7
	10	-3.9	0.5
Hilly and tussocky polygonal tundra covered with shrub, grass, and lichen; sand	3	-4.2	1.2
	6	-4.4	1.0
	10	-4.4	0.7

Table 16.2. Number of meteorological stations in each ACIA region (Instanes and Mjureke, 2002a).

	ACIA region	Number of stations
Iceland	1	1
Svalbard	1	1
Norway and Finland	1	2
Northwest Russia	1	1
Siberia	2	6
Alaska	3	5
Canada	4	3
Greenland	4	2

Region 1: Arctic Europe, East Greenland, European Russian North, North Atlantic
Region 2: Central Siberia
Region 3: Chukotka, Bering Sea, Alaska, western arctic Canada
Region 4: Northeast Canada, Labrador Sea, Davis Strait, West Greenland

Instanes A. (2003) presented a review of the use of air freezing and thawing indices for permafrost engineering design. The air thawing index (ATI) is a useful parameter to determine the "magnitude" of the thawing season and can be used to calculate active-layer thickness and maximum permafrost temperatures. The air thawing index is defined as the integral of the sinusoidal variation in mean daily or monthly air temperature (T) during one year for T >0 °C (the air freezing index, AFI, is defined as the integral of the sinusoidal air temperature variation during one year for T <0 °C).

Ground-surface temperatures differ from air temperatures. If observations of ground-surface temperatures are not available, they can be estimated from air temperatures using an empirically determined n-factor. Andersland and Ladanyi (1994) listed approximate n-factors for different types of surfaces. Variations in snow cover will also affect ground temperatures.

Instanes A. and Mjureke (2002a) carried out an extensive analysis of historic freezing and thawing indices for arctic meteorological stations. Many of these stations have more than 100 years of continuous temperature records. The data used in this analysis were mean monthly air temperatures from Russian datasets provided by O. Anisimov (State Hydrological Institute, St. Petersburg, Russia, 2001).

Twenty-one stations were chosen for this study, using the following criteria:

- the four ACIA regions (section 18.3) should be represented;
- station time series should be of considerable length (>30 years);
- station time series should be continuous;
- priority should be given to meteorological stations located near population concentrations and major infrastructure; and
- stations should be evenly distributed throughout the Arctic.

All the stations are north of 60° N and within the area covered by the Arctic Monitoring and Assessment Programme (AMAP) and the ACIA. Table 16.2 shows the number of stations in each ACIA region.

From an engineering point of view, current and past design levels of thawing and freezing indices are of interest mainly in comparison with values projected for the future. The impact of climate change on arctic

Table 16.3. Percentage of unusually warm summers and unusually warm winters between 1981 and 2000 (Instanes and Mjureke, 2002a).

Station	Location	Observed 1981–2000[a] (%)	Summer Expected[b] (%)	Trend[c]	Observed 1981–2000[d] (%)	Winter Expected[b] (%)	Trend[c]
Akureyri	Iceland	22	17	(+)	19	17	0
Ammassalik	Greenland	0	19	-	7	19	-
Anadyr	Russia	15	20	(-)	18	19	0
Barrow	Alaska	39	24	+	46	24	+
Bethel	Alaska	37	25	+	31	25	(+)
Kuglutuk (Coppermine)	Canada	60	32	+	49	32	(+)
Coral Harbour	Canada	29	35	(-)	37	35	0
Fairbanks	Alaska	52	21	+	28	20	(+)
Fort Smith	Canada	46	24	+	55	24	+
Naryan-Mar	Russia	33	26	(+)	18	26	(-)
Nome	Alaska	50	20	+	33	20	+
Nuuk	Greenland	0	14	-	5	14	-
Salekhard	Russia	35	17	+	29	16	+
Sodankylä	Finland	7	21	-	15	21	(-)
Svalbard Airport	Svalbard	47	25	+	26	25	0
Turukhansk	Russia	20	17	0	33	17	+
Valdez	Alaska	64	24	+	80	24	+
Vardø	Norway	9	13	0	21	13	(+)
Verkhoyansk	Russia	40	17	+	51	17	+
Vilyuysk	Russia	34	19	+	41	19	+
Yakutsk	Russia	22	16	(+)	50	16	+

[a]percentage of years with an air thawing index higher than the mean for the entire period of record plus one standard deviation; [b]the value from an ideal random data series, without trends; [c] + indicates warming, (+) indicates weak or possible warming, 0 indicates no trend, (-) indicates possible or weak cooling, and - indicates cooling; [d]percentage of years with an air freezing index lower than the mean for the entire period of record minus one standard deviation.

infrastructure will be very dependent on how future temperature levels relate to past design levels and historic variability at a specific site. However, an analysis of historic records in terms of freezing and thawing indices can provide indications of temperature increases between 1981 and 2000 (Instanes A. and Mjureke, 2002a).

Table 16.3 presents the percentage of unusually warm summers and unusually warm winters between 1981 and 2000. An unusually warm summer is defined as having an ATI higher than the mean value for the entire station record plus one standard deviation; an unusually warm winter is defined as having an AFI lower than the mean value for the entire station record minus one standard deviation.

Five stations exhibit both significant winter air temperature increases after 1970 and an unusually high frequency of warm winter events between 1981 and 2000. These stations are Fort Smith (Canada), Valdez (Alaska), and Verkhoyansk, Vilyuysk, and Yakutsk (central Russia). Four additional stations had a significantly high frequency of warm winter seasons between 1981 and 2000: Barrow and Nome (Alaska), and Salekhard and Turukhansk (central Russia). Two Greenland stations, Ammassalik and Nuuk, show clear evidence of a recent decrease in winter air temperatures.

Eight stations show evidence of significant recent summer air temperature increases in combination with a significantly high number of very warm summer seasons between 1981 and 2000 compared to the entire period of record. These are Barrow, Fairbanks, Nome, and Valdez (Alaska); Kuglutuk (Coppermine) and Fort Smith (Canada); Svalbard Airport; and Verkhoyansk (Central Russia). Bethel (Alaska) and Salekhard and Vilyuysk (central Russia) had a significantly high frequency of warm summer seasons between 1981 and 2000.

The two Greenland stations and Sodankylä (Finland) show evidence of recent air temperature decreases in both the mean ATI and the extreme summer values.

The results suggest a spatial pattern of recent climate change and are in agreement with results from other studies (e.g., AMAP, 1997). According to Table 16.3, temperatures have increased in central Russia, Alaska, and western Canada, while temperatures have decreased in southern Greenland. The trends are less clear in the Nordic countries and northwestern Russia.

16.2.2.3. Projected changes in permafrost

A constant rate of increase in air temperature is projected to have two related effects on ground temperature:

- an increase in the mean annual temperature at the ground surface, which will slowly propagate to greater depths and, depending on latitude, produce either a thinning or a complete disappearance of the permafrost layer; and
- changes in the annual amplitude of seasonal ground-temperature variation, damped with depth, and affected by related changes in precipitation (snow cover), groundwater hydrology, and vegetation. However, Riseborough (1990) pointed out that at temperatures close to 0 °C, latent heat effects may dominate and result in a smaller amplitude depending on the ice content of the soils.

Climate change is very likely to reduce the area occupied by frozen ground and to cause shifts between the zones of continuous, discontinuous, and sporadic permafrost. These changes can be projected using mathematical models of permafrost driven by scenarios of climate change. Projections of permafrost change in 2030, 2050, and 2080 using output from the five ACIA-designated climate models are presented in section 6.6.1.3.

The potential effects of increasing mean annual ground-surface temperature on permafrost will be very different for continuous and discontinuous permafrost zones. In the continuous zones, increasing air temperatures are very likely to increase permafrost temperatures and possibly increase the depth of the active layer (Burgess

Table 16.4. Comparison of maximum thaw depths simulated for 1990–1999 and 2090–2099 (Instanes A. and Mjureke, 2002b).

	Maximum thaw depth[a] (m)		Increase (%)
	1990–1999	2090–2099	
Barrow	0.7	1	43
Bethel	1.8	13	622
Naryan-Mar	1.4	1.8	29
Nuuk	1.1	1.7	55
Svalbard Airport	0.8	1.1[b]	38
Turukhansk	1.3	1.6	23
Verkhoyansk	1.4	1.5	7

[a]Thaw depths calculated for a theoretical sandy soil layer. Soil profile may not be representative for every location; [b]projection for 2040–2049.

Table 16.5. Mean ground-surface temperature for 1990–1999 (observed) and 2090–2099 (simulated) and the resulting loss in soil bearing strength at 10 m depth between the two periods (Instanes A. and Mjureke, 2002b; strength loss calculated after Ladanyi, 1996).

	Mean ground-surface temperature (°C)		Soil strength loss (%)
	1990–1999	2090–2099	
Barrow	-12	-7	23
Bethel	-2	0	40
Naryan-Mar	-4	-3	12
Nuuk	-2.5	-0.5	34
Svalbard Airport	-6	-4[a]	17
Turukhansk	-7	-5	15
Verkhoyansk	-16	-12	14

[a]Projection for 2040–2049.

et al., 2000a; Esch and Osterkamp, 1990; Osterkamp and Lachenbruch, 1990). In the discontinuous zone, the effects of a few degrees increase in the mean annual permafrost temperature are very likely to be substantial (Harris, 1986). Since the temperature of most of this permafrost is presently within a few degrees of the melting point, the permafrost is likely to disappear. Except for the southernmost zone of sporadic permafrost, many centuries will be required for the frozen ground to disappear entirely. However, increases in active-layer depth and thawing of the warmest permafrost from the top have already been observed (Burgess et al., 2000a; Esch and Osterkamp, 1990; Harris, 1986; Osterkamp and Lachenbruch, 1990).

Anisimov et al. (1997) used a permafrost model and climate scenarios for 2050 produced by general circulation models (GCMs) to project changes in active-layer thickness in the Arctic. The results of this study indicated that changes in active-layer thickness will vary by region, increasing by 10 to 15% to more than 50% between the mid-1990s and 2050. Instanes A. and Mjureke (2002b) used the ACIA-designated models (section 4.2.7) to project changes in active-layer thickness and maximum permafrost temperature for seven of the sites in Table 16.3: Barrow, Bethel, Naryan-Mar, Nuuk, Svalbard Airport, Turukhansk, and Verkhoyansk. The analysis used an identical soil profile with the same thermal properties for all the locations; therefore, it can only be used as an indication of relative climate differences between sites. The increase in maximum thaw depths between 1990–1999 and 2090–2099 and the changes in mean ground-surface temperature and soil bearing strength between 1990–1999 and 2090–2099 are presented in Tables 16.4 and 16.5, respectively.

The response of permafrost to climate change involves an important temperature threshold associated with

phase change beyond which future temperature increases will cause thawing of the frozen ground. The time required to reach this temperature threshold depends on the initial permafrost temperature and the rate of temperature increase. Table 16.6 presents projected changes in various types of permafrost soils for different rates of warming.

The projections discussed in this section suggest that a progressive increase in active-layer depth and temperature of the frozen ground is likely to be a relatively short-term reaction to climate change in permafrost regions. Changes in seasonal thaw depth are very likely to change the water-storage capacity of near-surface permafrost at local and regional scales, with substantial effects on vegetation, soil hydrology, and runoff, which will ultimately lead to changes in larger-scale processes such as landslides, erosion, and sedimentation.

With respect to cold-regions engineering and infrastructure in locations affected by permafrost, the temperature of the frozen ground and the depth of seasonal thawing is of critical importance for effective construction planning and the evaluation of potentially hazardous situations at existing facilities. Although the effects of an increase in mean annual air temperature on permafrost can be projected in a general sense, it is more difficult to project these effects for specific locations and regions. Factors such as microclimate, as well as soil type, ice content, and salinity will play a role, and may not necessarily be well known or readily projected (Riseborough, 1990; Smith M. and Riseborough, 1983, 1985).

16.2.2.4. Projected changes in freezing and thawing indices

Freezing and thawing indices were calculated using mean monthly air temperatures projected for 2000 to

Table 16.6. Projected changes in permafrost soils between 2000 and 2100 for different rates of increase in mean annual air temperature and different soil types (Parmuzin and Chepurnov, 2001).

Warming trend (°C/yr)	T_{ini} (°C)	T_{2100} (°C)			Year thawing begins			Thaw depth in 2100 (m)		
		0.06	0.03	0.01	0.06	0.03	0.01	0.06	0.03	0.01
Sands, loamy sands, loams	-7 to -9	-2 to -4	-5 to -7	-6 to -8	no permafrost thawing			<1.5	<1.2	<0.8
Sands								<6	<1.3	<0.8
Loamy sands, loams	-5 to -7	-1 to -2	-2 to -4	-4 to -6	2080–2090	no permafrost thawing		<3.5		
Peat								<1.5		
Sands								6–13	<6	<1.2
Loamy sands, loams	-3 to -5	-0.5 to -1	-1 to -2	-2 to -4	2050–2070	2080–2100	little thawing	3.5–8	<4	
Peat								1.5–3		
Sands								13–20	10–15	<1.2
Loamy sands, loams	-1 to -3	0 to -0.5	0 to -1	-0.5 to -2	2010–2040	2030–2080	little thawing	8–15	6–8	
Peat								3–5	<4	
Sands								15–25	13–20	10–15
Loamy sands, loams	0 to -1	~0			2000–2010	2010–2030	2060–2090	10–16	7–12	1.5–4
Peat								4–6	<5	<1.5

T_{ini}: initial mean annual temperature of permafrost soils; T_{2100}: mean annual temperature of permafrost soils projected for 2100.

2100 by the ACIA-designated climate models for the 21 stations shown in Table 16.3 (Instanes A. and Mjureke, 2002b). Output from four of the five ACIA-designated models (CGCM2 – Canadian Centre for Climate Modelling and Analysis, CSM_1.4 – National Center for Atmospheric Research, GFDL-R30_c – Geophysical Fluid Dynamics Laboratory, and HadCM3 – Hadley Centre for Climate Prediction and Research) was used for the analysis, along with a composite four-model mean (MEAN4).

In addition, results from empirical downscaling (Hanssen-Bauer et al., 2000) using the 2 m air temperature field from the ECHAM4/OPYC3 GSDIO integration (see section 4.6.2) were applied to Svalbard Airport.

Figures 16.3 and 16.4 show observed and projected freezing and thawing indices for Kugluktuk (Coppermine), Canada, from 1933 to 2100. The indices for 1933 to 2000 were calculated from meteorological observations, while the indices from 2000 to 2100 are based on output from the ACIA-designated models.

The figures show that the projections based on output from the different models "fit" the observed record to varying degrees. This is one of the major problems with using GCMs for impact studies. However, indices computed using output from the different models show generally similar trends. This suggests that the raw model output can probably be adjusted so that computed indices start where the observations leave off, providing better projections of future trends in freezing and thawing indices.

Plots similar to Figs. 16.3 and 16.4 showing observed and projected freezing and thawing indices for the 21 stations in Table 16.3 are reported by Instanes A. and Mjureke (2002b).

16.2.2.5. Engineering concerns

The physical and mechanical properties of frozen soils are generally temperature-dependent, and these dependencies are most pronounced at temperatures within 1 to 2 °C of the melting point. Esch and Osterkamp (1990) summarize most of the engineering concerns related to permafrost warming as follows.

- Warming of permafrost body at depth.
  a. Increase in creep rate of existing piles and footings.
  b. Increased creep of embankment foundations.
  c. Eventual loss of adfreeze bond support for pilings.
- Increases in seasonal thaw depth (active layer).
  a. Thaw settlement during seasonal thawing.
  b. Increased frost-heave forces on pilings.
  c. Increased total and differential frost heave during winter.
- Development of residual thaw zones (taliks).
  a. Decrease in effective length of piling located in permafrost.
  b. Progressive landslide movements.
  c. Progressive surface settlements.

Frozen-ground behavior

A constant rate of surface temperature increase due to projected climate change is very likely to lead to an increase in active-layer thickness. Woo et al. (1992), Kane et al. (1991), and Nakayama et al. (1993) attempted to simulate numerically the increase in active-layer thickness projected to result from climate change. Comparable simulations have been performed for three locations in the Mackenzie Basin, Canada (Burgess et al., 2000a).

In contrast to frozen rocks and dense gravels, whose strength depends mainly on mineral bonds and internal friction, the bulk of the mechanical strength of fine-grained frozen soils is due to ice bonding. Rising surface temperatures are likely to increase the unfrozen water content of fine-grained soils and decrease the ice bonding (cohesion) of soil particles, resulting in a gradual loss of strength in these soils.

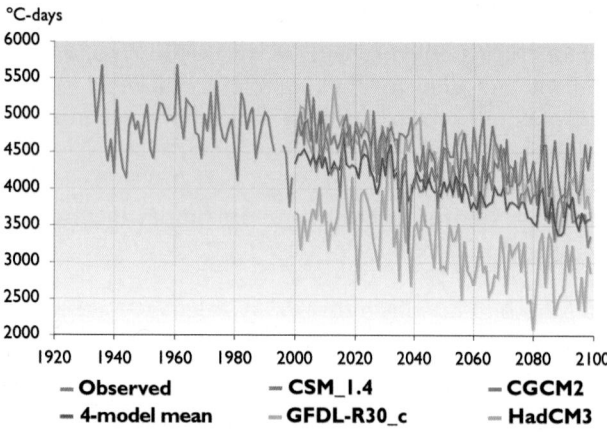

Fig. 16.3. Observed and projected freezing indices for Kugluktuk (Coppermine), Canada (Instanes A. and Mjureke, 2002b).

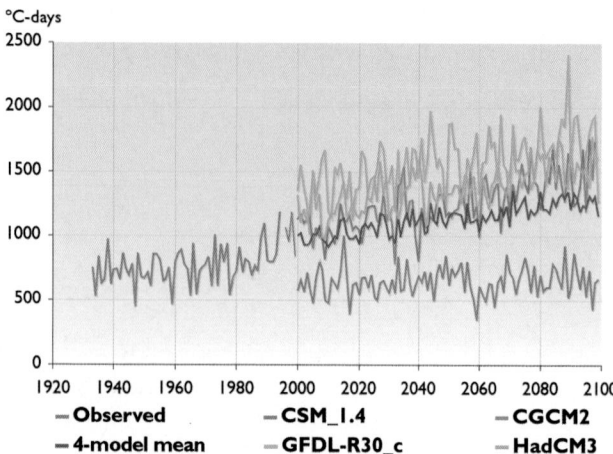

Fig. 16.4. Observed and projected thawing indices for Kugluktuk (Coppermine), Canada (Instanes A. and Mjureke, 2002b).

Soil and rocks can be classified by their sensitivity to climate change, similar to the classification normally used in permafrost engineering. In order of increasing sensitivity (defined by the potential impacts of climate change on strength and thaw settlement), geological materials are classified as follows.

1. Rocks
   - dense, with ice only in pores; and
   - shattered, with ice filling cracks and fissures (an existing rock mass classification system can be used for evaluating the degree of fragmentation and fissures).
2. Gravels and sands (according to their density and moisture content).
3. Silts (according to their density and moisture content).
4. Clays (according to their density and moisture content).
5. Organic soils and peat.
6. Ground ice.

Andersland and Ladanyi (1994) provided a more detailed classification of frozen soils.

Frozen soil will settle to a certain extent when completely thawed. For a given soil type, the amount of thaw settlement can be related to the increase in active-layer thickness, the soil bulk or dry density, and its ice saturation or total water content. Several correlations between the unit thaw settlement and the physical properties of frozen soils have been published (Haas and Barker, 1989; Johnson et al., 1984; Johnston, 1981; Ladanyi, 1994; McRoberts et al., 1978; Nixon, 1990a; Speer et al., 1973). One such correlation relates the percentage of thaw settlement to the frozen bulk density and is the preferred methodology for engineering purposes (first published by Speer et al., 1973, and completed by Johnston, 1981). In the last

20 years, several such correlations between thaw settlement and frozen bulk density for a wide range of frozen soils have been published (Haas and Barker, 1989; Leroueil et al., 1990; Nelson R. et al., 1983), and some have also been expressed by empirical equations. "Thaw sensitivity maps" for specific permafrost regions can be created using information from climate models, surficial geology maps, organic soil maps, ground temperature data, and the above-mentioned correlations. Smith S. et al. (2001), Smith S. and Burgess (1998, 1999, 2004) and Nelson F. et al. (2002) have constructed such maps for Canada and the circumpolar Arctic.

The strength of frozen soil depends not only on temperature, but also on soil density, ice content, and salinity. It is also affected by the degree of confinement and the applied strain rate.

The sensitivity of frozen-soil strength to a temperature increase can be expressed by the ratio:

$$S_T = \frac{\Delta q_{fi}}{q_{fi}}$$

Eqn. 16.1

where q_{fi} is the strength at temperature $\theta_i = -T_i$, and Δq_{fi} is its variation due to a temperature increase ΔT_i (see Fig. 16.5). The strength sensitivity index can also be expressed in terms of frozen soil creep parameters (Ladanyi, 1995, 1996, 1998).

The strength sensitivity index, S_T, defined in equation 16.1, may be a useful measure for evaluating the loss of strength in frozen soils in regions where climate change is not projected to cause complete permafrost thawing. The index requires information about the temperature sensitivities of strength and creep in typical arctic soils. Although some information already exists, further laboratory and field tests of permafrost soils are required. By combining information about permafrost occurrence, soil types and characteristics, and projected climate change, it may be possible to construct maps of projected effects on permafrost. Such maps would show not only the projected trends in active-layer depth and permafrost thawing, but also the projected reduction in permafrost strength. Permafrost sensitivity maps of this kind would be useful for projecting the effects of climate change on existing facilities in the Arctic, and for establishing guidelines for the design of new facilities. Vyalov et al. (1988) proposed the delineation of permafrost sensitivity zones in the Arctic, based on the mean annual ground temperature of permafrost (often measured at the level of negligible annual temperature amplitude, 10 to 20 m below the surface).

Frost heave is the result of ice lenses developing as soils freeze. Temperatures below 0 °C and frost-susceptible soils are required for frost heave to occur, while the availability of water and the freezing rate determine the degree of frost heave. The first two

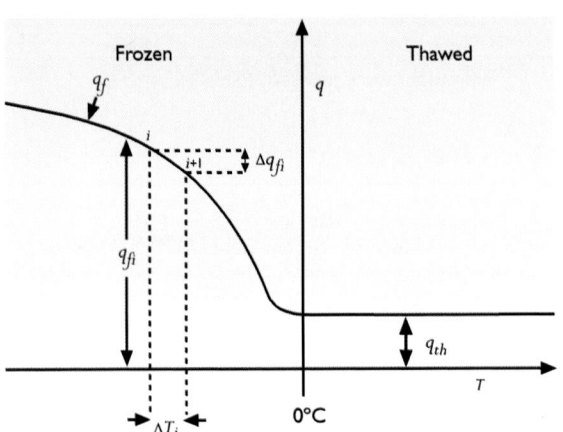

q = soil strength
q_f = soil strength of frozen ground
q_{th} = soil strength of thawed ground
q_{fi} = soil strength at a given temperature
$q_{fi} = q_{fi} - q_{f(i+1)}$ = change in soil strength for a specified temperature change, ΔT_i

Fig. 16.5. Variation in frozen-soil strength with temperature (Ladanyi, 1996).

conditions generally do not differ much between permafrost and seasonal freezing regions. However, the availability of groundwater for ice accumulation in the active layer is different in the two regions. The active layer is generally thinner in permafrost regions, thus the freezing rate is rapid and there is less time available for ice lens growth. In addition, the presence of nearly impermeable permafrost below the active layer may limit the water available for lens formation in permafrost areas, so that for comparable soil conditions there is less frost heave in permafrost regions than in regions of seasonal freezing. An increase in the mean annual air temperature is very likely to increase the thickness of the layer subjected to freeze-thaw cycles and subsequent frost heave.

Thaw settlement and pile creep

An increase in the mean annual air temperature in permafrost regions is very likely to lead to an increase in the thickness of the active layer, resulting in increased thaw settlement during seasonal thawing; and is very likely to lead to a decrease in frozen-ground creep strength (long-term strength of frozen soil), resulting in an increase in the creep settlement rate of existing piles and footings.

Numerical simulations that assume a specific rate of warming have been used to project the degree of settlement effects on existing and future structures in the Arctic. Nixon (1990a,b, 1994) used a one-dimensional geothermal model and assumed a mean surface temperature increase of 0.1 °C/yr for 25 years to examine the effects on thaw depth and pile creep settlement. The simulation of thaw depth below insulated surfaces in discontinuous permafrost projects a doubling of thaw depth after 25 years compared to a case with no temperature increase.

Thawing of permafrost soils can result in subsidence of the surface, thermokarst, and activation of freeze–thaw related processes such as solifluction. Parmuzin and Chepurnov (2001) projected soil subsidence in sandy loam soils by 2100 given different rates of warming and different soil ice content (Table 16.7). Other studies have projected the thaw settlement potential for Mackenzie Basin soils (Aylsworth et al., 2000; Burgess et al., 2000a; Burgess and Smith, 2003).

Such projections of the possible consequences of climate change may help inform the design of future facilities in permafrost regions.

16.2.2.6. Areas south of the permafrost border

In the Arctic and subarctic, there are large land areas south of the permafrost border that experience frost action during winter. Annual freezing of the top soil layer commonly causes frost heave of foundations and structures. Highway structures and embankments located above the frost-heave zone usually experience increased surface roughness and bumps (Andersland and Ladanyi, 1994). During the spring thaw, the bearing capacity of the structure may be considerably reduced, causing breakup of the pavement structure and failure of the embankment. It is possible that projected climate change will reduce the problems associated with winter frost action in these areas.

16.2.2.7. Summary

It is possible that projected climate change will be a factor in engineering projects if its effects go beyond those anticipated within the existing conservative design approach. Therefore, engineering design should take into account projected climate change where

Table 16.7. Projected soil subsidence between 2000 and 2100 due to the thawing of frozen deposits in sandy loam soils (Parmuzin and Chepurnov, 2001).

Initial (2000) permafrost soil temperature (°C)	Volumetric ice content of soil (%)	Air temperature increase (°C/yr)		
		0.06	0.03	0.01
		Subsidence by 2100 (m)		
-7 to -9	>40	no subsidence	no subsidence	no subsidence
-5 to -7	>40	<1.5		
	20–40	0.4–0.7	no subsidence	no subsidence
	<20	<0.5		
-3 to -5	>40	1.5–3.5	0.5–1.0	
	20–40	0.7–1.5	<0.5	no subsidence
	<20	<0.7	<0.1	
-1 to -3	>40	3.5–6.0	1.0–1.5	
	20–40	1.5–3.5	0.5–1.0	no subsidence
	<20	<1.5	<0.5	
0 to -1	>40	5.5–6.5	3.5–5.0	<0.5
	20–40	3.5–5.5	1.5–3.5	<0.3
	<20	<3.5	<0.5	<0.1

appropriate and where the potential effects represent an important component of the geothermal design.

The sensitivity of permafrost soil strength to projected climate change can be mapped using a simple strength sensitivity index, such as the one proposed in this section. A risk-based procedure for analyzing structures based on their sensitivity to the potential consequences of climate change is a reasonable approach to incorporating climate change concerns into the design process (section 16.4.1). The project-screening tool developed and currently in use in Canada is a very good guideline for such an approach (Bush et al., 1998).

16.2.3. Natural hazards

In some regions of the Arctic, climate change is projected to lead to increasing temperature and precipitation (sections 4.4.2, 4.4.3, and 6.2.3) and increasing storm frequency (Hanssen-Bauer and Forland, 1998). The type of precipitation is very likely to change as the temperature increases. Where the average winter temperature is close to 0 °C, a higher frequency of precipitation falling as rain instead of snow is expected. Runoff from the arctic river basins is likely to increase due to greater snow depth resulting from increased winter precipitation and to increased thawing of permafrost resulting from surface warming in summer. Greater winter snow depth coupled with rapid melting caused by higher spring temperatures is likely to increase the possibility of floods in the arctic river basins and increase erosion in thawing permafrost riverbanks. Thawing permafrost and increasing depth of the thawed layer are likely to make slopes vulnerable to slides caused by erosion, increasing pore water pressure, and earthquakes.

Floods and slides in soil, rock, and snow are directly or indirectly connected to weather phenomena. Slides in soil and rock can also be triggered by earthquakes. Most structures in the Arctic are located and designed based on historic observations of extreme weather events to meet defined criteria for acceptable risk. Climate change is very likely to change the probability of natural hazard occurrences. This implies that criteria for the location and design of infrastructure must be revised to keep risks at defined levels.

16.2.3.1. Infrastructure and natural hazards

Settlements are often located in areas of low hazard risk to avoid floods, mudflows, slides, and avalanches. River embankments are designed to control rivers during extreme flood events. The location and design of communities and structures are determined based on the risk of hazard occurrence (e.g., permanent settlements in Norway are only permitted in areas where the annual probability of natural hazards is less than 1×10^{-3}). Highways and railways crossing steep terrain are located where the risk of closure and accidents due to natural hazards is acceptable, or can be mitigated by protection facilities (e.g., snow sheds).

Houses, highways, roads, railways, transmission lines, and other infrastructure are sometimes located in areas exposed to snow accumulation and drifting. Highways and railways may be subjected to traffic restrictions or closure by high wind velocities and related snow drifting. To avoid dangerous snow accumulation, regulations in some areas dictate that houses and transmission lines are located in terrain where snow depths are acceptable or appropriate protection has been installed. At present, regulations governing the design of these facilities are based on acceptable risks of extreme snow depth, ice loads, wind forces, and storm frequencies.

16.2.3.2. Factors affecting slope stability and failure

Acceptable risk is directly related to the probability of slides and avalanches. Factors important to slope stability include the groundwater regime, and erosion caused by surface water flow, freeze–thaw processes, and human activity. The groundwater regime is affected by precipitation and meltwater infiltration. For a specific slope, the probability of slides can often be related to threshold values for water infiltration caused by the intensity and duration of rain and snowmelt.

Snow accumulation in avalanche release areas in mountainous regions of the Arctic (Scandinavia, Iceland, Russia, and North America) is dependent on wind velocity and duration in addition to the intensity of snowfall. Avalanche probability depends primarily on the rate of snow accumulation in the release area. The probability of slush avalanches (where water-saturated snow releases as a slide) is related to the porosity and permeability of the snow, which play a key role in snow stability. Slush avalanches release when the rate of water infiltration by rain and snowmelt reaches a threshold value for the specific type of snow on the slope.

Thawing of permafrost caused by climate change will possibly also influence the stability of a particular site. As shown in section 16.2.2.5, the strength of frozen soil drops rapidly as the temperature rises above 0 °C. The development of a weak saturated layer between frozen and unfrozen material can trigger landslides; slopes along arctic rivers are particularly sensitive to failure due to erosion of the toe of the slope (Dyke et al., 1997; Dyke, 2000). Such landslides are known as active-layer detachment slides or skin flows, which can also be triggered by forest fires that burn away the insulating organic layer, leading to increased absorption of solar radiation and more rapid thaw of the active layer.

Reservoirs are used to control flooding in some watersheds, but most arctic watersheds are unregulated. Flood intensity is dependent on precipitation and snowmelt rates and is tempered by the ability of the soil ability to absorb water.

Mudflows and debris flows are triggered as a consequence of a rapid increase in pore water pressure together with runoff-induced erosion (Sandersen et al., 1996). They occur during periods of intense rainstorms or as a consequence of rapid melting of snow and ice. Severe mudflows can also occur as a result of rapid drainage of glacier-dammed lakes due to glacial melting; examples of this phenomenon include the catastrophes in the Sima Valley (Norway) in 1893 and 1937, and in the city of Tyrnyauz (Caucasus) in 1977 and 1992 (Seinova, 1991; Seinova and Dandara, 1992).

16.2.3.3. Potential impacts of climate change on avalanche and slide activity

The ACIA-designated climate models forced with the B2 emissions scenario project that mean annual arctic temperatures (60°–90° N) will increase 1.2 °C by 2011–2030, 2.5 °C by 2041–2060, and 3.7 °C by 2071–2090 compared to the 1981–2000 baseline (5-model average, see section 4.4.2). The increase is projected to have an uneven spatial distribution, with the greatest increase in the Russian and Canadian Arctic, and the smallest increase in areas close to the Atlantic and Pacific coasts (IPCC, 2001). The changes are also projected to vary seasonally, with the greatest temperature increases occurring in winter.

The Norwegian Meteorological Institute projects that temperatures in Norway will increase by 0.2 to 0.7 °C between 2000 and 2010. The greatest increase is projected to occur in part of the Norwegian Arctic in winter (Haugen and Debenard, 2002). Regional downscaling of temperature projections for other regions of the Arctic has not been performed, but would be useful for assessing the potential impacts of climate change on natural hazards.

As a consequence of rising temperatures, the ACIA-designated models project average increases in precipitation of 4.3% by 2011–2030, 7.9% by 2041–2060, and 12.3% by 2071–2090 compared to the 1981–2000 baseline (section 4.4.3). Precipitation increases are projected to be greatest in winter, and smallest in summer (when a decrease is projected for some Russian watersheds).

Changes in the extent of snow cover in the Arctic are very likely to be influenced by both temperature and precipitation. Increasing temperature in a region is very likely to lead to earlier spring snowmelt and reduced snow cover extent at the end of the winter. Conversely, increasing precipitation is very likely to lead to greater snow depth in winter, especially in the coldest parts of the region. The ACIA-designated models project decreases in arctic snow-cover extent of 3–7%, 5–13%, and 9–18% by 2011–2030, 2041–2060, and 2071–2090, respectively, compared to the 1981–2000 baseline (section 6.4.3). The decrease in snow extent between the baseline (1981–2000) and 2071–2090 is projected to be greatest in spring (4.9 x 10⁶ km²) and winter (3.8 x 10⁶ km²), and lowest in summer (1.1 x 10⁶ km²) and autumn (3.3 x 10⁶ km²).

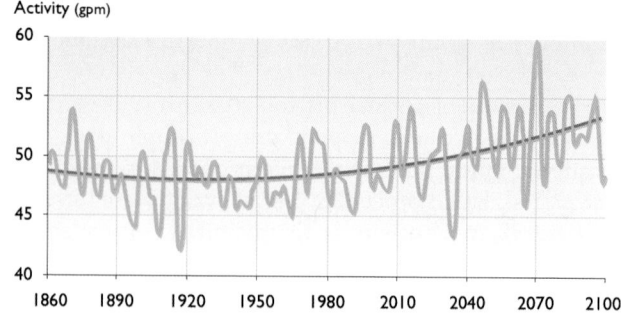

Fig. 16.6. Storm track activity (geopotential meters – gpm) over northwest Europe projected by the ECHAM4/OPYC greenhouse-gas scenario (4-yr running mean). The non-linear climate trend obtained from quadratic curve fitting is marked by the smooth curve (Ulbrich and Christoph, 1999, cited in IPCC, 2001).

Storms also affect avalanche and slide activity, and are projected to increase in frequency (5 to 10%) and amplitude over the 21st century (IPCC, 2001; see Fig. 16.6). For the west coast of Norway, the Norwegian Meteorological Institute projects a higher frequency of storms and greater amplitudes of storm activity over the next 50 years, combined with a 20% precipitation increase and a temperature increase of 2 to 3 °C, with the greatest change occurring in winter (Hackett, 2001; Haugen et al., 1999).

Avalanche activity depends on the rate of snow accumulation, which is dependent on temperature (<0 °C), precipitation rate, and storm frequency. A change in any of these factors is very likely to have an impact on avalanche activity. Increasing snow precipitation is likely to occur in areas with continental climates and high-altitude coastal regions, leading to an increase in avalanche activity. For example, an increase in precipitation rates and storm magnitude is likely to increase snow accumulation intensity in high-altitude avalanche release areas. Because avalanche run-out distance is related to the volume of snow released, it is possible that greater snow accumulation will cause longer run-outs than have historically occurred, resulting in increased risk to settlements and infrastructure.

As snow accumulation primarily occurs at temperatures below 0 °C, snow-cover extent and depth will depend on the duration of the frost period and the precipitation environment in any given area. In regions of the Arctic with long cold periods and low precipitation, changes in temperature and precipitation will have a negligible influence on avalanche activity. Conversely, in regions where winter temperatures are presently close to 0 °C and precipitation rates are high, the snow environment has a high sensitivity to changes in climate. For example, Naryan-Mar in northwest Russia (67.6° N, 53° E) has a typical arctic continental climate, with an average January temperature of -18.9 °C and a snow cover less sensitive to temperature change than Vardø in Norway (70.37° N, 31.1° E), which has a typical maritime climate and an average January temperature of -5.1 °C.

In the continental Russian, Canadian, and Alaskan regions of the Arctic, the winter is long and cold, with few periods of temperature above 0 °C. The coastal areas of Scandinavia and Iceland and the west coast of North America have shorter and warmer winters with a higher frequency of temperature fluctuations around 0 °C. A winter temperature change of a few degrees in cold continental regions of the Arctic is very likely to affect the duration of the winter, but not very likely to affect the snow environment.

As temperature is dependent on altitude, mountain areas with relatively low average temperatures are very likely to be less affected by temperature change than lower-elevation coastal areas. At low altitudes in a warmer maritime climate such as that of Scandinavia, the frequency of precipitation falling as rain is very likely to increase in the future. The frequency of snow avalanches with release areas at low altitudes (below 500 to 1000 m) is likely to be reduced due to this change in precipitation type. Increases in the frequency of rainstorms and intensity of storm precipitation are likely to lead to an increasing frequency of mudflows, as well as an increasing frequency of slush flows where the rate of thaw together with intense rain precipitation is the triggering mechanism (Hestnes, 1994). A higher frequency of winter rain events is very likely to increase the number of wet snow and slush avalanches. However, the duration of slush flow activity in the Arctic is projected to be shorter (Sidorova et al., 2001). The frequency of mudflows and debris flows is projected to increase, as the summer season is projected to be longer, with greater amounts of precipitation and a higher frequency of extreme events such as rainstorms and storm-induced flooding (Glazovskaya and Seliverstov, 1998). For Iceland, storm frequency and precipitation falling as rain are projected to increase along the east coast and decrease along the west coast, owing to a reduction in the ice sheet along the Greenland coast (Olafsson, pers. comm., University of Iceland, Reykjavik, 2003).

In maritime climates, the frequency of avalanches with long run-out distances is likely to decrease, owing to a projected change in snow type (from dry-snow to wet-snow and slush avalanches). This is very likely to have a positive effect on transportation routes in some areas. As the frequency of dry-snow avalanches with long run-out distances decreases, the exposure of highways and buildings to avalanches will be reduced.

Increased precipitation is projected to influence groundwater flow. Higher temperatures will probably also increase the thaw rate in spring and summer, increasing groundwater flow and flood potential. In low-altitude areas where snow is presently the predominant form of winter precipitation, an increase in winter rain events is likely to lead to a higher probability of slides in rock and soils (Fig. 16.7). Slopes that are stable under the current precipitation regime are likely to gradually become unstable if the frequency and magnitude of rainstorms increase (Sandersen et al., 1996), leading to a potential increase in rock and soil slide activity until a new equilibrium is established.

A change in the groundwater regime is also likely to affect the pore water pressure in quick clays (materials that can change rapidly from solid to liquid state) that are typical of some fjord districts in the Scandinavian and Canadian Arctic (Bjerrum, 1955; Janbu, 1996; Larsen et al., 1999), and may cause instability of these materials. Quick-clay slides (Fig. 16.8) have caused serious disasters with loss of lives and properties. Together with increased floods and erosion by rivers, higher amounts of groundwater are very likely to increase the risk of quick-clay slides in the future.

As shown in sections 6.6.1.2 and 6.8.2, observed air temperature changes in the Arctic have increased the thaw depth in permafrost areas and increased the discharge of water to the Arctic Ocean from Eurasia (Shiklomanov, et al., 2002). These changes are causing numerous slides in permafrost riverbanks (Fig. 16.9).

Fig. 16.7. Debris slide in a saturated moraine during spring thaw, Lofoten, northern Norway, 1998 (photo: Jan Otto Larsen, Norwegian Public Roads Administration, Oslo).

Fig. 16.8. Slide in a quick-clay deposit, Verdalen, Norway (photo: Jan Otto Larsen, Norwegian Public Roads Administration, Oslo).

Fig. 16.9. River erosion and slides in a permafrost riverbank with exposed ice wedge (photo: Edward C. Murphy, University of Alaska, Fairbanks).

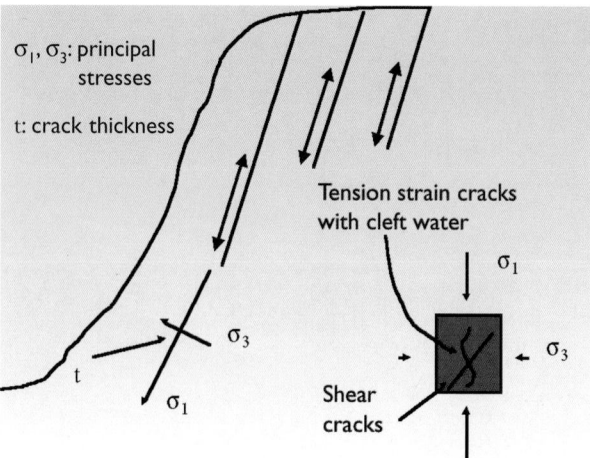

Fig. 16.10. Landslides in hard rock with joints parallel to the slope. Water accumulation in cracks is an important factor triggering failure (figure and photo provided by Jan Otto Larsen, Norwegian Public Roads Administration, Oslo, 2003).

Increasing storm frequencies are very likely to increase closure periods of wind-exposed roads, highways, railroads, and airports, and are likely to affect industries and other human activities dependent on transportation. For example, an increase in the frequency of closed roads is very likely to have an impact on the fishing industry in Norway where immediate transport of fresh fish to the European market is essential.

Greater amounts of precipitation combined with increased rates of snowmelt is likely to increase water infiltration in rock and influence the cleft water pressure in tension cracks in high mountain slopes (Terzaghi, 1963). The cleft water pressure depends on the permeability of the rock and the rate of infiltration; higher cleft water pressures can increase the probability of landslides in hard rock (Fig. 16.10).

16.2.3.4. Summary

Climate change is projected to increase precipitation frequencies and magnitudes, and it is possible that the frequency and magnitude of storms will increase in some regions. An increase in the frequency and magnitude of storms is very likely to lead to increased closures of roads, railways, and airports. Increases in temperature and precipitation together with increases in storm magnitude and frequency are very likely to increase the frequency of avalanches and slides in soil and rock. In some areas, the probability that these events will affect settlements, roads, and railways is likely to increase. Structures located in areas prone to slope failure are very likely to be more exposed to slide activity as groundwater amounts and pore water pressure increase. It is possible that floods of greater magnitude will occur due to greater amounts of precipitation and higher rates of snowmelt. Increased erosion due to higher river flows and thawing permafrost is very likely to initiate slope failure in riverbanks, exposing infrastructure such as buildings, harbors, and communication lines to potential damage. In low-altitude areas with maritime climates, increased temperatures and more precipitation falling as rain are likely to result in a higher frequency of wet-snow avalanches where dry-snow avalanches dominate at present. This is likely to reduce avalanche run-out distance and related problems on exposed traffic routes. The frequency and extent of slush-flow avalanches are likely to increase in the future. An increasing probability of slides coupled with increases in traffic and population concentrations is very likely to lead to expensive mitigation measures to maintain a defined risk level. The best way to address these problems is to incorporate the potential for increasing risk in the planning process for new settlements and transportation routes such as roads and railways.

16.2.4. Coastal environment

The Arctic has approximately 200 000 km of coastline, most of which is uninhabited. However, coastal development is critical to the economy and social well-being

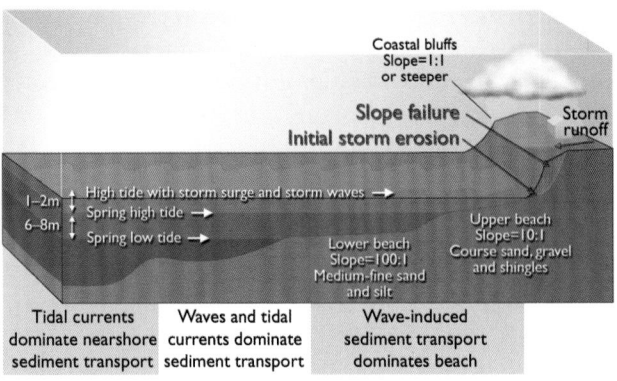

Fig. 16.11. Processes affecting bluff erosion in Cook Inlet, south-central Alaska (redrawn from Smith O. and Levasseur, 2002).

of nearly all arctic residents (see also Chapters 3, 12, and 15). Natural-resource development is concentrated along the coast, and the development of resources in these remote areas is constrained because of challenging transportation routes (Smith O. and Levasseur, 2002).

Arctic coastal dynamics are often affected directly or indirectly by the presence of permafrost. Permafrost coasts are especially vulnerable to erosive processes as ice beneath the seabed and shoreline melts from contact with warmer air and water. Thaw subsidence at the shore allows additional wave energy to reach unconsolidated erodible materials. Low-lying, ice-rich arctic permafrost coasts are the most vulnerable to thaw subsidence and subsequent wave-induced erosion.

The southern half of Alaska has coastal characteristics dominated by erodible glacial deposits and high tides. Cook Inlet, in south-central Alaska, has a 10 m tidal range at its northern extreme and an eroding shoreline of glacially deposited bluffs, as illustrated in Fig. 16.11 (Smith O. and Levasseur, 2002). Freezing of brackish water and ice deposition on broad tidal flats create huge blocks of "beach ice" (Fig. 16.12) that, when set afloat by higher tides, can carry coarse sediments for distances of over a hundred kilometers. Of all Cook Inlet sea ice, these sediment-laden ice blocks present

the greatest danger to ships in winter. The complex dynamics of bluff erosion and ice-borne sediment transport will become even more difficult to forecast with sea-level rise and a more erratic storm climate.

16.2.4.1. Observed changes in the coastal environment

Coastal erosion rates vary considerably across the Arctic. As indicated in Fig. 16.13, erosion rates are dependent on environmental forcing, sedimentology, geocryology, geochemistry, and anthropogenic disturbance of the coastline. In fine-grained icy–silty–clayey sediments, average erosion rates are typically 1 to 3 m/yr, while in silty–sandy sediments with high ice content that are directly exposed to waves and storm surges, erosion rates can be as high as 10 to 15 m/yr under extreme weather conditions. Frozen rock and sediments with low ice content may have erosion rates as low as 0.1 m/yr. Anthropogenic disturbance of the coastline can increase erosion rates, but there are also examples from Varandei in the Pechora Sea indicating that shore protection techniques can slow erosion rates. Table 16.8 presents examples of erosion rates along the arctic coast.

Erosion rates have increased along the arctic coast over the past 30 years. Coastal residents are concerned about the observed changes and the future of arctic coastal communities. However, arctic coastal survey data are often inadequate to reliably quantify accelerating shoreline retreat. Baseline surveys using satellite imagery will help the assessment of erosion rates and systematic planning of future responses significantly. Understanding of circumpolar coastal dynamics is also inadequate. Improved understanding of the physical processes involved in arctic coastal erosion will improve techniques for shore protection and other mitigation measures that may be necessary in the future.

The Arctic Coastal Dynamics project (ACD) is a recent international initiative, sponsored by the International Arctic Sciences Committee and the IPA, to address coastal change in the Arctic (Rachold et al., 2002,

Fig. 16.12. Beach ice in macro-tidal zones of Alaska (photo provided by Orson Smith, University of Alaska, Anchorage).

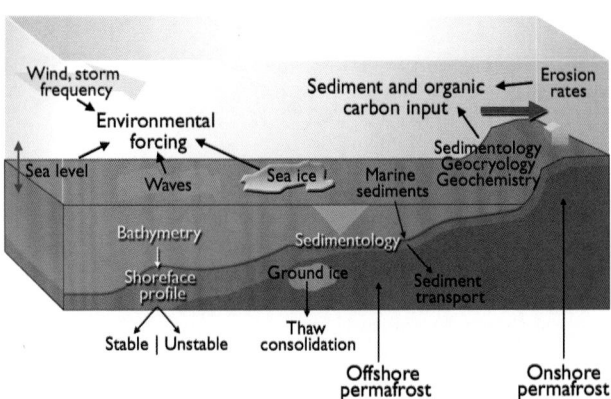

Fig. 16.13. Processes affecting coastal erosion in the Arctic (redrawn from Rachold et al., 2002).

2003). Its objective is to improve the understanding of coastal dynamics as a function of environmental forcing (including climate change), coastal geology and cryology, and changes in landforms (Fig. 16.13). The ACD has proposed:

- to establish the rates and magnitudes of erosion and accumulation along arctic coasts (e.g., Table 16.8);
- to develop a network of long-term monitoring sites including local community-based observation sites; and
- to develop empirical models to assess the sensitivity of arctic coasts to environmental variability and change and human impacts.

Output from studies such as those proposed by the ACD can be used for site-specific evaluation of arctic coastal areas. This type of research and monitoring is essential to document ongoing and future changes and to make policy recommendations.

16.2.4.2. Projected changes in the coastal environment

Arctic coastal conditions are likely to change as climate changes. Thinner, less extensive sea ice is very likely to improve navigation conditions along most northern shipping routes, such as the Northwest Passage and the

Northern Sea Route (see also section 16.3.7). However, decreasing sea-ice extent and thickness is very likely to affect traditional winter travel and hunting where sea ice has been used for these purposes.

Greater expanses and longer periods of open water are likely to result in wave generation by winds over longer fetches and durations. Wave energy is constrained by wind speed, duration of winds, extent of fetch, and water depth. Wave-induced coastal erosion along arctic shores is likely to increase as climate changes. Sea-level rise and thaw subsidence of permafrost shores are projected to exacerbate problems of increased wave energy at the coast. If more frequent and intense storms accompany climate change, these are also likely to contribute to greater wave energy.

Global sea level is rising, although the contribution of various factors to this rise is still being debated (section 6.9.2). The volume of water in the oceans is increasing due to thermodynamic expansion of seawater and melting of ice caps and glaciers. Rising sea level inundates marshes and coastal plains, accelerates beach erosion, exacerbates coastal flooding, and increases the salinity of bays, rivers, and groundwater. Some northern regions, such as the southern coasts of Alaska, have sea-level trends complicated by tectonic rebound of the land resulting from the retreat of continental glaciers at the end of the last glacial period. At Sitka, in south-

Table 16.8. Arctic coastal erosion rates measured at various sites (Brown et al., 2003; Rachold et al., 2002, 2003).

	Country	Time period	Average erosion rate (m/yr)
Yugorsky Peninsula, Kara Sea	Russia	1947–2001	0.6–1[a]
			1.3–1.6[b]
Herschel Island, Beaufort Sea	Canada	1954–1970	0.7
		1970–2000	1.0
		1954–2000	0.9
Barrow, Alaska, Chukchi Sea	United States	1948–1997	0.4–0.9
		1948–2000	1–2.5[c]
Pesyakov Island, Pechora Sea	Russia	2002	0.5–2.5[d]
Varandei, Pechora Sea	Russia	2002	1.8–2.0[e]
Varandei Island, Pechora Sea	Russia	1970–1980	7–10
		1980–1990	1.5–2[f]
		1987–2000	3–4
Maly Chukochiy Cape, East Siberian Sea	Russia	1984–1988	3.5
		1988–1990	1.5
		1990–1991	3
		1991–1994	5
		1994–1999	4.4
		1984–1999	4.3
		projected future	1.8[g]
Kolguev Island, Barents Sea	Russia	not available	1–2[h]
		not available	0.1–0.2
Beaufort Sea	Canada	not available	1–5

[a]scarp; [b]bluff; [c]Elson Lagoon; [d]Holocene terrace; [e]thermoabrasion coast; [f]shore protection; [g]based on complete melting of ice bodies; [h]includes thermo-erosion of ice-rich sediments.

east Alaska, the net effect is falling sea level. However, arctic coasts have a wide variation of tectonic trends. Low-lying coastal plains in the Arctic are generally not tectonically active, which is another reason why they are vulnerable to the adverse effects of sea-level rise. Global sea-level rise will possibly allow more wave energy to reach the coast and induce erosion as waves break at the shore.

Reduced sea-ice extent and thickness are very likely to provide opportunities for the export of natural resources and other waterborne commerce over new northern shipping routes. Reduced sea ice in the coastal regions of the Arctic Ocean is very likely to result in longer navigation seasons along the Russian Arctic coast, in the Canadian Arctic, along the eastern and western coasts of Greenland, and around Alaska. Enhanced regional arctic navigation, such between Europe and the Kara Sea, and potential trans-arctic voyages using icebreaking container ships and tankers, is very likely to shorten distances between markets and improve the delivery times for valuable products.

Climate change is also likely to change the use of arctic rivers for transportation routes, water sources, and habitat. Increased precipitation is very likely to result in higher stream flows and more flooding. Conditions for commercial river navigation will possibly improve the transport of minerals and bulk exports to tidewater. Erosion of thawing permafrost banks is very likely to accelerate, threatening the infrastructure of rural arctic river communities. River-ice breakup is very likely to occur earlier and ice jams and flooding risks are likely to be more difficult to project. The projection and prevention of ice-jam flooding warrant further study.

Higher sea levels at the mouths of rivers and estuaries are likely to allow salt to travel further inland, changing riparian habitats. Furthermore, climate change is projected to result in more frequent and intense storms accompanied by stronger winds. These winds are very likely to induce even higher water levels at the coast, accompanied by higher waves. Storms are also

likely to result in more intense rainfall at the coast, increasing runoff-related erosion and the mobile sediment in coastal waters.

Coastal communities are sensitive to climate change. Engineering solutions are available for shore protection (flood barriers, dikes, breakwaters, erosion control) but may not be able to reduce erosion rates sufficiently to save specific settlements. Moreover, while these protective measures may address one problem, they may create another by altering the dynamics of erosion and deposition processes.

16.2.5. Arctic Ocean

16.2.5.1. Observed changes in sea-ice extent

Climatic and environmental changes in the Arctic Basin include changes in air temperature, water temperature and salinity, and the distribution, extent, and thickness of sea ice. There is compelling empirical evidence of consistent environmental changes across the Arctic Ocean, including increases in air temperature, reductions in sea-ice extent, and freshening of the Beaufort Sea mixed layer (Maslanik et al., 1996; McPhee et al., 1998). Data from ice-floe measurements show a slight air temperature increase with statistically significant warming in May and June between 1961 and 1990. Air temperature anomalies in the Arctic Basin have been strongly positive since 1993. Between 1987 and 1997, the mean annual air temperature increased by 0.9 °C (Aleksandrov and Maistrova, 1998), comparable to temperature changes observed in the terrestrial Arctic.

The area of warm Atlantic waters in the polar basin increased by almost 500 000 km^2 over the past three decades (Kotlyakov, 1997), and the inflowing freshwater has warmed (Carmack, 2000; Carmack et al., 1995). Measurements from submarines indicate that surface waters in the Arctic Ocean basin warmed by 0.5 to 1 °C from the mid-1970s to the mid-1990s, with maximum warming observed in the Kara Sea (Alekseev et al., 1997).

One of the most valuable and graphic records of sea-ice extent changes in the Northern Hemisphere was produced by Walsh and Chapman (2001). Figure 16.14 shows a 103-year record (1900–2002) of sea-ice extent. A decreasing trend in sea-ice extent, starting in the mid-20th century, is evident for all four seasons. The greatest decreases have occurred in summer and spring: over the past 50 years, summer sea-ice extent has decreased by nearly 3 x 10^6 km^2. Although the decrease in sea-ice extent has been unevenly distributed around the coastal margins of the Arctic Ocean, it has provided greater marine access for ships.

Satellites have recorded increasing areas of open water along the Russian Arctic coast and in the Beaufort Sea. Figure 16.15 is a satellite passive microwave image showing the extent of arctic sea ice on 22 September

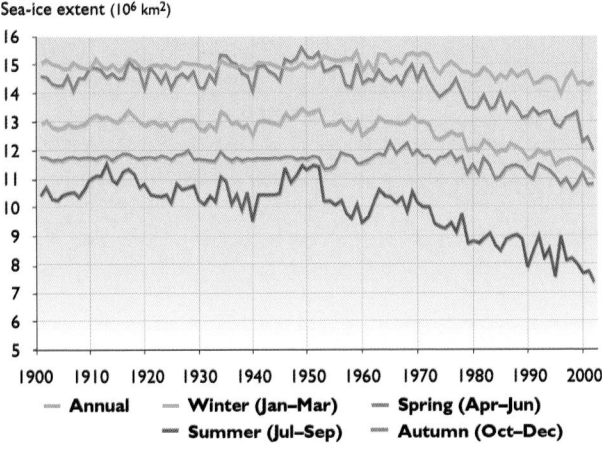

Sea-ice extent (10^6 km^2)

— **Annual**   — **Winter (Jan–Mar)**   — **Spring (Apr–Jun)**
— **Summer (Jul–Sep)**   — **Autumn (Oct–Dec)**

Fig. 16.14. Sea-ice extent in the Northern Hemisphere between 1900 and 2002 (Walsh and Chapman, 2001, updated).

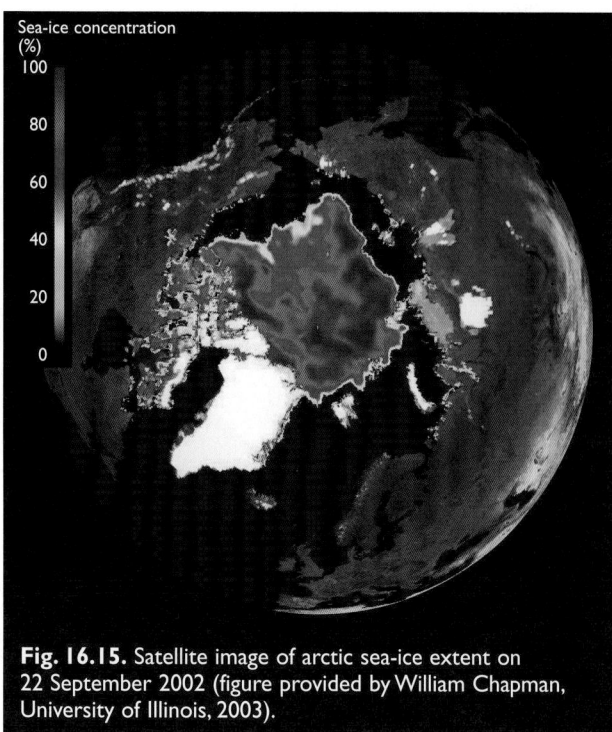

Fig. 16.15. Satellite image of arctic sea-ice extent on 22 September 2002 (figure provided by William Chapman, University of Illinois, 2003).

2002 – the date of the minimum observed extent in the 103-year record. This image illustrates the large areas of open water surrounding the Arctic Basin at the summer minimum extent of sea ice. Of significance for the Northern Sea Route, the only pack ice reaching the Russian Arctic coast is along the northern tip of Severnaya Zemlya; the sea ice in this image has also retreated record distances in the Beaufort, Chukchi, and East Siberian Seas.

Sea-ice conditions in the Canadian Arctic are very complex. Observations of minimum sea-ice extent in the

eastern and western regions of the Canadian Arctic between 1969 and 2003 (Fig. 16.16) illustrate the extraordinary interannual variability of the ice conditions. Although the trends in sea-ice extent are negative in both regions over the period shown, the year-to-year variability is extreme and sometimes differs between the two regions. For example, one of the largest observed minimum extents in the western region (for the period shown) occurred in 1991, while in the eastern region the minimum sea-ice extent that year was relatively low. While these observations indicate a recent overall decrease in the extent of sea ice in the Northwest Passage, the interannual and spatial variability is not conducive to planning a reliable marine transportation system.

16.2.5.2. Projected changes in sea-ice extent

Figure 16.17 shows the median of the sea-ice extents projected by the five ACIA-designated climate models for the three ACIA time slices. However, an important limitation of the ACIA-designated models is that they cannot resolve the complex geography of the Canadian Arctic and thus cannot provide adequate sea-ice projections for this region. In summer, the models project a substantial retreat of sea ice throughout the entire Arctic Ocean for each ACIA time slice, except for parts of the Canadian Archipelago and along the northern coast of Greenland. By mid-century (September 2041–2060), most of the alternative routes in the Northwest Passage and Northern Sea Route are projected to be nearly ice-free; three of the five models project open water conditions across the entire lengths of both. By the end of the 21st century, vast areas of the Arctic Ocean are projected to be ice-free in summer, increasing the possibility of shipping across the Arctic Ocean.

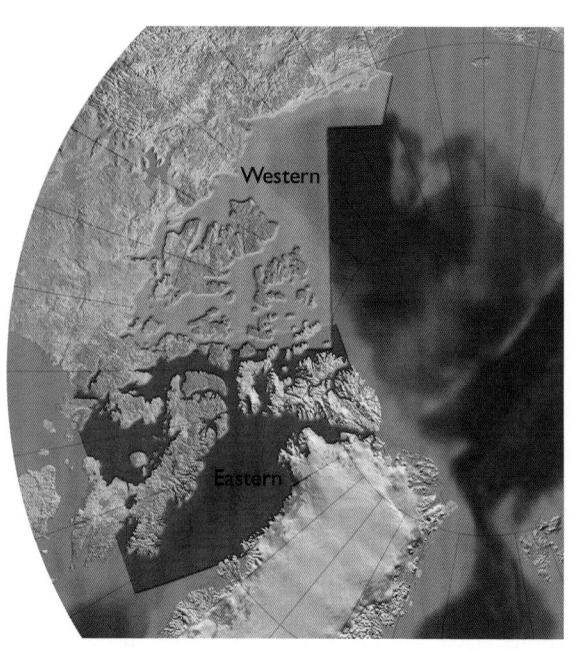

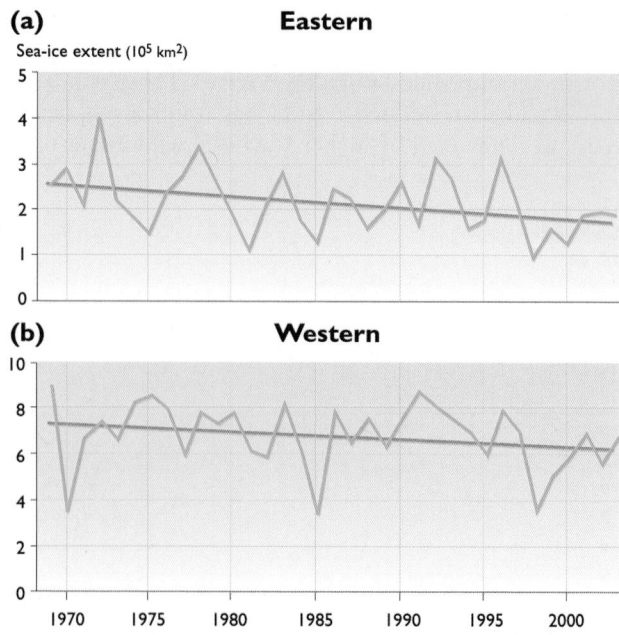

Fig. 16.16. Annual minimum sea-ice extent in (a) the eastern and (b) the western regions of the Canadian Arctic between 1969 and 2003, from a composite of remotely sensed and ground-based measurements. Lavender lines show linear trends (redrawn from figure provided by the Canadian Ice Service, 2004).

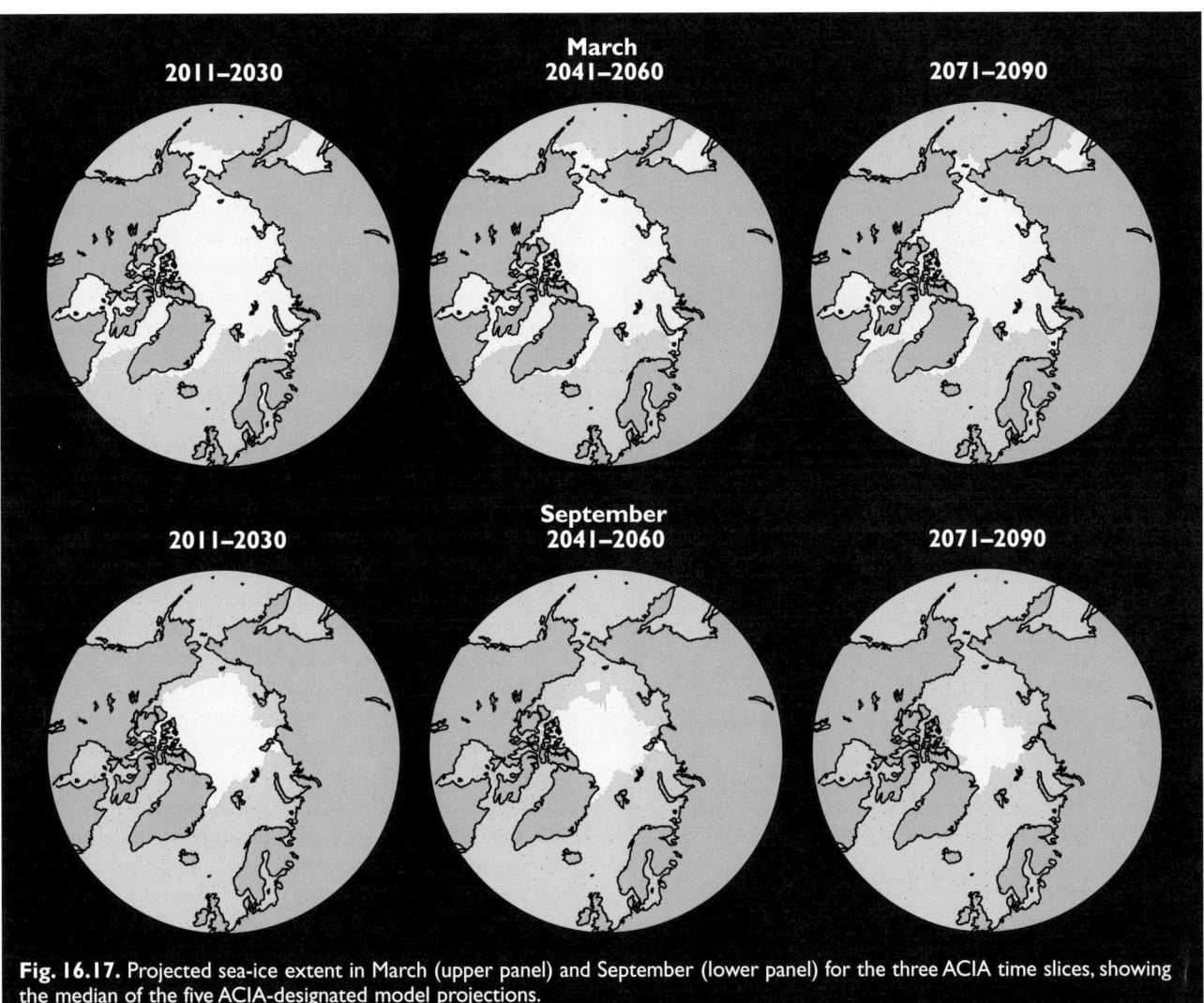

Fig. 16.17. Projected sea-ice extent in March (upper panel) and September (lower panel) for the three ACIA time slices, showing the median of the five ACIA-designated model projections.

Although there is some projected winter retreat of the sea-ice edge, particularly in the Bering and Barents Seas where ship access is likely to improve throughout the 21st century, most of the Arctic Ocean is projected to remain ice-covered in winter. However, sea-ice cover is likely to be on average thinner, and the area of multi-year sea ice in the central Arctic Ocean is likely to decrease through 2080. Winter navigation (throughout the year to Dudinka since 1978) along the western end of the Northern Sea Route, from the Barents Sea to the mouths of the Ob and Yenisey Rivers, will possibly encounter less first-year sea ice. Full transit of the Northern Sea Route through Vilkitskii Strait to the Bering Sea is very likely to remain challenging and require icebreaker escort. The only improvement in winter sea-ice conditions for the Laptev, East Siberian, and western Chukchi Seas is a possible reduction in the area and frequency of multi-year floes along the navigable coastal routes.

The significant interannual sea-ice variability in the arctic seas is also important to navigation. The projected increase in ice-free areas and reduced sea-ice concentrations will possibly lead to a more dynamic sea-ice cover under the influence of local and regional wind fields. These more variable environmental condi-

tions, combined with an increase in the number of ships making passages in the Arctic Ocean, will possibly lead to a higher demand for sea-ice information, long-range ice forecasting, and increased icebreaker support. While direct (across the North Pole) trans-arctic voyages are unlikely by mid-century, voyages may be possible north of the arctic island groups (such as those along the Russian Arctic coast) and away from shallow continental shelf areas that restrict navigation.

There is likely to be an increased requirement for real-time satellite imagery of sea-ice conditions. Increased ship access to the Arctic Ocean is very likely to require more resources to facilitate marine traffic and to support maritime safety and protection of the arctic marine environment.

The four ACIA regions are not conveniently drawn for assessment of future arctic marine transport routes. Therefore, as a case study, a new "sector" was defined between 60° and 90° N, and between 40° E and 170° W, which includes the coastal region from the eastern Barents Sea to Bering Strait. This sector also encompasses Russia's Northern Sea Route, defined by regulation as the routes or waterways between Kara Gate (at the southern end of Novaya Zemlya) and Bering Strait.

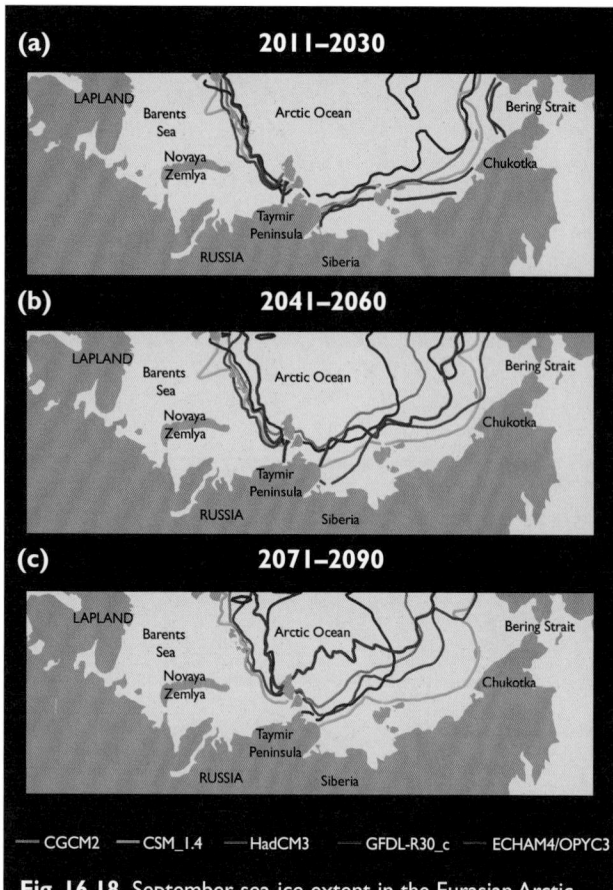

Fig. 16.18. September sea-ice extent in the Eurasian Arctic projected by each of the five ACIA-designated climate models.

As shown in Fig. 16.17, the median of the ACIA-designated model projections suggests that, in all the time slices, most coastal waters of the Eurasian Arctic (along the Northern Sea Route) will be relatively ice-free in September. Figure 16.18 illustrates the September sea-ice extents projected by each of the five models for the Eurasian Arctic sector used in this analysis. During September, the month of minimum sea-ice extent in the Arctic Ocean, the models consistently project sea ice in the vicinity of Severnaya Zemlya during each of the ACIA time slices. For ships to travel north of Severnaya Zemlya, highly capable, more expensive icebreaking ships would be necessary. However, it is more likely that ships would sail through the deep waters of Vilkitskii Strait to the south — between the Kara and Laptev Seas north of the Taymir Peninsula — where more open water is very likely to be found. Four of the five models also project open water to the east and north of the New Siberian Islands in September. Ships sailing along the Northern Sea Route are likely to take advantage of these ice-free conditions to avoid the shallow, narrow passages along the Eurasian coast. In addition, if there is a continued reduction in the proportion of winter multi-year sea ice in the Central Arctic Ocean (Johannessen et al., 1999), it is very likely that first-year sea ice will dominate the entire maritime region of the Northern Sea Route throughout the year, with a decreasing frequency of intrusions of multi-year sea ice into the coastal seas. Such changes in sea-ice conditions would have key

implications for ship construction (e.g., potentially lower construction costs) and route selection along the Northern Sea Route in summer and winter.

16.3. Infrastructure in the Arctic

Infrastructure is defined as facilities with permanent foundations or the essential elements of a community. It includes schools; hospitals; various types of buildings and structures; and facilities such as roads, railways, airports, pipelines, harbors, power stations, and power, water, and sewage lines. Infrastructure forms the basis for regional and national economic growth. In the Arctic, the largest population concentrations are located in North America and Russia (Freitag and McFadden, 1997), as is much of the existing infrastructure.

Most arctic facilities are connected with population concentrations, extraction of natural resources, or military activity. In the case of industrial or military developments, facilities typically include industrial buildings and warehouses, crew and worker quarters, embankments (roadway, airport, work pad), pipelines (both chilled and warm), and excavations of different types. Many industrial buildings must be designed to accommodate heavy equipment, increasing demands on the foundation system. In the case of human settlements, infrastructure includes public transportation systems, utility generation and distribution facilities, and buildings associated with residential or business activities. In addition to the complexity presented by the wide range in requirements of the different types of infrastructure, the problems associated with climate change are compounded by the range of environmental conditions over which human activities occur, extending from the sporadic permafrost/seasonal frost zone to the high Arctic with its cold, continuous permafrost layer.

Some of the engineering projects that are likely to be affected by climate change are as follows (see also Bush et al., 1998; Nixon, 1994).

- Northern pipelines are likely to be affected by frost heave and thaw settlement. Slope stability is also likely to be an issue in discontinuous permafrost (Nixon et al., 1990).
- The settlement of shallow pile foundations in permafrost could possibly be accelerated by temperature increases over the design life of a structure (~20 years). However, over the same period, there is very likely to be less of an effect on the deeper piles used for heavier structures (Nixon et al., 1990).
- Large tailings disposal facilities might be affected (negatively or positively) by climate change, due to the long-term effects on tailings layers. There is some chance that layers that freeze during winter deposition in northern seasonal-frost or permafrost areas would thaw out many years later, releasing excess water and contaminants into groundwater. There is some chance that

increasing temperature would significantly change the rate of thawing of such layers.

• The availability of off-road transportation routes (e.g., ice roads, snow roads) is likely to decrease owing to a reduction in the duration of the freezing season. The effect of a shorter freezing season on ice and snow roads has already been observed in Alaska and Canada (section 16.3.6).

• Climate change is likely to reduce ice-cover thickness on bodies of water and the resulting ice loading on structures such as bridge piers. However, until these effects are observed, it is unlikely that engineers will incorporate them into the design of such structures.

• The thickness of arctic sea-ice cover is also likely to change in response to climate change, and it is possible that this will affect the design of off-shore structures for ice loadings, and the design of ice roads used to access structures over landfast ice in winter.

• Precipitation changes are very likely to alter runoff patterns, and possibly the ice–water balance in the active layer. It is very difficult to assess the potential effects of these changes on structures such as bridges, pipeline river crossings, dikes, or erosion protection structures.

• The stability of open-pit mine walls will possibly be affected where steep slopes in permafrost overburden have been exposed for long periods of time. The engineering concerns relate to increased thaw depth over time, with consequent increased pore pressures in the soil and rock, and resulting loss of strength and pit-wall stability (Szymanski et al., 2003; Bush et al., 1998).

• The cleanup and abandonment of military and industrial facilities throughout the Arctic sometimes involves storage of potentially hazardous materials in permafrost or below the permafrost table. There is some chance that permafrost degradation associated with climate change will threaten these storage facilities. AMAP (1998) provides a detailed review of arctic pollution issues.

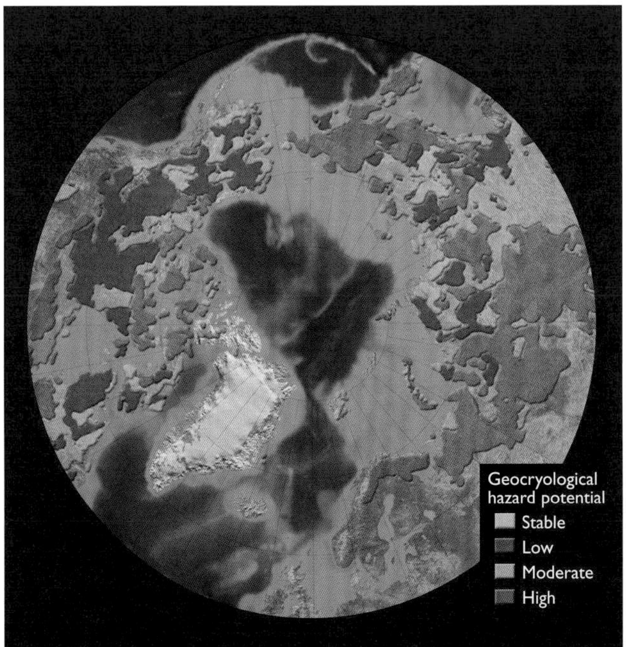

For all types of arctic infrastructure, the key climate-related concern is changes in the thermal state of the supporting soil layers. As described in section 16.2.2.5, changes in soil temperature can induce large variations in soil strength and bearing capacity and may also cause thaw settlement or frost heave. Many facilities and structures were designed for current climatic conditions, and it is possible that appreciable warming will introduce differential settlement beneath them. The susceptibility of permafrost to environmental hazards associated with thermokarst, ground settlement, and several other destructive cryogenic processes can be crudely evaluated using the geocryological hazard index, which is the combination of the projected percentage change in active-layer thickness and the ground ice content:

$$I_s = \Delta z_{al} \cdot V_{ice} \qquad \text{Eqn. 16.2}$$

where Δz_{al} is the percentage increase in active-layer thickness, and V_{ice} is the volumetric proportion of near-surface soil occupied by ground ice. The index provides a qualitative representation of relative risk, with lower values representing lower risk and vice versa.

Future geocryological hazards were projected using several scenarios of climate change (including those from the ACIA-designated models) as input to a permafrost model that included information about existing permafrost and ground ice distributions from the IPA. Figure 16.19 presents the geocryological hazard potential with respect to engineered structures projected for 2050, calculated using the HadCM3 scenario (Anisimov and Belolutskaia, 2002; Nelson F. et al., 2002). The map shows areas projected to have low, moderate, and high susceptibility to thaw-induced settlement, as well as areas where permafrost is projected to remain stable.

A zone of high and moderate risk potential is projected to extend discontinuously around the Arctic Ocean, indicating high potential for coastal erosion. North American population centers (e.g., Barrow, Inuvik) and river terminals on the arctic coast of Russia (e.g., Salekhard, Igarka, Dudinka, and Tiksi) fall within this zone. Transportation and pipeline corridors traverse areas of high projected hazard potential in northwestern North America. The area containing the Nadym-Pur-Taz natural gas production complex and associated infrastructure in northwest Siberia also falls in the projected high-risk category. Large portions of central Siberia, particularly the Sakha Republic (Yakutia) and the Russian Far East, have moderate or high projected hazard potential. These areas include several large population centers (Yakutsk, Noril'sk, and Vorkuta), an extensive network of roads and trails, and the Trans-

Fig. 16.19. Projected geocryological hazard potential in 2050, based on output from the HadCM3 model (Anisimov and Belolutskaia, 2002; Nelson F. et al., 2002).

Geocryological hazard potential
Stable
Low
Moderate
High

Siberian and Baikal–Amur mainline railroads. The Bilibino nuclear power station and its grid occupy an area of projected high hazard potential in the Russian Far East. Areas of lower projected hazard potential are associated with mountainous terrain and cratons (geologically stable interior portions of continents) where bedrock is at or near the surface.

Three main design approaches are employed when using permafrost soils as foundations for structures and infrastructure (Andersland and Ladanyi, 1994):

- to maintain the existing ground thermal regime (referred to as Principle I in Russia, and the passive method in North America);
- to accept changes in the ground thermal regime caused by construction and operation, or to modify foundation materials prior to construction (referred to as Principle II in Russia, and the active method in North America); and
- to use conventional foundation methods if the soils are thaw stable.

With the first two methods, it is necessary to estimate the maximum active-layer thickness and the maximum permafrost temperature as a function of depth that the structure will be subjected to in its lifetime. The air thawing index can be used to calculate active-layer thickness and maximum permafrost temperatures as a function of depth and time of year (see section 16.2.2.2). Other approaches can also be used to calculate active-layer thickness (e.g., using a full surface energy balance).

A significant consequence of permafrost degradation is likely to be a change in the maintenance conditions of many structures, especially for those that were designed without consideration of potential climate change. Projections of the change in bearing capacity and durability of foundations as temperatures change illustrate the potential for damage as a result of climate change. The results of such calculations for Yakutsk are presented in Table 16.9.

For structures utilizing Principle I (permafrost conservation), the table illustrates that foundation failures will possibly begin when the mean annual air temperature increases by a small amount (a few tenths of a

Table 16.9. Projected decrease in the bearing capacity and durability of foundations in Yakutsk for different increases in air temperature (Khrustalev, 2000).

	\multicolumn{5}{c}{Increase in mean annual air temperature (°C)}				
	0	0.5	1.0	1.5	2.0
Bearing capacity of structures built using Principle I (%)	100	93	85	73	50
Foundation durability of structures built using Principle II (%)	100	68	46	42	23

degree), and extend to all foundations when the increase exceeds 1.5 °C. This indicates that there is little margin for safety in the bearing capacity related to changes in air temperature. For structures utilizing Principle II (permafrost thawing), design is based on allowable deformations. As with structures utilizing Principle I, no factor of safety is included in the design. However, foundation failures will occur a relatively long time after the temperature changes. This means that the change in temperature affects only the durability of the foundations. Table 16.9 shows that a small increase in air temperature substantially affects the stability of the building, and the safety of the foundation decreases sharply with rising temperatures.

Temperature increases can result in a significant decrease in the lifetime and potential failure of the structure (Khrustalev, 2000, 2001). Permafrost engineers, therefore, face the problem of preserving infrastructure under projected future climate conditions. The compensation method (putting new buildings into operation as existing ones are damaged and abandoned) appears to be one of the possible ways to address this problem. However, Khrustalev (2000, 2001) states that this method will be inadequate, since the required rate of new construction rises exponentially from 5% per decade in 1980–1990 to 108% per decade in 2030–2040 (assuming a linear increase in mean annual temperature of 0.075 °C/yr).

Various methods have been suggested to address temperature-related foundation problems. Techniques to reduce warming and thawing, such as heat pumps, convection embankments, thermosyphons, winter-ventilated ducts, and passive cooling systems, are already common practice in North America, Scandinavia, and Russia (Andersland and Ladanyi, 1994; Couture et al., 2003; Goering and Kumar, 1996; Instanes, B., 2000; Khrustalev and Nikiforov, 1990; Smith S. et al., 2001).

16.3.1. Ultraviolet radiation and construction materials

Ultraviolet (UV) radiation adversely affects many materials used in construction and other outdoor applications. Exposure to UV radiation can alter the mechanical properties of synthetic polymers used in paint and plastics, and natural polymers present in wood. Increased exposure to UV radiation due to stratospheric ozone depletion is therefore likely to decrease the useful life of these materials (Andrady et al., 1998).

The impact of UV radiation on infrastructure in the Arctic is influenced by two compounding factors: the high surface reflectivity of snow or ice and long hours of sunlight. Both factors have strong seasonal components, generally resulting in increased UV radiation levels in the late spring. While the level of UV radiation incident on a horizontal surface (e.g., a flat roof) may be considerably lower in the Arctic than at mid-latitudes, the level of UV radiation incident on a vertical surface (e.g., a south-facing wall) may be higher than that on a horizontal surface at some times of the year, when reflection from snow augments the direct UV radiation incident on the surface. Materials degradation is often related to the total accumulated UV radiation exposure.

Long days during the arctic summer can result in large daily doses of UV radiation, even when noon levels remain moderate. If UV radiation levels increase as a result of ozone depletion or changes in cloud cover, the impacts on materials are likely to include earlier degradation and significant discoloration.

For natural polymers found in wood, exposure to UV radiation can lead to a decrease in the useful lifetime of the product. Even small doses of UV radiation may darken wood surfaces. Other effects of increased exposure to UV radiation on wood are less certain. The damage to finished wood products is limited primarily by protective surface coatings, but increased UV radiation levels will possibly lead to increased costs for more frequent painting or other maintenance.

The construction industry is increasingly turning to synthetic polymers for use in building materials. In the United States and Western Europe, the building sector uses 20 to 30% of the annual production of plastics (Mader, 1992). Plastics and other synthetic polymer products are used for a number of applications, including irrigation, water distribution and storage, and in fishing nets and agricultural films.

During the manufacture of virtually all polymer products, impurities can be introduced that make the end product susceptible to photodegradation by UV radiation. Although stabilizers may be added to retard photodegradation effects, their inclusion can substantially increase the cost of the final product. Unfortunately, much of the research on UV-induced degradation has been conducted on pure polymer resins, leading to problems in extrapolating the findings to processed products of the same polymer.

The effects of UV radiation on materials are closely tied to other environmental factors, including ambient temperatures. Polar regions have experienced the greatest ozone depletion and therefore the greatest potential increases in UV-B radiation levels. However, the cooler temperatures in these locations can help prevent rapid degradation of materials. Materials have different sensitivities that can depend on wavelength and dose. Some materials contain stabilizers designed to mitigate degradation, but the efficacy of those stabilizers under spectrally altered (e.g., higher than normal UV) conditions is not always known (Andrady, 1997).

UV-induced polymer deterioration has been widely observed. Polyvinyl chloride tends to undergo discoloration or yellowing, and to lose impact strength. This loss of impact strength can eventually lead to

cracking and other irreversible damage. Another polymer, polycarbonate, undergoes a rearrangement reaction when exposed to UV-B or UV-C radiation. When irradiated at longer wavelengths, including visible wavelengths, polycarbonates undergo oxidative reactions that result in yellowing (Factor et al., 1987).

Polystyrene, used as expanded foam in both building and packaging applications, also undergoes light-induced color changes. In polyethylene and polypropylene, which are used extensively in agricultural mulch films, greenhouse films, plastic pipes, and outdoor furniture, the loss of tensile properties and strength is a particular concern (Hamid and Pritchard, 1991; Hamid et al., 1995).

The cost of more frequent replacement of woods and polymers is likely to be higher in the Arctic than at lower latitudes because of the increased cost of shipping and placement. Environmental stresses in the Arctic, including high winds and repeated freezing and thawing, will possibly exacerbate minor materials problems that develop as a result of UV radiation damage.

16.3.2. Buildings

Several foundation systems are currently in use for industrial, commercial, and residential structures situated in the Arctic and subarctic. When building sites are underlain by permafrost, the foundation system must ensure that any warmth emanating from the structure does not induce thawing of the permafrost layer. For many structures, this is accomplished by elevating the building above the ground surface on a pile or adjustable foundation system. The resulting air space ensures that heat from the structure will not induce permafrost warming. Thousands of structures ranging from single-family residences to large living quarters and apartment blocks are currently supported on pile foundations, including many residential structures in the permafrost zones of Alaska, Canada, and Scandinavia, and many apartment buildings in Siberia. In most of these areas, existing structures are performing well, and there has been little evidence that climate change is inducing failures. However, as noted in section 16.3.8 many Siberian buildings are experiencing significant rates of structural failure that may be connected to increasing temperatures.

The bearing capacity of piles embedded in permafrost depends on the type of frozen soil (clay, silt, or sand), its temperature, and the length of pile embedment in the permafrost layer. A safe pile design is usually based on the calculated maximum temperatures of the frozen soil along the embedded pile length, determined from data on the mean annual temperature of the site and the seasonal temperature variation. Pile foundations are particularly sensitive to permafrost temperatures because of the large increase in creep rates as temperatures approach 0 °C (see section 16.2.2.5). For this reason, extra cooling measures, such as the use of thermopiles (thermosyphon-cooled pilings), are sometimes taken in the warmer discontinuous permafrost zone in order to lower temperatures and ensure a stable permafrost–piling bond.

An increase in ground temperature along an existing pile is very likely to reduce its bearing capacity or increase the rate of its settlement for two reasons: an increase in the active-layer thickness will reduce the effective embedment length of the pile; and increased temperature will reduce the strength of the frozen soil. As a result, if soil warming occurs, an existing structure founded on piles will experience an increasing settlement rate that is likely to lead to uneven settlement and damage to the structure.

The design of all future structures founded on piles embedded in permafrost soils should take into account projected future temperature increases. Depending on the estimated useful lifetime of the structure, the pile design should preferably be based on projected temperature conditions at the end of its lifetime. For any particular pile type, unless bearing on rock, this will result in longer pile length and increased cost.

Very light buildings in permafrost areas are often established directly on the ground surface and supported by a system of adjustable mechanical jacks providing a sufficient crawl space below the heated and insulated floor of the building and the ground surface. Although such buildings will not produce thaw settlement of underlying permafrost, they are likely to be subject to the effects of regional thaw settlement due to rising temperatures and the resulting increase in active-layer thickness. Although settlement in these types of buildings can be adjusted by the jacking system, the differential settlement of water supply and sewage evacuation pipes attached to the building must also be addressed.

Depending on local meteorological conditions, the foundation soils of buildings constructed on elevated foundation systems (either pile or adjustable supports) are likely to be less prone to temperature increases as climate change occurs. This is due to the combined influence of shielding the surface from solar radiative input (due to shading by the structure) and elimination of snow cover at the surface beneath the building, both of which have a significant cooling effect on ground-surface temperatures.

For industrial and equipment buildings that must support large floor loads, pile foundations are sometimes too costly. These buildings often have a slab-on-grade foundation with insulation installed beneath the floor to help protect the underlying permafrost. In addition to the insulation, some sort of cooling system under the slab is required to remove heat from beneath the structure. Both active and passive refrigeration systems have been employed for this purpose, but passive systems are generally preferred due to their lower operational and maintenance costs. Passive systems are based on either thermosyphon or air-duct cooling systems, and utilize low ambient temperatures during winter to refreeze a buffer layer of non-frost-susceptible material beneath the building. The buffer layer typically consists of a pad of granular material that is placed before building construction and sized to contain the seasonal thaw that develops beneath the building during summer months when the passive cooling system is inactive.

Increasing air temperature is very likely to have a detrimental effect on the operation of these foundation systems for two reasons. Higher air temperatures are likely to lengthen the thaw season and place increased requirements on the thickness of the buffer layer and/or the insulation system, and reduced air freezing indices (section 16.2.2.4) are very likely to decrease the capacity of the passive cooling system to refreeze the buffer layer material during winter.

As a result, existing buildings built on slabs are likely to experience an increasing failure rate as air temperatures rise and produce either significantly longer thaw seasons or a reduced freezing index. Future designs for such buildings will need to take into account temperature increases projected for the lifetime of the structure. This is very likely to increase costs due to the need for additional buffer layer material and higher cooling capacities.

16.3.3. Road and railway embankments and work pads

Transportation routes are likely to be particularly susceptible to destructive frost action under conditions of changing climate. Garagulya et al. (1997) developed a map showing areas with various probabilities of natural hazards. This map indicated that the regions of highest susceptibility to frost heave and thaw settlement are located along the Arctic Circle.

The design of road and railway embankments in the Arctic is complicated by the presence of underlying permafrost, due to the possibility of thaw settlement and significant permanent embankment deformation if thermal disturbance occurs. The situation is exacerbated by complex thermal interactions between the embankment and the surrounding environment. Embankment construction often produces a significant alteration of the surface microclimate that results in an increase in mean annual surface temperature of several degrees as compared to natural conditions. Precise temperature increases are a complex function of embankment surface conditions, maintenance operations (e.g., snow clearing patterns), and the pre-existing natural vegetation, and are sometimes difficult to project.

In the continuous permafrost zone, where the permafrost layer and surface conditions are generally colder, surface warming due to embankment construction can usually be accommodated using well-established design practices. In this case, the embankment thickness is adjusted to ensure that seasonal thawing is contained within the embankment itself, thus avoiding thawing of the underlying permafrost. The required embankment thickness is sensitive to climatic conditions, tending to increase significantly with warmer conditions.

In the discontinuous permafrost zone, permafrost and ground-surface temperatures are warmer, often within a few degrees of the melting point. In this zone, it is more difficult to accommodate surface warming due to embankment construction since the resulting mean surface temperatures are often above the melting point. In this case, long-term thaw of the permafrost layer can be expected and cost-effective design strategies are currently unavailable. In a limited number of cases, techniques such as thermosyphon, air duct, or convection cooling systems have been used to mitigate these problems (Andersland and Ladanyi, 1994; Goering, 2003; Goering and Kumar, 1996; Instanes, B., 2000; Phukan, 1985), however, the expense associated with these systems severely limits their utility. In practice, even under current climatic conditions, many road and railway embankments located in regions of warm discontinuous permafrost experience high failure rates and resulting high maintenance costs. Typical problems include differential thaw settlement and shoulder failure due to thawing permafrost, resulting in an uneven, cracked embankment surface (Fig. 16.20). The timescales associated with permafrost thaw beneath embankments are of the order of the embankment lifetime (20 to 30 years), thus necessitating a continuous maintenance program.

Increasing temperatures are very likely to affect embankment performance in both the continuous and discontinuous permafrost zones and should be considered in the design of future projects. In the discontinuous permafrost zone, the problems associated with permafrost thaw described above are very likely to increase as increasing air temperature adds to the warming influence of embankment construction. This is very likely to result in increased failure rates and higher maintenance costs. As climate change reduces permafrost extent in the southern discontinu-

ous permafrost zone, some reduction in these embankment problems is possible, although the timescales for projected warming and thaw are much longer than typical project lifetimes.

In the continuous permafrost zone, increasing temperatures are likely to have negative impacts on embankment performance for two reasons:

- increased surface temperatures will necessitate greater embankment thicknesses in order to contain the seasonal thaw depth. This is very likely to have a large impact on project costs due to the difficulty and expense associated with obtaining appropriate granular material for embankment construction; and
- as air and surface temperatures increase, a design regime shift will possibly occur in association with the northward movement of the boundary between the discontinuous and continuous permafrost zones. It is possible that increasing embankment thickness alone will no longer be sufficient to protect underlying permafrost and greater failure rates will occur, similar to those seen in the discontinuous permafrost zone.

As a result, it is likely that existing road, rail, or airport embankments will experience increasing failure rates both in the continuous and discontinuous permafrost zones. Future embankment designs should incorporate the effects of projected temperature increases over the lifetime of the project, which is likely to increase construction costs.

16.3.4. Pipelines

Many of the earth's remaining oil and gas reserves are located in regions of the Arctic far from population centers. These areas include the North Slope of Alaska, the Canadian Arctic, northwestern Russia, and Siberia. The limited exploitation of these resources to date has relied primarily on pipeline systems to transport products to market. Future expansion of these pipeline networks is likely given the increasing demand for fossil fuels worldwide. Examples include the large gas pipeline projects currently under consideration to connect natural gas reserves in the Alaskan and western Canadian Arctic to southern Canada and the continental United States. Many of the current and anticipated pipeline routes cross extensive areas of continuous and discontinuous permafrost and require special design considerations.

Oil and gas pipelines differ in their interactions with the surrounding environment because of variations in operating temperature. Transmission of oil through pipelines usually takes place at high temperatures because of high oil-well production temperatures and reduced pumping losses. Conversely, natural gas transmission through pipelines often takes place at temperatures below freezing in order to increase gas density

Fig. 16.20. Typical embankment cracking and differential thaw settlement in the discontinuous permafrost zone of Interior Alaska (Photo: Larry Hinzman, University of Alaska, Fairbanks).

and throughput. High- or low-temperature pipelines present different challenges to designers and will react in different ways to increased air temperatures.

Gas and oil pipelines are normally constructed below the ground surface, as this reduces construction costs and provides other benefits. In the case of warm-oil pipelines, this becomes problematic in areas where permafrost is encountered. The desire to keep the oil warm to limit viscosity and pumping costs is in direct conflict with the requirement to maintain the frozen state of the surrounding soil. If the pipeline is buried, no practical amount of insulation will prevent the warm oil from thawing surrounding permafrost, thus resulting in loss of strength, thaw settlement, and probable line failure. As a result, designers have relied on one of two methods to avoid permafrost degradation:

- an elevated oil pipeline that is supported above the ground surface on some sort of pile foundation, thus limiting the possibility of permafrost thaw; or
- a more conventional buried pipeline design, with the oil chilled to near-permafrost temperatures (typically below 0 °C for a large part of the year).

A major shortcoming of both methods, particularly with regard to potential climate change, is their reliance on the permafrost layer for structural support. In the first case, the piles are embedded in the permafrost, as for a building foundation, and the adfreeze bond between the pile and the permafrost supports the load. In the second case, the integrity of the pipe trench and support of the pipeline are dependent on the structural integrity of the underlying permafrost. Both methods result in increased cost; the first due to the large expense of constructing an elevated line, and the second due to the high pumping costs associated with moving chilled oil and related problems with potential wax formation in the line.

The best-known example of an existing elevated arctic oil pipeline is the Trans Alaska Pipeline System (TAPS), which stretches 1280 km from Prudhoe Bay to the ice-free port of Valdez in southern Alaska. This pipeline is elevated for just over half of its length in order to avoid potential permafrost problems. The northern sections, where permafrost temperatures are cold (lower than approximately -5 °C), utilize non-refrigerated pile supports and a work-pad embankment designed to protect underlying permafrost. In the more southerly sections, where warmer discontinuous permafrost is encountered, the piles utilize a passive refrigeration system consisting of pairs of thermosyphons installed in each piling. More than 120 000 thermosyphons are used (Sorensen et al., 2003). The thermosyphons are designed to ensure that any excess heat transported downward from the pipeline or entering the ground surface due to construction disturbance will not cause thawing of the permafrost where the piles are embedded.

Owing to the extensive use of pile supports, elevated oil pipelines are sensitive to increasing air and soil temperatures, as are building-support pilings. Increased soil temperatures are very likely to reduce the bearing capacity of these systems because of the reduction in the strength of the adfreeze bond between the frozen soil and the pile. In addition, increased air temperatures are very likely to result in a greater active-layer depth, which will reduce the effective embedment length of the pile in the frozen zone. Some of these effects can be countered by the use of refrigerated piles, as in the case of TAPS; however, increased air temperatures are also very likely to reduce the ability of these systems to provide adequate cooling. It is also possible that the pipeline right-of-way and work-pad embankment will begin to experience increased problems with thaw settlement due to the combination of surface disturbance and increasing air temperature. These factors should be considered during the design stage of future projects.

The Norman Wells Pipeline, which runs 896 km through the western Canadian Arctic from Norman Wells, Northwest Territories, to Zama City, Alberta, is an example of a chilled pipeline that is buried in permafrost terrain. The oil is chilled by a refrigeration system before it enters the line at Norman Wells, and operates at near-ambient permafrost temperatures. Even though the oil is chilled to minimize permafrost disturbance, the designers anticipated a significant amount of thaw settlement and/or frost heave along the route (Nixon et al., 1984). To resist the anticipated loading due to thaw strain or frost heave, a relatively high-strength system consisting of a small-diameter thick-wall pipe was used. Two major design/performance issues were identified as the most significant for this project: adequate thermal conditions must be maintained such that design loads due to thaw settlement or frost heave are not exceeded; and permafrost thaw within the pipeline trench and right-of-way must be limited in order to avoid slope instability (and potential landslides) in areas of sloped terrain.

Variations in line operating conditions have resulted in significant movement of the pipe due to thaw settlement and frost heave. In some places, thaw settlement near the pipeline trench has exceeded the design projections of 1 m (Burgess and Smith, 2003), and increases in thaw depth have reduced the factor of safety for slope stability (Oswell et al., 1998).

Many of the difficult operational issues identified above for buried ambient-temperature pipelines result from the thermal interaction between the pipe and the surrounding ground. These difficulties are likely to be exacerbated if air temperatures also change over time. Increased air temperatures are likely to aggravate problems with thaw settlement along the right-of-way and decrease slope stability. To some extent, it may be possible to reduce the severity of these problems by decreasing the operational temperatures of the pipeline, however this is not desirable because of the high pumping cost and wax formation issues mentioned previously. New projects should take projected temperature increases into account during the design stage and may have to increase measures designed to prevent slope instability and settlement associated with permafrost thaw.

Unlike oil pipelines, gas pipelines benefit from low-temperature operation and are often operated at temperatures significantly below the freezing point. When these pipelines are buried in continuous permafrost, they aid the maintenance of the permafrost layer and design is straightforward. On the other hand, where chilled buried pipelines must traverse zones of discontinuous permafrost, problems can be expected. In this case, the chilled pipeline will cause freezing of the thawed soil present along the route, some of which may be susceptible to frost. The resulting frost-heave loads on the pipe can be large and must be accounted for carefully. Previous studies have suggested that line operation temperatures should be kept only moderately below freezing in these areas in order to minimize frost-heave problems while, at the same time, avoiding thaw settlement in the permafrost portions of the route (Jahns and Heuer, 1983). Increased air temperatures will possibly expand the problematic portion of these pipeline routes as the boundary between continuous and discontinuous permafrost moves northward. However, the ability to control pipeline operating temperatures may help to adapt to changing climatic conditions.

16.3.5. Water-retaining structures

Water-retaining embankments in permafrost are discussed in detail by Andersland and Ladanyi (1994), and are generally one of two types: unfrozen embankments or frozen embankments. With unfrozen embankments it is assumed that the permafrost foundation will thaw during the lifetime of the structure. This type is limited to sites with thaw-stable foundation materials or bedrock, or cases where the water is retained for a short period of time. With frozen embankments it is assumed that the permafrost foundation will remain frozen during the lifetime of the structure. This type is suitable for continuous permafrost areas and other areas where the foundation materials are thaw-unstable.

The embankment design for a particular site must combine the principles of soil mechanics for unfrozen soils and the mechanical behavior of permafrost. The design should always include thermal and stability considerations, and for permanent structures, potential climate change should be taken into account. Sayles (1984, 1987) and Holubec et al. (2003) have summarized the factors that are relevant to embankment design.

Problems associated with water-retaining dams include seepage, frost heave (in areas of seasonal frost), settlement, slope stability, slope protection, and construction methods. Increased air temperatures are not likely to affect unfrozen embankments because the permafrost foundation is thaw stable. Frozen embankments usually require supplementary artificial freezing to ensure that the foundation and embankment remain frozen (Andersland and Ladanyi, 1994). Increased air temperatures are likely to increase the construction and operational costs of frozen embankments due to the increased energy demand required to keep the embankment frozen.

16.3.6. Off-road transportation routes

In recent years, temporary winter transportation routes have played an increasingly important role for community supply and industrial development in the permafrost zones of North America. These transportation corridors consist of ice roads that traverse frozen lakes, rivers, and tundra. In some cases, ice roads are constructed for one-time industrial mobilizations, such as oil and gas exploration activities. In other cases, permanent ice-road corridors have been established and are reopened each winter season. Winter ice roads offer important advantages that include low cost and minimal impact to the environment. Oil and gas exploration can be conducted from these road structures with very minimal ecological effects, and costs associated with construction and eventual removal of more permanent gravel roads or work pads can be avoided.

Winter ice-road construction is affected by a number of climatic factors, including air temperature, accumulated air freezing index, and snowfall. These roads depend on the structural integrity of the underlying frozen base material and, thus, a significant period of freezing temperature must occur each autumn before ice-road construction can begin. For water crossings, the critical factor influencing the start of the winter-road season is the rate and amount of ice formation. Ice thickness must reach critical minimum values before vehicles and freight can be supported safely. For tundra areas (particularly where temporary transportation routes are needed), a critical issue is protection of the existing vegetative cover. In this case, the active layer must be frozen to a depth that is sufficient to support anticipated loads and avoid damage to vegetation. Once a sufficient frozen layer has been established, the surface is covered with snow and water is applied and allowed to freeze in place. The result is a durable driving surface that can support significant loads without harming the underlying vegetation.

In North America, winter-road use and construction is regulated to avoid environmental damage to the tundra. Various inspection techniques are used to ensure adequate freezing before the winter-road season is opened. One technique employs a penetrometer that is pushed into the frozen active layer to measure the strength and thickness of the frozen zone. Based on these measurements, regulatory agencies make decisions regarding opening and closing dates for winter-road travel.

Climatic conditions play a strong role in determining the opening and closing dates for winter-road travel, although inspection techniques and load requirements, among other factors, are also important. Increased air temperature and reductions in the annual air freezing index are very likely to have a negative impact on the duration of the winter-road season. This will possibly become particularly problematic for oil and gas exploration because of the time needed at the beginning and end of the ice-road season for mobilization and demobilization.

Hinzman et al. (in press) present historic data for the opening and closing dates for tundra travel on the North Slope of Alaska that show a substantial reduction in the duration of the winter-road season (from over 200 days in the early 1970s to just over 100 days in 2002). The rate of reduction has been fairly consistent over the intervening years and is due primarily to delayed opening dates (from early November in the 1970s to late January in the 2000s), although closing dates have also been occurring earlier in the spring. Reductions in the duration of the winter-road season have also occurred in the Canadian Arctic, however the reductions are much smaller than those observed in Alaska, and in some cases the season length has increased.

Fig. 16.21 shows historic data for opening and closing dates of winter roads on the North Slope of Alaska and in Canada's Northwest Territories. The data for the North Slope are for temporary winter roads used primarily for oil and gas exploration, whereas the data for the Northwest Territories are for the winter roads between Inuvik and Tuktoyaktuk (187 km) and between the Yellowknife Highway and Wha Ti (103 km). The figure illustrates the reduction in season length for the North Slope of Alaska, and a trend of later opening

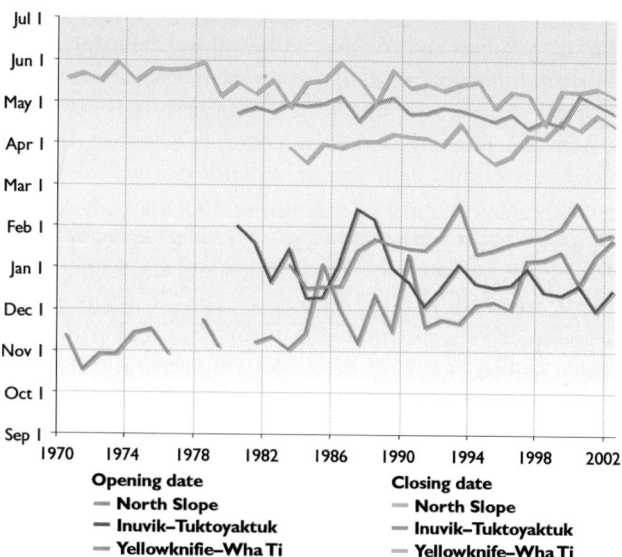

Fig. 16.21. Opening and closing dates for winter roads on the North Slope of Alaska and the Inuvik–Tuktoyaktuk and Yellowknife Highway–Wha Ti winter roads in western Canada (North Slope data after Hinzman et al., in press; Canadian data from the Government of the Northwest Territories, Department of Transportation, www.hwy.dot.gov.nt.ca/highways/, 2003).

dates for the Wha Ti road. Data for the Inuvik-Tuktoyaktuk road, however, indicate an increased season length.

The observed trend in Alaska shows that climate change is likely to lead to decreased availability of off-road transportation routes (ice roads, snow roads, etc.) due to reduced duration of the freezing season.

16.3.7. Offshore transportation routes

The global economy of the 21st century will need the natural resources of the Arctic and subarctic. Air transport remains unprofitable for mineral payloads and attention to arctic shipping is growing as a result. Road, rail, and pipeline routes are complicated in the far north by tectonically active glacier-contorted landscapes, low-lying frozen ground, and fragile ecosystems. Shorter routes from resource to tidewater minimize terrestrial complications only if a port can be built at the coast. New ice-breaking ship designs are continually improving the efficiency of arctic shipping. Growing evidence of climate change indicates that ice-free navigation seasons will probably be extended and thinner sea ice will probably reduce constraints on winter ship transits (section 16.2.5). Additional northern port capacity is critical to the success of arctic shipping strategies associated with northern resource development and potential climate change. Difficulties related to freezing temperatures, snow accumulation, and extended darkness compound the challenges of geotechnical, structural, architectural, mechanical, electrical, transportation, and coastal engineering in designing and operating sea and river ports.

The International Northern Sea Route Programme was a six-year (June 1993 – March 1999) international research program designed to create an extensive knowledge base about the ice-infested shipping lanes along the coast of the Russian Arctic, from Novaya Zemlya in the west to the Bering Strait in the east. This route was previously named the Northeast Passage, but is now more often known by its Russian name – the Northern Sea Route.

Projected reductions in sea-ice extent are likely to improve access along the Northern Sea Route and the Northwest Passage. Projected longer periods of open water are likely to foster greater access to all coastal seas around the Arctic Basin. While voyages across the Arctic Ocean (over the pole) will possibly become feasible this century, longer navigation seasons along the arctic coasts are more likely. Development of the offshore continental shelves and greater use of coastal shipping routes will possibly have significant social, political, and economic consequences for all residents of arctic coastal areas.

Output from the five ACIA-designated climate models was used to project the length of the navigation season along the Northern Sea Route based on the amount of sea ice present. Figure 16.22 shows the five-model mean for three conditions: 25, 50, and 75% open water across the Northern Sea Route. If ships sailing along the Northern Sea Route are designed for and capable of navigating in waters with 25% open water (75% sea-ice cover), the projected length of the navigation season is considerably longer than that for ships that are minimally ice-capable and can only navigate in 75% open water (25% sea-ice cover).

There are few days when, even at mid-century, the Northern Sea Route is covered by 75% open water (25% sea-ice cover). When days on which navigation is possible are defined by a higher ice-cover percentage, the length of the navigation season increases. In 2050, Fig. 16.22 shows a projected navigation season length of 125 days under conditions of 25% open water (75% sea-ice cover); conditions very favorable for the transit of ice-strengthened cargo ships. However, the ACIA-designated model projections provide no information on sea-ice thickness, a critical factor for ice navigation. Section 16.2.5.2 provides additional projections of future sea-ice conditions.

With increased marine access to arctic coastal seas, national and regional governments are likely to be called upon for increased services such as icebreaking assistance, improved sea-ice charts and forecasting, enhanced emergency response capabilities for sea-ice conditions, and greatly improved oil–ice cleanup capabilities. The sea ice, although thinning and decreasing in extent, will possibly become more mobile and dynamic in many coastal regions where fast ice was previously the norm. Competing marine users in newly open or partially ice-covered areas in the Arctic are likely to require increased enforcement presence and regulatory oversight.

A continued decrease in arctic sea-ice extent this century is very likely to increase seasonal and year-round

access for arctic marine transportation and offshore development. New and revised national and international regulations, focusing on marine safety and marine environmental protection, are likely to be required as a result of these trends. Another probable outcome of changing marine access will be an increase in potential conflicts between competing users of arctic waterways and coastal seas.

Based on the scenarios presented in this section and in section 16.2.5.2, a longer navigation season along the arctic coast is very likely and trans-arctic (polar) shipping is possible within the next 100 years.

16.3.8. Damage to infrastructure

Instanes A. (2003) pointed out that for structures on permafrost it is often difficult to differentiate between the effect of temperature increases and other factors that may affect a structure on permafrost. For example: the site conditions are different from the assumed design site conditions; the design of the structure did not take into account appropriate load conditions, active-layer thickness, and permafrost temperature; the contractor did not carry out construction according to the design; the maintenance program was not carried out according to plan; and/or the structure is not being used according to design assumptions. In addition, it is very difficult to find cost-effective engineering solutions for foundations or structures on warm (T >-1 °C), discontinuous permafrost.

Kronik (2001) summarized reports on damage to infrastructure in Russian permafrost areas from the 1980s to 2000. The reported deformations of foundations and structures were caused not only by climate change, but also by other factors such as those listed above, particularly the low quality and inadequate maintenance of structures. Unfortunately, it is difficult to distinguish between these factors. The analysis of deformation causes performed by Kronik (2001) for industrial and civil complexes showed that 22, 33, and 45% of the deformations were due to mistakes by designers, contractors, and maintenance services, respectively (see also Panova, 2003). Out of 376 buildings surveyed, 183 (48.4%) did not meet building code requirements in 1992, including 21 buildings (8.5%) that were unfit for use (Kronik, 2001). The percentage of dangerous buildings in large villages and cities in 1992 ranged from 22% in the village of Tiksi to 80% in the city of Vorkuta, including 55% in Magadan, 60% in Chita, 35% in Dudinka, 10% in Noril'sk, 50% in Pevek, 50% in Amderma, and 35% in Dikson.

The condition of land transportation routes in Russia is not much better. In the early 1990s, 10 to 16% of the subgrade in the permafrost areas of the Baikal–Amur railroad line was deformed because of permafrost thawing; this increased to 46% in 1998. The majority of runways in Norilsk, Yakutsk, Magadan, and other cities may be closed for shorter or longer periods due to lack

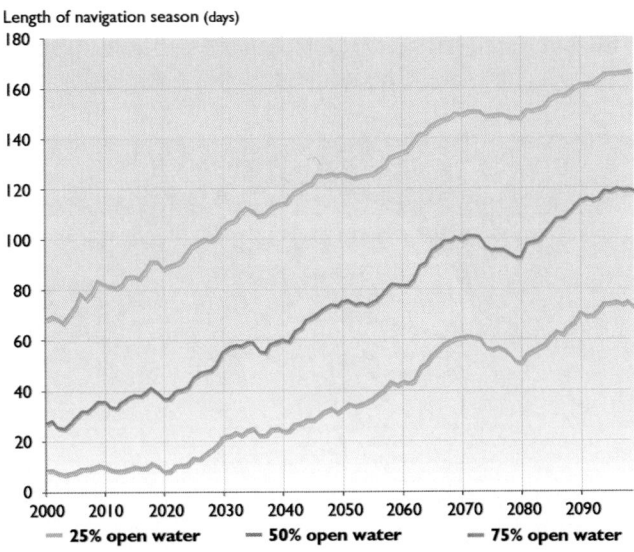

Length of navigation season (days)

Fig. 16.22. Length of the navigation season along the Northern Sea Route projected by the ACIA-designated climate models (five-model mean).

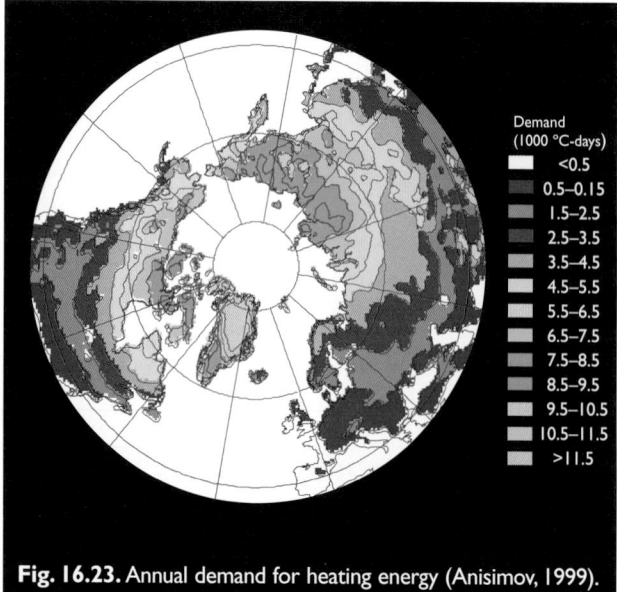

Fig. 16.23. Annual demand for heating energy (Anisimov, 1999).

of maintenance. The main gas and oil transmission lines in the permafrost region have also suffered damage related to permafrost thawing: 16 breaks were recorded on the Messoyakha–Noril'sk pipeline in 2000.

16.3.9. Energy consumption for heating

A reduction in the demand for heating energy is a potential positive effect of climate change in the Arctic and subarctic. The air-temperature threshold that defines the beginning of the cold period, when additional heating of living facilities, businesses, and industrial buildings is necessary, varies within and between countries. In North America and Western Europe, most civil buildings, private houses, apartment complexes, and even most buildings in large cities have local heating systems. In the United States the temperature threshold for heating is defined as 65 °F (17.8 °C), but because the local systems are very flexible and can be manipulated individually, evaluation of energy consumption is complex.

In Eastern Europe and Russia, most urban buildings have centralized heating systems. Under standard con-

ditions, such systems operate when the mean daily air temperature falls below 8 °C (47 °F). Because of the large thermal inertia of these centralized heating systems, comfortable indoor temperatures (e.g., 18 °C) are usually maintained throughout the winter.

Figure 16.23 shows the annual demand for heating energy (in 1000 °C-days) when building heating is required (mean daily air temperature below 8 °C) calculated for current climatic conditions. Annual heating degree-day totals characterize the demand for heating over the entire cold period. Daily heating degree-days are calculated by subtracting the mean daily temperature from the 8 °C threshold (e.g., a day with a mean daily temperature of 5 °C would result in three heating degree-days); days with a mean temperature at or above 8 °C result in zero heating degree-days.

Anisimov (1999) used the GFDL, ECHAM-1 (Max-Planck Institute for Meteorology), and HadCM transient climate scenarios for 2050 to calculate the reduction in the duration of the heating period and changes in the number of heating degree-days relative to 1999. Projected reductions in the number of heating degree-days (Fig. 16.24) can be used as a metric for the reduction in heating energy consumption. Figure 16.25 shows the percentage reduction in the duration of the heating period between 1999 and 2050; this decrease is projected to vary from a few weeks to more than a month, depending on the regional effects of climate change.

The energy savings from decreased demand for heating in northern regions are likely to be offset by increases in the temperature and duration of the warm period, leading to greater use of air conditioning.

16.3.10. Natural resources

The Arctic has large oil and natural gas reserves. Most are located in Russia: oil in the Pechora Basin, natural gas in the Lower Ob Basin, and other potential oil and gas fields along the Siberian coast. Canadian oil and gas fields are concentrated in two main basins in

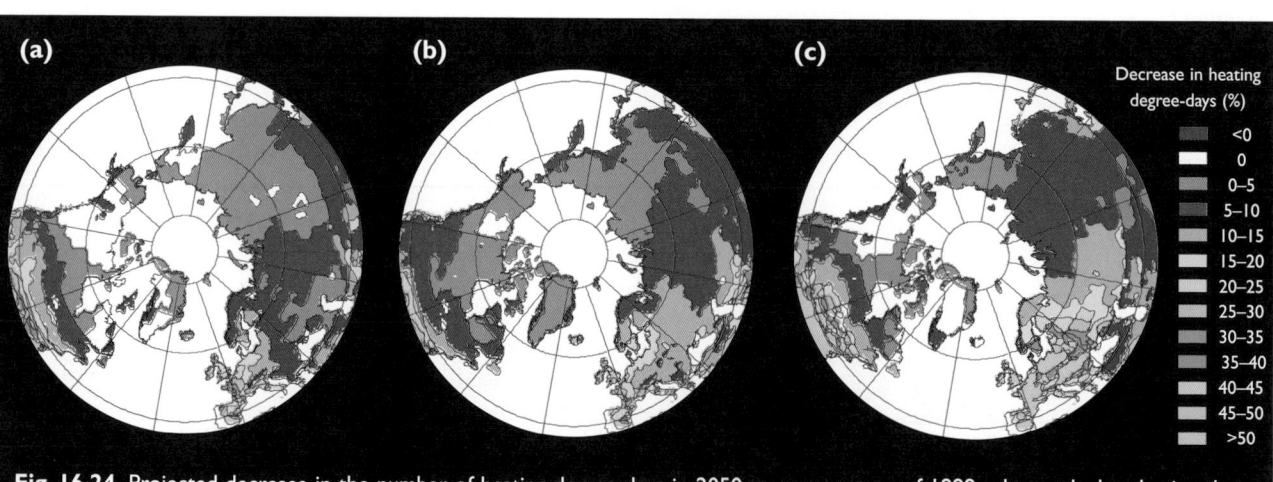

Fig. 16.24. Projected decrease in the number of heating degree-days in 2050 as a percentage of 1999 values, calculated using the (a) ECHAM-1; (b) GFDL; and (c) HadCM climate scenarios (Anisimov, 1999).

the Mackenzie Delta–Beaufort Sea region and in the high Arctic. Prudhoe Bay (Alaska) is the largest oil field in North America, and other fields have been discovered or are likely to be present along the Beaufort Sea coast. Oil and gas fields also exist in other arctic waters, for example, the Barents Sea and off the west coast of Greenland. The Arctic is an important supplier of oil and gas to the global market, and it is possible that climate change will have both positive and negative financial impacts on the exploration, production, and transportation activities of this industry.

Climate change impacts on oil and gas development have so far been minor, but are likely to result in both financial costs and benefits in the future. For example, offshore oil exploration and production is likely to benefit from less extensive and thinner sea ice because of cost reductions in the construction of platforms that must withstand ice forces. Conversely, ice roads, now used widely for access to offshore activities and facilities, are likely to be useable for shorter periods and less safe than at present. The thawing of permafrost, on which buildings, pipelines, airfields, and coastal installations supporting oil development are located, is very likely to adversely affect these structures and the cost of maintaining them (Parker, 1998; Weller and Lange, 1999; Weller et al., 1999).

The Arctic holds large stores of minerals, ranging from gemstones to fertilizers. Russia extracts the largest quantities, including nickel, copper, platinum, apatite, tin, diamonds, and gold, mostly on the Kola Peninsula but also in Siberia. Canadian mines in the Yukon and Northwest Territories supply lead, zinc, copper, and gold. Gold mining continues in Alaska, along with extraction of lead and zinc deposits from the Red Dog Mine, which contains two-thirds of US zinc resources. Coal mining also occurs in several areas of the Arctic. Mining activities in the Arctic are an important contributor of raw materials to the global economy and are likely to benefit from improved transportation conditions to bring products to market, due to a longer ice-free shipping season.

The coal and mineral extraction industries in the Arctic are important contributors to national economies, and the actual extraction process is not likely to be much affected by climate change. However, climate change will possibly affect the transportation of coal and minerals. Mines that export their products using marine transport are likely to experience savings due to reduced sea-ice extent and a longer shipping season. Conversely, mining facilities with roads on permafrost are likely to experience higher maintenance costs as the permafrost thaws (Parker, 1998; Weller and Lange, 1999; Weller et al., 1999).

Any expansion of oil and gas activities and mining is likely to require expansion of air, marine, and land transportation systems. The benefits of a longer shipping season in the Arctic, with the possibility of easy transit through the Northern Sea Route and Northwest Passage for at least part of the year, are likely to be significant. Other benefits are likely to include deeper drafts in harbors and channels as sea level rises, a reduced need for ice strengthening of ship hulls and offshore oil and gas platforms, and a reduced need for icebreaker support. Conversely, coping with greater wave heights, and possible flooding and erosion threats to coastal facilities, is likely to result in increased costs.

16.4. Engineering design for a changing climate

Climate change is likely to affect infrastructure in different ways, as has been described in previous sections of this chapter. Some infrastructure may be relatively insensitive to climate change or easily adaptable to changing conditions. In other cases, a large sensitivity may exist and/or the consequences of any failure may be high (Esch, 1993; Ladanyi et al., 1996, Khrustalev, 2001, 2000). For infrastructure in the first category, a detailed analysis of the potential impacts of climate change may not be important, particularly given the high level of uncertainty associated with local climate trends that will affect individual projects. Conversely, for projects where high sensitivity or large consequences are possible, a detailed analysis may be warranted.

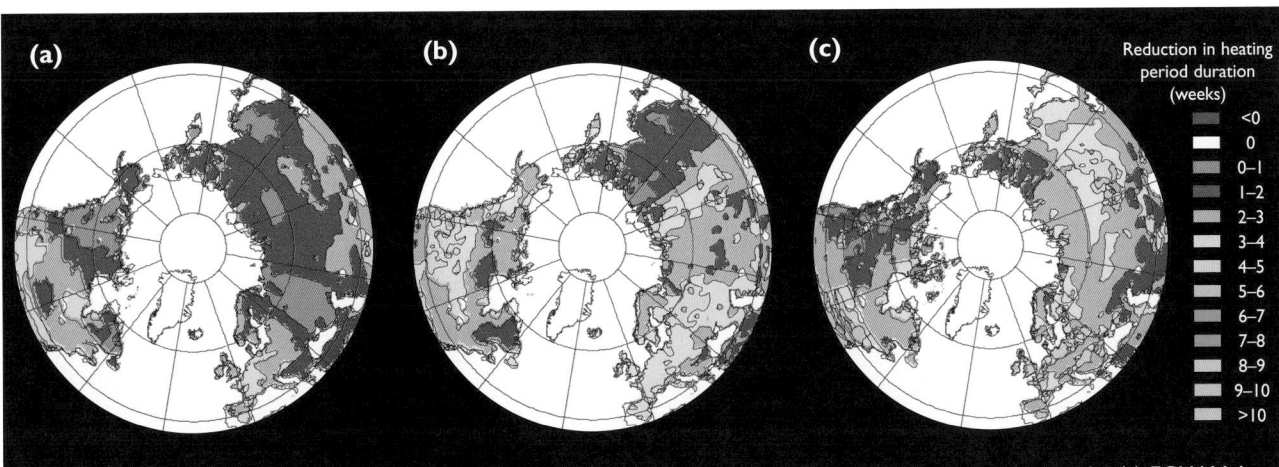

Fig. 16.25. Projected reduction in the duration of the heating period between 1999 and 2050, calculated using the (a) ECHAM-1; (b) GFDL; and (c) HadCM climate scenarios (Anisimov, 1999).

16.4.1. Risk-based evaluation of potential climate change impacts

Bush et al. (1998) presented a methodology for considering the impact of projected climate change within the framework of the engineering design process. They also explained how the same methodology can be utilized to identify and prioritize concerns about existing facilities with respect to climate change impacts. The method involves a multi-step approach that first assesses the sensitivity of a given project to climate change and then the consequences of any potential failures. The relationship between sensitivity and consequences defines the risk that climate change poses to the project. Finally, the degree of sensitivity and the severity of the consequences are used to determine what level of climate-change impact analysis should be carried out for a given project.

The sensitivity of a particular infrastructure project to climate change is determined by a number of factors, including the initial soil and permafrost temperatures, the temperature dependence of the material properties, the project lifetime, and the existing over-design or safety margin that might be included in the design for other reasons. Bush et al. (1998) included a procedure for categorizing these effects and determining the climate sensitivity of a project using a scale of high, medium, or low.

Any analysis of the consequences of failure begins with a determination of all relevant failure scenarios. These scenarios are then evaluated qualitatively using a scale of catastrophic, major, minor, or negligible, considering not only potential physical damage to the infrastructure but also socioeconomic or cultural impacts.

The final step in the screening process involves the determination of the level of climate-change impact analysis required at the design stage. Bush et al. (1998) included a table that suggests a level of analysis as a function of the degree of risk indicated by the project sensitivity and failure consequences. Level A requires a detailed quantitative analysis relying on numerical geotechnical models with refined input parameters, independent expert review, and a field-monitoring program with periodic evaluation of performance. Level B requires more limited quantitative analysis and also suggests a field-monitoring program. Level C suggests the use of qualitative analysis based on professional judgment, and Level D does not require any analysis of climate change impacts. This framework provides an organized approach to incorporating projected climate change into the design of arctic infrastructure projects.

16.4.2. Design thawing and freezing indices

Section 16.2.2.2 introduced the concept of air thawing and freezing indices. For design purposes, the design air thawing index (ATI) is commonly defined as the average ATI for the three warmest summers in the latest 30 years of record, or the warmest summer in the latest 10 years of record if 30 years of record are not available (Andersland and Anderson, 1978; Andersland and Ladanyi, 1994).

In Norway, statistical analysis of historic meteorological data is used to determine design air freezing indices (AFIs) with varying probabilities of occurrence (Heiersted, 1976). A similar approach can be used for determining design ATIs in permafrost areas (Instanes, A., 2003).

Table 16.10 shows different design ATIs and their probability of occurrence. The two-year design ATI (ATI_2) is approximately equal to the 30-year mean of the ATI. The average ATI for the three warmest summers in the latest 30 years of record usually lies somewhere between ATI_{20} and ATI_{50}. The magnitude of potential thawing incorporated in a design is dependent on the type of foundation or structure and the consequences of differential settlement or failure. For road embankments, ATI_2 to ATI_{10} is commonly used, while for buildings, ATI_{50} to ATI_{100} is used. For more sensitive structures, such as power plants and oil or gas pipelines, higher design ATIs should be considered.

In thermal analyses using advanced methods such as finite-element models, the design ATI is usually represented by a time series or sine curve that combines an average winter (AFI_2) and a design summer. Maximum thaw depth and permafrost temperature are usually caused by a combination of warm winters and summers. Therefore, combinations of warm winters (low AFI) and warm summers (high ATI) should also be considered.

Section 16.2.2.4 describes projected changes in freezing and thawing indices calculated using output from the ACIA-designated climate models. Figures 16.26 and 16.27 present an example of projected design freezing and thawing indices for Kugluktuk (Coppermine), Canada. The values from 1940 to 2000 are based on observations, while those from 2000 to 2100 are based on output from the CGCM2 model. The figures show curves for the 30-year mean value (probability of occurrence 50%); the mean of the warmest 3 years during the last 30 years; the warmest year during the last 10 years; AFI_{20}/ATI_{20} (probability of occurrence

Table 16.10. Design air thawing indexes (ATIs).

Magnitude of thawing	Probability of occurrence in any given year (%)	Predicted number of occurrences
ATI_2	50	once in 2 years
ATI_{10}	10	once in 10 years
ATI_{20}	5	once in 20 years
ATI_{100}	1	once in 100 years
ATI_{1000}	0.1	once in 1000 years
ATI_{10000}	0.01	once in 10000 years

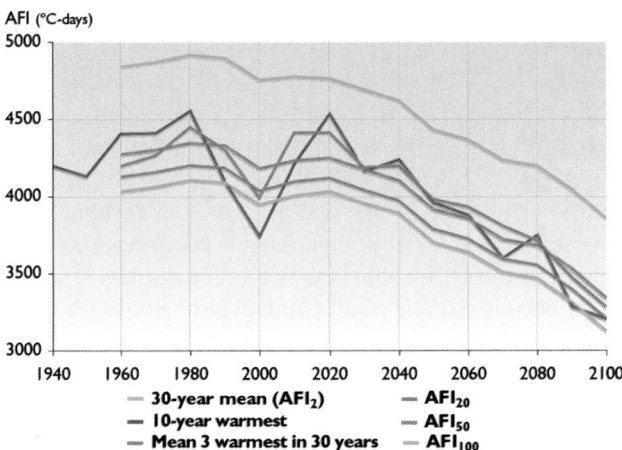

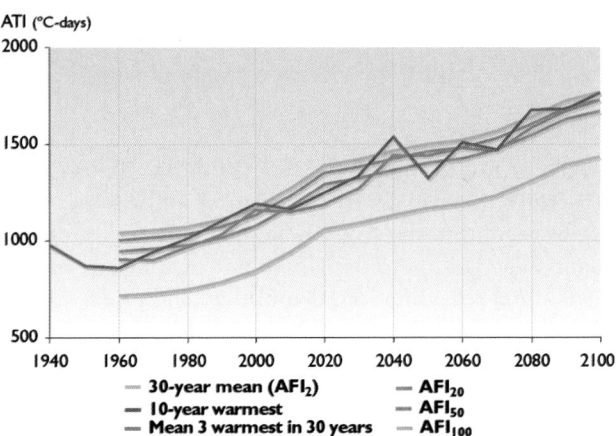

Fig. 16.26. Projected air freezing index (AFI) design levels for Kugluktuk (Coppermine), Canada (Instanes A. and Mjureke, 2002b).

Fig. 16.27. Projected air thawing index (ATI) design levels for Kugluktuk (Coppermine), Canada (Instanes A. and Mjureke, 2002b).

5%); AFI_{50}/ATI_{50} (probability of occurrence 2%); and AFI_{100}/ATI_{100}.

Figure 16.26 shows that the 30-year mean design AFI is projected to decrease from approximately 4850 degree-days in 1960 to 3850 degree-days in 2100. This represents an increase in winter (October–March) temperatures of approximately 5 to 6 °C. Figure 16.27 shows that the 30-year mean design ATI is projected to increase from approximately 720 degree-days in 1960 to 1430 degree-days in 2100. This represents an increase in summer (April–September) temperatures of approximately 4 °C. Instanes A. and Mjureke (2002b) presented similar plots of design freezing and thawing indices for the 21 stations listed in Table 16.3.

Using such projections, it is possible to estimate the potential impacts of climate change on specific structures in the Arctic while also including the effect of natural variability. Such an analysis has been carried out

for Svalbard Airport (Instanes A. and Mjureke, in press), using observed air temperature data for 1930 to 2000 and future temperature projections obtained by statistical downscaling of output from the ECHAM4/OPYC3 GSDIO integration (Hanssen-Bauer et. al, 2000, see also section 4.6.2). Figures 16.28 and 16.29 show the results of the analysis. Figure 16.28 shows that average thaw depths are projected to increase by as much as 50% between 2000 and 2050. As Fig. 16.29 shows, March ground temperatures at 2 m depth are projected to undergo the greatest increase (~4 °C) between 2000 and 2050, while the corresponding September temperatures are projected to increase by only 1 °C. At 40 m depth, mean annual temperatures are projected to increase by approximately 2 °C by 2050. At depths of 10 m and more, seasonal temperature variations are small. The climate scenario used in this study projects greater air temperature increases in winter than summer, which contributes to the strong positive trend in March temperatures at 2 m depth. The weak trend in September temperatures at 2 m is largely explained by the proximity of the 0 °C isotherm and the latent heat of the associated phase change. At shorter timescales, the large fluctuations in March ground temperatures reflect variable projected winter conditions, while summer conditions are

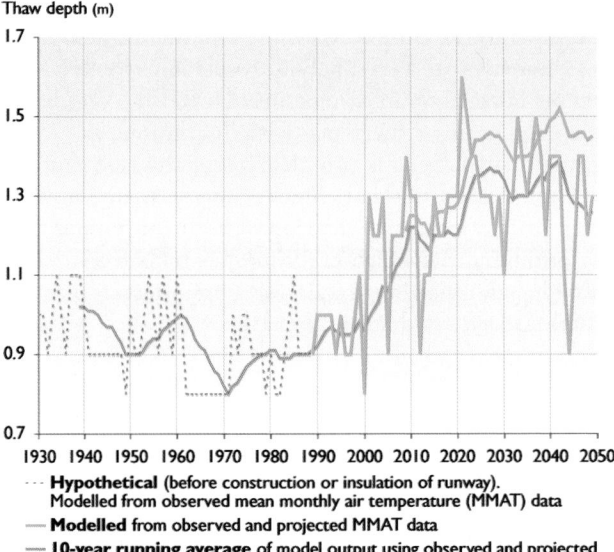

Fig. 16.28. Projected thaw depths below the runway at Svalbard Airport (Instanes A. and Mjureke, in press)

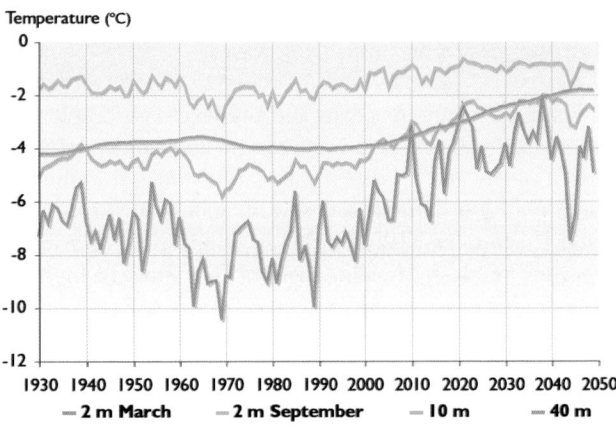

Fig. 16.29. Projected ground temperature changes at depths of 2, 10, and 40 m below the Svalbard Airport runway (Instanes A. and Mjureke, in press)

projected to be more uniform. This analysis suggests that the design of structures sensitive to thaw settlement should utilize the risk-based method described in section 16.4.1.

Instanes A. (2003) reported that the design lifetime for structures in permafrost regions is typically 30 to 50 years. Within this time frame, the structure should function according to design with normal maintenance costs. Total rehabilitation, demolition, and replacement of old structures must be expected and are part of sensible infrastructure planning and engineering practice. The effects of climate change on arctic infrastructure are, as previously indicated, difficult to quantify. Structural damage is often blamed on climate change, when a thorough investigation and case history indicates that the cause is either human error or the design lifetime being exceeded.

16.4.3. Coastal areas

Numerous arctic communities face the challenge of increased coastal erosion and warming of permafrost. The combined problems of increased wave action, sea-level rise, and thermal erosion have no simple engineering solutions. Two examples of erosion problems threatening communities and industrial facilities in Canada and Russia are discussed in the following sections. Coastal erosion is discussed in detail in section 6.9.

16.4.3.1. Severe erosion in Tuktoyaktuk, Canada

Tuktoyaktuk is the major port in the western Canadian Arctic and the only permanent settlement on the low-lying Beaufort Sea coast. The location of Tuktoyaktuk makes it highly vulnerable to increased coastal erosion due to decreased extent and duration of sea ice, accelerated thawing of permafrost, and sea-level rise. The Tuktoyaktuk Peninsula is characterized by sandy spits, barrier islands, and a series of lakes that have resulted from collapsed ground due to permafrost thawing ("thermokarst" lakes). Erosion is already a serious problem in and around Tuktoyaktuk, threatening cultural and archeological sites and causing the abandonment of an elementary school, housing, and other buildings. Successive shoreline protection structures have been rapidly destroyed by storm surges and accompanying waves.

As climate change proceeds and sea-level rise accelerates, impacts are likely to include further landward retreat of the coast, erosion of islands, more frequent flooding of low-lying areas, and breaching of freshwater thermokarst lakes and their consequent conversion into brackish or saline lagoons. The current high rates of cliff erosion are projected to increase due to higher sea levels, increased thawing of permafrost, and the increased potential for severe coastal storms during the extended open-water season. Attempts to control erosion at Tuktoyaktuk will become increasingly expensive as the surrounding coastline continues to retreat, and the site could ultimately become uninhabitable.

16.4.3.2. Erosion threatens Russian oil storage facility

The oil storage facility at Varandei on the Pechora Sea was built on a barrier island. Damage to the dunes and beach due to construction and use of the facility accelerated natural rates of coastal erosion. The Pechora Sea coasts are considered to be relatively stable, except where disturbed by human activity. Because this site has been perturbed, it is more vulnerable to damage from storm surges and the accompanying waves that will become a growing problem as climate continues to warm. As with the other sites discussed here, the reduction in sea-ice extent and duration, thawing coastal permafrost, and rising sea level are projected to exacerbate the existing erosion problem. This provides an example of the potential for combined impacts of climate change and other anthropogenic disturbances. Sites already threatened due to human activity are often more vulnerable to the impacts of climate change.

16.4.4. Summary

In areas of continuous permafrost, projected climate change is not likely to pose an immediate threat to infrastructure if the correct permafrost engineering design procedures have been followed; the infrastructure has not already been subjected to one of the factors mentioned at the beginning of this section or strains exceeding design values; and the infrastructure is not located on ice-rich terrain or along coastlines susceptible to erosion. Maintenance costs are likely to increase relative to those at present, but it should be possible to gradually adjust arctic infrastructure (through replacement and changing design approaches over time) to a warmer climate.

Projected climate change is very likely to have a serious effect on existing infrastructure in areas of discontinuous permafrost. Permafrost in these areas is already at temperatures close to thawing, and further temperature increases are very likely to result in extremely serious impacts on infrastructure. However, considerable engineering experience with discontinuous permafrost has been accumulated over the past century. Human activities and engineering construction very often lead to extensive thawing of both continuous and discontinuous permafrost. Techniques to address warming and thawing are already commonly used in North America and Scandinavia.

If the projections and trends presented in this assessment do occur over the next five to ten years, this is very likely to have a serious impact on the future design of engineering structures in permafrost areas. However, engineering design should still be based on actual meteorological observations and a risk-based method.

The most important engineering considerations related to projected climate change include: that risk-based methods should be used to evaluate projects in terms of potential climate change impacts; that design air thawing and freezing indices should be updated annually to account for observed climate variations and change; and that mitigation techniques such as artificial cooling of foundation soils should be considered as situations require.

In coastal areas, shore protection measures have to some extent reduced local erosion rates. However, thawing and erosion of ice-rich coastal sediments is a process that has been ongoing since the last glaciation and cannot be reversed given present climate trends.

16.5. Gaps in knowledge and research needs

The main gaps in knowledge are the lack of site-specific scenarios providing the probability of occurrence of various meteorological conditions (temperature, precipitation, wind, snow and sea-ice thickness and extent, waves, and erosion rates). Monitoring of infrastructure and the coastal environment is essential, as are climate sensitivity analyses.

It is also important to combine engineering knowledge with socioeconomic development scenarios and environmental impact assessments (see Chapters 3, 12, and 15) to evaluate how projected climate change may affect human lives in the Arctic in the future. Studies examining impacts and socioeconomic assessments have been performed in Canada (Couture et al., 2001). Studies are also underway for other regions of the Arctic, including Alaska (University of Alaska, Institute of Social and Economic Research) and northwest Russia (Centre for Economic Analysis, Oslo; Norwegian Institute of International Affairs, Oslo; and the Fridtjof Nansen Institute, Bergen).

References

Alekseev, G.V., Z.V. Bulatov, V.F. Zakharov and V.P. Ivanov, 1997. Influx of very warm water in the Arctic basin. Doklady Akademii Nauk SSSR, 356:401–403.

Aleksandrov, Ye.I. and V.V. Maistrova, 1998. Comparison of the air temperature measurements for the polar regions. The Antarctica, 34:60–72.

AMAP, 1997. Arctic Pollution Issues: A State of the Arctic Environment. Arctic Monitoring and Assessment Programme, Oslo, 188pp.

AMAP, 1998. AMAP Assessment Report: Arctic Pollution Issues. Arctic Monitoring and Assessment Programme, Oslo, xii+859pp.

Andersland, O.B. and D.M. Anderson, 1978. Geotechnical Engineering for Cold Regions. McGraw-Hill.

Andersland, O.B. and B. Ladanyi, 1994. An Introduction to Frozen Ground Engineering. Chapman & Hall, 352pp.

Andrady, A.L., 1997. Wavelength sensitivity in polymer photodegradation. Advances in Polymer Science, 128:49–94.

Andrady, A.L., S.H Hamid, X. Hu and A. Torikai, 1998. Effects of increased solar ultraviolet radiation on materials. Journal of Photochemistry and Photobiology B, 46:96–103.

Anisimov, O.A., 1999. Impacts of anthropogenic climate change on heating and air conditioning of buildings. Meteorology and Hydrology, 6:10–17. (In Russian)

Anisimov, O.A. 2001. Predicting patterns of near-surface air temperature using empirical data. Climatic Change, 50:297–315.

Anisimov, O.A. and M.A. Belolutskaia, 2002. Effects of changing climate and degradation of permafrost on infrastructure in northern Russia. Meteorology and Hydrology, 9:15–22. (In Russian)

Anisimov, O.A. and B. Fitzharris, 2001. Polar Regions (Arctic and Antarctic). In: J. McCarthy, O. Canziani, N.A. Leary, D.J. Dokken and K.S. White (eds.). Climate Change 2001: Impacts, Adaptation, and Vulnerability, pp. 801–841. Contribution of Working Group II to the Third Assessment Report of the Intergovernmental Panel on Climate Change. Cambridge University Press.

Anisimov, O.A., N.I. Shiklomanov and F.E. Nelson, 1997. Effects of global warming on permafrost and active-layer thickness: results from transient general circulation models. Global and Planetary Change, 61:61–77.

Aylsworth, J.M, M.M. Burgess, D.T. Desrochers, A. Duk-Rodkin, T. Robertson and J.A. Traynor, 2000. Surficial geology, subsurface materials, and thaw sensitivity of sediments. In: L.D. Dyke and G.R. Brooks (eds.). The Physical Environment of the Mackenzie Valley, Northwest Territories: a Baseline for the Assessment of Environmental Change. Geological Survey of Canada Bulletin, 547:41–47.

Bjerrum, L., 1955. Stability of natural slopes in quick clay. Geotechnique, 5(1):101–119.

Borzenkova, I.I., 1999a. Environmental indicators of recent global warming. In: Yu.A. Pykh (ed.). Environmental Indices, pp. 455–465. EOLSS Publishing.

Borzenkova, I.I., 1999b. About natural indicators of the present global warming. Meteorology and Hydrology, 6:98–116.

Brown, J., O.J. Ferrians, J.A. Heginbottom and E.S. Melnikov, 1997. International Permafrost Association Circum-Arctic Map of Permafrost and Ground Ice Conditions. Circum-Pacific map series, CP-45. U.S. Geological Survey.

Brown, J., K.M. Hinkel and F.E. Nelson, 2000. The circumpolar active layer (CALM) program: research designs and initial results. Polar Geography, 24:163–258.

Brown, J., M.T. Jorgenson, O.P. Smith and W. Lee, 2003. Long-term rates of erosion and carbon input, Elson Lagoon, Barrow, Alaska. In: M. Phillips, S.M. Springman and L.U. Arenson (eds.). Permafrost: Proceedings of the Eighth International Conference on Permafrost, pp. 101–106. A.A. Balkema.

Burgess, M.M. and S.L. Smith, 2003. 17 years of thaw penetration and surface settlement observations in permafrost terrain along the Norman Wells pipeline, Northwest Territories, Canada. In: M. Phillips, S.M. Springman and L.U. Arenson (eds.). Permafrost: Proceedings of the Eighth International Conference on Permafrost, pp. 1073–1078. A.A. Balkema.

Burgess, M.M., D.T. Desrochers and R. Saunders, 2000a. Potential changes in thaw depth and thaw settlement for three locations in the Mackenzie Valley. In: L.D. Dyke and G.R. Brooks (eds.). The Physical Environment of the Mackenzie Valley, Northwest Territories: a Baseline for the Assessment of Environmental Change. Geological Survey of Canada Bulletin, 547:187–195.

Burgess, M.M., S.L. Smith, J. Brown, V. Romanovsky and K. Hinkel, 2000b. Global terrestrial network for permafrost (GTNet-P): Permafrost monitoring contributing to global climate observations. Geological Survey of Canada, Current Research 2000-E14, 8pp.

Bush, E., D.A. Etkin, D. Hayley, E. Hivon, B. Ladanyi, B. Lavender, G. Paoli, D. Riseborough, J. Smith, and M. Smith, 1998. Climate Change Impacts on Permafrost Engineering Design. Environment Canada, Environmental Adaptation Research Group, Downsview, Ontario.

Carmack, E.C., 2000. Review of the Arctic Ocean's freshwater budget: sources, storage and export. In: E.L. Lewis, E.P. Jones, P. Lemke, T.D. Prowse and P. Wadhams (eds.). The Freshwater Budget of the Arctic Ocean. Kluwer Academic Publishers.

Carmack, E.C., R.W. Macdonald, F.A. Perkin, F.A. McLaughlin and R.J. Pearson, 1995. Evidence for warming of Atlantic water in the southern Canadian basin of the Arctic Ocean: Results from the Larsen-93 Exhibition. Geophysical Research Letters, 22:1061–1064.

Chapman, W.L. and J.E. Walsh, 2003. Observed climate change in the Arctic, updated from Chapman and Walsh, 1993. Recent variations of sea ice and air temperatures in high latitudes. Bulletin of the American Meteorological Society, 74(1):33–47. http://arctic.atmos.uiuc.edu/CLIMATESUMMARY/2003/

Couture, R., S.D. Robinson and M.M. Burgess, 2001. Climate change, permafrost degradation and impacts on infrastructure: two cases studies in the Mackenzie Valley. Proceedings of the 54th Canadian Geotechnical Conference, 4–9 September 2001, Calgary, 2:908–915.

Couture, R., S. Smith, S.D. Robinson, M.M. Burgess and S. Solomon, 2003. On the hazards to infrastructure in the Canadian North associated with thawing of permafrost. In: Proceedings of Geohazards 2003. Third Canadian Conference on Geotechnique and Natural Hazards, pp. 97–104.

Dyke, L.D. 2000. Stability of permafrost slopes in the Mackenzie valley. In: L.D. Dyke and G.R. Brooks (eds.). The Physical Environment of the Mackenzie Valley, Northwest Territories: a Baseline for the Assessment of Environmental Change. Geological Survey of Canada Bulletin, 547:177–186.

Dyke, L.D., J.M. Aylsworth, M.M. Burgess, F.M. Nixon and F. Wright, 1997. Permafrost in the Mackenzie Basin, its influences on land-altering processes and its relationship to climate change. In: Cohen, S.J. (ed.). Mackenzie Basin Impact Study (MBIS): Final report, pp. 112–117. Environment Canada.

Esch, D.C., 1993. Impacts of northern climate changes on Arctic engineering practice. In: Impacts of Climate Change on Resource Management in the North, 4th Canada-U.S. Symposium & Biennial AES/DIAND Meeting on Northern Climate, pp. 185–192.

Esch, D.C. and T.E. Osterkamp, 1990. Cold regions engineering: Climate warming concerns for Alaska. Journal of Cold Regions Engineering, 4(1):6–14.

Factor, A., W.V. Ligon, R.J. May and F.H. Greenburg, 1987. Recent advances in polycarbonate photodegradation. In: A.V. Patsis (ed.). Advances in Stabilization and Controlled Degradation of Polymers II, pp. 45–58. Technomic, Lancaster, Pennsylvania.

Freitag, D.R. and T. McFadden, 1997. Introduction to Cold Regions Engineering. ASCE Press, 738pp.

Garagulya, L.S., G.I. Gordeeva and L.N. Khrustalev, 1997. Zoning of the permafrost area by the impact of human-induced geocryological processes on environmental conditions. Kriosfera Zeml, 1(1):30–38.

Glazovskaya, T.G. and Yu.G. Seliverstov, 1998. Long-term forecasting of changes of snowiness and avalanche activity in the world due to the global warming. In: 25 Years of Snow Avalanche Research, Publ. No. 203. pp. 113–116. Norwegian Geotechnical Institute, Oslo.

Goering, D.J., 2003. Passively cooled railway embankments for use in permafrost areas. Journal of Cold Regions Engineering, 17(3):119–133.

Goering, D.J. and P. Kumar, 1996. Winter-time convection in open-graded embankments. Cold Regions Science and Engineering, 24:57–74.

Haas, W.M. and A.E. Barker, 1989. Frozen gravel: A study of compaction and thaw settlement behaviour. Cold Regions Engineering. Fifth International Conference, St. Paul, Minnesota, 6–8 February, 1989.

Hackett, B., 2001. Surge climate scenarios in the northern North Sea and along the Norwegian coast. Research Report No. 123. Norwegian Meteorological Institute, 51pp.

Hamid, S.H. and W. Pritchard, 1991. Mathematical modeling of weather-induced degradation of polymer properties. Journal of Applied Polymer Science, 43:651–678.

Hamid, S.H., M.B. Amin and A.G. Maadhah, 1995. Weathering degradation of polyethylene. In: S.H. Hamid, M.B. Amin and A.G. Maadhah (eds.). Handbook of Polymer Degradation. Marcel Dekker.

Hanssen-Bauer, I. and E.J. Forland, 1998. Long term trends in precipitation and temperature in the Norwegian Arctic: can they be explained by changes in atmospheric circulation patterns? Climate Research, 102:1–14.

Hanssen-Bauer, I., O.E. Tveito and E.J. Førland, 2000. Temperature scenarios for Norway: Empirical downscaling from the ECHAM4/OPYC3 GSDIO integration. Norwegian Meteorological Institute, report no. 24/00 KLIMA.

Harris, S.A., 1986. Permafrost distribution, zonation and stability along the eastern ranges of the Cordillera of North America. Arctic, 39:29–38.

Haugen J.E. and J. Debenard, 2002. Climate scenario for Norway in 2050 for the transportation sector. Internal document. Norwegian Meteorological Institute, Oslo, Norway (In Norwegian).

Haugen, J.E, D. Bjørge and T.E. Nordeng, 1999. A 20-year climate change experiment with HIRHAM, using MPI boundary data. RegClim Technical Report, 3:37–43. Norwegian Institute for Air Research.

Heiersted, R.S., 1976. Frost Action in Ground. Committee on Frost Action in Soils, Royal Norwegian Council for Scientific and Industrial Res., No. 17, Oslo. (In Norwegian)

Hestnes, E., 1994. Weather and snowpack conditions essential to slush flow release and down slope propagation. Norwegian Geotechnical Institute, Publication no. 582000-10.

Hinzman, L.D., N. Bettez, F.S. Chapin, M. Dyurgerov, C. Fastie, B. Griffith, R.D. Hollister, A. Hope, H.P. Huntington, A. Jensen, D. Kane, D.R. Klein, A. Lynch, A. Lloyd, A.D. McGuire, F. Nelson, W.C. Oechel, T. Osterkamp, C. Racine, V. Romanovsky, D. Stow, M. Sturm, C.E. Tweedie, G. Vourlitis, M. Walker, D. Walker, P.J. Webber, J. Welker, K. Winker and K. Yoshikawa, in press. Evidence and implications of recent climate change in terrestrial regions of the Arctic. Climatic Change.

Holubec, I., X. Hu, J. Wonnacott, R. Olive and D. Delarosbil, 2003. Design, construction and performance of dams in continuous permafrost. In: M. Phillips, S.M. Springman and L.U. Arenson (eds.). Permafrost: Proceedings of the Eighth International Conference on Permafrost, pp. 425–430. A.A. Balkema.

Instanes, A., 2003. Climate change and possible impact on Arctic infrastructure. In: M. Phillips, S.M. Springman and L.U. Arenson (eds.). Permafrost: Proceedings of the Eighth International Conference on Permafrost, pp. 461–466. A.A. Balkema.

Instanes, A. and D. Mjureke, 2002a. Design criteria for infrastructure on permafrost. SV4242-1. Instanes Svalbard AS (Report submitted to the Norwegian ACIA Secretariat, Norwegian Polar Institute, Tromsø).

Instanes, A. and D. Mjureke, 2002b. Design criteria for infrastructure on permafrost based on future climate scenarios. SV4242-2. Instanes Svalbard AS (Report submitted to the Norwegian ACIA Secretariat, Norwegian Polar Institute, Tromsø).

Instanes. A. and D. Mjureke, in press. Svalbard Airport runway - performance during a climate warming scenario 2000–2050. In: Proceedings of the International Conference on Bearing Capacity of Roads and Airfields, Trondheim, Norway, 27–29 June 2005.

Instanes, B., 2000. Permafrost engineering on Svalbard. In: Proceedings of the International workshop on Permafrost Engineering, Longyearbyen, Svalbard, 18–21 June, 2000, pp. 3–23. Tapir Publishers.

IPCC, 2001. Climate Change 2001: The Scientific Basis. Contribution of Working Group I to the Third Assessment Report of the Intergovernmental Panel on Climate Change. J.T. Houghton, Y. Ding, D.J. Griggs, M. Noguer, P.J. van der Linden, X. Dai, K. Maskell and C.A. Johnson (eds.). Cambridge University Press, 881pp.

Jahns, H.O. and C.E. Heuer, 1983. Frost heave mitigation and permafrost protection for a buried chilled-gas pipeline. In: Proceedings of the Fourth International Conference on Permafrost, Fairbanks, Alaska, July 1983, pp. 531–536. National Academy Press.

Janbu, N., 1996. Slope stability evaluation in engineering practice. In: Proceedings of the Seventh International Symposium on Landslides, Trondheim, Norway. Vol. 1, pp. 17–34. Balkema.

Johannessen, O., E. Shalina and M. Miles, 1999. Satellite evidence for an Arctic sea ice cover in transformation. Science, 286:1937–1939.

Johnson, T.C., E. McRoberts and J.F. Nixon, 1984. Design implications of subsoil thawing. In: R.L. Berg and E.A. Wright (eds.). Frost Action and its Control, pp. 45–103. American Society of Civil Engineers.

Johnston, G.H. (ed.), 1981. Permafrost, Engineering Design and Construction. J. Wiley & Sons, 352pp.

Jones, P.D., M. New, D.E. Parker, S. Martin and I.G. Rigor, 1999. Surface air temperature and its changes over the past 150 years. Reviews of Geophysics, 37:173–199.

Kane, D.L., L.D. Hinzman and J.P. Zarling, 1991. Thermal response of the active layer to climate warming in permafrost environment. Cold Regions Science and Technology, 19:111–122.

Khrustalev, L.N., 2000. Allowance for climate change in designing foundations on permafrost grounds. In: International Workshop on Permafrost Engineering, Longyearbyen, Norway, 18–21 June, 2000, pp. 25–36. Tapir Publishers.

Khrustalev, L.N., 2001. Problems of permafrost engineering as related to global climate warming. In: R. Paepe and V.P. Melnikov (eds.). NATO Advanced workshop on Permafrost Response on Economic Development, Environmental Security and Natural Resources, Novosibirsk, Russia, 12–16 November 1998, pp. 407–423. Kluwer Academic Publishers.

Khrustalev, L.N. and V.V. Nikiforov, 1990. Permafrost Stabilization in the Construction Basements. Nauka, Novosibirsk, 353pp.

Kotlyakov, V.M., 1997. Quickly warming Arctic basin. The Earth and the World, 4:107.

Kronik, Ya.A., 2001. Accident rate and safety of natural–anthropogenic systems in the permafrost zone. In: Proceedings of the Second Conference of Russian Geocryologists, vol. 4, pp. 138–146.

Ladanyi, B., 1994. La conception et la réhabilation des infrastructures de transport en régions nordiques. RTQ-94-07. Ministère des transport du Québec.

Ladanyi, B., 1995. Civil engineering concerns of climate warming in the Arctic. Transactions of the Royal Society of Canada, 6:7–20.

Ladanyi, B., 1996. A strength sensitivity index for assessing climate warming effects on permafrost. In: Proceedings of the Eighth International Conference on Cold Regions Engineering, Fairbanks, Alaska, 12–16 August 1996, pp. 35–45.

Ladanyi, B., 1998. Geotechnical microzonation in the Arctic related to climate warming. In: Proceedings of the Eleventh European Conference on Soil Mechanics and Geotechnical Engineering, Porec, Croatia, 25–29 May 1998, pp. 215–221. Balkema Publishers.

Ladanyi, B., M.W. Smith and D.W. Riseborough, 1996. Assessing the climate warming effects on permafrost engineering. In: Proceedings of the 49th Canadian Geotechnical Conference of the Canadian Geotechnical Society, St. John's, Newfoundland, 23–25 September 1996, pp. 295–302.

Larsen, J.O., L. Grande, S. Matsuura, S. Asano, T. Okamoto and S.G. Park, 1999. Slide activity in quick clay related to pore water pressure and weather parameters. In: Proceedings of the Ninth International Conference and Field Trip on Landslides, pp. 81–88.

Leroueil, S., G. Dionne and M. Allard, 1990. Tassement et consolidation au dégel d'un silt argileux à Kangiqsualujjuak. Fifth Canadian Permafrost Conference, Collection Nordicana No. 54, pp. 309–316. Laval University, Quebec.

Linell, K.A, 1973. Long-term effects of vegetative cover on permafrost stability in an area of discontinuous permafrost. In: North American contribution, Second International Conference on Permafrost, Yakutsk, USSR, pp. 688–693. National Academy of Sciences.

Mader, F.W., 1992. Plastics waste management in Europe. Macromolecular Symposia, 57:15–31.

Maslanik, J.A., M.C. Serreze and R.G. Barry, 1996. Recent decreases in Arctic summer ice cover and linkages to atmospheric circulation anomalies. Geophysical Research Letters, 23:1677–1680.

McPhee, M.G., T.P. Stanton, J.H. Morison and D.G. Martinson, 1998. Freshening of the upper ocean in the Central Arctic: Is perennial sea ice disappearing? Geophysical Research Letters, 25:1729–1732.

McRoberts, E.C., T.C. Law and E. Moniz, 1978. Thaw settlement studies in the discontinuous permafrost zone. Proceedings of the Third International Conference on Permafrost, Edmonton, Alberta, pp. 700–706.

Nakayama, T., T. Sone and M. Fukuda, 1993. Effects of climate warming on the active layer. In: Proceedings of the Sixth International Conference on Permafrost, Beijing, pp. 488–493.

Nelson, F.E., O.A. Anisimov and N.I. Shiklomanov, 2002. Climate change and hazard zonation in the circum-Arctic permafrost regions. Natural Hazards, 26(3):203–225.

Nelson, R.A., U. Luscher, R.W. Rooney and A.A. Stramler, 1983. Thaw strain data and thaw settlement predictions for Alaskan silts. In: Proceedings of the Fourth International Conference on Permafrost, Fairbanks, Alaska, pp. 912–917. National Academy Press.

Nixon, J.F., 1990a. Effect of climate warming on pile creep in permafrost. Journal of Cold Regions Engineering, ASCE, 4(1):67–73.

Nixon, J.F., 1990b. Seasonal and climate warming effects on pile creep in permafrost. In: Proceedings of the Fifth Canadian Permafrost Conference, Québec, July, 1990, pp. 335–340.

Nixon, J.F., 1994. Climate Change as an Engineering Design Consideration. Report prepared by Nixon Geotech Ltd. for the Canadian Climate Centre, Environment Canada, Downsview, Ontario, 11 pp.

Nixon, J.F., J. Stuchly and A.R. Pick, 1984. Design of Norman Wells pipeline for frost heave and thaw settlement. In: Proceedings of the Third International Offshore Mechanics and Arctic Engineering Symposium, American Society of Mechanical Engineers, New Orleans, Louisiana.

Nixon, J.F., N.A. Sortland and D.A. James, 1990. Geotechnical aspects of northern gas pipeline design. Collection Nordicana No. 54, Laval University, Québec, Canada, Proceedings of the Fifth Canadian Permafrost Conference, pp. 299–307.

Osterkamp, T.E. and A.H. Lachenbruch, 1990. Thermal regime of permafrost in Alaska and predicted global warming. Journal of Cold Regions Engineering, 4(1):38–42.

Osterkamp, T.E. and V.E. Romanovsky, 1999. Evidence of warming and thawing of discontinuous permafrost in Alaska. Permafrost and Periglacial Processes, 10:17–37.

Oswell, J.M., A.J. Hanna and R.M. Doblanko, 1998. Update on the performance of slopes on the Norman Wells Pipeline Project. In: Proceedings of the Seventh International Conference on Permafrost, Yellowknife, North West Territories.

Panova, Y.V., 2003. Permafrost processes inducing disastrous effects in engineering constructions, Medvezhje gas-field, Western Siberia. In: M. Phillips, S.M. Springman and L.U. Arenson (eds.). Permafrost: Proceedings of the Eighth International Conference on Permafrost, pp. 121–122. A.A. Balkema.

Parker, W.B., 2001. Effect of permafrost changes on economic development, environmental security and natural resource potential in Alaska. In: R.Paepe and V.P. Melnikov (eds.). NATO Advanced workshop on permafrost response on economic development, environmental security and natural resources, Novosibirsk, Russia, 12–16 November 1998. pp. 293–296. Kluwer Academic Publishers.

Parmuzin, S.Yu. and M.B. Chepurnov, 2001. Prediction of permafrost dynamics in the European North of Russia and Western Siberia in the 21st century. Moscow University Geology Bulletin, no. 4. (In Russian)

Pavlov, A.V., 1997. Cryological and climatic monitoring of Russia: methodology, observation results, and prediction. Kriosphera Zemli, 1(1):47–58.

Phukan, A., 1985. Frozen Ground Engineering. Prentice-Hall Inc. 336pp.

Rachold, V., J. Brown and S. Solomon (eds.), 2002. Arctic Coastal Dynamics. Report of an International Workshop, Potsdam, Germany, 26–30 November 2001. Reports on Polar Research 413, 103pp.

Rachold, V., J. Brown, S. Solomon and J.L. Sollid (eds.), 2003. Arctic Coastal Dynamics - Report of the Third International Workshop. Oslo, 2–5 December 2002. Reports on Polar and Marine Research 443, 127pp.

Riseborough, D.W., 1990. Soil latent heat as a filter of the climate signal in permafrost. Collection Nordicana No. 54, Laval University, Québec, Proceedings of the Fifth Canadian Permafrost Conference, pp. 199–205.

Sandersen, F., S. Bakkehøi, E. Hestnes and K. Lied, 1996. The influence of meteorological factors on the initiation of debris flows, rockslides and rockmass stability. In: Proceedings of the Seventh International Symposium on Landslides, Trondheim, Norway, pp. 97–114. Balkema.

Sayles, F. H. 1984. Design and performance of water-retaining embankments in permafrost. In: North American Contribution, Second International Conference on Permafrost, Fairbanks, Alaska, July 17–22 1983. Final Proceedings, pp. 31–42. National Academy Press.

Sayles, F.H. 1987. Embankment dams on permafrost: design and performance summary. US Army Cold Regions Research Engineering Laboratory, Special Report 87–11.

Seinova, I., 1991. Past and present changes of mudflow intensity in the Central Caucasus. Mountain Research and Development, 11(1):13–17.

Seinova, I. and N. Dandara, 1992. Hydro-technical arrangements for ecologically rational protection of the debris flow hazard. Internationales Symposium Interpraevent 1992 - Bern, Tagungspublikation, Band 3, pp. 315–326.

Serreze, M.C., J.E. Walsh, F.S. Chapin III, T.E. Osterkamp, V. Dyurgerov and D.M. Smith, 2000. Recent increase in the length of the melt season of perennial Arctic sea ice. Geophysical Research Letters, 25:655–658.

Shiklomanov, A.I., R.B. Lammers and C.J. Vorosmarty, 2002. Contemporary changes in Eurasian pan-Arctic river discharge. American Geophysical Union 83[47]. American Geophysical Union Meeting, 6–10 December 2002, San Francisco.

Sidorova, T., N. Belaya and V. Perov, 2001. Distribution of slushflows in northern Europe and their potential change due to global warming. Annals of Glaciology, 32:237–240.

Smith, M.W. and D.W. Riseborough, 1983. Permafrost sensitivity to climate change. In: Proceedings of the Fourth International Conference on Permafrost, Fairbanks, Alaska, pp. 1178–1183. National Academy Press.

Smith, M.W. and D.W. Riseborough, 1985. The sensitivity of thermal predictions to assumptions in soil properties. Proceedings of the Fourth International Conference on Ground Freezing, Sapporo, Japan, Vol. I, pp. 17–23. Balkema.

Smith, O.P. and G. Levasseur, 2002. Impact of climate change on transportation infrastructures in Alaska. Discussion paper. Department of Transportation Workshop, 1–2 October, 2002, Washington D.C. 11pp.

Smith, S.L. and M.M Burgess, 1998. Mapping the response of permafrost in Canada to climate warming. Current Research 1998-E, Geological Survey of Canada, pp. 163–171.

Smith, S.L. and M.M. Burgess, 1999. Mapping the sensitivity of Canadian permafrost to climate warming. In: M. Tranter, R. Armstrong, E. Brun, G. Jones, M. Sharp and M. Williams (eds.). Interactions between the Cryosphere, Climate and Greenhouse Gases, pp. 71–80. IAHS Publication No. 256.

Smith, S.L. and M.M. Burgess, 2004. Sensitivity of Permafrost to Climate Warming in Canada. Geological Survey of Canada Bulletin 579, 24 pp.

Smith, S.L., M.M. Burgess and J.A. Heginbottom, 2001. Permafrost in Canada, a challenge to northern development. In: G.R. Brooks (ed.). A Synthesis of Geological Hazards in Canada. Geological Survey of Canada Bulletin 548, pp. 241–264.

Sorensen, S., J. Smith and J. Zarling, 2003. Thermal performance of TAPS heat pipes with non-condensable gas blockage. In: M. Phillips, S.M. Springman and L.U. Arenson (eds.). Permafrost: Proceedings of the Eighth International Conference on Permafrost, pp. 1097–1102. Balkema.

Speer, T.L., G.H. Watson and R.K. Rowley, 1973. Effects of ground ice variability and resulting thaw settlements on buried warm oil pipelines. In: Proceedings of the Second International Conference on Permafrost, Yakutsk, Russia, pp. 746–752.

Szymanski, M.B., A. Zivkovic, A. Tchekovski and B. Swarbrick, 2003.
     Designing for closure of an open pit in the Canadian Arctic.
     In: M. Phillips, S.M. Springman and L.U. Arenson (eds.).
     Permafrost: Proceedings of the Eighth International Conference on
     Permafrost, pp. 1123–1128. A.A. Balkema.

Terzaghi, K., 1963. Stability of steep slopes on hard unweathered
     rock. Norwegian Geotechnical Institute, Publication no. 50.

Ulbrich, U. and M. Christoph, 1999. A shift of the NAO and
     increasing storm track activity over Europe due to anthro-
     pogenic greenhouse gas forcing. Climate Dynamics, 15:551–559
     (cited in IPCC, 2001).

Vyalov, S.S., A.S. Gerasimov, A.J. Zolotar and S.M. Fotiev, 1988.
     Ensuring structural stability and durability in permafrost ground
     areas at global warming of the Earth's climate. In: Proceedings of
     the Sixth International Conference on Permafrost, Beijing, pp.
     955–960.

Walsh, J.E. and W.L. Chapman, 2001. Twentieth-century sea-ice varia-
     tions from observational data. Annals of Glaciology, 33:444–448.

Weller, G. and M. Lange, 1999. Impacts of climate change in Arctic
     regions. Report from a workshop on the Impacts of Global Change,
     Tromsø, Norway, 25–26 April, 1999. International Arctic Science
     Committee, Oslo, 59pp.

Weller, G., P. Anderson and B. Wang, 1999. Preparing for a Changing
     Climate. The Potential Consequences of Climate Variability and
     Change. A Report of the Alaska Regional Assessment Group for the
     U.S. Global Change Research Program. Center for Global Change
     and Arctic System Research, Fairbanks, Alaska.

Woo, M.K., W.R. Rouse, A.G. Lewkowicz and K.L. Young, 1992.
     Adaptation of permafrost in the Canadian North: Present and
     future. Report No. 93-3. Canadian Climate Center, Environment
     Canada, Downsview, Ontario.

Chapter 17

Climate Change in the Context of Multiple Stressors and Resilience

Lead Author
James J. McCarthy, Marybeth Long Martello

Contributing Authors
Robert Corell, Noelle Eckley Selin, Shari Fox, Grete Hovelsrud-Broda, Svein Disch Mathiesen, Colin Polsky, Henrik Selin, Nicholas J.C. Tyler

Corresponding Author
Kirsti Strøm Bull, Inger Maria Gaup Eira, Nils Isak Eira, Siri Eriksen, Inger Hanssen-Bauer, Johan Klemet Kalstad, Christian Nellemann, Nils Oskal, Erik S. Reinert, Douglas Siegel-Causey, Paal Vegar Storeheier, Johan Mathis Turi

Contents

Summary

Climate change occurs amid myriad social and natural transformations. Understanding and anticipating the consequences of climate change, therefore, requires knowledge about the interactions of climate change and other stresses and about the resilience and vulnerability of human–environment systems that experience them. Vulnerability analysis offers a way of conceptualizing interacting stresses and their implications for particular human–environment systems. This chapter presents a framework for vulnerability analysis and uses this framework to illuminate examples in Sachs Harbour, Northwest Territories, Canada; coastal Greenland; and Finnmark, Norway. These examples focus on indigenous peoples and their experiences or potential experiences with climate change, organic and metallic pollution, and changing human and societal conditions. Indigenous peoples are the focus of these studies because of their (generally) close connections to the environments in which they live and because of the coping and adaptive strategies that have, for generations, sustained indigenous peoples in the highly variable arctic environment. The Sachs Harbour and Greenland examples are cursory since vulnerability field studies in these areas have yet to be undertaken. The Finnmark example provides a more in-depth analysis of Sámi reindeer herding developed through a collaborative effort involving scientists and herders, a subset of whom are authors of this chapter. These examples reveal a number of factors (e.g., changes in snow quality, changes in ice cover, contaminant concentrations in marine mammals, regulations, resource management practices, community dynamics, and economic development) likely to be important in determining the vulnerability of arctic peoples experiencing environmental and social change. The examples also illustrate the importance of understanding (and developing place-based methods to refine this understanding) stress interactions and the characteristics of particular human–environment systems, including their adaptive capacities. Moreover, meaningful analyses of human–environment dynamics require the full participation of local people, their knowledge, perspectives, and values.

Full vulnerability assessments for communities in Sachs Harbour and coastal Greenland, require in-depth investigations into what the people living in these areas view as key concerns and how these residents perceive the interrelations among, for example, natural resources and resource use, climate change, pollution, regulations, markets, and transnational political campaigns. This information will contribute to the identification of relevant stresses and to analysis of adaptation and coping, historically, presently, and in the future. For the Finnmark case study next steps should include attaining a more complete understanding of interrelations among reindeer herding, climate change, and governance and how reindeer herders might respond to consequences arising from changes in these factors. This case study highlights a number of other areas for future and/or continued investigation. These include analysis of the possibility that governmental management authorities or herders might respond to environmental and social changes in ways that enhance or degrade the reindeer herding habitat, and a more in-depth inquiry into extreme events and their implications for sustainable reindeer herding.

A comprehensive picture of the vulnerability of arctic human–environment systems to climate change and other changes will benefit from further development of case studies, longer periods of longitudinal analysis, and more comprehensive research with interdisciplinary teams that include local peoples as full participants. Case studies should be selected to provide information across a wide array of human–environment systems and conditions so as to enable comparative work across sites. This will lead to refinements in the vulnerability framework and improved understanding of resilience and vulnerability in this rapidly changing region.

17.1. Introduction

The impact assessments in the preceding chapters demonstrate significant effects that climate change and increases in ultraviolet (UV) radiation are now having and are expected to have on arctic peoples and ecosystems. These chapters also illustrate that (1) climate change and increases in UV radiation occur amidst a number of other interacting social and environmental changes, (2) the consequences of social and environmental changes depend on the interconnectedness of human and environmental systems and the ability of these coupled systems to cope with and otherwise respond to these changes, and (3) these changes and their consequences occur within and across scales from local to regional and even global dimensions (NRC, 1999). Assessments of potential impacts of social and environmental change in the Arctic will benefit from formalized frameworks for conceptualizing and analyzing these three characteristics and their implications for the dynamics of arctic social and biophysical systems. The fund of knowledge and learning that underpins these frameworks is based in risk–hazard and vulnerability studies, but only in recent years have these frameworks been applied in studies of arctic human and environment systems. Thus, unlike earlier chapters, this chapter does not have the benefit of a large body of published literature from which conclusions can be drawn regarding the resilience of arctic peoples and ecosystems in relation to future climate change and its interactions with other social and environmental changes.

This chapter develops the case for using a vulnerability framework to explore these interactions and ultimately to generate understanding as to where resilience, made possible through coping and adaptive strategies, could be effective in diminishing future climate change impacts in arctic coupled human–environment systems. "Coupled human–environment system" refers to the ensemble of inextricable relationships linking people and the environment within which they live. Use of the word "system" should not complicate this term, but rather it should communicate that various elements, from politics and history to the behavior of individuals and the ecology of plants and animals, form a complex whole. A vulnerability analysis that builds upon the assessment of climate impacts will consider a climate event in the context of other stresses and perturbations that together produce impacts of a compound character (Kasperson J. and Kasperson, 2001). Elements of a vulnerability approach are evident throughout preceding chapters of this assessment. The concept of vulnerability itself is noted in Chapters 1, 3, and 12, and adaptation and resilience are important themes in the overall assessment, particularly in Chapters 1, 3, 7, 11, 12, and 13.

This chapter uses the definitions of vulnerability and its elements that were adopted in the Third Assessment Report of the Intergovernmental Panel on Climate Change with vulnerability defined as the degree to which a system is susceptible to, or unable to cope with, adverse effects of stresses. Vulnerability is a function of the character, magnitude, and rate of change in stresses to which a system is exposed, its sensitivity, and its adaptive capacity. Exposure is the degree to which a system is in contact with particular stresses. Sensitivity is the degree to which a system is adversely or beneficially affected by stimuli. And adaptive capacity (or resilience) refers to a system's ability to adjust, to moderate possible harm, to realize opportunities, or to cope with consequences (IPCC, 2001b).

The presentation of vulnerability analysis in this chapter rests on three primary assumptions: (1) arctic human–environment systems are experiencing multiple and interacting stresses in addition to changes in climate and UV radiation; (2) consequences of social and environmental change depend upon how human–environment systems respond to such changes; and (3) the dynamics of changes, adaptations, and consequences span varied scales. Climate change and UV radiation increases trigger changes in ecosystems upon which arctic residents depend. For example, global warming is expected to increase net primary productivity in terrestrial and freshwater ecosystems (IPCC, 2001b; see also Chapters 7 and 8), but increased UV radiation penetration is likely to adversely affect productivity in aquatic ecosystems (AMAP, 1998). Although the Arctic is still a relatively pristine environment compared with many other areas, this region is experiencing significant problems associated with contaminants such as persistent organic pollutants (POPs) and heavy metals (AMAP, 1998, 2002). Climate change and exposure to pollutants interact, since changes in ice cover and runoff can cause lakes to become greater sinks for river-borne contaminants, and increased catchment rates and melting ice can lead to wider dispersion of pollutants. Moreover, sea-ice reductions can speed the entry of POPs trapped in Arctic Ocean ice into the food chain, posing risks to humans (AMAP, 2003; IPCC, 2001b). Linked human health effects of UV radiation, arctic diets, and pollutants have received little attention, but are plausible (De Fabo and

Björn, 2000). Clearly, an assessment of arctic vulnerabilities and the adaptive capacities that can modify vulnerabilities requires a holistic understanding of multiple drivers of change and their interactions.

Examples of resilience are also illustrated in the preceding chapters of this assessment. Consequences arising from climate change and increased UV radiation depend in large part both on the interconnectedness of human–environment systems and the capacities of these systems to respond to changes (see especially Chapters 1, 3, 7, 11, 12, 13; Freeman, 2000; Stenbaek, 1987). As noted by the authors of the Mackenzie Basin Impact Study:

> *Traditional lifestyles could be at risk from climate change, but this new challenge will not occur in a vacuum. Population growth and economic and institutional changes will influence the North's sensitivities and vulnerabilities to climate variability and climate change. They will also influence how regions and countries respond to the prospects of a global scale phenomenon that could affect their climate no matter what they do on their own.* Cohen, 1997

Studies of some regional arctic seas have also considered changes in factors that will interact with climate change. One such example is the Barents Sea Impact Study, which examines the possible mobilization of contaminants on the Kola Peninsula. The success of the Barents Sea Impact Study rests on a number of factors including place-based research that addresses socio-economic factors, the inclusion of indigenous knowledge, and attention to cross-scale interactions (Lange et al., 2003).

How arctic peoples experience, respond to, and cope with environmental phenomena will be shaped to some degree by the social changes they have experienced in the past (Freeman, 2000; Stenbaek, 1987; Chapters 1, 3, 11, 12, 13). Increasingly these changes concern relationships between local and central governments (Chapter 3), ties to a global economy and external markets and ways of life (Chapters 11 and 12), campaigns relating to animal rights and environmental issues (Chapter 12), resource management systems grounded in transnational as well as domestic policy fora (Chapters 11, 12, 13), habitat loss due to urbanization, industrial development, and agriculture (Chapter 11), and extraction of non-renewable resources (Chapters 11, 12, 16). Additional contemporary concerns of high priority for arctic peoples include poverty, domestic violence, substance abuse, inadequate housing, and substandard infrastructure (Chapters 3, 15, and 16).

Analysis of these and other changes and their implications for arctic human–environment systems must take account of dynamics at different scales. Some changes, such as those associated with climate change, for example, originate outside the Arctic, and arctic peoples contribute little to their sources. At the same time, the lives of many arctic peoples are closely interconnected with their environments through fishing, hunting, herding, and gathering (see Chapters 3 and 12). These relationships are also evolving, through, for example, technological changes, which can influence the future sustainability of arctic livelihoods. These close ties to transnational processes and intimate relationships between many arctic people and their environments underscore the importance of examining the vulnerability of particular arctic human–environment systems within the context of dynamics operating within and across local, regional, and global levels.

Social and environmental changes often yield benefits, as well as adverse effects for human–environment systems (Chapter 12). It is, therefore, appropriate to ask: in addition to the obvious desire to minimize future adverse effects of climate and other changes, in what ways might new opportunities be realized? Climate change could lead to increased vegetation growth/cover (Chapter 7), increased production of reindeer meat, new trade routes (Chapter 12), and new or intensified forms of commercial activity. Innovations in hunting equipment and practices might enable some hunters to hunt even more effectively and sustainably under snow and ice cover alterations brought about by climate change. Hunters may adapt to climate change by changing the type of species that they hunt and by altering the location, timing, and intensity of hunting. They may also take actions to minimize risk and uncertainty under unpredictable climate and ice conditions (e.g., by taking greater safety precautions or by electing not to hunt or fish) (Chapter 12).

The integrated vulnerability analysis described in this chapter begins with a general framework from Turner et al. (2003a). This framework provides a means of conceptualizing the vulnerability of coupled human–environment systems, under alterations in social and biophysical conditions arising from and interacting across global, regional, and local levels (e.g., NRC, 1999). Two examples are given where the extension of a climate impact analysis to a vulnerability analysis would be a logical next step. An example of a fully participatory exercise with a Sámi reindeer herding community in the Finnmark area of northern Norway is then used to explore aspects of vulnerability in their reindeer-herding livelihood. A full understanding of vulnerability in any of the systems examined is beyond the scope of this chapter. Such an analysis would require in-depth fieldwork and extensive participation of arctic residents (e.g., in planning and carrying out the assessment, in determining the stresses of greatest concern to them, in generating and disseminating results, etc.). The initial phase of work presented here illustrates, however, preliminary results of a conceptual and methodological approach to vulnerability analysis. These results offer insights into: the vulnerability of particular arctic human–environment systems to multiple human and environmental changes, how human and environmental conditions and behavior might attenuate or amplify these changes and their consequences, and what options exist to reduce vulnerability (see Turner et al., 2003a).

Examples used in this chapter focus on the experiences and likely future prospects for indigenous communities and the environments upon which they depend. Although non-indigenous populations far outnumber indigenous peoples in the Arctic, there are a number of reasons why a focus on indigenous livelihoods is particularly suited for initial analyses of interactions between climate and other factors that can contribute to the vulnerability of arctic residents. First, analyses of vulnerability require an understanding of human–environment interactions and their historical evolution. Such connections can be complex and difficult to discern. Indigenous ways of life, however, often offer ready insights into the ways in which people depend upon and adapt to their surroundings. Many indigenous peoples, for example, have livelihoods based partly or wholly on subsistence activities that entail strong human–environment relationships that have persisted through many generations. These activities include hunting, fishing, herding, and/or gathering, and their execution requires knowledge about the highly variable arctic environment, how to interact and cope with it, and how earlier generations adapted to past changes (Krupnik and Jolly, 2002). Second, analyses of vulnerability have the greatest potential for informing decisions regarding adaptation and mitigation when there is a distinct possibility of social and environmental loss. Arguably, the potential for such loss is particularly acute in indigenous arctic communities as they encounter varied forms of environmental and social change.

Rates of climate changes projected for some regions of the Arctic exceed, however, those likely to have been experienced during multiple past human generations. Thus, the resiliencies sufficient during the past may or may not suffice in the future. Moreover, while not all forms of likely future change portend likely negative consequences, climate change, UV radiation exposure, transboundary air pollution, and economic globalization, singly and in combination have the potential to adversely affect long-standing indigenous cultural practices, livelihoods, economies, and more. It is also noteworthy that among arctic residents a much larger body of literature is available on the resilience of indigenous peoples' livelihoods in response to climate change and in the context of multiple stressors.

The prospect of climate change in the Arctic has now begun to seriously influence planning in this region. Over an even shorter period researchers have begun to explore the degree to which likely future climate change will interact with other factors in the broader realm of human–environment interactions. At this early stage in the development of methodologies to quantitatively assess the vulnerability of different aspects of the human–environment system, studies of indigenous arctic communities are timely. Studies of indigenous peoples in other areas can now provide a common context within which to test characterizations of human–environment systems and their interactions, and to advance integrative data collection and analytical methodologies. Notable among these approaches is the absolute necessity of co-generating knowledge of exposures, sensitivities, and resiliencies inherent in these systems by involving indigenous peoples at the earliest stages of research planning and analysis.

Analyses of indigenous communities can also yield insights into the lives and livelihoods of non-indigenous arctic residents. However, without the same degree of historical and cultural ties to localities and ways of life, and with greater freedom to relocate, perhaps to an area outside the Arctic, non-indigenous residents will be vulnerable to likely future change in the Arctic in different ways. Eventually, suites of case studies focused on indigenous peoples and non-indigenous peoples and their environments will form useful comparative analyses from which questions regarding comparative resiliencies and ultimately their relative vulnerabilities can be assessed.

17.2. Conceptual approaches to vulnerability assessments

Large-scale studies of climate impacts have begun to examine the vulnerability of social and ecological systems to climate change. The seminal work of Timmerman (1981) provided intellectual underpinning for linking the concepts of vulnerability, resilience, and climate change. Examples of recent projects that incorporate these perspectives include the IPCC (particularly the contribution of Working Group II to the Third Scientific Assessment; IPCC 2001b), the Assessments of Impacts of and Adaptation to Climate Change in Multiple Regions and Sectors (AIACC) implemented by the United Nations Environment Programme, the Finnish global change research projects FIGARE and SILMU, the European Commission project on Tundra Degradation in the Russian Arctic (TUNDRA), the Norwegian project NORKLIMA, the US National Assessment of Climate Change Impacts on the United States (NAST, 2000), and the Regional Vulnerability Assessment (ReVA) Program under the United States Environmental Protection Agency (Smith, 2000). Some of these assessments were based on published research, and as such are limited in their completeness with respect to their spatial coverage, and especially to their inclusiveness of other stressors that can interact with climate to influence the vulnerability of human–environment systems. Other assessments are underway, and the surge in vulnerability research over the last few years will ensure that future climate impact assessments are more complete with respect to interactions with other stressors.

Vulnerability analysis is rooted in a long history (e.g., Cutter, 1996; Dow, 1992; Downing, 1992; Kates, 1971; Liverman, 1990; Turner et al., 2003a; White, 1974), and in research traditions (for recent reviews see Cutter, 1996; Golding, 2001; Kasperson J. et al., 2003; Polsky et al., 2003; Turner et al., 2003a) that encompass work on risk–hazards–disasters (Blaikie et al., 1994; Cutter, 1996), climate impacts (Cutter, 2001; IPCC, 1997; Kates et al., 1985; Parry 1978; Parry et

al., 1998), food security (Böhle et al., 1994; Downing, 1991; Easterling, 1996), national security (Bachler, 1998; Dabelko and Simmons, 1997; Gilmartin et al., 1996; Homer-Dixon and Blitt, 1998; Winnefield and Morris, 1994), and resilience (Berkes and Folke, 1998; Berkes et al., 2003; Turner et al., 2003a). Much of the applied hazards, climate impact, and food security research to date has focused on the source of and potential exposure to a hazard, and has sought to understand the magnitude, duration, and frequency of this hazard and the sensitivity of the exposed system (Burton et al., 1978; Cutter, 1996).

It is common to distinguish between impacts and vulnerability perspectives by saying that the former focuses more on system sensitivities and stops short of specifying whether or not a given combination of stress and sensitivity will result in an effective adaptation. The latter emphasizes the factors that constrain or enable a coupled human–environment system to adapt to a stress. Another distinction that has been drawn between climate impact and vulnerability assessments is that the former proceeds by examining a climate event and the stresses that are exerted upon an exposure unit to produce critical downstream outcomes. The latter, by contrast, considers the climate event in the context of other stresses and perturbations that together produce impacts from compound events (Vogel as quoted in Kasperson J. and Kasperson 2001).

These distinctions are, however, to some degree oversimplifications, since a lack of emphasis on adaptation applies more to past empirical studies of climate change impacts than to the conceptual underpinnings of such studies. Adaptation has long been at the heart of the debate on reducing vulnerability to environmental stresses (Turner et al., 2003a). Even the early models on climate change impacts (e.g., Kates et al., 1985) consider the importance of adaptation, and the same applies to the broader, related literature on risk/hazards (e.g., Burton et al., 1978; Cutter, 1996; Kasperson R. et al., 1988) and food security (e.g., Böhle et al., 1994; Downing, 1991). Parry and Carter (1998) also acknowledge the seminal ideas of Kates (1985) on this topic and go on to discuss the evolution from a climate impact approach to a climate interaction approach. They describe how the severe economic hardship experienced by Canadian prairie farmers in the 1930s arose as a result of interaction among multiple factors. "Economics, weather and farming technology interacted to create a severe economic and social impact that was perhaps preconditioned by the Depression but triggered by drought."

Thus, increasing interest in "global change vulnerability" is not so much the result of a revolution in ideas – although the theoretical bases are maturing (e.g., Adger and Kelly, 1999) – but more a response to a general dissatisfaction with the ways in which adaptive capacity has been captured in empirical research and the associated need to reconnect with this concept if climate impact and global change models are to improve.

Increasingly, studies of vulnerability go beyond understanding the behavior of a stress and the degree to which an exposed system reacts adversely or beneficially (Holling, 1996, 2001). These studies also investigate (1) ways in which the exposed system might respond to, intensify, and/or ameliorate the effects of multiple stresses; and (2) why the same hazard might affect different systems in different ways and what system characteristics (including political economy, social structures and institutions) help to explain this variation. The concept of resilience in ecological studies has also informed treatment of adaptive capacity in vulnerability assessment (Resilience Alliance, no date; Walker et al., 2002). Resilience generally refers to the ability of a system to return to a reference state or remain within a range of desirable states following a perturbation. Berkes and Jolly (2001) have pointed out that the concept of resilience has three defining characteristics. It is a measure of: the amount of change the system can experience and still retain the same controls on function and structure; the degree to which the system is capable of self-organization; and the systems' ability to sustain and increase its capacity for adaptation.

Similarly, adaptive capacity refers to ecosystem flexibility and social system responsiveness in the face of disturbances (Turner et al., 2003a). According to one line of thought in political ecology, for example, adaptive capacity derives from human ecology of production, entitlements pertaining to market exchanges, and political economy (Böhle et al., 1994). These factors depend, for example, on resources available to a social group, the ability to sell these resources, the selling price, and access to markets (Sen, 1981). In addition, social, institutional, and political conditions might affect the ability of a social system to utilize resources or make other adjustments in overcoming the effects of a disaster such as drought (Turner et al., 2003a). Initiatives such as the Management of Social Transformations Programme's Circumpolar Coping Processes Project (MOST CCPP) advances understanding of human responses to environmental and other forms of change. MOST CCPP is a cross-disciplinary network comprising participants from Norway, Finland, northwest Russia, Denmark, Faroe Islands, Greenland, Iceland, Canada, and Sweden. This project is a comparative research endeavor that examines ways in which local authorities, civil society actors, and enterprise networks cope locally and regionally with global technological, economic, and environmental changes.

Researchers are also increasingly attentive to the socio-ecological, multi-scalar, and dynamic nature of vulnerability. Studies aimed at understanding the vulnerability of particular places are forgoing the tendency to treat social and biophysical vulnerability as separate conditions (e.g., see Adger and Kelly, 1999; Kelly and Adger, 2000). They are instead examining the vulnerability of the coupled human–environment system with place-based approaches (Cutter, 1996; Turner et al., 2003a) (also see Berkes and Folke, 1998; Berkes et al., 2003).

In addition, conditions and phenomena spanning global, national, and local levels can have important implications for the vulnerability of specific people and areas. For example, the globalization of markets, technological innovations originating abroad, changes in national policy, and the condition of local infrastructure could all potentially increase or decrease the vulnerability of a particular household or community to drought or flood (Leichenko and O'Brien, 2002). The ever-changing character of biogeophysical, environmental, institutional, economic, and political processes that influence human–environment systems requires that vulnerability be treated as a process (Handmer et al., 1999; Leichenko and O'Brien, 2002; Reilly and Schimmelpfenning, 1999). In its simplest static state vulnerability can be seen as the residual of change after considering the resilience and adaptive capacity of a system. However, the dynamic nature of these processes requires that vulnerability also be considered as an integral part of the change rather than external to it.

17.2.1. A framework for analyzing vulnerability

Building on this history, the combined effects of climate and other stressors can be examined via the following questions:

1. How do social and biophysical conditions of human–environment systems in the Arctic influence the resilience of these systems when they are impacted by climate and other stressors?
2. How can the coupled condition of these systems be suitably characterized for analysis within a vulnerability framework?
3. To what stresses and combinations of stresses are coupled human–environment systems in the Arctic most vulnerable?
4. To what degree can mitigation and enhanced adaptation at local, regional, national, and global scales reduce vulnerabilities in these systems?

Answers to these questions require a holistic research approach that addresses the interconnected and multi-scale character of natural and social systems. A framework for this approach (Fig. 17.1) depicts a cross-scale, coupled human–environment system. The multiple and linked scales in each diagram are reflected in the nesting of different colors with blue (place), pink (region), and green (world). The place (whatever its spatial dimensions) contains the coupled human–environment system whose vulnerability is being investigated. Figure 17.2 presents a more detailed schematic of the place. The influences (including stresses) acting on the place arise from outside and inside its borders. However, given the complexity and possible non-linearity of these influences, their precise character (e.g., kind, magnitude, and sequence) is commonly specific to the place-based system. This system has certain attributes denoted as human and environmental conditions. These conditions can interact with one another and can enable or inhibit certain responses in, for example, the form of coping, adaptation, and impacts. Negative impacts at various scales result when stresses or perturbations exceed the ability of the place-based human–environment system to cope or respond. There are a number of feedbacks and interactions within and around the place-based system and these dynamics can extend across place-based, regional, and global levels. Impacts and mitigating and adaptive responses, for example, can modify societal conditions of the place and/or alter societal and environmental influences within the place and at regional and global scales.

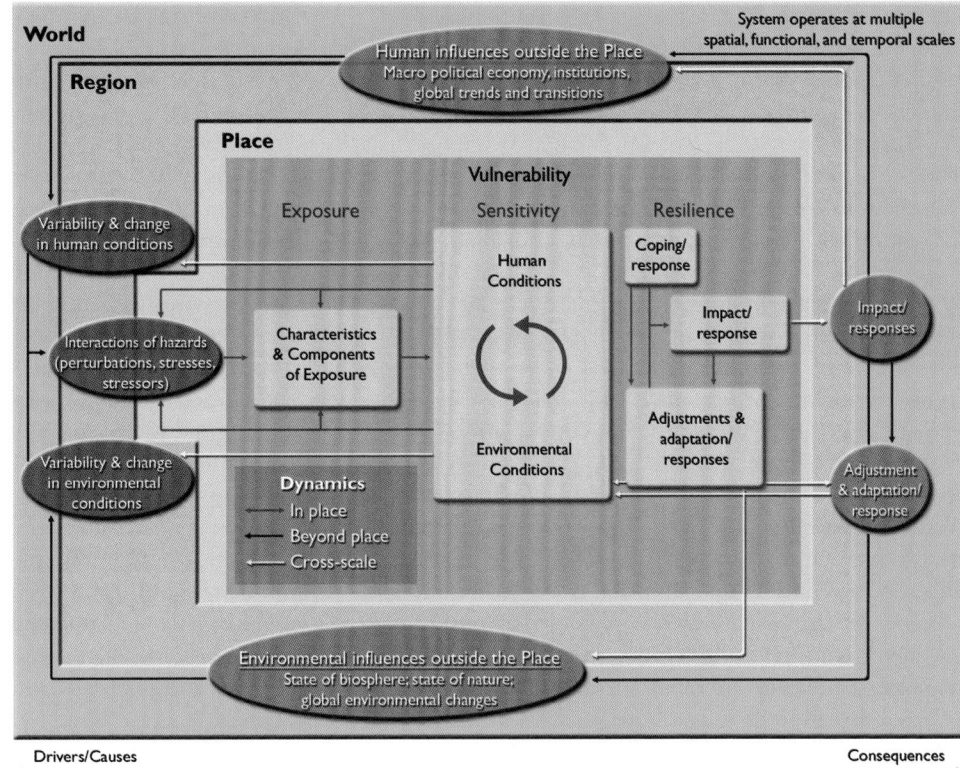

Fig. 17.1. Vulnerability framework (Turner et al., 2003a).

The vulnerability of the coupled human–environment system can be thought of as the potential for this system to experience adverse impacts, taking into consideration the system's resilience. Adverse impacts might arise from phenomena such as climate change, pollution, and social change. The system's resilience depends on its ability to counter sources of adverse change and to adapt to and otherwise cope with their consequences. It is important to note differences between mitigation and adaptation. Mitigation involves the amelioration of a stress at its source (e.g., changes in fossil fuel consumption resulting in reduced greenhouse gas (GHG) emissions). While the Arctic is experiencing the effects of climate change, actions to mitigate climate change through GHG reductions are largely dependent on the actions of people living at more southern latitudes. Adaptation (e.g., through mobility, new hunting or fishing practices, and/or the development or adoption of new technologies) requires that resources and other forms of capacity be accessible to the human–environment system in question. Such resources and capacity can take years, even generations to develop.

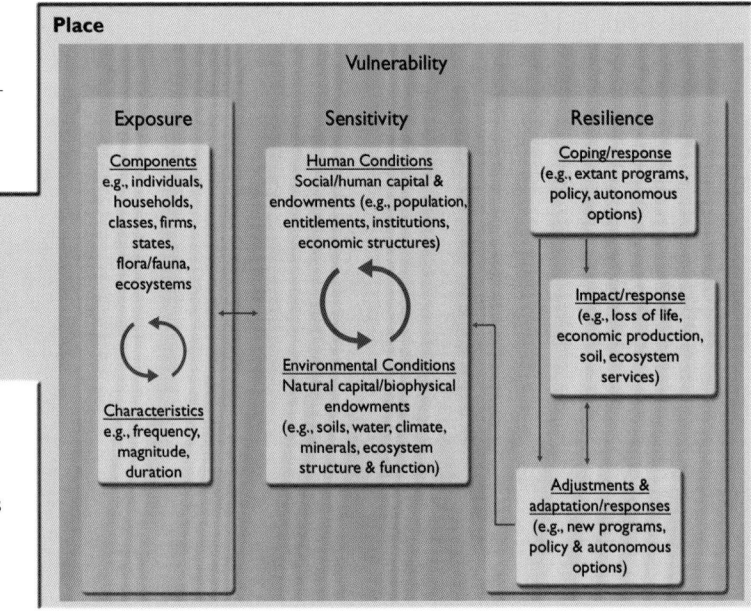

Fig. 17.2. Details of the exposure, sensitivity, and resilience components of the vulnerability framework (Turner et al., 2003a).

17.2.2. Focusing on interactive changes and stresses in the Arctic

The Arctic is experiencing a number of striking social and environmental changes and influences. While some are welcome, others (considered stresses) have adverse consequences. Given that vulnerability is highly complex and can vary significantly with location, it is essen-

tial to conduct "place-based" analyses, where "place-based" suggests a spatially continuous set of human–environment conditions or systems (Turner et al., 2003a). Since the vulnerability of a system is closely connected to the particular social and environmental conditions at a given location, the priorities and perspectives of people living in the location and those of other stakeholders are essential for identifying the key stresses, and understanding the exposure, sensitivity, and resilience of their coupled human–environment system (see section 17.3.4). Thus, knowledge, values, and understanding held by local residents are integral in determining which factors are most likely to test resilience of a system. Recent research on coupled human–environment systems (Turner et al., 2003a,b) demonstrates the efficacy of a participatory approach. The Sachs Harbour and Finnmark examples discussed in this chapter similarly evidence the importance of collaborating with local people and other stakeholders.

Preliminary consultations with arctic researchers and residents led the authors of this chapter to consider environmental pollution (POPs and heavy metals) and trends in human and societal conditions in addition to climate change and variability and UV radiation as potential stressors in arctic human–environment systems. The case for climate change is central to the Arctic Climate Impact Assessment, and so needs only a brief review in this section. Changes in UV radiation also receive considerable attention in the ACIA, but rather less is known about how possible harmful effects will be distributed in the Arctic. So little can be done at this time to assess vulnerability to this potential stress either in isolation or in combination with other factors. The comprehensive nature of the Arctic Monitoring and Assessment Programme (AMAP) studies allow much more to be said about organic and metal pollution as a potential stress in arctic systems, so more attention is given to pollution in this section.

It is important to note that, although they are often referred to here as stresses, climate change and trends in human and societal conditions can have positive as well as negative effects. Changes in climate and climate variability refer to changes in temperature, precipitation, snow cover, permafrost, sea ice, and extreme weather events. Major POPs include DDT (Dichlorodiphenyltrichloroethane), PCBs (polychlorinated biphenyls), HCH (hexachlorocyclohexane), and major heavy metals include lead (Pb), cadmium (Cd), and mercury (Hg) (AMAP, 1998, 2002). Human and societal trends of interest are consumption (especially pertaining to foodstuffs and technology), settlement patterns and demography, governance and regulation (particularly regarding natural renewable resources), connectivity (e.g., telephones, email, Internet), and markets and trade (Turner et al., 2003b) (see also Beach, 2000; Bjerregaard, 1995; Caulfield, 1997, 2000; Kuhnlein and Chan, 2000; Macdonald et al., 2003; Stenbaek, 1987; Svensson, 1987a,b; Wheelersburg, 1987). More in-depth fieldwork and analysis might reveal additional high priority stresses important for these sites.

17.2.2.1. Trends in human and societal conditions

Arctic peoples have experienced significant social changes over the past few human generations (Freeman, 2000; Stenbaek, 1987) as the Arctic's borders have become more permeable to southerners, material goods, and ways of thinking; as indigenous peoples have asserted their identity, rights, and culture in legal and policy forums; and as new relationships have formed

between local and national governments. The Arctic Council's Arctic Human Development Report (AHDR, 2004), especially its chapters on sustainable human development and economies is a logical and welcome next step in a synthesis of understanding in this area. Technology has been an important part of many such transformations. Satellites, television, the Internet, and telephones, for example, have revolutionized communication. Snowmobiles, all terrain vehicles, and more powerful small boats have brought new modes of transportation and recreation while accompanying changes in some hunting, herding, and fishing practices. The modernization of hunting equipment has also contributed to changes in approaches to whaling and marine mammal hunting. Individuals often have differing views about what types of social changes are beneficial and what types are unwanted. Some arctic residents, for example, might support the use of snowmobile technology in reindeer herding, while others might oppose it. Similarly, some people might view certain forms of human and societal change as adversely stressing a human–environment system, while others might view these changes as enhancing the resilience of that system. In seeking to understand how such changes bear on the vulnerability of arctic communities, this chapter examines a variety of human and societal factors including governance, population dynamics, migration, consumption, economies, markets and trade, and connectivity. These represent only a small subset of topics that constitute human and societal conditions. In some instances these factors are considered influences or stresses on the system (e.g., regulations limiting flexibility in reindeer herding). In other instances these factors can serve as both influences and part of the system's adaptive and coping responses (e.g., migration and changes in consumption).

17.2.2.2. Climate change

Projections of future climate change in the Arctic are documented in Chapter 4. Temperatures are projected to increase throughout the Arctic, even in sub-regions that have shown slight cooling trends in the latter half of the 20th century. Summer sea ice in the Arctic Ocean is projected to continue to decrease in area and thickness. The active layer of permafrost is projected to continue to deepen. Seasonal weather and precipitation patterns are likely to change, altering forms of precipitation between rain, freezing rain, and snow, and affecting snow quality. Recent evidence indicates that many of these changes are already affecting the distribution and abundances of terrestrial and marine species (see Chapters 7, 8, 9). Changes in temperature, precipitation, and storm patterns can affect the type, abundance, and location of animals and plants available to humans and may lessen the productivity of certain traditional forms of hunting and gathering. Decreases in the extent and thickness of sea ice can alter the distribution, age structure, and size of marine mammal populations, expose the arctic coast to more severe weather events, exacerbate coastal erosion, and affect modes of transportation and the ability of people to reach hunting locations and other villages. Changes in surface water budgets and wetlands can change coastal microclimates, alter the size and structure of peatlands, and result in pond drainage. In addition, damp, wet air during the traditional "drying season" makes it difficult to dry and preserve foods for winter months. These changes would, in turn, result in effects felt not only in human communities in the Arctic, but in other areas of the world as well (IPCC, 2001b).

17.2.2.3. UV radiation

Continued ozone depletion and the related problems of UV radiation exposure are likely to result in serious human and ecosystem impacts (Cahill and Weatherhead, 2001). UV radiation can harm humans directly via sunburn and skin cancer, immune system suppression, and eye damage, such as cataract photokeratitis (AMAP, 1998; De Fabo and Björn, 2000). The synergistic effects of UV radiation, climate change, and pollution could be more intense than the effects of any one of these stresses acting alone. For example, aquatic organisms that have assimilated UV-B absorbing polyaromatic hydrocarbons have shown phototoxic effects when exposed to UV-B radiation. Exposure to UV radiation has also been found to increase the toxicity of some chemicals, especially those associated with oil spills (Cahill and Weatherhead, 2001).

Adverse effects of UV radiation on arctic plants and animals can also indirectly affect humans. The vulnerability of arctic ecosystems to UV radiation is greatest in spring when ozone depletion is at its maximum and when new organisms are beginning life. Arctic plants have fewer protective pigments and are more sensitive to UV radiation than similar plants in other regions of the world, partly because at low temperatures plants are less able to repair UV radiation damage (AMAP, 1998; De Fabo and Björn, 2000).

Wildlife can experience UV radiation effects similar to those found in humans, although fur and plumage mean skin effects are less likely than eye damage (De Fabo and Björn, 2000). Increased UV radiation may affect fisheries through changes in planktonic food webs, but these changes are difficult to predict because they involve long-term alterations in species adaptation and community structure. If UV radiation were to change arctic aquatic ecosystems, this could in turn affect seabirds and land predators (e.g., seals, foxes, and bears) that feed on aquatic organisms (AMAP, 1998).

17.2.2.4. Pollution

AMAP concluded in both of its two recent assessments that pollution can pose problems in the Arctic (AMAP, 1998, 2002). Heavy metals and POPs are of particular concern, although there are important regional and local variations within the Arctic. Both heavy metals and POPs are transported to the Arctic via long-range air and water pathways and both bioaccumulate in food webs (see Fig. 17.3) (AMAP, 2002; Wania and Mackay, 1996). In addition to long-range transport, some pollutants originate from local sources such as the geology, industrial activities, pesticide use, and private use.

Heavy metals and POPs are associated with several environmental risks. These include estrogenic effects, disruption of endocrine functions, impairments of immune system functions, functional and physiological effects on reproduction capabilities, and reduced survival and growth of offspring (AMAP, 1998, 2002; UNECE, 1994). Data on human health effects suggest that human exposure to levels of POPs and heavy metals found in some traditional foods may cause adverse health effects, particularly during early development (AMAP, 2003; Ayotte et al., 1995; Colborn et al., 1996; Hild, 1995; Kuhnlein and Chan, 2000).

Traditional foods also provide health benefits, however, which need to be weighed against risks (see section 17.2.3.3). Many traditional foods are rich in vitamins and nutrients and low in saturated fats. Whale skin and blubber, for example, are a good source of vitamins A and C, thiamin, riboflavin, and niacin. They are also low in saturated fats and high in omega-3 polyunsaturated fatty acids that guard against cardiovascular diseases. Additional health benefits arise from the physical activity required to obtain traditional foods. Moreover, traditional harvesting, processing, and sharing of traditional foods serve important roles in the social, cultural, and economic life of many arctic inhabitants (AMAP, 2003; Freeman et al., 1998). In communities where contaminant levels are sufficiently high to prompt health concerns, balanced dietary advice is needed, especially for pregnant women and small children. Risk–benefit discussions have been most productive when they involve local communities, local public health authorities, and experts from a wide array of disciplines (AMAP, 2002, 2003).

Persistent organic pollutants that require special attention in arctic vulnerability studies include the industrial chemicals PCBs; the pesticide DDT; and the pesticide HCH, the most common form of which, γ-HCH, is the insecticide Lindane. These are well-known arctic pollutants of concern (AMAP, 1998, 2002) that are currently being addressed by national legislation and international agreements (Downie et al. 2004; Eckley, 2001; Selin, 2003; Selin and Eckley, 2003).

Many other POPs are known to be hazardous, as well as possibly other, lesser known organic substances that may have negative impacts. For example, levels of the flame-retardants polybrominated diphenyl ethers (PBDEs), polychlorinated naphthalenes (PCNs), and the pesticide endosulfan are increasingly found in the Arctic. Levels of PBDEs are increasing in the Canadian Arctic (AMAP, 2002; Ikonomou et al., 2002). Ikonomou et al. (2002) suggested that at current rates of bioaccumulation, PBDEs will surpass PCBs to become the most prevalent contaminant in ringed seals (*Phoca hispida*) in the Canadian Arctic by 2050.

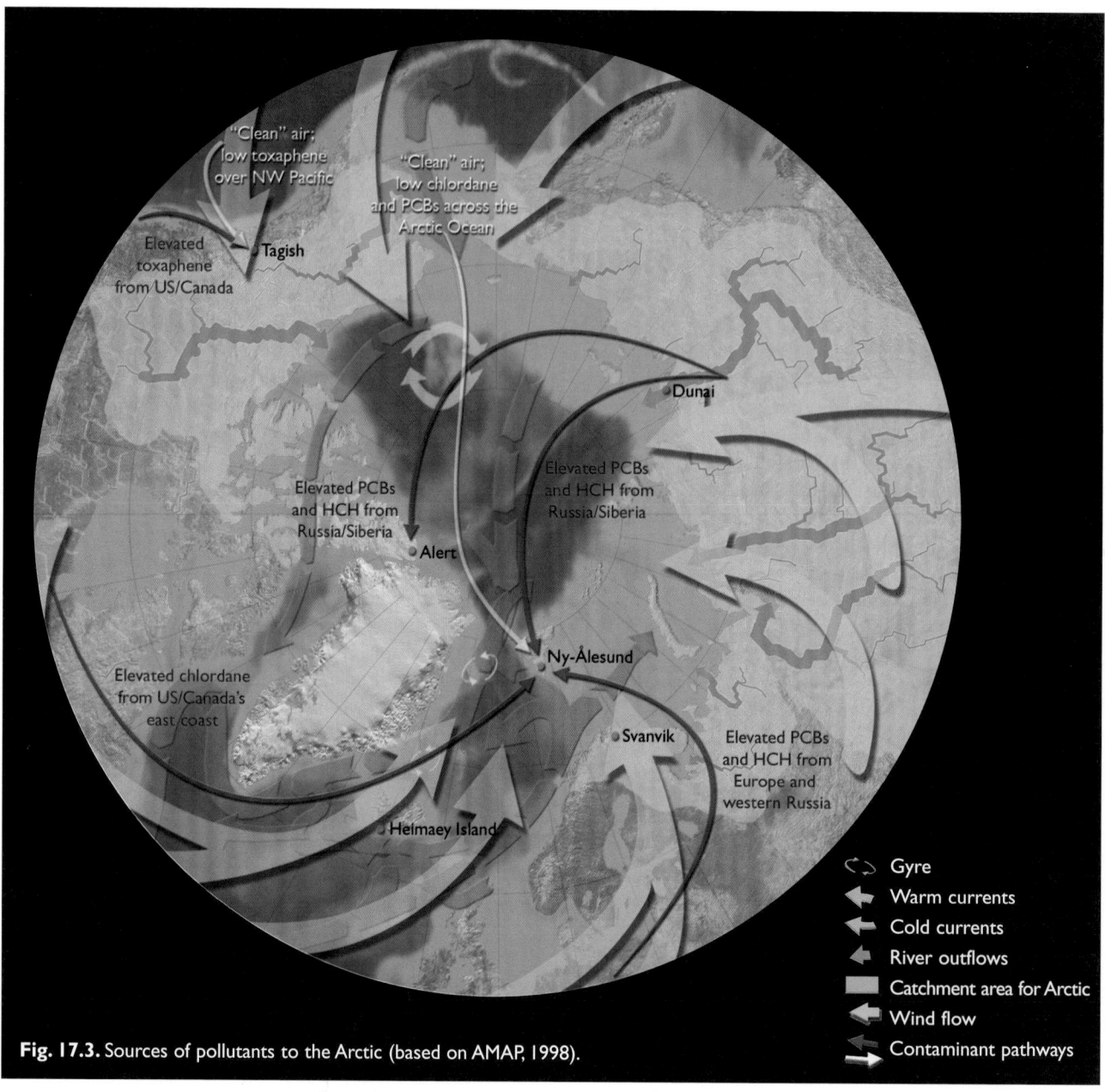

Fig. 17.3. Sources of pollutants to the Arctic (based on AMAP, 1998).

For heavy metals, special attention should be given to Cd, Pb, and Hg. The rationale for selecting these heavy metals is similar to the rationale for selecting the POPs; they are well known arctic pollutants that have been subject to much previous study (AMAP, 1998, 2002). They are also being addressed by national legislation and regional international agreements. Heavy metals are naturally-occurring environmental trace elements, and many are essential elements for living organisms. However, some have no known metabolic roles, and some are toxic even at low exposures. In the last 150 years there have been changes in the form in which these metals are released and dramatic increases in the quantity of these metals emitted to the environment. Anthropogenic emissions have altered the natural biogeochemical cycles of these elements (Nriagu, 1996). Anthropogenic sources of heavy metal pollution include industrial production, combustion processes, and waste incineration. These anthropogenic inputs add to the natural background levels

and can pose a toxic risk to environmental and human health (AMAP, 2002).

17.2.2.5. Pollutant interactions

Climate change, pollution, and human and societal conditions are interrelated and the consequences of these phenomena will depend largely on their interactions. It is becoming increasingly clear, for example, that climate change and pollution interact closely and that climate changes can affect the pollution transport chain (Alcamo et al., 2002). Air current changes affect pollutant transport patterns. Temperature changes affect which pollutants are deposited where, how they migrate, and which animals accumulate which pollutants. More extensive melting of multi-year sea ice and glacial ice can result in pulse releases of pollutants that were captured in the ice over multiple years or decades (AMAP, 2002; Macdonald et al., 2003).

Recently discovered mercury depletion events (MDEs) in the high Arctic (Schroeder et al., 1998) reflect additional ways in which pollution interacts with other factors. Levels of gaseous elemental mercury drop sharply each spring following polar sunrise, in a series of events that begin shortly after the first sunrise and continue until the snow melts. These MDEs are highly correlated with depletions in surface ozone, and appear to be caused by a reaction involving sunlight and bromine. The gaseous elemental mercury is transformed into reactive gaseous mercury, which is quickly deposited and can potentially enter food webs in a bioavailable form. Because MDEs occur at a time when biological productivity is increasing, the interactions between pollutant transport pathways, solar radiation, and climate can be extremely important (Lindberg et al., 2002; Lu et al., 2001). It is still unclear to what extent changes in climate and pollutant pathways may affect these events (AMAP, 2002).

Anthropogenic climate change and pollution are the products of societal activity and their consequences depend heavily on human and societal conditions. The effects of pollutants on human health are determined, in part, by regulations governing the use and disposal of hazardous chemicals, policies and public health guidance regarding human intake of potentially contaminated foods, public perceptions of and responses to such guidance, how much pollutant-contaminated food people ingest, cultural attitudes toward various types of food, and what access people have to these various foods.

17.2.3. Identifying coping and adaptation strategies

The Arctic has been inhabited by many diverse groups of people for several thousand years. Each group has its own distinct history, culture, language, and economic system. Despite the cultural and economic diversity found among arctic indigenous peoples, they have, through time, adapted to a number of similar conditions, such as a challenging and highly variable environment generally unsuited for most agriculture, severe climatic conditions, extended winter darkness, changes in wildlife populations, great expanses between settlements, and sparse populations (Chapter 3). The varied livelihoods of arctic indigenous peoples are examples of such adaptation. Reindeer herding in Finnmark and Russia, and fishing, sealing, and whaling in Greenland, Canada, Russia, and Alaska reflect the ability of arctic peoples to utilize and innovate with available resources, and to anticipate environmental and social changes in ways that enable people to take advantage of opportunities and guard against adverse effects.

Colonization has been another important source of change for arctic peoples. Prior to European contact, arctic indigenous peoples lived primarily in small settlements, and those dependent on terrestrial versus marine resources led nomadic lifestyles in order to follow the animals they relied upon for their livelihoods.

Historically, their cultures, identities, social organizations, and economies centered on these livelihoods, which represent successful adaptations to local environments. More recently, however, all arctic indigenous peoples, have, to greater or lesser extent, been colonized by outsiders interested in extracting and profiting from the Arctic's resources. In addition to centuries of European and Asian settlement, arctic indigenous peoples have also encountered missionaries and traders and, more recently social, economic, environmental, and political impacts and changes brought about by globalization (Freeman, 2000). In response, many indigenous peoples have developed mixed cash–subsistence economies. Yet, despite a number of challenges, these people continue to keep alive their traditional ways of life and in recent decades have acquired considerable authority in matters of governance. Arctic peoples have shown a remarkable resilience to extreme environmental conditions and profound societal change. At the same time, cultural change could reduce the adaptive capacity of arctic peoples (Chapter 3).

Adaptive responses to environmental changes are multi-dimensional. They include adjustments in hunting, herding, fishing, and gathering practices as well as alterations in emotional, cultural, and spiritual life. Arctic peoples change their hunting and herding grounds, become more selective about the quality of the fish they ingest, and build new partnerships between federal governments and indigenous peoples' governments and organizations. Adaptation can involve changing personal relationships between people and the weather and new forms of language and communication developed in response to novel environmental phenomena. Changes in knowledge and uses of knowledge can also constitute forms of adaptation. Altered weather prediction techniques are an example (Chapter 3).

In this chapter the term "adaptation" is used broadly, but in some instances it requires refinement. In their discussion on the term "adaptive" Berkes and Jolly (2001) apply terminology long used in anthropology (McCay, 1997) and the development literature (Davies, 1993), to distinguish between coping mechanisms and adaptive strategies. Coping responses are the ensemble of short-term responses to potential impacts that can be successfully applied season-to-season or year-to-year as needed to protect a resource, livelihood, etc. Some forms of coping are explicitly anticipatory and take the form of, for example, insurance schemes and emergency preparedness. Adaptive responses refer to the ways individuals, households, and communities change their productive activities and modify their rules and institutions to minimize risk to their resources and livelihoods. Depending on the frequency, duration, and suddenness in the onset of a stress, and on the resilience of a system, either coping or adaptive responses or both will come into play. With a progression of change in climatic conditions, coping mechanisms may at some point be overwhelmed, and by necessity supplanted by adaptive responses.

17.2.3.1. Governance, regulations, and subsistence

A number of changes in governance and regulation are transforming arctic governments and their relationships with the rest of the world. Since the early 1970s, authority has devolved from central governments to local and regional governing bodies in places like Greenland, Alaska's North Slope Borough, and northern Quebec's Nunavik region (Young, 2000). But while indigenous peoples in these communities have gained control over local affairs, external regulations have had considerable bearing on local ways of life. Seal harvest protests in Europe and the United States have affected seal hunting livelihoods in Greenland (e.g., Hovelsrud-Broda, 1997, 1999). Recently, proposals have been made to the International Whaling Commission (IWC) to deny the aboriginal subsistence hunters in Alaska and Chukotka in Russia a quota for bowhead whales (*Balaena mysticetus*). And Sámi reindeer herders must defend their practices against claims by others that they are allowing overgrazing (Beach, 2000).

For a long time, east–west tensions and core–periphery relationships (e.g., between Greenland and Copenhagen) kept arctic relations with the rest of the world connecting, for the most part, along north–south lines. Since the 1980s, however, arctic countries have become more open to pan-arctic cooperation with, for example, the thawing of the Cold War and the growing recognition of indigenous peoples' rights (Young, 1998a). Cooperative alliances include the Arctic Council, the Inuit Circumpolar Conference, and the North Atlantic Marine Mammal Commission. The Arctic Environmental Protection Strategy (AEPS), which provided a basis for the Arctic Council, was a pan-arctic initiative begun in 1991 when the eight arctic states signed the Declaration on the Protection of the Arctic Environment. A primary purpose of AEPS was a better understanding of environmental threats through a cooperative approach to these threats (Young, 1998a). There is also an increasing effort to link arctic initiatives with global regimes such as the Convention on Biological Diversity, the UN Framework Convention on Climate Change, ozone agreements, pollution-related agreements and initiatives, and the International Labour Organisation Convention 169 concerning Indigenous and Tribal Peoples in Independent Countries (Young, 2000).

Several factors are likely to characterize arctic international relations of the future. These include a greater role for non-state actors (especially indigenous peoples and environmental groups) in arctic affairs and a focus on sustainable development as a policy goal that means different things to different people. However, the future shape of environmental institutional arrangements (e.g., geographically broad with a narrow focus on an environmental program, or geographically limited, but encompassing a wide range of environmental issues) remains to be seen (Young, 1998a).

17.2.3.2. Settlements, population, and migration

Over the past decades indigenous populations in Greenland, Finnmark, and elsewhere have tended to migrate to towns and larger settlements. These movements have generally resulted in mixed economies where individuals are more likely to engage in wage labor and supplement their cash incomes with the sale of subsistence products. While these mixed economies can perpetuate traditional systems of land use and allow the use of cash to support household hunting and fishing (Caulfield, 1997), the diets of people who migrate from smaller settlements to larger towns tend to contain significantly less marine mammal and fish (Pars, 1997).

Indigenous peoples throughout the Arctic have often coped with and adapted to change via migration. Certain types of migration, however, can pose problems. People in general have responded to changes in animal populations and movements by altering their own locations and movement patterns and by varying the types of species hunted. Migration to towns might also serve as an adaptive strategy if, for example, economic trends, regulations and/or the effects of climate change and pollution make hunting or fishing in settlements impractical or unproductive. The movement of Greenlanders to permanent towns and settlements over recent decades restricted the ability of hunters to follow animals on their seasonal migrations, introduced more Greenlanders to wage labor, and helped to catalyze the indigenous political movement that culminated in Home Rule. Alternatively, certain economic conditions, regulatory policies, and/or the stresses of urban life could conceivably prompt people to move from towns to settlements. This type of coping through mobility is evident in Greenland over the past thirty years as the size, composition, and distribution of Greenland's population during this period has varied with changes in policies, economics, and educational and occupational opportunities.

Migration and settlement practices have also had implications for governance. Danish government policy encouraging a growth in town populations led many Greenlanders to concentrate in towns and major settlements in the 1960s. Migration from rural to more urban areas was part of Danish modernization programs of the 1950s and 1960s. These programs, for example, shut down a number of small settlements so that their inhabitants could work in fish-processing plants located in larger towns. Many Greenlanders were not in favor of these activities, believing that they were detrimental to Greenlandic culture and practices. Greenlandic resistance to forced migrations and other modernization initiatives eventually contributed to the establishment of Home Rule (Caulfield, 2000).

17.2.3.3. Consumption

Access to new foods and technologies have accompanied changes in diet and livelihood practices, respectively, and mark important ways in which consumption patterns are part of changing arctic lifestyles. The diets of indigenous peoples are changing as they use smaller amounts of traditional foods, and rely more on commercially available products and imports. These changes have implications for culture and health as traditional foods are closely tied to indigenous identity and offer significant nutritional benefits (Kuhnlein and Chan, 2000). A decrease in traditional foods combined with an increase in western foods in the diet of indigenous peoples increases the rate of western diseases such as heart disease. Examples of technological change include snowmobile use which has accompanied changes in transportation, hunting, trapping and fishing, and recreation and tourism among the Sámi in Norway (Muga, 1987). In addition, imported modern hunting equipment has made whaling and marine mammal hunting activities, in general, safer and more efficient in Alaska, Greenland, and Canada.

17.2.3.4. Economies, markets, and trade

Mixed economies based on wage labor and on subsistence activities are increasingly prevalent in arctic indigenous communities (Chapter 3) and broader trade and growing access to markets have innumerable implications for arctic indigenous peoples. Easier access to world markets continues to provide arctic inhabitants with increasingly better access to new material goods and new sources of income. At the same time, growing arctic-based businesses (e.g., tourism, see Chapter 12) can be sensitive to fluctuations in the distant economies to which they connect. An important question is whether, and if so how, this type of economic diversification affects resilience of local household and community economies.

17.2.3.5. Connectivity

A particular way in which technology is part of transformations in the Arctic is via the provision of new means of communication such as television, Internet, and telephones. The Anik satellites in Canada, for example, have been instrumental in exposing Inuit to outside cultures and in providing these peoples with a tool for asserting their own identity and culture (Stenbaek, 1987).

17.3. Methods and models for vulnerability analysis

A successful vulnerability assessment is one that prepares specific communities for the effects of likely future change. A vulnerability assessment should: draw upon a varied and flexible knowledge base; focus on a "place-based" study area; address multiple and interacting stresses; allow for differential adaptive capacity; and be both prospective and historical (Polsky et al., 2003). Data and methodologies to support such an assessment vary widely and any given vulnerability study is likely to involve a variety of quantitative and qualitative forms of data and methodological techniques. Interviews with "key informants" and surveys (Kelly and Adger, 2000) have been employed to obtain data on transience, immigration and education levels, income, education, age,

family structure (Clark et al., 1998), literacy, infant mortality, and life expectancy (Downing et al., 2001). Floodplain maps are important in analyzing the vulnerability of communities to extreme storm events (Clark et al., 1998). Agricultural vulnerability analyses often require information about extent of land degradation, crop type, soil moisture, runoff, and groundwater (Downing et al., 2001). As described by Cutter (1996) analytical techniques can include historical narratives, contextual analyses, case studies, statistical analyses and GIS approaches, mapping, factor analysis and data envelopment analysis, and vulnerability index development (see also Downing et al., 2001). Thus, what is novel about vulnerability assessments is not the individual techniques used to explore specific parts of a coupled human–environment system, but the integration of these techniques across varied intellectual domains.

A framework, such as that proposed by Turner et al. (2003a) enables at least two approaches for investigating vulnerability (see Fig. 17.1). One approach is to begin with knowledge about stresses and trace them through to consequences, while another is to begin with consequences and trace these back to stresses. It is also possible to work in both directions in an iterative fashion to yield a more comprehensive analysis. Figure 17.4 presents a research approach that allows for iterative analysis, in which (reading from left to right) information about stresses and their interactions are used both to develop scenarios and to project impacts. Impact projections can be used in conjunction with interviews, focus groups, workshops, and other means for engaging residents of the place of interest to explore coping strategies and adaptive capacities of a human–environment system. Knowledge of impacts and adaptive capacity can then be used to characterize site-specific vulnerabilities. Proceeding from consequences to stresses (right to left), researchers can work with residents of a particular locale to identify consequences experienced within a coupled human–environment system and then trace them back to identify the specific nature of the stresses.

Application of a framework to understand vulnerabilities within a coupled human–environment system requires different types of knowledge, as well as tools from a wide range of disciplines and from local and indigenous sources. For example, vulnerability analysis in the form presented here requires integration of natural science, social science, indigenous and local knowledge, cooperation among researchers and people who are part of the coupled human–environment system under study, and reliance on diverse techniques such as interviews, participant observation, focus groups, climate modeling, and climate downscaling. A proper vulnerability analysis will engage (1) a number of scientific disciplines (ecology, biology, climate and global change research, meteorology, social anthropology, sociology, political and policy science, economics, geography, ocean sciences, physiology and veterinary science, and environmental chemistry) and (2) local people with significant knowledge of their environment, of relevant

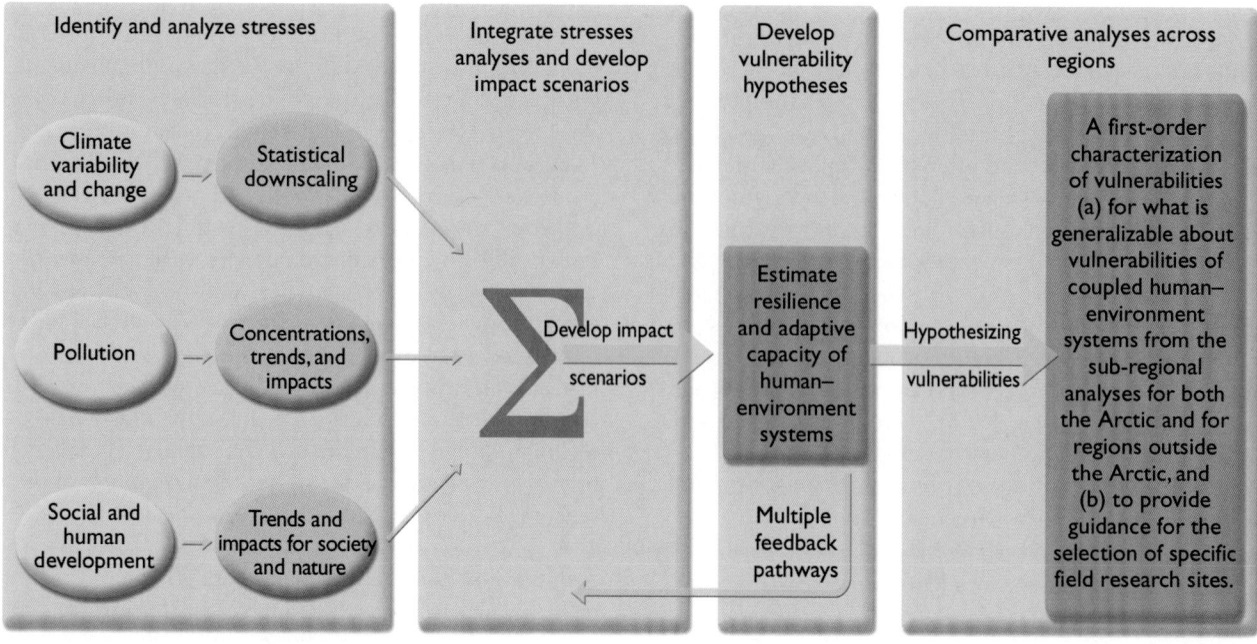

Fig. 17.4. Methodological framework.

social, political, and economic factors, and of human–environment interactions concerning, for example, hunting, herding, gathering, processing, and production. The success of a vulnerability assessment depends on the success of partnerships among various groups of stakeholders (Polsky et al., 2003).

17.3.1. Climate scenarios and downscaling to specific sites

An active area of climate change research is the translation of atmosphere–ocean general circulation model (AOGCM) projections calculated at large spatial scales to smaller spatial scales, a process termed "downscaling". In this way, selected study sites can be provided with customized climate projections. There are two principal approaches for downscaling: dynamical downscaling (also known as regional climate models) and statistical downscaling (also known as empirical downscaling). As described in Chapter 4, there are advantages and disadvantages to each approach. While both generate similar results for current climate, they have been known to generate different projections for future climates.

Greenhouse gas emissions scenarios used to drive the AOGCMs are based on projections of economic activity. In turn, the projected economic activity is a function of anticipated changes in global population, technology, and trends in international trade. As a result, each of the 40 scenarios used by the IPCC can be characterized by its anticipated trajectories of population, economy, environment, equity, technology, and globalization (IPCC, 2001a). It is impossible to assign likelihoods to these or any other GHG emissions scenario. Thus the IPCC emissions scenarios are individually equally plausible, but collectively represent only a subset of the possible futures.

Current arctic climate projections are limited in their utility for vulnerability analyses for two main reasons. First, the AOGCMs that produce these projections do not capture all important features of regional climate. For example, local ocean and atmospheric circulation patterns, and topographic relief are not well represented in AOGCMs. These factors often play a decisive role in determining local climate in the Arctic. As a result, additional analytical techniques are needed to produce local-scale climate projections. Chapter 4 reviews the various methods available for this task.

Second, for downscaling results to contribute to successful vulnerability assessments, local people must be involved in the planning and analysis of downscaling studies (Polsky et al., 2003). Otherwise, the downscaled climate projections may not reflect the climate factors relevant for decision-making to enable arctic residents to adapt or employ mitigation strategies. For example, one of the climate variables of concern for reindeer herders in northern Norway is snow quality. Too much snow hinders reindeer mobility and restricts their access to food on the ground, especially when the snow contains enough ice to mask the smell of the food. Too little snow, by contrast, makes it difficult for the herders to contain the reindeer (no restrictions on mobility) and to track the animals when they stray (no snow tracks). For these and other reasons, the Sámi employ many words to describe snow quality, as it relates to timing, amount, consistency, bearing, surface, trees, thawing, patches, accessibility, and other aspects (Ruong, 1967). The point here is that the way climate matters for any particular activity is specific to that activity. Thus a downscaled projection of, for example, mean monthly surface temperature may not be sufficient (or even necessary) information for contributing to the process of social adaptation to the effects of climate change for any group of arctic people.

Likelihoods of extreme conditions are difficult attributes to derive with confidence from climate scenarios, especially at the local scale, and this is the scale of most interest in assessing vulnerabilities in human–environment systems. As Smithers and Smit (1997) pointed out, "frequency, duration, and suddenness" of climatic events influence the character of successful adaptation strategies. Higher frequency of a potentially harmful event will heighten decision-makers' awareness of risk associated with a class of climate events. Greater duration of a climate event could inflict a correspondingly greater impact, or alternatively allow for more adaptation than would be possible with a shorter duration but otherwise similar event. Rapid onset of a particular climate event or condition, whether a specific flood or myriad aspects of the broader syndrome of climate change, will be much more limiting with respect to adaptation options than slowly rising water or very gradual climate change. Scales of social and political organization are important, since they reflect inertia in the response or adaptive capabilities of a human–environment system (Berkes and Jolly, 2001). Future projections of arctic climate change models (see Chapter 4) do not yet systematically resolve details in climate extremes that will be useful for assessing changes in frequency, duration, and suddenness of most climate events. Within the last few years, however, some progress has been made with ensemble simulations, especially for precipitation (Palmer and Räisänen, 2002).

17.3.2. Measurement and methodology for pollutant analyses

Data on POPs and heavy metal pollution and impacts for many arctic locations are now available. Although analytical methodology has advanced significantly for both since the late 1970s, some older and more recent data can be difficult to compare. For example, early PCB studies reported concentrations relative to industrial mixtures (e.g., Aroclor). Current studies report PCBs as a sum of individual congeners. In addition, levels of POPs and metals in biota may differ dramatically depending on the species sampled, the part of the individual animal from which the sample was taken, and the age and sex of the sampled individual. Shifts in diet can also affect exposure. Few studies have monitored the same species at the same site over a long timescale, a procedure necessary for making reliable comments on trends. Heavy metals pose an additional challenge for measurement and assessment because they are derived from both anthropogenic and natural sources. It is not always possible to determine whether a given concentration measured in biota originated from a natural or an anthropogenic source (Muir et al., 1999). In the last few years, the assessment and standardization efforts implemented by AMAP have dramatically improved knowledge of the pollutant situation in the Arctic. The data used here for analyzing interactions with multiple stresses and for identifying vulnerabilities draw heavily on the latest AMAP assessment (AMAP, 2002).

17.3.3. Analysis of human and societal trends

Data on trends in human and societal conditions can be obtained from published and unpublished institutional and governmental databases (e.g., Statistics Greenland) and from research in fields such as anthropology, sociology, political science, economics, and native studies. An in-depth analysis of human and societal trends requires a wide array of data sources, extensive statistical analysis, and knowledge shared and generated through interactions between researchers and arctic residents. Such an analysis might draw upon economic data pertaining to markets operating at various scales, employment, retail transactions, trade, imports, exports, and the processing and sale of natural and other resources. Public health information can also form an important part of human and societal trends analyses with information about contaminant levels in food and humans, diseases, and health care. Census data can provide information about general demographics, education, family structure, employment, and migration patterns. Election data can be useful in revealing trends in governance and the implementation and enforcement of regulations and policies emanating from transnational, national, and subnational decision-making bodies. Archives and other documentation that track negotiations and participation within such bodies are also useful. Surveys and interviews with local people are essential in ascertaining the views of individuals with, for example, respect to dietary practices, consumption patterns, values, and priorities.

17.3.4. Sources of local knowledge and stakeholders as participants

The vulnerability of a coupled human–environment system will be perceived differently across cultures, age groups, economic sectors, etc. A reindeer herder will most likely define the vulnerability of his/her human–environment system differently than would an outsider assessing the same system. There may also be diverging opinions within a community. There may well be a range of different perspectives on what constitutes a vulnerable condition and it is essential to recognize and address these perspectives in carrying out a vulnerability analysis. Evaluation of the exposure, sensitivity, and adaptive capacity of a coupled human–environment system will require the knowledge, observation, and participation of people who are part of the system. These people can, for example, identify important stresses, human–environment interactions, and outcomes that they seek to avoid. They can also identify changes in the human–environment system, describe coping and adaptive capacities, monitor environmental and social phenomena, and communicate research findings. The involvement of local people in research design, implementation, and the dissemination of research results should, therefore, be a central aspect of any comprehensive vulnerability study.

Participatory research has become increasingly common in arctic research on socio-ecological changes, as evident throughout this assessment and a number of other projects e.g., SnowChange (Mustonen, 2002), Scannet (Scannet, no date), the Mackenzie Basin Impact Study (Cohen, 1997), Voices from the Bay (McDonald et al., 1997), and Inuit Observations on Climate Change (IISD, 2001). Participatory research follows from a long tradition. From the time people in the "South" began to take an interest in the high latitudes, indigenous peoples of northern regions have often had a role to play in arctic research and exploration. Early anthropologists and archaeologists frequently used indigenous peoples as guides, laborers, informants, and/or interpreters (e.g., Boas, 1888; Rasmussen, 1908; Stefansson, 1941), as did arctic seamen and explorers (Peary, 1907). Since those days, indigenous peoples have continued to work with visiting researchers and explorers. Throughout, there have been a variety of reports on the positive and negative outcomes of these interactions and relationships (e.g., Harbsmeier, 2002). However, despite the many ways in which arctic indigenous peoples have contributed to arctic research, often there has been little mention of their role or acknowledgement of their efforts (Brewster, 1997).

The way outside researchers worked with indigenous peoples changed considerably in the mid-1980s. Coinciding with the settlement of land claims, the emergence of co-management regimes, and the ascendancy of indigenous peoples' power and influence in formal decision-making processes (Kuhn and Duerden, 1996), indigenous knowledge became a topic of interest for many researchers who worked in the Arctic. It was centered around a few key themes: documentation of indigenous knowledge about various aspects of the environment (Ferguson and Messier, 1997; Fox, 2004; Huntington and the communities of Buckland, 1999; Jolly et al., 2003; Kilabuck, 1998; McDonald et al., 1997; Mymrin and the communities of Novoe Chaplino, 1999; Reidlinger and Berkes, 2001); the increasing use of cooperative approaches to wildlife and environmental management (Berkes, 1998, 1999; Freeman and Carbyn, 1988; Huntington, 1992; Pinkerton, 1989; Usher, 2000); environmental impact assessment (Stevenson, 1996a); and collaborative research between scientists and indigenous peoples (Huntington, 2000; Krupnik and Jolly, 2002). This last theme has particular relevance for vulnerability studies. A brief literature review on the development of collaborative and participatory research in the Arctic follows.

Early efforts to involve arctic indigenous peoples in research and land management began with much discussion on the validity and utility of indigenous knowledge. Researchers who had worked with indigenous peoples for some time recognized that indigenous knowledge could reveal valuable information that could augment scientific understanding about many aspects of environment and ecology. Further, some researchers were beginning to recognize that indigenous knowledge holders needed to participate in the research process themselves (e.g., Wenzel, 1984). Seminal edited volumes on northern indigenous

knowledge, e.g., Johnson M. (1992) and Inglis (1993) use case studies from the Arctic to present a number of perspectives on the validity and utility of this knowledge. The case studies illustrate that indigenous knowledge can make a key contribution to resource management and sustainability. Arguing for the inclusion of indigenous peoples in research and decision making, several case studies present examples of how these initiatives were needed, or underway, in a variety of settings. For example Eythorsson (1993) explained why the knowledge of Sámi fishermen in northern Norway is integral to successful resource management there and Usher (1993) discussed some of the successes of the Beverly-Kaminuriak Caribou Management Board, one of the earliest examples of wildlife co-management in North America.

As interest in indigenous knowledge in the Arctic picked up through the 1990s, many people continued to focus on promoting the validity and utility of indigenous knowledge and the need to integrate it into research and management. However, critiques began to surface that questioned the methods behind the "integration", as well as the intentions. Bielawski (1996) examined interactions between scientists and indigenous land users involved in co-management systems and noted that, although co-management is supposed to combine scientific and indigenous expertise, the model and process for co-management is not integrative at all, but scientific and bureaucratic. Usher (2000) echoed this when stating that although indigenous knowledge (also called TEK, "traditional ecological knowledge") is required to be incorporated into Canadian resource management and environmental assessments there is little understanding of what TEK is and how to implement it in policy. This confusion was especially visible during 1996 when a senior policy advisor with the Northwest Territories (NWT) government wrote an article claiming that the inclusion of indigenous knowledge in environmental assessments not only hinders the scientific process, but is against the constitutional rights of Canadian citizens since indigenous knowledge is based on spiritual beliefs, not facts (Howard and Widdowson, 1996). The article created a heated debate (see Berkes and Henley, 1997; Stevenson, 1996b) and caused both researchers and managers to look more closely at the reasons and methods for incorporating indigenous knowledge into research and policy. As shown by Usher (2000), these reasons and methods remained unclear until the end of the 1990s. In 1999, others such as Nadasdy (1999) still believed that indigenous knowledge and the engagement of indigenous peoples in research were not taken seriously and were merely paid lip-service for political reasons. Nadasdy (2003) called for a more critical look at "successful" co-management efforts and the political, as well as methodological obstacles, to the integration of indigenous knowledge.

By 2000, a number of indigenous knowledge projects in the Canadian Arctic and Alaska had made advances in participatory methods for working with arctic communities (Krupnik and Jolly, 2002). For example the Tuktu (caribou) and Nogak (calves) Project (Thorpe et al.,

2001, 2002), which documented Inuit knowledge of Bathurst caribou and calving grounds in the Kitikmeot region of Nunavut from 1996 to 2001, established a local advisory board for the project and relied on trained local researchers to help with interviews and data analysis. Fox (1998, 2004), who has a long-term project with Nunavut communities regarding Inuit knowledge of climate and environmental change, has used an iterative approach to community work, incorporating community input and feedback in research methods. Jolly et al. (2002) also used an iterative approach over a one-year project in Sachs Harbour, NWT to collect Inuvialuit observations of climate change. In the Sachs Harbour project, scientific experts worked one-on-one with local experts to understand a variety of phenomena and community workshops were held to establish common goals for the research and to clarify information. A number of other projects and management systems have incorporated participatory approaches with much success in recent years (Huntington, 1998, 2000; Kofinas and the communities of Aklavik, 2002; Krupnik and Jolly, 2002). Common to many of these projects are some aspects of participatory research in the Arctic that have emerged as key including time, trust, communication, and meaningful goals and results. Many of these projects span multiple years, where researchers and community members form friendships and fruitful working relationships. In several projects, results were produced in forms that the community could use and found interesting. For example, the Tuktu and Nogak Project produced a community-directed book (Thorpe et al., 2001). Fox (2003) developed an interactive multi-media CD ROM and an Inuktitut book for participants, and the Sachs Harbour project created a documentary film (IISD, 2000).

Indigenous peoples themselves are also making an impact on participatory research in the Arctic. Many arctic communities and organizations are reaching out to scientists and decision-makers to set research priorities and form partnerships for investigations (Fenge, 2001).

17.4. Understanding and assessing vulnerabilities through case studies

Comparative vulnerability assessments at continental scales (IPCC, 2001b) can reveal regional differences in the vulnerability of human–environment systems. The differences reflect, for example, geographically uneven rates of climate change projected in regional climate scenarios and broad regional distinctions in the capacity of individuals and institutions to cope with and adapt to change.

The Third Scientific Assessment of the IPCC ascertained the following as likely projections for the Arctic using the IPCC scenarios for climate change (Anisimov and Fitzharris, 2001):

> *The Arctic is extremely vulnerable to climate change, and major ecological, sociological, and economic impacts are expected.*

> *Habitat loss [will occur] for some species, and apex consumers – with their low-reproductive outputs – are vulnerable to changes in the long polar marine food chains.*

> *Adaptation to climate change in natural polar ecosystems is likely to occur through migration and changing species assemblages but the details of these effects are unknown.*

> *Loss of sea-ice in the Arctic will provide increased opportunities for new sea routes, fishing and new settlements, but also for wider dispersal of pollutants.*

> *Although most indigenous peoples are highly resilient, the combined impacts of climate change and globalization create new and unexpected challenges. Because their livelihood and economy increasingly are tied to distant markets, they will be affected not only by climate change in the Arctic but also by other changes elsewhere. Local adjustments in harvest strategies and in allocation of labor and capital will be necessary. Perhaps the greatest threat of all is to maintenance of self-esteem, social cohesion, and cultural identity of communities.*

This chapter builds upon more general climate change vulnerability analyses by placing climate in the context of other factors that can enhance or diminish vulnerability of arctic systems to future climate change. In order to ensure that such an analysis realistically characterizes the perspectives of the people who will be making decisions to apply coping and adaptive strategies to increase resilience, and hence minimize vulnerability, a vulnerability analysis must focus on a particular place. The dynamic character of vulnerability requires that the human–environment system be represented on a scale that is meaningful to individual decision-makers.

Sections 17.4.1.1 and 17.4.1.2 provide examples for which a vulnerability assessment would be tractable, revealing, and ultimately useful to residents of the Arctic who will be making decisions in relation to future change. The first example (see section 17.4.1.1) is drawn from a case study on the Sachs Harbour community, NWT, Canada. This case study is one of several such cases presented in Chapters 3, 11, and 12 for which a vulnerability analysis would be illuminating. The study of Sachs Harbour, however, is particularly well suited because its design and development fully engaged the residents of this community, while detailing the ways in which local people and other stakeholders view, experience, and respond to climate change. The community has a mixed economy, with strong historical dependence on fish and wildlife, and lies in a region of the Arctic that experiences high rates of climate change. However, at this time it is not possible to go beyond the work that has already been conducted with respect to recent and likely future climate change. Without a new phase of research that fully involves the Inuvialuit people of Sachs Harbour, their resilience in accommodating interactive future impacts of changes in climate and other stressors cannot be assessed. A primary purpose of the description in section 17.4.1.1 is to signal the importance of the next phase in such an assessment in this community.

The second example (see section 17.4.1.2) focuses on the coastal communities of Greenland. These communities are attractive potential sites for vulnerability analysis because of their historical documentation of their strong dependence on marine living resource use, including fishing, sealing, and whaling for subsistence and income, and in particular their growing knowledge of the amounts of pollutants in the local marine food webs. In addition the governance of Greenland is evolving, and its institutions will continue to play important roles in shaping coping and adaptive capacities for Greenlanders. Research on this human–environment system is at a very early stage. In order to proceed, residents knowledgeable about the roles of past and likely future climate in local livelihood activities must be engaged.

Furthermore, assessing climate change in the context of multiple stresses and resilience in both Sachs Harbour and coastal Greenland will require scenarios, i.e., a range of plausible futures. These scenarios should reflect the combined effects of likely future changes in climate, in other environmental factors that could affect livelihoods (e.g., natural resources), and in human and societal conditions that influence resilience and coping strategies.

A third example (see section 17.4.2) focuses on reindeer populations and reindeer herding in Finnmark, Norway. Reindeer herding by Sámi takes place well inland in winter and on the coast and nearby islands in summer. This practice involves management of grazing grounds in both locations and migration of the herds between them in spring and autumn. The long history of these practices during periods of past climate change and the increasing role of governmental regulation raises interesting questions about the vulnerability of this system. Sámi reindeer herding represents a tightly coupled human–environment system in which indigenous peoples interact closely with an ecosystem upon which they depend for their way of life.

Two additional factors support the inclusion of the Finnmark case study. A vulnerability assessment has recently been conducted for Norway (O'Brien et al., 2004). It aptly demonstrates that the arctic region of Norway will be more vulnerable to climate change than more southern regions, and the importance of selecting the appropriate scale of analysis in vulnerability assessments. Secondly, work to date with the Sámi reindeer herding community in Finnmark offers an excellent example of the co-generation of knowledge involving academic scientists and indigenous peoples. A research effort is now underway to ascertain the resilience of this system to future change in climate and other factors, and the preliminary findings from this study are discussed below.

These three communities and livelihoods are not intended to be representative of the Arctic as a whole. They

have been selected because they present examples of tightly coupled human–environment systems in which indigenous peoples interact closely with local ecosystems and rely upon these ecosystems to support their ways of life. Yet, they also span different geographic settings, environmental conditions, governance systems, and socio-ecological dynamics. For example, neither Sámi reindeer herders nor reindeer herding itself are "typical" of anything beyond themselves. The system of which they form a part possesses unique ecological, sociological, and ethnological features. Therefore, the system represents a useful site in which to examine the plasticity and adaptability of a generalized methodological framework for vulnerability studies.

In each of these cases there lies potential to test concepts and methodologies described in this chapter and to present information about stresses, sensitivity, and resilience within the coupled human–environment system. Other examples could have been developed by building upon material presented in Chapters 3 and 12. For example, the Dene Nation, NWT, Canada, and the Yamal Nenets of Northwest Siberia. The Dene case study (Chapter 3) is of interest because of the success of the Dene in forming the Denendeh Environmental Working Group. The Dene culture emphasizes interconnectedness among all aspects of the environment, and this working group is observing, documenting, and communicating information related to climate change. However, there is not much information available to judge what factors in addition to climate change are contributing to the vulnerability and resilience of the Dene at this time, and the detailed content of their working group reports are not yet publicly available. The Yamal Nenets case study (Chapter 12) is of interest because it focuses, like the Finnmark study, on a reindeer herding livelihood with a history of adaptive management during times of change. The Yamal Nenet situation differs from that of the Finnmark herders in that the Yamal Nenets are experiencing stresses relating to oil and gas extraction, and might in the future experience stresses related to Northern Sea Route coastal development made possible by climate change. Krupnik (2000) and Krupnik and Jolly (2002) suggest that these other stresses may be straining the resilience of a livelihood that has sustained Nenet people for centuries. Although a comparative analysis of the Yamal and the Finnmark cases would certainly be instructive it is premature given the preliminary character of research on each of these at this time.

17.4.1. Candidate vulnerability case studies

The following are preliminary findings from the two sites for which assessments of climate change in the context of multiple stresses and resilience would be particularly timely. Knowledge about what makes a system either vulnerable to or resilient to change can be used to minimize risks and damage and to capitalize on opportunities. Regional scenarios for climate and other changes and field research conducted with the full participation of inhabitants of Sachs Harbour and coastal Greenland will be needed to develop an assessment of climate impacts and other changes beyond these initial findings.

17.4.1.1. Sachs Harbour

Berkes and Jolly (2001) have identified three reasons why the Arctic is a highly appropriate region in which to address questions relating to human adaptations to climate change. First, people living in the Arctic, particularly indigenous peoples with subsistence livelihoods, have historically experienced a high degree of climate variability, and their ability to adapt to varying climate, from seasonal to interannual, is part of their culture. Second, as is well documented in earlier chapters, the rate of climate change recently experienced in the Arctic, and likely to continue over the next several decades, may be exceeding the range of experience and hence capacity of arctic peoples to adapt. Third, there is a growing body of participatory research, with topics ranging from wildlife co-management to the use of traditional knowledge in environmental assessments (Berkes et al., 2001). The focus of the analysis by Berkes and Jolly (2001) is a Canadian Arctic community, Sachs Harbour, on Banks Island, NWT.

Sachs Harbour, a community of 150 Inuvialuit hunters and trappers was the subject of a two-year study (1999–2001) by the Canadian International Institute for Sustainable Development (Ashford and Castledown, 2001). This area is known to have been inhabited episodically, beginning with Pre-Dorset peoples over 3500 years ago. Traditional livelihoods (hunting, trapping, fishing) continue to thrive, and increasingly tourism, including guiding and the sale of arts and crafts, contributes to the local economy. The study became widely known through the distribution of an educational video, and several research papers (Ford, 2000; Fox et al., 2001; Riedlinger, 2000).

The report on this study describes a community at a crossroads. Climate has changed in recent decades and traditional ways of predicting weather are no longer reliable. Within the last few decades the later dates for autumn freeze-up, earlier dates for spring thaw, thinner winter ice, diminished extent of multi-year sea ice, thawing permafrost, and increased coastal erosion have altered abundances of and accessibility to fish and wildlife. The people of Sachs Harbour wonder whether they can maintain their traditional ways of life if these trends continue.

Berkes and Jolly (2001) analyzed the adaptive capacity of this community, considering a continuum of near-term coping responses to longer term cultural and ecological adaptations. Given the high degree of natural climate variability in the Arctic, coping strategies have always been essential for the success of indigenous peoples' livelihoods. These strategies include adjusting the timing of activities and switching between fished and hunted species to minimize risk and uncertainty in harvest. Waiting is also a coping strategy. "People wait for the geese to arrive, for the land to dry, for the weather to improve, or for the rain to end". But as annual climate cycles become more and more unfamiliar, new strategies are necessary. With changes in snow and ice cover, permafrost conditions, and coastal erosion, modes of transportation need to change. Greater unpredictability in weather also requires a greater caution for those who travel on ice. Coping with changes in harvest has in some regards become easier as alternatives to traditional diets have become more available with the growing reach of market economies.

Longer term adaptive responses that are considered central to the long-term success of indigenous peoples in the Arctic are categorized as follows: (1) mobility and flexibility in terms of group size; (2) flexibility with regard to seasonal cycles of harvest and resource use, backed up by oral traditions to provide group memory; (3) detailed local environmental knowledge and related skill sets; (4) sharing mechanisms and social networks to provide mutual support and minimize risks; and (5) intercommunity trade (Berkes and Jolly, 2001). The authors go on to suggest that the first response, mobility and group size, became much less relevant following the settlement of Inuit in permanent villages several decades ago. However, the remaining four responses have continuing potential to offer some adaptive capacity to deal with future climate change.

But will these time-proven strategies be sufficient for a future where factors in addition to climate change become increasingly important in this human–environment system? Are there other adaptive responses that need to be examined? Increasing dependence on cash economies and industries such as tourism raise new questions about the sustainability and overall vulnerability of this system.

The co-generation of knowledge in the IISD study, which fully involved the Inuvialuit people of Sachs Harbour, sets the stage for discussions about vulnerabilities in this location. However, in addition to participatory research methods to support the collaboration of researchers and local peoples, there are also institutional arrangements in this region that can facilitate the assessment of vulnerability in the context of multiple stresses, including climate change, pollution, and economic change. Over the last two decades co-management bodies have arisen that provide individual Inuit communities with formal mechanisms to interact with regional, territorial, and federal government institutions. These bodies provide greater local flexibility and response capacity in dealing with local uncertainties such as climate change (Berkes and Jolly, 2001). Moreover, they facilitate self-organization and learning across levels of organization, thus enhancing feedbacks across the levels. A community like Sachs Harbour can improve its understanding of risks and vulnerabilities and therefore better prepare itself for the future by examining possible effects of and responses to climate change in a historical perspective and within the context of other forms of social and environmental transformation.

17.4.1.2. Greenland

Communities in Greenland (and other similar communities elsewhere in the Arctic) that rely heavily on living natural resources, such as marine resources, might utilize vulnerability analysis in anticipating and planning for future social and environmental changes. Many such communities have a mixed subsistence/cash economy that involves a combination of commercial fishing, wage employment, and small-scale hunting and fishing activities. Commercial fishing (for shrimp, Greenland halibut (*Reinhardtius hippoglossoides*) and other species) is dominant in terms of monetary return. The residents of these communities and the environments with which they interact are affected by many factors including governance and market dynamics spanning local to global contexts, as well as climate and pollution. Recent decades have witnessed significant changes in these variables and the future may hold even more pronounced alterations. A vulnerability analysis could be useful for residents, other stakeholders, and decision-makers in identifying which social and environmental influences warrant their concern, the potentially advantageous or adverse consequences of these factors, and how human–environment systems could respond.

Climate change could have important consequences for Greenland's human–environment systems. Recent statistically downscaled temperature scenario results based on the Max Planck Institute climate model ECHAM4/OPYC3 for Greenland project a warming trend for the period 1990 to 2050 of 1.3 to 1.6 °C for West Greenland and around 0.4 °C for East Greenland (Førland et al., 2004). In West Greenland such a trend is likely to have significant implications for sea-ice cover, for ice-dependent marine mammal species such as the ringed seal, and for hunting and fishing activities that require secure ice cover.

Pollution is another factor that could bear on the vulnerability and resilience of Greenland's human–environment systems. Although no clear effects on human health are presently observed in Greenland, the occurrence of POPs and heavy metals in traditional foods is identified by some researchers as an important environmental threat to human health (AMAP, 2003; Bjerregaard, 1995). Health concerns have been expressed in particular for pregnant women and small children (AMAP, 2003). Greenlandic residents have in general a high consumption of marine mammals and fish and are exposed to POPs mainly through their marine diet. Data indicate that levels of some POPs in biota in Greenland have decreased over the past 20 years, although it is difficult to compare earlier and more recent data for methodological reasons (Frombert et al., 1999). Given international and national policies regarding these compounds, it is reasonable to expect that environmental levels of DDT, PCBs, and HCH in Greenland will decline toward 2020 (Macdonald et al., 2003). In contrast, PBDE levels are increasing in some parts of the Arctic (AMAP, 2002).

Data since the mid-1990s reveal that the estimated average human intake of both Hg and Cd from local marine food continues to greatly exceed the FAO/WHO limits (Johansen et al., 2000). The study that produced these data involved surveys in two towns and two settlements in the Disko Bay region where the main dietary source of Hg and Cd was seal liver. This study shows that Hg and Cd are still posing problems in arctic ecosystems and could affect humans whose diet results in high levels of exposure to these metals. Most Hg contamination arises from long-range transport and there are indications of increased Hg levels in seabirds and marine mammals in West Greenland (AMAP, 2002). Smoking is often the major source of human Cd contamination. Lead contamination, another heavy metal of concern in the Arctic, is linked to the use of lead shot for bird hunting (Johansen et al., 2001) and to continued use of leaded gasoline in parts of Russia and in some non-arctic countries.

Governance could shape the vulnerability and resilience of human–environment systems in marine resource use communities. The distribution of power among supranational, national, and sub-national decision-making bodies, for example, could help to create or ameliorate particular problems for a given human–environment system and could influence the ability of such a system to anticipate and react to stresses or potential stresses. Self-determination and self-government via Home Rule (established in 1979) allow Greenlanders a greater say in charting the country's economic and social development. However, some observers argue that the Greenlandic government, though supported by Greenland's inhabitants, has allowed for less autonomy at local levels than did its predecessor. According to this view, local people had more control over access to territory and other aspects of natural resource use and management under the more distant central government (Dahl, 2000). Currently, hunting methods and catches of a number of target species are influenced to a large extent by scientific data and by the management institutions that draft hunting and fishing regulations. Prior to the early 1990s, local communities granted territorial access for hunting and fishing to all members of a local community. This access (e.g., available for full-time hunters and fishers and fishing vessel owners) is now decided by the centralized Home Rule Government that manages hunting and fishing through regulations.

Natural resource management decisions are further influenced by international laws and policies and global markets. Greenland is heavily involved with transnational policymaking that has implications for domestic governance decisions regarding natural resource use and the environment. Consequently, Greenlandic hunters and fishers have to cope with and adapt to international politics and policymaking concerning species conservation and other matters. Greenland is represented in several multilateral fishery organizations and Greenland and the EU renegotiate a fishing agreement every five years (Nuttall, 2000; Statistics Greenland, 1997). The Home

Rule Government sets fishery quotas based on recommendations of biologists and international organizations (Statistics Greenland, 1997). In addition, global competition among commercial fisheries forces Royal Greenland (an independent limited company owned by the Home Rule Government that engages in the catching, processing, manufacture, and distribution of seafood products) to fish more efficiently (so affecting the nature of the fishery) and to cut costs (which can include shutting long-operating processing plants in some communities). This situation has important implications for households that engage in the fishing industry and rely to varying degrees on subsistence hunting and fishing.

Whaling of minke (*Balaenoptera acutorostrata*) and fin whales (*B. physalus*) in Greenland is subject to a variety of political pressures and regulations. Whales caught for subsistence purposes are to be used only for local consumption and may not be exported. The Greenland parliament regulates minke and fin whaling by, in part, requiring that whalers have a full-time hunting license, reside in Greenland, and have a "close affiliation" with Greenlandic society, a special whaling permit issued for each whale taken, and, at minimum, a 50 mm harpoon canon with a penthrite grenade (if a fishing vessel is used, the harpoon canon offers the best method for killing the animal). The Home Rule Government, in conjunction with the national hunters and fishers association (KNAPK) and the nationwide municipal government organization (KANUKOKA), allocates IWC quotas for minke and fin whales. After a municipality receives its annual quota consultations take place with the local hunter and fisher associations and quotas are assigned to vessels and collective hunters (Caulfield, 1994).

Marine mammal hunters have been subject to international protests and bans on marine mammal products since the early 1980s. The EU has maintained its 1983 trade ban on sealskins for certain species of seal pups, the 1972 Marine Mammal Protection Act remains in place, and the International Convention for the Regulation of Whaling (and its Commission – the IWC), sets quotas for aboriginal subsistence whaling. In addition, environmental and animal welfare organizations (e.g., Greenpeace, the International Fund for Animal Welfare, and the World Wildlife Fund) continue to criticize and protest against marine mammal utilization. Indigenous arctic peoples argue against restrictions on marine mammal hunting on the basis that the targeted animals are not endangered and that protests are not based on science, but on ethics particular to industrialized country politics (Caulfield, 2000).

Commercial and non-commercial fishing and hunting practices are inextricably linked and are integral to the social, economic, and cultural lives of Greenlanders. Marine resource use in general, entails cultural and social organization on many levels, through shared language, transmission of appropriate behavior, validation of identity, and reinforcement of social ties and kinship networks. Sealing and whaling, in particular, reflect the traditional social order of the communities and reinforce ties within and among families and households. The consumption, distribution, and exchange of marine resource products integrate the households and the community through a complex exchange network that reinforces cultural identity and social networks and provides important foodstuffs to households that are not able to hunt themselves.

Climate change, pollution, and governance are likely to be major factors in a vulnerability study of marine resource use communities in Greenland. Climate change and pollution could alter the availability, conditions, and health of animals such as seals and halibut, while changes in climate could affect the distribution and migration patterns of these animals, as well as ice and snow cover. Diminishing ice and snow cover could also have serious impacts on the mobility, hunting, and fishing activities of the residents of these communities. Climate alterations could further affect the ability of hunters and fishers to interpret and predict weather in planning safe and successful harvesting activities. Changes in politics, policies, and markets at local, national, and transnational levels could have negative or positive effects on the communities. Trade bans on marine mammal products, the increasing role that Home Rule and municipal governments play at the local level in towns and settlements, the growing importance of transnational policymaking forums, the financial support that individuals receive through transfer payments and subsidies, and consumption patterns of people near and far all have a bearing on the state of human–environment systems and their economies, social lives, and cultures. A vulnerability study would be useful in exploring how factors such as climate, pollution, and governance interact, their implications for human–environment systems, the resources Greenlanders might draw upon in reacting to social and environmental change, and the strategies that could be effective in guarding against negative consequences while capitalizing on opportunities.

17.4.2. A more advanced vulnerability case study

17.4.2.1. Reindeer nomadism in Finnmark, Norway

World reindeer herding

Reindeer herding is today the most extensive form of animal husbandry in the Eurasian Arctic and subarctic (also see Chapter 12). Some 2 million semi-domesticated reindeer (*Rangifer tarandus*) graze natural, contiguous mountain and tundra pastures covering an area of around 5 million km², which stretches from the North Sea to the Pacific Ocean (Fig. 17.5, Box 17.1). These reindeer provide the basis of the livelihood of herders belonging to some 28 different indigenous and other local peoples, from the Sámi of northern Fennoscandia (northern Norway, Sweden, and Finland) and the Kola Peninsula in northwest Russia, who herd approximately 500 000

Box 17.1. Biological adaptations by reindeer to life in the north (parts of this text have been published previously by Tyler and Blix, 1990)

Reindeer is one of only 13 out of 180 different species of ruminants that has been domesticated. The grazing areas of reindeer, however, cover almost 25% of land surface of the world (Turi, 1994). Reindeer inhabit a wide range of different biotopes. Like other species resident in the Arctic, reindeer are exposed to large seasonal variations in ambient light and temperature conditions and in the quality and availability of food.

The Arctic is a hostile place in winter, yet the cold, dark "polar wastes" sustain life. The environment is truly marginal and for this reason it might be thought that warm-blooded animals that spend the winter there must endure a truly marginal existence. However, most arctic animals usually neither freeze nor starve and it is therefore self-evident that they are well adapted to the several challenges of the natural environment in which they live.

Several species of monogastric mammals (i.e., those having a stomach with only one compartment) circumvent the problem of cold and the scarcity of food in winter by hibernating. Reindeer, however, are ruminants. Unlike monogastric species they have to remain active to feed continuously throughout winter. Moreover, they are truly homeothermic, requiring maintenance of a constant internal body temperature that is considerably above environmental temperature. For these, like other true homeotherms, the problem of survival becomes one of keeping warm. To do this they need both to reduce heat dissipation and to ensure an adequate supply of fuel, in the form of metabolites from food, for heat production. Therefore, adaptations for survival can be divided between those which help the animals to reduce their energy expenditure and those which help them to make best use of what little food they can find.

Reduction in energy losses

Reindeer and caribou have two principal defenses against cold. First, they are very well insulated by fur (e.g., Nilssen et al., 1984); second, they restrict loss of heat and water from the respiratory tract. In humans exposed to low ambient temperature but warmly dressed, the heat lost in exhaled air may account for more that 20% of metabolic heat production. In resting reindeer exposed to cold, by contrast, expired air is cooled and the animals are capable of conserving about 70% of the heat and 80% of the water added to the inspired air in the lungs (Folkow and Mercer, 1986).

Reduction in energy expenditure

Appetite and growth

Reindeer, like several other species of deer, show a pronounced seasonal cycle in appetite and growth which appears to follow an intrinsic rhythm entrained by photoperiod and associated with changes in levels of circulating hormones. In winter their appetite falls by as much as 70% of autumn values (Larsen et al., 1985; Mesteig et al., 2000; Tyler et al., 1999a). Growth slows or even stops (McEwan, 1968; Ryg and Jacobsen, 1982) and the animals begin to mobilize their fat reserves even when good quality food is freely available (e.g., Larsen et al., 1985). Intrinsic cycles of growth and fattening appear to be adaptations for survival in seasonal environments in which animals are confronted with long, predictable periods of potential under-nutrition. Slowed rate of growth and, to an even greater extent, actual loss of weight have the effect of reducing an animal's daily energy requirements (e.g., Tyler, 1987). This may be literally vitally important in winter when food is not only scarce and of poor quality but is also energetically expensive to acquire.

Activity

Besides minimizing heat loss in winter by means of increased insulation, reindeer and caribou can reduce energy expenditure by adopting appropriate behavior; in particular, by reducing the total daily locomotor activity. The nature of the surface over which animals travel is also very important. The relative net cost of locomotion in a caribou sinking to 60% of brisket height at each step is almost six times greater than the cost of walking on a hard surface (Fancy and White, 1985). The capacity of snow to support an animal depends on the hardness of the snow and the pressure (foot load) that the animal exerts on it. Thus, if snow hardness consistently exceeds foot loads, animals can walk on top of the snow or will sink to only a fraction of its total depth. The broad, spreading feet of reindeer and caribou, a well-known characteristic of this species, is clearly an adaptation to walking on snow, through minimizing the extent to which they break through the crust and sink in. Reindeer and caribou, with the exception of musk deer (*Moschus moschiferus*), have the lowest foot load measured in any ungulate

(Fancy and White, 1985). The potential significance of reducing locomotion as a means of saving energy is made clear from Fancy and White's (1985) calculation that the costs of locomotion for a 90 kg caribou breaking the trail at the head of the spring migration will represent an increment to its minimal metabolism of 82%. For the animals following the packed trail in its wake the incremental cost would be equivalent to about 33% of their minimal metabolism, a saving of more than half (Fancy and White, 1985).

Gathering and storing energy

Diet and digestion

Reindeer have an exceptional ability to cope with seasonal changes in the availability and quality of the different species of forage plants that they eat. This, together with the diversity of habitats in which the animals live, provides the basis for the capacity of reindeer to adapt toward climatic variability and change. Reindeer are highly adaptable intermediate mixed feeding types. They fall between true grazers that eat fibrous plants (25% of all species of ruminants) and concentrate selectors (40% of all species) that eat plants with low fiber content (Hofmann, 2000). By feeding selectively, they avoid highly fibrous plants and take, instead, the nutritious and easily digestible parts of a variety of different forage types including lichens, grasses, and some woody plants (Mathiesen et al., 1999, 2000; Storeheier et al., 2002a).

In some areas, the proportion of lichens in their diet increases in winter (Boertje, 1984; Mathiesen et al., 1999, 2000). Lichens are unusual as food for ruminants. They are rich in carbohydrate that is easily digestible in reindeer and are therefore also a good source of energy for the animals. However, they are deficient in nitrogen and minerals (Aagnes and Mathiesen, 1994; McEwan and Whitehead, 1970; Nieminen and Heiskari, 1989; Scotter, 1965; Storeheier et al., 2002a). Reindeer cannot, therefore, survive on lichens as their sole food supply. Ruminal fermentation of lichens has an important effect on ruminal absorption of energy rich volatile fatty acids in winter (Storeheier et al., 2003) and reindeer that eat lichens are better able to extract nitrogen from dietary vascular plants in winter (Storeheier et al., 2002b).

The consequences of increased temperatures over arctic ranges include an increase in the abundance of shrubs (Silapaswan et al., 2001; Strum et al., 2001) and a decrease in the abundance of lichens (Cornelissen et al., 2001). Reindeer herders report that the abundance and distribution of mountain birch (*Betula pubescens*) have increased and the abundance and distribution of mat-forming lichens have decreased in Finnmark over the last three to four decades. There are undoubtedly multiple causes underlying these changes. Thus, it is important to understand how reindeer can regulate their forage consumption to meet energy requirements under changing conditions. Though reindeer are able to survive without lichens in winter (Leader-Williams, 1988; Mathiesen et al., 1999; Sørmo et al., 1999) little is known about the level of production and the economy – and therefore, also, the vulnerability – of herding in lichen-free areas.

Fat

Many animals that live in highly seasonal environments store large amounts of energy as fat during summer and autumn in anticipation of food shortage during winter. In hibernating species, fat deposits may constitute up to 35% of the animals' total body weight. Ungulates, by contrast, usually store relatively little fat. The fat deposits of temperate and subarctic deer, for example, represent usually only between 4 and 10% of their total body weight in autumn (Tyler, 1987). Such low values cast doubt over the widely held view that fat is likely to be a major source of energy for deer and other ungulates in winter. Even using the most conservative models of energy expenditure it seems that the fat reserves of female Svalbard reindeer, the fattest of all reindeer, could contribute only between 10 and 25% of the animals' energy demands during winter (Mathiesen et al., 1984; Tyler, 1987). In practice, the contribution from fat is likely to be lower than these models predict because reindeer which survive winter do not normally use up all their fat (Tyler, 1987). Moreover, there is increasing evidence that the principal role of fat reserves in ungulates is to enhance reproductive success, rather than to provide a substitute for poor quality winter forage (although the very presence of fat will necessarily also provide insurance against death during periods of acute starvation). Substantial pre-rut fat reserves, for example, enable male deer to gather, defend, and serve their harems without being distracted by the need to feed and, in several species, males hardly eat at all for two or three weeks during the rut. It is more difficult to distinguish between alternative roles (reproduction and food supplement) for fat reserves in female ungulates because, in many species, these are pregnant throughout winter. Kay (1985) suggested that the principal role of fat reserves in females may be to supplement (but not to substitute for) their food intake during late pregnancy.

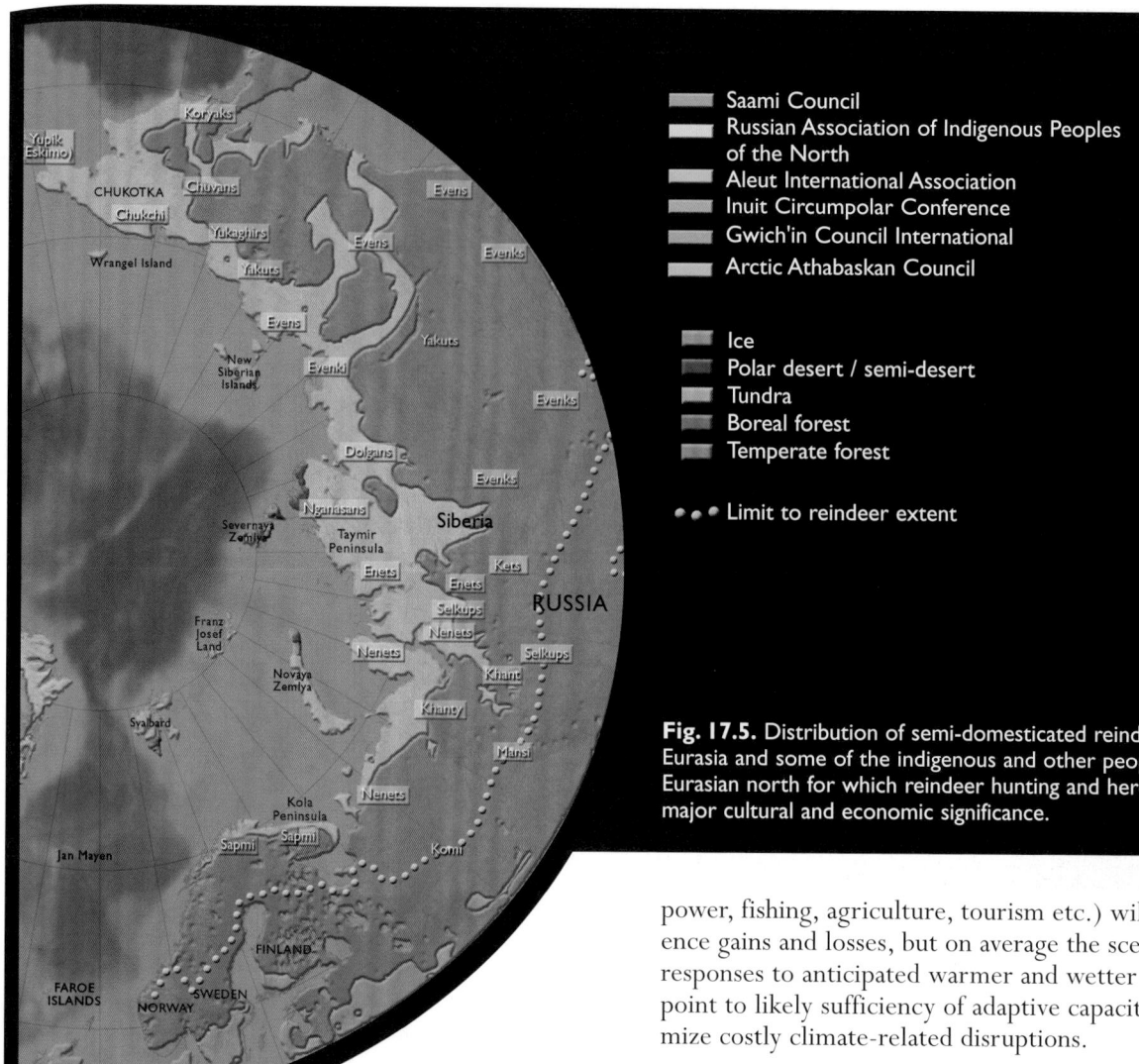

Fig. 17.5. Distribution of semi-domesticated reindeer in Eurasia and some of the indigenous and other peoples of the Eurasian north for which reindeer hunting and herding has major cultural and economic significance.

reindeer, to the Chukchi of the Chukotka Peninsula in the far east (Slezkine, 1994). The herding and hunting of reindeer has major cultural and economic significance for these people. Moreover, their herding practices, ancient in origin, represent models in the sustainable exploitation and management of northern terrestrial ecosystems that have developed and adapted *in situ* over hundreds of years to the climatic and administrative vagaries of these remote regions (Turi, 2002).

A Norwegian context

O'Brien, et al. (2004) asked whether Norway is vulnerable or resilient to future anthropogenic climate change (using projections from the ACACIA project, Parry, 2000). At a national level Norway can be considered relatively resilient and hence unlikely to be seriously affected by conditions forecast by climate scenarios over the next few decades. Its relative protection from hazards associated with sea-level rise, its weather-hardened architecture and infrastructure, its strong and equitable economy, its state of technological development, etc., all signal a good measure of resilience at the national scale. Certain economic sectors (oil and gas, hydro-

power, fishing, agriculture, tourism etc.) will experience gains and losses, but on average the scenarios for responses to anticipated warmer and wetter conditions point to likely sufficiency of adaptive capacity to minimize costly climate-related disruptions.

Through the application of multi-scale analyses, using dynamic and empirical downscaling techniques for regional and local climate scenarios, respectively, O'Brien, et al. (2004) were able to refine their assessments of vulnerability accordingly. Although climate extremes are not well captured in this analysis, it is clear that projections for differences in mean climate conditions vary greatly across Norway: northern, southwestern, and southeastern Norway fare quite differently. Only the first of these regions falls within the Arctic as defined in this chapter. Not surprisingly, it is this arctic portion of Norway that shows the greatest potential vulnerability to projected climate change; in large part due to the anticipated changes in natural ecosystems. The high dependence of human livelihoods on these resources, for economic and cultural reasons, contributes to a strong linkage between ecosystem changes and socio-economic consequences.

The primacy of fishing in many Norwegian coastal economies provides one example of such human–environment relationships. There is no historical analogue to allow confident predictions of fish stocks under a warmer coastal regime, and circulation changes in the North Atlantic may in fact be even more influential in determining the recruitment in key stock such as cod and herring. It is, however, likely that there will be changes in

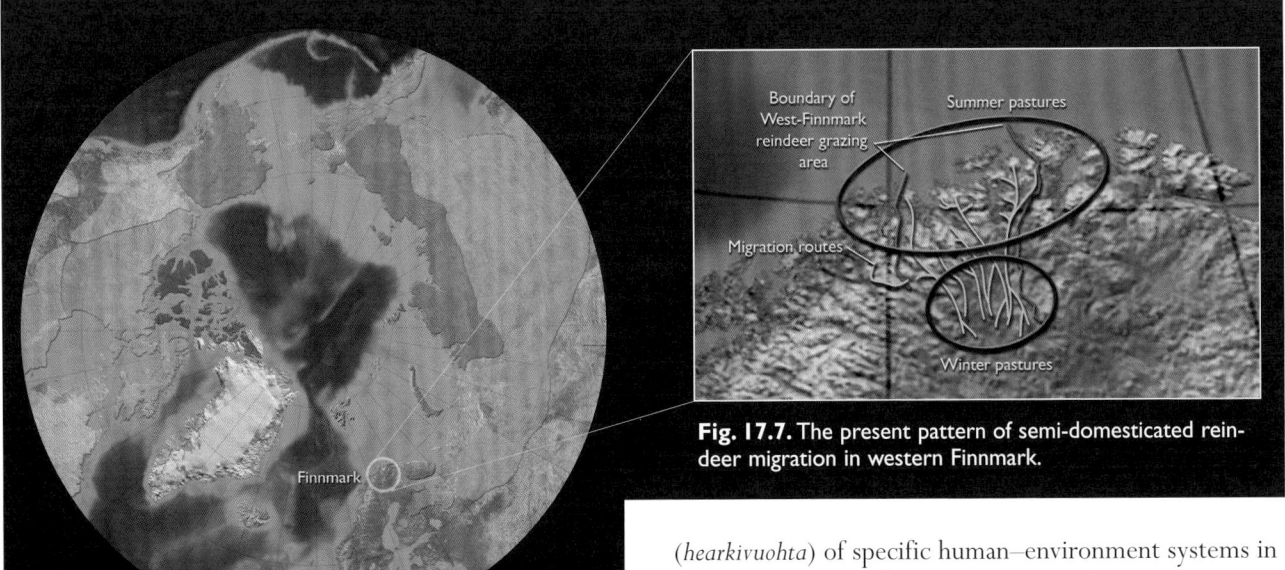

Fig. 17.7. The present pattern of semi-domesticated reindeer migration in western Finnmark.

Peary and Arctic-island caribou
Grant's caribou
Barren-ground caribou
Woodland caribou
Wild forest reindeer
Tundra reindeer
Svalbard reindeer

Fig. 17.6. World distribution of reindeer, showing Finnmark – the northernmost, largest, and least populated county in Norway (CAFF, 2001).

these marine ecosystems under projected climate regimes. Studies elsewhere reveal the difficulty that communities highly dependent on fishing have in adapting to alternative livelihoods when faced with permanently unfavorable changes in catch (Mariussen and Heen, 1998). Coastal areas in more temperate regions of North America and Europe contain many such examples.

Reindeer herding in northern Norway provides a similar example. Changes in temperature can affect vegetation and changes in the timing and form of precipitation can affect the animals' access to food. Either of these changes can influence the health and productivity of the herd, and hence the livelihoods and cultural practices of indigenous peoples who are highly dependent on this ecosystem.

O'Brien et al. (2004) also gave good examples of how the overall perspective on Norway's vulnerability could change with diminished importance of revenues from oil and gas over the next five decades (considered likely), and how climate impacts experienced in other nations can affect Norway via commerce, political relations, and movements of people. But an important underlying message is that for the foreseeable future the people most likely to be negatively affected by climate change are those whose lives are most intimately linked with terrestrial and marine ecosystems.

Finnmark Sámi reindeer herding

This analysis represents an interdisciplinary and inter-cultural approach to understanding the vulnerabilities

(*hearkivuohta*) of specific human–environment systems in the Arctic. As a work in progress it explores only some features of the human–environment system represented in reindeer nomadism. These features include climate and non-climate factors that impinge on, and may influence, the sensitivity and adaptive capacity of the system to environmental change. The perspective adopted here is that of members of local communities: the focus is on their interpretation of the concept of vulnerability analysis and on how it might usefully be applied to their situation. Thus, the information provided here is the result of a partnership between researchers and reindeer herders. The case study demonstrates how through active participation the reindeer herders modified and applied a general conceptual framework and interpreted research findings in a co-production of knowledge.

Finnmark is the northernmost, largest, and least populated county in Norway (Fig. 17.6). Within its 49 000 km² there live approximately 76 000 people, including a large proportion of Sámi. Populations of 114 000 reindeer and 2059 registered reindeer owners in Finnmark in 2000 represented 74 and 71% of semi-domesticated reindeer and Sámi reindeer owners in Norway, respectively (Reindriftsforvaltningen, 2002).

Reindeer in Finnmark are managed collectively in a nomadic manner rich in tradition. Herds of mixed age and sex, varying in size from 100 to 10 000 animals, are free-living and range in natural mountain pasture all year round. The herders typically make two migrations with their animals each year, moving between geographically separate summer and winter pastures. In spring (April and May), they and their animals generally move out to the mountainous coastal region where the reindeer are left on peninsulas or are swum or ferried across to islands where they feed throughout the summer, eating nutritious parts of bushes and shrubs, sedges, and grasses. In September the animals are gathered and taken inland to winter pastures in landscape typically consisting of open, upland plains of tundra and taiga birch scrub (Fig. 17.7, Paine, 1996; Tyler and Jonasson, 1993). The pattern of migration observed today is probably as much a legacy from earlier times, when Sámi moved to the coast to fish in the summer and retired inland to hunt game in winter, as a reflection of the natural behavior of their reindeer. The autumn migration inland is clearly an adaptation to climatic conditions. Winters are mild and wet near the coast but colder and drier inland (Fig. 17.8). Consequently, the climate is more continental inland and cycles of thawing and re-freezing (which increase both the density and the hardness of the snow making it increasingly difficult for the animals to dig to the plants beneath) occur less frequently than at the coast. Grazing (snow) conditions are generally better inland as a result.

Reindeer herding in northern Norway has many advantages over herding throughout much of the rest of the Eurasian Arctic and subarctic. First, although the absolute number of animals is small (the population in Finnmark, for example, represents approximately 4% of semi-domesticated reindeer in Eurasia), the density of reindeer is very high. This reflects, in part, the relatively high productivity of this region, which, in turn, is a consequence of the warming effect of a branch of the North Atlantic Current. The overall density of approxi-

mately 2 reindeer per km^2 in Finnmark is roughly four times greater than the density of reindeer in Russia. Second, reindeer meat is regarded as a delicacy in Norway and in many years production fails to meet demand. This, in combination with the richness of the Sámis' traditional gastronomic culture, provides opportunities for development of the economic basis of their industry through small-scale family-based productions focusing on the concept of adding value. Third, northern Fennoscandia possesses well developed infrastructure and transport and an electronic communication network superior to that in any other region of the circumpolar Arctic at similar latitude. These three factors form the basis of a potentially robust and vibrant form of cerviculture. They also represent features of both the natural and the social environments that potentially influence the vulnerability of reindeer herding to the effects of climate variability and change.

17.4.2.2. Modifying the general vulnerability framework

The first step in a vulnerability study is to evaluate the general methodological framework (Fig. 17.1) and modify it, where necessary, to suit the characteristics of the system of interest, in this case reindeer herding in Finnmark. A conceptual framework must be developed that focuses on the specific and, perhaps, even unique attributes of each particular case. Reindeer, reindeer herders, and the natural and social environments to which they belong represent a coupled human–environment system. Many of the components of this system, though only distantly related, are closely and functionally linked. Herders' livelihoods, for example, depend on the level of production of their herds. Production, in turn, depends on the size of herds and on the productivity of individual reindeer in them, which depend, again in turn, on the quantity and quality of forage available. The level of feeding the animals enjoy is determined in the short term by prevailing weather conditions including temperature in summer, which affects the growth and nutritional quality of forage plants, and by weather conditions in winter, in particular a combination of precipitation, temperature, and wind, which affect the quality of the snow pack and, hence, the availability of the forage beneath. In the medium and long term, however, feeding levels are also determined by a suite of non-climate factors all of which have a major influence on the level of production and, completing the circle, on the profitability of reindeer herding. These include the quality of pasture (in terms of the species composition and biomass of forage and the availability of other important natural resources), the area of pasture available, herders' rights of access to pasture, the level of competition between reindeer and other grazers, the level of predation to which herds are subjected, the monetary value of reindeer products and so on. Common to all these non-climate factors is that they are influenced by the decisions and policies of institutions far removed from Finnmark. Hence, it was clear at the outset that reindeer herding is a production system

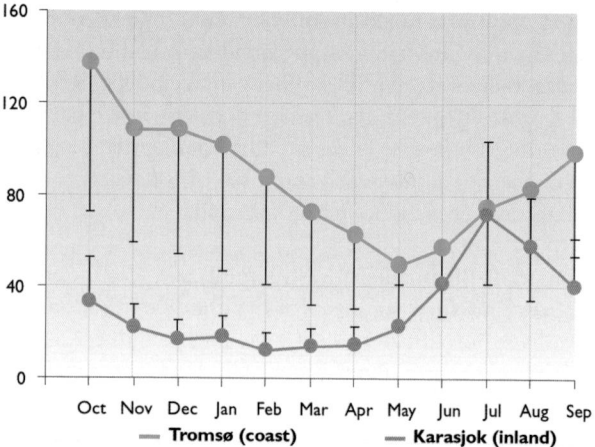

Mean monthly precipitation (mm)

— **Tromsø (coast)**   — **Karasjok (inland)**

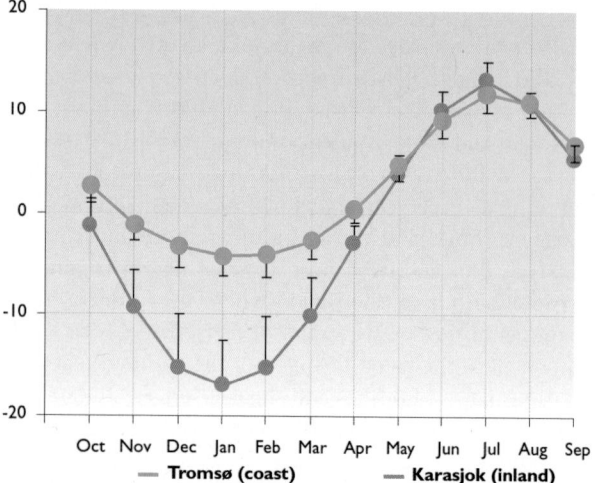

Mean monthly temperature (°C)

— **Tromsø (coast)**   — **Karasjok (inland)**

Fig. 17.8. Monthly mean precipitation and temperature at Tromsø and at Karasjok in Finnmark (26° E, 69° N). Data are for 1961 to 1990 and the bars indicate 1 standard deviation (data supplied by the Norwegian Meteorological Institute).

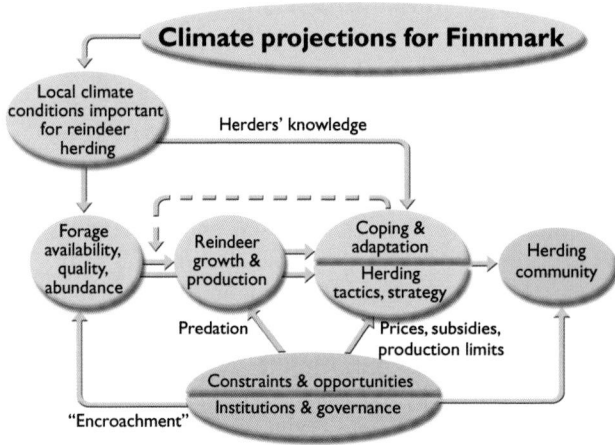

Fig. 17.9. Conceptual framework for the Finnmark case study.

affected not just by climate variability, and potentially also by climate change, but also potentially very strongly by the socio-economic environment in which it exists.

A conceptual model relevant for reindeer herding in Finnmark was developed at a five-day meeting held in Tromsø in August 2002. The president of the Association of World Reindeer Herders drew together a team of natural scientists, social scientists, administrators, and reindeer herders. All the participants were encouraged to emphasize their own particular perspectives and, working together in this way, the group then revised the generalized conceptual framework to suit the conditions prevailing in Finnmark. The herders, for example, were largely responsible for selecting the principal components included in the final model and upon which the study was based. The customized framework (Fig. 17.9) describes the perceived relationships through which (1) climate change influences the growth and productivity of herds of reindeer, (2) herders cope with climate-induced changes in the supply of forage and in the level of production of their herds, and (3) herders' ability to cope with climate-induced changes is constrained by extrinsic anthropogenic factors collectively called "institutions and governance". (These include "predation", the level of which is influenced by legislation that protects populations of predators.) Each part was tempered with herders' understanding of the dynamics of herding, of their society, and of the natural and social environments in which they live. Superficially the final model (Fig. 17.9) bore little resemblance to the general framework (Fig. 17.1) from which it evolved, yet key elements, including human and environmental driving forces, human and societal conditions, impacts, responses, and adaptation, all remain.

17.4.2.3. Climate change and climate variability in Finnmark: projections and potential effects

Climate change is one of a suite of factors that influence the physical environment, the biota and, ultimately, the cultures of indigenous and other arctic communities. Large-scale climate changes in the Arctic will influence

local climate (e.g., Bamzai, 2003), which, in turn, can possibly affect foraging conditions for reindeer, the productivity of herds and, ultimately, herders' income and livelihood.

Projections for Fennoscandia

The climate of northern Fennoscandia is milder than at similar latitudes in Russia or North America owing to the warming effect of a northeastern branch of the North Atlantic Current, which flows north along the coast of Norway. The mean July temperature at the coastal town of Vadsø (70° 05' N) in northern Norway, for example, is approximately 11 °C, while that at Point Barrow (71° 30' N) in Alaska is approximately 4 °C. Likewise, the mean January temperature inland at Kautokeino (68° 58' N) is approximately -16 °C compared to approximately -35 °C at Old Crow (67° 34' N) in Canada (both located at similar elevations).

These differences notwithstanding, recent modeling indicates that during the next 20 to 30 years the mean annual temperature over northern Fennoscandia is likely to increase by as much as 0.3 to 0.5 °C per decade (Christensen et al., 2001; Hanssen-Bauer et al., 2000; Hellström et al., 2001; see also Chapter 4). The projected rise in temperature is greater in the north than in the south of the region, greater inland than at the coast, and greater in winter than in summer.

Confidence in these projections is based on the trend in mean annual temperature for the period 1970 to 2000, generated retrospectively by the same models, corresponding reasonably well with empirical observations. Figure 17.10, for example, illustrates the observed mean annual temperature measured at Karasjok, a representative inland grazing region used in winter, between 1900 and 2000 and a modeled projection for mean annual temperature for the period 1950 to 2050 (Hanssen-Bauer et al., 2000). The trend in the projection from 1970 to 2000 compares well with, and is not significantly different from, the observed temperature trends (Hanssen-Bauer et al., 2003). The models do not, however, capture the changes in variability which

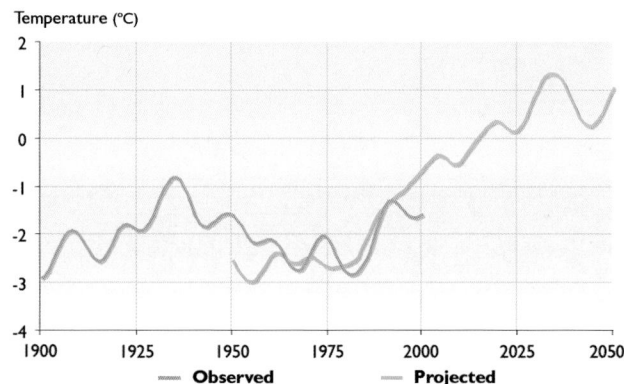

Fig. 17.10. Low-pass filtered series of observed and projected mean annual temperature in Karasjok, eastern Finnmark. The projected temperature is downscaled from the ECHAM4/OPYC3 global climate model, run with the IS92a emissions scenario (Hanssen-Bauer et al., 2000).

have been observed. At this stage, therefore, it is not possible to project with any degree of confidence to what extent the variability in mean annual temperature in northern Fennoscandia is likely to change over the next 50 to 100 years.

Global projections for the next 70 years or so indicate increased precipitation at high latitudes. These projections seem robust (e.g., Räisänen, 2001) and are qualitatively consistent with the expected intensification of the hydrological cycle caused by increased temperatures. Regional models for Fennoscandia project an increase in annual precipitation of between 1 and 4% per decade (Fig. 17.11). The regional precipitation scenarios are, however, generally less consistent than the regional temperature scenarios and their ability to reproduce the trends observed in recent decades remains limited.

Increases in temperature and precipitation can potentially affect snow conditions in a variety of ways that can influence foraging conditions for reindeer. Increased temperature in autumn might lead to a later start of the period with snow cover and increased temperature combined with more frequent precipitation may increase the frequency of snow falling on unfrozen ground. Furthermore, increased precipitation in winter would be expected to contribute to increased snow depth at the winter pastures of reindeer. With increased temperatures, the melting period in spring might start earlier but the last date of melting might be significantly delayed where the initial snow cover is deeper. The physical structure of the snow pack could also be affected by the projected changes in temperature and precipitation. No local projections for snow conditions in Finnmark have, however, yet been made. Their development would require an integration of the projections for temperature and precipitation, both of which are currently available only at a coarse scale of resolution. To be meaningful, models would have to be downscaled and would need to incorporate data on the physical structure of the landscape, especially altitude which influences local temperature profiles and, hence, the transition of precipitation from rain to snow (e.g., Mysterud et al., 2000; see also Chapter 7).

Downscaling global projections

The spatial resolution of the projections for temperature and precipitation over northern Fennoscandia is very coarse and, consequently, of limited use for projecting local trends in any but the most general terms. Downscaled scenarios, designed to improve the spatial resolution of the projections, have been developed for temperature and precipitation at selected stations in Finnmark. Projections for Karasjok in eastern Finnmark are shown in Figs. 17.10 and 17.11. The downscaled temperature scenarios show some of the same characteristics as the regional scenarios for Fennoscandia, including greater warming in winter than in summer and inland compared to the coast. However, the inland–coast gradient was in most cases greater in the downscaled projections than in the global scenarios. Downscaled projections for precipitation did not match the global projections for Fennoscandia well. This result was not unexpected and reflects the fact that downscaling, unlike global modeling, is sensitive to the effects of local topography on patterns of precipitation.

Local climate conditions important for reindeer herding

To be manageable, the models developed by downscaling analysis were necessarily made very simple. The weather patterns over reindeer pastures, by contrast, are highly complex and display a large degree of regional, local, and temporal variation. Some of the temporal variation is apparent from data for particular parameters. The observed winter precipitation in Karasjok, for example, has varied during the last five decades from less than half the 1961–1990 average to almost twice this value (Fig. 17.12). Likewise, at Tromsø at the coast, the date on which the last snow disappeared (between 1960 and 2000) has varied by as much as 60 days from year to year. There is also considerable spatial variability: the mean annual precipitation for Finnmark (1961–1990), for example, ranges from 325 mm at Kautokeino (inland) to 914 mm at Loppa (coast). The situation is, however, more complicated than these simple comparisons indicate owing to the many ways in which weather

(% of 1961-1990)

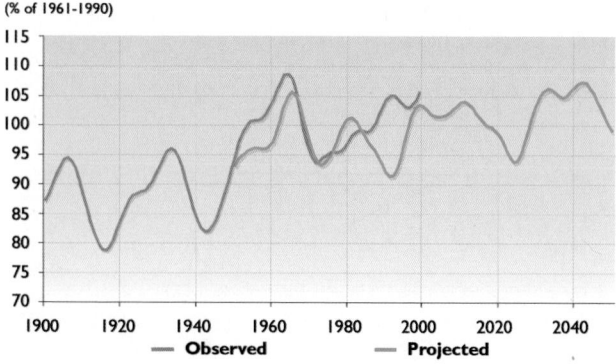

Fig. 17.11. Low-pass filtered series of observed and projected annual precipitation in Karasjok, eastern Finnmark. The projected precipitation is downscaled from the ECHAM4/OPYC3 global climate model, run with the IS92a emissions scenario (Hanssen-Bauer et al., 2000).

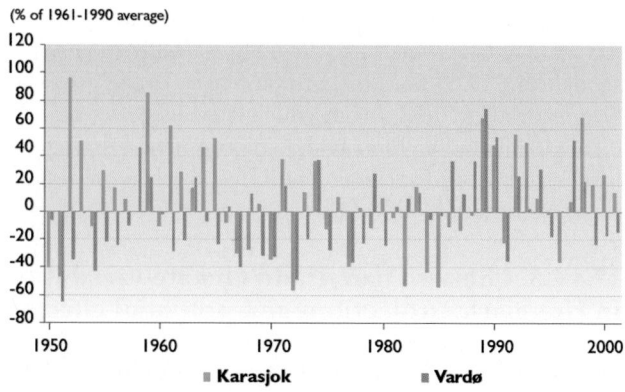

Fig. 17.12. Winter precipitation anomalies for Karasjok (eastern Finnmark) and Vardø (northeastern Norway), 1950 to 2001. The anomalies are given in percent relative to the 1961–1990 average.

can vary. Data collected over 35 years from two weather stations in central eastern Finnmark reveal, for example, that almost every year is a record year. Every year one parameter or another is colder, or warmer, or earlier, or deeper, and so on than ever before. There are, in effect, no "normal" years in Finnmark; instead, every year is exceptional. In herders' parlance: *Jahki ii leat jagi viellja* ("One year is not another's brother").

The challenge, therefore, is to extract data from global, regional, or local meteorological records in the form of selected parameters that, singly or in combination, represent useful proxies for those aspects of the weather in this complex system that significantly influence the growth and survival of reindeer and the work of the herders who look after them. Ecologists select proxies that are either as highly generalized or as highly specific as possible, including major atmospheric phenomena such as the North Atlantic Oscillation or monthly mean records of precipitation from a particular local weather station, respectively. The application of either highly

generalized or highly specific data can be useful and can yield robust results. The selection of proxies, however, is largely arbitrary and the results lack the sophistication that characterizes herders' understanding of the ways in which short-term variation in weather and longer term variation in climatic conditions affect their lives. Reindeer herders, like other people whose livelihood depends on close reading of the natural environment, have a deep understanding of the significance of the changing patterns of weather (Box 17.2). An important step in a vulnerability analysis of this kind, therefore, is to describe the effects of temporal and regional variation in weather on grazing conditions in terms of herders' experience and, hence, to identify climate phenomena and thresholds that are potentially important for reindeer production.

17.4.2.4. Ecological impacts

The ecological impact of large-scale climate variability and recent climate change on temperate species of

Box 17.2. The significance of snow

Reindeer herders, like other people whose livelihood depends on close reading of the natural environment, have a deep understanding of the significance of changing weather patterns. This knowledge is based on generations of experience accumulated and conserved in herding practice. Herders' understanding of snow (*muohta*) is one example.

In winter, each herding group grazes their animals in a defined area to which it has usufructory rights. In Finnmark, herds are typically tended continuously in winter, with herders taking watches that last seven to ten days. Their daily duties include maintaining the integrity of their herds – by preventing animals from straying and by keeping other reindeer away – and, most importantly, by finding fresh places for the animals to graze. "Good grazing" is a place where the snow is dry, friable, and not too deep to prevent the animals easily digging through it to reach the plants beneath. "Bad grazing" is a place where the snow is icy, hard, and heavy, or where a layer of ice lies over the vegetation on the ground beneath. "Exhausted grazing" is a place where reindeer have already dug and trampled the snow, consequently rendering it hard and effectively impenetrable.

Snow lies in the mountain pastures of northern Norway for up to 240 days per year and it is therefore not surprising that herders have learned to cope with varying snow conditions. The significance of snow for the lives of the people probably increased when they turned from hunting reindeer to herding them (Ruong, 1964). Winter grazing conditions must have become an important determinant of trade and, hence, an important topic of discussion. The distribution of snow and its physical characteristics such as its depth, hardness, density, structure, and variability all had to be expressed in a linguistic form. The Sámi recognize about 300 different qualities of snow and winter pasture – each defined by a separate word in their language (Eira, 1984, 1994; Jernsletten, 1997; Pruitt, 1979; Ryd, 2001).

A selection of Sámi words for snow

Čearga	Hard-packed drift snow "which one can't sink one's staff into" – impossible for reindeer to dig through.
Čiegar	Snow that has been dug up and trampled by reindeer, then frozen hard.
Fieski	Snow in an area where only a few reindeer have been, evidenced by few tracks.
Oppas	Thickly-packed snow through which reindeer can dig if the snow is of the *luotkku* (loose) or *seanas* type.
Sarti	A layer of frozen snow on the ground at the bottom of the snow pack that represents poor grazing conditions for reindeer.
Seanas	Dry, coarse-grained snow at the bottom of the snow pack. Easy for reindeer to dig through. Occurs in late winter and spring.
Skárta	A thin layer of frozen, hard snow on the ground that forms after rain. Also poor grazing conditions.

plants and animals is well documented (Ottersen et al., 2001; Post and Stenseth, 1999; Post et al., 2001; Stenseth et al., 2002; Walther et al., 2002). Among northern ungulates, variation in growth, body size, survival, fecundity, and population rates of increase correlate with large-scale atmospheric phenomena including the North Atlantic Oscillation (Forch-hammer et al., 1998, 2001, 2002; Post and Stenseth, 1999) and Arctic Oscillation (Aanes et al., 2002). Putative causal mechanisms underlying these correlations involve the climatic modulation of grazing conditions for the animals. The effects may be either direct, through the influence of climate on the animals' thermal environment or the availability of their forage beneath the snow in winter (e.g., Forchhammer et al., 2001; Mysterud et al., 2000), or indirect, through modulation, by late lying snow, of the phenological development and nutritional quality of forage plants in summer (e.g., Mysterud et al., 2001). The consequences for the animals may, in turn, be either direct, involving the survival of the current year's young, or indirect, whereby climate-induced variation in early growth influences the survival and breeding performance of the animals in adulthood (e.g., Forchhammer et al., 2001).

Some well-established reindeer populations characteristically display high-frequency, persistent instability (e.g., Solberg et al., 2001; Tyler et al., 1999b) indicating that their dynamics, and the dynamics of the grazing systems of which they are a part, may be strongly influenced by variation in climate (Behnke, 2000; Caughley and Gunn, 1993). However, despite a substantial volume of research related to the effects of snow on foraging conditions in tundra and taiga pastures (Adamczewski et al., 1988; Collins and Smith, 1991; Johnson C. et al., 2001; Miller et al., 1982; Priutt, 1959; Skogland, 1978) and, more recently, research related to the effects of variation in summer weather on forage (e.g., Lenart et al., 2002; Pentha et al., 2001; Van der Wal et al., 2000), only little evidence of a strong and pervasive influence of large-scale climate variation on the rate of growth of populations (Aanes et al., 2000, 2002, 2003; Post and Stenseth, 1999; Solberg et al., 2001; Tyler et al., 1999b) or the performance of individual reindeer (Post and Stenseth, 1999; Weladji et al., 2002) has yet emerged.

17.4.2.5. Coping with climate variability and change

The potential impact of climate variation and change on the productivity of herds can be ameliorated by tactical and strategic changes in herding practice. Herders' responses (feedback) represent coping (*birgehallat*), indicated by the dotted line in Fig. 17.9. The conceptual framework proposes that responses may be triggered at two levels. Ultimately, the herders respond to climate-induced changes in the performance of their animals. They also respond directly to the kinds of weather conditions that are important for suc-

cessful herding. This proximal response is indicated by the line marked "Herders' knowledge" in Fig. 17.9. The model makes no assumption about the extent or effectiveness of herders' ability to cope or the magnitude of the influence of climate change on the system.

A major point emphasized in this study is that climate change is not a new phenomenon in eastern Finnmark, even over the timescale of human memory. Systematic records of meteorological data have been made at Karasjok, close to the winter pastures, since 1870. These data provide clear evidence of climate change during the last 100 years. The dominant features of the temperature and precipitation records displayed in Figs. 17.10 and 17.11 are not the overall trends but, rather, the substantial decadal variation. Hence, although temperature displayed no statistically significant trend during the course of the last century, it is readily apparent that between 1900 and 2000 inner Finnmark experienced two periods with generally increasing temperatures. Between 1900 and 1935 and again between 1980 and 2000 the mean annual temperature at Karasjok increased by about 0.5 °C per decade. The observed rate of increase closely matches the projections for warming over Fennoscandia over the next 20 to 30 years that lie in the range of 0.3 to 0.5 °C per decade (see Chapter 4). Similarly, the modest net increase in precipitation during the last century, which occurred at a rate of 1.6% per decade, belies the observation that there were, in fact, three separate and substantial periods of increasing precipitation in those years. Between 1945 and 1965, for example, the mean annual precipitation at Karasjok increased by 20%. The rate of increase during this event greatly exceeds the current projections for precipitation increase over Fennoscandia of 1 to 4% per decade (see Chapter 4). Projections for future temperature and precipitation fail to capture these rapid changes and, instead, reflect only the modest trends observed across the 20th century as a whole.

Sámi reindeer herders have therefore, in the course of the last century, been exposed to climate change events of a magnitude at least as great as – and in some cases much greater than – those currently projected for northern Fennoscandia over the next 20 to 30 years. It needs to be noted, however, that a reoccurrence in the future of the large variations in climate experienced historically is certainly not excluded in the projections of climate change. Moreover, what is likely to be unprecedented historically is the level of mean climate around which these fluctuations will occur. One potentially useful approach to predicting the likely impact of, and herders' responses to, climate change, therefore, is to explore how they were affected in the past and what responses they displayed then. This kind of exploration requires the codification and analysis of herders' responses to weather-related changes in foraging conditions and of their perception and assessment of the risks associated with different coping options.

Strategic responses

Diversity in the structure of herds

Aboriginal production systems in extreme, highly variable, and unpredictable climates are based on the sequential utilization of, often, a large number of ecological or climatic niches (Murra, 1975). The essence of such systems is flexibility and the distribution of risk through diversity. Reindeer herders maintain high levels of phenotypic diversity in their herds with respect, for example, to the age, sex, size, color, and temperament of their animals (Oskal, 2000). A *čappa eallu* ("beautiful" herd of reindeer) is, therefore, highly diverse and, in this respect, is the antithesis of a purebred herd of livestock of the kind developed by careful selection to suit the requirements of a modern, high yielding agricultural ruminant production system.

The traditional diversity in the structure of reindeer herds is an example of a coping strategy aimed at reducing the vulnerability of the herd to the consequences of unfavorable – and unpredictable – conditions. Thus, in traditional reindeer herding, even apparently "non-productive" animals of either sex have particular roles which, when fulfilled, contribute significantly to the productivity of the herd as a whole. Traditionally, for example, reindeer herds in Finnmark typically consisted of as many as 40% adult males. Large numbers of large males were required for traction; they acted as focal points, helped keep the herd gathered, and reduced the general level of activity of the females: in modern jargon, the males contributed to energy conservation within the herd. Many were carefully castrated to this end (Linné, 1732). Their strength, moreover, enabled them to break crusted snow and to smash ice with their hooves, opening the snow pack to gain access to the plants beneath to the benefit of themselves and – incidentally – also for the females and calves in the herd. The modern agronomist, however, considers adult males unproductive and today few herds in Finnmark have more than 5% males. Males' role as draft animals and in gathering and steadying the herd has been largely superceded by snowmobiles – albeit at greatly increased economic cost. The reliance on snowmobiles, moreover, renders herding early in winter difficult in years when little or no snow arrives before the New Year. But old ways sometimes die out only slowly and there are ingenious solutions. When asked recently (in 2002) why he kept several heavy, barren females in his herd, Mattis Aslaksen Sara, a herder from Karasjok, replied "I have so few big males now – so who else will break the ice?" The decline in the diversity of the herd structure and, specifically, the increased proportion of females in today's herds is largely a result of government intervention. It reflects the influence of practices copied from sheep production systems that have been translated to reindeer herding by agronomists. The reduced heterogeneity of herds represents a reversal of the traditional approach; its consequences, in terms of the performance of the animals, remain largely unknown. The pattern of dispersion of female-dominated herds over the landscape is said to be different. The consequences of reduced heterogeneity in terms of changes in the vulnerability of the herding system to environmental change remain completely unknown.

Pastoral nomadism

The characteristic seasonal pattern of movement reflects herders' responses to the spatial and temporal heterogeneity and unpredictability of key resources, usually forage or water, whether these be for goats or cattle on a tropical savannah (Behnke et al., 1993) or for reindeer on northern taiga (Behnke, 2000). Nomadism is adaptive in the sense that, by moving his herd, the herder gains or averts what he anticipates will be the advantages or undesirable consequences of his doing so or not doing so, respectively.

Tactical responses

Movement

For Sámi nomads, one principal feature of the natural environment that influences the pattern of movement of herds into, within, and out of winter pastures is the condition of the snow pack. Snow determines the availability of forage (crusted snow is bad) and, in late winter, the mobility of herds (crusted snow is good). Skilled herders observe how the snow drifts, how it settles, and where conditions remain suitable for grazing and then make decisions about how and when to move after assessing the physical quality of the snow pack in relation to topography, vegetation, time of year, and condition of the animals. In the warm winters of the 1930s (see Fig. 17.10), for example, conditions were sometimes so difficult owing to heavy precipitation that herds spread out and moved to the coast earlier than normal in spring. Today, neighboring herding groups (*siida*) may even "trade" snow in the sense that one group may allow its neighbors to move their herd to an area of undisturbed snow (good grazing) on the former's land. In every case, success is contingent on the freedom to move.

Feeding

Reindeer husbandry in Norway is based on the sustainable exploitation of natural pasture. In winter, access to forage can be restricted by deep snow or ice and the animals have to cope with reduced food intake as a result. So extreme were snow conditions in the winter of 1917/18, with ensuing loss of animals, that Sámi herders in Norway employed Finnish settlers to dig snow to improve access to forage. Herders often provided small amounts of lichen both to reward animals they were in the process of taming and also as a supplement for draft animals or for hungry ones. Gathering lichens, however, is laborious and, instead, in addition to locally produced grass converted into hay or silage, several commercially available pelleted feeds have been developed (Aagnes et al., 1996; Bøe and Jacobsen, 1981; Jacobsen and

Skjenneberg, 1975; Mathiesen et al., 1984; Moen et al., 1998; Sletten and Hove, 1990). The provision of small amounts of supplementary feed can help to improve survival in winter (especially for calves), to increase the degree of tameness of the herd, and to improve the animal welfare image of reindeer herding in the eyes of the public. Negative effects include increased frequency of disease (Oksanen, 1999; Tryland et al., 2001) and increased cost. The use of pellets and locally harvested grass increased throughout Fennoscandia in the 1990s; reflecting this, many petrol stations in the reindeer herding areas of northern Finland now stock sacks of feed during winter. The use of pellets is less widespread in Norway owing in part to its high cost: the grain products in pelleted ruminant feeds are heavily taxed in Norway and the cost of providing artificial feed for reindeer is between four and six times higher than in Finland. In Norway, therefore, use of feed is generally restricted to periods of acute difficulty. This pattern might alter, however, in future should snowfall increase substantially.

17.4.2.6. Constraints on coping

The strategic and tactical decisions herders make in response to changes in pasture conditions represent aspects of coping. The success of the kinds of responses outlined in the previous section, however, depends to a large extent on herders' freedom of action. This section outlines five constraints or potential constraints on this freedom of action. The first four concern government policy (state, regional, and municipal) and present institutional arrangements that reduce the herders' ability to respond creatively to changing conditions, including climate variability and change. The fifth is pollution.

In Norway, Sámi reindeer herding takes place in a complex institutional setting heavily influenced by various forms of governance (see Fig. 17.1) that constrain herders' options. Constraints include the loss of habitat, predation (where the abundance of predators and, hence, the rates of mortality due to predation, is influenced by legislation protecting predators), and the governmental regulation of herding (including the regulation of rights of pasture, of the ownership of animals, and of the size and structure of herds) and of market- and price-controls. The effects of non-climate factors like these on the development of reindeer herding potentially dwarf the putative effects of climate change described previously. Institutions and governance have since the early 1980s demonstrably reduced the degree of freedom and the flexibility of operation under which reindeer herders traditionally acted. Their ability to cope with vagaries of climate may be reduced as a result. For these reasons, institutions and governance were included as a major element in the conceptual model (Fig. 17.9). The challenge remains to identify and quantify their impact on reindeer herding and to identify and understand the effects of this on herders' ability to cope with and adapt to changing environmental conditions. Of course, not all forms of governance and institutions are negative for reindeer herding: central administration also

provides important protection and opportunities for the industry and has supported both research and education. Interestingly, a major development in government support for reindeer herding was precipitated by an extreme climatic event. Severe icing over the pasture during the winter of 1967/68 resulted in substantial starvation and loss of reindeer in Finnmark. The government responded in an unprecedented manner and provided compensation equivalent, in today's monetary terms, to US$ 6.5 million. Out of this action arose a debate among the Sámi regarding the division and distribution of government funds within the reindeer industry, which continues, in one form or another, to this day. Loss of habitat, predation, the economic and socio-political environment, and law, however, were factors highlighted at the co-operative meeting in Tromsø (see section 17.4.2.2): their legitimacy and relevance lie in the fact that they are based on herders' subjective evaluation of their own situation.

Loss of habitat

Reindeer herding is a highly extensive form of land use. Roughly 40% (136 000 km^2) of Norway's mainland is designated reindeer pasture and within this area Sámi herders have – at least in principle – the right to graze their animals on uncultivated ground irrespective of land ownership. Herders' rights of usufruct, however, afford them neither exclusive access to the land nor protection from the interests of other land users. Conflicts of interest are common. For herders the principal issue is generally the securing of habitat in which to graze their reindeer. Indeed, the progressive and effectively irreversible loss of the uncultivated lands which reindeer use as pasture is probably the single greatest threat to reindeer husbandry in Norway today. Preservation of pastureland is, likewise, perhaps the single greatest priority for sustaining the resilience of reindeer herding confronted by changes in both the natural and the socio-economic environment.

Habitat loss occurs in two main ways: (1) through physical destruction and (2) through the effective, though non-destructive, removal of habitat or through a reduc-

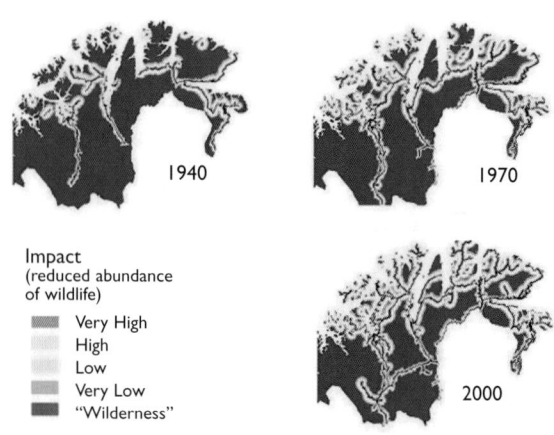

Fig. 17.13. Encroachment of roads in Finnmark 1940 to 2000, and the associated loss of reindeer pasture (UNEP, 2001).

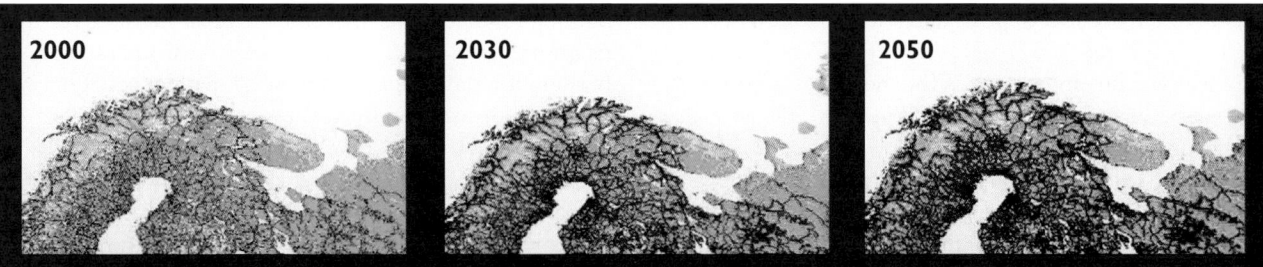

Fig. 17.14. Projected development of infrastructure (including roads, houses, military training areas) in the Barents Euro-Arctic region 2000–2050. This scenario is based on the historical development of infrastructure, the distribution and density of the human population, the existing infrastructure, the known location of natural resources, distance from the coast, and vegetation type (GloBio, 2002).

tion in its value as a resource. Physical destruction of habitat is chiefly a result of the development of infrastructure, including the construction of artillery ranges, buildings, hydro-electricity facilities, pipelines, roads, etc. The effective removal of habitat may result from disturbance (for example, by hunters, fishers, and walkers), local pollution, increased grazing pressure by potentially competing species (e.g., sheep, Coleman, 2000) or through loss of rights of access either locally (Strøm Bull, 2001) or as a result of the closure of regional or international borders (Hætta et al., 1994). Taking Norway as a whole, piecemeal development of infrastructure has resulted in an estimated loss of 70% of previously undisturbed reindeer habitat during the last 100 years (Nellemann et al., 2003); in Finnmark, the figure is close to 35% for the last 60 years alone (Figs. 17.13 and 17.14).

Predation

Fennoscandia is home to the last remaining sizeable populations of large mammalian predators in Western Europe, including bear (*Ursus arctos*), lynx (*Lynx lynx*), wolf (*Canis lupus*), and wolverine (*Gulo gulo*). These species are all capable of killing medium-sized ungulates like reindeer (although bears probably rarely do this). Wolverine, a major predator for reindeer, were completely protected in 1981 though limited hunting is now permitted. In Norway, very large numbers of domesticated animals range freely in the mountain areas in summer, including approximately 2 million sheep and 140 000 reindeer (which remain at pasture both in summer and winter) and these, not surprisingly, are potential prey. Reindeer herders in Finnmark, the county with the highest losses, estimate that between 30 and 60% of their calves are taken as prey each year (Anon., 2003); in some herds losses exceed 90% (Mathis Oskal, reindeer herder, pers. comm., 2003). Losses on this scale dwarf all other causes of mortality including climate-related deaths (Reindriftsforvaltningen, 2002) and are therefore a major determinant of levels of production in herds.

Norway's mountain pastures are an important renewable natural resource: their management as pasture, however, is clearly complicated by the presence of predators and the resulting predation on grazing animals. Intervention designed to ensure the sustained usefulness of mountain pastures as a resource for grazing animals by reducing the density of predator populations to levels at which they no longer represent a threat to the livelihood of the sheep farmers and reindeer herders, must select from among several unsatisfactory alternatives. Consequently, any solution is likely to be an unsatisfactory compromise. Alternative strategies range from implementing a general reduction in the density of predator populations, to establishing "predator-free zones" where grazing can continue uninterrupted while leaving the predators elsewhere undisturbed. Any course adopted must be commensurate both with Norway's commitment to the conservation of viable populations of mammalian predators under the terms of the Convention on the Conservation of European Wildlife and Natural Habitats (the "Bern Convention") and other international agreements and, at least as far as reindeer are concerned, by the country's commitment to safeguarding the special interests of the Sámi people. This commitment is enshrined in the terms of the International Labour Organisation (ILO) Convention No. 169 on Indigenous and Tribal Peoples. Moreover, it seems apparent, as reindeer herders argue, that obligations with respect to the intentions of ILO Convention No. 169 may take precedence over those of the Bern Convention (Schei, no date; Uggerud, 2001) and they press for the establishment of predator-free zones accordingly.

In practice, the situation remains unclear. No predator-free zones have been created. The culling of predators takes place only on a limited scale and herders – who have the best local knowledge about the predators – are not normally permitted to take part. Compensation for loss of animals is generally paid only in cases where claims are substantiated with unequivocal evidence such as post-mortem examination of carcasses. Herders, however, normally determine losses by observing the absence of particular animals and are only rarely able to support their claims by producing a carcass; the gathering up, transport, and delivery of carcasses is generally impracticable. Consequently, their claims are mostly unsubstantiated and usually rejected: in 2001–2002 herders in Finnmark were compensated for only one in four reindeer claimed lost (Lund, 2002). Loss of reindeer through predation, possibly exacerbated by increased snow, therefore, remains a major constraint on herd production levels and the herders, furthermore, remain largely powerless to tackle the situation owing to legislation that runs counter to their immediate interests.

Economic and socio-political environment

Reindeer herding in Norway is the most regulated reindeer husbandry in the world today. In 2000, the annual cost of its administration was US$ 21 million, which was more than twice the amount paid to reindeer herders for their meat. (This refers to the raw product; the market value of the reindeer meat sold is substantially greater.) The current high level of regulation of herding dates from 1978 when Sámi reindeer husbandry was brought more closely under the management of the Royal Norwegian Ministry of Agriculture where economic planning remained the policy-makers paradigm. This development reflected an earnest desire to improve the economic basis of Sámi reindeer husbandry and, hence, help herders to achieve the stable income indispensable in modern society. Its consequence was that central government became one of the most potent forces shaping the development of reindeer herding. The reigning paradigm of government policy was that of a modern, agricultural food-production system and an immediate consequence of its implementation was an increase in the number of reindeer and reindeer herders (Anon., 1992). Today, policies established by the central administration influence virtually every aspect of reindeer husbandry, from the granting of licenses to own reindeer and the allocation of grazing rights, to the monitoring and regulation of the size-, age-, sex-, and weight-structure of herds, the setting of production quotas, the influencing of both the age and sex composition of animals selected for slaughter, the timing of slaughtering, and determining to which slaughterhouses herders must sell their animals.

Some aspects of government intervention have been necessary and valuable. Once the aboriginal system of pasturing rights (the *siida* system) ceased to be recognized by Norwegian law (Strøm Bull, 2001), an alternative governance structure was needed. Other government interventions, such as a centralized regulation of the price of reindeer meat, have resulted in stagnation in herders' economy. Political and market power was lifted from the hands of the herders in 1978 and consolidated in early 2000 when an alliance took place between *Norsk Kjøtt* (a meat farmers' co-operative which controls 75% of slaughtering in Norway) and two large, private, reindeer slaughterhouses neither of which are Sámi owned. Sámi ownership and control is minimal: in Norway only a small proportion of reindeer are slaughtered by Sámi-owned enterprises compared to a large proportion in both Sweden and Finland (Reinert and Reinert, in press). Import tariffs and pricing policies have been used to protect and promote the interests of agricultural meat production at the expense of reindeer herding interests. The market mechanism has been eliminated as a price setting mechanism for reindeer meat and, instead, its price is negotiated annually by the herders' organization (the Sámi Reindeer Herders Association of Norway) and the government. In reality, the negotiating power of the herders is minimal because *Norsk Kjøtt* is responsible both for the marketing and the regulation of the reindeer meat market. For example, from 1976 to 1991 the net price paid to herders for their meat, corrected for inflation, fell by 45% largely in response to an increase in the level of production (Reinert and Reinert, in press). In the following decade the trend was reversed and the level of production was halved; yet, contrary to all normal practice, the real price paid to herders for their meat remained at the 1991 level. The absence of normal market mechanisms for price setting has been economically most disadvantageous for the herders. The fall in the price of reindeer meat over the last 25 years exemplifies the influence wielded over the economic development of reindeer husbandry by agricultural meat producers with vested interests. Lacking direct control over the slaughtering and marketing of reindeer meat, the Sámi of Norway became, *de facto*, an internal colony (Reinert and Reinert, in press). This recalls the term "Welfare Colonialism" coined by Paine (1977) to characterize culturally destructive colonialism in the Arctic.

Central administration, therefore, remains responsible for key aspects of the economic and socio-political environment in which herding exists and to which herders are obliged to adapt. The traditional fluidity and flexibility of practice that reindeer herding had developed to meet the vagaries of the natural environment of the north has been eroded. The exploration of the consequences of these developments for the adaptability, resilience, and vulnerability of Sámi reindeer herding under potential climate change remains, therefore, an important area of research.

Law

The elaborate legal structure upon which the regulation of reindeer husbandry is based is another aspect of the complex institutional setting in which Sámi reindeer herding is practiced in Norway. The law is comprehensive, complex and, occasionally, liberal to the point of ambiguity (Strøm Bull et al., 2001). It represents, therefore, a fourth non-climate factor that has a major influence on herding and which, by constraining herders' options, influences their ability to cope with changes in the natural environment.

Legislation governing reindeer husbandry is of considerable antiquity. A treaty agreed in 1751 between the respective joint kingdoms of Denmark/Norway and Sweden/Finland included the division along a common national border of hitherto undefined northern lands. This same border divides Norway and Finland today. The 18th century legislators realized that the creation of a border would potentially disrupt the lives of the nomads whose freedom of movement across the area had hitherto been unrestricted. An addendum was, therefore, included in the treaty confirming agreement between the two states that Sámi reindeer herders' customary utilization of the land should remain undisturbed notwithstanding the creation of a common border and the nomads' obligation to adopt one or other nationality.

This document, the Lapp Codicil, is the first formal legislation of reindeer husbandry. Crucially, it was built upon the principle of local self-government regarding the division of resources (Hætta et al., 1994).

The legislation of reindeer husbandry has evolved and increased in complexity since 1751. Successive statutes have been revised and new ones created to meet the challenges of changes in the economic and political climate, culminating in the Reindeer Husbandry Act of 1978 and its revision in 1996. Today's law includes provisions for the regulation of a wide range of issues. Section 2 alone includes rules for the designation of herding areas, the duration of grazing seasons within them, the size of herds, and the body mass of the animals in them. The level of detail of the legislation contrasts sharply with the lack of detail in the guidelines for its implementation. The Act is built on the premise that the organization of reindeer herding is best left in the hands of public administration. No groups have protected rights of usage. Instead, successive levels of the legislature – including the Ministry of Agriculture, the Reindeer Husbandry Board, Regional Boards, and Area Boards – determine, virtually unimpeded by legal barriers, the division of grazing districts, the allocation of herding franchises, and reindeer numbers. Regulation is achieved through rules, not statutes, as a consequence of which there remains considerable uncertainty among administrators and herders alike about the scope of the Act and a severe limitation of individual herder's opportunity to challenge administrative decisions.

The prevailing uncertainty is compounded by the fact that reindeer herding is regulated *de facto* by a Convention on Herding rather than through the provisions of the 1978 Act. The Convention is negotiated annually between the Government represented by the Ministry of Agriculture and the herders represented by the Sámi Reindeer Herders Association of Norway (NRL). The two parties are by no means equal. The Ministry is responsible both for drafting the regulations contained in each Convention, albeit in consultation with the NRL, and, ultimately, also for the interpretation and implementation of the final agreement. The regulations contained in the Convention are far more flexible than the Act but lack the legal checks and balances that the Act contains. The regulations agreed at each Convention, moreover, are frequently changed which only increases the level of uncertainty. Clearly, the complexities and ambiguities of the law contribute to the unpredictability of the administrative environment within which reindeer herding is practiced and, consequently, represent an important potential constraint on herders' ability to cope with changes in the natural environment.

Pollution

Pollution from sources outside the Arctic (AMAP, 2002) is another non-climate factor that can potentially influence the development of reindeer herding. Just as clean, local water is a fundamental human right, so also is the availability of uncontaminated food that can be gathered from traditional local sources. Imported agricultural food products are no substitute. Fortunately, chemical pollution is substantially less important for reindeer herding in Finnmark (generalizing from data for nearby regions in Russia) and, indeed, for all reindeer herding in Fennoscandia, than it is for marine resources (Bernhoft et al., 2002).

Most radioactive contamination on land in northern Fennocscandia is derived from fallout from atmospheric nuclear tests conducted up to 1980. Observed levels of contamination have not been considered hazardous for human health in Finnmark (Åhman, 1994; AMAP, 2004).

Radioactive contamination from the explosion at the Chernobyl nuclear power plant in 1986 is a major source of contamination in parts of southern Fennoscandia. This is a serious problem locally but is not directly a problem for reindeer herding as a whole because the majority of reindeer and reindeer herders live in the north of the region (Åhman, 1994). The radioactive pollution from Chernobyl has, however, been an indirect problem for the entire reindeer herding industry owing to negative focus in mass media, which failed to distinguish between those regions that received some fallout and those that were not affected at all. Misinformation of this kind can potentially turn consumers away and can encourage international food producers to step in and provide "clean", although non-traditional, substitute foods. The influence of effects of this kind on the vulnerability of small arctic enterprises like Sámi reindeer herding remains an important area of study.

Heavy metals accumulate in lichens (AMAP, 2002). Concentrations of heavy metals in reindeer meat, however, are no higher than in the meat of pigs, cattle, and poultry (Bernhoft et al., 2002). No data are available on concentrations of heavy metals in reindeer in Finnmark. Data on trends in heavy metals are available only for reindeer from Sweden where samples have been collected annually in three reindeer districts since the early 1980s (Swedish Museum of Natural History Contaminant Research Group, 2000). These data indicate that there have been no significant changes in the concentration of Pb, Cd, or Hg in reindeer meat for the period 1983 to 1998. In liver, the concentration of Pb decreased by 6.8% per year over this period, while the concentration of Cd showed a slight increase. Concentrations of Pb and Cd are very low (0.06 μmol/L) in blood among women in arctic Norway (Odland et al., 1999).

Cadmium is a potential problem owing to its tendency to accumulate in reindeer kidneys: people who consume these organs are exposed to this metal. AMAP (2002) reported that concentrations of Cd in reindeer kidneys in arctic Norway and Sweden are approximately three times higher than those in southern Norway and Sweden. Bernhoft et al. (2002) reported very low levels of Cd in the kidneys of reindeer from Kola in northwest Russia.

No data are available on concentrations of POPs in reindeer in Finnmark. In general, concentrations of POPs are lower in terrestrial mammals than in marine mammals (AMAP, 2002). Concentrations have been measured on an annual basis since 1983 in reindeer at Abisko, Sweden (AMAP, 1998). Current levels are not thought to represent a significant threat for reindeer (AMAP, 2002). Levels of two POPs in other species in the Swedish Arctic are declining, e.g., ΣDDT and ΣPCBs in otters in northern Sweden (Roos et al., 2001), and this trend is expected to continue.

17.4.2.7. Insights from the reindeer nomadism vulnerability case study

The reindeer nomadism vulnerability case study demonstrated the versatility of the general conceptual framework for vulnerability studies proposed in section 17.2. The development of a framework that was tailored specifically for reindeer herding in Finnmark also showed the diversity of the kinds of information that need to be included in an assessment of the vulnerability of a human–environment system in the Arctic. It illustrated the usefulness of reducing potential complexity to manageable proportions by developing a conceptual model containing just a few selected elements. It also showed the importance of collaborating with reindeer herders in a co-production of knowledge.

The validity and legitimacy of reducing an immensely complex system to something simple and, therefore, amenable to a vulnerability assessment depended wholly on the participation of herders themselves. Outsiders should not decide what factors, or suites of factors, influence reindeer herding: nobody, except for herders themselves, can legitimately make the required selection. The conceptual model, developed as a result of the interdisciplinary and intercultural effort, necessitated the integration of empirical data and herders' knowledge. The integration of different ways of knowing, called the "co-production of knowledge" (e.g., Kofinas and the communities of Aklavik, 2002), is not widely exploited in ecological research probably because aboriginal knowledge often does not lend itself to reductionist analysis and hypothesis testing. However, herders' knowledge of the impact of something as specific as climate variation on their way of life is based on an understanding resulting from generations of experience accumulated and conserved in herding practice and herders' specialized vocabulary. Consequently, in some instances herders can contribute knowledge gathered over a time span longer than the periods over which climate change has been documented by other means. The success of the approach outlined here was evident from the logical design and usefulness of the resulting conceptual model.

The joint effort in developing a conceptual model appropriate for a study of the vulnerability of reindeer herding in Finnmark to climate change quickly revealed that herding is affected by much more than just change in climate. Moreover, it is extremely likely that the effects on

reindeer herding of the non-climate factors introduced into the model potentially dwarf the putative effects of climate change on the system. Hence, the potential consequences of the projected increase in the average annual temperature at Karasjok over the next 20 to 30 years (Fig. 17.10) cannot meaningfully be considered independent of concurrent changes in the socio-economic environment for which, in some cases, clear predictions are already available (e.g., Figs. 17.13, 17.14).

Clearly, reindeer herding has been very resilient. The continued existence of nomadic reindeer herding by Sámi and other northern peoples in Eurasia today is evidence that these have, through the centuries, coped with and adapted to the vagaries and transitions of the socio-economic environment of the north. On one hand, it has not been overlooked that, if the marginalization of reindeer nomadism continues and if constraints on the freedom of action of the herders increase, new climatic conditions might threaten the resilience and increase the vulnerability of herding societies in ways that are without precedent. On the other hand, action provokes reaction: changes in climate and in the socio-economic environment might also create new opportunities for sustainable development in reindeer peoples' societies. Herders can be expected to grasp new opportunities, wherever they arise, and to take the initiative in improving the economy of their industry thereby reducing the vulnerability of their society.

17.5. Insights gained and implications for future vulnerability assessments

Arctic human–environment systems are subject to high rates of change in climate and/or other environmental and societal factors. Some changes emanate from outside the Arctic, while other changes arise from within the region. The vulnerability of human–environment systems in the face of such changes can vary widely with differences in the character and relative importance of environmental and societal changes across local settings. Vulnerability analysis offers an approach for exploring implications of environmental and social changes in a way that recognizes the interconnectedness of human and environment systems and the exposure, sensitivity, and adaptive capacity of these systems as they experience stresses or anticipate potential stresses arising from and interacting across local, regional, and global scales.

The Sachs Harbour and Greenland examples, plus the more developed case study on reindeer herding in Finnmark (section 17.4.2), reveal the importance of characterizing the place-based aspects of coupled human–environment systems, analyzing multiple and interacting stresses across multiple scales, accounting for adaptive capacity in assessing vulnerability, and incorporating varied forms of knowledge, analytical tools, and methodologies in vulnerability analysis. These case studies demonstrate that in their decision-making *arctic residents integrate their experiences and expectations of change in environmental and societal factors in addition to changes in climate and varia-*

tion in climate. They also illustrate that *vulnerability analysis can be applied in situations where the social and environmental changes important for a particular human–environment system operate at different scales.*

Given the close linkages between arctic peoples and the natural settings in which they live and on which they depend, a meaningful and useful analysis of arctic vulnerabilities requires the definition, characterization, and analysis of coupled human–environment systems. The human–environment systems at the heart of the case studies presented in this chapter are centered around human livelihoods (e.g., marine resource use). Seen from a human perspective, livelihoods are arguably the most salient aspect of a coupled human–environment system because the practices they entail involve close, fairly well-circumscribed, and critical interactions between social and natural systems in a particular locale, yet with discernible linkages to dynamics operating not only within, but also across local, regional, and global levels. Livelihoods are also the focal point of social organization, culture, and identity. The focus on reindeer herding in Finnmark, for example, enables the identification of specific climate-related changes (e.g., regarding snow pack and forage) and regulations (e.g., regarding land rights and predators) that affect this system. The identification of stresses, vulnerabilities, and response strategies for a more broadly defined system (e.g., for all indigenous peoples in Fennoscandia) would be more difficult and arguably less useful. The complex dynamics important for understanding vulnerability are apparent in all case studies. In Greenland and Sachs Harbour the size, health, and harvest of fish and marine mammals depend on climate, pollution, market factors, regulations, and technology. In Finnmark, climate changes and regulations have profound effects on reindeer, reindeer habitats, and herder practices and livelihoods. In Greenland with its market ties to distant localities via fish and fur products *it is evident that coupled human–environment systems in the Arctic are influenced by socio-political, socio-economic, and cultural factors originating outside as well as inside the region. Arctic residents accommodate this range of influences in their coping decisions.*

As these studies also show, primary stresses like climate change can have cascading and interacting impacts on many different aspects of the arctic physical and biological environment. Some factors, for example local climate shifts, can impact on many different components of the arctic system with differing magnitudes, timing, effects, and interactions. Thus, an increase in air and water temperature will probably affect the distribution of coastal winter sea ice and alter the access of local people to fishing and hunting areas; it may affect local oceanography and alter the habitats of marine mammals and their prey; it may increase the abundance of forage that reindeer eat; it may accelerate physical processes of pollutant transport and reactivity; and it may affect the health and well-being of arctic residents through decreased access to traditional foods and

increased incidence of disease, etc. Each of these effects can interact with others leading to more complex, higher-order effects. For example, the seasonal distribution and migratory routes of marine mammals may shift, forcing the hunters and their families either to follow the animals and relocate or to adopt new economies and lifestyles.

The case studies also illustrate *the importance of examining multiple, interacting stresses, operating within and across local, regional, and global scales, as well as the adaptive capacity of systems weathering these stresses.* Stresses (as well as potential opportunities) facing marine resource systems arise from interactions among, for example, climate, global markets, environmental and animal welfare campaigns, and changes in governance. Stresses (as well as potential opportunities) facing reindeer herding systems arise from interactions among changes in, for example, climate, forage, technology, and regulation. These factors do not, by definition, always lead to negative consequences. Changes in governance might be just as likely to reduce vulnerability as they are to contribute to vulnerability. A holistic understanding of vulnerability requires analysis of these many factors and their interactions, along with an understanding of how the coupled human–environment system in question might cope with or adapt to the changes brought about by these factors. *Coping and adaptation can diminish the vulnerability of certain components of the system and thereby offset adverse impacts. Vulnerability analyses reveal where actions can best be taken to enhance adaptive capacity, for example, via changes in public policy and new strategies in resource management, and anticipatory measures to prepare for adverse circumstances and mitigate their effects.* Arctic human and environment systems have a long history of coping with and adapting to social and natural changes. The resilience exhibited by arctic peoples provides insight into how these coupled human–environment systems might adapt in the future. Mobility, flexibility in livelihood (e.g., hunting, fishing, herding) practices, and a capacity for innovation all contribute to adaptive capacity, including a capacity to plan and prepare for contingencies. For example, the varied strategies that reindeer herders have developed for dealing with environmental and social changes are also strategies through which herders anticipate and prepare for future events. Nomadism itself is a way of anticipating future opportunities or adverse conditions. Because they are mobile, Sámi reindeer herders can respond quickly to unfavorable weather and/or snow conditions in one location by moving to another. "Trading" in snow is another practice that helps herders to successfully handle contingencies. An accounting of past and present adaptive measures is an important component of vulnerability assessment.

Vulnerability assessment also requires varied forms of knowledge and the development of new analytical tools and methodologies. Understanding the stresses facing place-based coupled human–environment systems and the adaptive measures they might take in response to these stresses necessitates novel modes of inquiry. *Involving indigenous peoples and other arctic residents in the research process is extremely important in developing such understandings.* Methods to integrate indigenous knowledge and scientific knowledge such as biology, climate science, political science, and anthropology are similarly important. Climate downscaling, pollutant modeling, scenario development for societal trends, environmental monitoring, interviews, focus groups, workshops, and ethnography comprise additional approaches that could be integral to vulnerability analysis.

The following sections contain general conclusions pertaining to the assessment of trends in climate, pollution, and human and societal conditions, and some comments on next steps.

17.5.1. Climate

The results of downscaling analyses reported for Finnmark provide preliminary insights into how temperature and precipitation may change in this region. The projections presented in the Finnmark case study were calculated using a single domain (20° W–40° E and 50° N–70° N for Karasjok, Norway). A more comprehensive downscaling program would provide projections using multiple downscaling domains. The models presented here also use a single predictor variable: large-scale temperature for projecting local temperature, and large-scale precipitation for projecting local precipitation. It remains to be seen how sensitive the results are to the selection of downscaling domain or predictor variables. Also important to include would be downscaling results for a number of additional variables such as snow and ice cover, permafrost conditions, and extreme events, as well as sensitivity analyses to examine the robustness of the various projections. A more comprehensive program would also involve residents more directly in research design, analysis, and dissemination.

Effective downscaling must engage local people. Snow quality, for example, is an extremely important factor for reindeer herding and reindeer herders have many words to describe snow quality. In contrast, climate downscaling provides information about a relatively limited number of variables. It is therefore not obvious how typical climate forecasting products and terminology might be made relevant for reindeer herders. Thus, in principle, analysts conducting downscaling for a vulnerability study should first assimilate the views and information needs of local people for the products of these analyses. In practice this will require creative ways for presenting results to non-climate specialists in order to address their needs and concerns and make most advantage of their local knowledge. As with any climate analysis, the models used in this study produce an enormous quantity of information – all of which is important for the analysis but most of which may not be useful for decision-makers. The risk of information overload is high. For example, at a minimum, for each downscaled climate variable, month, and general circulation model analyzed, vulnerability researchers should examine estimates of trend, variability, historical goodness-of-fit, and spatial distribution. Thus, climate analysts need to be willing to tailor their model products to the specific needs of local decision-makers.

17.5.2. Pollution

Information on POPs and heavy metals in the Arctic is widely available for the past two decades. Data on environmental concentrations for a number of chemicals exist for both western Greenland and Fennoscandia. These data, however, tend to be temporally and spatially dispersed. Data on local, long-term trends in environmental levels of POPs and heavy metals are much less abundant for both loci. There are, however, reliable time trends for certain species (e.g., reindeer and arctic char) in Fennoscandia. Data from the early 1980s to 2000 indicate generally declining environmental levels of POPs in both Disko Bay on the west coast of Greenland and in Fennoscandia. Trends in environmental heavy metal levels in western Greenland and Fennoscandia are less clear than for POPs. Some heavy metal levels have increased, while others show no change, or even a decrease.

Long-term local human trend data are even less available for western Greenland and Fennoscandia than environmental trend data. Available data suggest that observed human health problems relating to POPs and heavy metals are greater in western Greenland than in Fennoscandia. At the regional level, the greatest heavy metal threat to human health is due to Hg. Exposure to Hg in Greenland is at levels where subtle health effects can occur on fetal and neonatal development (AMAP, 2003). As in Fennoscandia, environmental heavy metal levels in the Disko Bay region show diverse trends. Daily human intake of Cd and Hg in the Disko Bay region is comparatively high.

Levels of POPs in both regions can be expected to decline toward 2020 due to increasing international regulation, although other POPs such as brominated flame retardants could become a growing problem (AMAP, 2002). Trends in environmental heavy metal levels to 2020 in both regions are more difficult to project than for POPs.

Future place-based pollutant research for vulnerability analysis would ideally consist of exposure and trend monitoring, human health and epidemiological analyses, and collection of other relevant data such as information about dietary intake, smoking, and other influences on pollutant burden. All these types of study are feasible and have been done at various sites; however, a vulnerability study necessitates that this information be available for a specific location. There is also a need to better understand local means of adaptation to problems with pollution, both in terms of what has been done and what could be done.

17.5.3. Trends in human and societal conditions

Several general trends (i.e., those concerning governance; population and migration; consumption; economy, markets and trade; and connectivity) are apparent in human and societal conditions throughout much of the Arctic. In recent decades, governing authority in Greenland and places within Canada and Alaska has rested increasingly with indigenous peoples. At the same time, regulations (particularly those pertaining to natural resource use) emanating from local, national, and international bodies are playing important roles in the lives of arctic peoples and the ways in which they are permitted to use land and to harvest fish and marine and terrestrial mammals. In addition, pan-arctic cooperation is increasing and transnational networks of indigenous peoples are growing. More people live in arctic urban areas than was the case thirty years ago, less traditional food is being consumed, a larger number and greater variety of imported technologies are employed, and people are more "wired" via the Internet, television, telephones, and satellites. Mixed economies have become more prevalent throughout the Arctic and the connections linking arctic economies with global markets are becoming stronger.

But while human and societal trends identified in this project are noteworthy for the Arctic as a whole, they by no means represent a complete inventory of such trends. Nor are they necessarily the most important trends for understanding the vulnerability of the case study sites. A more comprehensive and complete analysis of human and societal trends within the context of a fully-fledged vulnerability analysis would require the broad and systematic engagement of people living in the case study locations, and the use of tools such as surveys, participant observation, workshops, interviews, focus groups, and ethnography to ascertain what human and societal conditions are most relevant to a particular coupled human–environment system, how these conditions have changed over recent decades, and how they are expected to change in the future. The development of several alternative future societal scenarios would be useful in carrying out the difficult task of projecting future human and societal conditions and assessing their implications for coupled human–environment system vulnerabilities. The production and comparison of multiple scenarios could facilitate sensitivity analysis.

Oran Young (1998b) defined sustainable development as "…an analytic framework intended to provide structure and coherence to thinking about human/environment relations". Young calls for a sustainable development discourse that will facilitate efforts to identify and address arctic concerns. He adds that "To be useful in an arctic context, sustainable development must take into account the distinctive ecological, social, and cultural features of the region and offer an integrated approach to the endogenous and exogenous threats to sustainability peculiar to the circumpolar world". According to this view, vulnerability (and resilience) analysis as outlined in this chapter, can serve as a vehicle for conceptualizing and implementing sustainable development. Vulnerability analysis offers a holistic vision of human–environment systems and their dynamics at local to global scales. It recognizes that environmental changes are interactive, that ecology, culture, economics, history, and politics are interconnected, and that decisions about what to sustain and how must be made in particular social and ecological contexts. However, vulnerability analysis is more than a research strategy. It has the potential to provide processes in which people with varied backgrounds and interests can engage in dialogue, produce knowledge, and articulate values. Such processes can ultimately inform the ways in which communities and governments balance aspirations for human and societal development with those of environmental and social sustainability.

References

Aagnes, T.H. and S.D. Mathiesen, 1994. Food and snow intake, body mass and rumen function in reindeer fed lichen and subsequently starved for 4 days. Rangifer, 14:33–37.

Aagnes, T.H., A.S. Blix and S.D. Mathiesen, 1996. Food intake, digestibility and rumen fermentation in reindeer fed baled timothy silage in summer and winter. Journal of Agricultural Science (Cambridge), 127:517–523.

Aanes, R., B.-E. Sæther and N.A. Øritsland, 2000. Fluctuations of an introduced population of Svalbard reindeer: the effects of density. Ecography, 23:437–443.

Aanes, R., B.-E. Sæther, F.M. Smith, E.J. Cooper, P.A. Wookey and N.A. Øritsland, 2002. The Arctic Oscillation predicts effects of climate change in two trophic levels in a high-arctic ecosystem. Ecology Letters, 5:445–453.

Aanes, R., B.E. Sætre, E.J. Solberg, S. Aanes, O. Strand and N.A. Øritsland, 2003. Synchrony in Svalbard reindeer population dynamics. Canadian Journal of Zoology, 81:103–110.

Adamczewski, J.Z., C.C. Gates, B.M. Soiutar and R.J. Hudson, 1988. Limiting effects of snow on seasonal habitat use and diets of caribou (Rangifer tarandus groenlandicus) on Coats Island, Northwest Territories, Canada. Canadian Journal of Zoology, 66:1986–1996.

Adger, W. and P. Kelly, 1999. Social vulnerability to climate change and the architecture of entitlements. Mitigation and Adaptation Strategies for Global Change, 4:253–266.

AHDR, 2004. Arctic Human Development Report. Akureyri, Stefansson Arctic Institute.

Åhman, B. 1994. Radio cesium in reindeer (Rangifer tarandus) after fallout from Chernobyl accident. Swedish Agricultural University, Uppsala.

Alcamo, J., P. Mayerhofer, R. Guardans, T. van Harmelen, J. van Minnen, J. Onigkeit, M. Posch and B. de Vries, 2002. An integrated assessment of regional air pollution and climate change in Europe: findings of the AIR-CLIM Project. Environmental Science and Policy, 5:257–272.

AMAP, 1998. AMAP Assessment Report: Arctic Pollution Issues. Arctic Monitoring and Assessment Programme, Oslo, Norway, xii+859p.

AMAP, 2002. Arctic Pollution 2002: Persistent Organic Pollutants, Heavy Metals, Radioactivity, Human Health, Changing Pathways. Arctic Monitoring and Assessment Programme, Oslo, Norway. xii+112p.

AMAP, 2003. AMAP Assessment 2002: Human Health in the Arctic. Arctic Monitoring and Assessment Programme, Oslo, Norway. xiv+137p.

AMAP, 2004. AMAP Assessment 2002: Radioactivity in the Arctic. Arctic Monitoring and Assessment Programme, Oslo, Norway.

Anisimov, O. and B. Fitzharris, 2001. Polar Regions (Arctic and Antarctic). In: J.J. McCarthy, O.F. Canziani, N.A. Leary, D.J. Dokken, K.S. White (eds.). Climate Change 2001: Impacts, Adaptation, and Vulnerability, pp 801–842. Cambridge University Press.

Anon., 1992. En bærekraftig reindrift Nr. 28 (1991–1992). St. meld., Oslo. 134p. (In Norwegian)

Anon., 2003. Reindriftens rovdyrutvalg. Royal Ministries of Agriculture and the Environment, Oslo, Norway. (In Norwegian)

Ashford, G. and J. Castledown, 2001. Inuit Observations on Climate Change. Final Report. International Institute for Sustainable Development, 27p.

Ayotte, P., E. Dewailly, S. Bruneau, H. Careau and A. Vezina, 1995. Arctic air-pollution and human health – what effects should be expected. Science of the Total Environment, 160/161:529–537.

Bachler, G., 1998. Why environmental transformation causes violence. Environmental Change and Security Project Report 4.

Bamzai, A.S., 2003. Relationship between snow cover variability and Arctic oscillation index on a hierarchy of time scales. International Journal of Climatology, 23:131–142.

Beach, H., 2000. The Saami. In: M.M.R. Freeman (ed.). Endangered Peoples of the Arctic: Struggles to Survive and Thrive, pp. 223–246. Greenwood Press.

Behnke, R., 2000. Equilibrium and non-equilibrium models of livestock population dynamics in pastoral Africa: their relevance to Arctic grazing systems. Rangifer, 20:141–152.

Behnke, R.H., I. Scoones and C. Kerven (eds), 1993. Range Ecology at Disequilibrium: New Models of Natural Variability and Pastoral Adaptation in African Savannas. Overseas Development Institute, London.

Berkes, F., 1998. Indigenous knowledge and resource management systems in the Canadian subarctic. In: F. Berkes and C. Folke (eds.). Linking Social and Ecological Systems, pp. 98–128. Cambridge University Press.

Berkes, F., 1999. Sacred Ecology: Traditional Ecological Knowledge and Resource Management. Taylor and Francis, xvi+209p.

Berkes, F. and C. Folke (eds.), 1998. Linking Social and Ecological Systems: Management Practices and Social Mechanisms for Building Resilience. Cambridge University Press.

Berkes, F. and T. Henley, 1997. Co-management and traditional knowledge: threat or opportunity. Policy Options, 18:29–30.

Berkes, F. and D. Jolly, 2001. Adapting to climate change: Social-ecological resilience in a Canadian Western Arctic Community. Conservation Ecology, 5(2):18. http://www.consecol.org/vol5/iss2/art18

Berkes, F., J. Mathias, M. Kislalioglu and H. Fast, 2001. The Canadian Arctic and the Oceans Act: The development of participatory environmental research and management. Ocean and Coastal Management, 44:451–469.

Berkes, F., J. Colding and C. Folke (eds.), 2003. Navigating Social-Ecological Systems: Building Resilience for Complexity and Change. Cambridge University Press.

Bernhoft, A., T. Waaler, S.D. Mathiesen and A. Flaoyen, 2002. Trace elements in reindeer from Rybatsjij Ostrov, northwestern Russia. Rangifer, 22:67–73.

Bielawski, E., 1996. Inuit indigenous knowledge and science in the Arctic. In: L. Nader (ed.). Naked Science: Anthropological Inquiries into Boundaries, Power, and Knowledge. Routledge.

Bjerregaard, P., 1995. Health and environment in Greenland and other circumpolar areas. Science of the Total Environment, 160/161:521–527.

Blaikie, P., T. Cannon, I. Davis and B. Wisner, 1994. At Risk: Natural Hazards, People's Vulnerability, Disasters. Routledge.

Boas, F., 1888. The Central Eskimo. Smithsonian Institution, Washington, DC.

Bøe, U.B. and E. Jacobsen, 1981. Trials with different feeds to reindeer. Rangifer, 1:39–43.

Boertje, R.D., 1984. Seasonal diets of the Denali caribou herd, Alaska. Arctic, 37:161–165.

Böhle, H.G., T.E. Downing and M. Watts, 1994. Climate change and social vulnerability: the sociology and geography of food Insecurity. Global Environmental Change, 4:37–48.

Brewster, K. 1997. Native contributions to Arctic science at Barrow, Alaska. Arctic, 50:277–283.

Burton, I., R. Kates and G. White, 1978. The Environment as Hazard. Oxford University Press.

CAFF, 2001. Arctic Flora and Fauna: Status and Conservation. Conservation of Arctic Flora and Fauna, Helsinki.

Cahill, C. and E. Weatherhead, 2001. Ozone losses increase possible UV impacts in the Arctic. Witness the Arctic: Chronicles of the NSF Arctic Sciences Program, 8:1–2.

Caughley, G. and A. Gunn, 1993. Dynamics of large herbivores in deserts: kangaroos and caribou. Oikos, 67:47–55.

Caulfield, R.A., 1994. Aboriginal subsistence whaling in West Greenland. In: M.M.R. Freeman (ed.). Elephants and Whales: Resources for Whom? Gordon and Breach Science Publishers.

Caulfield, R.A., 1997. Greenlanders, Whales, and Whaling: Sustainability and Self-determination in the Arctic. University Press of New England.

Caulfield, R.A., 2000. The Kalaallit of West Greenland. In: M.M.R. Freeman (ed.). Endangered Peoples of the Arctic: Struggles to Survive and Thrive, pp. 167–186. Greenwood Press.

Christensen, J.H., J. Räisänen, T. Iversen, D. Bjørge, O.B. Christensen and M. Rummukainen, 2001. A synthesis of regional climatic change simulations. A Scandinavian perspective. Geophysical Research Letters, 28:1003–1006.

Clark, G.E., S.C. Moser, S.J. Ratick, K. Dow, W.B. Meyer, S. Emani, W. Jin, J.X. Kasperson, R.E. Kasperson and H.E. Schwarz, 1998. Assessing the vulnerability of coastal communities to extreme storms: the case of Revere, MA, USA. Mitigation and Adaptation Strategies for Global Change, 3:59–82.

Cohen, S.J. (ed.), 1997. Mackenzie Basin Study (MBIS) Final Report. Environment Canada, 372p.

Colborn, T., D. Dumanoski and J.P. Myers, 1996. Our Stolen Future. Dutton.

Coleman, J.E., 2000. Behaviour patterns of wild reindeer in relation to sheep and parasitic flies. PhD Thesis, University of Oslo.

Collins, W.B. and T.S. Smith, 1991. Effects of wind-hardened snow on foraging by reindeer (Rangifer tarandus). Arctic, 44:217–222.

Cornelissen, J.H.C., T. Callaghan, J.M. Alatalo, A. Michelsen, E. Graglia, A.E. Hartley, D.S. Hik, S.E. Hobbie, M.C. Press, C.H. Robinson, G.H.R. Henry, G.R. Shaver, G.K. Phoenix, D.G. Jones, S. Jonasson, F.S.I. Chapin, U. Molau, C. Neill, J.A. Lee, J.M. Melillo, B. Sveinbjornsson and R. Aerts, 2001. Global change and arctic ecosystems: is lichen decline a function of increases in vascular plant biomass? Journal of Ecology, 89:984–994.

Cutter, S.L., 1996. Vulnerability to environmental hazards. Progress in Human Geography, 20:529–539.

Cutter, S.L., 2001. American Hazardscapes: The Regionalization of Hazards and Disasters. Joseph Henry Press.

Dabelko, G. and P.J. Simmons, 1997. Environment and Security: Core Ideas and US Government Initiatives. SAIS Review, 71:127–146.

Dahl, J., 2000. Saqqaq: an Inuit hunting community in the modern world. University of Toronto Press.

Davies, S., 1993. Are coping strategies a cop out? Institute for Development Studies Bulletin, 24:60–72.

De Fabo, E. and L.O. Björn, 2000. Ozone depletion and UV-B radiation. In: M. Nuttall and T.V. Callaghan (eds.). The Arctic Environment, People, Policy, pp. 555–573. Harwood Academic Publishers.

Dow, K., 1992. Exploring differences in our common future(s): the meaning of vulnerability to global environmental change. Geoforum, 23:417–436.

Downie, D.L., J. Krueger and H. Selin, 2004. Global Policy for Hazardous Chemicals. In: R. Axelrod, D.L. Downie and N. Vig (eds.). Global Environmental Policy: Institutions, Law and Policy, pp. 125–145. CQ Press.

Downing, T., 1991. Assessing Socioeconomic Vulnerability to Famine and Country Studies in Zimbabwe, Kenya, Senegal and Chile. Research Report No. 1, Environmental Change Unit 2, University of Oxford.

Downing, T.E., 1992. Climate Change and Vulnerable Places: Global Food Security and Country Studies in Zimbabwe, Kenya, Senegal and Chile. Research Report No. 1, Environmental Change Unit, University of Oxford.

Downing, T.E., R. Butterfield, S. Cohen, S. Huq, R. Moss, A. Rahman, Y. Sokona and L. Stephen, 2001. Climate Change Vulnerability: Linking Impacts and Adaptation. Report to the Governing Council of the United Nations Environment Programme. United Nations Environment Programme and Oxford: Environmental Change Institute.

Easterling, W., 1996. Adapting North American agriculture to climate change in review. Agricultural and Forest Meteorology, 80:1–53.

Eckley, N., 2001. Traveling toxics: the science, policy, and management of persistent organic pollutants. Environment, 43:24–36.

Eira, N.I., 1984. Saami Reindeer Terminology. Diedut, 1:55–61.

Eira, N.I., 1994. Bohccuid luhtte: gulahallat ja ollásuhttit siidadoalu. Kautokeino, Norway. (In Sámisk)

Eythorsson, E., 1993. Sámi fjord fishermen and the state: traditional knowledge and resource management in northern Norway. In: J.T. Inglis (ed.). Traditional Ecological Knowledge: Concepts and Cases, pp. 132–142. International Program on Traditional Ecological Knowledge and International Development Research Centre, Ottawa.

Fancy, S.G. and R.G. White, 1985. Incremental cost of activity. In: R.J. Hudson and R.G. White (eds.). Bioenergetics of Wild Herbivores, pp. 143–159. CRC Press.

Fenge, R., 2001. The Inuit and climate change. Isuma, Winter:79–85.

Ferguson, M.A.D. and F. Messier, 1997. Collection and analysis of traditional ecological knowledge about a population of Arctic tundra caribou. Arctic, 50:17–28.

Folkow, L.P. and J.B. Mercer, 1986. Partition of heat loss in resting and exercising winter- and summer-insulated reindeer. American Journal of Physiology, 251:R32-R40.

Forchhammer, M.C., N.C. Stenseth, E. Post and R. Langvatn, 1998. Population dynamics of Norwegian red deer: density-dependence and climatic variation. Proceedings of the Royal Society of London B, 265:341–350.

Forchhammer, M.C., T.H. Clutton-Brock, J. Lindström and S.D. Albon, 2001. Climate and population density induce long-term cohort variation in a northern ungulate. Journal of Animal Ecology, 70:721–729.

Forchhammer, M.C., E. Post, N.C. Stenseth and D.M. Boertmann, 2002. Long-term responses in arctic ungulate dynamics to changes in climatic and trophic processes. Population Ecology, 44:113–200.

Ford, N., 2000. Communicating climate change from the perspective of local people: A case study from Arctic Canada. Journal of Development Communication, 1:93–108.

Forland, E.J., T.E. Skaugen, R.E. Benestad, I. Hanssen-Bauer and O.E. Tveito, 2004. Variations in thermal growing, heating, and freezing indices in the Nordic Arctic, 1900–2050. Arctic, Antarctic and Alpine Research, 36:347–356.

Fox, S.L., 1998. Inuit knowledge of climate and climate change. Department of Geography, University of Waterloo, 145p.

Fox, S.L., 2003. When the Weather is *Uggianaqtuq*: Inuit Observations of Environmental Change. A multi-media, interactive CD-ROM. Department of Geography, University of Colorado at Boulder. Distributed by the National Snow and Ice Data Center (NSIDC) and Arctic System Sciences (ARCSS), National Science Foundation.

Fox, S.L., 2004. When the Weather is *Uggianaqtuq*: Linking Inuit and Scientific Observations of Recent Environmental Change in Nunavut, Canada. PhD Thesis. Department of Geography, University of Colorado at Boulder.

Fox, S.L., D. Riedlinger and N. Thorpe, 2001. Inuit and Inuvialuit knowledge of climate change in the Canadian Arctic. In: J. Oakes (ed.). Native Voices in Research. University of Manitoba Native Studies Press.

Freeman, M.M.R. (ed.), 2000. Endangered Peoples of the Arctic: Struggles to Survive and Thrive. Greenwood Press.

Freeman, M.M.R. and L.N. Carbyn (eds.), 1988. Traditional knowledge and renewable resource management in northern regions. Boreal Institute for Northern Studies, Edmonton.

Freeman, M.M.R., L. Bogoslovskaya, R.A. Caulfield, I. Egede, I.I. Krupnik and M.G. Stevenson, 1998. Inuit Whaling and Sustainability. Sage Publications.

Frombert, A., M. Cleemann and L. Carlsen, 1999. Review on persistent organic pollutants in the environment of Greenland and Faroe Islands. Chemosphere, 38:3075–3093.

Gilmartin, T.J., B.R. Allenby and R.F. Lehman, 1996. Environmental Threats and National Security, An International Challenge to Science and Technology. University of California.

GloBio, 2002. The Arctic is getting more and more vulnerable, UNEP warns. UNEP GloBio: Mapping Human Impacts on the Biosphere. http://www.globio.info/press/2002-08-13.cfm. (14 Jan 2004)

Golding, D., 2001. Vulnerability. In: A.S. Goudie and D.J. Cuff (eds.). Encyclopedia of Global Change: Environmental Change and Human Society. Oxford University Press.

Hætta, J.I., O.K. Sara and I. Rushfeldt, 1994. Reindriften i Finnmark: lovgivning og distriktsinndeling. Reindriftsadminitrasjonen, Alta, Norway. 124p. (in Norwegian)

Handmer, J.W., S. Dovers and T.E. Downing, 1999. Societal vulnerability to climate change and variability. Mitigation and Adaptation Strategies for Global Change, 4:267–281.

Hanssen-Bauer, I., O.E. Tveito and E.J. Forland, 2000. Temperature scenarios for Norway. Empirical downscaling from ECHAM4/OPYC3. DNMI Klima Report 24/00, Norwegian Meteorological Institute, Oslo.

Hanssen-Bauer, I., E.J. Forland, J.E. Haugen and O.E. Tveito, 2003. Temperature and precipitation scenarios for Norway: comparison of results from dynamical and empirical downscaling. Climate Research, 25:15–27.

Harbsmeier, M., 2002. Bodies and voices from Ultima Thule: Inuit explorations of the Kablunat from Christian IV to Knud Rasmussen. In: M. Bravo and S. Sorlin (eds). Narrating the Arctic: a cultural history of Nordic scientific practices, pp. ix–373. Science History Publications.

Hellström, C., D. Chen, C. Achberger and J. Räisänen, 2001. A comparison of climate change scenarios for Sweden based on statistical and dynamical downscaling of monthly precipitation. Climate Research, 19:45–55.

Hild, C.M., 1995. The next step in assessing Arctic human health. Science of the Total Environment, 160/161:559–569.

Hofmann, R.R., 2000. Functional and comparative digestive system anatomy of Arctic ungulates. Rangifer, 20:71–81.

Holling, C.S., 1996. Engineering resilience versus ecological resilience. In: P. Schulze (ed.). Engineering Within Ecological Constraints, pp. 31–44. National Academy, Washington, D.C.

Holling, C.S., 2001. Understanding the complexity of economic, ecological and social systems. Ecosystems, 4:390–405.

Homer-Dixon, T.F. and J. Blitt (eds), 1998. Ecoviolence: Links Among Environment, Population, and Security. Rowman & Littlefield Publishers.

Hovelsrud-Broda, G.K., 1997. Arctic seal hunting households and the anti-sealing controversy. Research in Economic Anthropology, 18:17–34.

Hovelsrud-Broda, G.K., 1999. Contemporary seal hunting households: trade bans and subsidies. In: D.B. Small and N. Tannenbaum (eds.). At the Interface: The Household and Beyond. University Press of America.

Howard, A. and F. Widdowson, 1996. Traditional knowledge threatens environmental assessment. Policy Options, November:34–36.

Huntington, H.P., 1992. The Alaska Eskimo Whaling Commission and other cooperative marine mammal management organizations in Alaska. Polar Record, 28:119–126.

Huntington, H.P., 1998. Observations on the utility of the semi-directive interview for documenting traditional ecological knowledge. Arctic, 51:237–242.

Huntington, H.P., 2000. Using traditional ecological knowledge in science: methods and applications. Ecological Applications, 10:1270–1274.

Huntington, H.P. and the communities of Buckland, Point Lay, and Shatoolik, 1999. Traditional knowledge of the ecology of beluga whales (*Delphinapterus leucas*) in the eastern Chukchi and northern Bearing seas, Alaska. Arctic, 52:49–61.

IISD, 2000. Sila Alangotok: Inuit Observations on Climate Change, 2000. Videocassette. Directed by B. Dickie, produced by G. Ashford. International Institute for Sustainable Development.

IISD, 2001. Final Report: The Inuit Observations on Climate Change Project. International Institute for Sustainable Development, Winnipeg.

Ikonomou, M.G., S. Rayne and R.F. Addison, 2002. Exponential increases of the brominated flame retardants, polybrominated diphenyl ethers, in the Canadian Arctic from 1981 to 2000. Environmental Science and Technology, 36:1886–1892.

Inglis, J.T. (ed.), 1993. Traditional ecological knowledge: concepts and cases. International Program on Traditional Ecological Knowledge and International Development Research Centre, Ottawa.

IPCC, 1997. Summary for Policymakers. The Regional Impacts of Climate Change: An Assessment of Vulnerability. Intergovernmental Panel on Climate Change. Cambridge University Press.

IPCC, 2001a. Climate Change 2001: Synthesis Report. A Contribution of Working Groups I, II, and III to the Third Assessment Report of the Intergovernmental Panel on Climate Change. Watson, R.T., and the Core Writing Team (eds.). Cambridge University Press, 398 pp.

IPCC, 2001b. Climate Change 2001: Impacts, Adaptation, and Vulnerability: The Contribution of Working Group II to the Third Scientific Assessment of the Intergovernmental Panel on Climate Change. Cambridge University Press.

Jacobsen, E. and S. Skjenneberg, 1975. Some results from feeding experiments with reindeer. In: J.R. Luick, P.C. Lent, D.R. Klein and R.G. White (eds.). pp. 95–107, proceedings of the 1st International Reindeer/Caribou Symposium, Fairbanks, 1972. Biological papers of the University of Alaska, Special Report No.1. University of Alaska Fairbanks.

Jernsletten, N. 1997. Sámi Traditional Terminology: Professional Terms concerning Salmon, Reindeer and Snow. In: H. Gaski (ed.). Sami Culture in a New Era. Davvi Girji.

Johansen, P., G. Asmund and F. Riget, 2001. Lead contamination of seabirds harvested with lead shot – implications to human diet in Greenland. Environmental Pollution, 112:501–504.

Johansen, P., T. Pars and P. Bjerregaard, 2000. Lead, cadmium, mercury, and selenium intake by Greenlanders from local marine food. The Science of the Total Environment, 245:187–194.

Johnson, C.J., K.L. Parker and D.C. Heard, 2001. Foraging across a variable landscape: behavioural decisions made by woodland caribou at multiple spatial scales. Oecologia, 127:590–602.

Johnson, M. (ed.), 1992. Lore: Capturing Traditional Environmental Knowledge. Dene Cultural Institute and International Development, Ottawa.

Jolly, D., F. Berkes, J. Castleden, T. Nichols and the community of Sachs Harbour, 2002. We can't predict the weather like we used to: Inuvialuit observations of climate change, Sachs Harbor, Western Canadian Arctic. In: I. Krupnik and D. Jolly (eds). The Earth is Faster Now: Indigenous Observations of Arctic Environmental Change, pp. 92–125. Arctic Research Consortium of the United States, Fairbanks, Alaska.

Jolly, D., S. Fox and N. Thorpe, 2003. Inuit and Inuvialuit Knowledge of Climate Change in the Northwest Territories and Nunavut. In: J. Oakes and R. Riewe (eds.). Native Voices in Research: Northern and Native Studies, pp. 280–290. Native Studies Press.

Kasperson, J.X. and R.E. Kasperson, 2001. SEI Risk and Vulnerability Programme Report. 2001–01, Stockholm Environment Institute, Stockholm.

Kasperson, J.X., R.E. Kasperson, B.L. Turner, W. Hsieh and A. Schiller, 2003. Vulnerability to Global Environmental Change. In: A. Diekman, T. Dietz, C.C. Jaeger and E.A. Rosa (eds.). The Human Dimensions of Global Environmental Change, pp. 280–290. MIT Press.

Kasperson, R.E., O. Renn, P. Slovic, H. Brown, J. Emel, R. Goble, J.X. Kasperson and S. Ratick, 1988. The social amplification of risk: A conceptual framework. Risk Analysis, 8:177–187.

Kates, R.W., 1971. Natural hazards in human ecological perspective: hypotheses and models. Economic Geography, 47:438–451.

Kates, R.W., 1985. The interaction of climate and society. In: R.W. Kates, J.H. Ausubel and M. Berberian (eds.). Climate Impact Assessment, pp. 3–36. John Wiley & Sons.

Kates, R.W., J.H. Ausubel and M. Berberian (eds.), 1985. Climate Impact Assessment: Studies of the Interaction of Climate and Society. John Wiley & Sons.

Kay, R.N.B., 1985. Body size, patterns of growth, and efficiency of production in red deer. Bulletin of the Royal Society of New Zealand, 22:411–422.

Kelly, P.M. and W.N. Adger, 2000. Theory and practice in assessing vulnerability to climate change and facilitation adaptation. Climatic Change, 47:325–352.

Kilabuck, P., 1998. A Study of Inuit Knowledge of Southeast Baffin Beluga. Nunavut Wildlife Management Board, Iqaluit.

Kofinas, G. and the communities of Aklavik and Fort McPherson, 2002. Community Contributions to Ecological Monitoring: Knowledge Co-production in the U.S.-Canada Arctic Borderlands. In: I. Krupnik and D. Jolly (eds.). The Earth is Faster Now: Indigenous Observations of Arctic Environmental Change, pp. 54–91. Arctic Research Consortium of the United States, Fairbanks, Alaska.

Krupnik, I., 2000. Native perspectives on climate and sea ice changes. In: H. Huntington (ed.). Impact of changes in sea ice and other environmental parameters in the Arctic, pp. 25–39. Bethesda, MD: Marine Mammal Commission.

Krupnik, I. and D. Jolly (eds.), 2002. The Earth Is Faster Now: Indigenous Observations Of Arctic Environmental Change. Arctic Research Consortium of the United States, Fairbanks.

Kuhn, R.G. and F. Duerden, 1996. A review of traditional environmental knowledge: an interdisciplinary Canadian perspective. Culture, 16:71–84.

Kuhnlein, H.V. and H.M. Chan, 2000. Environment and contaminants in traditional food systems of northern indigenous peoples. Annual Review of Nutrition, 20:595–626.

Lange, M. and BASIS Consortium, 2003. The Barents Sea Impact Study (BASIS): methodology and first results. Continental Shelf Research, 23:1673–1684.

Larsen, T.S., N.Ö. Nilsson and A.S. Blix, 1985. Seasonal changes in lipotenesis and lipolysis in isolated adipocytes from Svalbard and Norwegian reindeer. Acta Physiologica Scandinavica, 123:97–104.

Leader-Williams, N., 1988. Reindeer on South Georgia: the ecology of an introduced population. Cambridge University Press, 319p.

Leichenko, R.M. and K.L. O'Brien, 2002. The dynamics of rural vulnerability to global change: the case of Southern Africa. Mitigation and Adaptation Strategies for Global Change, 7:1–18.

Lenart, E.A., R.T. Bowyer, J.V. Hoef and R. Ruess, 2002. Climate change and caribou: effects of summer weather on forage. Canadian Journal of Zoology, 80:664–678.

Lindberg, S.E., S.B. Brooks, C.J. Lin, K.J. Scott, M.S. Landis, R.K. Stevens, M. Goodsite and A. Richter, 2002. The dynamic oxidation of gaseous mercury in the Arctic atmosphere at polar sunrise. Environmental Science & Technology, 36. 1245–1256.

Linné, C.V., 1732. Linnæi iter Lapponicum Dei gratia institutum 1732:... avreste den 12 Maji, kom igen den 10 Oktober Uppsala, 1975 edition. Wahlström & Widstrand, 276p. (In Swedish)

Liverman, D., 1990. Vulnerability to global environmental change. In: J.X. Kasperson and R.E. Kasperson (eds.). Understanding Global Environmental Change: the Contributions of Risk Analysis and Management, pp.27–44. Clark University.

Lu, J.Y., W.H. Schroeder, L.A. Barrie, A. Steffen, H.E. Welch, K. Martin, L. Lockhart, R.V. Hunt, G. Boila and A. Richter, 2001. Magnification of atmospheric mercury deposition to polar regions in springtime: the link to tropospheric ozone depletion chemistry. Geophysical Research Letters, 28:3219–3222.

Lund, E., 2002. Rover018.xls. Norwegian Directorate for Nature Management. http://www.naturforvaltning.no/archive/attachments/01/36/rover018.xls. (In Norwegian). (14 Jan 2004)

Macdonald, R.W., T. Harner, J. Fyfe, H. Loeng, and T. Weingartner, 2003. AMAP Assessment 2002: The Influence of Global Change on Contaminant Pathways to, within, and from the Arctic. Arctic Monitoring and Assessment Programme, Oslo, Norway. xii+65p.

Mariussen, Å. and K. Heen, 1998, Dependency, uncertainty and climate change: the case of fishing. In: Barents Sea Impact Study (BASIS), Proceedings from the First International BASIS Research Conference, St. Petersburg, Russia, February 22–25, 1998.

Mathiesen, S.D., A. Rognmo and A.S. Blix, 1984. A test of the usefulness of a commercially available mill waste product (AB-84) as feed for starving reindeer (*Rangifer tarandus tarandus*). Rangifer, 4:28–34.

Mathiesen, S.D., T.H.A. Utsi and W. Sørmo, 1999. Forage chemistry and the digestive system in reindeer (*Rangifer tarandus tarandus*) in northern Norway and on South Georgia. Rangifer, 19:91–101.

Mathiesen, S.D., Ø.E. Haga, T. Kaino and N.J.C. Tyler, 2000. Diet composition, rumen papillation and maintenance of carcass mass in female Norwegian reindeer (*Rangifer tarandus tarandus*). Journal of Zoology (London), 251:129–138.

McCay, B.J., 1997. Systems ecology, people ecology and anthropology of fishing communities. Human Ecology, 6:397–422.

McDonald, M., L. Arragutainaq and Z. Novalinga, 1997. Voices from the Bay: Traditional Ecological Knowledge of Inuit and Cree in the Hudson Bay Bioregion. Canadian Arctic Resources Committee and Environmental Committee of the Municipality of Sanikiluaq, Ottawa.

McEwan, E.H., 1968. Growth and development of the barren-ground caribou. II. Postnatal growth rates. Canadian Journal of Zoology, 46:1023–1029.

McEwan, E.H. and P.E. Whitehead, 1970. Seasonal changes in the energy and nitrogen intake in reindeer and caribou. Canadian Journal of Zoology, 48:905–913.

Mesteig, K., N.J.C. Tyler and A.S. Blix, 2000. Seasonal changes in heart rate and food intake in reindeer (*Rangifer tarandus tarandus*). Acta Physiologica Scandinavica, 170:145–151.

Miller, F.L., E.J. Edmonds and A. Gunn, 1982. Foraging behaviour of Peary caribou in response to springtime snow and ice conditions. Occasional 48, Canadian Wildlife Service, Ottawa.

Moen, R., M.A. Olsen, Ø.E. Haga, W. Sørmo, T.H.A. Utsi and S.D. Mathiesen, 1998. Digestion of timothy silage and hay in reindeer. Rangifer, 18:35–45.

Muga, D.A., 1987. Effect of technology on an indigenous people: the case of the Norwegian Sámi. Journal of Ethnic Studies, 14:1–24.

Muir, D.C.G., B. Braune, B. DeMarch, R. Norstrom, R. Wagemann, L. Lockhart, B. Hargrave, D. Bright, R. Addison, J. Payne and K. Reimer, 1999. Spatial and temporal trends and effects of contaminants in the Canadian arctic marine ecosystem: a review. The Science of the Total Environment, 230:83–144.

Murra, J., 1975. Formaciones economicas y politicas del mundo andino. Instituto de Estudios Peruanos, Lima. (In Spanish)

Mustonen, T., 2002. Appendix: SnowChange 2000, Indigenous View on Climate Change, A Circumpolar Perspective. In: I. Krupnik and D. Jolly (eds.). The Earth is Faster Now: Indigenous Observations of Arctic Environmental Change, pp. 351–356. Arctic Research Consortium of the United States, Fairbanks.

Mymrin, N.S. and the communities of Novoe Chaplino, and Yanrakinnot. 1999. Traditional knowledge of the ecology of beluga whales (*Delphinapterus leucas*) in the northern Bering Sea, Chukotka, Russia. Arctic 52:62–70.

Mysterud, A., N.G. Yoccoz, N.C. Stenseth and R. Langvatn, 2000. Relationships between sex ratio, climate and density in red deer: the importance of spatial scale. Journal of Animal Ecology, 69:959–974.

Mysterud, A., N.C. Stenseth, N.G. Yoccoz, R. Langvatn and G. Steinhelm, 2001. Nonlinear effects of large-scale climatic variability on wild and domestic herbivores. Nature, 410:1096–1099.

Nadasdy, P., 1999. The politics of TEK: power and 'integration' of knowledge. Arctic Anthropology, 36:1–18.

Nadasdy, P., 2003. Reevaluating the co-management success story. Arctic, 56(4):367–380.

NAST, 2000. Climate Change Impacts on the United States: The Potential Consequences of Climate Variability and Change. A Report of the National Assessment Synthesis Team, US Global Change Research Program. Cambridge University Press.

Nellemann, C., I. Vistnes, P. Jordhøy, O. Strand and A. Newton, 2003. Progressive impacts of piecemeal development on wild reindeer. Biological Conservation, 101:351–360.

Nieminen, M. and U. Heiskari, 1989. Diets of freely grazing and captive reindeer during summer and winter. Rangifer, 9:17–34.

Nilssen, K.J., J.A. Sundsfjord and A.S. Blix, 1984. Regulation of metabolic rate in Svalbard reindeer (*Rangifer tarandus platyrhynchus*) and Norwegian reindeer (*Rangifer tarandus tarandus*). American Journal of Physiology, 247:R837–R841.

NRC, 1999. Our Common Journey: A Transition Toward Sustainability. National Research Council, National Academy Press, Washington, DC.

Nriagu, J.O., 1996. A history of global metal pollution. Science, 272:223–224.

Nuttall, M., 2000. Indigenous peoples, self-determination and the Arctic environment. In: M. Nuttall and T.V. Callaghan (eds.). The Arctic: Environment, People, Policy, pp. 377–410. Harwood Academic Publishers.

Nuttall, M., 2002. Barriers to Sustainability. WWF Arctic Bulletin, 202:10–14.

O'Brien, K.L., L. Sygna and J.E. Haugen, 2004. Vulnerable or resilient? Multi-scale assessments of the impacts of climate change in Norway. Climatic Change, 64 (1–2): pp. 193–225.

Odland, J.Ø., E. Nieboer, N. Romanova, Y. Thomassen and E. Lund, 1999. Blood lead and cadmium and birth weight among sub-arctic and arctic populations of Norway and Russia. Acta Obstetricia et Gynecologica Scandinavica, 78:852–860.

Oksanen, A., 1999. Endectocide treatment of the reindeer. Rangifer Special Issue:1–217.

Oskal, N., 2000. On nature and reindeer luck. Rangifer, Research, Management and Husbandry of Reindeer and other Northern Ungulates, 20:2–3.

Ottersen, G., B. Planque, A. Belgrano, E. Post, P.C. Reid and N.C. Stenseth, 2001. Ecological effects of the North Atlantic Oscillation. Oecologia, 128:1–14.

Paine, R. (ed.), 1977. The White Arctic. Memorial University of Newfoundland, St. John's, Newfoundland.

Paine, R., 1996. Saami reindeer pastoralism and the Norwegian state, 1960s–1990s. Nomadic Peoples, 38:125–135.

Palmer, T.N. and J. Räisänen, 2002. Quantifying the risk of extreme seasonal precipitation event: a changing climate. Nature, 415:512–514.

Parry, M.L., 1978. Climate Change, Agriculture and Settlement. Dawson UK Ltd.

Parry, M.L. (ed.), 2000. Assessment of Potential Effects and Adaptations for Climate Change in Europe: The Europe ACACIA Project. Jackson Environment Institute, University of East Anglia, 324p.

Parry, M. and T. Carter, 1998. Climate Impact and Adaptation Assessment: A Guide to the IPCC Approach. Earthscan Publications, Ltd., 166p.

Parry, M.L., M. Livermore, C. Rosenzweig, A. Iglesias and A. Fischer, 1998. The Impact of Climate on Food Supply. In: Climate Change and its Impacts: Some Highlights from the Ongoing UK Research Programme. Department of Environment, Transport, and Regions, with UK Public Meteorological Service Programme.

Pars, T., 1997. Dietary Habits in Greenland. In: E.B. Thorling and J.C. Hanson (eds.). Proceedings from Food Tables for Greenland Seminar, Copenhagen, January 7–8, 1997. Arctic Research Journal, Nuuk, Greenland. http://www.cam.gl/engelsk/inussuk/4foodtab/index.html.

Peary, R.E., 1907. Nearest the Pole: A narrative of the polar expedition of the Peary Arctic Club in the S.S. Roosevelt, 1905–1906. Doubleday, Page & Co.

Pentha, S.M., N.J. C. Tyler, P.A. Wookey and S.D. Mathiesen, 2001, Experimental warming of reindeer forage plants in situ increases their growth, 11th Nordic Conference on Reindeer Research, Kaamanen, Rangifer Report 5: 71 (Abstract).

Pinkerton, E. (ed.), 1989. Cooperative management of local fisheries. University of British Columbia Press.

Polsky, C., D. Schröter, A. Patt, S. Gaffin, M. Long Martello, R. Neff, A. Pulsipher and H. Selin, 2003. Assessing Vulnerabilities to the Effects of Global Change: An Eight-Step Approach. Research and assessment systems for sustainability program discussion paper 2003–05, Environment and Natural Resources Program, Harvard University.

Post, E. and N.C. Stenseth, 1999. Climate variability, plant phenology, and northern ungulates. Ecology, 80:1322–1339.

Post, E., M.C. Forchhammer, N.C. Stenseth and T. Callaghan, 2001. The timing of life-history events in a changing climate. Proceedings of the Royal Society of London B, 268:15–23.

Priutt, W.O., 1959. Snow as a factor in the winter ecology of the barren-ground caribou (Rangifer arcticus). Arctic, 12:158–179.

Pruitt, W.O., 1979. A numerical 'snow index' for reindeer (Rangifer tarandus) winter ecology (Mammalia, Cervidaea). Annales Zoologici Fennici, 16:271–280.

Räisänen, J., 2001. Intercomparison of 19 Global Climate Change Simulations from an Arctic Perspective. In: E. Källen, V. Kattsov, J. Walsh and E. Weatherhead (eds.). Report from the Arctic Climate Impact Assessment Modeling and Scenario Workshop, pp. 11–13. Stockholm, Sweden.

Rasmussen, K., 1908. The people of the polar north: a record. J.B. Lippincott Co.

Reidlinger, D. and F. Berkes, 2001. Contributions of traditional knowledge to understanding climate change in the Canadian Arctic. Polar Record, 37:315–328.

Reilly, J.M. and D. Schimmelpfenning, 1999. Agricultural impact assessment, vulnerability, and the scope for adaptation. Climatic Change, 42:745–748.

Reindriftsforvaltningen, 2002. Ressursregnskap for reindriftsnaeringen for reindrftsåret 1 April 2000–31 March 2001. Reindriftsforvaltningen, Alta. 143p. (In Norwegian)

Reinert, E. and H. Reinert, in press. The economics of reindeer herding: Saami entrepreneurship between cyclical sustainability and the powers of state and oligopolies. In: L.P. Dana, R. Anderson (eds.), Theory of Indigenous Entrepreneurship: The Edward Elgar Handbook of Research on Indigenous Enterprise. Edward Elgar.

Resilience Alliance. no date. http://www.resalliance.org/ev_en.php. (14 Jan 2004)

Riedlinger, D., 2000. Inuvialuit knowledge of climate change. In: J.E.A. Oakes (ed.). Pushing Margins: Northern and Native Research, pp. 346–355. Manitoba Native Studies Press.

Roos, A., E. Greyerz, M. Olsson and F. Sandegren, 2001. The otter (Lutra lutra) in Sweden – population trends in relation to sum-DDT and total PCB concentrations during 1968–1999. Environmental Pollution, 111:457–469.

Ruong, I., 1964. 'Jåhkåkaska sameby'. Svenska Landsmål och Svenskt Folkliv. (In Swedish)

Ruong, I., 1967. The Lapps in Sweden. Swedish Institute for Cultural Relations with Foreign Countries, Stockholm.

Ryd, Y., 2001. Snö : en renskötare berättar. Ordfront, Stockholm. 326p. (In Swedish)

Ryg, M. and E. Jacobsen, 1982. Effects of castration on growth and food intake cycles in young male reindeer (Rangifer tarandus tarandus). Canadian Journal of Zoology, 60:942–945.

Scannet. no date. Scandinavian l North European Network of Terrestrial Field Bases. http://www.envicat.co/scannet/Scannet. (14 Jan 2004)

Schei, A. no date. Norsk rovviltforvaltning og folkeretten. Norwegian Ministry of the Environment. http://odin.dep.no/md/rovviltmelding/hvaskjer/utredninger/099001-990034/index-dok000-b-n-a.html. (In Norwegian). (14 Jan 2004)

Schroeder, W.H., L.A. Anlauf, L.A. Barrie, J.Y. Lu, A. Steffen, D.R. Schneeberger and T. Berg, 1998. Arctic springtime depletion of mercury. Nature, 394:331–332.

Scotter, G.W., 1965. Chemical composition of forage lichens from northern Saskatchewan as related to use by barren-ground caribou. Canadian Journal of Plant Science, 45:246–250.

Selin, H., 2000. Towards International Chemical Safety: Taking Action on Persistent Organic Pollutants (POPs). Linköping Studies in Arts and Science, Linköping.

Selin, H., 2003. Regional POPs Policy: The UNECE/CLRTAP POPs Agreement. In: D.L. Downie and T. Fenge (eds.). Northern Lights against POPs: Combating Toxic Threats in the Arctic, pp. 111–132. Montreal: McGill-Queens University Press.

Selin, H. and N. Eckley, 2003. Science, politics and persistent organic pollutants: scientific assessments and their role in international environmental negotiations. International Environmental Agreements: Politics, Law, and Economics, 3:17–42.

Sen, A., 1981. Poverty and Famines: An Essay on Entitlements and Deprivation. Oxford University Press.

Silapaswan, C.S., D.L. Verbyla and A.D. McGuire, 2001. Land cover change on the Seward Peninsula: the use of remote sensing to evaluate the potential influences of climate warming on historical vegetation dynamics. Canadian Journal of Remote Sensing, 27:542–554.

Skogland, T., 1978. Characteristics of the snow cover and its relationship to wild mountain reindeer (Rangifer tarandus tarandus) feeding strategies. Arctic and Alpine Research, 10:569–580.

Sletten, H. and K. Hove, 1990. Digestive studies with a feed developed for realimentation of starving reindeer. Rangifer, 10:31–38.

Slezkine, Y., 1994. Arctic Mirrors: Russia and the Small Peoples of the North. Cornell University Press, 456p.

Smith, E.R., 2000. An Overview of EPA's Regional Vulnerability Assessment (ReVA) Program. Environmental Monitoring and Assessment, 64:9–15.

Smithers, J. and B. Smit, 1997. Human adaptation to climatic variability and change. Global Environmental Change, 7:129–146.

Solberg, E.J., P. Jordhøy, O. Strand, R. Aanes, A. Loison, B.-E. Sæther and J.D.C. Linnell, 2001. Effects of density-dependence and climate on the dynamics of a Svalbard reindeer population. Ecography, 24:441–451.

Sørmo, W., Ø.E. Haga, E. Gaare, R. Langvatn and S.D. Mathiesen, 1999. Forage chemistry and fermentation chambers in Svalbard reindeer (Rangifer tarandus platyrhynchus). Journal of Zoology (London), 247:247–256.

Statistics Greenland, 1997. Greenland 1997. The Greenlandic Society. http://www.statgreen.gl/english/publ/yearbook/1997/index.html (14 Jan 2004)

Stefansson, V., 1941. My Life with the Eskimo. Collier Books.

Stenbaek, M., 1987. Forty years of cultural change among the Inuit in Alaska, Canada and Greenland: some reflections. Arctic, 40:300–309.

Stenseth, N.C., A. Mysterud, G. Ottersen, J.W. Hurrell, K.-S. Chan and M. Lima, 2002. Ecological effects of climate fluctuations. Science, 297:1292–1296.

Stevenson, M.G., 1996a. Indigenous knowledge and environmental assessment. Arctic, 49:278–291.

Stevenson, M.G., 1996b. Ignorance and prejudice threaten environmental assessment. Policy Options, 18:25–28.

Storeheier, P.V., S.D. Mathiesen, N.J.C. Tyler and M.A. Olsen, 2002a. Nutritive value of terricolous lichens to reindeer in winter. The Lichenologist, 34:247–257.

Storeheier, P.V., S.D. Mathiesen, N.J.C. Tyler, I. Schjelderup and M.A. Olsen, 2002b. Utilization of nitrogen and mineral-rich vascular forage plants by reindeer in winter. Journal of Agricultural Science, 139:151–160.

Storeheier, P.V., J. Sehested, L. Diernaes, M.A. Sundset and S.D. Mathiesen, 2003. Effects of seasonal changes in food quality and food intake on the transport of sodium and butyrate across ruminal eupithelium of reindeer. Journal of Comparitive Physiology B doi 10.1007/s00360-003-0345-9.

Strøm Bull, K., 2001. Forslag til endringer i reindriftsloven. Norges offentlige utredninger 2001: 35. Statens forvaltningstjeneste. Statens forvaltningstjeneste, Oslo. 218p. (In Norwegian)

Strøm Bull, K., N. Oskal and M.N. Sara, 2001. Reindeer herding in Finnmark, history of law 1852–1960. Cappelen Akademic Press, 376 pp. (In Norwegian)

Strum, M., C. Racine and K. Tape, 2001. Climate change. Increasing shrub abundance in the Arctic. Nature, 411:546–547.

Svensson, T.G., 1987a. Industrial developments and the Sámi: ethnopolitical response to ecological crisis in the North. Anthropologica, 29:131–148.

Svensson, T.G., 1987b. Patterns of transformation and local self-determination: ethnopower and the larger society in the North, the Sámi case. Nomadic Peoples, 24:1–13.

Swedish Museum of Natural History Contaminant Research Group, 2000. Time trends of metals in liver and muscle of reindeer (*Rangifer tarandus*) from Lapland, north-western Sweden, 1983–1998: Swedish monitoring programme in terrestrial biota. Swedish Museum of Natural History, Stockholm.

Thorpe, N., S. Eyegetok, N. Hakongak and elders, 2001. Thunder on the Tundra: Inuit Qaujimajatuqangit of the Bathurst Caribou. Generation Printing.

Thorpe, N., S. Eyegetok, N. Hakongak and elders, 2002. Nowadays it's not the same: Inuit Qaujimajatuqangit, climate and caribou in the Kitikmeot Region of Nunavut, Canada. In: I. Krupnik and D. Jolly (eds.). The Earth is Faster Now: Indigenous Observations of Arctic Environmental Change, pp. 1998–1239. Arctic Research Consortium of the United States, Fairbanks, Alaska.

Timmerman, P., 1981. Vulnerability, Resilience and the Collapse of Society. Institute for Environmental Studies, University of Toronto, 42p.

Tryland, M., T.D. Josefsen, A. Oksanen and A. Ashfalk, 2001. Contagious ecthyma in Norwegian semidomesticated reindeer (*Rangifer tarandus tarandus*). Veterinary Record, 149:394–395.

Turi, J.M., 1994. A planet of reindeer. Reindriftsnytt, 2/3:32–37. (In Norwegian)

Turi, J.M., 2002. The world reindeer livelihood – current situation, treats and possibilities. In: S. Kankaanpaa, K. Muller-Willie, P. Susiluoto and M.-L. Sutinen (eds.). The Northern Timberline Forest: Environmental and Socio-Economic Issues and Concerns. Finnish Forest Research Institute, Helsinki.

Turner, B.L., II, R.E. Kasperson, P. Matson, J.J. McCarthy, R.W. Corell, L. Christensen, N. Eckley, J.X. Kasperson, A. Luers, M.L. Martello, C. Polsky, A. Pulsipher and A. Schiller, 2003a. A Framework for Vulnerability Analysis in Sustainability Science. Proceedings of the National Academy of Sciences, 100:8074–8079.

Turner, B.L., II, P. Matson, J.J. McCarthy, R.W. Corell, L. Christensen, N. Eckley, G. Hovelsrud-Broda, J.X. Kasperson, R.E. Kasperson, A. Luers, M.L. Martello, S. Mathiesen, R. Naylor, C. Polsky, A. Pulsipher, A. Schiller, H. Selin and N. Tyler, 2003b. Illustrating the coupled human-environment system for vulnerability analysis: Three Case Studies. Proceedings of the National Academy of Sciences, 100:8080–8085.

Tyler, N.J.C., 1987. Body composition and energy balance of pregnant and non-pregnant Svalbard reindeer during winter. Zoological Society of London Symposia, 57:203–229.

Tyler, N.J.C. and A.S. Blix, 1990. Survival strategies in Arctic ungulates. Rangifer Special Issue, 3:211–230.

Tyler, N.J.C. and M. Jonasson, 1993, Reindeer herding in Norway, Christchurch, New Zealand. A salute to world deer farming. Proceedings of the 1st World Deer Congress, Christchurch, New Zealand, February 1993, pp. 51–56.

Tyler, N.J.C., P. Fauchald, O. Johansen and H.R. Christiansen, 1999a. Seasonal inappetence and weight loss in female reindeer in winter. Ecological Bulletins, 47:105–116.

Tyler, N.J.C., M.C. Forchhammer and N.A. Øritsland, 1999b. Persistent instability in Svalbard reindeer dynamics: density-dependence in a fluctuating environment. 10th Arctic Ungulate Conference, 9–12 August 1999, University of Tromsø, (Abstract 94).

Uggerud, K. 2001. Rovdyrvern og menneskerettigheter i reindriftens områder. Betenkning utarbeidet etter oppdrag av Norske reindriftsamers landsforbund. Noørjen Båatsoesaemiej Rijhkesaervie. http://www.same.net/~nbr/betenkningen.doc. (In Norwegian). (14 Jan 2004)

UNECE, 1994. State of Knowledge Report of the UNECE Task Force on Persistent Organic Pollutants. United Nations Economic Commission for Europe.

UNEP, 2001. Global methodology for mapping human impacts on the biosphere. UNEP/DEWA/TR.01-3.

Usher, P., 1993. The Beverly-Kaminuiak Caribou Management Board: an experience in co-management. In: J.T. Inglis (ed.). Traditional Ecological Knowledge: Concepts and Cases, pp. 111–120. International Program on Traditional Ecological Knowledge and International Development Research Centre, Ottawa.

Usher, P., 2000. Tradition ecological knowledge in environment assessment and management. Arctic, 53:183–193.

Van der Wal, R., N. Madan, S. van Lieshout, C. Dormann, R. Langvatn and S.D. Albon, 2000. Trading forage quality for quantity? Plant phenology and patch choice by Svalbard reindeer. Oecologica, 123:108–115.

Walker, B., S. Carpenter, J. Anderies, N. Abel, G. Cumming, M. Janssen, L. Lebel, J. Norberg, G.D. Peterson and R. Pritchard, 2002. Resilience management in social-ecological systems: a working hypothesis for a participatory approach. Conservation Ecology, 6(1):14. [online] URL: http://www.consecol.org/vol6/iss1/art14

Walther, G.-R., E. Post, P. Convey, A. Menzel, C. Parmesan, T.J.C. Beebee, J.M. Fromentin, O. Hoegh-Guldberg and F. Bairlein, 2002. Ecological responses to recent climate change. Nature, 416:389–395.

Wania, F. and D. Mackay, 1996. Tracking the distribution of persistent organic pollutants. Environmental Science and Technology, 30:A390–A396.

Weladji, R.B., D.R. Klein, O. Holand and A. Mysterud, 2002. Comparative response of *Rangifer tarandus* and other northern ungulates to climatic variability. Rangifer, 22:33–50.

Wenzel, G., 1984. L'ecologie culturelle et les Inuit du Canada: une approche applique. Etudes Inuit Studies, 8:89–101. (In French)

Wheelersburg, R.P., 1987. New transportation technology among Swedish Sámi reindeer herders. Arctic Anthropology, 24:99–116.

White, G.F., 1974. Natural Hazards: Local, National, Global. Oxford University Press.

Winnefield, J.A. and M.E. Morris, 1994. Where Environmental Concerns and Security Strategies Meet: Green Conflict in Asia and the Middle East. RAND, Santa Monica.

Young, O., 1998b. Creating Regimes: Arctic Accords and International Governance. Cornell University Press.

Young, O.R., 1998b. Emerging priorities for sustainable development in the circumpolar north. The Northern Review, Special Issue, 18:38–46.

Young, O., 2000. The structure of Arctic cooperation: solving problems/seizing opportunities. A paper prepared at the request of Finland in preparation for the fourth Conference of Parliamentarians of the Arctic Region. Finnish chairmanship of the Arctic Council during the period 2000–2002, Rovaniemi. http://www.arcticparl.org/resource/images/conf4_sac.pdf

Chapter 18

Summary and Synthesis of the ACIA

Lead Author
Gunter Weller

Contributing Authors
Elizabeth Bush, Terry V. Callaghan, Robert Corell, Shari Fox, Christopher Furgal, Alf Håkon Hoel, Henry Huntington, Erland Källén, Vladimir M. Kattsov, David R. Klein, Harald Loeng, Marybeth Long Martello, Michael MacCracken, Mark Nuttall, Terry D. Prowse, Lars-Otto Reiersen, James D. Reist, Aapo Tanskanen, John E. Walsh, Betsy Weatherhead, Frederick J. Wrona

Contents

18.1. Introduction

This chapter provides a brief summary of the main conclusions of the seventeen preceding chapters of the Arctic Climate Impact Assessment (ACIA). The chapter has three main parts. The first part contains the conclusions of the assessment discussed chapter by chapter. Observed climate and ultraviolet (UV) radiation trends (Chapters 2 and 5) are summarized, using both scientific and indigenous (Chapter 3) observations, and information from the latest assessments by the Arctic Monitoring and Assessment Programme (AMAP, 1998) and the Intergovernmental Panel on Climate Change (IPCC, 2001). Projections of climate change over the 21st century, based on emissions scenarios and computer model simulations (Chapter 4) are described, as are the observed and projected changes in stratospheric ozone and UV radiation levels (Chapter 5). Next, the chapter summarizes arctic-wide consequences of climate change, by examining impacts on the environment (Chapters 6, 7, 8, and 9), on people's lives (Chapters 10, 11, 12, 15, and 17), and on economic sectors of importance in the Arctic (Chapters 13, 14, and 16). These impacts cut across the entire Arctic and are generally not dependent on resolving regional details. For example, the timing, intensity, and magnitude of the melting of snow and ice in a warmer climate will have widespread implications for the entire Arctic and the global environment, even if these changes vary regionally.

Projected major large-scale environmental changes in the Arctic are illustrated in Fig. 18.1, which shows the existing and projected boundaries for summer sea-ice extent, permafrost, and the treeline. The likely changes associated with these shifts are many and dramatic, as described in the preceding chapters of this assessment. For example, the map shows that the treeline is projected to reach the Arctic Ocean in most of Asia and western North America by the end of the century. This is likely to lead to a near total loss of tundra vegetation in these areas, with important consequences for many types of wildlife. The consequences of the permafrost thawing and sea-ice reductions shown in Fig. 18.1 are equally dramatic.

The second part of the chapter is a synthesis of impacts on a local and regional basis, providing details on four different regions of the Arctic. A regional emphasis is necessary because the Arctic covers a large area and so experiences significant regional variations in the changes in climate that will lead to different impacts and responses. Different regions also have different social, economic, and political systems, which will each be influenced differently, causing vulnerability and impacts to differ to a large extent on the basis of geopolitical and cultural boundaries. The four regions for which results are presented are:

- Region 1: East Greenland, the North Atlantic, northern Scandinavia, and northwestern Russia;
- Region 2: Siberia;

- Region 3: Chukotka, the Bering Sea, Alaska, and the western Canadian Arctic; and
- Region 4: the central and eastern Canadian Arctic, the Labrador Sea, Davis Strait, and West Greenland.

The rationale for selecting these four broad regions includes climatic, social, and other factors, and is discussed in section 18.3.

The final part of the chapter addresses cross-cutting issues that are important in the Arctic. These are discussed in several chapters of the assessment, although usually in the context of the main topic of the chapter, and include the carbon cycle, biodiversity, and extreme and abrupt climate change.

Changes in climate and UV radiation in the Arctic will not only have far-reaching consequences for the arctic environment and its peoples, but will also affect the rest of the world, including the global climate. The connections include arctic sources of change affecting the globe, e.g., feedback processes affecting the global climate, sea-level rise resulting from melting of arctic glaciers and ice sheets, and arctic-triggered changes in the global thermohaline circulation of the ocean.

The Arctic is also important to the global economy. There are large oil and gas and mineral reserves in many parts of the Arctic, and arctic fisheries are among the most productive in the world, providing food for millions (see section 18.2.2.3). Climate change is likely to benefit north–south connections, including shipping, the global economy, and migratory birds, fish, and mammals that are important conservation species in the south. The Arctic plays a unique role in the global context and climate change in the Arctic has consequences that extend well beyond the Arctic.

18.2. A summary of ACIA conclusions

18.2.1. Changes in climate and UV radiation

18.2.1.1. Observed climate change

The climate of the Arctic has undergone rapid and dramatic shifts in the past and there is no reason that it could not experience similar changes in the future. Past changes show climatic cycles that have occurred regularly on time scales from decades to centuries and longer and are most likely to have been caused by oceanic and atmospheric variability and variations in solar intensity. Examples of long-term cooler and warmer climates were the Little Ice Age and the Medieval Warm Period, respectively, while short-term decadal cycles like the North Atlantic Oscillation and Pacific Decadal Oscillation, among others, have also been found to affect the arctic climate. Since the industrial revolution in the 19th century, anthropogenic greenhouse gas (GHG) emissions have added another major climate driver. In the 1940s, the Arctic experienced a warm period, like the rest of the planet, although it did

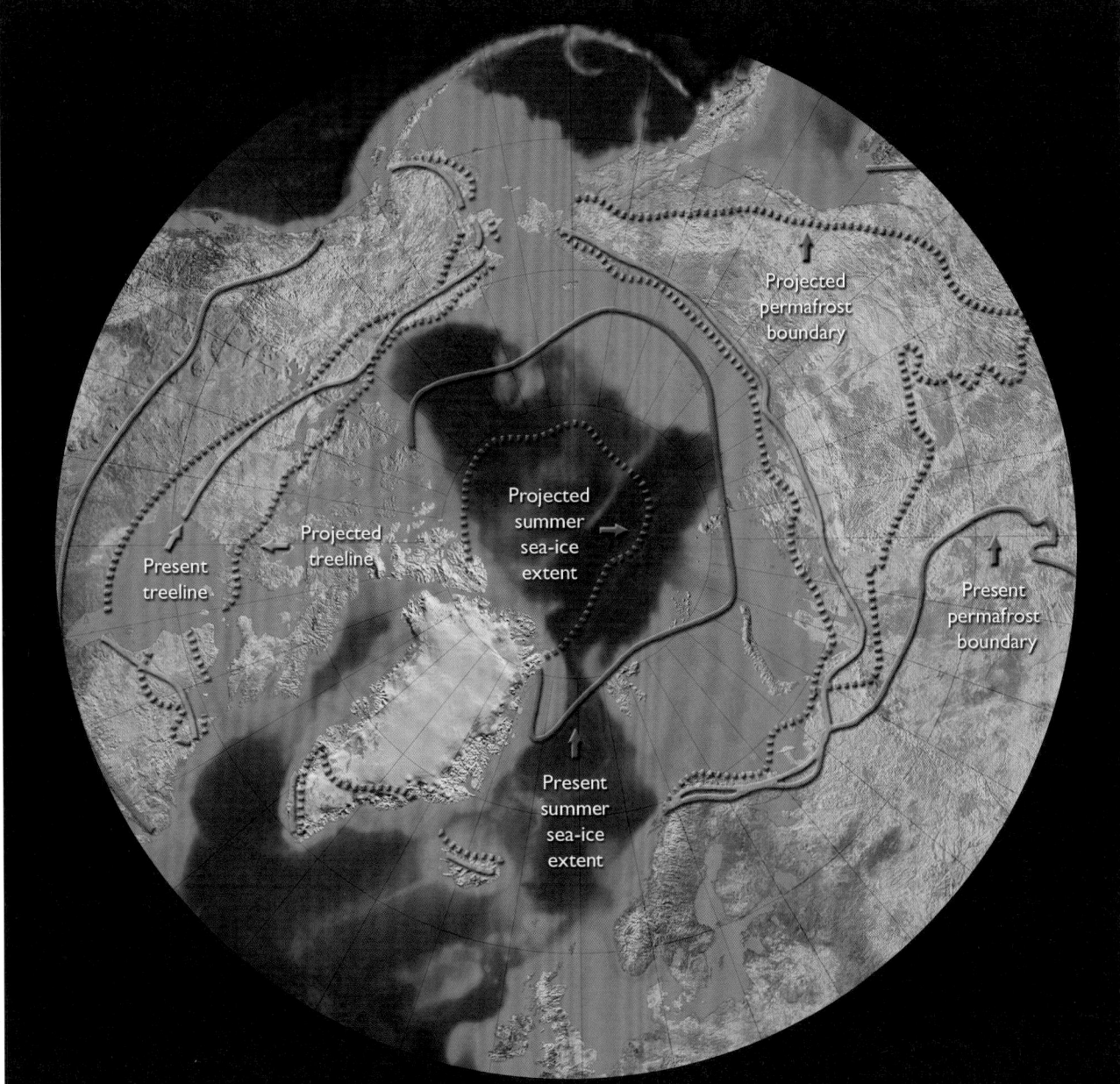

Fig. 18.1. Present and projected boundaries of summer sea-ice extent, permafrost, and the treeline. The changes are projected to occur over different time periods. Changes in summer sea-ice extent will occur by the end of the century, as projected by the five-model composite used by the ACIA (section 6.3, Figs. 6.3b and 6.9c). The projected changes in the treeline by the end of the century are from a vegetation model driven by output from the Hadley Centre model (section 7.1.1, Fig. 7.2 and section 7.5.3.2, Fig. 7.32). The change in the permafrost boundary assumes that the present areas of discontinuous permafrost (section 6.6.1, Fig. 6.21, although published sources differ) will be free of any permafrost in the future; this is likely to occur beyond the 21st century but it is not certain how long it will take.

not reach the level of the warming experienced in the 1990s. The IPCC stated that most of the global warming observed over the last 50 years is attributable to human activities (IPCC, 2001), and there is new and strong evidence that in the Arctic much of the observed warming over this period is also due to human activities.

Chapter 2 discusses the arctic climate system and observed changes in arctic climate over recent decades. Many types of observations indicate that the climate of the Arctic is changing. For example, air temperatures are generally warmer, the extent and duration of snow and sea ice are diminishing, and permafrost is thawing. However, there are also some regions where cooling has

occurred, and some areas where precipitation has increased. Reconstruction of the history of arctic climate over thousands to millions of years indicates that there have been very large changes in the past. Based on these indications that the arctic climate is sensitive to changes in natural forcing factors, it is very likely that human-induced factors, for example the rise in GHG concentrations and consequent enhancement of the global greenhouse effect, will lead to very large changes in climate, indeed, changes that will be much greater in the Arctic than at middle and lower latitudes.

The observed temperature changes in the Arctic over the five-decade period from 1954 to 2003 are shown in

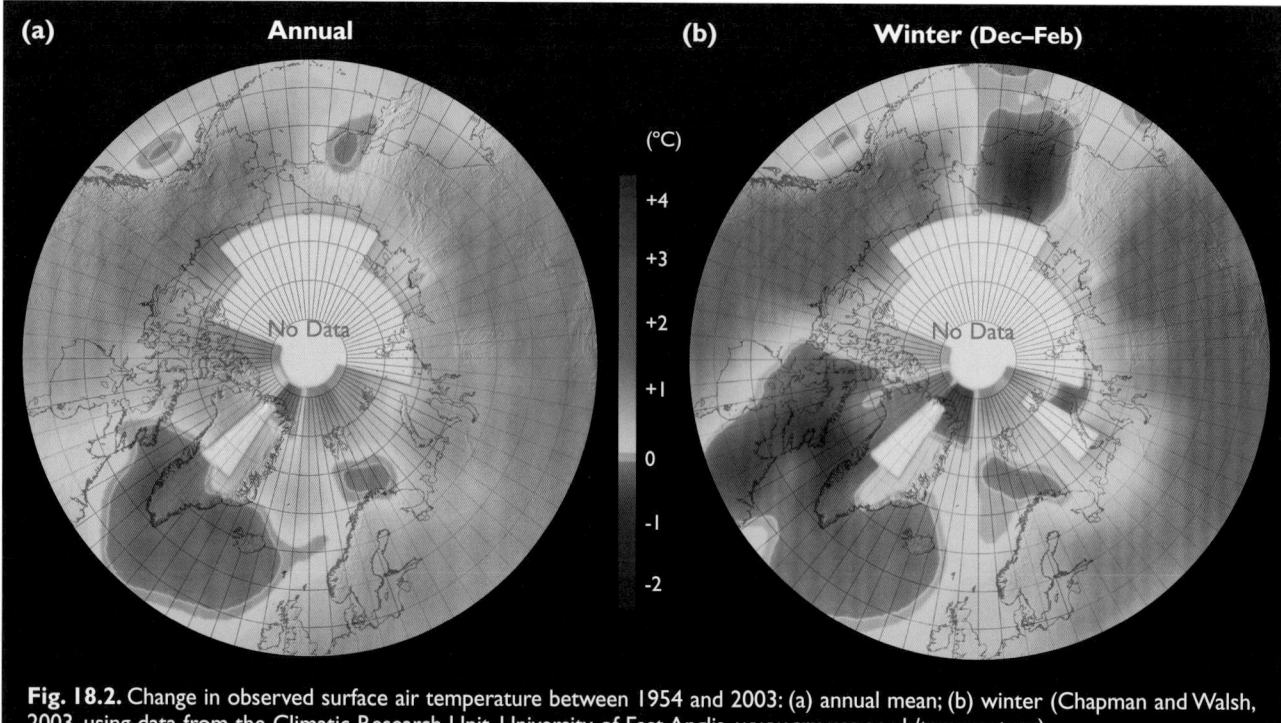

Fig. 18.2. Change in observed surface air temperature between 1954 and 2003: (a) annual mean; (b) winter (Chapman and Walsh, 2003, using data from the Climatic Research Unit, University of East Anglia, www.cru.uea.ac.uk/temperature).

Fig. 18.2. Owing to natural variations and the complex interactions of the climate system, the observed trends show variations within each region. Mean annual atmospheric surface temperature changes range from a 2 to 3 °C warming in Alaska and Siberia to a cooling of up to 1 °C in southern Greenland. Winter temperatures are up to 4 °C warmer in Siberia and in the western Canadian Arctic.

Although some regions have cooled slightly the overall trend for the Arctic is a substantial warming over the last few decades. For the Arctic as a whole, the 20th century can be divided into two warming periods, bracketing a 20-year cooling period (approximately 1945 to 1966) in the middle of the century. This pattern is less evident in northern Canada than in some other areas of the Arctic. The Canadian Archipelago and West Greenland did experience some cooling mid-century, although even then, there was substantial winter warming. The warming has been significant over the past few decades (1966 to 2003), particularly in the Northwest Territories – continuing the band of substantial warming across northwest North America that also covers Alaska and the Yukon – reaching an increase of 2 °C per decade. This warming is most evident in winter and spring. A more detailed description of observed climate change is given in Chapter 2. The climate change for each of the four ACIA regions is summarized in section 18.3.

Observations of arctic precipitation are restricted to a limited network of stations and are often unreliable since winter snowfall is not accurately measured by existing gauges, due to drifting snow. Available records indicate 20th-century increases in precipitation at high latitudes on the North American continent but little if any change in precipitation in the watersheds of the large Siberian rivers.

Rapid changes in regional climates (so-called regime shifts) are also evident in the climatic record. For example, in 1976 in the Bering Sea region there was a relatively sudden shift in prevailing climatic patterns, which included rapid warming and reduction in sea-ice extent. Such shifts have led to numerous, nearly instantaneous impacts on biota and ecosystems, as well as impacts on human communities and their interactions with the environment. Although such fluctuations are not fully understood and are therefore difficult to predict, regime shifts can be expected to continue to occur in the future, even as the baseline climate is also changing as a result of global warming (see section 18.4.3).

18.2.1.2. Indigenous observations of recent changes in climate

Indigenous observations of climate change, as discussed in Chapters 3 and 12, contribute to understanding of climate change and associated changes in the behavior and movement of animals. Through their various activities, which are closely linked to their surroundings, the indigenous peoples of the Arctic experience the climate in a very personal way. Over many generations and based on direct, everyday experience of living in the Arctic, they have developed specific ways of observing, interpreting, and adjusting to weather and climate changes. Based on careful observations, on which they often base life and death decisions and set priorities, indigenous peoples have come to possess a rich body of knowledge about their surroundings. Researchers are now working with indigenous peoples to learn from their observations and perspectives about the influences of climate change and weather events on the arctic environment and on their own lives and cultures. These studies are finding that the climate variations observed by indigenous people and by scientific observation are, for the most part, in good accord and often provide mutually reinforcing information.

The presently observed climate change is increasingly beyond the range experienced by the indigenous peoples in the past. These new conditions pose new risks to the lifestyles of the indigenous populations, as described in section 18.2.2. The magnitude of these threats is critically dependent on the rate at which change occurs. If change is slow, adaptation may be possible; if however, change is rapid, adaptation is very likely to be considerably more difficult, if possible at all in response to some types of impacts.

Recent observations by the indigenous peoples of the Arctic of major changes in the climate and associated impacts are summarized in Table 18.1. Taken together, the body of observations from people residing across the Arctic presents a compelling account of changes that are increasingly beyond what their experience tells them about the past.

Indigenous observations of climate and related environmental changes include many other effects on plants and animals that are important to them (Chapter 7). These observations provide evidence of nutritional stresses on many animals that are indicative of a changing environment and changes in food availability. New species, never before recorded in the Arctic, have also been observed. The distribution ranges of some species of birds, fish, and mammals now extend further to the north than in the past. These observations are significant for indigenous

communities since changes likely to occur in traditional food resources will have both negative and positive impacts on the culture and economy of arctic peoples.

18.2.1.3. Projections of future climate

In projections of future climate, uncertainties of many types can arise, especially for as complex a challenge as projecting ahead 100 years. The ACIA adopted a lexicon of terms (section 1.3.3) describing the likelihood of expected change (Fig. 18.3). These terms are used throughout this chapter, and in all ACIA documents.

Chapter 4 presents the ACIA projections of future changes in arctic climate. These projections extend the IPCC assessment (IPCC, 2001) by presenting regional (north of 60° N) climate parameters, derived from global model outputs. The ACIA used five different global climate models (CGCM2, Canadian Centre for Climate Modelling and Analysis; CSM_1.4, National Center for Atmospheric Research, United States; ECHAM4/OPYC3, Max Planck Institute for Meteorology, Germany; GFDL-R30_c, Geophysical Fluid Dynamics Laboratory, United States; and HadCM3, Hadley Centre for Climate Prediction and Research, United Kingdom) forced with two different emissions (GHG and aerosol) scenarios. The emissions scenarios are the B2 and A2 scenarios drawn from the IPCC Special Report on Emissions Scenarios (Nakićenović and Swart, 2000). The A2 emissions scenario assumes global emphasis on sustained economic development while the B2 emissions scenario reflects a world that promotes environmental sustainability. Neither scenario is considered an upper or lower

Table 18.1. Examples of indigenous observations of environmental change in the Arctic. This table is mainly based on Chapters 3 and 12.

	European Arctic	Canada and Greenland	Alaska
Atmosphere/ weather/winds	Weather patterns are changing so fast that traditional methods of prediction are no longer applicable. Winters are warmer. Seasonal patterns have changed.	Weather patterns are changing so fast that traditional methods of prediction are no longer applicable. Winters are warmer. There has been cooling in Hudson Strait/Baffin Island area, but greater variability.	Weather patterns are changing so fast that traditional methods of prediction are no longer applicable. There are more storms and fewer calm days. Winters are shorter and warmer, summers longer and hotter.
Rain/snow	Rain is more frequent in winter than before. There are more freeze–thaw cycles, thus more trouble for reindeer grazing in winter.	Snow is melting earlier and some permanent snow patches disappear. There is less snow and more wind, producing snow conditions that do not allow igloo building.	There is less snow.
Ocean/sea ice		Later freeze-up and earlier breakup of sea ice. Shore-fast ice is melting faster, creating large areas of open water earlier in summer.	Sea ice is thinner and is forming later. There is increased coastal erosion due to storms and lack of ice to protect the shoreline from waves.
Lakes/rivers/ permafrost	Ice on lakes and rivers is thinner.	Water levels in lakes and rivers are falling on the Canadian mainland. Thinner river ice affects caribou on migration (they fall through). Permafrost is thawing, slumping soil into rivers and draining lakes.	Lakes and wetlands are drying out. Permafrost thawing is affecting village water supply, sewage systems, and infrastructure.
Plants and animals	New species are moving into the region.	Caribou suffer from more insects; body condition has declined. Caribou migration routes have changed.	Trees and shrubs are advancing into tundra. There are die-offs of seabirds and marine mammals due to poor body condition. New species of insects are observed.

Fig. 18.3. Five-tier lexicon describing the likelihood of expected change.

bound on possible levels of future emissions. The climatic and environmental changes in the Arctic projected using the two scenarios are similar through about 2040, but diverge thereafter, with projections forced with the A2 emissions scenario showing greater warming.

These projections are not intended to capture a large range of possible futures for the Arctic under scenarios of continuing emissions of GHGs and other pollutants. For practical reasons, only a limited number of future change scenarios could be developed for this assessment. Nonetheless, while the ACIA used only two different emissions scenarios, five global climate models were used to project change under the two emissions scenarios, capturing a good range of the uncertainty associated with how different models represent climate system processes.

Under the A2 and B2 emissions scenarios, the models projected that mean annual arctic surface temperatures north of 60° N will be 2 to 4 °C higher by mid-century and 4 to 7 °C higher toward the end of the 21st century (Fig. 18.4), compared to the present. Precipitation is projected to increase by about 8% by mid-century and by about 20% toward the end of the 21st century. There are differences among the projections from the different models, however. Although the projected trends are similar for the next few decades, the scatter of results for either the A2 or B2 emissions scenarios is about 2 to 3 °C toward the end of the 21st century. The reasons for this scatter are differences in the rep-

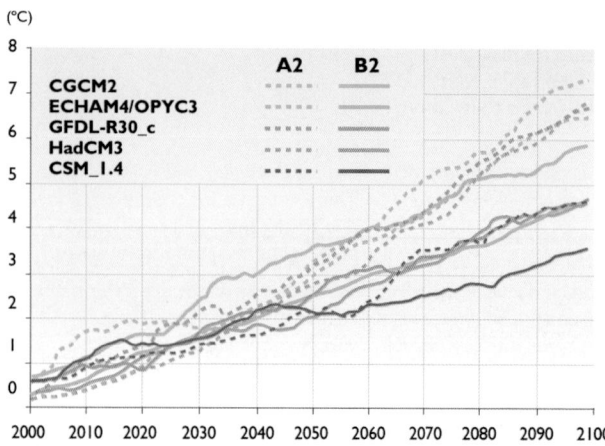

Fig. 18.4. Changes in surface air temperature north of 60° N between the 1981–2000 baseline and 2100 as projected by the five ACIA-designated models forced with the A2 and B2 emissions scenarios.

resentation of physical processes and feedbacks that are particularly important in the Arctic (e.g., changes in albedo due to reduced snow and ice cover, clouds, atmosphere–ocean interactions, ocean circulation), and natural variations that remain significant compared with the induced changes for at least the next few decades. While observations seem to indicate some increase in the frequency and severity of extreme events, these are very difficult to project.

Composite five-model projections for annual and winter mean surface air temperature changes in the Arctic between 1981–2000 and 2071–2090 using the B2 emissions scenario are shown in Fig. 18.5a (annual) and Fig. 18.5b (winter). Projected annual temperatures show a fairly uniform warming of 2 to 4 °C throughout the Arctic by the end of the century, with a slightly higher warming of up to 5 °C in the East Siberian Sea. Summer temperatures are projected to increase by 1 to 2 °C over land, with little change in the central Arctic Ocean, where sea ice melts each summer, keeping the ocean temperature close to 0 °C. Winter temperatures show the greatest warming: about 5 °C over land, and up to 8 to 9 °C in the central Arctic Ocean, where the feedback due to reduced sea-ice extent is largest. Regional and seasonal differences between the individual model results can be large, however, for the reasons previously discussed.

Changes in the Arctic affect the global system in several ways. Global climate change is influenced by feedback processes operating in the Arctic; these also amplify climate change in the Arctic itself. Apart from the well-known and often quoted ice- and snow-albedo feedback, another important arctic feedback is the thawing of permafrost, which is likely to lead to additional GHG releases. Arctic cloud feedbacks are also important but are still poorly understood. Some of these arctic feedbacks are adequately represented in general circulation models while others are not. Changes in the Arctic also affect the global system in other ways. Climate change is likely to increase precipitation throughout the Arctic and increase runoff to the Arctic Ocean; such a freshening is likely to slow the global thermohaline circulation with consequences for global climate. The melting of arctic glaciers and ice sheets is another effect of climate change that contributes to global sea-level rise with all its inherent problems. The global implications of these feedbacks are discussed in greater detail in section 18.2.2.

18.2.1.4. Ozone and UV radiation change

Atmospheric ozone is vital to life on earth. The stratosphere contains the majority of atmospheric ozone, which shields the biosphere by absorbing UV radiation from the sun. Anthropogenic chlorofluorocarbons are primarily responsible for the depletion of ozone in the stratosphere, particularly over the poles. Atmospheric dynamics and circulation strongly influence ozone

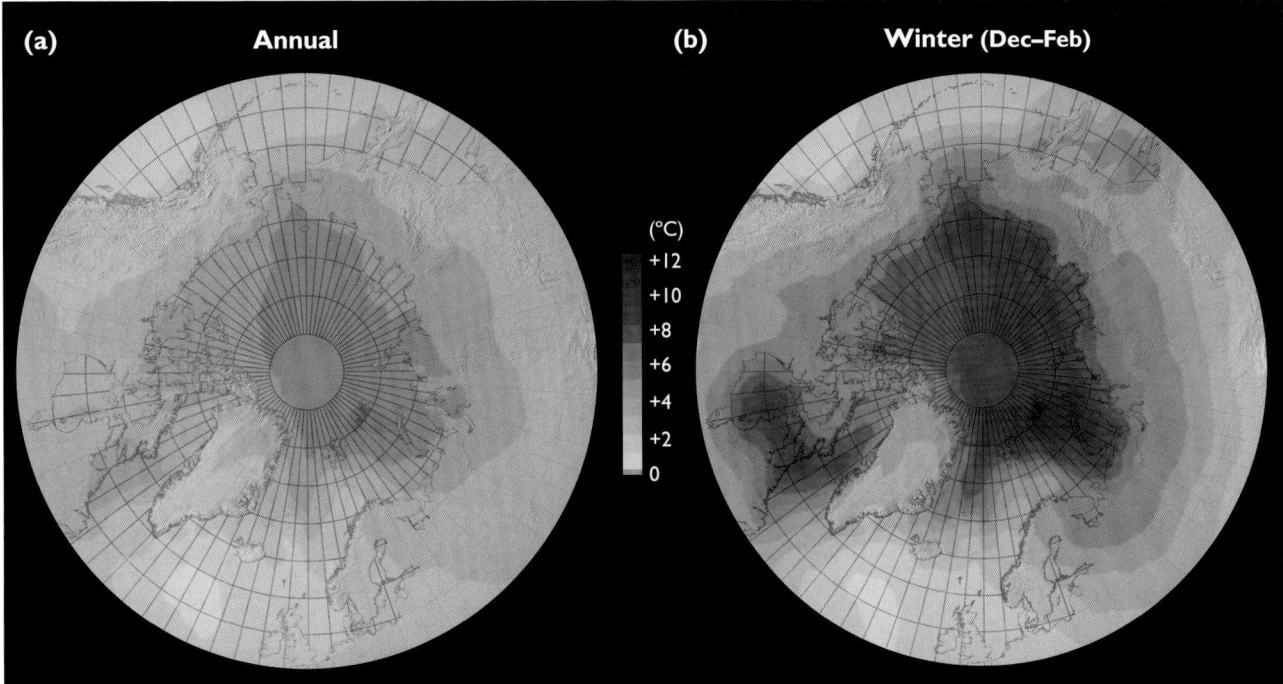

Fig. 18.5. (a) Projected annual surface air temperature change from the 1990s to the 2090s, based on the average change project-ed by the five ACIA-designated climate models using the B2 emissions scenario. (b) Projected surface air temperature change in winter from the 1990s to the 2090s, based on the average change projected by the five ACIA-designated climate models using the B2 emissions scenario.

amounts over the poles, which in normal conditions tend to be higher than over other regions on earth. During conditions of ozone depletion, however, ozone over the polar regions can be substantially reduced. This depletion is most severe in the late winter and early spring, when unperturbed ozone amounts are typically high. Ozone losses over the Arctic are also strongly influenced by meteorological variability and large-scale dynamical processes. Winter temperatures in the polar stratosphere tend to be near the threshold temperature for forming polar stratospheric clouds, which can accelerate ozone destruction, leading to significant and long-lasting deple-tion events. Because climate changes due to increasing GHGs are likely to lead to a cooling of the stratosphere, polar stratospheric cloud formation is likely to become more frequent in future years, causing episodes of severe ozone depletion to continue to occur over the Arctic.

Decreases in stratospheric ozone concentrations are very likely to lead to increased UV radiation levels at the sur-face of the earth. Clouds, aerosols, surface albedo, alti-tude, and other factors also influence the amount of UV radiation reaching the surface. Achieving an accurate pic-ture of UV radiation doses in the Arctic is complicated by low solar elevations and by reflectance off snow and ice. Ultraviolet radiation has long been a concern in the Arctic, as indicated by protective goggles found in the archaeological remains of indigenous peoples. Depletion of ozone over the Arctic, as has been observed in several years since the early 1980s, can lead to increased amounts of UV radiation, particularly UV-B radiation, reaching the surface of the earth, exposing humans and ecosystems to higher doses than have historically been observed. These higher doses are most likely to occur

during spring, which is also the time of year when many organisms produce their young and when plants experi-ence new growth. Because ecosystems are particularly vulnerable to UV radiation effects during these stages, increased UV radiation doses during spring could have serious implications throughout the Arctic.

Chapter 5 discusses observed and projected changes in atmospheric ozone and surface UV radiation levels. Satellite and ground-based observations since the early 1980s indicate substantial reductions in ozone over the Arctic during the late winter and early spring. Between 1979 and 2000, mean spring and annual atmospheric ozone levels over the Arctic declined by 11 and 7%, respectively. During the most severe depletion events, arctic ozone losses of up to 45% have occurred. Although international adherence to the Montreal Protocol and its amendments is starting to lead to a decline in the atmospheric concentrations of ozone-depleting substances, ozone levels over the Arctic are likely to remain depleted for the next several decades. For the Arctic, most models project little recovery over the next two decades. Episodes of very low spring ozone levels are likely to continue to occur, perhaps with increasing frequency and severity because of the stratospheric cooling projected to result from increasing concentrations of GHGs. These episodes of very low ozone can allow more UV radiation to reach the earth's surface, suggesting that people and ecosystems in the Arctic are likely to be exposed to higher-than-normal UV radiation doses for perhaps the next 50 years.

Table 18.2 summarizes some of the major aspects of changes in ozone and UV radiation levels.

Table 18.2. Observed and projected trends in ozone and ultraviolet radiation levels and factors affecting levels of ultraviolet radiation in the Arctic. This table is mainly based on Chapter 5.

	Observed	Projected
Ozone and UV radiation trends	Combined satellite and ground-based observations indicate that mean spring and annual atmospheric ozone levels over the Arctic declined by 11 and 7%, respectively, between 1979 and 2000. These losses allowed more UV radiation to reach the surface of the earth. Individual measurements suggest localized increases in surface UV radiation levels, although high natural variability makes it difficult to identify a trend conclusively.	Future ozone levels over the Arctic are difficult to project, partly owing to the link with climate change. Current projections suggest that ozone over the Arctic is likely to remain depleted for several decades. This depletion would allow UV radiation levels to remain elevated for several decades. The elevated levels are likely to be most pronounced in spring, when many ecosystems are most sensitive to UV radiation exposure.
Low ozone episodes	Multi-week episodes of very low ozone concentration (depletion of 25 to 45%) have been observed in several springs since the early 1990s.	Decreasing stratospheric temperatures resulting from climate change may cause low ozone episodes to become more frequent.
Seasonal variations in ozone and UV radiation levels	There is high seasonal and interannual variability in arctic ozone levels, due to atmospheric processes that influence ozone production and distribution. Over the Arctic, stratospheric ozone levels are typically highest in late winter and early spring, when ozone depletion is most likely to occur. Surface UV radiation amounts vary with solar angle and day length throughout the year. In general, UV radiation doses are highest in summer, but can also be significant in spring due to ozone depletion combined with UV radiation enhancements from reflection off snow.	
Cloud effects	Cloud cover typically attenuates the amount of UV radiation reaching the surface of the earth. When the ground is snow-covered, this attenuation is diminished and UV radiation levels reaching the surface may increase due to multiple scattering between the surface and cloud base.	Future changes in cloud cover are currently difficult to project, but are likely to be highly regional.
Albedo effects	Changes in snow and ice extent affect the amount of UV radiation reflected by the surface. Reflection off snow can increase biologically effective UV irradiance by over 50%. In addition, high surface albedo affects UV radiation amounts incident on vertical surfaces more strongly than amounts incident on horizontal surfaces. Snow-covered terrain can substantially enhance UV radiation exposure to the face or eyes, increasing cases of snow blindness and causing potential long-term skin or eye damage.	Climate changes are likely to alter snow cover and extent in the Arctic. Reduced snow and ice cover means less reflection of UV radiation, decreasing the UV radiation levels affecting organisms above the snow.
Snow and ice cover	Snow and ice cover shields many arctic ecosystems from UV radiation for much of the year.	Climate changes are likely to alter snow cover and extent in the Arctic. Reduced snow and ice cover will increase the UV radiation levels experienced by organisms that would otherwise be shielded by snow or ice cover.
Impacts of increased UV radiation levels	Ultraviolet radiation effects on organisms in human health and on terrestrial and aquatic ecosystems include skin cancer, corneal damage, immune suppression, sunburn, snow blindness. Ultraviolet radiation also damages wood, plastics, and other materials widely used in arctic infrastructure.	All the impacts noted in the "Observed" column are likely to worsen.

18.2.2. Arctic-wide impacts

18.2.2.1. Impacts on the environment

Changes in snow, ice, and permafrost

Recent observational data, discussed in detail in Chapter 6, present a generally consistent picture of cryospheric variations that are shaped by patterns of recent warming and variations in atmospheric circulation. Consistent with the overall increase in global temperatures, arctic snow and ice features have diminished in extent and volume. While the various cryospheric and atmospheric changes are consistent in an aggregate sense and are

quite large in some cases, it is possible that natural, low-frequency variations in the atmosphere and ocean have played at least some role in forcing the cryospheric and hydrological trends of the past few decades.

Model projections of anthropogenic climate change indicate a continuation of the recent trends through the 21st century, although the rates of the projected changes vary widely due to differences in model representations of feedback processes. Models project a 21st-century decrease in sea-ice extent of up to 100% in summer; a widespread decrease in snow-cover extent, particularly in spring and autumn; and permafrost degradation over 10 to 20% of the present permafrost area and a movement of the permafrost boundary northward by several hundred kilometers. The models also project river discharge increases of 5 to 25%; earlier breakup and later freeze-up of rivers and lakes; and a sea-level rise of several tens of centimeters resulting from glacier melting and thermal expansion, which is amplified or reduced in some areas due to long-term land subsidence or uplift.

Table 18.3 summarizes observed and projected trends in the snow and ice features of the Arctic, including snow cover, glaciers, permafrost, sea ice, and sea-level rise. Because the snow and ice features of the Arctic are not only sensitive indicators of climate change but also play a crucial role in shaping the arctic environment, any changes in these features are very likely to have profound effects on the environment, biota, ecosystems, and humans.

Changes on land

Climate change is also likely to have profound effects on the tundra and boreal forest ecosystems of the Arctic. Arctic plants, animals, and microorganisms adapted to climate change in the past primarily by relocation, and their main response to future climate change is also like-

ly to be relocation. In many areas of the Arctic, however, relocation possibilities are likely to be limited by regional and geographical barriers. Nevertheless, changes are already occurring in response to recent warming. Chapters 7, 8, and 14 provide details of the major conclusions presented here.

Some arctic species, especially those that are adapted to the cold arctic environment (e.g., mosses, lichens, and some herbivores and their predators) are especially at risk of loss in general, or displacement from their present locations. Present species diversity is more at risk in some ACIA regions than others; for example, Beringia (Region 3) has a higher number of threatened plant and animal species than any other ACIA region. While there will be some losses in many arctic areas, movement of species into the Arctic is likely to cause the overall number of species and their productivity to increase, thus overall biodiversity measured as species richness is likely to increase along with major changes at the ecosystem level.

Freshwater systems in the Arctic will also be affected due to changes in river runoff, including the timing of runoff from thawing permafrost, and changes in river-

Table 18.3. Observed and projected trends for the arctic cryosphere. This table is mainly based on Chapter 6.

	Observed	Projected for the 21st century
Snow cover	Snow-cover extent in the Northern Hemisphere has decreased by 5 to 10% since 1972; trends of such magnitude are rare in GCM simulations.	Snow-cover extent is projected to decrease by about 13% by 2071–2090 under the projected increase in mean annual temperature of about 4 °C. The projected reduction is greater in spring. Owing to warmer conditions, some winter precipitation in the form of rain is likely to increase the probability of ice layers over terrestrial vegetation.
Glaciers	Glaciers throughout the Northern Hemisphere have shrunk dramatically over the past few decades, contributing about 0.15 to 0.30 mm/yr to the average rate of sea-level rise in the 1990s.	The loss of glacial mass through melting is very likely to accelerate throughout the Arctic, with the Greenland Ice Sheet also starting to melt. These changes will tend to increase the rate of sea-level rise.
Permafrost	Permafrost temperatures in most of the Arctic and subarctic have increased by several tenths of a degree to as much as 2 to 3 °C (depending on location) since the early 1970s. Permafrost thawing has accompanied the warming.	Over the 21st century, permafrost degradation is likely to occur over 10 to 20% of the present permafrost area, and the southern limit of permafrost is likely to move northward by several hundred kilometers.
Sea ice	Summer sea-ice extent decreased by about 7% per decade between 1972 and 2002, and by 9% per decade between 1979 and 2002, reaching record low levels in 2002. The extent of multi-year sea ice has also decreased, and ice thickness in the Arctic Basin has decreased by up to 40% since the 1950s and 1960s due to climate-related and other factors.	Sea-ice extent is very likely to continue to decrease, particularly in summer. Model projections of summer sea-ice extent range from a loss of several percent to complete loss. As a result, the navigation season is projected to be extended by several months.
River discharge	River discharge has increased over much of the Arctic during the past few decades and the spring discharge pulse is occurring earlier.	Models project that total river discharge is likely to increase by an additional 5 to 25% by the late 21st century.
Breakup and freeze-up	Earlier breakup and later freeze-up of rivers and lakes across much of the Arctic have lengthened the ice-free season by 1 to 3 weeks.	The trend toward earlier breakup and later freeze-up of rivers and lakes is very likely to continue, consistent with increasing temperature. Breakup flooding is likely to be less severe.
Sea-level rise	Global average sea level rose between 10 and 20 cm during the 20th century. This change was amplified or moderated in particular regions by tectonic motion or isostatic rebound.	Models project that glacier contributions to sea-level rise will accelerate in the 21st century. Combined with the effects of thermal expansion, sea level is likely to rise by 20 to 70 cm (an average of 2 to 7 mm/year) by the end of the 21st century.

and lake-ice regimes. Changes in water flows as permafrost thaws are very likely to alter the biogeochemistry of many areas and create new wetlands and ponds, connected by new drainage networks. More water will alter the winter habitats in freshwater systems and increase survival of freshwater and sea-run fish. On hill slopes and higher ground, permafrost thawing is likely to drain and dry existing soils and wetlands. The productivity of these systems is likely to increase, as well as species diversity.

Table 18.4. Projected impacts on terrestrial ecosystems. This table is mainly based on Chapter 7.

	Projected impact
Ecotone transitions	Warming is very likely to lead to slow northward displacement of tundra by forests, while tundra will in turn displace high-arctic polar desert. Tundra is projected to decrease to its smallest extent in the last 21 000 years. In dry areas where thawing permafrost leads to drainage of the active layer, forests are likely to be replaced by tundra–steppe communities. Where thawing permafrost leads to waterlogging, forest will be displaced by bogs and wetlands.
Forest changes	Forests are likely to expand and in some areas, where present-day tundra occupies a narrow zone, are likely to reach the northern coastline. The expansion will be slowed by increased fire frequency, insect outbreaks, and vertebrate herbivory, as has already been observed in some parts of the Arctic.
Species diversity	Climate warming is very likely to lead to northward extension of the distribution ranges of species currently present in the Arctic and to an increase in the total number of species. Individual species will move at different rates and new communities of associated species are likely to form. Climate warming is also likely to lead to a decline or extirpation of populations of arctic species at their southern range margins. As additional species move in from warmer regions, the number of species in the Arctic and their productivity are very likely to increase.
Species at risk	Specialist species adapted to the cold arctic climate, ranging from mosses, lichens, vascular plants, some herbivores (lemmings and voles) and their predators, to ungulates (caribou and reindeer), are at risk of marked population decline or extirpation locally. This will be largely as a consequence of their inability to compete with species invading from the south. The biodiversity in Beringia is at risk as climate warms since it presently has a higher number of threatened plant and animal species than any other arctic region.
UV radiation effects	Increased UV radiation levels resulting from ozone depletion are likely to have both short- and long-term impacts on some ecosystem processes, including reduced nutrient cycling and decreased overall productivity. Many arctic plant species are assumed to be adaptable to high levels of UV-B radiation. Adaptation involves structural and chemical changes that can affect herbivores, decomposition, nutrient cycling, and productivity.
Carbon storage and fluxes	Over the long term, replacement of arctic vegetation with more productive southern vegetation is likely to increase net carbon storage in ecosystems, particularly in regions that are now tundra or high-arctic polar desert. Methane fluxes are likely to increase as vegetation grows in tundra ponds, and as wetlands become warmer (until they dry out). Methane fluxes are also likely to increase when permafrost thaws.
Albedo feedback	The positive feedback of albedo change (due to forest expansion) on climate is likely to dominate over the negative (cooling) feedback from an increase in carbon storage. The albedo reduction due to reduced terrestrial snow cover will be a major additional feedback.

Table 18.5. Projected impacts on freshwater ecosystems. This table is mainly based on Chapter 8.

	Projected impact
Lakes	Reduced ice cover and a longer open-water season are very likely to affect thermal regimes, particularly lake stratification. Permafrost thaw in ice-rich environments is very likely to lead to catastrophic lake drainage; increased groundwater flux will drain other lakes. A probable decrease in summer water levels of lakes and rivers is very likely to affect the quality and quantity of, and access to, aquatic habitats.
Rivers	A likely shift to less intense ice breakup will reduce the ability of flow systems to replenish riparian ecosystems, particularly in river deltas. Reduced climatic gradients along large northern rivers are likely to alter ice-flooding regimes and related ecological processes. A very likely increase in winter flows and reduced ice-cover growth will increase the availability of under-ice habitats.
Water quality	Enhanced permafrost thawing is very likely to increase nutrient, sediment, and carbon loadings to aquatic systems, with a mixture of positive and negative effects on freshwater chemistry. An earlier phase of enhanced sediment supply will probably be detrimental to benthic fauna but the balance will be ecosystem- or site-specific. Freshwater biogeochemistry is very likely to alter following changes in water budgets.
Wetlands	Changes in climate are very likely to lead to an increased extent of wetlands, ponds, and drainage networks in low-lying permafrost-dominated areas, but also to losses of wetlands on hill slopes and higher ground. Coastal erosion and inundation will generate new wetlands in some coastal areas. Conversely, increased evapotranspiration is likely to dry peatlands, particularly during the warm season.
Species diversity	Changes in the timing of freshwater habitat availability, quality, and suitability are very likely to alter the reproductive success of species. Correspondingly, the rate and magnitude of climate change and its effects on aquatic systems are likely to outstrip the capacity of many aquatic biota to adapt or acclimate. Climate change is very likely to act cumulatively and/or synergistically with other stressors to affect the overall biodiversity of aquatic ecosystems.
UV radiation effects	Reduced ice cover in freshwater ecosystems is likely to have a greater effect on underwater UV radiation exposure than projected levels of stratospheric ozone depletion. Little is known about the adaptive responses of aquatic organisms to changing UV radiation levels.

Changes in animal and plant populations are often triggered by extreme events, particularly winter processes. Weather extremes in winter are likely to have greater effects on the mammals and birds that remain active in winter, than on plants, insects, and other invertebrates that are dormant in winter. While some projections indicate a likely increase in the frequency and severity of extreme events (storms, floods, icing of snow layers, drought) the distribution of these events is very difficult to project. Rapid changes present additional stresses if they exceed the ability of species to adapt or relocate since they are likely to lead to increased incidence of fires, disease, and insect outbreaks, as well as to restricted forage availability.

The impact of changes in climate and UV radiation levels on species and ecosystems is likely to make the current use of many protected areas as a conservation practice almost obsolete. Although local measures to reduce hunting quotas might moderate impacts of climate change on wildlife species, habitat protection requires a new, more flexible paradigm. Comparison of areas in the Arctic in which vegetation is likely to dramatically change with the location of current protected areas shows that many habitats will be altered so that they will no longer serve to support the intended species or communities. These impacts will be reduced if simple measures are incorporated into the design of protected areas, for example, designating flexible boundaries that encompass extended latitudinal tracts of land and protect corridors for species movement.

As warming allows trees to grow, forests are projected to replace a significant proportion of the tundra.

This process is very likely to be slowed locally by natural barriers to movement, human activities, fires, insect attacks, browsing by vertebrate herbivores, and drying or waterlogging of soils, but the long-term effect on species composition will be significant. Displacement of tundra by forest will also lead to a decrease in albedo, which will increase the positive (warming) feedback to the climate system, especially during spring when snow melts, and amplify changes in the local climate. Warming and drying of tundra soils are likely to lead to an increased release of carbon, at least in the short term. However, current models suggest that the Arctic may become a net sink for carbon (although the uncertainties associated with the projections are high). There are also uncertainties about changes in methane (CH_4) fluxes (although current CH_4 emissions from arctic ecosystems are already forcing climate) from wetlands, permafrost, and CH_4 hydrates, so it is not known if the circumpolar tundra will become a carbon sink or carbon source in the long term.

Tables 18.4 and 18.5 summarize the most important impacts projected for terrestrial and freshwater ecosystems, respectively.

Changes in the ocean

Through its influence on the Atlantic thermohaline circulation, the Arctic plays a critical part in driving the global thermohaline circulation. It is possible that increased precipitation and runoff of fresh water and the melting of glaciers and ice sheets, and thawing of the extensive permafrost underlying northern Siberia, could

Table 18.6. Projected impacts in the Arctic Ocean and subarctic seas. This table is mainly based on Chapters 9 and 13.

	Projected impact
Ocean regime	Increased runoff from major arctic rivers and increased precipitation over the Arctic Ocean are very likely to decrease its salinity.
Thermohaline circulation	A slow-down of the global thermohaline circulation is likely as a result of increased freshwater input from melting glaciers and precipitation. This is likely to delay warming for several decades in the Atlantic sector of the Arctic as a result of reduced ocean heat transport.
Sea-ice regime	All the ACIA-designated models project substantial reductions in sea-ice extent and likely opening of the Northern Sea Route to shipping during summer. Some of the models project an entirely ice-free Arctic Ocean in summer by the end of the 21st century. Greater expanses of open water will also increase the positive feedback of albedo change to climate.
Marine ecosystems	Reduced sea-ice extent and more open water are very likely to change the distribution of marine mammals (particularly polar bears, walrus, ice-inhabiting seals, and narwhals) and some seabirds (particularly ivory gulls), reducing their populations to vulnerable low levels. It is likely that more open water will be favorable for some whale species and that the distribution range of these species is very likely to spread northward.
UV radiation effects	Ultraviolet radiation can act in combination with other stressors, including pollutants, habitat destruction, and changing predator populations, to adversely affect a number of aquatic species. In optically clear ocean waters, organisms living near the surface are likely to receive harmful doses of UV radiation. Sustained, increased UV radiation exposure could also have negative impacts on fisheries.
Fisheries	Changes in the distribution and migration patterns of fish stocks are likely. It is possible that higher primary productivity, increases in feeding areas, and higher growth rates could lead to more productive fisheries in some regions of the Arctic. New species are moving into the Arctic and competing with native species. The extinction of existing arctic fish species is unlikely.
Coastal regions	Serious coastal erosion problems are already evident in some low-lying coastal areas, especially in the Russian Far East, Alaska, and northwestern Canada, resulting from permafrost thawing and increased wave action and storm surges due to reduced sea-ice extent and sea-level rise. Ongoing or accelerated coastal-erosion trends are likely to lead to further relocations of coastal communities in the Arctic.

freshen arctic waters, causing a reduction in the over-turning circulation of the global ocean and thus affecting the global climate system and marine ecosystems. The IPCC 2001 assessment considers a future reduction of the Atlantic thermohaline circulation as likely, while a complete shutdown is considered as less likely, but not impossible. If half the oceanic heat flux were to disappear with a weakened Atlantic inflow, then the associated cooling would more than offset the projected heating in the 21st century. Thus, there is the possibility that some areas in the Atlantic Arctic will experience significant *regional* cooling rather than warming, but the present models can assess neither its probability, nor its extent and magnitude.

The most important projected trends for the marine systems of the Arctic are summarized in Table 18.6.

18.2.2.2. Impacts on people's lives

Several chapters address the impacts of climate change on people, including Chapters 10, 11, 12, 14, 15, and 16. The Arctic is home to a large number of distinct groups of indigenous peoples and the populations of eight nations. Between two and four million indigenous and non-indigenous people live in the Arctic, depending on how the Arctic is defined. Most live in cities; in Russia

large urban centers include Vorkuta and Norilsk with populations listed as exceeding 100000, and Murmansk with about 500000 people, although the population of these cities has decreased in recent years. Arctic towns in Scandinavia and North America are smaller; Reykjavik has around 110000 inhabitants and Rovaniemi about 65000. In total there are probably around 30 towns in the Arctic with more than 10000 inhabitants.

Table 18.7 summarizes the projected social impacts of climate change and UV radiation on the people of the Arctic. Climate change is only one, and perhaps not the most important, factor currently affecting people's lives and livelihood in the Arctic. For example, the people living in Russia's Far North have experienced dramatic political, social, and economic changes since the collapse of the former Soviet Union; and Europeans, Canadians, and Alaskans have experienced major changes resulting from the discovery of minerals, oil and gas reserves, and the declines or increases of some of the northern fisheries.

For the indigenous population, and particularly for those people who depend on hunting, herding, and fishing for a living, climate change is likely to be a matter of cultural survival, however. Their uniqueness as people with cultures based on harvesting marine mammals, hunting,

Table 18.7. Projected social impacts on arctic residents. This table is mainly based on Chapters 12, 15, and 16.

	Projected impact
Impacts on arctic residents	
Infrastructure	Permafrost thawing is very likely to threaten buildings, roads, and other infrastructure. This includes increases in the settling and breaking of underground pipes and other installations used for water supply, heating systems, and waste disposal, and threats to the integrity of containment structures such as tailing ponds and sewage lagoons.
Water	While increased river runoff is projected to occur mainly in winter and spring, lower water tables in rivers and lakes in summer will reduce available water and impede river travel in some areas (e.g., the Mackenzie River watershed).
Health	Circumpolar health problems such as those associated with changes in diet and UV radiation levels are likely to become more prominent. Increases in zoonotic diseases and injury rates are likely, due to environmental changes and climate variability.
Income	Impacts on the economy are expected as a consequence of climate change in the Arctic and will affect work opportunities and income of arctic residents. Expected increases in productivity and greater opportunity for settlement are also likely to benefit people within and beyond the region.
Impacts specific to indigenous communities	
Food security	Obtaining and sharing traditional foods, both cultural traditions, are very likely to become more difficult as the climate changes, because access to some food species will be reduced. The consequences of shifting to a more Western diet are likely to include increased incidence of diabetes, obesity, and cardiovascular diseases. Food from other sources may also be more costly.
Hunting	Hunter mobility and safety and the ability to move with changing distribution of resources, particularly on sea ice, are likely to decrease, leading to less hunting success. Similarly, access to caribou by hunters following changed snow and river-ice conditions is likely to become more difficult. Harvesting the threatened remaining populations of some marine mammals could accelerate their demise.
Herding	Changing snow conditions are very likely to adversely affect reindeer and caribou herding (e.g., ice layers and premature thawing will make grazing and migration difficult and increase herd die-offs). Shorter duration of snow cover and a longer plant growth season, on the other hand, are likely to increase forage production and herd productivity if range lands and stocking levels are adequately managed.
Cultural loss	For many Inuit, climate change is very likely to disrupt or even destroy their hunting culture because sea-ice extent is very likely to be reduced and the animals they now hunt are likely to decline in numbers, making them less accessible, or they may even disappear from some regions. Cultural adaptation to make use of newly introduced species may occur in some areas.

herding caribou and reindeer, or fishing, is at risk because climate change is likely to deprive them of access to their traditional food sources, although new species, as they move north, may become available in some regions. Indigenous peoples have adapted to changes in the past through careful observations and skillful adjustments of their traditional activities and lifestyles, but the addition of climate and UV radiation changes and impacts on existing social, political, and other environmental stresses is already posing serious challenges. Today, the indigenous peoples live in greatly circumscribed social and economic situations and their hunting and herding activities are determined to a large extent by resource management regimes and local, regional, and global economic market situations that reduce their ability to adapt and cope with climate variability and change. While they experience stress from other sources that threatens their lifestyles and cultures, climate change magnifies these threats.

Improvements in human health are very likely to continue through advances in technology, but the potential for emerging diseases (via the introduction of new insect and animal vectors) in northern communities makes it difficult to project how climate change is likely to affect the overall health of arctic residents. Several types of impacts seem likely. Because it will be more difficult to access marine animals when hunting, and because there is greater danger to the hunters when traveling over thinner sea ice, and in open water in less predictable weather conditions, direct health effects through a changing diet and increased accident rates are likely. Increased UV radiation levels are also likely to directly affect health, increasing incidences of skin cancer, cataracts, and viral infections, owing to effects on the immune system. Studies by the World Health Organization estimate that a person receives the majority of their lifetime UV radiation exposure before 18 years of age. An entire generation of people in the

Table 18.8. Projected impacts on important economic activities in the Arctic. This table is mainly based on Chapters 13, 14, and 16, but also draws information from other chapters.

	Projected impact
Non-renewable resources	
Oil and gas	
Exploration	Reduced sea ice is likely to facilitate some offshore operations but hamper winter seismic work on shore-fast ice. Later freeze-up and earlier melting are likely to limit the use of ice and snow roads.
Production	Reduced extent and thinner sea ice are likely to allow construction and operation of more economical offshore platforms. Storm surges and sea-level rise are likely to increase coastal erosion of shore facilities and artificial islands. The costs of maintaining infrastructure and minimizing environmental impacts are likely to increase as a result of thawing permafrost, storm surges, and erosion.
Transportation	Reduced extent and duration of sea and river ice are likely to lengthen the shipping season and shorten routes (including trans-polar routes). Permafrost thawing is likely to increase pipeline maintenance costs.
Coal and minerals	
Production	The costs of maintaining infrastructure and minimizing environmental impacts are likely to increase as a result of thawing permafrost, storm surges, and erosion.
Transportation	Reduced extent and duration of sea ice are likely to lengthen the shipping season. Thawing permafrost is likely to affect roads and infrastructure.
Renewable resources	
Fish, shellfish, freshwater fish	
Fish stocks	Temperature, currents, and salinity changes are likely to lead to changes in species availability (positive in some areas, negative in others).
Harvests	Changes in migration patterns are likely to lead to changes in distances to fishing grounds, and possible relocation of processing plants. Increased storms, and icing of ship superstructure are likely to increase risks and reduce catches.
Timber	Productivity is likely to increase if there is adequate soil moisture but decrease if there are summer droughts. Fire and insect outbreaks are likely to decrease productivity.
Agricultural products	A warmer climate is likely to lengthen the growing season and extend the northern range of agriculture. Increased insect problems are likely to decrease productivity.
Energy	
Hydropower	Precipitation changes are likely to affect the water supply. Melting glaciers are likely to reduce future seasonal water supply.
Power lines	Icing events, storms, and ground thaw are likely to affect power lines.
Wildlife	
Harvests	Changes in distribution and migration patterns are likely to affect access to wildlife and change harvests. Invasive species are likely to compete with existing populations.
Conservation	Habitat loss, longer seasons, and boat access are likely to lead to over-harvesting in protected areas and affect conservation.

Arctic is likely to continue to be exposed to above-normal UV radiation levels, and a new generation will grow to adulthood under increased UV radiation levels. Although behavioral adaptations can reduce the expected impacts, adequate information and education about these effects must be available.

18.2.2.3. Impacts on the economy

The three most important sectors of the commercial economy of the Arctic are oil and gas, fish, and minerals, each of which will be influenced by changes in the climate. There are also other economic sectors that will be affected by climate change, including forestry, agriculture, and tourism. Impacts on industry and commerce are described in greater detail in Chapters 13, 14, and 16. The use of local resources for traditional purposes, including fish, wildlife, plants, and wood for fuel and home construction, are also part of the arctic economy and have been addressed in Chapters 11, 12, and 17.

Oil and gas

The Arctic has large oil and gas reserves. Most are located in Russia: oil in the Pechora Basin, gas in the Lower Ob Basin, and other potential oil and gas fields along the Siberian coast. In Siberia, oil and gas development has expanded dramatically over the past few decades, and this region produces 78% of Russia's oil and 84% of its natural gas. Canadian oil and gas fields are concentrated in two main basins in the Mackenzie Delta/Beaufort Sea region and in the high Arctic. Oil and gas fields also occur in other arctic waters, for example the Barents Sea. The oil fields at Prudhoe Bay, Alaska, are the largest in North America, and by 2002, around 14 billion barrels had been produced at this site. There are also substantial reserves of natural gas and coal along the North Slope of Alaska. The Arctic is an important supplier of oil and gas to the global economy. Climate change impacts on the exploration, production, and transportation activities of this industry could have both positive and negative market and financial effects. These are summarized in Table 18.8.

Fish

The arctic seas contain some of the world's oldest and most productive commercial fishing grounds. In the Northeast Atlantic and the Bering Sea and Aleutian region, annual fish harvests in the past have exceeded two million tonnes in each of the two regions. In the Bering Sea, overall harvests have remained stable at about two million tonnes, but while some species like pollock (*Theragra chalcogramma*) are doing well, others like snow crab (*Chionoecetes opilio*) have declined. Important fisheries also exist around Iceland, Greenland, the Faroe Islands, and Canada. Fisheries are important to many arctic countries, as well as to the world as a whole. For example, Norway is one of the world's biggest fish exporters with exports worth US$ 4 billion in 2001. In some arctic regions aqua-

culture is a growing industry, providing local communities with jobs and income. Freshwater fisheries are also important in some areas. Changes in climatic conditions are likely to have both positive and negative financial impacts (see Table 18.8).

Minerals

The Arctic has large mineral reserves, ranging from gemstones to fertilizers. Russia extracts the greatest quantities of these minerals, including nickel, copper, platinum, apatite, tin, diamonds, and gold, mostly on the Kola Peninsula but also in the northern Ural Mountains, the Taymir region of Siberia, and the Far East. Canadian mining in the Yukon and Northwest Territories and Nunavut is for lead, zinc, copper, gold, and diamonds. In Alaska, lead and zinc are extracted at the Red Dog Mine, which sits atop two-thirds of US zinc resources, and gold mining continues in several areas. Coal mining occurs in several areas of the Arctic. Mining activities in the Arctic are an important contributor of raw materials to the global economy and are likely to expand with improving transportation conditions to bring products to market, due to a longer ice-free shipping season (Table 18.8).

Transportation industry

The cost of transporting products and goods into and out of the Arctic is a major theme of the potential impacts of climate change on many of the economic sectors described above. While climate change will affect many different modes of transport in the Arctic, the likelihood of reduced extent and duration of sea ice in the future will have a major impact. The projected opening of the Northern Sea Route (the opening of the Northwest Passage is less certain) to longer shipping seasons (Chapter 16) will provide faster and therefore cheaper access to the Arctic, as well as the possibility of trans-arctic shipping. This will provide new economic opportunities, as well as increased risks of oil and other pollution along these routes. Other regions of the Arctic will also benefit from easier shipping access due to less sea ice.

Projected climate-related impacts on the major economic sectors in the Arctic are listed in Table 18.8. This is a qualitative assessment only, since detailed financial estimates of economic impacts are not available at present, except in very few instances. Over the 21st century, new types of activities could arise (for example trans-arctic shipping) but there are likely to be others. This analysis focuses on how future climate change could affect the present economy, and is not based on projections of economic and demographic development in the Arctic over the 21st century.

Forestry, agriculture, and tourism

Forestry is an important economic activity in six of the eight arctic countries, and agriculture in its various forms also contributes to local economies in all eight countries. The basis for agricultural activities varies throughout the Arctic. In North America, the limited agriculture helps to meet the need for local fresh produce during the short summer. In northern Europe and across the Russian North, crop production along with reindeer husbandry and some other domestic livestock production serve traditional cultural needs and provide opportunities for income. Tourism is also becoming an increasingly important economic factor in many arctic regions. Impacts on these economic sectors in monetary terms are difficult to project and quantify since factors other than climate, including future regional economic development, play a major role.

Wildlife

Arctic wildlife resources support communities throughout the Arctic, through whaling, fishing, and hunting, and wildlife contributes to both the monetary and traditional economies of the Arctic. Climate change threatens the culture and traditional lifestyles of indigenous communities but is not discussed here. Likely economic impacts due to climate change are relevant here but are not easily quantified at present.

18.3. A synthesis of projected impacts in the four regions

This section examines impacts within a more regional setting. A spatial division is necessary because the Arctic is very large and different regions are likely to experience patterns of climate change in the coming decades that are significantly different. Different regions of the Arctic are also distinguished by different social, economic, and political systems, which will mediate the impacts of and responses to climate change. These distinctions are captured broadly by the four regions defined in this assessment (see Fig. 18.6).

Fig. 18.6. The four regions of the Arctic Climate Impact Assessment.

Differences in large-scale weather and climate-shaping factors were primary considerations in selecting the four regions. Observations also indicate that the climate is presently changing quite differently in each of these regions, and even within them, especially where there are pronounced variations in terrain, such as mountains versus coastal plains. There are also large north–south gradients in climate variability within each region. The scale was thought to be roughly appropriate given that a larger number of smaller regions would not have been practical for this assessment, or compatible with a focus at the circumpolar level.

Region 1 includes East Greenland, northern Scandinavia, and northwestern Russia, as well as the North Atlantic with the Norwegian, Greenland, and Barents Seas. This region is projected to experience similar types of changes because the entire area is under the influence of North Atlantic atmospheric and oceanic conditions, particularly the Icelandic Low.

Region 2 includes Central Siberia, from the Urals to Chukotka, and the Barents, Laptev, and East Siberian Seas. This region represents the coldest part of the Arctic and is under the influence of the Siberian high-pressure system during winter.

Region 3 includes Chukotka, Alaska, the western Canadian Arctic to the Mackenzie River, and the Bering, Chukchi, and Beaufort Seas. This region is largely under the influence of North Pacific atmospheric and oceanic processes and the Aleutian Low.

Region 4 includes the central and eastern Canadian Arctic east of the Mackenzie River, the Queen Elizabeth Islands south to Hudson Bay, and the Labrador Sea, Davis Strait, and West Greenland. The region's weather systems are connected to large-scale North American and western North Atlantic weather patterns.

Major impacts due to observed and projected climate change for each of the four regions are summarized in sections 18.3.1 to 18.3.4. Details, including relevant references and publications on earlier impact assessments in the four regions, can be found in the preceding 17 chapters. The previous regional impact assessments were useful source material and laid a critical foundation for the ACIA. Some impacts apply to more than one or to all of the regions; such impacts are described in section 18.2 and are not necessarily listed separately for each region.

While the importance of providing information at the regional level has been emphasized here, the focus of the ACIA was at the circumpolar scale. The following sections attempt to synthesize significant findings for each of the four regions where these have been provided by the relevant chapters of the assessment. In some cases, the evaluated literature did not support a very extensive assessment of the nature of changes in the four regions.

Table 18.9. Key consequences of climate change on the environment, the economy, and on people's lives in the four ACIA regions. Unless otherwise stated these key consequences are considered likely to occur.

	Environment	Economy	People's lives
Region 1 Eastern Greenland, North Atlantic, northern Scandinavia, northwestern Russia	• Northward shifts in the distribution ranges of plant and animal species (terrestrial, freshwater, and marine) • Many tundra areas disappear from the mainland, except in arctic Russia where bog growth prevents forest development • Carbon storage increases and albedo decreases, but less so than in other regions	• Change in the location of North Atlantic and Arctic fisheries due to warmer waters and change in yields of many commercial fish stocks • Improved access to oil, gas, and mineral resources in presently ice-covered waters and adjacent land areas • Rising sea level and more storm surges affect coastal facilities	• Reduced and changing snow cover, affecting reindeer herding and hunted wildlife • Traditional harvest of animals is riskier and less predictable • Emergence of zoonotic diseases as a threat to human health • Increased outdoor leisure and recreational opportunity, plus lower heating costs
Region 2 Siberia	• Northward shifts in the distribution ranges of plant and animal species (terrestrial, freshwater, and marine) • Changing forest character due to warmer climate and permafrost thawing, with possibly greater fire and pest threats • Tundra changing to shrub and forest, but northern tundra extension limited by the ocean • Increased river discharge, affecting sediment and nutrient fluxes	• Reopening of the Northern Sea Route (Northeast Passage), due to reduced extent and duration of sea ice, providing new economic possibilities, and also increased pollution risks (tankers) • Improved access to oil, gas, and mineral resources in presently ice-covered waters and adjacent land areas • Rising sea level and more storm surges affect coastal facilities	• Permafrost thawing, causing serious damage to buildings in Siberian cities and to houses and facilities in villages • Traditional harvest of animals is riskier and less predictable • Emergence of zoonotic diseases as a threat to human health • Lower heating costs
Region 3 Chukotka, Bering Sea, Alaska, western Canadian Arctic	• Northward shifts in the distribution ranges of plant and animal species (terrestrial, freshwater, and marine) • Forest disruption due to warming and increased pest outbreaks • Reduced sea ice and general warming, disrupting polar bears, marine mammals, and other wildlife • Low-lying coastal areas more frequently inundated by storm surges and sea-level rise	• Damage to infrastructure due to thawing permafrost • Improved access to oil, gas, and mineral resources in presently ice-covered waters and adjacent land areas • Change in recruitment, growth rates, abundance, and distribution of Bering Sea fish due to warmer waters • Rising sea level and storm surges, affect coastal facilities	• Retreating sea ice and earlier snowmelt, altering traditional lifestyle patterns and food security, increasing risks taken by hunters, further stressing nutritional status • Coastal erosion and flooding forcing relocation of villages • Emergence of zoonotic diseases as a threat to human health • Increased outdoor leisure and recreational opportunity, plus lower heating costs
Region 4 Central and Eastern Canadian Arctic, Labrador Sea, Davis Strait, West Greenland	• Thawing of warm permafrost and reduced river and lake ice, changing hydrological regimes • Northward shifts in the distribution ranges of plant and animal species (terrestrial, freshwater, and marine) • High-arctic polar desert replaced by tundra in some areas, leading to potential large carbon gains • Increased melting of the Greenland Ice Sheet, changing the coastal environment • Decreasing sea-ice extent; threatening the extinction of polar bears	• Potential increased shipping in the Northwest Passage as sea ice retreats, providing economic incentives such as cheaper transport of goods, but also increasing pollution and oil-spill risks • Shorter operating season for ice and snow roads • Changes in marine and freshwater fisheries with impacts on tourism and local economic development • Rising sea level and storm surges affect coastal facilities	• Traditional lifestyles and survival of indigenous hunting culture threatened by retreating sea ice and changing environment • Health impacts via stresses on food security and safety of travel conditions • Emergence of zoonotic diseases as a threat to human health • Increased outdoor leisure and recreational opportunity, plus lower heating costs

Table 18.9 summarizes the key consequences of climate change on the environment, the economy, and people's lives in each of the four regions. The following sections provide additional information by region. Projections of climate change for each region are based on output from the five ACIA-designated climate models.

18.3.1. Region 1

18.3.1.1. Changes in climate

Most of Region 1 experienced a modest increase in mean annual temperature (about 1 °C) between 1954 and 2003, with slightly higher winter temperature

increases over this period, except for Iceland, the Faroe Islands, and southern Greenland, where there has been some cooling (see Fig. 18.2 for regional details). From 1990 to 2000, greater warming was observed in northern Scandinavia, including Iceland, Svalbard, and East Greenland, but cooling was observed in other areas such as the Kola Peninsula.

Model projections (see Fig. 18.5 for regional details) indicate that this region is likely to experience additional increases in mean annual temperature of 2 to 3 °C in Scandinavia and East Greenland and up to 3 to 5 °C in northwestern Russia by the late 21st century. Although changes in atmospheric and oceanic circulation contributed to some cooling of the region during the 20th century, warming has occurred in recent decades and is projected to dominate throughout the 21st century. Precipitation has increased slightly and is projected to increase further by up to about 10% by the end of the century.

18.3.1.2. Impacts on the environment

The geography and environment of Region 1 are dominated by the North Atlantic Ocean, which has extensive connections with the Arctic Ocean via the Norwegian, Greenland, and Barents Seas. The North Atlantic Ocean separates Greenland in the west from the Fennoscandian, European, and western Russian landmasses in the east. Relatively isolated islands of Iceland, the Faroe Islands, and the Svalbard and Franz Josef archipelagos span the low to high arctic latitudes. The land areas are characterized by a north–south climatic contrast between low-arctic environments in the south isolated by ocean from the high-arctic environments of the high-latitude islands, and by an east–west climatic contrast between the Scandinavian landmass, which is uncharacteristically warm for its latitude, and East Greenland in the west of the region, which is heavily glaciated. The continuous south–north land corridors for movement of terrestrial and freshwater species and people found in Region 2, for example, are missing.

The Greenland, Iceland, Norwegian, and Barents Seas constitute a major part of this region. This vast oceanic area is influenced by the inflow of relatively warm Atlantic water, which enters along the coast of Norway and is the most northward branch of the Gulf Stream. Variability in the volume of this inflow, as experienced in the past and as projected by models for the future due to global climate change, is expected to have major consequences for the physical and biological regimes of the region. Sea surface temperature is expected to increase, and the Barents Sea is expected to be totally ice free in summer by 2080. Changes in the distribution of important fish stocks are expected to occur. Past integrative impact assessments of climate change in this region include publications by Lange et al. (1999, 2003) for the Barents Sea.

The arctic seas in Region 1 are projected to experience a temperature increase that will lead to a decrease in sea-ice cover, especially in summer, as well as earlier ice melt and later freeze-up. Unless compensated for by an increase in low-level cloudiness, decreases in sea-ice cover would reduce the overall planetary albedo of the region and provide a positive feedback to the global climate. The reduction in sea ice is likely to enhance primary productivity, lead to increases in zooplankton production, and possibly to increased fisheries production. Such changes would also lead to decreased natural habitat for polar bears (*Ursus maritimus*) and ringed seals (*Phoca hispida*) to an extent that is likely to threaten the survival of their populations in this region. Conversely, more open water is expected to favor some whale and seabird species.

Biodiversity is high in Region 1: around 6000 marine and terrestrial species have been recorded for Svalbard (Prestrud et al., 2004) and around 7200 species have been recorded for 22 705 km² between 68° and 70° N in northern Finland (Callaghan et al., 2004). The European Arctic and subarctic are important breeding areas for many bird species overwintering in more temperate regions. Excluding the Russian Arctic, over 43% of the European bird species pool occurs in Region 1.

Observations indicate very variable climate trends and ecological responses to them in Region 1 (see Chapter 7). Treelines in northern Sweden increased in altitude by up to 40 m during the first part of the 20th century, and a further 20 m during the warming of the past 40 years, giving recent rates of treeline increase of 0.5 m/yr and 40 m/°C (Callaghan et al., 2004). In northern Finland, the pine treeline is increasing in altitude and density, and in the Polar Ural Mountains, treeline has advanced. However, there is little evidence of a northward shift of the latitudinal treeline west of the Polar Ural Mountains. Unexpectedly, evidence shows a southward movement of the treeline in parts of the forest tundra of the Russian European Arctic, a change that appears to be associated with localized pollution, deforestation, agriculture, and the growth of bogs leading to tree death. In the Faroe Islands, there has been a lowering of the alpine altitudinal treeline in response to a cooling of 0.25 °C during the past 50 years. In some areas of Finland and northern Sweden, there is evidence of an increase in rapidly changing warm and cold episodes in winter that lead to increasing bud damage in birch. Recent warming in northern Sweden and Finland has led to a reduction in the extent of discontinuous permafrost in mires and a change in vegetation resulting in increased CH₄ flux to the atmosphere (Christensen et al., 2004).

Recent warm winters have resulted in unusual conditions (causing ice layers in the snow) unfavorable for reindeer and wildlife, and leading to an absence of lemming population peaks, and on Svalbard, a decline in wild reindeer (*Rangifer tarandus platyrhynchus*) through decreases in the availability of food resources. Changes in animal populations also include reductions in arctic fox (*Alopex lagopus*) and snowy owl (*Nyctea scandiaca*) (as well as several other bird species) populations on mainland Fennoscandia but a northward migration of larger butterflies and moths, the larvae of some being defoliators of trees and shrubs. Moose (*Alces alces*), red fox (*Vulpes vulpes*), and the invasive species mink (*Mustela vison*) are increasing in the east of the region and muskoxen (*Ovibos moschatus*), wolves, and pink-footed (*Anser brachyrhynchus*) and barnacle geese (*Branta leucopsis*) are increasing in northeast Greenland (Callaghan et al., 2004).

If warming occurs as projected, the deciduous mountain birch forest that forms much of the present treeline in the region, the boreal conifer forest and woodland, and the arctic and alpine tundra are very likely to begin shifting northward and upward in altitude. The potential for vegetation change within the region is perhaps greatest in northern Scandinavia, where large shifts occurred in the early Holocene in response to warming. Here, pine forest is projected to invade the lower belt of mountain birch forest. The birch treeline is projected to move upward and northward, displacing shrub tundra vegetation, which in turn is projected to displace alpine tundra. Alpine species in the north are expected to be the most threatened because there is no suitable geographic area for them to shift toward in order to avoid being lost from the Fennoscandian mainland. In Iceland, a warmer climate is likely to facilitate natural regeneration of the heavily degraded native birch woodland as well as aid current and future afforestation efforts (Chapter 14). Model projections suggest that arctic tundra will be displaced totally from the mainland by the end of the 21st century (Fig. 18.1), although in practice, the bogs of the western Russian European Arctic may prevent forests from reaching the coast. Model projections of change from tundra to taiga between 1960 and 2080 (5.0%), and of change from polar desert to tundra (4.2%), are the lowest of any of the four ACIA regions because of the lack of tundra areas and the separation of the high Arctic from the subarctic.

While the climate is changing, local forest damage is projected to occur as a result of winters that are warmer than normal. Warmer winters are likely to lead to an increase in insect damage to forests and decreases in populations of animals such as lemmings and voles that depend upon particular snow conditions for survival. These changes, in turn, are likely to cause decreases in populations of many existing bird species and other animals, with the most severe effects on carnivores, such as Arctic foxes, and raptors, such as snowy owls. Heathland and wetland areas are likely to be partially invaded by grasses, shrubs, and trees, and mosses and lichens are expected to decrease in extent. Unlike other arctic areas, fire is not likely to play a major role in controlling vegetation dynamics. Any changes in land-use patterns, including increased agriculture and domestic stock production in a warmer climate, will encroach on wildlife habitats and further threaten large carnivores.

Some areas in this region, such as East Greenland and the Faroe Islands, have experienced recent cooling, and future warming is expected to reverse the present downward vegetation shifts in the mountains of the Faroe Islands. The island settings in this region, particularly those of Greenland, Svalbard, Franz Josef Land, and Novaya Zemlya, are likely to delay the arrival of immigrant species and substantial change other than expansion and increased growth of some current species.

Changes in arctic ecosystems will not only have local consequences but will also have impacts at a global level because of the many linkages between the Arctic and other regions further south. For example, several hundreds of millions of birds migrate to the Arctic each year and their success in the Arctic determines their populations at lower latitudes. As previously noted,

excluding the Russian Arctic, over 43% of the European bird species pool is found in Region 1. Changes in their wintering areas far south of the Arctic also play an important part in the ecology of migratory birds and there are many important stop-over sites and over-wintering grounds in Europe. Birds that are already suffering major declines in the region include lesser white-fronted goose (*Anser erythropus*) and shore lark (*Eremophila alpestris*) that have almost become extinct in Fennoscandia, and snowy owls. In contrast, some southern bird species have become established. Examples include blue tit (*Parus caeruleus*) and greenfinch (*Carduelis chloris*) (Callaghan et al., 2004).

Changes in carbon storage and release from ecosystems also have potential global consequences. Christensen et al. (2004) estimated that CH_4 emissions have increased from between 1.8 and 2.2 mg $CH_4/m^2/hr$ to between 2.7 and 3.0 mg $CH_4/m^2/hr$ over the past 30 years in northern Sweden, as a result of permafrost thaw and vegetation change. Terrestrial carbon storage and net primary production are projected to increase, and albedo to decrease, but less than in any other region due to the ocean barriers and general lack of tundra: there is a transition from subarctic to high Arctic separated by seas where the mid-arctic tundra should be.

18.3.1.3. Impacts on the economy

Region 1 has abundant renewable and non-renewable resources (timber, fish, ore, oil, and natural gas). The highly productive marine life makes this region one of the most productive fishing grounds in the circumpolar North. Higher ocean temperatures are likely to cause shifts in the distribution of some fish species, as well as changes in the timing of their migration, possible extension of their feeding areas, and increased growth rates. The occurrence of several "warm years" or "cold years" in a row, which is a sequence that could occur more frequently as a result of continuing global climate change, seems likely to lead to repercussions on the major fish stocks and, ultimately, the lucrative and productive fisheries in the region. Provided that the fluctuations in Atlantic inflow to the area are maintained, along with a general warming of the North Atlantic waters, it is likely that annual recruitment in herring (*Clupea harengus*) and Atlantic cod (*Gadus morhua*) will increase from current levels and will be about the same as the long-term average during the first two to three decades of the 21st century. This projection is also based on the assumption that harvest rates are kept at levels that maintain spawning stocks well above the level at which recruitment is impaired.

Impacts of climate change on the fisheries sector of the region's economy are difficult to assess, however. A scenario of moderate warming could result in quite large positive changes in the catch of many species. A self-sustaining cod stock could be established in West Greenland waters through larval drift from Iceland. Past catches suggest that this could yield annual catches of about 300 000 t. Should that happen, it is estimated that catches of northern shrimp (*Pandalus borealis*) will decrease to around 30% of the present level, while those of snow crab and Greenland halibut (*Reinhardtius hippoglossoides*) might remain the same. Such a shift could approximately double the export earnings of the Greenlandic fishing industry, which roughly translates into the same amount as that presently paid by Denmark to subsidize the Greenland economy. Such dramatic changes are not expected in the Icelandic marine ecosystem. Nevertheless, there would be an overall gain through larger catches of demersal species such as cod, pelagic species like herring, and new fisheries of more southern species like mackerel (*Scomber scombrus*). On the other hand capelin (*Mallotus villosus*) catches would dwindle, both through diminished stock size and the necessity of conserving this very important forage fish for other species. Effective fisheries management will continue to play a key role both for Greenland and Iceland, however. Little can be said about possible changes under substantial climate warming because such a situation is outside any recorded experience.

Forestry and agriculture are important in Region 1; both have been affected by climate change in the past and impacts are likely to occur in the future. Longer growing seasons are likely to improve the growth of agricultural crops. While growth (net carbon assimilation) of forests and woodlands is likely to increase, this will not necessarily benefit the forestry industry as forest fires and pests will also increase. Forest pest outbreaks have been reported for the Russian part of the region, including the most extensive damage from the European pine sawfly (*Neodiprion sertifer*), which affected a number of areas, each covering more than 5000 ha. The annual number of insect outbreaks reported between 1989 and 1998 was 3.5 times higher than between 1956 and 1965. The mean annual intensity of forest damage increased two-fold between 1989 and 1998. Factors other than climate change are also important to forest-based economies. For example, while most of the region has seen modest growth in forestry, Russia has experienced a decline due to political and economic factors. These socio-economic problems are expected to be aggravated by global climate change, which in the short term will have negative effects on timber quality owing to fire and insect damage and on infrastructure and winter transport when permafrost thaws.

18.3.1.4. Impacts on people's lives

The prospects and opportunities of gaining access to important natural resources, both renewable and non-renewable, have attracted a large number of people to Region 1. The relatively intense industrial activities, particularly on the Kola Peninsula, have resulted in population densities that are the highest throughout the circumpolar North. Impacts of climate change on terrestrial and marine ecosystems and implications for the availability of natural resources may lead to major

changes in economic conditions and subsequent shifts in demography, societal structure, and cultural values.

Because they would affect food, fuel, and culture, changes in arctic ecosystems and their biota are particularly important to the peoples of the Arctic. Reindeer herding by the Saami and other indigenous peoples is an important economic and cultural activity and the people who herd reindeer are concerned about the impacts of climate change. Observations have shown that during autumn the weather in recent years has fluctuated between raining and freezing so that the ground surface has often been covered with an ice layer and reindeer in many areas have been unable to access the underlying lichen. These conditions are quite different from those in earlier years and have caused massive losses of reindeer in some years. Changes in snow conditions also pose problems. Since reindeer herding has become motorized, herders relying on snowmobiles have had to wait for the first snows to start herding. In some years, this has led to delays up to the middle of November. Also, the terrain has often been too difficult to travel over when the snow cover is light. Future changes in snow extent and condition have the potential to lead to major adverse consequences for reindeer herding and those aspects of health (physical, social, and mental) relating to the livelihood of reindeer herders.

The beneficial effects of a warmer climate on people's recreational and leisure activities (camping, hiking, and other outdoor activities) should not be overlooked. Even relatively modest warming will improve people's mental and physical health. A warmer climate is also likely to reduce heating costs.

18.3.2. Region 2

18.3.2.1. Changes in climate

Region 2, which experiences the coldest conditions in the Arctic, has experienced an increase in mean annual temperature of about 1 to 3 °C since 1954, and an increase of up to 3 to 5 °C in winter (see Fig. 18.2 for regional details).

Models project that the mean annual temperature of Region 2 is likely to increase by a further 3 to 5 °C by the late 21st century, and by up to 5 to 7 °C in winter (see Fig. 18.5 for regional details). In the far north, winter warming of up to 9 °C over the Arctic Ocean as a result of reduced sea-ice extent and thickness is projected. The summer warming over the land areas is projected to be 2 to 4 °C by the end of the century, but there is likely to be very little change in summer over the Arctic Ocean.

18.3.2.2. Impacts on the environment

This region has the largest continuous land mass, which stretches from the tropical regions to the high Arctic. It is very likely to experience major changes as the boreal forest expands northward, but tundra will persist, although with reduced area. For example, extensive tundra is likely to remain in the Taymir region but is likely to be displaced completely from the mainland in the Sakha region.

The large Siberian rivers draining into the Arctic Ocean are projected to experience major impacts. Projected increases in winter precipitation, and more importantly in precipitation minus evaporation, imply an increase in water availability for soil infiltration and runoff. The total projected increase in freshwater supplied to the Arctic Ocean could approach 15% by the latter decades of the 21st century. An increase in the supply of freshwater has potentially important implications for the stratification of the Arctic Ocean, for its sea-ice regime, and for its freshwater export to the North Atlantic. In addition, increased freshwater input into the coastal zone is likely to accelerate the degradation of coastal permafrost.

On land, the projected increase in precipitation is likely to lead to wetter soils when soils are not frozen, wetter active layers in summer, and greater ice content in the upper soil layer during winter. To the extent that the increase in precipitation occurs as an increase in snowfall during the cold season, snow depth and snow water equivalent will increase, although the seasonal duration of snow cover may be shorter if, as projected, warming accompanies the increased snowfall.

The projected changes in terrestrial watersheds will increase moisture availability in the upper soil layers in some areas, favoring plant growth in areas that are presently moisture-limited. The projected increase in river discharge during winter and spring is likely to result in enhanced fluxes of nutrients and sediments to the Arctic Ocean, with corresponding impacts on coastal marine ecosystems. Higher rates of river and stream flow are likely to have especially large impacts on riparian regions and flood plains in the Arctic. One important consequence is that the vast wetland and bog ecosystems of this region are very likely to expand, leading to higher CH_4 emissions.

18.3.2.3. Impacts on the economy

A potentially major impact on the economy of Region 2 and on the global economy could be the opening of the Northern Sea Route (Northeast Passage) to commercial shipping. Model projections of ice cover during the 21st century show considerable development of ice-free areas around the entire Arctic Basin. Most coastal waters of the Eurasian Arctic are projected to become relatively ice free during September by 2020, with more extensive melting occurring later in the century. Ships navigating the Northern Sea Route would clearly benefit from these ice-free conditions. In addition, if winter multi-year sea ice in the central Arctic Ocean continues to retreat, it is very likely that first-year sea ice will dominate the entire maritime Eurasian Arctic, with a decreasing frequency of multi-year ice intrusions into the coastal seas and more

open water during the summer. By 2100, one of the ACIA-designated models projects that the navigation season could be as long as 200 days, while the mean of the five ACIA-designated models projects a navigation-season length of 120 days (when defined as the period with sea-ice concentrations below 50%).

Such changes in sea-ice conditions are likely to have important implications for ship design and construction and route selection along the Northern Sea Route in summer and even in winter. The need for navigational aids, refueling and ship maintenance, and sea-ice monitoring will require major financial investment, however, to assure security and safety for shipping and protection of the marine environment.

The coal and mineral extraction industries in Region 2 are important parts of the Russian economy, but climate change is likely to have little effect on the actual extraction process. On the other hand, transportation of coal and minerals will be affected in both a positive and negative sense. Mines in Siberia that export their products by ship will experience savings resulting from reduced sea-ice extent and a longer shipping season. However, mining facilities relying on transport over roads on permafrost will experience higher maintenance costs as the permafrost thaws.

Forestry, another important sector of the Siberian economy, is likely to experience both positive and negative impacts. A potentially longer growing season and warmer climate are likely to enhance productivity. However, more frequent fires and insect outbreaks are likely as the climate warms and insects invade from warmer regions. Drying of soils as permafrost thaws is also likely to affect forest productivity in some areas. To meet the demands of the global economy, forestry is likely to become more important and transportation of wood and wood products to markets will improve as reduced sea-ice extent facilitates marine transport along the Siberian coast.

18.3.2.4. Impacts on people's lives

The change to a wetter climate is likely to lead to increased water resources for the region's residents. In permafrost-free areas, water tables are very likely to be closer to the surface, and more moisture is projected to be available for agricultural production. During the spring when enhanced precipitation and runoff are very likely to cause higher river levels, the risk of flooding will increase. Summer soil moisture changes remain an open question since the models do not give clear signals. It is possible that lower water levels will occur in summer, as projected for other regions (for example the Mackenzie River in Region 3), affecting river navigation in some areas, increasing the risk of forest fires, and affecting hydropower generation.

Other major environmental impacts projected for Region 2 are associated with thawing permafrost and melting sea ice. Warming during the 20th century produced noticeable impacts on permafrost, causing deeper seasonal thawing and changes in the distribution and temperature of the frozen ground. For example, from the late 1980s to 1998, temperatures in the upper permafrost layers increased by 0.1 to 1.0 °C on the western Yamal Peninsula. Permafrost degradation in the developed regions of northeast Russia, coupled with inadequate building design, has led to serious problems. For example, in 1966 a building affected by thermokarst and differential thaw settlement collapsed in Norilsk, killing 20 people. In Yakutsk, a city built over permafrost in central Siberia, more than 300 structures, including several large residential buildings, a local power station, and a runway at the airport, have been seriously damaged by thaw-induced settlement. Considerable advances in knowledge and technology for building on permafrost have been made in recent decades. Nevertheless, as global climate change continues to intensify changes in arctic climate, detrimental impacts on infrastructure and therefore on the economy, health, and well-being of the population throughout the permafrost regions are expected to increase.

18.3.3. Region 3

18.3.3.1. Changes in climate

Alaska experienced an increase in mean annual temperature of about 2 to 3 °C between 1954 and 2003. The temperature increase was similar in the western Canadian Arctic, but was only about 0.5 °C in the Bering Sea and Chukotka. Winter temperatures over the same period increased by up to 3 to 4 °C in Alaska and the western Canadian Arctic, but Chukotka experienced winter cooling of between 1 and 2 °C (see Fig. 18.2 for regional details).

The five ACIA-designated models project that mean annual temperatures will increase by 3 to 4 °C by the late 21st century (see Fig. 18.5 for regional details). All the models project that the warming is likely to be greater in the north, reaching up to 7 °C in winter. In the central Arctic Ocean, winter temperatures are projected to increase by up to 9 °C as a result of reduced sea-ice extent and thickness, but there is likely to be very little change in summer temperature. Trends in and future projections of ozone and UV radiation levels follow the Arctic-wide patterns.

18.3.3.2. Impacts on the environment

Two detailed assessments of the potential consequences of climate variability and change have been conducted in Region 3: one for the Mackenzie River watershed in Canada (Cohen, 1997a,b) and the other for Alaska and the Bering Sea (NAST, 2000, 2001; Weller et al., 1999) as part of the US Global Change Research Program. The Canada Country Study (Environment Canada, 1997) described impacts in the Yukon Territory. These assessments provided background information and input

for this assessment. No detailed impact studies have been conducted for Chukotka.

The entire region, but particularly Alaska and the western Canadian Arctic, has undergone a marked change over the last three decades, including a sharp reduction in snow-cover extent and duration, shorter river- and lake-ice seasons, melting of mountain glaciers, sea-ice retreat and thinning, permafrost retreat, and increased active-layer depth. These changes have caused major ecological and socio-economic impacts, which are likely to continue or worsen under projected future climate change. Thawing permafrost and northward movement of the permafrost boundary are likely to increase slope instabilities, which will lead to costly road replacement and increased maintenance costs for pipelines and other infrastructure. The projected shift in climate is likely to convert some forested areas into bogs when ice-rich permafrost thaws. Other areas of Alaska, such as the North Slope, are expected to continue drying. Reduced sea-ice extent and thickness, rising sea level, and increases in the length of the open-water season in the region will increase the frequency and intensity of storm surges and wave development, which in turn will increase coastal erosion and flooding.

Warmer temperatures have resulted in some northward expansion of boreal forest, as well as significant increases in fire frequency and intensity, unprecedented insect outbreaks, and a 20% increase in growing-degree days. The latter has benefited both agriculture and forestry. The expansion of forests in most areas and their increased vulnerability to fire and pest disruption are projected to increase. One simulation projects a three-fold increase in the total area burned per decade, destroying coniferous forests and eventually leading to a deciduous forest-dominated landscape on the Seward Peninsula in Alaska, after a warmer climate has led to forestation of the present tundra areas. Shrubbiness is already increasing in this area, a trend that is likely to continue.

Observations in the Bering Sea have shown abnormal conditions during recent years. The changes observed include significant reductions of seabird and marine mammal populations, unusual algal blooms, abnormally warm water temperatures, and low harvests of salmon on their return to spawning areas. Some of the changes observed in the 1997 and 1998 summers, such as warmer ocean temperatures and altered currents and atmospheric conditions, may have been exacerbated by the very strong El Niño event, but the area has been undergoing change for several decades. While the Bering Sea fishery has become one of the world's largest, the abundance of Steller sea lions (*Eumetopias jubatus*) has declined by between 50 and 80%. Northern fur seal (*Callorhinus ursinus*) pups on the Pribilof Islands – the major Bering Sea breeding grounds – declined by 50% between the 1950s and the 1980s. There have been significant declines in the populations of some seabird species, including common murre (*Uria aalge*), thick-billed murre (*U. lomvia*), and red- and blacklegged kittiwakes (*Rissa brevirostris* and *R. tridactyla*, respectively). Also, the number of salmon has been far below expected levels, the fish were smaller than average, and their traditional migratory patterns seemed to have altered.

Differentiating between the various factors affecting the Bering Sea ecosystem is a major focus of current and projected research. Well-documented climatic regime shifts occurred in the Bering Sea during the 20th century on roughly decadal time scales, alternating between warm and cool periods. A climatic regime shift occurred in the Bering Sea in 1976, changing the marine environment from a cool to a warm state. Information from the contrast between the warm and subsequent cool period forms the basis of projected responses of the Bering Sea ecosystem to scenarios of future warming. These projections show increased primary and secondary productivity with greater carrying capacity, poleward shifts in the distribution of some cold-water species, and possible negative effects in ice-associated species.

18.3.3.3. Impacts on the economy

Large oil and gas reserves exist in Alaska along the Beaufort Sea coast and in the Mackenzie River/Beaufort Sea area of Canada. To date, climate change impacts on oil and gas development in Region 3 have been minor but are likely to result in both financial costs and benefits in future. For example, offshore oil exploration and production is likely to benefit from less extensive and thinner sea ice, allowing savings in the construction of platforms that must withstand ice forces. Conversely, ice roads, now used widely for access to offshore activities and facilities, are likely to be less safe and useable for shorter periods; the same applies for over-snow transport on land given projected reductions in snow depth and duration. The thawing of permafrost, on which buildings, pipelines, airfields, and coastal installations supporting oil development are located, is very likely to adversely affect these structures and greatly increase the cost of maintaining or replacing them.

It is difficult to project impacts on the lucrative Bering Sea fisheries because many factors other than climate are involved, including fisheries policies, market demands and prices, harvesting practices, and fisheries technology. Large northward changes in the distribution of fish and shellfish are likely with a warmer climate. Relocating the fisheries infrastructure (fishing vessels, home ports, processing plants) may be necessary, and would incur substantial costs. Warmer waters are likely to lead to increased primary production in some regions, but a decline in cold-water species such as salmon and pollock.

Other economic sectors in this region, including forestry and agriculture, are far less developed and currently less important than oil and gas and fish and wildlife. Owing to this, economic impacts on forestry and agriculture resulting from climate change are unlikely to be significant, except locally. Impacts on tourism,

which is a large economic sector in this and other regions, are more difficult to assess, largely due to the relationship between tourism and economic conditions and social factors outside the Arctic. It is also unclear which features of Region 3 are primarily responsible for attracting tourists – large, undeveloped landscapes will not be directly affected by climate change, whereas marine mammal populations and accessible glaciers are likely to experience major changes. Whether such changes will reduce tourist interest is difficult to assess without more information.

18.3.3.4. Impacts on people's lives

Traditional lifestyles are already being threatened by multiple climate-related factors, including reduced or displaced populations of marine mammals, seabirds, and other wildlife, and reductions in the extent and thickness of sea ice, making hunting more difficult and dangerous. Indigenous communities depend on fish, marine mammals, and other wildlife, through hunting, trapping, fishing, and caribou/reindeer herding. These activities play social and cultural roles that may be far greater than their contribution to monetary incomes. Also, these foods from the land and sea make significant contributions to the daily diet and nutritional status of many indigenous populations and represent important opportunities for physical activity among populations that are increasingly sedentary.

Climate change is likely to have significant impacts on the availability of key marine and terrestrial species as food resources. At a minimum, salmon, herring, char, cod, walrus, seals, whales, caribou, moose, and various species of seabird are likely to undergo shifts in range and abundance. This will entail major local adjustments in harvest strategies and allocations of labor and equipment.

The following impacts on the lifestyles in indigenous villages and communities in Alaska and Canada, which depend heavily on fishing and hunting, have been observed in recent years:

- access to tundra and offshore food resources has been impeded by higher temperatures with milder winters, shorter duration of snow cover and sea ice, and less (or no) shore-fast ice and snow;
- recent decreases in anadromous fish stocks, which make up 60% of wildlife resources harvested by local residents, have directly affected their dietary and economic well-being;
- availability of marine mammals for local harvests has declined in some areas due to population declines associated with shifts in oceanographic and sea-ice conditions. Marine mammals are an important food source in many coastal communities;
- sea-level rise, permafrost thawing, and storm surges have triggered increased coastal erosion and threatened several villages along the Bering and Beaufort Sea coasts. The only long-term option

available has been to plan for relocation of villages, which will be very costly;

- storm surges have also reduced the protection of coastal habitats provided by barrier islands and spits, which are highly vulnerable to erosion and wave destruction; and
- infrastructure in villages constructed in or over permafrost has been affected by thawing permafrost and storm surges breaching coastlines into water supplies and sewage lagoons.

Changes in diet, nutritional health, and exposure to air-, water-, and food-borne contaminants are also likely. Adjustments in the balance between the "two economies" of rural areas (traditional and wage) will be accelerated by climate change. This suite of changes will be complex and largely indirect because of the mediating influences of market trends, the regulatory environment, and the pace and direction of rural development.

Other impacts are likely to occur in the future and have a substantial impact on people. For example:

- a decrease in the area of pack ice is projected to have important implications for primary productivity and the entire food chain. For example, walrus (*Odobenus rosmarus*) and bearded seals (*Erignathus barbatus*) require sea ice strong enough to support their weight, and ringed seals require stable shore-fast ice with adequate snow cover. Diving from ice over shallow waters allows walrus to reach the bottom to feed, and reductions in the extent and thickness of sea ice will adversely affect this species;
- as the boreal forest and associated shrub communities expand northward at the expense of tundra, changes in habitats, migration routes, ranges, and distribution and density of a number of wildlife species, particularly caribou and moose, are projected;
- a change in vegetation and landscape, affecting wildlife, is likely to change hunting practices, location of settlements, and local economic opportunities for people in many arctic regions;
- among wildlife species used as food, existing zoonotic diseases such as brucellosis and echinoccus are likely to become a greater threat to humans and wildlife. New diseases, such as West Nile virus, are likely to become established in a progressively warmer climate;
- lower water levels in some river basins, for example the Mackenzie River, cause increased erosion of riverbanks due to thawing permafrost. This erosion is very likely to increase the incidence of landslides, which have the potential to adversely affect community infrastructure; and
- more vigorous atmospheric and oceanic circulations are likely to increase the transport of contaminants from agricultural activities as well as military and industrial installations to arctic communities, both directly and via the food chain.

18.3.4. Region 4

A major integrative impact assessment for the region was published by Maxwell (1997) for the Canadian Arctic (encompassing the regions of the Northwest Territories and Nunavut). The Mackenzie River watershed assessment, mentioned in the summary for Region 3, also covers Region 4 (Cohen, 1997a,b). A number of studies of the traditional ecological knowledge of indigenous peoples of Region 4 have been conducted and are cited in earlier chapters. Detailed studies of integrated climate impacts in Greenland, on the other hand, have not been conducted prior to the ACIA.

18.3.4.1. Changes in climate

Temperature changes over the past decades have varied across Region 4. The amount of change depends on the time period chosen and, as shown in Chapter 2, the warming has been pronounced since 1966. Between 1954 and 2003, mean annual temperatures across most of arctic Canada increased by as much as 2 to 3 °C (see Fig. 18.2a), while temperatures in northeastern Canada, including Labrador and adjacent waters, showed little change. The southern part of West Greenland (including the surrounding ocean) cooled by about 1 °C while northern Greenland warmed by 1 to 2 °C. Winter temperature trends over the same period were noticeably warmer in the west and colder in the east than the annual trends (Fig. 18.2b). The landmass to the west of Hudson Bay warmed by up to 4 °C in winter while the area around Labrador, Baffin Island, and southwest Greenland experienced winter cooling of more than 1 °C.

Annual precipitation in Region 4 has increased over the past 50 years or so, and while seasonal differences were evident around the middle of the century, increases in precipitation have been evident in all seasons over the past few decades.

The physical complexity of the Queen Elizabeth Islands and the orography of Greenland create particular challenges for modeling past, present, and future arctic climate. Models project warming throughout Region 4 during the 21st century, with no cooling projected in any season. Figures 18.5a and 18.5b illustrate annual and winter temperature changes for the period 2071–2090 relative to the 1981–2000 baseline, projected by the five ACIA-designated models forced with the B2 emissions scenario. Projected winter warming in the Canadian areas of Region 4 ranges from about 3 °C up to about 9 °C, with the greatest warming projected to occur around southern Baffin Island and Hudson Bay, and substantially less warming projected in other seasons. Greenland is also projected to warm but the warming is weaker (up to about 3 °C by 2071–2090) and more consistent across seasons.

Precipitation increases are projected to be greatest in autumn and winter, and the areas of greatest increase (up to 30% by the end of the 21st century) generally correspond with the areas of greatest warming. Almost all areas of Region 4 are projected to experience some increase in precipitation after the first few decades of the 21st century.

18.3.4.2. Impacts on the environment

The Canadian part of Region 4 has significant areas of warm permafrost that are at risk of thawing with rising regional air temperatures. The boundary between continuous and discontinuous permafrost is projected to shift poleward, following, but with a lag in timing, the several-hundred-kilometer movement of the isotherms of mean annual temperature over the 21st century. This is likely to result in the disappearance of a substantial amount of the permafrost in the present discontinuous zone. Areas of warm permafrost are also likely to experience more widespread thermokarst development where soils are ice-rich, and increases in slope instability. In areas of remaining continuous and cold permafrost, increases in active-layer depth can be expected.

The maximum northward retreat of sea ice during summer is projected to increase from its present range of 150–200 km to 500–800 km. The thickness of fast ice in the Northwest Passage is likely to decrease substantially from its current value of 1 to 2 m.

The Greenland Ice Sheet is presently losing mass in its ablation zone and is likely to contribute substantially to sea-level rise in the future. While precipitation is projected to increase, it is possible that increased evaporation rates will lead to lower river and lake levels during the warm season.

In general terms, and consistent with results for other regions, the biomes of arctic Canada and Greenland are expected to change. Reductions in the area covered by polar deserts in Canada and Greenland are likely to result from the northward shift of the tundra, while reductions in the areas covered by arctic tundra are very likely to result from the northward shift of the treeline. Polar deserts in the region are extensive, and these areas could sequester large amounts of carbon dioxide (CO_2) if tundra vegetation displaces polar deserts. A reduction in polar desert area of about 36% by 2080 is projected, leading to the greatest projected carbon gain of any region. In contrast, increases in temperature and precipitation are likely to lead to relatively small increases in the area of taiga compared with other regions.

Many treelines, such as those in northeast Canada, have been relatively stable for the last few thousand years. A widespread and consistent observation from the late 20th century has been the infilling of sparse stands of trees near the tundra edge into dense stands that no longer retain the features of the tundra. Movement of the treeline northward is likely when climatic conditions become favorable, but the actual movement of trees will lag the climate warming considerably in time. The forests of northwestern Canada have recently experienced forest

health problems driven by insects, fire, and tree growth stress that are all associated with recent mild winters and warmer growing seasons. These findings for Region 4 complement those of Region 3 and accord with very large-scale environmental changes in the western North America subarctic. It is very likely that such forest health problems will become increasingly intense and widespread in response to future regional warming.

The Canadian High Arctic is characterized by land fragmentation within the archipelago and by large glaciated areas, leading to constraints on species movement and establishment. In West Greenland, loss of habitat and displacement of species in combination with time delays in species immigration from the south will ultimately lead to loss of the present biodiversity. However, Region 4 contains relatively few rare and endemic vascular plant species and threatened animal and plant species, compared with the other three regions, so biodiversity losses are likely to be less significant here.

Changes in timing and abundance of forage availability, insect harassment, and parasite infestations will increase stress on caribou, tending to reduce their populations. The ability of high-arctic Peary caribou and muskoxen to forage may become increasingly limited as a result of adverse snow conditions, in which case numbers will decline, with local extirpation in some areas. Direct involvement of the users of wildlife in its management at the local level has the potential for rapid management response to changes in wildlife populations and their availability for harvest.

Arctic freshwater systems are particularly sensitive to climate change because many hydro-ecological processes respond to even small changes in climatic regimes. These processes may change in a gradual way in response to changes in climate or in an abrupt manner as environmental or ecosystem thresholds are exceeded. Pronounced potential warming of freshwater systems in the autumn is particularly important because this is typically when these systems along the coastal margins currently experience freeze-up. Such warming is projected to delay freeze-up by up to 25 days in parts of Region 4. Also, high-latitude cold-season warming is likely to lead to less severe ice breakups and flooding as the spring flood wave pushes northward along arctic rivers. Hence, future changes in the spring timing of lake- and river-ice breakup and the export of freshwater to the Arctic Ocean are likely.

With respect to freshwater ecosystems, significant shifts in species range, composition, and trophic relations are also very likely to occur in response to the projected changes. Salmonids of northern Québec and Labrador, such as native Atlantic salmon and brook trout (*Salvelinus fontinalis*) and introduced brown trout (*Salmo trutta*) and rainbow trout (*Oncorhynchus mykiss*), are likely to extend their ranges northward. Because of these range extensions, the abundance of Arctic char (*Salvelinus alpinus*) is likely to be reduced throughout much of the southern part of Region 4, and brook trout are likely to become a more important component of native subsistence fisheries in rivers now lying within the tundra zone. Lake trout (*Salvelinus namaycush*) are also likely to disappear from rivers and the shallow margins of many northern lakes, and northern pike (*Esox lucius*) are expected to reduce in both numbers and size throughout much of their current range.

Marine mammal populations are likely to decline as the sea ice recedes but the populations of beluga and bowhead whales (*Delphinapterus leucas* and *Balaena mysticetus*, respectively) could increase (depending on the extent to which these whales become more vulnerable to predation as sea-ice cover decreases). If the Arctic Ocean becomes seasonally ice free for several years in a row, it is possible that polar bears would become extinct. Sea-level rise will change the location and distribution of coastal habitats for seabirds and some species of marine mammals (e.g., walrus haul-outs may become inundated).

18.3.4.3. Impacts on the economy

Oil and gas extraction and mining are active industries in Region 4. Diamond mining is underway in the Northwest Territories, and the development of a large nickel deposit in Voisey's Bay, Labrador, has recently been announced. Many rivers in the northern parts of Québec, Ontario, and Manitoba have been dammed for their hydroelectric potential. Roads, airstrips, and ports have been constructed and are essential to the economic infrastructure supporting these activities. Any expansion of oil and gas activities, mining, agriculture, or forestry is likely to require expansion of supporting infrastructure, including air, marine, and land transportation systems. Ice roads in nearshore areas and over-snow transport on land, systems that are important and are even now experiencing shorter seasons, are likely to be further curtailed in the future because of reduced extent and duration of sea ice and snow.

With reduced summer sea-ice extent, the shipping season in Canadian arctic waters is likely to be extended, although sea-ice conditions are likely to remain very challenging. Extension of the shipping season will result in costs and benefits, both of which are speculative. Benefits are likely to result from increased access to the natural resources of the region. As sea level rises, this will also benefit shipping by creating deeper drafts in harbors and channels. On the other hand, increased costs would result from greater wave heights, and possible flooding and erosion threats to coastal facilities. Increased rates of sediment movement during longer, more energetic open-water seasons are likely to increase rates of port and harbor infill and increase dredging costs. Increased ship traffic in the Northwest Passage will increase the risks and potential environmental damage from oil and other chemical spills.

Warmer air temperatures would be expected to reduce the power demand for heating, reduce insulation needs, and increase the length of the summer construction season. Other infrastructure likely to be affected by climate change includes northern pipeline design (negative); pile foundations in permafrost (negative but depending on depth of pile); bridges, pipeline river crossings, dikes, and erosion protection structures (negative); and open-pit mine wall stability (negative).

Impacts on marine fisheries in the eastern part of Region 4, under a moderate gradual warming, are likely to include a return to a cod–capelin system with a gradual decline in northern shrimp and snow crabs. Under more modest assumptions of ocean warming, the range of demersal species (those that tend to live near the bottom) are expected to expand northward. If ocean warming is more extreme, it is likely that the southern limit of the range of the demersal species would move northward. Many existing capelin-spawning beaches are likely to disappear as sea levels rise. If there is an increase in demersal spawning by capelin in the absence of new spawning beaches, capelin survival may decline. Seals may experience higher pup mortality as sea ice thins. Increases in regional storm intensities may also result in higher pup mortality. A reduction in the extent and duration of sea ice is likely to permit fishing further to the north and is likely to shorten the duration of Greenland halibut fisheries that are conducted through fast ice.

Impacts on freshwater and anadromous fisheries, and their economic benefits, such as tourism and local economic development, will vary across Region 4 and will depend on the local present-day and future species composition. Initially, local productivity associated with present-day freshwater and anadromous species is likely to increase, but as critical thresholds are reached (e.g., thermal limits) and as new species move in to the area, arctic-adapted species such as Arctic char are likely to experience declines in abundance and ultimately become extirpated. Loss of suitable habitat will result in decreased individual growth and declines of many populations, with resulting impacts on sport fisheries and local tourism.

18.3.4.4. Impacts on people's lives

The changes in climatic and environmental conditions projected for Region 4, and already being observed in some parts, affect people's lives in many ways. Seasonal unpredictability throughout Region 4 has already created dangerous environmental situations. For example, for the Inuit west of Hudson Bay, changing wind patterns and snow conditions make it difficult to build igloos as the snow is packed too hard. As a result, Inuit report increasing difficulty in building shelters during unexpected storms. In areas of Nunavik and Labrador the snow changes can differ, for example the type of snow now seen does not pack well enough. Changes in weather and ice conditions, such as earlier spring melt, later freeze-up, and formation of more cracks, such as those reported in the Kitikmeot region of Nunavut, result in increasingly difficult travel conditions and sometimes shifts in regular travel and harvesting times.

The changes in local environments experienced by the people in the Canadian part of Region 4 include thinner sea ice, early breakup and later freeze-up of sea ice and lake ice, sudden changes in wind direction and intensity, earlier and faster spring melt periods, decreasing water levels in mainland lakes and rivers, and the introduction of non-native animal and bird species. These changes affect lifestyles through changes in the timing of animal migrations as well as in the numbers and health of some animal populations, and in the quality of animal skins and pelts. The distribution and quality of animals and other resources will affect the livelihoods, and ultimately the health of northern communities in Region 4. For example, a shorter winter season with increased snowfall and less extensive and thinner sea ice is likely to decrease the opportunity and increase the risks for indigenous people to hunt and trap.

Other health impacts may arise from the introduction of new or increasingly present zoonotic and/or vector-borne diseases (e.g., potential spread of West Nile virus into warmer regions in the western Arctic), changes in exposure to UV radiation and contaminants that already threaten confidence in and safety of traditional diets, and the associated social and cultural impacts of this combination of changes. Relocation of low-lying communities may be forced by rising sea levels, with serious social impacts. Where these challenges to health already exist, and where infrastructure and support systems are stretched, the effects are likely be experienced to a greater extent and at a faster pace than elsewhere.

Many changes are reported and are currently experienced by Inuit, Dene, Gwitch'in, and other indigenous peoples in Region 4. These changes represent challenges to aspects of northern indigenous cultures and lifestyles that have existed for centuries. The ability of communities to cope with and adapt to climate-driven changes is also influenced by a number of other factors and is constrained by current social and economic aspects. For example, moving people to follow shifting resources is no longer an option with permanent settlements. Other factors complicating adaptations to change include regional resource regulations, industrial development, and global economic pressures. Climate change interacts with such forces and must be considered in assessing local risks and responses. As existing adaptation strategies become obsolete, new adaptations to climate impacts must develop as northern communities adjust to the many social, institutional, and economic changes related to land claim settlements, changes in job opportunities, and the creation of new political and social structures in the North.

18.4. Cross-cutting issues in the Arctic

This assessment has dealt with individual topics that reflect traditional academic and practical organization.

However, a strong thread running through the assessment is the interaction between the various topics and processes in the Arctic. This includes, for example, the *interactions* between physical atmospheric processes and biological processes in the major ecosystems, and the strong albedo and other feedback responses; the physical and biological *connections* between land, freshwater, and marine environments; and the integrity of the arctic system as a whole. Three important cross-cutting issues that illustrate the interactions within the Arctic and connections with the global system are carbon storage and carbon cycling, biodiversity, and abrupt climate change and extreme events.

18.4.1. Carbon storage and carbon cycling

18.4.1.1. Global importance of carbon in the Arctic

The Arctic contains large stores of carbon that have historically been sequestered from the atmospheric carbon pool. Estimates of arctic and boreal soil carbon (C) in the upper meter of soil vary considerably, ranging from 90 to 290 Pg C in upland boreal forest soils, 120 to 460 Pg C in peatland soils, and 60 to 190 Pg C in arctic tundra soils. There is also a general sparsity of high-latitude carbon data for aquatic ecosystems relative to arctic terrestrial systems, but some estimates from boreal lakes indicate that reserves can be significant (120 Pg C). An additional 450 Pg of organic C is stored as dissolved carbon in the Arctic Ocean (see Chapter 9). Estimates of carbon stored in the upper 100 meters of permafrost are as high as 10000 Pg C (Semiletov, 1999). In any case, the carbon stored in northern boreal forests, lakes, tundra, the Arctic Ocean, and permafrost is considerably greater than the *global* atmospheric pool of carbon, which is estimated at 730 Pg C (IPCC, 2001). In addition, up to 10000 Pg C in the form of CH_4 and CO_2 is stored as hydrates in marine permafrost below 100 m (Chapter 9), however, this figure is a maximum of estimates that span several orders of magnitude.

18.4.1.2. Spatial patterns of carbon storage

Within the Arctic, carbon storage generally decreases from south to north. On land, this represents parallel decreases from boreal forest to tundra to polar desert and from southern isolated, sporadic, and discontinuous permafrost to continuous permafrost in the north; in freshwater ecosystems there is a decrease from peatlands and lakes with high concentrations of dissolved organic carbon to tundra and high-arctic ponds with low dissolved organic carbon; and in the marine environment there is a decrease from areas of high organic matter production and sedimentation in the south and at the ice margin to relatively clear waters in the Arctic Ocean. Marine permafrost and gas hydrates show a different pattern in that they are concentrated in the area of continental shelves, which are particularly extensive along the northern coastlines of the arctic landmasses.

18.4.1.3. Processes involved in carbon storage and release

Over thousands of years, an imbalance between photosynthesis and decomposition has led to storage of carbon in lake and ocean sediments, and in forest and tundra soils (Chapters 7, 8, 9, and 14). On land, this imbalance was created because low temperatures, particularly when combined with high soil moisture, retarded microbial decomposition more than photosynthesis. In the marine environment, atmospheric carbon is dissolved as inorganic carbon in surface waters and stored at depth as a result of the physical pump; death and decomposition of organisms also lead to carbon storage in the form of dissolved and particulate organic carbon (Chapter 9). Low ocean temperatures have resulted in high solubility of carbon, while extensive sea-ice cover has reduced the duration and area for carbon exchange between air and surface waters (and thus photosynthesis).

Because low temperatures have been so important for the capture and storage of atmospheric carbon in the Arctic, projected temperature increases have the potential to lead to the release of old and more recently captured carbon to the atmosphere, although the older the stored carbon, the less responsive it will be to projected climate changes. The release of stored carbon will increase atmospheric GHG concentrations and provide a positive feedback to the climate system. However, increased temperatures are also likely to increase the photosynthetic capture of atmospheric carbon if other environmental conditions do not become limiting. On land, plants will grow faster and more productive vegetation will successively replace less productive vegetation at higher latitudes and altitudes (Chapters 7 and 14). In freshwater ecosystems, reduced duration of ice cover over lakes and ponds and increased temperatures are likely to increase primary production (Chapter 8). In the marine environment, primary production is expected to increase as areas where production has been limited by sea-ice cover become more restricted in extent. Also, it is likely that more carbon will be buried as deposition shifts from the continental shelves where primary production is currently concentrated to the deeper slope and basin region as the ice edge retreats (Chapter 9).

The balance between the opposing processes of increased carbon capture and release will determine future changes in the carbon feedback from the Arctic to global climate. However, there are great uncertainties in calculating this balance across permafrost, terrestrial soil, ocean, and freshwater systems and no quantitative integrative assessment has been performed to date.

18.4.1.4. Projected changes in carbon storage and release to the atmosphere

There is a consensus from the trace-gas measurement researchers that the terrestrial Arctic is presently a source of carbon and radiative forcing, but is likely to become a weak sink of carbon during future warming (Chapter 7). Modeling approaches suggest that circumpolar mean carbon uptake is likely to increase from the current 12 g $C/m^2/yr$ to 22 g $C/m^2/yr$ by 2100 and that carbon storage is likely to increase by 12 to 31 Pg C depending on the ACIA climate scenario used. However, the uncertainties are great: the projections are limited to terrestrial ecosystems and do not include carbon stored in permafrost and gas hydrates. Potential increases in human and natural disturbances are further uncertainties. The marine environment has been suggested as a weak sink, but the amount of carbon that the Arctic Ocean can sequester is likely to increase significantly under scenarios of decreased sea-ice cover, both through surface uptake and increased biological production, although there may be an abrupt release of CO_2 and CH_4 from thawing permafrost in marine sediments.

In the marine environment, there are vast stores of CH_4 and CO_2 (at least 10 000 Pg C in the form of gas hydrates in marine permafrost below 100 m; Semiletov, 1999). As there are currently about 4 Pg C in CH_4 in the atmosphere, even the release of a small percentage of CH_4 from gas hydrates could result in an abrupt and significant climate forcing (Chapter 9). The process of CH_4 release from gas hydrates under continental shelves could already be occurring due to the warming of earlier coastal landmasses during Holocene flooding. On land, however, natural gas hydrates are found only at depths of several hundreds of meters and are relatively inert.

18.4.2. Biodiversity

18.4.2.1. Background

The diversity of species in terrestrial, freshwater, and marine ecosystems of the Arctic is fundamental to the life support of the residents of the region and to commercial interests such as fishing at lower latitudes. Diversity is also important to the functioning of arctic ecosystems: productivity, carbon emissions, and albedo are all related to specific characteristics of current arctic species. While the Arctic contains some specialist species that are well adapted to the harsh arctic environment, it also contains species that migrate and contribute to the biodiversity of more southerly latitudes. Each year, whales, dolphins, and hundreds of millions of birds migrate from the Arctic to warmer latitudes. The Arctic is an area of relatively undisturbed and natural biodiversity because of generally lower human impacts than elsewhere on earth. However, at its southern border, human impacts are greater and particular areas, such as old growth forests on land, preserve biodiversity that is endangered in managed areas.

18.4.2.2. Patterns of diversity in the Arctic

The diversity of living organisms at any one time in the Arctic is a snapshot of complex, dynamic physical and biological processes that create habitats and opportunities or constraints for species, and genetically distinct populations of particular species, to colonize them.

The current diversity of organisms in the Arctic has been shaped by major climatic and associated changes in physical and chemical conditions of the land, wetlands, and oceans over past glacial and interglacial periods. Changes are presently occurring that are also driven by direct human activities such as fishing, hunting, and gathering, changes in land use, and habitat fragmentation, in addition to indirect human activities such as anthropogenic climate change, stratospheric ozone depletion, and transboundary movement of contaminants.

On land, and in freshwater and the marine environment, the fauna and flora are young in a geological context. Recent glaciations resulted in major losses of biodiversity, and recolonization has been slow because of the extreme environmental conditions and overall low productivity of the arctic system. On land, of at least 12 large herbivores and six large carnivores present in steppe–tundra areas at the last glacial maximum, only four and three respectively survive today and of these, only two herbivores (reindeer and musk ox) and two carnivores (brown bear, *Ursus arctos* and wolf, *Canis lupus*) presently occur in the arctic tundra biome. Arctic marine mammals to a large extent escaped the mass extinctions that affected their terrestrial counterparts at the end of the Pleistocene because of their great mobility. However, hunting in historical times had severe impacts on several species that were exterminated (great auk, *Pinguinus impennis*; Steller sea cow, *Hydrodamalis gigas*) or almost harvested to extinction (walrus; bowhead whale; sea otter, *Enhydra lutris*). Polar bears, all the Great Whales, white whales, and many species of colonially nesting birds were dramatically reduced.

The youth of arctic flora and fauna, together with the harsh physical environment of arctic habitats and to some extent over-harvesting, have resulted in lower species diversity in the Arctic compared to other regions. This results in arctic ecosystems, in a global sense, being "simple". Some of the species are specialists that are well adapted to the Arctic's physical environment; others were pre-adapted to the arctic environment and moved north during deglaciation. Overall however, many arctic species – marine, freshwater, and terrestrial – possess a suite of characteristics that allows them to survive in extreme environments. However, these characteristics, together with low diversity and simple relationships between species in food webs, render arctic species and ecosystems vulnerable to the environmental changes now occurring in the Arctic and those projected to occur in the future.

Although diversity of arctic species is relatively low, in absolute terms it can be high: about 6000 marine, freshwater, and terrestrial species have been catalogued in and around Svalbard (Prestrud et al., 2004) and about 7200 terrestrial and freshwater species have been recorded in a subarctic area of northern Finland (Callaghan et al., 2004). About 3% (around 5900 species) of the global flora occurs in the Arctic. The

diversity of arctic terrestrial animals beyond the latitudinal treeline (6000 species) is nearly twice as great as that of vascular plants and bryophytes. The arctic fauna accounts for about 2% of the global total. In the arctic region as defined by CAFF (Conservation of Arctic Flora and Fauna), which includes forested areas, some 450 species of birds have been recorded breeding, and around 280 species migrate. The diversity of vertebrate species in the arctic marine environment is less than on land. Species diversity differs from group to group: primitive species of land plants such as mosses and lichens are well represented in the Arctic whereas more advanced flowering plants are not; primitive species of land animals such as springtails are well represented whereas more advanced beetles and mammals are not. In contrast, although most taxonomic groups of freshwater organisms in the Arctic are not diverse, some groups such as fish have high diversity at and below the species level. One consequence of the generally low species diversity is that species will be susceptible to damage by new insect pests, parasites, and diseases. For example, low diversity of boreal trees together with low diversity of parasites and predators that control populations of insect pests exaggerates the impacts of the pests.

The number of species generally decreases with increasing latitude. The steep temperature gradient that has such a strong influence on species diversity occurs over much shorter distances in the Arctic than in other biomes. North of the treeline in Siberia, mean July temperature decreases from 12 to 2 °C over 900 km, whereas a 10 °C decline in July temperature is spread over 2000 km in the boreal zone, and July temperature decreases by less than 10 °C from the equator to the southern boreal zone. Patterns of species diversity in the Arctic also differ according to geography. With its complicated relief, geology, and biogeographic history, there are more species on land in Beringia at a given temperature than on the Taymir Peninsula. Taymir biodiversity values are intermediate between the higher values for Chukotka and Alaska, which have a more complicated relief, geology, and floristic history, and the lower values in the eastern Canadian Arctic with its impoverished flora resulting from relatively recent glaciation. Within any region, biological hot spots occur, for example below predictable leads in the sea ice, polynyas, oceanographic fronts, areas of intense mixing, and the marginal ice zone in the marine environment; in delta areas that lie at the interface between rivers and lakes or oceans; and at the ecotone between tundra and taiga on land where elements of both forest and tundra floras and faunas mix. Such hotspots are centers from which species with restricted distributions can expand during climatic warming.

An important consequence of the general decline in numbers of species with increasing latitude is an increase in abundance and dominance. For example, on land, one species of collembolan, *Folsomia regularis*, may constitute 60% of the total collembolan density in the polar desert.

In freshwater ecosystems, midge and mosquito larvae are particularly abundant but species-level diversity is low. These "super-dominant" species, such as lemmings in peak years of their population cycles, are generally highly plastic, occupy a wide range of habitats, and generally have large effects on ecosystem processes. Similarly, arctic fish communities of the marine environment are dominated by a few species, several of which are commercially important, while the abundance of fish, marine mammals, and birds attracted hunters and fishing enterprises in historical times. Loss/reduction of one or more of these species, particularly fish species, will have disproportionate impacts on economy and ecosystem function.

18.4.2.3. Characteristics of arctic species related to the arctic environment

Several physical factors make arctic marine systems unique from other oceanic regions including: a very high proportion of continental shelves and shallow water; a dramatic seasonality and overall low level of sunlight; extremely low water temperatures (but not compared with arctic terrestrial habitats); presence of extensive permanent and seasonal sea-ice cover; and a strong freshwater influence from rivers and ice melt. Arctic freshwater environments are also characterized by extreme seasonality and low levels of incident radiation, much of which is reflected due to the high albedo of snow and ice. In addition, the thermal energy of a substantive part of this incoming radiation is used to melt ice, rendering it unavailable to biota. However, large arctic rivers with headwaters south of the Arctic act as conduits of heat and biota.

On land, low solar angles, long snow-covered winters, cold soils with permafrost, and low nutrient availability in often primitive soils limit survival and productivity of organisms. Many species of marine and terrestrial environments migrate between relatively warm wintering grounds in the south and the rich, but short-lived, feeding and breeding grounds in the Arctic. In freshwater ecosystems, some fish are highly migratory, moving in response to environmental cues. Those species that do not migrate have a suite of characteristics (behavior, physiology, reproduction, growth, development) that allow them to avoid the harshest weather or to persist. Two characteristics common to marine, freshwater, and terrestrial arctic organisms are a protracted life span with slow development over several years to compensate for the brevity and harshness of each growing/feeding/breeding season, and low reproductive rates. These characteristics render arctic organisms in general vulnerable to disturbance and environmental change.

18.4.2.4. Responses of biodiversity to climate and UV-B radiation change

The past and present patterns of biodiversity in the Arctic, the characteristics of arctic species, and the experimental and modeling assessments described in Chapters 7, 8, and 9, together make it possible to infer the following likely changes to arctic biodiversity:

- The total number of species in the Arctic will increase as new species move northward during warming. Large, northward-flowing rivers are conduits for species to move northward. New communities will form.
- Present arctic species will relocate northward, as in the past, rather than adapt to new climate envelopes, particularly as the projected rate of climate change exceeds the ability of most species to adapt. However, some species, particularly freshwater species, may already be pre-adapted to acclimate successfully to rapid climate change.
- Locally adapted species may be extirpated from certain areas while arctic specialists and particular groups of species that are poorly represented in the Arctic – some through loss of species during earlier periods of climate warming – will be at risk of extinction. Examples of arctic specialists at risk include polar bears, and seals of the ice margins in the marine environment, and large ungulates and predators on land.
- Presently super-abundant species will be restricted in range and abundance with severe impacts on commercial fisheries, indigenous hunting, and ecosystem function. Examples on land include lemmings, mosses and lichens, and some migratory birds.
- On land, shifts in major vegetation zones such as forests and tundra will be accompanied by changes in the species associated with them. For example, tree seed-eating birds, and wood-eating insects will move northward with trees.
- Low biodiversity will render ecosystems more susceptible to disturbance through insect pest infestations, parasites, pathogens, and disease.
- At the small scale, changes will be seen in the representation of different genetically separate populations within species. In cases such as Arctic char, the species may remain but become geographically or ecologically marginalized with the potential loss of particular morphs.
- Changes in UV-B radiation levels are likely to have small effects on biodiversity compared with climate warming. However, UV-B radiation has harmful effects on some fish larvae, on those amphibians that might colonize the Arctic, and on some microorganisms and fungi. In freshwater ecosystems, increased UV-B radiation levels could potentially reduce biodiversity by disadvantaging sensitive species and changing algal communities.
- All the projected changes in biodiversity resulting from changes in climate and UV-B radiation levels are likely to be modified by direct human activities. Protection and management of some areas have led to the recovery of some previously declining species while deforestation, extractive industries, and pollution have prevented forests and associated species from moving northward during

recent warming in some areas. Protection of ecosystems from the impacts of changes in climate and UV-B radiation in the long term is difficult and perhaps impossible.

18.4.3. Abrupt climate change and extreme events

Human activities are causing atmospheric concentrations of CO_2 and other GHGs and aerosols to change slowly from year to year, thereby causing the radiative forcing that drives climate change to shift slowly. However, the resulting changes in climate and associated impacts do not necessarily have to change slowly and smoothly. First, the natural interactions of the atmosphere, oceans, snow and ice, and the land surface, both within and outside the Arctic, can cause climatic conditions to fluctuate. These variations can cause months, seasons, years, and even decades to be warmer or cooler, wetter or drier, and even more settled or more changeable than the multi-decadal average conditions. Intermittent volcanic eruptions and variations or cycles in the intensity of solar radiation can also cause such fluctuations. These types of fluctuations can be larger than the annual or even decadal increment of long-term anthropogenic global climate change. Present model simulations project a slow and relatively steady change in baseline climate while natural factors create fluctuations on monthly to decadal scales. As the baseline climate changes, the ongoing fluctuations are very likely to cause new extremes to be reached and the occurrence of conditions that currently create stress (e.g., summer temperatures greater than 30 °C) are likely to increase significantly.

The climatic history of the earth shows that instead of climate changing steadily and gradually, change can be intermittent and abrupt in particular regions – even very large regions. Reconstructions of climatic variations over the last glacial cycle and the early part of the current interglacial some 8000 to 20 000 years ago suggest that temperature changes of several degrees in the large-scale, long-term climate occurred over a relatively short period. For example, ice cores indicate that temperatures over Greenland dropped by as much as 5 °C within a few years during the period of warming following the last glacial. These changes were apparently driven by a sharp change in the thermohaline circulation of the ocean (also referred to as the Atlantic meridional overturning circulation), which probably also prompted changes in the atmospheric circulation that caused large climatic changes over land areas surrounding the North Atlantic and beyond. Over multi-decadal time periods, persistent shifts in atmospheric circulation patterns, such as the North Atlantic and Arctic Oscillations, have also caused changes in the prevailing weather regimes of arctic countries, contributing, for example, to warm decades, such as the 1930s and 1940s, and cool decades, such as the 1950s and 1960s.

A recent example of a rapid change in arctic climate was the so-called regime shift in the Bering Sea in 1976,

which had serious consequences and impacts on the environment. In 1976, mean annual temperatures in Alaska experienced a step-like increase of 1.5 °C to a lasting new high level, shown as the average of several measuring stations. Sea-ice extent in the Bering Sea showed a similar step-like decrease of about 5% (Weller et al., 1999). An analysis by Ebbesmeyer et al. (1991) gave statistical measures of deviation from the normal of 40 environmental parameters in the North Pacific region, as a consequence of this rapid change. The parameters included air and water temperatures, chlorophyll, geese, salmon, crabs, glaciers, atmospheric dust, coral, CO_2, winds, ice cover, and Bering Strait transport. The authors concluded that "apparently one of the Earth's large ecosystems occasionally undergo large abrupt shifts".

As anthropogenic climate change continues, the potential exists for oceanic and atmospheric circulations to shift to new or unusual states. Whether such changes, perhaps brought on when a temperature or precipitation threshold is crossed, will occur abruptly (i.e., within a few years) or more gradually (i.e., within several decades or more) remains to be determined. Such shifts could cause the relatively rapid onset of various types of impacts. A warm and wet anomaly might accelerate the onset of pests or infectious diseases. Warming exceeding about 3 °C might initiate the long-term deterioration of the Greenland Ice Sheet as temperatures above freezing spread across the plateau in summer. The tentative indication of an initial slowing of the thermohaline circulation could change into a significant slowdown, greatly reducing the northward transport of tropical warmth that now moderates European winters. The likelihood of any such shifts or changes occurring is not yet well established, but if the future is like the past, the possibility for abrupt change and new extremes is real.

18.5. Improving future assessments

A critical self-assessment of the ACIA shows achievements as well as deficiencies. Regional impacts were only covered in an exploratory manner and are hence a priority for future assessments. The ACIA did examine climate and UV radiation impacts in the Arctic on (1) the environment, (2) economic sectors, and (3) on people's lives. Impacts on the environment were covered very extensively, but the assessment has only qualitative information on economic impacts, and this must be a priority for future assessments. Impacts on people's lives covered indigenous communities but had little information concerning other arctic residents. Integrative vulnerability studies were only covered in an exploratory manner (in Chapter 17) and need attention in the future.

Regional impacts

While the ACIA was successful in many respects, it mostly addressed impacts at the large-scale circumpolar level. An attempt to differentiate between impacts within the four ACIA regions was exploratory and did not cover these regions in depth. There is a need to focus future assessments on smaller regions (perhaps at the landscape level) where an assessment of impacts of climate change has the greatest relevance and use for residents in the region and their activities.

Economic impacts

There are many important economic sectors in the Arctic, including oil and gas production, mining, transportation, fisheries, forestry and agriculture, and tourism. Some will gain from a warmer climate, others will not. In most cases, only qualitative information about the economic impacts (in monetary terms) of climate change is presently available. It is essential to involve a wide range of experts and stakeholders in future climate impact assessments to fill this gap and provide relevant information to users and decision makers.

Assessing vulnerabilities

Vulnerability is the degree to which a system is susceptible to or unable to cope with adverse effects of multiple and interacting stresses. Climate change occurs amidst a number of other interacting social and environmental changes across scales from local to regional and even global, and includes industrial development, contaminant transport and effects, and changes in social, political, and economic conditions. In this context it has become important to assess the vulnerabilities of coupled human–environment systems in the Arctic.

To undertake improved assessments on these three research topics a suite of improved observations and process studies, long-term monitoring, climate modeling, and impact analyses on society are necessary. These require new research efforts and studies funded by the various arctic countries.

Observations and process studies

To improve future climate impact assessments, many arctic processes require further study, both through scientific investigations and more detailed systematic documentation of indigenous knowledge. Priorities include collection of data ranging from satellite, surface, and paleo data on the climate and physical environment, to rates and ranges of change in arctic biota, and to the health status of arctic people.

Long-term monitoring

Long-term time series of climate and climate-related parameters are available from only a few locations in the Arctic. The need for continuing long-term acquisition of data is crucial, including upgrading of the climate observing system throughout the Arctic and monitoring snow and ice features, the discharge of major arctic rivers, ocean parameters, and changes in vegetation, biodiversity, and ecosystem processes.

Climate modeling

Improvements in numerical modeling of potential changes in climate are needed, including the representation in climate models of key arctic processes such as ocean processes, permafrost–soil–vegetation interactions, important feedback processes, and extreme events. The development and use of very high-resolution coupled regional models that provide useful information to local experts and decision makers is also required.

Analysis of impacts on society

Critical needs for improving projections of possible consequences for the environment and society include development and use of impact models; evaluating approaches for expressing relative levels of certainty and uncertainty; developing linkages between traditional and scientific knowledge; preparing scenarios of arctic population and economic development; and identifying and evaluating potential mitigation and adaptation measures to meet expected impacts.

18.6. Conclusions

With its almost continuous circle of land surrounding an ocean, which has a decadal circulation, the Arctic is a globally unique system and it is no accident that the Arctic Climate Impact Assessment is the first comprehensive regional assessment conducted to date. The ACIA is an authoritative synthesis of the consequences of changes in climate and UV radiation in the Arctic, involving hundreds of arctic experts. The assessment addresses the large climatic change that is very likely to occur over the 21st century and it concludes that changes in climate and in ozone and UV radiation levels are likely to affect every aspect of life in the Arctic.

However, assessment of the projected impacts of changes in climate and UV radiation is a difficult and long-term undertaking and the conclusions presented here, while as complete as present information allows, are only a first step in what must be a continuing process. There are likely to be future surprises, such as relatively rapid shifts in the prevailing trends in climatic regimes and in the frequency and intensity of extreme events; such changes, while likely, are expected to remain very difficult to project with high confidence. In future years, however, as additional data are gathered, a better understanding of the complex processes, interactions, and feedbacks will develop, and as model simulations are refined, findings and projections will be made with increasing confidence. As understanding of the climate system and its interactions with ozone amounts steadily improves, it will be possible to increase the usefulness of projections of the likely impacts in the Arctic, allowing more specificity in planning how best to adapt and respond.

An especially important task for future impact assessments will be to conduct comprehensive vulnerability studies of arctic communities, in which impacts modulated by adaptive capacity are examined in the context of both environmental and societal changes. The latter include changes in resource exploitation, human population, global trade and economies, introduction of new species, contamination, and new technologies. Chapter 17 points the way in this direction. It will be important to consider the interplay between impacts due to climate change and these other drivers. It is possible that many of the adverse impacts of variability and change can be moderated or even offset by implementing strategies for coping and adaptation, for example via changes in public policy and new strategies in resource management. The perspectives and concerns of local people will also be essential to consider more fully in future vulnerability analyses. To begin to address these policy-related issues, a separate process is ongoing to discuss mitigation and adaptation, as well as research, observation, and modeling needs, and communication and education issues pertaining to the Arctic Climate Impact Assessment.

Finally, it is important to re-emphasize that climate and UV radiation changes in the Arctic are likely to affect every aspect of human life in the region and the lives of many living outside the region. While more studies and a better understanding of the expected changes are important, action must begin to be taken to address current and anticipated changes before the scale of changes and impacts further reduces the options available for prevention, mitigation, and adaptation.

References

NOTE: This chapter is a summary based on the seventeen preceding chapters of the Arctic Climate Impact Assessment and a full list of references is provided in those chapters. Only references to major publications and data sources, including integrative regional assessments, and some papers reporting the most recent developments, are listed.

AMAP, 1998. Assessment Report: Arctic Pollution Issues. Arctic Monitoring and Assessment Programme, Oslo, 859p.
Callaghan, T.V., M. Johansson, O.W. Heal, N.R. Sælthun, L.J. Barkved, N. Bayfield, O. Brandt, R. Brooker, H.H. Christiansen, M.C. Forchhammer, T.T. Høye, O. Humlum, A. Järvinen, C. Jonasson, J. Kohler, B. Magnusson, H. Meltofte, L. Mortensen, S. Neuvonen, I. Pearce, M. Rasch, L. Turner, B. Hasholt, E. Huhta, E. Leskinen, N. Nielsen and P. Siikamäki, 2004. Environmental Changes in the North Atlantic Region: SCANNET as a collaborative approach for documenting, understanding and predicting changes. Ambio Special Report, 13:35–58.
Chapman, W.L. and J.E. Walsh, 2003. Observed climate change in the Arctic, updated from Chapman and Walsh, 1993: Recent variations of sea ice and air temperatures in high latitudes. Bulletin of the American Meteorological Society, 74(1):33–47. http://arctic.atmos.uiuc.edu/CLIMATESUMMARY/2003/
Christensen, T.R., T.R. Johansson, H.J. Åkerman, M. Mastepanov, N. Malmer, T. Friborg, P. Crill and B.H. Svensson, 2004. Thawing sub-arctic permafrost: Effects on vegetation and methane emissions. Geophysical Research Letters, 31:L04501,doi:10.1029/2003GL018680.
Cohen, S.J. (ed.), 1997a. Mackenzie Basin Impact Study (MBIS) Final Report. Environment Canada, 372p.
Cohen, S.J., 1997b. What if and so what in Northwest Canada: could climate change make a difference to the future of the Mackenzie Basin. Arctic, 50:293–307.

Ebbesmeyer, C.C., D.R. Cayan, D.R. McLain, F.H. Nichols, D.H. Peterson and K.T. Redmond, 1991. 1976 step in the Pacific climate: forty environmental changes between 1968–1975 and 1977–1984. In: J.L. Betancourt and V.L. Tharp (eds). Proceedings of the Seventh Annual Pacific Climate (PACLIM) Workshop, April 1990. California Dept. of Water Resources. Interagency Ecological Studies Program, Technical Report 26.
Environment Canada, 1997. The Canada Country Study: Climate Impacts and Adaptation. British Columbia and Yukon Summary, 8p.
IPCC, 2001. Climate Change 2001: Impacts, Adaptation, and Vulnerability. Contribution of Working Group II to the Third Assessment Report of the Intergovernmental Panel on Climate Change. Cambridge University Press, 1032p.
Lange, M.A., B. Bartling and K. Grosfeld (eds.), 1999. Global changes in the Barents Sea region. Proceedings of the First International BASIS Research Conference, St. Petersburg, Feb. 22–25, 1998, University of Münster, 470p.
Lange, M.A. and the BASIS consortium, 2003. The Barents Sea Impact Study (BASIS): Methodology and First Results. Continental Shelf Research, 23(17–19):1673–1694.
Maxwell, B., 1997. Responding to Global Climate Change in Canada's Arctic. Vol. II. The Canada Country Study: Climate Impacts and Adaptation. Environment Canada, 82p.
Nakićenović, N. and R. Swart (eds.), 2000. Intergovernmental Panel on Climate Change, Special Report on Emissions Scenarios. Cambridge University Press, 599p.
NAST, 2000. Climate Change Impacts on the United States: Overview. National Assessment Synthesis Team, US Global Change Research Program. Cambridge University Press, 153p.
NAST, 2001. Climate Change Impacts on the United States: Foundation. National Assessment Synthesis Team, US Global Change Research Program. Cambridge University Press, 612p.
Prestrud, P., S. Hallvard and H.V. Goldman, 2004. A catalogue of the terrestrial and marine animals of Svalbard. Skrifter 201. Norwegian Polar Institute, Tromsø, Norway, 137p.
Semiletov, I.P., 1999. On aquatic sources and sinks of CO_2 and CH_4 in the polar regions. Journal of Atmospheric Sciences, 56(2):286–306.
Weller, G., P. Anderson and B. Wang (eds.), 1999. Preparing for a Changing Climate: The Potential Consequences of Climate Change and Variability. A Report of the Alaska Regional Assessment Group for the U. S. Global Change Research Program, University of Alaska Fairbanks, 42p.

Chapter Authors

Chapter 1: An Introduction to the Arctic Climate Impact Assessment

Lead Authors
Henry Huntington, Huntington Consulting, USA
Gunter Weller, University of Alaska Fairbanks, USA
Contributing Authors
Elizabeth Bush, Environment Canada, Canada
Terry V. Callaghan, Abisko Scientific Research Station, Sweden; Sheffield Centre for Arctic Ecology, UK
Vladimir M. Kattsov, Voeikov Main Geophysical Observatory, Russia
Mark Nuttall, University of Aberdeen, Scotland, UK; University of Alberta, Canada

Chapter 2: Arctic Climate: Past and Present

Lead Author
Gordon McBean, University of Western Ontario, Canada
Contributing Authors
Genrikh Alekseev, Arctic and Antarctic Research Institute, Russia
Deliang Chen, Gothenburg University, Sweden
Eirik Førland, Norwegian Meteorological Institute, Norway
John Fyfe, Meteorological Service of Canada, Canada
Pavel Y. Groisman, NOAA National Climatic Data Center, USA
Roger King, University of Western Ontario, Canada
Humfrey Melling, Fisheries and Oceans Canada, Canada
Russell Vose, NOAA National Climatic Data Center, USA
Paul H. Whitfield, Meteorological Service of Canada, Canada

Chapter 3: The Changing Arctic: Indigenous Perspectives

Lead Authors
Henry Huntington, Huntington Consulting, USA
Shari Fox, University of Colorado at Boulder, USA
Contributing Authors
Fikret Berkes, University of Manitoba, Canada
Igor Krupnik, Smithsonian Institution, USA
Case Study Authors
Kotzebue:
    Alex Whiting, Native Village of Kotzebue, USA
    The Aleutian and Pribilof Islands Region, Alaska:
    Michael Zacharof, Aleutian International Association, USA
    Greg McGlashan, St. George Tribal Ecosystem Office, USA
    Michael Brubaker, Aleutian/Pribilof Islands Association, USA
    Victoria Gofman, Aleut International Association, USA
The Yukon Territory:
    Cindy Dickson, Arctic Athabascan Council, Canada
Denendeh:
    Chris Paci, Arctic Athabaskan Council, Canada
    Shirley Tsetta, Yellowknives Dene (N'dilo), Canada
    Chief Sam Gargan, Deh Gah Got'ine (Fort Providence), Canada
    Chief Roy Fabian, Katloodeeche (Hay River Dene Reserve), Canada
    Chief Jerry Paulette, Smith Landing First Nation, Canada
    Vice-Chief Michael Cazon, Deh Cho First Nations, Canada
    Diane Giroux, Sub-Chief Deninu Kue (Fort Resolution), Canada
    Pete King, Elder Akaitcho Territory, Canada
    Maurice Boucher, Deninu K-ue (Fort Resolution), Canada
    Louie Able, Elder Akaitcho Territory, Canada
    Jean Norin, Elder Akaitcho Territory, Canada
    Agatha Laboucan, Lutsel'Ke, Canada
    Philip Cheezie, Elder Akaitcho Territory, Canada
    Joseph Poitras, Elder, Canada
    Flora Abraham, Elder, Canada
    Bella T'selie, Sahtu Dene Council, Canada
    Jim Pierrot, Elder Sahtu, Canada
    Paul Cotchilly, Elder Sahtu, Canada
    George Lafferty, Tlicho Government, Canada
    James Rabesca, Tlicho Government, Canada
    Eddie Camille, Elder Tlicho, Canada
    John Edwards, Gwich'in Tribal Council, Canada
    John Carmichael, Elder Gwich'in, Canada
    Woody Elias, Elder Gwich'in, Canada
    Alison de Palham, Deh Cho First Nations, Canada
    Laura Pitkanen, Deh Cho First Nations, Canada
    Leo Norwegian, Elder Deh Cho, Canada
Nunavut:
    Shari Fox, University of Colorado at Boulder, USA
Qaanaaq, Greenland:
    Uusaqqak Qujaukitsoq, Inuit Circumpolar Conference, Greenland
    Nuka Møller, Inuit Circumpolar Conference, Greenland
Sapmi:
    Tero Mustonen, Tampere Polytechnic / Snowchange Project, Finland
    Mika Nieminen, Tampere Polytechnic / Snowchange Project, Finland
    Hanna Eklund, Tampere Polytechnic / Snowchange Project, Finland
Climate Change and the Saami:
    Elina Helander, University of Lapland, Finland
Kola:
    Tero Mustonen, Tampere Polytechnic / Snowchange Project, Finland
    Sergey Zavalko, Murmansk State Technical University, Russia
    Jyrki Terva, Tampere Polytechnic / Snowchange Project, Finland
    Alexey Cherenkov, Murmansk State Technical University, Russia
Consulting Authors
Anne Henshaw, Bowdoin College, USA
Terry Fenge, Inuit Circumpolar Conference, Canada
Scot Nickels, Inuit Tapiriit Kanatami, Canada
Simon Wilson, Arctic Monitoring and Assessment Programme, Norway

Chapter 4: Future Climate Change: Modeling and Scenarios for the Arctic

Lead Authors
Vladimir M. Kattsov, Voeikov Main Geophysical Observatory, Russia
Erland Källén, Stockholm University, Sweden
Contributing Authors
Howard Cattle, International CLIVAR Project Office, UK
Jens Christensen, Danish Meteorological Institute, Denmark
Helge Drange, Nansen Environmental and Remote Sensing Center and Bjerknes Centre for Climate Research, Norway
Inger Hanssen-Bauer, Norwegian Meteorological Institute, Norway
Tómas Jóhannesen, Icelandic Meteorological Office, Iceland
Igor Karol, Voeikov Main Geophysical Observatory, Russia
Jouni Räisänen, University of Helsinki, Finland
Gunilla Svensson, Stockholm University, Sweden
Stanislav Vavulin, Voeikov Main Geophysical Observatory, Russia
Consulting Authors
Deliang Chen, Gothenburg University, Sweden
Igor Polyakov, University of Alaska Fairbanks, USA
Annette Rinke, Alfred Wegener Institute for Polar and Marine Research, Germany

Chapter 5: Ozone and Ultraviolet Radiation

Lead Authors
Betsy Weatherhead, University of Colorado at Boulder, USA
Aapo Tanskanen, Finnish Meteorological Institute, Finland
Amy Stevermer, University of Colorado at Boulder, USA
Contributing Authors
Signe Bech Andersen, Danish Meteorological Institute, Denmark
Antti Arola, Finnish Meteorological Institute, Finland
John Austin, University Corporation for Atmospheric Research/ Geophysical Fluid Dynamics Laboratory, USA
Germar Bernhard, Biospherical Instruments Inc., USA
Howard Browman, Institute of Marine Research, Norway
Vitali Fioletov, Meteorological Service of Canada, Canada
Volker Grewe, DLR-Institut für Physik der Atmosphäre, Germany
Jay Herman, NASA Goddard Space Flight Center, USA
Weine Josefsson, Swedish Meteorological and Hydrological Institute, Sweden
Arve Kylling, Norwegian Institute for Air Research, Norway
Esko Kyrö, Finnish Meteorological Institute, Finland

Anders Lindfors, Finnish Meteorological Institute, Finland
Drew Shindell, NASA Goddard Institute for Space Studies, USA
Petteri Taalas, Finnish Meteorological Institute, Finland
David Tarasick, Meteorological Service of Canada, Canada
Consulting Authors
Valery Dorokhov, Central Aerological Observatory, Russia
Bjorn Johnsen, Norwegian Radiation Protection Authority, Norway
Jussi Kaurola, Finnish Meteorological Institute, Finland
Rigel Kivi, Finnish Meteorological Institute, Finland
Nikolay Krotkov, NASA Goddard Space Flight Center, USA
Kaisa Lakkala, Finnish Meteorological Institute, Finland
Jacqueline Lenoble, Université des Sciences et Technologies de Lille,
    France
David Sliney, U.S. Army Center for Health Promotion and Preventive
    Medicine, USA

Chapter 6: Cryosphere and Hydrology

Lead Author
John E. Walsh, University of Alaska Fairbanks, USA
Contributing Authors
Oleg Anisimov, State Hydrological Institute, Russia
Jon Ove M. Hagen, University of Oslo, Norway
Thor Jakobsson, Icelandic Meteorological Office, Iceland
Johannes Oerlemans, University of Utrecht, Netherlands
Terry D. Prowse, University of Victoria, Canada
Vladimir Romanovsky, University of Alaska Fairbanks, USA
Nina Savelieva, Pacific Oceanological Institute, Russia
Mark Serreze, University of Colorado at Boulder, USA
Alex Shiklomanov, University of New Hampshire, USA
Igor Shiklomanov, State Hydrological Institute, Russia
Steven Solomon, Geological Survey of Canada, Canada
Consulting Authors
Anthony Arendt, University of Alaska Fairbanks, USA
David Atkinson, University of Alaska Fairbanks, USA
Michael N. Demuth, Natural Resources Canada, Canada
Julian Dowdeswell, Scott Polar Research Institute, UK
Mark Dyurgerov, University of Colorado at Boulder, USA
Andrey Glazovsky, Institute of Geography, RAS, Russia
Roy M. Koerner, Geological Survey of Canada, Canada
Mark Meier, University of Colorado at Boulder, USA
Niels Reeh, Technical University of Denmark, Denmark
Oddur Sigurðsson, National Energy Authority, Hydrological Service,
    Iceland
Konrad Steffen, University of Colorado at Boulder, USA
Martin Truffer, University of Alaska Fairbanks, USA

Chapter 7: Arctic Tundra and Polar Desert Ecosystems

Lead Author
Terry V. Callaghan, Abisko Scientific Research Station, Sweden; Sheffield
    Centre for Arctic Ecology, UK
Contributing Authors
Lars Olof Björn, Lund University, Sweden
F. Stuart Chapin III, University of Alaska Fairbanks, USA
Yuri Chernov, A.N. Severtsov Institute of Evolutionary Morphology and
    Animal Ecology, RAS, Russia
Torben R. Christensen, Lund University, Sweden
Brian Huntley, University of Durham, UK
Rolf Ims, University of Tromsø, Norway
Margareta Johansson, Abisko Scientific Research Station, Sweden
Dyanna Jolly Riedlinger, Dyanna Jolly Consulting, Canada
Sven Jonasson, University of Copenhagen, Denmark
Nadya Matveyeva, Komarov Botanical Institute, RAS, Russia
Walter Oechel, San Diego State University, USA
Nicolai Panikov, Stevens Technical University, USA
Gus Shaver, Marine Biological Laboratory, USA
Consulting Authors
Josef Elster, University of South Bohemia, Czech Republic
Heikki Henttonen, Finnish Forest Research Institute, Finland
Ingibjörg S. Jónsdóttir, University of Svalbard, Norway
Kari Laine, University of Oulu, Finland
Sibyll Schaphoff, Potsdam Institute for Climate Impact Research,
    Germany
Stephen Sitch, Potsdam Institute for Climate Impact Research, Germany
Erja Taulavuori, University of Oulu, Finland
Kari Taulavuori, University of Oulu, Finland
Christoph Zöckler, UNEP World Conservation Monitoring Centre, UK

Chapter 8: Freshwater Ecosystems and Fisheries

Lead Authors
Frederick J. Wrona, National Water Research Institute, Canada
Terry D. Prowse, National Water Research Institute, Canada
James D. Reist, Fisheries and Oceans Canada, Canada
Contributing Authors
Richard Beamish, Fisheries and Oceans Canada, Canada
John J. Gibson, National Water Research Institute, Canada
John Hobbie, Marine Biological Laboratory, USA
Erik Jeppesen, National Environmental Research Institute, Denmark
Jackie King, Fisheries and Oceans Canada, Canada
Guenter Koeck, University of Innsbruck, Austria
Atte Korhola, University of Helsinki, Finland
Lucie Lévesque, National Water Research Institute, Canada
Robie Macdonald, Fisheries and Oceans Canada, Canada
Michael Power, University of Waterloo, Canada
Vladimir Skvortsov, Institute of Limnology, Russia
Warwick Vincent, Laval University, Canada
Consulting Authors
Robert Clark, Canadian Wildlife Service, Canada
Brian Dempson, Fisheries and Oceans Canada, Canada
David Lean, University of Ottawa, Canada
Hannu Lehtonen, University of Helsinki, Finland
Sofia Perin, University of Ottawa, Canada
Reinhard Pienitz, Laval University, Canada
Milla Rautio, Laval University, Canada
John Smol, Queen's University, Canada
Ross Tallman, Fisheries and Oceans Canada, Canada
Alexander Zhulidov, Centre for Preparation and Implementation of
    International Projects on Technical Assistance, Russia

Chapter 9: Marine Systems

Lead Author
Harald Loeng, Institute of Marine Research, Norway
Contributing Authors
Keith Brander, International Council for the Exploration of the Sea,
    Denmark
Eddy Carmack, Institute of Ocean Sciences, Canada
Stanislav Denisenko, Zoological Institute, RAS, Russia
Ken Drinkwater, Bedford Institute of Oceanography, Canada
Bogi Hansen, Fisheries Laboratory, Faroe Islands
Kit Kovacs, Norwegian Polar Institute, Norway
Pat Livingston, NOAA National Marine Fisheries Service, USA
Fiona McLaughlin, Institute of Ocean Sciences, Canada
Egil Sakshaug, Norwegian University of Science and Technology, Norway
Consulting Authors
Richard Bellerby, Bjerknes Centre for Climate Research, Norway
Howard Browman, Institute of Marine Research, Norway
Tore Furevik, University of Bergen, Norway
Jacqueline M. Grebmeier, University of Tennessee, USA
Eystein Jansen, Bjerknes Centre for Climate Research, Norway
Steingrimur Jónsson, Marine Research Institute, Iceland
Lis Lindal Jørgensen, Institute of Marine Research, Norway
Svend-Aage Malmberg, Marine Research Institute, Iceland
Svein Østerhus, Bjerknes Centre for Climate Research, Norway
Geir Ottersen, Institute of Marine Research, Norway
Koji Shimada, Japan Marine Science and Technology Center, Japan

Chapter 10: Principles of Conserving the Arctic's Biodiversity

Lead Author
Michael B. Usher, University of Stirling, Scotland, UK
Contributing Authors
Terry V. Callaghan, Abisko Scientific Research Station, Sweden; Sheffield
    Centre for Arctic Ecology, UK
Grant Gilchrist, Canadian Wildlife Service, Canada
Bill Heal, Durham University, UK
Glenn P. Juday, University of Alaska Fairbanks, USA
Harald Loeng, Institute of Marine Research, Norway
Magdalena A. K. Muir, Conservation of Arctic Flora and Fauna, Iceland;
    Arctic Institute of North America, Canada
Pål Prestrud, Centre for Climate Research in Oslo, Norway

Chapter 11: Management and Conservation of Wildlife in a Changing Arctic Environment

Lead Author
David R. Klein, University of Alaska Fairbanks, USA
Contributing Authors
Leonid M. Baskin, Institute of Ecology and Evolution, Russia
Lyudmila S. Bogoslovskaya, Russian Institute of Cultural and Natural Heritage, Russia
Kjell Danell, Swedish University of Agricultural Sciences, Sweden
Anne Gunn, Government of the Northwest Territory, Canada
David B. Irons, U.S. Fish and Wildlife Service, USA
Gary P. Kofinas, University of Alaska Fairbanks, USA
Kit M. Kovacs, Norwegian Polar Institute, Norway
Margarita Magomedova, Institute of Plant and Animal Ecology, Russia
Rosa H. Meehan, U.S. Fish and Wildlife Service, USA
Don E. Russell, Canadian Wildlife Service, Canada
Patrick Valkenburg, Alaska Department of Fish and Game, USA

Chapter 12: Hunting, Herding, Fishing, and Gathering: Indigenous Peoples and Renewable Resource Use in the Arctic

Lead Author
Mark Nuttall, University of Aberdeen, Scotland, UK; University of Alberta, Canada
Contributing Authors
Fikret Berkes, University of Manitoba, Canada
Bruce Forbes, University of Lapland, Finland
Gary Kofinas, University of Alaska Fairbanks, USA
Tatiana Vlassova, Russian Association of Indigenous Peoples of the North (RAIPON), Russia
George Wenzel, McGill University, Canada

Chapter 13: Fisheries and Aquaculture

Lead Authors
Hjalmar Vilhjálmsson, Marine Research Institute, Iceland
Alf Håkon Hoel, University of Tromso, Norway
Contributing Authors
Sveinn Agnarsson, University of Iceland, Iceland
Ragnar Arnason, University of Iceland, Iceland
James E. Carscadden, Fisheries and Oceans Canada, Canada
Arne Eide, University of Tromso, Norway
David Fluharty, University of Washington, USA
Geir Honneland, Fridtjof Nansen Institute, Norway
Carsten Hvingel, Pinngortitaleriffik, Greenland Institute of Natural Resources, Greenland
Jakob Jakobsson, Marine Research Institute, Iceland
George Lilly, Fisheries and Oceans Canada, Canada
Odd Nakken, Institute of Marine Research, Norway
Vladimir Radchenko, Sakhalin Research Institute of Fisheries and Oceanography, Russia
Susanne Ramstad, Norwegian Polar Institute, Norway
William Schrank, Memorial University of Newfoundland, Canada
Niels Vestergaard, University of Southern Denmark, Denmark
Thomas Wilderbuer, NOAA National Marine Fisheries Service, USA

Chapter 14: Forests, Land Management, and Agriculture

Lead Author
Glenn P. Juday, University of Alaska Fairbanks, USA
Contributing Authors
Valerie Barber, University of Alaska Fairbanks, USA
Paul Duffy, University of Alaska Fairbanks, USA
Hans Linderholm, Göteborg University, Sweden
Scott Rupp, University of Alaska Fairbanks, USA
Steve Sparrow, University of Alaska Fairbanks, USA
Eugene Vaganov, V.N. Sukachev Institute of Forest Research, RAS, Russia
John Yarie, University of Alaska Fairbanks, USA
Consulting Authors
Edward Berg, U.S. Fish and Wildlife Service, USA
Rosanne D'Arrigo, Lamont Doherty Earth Observatory, USA
Olafur Eggertsson, Icelandic Forest Research, Iceland
V.V. Furyaev, V.N. Sukachev Institute of Forest Research, RAS, Russia
Edward H. Hogg, Canadian Forest Service, Canada

Satu Huttunen, University of Oulu, Finland
Gordon Jacoby, Lamont Doherty Earth Observatory, USA
V.Ya. Kaplunov, V.N. Sukachev Institute of Forest Research, RAS, Russia
Seppo Kellomaki, University of Joensuu, Finland
A.V. Kirdyanov, V.N. Sukachev Institute of Forest Research, RAS, Russia
Carol E. Lewis, University of Alaska Fairbanks, USA
Sune Linder, Swedish University of Agricultural Sciences, Sweden
M.M. Naurzbaev, V.N. Sukachev Institute of Forest Research, RAS, USA
F.I. Pleshikov, V.N. Sukachev Institute of Forest Research, RAS, Russia
Ulf T. Runesson, Lakehead University, Canada
Yu.V. Savva, V.N. Sukachev Institute of Forest Research, RAS, Russia
O.V. Sidorova, V.N. Sukachev Institute of Forest Research, RAS, Russia
V.D. Stakanov, V.N. Sukachev Institute of Forest Research, RAS, Russia
N.M. Tchebakova, V.N. Sukachev Institute of Forest Research, RAS, Russia
E.N. Valendik, V.N. Sukachev Institute of Forest Research, RAS, Russia
E.F. Vedrova, V.N. Sukachev Institute of Forest Research, RAS, Russia
Martin Wilmking, Lamont-Doherty Earth Observatory, USA

Chapter 15: Human Health

Lead Authors
Jim Berner, Alaska Native Tribal Health Consortium, USA
Christopher Furgal, Laval University, Canada
Contributing Authors:
Peter Bjerregaard, National Institute of Public Health, Denmark
Mike Bradley, Alaska Native Tribal Health Consortium, USA
Tine Curtis, National Institute of Public Health, Denmark
Ed De Fabo, George Washington University, USA
Juhani Hassi, University of Oulu, Finland
William Keatinge, Queen Mary and Westfield College, UK
Siv Kvernmo, University of Tromso, Norway
Simo Nayha, University of Oulu, Finland
Hannu Rintamaki, Finnish Institute of Occupational Health, Finland
John Warren, Alaska Native Tribal Health Consortium, USA

Chapter 16: Infrastructure: Buildings, Support Systems, and Industrial Facilities

Lead Author
Arne Instanes, Instanes Consulting Engineers, Norway
Contributing Authors
Oleg Anisimov, State Hydrological Institute, Russia
Lawson Brigham, U.S. Arctic Research Commission, USA
Douglas Goering, University of Alaska Fairbanks, USA
Lev N. Khrustalev, Moscow State University, Russia
Branko Ladanyi, École Polytechnique de Montreal, Canada
Jan Otto Larsen, Norwegian University of Science and Technology, Norway
Consulting Authors
Orson Smith, University of Alaska Anchorage, USA
Amy Stevermer, University of Colorado at Boulder, USA
Betsy Weatherhead, University of Colorado at Boulder, USA
Gunter Weller, University of Alaska Fairbanks, USA

Chapter 17: Climate Change in the Context of Multiple Stressors and Resilience

Lead Authors
James J. McCarthy, Harvard University, USA
Marybeth Long Martello, Harvard University, USA
Contributing Authors
Robert Corell, American Meteorological Society and Harvard University, USA
Noelle Eckley Selin, Harvard University, USA
Shari Fox, University of Colorado at Boulder, USA
Grete Hovelsrud-Broda, Centre for International Climate and Environmental Research, Norway
Svein Disch Mathiesen, Norwegian School of Veterinary Science and Nordic Sámi Institute, Norway
Colin Polsky, Clark University, USA
Henrik Selin, Boston University, USA
Nicholas J.C. Tyler, University of Tromso, Norway
Consulting Authors
Kirsti Strøm Bull, University of Oslo and Nordic Sámi Institute, Norway
Inger Maria Gaup Eira, Nordic Sámi Institute, Norway
Nils Isak Eira, Fossbakken, Norway

Siri Eriksen, Centre for International Climate and Environmental Research, Norway
Inger Hanssen-Bauer, Norwegian Meteorological Institute, Norway
Johan Klemet Kalstad, Nordic Sámi Institute, Norway
Christian Nellemann, Norwegian Nature Research Institute, Norway
Nils Oskal, Sámi University College, Norway
Erik S. Reinert, Hvasser, Tønsberg, Norway
Douglas Siegel-Causey, Harvard University, USA
Paal Vegar Storeheier, University of Tromsø, Norway
Johan Mathis Turi, Association of World Reindeer Herders, Norway

Chapter 18: Summary and Synthesis of the ACIA

Lead Author
Gunter Weller, University of Alaska Fairbanks, USA
Contributing Authors
Elizabeth Bush, Environment Canada, Canada
Terry V. Callaghan, Abisko Scientific Research Station, Sweden; Sheffield Centre for Arctic Ecology, UK
Robert Corell, American Meteorological Society and Harvard University, USA
Shari Fox, University of Colorado at Boulder, USA
Christopher Furgal, Laval University, Canada
Alf Håkon Hoel, University of Tromsø, Norway
Henry Huntington, Huntington Consulting, USA
Erland Källén, Stockholm University, Sweden
Vladimir M. Kattsov, Voeikov Main Geophysical Observatory, Russia
David R. Klein, University of Alaska Fairbanks, USA
Harald Loeng, Institute of Marine Research, Norway
Michael MacCracken, Climate Institute, USA
Marybeth Long Martello, Harvard University, USA
Mark Nuttall, University of Aberdeen, Scotland, UK; University of Alberta, Canada
Terry D. Prowse, National Water Research Institute, Canada
Lars-Otto Reiersen, Arctic Monitoring and Assessment Programme, Norway
James D. Reist, Fisheries and Oceans Canada, Canada
Aapo Tanskanen, Finnish Meteorological Institute, Finland
John E. Walsh, University of Alaska Fairbanks, USA
Betsy Weatherhead, University of Colorado at Boulder, USA
Frederick J. Wrona, National Water Research Institute, Canada

Biographies

Lead authors

Dr. James Berner graduated from Oklahoma University Medical School in 1968 and spent three years in the U.S. Navy Medical Corps. He completed residency training and is board certified in Internal Medicine and Pediatrics. Dr. Berner has practiced medicine in the Alaska Native health care system since 1974 and has served as Director of Community Health for the Alaska Native health care system and part-time clinician since 1984. Dr. Berner currently directs the Alaska Native Traditional Food Safety Monitoring program, which assesses contaminant and micronutrient levels in pregnant Alaska Native women, and evaluates health effects in mothers and newborn infants. He has been the key national expert for the U.S. in the Human Health Expert Group of the Arctic Monitoring and Assessment Programme, a working group of the Arctic Council, since 1999.

Professor Terry V. Callaghan (B.Sc. Manchester University, Ph.D. Birmingham University, Ph.D. (honorary) Lund University, Ph.D. (honorary) Oulu University, D.Sc. Manchester University, and member of the Royal Swedish Academy of Sciences) has been involved in arctic ecological research for 37 years, and has worked in all eight arctic countries. He has been Director of the Royal Swedish Academy of Sciences' Abisko Scientific Research Station in the Swedish subarctic since 1996 and is concurrently Professor of Arctic Ecology at the University of Sheffield, UK, and the University of Lund, Sweden. His research focuses on the relationships between the arctic environment and the ecology of arctic plants, animals, and ecosystem processes, including ecological responses to changes in climate, atmospheric carbon dioxide concentrations, and ultraviolet (UV)-B radiation. Professor Callaghan is a member of the United Nations Environment Programme's expert panel on Stratospheric Ozone Depletion Effects, and was a lead author for the ecosystems chapter in the 1990 IPCC assessment of climate change, the polar chapter of the fourth IPCC assessment of climate change, and the polar chapter in the Millennium Ecosystem Assessment. Dr. Callaghan has initiated and chaired several international research groups within the International Arctic Science Committee, and is co-ordinator of SCANNET (Scandinavian and north European Network of Terrestrial Field Bases).

Dr. Shari Fox (BES, MES University of Waterloo, Canada; Ph.D. University of Colorado, Boulder) currently holds a post doctoral position at Harvard University as part of the U.S. National Oceanic and Atmospheric Administration Postdoctoral Program in Climate and Global Change (2003–2005). In 2006, she will join the National Snow and Ice Data Center at the University of Colorado as a research scientist. Dr. Fox has been working with Inuit in Nunavut, Canada, for over a decade. Her research focuses on Inuit knowledge of climate and environmental change and includes work on the documentation of Inuit observations, collaborative research approaches, finding linkages between Inuit and scientific knowledge, and experimenting with creative research products for use in local communities. Dr. Fox has been a consultant for local communities and the Government of Nunavut in efforts to develop a climate change strategy for Nunavut. In 2005, she was appointed to the National Academy of Sciences study committee on designing an Arctic Observing Network and is part of the Coastal Working Group for the second International Conference on Arctic Research Planning (ICARP II). Dr. Fox lives in Clyde River, an Inuit community on Baffin Island, Nunavut.

Dr. Chris Furgal (B.Sc. University of Western Ontario; M.Sc. and Ph.D. University of Waterloo) is a senior researcher at the Public Health Research Unit, Laval University Research Hospital and a research professor in the Department of Political Science at Laval University. For the past 13 years, he has been conducting multidisciplinary research in the biological, social, and health sciences on environmental health issues such as climate change and environmental contaminants, and their management and communication in the circumpolar North in cooperation with Inuit and other Indigenous organizations. He is a lead author for the polar chapter in the fourth IPCC assessment of climate change. Dr. Furgal is a member of the Canadian Federal Northern Contaminants Program Management Committee, the Nunavik Nutrition and Health Committee, and is Co-Director of the recently established Nasivvik Centre for Inuit Health and Changing Environments at Laval, one of eight federally funded centers for aboriginal health research and training in Canada.

Professor Alf Håkon Hoel (cand. polit. University of Oslo) teaches political science at the University of Tromsø, Norway. His research concerns international ocean governance issues and arctic affairs. He has published widely on the management of natural resources and the environment. Current projects include analyses of global change and fisheries, the experience of various countries with their resource management regimes, and the relationship between trade regimes and resource management regimes. He is also involved with the Arctic Monitoring and Assessment Programme's assessment on Oil and Gas in the Arctic. Professor Hoel has served as vice-president of the International Arctic Science Committee, and is a member of the Scientific Steering Committee of the Institutional Dimensions of Global Environmental Change program and the board of the Institute of Marine Research, Bergen.

Dr. Henry P. Huntington (A.B. Princeton University, M.Phil. and Ph.D. University of Cambridge) is an independent researcher in Eagle River, Alaska. His research has documented traditional ecological knowledge of beluga whales in Alaska and Russia, examined Iñupiat Eskimo knowledge and use of sea ice in Alaska, evaluated U.S. involvement in the Arctic Council, analyzed the co-management practices of the Alaska Beluga Whale Committee, studied the adaptation of wildlife management to incorporate subsistence hunting practices, and assessed the interactions of humans and forest fires in interior Alaska. This work has been funded by the National Science Foundation (NSF), the Trust for Mutual Understanding, the Exxon Valdez Oil Spill Trustee Council, the National Marine Fisheries Service, the Alaska Beluga Whale Committee, the Marine Mammal Commission, and other agencies and organizations. Dr. Huntington has also been involved as a researcher and writer in a number of international research programs, such as the Arctic Monitoring and Assessment Programme and the Program for the Conservation of Arctic Flora and Fauna. He was a member of the U.S. Polar Research Board from 1999 to 2005 and president of the Arctic Research Consortium of the United States from 2001 to 2003.

Dr. Arne Instanes (M.Sc. and Ph.D. Norwegian Institute of Technology, University of Trondheim) has worked with cold regions engineering for the last 15 years. He has work experience from research institutes and universities in Norway (SINTEF Geotechnical Engineering, Trondheim; Norwegian Geotechnical Institute, Oslo; University Centre in Svalbard, Longyearbyen) and Canada (University of Alberta, Edmonton). Dr. Instanes is currently vice-president of OPTICONSULT consulting engineers in Bergen, and director of Instanes Svalbard AS in Longyearbyen. His research on cold regions engineering includes work on stress-strain relationships in frozen soil, snow, and ice, thermal analysis of engineering structures, and the effect of pollution on the physical and mechanical properties of frozen soils. Dr. Instanes is co-chairman of the International Permafrost Association's Working Group on Permafrost Engineering and is a member of the International Society of Soil Mechanics and Geotechnical Engineering Technical Committee No. 8 on Frost.

Dr. Glenn Patrick Juday (B.S. Purdue University, Ph.D. Oregon State University) is currently Professor of Forest Ecology and Director of the Tree-Ring Laboratory in the Forest Sciences Department of the School of Natural Resources and Agricultural Sciences at the University of Alaska Fairbanks, where he has worked since 1981. His research specialties include climate change, tree-ring studies, biodiversity and forest management, and forest development following fire. He is currently a co-principal investigator in the NSF-supported Bonanza Creek Long-Term Ecological Research site, and has been on the science steering board of the Center for Global Change at the University of Alaska since its founding. Dr. Juday contributed to the U.S. National Climate Change regional assessment regional reports for Alaska and elsewhere. Dr. Juday teaches conservation biology and wilderness ecosystem management. He has served as science advisor for several television programs and in-depth news articles on climate warming in the United States, Europe, and Japan. He conducted research in the office of the vice president for science of The Nature Conservancy in 1988, and served as president of the Natural Areas Association for four years. Dr. Juday was recognized for outstanding accomplishments as the Chair of the Society of American Foresters Forest Ecology Working Group in 2000.

Professor Erland Källén (B.Sc. and Ph.D. Stockholm University, Sweden) is a professor of dynamic meteorology at the Department of Meteorology, Stockholm University and is presently head of department. His research areas are numerical weather prediction and climate modeling. He has contributed to the understanding of long wave dynamics in the atmosphere as well as methods for data assimilation in the field of numerical weather prediction. Dr. Källén contributed to the 2001 IPCC assessment of climate change, both as reviewer and participant in workshops and meetings. Dr. Källén was the first director of the Swedish Regional Climate Modelling Programme (SWE-CLIM) and his present activities include research on climate processes relevant to the Arctic. His positions on scientific bodies include: president of the Swedish Geophysical Society, chairman of the scientific advisory committee of the European Centre for Medium-Range Weather Forecasts, chairman of the Swedish committee to the World Climate Research Programme/ International Geosphere–Biosphere Programme, editorial board member of the journal Tellus, board member of the Swedish Meteorological and Hydrological Institute, member of a mission advisory group for the European Space Agency's Earth Explorer Atmospheric Dynamics mission. His main research interest is the large scale dynamics of the atmosphere and its applications to climate dynamics and weather prediction.

Dr. Vladimir M. Kattsov (M.Sc. Leningrad Hydrometeorological Institute; M.Sc. St. Petersburg State University; M.A. Kalinin State University, Ph.D. Leningrad Hydrometeorological Institute) is in his 17th year as a research scientist at the Voeikov Main Geophysical Observatory of the Russian Federal Service for Hydrometeorology and Environmental Monitoring, St. Petersburg. Since 2000, he has been head of the Department of Dynamic Meteorology. Dr. Kattsov's research includes global climate 3D modeling with a focus on polar climate dynamics. He was a lead author for the 2001 IPCC Working Group I report, and is currently a lead author for the chapter on model evaluation in the fourth IPCC assessment of climate change. Since 2000, Dr. Kattsov has been a member of the World Climate Research Programme's Working Group on Numerical Experimentation. He is a member of the Climate Commission of the Russian National Geophysical Committee, and a member of the Russian National Council on the WCRP project "Climate and Cryosphere".

Dr. David R. Klein (B.S. University of Connecticut, M.S. University of Alaska, Ph.D. University of British Columbia) was employed by the U.S. Fish and Wildlife Service prior to Alaskan statehood, and by the Alaska Department of Fish and Game immediately after statehood. He was leader of the Alaska Cooperative Wildlife Research Unit at the University of Alaska from 1962 to 1992 when he was appointed senior scientist with the Alaska Cooperative Fish and Wildlife Research Unit until his retirement in 1997. Dr. Klein spent sabbatical-type leaves undertaking research on roe deer in Denmark, wild reindeer in Norway (via a Fulbright Grant), and impala and blesbok in South Africa, and has been involved in other collaborative research in Canada, Greenland, Scandinavia, Siberia, and Portugal. Dr. Klein's research interests have focused on arctic and alpine ecology and habitat relationships of caribou, muskoxen, and other herbivores, assessment of impacts of northern development, and sustainability of arctic ecosystems. He serves on the Board of the Arctic Research Consortium of the United States and is currently Professor Emeritus with the Institute of Arctic Biology and the Department of Biology and Wildlife at the University of Alaska Fairbanks.

Professor Harald Loeng is head of the research group Oceanography and Climate at the Institute of Marine Research in Bergen and is adjunct professor at the University of Tromsø, Norway. He was responsible for the research program on Fish and Climate at IMR and has been head of the Norwegian Marine Data Centre. His main research interest has been climate change and variability and its impact on the marine ecosystem. His positions on scientific bodies include: chair of the Hydrography Committee, the Oceanography Committee, and the Consultative Committee of the International Council for the Exploration of the Sea (ICES); chair of the Norwegian National Committee on Polar Research under the Research Council of Norway; and vice-chair of the Arctic Ocean Science Board. Dr. Loeng has been the Norwegian member of the ICES Advisory Committee of Marine Environment. He was a lead author for the chapter on marine pathways in the 1998 assessment on Arctic Pollution Issues by the Arctic Monitoring and Assessment Programme. He has been an editorial board member for the Journal of Fisheries Oceanography since its beginning.

Dr. Marybeth Long Martello (B.S. and B.A. University of Connecticut, M.S. and Ph.D. Massachusetts Institute of Technology) is a research fellow in Harvard University's Science, Technology, and Society Program. Her research examines global change science and governance, and includes projects on scientific and political dimensions of vulnerability analysis, framing, analysis and representation of climate change impacts, scientific and intergovernmental efforts to address dryland degradation, local knowledge and traditional knowledge in the context of environmental science and policymaking, and corporate approaches to sustainability. She was formerly a research associate with the Kennedy School's Sustainability Systems Project, a policy fellow with the American Meteorological Society, a fellow with the Global Environmental

Assessment Project, and a fellow with an NSF-funded project on Sustainable Knowledge for the Global Environment. Dr. Martello is a contributor to the United Nations Environment Programme's fourth Global Environmental Outlook Report, and has worked as an environmental consultant. She has authored a number of journal articles and book chapters and is co-editor of Earthly Politics: Local and Global in Environmental Governance.

Professor Gordon McBean (M.Sc., McGill University; Ph.D., University of British Columbia) has been active in studies of the atmosphere and weather and climate systems for over 35 years. Dr. McBean was a scientist in Environment Canada and then moved to the Institute of Ocean Sciences. In 1988, he became Professor of Atmospheric and Oceanic Sciences at the University of British Columbia and chair of the WMO-IOC-ICSU Joint Scientific Committee for the World Climate Research Program. As chair, he initiated the Arctic Climate System Study (ACSyS), and other major programs. In 1994, he was appointed Assistant Deputy Minister for the Meteorological Service of Environment Canada, with overall responsibility for weather, climate, sea ice, and water sciences and services in the Canadian government. Since leaving government in 2000, he has been appointed professor in the Institute for Catastrophic Loss Reduction at the University of Western Ontario and Chair of the Board of Trustees for the Canadian Foundation for Climate and Atmospheric Sciences. Dr. McBean is theme leader for the Canadian ArcticNet research program, an integrated study of the coastal Canadian Arctic in the context of climate change. He has been elected a Fellow of the Royal Society of Canada, the American Meteorological Society, and the Canadian Meteorological and Oceanographic Society.

Dr. James J. McCarthy (B.S. Gonzaga University, Ph.D Scripps Institution of Oceanography) is Alexander Agassiz Professor of Biological Oceanography and Head Tutor for degrees in Environmental Science and Public Policy at Harvard University. He recently completed a two-decade term as Director of Harvard University's Museum of Comparative Zoology. His research interests concern the regulation of marine plankton productivity, and in recent years have focused on regions that are strongly affected by seasonal and interannual variation in climate. He has written many scientific papers, and currently teaches courses on biological oceanography, biogeochemical cycles, marine ecosystems, and global change and human health. Dr. McCarthy has served on many national and international planning committees, advisory panels, and commissions relating to oceanography, polar science, and the study of climate and global change. From 1986 to 1993, he chaired the International Geosphere–Biosphere Program. He was the founding editor for the American Geophysical Union's Global Biogeochemical Cycles. He was a convening lead author for the 1990 IPCC Working Group I report, and was co-chair of the 2001 IPCC Working Group II. He has been elected a Fellow of the American Association for the Advancement of Science, a Fellow of the American Academy of Arts and Sciences, and a Foreign Member of the Royal Swedish Academy of Sciences.

Professor Mark Nuttall (MA University of Aberdeen, Ph.D. University of Cambridge) holds the Henry Marshall Tory Chair of Anthropology at the University of Alberta and is Honorary Professor of Sociology at the University of Aberdeen. His work in the Arctic and North Atlantic is mainly concerned with environmental change and resource use issues in rural and coastal communities, depopulation and migration, climate change impacts on indigenous peoples and their livelihoods, the human dimensions of global environmental and sustainability issues, and historical ecology. He has worked extensively in Greenland, Alaska, Canada, and Scotland. Dr. Nuttall is author of Arctic Homeland: Kinship, community and development in northwest Greenland (University of Toronto Press, 1992), White Settlers: The impact of rural repopulation in Scotland (Routledge, 1996), and Protecting the Arctic: Indigenous peoples and cultural survival (Routledge, 1998); editor of the three-volume Encyclopedia of the Arctic (Routledge, 2005); and co-editor of The Arctic: Environment, people, policy (Taylor and Francis, 2000), Cultivating Arctic Landscapes: Knowing and managing animals in the circumpolar North (Berghahn, 2004), and The Russian North in Circumpolar Context (2003).

Dr. Terry D. Prowse (B.E.S. University of Waterloo, M.Sc. Trent University, Ph.D. University of Canterbury) holds an Environment Canada Research Chair and Professorship in Geography at the Water and Climate Impacts Research Centre, University of Victoria, investigating the impacts of climate on water resources. As a senior scientist with Environment Canada, he also heads a research program for the National Water Research Institute investigating the impacts of climate on hydrology and aquatic ecosystems. He was a lead author (chapters on the cryosphere, ecosystems, and polar regions) for the 1995 and 2001 IPCC assessments of climate change, and has a similar position for the 2007 IPCC assessment of the Arctic and Antarctic. His positions on scientific bodies include: President of the Canadian Geophysical Union, including the Hydrology Section; Canadian government representative for the UNESCO International Hydrologic Programme; Canadian Member of the International Association for Hydraulic Research-Ice; editorial board member for the journal Hydrological Processes, and associate editor for the Journal of Cold Regions Engineering. His main research interest is the impact of climate change on water resources and freshwater ecosystems, particularly in cold regions.

Dr. James (Jim) D. Reist (B.Sc. University of Calgary, M.Sc. University of Alberta, Ph.D. University of Toronto) is in his 22nd year as a research scientist in the Arctic Research Division at Fisheries and Oceans Canada, Central and Arctic Region, Winnipeg, where he has led the Arctic Fish Ecology and Assessment Research Section since 1989. His research addresses biodiversity of northern Canadian fishes using genetic, morphological, and ecological approaches with particular emphasis on chars and whitefishes. In addition to documenting fish diversity, biogeography, and understanding their roles in the structure and function of both arctic freshwater and marine ecosystems, his research addresses effects of anthropogenic activities such as exploitation, industrial development, and climate change. Dr. Reist has been active in the Conservation of Arctic Flora and Fauna working group of the Arctic Council, as well as in national and international programs to assess human impacts on fish in the Arctic. He has had adjunct status at several Canadian universities where he has supervised or co-supervised a number of graduate students researching northern fish biology and ecology. He has authored or co-authored over 80 scientific publications in both the primary literature and government publication series.

Amy J. Stevermer (M.S. Oregon State University) has been involved in research related to the transfer of radiation in the earth's atmosphere for more than a decade. Throughout her employment as an associate scientist at the University of Colorado at Boulder, she has contributed to projects focused on understanding the various parameters, including stratospheric aerosol and ozone, that affect UV radiation reaching the earth's surface. She has worked on data analysis and public outreach issues for the U.S Environmental Protection Agency's Ultraviolet Radiation Monitoring Network and has given talks on UV monitoring and effects studies at faculty workshops and national conferences.

Aapo Tanskanen (M.Sc. Lic.Tech. Helsinki University of Technology) is the head of the UV radiation research group at the Finnish Meteorological Institute. His research includes work on UV measurement techniques, radiative transfer modeling and development of methods for estimating surface UV irradiance using satellite data. He is a member of the Ozone Monitoring Instrument science team.

Professor Michael B Usher (B.Sc. and Ph.D. Edinburgh University, Honorary Doctorate University of Stirling) began his career at the University of York, with teaching and research interests in soil biodiversity and nature conservation. He undertook sabbatical periods to establish a termite research group in Ghana and to work on the soil mites and springtails in Antarctica. From 1991 until he retired he was chief scientist at Scottish Natural Heritage, the government's countryside and conservation agency in Scotland. In this role, as well as leading a large team of scientists, he was actively involved with advice to government ministers. Dr. Usher is a chartered biologist, a Fellow of the Institute of Biology, a Fellow of the Royal Entomological Society, and was elected a Fellow of the Royal Society of Edinburgh (Scotland's National Academy of Science & Letters) in 1999. Dr. Usher was awarded an OBE in the 2001 New Year Honour's List. Over the last few years he has chaired the U.K.'s Soil Biodiversity Research Programme, and has been active in the Scottish Biodiversity Forum and the Council of Europe; he is also a Trustee of the Royal Botanic Garden Edinburgh and the Woodland Trust, and continues to teach aspects of biodiversity conservation.

Dr. Hjálmar Vilhjálmsson (B.Sc. University of Glasgow, Ph.D. University of Bergen) has spent much of his career based at the Marine Research Institute in Reykjavík, initially working on operational fisheries research, namely the design and execution of surveys of pelagic fish migrations, abundance, and catchability, and on the environmental variables that affect them. The purpose of this work was to locate areas rich in target species, to predict future migrations and catchability, and to keep the fishing fleet informed. By necessity, the nature of his work changed to abundance assessments of these stocks and advising on their sustainable exploitation. He has been the senior pelagic fisheries biologist at the Marine Research Institute since 1990. Dr. Vilhjálmsson has been a member and vice-chairman of the board of the Icelandic Fisheries Fund, a long-term member of the ICES Northern Pelagic and Blue Whiting Working Group, and was appointed to serve on a special committee, organized by the National Research Council of Iceland, for evaluating existing fisheries science activities in Iceland and advising on research priorities. He is a member of the Icelandic Science Academy.

Dr. John Walsh (B.A. Dartmouth College, Ph.D. Massachusetts Institute of Technology) is a President's Professor of Global Change at the University of Alaska Fairbanks. He is also the Director of the Coopera-

tive Institute for Arctic Research and the Center for Global Change at the University of Alaska. His research has addressed arctic climate weather variability, with an emphasis on sea ice variability and the role of sea ice and snow cover in weather and climate. His work has also included evaluations of global climate model simulations of the Arctic. Dr. Walsh is a lead author for the polar regions in the fourth IPCC assessment. He is a member of the Polar Research Board and a panel chair for the Study of Environmental Arctic Change (SEARCH). Before joining the University of Alaska, he spent 30 years on the faculty of the University of Illinois, where he taught courses on weather and climate. He co-authored the textbook, Severe and Hazardous Weather, and is an associate editor of the Journal of Climate.

Professor Gunter Weller (Ph.D. University of Melbourne) is Professor of Geophysics Emeritus of the University of Alaska Fairbanks. His early research concerned climate change and its impacts in both the Arctic and the Antarctic. He has been program manager of the NSF's polar programs in meteorology, project manager of the NOAA-BLM Outer Continental Shelf Environmental Assessment Program in the Arctic, project director of the NASA-University of Alaska SAR Facility, and deputy director of the UAF Geophysical Institute. Among many scientific committee assignments he was the president of ICSU's International Commission on Polar Meteorology and chaired the U.S. National Research Council's Polar Research Board. He recently retired as the director of the Center for Global Change and Arctic System Research and director of the NOAA-UAF Cooperative Institute for Arctic Research and now lives in Australia. Dr. Weller was Executive Director of the ACIA for the last four years.

Dr. Frederick Wrona (B.Sc. and Ph.D. University of Calgary) is currently the Director of the Aquatic Ecosystems Impacts Research Branch, National Water Research Institute (Environment Canada) and is a professor in the Department of Geography, University of Victoria. Dr. Wrona has conducted and managed interdisciplinary aquatic ecosystem research for over 23 years, focusing on the ecology and eco-hydrology of cold-regions aquatic ecosystems. His research interests include understanding and predicting the impacts of climate variability/change on the structure and function of cold-regions aquatic ecosystems, identifying mechanisms responsible for the observed patterns of dynamics in aquatic predator-prey systems, assessing the ecotoxicology of aquatic organisms to contaminant stressors, and assessing the impacts of human developments on the health and sustainability of northern aquatic systems. He has served as the Science Director for the Northern River Basins Study and is currently involved with numerous national and international scientific and advisory committees related to the development and implementation of northern hydrological and ecological research programs (e.g., contributing author to the 2007 IPCC Working Group II assessment of the polar regions and the 2nd International Conference on Arctic Research Planning, Environment Canada's Northern Working Group). Dr. Wrona has a strong interest in science-policy linkages and is currently the Head and Chief Delegate for the Canadian National Committee for the UNESCO International Hydrological Programme.

Additional Members of the ACIA Implementation Team[5]

Dr. Robert W. Corell (B.S., M.S., Ph.D., Case Western Reserve University and MIT) joined the National Science Foundation (NSF) in 1987 as Assistant Director for Geosciences where for over 12 years he oversaw the Atmospheric, Earth, and Ocean Sciences and the global change programs of the NSF. While there, Dr. Corell chaired the National Science and Technology Council's committee that oversees the U.S. Global Change Research Program, and the international committee of government agencies funding global change research. He was also chair and principal U.S. delegate to many international bodies with interests in and responsibilities for climate and global change research programs. Dr Corell is currently a Senior Fellow at the Policy Program of the American Meteorological Society and is actively engaged in research concerned both with the science of global change and with the interface between science and public policy, particularly research activities focused on global and regional climate change and related environmental issues, and science to facilitate understanding of vulnerability and sustainable development. He co-chairs an international strategic planning group on harnessing science, technology, and innovation for sustainable development, and is the lead for an international partnership to better understand and plan for a transition to hydrogen for several nations, currently focused on Iceland, India, and the eight Arctic nations. He is leading a research project to explore methods, models, and conceptual frameworks for vulnerability research, analysis, and assessment – the current focus of which is on vulnerabilities of indigenous communities in the Arctic. Dr Corell was recently invited to join the Washington Advisory Group, LLC to work on the industry dimension of the climate issue.

---

[5]Entries for Terry Callaghan, Gordon McBean, and Gunter Weller may be found under "Lead authors"

Dr. Pål Prestrud (Master degree and Ph.D. University of Oslo) has been involved for the last 25 years in environmental research and management in the Arctic, with a special research interest in population dynamics of polar mammals and their physiological adaptation to the harsh polar conditions. Dr. Prestrud is currently Director of the Centre for Climate Research at the University of Oslo. He has been Director of Research for a number of years at the Norwegian Polar Institute, and has served as deputy director general in the Norwegian Ministry of Environment where he headed the Section on Polar Affairs and Cooperation with Russia. Dr. Prestrud has been involved in several environmental impact assessments conducted in the Norwegian Arctic over the last 20 years.

Lars-Otto Reiersen graduated in marine biology from the University of Oslo, Norway. He then worked at the University of Oslo for several years conducting research on basic processes in marine fish and the effects of oil and other contaminants. He later worked for the Norwegian State Pollution Control Authority dealing with the environmental regulation of shipping and oil and gas activities (exploration and exploitation) in the seas around Norway and at Svalbard. He was involved in the work of the Oslo and Paris Commissions and the London Dumping Commission, especially in relation to the testing of chemicals to be used offshore and chemical and biological monitoring of the marine environment. He chaired the group that made the assessment of the Pollution of the North Sea under the North Sea Task Force and was involved in the establishment and implementation of the Arctic Environmental Protection Strategy and the Arctic Council. Since 1992, he has been the Executive Secretary for the Arctic Monitoring and Assessment Programme.

Jan Idar Solbakken (M.Sc. University of Tromso) has worked at Saami University College, Norway for the last 13 years as an assistant professor in biology. From 2000 to 2003 he was Dean at Saami University College. He has represented the Saami Council, one of the Permanent Participants in the Arctic Council, within AMAP working groups since 1994. He also represented the Saami Council within the AMAP Assessment Steering Group during the first AMAP assessment of Arctic Pollution Issues.

Dr. Patricia A. Anderson (B.Sc. University of Iowa, M.A. Dalhousie University, Ph.D. New York University) has 18 years experience researching polar issues and managing polar science programs at the NSF and the University of Alaska Fairbanks. She was Executive Director of the U.S. Antarctic Program Safety Review Panel, an NSF activity that involved researching the history of U.S. exploration and science in Antarctica and the safety of these operations, and co-authoring the panel's report. She spent four years coordinating federal interagency programs on global climate change. Dr. Anderson's 12 years of experience at the University of Alaska Fairbanks involved the management of several arctic research and education activities, including the establishing of a competitive student research grant program, expanding participation of a wide range of stakeholder groups in assessing the impacts of climate change on Alaska, and facilitating interdisciplinary arctic system science through science management of the NSF Arctic System Science Land-Atmosphere-Ice Interactions program. In her capacity as Deputy Executive Director of the ACIA Secretariat, she has been responsible for coordinating all ACIA activities.

Elizabeth Bush (M.Sc. and M.A. University of Toronto) is a member of the Science Assessment and Integration Branch of Environment Canada, whose mandate it is to provide science advice and to coordinate science assessment activities on atmospheric issues. She has been involved in science assessment activities for many years, working first as an air quality advisor during which time she participated in Canadian national assessments of particulate matter and ground-level ozone. She currently works as a climate change science advisor and was the focal point in Canada for Canadian participation in the ACIA.

Paul Grabhorn is a communications consultant and photographer with 22 years of experience producing publications and campaigns on the subjects of global change, humanitarian action, and environmental research. A particular area of expertise is in the visual communication of complex subjects. His background in human ecology provides his work with a systems view and a synthesis perspective. Some of Grabhorn Studio's productions include: The U.S. National Assessment – Climate Change Impacts on the United States, Global Energy Technology Strategy: Addressing Climate Change, White House Conference on Science and Economics related to Global Change, National Energy Strategy, National Space Council Annual Reports, Coastal America campaign materials, Global Stewardship Brochure (White House), Army Corps of Engineers Environmental Stewardship Campaign, Defense and the Environment Initiative (US Army), GLOBE - Global Learning and Observations to Benefit the Environment (White House) – US Global Change Research Program – annual reports: Our Changing Planet, CIESIN; Understanding the Human Dimensions of Global Change, US Army Corps of Engineers recruitment and outreach materials, Government Buy Recycled Initiative, Technology for a Sustainable Future (White House), Bridge

to a Sustainable Future: National Environmental Technology Strategy (White House), Sustainable America: A New Consensus, Picturing Climate's Complexity, People on War campaign (International Committee of the Red Cross), So Why! Music goes to war campaign (ICRC). Paul Grabhorn has also undertaken photographic documentary missions for the International Committee of the Red Cross in many locations: Somalia, Bosnia, Rwanda, Azerbaijan, Georgia, Abkhazia, Armenia, Ngorni Kharabakh, Chechnya, Cambodia, Colombia, Philippines, Croatia, Nepal, Burundi, Guatemala, Mali, Angola, South Africa, Kenya, and Liberia among others.

Susan Joy Hassol is a researcher and writer with 20 years experience in global change science. Known for her ability to translate science into English, she synthesizes information from across the spectrum of scientific disciplines, and makes complex issues accessible to policymakers and the public. She was a lead author of Climate Change Impacts on the United States, the synthesis report of the U.S. National Assessment of the Consequences of Climate Change. Susan authored a chapter on energy efficiency in a book entitled Innovative Energy Strategies for CO_2 Stabilization (Cambridge University Press, 2002). She wrote a feature article entitled "A Change of Climate," in Issues in Science and Technology, a journal of the National Academy of Sciences, focusing on the actions of U.S. states, localities, and corporations in mitigating climate change. She has also written and edited numerous articles, papers, and books for organizations including the United Nations Environment Programme, the Scientific Committee on Problems of the Environment, and the Inter-American Institute for Global Change Research. She has served as Environment Fellow for the Aspen Institute and as Research Associate and Director of Communications for the Aspen Global Change Institute.

Dr. Michael C. MacCracken (B.S.E. Princeton University; Ph.D. University of California Davis) is Senior Scientist for Climate Change with the Climate Institute in Washington, DC. For 34 years, he was employed by the Lawrence Livermore National Laboratory, where his research included numerical modeling of various natural and anthropogenic causes of climate change and of factors affecting air quality in the Bay Area and northeastern United States. For the latter part of this period, Dr. MacCracken was on assignment with the interagency Office of the U.S. Global Change Research Program, serving for different periods as executive director of the Office and of its National Assessment Coordination Office. He also coordinated the U.S. Government technical review of the IPCC assessments. Dr. MacCracken is currently president of the International Association of Meteorology and Atmospheric Sciences and serves on the executive committees of the International Union of Geodesy and Geophysics and the Scientific Committee for Oceanic Research.

ACIA Science Editors

Lelani Arris (B.Sc. University of Vermont, M.Sc. Massachusetts Institute of Technology) has more than 14 years experience writing and editing technical and popular publications about climate change, ozone depletion, and other environmental science topics. She was editor of the bi-weekly newsletter Global Environmental Change Report for five years, senior editor of the quarterly magazine Global Change for three years, and has also written or edited publications for the Canadian Climate Impacts and Adaptation Research Network, the U.S. Global Change Research Program, and the British Columbia Ministry of Forests, among others.

Dr. Carolyn Symon (B.Sc. Loughborough University, M.Sc. Kings College London, Ph.D. Lancaster University and Proudman Oceanographic Laboratory) is a science editor specializing in multi-authored environmental assessments prepared by intergovernmental bodies. For the last ten years most of her work has focused on marine-related and polar-related issues. Dr. Symon has undertaken work for the OSPAR Commission, the British Antarctic Survey, the Secretariat for the Fifth North Sea Conference, ICES, CCAMLR, and AMAP.

Professor Bill (O.W.) Heal (BSc., Ph.D. Durham University, Honorary Professor Edinburgh University, Fellow Hatfield College, Durham) is now retired. His early research on protozoa expanded into soil biology and decomposition then into ecosystems. In the 1970s he led the UK International Biological Programme at the Moor House upland site. This linked naturally into the IBP Tundra Biome through its sub-Arctic climate and to his involvement in international co-ordination and synthesis. As Director of the Institute of Terrestrial Ecology he was responsible for a wide range of pure and applied national and international research. He led the EU Arctic Terrestrial Ecosystem Research project which helped to integrate arctic research and spawned a series of new Arctic–Alpine projects. He subsequently chaired the Polar Sciences Committee of the Natural Environment Research Council, helped to initiate the University of the Arctic, and participated in CAFF and AMAP and in the US NSF and LTER programs.

Appendix C

Reviewers

This appendix lists those international experts selected by ACIA that were willing to review one or more chapters of this assessment. Most of the experts listed below reviewed at least two related chapters and a few reviewed several chapters. Many additional reviewers, not listed here, were selected through national reviews conducted by the arctic countries. Reviews were received from about 200 individuals in total.

Hans Alexandersson, Swedish Meteorological and Hydrological Institute
Leif Anderson, Göteborg University, Sweden
Robert Barbault, Institut d'Ecologie Fondamentale et Appliquee, France
Roger Barry, National Snow and Ice Data Center, USA
Esfir G. Bogdanova, Voeikov Main Geophysical Observatory, Russia
Jerry Brown, International Permafrost Association, USA
Margo Burgess, Geological Survey of Canada
John Calder, National Oceanic and Atmospheric Administration, USA
JoLynn Carroll, Akvaplan-niva AS, Norway
Tim Carter, Finnish Environment Institute
Richard Caulfield, University of Alaska Fairbanks
Nataly Ye. Chubarova, Moscow State University, Russia
Stewart Cohen, University of British Columbia, Canada
Andre Corriveau, Government of the Northwest Territories, Dept. of Health and Social Services, Canada
Robert Crawford, University of St. Andrews (Emeritus), UK
Yvon Csonka, University of Greenland
Jens Dahl, International Work Group for Indigenous Affairs, Denmark
Klaus Dethloff, Alfred Wegener Institute for Polar and Marine Research, Germany
Mark Dyurgerov, University of Colorado at Boulder, USA
Michael A.D. Ferguson, Dept. of Sustainable Development, Government of Nunavut, Canada
Craig Fleener, Gwich'in Council International, USA
Sven Haakanson, Jr., Alutiiq Museum, Kodiak, Alaska, USA
Don Hayley, EBA Engineering Consultants Ltd., Canada
Bill Heal, University of Durham, UK
Raino Heino, Finnish Meteorological Institute
Annika Hofgaard, Norwegian Institute for Nature Research, Norway
Ad H.L. Huiskes, Netherlands Institute of Ecology
George Hunt, University of California, Irvine, USA
Ingvar Jarle Huse, Institute of Marine Research, Norway
Satu Huttunen, University of Oulu, Finland
Trond Iversen, University of Oslo, Norway
Robert Jefferies, University of Toronto, Canada
Peter Jones, Bedford Institute of Oceanography, Canada
Eigil Kaas, Danish Meteorological Institute
Anders Karlqvist, Swedish Polar Research Secretariat
Roy Koerner, Geological Survey of Canada
Pirkko Kortelainen, Finnish Environment Institute
Eduard Koster, Utrecht University, Netherlands
Peter Kuhry, Stockholm University, Sweden
Manfred Lange, University of Muenster, Germany
Donald S. Lemmen, Natural Resources Canada

Pentti Mälkki, Finnish Institute of Marine Research
Svend Aage Malmberg, Marine Research Institute, Iceland
Michael McGeehin, Centers for Disease Control, USA
Richard McKenzie, National Institute of Water and Atmospheric Research, New Zealand
Mark Meier, University of Colorado at Boulder, USA
Jamie Morison, University of Washington, USA
Lars Moseholm, National Environmental Research Institute, Denmark
Ted Munn, University of Toronto, Canada
Aynslie Ogden, Northern Climate ExChange, Canada
Erling Ögren, Swedish University of Agricultural Sciences
Mats Olsson, Swedish University of Agricultural Sciences
Olav Orheim, Norsk Polarinstitutt, Norway
Jim Overland, Pacific Marine Environmental Laboratory/NOAA, USA
Chris Paci, Dene Nation, Canada
Gísli Pálsson, University of Iceland
Walter Parker, Circumpolar Infrastructure Task Force of the Arctic Council; the Northern Forum, USA
Geoff Petts, University of Birmingham, UK
Henning Rodhe, Stockholm University, Sweden
Odd Rogne, International Arctic Science Committee, Norway
Ursula Schauer, Alfred Wegener Institute for Polar and Marine Research, Germany
Frank Sejersen, University of Copenhagen, Denmark
Stepan G. Shiyatov, Institute of Plant and Animal Ecology, Russian Academy of Sciences
Oddvar Skre, Norwegian Forest Research Institute
Kimberly Strong, University of Toronto, Canada
Thora E. Thorhallsdottir, University of Iceland
Darin Toohey, University of Colorado at Boulder, USA
Reidar Toresen, Institute of Marine Research, Norway
Adrian Tuck, NOAA Aeronomy Laboratory, USA
Jay Van Oostdam, Health Canada
Patrick J. Webber, Michigan State University, USA
Martin Weinstein, 'Namgis First Nation, Canada
Jan Weslawski, Institute of Oceanology, Polish Academy of Sciences
Ed Wiken, Wildlife Habitat Canada
Ming-Ko Woo, McMaster University, Canada
Oran Young, University of California, Santa Barbara, USA
T. Kue Young, University of Toronto, Faculty of Medicine, Canada
Alexander Zhulidov, South Russian Regional Centre for Preparation and Implementation of International Projects
Francis Zwiers, Meteorological Service of Canada

Species Names

Latin name	Common name/descriptor
Birds	
Aethia cristatella	crested auklet
Aethia psittacula	parakeet auklet
Aethia pusilla	least auklet
Alca torda	razorbill
Alle alle	little auk
Anas acuta	pintail
Anas americana	widgeon
Anas crecca	common teal
Anas penelope	Eurasian wigeon
Anas platyrhynchos	mallard
Anser albifrons	white-fronted goose or "yellow legs"
A. a. flavirostris	Greenland white-fronted goose
Anser anser	greylag goose
Anser brachyrhynchus	pink-footed goose
Anser caerulescens	snow goose
A. c. caerulescens	lesser snow goose
Anser canagicus	emperor goose
Anser erythropus	lesser white-fronted goose
Anser fabalis	bean goose
A. f. fabialis	taiga bean goose
A. f. rossicus	tundra bean goose
Arenaria interpres	ruddy turnstone
Asio flammeus	short-eared owl
Aythya affinis	lesser scaup
Aythya collaris	ring-necked duck
Aythya marila	greater scaup
Aythya valisineria	canvasback duck
Branta bernicla	Brent goose/ black brent
B. b. bernicla	dark-bellied brent goose
B. b. hrota	light-bellied brent goose
Branta canadensis	Canada goose
Branta leucopsis	barnacle goose
Branta ruficollis	red-breasted goose
Buteo spp.	raptors
Buteo lagopus	rough-legged buzzard
Calcarius lapponicus	Lapland longspur
Calidris acuminata	sharp-tailed sandpiper
Calidris alba	sanderling
Calidris alpina	dunlin
Calidris canutus	red knot
C. c. canutus	red knot (Taymir population)
C. c. islandica	red knot (Nearctic population)
Calidris ferruginea	curlew sandpiper
Calidris fuscicollis	white-rumped sandpiper
Calidris maritima	purple sandpiper
Calidris mauri	western sandpiper
Calidris melanotos	pectoral sandpiper
Calidris minuta	little stint
Calidris ruficollis	red-necked stint
Calidris tenuirostris	great knot
Carduelis chloris	greenfinch
Carduelis hornemanni	Arctic redpoll
Catharacta skua	great skua
Cepphus columba	pigeon guillemot
Cepphus grylle	black guillemot
Charadrius semipalmatus	semipalmated plover
Chen caerulescens	lesser snow goose
Clangula hyemalis	long-tailed duck/ oldsquaw
Corvus corax	raven/ common raven
Cygnus columbianus	tundra swan/ whistling swan
Emberiza pusilla	little bunting
Eremophila alpestris	shore lark
Eurynorhynchus pygmaeus	spoon-billed sandpiper
Falco peregrinus	peregrine falcon
Falco rusticolus	gyrfalcon
Fratercula arctica	Atlantic puffin
Fratercula cirrhata	tufted puffin
Fratercula corniculata	horned puffin
Fulmaris glacialis	northern fulmar
Gallinago gallinago	common snipe
Gavia adamsii	yellow-billed loon
Gavia arctica	Arctic loon
Gavia immer	common loon/ great northern diver
Gavia pacifica	Pacific loon
Gavia stellata	red-throated loon
Grus americana	whooping crane
Haematopus ostralegus	Eurasian oystercatcher
Hydrobates pelagicus	European storm petrel
Lagopus lagopus	willow grouse
Lagopus mutus	ptarmigan
L. m. hyperboreus	Svalbard ptarmigan
Larus argentatus	herring gull
Larus canus	mew gull
Larus fuscus	lesser black-backed gull
Larus glaucescens	glaucous-winged gull
Larus hyperboreus	glaucous gull
Larus marinus	great black-backed gull
Larus ridibundus	black-headed gull
Limnodromus scolopaceus	Long-billed dowitcher
Limosa lapponica	bar-tailed godwit
Limosa limosa	black-tailed godwit
Loxia spp.	crossbills
Melanitta fusca	white-winged scoter/ velvet scoter
Melanitta nigra	black scoter
Mergus merganser	common merganser
Mergus serrator	red-breasted merganser
Morus bassanus	northern gannet
Motacilla alba	white wagtail
Numenius borealis	Eskimo curlew
Nyctea scandiaca	snowy owl
Oceanodroma leucorhoa	Leach's storm-petrel
Oenanthe oenanthe	northern wheatear
Pagophila eburnea	ivory gull
Parus caeruleus	blue tit
Phalacrocorax aristotelis	European shag
Phalacrocorax carbo	great cormorant
Phalacrocorax pelagicus	pelagic cormorant
Phalacrocorax perspicillatus	Pallas's cormorant
Phalaropus fulicarius	red phalarope/ grey phalarope
Phalaropus lobatus	northern phalarope/ red-necked phalarope
Philomachus pugnax	ruff
Phylloscopus borealis	Arctic warbler
Phylloscopus inornatus	yellow-browed warbler
Pinguinus impennis	great auk
Plectrophenax nivalis	snow bunting
Pluvialis dominica	lesser golden plover
Pluvialis fulva	Pacific golden plover
Pluvialis squatarola	black-bellied plover
Polysticta stelleri	Steller's eider
Puffinus gravis	greater shearwater
Puffinus griseus	sooty shearwater
Rhodostethia rosea	Ross' gull
Rissa brevirostris	red-legged kittiwake
Rissa tridactyla	black-legged kittiwake
Somateria fisheri	spectacled eider
Somateria mollissima	common eider
Somateria spectabilis	king eider
Stercorarius longicaudus	long-tailed jaeger
Stercorarius parasiticus	Arctic skua/ parasitic jaeger
Sterna hirundo	common tern
Sterna paradisaea	Arctic tern
Sula bassana	northern gannet
Synthliboramphus antiquus	ancient murrelet
Tetrao urogallus	wood grouse
Tetrastes bonasia	hazel grouse
Tringa erythropus	spotted redshank

Tryngites subruficollis	buff-breasted sandpiper
Turdus iliacus	redwing
Turdus migratorius	American robin
Turdus pilaris	fieldfare
Uria aalge	common murre
Uria lomvia	Brünnich's guillemot/ thick billed murre
Vanellus vanellus	northern lapwing
Xema sabini	Sabine's gull

Fish

Abramis brama	carp bream
Alosa spp.	alewifes
Ammodytes americanus	sand lance
Ammodytes dubius	sand lance
Anarhichas lupus lupus	wolffish
Anguilla anguilla	eel
Arctogadus glacialis	Arctic cod
Atheresthes stomias	arrowtooth flounder
Berryteuthis magister magister	commander squid
Boreogadus saida	Arctic cod/ polar cod
Brosme brosme	tusk
Catostomus spp.	suckers
Cetorhinus maximus	basking shark
Chrosomus eos	northern redbelly dace
Clupea harengus	Atlantic herring
Clupea pallasi	Pacific herring
Coregonus albula	vendace
Coregonus artedi	lake cisco
Coregonus autumnalis	Arctic cisco
Coregonus clupeaformis	lake whitefish
Coregonus lavaretus	whitefish/ powan
Coregonus nasus	broad whitefish
Coregonus pidschian	Siberian whitefish
Coregonus sardinella	least cisco
Cottus spp.	sculpins
Cottus cognatus	slimy sculpin
Couesius plumbeus	lake chub
Dallia spp.	blackfishes
Dicentrachus labrax	sea bass
Eleginus gracilis	saffron cod
Engraulis mordax	northern anchovy
Esox lucius	northern pike
Gadus macrocephalus	Pacific cod
Gadus morhua	Atlantic cod
Gadus ogac	Greenland cod
Gasterosteus aculeatus	threespine stickleback
Glyptocephalus cynoglossus	witch
Gymnocephalus cernuus	ruffe
Haliotis rufuscens	abalone
Hemilepidotus jordani	yellow Irish lord
Hiodon alosoides	goldeye
Hippoglossoides elassodon	flathead sole
Hippoglossoides platessoides	long rough dab/ American plaice
Hippoglossoides robustus	northern flathead sole
Hippoglossus hippoglossus	halibut
Hippoglossus stenolepis	Pacific halibut
Hypomesus olidus	pond smelt
Icelus spiniger	thorny sculpin
Illex illecebrosus	squid
Lamna ditropis	salmon shark
Lepidopsetta bilineata	rock sole
Lepomis macrochirus	bluegill
Leuciscus idus	ide
Leuciscus leuciscus baicalensis	common dace
Limanda aspera	yellowfin sole
Lota lota	burbot
Mallotus villosus	capelin
Melanogrammus aeglefinus	haddock
Microgadus proximus	Pacific tomcod
Micromesistius poutassou	blue whiting
Micropterus dolomieu	smallmouth bass
Mola mola	ocean sunfish
Molva molva	ling
Myoxocephalus quadricornis	fourhorn sculpin
Noemacheilus barbatulus	stone loach
Notidanus griseus	bluntnose sixgill shark
Notropis atherinoides	emerald shiner
Oncorhynchus gorbuscha	pink salmon
Oncorhynchus keta	chum salmon
Oncorhynchus kisutch	coho salmon

Oncorhynchus mykiss	rainbow trout
Oncorhynchus nerka	sockeye salmon
Oncorhynchus tshawytscha	chinook salmon
Osmerus mordax	smelt/ rainbow smelt
Perca flavescens	yellow perch
Perca fluviatilis	European perch
Percopsis omiscomaycus	trout perch
Petromyzon marinus	lamprey
Pleurogrammus monoptergius	Atka mackerel
Pleuronectes asper	yellowfin sole
Pleuronectes bilineatus	rock sole
Pleuronectes ferrugineus	flounder
Pleuronectes glacialis	Arctic flounder
Pleuronectes platessa	plaice
Pleuronectes quadrituberculatus	Alaska plaice
Pollachius virens	coal fish/ saithe
Prosopium cylindraceum	round whitefish
Psetta maxima	turbot
Pungitius pungitius	ninespine stickleback
Reinhardtius hippoglossoides	Greenland halibut
Rutilus rutilus	roach
Salmo gairdneri	rainbow trout
Salmo salar	Atlantic salmon
Salmo trutta	brown trout
Salvelinus alpinus	Arctic char
Salvelinus confluentus	bull trout
Salvelinus fontinalis	brook trout
Salvelinus malma	Char/ Dolly Varden
Salvelinus namaycush	lake trout
Sander lucioperca	zander
Sander vitreus	walleye
Scomber japonicus	Pacific mackerel
Scomber scombrus	Atlantic mackerel
Scomberesox saurus	Atlantic saury
Sebastes marinus	redfish
Sebastes mentella	redfish
Sebastes viviparus	redfish
Squalus acanthias	spurdog
Stenodus leucichthys	inconnu
Thaleichthys pacificus	eulachon
Theragra chalcogramma	pollock
Thunnus thynnus	northern bluefin tuna
Thymallus arcticus	Arctic grayling
Trachurus trachurus	horse mackerel
Xiphias gladius	swordfish

Marine mammals

Balaena mysticetus	Greenland right whale/ bowhead whale
Balaenoptera acutorostrata	minke whale
Balaenoptera borealis	sei whale
Balaenoptera musculus	blue whale
Balaenoptera physalus	fin whale
Callorhinus ursinus	northern fur seal
Cystophora cristata	hooded seal
Delphinapterus leucas	beluga whale/ white whale
Enhydra lutris	sea otter
Erignathus barbatus	bearded seal
Eschrichtius robustus	grey whale
Eubalaena glacialis	right whales
Eubalaena japonica	North Pacific right whale
Eumetopias jubatus	Steller sea lion
Globicephala melaena	pilot whale
Halichoerus grypus	grey seal
Hydrodamalis gigas	Steller sea cow
Lagenorhynchus acutus	white-sided dolphin
Lagenorhynchus albirostris	white-beaked dolphin
Megaptera novaeangliae	humpback whale
Monodon monoceros	narwhal
Odobenus rosmarus	walrus
O. r. divergens	Pacific walrus
O. r. rosmarus	Atlantic wlarus
Orcinus orca	killer whale
Phoca fasciata	ribbon seal
Phoca groenlandica	harp seal
Phoca hispida	ringed seal
Phoca largha	spotted seal
Phoca vitulina	harbour seal
Phocoena phocoena	harbour porpoise
Phocoenoides dalli	Dall's porpoise/ Dahl's porpoise
Physeter catodon	sperm whale

Tursiops truncatus	bottlenose dolphin
Ursus maritimus	polar bear

Terrestrial mammals

Alces alces	moose
Alopex lagopus	Arctic fox
Bison bison	buffalo
Canis lupus	wolf
Castor canadensis	beaver
Cervus elaphus	red deer
Cervus nippon	Sika deer
Clethrionomys rufocanus	grey-sided vole
Clethrionomys rutilus	red-backed vole
Coelodonta antiquitatis	woolly rhinocerous
Dicrostonyx groenlandicus	collared lemming
Dicrostonyx torquatus	Arctic lemming
Erethizon dorsatum	porcupine
Gulo gulo	wolverine
Lemmus lemmus	Norway lemming
Lemmus sibiricus	brown lemming
Lepus americanus	snowshoe hare
Lepus arcticus	Arctic hare
Lepus timidus	hare/ mountain hare
Lynx lynx	lynx
Mammuthis primigenius	mammoth
Martes zibellina	sable
Megaloceros giganteus	giant deer/ Irish elk
Microtus abbreviatus	insular vole
Microtus gregalis	narrow-headed vole
Microtus middendorffi	Middendorf's vole
Microtus oeconomus	tundra vole
Microtus rossiaemeridionalis	sibling vole
Moschus moschiferus	musk deer
Mus musculus	house mouse
Mustela erminea	ermine
Mustela vison	mink
Mustela nivalis	least weasel
Odocoileus hemionus	mule deer
Ondatra zibethicus	muskrat
Ovibos moschatus	muskox
Ovis canadensis dalli	Dall sheep
Puma concolor	cougar
Rangifer tarandus	caribou/reindeer
R. t. pearyi	Peary caribou
Rattus norvegicus	Norway rat
Sorex spp.	shrews
Spermophilus parryii	ground squirrel
Tamiasciurus hudsonicus	red squirrel
Ursus arctos	grizzly bear/ brown bear
Ursus major	brown bear
Vulpes vulpes	red fox

Lower Animals

Acyrthosiphon spp.	aphids
Agrilus anxius	bronze birch borer
Alopecosa hirtipes	[spider]
Alvania	[gastropod]
Apherusa glacialis	[amphipod]
Asplanchna priodonta	[zooplankton]
Balanus balanoides	[barnacle]
Balanus balanus	[barnacle]
Bombus balteatus	[bumblebee]
Bombus cingulatus	[bumblebee]
Bombus hyperboreus	[bumblebee]
Bombus polaris	[bumblebee]
Bosmina longirostris	[zooplankton]
Calanus finmarchicus	[zooplankton]
Calanus hyperboreus	[zooplankton]
Calanus glacialis	[zooplankton]
Calanus marshallae	[zooplankton]
Calliopidae	[amphipod]
Cancer magister	Dungeness crab
Carabus truncaticollis	[ground beetle]
Ceriodaphnia quadrangula	[water flea]
Chaetozone setosa	[polychaete]
Chiloxanthus pilosus	[mite]
Chionoecetes bairdi	Tanner crab
Chionoecetes opilio	snow crab
Chlamys islandica	Iceland scallop

Chone paucibranchiata	[polychaete]
Choristoneura fumiferana	eastern spruce budworm
Choristoneura pinus pinus	jack pine budworm
Chrysaora melamaster	[jellyfish]
Clossiana sp.	fritillary butterfly
Cotesia jucunda	[parasitic wasp]
Curtonotus alpinus	[ground beetle]
Cyclops scutifer	[cladoceran]
Danaus plexippus	milkweed butterfly
Daphnia longiremis	[cladoceran]
Daphnia middendorffiana	[cladoceran]
Daphnia pulex	[cladoceran]
Daphnia pulicaria	[cladoceran]
Daphnia umbra	[cladoceran]
Dendroctonus rufipennis	spruce bark beetle
Dendrolimus sibiricus	Siberian silkworm
Dioryctria reniculelloides	spruce coneworm
Dreissena polymorpha	zebra mussel
Epirrita autumnata	autumnal moth
Eucalanus bungii	[zooplankton]
Euphausia pacifica	[krill]
Folsomia quadrioculata	[collembolan]
Folsomia regularis	[collembolan]
Folsomia sexoculata	[collembolan]
Gammarus lacustris	freshwater shrimp
Gammarus oceanicus	[amphipod]
Gammarus setosus	[amphipod]
Gammarus wilkitzkii	[amphipod]*
Gorgonocephalus caputmedusae	[brittle star]
Gynaephora groenlandica	[moth]
Harpinia spp.	amphipods
Heterocope spp.	copepods
Hiatella arctica	[bivalve]
Hormathia nodosa	[actinarian]
Hypogastrura tullbergi	[collembolan]
Hypogastrura viatica	[collembolan]
Ips typographus	spruce engraver beetle
Janira maculosa	[isopod]
Lepidurus	[benthic invertebrate]
Limacina helicina	[pteropod]
Lithodes aequispina	[crab]
Littorina saxatilis	[gastropod]
Lymantria dispar	gypsy moth
Lymnaea elodes	[snail]
Macoma spp.	clams
Malacosoma disstria	forest tent caterpillar
Maldane sarsi	[polychaete]
Metridia longa	[zooplankton]
Metridia pacifica	[zooplankton]
Munna	[isopod]
Mya arenaria	[clam]
Mya truncata	[clam]
Mysis relicta	opossum shrimp
Mytilus edulis	blue mussel
Neocalanus spp.	zooplankton
Neodiprion sertifer	European pine sawfly
Onisimus spp.	amphipods
Onychiurus arcticus	[collembolan]
Onychiurus groenlandicus	[collembolan]
Operophtera brumata	winter moth
Ophiopholis aculeata	[isopod]
Pandalopsis dispar	deepwater prawn
Pandalus borealis	deepwater prawn/ deepwater shrimp/ northern shrimp
Pandalus goniurus	humpy shrimp
Pandalus jordani	pandalid shrimp
Paragorgia arborea	red gorgonian
Paralithodes camtschatica	red king crab
Paralithodes camtschaticus	king crab
Parasyrphus tarsatus	[flower-fly]
Phyllocnistis populiella	aspen leaf miner
Pristiphora erichsonii	larch sawfly
Pterostichus costatus	[ground beetle]
Rana sylvatica	wood frog
Rana temporaria	common frog
Scoloplos armiger	[polychaete]
Spio filicornis	[polychaete]
Spiochaetopterus typicus	[polychaete]
Themisto libellula	[amphipod]
Thyasira	[bivalve]
Thysanoessa inermis	[krill]

Thysanoessa longicauda	[krill]
Thysanoessa longipes	[krill]
Thysanoessa raschii	[krill]
Tipula carinifrons	[crane fly]
Tonicella	[barnacle]
Umingmakstrongylus pallikuukensis	muskox lungworm
Urticina eques	[actinarian]
Vertagopus brevicaudus	[collembolan]

Higher plants

Abies spp.	firs
Abies sibirica	Siberian fir
Allium schoenoprasum	wild chive
Alnus spp.	alders
Alnus fruticosa	alder
Alopecurus alpinus	alpine foxtail
Andromeda polifolia	bog rosemary
Arctophila fulva	pendant grass
Artemisia spp.	sagebrushes
Betula ermanii	gold birch
Betula exilis	dwarf birch
Betula nana	dwarf birch
Betula papyrifera	paper birch
Betula pendula	silver birch
Betula pubescens	downy birch
B. p. czerepanovii	mountain birch
Calamagrostis lapponica	Lapland reedgrass
Calluna vulgaris	heather
Cardamine pratensis	cuckoo flower
Carex aquatilis	water sedge
Carex bigelowii	Bigelow's sedge
Carex bigelowii / arctisibirica	[sedge]
Carex chordorrhiza	creeping sedge
Carex duriuscula	needleleaf sedge
Carex ensifolia	[sedge]
Carex lugens	[sedge]
Carex stans	water sedge
Carex subspathacea	Hoppner's sedge
Carpinus spp.	hornbeams
Cassiope tetragona	white arctic mountain heather
Cerastium beeringianum	Bering chickweed
Cerastium regelii	Regel's chickweed
Chrysosplenium alternifolium	alternate-leaved golden-saxifrage
Corallorrhiza spp.	coralroots
Cortusa matthioli	bear's ear sanicle
Draba oblongata	Canadian arctic draba
Draba subcapitata	Ellesmereland whitlowgrass
Dryas integrifolia	Arctic dryad
Dryas octopetala / punctata	mountain avens
Dupontia fisheri	Fisher's tundra grass
Dupontia psilosantha	Fisher's tundra grass
Empetrum hermaphroditum	mountain crowberry
Empetrum nigrum	crowberry
Equisetum spp.	horsetails
Eriophorum angustifolium	tall cottongrass
Eriophorum scheuchzeri	white cottongrass
Eriophorum vaginatum	cottongrass
Eritrichium nanum	alpine cushion plant
Euphrasia frigida	cold eyebright
Fagus sylvatica	beech
Galium densiflorum	[herb]
Gentiana nivalis	snow gentian
Geum spp.	avens
Helictotrichon krylovii	[herb]
Juniperus communis	creeping juniper
Kobresia spp.	bog sedges
Koenigia islandica	Iceland purslane
Lagotis minor	little weaselsnout
Larix cajanderi	Cajander larch
Larix dahurica	Dahurian larch
Larix gmelinii	Gmelin larch
Larix laricina	tamarack
Larix sibirica	Siberian larch
Larix sukaczewii	Sukachev larch
Ledum spp.	Labrador teas
Lemna spp.	duckweeds
Lupinus spp.	lupines
Luzula confusa	northern woodrush
Menyanthes spp.	bogbeans
Oxycoccux spp.	cranberries

Oxyria digyna	Arctic sorrel
Papaver polare	Arctic poppy
Pedicularis hirsuta	hairy lousewort
Phippsia algida	ice grass
Phleum alpinum	alpine timothy
Picea abies	Norway spruce
Picea glauca	white spruce
Picea mariana	black spruce
Picea obovata	Siberian spruce
Picea sitchensis	Sitka spruce
Pinus contorta	lodgepole pine
Pinus pumila	Dwarf Siberian pine
Pinus sibirica	Siberian stone pine
Pinus sylvestris	Scots pine
Poa abbreviata	northern bluegrass
Polygonum amphibium	water smartweed
Polygonum viviparum	alpine bistort
Populus balsamifera	balsam poplar
Populus tremula	European aspen
Populus tremuloides	aspen
Populus trichocarpa	black cottonwood
Potamogeton spp.	pondweeds
Puccinellia phryganodes	creeping alkaligrass
Quercus spp.	oaks
Ranunculus glacialis	glacier buttercup
Ranunculus sabinei	Sardinian buttercup
Rhododendron spp.	rhododendrons
Rubus chamaemorus	cloudberry
Salix arctica	arctic willow
Salix glauca	glaucous willow
Salix herbacea	dwarf willow
Salix lanata	hairy willow
Salix myrsinifolia	dark-leaved willow
Salix myrsinites	whortle-leaved willow
Salix myrtilloides	[willow]
Salix phylicifolia	tea-leaved willow
Salix polaris	polar willow
Salix pulchra	tealeaf willow
Salix reptans	[willow]
Sanguisorba officinalis	official burnet
Saxifraga caespitosa	tufted alpine saxifrage
Saxifraga cernua	nodding saxifrage
Saxifraga hyperborea	pygmy saxifrage
Saxifraga nivalis	alpine saxifrage
Saxifraga oppositifolia	purple saxifrage
Silene acaulis	moss campion
Sorbus aucuparia	rowan
Taraxacum officinale	dandelion
Tilia spp.	lindens
Vaccinium myrtillus	blueberry/ bilberry
Vaccinium uliginosum	bog blueberry/ bog whortleberry/ bog bilberry
Vaccinium vitis-idaea	lingonberry

Lower plants

Achnanthes	[diatom]
Ahnfeltia plicata	[red algae]
Alaria esculenta	[kelp]
Arctocetraria nigricascens	[lichen]
Aulacomnium turgidum	[moss]
Bryum cyclophyllum	[moss]
Cetraria islandica	[lichen]
Cetrariella delisei	[lichen]
Cinclidium arcticum	[moss]
Cladina	[lichen]
Cladina rangiferina	[lichen]
Cladonia	[lichen]
Cladonia arbuscula	[lichen]
C. a. mitis	[lichen]
Cladonia pyxidata	[lichen]
Cladonia uncialis	[lichen]
Climacium dendroides	[forest moss]
Cyclotella	[diatom]
Dactylina madreporiformis	[lichen]
Dactylina ramulosa	[lichen]
Dicranoweisia crispula	[moss]
Drepanocladus intermedius	[moss]
Fragilaria	[diatom]
Fucus distichus	sea-tangle (seaweed)
Hylocomium splendens	[moss]

Laminaria digitata	[kelp]
Laminaria saccharina	[kelp]
Laminaria solidungula	[kelp]
Orthothecium chryseon	[moss]
Phaeocystis pouchetii	[flagellate]
Pleurozium shreberi	[forest moss]
Pogonatum alpinum	[bryophyte]
Polytrichum commune	[moss]
Polytrichum juniperinum	[moss]
Psora decipiens	[lichen]
Ptilidium ciliare	[liverwort]
Racomitrium lanuginosum	[moss]
Rhizoplaca melanophthalma	[lichen]
Rhytidiadelphus triquetrus	[forest moss]
Seligeria polaris	[moss]
Sphagnum fuscum	[moss]
Stereocaulon paschale	[lichen]
Thamnolia subuliformis	[lichen]
Tomentypnum nitens	[moss]
Tortula ruralis	[moss]
Xanthoria candelaria	[lichen]
Xanthoria parietina	[lichen]

Misc. fungi/bacteria etc.

Alternaria	[fungus]
Archaeoglobus	[bacterium]
Archaeoglobus fulgidus	[bacterium]
Arthrobacter	[bacterium]
Aspergillus	[fungus]
Azotobacter	[bacterium]
Bacillus	[bacterium]
Beijerinckia indica	[bacterium]
Botrytis	[fungus]
Clostridium	[bacterium]
Cortinarius	[fungus]
Cryptococcus laurentii	[yeast]
Exobasidium	[fungus]
Fusarium	[fungus]
Inocybe	[fungus]
Metarhizium	[fungus]
Metarhizium anisopliae	[fungus]
Methylocapsa	[bacterium]
Methylocella	[bacterium]
Microcoleus chthonoplastes	[cyanobacteria]
Mucor hiemalis	[fungus]
Nostoc spp.	[cyanobacteria]
Penicillium	[fungus]
Pseudomonas	[bacterium]
Pyrococcus	[bacterium]
Rhizopus	[fungus]
Thermococcus	[bacterium]
Truncatella truncata	[fungus]

Acronyms

ABL	Atmospheric boundary layer
ACIA	Arctic Climate Impact Assessment
ACD	Arctic Coastal Dynamics project
AEPS	Arctic Environmental Protection Strategy
AFI	Air freezing index
AGCM	Atmospheric general circulation model
AMAP	Arctic Monitoring and Assessment Programme
AO	Arctic Oscillation
AO$^-$	Low AO index
AO$^+$	High AO index
AOGCM	Atmosphere–ocean general circulation model
ARCMIP	Arctic Regional Climate Model Intercomparison Project
ATI	Air thawing index
AVHRR	Advanced Very High Resolution Radiometer
β-HCH	beta-Hexachlorocyclohexane
BOREAS	Boreal Ecosystem–Atmosphere Study
BP	Before present
C	Carbon
CAFF	Conservation of Arctic Flora and Fauna
CCSR	Center for Climate System Research (Japan)
Cd	Cadmium
CDOM	Colored dissolved organic matter
CDQ	Community Development Quota program (Alaska)
CFC	Chlorofluorocarbon
CGCM2	An AOGCM developed by the Canadian Centre for Climate Modelling and Analysis
CH_4	Methane
Chl-a	Chlorophyll-a
CMIP	Coupled Model Intercomparison Project
CO	Carbon monoxide
CO_2	Carbon dioxide
CPAN	Circumpolar Protected Area Network
CPUE	Catch-Per-Unit-Effort
CRU	Climatic Research Unit (University of East Anglia, UK)
CSM_1.4	An AOGCM developed by the National Center for Atmospheric Research (USA)
CTM	Chemical transport model
D-O	Dansgaard-Oeschger
DDT	Dichlorodiphenyltrichloroethane
DEWG	Denendeh Environmental Working Group
DIC	Dissolved inorganic carbon
DMS	Dimethyl sulfide
DO	Dissolved oxygen
DOC	Dissolved organic carbon
DU	Dobson unit
E	Evapotranspiration
ECHAM4/OPYC3	An AOGCM developed by the Max Planck Institute for Meteorology (Germany)
ECMWF	European Centre for Medium-range Weather Forecasts
EEZ	Exclusive economic zone
ENSO	El Niño–Southern Oscillation
FAO	United Nations Food and Agriculture Organization
fCO$_2$	Fugacity of CO_2
GCM	General circulation model
GDD	Growing degree-day
GDP	Gross domestic product
GEP	Gross ecosystem production
GFDL-R30_c	An AOGCM developed by the Geophysical Fluid Dynamics Laboratory (USA)
Gg	Gigagram (10^9 grams)
GHCN	Global Historical Climatology Network
GHG	Greenhouse gas
GPS	Global Positioning System
GWP	Global warming potential
H	Hydrogen
H_2	Molecular hydrogen

H_2S	Hydrogen sulfide
ha	Hectare
HadCM3	An AOGCM developed by the Hadley Centre for Climate Prediction and Research (UK)
HCB	Hexachlorobenzene
HCH	Hexachlorocyclohexane
Hg	Mercury
Hg^0	Elemental mercury
Hg^{2+}	Divalent mercury
IABP	International Arctic Buoy Programme
IASC	International Arctic Science Committee
IBA	Important Bird Area
IBP	International Biological Programme
ICC	Inuit Circumpolar Conference
ICES	International Council for the Exploration of the Sea
IGBP	International Geosphere–Biosphere Programme
IPA	International Permafrost Association
IPCC	Intergovernmental Panel on Climate Change
ITEX	International Tundra Experiment
ITQ	Individual Transferable Quota system
IUCN	World Conservation Union
IWC	International Whaling Commission
J	Joule
K	Potassium
Kg	Kilogram (10^3 grams)
LGM	Last glacial maximum
LIA	Little Ice Age
LPJ	Lund-Potsdam-Jena dynamic global vegetation model
MDE	Mercury depletion event
MeHg	Methyl mercury
MIP	Model intercomparison project
MPI	Max Planck Institute for Meteorology (Germany)
mwe	Meter water equivalent
MWP	Medieval Warm Period
My	Million years
N	Nitrogen
N_2O	Nitrous oxide
NADW	North Atlantic Deep Water
NAFO	Northwest Atlantic Fisheries Organization
NAMMCO	North Atlantic Marine Mammal Commission
NAO	North Atlantic Oscillation
NASA	National Aeronautics and Space Administration (US)
NCAR	National Center for Atmospheric Research (US)
NCARP	Northern Cod Adjustment and Recovery Program (Canada)
NCEP	National Centers for Environmental Prediction of NOAA
NDVI	Normalized Difference Vegetation Index
NEAFC	North East Atlantic Fisheries Commission
NEP	Net ecosystem production
NH_3	Ammonia
NH_4	Ammonium
NIES	National Institute for Environmental Studies (Japan)
nm	Nautical mile
NO_3	Nitrate
NOAA	National Oceanic and Atmospheric Administration (US)
NPP	Net primary production
NRL	Sámi Reindeer Herders Association of Norway
NSF	National Science Foundation (US)
NWT	Northwest Territories (Canada)
O	Oxygen
OUML	Ocean upper mixed layer
P	Phosphorus
P	Precipitation
PAH	Polycyclic aromatic hydrocarbon
PAR	Photosynthetically active radiation

Pb	Lead
PBDE	Polybrominated diphenyl ether
PCB	Polychlorinated biphenyl
PCN	Polychlorinated naphthalene
pCO$_2$	Difference in partial pressure of CO$_2$ (e.g., across the air–sea interface)
PDO	Pacific Decadal Oscillation
P-E	Precipitation minus evapotranspiration
Pg	Petagrams (10^{15} grams)
POC	Particulate organic carbon
POP	Persistent organic pollutant
ppmv	Parts per million by volume
PSC	Polar stratospheric cloud
PUFA	Polyunsaturated fatty acid
R	Runoff
R$_A$	Respiration, autotrophic
R$_E$	Respiration, ecosystem
R$_H$	Respiration, heterotrophic
RAF	Radiation amplification factor
RAIPON	Russian Association of Indigenous Peoples of the North
RCM	Regional climate model
RCS	Regional (age) curve standardization
RIMS	Rapid Integrated Monitoring System
RIVM	National Institute for Public Health and the Environment model
RUV	Remote underwater vehicle
S	Salinity
SBUV	Solar backscatter ultraviolet
SHEBA	Surface Heat Budget of the Arctic Ocean
SLP	Sea-level pressure
SO	Sulfur monoxide
SO$_2$	Sulfur dioxide
SO$_x$	Sulfur oxide
SRES	Special Report on Emissions Scenarios (by the IPCC)
SST	Sea surface temperature
Sv	Sverdrup (unit = 10^6 m^3/s)
SZA	Solar zenith angle
T	Tonne
TAC	Total allowable catch
TEK	Traditional ecological knowledge
Tg	Teragrams (10^{12} grams)
THC	Thermohaline circulation
TOMS	Total Ozone Mapping Spectrometer
TOPEX/ POSEIDON	joint French/US altimeter satellite
UIUC	University of Illinois at Urbana-Champaign (US)
ULAQ	Università degli studi dell'Aquila (Italy)
UNCED	United Nations Conference on Environment and Development
UNEP	United Nations Environment Programme
UV	Ultraviolet
UV-A	Ultraviolet-A radiation (315–400 nm)
UV-B	Ultraviolet-B radiation (280–315 nm)
W	Watt
WMO	World Meteorological Organization
WWF	World Wide Fund for Nature
Zn	Zinc
ΣDDT	Sum of DDT, DDD, and DDE (concentrations)
ΣPCBs	Sum of a number of individual polychlorinated (PCB) congeners

Glossary

Actinic flux

Radiation incident at a point, determined by integrating the spectral irradiance over all directions of incident light (units = W/m²/nm).

Action spectrum

A sensitivity function that describes the relative effectiveness of energy at different wavelengths in determining a biological response.

Active layer

The layer of ground that is subject to annual thawing and freezing in areas underlain by permafrost.

Adaptive ability or capacity

The ability of an organism, an ecosystem, or a human system (community, culture, enterprise) to adapt to environmental change.

Albedo

The fraction of solar radiation reflected by a surface or object, often expressed as a percentage. Snow covered surfaces have a high albedo; the albedo of soils ranges from high to low; vegetation covered surfaces and oceans have a low albedo. The earth's albedo varies mainly through varying cloudiness, snow, ice, leaf area, and land cover changes.

Allochthonous

Exogenous; originating outside and transported into a given system or area.

Anadromous

An adjective describing fish that exhibit migratory behavior between fresh and marine waters characterized by spawning (and in ice-covered arctic seas also by overwintering) in freshwater and summer feeding in marine water.

Aquaculture

Breeding and rearing fish and shellfish, etc.

Aquifer

A stratum of permeable rock that bears water. An unconfined aquifer is recharged directly by local rainfall, rivers, and lakes, and the rate of recharge will be influenced by the permeability of the overlying rocks and soils. A confined aquifer is characterized by an overlying bed that is impermeable and the local rainfall does not influence the aquifer.

Arctic

[See chapter 1, section 1.1, paragraph 4]

Athalassic

Used of waters or water bodies that have not had any connection to the sea in geologically recent times, all ions in solution are thus derived from the substratum or atmosphere.

Atmospheric boundary layer

The bottom layer of the troposphere that is in contact with the surface of the earth. It is often turbulent and is capped by a statically stable layer of air or temperature inversion. The atmospheric boundary layer depth (i.e., the inversion height) is variable in time and space, ranging from tens of meters in strongly statically stable situations, to several kilometers in convective conditions over deserts.

Benthic

Pertaining to the sea bed, river bed, or lake floor.

Biodiversity

The numbers and relative abundances of different genes (genetic diversity), species, and ecosystems (communities) in a particular area.

Biogeochemical cycle

The cyclical system through which a given chemical element is transferred between biotic and abiotic parts of the biosphere.

Biota

All living organisms of an area; the flora and fauna considered as a unit.

Bloom

A reproductive explosion of microscopic organisms in a lake, river, or ocean.

Catadromous

An adjective describing fish which exhibit migratory behavior between fresh and marine waters that is characterized by spawning in marine waters and feeding and early rearing in freshwaters.

Climate

Climate in a narrow sense is usually defined as the "average weather" or more rigorously as the statistical description in terms of the mean and variability of relevant quantities over a period of time ranging from months to thousands or millions of years. The classical period is 30 years as defined by the WMO. These relevant quantities are most often surface variables such as temperature, precipitation and wind. Climate in a wider sense is the state, including a statistical description, of the climate system.

Climate change

Climate change refers to a statistically significant variation in either the mean state of the climate, or in its variability, persisting for an extended period (typically decades or longer).

Climate feedback

An interaction between processes in the climate system, where the result of an initial process triggers a second process that in turn influences the initial one. A positive feedback intensifies the original process, and a negative feedback reduces it.

Climatological baseline

A period of years representing the current climate, the latter being understood as a statistical description in terms of the mean and variability over the period. A baseline period should: be representative of the present-day or recent average climate in the region considered; be of sufficient duration to encompass a range of climatic variations; cover a period for which data on all major climatological variables are abundant, adequately distributed in space, and readily available; include data of sufficiently high quality for use in evaluating impacts; and be consistent or readily comparable with baseline climatologies used in other impact assessments.

Co-management

A system for management of wildlife populations in which responsibility is shared between the users of the resource and government entities with legal authority for management of wildlife.

Conservation

The protection of environmental values associated with natural systems through planned management of natural resources to assure their continued viability and availability for human appreciation and use by preventing overexploitation, and protection from destruction or neglect.

Conspecific

Belonging to the same species.

Contaminant

A substance that is not naturally present in the environment or is present in unnatural concentrations that can, in sufficient concentration, result in potential negative effects on the health of humans, other organisms, and ecosystems.

Continental shelf

A shallow submarine plain of varying width forming a border to a continent and typically ending in a steep slope to the ocean abyss.

Cryosphere

The component of the climate system consisting of all snow, ice, and permafrost on and beneath the surface of the earth and ocean.

Demersal

Living at or near the bottom of a sea or lake but having the capacity for active swimming.

Diadromous

Migrating between fresh water and seawater.

Dose

Dose rate integrated over a time period of exposure (units = J/m² (effective)).

Dose rate

Spectral irradiance weighted by a biological action spectrum (units = W/m² (effective)).

Ecosystem

A system of interacting living organisms together with their physical environment. The boundaries of what could be called an ecosystem are somewhat arbitrary, depending on the focus of interest or study. Thus the extent of an ecosystem could range from very small spatial scales to, ultimately, the entire earth.

Ecosystem function

Ecosystem function includes carbon and nutrient cycling, soil processes, controls of trace gas exchange processes, primary and secondary productivity, and water and energy balance.

Ecosystem services

Ecological processes or functions that have value to individuals or society.

Ecosystem structure

The spatial structure of an ecosystem, trophic interactions, and community composition in terms of biodiversity.

Ecotone

A zone of transition from one major plant community to another. For example, the forest–tundra ecotone in high northern latitudes is a zone of patchy and often stunted tree growth intermixed with areas of tundra.

Emissions scenario

A plausible representation of the future development of emissions of substances that are potentially radiatively active (e.g., greenhouse gases and aerosols), based on a coherent and internally consistent set of assumptions about driving forces (such as demographic and socio-economic development, technological change) and their key relationships.

ENSO

El Niño in its original sense is a warm water current that periodically flows along the coast of Ecuador and Peru, disrupting the local fishery. This oceanic event is associated with a fluctuation in the intertropical surface pressure pattern and circulation in the Indian and Pacific Oceans, called the Southern Oscillation. This coupled atmospheric–oceanic phenomenon is collectively known as El Niño–Southern Oscillation, or ENSO.

Environment

The complex of climatic, edaphic, and biotic factors that act upon an organism or an ecological community and ultimately determine its form and survival. From the human perspective, also inclusive of the aggregate of social and cultural conditions that influence the life of an individual or community.

Erythema

Reddening of the skin. Commonly called sunburn, it is most effectively caused by UV-B radiation.

Evapotranspiration

The combined process of evaporation (the process by which a liquid becomes a gas) and transpiration (loss of water vapor from an organism through a membrane or through pores).

Extant

Existing or living at the present time.

Extinction

The complete disappearance of an entire species.

Extirpation

The disappearance of a species from part of its range; local extinction.

Fast ice (or land-fast ice)

Fast ice (or land-fast ice) is immobilized for up to 10 months each year by coastal geometry or by grounded ice ridges (stamukhi).

Finite-difference model

A model based on finite-difference approximations – the differences between the values of a function at two discrete points are used to approximate the derivatives of the function. Same as grid-point model.

Flux adjustment

To avoid the problem of coupled atmosphere–ocean general circulation models drifting into some unrealistic climate state, adjustment terms can be applied to the atmosphere–ocean fluxes of heat and moisture (and sometimes the surface stresses resulting from the effect of the wind on the ocean surface) before these fluxes are imposed on the model ocean and atmosphere.

Food chain

A sequence of organisms on successive trophic levels within a community, through which energy is transferred by feeding; energy enters the food chain during fixation by primary producers (mainly green plants) and passes to herbivores (primary consumers) and then to carnivores (secondary and tertiary consumers).

Food web

The network of interconnected food chains of a community.

Freshet

A rush of freshwater from rain or melted snow.

Gas hydrates or methane hydrates

In the presence of high concentrations of certain gases in the water, at low temperatures and high pressures, gas hydrates can form (i.e., open-structured water ice hosting gases such as methane, carbon dioxide or hydrogen sulphide). When the gas trapped in the icy compound is methane, this is known as methane hydrate. Methane hydrate is by far the most common naturally occurring gas hydrate. Other gases, including larger hydrocarbons and carbon dioxide, also form hydrate compounds.

Giardiasis

An infection caused by the parasite *Giardia lamblia*.

Greenhouse gases

Greenhouse gases are those gaseous constituents of the atmosphere, both natural and anthropogenic, that absorb and emit radiation at specific wavelengths within the spectrum of infrared radiation emitted by the earth's surface, atmosphere, and clouds. This property causes the greenhouse effect. Water vapor, carbon dioxide, nitrous oxide, methane, and ozone are the primary greenhouse gases in the earth's atmosphere. A major proportion of these gases derive from past and present life processes on the earth, including decomposition of organic matter, respiration of plants and animals, burning of forests and other plant material, and burning of coal, oil, and other fossil fuels.

Halocline

A zone of marked salinity gradient.

Indigenous people

People whose ancestors inhabited a place or a country when persons from another culture or ethnic background arrived on the scene and dominated them through conquest, settlement, or other means and who today live in more conformity with their own social, economic, and cultural customs and traditions than those of the country of which they now form a part. Such people are often referred to in the Arctic as "aboriginal", "Native", "first nations", or "tribal".

Irradiance

Radiant power per unit area (units = W/m^2).

Native

Official legal term used in Alaska for indigenous people as a result of wording in the Alaska Native Claims Settlement Act of 1971.

Net ecosystem production

Net gain or loss of carbon from an ecosystem. Net ecosystem production is equal to the gross primary production (carbon fixed by plants through the process of photosynthesis) minus the carbon lost through heterotrophic respiration.

Net primary production

The increase in plant biomass or carbon of a unit of a landscape. Net primary production is equal to the gross primary production (carbon fixed by plants through the process of photosynthesis) minus carbon lost through autotrophic respiration.

Nival

Pertaining to snow.

North Atlantic Oscillation

The North Atlantic Oscillation consists of opposing variations of barometric pressure near Iceland and near the Azores. On average, a westerly current between the Icelandic low pressure area and the Azores high pressure area carries cyclones and their associated frontal systems towards Europe. However, the pressure difference between Iceland and the Azores fluctuates on timescales of days to decades, and can be reversed at times. It is the dominant mode of winter climate variability in the North Atlantic region, ranging from central North America to Europe.

Northeast Passage (Northern Sea Route)

The route of potential ship transit through the Arctic Ocean north of Eurasia between the Barents and Bering Seas.

Northwest Passage

The route of potential ship transit north of North America between the Labrador and Bering Seas.

Nunatak

A mountain peak or rocky outcrop projecting above an ice cap.

Ocean outfall

A discharge pipe used for the final disposal of wastewater extending from a wastewater treatment works to the point of discharge in marine waters.

Ontogenetic migration

The occupation by an animal of different habitats at different stages of development.

Ozone layer

The stratosphere contains a layer in which the concentration of ozone is greatest, the so-called ozone layer. The layer extends from about 12 to 40 km. The ozone concentration reaches a maximum between 20 and 25 km. This layer is being depleted by human emissions of chlorine and bromine compounds. These compounds interact photochemically with the ozone to allow increased ultraviolet-B radiation to reach the earth's surface.

Pack ice

Ice formed on oceanic surfaces in polar regions, often encompassing ice bergs derived from calving of glaciers as glaciers enter the sea from land.

Paludification

The process of bog expansion.

Parameterization

In climate models, this term refers to the technique of representing processes that cannot be explicitly resolved at the spatial or temporal resolution of the model (sub-grid scale processes) by relationships between the area- or time-averaged effect of such sub-grid-scale processes and the larger scale flow.

Pathogen
A microbiological agent capable of causing disease.

Pelagic
Pertaining to the water column of the sea or lake; used of organisms inhabiting the open waters of an ocean or lake.

Permafrost
Ground (soil or rock and included ice and organic material) that remains at or below 0 °C for at least two consecutive years.

Phenology
The study of seasonal changes in plant and animal life and the relationships of these changes to weather and climate.

Phenotypic responses
Changes in the physical expression of a characteristic of an organism when experiencing a change in the environment and without genetic change.

Photokeratitis
Sunburn of the cornea resulting from overexposure to UV-B radiation that is usually reversible. It can occur after long periods on the snow, especially on bright, clear, sunny days, without adequate eye protection. It can be very painful for a couple of days and can result in transitory loss of vision.

Photoperiod
The relative lengths of seasonally alternating periods of lightness and darkness in the 24 hour day that affect the growth, activity, and reproductive timing in organisms.

Phytoplankton
The plant forms of plankton. Phytoplankton are the dominant plants in the sea, and are the basis of the entire marine food web. These single-celled organisms are the principal agents for photosynthetic carbon fixation in the ocean.

Piscivorous
Feeding on fish.

Pit privy (outhouse)
A structure that receives urine and excrement that is not water-borne; and is the final disposal site and not a temporary storage facility.

Planktivorous
Feeding on planktonic organisms.

Polar stratospheric clouds
Clouds that form at extremely low temperatures (below 195 K) in the stratosphere, mostly in the polar regions, and play a role in ozone depletion chemistry.

Polynya
A Russian term meaning an area of open water, possibly containing some thin ice, within the ice pack. A polynya is distinguished from a lead by being a broad opening rather than a long, narrow fracture.

Post and pad foundation
A building foundation system constructed with posts for vertical support and pads on the ground to distribute the load of each vertical support.

Prognostic variable
A variable that is described by an equation that contains a time derivative of this variable (a differential equation), and therefore its value can be determined at a later time when the other terms in the equation are known.

Proxy climate data
A proxy climate indicator is a local record that is interpreted, using physical and biophysical principles, to represent some combination of climate-related variations back in time. Climate-related data derived in this way are referred to as proxy data. Examples of proxies are tree ring records and various data derived from ice cores.

Quasi-biennial oscillation
An oscillation in the zonal winds of the equatorial stratosphere having a period that fluctuates between about 24 and 30 months. This oscillation is a manifestation of a downward propagation of winds with alternating sign. This phenomenon is sometimes referred to as the stratospheric quasi-biennial oscillation to distinguish it from other atmospheric features that also have spectral peaks near two years.

Refugium
An area that has escaped major climatic changes typical of a region as a whole and acts as a refuge for biota previously more widely distributed; an isolated habitat that retains the environmental conditions that were once widespread.

Regime shift
A rapid change in regional climate.

Resilience
Synonymous with "adaptive ability", or the ability of a system to undergo change without changing its state or identity.

Ruderal
Inhabiting disturbed sites.

Runoff
The water from rain or melted snow that travels over the ground surface.

Saline wedge
A salt-water layer flowing below a lower density freshwater layer that tends to form the shape of a wedge as it intrudes into a river system.

Species adaptation
Characteristics of an organism that have been selected by specific selection pressures exerted by other organisms or the physical environment and that have lead to a new genetic constitution.

Spectral irradiance
Radiant power per unit area (units $= W/m^2/nm$).

Spectral model
A model in which the prognostic field variables are represented as sums of a finite set of spectral modes rather than being given at grid points. The spectral modes may be Fourier modes in the one-dimensional case or double Fourier modes or spherical harmonics in the two-dimensional case. One advantage of a spectral model is that horizontal derivatives can be calculated exactly for the spectral modes represented in the model. Spectral models are, in general, computationally more efficient than a grid-point model with an equivalent resolution.

Stamukhi zone
The zone of heavily broken ice which marks the contact between land-fast ice and the moving pack-ice zones.

Stenothermal
A tolerance of a narrow range of environmental temperatures.

Storm surge
A temporary increase, at a particular locality, in the height of the sea due to extreme meteorological conditions (low atmospheric pressure and/or strong winds). The storm surge is defined as being the excess above the level expected from the tidal variation alone at that time and place.

Subpermafrost
Located beneath the permafrost.

Subsistence activity
An aspect of human existence involving derivation of food and other needs directly from the locally available natural resources.

Sustainability
The ability of a natural system (e.g., ecosystem, plant community, population of organisms) or a human-generated system (e.g., community, economy, culture) to maintain itself over time. Often used in reference to the ability of a renewable natural resource to yield a stable annual harvest over time.

Taiga
Russian term for the boreal or northern coniferous forest biome; the ecosystem adjacent to the arctic tundra.

Talik
A layer or body of unfrozen ground occurring in a permafrost area due to a local anomaly in thermal, hydrological, hydrogeological, or hydrochemical conditions.

Thermocline
A boundary region in water bodies (lakes or oceans) between two layers of water of different temperature, in which temperature changes sharply with depth.

Thermohaline circulation
Large-scale density-driven circulation in the ocean, caused by differences in temperature and salinity. In the north Atlantic, the thermohaline circulation consists of warm surface water flowing northward and cold deepwater flowing southward, resulting in a net poleward transport of heat. The surface water sinks in highly restricted sinking regions located in high latitudes.

Thermokarst
Irregular, hummocky topography in frozen ground caused by melting of ice.

Traditional knowledge
The accumulated knowledge of indigenous peoples about the environment in which they live that has been passed on via the elders of a community.

Trophic levels
The sequence of steps in a food chain; from producer to primary, secondary, or tertiary consumer.

Tundra
A type of ecosystem dominated by lichens, mosses, grasses, and dwarf woody plants. Tundra is found at high latitudes (arctic tundra) and high altitudes (alpine tundra). Arctic tundra is underlain by permafrost and is usually saturated.

Urocanic acid
A photoreceptor for the induction of UV immune suppression.

UV-A

The longest UV wavelengths (315–400 nm). Atmospheric gases absorb little UV-A radiation, so most reaches the earth's surface.

UV-B

Solar radiation within a wavelength range of 280–315 nm, the greater part of which is absorbed by stratospheric ozone. Enhanced UV-B radiation suppresses the immune system and can have other adverse effects on living organisms.

UV-C

The shortest UV wavelengths (100–280 nm). UV-C radiation is almost entirely absorbed by atmospheric oxygen and ozone.

UV index

A number reflecting the daily risk of overexposure (sunburning) to sunlight. Measured on a scale of 0 to >10, where 0 indicates minimal exposure and >10 indicates high to very high risk.

UV-induced immune suppression

A change in cell-mediated immunity induced by UV-B radiation. The result of UV-induced immune suppression is the production of regulatory T-cells (suppressor cells) as opposed to effector (antigen-attacking) T-cells.

Varve

A layer of sediment deposited in a lake during the course of a single year.

Vernal

Pertaining to the spring.

Wastewater

Waterborne human wastes or graywater derived from dwellings, commercial buildings, institutions, or similar structures; "wastewater" includes the contents of individual removable containers used to collect and temporarily store human wastes.

Weather

State of the atmosphere with regard to temperature, precipitation, wind, and degree of cloud cover.

Zooplankton

The animal forms of plankton. They consume phytoplankton or other zooplankton.

$\delta^{18}O$

An expression for the ratio of the ^{18}O to ^{16}O atoms (stable isotopes of oxygen) in a sample relative to a standard, used as an indicator of temperature change over time, and defined as: $\delta^{18}O = (^{18}O/^{16}O$ sample $- ^{18}O/^{18}O$ standard$)/(^{18}O/^{16}O$ standard$)$.